24th International Conference on Port and Ocean Engineering under Arctic Conditions (POAC'17)

Busan, Korea
11 - 16 June 2017

Volume 1 of 2

ISBN: 978-1-5108-5368-3
ISSN: 0376-6756

Printed from e-media with permission by:

Curran Associates, Inc.
57 Morehouse Lane
Red Hook, NY 12571

Some format issues inherent in the e-media version may also appear in this print version.

Printed by Curran Associates, Inc. (2018)

For permission requests, please contact Port and Ocean Engineering Under Arctic Conditions (POAC) at the address below.

Port and Ocean Engineering Under Arctic Conditions (POAC)
National Research Council of Canada
1200 Montreal Road, Bldg. M-32
Ottawa, Ontario
Canada K1A OR6

Phone: (613) 993-7695
Fax: (613) 952-7679

ivana.kubat@nrc.ca

Additional copies of this publication are available from:

Curran Associates, Inc.
57 Morehouse Lane
Red Hook, NY 12571 USA
Phone: 845-758-0400
Fax: 845-758-2633
Email: curran@proceedings.com
Web: www.proceedings.com

TABLE OF CONTENTS

VOLUME 1

SHIPS IN ICE #1

REMOTE SENSING AND ICE MODLLING #2

SHIPPING AND NSR NAVIGATION

ICE MANAGEMENT AND OPERATION

WAVE-ICE INTERACTION AND OIL SPILLS IN ICE WATER

PLENARY SESSION #2

RUSSIAN ARCTIC OFFSHORE

NUMERICAL MODELING AND SIMULATION #1

ICE MECHANICS #1

PLENARY SESSION #3

PLENARY SESSION #4

ICE TRIALS AND FIELD MEASUREMENT

VOLUME 2

NUMERICAL MODELING AND SIMULATION #2

POLAR CODES AND REGULATIONS

ARCTIC MARINE STRUCTURES AND CONSTRUCTION

STATION KEEPING IN ICE

ICE MECHANICS #2

SHIPS IN ICE #2

LABORATORY MODEL TEST

ICE-STRUCTURE INTERACTION #2

PLENARY SESSION #5

SHIPS IN ICE #3

ARCTIC ENVIRONMENTS AND SEA ICE PROPERTIES #3

WINTERIZATION AND ARCTIC OPERATION

POAC'17
Busan, KOREA

Proceedings of the 24th International Conference on
Port and Ocean Engineering under Arctic Conditions
June 11-16, 2017, Busan, Korea

Variability, Trends and Predictability of Arctic Sea Ice: Insights from Observations, Modeling and Remote Sensing at the National Snow and Ice Data Center

Mark C. Serreze, Director, National Snow and Ice Data Center and Professor, Department of Geography, University of Colorado Boulder

serreze@nsidc.org

Satellite data maintained at the National Snow and Ice Data Center (NSIDC) documents downward trends in Arctic sea ice extent in all months, strongest at the end of the summer melt season, with the largest summer retreats in the Beaufort and Chukchi seas. Data from various forms of remote sensing, including lidar, also documents a sharp decline in ice thickness and volume. Climate models project that a seasonally ice-free Arctic Ocean will be realized well within this century, perhaps as early as 2040. However, for years to come, sea ice conditions will be highly variable, especially on a regional basis. Regional variability in end-of-summer ice extent is closely tied to summer weather patterns; the prevailing weather patterns strongly determine whether or not the Northern Sea Route or the Northwest Passage will open in a given summer. Recent research at NSIDC has also shown that summer ice conditions in the Chukchi Sea – a key choke point for ships entering or leaving the Arctic Ocean, is strongly controlled by variability in the Bering Strait oceanic heat inflow. Knowing the oceanic heat flow provides a potential avenue via a simple linear model to predict the dates of seasonal sea ice retreat and advance several months in advance. Another source of predictability in this key area is seasonal ocean heat uptake and albedo feedback – early ice retreat means more absorbtion of heat in the upper ocean through summer, which then delays autumn freeze up. However, the vagaries of summer weather will always limit predictability. Along with conducting research on Arctic sea ice variability, trends, and projected future conditions, NSIDC maintains a number of satellite-based sea ice data sets that can support Arctic marine operations. MASIE (Multisensor Analyzed Sea Ice Extent), which mainly uses visible-band data, provides accurate information on the sea edge. MASIE products are developed in collaboration with the US National Ice Center. NSIDC also holds daily (and continuously updated) fields of sea ice concentration (October 1979 onwards) from the satellite passive microwave time series at a coarser 25 km resolution. A growing concern at NSIDC is a potential gap in data coverage; the record uses data from the US Defense Meteorological Satellite Program (DMSP) F-Series of satellites. F-18, the youngest and last in the series, is well past its design life, and the Advanced Microwave Sounding Radiometer 2 (AMSR-2) is also getting old. New platforms with appropriate passive microwave sensors are unlikely to launch until 2022.

Ridge Profile Measurements for Understanding Ridge Resistance

Mikko Suominen[1], Arttu Polojärvi[1], Annu Oikkonen[2]
[1] Aalto University, Department of Mechanical Engineering, Espoo, Finland
[2] Finnish Meteorological Institute, Helsinki, Finland

ABSTRACT

Ice ridges are the most difficult ice features that ships encounter in first-year ice conditions. Our understanding on ridge resistance on ships is currently limited, but we can assume that the keel of a ridge can have a significant impact to the resistance due to its large volume. In order to study the ridge resistance, we measured ridge profiles and recorded ship machinery and operational data during RV Aranda sea ice cruise in the Baltic Sea on spring 2016. This paper presents results from our ridge profile measurements made by drilling and with a sonar. We found a fairly good correspondence between these two methods. The maximum depth of the profiled ridges varied between 3 and 5 meters and the width of the cross sections from 10 to 25 meters.

KEY WORDS: Ridge profile; Drilling; Sonar; Ridge keel; Ridge sail.

INTRODUCTION

The ships are forced to navigate through the ice fields independently or in the assistance of ice breakers. In the first-year ice conditions, the thickness of the level ice is limited, but it can increase mechanically in a short time through ridge formation. Ridge formation leads to deformed ice features that may be even tens of meters thick (Timco and Burden, 1997; Strub-Klein and Sudom, 2012). Figure 1 illustrates how the first-year ridges consist of a sail, a refrozen consolidated layer, which is commonly thicker than the surrounding level ice (Kankaanpää, 1991; Veitch et al., 1991), and a keel, which can be very large in volume. Ridge properties and resistance are of interest, as the ridges are the most difficult ice features ships encounter in the first-year ice conditions.

In order to study the ridge resistance, a group of researchers participated to the Sea Ice cruise on board RV Aranda in March 2016. Extensive measurement campaign included, for example, surface laser profiling, electromagnetic measurements (EM), recording of ship machinery and navigational data, visual observations, ridge profiling, and snow thickness measurements. The profiles of six ridges were determined by drilling and with a sonar. After the ridge was profiled, the ship penetrated the ridge while the navigational and machinery data were recorded with high frequency. Machinery data enables the estimates of the used thrust, which together with measurements on the speed and location of the vessel with a differential GPS, allows the identification of the increased resistance due to the ridge.

In this paper we present the results from the ridge profiling measurements by drilling together with some of the results from the sonar measurements. While this data is beneficial for our future ridge resistance study, it can be also used in such transit, or ice mechanics, simulations as presented in Kuuliala et al. (2017), Polojärvi and Tuhkuri (2009; 2013) or Gong et al. (2017). We start the paper by introducing our measurements, and then proceed by describing the research voyage. After this, we introduce and briefly discuss our results before concluding the paper.

METHODS

Ridge Profiling by Drilling

We first measured the ridge thickness and profile using a standard way of profiling by drilling. When the drill head first goes through the level ice, the length of the drill inside of the level ice is measured. This gives the thickness of the ice. In a case of ridge keel, there are pores between the ice blocks that form the keel as Figure 1 illustrates. When the drill head goes through an ice block, the length of the drill inside the ice is again written down. Then the drill is pushed downwards to test for additional ice blocks under the most recent block that was drilled through. If an additional icc block is found, the drilling is continued. We continued drilling until we could not find additional blocks for, at least, 1 meter distance. The depth, where the drill went through the ice for the last time, was defined to be the depth of the keel at that point. For all profiled ridges, several boreholes in well-defined drill lines were used to measure the overall geometry of the ridge.

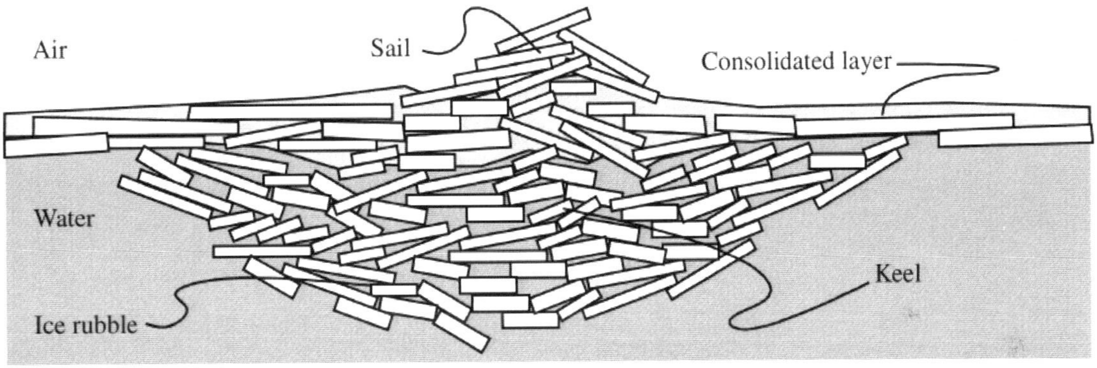

Figure 1. Cross section of a ridge with its main parts indicated: The sail above the waterline and the keel underwater. The keel comprises of a consolidated layer close to the waterline and the unconsolidated, or partly consolidated, ice rubble (Polojärvi, 2013).

Ridge Profiling using Sonar

The sonar head employed in the measurements was a Kongsberg 1071 series high resolution single-beam scanning sonar. The sonar enables imaging and profiling the keels even if not purposely built for it. The measurements were recorded and initially processed with Kongsberg MS 1000 software. The software records the distance of the sonar head from the echoes coming from ice for producing two type of results: (1) a cloud of data points describing the keel profile, and (2) a sonar image that can be used in the profiling.

The measurement of the distance with a sonar is based on the speed of the sound, which has to be determined. Here the water was estimated to be 1°C in temperature, the salinity was measured to be 3 ppt, and the average depth for the sonar head 3 meters. These were used as an input in the software, which gave the sound speed of 1411 m/s. The employed sonar head allows the use of a fan or cone-shaped beams. As the cone beam produces less noise in the measurements, it was deemed better and used in the measurements.

With the cone, two scanning frequencies, 900 and 1200 Hz, were allowed. Figure 2 presents example images from the sonar measurements with these frequencies used. As can be seen from Figure 2, the higher frequency produces considerably less background noise to the measurements. The images were processed using MATLAB. In brief, standard MATLAB image processing functions were used to first convert the original images to binary images shown in Figure 3. The binary images allowed us to simply scale the images (from image coordinates to the one used in the field) and to detect the features from them. These properties allowed a straightforward way of using the sonar images for verification and as a support for interpreting the sonar data.

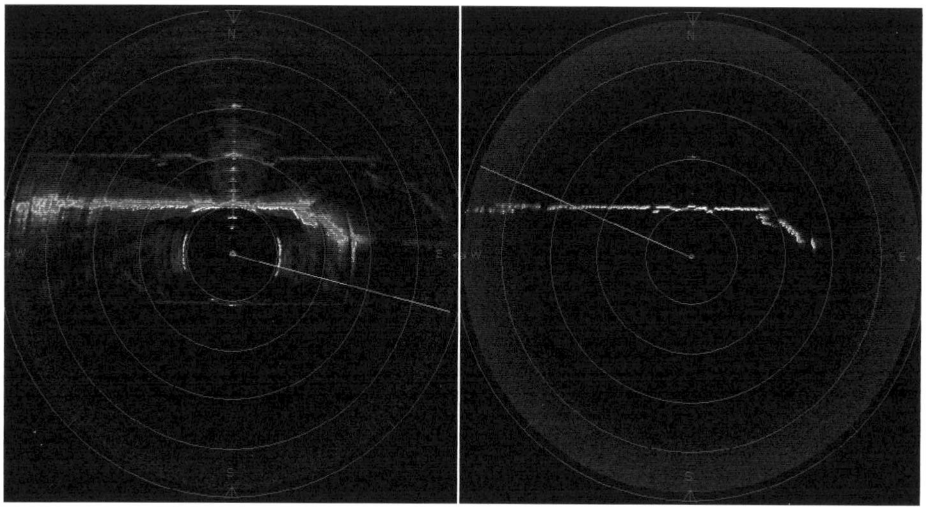

Figure 2. Raw images from the sonar measurements with a frequency of 900 Hz (left) and 1200 Hz (right) measured at the center line (CL) of ice station IS5. The circles indicate the distances from the sonar head with five-meter range.

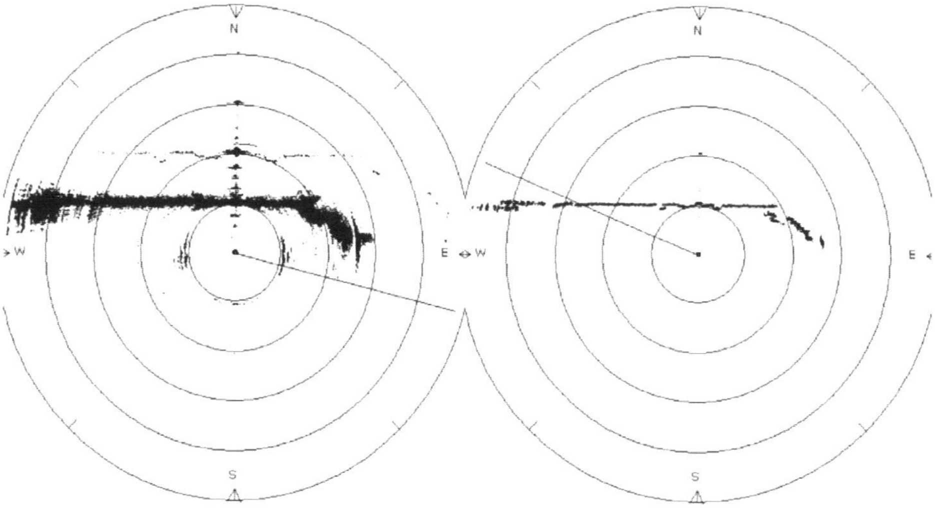

Figure 3. The binary images created from the sonar produced images of Figure 2. These images were created using standard MATLAB image processing functions for further processing of sonar data. The colors of the figures are here inverted.

DESCRIPTION OF THE VOYAGE

The voyage launched from Helsinki on February 29, 2016. Figure 4 presents the ship route during the expedition, and the general ice conditions in the Baltic Sea during the voyage. Due to the mild winter, sea ice existed only in the Northern part of the Baltic Sea, and thus the expedition headed to Bothnian Bay. The first ice was encountered near Tankar lighthouse (63.95° N, 22.85° E) on March 1. Some measurements were conducted in the early morning of March 2, after which the ship headed towards Oulu for researcher and crew exchange, and arrived there on March 3. The ice conditions were initially easy, but the severity of the conditions increased towards the north as the thickness and deformation level of the ice increased. The ship got stuck few times and had to ram through the heavier deformations, and circle around the heaviest conditions.

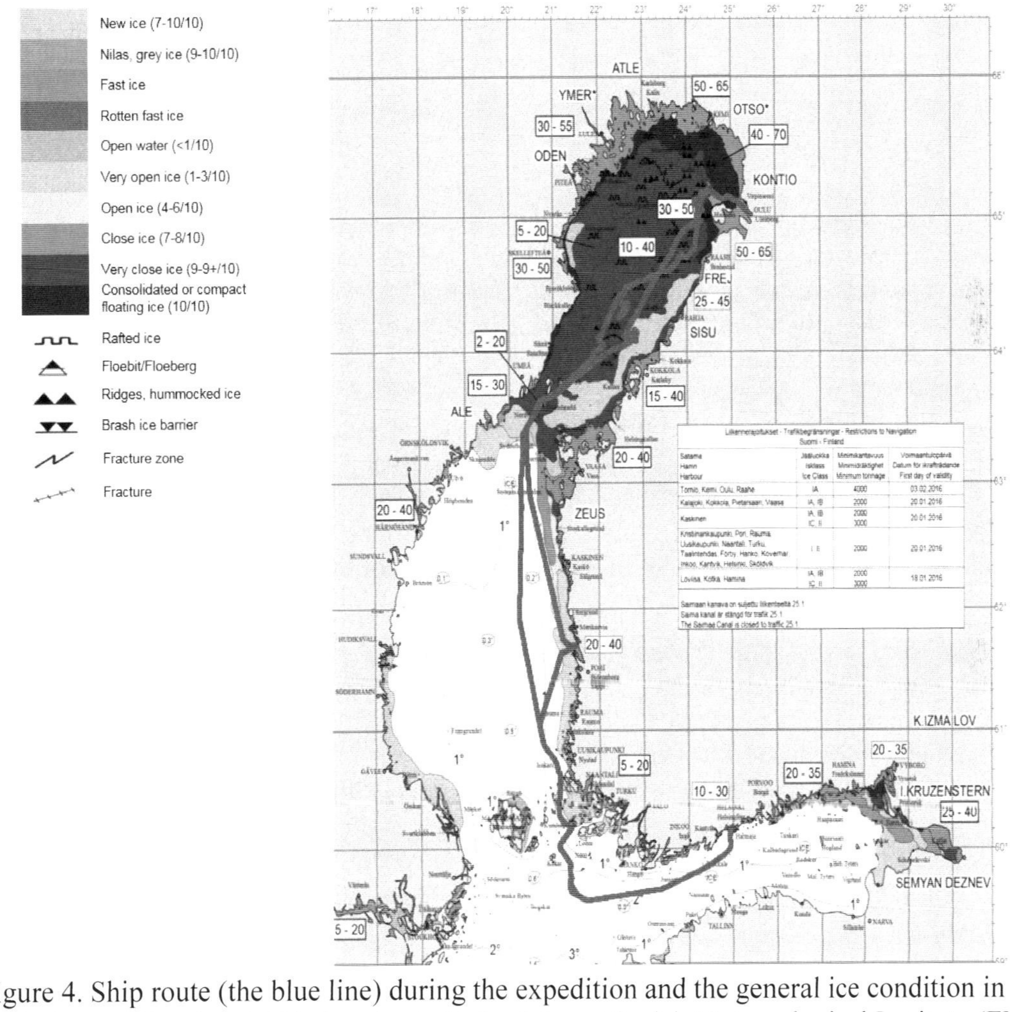

Figure 4. Ship route (the blue line) during the expedition and the general ice condition in the Baltic Sea on March 1, 2016. Ice chart obtained from Finnish Meteorological Institute (FMI).

After a short docking in the harbor, Aranda headed to the Northern part of Bothnian Bay. The following days followed the same pattern, the locations were searched from the drift ice areas from the evening to the morning, and the field measurements were conducted during the daylight time. All together, six ice stations IS1-IS6 were established at locations summarized by Table 1. Due to the mild winter, the level ice thickness was low for the time of the year. However, the ice stations were established in areas of deformed ice, where only little level ice existed (see Figure 4). After IS6, in the afternoon of March 9, the ship headed towards Pori where she arrived in the morning of March 10. The ship then continued towards Helsinki arriving to there in the evening of March 11.

Table 1. Times and locations of the Ice Stations.

Station	Date	Time (UTC)	Latitude	Longitude
IS1	Mar 4, 2016	11:09	65.11° N	24.19° E
IS2	Mar 5, 2016	7:54	65.05° N	24.12° E
IS3	Mar 6, 2016	8:38	65.09° N	24.18° E
IS4	Mar 7, 2016	7:28	64.41° N	23.52° E
IS5	Mar 8, 2016	7:15	64.21° N	22.29° E
IS6	Mar 9, 2016	8:03	63.58° N	22.06° E

RESULTS AND DISCUSSION

Figure 5 presents the results from the drilling measurements of ridge profiles at IS1-IS5. The sonar was employed in three stations, IS4-IS6. Initially we intended to perform sonar measurements at all of the stations, but due to an unfortunate accident, the sonar broke at IS1 and was under reparation at IS2 and 3. At IS1, the width of the ridge was actually greater than presented by Figure 5. These drilling measurements were conducted for validation of the sonar data, and the original plan was to measure the total width with the sonar before it broke. The measurements at IS6 are not reported here, as they included only sonar measurements, accompanied with a few additional drilling measurements for verification. At IS1 and IS2, the drilling included only the measurements on the keel profile. At IS3 – IS5 the measurements included both, the sail and the keel. As shown by Figure 5, the maximum depth of the keel was between three and four meters in all of the stations. The widths of these keels varied from 10 meters up to 25 meters.

The ridges were chosen so that the ship would be able to go through them with a single try. The ship moved to the positive x-direction (see Figure 5). This was achieved at IS2, IS4, IS5, and IS6. In IS3, there was thick layer of rafted ice in front of the ridge, which hindered the ship. In this case, the ridge resistance was thus not only dependent on keel thickness, but also due to these rafted layers, which are common for ice ridges (Tuhkuri and Lensu, 1999, 2002). As a result, she was not able to proceed through the actual ridge, but instead had to ram several times in order to get through it. Figure 6 shows a three-dimensional presentation of the ridges at IS3 – IS5. This type of data could be straightforwardly used as an input data for simulations of ice loads involving ice rubble or ridge keels.

Figure 7 shows the comparison between the sonar and drilling measurements, plotted on the top of the gray scale image of the ridge keel. As can be noted, the different measurements show fair correspondence, as was also seen in the similar measurements by Heinonen et al. (2015). The figures have the coordinate system translated so that the bottom of the intact ice sheet aligns with $y = 0$. This is due to the automated processing of the sonar images where the goal was to detect the bottom of the ice sheet and set it to $y = 0$. Hence, having the horizontal dark area aligning with $y = 0$ demonstrates that this type of automated image processing is fairly straightforward to implement with standard tools. Naturally, if the distance from the bottom of the ice sheet to water level is known, the translation back a coordinate system that has the water level at $y = 0$ is trivial. This distance can be always measured from the same borehole that is used for the sonar.

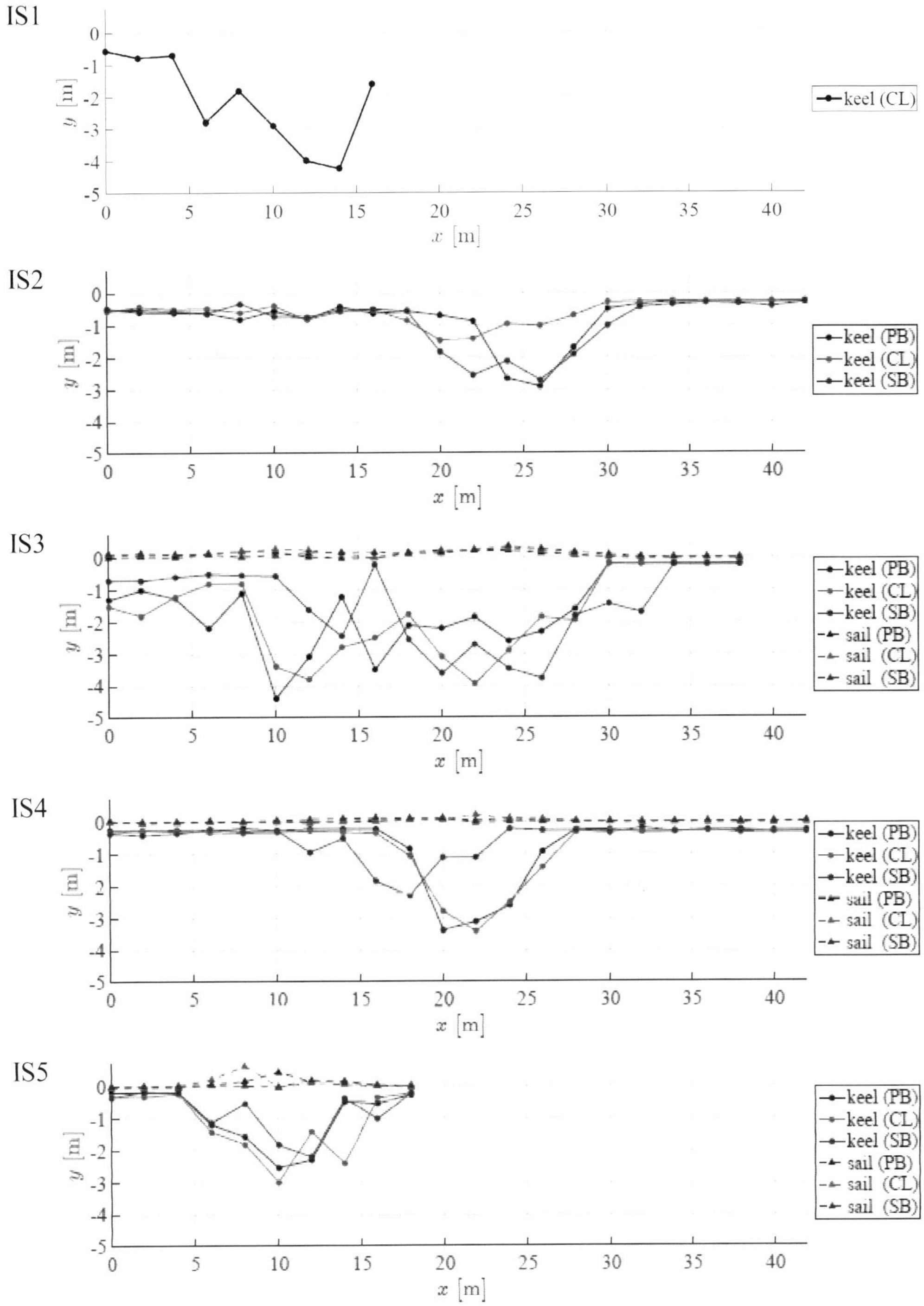

Figure 5. Results of the ridge keel profile measurements done by drilling measurements at ice stations IS1-5. Abbreviations used in the legend: PB = port side, CL = center line, and SB = star board. The horizontal distance between the drilling lines, SB-CL and CL-PB, was approximately 5 meters.

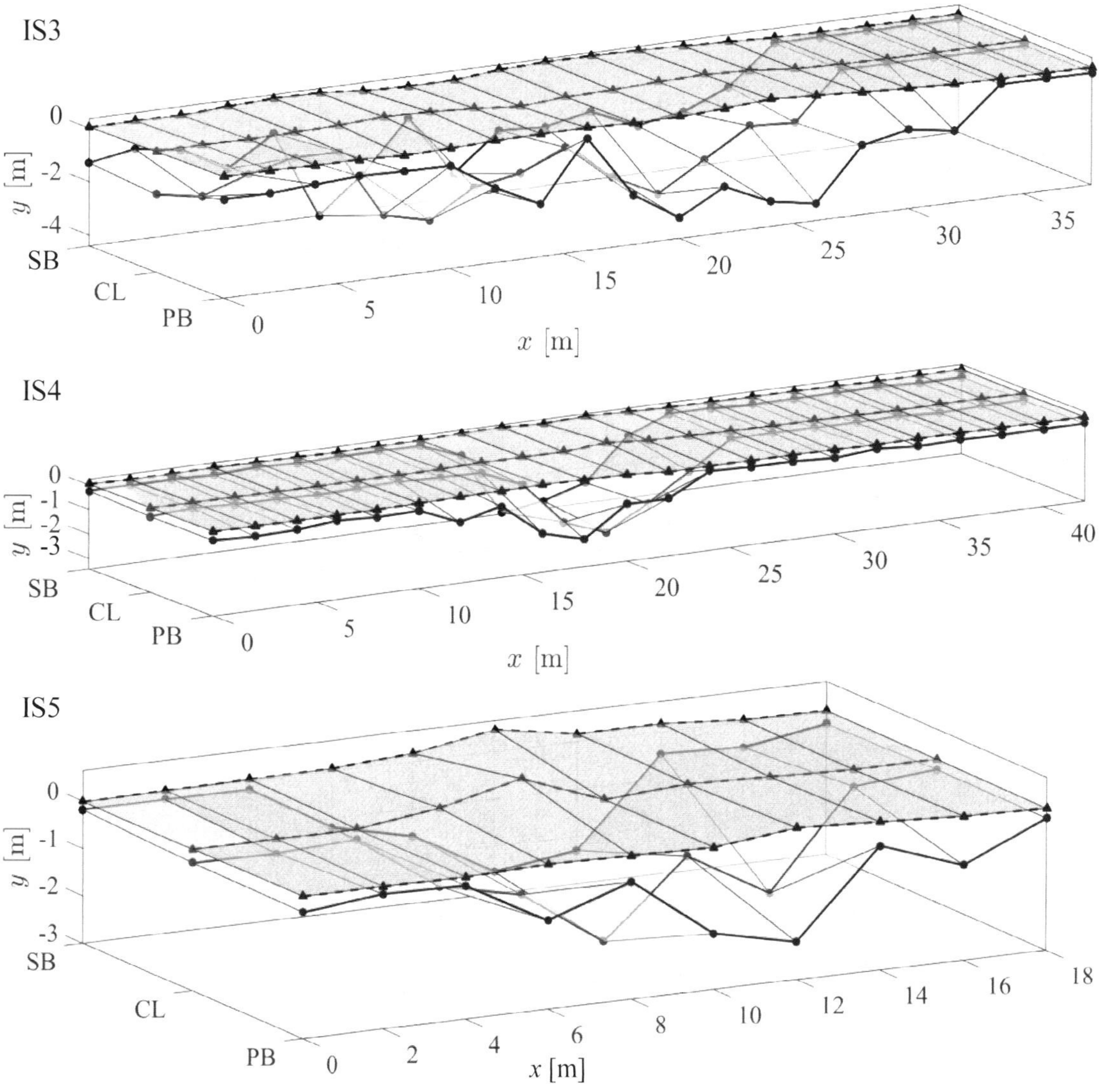

Figure 6. Data from drilling measurements at IS3-5 collected into three corresponding 3D surface plots.

As Figure 7 also illustrates, often it appeared that the sonar data would have given higher keel depth values than the drilling. For the star board (SB) line of keel in the figure, the drilling has likely bypassed an upwards-turned ice block close by. This block could lead to an over estimate of the keel depth (assuming that the drilling gives the ground truth data). We noticed, that in some cases the sonar data was difficult to interpret, and believe, that the properly processed image data, when available, should be used as support of the measurements. This was especially seen from the data from one of our ice stations, which showed that even small interferences in the echoes picked up by the sonar may lead to severe error in the measurements.

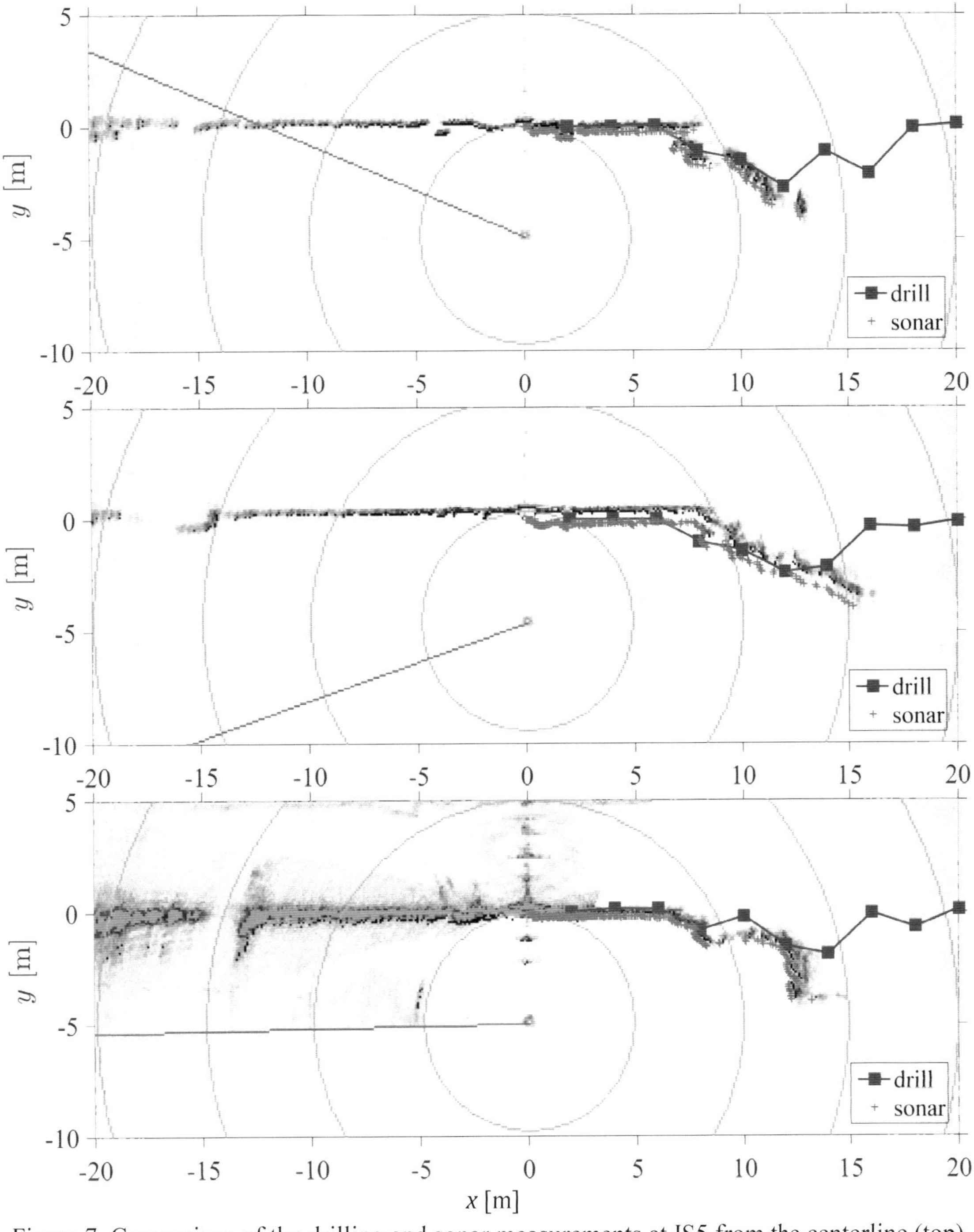

Figure 7. Comparison of the drilling and sonar measurements at IS5 from the centerline (top), port side (middle), and star board (bottom) drilling lines. The '+' markers are the data points given by the sonar software. The data is from the measurements with sonar head frequency 1200 Hz. Here the data is translated so that the bottom of the intact ice sheet aligns with $y = 0$.

CONCLUSIONS

The paper presented ridge profiles from five different locations in the Gulf of Bothnia determined by drilling. The maximum depth of the ridge keels varied from three to five meters, and the width of the cross sections from 10 to 25 meters. It should be note that ridges were chosen so that the ship would be able to get through them with a single try. Thus, the choosing was subjective and the ridges do not give typical representation for the ridge profiles in the area. The study showed a fairly good correspondence between the drilling and sonar measurements, but also suggests that the sonar image data, if available, should be used as a support when interpreting the sonar data. The remaining data will be processed and studied in the near future. The collected data is to be employed in the model scale measurements in Aalto Ice Tank when ridge resistance is studied in model scale.

ACKNOWLEDGEMENTS

We are grateful for the support from the Academy of Finland through the project Kara-Arctic Monitoring and Operation Planning Platform (KAMON) and Lloyd's Register Foundation through CEARCTIC project. The Lloyd's Register Foundation supports the advancement of engineering-related education and funds research and development that enhances the safety of life at sea, on land, and in the air. Authors would like to thank Jani Pötrönen and Juho Toijanaho from Meritaito for their help with practical matters related to operating the sonar. The help of Risto Haimelin (Aalto University), Eero Rinne (Finnish Meteorological Institute, FMI), Heidi Sallila (FMI), Henna-Reetta Hannula (FMI), Iiro Kokkonen (FMI), Rachel Tilling (University College London, UCL), Thomas Armitage (UCL), and Thsikana Rasehlomi (University of Cape Town) in the drilling is also acknowledged.

REFERENCES

Gong, H., Polojärvi, A., Tuhkuri, J., 2017. Preliminary 3D DEM Simulations on Ridge Keel Resistance on Ships. *To appear in the Proceedings of the 24th International conference on Port and Ocean Engineering under Arctic Conditions (POAC'17).*

Heinonen, T., Oikkonen, A., Pötrönen, J., 2015. Fast methods for measuring ice ridges in full-scale. *Proceedings of the 25th International Ocean and Polar Engineering Conference.*

Kankaanpää, P., 1991. Morphology of sea ice pressure ridges in the Baltic Sea. Licentiate thesis. University of Helsinki. Department of Geography. Helsinki. 119 p.

Kuuliala, L., Kujala, P., Suominen, M., Montewka, J., 2017. Estimating operability of ships in ridged ice fields, *Cold Regions Science and Technology*, 135, 51-61.

Polojärvi, A., 2013. Sea ice ridge keel punch through experiments: model experiments and numerical modeling with discrete and combined finite-discrete element methods. Doctoral Thesis. Aalto University publication series Doctoral Dissertations, 50/2013. 59 p.

Polojärvi, A., and Tuhkuri, J., 2009. 3D discrete numerical modeling of ridge keel punch through tests. *Cold Regions Science and Technology*, 56, 18-29.

Polojärvi, A., and Tuhkuri, J., 2013. On modeling cohesive ridge keel punch through tests with a combined finite-discrete element method. *Cold Regions Science and Technology*, 85, pp. 191-205.

Strub-Klein, L., Sudom, D., 2012. A comprehensive analysis of the morphology of first-year sea ice ridges, *Cold Regions Science and Technology*, 82, pp. 94-109.

Timco, G., Burden, R., 1997. Analysis of the Shapes of Sea Ice Ridges, *Cold Regions Science and Technology*, 25, pp. 65-77.

Veitch, B., Lensu, M., Riska, K., Kosloff, P., Keiley, P., Kujala, P., 1991. Field observations of ridges in the northern Baltic Sea. *Proceedings of the 11th International Conference on Port and Ocean Engineering under Arctic Conditions*. Vol. 1, pp. 381-400.

Tuhkuri, J., Lensu, M., 1999. Laboratory and Field Studies on the Mechanics of Ice Ridge Formation. *Proceedings of the 15th International Conference on Port and Ocean Engineering under Arctic Conditions*. Vol. 3, pp. 1118-1129

Tuhkuri, J., Lensu, M., 2002. Laboratory tests on ridging and rafting of ice sheets. *Journal Of Geophysical Research*, 107, No. C9, 3125

Proceedings of the 24th International Conference on
Port and Ocean Engineering under Arctic Conditions
June 11-16, 2017, Busan, Korea

Recent Ice Conditions in North-Norwegian Porsangerfjorden

Chris Petrich[1], Megan E. O'Sadnick[1], Line Dale[2]

[1]Northern Research Institute (Norut) Narvik, Narvik, Norway

[2]Norwegian Water Resources and Energy Directorate, Narvik, Norway

ABSTRACT

Ice is a common occurrence in mainland Norwegian fjords in the vicinity of river mouths in winter. Fjord ice provides a wide array of services to society, including (1) acting as a coastal buffer (e.g. from pollution) and marine hazard; (2) providing a platform for recreational activities (e.g., travel and fishing); (3) serving as a full-scale model for Arctic sea ice (e.g., response training); and (4) participating in the fjord and coastal ecosystems. In spite of the local importance of ice in mainland Norwegian fjords, systematic investigations of its seasonal development and geophysical properties appear to be scarce. This study presents a summary of ice conditions in Porsangerfjorden from satellite observations of the past two decades (2000-2016) and discusses these in the context of local air temperature data and regional measurements of flow rates in rivers. Maximum ice extent was observed in March and a useful first approximation of ice extent in March could be derived from freezing degree days of February. However, local distribution of ice appears to be governed by factors beyond air temperature.

KEY WORDS: Ice extent; Fjord; Freezing degree day; River discharge; Wind

INTRODUCTION

While the development of ice in Norwegian fjords has enjoyed considerable scientific attention in fjords of Svalbard (e.g., Gerland et al., 1999; Haarpaintner et al., 2001; Høyland 2009), there seems to be a lack of published studies on ice in mainland Norwegian fjords. Norway's climate is influenced by the North Atlantic (Seager et al., 2002), and ice conditions in fjords of Svalbard have been linked to water temperatures (Pavlov et al. 2013). Fjords are mostly ice-free along the coast of mainland Norway. However, according to local common lore, ice may be found in bays, coves, or at the end of fjords if there is a flow of freshwater from a river or creek leading into the fjord and if the winter is sufficiently cold. In various parts of the world local ice formation in larger water bodies has been attributed to a freshwater layer overlying saltwater that acts to suppress mixing and reduced heat exchange throughout the water column (e.g., Nof et al., 2006). In Norway, this form of ice is probably most often exploited as a platform for fishing, but it has also found entry into cultural entertainment (e.g., Markussen, 2001). Similar to sea ice, ice in fjords is of wide relevance to society (cf. Eicken et al., 2009). Fjord ice provides a wide array of services to society, including (1) acting as a coastal buffer (e.g. from pollution) and marine hazard; (2) providing a platform for recreational activities (e.g., travel and fishing); (3) serving as a full-scale model for Arctic sea ice (e.g., response training); and (4) participating in the fjord and coastal ecosystems.

In an attempt to provide a systematic assessment of system services of fjord ice, a preliminary analysis of ice conditions in one particular north Norwegian ford has been performed in this work. At this point, the analysis is based on remote sensing data and data archives with only limited input on specific local conditions. If successful, such an approach could be easily transferred to a larger number of fjords.

While ice formation and break-up of coastal sea ice has been linked to both generally warm weather and solar irradiance (e.g., Petrich et al., 2012 and references therein), predictive power is often found in simple freezing degree day assessments that only consider air temperature, or in statistical correlation with a wide range of potential parameters (e.g., Petrich et al. 2014 and references therein).

In this study, ice conditions in Porsangerfjorden were assessed from MODIS satellite images and correlations were attempted with a number of basic weather parameters.

METHODS

Ice conditions were investigated at the southern end of Porsangerfjorden, approx. 70.12° N, 25.0° E. Porsangerfjorden is split into two arms by Oldereidneset peninsula in the south, Vesterbotn in the West and Østerbotn in the East (Figure 1). Either or both arms may contain ice. The maximum northern ice extent is a few km north of Austmo. Two significant rivers end in Vesterbotn, one at its southern end at Banak airport and another on in Austmo. No information about significant river influx into Østerbotn is available.

Daily river flow data of the Norwegian Water Resources and Energy Directorate were available for Skoganvarre (through 2009 and in 2012) and Lombola measurement stations, representing the discharge at Banak and Austmo, respectively. Both measurement stations are located at fast-flowing points and do not usually freeze in winter. In addition, flow data were manually quality controlled by NVE to account for potential problems due to ice formation in winter. The rivers are fed by separate catchments. Since flow rates were measured upstream of the respective river mouths, absolute discharge rates into the fjord may be underestimated but are expected to be proportional.

Sea ice conditions were derived from data of the Moderate Resolution Imaging Spectroradiometer (MODIS) on board satellite TERRA. MOD09 Surface Reflectance product data were used from collection 6 (Vermote et al., 2015). The product is an estimate of the surface spectral reflectance for each band as it would have been measured at ground level as if there were no atmospheric scattering or absorption. It corrects for the effects of atmospheric gases and aerosols. False-color images were generated of bands 3-6-7 to facilitate the distinction between snow/ice (appearing red) and clouds (appearing white). For the present application, the product gives weekly ground conditions at a nominal pixel resolution of 500 m that are mostly cloud free. Light conditions are sufficient for reflective bands from early February. The ice season in Porsangerfjorden starts typically in February (earlier in some years) and ends in April. Maximum ice extent is observed in March.

Local air temperature data were obtained from three different sources: (1) 6-hourly eKlima weather station data of Banak airport provided by the Norwegian Meteorological Institute (mostly continuous record since 2005), (2) daily average gridded air temperature interpolation of seNorge (extracted at the eastern and western shores of Østerbotn and Vesterbotn, respectively), and (3) 3-hourly 0.75°x0.75° ERA40 Interim reanalysis, interpolated to Oldereidneset (71.12° N, 25.0° E), i.e. in the center of region of interest. Air temperatures were used to calculate monthly freezing degree days.

10 m surface wind data were analysed from ERA40 Interim since the airport weather station is located near a steep topographic rise (600 m). Monthly wind roses were generated from data interpolated at Oldereidneset.

Solar irradiance reanalyses of ERA40 Interim were investigated but will not be discussed due to a lack of local measurements for validation.

1°x1° gridded monthly measured (ship-borne) sea surface temperatures at the northern end of Porsangerfjorden were extracted from the International Comprehensive Ocean-Atmosphere Data Set (ICOADS). Data of two grid cells were averaged, representing 71°-72°N and 25°-27°E. The data are averaged ship-borne observations of unknown and varying accuracy.

RESULTS

Ice Conditions

Ice extent was classified in southern Porsangerfjorden separately for the eastern part (Ø) and western part (V) (Figure. 1). Distinction was made between essentially no ice (0), ice not extending Oldereidneset peninsula (1), ice not extending past Austmo (2), and ice extending past Austmo (3). An area of open water (>1 km^2) was visible offshore of Austmo when ice extended past Austmo – we cannot say whether this is a result of water from Lombola or effluent since area of open water existed each year. Maximum ice extent was observed in March although a few years had similar ice extent already in late February at the beginning of the period of satellite observation. The ice cover generally appeared to be solid (bright red in Figure. 1) although in particular the eastern section showed signs of thin ice or open ice cover (dark red as seen in section Ø3 in Figure. 1). The classification of ice conditions in Tab. 1 shows that maximum ice conditions range from as low as 1/0 (V/Ø) to as high as 3/3. Approximately half of the years saw very little or no ice in the eastern part (Ø), while the western part had ice (V).

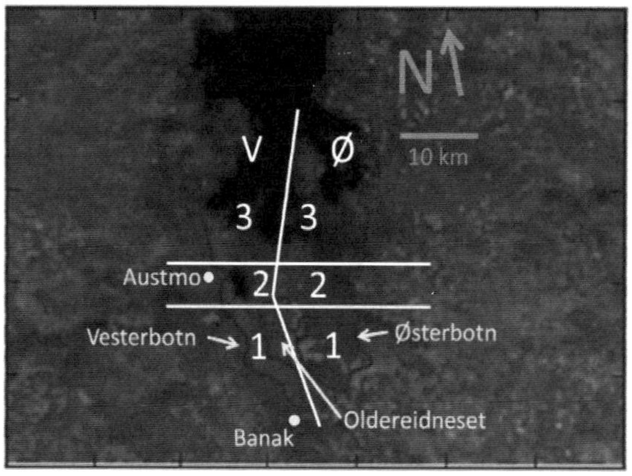

Figure 1. Classification of ice conditions at the southern end of Porsangerfjorden (blue outline) from MODIS false-color images used in Table 1. Red indicates snow or ice. Note the characteristic area of open water offshore Austmo. Image of 6 Mar 2001.

Freezing Degree Days

Calculated freezing degree days agree generally between data sources for all years (Tab. 1). Banak weather station data resemble seNorge Vesterbotn data, and seNorge Østerbotn data resemble ERA Interim data. seNorge Vesterbotn is consistently warmer than seNorge Østerbotn by an equivalent of about 1 °C per day. Notable disagreement exists only between seNorge Østerbotn data and ERA Interim in 2004 where seNorge suggested a notably colder month then ERA.

With few exceptions, ice extent seems to correlate with ERA Interim temperature data. Ice extent never exceeded category 1 in years with February FDD>-150 °C days, and only reached category 3 on either side of the fjord in years with FDD<-300 °C. Exceptions are 2008 which had ice extent lower than expected from February FDD, and 2003 which had an ice extent larger than expected from February FDD. 2004 and 2006 had almost identical ERA FDDs but considerably different ice covers. It appears as though either 2004 had too little ice given the February temperatures or 2006 had too much. A more close-up look at the temporal development of air temperatures and ice cover may explain the differences between 2004 and 2006.

Table 1. March ice extent in Østerbotn (Ø) and Vesterbotn (V), and freezing degree days (FDD) in February (°C days). In 2002, ice extent reached category 3 in Østerbotn but the ice cover appeared to be rather dispersed than continuous, hence the category is put in brackets. FDDs are given for Banak weather station (wx), seNorge (V, Ø), and ERA Interim (ERA). March flow rates at Lombola and Skoganvarre are indicated as low (-), medium (o), and high (+) as defined in the text.

Year	V	Ø	FDD wx	FDD V	FDD Ø	FDD ERA	Lom	Sko
2000	2	2		-205	-229	-245	+	+
2001	3	3		-292	-317	-334	-	-
2002	2	(3)		-216	-240	-243	+	+
2003	**2**	**3**		**-100**	**-120**	**-121**	+	**o**
2004	**2**	**0**		**-310**	**-347**	**-277**	**o**	**o**
2005	2	0	-168	-164	-183	-190	**o**	**o**
2006	**3**	**3**	**-235**	**-238**	**-266**	**-261**	-	**o**
2007	3	1	-396	-376	-408	-415	-	**o**
2008	**1**	**0**		**-164**	**-186**	**-205**	**o**	**o**
2009	3	3	-268	-294	-323	-316	-	-
2010	3	3	-400	-385	-415	-397	-	nd
2011	3	3	-372	-357	-389	-402	+	nd
2012	3	3	-266	-261	-292	-301	**o**	+
2013	2	0	-177	-184	-205	-197	+	nd
2014	1	0	-68	-71	-83	-108	+	nd
2015	2	0	-136	-139	-155	-178	+	nd
2016	2	0	-167	-177	-194	-196	nd	nd

River Discharge

Discharge rates measured in Lombola and Skoganvarre vary widely between years (Figure. 2). While flow rates are positively correlated between the stations in some years (2001), other years show anticorrelation (2003). Changes between positive correlation and anticorrelation can be found throughout the record since the 1920s (not shown) and may be related to wind and precipitation patterns across the respective catchments. Flow rates between months were generally correlated with low January flow rates corresponding to low flow rates in February and March.

Tab. 1 indicates whether March flow rates were ≤2.5 (-) or >3.0 m³/s (+) at Lombola, and ≤3.0 (-) or >4.5 m³/s (+) at Skoganvarre gauges. There is a general tendency for low flow rates to be associated with high ice extent in Vesterbotn.

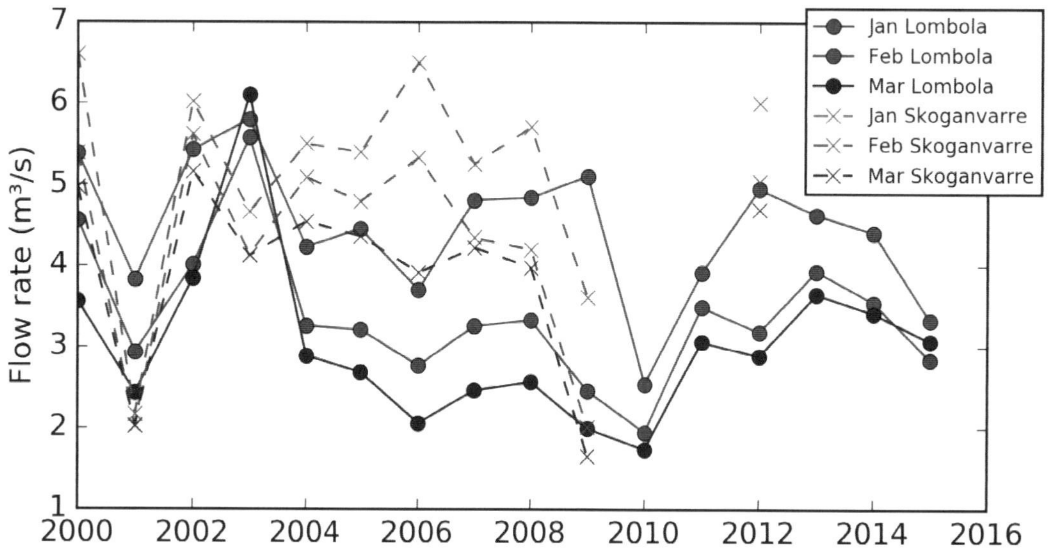

Figure 2. River discharge rates at the measurement stations averaged separately over months January, February, and March. Note that the vertical axis does not start at 0.

Winds

The local wind patterns have been assessed through monthly wind roses. Winds from south tend to dominate in the region through the years. No immediately obvious patterns were found explaining the years of no ice in Østerbotn. However, Figure 3 shows the winds for two years that had great ice extent (level 3), i.e. ice past Oldereidneset, on only one side of the fjord (Tab. 1, 2003 and 2007; Figure 4). In these years, there should have been sufficient ice available for drift. Figure 3 shows that February 2007 experienced winds from South-East while February 2003 experienced winds from West, which is consistent with ice distributed west and east in 2007 and 2003, respectively (Figure 4). Of all years investigated, 2003 had the strongest contribution from West in February in terms of both frequency and wind speed.

While years 2008 and 2014 had the smallest ice extent, 2008 was reasonably cold (Tab.1, Figure 6). Figure 5 shows that 2008 did not seem to have experienced persistently strong southerly winds that could have blown out the newly forming ice.

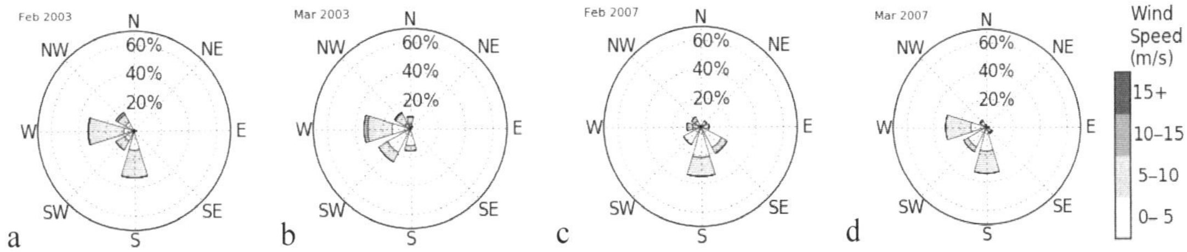

Figure 3. Wind conditions in February and March of two year that have large ice extent on predominantly one side of the fjord. 2003: East, 2007: West. (a) Feb 2003, (b) Mar 2003, (c) Feb 2007, (d) Mar 2007.

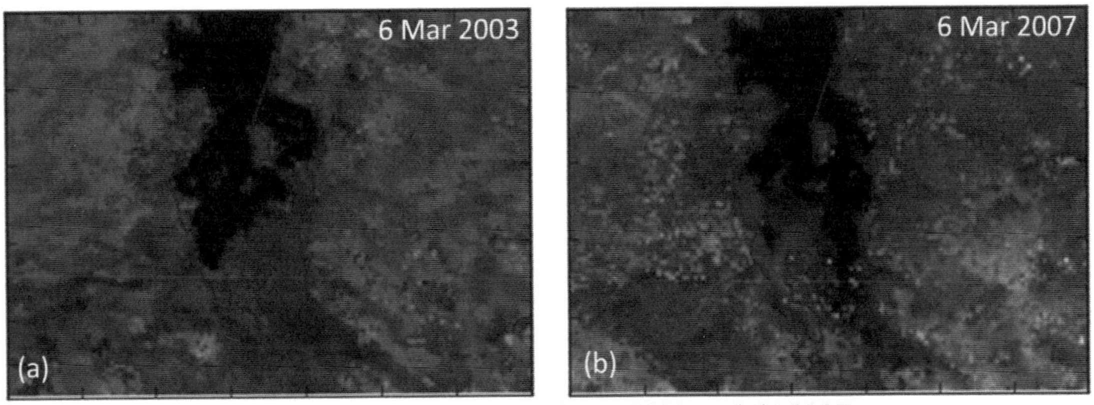

Figure 4. March ice extent in (a) 2003, and (b) 2007.

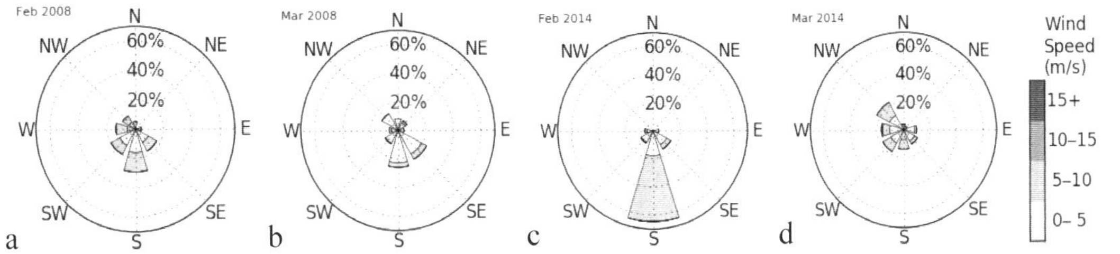

Figure 5. Wind conditions in February and March of the two years with the smallest ice extent. (a) Feb 2008, (b) Mar 2008, (c) Feb 2014, (d) Mar 2014.

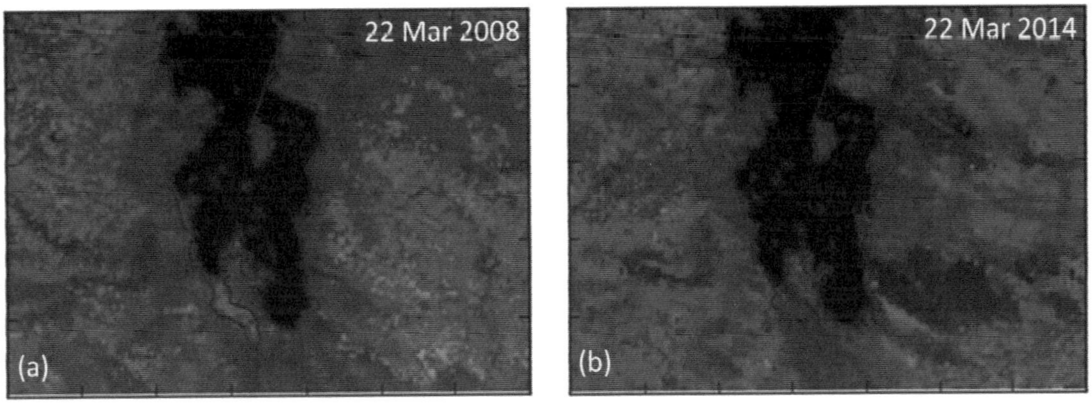

Figure 6. The two years of smallest March ice extent. (a) moderately cold year 2008, and (b) warm year 2014.

Sea Surface Temperature

The monthly variation in measured sea surface temperatures at the northern end of Porsangerfjorden is shown in Figure 7. While the data may enable the derivation of long term trends, they seem to be too noisy for comparisons between individual years. No further analysis of these data has been attempted.

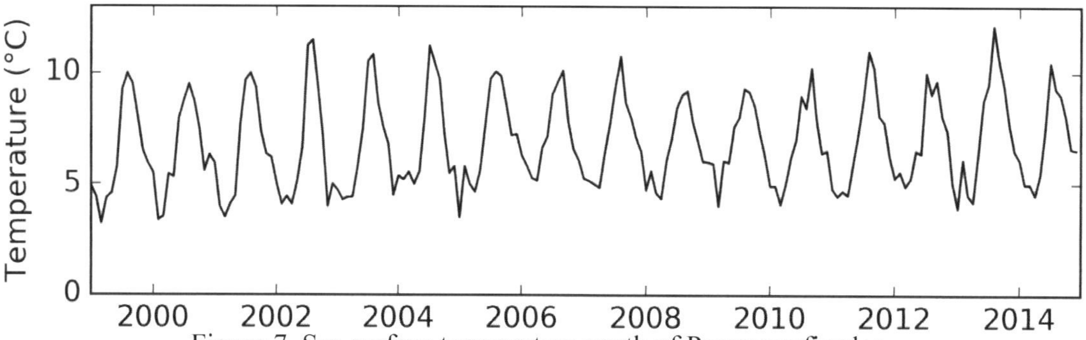

Figure 7. Sea surface temperature north of Porsangerfjorden.

DISCUSSION

Freezing degree days (FDD) can be directly related to ice thickness (Petrich and Eicken, 2017). However, the relationship to ice extent is more indirect. Extended periods of cold with a reasonable amount of wind increase heat removal from the water body, lowering its temperature closer to freezing and allowing ice to cover a larger area. Important parameters in such considerations are water depth and tides, two parameters that have not been addressed in this study. Earlier studies in a lagoon revealed a correlation between ice edge and bathymetry (Petrich et al., 2014), illustrating its relevance when quantitative relationships are being sought and when models developed for one area are to be applied to a different area.

Ice conditions were classified as 1/0 and 2/0 in 2008 and 2013, respectively, yet FDD were similar to each other in those years (Tab. 1). However, the difference between ice conditions of categories 1 and 2 can be quite small in Vesterbotn. It is conceivable that a short period of winds from the right direction leads to different assessments in MOD09 satellite products since floe size and separation may be smaller than image resolution. In that sense, ice conditions in 2008 and 2013 indicate that the monthly FDD approach has limitations in what can be resolved.

Ice extent in 2003 was considerable while February was similarly warm as 2008 (Tab. 1). Tab. 2 shows a more detailed FDD history going back to January. A significant difference between 2003 and 2008 was the very cold January in 2003, which was the coldest in the period studied. However, this alone cannot explain the large ice extent in 2003 since 2016 also had a very cold January but saw only a limited amount of ice in the fjord (Tab. 1, 2). The large ice extent in 2003 is of yet unexplained.

Years 2004 and 2006 had similar February FDDs yet different ice extent at the beginning of March. Tab. 2 shows that this difference may be due to extremely different temperatures during the first week of March. This indicates that monthly FDD are limited in their predictive power of ice extent since ice grows and decays over shorter characteristic time scales. This also indicates that ice has probably not reached particularly high thickness by 6 March 2006.

There was no correlation between river flow rates and ice extent apparent. This was somewhat surprising given the general importance of a freshwater layer to ice formation.

However, river flow rates are dependent on air temperature through snow melt and thus not independent of air temperature. Hence, to identify the impact of river flow rate, the flow rate data may have to be reduced by the effect of air temperature before a correlation with observations is attempted. To address the episodic absence of ice in Østerbotn, influx of water at the east side of the fjord should receive more scrutiny.

It was possible to find two examples where winds may have contributed significantly to the redistribution of ice in the fjord (2003/2007).

Geophysical factors that may be relevant for an understanding of ice extent but were not investigated include: solar radiation, snow cover, fjord bathymetry, tidal flow, ocean heat flux, river water temperature, and wind mixing of the water column. Also, short-term variation of parameters (weeks), in particular strong winds but also cold spells or hot spells, should be investigated in greater detail.

Table 2. Comparison of ERA Interim FDDs for selected years. Note the different lengths of the intervals.

Year	1-31 Jan	1-14 Feb	15-end Feb	1-7 Mar
2003	-455	-125	4.5	-39
2004	-270	-150	-127	-43
2006	-205	-144	-117	-144
2008	-196	-67	-138	-88
2014	-362	-58	-50	-16
2016	-418	-93	-103	-40

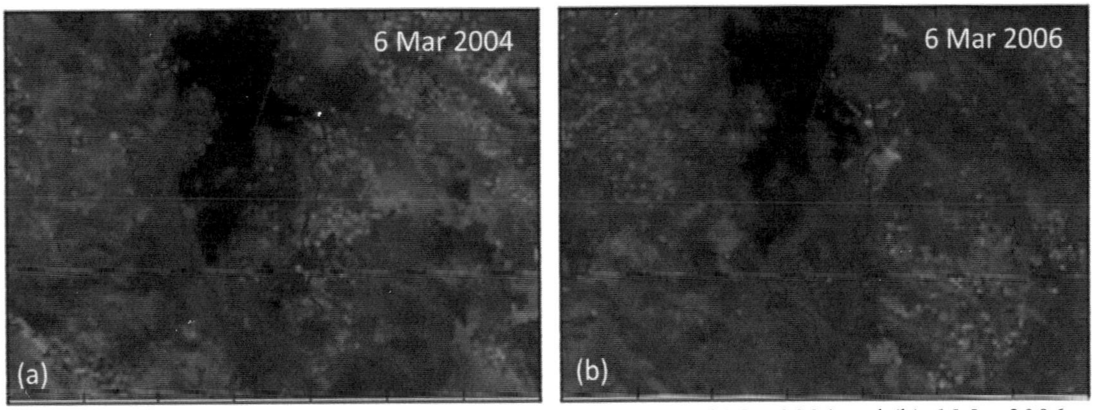

Figure 7. Ice extent in years of similar February FDD, (a) 6 Mar 2004 and (b) 6 Mar 2006. Same dates are shown for comparison but note that in 2006 ice extent was higher at the west side of the fjord and ice appeared more consolidated by 22 March. In 2004, (a) represented the greatest ice extent.

CONCLUSION

Reconstruction of local ice conditions becomes possible once the mechanisms governing ice conditions are understood and suitable environmental data are available. It has been shown that this is possible in principle for saltwater lagoons, and the present work suggests that ice conditions in fjords may also be amendable to prediction based on environmental conditions. Most of the variance of fjord ice extent in Porsangerfjorden in March seems to be explained by freezing degree days in February. Based on the investigated years 2000 through 2016, a relationship has been found that held in good approximation all winters except 2003. While the geophysical investigation into those years could easily be expanded, input from local residents should also be considered as a source of information.

To extend this work, the events that led to the comparatively high and low ice extent in 2003 and 2008, respectively, should be identified. Valuable insights may be gained by trying to address outliers. Also, ice conditions in other fjords should be investigated to assess to what degree results of this study can be generalized and applied on an operational level.

ACKNOWLEDGEMENTS

This work was supported by the Research Council of Norway center for research-based innovation (SFI) CIRFA, grant number 237906.

REFERENCES

Eicken, H., A.L. Lovecraft, M.L. Druckenmiller, 2009. Sea-Ice System Services: A framework to help identify and meet information needs relevant for observing networks. *Arctic*, 62(2), 119-136.

Gerland, S., J.-G. Winther, J.B. Ørbæk & B.V. Ivanov, 1999. Physical properties, spectral reflectance and thickness development of first year fast ice in Kongsfjorden, Svalbard. *Polar Research*, 18(2), 275–282, doi:10.1111/j.1751-8369.1999.tb00304.x.

Haarpaintner, J., J.-C. Gascard & P.M. Haugan, 2001. Ice production and brine formation in Storfjorden, Svalbard. *Journal of Geophysical Research – Oceans*, 106(C7), 14,001-14,013.

Høyland, K.V., 2009. Ice thickness, growth and salinity in Van Mijenfjorden, Svalbard, Norway. *Polar Research*, 23(8), 339-352, doi:10.1111/j.1751-8369.2009.00133.x.

Markussen, R.A., 2001. Jesus på Rombakens fjord. In *På plass* [CD], Groms Plass Brothers (artist). Tøtta Records, Narvik, Norway. (in Norwegian)

Nof, D., I. McKeague & N. Paldor, 2006. Is there a paleolimnological explanation for 'walking on water' in the Sea of Galilee? *Journal of Paleolimnology*, 35, 417-439, doi:10.1007/s10933-005-1996-1.

Pavlov, A.K., V. Tverberg, B.V. Ivanov, F. Nilsen, S. Falk-Petersen & M.A. Granskog, 2013. Warming of the Atlantic water in two West Spitzbergen fjords over the last century (1912-2009). *Polar Research*, 32, doi:10.3402/polar.v32i0.11206.

Petrich, C. & H. Eicken, 2017. Overview of sea ice growth and properties. In *Sea Ice*, 3rd ed., D. Thomas (ed.), Wiley-Blackwell, 664 pp.

Petrich, C., H. Eicken, J. Zhang, J.R. Krieger, Y. Fukamachi & K.I. Ohshima, 2012. Coastal sea ice melt and break-up in northern Alaska: processes and possibility to forecast. *Journal of Geophysical Research*, 117, C02003, 1–19, doi10.1029/2011JC007339.

Petrich, C., A.C. Tivy & D.H. Ward, 2014. Reconstruction of historic sea ice conditions in a sub-Arctic lagoon. *Cold Regions Science and Technology*, 98, 55-62, doi:10.1016/j.coldregions.2013.10.011.

Seager, R., D.S. Battisti, J. Yin, N. Gordon, N. Naik, A.C. Clement & M.A. Cane, 2002. Is the Gulf Stream responsible for Europe's mild winters? *Quarterly Journal of the Royal Meteorological Society*, 128(586), 2563-2586, doi:10.1256/qj.01.128.

Vermote, E.F., J.C. Roger & J.P. Ray, 2015. *MODIS Surface Reflectance User's Guide*. MODIS Land Surface Reflectance Science Computing Facility, 35 pages.

POAC'17
Busan, KOREA

Proceedings of the 24th International Conference on
Port and Ocean Engineering under Arctic Conditions
June 11-16, 2017, Busan, Korea

Wave Attenuation in the Barents Sea Marginal Ice Zone in the Spring of 2016

Andrei Tsarau[1], Aleksey Shestov[2, 1] and Sveinung Løset[1, 2]
[1]Sustainable Arctic Marine and Coastal Technology (SAMCoT), Centre for Research-based Innovation (CRI), Norwegian University of Science and Technology, Trondheim, Norway.
[2]University Centre in Svalbard (UNIS), Norway.

ABSTRACT

Simultaneous field measurements of the motion of several ice floes in the marginal ice zone near Svalbard were performed in April/May 2016 to study the process of wave attenuation in ice. The heave motion amplitudes registered by Inertial Measurement Units (IMUs) deployed on the ice were used to interpret the wave amplitude. Comparison of the wave amplitudes at two different locations gave the wave attenuation coefficient ranging from 0.1×10^{-4} to 3.5×10^{-4} m^{-1} for wave frequencies between 0.08 and 0.2 Hz, respectively. This is significantly lower than the attenuation coefficients reported from previous field experiments performed in the same area in 1990, which is perhaps due to the much thinner ice observed in this study compared with the ice in the previous survey.

KEY WORDS: Marginal Ice Zone; Field data.

INTRODUCTION

When ocean waves penetrate into an ice field a so-called Marginal Ice Zone (MIZ) is formed, which is the transition area where the ice cover is significantly affected by waves and open-ocean processes. Arctic shipping and engineering developments require knowledge on how ocean waves and ice may alter each other's properties. This knowledge is also needed for improving numerical models with respect to their capabilities to predict Arctic marine conditions.

It has been found in previous field studies (e.g., Wadhams et al. (1988)) that the wave amplitude decays exponentially as it travels further into an ice field and that the attenuation rate decreases with increasing wave period. However, the exact mechanism of wave attention and its dependence on the ice conditions are still not known as no theoretical model can fully explain the observations collected from various fields on both hemispheres. Squire et al. (1995) and Squire (2007) provide a comprehensive overview of the available experimental and theoretical studies related to the MIZ. One general conclusion from these studies is that wave attenuation strongly depends on the type of ice covers, and therefore measurements from one field may not be applicable to another. Another conclusion is that there are many effects that complicate field investigations and may lead to different results, e.g., wave breaking, wave generation by wind, oversplash, ice breaking and deterioration.

In the authors' opinion, there are two possible ways to overcome the above problems in MIZ studies: one is to minimise uncertainties in field measurements by focusing on one particular region, and the other is to carry out laboratory experiments in which ice effects on wave propagation would be isolated from other mechanisms. Hopefully, these two approaches can be combined at a later stage such that scale-independent relations will be established. This paper adopts the first approach with a focus on the Barents Sea MIZ and two other papers at POAC'17 (Cheng et al. (2017) and Li et al., (2017)) present experimental work following the second approach.

The first documented wave-attenuation measurements in the Barents Sea MIZ were performed in April 1990 as reported by Frankenstein et al. (2001). By measuring heave responses of individual ice floes under wave conditions, the wave attenuation rates were obtained. The present paper compares these results, along with the ice conditions at the study sites, with the most recent data from the same region.

ENVIRONMENTAL CONDITIONS AT THE STUDY SITE

The wave attenuation measurements presented in this paper were performed in the Barents Sea south-east of the Svalbard archipelago between Edgeøya and Hopen islands (Figure 1) as part of the R/V Lance research cruise that took place from 29 April to 7 May 2016. The R/V Lance was employed as the study platform for this cruise.

Ice conditions between Edgeøya and Hopen islands are determined by the East Spitsbergen Current which flows southward and brings ice formed locally in the Olga Strait and from the Arctic Ocean to eastern Svalbard (Løset and Carsten, 1996; Frankenstein et al., 2001). If the ice is compacted by wind, typically opposing the ice drift caused by current, a sharp ice edge may form. However, on the day of the experiment, May5th 2016, no distinct ice edge was observed and therefore it was decided to do measurements 15-30 km north-west from the imaginary ice edge where the ice was fully covering the water surface. Ice conditions in the area, as on April 28th 2016, are shown in the first Sentinel-1B satellite image (Figure 1b). Sentinel-1B is a European satellite carrying synthetic aperture radar, and it was launched on April 25th 2016.

The study site was obviously within the MIZ as the ice cover was fractured and pulverised by waves and consisted of approximately 70 % floes and 30 % brash ice (Figure 2). Regarding the floe size distribution, no significant variation was observed over the entire area where the sensors were deployed. The ice thickness measurements showed a variation from 0.21 to 0.44 m with an average of 0.3 m, and the ice was also rafted at some places.

The snow thickness on the ice ranged from 6-10 cm and four cantilever beam tests (Nesterov et al., 2016) in the area showed an average flexural strength of 278 ± 23 kPa and an elastic modulus of 1.10 ± 0.19 GPa. The average sea ice salinity was 6 ppt, the average ice temperature -2.2°C, average density 850 kg/m^3 and a sea water salinity of 34.5 ppt.

The visual observations of the sea state at the study site indicated that the wave height was up to 3 m; the dominant waves propagated towards the north-west at approximately 320° and this direction was nearly constant during the measurements. The water depth at the locations where the sensors were deployed and retrieved was 75 ± 5 m. Unfortunately, the wind data recorded by the on-board system are not reliable as the displayed wind direction was always biased; however, the recorded wind speed ranging between 18 and 28 knots seemed to be correct.

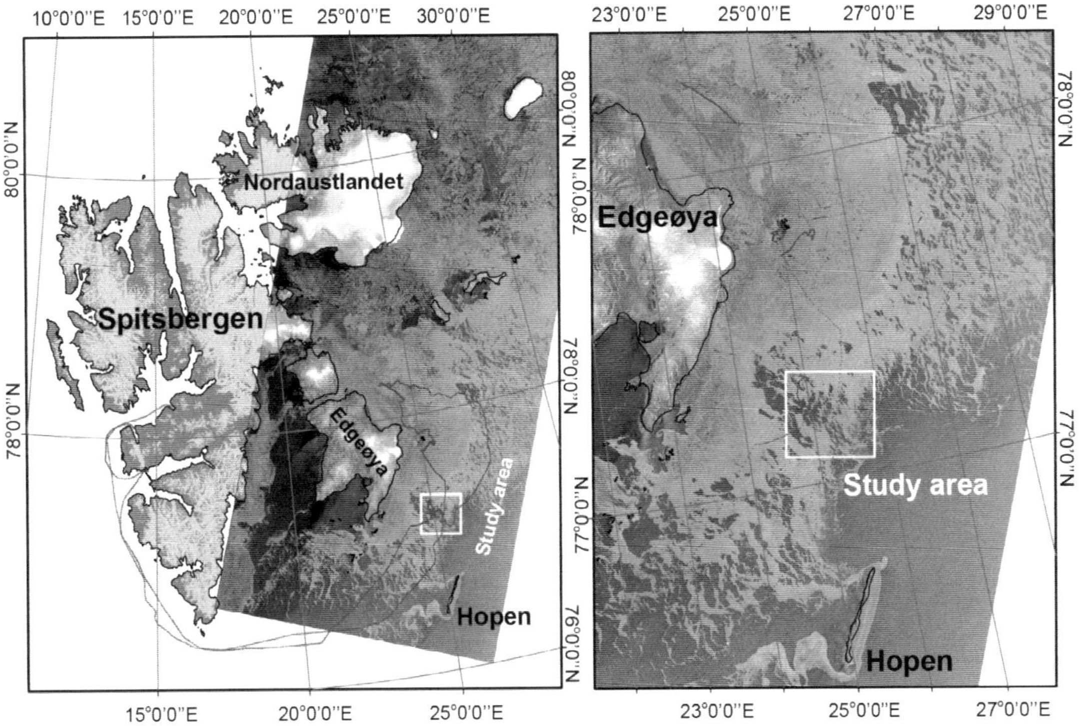

Figure 1. Track of the R/V Lance research cruise April 29th – May 7th 2016 (a). Detailed view of ice conditions in the study area (b). Ice conditions captured by Sentinel-1B satellite on April 28th. The experiment was carried out on May 5th.

Figure 2. The ice conditions at the study site (a typical floe size is 10 m).

MEASUREMENTS

Four IMU packages were deployed on individual ice floes in the MIZ 15-30 km from the imaginary ice edge in the direction of the wave propagation. Figure 3 demonstrates how the deployment was performed and Table 1 shows the measured physical parameters of each of the instrumented floes. Floe #1 was partly rafted on one side of the floe. The packages were fixed at the estimated centres of the nearly round floes to record the accelerations for approximately 10 hours. As the ice was drifting GPS trackers were employed to track, find and retrieve the IMUs at the end of the experiment. During this drift Floe #3 suffered from decay as can be seen in Figure 3 (right). This is due to the relatively thin and weak ice enabling the waves to split the floe in two pieces. The drift trajectories of the instrumented floes during the experiment are shown in Figure 4 where the instantaneous floe positions at 4:26 am and 1:22 pm (UTC), May 5th 2016, are given for calculating the distances between the floes.

Figure 3. Deploying sensors on Ice Floe #3 (left) and the same floe during retrieval of the sensors about 10 hours later (right). The latter photo shows that Floe #3 had split during the experiment.

The IMUs recorded motion data in six degrees of freedom; however, only the vertical accelerations (heave responses) were considered in this analysis. The lowest sampling frequency was 50 Hz and no filtering was applied. Figure 5 shows a typical signal from one of the IMUs (Ice Floe #4).

Table 1. Instrumented ice floes.

Ice Floe #	IMU #	Floe Diameter [m]	Floe Thickness [m]
1	6	10	0.21×2 (partly rafted)
2	2	9.5	0.27
3	5	9	0.30
4	13	12	0.32

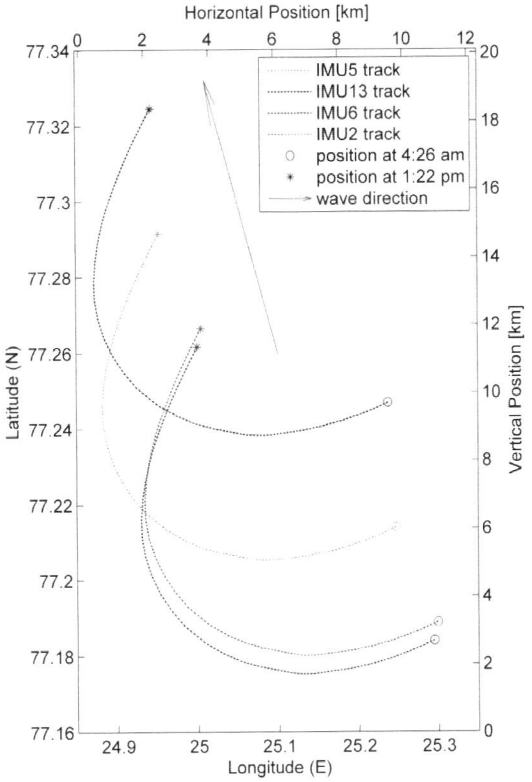

Figure 4. The drift trajectories of the instrumented ice floes.

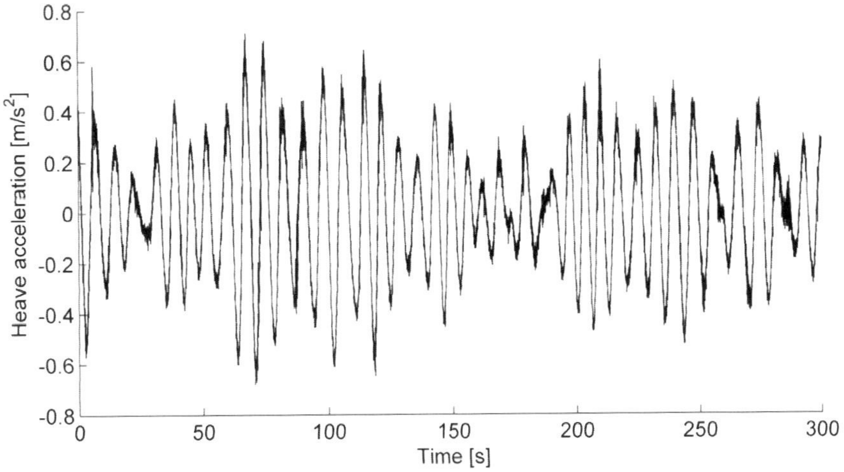

Figure 5. Heave response of Ice Floe #4 in waves.

DATA ANALYSIS

As mentioned, only the heave accelerations of the ice floes were analysed in this study. The recorded signals were trimmed to obtain 7-hour long overlapping data sets from each IMU. To obtain estimates of the power spectrum density (PSD) of the analysed data, the Matlab function pwelch was utilised with a window of 120 s. The estimated PSD of the floe heave responses from each IMU are presented in Figure 6.

The wave-induced heave motion of an ice floe can be presented in the form a non-dimensional response amplitude operator (RAO) defined as the ratio between heave amplitude and wave amplitudes. Masson and LeBlond (1989) showed that the RAO of an ice floe depends on the incident wavelength, the water depth, as well as the geometry and dimensions of the floe. Therefore, it is important to show that the measured heave responses can be used to identify the ratios between the actual wave heights at the sensor locations, which are necessary for calculating attenuation.

As seen from Figure 6, the wave spectrum is concentrated between 0.05 and 0.2 Hz, which corresponds to a wavelength band of 40 – 450 m for a nominal water depth of 75 m. According to Masson and LeBlond (1989), the heave RAO of a circular floe with a diameter of 10 m and a draft of 0.25 m is between 0.9 and 1.0 for this range of wavelengths. This RAO is a monotonic function of wavelength and can be used for the floes in this study, as they are approximately of the same size and shape. However, the exact values of the RAO are not necessary here because only ratios between wave amplitudes will be further analysed.

The wave attenuation coefficient can be calculated according to Frankenstein et al. (2001):

$$\alpha = -\ln(A_2/A_1)/(x_2 - x_1) \quad (1)$$

where α is the attenuation coefficient; A_2 and A_1 are the wave amplitudes at the two studied locations, the distance between which is $(x_2 - x_1)$. Assuming the same RAO for all instrumented ice floes, the ratio of the wave amplitudes A_2/A_1 can be obtained by taking the square root of the ratio between the areas under the corresponding PSD curve shown in Figure 6. The calculated attenuation coefficients for 5 different floe pairs, along with the average between them, are shown in Figure 7 as functions of wave frequency.

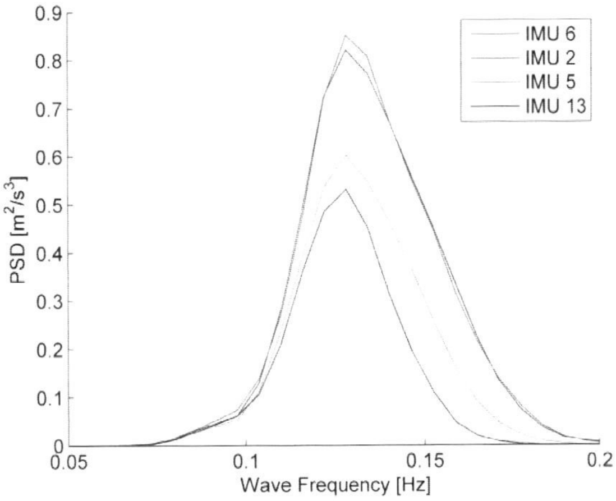

Figure 6. Heave response PSD of the ice floes versus wave frequency.

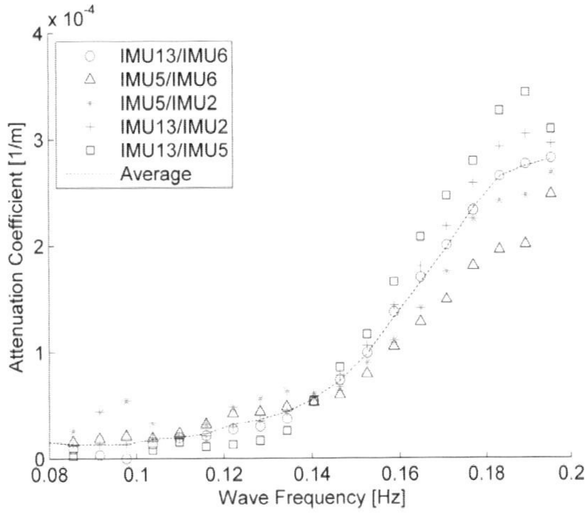

Figure 7. Wave attenuation coefficient versus wave frequency.

DISCUSSIONS

As highlighted by Cheng et al. (2017), wave attenuation depends on the composition of the ice field, e.g., ice type, concentration and floe size. Wadhams et al. (1988) indicated that the variation of the wave attenuation coefficient in the Arctic seas may be quite large; their measurements in the Bering and Greenland Seas showed a range of the wave attenuation coefficient between 0.2×10^{-4} and 8.7×10^{-4} m^{-1}. However, in these studies the ice covers consisted of floes from 5 to 50 m with a range of thickness from 0.4 to 1.2 m in the Bering Sea and 3.1 m in the Greenland Sea, which is very different from the ice conditions presented in this paper.

As it is rather difficult to find experiments with the same ice conditions as in this study, it may be interesting to compare the obtained results with the measurements reported by Frankenstein et al. (2001) which were performed in approximately the same region in April 1990. The authors reported wave attenuation coefficients ranging from 2.0×10^{-4} to 6.0×10^{-4}

m^{-1} over a corresponding frequency range of 0.04 – 0.16 Hz. The ice cover at the study site consisted of brash (35 – 60 %) and floes (40 – 65 %) from 2 to 20 m across. What is remarkable is that the ice thickness at that time was 1.3 m, which is almost five times the ice thickness in the recent study. The obvious decrease of the ice thickness in that area is most likely one of the reasons for the lower wave attenuation coefficient measured in the 2016 study compared to that from 1990. Another important parameter different in these two studies is the concentration of brash and very small ice floes; in the recent study most of the area, approximately 70 %, was covered by floes from 5 to 15 m and the rest was considered as brash, whereas in the 1990 study brash and floes smaller than 2 m were covering more than 50 % of the area. As noted by Cheng et al. (2017), smaller floes attenuate waves stronger than larger ones.

CONCLUSIONS

A field study on wave attenuation in the Barents Sea MIZ was performed in April/May 2016, based on which a range of wave attenuation coefficients from 0.1×10^{-4} to 3.5×10^{-4} m^{-1} was obtained for corresponding wave frequencies from 0.08 to 0.2 Hz. Comparison of these results with the results of a similar study conducted in 1990 shows that wave attenuation measurements from the same region in the Barents Sea MIZ, but conducted in different years, may significantly differ from each other. This should not be surprising if we take into account that ice conditions depend on climate changes and may also undergo seasonal variations; however, this conclusion questions repeatability of previously conducted measurements from various ice-covered seas.

Moreover, as mentioned by Cheng et al. (2017), a very wide range of wave attenuation coefficients has been reported for various ice conditions from different fields. A systematic approach to determine the effects of different parameters (e.g., floe size, as presented by Cheng et al. (2017)) on wave propagation is necessary, and laboratory experiments seem to be more feasible for this task than field studies.

ACKNOWLDEGEMENTS

This work was supported by the SAMCoT CRI through the Research Council of Norway and all of the SAMCoT Partners.

REFERENCES

Cheng, S., Tsarau, A., Li, H., Herman, A., Evers, K.U., and Shen, H., 2017. Loads on Structure and Waves in Ice (LS-WICE) project, Part 1: Wave attenuation and dispersion in broken ice fields. Proceedings of the 24th International Conference on Port and Ocean Engineering under Arctic Conditions, June 11-16, 2017, Busan, Korea.

Frankenstein, S., Løset, S., Shen, H.H., 2001.Wave–ice interactions in Barents Sea marginal ice zone. J. Cold Reg. Eng. 15 (2), 91–102.

Li, H., Tsarau, A., Shen, H., A., Herman, A., Evers, K.U., and Lubbad, R., 2017. Loads on Structure and Waves in Ice (LS-WICE) project, Part 4: Wave attenuation and dispersion in broken ice fields – Ice collisions under wave actions. Proceedings of the 24th International Conference on Port and Ocean Engineering under Arctic Conditions, June 11-16, 2017, Busan, Korea.

Løset, S. and Carstens, T., 1996. Sea Ice and Icebergs in the Western Barents Sea in 1987. Cold Regions Science and Technology, Vol. 24, No. 4, pp. 323-340.

Masson, D., and LeBlond, P. H., 1989. Spectral evolution of wind-generated surface gravity waves in a dispersed ice field. J. Fluid Mech., 202, 43–81.

Nesterov, A., Salganik, E., Vlasov, P., Bokovnya, S., and Lisova, S., 2016. Ice mechanics, Loads on Structures and Instrumentation. Unpublished report in student course AT-211, UNIS.

Squire, V.A., Dugan, J.P., Wadhams, P., Rottier, P.J., Liu, A.K., 1995. Of ocean waves and sea ice. Annu. Rev. Fluid Mech. 27, 115–168.

Squire, V. A. (2007). Of ocean waves and sea-ice revisited. Cold Reg. Sci. Technol., 49(2), 110–133.

Wadhams, P., Squire, V. A., Goodman, D. J., Cowan, A. M., and Moore, S. C., 1988. The attenuation rates of ocean waves in the marginal ice zone.'' J. Geophys. Res., 93(C6), 6799–6818.

Estimation of Lake Ice Thickness Using Measured Temperature Profile in Ice and Water

Xiaowei Cao[1], Zhijun Li[1], Qingkai Wang[1], Jiankang Wang[1], Guoyu Li[2]

[1] State Key Laboratory of Coastal and Offshore Engineering, Dalian University of Technology, Dalian 116024, China

[2] State Key Laboratory of Frozen Soil Engineering, Cold and Arid Regions Environmental and Engineering Research Institute, Chinese Academy of Sciences, Lanzhou 730000, China

ABSTRACT

Ice thickness is an important parameter in studies on engineering and climate change, and it is also difficult to obtain even by present methods of field investigation and remote sensing. Retrieval of ice thickness from measurement of ice temperature profile from buoy data is a potential method, but the retrieval method is still not so robust to different situations. To solve this problem, ice temperature profiles obtained by the temperature chain in Ulansuhai Lake from January to March 2016 were employed. The temperature profiles in ice were summarized into 6 typical types by analyzing the data and fitting the curve with different equations, such as the one or three order polynomial and the cubic spline function. Ice-water interface was estimated by the location of the 0 degrees Celsius on the temperature profile. Comparing the ice thickness obtained from the different fitting equations with that obtained from field measurements, it is showed that the difference between the ice thickness that estimated by the temperature profile and the measured data is less than the interval of temperature probe on the profile, namely, 5cm.

KEY WORDS: Ice thickness, Ice-water temperature profile, Fitting equation

INTRODUCTION

The sea-ice thickness distribution in the polar region has been shown to respond sensitively to global climate change (Mitchell et al., 1990). The ice hinders exchanges between the ocean and atmosphere, reflects solar radiation back to space, is an obstacle to ship travel, and has numerous impacts on plant and animal species (Haas et al., 2008; Parkinson, 2004). Recent years, the arctic sea ice extent in summer has famous reduced (Parkinson at al., 2008). The ice cover, extent and ice thickness are the most important parameters for the ice research. For these parameters, we can use different methods to obtain. The remote sensing can offer the highest spatial resolution for extraction of daily available, global sea ice concentration data (Spreen et al, 2008). It can also offer the ice thickness, but the accuracy of the result from remote sensing need to be further investigated (Laxon at el., 2013). Drilling is the most accuracy method to measure the ice thickness, but it can't be continue measurement in Arctic. The buoy is more and more widely used in Arctic research. We can get the ice-water temperature profile from the buoy data (Cheng at el., 2014). Retrieval of ice thickness from measurement of the ice-water temperature profile is a potential method, but the retrieval method is still not so robust to different situations .

In this paper, we measured ice-water temperature on Ulansuhai Lake, in Inner Mongolia China. We applied the temperature chain to measure ice-wate temperature and the underwater ultrasonic range finder to monitor the ice thickness. Our objectives are to: (1) Class the cure of the ice-water temperature profile by analyzing of the temperature profile data; (2) Use different equations to fit the temperature profile and retrieval of ice thickness; (3) Compare the ice thickness that estimated and measured data.

MEASUREMENT PLACE AND OBSERVATIONS

Measurement Place

Ulansuhai Lake is a freshwater lake in China, it is located in southwest of Inner Mongolia, situated in the north bank of the middle reaches of Yellow River (Figure 1). The lake is 233 km^2. The ice season lasts 4 months (from Dec. to Mar.), and the maximum ice thickness usually reaches more than 40 cm in late Junary, just before the onset of ice melt. We measured ice temperature profiles from January 12th to March 10th. The water depth of measurement place was 2.2 meters.

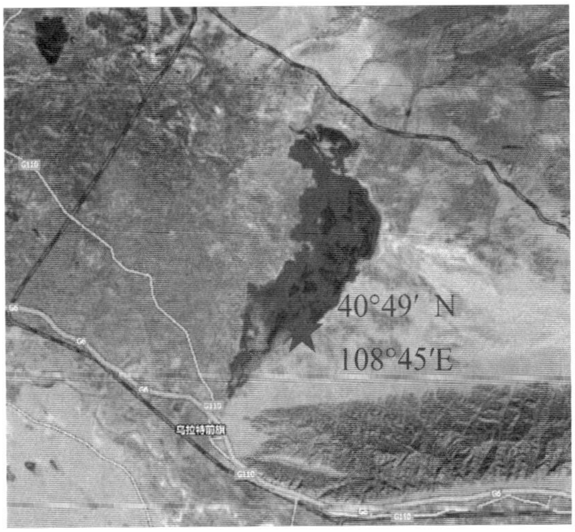

Figure 1. The Ulansuhai Lake and the measurement place

Measurement Method and Equipment

The temperature profile was measured by the platinum resistance temperature chain made by Dalian University and Technology. The interval of sensor is 5cm in the first 60cm, 40cm in the last 160cm. The accuracy of the temperature sensor is 0.01° C and the sampling interval is 1 min. The color map of temperature profiles are shown in Figure 2.

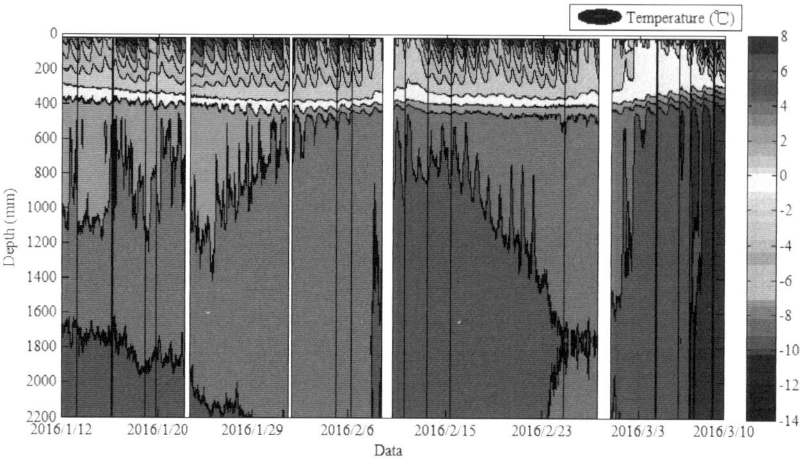

Figure 2. The color map of temperature profiles

The ice thickness was measured by an acoustic sensor in the interval 10 min. The measuring accuracy of the acoustic sensor is 0.1 mm. For recognizing the acoustic correction, there was thermal line method and drill method to measure ice thickness in random time. The three methods got fitted well data, see Figure 3.

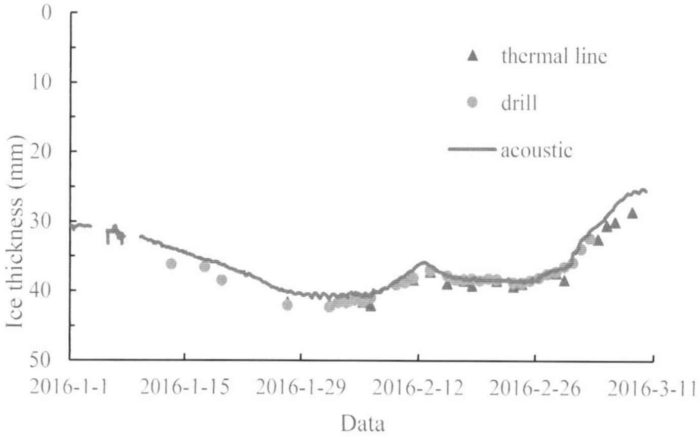

Figure 3. Ice thickness

DATA ANALYSIS

6 Typical Types of Temperature Profiles

By analyzing the data of the temperature profiles, we get the 6 typical types of temperature profiles (Figure 4). Type 1 and 5 always happened at the beginning freezing period. Type 2, 3, 4 always happened at the stable freezing period. Type 6 happened at the melting period.

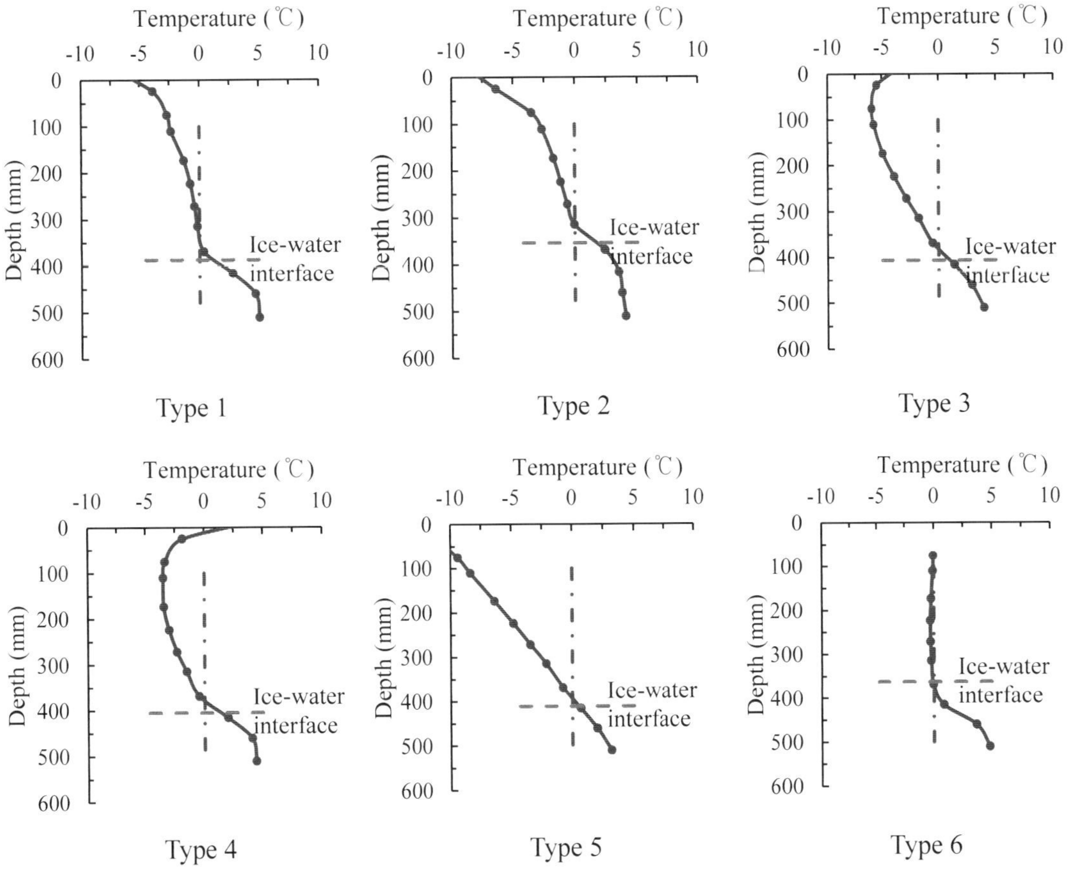

Figure 4. 6 typical types temperature profiles

Estimate Ice Thickness

First, we defined the location of 0° C for the ice-water interface. The shape of the temperature profiles of ice- water interface is different. Type 5 is nearly linear, the other are nearly the "S". We choose four points (two points in the ice and two points in the water) that different equations are used to fit, in order to obtain the location of the 0℃. Respectively using one or three order polynomial and the cubic spline function to fit these four points. The results that compare the ice thickness that estimated by the temperature profile and the measured data are shown in Tab. 1.

Table 1. Measure and estimate ice thickness

Type	Measure ice thickness (mm)	One order polynomial		Three order polynomial		Cubic spline function	
		Estimate ice thickness (mm)	Difference (mm)	Estimate ice thickness (mm)	Difference (mm)	Estimate ice thickness (mm)	Difference (mm)
1	386.63	318.53	-68.10	338.84	-47.79	338.82	-47.81
2	354.32	301.19	-53.13	316.11	-38.21	316.21	-38.12

3	406.20	374.17	-32.03	383.70	-22.50	384.22	-21.98
4	405.30	364.95	-40.35	383.42	-21.88	384.19	-21.11
5	411.20	391.64	-19.56	393.75	-17.45	393.84	-17.36
6	362.96	357	-5.96	375.83	12.87	375.99	13.03

CONCLUSIONS AND DISSCIONS

From the Tab. 1, we can see that different equations are applicable to different types of the ice-water temperature profiles. For the type 1, the difference of three order polynomial is the smallest. For the type 2 to 5, the applicable equation is cubic spline function. The one order polynomial is applicable to the type 6. The difference between the ice thickness that estimated by the temperature profile and the measured data is less than the interval of temperature sensor on the profile, namely, 5cm.

The temperature profile of sea ice is different from the fresh water ice. We will study the applicability for the sea ice-water temperature profiles in the next work, and look for other equation that is more suitable for fitting these points.

ACKNOWLEDGEMENTS

The in suit measurements on Ulansuhai Lake were supported by the Global Change Research National Major Scientific Research Projects (2015CB953901), National Natural Science Foundation of China (51579028, 41428603) and the Open Foundation of the State Key Laboratory of Frozen Soil Engineering (SKLFSE201604).

REFERENCES

Cheng, B., Vihma, T., Rontu, L., Kontu, A., Pour, H. K., & Duguay, C., et al. 2014. Evolution of snow and ice temperature, thickness and energy balance in lake orajavi, northern finland. *Tellus*, 66(1).

Haas, & Christian. 2008. Sea Ice: A Critical Habitat for Polar Marine Mammals and Birds. *Sea Ice: An Introduction to its Physics, Chemistry, Biology and Geology*.

Laxon, S. W., Giles, K. A., Ridout, A. L., Wingham, D. J., Willatt, R., & Cullen, R., et al. 2013. Cryosat-2 estimates of arctic sea ice thickness and volume. *Geophysical Research Letters*, 40(4), pp.732–737.

Mitchell, J., Manabe, S., Meleshko, V., & Tokioka, T. 1990. Equilibrium climate change -and its implications for the future. *Climate Change*.

Parkinson, C. L., & Cavalieri, D. J. (2008). Arctic sea ice variability and trends, 1979–2010. *Journal of Geophysical Research*, 113(C7), pp.341-355.

Parkinson, C. L., 2004. Southern Ocean sea ice and its wider linkages: Insights revealed from models and observations, *Antarctic Science*, 16, 4, pp.387–400.

Spreen, G, L. Kaleschke, and G. Heygster., 2008. Sea ice remote sensing using AMSR - E 89 - GHz channels. *Journal of Geophysical Research Atmospheres* 113.113(2008):C02S03.

An Experimental Study on Surface Wave Interactions with Viscoelastic Boundaries

Dharma Sree K. K.[1,2], Adrian W. K. Law[1,2*], Hayley H. Shen[1,3]

[1] Environmental Process Modelling Centre (EPMC), Nanyang Environment and Water Research Institute (NEWRI), Nanyang Technological University, 1 CleanTech Loop, Singapore

[2] School of Civil and Environmental Engineering, Nanyang Technological University, 50 Nanyang Avenue, Singapore

[3] 132 Rowley Laboratories, Civil and Environmental Engineering, Clarkson University, Potsdam, NY, USA

ABSTRACT

Recently, we performed an experimental study on wave interactions with floating viscoelastic covers that represent the varying sea ice conditions (Sree et al., 2016). In the present study, we continued to examine the wave interactions with a viscoelastic cover including the presence of a viscoelastic bottom in the laboratory. The topic has relevance in the port planning related to future Arctic shipping. The wave interactions with mud had been investigated earlier theoretically by MacPherson (1980), Hsiao and Shemdin (1980), Maa and Mheta (1990) and many others. However, no experimental verification of the theories using a homogeneous viscoelastic bottom has been reported in the literature, as far as we are aware. Our experimental approach utilized oil-doped PDMS (Polydimethylsiloxane) and glycerol-doped PDMS to create the top and bottom viscoelastic boundaries, respectively. Wave flume experiments with the viscoelastic boundaries were then performed in shallow water conditions. The surface wave profiles were measured using ultrasound sensors, and the changes in wave speed and wave height were compared to the reference condition of a rigid bottom. It was found that the wave length was modulated with the viscoelastic boundaries, and the attenuation of wave height along the region was different with and without the bottom layer. The results of the analysis shall be presented in the conference.

Keywords: viscoelastic material; floating cover; bottom layer; wave interaction

INTRODUCTION

Surface waves in the Arctic region experience the modulation in wavelength and attenuation in wave amplitude in the presence of ice covers and muddy bottoms, particularly in shallow waters. The analysis of three-layer system with ice covered shallow waters and muddy bottom, is necessary for port planning purposes in the Arctic region. It enables the determination of how the waves would transform near the port structures.

For wave attenuation with ice covers, both field studies and laboratory experiments had shown that the wave amplitude decays exponentially with distance (Wadhams et al., 1988; Newyear and Martin, 1999). The rate of decay depends on factors such as the thickness of the ice layer, its density, concentration as well as the size of ice floes. Many theoretical models had also been developed, the most generalized model of all ice types is the viscoelastic model (Squire and Alan, 1980; Robinson and Palmer, 1990 and Wang and Shen, 2010). The recent wave-ice experiments by Zhao and Shen (2015) showed that the viscoelastic model by Wang and Shen (2010) fit well for different types of ice covers.

The bottom mud also plays an important role in modulating the surface waves in the coastal regions (Jiang and Mehta, 1995, 1996; Mathew and Baba, 1995). Many theoretical studies had been performed modelling the mud as elastic, viscous as well as viscoelastic bottoms (MacPherson, 1980; Hsiao and Shemdin, 1980; Maa and Mehta, 1990). The related laboratory studies (Maa and Mehta, 1990; Zhao et al., 2006; Winterwerp et al., 2007) showed that the attenuation of wave height along the muddy bottom is frequency dependent. To the authors' knowledge, experiments on wave propagation over a homogeneous viscoelastic bottom have not been reported so far. This study focuses on the interactions of surface waves with different boundary conditions: viscoelastic cover only, viscoelastic bottom only, and a three-layer system with both viscoelastic boundaries.

MATERIAL PREPARATION AND EXPERIMENTAL PROCEDURES

The viscoelastic floating cover was made of oil-doped PDMS (Polydimethylsiloxane) as reported in Sree et al. (2016). The thickness of the cover was 1.0cm. The preparation included the mixing of the components (base, curing agent and white oil), degassing (to remove the trapped air bubbles) and curing. The mass percentage of the oil and curing agent used for the cover in the present study were 30% and 4%, respectively. Subsequently, the curing of the PDMS cover was conducted in a 2.0m long stainless steel tray for 7 days at 25°C.

The procedures for the preparation of the viscoelastic bottom were identical as the cover, except that the white oil was replaced with glycerol which is denser than water. The thickness of the bottom cover was 2.0cm. The mass percentage of the glycerol and curing agent used for the bottom were 10% and 2% respectively, and the curing was conducted in the 2.0m long stainless steel tray for a longer period of 15 days at 25°C. It should be noted that the rheological properties of the cover as well as the bottom can be varied by changing the mass percentage of curing agent, m_{CA} added at the time of preparation.

The rheological properties of the viscoelastic covers and bottoms were quantified using a rheometer (Anton Paar MCR 302, Germany) with parallel plate geometry (PP20). The rheometer tests included: (i) amplitude sweep, to determine the linear viscoelastic regime and (ii) frequency sweep, to determine the variation of storage modulus (G') and loss modulus (G'') with frequency of oscillation (ω) (Sree et al., 2016). The range of frequency of oscillation was determined by comparing with the required angular frequency ($\omega = 2\pi/T$). The variation of G' and G'' with ω of the viscoelastic cover and bottom are shown in Figure 1.

The experiments were conducted in a progressive wave flume at the Environmental Process Modelling Centre, NEWRI, Singapore. The flume was made of transparent tempered glass supported by stainless steel framework (8.0 m long, 0.30 m wide and 1.0 m deep). It was equipped with a paddle type wave generator at one end, and a dissipating beach made of wire mesh at the other end. The shallow water depth was set by allowing the viscoelastic bottom to rest on a raised platform (Figure 2).

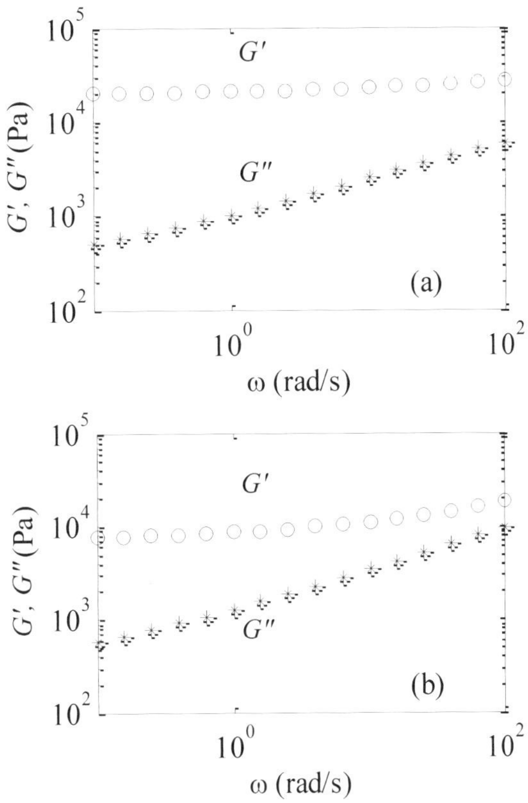

Figure 1. Viscoelastic properties against angular frequency for the (a) PDMS cover and (b) PDMS bottom. Circles = G' and asterisk =G''.

In the present study, the wave period was fixed at T = 0.7s, and the water depth above the viscoelastic bottom was 6.0cm. The wave height was set at 1.7cm. A capacitance wave gauge (WG_c) was installed near the wave maker that gave feedback to control the generation of the required wave profile.

Seven ultra sound sensors were installed along the length of the flume (Figure 2) to monitor the vertical surface displacement due to the progressive waves. The ultrasound sensor, US1, was kept at 20cm from the leading edge of the platform. All other ultrasound sensors, named in sequential order, were at 20cm interval from each other. Four resistance wave gauges were also installed near the wave generator to measure any possible reflection caused by the platform, viscoelastic bottom as well as viscoelastic cover. The sensors and wave gauges were connected to a data acquisition system, and the synchronized data were obtained through the LABVIEW software.

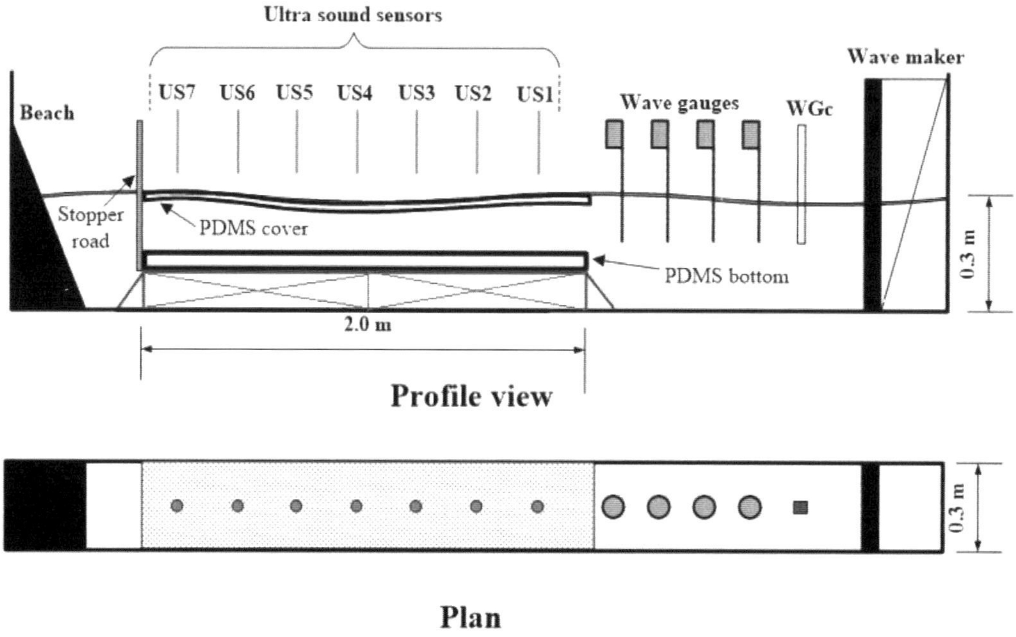

Figure 2. Schematic diagram of wave flume with viscoelastic boundaries.

RESULTS

The raw data acquired from LABVIEW was at 50 Hz, which was later increased to 200 Hz by interpolation using the *PCHIP* function in MATLAB (Figure 3). The location and value of peaks (crests) were then identified using the *findpeaks* function of MATLAB. Only the first three fully developed waves were considered for the analysis in order to avoid the effect of beach reflection.

The wave celerity in the viscoelastic region was determined by calculating the time taken for each of the fully developed waves to travel past the ultra sound sensors along the cover (Sree et al., 2016). The variation in celerity along the region is shown in Figure 4. The figure shows that the wave speed decreased with the presence of the viscoelastic boundaries. The wavenumber and attenuation coefficient obtained are shown in Table 1. The attenuation coefficient was determined by fitting an exponential curve ($\eta_a = \eta_0 \, e^{-\alpha x}$) to the measured data as illustrated in Figure 5. The attenuation coefficient obtained for different boundary conditions are provided in Table 1.

Overall, the experimental results showed that both the wavenumber and wave height were modulated by the presence of viscoelastic boundaries (Table 1). With the viscoelastic floating cover only, the wave attenuation increased significantly as expected, and wave length shortening was observed. With the viscoelastic bottom only, however, the attenuation reduced instead when compared to open water with a rigid bottom, while the wave length was preserved. With both, the effects were almost combined linearly.

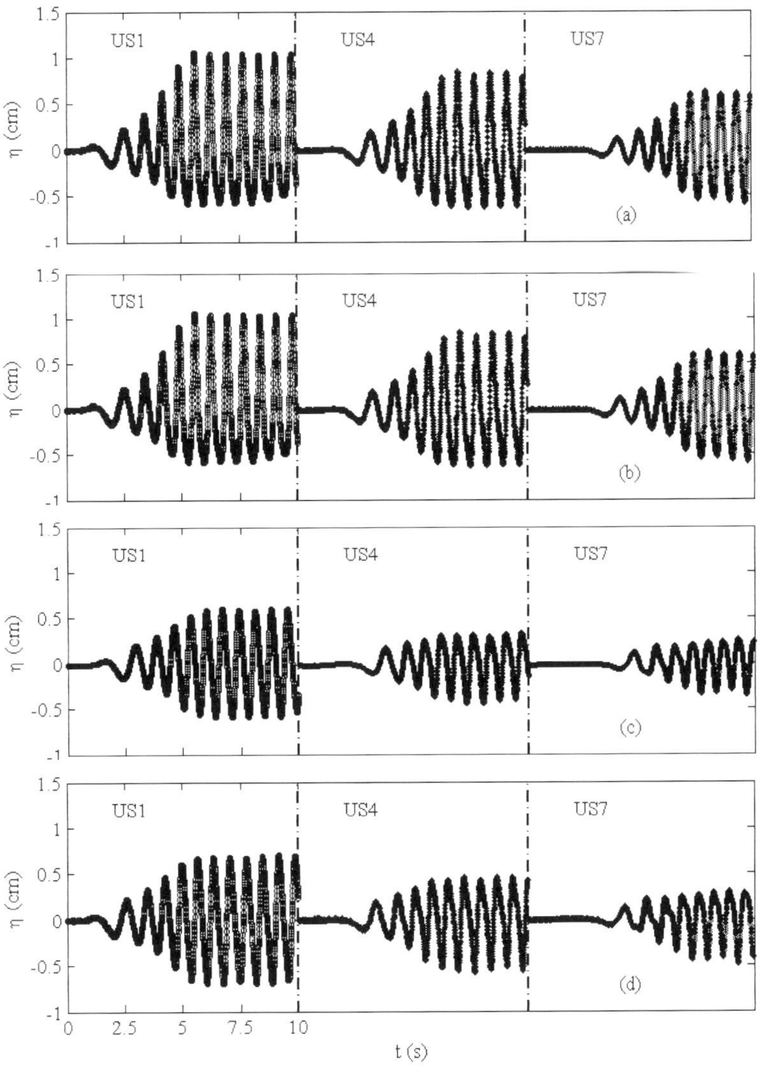

Figure 3. Time series of the wave record for US1, 4 and 7: (a) open water with rigid bottom, (b) viscoelastic bottom only, (c) viscoelastic cover only and (d) both viscoelastic cover and bottom.

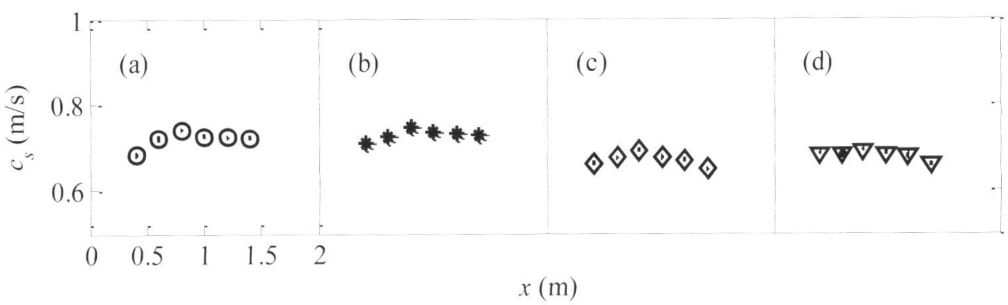

Figure 4. Celerity along the viscoelastic boundary region: (a) open water with rigid bottom, (b) viscoelastic bottom only, (c) viscoelastic cover only and (d) both viscoelastic cover and bottom.

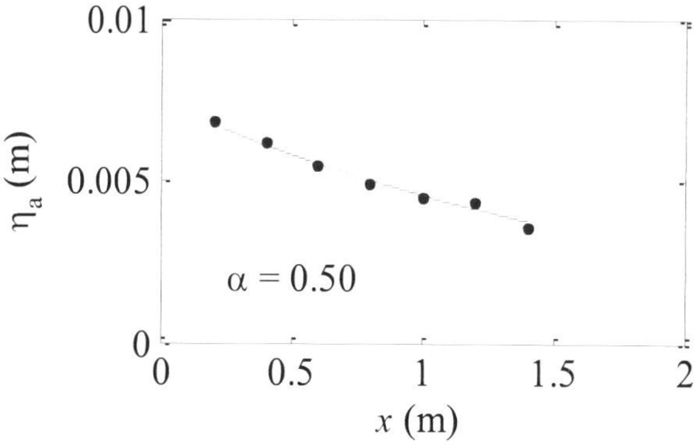

Figure 5. Wave amplitude along the region with both viscoelastic boundaries.

Table 1. Wave number and attenuation coefficient for different boundary conditions (T=0.7s)

Case	k (rad/m)	α (1/m)
Open water with rigid bottom	12.3	0.23
Viscoelastic bottom only	12.2	0.18
Viscoelastic cover only	13.3	0.56
Both viscoelastic boundaries	13.2	0.50

CONCLUSION

The present results thus demonstrated the feasibility of adopting the experimental approach with oil-doped and glycerol-doped PDMS materials to quantify the wave interactions with viscoelastic boundaries in the laboratory. We are currently conducting further experiments to cover a broader range of wave and viscoelastic characteristics.

ACKNOWLEDGEMENTS

The authors gratefully acknowledge the funding provided by Ministry of Education, Singapore through the AcRF Tier 2 Grant No. MOE2013-T2-1-054. This work is supported in part by the US Office of Naval Research Grant #N00014-13-1- 0294 and US Office of Naval Research Global Grant No: N62909-15-1-2069. The authors would like to thank Peh Zhisheng at the Nanyang Technological University for assisting in the experiments.

REFERENCES

Hsiao, S. V., & Shemdin, O. H., 1980. Interaction of ocean waves with a soft bottom. *Journal of Physical Oceanography*, 10(4), 605-610.

Jiang, F., & Mehta, A. J., 1995. Mudbanks of the southwest coast of India IV: mud viscoelastic properties. *Journal of Coastal Research*, 11(3), 918-926.

Jiang, F., & Mehta, A. J., 1996. Mudbanks of the southwest coast of India. V: Wave attenuation. *Journal of Coastal Research*, 12(4), 890-897.

Maa, J. P. Y., & Mehta, A. J., 1990. Soft mud response to water waves. *Journal of waterway, port, coastal, and ocean engineering*, *116*(5), 634-650.

Macpherson, H., 1980. The attenuation of water waves over a non-rigid bed. *Journal of Fluid Mechanics*, 97(4), 721-742.

Mathew, J., & Baba, M., 1995. Mudbanks of the southwest coast of India. II: Wave-mud interactions. *Journal of Coastal Research*, 11(1), 179-187.

Newyear, K., & Martin, S., 1999. Comparison of laboratory data with a viscous two-layer model of wave propagation in grease ice. *Journal of Geophysical Research: Oceans*, *104*(C4), 7837-7840.

Robinson, N. J., & Palmer, S. C., 1990. A modal analysis of a rectangular plate floating on an incompressible liquid. *Journal of sound and Vibration*, 142(3), 453-460.

Squire, V. A., & Alan A. J., 1980. Propagation of flexural gravity waves in sea ice, in *Sea Ice Processes and Models, Proceedings of the Arctic Ice Dynamics Joint Experiment*, edited by R. S. Pritchard, pp. 327–338, Univ. of Wash. Press, Seattle, Wash.

Sree, D. K., Law, A. W. K., & Shen, H. H., 2016. An experimental study on the interactions between surface waves and floating viscoelastic covers. *Wave Motion*. In press.

Wadhams, P., Squire, V. A., Goodman, D. J., Cowan, A. M., & Moore, S. C., 1988. The attenuation rates of ocean waves in the marginal ice zone. *Journal of Geophysical Research: Oceans*, 93(C6), 6799-6818.

Wang, R. & Shen, H.H., 2010. Gravity waves propagating into an ice-covered ocean: A viscoelastic model. *Journal of Geophysical Research: Oceans*, *115*(C6024), 1-12.

Winterwerp, J. C., De Graaff, R. F., Groeneweg, J., & Luijendijk, A. P., 2007. Modelling of wave damping at Guyana mud coast. *Coastal Engineering*, *54*(3), 249-261.

Zhao, Z. D., Lian, J. J., & Shi, J. Z., 2006. Interactions among waves, current, and mud: numerical and laboratory studies. *Advances in water resources*, *29*(11), 1731-1744.

Zhao, X., & Shen, H. H., 2015. Wave propagation in frazil/pancake, pancake, and fragmented ice covers. *Cold Regions Science and Technology*, *113*, 71-80.

Concrete-Ice Abrasion Laboratory Experiments

Guzel Shamsutdinova[1], Max A.N. Hendriks[1,2,] Stefan Jacobsen[1]
[1] Norwegian University of Science and Technology, Trondheim, Norway
[2] Delft University of Technology, Delft, Netherlands

ABSTRACT

Concrete-ice abrasion is a concrete surface degradation mechanism due to ice-structure interaction. The topic is especially relevant for concrete gravity-based structures in the Arctic offshore.

Experiments have become the main evaluation method for concrete durability under ice abrasion. The paper presents concrete-ice abrasion experiment between sawn concrete surface (cubic compressive strength is 90 MPa) and unidirectional fresh water ice. The experiment included two experimental conditions, with varied temperature of the concrete surface.

Results demonstrate the load cells response, coefficient of friction and abrasion depth. Based on our results and observations we identified wet and dry abrasion types. The first type caused most damage, whereas second one caused minor damage.

KEY WORDS: Concrete; Ice; Abrasion; Experiment.

INTRODUCTION

The concrete-ice abrasion process has been studied for the last 30 years, and has been defined as the surface degradation of concrete structures due to interaction with drifting ice floes. Several research groups have studied this topic through laboratory experiments (Huovinen, 1993, Bekker et al., 2011, Hanada et al., 1996, Fiorio, 2005, Møen et al., 2015, Tijsen et al., 2015) and field observations (Itoh et al., 1996, Huovinen, 1993). Hara et al. (Hara et al., 1995) recommended the concrete-ice sliding abrasion test, during evaluation of various test methods of concrete-ice abrasion resistance. And so far most experimental work has been based on the sliding interaction between ice and concrete, whether ice on concrete (Fiorio, 2005, Møen et al., 2015, Saeki et al., 1986) or concrete on ice (Hoff, 1989, Bekker et al., 2011, Tijsen et al., 2015, Itoh et al., 1988).

Our experimental method is based on the sliding of an ice sample along a fixed concrete sample, and controls exposure, measures relevant response parameters during concrete-ice interaction including concrete-ice abrasion with a non-contact laser scanner.

EXPERIMENT

This experimental studying included the abrasion machine for simulation of concrete-ice abrasion, and a laser scanner for further measuring of abrasion wear. The experimental equipment is described below together with the employed concrete and ice.

Abrasion Machine

The abrasion machine (Figure 1(a)) simulated concrete-ice abrasion according to the principle in Figure 1(b), with a sliding ice specimen along a fixed concrete sample.

The machine makes the ice sample holder move in repeated sliding movements in a horizontal direction, with the average velocity 0.16 m/s. The piston (Figure 1(a), no 3) continuously pushes the ice sample against the concrete surface with 1 MPa load. A feedback system keeps the loading as constant as possible during the test while moving back and forth. The temperature in the concrete-ice abrasion lab is kept at temperature of −10 °C.

The concrete sample temperature control goes through an aluminum heating plate that is below the concrete sample (Figure 1(a), no 7). The plate is connected to a controlled temperature liquid (alcohol) circulator. This means that the temperature of the concrete surface in the concrete-ice abrasion zone can be adjusted. The temperature of the concrete surface in the contact zone was measured with an infrared scanner. More detailed information about the components of the abrasion machine is given in (Shamsutdinova et al., 2015).

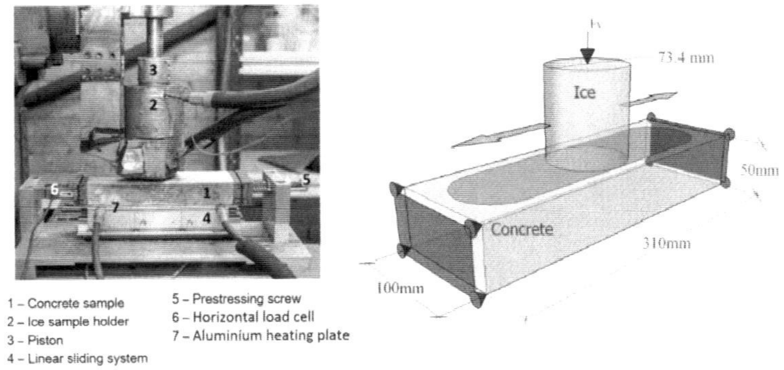

(a) (b)

Figure 1. The concrete-ice abrasion machine: (a) photo, (b) principal scheme.

Laser Scanner

Recently, we developed a non-contact Laser Scanning method (Figure 2(a)) to measure concrete-ice abrasion. It allows scanning of the concrete surface with accuracy 10 μm in reasonable time. The laser moves continuously along the sample according to a predefined "snaking" path. The measuring point distance is approximately 50 μm in the Y direction, and the step size in the X direction is 1 mm (Figure 2(b)). The measured data is transformed to a matrix of surface heights, with 1900x300 points.

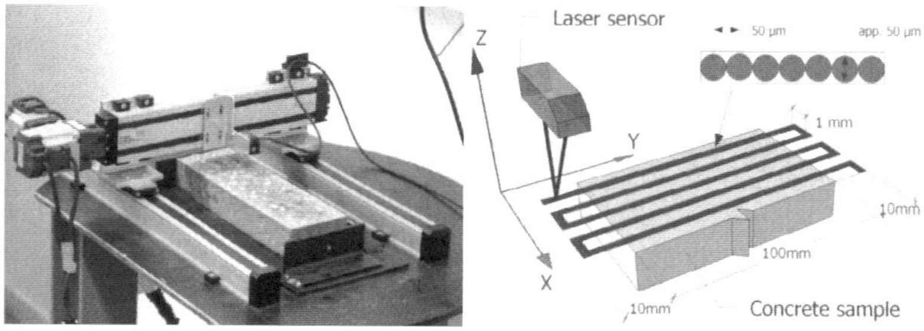

(a) (b)

Figure 2. Laser scanner: (a) photo; (b) simplified schematic of measurement process (not to scale).

Ice

Our abrasion machine is designed for fresh-water cylindrical ice samples. We used unidirectionally grown ice made from tap water, which can easily be reproduced in other laboratories.

An ice mould made of Polyoxymethylene (POM) 13.3 mm thick and 370 mm high (Figure 3(a)) and covered with thermo isolation on the sides and the bottom is filled with tap water and put in a freezer at −20 °C for 48-72 hours. The freezing of the water starts at the top of the mould, but later it also takes place from the bottom (Figure 3(b,c)). The upper part of the ice sample is transparent (unidirectionally grown ice) with very few air voids. The lower part of the ice sample contains a lot of air voids and unfrozen water. The ice sample is cut in two, and only the upper part is used for the test (Figure3(d).

(a) (b) (c) (d)

Figure 3. Schematic of ice production: (a) ice mould with isolation; (b) ice sample after 48 hours; (c) ice sample after 72 hours; (d) ice sample for the test.

Concrete

The tests were carried out with two concrete samples made of Norcem Anlegg (CEM I) cement (with 2% Elkem silica fume substitution) with fine aggregate (Årdal sand, 0-8 mm grain size) and coarse aggregate (Årdal, 8-16 mm grain size). The mix was made with the following proportion: W/(C+2S)=0.42, where W, C and S are the weight of water, cement and silica fume powder, respectively. The cement paste volume was 29.5%. Superplasticizing additive

Dynamon SX-23 from Mapei was used to achieve target workability.

The 28-day cube compressive strength of the concrete was 90 MPa. Samples were cured in water at +20°C for 11 months, reaching 110 MPa cube compressive strength by the start of the test.

Experimental Conditions

As was said above, the tests were performed in the abrasion laboratory at air temperature -10°C. The ice pressure was 1 MPa. The average sliding velocity was 0.16 m/s. The experiment was performed on two concrete samples with varied temperature of concrete surface in the sliding zone. The temperature variation went through aluminum heating plate (Figure 1(a), no 7). Sample 1 had surface temperature +2°C at the start of the test, and Sample 2 had surface temperature -4°C at the start of the test. Both samples had sawn concrete surface.

RESULTS

The data acquisition system logged horizontal and vertical load cells responses during the test at 500 Hz frequency. The absolute coefficient of friction (COF) is plotted in Figure 4 for two samples. The highest COF corresponds to the turning points of the ice specimen where the ice sample makes a full stop. We distinguished the coefficient of kinetic friction during sliding interaction, and the coefficient of static at the turning points. The coefficient of kinetic friction varied in the range 0 to 0.02, whereas the coefficient of static friction varied from 0.05 to 0.10. Figure 4 shows that Sample 2, with concrete surface temperature -4°C, resulted in the smallest COF.

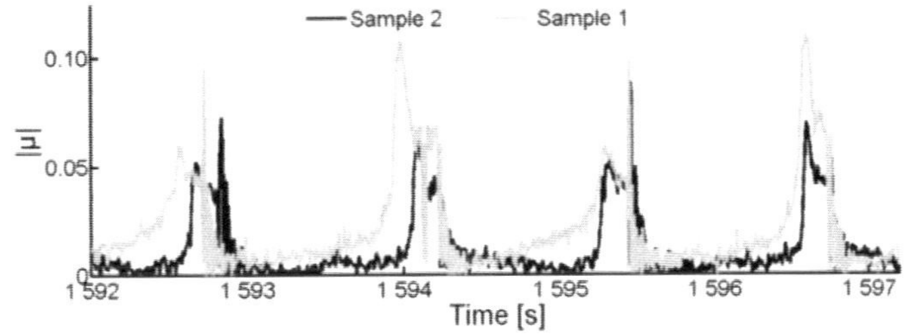

Figure 4. Coefficient of friction for two tested samples.

Figure 5 demonstrates the result of surface scanning before and after 4 km of abrasion test. On the left side, before the test, the concrete surface has diamond saw blade traces. On the right side, after 4 km of abrasion test, the concrete surface became rougher.

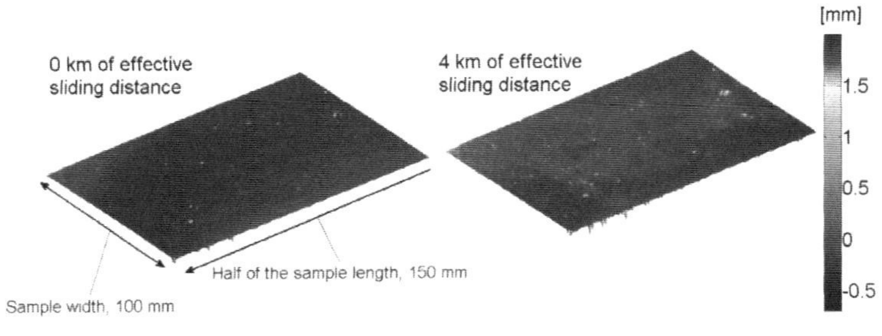

Figure 5. Scans of Sample 1, before abrasion test and after 4 km of abrasion test (half of the scanned sample).

Based on the surface measurements, abrasion was found as difference between the unabraded rim of the concrete sample and an abraded central band of 10 mm wide as done with mechanical measurements in (Møen et al., 2015). However, a much higher number of data points was collected with the laser scanner, so the calculation here was done for each millimeter of concrete sample length. Figure 6 shows the average profile of the abraded central band (10 mm wide) along the sample length, before and after 4 km abrasion test. The difference between dashed and solid lines shows the abrasion. The spikes, in Figure 6, corresponds to the air voids.

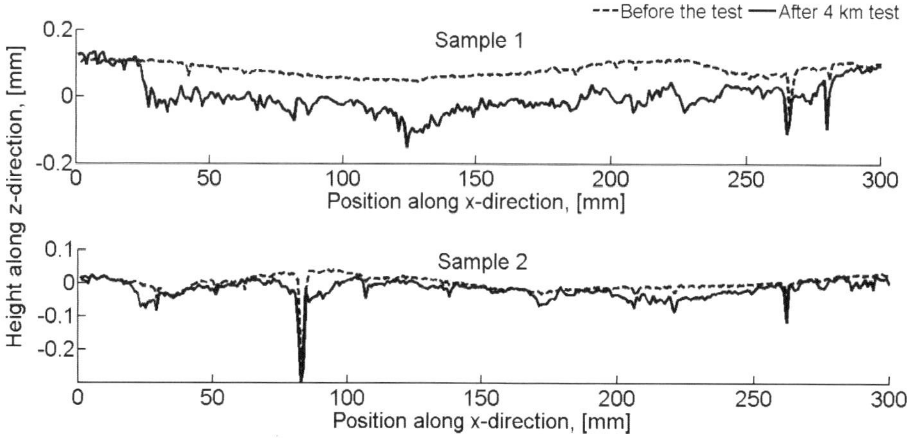

Figure 6. Average profile, of the central abraded band (10 mm wide), of sawn concrete surface along the sample length before and after 4 km abrasion test.

Figure 7 shows average abrasion rate for two samples. The abrasion for Sample 1 with the surface temperature +2°C at the start of the test is much higher than for Sample 2 with the surface temperature -4°C at the start of the test.

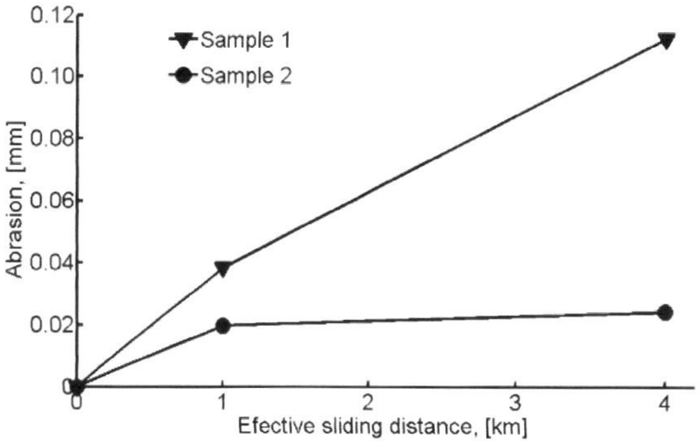

Figure 7. Abrasion of sawn concrete surface from zero to 1 and 4 km.

DISCUSSION AND CONCLUSION

The measured coefficient of kinetic friction is of the same magnitude as test results of other concrete-ice abrasion tests with the same sliding speed (0.16 m/s): 0.00 – 0.01 and 0.06 respectively (Møen et al., 2015, Itoh et al., 1988). Where Itoh et al. used ice pressure 0.2MPa, that is ¼ th of the present test.

The results of abrasion depth were found similar to previous study Møen et al. where concrete samples with cylindrical compressive strength from 72.8 to 147.8 MPa under similar experimental conditions had maximum abrasion rate 0.025 mm/km of effective sliding distance.

Based on our results and observations we distinguished between two types of concrete-ice abrasion exposure: wet and dry, corresponding to Sample 1 and Sample 2 respectively. Dry abrasion happened when the concrete sample had surface temperature -4°C at the start of the test. Ice filled the unevenness on the concrete surface during the first few sliding cycles, and all further sliding was between ice and ice.

In case of wet abrasion, the temperature of the concrete sample surface was +2°C at the start of the test, and it fluctuated between −1.8 and −2.6 °C during the abrasion test. There was a constant thin film of water on the concrete surface, with minor icing spots, during the test.

We conclude that wet abrasion test causes more wear than dry. It means that the concrete surface temperature control system is an important element of the experimental method.

ACKNOWLEDGEMENTS

This research forms part of the DACS (Durable Advanced Concrete Solutions) project. The financial contribution of the Norwegian Research Council is gratefully acknowledged. The DACS project partners are: Kværner AS (project owner), Axion AS (Stalite), AF Gruppen Norge AS, Concrete Structures AS, Mapei AS, Multiconsult AS, NorBetong AS, Norcem AS, NPRA (Statens Vegvesen), Norges Teknisk-Naturvitenskapelige Universitet (NTNU), SINTEF Byggforsk, Skanska Norge AS, Unicon AS and Veidekke Entreprenør AS.

Thanks to Giedrius Zirgulis for the photographs.

REFERENCES

Bekker, A.T., Uvarova, T.E., Pomnikov, E.E., Farafonov, A.E., Prytkov, I.G. & Tyutrin, R.S., 2011. Experimental study of concrete resistance to ice abrasion. Proceedings of the International Offshore and Polar Engineering Conference, pp. 1044-1047.

Fiorio, B., 2005. Wear characterisation and degradation mechanisms of a concrete surface under ice friction. Construction and Building Materials, 19, 366-375.

Hanada, M., Ujihira, M., Hara, F. & Saeki, H., 1996. Abrasion rate of various materials due to the movement of ice sheets. Proceedings of the International Offshore and Polar Engineering Conference, pp. 433-437.

Hara, F., Takahashi, Y. & Saeki, H., 1995. Evaluation of test methods of abrasion by ice movements on the surface of reinforced concrete structures. Concrete Under Severe Conditions Environment and Loading.

Hoff, G.C., 1989. Evaluation of ice abrasion of high-strength lightweight concretes for arctic applications. Proceedings of the International Offshore Mechanics and Arctic Engineering Symposium, pp. 583-590.

Huovinen, S., 1993. Abrasion of concrete structures by ice. Cement and Concrete Research, 23, pp.69-82.

Itoh, Y., Tanaka, Y., Delgado, A. & Saeki, H., 1996. Abrasion depth distribution of a cylindrical concrete structure due to sea ice movement. International Journal of Offshore and Polar Engineering, 6, pp. 144-151.

Itoh, Y., Yoshida, A., Tsuchiya, M., Katoh, K., Sasaki, K. & Saeki, H., 1988. An experimental study on abrasion of concrete due to sea ice. Offshore Technol. Conference, OTC 5687, pp.61-68.

Møen, E., Høiseth, K.V., Leira, B. & Høyland, K.V., 2015. Experimental study of concrete abrasion due to ice friction - Part I: Set-up, ice abrasion vs. material properties and exposure conditions. Cold Regions Science and Technology, 110, pp.183-201.

Saeki, H., Ono, T., Nakazawa, N., Sakai, M. & Tanaka, S., 1986. Coefficient of friction between sea ice and various materials used in offshore structures. Journal of Energy Resources Technology, Transactions of the ASME, 108, pp.65-71.

Shamsutdinova, G., Rike, P.B., Hendriks, M.A.N. & Jacobsen, S., 2015. Concrete ice abrasion rig and wear measurements. Proceedings of the International Conference on Port and Ocean Engineering under Arctic Conditions.

Tijsen, J., Bruneau, S. & Colbourne, B., 2015. Laboratory examination of ice loads and effects on concrete surfaces from bi-axial collision and adhesion events. Proceedings of the International Conference on Port and Ocean Engineering under Arctic Conditions.

POAC'17
Busan, KOREA

Proceedings of the 24th International Conference on
Port and Ocean Engineering under Arctic Conditions
June 11-16, 2017, Busan, Korea

A Continuum Model of Large Deformation Continuous Ductile Ice Crushing

Åse Ervik[1,2,3], Knut Vilhelm Høyland[1,2], Gustav Grimstad[1,2], Torodd Skjerve Nord[1,2,3]
[1]Sustainable Arctic Marine and Coastal Technology (NTNU, Trondheim, Norway)
[2]Department of Civil and Environmental Engineering (NTNU, Trondheim, Norway)
[3]UNIS (Longyearbyen, Norway)

ABSTRACT
This paper presents numerical modelling techniques for modelling of large deformation continuous ductile ice crushing. Two continuum formulations, the Coupled Eulerian Lagrangian (CEL) method and Smoothed Particle Hydrodynamics (SPH) method available in ABAQUS/Explicit 6-13 are used in combination with a modified version of a constitutive model originally developed for concrete. First, an analytical small strain problem is modelled in order to compare the accuracy of the two continuum formulations. With the CEL method the problem was solved exactly, while the SPH method produced an error of 42%. Finally, a medium-scale continuous ductile ice crushing indentation experiment conducted in the Van Mijen fjord in Spitsbergen 2003 is modelled with the CEL formulation. Required parameters like the initial uniaxial compressive strength and the elastic modulus was measured during the experiment, other required material parameters are estimated. The calculated global load on the cylindrical indenter was in good agreement with the load measured during the experiment.

KEY WORDS: CEL; SPH; Large deformations; Ductile ice crushing; Consolidated layer

INTRODUCTION
Ice ridges are accounted for in quasi-static design of structures in ice-infested waters. The combined force from the consolidated layer and the force from the ice rubble give the ice ridge design force. A relatively inexpensive means to assess ice ridge forces is by numerical modelling. In order to model the continuous failure of the consolidated layer and rubble accumulation, a suitable model must handle large deformations. This article focuses on crushing of the consolidated layer. First, two continuum formulations that allow for large deformations are presented: the Coupled Eulerian Lagrangian (CEL) method and the Smoothed Particle Hydrodynamics (SPH) method. Then a constitutive model for continuous ductile crushing of the consolidated layer is proposed. An analytical small strain problem is used to test the accuracy of the two continuum formulations. The medium-scale large deformation experiment presented in Moslet et al. (2004) and Moslet et al. (2005) is simulated. The results are also compared to the 2D finite difference results presented in Moslet (2008).

The type of ice failure is determined by the ice velocity, temperature and the structure shape and observed time histories of the measured forces and responses. In the modelled medium-scale indentation experiment, an ice sheet was towed against a vertical fixed cylindrical indenter with an average indentation speed of 3.5mm/s. According to Moslet et al. (2004) the ice failure mode was ductile crushing (see Figure 2a). However, because of the low stiffness of the system (indenter, steel wires etc.), the indentation speed was highly non-linear, resulting in cycles of ductile crushing. During force build up, elastic energy was stored in the system, and when the ice reached the failure limit, the elastic energy was released resulting in a high indentation speed. The cycles of ductile crushing lasted for approximately 50 s with an approximate indentation speed of 1 mm/s. According to Sodhi et al. (1998) this indentation speed is associated with ductile crushing. In the analysis, presented here, the experiment is simplified by assuming that the ice indentation speed is constant and that the cylindrical indenter is rigid. The purpose of this model is to predict the development in the peak force and therefore, the sudden force drops observed in the experiment force time history (Figure 2a) is outside the scope of this study.

Several constitutive models have been used to simulate ice crushing. E.g. the crushable foam model (Hallquist, 1998) was used by Gagnon (2011) to simulate brittle ice crushing including spalling. One problem with this model is the lack of physical explanation. Another model widely used to simulate brittle crushing during high velocity iceberg-ship interactions (Kirkegaard, 1993; Løset and Kvamsdal, 1994; Riska and Frederking, 1987), is the pressure dependent elliptical yield surface presented by Derradji-Aouat (2003). Liu et al. (2011) used this constitutive model to simulate a high velocity iceberg-ship impact and used element erosion to delete elements that exceeded an effective strain limit. For the current application (ductile crushing), element erosion as a failure criterion is not suited. Obtaining a physical consistent criterion for when to delete elements is difficult and additionally the energy balance is not satisfied when mass is removed from the analysis. In the end of the next section, we develop phenomenologically a constitutive model that captures the important physical behavior of ductile failure of ice in the consolidated layer.

BACKGROUND

From FEM to CEL

In the finite element (FE) Lagrangian formulation, expressed in the equations below, the material is fixed to the mesh, while in the CEL formulation the material moves independently of the mesh.

The concept of the FE method can be explained through kinematic-, constitutive- and equilibrium equations. First, the displacement vector $\mathbf{u}(x)$ is approximated by a linear combination of shape functions collected in a matrix $\mathbf{N}(x)$. Each shape function is associated with an unknown nodal displacement collected in a column vector $\mathbf{d}$, Equation 1.

$$\mathbf{u}(x) \approx \mathbf{N}(\mathrm{x}) \cdot \mathbf{d} \tag{1}$$

The kinematic equation is expressed in Equation 2, where ∂ is the Cartesian differential operator (with respect to the material configuration) and $\boldsymbol{\varepsilon}$ is the column strain vector.

$$\boldsymbol{\varepsilon}(x) \approx \partial\mathbf{N}(x)\cdot\mathbf{d} = \mathbf{B}(x)\cdot\mathbf{d} \tag{2}$$

The principal of virtual work in the weak form of the equilibrium equation, Equation 3-4, is used to derive the stiffness matrix $\mathbf{K}_e$. Where $\delta\mathbf{u}$ and $\delta\boldsymbol{\varepsilon}$ are arbitrary virtual displacements and strain fields, respectively.

$$\int_V \boldsymbol{\sigma}^T \delta\boldsymbol{\varepsilon} dV = \int_V \mathbf{b}^T \delta\mathbf{u} dV + \int_S \mathbf{t}^T \delta\mathbf{u} dS \tag{3}$$

$$\int_V \boldsymbol{\sigma}^T \mathbf{B}(x)\delta\mathbf{d} dV = \int_V \mathbf{b}^T \mathbf{N}(x)\delta\mathbf{d} dV + \int_S \mathbf{t}^T \mathbf{N}(x)\delta\mathbf{d} dS \tag{4}$$

For a linearly elastic material the stress ($\boldsymbol{\sigma}$) is expressed though the constitutive equation: $\boldsymbol{\sigma} \approx \mathbf{D}_e(x)\mathbf{B}(x)\mathbf{d}$, where $\mathbf{D}_e(x)$ is the elasticity matrix. For a linearly elastic material the element stiffness matrix (Equation 5) is used to evaluate the unknown displacement vector **d**.

$$\mathbf{K}_e\mathbf{d} = \mathbf{f}_{ext} \tag{5}$$

$$\mathbf{K}_e = \int_V \mathbf{B}^T(x)\mathbf{D}_e(x)\mathbf{B}(x)\,\mathrm{dV}$$

For non-linear materials (plasticity), the stress is a function of the $\mathbf{B}(x)$ matrix and x such that $\mathbf{K}_e$ becomes a nonlinear integral, which is evaluated through a numerical quadrature.

In ABAQUS, the CEL method is implemented through a Lagrangian-plus-remap algorithm, which consists of two steps for each explicit time increment. First, in a conventional Lagrangian increment, deformations, strains and stresses are computed. Second, in an advection remap increment, the material advection is computed and the mesh is brought back to its initial configuration.

SPH

The smoothed particle hydrodynamics (SPH) method is a mesh free Lagrangian formulation, first developed for astrophysical problems involving extreme deformations by Lucy (1977) and Gingold and Monaghan (1977) later extended to strength of materials by Libersky and Petschek (1991). There are four key ideas behind the SPH method. The material domain is discretized into a finite number of disengaged particles. Field functions (density, velocity, energy) are approximated by integral representations:

$$f(x) = \int_\Omega f(\tilde{x})\delta(x-\tilde{x})d\tilde{x}, \qquad \delta\begin{cases}1 & x=\tilde{x}\\ 0 & x\neq\tilde{x}\end{cases} \tag{6}$$

Where, $f(x)$ is a field function of the three dimensional position vector x, Ω is the volume of the support domain over which $f(x)$ is integrated, and δ is the Dirac delta function. The Dirac delta function is approximated by a smoothing kernel function W defined over a smoothing length h. The smoothing function should by a positive, even function, with compact support, and satisfy the Dirac delta function when the smoothing length approaches zero and unity over its support domain. The field function with the smoothing approximation accordingly reads:

$$\langle f(x)\rangle = \int_{\Omega} f(\tilde{x}) W(x-\tilde{x},h) d\tilde{x} \tag{7}$$

< > is the kernel approximation operator. Finally, through a particle approximation the integral in Equation 2, is discretized by summation over all particles within the support domain:

$$\langle f(x_i)\rangle = \sum_{j=1}^{N} \frac{m_j}{\rho_j} f(x_j) W(x_i - x_j, h) \tag{8}$$

Where, $d\tilde{x}$ is the volume of the support domain at the location of particle j that is replaced by a finite volume of the particle m_j/ρ_j.

Constitutive Model

The constitutive model presented in the following, will be used to simulate experiment no 2 presented in Moslet et al. (2004), material parameters and test setup is presented in the next section. The initial uniaxial compressive strength and the elastic modulus were measured during the experiment, other material parameters required for the model are estimated. The model was initially developed for modelling failure in concrete by Lubliner et al. (1989).

The elastic behavior of the consolidated layer is described though two isotropic elastic parameters, the elastic modulus and Poisson's ratio. Compared to orthotropic S2 level ice, the texture of the consolidated layer is isotropic (see e.g. Kankaanpää (1997)). Furthermore, Høyland (2007) and Poplin and Wang (1994) found little difference between the horizontal and vertical uniaxial compressive strength of ice from the consolidated layer. Therefore, the constitutive model of the consolidated layer is assumed isotropic.

The yield surface is pressure-dependent, and allow different strengths in compression and tension, expressed as follows:

$$\begin{aligned} F(\boldsymbol{\sigma},\boldsymbol{\kappa}) &= \frac{1}{1-\alpha}\left[\sqrt{3J_2} + \alpha I_1 + \beta(\boldsymbol{\kappa})\langle\sigma_{max}\rangle\right] - \sigma_c(\boldsymbol{\kappa}) \\ \alpha &= \frac{\sigma_b - \sigma_c}{2\sigma_b - \sigma_c} \\ \beta &= \frac{\sigma_c(\boldsymbol{\kappa})}{\sigma_t(\boldsymbol{\kappa})}(1-\alpha) - (1-\alpha) \end{aligned} \tag{9}$$

where J_2 is the second deviatoric invariant, I_1 is the first principal invariant, $\sigma_{\max}$ is the maximum principal stress, the Macauley bracket < > is defined as $\langle x\rangle = 0.5\cdot(x+|x|)$, $\boldsymbol{\kappa}$ is the hardening vector, σ_c is the uniaxial compressive strength, σ_t is the uniaxial tensile strength and σ_b is the biaxial compressive strength (tension is positive).

The initial ice strength depends on the state of the ice (temperature, strain rate, confinement) and the type of the ice (grain size and micro-porosity). Generally, the uniaxial compressive strength of an ice sample is approximately an order of magnitude greater than the uniaxial tensile strength of the same sample (Schulson and Duval, 2009). The tensile strength was not measured during the experiment (Moslet et al., 2004), we therefore assume that the uniaxial compressive strength was ten times that of the uniaxial tensile strength. Biaxial strength was neither measured and we assume that $\sigma_b = 1.2 \cdot \sigma_c$, which is in line with measurements of the biaxial strength of granular/discontinuous- columnar sea ice presented by Timco and Frederking (1984)

The main differences between the elliptical (Derradji-Aouat, 2003) failure surface and the present yield surface, is that the former is intended for brittle failure (without plasticity) while the latter is intended for ductile failure including plasticity. Additionally, the elliptical failure surface constrains the hydrostatic pressure. This implies that, upon a certain limit, increasing the hydrostatic pressure (confinement) results in a decrease in deviatoric strength. However, the data (Gagnon and Gammon, 1995; Gratz and Schulson, 1994; Jones, 1982; Rist and Murrell, 1994) used by Derradji-Aouat (2003), shows no decrease in deviatoric stress with increasing confinement. For polycrystalline ice at -11°C, Jones (1982) measured increasing deviatoric strength for hydrostatic pressures 0-20MPa and strain rates 10^{-6}-10^{-2}. In the problem of interest the confinement and the hydrostatic pressure is expected to be lower than 20MPa such that the shape of the failure surface at high confinement is irrelevant.

The non-associated flow rule is:

$$\mathrm{d}\boldsymbol{\varepsilon}^p = \mathrm{d}\lambda \frac{\partial G}{\partial \boldsymbol{\sigma}} \tag{10}$$

$$G = \sqrt{(\varsigma \sigma_{t0} \tan\psi)^2 + 3J_2} + \frac{I_1}{3}\tan\psi$$

Where, $d\boldsymbol{\varepsilon}^p$ is the plastic flow increment vector, $d\lambda$ is the plastic multiplier, G is plastic potential, σ_{t0} is the initial uniaxial tensile strength, ψ is the dilatancy angle and ς is the eccentricity.

Frictional materials, such as ice, concrete or rock, may contract or dilate during shearing, such that the plastic flow disobeys the normality condition. To the authors knowledge, there are no experiments which provide measurements of the dilatancy of isotropic ice. However, for concrete, rock and soils, the dilatancy angle was measured in the range 0°-20°, with decreasing values for increasing confinement (for loose soils negative dilatancy angles can occur). In test 2 (next section) a dilatancy of 15° was used.

The hardening/softening of the yield surface (expressed through $d\boldsymbol{\kappa}$) is generalized though the development of uniaxial strengths ($\boldsymbol{\kappa} = [\kappa_t\ \kappa_c]^T$). In a uniaxial test, a brittle failure is defined as a sharp drop in the stress- strain curve after the peak stress, whereas a ductile failure is defined as a continuous development in the stress-strain curve after the peak stress. The intention here is to model a ductile crushing failure and to study the effect of localized failure. Høyland et al. (2000) found that the residual uniaxial compressive strength of ice samples from the consolidated layer, tested at -10°C and at nominal strain of 10% was 25-50% of the peak stress. In the next section we will test the effect of a softening behavior compared to a perfectly plastic behavior. The expression for the hardening/softening is:

$$d\boldsymbol{\kappa} = \mathbf{h}(\boldsymbol{\sigma},\boldsymbol{\kappa})\, d\hat{\boldsymbol{\varepsilon}}^p \qquad (11)$$

$$\mathbf{h}(\boldsymbol{\sigma},\boldsymbol{\kappa}) = \begin{pmatrix} r(\boldsymbol{\sigma}) & 0 & 0 \\ 0 & 0 & (1-r(\boldsymbol{\sigma})) \end{pmatrix}$$

$$\mathrm{r}(\boldsymbol{\sigma}) = \frac{\sum_{i=1}^{3}\langle\sigma_i\rangle}{\sum_{i=1}^{3}|\sigma_i|}$$

Where σ_i are the principal stress components. The plastic strain increments in the modified vector $d\hat{\boldsymbol{\varepsilon}}^p$ are ordered from largest to smallest.

TEST SETUP

Small Strain Test

The first test is a small deformation (u<<L) and small strain (<1%) analytical problem. A linearly elastic cube (1x1x1 m^3) of material is tested in uniaxial compression. The cube is compressed 0.01m, against a frictionless rigid plate. The material parameters and the analytical force is given in Table 1. The cube is discretized by 10 elements/particles per 1m. For the CEL analysis, void elements are placed around the cube (Figure 1b). Since boundary conditions cannot be applied directly to particles in the SPH simulation (Dassault-Systèmes, 2013), a frictionless and rigid top plate is used to apply the displacement to the cube (Figure 1c). In the SPH simulation a cubic spline smoothing function is used (Equation 7). By default in the SPH formulation the smoothing length is computed in the beginning of the analysis containing on average 30-50 particles.

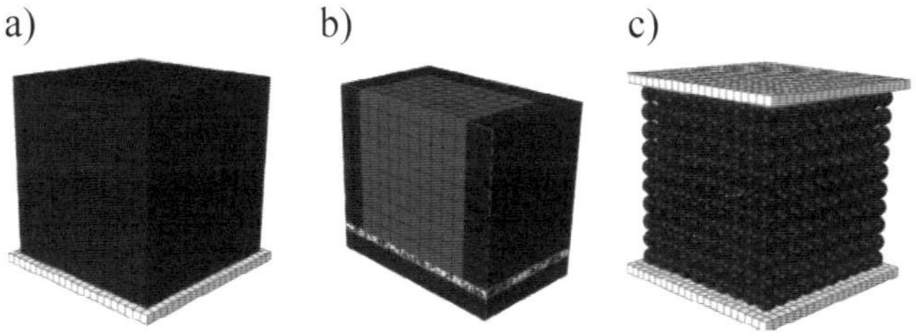

Figure 1. Uniaxial compression: the white plate is fixed and rigid a) FEM setup, blue is deformable cube b) CEL setup, red is the material and blue is the void elements c) SPH setup, particles with diameter of the parent element size additional top plate

Medium Scale Ice-Structure Indentation Test

In this test, we are modelling experiment number 2 described in Moslet et al. (2004). In the experiment, an ice sheet (0.16-0.26x7x8 m^3) was towed against a fixed cylindrical indenter (Ø0.665 m) in the Van Mijen fjord on Spitsbergen 2003. The setup is shown in Figure 2c. During the experiment the measured maximum ice force measured, was 161kN, the force time history is shown in Figure 2a. The predominant ice crystal structure was granular/isotropic (Moslet et al., 2004). The numerical setup is displayed in

a) b)

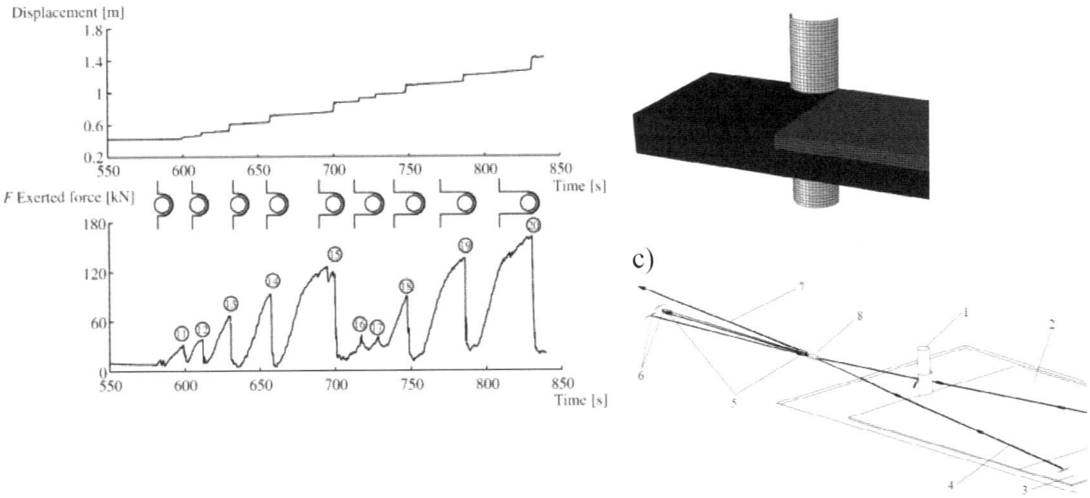

Figure 2b. Here, the cylindrical indenter is modelled as rigid and the ice sheet is fixed at the top surface and buoyancy is applied as an upward acceleration (g_B in Equation 12), according to this assumption the ice will accumulate below the ice sheet. This is done due to instabilities encountered when applying both buoyancy and gravity boundary conditions. Furthermore, the symmetry of the experiment is exploited by modeling only half of the cylinder, the ice sheet size is adjusted to 0.2x2x2 m^3. The material model, summarized above, is rate independent and therefore the velocity in the simulation is increased to 100mm/s to save computational time. Furthermore, the simulation is run until full contact between the ice sheet and the cylinder is reached. The simulations are carried out using the CEL formulation with the constitutive model described above, material parameters are summarized in Table 1. Two hardening/softening behaviors are tested, first, a perfectly plastic behavior and second a 50% reduction of κ_t and κ_c at 10% equivalent plastic strain (Table 1).

$$g_B = \frac{\rho_w - \rho_{ice}}{\rho_{ice}} g \qquad (12)$$

a) b)

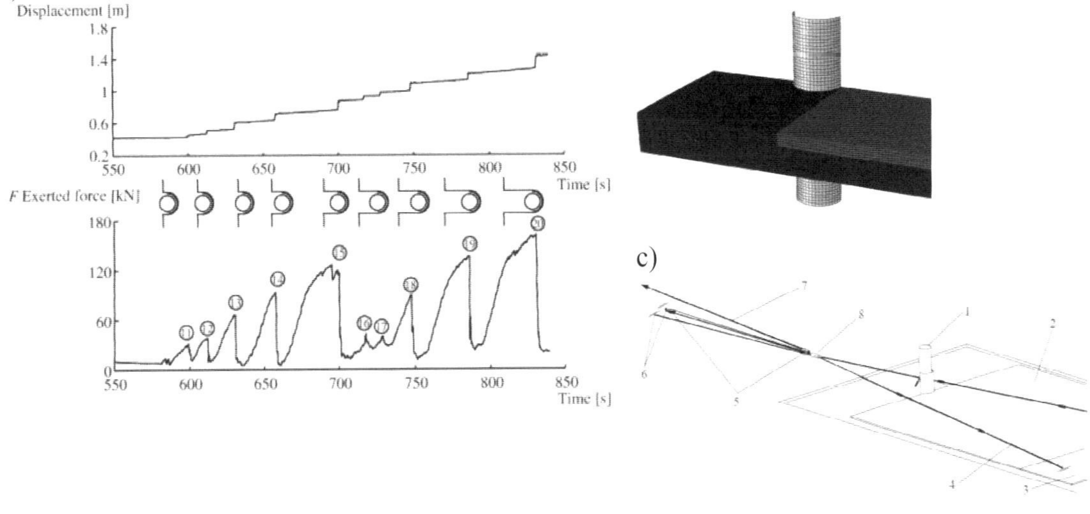

Figure 2. a) Force and displacement time histories (Moslet, 2008) b) CEL setup, the red shaped area is initially filled with material, the blue shaded area is void elements and the rigid cylinder is colored white c) Overview of the experiment setup. The arrow gives the pulling direction of the winch. 1. Cylinder structure, 2. Ice floe, 3. Parent ice piece used as towing floe, 4. Steel bars Ø35 mm, 5. Pulleys, 6. Ice anchors, 7. Chains connected to winch, 8. Load cell. (Moslet et al., 2004)

Table 1. Material properties, boundary conditions and test results

Material parameter	Test 1	Test 2 (perfectly plastic)	Test 2 (softening)
Ice density	900 kgm^{-3}	900 kgm^{-3}	900 kgm^{-3}
Water density	-	1000 kgm^{-3}	1000 kgm^{-3}
Ice elastic modulus	5 GPa	2.6 GPa	2.6 GPa
Poisson's ratio	0.3	0.3	0.3
Mean ice temperature	-	-4°C	-4°C
Uniaxial compressive strength	-	1.8 MPa	1.8 MPa (0.9 MPa ε^{p}_{c} 10%)
Uniaxial tensile strength	-	0.18 MPa	0.18 MPa (0.09 MPa ε^{p}_{t} 10%)
Biaxial compressive strength	-	2.4 MPa	2.4 MPa
Dilatancy	-	15°	15°
F_{max}	50 MN	161 kN	161 kN

RESULTS AND DISCUSSIONS

Small Strain Test

The rigid plate force is plotted in Figure 3 and the force at the end of the simulation is given in Table 2. The penalty contact stiffness was adjusted such that the contact penalty work was less than 1% of the external energy for all the formulations. The energy balance was checked and is summarized in Table 2. *Ae* is the artificial strain energy, *Vd* is viscous energy, *Pw* is penalty contact work, *Wk* is the external work and *Ie* is the internal energy. In an attempt to improve the results of the SPH simulation, the number of particles was doubled. This however had no effect on the results except increasing the computational time and causing instabilities that led to particles escaping the problem domain.

Table 2. Comparison of results

	CEL	FEM	SPH
$Ae / Ie \cdot 100\%$	2	0	0
$Vd / Ie \cdot 100\%$	0	0	0
$Pw / Wk \cdot 100\%$	1	0	-1
$Wk / Ie \cdot 100\%$	99	100	102
F_{end} [MN]	50	50	29

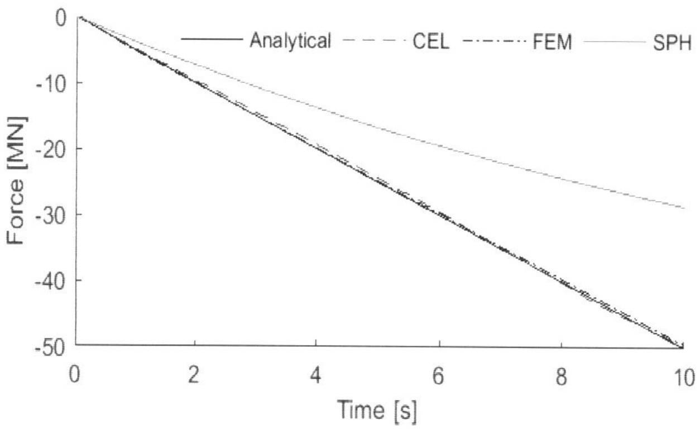

Figure 3. Forces on the bottom plate

With the element based methods (CEL, FEM) the problem was solved exact while the SPH formulation produced an error of 42%. In addition to the inaccuracy of the SPH formulation (in ABAQUS), it has several limitations. First, boundary conditions cannot be applied directly to particles. Second, different materials modelled with SPH particles do not interact with each other or Eulerian bodies (Dassault-Systèmes, 2013), such that the consolidated layer and the ice rubble modelled with SPH particles would not interact. Third, a positive (tensile) stress corresponds to a purely attractive potential in the SPH formulation, which leads to particle clumping together (tensile instability) (Liu and Liu, 2003).

Medium scale ice-structure indentation test

The result of the simulations are given in Figure 4a, where the total force on the indenter versus the ice sheet displacement is plotted. The grey line is the total force from the ice with a perfectly plastic behavior and the black line is the total force from the ice with softening behavior. The dashed line is the maximum total force measured during the experiment. Figure 4b and c shows the ice accumulation for the ice with the perfectly plastic behavior and the softening behavior, respectively. The energy balance was checked, and the contact energy diffusion was less than 4% and 8% of the external work for the perfectly plastic and softening case respectively. The viscus dissipation and artificial strain energy was less than 1% of the internal energy in both cases and the total energy was close to zero. The highest hydrostatic pressures reached during the simulations was 2MPa and 6 MPa for the perfectly plastic and the softening model respectively.

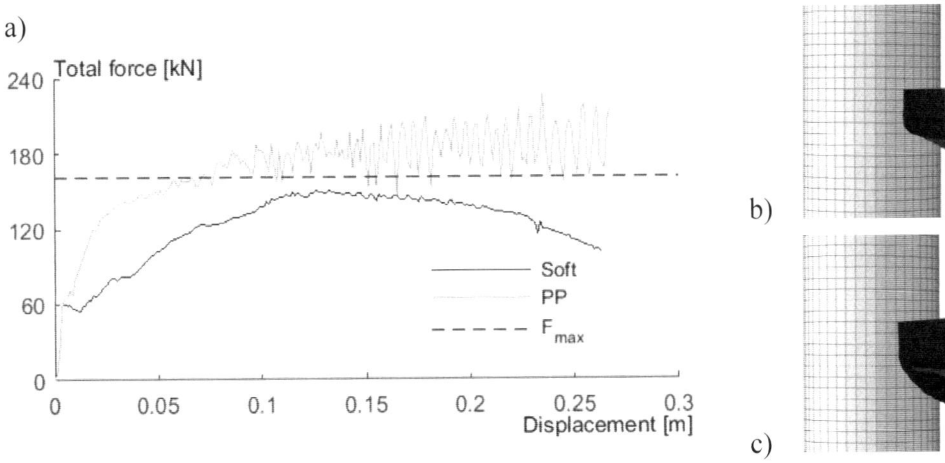

Figure 4. a) Total force on the indenter vs displacement of the ice sheet, the grey line shows the perfectly plastic behavior, the black line shows the softening (soft) behavior and the dashed line is the maximum force measured during the experiment b) is the ice accumulation with the perfectly plastic (PP) material behavior and c) is the ice accumulation with the softening behavior.

The simulation results show good agreement with the experimental measurements. The different hardening behaviors had a clear effect on the simulation results. The softening caused localized failure which increased the amount of local ice accumulation onto the structure. Furthermore, with the strength softening the total forces decreased as the plastic strain of the ice at the structure interface increased and the yield strength decreased. Therefore, the ice force decreased although the contact area increased. However, the peaks that occurred during the experiment increased in value as the ice contact area increased. Therefore, the development in the force record was best captured with the perfectly plastic behavior. The maximum force obtained during the simulation was 140% and 93% of the measured maximum force for the perfectly plastic and softening case respectively. The maximum hydrostatic pressure reached during the simulation was low (2-6MPa). Therefore, the shape of the yield surface at high hydrostatic pressures is irrelevant, as previously discussed in the section concerning the constitutive model.

To improve the results of the simulations the hardening behavior should be adjusted to the measured development in uniaxial strength, which was unfortunately not available in this study. Furthermore, for better results the system (chains, wires etc.) should be included in the simulations to study the effect of the elastic energy stored in the system. Alternatively, other data which is velocity controlled should be investigated, such as the Joia data presented in e.g. Sodhi et al. (1998).

CONCLUSIONS

The accuracy of two continuum formulations, the CEL and SPH methods, in ABAQUS/Explicit 6-13 was tested by modelling an analytical small strain problem. According to this test, the CEL method produced the highest accuracy between the two formulations. The CEL formulation and a concrete constitutive model was used to model a large deformation medium-scale indentation experiment conducted in the Van Mijen fjord in Spitsbergen in 2003. The elastic modulus and the initial uniaxial compressive strength was measured during the experiment, whereas the other material parameters were estimated. Two different hardening behaviours were tested in the material formulation. The perfectly plastic material model captured the development in the measured total force better that the softening model, whereas the softening model produced the maximum total force closest to the measured value. In future work the parameter sensitivity should be studied. Most of the parameters are uncertain and may vary significantly in nature. The influence of the variation in parameters on the response of the model is unknown and should be investigated further. Additionally, other data sets can be modelled to evaluate the applicability of the model, such as the JOIA experiments (Sodhi et al., 1998) or forces on the Nordstrømsgrund lighthouse (Sudom and Frederking, 2014).

ACKNOWLEDGEMENTS

The authors wish to acknowledge the support of the Research Council of Norway through the Centre of Research-based Innovation, SAMCoT, and the support of the SAMCoT partners.

REFERENCES

Dassault-Systèmes, 2013. Abaqus 6.13 User's manual

Derradji-Aouat, A., 2003. Multi-surface failure criterion for saline ice in the brittle regime. Cold Regions Science and Technology, 36(1–3): 47-70.

Gagnon, R.E., 2011. A numerical model of ice crushing using a foam analogue. Cold Regions Science and Technology, 65(3): 335-350.

Gagnon, R.E. and Gammon, P.H., 1995. Triaxial experiements on iceberg and glacier ice. Journal of Glaciology, 41(139): 528-540.

Gingold, R.A. and Monaghan, J.J., 1977. Smoothed particle hydrodynamics: theory and application to non-spherical stars. Monthly Notices of the Royal Astronomical Society, 181(3): 375-389.

Gratz, E. and Schulson, E., 1994. Preliminary observations of brittle compressive failure of columnar saline ice under triaxial loading. Annals of Glaciology(19): 33-38.

Høyland, K.V., 2007. Morphology and small-scale strength of ridges in the North-western Barents Sea. Cold Regions Science and Technology, 48(3): 169-187.

Høyland, K.V., Kjestveit, G., Heinonen, J. and Määttänen, M., 2000. LOLEIF ridge experiments at Marjaniemi; the size and strength of the consolidated layer., 15th International Symposium on Ice Gdansk, Polen, pp. 45-52.

Jones, J.S., 1982. The Confined Compressive Strength of Polycrystalline Ice. Journal of Glaciology, 28(98): 171-178.

Kankaanpää, P., 1997. Distribution, morphology and structure of sea ice pressure ridges in the Baltic Sea, Helsinki.

Kirkegaard, H., 1993. Ship Collisions with Icebergs. Doctoral thesis Thesis, Denmark Technology University, 149 pp.

Libersky, L.D. and Petschek, A.G., 1991. Smooth particle hydrodynamics with strength of materials. In: H.E. Trease, M.F. Fritts and W.P. Crowley (Editors). Springer Berlin Heidelberg, Berlin, Heidelberg, pp. 248-257.

Liu, G.R. and Liu, M.B., 2003. Smoothed Particle Hydrodynamics a meshfree particle method. World Scientific Publishing Co., 449 pp.

Liu, Z., Amdahl, J. and Løset, S., 2011. Plasticity based material modelling of ice and its application to ship–iceberg impacts. Cold Regions Science and Technology, 65(3): 326-334.

Lubliner, J., Oliver, J., Oller, S. and Oñate, E., 1989. A plastic-damage model for concrete. International Journal of Solids and Structures, 25(3): 299-326.

Lucy, L.B., 1977. Nummerical approach to testing the fission hypothesis. Astronomical Journal(82): 1013-1024.

Løset, S. and Kvamsdal, T., 1994. Iceberg/structure interaction-effects of ice temperature on the impact force, Proceedings of the IAHR Ice sympostium, Trondheim, Norway, pp. 205-218.

Moslet, P.O., 2008. Medium scale ice–structure interaction. Cold Regions Science and Technology, 54(2): 143-152.

Moslet, P.O., Bonnemaire, B., Valkonen, J., Høyland, K.V., Bjerkås, M., Dybdahl, J. and Løset, S., 2005. Sea ice- vertical pile interaction experiment, part 3: test results 2004, 18th International Conference on Port and Ocean Engineering Under Arctic Conditions, Portsdam, USA, pp. 471-480.

Moslet, P.O., Liferov, P., Nilsen, R., Høyland, K.V., Bjerkås, M., Bonnemaire, B., Dybdahl, J. and Løset, S., 2004. Sea ice- vertical pile interaction experiment, part 1: test set-up and ice properties, 17th International Symposium on Ice IAHR, St. Petersburg, pp. 11-16.

Poplin, J.P. and Wang, A.T., 1994. Mechanical properties of rafted annual sea ice. Cold Regions Science and Technology, 23(1): 41-67.

Riska, K. and Frederking, R., 1987. Ice load penetration modelling, Proceedings of the Port and Ocean Under Arctic Conditions Conference Fairbanks, Alaska, USA, pp. 317-327.

Rist, M.A. and Murrell, S.A.F., 1994. Ice triaxial deformation and fracture. Journal of Glaciology, 40(135): 305-318.

Schulson, E.M. and Duval, P., 2009. Creep and Fracture in Ice Cambridge University Press.

Sodhi, D.S., Takeuchi, T., Nakazawa, N., Akagawa, S. and Saeki, H., 1998. Medium-scale indentation tests on sea ice at various speeds. Cold Regions Science and Technology, 28(3): 161-182.

Sudom, D. and Frederking, R., 2014. First-year ice ridge loads on Norstrømsgrund lighthouse in 2002 and 2003, ICETECH 14.

Timco, G.W. and Frederking, R.M.W., 1984. An investigation of the failure envelope of granular/discontinuous-columnar sea ice. Cold Regions Science and Technology, 9(1): 17-27.

Is Linear Hydrodynamics Sufficient to Predict Breaking Lengths Correctly?

Chris Keijdener[1,2], Andrei Metrikine[1,2]
[1] Delft University of Technology, Delft, Netherlands
[2] SAMCoT, Trondheim, Norway

ABSTRACT

The bending failure of level ice in 2D is mainly characterized by its breaking length and interaction forces. In this paper a study is carried out to assert if a 2D ice-structure interaction model based on an incompressible, inviscid and irrotational fluid together with the *linearized* Bernoulli equation for ice can reproduce the dependence of the breaking length on the interaction velocity as observed in experiments. To this end a 2D model is composed in which the ice, modelled as a semi-infinite Euler-Bernoulli beam, is pushed into a rigid, immovable, downward-sloping structure with a fixed horizontal velocity. The beam rests on the finite depth, infinitely wide fluid layer.

To assert whether this linear description of the fluid suffices, the velocity dependence of the breaking length predicted by this model is compared with the experimentally validated 2D model of (Valanto 1992). The comparison shows that linear hydrodynamics gives significant error in certain velocity ranges. Recommendations are given as to a fluid model that would result in an improved prediction of the velocity dependence of the breaking length.

KEY WORDS: Bending failure, hydrodynamics, modeling, level ice.

INTRODUCTION

The breaking length of level ice is a widely studied topic in ice-structure interaction. A correct prediction is important for several reasons. Firstly, because the breaking length is directly linked to the time of failure, and therefore to the duration of the interaction, it governs the amount of work done by the ice on the structure and therefore governs the magnitude of the structure's response. Secondly, the breaking length, together with the interaction velocity, determines the period of ice failure and so is important for the dynamic response of the structure. A correct predictions of the breaking length is therefore imperative for a dynamic ice-structure interaction model.

To the authors' knowledge, only very few hydro-elastic ice-structure interaction model have been implemented to date, namely (Valanto 1992; Valanto 2001; Valanto 2006; Lubbad et al. 2008; Lu et al. 2012). Of these models only one is in 2D, namely (Valanto 1992). In this paper the dependence of the breaking length of level ice on the interaction velocity, i.e. the *dynamic breaking length*, is presented and shows good agreement with experimental results.

It remains unclear, however, which components of Valanto's model are essential in its correct prediction of the dynamic breaking length. As each component can incur a significant implementation cost, knowing which ones may be neglected is essential.

The goal of this paper is to determine whether a linear description of the fluid can replicate the dynamic breaking lengths observed by Valanto. This linear model is introduced in the next section. After this, the model's breaking lengths are compared with the experimental data by Valanto. This comparison will show that the proposed linear model has significant errors in certain velocities ranges. A discussion follows to estimate which assumption has led to this discrepancy.

In this paper 2D models are used for the sake of simplicity but the qualitative findings are postulated to also apply to 3D.

MODEL DESCRIPTION

In this section a brief description of the linear model is given. The purpose of this section is to highlight the assumptions on which the model is based so that any discrepancies in the breaking behavior can be traced back to a difference in the assumptions made. Because of this, details regarding the implementation are kept to a minimum for brevity. The model is depicted in the e below.

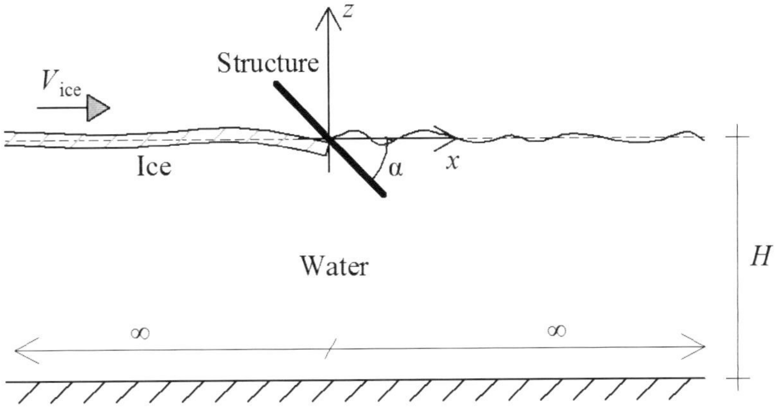

Figure 1. Problem overview of the linear 2D model

For simplicity the structure itself is considered in a simplified manner. Its geometry is not taking into account as a boundary condition for the fluid. However, for determining the ice-structure interaction loads the structure is modeled as a rigid, immovable, downward sloping wall with an angle α. The ice itself, located at $x \leq 0$, moves with a constant velocity V_{ice} towards the structure. The ice then fails dynamically because of the initial impact or quasi-statically as it is gradually pushed further down the wall. In both cases the ice fails in pure bending.

The ice rests a the fluid layer that has a depth H. The fluid layer is infinite in horizontal direction. As the ice is deflected, surface waves are generated that propagate away from the interaction point. These radiated waves act as a form of energy dissipation. Additionally, the fluid underneath the ice has to be mobilized when the ice deforms, resulting in additional resistance. Both effects thus resist the vertical motions of the ice. Since the ice is kinematically forced to move downward because of the structure, the extra resistance results in higher stresses in the ice and therefore in earlier times of failure and shorter breaking lengths. Note that these forms of resistance are not present in models that only capture hydrostatics. The mathematical description of the model is given next.

Mathematical Description of the Model

The fluid is assumed to be incompressible, inviscid and irrotational and so is governed by the Laplace equation:

$$\Delta\phi(x,z,t) = 0 \qquad \forall\, x \in (-\infty,0] \cap z \in [-H,0] \tag{1}$$

Its solution is sought in the form of the displacement potential $\phi(x,z,t)$. The boundary condition (BC) at the seabed prevents penetration of the fluid into the seabed:

$$\phi_z(x,-H,t) = 0 \tag{2}$$

where the subscript denotes derivatives. The fluid pressure is given by the *linearized* Bernoulli equation:

$$p(x,z,t) = -\rho_w\left(\ddot{\phi} + g(\phi_z + z)\right) \tag{3}$$

where the linearization removes the *dynamic pressure term* $1/2\left(\dot{\phi}_x^2 + \dot{\phi}_z^2\right)$, ρ_w is the fluid density, g the gravitational constant and the dot denotes time derivatives. At the surface the fluid pressure has to satisfy the following x-dependent BC:

$$p(x,0,t) = \begin{cases} \rho_i A\ddot{w} + EIw'''' & \forall\, x \in (-\infty,0] \\ 0 & \forall\, x \in (0,\infty) \end{cases} \tag{4}$$

where at $x \leq 0$ the ice is present and so the fluid pressure must balance with the stresses in the ice (modeled by a semi-infinite Euler-Bernoulli beam) and at $x > 0$ the fluid must satisfy the pressure release condition. Additionally, $w(x,t)$ is the displacement of the ice, ρ_i its density, A the area of its cross-section and I the second moment of its area. Continuity between ice and fluid dictates that their vertical displacement must be the same along their interface and so:

$$w(x,t) = \phi_z(x,0,t) \quad \forall\, x \in (-\infty,0] \tag{5}$$

Lastly, at the contact interface with the structure, the ice is assumed to crush and so the contact force is given by:

$$F_{contact}(t) = \begin{cases} \sigma_c A(t) & \text{when crushing (stage 1)} \\ \sigma_c^*(t)A^* & \text{when not crushing (stage 2)} \end{cases} \tag{6}$$

where σ_c is the crushing strength of the ice and $A(t)$ is the contact area. The contact model distinguishes between two stages. In the first stage the ice crushes and consequently the contact area is growing. Vibrations of the ice can cause a temporary drop in the required contact force and so crushing stops, entering stage 2. The time moment the contact force starts to drop is defined as t^*. During stage 2 the area remains constant at $A^* = A(t^*)$ and the pressure becomes variable within $\sigma_c^*(t) \in (0,\sigma_c)$. If at some point $\sigma_c^*(t) \leq 0$ contact is lost and if $\sigma_c^*(t) \geq \sigma_c$ crushing recommences (back to stage 1). $\sigma_c^*(t)$ is computed using

Lagrangian Multipliers as the contact is assumed to be stiff during the second stage. The force has an arm with respect to the neutral axis of the beam, generating a tip moment. Based on these loads the BCs for the beam are:

$$\begin{cases} EIw'''(0,t) = F_{contact}(t) \\ EIw''(0,t) = M_{contact}(t) \end{cases} \tag{7}$$

Further details of the contact model are given in (Keijdener & Metrikine 2014) but are omitted here as they are not relevant for this paper.

Solution Method

The defined problem is solved semi-analytically using an approach very similar to the Boundary Element Method. First, a Green's Matrix $\boldsymbol{G}(t)$ is computed that relates the excitation by contact loads $\boldsymbol{F}(t)$ to the vertical displacement and slope of the beam tip $\boldsymbol{u}(t)$. $\boldsymbol{G}(t)$ is thus a 2x2 matrix. Time-integration is then done based on the convolution of this Green's Matrix with the contact loads, so $\boldsymbol{u}(t) = \boldsymbol{G}(t) * \boldsymbol{F}(t)$, where $*$ denotes convolution.

The Green's Matrix is given by the numerically computed Inverse Fourier Transform of the frequency domain response $\widetilde{\boldsymbol{G}}(\omega) = \widetilde{\boldsymbol{T}}^{-1}(\omega)\widetilde{\boldsymbol{E}}(\omega)$ of the system to the elementary loading functions $\widetilde{\boldsymbol{E}}$. These loading functions act on both of its inputs (the tip force and moment) and are introduced to be able to perform the convolution numerically. They can be seen as a consequence of approximating the impulse (delta function) with a load that acts over a finite duration. The non-linear nature of the contact loads is handled using iterations at each time-step.

The transfer functions of the system $\widetilde{\boldsymbol{T}}$ are sought using two potentials: $\phi_{ice}(x,z,t)$ and $\phi_{ow}(x,z,t)$. ϕ_{ice} satisfies the ice-covered surface BC and the open water potential ϕ_{ow} satisfies the free-surface condition, see Eq. (4). Both potentials satisfy their respect BC along their entire surface, so $\forall\, x \in (-\infty, \infty)$. One half of each potential is then taken and these two halves are effectively 'glued' together using Eigenfunction Matching so that together they again span the full domain $(-\infty, \infty)$. This 'gluing' process assures that the fluid acts as a smooth continuum but, since each potential satisfies a different surface BC, one half is now covered with ice and the other half is not. An overview of the solution method is shown below in figure 2.

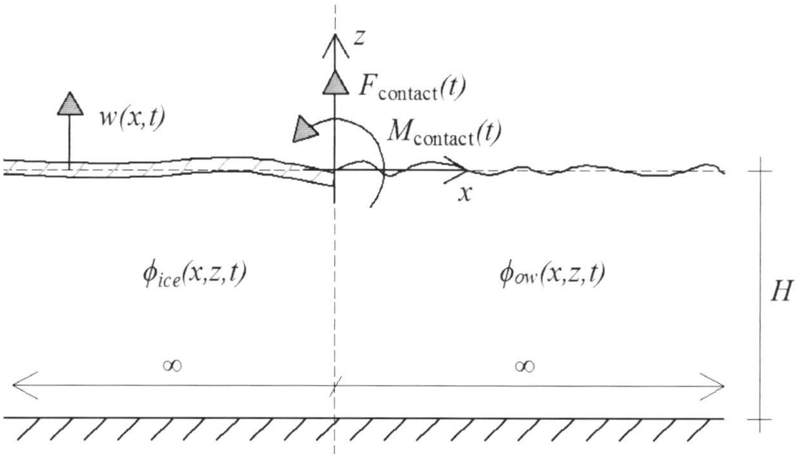

Figure 2. The solution is found by splitting the problem into two subdomains.

Using this model the breaking length of the level ice is computed for a range of velocities. The breaking length is defined by the location where the bending stress first exceeds the flexural strength σ_{fl} of the ice. The results are studied next. The parameters used for the model are the same as those used by Valanto and are: ice thickness $h = 1/33.33$ m, $\rho_i = 916$ kg/m^3, $\rho_w = 1025$ kg/m^3, $g = 9.81$ m/s^2, $H = 1$ m, $\sigma_c = 10.67$ kN/m^2, $\sigma_{fl} = 23.78$ kN/m^2, ice-steel friction coefficient $\xi = 0.1$ [-] and hull angle $\alpha = 15°$.

COMPARISON OF THE RESULTS

The graph below shows a comparison between the dynamic breaking length predicted by the herein proposed model and Valanto's model described in (Valanto 1992):

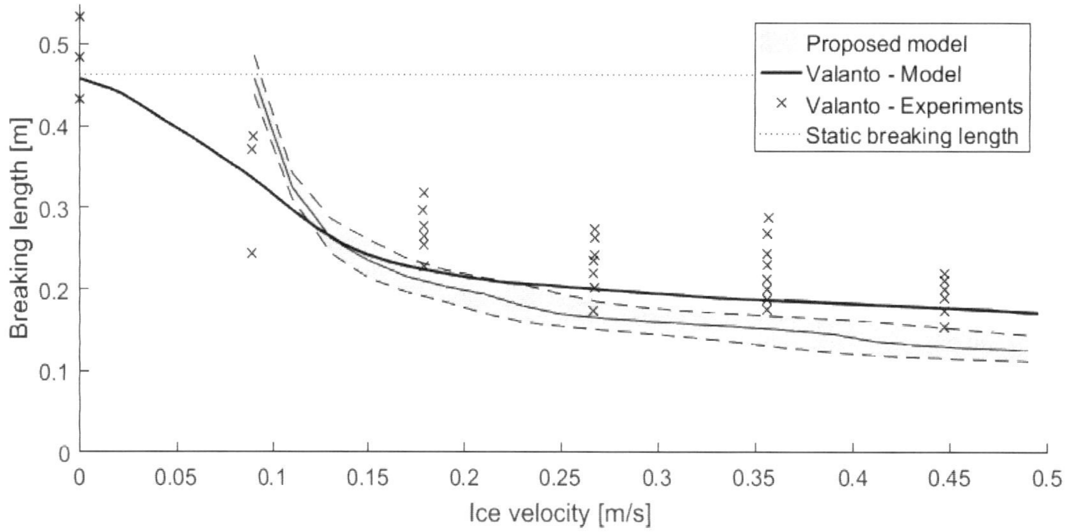

Figure 3. A comparison between the predicted dynamic breaking lengths. The Valanto data was directly extracted from figure 20 in the paper.

The figures shows the breaking lengths predicted by the proposed model, indicated by the gray area. A single value for the breaking length does not accurately portray the behavior seen by the model because at the time of failure a large portion of the ice is very close to failing. The gray area indicates the portion of the ice where the bending stress is higher than 99% of the yield stress at the moment of failure.

The graph shows there is a significant difference between the two models, especially at low velocities. This indicates that Valanto's model accounts for more forms of resistance causing the ice to fail with smaller breaking lengths and therefore earlier. The extra resistance even acts at very low velocities. This discrepancy must be caused by the different assumptions made by both models. Below is a list of the major items that are present in Valanto's model but are not present in the proposed model:

- the dynamic fluid pressure (see Eq. (4))
- rotational inertia of the ice
- steady-flow of the fluid around the vessel
- the effect of the vessel geometry on the fluid flow
- axial deformations of the ice
- ventilation effects

Of this list only two items were considered as a possible source of significant resistance: the rotational inertia and the dynamic pressure. To assess if rotational inertia can generate significant resistance it was incorporated by adding $\rho_i J \ddot{w}'$ to the equation of motion of the ice in Eq. (4), where J is the mass moment of inertia of the ice's cross-section. However, the rotational inertia has an insignificant effect $(< 1\%)$ on the breaking length and based on this was deemed not to be the source of the extra resistance.

Next the dynamic pressure was checked. Because of its non-linear nature incorporating it into the model is a non-trivial step. To gain insight into its effect without have to do a complete upgrade of the model to accommodate the nonlinearity, the dynamic pressure was evaluated using the solution of the linear model. To this end the transfer functions $\widetilde{\boldsymbol{T}}$ were analyzed and it was found that the magnitude of the dynamic pressure can be as much as 60% compared to the magnitude of all other forces acting on the ice combined (these are the ice inertia, bending stresses, hydrostatic pressure and linear hydrodynamic pressure). Additionally it was found that the dynamic pressure behaves very much like a rapidly decaying exponential, with a maximum magnitude at the contact and then rapidly decaying in space.

Based on this analysis it was concluded that the dynamic pressure is the most likely cause of the discrepancy given the that it can take on such large magnitudes in the vicinity of the contact. To verify this hypothesis the model is currently being upgraded to include the dynamic pressure. Future work based on the upgraded model will give insight into whether the dynamic pressure is indeed the source of the discrepancy and thus an essential component of dynamic ice-structure interaction.

CONCLUSIONS

This paper showed that a 2D ice-structure interaction model based on a linear description of the fluid, i.e. incompressible, inviscid and irrotational fluid together with the *linearized* Bernoulli equation for the fluid pressure, gives a significant error in its prediction of the dependence of the breaking length of level ice on the interaction velocity. The most likely cause for this discrepancy was identified as the omission of the dynamic pressure term that was neglected when the Bernoulli equation was linearized. Future work aims to verify whether this term is indeed the source of the discrepancy and whether this terms is essential for the correct prediction of the dynamic breaking length of level ice and should be included in ice-structure interaction models.

ACKNOWLEDGEMENTS

This work was supported by the SAMCoT CRI through the Research Council of Norway and all of the SAMCoT Partners. The authors would also like to thank the Wikiwaves.org website for providing a good open-source reference for hydro-elastic problems involving ice.

REFERENCES

Keijdener, C. & Metrikine, A. V, 2014. The effect of ice velocity on the breaking length of level ice failing in downward bending. *Proceedings of 22nd IAHR International Symposium on Ice.*

Lu, W., Løset, S. & Lubbad, R., 2012. Ventilation and Backfill Effect during Ice-sloping Structure Interactions. *Proceedings of the 19th IAHR Symposium on Ice, Dalian, China*, pp.826–841.

Lubbad, R., Moe, G. & Løset, S., 2008. Static and dynamic interaction of floating wedgeshaped ice beams and sloping structures. In *Proceedings of the 19th IAHR Symposium on Ice, Vancouver, Canada*. pp. 227–237.

Valanto, P., 2006. On the Ice Load Distribution on Ship Hulls in Level Ice. *Icetech 2006*, pp.1–10.

Valanto, P., 1992. The Icebreaking Problem in Two Dimensions : Experiments and Theory. *Journal of Ship Research*, 36(4), pp.299–316.

Valanto, P., 2001. The resistance of ship in level ice. *Transactions of the Society of Naval Architects and Marine Engineers*, 109, pp.53–83.

POAC'17
Busan, KOREA

Proceedings of the 24th International Conference on
Port and Ocean Engineering under Arctic Conditions
June 11-16, 2017, Busan, Korea

The Effects of Consolidation Time on the Strength and Failure Behavior of Freshwater Ice Rubble

Hamid Shayanfar[1, 2], Eleanor Bailey[1], Robert Pritchett[1], Rocky Taylor[1, 2]
[1] C-CORE's Centre for Arctic Resource Development, St. John's, Canada
[2]Memorial University of Newfoundland, St. John's, Canada

ABSTRACT

Ice ridges are common features in Arctic and sub-Arctic regions. They are large accumulations (typically 5-30 m) of ice rubble formed due to compressive or shear forces in ice cover. Understanding the deformation behavior and strength of ice rubble is key to estimating ice ridge loads on offshore structures. Medium-scale punch tests were conducted to measure and observe the strength and failure behavior of freshwater ice rubble. A custom-built punch box measuring 3.05 m in length and 0.94 m in width and height was used to perform the tests. The box walls were made from Plexiglas so that failure mechanisms could be observed. Ice rubble beams of nominal thickness 50 cm were produced by placing randomly sized ice pieces into the punch box, filled with water at its freezing temperature. After a specified consolidation time, the ice rubble beam was deformed by pushing a platen vertically downwards though the center of the beam until failure. Results showed that as the consolidation time varied from 0.2 to 70.5 hours, the failure behaviour changed considerably. For consolidation times less than 4 hours, the ice beam failed progressively and tended to fail by shearing; whereas, at times greater than 4 hours the beam failed by bending. The change in failure behaviour is controlled by the degree of bonding between ice blocks.

KEY WORDS: Ice rubble; Ridges; Consolidation time; Shear strength; Bending strength

NOMENCLATURE

η	Porosity
τ	Shear strength
σ	Bending strength
F_H	Hydraulic load
F_B	Buoyancy force
ρ_w	Density of water
g	Gravity
δ	Water level displacement
V_b	Nominal volume of submerged ice beam
A	Failure area
w	Box width
L	Box length

INTRODUCTION

Ice ridges are large accumulations of ice rubble that form due to compression or shear in the ice cover. They consist of two parts, a sail and a keel. The keel is the submerged part of an ice ridge, which is often frozen at the top forming what is often referred to as the consolidated or re-frozen layer. The non-submerged part of the ridge, the sail, is on average one fifth of the keel's depth (Timco et al, 2000). Ice ridges are complex features since they consist of individual ice blocks, random in size and orientation, and are bonded to each other with different degrees of strength from the consolidated layer to the base of the keel. Ridges are the thickest sea ice features in Arctic and sub-Arctic regions, and as such, must be considered during design loads estimation for vessels, offshore structures and subsea infrastructure.

A number of methods have been used to measure the mechanical properties of ice rubble. Initial tests began with the direct shear box test (Keinonen and Nyman (1978), Prodanovic (1979), Weiss et al., (1981), Hellmann (1984), Fransson and Sandkvist (1985)). Timco and Cornett (1999) reviewed these data and suggested that the wide range of data corresponded to the characteristics of the shear box setup, which induces non-uniform deformation and stress distribution in the ice rubble and forces the sample to fail along an induced failure plane (which may not be the weakest one). Other methods to investigate the mechanics of ice rubble included the shear box tests by Urroz and Ettema (1987), triaxial shear test by Gale et al. (1987) and Wong et al. (1990), biaxial shear tests by Timco et al. (1992), Sayed et al. (1992), Løset and Sayed (1993), Cornett and Timco (1995), and in situ punch tests conducted by Leppäranta and Hakala (1992), Croasdale and Associates Ltd. (1997, 1998) and Heinonen and Määttänen (2000).

Azarnejad and Brown (2001) and Lemee and Brown (2002) conducted laboratory scale punch tests on freshwater rubble to investigate the failure processes under more controlled conditions than are possible in the field. Azarnejad and Brown (2001) examined the influence that beam thickness (0.2 m-0.5 m), consolidation time (0-3 h) and hydraulic ram speed (9-120 mm s^{-1}) have on shear strength. They found that with higher consolidation time the cohesion and friction increased due to the development of freeze bonds between ice blocks. They also observed that at slow rates, ice beams behaved as a Mohr-Coulomb type material and the platen pushed a rectangular or trapezoidal plug out through the entire depth of the keel and failure occurred at the edges of the plug. At higher rates, failure occurred both under the platen as well as at a large surrounding area which had a triangular wedge. Lemee and Brown (2002) carried out similar tests using the same equipment but for larger ice rubble blocks and yielded comparable results. They also argued that ice rubble of larger dimensions fail in higher displacement. They observed that the lower the initial temperature of the ice blocks, the more bonded the rubble.

In this paper, initial results from a series of medium-scale punch tests are described. The focus of this work was to investigate the influence consolidation time had on the strength and failure behaviour of a 50 cm thick rubble beam. Tests consisted of deforming an ice rubble beam by pushing a platen vertically though its center until failure occurs (see Figure 1). In each case two tests were conducted; the first on the undeformed rubble beam and, the second, on the deformed beam to investigate frictional properties between the ice rubble blocks. The load applied to the platen was measured using a load cell and hydraulic ram displacement with a string pot. Video cameras were strategically positioned around the punch box to observe failure processes.

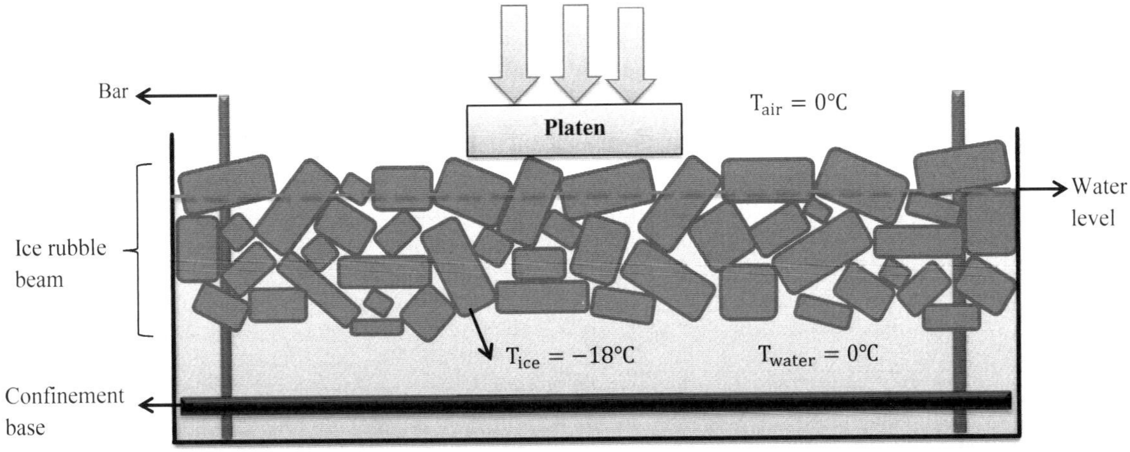

Figure 1. Schematic showing the punch test setup

APPARATUS AND INSTRUMENTATION

A custom-made punch box has been constructed for the test program that is 3.05 m in length and 0.94 m in width and height (see Figure 2). The walls are made from Plexiglas (with a grid drawn on) so that failure mechanisms and ice block motions can be observed and tracked. The punch platen is rectangular and spans the entire width of the box so that failure mechanisms can be observed. A platen width of 0.4 m was chosen based on careful consideration of how it will influence the failure of the beam and the block dimensions. Load was applied to the platen using a 20,000 lb (9 ton) hydraulic ram. To help make beams of even geometry, a bottom plate, mounted on 4 threaded rods, was brought up to produced beams that were 50 cm in thickness (see Figure 2).

During tests where the rubble was heavily bonded, the buoyancy of the non-loaded portion of the rubble was not sufficient to resist the applied load. As such, brackets had to be added to the end of the punch box to cause the beam to fail. This is similar to the procedure used by Azarnejad and Brown (2001) and Lemee and Brown (2002) where they glued small Plexiglas pieces on the tank walls to artificially increase friction between the tank walls and the ice rubble. The only difference being here that friction was only added to the ends of the box rather than the whole box to allow beams to fail in bending as well as shear.

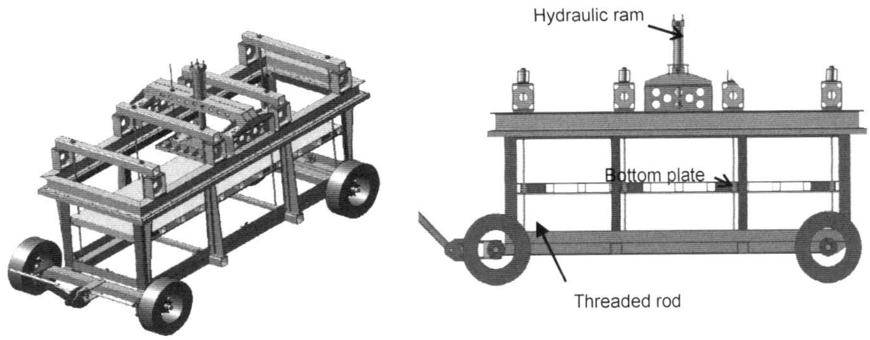

Figure 2. Computer aided design (CAD) images of the punch box assembly

A load cell was placed in line with the hydraulic ram to measure force applied to the rubble. A linear voltage displacement transducer (LVDT) was used to measure the vertical displacement and velocity throughout indentation. A total of six cameras were used to observe the punch box tests from different views. Four were placed side on to view the test through the Plexiglas window and two from the top (see Figure 3). A high speed video camera (HSV) was used to view the mid-point of the beam where the failure planes occurred, which is needed for calculating the shear strength. Three LED work lights were mounted on the opposite side of the punch box to illuminate the ice rubble beam. Cameras were synced with the load and displacement data through a noise trigger.

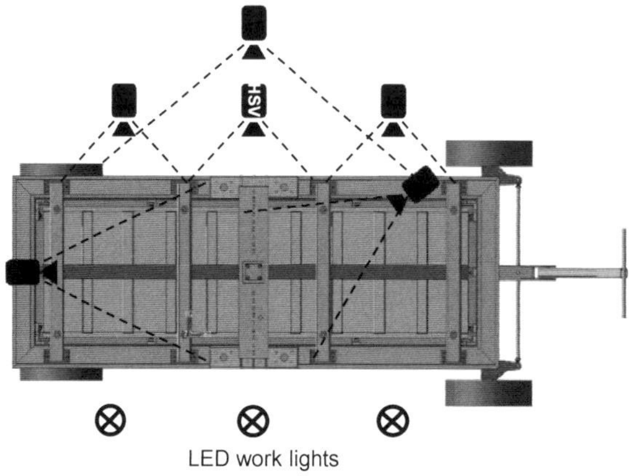

Figure 3. Camera and lighting locations

ICE RUBBLE PRODUCTION

Freshwater ice rubble was produced and stored in a refrigerated container. 1 m^2 pans were filled with tap water to an approximate depth of 10 cm and left to freeze at a nominal temperature of -18°C. Two dividers were placed in each pan prior to freezing to facilitate ice breakage. Once the water was frozen (approximately 5-6 days), pans were manually flipped over, allowing ice to break into randomly sized ice pieces. Larger sized ice pieces were broken with a hammer to give target lengths between 10 cm and 50 cm. Very small fragments and powdered ice which were less than 10 cm in thickness were swept away and thrown out. Figure 4 shows photographs of the reefer unit and the ice production methodology.

Figure 4. Ice rubble production and storage inside the refrigerated container

An analysis of the ice block dimensions was carried out to determine the ice block length, width and volume distributions. Photos of the ice blocks were taken during ice rubble fabrication and later analyzed using image processing software. Figure 5 shows that the length distribution of the ice blocks varied between 10 and 70 cm, with the majority falling in the range of 30 to 40 cm. The width distribution varied from 10 to 50 cm, with over 60% of the blocks being between the ranges of 20 to 30 cm. The length to width distribution shows that the ice blocks are generally square in shape with a maximum length to width ratio between 1 and 1.5.

Relatively large ice blocks were used in this test program so that the ice was 'thermodynamically scaled' to some extent. This we believe is particularly important to investigate the influence of consolidation time on the strength and failure behaviour of ice rubble as smaller ice blocks would not have sufficient cold reserves to form adequate freeze bonds. In addition, smaller ice pieces would result in a lower porosity and hence greater ice-ice contacts perhaps artificially increasing the rubble strength (Liferov and Bonnemaire, 2005).

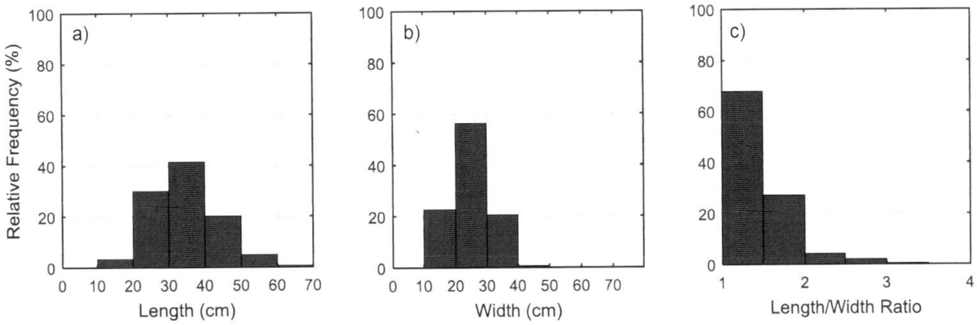

Figure 5. Histograms of the length (a), width (b) and length to width ratio (c) of the ice blocks

TEST PREPARATION & PLAN

The punch box tests were carried out in C-CORE's cold room which was set to a nominal temperature of 0°C to keep the water as close to its freezing point as possible without a consolidated layer forming. Approximately 750 kg of ice rubble were needed to produce a 50 cm thick ice rubble beam. The mass of each ice rubble bag was measured using the crane scale, which is needed to derive porosity. Whilst loading the ice rubble into the punch box the bottom plate was positioned 50 cm from the top of the box, which helped to produce a beam of even geometry.

Once the ice rubble beam was prepared, the bottom plate was lowered and the box moved back into the cold room where the ice rubble blocks were left to consolidate for a set period of time. The ice rubble beam was then deformed at a constant rate of 5 mm s^{-1} until the beam failed. After each punch test, the hydraulic ram was retracted causing the rubble beam to settle back to its pre-test position. A second test was then performed that was referred to as the friction test as it was assumed that all bonds would have failed in the previous test and as such would be a measure of the friction and interlocking between ice rubble blocks.

A total of seven punch tests were carried out to investigate the influence consolidation time had on the strength and failure behaviour of an ice rubble beam (see Table 1). The consolidation time was varied from 0 to 70.5 hrs, whilst the initial ice temperature was held constant at -18°C. Note that while best efforts were made to conduct the test in "0 hrs" the fastest test setup time that has actually been achieved was 0.2 hrs (12 minutes).

Table 1. Test matrix

Test No.	Consolidation time (hrs)	Initial ice temperature (°C)	Block size
1	0.2	-18	(Length 30-40 cm; Width 20-30 cm)
2	2.4		
3	4.2		
4	4.4		
5	10.1		
6	28.5		
7	70.5		

RESULTS

Analysis of the video and load data from the test carried out after 0.2 hours of consolidation (see Figure 6) showed that a shear plug failure took place, where the central part of the beam was punched through while the non-loaded portion of the beam remained in place. Failure of the shear plug was progressive, where the first failure took place to the right of the platen at 52 mm of displacement, followed by a second failure to the left of the platen at 85 mm of displacement. The load trace from the friction test showed that no load drops were observed, demonstrating that as expected, the loads were associated with frictional and buoyancy forces with little or no bond failure. The higher loads observed in the punch test are likely due to a combination of interlocking between ice blocks or potentially some initial freeze-bonds that may have formed in the time it took to setup and run the test. The increase in load observed with displacement in both the punch and friction tests is due to increasing buoyancy forces as the ice beam is submerged further below the water.

The failure behaviour observed in the test conducted after 70.5 hours of consolidation (see Figure 7), was noticeably different from that done at 0.2 hours, where the beam failed in bending as opposed to shear. The load trace measured during the punch test showed a single failure at 65 mm of displacement. The failure plane was located slightly to the left of the platen, as opposed to the center which would be the case if the beam was homogenous. This is not unexpected as the distribution of the ice blocks in the rubble beam is random resulting in a non-uniform stress distribution. The loads measured in 70.5 hour friction test were significantly lower than those measured in the punch test, especially during peak load. This is because the rubble beam was heavily bonded in the punch test which increased the strength of the beam, whereas in the friction test, the bonds along the failure plane had already been broken.

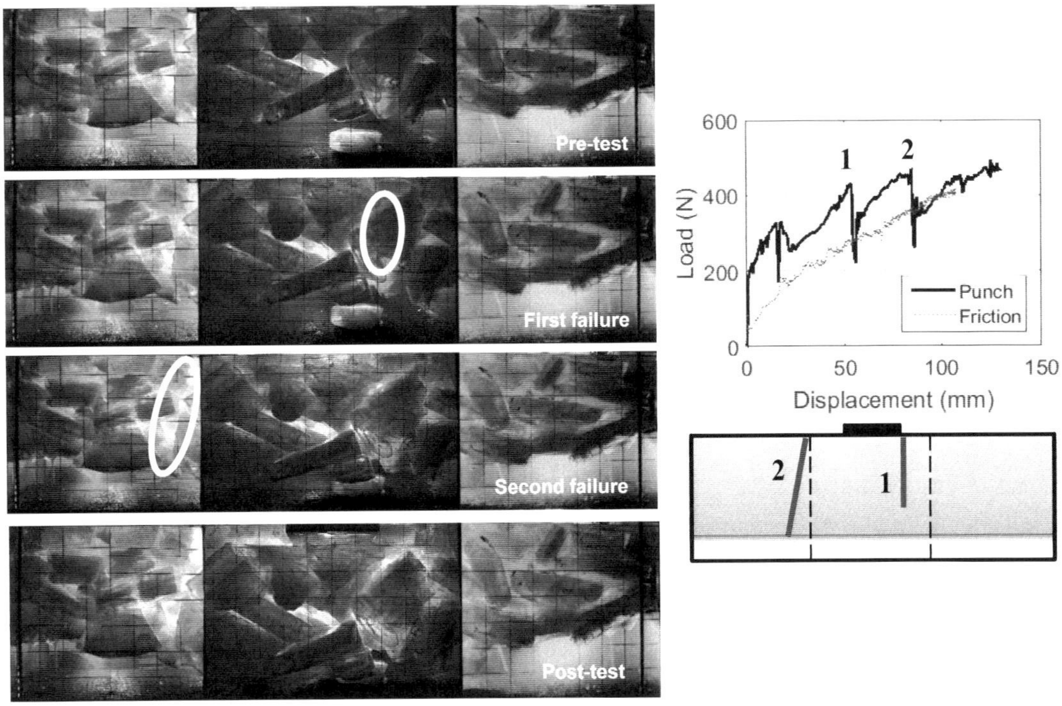

Figure 6. Analysis of video and load data from Test 1, which had a consolidation time of 0.2 hours, an initial ice temperature of -18°C, block size dimensions (length 30-40 cm, width 20-30cm) and an ice rubble beam thickness of 50 cm (see Table 1)

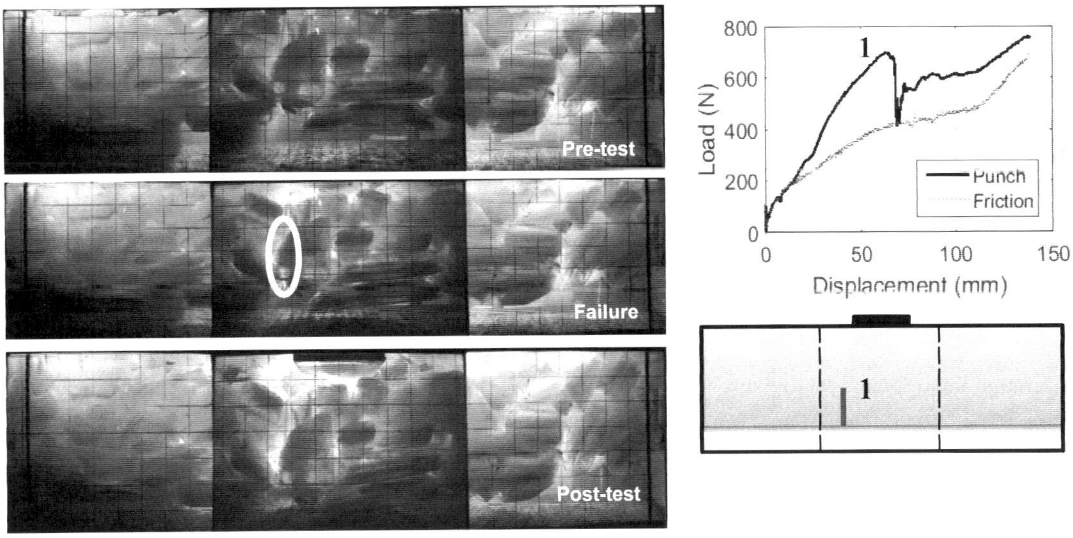

Figure 7. Analysis of video and load data from Test 7, which had a consolidation time of 70.5 hour, an initial ice temperature of -18°C, block size dimensions (length 30-40 cm, width 20-30cm) and an ice rubble beam thickness of 50 cm (see Table 1)

Figure 8 shows the force-displacement traces for all tests, as well as a schematic showing the observed failure behaviour of the beam. For consolidation times less than 4 hrs, the beam failed progressively in shear, where there were typically two load large drops observed before plug failure. For these shorter submersion times, since the bonds are expected to be weaker, a progressive sequence of block-block failures characteristic of macroscopic shear failure is more likely, which is analogous to shearing of a cohesive granular material. After 4 hrs of consolidation, the deformation behaviour appeared to change to a combination of bending and shearing or bending, where a single load drop resulted in failure of the beam. This behaviour may be attributed to increasing bonding which causes the rubble matrix to take on the behaviour of a brittle porous solid, in which higher stresses can be transmitted throughout the matrix resulting in macroscopic beam-like behaviour characterized by flexural failure. The peak load at failure increased with consolidation time from 460 N in the 0.2 hour test to 1400 N in the 4.4 hrs test, after which, it reduced to 700 N after 28.5 hrs of submersion. It is interesting that the 4.2 hrs test was much lower than the 4.4 hrs test (Tests 4 and 5 in Table 1), which was essentially a repeat test. This is possibly due to differences in setup times (where ice may have warmed up) or differences in the distribution of the blocks between the two tests. Additional repeat tests are needed to understand this discrepancy in more detail.

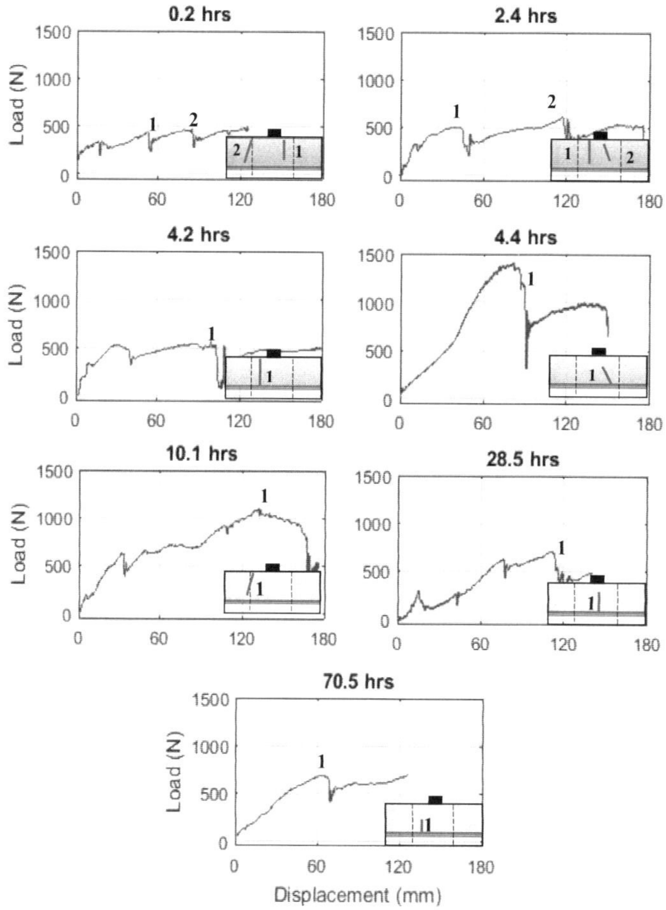

Figure 8. Force-displacement plots for all the punch box tests

DATA ANALYSIS

In order to measure the shear and bending strength of ice rubble, the buoyancy force must be calculated for each test. Two methods have been used to calculate the buoyancy force at the time of beam failure. In the first method, the rise in water level, δ, was used to estimate the buoyancy force at failure,

$$F_B = \delta w L \rho_w g \text{ [N]} \tag{1}$$

where ρ_w is the density of water, g is gravity and w and L are width and length of the box, respectively. δ was estimated from the height of the water level before the test (y_0) and at beam failure (y) i.e. $\delta = y - y_0$ (see *Figure* 9a). In the second method, the displacement observed at the underside side of the beam was used to estimate the buoyancy force at failure,

$$F_B = (1-\eta)\rho_w g V_b \text{[N]} \tag{2}$$

where V_b is the nominal volume of the displaced ice beam. V_b was determined from the video data where the position of the underside of the beam pre and post failure are shown in solid and lines, respectively (see *Figure* 9b). It is the product of the displacement-length plot area (S in Figure 9) and the box width. Note that this method assumes that the displaced volume at the underside of the beam is equal to the submerged volume at the top of the beam (i.e. no compaction of expansion took place).

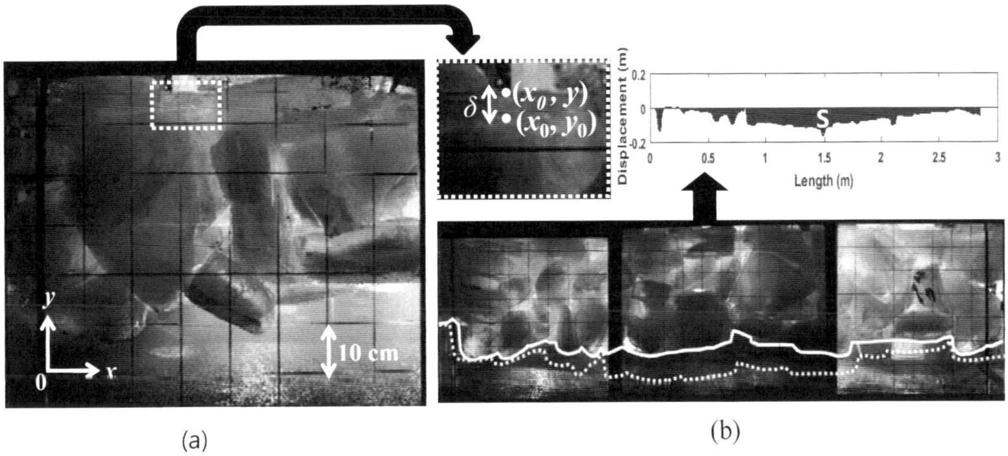

Figure 9. Snap shots of the video data showing how the water level displacement (δ) and the displaced beam volume (V_b) were estimated for the buoyancy force calculations used in method I (a) and method II (b), respectively. Note that the grid on the box is 10 cm^2.

The estimated buoyancy force at failure was calculated for each test using both method I and II (Table 2). Results for both methodologies are presented here, not only for comparison, but also because in Test 3 one of the cameras malfunctioned so we were not able to observe the full beam and in Test 6 the water level was above the viewing panel. The buoyancy forces estimated using both methodologies were similar for all tests with the exception of Test 5. We are uncertain why the buoyancy force calculated using method II in Test 5 was higher than that calculated using method I. It was, however, noted that in this test the beam failed at a much greater ram displacement. This could have resulted in greater tensile splitting at the bottom of the beam resulting in an increased porosity (which was assumed to be constant in

Eq. 2). Table 2 shows that with increasing hydraulic ram displacement the buoyancy force increases.

The shear and bending strength of the ice rubble beam was estimated from the following equations,

$$\tau_{shear} = \frac{F_H - F_B}{A} \text{ [kPa]} \tag{3}$$

$$\sigma_{bending} = \frac{3(F_H - F_B)L}{2wh^2} \text{ [kPa]} \tag{4}$$

where A is the observed failure area, F_H is the peak load at failure and L, w and h are length, width and thickness of the ice beam, respectively. The buoyancy force used to calculate the shearing/bending strength was taken as the average of both methodologies when possible. In Table 3 the shear and bending strengths for different consolidation times are given. No values was given for Test 3, as previously mentioned, the HSV camera malfunction so the failure area could not be determined.

Table 2. Estimated buoyancy force at peak load

Test No.	Consolidation time (hrs)	Hydraulic ram displacement (mm)	Porosity	Buoyancy force (N)		Average buoyancy force (N)
				Method I	Method II	
1	0.2	84.84	0.39	396	346	371
2	2.4	116.9	0.38	561	600	580.5
3	4.2	100.5	0.37	528	-	528
4	4.4	80.61	0.34	682	661	671.5
5	10.1	132.1	0.44	695	920	807.5
6	28.5	106.5	0.42	-	448	448
7	70.5	64.25	0.41	504	489	496.5

Table 3. Estimated shear/bending strength at peak load

Test No.	Consolidation time (hrs)	Observed mode of failure	Shear strength (kPa)	Bending strength (kPa)
1	0.2	Shear	0.24	-
2	2.4	Shear	0.1	-
3	4.2	-	-	-
4	4.4	Bending	-	14.3
5	10.1	Bending	-	5.5
6	28.5	Bending	-	4.94
7	70.5	Bending	-	3.82

DISCUSSION AND CONCLUSION

A total of seven punch box tests were conducted to investigate the influence that consolidation time has on the strength and failure behaviour of a 50 cm thick freshwater ice rubble beam. Results showed that as the consolidation time varied from 0.2 hrs to 70.5 hrs, the failure behaviour changed considerably. For consolidation times less than 4 hrs, the ice beam failed progressively and tended to fail by shearing, whereas at times greater than 4 hrs the beam failed in a combination of shearing and bending or pure bending. The change in failure behaviour is controlled by the degree of bonding between ice blocks, where at low consolidation times the rubble was less bonded and behaved more like a cohesive granular material. At greater consolidation times the increased bonding caused the rubble to behave like a porous solid.

Due to the different failure modes observed, the data were analyzed both in terms of a shear and bending strength. At low consolidation times (0 to 2.4 hrs), the shear strength of the rubble varied from 0.1 kPa to 0.24 kPa with an average of 0.17 kPa. These are similar to the values report by Azarnejad and Brown (2001), which is surprising considering the different test setup, block size dimensions, initial ice temperatures and ram speeds. At higher consolidation times (4.4 to 70.5 hrs), the flexural strength varied from 3.82 kPa to 14.3 kPa with an average of 7.14 kPa.

Given the results presented here it is clear that the failure behaviour of the rubble changes considerably with consolidation time. In such cases it may not be correct to assume the rubble fails in pure shear, as is currently done in most ice ridge loading models, but rather a combination of shear and bending. Further work is currently ongoing to investigate these processes in more detail.

ACKNOWLEDGEMENTS

The authors would like to thank Ken Croasdale for many useful discussions and the Hibernia Management and Development Company, Ltd. (HMDC) and the Research and Development Corporation of Newfoundland and Labrador (RDC) for their financial support.

REFERENCES

Azarnejad, A., & Brown, T., 2001. Ice rubble behavior in punch test. *Journal of Cold Regions Engineering*, 15(3), pp. 135-153.

Cornett, A., & Timco, G., 1995. Laboratory tests on the mechanical properties of saline ice rubble . NRC Report HYD-CTR-002, 171 pp.

Croasdale & Associates, 1997. In situ Ridge Strength Measurements. A study sponsored by NRC (PERD) and Exxon Production Research Co.

Croasdale & Associates, 1998. In situ Ridge Strength Measurements. A study sponsored by NRC (PERD) and Exxon Production Research Co.

Fransson, L., & Sandkvist, J., 1985. Brash ice shear properties - laboratory tests. *Proceedings of the 8th International Conference on Port and Ocean Engineering under Arctic Conditions*, Narssarssuaq, Greenland, vol. 1, pp. 75-87.

Gale, A., Wong, T., Sego, D., & Morgenstern, N., 1987. Stress-strain behavior of cohesionless broken ice. *Proceedings of the 9th International Conference on Port and Ocean Engineering under Arctic Conditions,* Fairbanks, AK, vol. 3, pp. 109-119.

Heinonen, J., & Määttänen, M., 2000. LOLEIF ridge-loading experiments-analysis of rubble strength in ridge keel punch test . *Proceedings of the 15th International Symposium on Ice*, Gdnask, Poland, vol.1, pp. 63-72.

Hellmann, J., 1984. Basic investigations of Mush Ice. *Proceedings of the 7th International Symposium on Ice*, Hamburg, Germany, vol.3, pp. 37-55.

Keinonen, A., & Nyman, T., 1978. An experimental model-scale study on compressible,frictional and cohesive behavior of broken ice masses. *Proceedings of the International Symposium on Ice*, Lulea, Sweden. Vol. 2, pp. 335-353.

Lemee, E., & Brown, T., 2002. Small-scale plane strain punch tests. *Proceedings of the 16th IAHR International Symposium on Ice*, Dunedin, New Zealand, vol. 2, pp. 1-8.

Leppäranta, M., & Hakala, R., 1992. The structure and strength of first-year ridges in the Baltic Sea. *Cold Regions Science and Technology* 20, pp. 295-311.

Liferov, P., & Bonnemaire, B., 2005. Ice rubble behaviour and strength: Part I. Review of testing and interpretation of results. Cold Regions Science and Technology 41, pp. 135-151.

Løset, S., & Sayed, M., 1993. Proportional strain tests of fresh water ice rubble. *Cold Regions Science and Technology* 7 (2), pp. 44-61.

Prodanovic, A., 1979. Model tests of ice rubble strength. *Proceedings of the 5th International Conference on Port and Ocean Engineering under Arctic Conditions*, Trondheim, Norway, pp. 89-105.

Sayed, M., Timco, G., & Lun, L., 1992. Testing ice rubble under proportional strains. *Proceedings of Offshore Mechanics and Arctic Engineering Conference*, Calgary, Canada. pp. 335-341.

Timco, G., & Cornett, A., 1999. Is ρ a constant for broken ice rubble? *Proceeding of the 10th Workshop on River Ice Management with a Changing Climate*, Winnipeg, Manitoba, Canada. *pp. 318-331.*

Timco, G., Croasdale, K., & Wright, B., 2000. An Overview of First-Year Sea Ice Ridges. NRC Publications Archive, Canada.

Timco, G., Funke, E., Sayed, M., & Laurich, P., 1992. A laboratory apparatus to measure the behavior of ice rubble. *Proceedings of Offshore Mechanics and Arctic Engineering Conference*, Calgary, Canada, pp. 369-375.

Urroz, G. E., & Ettema, R., 1987. Simple-shear box experiments with floating ice rubble. *Cold Regions Science and Technology*, 14, 185-199.

Weiss, R., Prodanovic, A., & Wood, K., 1981. Determination of ice rubble shear properties . Proceeding of the International Symposium on Ice, Quebec, Canada, pp. 860-872.

Wong, T., Morgenstern, N., & Sego, D., 1990. A constitutive model for broken ice. *Cold Regions Science and Technology* 17, 241-252.

Preliminary Analysis on the Drift Ice Floe Areas and Concentration near the Structures along Bohai Coastal based on Satellite Images

Zhijun Li[1], Jiankang Wang[1], Ge Li[2], Kecai Guo[2]

[1] State Key Laboratory of Coastal and Offshore Engineering, Dalian University of Technology, Dalian, China

[2] North China Sea Branch of State Oceanic Administration, Qingdao, China

ABSTRACT

With the development of wafts and harbors along the coastal of Bohai, China, the coastal structures along Bohai coastal increase fast. These structures affect the movement of current along the Bohai coastal. With the contributions from wind, tide, current speed and current direction, the drift ice distribution near the structures change under these environmental conditions. Although there were some literatures on the Bohai drift ice from satellite remote sensing, numerical modelling and some coastal observations, the drift ice characteristic floe area and concentration near structures were reported rarely. Recent years, the satellite productions with high-resolution increased fast and give choice to collect information from satellite images. About 44 coastal structures over 400 m in length along the Bohai coastal were taken from the collected satellite images of HJ-1A/B/C. The structure parameters and drift ice parameters were drawn. The structure parameters include dam length and principal axis direction. The drift ice parameters include floe areas, concentration near and far from the structures, and major and minor axis of equivalent ellipse. The natural environment parameters include wind direction, wind speed, air temperature, water temperature and tide status. This paper introduces the preliminary analysis methods on the drift ice floe areas and concentration based on these images parameters. We try to answer the possibility to determine designed drift ice criteria. Other parameters from the satellite image are discussed and will be analyzed in future.

KEY WORDS: Drift ice; Structures; Satellite images; Bohai Sea

INTRODUCTION

Bohai sea ice is as the south boundary of sea ice in northern hemisphere. There are some special characteristics. On some general sea ice conditions, they were summarized by Yang (2000). At that time, some ice conditions could not recognized because the shortage of data collection technology. With the Bohai industry developing, the structures along Bohai coastal were built more and more, the natural disasters from sea ice in winters need more and more detail ice data (Zhang et al, 2013). Except the local ice condition investigation (Wang et al, 2007)， the remote sensing technology has been very mature in sea ice monitoring. As one of the most important multispectral data source in China, HJ satellite data is commonly used in every aspect of marine, resources and environment(Zheng et al., 2014). The new satellite image made the past Bohai sea ice remote sensing application fast. In 2000, the Bohai sea ice remote sensing had a preliminary frame (Liu et al., 2000). Later different satellite productions were used, such as Liu et al., (2013) for engineering purposes. The ice criteria for engineering are ice period, level ice thickness, flexural strength, elastic modulus, compressive strength based on the Bohai history observed ice data and ice mechanical strength test results (Li and Wang, 2000). The air temperature, ice temperature, ice salinity, ice density, ice thickness and ice crystal were used to evaluate the porosity in ice, and then evaluated ice sheet strength in Bohai (Li et al., 2006). In fact, the sea ice in Bohai mostly is drift ice. The ice floe area, movement direction and speed also are necessary to study by different suitable methods (Lang and Meng, 2014). The wind and current effect on the floe movement is no double. However, the tidal effect on the floe was discussed rare (Li et al., 1999). On the ice-resistant structure design, the extreme upper load comes from the case of large area of ice sheet with much higher energy. In addition, there is a standard ice-grade for Bohai sea ice, which is 5 levels. Grade 5 is the heaviest ice situation and Grade 1 is the lightest ice situation. This national standard code is from the natural sea ice phenomes in Bohai. It is a geography code, not an engineering code. In fact, the Bohai sea ice is much different in Liaodong Bay, Bohai Bay and Laizhou Bay. Using the geography code, the same ice grade area has much different ice load on a same structure. From the view point of ice load controlling the structure manner, Li et al. (2006) developed a new ice design sub-zone by using designed ice thickness and ice strengths with different return period. In Bohai, most cases are small ice floe in the coastal dam structures. Therefore, the drift ice floe area and concentration, speed are necessary design criteria.

At present, the remote sensing image of HJ-1A/B/C data with high time-space resolution can provide powerful technical support for the analysis of the real-time sea ice monitoring and it's meaningful for data collection in port construction and shipping transportation. If we have longer enough time series of yearly drift ice data, the similar way for the level ice engineering design criteria methods can be used to determine drift ice design criteria. We select 8 of 44 structures along Bohai and the HJ-1A/B/C images in the winters from 2012-2016, the drift ice floe area and concentration with water temperature were analyzed because some area is melting zone in Bohai winter. If these methods and data are valid, they can be easy to be used for the sites near 44 structures and then detail drift ice feature in Bohai will be discussed. This paper is introduced the data collection from satellite images and meteriological and hydraulic conditions for the future analyzed. The simple relations between water temperature and drift ice floe area as well as between water temperature and drift ice concentration were shown.

BOHAI NATURAL ENVIRONMENT CONDITIONS

The drift ice movement depend on current, wind. Bohai currents have some special features, however the currents along the coastal mostly. Figure 1 shows the recent conclusions. In the figure, the red color means the Yellow Sea Warm Current and the green color means the cold flows. One is toward north and the other toward south. Black means closed loop current in Liaodong Bay, Bohai Bay and Laizhou Bay in winter. Also the currents made water movement in short time and some higher temperature water moves to the drift ice bottom (Li et al., 2007). Winds in winter come from west and north.

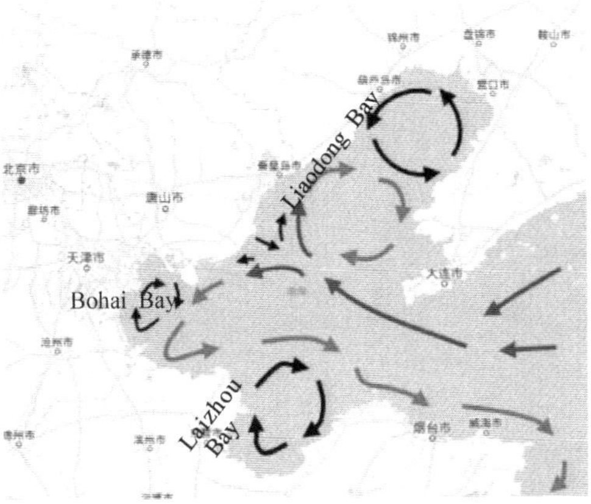

Figure 1. The density flow in Bohai in winter

(from: http://wenku.baidu.com/link?url=WKn8tEsdCOgBB1G6NShZxK8-TPNQVtWnytj 2lH1Xal33LdPGffC2URyHUNUgce7H7QygZgd95QEzvn6Jw7QUskbCLQchc6IKyv6R9MynHim)

Institute of Geography of Chinese Academy of Sciences (1990) summarized the general natural environmental conditions in Bohai (see Figure 2). The water temperature is not mentioned more on the drift ice in Bohai. Usually, the ice thermal dynamic process always assume the water temperature is same as the water freezing point under the ice. This assume is not correct for Bohai drift ice because Bohai water temperature varies in whole area. Most drift ice is not be frozen in the area, it just infested to the area. While ice floes move to the area, the ice temperature arise first, then maybe melt in longer time. The salinity decides the water freezing point, from Figure 2a the salinity is lower because the river water effect along the coastal zone. Figure 2b shows the water temperature. We can base on them to define the drift ice freezing or melting situation when it moves on the water surface. The tide is semi-diurnal tides, varies in 6 hr. and is big tidal range from 1 m to 5 m in most Bohai area (see Figure 2c and Figure 2d). In Bohai, tide also is an important factor while the ice frozen in shallow water area and broken, then moves to deep water area under offshore wind (Yang, 2000), also while the ice frozen in shallow water area and broken, then climb to beach or structure at the highest tide period under onshore wind (Wang et al., 2007).

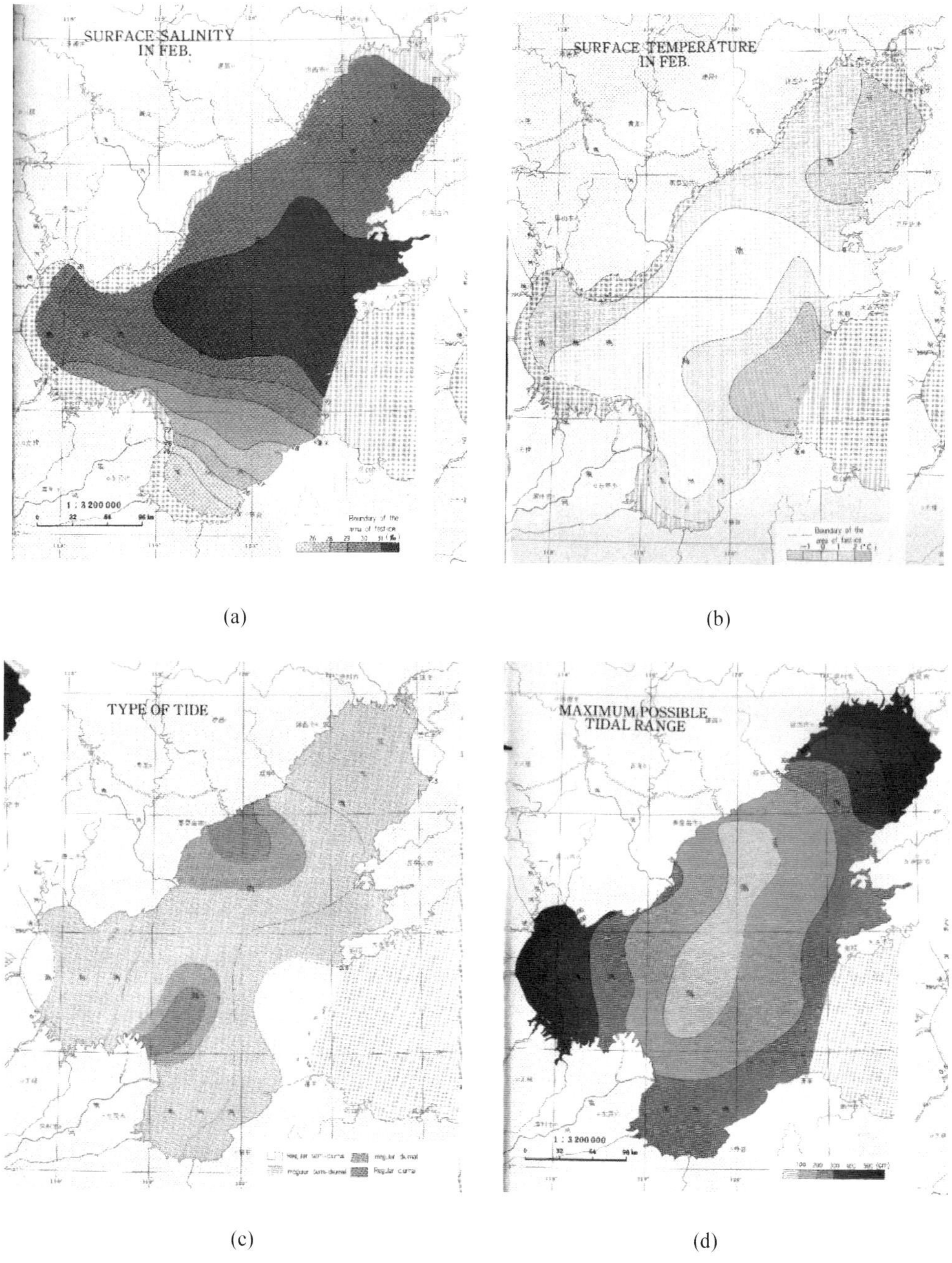

Figure 2. General natural environmental conditions in Bohai

BOHAI COASTAL STRUCTURES

With the Industry and fishery development, the coastal structures were built more and more along the coastal of Bohai. From the www.googlemap.com, we collected the structure information in Liaodong Bay (Figure 3), Bohai Bai (Figure 4) and Laizhou Bay (Figure 5).

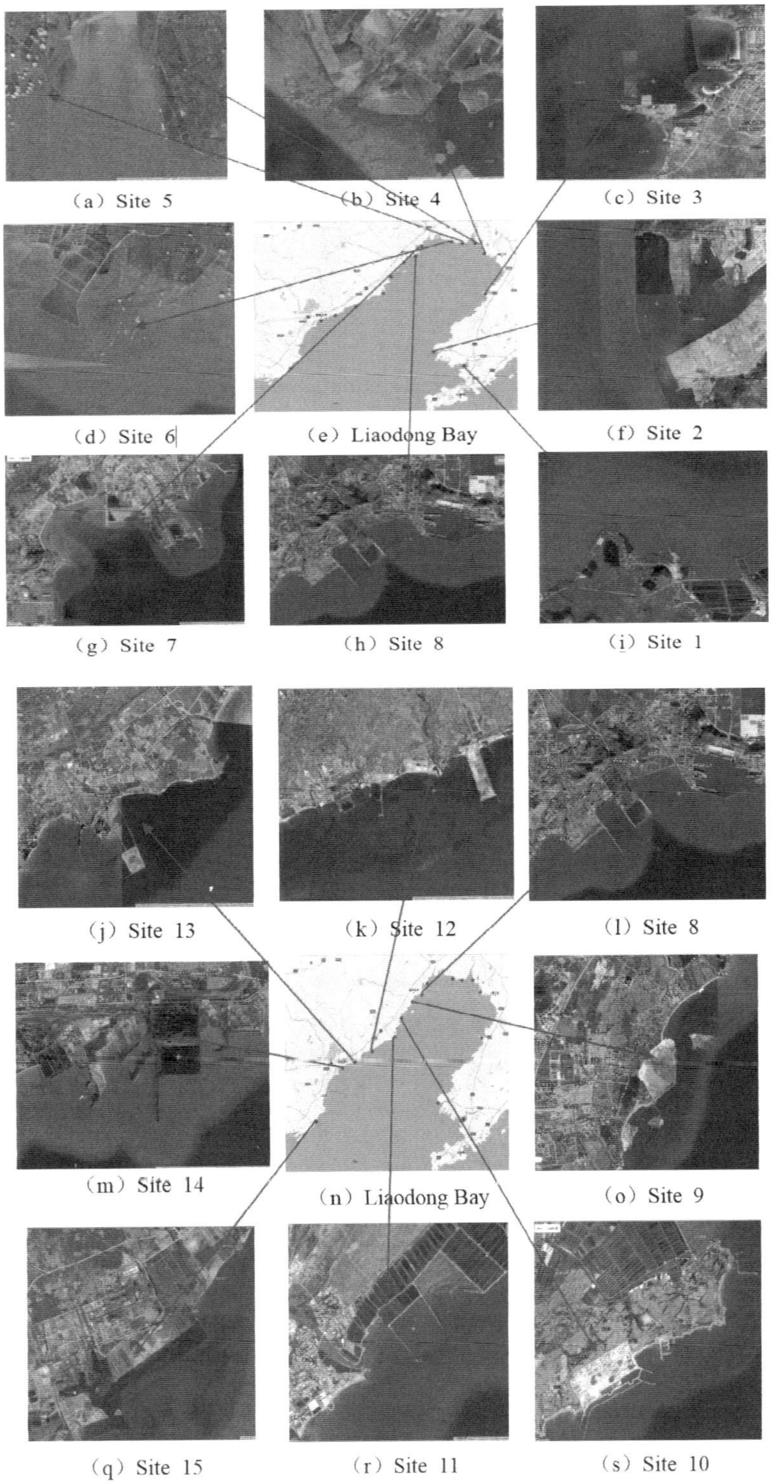

Figure 3. Locations of the selected coastal structures of Liaodong Bay

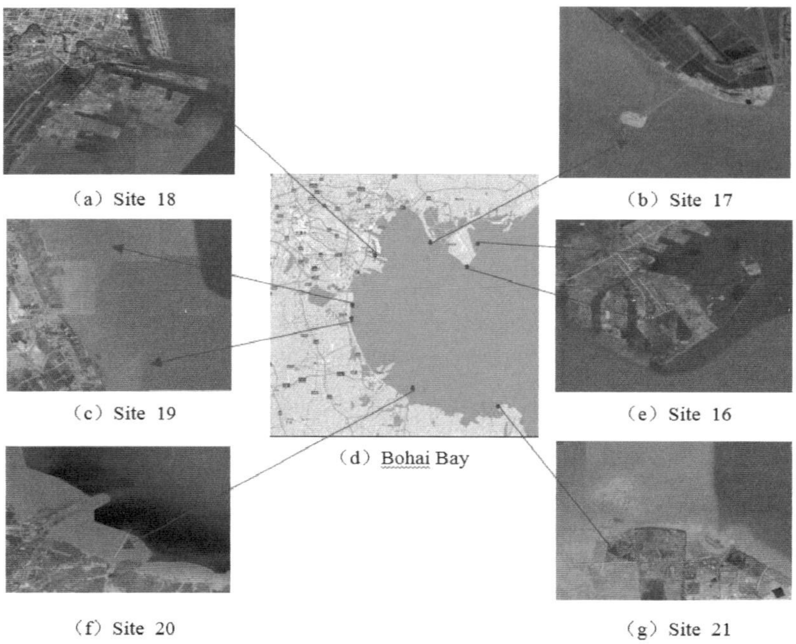

Figure 4. Locations of the selected coastal structures at Bohai Bay

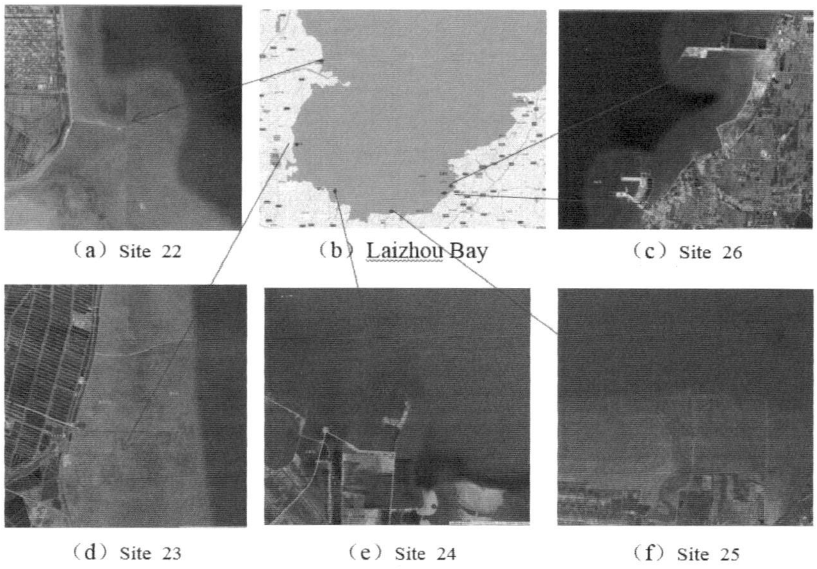

Figure 5. Locations of the selected coastal structures at Laizhou Bay

DATA RETRIEVE AND ANALYSIS

As a preliminary methodology analysis, 22 structures of the 44 structures along Bohai coastal were analyzed. These 22 structures were selected from 5 sites (see Figure 1) at Liaodong Bay, 4 sites at Bohai Bay and 4 sites at Laizhou Bay. Firstly, the drift ice floe area and concentration are analyzed. The relationships between hydrometeorological data (water temperature, tide, air temperature, wind direction and wind speed) and drift ice concentration, floe shape, floe area and distribution, and their effects on the structures will be studied in future.

The images were selected in the heaviest ice conditions in the winters between 2012 and 2016, see Figure 6. Each winter, 3 to 5 images were selected. The corresponding daily weather and ocean conditions were taken from Ocean Station records. Totally the weather data include wind speed and wind direction; ocean data include tide status and water temperature; structure data include the angle between structure axis and coastal, angle between wind direction and structure axis and structure length.

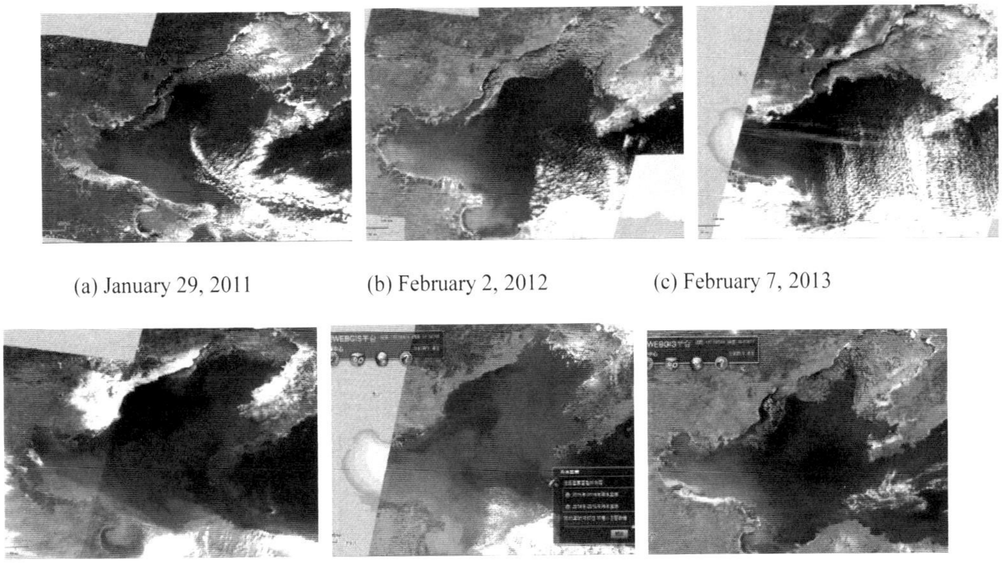

(a) January 29, 2011 (b) February 2, 2012 (c) February 7, 2013

(d) January 18, 2014 (e) January 17, 2015 (f) February 5, 2016

Figure 6. The selected satellite images in the heaviest ice conditions in winters

The drift ice data from image used the methods from Lu and Li (2010). The methods used by Lu and Li (2010) need to change the satellite images into black-white image (see Figure 7), and then use the computer program to calculate the each floe pixies area, boundary length. After these calculations and the scale of the reference object length, the drift ice floe areas and concentration are obtained.

For the data retrieved, we made two assumes. First, considering the largest ice floe was not nearby the structure, we statistics the each floe area and made the accumulated possibility distribution, the corresponding floe area at 90% is assumed as an index (see Figure 8).

Second, based on the Figures 2a and 2b, the coastal area is affected by river fresh water, its freezing point is assumed as -0.5°C. The water temperature below -0.5°C is the ice freezing case and the water temperature over -0.5°C is the ice melting case. Third, the blue line is an upper limit line for drift ice concentration or equal floe area. It decreases with water temperature increasing. The design criteria will be studied based on these data at the lowest water temperature (-1.6°C).

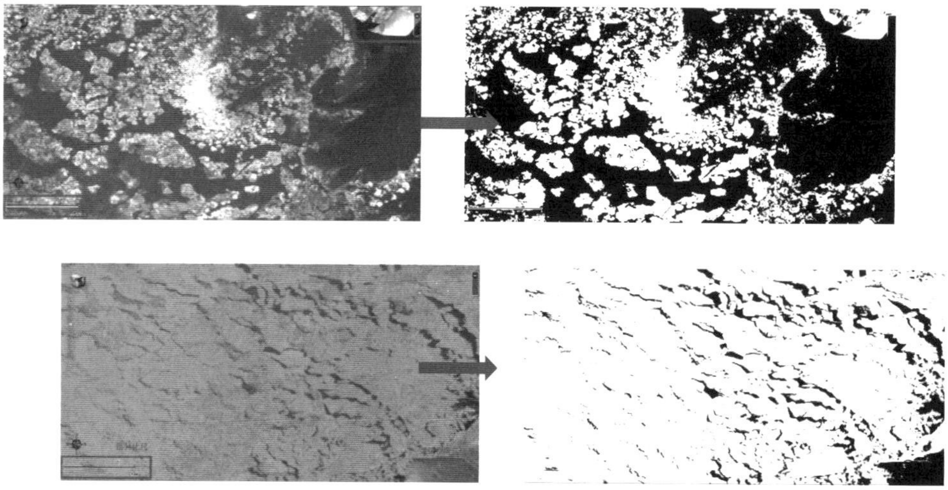

a) Satellite image b) Black-white image (white is ice & black is water)

Figure 7. The satellite image data retrieve method

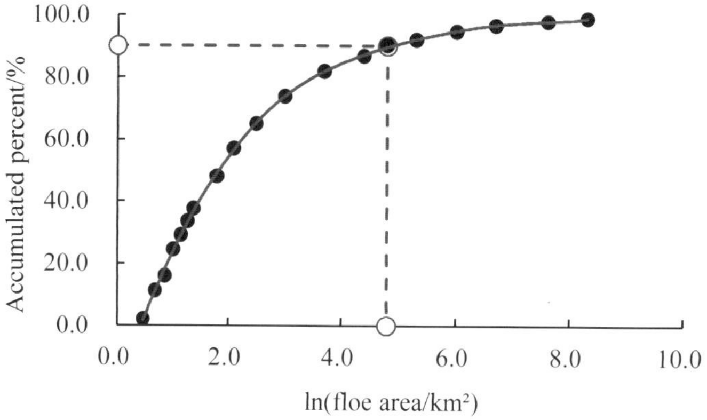

Figure 8. Obtaining the feature floe area from the satellite image near structure

As the examples, 3 sites from Liaodong Bay and 1 site from Laizhou Bay were selected. Figure 9 gives their floe areas determined curve and Figure 10 gives their floe concentration determined curve. Although the history data are not enough for different return period analysis, we can put them in the figure as different color. Following these data, an upper envelope curve can be drawn, see Figure 9. The upper envelope curve is regard as the design floe area and the curve increases with the water temperature decreasing.

Similarly Figure 10 gives the concentration upper envelope curve. The curves also inc reases with water temperature increasing to agree with the ice concentration at Liaodo ng Bay higher that that at Bohai Bay and Laizhou Bay.

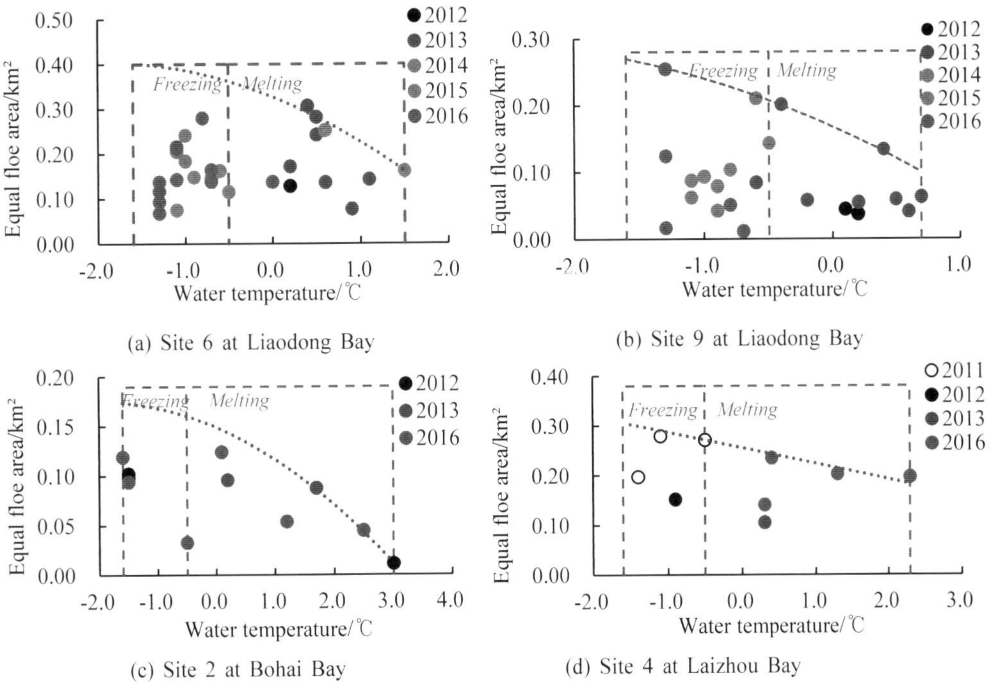

(a) Site 6 at Liaodong Bay (b) Site 9 at Liaodong Bay

(c) Site 2 at Bohai Bay (d) Site 4 at Laizhou Bay

Figure 9. The designed floe area at different sites along Bohai coastal

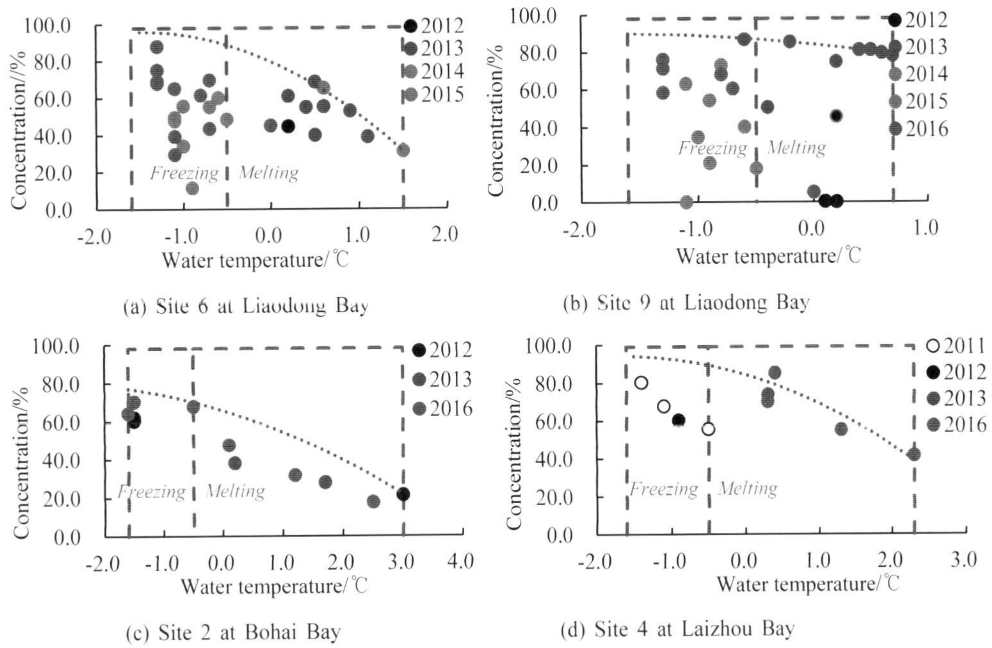

(a) Site 6 at Liaodong Bay (b) Site 9 at Liaodong Bay

(c) Site 2 at Bohai Bay (d) Site 4 at Laizhou Bay

Figure 10. The designed ice concentration at different sites along Bohai coastal

After collecting all selected 26 sites drift data, a table (see Table 1) is summarized. Based on Table 1, the drift ice design criteria can be determined if there are enough annual winter data. From Table 1, it is clear that the floe area is larger in Liaodong Bay than that in other two bays, and it is lager at the bay bottom than that near the bay mouth. Also the concentration is higher in northern and at bay bottom. The drifting ice in freezing case occupies higher near the larger river mouth areas.

Table 1. Ice floe parameters from satellite image analysis

Position	Site	Floe area/km^2	Concentration/%
Liaodong Bay	1	0.15	42
	2	0.35	82
	3	0.32	98
	4	0.36	87
	5	0.29	89
	6	0.41	93
	7	0.27	90
	8	0.36	80
	9	0.40	98
	10	0.20	90
	11	0.41	85
	12	0.40	81
	13	0.28	89
	14	0.36	81
	15	0.20	65
Bohai Bay	16	0.38	63
	17	0.14	78
	18	0.32	70
	19	0.30	60
	20	0.25	68
	21	0.12	36
Laizhou Bay	22	0.13	39
	23	0.18	61
	24	0.21	78
	25	0.30	92
	26	0.25	82

CONCLUSIONS AND DISSCIONS

As sea ice southern boundary, the water temperature in sea ice infested areas in Bohai is not same as the freezing point of the sea water. Somewhere, the water temperature is higher than freezing point. Therefore, sea ice drift to the area and melting. In this case, the higher the temperature and the longer the drifting ice staying, the thinner ice thickness, the lower concentration and the smaller floe area.

For the ice engineering design in Bohai, the design ice strength was the peak value at the strain rate of ice brittle-ductile behavior transfer zone because most drift ice in Bohai moves from lower speed to higher speed at the maximum 1.0 m/s. With the industry development along Bohai coastal, the structures maybe face drift ice with lower speed or quite wide caissons or dams, the 100% contact and upper limit ice broken load as design purpose is not suitable.

The drift ice floe area and concentration can be determined from high-resolution satellite productions. Based on the drift ice data, the design criteria for Bohai drift ice is possible based on the introduced methods in this paper.

ACKNOWLEDGEMENTS

The research were supported by Global Change Research National Major Scientific Research Projects (2015CB953901) and the National Natural Science Foundation of China (51639003 and 41376186).

REFERENCES

Institute of Geography of Chinese Academy of Sciences, 1990. *Atlas of Ecological Environment in the Beijing-Tianjin Area*. Science Press: Beijing.

Lang, W., Meng, J., 2014. Sea ice drift tracking in the Bohai Sea using geostationary ocean color imagery. *Journal of Applied Remote Sensing*, 8(1), doi: 10.1117/1.JRS.8.083650.

Li, H., Bai, S., Zhang, Z., et al., 1999. Numerical simulation of tidal effects on ice for the Bohai Sea. *Marine Forecasts*, 16(3), pp.39-47, in Chinese.

Li, Z. & Wang, Y., 2000. Design criteria of Bohai Sea ice-control engineering. *Ocean Engineering*, 18(1), pp. 62-65, in Chinese.

Li, Z., Ding, D., Yan, Q., et al., 1997. Preliminary study on the sea ice crystals and chlorophyll in Liaodong Bay. *Marine Environmental Science*, 16(3), pp.21-28, in Chinese.

Li, Z., Sodhi, D., Lu, P., 2006. Distribution of ice engineering design criteria of Bohai. *Engineering Mechanics*, 23(6), pp.167-172, in Chinese.

Liu, C., Gu, W., Chao, J., et al., 2013. Spatio-temporal characteristics of the sea-ice volume of the Bohai Sea, China, in winter 2009/10. *Annals of Glaciology*, 54(62), pp.97-104.

Liu, J., Huang, R., Jin, Z., et al., 2000. Bohai sea ice monitoring using satellite images. *Journal of Cold Regions Engineering,* 14(2), pp.93-100.

Lu, P. & Li, Z., 2010. A method of obtaining ice concentration and floe size from shipboard oblique sea ice images. *IEEE Transactions on Geoscience and Remote Sensing*, 48(7), pp.2771-2780.

Wang, X., Lei, R., Kong, X., et al., 2007. Investigation of topography along a beach in Liaodong Gulf. *Proceedings of 19th International Conference on Port and Ocean Engineering under Arctic Conditions (POAC'07)*, 2, pp.717-723.

Yang, G., 2000. Bohai Sea Ice Conditions. *Journal of Cold Regions Engineering*, 14(2), pp.54-67.

Zhang, X., Zhang, Z., Xu, Z., et al., 2013. Sea ice disasters and their impacts since 2000 in Laizhou Bay of Bohai Sea, China. *Natural Hazards*, 65(1), pp.27-40.

Zheng, G., Bai, J., Zhang, S., et al., 2014. Bohai bay sea ice monitoring based on the HJ satellite images. *International Congress on Image & Signal Processes*, pp. 558-562.

POAC'17
Busan, KOREA

Proceedings of the 24th International Conference on
Port and Ocean Engineering under Arctic Conditions
June 11-16, 2017, Busan, Korea

Mapping of summer sea ice in the Chukchi Sea from KOMPSAT-5 SAR

Hyangsun Han[1], Sang-Hoon Hong[2] and Hyun-cheol Kim[1]

[1] Unit of Arctic Sea-Ice Prediction, Korea Polar Research Institute

[2] Department of Geological Sciences, Pusan National University

ABSTRACT

Accurate sea ice mapping is important for providing information to guide human activities in polar regions and global climate change research. In this study, a sea ice mapping model based on Random Forest (RF) was developed for the Korea Multi-Purpose Satellite-5 (KOMPSAT-5) Synthetic Aperture Radar (SAR) data obtained from August to September 2015 in the Chukchi Sea. A total of 12 texture features derived from backscattering intensity and the gray-level co-occurrence matrix were used as input variables for sea ice mapping. The RF model produced a sea ice map with a grid size of 125 m, demonstrating excellent performance in the classification of sea ice and open water with an overall accuracy of 99.2% and a kappa coefficient of 98.5%. Sea ice concentration (SIC) retrieved from the RF-derived sea ice maps was compared with that from ice charts. The mean and median values of the differences between the SICs derived from the RF model and the ice charts were −8.85 and −8.38%, respectively. Such difference was attributed to the uncertainty in the ice charts rather than error in the RF model.

Very-High-Resolution Image Acquisition Over Arctic Sea Ice Using Multiple Remote Sensing Platforms

Chang-Uk Hyun[1], Hyun-cheol Kim[1]
[1] Korea Polar Research Institute, Incheon 21990, Republic of Korea

ABSTRACT

Remote sensing of arctic sea ice provides useful information on the status of sea ice such as concentration, extent, motion and surface composition. Very-high-resolution (VHR) imaging acquiring the images with a spatial resolution from few meters to few centimeters has become available from various remote sensing platforms. In this study VHR image acquisition over arctic sea ice using multiple remote sensing platforms, and image processing protocols for extracting information from the acquired images are demonstrated. The VHR images and products from the images can be applicable to validate sea ice concentration and motion data from coarse resolution remote sensing data, and can be used as an alternative reference data decreasing requirements for in-situ measurements. Furthermore, the VHR images can support safe research activities using icebreaker research vessel by providing the sea ice condition around the vessel.

KEY WORDS: Sea ice, Remote sensing, Very-high-resolution, Satellite image, Arctic ocean

INTRODUCTION

Sea ice and its temporal changes have been observed to understand dynamic climate system and to predict long-term changes, and low-resolution satellite remote sensing with a spatial resolution of 1 km or larger pixel size has provided useful information on the status of sea ice such as concentration, extent and motion (e.g. Heil et al., 2001, Holland and Kwok, 2012, Vincent et al., 2001). Relatively higher spatial resolution satellite remote sensing has been rarely used for the sea ice observation. Recently, very-high-resolution (VHR) imaging acquiring images with a spatial resolution from few centimeters to few meters has become available from various remote sensing platforms, e.g., satellite, helicopter and unmanned aerial vehicle. This study presents the feasibility of the VHR image acquisition over arctic sea ice using multiple remote sensing platforms, and image processing protocols for extracting information from the acquired images.

VERY-HIGH-RESOLUTION IMAGE ACQUISITION USING HELICIPTER

The VHR images over the arctic ocean were captured by helicopter camera on Aug 16, 2016 during 2-day sea ice camp supported by Korean icebreaker research vessel Araon. The geographical coordinates of the position of camera at each imaging site were recorded using GPS logger. The images were preprocessed and mosaicked using structure-from-motion (SfM) technique. The processing chain consists of image alignment, extraction of the dense cloud from overlapping regions between images, generation of a digital elevation model from the dense cloud, and mosaicking the orthorectified image. Detailed surface features of surrounding sea ice are clearly visible in the mosaicked image using 449 photos (Figure 1). The mosaicked photo includes target melt ponds for in-situ scientific activities with sufficient details and shows small-scale morphological features of sea ice around the icebreaker.

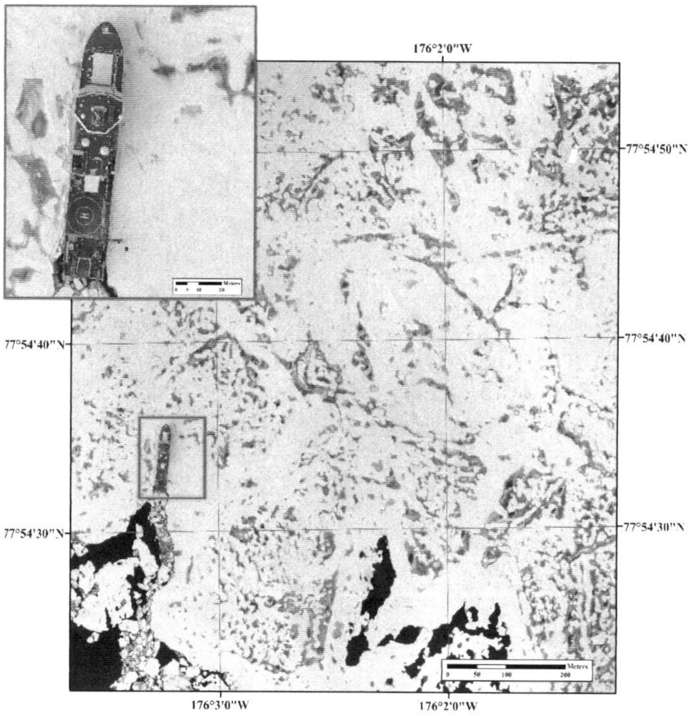

Figure 1. Arctic sea ice and Korean icebreaker research vessel Araon captured by helicopter during sea ice camp

VERY-HIGH-RESOLUTION IMAGE ACQUISITION USING SATELLITE

Sea ice motion from low-resolution satellite sensors (e.g. passive-microwave sensor and Advanced Very High Resolution Radiometer (AVHRR) sensor) with feature tracking methods, e.g. maximum cross-correlation (MCC), has been widely used to monitor changes in sea ice covers (e.g. Ninnis et al., 1986, Emery et al., 1991, Lavergne et al., 2010). This study presents measuring arctic sea ice motion using high-resolution optical satellite images acquired on August 2014. The high-resolution optical images were obtained from Korea Multi-Purpose Satellite-2 (KOMPSAT-2) and Korea Multi-Purpose Satellite-3 (KOMPSAT-3) with 1.0 m spatial resolution panchromatic (PAN) band and 4.0 m spatial resolution multi-spectral (MS) bands and 0.7 m spatial resolution PAN band and 2.8 m spatial resolution MS bands, respectively. The study area was around 77°20′N, 146°10′W in the Beaufort Sea.

The sea ice motion was retrieved using the MCC strategy, an automated block-based motion estimation technique by repeatedly evaluating the correlation between two image templates, i.e., reference block and candidate block. The velocity vector (or motion vector) can be calculated by measuring the displacement and the time gap between the two images. The motion measuring results can be represented as displacement vectors (Figure 2). The measurement results were able to be assessed by comparison with the track of the drifting buoy locating in the satellite images.

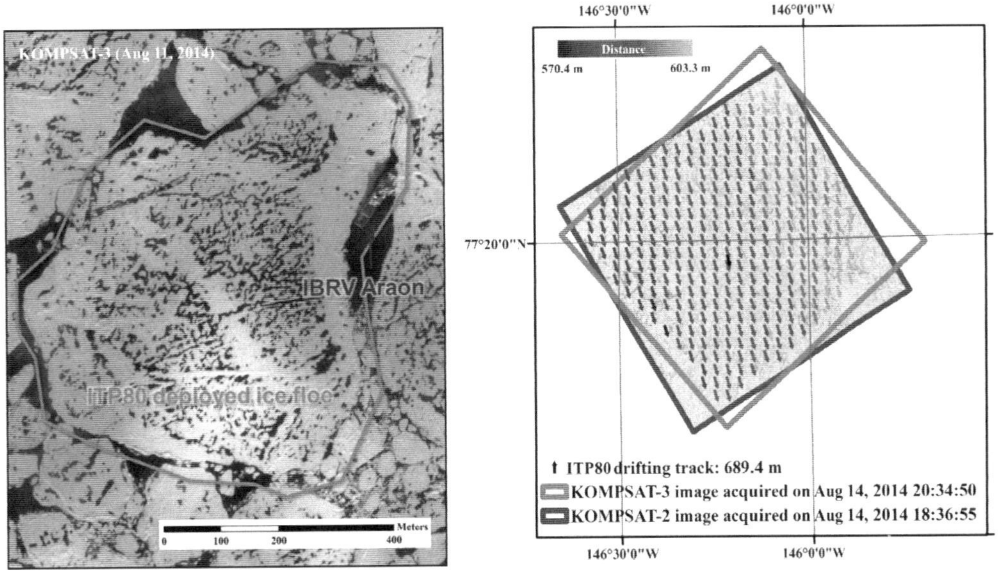

Figure 2. Arctic sea ice and Korean icebreaker research vessel Araon captured by satellite, and an example of the sea ice motions extracted from sequential VHR satellite images

The sea ice motions extracted from the sequential VHR satellite images are applicable to fill the gaps between sparse buoy data and the motions from low-resolution satellite data, so this sea ice motion from the high resolution optical satellite images can be used to expand reference dataset for arctic sea ice motion data validation.

CONCLUSIONS

The VHR image acquisition over arctic sea ice using multiple remote sensing platforms, and image processing protocols for extracting information from the acquired images are demonstrated. The VHR images and products from the images can be applicable to validate sea ice concentration data from coarse resolution remote sensing data and can be used as an alternative reference data decreasing requirements for in-situ measurements. Furthermore, the VHR images can support safe research activities using icebreaker research vessel by providing the sea ice condition around the vessel.

ACKNOWLEDGEMENTS

This study was supported by KOPRI project (PE17120).

REFERENCES

Emery, W. J., Fowler, C. W. Hawkins, J. & Preller, R. H. 1991, Fram Strait satellite image-derived ice motions, *Journal of Geophysical Research: Oceans*, 96(C3), pp.4751–4768.

Heil, P., Fowler, C. W., Maslanik, J. A., Emery, W. J., & Allison, I. 2001. A comparison of East Antarctic sea-ice motion derived using drifting buoys and remote sensing. *Annals of Glaciology*, 33(1), pp.139–144.

Holland, P. R., & Kwok, R. 2012. Wind-driven trends in Antarctic sea-ice drift. *Nature Geoscience*, 5(12), pp.872–875.

Lavergne, T., Eastwood, S., Teffah, Z., Schyberg, H., & Breivik, L. A. 2010. Sea ice motion from low-resolution satellite sensors: An alternative method and its validation in the Arctic. *Journal of Geophysical Research: Oceans*, 115(C10).

Ninnis, R. M., Emery, W. J., & Collins, M. J. 1986. Automated extraction of pack ice motion from advanced very high resolution radiometer imagery. *Journal of Geophysical Research: Oceans*, 91(C9), pp.10725–10734.

Vincent, R. F., Marsden, R. F., & McDonald, A. 2001. Short time-span ice tracking using sequential AVHRR imagery. *Atmosphere-Ocean*, 39(3), pp.279–288.

Proceedings of the 24th International Conference on
Port and Ocean Engineering under Arctic Conditions
June 11-16, 2017, Busan, Korea

Comparison of Satellite Imagery Based Ice Drift with Wind Model for the Caspian Sea

Yevgeniy Kadranov[1], Sigitov Anton[1], Sergey Vernyayev[1]

[1] LLP Iceman.kz, Shimkent, Kazakhstan

ABSTRACT

Many factors influencing the movement of ice such as wind, ice concentration, ice thickness, roughness, water currents, Coriolis force, bathymetry, artificial and natural obstacles in the area. Current speeds in the Caspian Sea are relatively small and so the main driving force for ice movements is wind. Therefore, main goal of this work was to study wind-ice movement velocities dependence in the region and check how ice concentration and thickness influence on the movement of ice.

A high number of measurements and observations was made to describe ice drift in the region, although the data was collected areas and usually not publicly available. In our work, we have used timely consequent optical and SAR satellite images to observe ice movements and its displacement over the area. Wind data for the same period and area was taken from wind models. Ice charts were prepared using visual interpretation of satellite imagery. Ice information (concentration, stage of development, floe size) were stored as vector data in SIGRID3 format. The described data has been correlated and analyzed.

The analysis provided in the work can be used for the forecast of short term ice drift on the operational basis and can be the first step for creation of ice drift forecast model for the region of North Caspian Sea. The used data, methods and results of the study are described in this paper.

KEY WORDS: Ice Charting, Ice behavior modeling, Ice drift, Caspian Sea, Optical and SAR Imagery interpretation, Weather model, GFS

INTRODUCTION

The area of North Caspian Sea is of great interest for Oil & Gas industry with the ongoing and projected offshore development of relatively complex oil fields. Ice detection and forecasting is needed for support of winter operations at site and along transit routes of marine supply lines. One of the major concerns is forecasting ice conditions in the region to plan logistics and avoid hazardous or time-consuming ice and weather impact on operations as discussed by Verlaan et al, 2011. As one of the major drivers of ice related hazards to the offshore installations and navigation is drift, it is useful to understand what the factors control the ice drift over the region. Developing the relationship between the forecast weather parameters, local conditions across region and the resulting drift is the way to build a model to predict ice drift in the region. Objective of this project is to study wind induced ice drift in the region and define to what extent ice concentration influence on the resulting displacements based on observations of the relatively mild ice season 2015-2016.

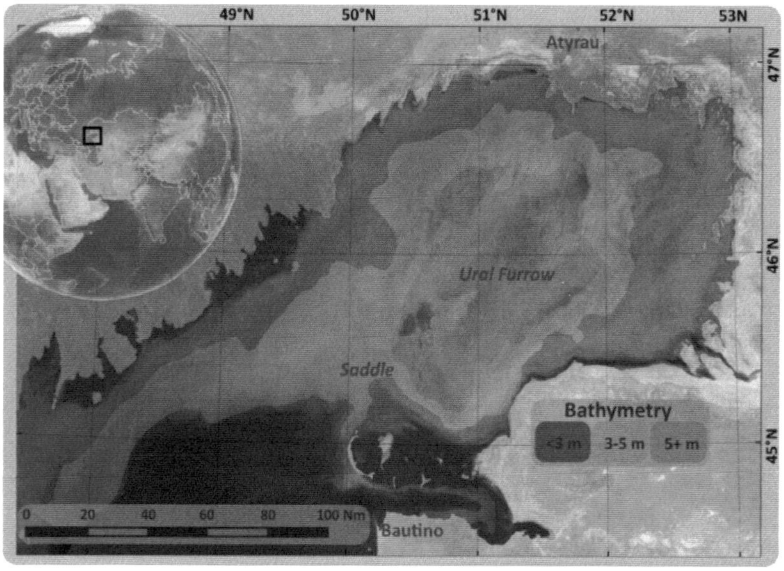

Figure 1. North-East Caspian Sea

North East Caspian Ice Conditions Summary

The North-East Caspian Sea normally stays ice covered from late November-Early December to Late March-Early April. Thermally grown ice thickness typically reaches 40-50 cm along the northern shoreline during an average winter. Ice drift is mostly wind driven and usually occurs over deeper parts of the NE Caspian in the central basin or south of the Saddle area (Figure 1), while ice sheet in the shallower area remains stationary during the major part of the winter. Ice compaction caused by strong wind events often lead to forming of ice features such as stamukhi (grounded rubble mounds) and pressure ridges as well as extensive rafted areas as observed by Crocker et al, 2011. Figure 2 illustrates seasonal variation of ice cover distribution over the last nine seasons 2007-2016.

Based on the analysis of remote sensing data it was noted that the warmer the winter is, more mobile areas are observed over the NE Caspian. Therefore, it was decided that the season 2015-2016 should be used for the comparison of ice drift/wind dependency. As the ice season lasted from late December to early March, it was shorter than usual. Only 221 FDD (as observed in Atyrau) were collected during this winter – nearly 3 times less than during the average Caspian ice season. Thickest level ice that has grown this season was about 25 cm as calculated by formula derived for this region by Jordaan et al, 2011. Ice conditions during the season 2015-2016 were more dynamic than usual as observed during execution of this project. This was mainly due to milder weather throughout the year (Figure 2). Most Ice in the NE Caspian never consolidated and remained mobile expect for the areas along the shore in the North and East. This created ideal conditions for ice drift analysis based on comparison of satellite images as same ice floe could be tracked for several weeks (while during other seasons, after several days of tracking floe would be deformed with compaction against stationary features or consolidates as part of stationary ice sheet).

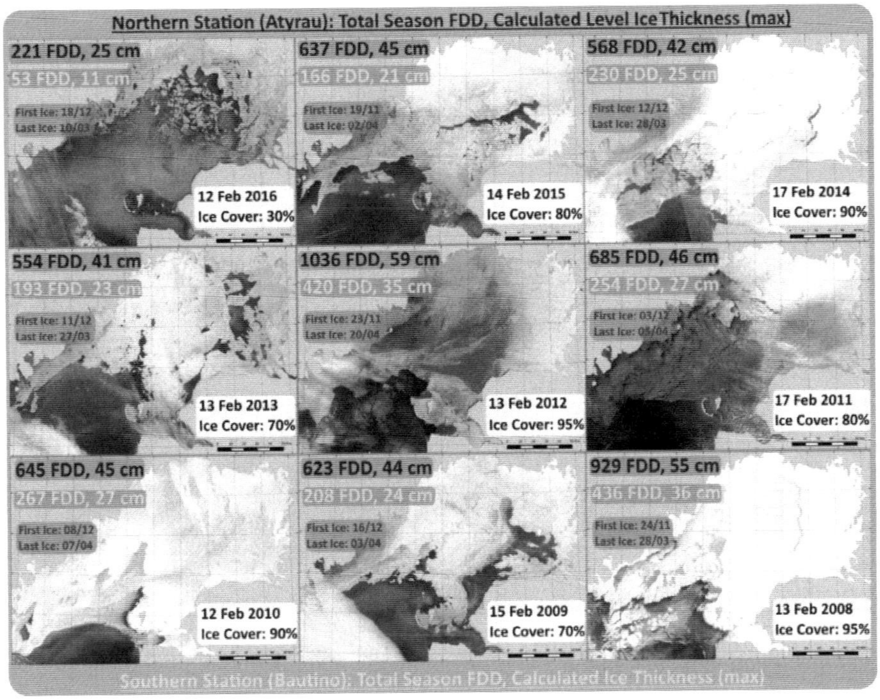

Figure 2. This picture compares 9 ice seasons (from 2007-2008 to 2015-2016). FDD data from NOAA National Centers for Environmental Information.

DATA SOURCES

Satellite Imagery

The major source of data for ice drift measurement within the scope of this project wass satellite images. Choosing between SAR and optical images, SAR is more reliable as it does not depend on cloud cover. However, due to geographical location of the area of interest SAR satellites have revisit time of 3-4 days per week and need to be supplemented with any available optical images.

The season had a fair amount of cloudless days (63% of MODIS images were either clear or partly cloudy as illustrated on Figure 3), which allowed to use optical MODIS Terra and Aqua data from NASA worldview portal intensively.

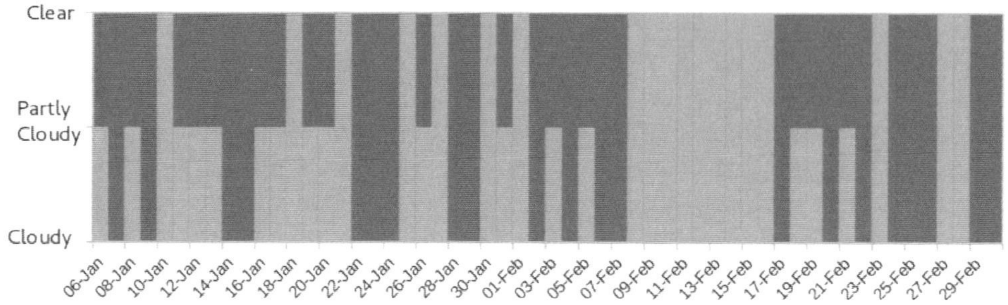

Figure 3. Cloudiness stats for Jan-Mar 2016

Publicly available Sentinel-1 and commercial RISAT-1, TerraSAR-X and Radarsat-2 SAR images were used to close some of the big gaps caused by cloudy sky conditions. Public SAR imagery were taken from Copernicus Science hub, while commercial SAR images were provided by Kongsberg Satellite Services (KSAT).

Ice Drift Data

Ice drift data over the period from 6 January 2015 to 2 March 2016 was gathered by identifying similar ice floes on subsequent images of the series and tracking their displacement. The start date of tracking is the day, when ice cover has established over the major part of the Central Basin of the NE Caspian and South of Saddle.

There were two automated steps of processing satellite images targeted to increase accuracy of the resulting displacement values. Both steps were performed within QGIS environment scripting specific plugins. During the first one, distinctive ice floes were identified on a pair of images and were polygonised as a geo-spatial feature with the following attributes assigned to each feature:

- Start and End Image date
- Floe ID
- Start and End ice concentration around the floe
- Area and perimeter values
- General Comments (rotation of floe, unique behavior, deviation from flow direction)

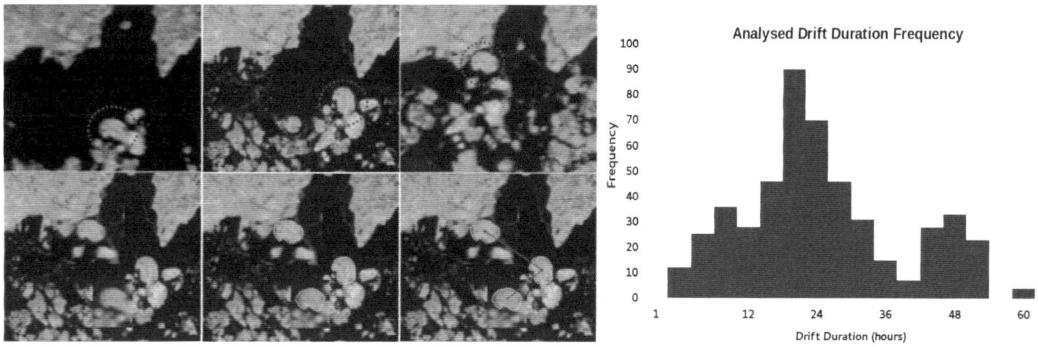

Figure 4. Example of digitizing same ice floe for 3 days (Left). Frequency distribution of drift observations intervals (Right)

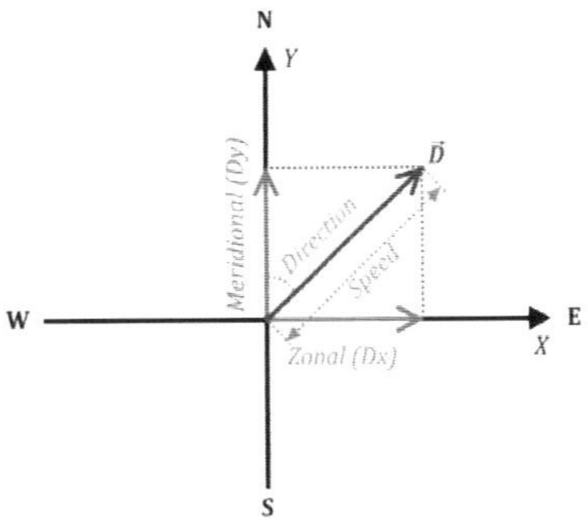

Figure 5. Drift vector and its components

As the second step the other QGIS plugin built vectors between centroids of floes using date and floe id attributes, and calculated vector displacement and direction between adjacent by date floes. Figure 4 (Left) illustrates the example of floes identification and resulting displacement vectors represented in graphical form.

In total for the whole season, 494 ice drift vectors were created and 63 pairs of satellite images were analyzed, making an average of about 8 drift vectors per analysis. The most frequent time interval between two images was about 24 hours, with minimum interval of around 2 hours (between MODIS Terra and Aqua from the same day) and rare maximum of 60 hours (Figure 4 Right).

Drift speed vectors consisting of vector magnitude, direction and two projections on x and y axes were derived from the displacement data. The magnitude of the average ice drift speed vector of the floe was found as an ice floe displacement distance divided by the time interval between the images. Direction of ice drift vector was measured as azimuth between displacement vector measured from true North. The zonal and meridional components of ice drift speed vector were found as orthogonal projections of ice drift speed vector on X (East-West) and Y (North-South) axes correspondingly (**Error! Reference source not found.**). The resulting values were stored in the database.

Ice Conditions Data

Ice conditions description such as total concentrations, partial concentrations, floe sizes and stages of development was prepared and stored as a geo-spatial database in standard SIGRID-3 format for the whole region. Ice conditions data has then been assigned to intersecting ice drift floes with the same date. The resulting ice concentrations were then joined with ice drift vector for each individual vector as attribute containing categories describing it at the start and the end of each track (Table 1). As the season was extremely mild the stage of development more than grey-white ice has never been grown. This has also allowed us to assume the ice thickness conditions were more or less uniform across the sea and neglect its effect on drift behavior.

Table 1. Ice drift concentration categories.

		Final		
		Low (1-3)	Medium (4-6)	High (7-10)
Initial	Low (1-3)	LL	LM	LH
	Medium (4-6)	ML	MM	MH
	High (7-10)	HL	HM	HH

Wind Data

GFS wind analysis model data (10m above Sea Level with spatial resolution of 30x20 nautical miles for the Caspian Sea) was taken as grib file from NOAA datahub, and converted into vector geo-spatial format. The database of wind speed vector components (meridional and zonal) for the season was derived based on four model runs per day, and contained two (+0 hrs and +3 hrs) output array data points from each model run. Considering the data points for resulting database were taken near the model run it was assumed that the dataset was close to real.

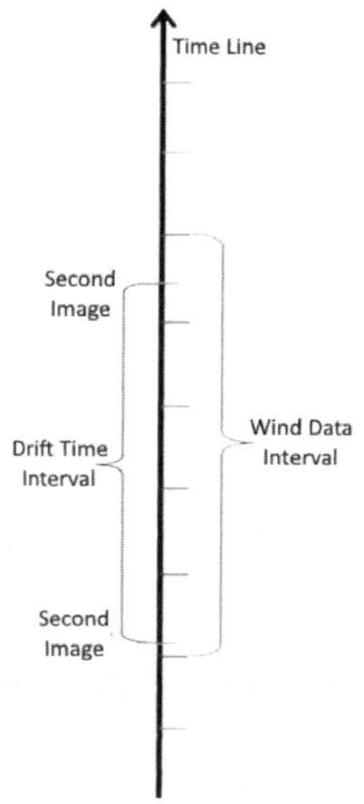

Figure 6. Selected wind data time interval for averaging for each drift interval.

FINAL DATABASE

The data described above was merged into a single database with each record containing corresponding information about ice drift data, wind data and surrounding ice conditions data to perform further analysis. To do that each drift vector's centroid was associated with the nearest GFS grid point. All the wind vectors at that point were averaged within time interval between two satellite images including one more wind vector before and after such interval. The following attributes were derived for each averaged wind vector:

- Averaged zonal (W-E) and meridian(N-S) components of wind vector

- Averaged wind speed (vector magnitude) and direction
- Maximum wind direction deviation from the averaged wind vector using Yamartino method (Yamartino, 1984).

Figure 7 shows wind data as per oceanographic convention (direction toward which the wind is blowing) for the period of observed drift intervals. Presentation of wind vectors with direction 'toward' as opposed to 'from' simplifies the calculus. The drift and wind values of compiled database are in form of drift speed distributions by direction. Most of the wind events have happened towards NW-W and S-SE directions as typical for the region as concluded from years of observations by Nilsen et al, 2011.

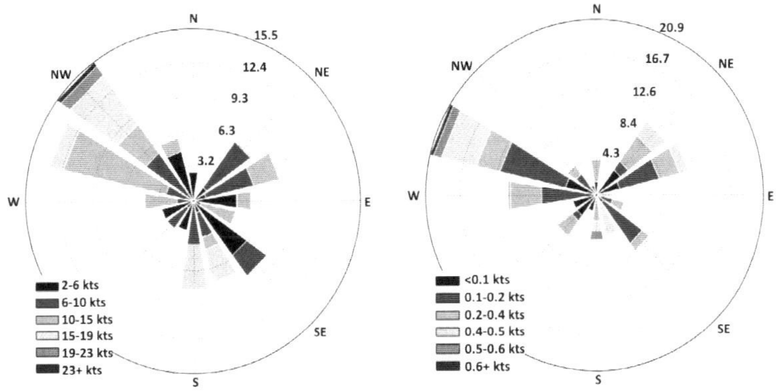

Figure 7. Wind speed and direction (toward) distribution for observed drift intervals (Left) and Ice drift speed and direction distribution (Right).

Flowchart illustrating all the steps of merging the data is schematically shown on Figure 8.

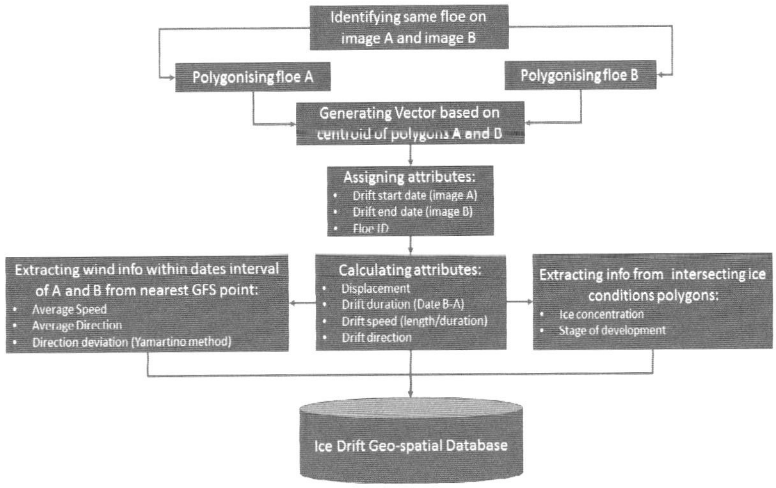

Figure 8. Data flow for Ice Drift geodatabase compilation

Resulting database revealed that ice drift vectors not always correspond to averaged wind vectors. Depending on time interval between the images ice floe could experience variable range of wind speeds with variable wind directions. For example, wind with even the constant speed can reverse its direction during the time interval, so the real track length of the floe will be much longer than the net floe displacement resulting in significant decrease of average drift speed. Figure 9 shows the wind and drift direction difference versus wind speed illustrating the point above. Most of the used data for wind speeds above 5 knots are within narrow interval of drift-wind difference values as it should be for wind driven drift as we expect in the Caspian.

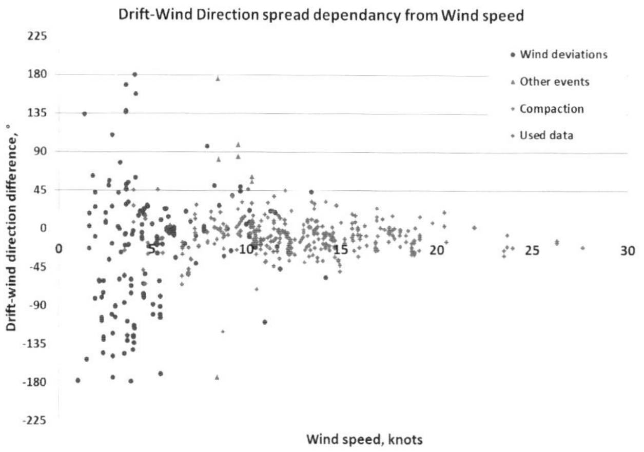

Figure 9. Scatter plot of Drift-Wind directions difference from wind speed

To clean the dataset from the discrepancies caused by the uncertainties of scarce data observations the whole database was filtered to exclude:

1. All drift vectors with the duration of less than 4 hours. It was found that even during homogeneous wind event, accuracy of drift vector length calculation can be significantly affected with image's resolution (250m for MODIS);
2. All the average wind vectors where wind direction deviation was more 60 degrees;
3. Most of the records where average wind speed was lower than 5 knots. As observed during the data compilation, such low values are normally associated with erroneous speed averaging process mentioned above. Low winds events resulting in small displacements with significant difference between drift and wind directions are also regulated by other factors such as currents, ice compaction, sea surface tilt during surge events that become noticeable, but could be neglected with higher winds;
4. Events when the compaction against coastlines or grounded ice features or diverging processes caused by non-homogeneous wind were observed. In this case the frictional forces and internal stresses between ice floes starts to play dominating role in floe behavior
5. Odd cases without reasonable explanation of drift behavior. Sometimes ice floe displacement was observed with incomparable wind speeds. It potentially could be caused by discrepancies in wind model output due to time lag or difficult synoptic situation that could not be handled within model.

As the dataset was cleaned of the cases listed above filtered data has shown clear relationship between wind and drift directions as can be seen on scatter (Figure 10 Right) as one would expect it to be in the region with prevailing wind driven drift as observed by Nilsen et al, 2011. The regression line for wind versus drift speed relationship (Figure 10 Left) showed a relatively big spread of data around the curve indicating that direct calculation of the drift speed based on one drift/wind ratio is not sufficient for modelling the ice drift in the region.

The distribution of the drift-wind speed ratios frequency shows that the range bin from 2 to 3% is dominant - 41% (Figure 11 Left) and the coefficients in range 1-4% appear for almost 90% of time. This has led us to conclusion we need to segregate these ratios based on concentrations and drift directions as these seemed to be the major factors affecting the coefficient. It can be seen from the Figure 11(Right) that, in general, the ratio coefficient tends to increase with lowering concentration.

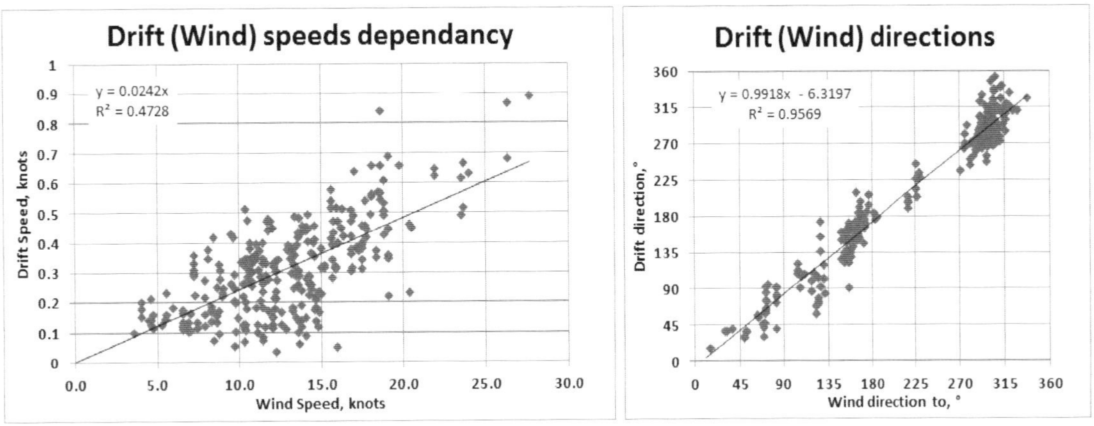

Figure 10. Scatter plot of data for Drift-wind speed (Left) and Drift-wind direction (Right)

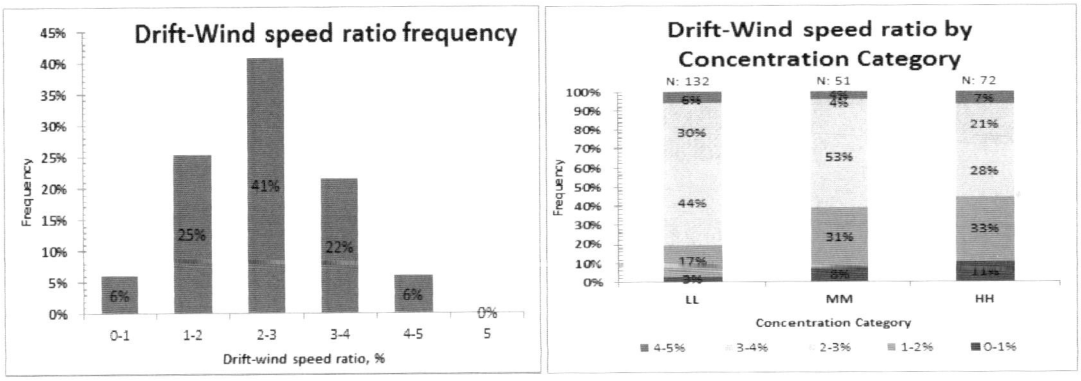

Figure 11. Drift-wind speed ratio frequency distribution (Left). Drift-wind speed ratio frequency by concentration category (Right).

DRIFT MODEL

General ice drift behavior in regards of its dependency to direction and concentrations became clear. Further on, equation (1) presented by Leppäranta (2005) and later simplified by Segboer and Verlaan (2007) was used to describe relationship between drift and wind. As currents measurements results are not publicly available they were accounted within this equation as residual variables described below.

$$\vec{D} = \mathbf{A}\vec{W} + \vec{E},$$

$$\begin{pmatrix} D_x \\ D_y \end{pmatrix} = \begin{pmatrix} a_{11} & a_{12} \\ a_{21} & a_{22} \end{pmatrix} \begin{pmatrix} W_x \\ W_y \end{pmatrix} + \begin{pmatrix} E_x \\ E_y \end{pmatrix} \quad (1)$$

where D_x, D_y are orthogonal zonal (West-East) and meridional (North-South) components of ice drift speed vector $\vec{D}$; W_x, W_y – components of synchronized wind speed vector $\vec{W}$, E_x, E_y – components of residual drift speed vector variation $\vec{E}$ that can't be explained by influence of the wind using linear transformations with matrix **A** with components a_{11}, a_{12}, a_{21}, a_{22}.

A linear regression analysis has been used to determine matrix coefficients a_{11}, a_{12}, a_{21}, a_{22} and *E* vector components from Eq. (1) based on records of drift vectors and corresponding averaged wind data. The results of the regression analysis are presented in Table 2 and Figure 12.

Table 2. Regression analysis outcome for different drift-wind records set

	All data	Filtered data			
		All	LL	MM	HH
Number of records	493	292	132	51	72
a_{11}	0.0254	0.0257	0.0286	0.0265	0.0219
a_{22}	0.0235	0.0254	0.0277	0.0237	0.0259
a_{12}	0.0009	0.0004	0.0010	0.0038	0.0008
a_{21}	0.0052	0.0070	0.0097	0.0050	0.0055
E_x	0.0152	0.0029	-0.0028	0.0466	0.0086
E_y	-0.0119	0.0038	0.0241	-0.0012	-0.0252
R^2 D_x	74.3%	82.8%	88.8%	82.4%	77.8%
R^2 D_y	75.2%	83.4%	84.2%	88.4%	77.0%

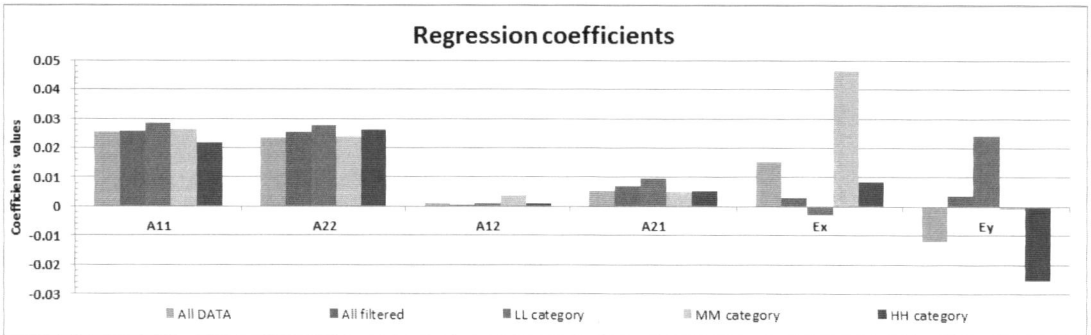

Figure 12. Regression coefficients comparison for various sets of data.

The set of computed coefficients above was subsequently used to hindcast drift for cases with known wind and drift conditions based on GFS forecast data archive. As an example, a single ice floe has been tracked for period of almost 2 weeks (30-Jan-2016 to 12-Feb-2016). The black arrows in Figure 13 (Left) indicate observed ice displacements over the imagery interval periods; green arrows are modeled ice displacement over the same periods using averaged wind and blue arrows are modeled ice displacement over 3-hours wind interval.

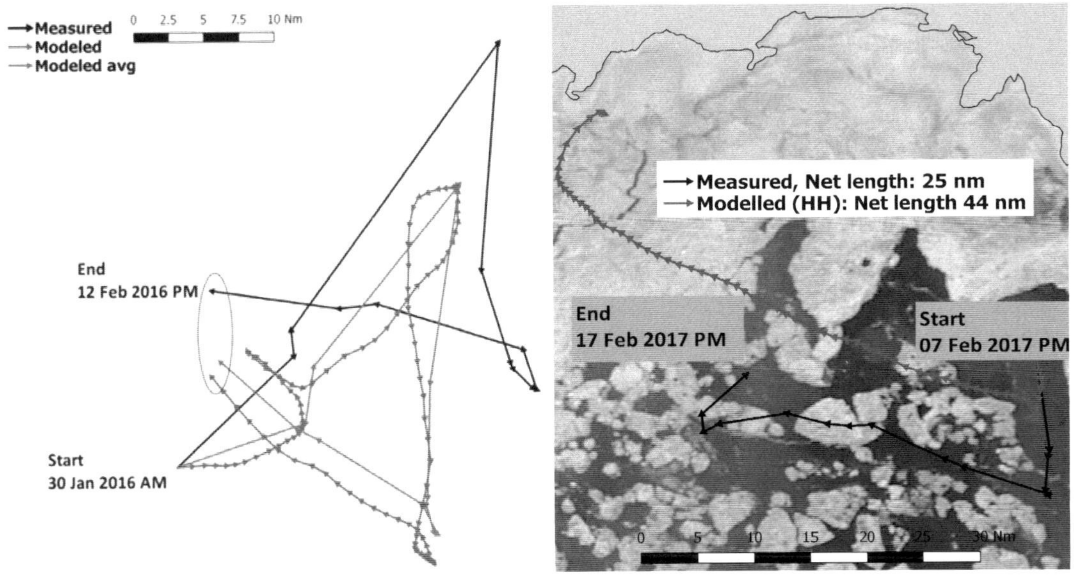

Figure 13. Comparison of measured and modeled ice drift for the periods 30/Jan/2016 to 12/Feb/2016 (Left) and 07/Feb/2016 – 17/Feb/2016 (Right)

Figure 14 shows two samples of modelled drift using GFS data and three different coefficients corresponding to different concentration ratings. These cases show better fit when corresponding set of coefficients is used in different by concentration conditions.

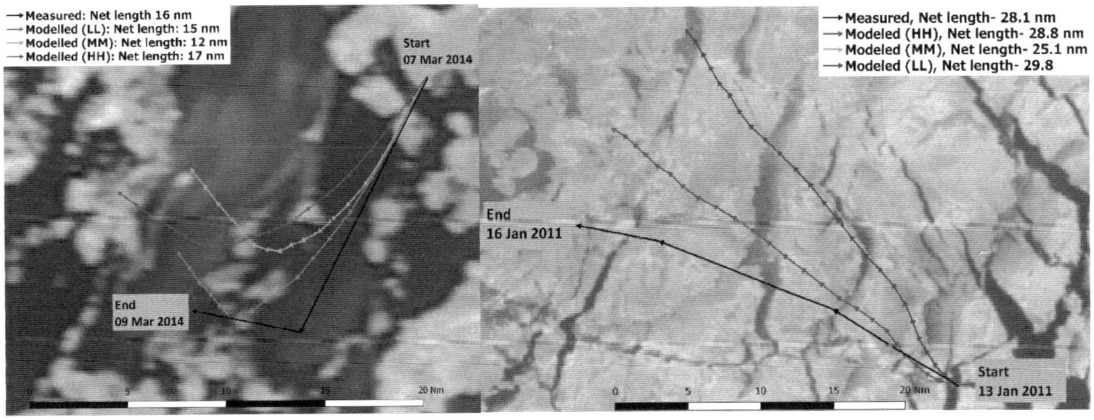

Figure 14. Results of verification in unconfined conditions of low (left) and high (right) concentration.

However, once there is an obstacle or confinement at coastline the model does not show any good fit at all as demonstrated on Figure 13 (Right), when the actual drift has finished in front of stable ice zone along the coastline and the model continued its progress.

DISCUSSION

The values of a_{11} and a_{22} for all set of data are between 0.02 and 0.03. The residual drift variation (E_x and E_y) values were significant for low wind speeds (especially for MM and HH concentration categories), but become negligible with increasing wind speeds as wind starts to dominate over other factors.

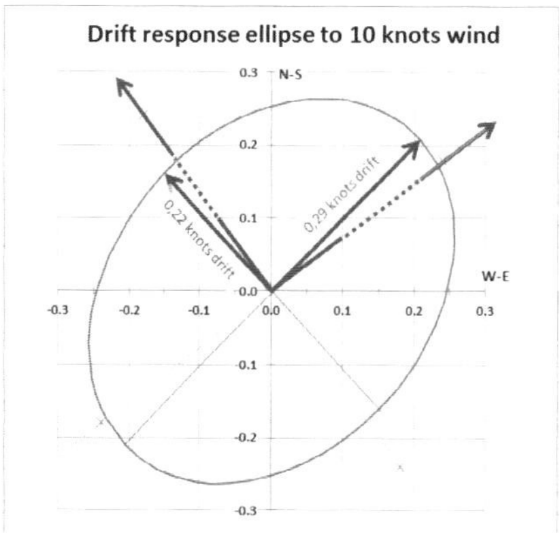

Figure 15. Drift response ellipse to wind speed circle of 10 knots.

The coefficients a_{12} and a_{21} are mostly much smaller than a_{11} and a_{22}, except for several cases when a_{21} almost reaches 0.01 for LL category, meaning that for Low concentrations wind zonal component (W-E) tends to influence drift meridional component (N-S) turning the drift vector a bit to the left from wind vector. Apparently positive a_{12} and a_{21} coefficients show that the response of ice depends not only on wind speed, but also on wind direction. To demonstrate this effect, the ice drift response ellipse was built (**Error! Reference source not found.**).

It shows the response of the ice to circumferential wind with constant speed of 10 knots. For wind of 10 knots in NE and NW direction (blue vectors) ice will tend to drift in the same direction with slight deviation to the left (red vectors). The general idea is that drift speed in NE or SW directions (big semi-axis), for example, tends to be almost 30% higher than for NW or SE (small semi-axis) for the same wind speed. These results make sense for the NE Caspian Sea (Figure 1), that has NE-SW stretched form and fast ice forming along coastlines creates confinement for mobile ice to easily drift in NE and SW directions.

CONCLUSIONS

The major achievement of this project is that a reliable dataset of ice drift in the region was compiled for the season 2015-2016. The dataset was used to gain initial understanding of how ice responds to wind in the conditions of the Caspian basin. Ice moves with 2 to 3% of wind speed for 41% of tracks and between 1-4% for about 90% of tracks. Correlation of wind direction to resulting drift direction was challenging due to many factors guiding the process. The method chosen to correlate wind and drift vectors could be used as a working solution for variable concentrations if there is no confinement of drift with coastline or landfast ice and obstacles in form of stationary ice features or artificial structures.

Introduction of Sentinel constellation with sufficient frequency of reattendance in the region starting last winter (2016-2017) the authors have received opportunity to grow the drift database and further improve the model. Introduction of additional set of coefficients and logic considering proximity to obstacles and coastline is the most obvious first step for improvement. This work is based on public domain data such as the MODIS Terra and Aqua satellite data and the GFS winds. With higher spatial resolution weather data (for example ECMWF) as well as more SAR satellite imagery to increase frequency of observations could improve the results. Coefficients can also be improved a lot if calibrated with field measurements (e.g. from drift buys) if they were publicly available.

ACKNOWLEDGEMENTS

Sean McDermott has inspired the whole development of the ice project in the history of its development. Authors are grateful to Paul Verlaan for giving his opinion on the work performed and advice on future development of this project. KSAT as the leading provider of SAR imagery in NRT operational mode has been the most significant facilitator of our Ice Charting operations in the Caspian and has contributed in our research with archive SAR images that helps us conduct our research and deliver better analysis to community. Carles Debart has spent significant time on technical discussions with us in regards of using SAR images and support in acquiring archive data.

REFERENCES

Crocker G., Ritch A., Nilsen R., 2011, Some Observations of Ice Features in the North Caspian Sea. *Proceedings of the 21st International Conference on Port and Ocean Engineering under Arctic Conditions,* July 10-14, 2011, Montréal, Canada, POAC11-118.

Jordaan I., Stuckey P., Bruce J., Croasdale K., Verlaan P., 2011, Probabilistic Modelling of the Ice Environment in the Northeast Caspian Sea and Associated Structural Loads. *Proceedings of the 21st International Conference on Port and Ocean Engineering under Arctic Conditions,* July 10-14, 2011, Montréal, Canada, POAC11-133.

Leppäranta, M. 2005. *The Drift of Sea ice*. Praxis Publishing Ltd., Chichester, UK.

Nilsen R., Verlaan P.A.J., 2011, The North Caspian Sea Ice Conditions and how Key Ice Data is Gathered. *Proceedings of the 21st International Conference on Port and Ocean Engineering under Arctic Conditions,* July 10-14, 2011, Montréal, Canada, POAC11-112.

Segboer T.J. and Verlaan P.A.J., 2007, Ice Drift Under the Influence of Winds and Currents along the Sakhalin Northeast Coast. 2007 Dalian University of Technology Press, Dalian, ISBN 978-7-5611-3631-7

Verlaan P.A.J., Croasdale K. 2011, Ice Issues Relating to the Kashagan Phase II Development, North Caspian Sea'. *Proceedings of the 21st International Conference on Port and Ocean Engineering under Arctic Conditions,* July 10-14, 2011, Montréal, Canada, POAC11-171.

Yamartino, R. J., 1984. A Comparison of Several "Single-pass" Estimators of the Standard Deviation of Wind Direction. J. Climate Appl. Meteor., Vol. 23, pp.1362-1366

Raw Data Sources: NASA worldview portal, NOAA National Centers for Environmental Information, GFS weather model data, Copernicus Science Hub

Arctic Lead Detection Using a Waveform Unmixing Algorithms from CryoSat-2

Sanggyun Lee[1], Jungho Im[1], Hyun-cheol Kim[2]

[1] School of Urban and Environmental Engineering Ulsan National Institute of Science of Technology (UNIST)

[2] Unit of Arctic sea- ice prediction, Korea Polar Research Institute (KOPRI)

ABSTRACT

Arctic areas consist of ice floes, leads, and polynyas. While leads and polynyas account for small parts in the Arctic Ocean, they play a key role in exchanging heat flux, moisture, and momentum between the atmosphere and ocean in wintertime because of their huge temperature difference In this study, a linear waveform unmixing approach was proposed to detect lead fraction. CryoSat-2 waveforms for pure leads, sea ice, and ocean were used as end-members based on visual interpretation of MODIS images coincident with CryoSat-2 data. The unmixing model produced lead, sea ice, and ocean abundances and a threshold (> 0.7) was applied to make a binary classification between lead and sea ice. The unmixing model produced better results than the existing models in the literature, which are based on simple thresholding approaches. The results were also comparable with our previous research using machine learning based models (i.e., decision trees and random forest). A monthly lead fraction was calculated, dividing the number of detected leads by the total number of measurements. The lead fraction around Beaufort Sea and Fram strait was high due to the anti-cyclonic rotation of Beaufort Gyre and the outflows of sea ice to the Atlantic. The lead fraction maps produced in this study were matched well with monthly lead fraction maps in the literature. The areas with thin sea ice identified from our previous research correspond to the high lead fraction areas in the present study. Furthermore, sea ice roughness from ASCAT scatterometer was compared to a lead fraction map to see the relationship between surface roughness and lead distribution. Active heat loss from ocean accelerates new ice in the leads. The leads weaken the internal strength

POAC'17
Busan, KOREA

Proceedings of the 24th International Conference on
Port and Ocean Engineering under Arctic Conditions
June 11-16, 2017, Busan, Korea

Evaluation of Suspended Sediment Retrieval from above Water and Satellite Reflectance in Fjord Waters of Svalbard

Young-Sun Son and Hyun-Cheol Kim[1]
[1] Unit of Arctic Sea-Ice Prediction, Korea Polar Research Institute (Incheon, South Korea)

ABSTRACT

The sky and sun glint correction of in situ measurements is most important in estimating R_{rs} from above water. If glint effect is not estimated accurately, significant errors can occur in the estimated R_{rs}. In this study, we evaluated the performance of five sky glint algorithms for above sea surface reflectance data collected from fjord waters of Svalbard. Based on the visual inspection of the R_{rs} spectra and the correlation with the SSC measurements, Ruddick's algorithm showed the best results in our evaluation for sky glint correction. The result of Ruddick's algorithm exhibited high correlation coefficient between R_{rs} and in situ SSC at 645 nm ($r^2 = 0.86$).

KEY WORDS: Remote sensing reflectance; Suspended sediment; Sky glint correction

INTRODUCTION

The use of ocean remote sensing has been focused on the water leaving reflectance (R_{rs}) properties associated with chlorophyll a, suspended sediment (SS) and colored dissolved organic matter (CDOM) in visible and near infrared (NIR) wavelength (Doxaran et al., 2002; Darecki and Stramski, 2004; Dall'olmo et al., 2005). The sky and sun glint correction of in situ measurements is most important in estimating R_{rs} from above water. Frequently, the NIR R_{rs} was assumed to be zero for the glint correction (Gordon and Wang, 1994; Quillion et al., 1996). However, this assumption is not true in turbid waters due to the back scattering on particles. The retrieval of suspended sediment concentration (SSC) can be measured in NIR if the R_{rs} intensity is strong enough to be measured (Gould et al., 2001; Ruddick et al., 2006). In this study, we investigate and assess which glint correction approach and wavelength region is effective to retrieval SSC from Arctic above-water reflectance.

METHODS

In situ data (above water reflectance and water samples) were collected at 28 stations in the Kongsfjorden of Svalbard between 06 and 11 August 2006. For SS measurements, 500 ml surface water samples are collected and filtered through a 25 mm GF/F filter. The filter weight was measured by drying in an oven at 60 ° C for 4 hours before filtering. After filtering, the filter was washed with distilled water to remove the salt remaining on the surface of the filter. The weight of filter after filtration was measured in the same way as the filter weigh measurement before filtering. The SSC was calculated by the filter weight difference before and after filtration. Above water reflectance measurements were collected using a ASD FieldSpec 512 channel radiometer covering spectral range of 350 -1050 nm. A dual spectrometer unit with two fiber optic cables was used to measure downwelling irradiance (E_d) and total upwelling radiance (L_t) simultaneously from above water. The L_t and L_s (sky radiance) were measured at a viewing direction of approximately 30° from nadir and zenith respectively. The azimuth angle to measure L_t and L_s was maintained at 135° from the Sun. The R_{rs} was calculated according to Eq. (1),

$$R_{rs} = \frac{L_t - \rho L_s}{E_d}, \quad (1)$$

where ρ is the sea surface reflectance factor on sky conditions, wind speed, solar zenith angle, and viewing geometry. If ρ is not estimated accurately, significant errors can occur in the estimated R_{rs}. Lee et al. (1997) proposed a value of ρ =0.021 for the removal of the Rayleigh component of the reflected skylight when the viewing angle is 30°. Mobley (1999) suggested that for viewing direction of 40° from the nadir and 135° from the Sun, a value of ρ = 0.028 is acceptable for wind speeds less than 5 m s^{-1} and overcast sky. The above corrections using a ρ value is often inaccurate due to the residual surface reflection, ε in Eq. (2),

$$R_{rs} = \frac{L_t - \rho L_s}{E_d} - \varepsilon. \quad (2)$$

Gould et al. (2001) presented a surface correction algorithm based on two assumptions which are at 715-735 nm that total absorption is due primarily to pure water absorption and that the surface reflected sky/cloud light and backscattering spectra are nearly flat. They calculated ε according to Eq. (3),

$$\varepsilon = \frac{R_t(735)a_w(735) - R_t(715)a_w(715)}{a_w(735) - a_w(715)} - \rho R_s(735), \quad (3)$$

where $R_t = L_t/E_d$, $R_s = L_s/E_d$ and a_w is the pure water absorption coefficient from Kou et al. (1993) and Pope and Fry (1997). With a similar approach, Ruddick et al. (2005, 2006) proposed an algorithm for the improvement of sky glint correction. They analyzed and tabulated as a similarity spectrum by normalization at 780 nm, and estimated the ε using measurements at 720 nm and 780 nm as follows

$$\varepsilon = \frac{\alpha(720,780)R_{rs}(780) - R_{rs}(720)}{\alpha(720,780) - 1}, \tag{4}$$

where $\alpha\,(720,780)$ can be read from the tabulated similarity spectrum from Ruddick et al. (2005). Kutser et al. (2013) recently developed a simple method for glint correction where reflectance values in the 350–380 nm range are used to determine the slope of the power function and reflectance values in the 890–900 nm range are used to determine the absolute value of glint. We tested above five approaches for sea surface reflectance correction and investigated the correlation between the sky glint corrected R_{rs} and in situ SSC.

RESULTS AND DISCUSSION

Figure 1 shows the in situ R_{rs} spectra corrected by the proposed algorithms for glint removal. A solid red line in Figure 1 indicates zero reflectance to comparison the performance of each correction algorithm. Dashed lines of different colors are representative in situ spectra to show the variations of R_{rs} according to SSC (red and blue: high SSC, green: moderate SSC, yellow: low SSC). The SSC of the black solid line spectra ranges from 0.5 to 6 mg/l.

The result for Mobley's correction showed that the R_{rs} values of some spectra are below zero at above 700 nm (Figure 1(a)). It seems to be due to overcorrection for glint. On the other hand, despite low SSC, some R_{rs} values are too high in NIR (700-900 nm). It may be because this correction does not consider the residual surface reflection (ε). Lee's algorithm exhibited a little improvement in overcorrection due to difference in ρ value, but did not show any significant differences from Mobley algorithm result (Figure 1(b)). Gould's algorithm exhibited better result than Mobley and Lee algorithms (Figure 1(c)). The residual surface reflection calibration resulted in significant differences at blue, green (400-500 nm) and NIR wavelengths (700-900 nm). Especially, the corrected spectra in the NIR clearly show the R_{rs} intensity difference due to SSC. However, some R_{rs} values were negative in the 800-900 nm. The steep decline of R_{rs} indicates that there is a polar environmental or instrumental error rather than an algorithmic problem. The Ruddick's algorithm result is similar to Gould's algorithm result (Figure 1(d)). This is because the two algorithms are based on similar assumptions. Kutser's algorithm did not produce good result in our experiment. They used R_{rs} values at two ranges of wavelengths (350–380 nm and 890–900 nm) for glint correction, but our in situ R_{rs} data were not good in that wavelengths because of lot of instrumental noise.

The correlation using linear regression between sky glint corrected R_{rs} and in situ SSC at wavelengths corresponding to MODIS band 1 (645 nm) and 2 (859 nm) is shown in Table 1. Table 1 shows that $R_{rs}(645)$ exhibits better results than $R_{rs}(859)$. The poor correlation result for $R_{rs}(859)$ may be because the SSCs of our field samples are too low. The R_{rs} of water with low SSC is generally close to zero in the NIR. Among five glint correction approaches, Ruddick's algorithm showed highest correlation coefficient between R_{rs} and in situ SSC at both 645 nm ($r^2 = 0.86$) and 859 nm ($r^2 = 0.68$).

CONCLUSIONS

This study evaluated the performance of sky glint algorithms for above sea surface reflectance data collected from fjord waters of Svalbard. Based on the visual inspection of the R_{rs} spectra and the correlation with the SSC observations, Ruddick's algorithm showed the best results in our evaluation for sky glint correction. However, our SSC measurements are too low and there are too few stations. Therefore future works need to further evaluation in various regions and periods with high SSC. We will analyze the long-term SSC variation in Svalbard fjord sea surface from 2006 to 2013 using in situ above-water surface reflectance and satellite reflectance data.

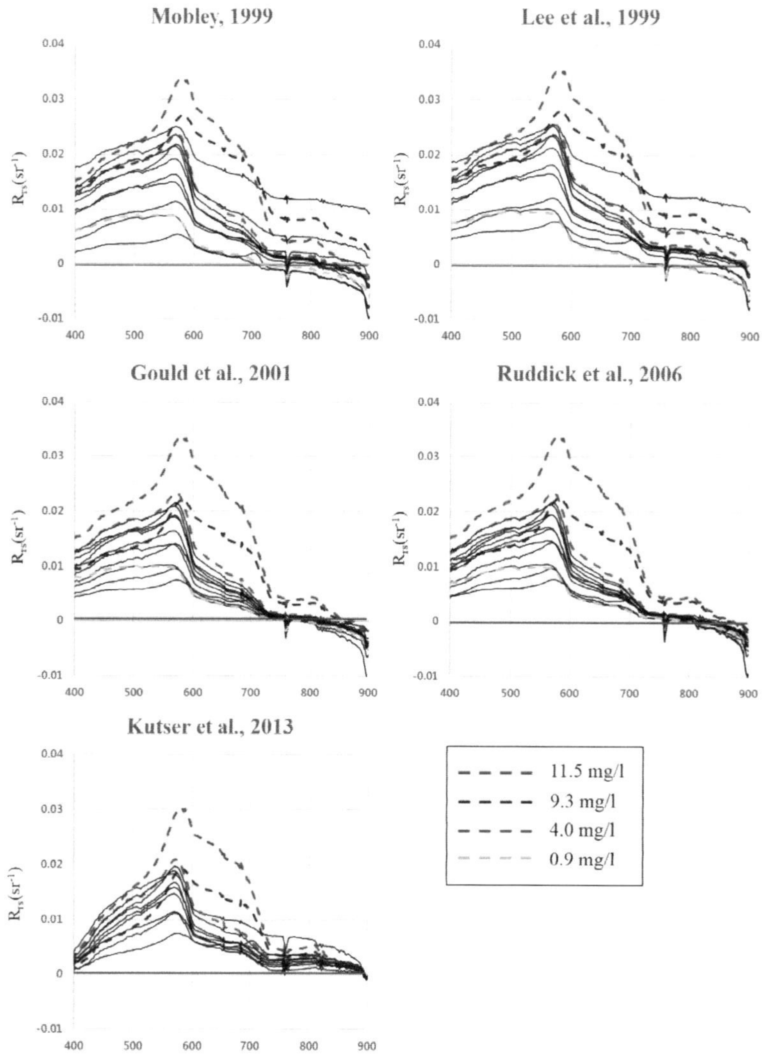

Figure 1. Above water R_{rs} spectra after sky glint correction using five approaches. Different colors of the dashed lines show SSC variation, representatively.

Table 1. Result of correlation coefficients using liner regression between in situ SSC data and above water R_{rs} after performing five glint correction approaches. The selected two wavelengths correspond to MODIS band 1 (645 nm) and band 2 (859 nm).

Wavelength (nm)	Mobley (1999)	Lee et al. (1999)	Gould et al. (2001)	Ruddick et al. (2006)	Kutser et al. (2013)

$R_{rs}(645)$	0.78	0.79	0.86	0.86	0.77
$R_{rs}(859)$	0.15	0.18	0.60	0.68	0

ACKNOWLEDGEMENTS

This study was supported by the Korea Polar Research Institute (grant number: PE17120).

REFERENCES

Dall'Olmo, G., Gitelson, A.A., Rundquist, D.C., Leavitt, B., Barrow, T. & Holz, J.C., 2005. Assessing the potential of SeaWiFS and MODIS for estimating chlorophyll concentration in turbid productive waters using red and near-infrared bands. *Remote Sensing of Environment*, 96, pp.176-187.

Darecki, M. & Stramski, D., 2004. An evaluation of MODIS and SeaWiFs bio-optical algorithms in the Baltic Sea. *Remote Sensing of Environment*, 89, pp.326-350.

Doxaran, D., Froidefond, J.-M. Lavender, S. & Castaing, P., 2002. Spectral signatures of highly turbid waters. Application with SPOT data to quantify total suspended matter concentrations. *Remote Sensing of Environment*, 81, pp.149-161.

Gordon, H.R. & Wang, M., 1994. Retrieval of water-leaving radiance and aerosol optical thickness over the oceans with SeaWiFS: A preliminary algorithm. *Applied Optics*, 33(3), pp.443-452.

Gould, R.W., Arone, R.A. & Sydor, M., 2001. Absorption, scattering and remote-sensing reflectance relationships in coastal waters: Testing a new inversion algorithm. *Journal of Coastal Research* 17(2), pp.328-341.

Kutser, T., Vahtmäe, E., Paavel, B. & Kauer, T., 2013. Removing glint effects from field radiometry data measured in optically complex coastal and inland waters. *Remote Sensing of Environment*, 133, pp.85-89.

Lee, Z., Carder, K.L., Steward, R.G., Peacock, T.G., Davis, C.O. & Mueller, J.L., 1997. Remote sensing reflectance and inherent optical properties of oceanic waters derived from above-water measurements. *Proceedings of the SPIE 2963*, Ocean Optics XIII, 160.

Mobley, C.D., 1999. Estimation of remote sensing reflectance from above-surface measurements. *Applied Optics*, 38(36), pp.7442-7455.

Quillon, S., Forget, P., Froidefond, J.-M. & Naudin, J.J., 1997. Estimating suspended matter concentrations from SPOT data and from field measurements in the Rhone River plume. *Marine Technology Society Journal*, 31(2), pp.15-20.

Ruddick, K.G., De Cauwer, V., Park, Y.-J. & Moore, G., 2006. Seaborne measurements of near infrared water-leaving reflectance: The similarity spectrum for turbid waters. *Limnology and Oceanography*, 51(2), pp.1167-1179.

Ruddick, K.G., De Cauwer, V. & Van Mol, B., 2005. Use of the infrared similarity reflectance spectrum for the quality control of remote sensing data. *Proceedings of the SPIE International Symposium on "optics and Photonics: Remote sensing of the coastal oceanic environment"*, paper #5885-1.

Proceedings of the 24th International Conference on
Port and Ocean Engineering under Arctic Conditions
June 11-16, 2017, Busan, Korea

Test on Friction of Sea Ice on Different Materials

Qingkai Wang, Zhijun Li, Xiaowei Cao, Zan Xiao, Jiankang Wang

State Key Laboratory of Coastal and Offshore Engineering, Dalian University of Technology, Dalian, China

ABSTRACT

Field experiments were conducted in the Bohai Sea to test the friction of sea ice on different materials, such as sea ice, steel and lacquered steel. The influences from different impact factors, for example, the sliding velocity (0.02 m/s to 3.96 m/s), interaction materials (sea ice, steel and lacquered steel), normal stress (3.10 kPa to 11.43 kPa), contacting surfaces of ice (top, bottom and lateral) and ice growth orientation with respect to the sliding direction (vertical and horizontal) on the friction coefficient of sea ice were thoroughly investigated. The results revealed that most of the static friction coefficients between ice and steel and lacquered steel firstly increased with velocity but then decreased at higher velocities with velocity increasing from 0.02 m/s to 3.96 m/s, compared with that of most of the kinetic friction coefficients between ice and steel and lacquered steel, which increased monotonously with increasing velocity. The static friction coefficients between ice lateral surfaces and ice lateral surface were the most of all the static coefficients, and the kinetic coefficients between steel/lacquered steel and ice lateral surfaces were the most of all the kinetic coefficients. Besides, sea ice friction coefficients were independent on normal stress. Finally, the ice growth orientation offered little contribution to the static friction and kinetic friction coefficients.

KEY WORDS: Sea ice; Field experiment; Friction coefficient

INTRODUCTION

In recent years, the decay of Arctic sea ice extent has become serious (Arrigo et al., 2008), making the Arctic shipping available (Smith and Stephenson, 2013). When a ship sails in the ice zone, the ice floes may slide along the hull of the ship (Kotras et al., 1984). The interaction between ice and the hull can not only hinder the voyage but also threaten the safety of navigation. Therefore, knowing the friction properties of sea ice is extremely urgent for the design of icebreaker. Previous tests have concentrated on the mechanics of ice friction. Saeki et al. (1984) tested the static and kinetic friction coefficients between sea ice and various materials and estimated the effect of velocity, normal stress and growth direction of sea ice and so on. Kenndey et al. (2000) studied the kinetic friction coefficient for ice sliding against itself and found that the variable decreased with increasing velocity. Pritchard et al. (2012) conducted friction tests by pulling sea ice blocks over the nature ice cover and proposed that static and kinetic friction coefficients did not occur at the times of maximum and minimum pulling forces, respectively. Sukhorukov and Løset (2013) suggested that the friction coefficient was independent on the velocity when sliding occurred between natural ice surfaces at velocity ranging from 6 mm/s to 105 mm/s. However, their experiments either were conducted in laboratory with processed specimens or tested at a very low velocity, which not aimed at the behavior of sea ice friction occurred during ice navigation. Therefore, we did a series of field friction tests by pulling sea ice blocks, steel and lacquered steel over the top, bottom and lateral surfaces of nature ice cover to simulate the friction occurred between ice and ship and between ice and ice. Moreover, according to the results this paper analyzed the effect of sliding velocity, normal stress, interaction materials and ice growth direction on the static and kinetic coefficients.

FIELD TESTS

Field experiments on the friction between sea ice and different materials were performed in January of 2015 in the Liaodong Bay (40°40.714′ N, 122°07.005′ E) of the Bohai Sea.

Four ice blocks were cut from the nature ice cover. Each block was selected one surface as the contacting surface (top, bottom and lateral). Moreover, the lateral surface was divided into two types: the ice growth orientation parallel to sliding direction (lateral//) and perpendicular to sliding direction (lateral⊥). Four 20 m-long ice tracks were processed at the ice cover, as shown in Figure1. The ice tracks consisted of ice blocks extracted from the ice cover. The bottom and lateral (lateral// and lateral⊥) surfaces of ice blocks were placed upward respectively to become the sliding plane of ice tracks. The top track used the nature top surface of ice cover.

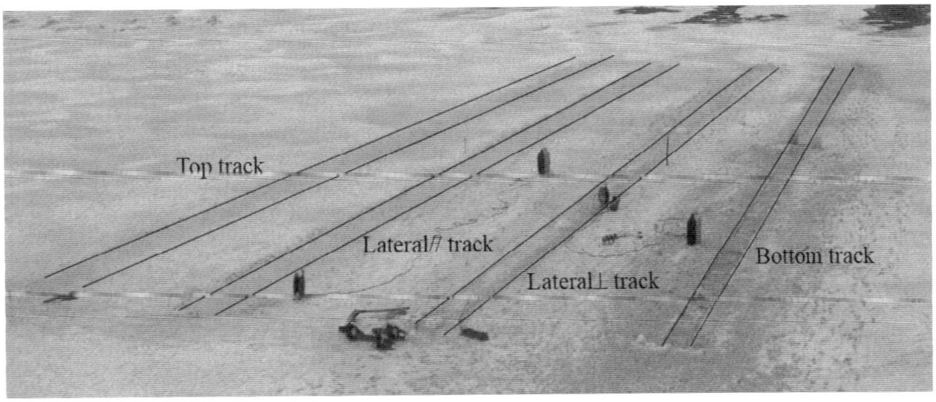

Figure 1. Ice tracks in the field

The experimental set-ups used in the tests were shown in Figure 2. The ice block slid along ice tracks using a pulling mechanism driven by a motor with a controlled velocity (Figure 2(b)). The motor was fixed on the ice cover with four ice screws. Depending on the expected load level, a load cell with a capacity of 2 kN was used in the tests. A pulling wirerope was connected between the load cell and the pulley of motor. As shown in Figure 2(d), the iron blocks were used to add the normal force of friction, and a steel and a lacquered steel plate were used to change interaction materials. A stainless steel basket was used to contain the ice blocks and iron blocks as shown in Figure 2(a) and (c), and the load cell was fixed on it.

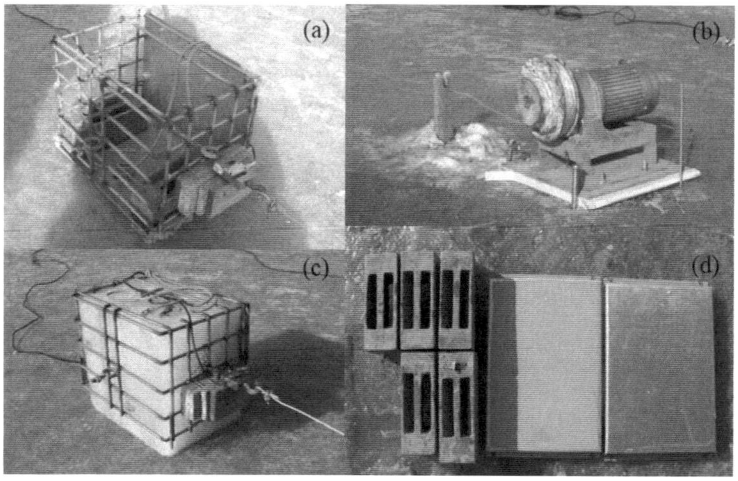

Figure 2. Friction tests: (a) lacquered steel slid on ice top track with iron blocks inside; (b) the motor used to pull the ice blocks; (c) ice lateral// surface slid on ice top track; (c) the iron blocks, the lacquered steel plate and the steel plate used in test

Friction between all of the surfaces of ice blocks and ice tracks were conducted, besides the steel plate and the lacquered steel plate slid on all of the ice tracks. The test conditions of sliding materials on ice tracks were list in Table 1. The sliding velocity ranged from 0 to 3.96 m/s for most of friction between steel/ lacquered steel and ice tracks, and 0 to 0.94 m/s for friction between ice and ice. The force sensor and the pulling mechanism worked simultaneously to record the whole process of friction.

Table 1. The conditions of sliding materials on ice tracks

	Ice top surface	Ice bottom surface	Ice lateral// surface	Ice lateral⊥ surface	Steel	Lacquered steel
Top track	✓	✓	✓	✓	✓	✓
Bottom Track		✓	✓	✓	✓	✓
Lateral// track			✓	✓	✓	✓
Lateral⊥ track				✓	✓	✓

RESULTS AND ANALYSIS

A force-time history was obtained during a friction process, as shown in Figure 3. The first peak load corresponds to the force needed to overcome the static friction. After sliding was initiated, the pulling force had to balance the kinetic friction force. The static and kinetic friction coefficients μ were calculated using the well-known expression:

$$\mu = \frac{F_f}{F_n} \quad (1)$$

where F_f is the static or kinetic friction force and F_n is the normal load. The kinetic friction force was obtained by taking an average of the measured friction force over the time when the ice block was steadily sliding. The test of each condition was repeated three times and the coefficients obtained were averaged.

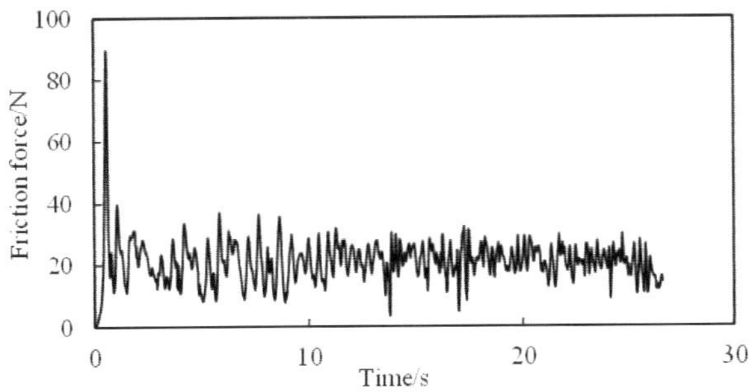

Figure 3. Temporal variations of the friction force during a friction of lacquered steel-top. The normal force F_n=292.70 N, the apparent contacting area S_a = 0.12 m^2, the sliding velocity v=0.19 m/s

Effect of the Velocity on the Static Friction Coefficient

The velocity for static friction refers to the steady sliding velocity after the sliding was initiated. The single logarithmic plots of the variation of static friction coefficients versus sliding velocity between steel and ice tracks are shown in Figure 4. There is a general trend that the static friction coefficients increase at low velocity and decrease at high velocity with increasing velocity, except that the static friction coefficients between steel and ice bottom track increase with increasing velocity without drop from 0 to 3.96 m/s . For steel-top friction, the static friction coefficients increase with velocity increasing from 0 to 0.38 m/s and decrease when velocity varies from 0.38 to 3.96 m/s. For friction between steel and ice lateral tracks (⊥ and //), the static friction coefficients increase with velocity increasing from 0 to 0.57 m/s, and drop at velocity ranging from 0.57 to 2.97 m/s. The static friction coefficients between steel and ice lateral surfaces are higher than other surfaces. The maximum static friction coefficient is 0.0763 at velocity of 0.75 m/s for steel-top friction, 0.1332 at velocity of 2.48 m/s for steel-bottom friction, 0.2272 at velocity of 0.94 m/s for steel-lateral⊥ friction and 0.2143 at velocity of 0.75 m/s for steel-lateral// friction.

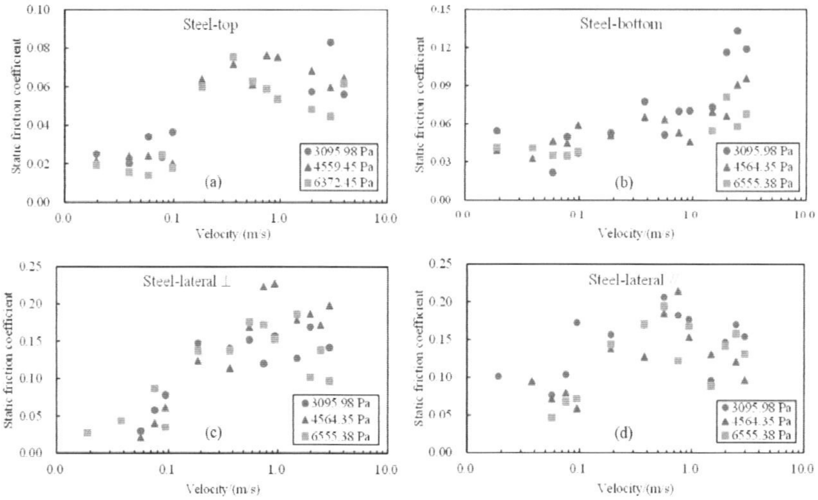

Figure 4. Single logarithmic plots of static friction coefficients versus sliding velocity between steel and ice tracks

Figure 5 shows the logarithmic plots of static friction coefficients between lacquered steel and ice tracks varying with respect to velocity. Similar to the trend of steel sliding on ice tracks, the static friction coefficients between lacquered steel and ice tracks increase at low velocity and decrease at high velocity. For friction between lacquered steel and ice lateral surfaces (⊥ and //) and ice bottom surface, the static friction coefficients increase with velocity increasing less than around 0.94 m/s and decrease with increasing velocity more than 0.94 m/s. For lacquered steel-top friction, the static friction coefficients show an abrupt rise and drop at velocity of 0.10 and 0.94 m/s respectively and the demarcation velocity is around 0.57 m/s. The maximum static friction coefficient is 0.3303 at velocity of 0.57 m/s for lacquered steel-top friction, 0.1184 at velocity of 1.49 m/s for lacquered steel-bottom friction, 0.2524 at velocity of 0.94 m/s for lacquered steel-lateral⊥ friction and 0.2353 at velocity of 0.75 m/s for lateral//-steel friction.

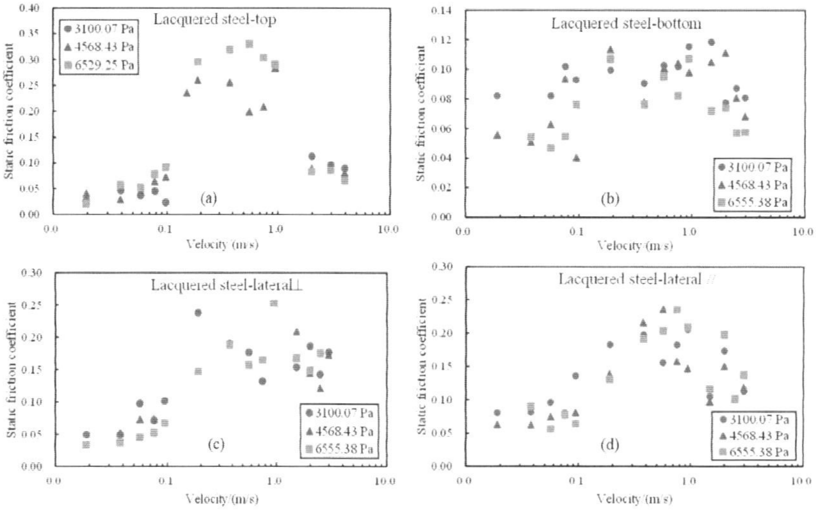

Figure 5. Single logarithmic plots of static friction coefficients versus sliding velocity between lacquered steel and ice tracks

Figure 6 shows the variation of static friction coefficients between ice and ice tracks with respect to velocity. As velocity increases from 0 to 0.94 m/s, the static friction coefficients increase slightly as shown in Figure 6(b), (c), (d), (e), (i) and (j) or remain constant as shown in Figure 6(f), (g) and (h). Besides, Figure 6(a) shows that the static friction coefficients for ice top-top friction are a little higher at velocity between 0 to 0.10 m/s and drop to a constant at velocity more than 0.10 m/s.

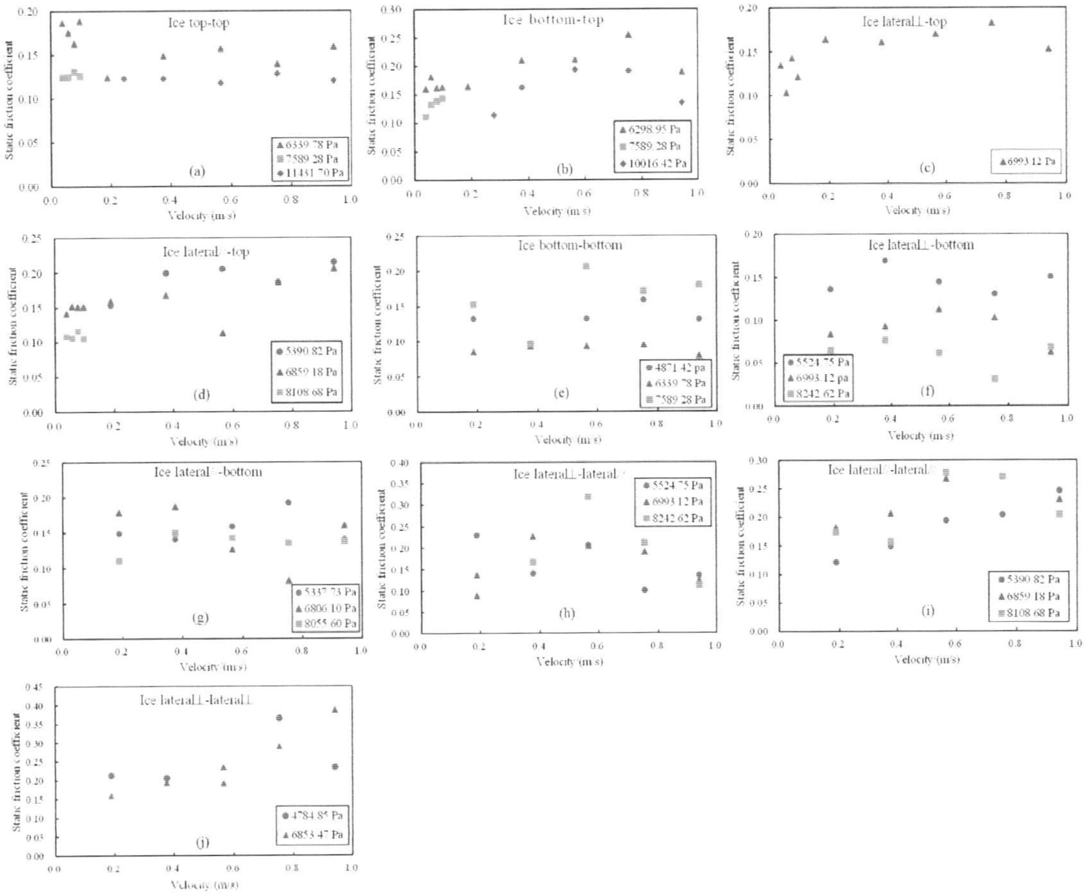

Figure 6. Static friction coefficients versus sliding velocity between ice and ice tracks

Effect of the Materials on the Static Friction Coefficient

Figure 4~6 show that the distribution of static friction coefficient has no correlation with respect to normal press. Table 2 summarizes the maximum static friction coefficients $\mu_{s,max}$ under all conditions. For friction between steel and ice, $\mu_{s,max}$ of steel-lateral⊥ and steel-lateral// friction are close and higher than others. For friction between lacquered steel and ice, $\mu_{s,max}$ of steel-lateral⊥ and steel-lateral// friction are also close and higher than that of lacquered steel-bottom friction. For friction between ice and ice, the $\mu_{s,max}$ of ice lateral surfaces sliding on ice lateral tracks is the highest among all conditions of coefficients. Comparing the $\mu_{s,max}$ of different materials sliding on ice top surface, the $\mu_{s,max}$ of lacquered steel is the highest and that of steel is the lowest. Among the $\mu_{s,max}$ of different materials sliding on ice bottom surface, the $\mu_{s,max}$ of ice is higher that of steel and lacquered steel. Among the $\mu_{s,max}$ of different materials sliding on ice lateral surfaces, the $\mu_{s,max}$ of ice lateral surfaces is the highest and that of ice top/bottom surfaces is the lowest. There is no differences between the $\mu_{s,max}$ of ice lateral//

surface and $\mu_{s,max}$ of ice lateral⊥ surface on various materials, therefore, the static friction coefficient is independent on ice growth direction.

Table 2. $\mu_{s,max}$ between different materials and ice tracks

Friction material		$\mu_{s,max}$	Friction material		$\mu_{s,max}$
Ice top	Ice top	0.1884	Steel	Ice top	0.0831
	Ice bottom	0.2546		Ice bottom	0.1332
	Ice lateral⊥	0.1828		Ice lateral⊥	0.2272
	Ice lateral//	0.2153		Ice lateral//	0.2143
Ice bottom	Ice bottom	0.2068	Lacquered steel	Ice top	0.3303
	Ice lateral⊥	0.1693		Ice bottom	0.1184
	Ice lateral//	0.1924		Lateral⊥	0.2524
Ice lateral//	Ice lateral⊥	0.3180		lateral//	0.2353
	Ice lateral//	0.2787			
Ice lateral⊥	Lateral⊥	0.3873			

Effect of the Velocity on the Kinetic Friction Coefficient

The single logarithmic plots of the variation of kinetic friction coefficients versus sliding velocity between steel and ice tracks are shown in Figure 7. The kinetic friction coefficients between steel and ice lateral surfaces (⊥ and //) remain at a low value when velocity varying from 0 to 0.09 m/s and increase with velocity increasing from 0.09 to 2.97 m/s. The kinetic friction coefficient between steel and ice top surface increases with increasing velocity while the kinetic friction coefficient between steel and ice bottom surface decreases with velocity increasing from 0 m/s to 0.18 m/s and increases when velocity increases from 0.18 m/s to 2.97 m/s.

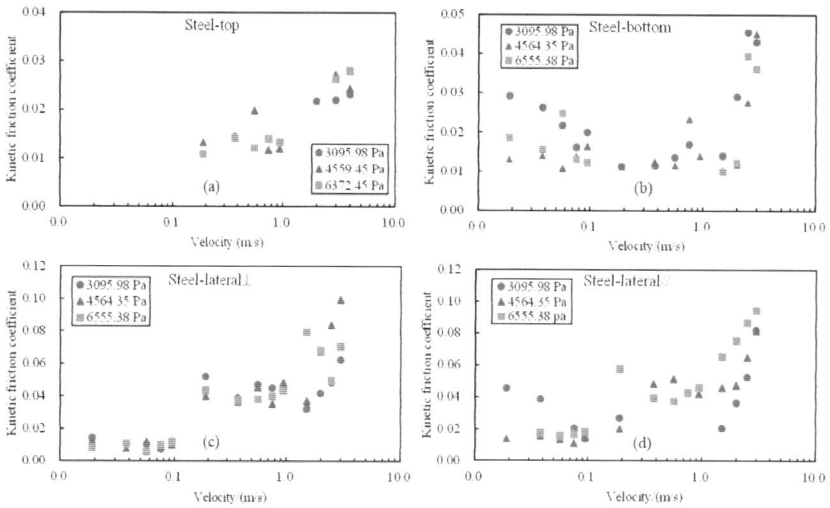

Figure 7. Single logarithmic plots of kinetic friction coefficients versus sliding velocity between steel and ice tracks

POAC17-032

Figure 8 shows the single logarithmic plots of the variation of kinetic friction coefficients between lacquered steel and ice tracks with respect to velocity. The kinetic friction coefficients between lacquered steel and ice lateral surfaces (⊥ and //) exhibit a good correlation with respect to velocity that the kinetic friction coefficients increase monotonically with velocity increasing from 0 to 2.97 m/s. The general variation of the kinetic friction coefficient of lacquered steel-bottom friction shows an increase trend with respect to increasing velocity. However, for lacquered-top friction, the kinetic friction coefficient is much higher than other conditions when velocity varies from 0.15 to 0.94 m/s and drops a lot at velocity more than 0.94 m/s.

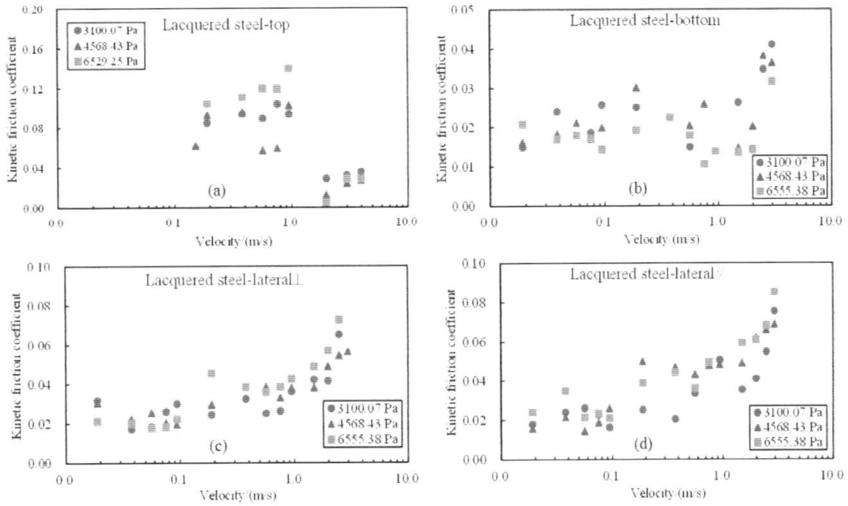

Figure 8. Single logarithmic plots of kinetic friction coefficients versus sliding velocity between lacquered steel and ice tracks

Figure 9 shows the kinetic friction coefficients between ice and ice tracks varying with velocity. The kinetic friction coefficients of lateral⊥-lateral⊥ and lateral//-lateral// friction increase with velocity increasing from 0 to 0.94 m/s. The kinetic friction coefficients of lateral⊥-bottom, lateral//-bottom and lateral⊥-lateral// friction are independent on velocity. The kinetic friction coefficients of top-top, bottom-top, lateral⊥-top and lateral//-top friction show a decrease at velocity less than 0.10 m/s and increase with increasing velocity at velocity ranging from 0.10 to 0.94 m/s.

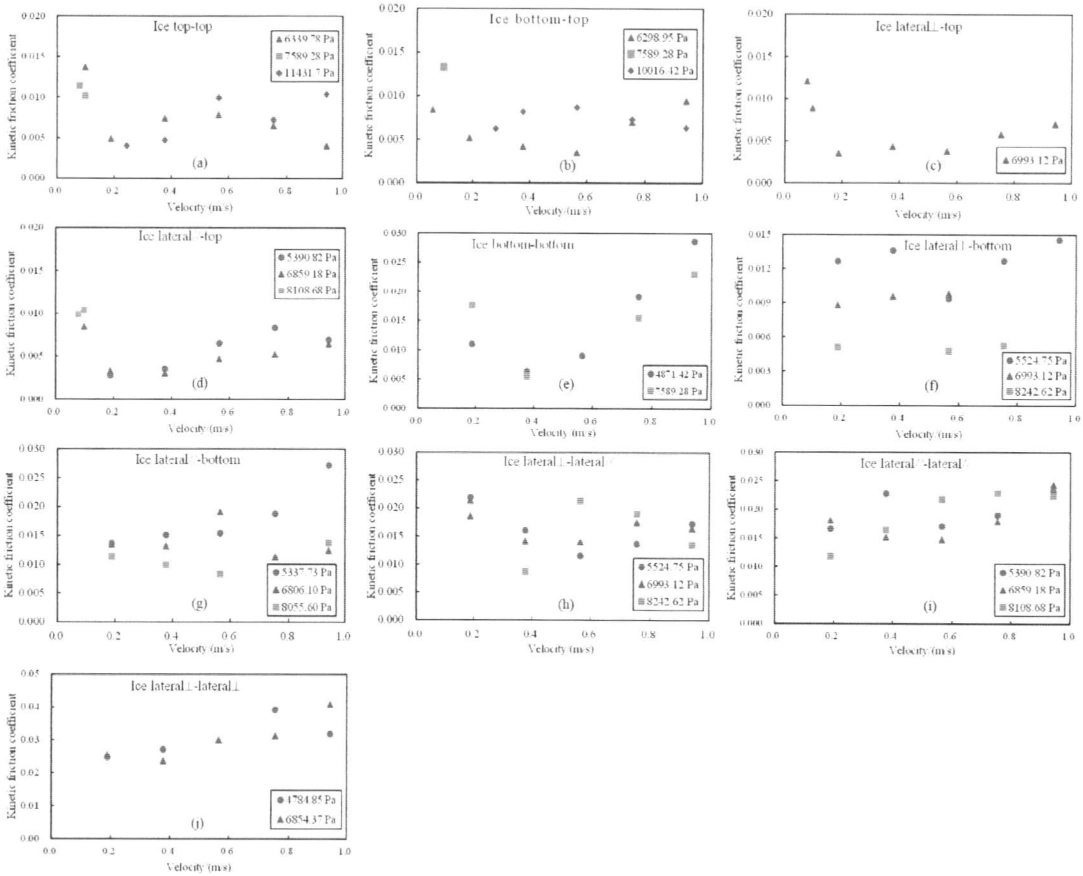

Figure 9. Kinetic friction coefficients versus sliding velocity between ice and ice tracks

Effect of the Materials on the Kinetic Friction Coefficient

The normal stress has no contribution to the kinetic friction coefficients according to Figure 7~9. Table 3 summarizes the maximum kinetic friction coefficients $\mu_{k,max}$ of all conditions. For the friction between steel and ice, the $\mu_{k,max}$ of steel-lateral⊥ and steel-lateral// friction are close and higher than that of steel-top/bottom friction. For the friction of lacquered steel sliding on ice, the $\mu_{k,max}$ of lacquered steel-lateral⊥ and lacquered steel-lateral// friction are close but less than that of lacquered steel-top friction. For the friction between ice and ice, the $\mu_{k,max}$ between lateral surfaces and lateral surfaces is higher than others. Among all of the coefficients, the $\mu_{k,max}$ of lateral⊥-lateral⊥ friction is the most. Comparing the friction coefficients of different materials sliding on ice tracks, the $\mu_{k,max}$ of steel and lacquered steel is higher than that of ice. It is found that the there is no difference between the $\mu_{k,max}$ of materials sliding on lateral⊥ track and lateral// track, therefore, the kinetic friction coefficients are independent on the ice growth direction.

Table 3. $\mu_{k,max}$ between different materials and ice tracks

Friction material		$\mu_{k,max}$	Friction material		$\mu_{k,max}$
Ice top	Ice top	0.0137	Steel	Ice top	0.0291
	Ice bottom	0.0133		Ice bottom	0.0454
	Ice lateral⊥	0.0121		Ice lateral⊥	0.0993
	Ice lateral//	0.0104		Ice lateral//	0.0943
Ice bottom	Ice bottom	0.0286	Lacquered steel	Ice top	0.1395
	Ice lateral⊥	0.0145		Ice bottom	0.0410
	Ice lateral//	0.0272		Lateral⊥	0.0726
Ice lateral//	Ice lateral⊥	0.0219		lateral//	0.0850
	Ice lateral//	0.0243			
Ice lateral⊥	Lateral⊥	0.0409			

DISCUSSION

Both static and kinetic friction coefficients have no correlation with normal stress, the average coefficients under different normal stress of a test condition can be used in discussion. Besides, both of them are independent on ice growth direction, therefore, the friction coefficients between materials and ice lateral⊥ and lateral// can be taken into consideration together.

Figure 10 shows the kinetic and static friction coefficients for friction between steel and ice lateral track and lacquered steel and ice lateral track in single logarithmic plots. The kinetic friction coefficients show a linear increasing relationship with increasing velocity which can be parameterized as Eq. (2) and (4) for steel-lateral and lacquered steel-lateral friction, respectively. While for static friction coefficients of steel/lacquered steel-lateral friction which increase firstly and then decrease with increasing velocity, nonlinear equations are used to describe the variation as Eq. (3) and (5).

Steel-lateral friction:

$$\mu_k = 0.02v + 0.02 \qquad r = 0.92 \tag{2}$$

$$\mu_s = \frac{v}{0.57 + 4.55v + 0.93v^2} \qquad r = 0.93 \tag{3}$$

Lacquered steel-lateral friction:

$$\mu_k = 0.02v + 0.023 \qquad r = 0.92 \tag{4}$$

$$\mu_s = \frac{v}{0.54 + 3.70v + 1.35v^2} \qquad r = 0.94 \tag{5}$$

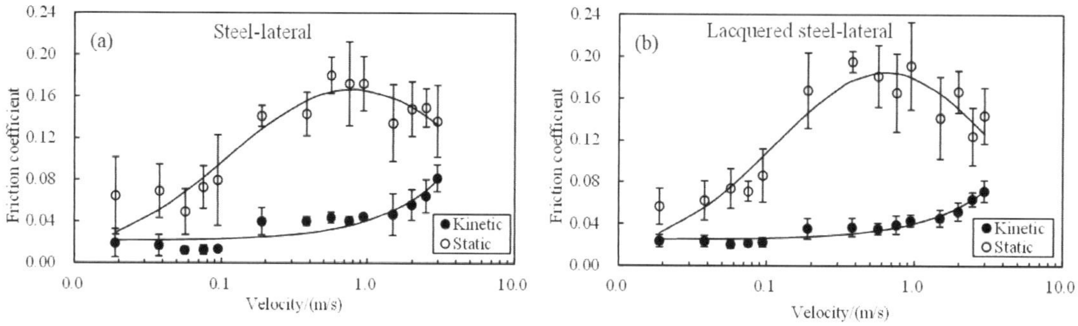

Fig. 10. Single logarithmic plots of kinetic (solid circles) and static (hollow circles) friction coefficients versus sliding velocity of (a) steel-lateral friction and (b) lacquered steel-lateral friction. Also shown are fitting curves (solid lines) and standard deviation (short lines across circles)

The relationships between friction coefficients and velocity for steel-bottom and lacquered steel-bottom friction are shown in Figure 11. The static friction coefficient of steel-bottom friction show an excellent linear relationship with velocity and can be expressed with Eq. (7), however, the kinetic friction coefficient decreases firstly and then increases, which is defined as Eq. (6). The kinetic friction coefficient of lacquered steel-bottom friction increases linearly with velocity, as parameterized with Eq. (8), however, the static friction coefficient climbs at first and then declines with increasing velocity, which is decribed as Eq. (9).

Steel-bottom friction:

$$\mu_k = \frac{v}{-0.72+83.93v+20.28v^2} \quad r=0.90 \tag{6}$$

$$\mu_s = 0.02v+0.04 \quad r=0.95 \tag{7}$$

Lacquered steel-bottom friction:

$$\mu_k = 4.50\times10^{-3}v+0.02 \quad r=0.65 \tag{8}$$

$$\mu_s = \frac{v}{0.24+9.25v+1.23v^2} \quad r=0.75 \tag{9}$$

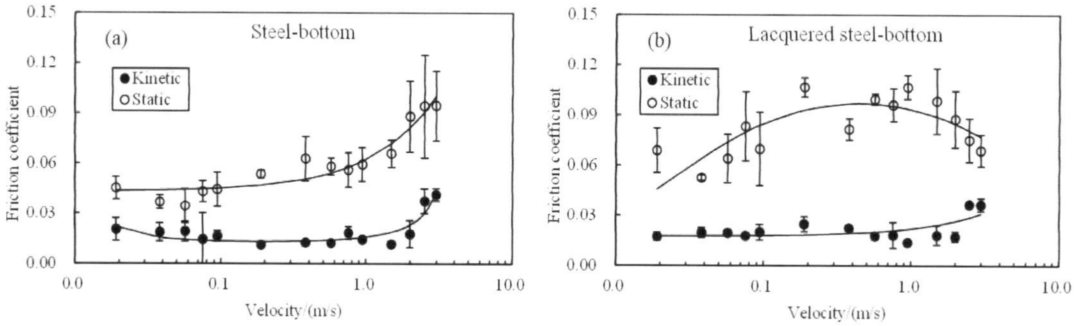

Figure 11. Single logarithmic plots of kinetic (solid circles) and static (hollow circles) friction coefficients versus sliding velocity of (a) steel-bottom friction and (b) lacquered steel-bottom friction. Also shown are fitting curves (solid lines) and standard deviation (short lines across circles)

Figure 12 shows the quantificational relationships between the coefficients of friction and velocity for steel-top and lacquered steel-top friction. The kinetic friction coefficient of steel-top increases linearly with velocity as Eq. (10). However, the static friction coefficient of steel-top friction and the kinetic and static friction coefficients of lacquered steel-top friction increase firstly and then decrease with increasing velocity, as described by Eq. (11), (12) and (13).

Steel-top friction:

$$\mu_k = 4.10\times10^{-3}v + 0.01 \qquad r=0.95 \tag{10}$$

$$\mu_s = \frac{v}{0.24+9.25v+1.23v^2} \qquad r=0.93 \tag{11}$$

Lacquered steel-top friction:

$$\mu_k = \frac{v}{0.24+9.25v+1.23v^2} \qquad r=0.88 \tag{12}$$

$$\mu_s = \frac{v}{0.89-0.52v-4.07v^2} \qquad r=0.95 \tag{13}$$

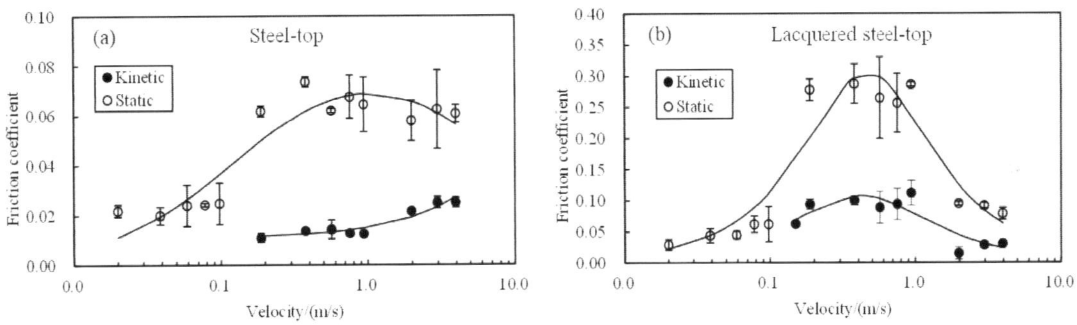

Figure 12. Single logarithmic plots of kinetic (solid circles) and static (hollow circles) friction coefficients versus sliding velocity of (a) steel-top friction and (b) lacquered steel-top friction. Also shown are fitting curves (solid lines) and standard deviation (short lines across circles)

CONCLUSIONS

The kinetic friction coefficients between ice and steel and lacquered steel increase linearly with velocity, except that the kinetic friction coefficient of lacquered steel-top friction increases at first and decreases afterwards with increasing velocity and the kinetic friction coefficient of steel-bottom friction decreases at first and then increases with velocity increasing. The variation of static friction coefficients with respect to velocity is divided into two parts: increasing with increasing velocity and decreasing with increasing velocity, except that the static friction coefficient of steel-bottom friction increases monotonously with velocity. For the friction between ice and ice, both static and kinetic friction coefficients increase with increasing velocity or remain constant.

Among the static friction coefficients, the coefficients between ice lateral surfaces and ice lateral surfaces are the most of all. Among the kinetic friction coefficients, the coefficients between steel and lacquered steel and ice lateral surfaces are the most of all. Both static and kinetic friction coefficients are independent on ice growth direction and normal stress.

ACKNOWLEDGEMENTS

The study is financially supported by the Global Change Research Program of China (No. 2015CB953901), the National Natural Science Foundation of China (No. 51639003 and No. 41376186) and the High Technology of Ship Research Project of the Ministry of Industry and Information Technology (No. [2013]417).

REFERENCES

Arrigo, K.R., van Dijken, G. & Pabi, S., 2008. Impact of a shrinking Arctic ice cover on marine primary production. *Geophysical Research Letters*, 35(19), pp.1-6.

Kotras, T.V., Baird, A.V. & Naegle, J.N., 1984. Predicting ship performance in level ice. *In Transactions - Society of Naval Architects and Marine Engineers. SNAME*, pp. 329-349.

Kennedy, F.E., Schulson, E.M. & Jones, D.E., 2000. The friction of ice on ice at low sliding velocities. *Philosophical Magazine A*, 80(5), pp.1093-1110.

Pritchard, R.S., Knoke, G.S. & Echert, D.C.S., 2012. Sliding friction of sea ice blocks. *Cold Regions Science and Technology*, 76-77, pp.8-16.

Saeki, H., et al., 1986. The Coefficient of Friction between Sea Ice and Various Materials Used in Offshore Structures. *Journal of Energy Resources Technology*, 108(1), pp.65-71.

Smith, L.C. & Stephenson, S.R., 2013. New Trans-Arctic shipping routes navigable by midcentury. *Proceedings of the National Academy of Sciences*, 2013(13), pp.6-10.

Sukhorukov, S. & Løset, S., 2013. Friction of sea ice on sea ice. *Cold Regions Science and Technology*, 94, pp.1-12.

New Insights about Ice Friction Obtained from Crushing-Friction Tests on Smooth and High-Roughness Surfaces

Robert E. Gagnon[1]
[1] Ocean, Coastal and River Engineering, National Research Council Canada,St. John's, NL, Canada

ABSTRACT

Ice crushing-friction tests were conducted in the lab at -10 °C using a set of acrylic ice-crushing platens that included a flat smooth surface and a variety of high-roughness surfaces with regular arrays of small prominences. The experiments were part of Phase II tests of the Blade Runners technology for reducing ice-induced vibration. Ice was crushed against the platens where the ice movement had both a vertical and a horizontal component. High-speed imaging through the platens was used to observe the ice contact zone as it evolved during the tests. Vertical crushing rates were in the range 10 – 30 mm/s and the horizontal sliding rates were in the range 4.14 – 30 mm/s. Three types of freshwater ice were used. Friction coefficients were extraordinarily low and were proportional to the ratio of the tangential sliding rate and the normal crushing rate. For the rough surfaces all of the friction coefficient variation was determined by the fluid dynamics of a slurry that flowed through channels that developed between leeward-facing facets of the prominences and the moving ice. The slurry originated from a highly-lubricating self-generating squeeze film of ice particles and melt located between the encroaching intact ice and the surfaces. These results are important because ice crushing occurs in many situations that involve a sliding frictional component such as sports involving ice-contact, ice interaction with bridges, piers, ship hulls, rock beds under glaciers and ice-on-ice sliding/crushing within glaciers and extraterrestrial ice masses (on Saturn's moon Enceladus) and between interacting sea ice floes.

KEY WORDS: Ice crushing-friction; Highly-lubricating ice-melt slurry; Solid-solid fluid interface friction mechanism; Surprising low friction on rough surfaces.

INTRODUCTION

As pointed out by Gagnon (2016), the frictional behavior of ice has been studied in many contexts, such as in relation to cold-environment sports involving friction on sled runners, skate blades and curling rocks, in the automotive industry concerning rubber tire performance on ice, in shipping regarding friction on hulls of icebreaking ships and other vessels in cold regions, in construction regarding the friction of moving ice on concrete such as piers, bridges and other fixed river and marine structures, in the movement of glaciers regarding ice-on-rock and ice-on-ice friction, and even in relation to extraterrestrial ice-on-ice friction in tectonically active regions of the icy Saturnian moon Enceladus. Over the past few decades ice crushing and indentation have also been studied intensively, mostly in relation to ice engineering associated with offshore oil and gas resources. The main characteristics of the ice behavior consistently recur during field and lab studies and at various scales (Gagnon, 1999). While ice friction in nature often involves some degree of crushing, as in most of the cases above, only a few studies have been conducted specifically on the topic of ice crushing friction where substantial crushing is an integral component (Gagnon and Mølgaard, 1989). Here we present results from ice crushing-friction tests, using novel apparatus, on high-roughness surfaces that reveal new friction mechanisms where the friction coefficient depends on both the normal crushing rate and the tangential sliding rate, and is extraordinarily low.

PRELIMINARY CONSIDERATIONS

To set the stage for discussing ice crushing friction we first consider an ice feature crushing against a surface where no lateral sliding of the ice feature is involved. Many studies (Gagnon, 1999; Riska et al., 1990; Fransson et al., 1991) have shown that during the crushing there are regions of relatively intact ice, small compared to the nominal contact area, in the contact zone that are surrounded by crushed ice which flows away from the intact zones (Fig. 1a). The peripheral crushed ice is essentially the debris of shattered spalls that have previously broken away sequentially from the intact hard zone. Fig. 1a shows a spall-creating fracture (dashed line) that is about to occur. When the spall forms (to the left of the fracture) and separates from the intact bulk ice it immediately shatters and pulverizes to become crushed ice. Thin section analysis from lab tests confirmed the intact nature of the ice (Gagnon, 1994) and thin and thick sections from the Hobson's Choice Ice Island indentation tests (Gagnon, 1998) showed similar features. The pressure on the intact ice zones (hard spots) is very high (30-70 MPa) and changes abruptly to low values (0-10 MPa) when crossing the boundary between intact ice and the crushed material at the perimeter. The high-pressure zones have been shown to be regions where a thin squeeze-film slurry layer of pressurized melt and ice particles is present between the intact ice and the surface (Fig. 1). The viscous flow of this slurry layer generates heat that accounts for the rapid melting component of the removal of ice from the hard zones during ice crushing. A similar process occurs at ice-on-ice contact (Gagnon, 2013) of ice fragments in the surrounding crushed ice matrix as it flows away from the high-pressure zones.

SLURRY LAYER DETAILS

Details of the thin layer have been the subject of investigation in four previous studies. Gagnon (2016) has summarized the results and provided a general view of the nature of the slurry layer during crushing for tests at a similar scale to that of the present ones. We can surmise that the slurry layer thickness in the present tests is somewhere in the approximate range 0.02 mm - 0.17 mm, and the liquid fraction of the layer is about 16%. We may think of the slurry layer as a self-generating squeeze film that is powered by the energy supplied by the loading system that causes the ice crushing.

EXPERIMENTAL SETUP

Fig. 2a shows a photograph of the test setup where the ice, confined in a rigid holder, is pressed in the vertical direction at a fixed rate against a transparent acrylic crushing platen (2.5 cm thickness) in a testing frame. Note that in Fig. 1 the ice is situated above the platen, for illustrative purposes, whereas in the actual tests (Fig. 2a) the ice was positioned below the platen. The crushing platen is backed by a secondary polished acrylic support plate (5 cm in thickness). The ice holder is attached to the top face of a 2.5 cm thick metal plate. The plate has four commercial 'frictionless' ball-bearing cars attached under it that ride on two metallic rails in the horizontal direction parallel to the long axes of the ice specimen and the crushing platen. An electric actuator is used to move the plate and ice holder at controllable fixed rates.

High-speed images of the ice-platen contact region are recorded by viewing through both the support plate and the crushing platen by means of a mirror situated between the supporting posts for the plate and platen. Each of the two platens primarily discussed here had either an array of small regular square pyramids machined onto its surface (Fig. 2b) or an array of

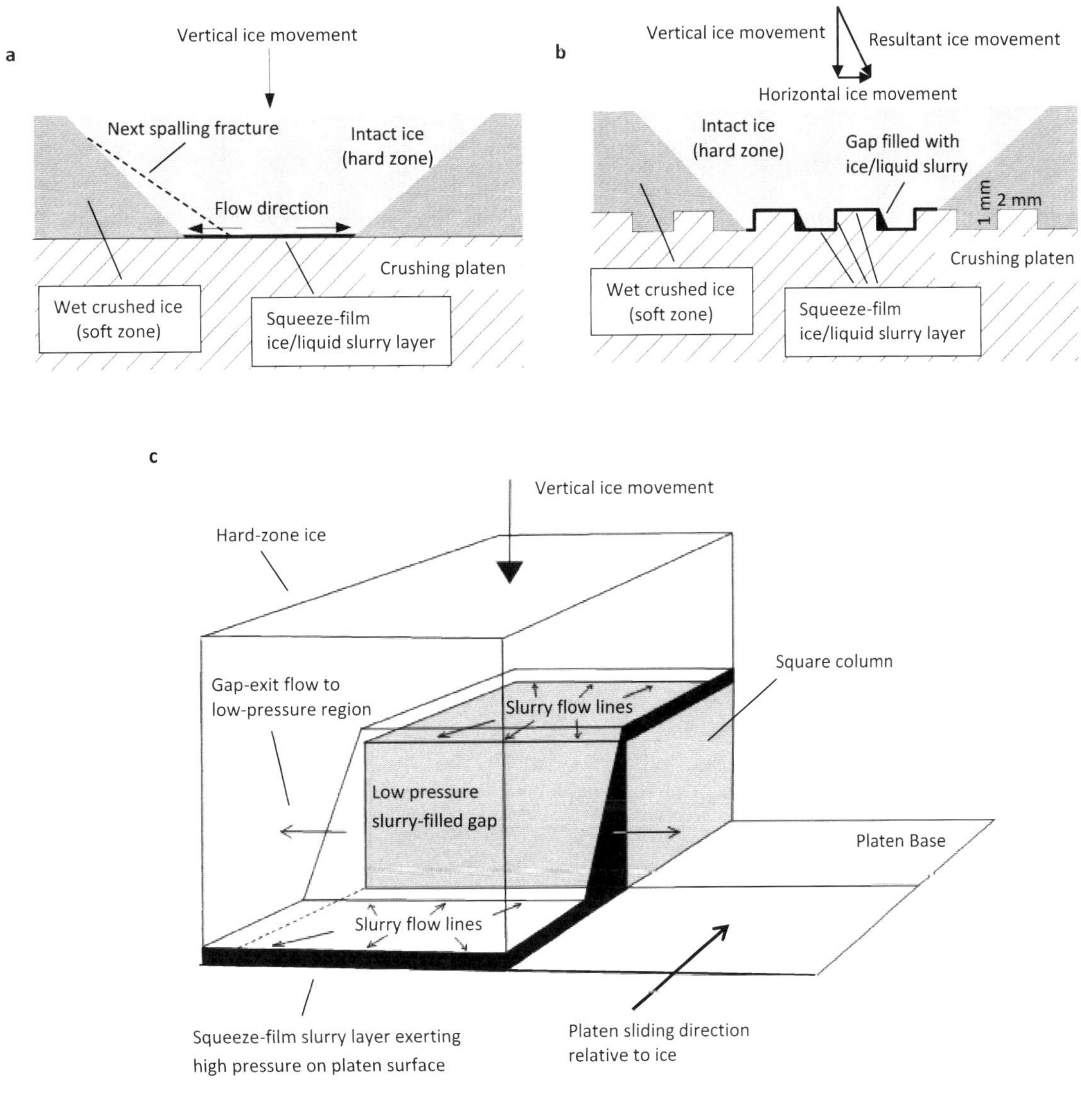

Fig. 1. Schematics showing aspects of the ice behavior during the crushing-friction experiments. **(a)** Schematic showing the essential characteristics of ice crushing against a flat rigid surface. **(b)** A 2D schematic depicting ice crushing against a platen surface with square columns. The ice has both a vertical and a horizontal component of movement relative to the crushing platen, where the resultant movement is as indicated. **(c)** A 3D schematic showing a small portion, a unit area containing one square column, of the view of the ice and platen shown in Fig. 1b. The schematic depicts the time-averaged general flow characteristics of the self-generating squeeze-film slurry as it moves from high-pressure regions, where it is generated, into the lower-pressure gap space and eventually out through the gap exits where low-pressure crushed ice is present. From Gagnon (2016).

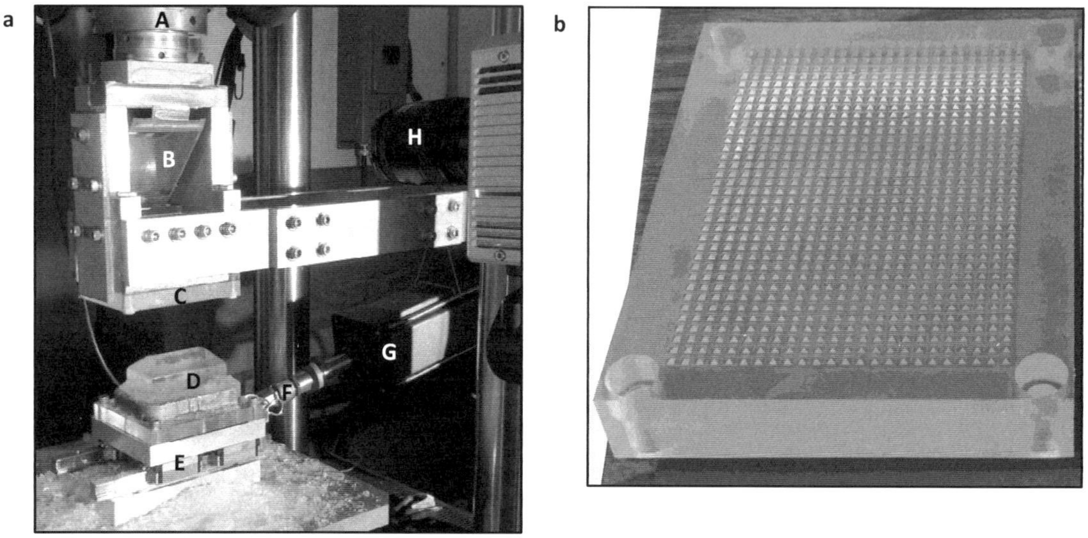

Fig. 2. Test apparatus photographs. **(a)** Photograph of the crushing-friction test setup. (A) Vertically-oriented test-frame load cell for measuring the normal load; (B) Mirror; (C) Acrylic crushing-platen; (D) Ice specimen in ice holder; (E) Rail-car assembly; (F) Load cell used to measure the horizontal friction force; (G) Linear actuator used to slide the rail-car and ice sample horizontally; (H) High-speed imaging camera. **(b)** Photograph of the acrylic platen with the array of small regular square pyramids on its surface (platen dimensions: 166 mm x 129 mm x 25 mm). The pyramids were 1 mm in height and 2 mm wide at the base. The space between each adjacent pyramid was 2 mm. The arithmetic average of the high-roughness profile for the surface of the platen was 0.075 mm. From Gagnon (2016).

square columns with the same base dimensions, height and spacing as the pyramids. The pyramids were 1 mm in height and 2 mm wide at the base. The space between each adjacent pyramid was 2 mm. The arithmetic average of the high-roughness profiles for the surfaces of the two platens were 0.075 mm (square pyramids) and 0.375 mm (square columns).

Most of the ice samples were prepared from freshwater granular ice blocks grown in the lab (grain size ~ 4 mm). Some ice samples were prepared from a columnar-grained freshwater ice sheet grown in a basin (column diameter ~ 5 mm). During tests the columnar grains were orthogonal to the vertical crushing direction and the long axis of the ice sample. Each ice sample was approximately 7 cm in height, and 12 cm by 6.5 cm at its base. The top of the samples was given a rounded-wedge shape. Each sample's base was freeze-bonded to an ice holder consisting of an acrylic plate with a rectangular band of steel (2 cm in height) attached to it that encompassed the base of the ice specimen. The ~ 1 cm gap between the ice sample and the confining steel band was filled with snow and then saturated with water near 0 °C so that, when frozen, it provided confinement at the base of the ice sample to prevent it from shattering during testing. Tests were carried out at –10 °C and the load data from the vertical crushing actuator and the horizontal sliding actuator were acquired at 6144 samples per second. The vertical crushing rates used for these tests were 10, 20 and 30 mm/s. The horizontal sliding rates used were 4.14, 10, 12.42 and 20 and 30 mm/s. All ice samples were crushed to a depth of ~ 35 mm.

TEST RESULTS AND DISCUSSION

Friction Data Characteristics

Fig. 3 (Top and Bottom) show the friction coefficient results for the tests conducted using the crushing platen with the array of pyramids. The reasonably good consistency of repeated test results is evident. Noteworthy observations from Fig. 3 (Top) are: (1) The friction coefficients are extraordinarily low given the high degree of roughness of the crushing platen. For comparison Fig. 3 (Top) includes two data points corresponding to two low-roughness unprepared flat steel plates, where each point represents the average friction coefficient value of two tests. (2) The friction coefficient increases with the horizontal sliding speed for any given vertical crushing rate. For the data corresponding to the 10 mm/s vertical crushing rate the friction coefficient dependence on sliding speed is roughly fitted by a line. (3) The friction coefficient is constant for the tests (Box A, excluding steel plate data) conducted at a constant ratio of vertical crushing rate and sliding speed, i.e. where the vertical crushing rate and the horizontal sliding rate ratios are (10 mm/s : 10 mm/s), (20 mm/s : 20 mm/s), and (30 mm/s : 30 mm/s). Below these data points on the chart are three other pairs (Box B) that demonstrate essentially the same thing, although the second pair of data points does not have exactly the same ratio as the first and last pair, (10 mm/s : 4.14 mm/s), (20 mm/s : 10 mm/s), and (30 mm/s : 12.42 mm/s). (4) The slopes of the three fit lines on the chart for the pyramids platen diminish as the vertical crushing rate increases. All of these observations may be understood in the context of the explanation of the ice behavior given below.

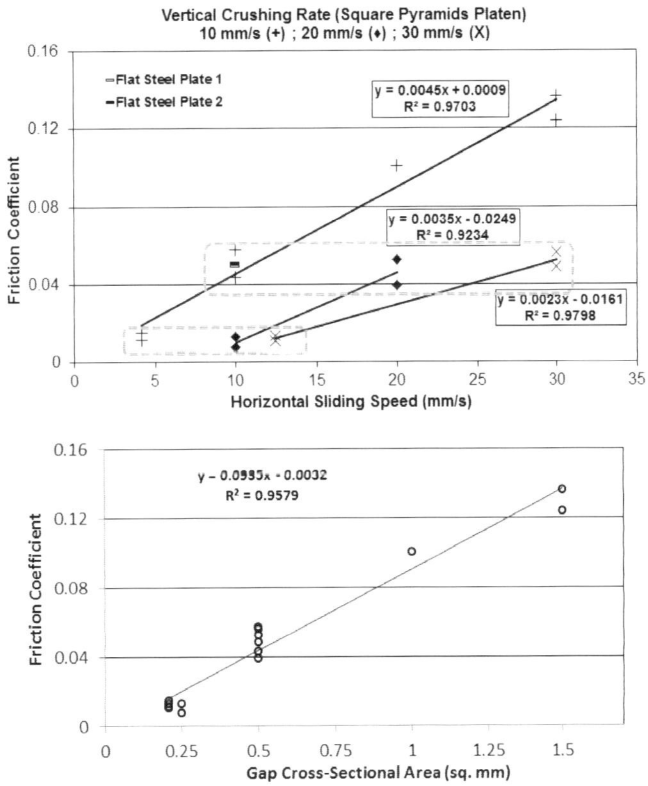

Fig. 3. Main friction coefficient results for the tests using the crushing platen with the array of square pyramids. (Top) Friction coefficient versus horizontal sliding speed. Tests corresponding to three vertical crushing rates and five horizontal sliding rates were conducted.

The included data points corresponding to the tests conducted using the two flat steel plates represent averages of two tests in each case, where the vertical crushing rate and horizontal sliding speed were both set at 10 mm/s. Scatter in the data, that is inherent in ice crushing and friction experiments, amounted to about ± 15%. (Bottom) Friction coefficient versus gap cross-sectional area. Granular ice was used for all tests shown in this figure. From Gagnon (2016).

Most of the experiments were conducted using the platen with the array of pyramids. Some tests were performed using the platen with the array of square columns (Fig. 4), including a few using columnar-grained freshwater ice that yielded very similar results. The friction values and slope of the fitted line are greater than that for the platen with the pyramid array (Fig. 3, Top). This is due to the higher roughness value for the platen with the square columns (0.375 mm) as opposed to 0.075 mm for the pyramids platen. Fig. 5 shows data from all platens used for friction tests. Table 1 describes these platens. The generally low friction coefficients and increase in values with sliding speed seen in Figs. 3 and 4 is evident for all platens with rough surfaces tested at two sliding speeds shown in Fig. 5.

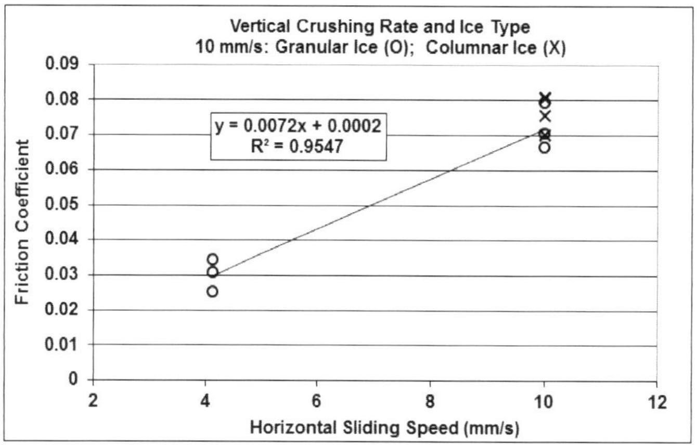

Fig. 4. Friction coefficient versus horizontal sliding speed for tests using the crushing platen with the array of square columns. Scatter in the data, which is inherent in ice crushing and friction experiments, amounted to about ± 15%. Granular ice was used for six tests and columnar ice was used for four tests. From Gagnon (2016).

Regarding ice crushing friction on high-roughness non-ice surfaces, the only data available are those presented in this paper. Similarly, the only analytical formulation to explain the data is found here, so comparisons with other studies are not possible as yet. On the other hand, the mechanisms described here, i.e. the inherent presence of the lubricating slurry and its fluid dynamic characteristics with respect to roughness features, are new and they may have some relevance to ice friction in cases where a degree of ice crushing (at some scale) occurs, such as on asperities under curling rocks, when skate blades plough/gouge the underlying ice and when rough edges of sea ice floes slide/grind against one another.

High-Speed Imaging Data of the Ice-Platen Interface

Fig. 6 shows an image from the high-speed image record of a test using the crushing platen with the array of square columns. The view is through the transparent acrylic crushing platen and the grid of small square columns is apparent. We see a roughly centrally-located horizontally-elongated dark hard zone that consists of relatively intact ice where the pressure is high (~ 53 MPa (Gagnon 2016)). The relatively intact hard zone appears dark because a thin sheet of black plastic was placed between the actuator head and the ice holder so that the black sheet was visible through the translucent ice sample and ice holder in areas where there was no opaque crushed ice (the white material). The elongated shape of the hard zone is due to the shape of the original ice sample (a horizontal wedge) that provides confinement governed by its geometry (Spencer and Masterson, 1993). The hard zone is surrounded by white material that consists of low-pressure (0-10 MPa) pulverized spall debris that is flowing away from the high-pressure region, generally extruding along the channels between the columns.

Table 1. Platen #, material, blade and array description/dimensions.

Platen #	Material	Blade and Array Description (all 'blades' are 1 mm in height)
2	Acrylic	Symmetric Crossed-Bars (45° wedged-shaped bars, 2 mm wide x 8.6 mm long); normal staggered array; 8.5 mm center-to-center spacing for rows and columns
5	Aluminum	Horizontal 45° Wedge Bar (8.6 mm long and 2 mm wide); normal staggered array; 8.5 mm center-to-center spacing for rows and columns; sliding perpendicular to bars
7	Acrylic	Horizontal 45° Wedge Bar (8.6 mm long and 2 mm wide); dense staggered array; 4.25 mm center-to-center spacing for rows and 8. 5 mm for columns; sliding along bars
11	Acrylic	Square Column (2 mm x 2 mm); square array; 4 mm center-to-center spacing
12	Acrylic	Square Pyramid (2 mm x 2 mm); square array; 4 mm center-to-center spacing
13	Acrylic	Square Column (2 mm x 2 mm); square array; 4 mm center-to-center spacing
14	Acrylic	Square Pyramid (2 mm x 2 mm); square array; 4 mm center-to-center spacing
16	Steel	Unprepared flat steel plate
20	Steel	Unprepared flat steel plate

FRICTION MECHANISMS AND FURTHER ANALYSIS

Schematics of the Physical Processes

In Fig. 6 the ice sample is sliding horizontally (to the right in the image) at the same rate (10 mm/s) as it is crushing vertically against the platen. The same processes described above for the ice crushing against the platen in the vertical direction apply to the ice crushing horizontally against the vertical walls of square columns facing the encroaching ice. For the intact hard zone to move horizontally to the right the thin squeeze-film slurry layer must be effectively removing hard-zone ice that is contacting the vertical walls of the square columns.

Fig. 1b shows a 2D schematic depicting ice crushing against a surface with square columns. Fig. 1c shows a 3D view of a smaller portion of that shown in Fig. 1b, containing only one square column. Although most of the data presented here is for the case of the platen with the square pyramids, we focus on the platen with square columns in the illustrations and associated discussion because the geometry and discussion are easier to depict and follow. The principles at play that are inculcated in the discussion, however, are the same in both cases. As discussed above the ice/surface contact consists of relatively-intact centrally-located hard-zone ice that is surrounded by crushed ice, i.e. shattered spall debris, at the sides of the hard zone. Between the hard zone and the underlying surface, either at the flat base of the platen or a vertical wall of a square column encroached by ice, is the heat-generating thin squeeze-film slurry layer that melts and erodes hard-zone ice as it flows. The slurry wets the crushed ice in its vicinity wherever it exits the high-pressure zone. The squeeze-film slurry dissipates the majority of the actuator energy supplied to the system because the load is mostly borne on the hard-zone ice (~ 88%, Gagnon (1994)), where the slurry is generated and flows. Consequently the energy dissipation rate of the slurry is directly proportional to the rate of crushing on whatever surface the ice crushes against, i.e. at horizontal surfaces or vertical surfaces of the platen.

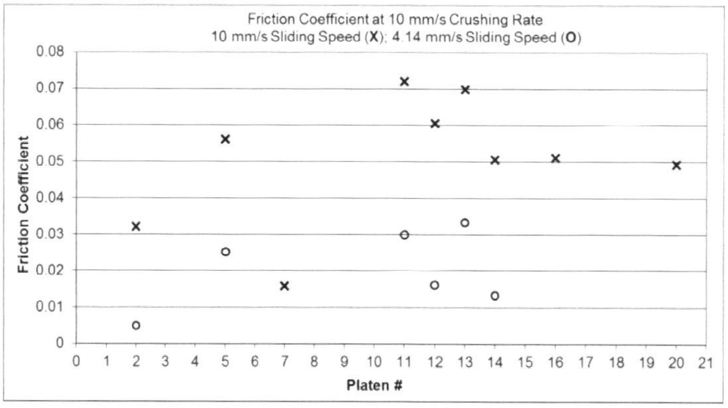

Fig. 5. Friction coefficient data for several of the platens used in the test program. Data points represent averages of multiple tests.

In Fig. 1b the ice crushes against the platen in the downward direction at a fixed rate and the ice is sliding horizontally to the right at a fixed rate. Since the actual movement of the ice relative to the platen is in the direction shown and since the ice is being melted and eroded on the tops of the square columns the shape of the ice in the space between two adjacent square columns will have a gap at the trailing left side, in the shape of a right wedge. The top angle of the right wedge is equal to the arctangent of the horizontal sliding speed divided by the vertical crushing rate. Hence the cross-sectional area of the gap is proportional to the horizontal sliding speed divided by the vertical crushing rate. The gap will be filled with flowing slurry since pressurized slurry will enter from the top of the square column and from the base of the platen in front of the gap, where the contacting hard-zone ice is melting/eroding, and will exit the gap at its sides where the pressure is lower, that is, in and out of the view of the figure. More detail on this is given below. In Fig. 1b the difference in pressure on the square column walls that face the oncoming ice and the lower pressure on the column walls on the leeward side of the oncoming ice, where the slurry-filled gaps are, accounts for the friction force. To give a better perspective of the gap, Fig. 1c shows a 3D view of a smaller portion of that shown in Fig. 1b, containing only one square column. This represents a unit area on the platen surface. Fig. 1c is a simplified schematic of what in reality is a highly dynamic situation. Gagnon (2016) noted that during a crushing-friction test a typical hard zone grows in size for certain time intervals and also suddenly, and frequently, diminishes in size by small amounts whenever small spalls break away from it at various locations. Furthermore, the hard zone is quite mobile (from the imaging records) in the vertical sense of Fig. 6 while remaining centrally located on average. In reality, due to these factors, the patterns and directions of flow of the pressurized slurry layer varies considerably both spatially and temporally, hence Fig. 1c represents a simplified time average of events.

The slurry fluid flows out of the sides of the gap where pressure is low because there is frequently low-pressure crushed ice in the immediate vicinity of one or both sides of the gap. This is due to the narrow width (a few times the width of a square column, or less) and elongated shape of the high-pressure zones, as seen in Fig. 6.

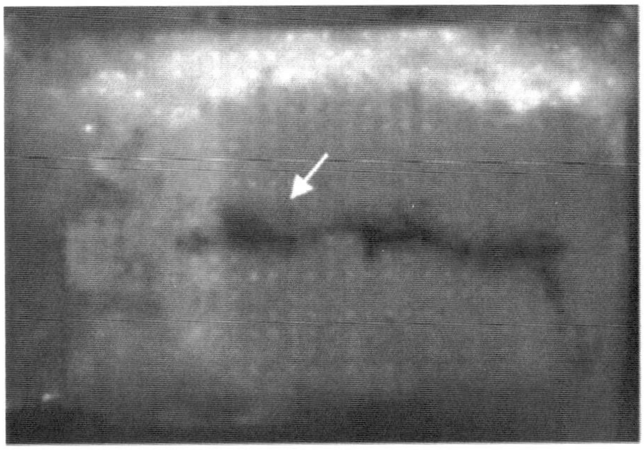

Fig. 6. An image from the high-speed image record of a test using the crushing platen with the array of square columns where the vertical crushing rate and horizontal sliding speed to the right were both set at 10 mm/s. The view is through the transparent acrylic crushing platen. A horizontally-elongated dark hard zone that consists of relatively intact ice, where the interface pressure is high, is visible. The hard zone is surrounded by crushed ice (the white material). The arrow points to a small region where a spalling event had occurred shortly before the image was captured. The width of the image is ~133 mm.

Analytical Description of the Friction Force

To analytically describe the friction force we need to know the parameters that determine the average pressure that the slurry exerts on the vertical column face in the gap. In the time-averaged sense, the gap acts somewhat as a short length of pipe or channel through which slurry flows. However, the geometry in Fig. 1c is clearly more complex than that of a simple channel and we would not expect that standard theory for flow within a channel would apply. Indeed in the present case all we know is that fluid enters the gap (channel) from the top of the column and base of the platen and exits at the sides of the gap. A complicating factor is that the pressurized slurry is entering the gap (channel) along its whole length, a very different scenario from fluid entering one end of a channel and exiting at the other end. For the average pressure of the gap slurry in the present scenario we might expect some sort of dependence on gap cross-sectional area. Due to the complexities of the geometry and flow entry characteristics, however, the analytical derivation of such a dependence from first principles is beyond the scope of this work.

Fortunately, all of the friction data in Fig. 4 (Top) for the platen with the square pyramids turns out to be proportional to the cross-sectional area (A) of the gap (Fig. 4, Bottom), where for geometrical simplicity we have assumed that the platen had square columns, i.e. where $A \approx \frac{1}{2}$ (horizontal sliding speed)/(vertical crushing rate) mm^2. While different in magnitude the cross-sectional areas of the gaps for the square pyramids would nevertheless be proportional to cross-sectional areas of gaps for square columns, as attested by the good fit in Fig. 4 (Bottom). To reiterate, the same proportional dependence of friction coefficient on gap cross-sectional area applies to both square pyramids and square columns. The above points imply that in these tests the average pressure that the slurry exerts on the column face within the gap is proportional to a constant maximal area minus the cross-sectional area of the gap. The friction coefficient is written as:

$$\mu = n\, w\, (P - P_{gap})/L = n\, w\, (P - P\, (A_{max} - A)/\, A_{max})/L \qquad (1)$$

where μ is the coefficient of friction, n is the number of square columns in the hard-zone area, w is the area of a column face, P (~ 55 MPa) is the slurry (squeeze-film) pressure on a column face on which ice is encroaching, P_{gap} is the average pressure of the flowing slurry on a column face in the gap, L is the normal crushing load, A_{max} is a certain constant area, and A is the cross-sectional area of the gap where $A_{max} > A$. The form of P_{gap} in equation (1), $P\,(A_{max} - A)/\, A_{max}$, derives from Fig. 3 (Bottom) that stipulates that the friction coefficient is proportional to A and tends to zero as A tends to zero, while P_{gap} tends to P as A tends to zero. Since we know that the gap cross-sectional area has the form stated above, equation (1) further simplifies and reduces to:

$$\mu = n\, w\, P\, A\, /\, (A_{max}\, L) = n\, w\, P\, H^2\, V_h\, /\, (2\, V_c\, A_{max}\, L) \qquad (2)$$

where H is the height of a column, V_h is the horizontal sliding speed and V_c is the vertical crushing rate.

The ice crushing-friction coefficient of a flat polished acrylic or aluminum surface is extremely low, virtually not resolvable with our apparatus, as indicated in the few tests that were conducted on flat acrylic and aluminum platens. Therefore equations (1) and (2) do not include any contribution to the friction force from ice crushing against the machined (near-polished) flat base of the platens in the areas between the square columns or square pyramids.

Equations (1) and (2) are essentially stating that the friction coefficient is directly proportional to the gap cross-sectional area A, as is evident in Fig. 4 (Bottom). Gagnon (2016) previously noted that the contribution to the friction force from the crushed ice (visible in Fig. 6) is negligible compared to that of the hard-zone ice. While equation (2) specifically relates to a platen with square columns, the patterns and trends of the data presented in Fig. 4 (Top) for the square-pyramids platen are nevertheless reflected in the equation. Gagnon (2016) investigated typical values for the two unknown quantities in equation (2), n and A_{max}, and obtained the following mutually reasonable estimates, n ~18 and A_{max} ~1.5 mm^2.

CONCLUSIONS

The generation and flow characteristics of a thin ice/liquid slurry layer, with respect to the regular array of prominences on the platen surfaces, explains the low magnitudes and the trends in the friction data. This viscous layer removes hard-zone ice from the contact interface through melting and erosion (Gagnon, 2016). The layer may be thought of as a self-generating squeeze film that is powered by the energy supplied by the loading system that causes the ice crushing. On flat relatively smooth surfaces the squeeze-film slurry layer causes extremely low friction forces. These friction mechanisms have not been observed before.

These results have beneficial consequences for the potential use of scaled-up panels with arrays of protrusions (Gagnon, 2015) that could be used to reduce ice-induced vibration of structures (e.g. offshore wind turbines and oil platforms) without causing undesirable high friction forces. Conversely, rough surfaces designed to have a high friction coefficient (e.g.

anti-slip surfaces on wheels or tracks) may have unexpectedly low friction coefficients when ice is crushed against them. The flow of glaciers may be influenced by the friction mechanisms described here at the glacier bases where ice slides over uneven rock surfaces. It was noted that the flowing slurry layer served to clean corrosion from the surfaces of unprepared steel plates. There may be specialized cleaning applications of natural or prepared hard surfaces, with low or high degrees of roughness, where ice crushing and sliding across the surface would be effective and non-contaminating, involving only short-duration exposure to tiny amounts of liquid water with no other cleaning agents.

The presence of a thin squeeze-film layer of ice/liquid slurry produced by ice crushing on flat aluminum and acrylic surfaces in this study specifically led to extremely low friction coefficients. This should be considered when attempting to explain the frictional behavior of skate blades, sled runners and curling stones where local crushing on ice asperities and/or small-scale ice unevenness, and due to gouging/plowing, occurs. By the term 'ice crushing' we include the production of the thin squeeze-film slurry layer both during spalling events and in the time intervals between spalling events.

ACKNOWLEDGEMENTS

The author would like to thank OCRE/NRC for its support of this research. The author is also grateful to Austin Bugden for technical assistance during the test program.

REFERENCES

Fransson, L., Olofsson, T., Sandkvist, J. 1991. Observations of the failure process in ice blocks crushed by a flat indentor. Proceedings of the 11th International Conference on Port and Ocean Engineering Under Arctic Conditions, 1, 501 514.

Gagnon, R.E., 1994. Generation of melt during crushing experiments on freshwater ice. Cold Regions Science and Technology, 22(4), 385-398.

Gagnon, R.E., 1998. Analysis of visual data from medium scale indentation experiments at Hobson's Choice Ice Island. Cold Regions Science and Technology, 28, 45-58.

Gagnon, R.E., 1999. Consistent observations of ice crushing in laboratory tests and field experiments covering three orders of magnitude in scale. *Proceedings of POAC-99*, **2**, 858-869.

Gagnon, R.E. and Mølgaard, J., 1989. Crushing friction experiments on freshwater ice. *Proceedings of the IUTAM/IAHR Symposium on Ice/Structure Interaction*, 405-421.

Gagnon, R.E., 2013. High-speed imaging of ice-on-ice crushing. *Proceedings of POAC 2013*, Paper No. 159.

Gagnon, R.E., 2015. Large scale spallation inducing ice protection. US Patent 9,181,670 B2, Nov. 10.

Gagnon, R.E., 2016. New friction mechanisms revealed by ice crushing-friction tests on high-roughness surfaces. Cold Regions Science and Technology, 131, 1-9.

Riska, K., Rantala, H., Joensuu, A., 1990. Full scale observations of ship ice contact. Laboratory of Naval Architecture and Marine Engineering, Helsinki University of Technology, Report M 97.

Spencer, P.A. and Masterson, D.M., 1993. A geometrical model for pressure aspect-ratio effects in ice-structure interaction. Proceedings of OMAE 1993, 4, 113-117.

POAC'17
Busan, KOREA

Proceedings of the 24th International Conference on
Port and Ocean Engineering under Arctic Conditions
June 11-16, 2017, Busan, Korea

Three Physical Mechanisms of Wave Energy Dissipation in Solid Ice

Aleksey Marchenko[1], David Cole[2]
[1] The University Centre in Svalbard, Norway
[2] CRREL, USA

ABSTRACT

In the present study we compare wave energy dissipation caused by three physical processes: anelastic and viscous deformations of ice due to the periodic bending, and migration of liquid brine in pore spaces, and friction on the ice-water interface. The wave damping due to anelastic and viscous deformations of ice is estimated using a model of ice behavior under cyclic loading developed by Cole (1995) and standard assumptions of thin plate theory. The wave damping due to ice-water friction is estimated using the theory developed by Liu and Mollo-Christensen (1988) and values of the eddy viscosity in ice adjacent layer calculated from field data (Marchenko et al., 2015). The wave damping due migration of liquid brine through the ice is estimated using a recently developed analytical model based on the Darcy's law that treats the ice as a permeable material and quantifies energy loss due to the periodic movement of liquid brine within the ice.

KEY WORDS: Wave, Sea ice, Cyclic loading, Deformations, Boundary Layer, Permeability

INTRODUCTION

Reduction of ice covered areas in the Arctic influences development of offshore activity, navigation and tourism in regions which were not considered earlier for these perspectives. Although the amount of sea ice in the Arctic becomes smaller it still exists in summer time and covers almost all Arctic Ocean in the winter. Depending on the weather conditions wind action on sea surface in ice free areas may influence an increase of storm activity there. Wind waves and swell may destroy drift ice over big areas in relatively short time and influence the formation of Marginal Ice Zone. Drift ice destroyed by waves into small floes becomes more dynamic, and increases energy exchange between the atmosphere and ocean. Therefore investigation of wave damping in ice covered areas is of interest for engineering applications and climate issues.

Wind action on sea surface creates waves in wide spectral range with periods from 5 s to 30 s. High frequency part of the wave spectrum is disappeared due to waves-floes interaction in the Marginal Ice Zone. Wave energy dissipates due to the waves scattering and water-ice friction. Longer waves penetrate below the ice consisting of big floes and solid ice. Measurements performed on solid drift ice in the Barents Sea demonstrate that waves with period about 10 sec penetrate on several tens of kilometers into the solid ice (Collins et al., 2015; Marchenko et al, 2015). Experimental data on the spectral composition of flexural–gravity waves in the Arctic were analyzed by Nagurny et al. (1994) and Wadhams et al. (1995) to show that a major portion of the wave energy corresponds to waves with periods greater than 16 s.

Capacity of waves to break up the ice depends on wave amplitude and wave length. Damping decreases the wave amplitude, and therefore influences ice breakup by waves. Wave damping is characterized by the ratio of energy dissipation rate due to some physical process to the wave energy. In the present paper we perform numerical estimates and compare energy dissipation rates for the three processes causing wave damping in solid ice including viscous and anelastic deforming of ice, migration of liquid brine through the ice, water-ice friction in ice adjacent boundary layer.

DISSIPATIVE PROCESSES IN SEA ICE AND ICE ADJUCENT BOUNDARY LAYER

Granular structure of sea ice is columnar with isotropic (S2) or anisotropic (S3) distribution of optical axes in the horizontal plane (see, e.g., Weeks and Hibler, 2010). Inclusions of liquid brine trapped inside the ice during sea water freezing are distributed between closed brine packets and permeable channels (see, e.g., Cole and Shapiro, 1998). Amount of liquid brine and its distribution between the packets and channels depends on the temperature and salinity of sea ice (Golden et al., 2007). The ice is permeable, and the brine is able to migrate through the matrix under the influence of pressure gradients. The permeability may be affected by thermal changes in the ice, as brine pockets separate and merge during freezing and melting cycles.

Cyclic bending deformation of ice influences cyclic tension/compression of horizontal layers in the ice, and creates cyclic changes of vertical pressure gradient inside the ice. In-plane deformation ε are of the order $h/2R$, where h is the ice thickness and R is the curvature radius of the ice surface. The curvature is approximated with the formula $1/R \sim ak^2$, when the ice is deformed by sinusoidal wave with amplitude a and wave length $\lambda=2\pi/k$. In-plane stresses are estimated as a product of the elastic modulus on the strain: $\sigma=E\varepsilon$. Dependence of the stress σ from the wave length λ is shown in Figure 1a, when E=1 GPa, h=1 m, and a=0.1 m. Period of gravity waves is greater 8 sec when the wave length is greater 100 m. From Figure 1a it follows that wave induced stresses in the ice are lower 0.2 MPa, and it is lower 0.4 MPa if the elastic modulus is equal to 2 GPa. In this case energy of anelastic and viscous deformation takes not more than 5% of the energy of elastic deformations (Cole, 1995). Anelastic and viscous deformations are related mainly with relative displacements of grain boundaries and density of dislocations is not growing. Anelastic and viscous deformations of ice dissipate the wave energy when wave propagates below the ice.

Vertical pressure gradient in the ice bended by waves is estimated with the formula ☆*p*/☆z~σ /*h*. It is smaller 0.2 MPa/m for the stresses shown in Figure 1a. The flux of the brine through the ice under the action of the pressure gradient is estimated with Darcy's formula *q*=-κ☆p/☆z/μ, where κ is the permeability of ice by brine, and μ is the dynamic viscosity of water. Assuming that $\kappa=10^{-12}$m^2 and $\mu=1.5\cdot10^{-3}$kg/(s·m) we estimate $q<1.3\cdot10^{-4}$m/s. Reynolds number is calculated with the formula $\mathrm{Re}=\rho q d/\mu$, where ρ is the brine density and *d* is the grain size. Assuming $\rho=10^3$kg/m^3 and *d*~1mm we estimate Re<0.086. It is smaller the critical values when the linear Darcy's law should be corrected (Polubarinova-Kochina, 1962). Thus we conclude that liquid brine inside ice will migrate in vertical direction through the ice when the ice is deformed by waves. Friction between moving brine and ice will dissipate the wave energy.

When wave propagate below the ice the surface particles of the water adhere to the ice bottom and lose their momentum. Effectivity of the momentum transfer from deeper particles to the surface particles depends on the turbulence in ice adjacent water level since fluctuations of wave induced water velocities are transported together with fluctuations induced by under ice turbulence. The process is characterized by the eddy viscosity which can be much greater the molecular viscosity of water. This mechanism of wave energy dissipation was considered by Liu and Mollo-Christensen (1988). The value of the eddy viscosity is most important for the wave damping.

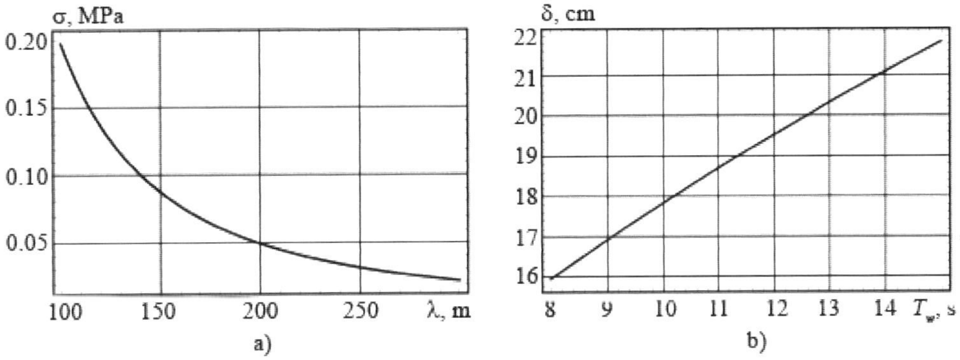

Figure 1. Bending stresses versus wave length (a). Thickness of wave induced boundary layer versus wave period (b).

WAVE DAMPING DUE TO VISCOUS AND ANELASTIC DEFORMATIONS OF ICE

Surface density of elastic energy of thin plate is calculated with the formula (see, e.g., Landau and Lifshitz, 1986)

$$F_{el} = D(\Delta\eta)^2/2,\quad D=\frac{Eh^3}{12(1-\nu_p^2)},\quad \Delta=\frac{\partial^2}{\partial x^2}+\frac{\partial^2}{\partial y^2}, \qquad (1)$$

where E is the effective elastic modulus, h is the ice thickness, and ν_p is the Poisson's ratio, x and y are the spatial coordinates.

When the plate has sinusoidal shape, $\eta=a\sin(kx-\omega t)$, the plate energy averaged over the wave length equals

$$\langle F_{el}\rangle = Da^2k^4/4. \tag{2}$$

Here a is the wave amplitude, k is the wave number, ω is the wave frequency, and t is the time.

According to the model of cyclic loading (Coal, 1995) the dissipated energy takes about 5% of the elastic energy when maximal load doesn't exceed 0.5 MPa and the loading frequency is around 0.1 Hz or higher. Thus the rate of energy dissipation due to the periodical bending of the plate is specified by the formula

$$\left\langle \dot{F}_v \right\rangle = \frac{0.05}{8\pi} D\omega a^2 k^4. \tag{3}$$

Energy of surface gravity wave is determined by the formula (Lamb, 1975),

$$\langle F_w\rangle = \frac{\rho\omega^2 a^2}{2k\tanh(kH)}, \tag{4}$$

where H is the water depth. The attenuation rate of the wave amplitude due to the viscous and anelastic deformations of ice is equal to

$$\gamma_v = \frac{\langle \dot{F}_v\rangle}{\langle F_w\rangle} = \frac{0.05}{4\pi}\frac{Dk^5\tanh(kH)}{\rho\omega}. \tag{5}$$

WAVE DAMPING DUE TO WAVE INDUCED MOTION OF BRINE

Vertical momentum balance of liquid brine in permeable channels inside the ice is expressed by the equation (Polubarinova-Kochina, 1962)

$$\frac{\rho}{\phi}\frac{\partial q}{\partial t} = -\frac{\partial p}{\partial z} - \frac{\mu}{\kappa}q. \tag{6}$$

where ϕ is the ice porosity. Pressure gradient is estimated from the theory of thin elastic plate

$$\frac{\partial p}{\partial z} \approx -\frac{1}{2}\frac{\partial \sigma_{xx}}{\partial z} = j_0\cos\theta,\quad j_0 = \frac{Eak^2}{2(1-\nu_p^2)},\quad \theta = kx-\omega t. \tag{7}$$

In case when the pressure gradient is expressed by formula (7) the solution of (6) has the form

$$q_z = \frac{j_0}{\Delta}\sin(\theta-\theta_0),\ \ \Delta^2 = \left(\frac{\rho\omega}{\phi}\right)^2 + \left(\frac{\mu}{\kappa}\right)^2,\ \ \theta_0 = \arccos\left(\frac{\rho\omega}{\phi\Delta}\right). \tag{8}$$

The permeability of saline ice by brine is estimated by the formula

$$\kappa = \kappa_0 e^{15\sqrt{v_b}}\ , \tag{9}$$

where v_b is the liquid brine content in ice, the coefficient κ_0=10^{-13}m^2 characterizes ice permeability at low values of the liquid brine content. Formula (9) approximates the data of the laboratory and numerical simulations shown in Figure 2 in Zhu et al (2006). Further we assume that $\phi = v_b$ and estimated with the formula of Frankenstein and Garner (1967)

$$v_b = \sigma_{si}\left(\frac{49.185}{|T|} + 0.532\right),\ \ -22.9^\circ\mathrm{C} < T < -0.5^\circ\mathrm{C}, \tag{10}$$

where σ_{si} and T_i are the sea ice temperature and salinity.

Assuming v_b=0.12 (warm sea ice with temperature -2C and salinity 5ppt) we find $\kappa \sim 1.8\cdot 10^{-11}$ m2, and $\Delta \sim \mu/\kappa >> 1$. Therefore solution (5) can be simplified to the formula

$$q \approx \frac{j_0\kappa}{\mu}\sin\theta\,. \tag{11}$$

Energy dissipation rate over unit area of the ice is calculated with the formula

$$\dot{F}_b = \frac{\mu h}{\kappa\phi} q^2\,. \tag{12}$$

Substituting formula (11) into formula (12) we find after the averaging over the wave period

$$\left\langle \dot{F}_b \right\rangle = \frac{\kappa h}{2\mu\phi} j_0^2\,. \tag{13}$$

The attenuation rate of the wave amplitude due to the vertical brine migration of brine through the ice is calculated with the formula

$$\gamma_b = \frac{\left\langle \dot{F}_b \right\rangle}{\left\langle F_w \right\rangle} = \frac{\kappa h}{4\rho\mu\phi}\frac{E^2 k^5 \tanh(kH)}{\omega^2(1-v_p^2)^2}\,. \tag{14}$$

WAVE DAMPING DUE TO WATER-ICE FRICTION

For the estimate of wave damping due to the friction on the ice-water interface we use the formula derived for the mean energy dissipation rate in oscillating water flow near a wall (formula (24.14) from Landau and Lifshitz, 1988),

$$\left\langle \dot{F}_f \right\rangle = \frac{\rho V^2}{2\sqrt{2}}\sqrt{\nu\omega}, \qquad (15)$$

where V is the amplitude of the wave induced water velocity outside the wave induced boundary layer below the ice, and ν is the kinematic eddy viscosity below the ice. The thickness δ of wave induced boundary layer is specified by the formula

$$\delta = \sqrt{2\nu/\omega}. \qquad (16)$$

The eddy viscosity below the drift ice is estimated from 100 cm^2/s to 200 cm^2/s (McPhee and Martinson, 1994). The thickness of the boundary layer calculated with ν=100 cm^2/s is shown versus wave period $T_w=2\pi/\omega$ in Figure 1b. High frequency measurements of under ice water velocity in the Barents Sea and data processing shown highest values of the eddy viscosity 140 cm^2/s. Locations of field measurements are shown in Figure 2. User setups of the deployments of the Acoustic Doppler Velocimeter SonTek 5 MHz are shown in Table 1. Calculated values of the eddy viscosities are shown in Table 2. Methodic of the eddy viscosity calculation is described by Marchenko et al (2015).

The amplitude of surface water velocity induced by wave is expressed via wave amplitude by the formula

$$V = a\omega / \tanh(kH). \qquad (17)$$

The attenuation rate of the wave amplitude due the water-ice friction is calculated with the formula

$$\gamma_f = \frac{\left\langle \dot{F}_f \right\rangle}{\left\langle F_w \right\rangle} = \frac{k\sqrt{\nu\omega}}{2\sqrt{2}\tanh(kH)}. \qquad (18)$$

Table 1. User setups of the ADV SonTek 5 MHz Ocean probe deployments in the Barents Sea.

N	Data	Sampling frequency, Hz	Burst interval, s	Samples per burst	Number of bursts	Depth, cm
1	April 24, 2006	5	1800	8000	2	190
2	April 25, 2006	10	1200	900	3	180
3	April 18-19, 2012	10	1200	1200	50	112
4	April 25, 2014	10	1200	3000	50	41
5	May 1-2, 2016	10	360	2400	94	80

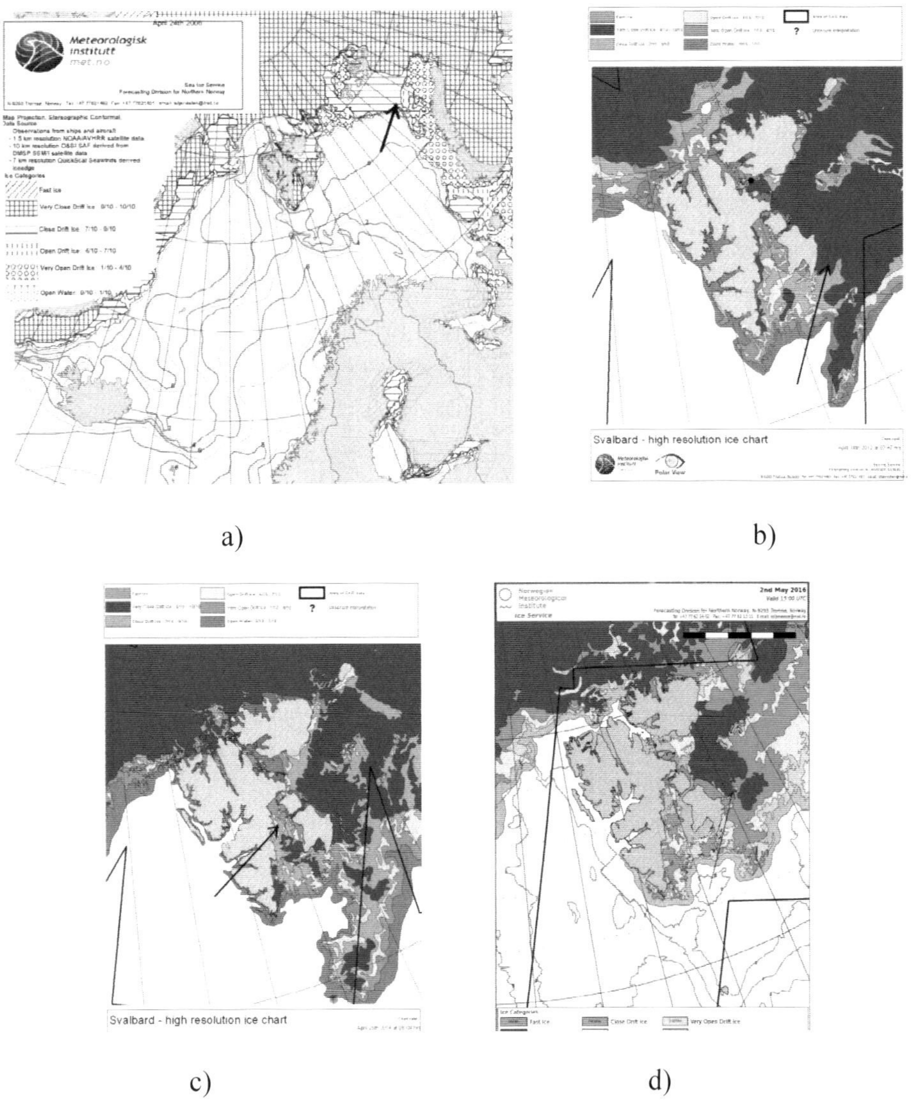

Figure 2. Locations of measurements of water velocity in ice adjacent boundary layer on April 24-25, 2006 (a); April 18-19, 2012 (b); April 25, 2014 (c); May 1-2, 2016 (d).

Table 2. Eddy viscosity below drift ice in the Barents Sea, cm^2/s.

1	2	3	4	5
0.25	0.002	96.07	0.1	140

DISCUSSION

Figure 3 shows the dependencies of the attenuation rates of the wave amplitude due the viscous and anelastic deformations of ice, due to the brine migration through ice, and due to the water-ice friction versus the wave length. Examples the sea ice salinity is 5 ppt. Figure 3a is constructed with sea ice temperature -5C and different ice thicknesses. Figures 3b,c are constructed with ice thickness equals 1 m. Figure 3c is constructed with ice temperature equals -2 C. The dependence of graphics shown in Figure 3 from the water depth is very small when H>20 m.

From Figure 3a and 3c it follows that the attenuation rates due the viscous and anelastic deformations of ice and due to the water-ice friction are of the same order when the eddy viscosity is relatively big. The attenuation rate due the viscous and anelastic deformations of ice dominates in rest water. Figure 3b demonstrates that the attenuation rate due to the brine migration through ice is much higher than the attenuation rates due the viscous and anelastic deformations of ice and due to the water-ice friction when the ice permeability is greater $10^{-13}m^2$. From Figure 3b it follows that waves with wave length 100-200 m practically don't penetrate into the ice covered regions. This effect is needed in future study. Probably wave induced brine migration influences micro-cracking in the ice. It can influence the reduction of filtration velocity and dissipated energy due to the friction between brine and ice. The micro-cracking may influence the increase of wave damping due to due the viscous and anelastic deformations of ice.

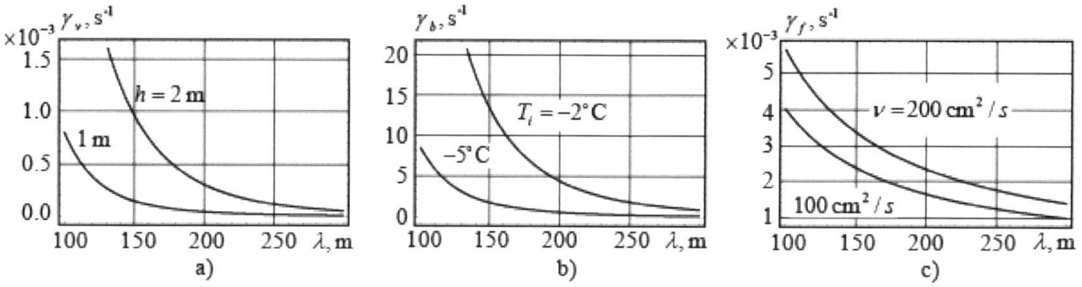

Figure 3. The attenuation rates of the wave amplitude due the viscous and anelastic deformations of ice (a), due to the brine migration through ice (b), and due to the water-ice friction (c) versus the wave length.

CONCLUSIONS

Three processes causing wave damping in solid ice are investigated. The attenuation rates of the wave amplitude due the viscous and anelastic deformations of ice, due to the brine migration through ice, and due to the water-ice friction are derived, analyzed and compared. It is discovered that the attenuation rates of the wave amplitude due the viscous and anelastic deformations of ice and due to the water-ice friction are of the same order when the eddy viscosity in ice adjacent water boundary layer is relatively high. In the rest water the attenuation rate of the wave amplitude due the viscous and anelastic deformations of ice dominates. The attenuation rate due to the brine migration through the ice is discovered very big. The estimates are based on Darcy's law and known values of sea ice permeability. The effect should be further investigated.

ACKNOWLEDGEMENTS

The authors wish to acknowledge the support of the Research Council of Norway through the SFI SAMCoT and PETROMAKS2 project Experiments on waves in oil and ice (WOICE).

REFERENCES

Cole, D.M.,1995. A model for the anelastic straining of saline ice subjected to cyclic loading. Philosophical Magazine A, 72(1), 231-248.

Cole, D.M., Shapiro, L.H., 1998. Observations of brine drainage networks and microstructure of first-year sea ice. Journal of Geophysical Research, 103(NC10), 21,739-21,750.

Collins, C,O., Rogers, W.A., Marchenko, A., Babanin, A.V., 2015. In situ measurements of an energetic waves event in the Arctic marginal ice zone. Geoph. Res. Letters, 42, 6, 1863-1870.

Frankenstein, G.E., Garner, R., 1967. Equations for determining the brine volume of sea ice from -0.5 to -22.9°C., J. Glaciol., 6(48), 943-944.

Golden, K.M., Eicken, H., Heaton, A.L., Miner, J., Pringle, D.J., Zhu, J. 2007. Thermal evolution of permeability and microstructure in sea ice. Geophysical Research Letters, 34, L 16501.

Landau LD and Lifshitz EM (1986) Theory of elasticity. Volume 7: Theoretical physics, 3rd edn. Butterworth-Heinemann, Oxford

Landau, L.D., Lifshitz,E.M., 1988. Fluid Mechanics, Second edition: Vol. 6 (Course of Theoretical Physics S), Butterworth-Heinemann, Oxford.

Liu, A. K. and E. Mollo-Christensen (1988), Wave propagation in a solid ice pack, J. Phys. Oceanogr., 18(11), 1702-1712.

Marchenko, A.V., Gorbatsky, V.V., Turnbull, I.D., 2015. Characteristics of under-ice ocean currents measured during wave propagation events in the Barents Sea. POAC15-00171.

McPhee, M.G., Martinson, D.G., 1994. Turbulent mixing under drifting pack ice in the Weddell Sea. Science, 263, 218-222.

Nagurny, A. P., V. G. Korostelev, and V. P. Abaza, 1994: Wave method for evaluating the effective thickness of sea ice in climate monitoring. (BRAS Physics/Supple), Phys. Vibrations, 58(3), 168–174.

Weeks, W.F., Hibler, W.D.III., 2010. On sea ice. University of Alaska Press, Fairbanks.

Wadhams, P., and S. C. S Wells, 1995: Ice surface oscillation measurements on SIMI using strain, heave and tilt sensors. Proc. Sea Ice Mechanics and Arctic Modeling Workshop, Anchorage, AK, Office of Naval Research, 176–189.

Polubarinova-Kochina, P.Ya., 1962. Theory of ground-water movement. Princeton Legacy Library.

Zhu, J., Jabini, A., Golden, K.M., Eicken, H., Morris, M. 2006. A network model for fluid transport through sea ice. Annals of Glaciology, 44, 129-133).

The Determination of Heat Transfer Coefficient on Water-ice Surface in a Free Convection

Xiaodong Chen[1,2], Knut Hoyland[2], Shunying Ji[1]
[1] Dalian University of Technology, Dalian, China
[2] Sustainable Arctic Marine and Coastal Technology (SAMCoT), Centre of Research-based Innovation (CRI). Norwegian University of Science and Technology, Trondheim, Norway

ABSTRACT

In ice ridges consolidation, the convective heat flux term comes critical due to the larger contact areas and surface temperature differences compare with those from level ice. In this paper, a submerging experiment was designed to determine the heat transfer coefficient (h) between fresh ice and fresh water in a free convection. A thermistor string was used to measure temperature changes while ice growth was recorded by photograph. To study the factor, the tests were carried on different ice thickness (4.9cm to 20.5cm) and initial temperatures (-20°C and -32°C). The result shows that the h exponential increased with temperature difference from 0.3 W/m^{2}K to 175 W/m^{2}K. On the other hand, the variation of initial thickness and temperature was not a direct influence for *h*. For convective heat transfer, the boundary layer condition is central for understanding the convection between ice surface and water flowing past it. From the governing equation, the water flow in a free convection is caused by density difference, which is driven by the thermal expansion. A large temperature difference between surface and environmental water creates a thicker boundary layer, which leads to a higher *h*.

KEY WORDS: Heat transfer coefficient; Consolidation; Free convection; Ice.

INTRODUCTION

In Arctic seas, ice ridges is one of the major concerns for oil exploration and shipping. Compare with ice floe, ridges exist in a more complicated structure, which includes sail, keel and consolidated layer. From previous work, it is believed that the latter two parts make most contribution to the ice-induced load (Timco et al., 2000). In ISO19906 standard, the keel porosity and consolidated layer thickness are key parameters for ice loads prediction. To study the mechanical properties, one needs to understand the formation of the ridges structure, which varies through seas and areas (Høyland, 2007a; Leppäranta and Hakala, 1992). Under the driven of winds and currents, the ice floes may move against each other and be end of small pieces, which are called ice rubble (Hopkins et al., 1999; Leppäranta et al., 1995). In a further stage, the new ice freezes to fill the gaps and the ice rubble becomes consolidated (Høyland, 2002). During three phases of consolidation, the strongest ridges are believed to form within the first two phases. In the study of Chen and Hoyland, the contribution of consolidation from the initial phase could be significant due to the saline properties (Chen and Høyland, 2016).

When the ice blocks fall into water, they experience a transient condition where the environment temperature is sudden changed. During this process, the cold ice is heated up by the energy released from phase change and convection, which is widely named oceanic flux. For the level ice, the oceanic flux varies in an order of 1-100W/m^2 and depends on the current condition (Høyland and Liferov, 2005; Leppäranta and Shirasawa, 2007; Shirasawa et al., 1997; Shirasawa et al., 2006). However, the water-ice temperature difference for level ice is much lower than that of a submerged ice. Mathematically, the convection is described by Newton's Law of cooling, which is a function of temperature difference and heat transfer coefficient. Therefore, this coefficient is critical in consolidation as well as its scaling since is it part of Biot number (Høyland, 2007b). Physically, the convection contains both conduction and advection, which is related to the fluid status(Bergman et al., 2011). When ice and water temperatures are closed, it is believed that the coefficient is a rather small value, which makes convection insignificant in level ice forming (Josberger, 1987; Leppäranta, 1993). However, the convection becomes essential under a large temperature gap, which leads to great thermal expansion and consequently fluid motion. Such a circumstance between water and ice has not been thoroughly studied.

In this paper, we designed and carried out a group of tests to investigate the heat transfer coefficient between ice and water in a free convection. To study the influence of initial condition, the experiments were done on varies thicknesses and initial temperatures.

THEORY

In this paper, the main purpose is to determine the heat transfer coefficient (h) between fresh water and fresh ice, where the ice block is submerged into water bath. To achieve this aim, related physical mechanism is needed to be understood in advance of experiments. Since the h is mainly associated with the boundary condition, which is dominated by the status of fluid for given solid and liquid material. It is reasonable to simplify the experiments to one-dimension.

Energy Contribution

When a cold ice piece falls into a water bath, where is much warmer than the ice, heat transfer starts under the driven of temperature difference. Such a case involves four energy contribution, which including conductive energy, convective energy, phase change energy and initial energy. Mainly, there is a warming up process inside the ice, where the inertial energy is increased by heat conduction. In the meantime, the heat from the warm water crosses water-ice surface, where ice freezing occurs. Among the energy terms, we are interested in the convection most since its association with heat transfer coefficient. However, it is very difficult to measure this part of energy directly. Therefore, in many studies, it is usually obtained by back calculation from energy conservation.

$$\Delta E_{in} = E_{lat} + E_{conv} \tag{1}$$

Where ΔE_{in} is the change of inertial energy; E_{lat} is the latent energy; E_{conv} is the convection energy.

$$\Delta E_{in} = \int_{T_0}^{T} mc_p dT = (T - T_0)\rho \cdot c_p \cdot L_0 \cdot A \tag{2}$$

Where T and T_0 are current and initial average temperature of ice block; L_0 is initial ice thickness; A is ice area exposed to water; ρ is density of ice; c_p is the specific heat capacity.

$$E_{lat} = \rho \cdot l \cdot \Delta L \cdot A \tag{3}$$

Where l is the latent hear; ΔL is the new ice thickness.

$$E_{conv} = \left(T_\infty - T_s\right) \cdot t \cdot A \cdot h \tag{4}$$

Where T_∞ is the water temperature; T_s is the temperature of ice at surface; t is time; h is the heat transfer coefficient.

Heat Transfer Coefficient

One way to study the heat convection is to find a solution of the correlation between thermal and velocity boundary layers. In this case, the governing equations of velocity layer is

$$u^* \frac{\partial u^*}{\partial x^*} + v^* \frac{\partial u^*}{\partial y^*} = \frac{g\beta\left(T_s - T_\infty\right)}{u_0^2} T^* + \frac{1}{Re_L} \frac{\partial^2 u^*}{\partial y^{*2}} \tag{5}$$

On the left-hand side of equation, the two terms are net rate of momentum flow from control volume; on the other side, the terms are buoyancy and viscous forces. Where g is gravity acceleration, β is thermal expansion coefficient.

The governing equation of thermal layer is

$$u^* \frac{\partial T^*}{\partial x^*} + v^* \frac{\partial T^*}{\partial y^*} = \frac{1}{Re_L \cdot Pr} \frac{\partial^2 T^*}{\partial y^{*2}} \tag{6}$$

If we take $u_0^2 = g\beta(T_s - T_\infty)L$, then we get Grashof number (Gr_L)

$$Gr_L = \frac{g\beta\left(T_s - T_\infty\right)L^3}{v^2} \tag{7}$$

Where g is gravity accelerate; β is thermal expansion coefficient; L is characteristic length; v is viscosity. Gr_L is a measure of the ratio of buoyancy forces to the viscous forces and its expression is

Based on the Eq. (5) and Eq. (6), we expect a correlation:

$$Nu_L = \frac{hL}{k} = f\left(Gr_L, Pr\right) = C_1 \cdot Gr_L^{\,n_1} \cdot Pr^{n_2} \tag{8}$$

Where k is thermal conductivity, C_1, n_1 and n_2 are constants.

$$Pr = \frac{v}{\alpha} \tag{9}$$

Where v is viscosity, α is diffusivity.

Substituting Eq. (9), Eq. (7) into Eq. (8), we obtain the expression:

$$h = C_1 \frac{k}{L}\left(\frac{v}{\alpha}\right)^{n_2}\left(\frac{g\beta L^3}{v^2}\right)^{n_1}\left(T_s - T_\infty\right)^{n_1} \tag{10}$$

Where h is a function of thermal properties, length and temperature difference, as shown in Eq.10.

EXPERIMENTAL SETUP

In this section, we designed and carried experiments to represent a free convection in transient condition. To achieve this process, cold ice blocks were submerged into warm water bath (0.2°C above freezing point). From analysis in section 2.1, one needs to measure latent energy and inertial energy in order to determine the convection. During each test, the temperature was measured by thermistor strings while the ice growth was visually recorded.

Setup Overview

Based on the theoretical analysis above, a group of experiments was carried out in NTNU cold laboratory to model the thermal process. During each test, the ice growth was recorded by camera while the ice temperature was measured by a thermistor string. The further description is introduced by Chen and Hoyland.(Chen and Høyland, 2016)

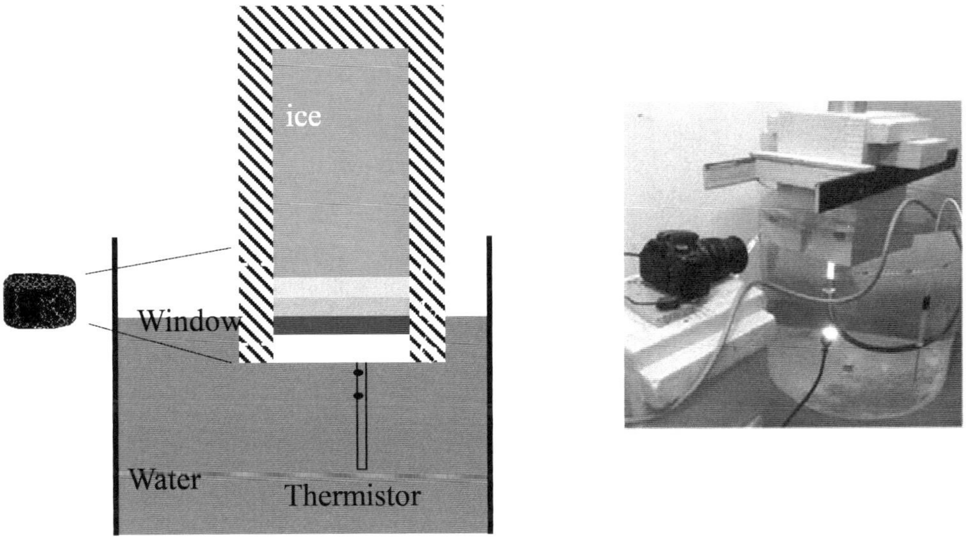

a. The setup sketch b. The setup photo

Figure 2. The overview of experiment setup

Experimental Matrix

In this work, the whole matrix is performed between fresh ice and fresh water. Since it is a model test, we are interested in size effect. Besides, the temperature different plays an important role in the coefficient. Therefore, the tests were done on the ice blocks with various initial temperatures and thicknesses as show in Table.1and Figure 3.

Figure 3. The ice samples with different thicknesses

Table 1. Experimental matrix

Experiment No.	Initial temperature (°C)	Initial thickness (cm)
01	-20±1	4.9±0.2
02	-20±1	6.4±0.2
03	-20±1	10.2±0.2
04	-20±1	20.5±0.2
06	-32±1	10.1±0.2

Determination of Heat Transfer Coefficient

As the derivation from Eq. (1)-(4), the coefficient can be back calculated as following.

$$h = \frac{\Delta E_{in} - E_{lat}}{\left(T_s - T_\infty\right) \cdot t \cdot A \cdot h} \tag{11}$$

In Eq. (11), the thermal properties do not change significantly in our interested temperature range while ice geometry can be measured directly (Sharqawy et al., 2010). The temperature difference is the trickiest one not only in this case but also in most of transient conditions. Because it is almost impossible to properly deploy a sensor at the ice surface during its growing, we make a statement as following equation:

$$T_s = \left(\frac{T_1 - T_2}{d}\right)\left(d + \Delta L\right) + T_2 \tag{12}$$

Where T1 and T2 are from two thermistors closest to the ice-water surface; d is the distance between neighboring sensors.

RESULT & DISCUSSION

As introduced in Sec.3, the temperature and thickness were hereby measured in each test.

Temperature

With the thermistor strings we built, it is able to see the temperature distribution within ice blocks. Figure 4 (a) shows the variation of temperature distribution of the sample with 10.2cm in thickness and -20°C in initial temperature. It is clear that the distribution had a non-linear beginning and linear ending. Therefore, it is a practical and reliable way to evaluate the surface temperature by the closest two sensor as we described in Sec 3.3.

In Figure 4 (b), the curves give us the information of averaged temperature. For each sample, the temperature climbed up due to the large temperature difference at beginnings and gently rose to freezing point. Since the thermal resistance is proportional to thickness, the warming up process proceeded slower in thicker ice samples. For instant, the curve from 4.9cm sample is steeper than that of 20.5cm sample. On the other hand, there was not much difference for the sample with lower initial temperature. For both samples with around 10cm thickness (10.2cm and 10.1cm), there is very little difference in the temperature curves.

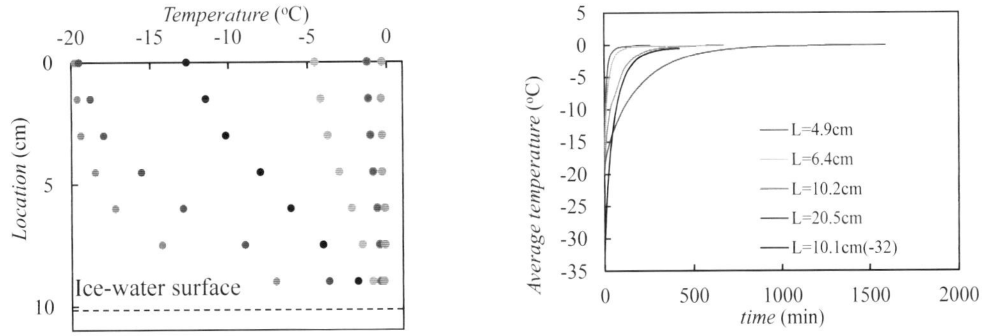

a. The temperature profile at 1st,5th, 30th, 100th, 200th, 500th minute (left to right)

b. The time series of average temperature

Figure 4. The temperature change of experiments

Ice Growth

The information of ice growth is given in Figure 5. The time series of thickness changes are plotted in (Figure 5 (a)), which is similar to the temperature, the ice grew faster at the beginning of each test and the growth rate slowed down until they reached the freezing point. That means, the temperature difference or gradient at the surface is the driven power to free new ice. Besides the common feature, the ice growth varied due to the different initial conditions. Apparently, there was more consolidation on colder and thicker ice pieces than others. It is because such samples has lower initial energy, which needs more energy to lift the temperature by freezing new ice.

If we leave a sight on the ratios between growth ice and initial ice as shown in Figure 5(b), the final percentage of thin and cold pieces are higher. Based on the energy conservation, the colder pieces absorb more energy, which is mainly released from consolidation. However, the convection leads to the difference for the samples with same initial temperature. For small ice block, the thermal resistance is smaller since it is a function of size and conductivity. Therefore, the test took longer time, which increased the convective part of energy. In another words, the small samples growth higher percentage since they obtained more energy from phase change rather than convection.

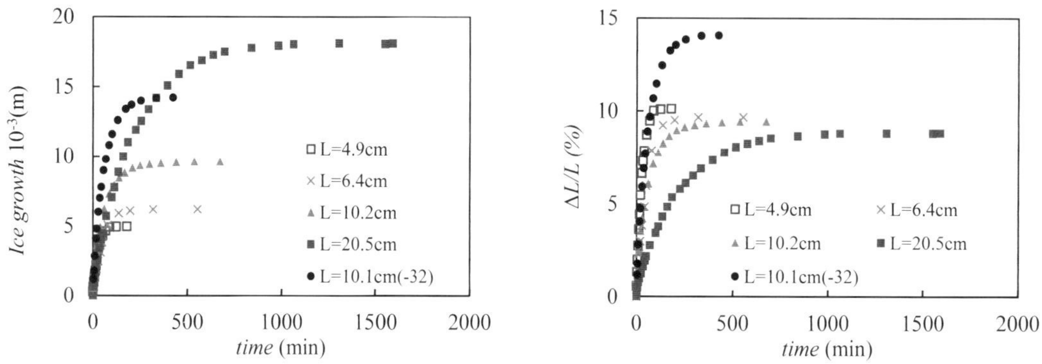

a. Time series of ice growth

b. Ratio between new ice thickness(ΔL) and initial ice thickness(L)

Figure 5. The ice growth

Heat Transfer Coefficient

Based on the data of thickness and temperature, the heat transfer coefficient is further calculated by Eq. (11). In the Figure 6, the relation between the coefficient and the temperature difference is plotted with both axis logged. Although the tests were done on varies initial conditions, it seems that those conditions did not lead to significant difference on the coefficient, neither thickness nor temperature. The only factor matters here is the temperature difference between ice surface and environmental water bath. This is because the convection is a combination of conduction and advection. While the conduction is mainly related to the material properties, the advection is dominated by the liquid motion. In our case, the water flow is driven by the thermal expansion, which is related with temperature difference. In other words, the temperature difference dominates the coefficient and hereby the convection. Furthermore, from the derivation in Sec.2.2 we learned that the coefficient exponential increase with temperature difference. From the equation of fitting curve, we can obtain the value of parameter n_l in Eq.10, which is closed to 1.537.

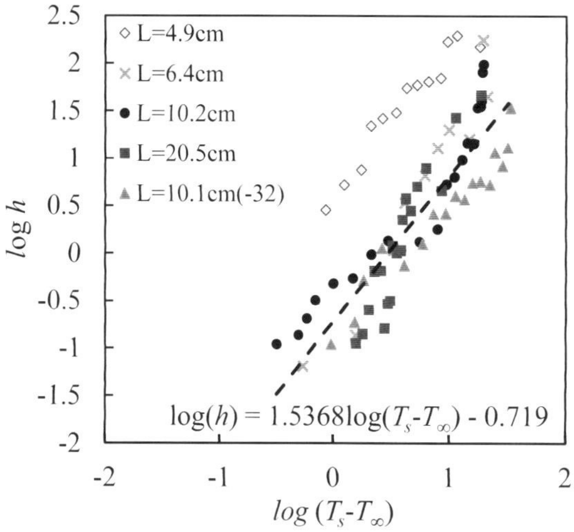

Figure 6. The heat transfer coefficient vs temperature difference

CONCLUSIONS

In this paper, a group of tests are done in NTNU cold laboratory to investigate the heat transfer coefficient of free convection. To represent the circumstance, an ice block was submerged into warm water bath. We varied the initial conditions of ice blocks to see their influences on the coefficient and conclude the information as following.

(1) When the ice is warming up, the thickness makes more influence to slow down the process than the initial temperature. (2) The initial temperature dominates the ice growth percentage while the thin ice grows slightly higher than thick ice. (3) The heat transfer coefficient is dominated by the temperature difference and independent with initial conditions. (4) The heat transfer coefficient exponentially increases with temperature difference and the power is 1.537.

ACKNOWLEDGEMENTS

Optional

The authors furthermore wish to acknowledge the support of the Research Council of Norway through the Centre for Research-based Innovation SAMCoT, and the support of all SAMCoT partners.

REFERENCES

Bergman, T.L., Incropera, F.P., DeWitt, D.P. and Lavine, A.S., 2011. Fundamentals of heat and mass transfer. John Wiley & Sons.

Chen, X. and Høyland, K.V., 2016. Laboratory Work on Heat Transfer in Submerged Ice, Theory, Experimental Setup and Results, Proceedings of the 23rd International Symposium on Ice (IAHR), pp. 4868615.

Hopkins, M.A., Tuhkuri, J. and Lensu, M., 1999. Rafting and ridging of thin ice sheets. Journal of Geophysical Research: Oceans, 104(C6): 13605-13613.

Høyland, K.V., 2002. Consolidation of first-year sea ice ridges. Journal of Geophysical Research: Oceans, 107(C6).

Høyland, K.V., 2007a. Morphology and small scale strength of ridges in the North-western Barents Sea. Cold Regions Science and Technology, 48(3): 169-187.

Høyland, K.V., 2007b. Thermal scaling of ice ridges, some dimensionless numbers, Proceedings of the International Conference on Port and Ocean Engineering Under Arctic Conditions.

Høyland, K.V. and Liferov, P., 2005. On the initial phase of consolidation. Cold regions science and technology, 41(1): 49-59.

Josberger, E.G., 1987. Bottom ablation and heat transfer coefficients from the 1983 marginal ice zone experiments. Journal of Geophysical Research: Oceans, 92(C7): 7012-7016.

Leppäranta, M., 1993. A review of analytical models of sea-ice growth. Atmosphere-Ocean, 31(1): 123-138.

Leppäranta, M. and Hakala, R., 1992. The structure and strength of first-year ice ridges in the Baltic Sea. Cold Regions Science and Technology, 20(3): 295-311.

Leppäranta, M., Lensu, M., Kosloff, P. and Veitch, B., 1995. The life story of a first-year sea ice ridge. Cold Regions Science and Technology, 23(3): 279-290.

Leppäranta, M. and Shirasawa, K., 2007. Influence of the ice-ocean heat flux on the ice thickness evolution in Saroma-ko lagoon, Hokkaido, Japan, Proceedings of the 16th

International Northern Research Basins Symposium and Workshop, 27 August-2 September, 2007, Petrozavodsk, Russia. Карельский научный центр РАН, pp. 83-88.
Sharqawy, M.H., Lienhard, J.H. and Zubair, S.M., 2010. Thermophysical properties of seawater: a review of existing correlations and data. Desalination and Water Treatment, 16(1-3): 354-380.
Shirasawa, K., Ingram, R.G. and Hudier, E.J.-J., 1997. Oceanic heat fluxes under thin sea ice in Saroma-ko Lagoon, Hokkaido, Japan. Journal of marine systems, 11(1): 9-19.
Shirasawa, K., Leppäranta, M., Kawamura, T., Ishikawa, M. and Takatsuka, T., 2006. Measurements and modelling of the water: ice heat flux in natural waters, Proceedings of the 18th IAHR International Symposium on Ice. IAHR, pp. 85-91.
Timco, G., Croasdale, K. and Wright, B., 2000. An overview of first-year sea ice ridges.

Proceedings of the 24th International Conference on
Port and Ocean Engineering under Arctic Conditions
June 11-16, 2017, Busan, Korea

Laser Scanning in Mechanical Ice Tests

Nataliya A. Marchenko[1], Alexander N. Sakharov[2]

[1] The University Centre in Svalbard, Longyearbyen, Norway
[2] Lomonosov Moscow State University, Moscow, Russia

ABSTRACT

Ice is interesting as natural phenomenon and in practical sense. Laser scanning is one of the numerous techniques to investigate ice properties, structure and dynamics.
Arctic Technology department of UNIS performs full-scale tests to investigate mechanical properties of sea ice. Laser scanning REIGL VZ-1000 was used to make 3D image of ice samples after testing and present the surface in details. Although this model of scanner was not specially designed for ice, the set of results was obtained. The scans were performed in spring 2016 on Svalbard (Van Mijen fjord) and in the Barents Sea during the cruise on RV Lance. Processing of 3-D images of ice surface of samples after full-scale indentation tests allows to see the fine structure, typical elements as splitting and tension crack, extrusion channel, etc. All these elements can be measured with high accuracy.
The main difficulties in scanning of ice are concern bad/strange scattering due to fractality of ice fracture surface. Scanning from several position gives the full 3D image, but it is complicated to obtain for small objects as the size of referencing point is large than scale of elements, which we would like to recognize on the point cloud (ice structure elements). The weather conditions during field work were quite warm (around 0° C), that influenced the tests and the scanning surface.
The advantages and disadvantages of various soft wares for processing of laser scans are discussed.

KEY WORDS: laser scanning; ice; mechanical tests; failure; crack

INTRODUCTION

On the way of mastering harsh polar environment, the knowledge about sea ice properties is vital, because ice can influence on constructions and transportation. Various methods and techniques are used to measure ice force and resistance. Full-scale mechanical test in natural environment give unique information about real sea ice and all additional means to get more data are valuables.
The international team of sea ice researchers performs full scale ice mechanical test in the

fjords of Svalbard and in the Barents Sea during last 7 years and essential experience are gained, reliable equipment is created, important data are obtained (Karulin, Marchenko et al. 2014, Karulin, Marchenko et al. 2015, Konstantinova, Marchenko et al. 2016, Murdza, Marchenko et al. 2016).

In 2016, laser scanning was firstly tried and tested with intention to find the way of saving the images of ice samples and deformation they received during testing, to understand what features of ice could be visible in the scans and what type of failure mode is due to mechanical behavior of ice in the tests.

In the course of ice testing we examine various ice properties and characterizations and the most important of them – strength under various type and rate of loading. Both brittle and ductile failure had been investigated in tension- compression tests, beam bending tests and the most interesting – indentation test in which the fracture surface can be scanned and investigated.

There was a lot of skepticism at the beginning, as it is known that ice has bad reflectivity. Even the idea to scan ice sample looked strange. But the possibility to see the fracture surface, cracks and crystals on 3D point cloud was very attractive. Due to the vector representation of the surface one can get quantitative characteristics. So various ice samples were scanned and processing gave interesting and encouraging results.

LASER SCANNING. DEVICES, TECNIQUE AND CHALLENGES

The laser scanner used for the measurements was a Riegl VZ-1000 which is part of Riegls V-line 3D Terrestrial Laser Scanners. It uses a narrow infrared pulsed laser beam in conjunction with fast rotating multi-facet polygonal mirror to acquire fast and precise laser ranging. It uses frequencies in the range from 70 to 300 kHz and achieves its maximum range of 1400 m at 70 kHz. The Riegl VZ-1000 is capable of detecting and processing multiple echo from the same direction, which means complex structures, fences, wires and vegetation can be handles. The minimum measurement distance is 1.5 m. Accuracy (conformity with actual value) is reported to be 8 mm and precision (degree to which further measurements show same results) 5 mm. A high-resolution, full-frame, calibrated Nikon D700 was used to automatically acquire RGB images for making textures during post-processing (Riegl 2017).

Riegl VZ-100 was not specially designed for ice and short distance scan. Ice test samples are usually has size 60-80 cm. The best results Riegl VZ-100 gives on the long range scanning (450-700 m distance) with opaque objects with size of several tens meters. Among Riegl VZ-400i, VZ-400 would be better. But it is too expensive equipment to have several units with various characteristics. So we used what we have. Furthermore the other models have own challenges. Designed for ice Riegl VZ-6000 is the best for glaciers and snow fields, but works on long distance (longer than 5 m) (Riegl 2017).

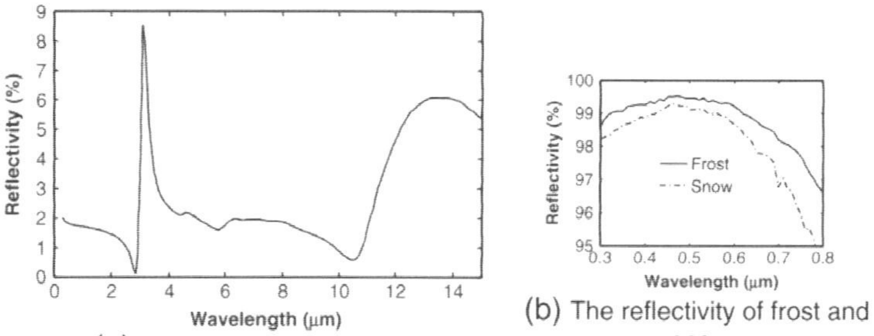

(a) The reflectivity of clear ice

(b) The reflectivity of frost and snow

Figure 1. The reflectivity of clear ice, frost and snow (directional (10°) hemispherical reflectance (Baldridge, Hook et al. 2009)

All scanners have a problem with ice. X. L. Gong and S. Bansmer made special investigation to find out the best way to scan ice and proposed a 3-D ice shape measurement technique through laser light sheet scanning in laboratorial condition (Gong and Bansmer 2015). They referred to A.M. Baldridge's and co-authors (Baldridge, Hook et al. 2009) investigation of ice reflectivity (Figure1) and use a thin layer of frost instead of opaque paints was used to increase the surface reflectance of clear ice. Fine snow layer gave also good results.
In our experiments we found that fresh ice samples, just after testing naturally covered by thin layer of ice powder and have sufficient reflectivity. After some time, for example keeping in cold storage room, samples lose that layer and not directly suitable for scanning. They looks nondiagnostic on scan, due to changing of reflectivity (Figure 2) and need to be covered by some powder - snow or hoarfrost. But it is difficult to perform on vertical surfaces.

Figure 2. Ice sample after storage on photos and on scans

During field work (see next section) we used the closest range scanning program 450 m (300 kHz) and tested it for various resolutions. The high resolution gives the large file (Table 1) and the hope to see the fine structure of surface, might be, large crystals, grain. But it comes that resolution higher than 0,01 deg gives much noise and hides the important features - the surface looks very fluffy. Actually resolution 0,02 is completely enough, and actually the best. Resolution 0,04 is not visible on illustration, for example in Word files, even it looks nice in RiScan program. So it is not presented in Table 1

Table 1. Appearance and size of scans with various resolutions.

Res-0,02. 27 MB	Res 0,01. 111 MB	Res-0,003 572 MB
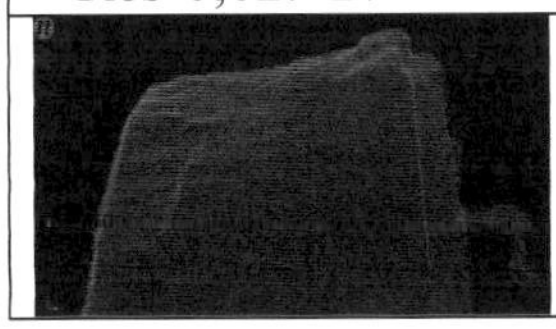	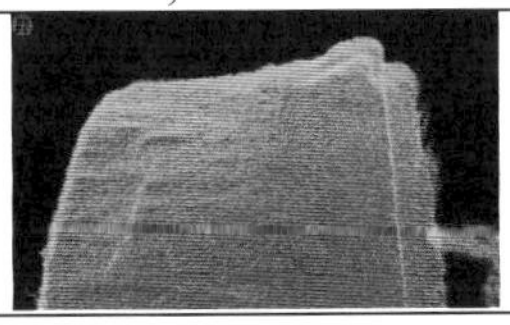	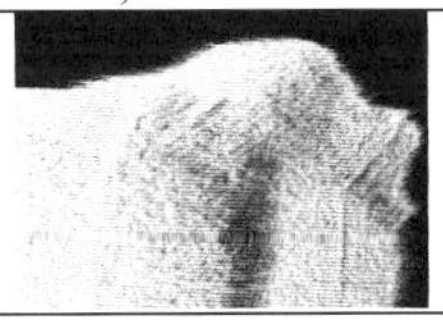

SCANNING OF SAMPLES IN THE FIELD. SPRING 2016

Laser scanning complain was performed during field work in Van Mijen fjord (Western fjord of Spitsbergen), performed by Arctic Technology department of UNIS and guest researches at the beginning of March 2016; and in western part of the Barents Sea while cruising on RV Lance in May 2016. The samples after several tests were scanned. These particular testes and results are described in details in (Konstantinova, Marchenko et al. 2016, Murdza, Marchenko et al. 2016). So here we only describe the scanning itself.

Indentation Test 1 (07.03.2016)

After indentation test, described in (Konstantinova, Marchenko) the sample of ice with splitting cracks was scanned with various resolutions from different points (Figure 3). The dimensions (depth – 0,368 m and opening at the mouth – 0, 036 m) of crack were visible and performed on scan image. The crack inclined in the angle of near 30^0 in vertical direction and propagated to warmer and weaker bottom layers of sample.

Figure 3. Test setup, place of sample, scanning process and result

Indentation Test 2 (08.03.2016)

The most impressive scan was performed after indentation test on 8 of March 2016.

Figure 4. Indentation test setup with ice sample for scanning and photo of sample

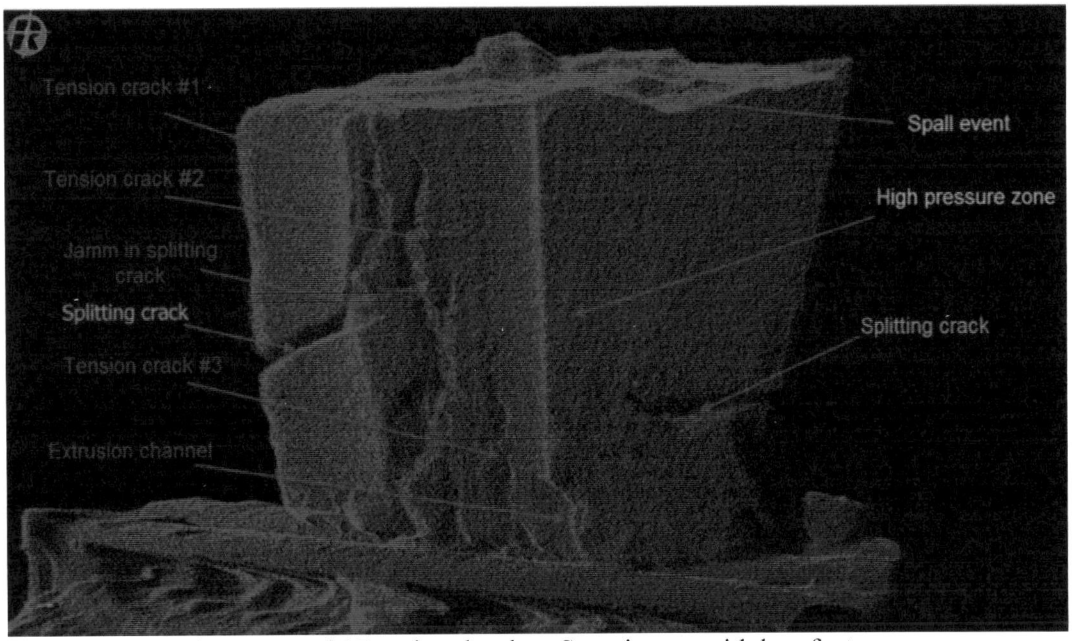

Figure 5. and 3D point cloud on Scan image with key features

Several splitting and tension cracks are visible. The result of indentation test in situ (Konstantinova, Marchenko et al. 2016) plot Force in load cylinder vs displacement (Figure 6) show the quasi periodical regime of ice failure during indentation. On the plot we can see the quasi period L_0 = 3-4 cm from one unloading event to another one. In complex sequences alternate the splitting and jam processes, spalling and other discontinuous events which are firing an unloading.

It's quite difficult to analyze the correlation between these elementary events and total force decreasing. We can only make a proposal that approximately half of these gash on load graph are in accordance of tension discontinuous, which formed at the back side of the semi cylinder of indentor (Figure 7).

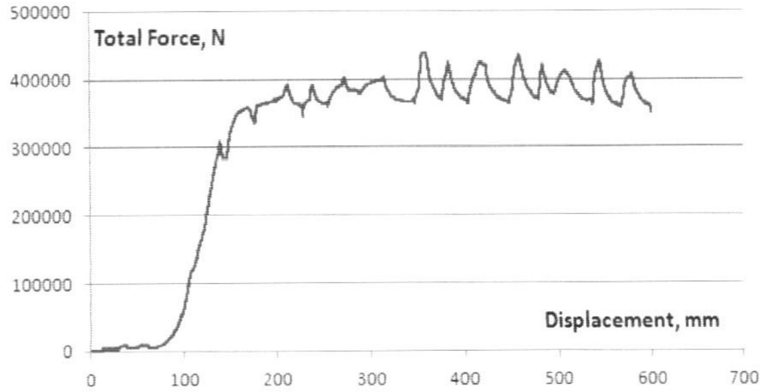

Figure 6.

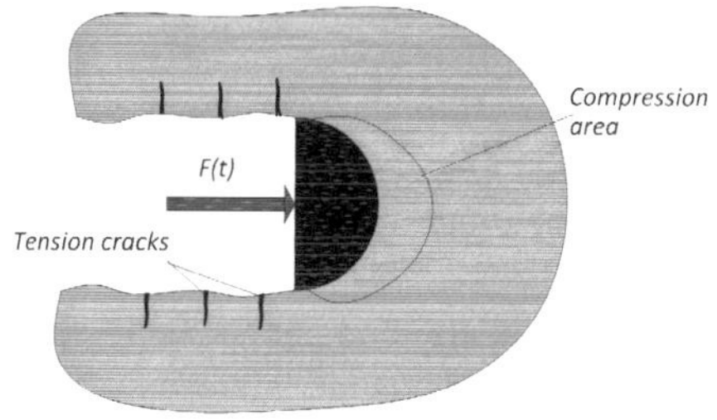

Figure 7.

L-shape Beam Test (10.03.2016)

Beam tests are used traditionally to deteminate flexural strength. In our modification test L-shape bending also torsion deformations and shear stress are included. Complex bending deformation of ice appears when the forces applied on the ice edge initiate torsion. It may happen when the ice edge is displaced upward and downward on different segments on the contact line with a structure. Complex bend is realized when surface waves propagating in different directions penetrate below the floating ice. Superposition of bend and plane stress state is realized when surface wave propagates in region covered by ice compressed in the direction not coinciding with the direction of wave propagation. Test with floating L-shape beam was firstly performed in Spring 2016 to formulate ice failure criterion in case when the principal stresses have different signs.

Figure 8. L-shape beam setup, sample place, photo and scan

Test with L-shape beam is describes in detail (Murdza, Marchenko et al. 2016). For laser scanning the sample cutting by crack had been taken. (Figure 8). So we actually scan and perform failure surface. It will be described in PROCESSING section

"Cheese ice" in the Ridge in the Barents Sea (03.05.2016)

During research cruise RV Lance in the Barents Sea, mechanical properties of ice in ridges and rubbles were investigated. The ice sample after indentation test had a loose structure with cylindrical holes, so we call it as "cheese ice" and captured by scanning (Figure 9). In the ice sample with size 0,563x0,398 m and approximately 0,3 m width, we discovered and imprinted 4 visible holes 0,069-0,079 m diameter and 0,007 m depth, that looked as thermokarst phenomena, because it difficult to imagine such holes created mechanically inside ridge.

We do not know exactly the genesis of such structure, but the object was recorded and can be further explored. So it is possible to investigate not only the fracture surface, that we get as the result of mechanical tests, but structures that are associated with thermodynamic processes within the ice and scan provides a good illustration of thermal phenomena in the hummocks (Figure 9).

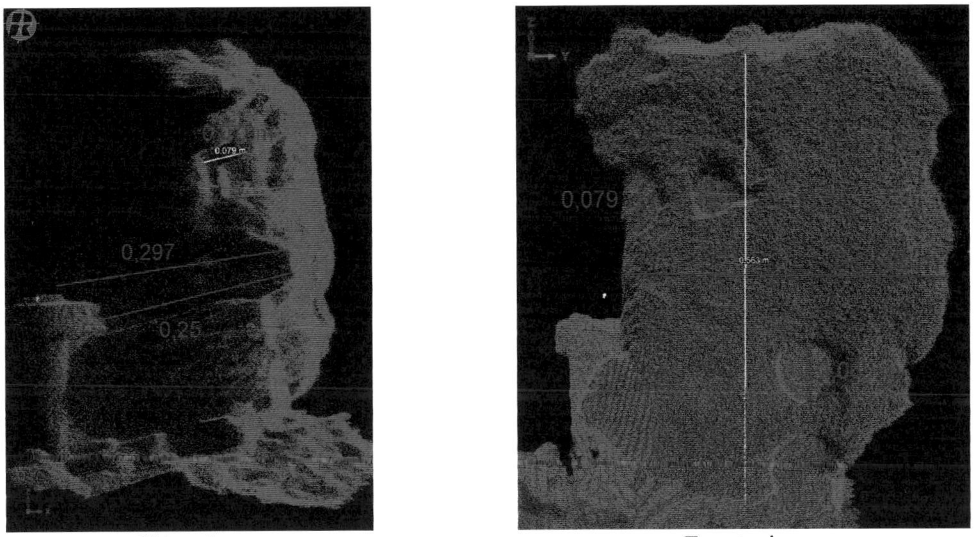

Side view Front view

Figure 9. "Cheese ice" sample after indentation test

PROCESSING

Let's consider various processing software and methods on the example of L-shape sample, that had size 0,62x0,68 m (Figure 10). We took scan with 0,02 deg resolution, that gave us after cleaning 215923 points in 3D point cloud of sample.

Processing of scanner data in RiScan program (Riegl 2016) (the very first step of working with scan data) allows to present 3D point cloud, clean from unnecessary points, apply various filters and export as ASCII code to TXT. file to process further in other more common program, for example ArcGIS (ESRI 2017)(Figure 10). RiScan program in special for Riegl scanner users, that has might be several thousand installations, rather expensive and unique. ArcGIS is very popular, widespread mapping software with millions of users and free viewers. So export to ArcGIS will make scanner data affordable for the masses. One can make profiles across ice samples in ArcMap (2D part of ArcGIS) (Figure 10. Middle)

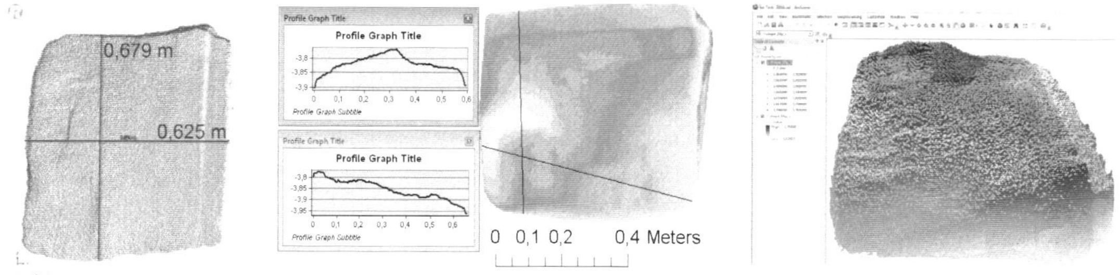

RiScan programm ArcMap ArcScene

Figure 10. L-shape sample images in various software

ArcMap allows to identify the geometric characteristics by the cross section. This is a way to describe the rough surfaces with a set of parallel lines and roughness of the ice can be studied with the help of profiles in ArcGIS. The other way is to analyze the surface by ArcGIS 3D Analyst tools, and as steepness representation, curvative map, hillshade (Figure 11).

Map of surface sectors steepness Curvature map Hillshade

Figure 11. Processing with 3D Analyst tools in ArcMAP

ArcScene (3D part of ArcGIS package) perform three-dimensional image that can be rotated, viewed from different angles and color, symbolized in accordance with different data/sample specific (Figure 10. Right).

Very important features of ArcGIS software (ESRI 2017) is openness, existing of viewers and online options. Publishing of surface maps (2D) and scene (3D) makes scanning data available for researchers and audience (students), who does not have RiScan and ArcGIS software.

Various filters and processing in RiScan

As it mentioned due to noise scan surface looks furry/fuzzy and could not be triangulated by plane triangulation method with good results (Figure 12)

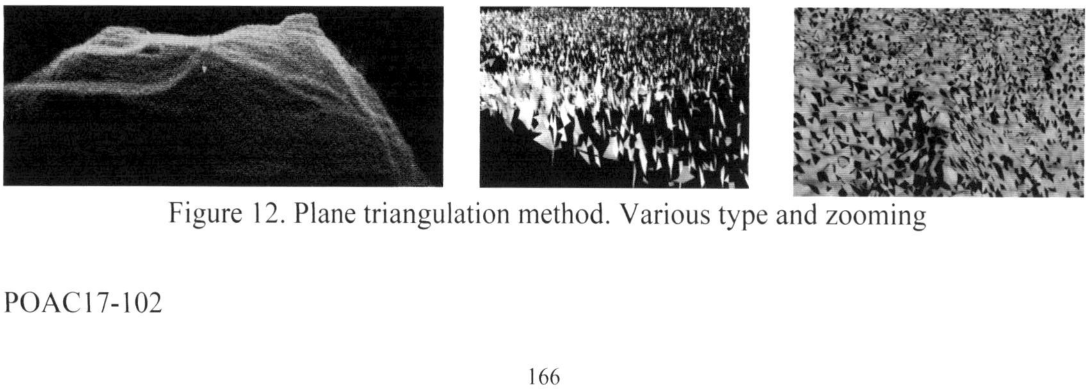

Figure 12. Plane triangulation method. Various type and zooming

But triangulation with icosahedron gives interesting results (Figure13). At level 9 you can see structures on ice fracture surface due to combine bending and torsion failure in L-shape beam bending test. Further we can export TXT file in other software, for example ArcGIS, where we can create elevation map and profiles (Figure 10).

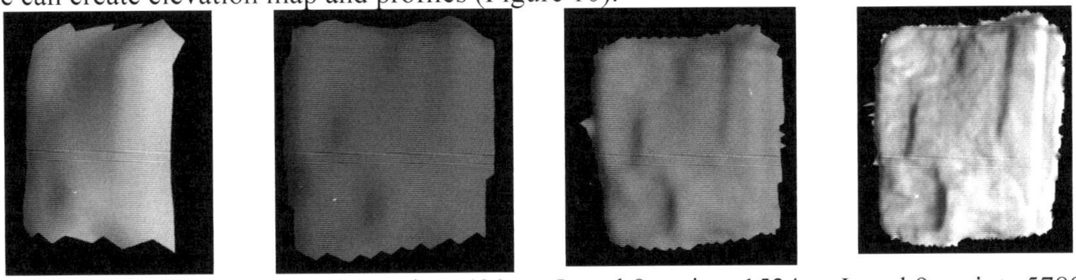

Level 6, points 124, polys 207 | Level 7, points 420, polys 763 | Level 8, points 1524, polys 2895 | Level 9, points 5788, polys 11256

Figure 13. Triangulation with icosahedron. Various levels

CONCLUSIONS

We would like to emphasize that there is no alternative for 3D laser scanning to reflect and store the image of ice, ice sample and visible results of mechanical (and other) tests. Since pictures can not allocate the structures, the work on the ice destruction are illustrated by the diagrams and plots/drawing, which are a generalization of what researchers have seen by their 3-dimensional vision. They could not fix on it in another way.

3D scan provides such an opportunity. It remembers the 3-dimensional image of the fracture surface and allows to decipher the structure of destruction not in the field (in harsh weather conditions and limited time), but in the course of a desk job, in office without losing of information.

It is not possible to perform the calculations associated with the analysis of the energy costs for the destruction taking into account the irregular nature of the surface by the methods which are used now widely. But the result of 3D scan and its processing provides such capability.

Moreover, the opportunity to receive a vector model of the fracture surface and to determine its properties, confronts engineers and programmers over the task of trying to build an algorithm for calculating the properties of the surface and to create the theory of destruction, which would be able to compute/evaluate the energy of destruction taking into account the structural features.

The results of 3D scanning give us the really new way to acquaint students with the structure and the show the nature of the destruction of ice. Students will be able to look into the cracks and cavities of ice that had melted long ago.

It is necessary to continue the work and create a bank of 3D models, which will be extremely useful in the future. Many mechanics agree that icc is a unique material from the standpoint of variety of forms and structures destruction. Therefore, such a bank model would also be in demand for the Materials Science. Even from a small number of scans can be concluded that the most interesting images of the structures were obtained in compression tests, as well as a crumpled ice on indentation tests. Moreover the little experience, which is now accumulated, gives us possibility to suggest that in situ mesoscale test is the most appropriate purpose for the laser scanning, allowing to perform not only a quantitative assessment, but also qualitative evaluation allowing you to allocate the structure. If vertical profiles of temperature and salinity of the ice would be added to a vector model of the scanned surface of the tested samples, it would have the necessary amount of raw data for measurement and analysis of test results.

On the way to creation of data bank and using in education of laser scanner point clouds, the combination RiScan and ArcGIS software is very important. ArcGIS viewers and online options make scanning data available for wide range of researchers and audience (students).

ACKNOWLEDGEMENTS

The authors would like to thank the SAMCoT project (funded by the Research Council of Norway), the SITRA project (funded by the Norwegian Centre for International Cooperation in Education) and the colleagues for supporting this study.

REFERENCES

Baldridge, A. M., S. J. Hook, C. I. Grove and G. Rivera (2009). "The ASTER spectral library version 2.0." Remote Sensing of Environment **113**(4): 711-715.

ESRI. (2017). "ArcGIS software." from http://www.esri.com/.

Gong, X. L. and S. Bansmer (2015). "Laser scanning applied for ice shape measurements." Cold Regions Science and Technology **115**: 64-76.

Karulin, E., A. Marchenko, M. Karulina, P. Chistyakov, A. Sakharov, A. V. Ervik and D. Sodhi (2014). Field Indentation Tests of Vertical Semi-Cylinder on First-Year Ice. 22nd IAHR International Symposium on Ice. Singapore: 8.

Karulin, E.B., Marchenko, A.V., Karulina, M.M., Sakharov, A.N., Chistyakov, P.V., Verbitskaya, M.Yu., Ignatieva, E.D., 2015. Compression testing of sea ice by penetrating the ice sheet with a semi-cylindrical indenter. Process numerical simulation. Transactions of the Krylov Shipbuilding Research Institute, 86 (370), 75-86. ISBN 0869-8422.

Konstantinova, M., A. Marchenko, M. Karulina, A. Sakharov, E. Karulin and P. Chistyakov (2016). In-situ Investigations of Ice Deformations and Loads in Indentation Tests. Proc. Of 23rd IARH International Symposium on Ice (IAHR-ICE 2016). AnnArbor, Michigan-US.

Murdza, A., A. Marchenko, A. Sakharov, P. Chistyakov, E. Karulin and M. Karulina (2016). Test with L-shaped Cantilever Beam for Complex Shear and Bending Strength. Proc. Of 23rd IARH International Symposium on Ice (IAHR-ICE 2016). AnnArbor, Michigan-US.

Riegl. (2016). "RiSCAN PRO. Operating and Processing Software for RIEGL 3D Laser

Scanners." from http://www.riegl.com/uploads/tx_pxpriegldownloads/11_DataSheet_RiSCAN-PRO_2016-09-19.pdf.

Riegl. (2017). "Terrestrial Laser Scanning." from http://www.riegl.com/nc/products/terrestrial-scanning/.

Evaluation Method Research of Power of Main Engine for Polar Carrier

Shaopeng Ji[1], Yukui Tian, Yinghui Wang[1], Baoshan Wu[1]
[1] China Ship Scientific Research Center, National Key Laboratory of Science and Technology on Hydrodynamics, Wuxi, China

ABSTRACT

In this paper, evaluation methods are established from theory to model test for polar carrier according to balance principle of resistance and thrust. The methods established above are verified for a polar carrier by conjunction with test results of ice resistance which come from the model test of ice tank. It is noted that the interaction between ice and propeller must be considered in the process of power estimation when the theory method is adopted by designer, and it mainly depends on experience of the designer. At the same time, the model test of low speed can also fulfill the requirement of the designer for the power evaluation of main engine in conventional towing tank. The evaluation results indicate that it is very effective to assess power of main engine by the theory and the model test, and it can supply reliable evaluation result for the designer.

KEY WORDS: Powering; Main engine; Theory evaluation; Model test; Polar carrier.

INTRODUCTION

The special requirement must be fulfilled for the polar carrier which voyage in the polar region and ice region. Firstly, the thruster of big power should be met to continuously break ice. Secondly, hull body can effectively resistant ice load. Thirdly, the equipment can adapt lower temperature at the polar region and ice region. So, it is very important for the polar carrier to evaluate power of main engine. However, it is very difficult to assess the power of main engine for the polar carrier. The first difficulty comes from reliability of ice resistance, the model test of ice tank is not suitable in the initial design phase thanks to time and cost, the numerical simulation is developed in recent years, and it needs to be demonstrated on reliability and stability of calculation by the model test, so, the empirical formulas are chosen by the designer in the initial design phase. Hyun-Soo kim (2014), Jian Hu (2015), Kyung-Duk Park (2014 & 2015), Seong-Rak Cho (2015), et al. study empirical formula and semi-empirical formula based on parameters of hull form. The research results show that some empirical formulas, such as Linqvist formula, Risk formula and Keinonen formula, can give reasonable results, but the obvious disadvantages are found in process of estimation thanks to limitation of ship parameters and sea ice conditions. The second difficulty comes from assessment of power of main engine based on ice resistance, and evaluation method for power of main engine is not published in any literature. At present, the direct estimation method of power of main engine in the ice class rules, such as RMRS (2010), CASPPR (2010) and ABS (2010), is suggested, but this method is not suitable for the polar carrier with high ice class condition. In view of mentioned above, it is necessary to develop a new method to evaluate power of main engine of polar carrier.

In this paper, the estimation methods of power of main engine are explored preliminarily based on ice resistance by the theory and the model test, respectively, and it is validated by the model test at ice tank for a polar carrier.

EVALUATION METHOD RESEARCH

Ice Resistance

Jungyong Wang (2011) gives the expression of ice resistance:

$$R_{ice} = C_{BR} S_N^{-\beta} \rho_i B h_i V^2 + C_B \Delta\rho g h_i BT + C_C F_h^{-a} \rho_i B h_i V^2 \text{ [kN]} \tag{1}$$

Where C_{BR} is icebreaking resistance coefficient, C_B is submersion resistance coefficient, C_C is sliding resistance coefficient. B is breath of ship at draught, T is draught, h_i is ice thickness, ρ_i is ice density, $\Delta\rho$ is density difference, g is gravity acceleration, V is ship speed. $S_N^{-\beta}$ is strength coefficient, F_h^{-a} is Froude number of ice thickness.

The ice resistance is influenced by the ship speed, ice thickness and flexural strength when the ship navigates in the polar region and ice region. It can be expressed quadratic polynomial equation between ice resistance and ship speed according to equation above when ice thickness and flexural strength are not changed in ice model test.

$$R_{ice} = a_0 + a_1 V_i + a_2 V_i^2 \text{ [kN]} \tag{2}$$

Where a_0 a_1 a_2 are polynomial coefficient, V_i is ship speed.

Stephen J. Jones (2005) gives nonlinear relationship between ice resistance and thickness with the growth of ice thickness when the ship speed and flexural strength are not changed, and the relationship can be expressed quadratic polynomial equation.

$$R_{ice} = b_0 + b_1 h_i + b_2 h_i^2 \text{ [kN]} \tag{3}$$

Where b_0 b_1 b_2 are polynomial coefficient, h_i is ice thickness.

Stephen J. Jones & Michael Lau (2005) give nonlinear relationship between ice resistance and flexural strength with the growth of flexural strength when the ship speed and ice thickness are not changed, and the relationship can be expressed quadratic polynomial equation.

$$R_{ice} = c_0 + c_1 \sigma_i + c_2 \sigma_i^2 \text{ [kN]} \tag{4}$$

Where c_0 c_1 c_2 are polynomial coefficient, σ_i is flexural strength of ice sheet.

Thruster Hydrodynamics

Theory Prediction of Thruster Hydrodynamics

The hydrodynamics of thruster in ice can be predicted by the lift surface and panel element methods.

It can be expressed higher degree relationship between thrust, torque and revolutions when the advance speed is not changed, as follows:

$$T = d_1 + d_2 N_s + d_3 N_s^{\,2} + d_4 N_s^{\,3} + d_5 N_s^{\,4} \text{ [kN]} \quad (5)$$

$$Q = e_1 + e_2 N_s + e_3 N_s^{\,2} + e_4 N_s^{\,3} + e_5 N_s^{\,4} \text{ [kN.m]} \quad (6)$$

Where $d_1\ d_2\ d_3\ d_4\ d_5\ e_1\ e_2\ e_3\ e_4\ e_5$ are biquadratic coefficient, N_S is propeller revolutions in full scale.

It must be considered for the suction effect of propeller and the interaction between ice and propeller when the thrust is used in the calculation for the power of main engine. The thrust deduction is used to assess thrust loss, and it depends on experience of the designer.

The net thrust of thruster in ice:

$$T_{net} = (1-t)T \text{ [kN]} \quad (7)$$

Where t is thrust deduction, T is total thrust.

Thruster Hydrodynamics behind the Ship in the Model Test

It can be expressed higher degree relationship between thrust, torque and revolutions when the ship speed is not changed in the model test, as follows:

$$T_{pm} = f_1 + f_2 N_m + f_3 N_m^2 + f_4 N_m^3 + f_5 N_m^4 \text{ [N]} \quad (8)$$

$$Q_m = g_1 + g_2 N_m + g_3 N_m^2 + g_4 N_m^3 + g_5 N_m^4 \text{ [N.m]} \quad (9)$$

Where T_{pm} is total thrust, Q_m is torque, N_m is revolution, $f_1\ f_2\ f_3\ f_4\ f_5\ g_1\ g_2\ g_3\ g_4\ g_5$ are biquadratic coefficient. The thrust deduction is expressed as:

$$t = 1 - \frac{T_{BPm}}{T_{Pm}} \quad (10)$$

Where T_{BPm} is bollard pull, T_{Pm} is thruster thrust.

Power Evaluation Based on Theory and Model Test

The revolutions of the propeller will be reduced when the work of main engine deviate the design point. It is very important to solve revolutions of propeller in ice.

Theory Assessment

According to balance of ice resistance and net thrust:

$$T_{net} = R_{ice} \text{ [kN]} \quad (11)$$

The following equation is acquired:

$$(1-t)(d_1 + d_2N_s + d_3N_s^{\ 2} + d_4N_s^{\ 3} + d_5N_s^{\ 4}) = a_0 + a_1V_i + a_2V_i^2 \quad (12)$$

Or

$$(1-t)(d_1 + d_2N_s + d_3N_s^{\ 2} + d_4N_s^{\ 3} + d_5N_s^{\ 4}) = b_0 + b_1h_i + b_2h_i^2 \quad (13)$$

Or

$$(1-t)(d_1 + d_2N_s + d_3N_s^{\ 2} + d_4N_s^{\ 3} + d_5N_s^{\ 4}) = c_0 + c_1\sigma_i + c_2\sigma_i^2 \quad (14)$$

The EQ. (12), (13) and 14 are solved by the method of newton iteration to obtain revolutions of propeller in ice.

The delivered power is:

$$P_D = 2\pi nQ \,[\text{MW}] \quad (15)$$

Eq.(6) into Eq.(15):

$$P_D = 2\pi n(e_1 + e_2(N_s) \ + e_3(N_s)^2 + e_4(N_s)^3 + e_5(N_s)^4) \,[\text{MW}] \quad (16)$$

The delivered power of thruster is obtained by actual propeller revolutions, and then the power of main engine is

$$P_{MCR} = \frac{P_D}{\eta_S} \,[\text{MW}] \quad (17)$$

Where P_{MCR} is power of main engine, η_S is the shaft efficiency, this value is usually equal to 0.99 in calculation for the conventional ship.

Model Test Assessment at Low Speed Condition in Conventional Towing Tank

The bollard pull at full scale:

$$T_{BPs} = \frac{\rho_s}{\rho_m}\lambda^3 T_{BPm} \,[\text{kN}] \quad (18)$$

The torque at full scale:

$$Q_S = \frac{\rho_s}{\rho_m}\lambda^4 Q_m \,[\text{kN.m}] \quad (19)$$

The revolutions of propeller at full scale:

$$N_s = \frac{N_m}{\sqrt{\lambda}} \,[\text{RPM}] \quad (20)$$

Where ρ_s is density of sea water, ρ_m is density of fresh water, λ is scale ratio, m denotes model, s denotes full scale.

The propeller revolutions and torque are acquired at work point in full scale according to balance of thrust and ice resistance. The delivered power of thruster at full scale is acquired by Eq.(15), and the power of main engine is acquired by Eq.(17).

EXAMPLE ANALYSIS

Solve Analysis Based on Theoretical Method

As an application, a polar carrier as a sample is analyzed to confirm evaluation method established above. The requirement of ice class with ARC6 in RMRS is fulfilled for this ship which can voyage in polar region with ice thickness of 1.5m and ship speed of 1.9kn in one year ice at ballast draught. The prediction of model test shows that the power of 32MW can meet requirement of ice class of ARC 6 for this ship at ballast draught.

The model test is conducted with condition of level ice at ice tank, as shown in Figure 1.

Figure 1. Model test in ice

The ice resistance is measured at the condition of ballast draught with ice thickness of 1.3m and 1.6m in model test according to the icebreaking requirement, the model test results of ice resistance are shown in Figure 2.

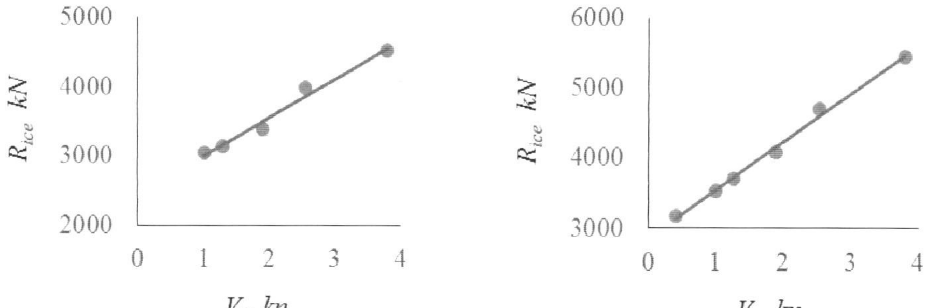

(a) Ice thickness of 1.3m (b) Ice thickness of 1.6m

Figure 2. Prediction of model test

The propeller diameter of 6.8m with twin screw type for this ship is estimated to meet propelled requirement in ice, the hydrodynamics of propeller are predicted by the lift surface method, and it is expressed four degree polynomial similar to Eq.(5) and (6), the results are shown in figure 3.

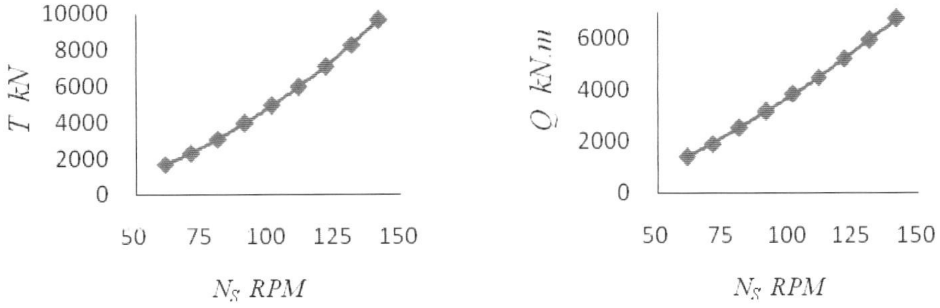

(a) Propeller thrust (b) Propeller torque

Figure 3. Prediction of lift surface

The thrust deduction of 0.1 is assumed in calculation according to experience of interaction between ice and propeller. The revolutions of propeller in ice are solved by the method of Newton iteration according to Eq.(13) at two different ice thickness with 1.3m and 1.6m, respectively. The delivered power of propeller and the minimum power of main engine are solved by Eq.(15) and (17), respectively, and the shaft efficiency of 0.99 is considered in calculation. The results of ice thickness of 1.5m are interpolated according to results of ice thickness of 1.3m and 1.6m, the detailed results are listed in table 1.

Table 1. The evaluation results with different ice thickness at the speed of 1.9kn

Ice thickness(m)	Revolutions(RPM)	Delivered power(MW)	Power of ME (MW)
1.3	83.4	26.419	26.686
1.5	88.6	32.115	32.439
1.6	91.2	34.962	35.315

Table 1 show that the minimum power of main engine is 32.439MW, which is close to the power requirement of 32MW. The evaluation method of theory indicates that it is reliable to calculate the power of main engine for the polar carrier.

Solve Analysis Based on Towing Test at Low Speed Condition

The polar carrier at low speed condition are tested in deep water towing tank of china scientific research center (CSSRC), the scale ratio of 35 is chosen in model test. The propeller diameter of 6.8m is adopted in model test according to result of theory evaluation above. The model test is shown in figure 4.

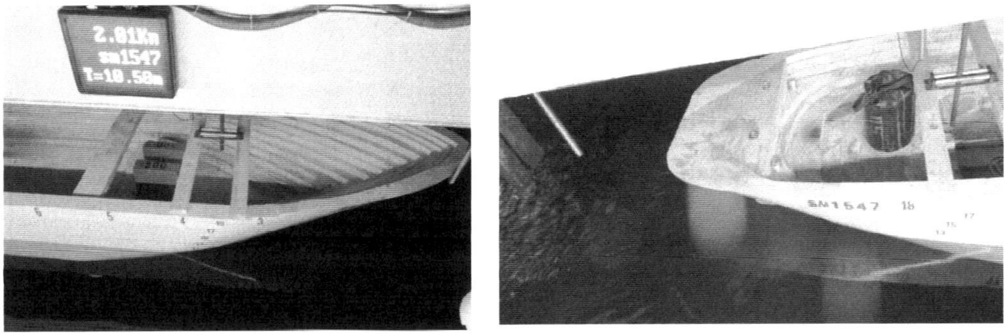

Figure 4. The towing test at speed of 2kn

The model test is conducted with towing speed of 1kn and 2kn at ballast draught, respectively. The average results of hydrodynamics between left and right propeller behind the ship model are given in figure 5 with speed of 1kn and 2kn, respectively.

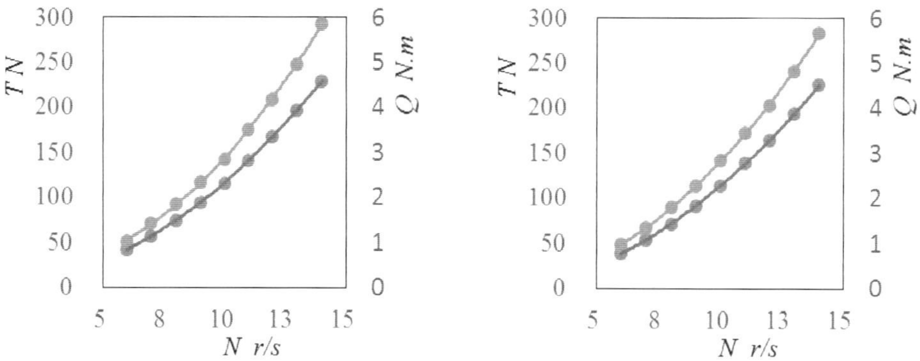

(a) Test results at speed of 1.0kn (b) Test results at speed of 2.0kn

Figure 5. The results of model test behind the ship model

The hydrodynamics results of propeller behind the ship model are fitted using method of least squares. The results at speed of 1.9kn are interpolated in calculation, and it is expressed four degree polynomial similar to Eq.(8) and (9), the results are shown in figure 6.

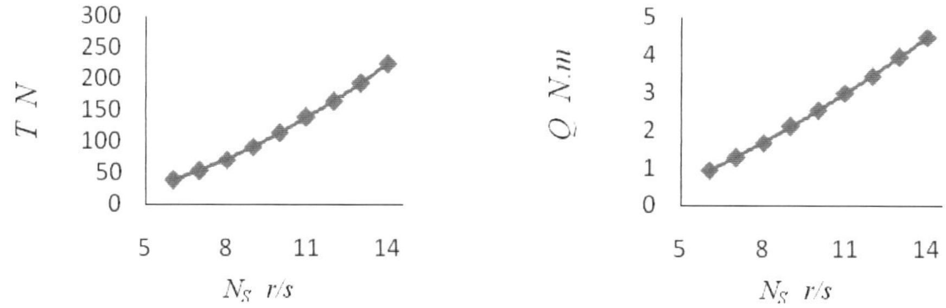

(a) Propeller thrust (b) Propeller torque

Figure 6. The results of model test at speed of 1.9kn

The propeller revolutions and delivered power are solved in full scale by Eq.(11), (15), (18), (19) and (20). The minimum power of main engine is solved by Eq.(17), the value of 0.99 is used for the shaft efficiency in calculation. The detailed results are listed in table 2.

Table 2. The evaluation results of power of main engine at the speed of 1.9kn

Ice thickness(m)	Thrust deduction	Revolutions(RPM)	Delivered power(MW)	Power of ME (MW)
1.5	0.107	89.1	32.120	32.444

Table 2 show that the minimum power of main engine is 32.444MW, which is close to the power requirement of 32MW.

The evaluation method of model test at condition of low speed indicates that the power of main engine can be evaluated precisely by the towing test at condition of low speed without interaction between ice and propeller in conventional towing tank.

CONCLUSIONS

Comparing to prediction results of model test of ice tank, it is feasible to evaluate the power of main engine by adoption theoretic method and model test of low speed in conventional towing tank. The important conclusions can be obtained:

(1) The propeller revolutions in ice and the value of thrust deduction are main influence factors in process of power assessment of main engine by theoretic method based on lift surface.

(2) Although there is not interaction of ice and propeller, the thrust deduction obtained can fulfill requirement of evaluation in conventional towing tank at low speed condition.

ACKNOWLEDGE

This paper is supported by the high technology ship project of Ministry of industry and information of china at 2014, and the author wants to appreciate advices of co-workers.

REFRENCES

Hyun-Soo kim, 2014. Development of estimation system of resistance with surface information of hull form. Ocean Engineering 92:PP.12-19.

Jian Hu, 2015. Experimental and numerical study on ice resistance for icebreaking vessels. ANAK Proceedings, PP.626-639.

Shaopeng Ji, 2016. Application research report of calculation requirement and test results for power of main engine. Technology report, pp.14-15.

Jungyong Wang, 2011. Resistance and Propulsion of CCGS Terry Fox in Ice from Model Tests to Full Scale Correlation. Test Report, pp.5-6.

Kyung-Duk Park, 2015. Calculation of ice clearing resistance using normal vector of hull form and direct calculation of buoyancy force under the hull. ANAK Proceedings, PP.699-707.

Kyung-Duk Park, 2014. Study on ship ice resistance estimation using empirical formulas. Proceedings of the international conference on OMAE, San Francisco, California, USA.

Rules for building and classing steel vessels ABS.2010

Rules for building and classing steel vessels RMRS.2010

Rules for building and classing steel vessels CASPPR.2010

Seong-Rak Cho, 2015. A Prediction method of icebreaking resistance using a multiple regression analysis. ANAK Proceedings, PP.712-719.

Seong-Rak Cho, 2015. The resistance of icebreaking ship using the regression analysis of model tests. Ocean Engineering, 108:PP.692-703.

Stephen J. Jones, 2005. Resistance and Propulsion Model Tests of the USCGC Healy (Model 546) in Ice. Test Report, LM-2005-02, pp.6-7.

Stephen J. Jones & Michael Lau, 2005. Propulsion and Maneuvering Model Tests of the USCGC Healy in Ice and Correlation with Full-Scale. Test Report, pp.5.2-4.

Measurement of Pack Ice Area and Concentration using Image Processing Package

Seok-Ho Byun[1], Jong-Ho Nam[1]
[1]Division of Naval Architecture and Ocean Systems Engineering, Korea Maritime and Ocean University, Busan, 49112, Korea

ABSTRACT

Due to the increasing possibility of passing the Arctic route, considerable works on the icebreaking vessels have been continuously published. The model ship test on the pack ice model in the ice tank is actively carried out as a way to investigate the performance of icebreaking vessels. The concentration as well as the individual area of pack ice are important since they directly affect the performance. However, it is not easy to obtain those properties because not only the pack ice has uneven shape but also it keeps floating around in the ice tank. In this paper, an algorithm to identify the concentration of pack ice is introduced. From a digital image of pack ice obtained in the ice tank, the goal is to measure the area of pack ice using an image processing technique. Instead of the general global thresholding that yields numerical errors in this problem, a local thresholding technique, coupled with image subdivision based on the quadtree structure, is developed. The individual area of each ice patch is calculated using the edge extraction and the watershed technique.

KEY WORDS: Pack ice; Concentration; Local thresholding; Individual area distribution; Segmentation.

INTRODUCTION

Recently, the frozen area of the Arctic has been decreasing and the duration of below freezing temperatures has been shortening. As a result, there is a growing need for basic research on icebreakers passing through the Arctic route. As a type of basic research, icebreaker model tests have been actively conducted by using integrated pack ice. In this experiment, the concentration and area distributions of each of the pack ice formations are the significant factors necessary to evaluate the performance of icebreakers. Pack ice is floating ice in a widely-bound state. Because pack ice floats freely in an ice tank, it has limited accessibility, and owing to its non-uniform shape, it is difficult to directly measure its concentration and area distribution. Therefore, we mainly use remote sensing to extract information on this object by using a sensor from a long distance (Rees, 2005). Among many methods, there is a method of analyzing the images of where pack ice is distributed, which yields lower initial investment requirements and operating costs than other methods. The images of a pack ice distribution were often analyzed via commercial software, which facilitated and simplified the measuring process (Cho, et al., 2013). Although it is suitable for general objects, the commercial software not only requires users to constantly engage with the software, but it has the drawbacks of not providing accurate results for specific measurements such as pack ice concentration measurement. Furthermore, this software is particularly not applicable when the pack ice is under irregular illumination.

An algorithm optimized for pack ice concentration measurement has been actively studied. As an example, a study analyzing an image of a large area of pack ice after manual subdivision of the image via preprocessing has been reported (Zhang, et al., 2015). An enhancement for a non-uniform light and color region using thresholding following manual subdivision was recently introduced by Byun, et al. (2016). Subsequently, an automatic subdivision algorithm using quadtree and a local thresholding result, which automatically solved the problem of light or color non-uniformity, were presented (Lee, et al., 2017).

In this paper, we describe the automatic subdivision algorithm using quadtree. The generated binary image is then applied to the watershed object segmentation algorithm to extract the separable pack ice formations and to calculate the pack ice distribution. We add a local maximum finder algorithm to the watershed object segmentation algorithm to search among pack ice of various sizes and to clearly distinguish proximity objects by applying morphology erosion. This algorithm is applied to the model pack ice image of an ice tank, and an example of deriving pack ice concentration and the area distribution result is demonstrated. Open-source digital image processing technology is used for image analysis (Itseez, 2014).

QUADTREE-BASED DEVELOPMENT SUBDIVISION ALGORITHM

Global thresholding using only one threshold per image can be used for ideal pack ice image analysis. Most images are, however, never perfect owing to irregularities in light or color, which may lead to inadequate image analysis results. This problem can be improved by using a local thresholding method of image thresholding followed by subdivision.

The method of image subdivision and frequency are determined via an automatic subdivision algorithm using quadtree. Fig. 1 shows the overall flow diagram of the development algorithm.

First, images are input into quadtree process for histogram analysis and subsequently classified into four groups (#1). If it is classified as a complete one-peak, the binary image is extracted via the one-peak thresholding algorithm and reflected in the accumulated binary image (#2). Alternatively, if it is classified as a complete two-peak, the binary image is extracted and reflected via the two-peak thresholding algorithm (#3).

An image classified as an incomplete one-peak is subjected to a quadtree subdivision process if it satisfies the minimum pixel number requirement for analysis (#4). Hence, four sub-images are generated and subsequently re-input for histogram analysis (#5). Additionally, for the image classified as two incomplete peaks, if the number of images is larger than the minimum number of pixels, the quadtree data is extracted and re-input (#6). At this point, after excluding the inaccurate image, a binary image of accurate image is extracted and reflected via a two-peak thresholding algorithm (#7).

Since the developed algorithm follows the recursive flow via peak analysis of the histogram, there is a high probability that the obtained peak is inaccurate. This inaccuracy is induced because the information needed to generate the peak is insufficient in images in which the number of pixels is below a certain value. Therefore, it is necessary to set the minimum pixel reference. In this study, it is set to the result of analyzing a large number of images with errors. Additionally, images with pixels below this number are excluded from the analysis.

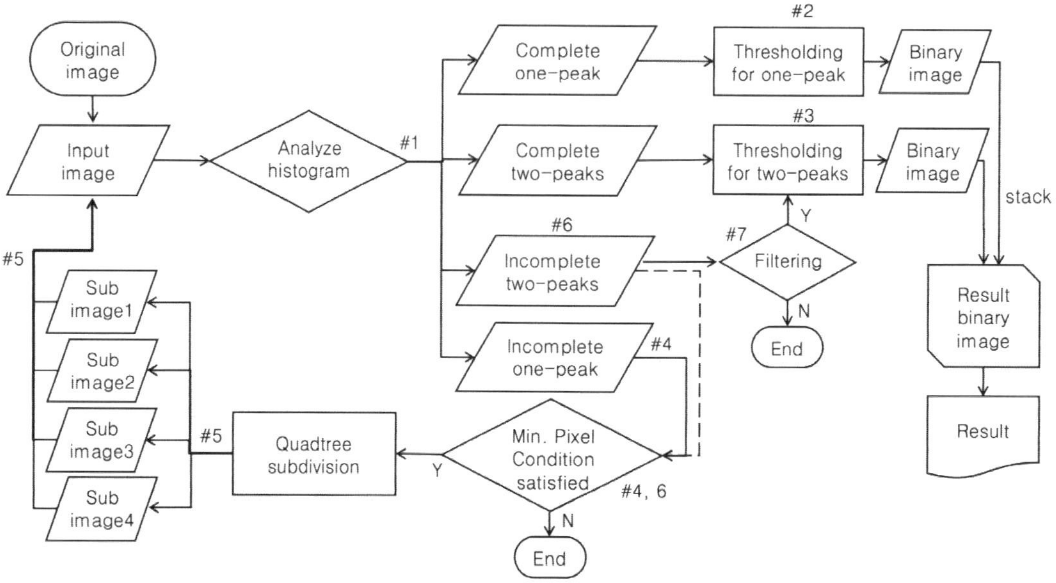

Figure 1. The overall flow diagram of the development algorithm

DETERMINING INDIVIDUAL AREA DISTRIBUTIONS USING WATERSHED OBJECT SEGMENTATION

Fig. 2 illustrates the watershed algorithm for object segmentation. The binary image is extracted via thresholding of the original input image. The edge and seed outline images are derived from the binary image via the edge detection and the seed outline extraction. The marker image merging these images and the original image are input as initial values of the watershed algorithm. Lastly the individual object segmentation images are derived.

This study utilizes a quadtree-based local thresholding development algorithm (#8), which minimizes the effects of color difference and light non-uniformity on the pack ice image. This enables the derivation of an optimized binary image that can distinguish between the pack ice and water. For the edge detection (#9), the Laplacian edge detection technique is utilized. This technique utilizes two-dimensional image differentials and is suitable for extracting pack ice edges with bidirectional distributions. Fig. 4 demonstrates the difference between the edges extracted in the first derivative and the second derivative of the original image. In the first derivative, only one directional edge is extracted, while in the second derivative both directional edges are extracted.

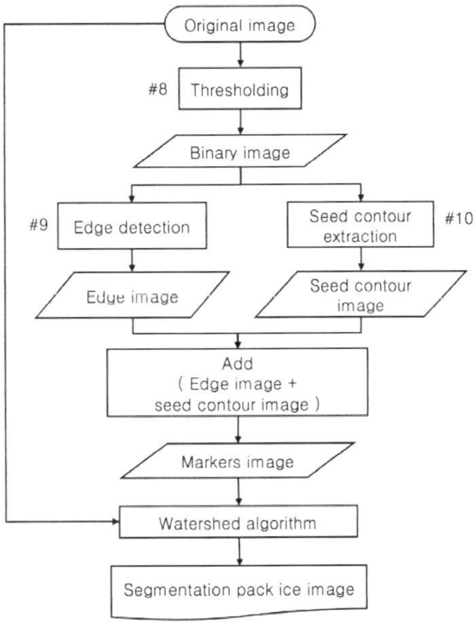

Figure 2. Watershed segmentation algorithm

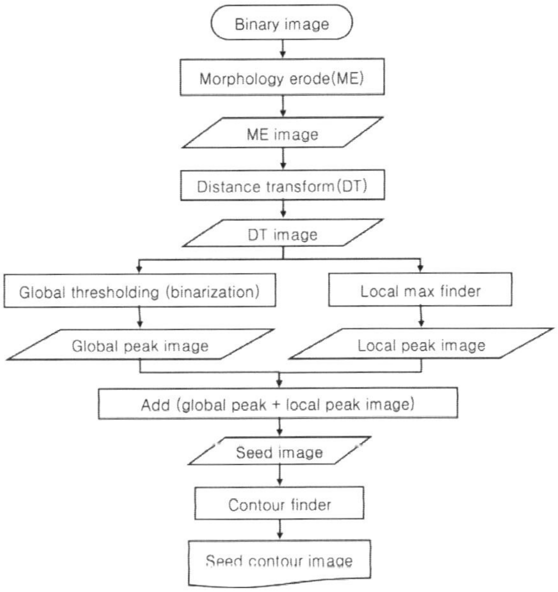

Figure 3. Seed contour extraction algorithm

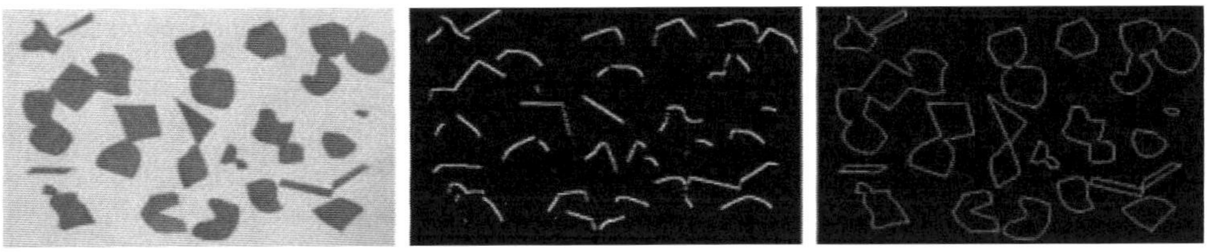

Figure 4. First (mid) and second derivative (right) edge of the original image (left)

Fig. 3 shows the seed contour extraction algorithm (#10). In this study, we applied morphology erosion, which reduces the white area representing the pack ice in the binary image, to enable clear distinguishability of the closely distributed pack ice. This method is applied to the distance transform. Fig. 5 depicts the before and after application of morphology erosion, where the proximity pack ice is distinctly separated after application. Fig. 6 shows the image obtained after applying the distance transform to the original image of Fig. 4.

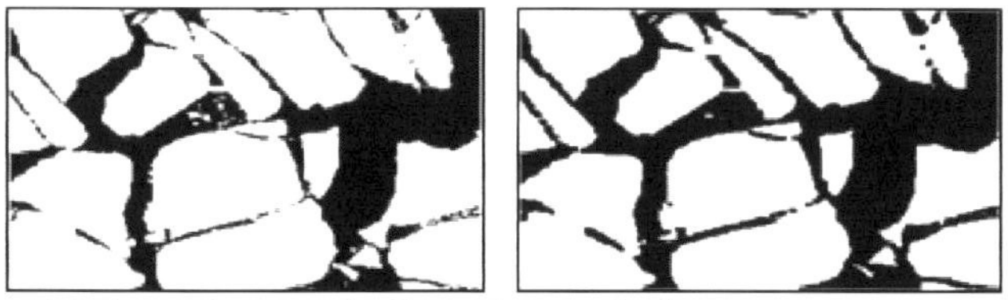

Figure 5. Binary image (left) and morphology erode image (right)

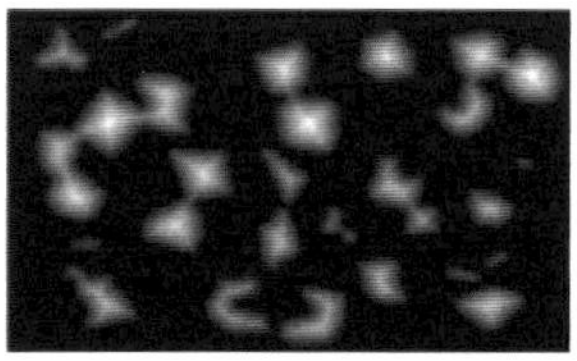

Figure 6. Example of distance transform

In general, global thresholding is used to extract only the global peaks located a sufficient distance from the outside. This method is only suitable for objects of a certain size. Since the size distribution of pack ice varies substantially, this study additionally incorporates a local maximum finder so that objects can be recognized regardless of size. Fig. 7 compares the segmentation between the global thresholding and the local thresholding, showing the segmentation of the original, via local thresholding, and global thresholding, respectively. Regarding distinguishability in global thresholding, small objects are not recognized. Conversely, local thresholding accurately recognizes small objects.

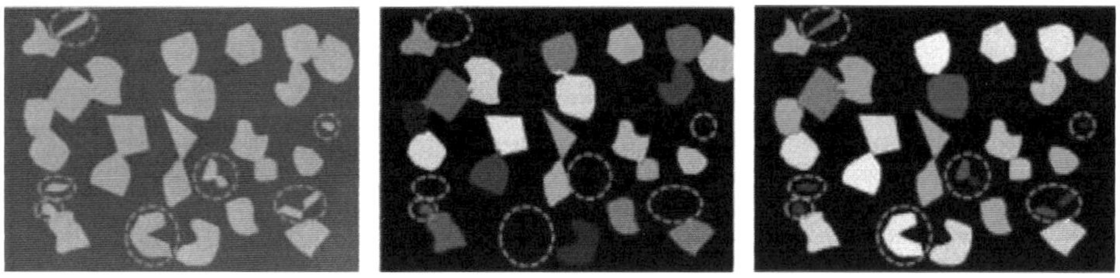

Figure 7. Images of original (left), segmentation with global (mid), and local thresholding (right)

Lastly, the local peak region is extracted and merged with the global peak to define the seed image. The seed outline image is thus extracted via the contour finder (Suzuki, 1985).

EXAMPLE

To verify the results of the local thresholding algorithm via quadtree, we compared the results to the global thresholding results. Fig. 8 shows the original and global thresholding results, in addition to the results obtained after applying the developed algorithm. The global thresholding results show that the lighting intensity of the right portion is not sufficient to detect pack ice. The developed algorithm, however, is observed to adequately handle this problem. Fig. 9 shows the area distributions of each of the actual pack ice formations. This figure provides the original image with color labeling and a number chart according by size.

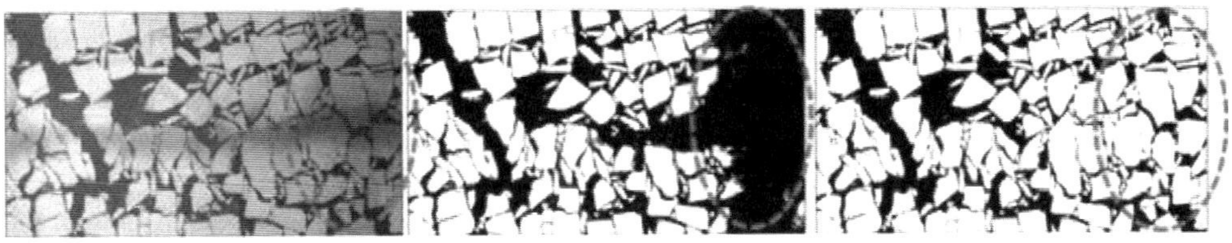

Figure 8. Original image (left), result of global thresholding (mid) and result of developed algorithm (right)

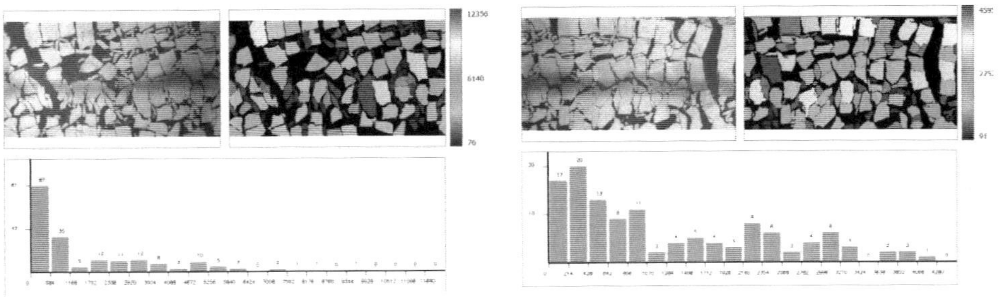

Figure 9. Example of individual pack ice extraction and area distribution chart

CONCLUSIONS

A localized thresholding algorithm using quadtree was developed to measure the pack ice concentration by distinguishing between pack ice and water. Additionally, the watershed object segmentation algorithm was modified to include pack ice extraction, yielding individual area distribution of pack ice.

Problems that may arise due to light non-uniformity or color difference are solved via local thresholding. The local thresholding method entails automatic extraction of small quadtree images and subsequent definition of each image into four categories according to the histogram results. It is confirmed that the development algorithm yields accurate binary results over the entire image independent of light irregularity or color difference.

The watershed algorithm is used to improve the sensitivity caused by the initial value inputs for pack ice extraction. For watershed-applied initial values, the edge was extracted via a Laplacian edge extraction technique that may be suitable for pack ice edge extraction of non-uniform shapes. Morphology erosion was applied to obtain clear distinction between the closely pack ice formations. We also added a local maximum finder to enable recognition of all sizes of pack ice.

Further research is needed to effectively apply an orthometric correction algorithm that is able to compensate for angle changes in angular images.

ACKNOWLEDGEMENTS

The part of this work has been supported by the projects "visualization of pack ice using image processing technique," and "specialized university of offshore plant."

REFERENCES

Byun, S.H., Nam, J-.H., Woo, J., & Lee.S-.J., 2016. An algorithm for analysis of pack ice concentration using image processing technique. *The 31st International Symposium on Okhotsk Sea & Sea Ice*, Mombetsu, Hokkaido, Japan, 21-24 February 2016, pp.17-20.

Cho, S.R., Jeong, S.Y. & Lee, S.S., 2013. Development of effective model test in pack ice conditions of square-type ice model basin. *Ocean Engineering*, 67, pp.35-44.

Itseez, 2014. *OpenCV 2.4.11.0 documentation*. Sphinx: United States.

Lee. J.H., Byun. S.H., Nam, J-.H. & Cho, S.R., 2017. An algorithm for measurement of pack ice concentration using localized binarization of quadtree-subdivided image. *Journal of the Society of Naval Architects of Korea*, 54(1), pp.49-56. (in Korean)

Rees, W.G., 2005. *Remote Sensing of Snow and Ice.*

Suzuki, S. & Abe, K., 1985. Topological Structural Analysis of Digitized Binary Images by Border Following. *Computer Vision Graphics and Image Processing*, 30(1), pp 32-46.

Zhang, Q., Skjetne, R., Metrikin, I. & Løset, S., 2015. Image processing for ice floe analyses in broken-ice model testing. *Cold Regions Science and Technology*, 111, pp.27-38.

POAC'17
Busan, KOREA

Proceedings of the 24th International Conference on
Port and Ocean Engineering under Arctic Conditions
June 11-16, 2017, Busan, Korea

Fatigue Damage Calculation for Ship Hulls Operating in Pack Ice

Yue Han, Junji Sawamura[1]
[1] Naval Architecture and Ocean Engineering, Osaka University, Osaka, Japan.

ABSTRACT

Fatigue damage induced by ice loads is an important issue for ships operating in ice-covered waters. This paper shows the fatigue damage calculation of a ship structure navigating in pack ice fields due to ice loads based on a numerical simulation. A discrete element method (DEM) is employed to simulate the interaction between drifting ice floes and a moving ship. The simulation domain contains hundreds of circular ice floes with random size and distribution. A Weibull model can be used to describe the ice load peaks. The cumulative density functions of ice load peaks in different ice conditions (ice thickness and ice concentration) can be determined from a series of numerical simulation. The structural fatigue stress is determined using structural beam theory. According to the joint probability distribution of ice thickness and ice concentration and a proper S-N curve, fatigue damage can be estimated based on the Palmgren–Miner's rule. An example of fatigue damage calculation is presented in this paper, and the calculated result reflects the fatigue damage due to ice-induced loads in pack ice is rather small.

KEY WORDS: Fatigue damage; Pack ice; Numerical simulation; Ice loads.

INTRODUCTION

Fatigue damage can be important for ships operating in ice-covered waters due to the harsh environment, which may lead to oil leaks, or even cause a catastrophic failure, threatening the overall safety of the structures. However, the research of fatigue damage due to ice action has not been developed well compared with wave action. At present, most studies on fatigue damage due to ice-induced loads are conducted by field measurements. Zhang and Bridges (2011) introduced a deterministic fatigue assessment procedure, the Ship Right FDA ICE Procedure, proposed by Lloyd's Register to assess fatigue damage of ship structure induced by ice loads. Suyuthi. A, et al. (2013) derived closed form expressions of the fatigue damage for several different statistical models of the stress amplitudes. However, the field measurements are usually quite limited and incomplete, so it is difficult to evaluate the fatigue damage correctly and to provide a guide for the design of new structural components or new ship routes. Compared to the field measurements, the ice conditions and ship hull can be easily varied in the numerical simulation, which can be used to complement the lack of data in some regions, or to predict the fatigue life for new structures. The numerical method seems quite promising to evaluate the fatigue damage.

Ships navigating in ice-covered waters can encounter a wide range of different ice conditions, including pack ice, level ice, ridged ice, ice in wave and so on. Therefore, the cases of fatigue damage calculation should also contain all kinds of ice conditions. In the authors' previous research (Han and Sawamura, 2016), fatigue damage calculation of ship operating in level ice has been proposed based on a numerical simulation. While in certain cases, such as managed ice fields and marginal ice zones, where a continuous ice sheet has been broken into smaller ice floes by icebreakers or wave actions, ships mainly operate in broken ice fields. In broken ice, occasional ship-ice collisions may occur, rather than continuous icebreaking process in level ice. Therefore, the fatigue damage in these pack ice cases will be studied in this paper.

Several numerical methods have been developed to simulate the interaction between ships and pack ice floes. Hansen and Løset (1999) proposed a DEM model for theoretical investigation of behavior of a mooring turret in broken ice. Daley et al. (2012) used a GPU-Event-Mechanics (GEM) approach to assess vessel performance in pack ice, in which the ice floes were all represented as convex polygons. Ji et al. (2012) modelled the ice floes with three-dimensional (3D) dilated disk elements, and the ship hull was modelled with 3D disks with overlaps.

In this paper, a 2D DEM numerical model is developed, with the ice floes represented as hundreds of circular disks, to produce ice loads data for fatigue damage calculation, instead of field measurement data. The ice loads in different ice conditions (ice thickness and ice concentration) can be determined from a series of numerical simulation. A Weibull statistical model is applied to represent the ice load process. Based on the structural beam theory, the distribution of ice load peaks can be converted into the distribution of stress amplitudes. The Palmgren-Miner cumulative damage rule is applied to calculate the fatigue damage. The calculated fatigue damage in broken ice is rather small, compared to the one in level ice.

MODELLING OF ICE-SHIP INTERACTION

Motion Equations

Motion of the Ship

Motion equations of ship are described by the equations of motion in three degrees of freedom and can be written as below:

$$(\mathbf{M}+\mathbf{A})\ddot{\mathbf{r}}(t)+\mathbf{B}\dot{\mathbf{r}}(t)+\mathbf{C}\mathbf{r}(t)=\mathbf{F}(t) \tag{1}$$

where **M**, **A**, **B**, and **C** are the mass, added mass, damping and restoring force matrices respectively, **F** is the excitation forces and moments, **r** is the displacement vector of ship. The damping term is not included in this simulation.

According to Newmark method and an assumption of linear acceleration (Su et al., 2011), three equations can be derived from the motion equation:

$$\dot{\mathbf{r}}(t_{k+1})=\dot{\mathbf{r}}(t_k)+\frac{1}{2}\ddot{\mathbf{r}}(t_k)\delta t+\frac{1}{2}\ddot{\mathbf{r}}(t_{k+1})\delta t \tag{2}$$

$$\mathbf{r}(t_{k+1}) = \mathbf{r}(t_k) + \dot{\mathbf{r}}(t_k)\delta t + \frac{1}{3}\ddot{\mathbf{r}}(t_k)\delta t^2 + \frac{1}{6}\ddot{\mathbf{r}}(t_{k+1})\delta t^2 \tag{3}$$

$$\mathbf{r}(t_{k+1}) = (\frac{6}{\delta t^2}(\mathbf{M}+\mathbf{A})+\mathbf{C})^{-1}(\mathbf{F}(t_{k+1})+(\mathbf{M}+\mathbf{A})(\frac{6}{\delta t^2}\mathbf{r}(t_k)+\frac{6}{\delta t}\dot{\mathbf{r}}(t_k)+2\ddot{\mathbf{r}}(t_k))) \tag{4}$$

where δt is the time interval of numerical integration, t_k represents the *kth* time step. The excitation forces and moments can be decomposed into four components:

$$\mathbf{F} = \mathbf{F}^{\text{ice}} + \mathbf{F}^{\text{p}} + \mathbf{F}^{\text{ow}} + \mathbf{F}^{\text{Euler}} \tag{5}$$

where the superscripts 'ice', 'p' , 'ow' and 'Euler' refer to ice-ship collision, propeller and rudder, open water, and a fictitious force induced by a non-uniformly rotating body-fixed frame respectively.

Motion of Ice Floes

Ice floes in this paper are modeled as hundreds of circular discs with random size and distribution. In each time step, the motion of every ice floe needs to be solved. The motions of ice floes follow the Newton's second law, and are solved by the assumption of linear acceleration as well. For the ice floes in the simulation domain, besides the ice-ice collision force and ship-ice collision force, they are also subjected to the water drag force. The added mass of ice floes should be taken into account (Ji et al., 2012):

$$M_{\text{a}} = C_{\text{m}}\rho_{\text{w}}V_{\text{sub}}\frac{d(|\boldsymbol{V}_i - \boldsymbol{V}_w|)}{dt} \tag{6}$$

where M_a is the added mass of floe ice, C_m is the added mass coefficient, ρ_w is the water density, V_{sub} is the submerged area of the floe, $\boldsymbol{V}_i$ and $\boldsymbol{V}_w$ are the velocity vectors of ice floe and water respectively. Here, for a circular disc the added mass is not included in the calculation of the moment of inertia, and it is also assumed that the water drag force act through the center of the disc and do hence not contribute to the torque of the disc floes.

Ice-Ice Contact Force and Ship-Ice Contact Force

The contact between two circular ice floes is presented in Figure 1. It is supposed that there are two components of contact force on the contact zone, i.e. the normal force and the tangential force. The **n-t** coordinate system with the origin O_i is defined as below:

$$\vec{n} = \frac{\overrightarrow{O_iO_j}}{|O_iO_j|} = (\cos\theta, \sin\theta) \tag{7}$$

$$\vec{t} = (-\sin\theta, \cos\theta) \tag{8}$$

where O_i and O_j are the centers of the two contact discs, $\vec{n}$ and $\vec{t}$ are the unit vectors of the normal axis and tangential axis respectively, θ is the angle from x-axis to the vector $\vec{n}$.

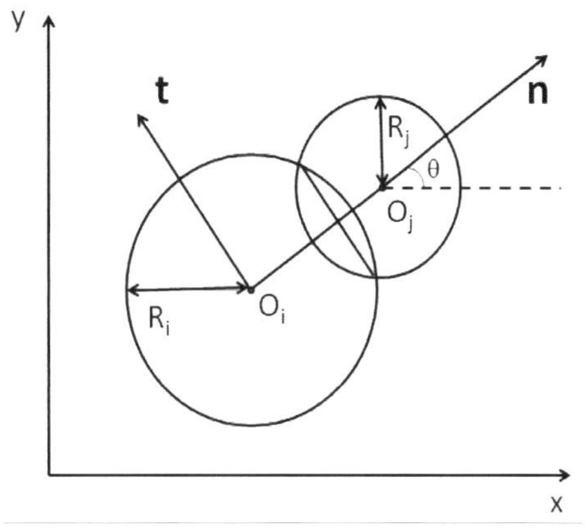

Figure 1. Collision between two circular ice floes.

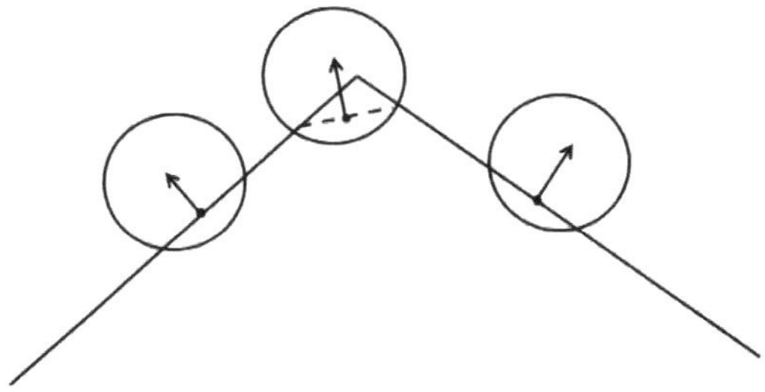

Figure 2. Collision between ship and ice floe.

The normal force is represented as a sum of elastic and damping terms:

$$F_{n}^{i} = -K_{ne}\delta_{n}^{i} - K_{nv}\dot{\delta}_{n}^{i} \qquad (9)$$

where the subscript n denotes the normal direction, the superscript i denotes the current time step, K_{ne} is the normal contact stiffness, K_{nv} is the normal contact viscosity, δ_n^i is the normal indentation of overlap, $\dot{\delta}_n^i$ is the relative velocity of the two disks at normal direction.

The tangential force is treated as linear-elastic:

$$F_{t}^{i*} = F_{t}^{i-1} - K_{te}\dot{\delta}_{t}^{i}\Delta t \qquad (10)$$

where the subscript t denotes the tangential direction, the superscript i-1 denotes the previous time step, K_{te} is the tangential contact stiffness, $\dot{\delta}_t^i$ is the relative velocity of the two disks at tangential direction, Δt is the time interval.

However, the upper limit of the tangential force is the Coulomb friction limit, so the tangential force can be expressed as:

$$F_{t}^{i} = \min(F_{t}^{i*}, \mathrm{sign}(F_{t}^{i*})\mu F_{n}^{i}) \qquad (11)$$

where μ is the ice–ice friction coefficient.

For ship-ice collision, the calculation method of the contact force is similar to the one of ice-ice collision. The ship waterline is represented as a polygon, including nodes and line segments. In the simulation, each segment has to be checked for contact with the discs. A case of collision between ship and ice floe is illustrated in Figure 2. In this paper, the middle point of the contact line is the reference contact point, and the contact normal direction is defined as perpendicular to the line that passes through the two intersecting points between the ship waterline and ice floe (Feng and Owen, 2004), by which no directional jump occurs at the corner when the ice floe continuously moves from the left position to the right. The total contact forces acting on ship hull and each ice floe are calculated as a sum of contact force induced by all the ship-ice collisions and ice-ice collisions.

FATIGUE DAMAGE CALCULATION

During operation in pack ice fields, a ship is expected to travel in a range of different stationary ice conditions (ice thickness and ice concentration), so the total fatigue damage D can be calculated by accumulating the fatigue damage contributions D_j in each stationary condition.

Ice-induced Loads Distribution

In each stationary ice condition, the time series of ice loads can be obtained from the numerical simulation. Generally, the distribution of load peaks may be approximately described by a Weibull distribution based on statistical analysis of field measurements data of ice loads. Its cumulative density function can be expressed as:

$$F(x) = 1 - \exp\left\{ (\frac{x}{\theta})^k \right\} \tag{12}$$

where θ is the scale parameter, k is the shape parameter of ice loads distribution. In order to determine the proper values for the parameters of Weibull distribution which is underlying the ice loads process, a probability paper can be employed.

Prior to the probability plotting, a proper separator is chosen and a Rayleigh separation is applied to identify the load peak values x_1, x_2,$\cdots$, x_n (Kujala et al., 2009). According to the empirical cumulative distribution function, the parameters of Weibull distribution can be estimated by fitting by means of the least square method.

Structural Response

The structural fatigue stress, which is induced by the ice loads, is determined using structural beam theory. As to transverse frame, the conversion from loads into the stress is a sort of linear transformation. Referring to Finnish Swedish Ice Class Rules (2010), the relationship between loads and stress is expressed as:

$$S = \gamma P_{\text{ice}} = \frac{P_{\text{ice}} s l}{m_{\text{t}} Z} \times 10^3 \tag{13}$$

where P_{ice} is the ice loads [kN/m], s is frame spacing [m], l is span of the frame [m] , Z is the section modulus [cm^3].

$$m_{\text{t}} = \frac{7 m_0}{7 - 5h/l} \tag{14}$$

where h is height of load area [m], m_0 takes the boundary conditions into account. The values of m_0 are given in the following table.

From the linear transformation, we can conclude that the ice-induced stress also follows the Weibull distribution, and the shape parameter p and the scale parameter q of stress distribution can be determined by $p = k$ and $q = \gamma\theta$.

Table 1. Boundary Conditions of the structural beam (Finnish Swedish Ice Class Rules, 2010).

Boundary Condition	m_0	Example
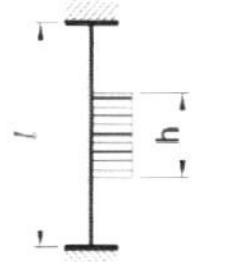	7.0	Frames in a bulk carrier with top wing tanks
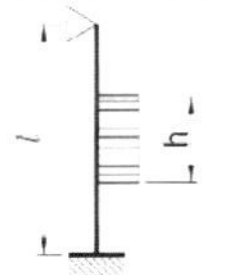	6.0	Frames extending from the tank top to a single deck
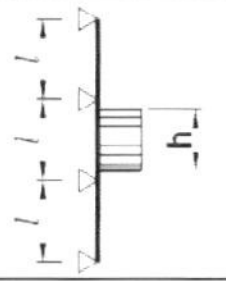	5.7	Continuous frames between several decks or stringers
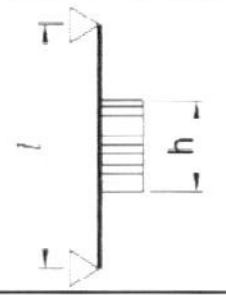	5.0	Frames extending between two decks only

Distribution of Ice Thickness and Ice Concentration

The thickness and concentration of the ice cover is highly variable, due to thermal and mechanical factors. In this paper, we use the ice conditions data, i.e. the mean value and standard deviation of ice thickness and ice concentration of Weddell region in Antarctic Sea as a calculation example (Worby et al., 2008). Worby et al. (2008) provides the histogram of ice thickness distribution, which is quite closed to lognormal model. Therefore, this study represents the ice thickness with a lognormal distribution. Worby et al. (2008) gives the mean value and standard deviation of ice concentration, but does not provide the statistic distribution. Here, the distribution of ice concentration is also assumed to follow a lognormal distribution in the calculation. A random variable of ice thickness or concentration is denoted as X, which follows a lognormal distribution, then $Y = ln(X)$ follows a normal distribution, i.e. $Y \sim N(\mu, \sigma)$. The relationship between the mean value and the variance of X and Y is:

$$E(X) = e^{\mu+\sigma^2/2} \tag{15}$$

$$D(X) = (e^{\sigma^2} - 1)e^{2\mu+\sigma^2} \tag{16}$$

The logarithmic function is a monotonic function, therefore the possibility of ice thickness or concentration $P(x)$ equals to the possibility of their logarithmic value, which can be determined by integration of the probability density function of the normal distribution between $ln(x-\Delta x/2)$ and $ln(x+\Delta x/2)$:

$$P_X(x) = P_Y(\ln(x)) = \int_{\ln(x-\frac{\Delta x}{2})}^{\ln(x+\frac{\Delta x}{2})} \frac{1}{\sqrt{2\pi}\sigma} \exp(-\frac{(y-\mu)^2}{2\sigma^2})dy \tag{17}$$

In the calculation of ice thickness probability, Δx is set as 0.1m, and in the calculation of ice concentration probability, Δx is set as a percentage of 10%. It is assumed that the random variables of ice thickness and ice concentration are independent of each other in this paper.

Fatigue Damage Expression

The Palmgren–Miner's linear damage hypothesis is applied for fatigue damage calculation in a particular stationary condition, D_j:

$$D_j = \sum_{i=1}^{n_s} \frac{n_i}{N_i} \tag{18}$$

where n_i is the number of stress amplitudes, N_i is number of amplitudes to failure for a constant stress S_i, n_s is the number of stress magnitudes.

S-N Curve

The relationship between S_i and N_i is given by a S-N curve, which can be expressed as:

$$N_i S_i^m = K \tag{19}$$

where K and m are the constants of S-N curve. The probability of the stress magnitude S_i can be written in the following two forms:

$$P(S_i) = \frac{n_i}{N_0} = f(S_i)\Delta S \tag{20}$$

where N_0 is the total number of stress amplitudes in each stationary condition, $f(S)$ is the probability density function of Weibull distribution of stress amplitudes:

$$f(S) = \frac{p}{q}\left(\frac{S}{q}\right)^{p-1} \exp\left\{-\left(\frac{S}{q}\right)^p\right\} \tag{21}$$

Insert Eqs.(19) , (20) and (21) into Eq.(18), the fatigue damage in a particular stationary condition can be translated as:

$$D_j = \frac{N_0}{K} q^m \Gamma\left(1+\frac{m}{p}\right) \tag{22}$$

where $\Gamma(\cdot)$ is the gamma function.

Impact Frequency

The number of ice impacts N_0 is variable and related to ice thickness and ice concentration. It can be determined by impact frequency per unit distance times sailed distance in each condition. Suyuthi. A, et al. (2013) proposed that the impact frequency in level ice can be estimated based on the inverse of the characteristic length of the ice plate:

$$\upsilon_d = \frac{1852}{13.3617 h_i^{3/4}} \tag{23}$$

where υ_d is the number of events per nautical mile. It should be noted that this estimation of the impact number is applied for the whole ship hull, so the impact frequency should be lower for a frame. The employment of Eq.(23) is over-estimated. The number of ice impacts N_0 in pack ice is determined by the calculated number in level ice times the percentage of ice concentration.

NUMERICAL CALCULATION RESULTS

Numerical Result of Ice Loads

A rectangular simulation domain is used to model the pack ice field with a dimension of 600 m×250 m. The radius of ice floes is set to be in the range from 2m to 10m randomly, with a specified ice concentration. Considering the effect of random position of ice floes on the ice-induced loads on a certain frame, five different random position distributions of ice floes with the same ice concentration are employed here. An example of random calculation domin with ice concentration of 60% is shown in Figure 3. In this paper, periodic boundary conditions are adopted, in case of which the discs leaving the ice domain will be re-introduced on the opposite boundary with their momentum unchanged, so as to ensure the ice concentration in the simulation domain constant. Ship navigates with a constant thrust, and its main characteristics is given in Table 2. Some computational parameters about ice properties are presented in Table 3.

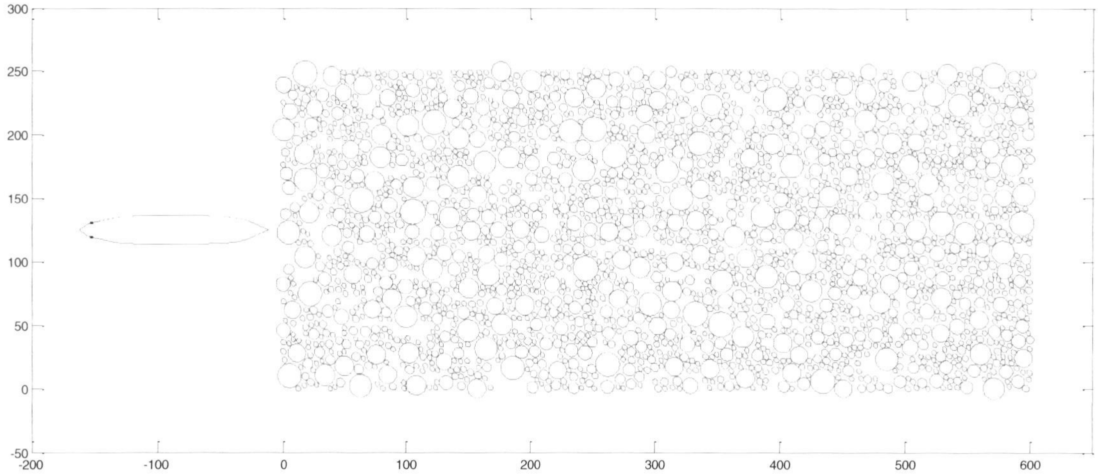

Figure 3. A random calculation domain with ice concentration of 60%

Table 2. Ship Characteristics

Length between perpendiculars	Breadth	Draught	Displacement	Moment of inertia I_{zz}
96.0m	18.4m	5.16m	4890.13t	2.57 ×10^9kg m^2

Table.3 Ice Properties

Normal contact stiffnes s K_{ne}	Tangential contact stiffness K_{te}	Normal contact viscosity K_{nv}	Friction coefficient μ	Added mass coefficient C_m	Water density ρ_w	Ice density ρ_i	Ice concentration Mean(std)	Ice thickness Mean(std)

587 KN/m	352 KN/m	5.87 KN·s/m	0.35	0.15	1000 kg/m^3	900 kg/m^3	38% (36%)	0.58m (0.55m)

An example of the time series of ice resistance obtained from numerical simulation is shown in Figure 4. The ice-induced load looks like a sequence of spikes, and the maximum ice load in this case is 1167 KN. Ji et al. (2012) simulated the ice loads of ship operating with a speed of 4.0 m/s under ice concentration of 60%, and the numerical maximum value was 1479KN. Daley et al. (2012) calculated the ice loads of a vessel with constant thrust under 35% ice coverage, and the simulated maximum value was about 1500KN. Compared to numerical results of ice loads in pack ice in these reference papers, the simulation values in this paper are in the same order of magnitude.

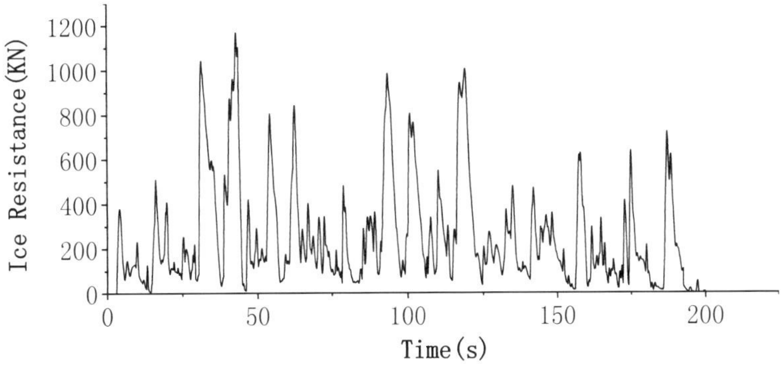

Figure 4. Example of the time history of ice load peaks calculated by the numerical simulation (Ice thickness = 0.5m, Ice concentration = 40%)

The ice load peaks on a specified frame are plotted in the Weibull distribution, as shown in Figure 5. The cumulative distributions of ice load peaks area plotted as a functions of the load level in logarithmic axis [ln(x)] and cumulative occurrence probability in twice logarithmic axis [-ln(-ln(1-F))]. The peak values forms nearly a straight line is observed, which means ice load peaks of the numerical simulation fits the Weibull distribution well.

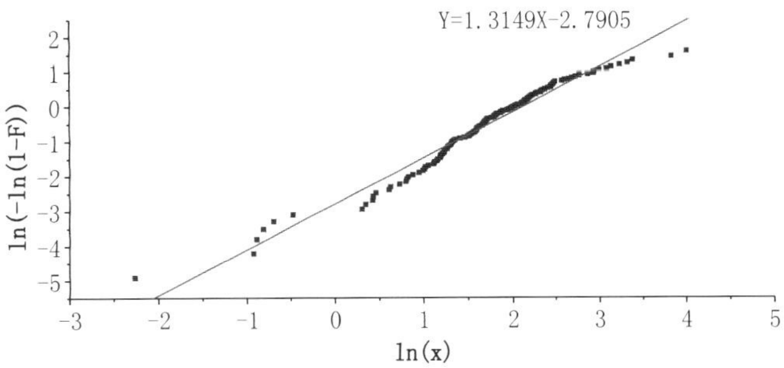

Figure 5. Example of the calculated local load peak distribution on a certain frame and fitted line of Weibull distribution (Ice thickness = 0.5m, Ice concentration = 40%).

Because nearly all the statistical results of ice-induced loads come from the field measurement data, and the field experiments contain all kinds of different ice conditions, including pack ice, level ice, ridged ice, etc. However, the ice condition studied in this paper is only pack ice, therefore, it is unreasonable to compare the statistical result of ice loads in this numerical simulation to those results in field measurements. Here, the global ice loads around the whole ship hull are plotted in the Weibull distribution as well, shown in Figure 6, in order to compare with the statistical result of Daley et al. (2014), which also focuses the study on pack ice, but with the ice floes represented as convex polygons. The case of Daley et al. (2014) is 40% ice coverage with different combination of ice thickness. The statistical result shows the shape parameter mainly concentrates in the range between 0.5 and 0.6. From the comparison, the shape parameter, i.e. the slope of the fitted line in this paper is 0.5929 (Ice thickness = 0.5m, Ice concentration = 40%), which is in the same range.

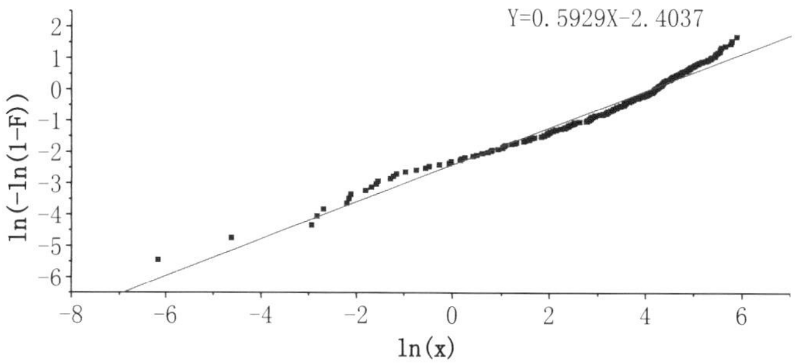

Figure 6. Example of the calculated global load peak distribution on a certain frame and fitted line of Weibull distribution (Ice thickness = 0.5m, Ice concentration = 40%).

Fatigue Calculation Result

The fatigue damage accumulates D_j in ice thickness from 0.2m to 1.0m and ice concentration from 10% to 60%. The constants which are needed in the fatigue calculation are shown in Table 4. The travelling distance per year is assumed to be 2500 nm. An example of fatigue damage calculation process for ice concentration 10%, 40% and 60% are presented in Table 5.

Table.4 Fatigue Calculation Constants

Frame Spacing s	Span of Frame l	Section Modulus Z	Height of Load Area h	Boundary Condition m_0	S-N curve parameter, K	S-N curve parameter, m
0.35m	1.5m	267cm^3	1.0m	5.0	1.0E+15.117	4.0

The total fatigue damage D of the calculated frame per year can be calculated by accumulating D_j in different ice conditions (ice thickness and ice concentration). The accumulation value is 1.788×10^{-4}, which is rather small, compared to the value in level ice in the authors' previous study (Han and Sawamura, 2016), which is 5.617×10^{-3}. It may be attributed to two factors: first, in pack ice fields, the distribution of ice floes is quite scattered, which will result in a remarkable decrease in the number of impacts between ship and sea ice. Then ice floes in pack ice field can be pushed away when the ship navigates through, therefore the magnitude of ice-induced loads is relatively lower. And the fatigue results based on numerical simulation seem reasonable preliminarily when compared to the fatigue damage value of Suyuthi. A, et al. (2013), which applied the field measurement data, although the ice data is not exactly the same, as shown in Table 6. The value of Suyuthi. A, et al. (2013) is 5.836×10^{-4}, which is between the above two values in pack ice and level ice based on numerical simulation respectively. It can be explained that the field measurements include all different kinds of ice conditions.

Table.5 Fatigue Damage Calculation

(1) Ice concentration = 10%

h_i	$P(h_i)$	N_0/year	p	q	D_j
0.2	0.158327347	7588.308373	2.6496	9.7441	6.99371E-08
0.3	0.151581688	5366.373432	2.4839	9.6343	5.08826E-08
0.4	0.124953085	3569.966254	1.5500	11.7473	1.88839E-07
0.5	0.098136737	2366.985408	1.8752	11.228	6.52014E-08
0.6	0.075940523	1595.797085	1.9527	10.5157	3.11866E-08
0.7	0.058670606	1099.277135	1.9790	10.328	1.94872E-08
0.8	0.045513197	772.6614357	1.8584	10.3881	1.58917E-08
0.9	0.035538153	551.8158333	1.8754	11.2128	1.51158E-08
1.0	0.027958931	402.2938638	1.4700	11.3122	2.15123E-08

(2) Ice concentration = 40%

h_i	$P(h_i)$	N_0/year	p	q	D_j
0.2	0.158327347	36520.56928	1.2338	8.7038	1.31244E-06
0.3	0.151581688	25826.97001	1.1166	10.5967	3.24644E-06
0.4	0.124953085	17181.326	0.911	9.5782	4.85455E-06
0.5	0.098136737	11391.68973	1.3149	8.3494	2.67554E-07
0.6	0.075940523	7680.159417	1.0518	8.427	5.30043E-07
0.7	0.058670606	5290.537072	1.4227	8.7389	1.12178E-07
0.8	0.045513197	3718.620028	1.1923	10.2582	2.99859E-07
0.9	0.035538153	2655.747155	1.1228	9.74613	2.32318E-07
1.0	0.027958931	1936.136515	1.0891	10.0265	2.21805E-07

(3) Ice concentration = 60%

h_i	$P(h_i)$	N_0/year	p	q	D_j
0.2	0.158327347	12679.82534	0.9966	22.0372	5.59571E-05
0.3	0.151581688	8967.041731	1.0621	19.458	1.66705E-05
0.4	0.124953085	5965.301666	1.0571	18.622	9.54759E-06
0.5	0.098136737	3955.158394	1.056	15.5344	3.08269E-06
0.6	0.075940523	2666.526889	1.068	17.346	3.03701E-06
0.7	0.058670606	1836.857622	1.0399	18.0731	2.8589E-06

0.8	0.045513197	1291.09303	0.936	19.1623	4.85725E-06
0.9	0.035538153	922.066954	0.9284	18.1544	2.95297E-06
1.0	0.027958931	672.2204316	0.9896	17.086	1.11888E-06

Table.6 Comparison of Fatigue Damage.

Published by	Sea are	Ice condition	Duration	Ship type	Fatigue damage
Suyuthi. A, et al. (2013)	Baltic Sea	Real sea ice (Pack ice + level ice + Ridge ice)	1 year (2500 nm)	Ice breaker	5.836×10^{-4}
Han and Sawamura, (2016)	Baltic Sea	Level ice	1 year (2500 nm)	Ice breaker	5.617×10^{-3}
Present paper	Antarctic Sea (Worby et al., 2008)	Pack ice	1 year (2500 nm)	Ice breaker	1.788×10^{-4}

CONCLUSIONS

A fatigue damage calculation method based on a numerical simulation has been introduced. The numerical simulation with a DEM model has been performed to obtain the time series of ice-induced loads in pack ice fields. The numerical result of ice loads has been compared with the existing numerical data preliminarily. Although the simulation model and the ship hull structure are not exactly the same, the magnitude of ice load and shape parameter of statistical Weibull model are quite consistent. The total fatigue damage can be an accumulation of damages in different ice conditions, according to the joint probability distribution of ice thickness and ice concentration and a proper S-N curve. The calculated fatigue result is much smaller than the one in level ice. The ice conditions and ship hull can be easily varied in numerical simulation, compared to field measurement, so the numerical method can be promising to calculate the fatigue damage. In order to evaluate the fatigue damage in real sea ice trial, the numerical method as a combination of simulation models in different ice conditions such as level ice, pack ice, ridge ice and also wave is needed. The proper ice conditions data, S-N data and structural response under ice load are also needed.

ACKNOWLEDGEMENTS

The author Yue Han warmly acknowledges the financial support of China Scholarship Council (CSC) for her PhD research.

REFERENCES

Daley, C., Alawneh, S., Peters, D., Quinton, B., & Colbourne, B., 2012. GPU modeling of ship operations in pack ice, *International Conference and Exhibition on Performance of Ships and Structures in Ice*, pp.20-23.

Daley, C., Alawneh, S., Peters, D., & Colbourne, B., 2014. GPU-event-mechanics evaluation of ice impact load statistics, *OTC Arctic Technology Conference*. Offshore Technology Conference.

Feng, Y. T., & Owen, D. R. J., 2004. A 2D polygon/polygon contact model: algorithmic aspects. *Engineering Computations*, 21(2/3/4), pp.265-277.

Finnish Maritime Administration, 2008. *Ice class regulations 2008 (finnish-swedish ice class rules)*, Bulletin 10 / 10.12.2010.

Han, Y., & Sawamura, J., 2016. Preliminary Fatigue Damage Calculation due to Ice-induced Loads based on Numerical Simulation, *The Japan Society of Naval Architects and Ocean Engineers Conference.*

Hansen, E.H., & Løset, S., 1999. Modelling floating offshore units moored in broken ice: model description. *Cold regions science and technology*, 29(2), pp.97-106.

Ji, S., Li, Z., Li, C., & Shang, J., 2013. Discrete element modeling of ice loads on ship hulls in broken ice fields. *Acta Oceanologica Sinica*, 32(11), pp.50-58.

Kujala, P., Suominen, M., & Riska, K., 2009. Statistics of ice loads measured on MT Uikku in the Baltic, *Proceedings of the International Conference on Port and Ocean Engineering Under Arctic Conditions*, No. PAOC09-51.

Su, B., Riska, K., & Moan, T., 2011. Numerical simulation of local ice loads in uniform and randomly varying ice conditions. *Cold Regions Science and Technology*, 65(2), pp.145-159.

Suyuthi, A., Leira, B. J., & Riska, K., 2013. Fatigue damage of ship hulls due to local ice-induced stresses. *Applied Ocean Research*, 42, pp.87-104.

Worby, A. P., Geiger, C. A., Paget, M. J., Van Woert, M. L., Ackley, S. F., & DeLiberty, T. L., 2008. Thickness distribution of Antarctic sea ice. *Journal of Geophysical Research: Oceans*, 113(C5)

Zhang, S., Bridges, R., & Tong, J., 2011. Fatigue Design Assessment of Ship Structures Induced by Ice Loading-An introduction to the ShipRight FDA ICE Procedure, *The Twenty-first International Offshore and Polar Engineering Conference.*

POAC'17
Busan, KOREA

Proceedings of the 24th International Conference on
Port and Ocean Engineering under Arctic Conditions
June 11-16, 2017, Busan, Korea

Estimation of Local Ice Load by Analyzing Shear Strain Data for the IBRV ARAON

Mincheul Jeon[1], Jung Ki Min[1], Kyungsik Choi[2], Jung Seok Ha[3]
[1] Ocean Science and Technology School, Korea Maritime and Ocean University, Busan, Korea
[2] Department of Ocean Engineering, Korea Maritime and Ocean University, Busan, Korea
[3] Korea Research Institute of Ship and Ocean Engineering, Daejeon, Korea

ABSTRACT

The icebreaking research vessel ARAON performed ice field tests during her 2016 Arctic voyage. The ship is subjected to ice loads through ice-ship interaction processes. Local ice load acting on ARAON's bow section was measured by using stain gauges installed on the inner hull plates and transverse frames of bow section. In this paper the local ice loads at transverse frames estimated from shear strain data were compared to ice loads from hull plate pressures by using the influence coefficient method. In addition to the analysis of local ice loads, the characteristics of peak ice loads with the ship speed is also discussed. It is recommended that the local ice loads estimated by calculating shear forces acting on transverse frames may be useful in estimating local ice loads on the hull of ship.

KEY WORDS: Full-scale Measurement; Local Ice Load; Shear Strain Data; Icebreaking Research Vessel ARAON.

INTRODUCTION

Icebreaking vessels in polar waters are subjected to various environmental loads such as wave loads and slamming forces, however, the largest one that acts on the ship hull is the ice load. To accurately estimate the ice load, the procedures for analyzing ice field data have been studied in various ways, but so far they were not defined well clearly (Cheon et al., 2014).

The Korean icebreaking research vessel, ARAON, has had the systematic ice field measurement program six times in the Antarctic and Arctic seas since 2010. This paper focuses on the last ice field measurement data in 2016. The ship seems to be subjected to mild to moderate ice loads through ice-ship interaction processes in 2016. Like the previous ice field measurements, the local ice load acting on ARAON's bow section was measured by using stain gauges installed on the inner hull plates and transverse frames of bow section. In this paper the local ice loads on the ARAON's hull were estimated from shear strain data recorded at transverse frames.

ARAON's 2016 Arctic voyage was carried out in the Chukchi and East Siberian Seas as shown in Figure 1. Transit route of the ARAON and research stations are marked in the same figure. The deformation of ship hull plating was measured through the strain gauges during entire ice transits. Ice conditions were moderate in general.

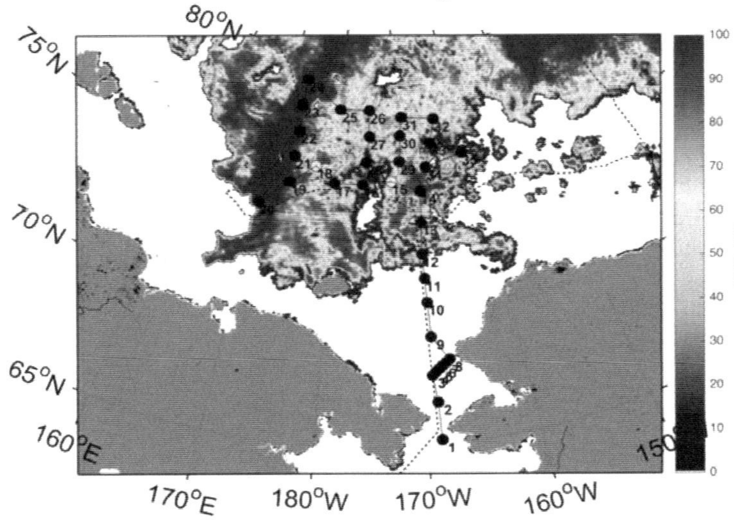

Figure 1. ARAON's voyage to the Chukchi Sea and East Siberian Sea (August/September 2016)

ARRANGEMENT OF STRAIN GAUGES

For the full-scale measurement of local ice load, the ARAON was instrumented with several different sensors which include strain gauges. During 2016 Arctic voyage of the ARAON, strain gauges were installed inside the shell plating and also at some transverse hull frames (FR.102~FR.110 at the port side and FR.106 ~ FR.110 at the starboard side). Lower part of second deck inside the bow thrust room was mostly instrumented. As shown in Figure 2, single axis gauges and 3-axis rosette gauges were used. Red circles are locations of strain gauges for measuring shear strains at the transverse hull frames. The location of strain gauges has been slightly adjusted in 2016 from previously designated location. In the port side, a total of 8 single gauges and 20 rosette gauges were installed inside shell plating and at the transverse frames as shown in the figure. In addition to strain gauges, 5 fiber optic sensors were additionally installed, but the analysis of fiber optic signals is not included in this paper. In the starboard side, only 12 rosette gauges were installed inside shell plating because of limited work space.

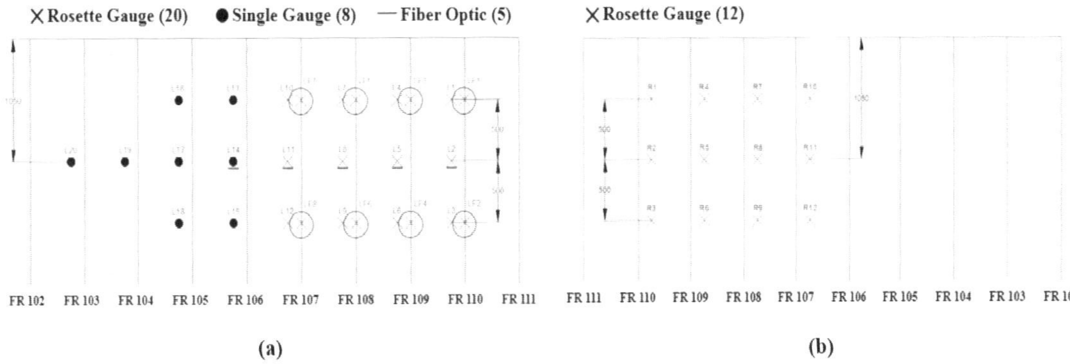

Figure 2. Locations of installed strain gauge (a) port side (b) starboard side

CONCEPT OF AN EVENT AND PEAK ICE LOAD

To understand the characteristics of a ship-ice interaction process, it is useful to adopt the concept of an 'event' from which the peak ice load and collision speed can be defined. This concept is well described in previous study (Min et al., 2016). When a ship collides with large ice feature, the speed of ship sharply decreases as ship penetrates into the ice. Contrary to decreasing ship speed, the ice load increases to a certain peak value and then ice fails. As shown in Figure 3, the time history of local ice load estimated by summing pressures at the strain gauge locations can be divided into a series of events. It is possible to select one peak ice load and one incoming ship speed during each event. The ship speed was recorded from a GPS onboard the ARAON.

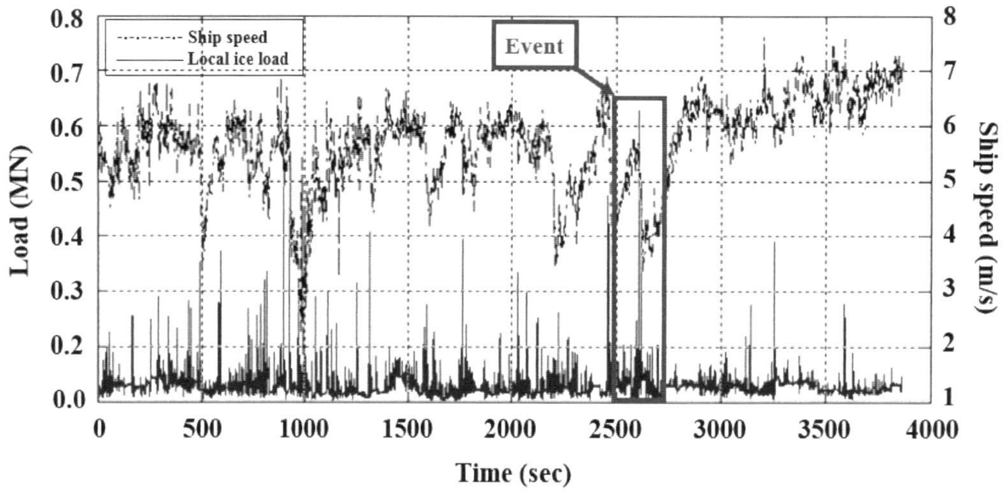

Figure 3. Typical time histories of local ice load and ship speed

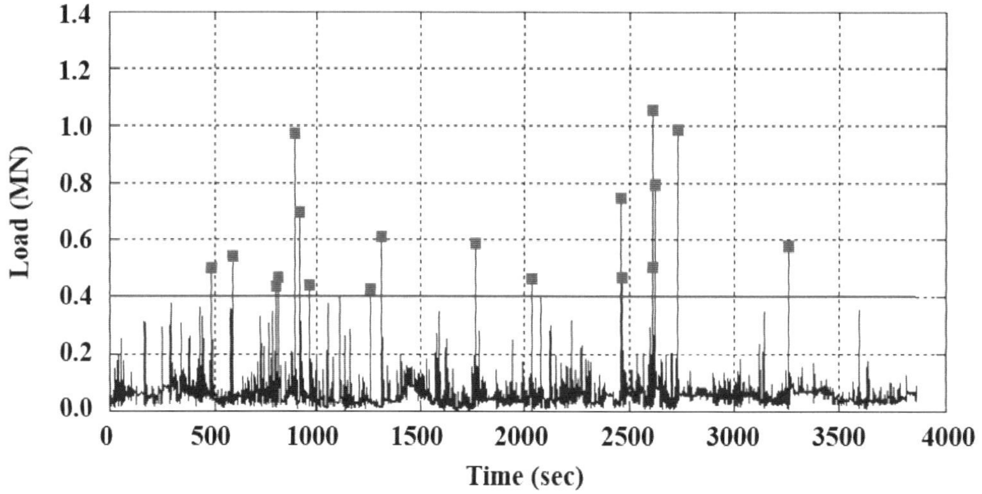

Figure 4. Extraction of peak ice loads and minimum threshold of 0.4MN

To obtain a meaningful data, it is needed to set a threshold for extracting a ship speed and peak ice load from an event. The threshold for peak ice load was set to 0.4MN for 2016 ice load data. The value of 0.4MN seem to be a little high while it is compared to 2015 ice load data (threshold value 0.04MN). The reason for this may be explained by comparing ice conditions in 2015 and 2016 seasons as shown in Figure 5. The speed of ship was selected as a speed of 0.5sec ahead the instance at which the peak ice load is achieved. Figure 4 shows the selected peak ice loads of each event.

(a) **(b)**

Figure 5. Typical ice conditions during (a) 2015 Arctic voyage (b) 2016 Arctic voyage

LOCAL ICE LOADS FROM SHEAR STRAIN DATA IN HULL FRAMES

During 2016 Arctic voyage, as mentioned, local ice load acting on ARAON's bow section was measured by using stain gauges installed on the inner hull plates and transverse frames of bow section. Two different approaches are possible to calculate local ice loads from strain gauge data. The first approach is to estimate ice pressures by analyzing stain gauge data measured on hull plating. The measured strain data were converted to equivalent von-Mises stresses from which ice pressures and local ice load were calculated by using a finite element analysis and an influence matrix. The procedure to calculate ice pressure and local ice loads on ship hull plating were described in previous studies (Choi et al., 2014; Min et al., 2016). An alternative approach is estimating shear forces and local ice loads by analyzing shear strain measurement at transverse frames. This paper focuses on this alternative approach. Local ice loads estimated from the analysis of shear strain data on frames can be compared to that from ice pressures.

Previous researches have reported that ice loads could be estimated by using shear gauges on the hull frame of ships such as CANMAR KIGORIAK, SOYA and S.A. AGULHAS II, but details of their procedures to analyze the data were not released (Ghoneim et al., 1984; Takimoto et al., 2006; Suominen et al., 2013).

In this study, the shear strain data gathered at the locations of frame is used for calculating the shear force along the frame. Two 3- axis rosette gauges were installed at the upper part and at the lower part of each frame. Shear forces can be calculated by the difference of shear strains in the frame. Schematic diagrams for analyzing shear forces on the frame are shown in Figure 6.

Shear strains are calculated from the measurement data using two-dimensional Mohr's circle diagram as in Eq.(1).

$$\gamma_a = -(2\epsilon_b - \epsilon_a - \epsilon_c) \tag{1}$$

Shear stresses are simply calculated by Eq.(2)

$$\tau = G \cdot \gamma_a = \frac{E \cdot \gamma_a}{2(1+\nu)} \tag{2}$$

where E is the elastic modulus, ν Poisson's ratio and γ_a is the measured shear strains. Finally shear forces are calculated by Eq.(3).

$$F_{shear} = k(\tau_1 - \tau_2)A \tag{3}$$

Where τ_1 and τ_2 are shear stresses at the upper and at the lower parts of the frame. A is the shear area of the frame shown in the figure. *k* is a factor that adjusts the shear area depending on shape of the frame. Temporarily 1.0 was used for *k* in this study.

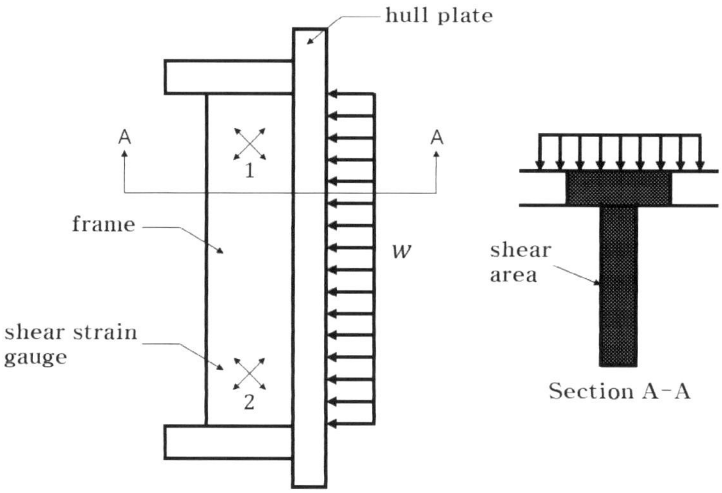

Figure 6. Location of strain gauges on the frame of ARAON's bow section

Table 1. Calculated shear areas

Frame No.	Area(mm^2)
106	23,940
107 ~ 110	15,640

COMPARISON OF LOCAL ICE LOADS BY TWO APPROACHES

In this paper two different approaches to estimate local ice loads from measured strain data were compared. Only the sub-region of the shell plating where shear strains were measured, was considered in the analysis to compare two approaches each other. As mentioned, the shear forces acting on the transverse frames were calculated by using shear strain data and shear stress/force equations. Hull stresses were calculated by using von Mises equivalent stress formula and ice pressures were derived by using the influence coefficient matrix. The local ice load can be calculated by multiplying corresponding areas.

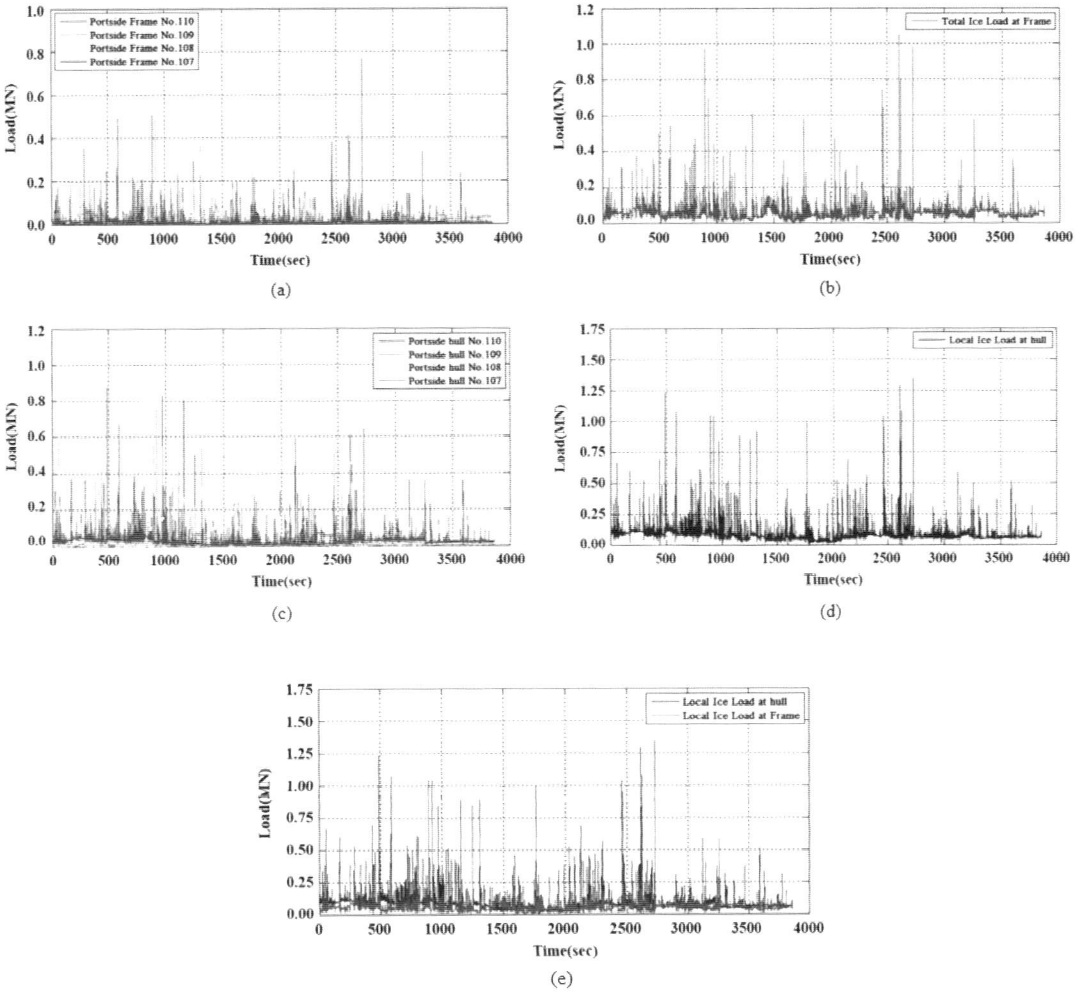

Figure 7. (a) Shear forces acting on each frame (b) Sum of shear forces acting on a frame (c) Ice loads on hull plates corresponding to each frame (d) Sum of ice loads on hull plating (e) Comparison of local ice loads acting on hull plating and transverse frame

A total of 32 data groups were analyzed for estimating the local ice loads from 2016 full-scale measurement data. Figure 7 shows the local ice loads estimated from the hull plating ice pressure and from shear forces along the transvers frames analyzing No.9 data group. The maximum load acting on hull plating is 1.34MN, which is a little higher than the maximum load acting on the frames (1.06MN). Local ice loads calculated by two approaches show a reasonable agreement in general as shown in Figure 7(e).

In order to understand the characteristics of peak ice loads acting on transverse frames and loads acting on hull plating, all the events were analyzed to extract the peak ice load and corresponding ship speed from the 2016 ice load data. Based on the threshold value of 0.4MN, a total of 267 events were found and the maximum value of peak ice loads was 1.45MN at 4.52m/s. By analyzing the events extracted from the local ice loads on hull plating, a total of 871 events were found and the maximum value of peak ice loads was 2.41MN at 4.52m/s. As shown in Figure 8, the peak ice loads acting on the hull plating section are a little higher than those on the transverse frames. The average value of peak ice loads acting on the frame is estimated at 85% of loading on the hull plating. Additionally, the peak ice loads data from the full-scale measurement shows accumulation within a particular speed range (3~7m/s).

The strain gauge data measured from 2015 Arctic voyages were compared to verify whether there is a similar trend of peak ice load vs. ship speed. By analyzing 2015 ice load data, based on the threshold value (0.04MN), a total of 120 events were found. The peak ice loads were distributed within 0.04~0.11MN and 3~7m/s range. Maximum value is 0.11MN at 4.64m/s (see Table 3).

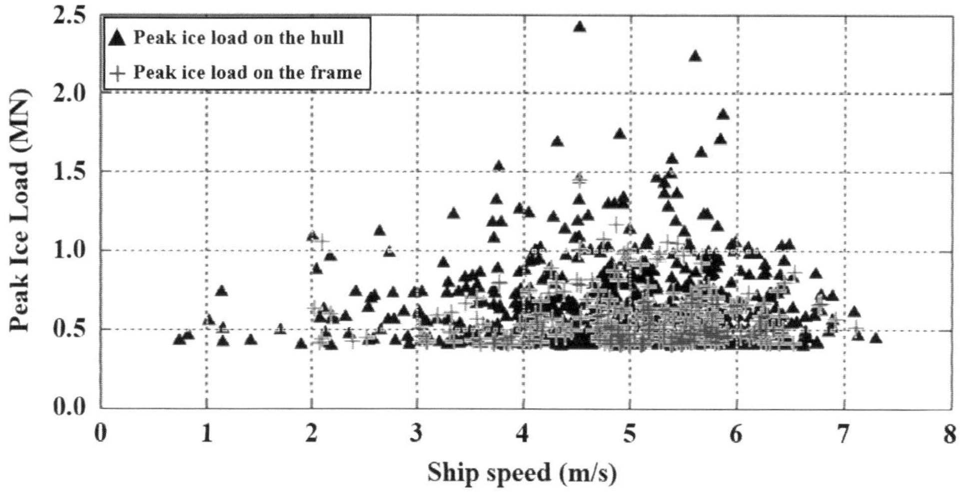

Figure 8. Peak ice loads vs. ship speed from hull plating and from transverse frames for all 32 data groups of 2016 ARAON's Arctic field measurement

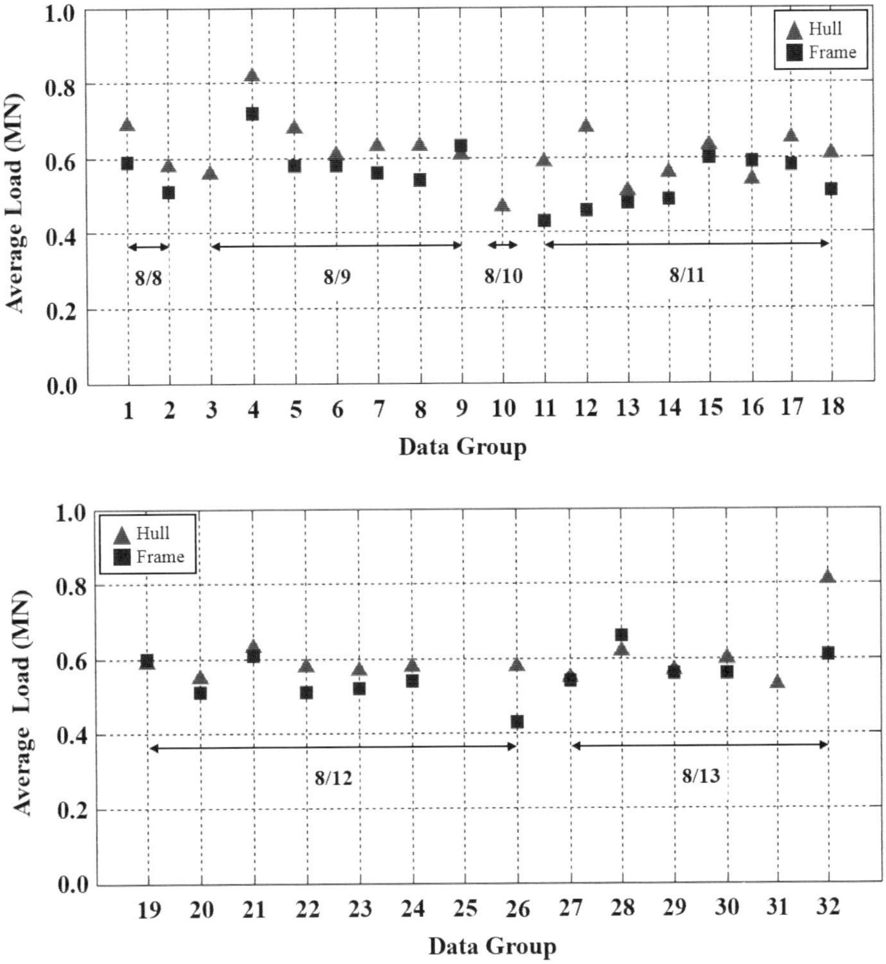

Figure 9. Average of peak ice loads from hull plating and from transverse frames for all 32 data groups of 2016 ARAON's Arctic field measurement

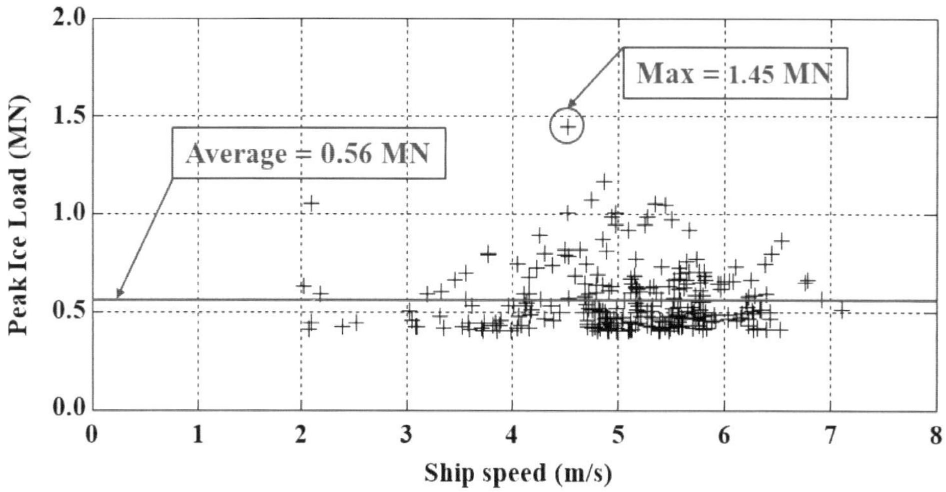

Figure 10. Peak ice loads vs. ship speed for 2016 ARAON's Arctic measurement

Table 3. Comparison of events extracted from shear data in 2015 and 2016 Arctic voyage

	2015 Arctic voyage	2016 Arctic voyage
Number of events	120	267
Average of peak ice loads	0.05MN	0.56MN
Maximum of peak ice loads	0.11MN	1.45MN

CONCLUSIONS

The ARAON had sea ice trails in the Arctic sea during 2016 summer season. This study focuses on estimating the local ice load, which is calculated by shear strains at transverse frames and comparing to local ice load calculated by the influence coefficient matrix acting on the ARAON's hull. In addition, the estimation of local ice load based on 2016 ice load data analyzed, the concept of the event is used for analyzing the relationship between the ship speed and peak ice load.

For the verification of the results, two approaches were used. According to the result of extracted peak ice loads, the peak ice loads on hull plates was a little higher than those of transverse frames. But the average of all peak ice loads on the frame was estimated at 85% of loading on hull plates. The alternate method of estimating local ice loads from shear strain measurement is recommended based on instrumentation cost and simple procedures. Use of strain gauges in the frames can not only estimate local ice loads on small areas but also can estimate global ice load acting on entire ship structure by summing all shear forces in frames.

Results show that peak ice load increases at the high speed. There is a large difference between hull plates and transverse frames in the number of extracted events. However, it is confirmed that there is a similar distribution between the ship speed and peak ice load and the peak ice loads are shown within particular speed ranges (3~7m/s).

ACKNOWLEDGEMENTS

Research funds supported by the Korea Research Institute of Ships & Ocean Engineering (Project No.PES3550) and also supported by Ministry of Trade, Industry and Energy of Korea (Project No.10063417) are greatly acknowledged.

REFERENCES

Cheon, E.J., Choi, K.S., Kim, H.Y. and Lee, T.K., 2014. Analysis of strain gauge data onboard the IBRV ARAON during icebreaking voyage in the Antarctic sea ice. *Journal of the Society of Naval Architecture of Korea*, 51(6), pp.489-494.

Choi, K., Cheon, E.J., Lee, T.K. and Kim, H.S., 2014, Analysis of ice load on the IBRV ARAON during icebreaking performance tests in the Arctic and Antarctic sea ice, *Proceedings of the 22nd IAHR International Symposium on Ice,* Singapore, Paper No.1115.

Ghoneim, G.A.M., Johansson, B.M., Smyth, M.W. and Grinstead, J., 1984. Global ship ice impact forces determined from full-scale tests and analytical modeling of the icebreakers Canmar Kigoriak and Robert LeMeur. *Society of Naval Architects and Marine Engineers Transactions*, 92, pp.253-282.

Jeong, S.Y., Choi, K. and Cheon, E.J., 2015. An analysis of characteristics of ice load distribution on model ship due to ship and ice interaction. *Journal of the Society of Naval Architecture of Korea*, 52(6), pp.603-610.

Min, J.K., Choi, K., Cheon, E.J. and Kim, J.M., 2016. Ice load estimation procedures for IBRV ARAON by analyzing shear strain data measured in Arctic sea. *Journal of Ocean Engineering and Technology*, 30(6), pp.468-473.

Suominen, M. et al., 2013. Full-scale measurements on board PSRV S.A. AGULHAS II in the Baltic Sea. *Proceedings of 22th International Conference on Port and Ocean Engineering under Arctic Conditions*, POAC13-148, Espoo, Finland.

Takimoto, T. et al., 2006. Measurement of ice load exerted on the hull of icebreaker SOYA in the southern Sea of Okhotsk. *Proceedings of the 18th IAHR International Symposium on Ice,* Sapporo, Japan, pp.245-262.

Initial results of a study into the estimation of the development of frequency lock-in for offshore structures subjected to ice loading

Hayo Hendrikse[1,4], Marc Seidel[2], Andrei Metrikine[1,4], Sveinung Løset[3,4]
[1] Delft University of Technology, Delft, the Netherlands
[2] Siemens Wind Power GmbH & Co. KG, Germany
[3] NTNU, Trondheim, Norway
[4] Sustainable Arctic Marine and Coastal Technology (SAMCoT), Trondheim, Norway

ABSTRACT

Ice-induced vibrations have to be considered in design of vertically sided offshore structures subjected to loading by sea ice, such as offshore wind turbines and oil- and gas platforms. The interaction between ice and structure may result in high global peak loads and the occurring structural vibrations can contribute significantly to the overall fatigue of the structure. A regime of particular interest is the frequency lock-in regime in which the interaction causes the structure to oscillate at high amplitude with a frequency close to one of its natural frequencies. Assessment of frequency lock-in in the design phase can be done based on simple approaches once for given ice conditions the natural modes to experience frequency lock-in and the range of ice drift velocities for which lock-in develops are known. Determining those modes and velocities is however challenging due to the nonlinear nature of the interaction between ice and structure and limited available reference data. In this paper two methods are applied to determine the structural modes and ice drift velocities required as an input for simplified design approaches. The first method is the application of design standards and estimation formulas available from literature. The second method is the application of a recently developed numerical model for simulation of the interaction. The methods are applied to two existing structures which have experienced frequency lock-in and an offshore wind turbine designed to be employed at a location with mild ice conditions. Results show that the estimation formulas do not match with full-scale observations of the existing structures and can therefore not be applied to obtain input for the simplified design approaches. The second method shows to give simulation results consistent with the full-scale observations. Application to the offshore wind turbine reveals that it is most susceptible to frequency lock-in in the second mode.

KEY WORDS: Ice-induced vibrations; frequency lock-in; offshore wind; structural design.

INTRODUCTION

Ice-induced vibrations of offshore structures in cold regions pose a design challenge which has been studied for decades. The Cook Inlet observations (Peyton, 1968, Blenkarn, 1970), the Molikpaq 'incident' in the eighties (Jefferies and Wright, 1988), and the large experimental and full-scale campaigns in the nineties (Saeki et al., 1996; Schwarz and

Jochmann, 2001) being of major importance to the developments. Recently the topic has gained new interest with respect to projected offshore wind developments at locations in the Baltic Sea and Bohai Sea.

Three regimes of interaction are typically distinguished (ISO19906, 2010), being intermittent crushing, frequency lock-in, and continuous brittle crushing. The frequency lock-in regime is of particular interest with respect to offshore wind developments. The typical support structures consisting of a tower and monopile result in low natural frequencies of the first modes and relatively low damping ratios which makes the structures theoretically very susceptible to frequency lock-in.

During frequency lock-in the structural oscillation shows some typical characteristics which allow for a simplified analysis of the impact for extreme (ULS) and fatigue (FLS) conditions (see for measured signals e.g. Huang et al., 2007; Izumiyama et al., 1994; Singh et al., 1990; Toyama et al., 1983). Typical structural displacement, structural velocity, and ice load signals illustrating those characteristics are shown in Figure 1. The signal is close to harmonic with a frequency often slightly below the natural frequency of the mode of the structure in which lock-in has developed. The deviation from the harmonic signal generally occurs at the time moment when the structure reverses its direction of motion and starts to move towards the ice as seen at the moments the velocity of the structure changes from positive to negative in Figure 1. The velocity amplitude of structural motion is found to be in the range of 1.0 up to 1.5 times the ice velocity in experiments and full-scale data (Toyama et al., 1983). Using these observations the oscillation of the structure during frequency lock-in can be approximated by:

$$u_s(t) = \frac{F_{mean}}{K_s} + \frac{\beta v_{ice}}{\omega_n} \sin(\omega_n t) \qquad (1)$$

, with u_s the structural displacement at the ice action point [m], F_{mean} the mean global ice load [N], K_s the structural stiffness at the ice action point [N/m], β a factor between 1.0 and 1.5 [-], v_{ice} the ice drift velocity [m/s], ω_n the natural frequency of the mode in which frequency lock-in develops [rad/s], and time t [s].

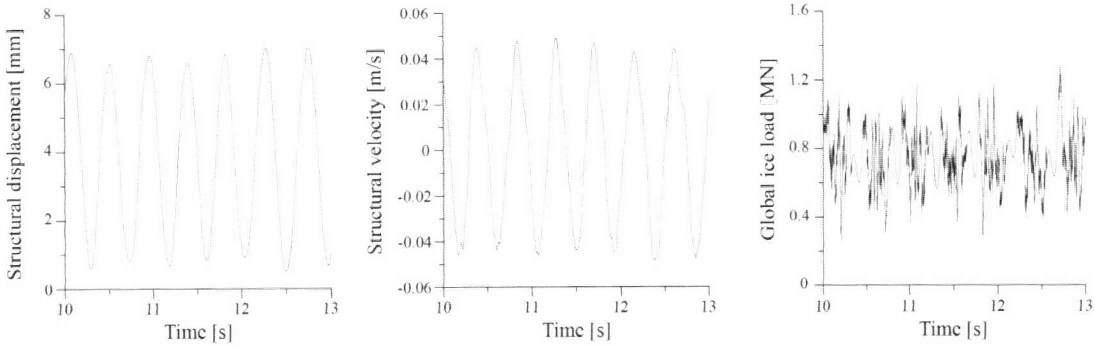

Figure 1. Typical (simulated) structural displacement, structural velocity, and global ice load during frequency lock-in of a structure with a natural frequency of 2.32 Hz at an ice drift velocity of 0.04 m/s.

The impact on design for ULS and FLS conditions can be verified by applying an oscillation pattern of the ice action point of the structure according to Eq. (1) in detailed numerical models of the structure. In order to do so it is required to know the global ice load, the modes to consider, and the ice drift velocities for which frequency lock-in can develop for relevant ice conditions. Note that this approach is valid for situations where other external loads are not causing significant structural motion. When significant interaction of the structure with wind or currents takes place, such that the structural motion is altered during interaction with the ice only a fully coupled numerical simulation of the interaction can be used to assess the development of frequency lock-in. It should additionally be checked whether limiting mechanisms such as ice buckling are expected to occur that would eliminate the possibility for frequency lock-in to develop.

In this paper we look into the challenge of estimating the structural modes and ice drift velocities for which frequency lock-in can develop in given ice conditions. Two methods are applied: the first method is based on ISO19906 (2010) and estimation formulas available in literature, and the second method incorporates a recently developed numerical model for simulation of the interaction (Hendrikse and Metrikine, 2016). Both methods are applied to two existing structures which have experienced frequency lock-in in order to evaluate their applicability. These structures are the Norströmsgrund lighthouse located in the Gulf of Bothnia and the Monopod MDP-1 located in the Bohai Sea. The methods are also applied to study an offshore wind turbine on monopile foundation designed for southern Baltic Sea conditions.

In Section 2 the relevant structural and ice parameters of the three structures are presented which are used as input for the analysis in this paper. In Section 3 an estimate of the modes in which frequency lock-in can develop and the velocity range over which it develops on the basis of design standards and estimation formulas from literature is presented. In Section 4 the same quantities are obtained from simulations with a recently developed numerical model. In Section 5 the results are discussed and in Section 6 the conclusions are presented.

STRUCTURES CONSIDERED IN THE ANALYSIS

For our analysis we take into consideration the first four modes of an offshore wind turbine on a cylindrical monopile foundation designed for southern Baltic Sea conditions. Higher modes are not included here as the analysis serves as an example, but for a real design all relevant modes have to be included. The modal properties of the first four modes are given in Table 1. Both the generalized mass M_g for eigenvectors scaled to one at maximum deflection, and the oscillating mass M_s used in a single-degree-of-freedom (scaled to modal deflection at the ice action point) representation are given. ϕ is the modal amplitude at the ice action point. The foundation has a diameter d of 6 m at the ice action point and the structure is envisaged to be employed at a location with a 50-year ice thickness h_{50} of 0.4 m.

Table 1. Modal properties of first four modes of the studied offshore wind turbine.

Mode	f_n [Hz]	ζ_n [-]	M_s [t]	ϕ [-]	M_g [t]
1	0.226	0.01	$2.9 \cdot 10^4$	0.147	615
2	0.626	0.01	$3.9 \cdot 10^3$	0.854	2850
3	1.546	0.01	$9.8 \cdot 10^6$	0.008	624
4	1.709	0.01	$1.5 \cdot 10^5$	0.059	513

We also consider two existing structures being the Norströmsgrund lighthouse located in the Gulf of Bothnia and the Monopod MDP-1 located in the Bohai Sea. For both structures only the first mode is considered as frequency lock-in has occurred mainly in mode one. It is again noted that in a real design scenario more modes should be included in the analysis. Table 2 shows the properties of the structures used in the analysis based on Guo (2012). Upper bound measured values for the ice thickness (h) and the reported range of velocities (v_{fli}) where frequency lock-in occurred are given (Yue et al., 2002; Ice Induced Vibrations JIP, 2011).

Table 2. Parameters used in analysis of the Norströmsgrund lighthouse and MDP-1 monopod.

Structure	f_n [Hz]	ζ [-]	M_s [t]	d [m]	h [m]	v_{fli} [m/s]
Norströmsgrund lighthouse	2.6	0.02	$3.5 \cdot 10^3$	7.5	1	0.02 – 0.12
Monopod MDP-1	2.32	0.03	$9.25 \cdot 10^2$	1.78	0.5	0.02 – 0.04

Estimates of the global ice load are required for the numerical simulations and are obtained based on formula A.8-21 in ISO19906 (2010):

$$F_{iso} = hdC_R \left(h\right)^{-0.5+h/5} \left(\frac{w}{h}\right)^{-0.16} \tag{2}$$

, where F_{iso} is an estimate of the maximum load during continuous brittle crushing [N], and w is the structure width or diameter [m]. Results for different C_R values are given in Table 3. We have chosen the C_R values to represent different estimations of ice load levels for continuous crushing based on ISO19906 and a discussion on ice load estimation by Gravesen and Kärnä (2009). The load levels presented in Table 3 are used as reference for input in the numerical simulations described in Section 4. It is noted that these estimates represent maxima during continuous brittle crushing which is a different interaction regime than frequency lock-in. During continuous brittle crushing the ice load is not affected by the structural motion.

Table 3. Maximum global load during continuous brittle crushing based on Eq. (2).

Structure	C_R = 1.0 MPa	C_R = 1.4 MPa	C_R = 1.8 MPa
Norströmsgrund lighthouse	$5.43 \cdot 10^6$ N	$7.61 \cdot 10^6$ N	$9.78 \cdot 10^6$ N
Monopod MDP-1	$9.58 \cdot 10^5$ N	$1.34 \cdot 10^6$ N	$1.73 \cdot 10^6$ N
Offshore wind turbine	$2.29 \cdot 10^6$ N	$3.20 \cdot 10^6$ N	$4.12 \cdot 10^6$ N

APPLICATION OF DESIGN STANDARDS AND ANALYTICAL FORMULAS

The applicable design standards IEC61400-3 and ISO19906 are soon to be revised, but the analysis method for frequency lock-in is not expected to change significantly from the method presented in the current version of ISO19906 (2010). We will therefore follow the approach presented in the current version of ISO19906 for our analysis.

The susceptibility of different modes of oscillation of a structure to frequency lock-in can be determined based on formula A.8-69 from ISO19906 (2010):

$$\zeta_n \geq \frac{\phi_{nc}^2}{4\pi f_n M_n} h\theta \qquad (3)$$

, with ζ_n the damping ratio as a fraction of critical of mode n [-], f_n the natural frequency of mode n [Hz], M_n the generalized mass for eigenvectors scaled to one at the maximum deflection of mode n [kg], h the ice thickness [m], ϕ_{nc} the modal amplitude at the ice action point of mode n [m], and θ an empirical coefficient equal to $40 \cdot 10^6$ [kg/ms]. Results of application of Eq. (3) to the structures considered are shown in Table 4. Frequency lock-in is expected for the lighthouse and monopod in mode one, and for the offshore wind turbine in mode one and two for the specified ice conditions.

Table 4. Susceptibility of modes based on Eq. (2).

Structure	ζ [-]	RHS Eq. (2)	FLI
Norströmsgrund lighthouse	0.02	0.35	Yes
Monopod MDP-1	0.03	0.74	Yes
OWT mode 1	0.01	0.19	Yes
OWT mode 2	0.01	0.52	Yes
OWT mode 3	0.01	0.00008	No
OWT mode 4	0.01	0.0051	No

The second step is to find the maximum ice drift velocity ($v_{maxfli,n}$), or the range of velocities (v_{fli}), at which frequency lock-in can develop for the identified susceptible modes. Several approaches have been proposed in literature to estimate the maximum velocity at which frequency lock-in can develop. ISO19906 (2010) gives formula (A.8-71):

$$v_{maxfli,n} = 0.06 f_n \qquad (4)$$

Guo (2013) suggests a dependence on the transitional strain rate between ductile and brittle ice behavior $\dot{\varepsilon}_{eq}$ and the structure diameter:

$$v_{maxfli} = 0.9 d \dot{\varepsilon}_{eq} \qquad (5)$$

, with $\dot{\varepsilon}_{eq}$ equal to 10^{-2} [s^{-1}]. Based on experimental and full-scale work Huang et al. (2007) define an estimate of the maximum velocity as:

$$v_{maxfli,n} = \frac{Df_n}{e^{-\frac{20h}{D \ln\left(\frac{K_n}{Eh}\right)}}} \qquad (6)$$

, with E the Young's modulus of ice [Pa], and K_n the stiffness used in a single-degree-of-freedom representation of mode n [N/m]. Palmer (2010) suggests a velocity range estimate based on:

$$v_{fli,n} = \alpha f_n h \tag{7}$$

, with α between 0.01 and 0.4. Application of these four approaches to the structures considered gives the estimates for the maximum velocity at which lock-in can develop shown in Table 5. For the Huang approach we have chosen a range of E of 1 GPa up to 5 GPa (Timco and Weeks, 2010)

Table 5. Estimates of maximum ice drift velocity for frequency lock-in based on analytical approaches.

Structure:	ISO19906 Eq. (4)	Guo (2013) Eq. (5)	Huang et al. (2007) Eq. (6)	Palmer et al. (2010) Eq. (7)
Norströmsgrund lighthouse	0.156	0.0675	0 – 3.98	0.026 - 1.04
Monopod MDP-1	0.1392	0.016	0.01 – 0.45	0.011 - 0.464
OWT mode 1	0.014	0.054	0.68 – 0.93	0.001 - 0.036
OWT mode 2	0.035	0.054	1.86 – 2.57	0.0025 - 0.1

The obtained results are discussed in Section 5 in comparison to the numerical approach applied in the following Section.

APPLICATION OF A NUMERICAL APPROACH

In this section we analyze the development of frequency lock-in for the different structures on the basis of numerical simulations with a recently developed numerical model (Hendrikse and Metrikine, 2016).

Model Input Parameters

The model requires eight input parameters to define the ice. Reference data from ice action on a rigid structure is used to define these parameters. As the ice behavior when crushing against a rigid structure is not influenced by structural motion pure ice related input parameters can be obtained. For flexible structures, such as analyzed in this paper, the interaction between ice and structure may lead to intermittent crushing or frequency lock-in in which case the data reflect the interaction rather than pure ice behavior. Only in the continuous brittle crushing range at high ice drift velocities is the behavior more or less similar for both flexible and rigid structures.

We use the best match currently available to Baltic Sea data derived from the observations on the Norströmsgrund lighthouse (Ice Induced Vibrations JIP, 2011). Seven of the input parameters used for simulations are given in Table 6 for reproducibility of the results presented in this Section. It is worth commenting on the two velocities used as model input. The value for v_{trans} is the velocity at which a transition from ductile to brittle ice behavior occurs in the model. The value of $v_{ref,high}$ is a reference velocity above which the ice fails in a completely brittle manner in continuous brittle crushing on a rigid structure. For velocities above this reference the mean global load and standard deviation are approximately constant as seen in the experiments by Sodhi and Morris (1984).

The remaining input parameter is the reference mean load level during continuous brittle crushing. This parameter is chosen such that the maximum simulated load levels during continuous brittle crushing correspond roughly to the load levels shown in Table 3. Table 7 shows the mean global ice load, maximum global ice load, and the amplitude during continuous brittle crushing as obtained from numerical simulation. For the monopod the ratio between amplitude and mean is much higher compared to the other structures as a result of the smaller ratio between structure diameter and ice thickness following the experimental results by Sodhi and Morris (1984).

Further discussion of the approach and definition of the parameters can be found in Hendrikse (2017). Simulations presented here were run for 100 cycles of oscillation of which the first 50 cycles were disregarded for the analysis to remove transient effects.

Table 6. Input parameters defining the ice behavior in the numerical model. For a detailed explanation see Hendrikse (2017).

Model input parameter	Magnitude
$v_{ref,high}$:	0.1 m/s
$f_{ref,peak}$:	20 Hz
v_{trans}	0.001 m/s
$F_{ref,max}/F_{ref,mean}$	5
$F_{ref,std}/F_{ref,mean}$	0.1
$F_{ref,mean,2}/F_{ref,mean}$	3
$T_{ref,peak}$	60 s

Table 7. Simulated global loads during continuous brittle crushing at high ice drift velocities for different load scenarios.

Structure and load scenario	Mean global load [MN]	Max global load [MN]	Amplitude	Amplitude / Mean [%]
Norströmsgrund lighthouse				
Load scenario 1	4.20	5.90	1.70	40%
Load scenario 2	5.90	8.30	2.40	41%
Load scenario 3	7.60	11.0	3.40	45%
Monopod MDP-1				
Load scenario 1	0.50	0.97	0.47	94%
Load scenario 2	0.71	1.30	0.59	83%
Load scenario 3	0.91	1.70	0.79	87%
Offshore wind turbine				

Load scenario 1	1.35	2.02	0.67	50%
Load scenario 2	1.90	2.80	0.90	47%
Load scenario 3	2.44	3.60	1.20	49%

Simulation Results

Simulation results for the Norströmsgrund lighthouse and the MDP-1 monopod are shown in Figure 2. The relation between the ice velocity and the maximum structural velocity is shown. The frequency lock-in range is given by the points which follow the relationship (Toyama et al., 1983):

$$v_{s,max} = \beta v_{ice} \tag{8}$$

, with $v_{s,max}$ the maximum structural velocity [m/s], and β a factor between 1.0 and 1.5.

The model predicts frequency lock-in for the Norströmsgrund lighthouse in the range of 0.02 m/s up to 0.09 m/s for the load scenarios considered. For the MDP-1 monopod frequency lock-in is predicted to occur between 0.02 and 0.06 m/s. It is seen that for higher load levels, corresponding to higher numbers of load scenario, the range where frequency lock-in is predicted to occur shifts to higher velocities. More energy is introduced by the ice at higher load levels resulting in this shift to higher velocities. For details see Hendrikse (2017).

Results for the first four modes of the offshore wind turbine are presented in Figure 3. Modes 3 and 4 show no frequency lock-in in any of the load scenarios considered. Mode 2 shows the largest range of frequency lock-in between 0.06 m/s and 0.11 m/s for the defined load scenarios. Mode 1 shows a transition in interaction between 0.04 m/s and 0.07 m/s for the considered load scenarios, but no pure frequency lock-in signal is obtained.

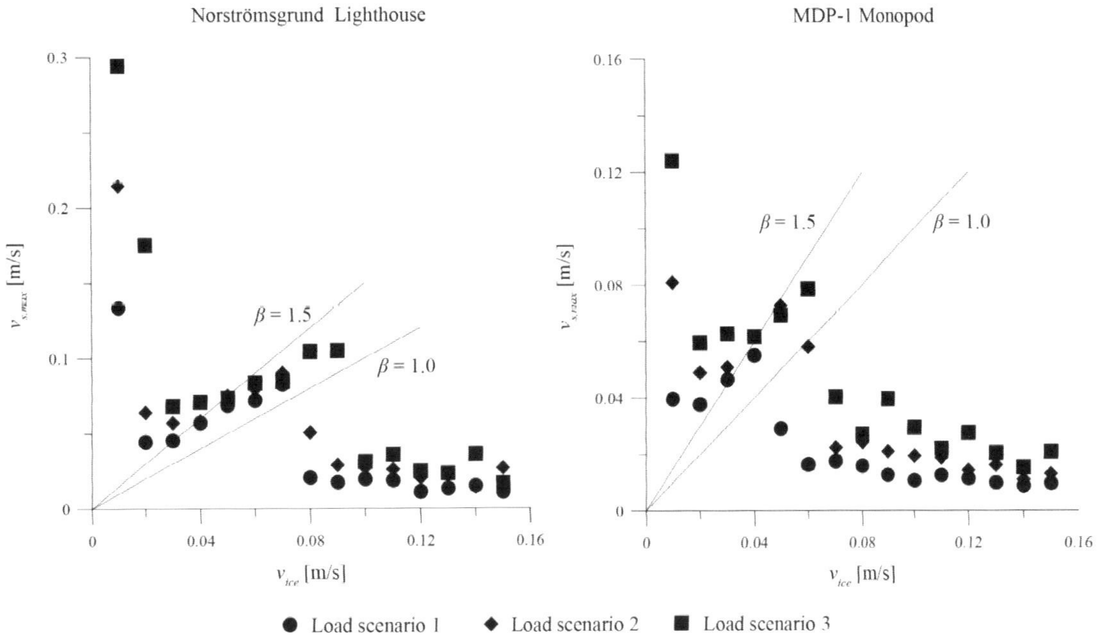

Figure 2. Left: Simulated dependence of maximum structural velocity on ice drift velocity for the Norströmsgrund lighthouse. Frequency lock-in is predicted to occur between 0.02 m/s and 0.09 m/s for the considered load scenarios. Right: Simulated dependence of maximum

structural velocity on ice drift velocity for the MDP-1 monopod. Frequency lock-in is predicted to occur between 0.02 m/s and 0.06 m/s for the different load scenarios considered. The solid lines represent a ratio of 1.0 and 1.5.

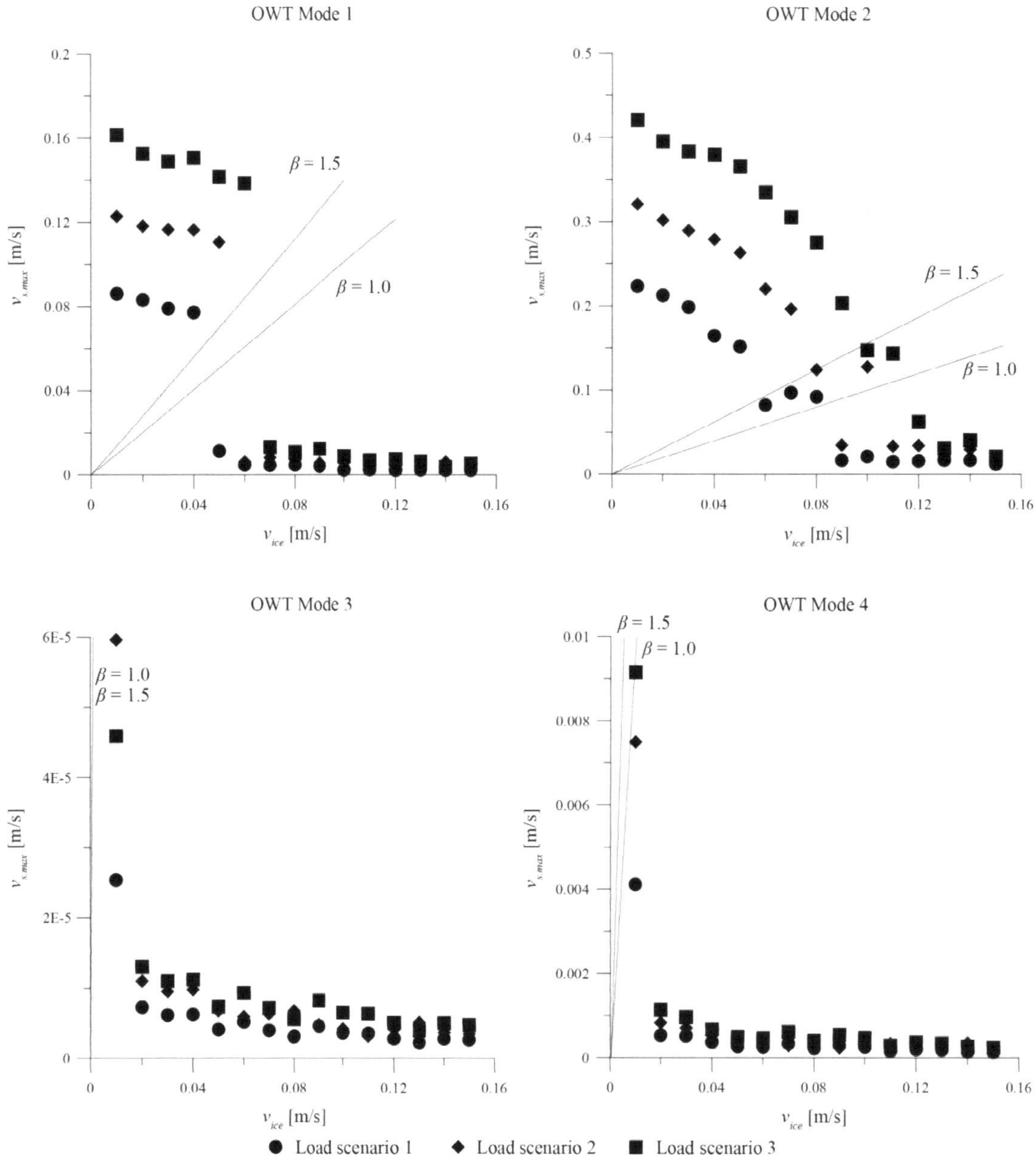

Figure 3. Simulated dependence of maximum structural velocity on ice drift velocity for the four modes of the offshore wind turbine. Modes 3 and 4 show no lock-in in any of the load scenarios. Mode 1 shows a transition in behavior in the range of 0.04 m/s up to 0.06 m/s, but no frequency lock-in is obtained. Mode 2 shows frequency lock-in between 0.06 m/s and 0.11 m/s for the different load scenarios considered. The solid lines represent a ratio of 1.0 and 1.5.

DISCUSSION

Modes Susceptible to Frequency Lock-in

With respect to the prediction of modes susceptible to frequency lock-in the analytical formula in ISO19906 (Eq. (2)) and the numerical approach used in Section 4 give the same results. Despite the fact that Eq. (2) does not reflect our current theoretical understanding of the interaction (Hendrikse, 2017) it gives a reasonable first indication of the susceptibility of a mode to frequency lock-in. Care must be taken with respect to the empirical parameter θ which incorporates effects of structure width and ice properties on the energy input from the ice and might not always be conservative.

Prediction of Ice Drift Velocities at which Frequency Lock-in Develops for Existing Structures

The predictions of the maximum velocity, or range of velocities, for frequency lock-in on the basis of the different estimation formulas (Eq. (3-6)) show a large scatter as illustrated in Figure 4. For the Norströmsgrund lighthouse only the formula used in ISO19906 predicts a reasonable upper bound estimate, but this is perhaps not surprising as the lighthouse were studied extensively as a basis for the standard. For the Monopod all formulas significantly over predict the maximum velocity compared to the observed velocity range, except for the estimation formula by Guo (2013) which results a value lower than the actual observations. None of the estimation formulas considered here gives a good estimate for the maximum velocities which could be used as input in Eq. (1) for design.

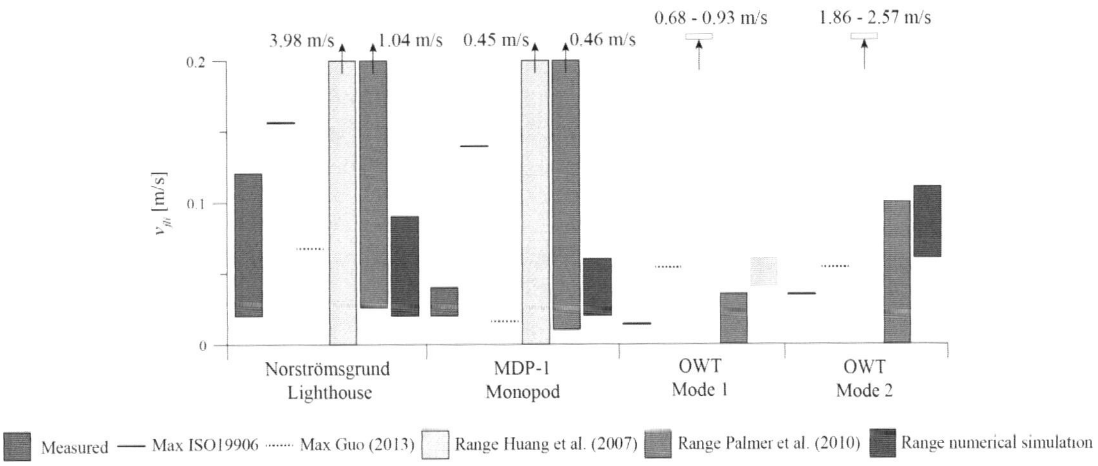

Figure 4. Summary of results for the predicted frequency lock-in velocity range or max velocity. For mode 1 of the offshore wind turbine the simulated possible range of frequency lock-in based on the behaviour transition illustrated in Figure 3 is shown.

The method using the numerical simulation model shows more consistent results for the lighthouse and the monopod as shown in Figure 4. The over prediction of the maximum velocity for the Monopod could be a result of the assumption of a too high global ice load in load scenarios 2 and 3. Load scenario one gives the best match with a range between 0.02 m/s and 0.04 m/s. There is not enough load data available to compare between simulated load levels and real load levels which have resulted in frequency lock-in of the monopod.

For the Norströmsgrund lighthouse a lower maximum is predicted in the considered ice load conditions. The load scenarios considered yield high load values compared to those actually measured on the lighthouse and therefore a similar over prediction of the maximum velocity as for the monopod would be expected. An effect which is not included here which is very relevant for stiffer structures, such as the Norströmsgrund lighthouse, is the importance of initial conditions. The interaction problem is to some extent sensitive to initial conditions as frequency lock-in may not be possible to develop from the initial equilibrium position, but the limit cycle can be reached with favorable initial conditions. At velocities above the range predicted as shown in Figure 4 short periods of frequency lock-in sometimes develop. An example is shown for the Norströmsgrund lighthouse in load scenario 2 at 0.12 m/s in Figure 5. Frequency lock-in has often been observed to develop for short amounts of time on the lighthouse (Bjerkås et al., 2014). This could indicate that the majority of events measured was actually driven by the initial conditions. A more detailed study into the effect of initial conditions on the predictions is ongoing.

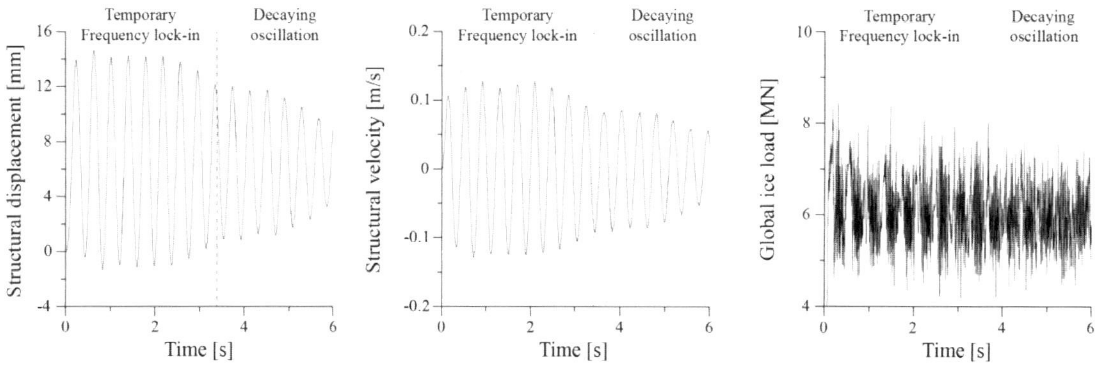

Figure 5. Short duration frequency lock-in of the Norströmsgrund lighthouse in load scenario 2 at a high velocity (0.12 m/s) as a result of initial conditions. Frequency lock-in is sustained for 7 cycles before the oscillations start to decay and the interaction is lost.

A degree of conservativeness has to be applied when using the maximum velocity from model simulations as input for simplified design approaches to account for the effect of initial conditions. For FLS predictions the effect is relatively small as initial condition driven frequency lock-in is generally not stable for long periods of time and changing ice conditions.

Prediction of Ice Drift Velocities at which Frequency Lock-in Dvelops for the Offshore Wind Turbine

Results for the offshore wind turbine shown in Figure 4 illustrate that in general the estimation formulas predict frequency lock-in at much lower velocities than for the lighthouse and the monopod, with the exception of the approach by Guo (2013). This is surprising as comparison between the second mode of the offshore wind turbine and the first mode of the lighthouse shows that the offshore wind turbine has lower damping and a lower natural frequency which on the basis of experiments would suggest frequency lock-in at higher velocities (Huang et al., 2007).

Based on the numerical approach the offshore wind turbine is shown to be susceptible to frequency lock-in primarily in the second mode. For the different load scenarios considered the range even extends to values higher than those simulated for the lighthouse. This is an important observation as it is often suggested in literature that sustained frequency lock-in cannot develop above 0.1 m/s (ISO19906, 2010). Full-scale observations on channel markers and model-scale experimental observations suggest that such theoretical limit does not exist and support the model predictions (Nordlund et al., 1988; Huang et al., 2007). Recent theoretical development (Hendrikse, 2017) also suggests that no such limit exists and that for structures with low damping in severe ice conditions the maximum velocity could increase significantly above 0.1 m/s.

With respect to the first mode it is observed that the natural period is so large such that during a single cycle of interaction a long time occurs during which the contact area between ice and structure increases in the model. This sometimes results in a large increase in the resistance of the ice. As a result the structure experiences a mix of frequency lock-in and intermittent crushing like behavior as illustrated in Figure 5 for load scenario 2 and a velocity of 0.04 m/s. There are no data available to validate this type of behavior. Experiments aimed at investigating frequency lock-in for structures with low natural frequencies are required.

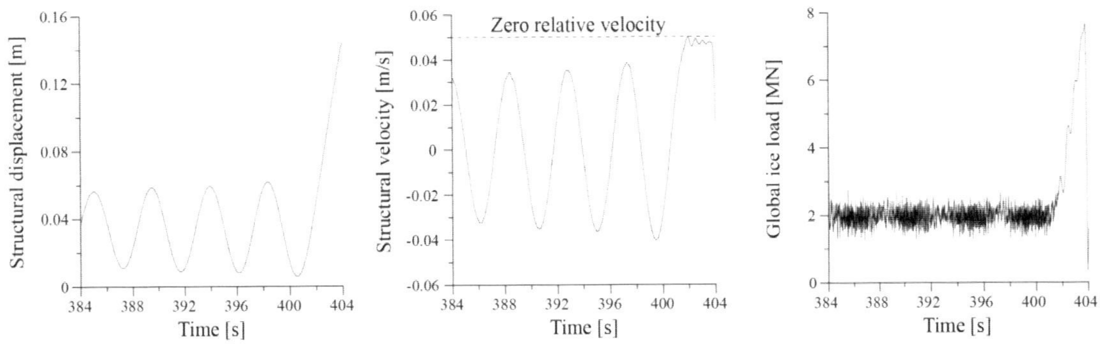

Figure 5 Simulated mixed frequency lock-in and intermittent crushing interaction at 0.04 m/s for load scenario 2 for mode 1 of the offshore wind turbine. A single mixed cycle is shown. At the moment when the amplitude of oscillation reaches close to zero relative velocity the ice resistance increases enough as a result of ductile ice behavior to prevent the structure from moving back and the oscillation pattern is interrupted by a typical intermittent crushing type of interaction.

CONCLUSION

The estimation of the susceptibility of structural modes to frequency lock-in and the range of ice drift velocities for which frequency lock-in can develop in given ice conditions has been studied. Three structures have been analyzed, being the Norströmsgrund lighthouse, MDP-1 monopod, and an offshore wind turbine designed for southern Baltic Sea ice conditions. The analysis has been conducted on the basis of both existing estimation formulas from literature and numerical modelling of the interaction with a recently developed model.

With respect to the prediction of the susceptibility of modes to frequency lock-in both the estimation formulas and the numerical approach show to be consistent with observations on the existing structures. The first mode of the lighthouse and monopod are correctly predicted to be susceptible to frequency lock-in. For the offshore wind turbine it has been found that the first and second mode are susceptible to frequency lock-in in the defined ice conditions based on both the estimation formulas and the numerical approach.

The estimation formulas from literature show a significant scatter in the predictions for the maximum velocity and velocity range at which frequency lock-in can develop. None of the formulas matches the observations of the lighthouse and monopod and they can therefore not be applied for simplified design approaches.

The numerical method shows results consistent with the observations of the lighthouse and monopod. A good estimate for the maximum velocity at which frequency lock-in can develop can be obtained with the numerical method allowing for a simplified design approach based on characteristics of the structural motion during frequency lock-in. Care should be taken to include a degree of safety in the estimate to account for the effect of initial conditions on the maximum predicted velocity at which frequency lock-in develops.

The results obtained with the numerical approach show that frequency lock-in is most likely to develop in the second mode of the offshore wind turbine considered. For moderate ice climates in the southern Baltic Sea, with ice thickness not exceeding 0.5 m and an ice load level given by a C_R value of 1.0, the example of the OWT structure calculated shows maximum ice drift velocities for frequency lock-in of 0.06 m/s for the first mode (in the region of 0.25Hz) and 0.1 m/s for the second mode (around 1.0Hz). These values can be used as an indication for conceptual design purposes, but validation with an interaction model is highly recommended at later design stages.

ACKNOWLEDGEMENTS

The authors acknowledge the support from the SAMCoT CRI through the Research Council of Norway and all the SAMCoT partners. Thanks to Chris Keijdener for his input during revision of the paper.

REFERENCES

Bjerkås, M., Alsos, H.S., & Wåsjø, K., 2014. Estimates of the number of vibration cycles from frequency locked-in ice loads. In *Proceedings of the ASME 2014 33rd International Conference on Ocean, Offshore and Arctic Engineering,* OMAE2014-23134.

Blenkarn, K. A., 1970. Measurements and analysis of ice forces on Cook Inlet structures. In *Proceedings of the Second Annual Offshore Technology Conference*, volume II, pp.365–378.

Gravesen, H., & Kärnä, T., 2009. Ice loads for offshore wind tutbines in Southern Baltic Sea. In *Proceedings of the 20th International Conference on Port and Ocean Engineering under Atctic Conditions,* POAC09-3.

Guo, F., 2012. Reanalysis of Ice Induced Steady State Vibration from an Engineering Perspective. In *Proceedings of the 21st IAHR International Symposium on Ice,* pp.1023-1034.

Guo, F., 2013. Analysis of the key parameters in ice induced frequency lock-in. In *Proceedings of the 22nd International Conference on Port and Ocean Engineering under Arctic Conditions*, POAC13-33.

Hendrikse, H., 2017. *Ice-induced vibrations of vertically sided offshore structures.* Ph.D. Delft: Delft University of Technology.

Hendrikse, H., & Metrikine, A., 2016. Ice-induced vibrations and ice buckling. *Cold Regions Science and Technology*, 121, pp.100-107.

Huang, Y., Shi, Q., & Song, A., 2007. Model test study of the interaction between ice and a compliant vertical narrow structure. *Cold Regions Science and Technology*, 49, pp.151-160.

Ice Induced Vibrations JIP, 2011. *Full-scale data analyses of frequency lock-in,OO-11179-1.2.2,* Norway: Dr.techn.OlavOlsen.

ISO19906 (2010). *Petroleum and natural gas industries – arctic offshore structures.*

Izumiyama, K., Irani, M.B., and Timco, G.W., 1994. Influence of compliance of structure on ice load. In *Proceedings of the 12th IAHR International Symposium on Ice*, volume 1, pp.229–238.

Jefferies, M.G., & Wright, W.H., 1988. Dynamic response of 'Molikpaq' to ice-structure interaction. In *Proceedings of the Seventh International Conference on Offshore Mechanics and Arctic Engineering*, volume 4, pp.201–220.

Nordlund, O.P., Kärnä, T., & Järvinen, E., 1988. Measurements of ice-induced vibrations of channel markers. In *Proceedings of the IAHR 9th International Symposium on Ice*, volume 1., pp.537-548.

Palmer, A., Yue, Q., & Guo, F., 2010. Ice-induced vibrations and scaling. *Cold Regions Science and Technology,* 60, pp.189-192.

Peyton, H.R., 1968. Sea ice forces. In *Ice Pressures Against Structures, compiled by L. Gold and G.Williams, NRC Techn.Memo No. 92.*

Saeki, H., Hirayama, K., Kawasaki, T., Akagawa, S., Kato, K., Kamesaki, K., Saka, K., & Kurokawa, A., 1996. JOIA project of study on ice load. In *Proceedings of the 13th IAHR International Symposiumon Ice*, volume 1, pp.17–27.

Schwarz, J., & Jochmann, P., 2001. Ice force measurements within the LOLEIF-project. In *Proceedings of the 16th International Conference on Port and Ocean Engineering under Arctic Conditions*, volume 2, pp.669–682.

Singh, S. K., Timco, G.W., Frederking, R.M.W., & Jordaan, I.J., 1990. Tests of ice crushing on a flexible structure. In *Proceedings of the Ninth Offshore Mechanics and Arcitc Engineering Symposium*, volume 4, pp.89–94.

Sodhi, D. S., & Morris, C. E., 1984. Ice forces on rigid, vertical, cylindrical structures. *Report 84-33*, US Army Cold Regions Research and Engineering Laboratory.

Timco, G.W., & Weeks, W.F., 2010. A review of the engineering properties of sea ice. *Cold Regions Science and Technology,* 60, pp.107-129.

Toyama, y., Sensu, T., Minami, M., & Yashima, N., 1983. Model tests on ice-induced self-excited vibration of cylindrical structures. In *Proceedings of the 7th International Conference on Port and Ocean Engineering under Arctic Conditions*, pp.834-844.

Yue, Q., Bi, X., Zhang, X., & Kärnä, T., 2002. Dynamic ice forces caused by crushing failure. In: Ice in the Environment, *Proceedings of the 16th IAHR International Symposium on Ice,* pp.231-237.

POAC'17
Busan, KOREA

Proceedings of the 24th International Conference on
Port and Ocean Engineering under Arctic Conditions
June 11-16, 2017, Busan, Korea

Arctic Sea Ice Sensitivity to Atmospheric Forcings in a Sea Ice Model

Su-Bong Lee[1], Beak-Min Kim[1], Joo-Hong Kim[1], Yoo-Geun Ham[2], Sung-Ho Kang[3]

[1] Unit of Arctic Sea-Ice prediction, Korea Polar Research Institute, Incheon, Korea

[2] Department of Oceanography, Chonnam National University, Gwangju, Korea

[3] Division of Polar Ocean Sciences, Korea Polar Research Institute, Incheon, Korea

ABSTRACT

Atmospheric thermodynamic/dynamic forcings play an important role for sea ice variability and contribute to sea ice melting as much as contribution from underlying oceanic forcings. In this study, we examine the sensitivity of a standalone sea ice model (CICE5) to different atmospheric forcing data. Downward shortwave radiation, downward longwave radiation, air temperature, wind, precipitation and specific humidity fields from the NCEP-NCAR reanalysis version 2 (NCEP-NCAR R2), ECMWF Interim Re-Analysis (ERA-Interim) and Japanese 55-year Reanalysis Project (JRA-55) are used to force the CICE5. Annual cycle of the Arctic sea ice extent in each experiment reproduces the observed sea ice extent retrieved by satellite reasonably well and resembles each other, while sea ice thickness exhibits different characteristics in each simulation. Simulation results forced by ERA-Interim show the largest melting at the top during summer. We find that this is due to the relatively large downward longwave radiation forcing. Furthermore, ERA-Interim data gives the largest incoming total energy fluxes among other products. During winter, the volume differences among reanalysis products are largely determined by the ice export. The largest amount of ice export is attained in NCEP NCAR R2 due to its strong wind speed. However, the largest uncertainties in the estimation of annual mean sea ice volume arise from the differences in the radiative forcings in the three reanalysis.

ICECAT-1-2-3: Operational Ice Chart for Seismic Planning

Sean M. McDermott[1], Yevgeniy Kadranov[2]
[1]Horizon Systems Group Inc, Halifax, Canada
[2] LLP Iceman.kz, Shymkent, Kazakhstan

ABSTRACT

2016 was the sixth and final season of TGS' 2D seismic acquisition program in the ice infested waters of North-East Greenland. Since the start of operations in 2011, it was clear that the standard WMO ice charts would not provide ice information in a form sufficient to support our efforts.

To work around this short-coming, an all-new ice chart product and procedure was developed, aimed directly at supporting ice management operations. Created in-situ, several iterations resulted in a simple and highly intuitive ice categorization product. Based on observed ice conditions data merged with satellite imagery and helicopter ice reconnaissance data, the information was then coalesced into an operational ice chart, relying heavily on the experience of the onboard ice management team. As well, the process weighted the capabilities and limitations of both the ice breaker and seismic vessel.

The target audience for this new ice chart was management… both onboard and shoreside. Its intuitive stop-light analogy made it obvious which areas would allow operations to occur, using a clear green-through-red color scheme. It readily showed areas of go/no-go ice conditions.

The result was an ice categorization product that was both process and output. Management saw none of the complexities of the development steps, just the clean result. This allowed operations people to focus more time ensuring safe and efficient operations and less interpreting a multitude of ice information sources. And while this process relied heavily on the experience of the ice management personnel, the result was well worth the extra effort invested.

TGS' multi-year operations in North-East Greenland wrapped up in the fall of 2016 and were by all measures, an outstanding achievement in one of the most challenging ice regimes in the world. The ice categorization process described herein was clearly part in achieving that success!

KEY WORDS: IceCat; Ice Categorization; Ice Charting; Polar Code; Greenland

INTRODUCTION

NE Greenland is one of the few remaining challenging areas for in-ice operations in Arctic regions and our plans would eventually take us into the most difficult areas within this region.

From 2006 to 2009, TGS-NOPEC Geophysical Company acquired 2D seismic data in NE Greenland on an as-able basis using an Akademik Class 2D seismic vessel, Shatskiy, and a small chase boat, Kvitbjorn. Both vessels had a light ice class rating which was sufficient for operations in open water and the occasional stray piece of small ice.

Ice information to support the project during this period came from public domain ice charts from various national ice services that cover the area of NE Greenland. The vessels acquiring seismic data simply used the charts to locate areas of open water and to follow along ice edges, something that large scale ice charts based on satellite imagery are very useful for.

The program to this point worked well enough until the need for seismic data from more challenging ice areas drove the requirement for a more formal ice management program that would provide higher resolution ice data to make better informed decisions. While the quality of public domain charts was good, the scale of these charts was insufficient to provide the small-scale ice information detail required to support seismic data collection in harsh areas.

It was knowing these limitations that saw us reintroduce ground-truthing, using ship borne and helicopter visual ice reconnaissance, and eventually (2016) even drone (UAV) data input. The ice management program would therefore require the ability to collect needed ice information on a frequent basis and augment it with satellite data of a higher-resolution than that used for public domain ice charts. This plan would necessitate the use of an ice reconnaissance helicopter and an ice management vessel capable of handling flight operations.

THE CHALLENGE

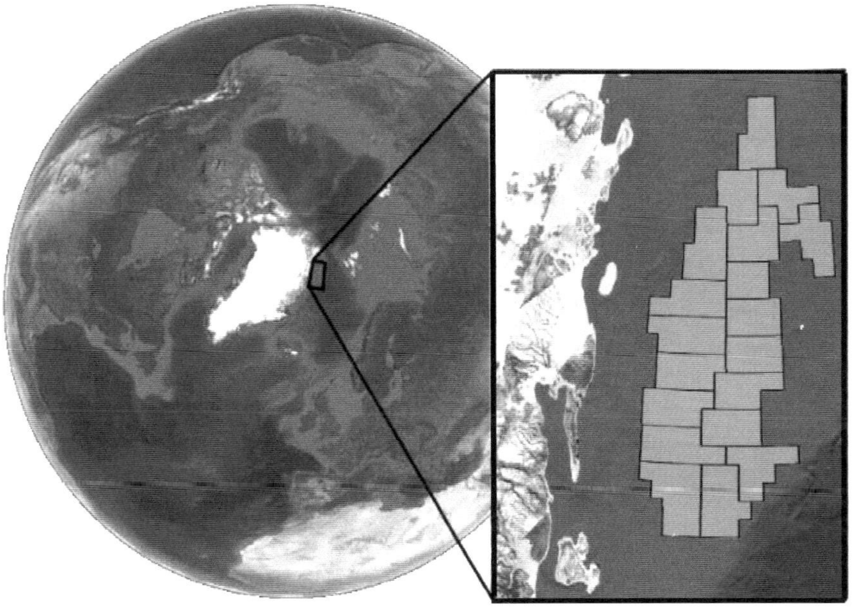

Figure 1. The area of interest (AOI) for the seismic acquisition program in NE Greenland is in one of the most ice-hostile regions of the Arctic. The AOI is divided into subsets called blocks. These blocks are leased by various oil exploration companies for seismic evaluation and possible future development.

Figure 2. (2016) Otso (icebreaker) leading Lazarev (seismic vessel) through heavy ice in the western extremes of the AOI, close to the NE Greenland coast. The six year program used the icebreaker Fennica in 2011 and 2014, the anchor handling ice breaker Balder Viking in 2012-13 and Otso in 2015-16. Lazarev replaced Shatskiy for the 2016 season.

The ice management program commenced with the 2011 summer season using the Finnish icebreaker Fennica, a 116m icebreaker with 15MW of power from 2 Aquamaster propulsion units. An A-Star helicopter was brought along for ice reconnaissance. Dedicated weather forecasting was added and the program launched northwards in late August of 2011. Initial ice charts from various ice services sources had shown a large percentage of multi-year ice existed throughout the block region AOI. As well, satellite imagery showed some areas of ice that could be open to interpretation. The first plan upon arrival at the southern ice edge was to ground truth the AOI using the helicopter and then bring in Fennica to test areas identified from the imagery and the heli-recce, to confirm what they were.

Figure 3. (2014) From 400m above, Fennica clearing the way for Shatskiy through light ice conditions. Note the clear wake aft of Fennica created by her powerful azimuth thrusters.

The format for the collection of seismic data was to have the lead icebreaker clear the way for the seismic vessel that followed just 200m behind, all while navigating carefully along the desired line of acquisition within a defined geographical region of interest or block. The acquisition line must be followed exactly, with little room for deviation, so knowing the exact ice conditions along the line was key. As well, the acquisition speed was just 4.5 knots and the 7km streamer must be kept moving forward or it will sink. When the streamer sinks, it triggers hydro-static floats that would bring it back to the surface, a surface often covered by ice. This occurrence would likely result in the total loss of the equipment and was something to be avoided at all costs. Further complicating these challenges was knowing what extreme ice features might be to the left or right of track out as far as 1km. This was because the streamer, following at a nominal depth of 15m was highly influenced by the unpredictable currents of NE Greenland. These currents would push the streamer left or right of track and potentially into the path of an extreme ice feature. If such an ice feature had a keel depth greater than 15m, the streamer would impact it causing damage and potential total loss. With currents playing such a strong role in moving the streamer off center it wasn't always the ice in front of the lead vessel that you had to worry about, but ice to the sides of the intended track out to several hundred meters. With often poor visibilities in snow and fog, this was a major challenge, even with shipborne radar in use.

And finally, the 4.5 knot acquisition speed is not best for an ice breaker that likes to break ice at 6-8 knots. There were several occasions when we would hit 100m floes but wouldn't break them due to our low speed and instead push them in front of the icebreaker. This floe would then impact a similar floe and so on until such time as we may have been pushing several of these small floes together, eventually slowing the ice breaker. We coined the phrase floe stacking where pushing one floe was fine but several stacked up together became problematic.

Figure 4. Otso (icebreaker) leads Lazarev (seismic vessel) through mixed ice conditions in NE Greenland. The black line shows the track that has been travelled while the red/yellow shading shows the potential offset of the seismic streamer should currents displace (in this concept example) the streamer to the right of track.

Figure 5. An extreme ice feature seen noted on a heli-recce. This is a small-medium (220m) size first-year floe with embedded heavy ridge elements having keel drafts estimated to be in excess of 15m. This ice feature would be a major challenge if encountered along a seismic line.

Our standard ice management technique was to bump floes off the track then flush them clear of the wake, something we intuitively termed "Bump & Flush". But for extreme ice features like icebergs and multi-year ridge remnants, something common during the 2016 program, we added "Avoidance" to our ice management techniques. If a track deviation was small enough we would simply avoid the ice feature ahead and then return to track once the streamer was clear of all possible danger. Through the six years of the program, "Wake Management" was coined as a term to capture the microscale level of ice management required to do the work we needed to do.

THE ICE MANAGEMENT PLAN

All of these challenges resulted in us acquiring as much data as we could to support our decision making. We developed an environmental database containing weather, sea conditions and ice data as well as vessel-in-ice performance characteristics. We acquired multiple daily satellite images which were managed by our ArcGIS based system that merged imagery with seismic planning and ship movements. We added high resolution dedicated weather forecasting and flew countless ice reconnaissance missions with a dedicated helicopter to look along planned lines to ensure they were clear or failing that, identify options if the original planned lines were ice covered. We used all available data to drive our understanding of drift patterns and then applied this to forecasting ice conditions along planned seismic lines.

With all the data at hand, it was possible to make informed operational decisions safely and efficiently. **What was lacking however was our ability to inform the shore based operations management about ice conditions now and what was expected in the next 24-48 hours.** They shared in most other data but for ice information they were still referencing broad scale public domain ice charts. The scale for these charts for NE Greenland considers vast areas of up to 300nm in a north to south orientation while we needed to consider ice conditions on a granular scale as small as 30nm. The public domain ice charts were coalesced from various satellite data and issued daily. We had access to the same base image data but the challenge for our operations was that currents moved ice about in random patterns that made satellite imagery older than 6 hours of academic interest only. Using the latest imagery and pre-flying lines planned for acquisition gave us a 6-7-hour look-ahead based on the vessel speeds of 4.5 knots.

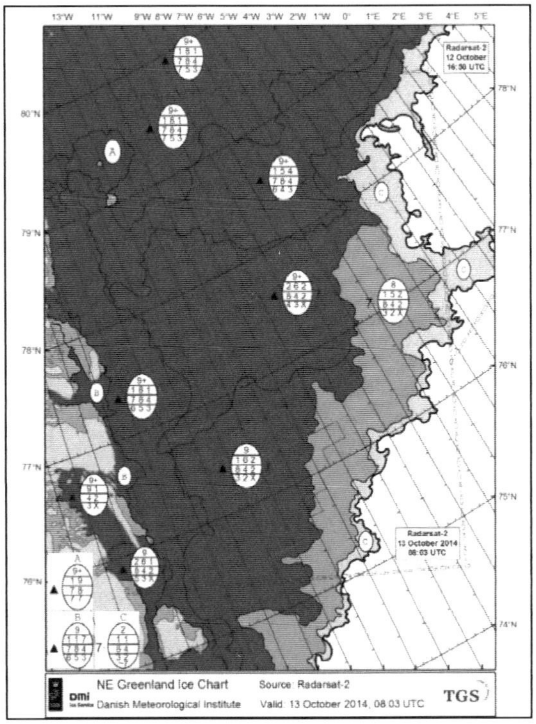

Figure 6. An example ice chart from the Danish national ice service (DMI) using the WMO color scheme for Ice Concentration

For example, at one juncture, we were getting questions from shore as to why we were focused on a certain area while onboard we were wondering why shore based operations wanted us to go to some other area. We were simply looking at different sources of ice information.

Intuitively, the chart on the right appears to indicate the possibility of penetrating the ice edge by 2-15nm but no further (at the boundary of red). The actual possibility for ice edge penetration is much greater if you consider the egg code classifications for the various ice boundaries.

Another example of where the color-coding doesn't intuitively help is a situation where 9+/10ths of Nilas would be color coded as red while 3/10ths of multi-year ice would be color coded as green. For this project we would easily be able to operate in 9+/10ths of Nilas (red) but not 3/10ths of multi-year (green).

With the shipboard ice management team using local knowledge and the shoreside offices using the broad scale ice charts from various ice services, it wasn't long before we knew we needed a product that suited both offshore and onshore operations.

The shoreside office used color coded WMO chart formats (Figure 7 Left) showing ice concentration by color. In one example, the office saw us working in a southern area and wanted us to move north to acquire seismic data in an area shown in yellow, or just 4-6/10ths of ice. This was an important area to acquire data from and given the short summer season, it was important to go there as soon as it was clear enough. The red color had recently changed to yellow as ice conditions eased and melted. But what changed was that the first-year ice melted leaving only multi-year & second-year ice remaining. This ice was impossible to work in for seismic acquisition but the change in colors from red to yellow had suggested a major improvement had taken place as far as the office was concerned.

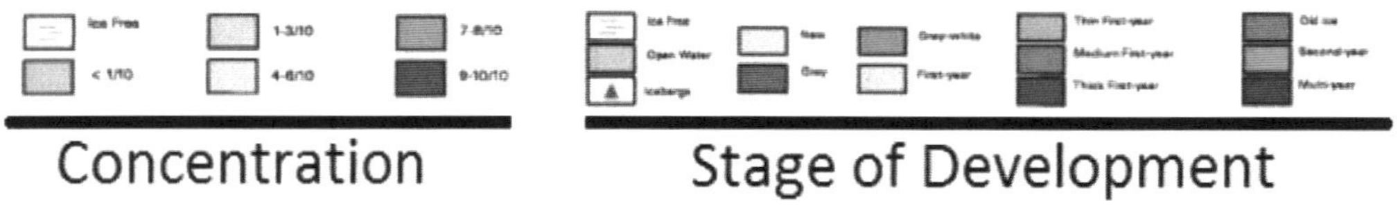

Figure 7. WMO Color code for Ice Concentration and Stage of Development

And this is where the second major challenge came in to play and the subject of this paper … How to capture all this high-resolution and verified ice conditions data in a form readily understood by marine and seismic operations, both onboard the ice breaker, the accompanying seismic vessel and shoreside management?

THE DEVELOPMENT OF ICECAT 1-2-3

With challenged bandwidth aboard the icebreakers, due to our far northern location, we knew we had to keep any product developed to be simple to reproduce and that answered the obvious question "Where can we work and for how long?". We originally experimented with a QuickLook format. The QuickLook came from the Dome Petroleum/Canadian Beaufort Sea days in the early 1980s and was a simple sketch taken from the latest airborne reconnaissance data and faxed to vessels offshore. It showed the current ice conditions in terms of ice edges, concentration boundaries and composition. What we needed was to take things one step further and identify areas where lines of acquisition were possible and what trends were occurring in terms of ice clearing in one area versus another.

The request from shoreside management was to produce a simple ice chart showing where we could work and where we couldn't and make it as intuitive as possible. This meant no egg codes and no underlying imagery that might require interpretation. It must show where it was possible to work for the pairing of the icebreaker being used and the seismic vessel it would escort.

We already knew the exact ice conditions based on our ground truthing efforts and this was aided by higher resolution imagery as well as understanding the limitations of our vessel pairing based on the ice trials we undertook. This then allowed us to develop a custom ice chart based simply on observed ice conditions, the most recent satellite data and the capabilities of the vessels involved. This saw the development of an operations-focused ice chart based simply on categorizing known ice conditions relevant to the operational limitations of the vessels involved.

Figure 8. An IceCat 1-2-3 product covering the entire AOI. Note Greenland coastline to west shown in gray.

Thus, the IceCat 1-2-3 product came into being. IceCat for ice categorization and 1-2-3 to symbolize the simplicity of interpretation. Using the universal stop light analogy where green is go, yellow is cautionary and red is danger, we added a few more categories based on our experience. Blue was readily understood to mean ice free or open water with no challenge to vessels, and orange was added for heavy ice conditions, one category higher than moderate (yellow) but not yet impassable (red).

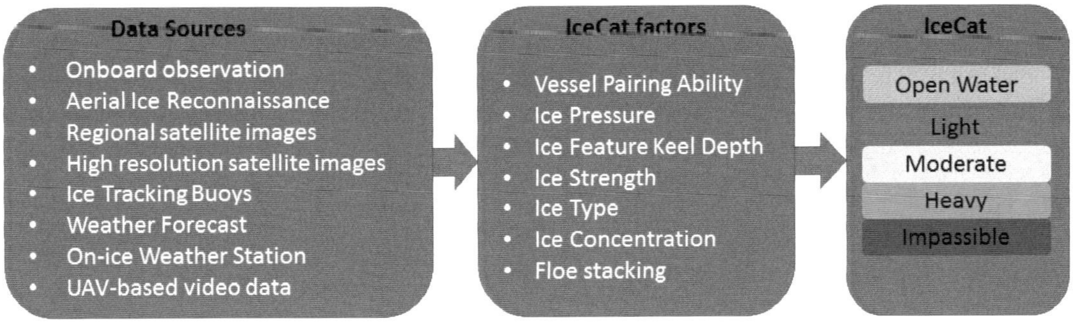

Figure 9. The data flow from input sources to factors considered in making an IceCat 1-2-3 chart to finally determine the appropriate color coding for the area of ice being investigated.

The chart covered the entire AOI and immediately showed where we could work (Blue through Orange) and where we couldn't (Red). A day to day comparison showed the trends in ice movement and decay/development.

Most importantly, and a key strength of this product was that it took the onus for chart interpretation from the end user (shoreside management & marine operations) and placed it solely with the ice management team onboard the icebreaker. Everything the ice team knew about the area ice was coalesced from various data sources, referenced to relevant IceCat factors such as vessel pairing, ice pressure, ice features, etc., and rendered into a colored chart called an IceCat 1-2-3. It was a daily snapshot of ice conditions that showed both line options as well as areas where seismic acquisition could be managed easier than other areas. The IceCat 1-2-3 didn't so much as embolden shoreside management to get involved in managing line choices as much as allow them to understand why we worked in areas where we did.

The IceCat 1-2-3 continued to develop through the six-year period of operations and improve with the addition of more and better data inputs.

For example, the 2015 season saw us 3D imaging some extreme ice features to calculate mass and then potentially infer danger to our operation. In 2016, we increased the resolution of our weather forecasting capabilities eventually adding a portable weather station to a grounded berg along the coastline due west of our AOI. This provided early warning of adverse weather and further strengthened the quality of our weather models and resultant forecast output. We started using ice tracking buoys that provided an accurate position every 30 minutes. We used these devices to track immediate threats by deploying and redeploying them on hazardous ice features, sometimes several redeployments per day when conditions were challenging.

Once we started to produce the IceCat 1-2-3 chart on a daily basis it became possible to monitor trends in ice cover changes. Our working area of interest was divided into many block areas that provided easy boundaries to use when investigating amounts and types of ice within each block and how the makeup of each block or region was changing. By clipping the boundaries either by block or region and using the inherent capabilities of the GIS-based IceCat 1-2-3 system, we could see if ice conditions were trending towards easier or more difficult. This provided TGS with the ability to monitor what trends were occurring and shift planning to more important areas as required. We had used a similar technique with our Caspian ice operations in previous years, so it made sense to port the technique for use in our NE Greenland operations.

THE ICECAT 1-2-3 SOFTWARE AND DATABASE

The IceCat 1-2-3 product is prepared within a GIS environment using Python plugins developed by the onboard ice management team. After gathering knowledge of current ice conditions from all available data sources, areas with different ice conditions were separated by lines and then converted to color-coded polygons using the IceCat 1-2-3 scheme.

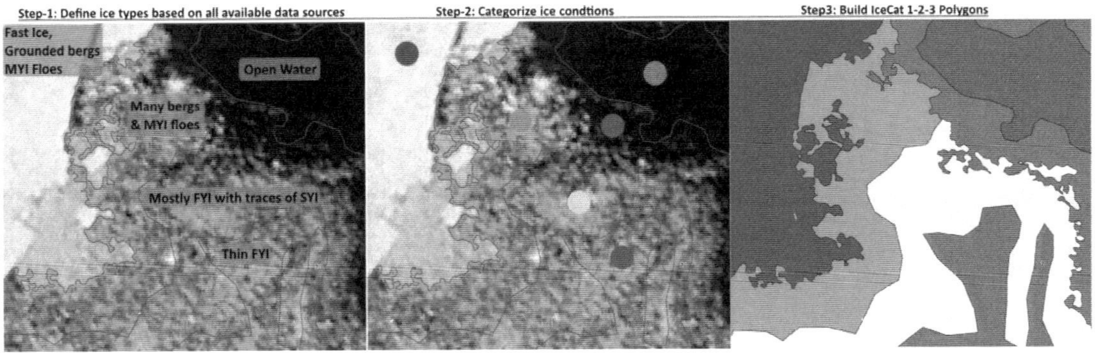

Figure 10. IceCat 1-2-3 product preparation steps.

These polygons were stored in the geo-spatial database, which allowed us to use the product for both operational (Figure 11) and post-seasonal analysis (Figure 12)

Block	OW	Light	Moderate	Heavy	Impassible
1	0.0%	65.3%	28.1%	6.7%	0.0%
2	0.0%	49.7%	27.5%	22.8%	0.0%
3	0.0%	64.7%	9.9%	19.4%	6.0%
4	0.0%	88.4%	5.5%	6.1%	0.0%
5	0.0%	18.9%	50.5%	22.1%	8.5%
6	0.0%	94.6%	2.7%	2.7%	0.0%
7	9.7%	86.2%	4.1%	0.0%	0.0%
8	12.3%	69.0%	10.4%	8.2%	0.1%
9	62.0%	33.9%	4.1%	0.0%	0.0%
10	83.6%	15.2%	0.0%	1.2%	0.0%
11	98.9%	1.1%	0.0%	0.0%	0.0%
12	94.0%	4.7%	1.3%	0.0%	0.0%
13	81.5%	18.5%	0.0%	0.0%	0.0%
14	85.2%	12.0%	2.8%	0.0%	0.0%
15	44.8%	51.7%	3.5%	0.0%	0.0%
16	100.0%	0.0%	0.0%	0.0%	0.0%
17	34.3%	56.3%	9.4%	0.0%	0.0%
18	98.5%	1.5%	0.0%	0.0%	0.0%
19	89.9%	10.1%	0.0%	0.0%	0.0%
All	49.8%	37.0%	8.0%	4.4%	0.7%

Data source: Radarsat-2
12 Oct 2015, 07:47 UTC

Statistical Summary

Current 12-Oct-2015:
Open Water: 7.4% (-2.7%)
Light: 17.8% (-11.0%)
Moderate: 17.5% (-2.1%)
Heavy: 51.3% (+14.2%)
Impassible: 6.0% (+1.5%)

Previous 11-Oct-2015:
Open Water: 10.1%
Light: 28.8%
Moderate: 19.6%
Heavy: 37.1%
Impassible: 4.5%

Conditions are:
Deteriorating

Data source: Radarsat-2
11 Oct 2015, 08:16 UTC

Statistical Summary

Current 11-Oct-2015:
Open Water: 10.1% (-2.6%)
Light: 28.8% (-2.4%)
Moderate: 19.6% (-7.0%)
Heavy: 37.1% (+9.9%)
Impassible: 4.5% (+2.2%)

Previous 10-Oct-2015:
Open Water: 12.7%
Light: 31.2%
Moderate: 26.6%
Heavy: 27.2%
Impassible: 2.3%

Conditions are:
Deteriorating

Figure 11. Example of statistics calculated for ice conditions occurring for 12 October 2015 for all blocks (left) and the comparison statistics calculated for ice conditions occurring for 11 October 2015, from two subsequent IceCat 1-2-3 charts covering all blocks (right). The timing of choice for this example covers the period where freeze-up is commencing with new ice forming throughout the AOI.

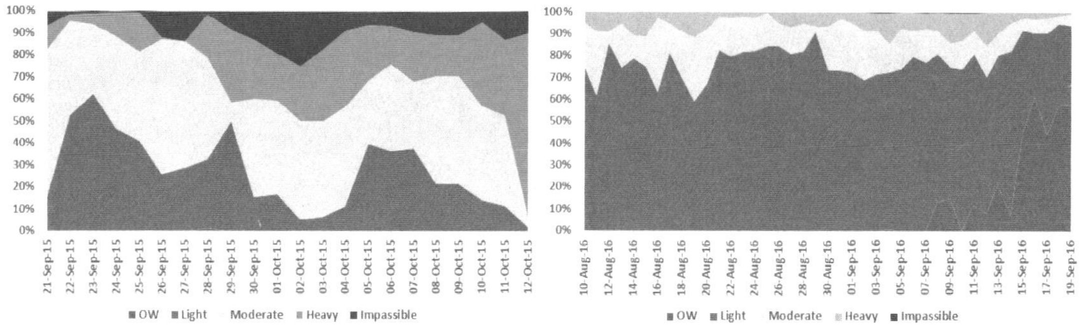

Figure 12. Seasonal comparison of the same license block for seasons 2015 (left) versus 2016 (right). 2015 showed substantially heavier (Orange) and impassable (Red) ice conditions in 2015 compared to the most recent season, 2016. This is in line with a generally observed trend of less heavy ice being noted throughout the AOI from project start in 2011 until project end in 2016.

Although the IceCat 1-2-3 database contains data for just six years, compared to decades of WMO ice chart databases, the recent rapid climate shift now taking place may make the 50-year data set less relevant for planning purposes. As well, and a key observation, is that ice chart interpretations made now are built from higher resolution imagery and much better data sources that were available even 10 years ago.

PORTABILITY OF THE ICECAT 1-2-3

The IceCat 1-2-3 can be ported to other geographic locations where a higher resolution of ice data may be required to drive specific operational or environmental reporting requirements. Such operations might include offshore oil exploration programs using floating or bottom founded rigs or even ice islands. It also shows promise for providing information on smaller or contained geographic areas such as the Great Lakes, Caspian Sea, navigable straits or even poorly sheltered harbors or transit routes in areas such as the northern sea routes. It would certainly work in areas like the southern Beaufort Sea, Chukchi Sea and along the Alaskan coast.

But there are challenges … the sheer volume of data required to produce the best IceCat 1-2-3 means that all inputs must be available in a timely fashion and that communications with all input sources such as satellite imagery be available via high-bandwidth.

It is very likely that in the not too distant future, clusters of satellites will provide frequent, high-resolution ice data imagery. Forecast weather models will continue to improve and the use of drones (UAVs) equipped with both optical and nano-SAR instrumentation requiring fast and reliable data streams will become standard. Upward-looking sonars deployed in arrays and connected via acoustic or fiber-optic networks will give early warning of heavy ice floes with substantial ridge element concentrations of encroaching the AOI. Real-time current meter data, from instruments deployed in wide areas at various depths will drive ice drift models with a much higher degree of accuracy than is available now.

To produce the best IceCat 1-2-3, all personnel involved must be very experienced and capable in ice management, marine operations, meteorology, oceanography, technically capable and have direct knowledge of the capabilities and limitations of the vessels or structure being supported as well as the skills of the personnel operating such vessels or structures.

And while quantifying vessel operations personnel and vessel capability is difficult to ascertain without a major time and testing investment, it is possible and is already being done to a limited degree. There is already a real push to qualify vessel management skills in ice. As well, through the use of the Polar Code, there is now a real effort to quantify a vessel's capability in ice.

That leaves just ground truthing missing from the mix of ingredients. Higher and higher resolution imagery will help but a standardized requirement for vessels to report ice conditions would certainly help. This is currently handled within National Coast Guard Fleets, ice breaker fleets, ferry services, the military and within other companies routinely operating in ice but a standardized reporting would help for vessels that follow. Weather reporting has been handled like this for many years by ships transiting the world's oceans so why not ice reporting? Airline pilots routinely report current turbulence for aircraft following behind so this is quite possible to handle with come cooperation and leadership.

All of this could lead to an ice chart, similar to an IceCat 1-2-3 product and specific to a vessel, its personnel and their geographic surroundings, being issued coincident with a change in any major input such as weather forecast, availability of newer ice imagery, other vessel's ice report, etc. Telling vessels where they can and can't go using an intuitive product put together by a capable ice service could increase safety of all operations carried out in remote Arctic water.

CONCLUSIONS

Having honed this product/service over several years in operational practice it is easy to see its possibilities for use with the new IMO Polar Code, where the code for each vessel would allow an IceCat 1-2-3 to be tailored for each vessels capability (polar code) based on ice conditions occurring in the planned area of operations. For multiple vessels intending to navigate through an area of ice, the IceCat 1-2-3 product would reflect the enhanced capabilities or limitations such as would occur when the vessel in question is being accompanied by a Polar 10 icebreaker (enhanced capability) or if the vessel in question picks up a barge for towing (degraded capability).

While this IceCat product is a large departure from traditional WMO egg code ice charts, it may be that adopting a simple product such as an IceCat 1-2-3 would allow all Arctic stakeholders from villagers to regulatory authorities and the public in general, to appreciate the ice challenges faced by vessels transiting through their waters, without the need for specialized knowledge.

ACKNOWLEDGEMENTS

The authors would like to thank TGS-NOPEC of Norway for inspiring the development of this paper, and in particular Sergej Usov, Manager Operations - Europe & Russia, and Reidun Myklebust, Senior Geologist Project Development, Europe & Russia.

We also thank the captains and crews of Balder Viking as well Arctia's Fennica and Otso.

REFERENCES

The Canadian Ice Service, *Interpreting Ice Charts - World Meteorological Organization Colour Code* [Online] Available at: https://www.ec.gc.ca/glaces-ice/default.asp?lang=En&n=D5F7EA14-1&offset=2&toc=hide#concentration-sea [Accessed 25 March 2017].

The WMO Polar Code, *Shipping in Polar Waters* [Online] Available at: http://www.imo.org/en/mediacentre/hottopics/polar/pages/default.aspx [Accessed 20 March 2017].

Viking Supply Ships, Viking Supply Ships develops the first Polar Code Course [Online] Available at: http://www.vikingsupply.com/newsarticle/viking-supply-ships-develops-the-first-polar-code-course%5B%5Bexternal%5D%5D/2142758 [Accessed: 15 March 2017]

World Meteorological Organization, 2004. WMO Sea Ice Nomenclature, WMO/OMM/BMO No. 259 & bull Edition 1970-2004.

Proceedings of the 24th International Conference on
Port and Ocean Engineering under Arctic Conditions
June 11-16, 2017, Busan, Korea

Arctic Sea Ice Sensitivity to Ice Thermodynamic Formulations in a Sea Ice Model

Joo-Hong Kim, Su-Bong Lee, and Baek-Min Kim

Abstract

We perform the sensitivity tests of Arctic sea ice to different thermodynamics components using the Los Alamos sea ice model version 5 (CICE5). The new mushy layer thermodynamics formulation (MUSHY) in which salinity evolves with time is compared to the previous thermodynamics (BL99) that assumes a fixed salinity profile. MUSHY exhibits larger simulated ice volume due to thicker ice. Comparison of the ice growth/melt rate reveals that MUSHY has the lower growth rate in cold season and the lower melt rate in warm season. The higher brine salinity in MUSHY may increase the thermal inertia of sea ice and ice temperature. Thus, the seasonal variation of sea ice volume is smaller in MUSHY. Although the change in thermodynamics formulation primarily modulate salinity and temperature of the ice, the other external processes become salient in warm season. In warm season, the lower melt rate in MUSHY is mostly due to the lower top melt rate, which originated from the more top snow volume persisting for the entire year, as well as the much less net shortwave flux and melt pond occurrence in MUSHY. The processes in warm season dominate the equilibrium ice thickness, leading to thicker ice in MUSHY.

POAC'17
Busan, KOREA

Proceedings of the 24th International Conference on
Port and Ocean Engineering under Arctic Conditions
June 11-16, 2017, Busan, Korea

Development of the Seasonal Forecast System of the Interannual Sea-ice Variability Using CICE5 Model

Yoo-Geun Ham[1], and Jeong-Gil Lee[1]

[1] Department of Oceanography, Chonnam National University, Gwangju, Korea

ABSTRACT

Using the CICE5 model, the initialization technique based on the nudging scheme is developed to assimilate the PIOMAS monthly sea-ice thickness, OISST daily sea-ice concentration, and OISST daily SST. The initialization using the observed SST and sea-ice concentration has a systematic problem that the sea-ice thickness has a serious positive bias from the beginning of the prediction. On the other hand, by assimilating the sea-ice thickness in addition to the sea-ice concentration and the SST, the bias in the initial condition is significantly reduced, and the prediction skill using the dynamical sea-ice model is also improved. By initializing the sea-ice thickness, sea-ice concentration, and SST, the correlation skill of total sea-ice extent in the forecast starting from September 1st, is over 0.5 up to 5-month lead forecast, even though the simple nudging scheme is utilized. In addition, the correlation skill over the Barent-Kara sea, which is important to predict the winter climate variability over the East Asia, is generally over 0.6 up to 6-month lead forecast. This implies that the sea-ice variables are well initialized in the developed dynamical prediction system to lead skillful forecast of the interannual sea-ice variation. The future development plan of the seasonal forecast system of the sea-ice variation will be also discussed.

Ice Intelligence Retrieval by Remote Sensing – Possibilities and Challenges in an Operational Setting

Runa A. Skarbø[1,2,3], Sveinung Løset[1,2]
[1] The Norwegian University of Science and Technology (NTNU), Trondheim, Norway
[2] Centre for Research-Based Innovation (CRI) SAMCoT (Sustainable Arctic Marine and Coastal Technology), NTNU, Trondheim, Norway
[3] CRI CIRFA (Centre for Remote Sensing and Forecasting for Arctic Operations), UiT The Arctic University of Norway, Tromsø, Norway

ABSTRACT

Detecting ice drift velocity when operating offshore in ice-covered waters is crucial during marine operations, as ice actions affect station keeping and ice management. Furthermore, other ice data/intelligence such as ice concentration and thickness are important parameters to determine ice resistance, evaluate performance of icebreakers and predict ice actions on structures. Different sensors are available and capable of providing ice intelligence; however, no single sensor is capable of providing all necessary ice intelligence alone. Thus, an operational scenario depends on combining ice intelligence from several sensors. Previous studies have assessed potential sensors that detect ice drift; however, the practical implications of applying these technologies in operational scenarios are often disregarded. This paper reviews the various sensors currently available for sensing ice drift and other ice intelligence, and their abilities to provide ice information for operational scenarios. The sensors satellite SAR, marine radars and optical cameras are assessed qualitatively in a case study. The study considers the scenarios of drilling and production of hydrocarbons at the Korpfjell prospect in the central eastern Barents Sea, where the Norwegian Ministry of Petroleum and Energy recently awarded a license. The case study shows that during an operational scenario, ice intelligence must be provided by a combination of regional and local sensors. Furthermore, great potential exists to combine intelligence from different sensors to form an operational monitoring, detection and surveillance tool for operational decision support.

KEY WORDS: Ice Intelligence; Ice Drift; Ice Management; Remote Sensing; Decision Support; Case Study; Barents Sea; Arctic Marine Operations.

INTRODUCTION

Offshore operations are demanding, even in open waters. Operations in ice-infested waters need to consider, identify and manage hazards posed by various ice features. In order to avoid and/or reduce the frequency and consequence of collisions with hazardous ice features, ice management may be performed. Eik (2008) defined ice management as the sum of all activities where the objective is to reduce or avoid actions from any kind of ice features.

Detection, tracking and forecasting of sea ice and glacial ice are considered being part of the ice management. Parameters such as ice thickness, concentration and drift need to be determined in order to assess that loads do not exceed design values for a structure and then if so envisage efficient ice management for load reduction.

Ice intelligence may be obtained in several ways. Traditional methods for providing information on ice drift velocity include deploying physical trackers on the ice, commonly by means of helicopters. This deployment method represents a risk for crew and equipment, has limited operability due to weather and light conditions and has a high carbon footprint. Ice concentration and thickness are normally determined manually by experienced ice observers on board. However, manual observations lack consistence and suffer from low log-in frequency (Hall et al., 2002). Therefore, the motivation for the use of remote sensing technologies operationally to automate ice intelligence retrieval is high.

Physical ice management requires tugboats or icebreakers to tow or break the ice. These vessels are expensive to have on standby on location, and operators may choose to rely on vessels that serve several fields in vicinity of each other or are on standby in nearby ports. However, in remote locations, the mobilization and transit times for such vessels may be substantial. In order to ensure safe operations and avoid over-specification of the available resources, automatic ice intelligence retrieval coupled with drift prediction models of a certain time period may ensure optimal configuration of resources.

Hydrocarbon exploration and production in Arctic conditions have been performed since the 1960s. However, there are only a few places around the world where year-round offshore hydrocarbon production in icy waters have taken place. Recent licenses awarded on the Norwegian Continental Shelf opened new locations for drilling in the central eastern Barents Sea, where the presence of sea ice and threat of icebergs has to be taken into account.

This paper considers an operational case study on the Korpfjell location, central east in the Barents Sea. Statoil plans to perform drilling at this site in 2017. Through a case study, the ice intelligence requirements on different planning time horizons are established for the scenarios of drilling and year-round production. Several remote sensing technologies are evaluated in terms of their performance and possibility to provide the required ice intelligence.

The structure of this paper is as follows: Remote sensing technologies and platforms currently available are introduced, and their application to the remote sensing of ice intelligence are discussed. The framework for the case study of the Korpfjell prospect is described, along with the parameters according to which the remote sensing techniques are evaluated. The remote sensing technologies chosen are compared and evaluated applied to the case study. A "three-layer concept" of ice intelligence is presented. Lastly, the results are discussed, and the conclusions are presented along with suggested further work.

REMOTE SENSING OF SEA ICE

Remote sensing is defined by Elachi and van Zyl (2006) as *the acquisition of information about an object without being in physical contact with it.* Information about the object is acquired by detecting and measuring changes that the object causes in its surroundings, by either electromagnetism, acoustics or potential.

There are several sensor types and platforms relevant for remote sensing of sea ice. Such sensor types include synthetic aperture radar (SAR) and other satellite based technologies, marine radar, thermal/infrared and optical cameras. These sensors can be deployed on platforms such as satellites, unmanned aerial vehicles (UAVs), ships, bottom-mounted buoys and autonomous underwater vehicles (AUVs). For this paper, only sensors capable of providing real-time or near real-time information are considered. This section introduces each platform and their relevant sensors, with focus on sensors relevant for the case study.

The Satellite Platform

The development of satellite technology has made it possible to gain detailed information about the planet, especially in remote and logistically challenging locations such as the Arctic. Imaging radars are especially suited for the task due to their all-day, all-weather capability.

Most satellites used in meteorology and earth observation are deployed at high inclination or *polar* orbits (Lubin and Massom, 2006), where their ground tracks converge spatially at high latitudes. Thus, polar areas have frequent temporal coverage by the sensors.

Due to technological limitations, trade-offs exist between coverage and spatial resolution. High-resolution sensors tend to offer limited spatial coverage over a narrow swath (≤100 km). Medium to coarse resolution (>250 m) sensors typically operate on much wider swaths (>1500 km) (Lubin and Massom, 2006). This is a major issue in operation scenarios that require a high temporal and spatial resolution, such as monitoring of highly dynamic sea ice.

Data from satellite imagery have operational value for planning operations at a given location, as they can provide information on the historical climatic conditions and the regional ice conditions. Furthermore, current satellite images may be used to detect and monitor hazardous ice features and conditions in near real-time. For example, several providers use satellite data to create near-real-time sea ice maps and drift charts, of which examples are illustrated in Figure 1. Such maps are typically updated once per day, produced for large areas and have low spatial resolution.

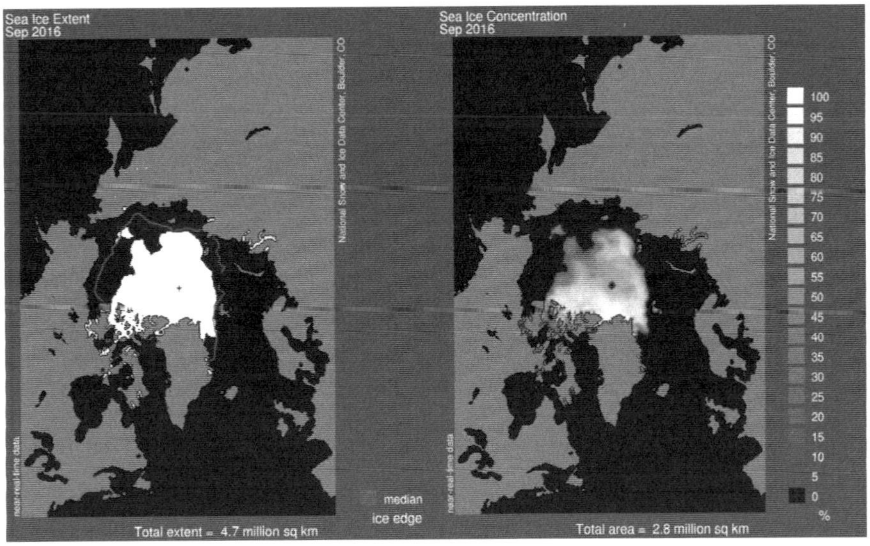

Figure 1. Monthly sea ice maps for September 2016, produced by the National Snow and Ice Data Center (NSIDC).

Several types of sensor technologies are used in satellite remote sensing. For the purpose of this study, the Synthetic Aperture Radar is of primary interest.

Synthetic Aperture Radar

Synthetic Aperture Radar (SAR) is an active microwave sensor system, unhindered by darkness or cloud cover. The radar waves may penetrate dry snow to provide volume structure information about the underlying ice. This property makes it a useful sensor to discriminate different types of ice, such as distinguishing the saline first-year ice (FYI) from the less saline and harder multi-year ice (FYI).

SAR products are high-resolution images where the grey scale is determined by the strength of the received radar return. The strength of the signal is proportional to the roughness of the surface, which is proportional to the wavelength of the radar pulse and the incidence angle of the radar. Because the incidence angle varies over the scene, materials with the same properties may appear different over the same scene. This makes automatic classification and analysis of SAR images challenging. Figure 2 shows a SAR image taken by Sentinel-1 over the marginal ice zone in the northern Barents Sea. The image illustrates well the difference in backscatter from the open sea surface, ranging from virtually no backscatter (black) to very strong backscatter (white), depending on the incidence angle.

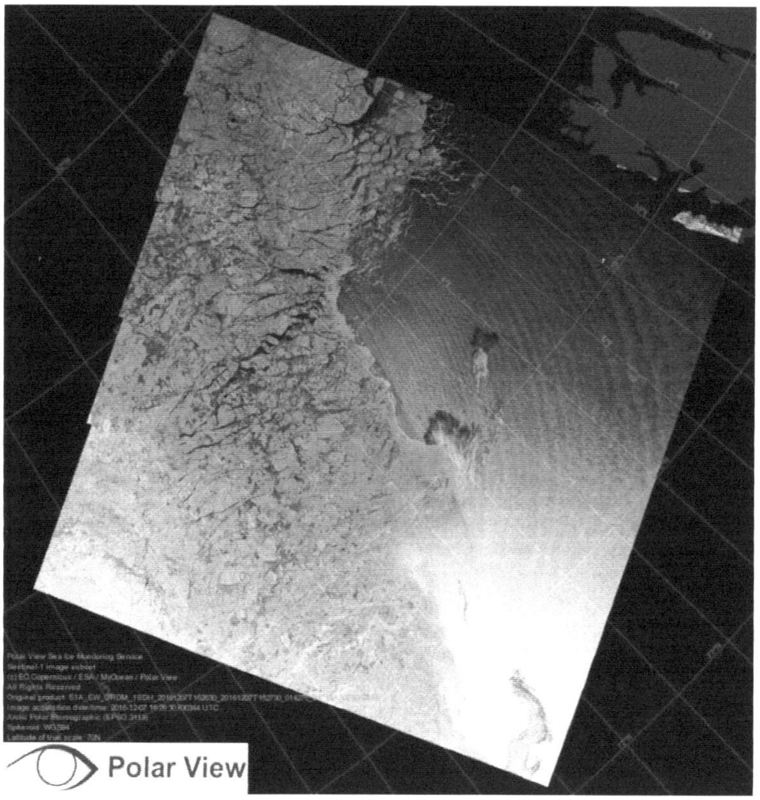

Figure 2. Example of a Sentinel-1 SAR extra wide scene from the ice edge north-west of Svalbard (masked in blue). In the image, sea ice can be seen to the left, while the sea closest to Svalbard is ice-free. Note the difference in ocean surface appearance over the incidence angle range of the image, ranging from almost white in the lower right corner to black at the top. (Image courtesy of Polar View).

Marine Radar

The International Conference on Safety of Life at Sea (SOLAS) of 1974 require all ships to be equipped with marine radars. Several commercial brands of dedicated ice radars exist, however their primarily use is for navigation in ice.

As all vessels are required to have marine radars, there has been an interest in using this sensor to retrieved ice intelligence, to avoid installation of proprietary and potentially costly additional sensors. Kjerstad et al. (2017) have proposed an automatic real-time algorithm that estimates the local ice drift near a vessel in real time, using the marine radar. Here, local refers to an area of a few (0.5 to 6) nautical miles in radius. The algorithm uses image processing to detect and track the motion of N distinct features (DFs), and two Kalman filters to select DFs and decouple the vessel motion. However, the algorithm is not capable of identifying ice features.

Unmanned Aerial Vehicles

With the improvement of remote controlled technology, there are several attempts of using UAVs for ice observations. Their flexibility in geographic coverage and low investment cost paired with high potential spatial and temporal resolution of collected data make this platform a strong candidate for ice observation. However, UAVs are sensitive to extreme environmental disturbances such as low temperatures, high wind speeds and atmospheric icing, which make reliability of these platforms an issue in an operational setting (Haugen et al., 2011).

Autonomous Underwater Vehicles

AUVs operate underwater below the ice, which makes it independent of surface conditions such as light, wind and temperature conditions. These properties make AUVs a good candidate for ice monitoring during harsh conditions in Arctic and cold climate environments.

Norgren and Skjetne (2014) investigated AUVs as a potential platform for ice monitoring. An Acoustic Doppler Current Profiler (ADCP) instrument can be used for measuring the ice-instrument relative velocity. However, limited range and deep waters put limitations on the applicability in the Arctic.

Infrared and Optical Cameras

Cameras may be mounted on platforms such as the ship itself or on UAVs.

Infrared sensors detect difference in thermal emissivity of objects. This is especially relevant for iceberg detection, as icebergs in general are much colder than the surrounding water. However, camera images are of low use in conditions with low visibility. Furthermore, maintenance of the cameras in atmospheric icing conditions and/or snow can be an issue.

Optical cameras are practical for monitoring the close surroundings of a vessel. Lu et al. (2016) have demonstrated use of optical cameras mounted on a vessel to identify nearby ice conditions. The algorithms developed by the authors were able to determine local ice concentration and thickness, traditionally determined manually by trained ice observers on board. A 180° camera looking obliquely towards the ship transit direction was used to provide information on ice concentration. The concept is illustrated in Figure 3. A downward-looking camera on the ship's side captured events of tilted pieces of broken ice, and was used to identify ice thickness.

Furthermore, floe and brash ice size distribution may also be retrieved from optical cameras, i.e. as presented by Zhang and Skjetne (2014).

Optical cameras have the same limitations as infrared cameras, with the addition of limited or no applicability in low visibility and darkness.

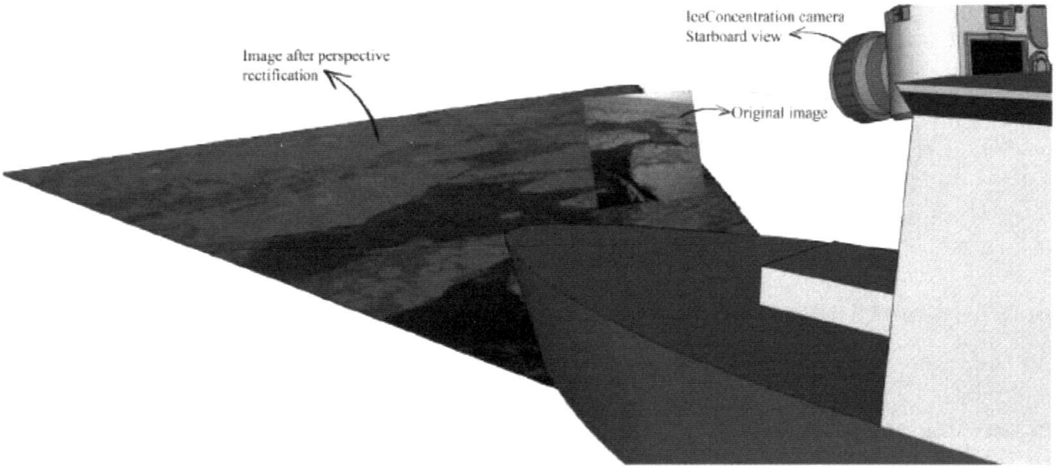

Figure 3. Illustration of the vessel-mounted forward-looking camera and perspective rectification in the method by Lu et al. (2016).

CASE STUDY: HYDROCARBON EXPLORATION AND PRODUCTION IN THE CENTRAL EASTERN BARENTS SEA

In 2015, the Norwegian Ministry of Petroleum and Energy (OED) announced the 23rd licensing round. Out of the 57 blocks announced, 54 of the blocks were in the Barents Sea. The Barents Sea licenses awarded in 2016 are illustrated in Figure 4. The most controversial blocks in the licensing round was the blocks PL 857, PL 858 and PL 859 in the central eastern Barents Sea, where sea ice may intrude during the winter months.

The location selected for the case study is the area PL 859 at 74°30'N 36°E, known as the Korpfjell prospect. Statoil plans exploration drilling at this location in Q2/Q3 of 2017 (Statoil, 2017).

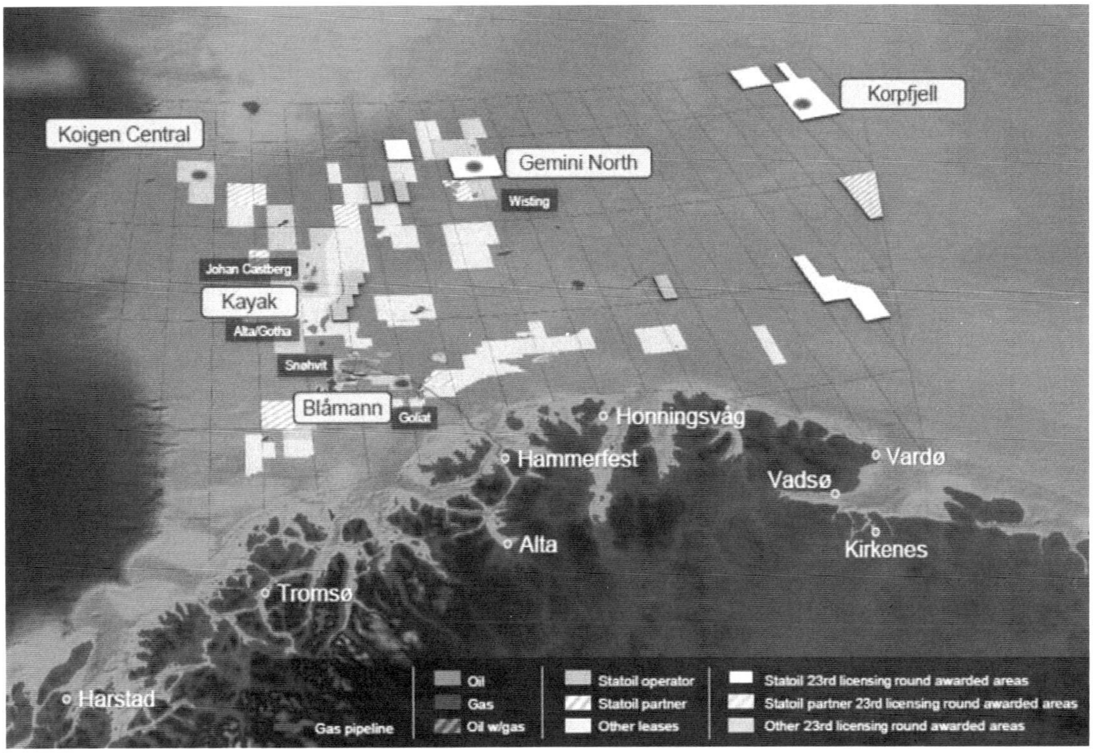

Figure 4. Barents Sea fields, discoveries, areas awarded and areas that have been opened for exploration activities. The locations of drilling in 2017 are indicated by name and red dots. (Image courtesy of Statoil.)

Due to the relatively warm water in the North Atlantic Coastal Current, the southern part of the Norwegian Barents Sea is usually ice-free year round. However, at 74°N, sea ice occurs in some years as well as drifting icebergs (Løset and Carstens, 1996).

The major source of icebergs in the Barents Sea are the glaciers on Franz Josef Land, Novaya Zemlya and to a smaller extent Nordaustlandet on Svalbard (Løset, 1993). There are generally more icebergs in the northern and eastern parts of the Barents Sea than the western parts (Abramov, 1996). However, limited data exist for the Barents Sea compared to i.e. the Grand Banks.

The NORSOK standard N-003 gives limits of annual probability exceedance of sea ice occurrence and iceberg collisions in the Barents Sea based on satellite data as shown in Figure 4. From Figure 4, it is evident that the Korpfjell prospect is located north of the 10^{-2} contour lines for both sea ice and iceberg occurrence. Thus, both presence of sea ice and potential iceberg collisions must be considered in both structural design and operations.

The water depth on the location is 253 metres (Statoil, 2017). The deep waters limit the application of bottom-founded structures.

The closest port on the mainland is Honningsvåg, 519 km away from Korpfjell. Today, this is the most important base in the Barents region for rescue operations and port for vessels to wait for weather. A ship transiting to Korpfjell from Honningsvåg travelling at 13 knots will be about 22 hours on the transit.

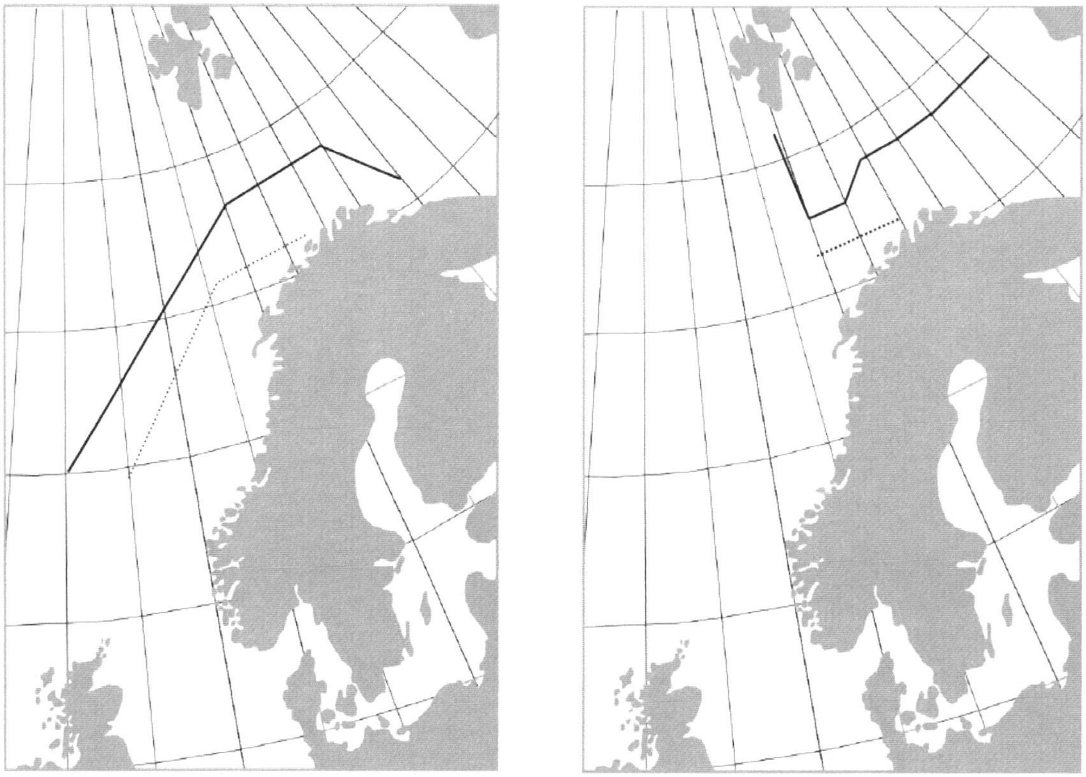

Figure 5. NORSOK N-003 southern limits of sea ice (left) and icebergs (right) with an annual probability of exceedance of 10^{-2} (solid line) and 10^{-4} (dotted line) (NORSOK N-003, 2007).

Drilling

Drilling operations are usually performed during open water and shoulder seasons, in order to reduce or eliminate the risk of operating in ice. Furthermore, drilling structures are usually temporarily present on the field, with a shorter disconnection time in case of emergency shutdown.

The semi-submersible rig Songa Enabler will be used for the exploration drilling in 2017 (Statoil, 2017).

In the licence requirements for the Korpfjell prospect, the Norwegian Petroleum Directorate (NPD) does not allow drilling if sea ice is observed closer than 50 km from the well location. In the emergency preparedness analysis for the Korpfjell prospect, Statoil states that they will perform ice surveillance before and during drilling operations and establish an ice management plan (Statoil, 2017).

Ice surveillance for a drilling operation in open water with possibilities for incoming ice features should be focused on detection of potential hazardous ice features that may drift into the location. Local ice surveillance is not necessary, as the rig would most likely disconnect and move off location in case of approaching ice features.

Production

Normally, a production structure will be present year-round and needs to deal with incoming sea ice and icebergs. Due to the water depth at the location, a production structure will most likely be a floater. An FPSO is one viable option, with other possible options such as a SPAR buoy or semi-submersible.

An ice intelligence retrieval system for a production scenario should be able to detect and monitor potentially hazardous ice features, as well as monitor the local ice conditions and drift for physical ice management on-site.

EVALUATION OF REMOTE SENSING TECHNOLOGIES

The remote sensing technologies satellite SAR, marine radar and optical cameras have been evaluated according to the criteria specified and described below.

Temporal and Spatial Resolution

As ice drift is highly dynamic, temporal resolution is of high importance to marine operations. Furthermore, spatial resolution of the sensor is of importance in order to evaluate at which scale ice features are identifiable. Table 1 lists the temporal and spatial resolution of the chosen sensor technologies, along with the coverage.

Table 1. Resolution and coverage of sensor technologies.

Sensor type	Temporal resolution	Spatial resolution	Coverage
Satellite SAR*	6-12 hours	>30 m	400 km
Marine radar	Second	Metres	1-3 km
Optical cameras	Sub-second	Sub-metre	100s of metres

* Parameters will vary depending on which satellite sensor is used. The data in the table are based on Sentinel-1 Extra Wide Swath mode, which is used for maritime, ice and polar zone operational services, where wide coverage and short revisit times is required.

As previously discussed, satellite SAR operates with a trade-off between spatial and temporal resolution and coverage. Furthermore, the highest temporal resolution available is on the few-times-per-day scale, and results in low spatial resolution due to a large swath width.

The marine radar has a high temporal resolution, in practice limited by the processing speed of the computer running the algorithm and the chosen frequency. Furthermore, the spatial coverage of the marine radar is limited by the operator setting, and can be in the range of 1 to 3 km (Kjerstad et al., 2017).

Optical cameras have high temporal and spatial resolution, but can only view the immediate surroundings of the vessel, and are limited in terms of areal coverage. They have high potential for identification of ice features, with sub-metre resolution.

Ice Feature Identification

The sensors ability to recognize hazardous ice features is of high importance for ice management operations. The assessment of the possibility to identify hazardous ice features from the sensor technologies are presented in Table 2.

Table 2. The sensor technologies' potential to identify ice features.

Sensor type	Ice floes	Icebergs	Ice ridges
Satellite SAR	Yes, large floes	Yes, free floating	Potentially
Marine radar	No	Potentially	Yes
Optical cameras	Yes*	Yes*	Yes*

* In light and fog-free conditions

Satellite SAR images are already widely used for sea ice monitoring on a large scale. Large, free-floating ice floes may be identified from satellite SAR images. The size of the floes depend on the spatial resolution of the image, at best 30 metres per pixel in wide swath mode. However, the operational resolution depends on the analysis post-processing of the images, and are generally lower than the pixel resolution.

The potential for identification of ice features from marine radars is generally lower than for SAR technology. The algorithm proposed by Kjerstad et al. (2017) is not capable of detecting ice features. However, commercial systems exist that claims to be able to identify and track ridges, icebergs and ice floes to some degree using the marine radar (FURUNO n.d., Rutter Inc. n.d.).

All ice features can be detected from optical camera images, given the right operating conditions for the camera. However, as discussed previously, the use of this technology may be limited by icing, fog and darkness.

Suitability of Sensor Technologies on Different Planning Time Scales

During an operation, different intelligence is needed on different timescales for ice management. In general, planning on the strategic time scale is for more than 48 hours ahead. On an operational time scale, planning between 2 to 48/72 hours ahead is performed. Real time planning is in the range of 2 to 10 hours ahead. The evaluation of the suitability of the different sensor technologies on different planning time scales is listed in Table 3.

Table 3. Suitability of the sensor technologies to provide ice intelligence on different time planning scales.

Sensor type	Strategic	Operational	Real time
Satellite SAR	Good	Intermediate	Poor
Marine radar	Intermediate	Good	Very good
Optical cameras	Poor	Intermediate	Very good

Satellite SAR images cover large areas, which makes it suited to provide intelligence on a strategic planning level with a long time horizon, and to monitor incoming ice. On an intermediate scale, the suitability of SAR images depends on when the image is received. As images are received potentially a couple of times per day, the images may or may not be useful, depending on current ice conditions and drift speeds. For real time monitoring, SAR images do not have an adequately high enough temporal or spatial resolution to monitor areas in close vicinity to the vessel.

The marine radar has limited range, and is thus not well suited to provide intelligence for long-term strategic planning. However, in the operational and real-time domains, with providing ice intelligence about conditions in proximity to the vessel in real-time, the marine radar has the possibility to provide valuable ice intelligence, especially on ice drift.

Optical cameras are mainly useful in the real time operational intelligence, for monitoring conditions close to the vessel. As previously stated, they are limited by factors such as darkness and visibility.

Availability and Communication

Availability is assessed in terms of how easily the operator can access the intelligence, how easily understood the intelligence is and potential delays of the information. The results of the analysis are listed in Table 4.

Table 4. The availability, ease of interpretation of the product and delay of ice intelligence collected by sensor technologies.

Sensor type	Availability	Interpretation	Delay
Satellite SAR	Poor to intermediate	Difficult	Hours/days
Marine radar	Excellent	Intermediate	None
Optical cameras	Excellent	Easy to intermediate	None

SAR products have to be downloaded from specific data services or sent by ice monitoring services. High volume data downloads may be challenging at high latitudes with limited data links. The interpretation process of the SAR images is not trivial, requiring expert analysis and preparation. Even then, non-experienced users may not be able to make use of the images without input from an experienced user. Furthermore, deriving additional information such as ice drift or feature tracking also require separate expert analyses. The delay may also be considerable, as the data have to be transferred from satellites to Earth, processed, analysed, and then transferred to the end user.

Marine radars have, like SAR, high availability due to all weather and light capabilities. Furthermore, the intelligence has no delay for the operator, as the technology is shipborne and produce images in real time. The images may require expert analysis in order to identify features, if even possible.

Camera images have limited availability, as they depend on weather and light conditions. However, the technology is, like the marine radar, shipborne, and may be available to the operator instantaneously. Camera images are easily interpreted by looking at them, given ideal operating conditions under high-visibility conditions. Furthermore, algorithms can be incorporated to automatically identify ice parameters of importance, such as ice concentration, floe sizes and ice thickness.

APPLICATION AND COMBINATION OF THE REMOTE SENSING TECHNOLOGIES DURING OPERATIONS

The findings in previous sections generally apply for all marine operations in ice prone waters. Offshore activities at the Korpfjell location, given that the area is not ice-covered all year, particularly depend on hazard detection and monitoring.

As discussed, none sensor is capable of providing all necessary ice intelligence. The three sensors satellite SAR, marine radars and optical cameras represent three layers of information, based on different time and spatial scales. The "three-layer concept" is illustrated in Figure 6.

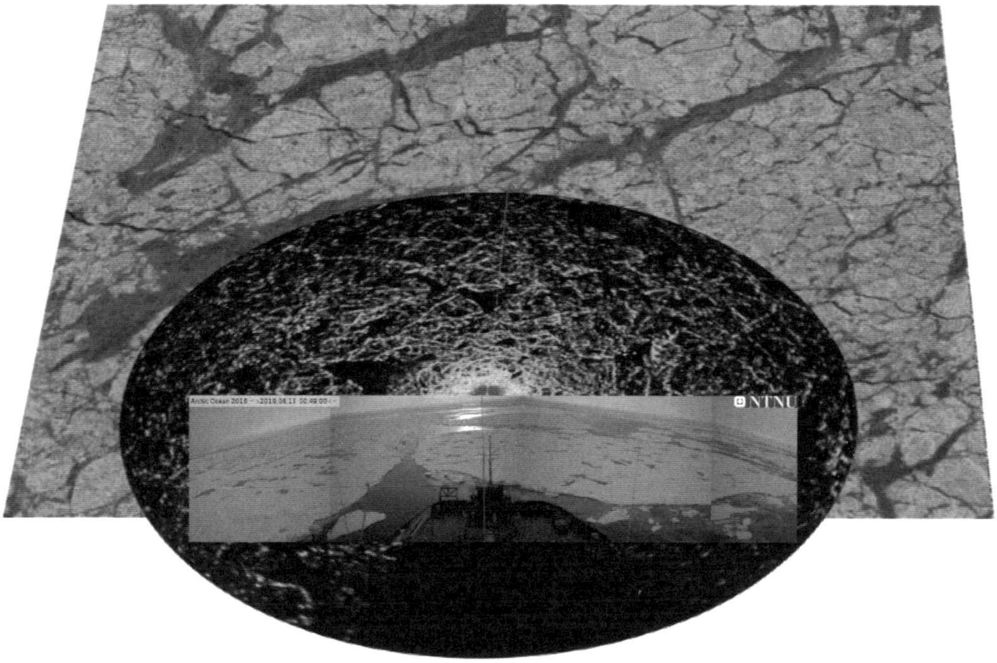

Figure 6. Visualization of the "three-layer concept" of ice intelligence; using satellite SAR, the marine radar and optical cameras mounted on the vessel can provide ice intelligence on different temporal and spatial scales to provide an overview of parameters of importance to the operation. (Satellite SAR image courtesy of Alaska SAR Facility.)

In the "three-layer concept", the satellite SAR sensor comprises the outer layer. As discussed in previous sections, satellite SAR images may provide information on the regional ice conditions; distinguishing ice types (first-year/multi-year), providing ice floe sizes and concentration, and retrieving large-scale regional ice drift on a once-per-day basis.

The second layer in the concept is the marine radar. Using i.e. the algorithm by (Kjerstad et al. 2017), local ice drift can be tracked and used in ice management operations planning and support.

The third layer represents optical images retrieved from cameras mounted on the vessel or structure. Cameras can be mounted in different configurations, depending on the floater concept selected. In case of a ship-shaped floater with upstream heading, front-looking 180° cameras facing the incoming ice would be most applicable. For an omnidirectional floater, cameras should cover a 360°-degree view. As previously discussed, cameras can be used to extract information about ice concentration, floe size distribution and ice thickness.

The "three-layer concept" used in Drilling at Korpfjell

Sea ice monitoring in a drilling operation at Korpfjell should mainly involve detection and monitoring of incoming hazardous ice features drifting in open water, such as large ice floes and icebergs. As no ice should be in the immediate surroundings of the rig (not closer to 50 km, according to the licence specifications), local surveillance sensors are of no use, and satellite SAR would be the most essential sensor for retrieving ice intelligence. Figure 7 displays an example of an operational ice surveillance scenario during drilling operations in open water at Korpfjell.

Satellite SAR images are used to monitor the pack ice and ice edge, generally to the north of the location. The chosen surveillance region will vary depending on the ice drift velocity, as the purpose is to monitor the potential incoming ice. However, due to "chaotic" current patterns in the Barents Sea, where several currents meet and interact, icebergs, which have previously drifted south of the location, may change direction and approach from the south. Thus, iceberg monitoring should be given close attention, and the satellite SAR surveillance region chosen accordingly.

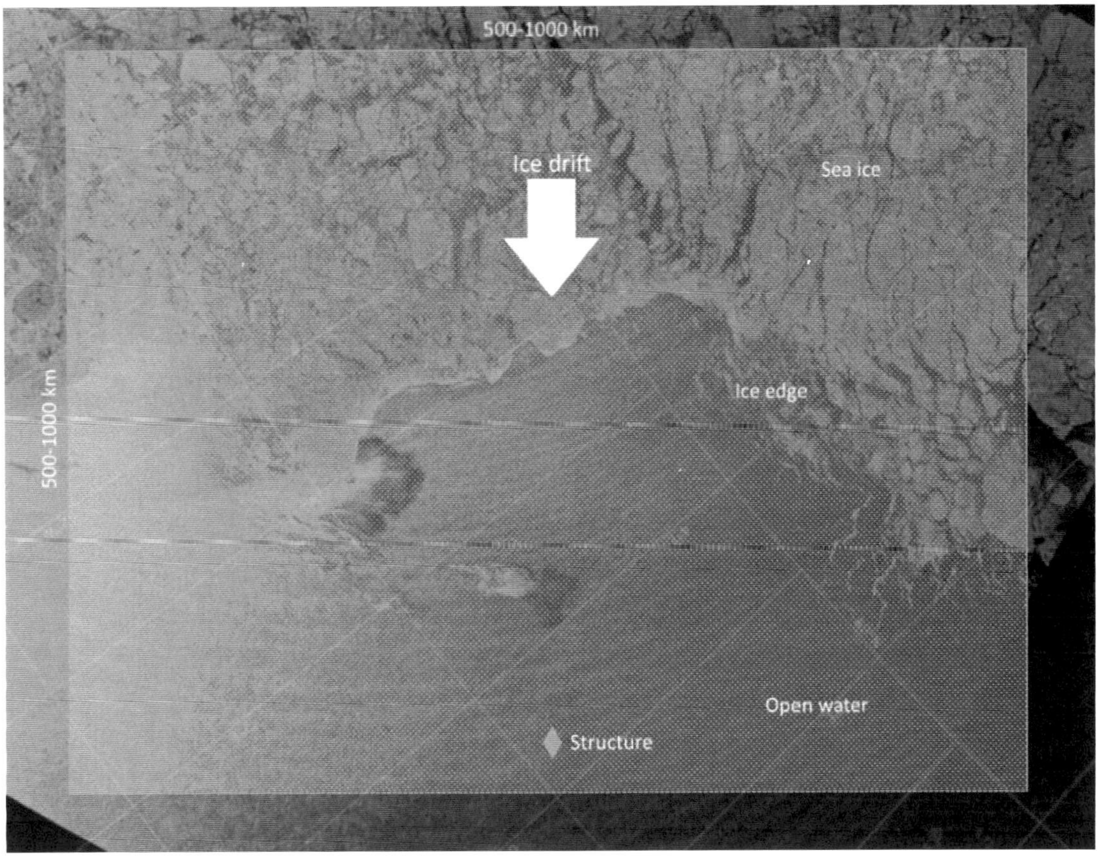

Figure 7. Surveillance region with satellite SAR represented in pink for a drilling scenario in the open water season on the Korpfjell prospect. The surveillance region will change depending on the ice drift velocity. (SAR image courtesy of Polar View).

The "three-layer concept" used in Production at Korpfjell

For a production case, which normally is year-round, both hazard detection during the open water season (as for the drilling case) and ice drift tracking and feature detection when sea ice is present at the site, is of relevance. Thus, all layers of the concept should be in place in order to have a proper operational picture of the ice conditions.

Figure 8 presents an illustration of a "three-layer surveillance concept" used for production at the Korpfjell prospect during a season with the ice edge extending south of the location. Satellite SAR is used for retrieving regional ice intelligence. The surveillance region is chosen depending on the current ice drift. As ice approaches the structure, local ice drift may be tracked using the marine radar. Furthermore, ice concentration and ice thickness can be monitored using onboard cameras facing upstream.

Note that for production in the open water season, the same surveillance configuration as for the drilling case, relying mostly on satellite SAR monitoring of the ice to the north, may be used (see Figure 7).

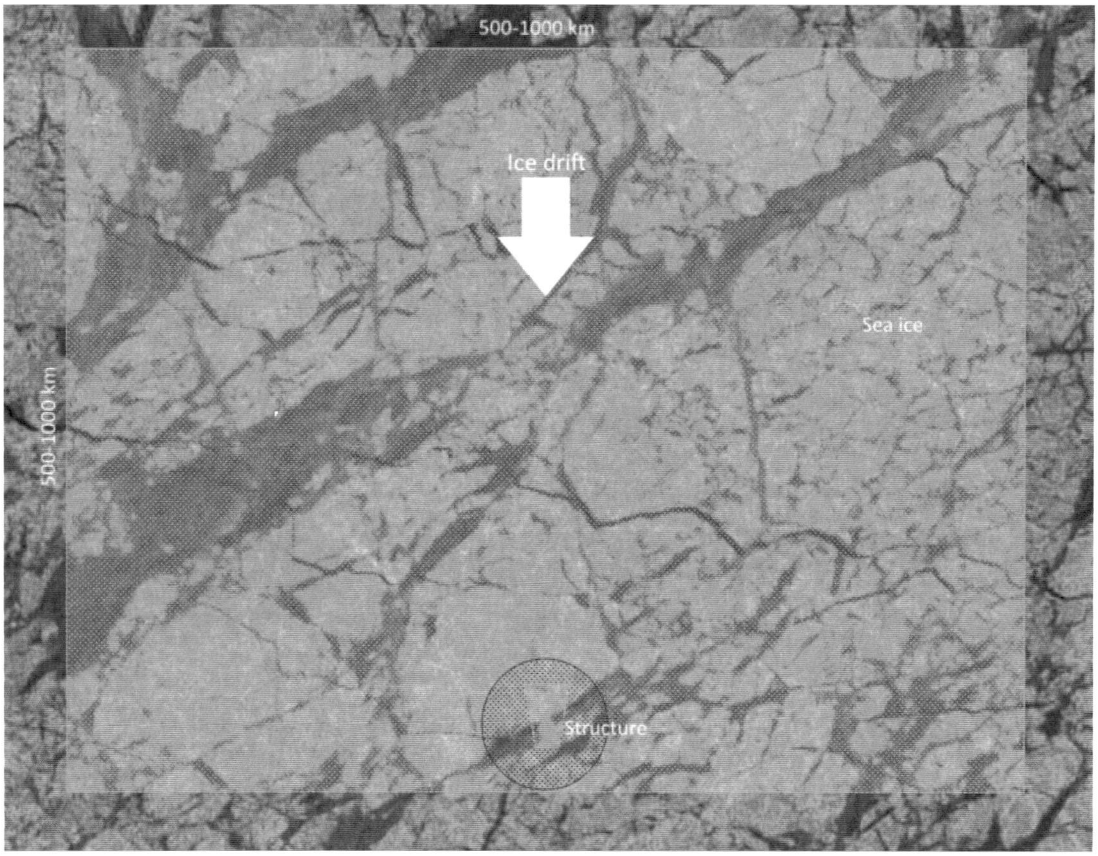

Figure 8. Ice surveillance regime for production at the Korpfjell prospect during a winter season with ice present. The pink area represents surveillance area with satellite SAR, blue indicates the marine radar surveillance region and green illustrates the camera surveillance region (not to scale). (Satellite SAR image courtesy of Alaska SAR Facility).

DISCUSSION

The evaluation of the different remote sensing technologies in previous sections shows that no single sensor is capable of providing all intelligence needed. Furthermore, all sensor technologies have trade-offs and limitations. Operational use of satellite SAR is challenging due to low temporal and spatial resolution. However, the sensor provide a good overview of the regional ice conditions and may be useful in planning operations on a longer time scale. The marine radar is currently capable of tracking ice drift, but no systems exist for feature detection.

The study evaluated and compared three different sensor technologies, satellite SAR, marine radars and cameras, for use in providing ice intelligence. These sensor types are considered most robust and viable for future development in ice intelligence retrieval. Sensors platforms such as UAVs and AUVs are promising, however at present they are not viable alternatives for operational ice intelligence retrieval.

Present-day operations would most likely use physical ice drift trackers to retrieve the local ice drift. However, with the advancement of remote sensing near-field technologies such as marine radar ice drift tracking, physical tracking technologies may be redundant in coming years.

In future scenarios, one could imagine development of a decision support tool, which incorporates ice intelligence from the different available sensors, and provided operators with detection, monitoring and forecasting of the physical environment. Such a tool, able to provide a common operational picture for the whole operation, would be of great support to operators in order to get a complete overview over the relevant parameters for different aspects of the operation. Furthermore, the information could easily be shared between the different decision makers involved, both on the floater but also for vessels supporting with ice management, land-based personnel and other relevant parties.

CONCLUSIONS AND FURTHER WORK

This paper presents different remote sensing techniques for retrieving ice intelligence in an operational setting. For all offshore marine operations, several sensors are needed to provide necessary ice intelligence on time planning horizons. No single sensor is capable of providing all intelligence required in an advanced marine operation.

In order to evaluate the potential of the remote sensing technologies in an operational setting, a case study was performed. The study considered the two operational scenarios, namely drilling and production on the newly awarded licence PL859 (also known as the Korpfjell prospect) in the Barents Sea. The conclusions of the case study are as follows:

- The main drilling operational window at the Korpfjell location is the open water season. The primary ice intelligence during drilling operations should be detection and tracking of hazardous ice features further north, which can pose a collision risk to the installation. The satellite SAR sensor is the best-suited provider of this intelligence.

- The Korpfjell location is prone to sea ice during some months of the year. Thus, year-round production at the Korpfjell location requires ice intelligence collected from several sensors;
    - In the open-water season, the ice intelligence is mainly the same as for the drilling scenario, with detection and tracking of potential hazardous ice features with satellite SAR technology as the key sensor to ice intelligence.
    - During the season when sea ice can be expected, ice intelligence from several sensors are needed to cover both regional and local ice conditions.

A "three-layer concept" of ice intelligence, using satellite SAR, marine radars and optical cameras on the vessel was proposed. Sensors and their intelligence are readily available and at low cost. Using ice intelligence from these sensors in combination will provide operators a holistic view of the ice conditions regionally and locally.

Combining ice intelligence from several sensors may provide possibilities for new tools for use in future operations. For instance, coupling the ice drift velocity retrieved with the marine radar with ice thickness and concentration could provide possibilities for tools predicting loads on the structure. Additionally, ice drift paired with ice thickness and floe size measurements could be incorporated in systems for planning and monitoring ice management performance.

Further work may be continued in forming an operational monitoring, detection and surveillance tool for use by the operators. Such a tool, which could provide a common operational picture that is available for key players in the operation, could provide major improvements to efficiency, safety and overview of the operations.

ACKNOWLEDGEMENTS

The authors wish to acknowledge the support from the Research Council of Norway through the Centres for Research-based Innovation *Sustainable Arctic Marine and Coastal Technology* (SAMCoT) and *Centre for Integrated Remote Sensing and Forecasting for Arctic Operations* (CIRFA). Furthermore, we also acknowledge the support from SAMCoT and CIRFA partners.

REFERENCES

Abramov, V. (1996), 'Atlas of Arctic icebergs: The Greenland, Barents, Kara, Laptev, East-Siberian and Chukchi Seas and the Arctic Basin', *Backbone Publ. Co.*

Eik, K. (2008), 'Review of Experiences within Ice and Iceberg Management', *Journal of Navigation* **61**(04), 557–572.

Elachi, C. and van Zyl, J. J. (2006), *Introduction to the Physics and Techniques of Remote Sensing*, Vol. 28, John Wiley & Sons.

FURUNO (n.d.), 'Ice Radar | Marine Equipment For Merchant Marine'. Retrieved 2016-12-13. http://www.furuno.com/en/merchant/iceradar/

Hall, R. J., Hughes, N. and Wadhams, P. (2002), 'A systematic method of obtaining ice concentration measurements from ship-based observations', *Cold Regions Science and Technology* **34**(2), 97–102.

Haugen, J., Imsland, L., Løset, S. and Skjetne, R. (2011), Ice Observer System for Ice Management Operations, *in* 'The Twenty-first International Offshore and Polar Engineering Conference, 19-24 June, Maui, Hawaii, USA', International Society of Offshore and Polar Engineers (ISOPE).

Kjerstad, Ø., Løset, S., Skjetne, R. and Skarbø, R. A. (2017), 'An Ice Drift Estimation Algorithm Using Radar and Ship Motion Measurements', *Manuscript submitted.*

Løset, S. (1993), 'Thermal Energy Conservation in Icebergs and Tracking by Temperature', *Journal of Geophysical Research: Oceans* **98**(C6), 10001–10012.

Løset, S. and Carstens, T. (1996), 'Sea ice and iceberg observations in the western Barents Sea in 1987', *Cold Regions Science and Technology* **24**(4), 323–340.

Lu, W., Zhang, Q., Lubbad, R. and Løset, S. (2016), A Shipborne Measurement System to Acquire Sea Ice Thickness and Concentration at Engineering Scale, *in* 'Arctic Technology Conference, 24-26 October, St. John's, Newfoundland and Labrador, Canada'.

Lubin, D. and Massom, R. (2006), *Polar Remote Sensing: Volume I: Atmosphere and Oceans*, Vol. I, Springer Science+Business Media.

Norgren, P. and Skjetne, R. (2014), 'Using Autonomous Underwater Vehicles as Sensor Platforms for Ice-Monitoring', *Modeling, Identification and Control* **35**(4), 263.

NORSOK N-003 (2007), 'Actions and action effects'. Edition 2, September 2007.

Rutter Inc. (n.d.), 'sigma S6 | Rutter'. Retrieved 2016-12-13. http://www.rutter.ca/sigma-s6

Statoil (2017), 'Beredskapsanalyse for letebrønn 7435/12-1 Korpfjell'. Distributed 2017-01-06. (In Norwegian).

Zhang, Q. and Skjetne, R. (2014), 'Image Techniques for Identifying Sea-Ice Parameters', *Modeling, Identification and Control* **35**(4), 293–301.

Proceedings of the 24th International Conference on
Port and Ocean Engineering under Arctic Conditions
June 11-16, 2017, Busan, Korea

Methods Development for Assessing Sea Depth on the Northern Sea Route Depending on the Recording of Hydrographical Relief Details

Andrei Afonin[1], Aleksandr Tezikov[1], Evgeniy Ol'khovik[1]
[1] Admiral Makarov State University of Maritime and Inland Shipping, Saint-Petersburg, Russia

ABSTRACT

In this article we presented the assessment of hydrographic study of routes of the Northern Sea Route (NSR) for safety of commercial shipping. Arctic Sea is mainly shallow, that is well known and demonstrated by the depth distribution and seas Square.

The main characteristics of high-altitude and traditional coastal routes are provided. In article data on distribution of depths on the water area of the Arctic seas are generalized. It should be noted that the difference between the minimum and the through depth increases with decreasing the details of mapping. It is noted that navigation on SMP is realized along the recommended routes that is connected with insufficient hydrographic study of a relief of a bottom. Severe ice conditions force vessels to deviate the recommended ways, both at independent swimming, and under conducting of the ice breaker because of risk of receiving ice damages. The extreme draft of the vessel calculated on the minimum depth found on the route at discrete measurements doesn't guarantee safe swimming on the route. At increase in discretization the level of credibility to the calculated value of extreme draft of the vessel decreases because the probability of presence of depth, which is smaller, than minimum depth found on the route, increases.

The original technique of an assessment of depths through passage of routes NSR is offered and examples of calculations are given. All studies and researches were carried out on the Arctic Faculty of Admiral Makarov state university

KEY WORDS: Northern Sea Route; Hydrographic study; Passing depth; Vessel draft; Minimum depth; Detail mapping.

INTRODUCTION

Northern Sea Route (NSR) passes through the Kara Sea, the Laptev Sea, the East Siberian sea and the Chukchi sea, and also northeast part of the Barents sea and northern part of the Bering sea. The length of the NSR tracks fluctuates from 2700 miles for circumpolar (high-latitude) routes to 3500 miles for the coastal routes. Total length of the NSR tracks exceeds 14 thousand miles. The access to the NSR tracks from the West is carried out through the Kara Gate strait or from the North from the Zhelanie cape. The distance from the port of Murmansk to the Kara Gate strait makes 528 nautical miles. The distance from the port of Murmansk to the Zhelanie cape is 758 nautical miles. The entrance on the NSR tracks from the East is carried out through the Dezhnyov strait. The distance from the Dezhnev Cape to the port of Petropavlovsk-Kamchatsky makes 1037 nautical miles.

Gavrilov (2015) described in detail the legal status of NSR and it is provided by the international marine legislation.

The Arctic seas are mainly shallow that is shown by distribution of depths and on the area of the seas in table 1 according to (Reshetnyak, 2006). In the Laptev Sea and the East Siberian Sea the depths to 20 meters occupy a half and more of the water area. In the Kara Sea this share reaches only 9%. The main traditional coastal and high-latitude NSR tracks pass on these shallow sites.

Table 1. Distribution of depths on the area of the Arctic seas

Range of depths, m	Share of area, %		
	The Kara Sea	The Laptev Sea	The East Siberian sea
0-10	3,5	11,1	14
10-20	5,3	37,4	47
20-30	10,1	22,7	25,5
30-40	9,3	17,8	13,1
40-50	3,8	11,0	-
50-100	18,3	-	-
100-200	146	-	-
> 200	214	-	-

The charts manufactured on materials of the hydrographic researches till 1990 make the basis of the collection of navigation sea charts. The detail of the performed mapping of bottom relief is shown in table 2 according to (Reshetnyak, 2006).

The charts made on materials of the hydrographic works performed by Federal State Unitary Enterprise "Hydrographic Enterprise" (Russia, official website – http://www.hydro-state.ru/) since 2010 till present are also included Into a collection. These charts cover parts of areas of the designed transit high-latitude routes and the water area of the Ob Bay and the Yenisei strait with passes (Federal State Unitary Enterprise "Hydrographic Enterprise", 2016)

Table 2. The detail of the mapping of bottom relief on the area of the Arctic seas

The detail (m)	Share of area, %			
	The Kara Sea	The Laptev Sea	The East Siberian sea	The Chukchi Sea
≤500	28	35	15	9
1000	38	39	20	-
2000-4000	16	11	14	21
≥4000	10	10	10	-
route	8	5	41	70

DEFINITION OF RESEARCH PROBLEM

NSR is mainly passing on sites of the Arctic seas with numerous underwater dangers through Severnaya Zemlya Archipelago and New Siberian Islands Archipelago on the Dmitry Laptev strait, Vilkitsky strait and Shokalsky strait. Navigation on the water area of NSR is carried out on the recommended ways. Significant long-term experience at the passage of these routes has not been accumulated yet; it belongs to the period of free navigation in clear water and at icebreaking assistance. The first options of the routes were mapped based on successfully completed navigations (Dmitriev, 2015; Pastusiak, 2016). The main criterion by which the routes for winter season were chosen was degree of landfast ice and adjacent areas, free from ice. During the summer season the routes were mapped so that to avoid the areas where there are ice massifs. Recently special attention is paid to the passing depths (Afonin, 2016; Batalin, 2008) because the commercial fleet has considerable draft.

Representation of an underwater relief of the water area of the Northern Sea Route (NSR) is based on discrete (in planned position) determination of depths and their coordinates. Such way of hydrographic mapping doesn't guarantee detection of all local raisings of a bottom which can be navigation dangers.

The extreme draft of the vessel calculated on the minimum depth found on the route at discrete measurements doesn't guarantee safe navigation on the route. At increasing of discretization (sampling) the level of credibility to the calculated value of extreme draft of the vessel decreases as the probability of depth presence smaller than minimum, found on the route, increases.

Study of these routes corresponded to requirements of the standards existing at the time of mapping and provided navigation of vessels with draft up to 6-7 meters in conditions of summer - autumn navigation. In winter navigation at icebreaking assistance the safety of transport vessels was provided by ice breaker which were going ahead and having the draft of 8 or 12 meters (Admiral Makarov State University of Maritime and Inland Shipping (SUMIS), 2016; Federal state Institution "The Northern Sea Route Administration", 2017).

Hydrographic study of the water area of NSR by modern criteria of the International hydrographic organization (IHO, S-44/5 (Standard of International Hydrographic Organization S-44/5, 2008)) is insufficient

According to these criteria it is necessary to perform the hydrographic areal mapping on the NSR, providing the identification of navigational hazards, which sizes are 1x1 and 2x2 meters, depending on the water reserve under the vessel keel (Admiral Makarov State University of Maritime and Inland Shipping (SUMIS), 2015).

Currently, transit high-latitude NSR tracks within a 2 kilometer strip (Federal State Unitary Enterprise "Hydrographic Enterprise", 2016; Federal state Institution "The Northern Sea Route Administration", 2017) and approaches to the ports of Dudinka and Sabetta correspond to such conditions.

The shallow-draft vessels pass on the coastal NSR tracks. Operation on coastal routes of courts with big draft without additional hydrographic inspection it is accompanied by risks of navigation incidents. Navigation on the recommended ways is applied in areas with insufficient hydrographic study according to rules of NSR Administration (Federal state Institution "The Northern Sea Route Administration", 2017).

Shallow water in a combination to difficult ice conditions of the Arctic are the major complicating factors of year-round navigation of heavy-tonnage vessels on the NSR.

Severe ice conditions force vessels to deviate the recommended ways, both at independent navigation, and at pilotage using icebreaking because of risk of ice damages receiving.

Considerable removal from the recommended ways in the conditions of shallow water increases risk of the accidents connected with grounding or contact of ground in the conditions of insufficient hydrographic study of adjacent water areas.

High variability in ice conditions means that vessels often cross a part of NSR along a route where ice conditions make impossible the further navigation, then there is a choice of other option at dynamically changing ice situation. All options available in the concrete seas are carefully discussed. Special attention is paid to the straits dividing certain seas on NSR.

DEVELOPMENT OF TECHNIQUE FOR ASSESSMENT OF THE PASSING DEPTHS ON NSR

The extreme draft of the vessel calculated on the minimum depth found on the route at discrete measurements doesn't guarantee safe navigation on the route. At increase in discretization the level of credibility to the calculated value of extreme draft of the vessel decreases as the probability of depth presence, smaller, than minimum depth found on the route, increases.

Assessment of the extreme draft of the vessel $d(L)$ guaranteeing safety of navigation on routes in shallow water in the conditions of insufficient hydrographic study ($L \neq 0$) is determined by a formula:

$$d(L) = f[Z_{\min}(p), \Delta, \Delta_L(p; L)], \quad (1)$$

where L - discretization of depths measurements; $Z_{min}(p)$ - the minimum depth determined with probability p; Δ - received reserve of water under the keel; Δ_L $(p;L)$ – the amendment for uncertainty of value of the minimum depth.

The following ratios are valid for areal mapping:

$$\Delta_L(p=1; L \to 0) = 0 \quad (2)$$

$$d(L \to 0) = f[Z_{\min}(p=1), \Delta] \quad (3)$$

The amendment for uncertainty Δ_L $(p;L)$ considers morphological and morphometric features of an underwater relief, his type and range of depths of the water area.

Formulas (1) – (3) are used for calculation of the extreme draft of the vessels on NSR.

The software (Afonin, 2016.) was developed for calculation of the extreme draft of the vessel. It is intended for the following types of hydrographic mapping of underwater relief: areal mapping, mapping with parallel tacks, ice measurement, route measurement.

Type of the realizing electronic computer is IBM PC compatible personal computer. Programming language is ObjectPascal (development environment is Lazarus, Delphi). The type and version of operating system is Windows 2003/XP/7, Linux(Ubuntu).

At calculating the program produces results of determination of passing depths corresponding to the extreme draft of the vessel in 6.0 meters, at the same time key parameters of the initial massif of depths and area of passing depths are displayed.

The given technique was used for assessment of passing depths (Batalin, 2008; Federal State Unitary Enterprise "Hydrographic Enterprise", 2016) on NSR. The table 3 shows the characteristics of high-latitude routes of NSR surveyed in 2010-2015 by detailed mapping. The detailed mapping provides sufficient hydrographic study of the water area of high-latitude routes. The first column shows direct and back courses of straight sections of the recommended way. The second column shows minimum depths found on the allocated straight sections of the recommended way.

The detailed mapping has allowed to find the minimum depths and all available underwater obstacles dangerous to navigation.

The values of passing depths are given in the third column. Passing depths differ from the minimum depths at a size of free reserve of water equal 2.0 meters.

Table 3. Characteristics of high-latitude routes

Course: direct - back	Minimum depth, m	Passing depth, m	Detail of mapping
West part			
36.2-216.2	44	42	areal mapping
74.5-254.5	25	23	areal mapping
90-270	45	43	areal mapping
56.4-236.4	27	25	areal mapping
65.6-245.6	35	33	areal mapping
Middle part			
53.1-233.1	24	22	areal mapping
90-270	19	17	areal mapping

61.3-241.3	21	19	areal mapping
94-274	23	21	areal mapping
53.1-233.1	24	22	areal mapping
East part			
106.2-286.2	22	20	areal mapping
132.5-312.5	22	20	areal mapping
137.7-317.7	20	18	areal mapping
119-299	28	26	areal mapping

The characteristic fragment of the coastal route of NSR and his key parameters are given in tab. 4.

Table 4. Characteristics of fragment of the coastal route

Exit from the Vilkitsky strait and Shokalski strait				
Course: direct - back	Length, km	Depths: min/max, m	Passing depth, m	Line spacing, *L*), m.
112.4-292.4	102	25/66	15.0	500
126.4-306.4	194	18/60	6.0	1000
0-180	104	20/185	2.0	2000

It should be noted that the divergence between the minimum and passing depths increases at reduction of detail of mapping (increase of parameter *L*).

DISCUSSION OF RESULTS

The developed technique and the software together with other forecasting models (Aksenov, 2017; Yamaguchi, 2016) and laying of safe routes in the water area of NSR will allow to provide navigation in the Arctic seas and reduce the economic risk connected with search of optimum routes and unpredictable change of time of voyage.

All data obtained in work are in good agreement with researches of other authors (Lee, 2015; Zhao, H., Hu, H., & Lin, Y., 2016; Zhao, H., & Hu, H., 2016) describing trial passes of the commercial fleet in the water areas of NSR by the Chinese and Korean companies in 2013-2014.

CONCLUSIONS

1. Techniques for assessment of passing depths on NSR tracks which consider both hydrographic study of routes, and morphological and morphometric features of underwater relief were developed.

2. Calculated data allow to estimate the possible minimum draft of vessel for the safety of navigation and identification of more optimum routes that is urgent in case of development of new tracks of NSR.

3. The developed technique and the software together with other forecasting models of safe conditions of navigation allow to find more optimum routes for high-latitude tracks of NSR

REFERENCES

Admiral Makarov State University of Maritime and Inland Shipping (SUMIS), 2016. *Assessment of the influence of navigation-hydrographic and hydrometeorological conditions on navigation along the Northern Sea Route*, SPb: SUMIS.

Admiral Makarov State University of Maritime and Inland Shipping (SUMIS), 2015. *Development of methods of the forecast of types of ice conditions in the southwest and northeast regions of the Kara Sea during the winter period*, SPb: SUMIS.

Afonin, A., Ol'khovik, E., & Tezikov, A., 2016. Development of the assessment methods of anadromous depths on the Northern Sea Route depending on the detail of survey of the bottom relief. *Vestnik Gosudarstvennogo universiteta morskogo i rechnogo flota imeni admirala S.O. Makarova*, 4(38), pp.62-68. (doi: 10.21821/2309-5180-2016-8-4-62-68). (rus)

Aksenov, Y., Popova, E.E., Yool, A., Nurser, A.J.G., Williams, T.D., Bertino, L., & Bergh, J., 2017. On the future navigability of Arctic sea routes: High-resolution projections of the Arctic Ocean and sea ice. *Marine Policy*, 75, pp.300-317. http://dx.doi.org/10.1016/j.marpol.2015.12.027.

Batalin, G., & Reshetnyak, S., 2008. The navigational and hydrographic support for shipping in the Arctic region. *Marine Fleet*, 3, pp.34-38. (rus)

Dmitriev, A.A., Gorbunov, Ju.A., & Sokolov, V.T., 2015. *Navigation history on the Northern Sea Route in XXth and the beginning of the XXIst century*. Marine encyclopedia: SPb. (rus)

Federal state Institution "The Northern Sea Route Administration". *Navigation rules on NSR*. [Online] (Updated 30 January 2017) Available at: http://asmp.morflot.ru/ [Accessed 12 December 2016].

Federal State Unitary Enterprise "Hydrographic Enterprise", 2016 [Online] (Updated 27 January 2017) Available at: http://www.hydro-state.ru/ [Accessed 12 December 2016].

Gavrilov, V.V., 2015. Legal Status of the Northern Sea Route of the Russian Federation. *Journal of Russian Law*, 3(2). DOI: 10.12737/7635 (rus)

Lee, T., & Kim, H. Ju., 2015. Barriers of voyaging on the Northern Sea Route: A perspective from shipping Companies. *Marine Policy*, 62, pp.264-270. (doi:10.1016/j.marpol.2015.09.006).

Pastusiak, T., 2016. Variants of the Northern Sea Route. *The Northern Sea Route as a Shipping Lane*, pp.131-152. (doi:10.1007/978-3-319-41834-6_6)

Reshetnyak, S.V., 2006. Hydrographic study of the Arctic secs underwater relief of Russia. *Geodesy and Cartography*, 4, pp.57-60. (rus)

Standard of International Hydrographic Organization S-44/5, 2008. Monaco.

Yamaguchi, H., & Nakano, Yu., 2016. Research on Navigation Support System and Optimum Route Search for the Northern Sea Route. *Proc. 23rd IAHR International Symposium on Ice*, pp.1-8.

Zhao, H., & Hu, H., 2016. Study on Economic Evaluation of Northern Sea Route: Taking Voyage of Yong Sheng as an Example. *Transportation Research Board 95th Annual Meeting*, №. 16-6420.

Zhao, H., Hu, H., & Lin, Y., 2016. Study on China-EU container shipping network in the context of Northern Sea Route. *Journal of Transport Geography*, 53, pp.50-60.

Proceedings of the 24th International Conference on
Port and Ocean Engineering under Arctic Conditions
June 11-16, 2017, Busan, Korea

A Method for Planning Arctic Sea Routes Under Multi-Constraint Conditions

Minyang Zhang [1,2], Di Zhang[1,2], Shanshan Fu[1,2] , Xinping Yan[1,2], Chi Zhang[1,2]
[1]Intelligent Transport Systems Research Center, Wuhan University of Technology, Wuhan, China
[2]National Engineering Research Center for Water Transport Safety, Wuhan, China

ABSTRACT
Recently, many ships are sailing for shorter distances by using the Northern Sea Route (NSR), rather than the Suez Canal Route (SCR). However, the NSR ships are exposed to the risk of ice-ship collisions when sailing in ice-covered waters. For this reason, a route optimization method is needed, in order to reduce the happening of collision accidents. In this paper, a route planning approach is proposed for ice-going ships, based on a modified A* algorithm under multi-constraint conditions. First, a grid-based geographical model using a high-precision electronic chart and ice data is developed. Second, multi-constraint conditions (physical constraints and operational constraints) are defined, and the optimum conditions are selected for a safe and economical route. Specifically, the ice class of ships, the conventional operation for crews and the avoidance of hazardous areas are taken into consideration in the proposed approach. Additionally, the approach integrates the grid model, the multi-constraint conditions, and the improved A* algorithm as a route design tool for ships sailing through ice-covered waters. The proposed route planning model is applied and validated in a case study of the YONGSHENG vessel.

KEY WORDS: Arctic waters; route planning method; multi-constraint conditions; A * algorithm

INTRODUCTION
As global warming accelerates the melting of polar ice, the extent and range of ice-covered waters are gradually decreasing. This, in turn, boosts the development of the shipping business in the Arctic region (Fu et al., 2016). In particular, the Russian government abolished the rule calling for compulsory icebreaker assistance on the Arctic Sea in 2014. While the Northern Sea Route (NSR) is regarded as the last route to be fully exploited, the safe navigation of ships through those waters is an issue that is becoming the focus of attention these years (Marken et al., 2015). Also, the Arctic Sea is located inside the sensitive and delicate polar circle. Once an accident occurs in the Arctic Sea, the accident will not only cause casualties and economic losses but may also have a massively negative impact on the fragile ecological environment in the Arctic region (Fu et al., 2016). As a consequence, coming up with a route planning method as a means to lower the risk when ships sail in Arctic ice-covered areas is very necessary (Somanathan et al., 2009).

The performance of computer technology has greatly improved the possibilities of carrying out ship transit modeling and transit simulations in ice (Kotovirta et al., 2009). In addition, it is a difficult task for ship crews to take into consideration all the ice environment information needed for safe route planning when already on board the ship. However, a computer algorithm could plan the best route option if given all the information available (Choi et al., 2014), for example, a system for route optimization in ice-covered waters. The route planning system contains an ice model, ship route planning model, and an end user system, all of which combine to optimize the route for ice-going ships in ice-covered waters was put forward (Kotovirta et al., 2008). Concerning the estimation in Nam et al., 2013, an enhanced algorithm, used to determine the optimal route in ice-covered waters, is introduced. Also, a route planning model is developed. The model is based on simulated sea ice and environment information in ice-covered waters. An Arctic route planning model based on graph algorithms is established (Zvyagin Petr et al. 2016). Some technical papers on the modeling of Arctic transportation systems are presented (Sören et al., 2014, 2015).

In all of the model route segments mentioned, the grid method is the main method used for the route planning algorithm which in turn is used to model the navigation environment. Recently, a wide range of route planning algorithms based on the grid has been used (Choi et al., 2014; Kotovirta et al., 2008; Zvyagin et al., 2016). Of all of those algorithms, the A* algorithm can efficiently figure out an optimal route from open waters to the Arctic Sea. However, some problems arise when using existing route planning algorithms based on the grid (Mei et al., 2012). Firstly, the search direction when using these algorithms is limited to an integer multiple of $\pi/4$. The routes we get from the route planning algorithms mentioned above are not the optimal routes. This is because the distance between the starting point and the goal is not the shortest. A lot of pivoting stations appear on the route, which poses a great danger to ships sailing on the Arctic Sea. In addition, the mutual restrictive relationships between the ice class of ship, the sailing speed, and the surrounding environment are not taken into account when utilizing the existing route planning algorithms. The existing papers reveal the lack of one route planning method, which takes a ship's ice class into consideration.

This paper aims to find an Arctic route planning algorithm under multiple restrictions and which selects the distance index and threat index as the main indicators. The two indexes represent the following parameters:

- the sailing time, distance and speed from the starting point to the goal point in the Arctic Sea;
- the navigation constraints in the Arctic region that stem from the ice thickness, degree of ice compression and ice type.

In order to propose an approach for Arctic Sea route planning under multi-constraint conditions, an A* algorithm will be combined with multiple constraints into a cost estimating function. Eventually, each of the ships with different ice classes will find a least-cost route (as the optimal route), according to the different weight values of the navigational constraint parameters.

The remainder of the paper is organized as follows: Section 2 presents the modeling of the Arctic environment by the grid method. Section 3 introduces the shortage of A* algorithm, the improved A * algorithm and the definition of optimal route planning. The simulation results are shown in Section 4. Our research findings are concluded in Section 5. The research framework is shown in Figure 1.

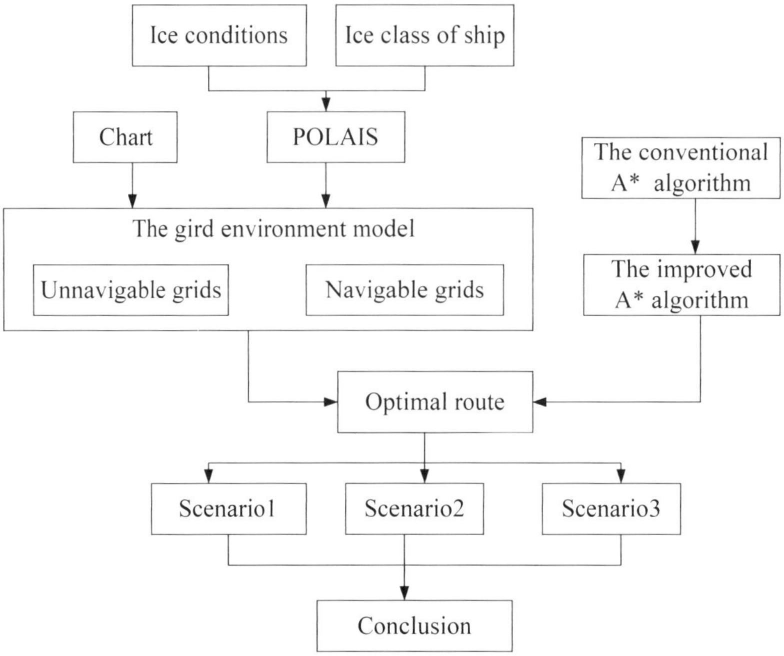

Figure 1. Research framework of the Arctic Sea route planning method

METHODOLOGY

The Grid Environment Model

The Grid Method

A model of the ice environment around a ship sailing on the Arctic Sea must be established before route planning in the Arctic area can be done. The aim of the ice environment model is to construct a representation scheme of navigation dangers, including shallow waters and severe ice environments. It is extremely important in Arctic Sea route planning to describe the environment around the ship and transform the actual environmental information (navigation dangers, shallow waters, severe ice environments, etc.) into human-readable data that is useful in route planning. What should be noted is that only reasonable environmental modeling is helpful regarding decreasing the search volume in the route planning and in finding a suitable route according to each ship's ice class. Different route planning methods are based on different types of environmental modeling.

The grid method was first proposed by W.E. Howden (Hart et al., 1968). The method divides the ice-covered waters around the ship into several grids. Every grid is linked to nearby grids (either four or eight grids), which forms a grid map. The grid map information directly corresponds to environmental information, making it easy to create, maintain and self-localize the grids. Because the route planning method based on an A-star algorithm is a route planning method based on picture research, this paper is based on a grid method used to establish environmental modeling.

The expression of the grid: Number all grids in sequence from left to right and from top to bottom. Every number N (N=0, 1, 2…) directly corresponds to one grid. The mapping relationship between the arbitrary points (x, y) with the number of grids is defined as follows:

$$N = x + \text{map} \times y$$
$$x = N \,\text{mod}/ \,map \tag{1}$$
$$y = \text{int}\, N \,/\, \text{map}$$

Here, "mod" represents taking remainders and "int" means rounding numbers. After the establishment of the environment model using the grid method, the quantization of such environmental modeling is required. In order to quantize above the grid, a matrix is introduced to establish the corresponding one-to-one relationship between the grid and the array element. This method adopts the grid position on the matrix as an index, which in turn contains the specific information of every grid. Also, N represents the number of the current grid. According to the information of every grid, the ice-breaking capability of a ship, an ice chart and the risk cost of each grid (as caused by relevant navigational rules in the Arctic area), our method can eventually figure out whether or not the ship can sail on the Arctic Sea. As for the navigation dangers, the cost value of obstacles like icebergs is equal to infinity. Meanwhile, once the extent of obstacles like navigation dangers, shallow waters and severe ice environments cannot fulfill one grid, the extent of those obstacles should be shown as one full grid, in order to ensure navigation safety.

The definition of cost of the grid

Ships must take into consideration risk factors like navigation dangers to navigation and shallow waters, and fully evaluate the risks caused by severe ice environments when sailing on the Arctic Sea. Every ship should also pay attention to the risks brought about by the quantization of the surrounding ice environment and the ship's ice class when planning a route in the Arctic via the grid model.

Risk management procedures for polar ship navigation in ice-covered waters are specified in the new International Maritime Organization (IMO) International Code for Ships Operating in Polar Waters (Polar Code) (Stoddard et al., 2016; Maritime Safety Committee 2014 POLARIS). The goal of the Polar Code is to provide safe ship operations by addressing the risks present in ice-covered waters. The Polar Operational Limit Assessment Risk Indexing System (POLARIS) also provides a risk assessment framework, which can be used to assess the degree of navigational safety in a given ice regime, using observed or historical ice conditions and concentration, as well as polar ship classification (International Maritime Organization, 2015).

The implementation of the Polar Operational Limit Assessment Risk Indexing System (POLARIS) takes into consideration both the Transport Canada Arctic Ice Regime Shipping System (AIRSS) and the rules of navigation in the water area of the Northern Sea Route. Great importance is attached to icebreaker assistance in the Arctic Sea regulations. In addition, POLARIS evaluates the risks of a variety of ice conditions by each ship's ice class.

Basically, POLARIS adopts a ship's ice-class classification and Risk Index Values (RIVs), which combines route evaluation with historical data and route evaluation with near-real-time data, in order to calculate the risk index value. The value can then be used to evaluate the risks and operational constraints when sailing on the Arctic Sea. The data of the risk index value are shown in the following Table 1.

Table 1. The risk index value on the Arctic Sea

Category	Ice class	Ice Free	New Ice	Grey Ice	Grey White Ice	Thin First-Year Ice 1st Stage	Thin First-Year Ice 2nd Stage	Medium First Year Ice	Medium First Year Ice 2nd	Thick First Year Ice	Second Year Ice	Light Multi-Year Ice	Heavy Multi-Year Ice
A	PC1	3	3	3	3	2	2	2	2	2	2	1	1
	PC2	3	3	3	3	2	2	2	2	2	1	1	0
	PC3	3	3	3	3	2	2	2	2	2	1	0	-1
	PC4	3	3	3	3	2	2	2	2	1	0	-1	-2
	PC5	3	3	3	3	2	2	2	1	0	-1	-2	-2
B	PC6	3	2	2	2	2	1	1	0	-1	-2	-3	-3
	PC7	3	2	2	2	1	1	0	-1	-2	-3	-3	-3
C	IA Super	3	2	2	2	2	1	0	-1	-2	-3	-4	-4
	IA	3	2	2	2	1	0	-1	-2	-3	-4	-4	-4
	IB	3	2	2	1	0	-1	-2	-3	-3	-4	-5	-5
	IC	3	2	1	0	-1	-2	-2	-3	-4	-4	-5	-6
	Not ice strengthened	3	1	0	-1	-2	-2	-3	-3	-4	-5	-6	-6

For each ice regime encountered, the Risk Index Values are used to determine a Risk Index Outcome (RIO) that forms the basis of the decision to operate or to limit operations. The RIO is determined by a summation of the RIVs for each ice type present in the ice regime, multiplied by the ice concentration (International Maritime Organization, 2015).

$$RIO = (C_1 \times RIV_1) + (C_2 \times RIV_2) + (C_3 \times RIV_3) + \cdots + + C_n \times RIV_n) \tag{2}$$

where, $C_1 \ldots C_n$ are the concentrations (in tenths) of ice types within the ice regime; $RIV_1 \ldots RIV_n$ are the Risk Index Values corresponding to the ship's ice class.

We utilize the Polar Operational Limit Assessment Risk Indexing System, A* algorithm and navigational risks in grid modeling that are defined by the RIO value, in order to transform the navigational risks or the Risk Index Values into the cost of navigation, as shown in Table 2.

Table 2. The navigational risks or the Risk Index Values in the Arctic Sea

RIO ship	Ice class PC1-PC7	Ice class below PC7 or ships not assigned an ice class	Navigation cost
RIO≥0	Normal operation	Normal operation	0
-10≤RIO<0	Elevator operational risk*	Operation subject to special consideration**	-RIO
RIO<-10	Operation subject to special consideration**	Operation subject to special consideration**	Unnavigable

This paper utilizes the grid method, together with the Polar Operational Limit Assessment Risk Indexing System (POLARIS), to establish an ice environment model. Our method's aim is to identify the characteristics of the Arctic's ice-covered waters according to the characteristics of the ice environment in each grid and for each ship's ice class. The RIO value can be determined as the cost of navigation of each grid. All of these steps can ultimately achieve the establishment of an ice grid environment model in the Arctic's ice-covered waters.

A* algorithm

The A* algorithm is a state-space heuristic algorithm, which is different from the conventional blind search (such as a depth-first search or a breadth-first search) (Hart et al., 1968). A heuristic search introduces heuristic information in the search process, evaluates each search location in the state space, obtains the optimal location, and then searches from this location until reaching the goal. This process eliminates a large number of invalid search routes and improves efficiency.

In the A* algorithm, a valuation function is needed to evaluate the cost value of each extended search grid. The general form of the evaluation function is as shown below:

$$f(n) = g(n) + h(n) \tag{3}$$

Here, n represents the grid to be extended; $f(n)$ is the estimated value of the minimum cost route from the starting point to the goal point via grid n; $g(n)$ is the navigation cost from the starting point to the current grid n, and $h(n)$ is the estimated cost from the current grid n to the goal point. Because $h(n)$ represents heuristic information in an A* algorithm, this function is called heuristic. In a two-dimensional search, the heuristic function $h(n)$ is usually the Euclidean distance between the two points. In order to guarantee the optimality of an A* algorithm, the heuristic function must satisfy the admissibility condition, that is, for all grids:

$$h(n) <= h^*(n) \tag{4}$$

where, $h^*(n)$ is the true minimum cost of the grid n to the goal grid. The A* algorithm's function is to select and extend the grid with the smallest f value from the candidate grids every time. When $h = 0$, A* algorithm degrades to a breadth-first search algorithm. When $g = 0$, A* algorithm degrades into a depth-first search algorithm. The A* algorithm firstly creates two tables, namely the OPEN table and the CLOSED table. Of these two tables, the OPEN table saves generated grids that have not been investigated. The CLOSED table records the grids that have been visited. The cost function f (n) evaluates the importance of each grid in the OPEN table. Both the paid cost and the cost to be paid must be considered, in order to determine the grids' order in the OPEN table. Also, g (n) points out the horizontal trend of search, which is beneficial to the completeness of search, but affects search efficiency. In determining f (n), we need to weigh the pros and cons of each grid, so that g (n) and h (n), respectively, occupy a certain proportion of the grid. In selecting the conditional grid type of grid, the cost of navigation in the grid can be defined according to the attributes of the defined grid. Therefore, the cost function f (n) is with the selectable condition.

In addition, f (n) is defined as the cost estimation function of the minimum cost route from the initial grid to the goal grid via n grid in the Arctic Sea. The minimum cost route is defined as follows:

$$f(n) = g(n) + h(n) \quad (5)$$

$$g(n)=g(n\text{-}1)+m(n) \quad (6)$$

$$h(n) = | x_n - x_{goal} | + | y_n - y_{goal} | \quad (7)$$

where, g(n) is the sum of the actual cost of the previous grid and the cost of moving to the current grid in the Arctic Sea; The Manhattan distance from grid n to the goal grid in the real environment is h(n), and the actual cost value of moving from the previous grid to the current grid is m(n). The environment grid is defined as the unit length of the dimension, which is set independently, according to different conditions. At the initial grid, g(n) = 0; (x_n, y_n) is the value of n order matrix, and (x_{goal}, y_{goal}) is the value of n order matrix. However, when defining route planning, an A* algorithm defines the center of each grid as a grid, which results in the mere capability to search adjacent areas when route planning and restricts movement direction to an integer multiple of π / 4. In this case, the acquired route is not the optimal route, as there are more turning points, as shown in Figure 2.

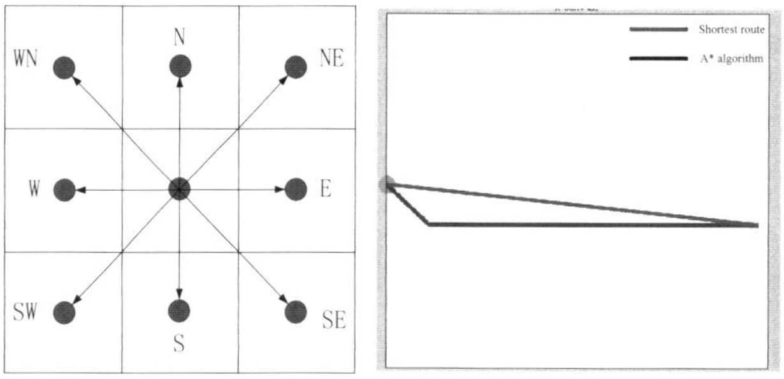

Figure 2. The eight neighborhood search process of an A * algorithm

To solve the problem whereby movement direction is restricted in the utilization of an A* algorithm, first of all, this paper makes a minor change in the heuristic algorithm grid map. More specifically, the center of the grid will not be defined as a nod, but the edge of each grid will be defined as a grid to be visited. As a result, every grid is surrounded by countless adjacent grids that can be researched, and each grid's sub-nod can be any one of those adjacent grids. This method leads to the possibility of a continuous movement direction angle in any direction when planning a route (Xin et al., 2014). The nodes on the edge of the grid can be classified as being in one of two categories: 1) the intersection of the edge, the amount of which is finite; 2) the nodes lying on the edge of the grid, as shown in Figure 3.

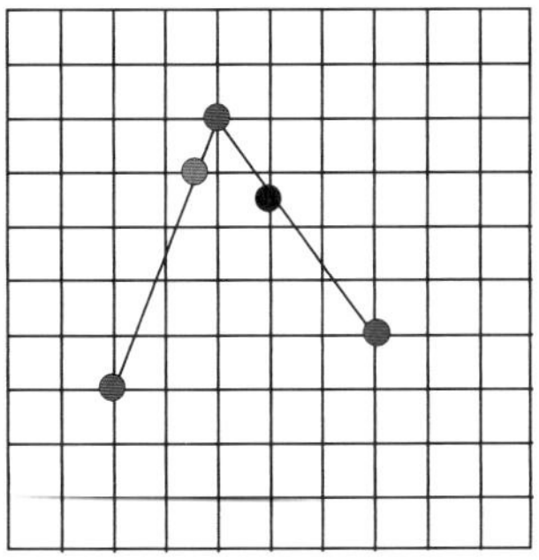

Figure 3. Redefining the position of a grid in a grid map

For the arbitrary point y on edge, we only need to figure out the minimum value of the valuation function, instead of finding the valuation function value f (n). Before evaluating the valuation function, nodes must be classified into two categories: 1) intersection points of the grid's edge; 2) nodes lying on the edges. However, when using the edge of the grid map as a nod, the cost of edge N-N1 should be the maximum value of grid a and grid b, and the cost of node Y, h(n) should be defined as follows:

$$\mathrm{h(n)} = d \times \mathrm{h}(N_2) + (1-\mathrm{d}) \times \mathrm{h(N_1)} \tag{8}$$

where, d is the distance from y to node N1.

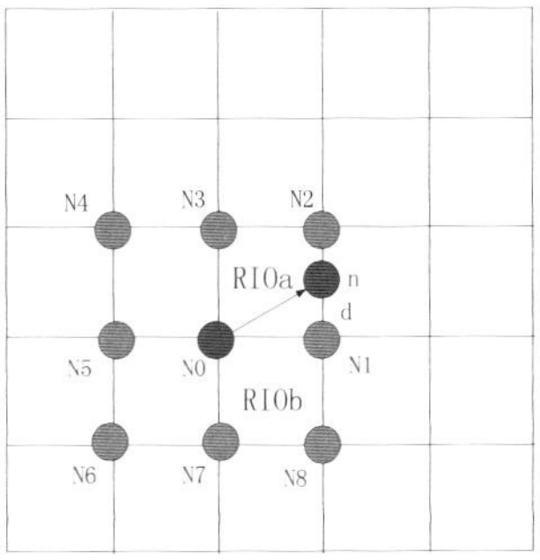

Figure 4. The current grid on the edge of a grid map

When the node is the intersection point of the grid map, the valuation function of the A* algorithm can be expressed as:

$$f(\mathrm{n}) = \begin{cases} g(\mathrm{N}) + VIO(a) \cdot \sqrt{t^2 + d^2} + d \cdot h(N_2) + (1-d) \cdot h(N_1) & \mathrm{n} \neq \mathrm{N}_1 \\ g(N) + \max\{VIO(a), VIO(b)\} \cdot t + h(N_1) & \mathrm{n{=}N}_1 \end{cases}$$
$$g(\mathrm{n}) = \begin{cases} g(N) + VIO(a) \cdot \sqrt{t^2 + d^2} & n \neq N_1 \\ g(N) + \max\{VIO(a), VIO(b)\} \cdot t & n = N_1 \end{cases} \tag{9}$$

When a node is not the intersection point of a grid map, the node can be placed into one of two categories. As shown in Figure 3, the black node is perpendicular to the edge, and the blue node is parallel to the edge. When the node is parallel to the arbitrary point y on the vertical edge, the cost value of y is defined as follows:

$$f(n)=\begin{cases} g(N)+VIO(a)\times\sqrt{1+d^2}+d\times h(N_2)+(1-d)\times h(N_1), n\neq N_1 \\ g(N)+\max\{VIO(a),VIO(b)\}+h(N_1), n=N_1 \end{cases}$$
$$g(n)=\begin{cases} g(N)+VIO(a)\times\sqrt{1+d^2}, n\neq N_1 \\ g(N)+\max\{VIO(a),VIO(b)\}, n=N_1 \end{cases} \quad (10)$$

When the node is perpendicular to the arbitrary point y on the parallel edge, the cost value of y is defined as follows:

$$f(\mathrm{n})=g(\mathrm{N})+VIO(a)\times\sqrt{1+(t-m)^2}+m\times h(N_3)+(1-m)\times h(N_2)$$
$$m\in[0,1] \quad (11)$$
$$g(n)=g(N)+VIO(a)\times\sqrt{1+(t-m)^2}$$

Here, t represents the horizontal distance from node N to arbitrary point y on the parallel edge.

We can find countless grid areas that can be researched around each node and improves the A* algorithm (in some aspects) by redefining the nod of the grid map. Also, by using MATLAB, the differences between the improved A* algorithm and the conventional A* algorithm are shown in the figure below. We can conclude that the improved A* algorithm is more suitable for route planning in the Arctic area.

Definition of Optimal Route

The improved methods in route planning are listed as follows: A route supposes to contain m route points (1, 2,…,m), and every route point is represented by a corresponding number on the grid. The first route point is the starting point, and the route point numbered m is the goal. The first route point is regarded as one end of a line segment. The route point numbered t (the number of which starts from 3) is regarded as the other end of the line segment. Figure out the distances from the line segment to the nearby navigation dangers. If all the distances are larger than the safe distance, which is determined by the ship's ice class, the next route point (numbered t+1) is regarded as one end of another line segment (one end of this line segment is the route point numbered t). Keep comparing the cost of distance and risk from the new line segment to the nearby navigation dangers with a safe distance, until all those distances are larger than the safe distance. Then, draw a new line segment from the route point numbered t+1 to the next route point and compare (following the above steps) until one end of the new line segment reaches the goal.

CASE STUDY

This paper builds a model for Arctic Sea route planning according to the Polar Operational Limit Assessment Risk Indexing System (POLARIS) and the improved A * algorithm. This method is employed to enable the safe planning of routes through ice-covered waters. The method uses the Risk Index Outcome (RIO) and the cost of navigation distance. Proper planning requires input data pertaining to ice conditions, as well as each ship's ice class and operational characteristics. In order to verify the validity of our method, we set up a grid model with the number 51*51, and we put the different ice factors into the grid model. Meanwhile, we created a simulation for ships of different ice classes, in order to achieve effective route planning in similar conditions. At the same time, the route results are compared. We can also verify that the results of the simulation are reliable.

Description of the Case Study

In 2013, 2015 and 2016, the Chinese merchant ship "YONG SHENG" successfully completed navigation of the Northern Sea Route (NSR) (Zhang et al., 2016). The ship is an Arc 4 general cargo vessel, according to the Russian Maritime Register of Shipping (RMRS) and the Rules of Navigation in the Northern Sea Route Water Area (which is nominally equivalent to the IMO PC6). In 2013, the ship was escorted by the "50 LET POBEDY" ship, a Russian nuclear-powered icebreaker. The escort was necessary because the "YONG SHENG" could not independently plan a safe route.

Our method for Arctic Sea route planning has been validated using the ice condition data and the experience of the ship. In addition, the ice condition data were analyzed based on the Polar Operational Limit Assessment Risk Indexing System (POLARIS), in order to obtain the characteristic parameters for validation, as shown in Figure 5.

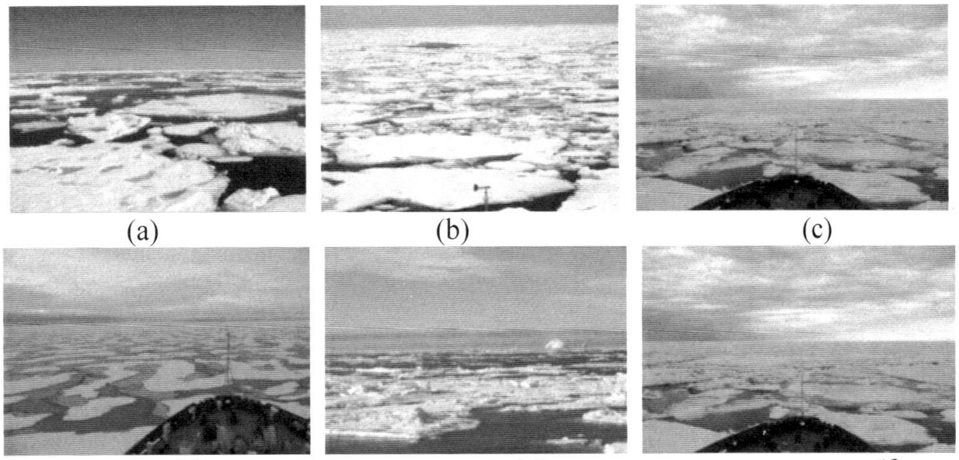

(a) (b) (c)

(d) (e) (f)

Figure 5. The ice environment in the Arctic

We analyze the Arctic ice information according to POLARIS. We first calculate the RIO value based on each ship's ice class and the prevailing ice conditions. Then, we use each ship's the ice class and RIO value to determine which class of ships can pass through a given area.

Table 3. Definitions of no navigation parameters, according to the ice conditions

	Ice conditions	No navigation ship
a	4/10th of thick first-year ice, and 1/10th of medium first-year ice	Ice class below IC
b	2/10th of heavy multi-year ice, and 7/10th of thick first-year ice	Ice class below PC7
c	1/10th of multi-year ice, and 8/10th of new ice	Ice class below PC7
d	6/10th of thick first-year ice, and 4/10th of new ice	Not ice-strengthened
e	6/10th of medium first-year ice	Ice class below IC
f	1/10th of heavy multi-year ice, and 8/10th of thick first-year ice	Ice class below PC6

Comparing a conventional A * algorithm and improved A * algorithm
In Scenario 1, we built the grid model under the same ice conditions and set up a simulation to plan routes for ships of the same ice class, in the grid model based on the conventional A * algorithm and an improved A * algorithm. For verification purposes, two routes with the same starting point and the same goal point were compared. As shown in Figure 6, the black route abandons the shortcomings of the conventional A * algorithm and is an optimized route in terms of distance.

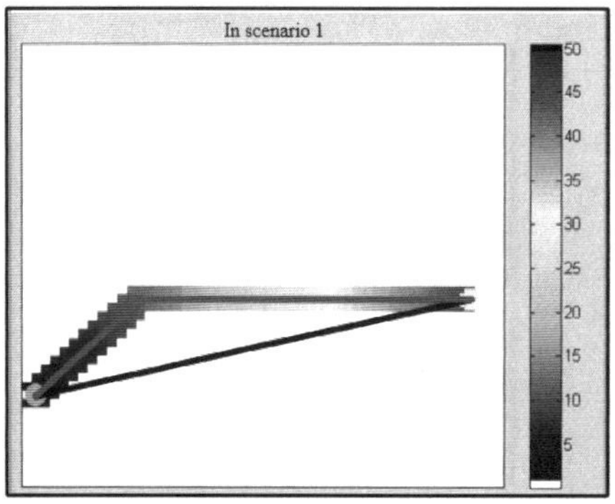

Figure 6. Comparison of route planning

In the case of the black route created by the improved A * algorithm and the red route created by the conventional A * algorithm, it is very clear that the black route is comparatively shorter than the red route. This result indicates that this improved A * algorithm is effective when planning a route in the Arctic Sea.

Comparing a Ship's Navigation in Different Ice Conditions
In Scenario 2, we built the grid model using Matlab. We introduced the different ice conditions that we discussed in Section 4. We also created a simulation, with the intention of planning a route for the "YONG SHENG". This route was based on the improved A * algorithm. For verification, we compared three routes with the same starting point and the same goal point. As shown in Figure 7, the black area represents an "unnavigable area". We placed our "unnavigable areas" together at the center of the chart, in order to validate the reliability of the algorithm.

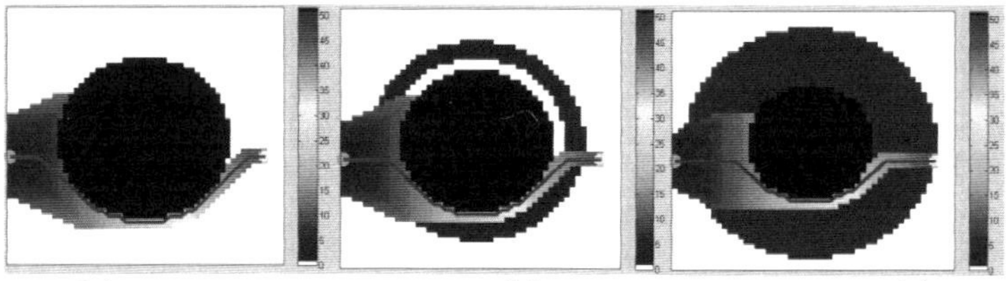
(a) (b) (c)
Figure 7. Comparison of route planning in different ice conditions

Table 4. Route planning results in ice-covered waters

Ice condition	5/10th of heavy multi-year ice	3/10th of heavy multi-year ice, and 2/10th of medium first-year ice, 2/10th of second-year ice	2/10th of heavy multi-year ice, and 2/10th of medium first-year ice
Ship ice class	PC6	PC6	PC6
Cost of distance*	56	47	35
Cost of ice-covered	0	21	25

waters**			
Total	56	68	60

The cost of distance* stands for the distance when we plan routes in ice conditions.
The cost of ice-covered waters** stands for the risk according to the RIO value.

We analyzed the cases of the three routes created by the improved A * algorithm in different ice conditions but for the same ship. We also calculated the cost of distance and the risk of navigation through ice-filled waters, as shown in Table 4. Clearly, all routes avoid the unnavigable areas in different ice conditions, and all three routes created are the shortest possible safe route. Our results show that the proposed Arctic Sea planning algorithm was able to conduct the search of the safety area and plan the lowest-cost routes according to each ship's ice class.

Comparing Different Ship Navigation Plans in Similar Ice Conditions
In Scenario 3, we built the grid environment model using the same ice conditions. We also created a simulation, in order to plan routes for different ships, based on the improved A * algorithm. For verification, the three routes with the same start point and the same goal point are compared. As shown in Figure 8, the black areas again represent "unnavigable areas" that are different, depending on the ship's ice class. We placed our "unnavigable areas" together at the center of the chart, in order to validate the reliability of the algorithm, as shown in Figure 8.

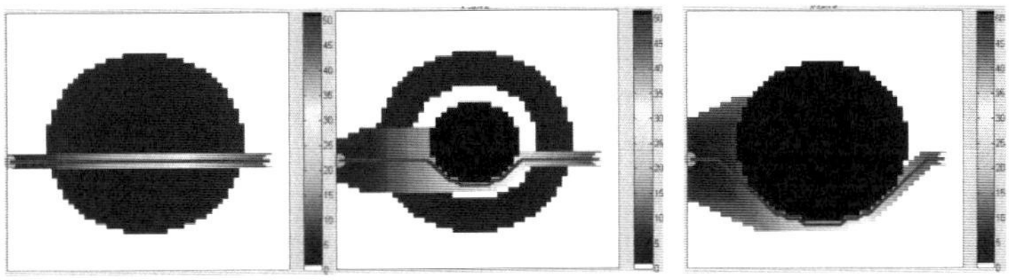

(a) (b) (c)
Figure 8. Comparison of route planning for different ships

Table 5. Results of route planning in ice-covered waters

Ice condition	1/10th of heavy multi-year ice, 1/10th of medium first-year ice, 1/10th of thick first-year ice, and 1/10th of new ice		
Ship's ice class	PC1	PC6	Not ice-strengthened
Cost of distance*	28	38	46
Cost of ice-covered waters**	40	22	0
Total	68	60	46

The cost of distance* stands for the distance when we plan routes in ice conditions.
The cost of ice-covered waters** stands for the risk according to the RIO value.

We analyzed the three routes created for the same ship by the improved A * algorithm. We also calculated the cost of distance and the risk of navigation through ice, as shown in Table 5. Once again, all of the routes clearly avoid the unnavigable area (which is different for each ship's ice class) in similar ice conditions. Again, all three of the analyzed routes are the shortest routes. This finding indicates that the algorithm can analyze the difference between high ice class ships and low ice class ships.

CONCLUSION

This paper introduced the recent studies pertaining to route planning in the Arctic region. We then analyzed the deficiency of the A* algorithm in terms of route planning. By using an improved A* algorithm, the number of search neighbors in the ice environment can be adjusted from eight to infinite. Also, the improved algorithm not only independently adapts to the needs of ice-going ships and reduces the ice-going ship's steering changes and steering angle frequency, but the improved A* algorithm also finds the shortest route in terms of distance in the ice grid environment. In the Arctic Sea route design, this paper established the ice class of different ships and the navigable relationship between the sea ice areas according to the Polar Operational Limit Assessment Risk Indexing System (POLARIS). At the same time, we established the ice route planning algorithm by using "the Risk Index Outcome" (RIO) as the navigable ice navigation cost. This algorithm not only considers the cost of distance between the start and the goal, but the improved algorithm also fully considers the impact of sea ice on the different ice classes of ships. We were able to conclude that ships of a low ice class would incur a higher cost when navigating in a harsh ice environment.

The method of polar navigation planning as proposed here was implemented using MATLAB, and was verified by setting up the simulation scenarios. First of all, we used the conventional A* algorithm and then the improved A* algorithm, respectively, in the grid model of ice-free route planning. We did this by setting the grid model of an open water environment. Our results show that the improved A* algorithm can achieve arbitrary ship courses (rather than eight discrete directions) and is also able to find the shortest route. Secondly, we set up different ice grid models and used the improved A* algorithm for route planning for ships of the same ice class. Our results show that the proposed polar ice navigation planning algorithm was able to conduct a search of the safest area at the lowest cost for each ship of each ice class. Finally, the improved A* algorithm was used for route planning by setting up the same ice grid model but using ships of different ice classes. We were able to show that the method can analyze the difference between high ice class ships and low ice class ships.

In short, we put forward a new type of route planning method for Arctic Sea navigation. Our method is based on each ship's ice class. The feasibility and validity of our method have been verified by the use of simulation scenarios. This method could be used for effective navigation planning for ships on the Arctic Sea.

ACKNOWLEDGEMENTS

This study was supported by National Science Foundation of China (NSFC) under Grant No. 51579203 and 51711530033 and The Excellent Dissertation Cultivation Funds of Wuhan University of Technology under grant No. 2016-YS-041. The authors would also like to thank Dr. Jakub Montewka for his valuable suggestions and comments on this work.

REFERENCES

Choi M, Chung H, Yamaguchi H, et al. 2014. Arctic sea route path planning based on an uncertain ice prediction model. *Cold Regions Science & Technology*, 109(109), pp. 61-69.

Bergström M, Ehlers S, Erikstad S O, et al. 2014. Development of an Approach Towards Mission-Based Design of Arctic Maritime Transport Systems. *ASME 2014 33rd International Conference on Ocean, Offshore and Arctic Engineering. American Society of Mechanical Engineers.*

Bergström M, Erikstad S O, Ehlers S. 2015. Applying risk-based design to arctic ships. *ASME 2015, International Conference on Ocean, Offshore and Arctic Engineering.*

Fu S, Zhang D, Montewka J, et al. (2016). A probabilistic model for consequences analysis of ship stuck in ice in arctic waters. *ASME 2016 35th International Conference on Ocean, Offshore and Arctic Engineering, Busan.*

Fu S, Zhang D, Montewka J, et al. 2016. Towards a probabilistic model for predicting ship besetting in ice in arctic waters. *Reliability Engineering & System Safety*, 155. pp. 124-136.

Hart P E, Nilsson N J, Raphael B. 1968. A Formal Basis for the Heuristic Determination of Minimum Cost Paths. *IEEE Transactions on Systems Science & Cybernetics*, 4(2), pp. 100-107.

International Maritime Organization 2015 *International Code for Ship Operating in Polar Waters (POLAR CODE).* https://edocs.imo.org/FinalDocuments/English/MEPC68-21-ADD.1(E). [Accessed 17 July 2016].

Kotovirta V, Jalonen R, Axell L, et al. 2009. A system for route optimization in ice-covered waters. *Cold Regions Science & Technology*, 55(1), pp. 52-62.

Maritime Safety Committee 2014 POLARIS: Proposed system for determining operational limitations in ice (International Maritime Organization).

Marken V B, Ehlers S, Khan F. 2015. Delay risk analysis of ship sailing the northern sea route. *Ship Technology Research*, 62(1), pp. 26-35.

Mei T, Liang H, Kong B, et al. 2012. Development of 'Intelligent Pioneer' unmanned vehicle. *Intelligent Vehicles Symposium. IEEE*. pp.938-943.

Nam J H, Park I, Lee H J, et al. 2014. Simulation of optimal arctic routes using a numerical sea ice model based on an ice-coupled ocean circulation method. *International Journal of Naval Architecture & Ocean Engineering*, 5(2), pp. 210-226.

Somanathan S, Flynn P C, Szymanski J. 2009. The Northwest Passage: A Simulation. *Transportation research part A*. pp. 127-139.

Stoddard M A, Etienne L, Fournier M, et al. 2016. Making sense of Arctic maritime traffic using the Polar Operational Limits Assessment Risk Indexing System (POLARIS).34, pp. 012-034.

Xin Y, Liang H, Du M, et al. 2014. An improved A* algorithm for searching infinite neighbourhoods. *Robot*.

Zhang M, Zhang D, Fu S, Yan X, Goncharov V. 2017. Safety distance modeling for vessels under the icebreaker assistance： taking "YONG SHENG" and "50 let pobedy" as an example. *In Transportation Research Board 96th Annual Meeting*, Washington, D.C.

Zvyagin P, Voitkunskaia A. 2016. Model of Transit Transport in Arctic Based on Graph Algorithms. *ASME 2016, International Conference on Ocean, Offshore and Arctic Engineering.*

POAC'17
Busan, KOREA

Proceedings of the 24th International Conference on
Port and Ocean Engineering under Arctic Conditions
June 11-16, 2017, Busan, Korea

The Concept of the shared Multi-Users Transportation System in the Northern Sea Route Area

Alexander V. Kirichenko[1], Alexander L. Kuznetsov[1]
[1] Admiral Makarov State University of Maritime and Inland Shipping, Saint Petersburg, Russia

ABSTRACT

Today the aquatory along the Northern Sea Route (NSR), as well as the whole Arctic zone of Russian Federation, witnesses the growth of various activities: merchant shipping, industrial fishing, geological explorations and carbohydrates production on the polar shelf. The government policy articulates the support of the further development of the Arctic region. Consequently, one can observe a remarkable intensification of transportation supporting the everyday activities of industrial enterprises, private and state companies and their departments, directly involved in works in Arctic and socially important for the local population. In the same time, this activity generates the growing demand for the transport and supporting infrastructure's development. The specific conditions of Polar Regions make it economically unwise to create any redundant facilities. In the same time, the current studies revealed that there are separate shipping parties of supply cargo (including those bound for open roadsteads) consisting of homogenous issues in many numbers, but relatively small by individual size. Simultaneously, many shippers (belonging to different administrative bodies) send similar cargo quite independently, thus providing the fleet overcapacity. All these factors drive to conclusions that a more rational usage of the infrastructural objects and transportation vehicles by the federal executive authorities and commercial companies in Russian Arctic would minimize the total budget expenses allocated for materials and supplies transportation.

KEY WORDS: Shared multi-users transportation system; Northern Sea Route; Arctic.

HISTORY AND CURRENT STATE

In the era of the administrative economy, the central authorities solved all the questions of utilization of the NSR. For this purpose in 1932 the Main Directorate of the NSR ("GlavSevMorPut"), – a governmental organization for national economic development and provision of navigation along the NSR, was created. Until 1946 it belonged to the Council of People Commissars of USSR, from 1946 until 1953 – to the Council of Ministers of USSR. The procedure of tonnage allocation for the supply transportation under the "Northern (expedition) delivery" program assumed the submission of administrative requests through the State Planning Committee of the Council of Ministers of USSR ("GosPlan"). GosPlan was a governmental body responsible both for overall national planning of the country's economy and the control of these plans' implementation. In addition, the Council on problems of the North and Arctic at the Government of USSR, and the State Committee on the affairs of the North existed to deal with the implementation of the state policy in Arctic.

Today the inherited organization is the Federal Governmental State enterprise "The Administration of the NSR". It was created by the Resolution of the Government of Russian Federation by March, 15th of 2013 № 358-p that was based on item 3 of Article 5.1 of the Federal Low by April, 30th 1999 № 81-FZ «Code of Maritime shipping of Russian, responsible solely for the organization of navigation in the water areas of the NSR». As the main goals of this organization, «the provision of safety of navigation and environmental protection in the water areas of the NSR. (Scope of activities, 2015)» were declared (The principles...,2008; The Strategy..., 2013; Resolution of the Government..., 2014; The scope of responsibility..., 2015).

The academic researches of the Russian Artic area are numerous. Still, the bibliographic study shows that by now those researches are fragmentary and inner inconsistent (Gorodetskiy, A.E., Ivanov, V.V., Filin, B.N., 2014; Izotov, O.A., 2005; Konovalov, A.M., 2013). In the studies (Izotov, O.A., 2005; Tolstych, D.A., 2009), most closely related to the offered hypothesis, the idea of formation of a general shared transportation systems was firstly introduced. Still, the development of a solid methodological foundation for rational organization, management and control of the multi-users shared transportation system in Russian Arctic region remains an actual scientific problem. The solution of this problem would minimize the total costs of supply material transportation, including the expeditional supply and delivery of material resources.

GOALS, METHODOLOGY AND STAGING

The preliminary stage of the concept's development revealed three main components of the maritime transportation in Arctic:

transit (including the cross-trading);

import-export, mainly connected with the world trade;

short-sea (inner maritime economy), defined by domestic needs, including social ones.

Accordingly, the following methodology was adopted:

First stage: in every sector of the transportation defined above, the identification of the key participants involved in planning and transportation, as well as other relevant national economic activities in Arctic; their specific demands and features perception, an arbitrary clusterization of these participants.

The cluster of "external" participants includes those operating beyond the boundaries of the system's control and permitting only reactive and/or proactive response.

The cluster of "internal participants" includes the players whose behavior to a certain extent is the subject of regulation, control and planning. For these participants, deeply and widely involved in the sea transportation and auxiliary economic activity, a preliminary analysis is needed, the analysis of their needs and demands for transport infrastructure (including ports) and technologies. The goal here is to assess the possibility of joint exploitation and competition, both economic and technological.

The core hypothesis of this stage is that the transport operations based on the same or similar technology could be performed complexly, for the interests of several participants, thus resulting in the economy of scale. The methodology implies that these complex analyses should take into account not only pure transport and national economic criteria, but also technological features and geographic location criteria.

Accordingly, maritime shipping routes break into following categories:

- transportation of import/export multi-issues cargo between the general public ports;
- import of bulk cargoes from Northern ports and port points;
- regular supply deliveries for the Northern ports, port point and hydrotechnical structures;
- expeditional supply deliveries to the port points and unequipped shores;
- construction material supply for port points;
- construction material and supply deliveries to unequipped shores and facilities;
- expedition, including ecological monitoring and cleaning;
- transit transportation;
- passenger transportation;
- passenger cruises.

This approach would enable to percept a general outlook of the shared transportation systems for many users. In addition to pure transportation, it also could be extended over some other maritime economic activities in Arctic.

This analysis will result in formulating criteria for matching every Arctic port and port point to one of the technologic development stage (under the classification of UNCTAD). Based on this classification, certain recommendations could be derived, taking into account types and quantities of technological equipment in ports, demands for required territories, buildings, constructions, facilities and water areas of ports and port points.

It is expected that in some cases there will be a necessity to shift from conventional sea port to modern cargo handling complexes, among other features including logistic infrastructure for consolidation and distribution of multi-issue supply and transit shipment.

Second stage – the potential aggregated joint transportation systems are identified, among them large unified inter-department sub-systems for transportation of multi-item cargo and supplies are selected, those which used for transportation aimed for development of the infrastructure and expeditional delivery.

Third stage – complex mathematical models (optimizational, imitational) of separated different transportation sub-systems' operations providing their rational planning and exploitation.

Fourth stage – the complex integrated simulation model for the sub-systems joint operations enabling to perform an adequate and accurate prognoses of general and particial parameters of single interacting systems aiming both for rational regional development of separate components and coordination of the component's functions within the external system environment. Different scenarios of development are to be studied.

ANALYSES AND SOLUTIONS

In Russian Federation there is a group of systematically performing transportations demanding for out-of-port handling of ships. This group consists of expeditional (navigational, or so called "Northern") delivery. This regular navigational delivery is a complex of annual governmental measures aimed for supply of the Russian Northern territories with vitally important goods, mainly food and fuel, in anticipation of the winter season.

The phenomenon of the Northern delivery can be explained by the following reasons:

- the absence on the Russian Northern territories the own production base for the majority of manufactured goods and food staffs;

- the remoteness by thousand kilometer from main industrial areas, which makes very difficult and expensive for persons and companies their own transportation of goods;

- the total absence of the transport infrastructure, with only rudiment elements for air and inner waterways modes.

A more generic reason is in extreme natural and climatic conditions in Russian Northern areas, which occupy over two third part of the Russia's territory. Under these conditions, the only solution is the centralization of the purchase, transportation and distribution of the goods from Southern areas of the country. This responsibility both in the USSR and in Russian Federation bears the state, sponsoring the task from the federal budget and performing it via regional and local authorities. This activity assumes out-of-port handling of ships, which makes containerization quite difficult task. The opportunities offered by existing out-of-port handling technologies are shown in Figure. 1. Accordingly, an important problem today is the development of the pool of out-of-port container handling equipment, like rough terrain container handlers (RTCH) etc.

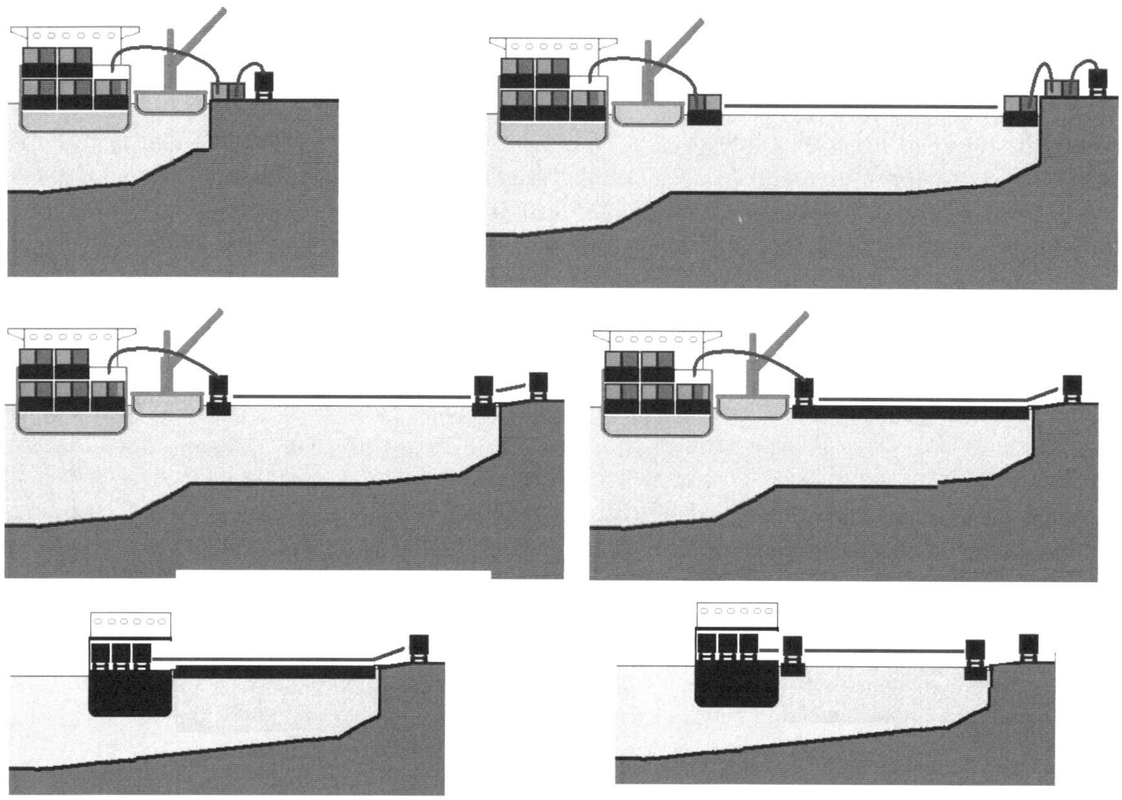

Figure 1. General variants for out-of-port container handling

While the range of the specialized handling equipment for "civilian" transportation is relatively stable and well-studied, the out-of-port container handling equipment need a lot of researches and studies to find the boundaries of applications for existing types and possibly developing of new prototypes, t.t. like on Figure. 2.

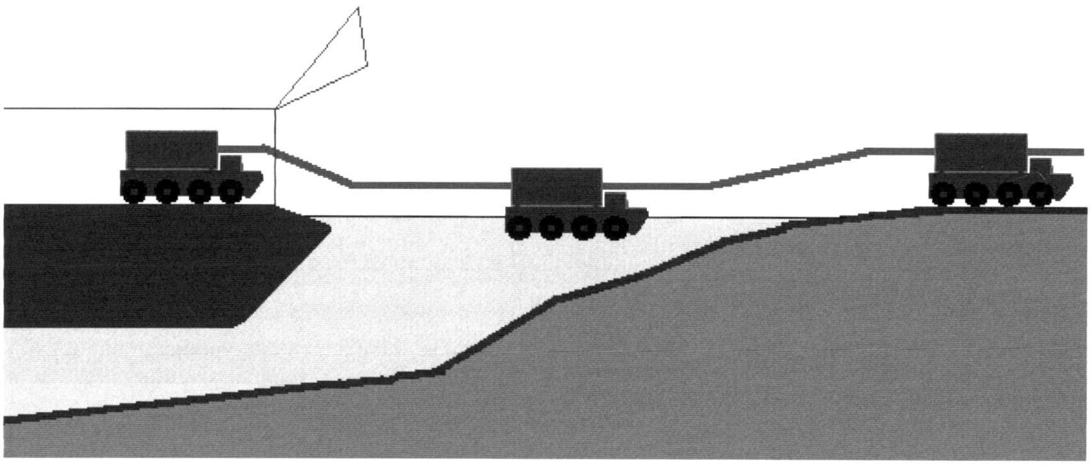

Figure 2. Example of amphibian technology for container handling

For the container handling in out-of-port conditions there might be finger-pierce floating berths arranged formed out of lighters or pontoons, with the last one equipped with a ramp. Another variant is to deliver ashore by a lighter a container handling equipment specimen, e.g.

RTCH, for further handling of containers shuttled between ship and shore by lighters. There might be a lot of combinations of similar schema.

In the same time, all the advantages of above mentioned variants could be integrated into an unified transportation/handling solution base on the lighters-abroad-ship system (LASH). These systems already proved their flexibility and efficiency, being intensively used in many countries. In Russian Federation (in contrast with the former USSR) these systems do not exist. The only exemption is quite unique super-powerful LASH system built around nuclear-powered ice-breaking LASH vessel "Sevmorput" (or «Northern Sea Route" in Russian).

The ship was built in 1988 and totally rehabilitated in 2016. Originally, the ship was designated to deliver goods on lighters and in containers for distant areas of the Russian Arctic. The ship can navigate through the ice up to 1 m thick, she carries 74 dedicated lighters or 1336 TEUs (Figure. 3).

Figure 3. Nuclear powered ice-breaker LASH ship "Sevmorput"

The existence of this LASH vessel with these unique features provides a scaled logistical advantage both for defense transportation system and all Arctic cargo flows support.

For the full realization of this potential advantage, it is required to develop a lighter set very well-balanced functionally. Its functional profile should be defined by taking into account the interests of all participants involved in transportation activity in Arctic region.

At this stage the authors propose the following content of the functional components of the system:

- nuclear-powered ice-breaking LASH vessel, equipped with auxiliary lifting cranes;

- specialized LASH standard modules: pontoon-berths, hold-types, platforms, ramps (Figure. 4);

- self-propelled barges with ramps;

- rough terrain container handlers.

Figure. 4–5 give examples of cargo handling based upon this system.

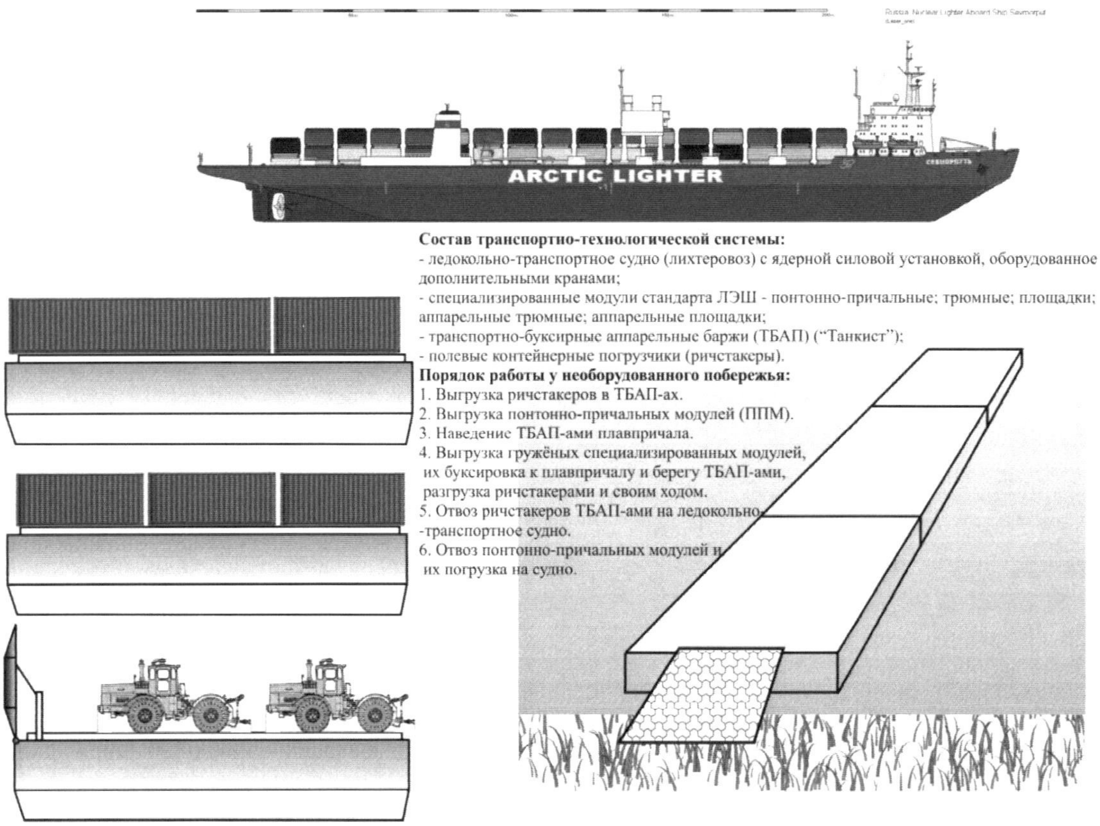

Figure 4. Components of the LASH system

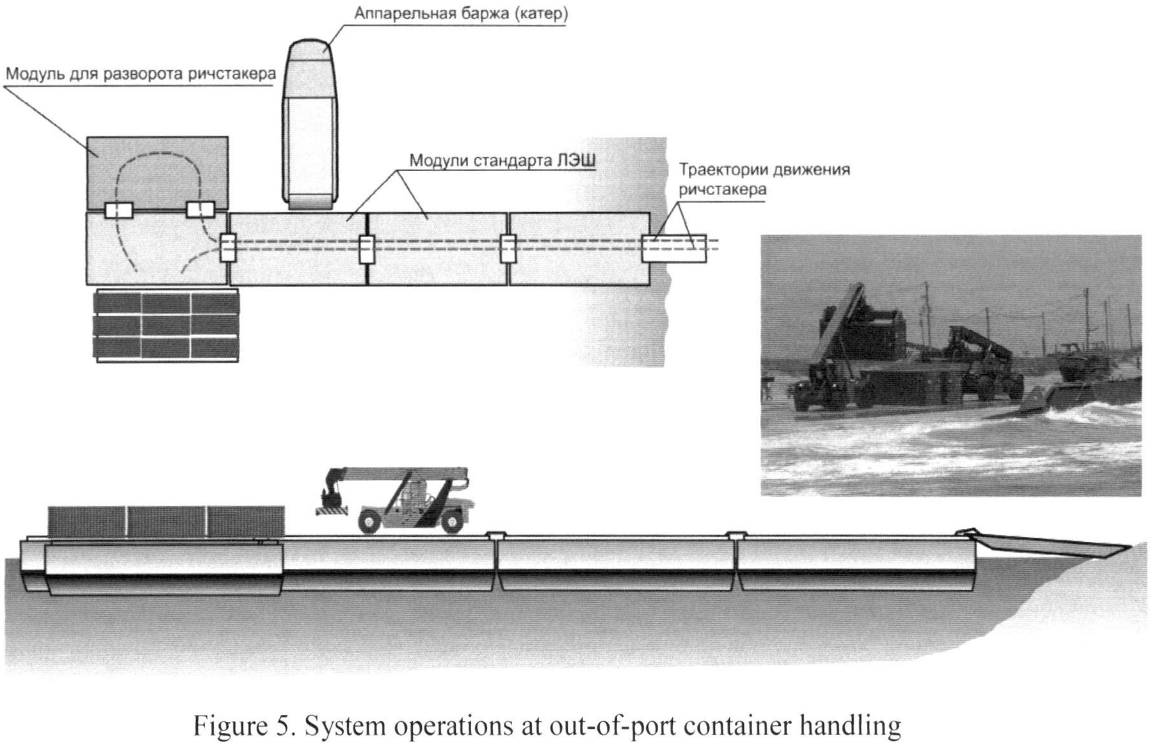

Figure 5. System operations at out-of-port container handling

CONCLUSIONS

The proposed methodology provides the results of investigations which help to optimize the industrial activities of separate transportation and national economic bodies, i.e. directly providing the improvements in efficiency of different departmental and administrative industrial systems – users of sea areas and infrastructure of Arctic.

In respect of the Federal Executive government bodies, who are the key stakeholders of transportation and economic activities in the sea areas of Arctic , the expected result could be the creation of the governmental unified integrated management system, providing the joint exploitation of transportation vehicles and infrastructural objects, minimizing the total budget investment planning and execution.

The proposed hypothesis and methodology assume also investigations of legislative and regulation base with the aim of creation of unified information management system for transportation and national economic activities, bases upon the positive experience of similar Federal Executive bodies of state power in 2000th.

REFERENCES

Gorodetskiy, A.E., Ivanov, V.V., & Filin, B.N., 2014. Legislative and methodological problems of strategic planning of development of the Russian Federation Arctic zone. *Ecology and economy,* 4(16), pp.4–13.

Izotov, O.A., 2005. *Optimization of the process of exploitation of the objects of maritime and river transport in the context of joint utilization systems.* PhD. St. Petersburg: Admiral Makarov State Maritime Academy.

Konovalov, A.M., 2013. The system of state strategic planning of the Russian Federation Arctic zone. *Science and transport. Maritime u river transportation,* 1(5), pp.57–63.

Resolution of the Government of Russian Federation by 21/04/2014 № 366 "On approval of the State program of Russian Federation "Social - economic development of Russian Federation Arctic zone".

The principles of Russian Federation state policy in Arctic for the period until 2020 and further future. Approved by the President of Russian Federation 18/09/2008 Act №1969.

The scope of responsibility and functions of the Administration of the Northern Sea Routs. Official website of the Federal governmental state organization "the Administration of the Northern Sea Routs". (Updated 12 April 2015) Available at: http://asmp.morflot.ru/ru/celi_funktsii/

The Strategy of Russian Federation Arctic zone development and provision of national security for the period until 2020. Approved by the President of Russian Federation 18/02/2013.

Tolstych, D.A., 2009. Regional industrial-logistic terminals as the element of formation of the integrated transport infrastructure. PhD. St. Petersburg: Admiral Makarov State Maritime Academy.

POAC'17
Busan, KOREA

Proceedings of the 24th International Conference on
Port and Ocean Engineering under Arctic Conditions
June 11-16, 2017, Busan, Korea

Research of Quantitative Indicators of Tightness of the Northern Sea Route (NSR)

Aleksandr Tezikov[1], Andrei Afonin[1], Vitaly Kljuev[2]
[1] Admiral Makarov State University of Maritime and Inland Shipping, Saint-Petersburg, Russia
[2] The Ministry of Transport of the Russian Federation, Moscow, Russia

ABSTRACT

The paper discusses the features of the Arctic shipping and the main sources of navigational accident risk in the Arctic seas in the Northern Sea Route (NSR).
The results of the analysis of risk assessment techniques of navigation accidents, based on the use of the global and regional statistics of marine accidents are considered. In addition, we used data from the Russian Northern Sea Route Administration. Statistical representativeness of navigational accidents related to the seas of the Arctic basin was estimated.
The study found an association between the frequency of incidents of navigation and tightness of aquatory.
The term "tightness of aquatory", which is usually used in connection with the navigation of vessels in narrow waters and in shallow water, is proposed to use as the generalized characteristics of the entire water area of the NSR and its individual parts including for seaports.
To determine the degree of tightness of aquatory we suggested to use a method, based on the theory of geometric probability.
The examples of the effective use of geometric probability theory to solve practical problems of hydrography and decoding images are presented.
The paper presents the basic scientific principles, formulas and ratios that allow to calculate the required probability analytically for two variants of the direction of flow of ship, including isotropic and anisotropic many routes.
All studies were conducted in the Arctic Faculty of the Admiral Makarov State University of maritime and inland shipping. In conclusion, the article gives a general evaluation of the prop osed method and the directions of its perfection.

KEY WORDS: Polar Code; FSA; Quantitative assessment; Geometrical probability; Navigational hazard.

INTRODUCTION

The principles of Formal Safety Assessment (FSA) stated in the Temporary manual were developed and approved by Maritime Safety Committee and Marine Environment Protection Committee of the International Maritime Organization (IMO) (Revised guidelines for formal safety (FSA) for use in the IMO rule making process, 2014) since 1997. The Polar Code, 2015 which is also in many respects based on the principles of FSA came into force in 2017.

The essence of the FSA concept is aprioristic risk assessment of marine navigational and environmental incidents and development of the actions directed at decreasing these risks (Chen, 2001). The world or regional statistics of accidents is used, as a rule, in case of risks assessment of navigation incidents (Pastusiak, 2016). Statistics allows to select some general reasons of accidents by their types, for example ecological - (Moe, 2000), on the frequency and conditions and to use the obtained data as the initial aprioristic estimates at FSA development.

The available statistics of navigation incidents on the water area of NSR (Federal state Institution "The Northern Sea Route Administration", 2017) can't be objective for the existing and perspective year-round Arctic transport projects because generally it was created in the conditions of extremely low navigation intensity, only for the summer navigation period and as a rule, for vessels with insignificant draft and not having ice strengthening.

The traditional method of aprioristic assessments of probabilities of navigation incidents can be applied to the water area of NSR only partially because the conditions of the Arctic navigation and its risks substantially differ from the navigation conditions and the corresponding risks on not Arctic water area of the World Ocean.

DEFINITION OF RESEARCH PROBLEM

Aprioristic assessment of probability can be carried out on the basis of the general assumptions of adverse events in the absence of reliable statistical data. It is established (Kljuev, 2016) that navigation incidents happen on the "constrained" sites of water areas more often. At the same time degree of "tightness" of the water area is defined by sizes and quantity of its sites dangerous to navigation. It should be noted that the term "tightness" of the water area is usually used for vessels navigation in narrow waters and in shallow water where tortuosity of fairway and proximity to surface and underwater navigation hazards significantly influence maneuverable characteristics of the vessel.

Results of hydrographic research of the Arctic seas (Batalin, 2008; Federal State Unitary Enterprise "Hydrographic Enterprise", 2016; Kastner, 2015; Reshetnyak, 2006; Svahn, 2015) show that the water area of NSR is mainly shallow, abounds with a large number of underwater navigational hazards which are shallows, banks, reefs and rocks and therefore can be characterized as "constrained". Comparative assessment of "tightness" degree of the water area of NSR and its separate parts is required to perform for definition of areas potentially dangerous for navigation and implementation of the complex actions directed to decreasing of incidents risks (Revised guidelines for formal safety (FSA) for use in the IMO rule making process, 2014). It is expediently to carry out the comparative assessment using some generalized quantitative indicators of the water area. Obviously that the calculation method of such indicators and the indicators themselves have to be justified, realized in practice and objective.

The quantitative assessment of "tightness" degree of separate parts of the NSR water area is offered to perform by the methods developed in the theory of geometrical probabilities and corresponding the listed requirements. The example of use of the theory of geometrical probabilities at the solution of practical tasks is use during the performing of shooting of underwater relief and search of underwater objects and also recognition of the scanned images. Use of the theory of geometrical probabilities allows to reduce a task of quantitative assessment of the water area "tightness" to a task about crossing of straight lines with curves on the plane. At the same time the probability of crossing of straight lines (vessel way) with curves (contours of navigation hazards) P_o is accepted as a quantitative measure of "tightness" of the water area M_{TWA}.

DEVELOPMENT AND JUSTIFICATION OF METHOD OF QUANTITATIVE ASSESSMENT OF THE WATER AREA "TIGHTNESS"

Local rise of bottom which is dangerous to the vessel with draft d is given as

$$R_{hi}^{d} = R_{hi}\left[x, y, z < z_{ex}(d)\right], \quad (1)$$

where the lower index "h" designates "hazard"; i — the number of hazard; x, y — planned coordinates of a local rise; z — depths in the internal area of local rise of bottom; $z_{ex}(d)$ — extreme passing depths for the vessel with draft d.

Vessel routes are set by array of lines as

$$G = \{G(d)\}. \quad (2)$$

The consideration of draft d in models (1) and (2) is fundamental because passing depths and sizes of that part of local rise of bottom which has to be accepted as dangerous depend on her value.

The probability P_o of crossing of routes of the vessel G with danger R_{hi}^{d} represents a measure of undesirable event $M_{TWA}[G \cap R_{hi}^{d} \neq 0]$ that is written as

$$P_o = P\left[G \cap R_{hi}^{d} \neq 0\right] = M_{TWA}\left[G \cap R_{hi}^{d}\right]. \quad (3)$$

The probability of safe navigation P_S is defined by a condition under which any of routes G won't cross hazard R_{hi}^{d}. This condition is established by ratio

$$P_S = P\left[G \cap R_{hi}^{d} = 0\right]. \quad (4)$$

The events set by conditions (3) and (4) make full group of events, i.e.

$$P_o + P_S = 1.$$

Algorithms of probabilities (3) and (4) calculation depend on the choice of the direction of routes G for which the assessment of " tightness " of the water area is carried out. It is possible to allocate two main ways of setting of routes directions: isotropic and anisotropic. At the isotropic method the routes are arranged in all possible directions. At the anisotropic method the priority direction of routes, for example, "East" - "West" is allocated. The routes which are different from the routes set in the direction are considered forbidden at the anisotropic method of the routes direction.

ISOTROPIC SET OF ROUTES

We consider a part of the water area which is limited to a quadrate of the square S and perimeter $L(S)$. We consider that there is one local rise of bottom R_{hi}^{d}, where $i = 1$, on the water area

If we assume that routes G can cross the allocated part of the water area in any direction that corresponds to the isotropic set of straight lines, then the probability of undesirable event (3) equal to measure of "tightness" of the water area $\mathrm{M}^1_{\mathrm{TWA}}$ can be calculated on a formula

$$P_{o} = \mathrm{M}^{1}_{\mathrm{TWA}} = \frac{L(R^{d}_{h1})}{L(S)}, \tag{5}$$

where $L(R^d_{o1})$ — perimeter of contour of local rise R^d_{h1}; superscript "1" indicates the number of local rises on the water area.

The equation (5) implies the probability of undesirable event doesn't depend on the position of the local rise R^d_{h1} in the allocated part of the water area.

Let's consider more general case when on the allocated part of the water area there are two dangerous rises R^d_{h1} and R^d_{h2} removed from each other so that the safe route can be laid between them.

We will unite the rises R^d_{h1} и R^d_{h2} by the general contour which perimeter we will designate as $L(\oplus) = L\left(R^d_{h1} \oplus R^d_{h2}\right)$. The probability that routes at their isotropic arrangement will cross the general contour can be calculated by a formula

$$P_{o} = \frac{L(\oplus)}{L(S)}. \tag{6}$$

The probability that any of routes won't cross the general contour will be equal

$$P_S = 1 - P_o. \tag{7}$$

Expression (7) is received on the basis that probabilities P_o and P_S form full group of casual events.

The routes passing through the general contour are divided into four groups:

- $G \cap R^d_{h1} \neq 0$ — the routes crossing the rise R^d_{h1} (an adverse event);
- $G \cap R^d_{h2} \neq 0$ — the routes crossing the rise R^d_{h2} (an adverse event);
- $G \cap \left(R^d_{h1}, R^d_{h2}\right) \neq 0$ — the routes which are at the same time crossing both rises (an adverse event);
- $G \cap \left(R^d_{h1}, R^d_{h2}\right) = 0$ — the routes passing between rises R^d_{h1} and R^d_{o2} (a favorable event).

From the listed groups of routes the fourth group characterizes safe navigation whereas the first three define degree of "tightness" of the water area.

The measurement of length of the "twisted" contour encompassing rises R^d_{h1} and R^d_{h2} is required for determination of probability of each of the listed four events. The length of the contour is denoted as $L(\otimes) = L\left(R^d_{h1} \otimes R^d_{h2}\right)$.

Taking into account the designations presented above the probabilities of each of the listed four events are calculated by the following formulas:

$$P\left(G\cap R_{h1}^{d}\neq 0\right)=\frac{L(\oplus)-L(\otimes)+L\left(R_{h1}^{d}\right)}{L(S)};\qquad(8)$$

$$P\left(G\cap R_{h2}^{d}\neq 0\right)=\frac{L(\oplus)-L(\otimes)+L\left(R_{h2}^{d}\right)}{L(S)};\qquad(9)$$

$$P\left(G\cap\left(R_{h1}^{d},R_{h2}^{d}\right)\neq 0\right)=\frac{L(\otimes)-L(\oplus)}{L(S)};\qquad(10)$$

$$P\left(G\cap\left(R_{h1}^{d},R_{h2}^{d}\right)=0\right)=\frac{L(\otimes)-L\left(R_{h1}^{d}\right)-L\left(R_{h2}^{d}\right)}{L(S)}.\qquad(11)$$

The sum of the probabilities calculated by formulas (8) – (11) is equal to the ratio $L(\oplus)/L(S)$, that corresponds to the probability calculated by formula (6).

Quantitative assessment of "tightness" of the water area at the isotropic arrangement of routes on the basis of formulas (6) – (11) for two local spaced dangerous rises, equal to the probability of crossing at least of one of hazards, is determined by formula

$$\mathrm{M}_{\mathrm{TWA}}^{2}=\frac{L\left(R_{h1}^{d}\right)+L\left(R_{h2}^{d}\right)}{L(S)}.\qquad(12)$$

Generalizing the received result on n dangerous rises of bottom at the isotropic arrangement of routes, we will receive expression for quantitative assessment of " tightness " of the water area as

$$\mathrm{M}_{\mathrm{TWA}}^{n}=\frac{\sum_{1}^{n}L\left(R_{hi}^{d}\right)}{L(S)}.\qquad(13)$$

The formula (13) shows that the indicator of "tightness" of the water area in the absence of any preferred direction of the chosen routes is defined by a ratio of the sizes of the water area and dangers.

ANISOTROPIC SET OF ROUTES

Let's consider the part of the water area limited to a rectangle with square S and perimeter $L(S)$.

We assume that there are no local rises of bottom on the water area. Routes belong to a set of straight lines which at the same time cross the right and left vertical framework of the area in any their point. At the same time any of them can't cross either the top horizontal frame, or lower. Each of horizontal frameworks, by analogy with (1), can be interpreted as some dangerous object therefore for their description we use following expression

$$R_{Ei}^{d}=R_{Ei}\left[x,y,z<z_{ex}(d)\right],\qquad(14)$$

where the lower index «E» is border of forbidden area; index «i» is number of a framework which is forbidden to be crossed.

Replacement of object (1) on the object (14) allows the task of calculation of "tightness" indicator of the water area for the anisotropic set of routes to reduce to the task of calculation of "tightness" indicator for the isotropic set of routes considered earlier.

The full set of routes G crossing the allocated part of the water area is divided into four groups:

- $G \cap R_{E1}^{d} \neq 0$ — the routes crossing framework R_{E1}^{d};
- $G \cap R_{E2}^{d} \neq 0$ — the routes crossing framework R_{E2}^{d};
- $G \cap \left(R_{E1}^{d}, R_{E2}^{d}\right) \neq 0$ — the routes which are crossing both horizontal frames at the same time;
- $G \cap \left(R_{E1}^{d}, R_{E2}^{d}\right) = 0$ — the routes passing between frameworks R_{E1}^{d} and R_{E2}^{d}.

From the listed groups of routes the fourth group characterizes safe navigation whereas the first three groups define degree of "tightness" of the water area. Probabilities of each of the listed four groups are calculated by formulas:

$$P\left(G \cap R_{E1}^{d} \neq 0\right) = \frac{L(\oplus) - L(\otimes) + L\left(R_{E1}^{d}\right)}{L(\oplus)}; \tag{15}$$

$$P\left(G \cap R_{E2}^{d} \neq 0\right) = \frac{L(\oplus) - L(\otimes) + L\left(R_{E2}^{d}\right)}{L(\oplus)}; \tag{16}$$

$$P\left(G \cap \left(R_{E1}^{d}, R_{E2}^{d}\right) \neq 0\right) = \frac{L(\otimes) - L(\oplus)}{L(\oplus)}; \tag{17}$$

$$P\left(G \cap \left(R_{E1}^{d}, R_{E2}^{d}\right) = 0\right) = \frac{L(\otimes) - L\left(R_{E1}^{d}\right) - L\left(R_{E2}^{d}\right)}{L(\oplus)}. \tag{18}$$

In formulas (15), (16) and (18) the summand $L\left(R_{E1}^{d}\right)$ and $L\left(R_{E2}^{d}\right)$ representing perimeters of segments of straight lines, are equal to the doubled length of each of segments.

Based on the fact that the events connected with crossing by routes of all four borders of the allocated area make full group

$$P\left(G \cap R_{E1}^{d} \neq 0\right) + P\left(G \cap R_{E2}^{d} \neq 0\right) + P\left(G \cap \left(R_{E1}^{d}, R_{E2}^{d}\right) \neq 0\right) + P\left(G \cap \left(R_{E1}^{d}, R_{E2}^{d}\right) = 0\right) = 1, \tag{19}$$

the "tightness" indicator of the water area has the following form:

$$\mathrm{M}_{\mathrm{TWA}}^{2} = 1 - P\left(G \cap \left(R_{E1}^{d}, R_{E2}^{d}\right) = 0\right). \tag{20}$$

Taking into account the expression (18), we will receive calculated formula for the indicator of the water area constrained by two borders:

$$\mathrm{M}_{\mathrm{TWA}}^{2} = 1 - \frac{L(\otimes) - L\left(R_{E1}^{d}\right) - L\left(R_{E2}^{d}\right)}{L(\oplus)}. \tag{21}$$

The analysis of formula (21) shows that at the anisotropic choice of routes the "tightness" indicator of the water area depends not only on a ratio of the sizes of the water area and hazards, but also on their mutual arrangement.

The "tightness" indicator of the water area with other things being equal always has bigger numerical significance for areas where the priority direction of the vessels movement is allocated in comparison with areas where the priority direction of the movement isn't established. With use of the developed technique there is a possibility of laying of the most optimum routes on NSR, including and with use of the software developed earlier (Afonin, 2016).

CONCLUSION

It is offered to use probability of crossing of vessel routes with navigation hazards as a quantitative measure of "tightness" of the water area. The article presents the basic formulas and ratios allowing to calculate analytically the required probability for two options of the direction of the vessels streams including the isotropic set of routes and anisotropic set of routes. The general assessment of the offered method and the directions of his improvement:

1. The quantitative indicator of "tightness" of the water area can be used in FSA for polar and Arctic navigation.

2. Use of the theory of geometrical probabilities allows to receive the quantitative indicator of "tightness" of the water area having obvious geometrical interpretation of the events leading to collision with hazards.

3. Algorithm of calculation of "tightness" indicators of the water area is rather simple in realization because it is based on measurements of lines lengths on navigation charts and calculation of their ratios. The algorithm is characterized by sufficient simplicity.

4. "Tightness" of the water area, apparently, has to be estimated by some set of quantitative indicators because it depends on the main vessels sizes, the established directions of their movement, borders of the water area and other factors. The question of structure of such indicators needs additional justification.

5. The approach to assessment of "tightness" of the water area used in this work can be extended to the dynamic objects influencing risk of navigation incidents in the Arctic seas.

All researches within this article have been executed at the Arctic faculty of Admiral Makarov State University of Maritime and Inland Shipping (Russia, Saint-Petersburg).

REFERENCES

Revised guidelines for formal safety (FSA) for use in the IMO rule making process - MSC-MEPC.2/Circ.12/Rev.1. IMO: London, 2014.

Polar Code, Resolution MSC/385(94), MEPC.264 (68). IMO, 2015.

Chen, Y. & Lin, S., 2001. Comprehensive Fuzzy Approach in Hazard Identification of Formal Safety Assessment (FSA). *Practical design of ships and other floating structures*, pp.245-251. (doi:10.1016/b978-008043950-1/50031-4).

Pastusiak, T., 2016. Variants of the Northern Sea Route. *The Northern Sea Route as a Shipping Lane*, pp.131-152. doi:10.1007/978-3-319-41834-6_6.

Moe, K.A., 2000. Northern Sea Route Activities — Environmental Assessments. *The 21st Century — Turning Point for the Northern Sea Route?*, pp.63-69. doi:10.1007/978-94-017-3228-4_13.

Federal state Institution "The Northern Sea Route Administration", 2017. Navigation rules on NSR. [Online] (Updated 30 January 2017) Available at: http://www.nsra.ru/ [Accessed 12 December 2016].

Kljuev, V., 2016. Quantitative assessment of constraint of the Northern Sea Route. *Vestnik Gosudarstvennogo universiteta morskogo i rechnogo flota imeni admirala S.O. Makarova,* 5(39), pp.109-117. DOI: 10.21821/2309-5180-2016-8-5-109-117. (rus)

Kastner, P., 2015. International Legal Dimensions of the Northern Sea Route. *The Northern Sea Route*, pp.39-52. doi: 10.1007/978-3-658-04081-9_3.

Svahn, P., 2015. Through the Northern Sea Route by Stena Polaris: A logbook. *The Northern Sea Route*, pp.69-88. (doi: 10.1007/978-3-658-04081-9_5).

Reshetnyak, S.V., 2006. Hydrographic study of the Arctic sees underwater relief of Russia. *Geodesy and Cartography,* 4, pp.57-60. (rus)

Federal State Unitary Enterprise "Hydrographic Enterprise", 2016 [Online] (Updated 27 January 2017) Available at: http://www.hydro-state.ru/ [Accessed 12 December 2016].

Batalin, G., & Reshetnyak, S., 2008. The navigational and hydrographic support for shipping in the Arctic region. *Marine Fleet,* 3, pp.34-38. (rus)

Afonin, A., Ol'hovik E., & Tezikov, A., 2016. Development of the assessment methods of anadromous depths on the Northern Sea Route depending on the detail of survey of the bottom relief. *Vestnik Gosudarstvennogo universiteta morskogo i rechnogo flota imeni admirala S.O. Makarova,* 4(38), pp.62-68. doi: 10.21821/2309-5180-2016-8-4-62-68. (rus)

Interpretative Structural Modeling of Ship Energy Efficiency in Arctic Navigation

Chi Zhang[1,2], Di Zhang[1,2], Shanshan Fu[1,2], Mingyang Zhang[1,2]
[1]Intelligent Transport Systems Research Center, Wuhan University of Technology, Wuhan, China
[2]National Engineering Research Center for Water Transport Safety, Wuhan, China

ABSTRACT

For safe and efficient exploitation of ice-covered waters, knowledge about ship performance in ice is crucial. While the existing literature describes various models that interpret ship risk in ice, few studies have focused on the aspect of efficiency. This research studies the energy efficiency of Arctic ships via the Energy Efficiency Operational Indicator (EEOI), introduced by the International Maritime Organization (IMO). First, the factors influencing energy efficiency are identified. Second, the model of energy efficiency in the Arctic is built based on interpretative structural modeling (ISM) method. Finally, the paper analyses the model and provides suggestions for the improvement of the energy efficiency. The proposed research would help crews and maritime managers understand the factors influencing energy efficiency of Arctic navigation.

KEY WORDS: Energy efficiency; Ice conditions; Energy efficiency operational indicator; Arctic navigation

INTRODUCTION

Given the ongoing effects of climate change, the Arctic sea ice has begun to melt rapidly. It has been predicted that navigation will be entirely possible throughout the Arctic Sea Route during the summer. The Northeast Passage is a new route, and a shorter shipping line compared with the traditional Suez canal (Schøyen et al., 2011). This international route connects the Far East with North European countries. Exploiting this route will thus enable the rapid development of northern Chinese ports. The Chinese merchant vessel, the Yongsheng, went through this route successfully in 2013, which became a key moment in Chinese marine navigation. With reference to estimations about Arctic routes (Zhang et al. 2013), in 2030 natural gas and container/freight transport from the Arctic regions to East Asia are set to rise to 10 and 17.43 million tonnes, respectively.

According to the International Marine Organization's (IMO) second greenhouse gas (GHG) study (2010), shipping emitted 1046 million tonnes of CO_2 in 2007, representing approximately 3.3% of global emissions. For many existing ships, reducing fuel consumption to reduce these types of emissions involves applying operational measures rather than adopting alternative technologies, owing to the long payback periods associated with a new investment (Fan et al., 2016).

As of January 2013, new ships are required to follow the Energy Efficiency Design Index (EEDI) stipulations, and to maintain a Ship Energy Efficiency Management Plan (SEEMP) (IMO, 2011). While the EEDI puts forward minimum technological standards with long-term impacts for all new ships, SEEMP is an operational implementation that seeks to enhance ships' energy efficiency through using the already available technologies in ships (Beşikçi et al., 2016). The Energy Efficiency Operational Indicator (EEOI) was put forward as a monitoring tool of the SEEMP (IMO, 2009). The IMO recommends that a straightforward EEOI calculation, first described in MEPC.1/Circ.684, should be used to evaluate the efficiency of performance according to a ship's SEEMP, with the EEOI reflecting the emission of CO_2 per unit of transport work. Energy conservation and emissions reduction are important measures for solving both the problem of environmental pollution and increasing operating costs for the shipping industry.

As far as operation energy efficiency is concerned, there are differences between ships sailing the Arctic region and others. The Arctic environment is more complex and sensitive than others. In addition, the factors which influence energy efficiency in Arctic waters are different to those influencing other regions. For example, ice injects considerable uncertainty into Arctic navigation — the biggest difference to navigating inland rivers. For instance, when ships are sailing in an ice formation without piloting, they need to consume more energy to break the ice while at the same time sustaining their speed to avoid getting stuck in the ice.

While navigational accidents occur more frequently in Arctic navigation, such as in ice conditions than in open water, they typically lead to less serious consequences (Goerlandt et al., 2016). However, the growing changes in Arctic environmental conditions have made more of this region ice-free in the summer season, thus reducing the difficulty of marine transport in Arctic waters (Khan et al., 2014). This now means that energy efficiency has become a priority issue even though, to date, there are very few studies considering the energy efficiency of ships on an Arctic voyage.

The current paper aims to fill this gap by analyzing the energy efficiency of ships sailing in the Arctic. To do so, we employed interpretative structural modeling (ISM), which can be used to establish levels to help describe problem structures and factor priorities; this process enabled the study to analyze the relationship between the different factors exerting an impact on the energy efficiency of Arctic navigation.

INTERPRETATIVE STRUCTURAL MODELING (ISM)

Interpretive structural modeling (ISM) is an effective methodology for dealing with complex issues, first proposed by J. Warfield in 1973. It enables individuals or groups to develop a map of the complex relationships between the diverse elements involved in a complex situation. ISM is often used to provide a fundamental understanding of complex situations, as well as to put together a course of action for problem-solving. It has been used worldwide by many prestigious organizations, including NASA.

ISM is a multivariate analysis method that can be used to analyze the causal relationships between variables in the social sciences. ISM analyses the structure of the selected characteristics. ISM has gradually come to be used in the structure analysis of complicated systems and some engineering fields, where it is typically employed to seek out complex relationships. The review clearly brings out the versatility of ISM as a tool capable of modeling a diverse range of complex issues. The tool brings structural clarity and establishes a hierarchical order for prioritization and consequent action.

ISM uses experts' practical experience and knowledge to deconstruct a complicated system into several subsystems and construct a multilevel structural model. The fundamental concepts of the process are those of an 'element set' and a 'contextual relationship'. The element set is identified within a particular situational context, and the contextual relationship is selected as a possible statement of the relationship between the elements, in a manner that is contextually significant for the purposes of the inquiry. The elements correspond to the nodes of a network model, and the presence of the relationship between any two elements is denoted by a directed line connecting those two element nodes. In an equivalent binary matrix representation, the elements represent the contents of the index set for the rows and columns of the matrix and denote the presence of a relationship.

An adjacency matrix is employed to demonstrate the relationship between the elements in ISM and can be obtained from the directed ISM graph (Zhang et al., 2015). The element e_{ij} of the adjacency matrix is defined as:

$$e_{ij} = \begin{cases} 1 & \text{there is a directed line from } s_i \text{ to } s_j \\ 0 & \text{there is no directed line from } s_i \text{ to } s_j \end{cases} \quad (1)$$

Thus, the adjacent matrix of the system shown in Figure 1. Is

$$E = \begin{bmatrix} 0 & 0 & 0 & 0 \\ 1 & 0 & 0 & 0 \\ 1 & 0 & 0 & 0 \\ 0 & 1 & 1 & 0 \end{bmatrix} \quad (2)$$

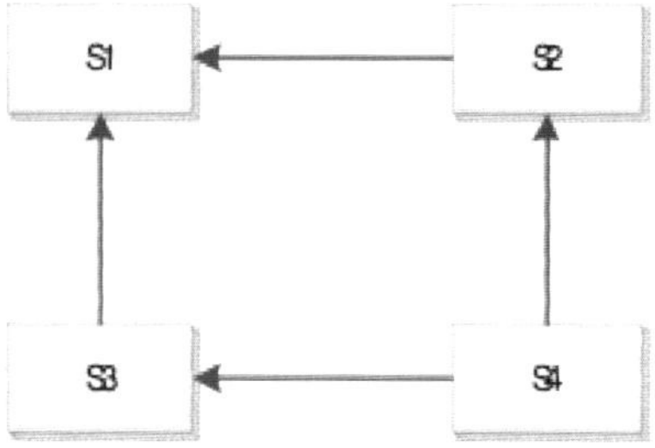

Figure 1. The directed digraph of 4 elements

The reachability matrix is employed to demonstrate the capability of one element having impacts on the others. The sum of E and the unit matrix I lead to a new matrix, E_1, and the square of the new matrix E_1 can be calculated using Boolean rules ($0 + 0 = 0$, $0 + 1 = 1$, $1 + 1 = 1$, $0\times0 = 0$, $0\times1 = 0$, $1\times1 = 1$). The result is as follows: $E_2 = (E_1)^2 = (E + I)^2 = E^2 + E + I$. Through sequential calculations, it can be obtained that $E_1 \neq E_2 \neq \ldots \neq E^{q-1} = E^{q}$; this means that $(E + I) \neq (E + I)^2 \neq \ldots \neq (E + I)^{q-1} = (E + I)^q = M$, and K is defined as the reachability matrix.

The reachability set, antecedent set and commonality set are defined as follows: (1) the reachability set $R(s_i)=\{s_j\in\Omega \mid k_{ij} = 1\}$; (2) the antecedent set $A(s_i)=\{s_j\in\Omega \mid k_{ij} = 1\}$; and (3) the commonality set $T(s_i) = R(s_i) \cap A(s_i)$ $(0 < i, j < Z)$. Z refers to the number of the elements considered in ISM, and k_{ij} is the i-th row and j-th column element of K. The reachability set, antecedent set, and commonality set can be obtained by K. If $R(s_i) = R(s_i) \cap A(s_i)$ holds, $R(s_i)$ is the set of elements of the top level. Then, the rows and columns related to the elements of $R(s_i)$ are removed from K, in order to obtain the reachability matrix of the next level, and the elements of the other levels are consequently obtained. Finally, the structure model is built with the elements of all the levels, and the relationship between the elements is represented by the reachability matrix.

CASE STUDY

We chose the polar voyage of the ship, Yong Sheng, in the Arctic region as a case study to analyze the energy efficiency of Arctic navigation. Yong Sheng completed its voyage of the Northeast Sea Route from August 28 to September 2 in 2013 (Fu et al., 2015). In total, 179 sets of data were obtained, including air temperature, relative wind speed, visibility, water temperature, ice concentration, GPS, gyrocompass, magnetic compass, navigational speed and engine power during navigation. According to these data, we screened out some useful message about energy efficiency analysis. The main characteristics of the ship are listed in table 1.

Table 1. Characteristics of the ship Yong Sheng

Name	Yong sheng
Type	General cargo
Ice class	Arc 4 (Russian ice class)
DWT (tonnage)	14357
Length (m)	159.985
Breadth (m)	23.70
Draught (m)	41.10
Operational power (kw)	7074

Factors influencing energy efficiency

The factors influencing energy efficiency are shown in table 2; the selection of these factors was conducted based on a literature review and the data drawn from Yong Sheng.

Table 2. Factors influencing energy efficiency

Number	Factors	Description/reason for consideration
1	Wind speed	The wind speed of the voyage

2	Air temperature	A low air temperature may lead to a freeze of cold air, which influences equipment conditions
3	Ice concentration	Ice condition
4	Ice thickness	Ice condition
5	Visibility	Visibility is a variable expressing the impact of fog and aurora borealis during the voyage
6	Sea temperature	Sea temperature affects the formation of sea ice
7	Ship speed	The ship's speed has a major impact on reducing fuel consumption
8	Route selection	The route selection mainly influences the distance of the voyage
9	Oil consumption	The oil consumption of the whole voyage
10	Energy efficiency	The energy efficiency of the ship

Analyzing energy efficiency using ISM

Employing the aforementioned factors, adjacency matrix A was as follows:

$$A = \begin{bmatrix} 0 & 0 & 1 & 0 & 0 & 0 & 0 & 0 & 0 & 0 \\ 0 & 0 & 1 & 0 & 0 & 0 & 0 & 0 & 0 & 0 \\ 0 & 0 & 0 & 0 & 0 & 0 & 1 & 0 & 0 & 0 \\ 0 & 0 & 0 & 0 & 0 & 0 & 1 & 0 & 0 & 0 \\ 0 & 0 & 0 & 0 & 0 & 0 & 1 & 0 & 0 & 0 \\ 0 & 0 & 1 & 1 & 0 & 0 & 0 & 0 & 0 & 0 \\ 0 & 0 & 0 & 0 & 0 & 0 & 0 & 0 & 1 & 0 \\ 0 & 0 & 0 & 0 & 0 & 0 & 0 & 0 & 0 & 1 \\ 0 & 0 & 0 & 0 & 0 & 0 & 0 & 0 & 0 & 1 \\ 0 & 0 & 0 & 0 & 0 & 0 & 0 & 0 & 0 & 0 \end{bmatrix} \quad (3)$$

The calculation result based on Boolean rules was: $(E + I) \neq (E + I)^2 = (E + I)^3 = M$, by which the reachability matrix could be obtained. The reachability matrix indicates the relationship between the elements arrived at in binary form. With the help of MATLAB, reachability matrix M was developed, as follows:

$$M = \begin{bmatrix} 1 & 0 & 1 & 1 & 0 & 0 & 1 & 0 & 1 & 1 \\ 0 & 1 & 1 & 1 & 0 & 0 & 1 & 0 & 1 & 1 \\ 0 & 0 & 1 & 0 & 0 & 0 & 1 & 0 & 1 & 1 \\ 0 & 0 & 0 & 1 & 0 & 0 & 1 & 0 & 1 & 1 \\ 0 & 0 & 0 & 0 & 1 & 0 & 1 & 0 & 1 & 1 \\ 0 & 0 & 1 & 1 & 0 & 1 & 1 & 0 & 1 & 1 \\ 0 & 0 & 0 & 0 & 0 & 0 & 1 & 0 & 1 & 1 \\ 0 & 0 & 0 & 0 & 0 & 0 & 0 & 1 & 0 & 1 \\ 0 & 0 & 0 & 0 & 0 & 0 & 0 & 0 & 1 & 1 \\ 0 & 0 & 0 & 0 & 0 & 0 & 0 & 0 & 0 & 1 \end{bmatrix} \quad (4)$$

Partitioning the Reachability Matrix

The reachability and antecedent sets for each barrier were derived from the final reachability set. The reachability set consists of the element itself and other elements which the element itself may help achieve, while the antecedent set consists of the element itself and other elements, which may help to achieve it. The intersection of these two sets is termed the intersection set for each barrier. The top level barrier of the table will not help to achieve any other element (Digalwar, 2015). Once the top-level element was found, it was separated out from the other elements. The same method was then used to determine the next level barrier. The process was repeated until the level of each barrier was determined. The levels were then used to construct the ISM model for the energy efficiency of Arctic navigation.

Reachability set $R(s_i)$, antecedent set $A(s_i)$ and commonality set $T(s_i)$ at the top level were obtained using M, as shown in table 3.

Table 3. Reachability set, antecedent set and commonality set of level 1

s_i	Set $R(s_i)$	Set $A(s_i)$	Set $T(s_i)$
1	1, 3, 7, 9, 10	1	1
2	2, 3, 7, 9, 10	2	2
3	3, 7, 9, 10	1, 2, 3, 6	3
4	4, 7, 9, 10	4, 6,	4
5	5, 7, 9, 10	5	5
6	3, 4, 6, 7, 9, 10	6	6
7	7, 9, 10	1, 2, 3, 4, 5, 6, 7	7
8	8, 10	8	8
9	9, 10	1, 2, 3, 4, 5, 6, 7, 9	9
10	10	1, 2, 3, 4, 5, 6, 7, 8, 9, 10	10

These three sets denote the set of the elements that can be influenced by s_i, the set of the elements that influence s_i, and the intersection of these two sets, respectively.

As can be seen from table 3, R(10)= R(10) ∩ A(10), meaning that the s_1 was set at the top level. Next, we removed the row and column related s_1, so that the reachability matrix of level 2 could be obtained, as follows:

$$M_2 = \begin{bmatrix} 1 & 0 & 1 & 1 & 0 & 0 & 1 & 0 & 1 \\ 0 & 1 & 1 & 1 & 0 & 0 & 1 & 0 & 1 \\ 0 & 0 & 1 & 0 & 0 & 0 & 1 & 0 & 1 \\ 0 & 0 & 0 & 1 & 0 & 0 & 1 & 0 & 1 \\ 0 & 0 & 0 & 0 & 1 & 0 & 1 & 0 & 1 \\ 0 & 0 & 1 & 1 & 0 & 1 & 1 & 0 & 1 \\ 0 & 0 & 0 & 0 & 0 & 0 & 1 & 0 & 1 \\ 0 & 0 & 0 & 0 & 0 & 0 & 0 & 1 & 0 \\ 0 & 0 & 0 & 0 & 0 & 0 & 0 & 0 & 1 \end{bmatrix} \quad (5)$$

We thus obtained the reachability set, antecedent set and commonality set of level 2, which is shown in table 4.

Table 4. Reachability set, antecedent set and commonality set of level 2

s_i	Set R(s_i)	Set A(s_i)	Set T(s_i)
1	1, 3, 7, 9,	1	1
2	2, 3, 7, 9,	2	2
3	3, 7, 9,	1, 2, 3, 6	3
4	4, 7, 9,	4, 6,	4
5	5, 7, 9,	5	5
6	3, 4, 6, 7, 9,	6	6
7	7, 9,	1, 2, 3, 4, 5, 6, 7	7
8	8,	8	8
9	9,	1, 2, 3, 4, 5, 6, 7, 9	9

As shown in table 4, R(8)= R(8) ∩ A(8) and R(9)= R(9) ∩ A(9), meaning that the s_8 and s_9 were set at level 2. We then removed the row and column related y_8 and y_9, obtaining the reachability matrix of level 3, as follows:

$$M_3 = \begin{bmatrix} 1 & 0 & 1 & 0 & 0 & 0 & 1 \\ 0 & 1 & 1 & 0 & 0 & 0 & 1 \\ 0 & 0 & 1 & 0 & 0 & 0 & 1 \\ 0 & 0 & 0 & 1 & 0 & 0 & 1 \\ 0 & 0 & 0 & 0 & 1 & 0 & 1 \\ 0 & 0 & 1 & 1 & 0 & 1 & 1 \\ 0 & 0 & 0 & 0 & 0 & 0 & 1 \end{bmatrix} \quad (6)$$

We then performed the same steps as previously in order to obtain the reachability set, antecedent set and commonality set of level 3.

Table 5. Reachability set, antecedent set and commonality set of level 3

s_i	Set $R(s_i)$	Set $A(s_i)$	Set $T(s_i)$
1	1, 3, 7,	1	1
2	2, 3, 7,	2	2
3	3, 7,	1, 2, 3, 6	3
4	4, 7,	4, 6,	4
5	5, 7,	5	5
6	3, 4, 6, 7,	6	6
7	7,	1, 2, 3, 4, 5, 6, 7	7

According to table 5, it is evident that $R(7) = R(7) \cap A(7)$. So, we removed the row and column related s_7, obtaining the reachability matrix of the next level, as followed:

$$M_4 = \begin{bmatrix} 1 & 0 & 1 & 0 & 0 & 0 \\ 0 & 1 & 1 & 0 & 0 & 0 \\ 0 & 0 & 1 & 0 & 0 & 0 \\ 0 & 0 & 0 & 1 & 0 & 0 \\ 0 & 0 & 0 & 0 & 1 & 0 \\ 0 & 0 & 1 & 1 & 0 & 1 \end{bmatrix} \quad (7)$$

Following this, we used the same aforementioned method to partition the reachability matrix, enabling us to obtain the reachability sets, antecedent sets and commonality sets of levels 4 and 5.

Table 6. Reachability set, antecedent set and commonality set of level 4

si	Set R(si)	Set A(si)	Set T(si)
1	1, 3,	1	1

2	2, 3,	2	2
3	3,	1, 2, 3, 6	3
4	4,	4, 6,	4
5	5,	5	5
6	3, 4, 6,	6	6

Table 7. Reachability set, antecedent set and commonality set of level 5

si	Set R(si)	Set A(si)	Set T(si)
1	1	1	1
2	2	2	2
6	6	6	6

ISM-based model formation for ship energy efficiency in the Arctic region

After the reachability matrix was totally partitioned, we obtained the ISM hierarchy and constructed the final graphical model, which is shown in figure 2.

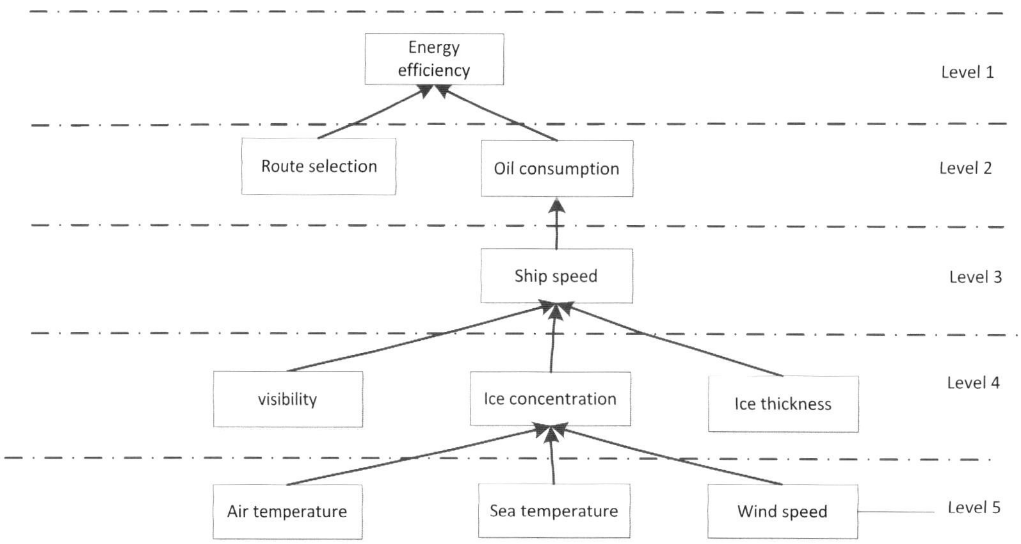

Figure 2. Interpretive structural modeling (ISM) of the elements in the energy efficiency of Arctic navigation

As can be seen from figure 2, the factors influencing energy efficiency are divided into five levels. In this study, energy efficiency is set at the top level as it cannot help to achieve any other elements. The elements at level 2 directly influence energy efficiency with the help of other elements at level 3.

At level 3, we find ship speed, which can determine the oil consumption. Ship speed optimization may be an effective method of improving energy efficiency. Indeed, it has become a major research subject from both an environmental and economic point of view, particularly in the Arctic region where the conditions are more complex than in others.

The elements at level 4 are ice condition and visibility, which influence ship speed most than the other elements in Arctic area.

The elements at level 5 were set as environmental conditions. The element of 'ice concentration' at level 4 was investigated based on its associated factors given at level 5.

To be specific with regard to environmental conditions, ice conditions were considered and set at level 5. As for the elements at level 5, they are the basic elements present in Arctic navigation. When strong winds appear and are going in the same direction, if there is a lot of trash ice in the upstream direction, the ice will quickly gather with the wind. Similarly, if strong currents appear in a particular area, ice will quickly gather there owing to the strong flow. This will be followed by the appearance of much floating ice, thus influencing the ship's speed. These changes in the ice conditions mean that ship speed needs to be controlled and the route optimized.

In the Arctic region, especially along the coast where the weather is unpredictable, strong winds of class 9-10 can suddenly appear without warning. This is a significant threat to ships sailing in this area. When ships meet floating ice, they must take certain measures in order to avoid a collision, meaning that their energy efficiency would decrease.

From what has been discussed above, we obtain the hierarchical structure of factors influencing energy efficiency of Arctic navigation and analyze them in each level. The interpretative structural modeling presented in this paper is consistent with an actual situation of a ship navigating ice in Arctic waters.

DISCUSSION

By analyzing and calculating the model of energy efficiency in Arctic navigation, the current study was able to put forward the following recommendations to improve this efficiency.

The National Marine Environmental Forecasting Center (NMEFC) should strengthen research on the meteorological and hydrological characteristics of the polar region. The collection of data, research, and development technology is also important. Systems enabling real-time observation and forecasting the Arctic weather environment should also be built to improve the energy efficiency of Arctic navigation.

Considering the particularity of polar navigation, we should protect or strengthen the key equipment on board. A ship's propeller, rudder, and other vulnerable parts need to be considered in detail. Clogging and freezing need to be prevented in the water filter. Deck equipment and parts vulnerable to low temperature damage should undergo heat preservation processing and simultaneously be provided with melting equipment. The location of the tank and the corresponding pipeline should be chosen in which damage caused by freezing expansion can reduce. Finally, making sure that the equipment operates well is basic to improving a ship's energy efficiency.

Regarding the choice of routes, more attention should be paid to the influence of energy efficiency. On the premise of guaranteeing safety, the most efficient route should be selected as much as possible.

China should strengthen its international cooperation, and undertake academic research as well as further infrastructure construction in the polar region. While it is important that ship training crews are prepared for t Arctic navigation, at the same time China should develop ships that can adapt well to the new Arctic conditions.

CONCLUSION

As the safety of Arctic navigation continuously improves, more attention should be paid to energy efficiency. The current paper analyzed the factors influencing the energy efficiency of Arctic navigation. Firstly, we selected these influencing factors. Next, an interpretative structural modeling (ISM) was built to provide a hierarchical analysis about energy efficiency in Arctic navigation. This enabled us to derive the direct or indirect relationships between the factors. Last but not least, and based on the model constructed, we put forward several suggestions about how to improve the energy efficiency of ships sailing in the Arctic region.

ACKNOWLEDGEMENTS

This study was supported by National Science Foundation of China (NSFC) under grant No. 51579203 and 51711530033 and The Excellent Dissertation Cultivation Funds of Wuhan University of Technology under grant No. 2016-YS-041.

REFERENCES

Ansari, M.F. , Kharb, R.K. , Luthra, S. , Shimmi, S.L. , Chatterji, S., 2013. Analysis of barriers to implement solar power installations in India using interpretive structural modeling technique. *Renewable and Sustainable Energy Reviews*, 27(6):163-174.

Digalwar, A. K., Giridhar, G.2015. Interpretive Structural Modeling approach for development of electric vehicle market in India. *12th Global Conference on Sustainable Manufacturing*, Procedia CIRP 26, 40 – 45

E. Bal Beşikçi, E., Kececi, T., Arslan, O., Turan, O., 2016. An application of fuzzy-AHP to ship operational energy efficiency measures. *Ocean Engineering*, 121, 392–402.

Fu, S. S., Zhang, D, Montewka, J., Yan, X. P., Zio, E., 2016. Towards a probabilistic model for predicting ship besetting in ice in Arctic waters. *Reliability Engineering & System Safety*, 155, 124-136.

Fan, A.L, Yan, X. P., Yin, Q.Z., 2016. A Multisource Information System for Monitoring and Improving Ship Energy Efficiency. *Journal of Coastal Research*, 32(5), 1235–1245.

Fu, S., Zhang, D., Yan, X. P., Shi, J., Xu, L. 2015. Risk factors analysis of Arctic maritime transportation system using structural equation modeling. *Proceedings of the 23rd international conference on port and ocean engineering under arctic conditions*.

Goerlandt, F., Montewk, J., Zhang, W. B., Kujala, P. 2016. An analysis of ship escort and convoy operations in ice conditions. *Safety Science*.

IMO, 2009. MEPC.1Circ.684, Guidelines for the voluntary use of the ship energy efficiency operational indicator (EEOI).

Khan, F. , Yang, M. , Veitch, B. , Ehlers, S. , Chai, S. 2014. Transportation risk analysis framework for Arctic waters. *Proceedings of the ASME 2014 33rd International Conference on Ocean, Offshore and Arctic Engineering.*

Schøyen, H., & Bråthen, S. (2011). The Northern Sea Route versus the Suez Canal: cases from bulk shipping. *Journal of Transport Geography*, 19(4), 977-983.

Sun, X., Yan, X. P., Wu, B., Song, X., 2013. Analysis of the operational energy efficiency for inland river ships. *Transportation Research Part D*, 22 (2013) , 34–39.

Zhang, C., Sun, L., Wen, F. S., Lin, Z. Z., Gerard L., Xue, Y. S., 2015.An interpretative structural modeling based network reconfiguration strategy for power systems. *Electrical Power and Energy Systems*, 65, 83–93.

Zhang M, Zhang D, Fu S, Yan X, Goncharov V. 2017. Safety distance modeling for vessels under the icebreaker assistance : taking "YONG SHENG" and "50 let pobedy" as an example. *In Transportation Research Board 96th Annual Meeting*, Washington, D.C.

Zhang, X., Shou, J. M., Zhou, H. J. 2013. The research on the scale and shipping cargo type of the Arctic waterway. *Polar research*, 2, 66-74.

Identification of Hazards during Search and Rescue Operation s in Cold Climate

Nese, T.[1], Dalsand, R.[1], Gudmestad, O. T.[1,2], Solberg, K. E.[2,3], Barabady, J.[1], Barabadi, A.[1], Ayele, Y. Z.[1]
[1] UiT The Arctic University, Tromsø, Norway
[2] University of Stavanger, Stavanger, Norway
[3] GMC Maritime AS, Stavanger, Norway

ABSTRACT

A search and rescue exercise simulating the evacuation from a cruise ship in Arctic conditions was carried out jointly between the Norwegian Coast Guard, academic institutions, and participants from the industry, in the north of Spitzbergen in April 2016. Reference for the exercise was the new Polar Code, which came into force in January 2017. This Code is functional, however, it prescribes that survival shall be ensured after 5 days in the evacuated craft. An important part of the exercise was an evaluation of the risks during the use of lifeboats and life rafts; the transfer to the rescue means, the stay in the lifeboats and life rafts and the rescue operation.
The aim of this paper is to present the risks identified for these operations and, their mitigating measures. The focus is to present a review of risks during evacuation and rescue operations in cold climate, evaluate the findings from the exercise, and give a summary of gaps to be closed to meet the requirements of the Polar Code.

KEY WORDS: Arctic, cold climate, hazard identification, Polar Code, search and rescue

INTRODUCTION

The International Maritime Organization (IMO, 2016a), has developed the functionally based Polar Code, which came into force on 01 January 2017. The code requires marine operators to provide lifesaving equipment that ensures a minimum of five days' survival time. This requirement puts additional strain on existing lifesaving appliances. The objective of a full-scale rescue exercise, the SARex Spitzbergen, April 2016, Solberg et al. (2016), was to identify and explore the gaps between the functionality provided by existing SOLAS (The International Convention for the Safety of Life at Sea) approved safety equipment (IMO, 2016b) and, the functionality required by the Polar Code. This was performed through an exercise conducted jointly by the Norwegian Coast Guard and leading experts from industry, governmental organizations and academia. The exercise was to be along the lines of a "*Maxim Gorkiy* scenario", where an expedition cruise ship was about to sink in the marginal ice zone off the coast of Svalbard (Hovden, 2012). Figure 1 illustrates the SARex exercise location.
The polar conditions generate additional polar-specific challenges for exercise's participants

and for the lifesaving equipment; these were identified and assessed. In general, the functional requirements, for escape routes states that exposed escape routes shall remain accessible and safe, taking into consideration the potential for icing of structures and snow accumulation. In addition, the survival craft and muster and, embarkation arrangements shall provide safe abandonment of ship, taking into consideration the possible adverse environmental conditions during an emergency. On the other hand, the functional requirements, regarding evacuation states that all life-saving appliances and associated equipment shall provide safe evacuation and be functional under the possible adverse environmental conditions during the maximum expected time of rescue.

Hence, the SARex was aiming to simulate the relevant polar conditions, incorporating sea ice, sea swell, low air and water temperatures and remoteness. Further, SARex was aiming on assessing the possible gaps in the functionality requirements, such as:

i) exposed escape routes may not be accessible or safe, taking into consideration the potential for icing of structures and snow accumulation;

ii) all life-saving appliances might not provide safe evacuation and be functional under the possible adverse environmental conditions and, adequate thermal protection might not be provided for all persons on board;

iii) escape routes and embarkation arrangements might not be arranged for and adequate for persons wearing additional polar clothing;

iv) the lifeboats and life rafts might not have enough space to accommodate persons equipped with thermal protection adequate for the environment, etc.

About SARex

- SARex was a full-scale exercise that sought:
    - To identify and explore the gaps between the functionality provided by existing SOLAS approved safety equipment and the functionality required by the Polar Code.
- The full-scale exercise was held in Woodfjorden in northern Svalbard in late April 2016.

Aim of SARex

- The exercise aimed to simulate relevant polar conditions as well as incorporate:
    - sea ice,
    - sea swell,
    - low air and
    - water temperatures, and remoteness

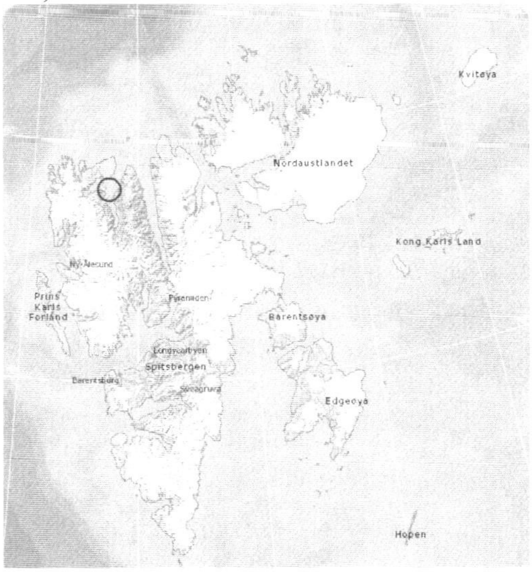

Figure 1. SARex test location in Woodfjorden, indicated with a red circle. Map © Norwegian Polar Institute

POAC17-122

The full report regarding the exercise can be found, Solberg et al. (2016). The rest of the paper is organized as follows: a brief description about the main aim of the SARex research exercise is presented in Section 2. Thereafter, in Section 3, the main findings of the Preliminary Hazard Analysis (PHA) are summarized. Finally, concluding remarks are presented in Section 4.

CASE DESCRIPTION – SARex RESEARCH EXERCISE

Prior to the SARex research exercise, a cross-disciplinary team comprised of medical doctors, suppliers, researchers, academicians, regulators and users, assessed the Polar Code, especially focusing on the interpretation of Chapter 8: Lifesaving Appliances and Arrangements. The main goal of the Chapter 8 of the Polar Code is:

> *"The equipment required by the Polar Code is to provide functionality that enables the individual safety, which means to maintain cognitive abilities, body control and fine motor skills for the maximum expected time of rescue."*

In order to address and investigate the potential gaps, the SARex research exercise was carried out jointly between the Norwegian Coast Guard, academic institutions, and participants from the maritime as well as oil and gas industry. SARex was a full-scale exercise that was held in Woodfjorden in northern Svalbard in late April 2016. In general, the main objectives of the exercise and the associated research trip were to:

- Assess the adequacy of the lifesaving appliances as required by the IMO Polar Code.
- Identify the gaps between SOLAS approved rescue craft (lifeboats and life rafts) and the requirements defined in the IMO Polar Code.
- Identify the gap between SOLAS approved personal protective equipment (PPE) and the requirements defined in the IMO Polar Code.
- Assess the personal/group survival kits as defined by the IMO Polar Code.
- Train Norwegian Coast Guard personnel in emergency procedures in ice-infested waters, with particular reference to evacuation and rescue from cruise ships.

Moreover, the exercise addressed the following topics (Solberg et al., 2016):

- functionality of life raft/lifeboat under polar conditions;
- functionality of personal protective equipment (PPE) (e.g. thermal protection/survival suits);
- additional training requirements for crew and passengers;
- evaluation of Coast Guard's search and rescue procedures, including handling of mass evacuations in Polar Regions.

The exercise was carried out in three phases:

- *Phase one:* Testing survival times in lifeboat and life raft with participants wearing different kinds of personal protective equipment.
- *Phase two:* Training for the Coast Guard personnel on mass evacuation from lifeboat and triage of passengers.
- *Phase three:* Testing the lifeboat and life raft in ice-infested waters as well as testing personal/group survival equipment and personal protective equipment on ice/close to ice.

Figure 2. Lifeboat being lowered into the water with the deck crane. Photo © Jan Erik Jensen

As a part of the preparations for the exercise, the lifeboat was launched from the deck of KV Svalbard using the deck crane (Figure 2). In addition, transport between KV Svalbard and the survival crafts was carried out using the two MOBs (Man Overboard Boats), and these were also utilized for toilet breaks. Moreover, there were at least one MOB boat stationed close to the survival crafts during the exercise for safety reasons. The exercise was started Sunday 24th of April, at approximately 09:40, when all participants had been transported to the lifeboat/ life raft with the MOBs. The weather during the exercise was ideal for performing the exercise, with an ambient air temperature of about –9°C, a water temperature of about –1°C and little wind. These are considered representative weather conditions for the cruise ship season in Svalbard. Due to the favorable weather conditions, seasickness was not an issue for any of the participants.

The lifeboat and life raft employed for the exercise as well as PPEs, which the participants wore during the exercise are depicted in Figures 3 to 6. In phase 1, each participant wore standard SOLAS approved PPEs, ranging from life jackets to insulated survival suits. The goal was to see how long the participants would stay in the lifeboat before they experience the cold stress – the response of the body to cold temperatures resulting from heat loss. The heat loss can be due to whole-body cooling and/or local cooling, including extremity cooling, convective skin cooling, conductive skin cooling (contact cooling), and cooling of respiratory tract.

Figure 3. The lifeboat and the raft © Trond Spande

Figure 4. Lifeboat drifted into a belt of ice during the exerciseoto © Trond Spande

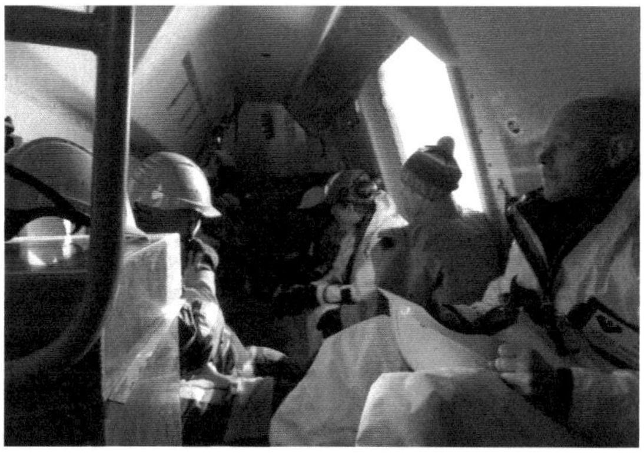

Figure 5. Situation inside the lifeboat © Bjarte Odin Kvamme

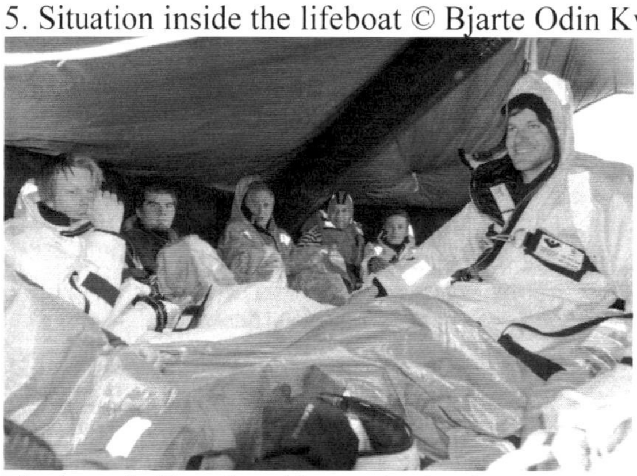

Figure 6. Situation inside the raft. People with survival suits are warm, and had to open their suit to avoid sweating. © Jan Erik Jensen

PRELIMINARY HAZARD ANALYSIS

In general, in polar waters – in Arctic water – it is not enough to keep people afloat with either life jackets, lifeboats or life rafts. People must also be kept dry and warm to be able to survive in such a harsh climate. With a maximum expected time of rescue of five days, it is also important that people have access to food and water to sustain them for this time. The most important life-saving appliances in polar waters is the immersion suit and, a lifeboat/ life raft stocked with the necessary equipment. In addition, in case of a wet evacuation, an immersion suit together with personal survival equipment is crucial to survival, but even then, people cannot survive for long if they cannot get into a lifeboat/ life raft or onto land or ice.

Further, in an evacuation situation, in Arctic water, getting to the lifeboat/ life raft can be hazardous if the walkways are unsheltered and slippery due to ice and snow accumulation. For instance, cruise passengers are often elderly persons that can have a difficult time moving around and, they can be in need of some sort of assistance. In these cases, a slippery walkway can be a daunting task to maneuver through safely. The Polar Code requires exposed escape routes to be accessible and safe when taking into consideration the potential for icing of structures and snow accumulation. This can be achieved with sheltered walkways. Hence, to ensure that the life-saving appliances are functional under any circumstance in the Arctic, they have to be designed to work with the presence of low temperatures, snow and ice. Further, the Arctic cruise ship industry should work towards fulfilling the goals with life-saving appliances that meets all the functional requirements of the Polar Code.

To maximize the chances of prolonged survival in polar waters, the potential hazards and risks in case of accident should be investigated and, risk reduction as well as mitigation measures should be assessed. In order to fulfil these aims, during SARex efforts were done to evaluate the potential hazards and risks under Arctic conditions. Further, for ensuring the safety of the participants, after the exercise, all those participating in the exercise were taken back to KV Svalbard as soon as one or more of the predefined abortion criteria was met. The predefined abortion criteria were defined as:

- Loss of cognitive abilities
- Loss of body control (e.g. uncontrollable shivering)
- Loss of fine motoric skills

Further, since the participants aborted at different times, the ship's medical doctor was on stand-by during the entire exercise period. Immediately after returning to KV Svalbard, the following medical parameters was checked on all participants: body temperature, pulse, and blood pressure.

Functionality of Life Raft/Lifeboat Under Cold Climate Conditions

The evacuation from the Coast Guard vessel *KV Svalbard* was performed with the help of the vessel's man overboard boats (MOBs). The transfer of the survivors from the lifeboat and life raft to the vessel was also carried out by MOBs. The hazard prior to the evacuation of personnel to the lifeboat and life raft were identified and, the potential consequences are discussed.

Figure 7. The lifeboat experienced snow accumulation during the exercise. © Trond Spande

Thereafter, risk-reducing measures were suggested. For instance, when launching a lifeboat under Arctic conditions, one of the main risk factors are icing on the parts and components (Figure 7). In cases of icing, a lifeboat is unable to launch properly, which consequently leads to drastically reducing the passengers' survival chances. To reduce the impact of icing, employing risk-based inspection procedures and manual ice removal as a risk reduction measures should be considered. Further, enclosing life rafts and lifeboats as well as implementing heating systems can be effective risk reduction measures.

Table 1 illustrates an example of the risk assessment for the transfer to the lifeboat. The hazard identification was considered essential for the safety of those involved in the exercise and the complete assessments are presented in Appendix (Table A1 and Table A2). The lifeboat, which was employed for testing in the SARex project was a conventional Totally Enclosed Lifeboat (TELB) with model name "Miriam 8.5", and was designed and manufactured in accordance to the latest SOLAS, Classification Society and National Authority requirements. In addition, the lifeboat was equipped with a compressed air system and exterior deluge system and, seating inside the lifeboat consisted of GRP (glass fiber reinforced plastics) benches with safety belts. Moreover, the lifeboat was launched and retrieved using a davit, with two lifting wires, which connected to the lifeboats' two release hooks. Further, when installed on a ship's side, the lifeboat is equipped with skates that allows it to slide easily on the ships' side, along with shock-absorbing fenders.

Table 1. An example of SARex Phase 1 Risk assessment – lifeboat

Potential hazard	Cause	Consequence	Risk-reducing measures
Rescuing passengers from sea to lifeboat	Insufficient arrangements on lifeboat for extraction of people. Mob boat far away.	Difficult/impossible to rescue survivors from sea	Mob boat nearby.
Transfer of persons from lifeboat to rescue vessel	-Insufficient/no arrangements for moving passengers from one craft to another -Passengers have physical problems after long time in lifeboat -Heavy seas/strong winds	Complicated rescue process, potential injuries/fatalities Time consuming 30 minutes?	Helmet Evenly distribute the weight in the mob boat Hold onto the rail in the boat
Danger getting down to lifeboat	Hoisted down in mob boat	Falling into sea/injury	Helmets, life-vest or survival suit, instructions on holding on while lowering.
Falling into sea	Transfer of persons between lifeboat and mob boat. Trying to urinate from either lifeboat or mob boat.	Becoming soaking wet and very cold. Exercise over for that person.	Assistance from mob boat crew. Mob boat always nearby, *KV Svalbard* also nearby.
Tripping and falling	-Passengers not used to heavy suits/equipment -Slippery surfaces	-Fall damage (injuries, broken bones) -Ending up in the water	-Be cautious -Follow instructions -Detailed safety information and procedures from *KV Svalbard* crew
MOB boat lifting hook swinging after release	-Hook operator error -Rough seas	-Hook arrangement hits passengers in MOB boat, leading to injuries.	-Follow instructions -Detailed safety information and procedures from *KV Svalbard* crew
Exercise participant becomes ill/injured needs immediate assistance	Decease, accident, medical issues.	They are in need of medical assistance.	Mob boat nearby, *KV Svalbard* nearby. Helicopter from Longyearbyen next option.
Personal protective equipment not functioning	Damaged, production error, not maintained correctly.	PPE does not work as intended.	Maintenance &functionality check before use.

The lifeboat exercise also highlighted several potential hazards and possible gaps. For instance, one of the highlighted problems was mist build-up that led to poor sight, which was due to mist building up on the inside of the lifeboat windows. This mist build-up forced the coxswain (the person in charge of the navigation) to wipe the windows continuously to be able to see the external environment. Further, inside the lifeboat, the air quality and low oxygen levels were issues, as the ventilation system required the engine to be operating. In addition, the participants experienced a significant heat loss from the structure (floor, seat and backrest) of the lifeboat. Hence, for improving the performance of the lifeboat, various measures should be implemented. For instance, to reduce the mist build-up, heating in the windows and air vents should be considered. Further, to obtain a survivability rate in accordance with the minimum five-day requirement set by the Polar Code, the insulation system of the lifeboat in combination with the insulating capabilities of the personal protective equipment should be improved.

In addition to evaluating the functionality of lifeboats, during the SARex exercise, the potential hazards and possible gaps that are associated with life rafts were investigated. Table 2 illustrates an example of risk assessment for life rafts. Several potential hazards and possible gaps were highlighted during the SARex exercise. One of the main issues was that the participants in the life raft experienced a significant heat loss, especially through the bottom of the life raft. This became especially evident when sitting or lying down, regardless of the type of personal protective equipment used. To reduce the heat loss, the life raft canopy was kept closed to retain heat. Consequently, the lack of ventilation caused the air to be moist, and extensive condensation developed. Thereafter, due to the combination of condensation and sweat, survivors suffered from wet insulation layers in their PPEs, followed by loss of insulation and freezing. Moreover, low oxygen levels were also an issue, and the raft had to be vented frequently, losing a significant amount of heat in the process. Furthermore, the congestion inside the raft was a problem, causing reduced ability to move, triggering reduced blood circulation in the body's extremities, resulting in freezing hands and feet.

Table 2. An example of SARex Phase 1 Risk assessment – Life raft

Potential hazard	**Cause**	**Consequence**	**Risk-reducing measures**
Life raft integrity compromised	-Collision with ship during/after launch -Collision ice floes	-Water intrusion, exercise stopped	-Abortion if the ice conditions gets too severe -Procedures for rapid evacuation of all participants
Transfer of persons from life raft to rescue vessel	-Insufficient/no arrangements for moving passengers from one craft to another -Passengers have physical problems after long time in life raft -Heavy seas/strong winds	-Complicated rescue process -Potential injuries/fatalities -Time consuming	-MOB boats will be used for transfer of passengers from life raft to *KV Svalbard*, piloted by experienced crew.
Tripping and falling	-Passengers not used to heavy suits/equipment -Slippery surfaces	-Fall damage (injuries, broken bones) -Ending up in the water	-Be cautious -Follow instructions -Detailed safety information and procedures from *KV Svalbard* crew
Communication difficulties	-Wind noise -Many people talking at the same time -Routines for how to communicate -Radio equipment failure	-Important messages cannot be communicated via radio	-Backup radio equipment/battery -Clarify communication routines prior to test
Sea spray into life raft	-The life raft canopy are open (ventilation, toilet breaks, extracting persons from sea, etc.)	-People get wet and cold -Water enters the life raft.	-Keep canopy closed whenever possible
Sea spray icing on life raft	-Sea spray combined with low temperatures	-Change in life raft buoyancy qualities -Zippers and other small details frozen stuck	-Shaking canopy from the inside to loosen any ice
Clogging/blocking of ventilation	-Warm and moist air from inside the life raft condensates and freezes around/in the ventilation outlet	-Reduced ventilation -Deterioration of air quality	-Opening canopy for ventilation

From the risk assessment results of the SARex exercise, it can be deduced that it is unlikely that the majority of those evacuated to a life raft and lifeboat (engine shut off) would survive for a minimum of five days according to the Polar Code criteria. In addition, the presence of a well-trained lifeboat/life raft captain proved very important for maintaining both the safety and, the morale of the personnel on board. This is especially important when the duration of the stay is long (a minimum of five days, along the lines defined in the Polar Code). Further, due to the high risk of cardiac arrest connected to hypothermia, having a defibrillator onboard is recommended. However, this is costly and someone on board the lifeboat would have to know how to use it. The Polar Code does not address the medical problem, but the cruise operators and their organizations will need to consider it carefully.

In addition, when it comes to having a habitable environment, seasickness and hygiene issues have to be addressed. For instance, seasickness can make the lifeboat an uncomfortable place to be, as well as cause dehydration and starvation. Especially water is crucial; little water in combination with seasickness can easily lead to dehydration. Further, it should be noted that most of the people involved in the SARex exercise were either physically fit young persons or mature persons with good physical health. The lack of elderly or disabled persons involved in the exercise renders the results on the positive side, as the participants were apt and, in better physical and physiological shape than the average seafarer/ passenger. Seasickness was not an issue in this exercise, as the waters were calm, and all participants were given seasickness pills to prevent any occurrence of vomiting.

Handling of Mass Evacuations in Polar Regions

One element of the SARex was to require the Coast Guard staff to conduct a mass evacuation from the rescue craft onto the Coast Guard vessel. A large number of the evacuated personnel simulated a hypothermia state. Establishing, implementing, and conducting regular training on the procedures for handling disabled, wounded, and immobile passengers when evacuating a large group of people is of great importance for ensuring an efficient evacuation. Evacuating a large number of immobile casualties takes an excessive amount of time and, puts a great strain on the staff on board the Coast Guard vessel. The medical state of the casualties is of key importance in determining the time required to evacuate personnel from a lifeboat/life raft to a rescue vessel. The potential of involving those casualties who are in good condition in monitoring/aiding the caretaking process of the patients should be emphasized.

CONCLUSIONS

This paper has presented results from the SARex research exercise, which was carried out to identify and explore the gaps between the functionality provided by existing SOLAS approved safety equipment and, the functionality required by the new Polar Code. The exercise demonstrated that the peculiar Arctic operating environment will have a significant impact on the functionality of the lifeboats and life rafts. In addition, high wind speeds, low temperatures, fog and large waves all create their own set of problems. Moreover, the available radio links and satellite communication in the Arctic lacks in reliability, and therefore presents a safety hazard in major accident scenarios. Further, SARex demonstrated that mass evacuation can be demanding and, in cases of a crowded lifeboat and life raft, it became challenging for the rescue personnel to evacuate the most injured people first, due to the limited available space in the lifeboat/ life raft. In such scenarios, the rescuers were forced to get the people that were able to help themselves off first so they had room to handle the people that required assistance. Furthermore, abandoning a ship in Arctic will require that every person on board have access to life-saving appliances designed for polar waters. Hence, to maximize chances of survival: all passengers will want to be in an enclosed lifeboat if they manage a dry evacuation, or in case of a wet evacuation, an insulated immersion suit designed for polar water survival. The chances of prolonged survival in polar waters will be drastically reduced without these life-saving appliances.

Our conclusion is that in order to fulfil the minimum functional requirement, especially the requirement of Polar Code Chapter 8, which states that a vessel is to provide equipment that enables the passengers to survive a minimum of five days or the anticipated time of rescue, a holistic safety management approach is required. This ensures that the vessel owner considers all relevant conditions, factors and parameters. In general, when assessing the probability of survival, it is recommended to consider the following list of conditions, factors and parameters:

- governing metocean (meteorological and oceanographic) conditions for the area of operation,
- remoteness,
- available SAR infrastructure,
- performance of SAR operators,
- energy required to maintain the core temperature of the persons,
- water/food required to maintain an adequate metabolism for human heat generation,
- safe transfer and stay in the rescue means,
- insulating properties of the rescue craft,
- insulating properties of the PPEs,
- number of passengers and physical condition of the passengers,
- cumulative weight of group and personal survival equipment and carrying capacity of survival craft,
- abandon ship activities,
- survival strategies onboard the evacuation vessel,
- survival craft management, etc.

ACKNOWLEDGEMENTS

The SARex exercise conducted north of Spitzbergen in April 2016 was made possible through the cooperation between the Norwegian Coast Guard and Norwegian official and academic institutions as well as industrial companies. We appreciate in particular the positive response of the Norwegian Coast Guard to participating in the exercise. In particular we appreciate the attitude and interest of all the personnel on board the Coast Guard vessel *KV Svalbard* during the planning and execution of the exercise, The involvement of the 23 scientists, professionals and students on board, representing scientific institutions and industrial companies, furthermore, made the exercise a success by meeting all its planned objectives. We also appreciate the full support of the Faculty of Science and Technology at the University of Stavanger during the planning of the exercise.

REFERENCES

Hovden, S. T. 2012. *Redningsdåden (om Maksim Gorkiy-havariet utenfor Svalbard i 1989)*, Commentum Publishing.
IMO 2016a. International code for ships operating in polar waters (polar code). The marine environment protection committee of the International Maritime Organization.
IMO 2016b. International Convention for the Safety of Life at Sea (SOLAS), 1974. International Maritime Organization.
Solberg, K. E., Gudmestad, O. T. & Kvamme, B. O. 2016. SARex Spitzbergen - Search and rescue exercise conducted off North Spitzbergen : Exercise report, University of Stavanger, http://hdl.handle.net/11250/2414815

Appendix

Table A1. SARex Phase 1 Risk assessment – Lifeboat

Potential hazard	Cause	Consequence	Risk-reducing measures
Maneuvering difficulties	Heavy sea, sea ice, wind	Little or no control of lifeboat position/heading Lifeboat cannot be positioned correctly in the waves, heavy lifeboat motion (uncomfortable for passengers)	-Lifeboat pilot training/experience
MOB boat occupied when an accident occurs	-MOB boat have many tasks	-Long time to rescue -People getting seriously chilled	-Use both MOB-boats for redundancy
Passengers not noticing getting severely cold (core and extremities)	-Individual differences -Little or no experience with being cold -Sleeping	-Risk of injuries/fatalities	-Have buddies near you which can check on you -Leader onboard raft should keep overview.
Freezing body extremities	-Getting wet -Little clothing	-Frost bite leading to injuries	-Bring hats, gloves, etc. for backup in case. -Low threshold for returning people to *KV Svalbard* -Additional immersion suits in lifeboat for emergencies (for those not wearing suits during tests)
Pilot is incapable of leading	Injury, death, pilot has to abort exercise and return to ship.	Anarchy? Without leadership, people might not survive as long.	Find a new leader/next in command
Immersion suit integrity compromised	-Improper entering of suit -- Openings not properly closed (zippers), etc.	-Exposure to cold water with potential injuries/chilling of body	-Buddy check on suit after putting it on, prior to test
Lack of sleep	Uncomfortable seating, stressful situation (physical and psychological)	Sleep deprivation	
CO and CO2 build-up inside lifeboat	Insufficient ventilation, many people breathing. Leak from exhaust system	Headaches, sleepiness, poor concentration, loss of attention, increased heart rate, slight nausea, oxygen deprivation.	-Detectors will measure CO and CO2 build-up and give alarms. -Opening hatches
High temperature inside lifeboat	Insufficient ventilation, many people generating heat	Dehydration caused by perspiration. Nausea, which can lead to vomiting, causing further dehydration.	-Opening hatches
Low temperature in lifeboat	Outside temperature.	Core body temperature of passengers dangerously low (hypothermia)	-Passengers wearing warm (waterproof) clothing
Hygiene	No toilet available	Insanitary conditions	Bucket or other solutions? -Bottles used in hospitals/small aircraft.
Clogging/blocking of ventilation	Warm and moist air from inside the lifeboat condensates and freezes around the ventilation outlet	Reduced ventilation, rapid deterioration of air quality	Opening hatches
Not enough food	Lifeboat not stocked	Hunger	-Ensure lifeboat carries enough food for exercise duration
Not enough water	Lifeboat not stocked	Thirst	-Ensure lifeboat carries enough water for exercise duration
Seasickness	Lifeboat movements	Vomiting, inducing dehydration	-Anti-seasickness medicine

		and starvation.	How to handle this if exercise participants starts vomiting?
Sea spray into lifeboat	The lifeboat hatches are open (ventilation, extracting persons from sea, etc.)	People get wet and cold. Water inside the lifeboat.	-Close hatches
Poor sight	Fog, snow squalls. Icing (sea spray) on windows	Navigation difficulties Possible collision	-Lifeboat pilot training
Sea spray icing on lifeboat	Sea spray combined with low temperatures	Skew loads, hinges and locks on hatches stuck, ventilation compromised.	
Rapid weather changes	-Weather in this region can change in minutes	-Exercise gets much more difficult -Stopping the might be necessary	-Check weather report prior to/during test -Procedure for rapid evacuation of all participants -Ensure that MOB boat is close to lifeboat during test, for emergency preparedness
Communication difficulties	-Wind noise -Many people talking at the same time -Routines for how to communicate -Radio equipment failure	-Important messages cannot be communicated via radio	-Backup radio equipment/battery -Clarify communication routines prior to test
Disturbance from other vessels in area, not part of exercise	-Nearby vessels not informed of test	-Interruption of test -Possible collisions and hazard for participants	-Notify any nearby vessels of the test -Establish safety zone around test area

Table A2. SARex Phase 1 Risk assessment – Life raft

Problem	Cause	Consequence	Risk-reducing measures
Life raft damaged	-Production error	-Equipment unusable	-Check prior to launch (Viking)
Disturbance from other vessels in area, not part of exercise	-Nearby vessels not informed of test	-Interruption of test -Possible collisions and hazard for participants	-Notify any nearby vessels of the test -Establish safety zone around test area
MOB boat lifting hook swinging after release	-Hook operator error -Rough seas	-Hook arrangement hits passengers in MOB boat, leading to injuries.	-Follow instructions -Detailed safety information and procedures from *KV Svalbard* crew
Falling into water during transfer between MOB boat and life raft	-Slippery surfaces -Distance between raft and MOB boat (e.g. due to poor mooring) -Rough seas	-Rapid cooling of persons in the sea	-*KV Svalbard* personnel entering life raft first, to assist with keeping the life raft and MOB boat close, and to help passengers from one vessel to the other
Many people ending up in the water at the same time	-Life raft integrity compromised -Capsizing -Etc.	-Mass injuries/hypothermia -Possible fatalities	-Establish procedures for rapid evacuation of all participants it necessary
MOB boat occupied when an accident occurs	-MOB boat have many tasks	-Long time to rescue -People getting seriously chilled	-Use both MOB-boats for redundancy
Long distance from MOB boat to life raft during test	-MOB-boat have many tasks -MOB-boat not on the water during test	-Long time to rescue -People getting seriously chilled -Long rescue time if people fall into the sea	-Ensure that MOB boat is close to raft during test for emergency preparedness -Use both MOB-boats for redundancy
Rescuing passengers from sea to life raft	-Insufficient arrangements on life raft for extraction of people	-Difficult/impossible to rescue survivors from sea	-Ensure that MOB boat is close to raft during test for emergency preparedness
Insufficient observation during test (of the entire area)	-Poor overview	-People fall into water without someone noticing	-Crew onboard *KV Svalbard* and MOB boats ensures lookout
Immersion suit integrity compromised	-Improper entering of suit -- Openings not properly closed (zippers), etc.	-Exposure to cold water with potential injuries/chilling of body	-Buddy check on suit after putting it on, prior to test
Poor sight	-Fog -Snow squalls -Sea spray	-Impact with drift ice -Difficulties with keeping an overview (polar bear lookouts spot/locating participants falling into sea, etc.)	-Abort test if weather conditions gets too severe -Ensure that MOB boat is close to raft during test for emergency preparedness

CO_2 build-up inside life raft	-Insufficient ventilation, combined with many people breathing.	-Headaches, sleepiness, poor concentration, loss of attention, increased heart rate, slight nausea, oxygen deprivation.	-Opening the canopy -Air quality measurement instruments onboard life raft -Ensure control of air vents.
High temperature inside life raft	-Insufficient ventilation, combined with many people generating heat	-Heat stress: Dehydration caused by perspiration. Nausea, which can lead to vomiting, causing further dehydration.	-Opening canopy when necessary
Low temperature in life raft	-Outside temperature.	-Core body temperature of passengers drops dangerously low (hypothermia)	-Bring hats, gloves, etc. for backup in case. -Low threshold for returning people to *KV Svalbard* -Ensure that MOB boat is close to raft during test for emergency preparedness
Rapid weather changes	-Weather in this region can change in minutes	-Exercise gets much more difficult -Stopping the might be necessary	-Check weather report prior to/during test -Procedure for rapid evacuation of all participants -Ensure that MOB boat is close to raft during test, for emergency preparedness
Passengers not noticing getting severely cold (core and extremities)	-Individual differences -Little or no experience with being cold -Sleeping	-Risk of injuries/fatalities	-Have buddies near you which can check on you -Leader onboard raft should keep overview.
Freezing body extremities	-Getting wet -Little clothing	-Frost bite leading to injuries	-Bring hats, gloves, etc. for backup in case. -Low threshold for returning people to *KV Svalbard* -Additional immersion suits in life raft for emergencies (for those not wearing suits during tests)
Medical problems of passengers	-Latent health issues -Other medical condition factors	-Possible medical emergencies for participants	-Low threshold for returning people to *KV Svalbard*
Seasickness	-Lifeboat movements -Seasickness medicine not effective immediately	Vomiting, inducing dehydration and starvation.	-Anti-seasickness medicine -Take medicine prior to test -Check with KV personnel on advice
Not enough food	-Lifeboat not stocked	-Hunger -Exercise must be stopped	-Ensure lifeboat carries enough food for exercise duration
Not enough water	-Lifeboat not stocked	-Thirst -Exercise must be stopped	-Ensure lifeboat carries enough water for exercise duration
Hygiene	-No toilet available	-Insanitary conditions	-Bucket or other solutions? -Transport people with MOB boat to *KV Svalbard* for toilet visits
Lack of sleep	Uncomfortable seating, stressful situation (physical and psychological)	-Sleep deprivation: Reduced cognitive abilities Ability to take care of yourself reduced, etc.	-Try to sleep when possible
Polar bear attack	-Animal curiosity/hunger/threatened	-Raft puncture -Injuries/fatalities	-Polar bear guard (*KV Svalbard*) -Armed personnel onboard MOB-boat -Flares/signal rockets -Situation awareness
Walrus/orca attack	-Animal curiosity/hunger/threatened	-Raft puncture -Injuries/fatalities	-Armed personnel onboard MOB-boat -Flares/signal rockets -Situation awareness

POAC'17
Busan, KOREA

Proceedings of the 24th International Conference on
Port and Ocean Engineering under Arctic Conditions
June 11-16, 2017, Busan, Korea

Investigation of Secondary Icebreaker Performance during an Ice Management Operation

Yujian Huang[1], Freeman Ralph[1]
[1] C-Core Centre for Arctic Resource Development (CARD), St. John's, Newfoundland and Labrador, Canada

ABSTRACT

Depending on ice conditions, ice management operations may involve two or more icebreakers, known as primary and secondary icebreakers. Whether developing an ice management strategy, or executing operations, understand the performance of the icebreaker fleet is most important. The Lindqvist model of ice resistance is chosen to predict icebreaker performance in level ice with estimated hydrodynamic resistance. While one may assume that a broken ice regime (following primary icebreaker activity) may have less resistance, ice management operations for a secondary icebreaker will require continuous turning /maneuvering as it interacts with a more complex ice regime. Therefore, it is important to look into secondary icebreaker performance with different operating situations. Recent field observations suggest that existing models may over predict secondary vessel performance. Understanding the influences of the performance reduction for the secondary icebreaker is the key to successful develop of performance model for icebreakers during ice management operations.

This paper presents an overview of the secondary icebreaker performance during an ice management operation and a new analytical model that is compared with field trials data. The resulting model demonstrates reduced icebreaker capabilities with an example.

KEY WORDS: Secondary Icebreaker; Ice Management Operation.

INTRODUCTION

The increased activity in the high Arctic by industry is driven by the world's remaining undiscovered hydrocarbons. The capability of station-keeping platforms (mooring and hull) in Arctic and subarctic waters is one of the main challenges particularly given the desire and benefit of safely extending the open water season. To reduce/mitigate ice loads on station-keeping platforms, ice management operations are usually involved. The required ice management operations are reviewed by many researchers, such as (Browne et al., 2014; Hamilton, 2011; Hamilton et al., 2016, 2014, 2013, Hamilton et al., 2011, 2011a, 2011b; Wright et al., 2014; Younan et al., 2012). Also, Kubat and Sayed (2014) have a detailed review of station keeping and ice management.

An ice management operation usually involves two or more icebreakers. The leading or primary icebreaker manages unbroken or large ice floes, and the secondary icebreaker further manages these floes into smaller sizes based on broken ice forces on the station-keeping vessel and its mooring/hull structure capacity. In modeling icebreaker performance, this paper focuses on bow breaking technique only. The floe size, thickness, and properties of the ice encountered by the icebreaker influence the icebreaker performance.

Secondary icebreaker performance in broken ice with reduced concentrations will consider open water resistance. It may seem reasonable to model icebreaker performance based on equivalent resistance that considers hydrodynamic resistance and ice resistance. However, it is recognized that the performance of an icebreaker in the water with ice present is much more complicated than a vessel in the open water only, and uncertainty exists.

Modelling secondary icebreaker performance as equivalent resistance would suggest enhanced performance as floe sizes and equivalent concentration of the pre-managed or broken ice field following primary icebreaker activity are reduced. However, recent field observations suggest that current models may over predict secondary vessel performance. Added maneuvering required to maintain course or deviate from a track to smaller floes, which are targeted by the secondary vessel, may lead to reduced performance. The degree of speed reduction for the secondary icebreaker is the main driver of this study.

This study assumes the predicted secondary icebreaker performance will be a function of level ice performance, turning performance, continuous ship-ice interactions with broken floes, and on-demand maneuvering. It is recognized that variability exists (floe size, current and wind conditions, etc.) and modeling performance probabilistically to satisfy some reliability target is desirable. Such analysis allows one to also consider extreme events that may not yet have been observed in the field, but models suggest they may occur.

A balance between system design (more capable and reliable systems cost more but will reduce downtime) and potential downtime must be achieved. For success, the icebreaker must break the ice floes into the target floe size, in a time that is less than the time for the unmanaged ice floes to arrive at the vessel. If this is not satisfied, operations will suspend to make more time available, until either the threat is gone, or the vessel leaves location.

ICEBREAKER LEVEL ICE PERFORMANCE MODELING

For ships traversing ice covered waters, the main contributing resistances are hydrodynamic (drag) resistance (R_{ow}) and ice resistance (R_{ice}). Ignoring other minor resistances, such as the resistances from wind and wave effect, the total resistance (R) could be represented by:

$$R = R_{ow} + R_{ice} \tag{1}$$

Hydrodynamic hull resistance could be estimated through towing tank tests or full scale sea-trials. The common open water resistance model is represented as

$$R_{ow} = \frac{1}{2} \cdot \rho \cdot V_i^2 \cdot C_D \cdot A \tag{2}$$

where,

C_D = Drag coefficient, estimated through tests or trials,
ρ = Water density,
A = Wetted area = $L \cdot (B + T)$,
L = Length of the ship at waterline,
B = Bean of the ship, and
T = Draft of the ship.

Also, field trial data suggested similar relation with additional coefficient term as shown in Figure 1.

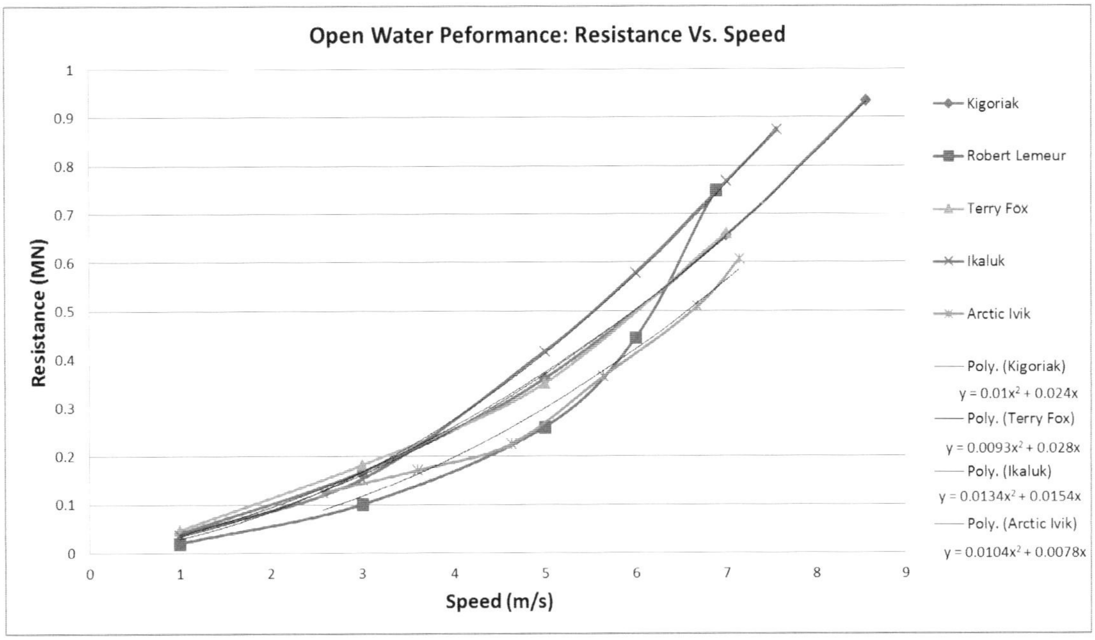

Figure 1. Open water performance (after Keinonen et al., 1989)

Therefore, the open water resistance for these icebreakers could also be estimated as

$$R_{ow} = 0.01 \cdot V_i^2 + 0.03 \cdot V_i \qquad (3)$$

where,

V_i = Icebreaker initial achieved speed.

There are three common analytical models for ice resistance, including the Riska model (Riska et al., 1998), the Keinonen model (Keinonen et al., 1996, 1991, 1989), and the Lindqvist model (Lindqvist, 1989). To compare, the Riska model only involves ice thickness in its formulas, the Keinonen model considers ice thickness, flexural strength, temperature, and salinity, and the Lindqvist model considers ice thickness, flexural strength, ice friction, shear strength, ice density, Poisson's ratio, and elastic modulus of ice. The Lindqvist model is considered the preferred model to estimate icebreaker ice resistance in level ice due to its comprehensiveness of ice properties.

Lindqvist model is comprised of three resistance components, the crushing resistance (R_C), the bending (flexural) resistance (R_B), and the submergence resistance (R_S), and given as

$$R_{ice} = (R_B + R_C) \cdot \left(1 + 1.4 \cdot \frac{v}{\sqrt{g \cdot H_{ice}}}\right) + R_s \cdot \left(1 + 9.4 \cdot v/\sqrt{g \cdot L}\right) \qquad (4)$$

where,

$$R_C = F_v \cdot \frac{\tan\phi + \mu \cdot \cos\phi / \cos\psi}{1 - \mu \cdot \sin\phi / \cos\psi} \qquad (4)$$

$$F_v = 0.5 \cdot \sigma_f \cdot H_{ice},$$

$$\psi = \arctan(\tan\phi / \sin\alpha),$$

$$R_B = 0.003 \cdot \sigma_f \cdot B \cdot \frac{H_{ice}^{1.5}}{\sqrt{m}} \cdot \left(\frac{\tan\bar{\psi} + \mu\cos\bar{\phi}}{\sin\bar{\alpha} \cdot \cos\bar{\psi}}\right) \cdot \left(1 + \frac{1}{\cos\bar{\psi}}\right), \quad (5)$$

$$R_s = \delta\rho \cdot g \cdot H_{ice} \cdot B \cdot \left(T\frac{B+T}{B+2T} + \mu(0.7 \cdot L - \frac{T}{tan\bar{\phi}} - \frac{0.25 \cdot B}{tan\bar{\alpha}} + T \cdot cos\bar{\phi} \cdot cos\bar{\psi} \cdot \sqrt{\frac{1}{\sin^2\bar{\phi}} + \frac{1}{\tan^2\bar{\alpha}}}\right), \quad (6)$$

v = Icebreaker speed in level ice = V_i ,
H_{ice} = Ice thickness,
L = Length of the ship at waterline,
F_v = Vertical force needed to fail an ice edge in flexure,
μ = Friction factor (between ice and ship hull),
σ_f = Flexure strength of ice,
m = 1,
$\delta\rho$ = Density difference between ice and water,
B = Beam of the ship,
T = Draft of the ship,
ϕ, α, ψ = Refers to the angle at the stem, and
$\bar{\phi}, \bar{\alpha}, \bar{\psi}$ = Average angles over the bow sides.

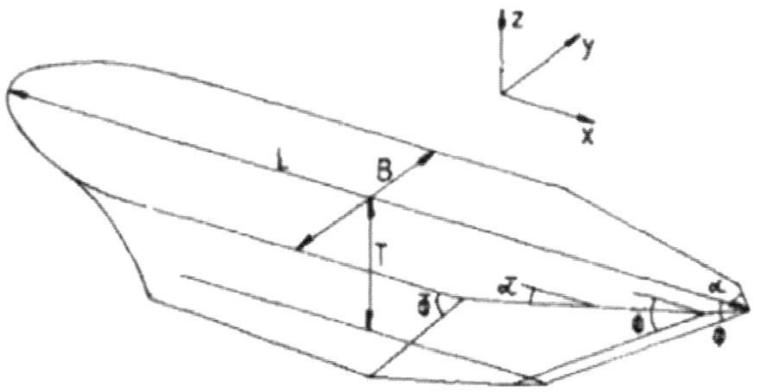

There are many investigators who have studied ice microstructure and its mechanics. For example, Timco and Weeks (2010) provides a good review of engineering properties for first year sea ice and old ice, including growth and microstructure, thickness, salinity and porosity, density, tensile strength, flexural strength, shear strength, compressive strength, multi-axial loading, borehole strength measurements, creep, elastic and strain modulus, Poisson's ratio, fracture toughness, and friction.

With the purpose to determine ice resistance through analytical modelling, ice engineering properties, including thickness, density, friction, flexural strength, shear strength, Poisson's ratio, and elastic modulus, have been reviewed as illustrated in Table *1*.

Table 1. Ice Property Review

Ice Property	Literature Review	Assumed Value
Thickness	Estimated through local in-situ measurements, satellites, marine radar, aircraft, helicopter, or icebreaker reconnaissance, as well as freezing degree day analytical modeling	-
Density	Reported to vary in the range 720 kg/m^3 to 940 kg/m^3. The density is also different for ice above or below waterline. For first year sea ice, an estimated value of 920 kg/m^3 is commonly used.	920 kg/m^3
Friction	Liu (2009) summarizes the model tests and full-scale experiments work from IOT-NRC and states the ice-hull friction coefficient is usually between 0.01 and 0.2. Timco and Weeks (2010) reviewed the ice friction as static friction and kinetic friction.	Determined by hull conditions
Flexure Strength	(Timco and S. O'Brien, 1994) equation: $\sigma_f = 1.76 \cdot e^{-5.88\sqrt{v_b}}$ Frankenstein and Garner (1967) derived brine volume, from Assur's brine volume table (Assur, 1960), for associated ice salinity and ice temperature between -0.5 °C and -22.9 °C. $v_b = S_i \left[\frac{49.185}{\lvert T_i \rvert} + 0.532\right]$	Equations with environmental data
Shear Strength	Timco and Weeks (2010) listed shear strength for granular ice are ranged from 400 kPa to 700 kPa and for columnar ice are ranged from 550 kPa to 900 kPa. Lindqvist (1989) assumed the shear strength equals to the bending strength (flexural strength).	Lindqvist (1989) assumption
Poisson's Ratio	Lindqvist (1989) assumed the Poisson's ratio equal to 0.3 for sea ice. Langleben and Pounder (1963) determined the mean value of Poisson's ratio for ice is 0.295±0.009. (Weeks and Assur, 1968, 1967) proposed $\mu_D = 0.333 + 0.06105 \cdot e^{\frac{T_i}{5.48}}$	Lindqvist (1989) assumption
Elastic Modulus	Timco and Weeks (2010) briefed the effective static elastic modulus of ice is in the range of 1.7 to 5.7 GPa, and the effective dynamic elastic modulus of ice is in the range of 1.7 to 9.1 GPa. Lindqvist (1989) used an elastic modulus equals to 2 GPa for his model.	Lindqvist (1989) value

STEADY TURNING PERFORMANCE (SPEED)

The reduction in ship speed while turning depends on many factors, both the ship and the ocean. For example, House (2007) points out that the ship's ability to turn is decreased significantly in shallow water, comparing with deep water. More importantly, he indicates the ship speed during hard over turn is decreased by a considerable amount, 30 to 40 percent from the full speed. In addition, he summarizes other factors which include:

- Ship length and superstructure arrangements;
- Distribution of any cargo or weight;
- Draft, trim, and heel;
- The relationship between power and ship displacement;
- Number and type of propellers;
- Rudder angle; and
- External forces that cause heading drift.

While a general understanding of the factors that influence ship turning exists, data to quantify these influences are limited. Even though all factors listed are significant, this study focuses on ship length, heel, and turning diameter in modeling performance.

Ship Turning in Relation to Turning Diameters and Ship Length

Among those factors described by (House, 2007), the most significant influence is the speed drop while turning, which is related to the ship length and sharpness of the turn. Shiba, et al. (1959) reports experimental findings of ship turning through three ship models. Their main objective is to model the rudder area effect on ship turning, although application may be limited as the rudder properties are not often published. They do however demonstrate a relationship between steady turning speeding over approaching speed and turning diameters over ship length, as shown in Figure 2. Halpern (2007) compares ship turning by Shiba, et al., (1959) with Davidson, (1944) and finds that they have good agreement, as shown in Figure 3.

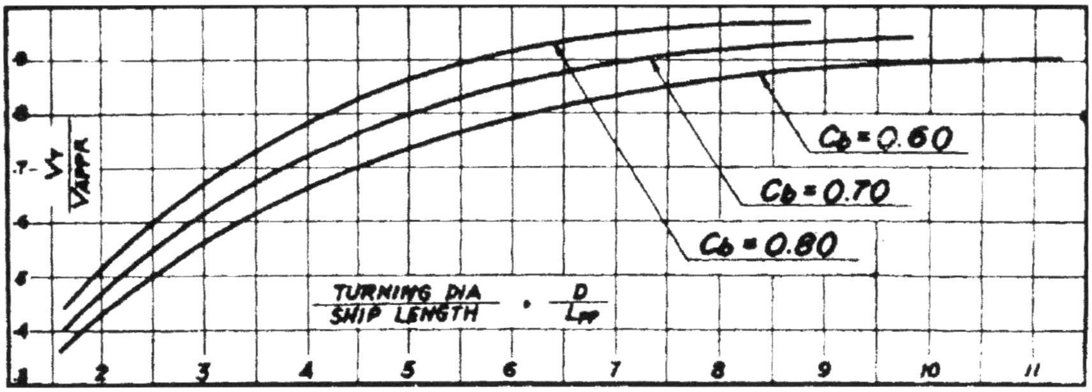

Figure 2. Ship turning performance test (Shiba et al., 1959)

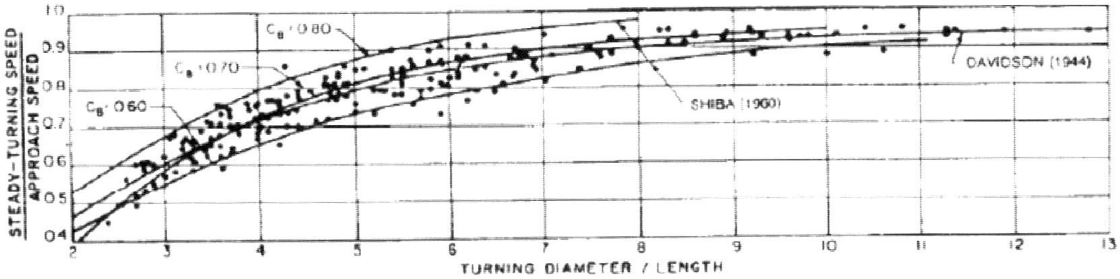

Figure 3. Ship turning performance test: compare (Shiba et al., 1959) with (Davidson 1944)

With this agreement, these curves can be extended to predict turning performance in open water for different conventional propelled ships.

There are many studies on ship turning performance in ice, such as (Brown, 2002; Glen et al., 1991; Lau, 2011; Liu, 2009; Martio, 2007; Quinton and Lau, 2006; Shi, 2002; Zhan et al., 2011; Zhou et al., 2016). Instead of applying a comprehensive model for icebreaker maneuvering, this study takes a simplified approach that models the speed reduction during turning in ice. Assuming the icebreaker turning in ice has a consistent speed reduction as the icebreaker maneuvering in the open water, an icebreaker turning speed coefficient ($C_{Turning}$) for level ice is model as

$$C_{Turning} = \frac{V_{Turning}}{V} \tag{7}$$

where,

V = Icebreaker speed in ice, and

$V_{Turning}$ = Icebreaker turning speed in ice.

For icebreakers with conventional propellers, the smaller the turning diameter, the slower the icebreaker can advance in certain level ice conditions. The secondary icebreaker in particular, has smaller turning diameter requirement. Given that secondary floes are not aligned with an idealistic path or track of the secondary icebreaker, performance will be degraded as 1) frequent maneuvering will be required (i.e. break floes in half), and 2) interference from bow collisions with small floes.

Active Heeling System

Active rapid heeling systems are sometimes used on ships to enhanced maneuvering and improve ship turning performance. Table 2 demonstrates the effectiveness of heeling degrees on icebreaker turning with or without rudder engagement.

Table 2. Study of Heeling Effect for Icebreaker Maneuvering (after Keinonen et al., 1989)

Vessel	Ice Thickness (m)	Heeling (Deg.)	Rudder	Turning Diameter/ Diameter at Zero Heel, Full Rudder in Same Ice
Kigoriak	1.25	6	Full	7.9/18.7 = 0.42
	1.25	3	Full	13/18.7 = 0.7
	1.25	6	0	19/18.7 = 1.02
	1.25	3	0	33/18.7 = 1.76
	1.25	6	Full Opposite	63/18.7 = 3.37
	1.4	6	Full	5.6/18.7 = 0.32
Robert Lemeur	0.6	6	Full	12.8/16 = 0.8
	0.9	3	Full	11.6/13 = 0.91
	0.8	4	Full Opposite	9.1/14.5 = 0.63

SECONDARY ICEBREAKER IN BROKEN ICE FIELD

For large level ice floes or continuous level ice, the resistance to the primary icebreaker is rather consistent (recognizing that ice thickness has natural variability). However, resistance modeling for the secondary icebreaker is more complex.

Besides breaking larger floes into smaller ones, interactions with small floes in a broken ice channel or regime include floe clearing and floe submergence. These contribute to added resistance and loss of heading control.

Submerging Floes

Depending on the floe size and bow geometry, one component of resistance influencing the secondary icebreaker performance in broken ice channels is floe submergence. With the water drag resistance, the combined resistance R_{t_sub} could be estimated as described by Eq. (6) and (2) or (3) as

$$R_{t\_sub} = R_s + R_{ow} \tag{8}$$

Floe Clearing or Ramming

Due to the geometry of the bow and vessel mass relative to the ice floe mass, ice floes that do not submerge may be cleared to the side. If the ice floes have sufficient mass, the energy used to clear the ice floes could be significant. The vessel may also be pushed off course. The energy transfer can be described as

$$E_{IB\_in} + \sum E_{IF\_in} = E_{IB\_out} + \sum E_{IF\_out} \tag{9}$$

where,

E_{IB_in} = Energy of the icebreaker before impact,

$\sum E_{IF\_in}$ = Sum of energy of the ice floes before impact,
E_{IB_out} = Energy of the icebreaker after impact, and
$\sum E_{IF\_out}$ = Sum of energy of the ice floes after impact from the icebreaker.

Combining with icebreaker position, ice drift, and surrounding environment, estimating ice floe energy after any interactions with the icebreaker bow may be challenging. However, it is reasonable that the speed of the ice management icebreaker will be reduced. For modeling the degrees of reduction, different analytical models could be considered with reasonable assumptions, such as the Popov model (Popov et al., 1969). At the initial stage of this study, the reduced icebreaker speed could be represented as

$$\frac{1}{2} \cdot m_{IB} \cdot V_{IB\_in}^2 = \frac{1}{2} \cdot m_{IB} \cdot V_{IB\_out}^2 + \frac{1}{2} \cdot m_{IF} \cdot \left(C_p \cdot V_{IB\_out}\right)^2 \quad (10)$$

where,

m_{IB} = Mass of the icebreaker,
V_{IB_in} = Initial icebreaker speed,
V_{IB_out} = Speed of the icebreaker after impact,
m_{IF} = Mass of the ice floe, and
C_p = Coefficient of the potential ice floe speed relative to the icebreaker (e.g. $C_p = 1$).

An estimate of floe mass assuming a plan view can be approximated by an equivalent circle as

$$m_{IF} = \rho_{ice} \cdot H_{ice} \cdot \pi \cdot R_{ice}^2 \quad (11)$$

where,

ρ_{ice} = Density of the ice floe,
H_{ice} = Thickness of the ice floe, and
R_{ice} = Radius of the ice floe.

The resulting speed coefficient model (C_{Broken}) for an icebreaker in a broken ice field can be estimated as

$$C_{Broken} = \frac{V_{IB\_out}}{V_{IB\_in}} = \sqrt{\frac{m_{IB}}{m_{IB} + m_{IF}}} \quad (12)$$

The total length of the broken ice field is determined by the beam of the icebreaker and number of ice channels encountered. Hamilton et al. (2011) experimentally demonstrated that an icebreaker forms a channel having a width equal to its beam that is filled with brash ice and will break ice between channels with a nominal 1 to 1 aspect ratio as shown in Figure 4. Hisette et al. (2014) suggested the formed ice channel equal to 1.3 times of the beam width during ice management operations. A secondary icebreaker would typically run into broken ice field at each loop of ice management operation. A theoretical example is shown in Figure 5a.

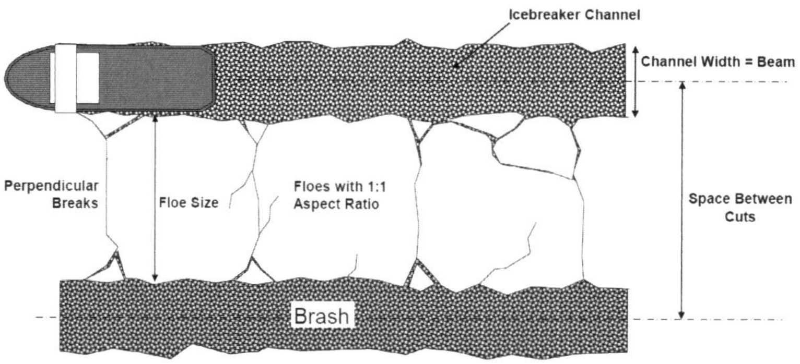

Figure 4. Ice break-up characteristics (Hamilton et al., 2011c)

If the ice floe has sufficient mass and sufficient thickness to prevent flexure failure, the icebreaker may require multiple attempts (i.e. rams) to transect the floe. For scenarios where continuous ramming may be required, comprehensive ship-ice interaction models (e.g. F_{MAX} software by Carter et al. 1995) could be used to estimate required time for the icebreaker to ram through the tough ice field. Field experience from Wright (1999) could also be referenced to estimate time requirement for an icebreaker to break a ridge in general.

CONTINUOUS MANOEUVRING

For a drifting large ice field, icebreakers need to adjust speed to achieve advance distance and target managed floe sizes.

Deflecting and On-Demand Manoeuvring

Theoretically, idealized icebreaking patterns may be chosen to reduce the size of oncoming ice floes based on a target managed floe size. The actual track however, as illustrated in Figure 5b, may be quite different as the location of the floes will not likely be aligned with the idealized track. The maneuvering associated with this actual track will reduce performance.

If the secondary icebreaker gets deflected off course, the vessel has to continuously maneuver to correct the heading to maintain some target track. The vessel may also have to continually chance course or maneuver to align with an ice floe that is off its track to ensure the floes are managed to achieve the target floe size. These are conceptually illustrated in Figure 6 and each will reduce speed and time required to achieve performance targets.

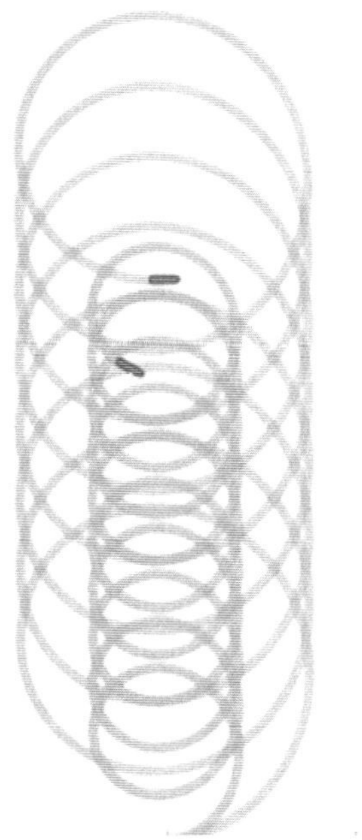

(a) Ideal planned

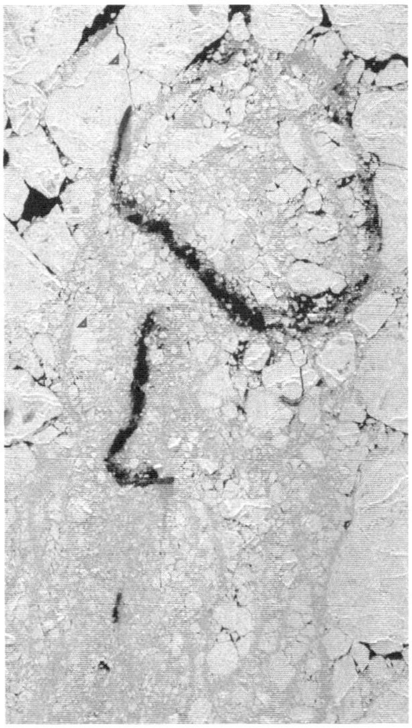

(b) Actual
(Shipboard Scientific Party, 2005)

Figure 5. Comparing (a) ideal planned to (b) actual ice management operation from Arctic Coring Expedition 2004

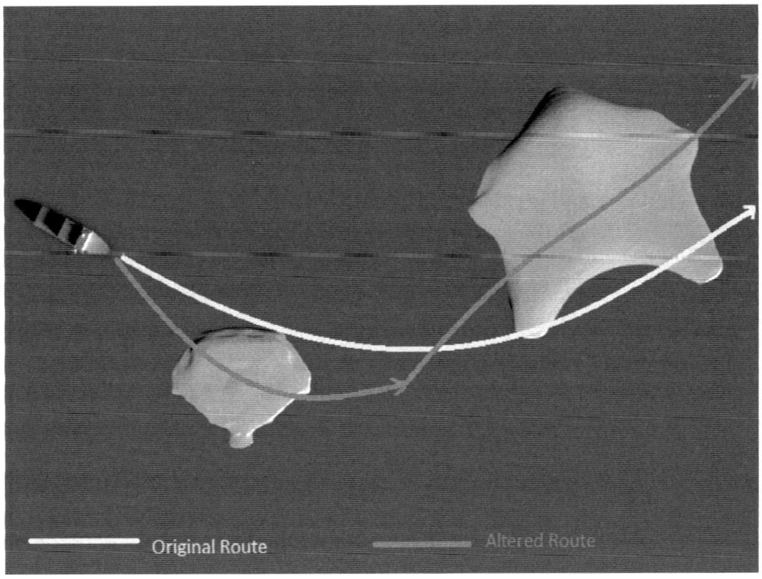

Figure 6. Original route vs. altered route

Ice Drift

Ice floes drift based on wind, current, and earth rotation effects. While ultimate management performance will depend on the accuracy of forecasting, the focus here is on performance given ice floes are present.

With respect to ice drift, frequent changes in drift speed and direction will require active changes to maneuvering parameters, which will slow ice management progress requiring more time.

To account for frequent turning requirements relative to planned turning, on-demand maneuvering coefficient $C_{Maneuvering}$ is defined as

$$C_{Maneuvering} = \frac{V_M}{V_{in}} \quad (13)$$

where,

V_M = Icebreaker on-demand maneuvering speed, and
V_{in} = Icebreaker speed before on-demand maneuvering.

SECONDARY ICEBREAKER PERFORMANCE MODELING

The secondary icebreaker performance model is comprised of icebreaker level ice performance, icebreaker steady turning performance, icebreaker performance in broken ice fields, and icebreaker on-demand maneuvering performance. Assuming an ice management loop to be an idealized circle, the effective secondary icebreaker operational speed (V_{E_SIB}) can be estimated

$$V_{E\_SIB} = \frac{L_{LI}}{\pi \cdot D} \cdot V_{LI\_SIB} \cdot C_{Turning} \cdot C_{Broken} \cdot C_{Maneuvering} + \frac{L_{BI}}{\pi \cdot D} \cdot V_{BI\_SIB} \cdot C_{Turning} \cdot C_{Broken} \cdot C_{Maneuvering} \quad (14)$$

where $L/(\pi \cdot D)$ represents the ratio of advance relative to an idealized planned circle and V_{LI_SIB} and V_{BI_SIB} are the secondary icebreaker speeds in level and broken ice respectively.

Comparison between the Model and Field Trial Data

Keinonen et al. (1989) reported some field trial data of icebreaker performance for different icebreakers in different ice and environmental conditions (i.e. icebreaker full power performance, and icebreaker turning circle performance at effective turning diameters). Based on the data, the level ice part of the performance model is tested. Since the turning circle trials were carried out with full or close to full power, full rudder, and no heel conditions at continuous turning, coefficients from broken ice and maneuvering (C_{Broken} and $C_{Maneuvering}$) in Eq. (12) and (13) are set to 1 respectively. Estimated turning speeds with recorded turning speeds for given ice and environmental conditions, as shown in Table 3. Due to relative large turning diameters, some estimated turning speeds are close or equal to estimated level ice speed. While comparison data are limited, reasonable consistency is achieved.

Table 3. Comparing Modeled Performance with Data

Vessel	Ice Thickness (m)	Power (MW)	Hull Condition	Turning Circle Diameter (m)	Recorded Turning Speed (m/s)	*Estimated Level Ice Speed (m/s)	Estimated Speed with $C_{Turning}$ (m/s)
Terry Fox	Open Water	14	Inerta Coating	329	-	9.0	5.00
	0.22			366	4.98	8.03	4.88
	0.75			2100	-	4.94	4.94
	1.5	12.8		3744	-	1.22	1.22
Kigoriak	Open Water	12	Bare, smooth, Lube on	438	-	8.70	5.28
	0.35			455	-	7.23	4.52
	0.67			775	-	5.80	3.62
	1.22	11.2		1196	2.30	2.65	2.29
	1.26	10.2	Bare, rough	1579	1.58	1.70	1.57
	1.43	11.2	Lube on	1423	1.50	1.70	1.53
Robert Lemeur	Open Water	7.1	Inerta Coating	198	-	7.1	2.71
	0.63			1369	-	4.57	4.57
	0.64			1266	-	4.53	4.49
	1.1			1543	-	2.09	2.09

*Note: Ice properties are based on average values from the data;
Ice-hull frictions are based on hull condition.

Application

For example, let us assume a Kigoriak type vessel is operating at full power managing 1 m thick ice at 500 m turning diameter ($C_{Turning} = 0.67$) to produce 50 m diameter floes. For each loop (see Figure 5a), it is assumed that:

- the secondary icebreaker encounters broken ice channel an average of 12 times where the width of each channel $L_{BI} = 25\ m$,
- one small floe fragment occurs for each broken channel width that deflect the vessel from its planned course,
- nine tenths of the pre-managed floes are typical level ice with one tenth of the floes having a small diameters approximately one half of the planned ice floe size diameter (i.e. $D_{UB} = 25m$), such compliance of the floe and loss of energy reduces the icebreaking capability; and
- one unexpected maneuvering is required within each planned loop at the vessel's maximum maneuverability (e.g. maneuver back to original route within about 3 ship length).

Based on these assumptions for each loop, the following are considered in modeling the secondary icebreaker performance:

- icebreaker in large ice floes ($V_{LI\_SIB} = 3.3\ m/s, C_{Broken} = 1\ and\ C_{Maneuvering} = 1$),
- icebreaker between two large ice floes with one smaller unbreakable/un-submergible floe fragment ($C_{Broken\_1} = 0.67\ and\ C_{Maneuvering} = 1$),
- icebreaker maneuvering between two large ice floes ($C_{Broken} = 1\ and\ C_{Maneuvering\_1} = 0.36$),
- and icebreaker in broken ice channel with one smaller unbreakable/un-submergible floe fragment ($V_{BI\_SIB} = 6.5\ m/s, C_{Broken\_2} = 0.67\ and\ C_{Maneuvering} = 1$).

Hence, the effective speed of the second icebreaker can be estimated using Eq. (14) as

$$V_{E\_SIB} = \frac{10}{12} \cdot \frac{L_{LI}}{\pi \cdot D} \cdot V_{LI\_SIB} \cdot C_{Turning} \cdot \left(\frac{9}{10} + \frac{1}{10} \cdot C_{Broken\_1}\right) + \frac{2}{12} \cdot \frac{L_{LI}}{\pi \cdot D} \cdot V_{LI\_SIB} \cdot C_{Turning} \cdot C_{Maneuvering\_1} + \frac{L_{BI}}{\pi \cdot D} \cdot V_{BI\_SIB} \cdot C_{Turning} \cdot C_{Broken\_2} \quad (15)$$

$$V_{E\_SIB} = 2.1\ m/s$$

The resulting estimated performance illustrates that the vessel could only maintain about 60% of its level ice capability as a secondary icebreaker. The assumptions are considered conservative. If there are more unbreakable/un-submergible/embedded multi-year ice floes, more requirements on maneuvering, or the vessel's maneuverability is over-predicted, the secondary icebreaker might be less than 50% of her level ice capability. Additional test data (field and model) will help clarify these uncertainties.

CONCLUSIONS

Understanding the factors that influence reduced secondary icebreaker performance are important for ice management operational planning.

Compared with the primary icebreaker, the secondary icebreaker may require increased icebreaking and turning capabilities depending on ice conditions. To model the speed reduction of an icebreaker while turning, a simple approach has been developed and compared with field trial data. With the complexity of secondary icebreaker operations, two extra influence factors are considered, including interactions with smaller ice floes and on-demand maneuvering for unexpected situations.

Consistent with field observations, the resulting model and illustrative example demonstrates the speed reduction in secondary icebreaker performance compared with idealistic modeling. This reduction in performance is important when developing ice management plans and requirements for icebreaker support.

While the performance model compares well with available operational field data, additional verification with more recent field trial data are recommended. The resulting performance model will be implemented in C-CORE's ice management analysis and planning software which has a probabilistic modeling capability to assist with designing an ice management fleet to support station-keeping operations in specific regions having particular ice conditions. Modeling the variability that exists with ice and environmental data, the probabilistic approach allows one to base decisions on a target annual probability of exceedance (e.g. 10^{-2}).

ACKNOWLEDGEMENTS

The authors gratefully acknowledge core funding for C-CORE's CARD program by the Hibernia Management and Development Company Ltd. (HMDC), the Terra Nova Development (Suncor Energy Inc. – Operator), and the Research and Development Corporation of Newfoundland and Labrador (RDC).

REFERENCES

Assur, A., 1960. Composition of sea ice and its tensile strength (Research Report No. 44). U.S. Army Corps of Engineers.

Brown, R.C., 2002. An Experimental Investigation of Ship Manoeuvrability in Pack Ice. Memorial University of Newfoundland, St. John's, NL.

Browne, R., Wright, B., Connelly, D., 2014. The Question of Ice Management and Some Associated Realities. Presented at the OTC Arctic Technology Conference, Offshore Technology Conference. doi:10.4043/24655-MS

Davidson, K.S.M., 1944. Turning Steering of Ships.

Frankenstein, G., Garner, R., 1967. Equation for Determining the Brine Volume of Sea ice From -0.5° to −22.9 °C. J. Glaciol. 6, 943–944.

Glen, I.F., Menon, B., Steele, M., Hardiman, K., Brennan, W., 1991. Investigation of Ship Manoeuvrability in Ice Phase 1 (Technical Report No. TP 10922E). Fleet Technology Limited.

Halpern, S., 2007. She Turned Two Points In 37 Seconds.

Hamilton, J.M., 2011. The Challenges of Deep-Water Arctic Development. Int. J. Offshore Polar Eng. 21.

Hamilton, J.M., Fenz, D.M., Kokkinis, T., Holub, C.J., 2013. Aligning the Needs of Floating Drilling and the Capabilities of Ice Management, in: Proceddings of the 22nd International Conference on Port and Ocean Engineering Under Arctic Conditions (POAC). Presented at the POAC 13, Espoo, Finland.

Hamilton, J.M., Holub, C., Blunt, J., 2011a. Simulation of Ice Management Fleet Operations Using Two Decades of Beaufort Sea Ice Drift And Thickness Time Histories. Presented at the The Twenty-first International Offshore and Polar Engineering Conference, International Society of Offshore and Polar Engineers.

Hamilton, J.M., Holub, C., Blunt, J., 2011b. Simulation of Ice Management Fleet Operations Using Two Decades of Beaufort Sea Ice Drift And Thickness Time Histories. Presented at the The Twenty-first International Offshore and Polar Engineering Conference, International Society of Offshore and Polar Engineers.

Hamilton, J.M., Holub, C., Blunt, J., Mitchell, D., Kokkinis, T., 2011c. Ice Management for Support of Arctic Floating Operations. Presented at the OTC Arctic Technology Conference, Offshore Technology Conference. doi:10.4043/22105-MS

Hamilton, J.M., Holub, C.J., Shafrova, S., Blunt, J., Foltz, R., Ritch, R., 2014. A Case Study of Ice Management for Exploration Floating Drilling, in: ICETECH. Presented at the ICETECH 14.

Hamilton, J.M., Kokkinis, T., Holub, C., Matskevitch, D., Cheng, T., Harris, M., Shafrova, S.,

2016. Near-Field Ice Management Tactics for Floating Drilling in Arctic Pack Ice. Presented at the Arctic Technology Conference, Offshore Technology Conference. doi:10.4043/27341-MS
Hisette, Q., Jochmann, P., Bronsart, R., 2014. Simulation of Ice Management Operations. Presented at the The Twenty-fourth International Ocean and Polar Engineering Conference, International Society of Offshore and Polar Engineers.
House, D.J., 2007. Ship Handling: Theory and Practice. Elsevier.
Keinonen, A., Browne, R.P., Revill, C., 1991. Icebreaker Design Sythesis Phase 2 Analysis of Contemporary Icebreaker Performance (Technical Report No. TP 10923E).
Keinonen, A., Browne, R.P., Revill, C., 1989. Icebreaker Design Synthesis Analysis of Contemporary Icebreaker Performance (Technical Report No. TP 9992E). Arno Keinonen Arctic Consulting Inc., Calgary, Alberta, Canada.
Keinonen, A., Browne, R.P., Revill, C., Reynolds, A., 1996. Icebreaker Characteristics Synthesis (Technical Report No. TP 12812E).
Kubat, I., Sayed, M., 2014. Literature survey of station keeping and ice management, in: ICETECH. Presented at the ICETECH 14.
Langleben, M.P., Pounder, E.R., 1963. Elastic Parameters of Sea Ice, Reprinted from Ice and Snow edited by W.D. Kingery. MIT Press, USA.
Lau, M., 2011. Ship Manoeuvring-in-Ice modeling software OSIS-IHI, in: POAC 11. Presented at the POAC 11, Montreal, Canada.
Lindqvist, G., 1989. A Straightforward Method for Calculation of Ice Resistance of Ships. POAC 1989 2, 722–735.
Liu, J., 2009. Mathematical Modeling Ice-Hull Interaction for Real Time Simulations of Ship Manoeuvring in Level Ice. Memorial University of Newfoundland, St. John's, NL.
Martio, J., 2007. Numerical Simulation of Vessel's Manoeuvring Performance in Uniform Ice. Helsinki University of Technology, Helsinki.
Popov, Y.N., Faddeev, O.V., Kheisin, D.E., Yakovlev, A.A., 1969. STRENGTH OF SHIPS SAILING IN ICE.
Quinton, B.W., Lau, M., 2006. Manoeuvring in Ice: A Test/Trial Database (Technical Report). Institute for Ocean Technology, National Research Council.
Riska, K., Wilhelmson, M., Englund, K., LEIVISKÄ, T., 1998. Performance of Merchant Vessels in Ice in the Baltic (Research Report No. 52). Finnish Maritime Administration.
Shi, Y., 2002. Model test data analysis of ship maneuverability in ice. Memorial University of Newfoundland, St. John's, NL.
Shiba, H., Mizuno, T., Tomita, S., Eda, H., 1959. Turning Model Experiments on Optimum Rudder Area of Ships.
Shipboard Scientific Party, 2005. Arctic Coring Expedition (ACEX): Paleoceanographic and tectonic evolution of the central Arctic Ocean, IODP Preliminary Report. Integrated Ocean Drilling Program.
Timco, G.W., S. O'Brien, 1994. Flexural strength equation for sea ice. Cold Reg. Sci. Technol. 22, 285–298. doi:10.1016/0165-232X(94)90006-X
Timco, G.W., Weeks, W.F., 2010. A review of the engineering properties of sea ice. Cold Reg. Sci. Technol. 60, 107–129. doi:10.1016/j.coldregions.2009.10.003
Weeks, W.F., Assur, A., 1968. The Mechanical Properties of Sea Ice, in: Proceedings, Conference on Ice Pressures Against Structures. National Research Council of Canada, Quebec City, Laval University, pp. 25–78.
Weeks, W.F., Assur, A., 1967. The Mechanical Properties of Sea Ice. U.S. Army CRREL Monograph II-C3, Hanover, NH, USA.
Wright, B., 1999. Evaluation of full-scale data for moored vessel stationkeeping in pack ice

(Technical Report No. PERD/CHC Report 26-200).
Wright, B., McKenna, R., Browne, R., 2014. A Tactical Ice Management Simulation Methodology and Approach. Presented at the OTC Arctic Technology Conference, Offshore Technology Conference. doi:10.4043/24654-MS
Younan, A.H., Hamilton, J.M., Garas-Yanni, V.Y., Blunt, J., Holub, C.J., Kokkinis, T., 2012. An Integrated Ice Management Alert System. Presented at the The Twenty-second International Offshore and Polar Engineering Conference, International Society of Offshore and Polar Engineers.
Zhan, D., Agar, D., He, M., Spencer, D., Molyneux, D., 2011. Numerical Simulation of Ship Maneuvering in Pack Ice, in: OMAE 2010. Presented at the OMAE 2010.
Zhou, Q., Peng, H., Qiu, W., 2016. Numerical investigations of ship–ice interaction and maneuvering performance in level ice. Cold Reg. Sci. Technol. 122, 36–49. doi:10.1016/j.coldregions.2015.10.015

POAC'17
Busan, KOREA

Proceedings of the 24th International Conference on
Port and Ocean Engineering under Arctic Conditions
June 11-16, 2017, Busan, Korea

Automated Monitoring System of Planned Position of Arctic Quay Walls

Pavel Andreevich Garibin[1], Evgeniy Olegovich Ol'khovik[1]
[1] Admiral Makarov State University of Maritime and Inland Shipping, Saint-Petersburg, Russia

ABSTRACT

In current practice of construction projects Arctic offshore pier may be deviations from the planned situation building structures as during the construction phase and during operation. In winter period there is a powerful influence ice loads, measuring and control that with traditional method of surveying extremely difficult. The objective of this research is to automate monitoring procedure of building structures. Article suggests a new method and apparatus for automated health monitoring of construction elements of geotechnical engineering structures of maritime transport in the Arctic port piers. Maritime port located on the Arctic shelf is exposed to difficult environmental impact, the study and the characteristics of which are not fully understood. High freight load, a limited period of use in the summer navigation, the lack of scientific support for the project, all this lead to the need to perform regular observations of engineering.

KEY WORDS: Arctic port pier; Quay walls; Monitoring system; Technical state.

INTRODUCTION

Many Arctic offshore piers have a popular structure as quay walls construction, because it is easier and faster than building of gravitational structures, which is more difficult to deliver bulky and heavy construction materials. Sheet piling (or pipe piling) installation in permafrost and Arctic conditions is also challenging, but it is still easier than building of large gravitational structures. Coastal structures are experienced high loads from naval cargo. In recent years, there is the problem of structural integrity of harbor. It becomes more important question in connection with various environmental constraints and extreme phenomena such as typhoon or collision of the vessel. Common failures such as deformation of quay wall or its movement affect the stability of whole structure. Local faults affect the structural stability of entire structure and may cause the total destruction or restrict performance. Therefore, the integrity monitoring of the existing sheet piling structures is absolutely necessary.

Currently, structural technology of the technical condition monitoring of building constructions is actively developed. Existing inclination control devices have some drawbacks, the main of them is the high cost and secrecy of data processing technology. Also, most of measuring systems don't allow to determine spatial structure of individual elements of situation as part of whole buildings complex, those, they don't have a linked network structure for processing and analyzing data about change the geometric parameters.

Metzger, Hutchinson and Kwiatkowski (2014) proposed the measurement system for berthing parameters of marine vessel, it's the most complete and perfect system, but it requires a special connection with berthing ships. Lee and Kim (2015) reviewed the water level influence on the vibration characteristics of gravitational structures by analyzing of the modal parameters.

Van der Linden, Emami-Naeini, Kosut, Sedarat and Lynch (2011) describes an analysis to determine optimal placement of inclinometers for deformations evaluating of operated bridge, examines the results of model tests, but definitive conclusions about optimal structure of sensor network cannot be moved to a random construction structure.

Ichii, Kitade, Kawano and Taguchi (2014) discuss cases of port facilities worsening. Method is used for monitoring of the state of backfill sheet pile quay wall which is the main structural component of construction. It was found that as a result of corrosion interlocks discontinuity of some of sheet piles occurs and as a result, fine particles of backfill are washed out, which is a serious problem that must be considered in order to maintain security of the harbor wall. According to the results of numerical simulation with using the finite element method (FEM) in the dynamic analysis was formulated the deformation changes acting in pier. The results of analysis were compared with results of direct measurements in natural conditions.

METHODOLOGY AND APPARATUS

In this study we suggested a structural method for measuring of local deformations (angles of inclination) to assess the technical condition of quay wall construction. We have proposed a multiple analysis of variance design line cordon berth. To achieve this goal, the following approaches were implemented. On first phase we developed a simplified structural model of the sheet pile wall, which has a kinematic degree of freedom, at the expense of the formulated structure for measuring of network planning provisions which, in turn, corresponds to a set of sensors. On second phase we make the algorithm is based on the identification and use of the obtained inclination angle measurement sheet pile wall elements. This makes it possible to formulate a more accurate model by additional analysis of the elastic properties and structural rigidity of the quay wall. On third phase, the forecast of technical condition of sheet pile wall design with the refinement of the damping parameters was obtained as a result of in-situ measurements. At each stage it is possible to take into account additional parameters that determine the arctic conditions, if possible to fix instrumental observations.

Network sensor implementation can continuously monitor technical condition of building structure in automatic mode, which will reduce the cost of engineering staff. It is important for the Arctic port facilities under a long winter period, when surveying is complicated because of the severe winter conditions. With this system, continuous monitoring of quay walls parameters can be in real-time. Developed project of monitoring system is presented in this article and demonstrates the usefulness of automated system. The article provides reasonable scheme for installation of measurement sensors and examples of data obtained from test measurements.

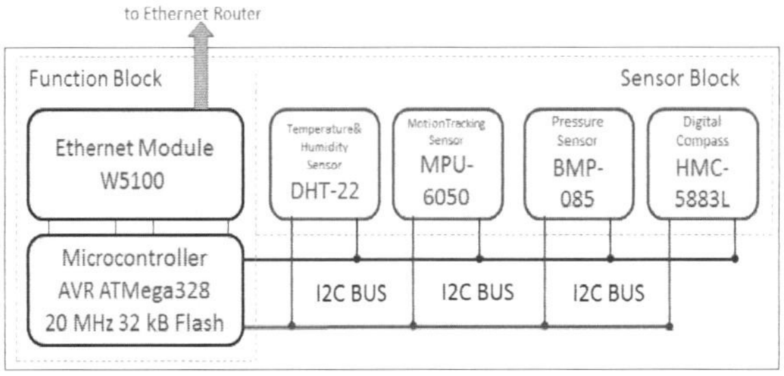

Figure 1. Structural diagram of the sensor for controlling of the inclination angle of the structural element of quay walls

The developed sensor is arranged in a sealed housing from non-magnetic material (aluminum cast alloy) inside which is a three-axis accelerometer, gyroscope and magnetometer (compass). Each of sensing axes of accelerometer is collinear of magnetometer axis and gyroscope axis. Temperature, humidity and barometric atmospheric pressure sensors are additionally installed in the device, these sensor allow to adjust correcting the measuring data depending of meteorological conditions. A detailed diagram is shown in fig.1, which discloses a composition of all components. The developed sensor has a standard network interface and can be connected via a router to Internet for next data transfer to server. Technical characteristics of server for receiving and collecting data are minimal, can be used as a local or remote server. Photo of the developed sensor is shown in fig.2.

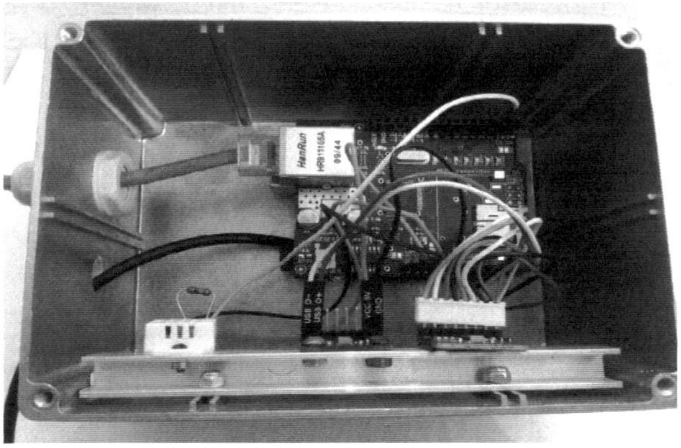

Figure 2. Photography of measuring sensor of automated system for control of planned position

Developed automated system identifies hazardous angular displacements of sheet pile wall in real time and storing all data about deformation on special remote server, which is important information for further calculations and evaluation of safety of operation on berth. Developed equipment includes a microcontroller AVR ATmega328 and some measuring sensors, which accurately detect angular deflection of structural elements. One measurement module contains a micromechanical accelerometer, magnetic compass, temperature sensor, humidity and atmospheric pressure, together they provide objective information about majority of phenomena that affect on structure of pier.

THE MAIN TASKS FOR USE OF DEVELOPED MONITORING SYSTEM FOR MARITIME TRANSPORT PORT'S PIERS IN ARCTIC

Most of the regulations and rules suggest that offshore facilities are potentially dangerous as technical constructions. Then, for designs of high hazard class it is recommended to use the automated monitoring system. This article discusses the berths with varying depth at the quay wall. However, for arctic conditions (shift method of work) and a long period of people absence on facility, requirements for the regulations execution for conducting of periodic inspections of technical condition is almost impossible, and due diligence requires significant capital expenditures. Similar conditions correspond to stationary and gravity platforms on the Arctic shelf.

The scientific methodology for organization of the system for continuous automated monitoring of some operation on hydraulic structures of water transport, which interact with the environment was proposed (Garibin and Ol'khovik, 2015; Oleynik, Makshanov and Marley, 2015) and now we're developed the new strategy for the technical state monitoring is a series of elaborated structural schemes for automation of measurements. The essence of proposed deterministic model for assessing of technical state of arctic port pier is disclosed in that form: "The automated monitoring of the technical state" - "The complex model of assessment of the technical state" - "The predictive model".

New developed model allows providing the safety of operation of the hydraulic structures already on a completely new practical level and gives the qualitative prediction on the predetermined time period, i.e. fully implementing of the concept of the life-cycle analysis. The methods of joint inclinometric and dynamic instrumental observations are offered new integrated approach that in complex implement for quay walls monitoring and give a possibility in future to get safe and justified information about the technical state.

MODEL OF THE TECHNICAL STATE ASSESSMENT

We used the standard approach for evaluating of technical condition - the calculation of stress-strain state under load and natural factors. For this we considered the classic type of revetment structure form tongue with anker system and distribution belt. For the stress-strain state calculations we used the software GeoWall (freeware demo version from www.geo-soft.ru) and some own research model. The results are present in fig.3, 4.

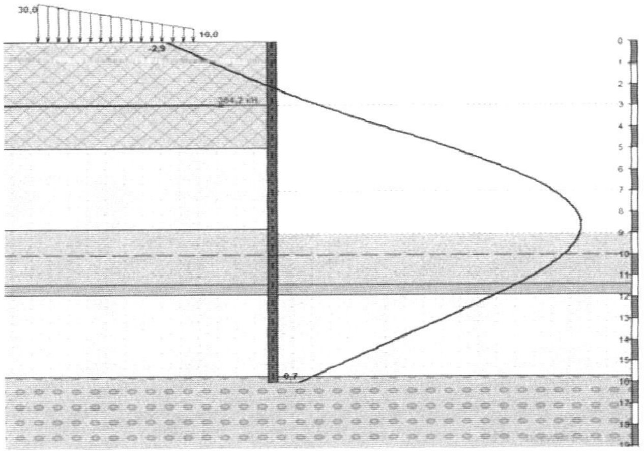

Figure 3. Design of sheet pile quay wall with one horizontal anchor

The simulated displacement and deformation of quay walls can be calculated with sufficient accuracy by using software (GeoWall or other) – in fig.3. Further, it is possible to determine the parameters of critical strain as a deviation in plane of geometrical dimensions or wall angle from horizontal (fig.4). Also, the calculated displacement and deformation in the mathematical model can be defined for the exact values of the stress-strain state. Then, all measurements obtained from the proposed equipment (fig.1, 2) as a result of long-time monitoring can be interpreted as the variable load on the building structure of sheet pile wall.

Thus, the relationship can be established for controlled planned position of berthing wall with various load and environmental factors in the form of constant monitoring of technical condition. That is especially important for the Arctic designs construction, which is impossible for periodic inspections due to adverse weather conditions and the absence of special service personnel.

CALCULATION MODEL

The bending profile (fig.4) of sheet pile wall for most of load corresponds to the quadratic parabola with zero angle of rotation in the base soil.

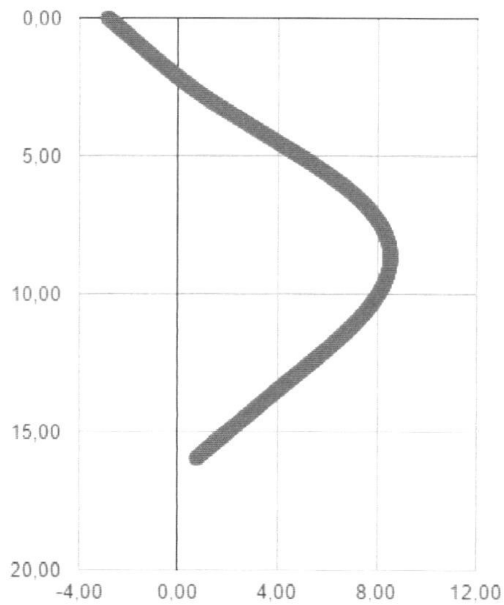

Figure 4. Calculation of the horizontal displacement of sheet pile wall

Also, power of marine ships which alone maneuver in water area is constantly increasing. That is cause of destruction of shore protection structures in ports. To evaluate the effects of using empirical methods and periodic inspections, Abramowicz-Gerigk (2014) proposed some ways of automatic identification of degradation processes of shore protection. However, these methods are based on measurement of ship propeller jets velocities and have some limitation.

Network from the measurements sensors are mounted at top of well head harbor wall have full control of displacements and rotations, as shown in Figure 5. Movement in X-Y plane (planned situation) is in Fig. 5(a) or rotation in Y-Z plane is in fig. 5(b). For this purpose, data from accelerometer and magnetic field sensor used in the automated monitoring system of technical state.

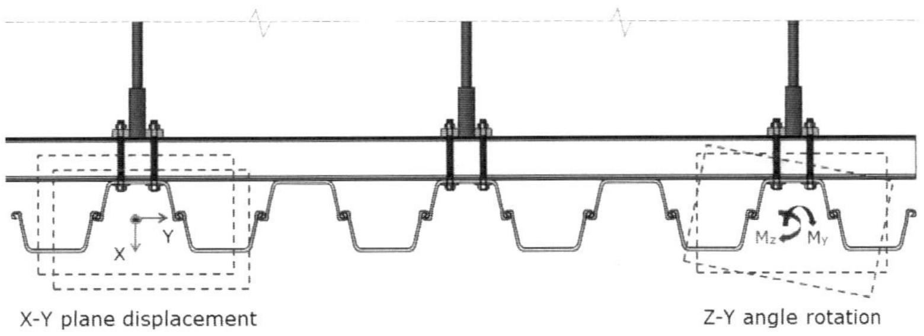

a) b)

Figure 5. The scheme for measuring the angular deviation of sheet pile wall.
a) Displacement through parallel shift in horizontal plane;
b) Displacement through rotate or bend deformation in vertical plane

We have developed software, which analyzes incoming data from sensor network. Data updating is provided every minute that is sufficient. The signal system is realized depending on the angle of deformation, for example: 0÷1 ° – a normal state, 1÷2 ° – warning, 2÷3 ° – danger. Such control allows estimating loadings more precisely at cargo operation of pier, including processes of freights unloading in various pier zones.

На рисунке 6 представлены тестовые измерения для угла отклонения sheet pile wall of небольшого грузового причала (протяженность около 100 метров) в зимний период в Санкт-Петербурге, метеорологические условия были максимально приближены к арктическим (отрицательная температура, лед в акватории). В условиях швартовки и отхода судна (буксира) были зафиксированы небольшие отклонения от вертикальной оси (с 22 по 55 минуту), что подтверждает работоспособность разработанной системы мониторинга.

Figure 6 shows the test measurements for the angle of deflection of sheet pile wall of a small cargo quay (about 100 meters long) in winter period in St. Petersburg, meteorological conditions were as close as possible to the Arctic (negative temperature, ice in the water area).

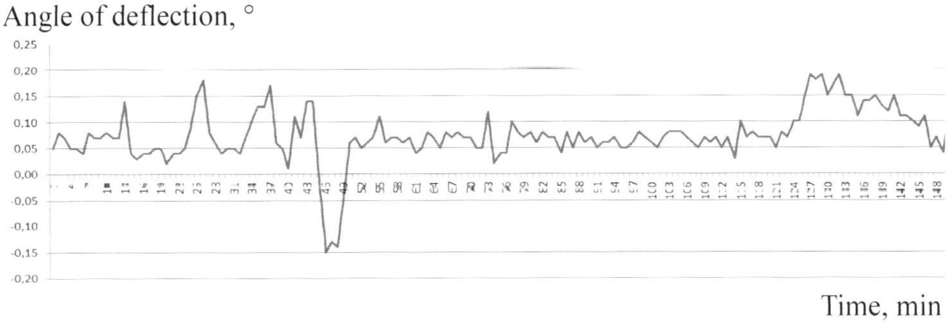

Figure 6. Results of monitoring (measurements) of local deflection of sheet pile wall in X-Y displacement

CONCLUSIONS

Continuous geodetic monitoring of planned position of elements of berthing constructions is an indispensable condition for their safe operation, it belongs also to the Arctic ports with severe climatic conditions and the minimum personnel during the winter period.

In perspective the developed automatic monitoring system allows to solve some problems with definition of current technical condition of an arctic quay walls. Applied algorithm detects the slightest position deviations within 0.15 angle degrees, which is sufficient to detect significant deviations of planned position of quay wall.

The developed monitoring system can be applied together with other control systems, for example, with control system described by Ruggeri, Segato and Scarpelli (2013) where the technique for a tension control in anchor devices is presented. Also, as some application is control of vessel mooring on a wall, the sufficient mathematical apparatus for similar calculations is presented (Liu, Wang and Zhu, 2015; Sato and Tabata, 2009), it can be confirmed with data directly from a network of sensors installed on a quay wall.

Sharapov, Shkhinek and Delvalls (2015) proposed technology that due to heating of the inner site of the sheet pile wall allows to significantly reduce the ice load on the design of pier and remove ice accumulations. Lack of this technology is high electricity consumption. Joint use of our monitoring system and heating system can provide smooth energy consumption due to regulation of its consumption only at the necessary moment when there is a significant force impact on the anchor tractions and the quay wall.

Wrangborg, Marchenko & Murashkin (2015) in research cycle during the winter season 2013 and 2014 using Geokon Earth Pressure cells, fiber optics and water pressure sensors (Sinitsyn, A. (2012) used 3D Scanner) made measurements on both the inside and outside of one of the open cofferdams supporting the quay. The systematic studies of ice impacts proposed by them are unique, but the complexity of the equipment will not allow the introduction of these methods for mass use. Автоматизированная система мониторинга, разрабатываемая нами, обладает большей универсальностью, поскольку выполнена на основе единого сенсора, также имеет преимущество за счет низкой стоимости конечного продукта. The automated monitoring system which is being developed by us is more versatile, because it is based on a single sensor. It also has the advantage of low cost of the final product.

In future articles we will discuss the details of use of the developed monitoring system of the quay walls state.

REFERENCES

Abramowicz-Gerigk, T., 2014. Identification of Degradation Processes of Seabed Protection in Ports. *Safety and Reliability: Methodology and Applications*, CRC Press, pp. 889–894. Available from: doi: 10.1201/b17399-126.

Garibin, P.A. & Ol'khovik, E.O., 2015. The methodology of organization of continuous monitoring of technical state of hydraulic structures of water transport. *News of Higher Educational Institutions. Construction*, 2(674), pp.58-72. (rus)

Garibin, P.A. & Ol'khovik, E.O., 2016. Development of the method and apparatus for automated monitoring of the planned position of marine berthing facilities. *Vestnik of the Siberian State University of Geosystems and Technologies (SSUGT)*, 2(34), pp.55-64. (rus)

Ichii, K., Kitade, K., Kawano, M. & Taguchi, I., 2014. Surface Wave-Based Health Monitoring Method for a Sheet Pile Quay Wall. *International Journal of Structural Stability and Dynamics*, 14(05), 1440009. Available from: doi:10.1142/s0219455414400094.

Lee, S.Y. & Kim, J.T., 2015. Effects of foundation damage and water-level change on vibration modal parameters of gravity-type caisson structure. *Science China Technological Sciences*, 58(2), pp.316-329. Available from: doi:10.1007/s11431-014-5748-1.

Liu, P., Wang, H. & Chao, Z., 2015. Numerical Simulation Analysis of Collision Between Ship and Steel Sheet Pile Quay-Wall. *Applied Mechanics and Materials*, 744-746, pp.1175–1179. Available from: doi: 10.4028/www.scientific.net/amm.744-746.1175.

Metzger, A.T., Hutchinson, J. & Kwiatkowski, J., 2014. Measurement of marine vessel berthing parameters. *Marine Structures*, 39, pp.350-372.

Oleynik, A.Yu., Makshanov, A.V. & Marley, V.E., 2015. Development monitoring system strain engineering at the web-based and microcontrollers. *Vestnik gosudarstvennogo universiteta morskogo i rechnogo flota imeni admirala S.O. Makarova*, 1(29), pp.215-222. (rus)

Ruggeri, P., Segato, D. & Scarpelli, G., 2013. Sheet Pile Quay Wall Safety: Investigation of Posttensioned Anchor Failures. *Journal of Geotechnical and Geoenvironmental Engineering*, 139(9), pp.1567–1574. Available from: doi: 10.1061/(asce)gt.1943-5606.0000886.

Sato, M. & Tabata, K., 2009. Shaking Table Test of a Large-Size Model on Failure Mechanism of Sheet-Pile Quay Wall and Pile Foundation Due to Lateral Spreading. *Jiban Kogaku Janaru (Japanese Geotechnical Journal)*, 4(4), pp.259–271. Available from: doi: 10.3208/jgs.4.259.

Sharapov, D., Shkhinek, K. & Delvalls, T.Á., 2015. An Estimation of the Amount of the Thermal Energy for the Moorage Wall Heating in the Arctic Harbors to Avoid Ice Accumulation. *Ocean Engineering*, 100, pp.90-96. Available from: doi: 10.1016/j.oceaneng.2015.03.016.

Sinitsyn, A., Wrangborg, D., Yulmetov, R., Sund, A.T., & Marchenko, A., 2012. Measurements of Deformations and Displacements of Stationary Quays in Svalbard with 3D Laser Scanner Riegl VZ-1000. *21st IAHR International Symposium on Ice "Ice Research for a Sustainable Environment",* Dalian University of Technology Press, Dalian,. Available from: doi: 10.13140/2.1.4118.8483

Van der Linden, G.W., Emami-Naeini, A., Kosut, R.L., Sedarat, H. & Lynch, J.P., 2011. Optimal sensor placement for health monitoring of civil structures. *Proceedings of the 2011 American Control Conference*. Available from: doi:10.1109/acc.2011.5991121.

Wrangborg, D., Marchenko, A. & Murashkin, D., 2015. Measurement of loads exerted by sea ice on the quay at Kapp Amsterdam on Svalbard. *Proceedings of the 23rd International Conference on Port and Ocean Engineering under Arctic Conditions, POAC.*

POAC'17
Busan, KOREA

Proceedings of the 24th International Conference on
Port and Ocean Engineering under Arctic Conditions
June 11-16, 2017, Busan, Korea

A Study for the State-of-the-Art on Arctic Issues for Floating Ices and Subsea Operations

Byungmo Kim[1], Hyungju Kim[2], Seunghyun Ha[3], Yiliu Liu[2] and Kyungsik Choi[3]
[1] Department of Convergence Study on the Ocean Science and Technology, Ocean Science and Technology School of Korea Maritime and Ocean University, Busan, Republic of Korea
[2] Department of Mechanical and Industrial Engineering, Norwegian University of Science and Technology, Trondheim, Norway
[3] Department of Ocean Engineering, Korea Maritime and Ocean University, Busan, Republic of Korea

ABSTRACT

While there are huge energy potentials in the Arctic, latent risks for the recovering of hydrocarbons still remain, as well. One of the main causes of the risks is harsh environmental conditions, for instance, geographic remoteness, ice-covered areas, icing and ices. In particular, various types of floating ices, such as icebergs, can directly and immediately damage offshore and subsea facilities, and this can lead to serious consequences to the environment. Thus, this paper aims to review literatures and define research agenda on arctic issues for floating ices and subsea operation.

KEY WORDS : Arctic; Floating Ice; Subsea; Safety System; Structural Interaction.

INTRODUCTION

There are huge potential and opportunity of the Arctic in present. As one of underlying strengths, the length of shipping route and the shipping period from Asia to Europe through the Arctic are approximately forty percent and twenty five percent shorter, respectively, than those of existing routes through the Suez Canal (Baek, et al., 2016). Meanwhile, in terms of energy resources, it is expected that twenty five percent of undiscovered deposit of worldwide remaining petroleum resources is placed in the Arctic (Norway's technology strategy for the petroleum sector (OG21, 2006). According to the United States Geological Survey, moreover, it is assessed that almost thirty percent of the world's undiscovered natural gases are in the Arctic (King, N.D.).

In the exploitation of petroleum resource, development history of offshore industry will be looked over to find which part will receive attention in the future, and it will be analyzed whether the part is well-matched with the characteristics of the Arctic development.

Table 1. Futurity based on history of offshore industry

	1platform / 1well	Subsea production	Multi-well production	Subsea processing	Minimized platform and all subsea
Onshore	1887				
Shallow water	1947	1961	1973	2007	
Deep water				2010	Future
The Arctic and The Subarctic		1963			Future

In 1887, the first onshore well connected to sea was developed by H.L. Williams. The Christmas tree system was invented in 1922 and the first offshore well from a fixed platform was drilled in 1947, which is the start of the modern offshore system (NOIA, N.D.). The world's first subsea well production began in 1961 (UTF, N.D.). Meanwhile, the initial offshore development in ice-covered areas is recorded in 1963 and the regions have been expanded (SAMCoT, 2012; Shell, N.D.). The first multi well subsea template was applied in 1973. The first practical subsea processing system called Tordis, including a separator, a booster, and an injector, was installed in 2007. (Gundersen, et al., 2014) The deepest subsea field was exploited in 2010, which is located in the Gulf of Mexico and of which depth is 2934 m (DNV GL, 2014). Recently, management plan in two shifts, four weeks unmanned and two weeks manned, is being applied to the Valemon gas and condensate field in Norway. Besides, oil majors have started to introduce the concept that all functions of an offshore are intensively focused on subsea system, for instance, Subsea Factories of STATOIL, Subsea to Market of ABB and All Subsea of DNV GL. Based on this development history of offshore industry, it sounds clarified that the futurity would direct unmanned and minimized platforms and further all subsea system.

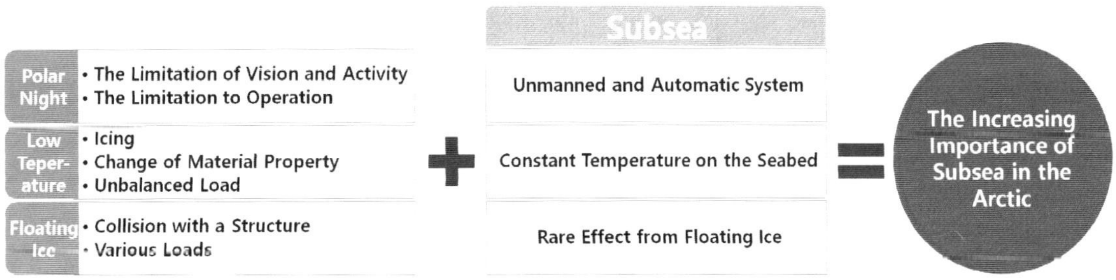

Figure 1. Subsea in the Arctic

Such tendency is expected to ultimately reach up to arctic region because the importance of subsea in the Arctic would increase. The reason for this is that it would be a key to solve three major obstacles in the area, which are polar nights, extremely low temperature, and floating ices (Figure 1). Firstly, it is clear that the polar night blocks vision and activity of workers, thereby interfering with the operation. Secondly, the extremely low temperature makes troubles, such as icing shown in Figure 2, triggering operation problems, the change of material property and unbalanced weight which contributes to capsizing a structure. For instance, icing is considered as one of causes of capsized Koskaya jack-up (Efimov & Kornishin, 2012). Thirdly, the collision with floating ices could damages a structure, but it is still hard to evaluate the exact impact because the physical characteristics of the ices is obscured by reason of the various kinds of the ices depending on development process, shape, and so on. For these matters, the unmanned and automatic system of subsea operations, constant water temperature on seabed, and depth of seabed would respectively help to overcome.

Figure 2. Examples of an icing matter (L: Davies, 2012; R: Efimov & Kornishin, 2012)

Nevertheless, several unsolved challenges still remain. One of them is a matter of interaction between a large floating ice and subsea operations. To explain, a floating ice can cause serious damage to subsea structure. The extent of damage varies depending on a variety of factors such as properties of ice, ice drift, strength of structure, structural protection, and so on. The structural damage can lead to various consequences of subsea operation (e.g., major oil spill, minor oil spill, no oil spill, etc.). This depends on the magnitude of the impact, the overall configuration of subsea facilities, and subsea safety systems. To prevent this difficulty, on the other hand, lots of efforts would be made, for instance, observation, forecast, towing, and destruction, and design criteria would be overestimated, as well. All of these will give rise to huge financial burdens. Hence, it is important to analyze in order to get over the threat and to reduce the financial burden.

Therefore, the main objective of this paper is to identify future direction of research on the interaction between floating ices and subsea operations through literature review. In details, it will be conducted to review literature on structural interaction between floating ice and subsea facility and against floating ice. Lastly, an appropriate research agenda for the future will be proposed.

OVERVIEW OF SUBSEA RISK CAUSED DUE TO FLOATING ICES

For systematic literature review on Arctic issues for floating ices and subsea operation, an overview of subsea risk caused by floating ices is provided in this section, using a bow-tie diagram.

A bow-tie diagram is an illustration that shows the relationship between a hazardous event, its causes and consequences, and the barriers that can reduce the probability of the hazardous event and/or mitigate the consequences (Rausand, 2011). A bow-tie diagram connects the hazards, the hazardous event and the consequences through a series of event lines and depicts the routes to accidents (Rausand, 2011).

A floating ice is one of the various hazards that may lead to the damage of subsea facilities due to external impacts, which is the hazardous event. The protection structure against floating ices is the barrier to reduce the probability of the hazardous event. The other hazards associated with external impact, which are not the scope of this study, are trawling, dropped object, subsea landslide, earthquake, etc. (Brand, 2004; ISO 13628-1, 2006; Gundersen, et al., 2014).

Once the subsea facilities are damaged by an external impact (in other words, after the occurrence of the hazardous event), subsea safety systems prevent or mitigate the consequences. If these barriers fail to respond to the hazardous event, then the worst consequence, large amount of hydrocarbon release, may occur.

The relationship between the hazards, hazardous event, consequences and barriers for subsea risk associated with floating ices is illustrated as a bow-tie diagram in Figure 3.

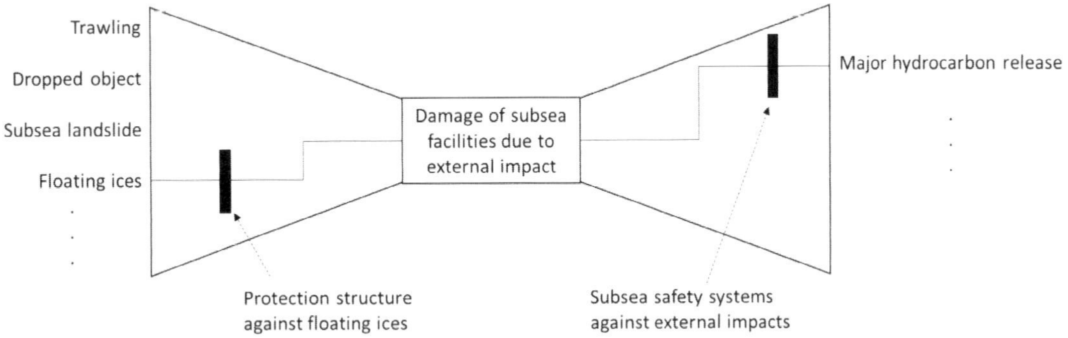

Figure 3. A bow-tie diagram for subsea risks caused due to floating ices

The barrier on the left side of the bow-tie diagram is a proactive barrier that prevents or reduces the probability of the hazardous event, while the barrier on the right side is a reactive barrier that is installed to avoid or reduce the consequences of the hazardous event (Rausand, 2011).

In the bow-tie diagram in Figure 3, protection structure against floating ices is the proactive barrier, and subsea safety systems are the reactive barrier. In other words, the occurrence of the external damage of subsea facilities is relevant to the structural interaction between floating ices and subsea facility, while the mitigation of the consequence is closely related with response of subsea safety systems against external impacts. Based on this result, following literature review will be conducted in two parts: 1) structural interaction between floating ice and subsea facility, and 2) subsea safety systems against external impacts.

LITERATURE ON STRUCTURAL INTERACTION BETWEEN FLOATING ICE AND SUBSEA FACILITY

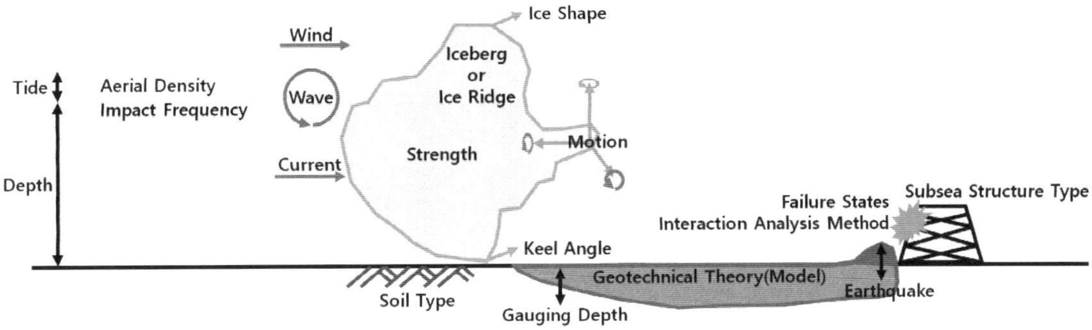

Figure 4. Considerations for interaction between a floating ice and a subsea facility

In the Arctic area, there are three kinds of big floating ice, which are ice ridges, icebergs, and ice islands. While ice ridges originate from accumulated sea ices, others come from glaciers, so they are likely to be harder than ice ridge. Plus, draughts of icebergs are deeper, in general, than keel depths of ice ridges and thicknesses of ice islands. These means that the effect and possibility of iceberg interaction with subsea operations would be prior to those of others. This is the reason why the focus will be on icebergs.

Studies about iceberg collision with GBS (Ground Based Structure) and Floating Structure like FPSO have considerably accumulated. In the case of subsea system, however, some of the knowledge would not be applicable on account of dissimilarities of the systems. Figure 4 introduces expected factors associated with interaction of a floating ice and a subsea structure based on the established research results. These factors would be categorized into two groups relying on whether methodology or tendency are resemble to the accumulated research or not. As an example, it is presumed that areal density, iceberg drift arising from wind, wave, and current, and uncertainty of iceberg shapes would be dealt with by the same way whereas subsea structure types, possible sites relevant with water depth, and interaction mechanism would be differently addressed.

Subsea Structures

A subsea system has been complicatedly developed for recent decades. In the beginning, production system including wellhead, X-mas tree, manifold, pipeline, and riser was necessarily applied. And then, processing system, such as subsea separator, injector and booster (SSIB) and gas compressor, has been added so as to improve effectiveness. All these facilities might be classified into three types regarding iceberg problem; subsea structures on seabed, pipelines on/under seabed, and risers. In spite of a cylindrical shape similar with a GBS, first of all, a riser has totally different structural characteristics owing to its slenderness. Further, differently from other two subsea structure types, its collision is nothing to do with water depth because it reaches from sea surface to seabed. Secondly, it is surmised that pipeline may be involved in the effect of interaction with soils and an iceberg called scouring, as well as direct contact with an iceberg. Lastly, other subsea operations on seabed have to be considered depending on their types, size and functions. This paper will mainly concentrate on this. The Figure 5 describes overall outline of a subsea system.

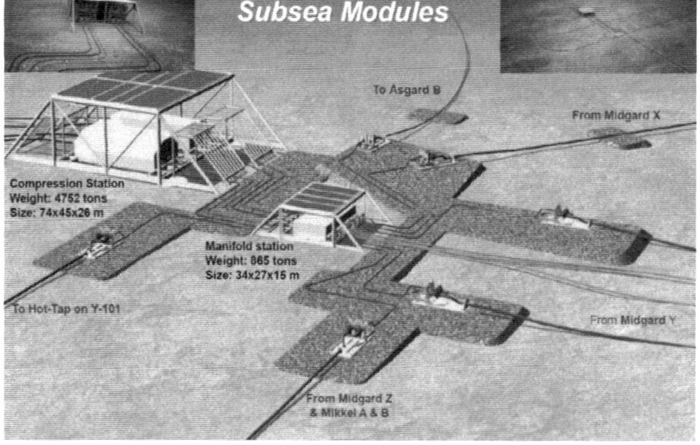

Figure 5. Outline of subsea modules (Vinterstø, N.D.)

As shown in Figure 5, lots of equipment ranging from a small power unit to a huge station template are applied to a subsea system in these days. When it comes to collision with an iceberg and a subsea module, it is plausible that bigger and taller structures would be more risky. The table 2 indicates specification of several subsea installations relatively larger than others.

Table 2. Size and weight of several subsea operations (Aguilera, 2013; Vinterstø, N.D.; Jahanshahi, 2013)

Project	Åsgard	Ormen Lange	Gulfaks	Pazflor
Country	Norway	Norway	Norway	Angola
Type	Compression Station	Compression Station	Compression Station	Separation
Dimension	75×45×26 m^3	60×38×12 m^3	34×20×12 m^3	21×21×19 m^3
Weight	4800 tons	3300 tons	950 tons	900 tons

The biggest type of operations is a compression station mainly targeted at a natural gas field. The dimensions of a compressor station relies on the pipe diameter, the volume of gas, and the differences of gas elevation. (Rover Pipeline, 2016) Hence, it seems convincing that the priority of considerations for structure type would be given to the compression station due to its size and role.

Sites and Areal Density

Several conditions are fundamentally required for sites of this collision scenario. To start with, submarine reservoirs have to be located in the sites. In particular, natural gas fields are more to do with this scenario because of the compression station as mentioned previously. The Figure 6 shows main resource regions in the Arctic circle. (EIA, 2012)

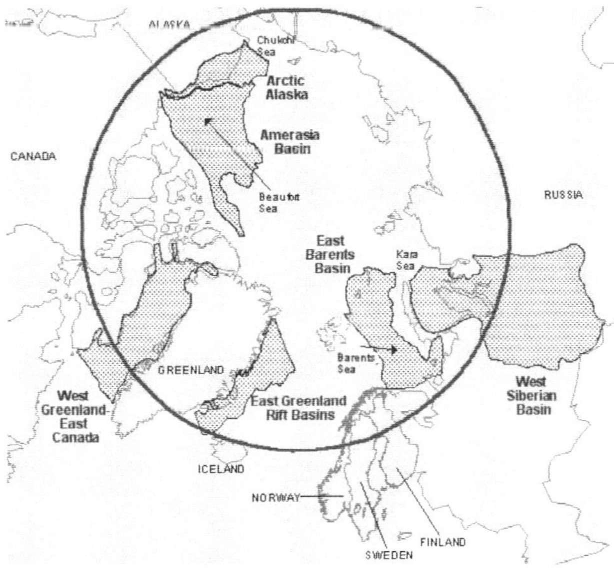

Figure 6. Arctic resource province map

In addition, icebergs ought to appear at least intermittently in the above marked sites, Beaufort Sea, Kara Sea, Barents Sea, Baffin Bay(West Greenland, East Canada), and East Greenland in Figure 6. Concerning these seas, Table 3 presents average annual values of icebergs. (ISO 19906, 2010; Diemand, 2001) In accordance with the Table 3, the frequency of iceberg appearance is comparatively higher for Baffin Bay, Greenland, Western and North-eastern Barents Sea than Beaufort and Pechora sea in Barents. Considering the maximum size of subsea facility (compression station : 26 m) and the maximum iceberg draft (130 m) in Table 3, the most possible sites in the four locations would be inferred as a place with natural gas reservoirs not deeper than the sum of the height and the iceberg draft.

Table 3. Average annual value of icebergs (ISO 19906, 2010)

		Beaufort	Western Barents	North-eastern Barents	Pechora Sea in Barents	Greenland	Baffin Bay
Size	Mass [10^6 ton]	10	~ 6	~ 4	ND	0.5 ~ 1 (Max. 8)	5 ~ 10
	Mean draught [m]	-	-	-	-	60 ~ 80	100
	Maximum draught [m]	-	-	-	-	120 ~ 130	-
Frequency	Months present	Poorly known	1 ~ 6	All year	Infrequent occurrence	All year	12
	Number / year	Poorly known	10 ~ 40	ND	ND		1000
	Maximum number / month	Rare	30	ND	ND		Numerous

Concerning West Greenland waters, according to a research article (Chumakov, 2011), it has been known that there was no major discovery for petroleum resource. On the other hand, a twenty-year plan for offshore development around Svalbard was declared in 2011 (Gibbs & Koranyi, 2011), and seven test drillings near Svalbard were arranged and financial approval was also decided by Norwegian Parliament. (Leiendecker, 2015) Although economical reservoir has not been found yet, it seemed highly positive for the potential to be revealed soon. The rationale behind this is that, as shown in Figure 7 (Fadeyev, 2012), it is obvious that gas condensate, oil and even natural gas are widely exist in the Barents and Kara region.

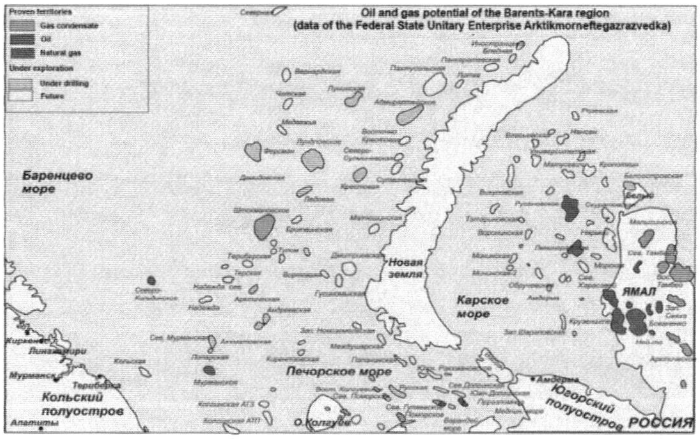

Figure 7. Oil and gas potential of the Barents-Kara region

For these reasons, the sites are selected by combining Arctic resource province map (Figure 6) and Bathymetry of the Barents and Kara (Foraminiferal Research at Byrd Polar Research Center, N.D.) based on boundaries and regions of the Barents sea seen in the Table 3 (ISO19906, 2010). To explain, the circled spots in Figure 8 are shallower than 100 m. Concurrently, they are included not only in the shadow areas in Figure 6 but also in the region that icebergs quite frequently appear shown in Table 3.

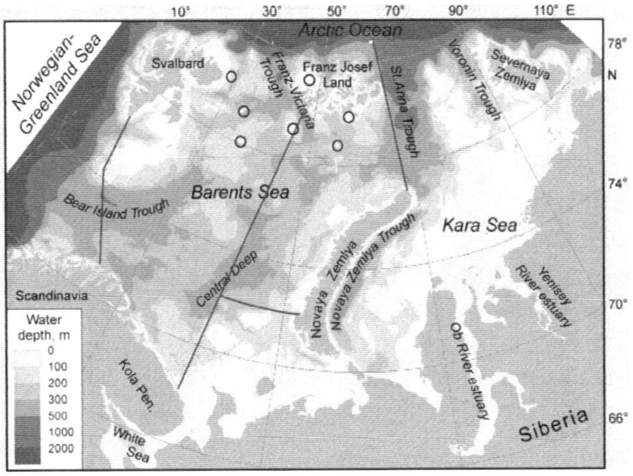

Figure 8. Expected sites for interaction between an iceberg and a subsea operation

It is evident that all environmental factors shown in Figure 4 would be actually involved in sites and have their own uncertainties. Moreover, the iceberg collision is not thought to be prevalent. Hence, it would be more rational that the factors are treated as probabilistic parameters, if possible. The representative stochastic parameter depending on the site is encounter frequency, save for impetuses of iceberg movement, for example wave. It is the rate of interactions per unit time and calculated from areal density which is the average annual number of features per unit area. (ISO 19906, 2010) The history of major research tendency on this topic is well-summarized on the whole in the most recent study. (Habib, et al., 2016) According to the summary, since IIP(International Ice Patrol) started to observe icebergs, the IIP data has been a basis for estimation of areal density. Deterioration and drift models were employed to anticipate more exact areal density. (Anderson, 1971) The observation of the areal density has been improved by using more various ways for iceberg detection to collect more data. Extensive dataset since 1960 by 1992 were added to the IIP data in order to study the characteristics of iceberg frequency in the degree square (Fuglem, et al. 1996, 1996a). Comprehensive database for icebergs has been provided by the PERD(Program on Energy Research and Development) since 1998. The database consists of the data mostly from the IIP and Province Airlines, and additionally from ships and satellites. (Trott & Comfort, 2007) In order to utilize data from the IIP and the Envisat ASAR satellite at the same time, a formula to mutually compensate for upside and downside of the each was proposed. (Habib, et al., 2015) To be specific, the IIP data could be more trustful in certain fields but the coverage would be restricted owing to the limited number of observing flights whereas the satellite data could cover a target range completely, but not thoroughly due to the limitation of imagery resolution. Lastly, Habib et al. also compared the results of areal densities calculated from aerial reconnaissance data and from bi-weekly iceberg charts. The study represented that the areal densities from different data sources in the same regions could be unmatched for each other, therefore ultimately leading to exaggerate or belittle the iceberg impact load.

Iceberg Drift

Iceberg drift is one of the most important things for this issue. This is because it plays vital roles in evaluation for iceberg design load as well as in risk assessment of iceberg and iceberg management which, in turn, could become probabilistic factors of the design load evaluation. Iceberg drift model includes trajectories of iceberg movements and drift speeds. Hence, the iceberg drift model fairly depends on weather forecasts such as winds, currents, and waves, and the simulation of the iceberg drift normally regards an iceberg as a point mass (SAMCoT, 2012).

In order to calculate iceberg impact, speed of an iceberg is essential because the kinematic energy, deemed as the governing factor of the limit energy mechanism which is usually applied to analysis of a huge floating ice, originates from the velocity of the iceberg. Basically, the iceberg speed would arise from the interaction between the iceberg and fluids like air and water. Such a fluid-structure interaction is fundamentally affected by a projection area and speed of the flows like current, wind, and wave. On this point, the drift speed model accepted in general nowadays has been developed as a function of significant wave height and waterline length of the iceberg. (Fuglem, 1997; Stuckey, 2008)

First of all, waterline length is the representative parameter of the projection area, which is the longest width of the projection area of an iceberg and easiest to be measured. In detail, iceberg draft could be deduced from a waterline length with a power law relation (Stuckey et al., 2016(a)), and the projection area, in turn, could be calculated. When it comes to the velocity of flows, on the other hand, the significant wave height plays a key role because of the well-known relationships between wind speed and significant wave height and between wind speed and current speed from Ekman theory. The relationships are based on the fact that wave and current are triggered by wind in nature.

So as to infer the iceberg velocity, a basic assumption is that the iceberg is supposed to show relatively steady movement, so considerations involved in acceleration and inertia would be ignored. This means that all external environment forces would be kept in equivalence one another. Therefore, consequent velocity of iceberg drift would be decided by the total environment load.

Based on this, probabilistic drift model was proposed (Stucky, 2008). Sensitivity analysis was, firstly, performed for iceberg interaction with FPSO and GBS. Then, lots of data containing waterline lengths and locations of icebergs with significant wave heights were analyzed on the Grand Bank in Canada. Finally, the result was compared with the deterministic result.

Prevention

Generally, the definition of ice management is all actions to diminish or avoid any types of ice features (SAMCoT, 2012). Because of immovable subsea structures including pipelines and risers, the only way for avoidance would be to get rid of an approaching iceberg before impact. The conventional method starts from forecasting approaches of the potentially risky icebergs based on the information about traveling icebergs detected by IIP stated previously. If an iceberg comes in the immediate vicinity of an offshore, a tug vessel lets it towed away. There are several towing methods in history; predominantly single vessel rope, prop washing, net, water cannon, two vessel tow, ramming, and others arranged in order of the number of towing operations (Young & Rudkin, 2006). The maximum speed of an iceberg is nearly 2.7 kilometer per an hour (SAMCoT, 2012). Additionally, it may take ten hours for the towing vessel to reach a speed of just one knot during towing owing to the massive weight of the iceberg, so the towing can need up to three days (Kaushik, 2014). For these reasons, the task would have to complete within a time not so long.

Between the forecast and the removal action, a supplementary step may be needed to calculate accurate probability of iceberg collision with subsea modules. The precise prediction would help to improve cost-effectiveness and operation efficiency since it does not necessarily mean colliding each other just because an adjacent iceberg approaches to subsea operations or floats above them, due to the fact that the iceberg keel might not be reachable up to structures on seabed. However, it is not easy at all to make out the submerged part of icebergs by reason of the invisibility of the underwater and the randomness of iceberg shapes. The 2012 Iceberg Profiling Program can give a direction for this. Twenty nine icebergs were measured, using optical cameras, laser finder, and GPS for parts above water and multi-beam sonar installed in ROV for submerged parts, along the coastal line of Newfoundland and Labrador. (Younan, et al., 2016) This database contributed to upgrade various related research, such as iceberg impact modelling, iceberg towing, iceberg hydrodynamics, iceberg risk, iceberg design load. (Fuglem & Younan, 2016; Bruce, et al., 2016; Talimi, et al., 2016; King., et al., 2016; Stuckey, et al., 2016(b))

Compared to the surveyed field, however, the selected regions in the Barents sea are much closer to the origin of icebergs as seen in Figure 10. Thus, more immediate system, for example, on-the-spot iceberg observation and profiling sonar system on seabed, may be needed in the future to develop offshore and subsea system in the Arctic region to prevent the crash of icebergs.

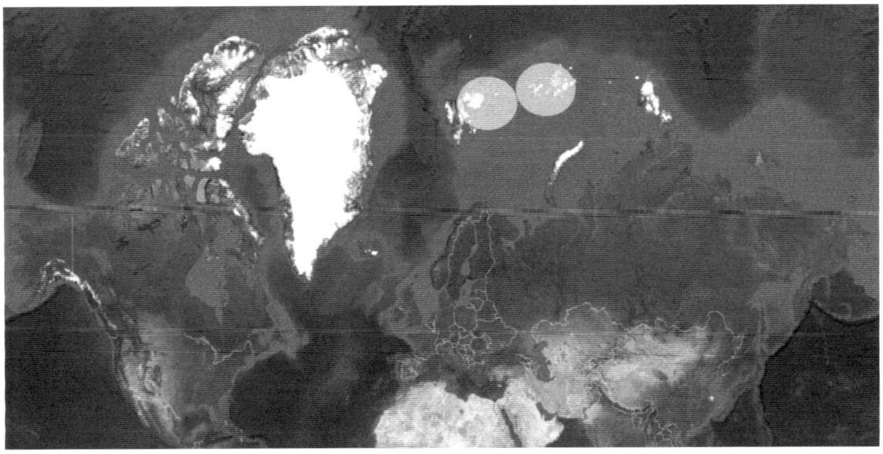

Figure 10. The selected sites (yellow mark) and the surveyed field of the 2012 Iceberg Profiling Program (red mark)

LITERATURES ON SUBSEA SYSTEMS AGAINST EXTERNAL IMPACTS

The first and most important barrier that is installed to stop the unwanted hydrocarbon release is the downhole safety valve (DHSV) (Torbergsen, et al., 2012; Kim, et al. 2016). The DHSV closes automatically when the hydraulic pressure is lost due to a command from the topside emergency shut down (ESD) system, or due to an abnormal situation, like rupture of hydraulic supply lines (Bai & Bai, 2012; Kim, et al. 2016).

The second pressure barrier to control the flow of hydrocarbons is the production master valve (PMV) and production wing valve (PWV) installed in the subsea X-mas tree (Bai & Bai, 2012; Kim, et al. 2016). The PMV and PWV closes automatically when the hydraulic pressure is lost, and the loss of hydraulic pressure can be initiated by a command of the topside ESD, production shutdown (PSD) system and subsea PSD system.

A brief overview of subsea facilities and the locations of DHSV, PMV, and PWV are illustrated in Figure 9. If a large floating ice damages subsea facilities and/or pipelines, then DHSV, PMW, and PWV can shut down the hydrocarbon flow to prevent unwanted hydrocarbon release to the environment.

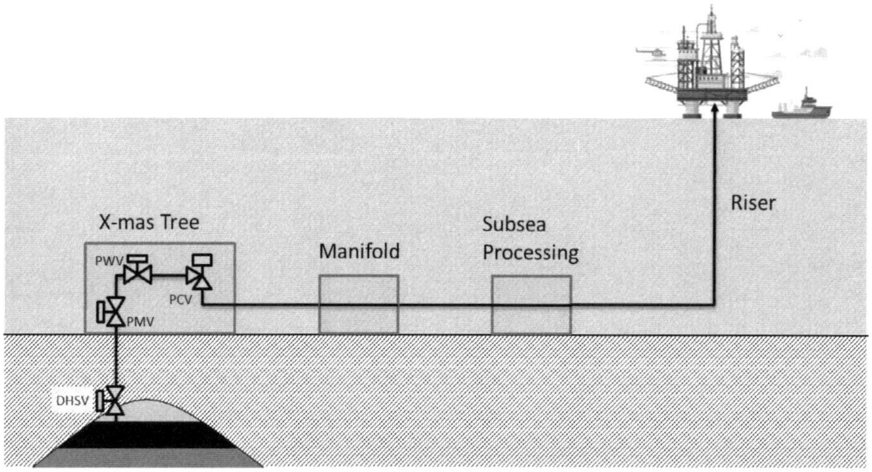

Figure 9. DHSV, PMW, and PWV

These subsea safety systems are installed not only for floating ices, but also for all the other kinds of subsea hazardous events, such as hydrocarbon release due to critical loads from trawling, dropped objects, material defect, erosion/corrosion, rupture of riser, etc. Therefore, there have been a high number of studies and efforts on theses subsea safety systems, including standards and regulations such as ISO 13628-6 (2006), NORSOK S-001 (2008), NOG GL 070 (2001), API RP 17A (2010), The Facilities Regulations (PSA, 2015), etc. For instance, NORSOK S-001 (2008) suggests ESD shut down principle hierarchy as shown in Figure 10, and the reliability analysis of ESD systems have been studied by Farwana (2014), Hauge, et al. (2010), Byrne (1994), Lundteigen and Rausand (2007), Hauge et al. (2015), and by various reliability researchers. In addition, NOG GL 070 (2001) provides guidelines for quantitative reliability analysis of subsea safety systems, based on IEC61508 (2010) and IEC61511 (2003).

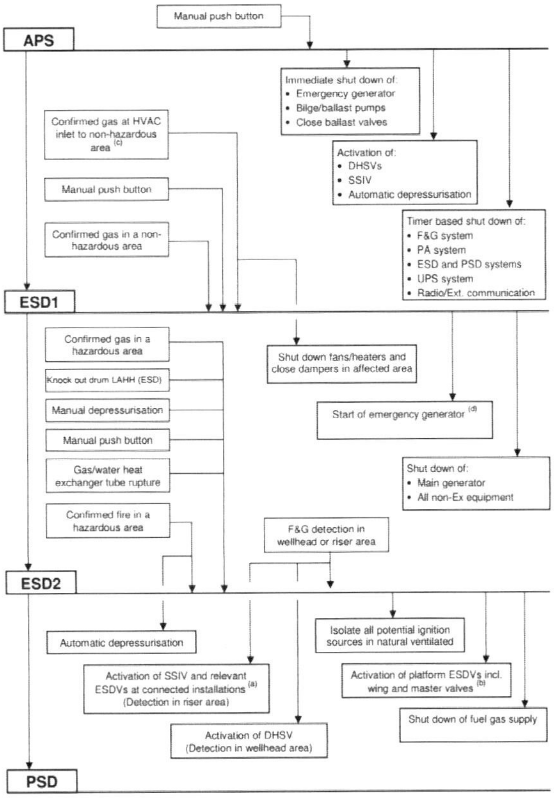

Figure 10. Emergency shut down (ESD) principle hierarchy (NORSOK S-001, 2008)

However, there are limited studies on specific consideration of subsea safety systems against floating ices. While several studies discuss the challenges and solutions of Arctic subsea operation (Lanan, et al., 2001; Aggarwal & D'Souza 2011, El-Wardani 2013), interaction between subsea safety systems and floating ices is still an unexplored field. In addition, the overall risk model associated with floating ices, including both subsea structure and safety systems, is not well investigated yet.

CONCLUSIONS

In this paper, a wide range of recent Arctic issues related to interaction between an iceberg and a subsea structure are briefly introduced. Firstly, it was demonstrated that a role of subsea system in the Arctic would increase more and more. Thus, subsea risks from floating ices were arranged by a Bow-tie diagram. Secondly, factors linked to the interaction between subsea and an iceberg are arrayed based on the existing research. Thirdly, a reasonable structure type on seabed in iceberg impact was clarified, and possible sites are specified based on the type. Finally, assessment methodology of areal density and of iceberg drift were simply summarized respectively, and safety systems applied to subsea operation against external impacts were summed up, as well. As a result, this paper would contribute to identify research agenda and develop future studies on interaction with a floating ice and subsea operations in that this contents could be a base of such relevant issues because there has seldom been research exactly fitted with this topic.

In the future, other possible sites could be searched broadly regarding similar problems, such as ice ridges in Sakhalin Sea and icebergs in Baffin Bay. Furthermore, effect of variation of keel angle attributed to rotation of a floating ice like rolling would be identified as a key point. Furthermore, geotechnical mechanism between soils and pipelines under iceberg impact, if possible including earthquake, could be an interesting problem. Also, it may be needed to research thoroughly if it is right or not to employ the existing design load estimation to a subsea structure due to the fact that the specification of the subsea facility is completely different from a huge GBS and/or FPSO. Lastly, all these further research would be connected with risk assessment and development of subsea protection and/or prevention system against iceberg impact

ACKNOWLEDGEMENTS

Research funds supported by the Norwegian Centre for International Cooperation in Education (High North Programme, Project No.HNPla-2016/10059) and also supported by Ministry of Trade, Industry and Energy of Korea (Project No.10063417) are greatly acknowledged.

REFERENCES

Aggarwal, R. & D'Souza, R., 2011. Deepwater Arctic-Technical Challenges and Solutions, Proceedings of Arctic Technology Conference (2011), 7-9 Feb. 2011, Houston, Texas, USA, Offshore Technology Conference, OTC-22155-MS.

Aguilera, L.C.P., 2013. Subsea Wet Gas Compressor Dynamics, Norwegian University of Science and Technology, Trondheim, Norway.

Anderson, C., 1971. The Flow of Icebergs along the Canadian East Coast, Proceedings of the Canadian Seminar on Icebergs, 6-7 Dec. 1971, Halifax, Nova Scotia, Canada. Quoted in Habib, et al., 2016.

API RP 17A, 2010. Design and Operation of Subsea Production Systems-General Requirements and Recommendations, American Petroleum Institute, USA.

Baek, A., lee, S. & Joh, C.H., 2016. The possibility and Analysis for Northern Sea Route, Journal of Climate Research, 11(2), pp.121-130, KU Climate Research Institute.

Bai, Y. & Bai, Q., 2012. Subsea Engineering Handbook, Gulf Professional Publishing.

Brand, H., 2004. Risk Comparison Subsea vs. Surface Processing, Houston, USA, Det Norske Veritas (USA) Inc.

Bruce, J., Younan, A., & Macneil, A., 2016. Applications of Iceberg Profiling Data to Improve Iceberg management Success, Proceedings of Arctic Technology Conference (2016), 24-26 Oct. 2016, St. Jonhn's, Newfounland and Labrador, Canada, Offshore Technology Conference, OTC-27470-MS.

Byrne, S., 1994. Subsea Well Control Systems The Specification of Reliability, Availability and Maintainability, Subsea Control and Data Acquisition: Proceedings of an international conference, 20-21 Mar. 1994, London, UK, Society of Underwater Technology, SUT-AUTOE-v32-187.

Chumakov, D.S., 2011. Main Vectors of International Cooperation in the Arctic, International Relations and World Politics, 25(2), pp.41-61, Bulletin of Moskow State University, Quoted in ФАДЕЕВ, А.М. & ЛАРИЧКИН, Ф.Д., 2012.

Davis, G., 2012. The High North Visions and Strategy, Meld. St. 7, Norwegian Ministry of Foreign Affairs, Norway.

Diemand, D., 2001. Icebergs, pp.1255-1264, 2001 Academic Press.

DNV GL, 2014. Introduction to Subsea Production Systems, [Online] (Updated 28 Aug. 2014) Available at: http://www.uio.no/studier/emner/matnat/math/MEK4450/h14/undervisningsmateriale/module-2/mek4450-dnvgl-02-what-is-subsea.pdf [Accessed 3 Jan.2017].

Efimov, Y. & Kornishin, K., 2012. Vessel Icing on the Shtokman FPSO, Proceedings of Arctic Technology Conference (2012), 3-5 Dec. 2012, Houston, Texas, USA, Offshore Technology Conference, OTC-23718.

EIA(U.S. Energy Information Administration), 2012. Arctic Oil and Natural Gas Resources, Today in Energy, [Online] (Updated 20 Jan. 2012) Available at: https://www.eia.gov/todayinenergy/detail.php?id=4650 [Accessed 20 Jan. 2017].

El-Wardani, R., 2013. Challenges and Solution in Subsea Field Development for the High North and Arctic, University of Stavanger, Stavanger, Norway.

Fadeyev, A., 2012. International Cooperation in the Exploration of the Arctic, RIAC (Russian International Affairs Council), [Online] (Updated 7 Aug. 2012) Available at: http://russiancouncil.ru/en/inner/?id_4=682#2 [Accessed 8 Feb. 2017].

Farwana, A.O., 2008. Managing Well Integrity using Reliability Based Models, Imperial College London, London, UK.

Foraminiferal Research at Byrd Polar Research Center (Digital source: GEBCO), N.D. Bathymetry of the Barents and Kara, [Online] Available at: http://research.bpcrc.osu.edu/foram/maps.htm [Accessed 1 Jan. 2016].

Fuglem, M., 1997. Dicision-Making for Offshore Resource Development, Memorial University of Newfoundland, St. Jonhn's, Newfounland and Labrador, Canada.

Fuglem, M. & Younan, A., 2016. A 3D Time-Domain Model for Iceberg Impacts with Offshore Platforms and Subsea Equipment, Proceedings of Arctic Technology Conference (2016), 24-26 Oct. 2016, St. Jonhn's, Newfounland and Labrador, Canada, Offshore Technology Conference, OTC-27359-MS.

Fuglem, M., Jordaan, I., & Coaker, G., 1996(a). Iceberg-Structure Interaction Probabilities for Design, Canadian Journal of Civil Engineering, 23(1), pp.231-241, Canadian Journal of Civil Engineering.

Fuglem, M., Jordaan, I., Coaker, G., Cammaert, C., & Berry, B., 1996(b). Environmental Factors in Iceberg Collision Risks for Floating Systems, Cold Regions Science and Technology, 24(3), pp.251-261, Elsevier.

Gibbs, W. & Koranyi, B., 2011. Norway Mobilises for Oil Push into Arctic, Reuter, (Ed. by) William Hardy [Online] (Updated 18 Nov. 2011) Available at: http://www.reuters.com/article/norway-highnorth-IdUSL5E7MI1GK20111118 [Accessed 10 Nov. 2016].

Gundersen, S., Lønvik, K., Søgård, B., Markussen, C., & Wallin, P., 2013. Subsea Facilities-Technology Developments, Incidents and Future Trends, Report No. 2014-0113, Rev. 03, Ice-Structure Interaction Activity of the Program of Energy Research and Development(PERD), Project No. PP088320, DNV GL, 14 Mar. 2014, Norway.

Habib, K., Hicks, M., Stuckey, P., & King, T., 2016. A Revised Basis for Iceberg Areal Density Values for Risk Analysis, Proceedings of the Arctic Technology Conference, 24-26 Oct. 2016, St. Jonhn's, Newfounland and Labrador, Canada, Offshore Technology Conference, OTC-27369-MS.

Habib, K.B., Cuff, A., & King, T., 2015. Analysis of Iceberg Frequency in Labrador Sea Using Aerial Reconnaissance Flight Surveys and Satellite Radar Data, Proceedings of the ASME 2015 34th International Conference on Ocean, Offshore and the Arctic Engineering(OMAE 2015), 31 May - 5 Jun. 2015, St. Jonhn's, Newfounland and Labrador, Canada, ASME.

Hauge, S., Hoem, Å.S., Hokstad, P., Håbrekke, S., & Lundteigen, M.A., 2015. Common Cause Failures in Safety Instrumented Systems, Report No. SINTEF A26922, Project No. 102001186, 20 May 2015, Norway.

Hauge, S., Lundteigen, M.A., Hokstad, P.R., & Håbrekke, S., 2010. Reliability Prediction Method for Safety Instrumented Systems - PDS Method Handbook, 2010 Edition, Report No. SINTEF report STF50 A, 6031, SINTEF Technology and Society, Norway.

IEC61508, 2010. Functional Safety of Electrical/Electronic/Programmable Electronic Safety-Related Systems, (Eds.), 2, (Published) 30 Apr. 2010, Geneva, International Electrotechnical Commission (IEC).

IEC61511, 2003. Functional Safety-Safety Instrumented Systems for the Process Industry Sector, Eds. 1, (Published) 30 Jan. 2003, Geneva, International Electrotechnical Commission (IEC).

ISO 13628-1, 2006. Petroleum and natural gas industries - Design and operation of subsea production systems - Part 1: General requirements and recommendations, International Organization for Standardization.

ISO 13628-6, 2006. Petroleum and natural gas industries - Design and operation of subsea production systems - Part 6: Subsea production control systems, International Organization for Standardization.

ISO 19906, 2010. Petroleum and Natural Gas Industries-Arctic Offshore Structures, International Organization for Standardization.

ISO 19906, 2010. Petroleum and natural gas industries – Arctic offshore structures, International Organization for Standardization.

Jahanshahi, E., 2013. Challenges in Design, Operation and Control of Subsea Separation Processes, Public Trail Lecture of Norwegian University of Science and Technology(NTNU), 12 Oct. 2013, Trondheim, Norway.

Kaushik, 2014. Towing Icebergs Away From Oil Platforms, [Online] (Updated 2 May 2014) Available at: http://www.amusingplanet.com/2014/05/towing-icebergs-away-from-oil-platforms.html#comment-form [Accessed 27 Feb.2017].

Kim, H., Lundteigen, M.A., & Holden, C., 2016. A Gap Analysis for Subsea Control and Safety Philosophies on the Norwegian Continental Shelf, Proceedings of 13th International Conference on Probabilistic Safety Assessment and Management(PSAM 13), 2-7 Oct. 2016, Seoul, South Korea, PSAM13.

King, H., N.D. Oil and Natural Gas Resources of the Arctic, Geology.com, [Online] Available at: http://geology.com/articles/arctic-oil-and-gas [Accessed 2 Feb. 2017].

King, T., Younan, A., Richard, M., Bruce, J., Fuglem, M., & Phillips, R., 2016. Subsea Risk Update Using High Resolution Iceberg Profiles, Proceedings of Arctic Technology Conference (2016), 24-26 Oct. 2016, St. Jonhn's, Newfounland and Labrador, Canada, Offshore Technology Conference, OTC-27358-MS.

Leiendecker, J., 2015. Oil and gas from the Arctic? Test Drillings northeast of Svalbard, [Online] (Updated 1 Nov. 2015) Available at: https://www.spitsbergen-svalbard.com/2015/11/01/oil-and-gas-from-the-arctic-test-drillings-northeast-of-svalbard.html [Accessed 1 Jan. 2016].

Lenan, G.A., Ennis, J.O., Egger, P.S., & Yockey, K.E., 2001. Northstar Offshore Arctic Pipeline Design and Construction, Proceedings of Offshore Technology Conference (2001), 30 Apr. - 3 May 2001, Houston, Texas, USA, Offshore Technology Conference, OTC-13133-MS.

Lundteigen, M.A., & Rausand, M., 2007. The effect of partial stroke testing on the reliability of safety valves, Proceedings of ESREL '07, 25-27 Jun. 2007, Stavanger, Norway, European Safety and Reliability Associaion (ESRA), pp.1-15.

NOG GL 070, 2001. Norwegian oil and gas application of IEC 61508 and IEC 61511 in the Norwegian petroleum industry, Norsk Olje & Gas, Norway.

NOIA(America's Offshore Energy Industry), 2006. History of Offshore, [Online] (Updated Sep. 2016) Available at: http://www.noia.org/history-of-offshore [Accessed 18 Dec. 2016].

NORSOK S-001, 2008. Technical Safety, Standards, Norway.

OG21(Norway's Technology Strategy for the Petroleum Sector), 2006. Technology Strategy for The Arctic.

PSA, 2015. Regulations relating to design and outfitting of facilities, etc. in the petroleum activities (The Facilities Regulations), Petroleum Safety Authority, Norway.

Rausand, M., 2011. Risk Assessment; Theory, Methods, and Applications, Hoboken, NJ, Wiley.

Rover Pipeline, 2016. Compressor Stations, [Online] (Updated 18 Feb. 2016) Available at: http://www.roverpipelinefacts.com/documents/02172016/Rover_Compressor_Stations_0218 2016.pdf [Accessed 23 Feb. 2017].

SAMCoT(Sustainable Arctic Marine and Coastal Technology), 2012. SAMCoT Annual Report 2012, Norwegian University of Science and Technology, Norway.

Shell, N.D. Shell Offshore Alaska, [Online] Available at: http://www.shell.us/content/dam/shell/static/u sa/downloads/alaska/alaska-022510.pdf [Accessed 23 Feb. 2017].

Stuckey, P., Younan, A., Burton, R., & Alawneh, S., 2016(a). Modelling Iceberg-Topsides Impacts Using High Resolution Iceberg Profiles, Proceedings of Arctic Technology Conference (2016), 24-26 Oct. 2016, St. Jonhn's, Newfounland and Labrador, Canada, Offshore Technology Conference, OTC-27429-MS.

Stuckey, P., Younan, A., Parr, G., & Fuglem, M., 2016(b). Updating the Iceberg Design Load Software Using High Resolution Iceberg Profiles, Proceedings of Arctic Technology Conference (2016), 24-26 Oct. 2016, St. Jonhn's, Newfounland and Labrador, Canada, Offshore Technology Conference, OTC-27492-MS.

Stuckey, P.D., 2008. Drift Speed Distribution of Icebergs on the Grand Banks and Influence on Design Loads, Memorial University of Newfoundland, St. Jonhn's, Newfounland and Labrador, Canada.

Talimi, V., Ni, S., Qiu, W., Fuglem, M., Macneil, A., & Younan, A., 2016. Investigation of Iceberg Hydrodynamics, Proceedings of Arctic Technology Conference (2016), 24-26 Oct. 2016, St. Jonhn's, Newfounland and Labrador, Canada, Offshore Technology Conference, OTC-27493-MS.

Torbergsen, H.B., Haga, H.B., Sangesland, S., Aadnøy, B.S., Sæby, J., Johnsen, S., Rausand, M., & Lundeteigen, M.A., 2012. An Introduction to Well Integrity, Norsk Olje & Gas, 4 Dec. 2012, Norway.

Trott, B., & Comfort., G., 2007. PERD Iceberg Sighting Database Update & Quality Assurance: 2006~2007, PERD/CHC Report 20-86-6143C.DFR, Ice-Structure Interaction Activity of the Program of Energy Research and Development(PERD), BMT Fleet Technology Limited, 30 Mar. 2007, pp.19-41, Canada.

UTF(Underwater Technology Foundation), N.D. Subsea History, [Online] Available at: http://www.utc.no/utf/om_subsea/subsea_history [Accessed 3 Jan. 2017].

Vineterstø, T., N.D. Åsgard Subsea Gas Compression-STATOIL, [Online] Available at: http://hniforum.no/cmsAdmin/uploads/statoil.pdf [Accessed 2 Feb. 2017].

Younan, A., Ralph, F., Ralph, T., & Bruce, J., 2016. Overview of the 2012 Iceberg Profiling Program, Proceedings of Arctic Technology Conference (2016), 24-26 Oct. 2016, St. Jonhn's, Newfounland and Labrador, Canada, Offshore Technology Conference, OTC-27469-MS.

Young, C.J.L. & Rudkin, P., 2006. Statistical Analysis of the Optimum Amount of Bollard Pull Required for Towing and Iceberg, Proceedings of the Seventh International Conference and Exhibition on Performance of Ships and Structures in Ice 2006 (ICETECH 2006), Jul. 2006, Banff, Alberta, Canada, ICETECH, ICETECH06-121-RF.

ФАДЕЕВ, А.М. & ЛАРИЧКИН, Ф.Д., 2012. Effective International Cooperation in Industrial Supplying of the Arctic Shelf Projects, Proceedings of 20th International Scientific Conference CO-MAT-TECH 2012, 11-12 Oct. 2012, Trvana, Trvana, Slovakia, 20th International Scientific Conference CO-MAT-TECH 2012, [Online] Available at: http://www.scss.sk/cd_ap vv_ lpp.../Fadejev,%20Larichkin.pdf.

POAC'17
Busan, KOREA

Proceedings of the 24th International Conference on
Port and Ocean Engineering under Arctic Conditions
June 11-16, 2017, Busan, Korea

The Use of Modern Communications Technology to Ensure the Safe Navigability of the River Estuaries of the Arctic Basin

V. V. Karetnikov[1], S.V. Rudy'kh[1], A. A. Sikarev[1]
[1] Admiral Makarov State University of Maritime and Inland Shipping, Saint-Petersburg, Russia

ABSTRACT

The constructive decisions allowing to count range of an area of coverage of the modern telecommunication systems are considered in the article. These systems functioning in VHF range are intended for safety of navigation in coastal zone and estuarial areas of rivers of the Arctic basin. The specified decisions allow to find solutions for the VHF range taking into account the actual parameters of transceiver equipment. Moreover influence of the blocking land relief, and also relocation of the ship transponder relative to the base station of Automatic identification system (AIS) are possible to take into consideration. The results of coverage calculations of such systems for the estuarial area of river of the Arctic basin are given in the final part.

KEY WORDS: Estuarial area of river of the Arctic basin; Automatic identification system (AIS); Modern telecommunication technologies; blocking relief; Relative movement of the ship transponder.

FEATURES OF PROVIDING OF SAFE NAVIGABLE CONDITIONS ON ESTUARIAL AREAS OF RIVERS OF THE ARCTIC BASIN

Creating and maintaining of safe navigational conditions in the estuarial areas of the Arctic rivers of the Russian Federation to ensure the timely transportation of cargoes on the rivers of the Arctic basin with access to the Northern Sea Route (Pastusiak, 2016) is a very important and urgent task, because the transport infrastructure in the region is still undeveloped and water transport remains uncontested. This problem is particularly acute in the conditions of intensification of oil and gas mining in the region of the West Siberian Plain and the Arctic shelf, as well as increasing of exploitation of mineral resources of the Republic of Sakha (Yakutia).

The main problem to ensure the navigable depth on the estuarial areas is shallow water on estuarial bars. For maintenance of navigable depths on bar areas it is required to produce annual dredging works, volume of which is estimated in millions of cubic meters of soil. This significant amount of dredging must be produced in a short time, because for a short period the Arctic navigation is required not only to prepare safe navigable waterway, but also provide a significant amount of cargo transportation.

The systematic underfunding of dredging works and the lack of sufficient replenishment of vessels of technical fleet which are decommissioned because of physical and mental deterioration has led to the fact that the depth on estuarial bars decreased significantly, almost reaching their natural values, currently not allowing the vessels of "river-sea" type to pass freely. As a result, ship owners have to bear substantial losses because of downtime of vessels waiting at anchor close to shore line or rising of water levels from the effect of wind surges.

Modernization and acquisition of additional dredging technique, implantation of automated systems for monitoring and control of the technical and auxiliary fleet (ASC TAF) and monitoring of navigation means (SM NM) (Park, et al., 2009) are required in such circumstances to recover the lost dimensions of the ship moves on estuarial areas for the passage of large vessels and qualitatively new navigation and hydrographic support of navigation. Such automated systems based on the use of global navigation systems (GNSS) GLONASS / GPS (Bruyninx, 2006 and Soler, 2004) and modern transceiver systems will allow the technical and auxiliary fleet to carry out their tasks more efficiently and expeditiously. It is necessary to equip the technical fleet by the modern navigation and communication equipment, including electronic chart display and information system (ECDIS) with the possibility of imposing the radar picture on the ECDIS. Under certain conditions it allows to perform track work in conditions of reduced visibility. Equipment of floating beacons and land marks by monitoring system of navigation means in combination with implementation of automated systems of statement of floating navigation beacons will give the chance to reduce operating costs on the fleet. The implementation of the differential mode of GNSS GLONASS, AIS (The British Standards Institution, 2017) and communication systems, modernization of communication systems will accurately track and manage the vessels.

Naturally, the harmonious functioning of these areas of improvement of monitoring and increase of work efficiency of technical and auxiliary fleet is only possible in the structural framework of such ergatic system as ASC TAF which complements the well-known hierarchical infocommunication triad "Corporate river information system (CRIS) - River Information Services (RIS) – vessel traffic service (VTS)".

At the same time it is necessary to consider that before making decision on quantity and location of the base stations (BS) of AIS the carrying out serious scientific and engineering researches is necessary for determination of their topology and range of their action, taking into account architecture and methods of creation of the modern automated networks of safety and communication.

The task of optimization of zones and radius of action of BS AIS defining, eventually, architectonic, and also technological features of their construction is very important for development of topological structure of AIS.

Already created general approach to the solution of task of optimization is based on use of stochastic models of information channels considering the probabilistic nature of three main groups of the facts influencing fidelity of messages transmission.

DISTRIBUTION OF VHF RADIO WAVES TAKING INTO ACCOUNT PARAMETERS OF TRANSCEIVER TRACTS OF APPARATUS

The signals of VHF range used in AIS (Nelson, et al., 2011) can be generally presented as amplitude coefficient of transfer of information channel μ, which is Gaussian independent random variable with mean values m_c and m_s, and also dispersions σ_c^2 and σ_s^2 of orthogonal components.

The one-dimensional density of distribution of transfer coefficient $\mu(t)$, depending on these four parameters, had received the name of four-parameter distribution

$$W(\mu)=\frac{\mu}{\sigma_c\sigma_s}\exp\left(\frac{\mu^2-m_s^2}{2\sigma_s}-\frac{m_c^2}{2\sigma_c^2}\right)\sum_{k=0}^{\infty}\frac{H_{2k}^{(\alpha)}}{(2k)!!2^k}\mu^k\left(\frac{\sigma_s}{m_s}\right)^k\left(\frac{1}{\sigma_c^2}-\frac{1}{\sigma_s^2}\right)^k I_k\left(\frac{m_s}{\sigma_s^2}\right), \quad (1)$$

where

$$\sigma_c^2\le\sigma_s^2;\ \frac{\sigma_c^2}{\sigma_s^2}<1;\quad \alpha=\frac{m_s}{\sqrt{2\sigma_s^2}\sqrt{\frac{1}{\sigma_c^2}-\frac{1}{\sigma_s^2}}}, \quad (2)$$

$H_{2k}(.)$ – Hermite polynomials of degree $2k$ (Society of Photo-Optical Instrumentation Engineers (SPIE), 2011).

$I_k(.)$ – modified Bessel function of the first kind of order k (Balogh, 1966).

At certain phasing of regular component μ_p of signal the following takes place

$m_s=0;\ \mu_p=m_c$ (3)

Then three-parameter Beckmann distribution follows from (1):

$$W(\gamma)=\frac{\mu}{\sigma_c\sigma_s}\exp\left(-\frac{\mu^2\mu_p^2}{2\sigma_c^2}\right)\sum_{k=0}^{\infty}\frac{(2k-1)!!\left(\sigma_s^2-\sigma_c^2\right)^k}{k!2^k\cdot\sigma_s^{2k}\cdot\mu_p^k\sigma}\cdot I_k\left(\frac{\mu\,\mu_p}{{}_c^2}\right).$$

(4)

At symmetry of information channel on dispersions of quadrature components when the following condition is satisfied:

$$\sigma_c^2=\sigma_s^2=\sigma^2$$

(5)

Rice-Rayleigh distribution (Donato, et al., 1967) or generalized Rayleigh distribution (Raqab, et al., 2011) which are very extended in practice follow:

$$W(\gamma)=\frac{\mu}{\sigma^2}\exp\left(-\frac{\mu^2+\mu_p^2}{2\sigma_c^2}\right)I_k\left(\frac{\mu\,\mu_p}{\sigma^2}\right). \quad (6)$$

In the absence of regular component of signal ($\mu_r=0$) the Rayleigh distributions for the coefficient of transfer of information channel follows:

$$W(\gamma)=\frac{\mu}{\sigma^2}\exp\left(\frac{\mu^2}{2\sigma^2}\right). \tag{7}$$

In the absence of random changes of quadrature signal components when $\sigma^2 \to 0$ and $\mu = \mu_p$ it is necessary to use model of the channel with constant parameters.

CHANGE OF THE BLOCKING RELIEF BETWEEN THE BS AIS AND THE SHIP TRANSPONDER (ST) AIS

The models for description of analytical probable dependences of relief heights can be used for taking into account of the influence of the blocking relief between BS AIS and ST AIS.

The normal (Gaussian) law of distribution of the relief heights is widely used in case of approximation of various reliefs by methods of mathematical statistics.

$$W(H)=\frac{1}{\sqrt{2\pi\sigma_H}}\cdot\exp\left[-\frac{(H-m_H)^2}{2\sigma_H^2}\right], \tag{8}$$

where m_H и σ_H^2 – mean value and dispersion of relief of the route between correspondents.

For the law of the even area it is possible to use the following ratio with a one-dimensional probability density

$$W(H)=\frac{1}{(H_{\max}-H_{\min})},\ H_{\min}\le H\le H_{\max}, \tag{9}$$

The following expression is used for exponential distribution law :

$$W(H)=\alpha\, e^{-\alpha H}\ H\ge 0, \tag{10}$$

where α– parameter of heights distribution.

PROCESS OF MOVEMENT OF THE ST AIS RELATIVE TO THE BS AIS

Random distances R between the vessel (ST AIS) and the shore (BS AIS) station have significant effect on fidelity of messages transmission in the information channels of AIS.

For the description of such distances it is possible to use, at first, Maxwell's law with one-dimensional probability density:

$$W(R)=\sqrt{\frac{2R^2}{\pi\sigma^3}}\exp\left(-\frac{R^2}{2\sigma^3}\right),\ R\ge 0, \tag{11}$$

where $\sigma^2 = \pi / 8R_{av}^2$, R_{av} – the average value of R.

Secondly, the models on the basis of the Rayleigh law:

$$W(R)=\frac{R}{\sigma^2}\exp\left(-\frac{R^2}{2\sigma^2}\right), R\ge 0, \tag{12}$$

where $\sigma^2 = \frac{2}{\pi} R_{av}^2$,

In case the static data on movement of vessels in transport process are absent, then the uniform law is used:

$$W(R) = \frac{1}{R_{\max} - R_{\min}}, \; R_{\min} \le R \le R_{\max} \tag{13}$$

Also the consideration of electromagnetic hindrances influence is very important, however it doesn't keep within a framework of the current article and demands separate consideration.

CONCLUSIONS

The technique offered in the article allows to model areas of coverage of the modern telecommunication systems functioning in VHF range, for example AIS which is intended for safety of navigation in estuarial areas of rivers of the polar basin. The result of such simulation for the estuarial area of the Yana river is presented in the figure 1. Use of expressions 1 – 13 has allowed to choose the quasioptimum place for the installation of BS AIS. In this case the installation of BS AIS is supposed on the Yuedey cape (0th kilometer).

The installation of BS AIS on the Yuedey cape allows to provide by the telecommunication field of AIS the access to the estuary of the Yana river on remoteness more than 20 kilometers from the coastal line with probability of reception of discrete messages 10^{-3}.

A choice of the installation of BS AIS had been made taking into account, at first, quantitative estimates of range of actions of BS AIS received taking into account the blocking relief and the vessel movement relatively BS AIS, secondly, taking into account existence of elements of infrastructure necessary for the installation of BS of AIS. In case of placement of BS of AIS on the Yuedey cape the automated control of movement of the vessels equipped with the AIS equipment will be possible not only during passing all site of shallow estuarial bar from the receiving buoy to the Yuedey cape, but also to the river port of Nizhneyansk. It should be noted that in certain cases for estuarial areas of the Arctic rivers the impact of industrial noises on the coverage area of BS of AIS is possible not consider in connection with rather low intensity or even lack of the specified class of noises.

As the conclusion it should be noted that the offered technique can be also used rather successfully for other rivers of the Arctic region.

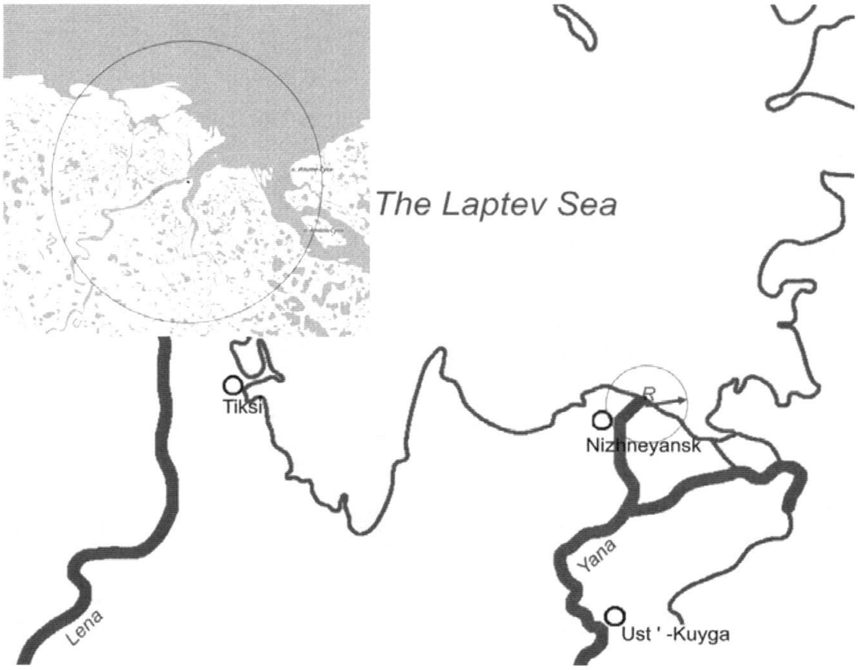

Figure 1. Estimated range of the coverage area of base station (BS) of Automatic identification system (AIS) in the estuary of the Yana River

REFERENCES

Pastusiak, T., 2016. Variants of the Northern Sea Route. *The Northern Sea Route as a Shipping Lane*, pp.131–152. Available at: http://dx.doi.org/10.1007/978-3-319-41834-6_6.

Park, I.-H., Hwang, S.-W. & Lee, S.-J., 2009. The Extention and Implementation of User-defined AIS AtoN for Marine Safety Information Service. *Journal of Korean navigation and port research*, 33(6), pp.423–428. Available at: http://dx.doi.org/10.5394/kinpr.2009.33.6.423.

Soler, T., 2004. GPS/GNSS Current Bibliography. *GPS Solutions*, 8(2), pp.115–117. Available at: http://dx.doi.org/10.1007/s10291-004-0097-3.

Bruyninx, C., 2006. Comparing GPS-only with GPS + GLONASS positioning in a regional permanent GNSS network. *GPS Solutions*, 11(2), pp.97–106. Available at: http://dx.doi.org/10.1007/s10291-006-0041-9.

The British Standards Institution, 2017. *Maritime navigation and radiocommunication equipment and systems. Automatic identification system (AIS)*. Available at: http://dx.doi.org/10.3403/bsen62320.

Nelson, D.J., Hopkins, J.R. & Bartos, A.L., 2011. Coherent demodulation of AIS-GMSK signals in co-channel interference. *2011 Conference Record of the Forty Fifth Asilomar Conference on Signals, Systems and Computers (ASILOMAR)*, pp.295–299. Available at: http://dx.doi.org/10.1109/acssc.2011.6190005.

Society of Photo-Optical Instrumentation Engineers (SPIE), 2011. *Hermite Polynomial Example. Field Guide to Special Functions for Engineers*, pp.48–48. Available at: http://dx.doi.org/10.1117/3.885484.ch48.

Balogh, C.B., 1966. Uniform asymptotic expansions of the modified Bessel function of the third kind of large imaginary order. *Bulletin of the American Mathematical Society*, 72(1), pp.40–44. Available at: http://dx.doi.org/10.1090/s0002-9904-1966-11408-8.

Donato, A.R.D. et al., 1967. Tables of the Rayleigh-Rice Distribution. *Mathematics of Computation*, 21(98), pp.267–269. Available at: http://dx.doi.org/10.2307/2004186.

Raqab, M.Z. & Madi, M.T., 2011. Generalized Rayleigh Distribution. *International Encyclopedia of Statistical Science*, pp.599–603. Available at: http://dx.doi.org/10.1007/978-3-642-04898-2_275.

Numerical Study of Oil Spilled Behavior under Ice–Covered Area in the East Siberian Sea

Liyanarachchi W. A. DE SILVA[1], Hajime YAMAGUCHI [1]
[1] The University of Tokyo, Tokyo, Japan

ABSTRACT

The Recent demands of commercial shipping through the Arctic and exploit of natural resources increase the risk of an oil spill in the Arctic Ocean. State-of-the-art modeling techniques are necessary for prevention measures and oil response and determine the fate and effect of oil in the environments. So we developed a numerical simulation model for spilled oil forecasting under ice-covered areas. This simulation model can perform high-resolution computations at low computational cost. Numerical model results of oil spreading over open water, under fully ice-covered water, and under a broken ice field are agreed reasonably well with the previous experimental results. Results of oil spilled behavior under the ice-covered area in the East Siberian Sea have shown the possibility of establishing an accurate and practical forecast system for oil spreading under a marginal ice covers.

KEY WORDS: Oil spill; Ice cover; Numerical modeling of oil.

INTRODUCTION

The retreat of summer sea ice in the Arctic Ocean continues to attract interest in exploring Arctic areas in order to locate and exploit natural resources and to establish shorter commercial shipping routes. Increasing those activities in the Arctic have increased the risk of oil spill accidents in ice-covered areas. If oil spill accident happens in this area, it could impact the marine eco systems and regional economy of coastal areas.

Oil spill behavior is depending on two main processers. First process is weathering, the physical and chemical properties of the oil is changed due to the environmental conditions (eg. evaporation and emulsification)(Díez *et al.*, 2007; Boehm *et al.*, 2008). Second process is movement of oil in the environment (eg. Wind and ocean currents, sea ice conditions)(French-McCay, 2004). Weathering and movement process can also overlap on each other.

Many researchers have studied the spreading of oil spill in open water and many numerical models have been introduced to simulate the process(Fay, 1971; Fannelop and Waldman, 1972). In general, numerical prediction of oil spill spreading under the ice-covered water is more complex than the open water spreading. Therefore, research on oil spreading under ice-covered water has been very limited.

However, when oil spill accident happens in ice-covered area, the ability to predict oil spreading is important in assisting cleanup operation, evaluating the degree of environmental impact and spills prevention planning. Therefore, high accurate modeling technique is necessary for predict the oil spill behavior in order to spill preventions and spill responses.

Yapa and Chowdhury (1990) introduced the theoretical equations, based on Navier-Stokes equations, for spreading of oil under the ice. Izumiyama *et al.* (1998) improved the Yapa and Chowdhury (1990) oil spreading theory by introducing the net interfacial tension. Izumiyama *et al.* (2002) validated their previous results by conducting series of experiments. Those results are very important for numerical modeling.

In 2006, Engineering Advancement Association of Japan conducted the three-year research program to develop a prototype model for numerical prediction of spilled oil diffusion and advection in the Sea of Okhotsk (Terashima, 2003; Rheem and Yamaguchi, 2004; Hara, et al., 2008). In this study we used the same oil spill simulation model to predict the oil trajectories, weathering of oil spill and also simulate the variety of scenarios for the oil spill in the East Siberian Sea using different ensembles to create the hazards map. We also validated the numerical simulation under the ice-coved area comparing with the laboratory experiment data.

MODEL DESCRIPTION

Oil spill model used in this study is based on the model developed by Terashima (2003) and Rheem and Yamaguchi (2004). Vertical diffusion of oil on ice-covered area is defined on four stages as shown in Figure 1.

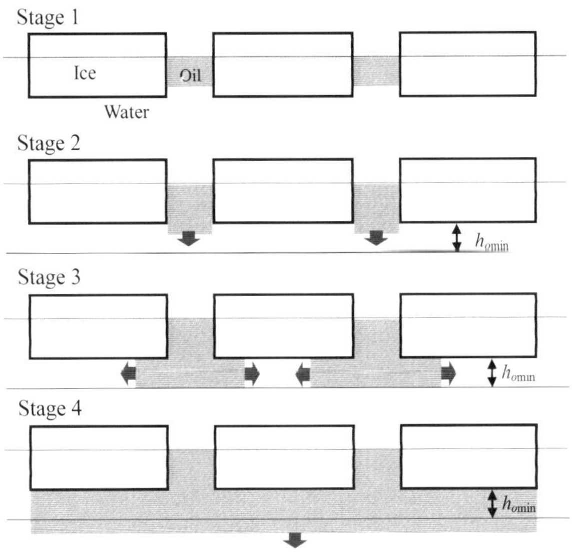

Figure 1. Schematic diagram of vertical spreading of oil under ice-covered water. h_{omin} is the minimum oil thickness for spreading under ice define in Eq.(1).

In the stage 1, spilled oil spreads horizontally through the gaps between ice floes. After stage 1, if oil contacts with land or surrounded by dense sea ice, spilled oil is thickened until the oil thickness below the ice bottom and until the minimum oil thickness (Mackay, Medir and Thornton, 1976)(Mackay et al., 1976) for spreading under ice (h_{omin}), which we call the stage

2. The minimum oil thickness for spreading under ice is defined by Eq.(1)

$$h_{o\min} = \sqrt{\frac{2\sigma_{o-i-w}}{g(\rho_w - \rho_o)}} \tag{1}$$

where, σ_{o-i-w} denote surface tension between ice, water and oil. g is a gravitational acceleration and ρ_w, ρ_o denotes the density of water and oil respectively. If oil further thickened and reaches the minimum oil thickness level, oil starts spreading horizontally under the ice in stage 3 while maintaining a constant thickness (h_{omin}). In stage 3 surface tension forces between ice, water and oil dominate the gravity force. In final stage (stage 4), when the oil spread whole sea ice bottom, then oil begins to further thickening due to the supply of oil.

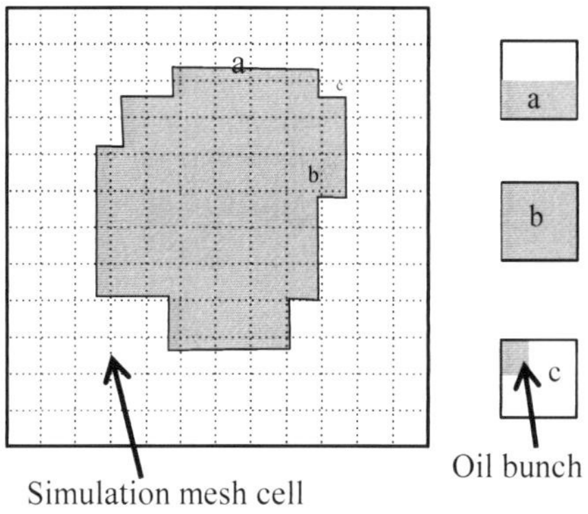

Figure 2. Simulation mesh and oil bunch locations

To improve the ability of the oil spill model to locate the oil edges accurately, we implemented the Eulerian–Lagrangian method for advection (Rheem et al., 1997) of the oil spill (the so-called subgrid-scale oil motion). First, we solve the momentum equation in a Eulerian grid, and then we solve the oil conservation laws in a Lagrangian grid. Spilled oil is simplified to one rectangular oil bunch at each simulation mesh as shown in Figure 2.

On the other hand, the ice field is represented by rectangular floe particles (side length D) with a given thickness and bundle to the rectangular bunches on top of the oil bunch as shown in Figure 3. We assumed ice floes are homogenously distributed among the cells and distance between two ice floes is l and number of ice floes in X direction is *nx* and Y direction *ny*. Sea ice concentration (*A*) can be written by the Eq.(2)

$$A = \frac{(Dnx)(Dny)}{bxdx \cdot bydy} \tag{2}$$

where, *bx* and *by* are oil bunch locations (varied from 0-1). *dx* and *dy* are cell size. Eq.(3) and 4 show the number of ice floes in X and Y directions.

$$nx = \frac{\sqrt{A}bxdx}{D} \tag{3}$$

$$ny = \frac{\sqrt{A}bydy}{D} \tag{4}$$

Eq.(5) shows the distance between two ice floes

$$l = \frac{\left(1-\sqrt{A}\right)D}{\sqrt{A}} \tag{5}$$

In each Eulerian grid cell facers, the momentum equation (Eq.(6)) is solved for diffusion oil velocities.

$$m_o\left[\frac{\partial \bar{u}_o}{\partial t} + f\bar{k}\times\bar{u}_o\right] = \bar{F}_p + \bar{F}_f + \bar{F}_m + \bar{F}_{st} \tag{6}$$

where m_o is oil mass, u_o is oil velocity, f is Coriolis parameter, k unit vector in vertical direction and F_p, F_f, F_m, F_{st} are pressure force, frictional force, added mass force and surface tension force respectively. Over bar denotes the vectors.

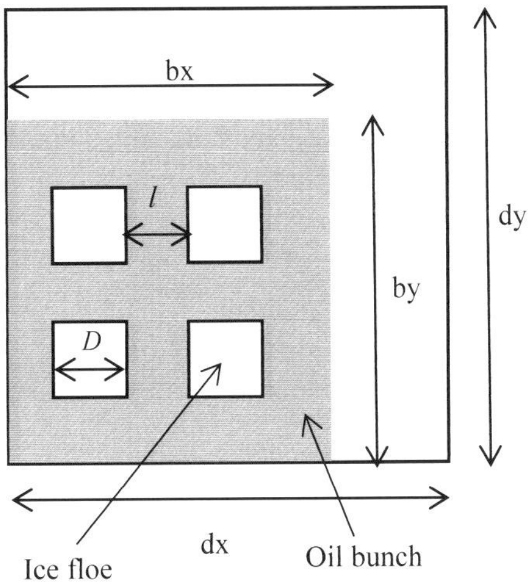

Figure 3. Simulation mesh, ice floe distribution and oil bunch location. Where dx, dy are cell size and bx, by are oil bunch size. D is length of rectangular ice floe size and l is distance between two ice floes

The external forces change in different sea surface conditions. When oil spills on sea surface where is no sea ice, forces acting on oil at the direction of diffusion are the surface tension, pressure force from gravity. The forces acting on oil at the direction of anti-diffusion are force of friction on sea current and forces by added mass as shown in Figure 4(a). On the other hand, when oil spills under sea ice, the force acting on oil at the direction of diffusion is only the pressure force from gravity, and the forces acting on oil at the direction of anti-diffusion are the surface tension, the forces by added mass, and the frictional force on sea ice and sea current as shown in Figure 4(b).

After calculating the cell each cell face velocities, each Eulerian cell oil bunch is advected in space creating new configurations. Finally, a new oil state is obtained for the Eulerian grid by summing and redistributing the advected configurations. More details about redistribution can be find in (Rheem et al., 1997).

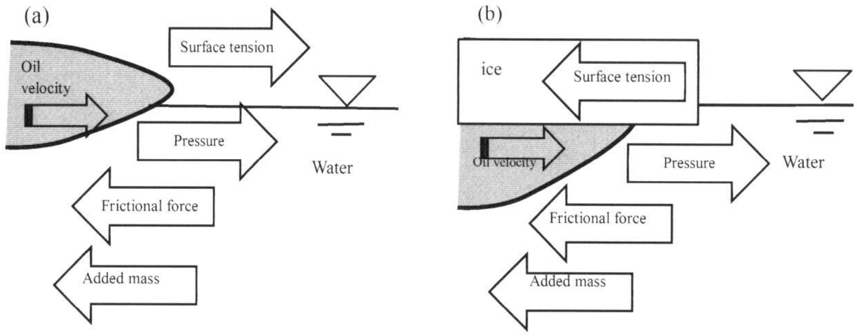

Figure 4. External forcers on oil (a) On water surface (b) under sea ice

ACCURACY VERIFICATION OF OIL SPILL DIFFUSION MODEL

In this section, oil spill experiments conducted in the past with different sea ice concentrations are compared with the oil spill numerical model. Three patterns of spilled oil diffusion radius were computed. First, spilled oil radius on open water this experiment is conducted by Yamaguchi, (2006). Second, spilled oil on fully ice covered (100% sea ice concentration) area experiment conducted by Izumiyama et al., (2002). Third, spilled oil under partially ice covered (sea ice concentration 65%) surface with the effect of ice floe motion experiment conducted by Gjosteen and Loset,(2002). In each experiment physical properties of the oil and experimental conditions are shown in Table 1.

Table 1. Experimental and calculation conditions of oil spill accuracy verifications

Experiment	Open water	Fully ice covered	Partially ice covered
Oil name	Idemitsu Dafny #32	Lubrication oil	IFO 30 bunker oil
Density [kg/m^3]	863	890	938.63
Water temperature [°C]	12	-2	-2
Viscosity [Pa s]	1.1×10^{-1}	1.23×10^{-1}	6.87×10^{-1}
Oil-water surface tension [N/m]	1×10^{-2}	2.56×10^{-2}	2.32×10^{-2}
Spilled oil volume [m^3]	1.0725×10^{-3}	16.32×10^{-3}	9.8×10^{-3}
Oil spilled time [s]	205.6	600	57
Sea ice concentration	-	100%	65%
Sea ice thickness [m]	-	0.01	0.038
Ice bottom	-	Flat	Flat
Sea ice rotational velocity [rad/s]	-	-	0.003
Time step Δt [s]	0.1	0.1	0.1
Grid resolution [m]	0.03	0.03	0.04
The number of grids	100×100	100×100	100×100

Accuracy Verification of Oil Diffusion Model in Open Water Surface

The oil spill experiment on open water is carried out in the Institute of Industrial Science, University of Tokyo, tsunami and storm surge water tank by Yamaguchi,(2007). Physical properties of the oil and experimental conditions are shown in Table 1. Figure 5 shows the comparison of the diffusion radius of the oil obtained in the experiment (blue dots) and numerical model (black line). Present numerical model shows good agreement with experimental results.

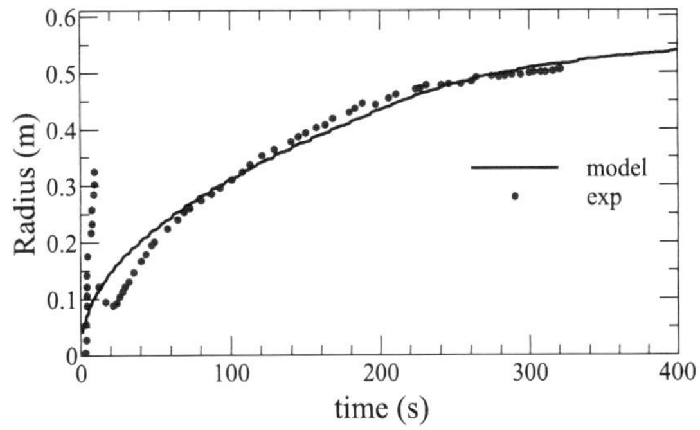

Figure 5. Accuracy verification of oil diffusion radius in open water. Blue dots denote the experimental results and black line denotes the numerical results.

Accuracy Verification of Oil Diffusion Model under Fully Ice Covered Surface

The oil spill experiment in the fully ice covered surface is carried out in the ice tank of National Maritime Research Institute, Japan by Izumiyama et al., (2002). The test was performed for level ice sheets with flat bottom and without the ice sheet movement. Physical properties of the oil and experimental conditions are shown in Table 1.

Figure 6 shows the comparison of diffusion oil slick radius of the experimental results and numerical model results. Distribution of oil radius with respect to the time shows the similar pattern compared with experiment and simulations. However, model over predict the radius about 0.05m.

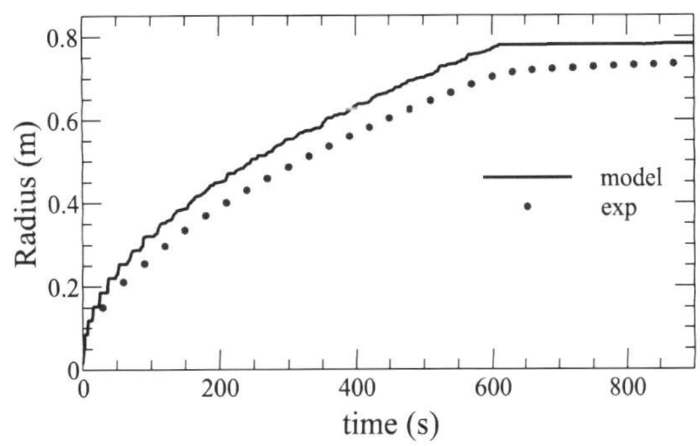

Figure 6. Accuracy verification of oil diffusion radius under the fully ice covered surface. Blue dots denote the experimental results and black line denotes the numerical results.

Accuracy Verification of Oil Diffusion Model under Partially Ice Covered Surface

Gjosteen and Loset,(2002) conducted the laboratory experiment of oil spill under the broken ice field at the Hamburgische Schiffbau Versuchsanstalt ice tank. Oil spreading and ice floe motion were monitored by video cameras. The results they have obtained are very useful for validate the oil spreading model under the influence of ice flow dynamics. Physical properties of the oil and experimental conditions are shown in Table 1.

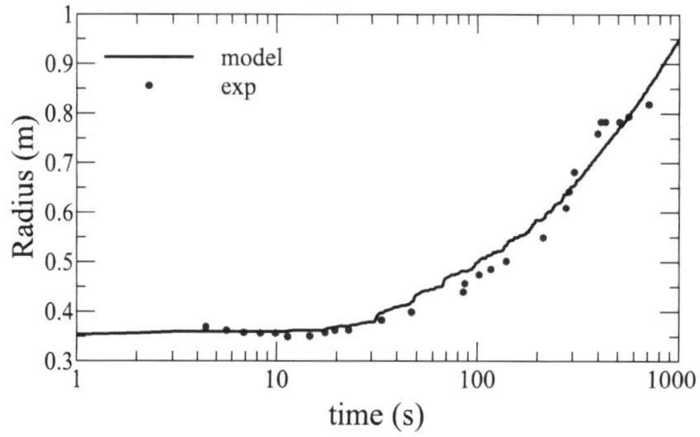

Figure 7. Accuracy verification of oil diffusion radius under partially ice-covered surface. Blue dots denote the experimental results and black line denotes the numerical results.

Figure 7 shows the comparison of oil diffusion radius between experimental results of (Gjosteen and Loset, 2002) and numerical model results. Numerical model has reproduced the experimental results reasonably. In this experiment ice floes had a clockwise, circular motion. These ice floe motions exert th e drag force on the oil and enhance the spreading further. In this experiment oil is place near the center of the domain and constantly spilled the oil for 57s.

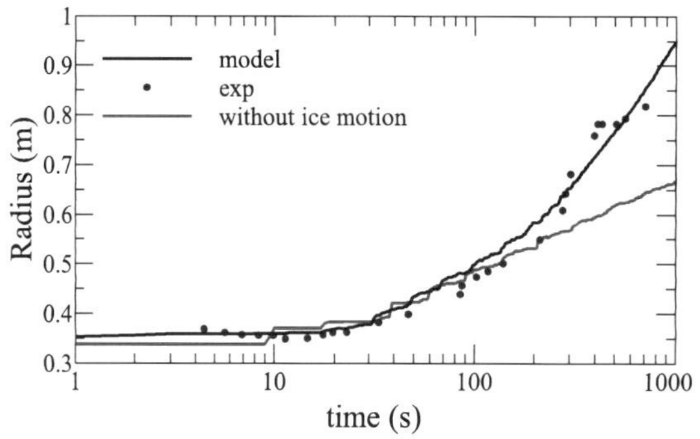

Figure 8. Accuracy verification of oil diffusion radius under partially ice-covered surface with and without ice motion. Blue dots denote the experimental results and black line denotes the numerical result with ice motion and read line denotes the numerical results without ice motion.

Short time after the injection was completed, ice cover was brought in to motion. At first oil is spread horizontally along the ice floes and very slowly spreading on vertical direction. After the oil thickness reach into the sea ice thickness oil begins to spread under the ice floes while receiving frictional force from ice bottom. Our model is able to reproduce those treads accurately. We also tried the effect of ice motion on oil spill simulation. Figure 8 shows the numerical results with and without ice rotational motion. Before the 200s both with and without rotational cases oil is spreading in similar radius. But after 200s with ice rotational experiment accelerate the oil spreading exponentially compared to the without rotational experiment.

COMPUTATION OF SPILLED OIL BEHAVIOR IN THE EAST SIBERIAN SEA

We carried out a numerical computation of sea ice and spilled oil behavior in the East Siberian Sea. Initial sea ice distribution and ocean conditions are given by high-resolution computation of ice-POM model (De Silva, Yamaguchi and Ono, 2015). It is assumed that the oil spilled at a rate of 2 m^3/s for 3000 seconds. Oil density is 890 kg/m^3 and viscosity 0.123 Pa s. Oil is spilled at the south of the Novaya Sibir island on 2015 July. Figure 9 shows the ice and oil distribution on initial 3,7 and 10 days later. It is demonstrated that the oil spilled near the East Siberian Sea can reach the coastal areas of the Novaya Sibir islands within few days.

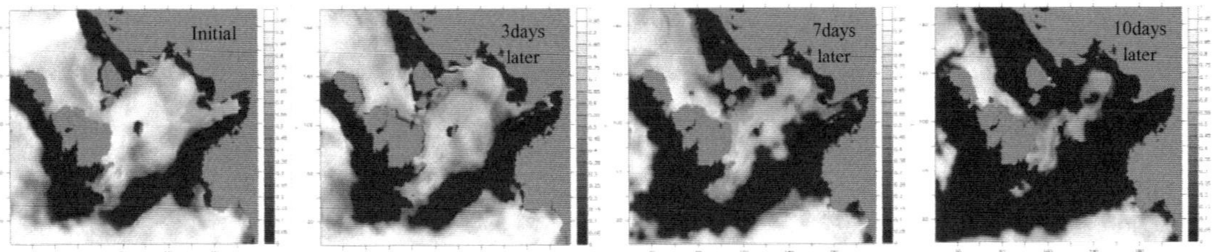

Figure 9. Sea ice distribution and oil spilled behavior in the first 10 days of computation in 2015 July. Red area shows the spilled oil in the computational grid

10-days and 15-days computations similar to the Figure 9 are made with the same oil spill conditions in July 2015.

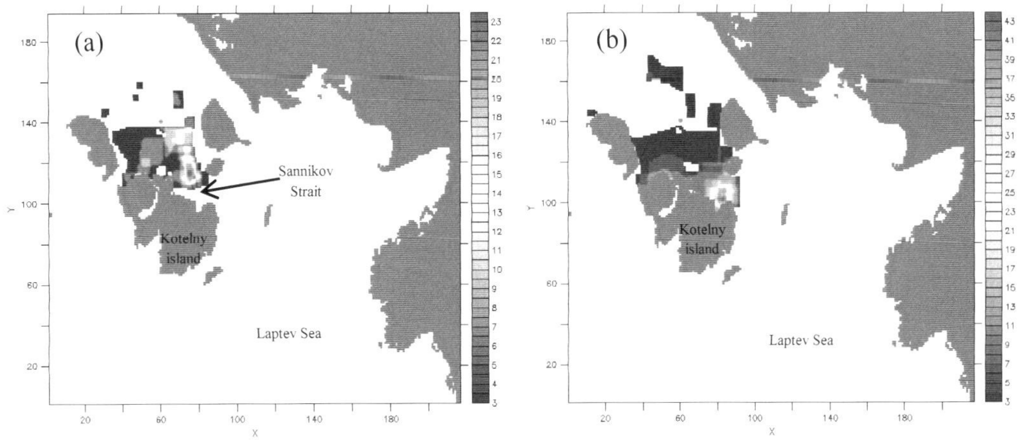

Figure 10. Probability distribution of oil presence (a) after 10days and (b) after 15 days later. Calculated by 30 ensembles from 1-30 July 2015. Note that color scales are not same.

30 ensemble computations are carried out; by changing initial oil spill dates. Figure 10 shows the probability distribution where spilled oil reaches within 10 days and 15days. Both figures shows widespread of probability distribution in the area and within 10 days 23% probability that oil may reach in the Sannikov Strait and after 15 days 43% probability that oil reach the east coast of Kotelny island.

CONCLUSIONS

Oil spill experiments conducted in the past with different sea ice concentrations are compared with the oil spill numerical model. Numerical model shows the good agreement with the laboratory experiments. Next, we have carried out a numerical computation of sea ice and spilled oil behavior in ice-covered East Siberian Sea. We have shown if oil spills in East Siberian Sea, the oil can reach coastal area of Kotelny Island in a few days. To accomplish the high accuracy spilled oil forecasting, we have to carry out high accuracy sea ice forecasting using ice-POM model.

REFERENCES

Boehm, P. D., Page, D. S., Brown, J. S., Neff, J. M., Bragg, J. R. and Atlas, R. M. (2008) 'Distribution and weathering of crude oil residues on shorelines 18 years after the Exxon Valdez spill.', *Environmental science & technology*, 42(24), pp. 9210–9216. Available at: http://www.ncbi.nlm.nih.gov/pubmed/19174894 (Accessed: 13 September 2016).

Díez, S., Jover, E., Bayona, J. M. and Albaigés, J. (2007) '*Prestige* Oil Spill. III. Fate of a Heavy Oil in the Marine Environment', *Environmental Science & Technology*. American Chemical Society, 41(9), pp. 3075–3082. doi: 10.1021/es0629559.

Fannelop, T. K. and Waldman, G. D. (1972) 'Dynamics of Oil Slicks', *AIAA Journal*, 10(4), pp. 506–510. doi: 10.2514/3.50127.

Fay, J. A. (1971) 'Physical processes in the spread of oil on a water surface', *International Oil Spill Conference Proceedings*. American Petroleum Institute, 1971(1), pp. 463–467. doi: 10.7901/2169-3358-1971-1-463.

French-McCay, D. P. (2004) 'Oil spill impact modeling: development and validation.', *Environmental toxicology and chemistry / SETAC*, 23(10), pp. 2441–2456. Available at: http://www.ncbi.nlm.nih.gov/pubmed/15511105 (Accessed: 13 September 2016).

Gjosteen, J. K. O. and Loset, S. (2002) 'Laboratory studies of oil spreading in broken ice', in *Proceedings of the 16th IAHR International Symposium on ice*. Dunedin, New Zealand, pp. 110–119.

Hara, K., Yamaguchi, H. and Sagawa, G. (2008) 'A numerical simulation of spilled oil behaviors of ice-covered sea in the Sea of Okhotsk', in *Proceeding of the 23rd International Symposium on Okhotsk Sea and Sea ice*. Mombetsu, pp. 108–113.

Izumiyama, K., Konno, A. and Sakai, S. (2002) 'Experimental Study On Spreading of Oil Under Ice Covers', in *Proceedings of The Twelfth International Offshore and Polar Engineering Conference*. Kitakyushu, Japan: International Society of Offshore and Polar

Engineers, pp. 821–826.

Izumiyama, K., Uto, S., Narita, S. and Tasaki, R. (1998) 'Effect of interfacial tension on the spreading of oil under an ice cover', *Ice in Surface Waters*, Shen (ed.), pp. 419–426.

Mackay, D., Medir, M. and Thornton, D. E. (1976) 'Interfacial behaviour of oil under ice', *The Canadian Journal of Chemical Engineering*. Wiley Subscription Services, Inc., A Wiley Company, 54(1–2), pp. 72–74. doi: 10.1002/cjce.5450540110.

Rheem, C.-K. and Yamaguchi, H. (2004) 'A Numerical Simulation Model for Oil Spill in Ice Covered Sea', *SEISAN KENKYU*. 東京大学生産技術研究所, 56(6), pp. 483–486. doi: 10.11188/SEISANKENKYU.56.483.

Rheem, C. K., Yamaguchi, H. and Kato, H. (1997) 'Distributed mass/discrete floe model for pack ice rheology computation', *Journal of Marine Science and Technology*, 2(2), pp. 101–121. doi: 10.1007/BF02491524.

De Silva, L. W. A., Yamaguchi, H. and Ono, J. (2015) 'Ice–ocean coupled computations for sea-ice prediction to support ice navigation in Arctic sea routes', *Polar Research*, 34. doi: 10.3402/polar.v34.25008.

Terashima, K. (2003) *A study of numerical model for oil diffusion in ice-covered sea (in Japanese)*. The University of Tokyo.

Yamaguchi, D. (2007) *Research on Numerical Simulation Model of Spilled Oil Behavior in Ice Covered Sea*. Nihon University, JAPAN.

Yapa, P. D. and Chowdhury, T. (1990) 'Spreading of Oil Spilled under Ice', *Journal of Hydraulic Engineering*, 116(12), pp. 1468–1483. doi: 10.1061/(ASCE)0733-9429(1990)116:12(1468).

POAC'17
Busan, KOREA

Proceedings of the 24th International Conference on
Port and Ocean Engineering under Arctic Conditions June 11-16, 2017, Busan, Korea

Verification and Validation of An in-Ice Oil Spill Trajectory Model Based on Satellite-Derived Ice Drift Data

Hossein Babaei[1], David Watson and Richard Burcher[1]
[1] Ocean, Coastal and River Engineering, National Research Council Canada, Ottawa, Ontario

ABSTRACT

A future increase in hydrocarbon exploration and development activities driven by the probable existence of hydrocarbon reserves and an expected increase in shipping activities due to less severe ice conditions, pose a risk of potential oil spills in the offshore Arctic. Estimating oil spill trajectories is essential in quantifying risks and planning an effective spill response.

An in-ice spill trajectory modelling, analysis and visualization tool suitable for spills in highly ice-infested waters has been previously developed at NRC. The source data is historical satellite-derived ice drift. The model has been enhanced by including time dependent land-fast ice extent to better estimate coastal spill trajectories.

Two hypothetical in-ice spill scenarios in the Canadian Beaufort Sea were modelled based on 34 years of ice velocity data. In four months starting in November, a deep water spill in ice could travel over 700 km, while for a shallow water spill in ice, the travel distance could exceed 400 km. Depending on how fast an in-ice spill could be cleaned, both investigated deep water and shallow water spills could be an international issue, particularly the deep water spill scenario.

Present model results were compared with an observed in-ice spill trajectory in the Barents Sea. Because of an underestimation of ice speeds in the input satellite-derived ice drift dataset, the present model underestimates the extent of the trajectory. However, the model estimated the trajectory of an observed buoy well.

Present model results were also compared with an independent numerical study of oil spills in the Beaufort Sea. Coastward motions of an in-ice spill are found to be generally similar, however, along the coast, motions deviate after a certain time in the modelled period. Both models are based on data that are expected to be less accurate in the nearshore zone. We did not investigate what caused this deviation or whether the present model or the independent study is a better representation of reality.

KEY WORDS: Oil spill; Trajectory modelling; Ice; Validation; Verification; Satellite-derived ice drift

INTRODUCTION

A large proportion of the world's undiscovered hydrocarbon reserves are expected to be found in the offshore Arctic (Eurasia Group, 2014) where the average monthly ice extent has been declining since 1978 (National Snow and Ice Data Centre, 2016). This means a probable increase in the exploration and development of hydrocarbon resources and an increase in shipping if the decline of ice extent continues. More activities in the offshore Arctic pose a higher risk of large spills of chemicals, particularly crude oil with potentially significant negative socio-economic and environmental impacts. The trajectory estimation of such spills is crucial for timely, optimized and successful clean-up efforts, particularly in colder seasons when cleanup operations are more challenging because of shorter or no daylight periods and harsher ice conditions.

Trajectory of an oil spill in the presence of ice depends on several factors including the concentration of ice. In high ice concentrations, oil tends to move with the ice and not with the water current. Oil follows the trajectory of ice for several reasons: it can be trapped in leads between ice floes, absorbed in snow on ice (for an on-ice spill), encapsulated in ice brine channels (for an under-ice spill), trapped in under-ice depressions, and adhere to ice floe edges. An under-ice spill could also move relative to the ice when currents are high, however this is not generally expected except in river and fjords and where tides are strong (MAR Incorporated et al., 2008). The ice concentration required for oil to move with the ice is not fully known; values as low as 30% (Venkatesh et al., 1990) and as high as 70% (LOOKNorth, 2014) have been mentioned in literature.

Computer modelling of different aspects of in-ice oil spills including trajectory estimation is under active development. A review of some of the models can be found in LOOKNorth, 2014 and Drozdowski et al., 2011. To model the trajectory of a spill in highly ice covered waters, the motion and deformation of the ice field is required. To our knowledge, all numerical oil spill trajectory models rely on numerical estimations of ice motion and deformation computed by different numerical ice-ocean models with different length scale applicability, levels of sophistication and accuracy. Satellite-derived ice motion data may potentially be used to estimate the trajectory of in-ice spills as an alternative to numerically generated ice motion data or to validate it. Several publicly available satellite-derived ice drift datasets exist (see Tschudi et al., 2016; Muir et al., 2011; Kimura et al., 2013; Lavergne et al., 2010, and www.seaice.dk) which could be considered for trajectory estimation, however the literature on this subject is not fully developed.

We have previously developed a model based on a satellite-derived ice drift dataset for in-ice trajectory modelling, analysis and visualization and studied several hypothetical spill scenarios in the Beaufort Sea during fall and winter (Babaei et al. 2016) when usually the ice concentration is high and hence oil spills are expected to move mainly with ice. In that study, the trajectory algorithm of the model was successfully validated by comparing the model trajectory with an observed buoy trajectory for a 5-month-long period from November to March. The present paper reports about further developments, validation and verification of the model. To improve the reliability of the model in coastal regions of the Beaufort Sea, observed time-dependent landfast ice extents from ice charts were included in the model. We have verified results of the present model with results of an independent numerical oil spill study. We have also compared the present model results with an actual in-ice oil spill trajectory in the Barents Sea and with an observed buoy trajectory in the Arctic Ocean.

MODEL DESCRIPTION AND TWO HYPOTHETICAL IN-ICE SPILLS IN THE BEAUFORT SEA

Model Description

The core of the model simply calculates trajectory of particles in a gridded time dependent ice velocity field. An oil slick is numerically represented by a cluster of points that are moved separately. This movement of the points is calculated using a spatially and temporally gridded ice velocity field. The velocity of each point is interpolated using the three nearest nodes on the grid. To model a spill lasting for a given period, a point is introduced at the location of spill with a frequency equal to the temporal resolution of the ice velocity dataset. Each point is then moved and tracked based on the above algorithm to the end of the simulation. Weathering of oil is not modelled.

Generally, both numerical and satellite-based ocean ice drift data are less accurate for coastal regions than farther offshore. For numerical ice drift data, a reason could be the challenge involved to capture current features caused by typical high coastal bathymetric gradients and by rivers flows at their mouth. For satellite-derived ice drift data, there is the challenge to correctly distinguish land from sea ice depending on the spatial resolution of the satellite sensors.

In the present study, to model in-ice coastal spills based on satellite-derived ice drift data, time-dependant landfast ice extents from historical ice charts have been incorporated into the model; ice velocities within landfast ice are set to zero (if not already zero in the raw dataset) and when and if spill points are in landfast ice, spill points remain stationary until the landfast ice recedes and is replaced by drift ice. Coastlines are also included in the model to distinguish land from ocean. The present model is also capable of Monte-Carlo modelling of in-ice spills when uncertainties in the input satellite-derived ice drift dataset are known.

The third version of the Polar Pathfinder satellite-derived ice drift dataset (Tschudi et al., 2016) was selected for the present study because of the dataset's long term data availability spanning from 1978 to 2015 and the dataset's spatial coverage availability for our locations of interest. The spatial resolution of the dataset is 25 km and the selected temporal resolution is 1 day. The dataset is based on several different data sources including several satellite sensors and it is improved by observed buoy velocities collected under the International Arctic Buoy Program (IABP).

Model results include contamination probability (the probability that oil reaches a location), contamination intensity (how frequently a location is contaminated), contamination probability for water and land boundaries including territorial water boundary and coastlines, and residence time of spill in different locations.

Two Hypothetical In-ice Spills in the Beaufort Sea

Two hypothetical spill scenarios were selected to briefly show some of the model outputs, capabilities, and applications for spill analysis. Scenarios only differ in the spill location. We have modelled a deep (location: 71.036N, 136.507W; water depth: ~ 1170 m; offshore distance: ~ 180 km) and a shallow (location: 70.174N, 133.357W; water depth: ~30 m; offshore distance: ~ 70 km) water spill starting on November 1st lasting for 42 days. The location of the spills is the Canadian Beaufort Sea and the total model duration is 120 days.

For each scenario, spill was modelled for each year from 1978 to 2012 (thirty four years) and trajectories were combined and analyzed. Canadian Ice Service (CIS)'s Regional Western Arctic ice charts were used to extract the extent of landfast ice for the above period. The frequency of ice charts for the period of interest is different from year to year and month to month with an average of 2 weeks.

Figure 1 shows the contamination probability for the deep water spill scenario. The dominant direction of motion is towards the west. This dominant direction is consistent with the general ice drift direction in the Beaufort Sea known as the Beaufort Gyre. In 120 days, the spill could move west more than 700 km and there is a 40% to 50% chance that the spill moves west more than 200 km reaching to the US Beaufort Sea. The spill does not reach the coastline although it can reach the Alaskan territorial waters (dashed black line in the figure) close to Utqiagvik (Barrow) (<10% chance, shown by dark blue in Figure 1).

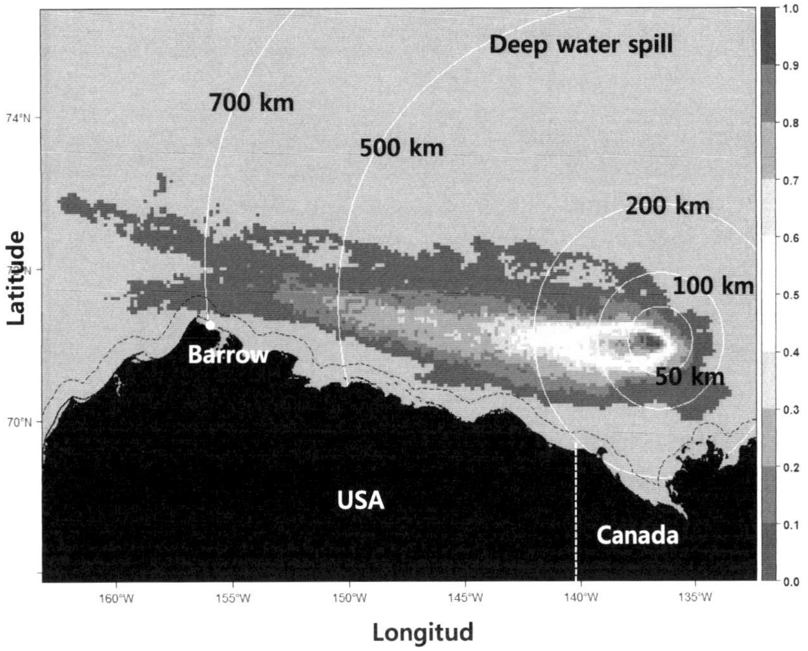

Figure 1. Probability of contamination (the chance that the spill reaches a location) for the deep water in-ice spill scenario. In 120 days, the spill could move west more than 700 km, but coastlines are not impacted. The dashed line shows the offshore distance of 25 km.

Figure 2 shows the contamination probability for the shallow water spill scenario. In 120 days, the spill could move west more than 400 km which is considerably less than the deep water spill scenario. As expected, the shallow water spill scenario impacts coastlines and territorial waters much more than the deep water spill scenario; the spill could reach the Alaskan coastline east of Kaktovic (<10% chance, shown by dark blue in Figure 2).

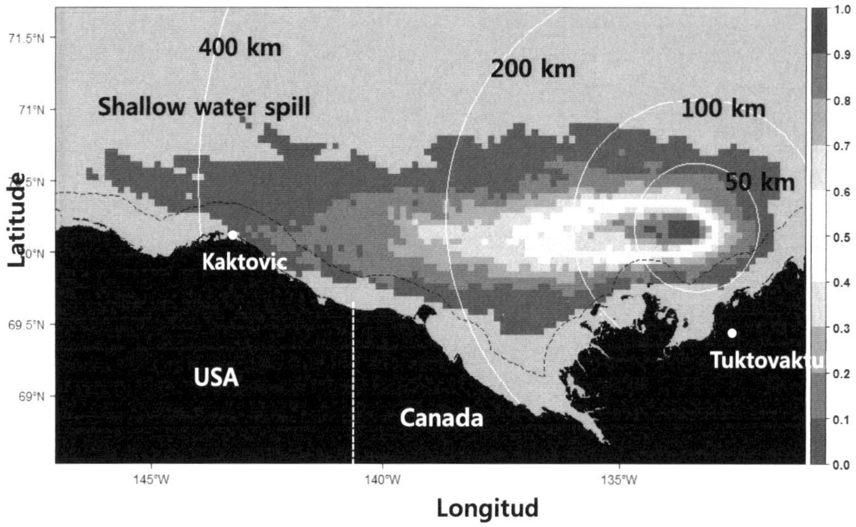

Figure 2. Probability of contamination (the chance that the spill reaches a location) for the shallow water in-ice spill scenario. In 120 days, the spill could move west more than 400 km and coastlines could be impacted. The dashed line shows the offshore distance of 25 km.

VALIDATION OF THE MODEL WITH AN OBSERVED IN-ICE OIL SPILL AND A BOUY TRAJECTORY

Validation of the Model with an Observed In-ice Oil Spill

As part of a SINTEF-led Joint Industry Partnership to study fate and behavior of spilled oil and the effectiveness and applicability of different cleanup technologies in ice-covered waters, oil was released between ice floes in the Barents Sea southeast of Svalbard in May 2009 (Sorstrom et al., 2010). The spill was tracked for 6 days (from May 15th to 20th) and reported to move with and remain between floes of ice, with concentrations ranging from 70% to 90% (Sorstrom et al., 2010). Figure 3a through 3d show the observed oil spill trajectory (reproduced from Sorstrom et al., 2010 and Ragnhild et al., 2011) on four available ice charts for the study period. The time-dependent location of the spilled oil is shown in Figure 3e.

Figure 4 shows the present NRC model results and the SINTEF observed trajectory. Although the present model predicts the general shape of the trajectory correctly, the extent of the trajectory is significantly underestimated. This underestimation is more pronounced along the east-west direction (~ 3 times) than along the north-south direction (~ 1.6 times). The final location of the modelled spill is approximately 23 km away from the start location of the spill while the observed value is approximately 37 km away. Table 1 lists the approximate observed and modelled movement of the oil for each day from May 15th to 20th 2009.

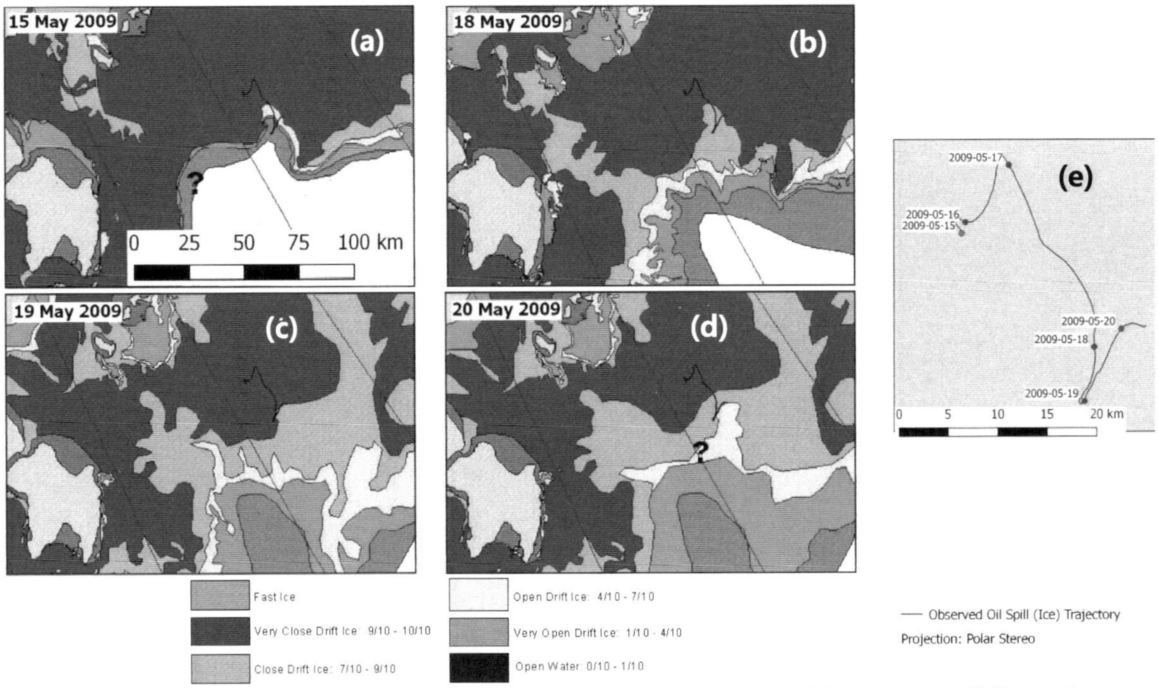

Figure 3. The relative location of the observed oil spill trajectory with respect to different ice categories for four days from the observation period in 2009. (a) 15 May, (b) 18 May, (c) 19 May, (d) 20 May. Ice charts are from the Norwegian Meteorological Institute. Observed oil is within 50 km from open water. (e) Observed trajectory (Sorstrom et al., 2010; Ragnhild et al., 2011) with dates.

Table 1. Approximate daily displacement of observed and modelled oil

Time period in 2009	Approximate daily movement of oil, km	
	Observed (Sorstrom et al., 2010 ; Ragnhild et al., 2011)	Modelled
May 15 to 16	3	3.5
May 16 to 17	13.5	2.5
May 17 to 18	42	16
May 18 to 19	10	8
May 19 to 20	17	5

The difference between the modelled trajectory and observation is attributed to the underestimation of ice speeds in the Polar Pathfinder dataset. It is known that the accuracy of the Polar Pathfinder data is higher when the concentration of ice and/or ice thickness is high (Sumata et al., 2014). Figure 3a through 3d show that the spill is within one to two Polar Pathfinder grid cell sizes from open water. We do not explore further whether this proximity to open water is the reason for the underestimation of ice drift by the Polar Pathfinder. We however briefly studied how the Polar Pathfinder data uncertainty impacts the modelled spill trajectory. The Polar Pathfinder dataset reports root-mean-square of error in the estimation of ice drift as a function of time and location. This error estimate with the mean value of error reported for the entire temporal and spatial coverage of the dataset (Miller et al. 2006) was the basis of a previous Monte-Carlo modelling whose details is given in (Babaei et al. 2016). Figure 4 shows the Monte-Carlo results as the probability of the oil reaching to any location. Although the results explain some of the differences between the modelled and observed trajectories, the modelled extent of the spill is still underestimated even if the uncertainties are included in the trajectory modelling.

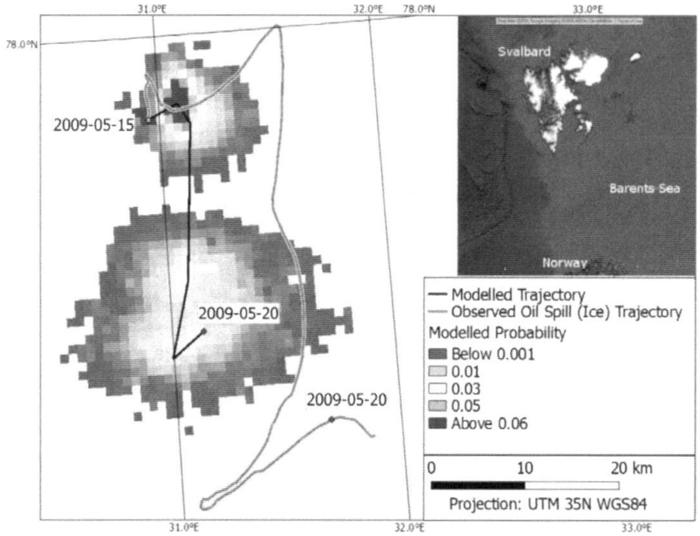

Figure 4. Observed oil moved more than the modelled oil. This difference is in part explained by considering the uncertainty in the Polar Pathfinder data input to the present model. The observed trajectory is reproduced from Sorstrom et al., 2010 and Ragnhild et al., 2011.

Validation of the Model with a Buoy Trajectory

Buoys of the International Arctic Buoy Program (IABP) are used to improve the accuracy of the Polar Pathfinder dataset (Fowler et al. 2013). It is then expected that the Polar Pathfinder data to be closely correlated to IABP data. We have previously validated the trajectory algorithm of the present model by comparing a modelled trajectory in the Beaufort Sea with an IABP buoy trajectory (Babaei et al. 2016). To further validate the present trajectory algorithm and to investigate how closely the Polar Pathfinder data is correlated to IABP data, we selected the trajectory of an IABP buoy in the Arctic Ocean north of Svalbard and compared it with the present model trajectory. Figure 5 shows the observed and modelled trajectories for the 20 Feb. 2009 to 8 Mar. 2009 period. The modelled trajectory closely follows the observed trajectory. The total observed travel distance is approximately 206 km while the modelled travel distance is approximately 188 km. The maximum difference between the location of the modelled and observed trajectory is approximately 15 km. The slight difference between the modelled and observed trajectories could be partially attributed to the large spatial (25 km) and temporal (1 day) resolution of the Polar Pathfinder dataset which are not enough to fully resolve sub-resolution events.

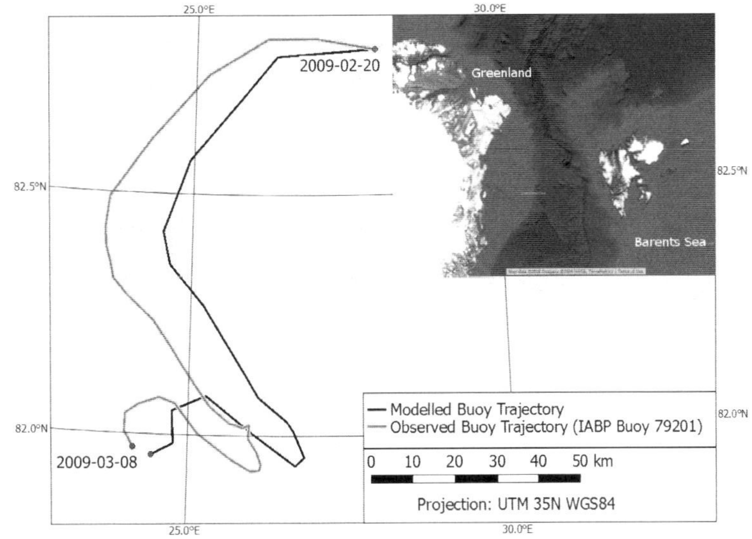

Figure 5. The present model closely predicts the trajectory of a buoy in the Arctic Ocean. The observed buoy moves approximately 206 km and the maximum difference between the model and observed trajectory is approximately 15 km.

VERIFICATION OF THE MODEL WITH AN INDEPENDENT NUMERICAL MODEL RESULTS FOR THE BEAUFORT SEA

The World Wildlife Fund (WWF) Canada sponsored a numerical study (Gearon et al., 2014) to model the fate, effects and trajectory of different hypothetical oil spill scenarios in the Beaufort Sea and their impact on wildlife. Spill scenarios included shallow and deep water blowouts and transportation and pipeline-based spills with different occurrence probabilities, locations, and onsets. Ice velocities for the modelling were based on a numerical ice-ocean data assimilation system named TOPAZ4 as a driver of a commercial oil spill modelling suite. When ice concentrations were more than 30%, the surface oil plume was modelled to move with ice. Modelled weekly locations of the surface oil plume are publicly available online through an interactive map at http://arcticspills.wwf.ca/. In our case, we selected an in-ice period of one of the modelled scenarios named Eastern Shipping-Bulker from the WWF study, to compare with the present model results. We modelled the trajectory of the in-ice plume from 27 Oct. to 15 Dec. 2009 and compared it with the WWF plumes reproduced from the online interactive map.

The Polar Pathfinder dataset, which is the input to the present model, does not have ice drift data for a region associated with a portion of the WWF plume on 27 Oct. 2009. This portion overlaps an open water region identified on the CIS ice chart for 26 Oct. 2009 (no ice chart was available for 27 Oct. 2009) shown in Figure 6a. The existence of the open water region explains the unavailability of ice drift data in the Polar Pathfinder dataset. We then chose to model only the trajectory of the in-ice portion of the WWF plume on 27 Oct. 2009.

Figure 6a through 6c compares the present modelled plume locations with those of the WWF study. WWF plumes reach closer to the coastline than the present modeled plumes by 40 km or less. Along parallels to the coastline, the difference between the location of the west edge of the plumes is less than 15 km until 17 Nov. 2009 after which the present NRC model plume moves faster than the WWF plume. This difference in the parallel to the coastline component of the plume (ice) velocity leads to an approximately 130 km difference in the locations of the west edges of the plumes on 15 Dec. 2009. Landfast ice is treated similarly in the present model and in the WWF study. However, the landfast ice extents used for the present model and in the WWF study were different; for the Alaskan Beaufort, the WWF study used a monthly average landfast ice extent based on observations between 1996 and 2008, and the present model uses time varying landfast ice extents extracted from the CIS's Regional Western Arctic ice charts. Figure 6c shows the average landfast ice extent used in the WWF study for the month of Dec. and one of the weekly CIS landfast ice extents for the month of Dec. 2009. We investigated whether this difference in the landfast ice extents caused the difference in the location of the plume; neither the present model plume nor the WWF plume is impacted by the existence of landfast ice because both plumes are away from the landfast ice edge. Note that we have not shown landfast ice extents in Figures 6a and 6b because plumes are far from the land for the dates associated with these two figures.

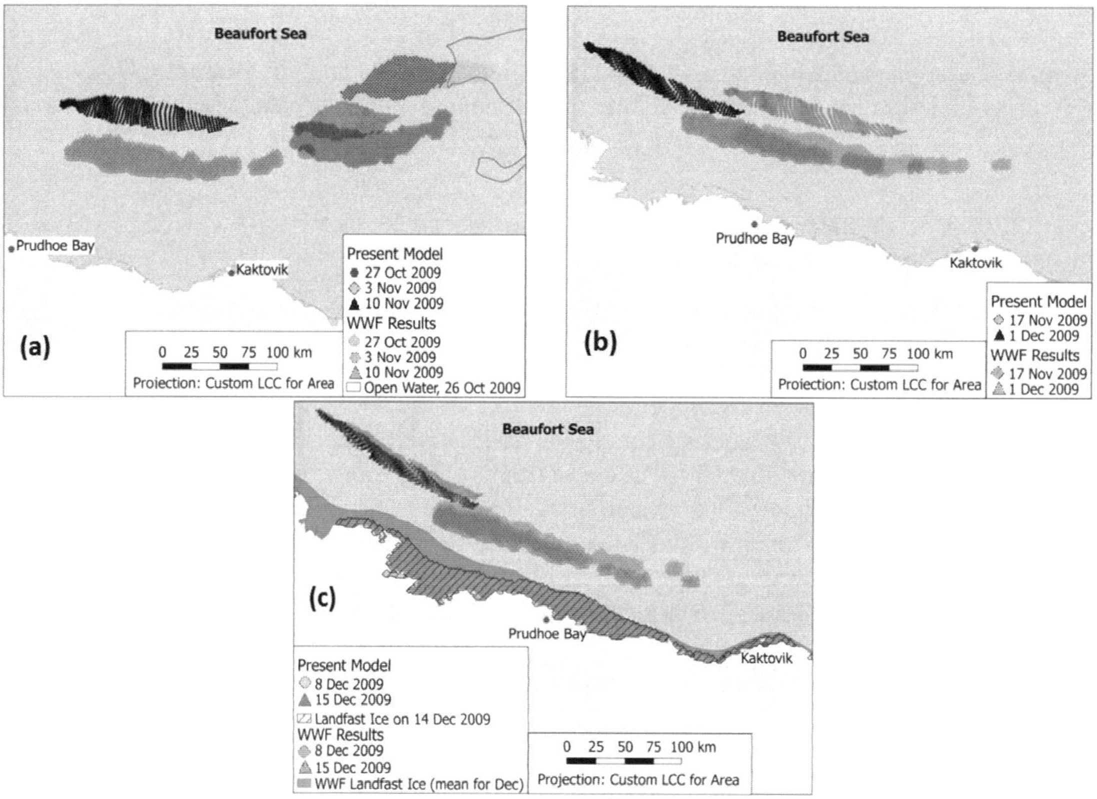

Figure 6. Comparison of the present model results with the WWF results reproduced from http://arcticspills.wwf.ca/. The difference in the coastward motion of the modelled and WWF plumes is within 40 km. The difference between the along the coast motion is within 15 km until Nov. 17 2009 after which the present model predicts faster motions.

It is challenging to find out whether the WWF or the present model results is a closer estimate of where the plume would go in reality. This is particularly true for the modelled nearshore plumes because both models are based on data known to have higher uncertainty levels closer to coastlines. For example, the Polar Pathfinder dataset, used in the present model, is expected to be less accurate close to coastlines and possibly in areas and times that no IABP buoys exists. To the best of the authors' knowledge, no IABP buoy exists in the space-time window of interest for the modelled scenario. The TOPAZ4 dataset, used for the WWF study, is expected to be less accurate nearshore (Gearon et al., 2014; Xie et al., 2016; Sakov et al., 2012) or even unavailable in some coastal regions of interest (Gearon et al., 2014).

CONCLUSIONS

Satellite-derived ice drifts were the basis of an in-ice oil spill trajectory model developed based on the assumption that oil moves with ice. To enhance the model, time dependent, spatially varying landfast ice extents were included in the model to reduce the uncertainty of satellite-derived ice drifts in coastal regions. Two hypothetical in-ice spill scenarios in deep and shallow waters of the Beaufort Sea were modelled and analyzed. The present model was used to estimate an observed spill trajectory that moved with ice in the Barents Sea; the model underestimated the extent of the trajectory because of the underestimation of ice drift in the input satellite-derived ice drift dataset. The trajectory of a buoy in the same general area was, however, closely estimated. The present model results were also compared with an independent numerical model results for in-ice oil spills in the Beaufort Sea. Coastward motions were generally similar but deviate after some time. Both the present model and the independent study results could be more uncertain nearshore and it is then challenging to know whether the present model or the independent study is a better representation of reality.

ACKNOWLEDGMENTS

We thank NRC's Philippe Lamontagne for processing the latest version of the Polar Pathfinder dataset and the CIS ice charts. We also thank the NRC's Arctic Program for their financial support.

REFERENCES

Babaei, H., Burcher, R., Watson, D., and Tivy, A., 2016. "In-ice Oil Spill Trajectory Modelling based on a Satellite-derived Ice Drift Dataset for the Beaufort Sea", *Proceedings of the thirty-ninth AMOP Technical Seminar*, Environment and Climate Change Canada, Ottawa, ON, pp.1-21.

Drozdowski, A., Nudds, S., Hannah, C.G., Niu, H., Peterson, I.K., and Perrie, W.A., 2011. *Review of Oil Spill Trajectory Modelling in the Presence of Ice*, Canadian Technical Report of Hydrography and Ocean Sciences 274, Fisheries and Oceans Canada, 92 p.

Eurasia Group, 2014. *Opportunities and Challenges for Arctic Oil and Gas Development*, The Wilson Center, Washington, D.C., 29 p.

Gearon, M. S., McCay, D. F., Chaite, E., Zamorski, S., Reich, D., Rowe, J., and Schmidt-Etkin, D., 2014. *SIMAP Modelling of Hypothetical Oil Spills in the Beaufort Sea for World Wildlife Fund (WWF)*, Project Number, ASA 13-235, 311 p.

Kimura, N. Nishimura, A., Tanaka, Y., and Yamaguchi, H., 2013. Influence of Winter Sea Ice Motion on Summer Ice Cover in the Arctic, *Polar Research*, 32, 20193.

Lavergne, T., Eastwood, S., Teffah, Z., Schyberg, H. and Breivik, L.-A., 2010. Sea Ice Motion from Low Resolution Satellite Sensors: An Alternative Method and its Validation in the Arctic, *Journal of Geophysical Research, Oceans*, 115, C10032.

LOOKNorth, 2014. *Oil Spill Detection and Modeling in the Hudson and Davis Straits*, Prepared for Nunavut Planning Commission, Report R-13-087-1096, 274 p.

MAR Incorporated, SL Ross Environmental Research Ltd., DF Dickins Associates Ltd., and Emergencies Science and Technology Division, 2008. *Empirical Weathering Properties of Oil in Ice and Snow*, Project Number 1435-01-04-RP-34501, Prepared for U.S. Department of the Interior Minerals Management Service Alaska Outer Continental Shelf Region, 170 p.

Miller, P.A., Laxon, S.W., Feltham, D.L., and Cresswell, D.J., 2006. Optimization of a Sea Ice Model Using Basinwide Observations of Arctic Sea Ice Thickness, Extent, and Velocity, *Journal of Climate*, 19, pp.1089–1108.

Muir, A., Baker, S., Laxon, S.W., Ridley, J., and Kwok, R., 2011. Arctic Sea Ice Dynamics for Climate Models: GlobICE Project Results and Products, *American Geophysical Union*, abstract #C52B-04.

National Snow and Ice Data Centre, 2016. *Arctic Sea Ice News & Analysis* [Online] Available at: http://nsidc.org/arcticseaicenews [Accessed 21 Nov. 2016].

Ragnhild, L. D., Faksness, L.G., Durgut, I., Brandvik, P. J., Leirvik F., 2011. *Modelling of Oil in Ice with OSCAR*, SINTEF A19804.

Sakov, P., Counillon, F., Bertino, L., Lisæter, K. A., Oke, P.R., Korablev, A., 2012. TOPAZ4: An Ocean-sea Ice Data Assimilation System for the North Atlantic and Arctic, *Ocean Science*, 8, pp. 633–656, doi:10.5194/os-8-633-2012.

Sorstrom, S., K., Brandvik, P., J., Buist, I., Daling, Per., Dickins, D., Faksness, L. G., Potter. S., Rasmussen, J.,F., Singsaas, I., 2010. *Joint Industry Program on Oil Spill Contingency for Arctic and Ice-covered Waters*, Summary Report: SINTEF A14181. Open.

Sumata, H., Lavergne, T., Girard-Ardhuin, F., Kimura, N., Tschudi, M.A., Kauker, F., Karcher, M., and Gerdes, R., 2014. An Intercomparison of Arctic Ice Drift Products to Deduce Uncertainty Estimates, *Journal of Geophysical Research, Oceans*, 119, 8, pp.4887-4921.

Tschudi, M., C. Fowler, J. Maslanik, J. S. Stewart, and W. Meier. 2016. *Polar Pathfinder Daily 25 km EASE-Grid Sea Ice Motion Vectors. Version 3. Boulder, Colorado USA: National Snow and Ice Data Center* [Online] Available at: http://dx.doi.org/10.5067/O57VAIT2AYYY [Accessed September 2016].

Venkatesh, S., El-Taha, H., Comfort, G. and Abdelnour, R., 1990. Modeling the Behaviour of Oil Spills in Ice-infested Waters. *Atmosphere-Ocean*, 28, pp. 303-329.

Xie, J., Bertino, L., Counillon, F., Lisæter, K. A., Sakov, P., 2016. Quality Assessment of the TOPAZ4 Reanalysis in the Arctic over the Period 1991-2013, *Ocean Science Discussions*, doi:10.5194/os-2016-38, 53 p.

POAC'17
Busan, KOREA

Proceedings of the 24th International Conference on
Port and Ocean Engineering under Arctic Conditions
June 11-16, 2017, Busan, Korea

The Entrainment and Migration of Crude Oil in Sea Ice, the Use of Vegetable Oil as a Substitute, and Other Lessons from Laboratory Experiments

Megan O'Sadnick[1], Chris Petrich[1], Nga Dang Phuong[1]
[1] Northern Research Institute (Norut) Narvik, Narvik, Norway

ABSTRACT

Understanding the interaction between oil and sea ice is essential in the development of oil spill detection and response technology in the Arctic. Laboratory experiments were performed to examine oil migration in sea ice during the cold phase and during warming, and investigate the behavior of different oils in sea ice to identify suitable substitute oils for crude oil. Vegetable oil and Troll B crude oil were injected underneath laboratory-grown sea ice with oil lenses either 1 cm or 3 cm in thickness. Results show similar behavior of vegetable oil and crude oil in sea ice both shortly after oil injection and during the warm phase. Further, results are independent of lens thickness. The implications of this result are discussed. In addition, the impact of cylindrical confinement of ice is shown on crystal and brine channel structure, and the energy balance of the tank are discussed.

KEY WORDS: Sea ice; Oil; Microstructure

INTRODUCTION

Research into the interaction between crude oil and sea ice began in earnest in the 1970s with both large scale field studies designed to gained a stronger overall understanding of the topic (NORCOR, 1975; Martin, 1979; Deslauriers, 1978) and smaller scale laboratory experiments which focused on specific aspects of the interaction between oil and sea ice (Keevil and Ramseier, 1975, Mackay et al., 1976) Since that time, research into the oil and ice has expanded with an increasing number of studies focused on the microscale interaction of oil and sea ice including the influence of pore structure on the rise of oil through ice and the associated parameters that may impact this process (Karlsson et al, 2011; Petrich et al., 2013; Maus et al., 2015; Salomon et al., 2017).

Due to current environmental regulations, it is difficult to perform experiments in the field examining the interaction between crude oil and sea ice. This concern provides the first objective for our experiment:

1) *The comparison of crude oil to vegetable oil* to determine the suitability of the latter as a replacement in experiments.

The three additional objectives of this study are:

2) *The examination of the parameter space including the influence of lens thickness without added hydrostatic pressure from encapsulation and time elapsed on the migration of oil upward.* To ensure only lens thickness influences movement of oil upwards, ice melt is initiated before an ice cap can form beneath the lens. The pressure potentially created as ice encapsulates the oil lens is therefore prevented simplifying analysis of results from lenses of differing thicknesses.
3) *The influence of cardboard cylinders on ice growth and oil migration.* Use of cylinders allows for the easy containment of oil lenses and ability to control lens diameter and thickness independent of under-ice topography. However, ice structure and brine channels in cylinders have not yet been characterized.
4) *The quantification of the energy balance of ice tanks.* Quantifying growth conditions in ice tanks allows for more accurate and detailed comparisons to be made across experiments, in both the laboratory and field, and enhance our ability to alter conditions to suit the needs of an experiment.

METHODS

To prepare samples of artificial sea ice, two tanks with slanted walls, each having the dimensions of 0.50 x 0.50 x 0.50 m, were first filled with a mixture of fresh water and Blue Treasure Synthetic Salt to a level of 0.41 m (Salomon et al., 2016). Starting salinity in both tanks was 32.9 ppt. A probe measuring temperatures in 1 cm vertical intervals was installed in each tank. Thermocouple wires ran horizontally along the width of the tank in an effort to decrease heat conduction and the buildup of ice at the temperature probe (Figure 1b). Cardboard tubes to confine oil during injection were suspended from metal rods above the water surface and reached to a depth of about 0.16 m below the water surface (Figure 1a & 1b).

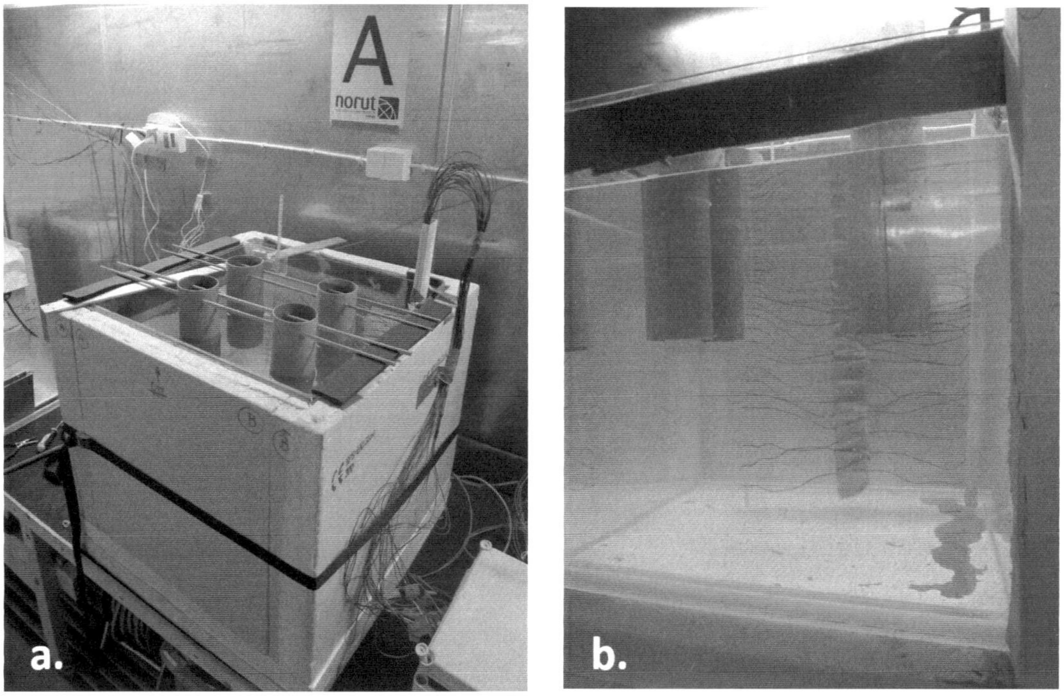

Figure 1. Experimental setup a) Overview of tank setup showing placement of thermocouples and cardboard tubes for oil injection; b) View of tubes and thermocouples from below the water surface.

Once setup was complete, the ambient temperature in the cold room was lowered to 0 °C to allow for the even cooling of the salt water prior to onset of ice formation. After two days, the ambient temperature was set to -15 °C to initiate ice growth. Due to the slanted design of the tank, the ice grew free-floating and developed a natural freeboard.

Once the ice reached a thickness of approximate 0.10 m, oil was injected. In Tank A, a mixture of rapeseed oil dyed with Sudan III at a concentration of 0.03% (w/w) was used. Of a range of store-brand vegetable oils tested (olive, sunflower, soybean, rapeseed, and fry oil),rapeseed oil was chosen as it was the oil with the lowest pour point of around -15 °C. Sudan III is a lysochrome, i.e, an oil soluble dye used to improve visualization of oil movement in the ice. In Tank B, North Sea crude oil from the Troll B field was used. Troll B is a sweet, naphthenic crude oil with a low pour point. A syringe having volume of 60 ml and a teflon tube were used to inject the oil. Care was taken to avoid the presence of air bubbles in the tube and subsequently the oil lens. Given the presence of cardboard tubes, injection was easy to control with spillage outside of the tubes being negligible. Four oil lenses were injected into each tank, with two lenses having a thickness of approximately 0.01 m and the other two lenses a thickness of 0.03 m - volumes of $2.8 * 10^{-5}$ m^3 and $8.5 * 10^{-5}$ m^3, respectively.

After four hours, one lens of each thickness was removed from both tanks using a saw to cut around the cardboard tubes. Cores were held vertically upright to prevent oil moving higher up into the ice than it had already. Since the oil lenses were not overgrown yet, most of the oil held in the lens was lost. Immediately after core removal, dry ice was packed upward into the cardboard tube to cool ice and oil and retain oil in the pore space (and potentially at the lower ice interface). The samples were quickly transported to a -80 °C freezer to stop any further movement of oil and prepare for future analysis. Samples were stored vertically. As a substantial volume of ice was removed from the tanks during this process, additional ice was placed into the tanks to ensure freeboard, similar to before removal, was maintained.

To examine the migration of oil upwards without the potential impact of pressure arising from the encapsulation of the oil lens, melt was initiated immediately after ice removal by increasing the air temperature to -2 °C. All four remaining cores were removed at the first observed surfacing of oil, which occurred about 30 hours after oil injection. In an attempt to preserve the location of oil in brine channels, cores were laid horizontally immediately after removal and dry ice was packed into the bottom and top of the cardboard tube. All samples were next placed in the -80 °C freezer horizontally.

Cores were stored in the -80 °C freezer between 8 and 17 days before being sectioned and photographed in the cold room at -15 °C to document any migration of oil upwards. Careful sectioning and realignment enabled us to take both vertical and horizontal sections of the ice cores. Horizontal sections between 5 and 20 mm thickness were melted to measure oil concentration and salinity. To measure the concentration of vegetable oil in cores removed from Tank A heptane was next added to the sample in a beaker of known weight in an amount at least equal in volume to the brine and oil mixture. Through vigorous mixing, the oil was extracted into heptane. Once a clear separation of heptane and brine appeared, brine was separated with a pipette. The remaining mixture of oil and heptane was left for heptane to evaporate, leaving behind only vegetable oil. The beaker was again weighed and the difference taken to be the mass of oil present in that slice of ice.

For melted crude oil samples, the oil was extracted with heptane. The exact amount of heptane used was recorded. The oil concentration in the heptane was then measured using a UV-fluorescence meter TD500™ (Turner Designs Hydrocarbon Instruments, Inc) which was calibrated to the crude oil prior to analyses according to standard procedures.

For both vegetable oil and crude oil samples, measurements of conductivity were taken on the remaining brine to determine sea ice bulk salinity. Temperature and salinity measurements of the ice were used to calculate brine volume fraction using equations derived from Cox and Weeks (1983).

RESULTS

In Figure 2, the evolution of ice and water temperature during the ice growth phase is presented. The freeze-in of each thermocouple is represented as a clear deviation from the average sea water temperature located around -2 °C. This thus provides a view of the rate of ice growth in each tank.

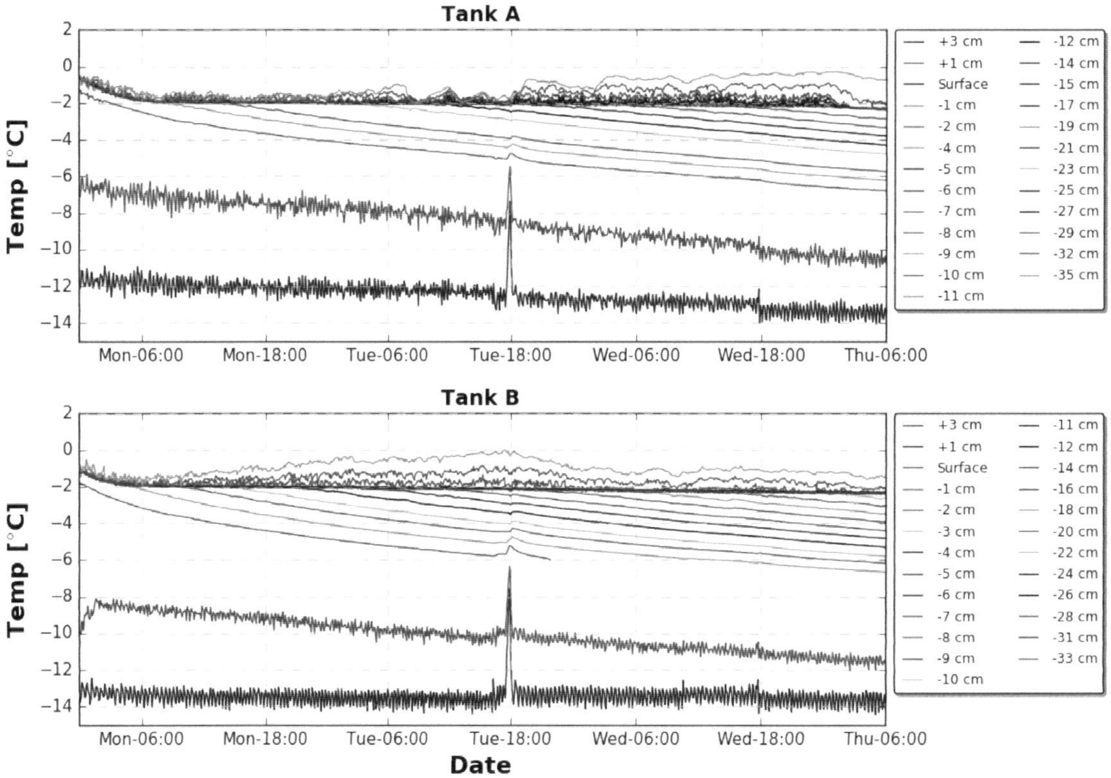

Figure 2. Evolution of temperature in Tanks A and B.

To examine the geometry of brine channels and the path oil traveled upwards, vertical thick sections were photographed of all cores (Figure 3). The vertical section was cut near brine channels visible in the bottom of the core. The temperature, salinity, and brine volume fraction profiles are presented in Figure 4 along with measurements of oil concentration.

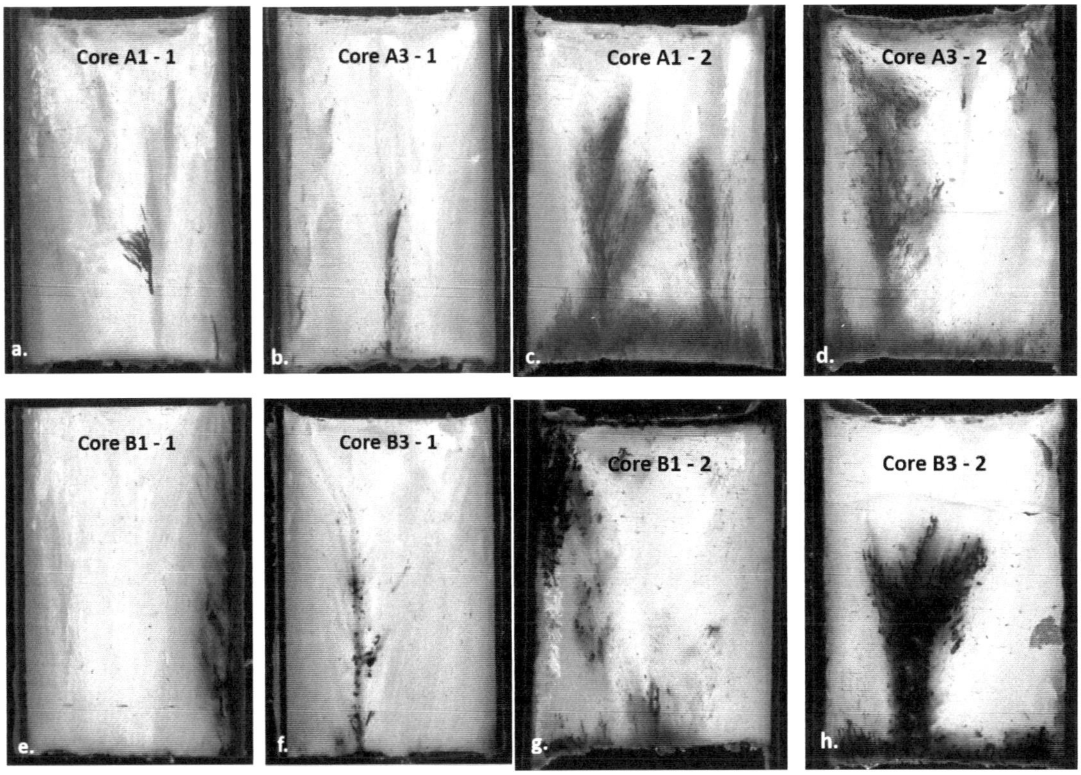

Figure 3. Vertical thick sections (approx. 10 mm thick) obtained from each core removed. (a,b,c,d) dyed vegetable oil and (e,f,g,h) crude oil, (a,c,e,g) oil lens thickness 0.01 m and (b,d,f,h) 0.03 m, and (a,b,e,f) cold ice after 4 hours, and (c,d,g,h) warm ice after 30 hours. The lower part of the channel in (a) was outside of the plane of the thick section.

Ice Properties

Four hours after injection, ice in Tank B was approximately 0.5 °C colder than in Tank A (Figure 2). This is likely due to circulation patterns in the cold room with Tank B being closer to the air outlet and thus exposed to drafts that led to more convective heat loss. As a result, ice thickness was slightly greater and porosity lower throughout the ice in Tank B in comparison to Tank A. However, bulk salinity was relatively consistent between the two tanks (Figure 4). After 30 hours of warming, ice temperature had evolved to be homogenous in both ice tanks. As expected, salinity generally decreased as brine drained resulting in an increased salinity only in the bottom 0.01 – 0.03 m. Porosity was below 20% in the cold ice and increased above 20% throughout the ice during melt (Figure 4). Inside the cardboard tubes, the lamellar structure at the ice–water interface and the crystal structure in horizontal sections (Figure 5) showed axial-symmetric alignment, i.e. horizontal *c*-axes aligned radially from the center.

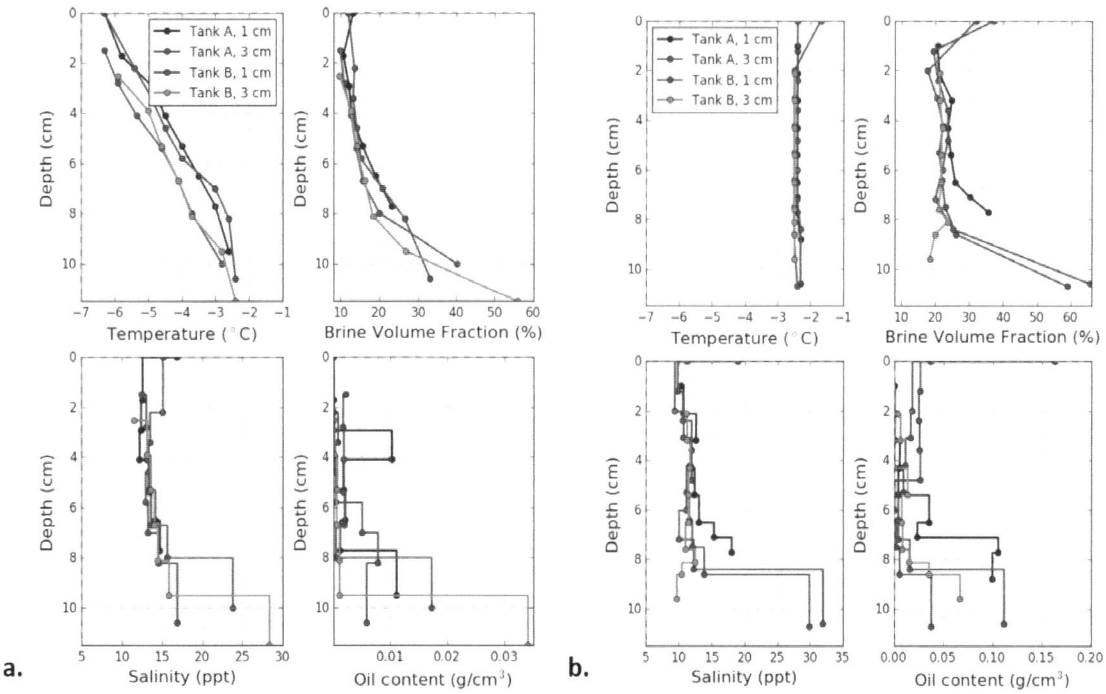

Figure 4. Profiles of ice properties and oil concentration from cores removed at 4 hours (a) and 30 hours (b). Note the different scales of oil content.

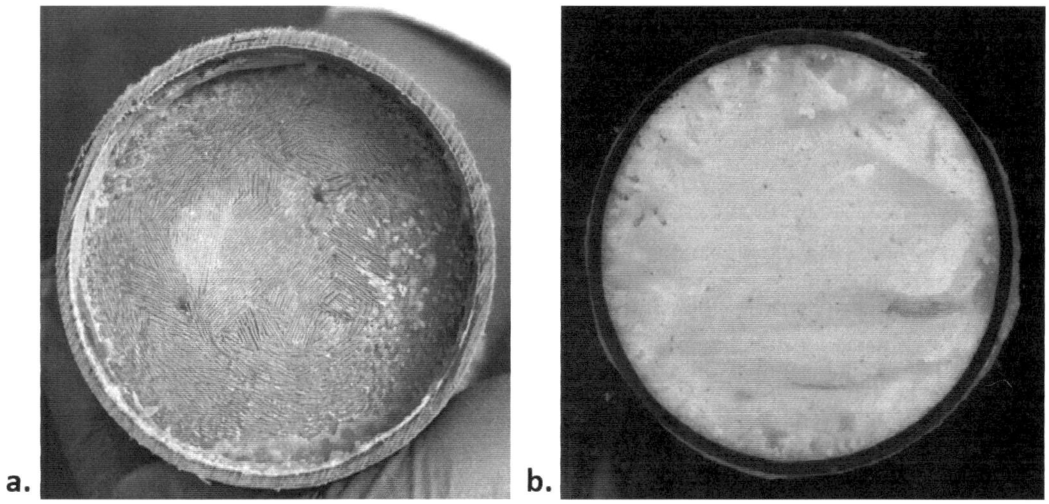

Figure 5. a) Lamellae structure of Core A1-2 at ice-water interface. Note the circular pattern of lamellae around the perimeter, and two brine channels (circular holes). b) Horizontal thick section showing crystal structure of Core B1-1, through cross-polarized filters, at approximately 6 cm depth.

Migration during the initial 4 hours

Figures 3a, 3b, 3e, and 3f as well as plots of oil concentration in Figure 4a, reveal migration only 4 hours after injection in all cores. In one of the two crude oil samples (Core B1-1, Figure

3f), oil travelled up only along the cardboard tube to reach the surface. It is presumed that this extreme migration is due to a weak bond or high porosity at the interface between ice and the cardboard. In the remaining three samples, migration occurred primarily through a brine channel located in the center of the core.

Of all samples taken in cold ice, one vegetable oil sample showed particularly pronounced top-heavy fanning of an oil-filled channel (Figure 3a, Core A1-1). This fanning of the brine channel is reflected in a peak of oil concentration 0.04 to 0.05 m below the ice–air interface (Figure 4a, Tank A 1 cm). Neither such peak (Figure 4a), nor such pronounced fanning (Figure 3) have been found at the upper ends of the channels in either Core B3 - 1 and Core A3-1. In fact, both cores show a trend toward decreasing oil content with distance above the oil lens.

The amount of oil entrained into the skeletal layer is small and limited to the bottom-most millimeters in Cores 3A-1 and 3B-1 (Figure 3). Entrainment appears to be in droplet form and more concentrated around the brine channel. Measurements of these small amounts of entrainment into the skeletal layer are difficult to perform quantitatively since the separation between oil lens and skeletal layer may not have been clean and some oil from the lens may have held on to the interface before it got immobilized by dry ice. As a result, all cores have a measureable amount oil at this depth despite little visual evidence inside the lamellar layer.

Migration over 30 hours

The remaining cores were all removed upon the first sight of oil surfacing. Vertical thick sections show more diffuse oil distribution compared to the distinct channel features in cold ice (Figure 3). Surfacing of oil occurred in Cores A3-2 (vegetable oil) and B1-2 (crude oil) 30 hours after injection. In both cores, the migration reaching the surface seemed to have occurred along the interface between the ice and cardboard tube. In Core A3-2, a large oil-filled brine channel was also present at the center of the core. The combined contributions may explain the high oil content from the surface down to 0.05 m depth (Figure 4b, Tank A, 3 cm).

In Cores B3-2 and A1-2, migration of oil occurred primarily through large brine channels close to the center of the core (Figure 3c, h). Two oil-filled channels developed in Core A1-2, neither leading to the surface but with one branching substantially outward toward the top. In Core B3-2, this branching occurred to the greatest degree of any of the cores. However, the oil profile in Figure 4b shows that the oil concentration decreased monotonously with distance from the oil lens, indicating that the increase in penetrated horizontal cross-sectional ice area was accompanied by a decrease of oil concentration within that penetrated cross-section.

Evident in all warm cores is entrainment of oil into the skeletal layer between 10 and 20 mm. Oil appears to have travelled the greatest distance in Core A1-2. This behavior is seen in measurements of oil concentration with Core A1-2 having 0.05-0.10 g/cm^3 in the 20 mm above the oil lens (i.e., from 0.07-0.09 m), equivalent to 10% oil by volume. In the other cores, high oil concentration in the skeletal layer is only seen in the bottom 0.01 m.

DISCUSSION

Comparing Vegetable Oil to Crude Oil

In comparing the behavior of vegetable oil to crude oil over 4 hours and 30 hours, many similarities are apparent between the oils. After 4 hours, little to no infiltration of the skeletal layer occurred in samples containing both kinds of oil (Figure 3). The migration of oil upwards was largely dependent on the presence of a main brine channel in the center of the core. When present, oil in all cores travelled up to 0.04 to 0.03 m below the ice surface. While a higher concentration of oil was present in cores with vegetable oil, this may be due to the branching away from the main brine channel observed, particularly apparent in Core A1-1. Possibly the most plausible explanation for this was that the brine channel in Core A1-1 fanned out more than in other cores. While no measurements of viscosity were obtained on vegetable or crude oil samples at relevant temperatures, at room temperature, the viscosity of vegetable oil is about 10 times higher than the viscosity of Troll B crude oil. Measurements of the respective interface tension are not available. One could surmise from our findings however, that the vegetable oil did not flow much easier through cold ice than crude oil. Examination of cores gathered after 30 hours provides some insight into these findings.

Cores A3-2 and B1-2 both had oil surface due to oil traveling up along the side of the core (Figure 3). Cores A1-2 and B3-2 however, had the majority of oil migration occur through brine channels located near to the center of the core. Oil lens thickness did differ between these two cores, however the height to which oil travelled upward is comparable between the two cores. The largest difference in oil concentration was observed near the bottom of the core, 0.08 to 0.10 m from the top. In Core A1-2, oil concentration reached over 0.10 g/cm^3 due to the high level of oil infiltration into the skeletal layer. From these results, it appears the vegetable oil and crude oil exhibit similar behavior in ice when large brine channels exist. Where differences occur and where further investigation is required is the ability of both to travel into the skeletal layer as well as into narrower channels that branch from the main brine channel. The low entrainment depth is particularly noteworthy in cold ice compared with expectations from earlier experiments and field work (Karlsson et al. 2011; Petrich et al. 2013). Differences between those experiments and the experiments reported here are the absence of overgrowth of the oil lens. It had been hypothesized that the overgrown ice cap pressurizes the oil due to buoyancy (Martin, 1979), potentially explaining the missing observation of entrainment.

Impact of Lens Thickness, Time Elapsed, and Oil Movement

In examining cores to determine the impact of lens thickness and time elapsed, perhaps the most evident differences occur in the skeletal layer. The skeletal layer is composed of freshly formed platelets slowly protruding themselves into a thin layer of supercooled sea water below. It is highly porous and in both this and previous laboratory experiments performed at Norut Narvik, has been found to be typically 0.01m thick in cold ice. Therefore, unlike brine channels, the structure of the skeletal layer is relatively consistent across experiments and tanks. Examining entrainment of the skeletal layer as a potential function of lens thickness, no strong pattern emerges. At 4 hours, all cores of both differing oil type and lens thickness display little to no oil in the skeletal layer. It is the two cores with 0.01 m lens thickness that held the highest concentration of oil, between 0.01 and 0.02 g/cm^3 – an amount lower by an order of magnitude from cores collected at 30 hours. In cores with a lens of 0.03 m, small droplets of oil are visible particularly centered nearer to the brine channel. While it was not possible to quantify that oil with our methods, it suggests that lens thickness may impact how quickly oil settles into the skeletal layer initially. Closer examination with techniques such as micro tomography (uCT) may shed light on the entrainment process (Salomon et al., 2016; 2017).

Despite little movement into the skeletal layer, both vegetable oil and crude oil migrated through established brine channels after only four hours. The length of these channels varied from approximately 0.04 m to 0.07 m. This finding compares nicely to that of the 1975 NORCOR crude oil in ice experiment where oil spilled beneath cold natural ice in late fall also migrated on average between 0.04 – 0.06 m up brine channels. A greater amount of oil was observed in the skeletal layer during the NORCOR experiment, however, this may possibly be related to the encapsulation of the oil lens that occurred.

In cores removed after 30 hours, no dominant pattern emerged between cores of differing lens thickness. The largest entrainment depth was observed in vegetable oil of 0.01 m lens thickness, essentially the one expected to develop the least entrainment. Reduced oil entrainment may be attributed to the presence of brine channels for oil to travel upward. The two cores with the greatest concentration of oil in the skeletal layer had little to no movement of oil up along the cardboard tube. It is therefore surmised that the channels that formed at the cardboard tube acted to drain oil from the lens and reduce hydrostatic pressure before oil was able to fully entrain the skeletal layer.

Given the difference in oil movement above the skeletal layer up into channels, it is harder to draw conclusions on the impact of lens thickness on oil movement through channels and branches leading away from the main channel. Core A3-2 displayed a large broad channel with oil branching from the main trunk in addition to oil between the core and cardboard. In contrast, Core B1-2 showed greater migration of oil upwards than Core B3-2 due to a brine channel along the edge of the core. Oil is also visible within the core, possibly from movement up other small channels. In the sample having a 0.03 m crude oil lens, oil did not reach the surface but did branch outward extensively. Thus, from these findings, it appears that lens thickness did not play a dominant role in the migration of oil upward.

Further Lessons for Future Experiments

Cardboard Cylinders for Oil Containment

The use of cardboard cylinders to contain oil has its downfalls as displayed particularly in Figure 3d, 3e, and 3g. In these cores, oil traveled primarily up through the pore space at the ice-cardboard interface, the result likely of a weak bond between the two. In addition, a clear pattern arose in the ice crystal structure and lamellae somewhat circular in shape and reminiscent of tree rings (Figure 5). This pattern is believed to be caused by convection patterns formed within the tube. Convection has been shown to drive crystal alignment, including at the scale of 10-cm relevant for this experiment, and it is the reason for the development of brine channels (Langhorne and Robinson, 1986; Worster 1997; Petrich et al. 2007). As the development of brine channels in sea ice is often unpredictable, this finding may aid in the design of future experiments involving brine channels. Measures to avoid migration along the cardboard-ice interface are needed however. One simple solution is to prevent contact between cardboard and oil entirely by placing small pieces of insulation at the surface. This will create thicker ice at the walls of the cardboard thus creating a small 'pocket' at the center of the core.

The Balance of Heat in Ice Tanks

Ice, air and water temperature data from the temperature probe were evaluated to examine the energy balance of both tanks during the ice growth phase. Ice growth took place in darkness, reducing the energy balance equation to

$$\dot{Q}_H = \dot{Q}_C + \dot{Q}_I + \dot{Q}_L, \tag{1}$$

where $\dot{Q}_H$ is the rate of change of enthalpy of the sea ice and water, $\dot{Q}_C$ is the conductive heat flow across the ice–air interface, $\dot{Q}_I$ is the sensible heat input into the system at the bottom of the tank, and $\dot{Q}_L$ is the heat loss through the sides of the tank and at the bottom. The heat loss terms are calculated as the residual of Eq. (1), all other terms are calculated or known.

In Figure 6a, ice temperature is used to model depth-average brine salinity within the ice, depth-averaged porosity, and enthalpy of the tank. Results for porosity show higher porosity in Tank A than in Tank B, consistent with Figure 4a. Also, the enthalpy stored in Tank B was larger than in Tank A.

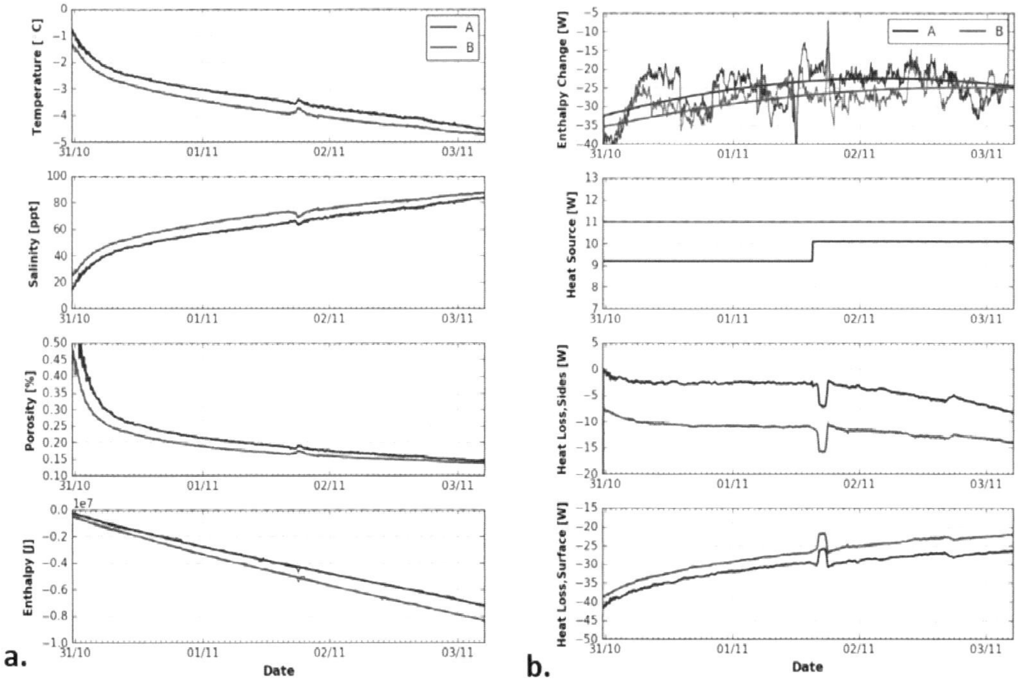

Figure 6. a) Calculation of enthalpy from modeled values of porosity and salinity derived from measurements of temperature; b) Heat balance with Tanks A and B. In (b), the rate of enthalpy change was calculated over 5 minute intervals with a third-order polynomial fit applied over measurements every 6 hours. The cross sectional area of the tank was 0.25 m^2, suggesting initial conductive heat loss rates through the surface of up to 160 W/m^2.

Figure 6b compares the derived heat flow rates. At the beginning of the experiment, the energy balance is dominated by the conductive heat flow and enthalpy change, indicating that ice formation is dominated by 1-dimensional heat loss to the surface. However, the heat loss to the sides and bottom increases in magnitude toward the end of the ice growth period while the surface heat flow decreases in response to thickening ice. The former is likely due to an increase in the difference between tank and air temperature which will continue to grow in magnitude as ice thickens. In Tank A, the magnitude of the surface flow is shown to be consistently larger than in Tank B due to ice being 0.01 m less in thickness. The difference in thickness is further reflected in measurements of enthalpy and enthalpy change. Values represent enthalpy in both ice and water and were derived using values of average temperature found in both. The greater the magnitude of enthalpy change, the greater amount of ice present in the system. The heat loss to sides and bottom of the system are consistently higher in Tank B than in Tank A, which is ascribed primarily to the placement of Tank B in the direct draft of the cooling system.

In the second panel of Figure 6b, heat input is presented. While both tanks were set up similarly, a higher heat flux was initially applied to Tank B in an effort to counter the greater amount of heat loss. In Tank A, heat input was increased midway through the ice growth process due to the formation of platelet ice along the sides of the tank and at the ice–water interface. This action was helpful in avoiding platelet growth and did not impact total ice growth substantially. This finding is useful in designing future experiments where one aims for consistent ice growth, a flat ice/water interface, and little to no platelet formation.

Heat flux through the side was determined as the residual of the other three terms. This is the flux through the four sides of the tank and the bottom, each having approximately 0.05 m of insulation. As time progressed, ice thickened, and average temperature in the two tanks dropped, heat flux through the sides increased in magnitude. The difference between Tank A and Tank B is particularly evident in this plot with Tank B having often greater than double the flux through the tank sides than in Tank A. However, distributing the heat flow through the sides between four or five sides, it is evident that the sides were insulated and ice growth is still dominated by flux through the surface in both tanks. The flux through the surface was often more than ten times greater than flux through any one side of the respective tank.

CONCLUSIONS

To develop a complete understanding of the migration of oil through sea ice it is at times advantageous to downscale experiments to constrain the impact of various parameters and focus on specific aspects of the process. In doing so, useful data can be collected while also often saving on both cost and time. Using this approach, we were able to reach the following conclusions that both add to current knowledge of oil in sea ice and will assist in the design of future laboratory experiments.

1) Vegetable oil that is known to still flow at -15 °C displays similarities to Troll B crude oil in its movement through brine channels and the rate of migration to the surface. Hence, vegetable oil can be a suitable alternative to study the general process of oil migration through sea ice. Further studies need to be conducted quantifying the crude oils and circumstances where vegetable oil cannot serve as a substitute.
2) The thickness of the oil lens did not have a noticeable impact on the migration of oil upward in this experiment (comparing 0.01 m with 0.03 m lens thickness under 0.1 m ice). The dominant factor was instead the presence and placement of brine channels. Entrainment in the skeletal layer after four hours was low or nonexistent for lenses of either thickness. Presumably, this was because an ice cap had not yet formed beneath the oil lens, limiting the magnitude of hydrostatic pressure driving oil into the ice.
3) Cardboard cylinders do have an impact on the crystal structure and brine channel morphology of the sea ice formed within due likely to patterns in fluid flow created by the cylinder (cf. Petrich et al., 2007). Such cylinders therefore provide a potential method to generate brine channels above the oil lens for a targeted study. Much care needs to be taken, however, to limit migration at the interface between the cardboard and ice.
4) The heat balance within the tanks shows the greatest amount of heat loss is through the surface with loss through the sides being an order of magnitude lower. Heat flux in to the tank from resistors placed below, assisted in the prevention of platelet formation while not impacting overall ice growth.

ACKNOWLEDGEMENTS

Funding was provided by the Research Council of Norway (RCN) PETROMAKS2 program through project MOSIDEO (project number 243812), RCN Center for Research-Based Innovation (SFI) CIRFA (grant number 237906), and the European Union's Horizon 2020 research and innovation program under grant agreement No 679266 (GRACE). In addition, we would thank Statoil for supplying crude oil from the Troll B field.

REFERENCES

Deslauriers, P.C. (1978). The Bouchard No. 65 Oil Spill in the Ice-Covered Water of Buzzards Bay. *Journal of Petroleum Technology,* pp. 1092- 1100.

Karlsson, J., Petrich, C., & Eicken, H. (2011). Oil entrainment and migration in laboratory-grown saltwater ice. *Proceedings of the 21st International Conference on Port and Ocean Engineering under Arctic Conditions (POAC)*, Montreal, Quebec, Canada, 10–14 July 2011. Paper POAC11-186, 10 pp.

Keevil, B.E., & Ramseier, R.O. (1975). Behavior of Oil Spilled Under Floating Ice. *Conference on Prevention and Control of Oil Pollution*, pp. 497-501.

Langhorne, P. J., & Robinson, W. H. (1986), Alignment of crystals in sea ice due to fluid motion, *Cold Regions Science and Technology*, 12, 197–214.

Mackay, D., Medir, M., & Thornton, D.E. (1976). Interfacial Behavior of Oil Under Ice. *The Canadian Journal of Chemical Engineering*, 54, pp.72-4.

Martin, S. (1979). A Field Study of Brine Drainage and Oil Entrainment in First-Year Sea Ice. *Journal of Glaciology*, 22(88), 473-502.

Maus, S., Becker, J., Leisinger, S., Matzl, M., Schneebeli, M., & Wiegmann, A. (2015). Oil Saturation of the Sea Ice Pore Space. *Proceedings of the 23rd International Conference on Port and Ocean Engineering under Arctic Conditions*, Trondheim, Norway.

NORCOR, (1975). The Interaction of Crude Oil With Arctic Sea Ice. *Beaufort Sea Project, Department of the Environment*

Petrich, C., & Arntsen, M. (2013). Laboratory studies of oil encapsulation under growing sea ice. *Proceedings of the 22nd International Conference on Port and Ocean Engineering under Arctic Conditions (POAC)*, Espoo, Finland, 9–13 June 2013. Paper POAC13-014, 10 pp.

Petrich, C., Karlsson, J., & Eicken, H. (2013). Porosity of growing sea ice and potential for oil entrainment. *Cold Regions Science and Technology*, 87, 27–32, http://dx.doi.org/10.1016/j.coldregions.2012.12.002.

Petrich, C., Langhorne, & Haskell, P.J. (2007). Formation and structure of refrozen cracks in land-fast first-year sea ice, *Journal of Geophysical Research*, 112, C04006, 1–13, http://dx.doi.org/10.1029/2006JC003466.

Salomon, M. L., Maus, S., Arntsen, M., O'Sadnick, M.,Petrich, C., & Wilde, F. (2016). Distribution of oil in sea ice: Laboratory experiments for 3-dimensional micro-CT investigations. Presented at the *2016 Conference of the The International Society of Offshore and Polar Engineers (ISOPE),* Rodos, Greece, 26 June-1 July 2016.

Salomon, M.L., Arntsen, M., Dang, N.P., Maus, S., O'Sadnick, M., Petrich, C., Schneebeli, M., Weise, M. (2017). Experimental and micro-CT study on the oil distribution in laboratory grown sea ice. *Proceedings of the 24th International Conference on Port and Ocean Engineering Under Arctic Conditions*, Busan, Korea,11-16 June 2017.

Worster, M.G. (1997), Convection in mushy layers, *Annual Review of Fluid Mechanics*, 29, 91–122.

POAC'17
Busan, KOREA

Proceedings of the 24th International Conference on
Port and Ocean Engineering under Arctic Conditions
June 11-16, 2017, Busan, Korea

Experimental and Micro-CT Study on the Oil Distribution in Laboratory Grown Sea Ice

Martina Lan Salomon[1], Martin Arntsen[2], Nga Dang Phuong[2], Sönke Maus[1],
Megan O' Sadnick[2], Chris Petrich[2], Martin Schneebeli[3], Mareike Wiese[3]

[1] Norwegian University of Science and Technology, Trondheim (NTNU), Norway
[2] Northern Research Institute Narvik (NORUT), Narvik, Norway
[3] WSL Swiss Federal Institute for Snow and Avalanche Research SLF, Davos, Switzerland

ABSTRACT

Increasing activities in Arctic waters bare a risk of oil spills under ice-covered conditions and afford sound understanding of the interplay between sea ice and oil. Towards better knowledge, this study focuses on X-ray- micro computed-tomography (μ-CT) investigations of a laboratory oil- in ice experiment. The 3-dimensional distribution of oil in the porous space of 11- 13 cm columnar ice grown in a laboratory was investigated. Two different oil content measurement methods are discussed. (i) The first method quantifies the oil volume fraction based on μ -CT-scan investigations, allowing spatial oil distribution analysis in the porous space of sea ice. Oil inclusions were mapped manually over the acquired CT-scans with a resolution of 18 μm and 25 μm, respectively. Results give higher oil contents for smaller resolutions. Oil migration of 4 cm was observed. (ii) The second method quantifies the present oil concentration with fluorescent measurements. CT- scans give in comparison to fluorescent measurements a root mean square error of 1.27 % (18 μm) 0.76 % (25 μm), respectively. Bulk salinity determined from melted samples is compared with salinity estimated from μ –CT data.

KEY WORDS: Sea ice, Computer Tomography, Microstructure, Oil, Porosity

INTRODUCTION

Activities in polar marine areas are increasing. Including interest in Arctic resources. Oil exploration in potentially ice-covered regions and growing shipping activities in high latitudes, bare a risk of accidental oil spills. Giving objective to study the interaction of sea ice and oil for environmental and sustainable risk management.

Networks of brine channels and air inclusions make sea ice to a permeable material and a potential entrapment medium for oil. Investigations on the interaction of oil and sea ice have been published since the 1970's (Glaeser & Vance, 1971; NORCOR, 1975). Several field studies and experimental work were conducted since then. Studies were mainly based on destructive two- dimensional observations of thin sections and macroscopic descriptions, characterizing processes from oil encapsulation over oil migration to oil surfacing on sea ice (c.f Martin, 1979; Nelson, 1981; Buist & Pistruzak, 1981; Buist et al., 1983; Otsuka et al. 2004; Karlsson, 2009; Karlsson et al., 2011). Experiments quantifying the oil content in sea ice were conducted by Karlsson et al., (2011) and Otsuka (2004), reporting that the ability of oil to penetrate sea ice is limited by its porosity. A porosity threshold of 0.1 to 0.15 was determined and an oil saturation up to 4.5 to 7 mass percent oil by mass of sea ice was reported. Based on these observations and model assumptions (Petrich, Karlsson, & Eicken, 2013)) computed the potential oil penetration depth and concluded on oil entrainment volumes from less than 1 L/ m^2 as high as 5 to 10 L/ m^2 with increasing ice temperatures. 3-dimensional insights and application of X-ray tomography (XRT) to sea ice samples were first given by Kawamura (1988) with a resolution of 1 mm. Maus, 2009, 2013, 2015 applied XRT and synchrotron- XRT on sea ice with a resolution of 18 to 25 μm, respectively a resolution of 2 x 5.6 μm. Based on 3- dimensional micro- CT data, oil infiltration rate was computed, concluding on entrainment volume not solely depending on porosity. Furthermore, the control mechanism are described by parameters such as, pore size, pore necking and convection driven desalinization. (Maus et al., 2013) Salomon et al. (2016) give preliminary results on the detection of oil in sea ice based on synchrotron- XRT.

The present experiment was designed to detect the 3-dimensional distribution and the concentration of oil in the porous space of sea ice based on μ -CT investigation. Oil concentration measurements were achieved with two different approaches. (i) First method is based on μ-CT-scans and mapping the oil content over the 2- dimensional μ-CT sections. (ii) Second approach focused on UV-light stimulation and emitted fluorescence measurements of hydrocarbons. Resulted brine volume fraction from μ-CT-scans were converted into bulk salinity and are compared with measured bulk salinity.

METHODS

Experimental Setup

Ice growth was performed in a cold lab at NORUT Narvik in slightly conical double-walled plexiglas container with outer-dimensions of 51 x 51 x 51 cm and inner-dimensions of 50 x 50 x 50 cm. Approximately 100 liter of artificial saltwater (Blue Treasure Synthetic Salt) with an initial salinity of 31.5 ppt was filled into the tank. The plexiglas tank allowed visual access to the freezing process and facilitated the oil injection. The double-tank arrangement, surrounded by a 5 cm thick Styrofoam- isolation and air-cooling allowed one-dimensional ice growth. Heat supply from the bottom of the tank avoided supercooling of the water column and led to columnar ice growth. Fans beneath the tank assured homogenous heat distribution.

Prior to ice growth initiation, artificial seawater had a water temperature of approximately 0 °C. Referred starting point of the experiment is the set of ambient temperature to –15 °C. After 64 hour ice reached a thickness of 8 cm and oil was injected below the ice with a silicon syringe of 5 mm diameter. 28 h hours after oil release, room temperature was raised to -10 °C, to facilitate oil migration, to ensure that the surface temperature well above the nominal pour point of crude oil at -15 °C. Sampling commenced 91.5 hours after experimental start when the ice thickness reached 11-13 cm. Cores were taken with a drill 50 mm in diameter. Samples were stored over night at -80 °C and than transported on dry ice (-78.5 °C) from NORUT Narvik (Norway) to NTNU Trondheim (Norway). After storage at - 50 °C over a period of 9.5 months they were prepared for imaging with X-ray- micro computed-tomography (μ-CT). The cores were cut into subsamples of 2 to 3.5 cm thickness and a diameter of 35 mm. During the cutting process it was attempted to preserve the oil lenses.

During transport from and to WSL Swiss Federal Institute for Snow and Avalanche Research SLF (Davos,Switzerland), where the first μ-CT measurements were performed at -15 °C, sample temperature was kept at the same temperature by eutectic cooling elements. Back to NTNU samples were stored for another 2 months at -50 °C. Imaging of the same samples was conducted at the RECX, Norwegian Centre for X-Ray diffraction, scattering and imaging at the Norwegian University of Science and Technology (NTNU, Trondheim, Norway). Before imaging was performed samples were warmed up to the imaging temperature of -15 °C over night. Prior to fluorescent measurements sample were molten in airtight glass containers and shipped to NORUT Narvik.

X-ray Micro Computed- Tomography (μ-CT) Imaging

μ-CT measurements were conducted with three instruments (i) ScancoMedical AG micro-CT 40 at SLF, (ii) SacancoMedical AG micro-CT 80 at SLF and (iii) XT H 225 ST micro-CT system from Nikon Metrology NV at NTNU. (i & ii) were operated with a micro focus X-ray source (7 μm diameter) current of 177 μA and an acceleration voltage of 45 kV. Detector panels of 2048 x 256 and 2048 x 128 pixels, respectively were used. Tomographic scans were performed in high resolution mode with 2000 rotation per 360°. The height of scans could be chosen according to the sample height. Scans were operated with a field of view (FOV) of 37 mm, resulting in a pixel size of 18 μm. Samples were acquired in a cold- room at ambient temperatures of -15 °C at SLF. Acquisition time at SLF was limited to a window of 5 days, therefore two CTs were used simultaneously. Settings were the same and a difference at the acquisition is not assumed. (iii) a current source of 250 μA and an acceleration voltage of 150 kV. Scans were performed with 3142 rotation per 360°. A Perkin Elmer 1620 with a pixel size of 200 x 200 μm and 2048 x 2018 pixels operated as flat panel detector. The FOV of 50 mm corresponds to a pixel size of 25 μm. Scanning temperature (-15 °C) was controlled with a thermoelectric assembly (www.lairdtech.com) by top and bottom sample cooling.

CT- data provide a 2-d map of the variation of X-ray absorption of substructures within the imaged sea ice. Those 2-d maps are called radiographs and are acquired over a finite number of radiographic viewing directions to reconstruct the distribution of X-ray absorptivity within a horizontal cross-section of the imaged sample. The cross sections are stored as 16-bit grey value stacks and give a 3- dimensional view of the scanned object.The resulting 3-d images of ice cores were processed in ImageJ (rsb.info.nih.gov/ij). First a rectangular cylinder with the largest possible horizontal cross-section was cropped. After cropping a 2 pixel median and 1.5 standard deviation Gaussian blur filter were applied. Segmentation of air and ice was based on using Otsu's algorithm (Otsu, 1979; Maus, 2015) For segmentation of brine from ice the Triangle algorithm was applied (Zack et al., 1977). A different segmentation method was chosen for setting the brine threshold, as the Otsu algorithm was found to give too high brine volumes (Hullar et al., 2016).

Absorption contrast between ice and oil was insufficient for automatic segmentation. Segmentation was performed manually using the Wand Tool in ImageJ. Oil inclusions were selected by tracing objects. Therefore an initial pixel (iP) value was chosen for each oil bubble in the acquired 2-D slices. The Wand Tool checked the 4 neighbouring pixels of the iP. If the adjoining pixel was within the tolerance range of 3-5 of the iP value, it was detected as oil and its neighbouring pixels were checked in the same manner. If one pixel did not fulfil the criterion of falling into the range of the iP value, it was not accounted longer for oil and a border was set. Manual selection of oil was repeated on every 5th to 20th slice of an image stack. The selected areas were interpolated between the 2-D slices with the Region of Interest (ROI) manager and edited with the Segmentation Editor (Schindelin et al., 2012) giving a 3-D view of the oil inclusion. Segmentation results in 3-d images, where any voxel is either air, ice, oil or brine. The respective volume fraction were computed by GeoDict(2017).

The brine volume fraction Vb_{CT} was used to estimate the bulk sea ice salinity S_{CT} from

$$S_{CT} = \frac{S_b * \rho_b * Vb_{CT}}{\rho_i} \tag{1}$$

where S_b is the brine salinity at imaging temperature, ρb is the brine density at imaging temperature and ρi is the bulk sea ice density. Sb 177.96 ppt was interpolated from the table given by Cox and Weeks (1983), ρb was taken as $1000+0.8Sb$ and bulk sea ice density ρi $[kgm^{-3}]$ was computed from CT- data, knowing the analyzed volume as well as the ice, air, oil and brine fractions. For pure ice the density of 920 kgm^{-3} was assumed.

Oil density was estimated as 877 kgm^{-3} from measured Crude oil Troll B density of 0.879 Mgm^{-3} at a temperature of 20 °C, assuming a linear increase of oil density with 0.01 Mgm^{-3}/15 °C. The oil volume fraction Vo_{CT} was converted to oil mass fractions by multiplication with $\rho oil/\rho i$.

Salinity and Oil Content of Melted Samples

Bulk salinity S of melted samples is based on electric conductivity measurements. Conversion from conductivity to ppt were conducted with the Gibbs Sea Water (GSW) Oceanographic Toolbox of TEOS- 10 (Dougall et al. 2011).

The mass fraction $Mo_{Fluroescent}$ oil in the melted samples was extracted with heptane by adding an equal volume of heptane to oil-containing samples. The concentration of oil in the heptane was subsequently analysed using a UV-fluorescence meter TD500™ (Turner Designs Hydrocarbon Instruments, Inc). Under UV or near UV light, aromatic hydrocarbons in the oil are stimulated or excited to fluoresce. The intensity of fluorescent emission reflects the concentration of aromatic hydrocarbons in the oil. Since the UV-fluorescence meter was calibrated with the same crude oil, the intensity of fluorescent emission corresponds to the concentration of the oil (Brost et al., 2011).

RESULTS

In our laboratory experiment columnar sea ice grew over the course of 91.5 hours to a thickness of 11-13 cm at an air temperature of -15 °C. 63.75 hour after the experiment start, ice thickness reached 8 cm and oil was released. Four lenses of Troll B crude oil were injected under the ice, each with a volume of 20-25 ml and a thickness of approximately 4 mm. Injected oil migrated along the ice-water interface before it collected at the highest points of small under-ice undulations.[1] Within 4 hours, a thin rim of horizontal platelets formed around the oil lenses. After 5.5 hours, a thin layer of ice platelets had formed beneath all four oil lenses. Ice cores were taken was taken when the ice layer underneath the oil lenses was 2-3 cm and an overall thickness of 11- 13 cm. Table 1 summarizes the analysis of cores taken as described in the method section and presented in the following figures.

Air Fraction

CT- based results for total and closed air volume are given for both acquisitions at SLF and at NTNU and are compared with each other in Figure 1. Open porosity is defined by a continuous path from the pore to one of the surface edges in x-, y-, or z- direction, whereas closed pores

[1] Phenomena of oil collecting in concave under- ice cavities and encapsulation process are for e.g. described by Glaeser & Vance (1971), NORCOR (1975), Goodman and Fingas (1983) Buist et al. (1983) and Wilkinson et al. (2007).

are not connected to any surface.

Total and closed air volume fraction Va_{SLF} and Va_{NTNU} are plotted in % over the ice thickness [cm] for cores C,D,E and F each with a length of 13, 12,11 and 12 cm. Air porosity for sample C4 is not shown in Figure 1 as Va_{SLF} and Va_{NTNU} the sample broke during transport. Therefore different areas were analysed resulting in a significant variance for Va_{SLF} and Va_{NTNU}. Results from Closed air porosities in NTNU-acquisition excluding samples with oil lenses tend to be with a mean of 0.60% and a standard deviation of 0.23% higher than results from SLF with a mean of 0.52% and a standard deviation of 0.16%.

Brine Fraction and Salinity

Comparison of estimated bulk salinity S_{SLF} and S_{NTNU} determined from Vb_{SLF} and Vb_{NTNU} with measured bulk salinity S are shown for each core over the ice thickness in Figure 2. Salinity profiles show a characteristic C-Shape profile (Eicken, 1992). All salinity derived from μ-CT measurements is compared with direct measurements in Figure 4(a). Salinity calculation for S_{SLF} differs from direct measurements with a root mean square deviation of 1.64 %. S_{NTNU} differs from measurements with a root mean square error of 2.10 % from measured salinity.

Oil Fraction

Oil content for samples imaged at the SLF and NTNU are expressed in oil mass fraction [%] to be comparable with fluorescent measurements (Figure 3 & 4). Oil injection level is highlighted in the plots at an ice thickness of 8 cm. Measurements displayed with a black circle, correspond to highlighted results in Table 1. The highest oil volume was found in sections close to the injection levels, i.e. at and near the location of the oil lenses. Results of oil content Mo_{SLF} and Mo_{NTNU} are compared with fluorescent oil content measurements in Figure 4. Differences between Mo_{NTNU} and $Mo_{Fluorescenct}$ are given by a root mean square deviation of 0.76%. Mo_{SLF} deviates from fluorescent measurements $Mo_{Fluorescence}$ with a root mean square error of 1.27%. While μ -CT measurements seem to be biased slightly high or low, agreement with fluorescent measurements is very good. With the notable exception of C4 acquired with a resolution of 18 μm. Oil and ice absorption contrast was too low in the mentioned sample for visual detection.

Table 1. Summary of key parameters and experiment results.

Sample	Sample Depth from air-ice interface	Air Volume Fraction (SLF) Va_{SLF}		Air Volume Fraction (NTNU) Va_{NTNU}		Brine Volume Fraction (SLF) Vb_{SLF}		Brine Volume Fraction (NTNU) Vb_{NTNU}		Bulk Salinity S_{SLF} from Vb_{SLF}	Bulk Salinity S_{NTNU} from Vb_{NTNU}	Bulk Salinity S	Oil Mass Fraction (SLF) Mo_{SLF}	Oil Mass Fraction (NTNU) Mo_{NTNU}	Oil Mass Fraction (Fluorescent) $Mo_{Fluorescent}$
-	cm	%		%		%		%		ppt	ppt	ppt	%	%	%
-	-	total	closed	total	closed	total	closed	total	closed			-	-	-	-
C1	0-2.5	-	-	3.42	0.99	-	-	4.94	3.69	-	10.9	6.8	-	0.00	-
C2	2.5-3.5	-	-	0.57	0.51	-	-	3.34	2.34	-	7.4	5.1	-	0.00	-
C3	3.5-6	0.89	0.61	0.95	0.58	2.61	2.30	2.18	1.97	5.8	4.8	6.4	0.44	0.47	0.57
C4	6-8.5	26.45	0.19	0.65	0.12	2.54	1.99	4.09	2.27	5.6	9.0	5.3	0.98	0.00	1.19
D1	0-2.5	-	-	0.79	0.71	-	-	3.54	2.5	-	7.8	6.7	-	0.00	-
D2	2.5-5	0.63	0.48	0.60	0.50	3.66	2.77	3.04	2.44	8.6	6.7	6.4	0.008	0.01	0.02
D3	5-7.5	1.76	0.85	4.54	3.75	3.49	2.99	1.94	1.63	7.7	4.3	5.9	10.09	8.61	6.74
D4	7.5-11	-	-	0.48	0.44	-	-	4.72	4.22	-	10.4	9.1	-	0.00	0.56
E1	0-2.5	-	-	1.26	0.95	-	-	3.47	2.71	-	7.7	6.8	-	0.00	
E2	5-7.5	0.61	0.52	0.66	0.58	4.01	3.30	2.85	0.62	8.9	6.7	5.8	0.09	0.04	0.12
E3	7.5-10	0.97	0.67	1.04	0.55	3.10	2.75	1.61	1.53	6.9	3.6	5.3	1.39	0.72	0.73
E4	10-11.5	5.81	0.78	-	-	2.35	1.98	-	-	5.2	-	6.9	4.77	-	3.09
F1	0-2.5	-	-	6.38	0.78	-	-	2.79	2.07	-	6.2	7.1	-	0.00	-
F2	2.5-5	0.86	0.66	0.84	0.72	2.85	2.33	1.36	1.23	6.3	3.0	5.5	0.12	0.02	0.10
F3	5-8	4.47	1.05	3.60	3.52	2..40	2.18	1.85	1.54	5.3	4.2	4.5	6.88	5.07	-
F4	8-10	-	-	0.59	0.36	-	-	4.69	3.26	-	10.4	7.1	0.00	0.00	-

Temperature [°C]	Brine Density ρb [Mgm^{-3}]	Brine Salinity Sb [ppt]	Oil Density [Mgm^{-3}]
-15	1.14236	177.95	0.877

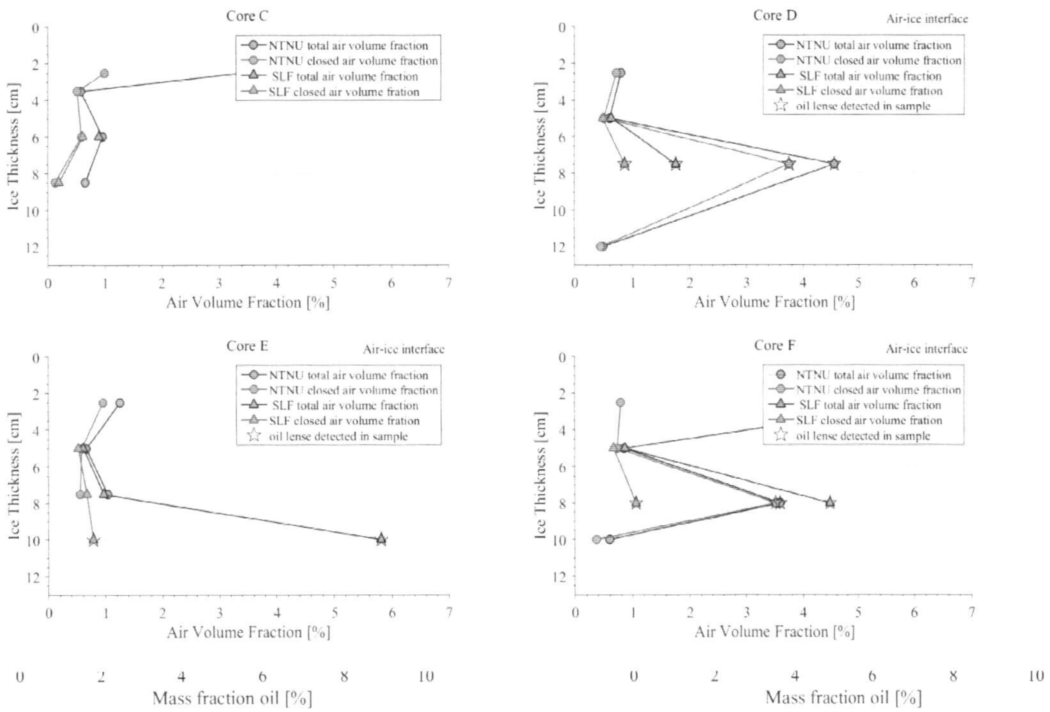

Figure 1. Closed and total air volume fraction from μ-CT- scans at SLF and NTNU in comparison.

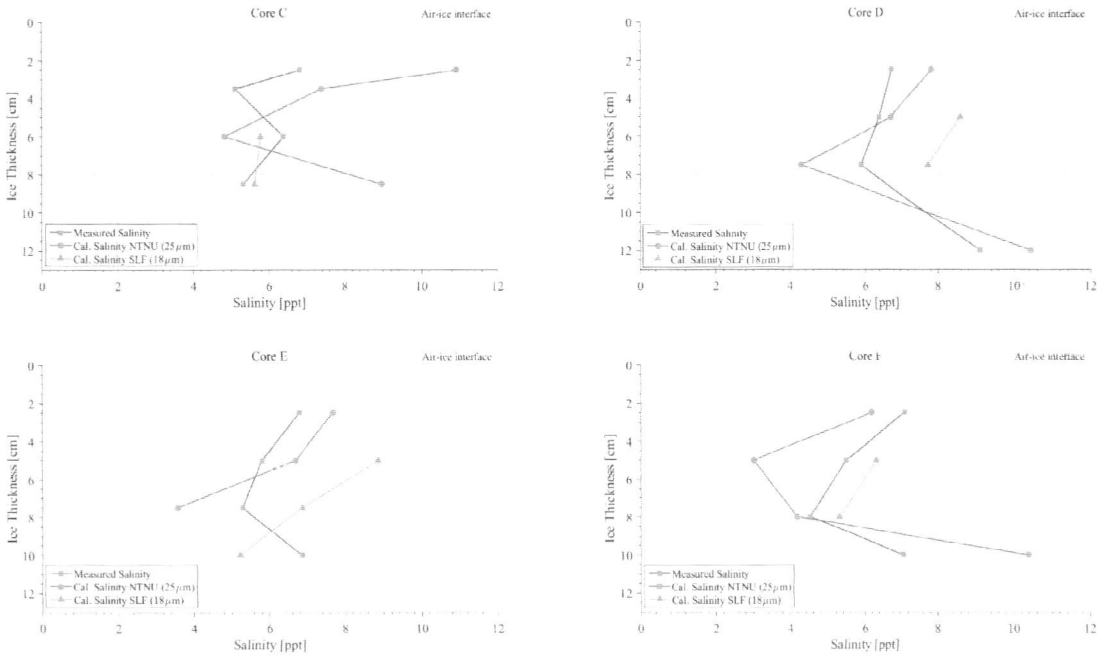

Figure 2. Measured bulk ice salinity in comparison with approximated bulk salinity based on segmented brine volume fractions of SLF-scans and NTNU-scans.

Figure 3. Oil content expressed in mass fraction in comparison between fluorescent measurement, μ-CT images from SLF and NTNU. Results highlighted with a black circle are averaged over the whole volume.

Figure 4. Bulk salinity of three different sources CT-scans (NTNU), CT-scans (SLF) and measured in comparison; Oil content from CT-scans (NTNU), CT-scans (SLF) and fluorescent in comparison.

DISCUSSION

Air Fraction

Air segmentation with Otsu algorithm gives reasonable and consistent results for SLF and NTNU scans in comparison (Figure 1). Total porosity is influenced by the open porosity, which might result from brine drainage. Values for closed porosity are there for more reliable. Spikes in air volume fraction are observed for samples containing oil concentrated in lenses (Figure 1), with one exception for sample F1. Sample F1 shows higher air content in the top most part. Observed air volume are open pores, indicating that potentially brine was leaking

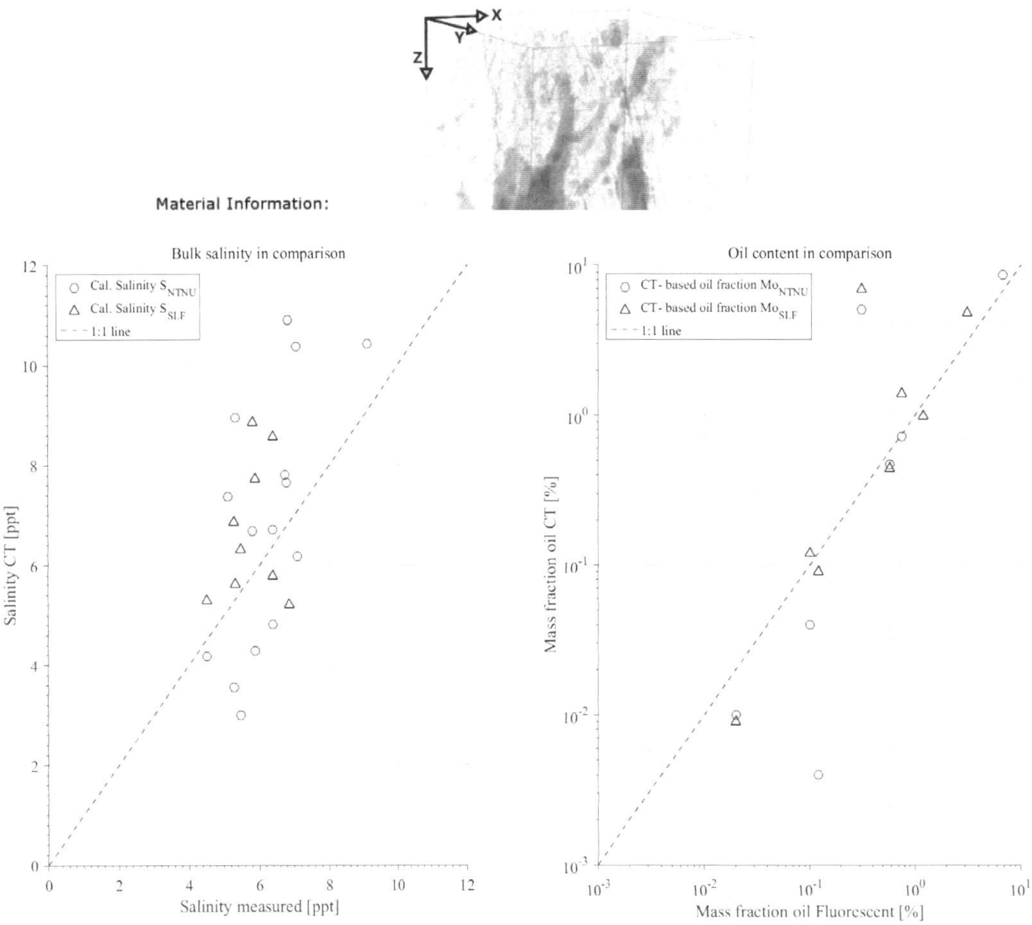

during sampling, storage and transportation.

We noticed larger air volumes above all oil lenses. Samples F3 and D3 where sectioned in a way, that preserved the closed air volume forming channels above the oil lens (Figure 6). Air volumes of 73.6 and 226.3 mm^3, respectively were found. Additionally smaller air bubbles were detected within the lenses of oil with a volume of 1 to 3 mm^3. However, we cannot be sure whether these are real due to air content of the oil or resulting from thermal cycling.

Figure 5. 3-D image of segmented oil (red) air (blue) and brine (green) for sample F3 resulting from μ-CT at SLF. Ice is invisible, except of the last image where ice is shown in white.

Too high closed air porosities in sample F3 and D3 from images at NTNU are observed in comparison to SLF data. Short thermal cycles above –15 °C might have occurred during scanning at SLF. As the temperature in the CT was controlled over the room temperature in the cold lab. Exposure to temperatures above the crude oil pour point (-15 °C) might lead to oil migration and drainage within the sample. Consequently the initial oil filled space is replaced by air. By cooling the sample afterwards down to the transport temperature, refreezing processes start and leave a closed pore.

While the source of these volumes could not been established unambiguously candidates include:

a. Unintended air release during first oil injection.
b. Air collected by the oil droplets migrating beneath the ice prior to freeze-in leading to an estimated air volume of 300 mm^3, for a prior freeze migration length of 10 mm by an oil lens diameter of 20 mm;
c. fluid loss during repeated warming and cooling cycles.

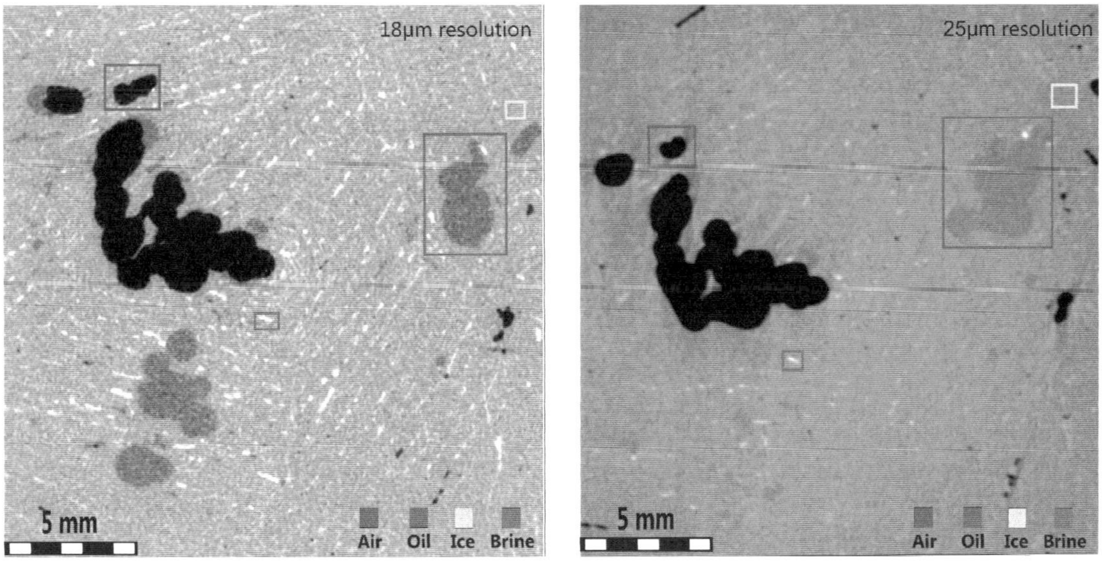

Figure 6. 2-D sections 18 μm respectively 25 μm resolution. Showing typical grey values for air, oil, ice and brine.

Brine Fraction and Salinity

CT- based salinity evaluation results in comparison to measured salinity in an error of approximately 2% (Figure 4). Where S_{SLF} shows a slightly higher salinity than S_{NTNU}. Error can possibly result from different:

(a) chosen energy setting for SLF and NTNU acquisition. Higher energy resolve the ice to brine contrast better, resulting in higher evaluated salinity. Furthermore, plastic sample holder (SLF) with a lower X-ray absorption in comparison to aluminum holder (NTNU) were chosen, having a possible effect on the results. Systematically evaluation on a bigger data set is needed to determine influences.
(b) pixel size of 25 μm and 18 μm results in resolving objects ≥ 50 μm and 36 μm respectively, i.e. 2x pixel size. This assumption would lead to an higher estimated salinity for smaller pixel sizes. Relevance of resolution for brine volume segmentation is not certain.
(c) imaging temperature was chosen to be -15°C for both acquisitions, the actual temperature in the CT at the SLF, was controlled over the room temperature and might therfore be slightly different.

Oil Fraction

Manual oil segmentation is time consuming and subjective. Figure 5 shows the air, brine, and oil distribution in sample F3 acquired at the SLF. In general, most oil was found a few mm underneath and above the respective oil injection level. Above the oil lenses, oil was found in channels with a similar orientation to the brine layer direction and as individual, tiny oil pockets. Oil was detected in inclusions as small as 0-36 μm in images acquired at SLF. Whereas observation of acquisition NTNU show that the smallest oil inclusions was found in pockets with a diameter in between 0- 50 μm, which is actually 2x the pixel size. Largest inclusion was found in a pore with 19 mm in diameter. Oil migration was observed up 2.5 to 5 cm above the respective lenses. Where oil was detected in the lower third up to 2/3 of the sample, approximately to a level 4 cm from the air- ice interface. I.e., within 46 hours oil migrated approximately 4 cm upwards from the oil injection level.

Oil content in CT- data in comparison with fluorescence measurements tend to be higher for samples with oil lenses and underestimated for samples with smaller oil inclusions.Inclusions were classified as oil only if they appeared over more than one slice. More objective results could be expected from automated segmentation. However, this was not possible in the current experiments since the oil and ice absorption contrast was too low. Enhancement of the oil-ice contrast could be achieved by doping the oil, e.g. with iodoheptane (Brown et al., 2014).

The oil content compared in Figure 3 shows that trends were consistent. However, Vo_{SLF} tends to be higher than Vo_{NTNU}. Observed trend can be explained by several factors:

a. due to subjective manual segmentation and the tendency to discard doubtful oil signals, just appearing in a single slice
b. Lower resolution data do show a coarser vertical grid, smaller inclusions were therefore easier excluded and accounted for ice;
c. choice of energy during CT-acquisition show a significant effect on the oil detectability (Figure6), lower energies used for acquisition at 18 μm result in a higher absorption contrast between oil and ice, and are easier to detect.
d. oil loss and drainage during transport and acquisition at SLF.

CONCLUSIONS

μ-CT is a powerful instrument to obtain non-destructive 3-dimensional insights into the interior of sea ice. The employed laboratory CTs provide satisfying results on low absorption contrast materials such as sea ice and oil (Figure 6) and reveal their spatial distribution (Figure 4).

Automated segmentation for air with the Otsu gave reasonable results for oil free samples with small bias in vertical profiles. Comparison of salinity evaluated from CT-data with measured salinity show a slight overestimation of brine volume fractions.

Spatial distribution of oil can be studied with significant investment of time into manual segmentation. Oil was found in inclusions with a diameter as small as 0- 36 μm, i.e actually 2 times pixel size. Comparisson of oil content from μ-CT with fluorescent measurements show consistent trends.Future motivation is given by the improvements in segmentation to open the possibility for quantitative oil and ice studies.

ACKNOWLEDGEMENTS

We gratefully acknowledge Matthias Jaggi for assistance and expertise at WSL Swiss Federal Institute for Snow and Avalanche Research SLF, as well as Ole Tore Buset and Leander Michels for guidance and help during μ-CT measurements at the Norwegian University of Science and Technology (NTNU). Special thanks to Math2Market, GeoDict for providing us with the imaging analysis software. Much appreciation to Statoil for providing Troll B crude oil. This work was supported by the Research Council of Norway PETROMAKS2 program, project number 243812 (MOSIDEO).

REFERENCES

Buist, I. A., Potter, S., & Dickins, D. (1983). Fate and Behaviour of Water-in-Oil Emulsion in Ice. *Proceedings of the 6th Annual AMOP Technical Seminar, 263-279.*

Brost, D. F., Foster, A. E., Holmes, M., Designs, T., & Instruments, H. (2011). No-Solvent Oil-in-Water Analysis – A Robust Alternative to Conventional Solvent Extraction Methods. *Offshore Technology Conference*, (January 2011), 1–18. http://doi.org/10.4043/21695-MS

Brown, K., Schlüter, S., Sheppard, A., & Wildenschild, D. (2014). On the challenges of measuring interfacial characteristics of three-phase fluid flow with x-ray microtomography. *Journal of Microscopy*, *253*(3), 171–182.

Buist, I. A., & Pistruzak, W. M. (1981). Dome Petroleum's Oil and Gas Undersea Ice Study. In *Proceedings of the Arctic Marine OIl Spill Programm Technichal Seminar, Edonton, Alberta*.

Cox, G. F. N., & Weeks, W. F. (1983). Equations for determining the gas and brine volumes in sea - ice samples. *Journal of Glciology*, *29*(102), 306–316.

Eicken, H. (1992). Salinity Profiles of Antarctic Sea Ice : Field Data and Model Results. *Journal of Geophysical Research,* 97(C10), 15545-15557

Glaeser, J. L., & Vance, G. P. (1971). *A Study of Behaviour of Oil Spills in the Arctic*.

Hullar, T., & Anastasio, C. (2016). Direct visualization of solute locations in laboratory ice samples. *The Cryosphere*, 10(5), 2057-2068

Karlsson, J. (2009). Oil Movment in Sea Ice, A laboratory study of fixation , release rates and small scale movement of oil in. *University of Copenhagen, Copenhagen,Denmark.*

Karlsson, J., Petrich, C., & Eicken, H. (2011). OIL ENTRAINMENT AND MIGRATION IN

LABORATORY- GROWN SALTWATER ICE. In *Proceedings of the 21st International Conference on Port and Ocean Engineering under Arctic Conditions, Montreal ,Canada.*

Kawamura, T. (1988). Observations of the internal structure of sea ice by X ray computed tomography. *Journal of Geophysical Research*, *93*(C3), 2343.

Martin, S. (1979). A field study of brine drainage and oil entrainment in first-year sea ice. *Journal of Glaciology*, *22*(88), 473–502.

Maus, S. (2009). Synchrotron-based X-ray micro-tomography: Insights into Sea Ice Microstructure.*Report Series in Geophysics, Proceedings of The Sixth Workshop on Baltic Sea Ice Climate.*

Maus, S., Becker, J., Leisinger, S., Matzl, M., Schneebeli, M., & Wiegmann, A. (2015). Oil Saturation of the Sea Ice Pore Space. In *Proceedings of the 23rd International Conference on Port and Ocean Engineering under Arctic Conditions, Trondheim, Norway.*

Maus, S., Leisinger, S., Matzl, M., Schneebeli, M., & Wiegmann, A. (2013). Modelling oil entrapment in sea ice on the basis of 3d micro-tomographic images. In *Proceedings of the 22nd International Conference on Port and Ocean Engineering under Arctic Conditions, Espoo, Finnland.*

McDougall, T. J., & Barker, P. M. (2011). Getting started with TEOS-10 and the Gibbs Seawater (GSW) oceanographic toolbox. SCOR/IAPSO WG, 127, 1-28.

Nelson, W. G. (1981). Oil Migration and Modification Processes in Solid Sea Ice. *International Oil Spill Conference, American Petroleum Institute, 1.*

NORCOR, (1975). The Interaction of Crude Oil With Arctic Sea Ice.*NORCOR Engineering and Research Limited Yellowknife, N.W.T.*, Beauforts Sea Technical Report 27

Otsu, N. (1979). A Threshold Selection Method from Gray-Level Histograms. *IEEE Transactions on Systems, Man, and Cybernetics*, *9*(1), 62–66.

Otsuka, N., Kondo, H., & Saeki, H. (2004). Experimental study on the characteristics of oil ice sandwich. In *Oceans '04 MTS/IEEE Techno-Ocean '04 (IEEE Cat. No.04CH37600)* (Vol. 3, pp. 1470–1475).

Petrich, C., Karlsson, J., & Eicken, H. (2013). Porosity of growing sea ice and potential for oil entrainment. *Cold Regions Science and Technology*, *87*, 27–32.

Schindelin, J., Arganda-Carreras, I., Frise, E., Kaynig, V., Longair, M., Pietzsch, T., … Cardona, A. (2012). Fiji: an open-source platform for biological-image analysis. *Nature Methods*, *9*(7), 676–682. http://doi.org/10.1038/nmeth.2019

Timco, G. W., & Weeks, W. F. (2010). A review of the engineering properties of sea ice. *Cold Regions Science and Technology*, *60*(2), 107–129.

Wilkinson, J. P., Wadhams, P., & Hughes, N. E. (2007). Modelling the spread of oil under fast sea ice using three-dimensional multibeam sonar data. *Geophysical Research Letters*, 34(22).

Zack, G. W., Rogers, W. E., & Latp, S. A. (1977). Automatic Measurement of Sister Chromatid Exchange Frequency. *The Journal of Histochemistry and Cytochemistry*, *25*(7), 741–753.

POAC'17
Busan, KOREA

Proceedings of the 24th International Conference on
Port and Ocean Engineering under Arctic Conditions
June 11-16, 2017, Busan, Korea

Road to the North – DSME efforts to YAMAL

Sungkon Han[1], Sung-Pyo Kim[1], Joonghyo Choi[1]
[1] Daewoo Shipbuilding & Marine Engineering. Ltd. Co., Seoul, Korea

ABSTRACT

This key note speech addresses all the involved parties' effort not limited to DSME's initiative since the achievement should not be claimed as a small group's excellence but as a group of passionate frontiers' efforts. Project Overview offers a bird view to the overall picture of YAMAL LNG Carrier Project in order to attract readers' attention to an epic full of exciting stories. The first main story covers engineering activities to overcome challenges from the unexplored areas and to achieve the proclaimed goal as well. The second story presents preparation and experience for the ice trial of the world's first icebreaking LNG carrier at the North, which was performed for the first vessel out of fifteen ships.

KEY WORDS : Arctic class LNG carriers; Ice-Structure interaction; Ice trial; Winterization; D-VACO.

PROJECT OVERVIEW

YAMAL project involves development of fifteen LNG carriers operating from Sabetta on the Yamal peninsula to worldwide major ports to transport LNG cargo. The vessels are to be designed complying with the requirements of Icebreaking ARC7 LNG carrier to be able to operate in the harsh ice field as well as open sea without ice. Optimization and/or compromising were inevitable to meet the two incompatible requirements using one design.

CHALLENGE AND RESPONSE

This section addresses designers' efforts from exploring challenges during the targeted operation to seeking design solutions. The contents cover the following:

- CCS Selection considering Ice-Structure Interaction
- Hull form development
- Hull structure design and construction
- Winterizations
- LNG as fuel gas – D-VACO

BEAUTIFUL BUT CROUCHING WHITE PLANE

The first vessel is launched and sailed away to the North after successful sea trial in an open sea. It still needs ice trial, which eventually proves all the equipped devices, as a whole, will work fine in the ice field. The section covers the following:

- Road to Yamal Peninsula
- Ice Trial that the world has never dared to attempt

A LIGHT ON THE FROZEN FIELD

No one dares to say that human beings live with the Nature but struggle to live in the Nature. This section concludes our story about how to come to this place but leaves awe to the Mother Nature at the same time.

- A small step to the unforgiving harsh land
- Humble Efforts to live in the Nature

POAC'17
Busan, KOREA

Proceedings of the 24th International Conference on
Port and Ocean Engineering under Arctic Conditions
June 11-16, 2017, Busan, Korea

Ice Collision Analysis and Alternative Full Scale Impact Test for ARC 7 LNG Carrier

Hyeok Geun Ki[1], Joong Hyo Choi[1], Sung Gun Park[1], Sung Kon Han[1]
[1] Ship & Ocean R&D Institute (Daewoo Shipbuilding and Marine Engineering Co., Ltd., Seoul, Korea)

ABSTRACT

The structural strength of ARC7 LNG carrier was guaranteed in consideration of reinforcement according to class rule. But more detailed strength estimation based on directly calculated ice loads followed by FEA was requested to secure the structural strength under the unexpected ice conditions. For ice conditions from ARC7 requirement, detailed ice collision analysis was carried out to verify hull strength as well as safety of CCS (Cargo Containment System). Based on ice load from direct calculation, the nonlinear static analysis was carried out according to ice collision scenarios defined by ship speed and ice thickness.

The unexpected thick ice may cause high stress in local ice belt zone even though Arctic class vessel has strongly reinforced structure against ice loads. Therefore, the full scale impact test was also performed to check the soundness of hull structure and weld strength against dropping steel object. The high local strain was measured by means of strain gauges. The measured structural response was compared to FE analysis result.

KEY WORDS: Ice Collision, Arctic LNG, Ice Load, Plasticity

INTRODUCTION

An arctic LNG carrier was newly constructed according to ARC 7 class of RMRS(Russian Maritime Register of Shipping). Even though she is designed in accordance with RMRS rule to have enough strength against ice loads, detailed strength assessment based on directly calculated ice loads was requested to check structural capacity of hull as well as cargo containment system against ice loads.

For ice loads from level ice, the ARC7 requirement for ice conditions (ice thickness and ship speed combination) was considered. First ice condition is based on ARC7 requirement of RMRS, which defines structural capacity considering local plastic deformation of hull structure. All of ice pressures were calculated using 'Ice Load+' developed by KSRC(Krylov State Research Center). Hull form angles, ship speeds and ice thicknesses related to requirements were used as the input data of 'Ice Load+'. According to ice collision analysis, it was verified that no additional reinforcement is required and hull design has enough strength against ice load.

The full scale impact test was performed to check the soundness of hull structure and weld strength against impact load. The dropping steel of 7 ton with controlled dropping height was used to realize impact load equivalent to encountered ice load. The drop height was determined based on the iterative drop object analysis result which is similar with the nonlinear analysis result for ARC 7.

The high local strain induced from impact load was measured by means of strain gauges. From the measured strain data, the plasticity could be founded. However, due to absence of available dynamic material property of high tensile steel in the FE analysis, it was founded that the stress response during the impact test was not identical to that from dropped object simulation.

ICE LOAD CALCULATION FOR HULL STRENGTH EVALUATION

Ice loads were calculated for the NO.1 cargo hold region described in Figure 1. Inclined angles (α and β) of each location were measured and used as input data for ice load calculation. To determine the ice thicknesses and ship speeds inducing ice loads, the BDSC curve was introduced and ARC 7 curve among BDSC curves described in Figure 2 was selected to calculate ice pressure and related patch size. Ice pressure and its patch sizes (width and length) were obtained from 'Ice Load+' program. 'Ice Load+' calculates ice loads for ten points specified in BDSC curve, i.e., ten combination of ship speed and ice thickness. Among calculated ice loads, the maximum values were used in ice collision analysis.

Figure 1.NO.1 Cargo Hold Region Model

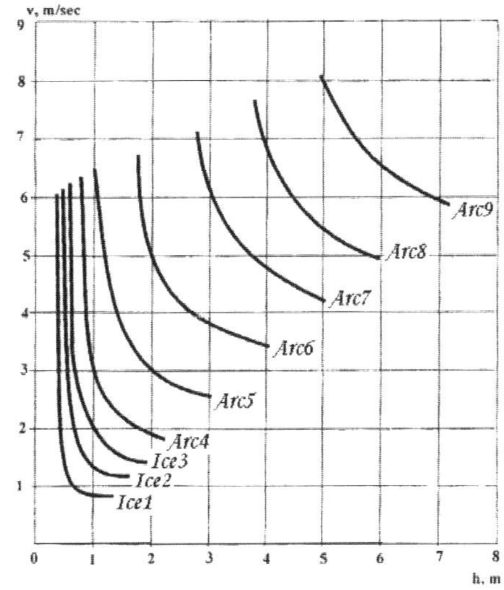

Figure 2.Recommended format for paragraphs

CONSIDERED ADDITIONAL SCENARIOS AND CALCULATED ICE LOADS

When a ship is escorted by an icebreaker of less width, the canal width is increased through interaction of its edges with the large ship hull. This interaction in the form of periodic impacts resulting in bending failure of canal edge portions is called as the finish breaking. From increased impact number and edge failure, the ice force is to be increased by 1.16 times in A1I area specified in RMRS rule. The relationship between original and increased ice pressure is as follows;

$$p_{A_1I}^{KT} = p_{A_1I} \cdot k_2; \qquad (1)$$

Where, p_{A_1I} = original ice pressure in A_1I area ignoring large ship features;

$p_{A_1I}^{KT}$ = increase ice pressure for large ship;

k_2 = enlargement coefficient that could be taken equal to $k_2 = 1.16$.

Based on additional scenarios, the ice load were calculated using 'Ice Load +'. Calculated pressure is summarized in Table 1. For strength verification, the trapezoidal ice load pattern described in Figure 3 was applied in the direct strength evaluation.

Table 1.The Ice Load Calculated for Strength Evaluation

BDSC 10 Condition		Ice Pressure	Ice Pressure Patch Size	
Ship Speed [m/s]	Ice Thickness [m]	P [MPa]	b [m]	l^H [m]
4.22	5	9.78	1.17	5.97

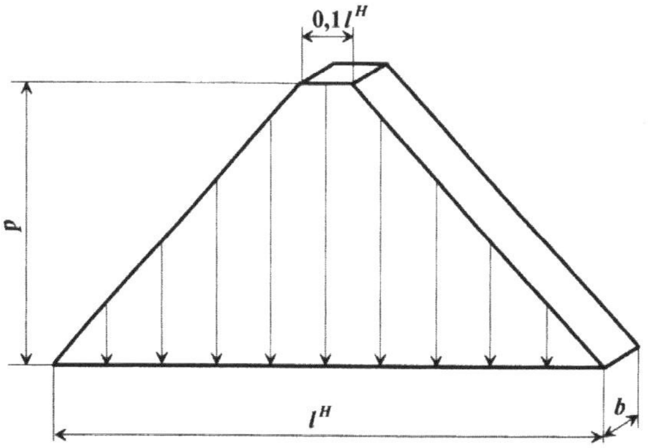

Figure 3.Ice Load Pattern

ALLOWABLE CRITERIA

The several design rules for Arctic vessels including RMRS rule adopt plastic design concept against ice load. For loads corresponding to ARC7 class, the structural capacity was calculated considering slight plastic deformation. To confirm the hull safety against ice load, the strength capacity should be calculated. Using the relation between load and deflection, strength capacity can be evaluated. Strength capacity, i.e., limit load calculation method is shown in Figure 4. Ultimate strength (limit load P_{ult}) can be defined as the intersection point of the tangent line at 5 times of the yield point and tangent line in the elastic region.

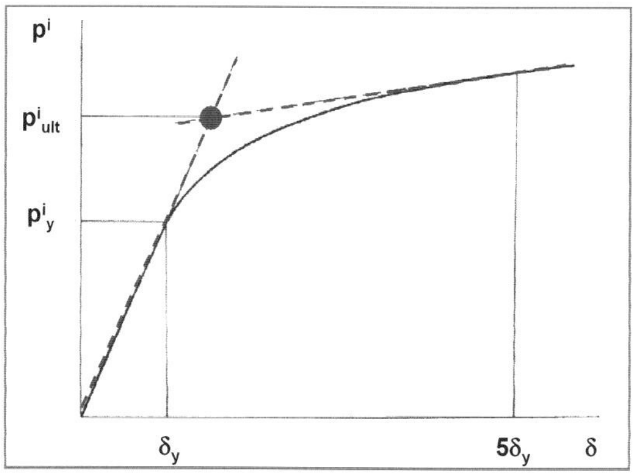

Figure 4.Limit Load Calculation Method

In case of GTT NO.96 containment system, the coupler provides the insulation boxes with coercing forces to remain in erection position and flexibility that can absorb any accidental deflection damages. According to GTT, the operating limit of coupler base with flexibility is 4.6mm/m which is a kind of relative deflection.

STRENGTH EVALUATION RESULT

FE analysis was carried out to verify hull strength according to ARC7. Nonlinear static analysis was adopted to check structural capacity and the inner hull deformation for operational limit of CCS(Cargo Containment System) strength. NASTRAN solution sequence 600 was used for nonlinear static analysis.

As shown in Table 2, it is concluded that structural capacities of reinforced structures exceed the demand which is defined as the ice load at concerned region. The load deflection curve and von-Mises stress contour were described in Figure 5.

The relative deflection for operational limit check was obtained from deflection per 1m distance based on length of plywood box. Figure 6 shows detailed description for relative deflection calculation methodology. As this result, 0.3 mm/m of maximum relative deflection, shown in Figure 7, was calculated for No.1 cargo hold. The resultant relative deflection is much smaller than operational limit of 4.6 mm/meter.

Table 2.Strength Evaluation Result

Hull Strength Evaluation		CCS Strength Check	
Demand [MPa]	Capacity [MPa]	Relative Deflection [mm/m]	Operational Limit [mm/m]
9.78	10.06	0.3	4.6

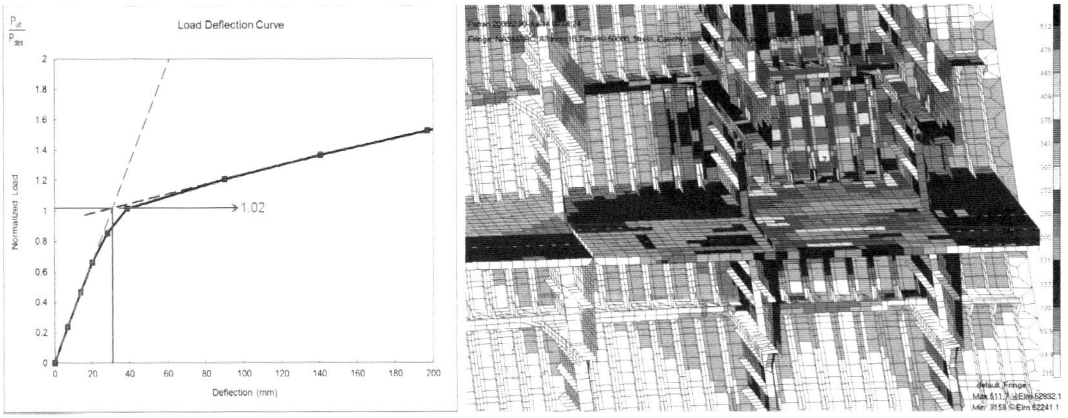

Figure 5.Normalized Load Deflection Curve and von-Mises Stress Contour

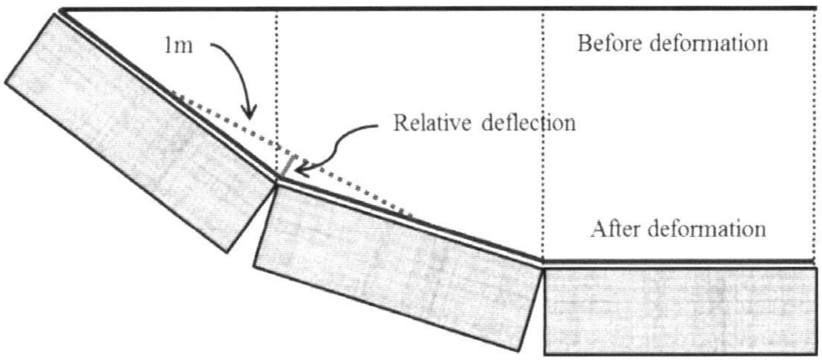

Figure 6.Relative deflection calculation methodology

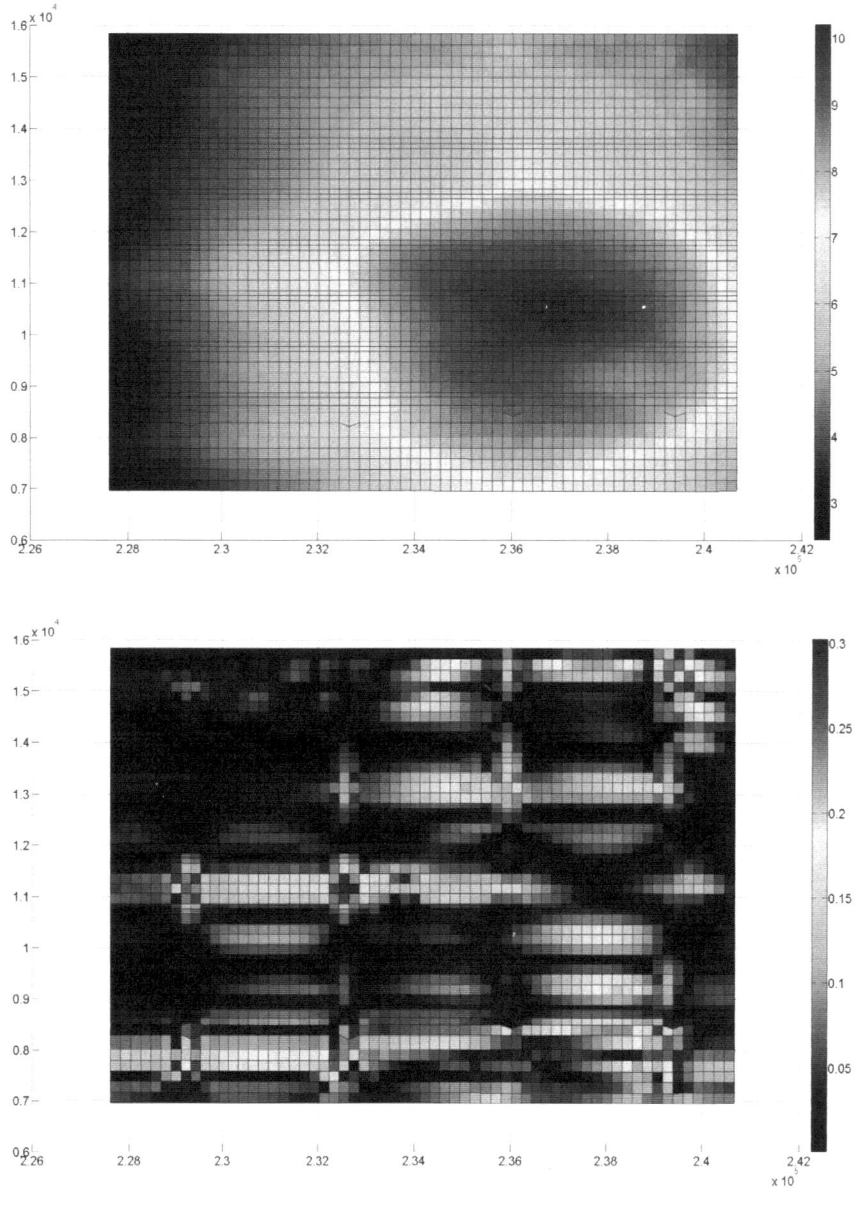

Figure 7.Absolute (Upper) and Relative (Lower) Deflection of NO.1 Hold L.BHD

FULL SCALE IMPACT TEST

The local high stresses against the ice load would occur at the ice strengthened structures. Even though the reinforcement for ARC class ship was regulated in RMRS rule, it is necessary to verify the soundness of hull structure and weld strength. Therefore, without the scale down, the full scale impact test was performed in the real block of A_1I region.

The weight bar of 7 ton with controlled dropping height was used to realize impact load equivalent to encountered ice load. The drop height was determined based on the iterative drop object analysis result which is similar with the nonlinear analysis result for ARC 7.

Figure 8 shows the test setting and weight bar of 7 ton. In order to drop on the exact location, the casing, such as guiding structure for weight bar, was installed on the concerned location of block. Firstly the weight bar was inserted into the casing by the crane. After adjusting drop height of 2m, the weight bar was dropped freely through the casing.

Figure 8.Test Setting (Left) and Weight Bar (Right)

The 3-axial and 1-axial strain gauges were installed at the each impact location in order to record the behavior of the impacted structure against impact load. For each impact point, four 3-axial gauges and one axial gauge were used as shown in Figure 9.

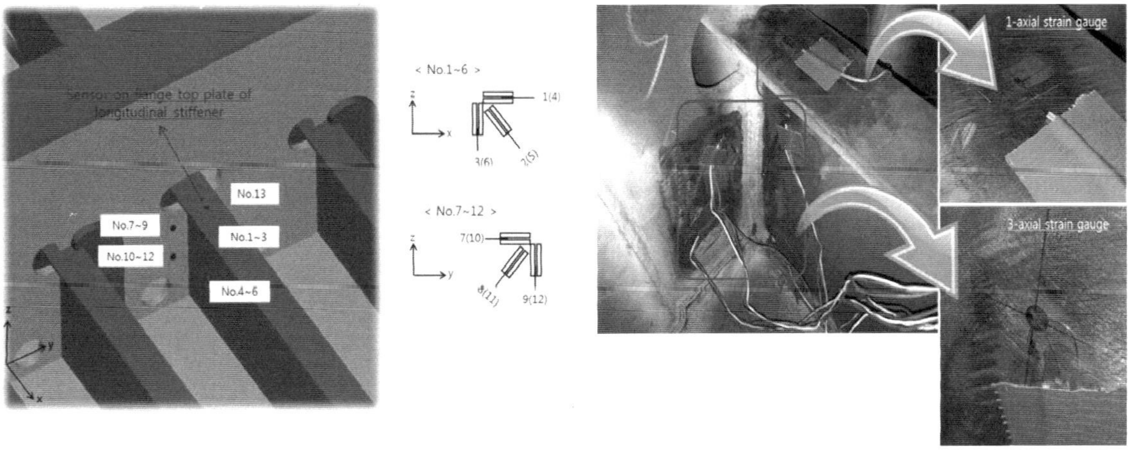

Figure 9.Strain Gage Installation at Joint between Intermediate Frame and Longi. Stiffener

STRAIN HARDENING EFFECT

In the dropped object simulation and the post processing of impact test data, the large deformation of structures and elasto-plastic material property were adopted to consider strain hardening effect. The strain hardening effect as well as ultimate stress was considered as a bi-linear strain-stress curve according to material grades as shown in Figure 10. The fracture was determined on the basis of the critical plastic strain of the steel material used as per the NORSOK document. In case of steel material, the critical strain for rupture was determined from data of mild steel and HT 32 only according to NORSOK standard. For material not shown in NORSOK standard, critical strain was calculated through linear extrapolation based on known data such as material data of HT 32. The concerned locations were designed by S500 steel.

Figure 11 shows material properties used in nonlinear simulation. The Cowper-Symonds rate enhancement formula was used to consider the effect of strain rate on material properties as shown in Figure 12.

To realize strain hardening effect in FE analysis, the commercial software LS-Dyna was used. As a material card, in flexible body, *MAT_PIECEWISE_LINEAR_PLASTICITY was used to simulate rupture and strain hardening of hull structure.

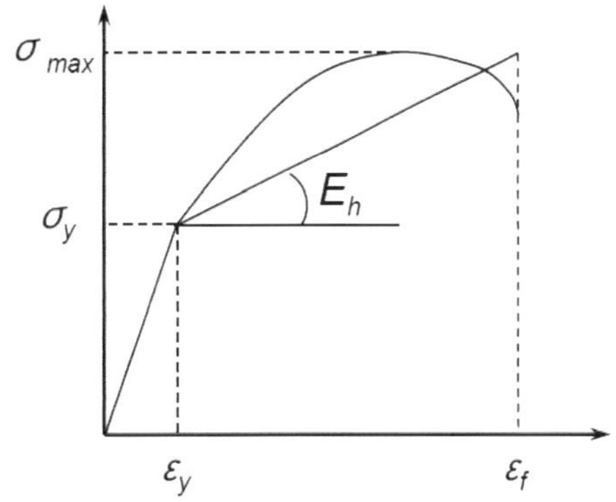

$$\sigma_p = \sigma_y + \frac{EE_h}{E - E_h}\varepsilon_p$$

$$E_h = \frac{\sigma_{max} - \sigma_y}{\varepsilon_f - \varepsilon_y}$$

σ_y = Yield stress

E_h = Young's modulus

E_h = Hardening modulus

σ_p, ε_p = Plastic stress & Plastic strain

Figure 10.Stress-strain Curve for Bi-linear Material

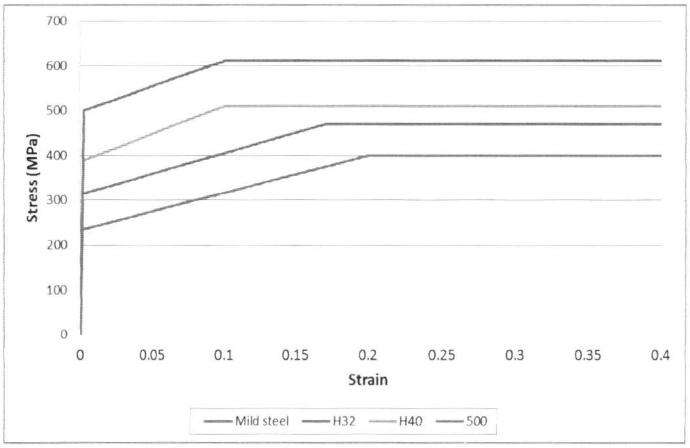

Figure 11.Stress and Strain Curves for FE Analysis

POAC17-150

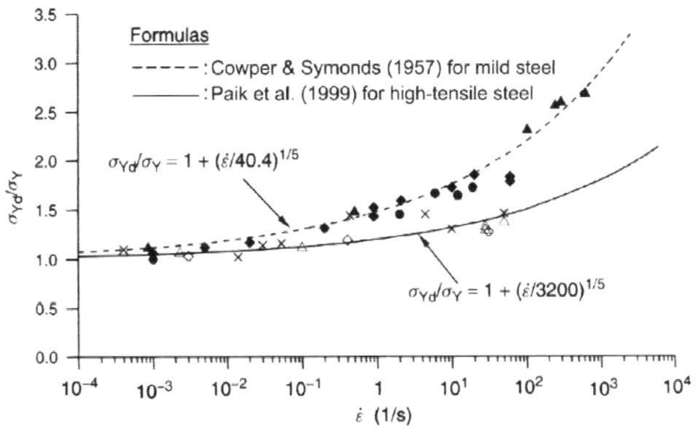

$$\frac{\sigma_{yd}}{\sigma_y} = 1 + \left(\frac{\dot{\varepsilon}}{C}\right)^{1/p}$$

Mild Steel: C= 40.4, p=5
HT Steel: C=3200, p=5

Figure 12.Strain Rate Effect of Steel

COMPARISON OF IMPACT TEST AND FE ANALYSIS RESULT

The total equivalent strains at concerned locations same in impact test were obtained from dropped object simulation. The plasticity was not founded at concerned locations. Total equivalent strains were smaller than the equivalent yield strain and the value at intermediate frame, location3 and 4, is bigger than at longi. stiffener. However, in full scale impact test, plasticity was clearly identified as shown in Table 3 and Figure 13 which represent total strain value at concerned locations for comparison.

As the result shown in this difference, it is necessary to verify the contact condition and the strain hardening effect in the FE analysis. The mesh shape of slave and mister part should be adjusted until the plastic strain at intermediate frame is identified. Moreover, the parameters for stain hardening effect, such as in Cowper Symonds, should be iterated to decrease gap with impact test result.

Table 3.Maximum Total Strain

		Max. Equivalent Strain (mm/mm)	
	Strain Gage No.	FE analysis	Impact Test
Location01	No.01~03	1.79E-03	2.27E-03
Location02	No.04~06	1.50E-03	1.59E-03
Location03	No.07~09	2.10E-03	1.17E-02
Location04	No.10~12	1.87E-03	7.61E-03

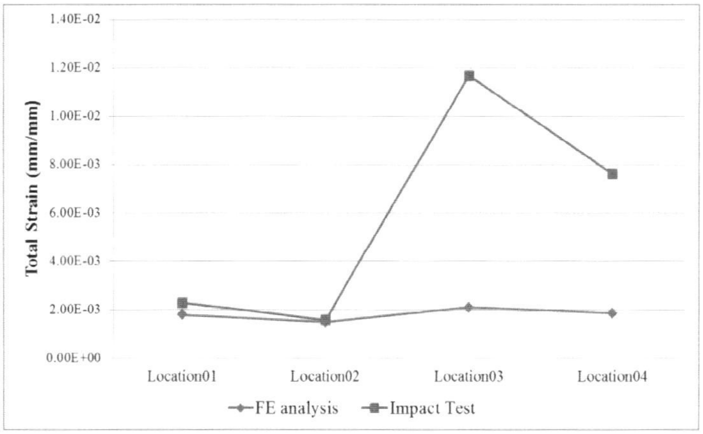

Figure 13.Maximum Total Strain

The weld crack against impact load was not identified according to NDT (Non-destructive testing), after impact test. The plasticity was not founded in Location 1 and 2. In measured effective plastic strain and total strain for von-Mises criteria as shown in Figure 14, the soundness of hull structure and weld strength was identified because measured values were below than the critical strain of concerned material (S500 steel: 0.1).

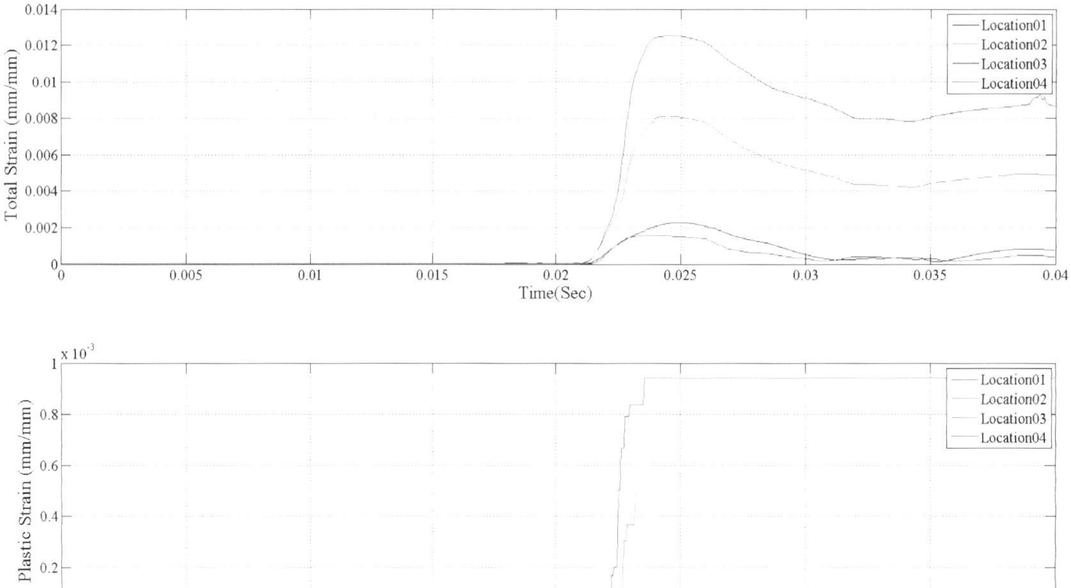

Figure 14.Effective Plastic Strain and Total Strain for von-Mises Criteria

CONCLUSIONS

According to direct ice load calculation introducing the finishing breaking scenario, the ice load for ARC 7 requirement was calculated based on real LNGC design. The nonlinear analysis was performed to guarantee the hull and CCS strength against ice load. The requirement was satisfied as the FE analysis result. Especially, the enough margin of CCS strength was identified in relative deflection check.

From the full scale impact test, the designed structure has enough impact strength, even though the high local strain induced from impact load was measured. However, it was founded that the impact response in FE analysis was not matched with in impact test. In future study, the contact option and the parameters for stain hardening effect in FE analysis will be deeply investigated to simulate realistic impact phenomenon.

REFERENCES

Daewoo Shipbuilding and Marine Engineering (DSME), 2016. *Ice Collision Analysis Report*, Seoul: DSME.

Hallquist, J.O., 2007. *LS-DYNA Keyword User's Manual Version 971*, Livermore Software Technology Corporation.

Krabbenhøft, K., 2002. *Basic Computational Plasticity*. Department of Civil Engineering Technical University of Denmark

Paik, J.K. & Thayamballi, A.K., 2006. *Ultimate Limit State Design of Steel-Plated Structures*. John Wiley & Sons:Chichester

Numerical Study on the Structural Response of Energy-Saving Device of Ice-Class Vessel due to Impact of Ice block

Sadaoki Matsui[1], Shotaro Uto[1], Yasuhira Yamada[1], Shinpei Watanabe[2]
[1] National Maritime Research Institute, Japan
[2] Nippon Kaiji Kyoukai, Japan

ABSTRACT

The present paper considers the contact between energy-saving device of ice-class vessel and ice block and clarifies the ice impact force and the structural response from various points of view. The contact analysis is performed by using LS-DYNA finite element code. The main collision scenario is based on Finnish-Swedish ice class rules and a stern duct model is used as an energy-saving device. For the contact force, two modelling approaches are adopted. One is dynamic indentation model of ice block based on the pressure-area curve. The other is numerical material modelling by LS-DYNA. The authors investigated the sensitivity of the structural response against the ice contact pressure, the interaction effect between structure and ice block, and the influence of eccentric collision. Furthermore, the simplified mechanical model was proposed to evaluate the structural response. The results of these simulations are presented and discussed with respect to structural safety.

KEY WORDS: Ice-class Vessel; Energy-saving Devise; Ice collision

INTRODUCTION

An energy-saving device is one of the most effective measures to improve propulsive efficiency in ice-free water. It is expected to be effective for ice-class vessels. However, its strength due to ice contact should be carefully considered in its design, and it is a significant technological issue to evaluate structural response due to impact of ice. Since some types of energy-saving devices are protuberant structure, they are strongly affected by dynamic factor such as oscillation. Therefore, "Time-domain" analysis is necessary instead of conventional "Static" estimation based on energy theory. Kinnunen et al (2013, 2015) implemented the contact analysis between ice and the azimuthing thruster. They indicated that the dynamic behavior of the structure affected the impact load significantly.

The purpose of the present study is to investigate the ice contact load and the structural response of the stern duct equipped for ice-class vessels. The main contact scenario is based on Finnish-Swedish ice class rules, and the used stern duct structure is Weather Adapted Duct (Kawasshima, et al. 2014) developed in National Maritime Research Institute, Japan. In this study, the authors carried out a series of non-linear dynamic finite element simulations to evaluate the structural safety and estimate the ice contact load, including the effect of ice-structure interaction. For the ice contact load, two modelling approaches are adopted. One is dynamic indentation model of an ice block based on the pressure-area curve. The other is numerical material modelling by LS-DYNA to consider ice-structure interaction. For the evaluation of structural safety, the sensitivity analysis against the ice contact pressure was performed and thereby the simplified mechanical structural model was proposed. In addition to the structural response, the ice contact load was investigated in detail such as the effect of the structural interaction and the influence of eccentric collision.

ICE LOAD SCENARIOS

It is assumed that the ice block contacts to the front end of stern duct as shown in Figure 1. According to Finnish-Swedish Ice Class Rule (2010), the maximum design ice block entering the propeller is a rectangular ice block with the dimensions of $H_{\mathrm{ice}} \cdot 2H_{\mathrm{ice}} \cdot 3H_{\mathrm{ice}}$. The thickness of the ice block H_{ice} is given as 1.5[m] for ice-class 1A. The density of ice block ρ_{ice} was assumed to be 880[kg/m^3], and the added mass was given as the following equation. Only the added mass was considered as the fluid force effect.

$$M_{\mathrm{a}} = 1.82\rho H_{\mathrm{ice}}^3 \tag{1}$$

The initial velocity of the ice block is assumed to be 5[knot] that is required as a vessel speed in the brash ice channels for the powering requirement of Finnish-Swedish ice class rules. These values are summarized in Table 1.

Table 1. Ice block basic parameter

Ice thickness H_{ice}	1.5[m]
Ice mass M_{ice}	17820[kg]
Ice added mass M_{a}	6296[kg]
Initial velocity v_0	2.572[m/s]

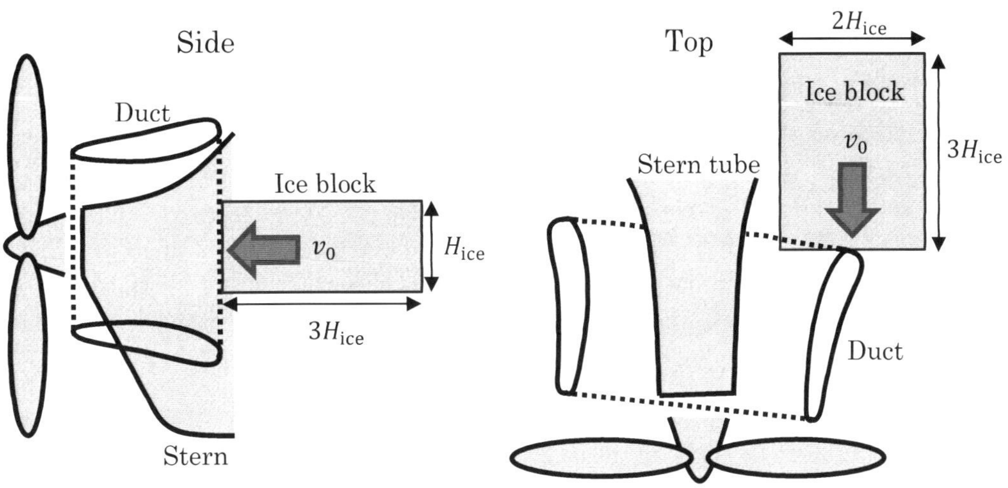

Figure 1. Contact between stern duct and ice block

ICE MODELLING

To perform safety evaluation, it is one of the most important processes to estimate an ice contact load. In this study, two load modelling approaches are adopted. One is the simple estimation model based on pressure-area curve (Daley, 1999). The ice load for structural requirement in UR-I (IACS, 2007) is based on this model. The other is FEM modelling. The dynamic contact simulation between ice block and structure was performed.

Load Model Based on Pressure-Area Curve

The general principles of the model are shown in Figure 2. The fore part of the stern duct is modelled to be rigid cylinder, and the indentation depth ζ_{n} can be obtained by the motion equation of an ice block.

$$(M_{\mathrm{ice}} + M_a)\ddot{\zeta}_{\mathrm{n}} = -F_{\mathrm{n}} \tag{2}$$

Here, the ice contact force F_{n} is expressed as a product of average contact pressure P and nominal contact area A:

$$F_{\mathrm{n}} = PA \tag{3}$$

Here, P is expressed as a function of A:

$$P(A) = P_0 A^{\mathrm{ex}} \tag{4}$$

Here, P_0 is average contact pressure at $A = 1[\mathrm{m}^2]$, and ex is an exponent constant. In this paper, the following equations are mainly used.

$$P(A) = 2.2A^{-1/3} \tag{5}$$

$$P(A) = 7.4A^{-0.7} \tag{6}$$

The equation (5) is proposed by Frederking and Ritch (2009) based on impact tests of bergy bit and ship. The equation (6) is proposed by Masterson (Palmer et al. 2009) for isolated small areas (<10 m^2), which is also recommended by the ISO code (ISO 19906, 2010). These relations are plotted with experimental data in Figure 3. The contact area is at most 0.5[m] in this simulation, hence the equation (6) gives much higher pressure. A is expressed as a function of indentation depth ζ_{n}:

$$A(\zeta_{\mathrm{n}}) = 2H\sqrt{2R\zeta_{\mathrm{n}} - \zeta_n^2} \tag{7}$$

by geometric consideration of indentation, regarding the fore part of the stern duct as a cylinder with a radius of R=0.13[m] as shown in Figure 2. Thus, F_{n} can be expressed by ζ_{n} and the load history $F(t)$ is obtained by solving the equation (2). To solve the differential equation, Runge-Kutta method was used. Both load curves based on equation (5) and (6) are called as "low" and "high" respectively and indicated in Figure 4. The termination of contact load is the moment when the indentation velocity is equal to zero. These areas are equal to the impulse of ice : $(M_{\mathrm{ice}} + M_{\mathrm{a}})v_0$.

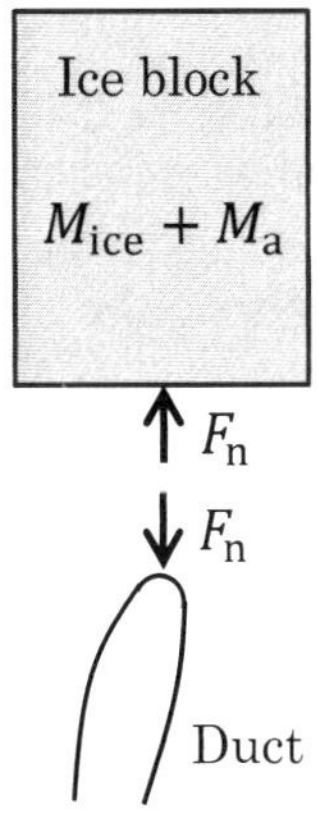

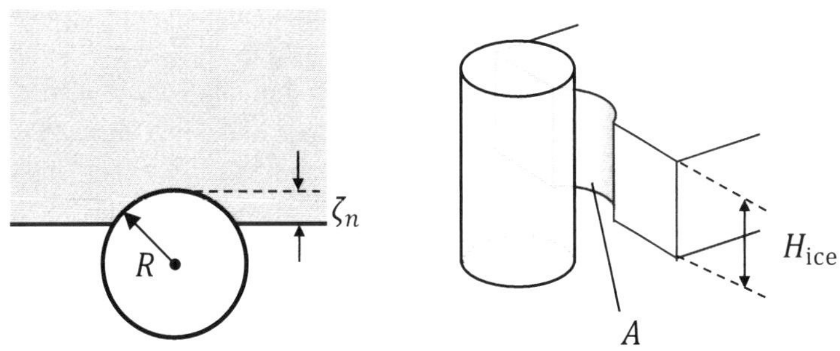

Figure 2. Indentation of vertical cylinder

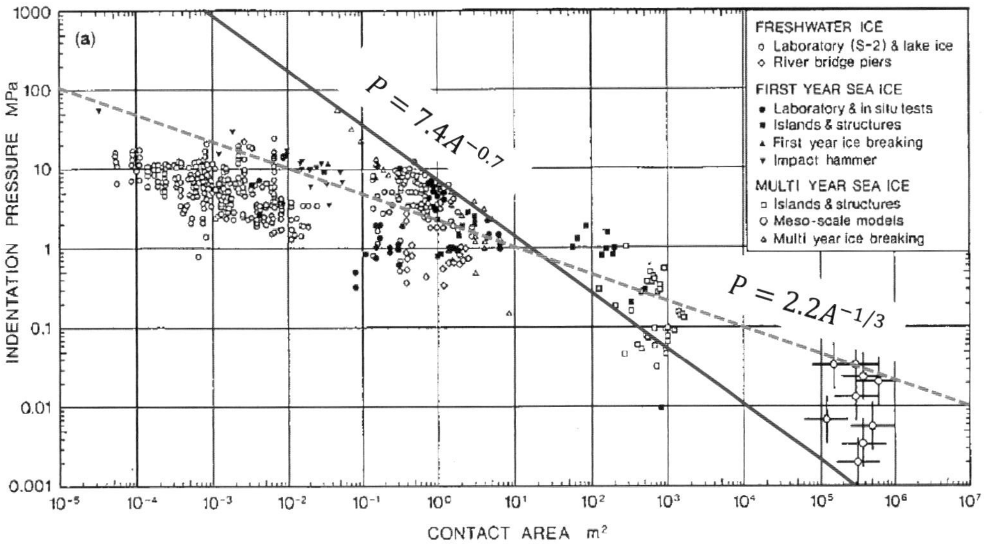

Figure 3. Pressure-area curve (Sanderson J.O.,1988)

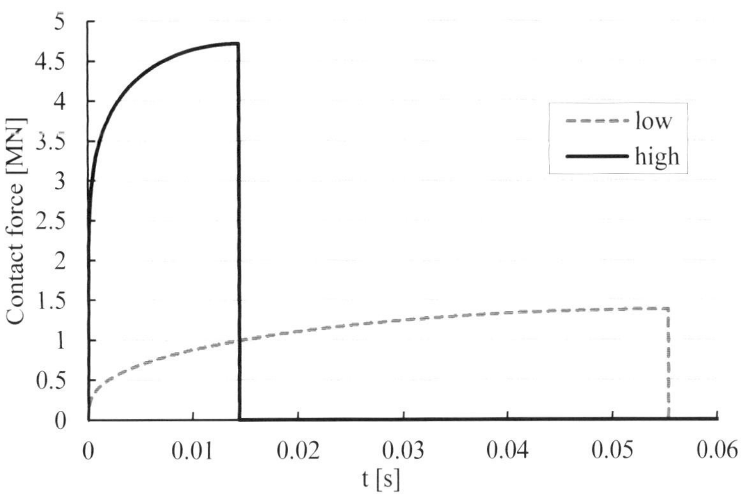

Figure 4. Time history of ice load

Finite Element Model

Several FEM modelling of ice have been tried and it was reported that they have a certain level of accuracy to estimate the contact force. The dynamic FEM analysis has been performed by using some material properties of ice calibrated already. To validate such theoretical models, the contact simulation between rectangular ice block and rigid cylinder with a radius of R=0.13[m] was performed. For the numerical simulation, LS-DYNA was chosen. The FEM models of the ice block and the cylinder are indicated in Figure 5. The ice material property was applied to only fine mesh zone and the remaining region was assumed to be a rigid body.

Gagnon (2011) calibrated volumetric strain-stress relationship based on MAT63 CRUSHABLE_FORM through the bergy bit field trial data. He carried out FEM simulation of the compression of the pyramid-shaped ice. He modelled several thin ice layers to express spalling behavior. The volumetric strain-stress relationships of the high and low pressure zone are expressed as shown in Figure 6. It should be noted that the high and low curves are considered separately although he combined them in his simulations. The young's modulus and Poisson's ratio is assumed to be 8[GPa] and 0.003, respectively.

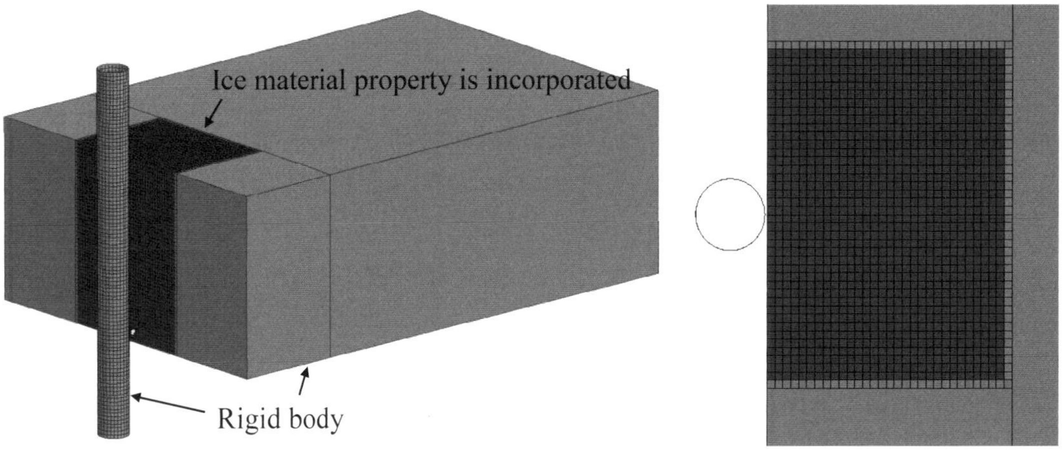

Figure 5. FEM model (Left: 3D view, Right: Top view)

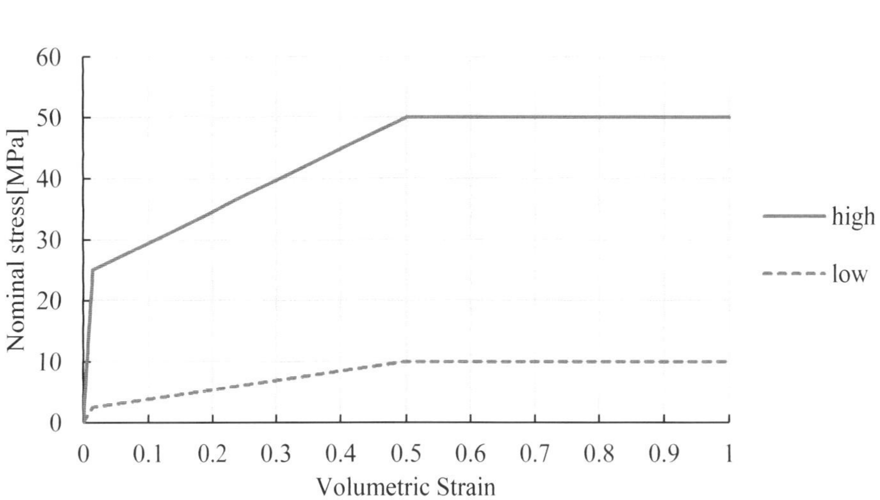

Figure 6. Relationship between nominal stress and volumetric strain

Liu (2011) used "Tsai-Wu" model as the yield criterion of ice:

$$f = J_2 - (a_0 + a_1 p + a_2 p^2) = 0 \quad (8)$$

where J_2 represents the second invariant of the deviatoric stress tensor s_{ij}, p is the hydrostatic pressure expressed in equation (9), and a_0, a_1 and a_2 are constants that require fitting to the triaxial experimental data. The equation (8) denotes the ellipsoid yield surface in the meridian plane. In this paper, the constant values of $a_0 = 22.93[\mathrm{MPa}^2]$, $a_1 = 2.06[\mathrm{MPa}]$, $a_2 = -0.023$ recommended by Derradji-Aouat (2000) are adopted.

$$J_2 = \frac{1}{2} s_{ij} s_{ji}, \; p = -\frac{1}{3}\sigma_{ii} \quad (9)$$

Liu also defined failure criterion as follows.

$$\bar{\varepsilon}^p > \varepsilon_f \quad (10)$$

$$\bar{\varepsilon}^p = \sqrt{\frac{2}{3}\varepsilon_{ij}^p : \varepsilon_{ij}^p} \quad (11)$$

$$\varepsilon_f = \varepsilon_0 + \left(\frac{p}{p_2} - 0.5\right)^2 \quad (12)$$

where $\bar{\varepsilon}^p$ is the effective plastic strain, ε_f is the element failure strain, p_2 is the larger root of the yield function, ε_0 is the initial failure strain, which is assumed to be 0.01. Once the failure condition (10) is satisfied, the element fails and its stiffness is immediately set to zero in the numerical simulation. These material properties are defined in LS-DYNA using a MAT41_USE_DEFINED_MATERIAL_ MODELS which is defined by user subroutine. As a return mapping method for the yield surface, the cutting-plane algorithm (E. A. de Souza Neto, et. al, 2008) was used. The effect of strain rate and work hardening were not considered.

The time histories of the ice load are obtained by MAT 63 and MAT 41 and shown in Figure 7. In Figure 7, "MAT63 (high)" and "MAT63 (low)" correspond to the high and low stress model in Figure 6, respectively. For MAT 41, two cases that failure criterion of equation (10) is considered (Failure) or not (No Failure) are plotted. The two curves calculated by the pressure-area curve in Figure 4 are also indicated.

Two load curves of MAT 63 are consistent with the two load curves calculated by the pressure-area curve. On the other hand, "MAT41 (No Failure)" is consistent with the curve of high pressure model of pressure-area curve, whereas "MAT41 (Failure)" indicates the irrational intermittent behavior. This is because elements which satisfy the failure criterion are vanished immediately in the numerical calculation. In reality, crushed ice fragments also contribute to the compression stress in the contact area. Such the failure behavior strongly depends on the size of finite element and stress condition. The equation (10) is just pure empirical formula calibrated by Liu (2011), hence the failure criterion should be revised by the experiment in present condition. The indentation of the ice is treated by only continues behavior for such a reason, although the ice should behave in brittle failure mode in the present high strain rate condition. Further examination should be considered through an experiment in future.

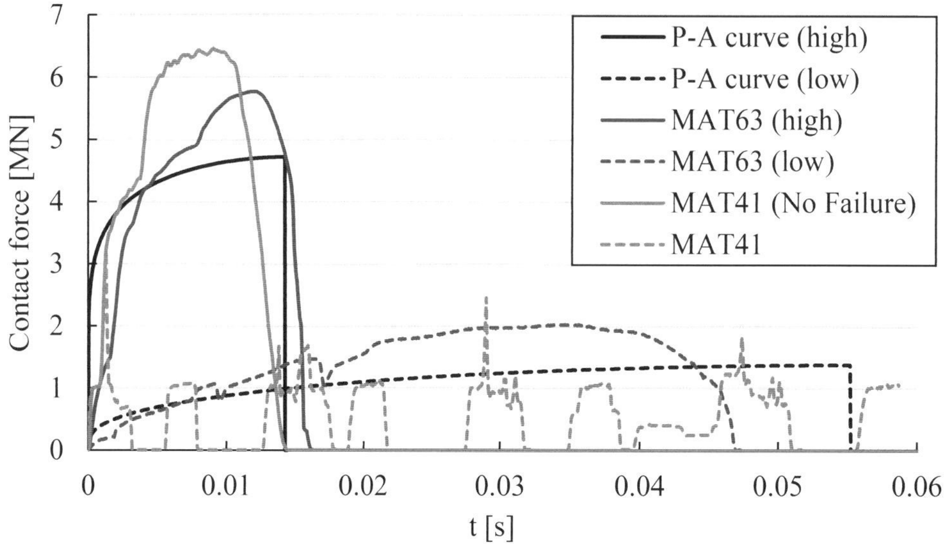

Figure 7. Time history of ice load

STRUCTURE MODEL

In this section, the FEM model of the stern and the duct structure and the calculation conditions are described. The detail of structural model and boundary conditions are described in Table 2, and the stern and the duct FEM model are shown in Figure 8. The junction between solid and shell element is modelled by Shell-Solid coupling model (Osawa, 2007).

Table 2. FEM analysis condition

Modelling region	Duct and stern model (35[m] from A.E.)
Element type	Bearing … solid element Another member … shell element
Thickness of duct	Inside web … 20[mm] Outside shell … 25[mm]
Material	Mild steel (MAT24) Young modulus … 206[GPa] Yielding stress … 235[MPa] Strain rate dependency … Cowper Symonds low
Boundary condition	Tight bulkhead of stern is fixed

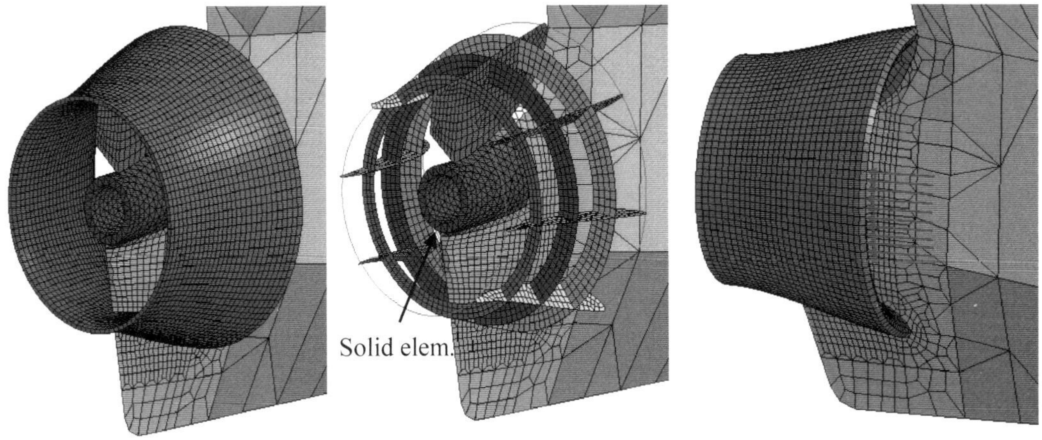

Figure 8. Structure FEM model

STRUCTURAL SENSITIVITY ANALYSIS

Since the contact pressure of ice has large uncertainty, the maximum contact load may have great variability even though the other conditions, such as mass and initial speed of the ice block, are the same. Accordingly, the structural sensitivity analysis against contact pressure is needed to evaluate the structural safety of the stern duct. Ice contact load based on the pressure-area curve are exerted at the fore part of the stern duct as nodal force as shown in the right figure of Figure 8. Those curves were calculated by equation (4), which ex is set to -1/3 and P_0 is changed, in the same manner as in Figure 4. Moreover, those were modified to sinusoidal curves while keeping each maximum force and area (impulse) the same. This is because of avoiding the irrational oscillation in the structure due to the step of load curve at the end of indentation, which is generated by the absence of ice elasticity. An actual load curve might take the medium form of the both curves. Some examples of used load curves are shown in Figure 9.

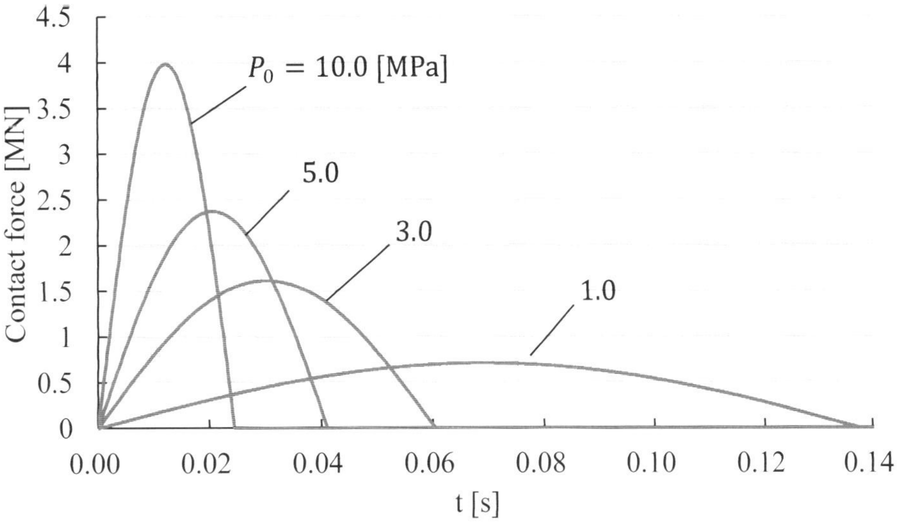

Figure 9. Load curves under changing contact pressure (ex $= -1/3$)

Figure 10 shows an example of Von-Mises stress distribution on the stern duct at the moment when the maximum stress occurs. As the general response of the stern duct, it was found that the relatively high stress occurs in connection of duct and hull, X and Y in Figure 10. The sensitivity of the maximum Von-Mises stress value at X and Y are shown in Figure 11. The broken line indicates the tolerance stress value in CSR-B (IACS, 2006) and is 280[MPa] when the element size is less than 200[mm]. Figure 11 indicates that the larger response occurs as the nominal pressure increases even though the impulses of curves are the same value. This is because that the maximum force becomes large as P_0 becomes large and the acting time approaches to the quarter of the eigenperiod of the stern duct, which is about 0.05[s]. In the case that the acting time is enough long, the external force does negative work while the stern duct turns back by elastic behavior. It is also found in Figure 11 that the maximums stress occurs at X when P_0 is smaller than 4.5[MPa] and stress at Y is larger beyond this value. This is because the static horizontal bending is dominant when P_0 is small and the position of X is affected by this effect mainly. Meanwhile, the dynamic effects such as free oscillation are remarkable when P_0 is large and the extremely large tensile stress is induced at Y which is the connection between duct and hull.

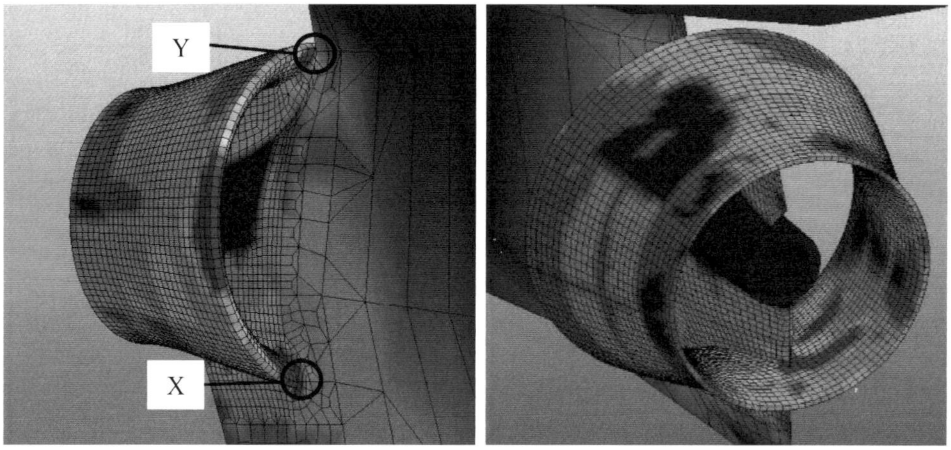

Figure 10. Von-Mises stress distribution (Red : High ↔ Blue : Low)

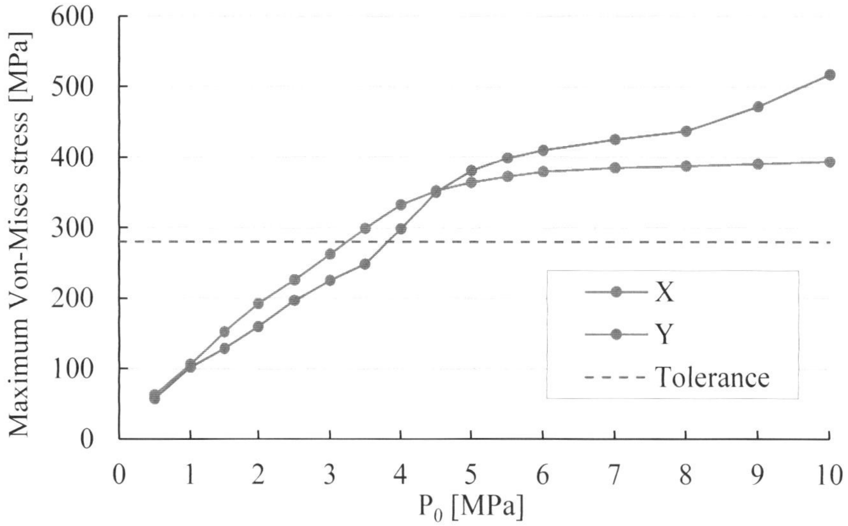

Figure 11. Sensitivity of the maximum Von-Mises stress on the stern duct against nominal contact pressure

The maximum displacement of the point where load acts in aft direction is shown in Figure 12. In addition to the result by LS-DYNA, the result by proposed simplified model "MCK model" is also indicated. MCK model means single-degree-of-freedom spring-dumper-mass model:

$$M\ddot{\xi} + C\dot{\xi} + K\xi = F_n \tag{13}$$

Here, ξ is the displacement of the stern duct at the contact point in aft direction, M, C, K are mass, damping and spring coefficient of the stern duct, respectively. These responses correspond reasonably, although the stern duct is significantly complex structure which consists of duct, stern tube, and other local members. This is because the maximum displacement occurs in the first amplitude of oscillation, and the value is determined by the most dominant component of oscillations. For this reason, the MCK model can express the maximum response of the structure accurately.

For the simplification into MCK model, it is necessary to determine the constant of mass, damper and spring coefficients. These values should be estimated by moment of inertia, structural damping, and horizontal bending stiffness of the stern duct respectively. In this paper, these constants are calibrated by results of LS-DYNA. If the relationship between the displacement and the local stress becomes clear, the structural response of the stern duct would be able to estimate by using MCK model.

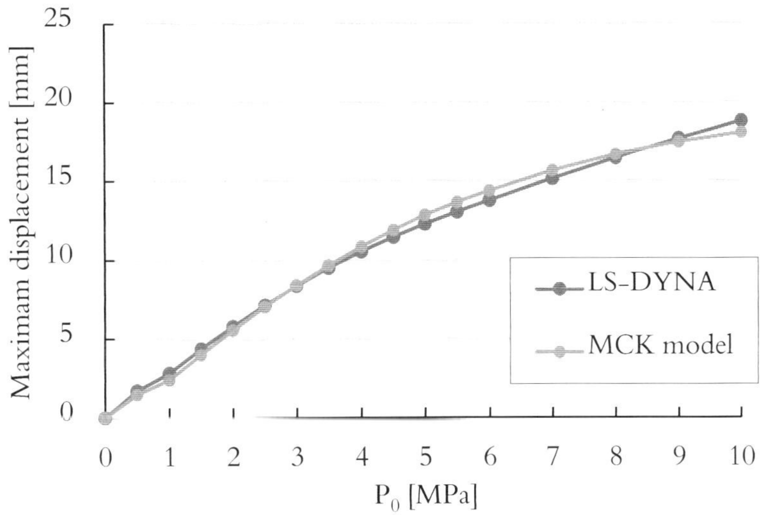

Figure 12. Maximum displacement of the point where load acts

EFFECT OF SHIP-ICE INTERACTION ON THE ICE LOAD

To obtain the ice contact load, steel structures can be assumed to be a rigid body in general because there is a large difference of rigidity between the ice and the steel structure. Meanwhile, in case of the protuberant structure such as a stern duct, oscillation occurs significantly by an impact force and the structural response may affect the impact load. To investigate such influence quantitatively, the simulation were performed with a stern duct of rigid (hereafter called as decouple) and compliant body (couple) and both contact loads were compared. The used FEM model is shown in Figure 13. MAT63 (high) and MAT63 (low) were applied to only fine mesh zone and the remaining region was assumed to be rigid body.

The contact point and the center of gravity of the ice block are in the straight line in the direction of contact velocity to avoid rotation of the ice. Each load history is shown in Figure 14. The reason why the peak load of the decouple simulations are smaller than the results of MAT 63 in Figure 7 is that the collision angle must be applied because of the stern shape, and the contact area is smaller than the case that regards the stern duct as a cylinder.

Figure 14 shows that the contact force by the couple analysis is smaller than that by the decouple analysis in high-stress model but the little difference is seen in low-stress model. The difference of both tendencies can be explained by the deformation of the stern duct. The history of displacement of the stern duct at the contact position and the center of gravity of the ice block in collision direction by the couple analysis is shown in Figure 15. The contact load is primarily determined by the indentation depth, e.g. the difference of the displacements between the stern duct and the ice block. Consequently, the contact load is reduced by the displacement of the stern duct in the couple analysis. The influence of the displacement of the stern duct is relatively large in the case of high-stress model, on the other hand, that is relatively small in the case of low-stress model. From these facts, it is presumed that the ice-structure interaction effect becomes greater as the contact pressure is higher. However, if the acting time is much less than the eigenperiod of the structure, the structural displacement hardly occurs and the interaction effect may become small conversely. In the present analysis, the influence of the structural response was at most 20% and it is smaller than the uncertainty of ice property.

ECCENTRIC CONTACT ANALYSIS

In the actual collision, the contact position and the center of gravity of ice are not always in line. Thus the initial kinematic energy is absorbed partly into the rotational energy of the ice block by eccentric collision and the contact load should be somewhat reduced. To clarify the reduction of the contact load against the eccentricity, the eccentric contact analysis was performed by LS-DYNA. The material property of ice is assumed to be MAT63 (high), and the structure is assumed to be a rigid body. The six cases in different eccentricities, 0, 0.3, 0.6, 0.9 1.2, 1.45[m] (The half breadh of ice block is 1.5[m]) were performed. The contact load histoiry is shown in Figure 16. Reduction of the contact load by eccentricity can be seen clearly. Since no marked differences are found in the shape of these curves, the magnitude of the load may be estimated by only its impulse. The impulse due to collision can be derived from momentum theory, and the following relations are derived in the case that the slip at the contact point is allowed or not :

$$P_{\text{slip}} = \frac{I}{I + e_y^2 m} m v_0 \tag{14}$$

$$P_{\text{non-slip}} = \frac{I + e_x^2 m}{I + \left(e_x^2 + e_y^2\right) m} m v_0 \tag{15}$$

where m, I, v_0 are the mass, the moment of inertia and the initial velocity of the ice block, respectively, e_y is the eccentric distance, and e_x is the distance between the contact position and the center of gravity of the ice block in the collision direction. The completely plastic collision is assumed because the plastic behavior is dominant in the ice property. The derivation of the above formulae are described in APPENDIX in detail. The influence of the eccentricity against the impulse is shown in Figure 17. Ecah value is normalized by the value

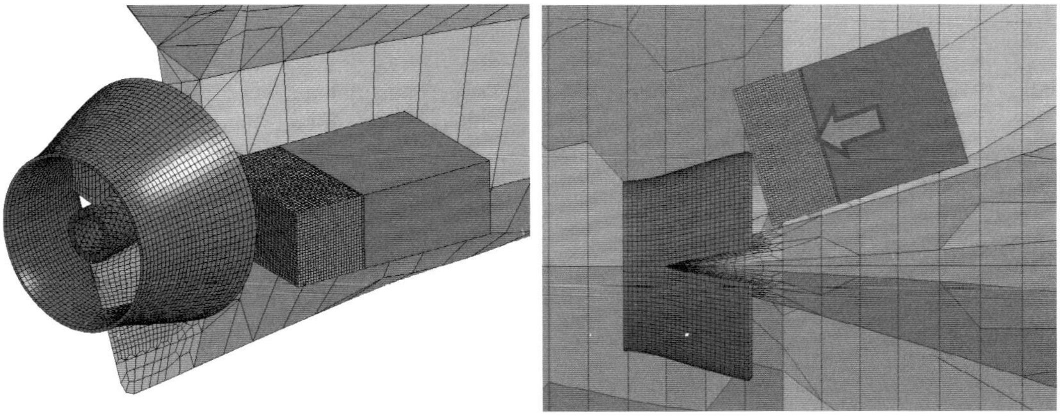

Figure 13. FEM model of stern duct and ice block (Left: 3D view, Right: Bottom view)

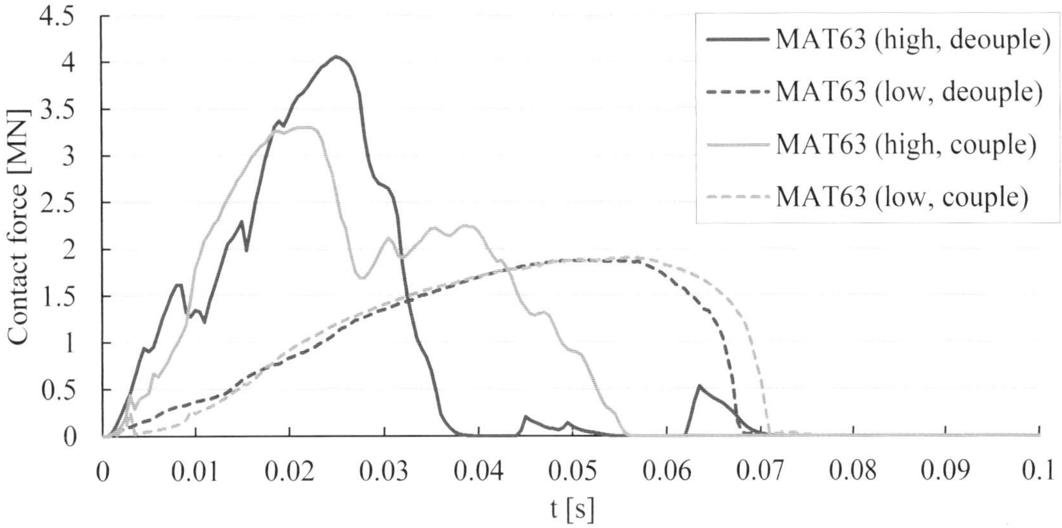

Figure 14. Load history of couple and decouple analysis

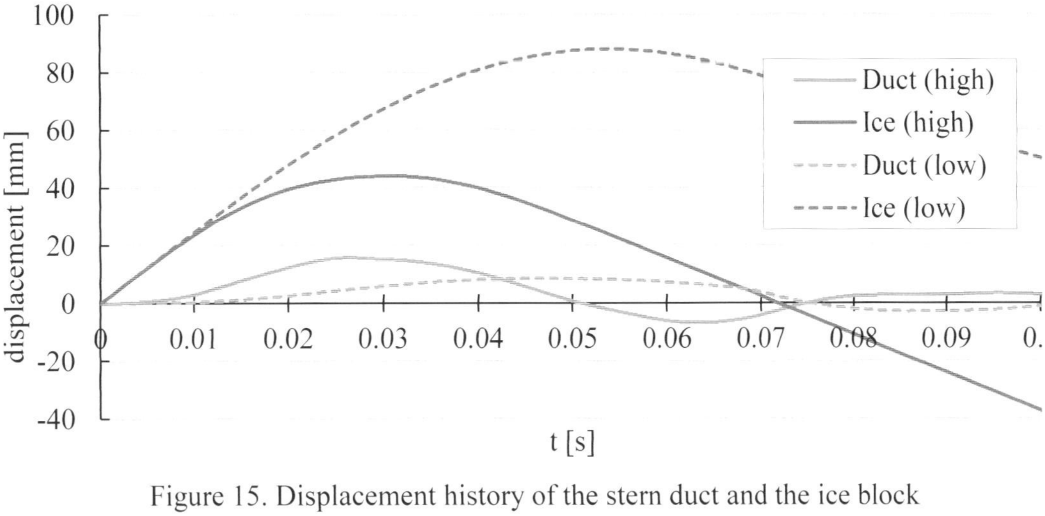

Figure 15. Displacement history of the stern duct and the ice block

of non-eccentric collision. In addition to the results of LS-DYNA, the equation (14) and (15) are also indicated. The results of LS-DYNA almost correspond with the momentum theory of the non-slip case, and these reductions are small in comparison with the momentum theory of the slip-allowed case. This is because the ice block is indented into the edge of the stern duct and the center of rotation corresponds to the contact position, and thus the rotational motion of the ice block is hard to occur in comparison with the slip-allowed case. On the other hand, for example, the ice contact with hull does not cause indentation, hence its tendency may be close to slip-allowed case.

It is indicated from the result that the load reduction by the eccentric collision is not so large against the structure such as the stern duct, and to simulate under non-eccentric collision scenario does not result in overestimation for the structural safety. Such tendency becomes remarkable when the distance between the contact position and the center of gravity of the ice block e_x is large.

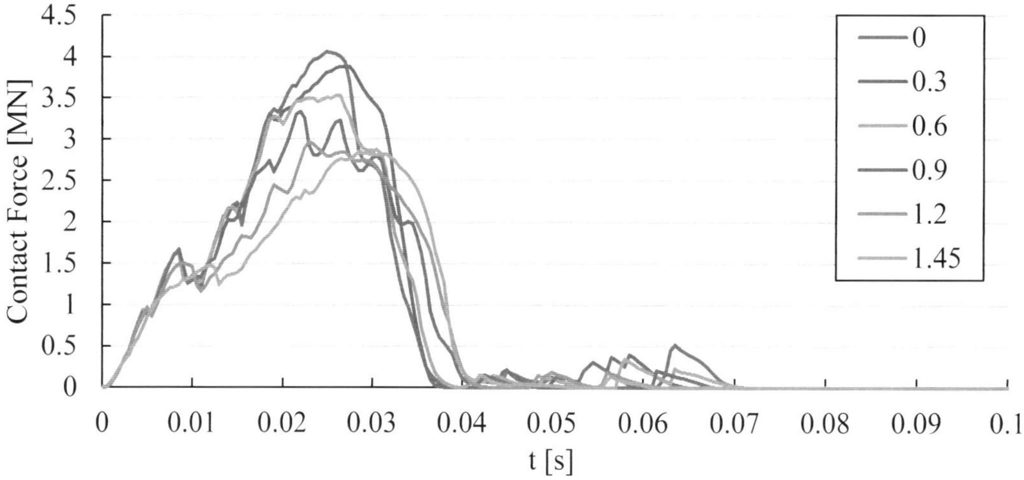

Figure 16. Time history of eccentric contact load

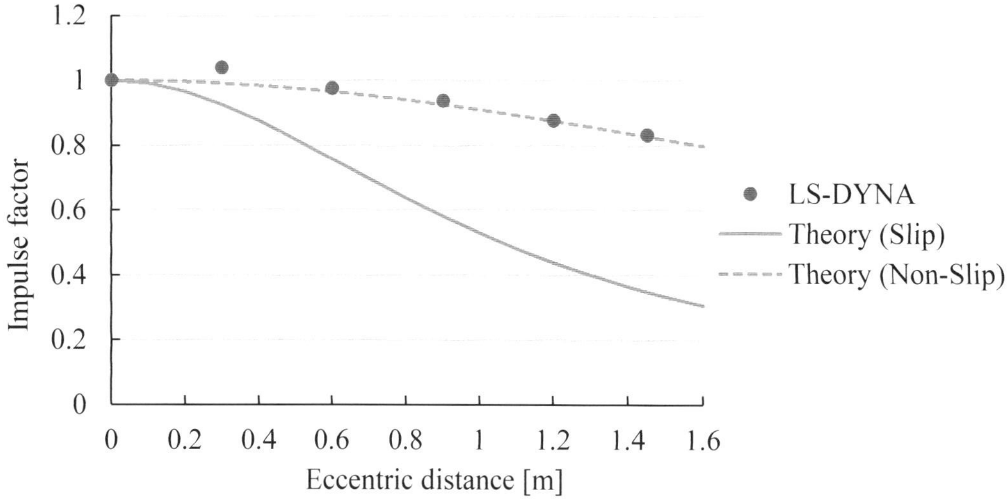

Figure 17. Eccentric effect on contact force

CONCLUSIONS

The evaluation of structural safety of the stern duct due to the impact of ice block was performed through the series of dynamic FEM analysis. Several conclusions are described as follows.

About the structural response:

- The structural response becomes large as contact pressure of ice becomes high even though the impulses are the same.
- Relatively high stress occurs in the connection of the duct and the hull. There is a possibility that the maximum stress in the structure exceeds the permissible stress depending on the contact pressure.
- The maximum deformation of the stern duct due to impact force can be expressed by the single-degree-of-freedom spring-damper-mass model accurately. This model may apply the simple estimation for occurring stress in the structure.

About the contact load:

- The ice contact load reduces by considering the ice-structure interaction. The reduction is large as the contact pressure becomes high, but the influence of the interaction was at most 20% in the present case and is smaller than the uncertainty of ice property.
- The reduction of ice contact load due to the eccentric effect is relatively small because the contact point becomes the center of rotation of the ice block because of indentation. Accordingly, the safety evaluation based on non-eccentric collision analysis is not over-estimation for thin column structure such as the present structure.

In this study, the incorporated material model of ice in FEM analysis was validated indirectly comparing with the theoretical model based on the pressure-area curve. However, only the yield criterion is defined as continuous behavior, and it can't express the tendency of real ice that the contact pressure becomes large as the contact area becomes small due to the non-simultaneity of contact between ice and structure. This tendency can be imitated to some extent by introducing the failure criterion properly, as mentioned by Liu (2011) and Gao (2015). In order to carry out such modelling, the examination of the criteria and the mesh size in detail should be performed and verified by experiments.

ACKNOWLEDGEMENTS

This research was carried out jointly by Oshima Shipbuilding Co., Ltd., and ClassNK as part of the ClassNK Joint R&D for Industry Program.

REFERENCES

Aki Kinnunen, Pekka Koskinen, Maria Tikanmaki, 2015. ICE-STRUCTURE IMPACT CONTACT LOAD CALCULATION WITH DYNAMIC MODEL AND SIMPLIFIED LOAD FORMULA, *Port and Ocean Engineering under Arctic Conditions,* June 14-18 2015, Trondheim, Norway

Aki Kinnunen, Maria Tikanmaki, Jaakko Heinonen, Juha Kurkela, Pekka Koskinen, Matti Jussila, 2013. Azimuthing thrusters ice impact load calculation, *Azimuthing thrusters ice class rule development Steering group meeting,* 2013-10-31

Claude Daley: ENERGY BASED ICE COLLISION FORCES, Expanded version (full derivations) of paper for POAC 99, Proc. of the 15th International Conference on Port and Ocean Engineering under Arctic Conditions, Helsinki University of Technology in Espoo, Finland on August 23-27, 1999.

Derradji-Aouat, A., 2000. A unified failure envelope for isotropic fresh water ice and iceberg ice, *Offshore mechanics and arctic engineering : presented at ETCE/OMAE 2000 Joint Conference*, Vol. 2, Part B, pp. 1241-1248.

E.A. de Souza Neto, D. Peric, D. R. J. Owen, 2008. Computational Method for Plasticity : Theory and Applications

Frederking, R. & Ritch, R. 2009. The nature of the process pressure-area relation from CCGS Terry Fox bergy bit impacts. 19th (2009) International Offshore and Polar Engineering Conference, June 21, 2009 – June 26, 2009. Osaka, Japan: *International Society of Offshore and Polar Engineers*. pp. 608-613. ISSN 10986189

Gagnon, R.E., 2011. A numerical model of ice crushing using a foam analogue, *Cold Region Science and Technology*, Vol. 65, pp.335-350.

H. Kawashima, K. Kume, N. Sakamoto, 2014. Study of Weather Adapted Duct, *Reports of National Maritime Research Institute, 2014,* Vol. 14-2, pp. 19-34.

ISO/CD 19906, 2010. Petroleum and Natural Gas Industries-Arctic Offshore Structures, ISO TC 67/SC 7/WG 8, *Final Draft International Standard. International Standardization Organization*, Geneva, Switzerland p. 434.

IACS, 2007, Requirements concerning POLAR CLASS, Unified requirements

IACS, 2006, COMMON STRUCTURAL RULES FOR BULK CARRIERS

IMO, 2012. DEVELOPMENT OF MANDATORY CODE FOR SHIPS OPERATING IN POLAR WATERS, *SUB-COMMITTEE ON SHIP DESIGN AND EQUIPMENT*, 57th session Agenda item 11,

Liu, Z., et al., 2011. Plasticity based material modelling of ice and its application to ship-iceberg impacts, *Cold Regions Science and Technology*, Vol. 65, pp. 326-334.

N. Osawa, K. Hashimoto, J. Sawamura, T. Nakai, S. Suzuki, 2007. Study on shell-solid coupling FE analysis for fatigue assessment of ship structure, *Marine Structures*, Vol. 20, pp. 143-163.

Palmer, A. C., Dempsey, J.P. & Masterson, D. M. 2009. A revised ice pressure-area curve and a fracture mechanics explanation. *Cold Regions Science and Technology*, Vol. 56, pp. 73-76.

Sanderson J.O., Ice Mechanics, Risks to Offshore Structures

TRAFI, 2010. *ICE CLASS REGULATIONS 2010 "FINNISH-SWEDISH ICE CLASS RULES 2010"*, TRAFI/31298/03.04.01.00/2010

Yan Gao, Zhiqiang Hu, et al, 2015. An elastic-plastic ice material model for ship-iceberg collision simulations, *Ocean Engineering,* Vol. 102, pp. 27-39

APPENDIX

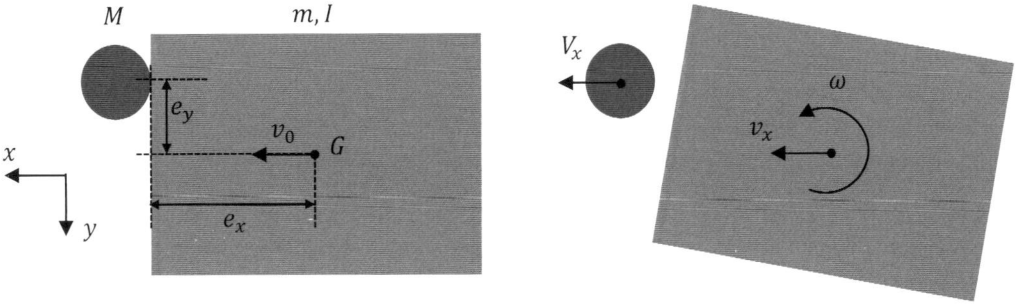

Let us assume that an ice block which has mass m and moment of inertia I collides against a point mass in an initial velocity v_0 and an eccentiricity e_y, and after collision, the velocity of the point mass V_x and the ice block v_x and angular velocity of the ice block ω are generated. The difinition of the direction is shown in the above figure.

In the case of completely plastic collision, the relative velocity in x direction between the ice block and the point mass after collision is equal to zero. Consequently, V_x, v_x, ω can be obtained by the law of conservation of momentum and angular momentum and the equation of the relative velocity after collision.

$$\begin{cases} mv_0 = MV_x + mv_x \\ 0 = Me_yV_x + I\omega \\ V_x = v_x + e_y\omega \end{cases} \tag{16}$$

Accordingly, V_x, v_x, ω can be written as follows.

$$\begin{cases} V_x = \dfrac{\mathscr{m}}{1+\mathscr{m}+\mathcal{J}} v_0 \\ v_x = \dfrac{1}{1+\mathscr{m}}\left(\mathscr{m} + \dfrac{\mathcal{J}}{1+\mathscr{m}+\mathcal{J}}\right) v_0 \\ \omega = -\dfrac{\mathcal{J}}{1+\mathscr{m}+\mathcal{J}} \dfrac{v_0}{e_y} \end{cases} \tag{17}$$

where

$$\mathscr{m} = \frac{m}{M}, \quad \mathcal{J} = \frac{e_y^2 m}{I}$$

The impulse of collision is described as follows.

$$P_{\text{slip}} = M(V_x - 0) = \left(\frac{1}{1+\mathscr{m}+\mathcal{J}}\right) mv_0 \tag{18}$$

If M is infinity, the impulse approaches to the following value.

$$P_{\text{slip}} = \frac{I}{I + e_y^2 m} mv_0 \tag{19}$$

In the case of completely plastic collision and non-slip condition at the contact point, center of the rotation of the ice block corresponds to the contact point. If the mass is infinite, the law of conservation of angular momentum can be described:

$$-e_y m v_0 = \{I + (e_x^2 + e_y^2)m\}\omega \tag{20}$$

and v_x is written as:

$$v_x = -\omega e_y = \frac{e_y^2 m}{I + (e_x^2 + e_y^2)m} v_0 \tag{21}$$

Finally, the impulse of collision in non-slip condition is obtained as follows.

$$P_{\text{non-slip}} = -m(v_x - v_0) = \frac{I + e_x^2 m}{I + (e_x^2 + e_y^2)m} m v_0 \tag{22}$$

Proceedings of the 24th International Conference on
Port and Ocean Engineering under Arctic Conditions
June 11-16, 2017, Busan, Korea

Bow Optimization of A Polar Tanker Based on Approximate Technology

Zhailiu Hao[1], Zhipeng Wang[1], Chengsheng Wu[1], Yukui Tian[1]
[1] China Ship Scientific Research Center, National Key Laboratory of Science and Technology on Hydrodynamics, Wuxi, China

ABSTRACT

The hydrodynamic performances in both of the ice area and the open water area should be considered in the design optimization of polar ship. In this study, the bow of a polar tanker is optimized to reduce the resistance in open water while the ice resistance is being kept. To solve the problem of consuming too much calculation time, the optimization method based on approximate technology is established. The parametric modeling method based on characteristic parameter is used to generate the parametric model of the bow. In approximate technology the uniform experimental design is improved by cutting method and genetic algorithm to improve the approximation accuracy. In optimization the resistance in open water is calculated by CFD numerical simulation based on viscous theory, and the ice resistance is predicted by discrete element method. After optimization, the resistance in open water of model ship and effective power of full scale ship are respectively decreased by 4%~6% and 7%~10% while the ice resistance is being improved, and the principal parameters, such as the displacement, almost keep the same.

KEY WORDS: Polar tanker; Bow optimization; Ice resistance; Resistance in open water; Approximate technology.

INTRODUCTION

Polar ship is the key equipment for shipping and resource exploitation in polar region. With the progress of the arctic route opening and the polar resource exploitation, many countries are focusing on the relevant key technology research and building a lot of polar ships.

As a primary technology for ship development, the ship design optimization is being widely studied. But there are few researches on design optimization of the polar ship recently. The optimization mostly focuses on the ship navigating in open water area. Peri, D. (2010), Jiankui Qian (2012), Zhailiu Hao (2015) et al. research the parametric modeling method, approximate technology, multi-objective optimization algorithm and multidisciplinary design optimization method to improve the hydrodynamic performance in open water. But for the polar ship, the related research mainly concentrates on ice resistance calculation. Shunying Ji

(2011), Konuk, I. (2009), Belytschko, T. (1999), Gürtner, A. (2008) et al. study the discrete element method (DEM), computational cohesive element method(CCEM) and finite element method (FEM) to improve the calculation accuracy of ice resistance. The researches on design optimization of polar ship are few. Kazuo Nozawa (2009) studies the parametric design method and optimizes the bulbous bow to improve the icebreaking ability.

The polar ships navigate not only in ice area, but also in open water area. So the hydrodynamic performances in both of the ice area and the open water area should be considered in the design optimization of these ships. In this study, the bow of a polar tanker is optimized to reduce the resistance in open water while the ice resistance is being kept. The optimization method based on approximate technology is established to obtain the optimized hull. Firstly the optimization process is established. Then the parametric modeling method based on characteristic parameter is used to generate the parametric model of the bow. After that, the approximate model of resistance in open water is established. In approximate technology, the uniform experimental design is improved by cutting method and genetic algorithm to improve approximation accuracy. In optimization process, the resistance in open water is calculated by CFD numerical simulation based on viscous theory, and the ice resistance is predicted by discrete element method. Based on the above works, the approximate model is optimized by multi-island genetic algorithm (MIGA) to obtain the optimized hull. Finally the resistances in open water and the ice resistances of the initial ship and the optimized ship are compared to verify the effectiveness of the bow optimization.

OPTIMIZATION MODEL

The optimization object is a polar tanker with ship length between perpendiculars 259.2m, breath 44.0m, draft 15.89m and volume 145663m^3. Its geometric shape is shown in Figure 1, and the optimization part is the bow. The optimization speed for resistance in open water is the design speed 15kn, and the design icebreaking speed is 2.54kn.

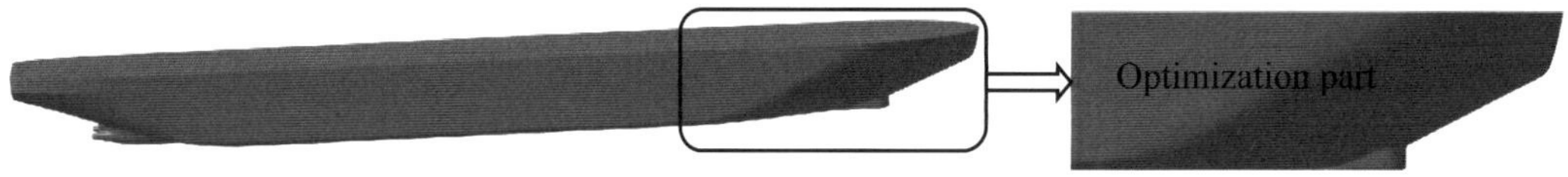

Figure 1. Hull geometry of the polar tanker

OPTIMIZATION METHOD

Optimization Process

To solve the problem of consuming too much calculation time, the optimization method based on approximate technology is established, and the corresponding optimization process is specified as follows. Step 1, the characteristic parameters are selected and their ranges are determined, then the bow of the polar tanker is parametric modeled. Step 2, the uniform experimental design is improved to generate a series of hulls. Step 3, the resistances in open water of these hulls are calculated. Step 4, the approximate model of resistance in open water is established by using neural network model. Step5, the approximate model is optimized by MIGA algorithm to obtain the optimized hull. Step6, the resistance in open water and ice resistance of optimized ship are predicted and verified. Step7, if the resistance is satisfied with design requirement, output the optimized hull, otherwise go back to step 1 to adjust the characteristic parameters and re-optimize the bow until the satisfied hull is obtained.

Parametric Modeling

The parametric modeling method based on characteristic parameter is used to generate the parametric model of the bow. Considering resistance in open water, ice resistance and smoothness of hull surface, 5 characteristic parameters shown in Figure 2 are selected to obtain the parametric model of the bow, which are water line curvature of forward shoulder, water line angle, curvature of bow section lines, stem angle and width of forward skeg. After the parametric model is generated, different bow lines can be obtained by changing the value of the 5 characteristic parameters.

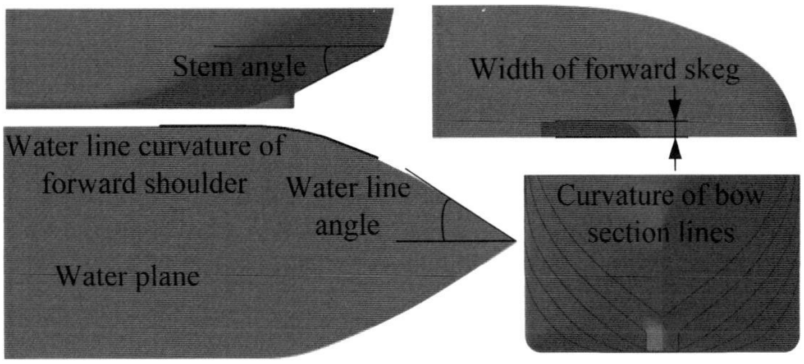

Figure 2. Characteristic parameters

In this optimization the following limiting conditions need to be considered for determining the ranges of characteristic parameters. (1) The ranges of water line angle and stem angle should be limited to keep the ice resistance. (2) The change rates of displacement and longitudinal center of buoyancy need to be less than 1%. (3) Considering the separation of crushed ice, the longitudinal projection contour of forward skeg should remain the same.

Approximate Technology

The parametric hull transformation and the calculation of resistance in open water consume too much time, which greatly influences the optimization efficiency and quality. Therefore, the approximate technology is introduced to generate approximate model for optimization.

The successful application of approximate model depends on its approximate accuracy, and the experimental design has great influence on the approximate accuracy. For this purpose, the uniform experimental design is chosen to generate the approximate model to improve approximate accuracy in this paper. As for the uniform experimental design, it is difficult to satisfy the requirement of uniformity and computational efficiency simultaneously in the present study, which restricts its application in engineering practice. So the cutting method and the genetic algorithm are introduced in this paper to solve this problem to efficiently generate the uniform experimental design with arbitrary number of variables and samples.

The uniformity measure of uniform experimental design in this study is the center deviation L_2- $CD_2(P_n)$ which has the expression:

$$CD_2(P_n)=\left[\left(\frac{13}{12}\right)^s-\frac{2^{s-1}}{n}\sum_{k=1}^{n}\prod_{i=1}^{s}\left(2+\left|x_{ki}-\frac{1}{2}\right|-\left|x_{ki}-\frac{1}{2}\right|^2\right)\right.$$
$$\left.+\frac{1}{n^2}\sum_{k,l=1}^{n}\prod_{i=1}^{s}\left(1+\frac{1}{2}\left|x_{ki}-\frac{1}{2}\right|+\frac{1}{2}\left|x_{li}-\frac{1}{2}\right|-\frac{1}{2}\left|x_{ki}-x_{li}\right|\right)\right]^{1/2} \quad (1)$$

Where $P_n = \{x_k = (x_{k1},\ldots,x_{ks}), k = 1,\ldots,n\}$ is the design matrix in the test area C^s, n is the number of samples, and s is the number of variables.

In uniform experimental design the U matrix is generated by good lattice point method in this paper. When sample number is even, the uniformity of the generated schemes is hard to meet the requirement. Therefore, the cutting method proposed by Changxing Ma (2004) is introduced in this paper to improve the uniformity, and the steps are as follows:

Step 1, for the uniform experimental design $U_n(n^s)$, select a prime p where $p>n$, and obtain the uniform experimental design $U_p(p^s)$ with the design matrix $P_u^{'} = \{c_1,\ldots,c_s\}$ by good lattice point method. Step 2, denote $P_u^{'}$ as $C=(c_{ij})$, and reorder C according to the sample value in its lth column to generate the reordered matrix $C^{(l)} = (c_{kj}^{(l)})$ where $l = 1,\ldots,s$. Step 3, for $m = 1,\ldots,p$, $C^{(l,m)} = (c_{kj}^{(l,m)})$ is computed as follows:

$$c_{kj}^{(l,m)} = \begin{cases} c_{k+m-n-1,j}^{(l)} & m>n, k=1,\ldots,n, j=1,\ldots,s \quad 1 \\ c_{k,j}^{(l)} & m \le n, k=1,\ldots,m-1, j=1,\ldots,s \quad 1 \\ c_{k+p-n,j}^{(l)} & m \le n, k=m,\ldots,n, j=1,\ldots,s \quad 1 \end{cases} \tag{2}$$

Step 4, for each m and l, a $C^{(l,m)}$ with n rows and s columns can be obtained, and the total number of $C^{(l,m)}$ is ns. The uniform experimental design $U_n(n^s)$ with the minimum CD_2 in the uniform experimental designs corresponding to all of the $C^{(l,m)}$ is expected.

The uniform experimental design can be obtained by the cutting method when the numbers of variables and samples are small. But as the numbers increasing, the computational efficiency will be reduced significantly. Actually, experimental design is an optimization problem with certain uniformity measure as the objective. So the genetic algorithm with great search ability is able to solve the computational efficiency problem.

The uniform experimental design $U_n(n^s)$ generated by genetic algorithm is executed in two steps. The first step is to generate $U_p(p^s)$ where p is prime. The U-type design with p rows and $p-1$ columns can be obtained by good lattice point method. Then the s columns from the $p-1$ columns with the minimum CD_2 are selected to generate $U_p(p^s)$. When executing genetic algorithm, there are s variables and the objective is CD_2. Each variable represents each column of the U-type matrix, and is natural number with the value from 1 to $p-1$. The second step is to select $U_n(n^s)$ in $U_p(p^s)$ by the cutting method. When executing genetic algorithm, there are two variables and the objective is CD_2. Both of the variables are natural numbers. One is related to the row of $U_p(p^s)$ with the value from 1 to p, and the other is related to the column of $U_p(p^s)$ with the value from 1 to s. By executing genetic algorithm two times, the uniform experimental design with arbitrary number of samples and variables is able to be generated efficiently.

In the bow optimization of polar tanker, comprehensively considering calculation work and approximate accuracy, 125 hulls are generated in this study, and their projection of 3 variables is shown in Figure 3.

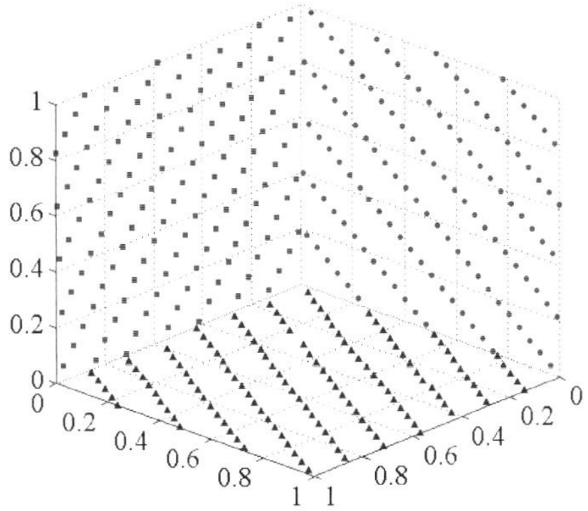

Figure 3. Projection of 3 variables

RESISTACE CLACULATION METHOD

The resistance in open water is calculated by CFD numerical simulation based on viscous theory, as is introduced by Chengsheng Wu (2010). The control equation is continuity equation and RANS equation, which are expressed as follows:

$$\frac{\partial u_i}{\partial x_i} = 0 \tag{3}$$

$$\frac{\partial u_i}{\partial t} + \frac{\partial u_i u_j}{\partial x_j} = -\frac{1}{\rho}\frac{\partial P}{\partial x_i} + \frac{\partial}{\partial x_j}\left[\nu \left(\frac{\partial u_i}{\partial x_j} + \frac{\partial u_j}{\partial x_i} \right) \right] - \frac{\partial \left(\overline{u_i' u_j'} \right)}{\partial x_j} + f_i \tag{4}$$

The SST k–ω turbulence model is used to close the RANS equation. The free surface is captured by VOF (volume of fluid) method, which introduces volume fraction to simulate two or more kinds of fluids.

The ice resistance is calculated by discrete element method. The level ice unit is simulated by particle parallel bond model, and the hull structure is expressed by triangular element with certain arrangement mode. The parallel bond model sets an elastic bond disk between two particle bond units, and the disk can transfer force and moment between the two units, including tension, shear, bending moment and torque. The ice resistance is able to be obtained by using contact model to calculate contact force and simulating interaction of friction and collision between ship and ice or among ice units.

The particle unit force can be divided into normal force and tangential force. The normal force F_n and the tangential force F_s are defined as:

$$F_n = K_n x_n - C_n x_n \tag{5}$$

$$F_s^* = K_s x_s - C_s x_s = F_s \quad \min\left[F_s^*, sign(F_s^*) \right] \mu F_n \tag{6}$$

Where K_n, C_n, x_n and x_n are the stiffness coefficient, damping coefficient, deformation and strain rate of particle in the normal direction, K_s , C_s , x_s and x_s are the corresponding parameters in the tangential direction, and μ is friction coefficient.

The frost action among particles is realized by establishing the parallel bond model. The force

$\overline{F}_i$ and moment $\overline{M}_i$ on the bond disk are:

$$\overline{F}_i = \overline{F}_i^s + \overline{F}_i^n \quad \overline{M}_i = \overline{M}_i^s + \overline{M}_i^n \tag{7}$$

Where $\overline{F}_i^n$ and $\overline{F}_i^s$ are the normal and tangential force, $\overline{M}_i^n$ is the torque, and $\overline{M}_i^s$ is the moment on the bond disk.

OPTIMIZATION RESULT AND ANALYSIS

Optimization Result of Resistance in Open Water

The optimization objective is the model resistance in open water at the design speed 15kn. After the approximate model of resistance in open water is established, the optimized ship is obtained by optimizing this approximate model with MIGA algorithm, which has 5 islands with 20 populations in each island and 200 generations.

According to CFD calculation, the comparison of resistances between initial ship and optimized ship is shown in Table 1, where R_m is the model resistance in open water, R_s and P_E are the resistance and effective power of full scale ship, $CHAR_m$ is the change rate of R_m, and the $CHAP_E$ is the change rate of P_E. As for the optimized ship, its R_m and P_E are respectively decreased by 4%~6% and 7%~10%, and especially at the design speed its R_m and P_E are respectively decreased by 5.08% and 8.27%.

Table 1. Comparison of total resistance and effective power

V_s(kn)	Fr	R_m(N)			R_s(kN)		P_E(kW)		
		Initial ship	Optimized ship	$CHAR_m$	Initial ship	Optimized ship	Initial ship	Optimized ship	$CHAP_E$
10	0.099	23.38	22.40	-4.18%	567.5	528.0	2919.2	2715.8	-6.97%
12	0.123	32.98	31.63	-4.08%	811.5	757.0	5009.1	4672.8	-6.71%
14	0.143	44.55	42.64	-4.30%	1116.5	1038.5	8040.5	7478.9	-6.98%
15	0.153	51.16	48.57	-5.08%	1295.8	1188.7	9998.2	9171.7	-8.27%
16	0.163	59.29	56.08	-5.42%	1535.6	1402.3	12638.7	11541.5	-8.68%
18	0.184	77.02	72.09	-6.40%	2062.1	1855.1	19093.3	17176.3	-10.04%

The ship length, breath and draft of the optimized ship are not changed, and the other principal parameters, such as the displacement and the longitudinal center of buoyancy, almost keep the same with the initial ship. After optimization, it is obtained that: (1) the water line angle is increased; (2) the stem angle is decreased; (3) the transition between bow and midship area is more smooth; (4) as for the forward skeg, its longitudinal projection contour remains the same and its width is decreased; (5) the radian of section lines become larger when approaching the midship; (6) and as the waterlines going down, its entrance radian is growing.

Figure 4 compares the wave-making between initial ship and optimized ship. It is observed from the figures that the wave-making of optimized ship is obviously improved.

Optimization Result of Ice Resistance

After the optimization of resistance in open water, the ice resistance of the optimized ship is verified by discrete element method. The resistances when breaking level ice with thickness

1.60m at 3 speeds are predicted and shown in Table 2 and Figure 5, where V_S represents ship speed and R_I represents ice resistance. The collision of level ice with ship hull is shown in Figure 6. It is obtained that at low speed the ice resistances of optimized ship and initial ship are similar, and as the speed increasing the ice resistance of optimized ship is being obviously improved. At design icebreaking speed 2.54kn, the ice resistance of optimized ship is decreased by 3.74%. So the ice resistance is improved and the precondition of the bow optimization in this study is satisfied.

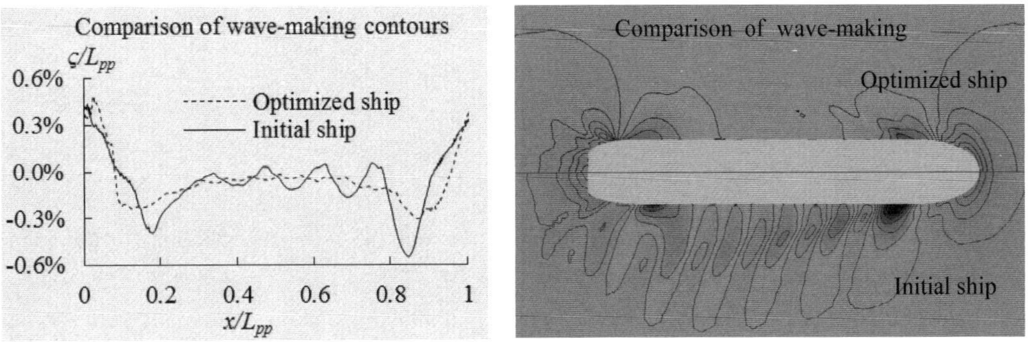

Figure 4. Comparison of wave-making

Table 2. Comparison of ice resistance result

Speed (kn)	Ice resistance of initial ship (kN)	Ice resistance of optimized ship (kN)	Change rate
1.90	4558.9	4510.3	-1.07%
2.54	5004.5	4817.2	-3.74%
3.80	6074.0	5374.4	-11.52%

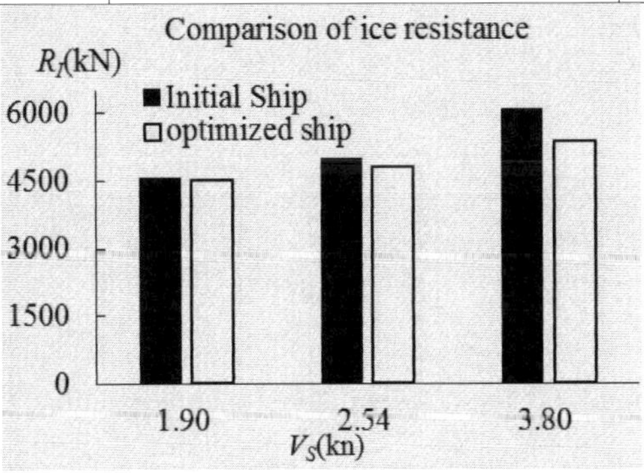

Figure 5. Comparison of ice resistance

Figure 6. Collision of level ice with ship hull

CONCLUSIONS

The bow of a polar tanker is optimized to reduce the resistance in open water while the ice resistance is being kept. To solve the problem of consuming too much calculation time, the optimization method based on approximate technology is established to obtain the optimized hull. The water line curvature of forward shoulder, water line angle, curvature of bow section lines, stem angle and width of forward skeg are selected as the characteristic parameters to generate the parametric model of the bow. In approximate technology, the uniform experimental design is improved by cutting method and genetic algorithm to improve the approximate accuracy. In optimization, the resistance in open water is calculated by CFD numerical simulation based on viscous theory, and the ice resistance is predicted by discrete element method. After optimization, the resistance in open water of model ship and effective power of full scale ship are respectively decreased by 4%~6% and 7%~10% while the ice resistance is being improved, and the principal parameters, such as the displacement, almost keep the same.

Having high optimization efficiency by introducing approximate technology, and being able to optimize the detail flow field based on CFD simulation, the optimization method established in this study can be effectively used for polar ship optimization in engineering practice.

ACKNOWLEDGEMENTS

The author is gratefully thankful to the research team members, and the support of High-tech Ship Scientific Research Project (Key Technology Research of Polar Tanker).

REFERENCES

Belytschko, T. & Black, T, 1999. Elastic Crack Growth in Finite Elements with Minimal Remeshing. *International Journal for Numerical Methods in Engineering*, 45(5), pp. 601-620.

Changxing Ma & Kai-Tai Fang, 2004. A new approach to construction of nearly uniform designs. *International Journal of Materials and Product Technology*, 20(1), pp. 115-126.

Chengsheng Wu, Xiong Chen et.al, 2010. Numerical simulation of free ship model towed in still water. *Journal of Ship Mechanics*, 14(8), pp. 823-833.

Gürtner, A., Konuk, I., Gudmestad, O.T. & Liferov, P, 2008. Innovative Ice Protection for Shallow Water Drilling - Part III: Finite Element Modelling of Ice Accumulation. *Proc. of the 27th International Conference on Offshore Mechanics and Arctic Engineering*, Estoril, Portugal.

Jiankui Qian, Xiaofei Mao, Xiaoyi Wang & Qiuqin Yun, 2012. Ship hull automated optimization of minimum resistance via CFD and RSM technique. *Journal of Ship Mechanics*, 16(1), pp: 36-43.

Kazuo Nozawa, 2009. A Consideration on Bow Design of Arctic Tanker Transiting in Thin Level Ice and in Broken Ice Channel. *Proc. of the 19th International Offshore and Polar Engineering Conference*, Osaka, Japan.

Konuk, I., Gürtner, A. & Yu, S, 2009. Study of Dynamic Ice and Cylindrical Structure Interaction by the Cohesive Element Method. *Int. Conf. on Port and Ocean Engineering under Arctic Conditions 2009.*

Peri, D., Campana, E.F., Tahara, Y., Kandasamy, M. & Stern, F, 2010. New developments in Simulation-Based Design with application to High Speed Waterjet Ship Design. *Proc. of the 28th Symposium on Naval Hydrodynamics*, Pasadena, USA.

Shunying Ji & Qianjin Yue, 2011. *Engineering sea ice model and its application.* Science Press: Beijing.

Zhailiu Hao, Baiwei Feng & Zu Yuanliu, 2015. *Physical programming and its application in ship design optimization.* National Defense Industry Press: Beijing.

Sensor Arrangement for Ice Load Monitoring to Estimate Local Ice Load in Arctic Vessel

Young-Chun Jo[1], Joong-Hyo Choi[1], Sung-Gun Park[1], Sungkon Han[1]
[1] Ship & Ocean R&D Institute (Daewoo Shipbuilding & Marine Engineering, Seoul, Korea)

ABSTRACT

The more opportunities for exploitation and transportation are given in the Arctic area, the larger commercial vessels are needed for mass transportation. The environmental loads due to the interaction between ice and hull are increased to enlargement of arctic vessels. The ice load monitoring system is recommended for the research of ice load estimation and the safety of hull structure. This paper is concerned with sensors arrangement for ice load monitoring in arctic vessel. The number and arrangement of sensors were determined for the estimation of ice induced loads acting on the hull. FE analyses were carried out to verify the estimation of ice loads. The alternative arrangement of sensors was proposed considering complex structural details which cause the stress concentrations and discontinuities.

KEY WORDS: Sensor; Sensor arrangement; Ice load; Ice load monitoring; Arctic vessel.

INTRODUCTION

The activities related to exploitation for oil and gas in the Arctic areas increase significantly. In order to transport increased resources in the Arctic areas, large Arctic commercial vessels such as gas carriers, oil tankers, bulk carriers, etc. are needed for mass transportation. In Arctic area, ice is the main factor of environmental load acting on Arctic vessel. The ice load is increased with the enlargement of vessel.

The largest Arctic commercial vessel was built by DSME. Table 1 shows main dimension of Arctic LNG carrier. The size of Arctic LNG carrier is larger than any other Arctic vessels have been constructed so far. Ice load monitoring system was installed for safety of ice navigating and hull structure of this large LNG carrier.

Table 1. Main dimensions of Arctic LNG carrier

	Length (LOA)	Breadth	Height
Dimension(m)	Abt. 300	50	26.5

This paper is concerned with sensors arrangement for ice load monitoring in arctic vessel. The number and arrangement of sensors are determined for the estimation of ice induced loads acting on the hull. FE analyses were carried out to verify the estimation of ice loads. The alternative arrangement of sensors was proposed considering complex structural details which cause the stress concentrations and discontinuities.

ICE LOAD MONITORING SYSTEM (ILM)

Ice load monitoring system consists of sensors, wire, data acquisition device, data process device and so on. Because sensors are installed directly on the structure at measuring location and ice load on hull structure is estimated based on measured electrical signal of these sensors, the number and arrangement of them are crucial in ILM.

Installed Locations of Sensors

The installed locations of strain sensors were determined considering expected high ice loaded area. Figure 1 shows the sensors installed locations for ILM.

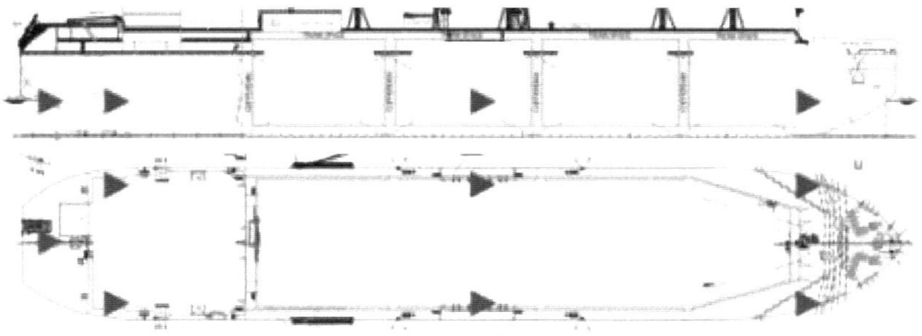

Figure 1. Sensors installed locations for ILM

Sensors Arrangement

The arrangement of strain sensors at each location is shown as figure2. The type and purpose of arranged sensors are described as follows.

- The sensors on flange of longitudinal/transverse frame have 1-axis filament. They measure bending stress of frame.
- The sensors on both web ends of longitudinal/transverse frame have 2-axis filament.

They measure shear stress of frame.

- The sensors on side shell have 3-axis filament. They assess the damage of hull plate.

Finally, 49 strain gauges were installed in Arctic LNG carrier.

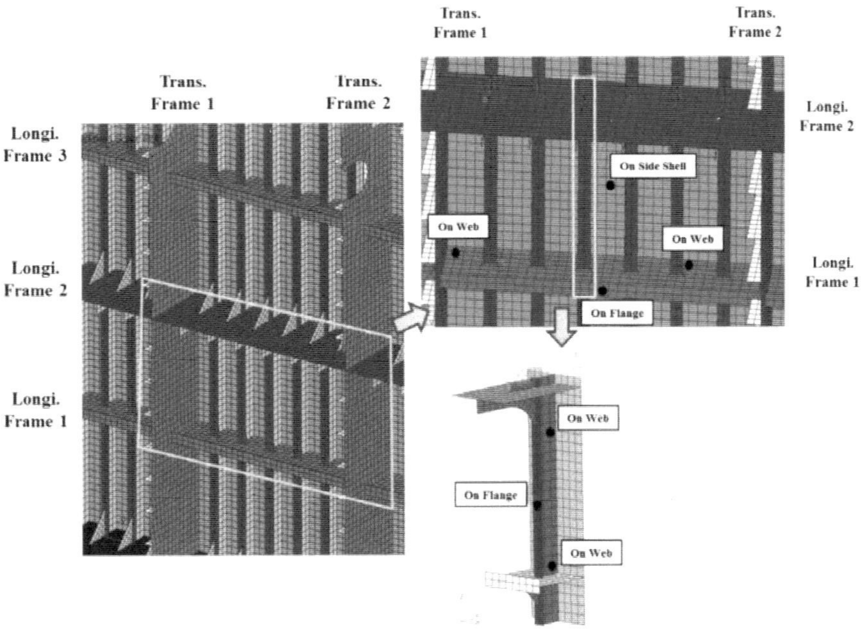

Figure 2. Details of Sensors Arrangement in Measuring Area

ICE LOAD ESTIMATION

In this section, the procedure of stress calculation from measured strain value at sensors and estimation of ice load which is derived from stress components were described.

Calculation of Stress Components

The stress in the cross section of the stiffener was divided into axial, transverse and shear stress. Eq. (1) is the general expression for principal strain directions at any location.

$$\varepsilon(\phi)=\frac{\varepsilon_x+\varepsilon_y}{2}+\frac{\varepsilon_x-\varepsilon_y}{2}\cdot\cos(2\phi)+\frac{1}{2}\cdot\gamma_{xy}\sin(2\phi) \tag{1}$$

Where, ϕ is the degree of angle about x-axis with 2 dimensional Cartesian coordinate system.

Because the strain sensor with 3-axis filament can measure 3 principal strain directions, stress components can be obtained by stress-strain relation. But, the study result proposed by Holm (2012) was applied to prevent increase of the number of sensor. Holm (2012) showed that the strain sensor with 2-axis filament can calculate the stress components in a particular case.

The strain sensors with 2-axis filament were installed near the both ends of stiffener avoiding bracket as shown in Figure 3. The sensors were located on neutral axis.

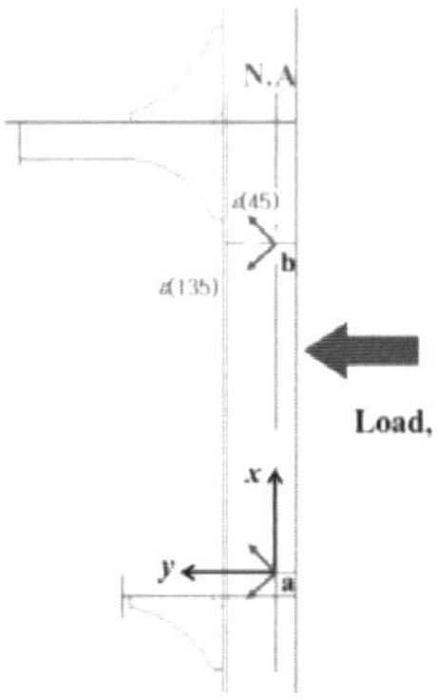

Figure 3. Sensors Arrangement of Frame

X axis is parallel with length direction of stiffener. The filaments of each strain sensor were attached at angles of 45 degrees and 135 degrees, respectively.

Eq. (2) is the matrix form of Eq. (1) considering angles of filament arrangement. $\varepsilon(0)$ is assumed as the dummy strain.

$$\begin{bmatrix} \varepsilon(0) \\ \varepsilon(45) \\ \varepsilon(135) \end{bmatrix} = \begin{bmatrix} 1 & 0 & 0 \\ \frac{1}{2} & \frac{1}{2} & \frac{1}{2} \\ \frac{1}{2} & \frac{1}{2} & -\frac{1}{2} \end{bmatrix} \cdot \begin{bmatrix} \varepsilon_x \\ \varepsilon_y \\ \gamma_{xy} \end{bmatrix} \tag{2}$$

By solving Eq. (2), the strain components of Cartesian coordinate is obtained as Eq. (3)

$$\begin{bmatrix} \varepsilon_x \\ \varepsilon_y \\ \gamma_{xy} \end{bmatrix} = \begin{bmatrix} 1 & 0 & 0 \\ -1 & 1 & 1 \\ 0 & 1 & -1 \end{bmatrix} \cdot \begin{bmatrix} \varepsilon(0) \\ \varepsilon(45) \\ \varepsilon(135) \end{bmatrix} \tag{3}$$

On beam bending problem, the axial strain along neutral axis is 0 (zero).

$$\varepsilon_x = 0 \tag{4}$$

Strain components can be obtained from measured strain $\varepsilon(45)$, $\varepsilon(135)$ using Eq. (3).

$$\varepsilon_y = \varepsilon(45) + \varepsilon(135), \quad \gamma_{xy} = \varepsilon(45) - \varepsilon(135) \tag{5}$$

Stress components can be calculated as stress-strain relation.

$$\sigma_x = E\varepsilon_x = 0, \quad \sigma_y = E\varepsilon_y, \quad \tau_{xy} = G\gamma_{xy} \tag{6}$$

Load Estimation

When load P_{app} is applied between sensor positions, shear stresses at each gauge are respectively obtained as τ_a and τ_b according to the previous section as Figure 4.

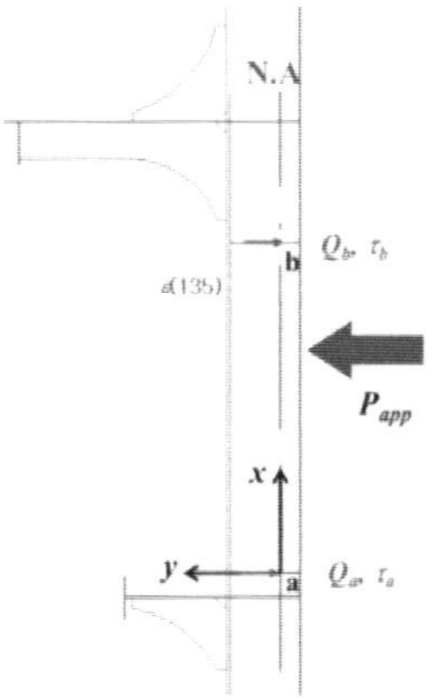

Figure 4. Frame acting on applied load, P_{app}

Estimated load, P_{est}, can be derived from shear forces at the sensor positions.

$$P_{est} = Q_a - Q_b = f_Q \cdot (\tau_a - \tau_b) \cdot A_w \tag{7}$$

Where, f_Q is the correction factor and A_w is the cross-sectional area of web.

For accuracy of estimation, estimated load should be same as applied load. If applied load, P_{app}, is known and shear stresses can be calculated by FEA, correction factor f_Q can be obtained as Eq. (8).

$$f_Q = \frac{P_{app}}{(\tau_a - \tau_b) \cdot A_w} \tag{8}$$

FE ANALYSIS

FE analysis was carried out in order to verify the arrangement of strain sensors. MSC PATRAN/NASTRAN was used for FE analysis.

FE Models

While the number of sensors installed locations was seven (7), the number of models was reduced to four (4) considering symmetry. Figure 5 shows FE models and detail in each sensor installed location. Very fine mesh for more detailed distribution of stress components was used in the models. Figure 6 shows very fine meshed details of transverse and longitudinal frames in fore body model.

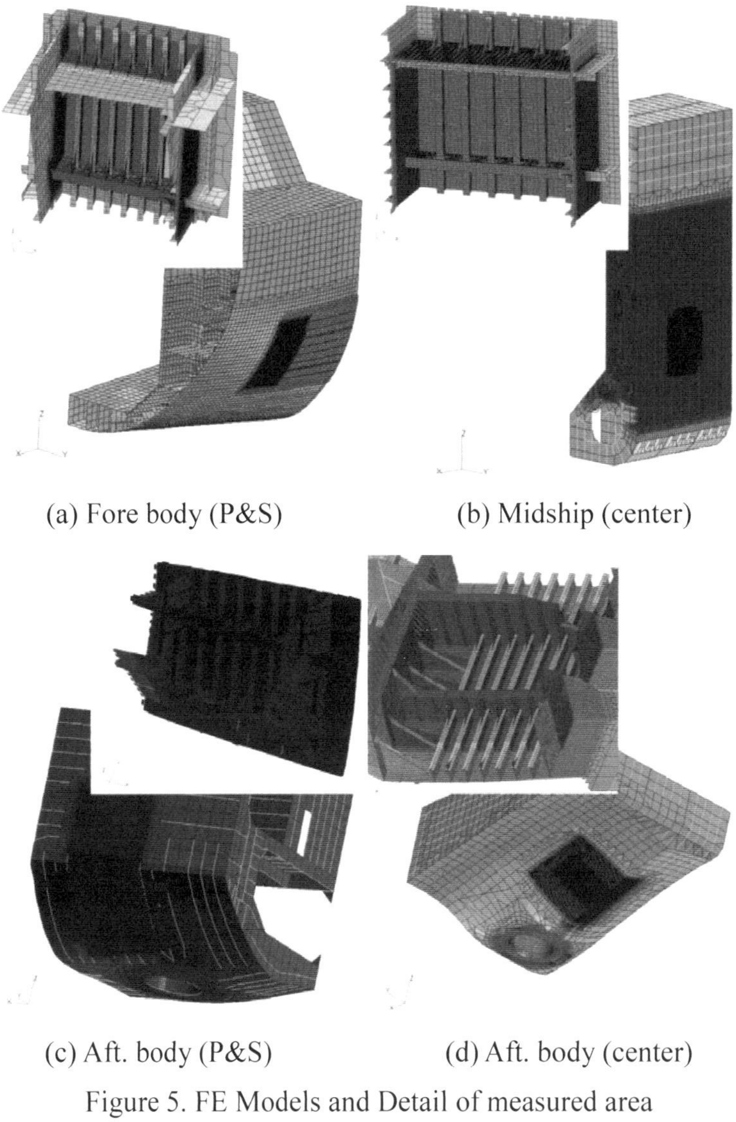

(a) Fore body (P&S) (b) Midship (center)

(c) Aft. body (P&S) (d) Aft. body (center)

Figure 5. FE Models and Detail of measured area

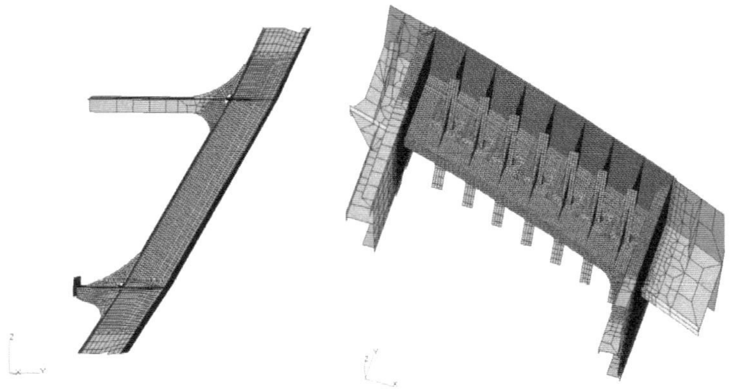

(a) transverse frame (b) longitudinal frame

Figure 6. Details of Frames (fore body)

Loading & BCs

It was assumed that relationship between applied load and structural response is linear. Unit load (1kN) was applied to FE model. Area subject to ice load was determined considering ice navigating draft of Arctic LNG carrier. In order to consider the vessel's movement, 17 load steps were created as Figure 7.

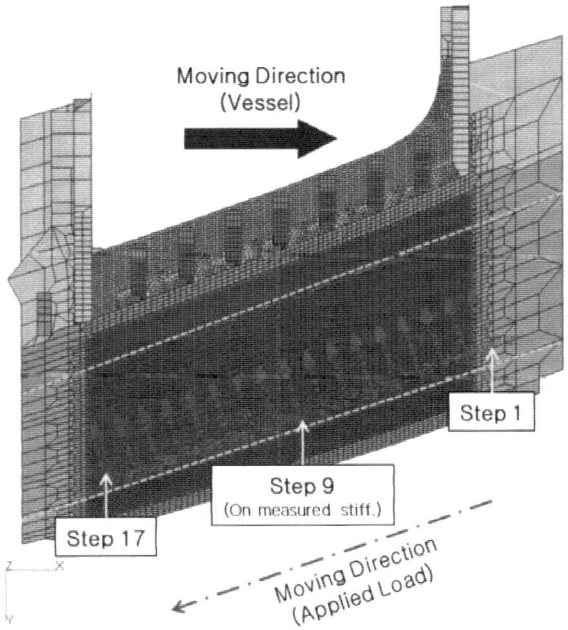

Figure 7. Scheme of Load Step

Results

Figure8 shows shear stress distribution in transverse frame and the graph of shear stress distribution on neutral axis. Load estimation results of transverse frame were shown in Table 2.

Table 2. Load Estimation Result of Transverse Frame

Position	Shear force (kN)	Est. load (kN) a – b	Applied load (kN)	Correction factor, f_Q
Sensor a	0.517	0.970	1.00	1.03
Sensor b	-0.453			

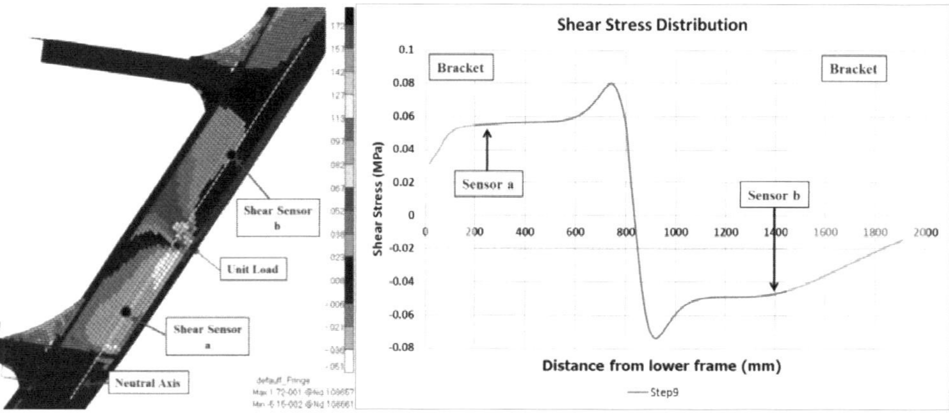

Figure 8. Shear Stress Distribution in Transverse Frame

Figure9 shows shear stress distribution in longitudinal frame and the graph of shear stress distribution on neutral axis. At connections with transvers frames, stress concentrations and discontinuities were shown.

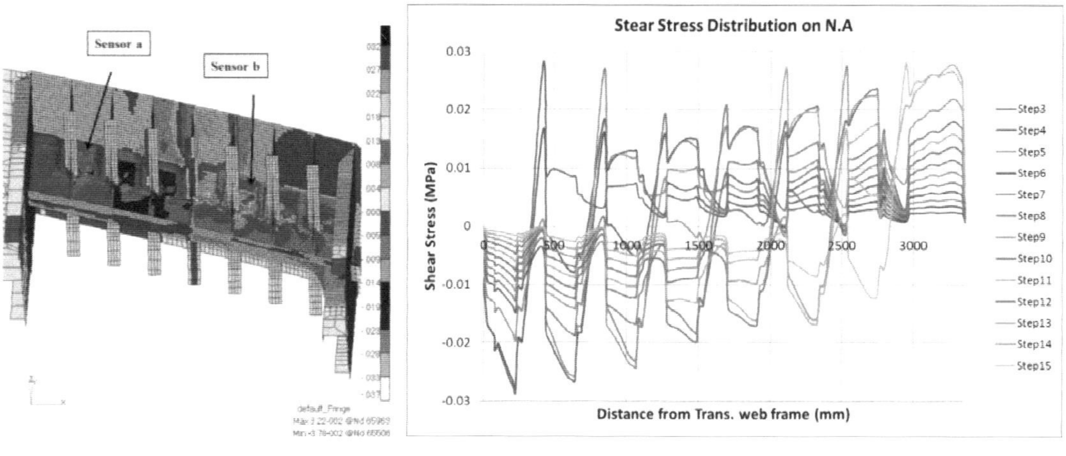

Figure 9. Shear Stress Distribution in Longitudinal Frame

POAC17-147

Load estimation result from shear stress at positions of sensors is shown in Figure 10. Estimated values from load step 5 to load step 11 were stable. Correction factor was able to be obtained using average estimated value of stable load steps as table 3. In spite of structural discontinuity, load could be estimated from strain sensors.

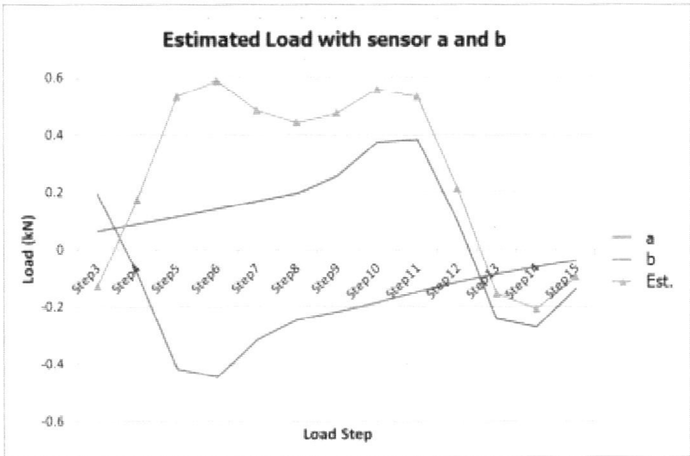

Figure 10. Estimated Load at Load Steps

Table 3. Load Estimation Result of Longitudinal Frame

Position	Avg. Est. load (kN)	Applied load (kN)	Correction factor, f_Q
Sensor a	0.516	1.00	1.94
Sensor b			

Modification of Sensors Arrangement

In longitudinal frame, strain sensors were not able to be installed because the residual area considering transverse frame connections, such as slot hole and collar plate, is smaller than the required area for sensors installation. Positions of strain sensors in longitudinal frame were modified out of neutral axis. Shear stress distribution on cross-sectional area at each sensor's position was reviewed above all as Figure 11. Although stress at upper part of cross-section was smaller than lower part, distribution trends were similar with each other. Load estimation with strain sensors out of neutral axis was able to carry out through the recalculation of correction factor.

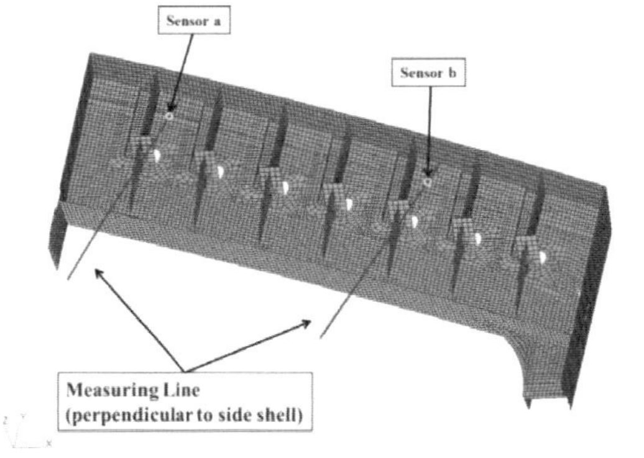

(a) Reviewed Line at Each Sensor's Position

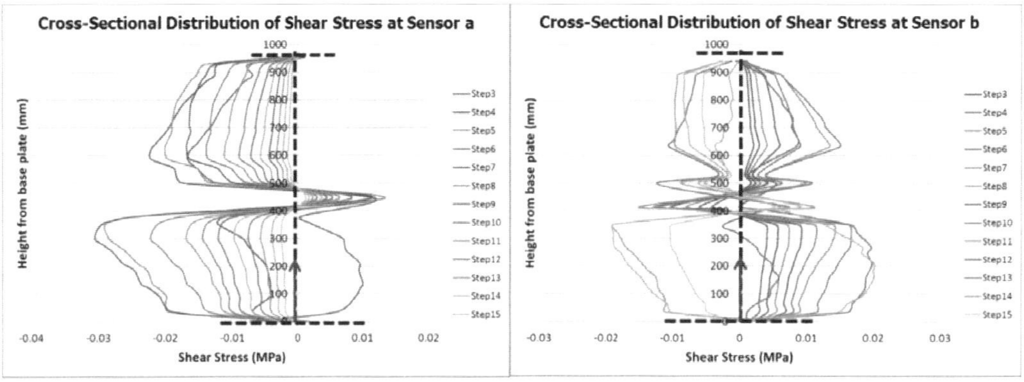

(a) sensor a (b) sensror b

Figure 11. Cross-Sectional Distribution of Shear Stress in Longitudinal Frame

Load estimations were carried out with candidates (a', a'', b') for modified sensors positions as Figure 12. Estimation results are shown in Figure 13. The result with a'' and b' was more stable, alternative positions of strain sensors were determined. Table 4 shows estimation result and correction factor of modified positions.

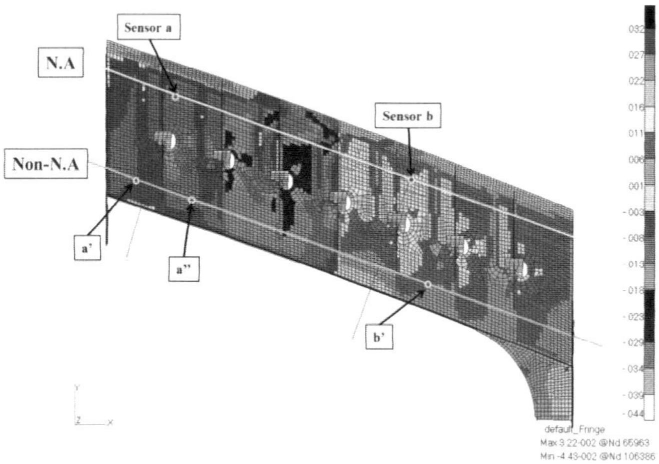

Figure 12. Modified Arrangement of Strain Sensors in Longitudinal Frame

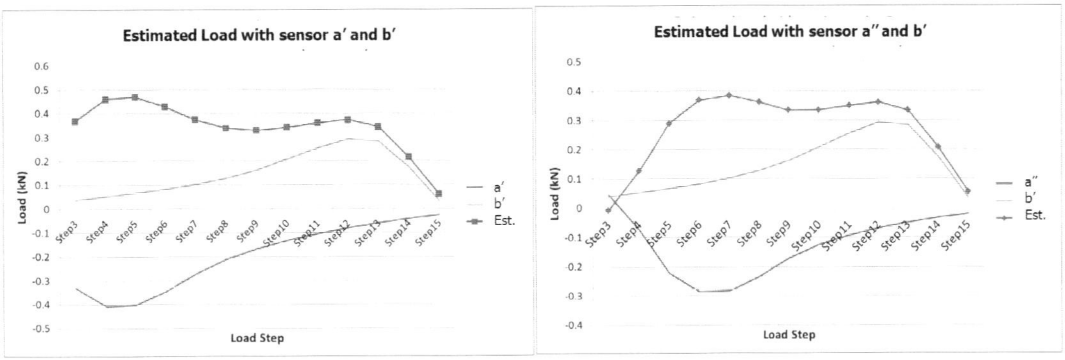

Figure 13. Load Estimation Results

Table 4. Load Estimation Result of Modified Position

Position	Avg. Est. load (kN)	Applied load (kN)	Correction factor, f_Q
Sensor a''	0.356	1.00	2.81
Sensor b'			

CONCLUSIONS

- The seven (7) locations for the installation of strain sensors were determined in Arctic LNG carrier built by DSME. Seven (7) strain sensors are required at each location. Total 49 strain sensors were used for ice load monitoring in Arctic LNG carrier.
- Strain components could be derived from 2-axis strain sensors on neutral axis of longitudinal/transverse frames. Stress component could be obtained with stress-strain relationship and load estimation could be performed.
- FE analyses were carried out in order to verify sensors arrangement and load estimation. Load could be estimated with proposed sensors arrangement despite structural discontinuity and stress concentration of complex stiffened structure.
- Sensors in longitudinal frame were moved out of neutral axis due to installation problem. Load estimation could be carried out.

ACKNOWLEDGEMENTS

This work was supported by the Industrial Convergence Strategic technology development program (10063417, Development of basic design technology for ARC7 class Arctic offshore structures) funded by the Ministry of Trade, industry & Energy (MI, Korea).

REFERENCES

Choi, J., Park, G., Kim, Y., Jang, K. Park, S., Ha, M., Yu, H., Iyerusalimskiy, A. & St. John, J., 2009. Ice Load Monitoring System for Large Arctic Shuttle Tanker, Proceedings of International Conference on Ship and Offshore Technology,

Holm, H., 2012. *Ice-Induced Loads on Ship Hulls*. Master Thesis: Norwegian University of Science and Technology.

Iyerusalimskiy, A., Choi, J., Park, G., Kim, Y. & Yu, H., 2011. The Interim Results of the Long-term Ice Loads Monitoring on the Large Arctic Tanker, *Proceedings of the 21st International Conference on Port and Ocean Engineering under Arctic Conditions*, paper No. POAC11-088

Leira, B., Børsheim, L., Espeland, Ø. & Amdahl, J., 2009. Ice-Load Estimation for a Ship Hull based on Continuous Response Monitoring, *Proceedings of the Institution of Mechanical Engineers*, Vol. 223 pp.529-540.

Yu, H., Iyerusalimskiy, A., Kim, Y. S. & St. John, J., 2012. Hull Structural Performance Monitoring System for Ships Operating in Ice-Covered Waters, *Proceedings of the Arctic Technology Conference*, paper No. 1368063.

POAC'17
Busan, KOREA

Proceedings of the 24th International Conference on
Port and Ocean Engineering under Arctic Conditions
June 11-16, 2017, Busan, Korea

Bond Model of Dilated Polyhedral DEM and its Applications in Sea Ice Simulation

Lu Liu, Shunying Ji [1]
[1] State Key Laboratory of Structure Analysis of Industrial Equipment, Dalian University of Technology, Dalian 116023, China

ABSTRACT

The simulation of breaking process of continuum is an important issue for discrete element method. An explicit bonding method is developed based on Rigid Body Spring Method (RBSM) to simulate fracture with Dilated-polyhedron DEM (DP-DEM) in this paper. Minkowski sum of basic polyhedron and sphere is used to generate the dilated polyhedral element. Bond points are fixed on the common face between elements. Each bond point represents average area of the common face. The strain on every bond point is calculated by the division of distance between bond points and characteristic length. Then the stress can be determined according the stiffness in RBSM. The bond force at bond point is evaluated by stress and average area of every bond point. So the resultant force and moment of each element could be obtained by summation of force at each bond point and its moment to mass center. The complex matrix calculation and the integral for stiffness in RBSM could be avoided to save computational resources. The fracture on the interface is decided by Mohr-Coulomb criterion. After the fracture, DEM would take charge of the broken element in contact computation. This method is built to simulate the fracture of sea ice interacting with structure. The ice loads on structure and broken ice shape is analyzed to measure the applicability of this method in sea ice simulations.

KEY WORDS: Bonding model, Dilated polyhedral DEM, Sea ice, Breaking process, Ice loads

INTRODUCTION

Since the discrete element method (DEM) has been acknowledged to be a valid method to study granular matter, the studies of simulating fragmentation of brittle material with DEM have obtained a lot of attentions for its simple mathematical model and parallel algorithm (Potyondy and Cundall, 2004). However, sphere-based DEM has drawbacks because of the intervals inside the assembly of spheres. Therefore, the polygonal/polyhedral element is still widely used in DEM.

Minkowski sum is introduced to generate dilated polyhedron with tunable grain roundness.

This method avoids the singularity of corner contact pattern and the choice of contact point. Due to the dilated part in geometry, the contact detection could overcome the unphysical behavior of multiple simultaneous contacts. Based on dilated polyhedral element, Galindo-Torres (2012) developed a bond model to simulate fracture of material by an appropriate failure criterion. However, the stress and strain on the cohesive interface are evaluated inappropriately in the bond model. Besides, the parameters used in this model can only be determined by a series of numerical tests.

Finite Element Method (FEM) is employed with DEM to analyze breaking process named as coupled FEM-DEM model due to the calculative precision of stress and strain. The Rigid Finite Element Method (RFEM), or called as Rigid Body-Spring Method (RBSM) (Kawai, 1978) builds cohesion on the interface between adjacent elements. This method uses displacement of every element mass center as the degree of freedom, while the stress precision can be maintained at the same level with displacement precision. The coupled RBSM and DEM method is successful in simulating concrete failure process (Nagai et al., 2005). Thus RBSM is an effective numerical way to study failure process.

DEM and FEM, and their coupling method are valid ways to study ice failure and ice loads. The dilated disk element is used to simulate floe ice jam in a river, which is similar with dilated polyhedral element (Hopkins, 2004). The coupled FEM-DEM method is adopted in ridge keel punch through tests by modelling freeze bonds between ice blocks (Polojärvi and Tuhkuri, 2013). The crack propagation by wing crack model in ice is studied by FEM (Kolari, 2015). Through numerical tests of compression and bending, the parameters of parallel-bond model of spherical element in DEM are studied to analyze the ice loads on structure (Ji et al., 2016). These numerical approaches become more and more important in engineering.

RBSM is modified as an explicit algorithm in this paper. Then it is adopted in dilated polyhedral DEM (DP-DEM) to simulate interaction between ice and conical structure. The ice loads on structure is analyzed and compared with field data of Bohai Sea.

NUMERICAL METHODOLOGY

Minkowski sum theory is used to generate the dilated polyhedral element (Ji et al., 2016). DEM with the dilated polyhedral element is established. Hereby RBSM is modified by abandoning matrix calculation, which fits to insert to the data structure in DEM.

Dilated Polyhedral DEM

The nature of Minkowski sum theory means sweeping one geometric shape around the profile of the other. The dilated polyhedron is the result of Minkowski sum by a basic polyhedron and a sphere, i.e., the sphere sweeps on the profile of the basic polyhedron, shown in Figure 1.

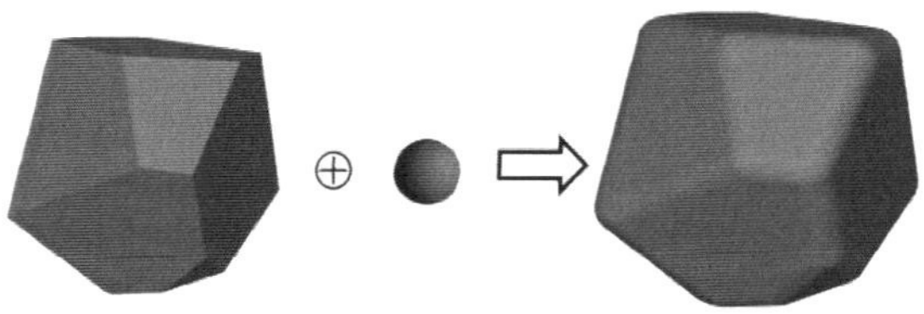

Figure 1. The dilated polyhedron generated by a basic polyhedron and a sphere.

Due to the dilated part in geometry, the profile of dilated polyhedron is smooth curve, including sphere, cylinder and flat face. The contact force model could be built by Hertz Model (Liu et al., 2016).

The normal force between them can be modelled by considering the elastic force caused by normal elastic deformation and the normal viscous force caused by relative velocity as viscoelasticity model:

$$F_{\mathrm{n}} = k_{\mathrm{n}}\delta_{\mathrm{n}}^{\kappa} + C_{\mathrm{n}}\delta_{\mathrm{n}}^{\kappa-1}\dot{\delta}_{\mathrm{n}} \tag{1}$$

where k_{n} is normal contact stiffness which depends on specific contact pattern; κ is usually 1.5 in Hertz model; δ_{n} and $\dot{\delta}_{\mathrm{n}}$ are normal deformation and normal relative velocity; C_{n} is normal damping coefficient.

Likewise, ignored tangential viscous force, tangential force is defined by tangential elastic force, and related with friction force:

$$F_{\mathrm{s}}^{*} = k_{\mathrm{s}}\delta_{\mathrm{n}}^{\kappa-1}\delta_{\mathrm{s}} \tag{2}$$

$$F_{\mathrm{s}} = \min(F_{\mathrm{s}}^{*}, \mathrm{sign}(F_{\mathrm{s}}^{*})\mu F_{\mathrm{n}}) \tag{3}$$

where k_{s} is tangential stiffness, and $k_{\mathrm{s}} = r_{\mathrm{sn}}k_{\mathrm{n}}$. Here r_{sn} is defined as $1/2(1+\nu)$ according to the relationship between elastic modules and shear modules of isotropic material, ν is the Poisson ratio. δ_{s} is the tangential elastic deformation; μ is friction coefficient.

Modified RBSM

The integral equilibrium equation of RBSM can be obtained by Galerkin theory, written as

$$\int_{S} \delta\boldsymbol{u}^{\mathrm{T}}(\boldsymbol{\sigma n})dS + \int_{V} \rho\delta\boldsymbol{u}^{\mathrm{T}}(\boldsymbol{f} - \boldsymbol{a})dV = 0 \tag{4}$$

where $\boldsymbol{\sigma}$ is the Green stress tensor, $\boldsymbol{n}$ is the outer normal vector of the interface between two adjacent elements, S is the area of interface, ρ is the mass density, $\boldsymbol{f}$ is the mass force, $\boldsymbol{a}$ is the acceleration, V means the continuum volume, $\boldsymbol{u} = \{u \ v \ w\}^{\mathrm{T}}$ is the displacement of the mass center.

As shown in Figure 2, the bonds (springs) are established on the point of the common face between the two adjacent elements. The normal and shear force are considered between the two bonded points. In this paper, the bonded points are chosen as the corner points of the common face. The normal strain is evaluated on every bonded point by the following formula

$$\varepsilon_{\mathrm{n}} = \frac{\boldsymbol{d} \cdot \boldsymbol{n}}{C} \tag{5}$$

where $\boldsymbol{n}$ is the normal vector, which is defined as the normal of the angle bisecting plane between the bond faces of the two adjacent elements; C is a scalar named as characteristic length, which is the summation of the distances from element centers to the interface of the two elements, as

$$C = h_m + h_n \tag{6}$$

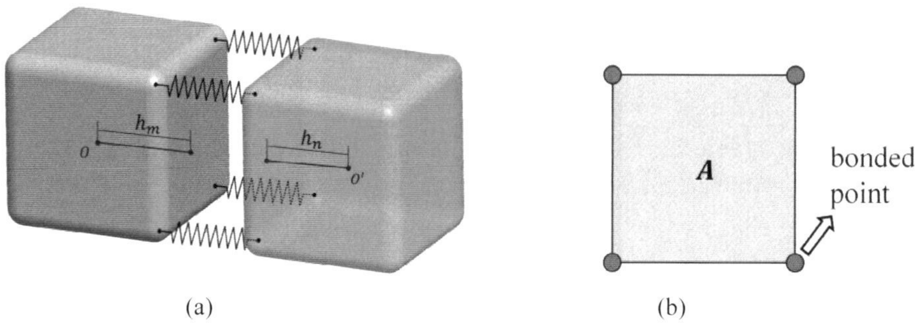

Figure 2. The common face of the adjacent elements: (a) Two adjacent elements; (b) The common face of the two adjacent elements, A is the area of the face.

The shear strain is obtained by

$$\varepsilon_{\mathrm{t}} = \frac{|\boldsymbol{d} - \boldsymbol{d} \cdot \boldsymbol{n}|}{C} \tag{7}$$

The two directions of shear force are considered to compose in only one direction $\boldsymbol{t} = \mathrm{unit}(\boldsymbol{d} - \boldsymbol{d} \cdot \boldsymbol{n})$, because the shear force would finally be transformed on the mass center. Based on the elastic matrix in RBSM, the stress is calculated by

$$\boldsymbol{\sigma} = \frac{E(1-\mu)}{(1+\mu)(1-2\mu)} \begin{bmatrix} 1 & 0 \\ 0 & \dfrac{1-2\mu}{2(1-\mu)} \end{bmatrix} \boldsymbol{\varepsilon} \tag{8}$$

where $\boldsymbol{\sigma} = \{\sigma \ \tau\}^{\mathrm{T}}$, $\boldsymbol{\varepsilon} = \{\varepsilon_{\mathrm{n}} \ \varepsilon_{\mathrm{t}}\}^{\mathrm{T}}$. The bond force between two bonded points is

$$\boldsymbol{F} = \boldsymbol{\sigma} \cdot \frac{A}{n} \tag{9}$$

All bond force between bonded points on the interface would be determined by Eq. (9) and transformed on the mass center. With contact force together, the motion of every element is obtained by explicit integral algorithm.

Fracture Criterion

Mohr-Coulomb criterion is adopted to detect the failure which is divided in the normal and shear direction separately, see Figure 3. In normal direction, the tensile strength is a given threshold $\sigma_{\mathrm{b}}^{\mathrm{n}}$, and the failure in tension is determined by

$$\sigma < \sigma_{\mathrm{b}}^{\mathrm{n}} \tag{10}$$

The shear strength is controlled by the following equation

$$\tau_{\mathrm{b}} = C + \sigma_{\mathrm{n}} \tan\theta \tag{11}$$

where C is cohesion; θ is internal friction angle, $\mu_{\mathrm{b}} = \tan\theta$ is internal friction coefficient; σ_{n} is the normal stress. Note that tension is negative and compression is positive. With Eq. (11), the compressive stress is related with shear strength. The failure in shear direction is determined by

$$\tau < \tau_{\mathrm{b}} \tag{12}$$

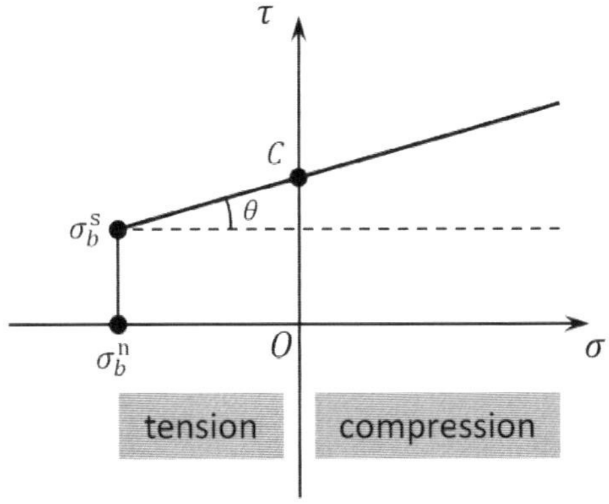

Figure 3. Failure criterion. The normal strength is a given threshold, while the shear strength is related to the compression stress determined as Eq. (11).

Mohr-Coulomb criterion is not quite appropriate for the compressive failure. Other strength criterions, related to stress intensity factor, fracture toughness, etc., could be adopted here to improve the computational accuracy.

Contact Detection

Contact detection is the most cumbersome part which affects the computational efficiency seriously in polyhedral DEM. The detecting time interval is taken into the simulation which means a potential contact list is built and updated at every detecting time interval. The detecting time interval is usually 10~20 times as much as simulating time interval. Besides, because the contact detection is not implemented every time step, the detecting terms need to be softened a little to ensure the accuracy during detecting time interval.

The contact detection is divided into three steps: uniform grid method, filter contact pairs, and detecting contact pattern and contact point. Uniform grid method is widely used in collision detection in game programming, and it is quite efficient in spherical DEM. Filtering the possible contact pair generated in the upper step is the most important step here. The main object of the filtering process is to find a face of a polyhedron which all the vertices of the adjacent polyhedron are outside of. Different from traversing all vertices and faces of polyhedra in SAA, this algorithm would break out if the object has already been achieved, which aimed at convex polyhedron technically.

Although a remarkable improvement is provided in the detection part, we still pursue a higher efficient way for contact detection and even the whole simulation process. Thus, the OpenMP, an effective parallel computing technology, is adopted here for element-level parallization which improve the computational efficiency on the basis of the modified detection method. This parallel level is coarse-grained, and fine-grained parallelization could be built at contact-type-level actually. The fine-grained parallelization would provide a more efficient way, but it requires a more complex data structure to avoid the access conflict in threads (Nishiura et al., 2011), which would also occupy more memory.

DP-DEM SIMULATION OF ICE-STRUCTURE INTERACTION

The ice load on conical structure interacting with level ice is simulated by DP-DEM with bond model. The structure is modelled according to the cone pile in Bohai Sea for the comparison between simulation and field data. Figure 4 is a snapshot of the ice breaking process.

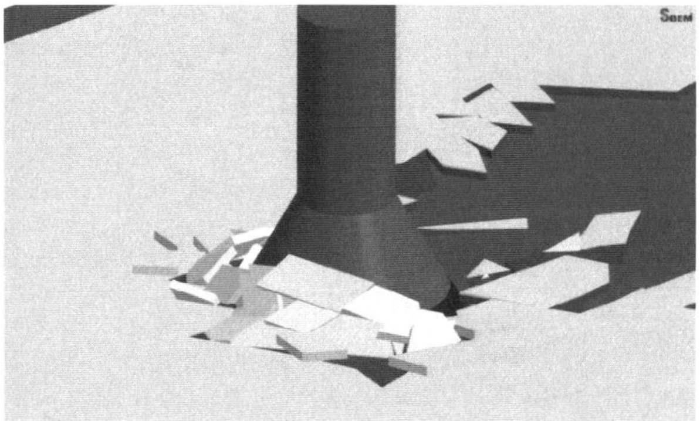

Figure 4. A snapshot of level ice simulation. Obvious bending failure is observed during the ice breaking process in this picture.

Figure 5 is the ice load in x direction when $\sigma_b^n = 1.0$ MPa and $\sigma_b^s = 0.5$ MPa. Comparing the simulation result and the field data in Bohai Sea, the data is shown in Table 2. The ice load exhibits the feature of bending failure, which conforms to Figure 4. The comparison with field data of Bohai Sea is shown in Table 2.

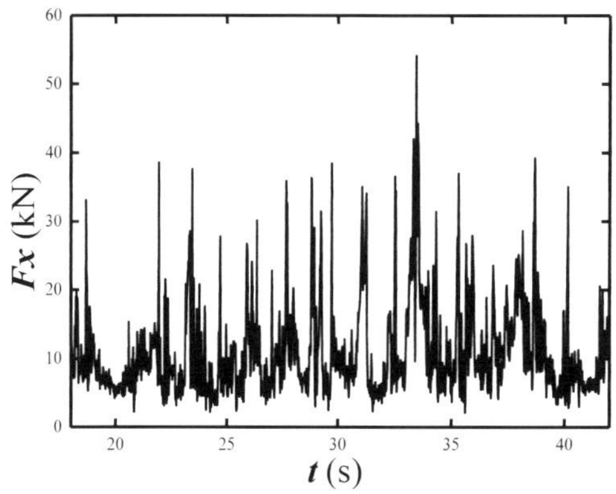

Figure 5. Ice load in x direction: $\sigma_b^n = 1.0$ MPa and $\sigma_b^s = 0.5$ MPa.

The comparison with field data of Bohai Sea is shown in Table 1. The average mean forces are close, while the standard deviation has bigger difference. Obviously bond strengths and their ratio have important effects on the ice load. The approach to choose appropriate strengths requires more specific study on this numerical method. Usually the strengths used in simulation are based on bending and compressive experiments of ice. However, the strengths used in this method have deviation with practical strengths due to the numerical approximation of stress determination. The validation of bond strengths is necessary to evaluate the error of this method.

Table 1. Comparison between DEM results and field data

Definitions	Simulation (kN)	Field data in Bohai (kN)
Average mean force	21.78	29.3
Standard deviation	9.63	21.3

On the other hand, the ice load on conical structure is affected by ice thickness and cone diameter. The ratio between ice thickness (h) and cone diameter (D) is studied by comparing with field data of Bohai Sea, see Figure 6. Numerical results are bigger but close to field results. The ice load in this figure exhibits in linear. However, it cannot be summarized that the ice load is linear with h/D. The ratio changes in the simulation are all due to the change of ice thickness, while the cone diameter keeps the same. In fact, the cone diameter also changes little in the field data. So the cone diameter is not considered severely. But the comparison shows DP-DEM is an effective way to study the ice load on structure.

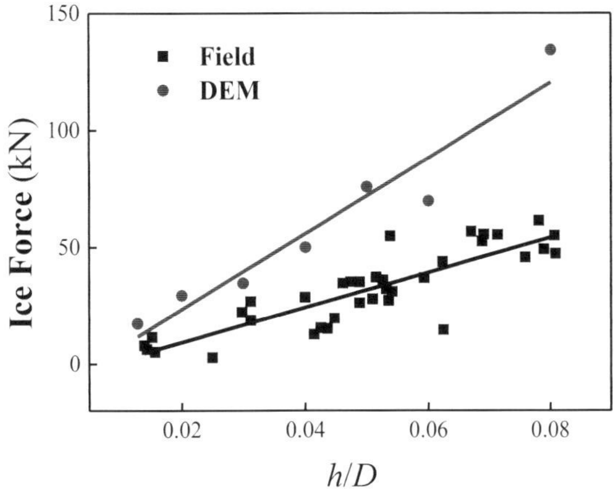

Figure 6. The ice load affected by the ratio between ice thickness and cone diameter. The ice loads in this figure are all maximum value. The field data of Bohai Sea is from Qu et al. (2006).

CONCLUSIONS

A bond model modified according to RBSM is introduced with DP-DEM. This method is based on building the cohesion of point pairs on the common face. The elastic matrix in RBSM is employed to determine the bond force between bonded points. Meanwhile, Mohr-Coulomb criterion is adopted to detect the failure process. The stress is detected separately in normal and shear with different strength thresholds. The contact detection of dilated polyhedron is introduced to improve the computational efficiency. This method is used to simulate the interaction between level ice and conical structure. The result is compared with field data of Bohai Sea. Generally, the parameters, especially bond strengths, need more sensibility analysis on ice failure modes.

ACKNOWLEDGEMENTS

The authors gratefully appreciate the financial support provided by the National Natural Science Foundation of China (Grant No. 41576179, 51639004), and the Key Research and Development Program of China (Grant No. 2016YFC1401505).

REFERENCES

Galindo-Torres, S.A., Pedroso, D.M., Williams, D.J., Li, L., 2012. Breaking processes in three-dimensional bonded granular materials with general shapes. *Computer Physics Communications*, 183(2), pp.266-277.

Hopkins, M.A., 2004. Discrete element modeling with dilated particles. *Engineering*

Computations, 21, pp.422-430.

Ji, S., Di, S., Long, X., 2016. DEM Simulation of Uniaxial Compressive and Flexural Strength of Sea Ice: Parametric Study. *Journal of Engineering Mechanics*, C4016010.

Ji, S., Sun, S., Yan, Y., 2016. Discrete element modeling of dynamic behaviors of railway ballast under cyclic loading with dilated polyhedra. *International Journal for Numerical & Analytical Methods in Geomechanics*.

Kawai, T., 1978. New discrete models and their application to seismic response analysis of structures. *Nuclear Engineering & Design*, 48(1), pp.207-229.

Kolari, K., 2015. Simulation of the temperature and grain size dependent uniaxial compressive strength using 3D wing crack model. In: Throndheim, Norway, *Proceedings of the 23rd International Conference on Port and Ocean Engineering under Arctic Conditions*.

Liu, L., Sun, S., Ji, S., 2016. Interaction Between Floater and Sea Ice Simulated with Dilated Polyhedral DEM. In: Dalian, China, *Proceedings of the 7th International Conference on Discrete Element Methods*, Volume 188 of the series Springer Proceedings in Physics, pp.1065-1074.

Nagai, K., Sato, Y., Ueda, T., 2005. Mesoscopic Simulation of Failure of Mortar and Concrete by 3D RBSM. *Journal of Advanced Concrete Technology*, 3(3), pp.385-402.

Nishiura, D., Sakaguchi, H., 2011. Parallel-vector algorithms for particle simulations on shared-memory multiprocessors. *Journal of Computational Physics*, 230, pp.1923-1938.

Polojärvi, A., Tuhkuri, J., 2013. On modeling cohesive ridge keel punch through tests with a combined finite-discrete element method. *Cold Regions Science & Technology*, 85, pp.191-205.

Potyondy, D.O., Cundall, P.A., 2004. A bonded-particle model for rock. *International Journal of Rock Mechanics & Mining Sciences*, 41(8), pp.1329-1364.

Qu, Y., Yue, Q., Bi, X., Karna, T., 2006. A random ice force model for narrow conical structures. *Cold Regions Science & Technology*, 45(3), pp.148-157.

POAC'17
Busan, KOREA

Proceedings of the 24th International Conference on
Port and Ocean Engineering under Arctic Conditions
June 11-16, 2017, Busan, Korea

A Three-Dimensional FEM-DEM Model of an Ice Sheet

Ville-Pekka Lilja[1], Arttu Polojärvi[1], Jukka Tuhkuri[1], Jani Paavilainen[2]
[1] Aalto University, School of Engineering, Department of Mechanical Engineering, Finland
[2] Rand Simulation Ltd., Vantaa, Finland

ABSTRACT

A three-dimensional combined finite-discrete element (FEM-DEM) model of an ice sheet is presented. Centroidal Voronoi tessellation with a random generating point set is used to produce an unstructured mesh. Mesh consists of convex polyhedra as the discrete elements and a Delaunay-triangulated network, or a lattice, of Timoshenko beam elements connecting their mass centroids. We investigate, by performing simple in-plane and out-of-plane mechanical tests, how the model behaves while the sheet is still intact. We compute the effective Young's modulus and study the bending of an ice sheet on an elastic foundation for several different specimen sizes.

KEY WORDS: Ice; Combined FEM-DEM; Beam lattice network; Size effect; Bending

INTRODUCTION

A sea ice sheet interacting with a structure deforms and eventually breaks into a large number of distinct fragments. Broken fragments accumulate in front of the structure and affect the subsequent failure process. Modeling the interaction process is computationally demanding and requires the development of efficient methods.

This paper presents a three-dimensional combined finite-discrete element (FEM-DEM) model of an ice sheet meant to be used for conducting such ice-structure interaction simulations. We investigate, by performing simple in-plane and out-of-plane mechanical tests, how the model behaves while the sheet is still intact. We compute the effective Young's modulus and study the bending of an ice sheet on an elastic foundation for several different specimen sizes.

Similar, namely discrete element, combined finite-discrete element, finite element, or lattice-based numerical methods to model a floating ice sheet and its interaction with an offshore structure have been used in the past by for example Hocking (1992), Jirásek & Bažant (1995), Hopkins (1998), Sayed & Timco (1998), Selvadurai & Sepehr (1999), Sand (2008), Dorival et al. (2008), Konuk et al. (2009), Gürtner (2009), Paavilainen (2013), Kuutti et al. (2013), Lu et al. (2014), Herman (2016) and van den Berg (2016).

CONSTRUCTION OF THE NUMERICAL MODEL

The modeled ice sheet consists of a network of two-noded, three-dimensional, co-rotational Timoshenko beam elements connected with the mass centroids of the overlaid undeformable discrete elements, Figure 1. Centroidal Voronoi tessellation (CVT) with a random generating point set and Lloyd's algorithm, (Du et al., 1999), is used to produce an unstructured mesh, Figure 2. Mesh consists of convex polyhedra as the discrete elements and a Delaunay-triangulated network, or a lattice, of Timoshenko beam elements.

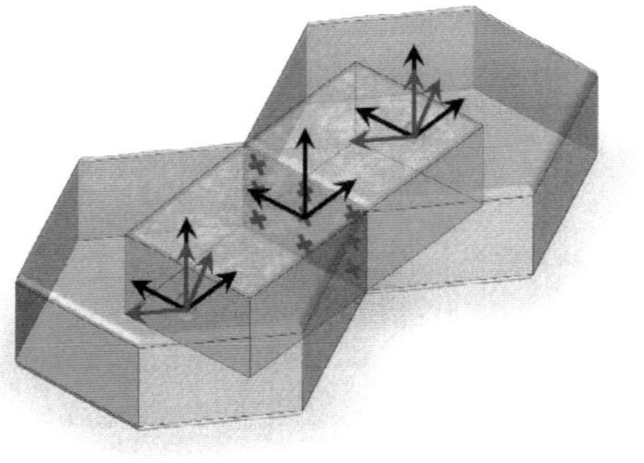

Figure 1. Two discrete elements connected with a Timoshenko beam element.

Elasticity of an ice sheet, prior to fracture, is modeled solely with the beams, contacts between the broken pieces of ice by the discrete elements. The beam formulation follows closely Crisfield (1990,1997) while the contacts between the discrete elements are resolved as in Polojärvi et al. (2012). For the beams, Hooke's linearly elastic constitutive law, with a viscous damping model, is used. To model damage, the beams are allowed to fracture allowing a crack to propagate along an interface separating two adjacent discrete elements. This paper does not, however, consider fracture. For a description of fracture, see Lilja et al. (2017). In esse nce, the model is a three-dimensional extension of our earlier two-dimensional FEM-DEM model by Paavilainen et al. (2009) and was introduced by Paavilainen (2010).

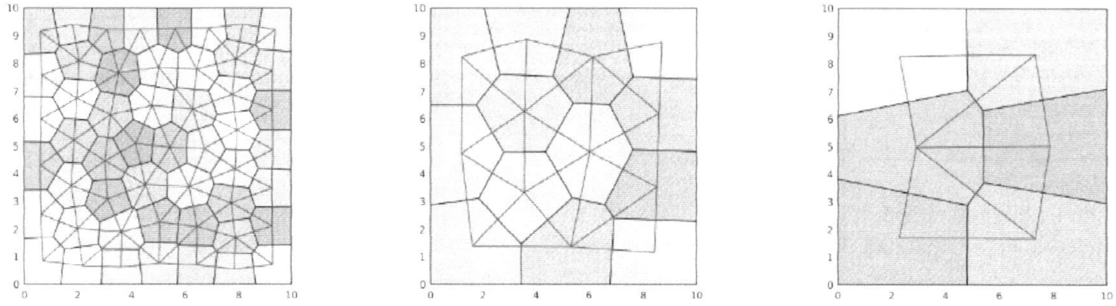

Figure 2. Three CVT-tessellated square-shaped ice sheets with a side length of $L = 10$ m. From left to right, sheets having a thickness of $h = 0.5$, 1.0 and 1.5 m and a discrete element size of $l = 3h$. Cross-sectional area of each beam element gets determined by the area of the joint surface separating two adjacent discrete elements. Red lines denote the longitudinal axes of the beam elements.

A sea ice sheet modeled as described above behaves mechanically as a cellular solid. Effective macroscopic material properties behave as functions of the ratio between the macroscopic size of an ice sheet and the size of a unit cell of the beam lattice network in use: there is a size effect. It can be also shown that a beam element network is a discrete equivalent of a micropolar continuum (Bažant & Christensen, 1972), and that on a large enough scale a CVT-tessellated mesh behaves like a transversely isotropic medium, as the mesh consists of beams forming (nearly) equilateral triangles.

Equations to be solved for each discrete element, discretely in time, are the three translational equations of motion and the three rotational equations of motion. Internal forces and moments inside a sheet consist of the resultants of all the nodal force and moment components of all the deformable beam elements sharing nodes with the discrete elements. Simulations are explicit.

SIMULATIONS

Main goals of our simulations are: (1) to compute the effective Young's modulus E of an ice sheet for several different specimen sizes and (2) to test how well our model describes a floating ice sheet in bending. Values of the computed effective Young's moduli E are compared with the Young's modulus E^b as given to the beams and with the Young's modulus E^* of an infinite lattice composed of Timoshenko beams and having a regular topology consisting of perfectly hexagonal unit cells (beams forming equilateral triangles) using the theory of a micropolar continuum (Ostoja-Starzewski, 2002; Karihaloo et al., 2003). Bending of an elastic ice sheet on an elastic foundation having a concentrated vertical point load is investigated subsequently. Deflections are compared with the deflections of a point-loaded infinite ice sheet floating on water, as presented by Wyman (1950).

Mesh Creation

Figure 3 shows the sheets used in the tests. Sheets were square-shaped and had the side lengths of $L = 10, 20, 40, 80$ and 160 m and the thicknesses of $h = 0.5, 1.0$ and 1.5 m. For each sheet, we used two mesh densities: a mesh with an average discrete element size of $l = 2h$ and a mesh with an average discrete element size of $l = 3h$. In the case of a possible fracture, each broken fragment should have, thus, in minimum, a size that is comparable with actual block sizes measured from ridge sails (Kankaanpää, 1988; Høyland, 2007; Kulyakhtin, 2014). Ten randomized CVT-meshes were created for all of the sheets, excluding the sheet with L=160 m, h = 0.5 m and $l = 2h$, for which six meshes were produced.

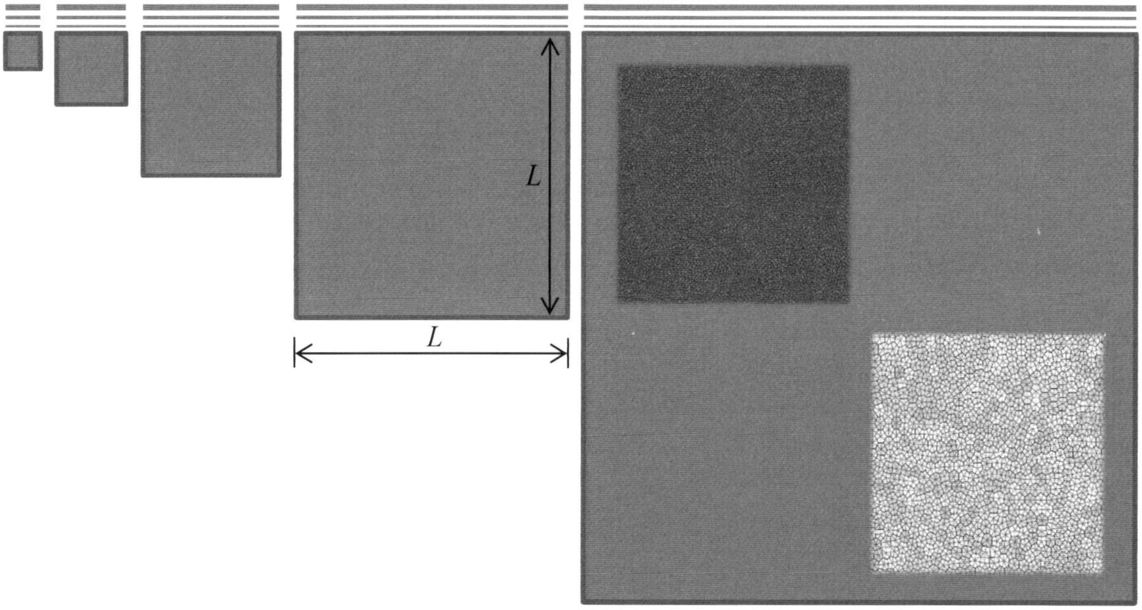

Figure 3. Size range (1:16) of the ice sheets considered. From left to right: $L = 10, 20, 40, 80$ and 160 m. The largest sheet on the right shows also sections of the two meshes that have the most and the least amount of discrete elements for that size: 29561 discrete elements ($h = 0.5$ m, $l = 2h$) and 1460 discrete elements ($h = 1.5$ m, $l = 3h$).

Effective Young's Modulus

Effective Young's modulus E was computed by carrying out a simple uniaxial tensile test. Each sheet was loaded under displacement control, as shown in Figure 4. Nodes on the right boundary were pulled with a constant velocity $v = 0.1$ m/s into the positive X-direction while the X-direction motion of the nodes on the left boundary were fixed. There were no other constraints in the system. Beams were not allowed to fracture.

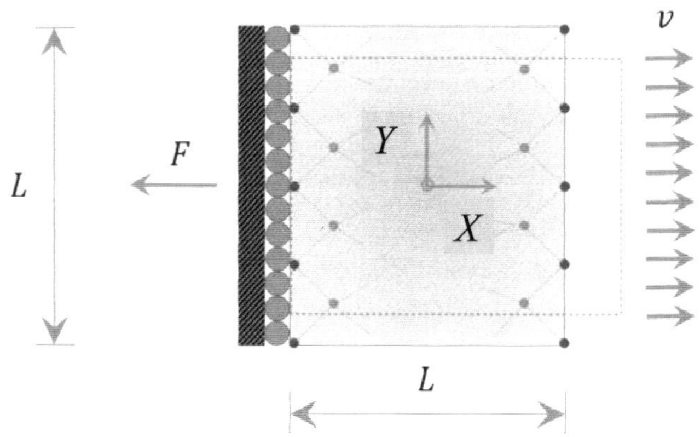

Figure 4. Boundary conditions and the reaction force F used in the computation of the effective Young's modulus E.

Effective Young's modulus E was evaluated through Hooke's law as $E = F/(A_0\varepsilon_X)$, where F is the computed resultant boundary reaction, $A_0 = Lh$ is the initial cross-sectional area of an ice sheet and ε_X is the applied axial strain (a value of 0.001 was used). Reaction F was calculated for each sheet by summing up the X-direction components of the internal nodal force vectors of the beams having nodes on the left boundary. Table 1 shows the main parameters of the simulations.

Table 1. Main simulation parameters.

Parameter	Symbol	Unit	Value
Young's modulus	E^b	GPa	4
Poisson's coefficient	ν		0.3
Density of ice	ρ_{ice}	kg/m^3	920
Density of water	ρ_{water}	kg/m^3	1010
Damping constant	c		critical[1]
Time step	Δt	s	$1.0\times10^{-5}...5.0\times10^{-5}$
Acceleration of gravity	g	m/s^2	9.81

[1] $c = \sqrt{m_{eff}E^b}$, where m_{eff} is the average of the masses of the two discrete elements a beam connects.

Results

Figure 5 shows the computed effective average Young's moduli E and their standard deviations as functions of sheet size L, sheet thickness h and discrete element size l. Results shown are the averages calculated using ten randomized CVT-meshes, except for the case with L=160 m, h = 0.5 m and $l = 2h$, for which six meshes were used. For each L the results are arranged in an order of ascending h and l (for each h). Modulus E grows as a function of h and decreases as a function of L. For the smallest sheet size the differences between the results, whether a discrete element size of either $l = 2h$ or $l = 3h$ is used, are large. Sheets that are smaller, thicker or meshed with a discrete element size of $l = 3h$ have a higher E.

Figure 6 shows the ratios E/E^* and E/E^b as functions of the relative sheet size $L_{rel} = L/l$. The upper curve corresponds to the ratios E/E^* between the effective and the micropolar Young's moduli and the lower curve to the ratios E/E^b between the effective and the material (as given to the beams) Young's moduli. Ratio E/E^* converges approximately to a value of 1.0. Ratio varies from approximately 1.94 for the smallest and the thickest sheets to approximately 1.0 for the largest sheets. Ratio E/E^b converges approximately to a value of 0.8. Ratio varies from approximately 1.53 for the smallest and the thickest sheets to approximately 0.8 for the largest sheets. Size effect vanishes (practically) for $L_{rel} \geq 25$. Except for the smallest and the thickest sheets, having only a few elements across a sheet, deviations are observed to be small.

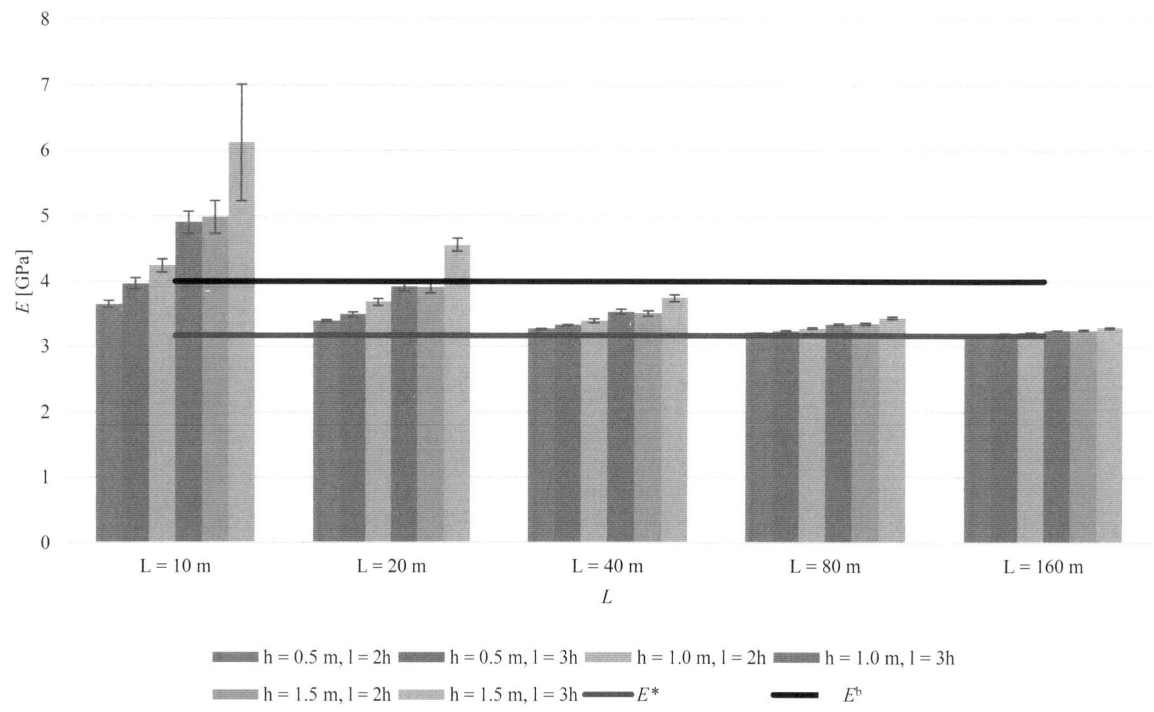

Figure 5. Effective Young's modulus E as a function of sheet size L, sheet thickness h and discrete element size l. Young's modulus E^b of the beams is indicated with a black color and the Young's modulus E^* as predicted by using the theory of a micropolar continuum with a red color.

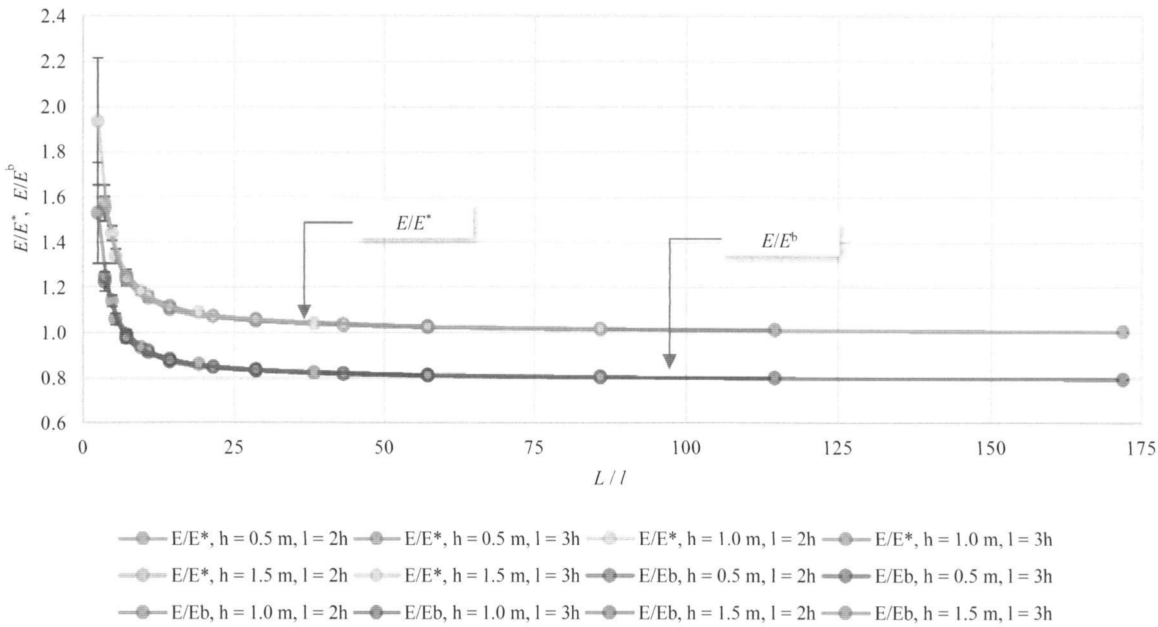

Figure 6. Ratios E/E^* and E/E^b as functions of the relative sheet size $L_{rel} = L / l$.

Bending of an Ice Sheet

Bending tests were done on an elastic foundation modeled by using both buoyancy and drag. Stoke's drag (a drag model giving a drag force that is linearly proportional to the absolute velocity) was used to reach a quasi-static equilibrium. A vertical point load $F_Z = -10$ kN (acting towards water, see Figure 4) was applied to a node that was closest to the geometrical center of each sheet by using a statically equivalent force system. Deflections were measured along a path: $X = 0$, $0 \leq Y \leq L/2$, corresponding to the edge of a quadrant of each sheet. Gravitation was taken into account, no displacement constraints were applied, beams were not allowed to fracture and the smallest sheets were excluded from the analysis. Sheets did not submerge under the loading.

Results

Figure 7 shows the average deflection profiles (with their standard deviations) of all the sheet sizes as functions of their normalized half-widths. Deflection profiles represent the average deflection profiles calculated using ten randomized CVT-meshes (for the case with L = 160 m, h = 0.5 m, l = $2h$, with six meshes). Deflections are normalized with the maximum deflection of an infinite ice sheet having a thickness h, loaded by a concentrated vertical point load F_Z and floating on water (Wyman, 1950). Deflection profiles of the L = 80 m and the L = 160 m sheets are shown, again, in Figure 8.

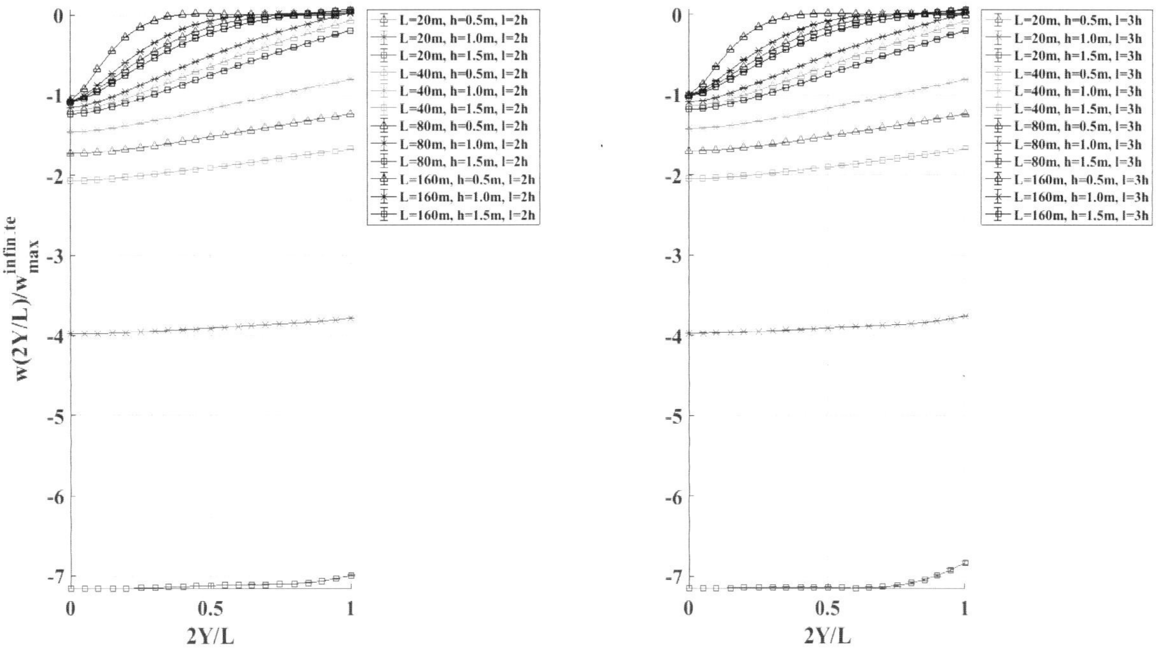

Figure 7. Normalized deflection profiles of all the sheet sizes as functions of their normalized half-widths. Discrete element size on the left: $l = 2h$, discrete element size on the right: $l = 3h$.

As the sheet size grows, the maximum deflections approach those by Wyman (1950). As the sheets get smaller or thicker, sheets tend to deform less and sink. Differences between the deflection profiles, whether a discrete element size of either $l = 2h$ or $l = 3h$ is used, are small. The maximum deflections, however, agree slightly better with Wyman (1950) when a discrete element size of $l = 3h$ is used.

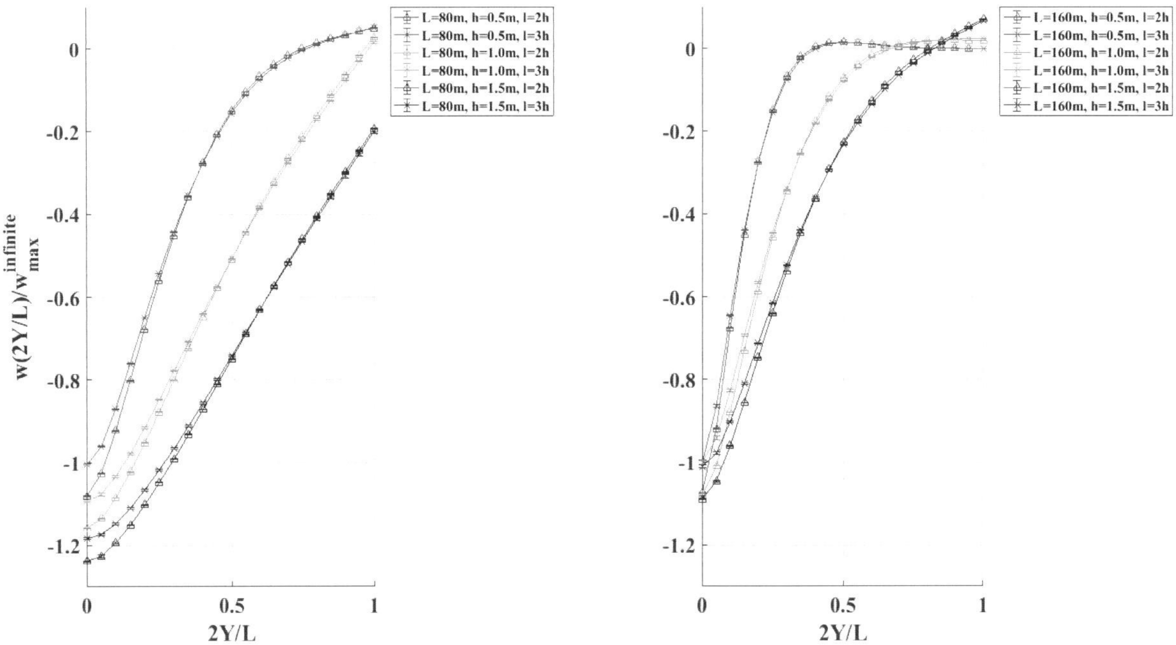

Figure 8. Normalized deflection profiles of the L = 80 m and the L = 160 m sheets as functions of their normalized half-widths.

Figure 9 shows the deflection profiles of the largest L = 160 m sheets as functions of characteristic length ($l_{ch} = \sqrt[4]{D/k}$, where $D = E^b h^3 / 12(1-\nu^2)$ is the flexural stiffness of an ice sheet and $k = \rho_{water} g$ is the specific weight of water). Deflections are, again, normalized with the maximum deflection of an infinite ice sheet having a thickness h, loaded by a concentrated vertical point load F_Z and floating on water (Wyman, 1950). Deflection of an infinite ice sheet is shown with a dashed line.

Deflections follow the corresponding normalized analytical deflection profiles well, especially for the two thinnest sheets. They, thus, seem to be able to represent point-loaded infinite ice sheets floating on water (with corresponding thicknesses) fairly well.

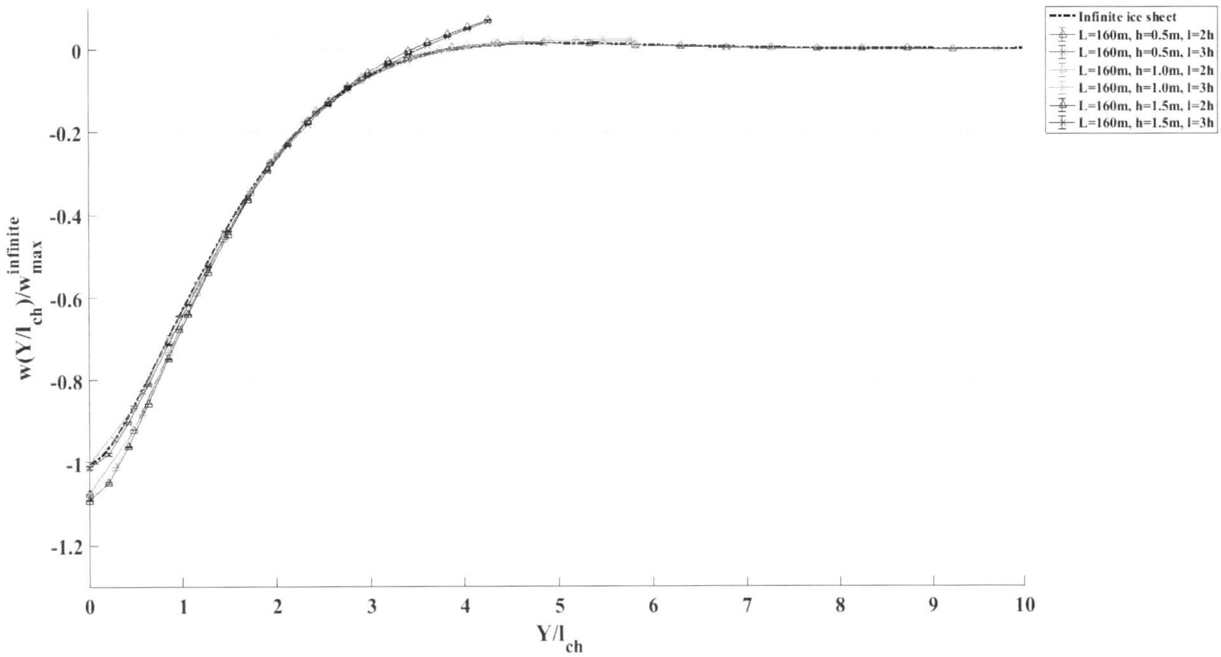

Figure 9. Normalized deflection profiles of the largest L = 160 m sheets as well as the analytical deflection profile of an infinite ice sheet loaded by a concentrated vertical point load and floating on water (Wyman, 1950).

CONCLUSIONS

A three-dimensional FEM-DEM model of an ice sheet was presented. Model consists of rigid discrete elements joined by two-noded, three-dimensional, co-rotational Timoshenko beam elements. Centroidal Voronoi tessellation was used to create the meshes.

Effective Young's modulus E was computed as a function of sheet size L, sheet thickness h and discrete element size l by carrying out a simple uniaxial tensile test. Sheets that are smaller, thicker or have a coarser mesh have a higher effective Young's modulus E. Size effect vanishes (practically) for $L_{rel} \geq 25$.

Ratio between the effective Young's modulus E and the Young's modulus E^* as calculated by using the theory of a micropolar continuum (using as a reference topology an infinite regular lattice composed of Timoshenko beams forming perfectly equilateral triangles) converges approximately to a value of 1.0. Ratio between the effective Young's modulus E and the Young's modulus E^b of the beams converges approximately to a value of 0.8.

Bending of an elastic ice sheet on an elastic foundation was studied in the case of central loading. As the sheet size grows, the maximum deflections approach those by Wyman (1950). As the sheets get smaller or thicker, the sheets tend to deform less and sink. The sheets having a size of L = 160 m and the thicknesses of either h = 0.5 m or h = 1.0 m seem to be able to represent point-loaded infinite ice sheets floating on water (with the same thicknesses) fairly well.

ACKNOWLEDGEMENTS

Authors gratefully acknowledge the financial support by the Academy of Finland through the project: "Discrete Numerical Simulation and Statistical Analysis of the Failure Process of a Non-homogeneous Ice Sheet Against an Offshore Structure."

REFERENCES

Bažant, Z. P. & Christensen, M., 1972. Analogy between micropolar continuum and grid frameworks under initial stress. *International Journal of Solids and Structures,* Volume 8, pp. 327-346.

Crisfield, M. A., 1990. A consistent co-rotational formulation for non-linear, three-dimensional, beam-elements. *Computer methods in applied mechanics and engineering,* Volume 81, pp. 131-150.

Crisfield, M. A., 1997. chs. 17.1-2. In: *Non-linear Finite Element Analysis of Solids and Structures, volume 2: Advanced Topics.* John Wiley & Sons Ltd.: New York, pp. 213-232.

Dorival, O., Metrikine, A. & Simone, A., 2008. *A lattice model to simulate ice-structure interaction.* ASME 2008 27th International Conference on Offshore Mechanics and Arctic Engineering, Estoril, Portugal, pp. 989-996.

Du, Q., Faber, V. & Gunzburger, M., 1999. Centroidal Voronoi Tessellations: Applications and Algorithms. *SIAM Review,* Volume 41, p. 637–676.

Gürtner, A., 2009. *Experimental and Numerical Investigations of Ice-Structure Interaction*, Doctoral thesis, Norwegian University of Science and Technology, NTNU: Trondheim.

Herman, A., 2016. Discrete-Element bonded-particle Sea Ice model DESIgn, version 1.3a – model description and implementation. *Geoscientific Model Development,* Volume 9, p. 1219–1241.

Hocking, G., 1992. The discrete element method for analysis of fragmentation of discontinua. *Engineering Computations,* Volume 9, pp. 145 - 155.

Hopkins, M., 1998. Four stages of pressure ridging. *Journal of geophysical research,* Volume 103, pp. 21883-21891.

Høyland, K. V., 2007. Morphology and small-scale strength of ridges in the North-Western Barents Sea. *Cold Regions Science and Technology,* Volume 48, pp. 169-187.

Jirásek, M. & Bažant, Z. P., 1995. Particle Model for Quasibrittle Fracture and Application to Sea Ice. *Journal of Engineering Mechanics,* Volume 121, pp. 1016-1025.

Kankaanpää, P., 1988. Morphology of sea ice pressure ridges in the Baltic Sea. *Geophysica,* Volume 24, pp. 15-44.

Karihaloo, B. L., Shao, P. F. & Xiao, Q. Z., 2003. Lattice modelling of the failure of particle composites. *Engineering fracture mechanics,* Volume 70, pp. 2385-2406.

Konuk, I., Gürtner, A. & Yu, S., 2009. *A Cohesive Element Framework for Dynamic Ice-Structure Interaction Problems - Part I: Review and Formulation.* ASME 2009 28th International Conference on Ocean, Offshore and Arctic Engineering, Honolulu, Hawaii, USA, May 31 - June 5, pp. 33-41.

Kulyakhtin, S., 2014. *Distribution of Ice Block Sizes in Sails of Pressure Ice Ridges.* 22nd IAHR International Symposium on Ice, Singapore, August 11 to 15.

Kuutti, J., Kolari, K. & Marjavaara, P., 2013. Simulation of ice crushing experiments with cohesive surface methodology. *Cold Regions Science and Technology,* Volume 92, pp. 17-28.

Lilja, V. P., Polojärvi, A., Tuhkuri, J. & Paavilainen, J., 2017. *Effective tensile strength of an ice sheet using a three-dimensional FEM-DEM approach.* Proceedings of the 24th International Conference on Port and Ocean Engineering under Arctic Conditions, Busan, Korea, June 11-16.

Lu, W., Lubbad, R. & Løset, S., 2014. Simulating Ice-Sloping Structure Interactions With the Cohesive Element Method. *Journal of Offshore Mechanics and Arctic Engineering,* Volume 136, pp. 1-16.

Ostoja-Starzewski, M., 2002. Lattice models in micromechanics. *Applied Mechanics Reviews,* Volume 55, pp. 35-60.

Paavilainen, J., 2010. Jäälautan murtuminen kartiorakennetta vasten. In: J. Heinonen, ed. *Jatkuvan murtumisprosessin mallinnus jää-rakenne vuorovaikutusprosessissa*, research report (in Finnish), project: "STRUTSI". Technical Research Center of Finland, Espoo: VTT, pp. 25-27.

Paavilainen, J., 2013. *Factors affecting ice loads during the rubbling process using a 2D FE-DE Approach*, Doctoral thesis, Aalto University: Espoo.

Paavilainen, J., Tuhkuri, J. & Polojärvi, A., 2009. 2D combined finite-discrete element method to model multi-fracture of beam structures. *Engineering computations,* Volume 26, pp. 578-598.

Polojärvi, A., Tuhkuri, J. & Korkalo, O., 2012. Comparison and analysis of experimental and virtual laboratory scale punch through tests. *Cold Regions Science and Technology*, Volume 81, pp. 11-25.

Sand, B., 2008. *Nonlinear finite element simulations of ice forces on offshore structures*, Doctoral thesis, Luleå University of Technology: Luleå.

Sayed, M. & Timco, G., 1998. *A Lattice Model of Ice Failure,* National Research Council of Canada: Ottawa.

Selvadurai, A. P. S. & Sepehr, K., 1999. Two-dimensional discrete element simulations of ice-structure interaction. *International Journal of Solids and Structures,* Volume 36, pp. 4919-4940.

van den Berg, M., 2016. *A 3-D Random Lattice Model of Sea Ice.* Arctic Technology Conference, St. John's, Newfoundland and Labrador, Canada, 24-26 October.

Wyman, M., 1950. Deflections of an infinite plate. *Canadian Journal of Research,* Volume 28, pp. 293-302.

POAC'17
Busan, KOREA

Proceedings of the 24th International Conference on
Port and Ocean Engineering under Arctic Conditions
June 11-16, 2017, Busan, Korea

Brittle Failure Simulation of Ice Beam Using a Fully Lagrangian Particle Method

D. Ren[1], J.-C. Park[1*], S.-C. Hwang[2], S.-Y. Jeong[3], H.-S. Kim[4]
[1] Dept. of Naval Architecture & Ocean Engineering, Pusan National University (PNU), Busan, Korea
[2] Offshore Plant Research Division, Korea Research Institute of Ships and Ocean Engineering (KRISO), Daejeon, Korea
[3] Ship Hydrodynamics Research Group (Ice Model Basin), Korea Research Institute of Ships and Ocean Engineering (KRISO), Daejeon, Korea
[4] Dept. of Naval Architecture & Ocean Engineering, Inha Technical College, Incheon, Korea

ABSTRACT

A realistic numerical simulation technology using a particle-based fluid-structure interaction (FSI) model is combined with a brittle fracture algorithm to predict the fluid-ice-structure interaction. With respect to the failure of ice, it is modeled as a brittle fracture of elastic body by applying a novel FSI model based on the Moving Particle Semi-implicit (MPS) method. For verifying the developed fracture algorithm, a series of numerical simulation for 3-point bending tests with ice beam are performed and compared with the experiments carried out in the ice room. And then, for the application of the developed FSI model, a dropping water droplet interacting with a cantilever ice beam is simulated with and without the fracture algorithm. As the results from the simulation, it is seen that the consequent effects of fracture that might be occurred in the process of FSI simulation is available to be studied.

KEY WORDS: Brittle fracture, 3-point bending problem, Ice fracture, Fluid-ice-structure interaction, MPS (Moving Particle Semi-implicit) method

INTRODUCTION

In Arctic environment, ice accumulation plays a critical role as extremely high loads acting on the moored structures or the structures operating in ice-covered waters. The floating ice loads on the Arctic structures affect serious structural damages and safety for not only stationary operating offshore platforms but also sailing vessels in icy environments. Especially, when the icebreaker conducting icebreaking operations, the interaction of the crushed ice and marine propulsor has become an important factor affecting the performance and safety of the ship. Primary consideration for the Arctic transportations will be safety, effectiveness and cost. In order to evaluate a new design of ice strengthen vessel, physical model testing or numerical simulation should be required. To perform such simulations, the simulation technology of fluid-ice-structure interaction has become a necessary prerequisite.

Most of the fluid-structure interaction (FSI) simulation was performed by grid system (Hübner, et. al., 2004). When fracture is engaged, however, the grid system is interfering and the meshes need to be separated or decomposed to represent numerically the crack propagation. To solve these problems, therefore, a relatively complex algorithm should be required. In this respect, the meshless method is less restrictive from such matters.

Up to now, there are many researchers studied on the fracture by mesh or meshless method. Chan (Chan, 1981) studied the fracture toughness and creep behavior of ice by Finite Element Method (FEM). Sakharov (Sakharov et. al., 2015) carried out experiment of fixed end beam bending test with lake ice and compared the experimental results with the finite element simulation. Abbas (Abbas et. al., 2010) proposed two model-independent approaches based on Extended Finite Element Method (XFEM), which the author claim to be independent of the fracture model consideration. Sepehri (Sepehri, 2014) studied the hydraulic fracture propagation pattern used XFEM. Peixiang (Peixiang et al., 2013) studied he dynamic fracture problem in functional graded material based on Element-Free Galerkin Method (EFGM). Bui (Bui, 2008) presented a study of large deformation and failure flows of geo-material use Smoothed Particle Hydrodynamics (SPH). Tan (Tan et al., 2009) simulate the microscopic machining process of ceramics by considering the fracture and damage. Beckmann (Beckmann et. al., 2014) studied concrete fracture phenomena. Both of them used DEM model to achieve that. In recent, especially, there are also some research activities for numerical simulations to handle the ice breakup feature during the ice-structure interaction and treat broken ice as discrete-continuum material using the discrete element method (DEM). (Shen et al., 1987; Lepparanta et al., 1990; Hopkins, 1998; Hansen and Løset, 1999a, b; Selvadurai and Sepehr, 1999; Hopkins and Shen, 2001; Dai et al, 2004; Polojarvi and Tuhkuri, 2009; Zhan et al., 2010; Karulin and Karulina, 2011; Lau et al., 2011; Sun and Shen, 2012; Xu et al., 2012; Ji et al., 2013; Metrikin and Løset, 2013; Morgan, et al., 2015). However, most of DEM-based simulations concerned with ice floes have performed under no consideration of fully interacting with sea water.

In the present study, a realistic numerical simulation technology for predicting the fluid-ice-structure interaction has been newly developed using a particle-based FSI model which is combined with a brittle fracture algorithm. With respect to failure of ice, it is modeled as a brittle fracture of elastic by applying a novel FSI model based on the Moving Particle Semi-implicit (MPS) method (Hwang et al., 2014; 2016). To validate the developed fracture algorithm, the 3-point bending simulation with ice beam is performed and compared with the experiments which was carried out in a cold room. And then, for the application of FSI model, the dropping water droplet interacting with a cantilever ice beam is simulated comparatively with and without the fracture algorithm.

GOVERNING EQUATIONS

The momentum conservation equation for both of fluid and structure can be expressed as:

$$\frac{D\vec{u}}{Dt} = -\frac{1}{\rho}\nabla\cdot\sigma + \vec{f} + \vec{g} \tag{1}$$

where, ρ, t, $\vec{u}$, σ, $\vec{f}$, $\vec{g}$ indicate the density of fluid, time, velocity vector, stress tensor, external force and the gravity acceleration, respectively. The stress tensor σ is defined for fluid and structure analysis differently. For fluid, the stress tensor consists of a gradient of pressure and a viscosity term. For structure, the stress tensor is composed of normal stress and shear stress.

$$\sigma_{ij}^{F} = -P\delta_{ij} + \frac{\mu^{F}}{2}\left(\frac{\partial u_i}{\partial x_j} + \frac{\partial u_j}{\partial x_i}\right) \tag{2}$$

$$\sigma_{ij}^{S} = \lambda^{S}\frac{\partial \Delta x_k}{\partial x_k}\delta_{ij} + \frac{\mu^{S}}{2}\left(\frac{\partial \Delta x_i}{\partial x_j} + \frac{\partial \Delta x_j}{\partial x_i}\right) \tag{3}$$

where, P, μ^{F}, λ^{S} μ^{S} u Δx indicate the pressure and viscosity coefficient of fluid, the first and second Lamé's parameters, and the components of velocity and displacement.

Governing Equations of Fluid

By introducing Eq. (2) into Eq. (1), we obtain the momentum conservation equations of fluid, also known as Navier-Stokes equation. And in order to solve the pressure term by the relationship between the density and the divergence of the velocity, we also introduced the mass conservation equation as governing equation:

$$\frac{D\vec{u}}{Dt} = -\frac{1}{\rho_F}\nabla P + \nu\nabla^2\vec{u} + \vec{f}_{S_{to}F} + \vec{g} \tag{4}$$

$$\frac{D\rho_F}{Dt} + \rho_F\left(\nabla\cdot\vec{u}\right) = 0 \tag{5}$$

where, ρ_F, t, $\vec{u}$, $\vec{f}_{S_{to}F}$, $\vec{g}$, ν indicate the density of fluid, time, velocity vector, coupling force acting on the fluid boundary, gravity acceleration and the kinematic viscosity, respectively.

Governing Equations of Structure

By introducing Eq. (3) into Eq. (1), we obtain the momentum conservation equations of structure as:

$$\frac{D\vec{u}}{Dt} = \frac{1}{\rho_s}\nabla\cdot\left(\lambda_s tr(\varepsilon) + 2\mu_s\varepsilon\right) + \vec{f}_{F_{to}S} + \vec{g} \tag{6}$$

where, ρ_s, $\vec{f}_{F_{to}S}$, $\vec{g}$, ε imply the density of structure, stress tensor, load acting on the structure and the gravity acceleration, respectively.

And the Lame's constants λ_s and μ_s can be calculated by Young's modulus E and Poisson ratio v_s as:

$$\lambda_s = \frac{Ev_s}{(1+v_s)(1-2v_s)} \quad \mu_s = \frac{E}{2(1+v_s)} \tag{7}$$

To consider the rotation of structure particles, the angular momentum conservation equation was introduced as:

$$I\frac{D\vec{\omega}}{Dt} = \vec{r} \times \vec{f}_{shear} \quad (8)$$

where, I, $\vec{\omega}$, $\vec{r}$, $\vec{f}_{shear}$ indicate the moment of inertia, angular velocity, position vector and shear force, respectively.

Fluid-Structure Interaction Algorithm

A concept diagram of the FSI coupling system which is based on the improved MPS-based model (Hwang et al., 2014; 2016) is illustrated in Figure 1. After the fluid analysis is completed in a time step, the coupling force to structure analysis is calculated by integrating the pressure of fluid on the surface of structure. And then, the structure analysis is performed considering the coupling force from the fluid. Consequentially, the updated position of structure particles acts a role of new boundary condition and acceleration at the new interface is given as the coupling force to fluid analysis in next time step.

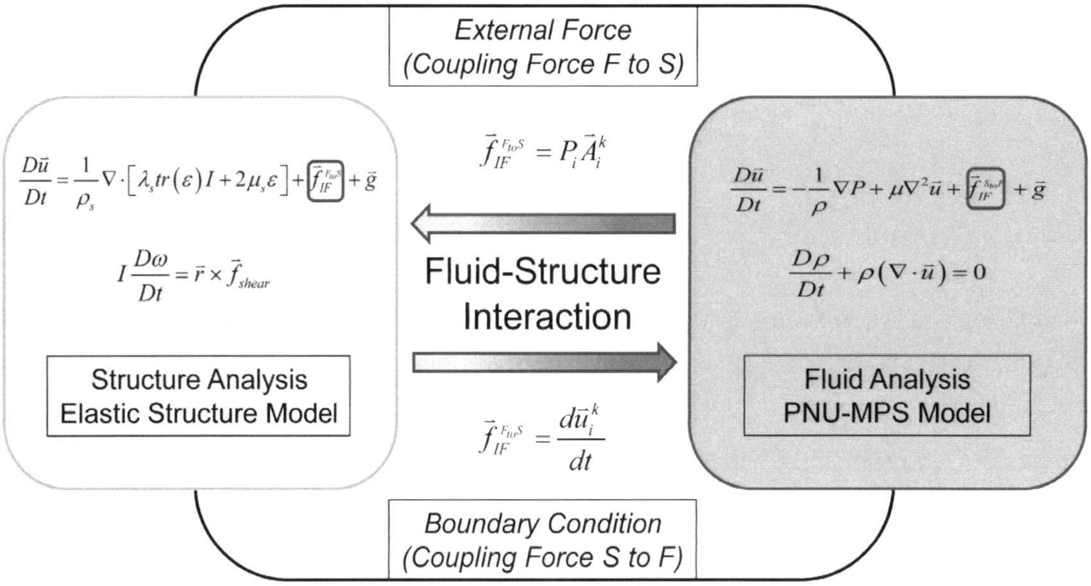

Figure 1. Concept diagram of FSI coupling procedure

In the calculation procedure, the prediction–correction solution algorithm is introduced for fluid analysis based on the PNU-MPS method (Lee et al., 2011). Firstly, the intermediate velocity is obtained from the viscous, coupling force and gravity terms explicitly as below:

$$\vec{u}_i^* = \vec{u}_{i,F}^k + \Delta t\left(\nu^F \nabla^2 \vec{u}_i^k + \vec{f}_{i,S_{to}F}^k + \vec{g}\right) \quad (9)$$

where Δt is the time step and the superscript k implies time step number. Also, the coupling force $\vec{f}_{i,S_{to}F}$ can be defined by the acceleration of the interface of structure domain as:

$$\vec{f}_{i,S_{to}F} = \frac{d\vec{u}^{k}_{IF,F}}{dt} \tag{10}$$

After calculating the intermediate velocity by Eq. (9), the pressure field of fluid at time step $k+1$ is obtained implicitly by solving the Pressure Poisson Equation (PPE) as follows:

$$\nabla^2 P^{k+1}_{i,F} = \frac{\rho_F}{\Delta t}\left(\nabla \cdot \vec{u}^{*}\right)_i + \gamma \frac{\rho_F}{\Delta t^2}\frac{n^0 - n^k_i}{n^0} \tag{11}$$

where γ is a blending parameter considered as 0.03 (Lee et al., 2011), n^0 is the particle number density of initial arrangement and n^k is the particle number density at the k-th time step obtained as:

$$n_i = \sum_{j \neq i} w_F\left(\left|\vec{r}_j - \vec{r}_i\right|\right) \tag{12}$$

$$w_F = \begin{cases} \frac{r_e}{r} - 1 & 0 \leq r \leq r_e \\ 0 & r > r_e \end{cases} \tag{13}$$

In the correction step, finally, the new velocity at the time step $k+1$ is updated by taking the summation of intermediate velocity and pressure gradient term in Eq. (14).

$$\vec{u}^{k+1}_{i,F} = \vec{u}^{*}_i - \frac{\Delta t}{\rho_F}\nabla P^{k+1}_i \tag{14}$$

$$\vec{r}^{k+1}_{i,F} = \vec{r}^{k}_{i,F} + \Delta t \vec{u}^{k+1}_{i,F} \tag{15}$$

In Eq. (14), the Gradient Correction (GC) scheme (Khayyer and Gotoh, 2011) is used for calculating the pressure gradient term.

After completing the fluid analysis, the structure analysis is performed according to Hwang et. al.(2014; 2016). as:

$$\frac{D\vec{u}^{k+1}_{i,S}}{Dt} = \frac{1}{\rho_S}\nabla \cdot \left(\lambda_S \varepsilon^k_{ll}\right) + \frac{1}{\rho_S}\left[2\mu_S \nabla^2 \left(\Delta \vec{x}\right)^k_{i,S}\right] + \vec{f}^{k}_{F_{to}S} + \vec{g} \tag{16}$$

where $\left(\Delta \vec{x}\right)^k_{i,S}$ is the displacement of the particle i from its initial location and ε_{ll} the strain rate represented as follows:

$$\varepsilon_{ll} = \varepsilon_{ii} + \varepsilon_{jj} + \varepsilon_{kk} = \frac{d}{n^0_i}\sum_{j \neq i}\frac{\left|\vec{x}^k_{j,S} - \vec{x}^k_{i,S}\right| - \left|\vec{x}^0_{j,S} - \vec{x}^0_{i,S}\right|}{\left|\vec{x}^0_{j,S} - \vec{x}^0_{i,S}\right|} w\left(\left|\vec{x}^0_{j,S} - \vec{x}^0_{i,S}\right|\right) \tag{17}$$

And a coupling force $\vec{f}^{k}_{F_{to}S}$ in Eq. (16) can be calculated by integrating the pressure on the interface as:

$$\vec{f}^{k}_{F_{to}S} = \int_{IF} P^{k+1}_{i,F} d\vec{s} = \frac{\sum_j P_{j_F} w\left(\left|\vec{x}_{j,F} - \vec{x}_{i,IF}\right|\right)}{\sum_j w\left(\left|\vec{x}_{j,F} - \vec{x}_{i,IF}\right|\right)} \quad (18)$$

where $\vec{x}_{i,IF}$ represents the position vector of the fluid-solid interface corresponding to the structure particle i and is calculated by the following formula (Antoci et. al., 2007).

$$\vec{r}_{i_{surf}} = \vec{r}_i + 0.5 l_0 \hat{n}_i \quad (19)$$

where l_0 is the initial particle length and $\hat{n}_i$ the normal vector of structure particle i.

For more detail description on the FSI algorithm, refer to Hwang et al. (2016; 2016).

Figure 2 shows the computational procedure of numerical algorithm for FSI simulation

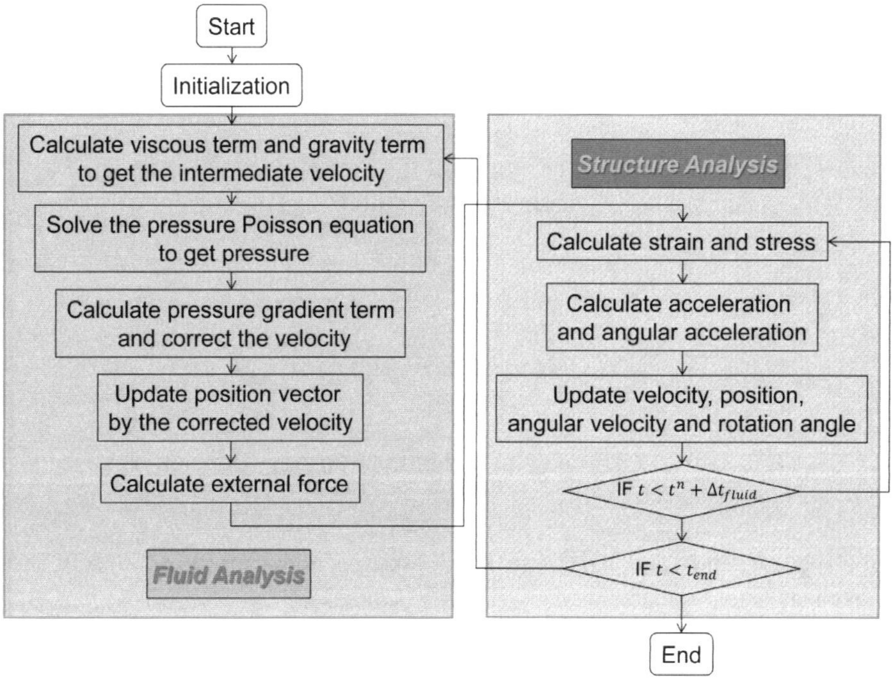

Figure 2. Computational procedure of numerical algorithm for FSI simulation

Brittle Failure

In the present study, the maximum normal stress criterion is used for a failure criteria of brittle fracture (Juvinall & Marshek, 2006), i.e. a brittle material will fail when the maximum normal stress, σ_1, exceeds the uniaxial tensile strength of the material, σ_f, as:

$$\sigma_1 \geq \sigma_f \quad (27)$$

Since the maximum normal stress that causes the failure may occur in any direction, we consider the rotation of stress tensor to obtain the maximum normal stress as follows:

$$\tilde{\sigma} = R^T \sigma R = \begin{bmatrix} \cos\alpha & \sin\alpha \\ -\sin\alpha & \cos\alpha \end{bmatrix} \begin{bmatrix} \sigma_{xx} & \sigma_{yx} \\ \sigma_{xy} & \sigma_{yy} \end{bmatrix} \begin{bmatrix} \cos\alpha & -\sin\alpha \\ \sin\alpha & \cos\alpha \end{bmatrix}$$
$$= \begin{bmatrix} \cos^2\alpha\sigma_{xx} + 2\cos\alpha\sin\alpha\sigma_{xy} + \sin^2\alpha\sigma_{yy} & \left(\cos^2\alpha - \sin^2\alpha\right)\sigma_{xy} + \cos\alpha\sin\alpha\left(\sigma_{yy} - \sigma_{xx}\right) \\ \left(\cos^2\alpha - \sin^2\alpha\right)\sigma_{xy} + \cos\alpha\sin\alpha\left(\sigma_{yy} - \sigma_{xx}\right) & \sin^2\alpha\sigma_{xx} - 2\cos\alpha\sin\alpha\sigma_{xy} + \cos^2\alpha\sigma_{yy} \end{bmatrix} \quad (28)$$

From the above equation, the first normal components of stress, $\tilde{\sigma}_{n1}$, can be written as:

$$\tilde{\sigma}_{n1} = \tilde{\sigma}\left(\cos\alpha\right) = \cos^2\alpha\sigma_{xx} + 2\cos\alpha\sqrt{1-\cos^2\alpha}\,\sigma_{xy} + \left(1-\cos^2\alpha\right)\sigma_{yy} \quad (29)$$

In Eq. (29), the stress tensor is a function of rotation angle and the maximum stress in the normal direction should be satisfied to the following equation as:

$$\frac{\partial\sigma_{n1}}{\partial\cos\alpha} = 2\cos\alpha\left(\sigma_{xx} - \sigma_{yy}\right) + 2\frac{1-2\cos^2\alpha}{\sqrt{1-\cos^2\alpha}}\sigma_{xy} = 0 \quad (30)$$

The above equation can be rewritten as:

$$\cos\alpha\left(\sigma_{xx} - \sigma_{yy}\right) + \frac{1-2\cos^2\alpha}{\sqrt{1-\cos^2\alpha}}\sigma_{xy} = 0 \quad (31)$$

In general, Eq. (31) becomes a quadratic equation as:

$$\chi^2 - \chi + \frac{B}{\left(A+4B\right)} = 0 \quad (32)$$

where $\chi = \cos^2\alpha$, $A = \left(\sigma_{xx} - \sigma_{yy}\right)^2$, $B = \sigma_{xy}^2$.

Finally, the following equation can be obtained as the solution of Eq. (32):

$$\chi = \cos^2\alpha = \frac{1}{2} \pm \frac{1}{2}\sqrt{1 - \frac{4B}{A+4B}} \quad (33)$$

If the direction angle α for the maximum normal stress is obtained from Eq. (33), therefore, the corresponding maximum normal stress $\tilde{\sigma}_{max} (= \sigma_1)$ can be determined by substituting α into Eq. (29).

In addition, there is another way to consider the maximum normal stress, so-called as the first principal stress. Because the maximum normal stress is indicated over the crack propagation direction, all components of shear stress should be zero, which means the first principal stress becomes the largest eigenvalue of the stress matrix. Hence, the direction of the maximum normal stress is the eigenvector corresponding to the eigenvalue of the stress matrix.

In the numerical process, when a particle i is judged that the crack occurs locally, a spring-like relationship will be disconnected with the neighboring particle j which is located opposite to the crack plane within the effective radius r_e. For any pair of two particles i and j in the vicinity of the crack, as long as they meet the following two conditions, the mutual influence each other will be eliminated by setting the value of weight function zero:

i) A pair particle needs to locate on both sides of the crack plane and can expressed as:

$$\left[(\vec{r}_i-\vec{r}_c)\cdot\hat{n}_c\right]\left[(\vec{r}_j-\vec{r}_c)\cdot\hat{n}_c\right]<0 \tag{34}$$

where, $\vec{r}_c=(\vec{r}_i+\vec{r}_j)/2$ implies the midpoint between two particles i and j, and $\hat{n}_c$ the direction of crack propagation as given in $\hat{n}_c=(\cos\alpha,\, \sin\alpha)$.

ii) One of two particles needs to be located opposite to the crack as:

$$\left|\left\{(\vec{r}_i-\vec{r}_c)-\left[(\vec{r}_i-\vec{r}_c)\cdot\hat{n}_c\right]\hat{n}_c\right\}\right|\le C_w l_0$$
$$\text{or} \tag{35}$$
$$\left|\left\{(\vec{r}_j-\vec{r}_c)-\left[(\vec{r}_j-\vec{r}_c)\cdot\hat{n}_c\right]\hat{n}_c\right\}\right|\le C_w l_0$$

where, $C_w l_0$ should be determined in a range of $\frac{l_0}{2}\le C_w l_0\le r_e$ and C_w can be set at $\sqrt{d}/2$ to avoid discontinuous crack capturing especially in diagonal direction.

SIMULATION RESULTS

Simulation of Three-Point Bending Test

To validate the developed fracture algorithm, a series of numerical simulation of 3-point bending test with ice beam made of fresh water is performed and compared with the experimental data obtained from the experiment which was independently conducted in a cold room at Korea Maritime and Ocean University.

Figure 4 shows the schematic view of the three-point bending test. The distance between two simple support points is 180mm and the thickness of ice specimen is 40mm. A concentrated load with a constant speed of 0.0025m/s is applied to the midpoint of the ice beam.

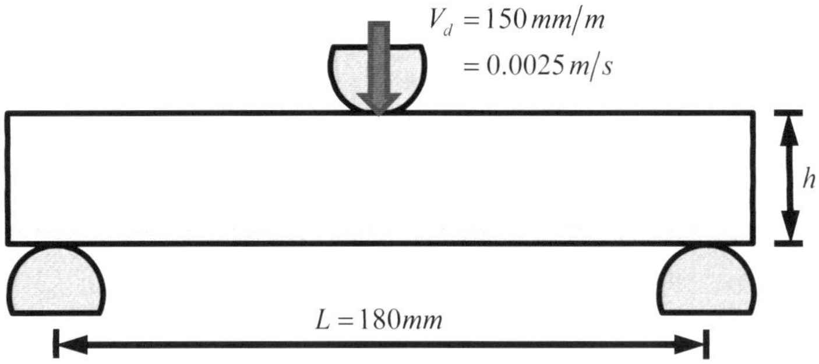

Figure 4. Schematic view of three-point bending test with ice beam

Photo 1 shows the typical snapshots of three-point bending test before and after crack of ice beam. It is seen clearly that a vertical crack appears just below the point where the load is acting when the load is large enough, which directly leads to the breakage of the ice beam.

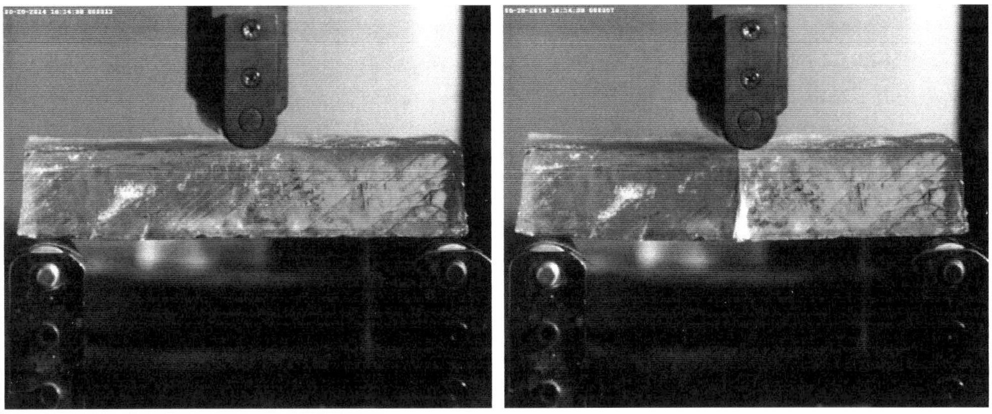

Photo 1. Experimental photos of three-point bending test with ice beam before (left) and after (right) cracking

From the experiment, the failure load $P_{\max}$ and deformation on the loading point $\Delta\delta$ were measured and the material properties of ice beam, such as Young's modulus E, flexural strength σ_f and flexural strain ε_f, were calculated as:

$$E = \frac{1}{4w}\left(\frac{L}{h}\right)^3 \frac{P_{\max}}{\Delta\delta} \tag{36}$$

$$\sigma_f = \frac{3}{2}\frac{P_{\max}L}{wh^2} \tag{37}$$

$$\varepsilon_f = \frac{6h\Delta\delta}{h^2} \tag{38}$$

where, w and h are the width and thickness of specimen.

The summarized experimental results are listed in Table 1.

Table 1. Summary of experiments for three-point bending test with ice beam

	Raw data				Calculation results		
Exp. No.	$w(m)$	$h(m)$	$P_{\max}(N)$	$\Delta\delta(m)$	$E(MPa)$	$\sigma_f(MPa)$	ε_f
02	0.082	0.038	500.1	6e-4	270.084	1.140	0.0042
03	0.084	0.044	411.9	4e-4	209.822	0.684	0.0032
04	0.081	0.040	588.4	9e-4	183.875	1.226	0.0067
05	0.079	0.039	441.3	1e-3	137.300	0.992	0.0072
10	0.082	0.037	323.6	1.15e-3	98.775	0.778	0.0079

In the simulation, the Young's modulus and flexural strength are used as the input data for material properties of ice beam. The Poisson's ratio is assumed to be 0.3. The time step and particle length are set at 10^{-6} s and 2×10^{-3} m, respectively.

The time-sequential snapshots of crack path and distribution of horizontal normal stress are shown in Figure 5. Before cracks appeared at t=0.0334s, the tensile stress is observed at lower part and compressive stress separated near central position where the concentrated load acts at upper part. As the crack evolves at t=0.0335s and 0.0339s, the concentration of horizontal stress is evident. Eventually the ice beam is completely broken at t=0.0342s and the corresponding stress is relived.

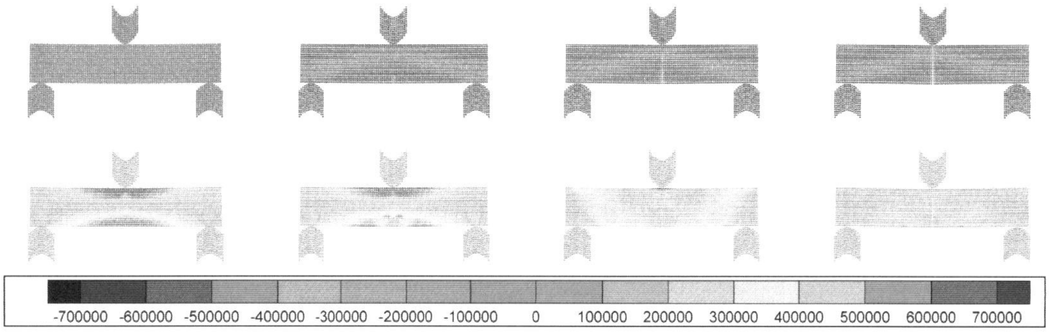

(a) t=0.0334s (b) t=0.0335s (c) t=0.0339s (d) t=0.0342s

Figure 5. Snapshots of crack path (upper) and normal stress in *x*-direction (lower)

Figure 6 shows the time history of horizontal normal stress at two middle points on top and bottom surface of the ice beam. The stress curves increase gradually with small elastic vibration before fracture as time increasing due to the assumption that the structure is considered as elastic material. And then after the fracture, the stresses dramatically decrease near zero.

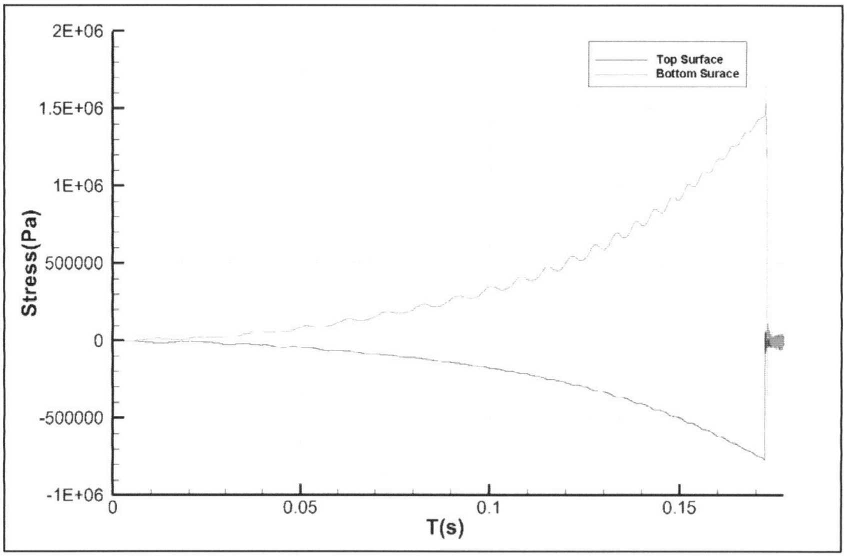

Figure 6. Time history of stress at two mid-points on top and bottom surface of beam

Finally, the flexural strains of the present simulation are compared with those of experiment in Figure 7. Overall agreements are quite good, but it tends that the deviation with the experiment becomes larger if the Young's modulus of the materials becomes smaller. The maximum error between both is below 7.5 percent.

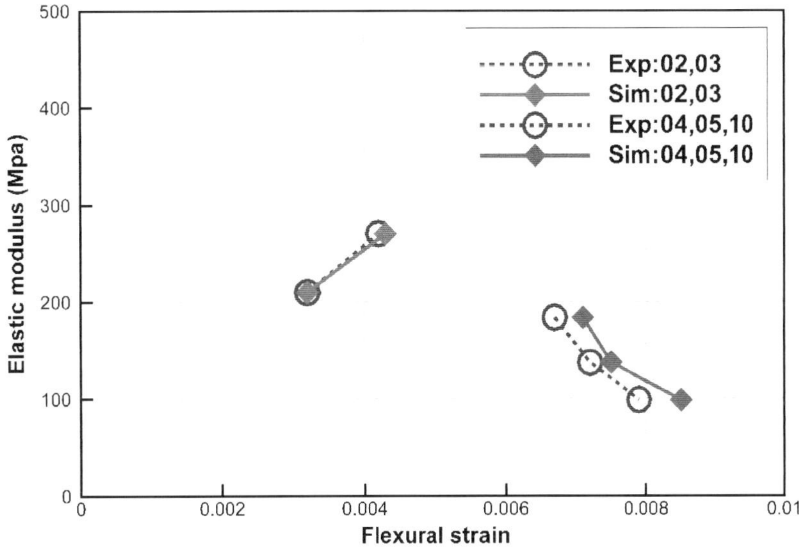

Figure 7. Comparison of flexural strain with experiments

FSI Simulation with Fracture Model

A dropping water droplet interacting with a cantilever ice beam is carried out for the application of FSI model. Figure 10 shows an initial setup for FSI simulation. The length and thickness of the ice beam are 0.5m and 0.05 m, respectively. The left end is fixed and the square water droplet with 0.1 m is falling down with the initial velocity of 9 m/s above the free end of the beam. The material properties of the cantilever ice beam are set the same as case 02 of 3-point bending test with the Young's modulus of 270.084 MPa, the Poisson's ratio of 0.3 and the flexural strength of 1.140 MPa. The time step for fluid and structure analysis are 1×10^{-5} s and 1×10^{-7} s, respectively, and the particle size is 2.5×10^{-3} m.

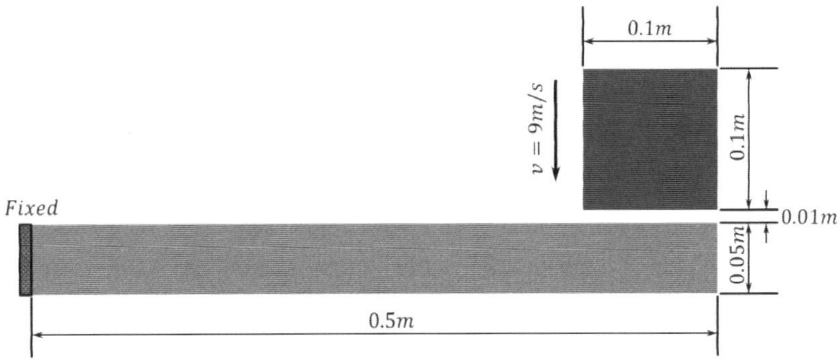

Figure 10. Schematic view and dimension of FSI simulation

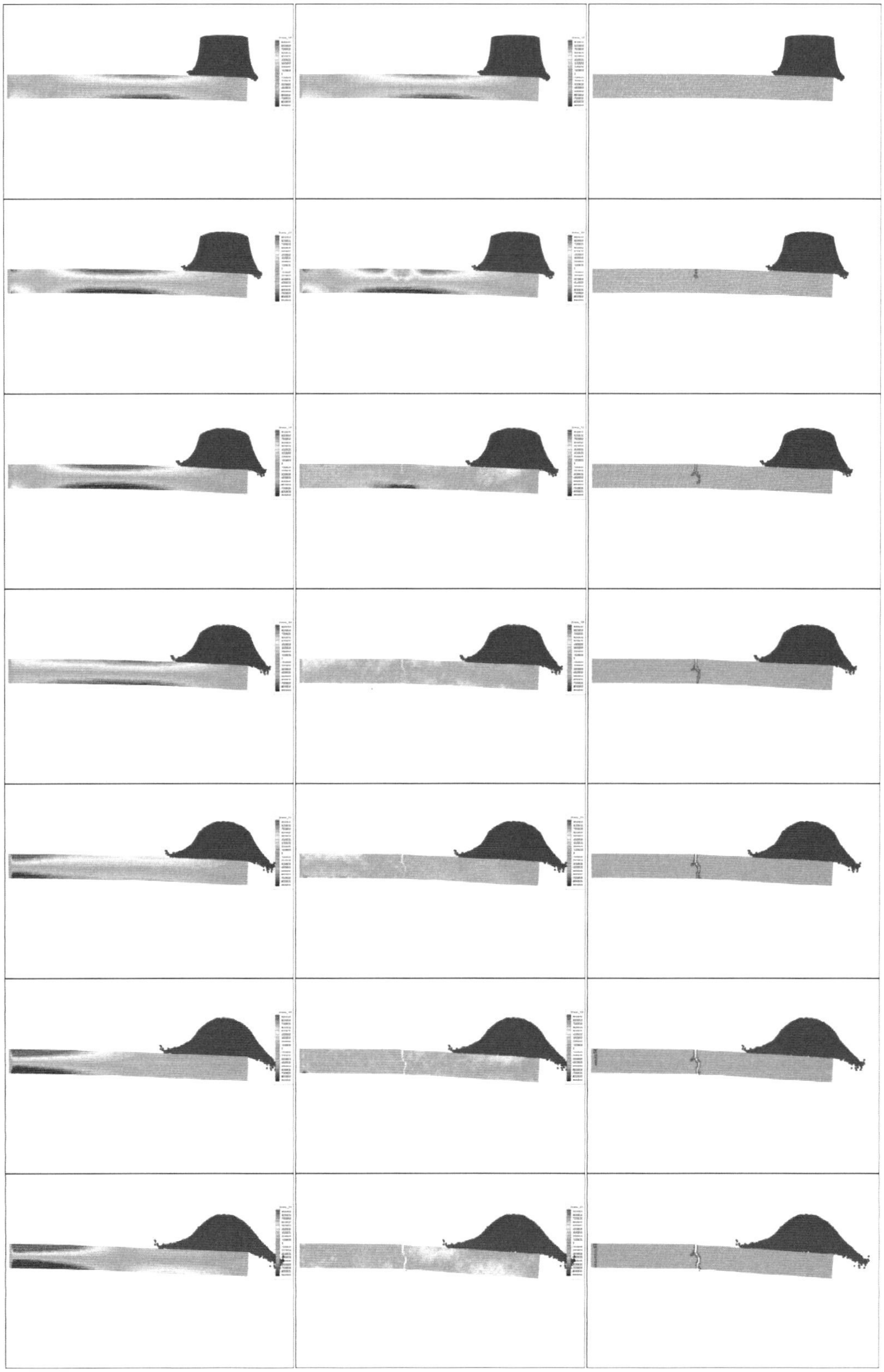

Figure 11. Snapshots of FSI simulation without failure (left), with failure (middle) and the crack path (right) from t=0.004s to t=0.01s with the step of 0.001s

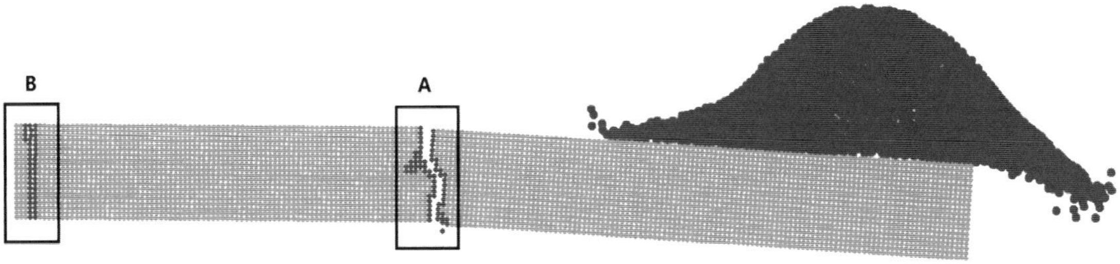

Figure 12. Definition of corresponding positions A and B

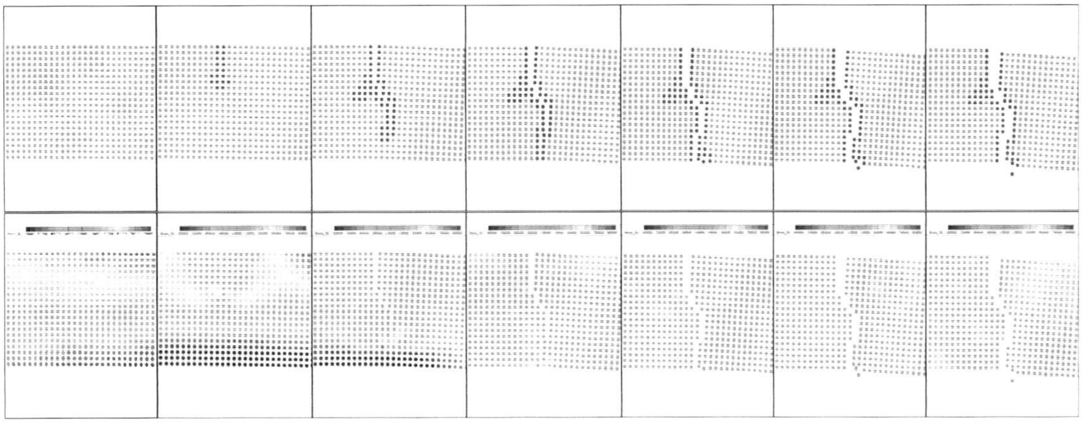

Figure 13. Snapshots of the failure path (upper) and horizontal normal stress distribution (below) at position A from t=0.004s to t=0.01s with interval of 0.001s

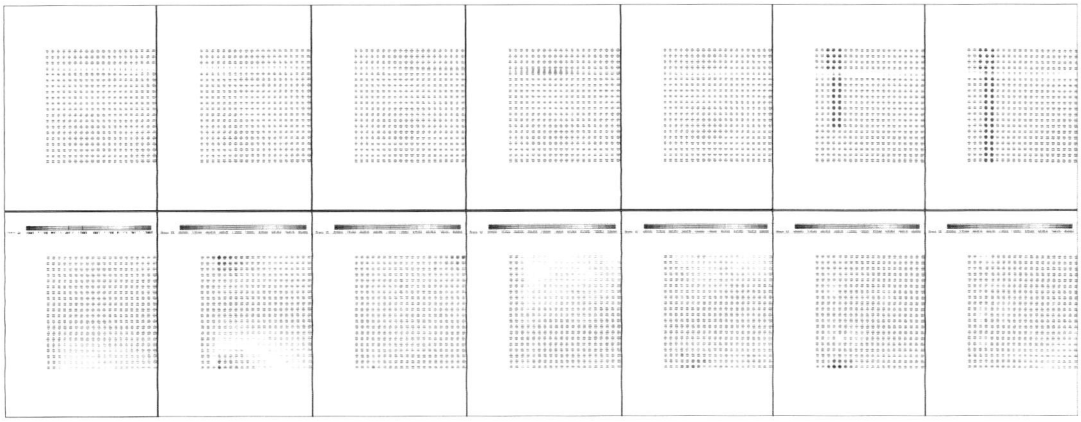

Figure 14. Snapshots of the failure path (upper) and the horizontal normal stress distribution (below) at position B from t=0.004s to t=0.01s with interval of 0.001s

Figure 11 illustrates the time-sequential snapshots for FSI simulation from t=0.004s to 0.01s with interval of 0.001s. The first and second column are the cases without and with the brittle failure algorithm, respectively, and the last column the crack path of ice in the FSI simulation with the failure algorithm. It can be shown that the failure occurs from where the local stress exceeds the critical value, and the dynamic behavior and stress distribution over the beam are completely different from the simulation without the failure algorithm.

A series of snapshots for local stress distribution and crack propagation around two different positions A and B marked in Figure 12 are depicted in Figures 13 and 14. It is observed the crack occurs where the stress is concentrated and location of stress concentration moves according to the front of crack propagation. At the same time, the corresponding stress distribution becomes unstable, which is whereby resulted in the direction of crack propagation unpredictably changed. Thus, small differences in conditions would reproduce completely different simulation results. It makes to predict accurately these kinds of problems difficult.

CONCLUSIONS

A realistic numerical simulation technology using a particle-based Fluid-Structure Interaction (FSI) model combined with a brittle fracture algorithm has been developed to simulate the fluid-ice-structure interaction problems. The failure of ice is modeled as a brittle fracture of elastic by applying a novel FSI model based on the Moving Particle Semi-implicit (MPS) method. The developed numerical algorithm of fracture was validated through the comparison with the experiments in a cold room for 3-point bending tests with ice beam. It was seen that the brittle failure was successfully simulated and the acceptable range of error was obtained within 7.5% compare to the experiments. After then, for the application of FSI model associated with the fracture algorithm, the numerical simulation of the dropping water droplet interacting with a cantilever ice beam was carried out comparatively with and without the fracture algorithm. From the simulation, it can be said that the consequent effects of fracture that might be occurred in the process of FSI simulation are available to be studied.

ACKNOWLEDGEMENTS

This work was supported by the IT R&D program of MOTIE/KEIT. [10060329, Development of numerical simulator for environmental load on ship and offshore in polar region]

REFERENCES

Antoci, C., Gallati, M., & Sibilla, S., 2007. Numerical simulation of fluid–structure interaction by SPH. *Computers & Structures,* 85(11), pp. 879-890.

Abbas, S., Alizada, A., & Fries, T. P., 2010. Model-independent approaches for the XFEM in fracture mechanics. *Int J Numer Methods Eng*, 1, pp. 0-10.

Beckmann, B., Schicktanz, K., & Curbach, M. 2014. *DEM Simulation of Concrete Fracture Phenomena.*

Bui, H. H., Fukagawa, R., Sako, K., & Ohno, S. (2008). Lagrangian meshfree particles method (SPH) for large deformation and failure flows of geomaterial using elastic–plastic soil constitutive model. *International Journal for Numerical and Analytical Methods in Geomechanics,* 32(12), pp. 1537-1570.

Chan, H. K. D., 1981. *Creep and fracture simulation of ice using the finite element method.* Ph.D. McMaster University.

Dai, M., Shen, H.H., Hopkins, M.A., Ackley, S.F., 2004. Wave rafting and the equilibrium pancake ice cover thickness. *Journal of Geophysical Research*, 109, C07023.

Hansen, E.H., Løset, S., 1999. Modelling floating offshore units moored in broken ice: model description. *Cold Regions Science and Technology*, 29, 97–106.

Hopkins, M.A., 1998. Four stages of pressure ridging. *Journal of Geophysical Research*, 103, 21883–21891.

Hopkins, M.A., Shen, H.H., 2001. Simulation of pancake-ice dynamics in wave field. *Annals of Glaciology*, 33, 355–360.

Hwang, S.C., Khayyer, A., Gotoh, H., & Park, J.C., 2014. Development of a fully lagrangian MPS-based coupled method for simulation of fluid–structure interaction problems. *Journal of Fluids and Structures,* 50, pp. 497-511.

Hwang, S.C., Park, J.C., Gotoh, H., Khayyer, A., & Kang, K.J., 2016. Numerical Simulation of Sloshing Flows with Elastic Baffles by Using a Particle-based Fluid–Structure Interaction Analysis Method. *Ocean Engineering,* 118, pp. 227-241.

Hübner, B., Walhorn, E., & Dinkler, D., 2004. A monolithic approach to fluid–structure interaction using space–time finite elements. *Computer methods in applied mechanics and engineering,* 193(23), pp. 2087-2104.

Ji, S., Li, Z., Li, C., Shang, J., 2013. Discrete Element Modeling of Ice Loads on Ship Hulls in Broken Ice Fields. *Acta Oceanologica Sinica*, 32(11), 50-58.

Juvinall, R. C., & Marshek, K. M., 2006. *Fundamentals of machine component design (Vol. 83).* John Wiley & Sons: New York.

Karulin, E.B., Karulina, M.M., 2011. Numerical and physical simulations ofmoored tanker behaviour. *Ship and Offshore Structures*, 6(3), 179–184.

Khayyer, A., & Gotoh, H., 2010. A higher order Laplacian model for enhancement and stabilization of pressure calculation by the MPS method. *Applied Ocean Research,* 32(1), pp. 124-131.

Khayyer, A., & Gotoh, H., 2011. Enhancement of stability and accuracy of the moving particle semi-implicit method. *Journal of Computational Physics,* 230(8), pp. 3093-3118.

Koshizuka, S., & Oka, Y., 1996. Moving-particle semi-implicit method for fragmentation of incompressible fluid. *Nuclear science and engineering,* 123(3), pp. 421-434.

Lau, M., Lawrence, K.P., Rothenburg, L., 2011. Discrete Element Analysis of Ice Loads on Ships and Structures. *Ships and Offshore Structures*, 6(3), 211-221.

Lee, B.H., Park, J.C., Kim, M.H., & Hwang, S.C., 2011. Step-by-step improvement of MPS method in simulating violent free-surface motions and impact-loads. *Computer methods in applied mechanics and engineering,* 200(9), pp. 1113-1125.

Lepparanta, M., Lensu, M., Lu, Q.M., 1990. Shear flow of sea ice in the Marginal Ice Zone with collision rheology. *Geophysica*, 25(1–2), 57–74.

Metrikin, I., Løset, S., 2013. Nonsmooth 3D Discrete Element Simulation of a Drillship in Discontinuous Ice. *Proceedings of the 22nd International Conference on Port and Ocean Engineering under Arctic Conditions*, Espoo, Finland.

Morgan, D., Sarracino, R., KcKenna, R., Thijssen, J.W., 2015. Simulation of Ice Rubbling against Conical Structures Using 3D DEM. *Proceedings of the 23rd International Conference on Port and Ocean Engineering under Arctic Conditions*, Trondheim, Norway.

Polojarvi, A., Tuhkuri, J., 2009. 3D discrete numerical modelling of ridge keek punch through tests. *Cold Regions Science and Technology*, 56, 18–29.

Peixiang, H., Ziran, L., & Changchun, W., 2013. An element-free Galerkin method for dynamic fracture in functional graded material. *In ICF10*, Honolulu (USA) 2001.

Sakharov, A., Karulin, E., Marchenko, A., Karulina, M., Sodhi, D., & Chistyakov, P., 2015. Failure Envelope of the Brittle Strength of Ice in the Fixed-End Beam Test (Two Scenarios). *In 23rd International Conference on Port and Ocean Engineering Under Arctic Conditions,* Trondheim, Norway.

Selvadurai, A.P.S., Sepehr, K., 1999. Two-dimensional discrete element simulations of ice-structure interaction. *International Journal of Solids and Structures*, 36, 4919–4940.

Sepehri, J., 2014. *Application of extended finite element method (XFEM) to simulate hydraulic fracture propagation from oriented perforations.* Ph.D. Texas Tech University.

Shen, H.H., Hibler, W.D., Lepparanta, M., 1987. The role of floe collisions in sea ice rheology. *Journal of Geophysical Research*, 92(C10), 7085–7096.

Sun, S., Shen, H.H., 2012. Simulation of pancake ice load on a circular cylinder in a wave and current field. *Cold Regions Science and Technology*, 78, 31–39.

Tan, Y., Yang, D., & Sheng, Y., 2009. Discrete element method (DEM) modeling of fracture and damage in the machining process of polycrystalline SiC. *Journal of the European ceramic society*, 29(6), pp. 1029-1037.

Tanaka, M., & Masunaga, T., 2010. Stabilization and smoothing of pressure in MPS method by Quasi-Compressibility. *Journal of Computational Physics,* 229(11), pp. 4279-4290.

Xu, Z., Tartakovsky, A.M., Pan, W., 2012. Discrete-element model for the interaction between ocean waves and sea ice. *Physical Review E*, 85, 016703.

Zhan, D., Agar, D., He, M., Spencer, D., Molyneux, D., 2010. Numerical simulation of ship maneuvering in pack ice. *Proceedings of the ASME 2010 29th International Conference on Ocean, Offshore and Arctic Engineering*, Shanghai, China, OMAE2010-21109.

POAC'17
Busan, KOREA

Proceedings of the 24th International Conference on
Port and Ocean Engineering under Arctic Conditions
June 11-16, 2017, Busan, Korea

Simulating Interaction Between Level Ice and Conical Structures with a 2D Lattice Model

Renate van Vliet[1,2], Andrei V. Metrikine[2]
[1] Shell Projects & Technology, Rijswijk, The Netherlands
[2] Delft University of Technology, The Netherlands

ABSTRACT

Structures with a downward sloping waterline shape in Arctic or sub-Arctic regions experience reduced loading from ice on the structure and its foundation compared to vertical structures as the slope causes the ice to fail in bending. For the design of these structures, a numerical model is desired that can predict the loading from the ice on the structure by simulating the physical processes that occur when the ice fails.

A numerical 2D lattice model has been developed to simulate level ice behavior. The model is composed of masses and interconnecting springs, taking into account deformation in tension, compression, bending, shear and torsion in the ice sheet. Deformation and failure criteria in the model are based on first principles, enabling a physically sound simulation of breaking processes. In addition, the discrete lattice model avoids stress singularities in fracture modeling since the model only describes displacements and forces in the connections and no stresses are present.

In this paper interaction between ice and a downward sloping conical structure is simulated with the lattice model and compared with data from scale model tests. From the model test results it was observed that level ice and larger ice floes fail in sequential bending during interaction with a structure, giving a repetitive load pattern. Smaller ice floes split and break in bending into smaller parts and a sequential breaking pattern does not develop. The lattice model is capable of simulating failure in bending, splitting and combinations of these. In contrast, predictions of the model presented herein suffer from limitations of the contact model between the ice and the structure, the lack of a clearing mechanism and the use of a regular lattice mesh.

KEY WORDS: Ice; Failure; Lattice model.

INTRODUCTION

To build structures in Arctic and sub-Arctic regions, assessment and minimization of loading from ice on structures is required. Structures with a downward sloping waterline shape experience reduced loading from ice on the structure and its foundation compared to vertical structures as the slope causes the ice to fail in bending. The manner in which the ice floes fail when encountering the structure depends on the size and material parameters of the floes; most observed failure modes are in bending, splitting or a combination of these. By means of numerical assessment, a multitude of structures and loading conditions can be analyzed and structural shapes can be optimized for ice load reduction. Ice management processes can be optimized as well in case the effects of managed floe sizes on the loading of a structure are known. A numerical model is used in this paper that can predict the loading from the ice on the structure by simulating the physical failure processes and effects of floe parameters.

Several research institutes and companies have developed numerical assessment tools to study ice-sloping-structure interaction, in the form of 1D, 2D or 3D models. Aksnes (2010) and Wille, Kuiper, & Metrikine (2011) simulated bending failure of ice in interaction with a downward sloping conical structure using a beam model. Beam models are limited in the prediction of loading frequencies and breaking length of the ice in case of circular or ship-shaped structures as 2D effects have significant influence on the breaking pattern for these structures. To simulate real-time ice-structure interaction Lubbad and Løaset (2011) adopted a model of wedge-shaped beams failing adjacently based on theory of a semi-infinite plate resting on an elastic foundation, for which a solution was presented by Nevel (1992). Ice failure was modeled in a similar manner to simulate a floating drillship in level ice with SIBIS software by Metrikin et al. (2015). The discrete element tool described by Lau et al. (2008) has a module that allows for pre-defined crack patterns in simulations for ships in level ice. Bonnemaire et al. (2014) used a semi-empirical model to simulate the response of a moored floating structure in ice. The discrete element method was used to analyze the breaking process of sea ice by Ji et al. (2015) by connecting particles with parallel-bonds with associated moment and force and breaking the bonds when internal stresses exceeded a certain threshold.

In this paper a numerical lattice model developed by the authors is employed to simulate interaction between ice and a conically shaped structure. The lattice model is composed of masses and springs of various types, taking into account deformation in tension, compression, bending, shear and torsion in the ice sheet. Lattice models have been used in the past to simulate ice-structure interaction. Sayed (1997) used a lattice model for fracture of sea-ice, using elastic and visco-elastic bonds in normal direction between the masses in the model. Dorival, Metrikine, & Simone (2008) developed a lattice model to simulate ice crushing against a structure. Van den Berg (2016) presented a 3-dimensional lattice model to simulate ice-structure interaction. The 2-dimensional model of an ice plate that is used in this paper applies deformation and failure criteria that are based on first principles, enabling physically sound simulation of failure processes that occur in ice-structure interaction. Another advantage of the lattice model is that stress singularities in fracture modeling are avoided as the model only describes displacements and forces in the connections instead of strains and stresses inherently present in any continuum model.

The lattice model is validated in this paper against model test results for a rigid downward breaking conical structure interacting with ice floes of various sizes. These tests were performed as part of a Joint Industry Project and described by Bruun et al. (2011). Focus in this validation study is on combined bending and splitting behavior of the ice, which depends on the floe size. It is shown that the physical failure processes of combined bending and splitting can be simulated with the lattice model.

Although fracture in bending and splitting can be predicted with the lattice model, there are limitations to the fracture patterns that can be simulated as the square mesh that is applied in the current study causes the fractures in the ice plate to follow mesh-related patterns. This is partly resolved by application of local variations to spring stiffness and nodal locations in the model. Further improvement could be achieved in the future by further study of local variations, improvement of the contact model, adding ice clearance and local mesh refinement.

The paper is structured as follows. In the next section the numerical lattice model is described. The third section presents results of experimental measurements that are used for validation of the model predictions. This is followed by a comparison is made between experimental results and simulations in section 4. In the last section conclusions are listed.

DESCRIPTION OF THE LATTICE MODEL

A lattice model has been developed to simulate ice-structure interaction. This 2-dimensional model is composed of masses and springs, taking into account deformation in tension, compression, bending, shear and torsion in the ice sheet. The lattice, which has a square mesh, and its connections are depicted in Figure 1. The model can be applied in both time- and frequency-domain simulations. Due to strongly non-linear fracture modeling, in this paper simulations are carried out in the time domain.

In axial directions, parallel with the x- and y- axes, the lattice model has more types of connectors than in diagonal directions. The connectors in axial and diagonal directions have different stiffness and their stiffness is assigned such that the global bending deformations of the plate match the Mindlin-Reissner equations in out-of-plane and those of the classical elastic continuum in in-plane direction in the long-wave approximation. The method used to derive and verify stiffness of the springs in the lattice model for in-plane-deformation are described in Suiker et al. (2001). Linear viscous elements are added to all connectors, to represent material damping.

The Mindlin-Reissner equations, to which the out-of-plane deformations of the lattice model are matched in the long-wave limit, describe the dynamic out-of-plane behavior of shear-deformable plates. The lattice model deformations are based on first principles and not limited in width-to-thickness ratio of the ice plate deformations that are simulated with it.

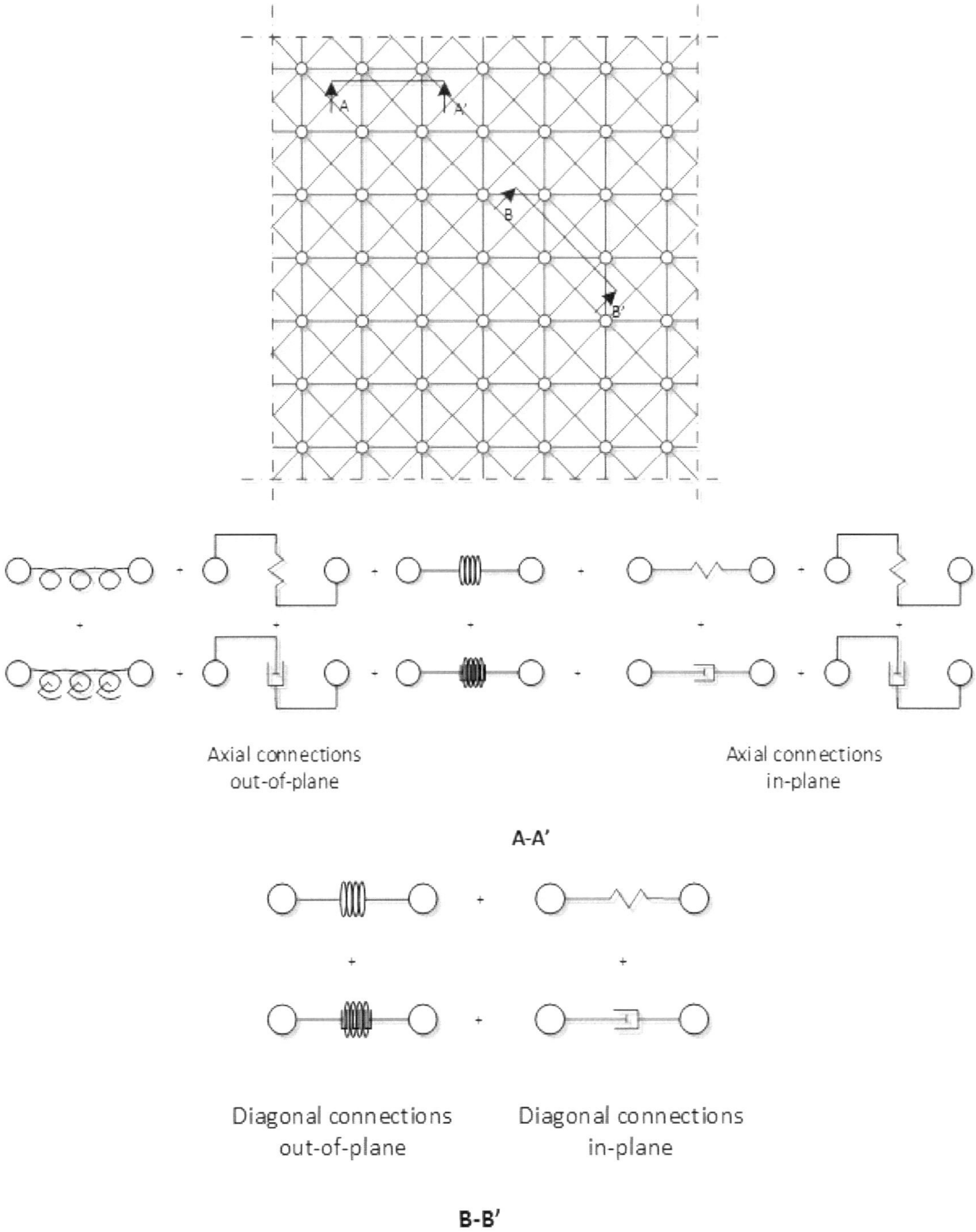

Figure 1. The lattice model and its connections working in-and out-of-plane direction. For section A-A' from left to right out-of-plane-bending, out-of-plane-shear, torsion, axial deformation and in-plane shear. For section B-B' from left to right torsion and axial deformation.

A Kelvin foundation is applied to the model as a simplistic manner to account for stiffness and damping associated with the fluid the ice plate is assumed to float on. Friction working parallel to the ice sheet is included by applying form friction and skin friction drag loads that are related to the squared velocity of the ice plate.

An algorithm for connection failure is included in the lattice model to study fracture. The criterion for failure of the connections is based on relative displacement and rotation of the springs. When the combined bending and tensile deformation in a connection reaches a critical value, the stiffness of all springs forming that connection are set to zero. Critical deformation is linked to the flexural strength of the ice via linear stress-strain relation accounting for the varying distance between the masses for different lattice cell sizes.

The advantage of the lattice model is that stress singularities in fracture modeling are avoided as the model only describes displacements and forces. Other benefits are that the shape of the model can easily be varied by cutting-out the required shape and size along the mesh. Natural variations of a material, which are of significant relevance in ice-mechanics, can be easily implemented by varying the spring stiffnesses in the different connections or displacing the nodes in the mesh of the lattice model.

To simulate contact load between the lattice and a conical structure, a simplified contact algorithm is applied. At each time step of the simulation the amount of overlap between the ice and the structure is determined, based on the structural shape, its indentation and the deflection and tip rotation of the ice. From this overlap the contact area, a_{crush} in Figure 2, is determined. Please note that Figure 2 is a side view of the overlap and that contact forces are calculated in three dimensions. The force from the structure on the ice, which is a linear function of the contact area and the crushing strength of the ice, and works perpendicular to the structure, is calculated and applied to the ice. A friction force is applied in opposite direction to the motion of the ice along the structure and is related to the contact force via a friction coefficient.

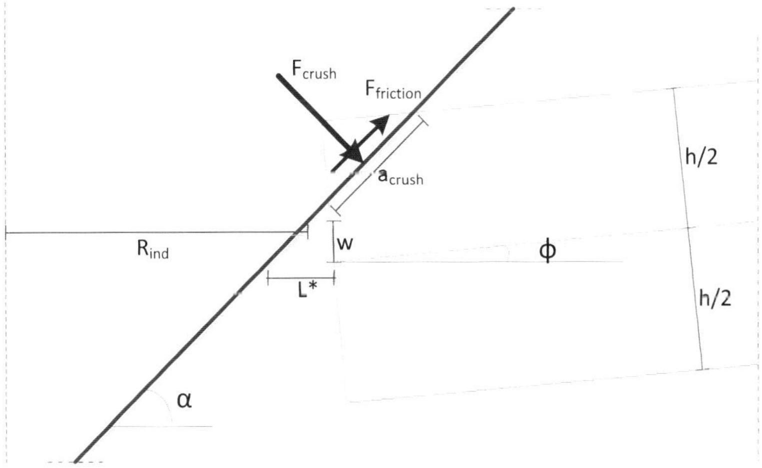

Figure 2. Side view indicating important parameters in determining the contact load: R_{ind} is the radius of the cone, α is the slope of the cone, L^* is the distance between the center of the ice edge and the cone, w is the deflection of the ice, ϕ is the tip rotation of the ice, h is ice thickness, a_{crush} is the contact length between the ice and the structure, F_{crush} is the crushing load, perpendicular to the structure and $F_{friction}$ is the friction load, directed opposite to the motion of the ice along the structure.

OBSERVATIONS FROM MODEL TESTING

Three tests that were executed as part of the JIP model testing campaign, of which an overview is presented by Bruun et al. (2011), are studied in this paper. All of the tests were executed using the same fixed structure, depicted in Figure 3, which was moved through ice with a thickness of 3m at a velocity of 0.1 m/s. All reported values in this paper are in full scale.

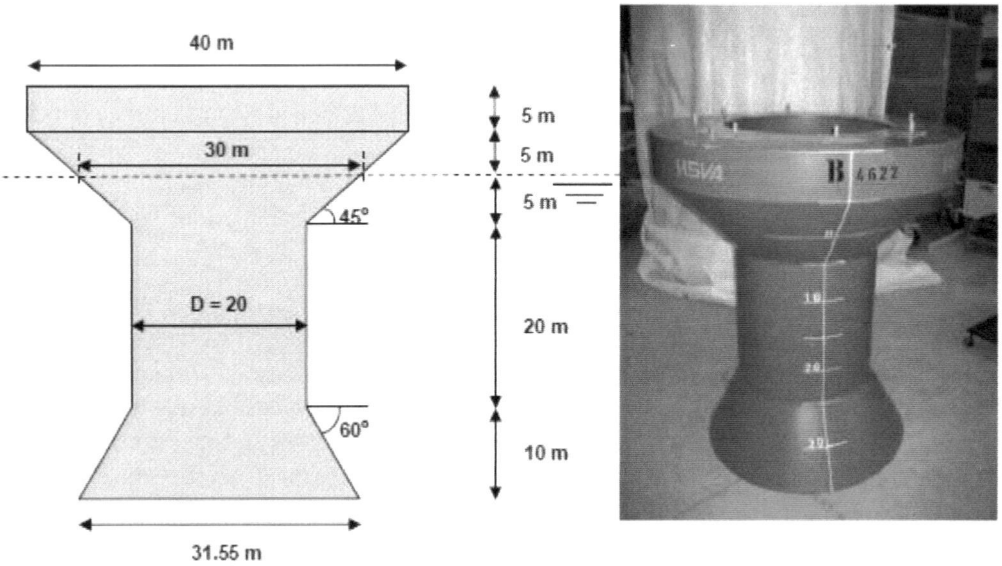

Figure 3. Dimensions of the structure that was used during the ice basin tests. The waterline diameter was 30m. (HSVA 2010a)

a. b.

Figure 4. Overview of the floe distribution in the ice basin a) large floes and b) smaller floes (HSVA 2010a)

In the three tests the structure encountered ice plates of various sizes: one test was executed in level ice, one test in large ice floes and one test in smaller ice floes. Figure 4 presents an overview of the ice floe distribution for the two tests in broken ice.

Figure 5 shows the structure during the test in level ice. Only bending failure can be observed when looking at images from the tests, but the failure pattern is not very clear. For tests in broken ice, the failure patterns are clearer as the ice floes drift apart and reveal fractures. Figure 6, Figure 7 and Figure 8 show images from test video's for interaction between the structure and finite size ice floes. Events were selected where the structure encounters the ice floe edge in near the centre. Failure occurs as a combination of bending and splitting.

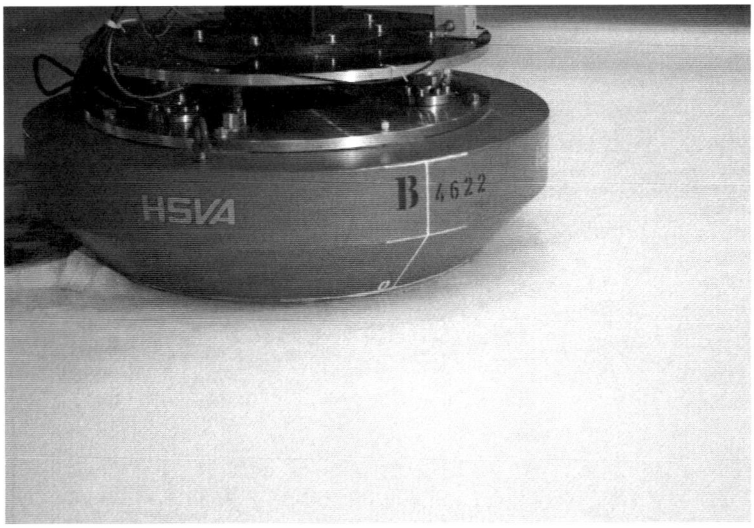

Figure 5. Structure in level ice during model test (HSVA 2010b)

During interaction with a large ice floe of 165x165m, first a piece of ice fails in bending, showing a circumferential breaking pattern. Shortly after the structure encounters the new broken edge of the ice, the ice plate splits in two. Then the ice fails close to the structure in combined radial and circumferential fracture as can be seen on the right hand side of Figure 6. During the remaining encounters between the structure and the ice floe, splitting and circumferential and radial fracture alternate.

During interaction with smaller ice floes of 69x34m and 55x55m in both cases first a piece of ice breaks off circumferentially. The rectangular floe immediately splits apart during this first interaction. The square floe splits apart at the second encounter. The remaining ice pieces experience confinement from the surrounding pieces of ice, loading the structure during the clearing process. Since all ice floes in the basin are of different size and the structure approaches each ice floe at a different angle, breaking patterns of each ice floe are different. In case an ice floe is encountered at the edge, typically the edge breaks off before the ice floe splits during further encounter with the structure.

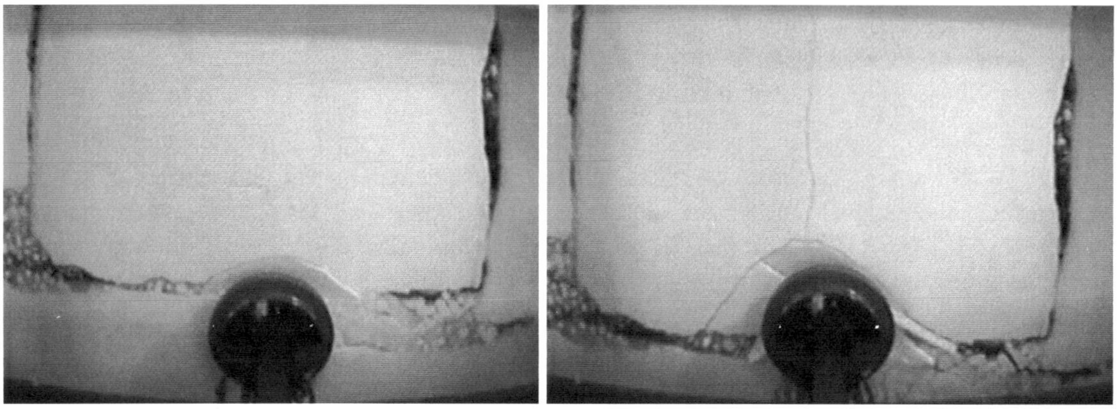

Figure 6. Image from model test video. Encounter between the structure and a large ice floe of 165x165m (HSVA 2010b)

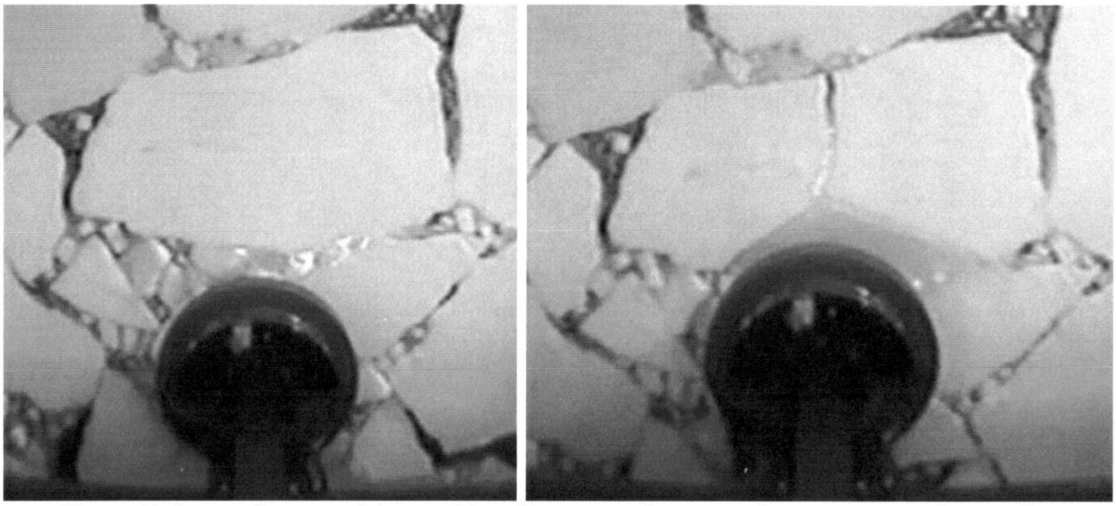

Figure 7. Image from model test video. Encounter between the structure and a small rectangular ice floe of 69x34m (HSVA 2010b)

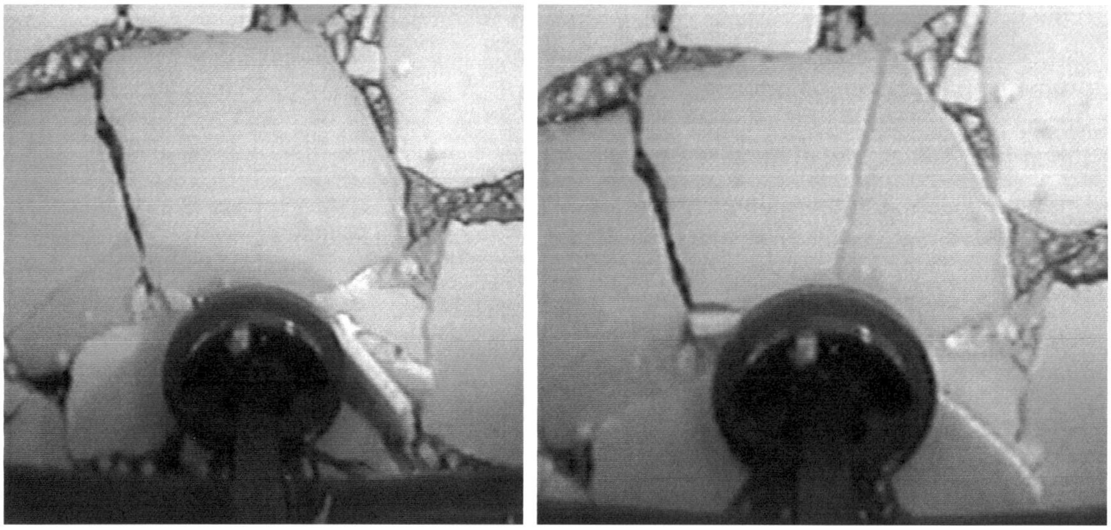

Figure 8. Image from model test video. Encounter between the structure and a small square ice floe of 55x55m (HSVA 2010b)

The measured force on the structure as function of time is presented in Figure 9 for vertical Z-direction and in Figure 10 for horizontal X-direction, in-line with the motion of the structure. For the tests in level ice (blue curve) there is a clear sequential pattern of ice loading building up until the ice fails and reducing again as the ice clears around the structure, both for loading in Z-direction and for loading in X-direction. The difference in height between the peaks indicates natural variations in the failure process. For the large ice floes, these sequential peaks are also observed, but not continuously during the tests. The peaks coincide with events of bending failure and associated clearance during encounters with larger ice floes. Circumferential breaking lengths were approximately 6-10m. Despite the splitting fracture of the ice floe, similar load levels and failure periods as in level ice are achieved, both in X-and Z-direction. It should be noted here that the three tests were executed at different timing, such that the ice properties in the basin were not exactly the same. For the smaller ice floes the sequential failure patterns are not observed. The smaller ice floes break apart almost immediately after encounter with the structure, such that sequential bending failure will not occur. The higher horizontal loading peaks in the plots for the structure in broken ice, are associated with pushing broken pieces of ice against other ice pieces in the basin. The horizontal loading peaks can reach similar levels for all floe sizes. Periodic failure is less likely to occur as the floe sizes get smaller.

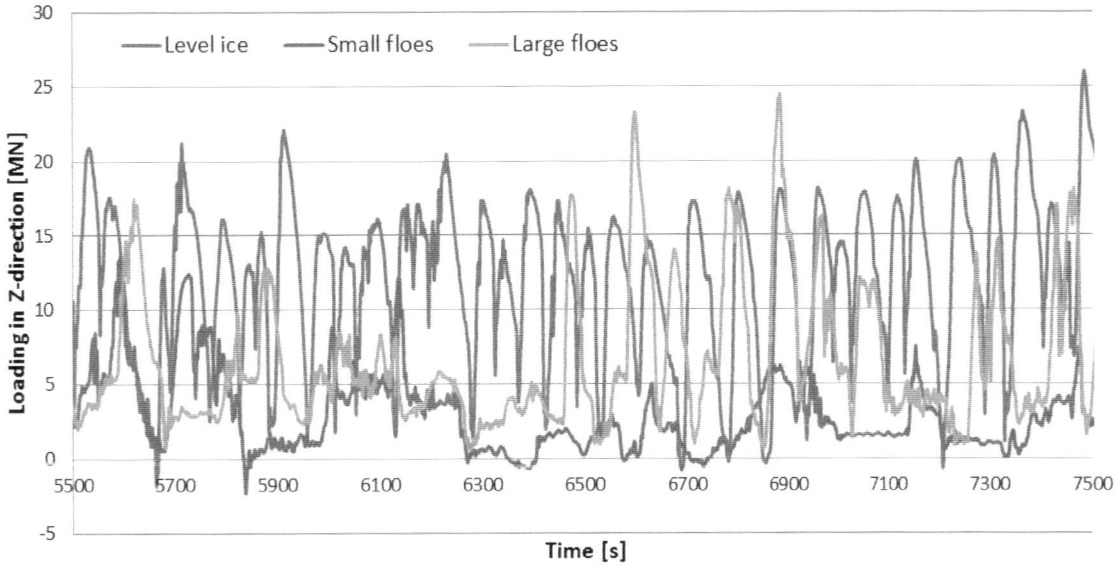

Figure 9. Loading in Z-direction, measured during the model testing campaign for a sloping structure interacting with level ice and ice floes of different sizes

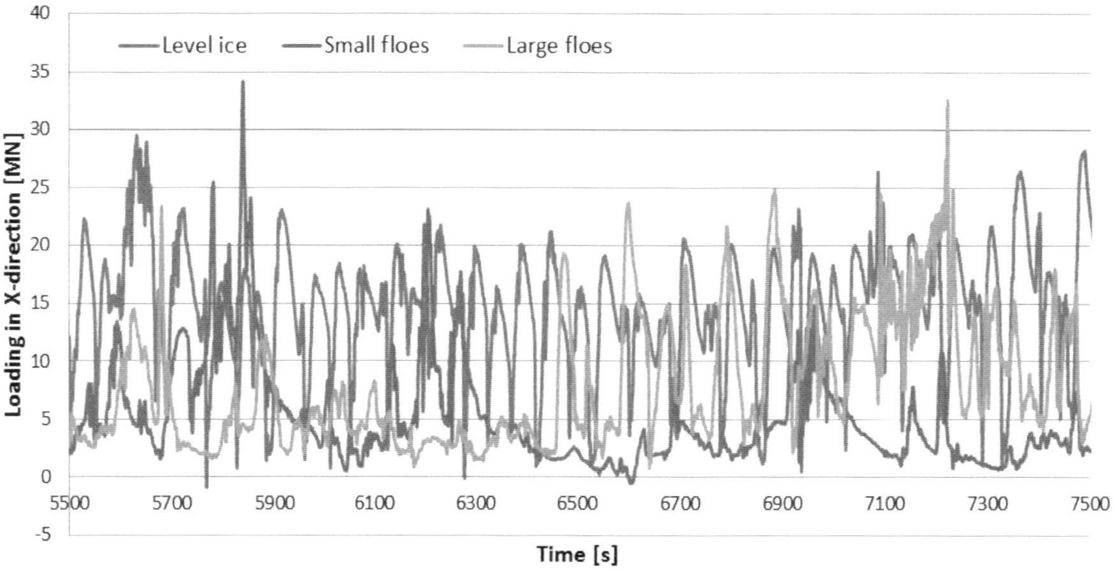

Figure 10. Loading in X-direction, measured during the model testing campaign for a sloping structure interacting with level ice and ice floes of different sizes

The load levels in Figure 9 and Figure 10 are achieved after running through the ice for a longer period of time. At initial encounter with the ice, the load levels are lower, as shown in Figure 11 for level ice. This is due to: 1) at first encounter the structure hits the free ice edge whereas at steady-state, the structure is surrounded by ice, except for its wake and 2) the ice rubble that accumulates underneath the ice near the structure increasing the load that is required to push the ice downward and break in bending, as shown in Figure 12.

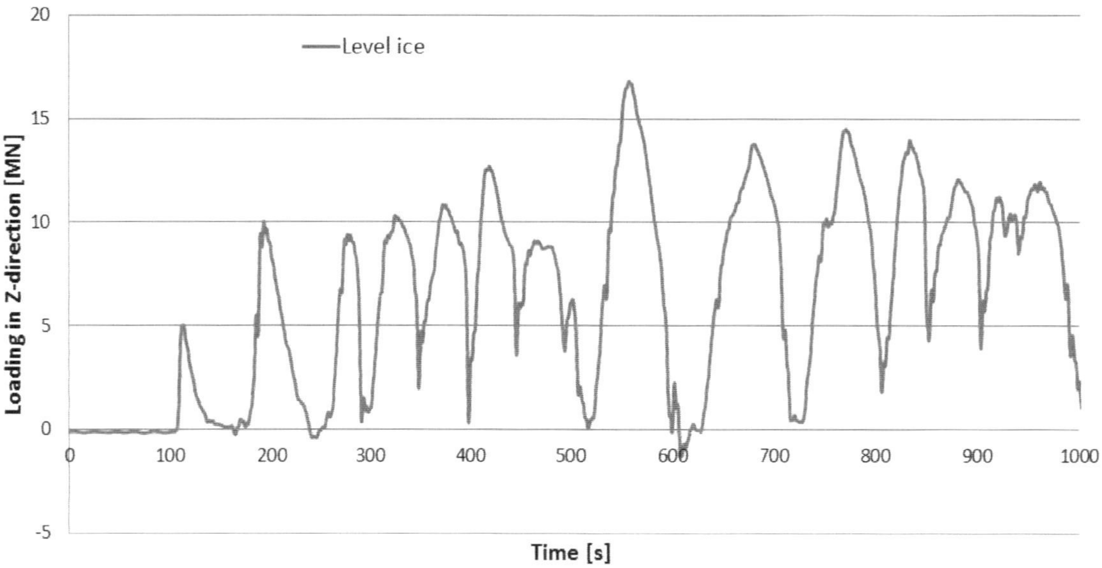

Figure 11. Loading in Z-direction, measured during the model testing campaign for a sloping structure interacting with level ice during the first stage of the test

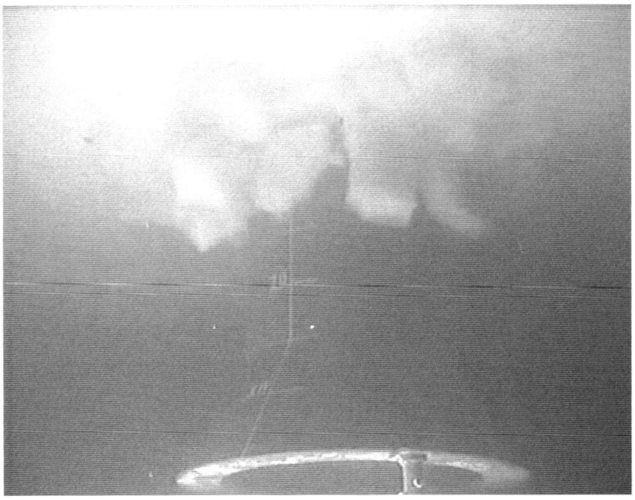

Figure 12. Ice rubble accumulates near the structure, beneath the ice, still from model test video (HSVA 2010a)

COMPARISON OF LATTICE MODEL AND MODEL TEST RESULTS

Interaction between the conical structure and level ice and ice floes of 165x165m, 55x55m and 34x69m is simulated with the lattice model, focusing on the moment of first encountering the ice. The input parameters that are used are the same as in the model tests and presented in Table 1. It is assumed that the structure hits the ice floes at the mid positon of the edge. Confinement is applied at the back of the ice floes, representing confinement from the other ice floes in the basin. A square mesh was selected for the lattice model in this study to keep the model as simple as possible.

Table 1. Input parameters for ice-structure interaction simulations with the lattice model, based on model test input parameters.

Parameter	Quantity
Ice thickness	3 m
Flexural strength	750 kPa
Young's modulus	8.7 GPa
Ice density	846 kg/m3
Waterline diameter of the cone	30 m
Velocity of the cone	0.1 m/s

The resulting fracture patterns of the simulations with the lattice model are presented in Figure 13 with light blue lines. In interaction with the level ice (Figure 13a) bending fracture only occurs near the structure. This is in line with observations from the model tests. A split runs into the ice sheet in the numerical simulations, which was not seen in the modeltests. It could be that the split in the numerical simulations is caused by effects from the boundary or simplifications in the contact model. For the finite size ice floes (Figure 13 b-d) the fracture patterns consist of splitting of the ice floe and bending fracture near the structure, this is in line with observations from the model tests. For the 165x165m ice floe, first a small piece breaks off during the model tests, before the floe splits during the second encounter. In the numerical simulations a split occurs, which does not reach the end of the ice-sheet at the first encounter, but is likely to reach the end of the sheet at the second encounter between the ice edge and the structure. In interaction with the 34x69m ice floe, the ice floe splits apart and two triangular-shaped pieces of ice break off. In the numerical simulation a similar failure pattern is observed. The numerical pattern is slightly mesh-dependent as can be seen by the straight fracture lines along the mesh near the structure. This could be avoided by applying randomness in nodal locations or local strength or stiffness properties in the mesh. For the 55x55m ice floe the ice fails near the structure and splits apart. In the model tests the structure hits the ice floe to the side and under an angle with the x-axis, which causes a piece of ice to break off first, before the ice splits.

The maximum load the structure imposes on the ice for the case with level ice and the 165x165m ice floe is 3.6MN. This is of the same order of magnitude, but lower than the loading of 5MN that was observed during the model testing for the first encounter with the ice, where no rubble was present ahead of the structure and the structure encountered the ice at the edge. Uncertainties and local variations in ice strength and contact simulation may be a source of these differences. A lower load of 1.6MN for the 34x69m ice floe and 1.3MN for the 55x55m ice floe are simulated. This is in line with the model tests, where lower loads were found for encounter with smaller ice floes. The build-up time of the load in the simulations is similar to the build-up time in the model tests; a few seconds. The ice clearance phase around the structure is not simulated.

The breaking radius of circumferential fracture in the level ice and 165x165m ice floe is approximately 11m in the simulations with the lattice model. This is a bit higher than the 6-10m that was observed in model testing. Including local imperfections (lower strength of connections or weaker connections) could make the breaking length estimate more accurate. Mesh refinement near the structure may also lead to more accurate failure length calculation. The triangular piece that breaks of in the 34x69m ice floe is a bit wider in the model tests than in the numerical simulation. The ice rubble that was present under the ice in the model test, but was not simulated numerically, could play a role in discrepancies between numerical results and model tests.

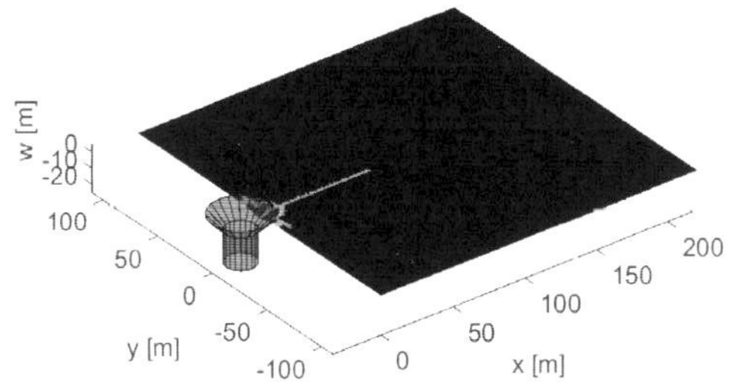

a.

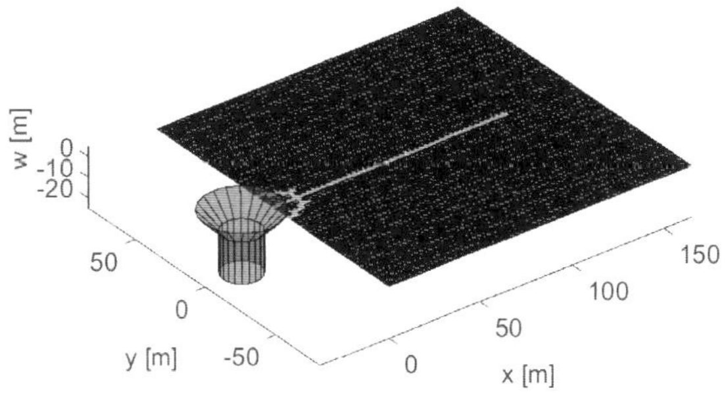

b.

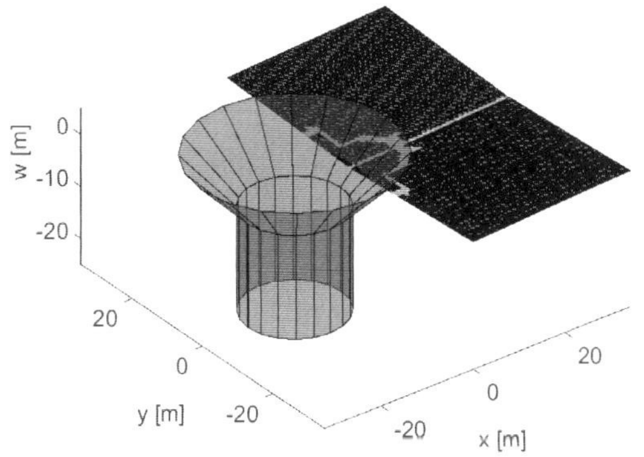

c.

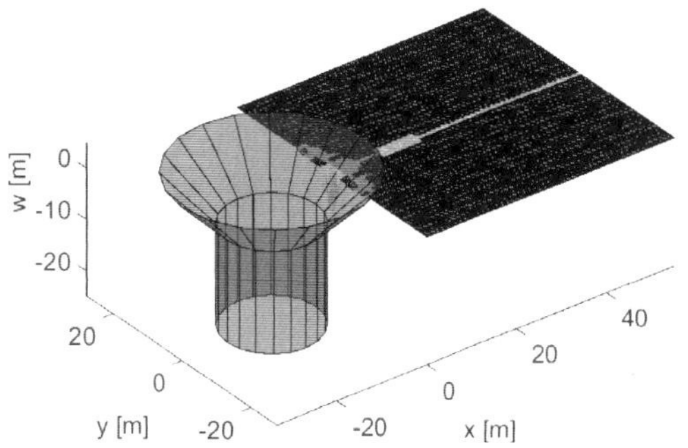

d.

Figure 13. Fracture patterns of interaction between the lattice model and the structure. a) level ice b) 165x165m ice floe c) 34x69m ice floe d) 55x55m ice floe

To simulate natural variations in the ice, the stiffness of each spring in the mesh was changed by adding random fluctuation using a normal distribution with zero mean and unit variance, multiplied with 1/10th of the original stiffness. The nodes of the lattice were displaced by adding random fluctuation using a normal distribution with zero mean and unit variance, multiplied with 1/16th of the distance between the nodes. Results are presented in Figure 14 for the 69x34m ice floe. The fracture pattern is now influenced by the irregular mesh and initially a piece of ice breaks off towards one side. The failure load in the numerical simulation is 1.7MN. The randomness in the model could be further optimized to improve accuracy of failure predictions with the lattice model.

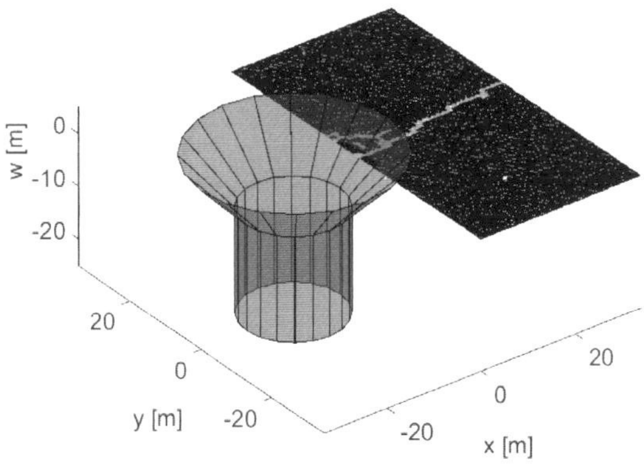

Figure 14. Fracture patterns of interaction between the lattice model for a 34x69m ice floe and the structure with randomized stiffness of the connections.

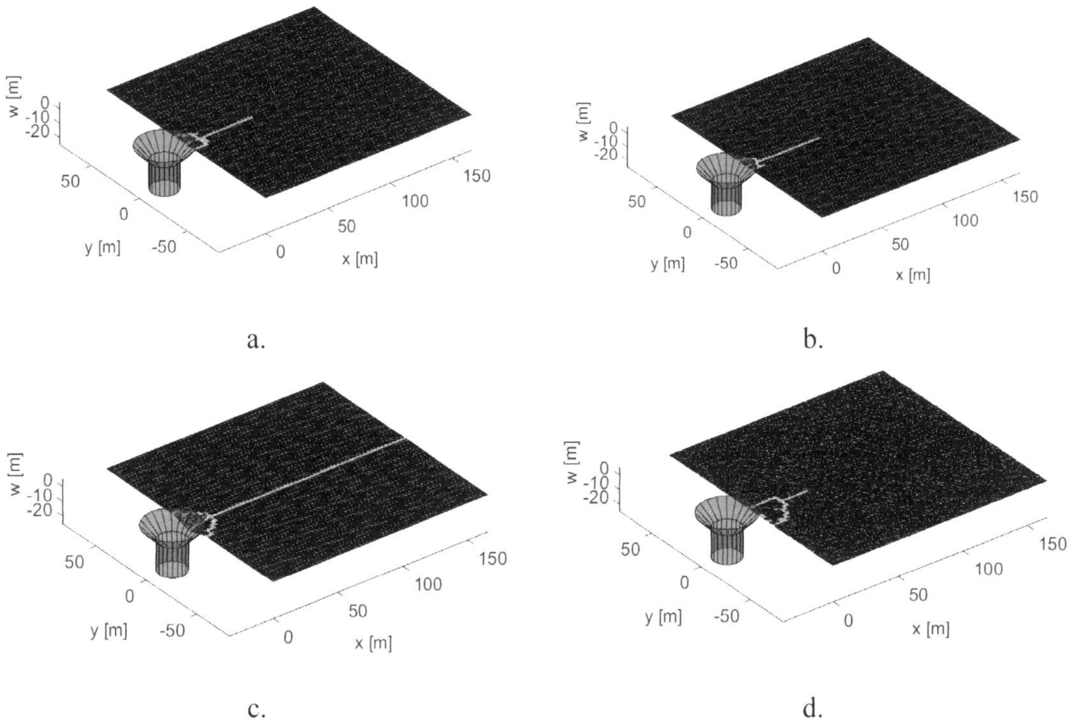

Figure 15. Fracture patterns of interaction between the lattice model and the structure for 165x165m ice floe. Sensitivity checks for a) Young's Modulus 2GPa b) Ice thickness 2m c) Crushing strength 0.1MPa d) Randomized mesh

Four sensitivity checks are done for the 165x165m ice floe. The fracture patterns are presented in Figure 15. First the Young's Modulus is reduced to 2GPa, which reduces the length of the split. The maximum load on the structure remains 3.6 MN. Next, the ice thickness is reduced to 2m and again reduction the length of the split is observed, and the maximum load reduces to 1.5MN. Then the crushing strength of the ice is reduced from 2MPa to 0.2MPa causing the ice plate to split over its full length. The failure load slightly increases to 3.9 MN and the failure process takes longer. Finally, the effect of randomization of the ice is tested for the large ice floe, resulting in a different fracture pattern; the split does not follow the mesh in a straight line. The failure load for this realization is 2.1MN.

The lattice model is capable of simulating failure in bending, splitting and combinations of these. The deformation and fracture algorithm are based on first principles, enabling correct simulation of physical processes without any calibration to the model. Currently sequential failure cannot be simulated with the model as a clearing mechanism of the ice blocks after failure has not been developed. Remaining differences in failure pattern and ice load between model test results and lattice model simulations could be reduced by making further improvements such as local mesh refinement, optimized mesh randomization and an improved contact model. The numerical lattice model could be linked to existing numerical models, for example discrete element method, to make optimum use of existing codes and developments.

CONCLUSIONS

A numerical lattice model has been used to simulate level ice behavior and its capabilities of predicting fracture patterns and loads have been compared to results from model testing. Deformation and failure criteria in the 2-dimensional lattice model are based on first principles, enabling sound simulation of physical processes. In addition, the discrete lattice model avoids stress singularities in fracture modeling since the model only describes displacements and forces in the connections and no stresses are present. By means of numerical assessment, a multitude of structures and loading conditions can be analyzed and structural shapes can be optimized for interaction with ice. Ice management processes can be optimized in case the effects of managed floe sizes on the loading of a structure are known.

In model tests with a conical structure in 3m thick level ice and ice floes of different sizes it was found that for level ice and larger ice floes the floes fail in bending near the structure, whereas the smaller ice floes split apart in interaction with the structure. Sequential loading peaks are observed in interaction with level ice and larger floe sizes, but not for smaller floe sizes. Load on the structure in vertical direction is smaller in interaction with smaller floes. Horizontal loading, however, can reach similar levels for all floe sizes, in case of high confinement, as pieces of ice are pushed against other ice pieces forming load bridges.

Interaction between ice and a downward sloping conical structure is simulated with the lattice model and compared with observations from the model tests, focusing on the first encounter with the ice floe. The lattice model is able to predict whether an ice floe splits or whether failure occurs near the structure in bending. Slightly lower load levels are found in simulations with the lattice model. A sensitivity study is done to analyse the influence of heterogeneities by varying the stiffness of the connections in the lattice model and nodal locations, which could make the lattice model less mesh sensitive. The fracture pattern becomes less senstitive to the mesh, but further improvements to the randomization process are required. Other additions such as ice clearance mechanisms, local mesh refinement, optimized randomization and an improved contact model would enable more accurate fracture pattern predictions and maximize the efficiency of the lattice model. The numerical lattice model could be linked to existing numerical models, for example a discrete element method, to make optimum use of existing codes and developments.

ACKNOWLEDGEMENTS

The authors would like to thank all partners in the "Ice model testing of structures with a downward breaking cone in the waterline" JIP. Initiated and led by Aker Solutions, these tests took place at HSVA in Hamburg, Germany in 2008 and 2009. The project was funded by the Norwegian Research Council (DEMO2000), Aker Solutions, ExxonMobil, Chevron, StatoilHydro and Shell. It was carried out in cooperation with the Norwegian University of Science and Technology (NTNU). Furthermore, the authors are grateful for the support of this research project by Shell International Exploration and Production Inc. and Delft University of Technology as well as the support by Dr. G.L. Kuiper as reviewer.

REFERENCES

Aksnes, V., 2010. A simplified interaction model for moored ships in level ice. *Cold Regions Science and Technology*, 63(1-2), pp.29–39.

van den Berg, M., 2016. A 3-D Random Lattice Model of Sea Ice. In *Arctic Technology Conference*.

Bonnemaire, B. et al., 2014. Ice breaking and accumulation around a moored structure: Ice basin tests and numerical simulations. In *Society of Petroleum Engineers - Arctic Technology Conference 2014*.

Bruun, P.K. et al., 2011. Ice Model Testing of Structures With a Downward Breaking Cone. In *OMAE*.

Dorival, O., Metrikine, A.V. & Simone, A., 2008. A Lattice Model to Simulate Ice-Structure Interaction. In *ASME 2008 27th International Conference on Offshore Mechanics and Arctic Engineering, Jun 2008, Estoril, Portugal*. pp. 989–996.

HSVA, 2010a. *JIP Phase II Ice Model Testing of Structures in Ice*,

HSVA, 2010b. *JIP Phase II Model test videos*,

Ji, S., Di, S. & Liu, S., 2015. Analysis of ice load on conical structure with discrete element method. *Engineering computations: International Journal for Computer-Aided Engineering and Software*, 32(4), pp.1121–1134.

Lau, M., Lawrence, K. & Rothenberg, L., 2008. Discrete element analysis of ice loads on ships and structures. In *ICETECH*.

Lubbad, R. & Løset, S., 2011. A numerical model for real-time simulation of ship-ice interaction. *Cold Regions Science and Technology*, 65(2).

Metrikin, I. et al., 2015. Experimental and numerical investigation of dynamic positioning in level ice. *Journal of Offshore Mechanics and Arctic Engineering*, 137(3).

Nevel, D.E., 1992. Ice forces on cones from floes. In *Proceedings of IAHR '92, 11th International Symposium on Ice Problems 3*. pp. 1391–1404.

Sayed, M., 1997. Discrete And Lattice Models of Floating Ice Covers. In *The Seventh International Offshore and Polar Engineering Conference*. pp. 428–433.

Suiker, A.S.J., Metrikine, A.V. & de Borst, R., 2001. Dynamic behaviour of a layer of discrete particles, part 1: analysis of body waves and eigenmodes. *Journal of Sound and vibration*, 240(1), pp.1–18.

Wille, S.F., Kuiper, G.L. & Metrikine, A.V., 2011. On the dynamic interaction between drifting level ice and moored downward conical structures: A critical assessment of the applicability of a beam model for the ice. In *Proceedings of the ASME 2011 30th International Conference on Ocean, Offshore and Arctic Engineering*..

.

The Attributes of Local Ice Pressure Analyzed by Discrete Element Method

Xue Long[1], Shunying Ji[1]
[1] State Key Laboratory of Structural Analysis for Industrial Equipment, Dalian University of Technology, Dalian 116023, China

ABSTRACT

The local ice pressure of the ice-structure interaction plays an important role in structure design. The discrete element method (DEM) with bonded particles is adopted to simulate ice pressures during the crushing failure of level ice interacting on the vertical structure. The major attributes of ice pressure are explored to compare with existing empirical models, including pressure distribution and the relationship between the global and local pressures. The results show that the high-pressure zones (HPZs) always concentrate on the horizontal middle line of contact area. In this case, the maximum local pressure exceeds 100 MPa, while the maximum global one is only 3 MPa. The influence of structure width on the local pressure is analyzed to compare with the ISO. The ice pressure always decreases with the increasing of contact area which can be expressed as a pressure-area plot. Moreover, as the loading rate increases, the failure mode of sea ice changes from ductile failure to brittle failure in the numerical simulations. Therefore, the numerical analysis of local pressure provides a reference for stiffening elements of vertical structures.

KEY WORDS: local pressure; sea ice; DEM; crushing failure.

INTRODUCTION

The analysis of ice pressure during the action between level ice and vertical structure is of great significance to the structural design. Generally, the ice pressure is divided into the global ice pressure and the local ice pressure. The global pressure (P_G) is the ice pressure averaged over the nominal contact area associated with the global action which is important for overall structure design and stability. The local pressure (p) is the pressure acting on some local and small zones relative to the whole contact area. These zones with stress concentrations are produced by the non-simultaneous destruction of ice cover interacted with the vertical structure, which is called the high-pressure zone (HPZ) (Masterson et al., 1993). Many studies have shown that the local pressure is much greater than the global pressure (Frederking, 2004). It indicates that local ice pressure plays an important role in structural failure, which could be adopted to strengthen the local structural elements, for instant the plating between frames.

It is found that the high pressure zone is mainly distributed along the centerline of the ice-structure contact area which is called 'line-load' contact geometry during the crushing failure of ice (Dempsey et al., 2001). The contact area between ice and structure is related to the ice pressure directly. It has been widely recognized that the ice pressure decrease with the increasing contact area in the research field of sea ice characteristics. Masterson et al. (2007) proposed a pressure-area plot for local pressures based on the data from the Molikpaq offshore platform, indenter tests and some flat jack tests by $p=7.4A^{-0.7}$ for areas up to 10 m^2, where p is the pressure (in MPa) and A is the contact area (in m^2). This relationship has been employed into the ISO 19906 (2010).

Most of the researches on ice pressure are based on the measured and experimental results, but the related numerical simulation methods are less applied. The finite element method is also used in the ice pressure distribution of the slope structure by Paavilainen and Tuhkuri (2013). Considering that the numerical simulation is required by the accuracy and time of calculation, the discrete element method (DEM) based on GPU high performance algorithm has certain advantages in simulating ice-structural interactions. This method with a high computational efficiency has an advantage on describing the failure process of sea ice realistically (Ji et al., 2015).

In this paper, the discrete element method with the bonding and crushing functions is adopted to simulate the ice-structural interaction. The simulated results of local pressure distribution and ice load are compared with the Frederking (2004)'s filed experiments to verify the rationality of the DEM. The influences of contact areas and load speeds on local ice pressures were analyzed. The comparison with ISO standard indicates that DEM is suitable to study the attributes of ice pressure.

PARALLEL-BONDING MODEL

The DEM with the parallel-bonding model is adopted to simulate sea ice as shown in Figure 1 (Potyondy and Cundall, 2004). An elastic bonding disk between two particles is set to transfer both forces and moments. The parallel bond can be regarded as elastic springs with constant normal and shear stiffness, which is distributed over a circular disk lying on the contact plane and centered at the contact point.

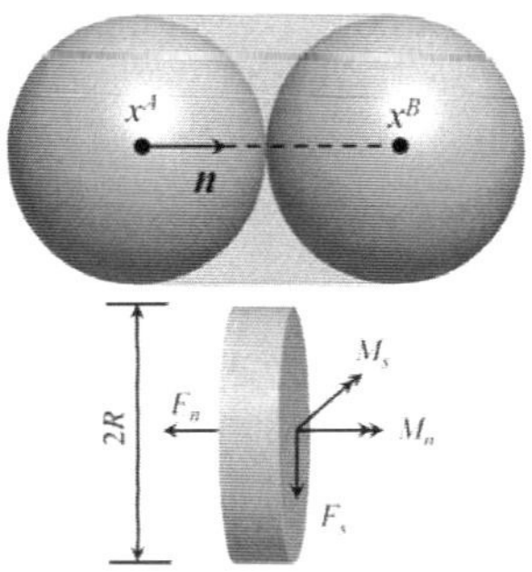

Figure 1. Bonding model between two spherical particles

The force and moment in normal direction and shear direction associated with the parallel bond are denoted by F_n, F_s, M_n and M_s. The tensile and shear stresses acting on the bonding disk are calculated based on the beam theory as

$$\sigma_{max} = \frac{F_n}{A} + \frac{|M_s|}{I} R, \quad \tau_{max} = \frac{|F_s|}{A} + \frac{|M_n|}{J} R \tag{1}$$

where the variables A, I and J are given by

$$A = \pi R^2, \quad J = \frac{1}{2}\pi R^4, \quad I = \frac{1}{4}\pi R^4 \tag{2}$$

If tensile stress exceeds the normal strength, or shear stress exceeds the shear strength, then the parallel bond breaks.

COMPARISON WITH A FIELD TES

Ice-structure Interaction

A mass of field measures and tests show the existence of high pressure zones. The field test of Frederking (2004) is simulated to validate the DEM mode. In the field test, an indenter with 1.5m width and 0.5m height is driven to act with a sea sheet with 168 mm thickness. As the test data, the strength of ice is about 2.46 MPa. The sensor elements over the contact face are adopted to measure the ice load and local pressures. A DEM simulation is designed according to the test data and the major DEM parameters are listed in table 1. In this simulation, the ice cover with 0.8 m length and 2 m width is composed of regularly arranged spherical particles. The indenter is regarded as a rigid body with a load speed of 3 mm/s.

The failure process of ice acted on the indenter is presented in Figure 2 which indicates that the sea ice will be evidently accumulated in front of the structure. These accumulated ices have little direct influence on the ice load on indenter because of the lower compressive strengths. However they have a vertical squeezing effect on the ice cover that can make the particles in middle cover bonded stronger during the crush.

Table 1. Major parameters in DEM

Definition	Symbol	unit	value
Ice elastic modulus	E	MPa	10^9
Normal stiffness	k_n	N/m	2.5×10^6
Shear Stiffness	k_s	N/m	2.5×10^5
Bonding strength	σ_b	MPa	0.6
Particle friction	μ_p		0.2
Particle restitution	e_p		0.3
Structure friction	μ_w		0.15
Structure restitution	e_w		0.3

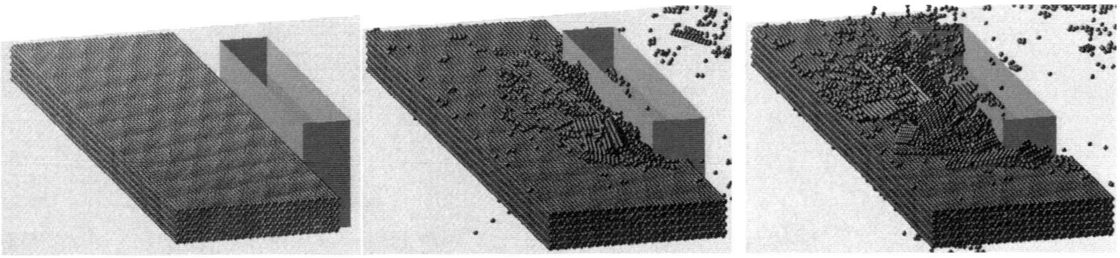

(a) $t = 0$s (b) $t = 50$s (c) $t = 100$s

Figure 2. The failure process of ice-structure interaction at different time

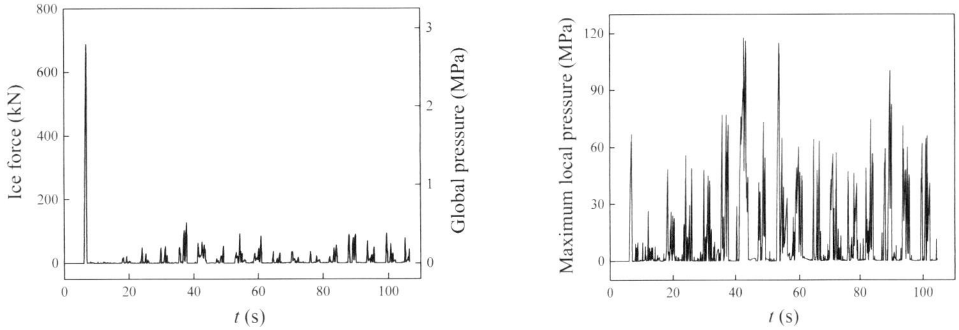

(a) (b)

Figure 3. The ice-force curve of ice-structure interaction

At the early stage of the ice-structure interaction, the level ice constituted by bonding particles has a great compressive strength to prevent the crushing failure. Hence the first force peak is about 710 kN in Figure 3(a) which is evidently larger than others. This strong impact makes the ice break into small pieces, so that the ice force keeps quiet low in later 15s. Then, the ice force gradually increases and changed by a stable period of 5s. During this period the maximum ice force is 125 kN. These attributes of ice force curve in the simulation are in good agreement with the experimental data. Meanwhile the global ice pressure as the ratio of ice force to contact area has the same trend with the ice-force curve.

Local Pressure Distribution

It is clearly observed the crushing process of sea ice in the interface between sea ice and structure by the numerical simulations. Figure 4 shows the failure process in detail by a vertical cross section of sea ice and structure. During this interaction, the particles will be crushed out of the contact area from up and down edge of ice sheet. In Figure 4(a), the different particle velocities which are represented by the colors indicate the trend of particles' movement. The cracks of ice sheet are always along the slops (two red lines) after the first action on the structure. Then the broken ice pieces accumulate before the vertical structure gradually at t = 100s as shown in Figure 4(b) where the colors mean particle forces. It is found that the inter forces of particles at the middle of ice sheet are higher than other places and acted on the structure more strongly. Hence these simulated results indicate that the high pressure zones on structure are caused by the non-simultaneous destruction of ice sheet during the interaction.

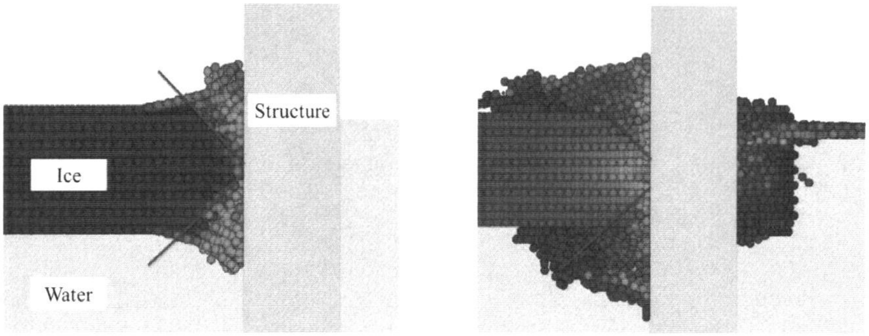

(a) t = 10s (b) t = 100s

Figure 4. The vertical profiles of ice-structure interaction at different time

All the particle forces acting on the structure could be recorded based the pre-draw grids on contact area. The grid size is same with the sensor elements of indenter in the test as 0.01m × 0.01m. The pressures on each element are defined as the local pressures as Figure 5(a) (in MPa). The elements in red are regarded as high pressure zones with a 'line-like' distribution. The distribution along thickness direction in Figure 5(b) also indicates that the ice pressure is mainly provided by the middle of the ice layer, instead of the upper and lower edges.

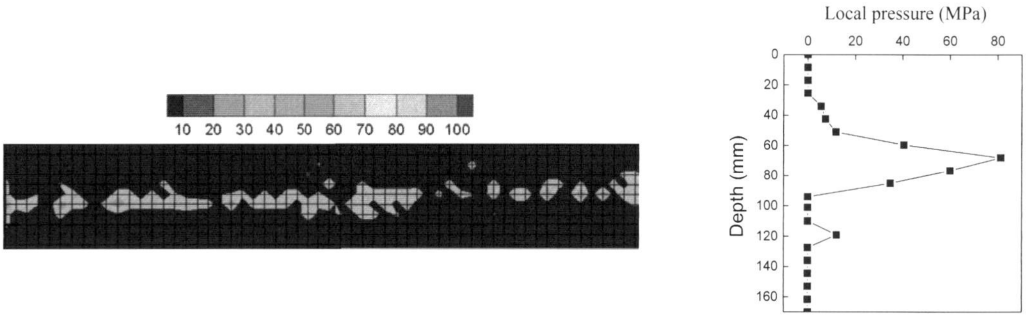

(a) (b)

Figure 5. The HPZs distribution on contact area at 100s

There is a big difference between the local pressure and average pressure as shown in Figure 3. The local pressure in Figure 3(b) is the maximum value of the whole contact area at every time step which is much higher than the globe average pressure in Figure 3(a). The max global pressure is corresponded to the first load peak at 6.7s, while the max local pressure is at 42.6s. There is no local failure for ice until the first crushing action, so the local pressure at this moment is lower relatively.

INFLUENCING FACTORS ON THE LOCAL PRESSURE

Two major factors are considered in this paper, including structure width and ice velocity. It is widely agreed that the ice local pressure (p) always decreases as the area of contact (A) increases which has been adopt in ISO by a curve of $p=7.4A^{-0.7}$. To research this relationship, a comparison is proposed in this paper by DEM simulations where the same ice cover is acted on the structures with different width from 0.5m ~ 1.5m. The local pressures are analyzed by both maximum and mean peak pressure as shown in Figure 6(a). The simulated curve has a similar trend with the ISO as Equation 3, but the local pressures in simulations are about two times of the standard value which is not considered the effect of other factors, such as load speed. The equation in the ISO is not applicable to all ice conditions, especially for thin ice.

$$p = 16.5A^{-0.6} \tag{3}$$

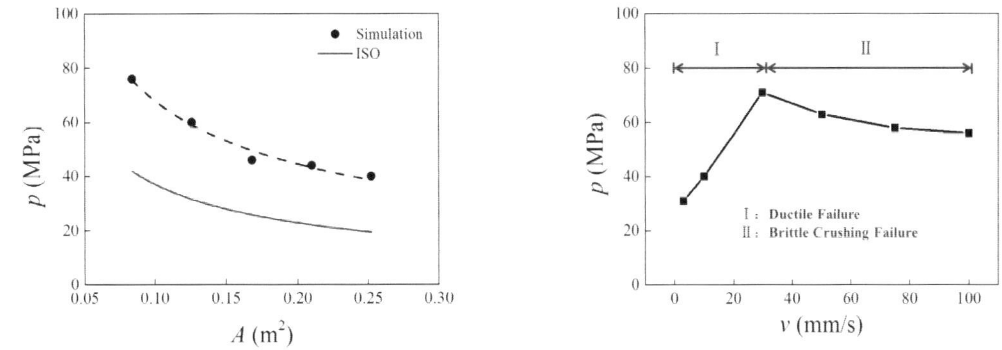

(a) contact area (b) load speed

Figure 6 The contact area and load speed effect on local ice pressure

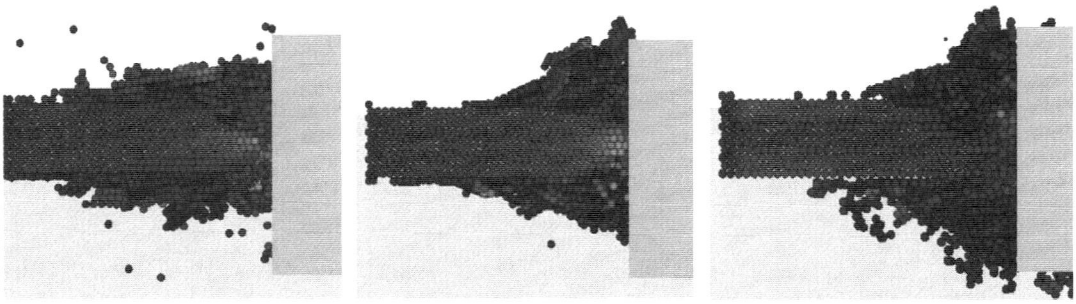

(a) v = 10 mm/s (b) v = 30 mm/s (c) v = 75 mm/s

Figure 7 The failure process of ice with different load speeds

On the other hand the load speeds of ice-structure interactions which are not considered in ISO also affect the ice pressures by changing the failure modes of sea ice (Sodhi, 2001; Jordaan et al., 2010; Timco and Sudom, 2013). Hence some simulated cases in this paper with different velocities of ice (3mm/s ~ 100mm/s) are provided to the effect on local pressures. The result in Figure 6(b) indicates that ice velocity is related to the local pressure. The failure processes in Figure 7 at different load speeds and the same moment, illustrate the effect of the load speed on failure modes of ice. When the load speed is low, the upper and lower surfaces of the ice sheet begin to be broken and the middle particles of ice are still bond and solid. Contrarily, when the load speed is high, all the particles on the interacted face are almost broken at same time. Therefore, the results show that brittle crushing of ice occurs at higher speed and ductile failure occurs at lower speed. The local pressure increases with the velocity during ductile failure linearly. Moreover, ice velocity has little effect on local pressure during brittle failure.

CONCLUSIONS

In this paper, the interaction between sea ice and vertical structures is simulated by constructing the spherical DEM model with the functions of bonding and breaking. A

comparison with field test from Frederking (2004) is provided to validate the rationality of this numerical model. From the DEM results, both the ice force and the distribution characteristics of local pressures reach an agreement the experimental data. Moreover a 'line-like' distribution of the high-pressure zones also could be obtained in this simulation.
Two factors influencing on the local pressures are researched, including contact area and load speed. The local pressure decreases with the increasing area which is consistent with the relationship of ISO. On the other hand, the load speed has a positive influence on local ice pressures which could lead to the change of the failure modes. The DEM simulations could be applied in the study of ice pressure to reinforce the structures reasonably. Meanwhile some other factors such as the ratio of structure width and ice thickness, and ice strength, are necessary for detailed analysis of ice pressure.

ACKNOWLEDGEMENTS

The authors gratefully appreciate the financial support provided by the National Natural Science Foundation of China (Grant No. 41176012, 41576179).

REFERENCES

Dempsey, J.P., Palmer, A.C., Sodhi, D.S., 2001. High pressure zone formation during compressive ice failure. *Engineering Fracture Mechanics*, 68(17-18), pp.1961-1974.

Frederking, R.M.W., 2004. Ice pressure variations during indentation. In: Saint Petersburg, Russia, *Proceedings of the 17th IAHR International Symposium on Ice*, pp. 307-314.

ISO, 2010. Petroleum and Natural Gas Industries—Arctic Offshore Structures. International Organization for Standardization, ISO 19906, Geneva, Switzerland.

Ji, S., Di, S., Liu, S., 2015. Analysis of ice load on conical structure with discrete element method. *Engineering Computations*, 32(4), pp.1121-1134.

Jordaan, I., Bruce, J., Dan, M., Frederking, R., 2010. Local ice pressures for multiyear ice accounting for exposure. *Cold Regions Science and Technology*, 61(2-3), pp.97-106.

Liu L., Sun S., Ji S., 2016. Interaction Between Floater and Sea Ice Simulated with Dilated Polyhedral DEM. In: Dalian, China, *Proceedings of the 7th International Conference on Discrete Element Methods*, Volume 188 of the series Springer Proceedings in Physics, pp.1065-1074.

Masterson, D.M, Frederking, R.M.W., Wright, B., et al., 2007. A revised ice pressure-area curve. In: Dalian, China, *Proceedings of the 19th International Conference on Port and Ocean Engineering under Arctic Conditions*, pp.305-314

00Masterson, D.M., Frederking, R.M.W., Jordaan, I.J., Spencer, P.A., 1993. Description of multi-year ice indentation tests at Hobson's Choice Ice Island — 1990. In: Glasgow, Scotland, *Proceedings of the 12th OMAE Conference*, Volume 4 of American Society of Mechanical Engineers, pp.145-155.

Paavilainen, J., Tuhkuri, J., 2013 Pressure distributions and force chains during simulated ice rubbling against sloped structures. *Cold Regions Science and Technology*, 85, pp.157-174.

Potyondy, D.O., Cundall, P.A., 2004. A bonded-particle model for rock. *International Journal of Rock Mechanics and mining Sciences*, 41(8), pp. 1329-1364.

Sodhi D S. Crushing failure during ice-structure interaction. *Engineering Fracture Mechanics*,

2001, 68(17), pp.1889-1921.

Timco, G.W., Sudom, D., 2013. Revisiting the Sanderson pressure–area curve: Defining parameters that influence ice pressure. *Cold Regions Science and Technology*, 95(11), pp.53-66.

Some Limitations of Using Saline Ice in Lab Tests to Study Structure Interaction with Sea Ice at Full Scale

Robert E. Gagnon[1]
[1] Ocean, Coastal and River Engineering, National Research Council Canada, St. John's, NL, Canada

ABSTRACT

Studies show that crushing of small-scale freshwater ice samples against smooth platens exhibits high-amplitude sawtooth load patterns (HASLP), due to spalling. For comparable ice temperatures and crushing rates it is known that sea ice sheets encroaching on structures (e.g. Molikpaq) can produce HASLP. Hence, small-scale freshwater ice tests are useful for studying the large-scale phenomenon. The present small-scale crushing tests on lab-grown saline ice, however, produced no HASLP at any crushing rate for temperatures higher than –16 °C. Therefore small-scale saline ice tests may not be suitable to study large-scale HASLP. Some differences in behaviour of the two ice types is attributable to brine channels/pockets within the saline ice matrix that tend to make the ice more compliant at the small scale, where there is less confinement, than at field scale. Similarly, when Blade-Runners platens, that had regular arrays of small prominences, were used for the saline ice tests no ice spallation at individual blades was observed, whereas for previous freshwater ice tests small spalling events at the blades were prevalent and were the reason why the technology reduces HASLP. At the lab scale, where the individual blade dimensions and spacing were comparable to that of the size and spacing of brine channels/pockets in the saline ice, the compliance of the ice in the vicinity of blades may limit non-uniform local stresses to values lower than that needed to nucleate the small fractures/spalls at the blades. This suggests that full-scale experiments are required to properly test the technology for sea ice.

KEY WORDS: High-amplitude sawtooth load patterns; Spalling; Saline ice crushing tests; Scale effects; Blade Runners technology.

INTRODUCTION

Many studies have been successfully performed in the lab to gain insight into the behaviour of ice when it interacts with structures at full scale, including those focused on ice crushing characteristics and ice induced vibration (IIV) (e.g. Määttänen et al., 2011; Gagnon, 2008; Sodhi, 1991; Evans et al., 1984). In the latter case the most widely known and studied events are those associated with the Gulf Canada Resources Ltd. Molikpaq caisson facility that occurred in 1986 during operations at the Amauligak I-65 site in the Canadian Beaufort Sea. IIV is caused by cyclic loading that is due to repetitive spalling from hard-zone regions, relatively intact ice, at the ice contact zone (Gagnon, 2012). While much has been learned about the full-scale phenomena from small-scale laboratory experiments using freshwater ice samples, we have seen in the present saline ice crushing experiments that considerable caution should be exercised before assuming that results from small-scale saline-ice lab tests will also manifest at full scale for sea ice at similar temperatures and crushing-rates. This paper discusses scale and geometry related reasons for the differences in ice behavior and further points out why a new technology for reducing ice-induced vibration, known as Blade Runners, should be tested under full-scale sea ice conditions rather than in small-scale lab tests using saline ice.

Figure 1. Photograph of the acrylic platen with the array of small regular square pyramids on its surface (platen dimensions: 166 mm x 129 mm x 25 mm). The pyramids were 1 mm in height and 2 mm wide at the base. The space between each adjacent pyramid was 2 mm. The arithmetic average of the high-roughness profile for the surface of the platen was 0.075 mm. From Gagnon (2016).

EXPERIMENTAL SETUP

To test the Blade Runners concept two styles of crushing platens were fabricated, with square columns and square pyramids blade shapes. Figure 1. is a photograph of the crushing platen with square pyramids. The pyramids and columns were 1 mm in height and 2 mm wide at the base. The space between adjacent blades was 2 mm. In previous tests using freshwater ice samples (Gagnon, 2016) these two platens were the best performers in terms of reducing large-amplitude cyclic loading due to spallation. All platens, except one made of aluminum, were made of acrylic. The ice crushing tests were performed in NRC-OCRE's Cold Room facility. The initial plan was to crush samples of ice against platens with blade arrays and compare those results with those from crushing experiments using a flat platen with no blade array.

2 shows the essentials of the test setup where the ice, confined in a rigid holder, is pressed in the vertical direction at a fixed rate against a transparent acrylic crushing platen (2.5 cm thickness) in a testing frame. The crushing platen is backed by a secondary polished acrylic support plate (5 cm in thickness). The ice holder is attached to the top face of a 2.5 cm thick metal plate. For the acrylic platens high-speed images of the ice-platen contact region were

recorded by viewing through both the support plate and the crushing platen by means of a mirror situated between the supporting posts for the plate and platen.

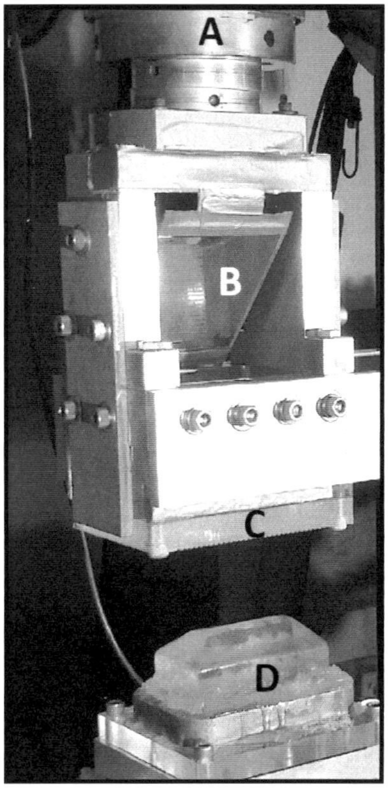

Figure 2. Essentials of the ice-crushing test setup, from an earlier program. (A) Vertically-oriented test-frame load cell for measuring the normal load; (B) Mirror; (C) Acrylic crushing-platen; (D) Ice specimen in ice holder.

Ice samples were prepared from the saline ice grown in the lab. During tests the columnar grains (roughly 3 mm in diameter) were orthogonal to the vertical crushing direction and the long axis of the ice sample. Each ice sample was approximately 7 cm in height, and 12 cm by 6.5 cm at its base. The top of the samples was given a rounded-wedge shape. Each sample's base was freeze-bonded to an ice holder consisting of an acrylic plate with a rectangular band of steel (2 cm in height) attached to it that encompassed the base of the ice specimen. The ~ 1 cm gap between the ice sample and the confining steel band was filled with snow and then saturated with water near 0 °C so that, when frozen, it provided confinement at the base of the ice sample to prevent it from shattering during testing. Tests were carried out at –10 °C, –16 °C and –21 °C. The load data from the vertical crushing actuator were acquired at 6144 samples per second. The vertical crushing rates used for these tests were 5, 10, 20 and 40 mm/s. Ice samples were crushed to a depth of ~ 35 mm.

Figure 3. Thin section of the ice showing the initial vertical columnar structure that became disrupted and somewhat jumbled (about half-way down the image). The markings on the

Figure 4. Same thin section as in Figure 3., with side lighting and without using polaroid filters, showing brine and air filled channels and pockets in the ice. The section is 1.0 mm thick. The markings on the scale at the left are mm
scale at the right are mm. The section is 0.5 mm thick.

The purpose of the test program was to test the efficacy of the Blade Runners technology for reducing ice induced vibration of structures in small-scale experiments using lab-grown saline ice. The ice was grown in a cold room in a basin that was initially filled with fresh water and cooled to ~ 2 °C. Then commercial 'sea salt' was added to obtain a concentration of ~16 ppt. The basin and contents were further cooled by setting the air temperature to – 16 °C. Just prior to the initiation of freezing, the surface of the liquid in the basin was seeded with granular freshwater ice with a grain diameter of approximately 3 mm. Columnar-grained ice started to grow within minutes of seeding. After a certain period of freezing (approximately 24 hours), when the ice was about 6.0 cm thick, the basin was inadvertently disturbed/shaken somewhat during ice thickness measurements. Consequently, the uniform columnar ice structure changed to a more jumbled structure because vertically-oriented flakes at the bottom of the growing ice were dislodged and changed their orientation thereby altering the fabric and texture of the ice that grew following the disturbance (Figure 3.). This attribute of the ice proved to be beneficial, as will be seen below. The salinity of the ice was approximately 5.2 ppt, that is typical for first year sea ice (e.g. Dempsey et al., 1999). Figure 4. shows brine and air filled channels/pockets in the same ice section shown in Figure 3.

RESULTS AND DISCUSSION

As mentioned, the original intention of the test program was to investigate the efficacy of the Blade Runners technology using saline ice. At the start it was decided to test some saline ice samples by crushing them against a flat platen at –10 °C and rate of 10 mm/s (typical conditions for previous testing of the technology (Gagnon, 2016) to establish a baseline of high-amplitude cyclic sawtooth load patterns (HASLP) that could be used for later comparisons with records using the Blade Runners platens. But these initial tests did not exhibit the expected HASLP (see Figure 5.), in spite of the fact that sea ice sheets crushing against structures (e.g. Molikpaq, Figure 6.) at similar temperatures and rates do exhibit HASLP. However, when tests were conducted at –16 °C HASLP did occur (Figure 5.). Fortunately the high-speed imaging system captured visual data (500 – 1500 images/s) that helped explain the difference in results at the two temperatures. This will be discussed below. Having established conditions where HASLP would occur for crushing against a flat platen, the next tests were conducted at –16 °C using the Blade Runners platens, where one platen had square-column blades and the other had square pyramids (described above). In these experiments the HASLP was not affected, indeed the load patterns looked the same as that obtained with a flat platen, i.e. as though the blades were not present (Figure 7.). Subsequently, in attempts to get the desired effect from the blades, i.e. nucleation of small-scale spalls at the blades (the key mechanism of the technology), the temperature was lowered to –21 °C and more experiments were performed for a variety of crushing rates (5, 10, 20 and 40 mm/s). HASLP persisted for all these tests. Even when platens (one acrylic and one aluminium) with blades that had twice the height of the original blades were used the results still showed HASLP, that is, the blades were not performing as anticipated. We note that with respect to the Blade Runners technology the ice behaviour did not appear to depend on whether the hard-zone region, that wandered upwards and downwards in the view during the course of an experiment, encompassed the columnar structure of the ice or the jumbled structure mentioned above (Figure 3.). A similar observation was made during earlier tests using columnar and granular freshwater ice samples (Gagnon, 2016).

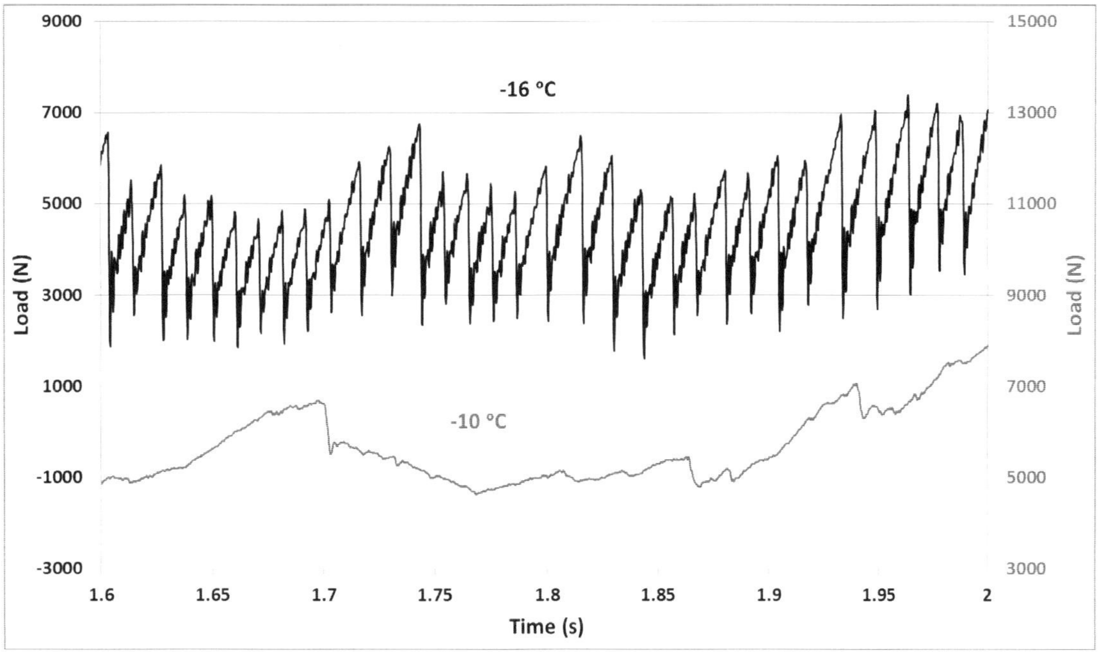

Figure 5. Comparison of load data from two tests using saline ice samples that were crushed against a flat platen where one test was conducted at –10 °C and the other at – 16 °C.

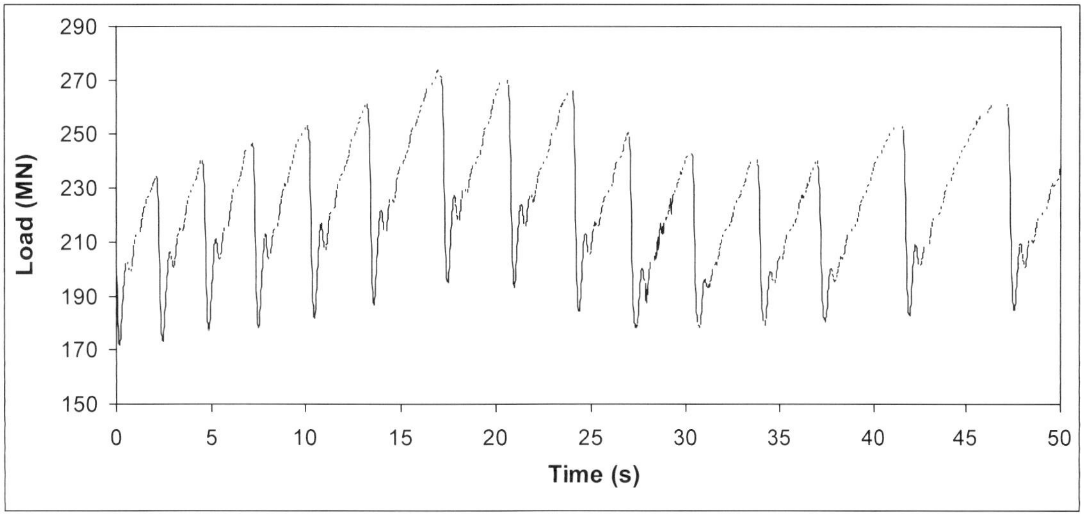

Figure 6. Full-scale load record illustrating a high amplitude sawtooth load pattern (HASLP) that occurred when a sea ice sheet encroached on the Molikpaq structure in 1986. The cyclic pattern is due to ice spalling events that regularly occur in the ice/structure contact zone, as described by Gagnon (2012).

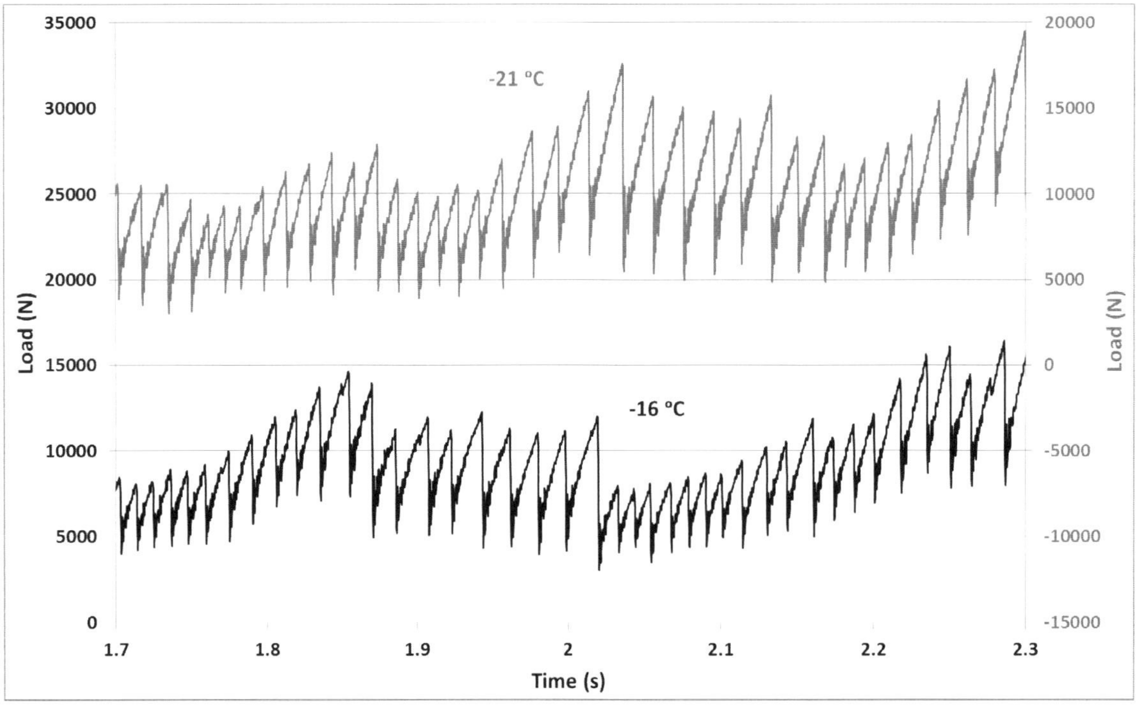

Figure 7. Comparison of load data from two tests using saline ice samples that were crushed against a Blade-Runners platen with an array of square-columns blades where one test was conducted at –16 °C and the other at – 21 °C.

EXPLANATION OF THE SALINE ICE CRUSHING BEHAVIOR AT SMALL-SCALE

We now present an explanation (in point form) of the initially perplexing observations, above, that not only relates to the Blade Runners technology but that also has implications regarding the use of saline ice in small-scale dynamic crushing tests in general. The explanation is based on the size and spacing of the brine and air filled channels/pockets in the ice, the similar size and spacing of the blades on the crushing platens and the shape and dimensions of the ice hard zone (relatively intact ice) that is characteristically present during crushing. Since we make reference to earlier crushing experiments using freshwater ice in the lab to test the Blade Runners technology that were successful in reducing IIV, we include an example (Figure 8.) for comparison with the present results.

Point 1. For saline ice crushing against a flat surface at -10 °C the interfacial pressure in the hard-zone area is lower than the value for freshwater ice (~55 MPa; Gagnon, 2016), but it is high enough to nucleate small

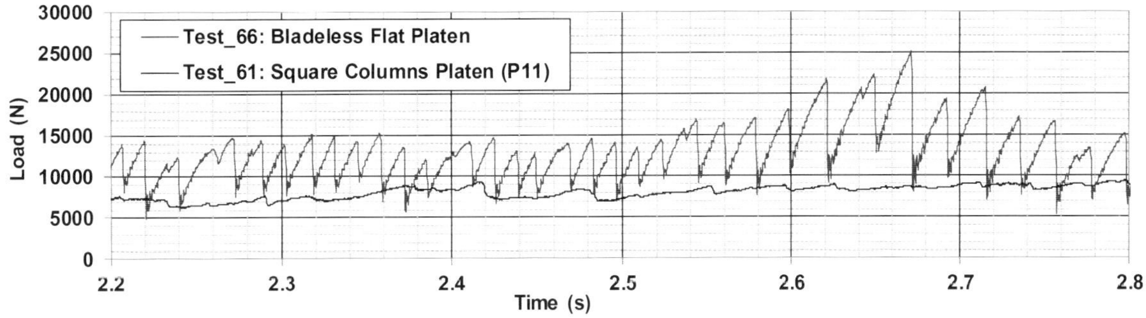

Figure 8. Partial load records from one of the best performing Blade Runners crushing platens (P11 – with square columns blades) and from a 'bladeless' flat acrylic platen. A typical high-amplitude sawtooth load pattern (HASLP) is evident for the test using the flat platen whereas the Blade Runners platen shows a dramatic reduction in HASLP. From Gagnon (2016).

uncoordinated spalls on the scale of the size and spacing of the brine channels/pockets that act as flaws in the ice (see Figure 8.). The lower interfacial pressure in the hard-zone region is due to the brine pockets and channels in the ice that reduce actual ice contact on the surfaces (i.e. lowering the average pressure) and that also make the ice more compliant due to grain-boundary sliding and reduced solid ice density. Small-scale spallation is facilitated by the fact that the hard-zone lateral span (its vertical extent in Figure 9.) is on the scale of the brine channels/pockets size and spacing that, as mentioned above, act as flaws.

The differences in hard-zone interface pressure between freshwater ice and saline ice were confirmed from observations of damage to the blades on acrylic platens. Several square-column blades (with 1 mm height) were partly or completely broken off at their base during the course of a few crushing tests for the freshwater ice case whereas no blades of that height were damaged when crushing saline ice. The cleavages were obviously from side forces on the blades that cause fracture at the base of the blade that expands outwards and away (plumose style) from the higher pressure side of the blade (i.e. generally upwards or downwards when viewing the high-speed images (Figure 9.), that is, perpendicular to the long axis of the linear hard zone, except at its ends. For saline ice the differential pressure on the opposing faces of a blade may be roughly ½ to ¼ that for freshwater ice from considering the moment exerted by a uniformly-distributed load on a cantilever beam, since we know that extensive blade damage did occur for square–column blades that were 2 mm in height while no damage occurred for blades that were 1 mm in height.

Point 2. Lowering the temperature to – 16 °C strengthens the ice enough to inhibit small-scale crack nucleation so that small spalls, and associated stress relief, do not occur. Consequently stress builds up in the bulk ice until a large spall occurs that is many times the size and spacing of the brine channels/pockets (Figure 10.), that is, HASLP occurs.

Point 3. Since the saline ice is more compliant and the hard-zone interface pressure is lower, as stated in point 1 above, in penetrating the hard-zone ice during crushing the blades do not create enough local differential pressure at the sides and/or tops of the blades to nucleate small spalls on the scale of the blades. Only large spalls that are many times the size of the blades are produced due to the large build ups of stress in the ice (i.e. HASLP), as though the blades were not present.

Point 4. It is anticipated that for full-scale testing of the Blade Runners technology in sea ice, or in large-scale tests in the lab using big saline ice samples, that the difference in scale will eliminate the peculiar behaviors observed during the small-scale tests on saline ice. That is, in the case of large lab-grown saline ice samples (or typical sea ice in the field), the size of the hard zones will be much greater than the size and spacing of brine channels/pockets so that the ice will exhibit HASLP throughout the typical range of ice temperatures (- 4 to -12 °C) when crushed against a flat surface. Furthermore, when the Blade Runners technology is used the blades, blade spacing and hard-zone size will be much greater than brine channels/pockets size and spacing. Hence, behavior related to similarity of scale of the blades and brine channels/pockets in the small-scale saline ice tests should be eliminated. In that case the blades will nucleate many relatively small uncoordinated spalls (on the scale of the blades) rather than repetitive large-scale spallations that cause HASLP. Under these realistic scaling conditions the ice presents itself as a uniform material.

Point 5. The interfacial pressure in the hard-zone regions may be higher for large-scale tests than small-scale tests for the same reasons outlined in Point 4, that is, the hard-zone size is much greater in size than the brine channel/pocket sizes and spacing so that confinement in the hard zone is greater. As mentioned above, the ice essentially presents itself as a uniform material at the large scale.

CONCLUSIONS

At –10 °C saline ice crushing against a flat platen leads to similar behavior that Blade Runners platens create when freshwater ice crushes against them. That is, the hard-zone ice region experiences many small uncoordinated spall events. In the latter case of the freshwater ice it is the blades that nucleate the tinny uncoordinated spalls whereas in the former case of the saline ice crushing against a flat platen it is the small closely-spaced flaws (brine channels/pockets) that serve as weak spots in the ice for small-spall cracks to nucleate.

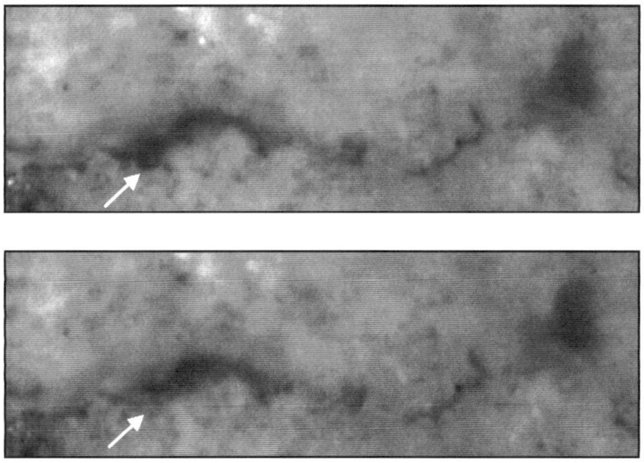

Figure 9. Two images from the high-speed imaging record, as viewed through the flat acrylic crushing platen, of a test where saline ice was crushed against the platen at a temperature and rate of -10 °C and 10 mm/s. A hard zone, the dark snake-like shape of varying thickness that spans the images, is visible. The white material above and below the hard zone is crushed ice, i.e. shattered spall debris. The top image was taken just prior to a small-scale spalling event (indicated by the arrow) and the bottom image was captured just afterwards. The width of the images is ~ 5.8 cm.

Considerable caution has to be exercised when using small-scale saline ice samples in lab tests to study the crushing behavior of full-scale sea ice. For example, high amplitude sawtooth load patterns, that typically occur when actual sea ice sheets crush against structures, will not occur unless the temperature for the small-scale lab tests is sufficiently low. Alternatively, it may be advisable in such studies to use freshwater ice samples instead since the ice's characteristic crushing behaviors, as observed in previous studies, reflect the behavior of full-scale sea ice for similar temperatures and crushing rates.

The difficulties that arise when using small-scale saline ice samples for crushing tests are due to the similarity of scale of the hard-zone region and the flaws (brine channels/pockets) in the ice. That is, at certain temperatures spalling behavior is highly influenced by the flaws since the vertical width of the elongated horizontal hard zone is roughly on the scale of the size and spacing of the inclusions. In the same vein, if the efficacy of a technology such as Blade Runners is being investigated in small-scale crushing experiments using saline ice the size and spacing of the blades will be unrealistically similar to that of the flaws in the ice (Figure 11.) so that results, whether favorable and unfavorable, may not be valid. To avoid this situation large-scale experiments, either in the lab or the field, are required to observe the actual behavior of sea ice and to properly test the Blade Runners technology.

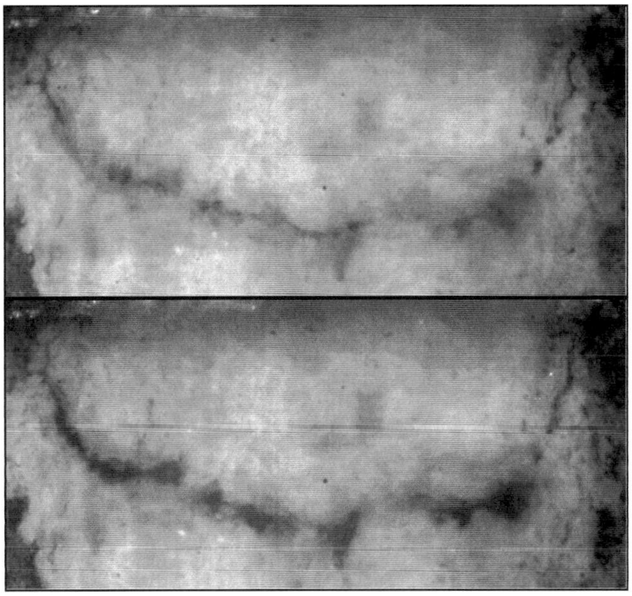

Figure 10. Two images from the high-speed imaging record, as viewed through the flat acrylic crushing platen, of a test where saline ice was crushed against the platen at a temperature and rate of -16 °C and 10 mm/s. A hard zone, the dark catenary-like shape of varying thickness that spans the images, is visible. The white material above and below the hard zone is crushed ice, i.e. shattered spall debris. The top image was taken just prior to a large-scale spalling event that spanned the full length of the hard zone and substantially reduced its thickness. The bottom image was captured just after the event. The width of the images is ~ 10.5 cm.

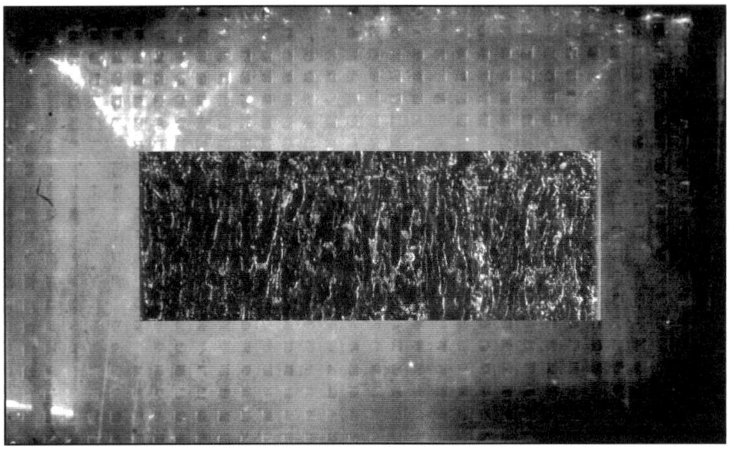

Figure 11. View through the acrylic crushing platen that had the square-columns blade array (square tops of the columns are visible in the image) just prior to a test. The wedge-shaped ice sample (light-colored object) is visible. A second image, the dark centrally-located rectangle, has been inset (at the same scale) on the central portion of the ice sample image. The inset is a photo of the brine and air filled channels/pockets in the ice obtained from a thin section. The size and spacing of the brine/air inclusions are on the scale of that of the square columns. The width of the image is ~13.2 cm.

ACKNOWLEDGEMENTS

I would like to thank Austin Bugden for technical assistance during the test program and analysis of results, and I am grateful to NRC for its support of this work.

REFERENCES

Dempsey, J.P., Adamson, R.M. and Mulmule, S.V., 1999. Scale effects on the in-situ tensile strength and fracture of ice. Part II: First-year sea ice at Resolute, N.W.T. International Journal of Fracture, **95**, pp. 347-366.

Evans, A.G., Palmer, A.C., Goodman, D.J., Ashby, M.F., Hutchison, J.W., Ponter, A.R.S., Williams, G.J., 1984. Indentation spalling of edge-loaded ice sheets. Proceedings of IAHR 1984, Hamburg, pp. 113–121.

Gagnon, R.E., 2008. High-speed imaging of mechanisms responsible for sawtooth cyclic loading during ice crushing. Proceedings of IAHR 2008, Vancouver, Canada, Vol. 2, pp. 983–991.

Gagnon, R., 2012. An Explanation for the Molikpaq May 12, 1986 Event. Cold Regions Science and Technology 82 (2012) 75-93.

Gagnon, R., 2016. Phase II Lab Tests of the Blade Runners Concept for Reducing Ice-Induced Vibration of Structures. Proceedings of the 23nd International Symposium on Ice, IAHR 2016, Ann Arbor, Michigan, Paper #494880.

Määttänen, M., Marjavaara, P., Saarinen, S., 2011. Ice crushing pressure distribution against a compliant stiffened panel. Proc. 21st Int. Conf. On Port and Ocean Engineering under Arctic Conditions, Montreal, Canada, POAC 2011, paper #038.

Sodhi, D.S., 1991. Ice-structure interaction during indentation tests. In Ice-Structure Interaction: Proceedings of IUTAM-IAHR Symposium 1989, edited by S. Jones et al., Springer-Verlag, pp. 619–640.

Loads on Structure and Waves in Ice (LS-WICE) Project, Part 1: Wave Attenuation and Dispersion in Broken Ice Fields

Sukun Cheng[1], Andrei Tsarau[2], Hongtao Li[2], Agnieszka Herman[3], Karl-Ulrich Evers[4], Hayley Shen[1]
[1] Department of Civil & Environmental Engineering, Clarkson University, Potsdam, NY 13699 USA
[2] Norwegian University of Science and Technology, Trondheim, Norway
[3] Institute of Oceanography, University of Gdansk, Poland
[4] Arctic Technology, Hamburgische Schiffbau-Versuchsanstalt GmbH, Hamburg Ship Model Basin, Bramfelder Straße 164, D-22305 Hamburg, Germany

ABSTRACT

A multi-group investigation was conducted at Hamburgische Schiffbau-Versuchsanstalt GmbH (HSVA) from Oct. 24 to Nov. 11, 2016 under the Hydralab+ Transnational Access project: Loads on Structure and Waves in Ice (LS-WICE). There are three parts to this investigation: ice fracture under wave actions, wave attenuation/dispersion in broken ice covers, and ice-structure interaction under wave conditions. This paper focuses on the wave attenuation/dispersion part of the investigation. A level ice sheet was produced first. The physical and mechanical properties were measured in the Large Ice Model Basin (LIMB) before passing through a range of waves to obtain the attenuation/dispersion relation. Wave measurements were monitored with pressure transducers and ultrasound sensors. The test sequence began with a continuous ice sheet followed by broken ice sheet. The broken ice sheet was produced by cutting the continuous ice sheet into uniform size floes. A range of floe sizes and wave periods were tested. Results of the data obtained and their implications to wave propagation in the marginal ice zone populated with different broken ice floes will be presented.

KEY WORDS: Ice; Waves; Attenuation; Dispersion.

INTRODUCTION

Reduction of sea ice coverage in the Arctic has opened up shipping and offshore engineering developments. The safe operation of these developments and effective environmental protection strategies rely on better predictions of the Arctic marine condition. One of the most obvious needs is a better wave model, which can take into account the dynamically changing ice cover, in particular in the marginal ice zone. After several decades of intensive development, global wave models such as WAVEWATCH III (Tolman, 2014) have reached a satisfactory level for open water regions. However, under partially and fully ice covered seas, these models need much more work.

The project LS-WICE "Loads on structures and waves in ice" was granted by the Hydralab+ project under the Horizon 2020 EU-Framework programme for research and innovation (H2020-INFAIA-2014-2015). The current Hydralab+ project runs from September 2015 until August 2019, in which 24 partners and 9 associated partners form a network of research institutes. The goal of this network is to strengthen the experimental hydraulic and ice engineering research to better address climate change adaptation issues.

The LS-WICE experiments were carried out by an international group of researchers from Norway, Poland and USA in the Large Ice Model Basin (LIMB) at the Hamburg Ship Model Basin (HSVA).The test facility in the LIMB was a 72 m long, 10 m wide and 2.5 m deep ice tank. Four flap type mobile wave generator modules cover the total 10 m width of the ice tank. The maximum wave height of this facility is limited to 0.25 m and the maximum wave period is 3 s. At the end of the basin a parabolic shaped beach especially designed for this project was installed in order to minimize wave reflection. Ice sheets were produced under air temperature of about -22 deg Celsius.

Figure 1 shows the elevation and the plan view of the LIMB and Figure 2 shows the mobile wave generator modules and the parabolic shaped beach at the end of the ice tank.

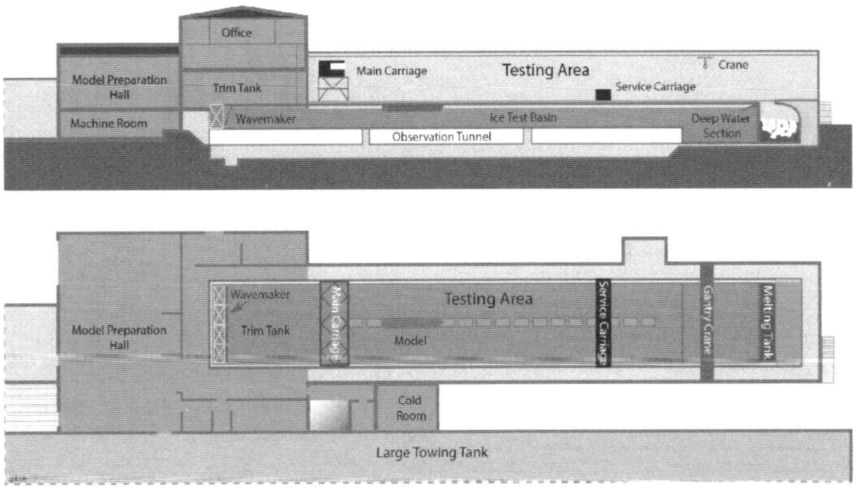

Figure 1. Elevation view (top) and plan view (bottom) of the LIMB.

Figure 2. Mobile wave generator modules (left) and parabolic shaped beach (right).

The test program was executed in open water and in 4 ice sheets. A cylindrical structure (0.69 m diameter) was stationarily anchored in ice sheet no. 4 in order to measure the ice and wave forces acting on the cylinder as shown in Figure 3. Each of these tests had different purposes. The open water case was performed to characterize the wave tank behavior and to calibrate all sensors. Ice sheet no. 1 was constructed to perform the break-up test. Ice sheets no. 2 and no. 3 were used to test wave attenuation and dispersion under different monochromatic waves. Ice sheet no. 4 was produced to test ice-structure interaction in the presence of a wave field.

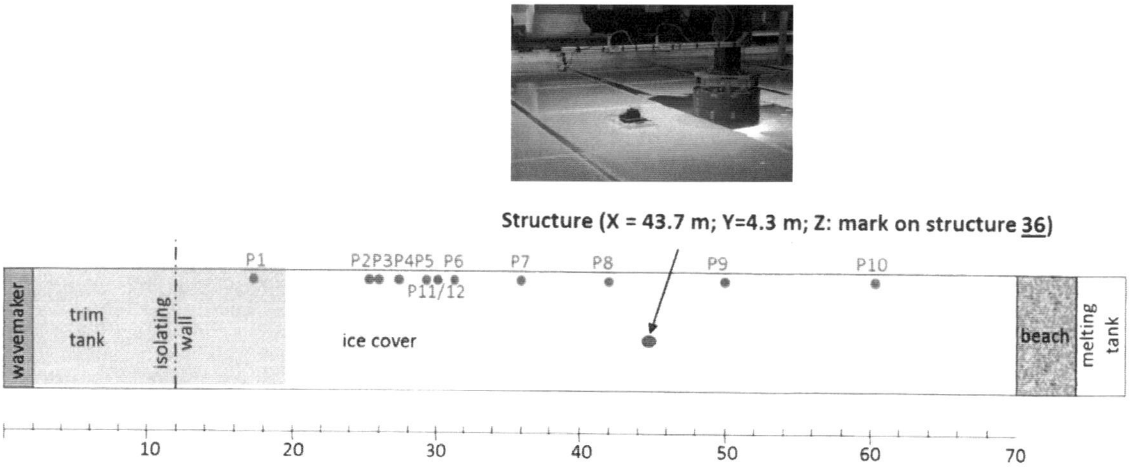

Figure 3. Layout of the wave tank and location of pressure sensors in open water tests (note: locations of sensors changed everyday.)

The wave field was monitored by two types of sensors: ultrasound and pressure, one determined directly the surface elevation and the other the dynamic pressure which was related to the surface elevation. The ultrasound sensors were UltraLab (model ULS Advanced USS 13-HF, IP65, M30x1.5). The pressure transducers included ten BD (model LMP307, 0.1 bar range), and two Omega (PX437 and PX438, range 0.14 bar range). A Qualisys Motion Capture System together with markers that were placed on the ice along the central axis of the tank to monitor the motion of ice surface. In order to record the time evolution of the ice breaking pattern, large parts of the ice sheet were continuously monitored with an AXIS camera mounted at the ceiling and two sideward-looking GoPro cameras mounted at the walls. Additionally, static images of the entire ice sheet were taken with a Canon EOS 550D digital reflex camera after two tests in which major ice breaking was observed in order to determine the resulting floe-size distribution.

The study presented in this paper concerns only the wave attenuation and dispersion part of the LS-WICE project. There are three accompanying papers at this conference each addresses a different topic (Herman et al., Tsarau et al., and Li et al., see this Proceedings). This paper is motivated by this purpose: to determine the effect of ice in the marginal ice zone on wave propagation.

BACKGROUND OF THE STUDY

Sakai and Hanai (2002) reported an interesting experiment that showed floe size effect on wave dispersion. The experiment used a polyethylene sheet floating on water. The phase speed of a range of wave period propagating over this sheet was measured using ultrasound sensors. The first set of measurements was conducted using an intact sheet. The sheet was then cut successively into smaller and smaller strips. After each reduction of strip size, keeping the entire cover the same length, the same wave tests were conducted. The results showed clearly a decreasing wave speed as the strip size reduced, suggesting a reduction of the "equivalent" elastic modulus of the fragmented floating cover. This study inspired a couple of subsequent laboratory investigations. However, instead of focusing only on the floe size, these later studies included other effects. In Wang and Shen (2010), pancake ice mixed with frazils was used. In Zhao and Shen (2015) grease ice, pancake ice, and randomly fragmented ice covers were used. These tests showed that wave speed might change in different ice covers, particularly in fragmented ice covers. Furthermore, wave attenuation was clearly detected in all cases. However, due to the size of the wave tank, only very narrow range of wave period could be tested. These past studies encouraged a systematic investigation using real ice cover but in such a way that only floe size was allowed to change, as reported in this paper.

DESCRIPTION OF THE EXPERIMENT

We focused on the effect of floe size (D) relative to wavelength (λ) on wave propagation. Parameters to be determined from these two inputs include the wavelength and the attenuation rate.

The procedure of producing the ice cover with different floe sizes is described below. A level ice sheet was formed by seeding first to create a granular layer. Sustained cooling thickened this top layer by columnar growth until the whole sheet reaches about 0.035 m thickness. The Young's modulus, flexural strength, ice thickness, density and salinity were then obtained.

The intact ice cover was first freed by cutting a narrow strip about 0.1 m from the walls. A sequence of wave tests with different wave periods was first applied to the intact level ice cover. The intact level ice was then cut longitudinally into six 1.5 m wide strips parallel to the wave tank wall. Transverse cuts were applied to create floes of uniform sizes. We started with cuts of 6 m long floes. After applying a set of wave tests with different periods, these floes were further cut into 3 m, then 1.5 m, and finally 0.5 m floes. Each time a pattern of uniform size floes was produced. Figure 2 shows the floe field for the 6 m case and the 0.5 m case.

Figure 4. The regular pattern of ice floes tested were 1.5 m wide. The length of the floes started with 6 m, as shown in the left side photo and ended with 0.5 m, as shown in the right side photo, both taken just after the cutting prior to the wave tests.The whole ice cover was refrained from spreading by the boom at the leading edge. The first row of the 6 m long floes was broken by waves as evident from the right side photo.

The wave tests included wave periods between 0.9 to 2 s, starting from long wave to minimize the risk of breaking floes. However, the 6 m floes were proven to be too long to keep their integrity. The first row of these floes did break, as can be seen from Figure 4 (right side photo).

A stop-go wave test was performed. Each run lasted between 80 to 90 s, sufficient for each of the wave tested to travel the entire wave tank and return to ice edge. However, because no beach could perfectly absorb wave energy, to avoid effects from reflection, only a portion of the wave record before any wave reflection will be used for further analysis.

RESULTS

A typical set of pressure sensor data are shown in Figure 5. In Figure 5(a). from bottom to top the curves represent measured dynamic pressure from P1 to P9, respectively. Wave height in the open water (measured by P1) is clearly higher than that by the first sensor under ice cover (P2), suggesting some wave energy was reflected by the ice cover. Attenuation is clear by comparing the wave height from successive sensors. However, dynamic pressure defined as $\rho g \eta \frac{\cosh k(h+z)}{\cosh kh}$ is influenced by the sensor location z (with origin at the calm water level and positive upward), where η is the water surface elevation, h the water depth, ρ the water density and g the gravitational acceleration (Dean and Dalrymple, 1984). Data presented in Figure 5(a). need to be corrected by the small but not negligible differences between sensors before attenuation may be correctly measured. Phase shift between sensors can be seen from Figure 5(b). The time lag and distance between a pair of sensors will be used to determine the wave phase velocity first. From which the wavelength can be calculated using the relation $c = \lambda/T$, where c is the wave phase velocity and T is the wave period.

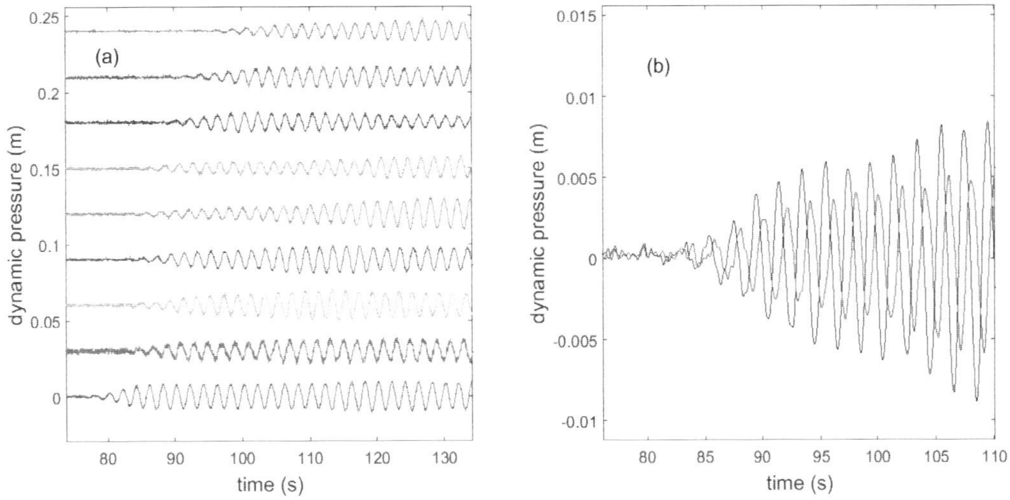

Figure 5. A sample test data. The vertical axis is the dynamic pressure divided by ρg. (a) All pressure data were vertically shifted to aid visualization; (b) A pair of pressure sensors, where both wave height and phase difference are clearly seen. The wave period in this case was 2 s and floe size 3 m.

In Figure 6. we compare the wave speed between two cases of different floe sizes, both are from $T = 1.4$ s. The left panel is from floe size 1.5 m and the right panel is from floe size 0.5 m. The arrival of the wave from P2 to P9 is faster for the ice cover consisting floes of 1.5 m than that from floes of 0.5 m. Attenuation is also observed to be greater in the case with smaller floes.

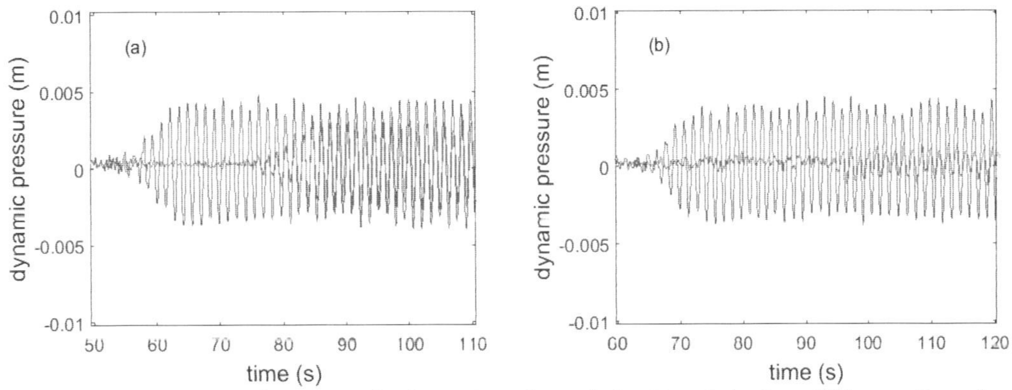

Figure 6. Comparison of wave arrival at P2 and at P9 for $T = 1.4$ s but different floe sizes: (a) 1.5 m (b) 0.5 m.

DISCUSSION AND CONCLUSIONS

In addition to laboratory studies mentioned earlier, a recent field experiment (Rogers et al., 2016) also showed that wave attenuation depended on the type of ice covers. Theoretical studies of how an ice cover might attenuate waves and change its dispersion relation are many. Squire (2007) provided a review of the theories up to that publication time. These theories have only been partially confirmed by a number of field experiments. Meylan et al. (2015) reported attenuation in the order of 10^{-5} (m^{-1}) in the Southern Ocean marginal ice zone with floes that were 2-3 m near the ice edge to 10-20 m about 200 km from the ice edge. The floes were 0.5-1 m thick first year ice. Doble et al. (2015) reported attenuation in the order of 10^{-3} (m^{-1}) in the Weddell Sea. The floes were newly formed pancake ice with size ~0.7 m and thickness 0.05-0.1 m. The space between floes was filled with frazil ice. The ice conditions of these two studies were very different from each other, and also very different from the much earlier study of Wadhams et al. (1988), in which they reported several field studies in the Arctic marginal ice zone including Bering and Greenland Seas. The range of attenuation was from 10^{-5} – 10^{-4} (m^{-1}). The ice covers consisted floes from 5-50 m with a range of thickness from 0.4 to 1.2 m in the Bering Sea and 3.1 m in the Greenland Sea. These studies strongly suggested that wave attenuation is a result of the combination of ice type and floe sizes.

To isolate ice effect on wave propagation from other mechanisms such as wind wave generation, nonlinear wave-wave energy transfer, and wave breaking, laboratory studies are necessary. The present study demonstrated a systematic approach to determine floe size effect on wave propagation. To scale up the results from a laboratory size wave and floes to the field, we need to determine the dominant length scales that can link the laboratory to the field conditions.

Detailed results of this study will be reported later in a journal paper. Wave-ice interaction may include many physical mechanisms as discussed in Shen and Squire (1998). In this experiment, the total attenuation from all processes present in the experiment is measured. Likewise, wave dispersion data are also the result of coexisting mechanisms. The dataset can potentially be used in many future studies.

ACKNOWLEDGEMENTS

The work described in this publication was supported by the European Community's Horizon 2020 Programme through the grant to the budget of the Integrated Infrastructure Initiative HYDRALAB+, Contract no. 654110. The author(s) would like to thank the Hamburg Ship Model Basin (HSVA), especially the ice tank crew, for the hospitality, technical and scientific support and the professional execution of the test programme in the Research Infrastructure ARCTECLAB, and the instrumentation help from Meleta Truax of the Clarkson University. Authors SC and HS are also funded in part by the Office of Naval Research grant number #N00014-13-1-0294.

REFERENCES

Dean, R. G. and Dalrymple, R. A. (1984) Water wave mechanics for engineers and scientists. World Scientific, ISBN: 978-981-02-0421-1, 353p.

Doble, M. De Carolis, J., G. Meylan, M. H. Bidlot, J.-R., and Wadhams P. (2015) Relating wave attenuation to pancake ice thickness, using field measurements and model results, Geophys. Res. Lett., 42, 4473–4481, doi:10.1002/2015GL063628.

Herman, A., Tsarau, A., Evers, K.-U., Li, H., and Shen, H.H. (2017) Loads on Structure and Waves in Ice (LS-WICE) project, Part 2: Sea ice breaking by waves, Proc. 24th Int.Conf. Port Ocean Eng. Under Arctic Conditions, June 11-16, 2017, Busan, Korea.

Li, H., Tsarau, A., Shen, H.H., Herman, A., Evers, K.-U., and Lubbad, R. (2017) Loads on Structure and Waves in Ice (LS-WICE) project, Part 4: Wave attenuation and dispersion in broken ice fields – Ice floes collisions under wave actions

Meylan, M. ., Bennetts, L. G. and Kohout A. L. (2014) In situ measurements and analysis of ocean waves in the Antarctic marginal ice zone, Geophys. Res. Lett., 41, 5046–5051, doi:10.1002/2014GL060809.

Rogers, W. E., Thomson, J., Shen, H.H., Doble, M.J., Wadhams, P., and Cheng, S. (2016) Dissipation of wind waves by pancake and frazil ice in the autumn Beaufort Sea, J. Geophys. Res. – Oceans, DOI 10.1002/2016JC012251.

Sakai, S., and Hanai, K. (2002) Empirical formula of dispersion relation of waves in sea ice, In: Squire, V. A. and P. J. Langhorne (Eds), Ice in the Environment: Proc. 16th Int. Symp. Ice, V. 2. Int. Asso. Hydraulic Eng. Res., Dunedin, New Zealand, 327-335.

Shen, H.H. and Squire, V.A. (1998) Wave damping in compact pancake ice fields due to interactions between pancakes, AGU Antarctic Research Series 74, *Antarctic Sea Ice: Physical Processes*, Interactions and Variability, (Ed. Dr. Martin O. Jeffries): 325-342.

Squire, V.A. (2007) Of ocean waves and sea-ice revisited, Cold Reg. Sci. Technol., 49(2), 110–133.

Tolman, H. L. and the WAVEWATCH III® Development Group (2014) User manual and system documentation of WAVEWATCH III® version 4.18, NOAA Technical Note, MMAB Contribution No. 316, 311p.

Tsarau, A., Sukhorukov, S., Herman, A., Evers, K.-U., and Løset, S. (2017) Loads on Structure and Waves in Ice (LS-WICE) project, Part 3: Ice-structure interaction under wave conditions, Proc. 24th Int.Conf. Port Ocean Eng. Under Arctic Conditions, June 11-16, 2017, Busan, Korea.

Wadhams, P., Squire, V.A., Goodman, D.J., Cowan, A.M. and Moore, S.C. (1988) The Attenuation Rates of Ocean Waves in the Marginal Ice Zone, J. Geophys. Res., 93(C6), 6799-6818.

Wang, R. and Shen, H.H. (2010) Experimental study on surface wave propagating through a grease-pancake ice mixture. Cold Reg. Sci. & Tech. doi:10.1016/j.coldregions.2010.01.011.

Zhao, X and Shen, H.H. (2015) Wave propagation in frazil/pancake, pancake, and fragmented ice covers, Cold Regions Science and Technology, doi:10.1016/j.coldregions.2015.02.007.

Proceedings of the 24th International Conference on
Port and Ocean Engineering under Arctic Conditions
June 11-16, 2017, Busan, Korea

Loads on Structure and Waves in Ice (LS-WICE) Project, Part 2: Sea Ice Breaking by Waves

Agnieszka Herman[1], Andrei Tsarau[2], Karl-Ulrich Evers[3], Hongtao Li[2], and Hayley Shen[4]
[1] Institute of Oceanography, University of Gdansk, Poland
[2] Norwegian University of Science and Technology, Trondheim, Norway
[3] Arctic Technology, Hamburgische Schiffbau-Versuchsanstalt GmbH (Hamburg Ship Model Basin), Hamburg, Germany
[4] Department of Civil & Environmental Engineering, Clarkson University, Potsdam, NY, USA

ABSTRACT

A multi-group investigation was conducted at Hamburgische Schiffbau-Versuchsanstalt GmbH (HSVA) from Oct. 24 to Nov. 11, 2016 under the Hydralab+ Transnational Access project: Loads on Structure and Waves in Ice (LS-WICE). There are four parts to this investigation: wave attenuation/dispersion in broken ice covers, wave-induced ice-floe collisions, ice fracture under wave action, and ice-structure interaction under wave conditions. This paper focuses on the part of the experiment related to wave-induced ice breaking. During the preparatory stage of the tests, a continuous ice sheet was produced, and the mechanical and physical ice properties were determined at a number of selected locations in the Large Ice Model Basin (LIMB). Pressure transducers and ultrasound sensors located at different positions along the ice tank were used to monitor the wave propagation through the ice. Additionally, 3D motion of a few markers placed on the ice was tracked with an infrared camera system (Qualisys Motion Capture System). The progress of breaking was recorded with cameras placed above as well as sideways from the tank. The measurements consisted of two groups of tests, with a constant wave period in each group. The wave amplitude was increased stepwise, starting from a value too low to break the ice, until a major fragmentation of ice occurred. After each group of tests, the locations of cracks were determined and the floe-size distribution was estimated. The paper presents the data collected during the experiment, together with a short discussion of the observed breaking patterns and processes that created them.

KEY WORDS: Ice; Waves; Ice breaking; Floe-size distribution.

INTRODUCTION

Sea ice–waves interactions belong to the dominating processes shaping the features and behavior of the marginal ice zone (MIZ). In fact, mutual wave–ice interactions are a defining characteristic of this type of the ice cover, occurring within a certain distance from the ice edge. The details of these interactions depend on wave parameters, ice thickness and material properties (density, elastic modulus, etc.), but they also influence and are influenced by the floe-size distribution (FSD). The scattering and attenuation of the wave energy – and thus the width of the MIZ – very strongly depend on the sizes of ice floes, which in turn are a product of wave-induced breaking. In general, FSD is only rarely taken into account in sea ice models, and only very recently equations have been proposed that describe the evolution of FSD in time (Horvat & Tzipermann, 2015) in a manner analogous to the evolution of the ice-thickness distribution, for which a mathematical description suitable for continuum sea ice models is well established. Although substantial effort has been put recently in developing parameterizations of sea ice–wave interactions for numerical models (e.g., Dumont *et al.*, 2011; Williams *et al.* 2013), our understanding of the details of many important aspects of these interactions still remains unsatisfactory. This is true for wave-induced ice breaking, for which only very rudimentary, approximate models exist (e.g., Langhorne *et al.*, 1998; Squire, 2007; Squire *et al.*, 1995; Kohout & Meylan, 2008). Moreover, lack of validation data restrains further development. The Hydralab+ Transnational Access project "Loads on Structure and Waves in Ice" (LS-WICE) belongs to the first experiments in which a group of tests was devoted to measuring and observing ice breaking by waves under controlled, laboratory conditions. Apart from measuring the wave characteristics at several locations along the ice tank, and the motion of the ice itself, we recorded the progress of breaking, making it possible to analyze the relationships between the wave propagation and the temporal evolution of the crack pattern. Other LS-WICE tests, devoted to wave attenuation in broken ice, kinematics of floes' collisions, and ice–structure interactions are described in accompanying papers in this volume (Cheng *et al.*, 2017; Li *et al.*, 2017; Tsarau *et al.*, 2017).

DESCRIPTION OF THE EXPERIMENT

An overview of the whole LS-WICE project, including the description of the facility (Large Ice Model Basin at HSVA) and the measuring equipment, can be found in the accompanying paper by Cheng *et al.* (2017) and will not be repeated here. For the ice-breaking group of tests, a continuous ice sheet was prepared and positioned as shown in Figure 1. by cutting it from the side walls of the tank (the ice strips at both sides, ~10 cm wide, have been removed in order to prevent affecting the ice motions by the interactions with wall and thus the breaking patterns obtained) and by pushing it by 6 m towards the melting tank so that the final ice edge position was at $x = 20$ m, approximately 18 m from the wavemaker. The thickness of the ice, measured at a number of locations in the tank, varied between 32.5 and 38.5 mm, with an average of 34.8 mm; the ice elastic modulus equaled 16 MPa; the bending strength varied from 41.5 kPa close to the ice edge to 67.1 kPa in the area close to the beach.

The locations of the pressure and ultrasound sensors used in this group of tests is shown in Figure 1, together with the locations of five markers of the Qualisys Motion Capture System that were placed on the ice along the central axis of the tank, ~1.5 m apart from each other. In order to record the time evolution of the ice breaking pattern, large parts of the ice sheet were continuously monitored with an AXIS camera mounted at the ceiling and two sideward-looking GoPro cameras mounted at the walls. Additionally, static images of the entire ice sheet were taken after two tests in which major ice breaking was observed in order to determine the resulting floe-size distribution.

The list of tests performed in this test group is given in Table 1. The purpose of tests 1100, 1200 and 1300, performed with a small wave height (H = 10 cm) for three different wave periods, was to analyze wave attenuation within the continuous, unbroken ice cover – thus, these tests are complementary to those described in Cheng *et al.* (2017) and devoted

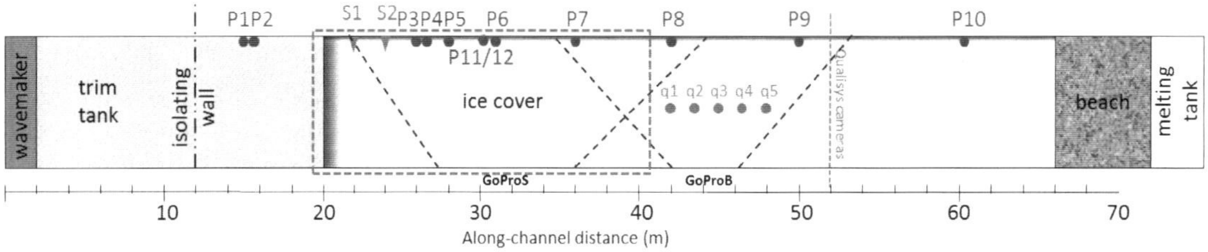

Figure 1. Instrument setup during the ice breaking test: P1–P10 are single pressure sensors, P11/12 – a double pressure sensor, S1 and S2 – ultrasound sensors, q1–q5 – Qualisys markers, dashed black lines – fields of view of sideward-looking GoPro cameras (Silver and Black), dashed blue lines – field of view of the AXIS camera mounted on the ceiling.

Table 1. Summary of test runs performed during the ice-breakup experiment

Run No.	Wave height (mm)	Wave period (s)	Test duration (s)	Remarks
1100	10.0	2.0	90	no breaking observed
1200	10.0	1.6	90	no breaking observed
1300	10.0	1.2	90	no breaking observed
1400	20.0	2.0	90	no breaking observed
1410	30.0	2.0	90	no breaking observed
1420	40.0	2.0	90	no breaking observed
1430	50.0	2.0	90	no breaking observed
1440	70.0	2.0	115	first major crack at x ≈ 44 m
1450	90.0	2.0	several minutes	major breaking; test continued until no breaking occurred
photos from the crane camera for FSD				
1500	50.0	1.6	90	only a few new cracks observed
1510	70.0	1.6	several minutes	major breaking; test continued until no breaking occurred
photos from the crane camera for FSD				

specifically to measuring wave attenuation in broken ice fields. Subsequently, in tests 1400–1450, the wave period was kept constant at $T = 2$ s, corresponding to deep-water wavelength $L = 6.24$ m, and the wave amplitude was increased stepwise until major breaking of the ice was observed. Finally, two additional tests with shorter waves were performed ($T = 1.6$ s, $L = 3.99$ m) that led to further breaking of the ice into smaller floes. In all tests, the wave steepness was small enough to justify the use of the linear wave theory (the ratio H/L was largest in test 1510 and equaled 0.0175).

RESULTS

Stages of Ice Breaking

For long waves used in tests 1400–1450 (Table 1), no breaking of the ice was observed for wave heights up to 50 mm (test 1430). Remarkably, the first major crack, which appeared in test 1440, did not develop in vicinity of the ice edge, but approximately in the middle of the ice sheet, i.e., at x ≈ 44 m (Figure 2a.), after 93 seconds from the start of the wavemaker. In the following test 1450, which led to a fragmentation of the entire ice sheet, this first crack acted

(a) (b) (c)

Figure 2. Snapshots of the ice sheet from the GoPro Silver camera after test 1440 (a), 1450 (b) and 1510 (c), after removal of fish-eye effects. The Qualisys markers are highlighted with yellow circles for better visibility. In (b) and (c), the location of the initial crack is marked with a dashed blue line. Additionally in (b), the time sequence of the crack formation in the initial phases of test 1450 is shown with lines of different colors (according to the color code in the upper left corner of the image).

as a "secondary" ice edge, from which breaking advanced in a down-wave direction in a manner similar to that at the "proper" ice edge. This is illustrated in Figure 2b, in which the sequence of crack formation is shown with lines of different color. Notably, whereas all cracks numbered 2–6 in Figure 2b. formed during the first 40 seconds of test 1450, no single crack could be recognized on the up-wave side of the first crack during the first 60 seconds. Consequently, at the end of this test, the floes on the down-wave side of this initial crack were smaller than those on its up-wave side (Figure 2b.) and they experienced more intense flooding. Interestingly, a similar breaking sequence could be observed for the part of the ice sheet between the ice edge and the first crack, i.e., after a few floes separated from the main sheet in the ice edge region, a transverse crack formed roughly in the middle, and further breaking took place down-wave from this crack (not shown). In the subsequent test 1500, only a few new (mostly short) cracks developed, even though the overall behavior of the ice cover was visibly different than in previous tests, primarily due to the drift of the ice floes close to the ice edge in the up-wave direction, which led to a decrease of the ice concentration and thus changed the motion of individual floes (larger amplitudes of horizontal oscillatory motion, more frequent and more energetic collisions). Finally, in test 1510, strong breaking took place throughout the ice sheet, accompanied by a strong flooding (Figure 2c.) and progressive deterioration of the floes' edges due to collisions – a process that produced many very small ice fragments filling spaces between larger floes.

Ice Motion in Vicinity of the First Major Crack

As can be seen in Figure 2, the Qualisys markers were placed along a line that was crossed by the first crack in test 1440, and remained in the zone of major crack formation throughout subsequent tests. (Note that although Figures 2 b,c suggest that the markers were directly at the edges of the floes, they remained firmly placed on the ice until the end of the last test.) A detailed analysis of the data from the pressure sensors and the Qualisys system is beyond the scope of this paper, but it is worth mentioning here some overall behavior of the ice sheet, visible in the time series of the vertical displacements of the Qualysis markers q1–q5 (Figure 3.). During the first ~55 s of test 1440, the amplitude of these displacements was nearly identical at all five locations (Figure 3a; note that the distance between q1 and q5 equaled approx. 6 m, i.e., one wavelength). The situation changed during the following ~20 s, after which a new pattern established (Figure 3b.), with increased amplitudes at q2 and q4 and decreased amplitudes at the remaining three markers – presumably due to wave reflection from the beach. In the videos, the first major crack became visible at time $t = 93$ s, i.e., the reflected waves presumably contributed to its formation.

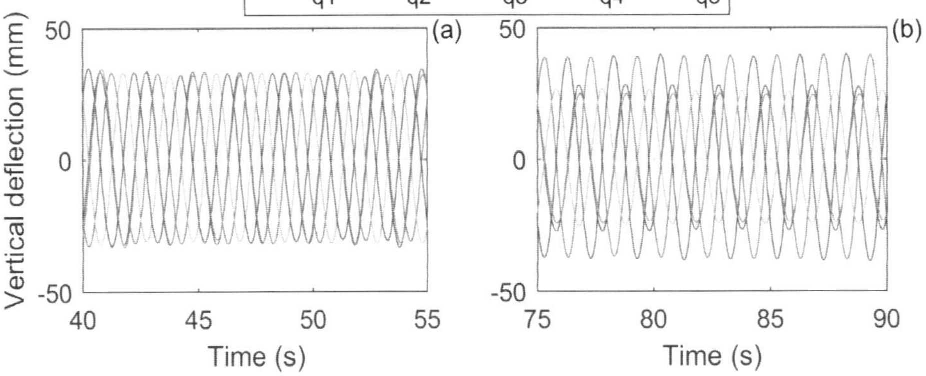

Figure 3. Fragments of time series of the vertical deflections (in mm) of Qualisys markers q1–q5 (see Figure 1. for their locations). Time is measured from the start of the wavemaker.

Floe-Size Distribution

Figure 4. shows a fragment of an image of the ice sheet obtained by: (i) stitching together overlapping images of the ice sheet taken with a downward-looking crane camera, (ii) transforming the result to a binary image of ice and water, and (iii) identifying floe boundaries. The floe-size distributions determined after tests 1450 (N_f = 2269 identified floes) and 1510 (N_f = 3422) are shown in Figure 4. A rank-order statistics is shown rather than binned histograms in order to avoid binning effects that would strongly influence the results, especially in the range of large floe sizes (as can be seen in Figure 5, in both cases less than 100 floes have surface areas larger than 1 m^2, and less than 10 floes have areas larger than 10 m^2). Moreover, because the shapes of the floes, especially the larger ones, tend to be irregular and elongated, the results are analyzed in terms of the floes' surface areas instead of their radii? (obviously, this choice does not affect the shape of the curves in a log–log plot in Figure 5.). As can be seen, the floe sizes have an upper bound dependent on the wavelength, and the slopes of the distributions change gradually between the large and small floe regime. In the latter case, as can be expected, the slope is largely independent of the wave properties.

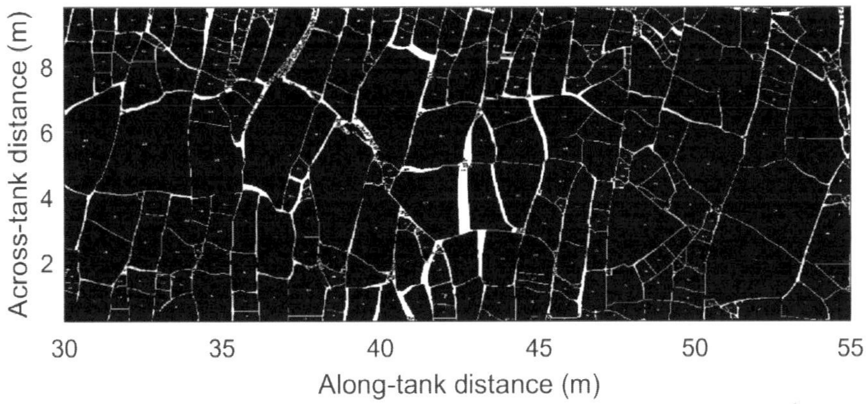

Figure 4. A fragment of a binary image of the ice sheet after test 1510, with identified ice floes' boundaries.

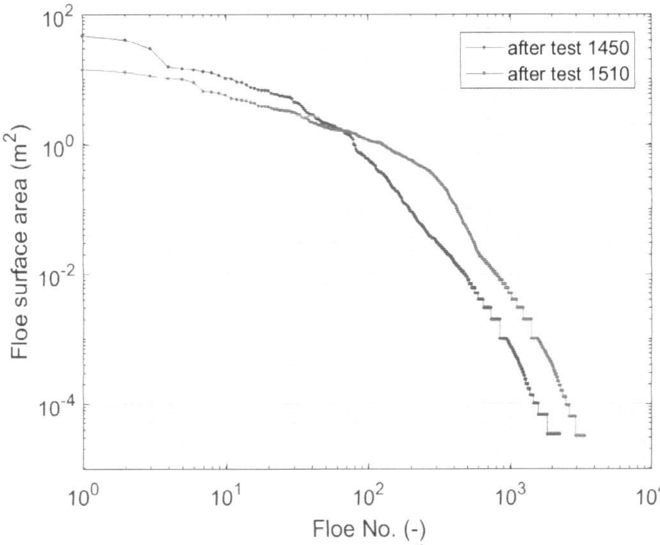

Figure 5. Rank-order statistics of the surface areas (m^2) of ice floes after tests 1450 and 1510.

DISCUSSION AND CONCLUSIONS

Contrary to the expectations, we did not observe progressive breaking starting from the ice edge. Instead, the ice sheet first broke approximately in the middle of its length, and this first major crack presumably had a profound influence on the subsequent development of fractures. Thus, the breaking pattern in this experiment was very different, e.g., from that observed previously in preliminary tests at the same facility (Nils Reimer, HSVA – personal communication) or in the field (Squire *et al.*, 1995; see also https://vimeo.com/106835989). Obviously, with just one case, it is impossible to determine whether this kind of breaking behavior is a repeatable feature or just a singular event caused, e.g., by unrecognized defects in the ice sheet that may have been created during ice preparation. The data from the pressure and Qualisys sensors suggest the role of reflected waves in the process. Moreover, it should be remembered that the ice sheet was not frozen to the beach at its down-wave end, i.e., it was a freely floating floe roughly ten wavelengths long rather than shore-fast ice, for which most models are developed.

A more in-depth analysis of the experiment described here will be presented in a subsequent paper, together with results of numerical modeling with a new version of the Discrete-Element bonded-particle Sea Ice model (DESIgn; Herman, 2016), extended to include a wave model coupled with an ice model.

ACKNOWLEDGEMENTS

The work described in this publication was supported by the European Community's Horizon 2020 Programme through the grant to the budget of the Integrated Infrastructure Initiative HYDRALAB+, Contract no. 654110. The author(s) would like to thank the Hamburg Ship Model Basin (HSVA), especially the ice tank crew, for the hospitality, technical and scientific support and the professional execution of the test programme in the Research Infrastructure ARCTECLAB.

REFERENCES

Cheng, S., Tsarau, A., Li, H., Herman, A., Evers, K.-U. & Shen, H., 2017. Loads on Structure and Waves in Ice (LS-WICE) project, Part 1: Wave attenuation and dispersion in broken ice fields, *Proc. 24th Int. Conf. Port and Ocean Engineering under Arctic Conditions (POAC)*, 11–16.06.2017, Busan, Korea. (*this volume*)

Dumont, D., Kohout, A. & Bertino, L., 2011. A wave-based model for the marginal ice zone including floe breaking parameterization, *J. Geophys. Res.*, 116, C04 001.

Herman, A., 2016. Discrete-Element bonded-particle Sea Ice model DESIgn, version 1.3a – model description and implementation, *Geosci. Model Dev.*, 9, 1219–1241.

Horvat, C. & Tziperman, E., 2015. A prognostic model of the sea-ice floe size and thickness distribution, The Cryosphere, 9, 2119–2134.

Kohout, A. & Meylan, M., 2008. An elastic plate model for wave attenuation and ice floe breaking in the marginal ice zone, *J. Geophys. Res.*, 113, C09 016.

Langhorne, P., Squire, V., Fox, C. & Haskell, T., 1998. Break-up of sea ice by ocean waves, *Annals Glaciology*, 27, 438–442.

Li, H., Tsarau, A., Shen, H., Herman, A., Evers, K.-U., 2017. Loads on Structure and Waves in Ice (LS-WICE) project, Part 4: Ice collisions under wave actions, *Proc. 24th Int. Conf. Port and Ocean Engineering under Arctic Conditions (POAC)*, 11–16.06.2017, Busan, Korea. (*this volume*)

Squire, V., Dugan, J., Wadhams, P., Rottier, P. & Liu, A., 1995. Of ocean waves and sea ice, *Annu. Rev. Fluid Mech.*, 27, 115–168.

Squire, V., 2007. Of ocean waves and sea-ice revisited, *Cold Regions Sci. Tech.*, 49, 110–133.

Tsarau, A., Sukhorukov, S., Herman, A., Evers, K.-U. & Løset, S., 2017. Loads on Structure and Waves in Ice (LS-WICE) project, Part 3: Ice-structure interaction under wave conditions, *Proc. 24th Int. Conf. Port and Ocean Engineering under Arctic Conditions (POAC)*, 11–16.06.2017, Busan, Korea. (*this volume*)

Williams, T., Bennetts, L., Squire, V., Dumont, D. & Bertino, L., 2013. Wave-ice interactions in the marginal ice zone. Part 1: Theoretical foundations. *Ocean Modelling*, 71, 81–91.

Thermodynamics and Consolidation of Ice Ridges for Laboratory Scale

Evgenii Salganik[1,2], Knut Vilhelm Høyland[1], Aleksey Shestov[1,2]
[1] Sustainable Arctic Marine and Coastal Technology (SAMCoT), Centre for Research-based Innovations (CRI), Norwegian University of Science and Technology, Trondheim, Norway
[2] The University Centre in Svalbard, Longyearbyen, Norway

ABSTRACT

First-year ice ridge interaction with structures often gives highest loads and can be modelled in controlled environment in ice basins. Five laboratory experiments were performed to study model-scale first-year ice ridge development. Effect of initial rubble temperature on consolidated level growth was observed. For both ridges with low and high initial rubble temperatures, consolidated layer was 2–4 times thicker than surrounding level ice at the initial phase of experiment. At the main phase of consolidation this ratio approaches lower equilibrium value of 1.2–1.7 of level ice thickness that is also depends on initial rubble temperature. Non-linear sea ice specific heat capacity can change consolidation development that results in sufficient difference from ice thickness prediction using Leppäranta (1993) and Ashton (1989) approaches.

Observed ratios of air, ice top and bottom surface temperatures can be used for consolidated layer thickness predictions in laboratory conditions using obtained heat transfer coefficient H_{ia}.

During the main phase vertical conductive heat flux at the top of consolidated layer was about two times higher than heat flux at the bottom part due to sea ice cooling. Latent heat flux was slightly lower than vertical conductive heat flux at the bottom of consolidated layer due to natural water convection.

Consolidated layer bulk salinity was always lower than salinity of surrounding level ice for provided experiments. This difference was becoming larger after melting phase.

This study can be approach for better understanding of the main differences between thermodynamics of model-scale and full-scale ice ridges.

KEY WORDS: Ice, ridges, thermodynamics, consolidation, laboratory.

INTRODUCTION

Increasing level of transportation and exploration in the Arctic enhance the significance of ice loads on coastal and offshore structures. Loads from ice ridges often give highest quasi-static loads. In contrast with level ice, loads from ice ridges depends on several that are hard to measure directly in field.

The thickness of consolidated layer h_c is one of these and because of uncertain data the load estimation may become inaccurate. Analysis of mesoscale experiments requires good understanding of aspect ratio effect on stress distributions. However, it is possible to model ice-structure interaction with controlled key parameters in ice basins.

Mechanical scaling traditionally involves decreasing of both ice strength and elastic parameters. This originates from studies of icebreakers advancing through level ice where inertia and gravity forces both play key roles. It is not obvious that the gravity contribution is necessary for relatively slow interaction between ice ridges and fixed structures. One of the main problems connected to mechanical scaling is that in vicinity of the melting point not only strength and elastic parameters are changing but also mechanisms of ice failure and applicability of elastic material model.

Thermodynamics of ice ridge governs both the thickness and the strength of consolidated layer, two key parameters for ice ridge load determination: ISO/FDIS/19906 (2010) recommends modelling the consolidated layer of ice ridges as thick level ice, even though it may have different salinity, ice texture and temperature profile.

Laboratory experiments were provided to understand how controlled consolidation parameters (air and water temperature, initial ice temperature, dopant fraction and time) could affect both consolidated layer thickness and salinity for laboratory scale.

The main goal of this study is to investigate ridge consolidation process because ratios of different thermal processes (conduction, convection, solidification, salt expulsion and initial rubble sensible heat at temperature T_0) is different for different scales while laboratory scale is used for basin tests and full scale is used for collecting and verification of ice ridge thermal, mechanical and geometrical parameters.

RIDGE CONSOLIDATION THEORY

Full-scale ridge development usually consists of three main phases: initial, main and decay (Høyland and Liferov, 2005). For laboratory scale, initial phase (when ice rubble temperature T_0 is lower than freezing point of surrounding water T_f) can continue during significant part of the whole experiment time (Chen and Høyland, 2016). For adiabatic conditions, realized in an ice rubble, the change of initial keel macro-porosity $\Delta\eta$ for fresh ice is equal to:

$$\Delta\eta = (1-\eta)\frac{c_i\left(T_f - T_0\right)}{L_i} = (1-\eta)Ste, \tag{1}$$

where η is the initial rubble macro-porosity, c_i is the fresh ice specific heat capacity, L_i is the latent heat of ice, Ste is the Stefan number.

Change of keel macro-porosity $\Delta\eta$ for saline ice is higher due to change of ice micro-porosity during temperature change and so that strongly depends on freezing temperature of surrounding water (Schwerdtfeger, 1963):

$$\Delta\eta = (1-\eta)\frac{\int_{T_0}^{T_f} c_{si}(T)dT}{L_{si}}, \tag{2}$$

where c_{si} is the sea ice specific heat capacity, L_{si} is the latent heat of sea ice.

Heat convection in the water initiated by solidification can decrease saline ice growth rate providing heat flux around 280 W/m^2 in the beginning of initial phase and around 90 W/m^2 in the late phase for ice initial temperature of 35°C (Chen and Høyland, 2016). That corresponds to heat transfer coefficient H_{iw} of 8 W/m^{2}K for the initial phase.

Forced and natural air convection affects consolidation rate by governing ice top surface temperature T_s depending on ice thickness h and ice surface roughness, air temperature T_a and circulation, and water freezing temperature T_f. Biot number represents ratio of conduction resistance within a solid ice to the external convection resistance offered by the surrounding air (Bergman, et al., 2011):

$$Bi = \frac{H_{ia}h}{k_i} = \frac{T_f - T_s}{T_a - T_s}, \tag{3}$$

where H_{ia} is the heat transfer coefficient, h is the ice thickness, k_i is the ice thermal conductivity.

Recommended values of heat transfer coefficient H_{ia} are in the range of 10–30 W/m^{2}K for still air and 6.7 m/s wind speed correspondingly (Ashton, 1989).

Ice thickness under assumption of linear temperature profile and no water convection can be calculated from the top ice surface T_s and the water freezing temperature T_f as (Stefan, 1891):

$$h(t) = \sqrt{h(t_0)^2 + \frac{2k_i}{\rho_i L_i}\int_{t_o}^{t}(T_f - T_s)dt}, \tag{4}$$

where k_i is the ice thermal conductivity, ρ_i is the ice density, L_i is the latent heat of ice.

The surface temperature T_s can be significantly higher than the air temperature T_a for thin ice growth. Under assumption of equal convective heat flux to the atmosphere and conductive heat flux through the ice Ashton (1989) derived an equation for the ice thickness, based on values of air temperature T_a, water freezing point T_f, and heat transfer coefficient H_{ia}:

$$h(t) = \sqrt{\frac{2k_i}{\rho_i L_i}(T_f - T_a)t + \left(\frac{k_i}{H_{ia}}\right)^2} - \frac{k_i}{H_{ia}} \tag{5}$$

Consolidated layer growth can be calculated assuming reduced value of the ice latent heat multiplied by the value of macro-porosity η (Leppäranta, 1993):

$$L_c = \eta L_i \tag{6}$$

This equation is valid only under assumption of constant macro-porosity in the ridge keel part that will consolidate.

EXPERIMENTAL SETUP

A series of laboratory experiments were performed with different initial rubble temperature and different thermal boundary conditions: 2D and 3D configurations (Figure 1). One vertical layer of ice rubble partly insulated from sides and from the bottom by acrylic walls was used at 2D configuration. Plastic net with 30x30 cm horizontal cross-section was filled with ice rubble at 3D configuration. Two sides of the plastic net were thermally insulated during run 4 and there was no insulation of model ridge at run 5.

Ice for model ridges was prepared in 1000 L water tank and cut into pieces of 8x4x4 cm (run 1, 2, 3 and 5) and 7x5x3 cm (run 4). Then transparent acrylic box (2D) or plastic net (3D) with ice rubble at initial temperature was placed in the middle of the tank with water at the freezing point. Two of the models (run 1 and 5) went through transition zone from consolidation to melting. Air temperatures were in the range of -(4–15)°C during initial and main phases of consolidation and around +5°C during decay phase. Amount of freezing degree-days ($FDD_{air} = \int(T_a - T_w)dt$) was in the range of 11–30°Cd. Initial water salinity $S_{w,0}$ was in the range of 20–34 ppt, rubble salinity $S_{i,0}$ was 3.8–7.0 ppt for different experiment runs.

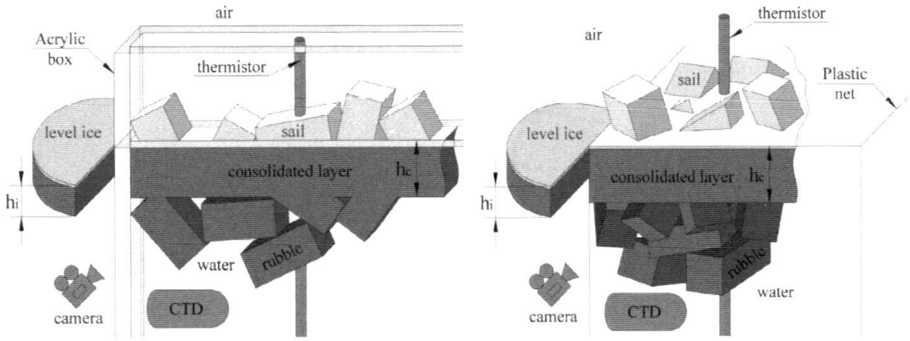

Figure 1. Experimental setup for 2D (left) and 3D (right) configurations

Thermistor strings and CTD (electrical conductivity, temperature and depth) sensors measured vertical temperature profile in air, consolidated layer, rubble, water, and water salinity and freezing temperature. Initial macro-porosity and consolidated layer thickness of 2D model ridges were obtained by underwater camera image processing. Freezing and upper surface temperatures of ridges were assumed equal to the temperature values measured by thermistor string at the bottom and the top surface of consolidated layer. Level ice thickness, water freezing temperature and salinity of water below level ice bottom surface were manually measured by ruler, thermometer (Ebro TFX 410-1) and conductivity meter (Mettler Toledo SG7-FK2).

Consolidated layer thickness was measured manually after each experiment. Initial macro-porosity of 3D model ridges was obtained from manually measured keel volume and number of rubble blocks. Evolution of consolidated layer and corresponding freezing temperatures were obtained from analysis of vertical temperature profiles from thermistors.

Two types of thermistor strings were used: 100 cm length thermistor string with metal cover and 15 cm length negative temperature coefficient thermistor string with plastic cover (Chen and Høyland, 2016).

After experiments, model-scale ridges were taken from water tank for geometrical, temperature, density and salinity measurements. Sea ice density ρ_{si} was measured by hydrostatic weighing in paraffin (Pustogvar and Kulyakhtin, 2016).

RESULTS

Key initial parameter's values and experiment results after a main consolidation phase are presented in Table 1. The ratio of consolidated layer h_c and level ice thickness h_i is called degree of consolidation $R = h_c/h_i$. It can be estimated from experimental data and also can be derived using Stefan's predictions $R^{Ste} = h_c^{Ste}/h_i^{Ste}$ based on measured consolidated layer surface temperatures and initial macro-porosity of consolidated layer η.

For lower rubble initial temperature for both 2D and 3D experiments give higher degree of consolidation in comparison to Stefan's predictions. Average values of heat transfer coefficient

H_{ia}, derived from consolidated layer top and bottom surfaces and air temperatures using equation (3), were higher for experiments with colder initial rubble. At the same moment heat transfer coefficient H_{ia}, based on direct level ice thickness and air temperature values, was around 20 W/m^{2}K for all provided experiments.

Table 1. Consolidation experiments summary

N	Type	T_0	η	$\Delta\eta$	$S_{i,0}$	$S_{w,0}$	FDD_{air}	H_{ia}	h_i	h_i^{Ste}	h_c	h_c^{Ste}	R	R^{Ste}	R/R^{Ste}
-	-	°C	-	-	ppt	ppt	°Cd	W/m^{2}K	cm	cm	cm	cm	-	-	-
1	2D	-2.7	0.41	0.20	7.0	20.2	15.4	7.0	6.5	4.2	7.5	6.6	1.15	1.57	0.73
2	2D	-7.9	0.49	0.25	7.0	22.0	11.0	15.0	4.6	6.1	8.0	8.7	1.74	1.43	1.22
3	2D	-6.4	0.38	0.25	7.0	24.1	29.3	22.0	9.5	13.7	12.9	20.5	1.36	1.50	0.91
4	3D	-17.6	0.47	0.21	5.0	24.0	15.1	20.0	9.0	9.3	11.9	13.6	1.32	1.46	0.90
5	3D	-3.0	0.31	0.04	3.8	34.2	29.5	7.0	13.5	18.1	18.0	32.3	1.33	1.78	0.75

Both ice top and bottom surface temperatures of consolidated layer are different from level ice temperatures due to different ratios of conduction and convection, and due to different growth rate and salt expulsion for 2D and 3D configurations. Consolidated layer average vertical temperature gradient was almost constant during experiment time while values of the gradient were significantly higher at the top ice surface than at the bottom surface. For typical vertical temperature profile, the air temperature was significantly lower than ice top surface temperature while ice bottom temperature was slightly lower than water temperature below (Figure 2).

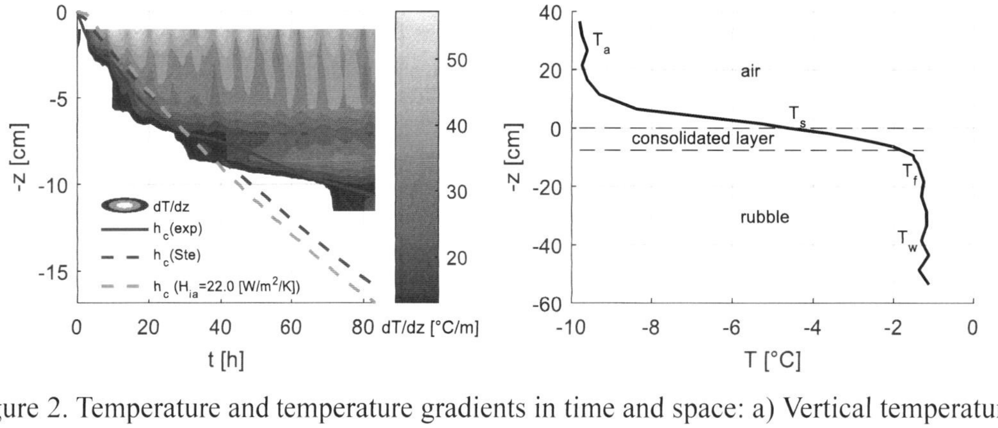

Figure 2. Temperature and temperature gradients in time and space: a) Vertical temperature gradient dT/dz vs time t and water depth z. Red solid curve is h_c from video analysis, red dashed curve represents h_c prediction from Stefan's equation and green dashed curve represents h_c prediction from Ashton's equation using heat transfer coefficient of 22 W/m^{2}K and b) temperature profile of model ridge after t = 40 h (right) for run 3 (2D)

Macro-porosity measurements are complex for volumetric (3D) ice ridges while image processing can give its values for planar (2D) consolidation experiments. Initial macro-porosity vertical distribution is important for ice growth analysis and its values could have significant deviations from average values. Without taking into account low values at the box corners macro-porosity for run 3 was in the range of 0.12–0.26 (Figure 3).

For thin ice growth, difference between air and water freezing temperature is usually much higher than difference between ice top and bottom surface temperatures. For provided experiments ratio between accumulated $FDD_{air} = \int(T_a - T_w)dt$ and $FDD_{ice} = \int(T_s - T_f)dt$ was in the range of 1.7–2.4 during experiments.

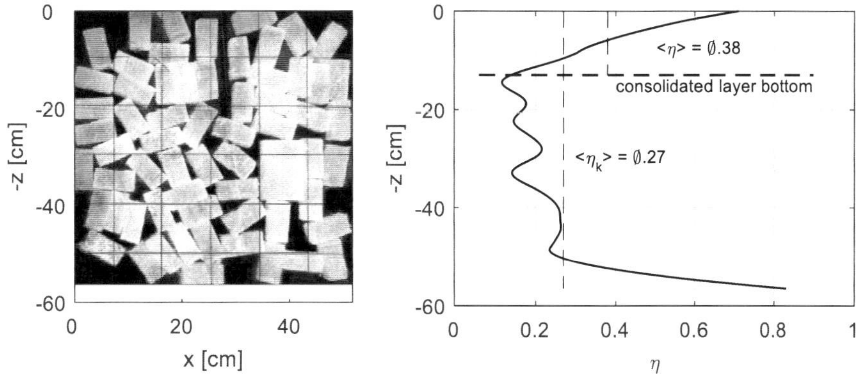

Figure 3. Initial rubble configuration (left) and macro-porosity distribution η vs water depth z (right) for run 3

Ratio of consolidated layer and level ice thicknesses, called degree of consolidation R, is a convenient way to represent ridge development. To neglect different ice growth rate for different macro-porosities η for different experiments and their stages, ratios of experimental and analytical degree of consolidation are presented at the Figure 4. Analytical values are based on Stefan's equation for ice growth, measured surface and water freezing temperatures and latent heat of fusion for consolidated layer $L_{i,c} = \eta L_i$.

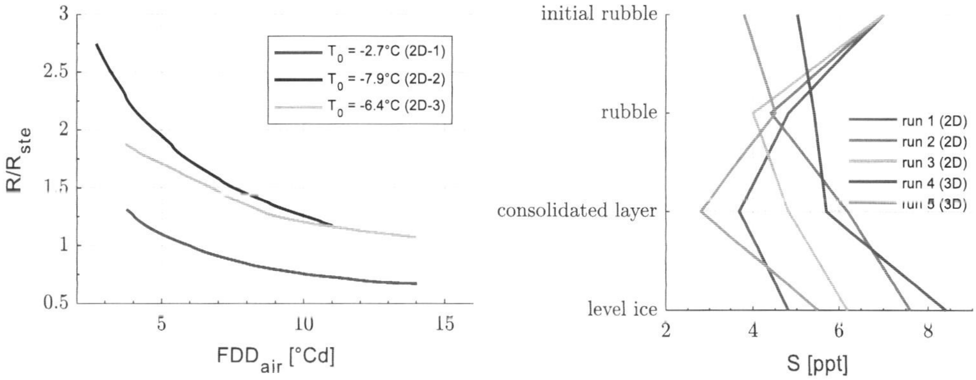

Figure 4. Ratio of experimental and analytical degree of consolidation vs FDD_{air} for 2D experiments with different initial rubble temperatures T_0 (left) and salinity of level ice, initial ridge rubble, consolidated layer and surrounding level ice (right)

Ice salinity usually depends on water salinity and ice growth rate. Consolidated layer was more saline than surrounding level ice (Figure 4). After decay phase, consolidated layer was becoming less saline than rubble and almost two times less saline than level ice. That means that ice consolidated layer had almost two times less micro-porosity than level ice around model ice ridge after the main phase. Bulk ice salinity was linearly decreasing during freezing.

Air content of sail is close to air content of consolidated layer top while air content of consolidated layer bottom is close to air content of rubble. Average air contents of level ice and consolidated layer are close. Density of model ridge rubble was slightly lower than density of level ice and consolidated layer.

DISCUSSION

After short time of consolidation, less than 5°Cd, consolidated layer thickness was almost 2 times higher than thickness value predicted only by initial rubble macro-porosity in comparison to level ice thickness (Figure 4). This effect was decreasing in time, and consolidated layer thickness started approaching equilibrium value that also depends on initial rubble temperature. For initial rubble temperatures close to water freezing point, this value is around 0.7 for both 2D and 3D experiments. For lower initial temperatures values of R/R^{Ste} are higher for 2D configuration than for 3D.

It is possible to estimate change of initial macro-porosity $\Delta\eta$ based on values of the rubble initial temperature, the sea ice specific heat capacity, the sea ice latent heat and initial rubble porosity η using equation (2). Ratio of R/R^{Ste} can be estimated from macro-porosity as:

$$\frac{R}{R^{Ste}} = \sqrt{\frac{\eta}{\eta - \Delta\eta}} \qquad (7)$$

It is assumed that heat diffusion in the rubble is fast enough to form new ice before consolidated layer significant growth. For provided experiments, values of estimated ratios of R/R^{Ste} were in the range of 1.1–1.7. However, this assumption cannot explain higher R/R^{Ste} experimental values during initial phase. It also does not account heat loss due to water convection and ice cooling that can explain slower ice growth at the late stages of experiments.

Two main processes govern consolidated layer growth rate: extraction of initial sensible energy $\int cdT$ due to temperature increase and upper surface cooling by air convection. First process strongly depends on ice rubble salinity. New ice is forming faster around fresh ice (for several hours depending on ice rubble size) but saline ice can store higher amount of energy to form new ice around it due to non-linear sea ice specific heat capacity c_{si} and sea ice thermal conductivity k_{si}. Fourier number can be used for dimensionless analysis of described heat transfer.

Second governing process is consolidation due to cooling from above. This process also depends on ice salinity because for high values of saline ice specific heat capacity significant amount of heat should be spent on sea ice cooling so less heat can be available for ice formation. It can be seen from non-linear ice vertical temperature gradient profile during provided experiments.

Cooling can decrease effect of initial macro-porosity change because ice at freezing temperatures of surrounding contains sufficient amount of brine that should be partly frozen after cooling. It should be said that for engineering or basin test application consolidated layer thickness at the air temperature equal to the surrounding water freezing temperature is a value of interest. It is also a convenient value for experimental analysis because all the energy stored in initial rubble should be spent on consolidation process not on cooling.

Another complication of consolidated layer formation analysis is the effect that due to lower permeability and slow solute diffusion water salinity near consolidation surface is significantly higher than initial water salinity that leads to less amount energy that could be extracted from cold rubble.

According to Chen and Høyland (2016) for ice surrounded by fresh water, amount of new formed fresh ice around saline ice $\Delta V/V$ is 48 % higher than for fresh ice: 26 % and 16 % respectively. This value corresponds to the volume change $\Delta V/V$ calculated from sea ice specific heat capacity values by Schwerdtfeger (1963) including total melting of initial saline ice:

$$\frac{\Delta V}{V} = \frac{\int_{T_0}^{T_f} c_{si}(T)dT}{L_i} = \frac{\int_{-35}^{-0.13} c_{si}(T)dT}{L_i} = 1.34, \tag{8}$$

where $T_f = -0.13°C$ is the freezing temperature of sea ice with salinity of 2.65 ppt.

The test results prove that initial rubble temperature changes not only initial degree of consolidation values but also its values during cooling. This degree of consolidation value depends on ratio of initial rubble temperature and consolidated layer temperature at the end of experiment. It can be confirmed by the fact that degree of consolidation values after decay phase (when consolidated layer temperature is close to water freezing point) were approaching $1/\sqrt{\eta}$ values (and $R/R^{Ste} = 1$) for experiments with high rubble initial temperature.

For experiment runs 2 and 3 with rubble initial temperatures of -7.9°C and -6.4°C degree of consolidation values are approaching values $1/\sqrt{\eta}$ because these temperatures are relatively close to final consolidated layer surface temperatures of -4.0°C and -5.4°C respectively.

These assumptions can explain why consolidated layer growth is faster at the beginning and slower after initial phase (Figure 2).

During the main phase vertical conductive heat flux at the top of consolidated layer $(k_{si}dT/dz|_{z=0+})$ is about two times higher than heat flux at the bottom part $\left(k_{si}dT/dz|_{z=h_c-}\right)$ while this difference accounts for ice cooling. Latent heat flux $(\rho_{si}L_{i,c}\, dh_c/dt)$ is at the same moment slightly lower than vertical conductive heat flux at the bottom of consolidated layer. This difference accounts for water convection that is not considered at Stefan's condition:

$$\rho_{si}L_{i,c}\, dh_c/dt = k_{si}dT/dz|_{z=h_c-} \tag{9}$$

Lower initial rubble temperature also leads to higher ratio of conduction resistance to the external convection air resistance. While thin level ice growth is governed by air convection, consolidated layer growth is also controlled by initial sensible energy. However, for full-scale first-year ice ridges difference between sail surface temperature and air temperature is insignificant (Shestov and Ervik, 2016).

2D experiments give easily accessible data of macro-porosity distribution that is very valuable for consolidation analysis. 3D configuration provides realistic thermal boundary conditions with insulation along imaginable model ridge axis, without insulation at water-keel boundaries and realistic permeability of rubble. There is no perfect insulation at ridge sides for 2D configuration. It creates addition heat fluxes oriented not in vertical direction. Permeability difference leads to higher difference in level ice and consolidated layer bottom surface temperatures due to higher water bulk salinity around rubble inside 2D box.

Consolidated layer bulk salinity was always lower than salinity of surrounding level ice. This difference was becoming larger after melting phase. It could be critical for scale-model mechanical experiments because ice strength is governed by salinity and temperature while thinner level ice is more saline and its temperature have to decrease faster than for thicker consolidated layer.

CONCLUSIONS

Five laboratory experiments were performed to study model-scale ridge development. Effect of initial rubble temperature on consolidated level growth was observed during initial and main phases of ridge formation. For both ridges with low and high initial rubble temperatures, consolidated layer was 2–4 times thicker than surrounding level ice at the initial phase of experiment. At the main phase of consolidation this ratio approaches lower equilibrium values in the range of 1.2–1.7 that is also depends on initial rubble temperature. Effect of non-linear sea ice specific heat capacity on degree consolidation was described in order to explain sufficient difference from ice thickness prediction using Leppäranta (1993) and Ashton (1989) approaches.

Difference in ratio of conduction resistance within a solid ice to the external convection resistance offered by the surrounding air was observed for level ice and consolidated layer formed from a rubble of different initial temperatures. Observed ratios of air, ice surface and water freezing temperatures can be used for consolidated layer thickness predictions for laboratory conditions using obtained heat transfer coefficient H_{ia}.

During the main phase vertical conductive heat flux at the top of consolidated layer was about two times higher than heat flux at the bottom part due to ice cooling and high non-linear values of sea ice specific heat capacity. Latent heat flux was slightly lower than vertical conductive heat flux at the bottom of consolidated layer due to natural water convection.

2D and 3D experimental configurations and their advantages and potential method uncertainties of usage for consolidation process study were described.

Effect of different consolidation conditions for level ice and rubble was observed during experiments. Consolidated layer bulk salinity was always lower than salinity of surrounding level ice for provided experiments. This difference was becoming larger after melting phase that could be critical for scale-model mechanical experiments because ice strength is governed by salinity and temperature while thinner level ice is more saline and its temperature have to decrease faster than for thicker consolidated layer.

This study can be approach for better understanding of the main differences between thermodynamics of model-scale and full-scale ice ridges and for the development of ridge consolidation model. Future consolidation studies have to take into account main effects governing this process: air convection above ice surface, salt expulsion and diffusion below consolidated layer, initial sensible energy, macro-porosity distribution and non-linear specific heat capacity of sea ice during bulk salinity development.

ACKNOWLEDGEMENTS

The authors wish to acknowledge the support of the Research Council of Norway through the Centre for Research-based Innovation, SAMCoT, and the support of the SAMCoT partners. The authors wish to acknowledge the economic support of the Research Council of Norway through the project “Field studies and modelling of sea state, drift ice, ice actions and methods of icebergs management on the Arctic shelf, 2015-2017 (FIMA)”. The logistics and Arctic Technology department at UNIS are acknowledges for helping during laboratory experiments.

REFERENCES

Ashton G.D., 1989. Thin Ice Growth. *Journal of Water Resources Research.* Vol. 25, No. 3. pp. 564–566.

Bergman, T.L., Incropera, F.P. and Lavine, A.S., 2011. *Fundamentals of heat and mass transfer*. John Wiley & Sons.

Chen, X. and Høyland, K.V., 2016. Laboratory Work on Heat Transfer in Submerged Ice, Theory, Experimental Setup and Results. *Proceedings the 23rd IAHR International Ice Symposium on Ice*, Ann Arbor, USA.

Høyland, K.V., Liferov, P., 2005. On the initial phase of consolidation. *Cold Regions Science and Technology*. (41, 1), pp. 49-59.

Høyland, K.V., 2002. Consolidation of first-year sea ice ridges. *Journal of geophysical research*. 107(C6): 3062.

ISO/FDIS/19906, 2010. Petroleum and natural gas industries - Arctic offshore structures. Standard.

Leppäranta, M., 1993. A review of analytical models of sea- ice growth. *Atmosphere-Ocean*, 31, pp. 123-138.

Pustogvar, A. and Kulyakhtin, A., 2016. Sea ice density measurements. Methods and uncertainties. *Cold Regions Science and Technology*. vol. 131.

Schwerdtfeger, P., 1963. The thermal properties of sea ice. *Journal of Glaciology*, 4: pp. 789-807.

Shestov, A. and Ervik, Å., 2016. Studies of Drifting Ice Ridges in the Arctic Ocean during May-June 2015, Part II. Thermodynamic properties and melting rate. *Proceedings the 23rd IAHR International Ice Symposium on Ice*, Ann Arbor, USA.

Stefan, J., 1891. Uber die Theorie der Eisbildung, insbesondere uber die Eisbildung im Polarmeere, *Annalen der Physik und Chemie*, 42, pp. 269-286.

Loads on Structure and Waves in Ice (LS-WICE) project, Part 3: Ice-structure interaction under wave conditions

Andrei Tsarau[1], Sergiy Sukhorukov[2], Agnieszka Herman[3], Karl-Ulrich Evers[4], and Sveinung Løset[1]
[1] Sustainable Arctic Marine and Coastal Technology (SAMCoT), Centre for Research-based Innovation (CRI), Norwegian University of Science and Technology, Trondheim, Norway
[2] KvaernerAS, Lysaker, Norway
[3] Institute of Oceanography, University of Gdansk, Poland
[4] Arctic Technology, Hamburgische Schiffbau-Versuchsanstalt GmbH, Hamburg Ship Model Basin, Bramfelder Straße 164, D-22305 Hamburg, Germany

ABSTRACT

A multi-group investigation was conducted at Hamburgische Schiffbau-Versuchsanstalt GmbH (HSVA) from Oct. 24 to Nov. 11, 2016 under the project: Loads on Structure and Waves in Ice (LS-WICE). There are three parts to this investigation: ice fracture under wave actions, wave attenuation/dispersion in broken ice covers, and ice-structure interaction under wave conditions. This paper focuses on the last part of the investigation in which a cylindrical structure was subjected to impact loads due to ice floes in regular waves.

The global loads from both only waves and combined ice/wave action were obtained. Waves with periods 1.5 s – 2.0 s and heights 25 mm – 75 mm were used in the tests. Two additional runs were performed with wave heights 100 mm and 200 mm and wave period 1.6 s; major fractures of the ice floes were observed in these test runs. All tests were well documented with help of pressure transducers, ultrasound sensors, accelerometers on the ice, load cells on the structure, an optical system and several cameras, which were covering the ice-structure interaction zone both above and under the water level.

KEY WORDS: Marginal Ice Zone; Waves; Loads on Structure; Model Tests.

INTRODUCTION

In the light of the observed climate change, Arctic marine structures may have to operate in Marginal Ice Zones (MIZ), in which the ice cover typically consists of individual ice floes formed by ocean waves penetrating into the ice field. Both vessel traffic and offshore structures may be subject to MIZ conditions, at least during a part of the ice season. Therefore, studying wave-ice interactions and their effects on a marine structure in such conditions is of practical importance.

A multi-group investigation was conducted at Hamburgische Schiffbau-Versuchsanstalt GmbH (HSVA) from Oct. 24 to Nov. 11, 2016 under the Hydralab+ Transnational Access project: Loads on Structure and Waves in Ice (LS-WICE). There are three parts to this investigation: ice fracture under wave actions, wave attenuation/dispersion in broken ice covers, and ice-structure interaction under wave conditions. This paper focuses only on the third part. Several accompanying papers at this conference present the other parts of the LS-WICE project (scc Chcng et al., 2017; Herman et al., 2017; and Li et al., 2017).

DESCRIPTION OF THE EXPERIMENT

The experiment focused on studying combined wave and ice actions on a fixed structure represented by a cylinder with a diameter of 0.69 m. The structure was located approximately in the middle of the ice tank at 43.7 m from the wave maker as shown in Figure 1. The ice concentration near the structure was approximately 100%. Apart from the structure and the ice field, Figure 1 shows a number of sensors and video cameras that were used in the tests: 12 pressure sensors and 2 ultrasound sensors to measure wave height; Qualisys cameras and markers to monitor ice deflection; an Inertial Measurement Unit (IMU) to record the accelerations of an ice floe; video cameras on the ceiling, side walls, above the structure and under the water surface to monitor the ice cover and the ice-structure interaction area. A set of load cells that were installed inside the structure and fixed to a rigid carriage was used to measure loads on the structure. The IMU was installed at the centre of an ice floe adjacent to the structure to obtain the response in waves and impact accelerations (Figure 2).

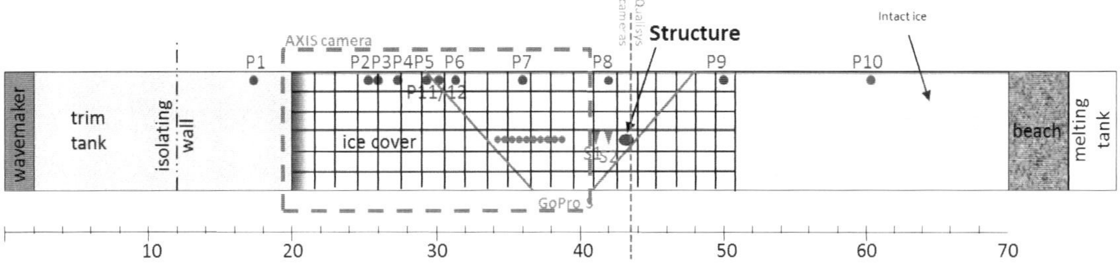

Figure 1. Layout of the wave tank and sensors. P1 – P10 are single pressure sensors, P11/12 – double pressure sensor, S1 and S2 – ultrasound sensors, continuous blue lines – fields of view of GoPro camera, dashed blue lines – field of view of the AXIS camera, continuous black lines – locations of longitudinal and transverse cuts (i.e., floe boundaries), yellow rectangle – a region, where no transverse cuts were done during the tests, green points – location of the Qualisys markers, dashed green line – approximate location of the Qualisys cameras and the IMU.

Figure 2. Structure and an adjacent ice floe equipped with an IMU. X shows the surge direction.

The procedure of producing the ice cover is described by Cheng et al. (2017). A level ice sheet was formed by seeding first to create a granular layer. Sustained cooling thickened this top layer by columnar growth until the whole sheet reaches about 0.035 m thickness. The Young's modulus, flexural strength, ice thickness, density and salinity were then obtained (Table 1). The choice of the ice properties was defined by the intention to have similar ice properties as in the tests of wave attenuation in broken ice (Cheng et al., 2017). The intact level ice was then cut longitudinally into six 1.6 m wide strips parallel to the wave tank wall. Transverse cuts were applied to create square floes of a uniform size.

Table 1. Ice properties.

Parameter	Unit	Mean Value
Thickness	m	0.035
Density	kg/m^3	924.8
Salinity	‰	4.0
Elastic Modulus	MPa	40.3
Flexural Strength	kPa	67.5
Compressive Strength*	kPa	200

* The compressive strength was not measured directly. It was obtained using a relationship between the flexural strength and the compressive strength for model ice as provided by HSVA.

Three test series were conducted with different wave periods between 1.5 – 2.0 s and three controlled wave heights between 25 and 75 mm at the wave maker. Two additional runs were performed with wave heights of 100 and 200 mm and a wave period of 1.6 s, during which major ice breaking was observed and therefore the IMU was removed. The water depth in the tank was 2.48 m (the deep-water section did not affect this experiment).

TEST SERIES AND RESULTS

This paper presents only a selected set of data obtained during the experiment. We mainly focus on the following parameters: the surge acceleration of the ice floe adjacent to the structure, a, and its oscillation amplitude due to waves, a_{amp}; additionally, the peak acceleration of the floe when it was impacting the structure, a_{peak}; the amplitude of the hydrodynamic wave force on the structure in the surge direction, F_{hd}; the surge force between the structure and the rigid carriage when the ice floe was impacting, R, and its peak value R_{peak}. The latter parameter is only used to indicate the impact load and cannot be directly interpreted as the total force on the structure due to waves and ice impacts; this is because the structure often responded to impacts at its natural frequency at approximately 10 Hz. This issue did not affect the measurements of F_{hd} as the wave frequencies were much below 10 Hz. Instead of using an inverse analysis of R to identify the actual impact force between the structure and the ice floe, this force was approximately calculated as $F_{imp} = -m*a_{peak}$, where m is the mass of the ice floe. F_{imp} also includes the added-mass effects on the ice floe.

Table 2 summarises all parameters described above for the test series in which the ice floe adjacent to the structure remained intact, except for a small part of the floe crushed by the structure in the interaction zone. The controlled parameters in each test were the wave period T and the wave height at the wave maker. Instead of the latter, Table 2 presents the wave height H_{P8} which was measured by the pressure sensor located closest to the structure (see P8 in Figure 1); this is because of the fact that waves undergo attenuation in broken ice fields. Note that all parameters in Table 2 are related to the model scale.

Table 2. Controlled and measured parameters.

Run Number	T [s]	H_{P8} [mm]	a_{amp} [m/s^2]	F_{imp} [N]	R_{peak} [N]	F_{hd} [N]
4110	2.0	25	0.06	126	186	21
4120	1.8	22	0.2	122	169	20
4130	1.6	18	0.09	141	167	16
4140	1.5	16	0.14	110	159	21
4210	2.0	45	0.14	270	341	35
4220	1.8	40	0.13	289	360	35
4230	1.6	28	0.17	361	353	39
4240	1.5	27	0.26	379	276	41
4310	2.0	67	0.32	--	--	52
4320	1.8	55	0.26	319	438	55
4330	1.6	36	0.26	249	371	56

Typical time series of the forces obtained from the experiment are shown in Figures 3 and 4. The red curve is the surge force on the structure measured by the load cells, and the black force is the total force on the ice floe in the inverse surge direction, which was calculated as -

m^*a. In these figures, the impact forces due to ice-structure collisions appear as long vertical spikes with positive peaks; spikes with negative peaks appear only on the black curves and are attributed to floe-floe collisions.

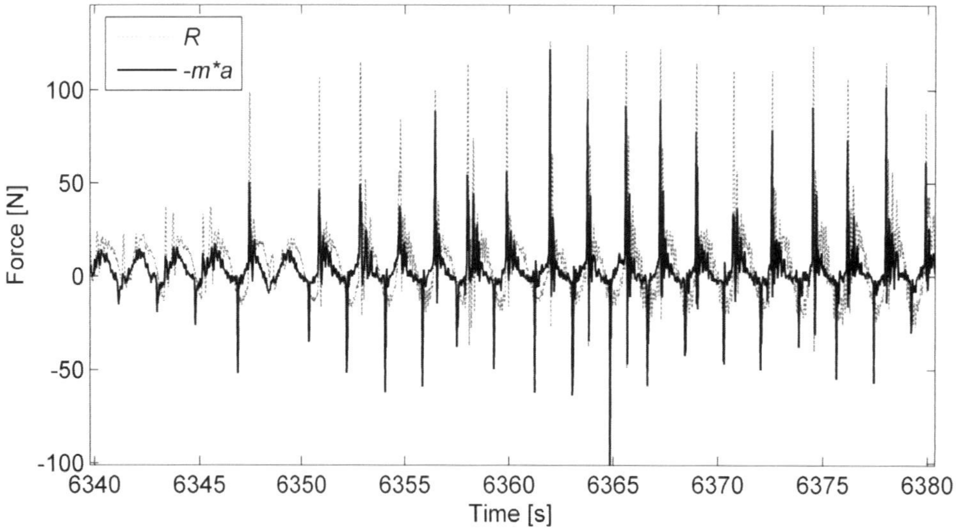

Figure 3. Time series of the surge force on the structure (red) and the inverse surge force on the ice floe (black) in Test 4120.

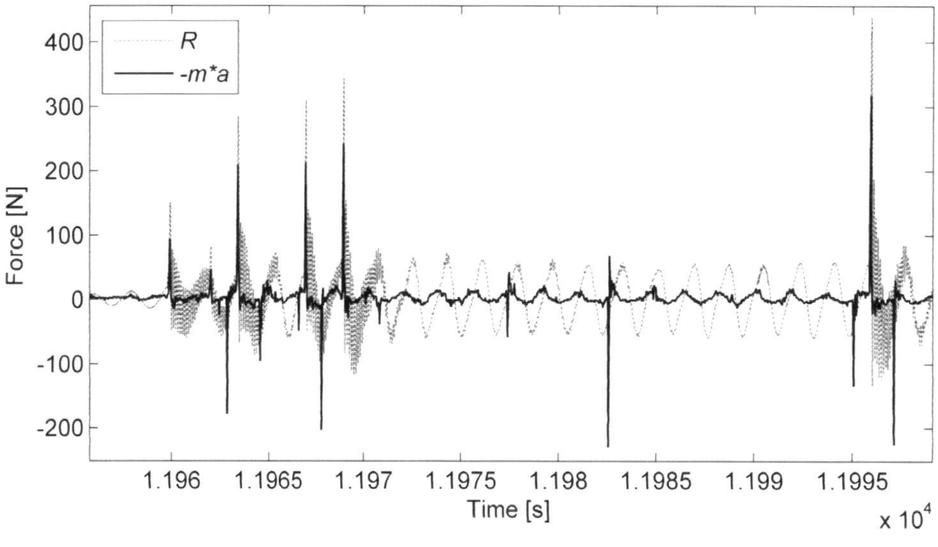

Figure 4. Time series of the surge force on the structure (red) and the inverse surge force on the ice floe (black) in Test 4320.

DISCUSSION ON THE FLOE RESPONSE NEAR THE STRUCTURE

The wave-induced motion of a floating body is usually presented in the form of non-dimensional response amplitude operators (RAO). The surge RAO of an ice floe is defined as

$x/(\frac{1}{2}H)$, where x is the floe motion amplitude in surge and H is the wave height. As the wave frequencies in the tests are known, x can be obtained by analytically integrating a_{amp} from Table 2. Figure 5 presents the calculated RAOs for the ice floe adjacent to the structure based on the acceleration data prior or between any collisions. The results are averaged for different wave heights and the bars show scatter around mean value.

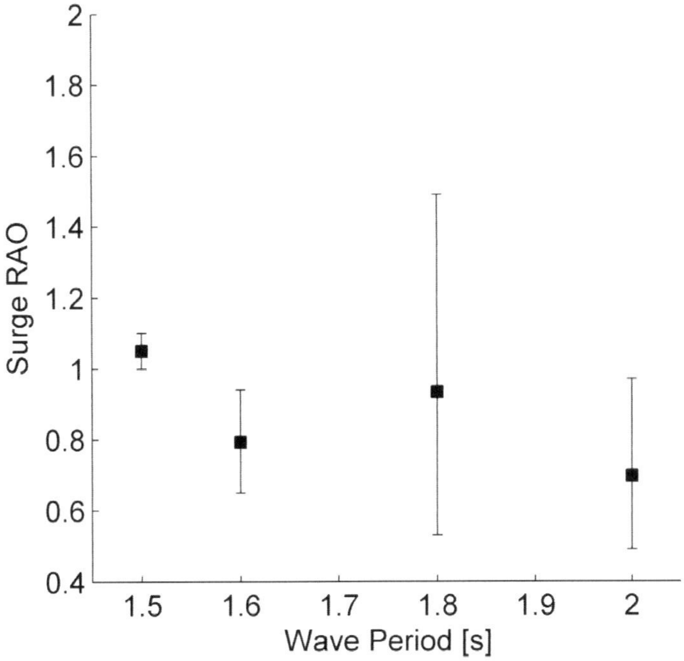

Figure 5. Surge RAO of the ice floe near the structure as a function of wave period.

According to Masson and LeBlond (1989), when the floe size is small compared to the wavelength, the floe tends to behave essentially as a fluid particle. For deep water waves, the water particle travels along closed circular orbits of radius $\frac{1}{2}H$, resulting in a surge RAO of 1. In our case, the obtained RAOs differ from unity because of several reasons: first, the wavelengths were only 2 – 4 times the floe size and therefore the long-wave assumption did not apply; second, the floe responses were also influenced by the presence of the structure due to the hydrodynamic interaction. The latter effect was thoroughly investigated by McGovern and Bai (2014) in a number of wave-tank experiments with a cylindrical structure and floes of different sizes and shapes. Their results showed that when an ice floe was in the vicinity of the structure, the floe's response in surge, regardless of whether there was an eventual impact, was markedly reduced compared to that in the far-field region. The surge RAO of the floe as it approached the cylinder decreased from 1.2 (in the far-field) to approximately 0.8 – 1 just before the impact. Similar results are also demonstrated in Figure 5 for the near-filed RAO.

DISCUSSION ON THE IMPACT FORCES

The experimental setup (Figure 2) ensured that all impacts of the upstream floe on the structure were nearly head-on. However, the impact occurrence in the experiment was not regular: in some tests impacts occurred almost at every wave cycle (e.g., Test 4120, see

Figure 3); in some tests impacts were rare (as shown in Figure 4 for Test 4320); in Test 4310 no impact was detected at all. As it was observed during the tests, impact occurrence seemed to be dependent on the initial separation distance between the floe and the structure and floe-floe interactions. No exact parametric dependence of impact occurrence was found.

As Table 2 presents, the impact forces F_{imp} were more than 5 times higher than the corresponding hydrodynamic forces F_{hd}. This was observed in all tests until the floe finally broke apart due to the wave action and impacts on the structure during Test 4330. There was a clear increase in F_{imp} with increasing wave height in all tests; however, the effect of wave period on F_{imp} is unclear. The increase in F_{imp} from Test 4210 to Test 4240 is most likely attributed to the increase of the interaction area between the floe and the structure due to crushing. Figure 6 shows how the crushed area was developing.

Figure 6. Increasing interaction area in subsequent tests.

As this paper presents only a preliminary analysis of the obtained data, no functional relationships were intended to be established between the impact force and the identified influencing parameters such as wave height and period, floe response in waves, floe-floe interaction prior to impact, penetration depth or interaction area. However, the range of data from the LS-WICE project allows a more thorough analysis that would result in such relationships.

CONCLUSIONS

A preliminary analysis of the LS-WICE data on wave-ice-structure interaction was presented. This analysis focused on both the response of a wave-driven ice floe near the structure and the forces on the structure due to both waves and ice impacts. Regarding the surge RAO of the ice floe, its mean values were not significantly different from 1 for all considered wavelengths. However, it was found that this RAO does not solely account for the variation of the impact forces on the structure and impact occurrence in the experiment. Floe-floe collisions seemed to affect impact occurrence as well.

Among the parameters influencing the impact force, we identified the following:

- wave height and period (also wavelength for shallow water);
- ice-floe kinematics in waves, including momentum exchange due to floe-floe interaction;
- interaction area between the structure and the ice floe.

The effect of ice properties was not assessed in this study. A major difference of this experiment compared to previous experiments on impacts of a wave-driven ice mass on a structure is the utilisation of a wave tank fully covered with ice floes instead of considering only one isolated ice mass and a structure. Full ice coverage ensures a better representation of MIZ conditions and allows taking into account floe-floe interactions and the wave dispersion effects in ice-covered water.

ACKNOWLEDGEMENTS

The work described in this publication was supported by the European Community's Horizon 2020 Programme through the grant to the budget of the Integrated Infrastructure Initiative HYDRALAB+, Contract no. 654110. The author(s) would like to thank the Hamburg Ship Model Basin (HSVA), especially the ice tank crew, for the hospitality, technical and scientific support and the professional execution of the test programme in the Research Infrastructure ARCTECLAB, and the instrumentation help from Meleta Truax of the Clarkson University. Authors AT and SL are also funded in part by the SAMCoT CRI through the Research Council of Norway and all of the SAMCoT Parties.

REFERENCES

Cheng, S., Tsarau, A., Li, H., Herman, A., Evers, K.U., and Shen, H. (2017). Loads on Structure and Waves in Ice (LS-WICE) project, Part 1: Wave attenuation and dispersion in broken ice fields. Proceedings of the 24th International Conference on Port and Ocean Engineering under Arctic Conditions, June 11-16, 2017, Busan, Korea.

Herman, A., Tsarau, A., Evers, K.-U., Li, H., and Shen, H.H. (2017) Loads on Structure and Waves in Ice (LS-WICE) project, Part 2: Sea ice breaking by waves, Proceedings of the 24th International Conference on Port and Ocean Engineering under Arctic Conditions, June 11-16, 2017, Busan, Korea.

Li, H., Tsarau, A., Shen, H., A., Herman, A., Evers, K.U., and Lubbad, R. (2017). Loads on Structure and Waves in Ice (LS-WICE) project, Part 4: Wave attenuation and dispersion in broken ice fields – Ice collisions under wave actions Proceedings of the 24th International Conference on Port and Ocean Engineering under Arctic Conditions, June 11-16, 2017, Busan, Korea.

Masson, D., and LeBlond, P. H. (1989). Spectral evolution of wind-generated surface gravity waves in a dispersed ice field. J. Fluid Mech., 202, 43–81.

McGovern, D.J., Bai, W., 2014. Experimental study of wave-driven impact of sea ice floes on a circular cylinder. Cold Reg. Sci. Technol. 108, 36–48.

POAC'17
Busan, KOREA

Proceedings of the 24th International Conference on
Port and Ocean Engineering under Arctic Conditions
June 11-16, 2017, Busan, Korea

Sea Ice as an Essential Climate Variable from Remote Sensing and Modelling

Prof. Stein Sandven[1]

[1] Nansen Environmental and Remote Sensing Center, Thormohlensgate 47, N-5006 Bergen, Norway

Sea ice covers the polar oceans on both hemispheres and it has a large seasonal variability. Sea ice is an important component of the climate system because it has a high surface albedo compared to open water, together with the polar surface water it insulates the relatively warm ocean from the cold atmosphere, and it forms a barrier to the exchange of momentum and gases such as water vapor and CO_2 between the ocean and atmosphere. Regional climate changes affect the sea ice characteristics and those changes can feed back on the climate system, both regionally and globally.

Systematic and long-term observations of the major sea ice variables is only possible using past and present satellite Earth Observation (EO) data. Sea ice data from satellites has been collected for more than four decades and sea ice mapping is one of the most successful applications of EO data in climate change studies. Several sensors and retrieval methods have been developed and successfully utilized to measure sea ice area, concentration and drift. There are also other sea ice parameters of importance for climate research such as thickness, albedo, snow cover, temperature, duration of the melting season, the density of leads/polynyas and the volume of ridges. [e.g. GCOS, 2010; IGOS, 2007]. Satellite remote sensing can contribute to retrieving quantitative measurements of most of these variables, even though GCOS defines sea ice in general as one **essential climate variable** (ECV). In order to provide quantitative data on sea ice it is necessary to define the variables that can be measured. For climate change studies it is generally accepted that the most important and mature variables, where quantitative data have been obtained over several decades, are ice extent (area), thickness, and drift.

There is evidence that the polar amplification of global climate change affects the sea ice covers of the Arctic and the Antarctic in different ways – in line with contrasting observations of climate relevant parameters during the last decades. The most pronounced change in the Arctic sea ice over the last three decades is the reduction of the sea ice extent observed from time series of passive microwave data in particular the reduction of the summer ice (Fig. 1). This change is also observed in reduction of multiyear ice fraction (Fig. 2), the increase of the length of melt season and as well as in reduction of the ice thickness [e.g. Vaughan et al. 2013, Lindsay and Schweiger, 2015]. The reduction in Arctic ice thickness has been documented by combined observations from submarine sonar data, airborne surveys, in situ measurements and recently by satellite altimeter data from ICESat-1 and CryoSat-2 [e.g. Fig. 3, Kwok and Cunningham, 2015].

The integrated estimate of ice thickness reduction reported by IPCC is 0.62 m per decade, corresponding to about -19.4 % per decade (Table 1). An important aspect for the Arctic is that the thickness reduction is closely linked to the decline of the multiyear ice cover [Vaughan et al. 2013]. While ice extent has decreased at a rate of -3.8 % per decade, the multiyear ice cover has decreased by -13.5 % per decade (Table 1).

Table 1. Trends in Arctic sea ice.

Parameter	Change per decade	Parameter	Change per decade
Ice extent: annual mean	-3.8±0.3 %	Ice thickness (1980-2000, submarine)	-16.5 %
Ice extent: winter	-2.3±0.5 %	Ice thickness (2004-2008, IceSat)	-22.7 % per 5 years
Ice extent: spring	-1.8±0.5 %	Ice thickness (Integrated)	-19.4 %
Ice extent: summer	-6.1±0.8 %	Ice drift (winter average)	+ 10.6 ± 0.9 %
Ice extent: autumn	-7.0±1.5 %	Length of melt season (total)	+ 5.7 days/decade
Ice extent: MY fraction	13.5±2.5 %	Length of melt season (margins)	+10 days/decade

(ref. IPCC, Vaughan et al., 2013)

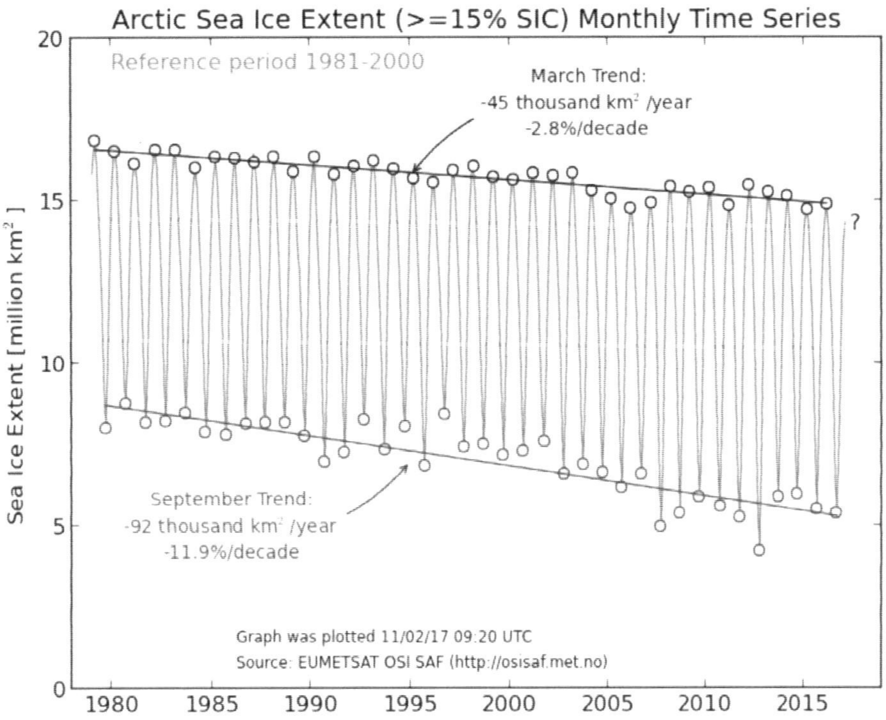

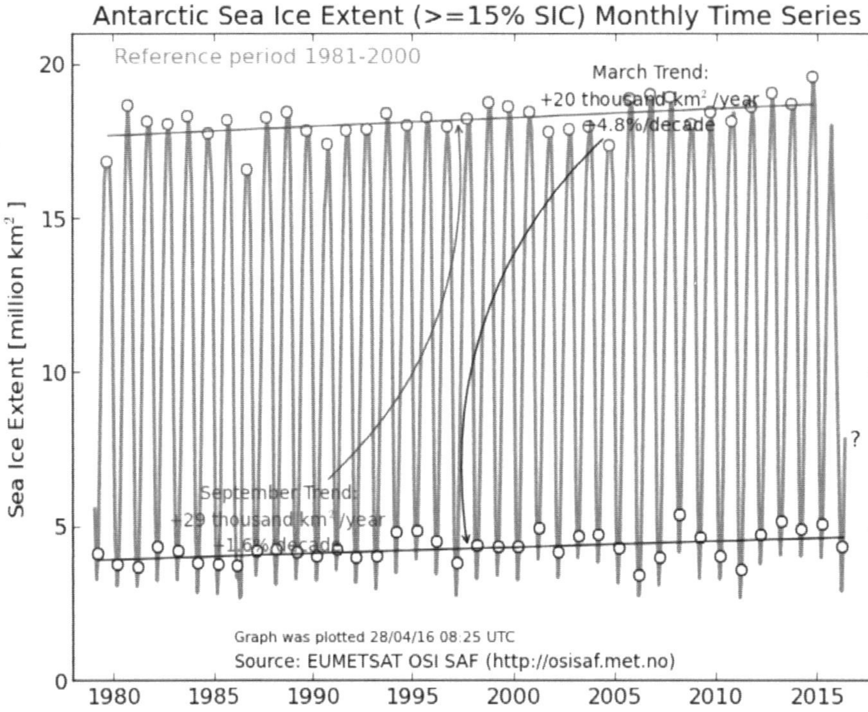

Figure 1. The trend and seasonal cycle of sea ice extent in the Arctic (upper graph) and Antarctic (lower graph) based on passive microwave satellite data from 1978 to present. The data are provided by the Eumetsat Ocean and Sea Ice Application Facility (http://osisaf.met.no).

The multiyear ice extent is a very sensitive climate variable that is not yet established as an ECV. The amount of multiyear ice is important to quantify because multiyear ice is thicker, it has thicker layer of snow and has different physical properties compared to first-year ice. Methods to derive multiyear ice fraction exist but a thorough investigation and quantification of the uncertainties involved has not been undertaken yet. Algorithms combining radiometer and scatterometer data have the potential to improve current time series of the multiyear ice extent. A longer high-quality time series of the multiyear ice extent is also required for improved sea ice thickness retrieval because it permits an improved choice of sea ice densities [Kern et al., 2015].

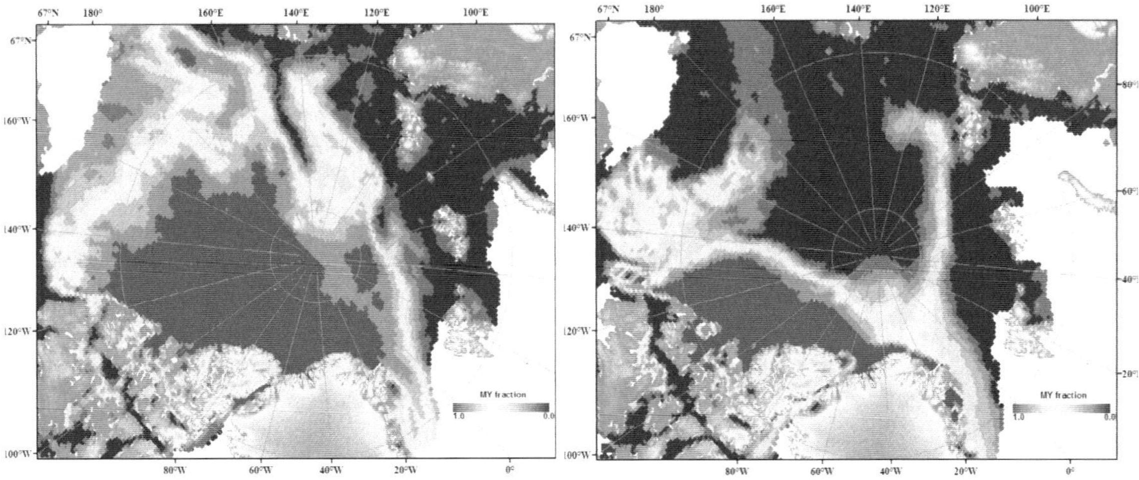

Figure 2. Multiyear fraction of sea ice concentration in the Arctic for March 2002 (left graph) and March 2008 (right graph). Red and yellow areas indicate where multiyear ice is dominating , while ble areas indicate firstyear ice. The maps are produced from scatterometer data from Quikscat available at http://nsidc.org.

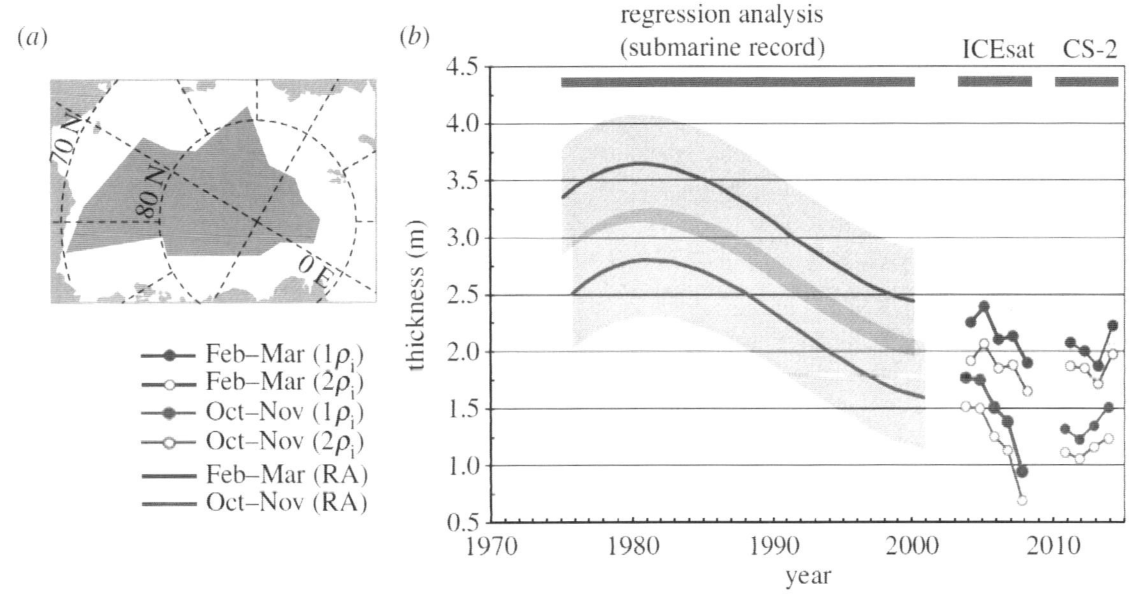

Figure 3. Average winter and autumn ice thickness from regression of submarine data, ICESat, and CryoSat-2 data. (a) data coverage area, (b) ice thickness decline between 1980 and 2014 (Kwok and Cunningham, 2014).

In the Antarctic, there is an increase in sea ice extent (Fig. 1), however, regional trends are diverse [Parkinson and Cavalieri, 2012]. Long, consistent, and error-characterized time series for the other parameters like thickness and drift as well as onset and length of melt season are not available. These are often less thoroughly analyzed and far less mature than their Arctic counterparts. The Antarctic sea ice is, on average, younger and hence thinner than Arctic sea ice while its snow load is, on average, thicker than on Arctic sea ice. As a consequence the large areas of the seasonal Antarctic sea ice have a sea ice freeboard, i.e. the part of the sea ice above the water line, which is close to zero if not even negative. Only the old, thick Antarctic sea ice or deformed seasonal sea ice can exhibit substantially larger sea ice freeboards.

Today operational sea ice monitoring and analysis is fully dependent on use of satellite data. However, new and improved satellite systems, such as multi-polarisation SAR, radar and laser altimeters, require further studies to develop more advanced sea ice remote sensing methods. In climate change studies based on satellite data, it is a major challenge to construct homogeneous time series from a series of consecutive satellite sensors needed for detection of changes over several decades [e.g. Meier et al. 2013]. At the same time there is progress in sensors and observation technology, which makes it possible to observe new parameters in the future.

It is important that the observational community works closely with the modeling community in order to communicate caveats and usefulness of satellite data products from the observational side and requirements to data and their importance from the modeling side. Available sea ice drift data are not necessarily free of inconsistencies due to changes in sensor technology used. Available sea ice thickness data may be based on sub-optimal assumptions.

The ESA CCI Sea Ice project, running from 2012-2018, represents an extensive effort to combine and extend ongoing research to develop improved and validated timeseries of ice concentration and ice thickness for use in climate research (http:// http://esa-cci.nersc.no/). Since sea ice is a sensitive climate indicator with large seasonal and regional variability, the climate research community requires long-term and regular observations of the key ice parameters in both Arctic and Antarctic. The project provides global data sets on ice concentration for Arctic and Antarctic, and ice thickness data sets for the Arctic, to support climate research and monitoring according to the GCOS requirements for generation of satellite-based data sets and products. This implies provision of data sets with associated metadata, software systems, technical documentation and scientific reports/publications.

A prototype ECV production system for ice concentration has been developed, which has been integrated in to the Eumetsat OSI SAF system by Met.no (http://osisaf.met.no). For ice thickness from radar altimetry, an operational processing system has been developed by AWI for ENVISAT and CryoSat data, as well as for Sentinel-3 data. The processing system is described in detail by Hendricks et al., 2016.

Improved satellite observations of the global sea-ice cover are required by the GCM mainly for two applications: First, the improved observations allow researchers to more accurately monitor and quantify past and ongoing changes in the ice cover. Second, these observations are needed to improve the quality of numerical models that aim both at understanding the drivers for past sea-ice changes and at estimating the future evolution of the sea-ice pack (Fig. 4). The main objective of involving the global climate modeling community deals with this second application: we will answer the question to what extent the improved satellite observations allow us to also improve the representation of sea-ice in coupled Earth-System Models (ESMs) and in turn to what extent modern Earth-System Models allow us to improve the quality of the observational record

In addition to the climate modelling community, there are other climate research groups who need sea ice ECV data. These include scientists working with sea ice data assimilation in regional models, scientists working with creation of longer time series of sea ice data, by combining historical records with modern satellite data, and scientists developing new observing systems using aircraft, in situ and under-ice platform to collect various sea ice data that are complementary to EO data and can be used for validation of the EO-retrievals. The polar regions are severely undersampled regarding observational data for climate research. It is therefore of high priority to improve the collection and retrieval of climate parameters from both EO and non-EO observations. With access to regular sea ice data in the polar regions, it will be possible to use data assimilation for reanalysis and improvement of the climate models in the polar regions. Data assimilation has been proven to be a useful tool in atmospheric and ocean modeling for many years, but assimilation of sea ice data has only started recently.

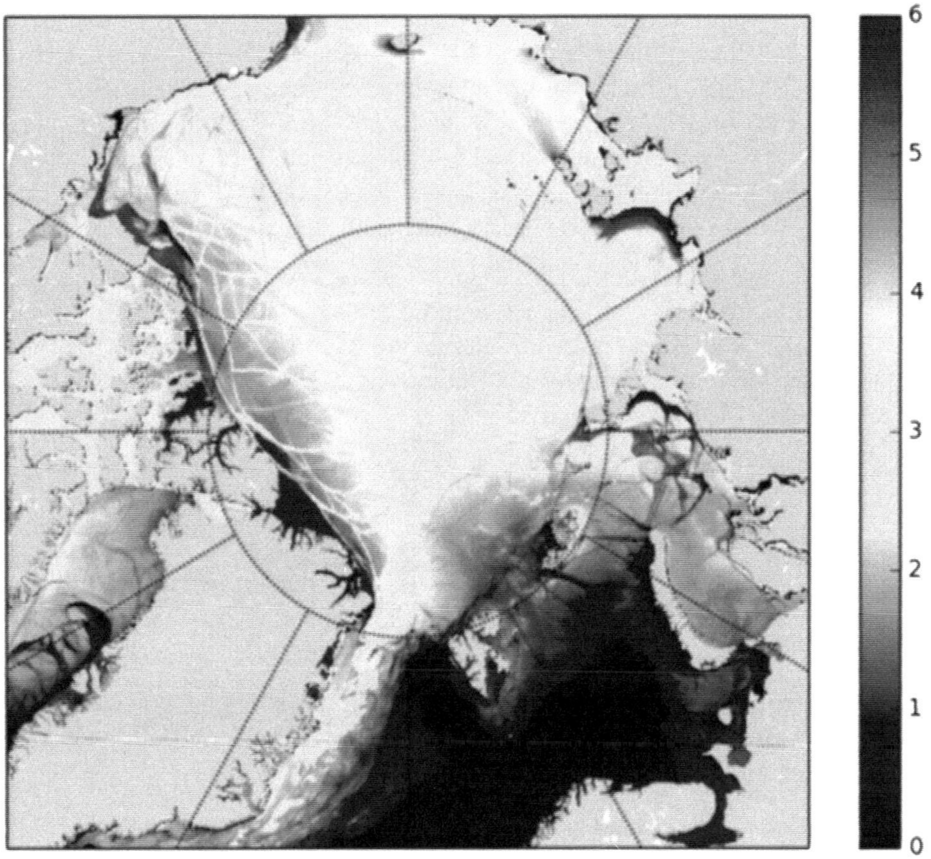

Figure 4. Sea ice thickness (m) on 30 Mar 2001 as simulated by the MITgcm (sea ice–ocean model forced with reanalysis data) at a horizontal resolution of about 4 km (Jung et al., 2016). The simulation is very similar to the one described in Nguyen et al. (2012).

In recent years, Arctic sea ice has been retreating at a rate much faster than that projected by the Earth-System Models (ESMs) used for the last IPCC report. Overcoming this discrepancy is crucial for an improved modelling of the future evolution of the sea-ice cover and will therefore be the ultimate aim for involvement of the GMC.

It will be particularly important to be in dialogue with the group of scientists working in the Climate Model Intercomparison Project. This group plays a key role to analyze climate model results for the Arctic and Antarctic and provide input to the IPCC reports. The ESA CCI Sea Ice project will be completed in early 2018.

REFERENCES

Hendricks, S., Ricker, R., and Helm, V.: User Guide - AWI CryoSat-2 Sea Ice Thickness Data Product (v1.2), 2016.

Jung, T., N. Gordon, P. Bauer, D. Bromwich, M. Chevallier, J. Day, J. Dawson, F. Doblas-Reyes, C. Fairall, H. Goessling, M. Holland, J. Inoue, T. Iversen, S. Klebe, P. Lemke, M. Losch, A. Makshtas, B. Mills, P. Nurmi, D. Perovich, P. Reid, I. Renfrew, G. Smith, G. Svensson, M. Tolstykh, and Q. Yang, 2016: Advancing polar prediction capabilities on daily to seasonal time scales. Bull. Amer. Meteor. Soc. doi:10.1175/BAMS-D-14-00246.1, published online 18 Oct 2016.

Kern, S., Khvorostovsky, K., Skourup, H., Rinne, E., Parsakhoo, Z. S., Djepa, V., Wadhams, P., and Sandven, S.: The impact of snow depth, snow density and ice density on sea ice thickness retrieval from satellite radar altimetry: results from the ESA-CCI Sea Ice ECV Project Round Robin Exercise, The Cryosphere, 9, 37-52, doi:10.5194/tc-9-37-2015, 2015

Kwok R, Cunningham GF. 2015 Variability of Arctic sea ice thickness and volume from CryoSat-2. Phil. Trans. R. Soc. A 373: 20140157. http://dx.doi.org/10.1098/rsta.2014.0157

Lindsay, R. and A. Schweiger, Arctic sea ice thickness loss determined using subsurface, aircraft, and satellite observations. The Cryosphere, 9, 269–283, 2015, www.the-cryosphere.net/9/269/2015/, doi:10.5194/tc-9-269-2015

Meier, W. N., J. Stroeve, A. Barrett, and F. Fetterer (2012), A simple approach to providing a more consistent Arctic sea ice extent time series from the 1950s to present, The Cryosphere, 6, 1359-1368.

Parkinson, C. L., and D. J. Cavalieri (2012), Antarctic sea ice variability and trends, 1979-2010, The Cryosphere, 6, 871-880, doi:10.5194/tc-6-871-2012.

Vaughan, D.G., J.C. Comiso, I. Allison, J. Carrasco, G. Kaser, R. Kwok, P. Mote, T. Murray, F. Paul, J. Ren, E. Rignot, O. Solomina, K. Steffen and T. Zhang, 2013: Observations: Cryosphere. In: Climate Change 2013: The Physical Science Basis. Contribution of Working Group I to the Fifth Assessment Report of the Intergovernmental Panel on Climate Change [Stocker, T.F., D. Qin, G.-K. Plattner, M. Tignor, S.K. Allen, J. Boschung, A. Nauels, Y. Xia, V. Bex and P.M. Midgley (eds.)]. Cambridge University Press, Cambridge, United Kingdom and New York, NY, USA.

POAC'17
Busan, KOREA

Proceedings of the 24th International Conference on
Port and Ocean Engineering under Arctic Conditions
June 11-16, 2017, Busan, Korea

Local Ice Load on Ship Hulls
- An Approach from Model Testing -

Koh Izumiyama[1]
[1] Hokkaido University (Sapporo, Japan)

ABSTRACT

This paper discusses local ice load measurement in model testing. Studying local ice load is important in terms of both safety and performance of ships in ice. A tactile sensor system is used in model tests to measure local ice loads on the model. Model test results at NMRI ice tank are presented. The test was performed to study local ice loading from a viewpoint of ship safety in ice. A model ship with podded propulsors and a model ship with a conventional rudder and propeller system were tested in a free-running mode. Discussion is made of the calibration of the tactile sensor, ice load distributions, and comparison of the test results with full-scale data.

KEY WORDS: Local Ice Load; Tactile Sensor; Ice Tank Test

INTRODUCTION

Ships sailing in ice-covered waters are exerted by loads from the ice. Ice load is characterized as localized high load in comparison with hydrodynamic loads. Such local ice load is the foremost concern for the safety of ships in ice. Ship hull structures should be designed and constructed to withstand local ice loads to safeguard the people and cargoes on-board. Local ice load is also significant from the viewpoint of ship performance in ice. Ship resistance in ice can be given as a vector sum of local ice loads on the hull. Knowing the behavior of local ice load will deepen the understanding of ice resistance. In this way, local ice load is important in terms of both safety and performance of ships in ice.

Model testing in ice tanks provides useful data for ship-in-ice research. It is an indispensable tool for designing ice-going ships today. There have been some model studies done with local ice load measurements (e.g.: Liukkonen and Nortala-Hoikkanen, 1992; and Kayo, 1993). In these studies segmented models were tested. However, a segmented model with many load cells installed is a complicated model to manufacture. It is difficult to have many small segments in the model. This limits the spatial resolution of the local ice load measurement.

Recently, a tactile sensor system has been used for ice tank tests to measure local ice loads on the model. National Maritime Research Institute (NMRI) of Japan pioneered the use of tactile sensor for model testing in ice. A wide range of tests have been made at NMRI ice tank with tactile measurements. This includes an indentation test of a flat indentor (Izumiyama et al., 1999a), resistance test of a ship (Izumiyama et al., 1999b and Izumiyama et al., 2001), free-running test of ships (Matsuzawa et al., 2006 and Izumiyama et al., 2007) measurement of ice load on a conical structure and towing test of a drill ship in broken ice (Izumiyama et al., 2016). Model tests with the tactile sensor have also been performed at ice tanks of Aalto University (Valkonen et al., 2007 and Kujala and Arughadhoss, 2012) and Korea Research Institute of Ships and Ocean Engineering (Jeong et al., 2015).

This paper discusses the model test measurement of local ice load by way of the tactile sensor. From the above-mentioned model tests performed at NMRI ice tank, free-running test of ship models is presented. This test was made to study local ice load from a viewpoint of ship safety in ice. There have been full-scale measurements made on this issue for various ships and sea areas (e.g.: Hanninen, 2007). However, full-scale measurements are usually made in relatively small areas in the hull. It is difficult to see a whole picture of local ice loading over a ship hull. The model test was performed to have such a picture. In the following, the sensor system is briefly described first. This is followed by the presentation of the method and results of the model test. Finally discussion is made of the calibration of the tactile sensor, ice load distributions along the hull, and comparison of test results with full-scale data.

TACTILE SENSOR

Sensor Description

A tactile sensor system I-Scan is used for the model tests at NMRI ice tank. The system is composed of sensor films and software to acquire and process the data by a PC. Sensor films are connected to PC via handles. The sensor film is thin and flexible. This allows the film to be installed on the model hull surface by adhesive tapes without creating roughness that can affect model-ice interaction. The software displays in real time the measured pressure data. It also has functions to store the data as files and to export them as ASCII text files.

In a sensor film there are in total 1936 pressure sensing cells locating in a square latticed pattern of 44 rows by 44 columns. The manufacturer provides sensor films of different sizes. Sensor films with a sensing area of 238 mm square (I-Scan 210) are used for the ship model tests at NMRI ice tank. In this sensor film, sensing cells are located at 5.4 mm intervals at the center. Each sensing cell is of 3.5 mm square area and there are non-sensing gaps of about 2 mm between cells.

The output from the I-Scan system is not pressure but a digital counter called a "Raw" from 0 (no pressure) to 255 (maximum pressure). A "Row Sum" is a sum of Raws from sensing cells in the whole or a part of a sensor film. Calibration needs to be made for converting Raw into a pressure or force value. The method to calibrate the I-Scan system for the purpose of ice model tests will be discussed later in this paper.

Basic Characteristics of the Tactile Sensor

Prior to the use of I-Scan sensor system for the model testing in the ice tank, a series of trial tests were performed to obtain basic characteristics of the sensor. In the test, a sensor film with a 112 mm square sensing area (I-Scan 100) was loaded in three different ways of 1) stepwise loading, 2) sinusoidal loading and 3) creep test. Loads were applied in a 70 mm square area in the center of the sensor film via a rubber sheet. Raw Sum from the loaded area was analyzed. Test results are shown in Figure 1. Loads applied and I-Scan responses are summarized below.

- Stepwise Loading Test: The sensor film is loaded stepwise loads up to 10 kN. The response of I-Scan is somewhat non-linear in the high load region. There is slight hysteresis seen between the loading and unloading phases.
- Sinusoidal Loading Test: The sensor film is loaded by loads fluctuating between 2 kN and 8 kN in a sinusoidal pattern. Tests were made at different loading frequencies in a range from 0.1 Hz to 35 Hz. I-Scan shows a flat response to the loads through the frequency range of up to 35 Hz.
- Creep Test: The sensor film is loaded by static loads of 5 kN and 10 kN for a duration of about 300 seconds. Output Raw Sum shows some drift with time. Drifting rate is approximately 5%/min.

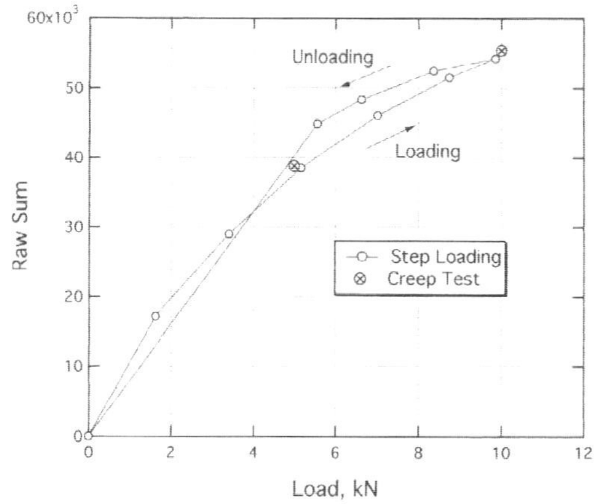

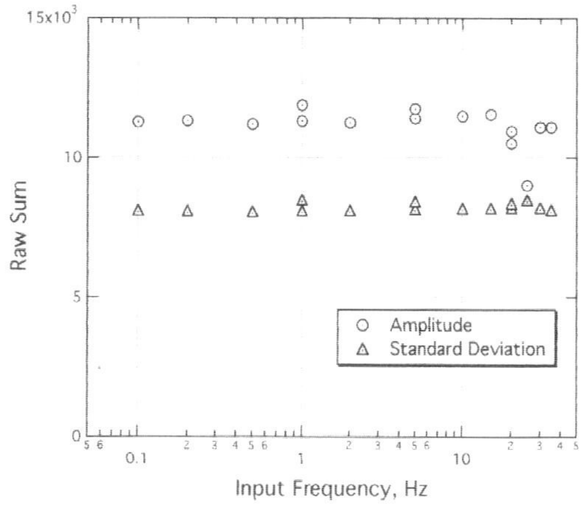

(a) Stepwise Loading (b) Sinusoidal Loading

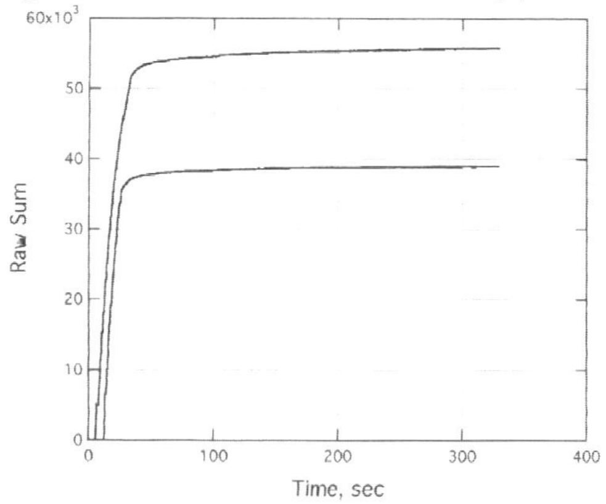

(c) Creep Test

Figure 1. Results of Trial Tests

MODEL TEST

Model Ships and Test Method

Two model ships were tested at NMRI ice tank. One is a 1:16 scaled model of an icebreaker (Shimoda et al., 2002) and the other is a 1:36 scaled model of an cargo ship (Izumiyama, 1995). Main dimensions of these models and ships are shown in Table 1. These model ships were designed for research projects and there are no actual existing ships for them. Both model ships have a conventional wedge-shaped bow with V-shaped frame lines. The icebreaker model is equipped with twin podded propulsors ( ), while the cargo ship model has a conventional propulsion system with a single propeller and a rudder.

Table 1. Main Dimensions of Models and Ships

(a) Icebreaker

	Model	Ship
Length, m	4.688	75.0
Width, m	0.857	14.0
Draft, m	0.250	4.0

(b) Cargo Ship

	Model	Ship
Length, m	4.861	175.0
Width, m	0.667	24.0
Draft, m	0.222	8.0

Eight tactile sensor films were installed on the model ship hulls along the water line from the bow to the stern. Figure 3 shows tactile sensor locations for the icebreaker model. Model ships were tested in in a free-running mode where they had six degrees of freedom in motion. Tests were performed in level ice with thickness of 0.50 m and 0.8 m for the icebreaker and 1.0 m for the cargo ship in full-scale. Turning tests as well as straight-going tests were performed. In some tests an S-shaped turning was made. Figure 4 shows an S-shaped turning track by the icebreaker model.

Figure 2. Podded Propulsors

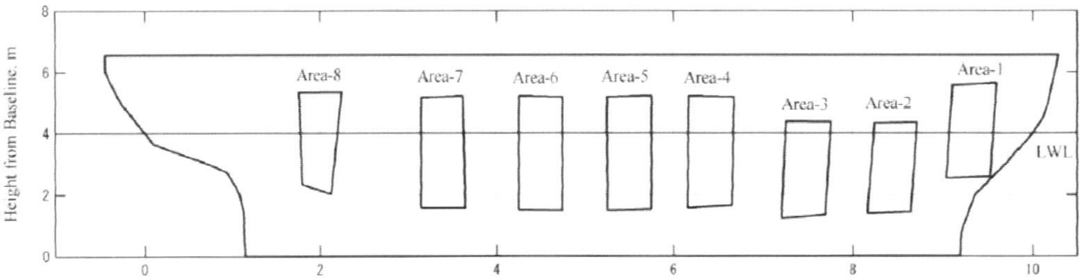

Figure 3. Tactile Sensor Areas on Icebreaker Model

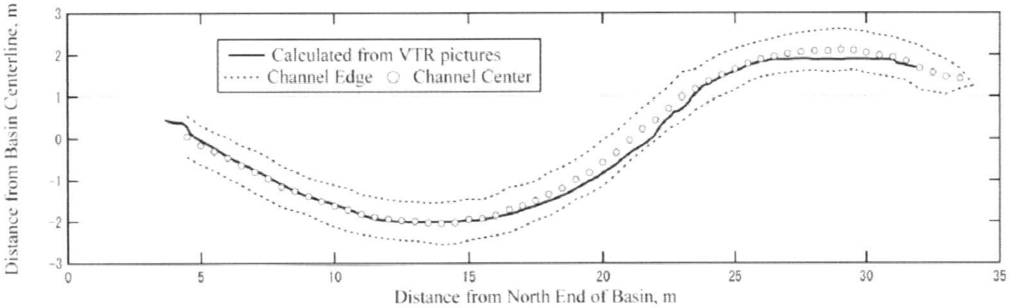

Figure 4. Model Ship Track in S-shaped Turning Test

Analysis

In a full-scale measurement, local ice loads on the hull are usually measured by strain gauges installed on transvers frames. Ice load on each frame is calculated from the strain measured. Analysis of the tactile sensor data was done so that it can be comparable with this full-scale measurement method. For this purpose the sensor was divided into 11 narrow vertical segments. Each segment is of 4 sensing cells in width and of full sensor height (Figure 5). The width of a segment is 0.35 m and 0.79 m in full-scale for the icebreaker and cargo ship, respectively. Ice load on each segment was calculated. The ice load value of non-exceedence probability of 99% was taken as the representative value for the local ice load on the segment.

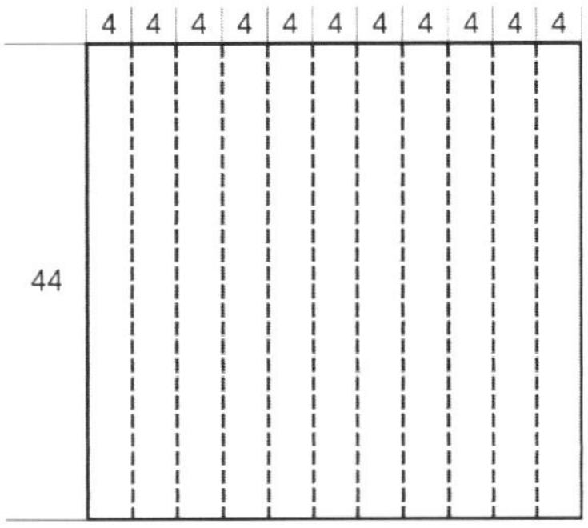

Figure 5. Data Analysis Segments

Results

The I-Scan system has a function to display measured pressure as a two-dimesional distribution map.

Figure 6 is an example of such display window. Loaded cells are colored. Data measured in the bow of a model ship is shown. Load patches of short length are aligned in a line forming a "broken-line-like" pattern. The load patch line moves sternward and downward as the model moves in ice.

Figure 7 (a) shows a time curve of local ice load acting on a data analysis segment. Ice load shows a spike-like pattern with many peaks. These peaks were recorded as a load patch passes over the data analysis segment. Figure 7 (b) shows a time curve of local ice load measured in a full-scale test performed in the Sea of Okhotsk. (Uto et al., 2005). Full-scale ice load also shows a spike-like pattern with many peaks as observed in the model test result.

This similarity indicates that full-scale ice load also acts on load patches of short length moving sternwards.

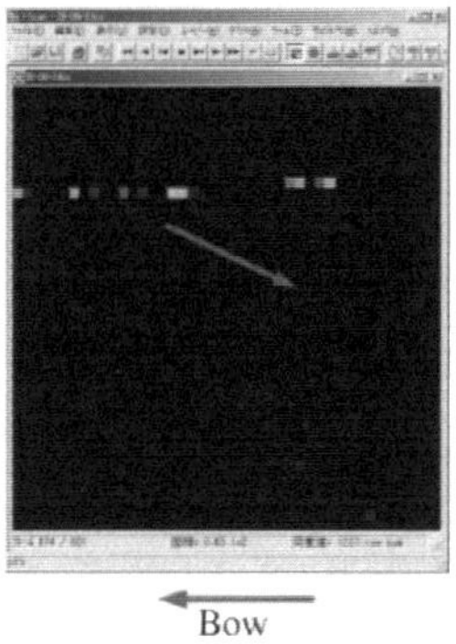

Figure 6. Ice Load Patch Pattern

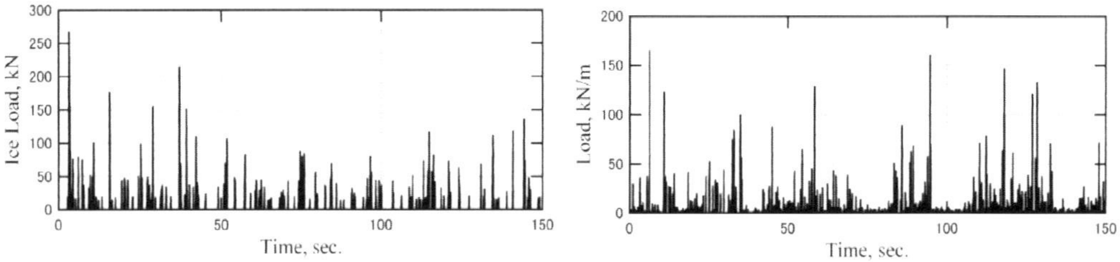

(a) Model Test Result (b) Full-Scale Data

Figure 7. Time Curves of Local Ice Load

Figure 8 shows ice load distribution along the hull in the straight-going tests for icebreaker and cargo ship. Ice load data measured by eight sensor films were plotted against the locations of the films on the model hull. These data are mean of representative values of ice load from 11 data analysis segments in a sensor film. Figure 9 shows ice load distribution in the turning tests. Ice load data in the inside of the turn are shown.

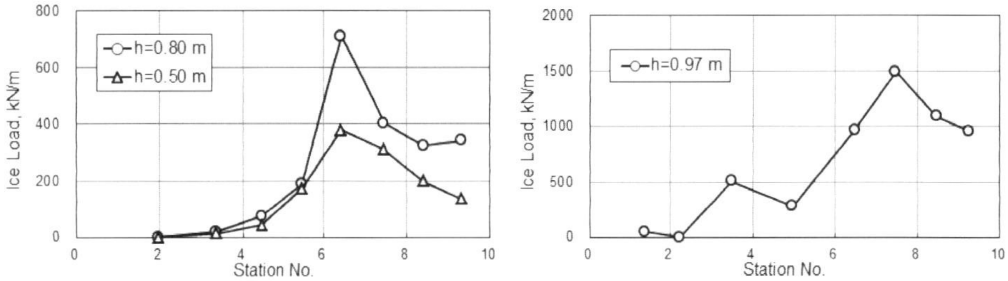

(a) Icebreaker (b) Cargo Ship

Figure 8 Ice Load Distribution in Straight-going Mode

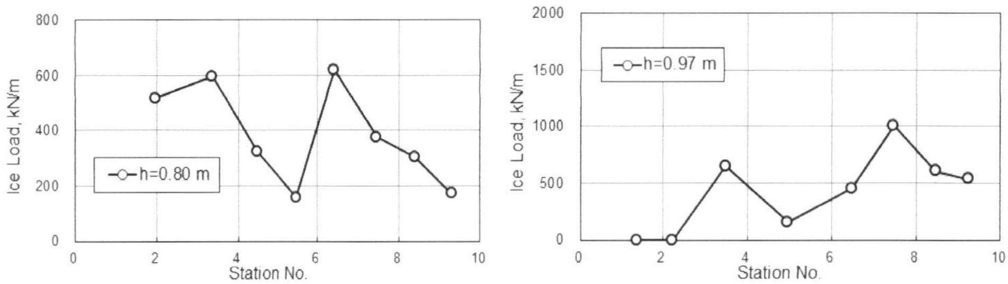

(a) Icebreaker (b) Cargo Ship

Figure 9. Ice Load Distribution in Turning Mode

DISCUSSION

Sensor Calibration

I-Scan sensor needs to be calibrated to obtain load value from digital counter Raw. However, calibrating the sensor for the purpose of ice model testing is not straightforward. Figure 10 shows the reason for this (Izumiyama et al., 1999a). The figure shows the result of a test in which loads are applied on individual sensing cells. Six cells were loaded over the entire cell area of 3.5 mm square (open circles). Outputs from these cells agree well each other indicating good spatial uniformity of the sensor film. Another cell was loaded only in a quarter of the cell area (full circles). Outputs from this cell significantly differ from those of entirely loaded cells for same loads. The test shows that the output from the I-Scan sensor system depends not only on the load applied but also on the area on which load acts, or more generally, contact conditions at the interface.

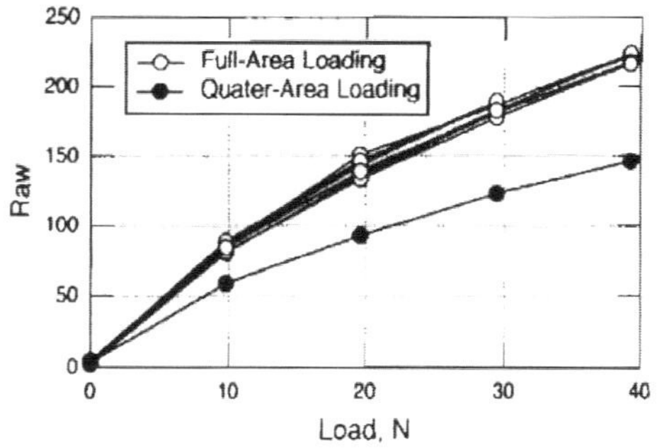

Figure 10. Loading on Individual Cells

However, it is difficult to know the actual contact conditions at the interface for solid-solid contact problems including hull-ice contact. Actual contact area may differ from the virtual one. Pressure cannot be uniform but vary within a contact area. In the case of hull-ice contact, local ice failure at the interface will make the problem more complicated. Based on this consideration, it was decided to calibrate the I-Scan system by using ice load acting on the model. In this way, the effects of the above-mentioned complexity related

to the hull-ice contact conditions can be included in the calibration.

In the present model test, resistance is used for the calibration of the I-Scan system (Matsuzawa et al., 2006). Resistance of the model is given as a vector sum of local ice loads on the model hull. Sensor films gives Raw Sum corresponding to local ice loads normal to the model surface. Longitudinal component of those Raw Sums are calculated using hull angles at the location of sensor films and summed up over the bow region of the model. Raw Sum for the hull areas between sensor films are interpreted from data of neighboring films. Calculated value was doubled to account for resistance on the other side of the model where sensor films are not installed. Resistance was calculated from propeller thrust assuming a thrust deduction factor. Figure 11 shows the correlation of Sum of longitudinal Raw Sums and resistance of the model. Dashed line in the figure was used for the sensor calibration in the present model test.

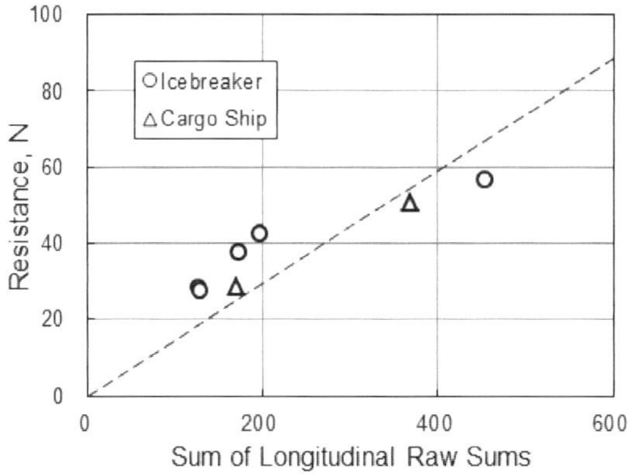

Figure 11. Calibration Line

Ice Load in Straight-going Mode

In the straight-going test, ice loading is predominant in the bow region as shown in Figure 8. This ice load distribution results from the way the bow breaks the ice (Figure 12). With a very few exceptions, ice-going ships are designed to break ice in bending by an inclined bow. When a ship sails in ice, the inclined bow pushes the ice downwards. Ice then breaks with a crack running some distance away from the contact point forming a broken ice piece. The ice piece is further pushed down to be submerged and a new contact takes place. The process is repeated to form an open channel in the ice. The bow is exerted by loads when it bends, breaks and submerges the ice. Hull-ice contacts and ice loading take place at many points on the bow. Ice loading in the ship region after the bow is much more moderate than that in the bow as the channel formed by the bow is wider than the ship beam and there is less chance for the hull to contact the ice.

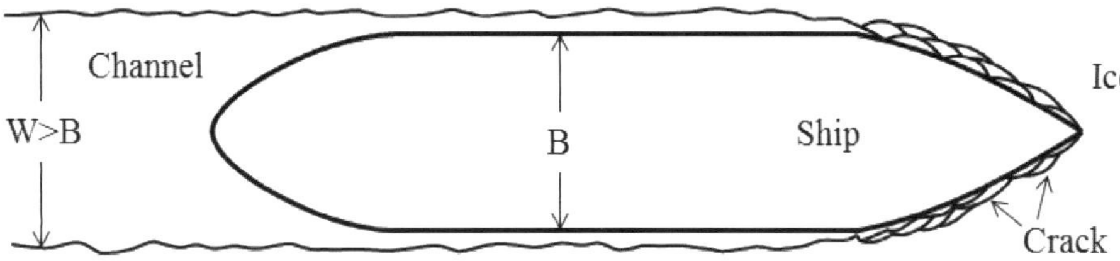

Figure 12 Schematic of Ship and Channel in Ice

Difference in ice load intensity in different hull regions is taken into account in the design ice loads calculation in ship design rules. Figure 13 shows a factor defined in Finnish-Swedish Ice Class Rules (FSICR), a *de facto* standard for ship design for first-year ice conditions, to calculate design ice loads. The factor "takes account of the probability that the design ice pressure occurs in a certain region of the hull for the ice class in question". The factor gives design ice loads for the midbody and stern regions relative to the bow (factors for the bow is always unity). FSICR defines four ice classes of IA Super, IA, IB and IC with IA Super being the highest and IC the lowest. Figure 13 shows factors for these ice classes. The factor decreases from the bow toward the stem. The lower the ice class is the more the factor is reduced for the midbody and the stern.

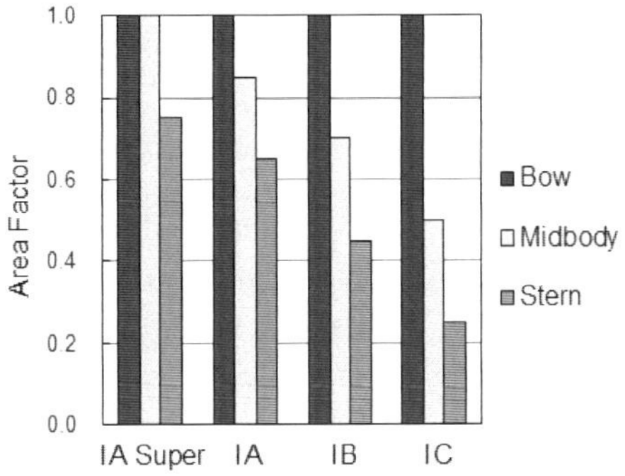

Figure 13. Factors defined in FSICR

In the bow region, the ice load increases from the stem toward the fore shoulder. It is interesting to see this ice load distribution in regard with hull angles of the bow. Varsta (1983) performed a numerical calculation of ice load acting on a ship bow. His calculation showed the ice load is a function of the inclination angle (angle off the vertical) of the hull as shown in Figure 14 (a). Ice load increases with the decrease of the angle. Result of the model test is shown in Figure 14 (b). In the figure, non-dimensional ice load (ice load divided by resistance) is plotted against the spread angle of the model hull. Model test result also shows that ice load increases with the decrease of the hull angle.

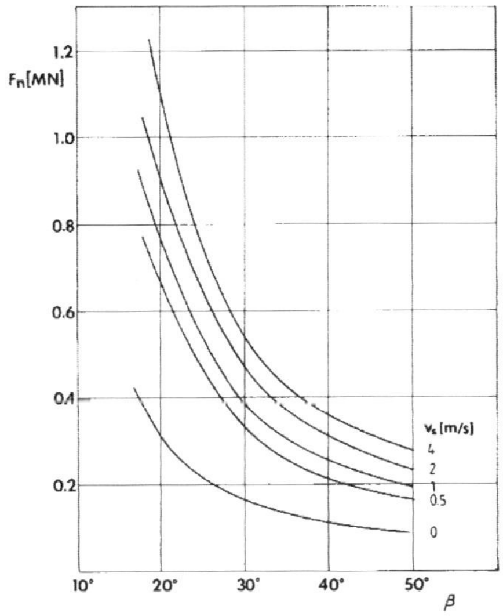

(a) Ice Load Calculation by Varsta (1983)

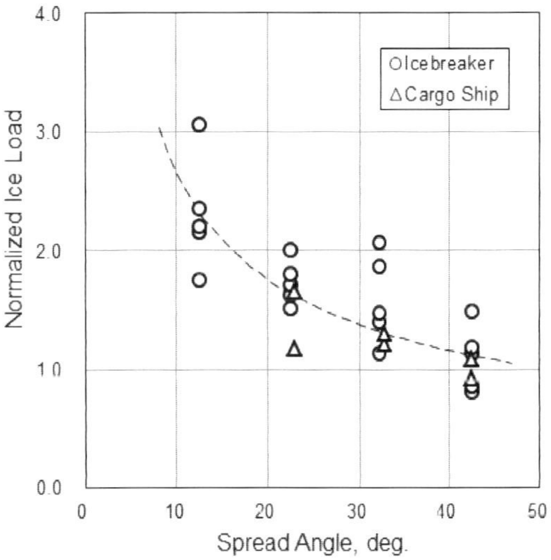

(b) Model Test Result

Figure 14. Effects of Hull Angle on Ice Load

Ice Load in Turing Mode

In the turning mode, ice load distribution along the hull is totally different from that in the straight-going mode (Figure 9). The stern, not only the bow, is exerted by significant ice loading. This can also be explained by ship motion in the open channel (Figure 15). When a ship turns in ice, force from the rudder or propulsor pivots the ship in the channel. The aftbody of the ship is pressed against the edge of the channel and thereby exerted by loads from the ice.

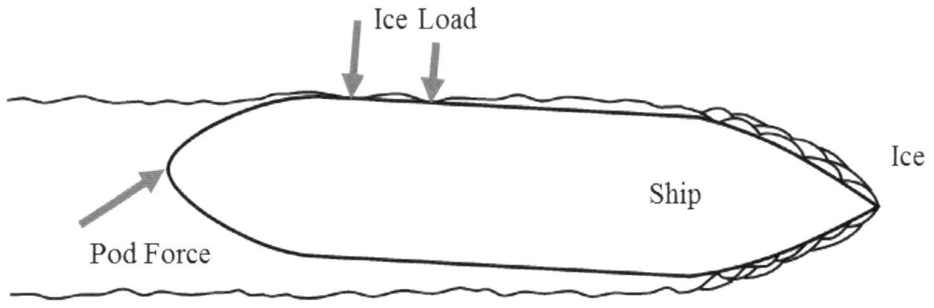

Figure 15. Schematic of Ship Pivoting in Channel

Ice loading in the stern is more notable for the icebreaker model with podded propulsors than the cargo ship model with the conventional rudder and properllar system. This is due to high lateral force from the propulsors to turn the model. Figure 16 shows correlation of ice load measured in sensor films located in the stern of the icebreaker model and lateral force of the pods. Stern ice load increases as the pod force increases.

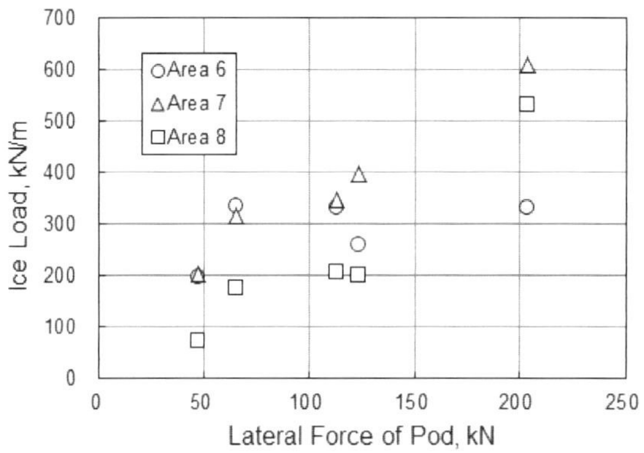

Figure 16. Correlation of Stern Ice Load and Lateral Force of Propulsors

Podded propulsors are increasing utilized in ice-going ships in recent years. Those ships show superb manoeuvrability in ice (e.g.: Nyman et al., 1999). Test results shown in Figure 16 indicate that such ships with podded propulsor could receive high ice loads in the aft region as well as the bow. It may not be relevant to use factors shown in Figure 13 to calculate design ice loads for those ships. Regarding this matter, FSICR states "The introduction of new propulsion arrangements with azimuthing thrusters or "podded" propellers, which provide an improved manoeuvrability, will result in increased ice loading of the Stern region and the stern area. This fact should be considered in the design of the aft/stern structures." However, no rule is given for how it should be "considered" in the ship design. This may be attributed to the lack of full-scale data from ships with podded propulsors. Accumulation of such data is needed.

Comparison with Full-scale Data

Model testing is performed so that the similarity law holds between the model- and full-scale conditions. For ship model testing in ice, the similarity law requires the followings.

$$l_S = \alpha l_M, \quad \sigma_S = \alpha \sigma_M, \quad t_S = \alpha^{1/2} l_M, \qquad (1)$$

where α is the scale ratio, l, σ and t are dimensions of ship and ice, ice strength and time, respectively. Suffices M and S refer to mode- and full-scale, respectively. For force, F, the cube law holds as

$$F_S = \alpha^3 F_M . \quad (2)$$

In the present analysis, full-scale local ice load was calculated from measured tactile data based on the above cube law.

As there are no real existing ships for the models tested, it is difficult to compare the ice load calculated from model test results with full-scale data directly. Instead, test results are compared with design ice loads calculated based on the ice load formula of FSICR. FSICR assumes ships of each ice class to operate in conditions corresponding to level ice thickness not exceeding the value shown in Table 2. Design ice load for the bow was calculated based on ice thickness of the model test and ice class corresponding to the thickness.

Table 2. Level Ice Thickness for FSICR Ice Classes

Ice Class	Ice Thickness
IA Super	1.0 m
IA	0.8 m
IB	0.6 m
IC	0.4 m

Figure 17 compares calculated ice load with design ice load. The figure shows frequency distribution of the ratio of measured and design ice loads. Although most of measured ice loads are lower than the design ice load, there are a few data higher than it. Design loads are loads expected to occur during a long return period, whereas the length of each run is only about 30 m that is 0.5 to 1 km in full-scale. If one takes this into account, it may be interpreted form the result shown in Figure 17 that the model test tends to overestimate the local ice loads on ship hulls.

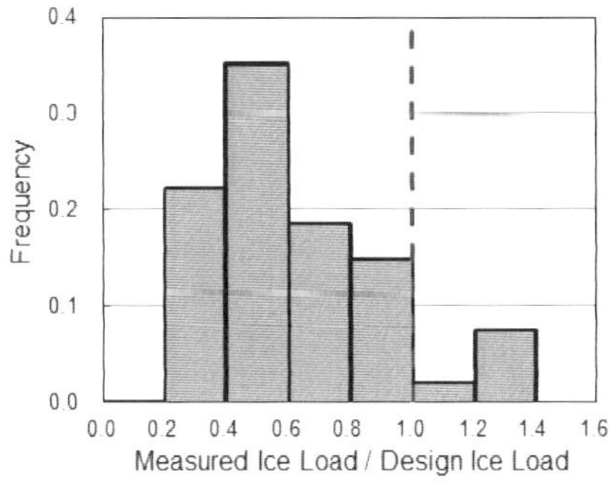

Figure 17. Comparison of Measured Ice Load and Design Ice Load

Local ice load occurs as a reaction force against the downward force to break ice in bending. Tactile measurement in the model test showed that hull-ice contact is "broken-line-like" consisting of short load patches (

Figure 6). Similarity between model- and full-scale load time curves (Figure 7) indicates that full-scale load patch is also broken-line like. It is reasonable to assume that the local ice load occur in individual load patch depends not only on the total load required to break ice but also on the contact condition between hull and ice. Figure 18 schematically shows ice loading in different contact conditions. Large local ice loads occur in a contact points (left), while the load is low when there are many contact points (right).

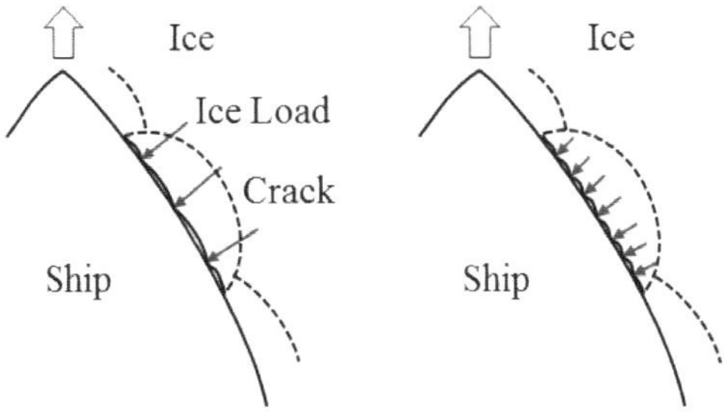

Figure 18. Schematics of Hull-Ice Contact and Local Ice Loads

Model ice has been developed mainly for the purpose of ship model testing. Mechanical properties of model ice are adjusted based on the similarity law so that it can properly simulate icebreaking by a ship including the downward force required to break the ice in bending and the way ice breaks. However, no study has been made of local contact conditions between ship hull and ice as discussed above to examine how model ice can be compared with sea ice in this regard. More study is required of this issue for ice model testing with tactile measurement to be a tool for estimating full-scale local ice loads as it is so for ship performance in ice.

There are two types of model ice in terms of the crystal structure – columnar and granular. Difference in model ice structure may lead to different contact conditions with models. The present test was performed in model ice with a columnar structure. Test in granular-structured model ice may give different results in regard with full-scale ice load estimation.

SUMMARY

This paper discussed local ice load measurement in model testing. A tactile sensor system is used for the model test at NMRI ice tank. Prior to the use for model test, the sensor system was tested by three different ways of loading to obtain basic characteristics of the sensor. Results of model test at NMRI ice tank were presented. Two model ships, one with podded propulsors and the other with conventional propeller and rudder system, were tested in a free-running mode. Discussion was made of the calibration of the tactile sensor, ice load distributions, and comparison of the test results with full-scale data. Output of the tactile sensor depends on contact conditions at the interface as well as load applied on it. Calibration of the tactile sensor needs considerations on this point. Local ice load is predominant in the bow when a ship is sailing straight. Stern is exerted by significant ice loading when a ship turns. Stern ice load correlates well with the lateral force of pods. Full-scale ice load is calculated based on the cube law. Calculated ice load is compared with design ice load. More study is required of local contact conditions between ship hull and ice to examine how model ice can be compared with sea ice in this regard.

REFERENCES

Hanninen, S., 2007. Inventory of Data Sources Available on Ship Hull Ice Loads and Damages. In: *Increasing the Safety of Icebound Shipping Final Scientific Report: Volume 1*, Helsinki University of Technology Ship Laboratory M-302, pp.22-33.

Izumiyama, K. and Uto, S., 1995. Ice Resistance of Three Bow Forms for the NSR Cargo Ship. Northern Sea Route; Future & Prospective, *The Proceedings of INSROP Symposium Tokyo '95 (1-6 October 1995)*, pp. 459-467.

Izumiyama, K., Wako, D. and Uto, S., 1999a. Ice Force Distribution on a Flat Indentor. In: Ice in Surface Waters, Shen (ed.), Vol. 2, pp. 917-922.

Izumiyama, K., Wako, D. and Uto, S., 1999b. Ice Force Distribuion around a Ship Hull. *Proceedings of the 15th International Conference on Port and Ocean Engineering under Arctic Conditions*, Vol. 2, pp. 707-716.

Izumiyama, K., Wako, D. and Uto, S., 2001. Ice Pressure acting over a Model Ship Hull. *Proceedings of the 16th International Conference on Port and Ocean Engineering under Arctic Conditions*, Vol. 2, pp. 793-802.

Izumiyama, K., 2007. Description of Local Ice Loading and Hull Area Factor on Model Test Results. In: *Increasing the Safety of Icebound Shipping Final Scientific Report: Volume 1*, Helsinki University of Technology Ship Laboratory M-302, pp.115-134.

Izumiyama, K., Uto, S., Otsubo, K., Matsuzawa, T., Shimoda, H., Wako, D., Asanuma, T. and Yamamoto, J., 2016. Ice Load Distribution on a Model Ship Station-keeping in Broken Ice. *Proceedings of the 31st International Symposium on Okhotsk Sea & Sea Ice*, pp. 92-95.

Jeong, S., Choi, K., Ha, J., Kang, K. and Cheon E., 2015. Distribution of Ice Load acting on Model Hull due to Ship-Ice Interaction. *Proceedings of the 23rd International Conference on Port and Ocean Engineering under Arctic Conditions*.

Kayo, Y., 1993. Measurement of Ice Load Distribution on an Icebreaker Bow Model composed by Loade Panels. *Proceedings of the 12th International Conference on Port and Ocean Engineering under Arctic Conditions*, Vol. 1, pp. 328-337.

Kujala, P. and Arughadhoss, S.., 2012. Statistical Analysis of Ice Crushing Pressure on a Ship's Hull dring Hull-Ice Interaction. *Cold Regions Science and Technology*, Volume 70, January 2012, pp. 1-11.

Liukkonen, S. and Nortala-Hoikkanen, A., 1992. Ice Resistance Tests on a Segmented Icebreaker Model. *Proceedings of the 11th IAHR International Symposium on Ice*, Vol. 1, pp. 296-308.

Matsuzawa, T., Wako, D. and Izumiyama, K., 2006. Local Ice Load on a Ship with Podded Propulsors. *Proceedings of the 18th IAHR International Symposium on Ice (2006)*, Vol. 2, pp. 33-40.

Nyman, T., Riska, K., Soininen, H., Lensu, M., Jalonene, R., Lohi, P. and Hrjula, A., 1999. The Ice Capability of the Multipurpose Icebreaker Bothnica – Full Scale Results. *Proceeding of the 15th International Conference on Port and Ocean Engineering under Arctic Conditions*,

Vol. 2, pp. 631-643.

Shimoda, H., Wako, D., Takimoto, T., Tamura, K. and Uto, S., 2002. Development of an Icebreaking Patrol Vessel with Podded Propulsors – Basic Design and Open Water Performance – (in Japanese). *Proceedings of 2002 Cold Region Technology Conference*, pp. 204-210.

Transport Safety Agency of Finland, 2010. Finnish-Swedish Ice Class Rules 2010. p. 48.

Valkonen, J. Izumiyama, K. and Kujala, P., 2007. Measuring of Ice induced Pressures and Loads on Ships in Model Scale. *Proceedings of the 10th International Symposium on Practical Design of Ships and other Floating Structures*.

Uto, S., Murakami, C., Takimoto, T. and Izumiyama, K., 2005. Ice Load exerted on the Hull of Icebreaker PM Teshio in the Southern Sea of Okhotsk. *Proceeding of the 18th International Conference on Port and Ocean Engineering under Arctic Conditions*, Vol. 2, pp. 683-692.

Varsta, P., 1983. On the Mechanics of Ice Load on Ships in Level Ice in the Baltic Sea. *Technical Research Center of Finland, Publication 11*, p. 91.

POAC'17
Busan, KOREA

Proceedings of the 24th International Conference on Port and Ocean Engineering under Arctic Conditions
June 11-16, 2017, Busan, Korea

Bi-Polar Full-Scale Measurements of Operational Loading on Polar Vessel Shaft-Lines

Rosca J.O. de Waal[1], Anriëtte Bekker[1], P. Stephan Heyns[2]

[1] Sound and Vibration Research Group, Stellenbosch University, Stellenbosch, South Africa

[2] Centre for Asset Integrity Management , University of Pretoria, Pretoria, South Africa

ABSTRACT

The port side shaft-lines of two polar research vessels, the SA Agulhas II and the FS Polarstern, were instrumented with in-board strain gauges to capture operational measurements of shear strain and axial strain. These measurements were used to calculate the instantaneous torque and thrust in the respective shafts. Vessel operations were categorized into open water navigation, ice passage and stationary operations. Subsequent to this rainflow counting of the transient torque and thrust cycles was performed in order to gain insight into the respective demands of different operational environments. During the 2015/2016 Antarctic relief voyage of the SA Agulhas II the vessel spent 16% of its voyage duration in ice and a further 54% in open water navigation. In comparison the FS Polarstern was engaged in ice passage for 24% of its Arctic voyage from Tromsø and spent the majority of her time performing stationary operations. Although both vessels experienced similar levels of torsional loading, the SA Agulhas II shaft line was more critically loaded relative to its Maximum Continuous Rating.

KEY WORDS: Shaft loads; Rainflow counting; Torque; Thrust; Ice.

INTRODUCTION

Efficient and safe shipping in Arctic regions is an increasing requirement as maritime transport in ice-covered seas is expected to increase in future decades (Ikonen et al. 2014). Vessel passage through icy waters entails exposure to additional ice-related loads on the propulsion system. Ice-related loads furthermore affect the efficiency and safety of vessel operations (Polić et al. 2014) and could result in failure of other shaft line elements if not accounted for. A study by the Transportation Safety Board of Canada (2010) found that propulsion system failure is still one of the greatest contributors to vessel failure.

Shaft-line components, including the propeller, are therefore required to operate efficiently in ice and open water while withstanding both extreme loads and fatigue loads (Huisman et al. 2014). Extreme and cyclic moderate loading of the propeller is increased during propeller-ice interaction. Extreme loading dictates the ultimate strength of propeller design and cyclic moderate loading determines the fatigue life of the propeller (Huisman et al. 2014). These loads are transmitted to other shaft-line elements which should not deform plastically in the event of propeller failure. The propeller and shaft line elements of ice going vessels are exposed to transient torsional and axial vibration induced by propeller-ice interaction which leads to the interest in the effect of ice loads on propulsion designs (Batrak et al. 2014).

During both full-scale trials and model-scale investigations, ice-related propeller loads are typically measured between the engine and the propeller on the shaft (Polić et al. 2014). These loads are not measured directly and include the dynamic response of the mechanical transmission components (Polić et al. 2014). The use of numerical methods is useful to predict the global loads at the propeller but is not adequate to optimize the efficiency and reliability design of the propeller (Huisman et al. 2014). Tang & Brennan (2013) recommend that multi-point measurements are more reliable than conventional single measurements, especially for shaft lines containing highly flexible components. To date, there exists no in-depth, definitive study on propeller-ice interaction (Sampson et al. 2009) and there is a lack of knowledge regarding the physical processes during propeller-ice interaction which cause ice-loading (Huisman et al. 2014).

Shipping registers, such as Korean Register (Korean Register 2015), American Bureau of Shipping (American Bureau of Shipping 2006), Det Norske Veritas (Det Norske Veritas 2011a) and Lloyd's Register (Germanischer Lloyd 2007) specify the requirement for calculation of torsional shaft line vibration during the design stage. Full-scale measurements are subsequently performed during sea trials in order to ensure safe vessel operation (Tang & Brennan 2013). Although full-scale measurements are time-consuming and costly, such measurements offer advantages over a model-scale approach. The design and extensive calibration of a model test rig is avoided (Brouwer et al. 2013). Furthermore, full-scale measurements of vessels provide accurate performance data of the vessel relative to the environmental conditions (Dinham-Peren & Dand 2010).

It is required to design the propulsion line components with sufficient strength to withstand the maximum induced loads on the propeller. At the same time, damage to other propulsion line components is to be prevented in the event of plastic bending of a propeller blade (Det Norske Veritas 2011a). A further requirement is that fatigue strength is to accommodate the load distribution as shown in Eq. 1 (Det Norske Veritas 2011a):

$$Q_A(N) = Q_{A,max}\left[1-\frac{log(N)}{log(ZN_{ice})}\right]^{1/k_w} \quad [Nm] \tag{1}$$

where Q_A is the response torque amplitude on the shaft during a sequence of ice impacts on the propeller and $Q_{A,max}$ is the highest response torque amplitude on the shaft during a sequence of ice impacts on the propeller. N_{ice} is the number of ice impacts, N is the number of cycles analyzed and Z is the number of propeller blades. The Weibull shape parameter is assumed to have a value of $k_w = 1$ (Det Norske Veritas 2011a).

Cycle counting is a convenient method to summarize irregular loading histories by determining the number of times certain cycles of various magnitudes occur (ASTM International E1049-85 2011). Many different cycle counting methods, of which level-crossing counting, peak counting, simple-range counting and rainflow counting are the most well-known (ASTM International E1049-85 2011). Of these methods, rainflow counting has been proven to be the better method for irregular loads (Connor et al. 2010). Rainflow counting was also used by Myklebost & Dahler (2013) during the analysis of shaft line torque. Rainflow counting enables one to extract cycles from randomly loaded data, with amplitudes defined as half the difference between two consecutive peaks and troughs.

Vessel propulsion systems are exposed to a wide spectrum of loads during operation. Only the dominating cases are considered during fatigue analysis as loads below the Mean Continuous Rating (MCR) in bollard condition theoretically do not result in shaft line failure, with MCR defined as the maximum rated torque Q_{MCR} the propulsion system can safely operate at continuously (Det Norske Veritas 2011a).

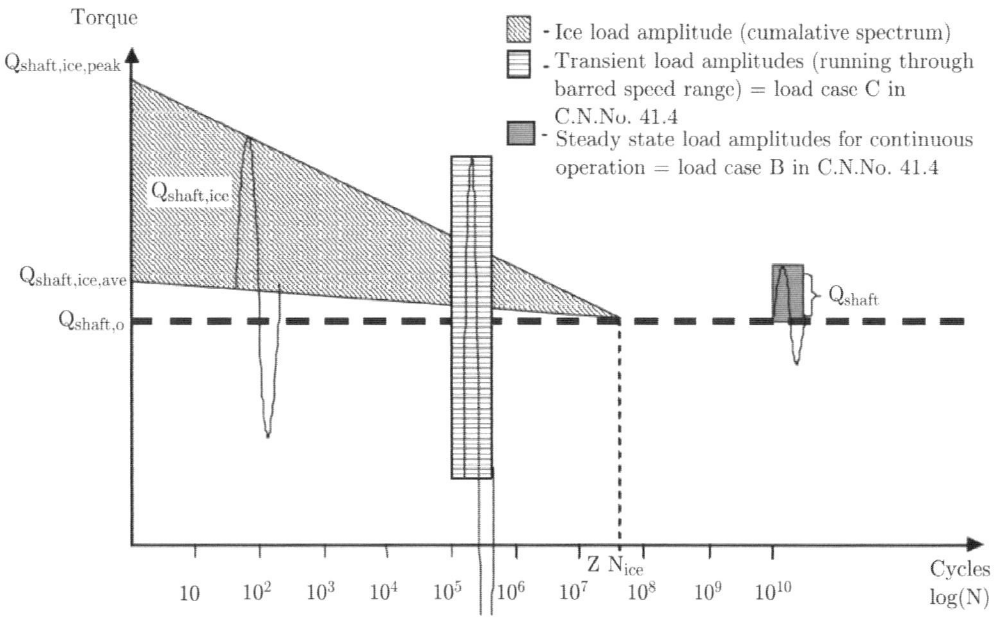

Figure 1. Typical load cases to be assessed for directly coupled two-stroke plant ice class vessels. Adapted from (Det Norske Veritas 2011a)

Figure 1 represents the typical load cases that need to be assessed for vessels navigating in ice with a directly coupled two-stroke plant. $Q_{shaft,ice,peak}$ is the maximum response peak torque measured in the shaft due to ice impacts on the propeller, $Q_{shaft,ice,ave}$ is the average torque in the shaft during an ice milling sequence, $Q_{shaft,o}$ is the shaft torque at maximum continuous power in bollard condition, $Q_{shaft,ice}$ is the response torque in the shaft due to ice impacts on the propeller and Q_{shaft} is the total torque response in the shaft due to external loading on the propeller.

In light of the sparsity of public domain operational data, full-scale measurements of axial- and shear strain on two polar research vessels were undertaken. Measurements on the shaft-line have a two-fold purpose: Firstly, to obtain an indication of the transient torque and thrust in the shaft and secondly, to provide a means through which propeller loads can be estimated by inverse methods (Ikonen et al. 2014). The present study focuses on the former purpose, whereby the operational loading on the propulsion components is to be informed and compared to existing ice class rules (Det Norske Veritas 2011b). The proportional time of vessel activities is categorized from response data. Rainflow counting is performed on the transient thrust and torque histories which are calculated from strain measurements for each operational category. The measurement data is extrapolated over the full voyage durations and this information is subsequently used to predict the operational load profiles for the investigated vessels.

FULL-SCALE MEAUREMENTS

Vessel Descriptions

The SAA II, depicted in Figure 2, was manufactured in Rauma shipyard in 2012 by STX Finland (Bekker et al. 2014). The hull was strengthened in accordance with DNV ICE-10 and the vessel classified to Polar Ice Class PC-5 (Bekker et al. 2014). She is therefore rated for year-round operations in medium first-year ice containing old ice inclusions (International Association of Classification Societies 2011). The ship is propelled by four 3 MW diesel generators which power two Conver Team electric motors of 4.5 MW each. She is equipped with two four-bladed variable pitch propellers with individual shaft lines (STX Finland Oy 2012).

Figure 2. SAA II vessel instrumented for the 2015/2016 relief voyage Antarctica.

Figure 3. The FS Polarstern was instrumented for the PS-100 Arctic voyage, 2016.

The FS Polarstern (Figure 3) was manufactured in 1982 by Howaldtwerke-Deutsche Werft AG in Kiel and outfitted by Werft Nobiskrug GmbH in Rendsburg (Grobe & Alfred Wegener Institute 2007). The hull was strengthened in accordance with Germanischer Lloyd Ice Class E3, which is the equivalent of Polar Ice Class PC-5 (Transport Safety Agency 2010). Her propulsion design comprises four diesel engines, each 3.5 MW, which in turn powers two shaft lines with four-bladed variable pitch propellers.

A key difference between the FS Polarstern and the SAA II is that the SAA II has open propellers and a direct diesel to electric drive to the shaft line as shown in Figure 4. The FS

Polarstern design includes ducted propellers (*Figure 5*) as well as a direct diesel engine drive through a gearbox. Principal specifications of the vessels are presented in Table 1.

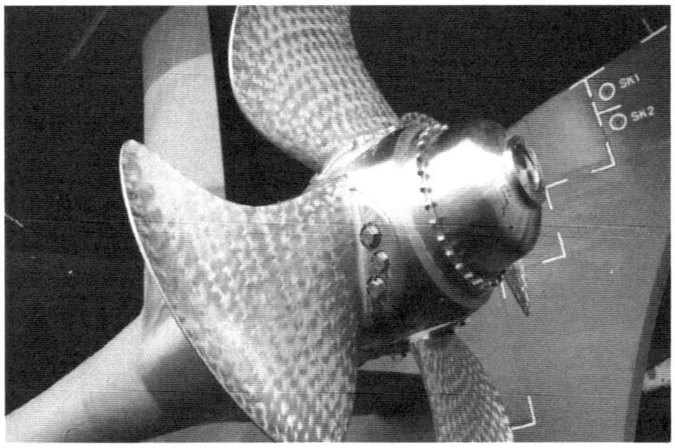

Figure 4. The four-bladed propeller of the SAA II.

Figure 5. The FS Polarstern has a ducted propeller design.

Table 1. Vessel specifications (STX Finland Oy, 2012; The Alfred Wegener Institut, 2016).

	SA Agulhas II	FS Polarstern
Gross tonnage	12 897 tons	12 614 tons
Length / Breadth	134 m / 22 m	118 m / 25 m
Classification	Det Norske Veritas	Germanischer Lloyd
Class notation	1A1 PC-5/ICE-10	100 A5 ARC 3
Yard	STX Finland, Rauma, Finland	Howaldtswerke-Deutsche, Hamburg & Kiel, Germany
Year built	2012	1982
Main engine maker	Wärtsilä	Klöckner-Humboldt-Deutz
Diesel engine type	6L32	KHD RBV 8 M 540
Electric motor type	N3 HXC 1120 LL8	-
Speed / Power at MCR	140 rpm / 4500 kW	182.4 rpm / 7765 kW
Nominal torque	$307\,k \cdot Nm$	$407\ k \cdot Nm$
Propeller maker	Rolls-Royce	Vereinigte Edelstahlwerke
No. of blades / Diameter	4 / 4.3 m	4 / 4.2 m
Shaft characteristics	Direct drive	1:3.563 gearbox ratio
No. of motors / propellers	2 / 2	4 / 2

Arctic and Antarctic Voyages

The global positioning system (GPS) track of the 2015/2016 Antarctic relief voyage of the SAA II presented in Figure 6(a). The vessel departed Cape Town Harbour (1) and headed south along the Greenwich Meridian. Ice was encountered prior to reaching the ice shelf at

Penguin Bukta (3). From here the vessel departed for Akta Bukta near the German Antarctic Research Station, Neumayer and continued through heavy pack ice towards the South Sandwich Islands towards South Thule (4). She exited the ice field and reached South Georgia (5) and returned to Penguin Bukta (3) before heading back to Cape Town (1).

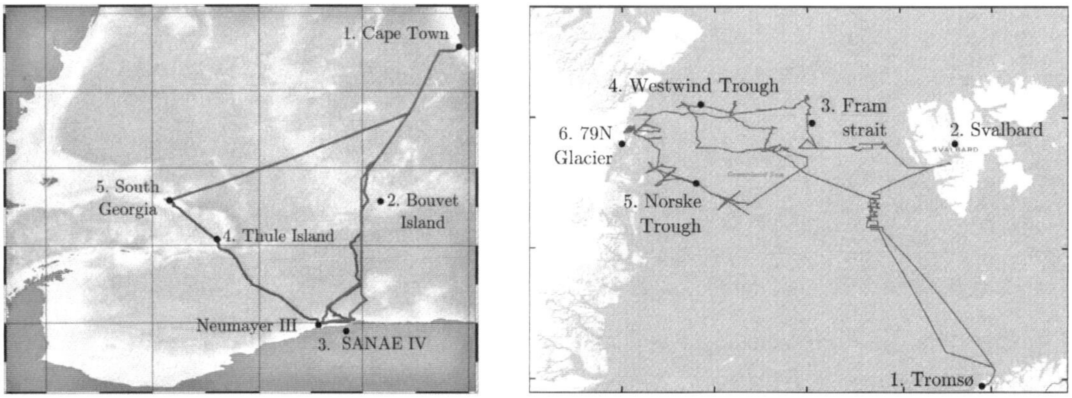

(a) Antarctic voyage (b) Arctic voyage

Figure 6. Round voyages for (a) SAA II to Antarctica (Red - outbound, blue - return voyage) and (b) FS Polarstern to the Arctic. Background for Antarctica adapted from AWI (2015) and for Arctic from Google (2016).

For the PS-100 Arctic voyage on-board the FS Polarstern (Figure 6(b)) the FS Polarstern departed from Tromsø (1), Norway and sailed for Longyearbyen, Svalbard (2). From there she headed West North West towards the zero degree meridian and sailed along the Fram Strait (3). Ice was encountered during navigation along the zero meridian to just below 81 degrees North. The voyage continued Westwards towards the West wind Trough (4) where ice cover was less concentrated. The ship re-entered open water on her southerly journey to the Norske Trough (5) where she re-encountered significant ice cover. She proceeded towards the 79 North Glacier (6) at Greenland before exiting the ice filed and returning to Tromsø (1). The total duration of the voyage was 51 days, of which an estimated 70 hours was spent navigating in thick Arctic ice.

Instrumentation

Strain gauges were installed on the port side shaft lines of each of the SAA II and FS Polarstern to determine the shear and axial strain. This allows for the determination of the instantaneous torque and thrust at the measurement locations. For shear strain measurements T-rosette gauges were installed at a 45° angle to the shaft mid-plane in a Wheatstone bridge configuration. Figure 7(a) provides a diagram of the Wheatstone bridge layout, indicating the supply voltage, U_E, and output voltage, U_A, as well as the strain gauge resistances (R_1 to R_4) for the four gauges in a full bridge. Figure 7(b) depicts the orientation of the strain gauges for shear strain measurement on the shaft. The bridge was set up to reject both axial- and bending strain whilst compensating for temperature variations.

A further set of T-rosette strain gauges were installed on either side of the shaft line to enable axial strain measurements. These gauges were oriented parallel to the shaft mid-plane as shown in Figure 7(c). The full bridge configuration is robust for axial strain measurement as it is compensated for temperature variation and bending strain.

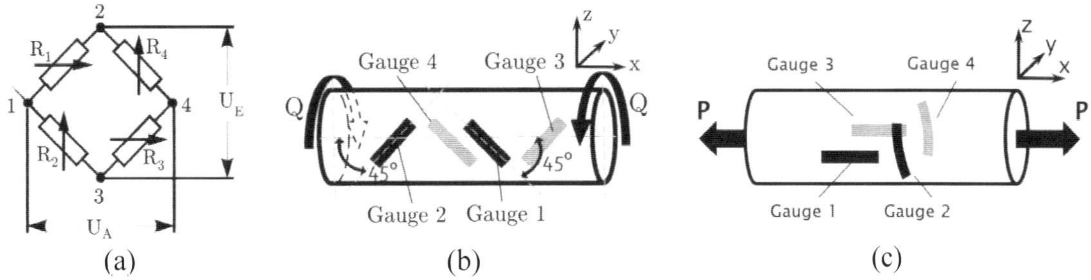

Figure 7. Strain gauge placement on shaft for (a) torque and (b) thrust measurements. Adapted from Hoffmann (Hoffmann 2001).

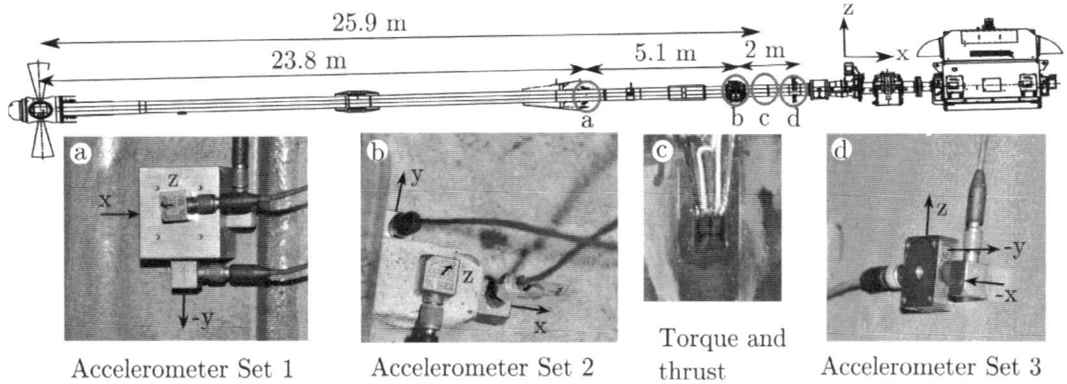

Figure 8. Accelerometers and strain gauges mounted along the shaft line for the SAA II (Adapted from technical ship drawings (STX Finland Oy 2012))

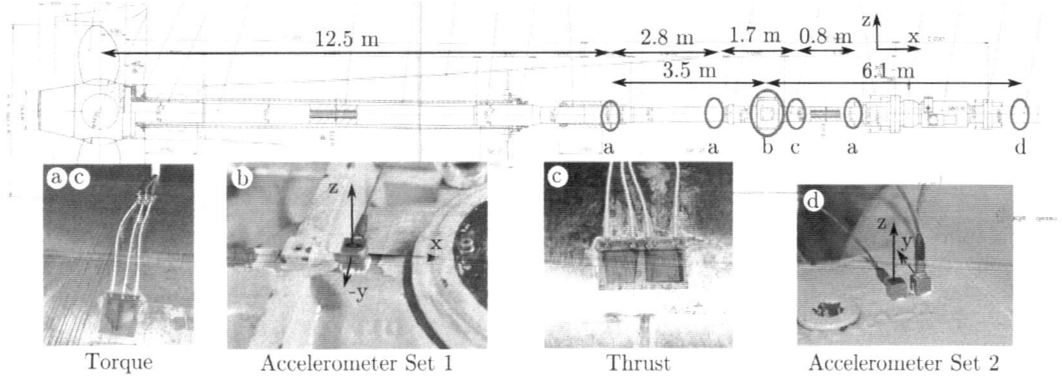

Figure 9. Accelerometers and strain gauges mounted along the shaft line for the FS Polarstern (Adapted from technical ship drawings (Werft Nobiskrug GmbH 1980))

The strain gauge installations on the SAA II were located at 25.9 m inboard from the centre of gravity of the port side propeller as shown in Figure 8. The measurements were wirelessly transmitted via wi-fi from a V-link system to a WSDA-Base data gateway, managed through Node Commander software. Data was acquired with a Hottinger Baldwin Messtechnik (HBM) acquisition system. In addition, three accelerometer sets (PCB, 100 mV/g) were mounted at three locations on the shaft support structures. The orientations of the accelerometers are indicated in Figure 8. Data was acquired through a HBM Quantum mobile data acquisition system and transmitted to a laptop via an ethernet cable and recorded through Catman AP V3.5 software at a sample rate of 600 Hz.

Similar measurements were performed on the FS Polarstern as shown in the instrumentation layout presented in Figure 9. The power source in the shaft line room originated directly from the generator, resulting in unstable power. Two sets of accelerometers were installed on the FS Polarstern on the available shaft support structures within the intermediate shaft line room. One x-direction (fore-aft) measurement was recorded as the bearings on which accelerometer measurements were performed are radial bearings that do not transfer axial vibration well.

METHODS

Calculation of Shaft Line Torque and Thrust

The torque in the shaft can be determined from the output voltage of the Wheatstone bridge through:

$$Q_{shaft} = U_A \frac{\pi E(d_o^4 - d_{in}^4)}{16U_E k d_o (1+\nu)}_{[Nm]} \tag{2}$$

Here, E is the Young's modulus, ν is the Poisson's ratio and d_o and d_{in} respectively reflect the outer and inner diameters of the hollow shaft. The strain gauge factor, k, is directly obtained from manufacturer specifications. The thrust was calculated from the T-rosette Wheatstone bridge output voltage with Eq, (3):

$$T_{shaft} = \frac{U_A E\pi(d_o^2 - d_{in}^2)}{2U_E k(1+\nu)}_{[N]} \tag{3}$$

Table 2. Shaft line dimensions, material properties and shaft related variables for measurement location (STX Finland Oy 2012; Rolls-Royce AB 2010; Escher Wyss 1980; Metallurgica Veneta 2004; Det Norske Veritas 2011a).

Description	Symbol	SA Agulhas II	FS Polarstern
Modulus of elasticity	E	210 GPa	220 GPa
Shear modulus	G	81 GPa	84 GPa
Outer diameter	d_o	0.5 m	0.39 m
Inner diameter	d_{in}	0.175 m	0.213 m
Hub diameter	d_h	1.32 m	1.60 m
Maximum ice thickness	H_{ice}	2.0 m	2.0 m
Ice strength index	S_{ice}	1.1 m	1.1 m
Pitch at $0.7 \cdot$ radius	$P_{0.7}$	5.15 m	4.42 m
Expanded blade area ratio	EAR	0.51	0.55
Depth of propeller centerline	h_o	3.75 m	6.92 m

The shaft line dimensions for the SAA II were obtained from engineering drawings by (STX Finland Oy 2012). The material specifications were sourced from manufacturer documentation (Rolls-Royce AB 2010). Polarstern shaft line dimensions were sourced from engineering drawings by (Werft Nobiskrug GmbH 1980) and the material properties from a test certificate of materials (Germanischer Lloyd 1981). As the FS Polarstern was built in

1982, old German material specifications were used. The current equivalent classification was obtained from a catalogue (Metallurgica Veneta 2004), in which the new material grade equivalent for St52-3N was found to be S355J2. The dimensions, material properties and shaft related variables for the two vessels are presented in Table 2. The depth of the propeller centreline, h_o, was not directly obtainable and was extracted by scaling a measurement from the vessel drawings.

Determination of Representative Loading Profiles

As data recordings were not possible throughout the voyages, it was required to classify vessel activities along with their associated load histories from the recorded data. This was achieved by categorizing the data into three categories namely: open water, ice navigation and stationary vessel operations. The appropriate categories could be determined by consideration of the vessel location, speed, shaft line data and logbook information. Vessel operations are recorded in the ship logbooks which assisted with the determination of stationary periods. The boundary between ice navigation and open water operational categories was more difficult to distinguish precisely as low ice concentrations and thickness do not prominently affect the shaft line response. The threshold between ice and open water navigation was distinguished based on three indicators: i.) By classification of the corresponding data in accordance with continual ice observation records from the bridge. ii.) Investigation of the shaft line torque and thrust, which showed a more oscillatory, higher amplitude response in ice compared to open water. iii.) Vessel speed is commonly reduced during navigation in thick ice.

Rainflow Counting

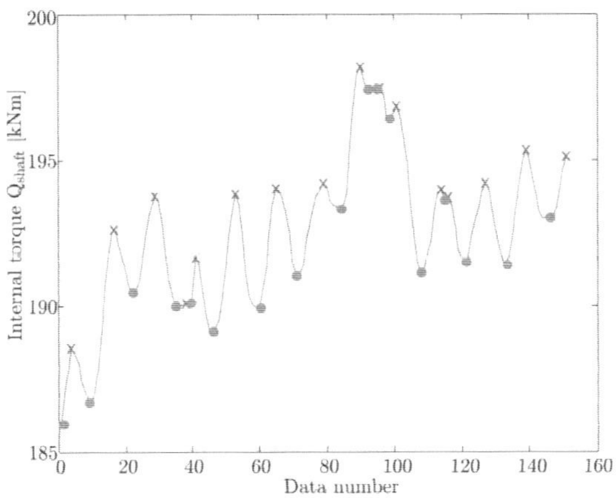

Figure 10. The identification of peaks(crosses) and troughs (circles) by the `sig2ext.m` (Nieslony 2003) algorithm.

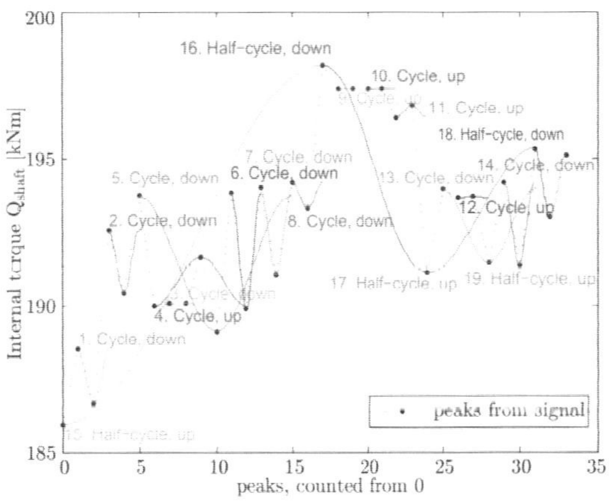

Figure 11. A diagram which demonstrates the extraction of loading cycles by the `rainflow.m` (Nieslony 2003) algorithm.

Rainflow counting involved peak and cycle counting on the calculated torque (Eq.1) and thrust (Eq. 2) from shaft line measurement data of the two vessels. The local maxima and minima (turning points) of the signals were first identified using a MATLAB algorithm, `sig2ext.m` (Nieslony 2003). A graphical result for a half second time signal is presented in Figure 10 where red crosses indicate the peaks and green dots the troughs, which were identified by the algorithm. To obtain the amplitudes of the data signals, the identified maxima and minima are processed by a second algorithm `rainflow.m` (Nieslony 2003) which extracts and counts the identified cycles as demonstrated in Figure 11.

RESULTS

Vessel Loading Profiles

The total time spent in ice, open water and stationary operations was determined for each voyage as shown in Figure 12 and Figure 13. The actual duration was compared to the time of the available full-scale recordings to determine scaling factors for the estimation of the expected load profile for each vessel (Table 3).

Table 3. Duration of voyage spent in open water, ice and stationary with recorded times for each of these conditions.

	SA Agulhas II			FS Polarstern		
	Actual [days]	Recorded [days]	Scaling factor [days]	Actual [days]	Recorded [days]	Scaling factor [days]
Ice navigation	10.67	8.54	1.25	11.69	9.15	1.27
Open water	39.96	5.42	7.38	17.31	5.03	3.44
Stationary	17.46	1.58	11.03	19.97	10.95	1.82
Total	68.09	15.54	4.38	48.97	25.13	1.95

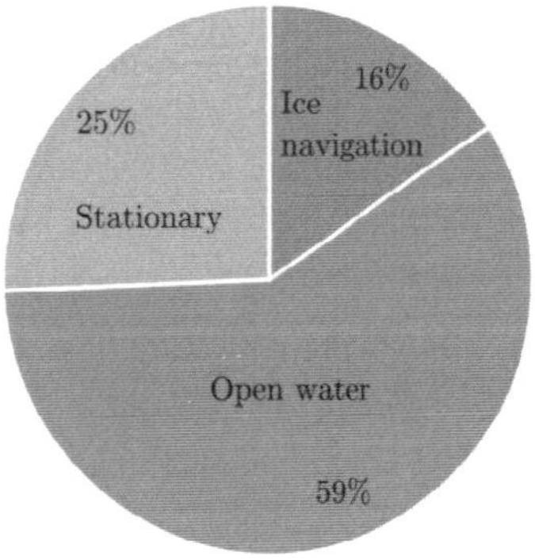

Figure 12. A classification of the activities of the SAA II during the 2015/2016 Antarctic voyage.

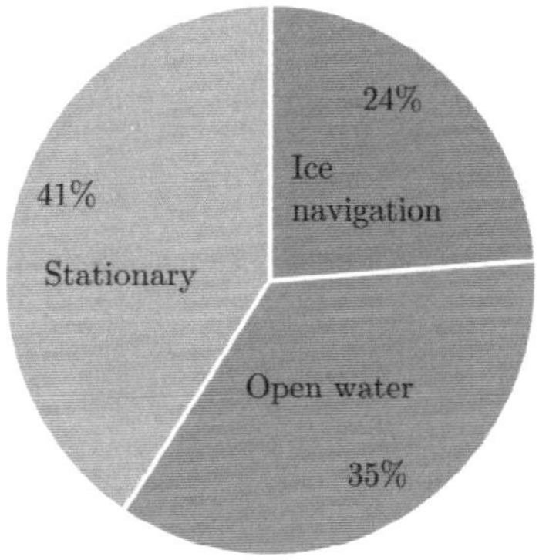

Figure 13. A classification of the activities of the FS Polarstern for the PS-100 voyage to the Arctic, 2016.

Rainflow Counting

Rainflow counting analyses of peak torque values are presented in Figure 14 for both vessels. The plots only include significant peaks, as there are many high magnitude peaks occurring at low cycle magnitudes. The highest peak values and total cycles recorded during these conditions are presented in Figure 16. The bin size for the histograms was selected to be 1 kN·m for torque and 1 kN for thrust. When comparing the results from the operational categories for the two vessels, the following was observed for ice navigation in Figure 14 (a) and (b):

i.) Ice navigation results in higher peak values, with the SAA II having a higher distribution overall.

ii.) The most peak torsional loads in ice occurred around 100 kN · m for the SAA II and around 60 kN · m and 80 kN · m for the FS Polarstern.

iii.) During ice navigation both vessels display a secondary region of increased loading cycles at higher torque magnitudes of 310 kN · m for the SAA II and 250 kN · m for the FS Polarstern, whereby these high magnitude peaks were not as evident in open water or stationary operations.

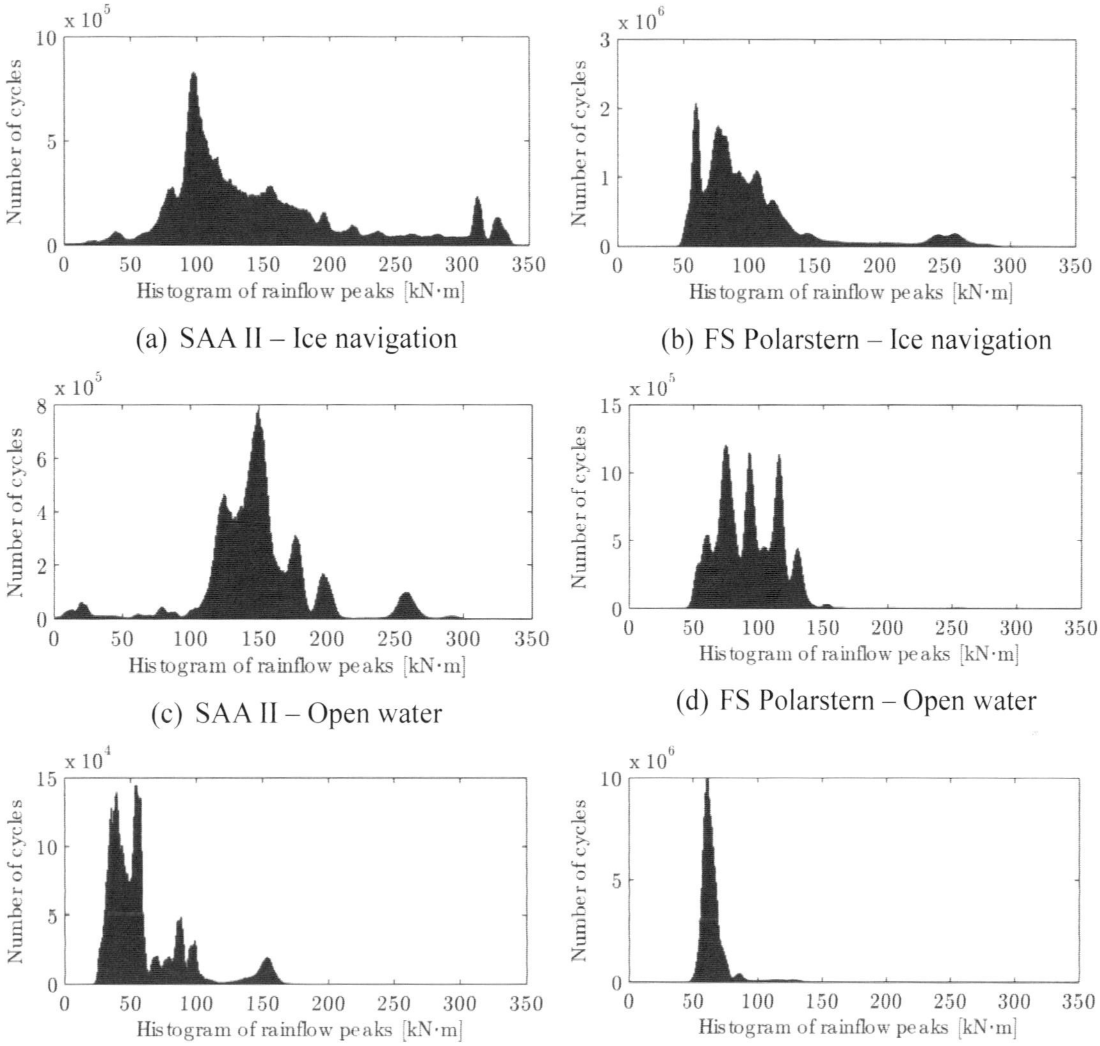

(a) SAA II – Ice navigation

(b) FS Polarstern – Ice navigation

(c) SAA II – Open water

(d) FS Polarstern – Open water

(e) SAA II - Stationary

(f) FS Polarstern – Stationary

Figure 14. Comparison of torque peak rainflow cycles for the SAA II (left) during the 2015/2016 Antarctica voyage and for the FS Polarstern (right) during the 2016 Arctic voyage.

Rainflow counting for open water passage in Figure 14(c) and (d) show that:

i.) The greatest number of torque peak cycles for the SAA II occurred at 150 kN · m which amounts to more cycles at a higher cycle count than for ice navigation. This is explained by considering that the pitch of the propeller and shaft line speed is lowered during ice navigation to reduce the speed of the vessel while maintaining enough torque on the shaft line to allow the propeller to chop through ice. In open water, the SAA II generally operated above 7 knots at a pitch of 87% and shaft line speed of 135 rpm, compared to a speed below 6 knots, pitch of 74% and shaft line speed of between 90 and 110 rpm during ice navigation.

ii.) The high cycle loading just above 250 k·Nm could be a result of vessel operations just shy of the maximum pitch of 99.5% in open water conditions upon the return leg from Antarctica to Cape Town.

iii.) The FS Polarstern data presented a much narrower spread of high cycle peak torque values, ranging from 50 to 150 k·Nm. Within this range there were three distinct high cycle peaks, which were also apparent for ice navigation condition. The FS Polarstern generally operated below 7 knots during ice navigation, however the shaft line speed remained relatively constant around 174 rpm throughout the voyage, as they were directly driven by the diesel generators.

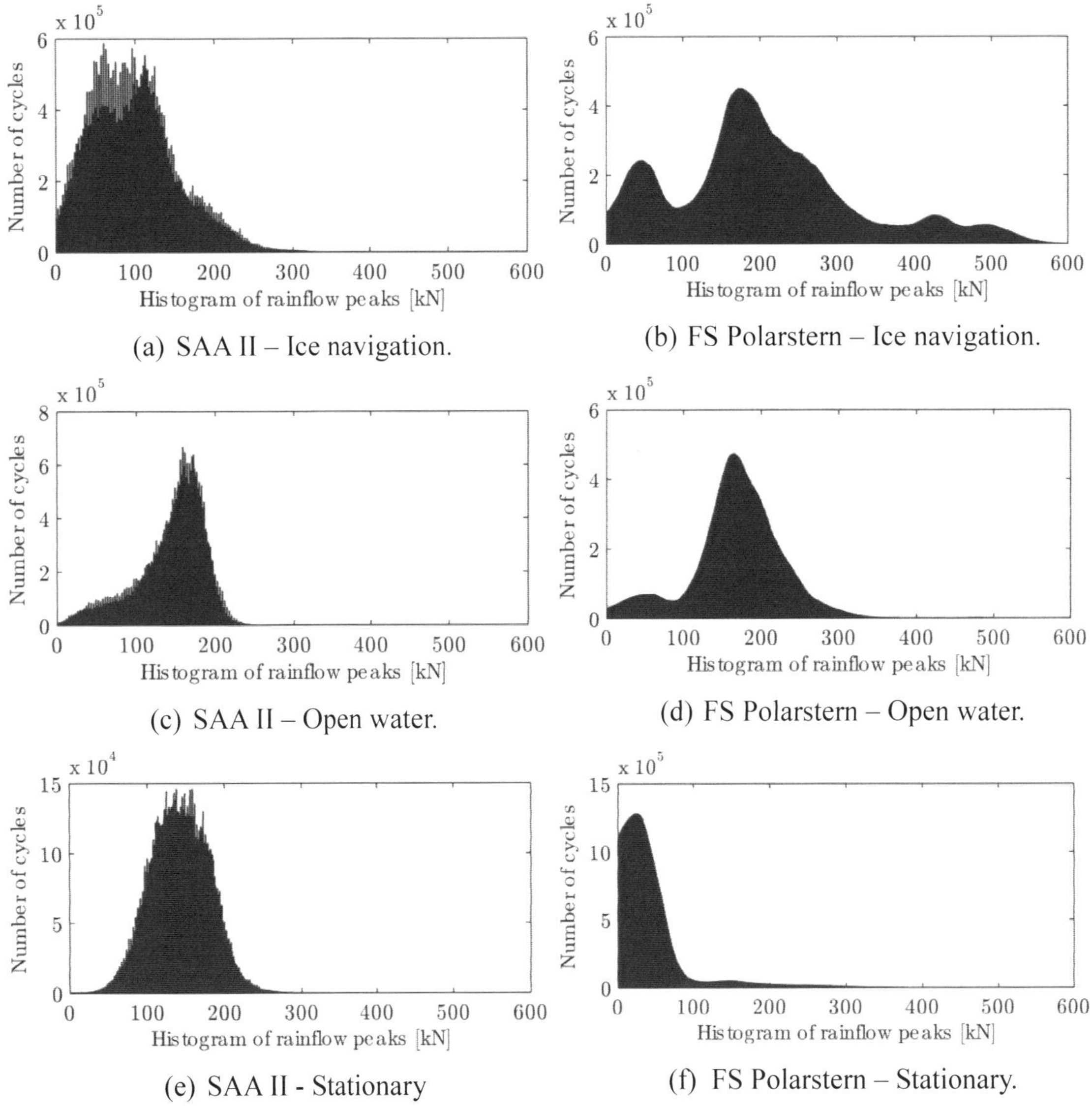

(a) SAA II – Ice navigation. (b) FS Polarstern – Ice navigation.

(c) SAA II – Open water. (d) FS Polarstern – Open water.

(e) SAA II - Stationary (f) FS Polarstern – Stationary.

Figure 15. Comparison of thrust peak rainflow cycles for the SAA II (left) during the 2015/2016 Antarctica voyage and for the FS Polarstern (right) during the 2016 Arctic voyage.

For the stationary case (Figure 14(e) and (f)):

i.) Low torque peak cycles are evident as the pitch of the propeller was set to approximately zero during these conditions.

i) The maximum cycles for the FS Polarstern occur above 50 kN·m. This could be due to the direct diesel engine drive which induces a torque greater than 50 kN·m, whereas the SAA II has a direct electric drive which is not subjected to gearbox and combustion engine vibrations.

There are several high peak torque cycles evident for the stationary case for both vessels. The vessels use their respective propellers to operate on station during deployment of equipment overboard or retaining constant thrust on the propeller to maintain the vessel stationary against the shelf or bay ice. Histograms of the rainflow counting analyses of thrust peaks are presented in Figure 15. The following observations were made:

i.) The FS Polarstern experienced a greater range of significant thrust loading peaks during ice navigation, ranging from 0 to 600 kN.

ii.) The largest number of cycles for the FS Polarstern shaft occurred at 180 kN for both ice and open water conditions and at 25 kN for the stationary conditions.

iii.) For the SAA II, the largest cycles were distributed between 50 and 150 kN for the ice condition, 170 kN for the open water condition and 140 kN for the stationary condition.

iv.) The reason for the larger significant peaks during open water compared to ice navigation is attributed to the greater propeller pitch under these conditions.

The maximum bin size for each operational category is presented in Figure 16 from which it is evident that the FS Polarstern measurements presented greater torque peaks, torque amplitudes and thrust amplitudes during the stationary condition compared to the open water condition. This is explained by the incidence of rapid shaft line loading during stations whereby the vessel is required to remain positioned in the same location. Such operations often require pushing back against floe ice.

The FS Polarstern experienced higher peak loads compared to the SAA II. This could be due to the Arctic containing harder, multi-year ice compared to the Antarctic. Alternatively the design of the FS Polarstern induces higher loads compared to the SAA II due to shaft line geometry. The location of measurements along the shaft is also a factor as torque fluctuates throughout the shaft. For a fair comparison of load conditions shaft line loads need to be converted to propeller loads through inverse methods (Ikonen et al. 2014) whereby ice induced loading on the propeller could be compared.

The predicted load profiles, which were derived from the three operational categories, are compared in Figure 17 on logarithmic scales. The torque peak values were divided by the nominal rated torque, Q_{MCR}, to obtain the load response factor, K_a which is used for vessels exposed to varying loads. This allows for the safe loading on components along the shaft line, specifically the propeller, to be determined.

Raw load response data typically exhibits a change in slope of which the cause is ascribed to one of two possibilities Myklebost & Dahler (2013): either the statistical distribution differs from medium to high impact values, or the distribution below the change in slope is a result of control system or motor excitations and the distribution above the change in slope is as a result of propeller-ice impacts. This corresponds to the typical load cases in Figure 1 (Det Norske Veritas 2011a) in which ice load amplitudes occur above the distinct change in slope.

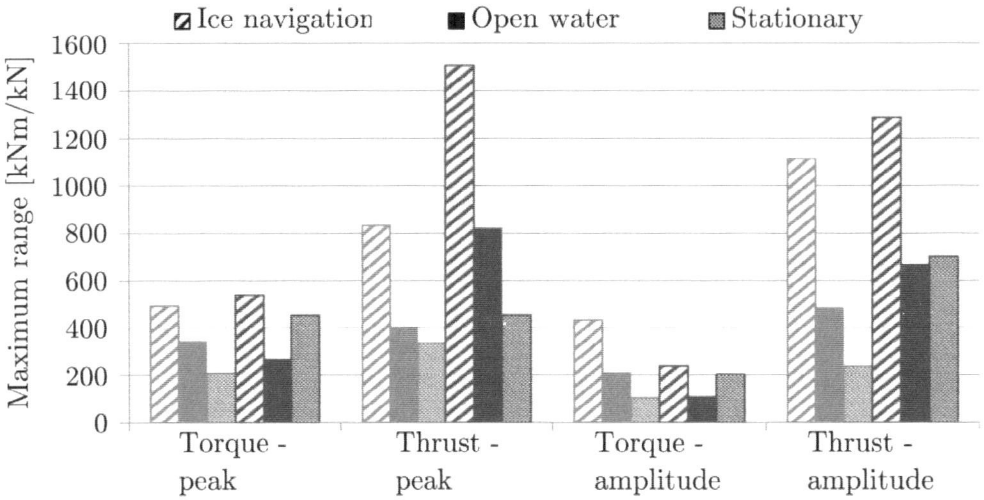

Figure 16. Maximum recorded bin size for the SAA II (blue) and FS Polarstern (red).

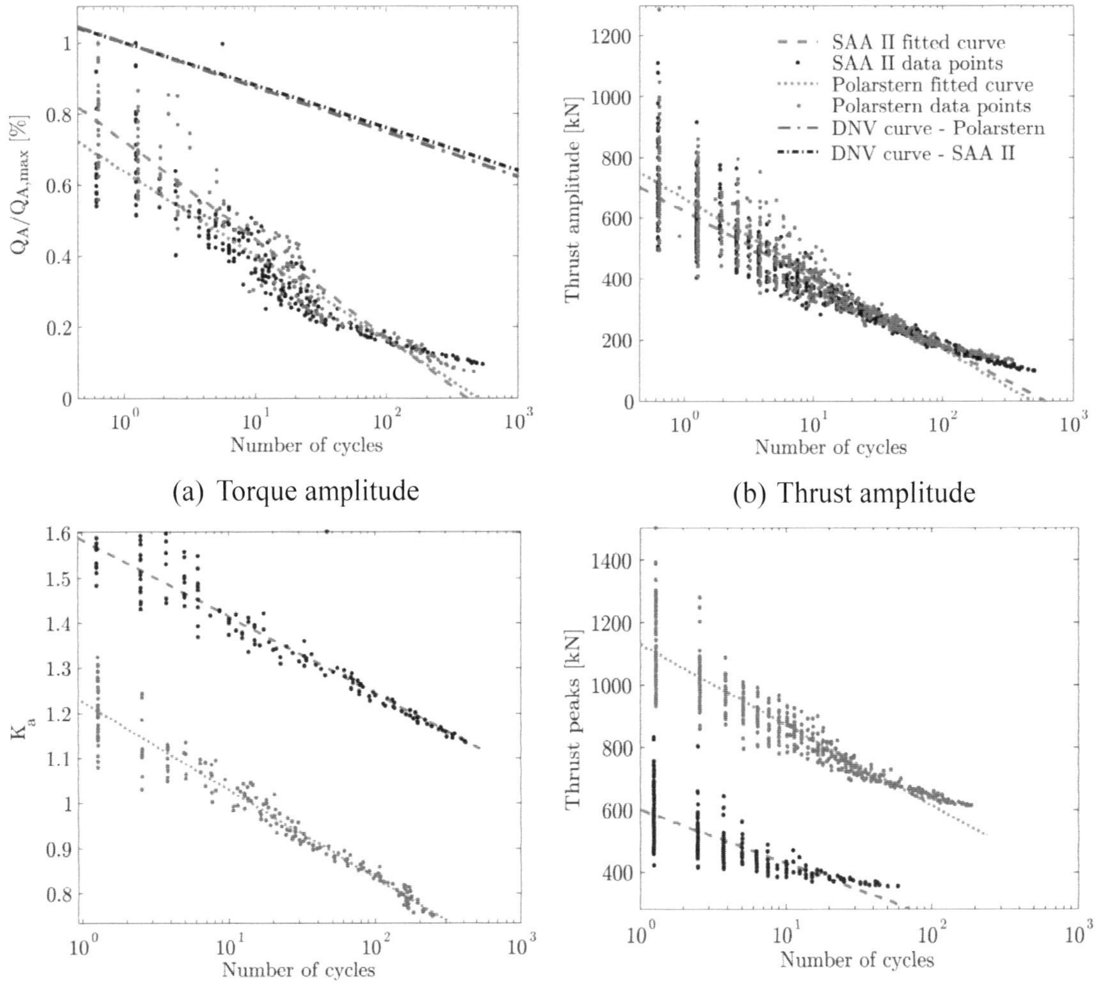

(a) Torque amplitude

(b) Thrust amplitude

(c) Torque peaks

(d) Thrust peaks

Figure 17. Comparison between two vessels of predicted load profiling during a voyage to the Arctic and Antarctica.

DNV (Det Norske Veritas 2011a) provide a relationship to predict the cumulative shaft torque distribution Eq.(1) through torque amplitudes as a function of N_{ice}. This equation was presented as an accumulated load spectrum in Figure 17 (a). To obtain this curve, the procedures described in the DNV ice rules (Det Norske Veritas 2011b) were used with variables from Table 2. These curves are presented in Figure 17(a) for the SAA II and the FS Polarstern. The representative curves for the thrust and torque peak and amplitude analyses are recorded in Table 4, with regression values and total cycles for each of the four cases for each vessel.

Table 4. Regression, total cycles and curve fit equations for rainflow counting data of the form $y = A\log x + B$.

	A	B	R^2	Total cycles
SAA II				
Torque amplitude	-102.47	275.81	0.82	8.73×10^7
Torque peak	-0.17	1.58	0.90	8.72×10^7
Thrust amplitude	-223.87	623.52	0.86	1.27×10^8
Thrust peak	-171.93	600.08	0.60	1.24×10^8
DNV curve	-0.12	1		
FS Polarstern				
Torque amplitude	266.41	173.21	0.89	2.72×10^8
Torque peak	-0.20	1.22	0.94	2.72×10^8
Thrust amplitude	-249.58	665.97	0.83	2.63×10^8
Thrust peak	-258.60	1131.57	0.85	2.21×10^8
DNV curve	-0.13	1		

Myklebost & Dahler (2013) determined that this change in slope occurred at $1.4\cdot Q_{MCR}$ during the 2012/2013 voyage for the SAA II. For the 2015/2016 voyage, it was determined to be $1.14\cdot Q_{MCR}$ for SAA II and $0.75\cdot Q_{MCR}$ for the FS Polarstern during the 2016 PS-100 voyage. The change in slopes for peak torque, peak thrust, torque amplitude and thrust amplitude were determined and the data above these points presented in Figure 17. This was done to focus the analysis on the ice related loads. The following observations are made from Figure 17:

i.) The SAA II experienced greater torque peaks relative to Q_{MCR}, however similar magnitudes were experienced by the two vessels overall. The plot is deceiving as it appears the SAA II experienced greater torque peaks. This is due to the 100 kN·m difference between the Q_{MCR} of the two vessels.

ii.) Significantly smaller thrust peaks were experienced by the SAA II throughout its voyage compared to the FS Polarstern. One possible reason could be the smaller diameter shaft of the FS Polarstern, thus experiencing a greater thrust value at the location of the strain gauge measurement.

CONCLUSIONS

Through rainflow counting analyses, the cyclic torsional and thrust loading of two polar vessel shaft lines have been highlighted for operations in open water, ice and stationary work. In the Arctic, the FS Polarstern was exposed to much greater thrust peaks in comparison to the SAA II, during her Antarctic voyage. Comparison of the peak torque cycles of the two vessels showed that the SAA II experienced greater torque peaks relative to its MCR compared to the FS Polarstern. The two vessels did however experience the same order of magnitude peak torques during their respective voyages. This does not necessarily imply that vessels operating in the Arctic and Antarctica will be exposed to the same magnitude of peak torque. Shaft geometry and measurement location do influence the obtained dynamic responses. An objective comparison would include inverse force estimation to eliminate the dynamic response of the shaft and propulsion components and subsequent comparison of the applied ice loads at the propeller. This analysis does however emphasize the importance of shaft line torsional analysis, which could identify the operation of the vessels above its MCR.

ACKNOWLEDGEMENTS

The financial assistance of the National Research Foundation (NRF) towards this research is hereby acknowledged. Captain Gavin Syndercombe and the SMIT Amandla crew and Kapitän Stefan Schwarze and his crew on-board the Polarstern is appreciated during the respective Antarctic and Arctic voyages. The collaboration of Aalto University, Finnish Meteorological Institute, the Alfred-Wegener-Institut and the Department of Environmental Affairs of South Africa is gratefully acknowledged.

REFERENCES

American Bureau of Shipping, 2006. Guidance notes on ship vibration.

ASTM International E1049-85, 2011. ASTM E1049-85: Standard practices for cycle counting in fatigue analysis. *Designation: E1049 - 85*.

Batrak, Y.A., Serdjuchenko, A.M. & Tarasenko, A.I., 2014. Calculation of torsional vibration responses in propulsion shafting system caused by ice impacts. *Torsional Vibration Symposium*, (January 2011).

Bekker, A. et al., 2014. OMAE2014-24128. In *ASME 33rd International Conference in Ocean, Offshore and Arctic Engineering*. San Francisco, California, USA.

Brouwer, J., Hagesteijn, G. & Bosman, R., 2013. Propeller-ice impacts measurements with a six-component blade load sensor. In *Third International Symposium on Marine Propulsors*. Tasmania, pp. 47–54.

Connor, S.O. et al., 2010. Fatigue life monitoring of metallic structures by decentralized rainflow counting embedded in a wireless sensor network. In *Proceedings of the ASME 2010 Conference on Smart Materials, Adaptive Structures and Intelligent Systems*. Philadelphia, PA, USA.

Det Norske Veritas, 2011a. Classification of ships for navigation in ice. *Rules for Classification of Ships*, 5(1322), pp.1–142.

Det Norske Veritas, 2011b. Ice strengthening of propulsion machinery. *Classification Notes*, 1(51).

Dinham-Peren, T. & Dand, I., 2010. The need for full scale measurements. In *William Froude Conference: Advances in Theoretical and Applied Hydrodynamics - Past and Future*. Portsmouth. Available at: http://media.bmt.org/bmt_media/resources/33/RINA Froude Conf - The Need for Full Scale Measurements.pdf.pdf.

Escher Wyss, 1980. Gesamtanordnung VP Anlage.

Germanischer Lloyd, 2007. Rules for classification and construction. *Part 6 Offshore Technology, Chapter 5 Machinery Installations, Section 7 Torsional Vibrations*.

Germanischer Lloyd, 1981. Test certificate for shaft line materials.

Grobe, H. & Alfred Wegener Institute, 2007. Photo from the German research vessel Polarstern port propeller. *Wikimedia Commons*. Available at: https://commons.wikimedia.org/wiki/File:Polarstern_propeller-bb_hg.jpg [Accessed October 28, 2016].

Hoffmann, K., 2001. Applying the Wheatstone bridge circuit. *HBM W1569-1.0 en, HBM, Darmstadt, Germany*, pp.1–28.

Huisman, T.J. et al., 2014. Interaction between warm model ice and a propeller. In *Proceedings of the ASME 2014 33rd International Conference on Ocean, Offshore and Arctic Engineering OMAE2014*. San Francisco, California, USA.

Ikonen, T., Peltokorpi, O. & Karhunen, J., 2014. Inverse ice-induced moment determination on the propeller of an ice-going vessel. *Cold Regions Science and Technology*, 112, pp.1–13.

International Association of Classification Societies, 2011. Requirements Concerning Polar Class. , pp.1–26.

Korean Register, 2015. Guidance for ship for navigation in ice. Available at: http://www.krs.co.kr/KRRules/KRRules2015/KRRulesE.html [Accessed November 10, 2016].

Metallurgica Veneta, 2004. Material specifications for S355J2.

Myklebost, M.R. & Dahler, G., 2013. Agulhas II data analysis of torsional shaft responses from operation in Arctic waters - Technical summary report. *Det Norske Veritas AS*, pp.1–40.

Nieslony, A., 2003. Rainflow counting algorithm. *MathWorks File Exchange*. Available at: https://www.mathworks.com/matlabcentral/fileexchange/3026-rainflow-counting algorithm [Accessed October 26, 2016].

Polić, D. et al., 2014. Shaft response as a propulsion machinery design load. In *Proceedings of the ASME 2014 33rd International Conference on Ocean, Offshore and Arctic Engineering OMAE2014*. California,.

Rolls-Royce AB, 2010. Whirling calculation. *Report SA1090*, pp.1–15.

Sampson, R., Atlar, M. & Sasaki, N., 2009. Propeller ice interaction - effect of blockage proximity. In *First International Symposium on Marine Propulsors*. Trondheim, Norway.

STX Finland Oy, 2012. Polar supply and research vessel shaft line arrangement.

Tang, B. & Brennan, M.J., 2013. On the influence of the mode-shapes of a marine propulsion shafting system on the prediction of torsional stresses. *Journal of Marine Science and Technology*, 21(2), pp.209–214.

The Alfred Wegener Institut, 2016. Technical data. Available at: http://www.fs-polarstern.de/index.php?lang=EN&aktion=anzeigen&rubrik=015006 [Accessed August 16, 2016].

Transport Safety Agency, 2010. Finnish ice classes equivalent to class notations of recognized classification societies. *Maritime Safety Regulations*.

Transportation Safety Board of Canada, 2010. Statistical summary marine occurences 2010.

Werft Nobiskrug GmbH, 1980. Wellenleitung, Hauptmasse.

POAC'17
Busan, KOREA

Proceedings of the 24th International Conference on
Port and Ocean Engineering under Arctic Conditions
June 11-16, 2017, Busan, Korea

Global Ice Load Prediction for the Icebreaker Using 6-DOF Motion Measurement Method

Sang Chul Lee[1], Sunho Park[1], Kyungsik Choi[1], and Seong-Yeob Jeong[2]
[1] Korea Maritime and Ocean University, Busan, Korea
[2] Korea Research Institute of Ships and Ocean Engineering, Daejeon, Korea

ABSTRACT

A global ice load acting on a hull of an icebreaker with a six degrees of freedom (6-DOF) inertial measurement system method was described in this paper. The 6-DOF inertial measurement system treat the ship as a rigid body and measures whole ship motions, and the global ice loads were calculated with the data measured by full-scale ice sea trials of Korean ice breaking research vessel (IBRV) ARAON during her voyage in 2015. To calculate the maximum global ice load, the point of impact (POI) approach and the center of gravity (COG) approach were used, respectively. The differences of the calculated results using the POI and COG approaches were from 254% to 1,794%. Comparing with the resultant forces from strain gauges which ranged from 0.11 to 0.8 MN, the global ice loads from the POI approach were from 59% lower to 239% higher than forces from the strain gauges, and the global ice loads from the COG approach were up to 1,907% higher than the forces from the strain gauges. The calculated global ice loads from both the POI and COG approaches were within the range of an empirical ice load estimation formula under normal operating conditions.

KEY WORDS: Korean IBRV ARAON; Global ice load; 6-DOF motion measurement method; COG approach; POI approach.

INTRODUCTION

Exact ice load estimation is very important for icebreakers that operate in the ice seas. This is because fundamental design factors, such as a hull structure, hull form, and thruster type, are determined based on it.

To measure the global ice loads acting on the hull of an icebreaker, strain gauges were used traditionally. In this method, the hull of an icebreaker is treated as an elastically deformable body, and the ice load is measured by the deformation of the hull during it interacts with the sea ice. Although this method has been proved as an effective and a reliable method, it needs much time and cost for installation and operation. In addition, every time when it is applied on another hull, same amount of time and cost have to be spent repeatedly.

Then another method which uses the inertial measurement system named as MOTAN (MOTion-ANalysis) was introduced by the Canadian Hydraulic Centre (CHC) of the National Research Council of Canada (Johnston et al., 2003). In this method, the hull of an icebreaker is treated as a rigid body, and the global ice load is measured by the whole-ship motions in the six-degrees of freedom. It is economical for installation and operation, and simple for applying on another icebreaker to predict the global ice loads on a hull.

In this study, global ice loads on the Korean ice breaking research vessel (IBRV) ARAON was predicted using the data measured by an inertial measurement system during her Arctic Ocean voyage in 2015. This Arctic Ocean voyage allocation was carried out from July 30 to August 23, 2015 in the Chukchi and East Siberian Seas.

The motion data in six-degrees of freedom were measured by an inertial measurement system "MotionPak Ⅱ" of SYSTRON DONNER INERTIAL, Inc., which shown in Figure 1, and it consists of three translational accelerometers and three angular rate sensors then each sensor measures the total ship accelerations and the angular rotational rates of the ship in the directions of x-, y-, and z-coordinates, respectively.

Figure 1. Inertial measurement system "MotionPak Ⅱ" of SYSTRON DONNER INERTIAL, Inc.,

DATA PROCESS

Data Calibration and Filtering

For calculating the global ice loads using the measured raw data from the inertial measurement system, two-stepped data processing procedures are required. The first step is data calibration. Since there is no six-degrees of freedom motion before the ice breaker collides with the sea ice, so all the initial values of the raw data have to be shifted to the base level, i.e., set to zero. The second step is data filtering with a low pass filter (LPF). Since the raw data of the inertial measurement system contain motions that are related with modal vibrations and not with sea ice interactions, those motions have to be removed by filtering. It was well known that whole-ship motions commonly occur at the lower frequencies than 5 Hz, and the higher frequencies than 2 Hz are related with the secondary ship vibrations that are quite separated from the global ship accelerations. Therefore the frequencies higher than 2 Hz must be removed (Chen et al., 1990; Johnston et al., 2004).

As the results after the calibration and filtering procedures, the comparison of acceleration in the y-direction between the raw and processed data was shown in Figure 2, and it was found that the amplitude of the processed data becomes smaller than the raw data because the components of the higher frequencies were removed. Also, The result of the analysis of signal intensity using the fast Fourier transform (FFT) shows that the signals of the higher frequencies than 2 Hz were attenuated rapidly, as shown in Figure 3, by the effect of the filter.

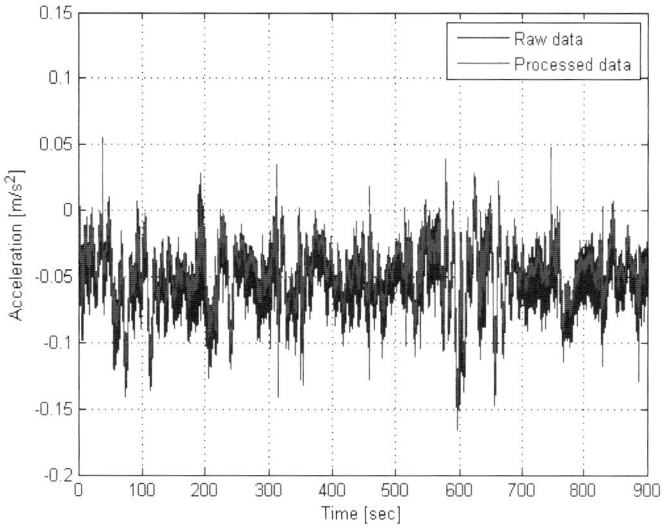

Figure 2. Comparison of amplitudes between raw and processed accelerations in y-direction

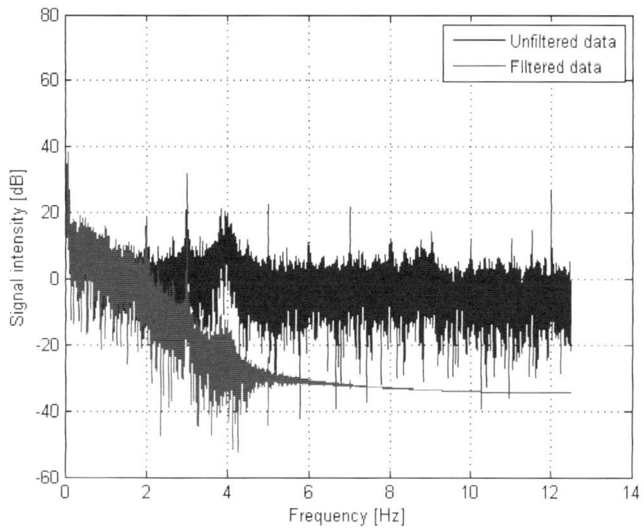

Figure 3. Comparison of signal intensities between unfiltered and filtered accelerations in y-direction

Numerical Calculation

For convenience, the subscripts 1, 2, and 3 represent the translational components of x-, y-, and z-coordinate axes, and the subscripts 4, 5, and 6 represent the rotational components of x-, y-, and z-coordinate axes, respectively. The capital letters *A, V*, and *D* represent the acceleration, velocity (or rate), and displacement, respectively. When using above notations, consequently, all measured variables by the inertial measurement system are A_1, A_2, A_3, V_4, V_5, and V_6.

The unknown angular accelerations A_4, A_5, and A_6 were calculated by the numerical differentiation using the known angular rates V_4, V_5, and V_6, respectively, and the numerical differentiation can be written

$$f'(t) = \{f(t + \Delta t) - f(t)\}/h \tag{1}$$

where *t* is time, and Δt is measuring time step.

The other unknown variables were calculated by the numerical integration, i.e., V_1, V_2, and V_3 were from A_1, A_2, and A_3.. The trapezoidal rule was used for the numerical integration (Atkinson, 1993), and it can be written as

$$\int_{t_o}^{t_o+\Delta t} f(t)dt = \Delta t[\{f(t_o + \Delta t) + f(t_o)\}/2] \tag{2}$$

The relations between the variables, and numerical differentiation and integration are shown in Figure 4.

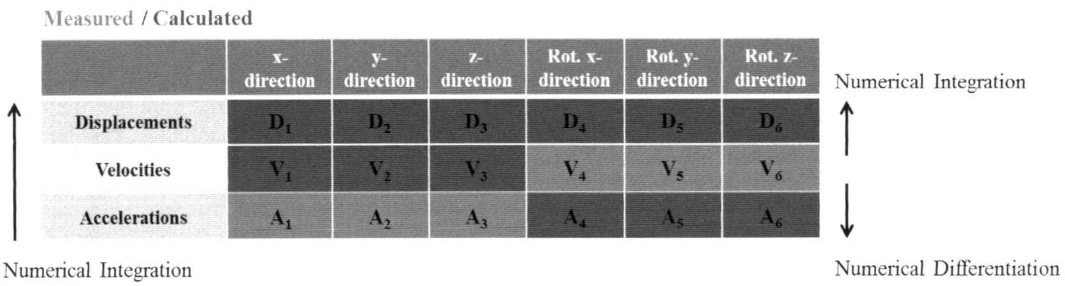

Figure 4. Relations between variables and numerical methods

GLOBAL ICE LOAD

Equations of Motion and Coefficients

The external forces and moments which will be used to calculate the global ice loads can be obtained by the equations of motion in six-degrees of freedom, and the equations of motion can be written as

$$\left[M_{ij} + M_{a_{ij}}\right]\{A_j\} + [B_{ij}]\{V_j\} + [C_{ij}]\{D_j\} = \{F_i\} \tag{3}$$

where M_{ij} is the mass matrix, M_{aij} is the added mass matrix, B_{ij} is the damping coefficient matrix, C_{ij} is the restoring coefficient matrix, A_j is the acceleration vector, V_j is the velocity vector, D_j is the displacement vector, and F_i is the external force and moment vector. The coefficient matrices such as the mass, added mass, and restoring coefficient were calculated using the potential flow-based simulation code.

Global Ice Load Calculations

The external forces and moments obtained by the equations of motion were generated by the interactions between the ship and the sea ice, and they are acting on the center of gravity. F_1, F_2, and F_3 means the external forces acting on the center of gravity in the directions of x-, y-, and z-coordinate axes, i.e., surge, sway, and heave forces, and F_4, F_5, and F_6 means the external moments acting on the same position in the rotational directions about x-, y-,and z-coordinate axes, i.e., roll, pitch, and yaw moments, respectively.

To calculate the maximum global ice load, two approaches, i.e., the point of impact (POI) approach that can be used when the location of impact with ices is exactly known and while the center of gravity (COG) approach that can be used when the impact location is not known, were used, respectively (Johnston et al., 2008a; Johnston et al., 2008b).

For the COG approach, the global ice load was calculated at the center of gravity, and the surge, sway, and heave forces acting on the center of gravity were used. The resultant global ice load is calculated as

$$F_{COG} = \sqrt{(F_1)^2 + (F_2)^2 + (F_3)^2} \tag{4}$$

where F_1, F_2, and F_3 are the surge, sway, and heave forces acting on the center of gravity, respectively.

For the POI approach, the global ice load was calculated at the point of impact, and this approach can be used only when the exact ice impact locations are known. In this approach, the surge force, and yaw and pitch moments acting on the center of gravity were used, and the resultant global ice load is calculated as

$$F_{POI} = \sqrt{(F_1)^2 + (F_6/X_{ab})^2 + (F_5/X_{ab})^2} \tag{5}$$

where F_6 and F_5 are the yaw and pitch moments acting on the center of gravity, and X_{ab} is the longitudinal distance in the direction of x-coordinate axis from the location where the inertial measurement system is installed to the point of impact.

RESULTS AND DISCUSSION

Motion Data Analysis

The translational accelerations and rotational angular rates measured by the inertial measurement system were shown in Figures 5 and 6. In the general icebreaking process with the level ice, the surge motions which appear during the ship collides with the ice and the pitch motions which appear in the process of ice breaking that the ice breaker glides over the ice and crushes it downward with her weight are dominant. However, in the data of year 2015, the sway motions in the translational accelerations, and the roll and yaw motions in the angular rates are dominant as shown in Figures 5 and 6. This is because the ice conditions during the Arctic Ocean voyage in 2015 were not the level ice but small pack ices.

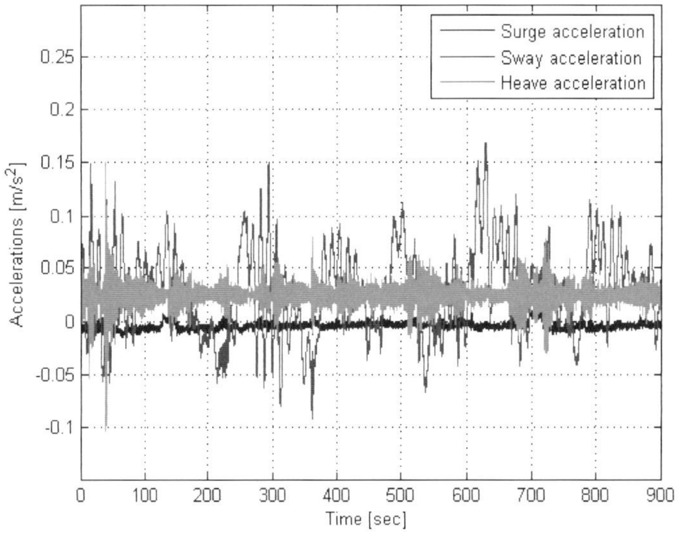

Figure 5. Translational accelerations measured by inertial measurement system

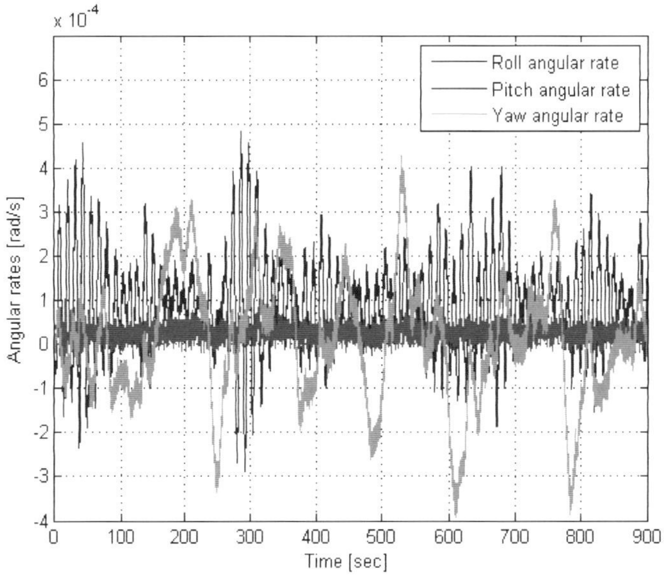

Figure 6. Angular rates measured by inertial measurement system

Global Ice Loads

The global ice load calculations were performed for the total of 138 tests in 30 cases. The resultant global ice loads ranged from 0.11 to 1.06 MN using the POI approach and from 1.41 to 4.87 MN using the COG approach, respectively. At this time, the ship speed corresponding to each test was calculated by using a global positioning system (GPS) device information and the range was from 0.3 to 7.4 *m/s*. The maximum deviation between the POI and COG approaches was 1,794% for the case of the Official 26_2, i.e., the second test in the 26th official case, and the minimum one was 254% for the case of the Official 28_1. The global ice loads for the cases of Official 26-2 and 28_1 were shown in Figures 7 and 8.

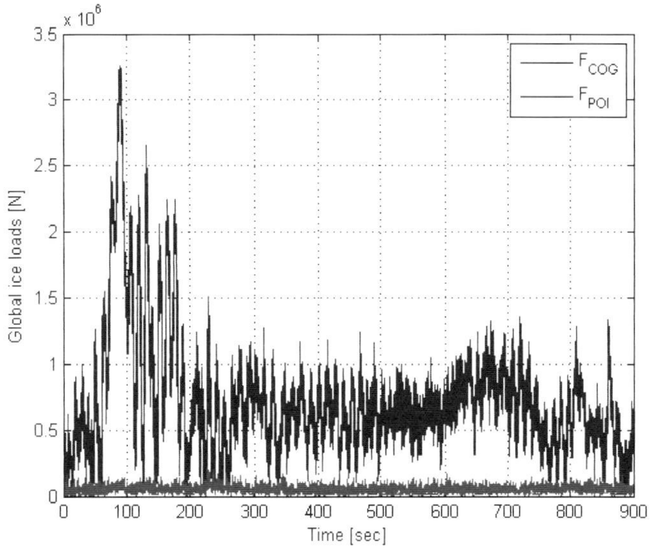

Figure 7. Global ice loads for case of Official 26_2

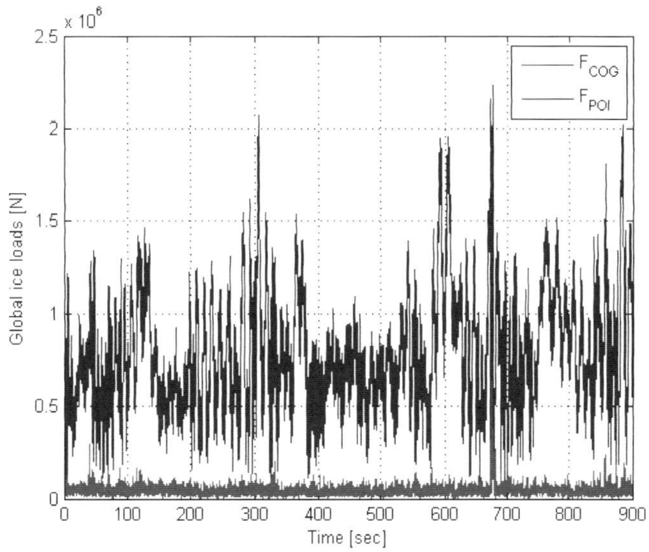

Figure 8. Global ice loads for case of Official 28_1

Comparison with Results of Strain Gauges

The resultant global ice loads calculated from the inertial measurement system were compared with the forces measured from the strain gauge (Min et al., 2016) for the cases of the Official 5, 14, 19, 23, and 28 that contain four tests in each case, and the results were shown in from Figures 9 to 13. The ice loads calculated from the COG approach were consistently much higher than those of the strain gauges, but the ice loads from the POI approach were lower or higher than those of the strain gauges and showed the better agreement with the result of the strain gauges. On the other hand, when considering the ship speed, there is no clear relationship between the ship speed and the ice load. It seems that the weight of each pack ice was more important to generate ice loads than the ship speed.

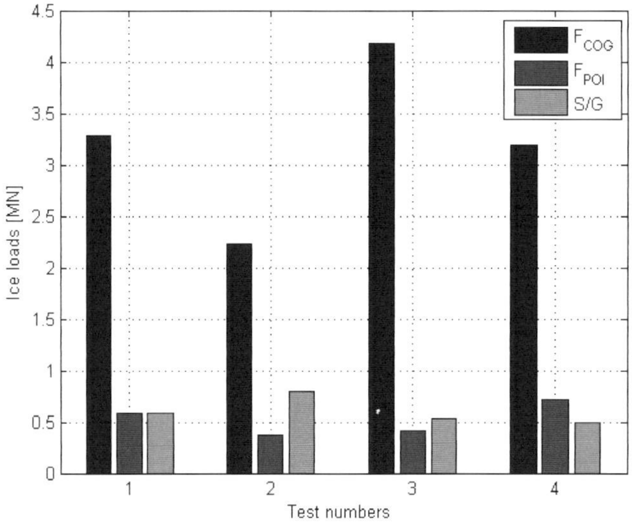

Figure 9. Resultant ice loads from COG and POI approaches and strain gauge for the case of Official 5

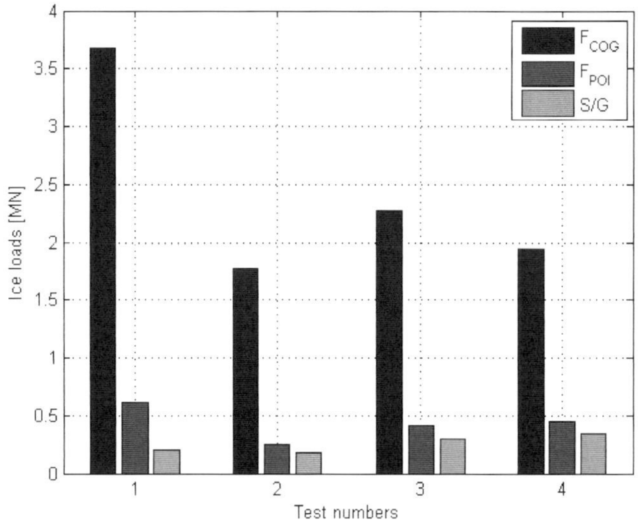

Figure 10. Resultant ice loads from COG and POI approaches and strain gauge for the case of Official 14

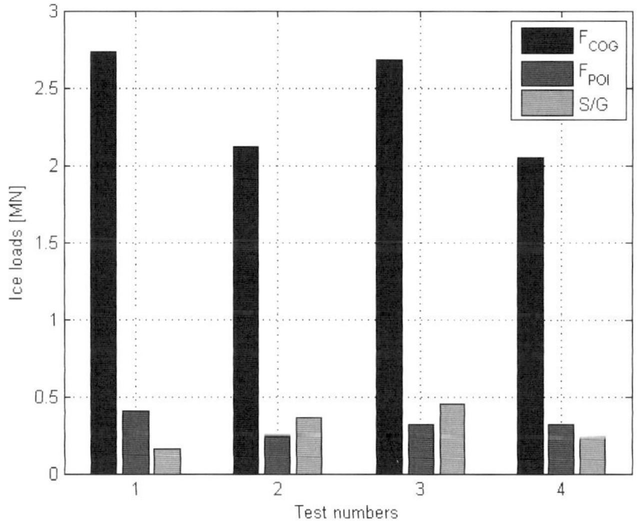

Figure 11. Resultant ice loads from COG and POI approaches and strain gauge for the case of Official 19

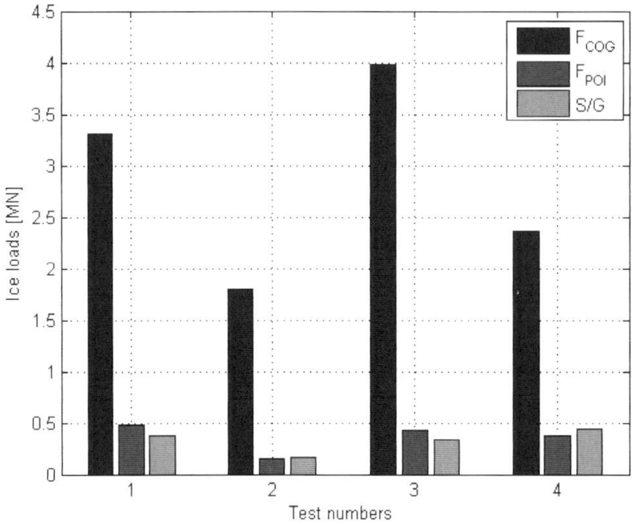

Figure 12. Resultant ice loads from COG and POI approaches and strain gauge for the case of Official 23

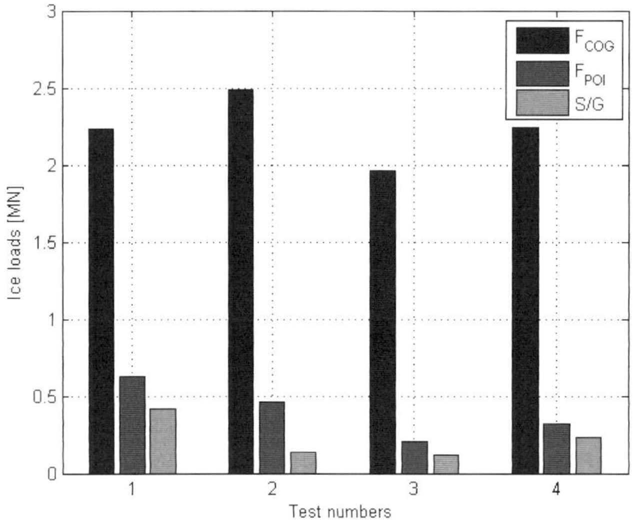

Figure 13. Resultant ice loads from COG and POI approaches and strain gauge for the case of Official 28

Comparison with the Empirical Ice Load Formula

The resultant global ice loads calculated from the inertial measurement system were compared with the empirical ice load estimation formula under the normal operating conditions (Choi et al., 2009) and the formula is expressed as

$$F = 0.824\,\Delta^{0.4}\left(\sigma_f\,h^2\,V\,\cos\alpha\right)^{0.283} \tag{6}$$

where Δ (10^3ton) is the displacement of a ship, σ_f (kPa) is the flexural strength of ice, h (m) is the ice thickness, V (m/s) is the ship speed, and α (degree) is the stem angle of ship. The values of 1 m and 315 kPa were used for the ice thickness and the flexural strength of ice.

The comparison of the calculated ice loads from the inertial measurement system and the strain gauge for the eight cases in the previous section with the estimated ice loads under the normal operating conditions was shown in Figure 14. All the calculated ice loads located below the ice loads estimation formula curve hence they were within the range of the empirical ice load estimation formula under the normal operating conditions. Additionally, all the calculated ice loads showed much lower values than the estimation formula, and there might be several reasons for this. The estimation ice load formula is derived from the conditions with a continuous icebreaking of the level ice with the same material properties, such as the thickness and the flexural strength, while the ice conditions in 2015 were small pack ices. And since the small pack ices are not fixed at one end like a cantilever, the actual flexural strength was considered to be much smaller than the value of 315 kPa used in the calculation. Finally, when the video taken during the voyage in 2015 was investigated (Min et al., 2016), it was estimated that the thickness of the actual sea ice is thinner than 1 m used in the calculation.

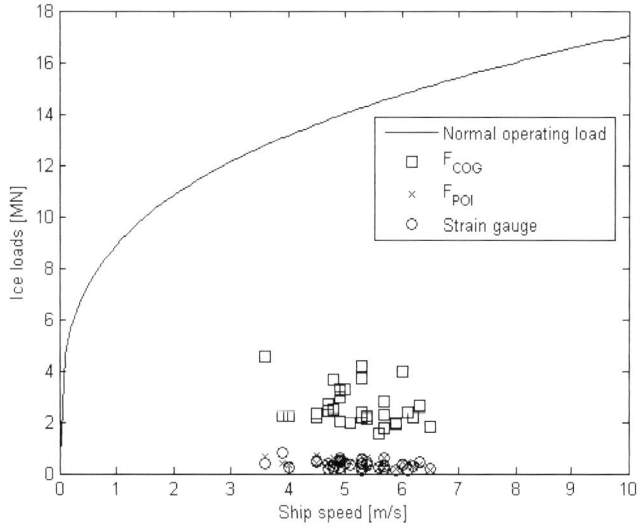

Figure 14. Comparison of calculated ice loads with the empirical ice load estimation formula under normal operating conditions

CONCLUSIONS

To solve the equations of motion, and calculating the global ice loads, procedures, such as calibration, filtering, numerical differentiation and integration were presented. The global ice loads were calculated with the assumption that the ship hull is the rigid body from the motion data. The center of gravity (COG) approach and the point of impact (POI) approach were used to calculate the global ice loads, respectively. From the motion data, the sway, roll and yaw motions were dominant in the collision with the small pack ices because the pack ices were pushed to the side. The ice loads from the COG approach was calculated larger than the those from the POI approach. The ice loads from the strain gauges was similar to those from the POI approach. On the other hand, when considering the ship speed, there is no clear relationship between the ship speed and the ice load.

All the calculated ice loads were within the range of the empirical ice load estimation formula under the normal operating conditions, but their values were very small compared to the estimation formula. This is due to the fact that the ice conditions were small pack ices, whereas the estimation formula is originally to estimate of the ice load for continuous icebreaking in level ice.

ACKNOWLEDGEMENTS

Our research was supported by the National Research Foundation (2015037577) and the IT R&D Program of MOTIE/KEIT (10060329) funded by the Korea government, and carried out within the scope of the primary program ("Establishment of experimental foundation for unconventional test to evaluate hydrodynamic performance of ships," Project No. PES3550), which is funded by KRISO.

REFERENCES

Atkinson, K., 1993. Elementary Numerical Analysis. John Wiley & Sons, pp.161-204.

Butterworth, S., 1930. On the theory of filter amplifier. Experimental Wireless & the Wireless Engineer, 7, pp.536-541.

Chen, Y., Tunik, A., & Chen, A., 1990. Global Ice Forces and Ship Response to Ice – Analysis of Ice Ramming Forces. Report submitted to Maritime Administration by American Bureau of Shipping, Report SSC-342, pp.104.

Choi, K., Jeong, S.Y., & Nam, J.H., 2009. Prediction of Design Ice Load on Icebreaking Vessels under Normal Operating Conditions, Proceedings of the 20th Int Conf on Port and Ocean Engineering under Arctic Conditions, Lulea, Sweden, 9-12 June, 2009.

Johnston, M., Frederking, R., Timco, G., & Miles, M., 2003. MOTAN: A Novel Approach for Determining Ice-induced Global Loads on Ships, Proceedings of MARI-TECH 2003, Montreal, Canada, 28-30 May, 2003.

Johnston, M., Frederking, R., and Timco, G., and Miles, M., 2004. Using MOTAN to Measure Global Accelerations of the CCGS Terry Fox During Bergy Bit Trials, Proceedings of 23rd Int Conf on Offshore Mechanics and Arctic Engineering, Vancouver, Canada, 20-25 June, 2004.

Johnston, M., Ritch, R., & Gagnon, R., 2008a. Comparison of Impact Forces Measured by Different Instrumentation systems on the CCGS Terry Fox During the Bergy Bit Trials. Cold Regions Science and Technology, 52, pp83-97.

Johnston, M., Timco, G., Frederking, R., and Miles, M., 2008b. Measuring Global Impact Forces on the CCGS Terry Fox with an Inertial Measurement System Called MOTAN. Cold Regions Science and Technology, 52, pp67-82.

Min, J.K., Cheon, E.J., Kim, J.M., Lee, S.C., & Choi, K., 2016. Comparison of the 6-DOF Motion Sensor and Strain Gauge Data for Ice Load Estimation on IBRV ARAON. Journal of the Society of Naval Architects of Korea, 53(6), pp.529-535.

Newman, J.N., 1997. Marine Hydrodynamics. The MIT Press, pp.295-300.

Salvesen, M., Tuch, E.O., & Faltinsen, O., 1970. Ship Motions and Sea Loads. Transactions of Society of Naval Architects and Marine Engineers, 78, pp.250-287.

Timco, G., & O.'Brien, S., 1994. Flexural Strength Equation for Sea Ice. Cold Regions Science and Technology, 22(3), pp.285-298.

Development of Accurate Ice Observation Technologies for Arctic Oil and Gas Developments and Verification by Field Campaigns

Takatoshi Matsuzawa[1], Kenta Hasegawa[1], Shotaro Uto[1], Koh Izumiyama[1]
Kazutaka Tateyama[2], Kazuki Nakamura[3], Chang-kyu Rheem[4], Akira Asada[4]
Joji Yamamoto[5] and Takayuki Asanuma[5]

[1] National Maritime Research Institute, Tokyo, Japan
[2] Kitami Institute of Technology, Hokkaido, Japan
[3] Nihon University, Fukushima, Japan
[4] Institute of Industrial Science, the University of Tokyo, Tokyo, Japan
[5] Japan Oil, Gas and Metals National Corporation, Tokyo, Japan

ABSTRACT

Arctic is one of the most important areas for the future oil and gas development. However presence of ice is a serious hazard for safe, efficient and sustainable developments. The authors developed accurate ice observation technologies and verified their functions by field campaigns in Japan and Greenland. The "EM-BIRD" (airborne Electro-Magnetic induction sensor of ice thickness) was improved by installing a portable microwave radiometer (PMR) and a laser scanner for observing more comprehensive information including snow depth and surface geometry of ice. The authors developed the on-board pulse-doppler ice radar and the parametric sub-bottom profiler which are capable of measuring motions and distributions of sea ice, and gouging depth by massive ice features with high accuracy. The authors also developed ice database system containing satellite remote-sensing and field data measured by the developed sensors for the operability analysis of arctic facilities.

KEY WORDS: Ice observation; EM-BIRD; Pulse doppler radar ; Sub-bottom profiler

INTRODUCTION

Arctic is one of the most important areas for the future oil and gas development. For its safe, efficient and environmental-friendly developments, it is indispensable to accumulate and analyse ice information with high accuracy. Nowadays, satellite remote sensing by passive microwave radiometers is capable of providing sea ice information such as concentration, thickness and motion of sea ice on a regular basis. However, from the engineering point of view, sea ice information with higher special resolution and higher accuracy is required for the rational design and safe operation of arctic facilities. JOGMEC (Japan Oil, Gas and Metals National Corporation) conducted an R&D project on accurate ice observation technologies and ice database for offshore oil and gas development. Three Japanese universities, one research institute and one company joined the project. The authors established advanced sensor technologies for observing ice geometry and motion, and ice gouging on seabed with high accuracy. The authors also developed an ice database system to support the decision-making for industries to go into Arctic oil and gas developments. The main results of this project are summarized in the paper.

HIGH-ACCURATE ICE OBSERVATION SENSORS

Development of Advanced EM-BIRD

EM-BIRD (airborne Electro-Magnetic induction sea ice thickness profiler) is one of the most sophisticated technologies for measuring sea ice thickness with high accuracy in wide area (Haas and others, 2009). The first EM-BIRD in Japan was developed by Kitami Institute of Technology (KIT) in FY2011 under the support from JOGMEC. In the present R&D project, KIT and National Maritime Research Institute (NMRI) further improved the functions of the EM-BIRD by 1) installing a portable microwave radiometer (PMR) and a 2D laser scanner for observing more comprehensive information including snow depth and surface geometry of ice, 2) supplying electric power via a battery system, while it is supplied from helicopter for the original EM-BIRD. Figure1 shows the sensor part of the advanced EM-BIRD, the portable microwave radiometer stored inside the EM-BIRD and the test flight using the battery system.

Figure1. Development of Advanced EM-BIRD

KIT and NMRI conducted the field campaigns at Saroma Lagoon in the northern Japan and Qaanaaq in Greenland. From the field campaign in Saroma Lagoon performed from 25th February to 6th March in 2016 (right hand side of Figure2), it was verified that the EM-BIRD could measure the total thickness (ice plus snow thickness) of first-year ice ranged from 0.3 to 0.8m with a root-mean-square error (RMSE) of 0.05m against drill-hole measurements. It means that accuracy is about 10% of the total thickness of ice.

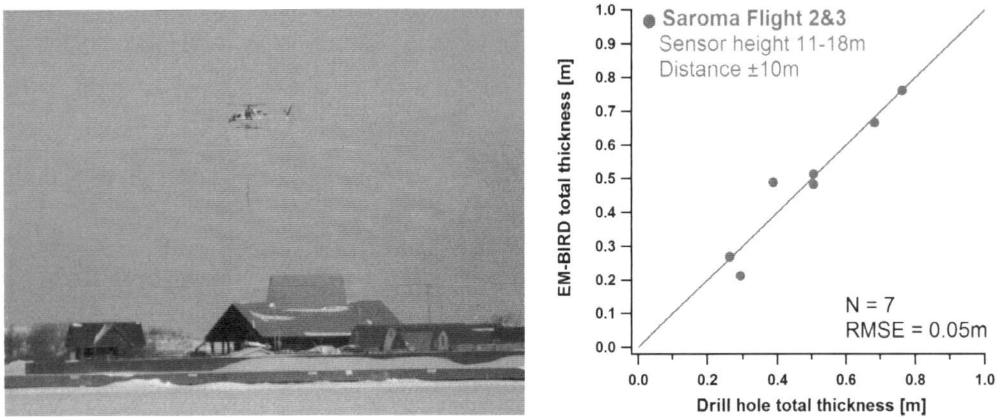

Figure 2. Field campaign at Saroma Lagoon

Helicopter flight (left), comparison of total thickness between EM-BIRD and in-situ measurement (right)

The field campaign was also conducted in Qaanaaq, northwest coast of Greenland from 21st to 26th April in 2016. Comparison with drill-hole measurements revealed that the RMSE of the total thickness was 0.07m for ice of 1.2 to 1.4m thick (Figure 3 left). Figure 3 right shows the comparison of snow depth between PMR and in-situ measurement indicating a fairly good agreement (RSME=0.06m). Advanced EM-BIRD has potential to measure ice thickness and snow depth separately.

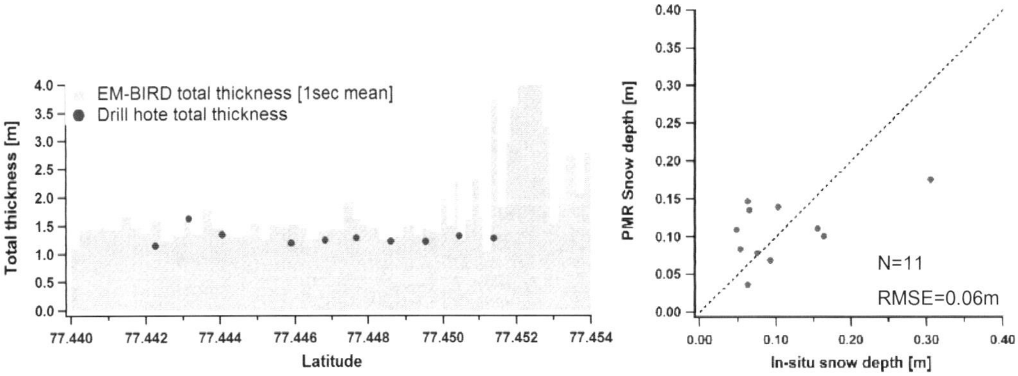

Figure 3. Field campaign at Qaanaaq, Greenland

Comparison of total thickness between EM-BIRD and in-situ measurement (left)

Comparison of snow depth between PMR and in-situ measurement (right)

Hybrid use of EM-BIRD and satellite SAR (Synthetic Aperture Radar) image is one of the key issues for obtaining high-resolution ice information around arctic oil and gas facilities. By comparing with the sea ice thickness by the EM-BIRD, the authors found a decreasing tendency of the back scatter coefficient of Synthetic Aperture Radar (SAR) against ice thickness. The authors proposed an algorithm for discrimination between un-deformed and deformed first year ice.

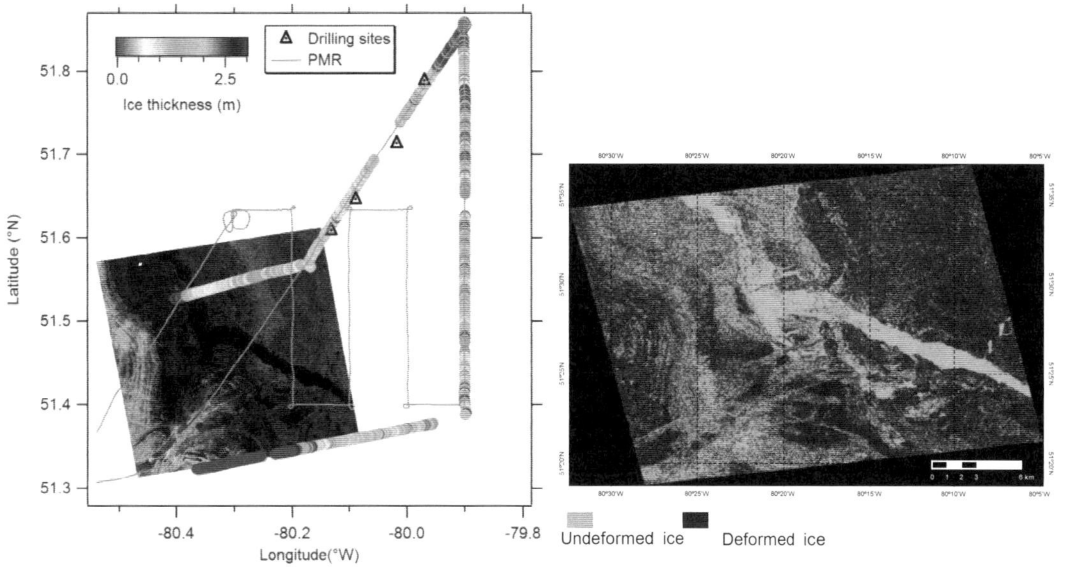

Figure 4. Flight path and total thickness by EM-BIRD and SAR image (left)
Discrimination of un-deformed and deformed ice (right)

Development of Onboard Pulse-doppler Radar

An X-band, pulse-doppler radar is one of the promising technologies for detecting motions of ice with higher accuracy than the existing marine radar. The University of Tokyo developed a prototype of a pulse-doppler radar and conducted a field campaign at the coast of the Sea of Okhotsk, Japan (Figure 5).

Figure 5. Ice radar at Okhotsk tower (left) and icebreaker Garinko (right)

A new algorithm was developed to detect presence and types of thin ice such as nilas and young ice in the range of 4 km (Figure 6). It is estimated that the ice radar system is capable of detecting massive ice feature 10km away from the radar.

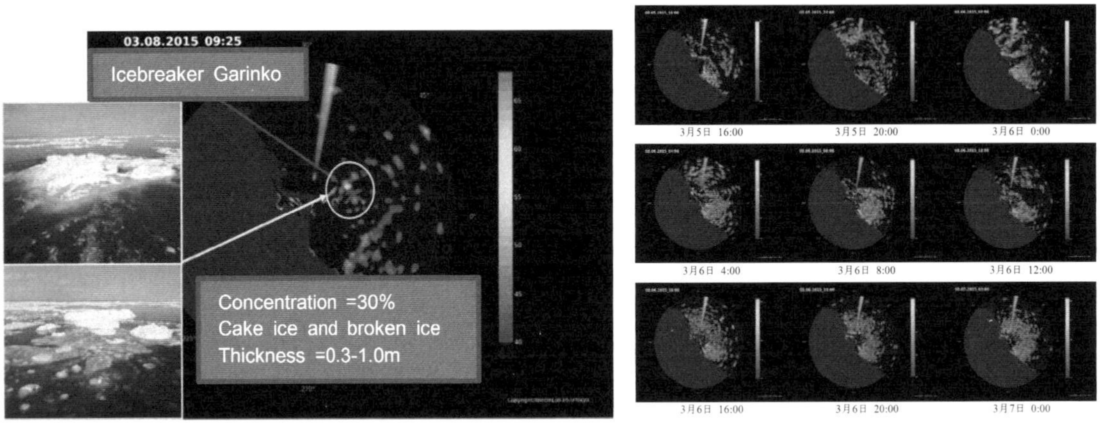

Figure 6. Matching radar image and ice observations (left) and four-hour interval images from 5th to 6th March, 2015 (right)

Development of Parametric Sub-bottom Profiler

Accurate measurement of gouge depth is important to protecting subsea equipment and pipelines from massive ice features such as icebergs. However, it is difficult to estimate its true depth if gouge is infilled by seabed sediment transport (ex. Palmer and Croasdale, 2012). The University of Tokyo developed a parametric sub-bottom profiler (PSBP) for detecting ice gouging depth including sub-bottom structure of sediment layers.

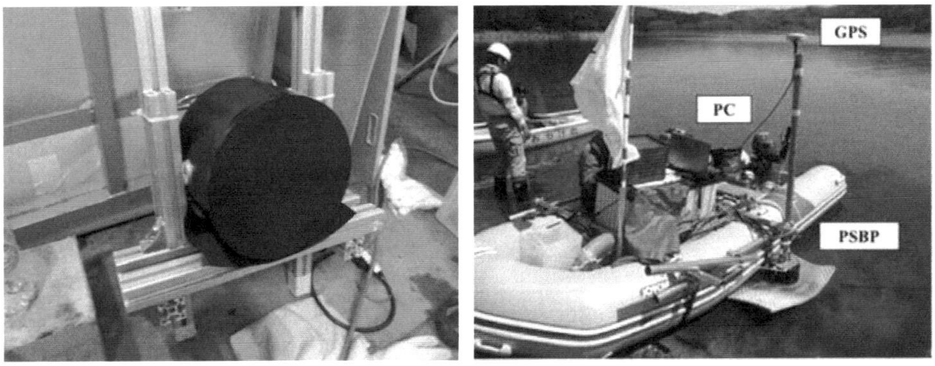

Figure 7. Sensor probe of PSBP (left) and experimental setup (right)

Field campaigns were conducted at Lake Aoki, Nagano and Saroma Lagoon, Hokkaido, Japan. The PSBP was able to have a finer view of sediment structure which was identified as a single layer by the boring survey. The PSBP has a potential to detect the original gouge depth infilled by sediment layers.

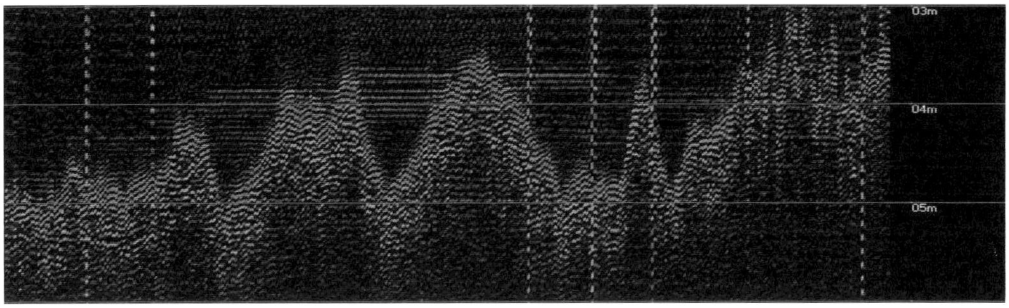

Figure 8. Sediment layer structure obtained by PSBP

ICE DATABASE SYSTEM

An ice database system was developed for supporting designs and operations of Arctic oil and gas developments. It consists of ice data covering sea ice regions in the Northern Hemisphere and the interface enabling for users to browse and analyze data through the network as a web application (Figure 9). It has functions of spatial, temporal and operability analyses of ice conditions for design purposes, and those of displaying in-situ data measured by sensors such as the Advanced EM-BIRD, the on-board pulse doppler radar and the PSBP for the operation support. Figure 10 shows the result of operability analyses. Based on the limiting ice thickness derived from the experinences of operations in Beaufort Sea and using ice thickness information, it is capable of analyzing the operation window, a time window in which ice condition allows a given operation, from 2002 to 2011 for the drilling operation in North East Greenland (NEG).

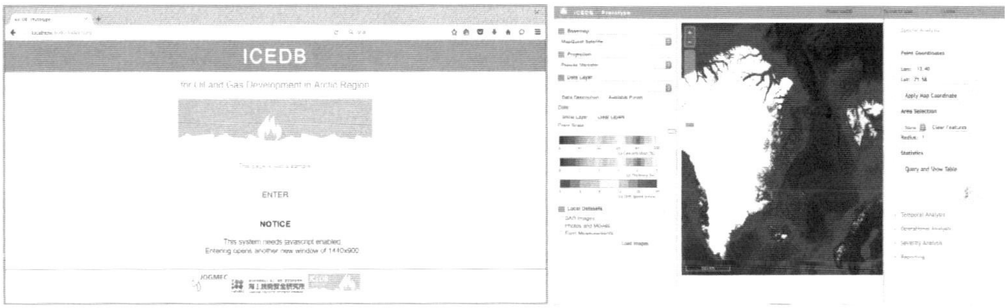

(a) Top page **(b) Interface**

Figure 9. Example views of the ice database system

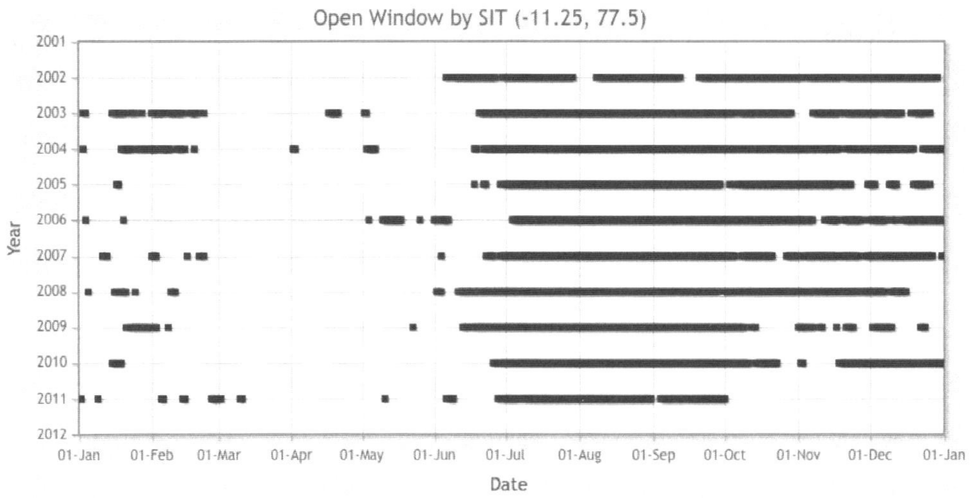

(a) Operation window for drilling in NEG based on the limiting ice thickness condition

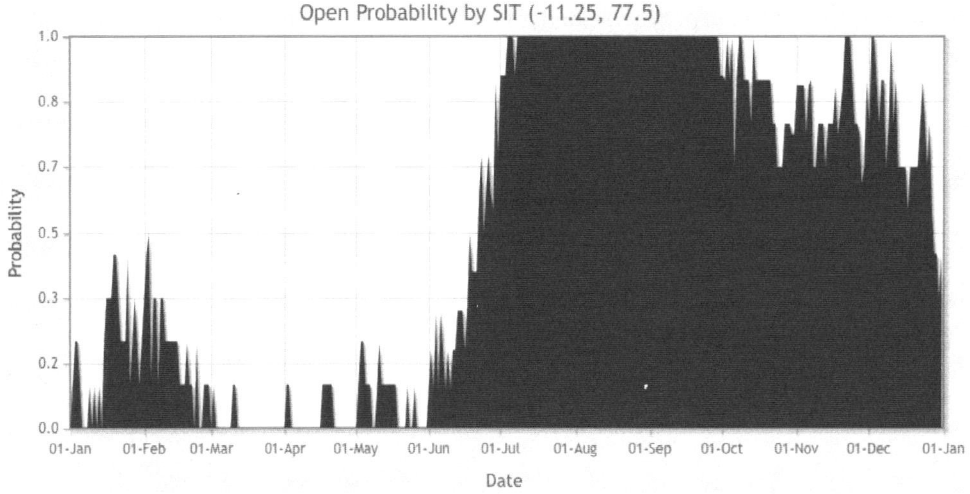

(b) Probability of operation windows

Figure 10. Operation window for the drilling operation in NEG

CONCLUSIONS

Arctic is one of the most important areas for the future oil and gas development. However presence of ice is a serious hazard for safe, efficient and sustainable developments. From the engineering point of view, sea ice information with high resolution and accuracy is required for the rational design and safe operation of arctic facilities. JOGMEC conducted the R&D project on accurate ice observation technologies and ice database for oil & gas development. Three Japanese universities, one research Institute and one company joined this project and developed the Advanced EM-BIRD, pulse doppler ice radar and sub-bottom profiler for observing ice and snow thickness, ice motion and ice-type discrimination, and gouging depth with layered sub-bottom sediment structure, respectively. Extensive field campaigns were conducted in Japan and northwest Greenland to verify their functionality and measurement accuracy. Ice database system was also developed containing satellite remote-sensing and field data measured by the developed sensors for the operability analysis of arctic facilities.

REFERENCES

Haas, C., Lobach, J., Hendricks, S., Rabenstein, L. and Pfaffling, A.,"Helicopter-borne measurements of sea ice thickness, using a small and lightweight, digital EM system" Journal of Applied Geophysics, 67, 234-241, 2009.

Palmer, A. and Croasdale, K., "Arctic Offshore Engineering", World Scientific, 2012

AUTHOR INDEX

AUTHOR INDEX

AUTHOR INDEX

Printed from e-media with permission by:

Curran Associates, Inc.
57 Morehouse Lane
Red Hook, NY 12571

Some format issues inherent in the e-media version may also appear in this print version.

Printed by Curran Associates, Inc. (2018)

For permission requests, please contact Port and Ocean Engineering Under Arctic Conditions (POAC) at the address below.

Port and Ocean Engineering Under Arctic Conditions (POAC)
National Research Council of Canada
1200 Montreal Road, Bldg. M-32
Ottawa, Ontario
Canada K1A OR6

Phone: (613) 993-7695
Fax: (613) 952-7679

ivana.kubat@nrc.ca

Additional copies of this publication are available from:

Curran Associates, Inc.
57 Morehouse Lane
Red Hook, NY 12571 USA
Phone: 845-758-0400
Fax: 845-758-2633
Email: curran@proceedings.com
Web: www.proceedings.com

TABLE OF CONTENTS

VOLUME 1

SHIPS IN ICE #1

REMOTE SENSING AND ICE MODLLING #2

SHIPPING AND NSR NAVIGATION

ICE MANAGEMENT AND OPERATION

PLENARY SESSION #3

PLENARY SESSION #4

ICE TRIALS AND FIELD MEASUREMENT

VOLUME 2

NUMERICAL MODELING AND SIMULATION #2

POLAR CODES AND REGULATIONS

ARCTIC MARINE STRUCTURES AND CONSTRUCTION

STATION KEEPING IN ICE

ICE MECHANICS #2

SHIPS IN ICE #2

LABORATORY MODEL TEST

ICE-STRUCTURE INTERACTION #2

PLENARY SESSION #5

SHIPS IN ICE #3

ARCTIC ENVIRONMENTS AND SEA ICE PROPERTIES #3

WINTERIZATION AND ARCTIC OPERATION

Proceedings of the 24th International Conference on
Port and Ocean Engineering under Arctic Conditions
June 11-16, 2017, Busan, Korea

An Analysis on Strain Gauge Signal Measured from Repetitive Ramming in Heavy Ice Condition

Se-Jin Ahn[1], Woo-Seong An[1], Tak-Kee Lee[2] and Kyungsik Choi[3]
[1] Department of Ocean System Eng., Graduate School, Gyeongsang National University, Tongyoung, Korea
[2] Department of Naval Archi. & Ocean Eng., Gyeongsang National University, Institute of Marine Industry, Tongyoung, Korea
[3] Department of Ocean Engineering, Korea Maritime and Ocean University, Busan, Korea

ABSTRACT

To navigate ice-covered waters, ice-breaking process is required. Ice-breaking mode depends on material properties of sea ice and ice conditions. Ice-breaking mode is classified into ramming and continuous ice-breaking. Ramming has a great effect on large ice features, such as thick ice ridge and icebergs, and continuous ice-breaking does on level ice. In general, the impact time duration in crushing or bending on ice sheets ranges from 0.2 to 1.0 second. However, impact duration in ramming will be increased.

The Korean ice-breaking research vessel ARAON conducted her research voyage in the Antarctic sea ice during the winter of 2012. The IBRV ARAON measured strain in ramming and continuous ice-breaking. Strain gauge signals were recorded during the planned ice-breaking performance and the unplanned ice transits in heavy ice conditions.

The aim of this study is to investigate the strain gauge signals measured in ramming processes under the heavy ice condition. Based on the time history of the signals, a raising and a half-decaying time were investigated and compared with the previous study which was suggested the five profiles on the strain gauge signals.

KEY WORDS: Ice-breaking process, Ramming, Strain gauge signal, Time-history, Raising time, Half-decaying time

INTRODUCTION

As the Arctic ice was reduced to less than 4 million km^2 in the recent 30 years due to global warming, it was reduced to the level allowing free voyage of vessels. Due to climate change, the Northern Sea Route (NSR) came to be provided with conditions allowing voyage during summer period. The NSR means the sea route connecting Europe, Asia and America through the Arctic sea. It also collectively refers to the northwest sea route and the northeast sea route. According to the demonstration voyage of NSR by Stena Bulk Company, about 48 days of voyage is required when the existing sea route is used from Busan to Ust-Luga of Russia through Suez Canal. However, when NSR along the north coast of Siberia through Bering Strait is used, about 35 days of voyage is required (Stena Bulk, 2013). When compared with the case of using south sea route, The NSR shortens the sea route by about 7,200km, which is the greatest advantage of its route.

Such construction of sea route leads to the advantages of allowing commercial use of various natural resources such as enormous crude oil and natural gas, etc. of the Arctic, drawing worldwide attention due to interests in the Arctic circle. Also, it is being expanded to an increase in the demand for icebreaking cargo ships in the market for ice-going ships which include oil tankers containing icebreaking capability in the features of the existing ships, icebreaking LNG carriers, icebreaking container carriers, etc. in addition to icebreakers.

Icebreaking capability is essential requirement for voyage of ships in the ice field. In general, icebreaking is divided into ramming and continuous icebreaking. The ramming has a great effect for thick ice or ice ridge, while continuous icebreaking is greatly effective for level ice. When icebreaking operation is performed, the ship is subjected to forces as a global or locally due to collision or contact with sea ice. Such force is called ice load or ice pressure. Ice pressure is the expression in a general form of local ice load acting only on a local area of a particular part. Since ice loads acting on the icebreaking ship are larger than the wave loads produced in general shore regions or the forces due to slamming, the magnitude of ice load is given a high importance upon design (Kim, 2014).

The ARAON as the Korean first icebreaking research vessel sailed Amundsen Sea of the Antarctic between January 31 and March 20, 2012 (Choi et al., 2012). The ARAON conducted full scale measurement tests of the bow side part by division into icebreaking voyage and general voyage in the ice field. For ice load measurements, 21 rosette gauges were attached to inside of the side shell in the bow side part. The data measured through this was recorded during two official icebreaking performance tests and general ice transit with breaking of ice in the ice fields with various ice concentrations.

Since the amount of impact was proportional to time duration in ramming and impact load, in this study focused on the fact that damages to the hull produced by ramming could be increased, the longer the time duration in ramming. Prior to data analysis, strain data measured upon ramming was converted to stress. The measurement data when peak stress of higher than 20MPa was recorded was analyzed. After dimensionless conversion of the time-history by using the maximum stress, raising time and half-decaying time were checked by using this graph, and impact time duration calculated. Also, classification was made with application of the criteria of the previous study (Lee et al., 2016) where the measured signals were represented by 5 types, and the characteristics were examined and analyzed.

ANTARCTIC VOYAGE OF THE IBRV ARAON IN 2012

The ARAON has conducted various full scale measurement tests such as ice load and ice resistance, etc. in Antarctic Ocean and Arctic Ocean from January, 2010 through recent times since its construction. Between January 31 and March 20, 2012, the ARAON conducted full scale measurement and icebreaking performance tests in the ice field of Amundsen Sea, Antarctic. The place where ramming was conducted is displayed by (X) in Figure 1.

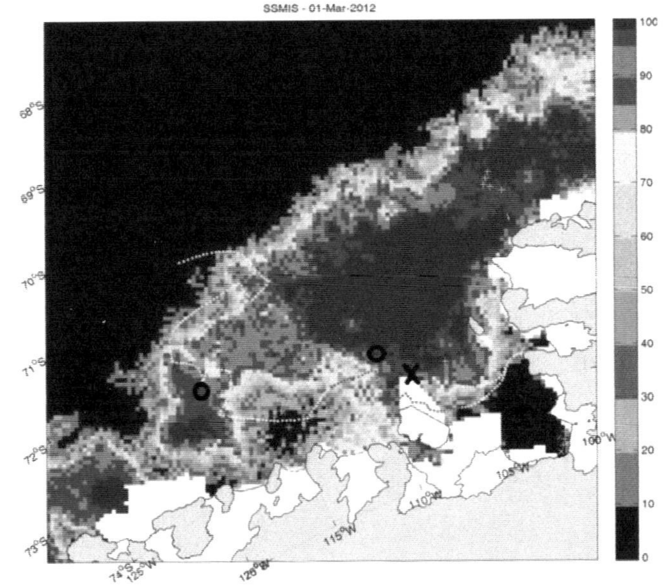

Figure 1. Ice concentration in the Amundsen Sea, Antarctic
(O) Official ice performance test sites; (X) Location where ARAON was stuck on heavy ice (Choi et al., 2015)

A total of 21 rosette gauge including 10 ea. on the port and 11 ea. on the starboard were attached around the draft part inside the side shell between FR.106 and FR.111 of the lower part of second deck inside the bow thrust room. It was configured so as to allow simultaneous measurement for a total of 63 channels in rosette form. The parts instrumented by rosette gauges are shown in Figure 2. Shell material is EH36 steel with 355MPa for yield strength, 200GPa for elastic modulus, and 0.3 for Poisson's ratio (Kim, 2014).

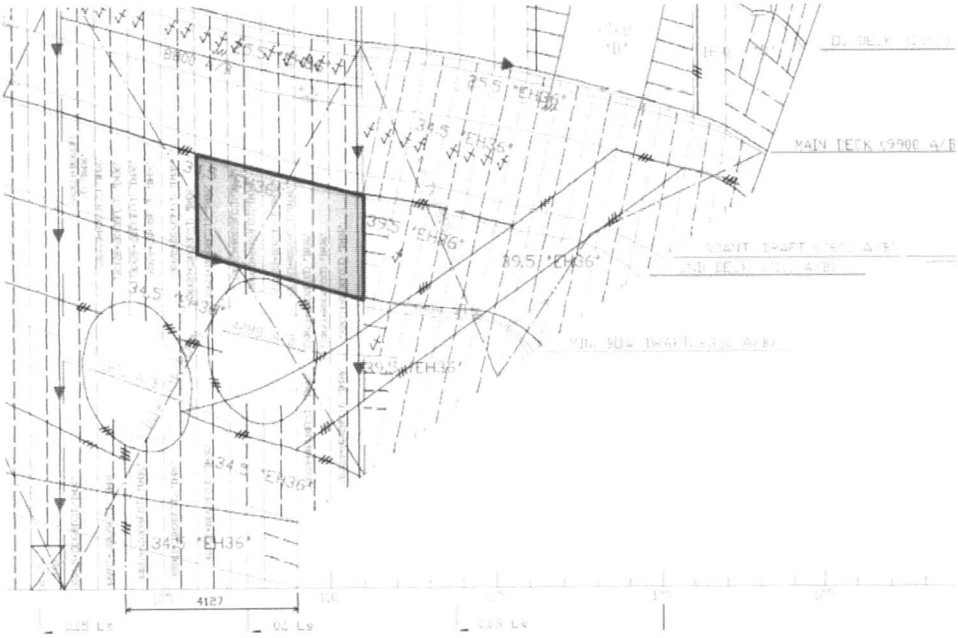

Figure 2. Area instrumented by strain gauges in the IBRV ARAON (Kim, 2014)

POAC17-108

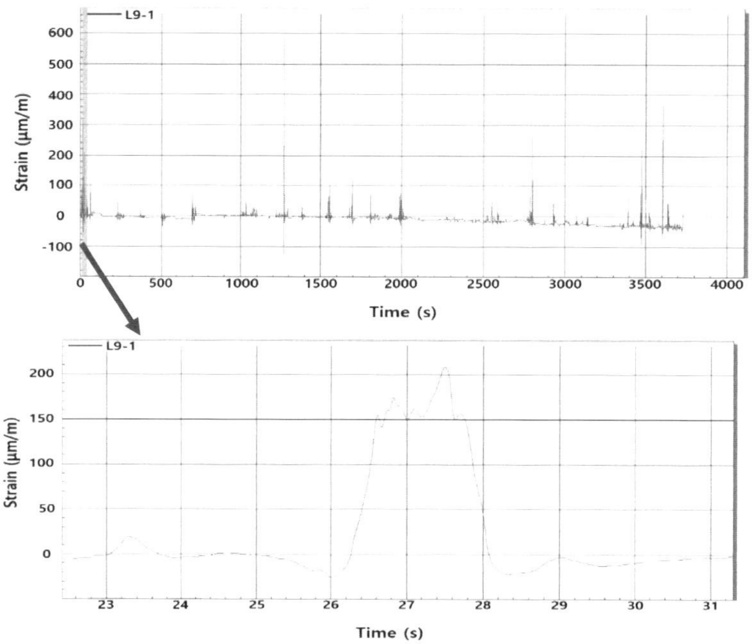

Figure 3. Example of measured strain gauge data

LOCATION AND SITUATION IN REPEATED RAMMING

Table 1 shows a summary of information related to data involving ramming among the data measured during general ice transit of the ARAON. Here, the test number represents the number continuously granted for each measurement by the measurement team during general voyage. In this study, it is focused on two times of measurement for nos. 9 and 10 where ramming occurred. Also, measurement times are arranged based on UTC (universal time coordinated). Ramming occurred when moving from the position of 73°37'33015"S, 109°29'64878"W to that of 73°34'44119"S, 109°38'61929"W and while moving from 73°33'20699"S, 109°40'98492"W to 73°31'95289"S, 109°45'51126"W. No measurements with R5 gauge were made since the gauge was defective at the time of measurement.

Description of the sailing situation with the ARAON on the relevant day is as follows. On February 28, 2012, the open channel was closed due to gale. The ARAON became trapped in the middle of heavy ice. On that day, thick ice floe moved by more than 20km. Subsequently, the front side of open channel completely vanished. For 20 hours, the ARAON made efforts to find a way out of the thick ice flow by enforcing repeated ramming at the full engine power. As a result, the ARAON successfully escaped, and a considerably large amount was recorded for the strain gauge signals (Choi et al., 2015).

Table 1. Summary of information related to data involving ramming

Test no.	Date	Time(UTC)	Measuring site	Measuring time(sec)
9	2012.02.28	14:35 ~ 15:40	73° 37' 33015"S-109° 29' 64878"W~ 73° 34' 44119"S-109° 38' 61929W	3,900
10	2012.02.28	15:45 ~ 18:00	73° 33' 20699"S-109° 40' 98492"W~ 73° 31' 95289"S-109° 45' 51126"W	8,100

Figure 4. Heavy ice condition in repeated ramming (Choi et al., 2015)

ANALYSIS OF ICE LOAD SIGNALS MEASURED IN RAMMING

Data measured from rosette gauge was converted to stress by application of Hooke's law. Among the signals measured from rosette gauge, the value in the direction of bow & stern was taken as the representative value, which considered that the ice broken by bow icebreaking flowed along the side shell according to advancement of the ship (Jeon et al., 2013). In Tables 2, S_{max} shows the maximum value of peak stress, T_r the raising time, T_{hd} the half-decaying time, and T_d the time duration. Here, the raising time means the time taken from the time where signal passed through zero point to the time of arriving at peak point, while the half-decaying time is the time taken by dimensionless stress level moving from the peak point to the 0.5 point (Lee et al., 2016). And, time durations of Type Ⅰ and Ⅱ were taken as the value of doubling the half-decaying time with the raising time added. However, time duration values of Type Ⅲ, Ⅳ and Ⅴ were taken as the time difference from the raising time point to the calculated point by calculation of the time when the stress became after decaying time had ended.

In the ramming on February 28, 2012, the maximum peak stresses higher than 20MPa for a total of 12,000 seconds were recorded for a total of 347 times. Data was classified into 5 types of signals based on the presence status of intermediate peak (Lee et al., 2016). Maximum value, minimum value, and average were arranged for peak stress, raising time, half-decaying time, time duration in Table 2, respectively. The maximum value for measured peak stress was 159.6MPa. Also, the average of raising time was 0.540 second, the average of half-decaying time 0.326 second, and the average of time duration 1.173 seconds.

Table 2. Summary on peak data in ramming

Type	No. of peak over 20MPa		S_{max} (MPa)	T_r (sec)	T_{hd} (sec)	T_d (sec)
I	185	Max.	159.6	7.231	1.854	9.750
		Min.	20.1	0.050	0.013	0.128
		Avg.	37.8	0.286	0.138	0.562
II	43	Max.	111.3	4.863	1.781	8.425
		Min.	20.3	0.124	0.042	0.225
		Avg.	32.1	0.516	0.181	0.879
III	43	Max.	119.4	3.237	3.733	9.952
		Min.	20.1	0.077	0.066	0.242
		Avg.	39.6	0.305	0.333	0.986
IV	16	Max.	83.2	1.308	0.812	3.326
		Min.	20.9	0.188	0.048	0.467
		Avg.	39.4	0.557	0.337	1.223
V	60	Max.	90.9	6.470	5.698	14.077
		Min.	20.3	0.174	0.108	0.497
		Avg.	35.4	1.507	0.999	3.387
Total	347	Max.	159.6	7.231	5.698	14.077
		Min.	20.1	0.050	0.013	0.128
		Avg.	37.0	0.540	0.326	1.173

Figure 5 is the graph showing relationship between the raising time and the peak stress among data measured in ramming. Peak stresses measured in general ice transit were shown to be high in the section with a short raising time as analyzed by Lee et al. (2016). However, some data was recorded to have relatively high peak stresses in the sections with a long raising time, which is attributed to the situation at the time of icebreaking of relatively thick ice at a low speed (Kim, 2014). When peak stress was high, signals of Type Ⅰ were recorded most frequently. And, when the raising time is more than 2 seconds, the majority was classified as Type Ⅴ. Such data with a long raising time was not observed at all in the data for analysis of ice load signals of Arctic Ocean in 2010 (Lee et al., 2016). Also, much of the data for Type Ⅰ, Ⅲ and Ⅳ is distributed before about 0.4 second, while much of the data for Type Ⅱ and Ⅴ was distributed after about 0.4 second.

Figure 6 shows the relationship between half-decaying time and peak stress. When the half-decaying time is more than 2 seconds unlike the raising time, only signals of Type Ⅴ were recorded. However an exception of Type Ⅲ was recorded. Because bending strength of ice is higher than other, the bow has a long time on ice in ramming. Also, with 0.4 second as the reference, much of the data for Types Ⅰ, Ⅱ, Ⅲ and Ⅳ can be seen to be distributed before, while much of the data for Types Ⅴ are distributed after.

Figure 7 shows the relationship between time duration and peak stress. Hysing et al. (1979, in Choi, 2015) presented that time duration was in the range of 0.3~0.5 second when ramming by an icebreaking cargo ship. On the other hand, the average of time duration in ramming measured while ramming was actually executed in the Antarctic was 1.173 seconds as seen from ramming. Also, the range of distribution was varied with types. In the case of dominant Type Ⅰ among the entire data, it was distributed in a wide range from 0.1~10 second. Type's Ⅱ and Ⅲ were distributed in 0.2~10 seconds, while Type Ⅳ was distributed in 0.4~4 seconds. In the case of Type Ⅴ having the longest time duration, it lasted up to the maximum of about

14 seconds. However, it can be seen to be distributed between 0.8~10 seconds through the graphs in most cases.

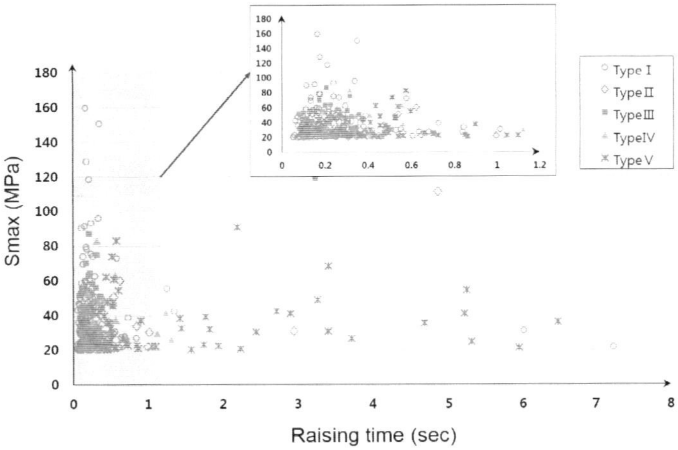

Figure 5. Relationship between raising time and peak stress

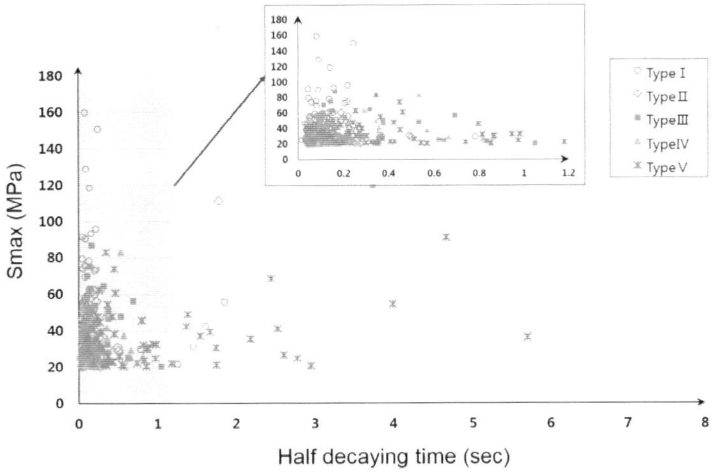

Figure 6. Relationship between half-decaying time and peak stress

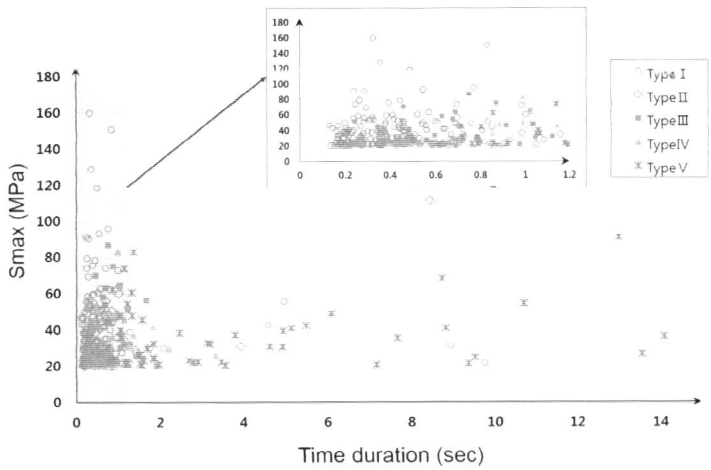

Figure 7. Relationship between time duration and peak stress

Table 3 summarizes peak stress, raising time, half-decaying time, and time duration by extraction of 3 data with the longest time duration in ramming from ramming. Average of time duration for ramming was shown to be 13.537 seconds. Also, while all case shown to be Type V for ramming. The top 3 on time duration appeared at only test no.9.

Since the ice as level ice or pack ice is mainly distributed when the icebreaker conducts continuous icebreaking, icebreaking can be executed continuously while sailing at a speed higher than a given value. However, since ice ridge or thick ice is broken when ramming is conducted, there is a feature that icebreaking occurs at a relatively higher engine power in a low-speed. The fact that time duration is observed to be long upon ramming can be considered attributable to such characteristics of icebreaking.

Table 3. Summary of the top 3 on time duration

No.	Type	S_{max} (MPa)	T_r (sec)	T_{hd} (sec)	T_d (sec)
R6 in Test no.9	V	36.3	6.470	5.698	14.077
R7 in Test no.9	V	26.8	3.709	2.612	13.543
R7 in Test no.9	V	90.9	2.197	4.664	12.992
Avg.		51.3	4.125	4.325	13.537

Figure 8 shows the data of Table 3. When a ship conducts icebreaking of such as ice ridge, ramming is conducted. Since the ramming is characterized by low-speed execution, the trend can be observed where the stresses are steadily increased without intermediate peaks upon raising due to continuous contact with ice. Here, the sizes of ice and ship are also considered to have an effect to a certain extent. Namely, when the ice is relatively very large with the ship, the ice will be resisted against the ship's motion. Here, a continuous increase in stresses can occur due to withstanding by ice.

The fact that small peaks occur for few times upon decaying as in the signal profile of R7 shown by a dotted line might be caused by the contact with other surfaces immediately after the surface of ice having come into contact with the hull is broken. And, when stresses higher than a given level are continued as R6 shown by the thick solid line, it is presumably due to the fact that the ice withstands without being pushed away upon ramming as the size of the ice is relatively large compared with that of the ship.

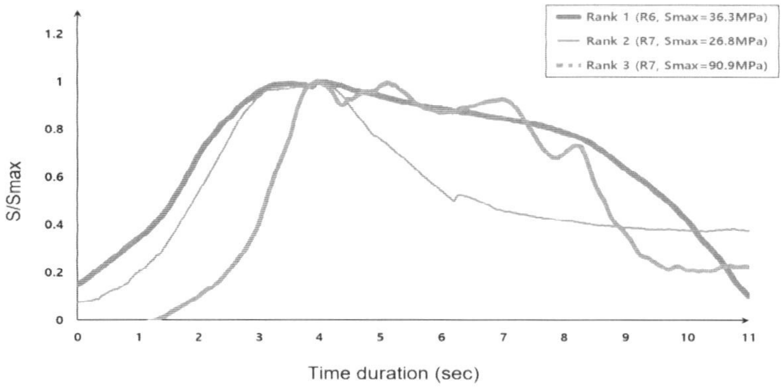

Figure 8. Comparison of the top 3 on time duration profile

COMPARISON WITH PREVIOUS STUDIES

Ritch et al. (1999) presented ice load data measured during the voyage of the Louis S. St. Laurent as a Canadian icebreaker through the Arctic in the summer of 1994. Data was examined after converting the strains inputted in the strain gauge into the stresses. Time history among the measured data was comparatively analyzed in terms of 'sharp impact' event and 'sliding' event. First, considering the 'sharp impact' event, time duration was shown to be about 0.4 second, and can be seen to be similar to Type Ⅰ among 5 signal types classified in the previous study (Lee et al., 2016).

The graph of Figure 8 is compared with an example of the 'sliding' event where time duration is 5 seconds reported by Ritch et al. (1999). The 'sliding' event is presumed to be similar to the signal with impulsive tail in the previous study (Jeon et al., 2013). These two types of signals have longer decaying time than normal one. In this study, the half-decaying time is characterized as being continued for long like in the 'sliding' event. This is presumed to originate from the execution of ramming by the ARAON in a low-speed as well as the relative size difference between the ship and the ice. A ship in the low-speed will be in continuous contact after the first collision with ice to push away the ice. Here, if the ice is relatively small, it will be immediately pushed away. However, when the ice is larger than a given scale, it will withstand rather than being pushed away to extend the time duration in ramming, and the tendency toward a gradual decrease in stresses has been presumably exhibited.

Hysing et al. (1979, in Choi, 2015), has reported that the time duration is increased upon ramming and the average of time duration in ramming of icebreaking cargo ship on 40cm-thick ice is between 0.3~0.5 second. Time duration in ramming of an icebreaking cargo ship was estimated through the measurement data for an icebreaker. While the length of the crushing zone was increased to 3m when icebreaker pass through thick ice or ice ridge that it was reported that the length may be 5~6m in the case of icebreaking cargo ship (Choi, 2015). According to Kim (2014), the ship speed in ramming was reported to be less than 0.5 m/s. Upon calculation with application of 0.5 m/s for the ship speed, time duration in ramming for an icebreaking cargo ship can be shown to be longer by 4~6 seconds compared with an icebreaker due to the difference of that length. In view that the average of time duration in ramming from Table 3 was 13.537 seconds, it is presumed that about 17.537~19.537 seconds can also be measured for an icebreaking cargo ship. These values are very longer them that of Hysing et al. (1979). However, it is necessary to keep in mind that such consideration did not take into account the relative size difference between the ship and ice.

CONCLUSIONS

In this study, data with the relatively high stress level of 20MPa exceeded was extracted based on the data measured upon ramming in the ice field of Amundsen Sea in 2012 using the ARAON as the icebreaking research vessel, and waveform analysis of the signals has been attempted. A summary of conclusions obtained from the present study is as follows.

1. 347 data for test no. 9 and test no. 10 measured from the Amundsen Sea in 2012 was analyzed and classified into 5 signal types as presented in the previous study (Lee et al., 2016).
2. According to the data classification results, Type Ⅰ as the most general form was measured most frequently, accounting for 53.3%. In the case of Type Ⅱ, it accounted

for about 12.4%, about 12.4% in the case of Type Ⅲ, about 4.6% in the case of Type Ⅳ, and about 17.3% in the case of Type Ⅴ, showing much difference from the measurements of the previous study (Lee et al., 2016). Also, when time duration is long while executing ramming, Type Ⅴ is analyzed to have been frequently measured.

3. According to the analysis results of 3 data with the longest time duration in ramming, the signal type of Type Ⅴ were only observed. Since collision was executed at a low speed, impacts with continuous increase in stresses occurred and the phenomenon with extension of raising time has been discovered.
4. In ramming, the signal types of Type Ⅰ and Ⅴ were most frequently observed, where Type Ⅰ can be seen to be dominant when the time duration was short, and Type Ⅴ dominant when the time duration was long. Unlike the previous study (Hysing, 1979 in Choi, 2015) where the time duration in ramming was 0.3~0.5 second, it was diversely distributed in 0.1~14 seconds, with the average of time duration calculated to be 1.173 seconds. Also, the average duration for top 3 in ramming was calculated to be 13.537sec. Such measurement signals are considered to be precious data which are not so frequently discovered in the previous studies.

ACKNOWLEDGEMENT

This study was supported by the BK21 Plus project. The authors gratefully acknowledge this support. Also, thanks to Tae-Hoon Bae and Ju-Young Lee for helping we organize our research data.

REFERENCES

1. Choi, K., 2015. *Ice Loads on Icebreaking Vessels*, Munundang.
2. Choi, K., Cheon, E.J., Kim, H.Y., Nam, J.H., and Lee, T.K., 2015. Comparison of Peak Ice Pressures on the IBRV ARAON during the Planned and the Unusual Ice Transits. The Twenty-third (2015) Port and Ocean Engineering under Arctic Conditions. Trondheim, Norway.
3. Choi, K., Kim, H.S., Choi, G.G., Lee, J.M and Ha, J.S., 2012. Sea Ice Field Test Report: Measuring Ice Properties during 2012 Antarctic Voyage of IBRV ARAO, KMU-ARL.
4. Hysing, T. et al., 1979, Ice Loads on Marine Structures, Marine Structures and Ships in Ice, a Joint Norwegian Research Project Report No.79-02, in Choi, K.S., 2015. *Ice Loads on Icebreaking Vessels*, Munundang.
5. Jeon, Y.J., Rim, C.W., Lee, T.K., 2013. Profile analysis on signal measured local ice load during icebreaking in Arctic Sea. J. Navig. Port. Res. 37 (2), pp. 143-148
6. Kim, H.Y., 2014. "Analysis of Strain Gauge Data and Estimation of Ice Load on the IBRV ARAON in the Antarctic Sea", Master's Thesis, Korea Maritime & Ocean University, Department of Ocean Engineering.
7. Lee, J.H., Kwon, Y.H., Rim, C.W., and Lee, T.K., 2016. Characteristics Analysis of Local Ice Load Signals in Ice-covered Waters. International Journal of Naval Architecture and Ocean Engineering 8, pp. 66-72

8. Ritch, R., John, J.S., Browne, R., Sheinberg, R., 1999. Ice Load Impact Measurements on the CCGS Louis S. St. Laurent during the 1994 Artic Ocean Crossing. Offshore mechanics and Artic engineering; OMAE 1999. St John's, Canada, pp. 39-46.
9. Stena Bulk, 2013. Northern Sea Route Map, Online Available at: http://www.stenanorthernsearoute.com/northern-sea-route/

Preliminary 3D DEM Simulations on Ridge Keel Resistance on Ships

Hanyang Gong[1], Arttu Polojärvi[1], Jukka Tuhkuri[1]
[1] Aalto University, School of Engineering, Department of Mechanical Engineering, P.O. Box 14300, FI-00076 Aalto, Finland

ABSTRACT

Ice ridge resistance of ships operating on Arctic seas is of importance when estimating the ship traffic emissions and the efficiency of ship transport systems. Here we use 3D discrete element method (DEM) to simulate the interaction between a ridge keel and a ship. In this preliminary study, we will present results from simulations with various ridge keel lengths and accounting for the symmetry of the interaction scenario. We also study the effect of ridge keel widths on the keel resistance in our simulations.

KEY WORDS: Ice ridge; Keel resistance; Ship; Keel width; 3D DEM.

INSTRUCTION

Ice ridge resistance on ships has a central role in route planning and optimization of marine transport on Arctic seas. Ridge resistance is usually divided into components related to the structure of a ridge: keel (underwater pile of loose or partly cohesive ice rubble), consolidated layer (refrozen layer close to water line), and sail (rubble pile above water line). In recent transit models (Ehle, 2012; Kuuliala et al., 2016), the consolidated layer resistance is assumed equal to that of level ice (Riska, 2014), while the keel, or ice rubble, resistance is derived based on the methods presented by Mellor (1980) and Malmberg (1983). (Sail is small in volume and often neglected.) This paper focuses on three-dimensional discrete element method (DEM) study on rubble resistance, and on the effect of rubble pile geometry on the resistance.

The only analytical methods for estimating the keel resistance were developed by Keinonen (1979) and Malmberg (1983). They assumed that the rubble in the keel behaves like a body of homogeneous and isotropic granular material, which has a semi-infinite boundary, and applied Coulomb failure criterion together with Rankine theory on its failure. Many analytical approaches on the ridge-structure interaction follow similar ideas (Dolgopolov, 1975; Cammaert et al., 1993; Croasdale, 2012; Palmer and Croasdale, 2013). Modelling of the ice rubble has been performed using continuum models (Heinonen, 2004; Serre, 2011 a, b) and discrete approaches that account for the individual ice blocks in the rubble (Hopkins, 1992; Hopkins, 1998; Hopkins and Hibler, 1991; Polojärvi and Tuhkuri, 2009, 2012; Polojärvi et al., 2012). Liu et al. (2016) recently used dilated polyhedron-based DEM to study ice floe-ship hull interaction. DEM enables studies on the rubble resistance without choosing a rubble material model and making a priori decision on rubble pile failure criterion.

The main finding of this paper is that the rubble pile width has a strong effect on the resistance. This effect is not accounted for in the analytical models. We will first introduce the simulation set-up. Then the keel resistance records are presented and assumptions used in the modelling are addressed. After this, the results on the effect of rubble pile width on the resistance are presented. Before concluding our paper, we compare the simulation-based resistance values to those measured in the field and given by the analytical formulas.

SIMULATION SET UP

The 3D DEM code we use is an in-house code of Aalto University Ice Mechanics Group (Polojärvi and Tuhkuri, 2009, 2012). In brief, the simulations are explicit and we model the ship and all the blocks in the rubble as polyhedral rigid bodies. The ship-to-block and block-to-block contacts cause forces, which compose of a normal component due to elastic and viscous damping forces, and a tangential component due to friction. In addition to contact forces, buoyant, gravitational and drag forces act on the blocks. The block sizes vary randomly with a distribution given by Kulyakhtin (2014). The geometry of the modelled ship is that of M/T Uikku. All simulation parameters are given in Table 1.

The simulation basin is illustrated by Figure 1. The whole basin was covered by a rigid plate mimicking the ice cover on top of the ice rubble. The basin extended far enough into the x- and z-directions to keep any of the blocks, or the ship, from interacting with the basin walls in these directions. As will be explained below, the basin extent into the y-direction was varied, as in this direction the basin walls were in contact with the rubble pile. The centerline of the ship was parallel to the x-axis and aligned with the centerline of the basin. The model has a plane of symmetry on xz-plane and we could use a half model in our simulations (rigid wall at $y = 0$ and rubble in the half domain $y < 0$ only).

We tested a number of pile widths (into the x-direction) and lengths (into the y-direction), and ran simulations with two different ice-ice friction coefficient μ_i values as summarized in Table 2. Figures 2a and b describe the pile geometries used. The narrowest of the modelled piles had a triangular shape (Figure 2a) while the shape of the wider keels was trapezoidal (Figure 2b). All piles had a depth h of 5 m and pile angle of repose of 30° (Timco and Burden, 1997). We also justified the use of the half model by running simulations with full and half models. Naturally, the resistance of a full rubble pile model having the length l_f should be the double of that yielded by a corresponding half model with length $l_h = 0.5 l_f$.

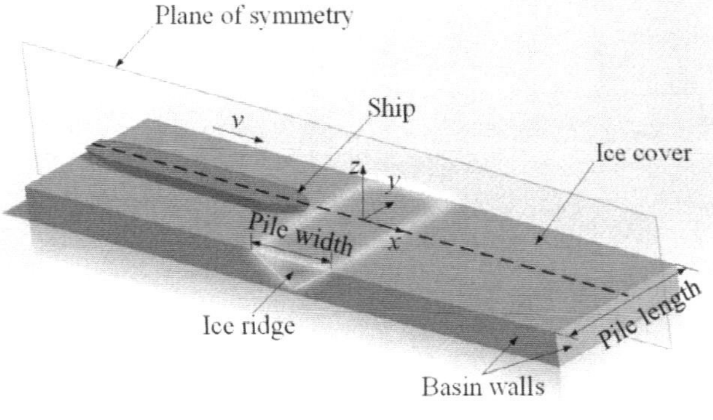

Figure 1. A sketch of the simulation basin and the coordinate system used in this paper.

Table 1. The main parameters used in simulations. The geometry of the modelled ship was that of M/T Uikku. Figure 2 illustrates the pile dimensions given in the Table.

	Parameter	Unit	Value
General	Gravitational acceleration	ms^{-2}	9.81
Contact	Penalty term	-	$1 \cdot 10^7$
	Damping constant	-	$5 \cdot 10^5$
	Time step	s	$5 \cdot 10^{-5}$
Ice blocks	Thickness	ms^{-2}	0.3
	Block aspect ratio	-	0.6…15
	Ice-ice friction coefficient μ_i	-	0.3, 0.6
	Mass density	kgm^{-3}	920
Water	Mass density	kgm^{-3}	1010
Rubble piles*	Pile depth h	m	5
	Pile angle θ	°	30
	Pile width w	m	17.3…80
	Pile length l_f (= $2l_h$)	m	60…100
Ship	Waterline length	m	150
	Parallel midbody L_m	m	67.5
	Breadth B	m	22.2
	Draft T	m	9.5
	Waterline half-angle α	°	25.2
	Stem angle φ	°	29.7
	Ship-ice friction coefficient μ_s	-	0.1
	Ship velocity v_{ship}	ms^{-1}	1

*all values refer to the initial pile geometry. Indices f and h respectively refer to full and half model.

Table 2. Summary of the simulations used in this paper. All of these simulations were ran with ice-ice friction coefficient μ_i values 0.3 and 0.6, so that the total number of simulations was 20. Figure 2 illustrates the pile dimensions given in the Table.

	Pile length* l_f	Pile length* l_h	Pile width w
Full model	100	-	17.3
	80	-	17.3
	60	-	17.3
Half model	-	50	17.3
	-	40	17.3
	-	30	17.3
	-	30	30
	-	30	40
	-	30	60
	-	30	80

*all values refer to the initial pile geometry. Indices f and h respectively refer to full and half model.

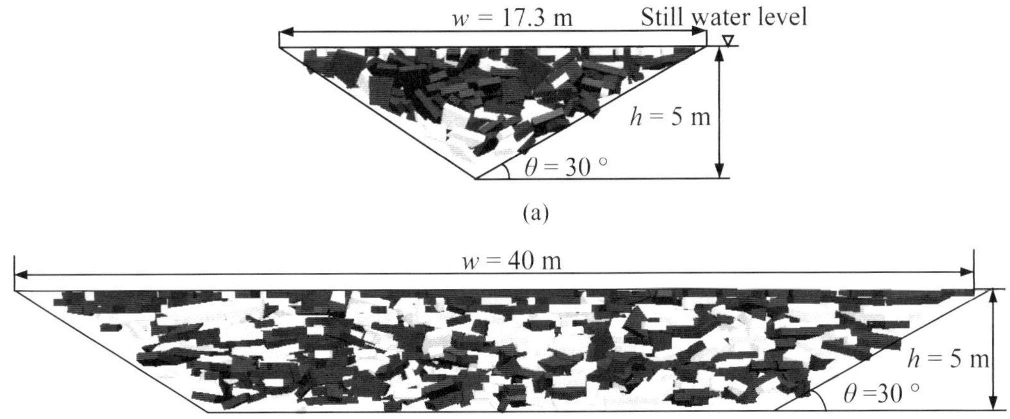

(b)

Figure 2. The pile geometries used in our simulations: (a) is the narrowest modelled pile with a triangle shape and (b) has a trapezoidal shape for the wider pile.

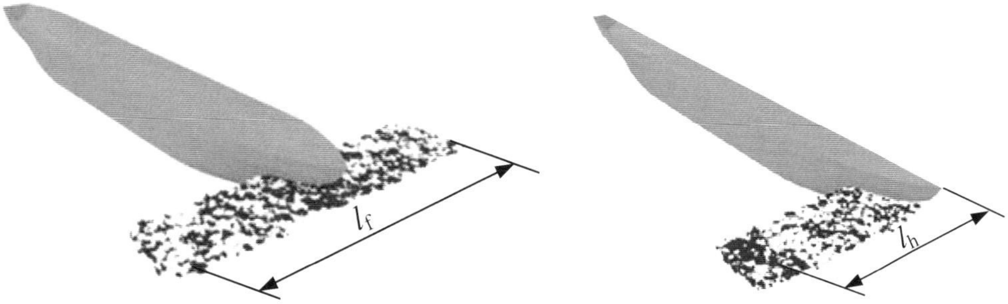

(a) (b)

Figure 3. Simulation snapshots of the actual ship-rubble interaction: (a) to illustrate the ship penetration through the full model with rubble pile length of l_f and the half model with l_h (b).

The simulations were performed in two stages: first the ice rubble pile was generated and then the ship was driven through the rubble. The ice rubble was generated by releasing ice blocks underwater, and by letting the simulation run until the viscous damping and frictional forces had dissipated virtually all of the kinetic energy of the system (Polojärvi and Tuhkuri, 2009). The basin cover on the top of the rubble provided frictional resistance on the rubble pile so that it would not dissolve. After the rubble mass had become to rest, the blocks that did not belong to the wanted pile geometry were removed from the simulation.

Figures 3a and b show two snapshots from a simulation of the actual ship-rubble interaction. In these simulations, the ship moved with a constant velocity v_{ship} into the positive x-direction and penetrated the rubble pile. The basin cover did not interact with the ship (no contact forces between the ship and the cover) nor did the cover fracture. Hence, the cover provided just frictional resistance for the rubble pile. On each time step, all of the ship-block contact forces were superposed into a force resultant $\mathbf{f}_S = [f_{sx} \quad f_{sy} \quad f_{sz}]^T$ acting on the ship. The rubble resistance was defined as $R = -f_{sx}$.

RESULTS

The Keel Resistance Records

Figure 4 shows ridge keel resistance - ship penetration (R-δ) records from two simulations: one with a full model (l_f = 60 m), and another with a half model (l_h = 30 m). The pile width w and the ice-ice friction coefficient μ_i were 17.3 m and 0.3, respectively, and the half model resistance data is multiplied by a factor of two. The noise in the force record is due to the individual ice blocks within the rubble, which had a random initial configuration, impacting the ship hull. While the initial configuration of the rubble affected this noise, the general features of the R-δ records were similar for all configurations.

Figure 4 describes the main features of the R-δ records. R initially increases up to a peak value R^p, which is reached when the ship bow has nearly penetrated the whole ridge. Subsequently, R decreases and reaches an approximately constant value for a period, during which only the midbody of the ship interacts with the rubble. During this period, R is due to the frictional forces acting on the midbody of the ship. (With large w, R did not show this period of a constant value, as seen below.) When the whole ship has passed through the rubble, R decreases to zero. The change in ice-ice friction coefficient μ_i did not affect the main features of the R-δ records, but only increased the R throughout the interaction process.

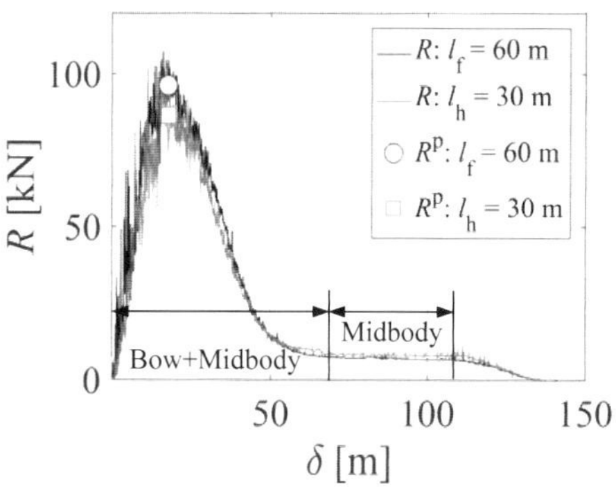

Figure 4. The resistance (R) - ship penetration (δ) records for a full model (l_f = 60 m) and a half model (l_h = 30 m). The peak resistance R^p is indicated with a marker. The rubble width w = 17.3 m and the ice-ice friction coefficient μ_i = 0.3. Half model R is multiplied by two.

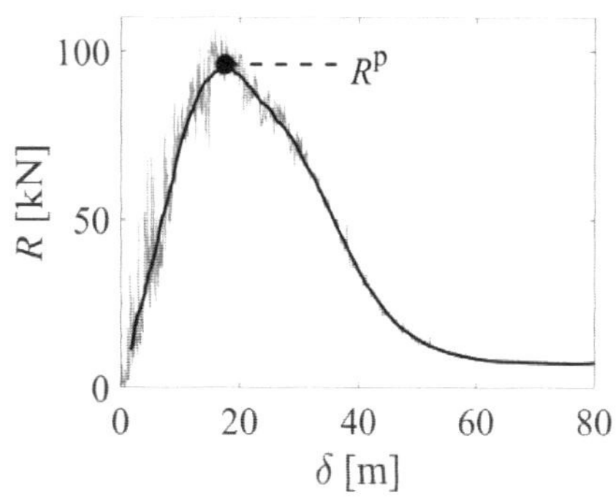

Figure 5. The resistance-penetration record (gray), the running average of it (black) and the peak resistance R^p. The data is from a simulation with full model, window size for running average 6 m, pile length l_f = 60 m, pile width w = 17.3 m, and ice-ice friction coefficient μ_i = 0.3.

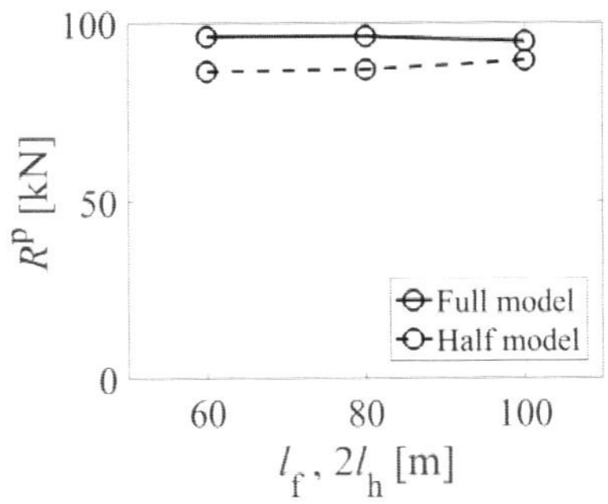

Figure 6. Peak resistance R^p from simulations with the full models and the half models. R^p is plotted as a function of pile length. Half model R^p values are multiplied by a factor of two.

Figure 4 further indicates the peak resistance R^p values of the simulations with two markers. The R^p values were defined as illustrated in Figure 5: R^p was the maximum of the running average on the R-δ data. The running average enabled us to study the general trends in R^p data as it filtered out the configuration-depended noise in the R-δ records. The window size for the running average was 6 m. As Figure 5 shows, the general features of the R-δ records are well preserved when the running average with this window size is used.

Figure 6 compares the R^p values from the simulations with the full model to those with the half model. The data is from simulations where the pile width w and the ice-ice friction coefficient μ_i were 17.3 m and 0.3, respectively. The R^p values are plotted against pile length l, and it can be seen that the pile length has virtually no effect on the R^p values. The figure also shows that the R^p values in the simulations with the full model are 6-10 % higher than in the simulations with the half model. Based on this, and the data in Figure 4, we use the half model with pile length l_h = 30 m in the further studies.

The Effect of Keel Width on the Resistance

Figure 7 presents the R-δ records from five simulations that had different pile widths w (see Table 2 for the simulation matrix). It is interesting that w has a strong effect on the R-δ record: as w increases, so does the peak resistance value R^p. Further, the peak resistance value R^p is reached later in wider rubble piles.

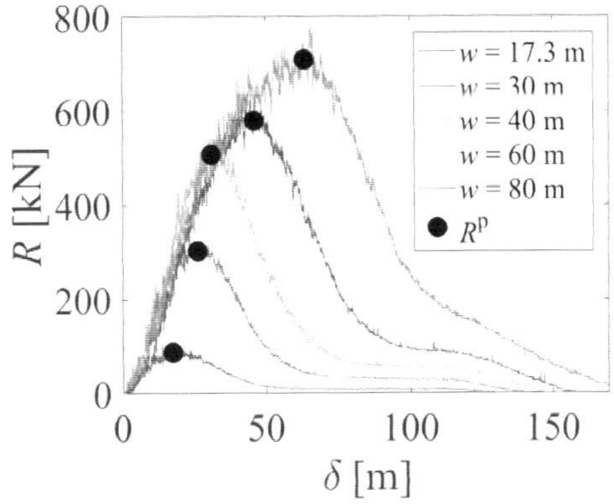

Figure 7. The keel resistance-ship penetration records from simulations with different pile widths w. The figure also indicates the peak resistance R^p values. In these simulations, a half model with l_h = 30 m and μ_i = 0.3 is used. All values are multiplied by a factor of two.

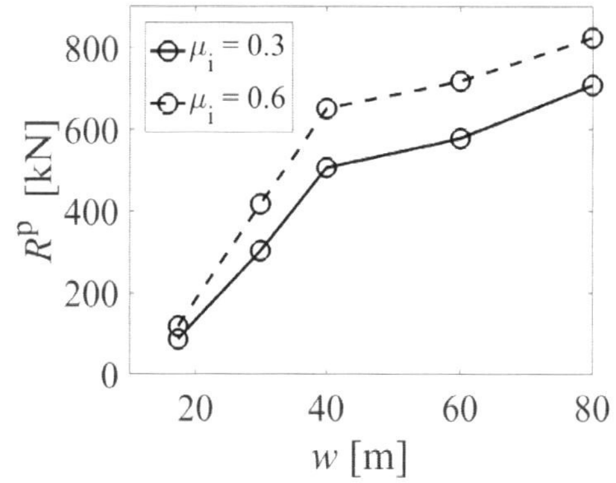

Figure 8. The peak resistance-pile width (R^{p}-w) records from simulations for two ice-ice friction coefficients μ_{i}. All values are multiplied by a factor of two.

Additionally, for $w < 60$ m, the R value for the interval of constant resistance increases with w (see also Figure 4). This is another manifestation that this interval of constant R is due to the frictional resistance acting on the midbody of the ship: the total contact area between the midbody and the rubble mass increases with w. Further, the interval of constant R starts later during the interaction process as w is increased. However, once w is large enough (80 m), the stern of the ship enters the rubble before the interval of constant R would start, and no period of constant R occurs.

The values for peak resistance R^{p} are plotted against w and μ_{i} in Figure 8. The effect w on R^{p} is drastic. R^{p} shows an eight-fold increase with the w values used (17.32 … 80 m). The R^{p}-w data show two distinct regimes. When $w < 40$ m, R^{p} shows a strong dependency on w, whereas when $w > 40$ m, the increase in R^{p} is slower. Ice-ice friction coefficient μ_{i} has a clear, but less pronounced effect on R^{p} values. Increasing μ_{i} from 0.3 to 0.6 increased R^{p} by 16-38 %, depending on w. The effect of μ_{i} was strongest with low w and then got weaker as w increased. The study on the reasons behind these results is out of the scope of this paper.

DISCUSSION

We compared the simulation results to field data and analytical formulas. The field-measured ridge resistance of a ship having close to same dimensions as the one modelled here was about 1900 kN for a ridge with a depth of 5 m (Keinonen, 1979). This is much higher than the peak keel resistance R^{p} yielded by our simulations. We are looking into this discrepancy, but unfortunately, there is a lack of a detailed description on the geometry (especially its width), and even on the existence of a potential consolidated layer of the full-scale ridge.

We also compared the results to the analytical formula for rubble resistance R_{A} by Malmberg (1983), which is recommended by a report for Finnish-Swedish ice class rules (Riska, 1997). The formula reads (Table 1 describes most of symbols)

$$R_{\mathrm{A}} = C_1 Th\left(\frac{B}{2} + h\tan\psi\cos\alpha\right)(\mu_{\mathrm{s}}\cos\alpha + \sin\psi\sin\alpha) + C_2 TL_{\mathrm{m}}\left[K_0 h + \left(\frac{h}{T} - \frac{1}{2}\right)B\right], \quad (1)$$

where L_{m} is the length the part of the midbody that is parallel to the centerline of the ship and the term h/T-1/2 should always be positive or it is to be set to zero. The angle ψ is the normal angle from horizontal at the bow, and is obtained from $\psi = \arctan(\tan\varphi / \sin\alpha)$, where φ and α are the stem angle and waterline half-angle of the ship, respectively. Further, K_0 is the coefficient of active pressure at rest, which describes the pressure due to the static ice rubble on the ship, and should be 0.27 according to (Riska, 1997). The constants C_1 and C_2 are

$$C_1 = (1-p)(\rho_w - \rho_i)gK_p \quad \text{and} \quad C_2 = (1-p)(\rho_w - \rho_i)g\mu_s \tag{2}$$

where p is the rubble porosity, μ_s is the ship-ice friction coefficient and g is the gravitational acceleration. Further, $K_p = \tan^2(45+\phi/2)$ is the coefficient of passive pressure, where ϕ is the internal friction angle of the rubble pile.

With the dimensions and parameters of our simulations, the resistance R_A given by Equation 1 is about 1000 kN, when we set $\phi = 40°$ and $p = 0.3$. These are typical ϕ and p values assumed for ice rubble (Kulyathkin, 2015). (Porosity p in our simulations was about 0.3 (Polojärvi and Tuhkuri, 2009) but we have not tested the simulated rubble for ϕ.) Peak resistance R^p in our simulations with the keel width w = 80 m was about 700-800 kN depending on μ_i (see Figure 8). The peak resistance values for wide rubble piles in our simulations compare well with R_A values given by Equation 1 when we use ϕ and p values typical for natural ice rubble.

The R^p values however vary with w as Figure 8 showed. We believe that this is the first time the dependency of R on w is discussed, and that this is a finding of potential importance. Route planning and optimization for Arctic transit often rely on estimates on ridge resistance. The often-used analytical methods (Keinonen, 1979; Mamlberg, 1983; Riska, 1997) for ridge resistance do not account for the effect of w. They assume that the pile width is infinite, and effectively then, apply to rubble fields rather than individual ridge keels. In other words, it is possible, that the resistance for ridge keels having limited w is often overestimated (or, at least, that the contribution of the ice rubble on ridge resistance is overestimated).

CONCLUSIONS

We simulated ship-ice rubble interaction process using 3D DEM. In these simulations, a ship penetrated through a non-cohesive rubble pile of width w with a constant velocity. As a result, rubble resistance R curves were achieved. Our main finding was that the R depends on w. This finding may have implications on, for example, marine traffic route optimization and transit simulations.

ACKNOWLEDGEMENTS

We are grateful for the support from the Academy of Finland through the project Kara-Arctic Monitoring and Operation Planning Platform (KAMON).

REFERENCES

Croasdale, K. R., 2012. A simple model for first-year ridge loads on sloping structures. In: *Proc. of the Int. Conf. and Exhibition on Performance of Ships and Structures in Ice*.

Dolgopolov, Y., Afanasiev, V. P., Koren'Kov, V. A., & Panfilov, D. F., 1975. Effect of hummocked ice on the piers of marine hydraulic structures. In: *International Symposium on Ice Problems, 3rd, Proceedings*, pp.460-477.

Ehle, D., 2012. Ships Breaking through Sea Ice Ridges. In: *Proceedings of the Twenty-second International Offshore and Polar Engineering Conference*, pp.1188-1193.

Heinonen, J., 2004. *Constitutive modeling of ice rubble in first-year ridge keel.* Doctoral thesis. Helsinki University of Technology. VTT Publications 1235-0621 536. VTT Technical Research Centre of Finland, Espoo, Finland.

Riska, K., Wilhelmson, M. & Englund, K. and Leiviskä, T., 1997. *Performance of Merchant Vessels in Ice in the Baltic.* Finnish Maritime Administration, Winter Navigation Research Board, Rpt. No. 52, Helsinki, 73 p.

Hopkins, M.A., Hibler, W.D. and Flato, G.M., 1991. On the numerical simulation of the sea ice ridging process. *Journal of Geophysical Research: Oceans*, 96(C3), pp. 4809-4820.

Hopkins, M. A., 1992. *Numerical simulation of systems of multitudinous polygonal blocks.* Technical Report 92-22, Cold Regions Research and Engineering Laboratory, CRREL. 69 p.

Hopkins, M.A., 1998. Four stages of pressure ridging. *Journal of Geophysical Research: Oceans,* 103(C10), pp.21883-21891.

Keinonen, A., 1979. *An analytical method for calculating the pure ridge resistance encountered by ships in first year ice ridges.* Doctoral thesis. Helsinki University of Technology.

Kulyakhtin, S. and Høyland, K.V., 2014. Study of the Volumetric Behaviour of Ice Rubble Based on Bi-Axial Compression Data. In: *22nd IAHR Int. Symposium on Ice*, pp.235-240.

Kulyakhtin, S. and Høyland, K.V., 2015. Ice rubble frictional resistance by critical state theories. *Cold Regions Science and Technology*, 119, pp.145-150.

Kuuliala, L., Kujala, P., Suominen, M. and Montewka, J., 2016. Estimating operability of ships in ridged ice fields. *Cold Regions Science and Technology*, 135, pp.51-61.

Liu L., Sun S., Ji S., 2016. Dilated Polyhedra-based Discrete Element Simulation of Ice Loads on Ship Hull. In: *23rd IAHR International Symposium on Ice*, pp.1-8.

Malmberg, S., 1983. *Om fartygs fastkilning i is.* Master's thesis. Helsinki University of Technology.

Mellor, M., 1980. Ship resistance in thick brash ice. *Cold Regions Science and Technology*, 3(4), pp.305-321.

Palmer, A. and Croasdale, K., 2013. *Arctic offshore engineering.* World Scientific: Singapore.

Polojärvi, A. and Tuhkuri, J., 2009. 3D discrete numerical modelling of ridge keel punch through tests. *Cold Regions Science and Technology*, 56(1), pp.18-29.

Polojärvi, A., and Tuhkuri, J., 2012. Velocity effects in laboratory scale punch through experiments. *Cold Regions Science and Technology*, 70, pp.81-93.

Polojärvi, A., Tuhkuri, J. and Korkalo, O., 2012. Comparison and analysis of experimental and virtual laboratory scale punch through tests. *Cold Regions Science and Technology*, 81, pp.11-25.

Timco, G.W. and Burden, R.P., 1997. An analysis of the shapes of sea ice ridges. *Cold regions science and technology*, 25(1), pp.65-77.

Serré, N., 2011a. Mechanical properties of model ice ridge keels. *Cold Regions Science and Technology*, 67(3), pp.89-106.

Serré, N., 2011b. Numerical modelling of ice ridge keel action on subsea structures. *Cold Regions Science and Technology*, 67(3), pp.107-119.

Resistance and Seakeeping Performance Analysis of an Icebreaker Vessel in Oblique Bow Waves Using SNUFOAM

Woo-Chan Seok[1], Hak-Kyu Choi[1], Shin Hyung Rhee[1,2]
[1] Department of Naval Architecture and Ocean Engineering, Seoul National University, Seoul, Korea
[2] Research Institute of Marine Systems Engineering, Seoul National University, Seoul, Korea

ABSTRACT

The present study is concerned with computational fluid dynamics (CFD) analysis of resistance and seakeeping performance of an icebreaker vessel. SNUFOAM, a CFD code developed by Seoul National University towing tank laboratory and specialized in fluid dynamics problem in naval architecture and ocean engineering, was used for this study. In SNUFOAM, the dynamic interface compression method was applied to obtain precise and sharp free-surface behavior. A mesh deforming technique was used to enable six degrees-of-freedom motion. Resistance performance in calm water were analyzed first. By comparison with experimental data, it was confirmed that SNUFOAM accurately predicts the ship resistance. Using SNUFOAM, an analysis procedure on seakeeping performance in regular wave was established and applied to oblique bow wave conditions, to acquire the resistance and motion responses. Test conditions were selected based on Arctic Ocean condition. It was found that the icebreaker vessel satisfies required resistance and seakeeping performance at the design speed.

KEY WORDS :Computational Fluid dynamics; Arctic Region; Seakeeping; Icebreaker.

INTRODUCTION

Operation of vessels and offshore plants in polar region are exposed to harsh environments compared to ships operating on conventional routes. The sea ice, snow storm, and ice fog, due to low temperature below -10 ° C are present and there is a possibility of damages to the ship due to various ice conditions such as drift ice and flat ice. It contributes to increase the design costs of the ships.

These vessels for arctic region are experimentally studied in special facilities such as ice tanks that can reproduce the polar environment (Lee, et al., 2006), but it takes a lot of time and resources to conduct experiments with ice. Therefore, to reduce the design cost, it is essential to develop and utilize a simulation program specialized for polar environment conditions (Hansen and Loset, 1999; Lubbad and Loset, 2011).

Computational fluid dynamics (CFD) analysis is a feasible and useful simulation method for designing polar region application. The CFD results can be combined with other computer aided engineering applications, such as structural analysis and design optimization. To utilize CFD analysis in polar environments case, simulation of ship motions coupled with external causes, such as ice behavior, is required.

In the present study, as preliminary study for estimating the ice load of a ship and offshore plant in the polar environment and carrying out the analysis of ice-fluid interaction, resistance and seakeeping performance of an icebreaker ship was analyzed using SNUFOAM, a CFD code developed by Seoul National University towing tank laboratory and specialized in fluid dynamics problem in naval architecture and ocean engineering. To perform the seakeeping analyses, a dynamic interface compression, grid morphing, and numerical towing tank technique were applied to SNUFOAM.

The resistance analysis results were compared with the experiments in the same condition, to validate the CFD code. In addition, by seakeeping analyses, six degrees-of-freedom (6DOF) motion response of the icebreaker in regular wave was obtained.

TEST MODEL, CONDITON AND COMPUTATIONAL METHODS

The test model is ARAON, an icebreaker designed and built in Korea. To be coherent with experiments, the scale ratio was 1/18.667 and corresponding model ship length between perpendiculars was 5.09 m. Figure 1 shows the design and principal dimensions of ARAON.

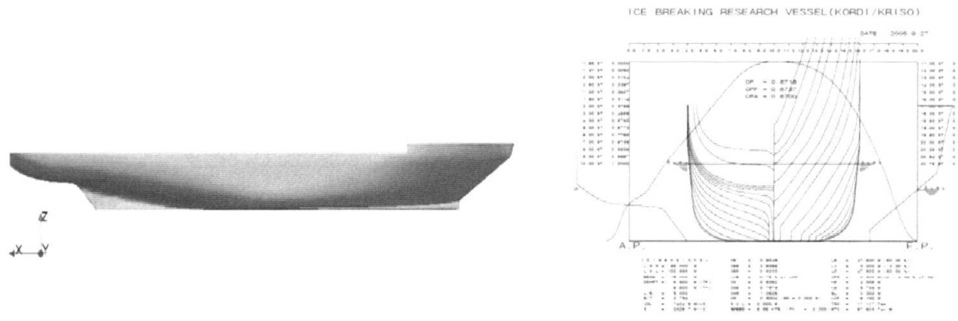

	Lpp (m)	B (m)	T (m)	Disp. (m^3)	Scale Ratio
Prototype	95	19	6.8	7424.8	18.667
Model	5.09	1.017	0.364	1.1414	

Figure 1. Design and principal dimensions of ARAON

Resistance and seakeeping tests were performed using SNUFOAM (Lee, 2015). The continuity and Reynolds-averaged Navier Stokes equations were used as the governing equations. In addition, a volume of fluid (VOF) method was used to enable free-surface behavior. The ship motion in waves was reproduced by mesh deformation, which was used for performing body motions with considerable boundary deformation. (Jasak, H., 2009).

The resistance test was conducted in the calm water condition, with advance speed variations, as shown in Table 1. Seven ship speeds in the full scale were chosen and corresponding advance speeds in the model scale were derived, following Froude's scaling law. Seakeeping simulation were conducted in regular wave condition, considering the wave condition in Arctic Ocean. The period and wave height in the model scale were 1.80 s and 0.1 m, respectively. In the wave condition, wave length was equivalent to the ship length. The wave has encounter angle of 170°, thus the model ship was in the oblique bow wave condition. Three Froude number conditions were selected: 0, 0.067, and 0.202.

Table 1. Resistance test condition

Vs (knots)	Froude number	Vm (m/s)
4	0.067	0.476
8	0.134	0.953
10	0.168	1.191
12	0.202	1.429
14	0.236	1.667
15	0.253	1.786
16	0.270	1.905

A right-handed coordinate system was used. The origin located at the intersection of station 10 and calm water surface. The size of the computational domain was $-2.36L < X < 2.36L$, $0 < Y < 1.2L$, $-1.3L < Z < 0.52L$. The computational mesh consisted of hexagonal cells and was generated by snappyHexMesh. The number of cells were around 1,300,000. Dirichlet boundary condition was applied to the inlet and hull surface. Neumann boundary condition was used to the outlet. Time and space were differentiated by second order scheme.

RESULTS AND DISCUSSIONS

Figure 2 shows resistance analysis results of ARAON and comparisons with experiments. By the comparison, it was shown that SNUFOAM has error under 3% and is feasible for analyzing resistance performance. The total resistance increased in high Froude number conditions, but in terms of total resistance coefficient, the ship showed good resistance performance around the design speed with low resistance coefficient.

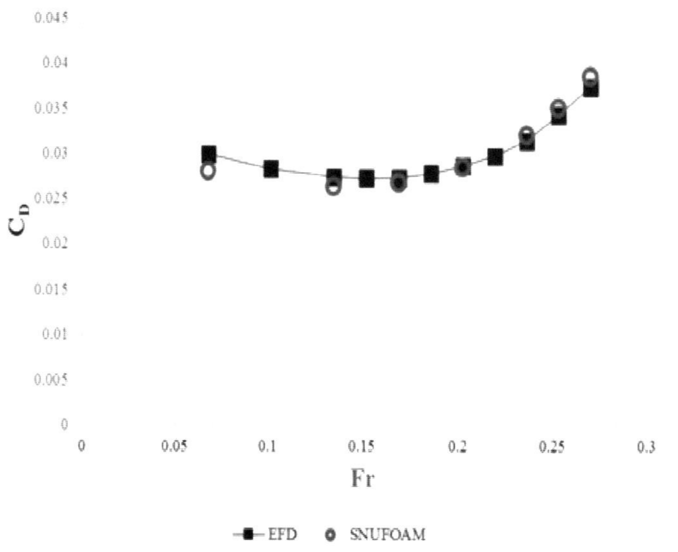

Figure 2. Resistance coefficient of ARAON model in calm water condition

Figure 3 shows surface pressure distribution in waves. In oblique bow waves, the pressure distribution developed with wave profile, but 6DOF motion of the ship also affected the surface pressure. Maximum local pressure was found when the bow impact water surface.

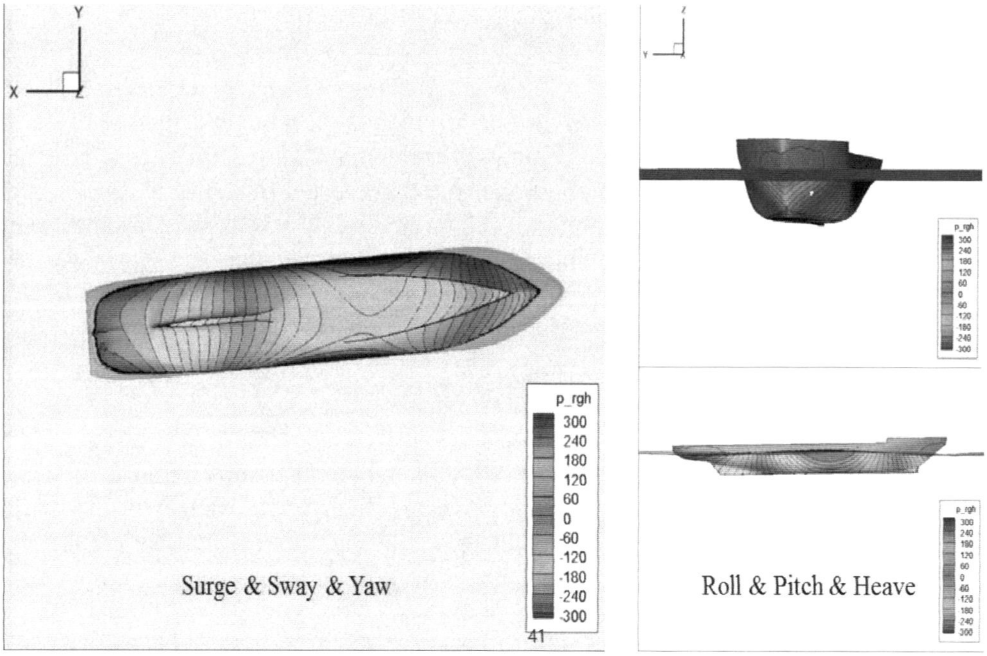

Figure 3. Surface pressure distribution of ARAON model in oblique bow wave condition

Figure 4 shows time-history of 6DOF motion of ARAON model in oblique bow waves and zero advance speed. Surge and sway motion oscillated with wave encounter. The model drifted in waves, thus the wave encounter frequency increased. Like surge motion, the yaw angle also increased while pitch and roll have regular response.

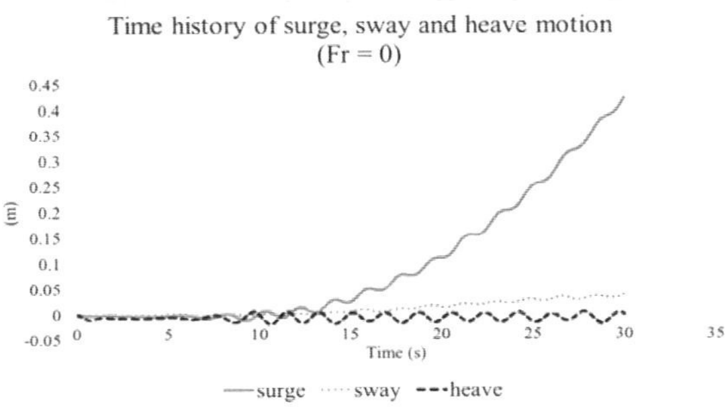

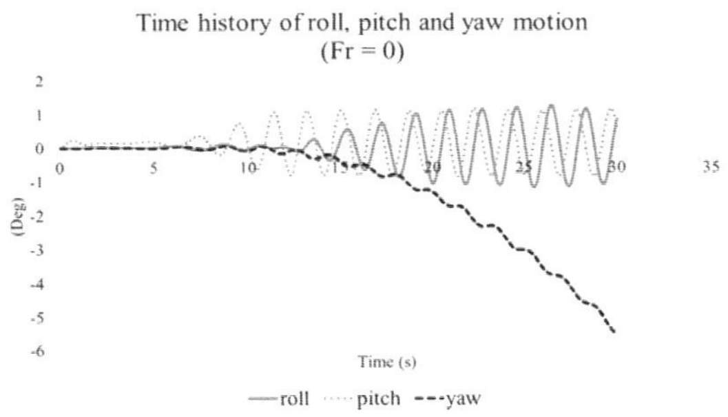

Figure 4. Time history of 6DOF motion of ARAON model in zero advance speed

CONCLUSIONS

In the present study, resistance and seakeeping performance of ARAON was analyzed using SNUFOAM, to analyze resistance and 6DOF motion. The calm water resistance showed error under 3%, compared with experiments. A 6DOF motion analysis process was established and applied to oblique bow waves. By the CFD analyses, it was shown that ARAON has reasonable resistance and seakeeping performance.

ACKNOWLEDGEMENTS

This research was supported by the IT R&D program of MOTIE/KEIT (Grant No. 100660329) and National Research Council of Science & Technology (CMP-16-03-KISTI), grant by the Korea government.

REFERENCES

Hansen, E.H. & Loset, S., 1999. Modelling floating offshore units moored in broken ice: comparing simulations with the ice tank tests. *Cold Regions Science and Technology,* 29(2), pp.107-119.

Jasak, H., 2009. Dynamic Mesh Handling in OpenFOAM. Aerospace Sciences Meeting Including The New Horizons Forum and Aerospace Exposition.

Lee, C.J., Koh, C.D., Lee, Y.Y., & Park, I.R., 2006. A development of icebreaking hull form for Antarctic research vessel. *Proceedings of the 16th Annual ISOPE conference.*

Lee, H., 2015. *A study of computational schemes for six degree-of-freedom motion of a ship in waves.* Ph.D. thesis, Seoul National University.

Lubbad, R. & Loset, S., 2011. A numerical model for real-time simulation of ship-ice interaction. *Cold Regions Science and Technology,* 65(2), pp.111-127.

POAC'17
Busan, KOREA

Proceedings of the 24th International Conference on
Port and Ocean Engineering under Arctic Conditions
June 11-16, 2017, Busan, Korea

A Discrete Element Model of Ice Ridge Interaction with a Conical Structure

Renat Yulmetov[1], Eleanor Bailey[2], Freeman Ralph[2]
[1] Memorial University of Newfoundland, St. John's, NL, Canada
[2] C-CORE's Centre for Arctic Resource Development, St. John's, NL, Canada

ABSTRACT

A numerical model of ice ridge interaction with narrow sloped structure is being developed to examine possible failure scenarios and loads it may impart on the structure. In reality there is no dominant interaction scenario and the roles of the consolidated layer, keel rubble and rubble pile accumulation on the slope still remain uncertain. Bonded-particle discrete element models are capable of simulating both the failure of continuous layers of ice and the granular behaviour of ice rubble. The computational complexity of the model is high; therefore, the simulations are run on high-performance computers. Ice ridge interactions with a Confederation Bridge pier have been simulated and the results compared with analytical estimates. The results show that the rubble pile heights and load levels obtained in the simulations are in agreement with observations indicating that the model performs well. The model will be used to further study ridge-structure interactions and hopefully help reduce uncertainties related to ridge properties and load determinations.

KEYWORDS: Ice; Ridge; Rubble; DEM; Failure

INTRODUCTION

Mechanical interaction between an ice ridge and a sloping offshore structure is extremely complex. It involves failure of ice (mostly in bending; however, all modes are possible), formation and evolution of a rubble pile, interaction between the rubble pile and approaching ice, interaction between the ridge keel rubble and the structure, hydrodynamic effects, etc. In addition, the mechanical properties of the ice and geometry of ice ridges are highly variable. This makes the process very difficult to model mathematically. As a consequence, the overall uncertainty becomes the major challenge for design load determination.

The analytical estimate of the total load from an ice ridge on a sloping structure currently proposed by ISO 19906: 2010 (ISO, 2010) is the sum of the loads caused by the consolidated layer and the rubble keel. The first term is found by considering the breaking process of level ice of equivalent thickness on an elastic foundation with consequent interaction with a rubble pile on the cone (Croasdale & Cammaert 1994), and the second term is calculated assuming plug failure of cohesive Mohr-Coulomb rubble (Dolgopolov et al. 1975). This approach often overestimates the real load, and it is said to give an upper-bound load estimate. Recent study by Croasdale et al. (2016) improves the model by introducing secondary breaking of the ice and better rubble pile geometry representation. Worthy of note, is the model of Mayne (2007) that considers the geometry of rubble pile in more detail and calculates the total load using "slices" through the pile. While these methods are very robust, they do not provide transient solutions and do not provide an opportunity to examine the processes inside the ridge. In addition, the mutual effect of the rubble keel and the consolidated layer on their strength is unknown.

Recent numerical models enable time-domain analysis of ice motion and dynamics, i.e. evolution of ice loads. A variety of methods have been developed to address ridge-structure interaction: finite-element method (Heinonen 2004), finite-discrete element method (Paavilainen et al. 2011), discrete element method (Liu, J. et al., 2016), and particle-in-cell (Barker et al. 2014). For the discrete element method (DEM), the physical size of the problem often prohibits reaching the required resolution. Domain decomposition techniques may be used to effectively parallel the computations. Thus, the model may be run on a graphics processing unit (GPU) or on high-performance computers (HPC). This helps to overcome the limitation and accelerate the simulations.

This study presents a DEM-based model of ridge-sloping structure interaction, which is currently under development, and demonstrates its performance. Ice ridge-interactions with a Confederation Bridge pier are simulated because some data are available to help validating the numerical model. The influence keel depth has on the failure mechanics and simulated loads are investigated. In particular, the influence of the truncated cone on deep ridge keels is discussed. Simulated pile heights are compared with the work of Mayne (2007) and total loads with ISO analytical solutions.

NUMERICAL METHOD

Bonded-particle model for discrete element method (Potyondy and Cundall 2004) was used to simulate the ice in the model. As in conventional discrete element method, the material consisted of spherical particles that had visco-elastic frictional contacts. In addition, the particles were bonded through visco-elastic contacts, which produced tension. If the bond stresses exceed a certain critical value, the bonds would break and the applied tension force removed.

The solid ice structure was produced using spherical particles that were the same size and arranged in hexagonal close packing (HCP). This arrangement of particles created anisotropy, which, is often observed in columnar ice. The relation between the normal and tangential stiffness and material properties was estimated by considering the dynamics of a particle in a unit HCP-cell. The normal stiffness was set to be equal to the tangential stiffness and was expressed as

$$K_n = K_\tau = \frac{1}{12}\pi E D \tag{1}$$

where E is the elastic modulus of ice and D is the particle diameter. The time step Δt in the simulations was chosen as a small fraction of a characteristic period of oscillation

$$\Delta t = \alpha D \sqrt{\frac{\rho_i}{E}} \quad (2)$$

where ρ_i is the density of ice and $\alpha = 0.1$ is a constant maximizing time step that provides stability to the simulation.

The model was implemented using an open-source package for granular materials called LIGGGHTS (Kloss et al. 2012), which can be run on multi-core distributed systems. This is extremely advantageous in that it allows large numbers of particles to be simulated, which is needed to consider the three-dimensional geometry of ice ridges and the failure of solid ice blocks in the keel.

The numerical set-up consisted of a Confederation Bridge pier imported as a triangular mesh and an ice ridge embedded in level ice. The motion of the ridge was limited by a confining wall at one end, and the pier was displaced at a constant velocity of 0.4 m/s from the opposite end. The velocity magnitude was chosen as certain representative value for the ice floe velocities measured at the bridge (Brown et al., 2010). No confinement was applied on the sides to see if splitting was possible.

It was assumed that the hydrodynamic force caused by the water flow around the pier was significantly less than the total force produced by ice. Yet, the water flow may influence the clearing of the ice blocks in the vicinity of the structure. However, this hydrodynamic effect was not implemented in the current version of the model.

Generation of Ice Ridges

Ice ridges for the simulations were constructed in several stages. First, the keel was formed by emerging cuboid-shaped blocks under level ice with a consolidated layer using a rigid body engine. It was more efficient than emerging blocks consisting of large number of particles, obviously because of reduced number of bodies and much larger time steps used in rigid body dynamics. When the blocks had settled and total kinetic energy of the system became low, each block was filled with spherical particles forming a lattice oriented in accordance with the corresponding block orientation. The ridge sail was formed in similar way. Next, all the particles in the keel and sail were exported and joined with the level ice and consolidated layer made of latticed particles, and the block-block contacts added. The obtained structure was released into the water and simulated for a short period with no forces applied except the gravity force and buoyancy, to dissipate any remaining pre-stresses between the blocks.

The size of the generated ice rubble blocks was random. The ratio between the length and the thickness for the blocks was made to follow lognormal distribution with the mean equal to 1.06 and the standard deviation equal to 0.5 according to (Kulyakhtin, 2014). The blocks were rectangular in shape so that interlocking between blocks is possible, which as Liu et al. (2015) showed is important when considering any kind of compaction build up around a structure.

A newly created ice ridge is shown in Figure 1. For this particular ridge the height, width and length of the ridge were 9.7 m, 100 m and 200 m, respectively, where the length is perpendicular to the interaction direction. It consists of more than half a million particles. The shoulders of the ridge have slope angles of approximately 6° which is in accordance with (Wadhams & Doble 2014). All of the ridges in this study were embedded into a level ice of 0.6 m thickness with a 20-m-wide and 0.9-m-thick consolidated layer in the middle.

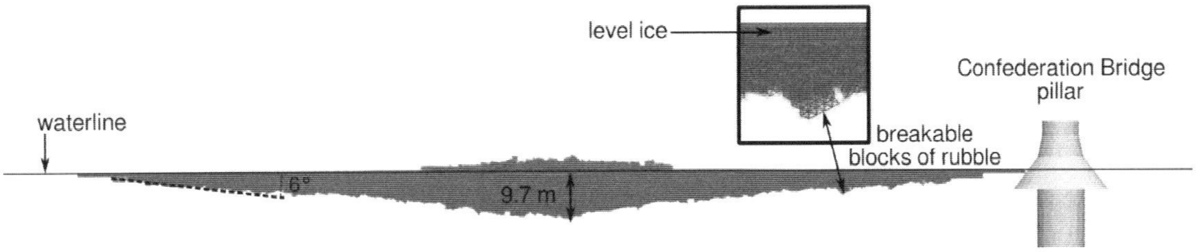

Figure 1. Side view of a generated ice ridge. The half-width is approximately 90 m and the slope angle is approximately 6°.

RESULTS

Ice Failure Behaviour

In the model the interaction with an advancing ice ridge started from level ice breaking on the slope of the cone. The particles contacting the cone experienced an upward contact force component. While the contact area was small, the bonds supporting these particles became broken and many small pieces of ice became free; this may be considered as crushing at the interface. When the interaction area expanded, the ice sheet could get radial cracks followed by flexural failures of newly formed wedges (Figure 2a).

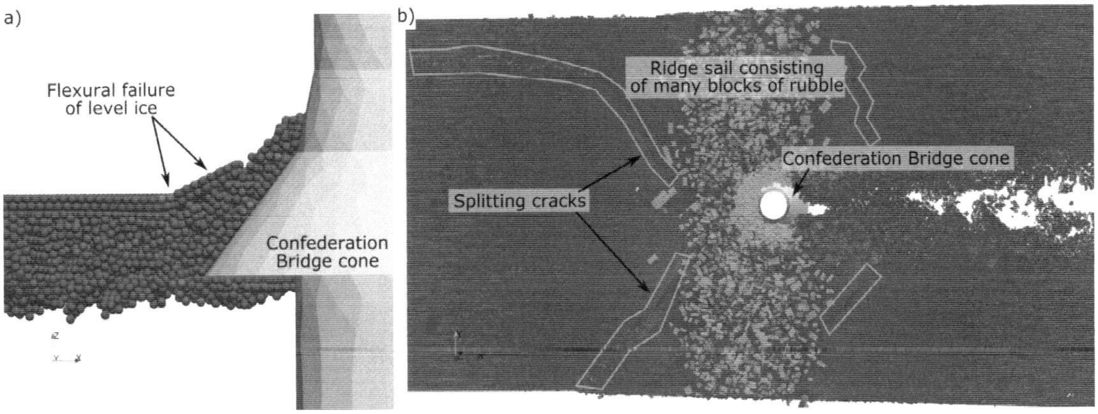

Figure 2. a) Cross-section of an ice ridge shoulder interacting with the Confederation Bridge cone; b) Top view snapshot showing formation of long splitting cracks occurring when the cone goes through a 100-m-wide section of an ice ridge.

The characteristic length of a 0.6-m-thick, quasi-infinite, ice sheet should be approximately 10.8 m; however, in our simulations the ice broke in smaller pieces. Similar behaviour has been observed by Mayne (2007); he reports the average of the maximum observed piece length of 4 m for 1-m-thick level ice. While being pushed through the pile, the blocks in the simulation were broken down to single particles. The blocks size observed in the reality was on the order of 0.5 m, which is the size of the spherical particles used in the simulations. The rubble blocks in the ridge keel often remained intact and emerged, once freeze bonds were broken.

Longitudinal cracks appeared later in the simulation, when the cone started to penetrate the consolidated layer (Figure 2b). The ice, however, continued to interact with the cone and no abrupt force reduction was observed. Interestingly, the longitudinal cracks occurred only for large ridges.

Rubble Pile Characteristics

During the simulated ridge-structure interaction, the ice formed a rubble pile on the cone. The pile was deforming, and its shape and height constantly varied. The shape of the pile cross-section on the cone can be described as linear mostly for level ice interactions, when ice accumulation is weak. The cross-section becomes parabolic when ice accumulates while interacting with sail of the ridge supported by consolidated layer and large keel. As expected, the highest point of the pile is usually located at the upstream surface of the cone, whereas rubble pile height is lower on the sides of the cone (Figure 2b).

The variation of rubble pile height with time is shown in Figure 3. It can be seen that after approximately 10 m of penetration into the ice floe (or after approximately 25 s) the pile height remains on the same level. No significant change in pile height was observed in the simulations when the pier interacted with the consolidated layer.

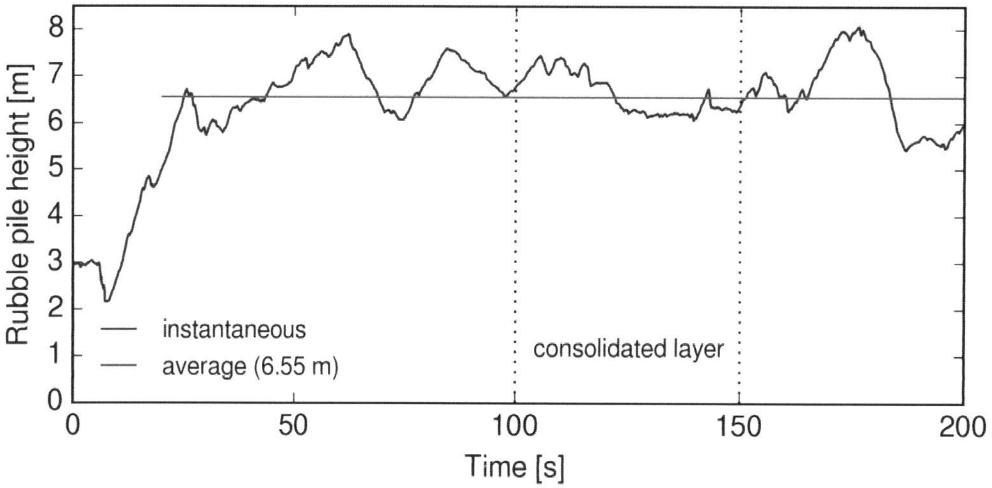

Figure 3. Evolution of a rubble pile height during interaction with a ridge having 7.9-m-thick keel. No significant changes in rubble pile height has been observed during the interaction with the consolidated layer.

Total Force

The evolution of the x-component of the total force acting on the structure is shown in Figure 4. The force generally follows the profile of the ridge with an additional increase when the consolidated layer is encountered. However, when the ridge splits, the force does not drop instantly. The remaining pieces of ice continue to break on the cone, but under less confinement.

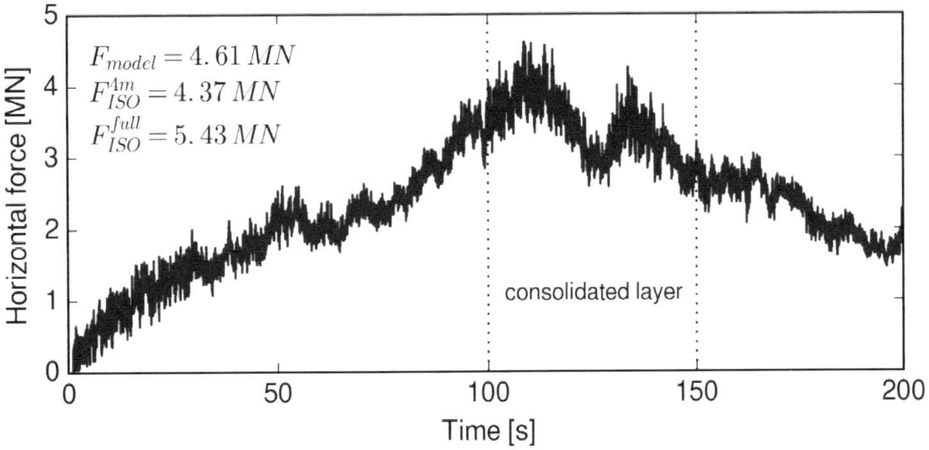

Figure 4. Evolution of the x-component of the total force obtained for an ice ridge with 7.9-m-deep keel. Estimates according to ISO are given for a 4-m-deep keel equivalent and full keel depth.

DISCUSSION

A scenario when an embedded ice ridge interacts with a Confederation Bridge pier has been simulated numerically. The numerical model employs strong assumptions regarding ridge geometry, mechanical properties and hydrodynamics, while the real environment has more natural variability particularly in ridge profile geometry. Nevertheless, the numerical model provides control over the parameters and helps to capture the ice behaviour which may be hard to observe in reality.

Simulations showed that the rubble pile geometry varied constantly, but the pile height remained approximately on the same level. The rubble accumulates due to incoming ice sheet breaking on the cone, and it is eventually spread on both sides of the cone. The rubble pile also collapses downwards, from time to time, if the pileup becomes too heavy. Both linear (as in Figure 2a) and more complex (Figure 5a) pile cross-sections were observed. The thickest profiles were observed when crossing the crest of the ridge and the keel was at its maximum depth. This is as expected as there would have been the greatest buoyancy forces preventing the rubble pile from collapsing. Therefore, more rubble was able to accumulate and consequently impart higher loads on the structure.

The rubble pile heights obtained from the model and measured in reality (Mayne, 2007) are compared in Figure 6. The numerical model slightly overestimates the pile height, because the simulated pile porosity is higher than the real one. This occurs due to equal-sized spherical particles representing the material. The minimum porosity that an assembly of these particles may have is 26%, if they are closely packed in a lattice. Assuming the packing in the pile is random during the rubbling process, the porosity is likely to be on the order of 40%, while the porosity values between 10% and 30% are often assumed when using ISO. This discrepancy in porosity calculations will likely not have much effect on the simulations, as the breaking of the incoming ice sheet is determined from the mass of the rubble pile. It is the mass that has to overcome the bearing capacity of the ice sheet in order to collapse the pile. More simulations are planned for the range of velocities to find the upper bound pile height and compare it with the results of Mayne (2007).

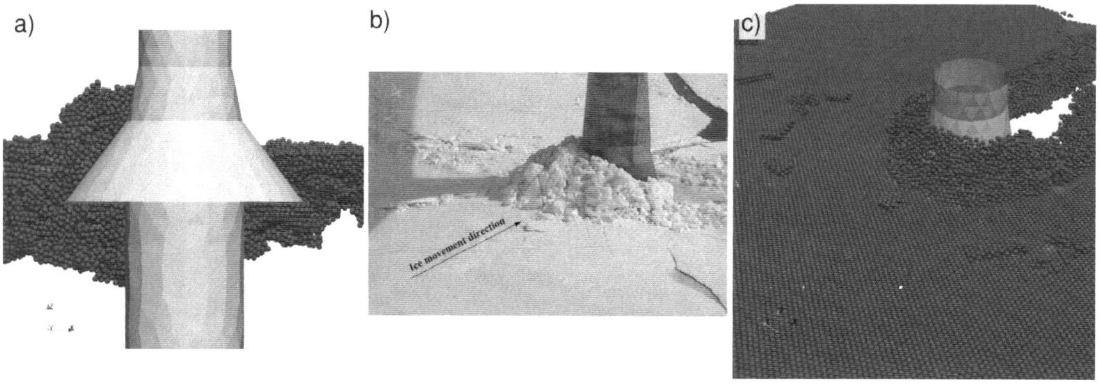

Figure 5. a) Parabolic rubble pile profile; b) observed rubble pile from Brown et al. (2010); c) snapshot from the simulation with similar geometry.

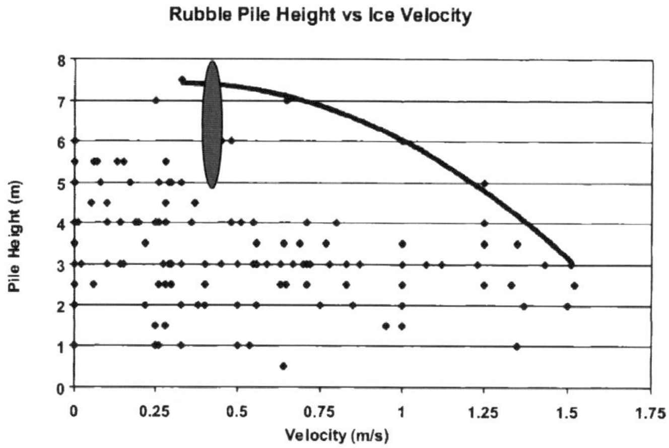

Figure 6. The rubble pile heights measured by Mayne (2007) compared to those obtained during the simulations which were done at 0.4 m/s (the set is denoted by red circle).

The simulations have been performed for several ridges of different keel depths. In one test only the consolidated layer was considered (i.e. zero keel depth). Overall, the ice loads in the simulations approximately followed the ridge cross-section profiles. Peak ice loads obtained in the simulations are compared with estimates provided by ISO 19906: 2010 (Figure 7). Rubble pile heights and porosity from the numerical simulations have been used in the analytical equations.

The peak load for only the consolidated layer is very close to that of Croasdale and Cammaert (1994). Once the keel is considered, the analytical solutions used in ISO are systematically higher than those simulated by the numerical model. This may be because they are upper bound solutions, but also there are other sources of error in the model. Firstly, the ice ridge in the simulation was not confined laterally which could have caused a load reduction. Longitudinal splitting was only observed in one simulation, when the ice ridge was 8 m deep and it took place at the moment the pier was crossing the consolidated layer. Secondly, for unconsolidated ice ridges, the cohesion of the rubble keel may be too low. A cohesion of 3 kPa and friction angle of 30° have been used to give the estimates shown in Figure 7.

Although, the blocks in the keel were frozen in the simulation, they may not have been sufficiently bonded to simulate the correct behavior. The effects of freeze bond strength will be studied in the future.

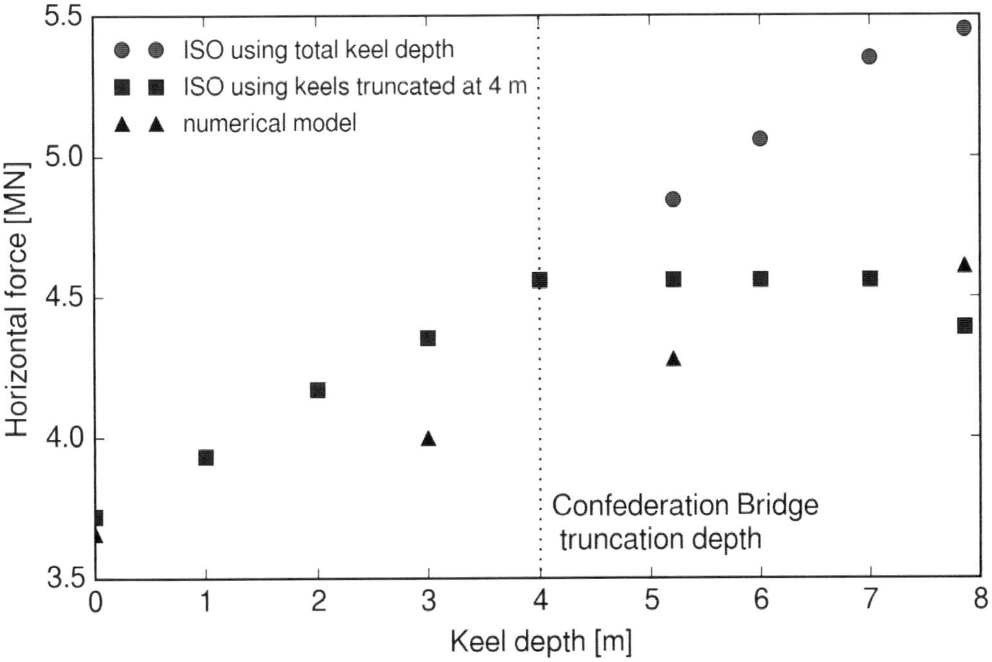

Figure 7. Results of the simulations compared with ISO 19906: 2010.

The effects of the truncated cone are also depicted in Figure 7. It has been reported previously, that marine life was present on a pier shaft under the cone (Lemee and Brown, 2005) meaning that ice interaction from the rubble keel is minimal if not absent at all. In addition, data provided by ElSeify and Brown (2006) confirm that the variation in keel depth appeared to have no effect on the peak loads. In the numerical model, the keel rubble interacts with the leading side of the shaft. It can be seen in Figure 7 that the loads from the model are growing monotonously with increasing keel depth. The ISO estimates using load levels "frozen" at 4 m keel depth are shown in the same figure. The numerical model lacking accurate hydrodynamics may cause some overestimate of the keel forces. The flow of water around the shaft likely facilitates the rubble transport and prevents violent interaction between the blocks and the shaft. On the other hand, the data presented by ElSeify and Brown (2006) has considerable variability and given for various velocities, confinements and level ice thicknesses. Further investigation is warranted to clarify the effect of the truncated cone.

CONCLUSIONS

The preliminary results of the numerical model testing clearly demonstrate that it is capable to simulate ice failure in different modes on a conical surface, and it captures some features of rubble behaviour, both on the slope of the cone and in the keel. The model is planned to be applied to simulate interaction using actual ice ridge profiles obtained by sonars and implementing limit momentum scenarios to improve reliability. This will hopefully result in improved accuracy for the design values for sloped structures. It is also worth noting that the model may be potentially applied to simulate ridge-ship interactions during station-keeping operations, which is likely to be the next phase of modelling.

ACKNOWLEDGEMENTS

The authors gratefully acknowledge financial support from the Research and Development Corporation (RDC) of Newfoundland and Labrador. Computational facilities are provided by ACENET, the regional high performance computing consortium for universities in Atlantic Canada. ACENET is funded by the Canada Foundation for Innovation (CFI), the Atlantic Canada Opportunities Agency (ACOA), and the provinces of Newfoundland and Labrador, Nova Scotia, and New Brunswick.

REFERENCES

Barker, A., Sudom, D. & Sayed, M., 2014. Conical Structures in Ice: The Roles Friction, Slope and Shape Play. In *Proceedings of the Arctic Technology Conference*, Houston, TX, USA, February 10-12, 2014.

Brown, T., Tibbo, S., Tripathi, D., Obert, K. & Sherstra, N., 2010. Extreme ice load events on the Confederation Bridge, *Cold Regions Science and Technology*, 60(1), pp.1–14.

Croasdale, K, Brown, T., Wong, C., Shrestha, N., Li, G., Spring, W., Fuglem, M. & Thijssen, J., 2016. Improved Equations for the Actions of Thick Level Ice on Sloping Platforms. In *Proccedings of the Arctic Technology Conference*, St. John's, Canada, October 24-26, 2016.

Croasdale, K.R. & Cammaert, A.B., 1994. An improved method for the calculation of ice loads on sloping structures in first-year ice. *Hydrotechnical Construction*, 28(3), pp.174–179.

Dolgopolov, Y., Afanasiev, V., Koren'kov, V. & Panfilov, D., 1975. Effect of Hummocked Ice on the Piers of Marine Hydraulic Structures. In *Proceedings of the 3rd International Symposium on Ice Problems*, Dartmouth College, Hanover, NH, August 18-21, 1975.

ElSeify, M.O. & Brown, T.G., 2006. Ice ridges interaction with Confederation Bridge piers, results of detailed video and sonar data analysis. In *Proceedings of the 18th International Symposium on Ice*, Sapporo, Japan, August 28 – September 1, 2006.

Heinonen, J., 2004. Constitutive modeling of ice rubble in first-year ridge keel. *Doctoral thesis,* VTT Technical Research Centre of Finland.

ISO 19906, 2010. ISO 19906: 2010 Petroleum and natural gas industries - Arctic offshore structures.

Kloss, C., Goniva. C., Hager, A., Amberger, S. & Pirker, S., 2012. Models, algorithms and validation for opensource DEM and CFD–DEM. *Progress in Computational Fluid Dynamics, an International Journal*, 12(2–3), pp.140–152.

Kulyakhtin, S., 2014. Distribution of Ice Block Sizes in Sails of Pressure Ice Ridges. In *Proceedings of the 22nd IAHR Internations Symposium on Ice*, Singapore, August 11-15, 2014, Singapore.

Lemee, E. & Brown, T., 2005. Review of ridge failure against the confederation bridge, *Cold Regions Science and Technology*, 42(1), pp.1-15.

Liu, J., Liu, X., Chen, Y., Long, X., & Ji, S., 2016. Numerical Simulations of Ice Loads on Fixed or Floating Offshore Structures Using Discrete Element Method. In *Proceedings of the Arctic Technology Conference*, St. John's, Canada, October 24-26, 2016.

Liu, L., Bailey, E., Sarracino, R., Taylor, R., & Power, C., 2015. Numerical Simulation of Ice Ridge Gouging. In *Proceedings of the ASME 2015 34th International Conference on Ocean, Offshore and Arctic Engineering*, May 31-June 5, 2015, St. John's, NL, Canada.

Mayne, D.C., 2007. Level Ice and Rubble Actions on Offshore Conical and Sloping Structures. *Doctoral thesis*, Calgary: University of Calgary.

Paavilainen, J., Tuhkuri, J. & Polojärvi, A., 2011. 2D numerical simulations of ice rubble formation process against an inclined structure. *Cold Regions Science and Technology*, 68(1–2), pp.20–34.

Potyondy, D. & Cundall, P., 2004. A bonded-particle model for rock. *International Journal of Rock Mechanics and Mining Sciences*, 41(8), pp.1329-1364.

Wadhams, P. & Doble, M., 2014. Report on ridge shapes and distributions for extreme value analyses in WP2, WP4 and under-ice ecosystems in WP3, Arctic Climate Change, Economy and Society program.

Proceedings of the 24th International Conference on
Port and Ocean Engineering under Arctic Conditions
June 11-16, 2017, Busan, Korea

Effective Tensile Strength of an Ice Sheet Using a Three-Dimensional FEM-DEM Approach

Ville-Pekka Lilja[1], Arttu Polojärvi[1], Jukka Tuhkuri[1], Jani Paavilainen[2]
[1] Aalto University, School of Engineering, Department of Mechanical Engineering, Finland
[2] Rand Simulation Ltd., Vantaa, Finland

ABSTRACT

A sea ice sheet may fail through several mechanisms when interacting with an obstacle. Susceptibility and mode of failure behave as functions of the physical size of a sheet. In this paper we compute the effective uniaxial tensile strength of an ice sheet for several different specimen sizes by applying a three-dimensional combined finite-discrete element approach. We evaluate the effect of the loading rate by using two displacement rates. Although a rate-independent cohesive formulation is used and the displacement rates applied are low, a significant rate effect emerges.

KEYWORDS: Ice; Combined FEM-DEM; Beam lattice network; Rate/size effect; Effective tensile strength

INTRODUCTION

Operating safely on ice-infested waters requires reliable knowledge of ice loads on structures. Information of ice-exerted loads can be obtained either through experimental campaigns or by using analytical or numerical methods. This paper investigates how an ice sheet modeled by using a three-dimensional combined finite-discrete element method (FEM-DEM) and meant to be used when conducting ice-structure interaction simulations, fractures while under uniaxial tension. We compute the effective uniaxial tensile strength of an ice sheet for several different specimen sizes. We evaluate the effect of the loading rate by using two displacement rates.

Failure of an ice sheet, using similar numerical methods, has been treated in the past by for example Hocking (1992), Jirásek & Bažant (1995), Hopkins (1998), Sayed & Timco (1998), Selvadurai & Sepehr (1999), Sand (2008), Paavilainen et al. (2009), Dorival et al. (2008), Konuk et al. (2009), Gürtner (2009), Kuutti et al. (2013), Lu et al. (2014), Herman (2016) and van den Berg (2016).

CONSTRUCTION OF THE NUMERICAL MODEL

The modeled ice sheet consists of a network of two-noded, three-dimensional co-rotational Timoshenko beam elements connected with the mass centroids of the overlaid undeformable discrete elements, Figure 1. Mesh is constructed by using centroidal Voronoi tessellation (CVT)

(Du et al., 1999). Mesh consists of convex polyhedra as the discrete elements and a Delaunay-triangulated network, or a lattice, of Timoshenko beam elements. Elasticity of an ice sheet is modeled solely with the beams, contacts between the broken pieces of ice by the discrete elements. The beam formulation follows closely Crisfield (1990,1997) while the contacts between the discrete elements are resolved as in Polojärvi et al. (2012). In essence, the model is a three-dimensional extension of our earlier two-dimensional FEM-DEM-model by Paavilainen et al. (2009) and was introduced by Paavilainen (2010). See Lilja et al. (2017) for additional details.

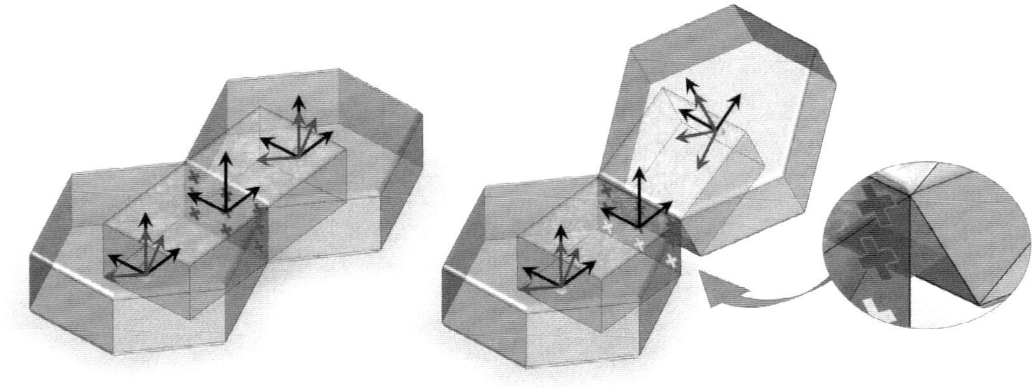

Figure 1. On the left: two discrete elements connected with a Timoshenko beam element. On the right: a partially fractured interface.

Description of Fracture

We use cohesive beam elements to model fracture. Cohesive sections are located at the mid-span of each beam element. Hooke's linearly elastic constitutive law, with a viscous damping model, is followed up until the onset of damage. When the cohesive cracking starts, a rate-independent linearly softening traction-separation law is activated and applied integration point wise to allow a crack to propagate along an interface between two adjacent discrete elements. As the constitutive model gets changed once the cohesive cracking initiates, approach is extrinsic (Seagraves & Radovitzky, 2010).

Numerical treatment of the softening process follows the model by Paavilainen et al. (2009). In their model a crack was able to propagate only vertically due to the modeling space being two-dimensional. As the model here is three-dimensional a crack is able to propagate both horizontally and vertically. Mixed mode fracture gets introduced by using the concept of an effective traction vector and representation of the state of stress with respect to a failure surface (Schreyer et al., 2006). Stress state calculation is done as with the numerical treatment of softening plasticity and includes a radial return mapping algorithm. Simpson's two-dimensional (composite) numerical integration rule is used in integrating the internal force and moment resultants.

Description of fracture using cohesive elements, when the crack path is not known in advance, is not energetically convergent with any reasonable number of elements used (Seagraves, 2013). If a mesh is structured, crack growth is also prone to preferred directions as the artificial toughness induced by the mesh varies with direction. Using a CVT-based mesh renders the crack growth isotropic. CVT-tessellated meshes are isotropic although the mesh induced toughness is still artificially high (Leon et al., 2014; Spring et al., 2014; Rimoli & Rojas, 2015).

SIMULATIONS

Main goals of our simulations are: (1) to compute the effective uniaxial tensile strength of an ice sheet for several different specimen sizes and (2) to evaluate the effect of the loading rate. A set of square-shaped sheets having a size range of 1:16 was, thus, created for the tests, Figure 2. Sheets had the side lengths of L = 10, 20, 40, 80 and 160 m and the thicknesses of h = 0.5, 1.0 and 1.5 m. For each sheet, we used two mesh densities: a mesh with an average discrete element size of $l = 2h$ and a mesh with an average discrete element size of $l = 3h$. Each broken fragment should have, thus, in minimum, a size that is comparable with actual block sizes measured from ridge sails (Kankaanpää, 1988; Høyland, 2007; Kulyakhtin, 2014). Ten randomized CVT-meshes were created for all of the sheets, excluding the sheet with L=160 m, h = 0.5 m and $l = 2h$, for which six meshes were produced.

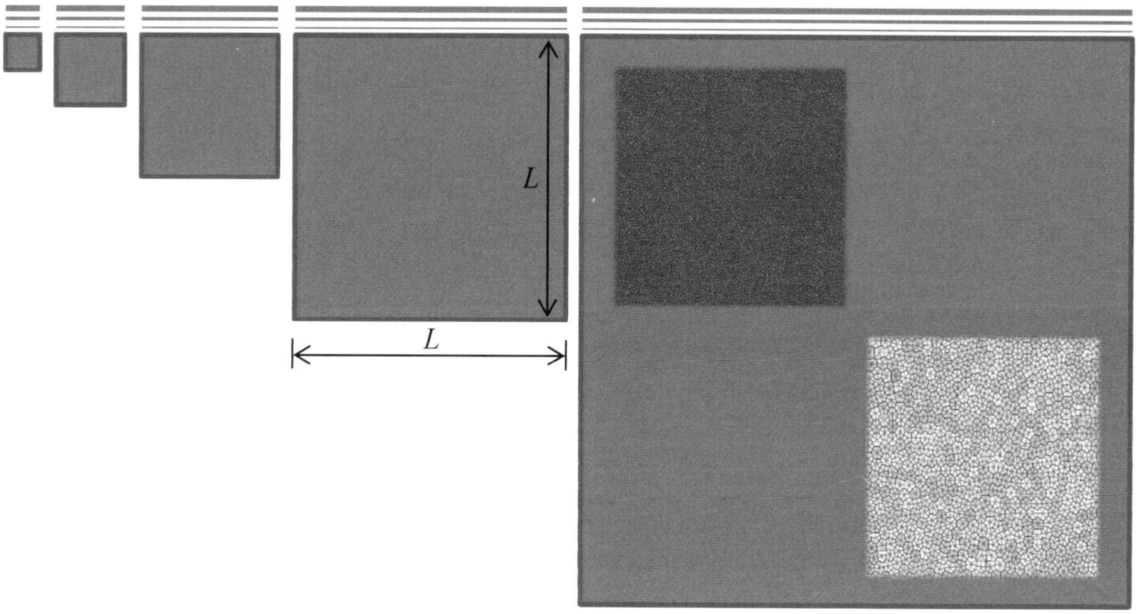

Figure 2. Size range (1:16) of the ice sheets considered. From left to right: L = 10, 20, 40, 80 and 160 m. The largest sheet on the right shows also sections of the two meshes that have the most and the least amount of discrete elements for that size: 29561 discrete elements (h = 0.5 m, $l = 2h$) and 1460 discrete elements (h = 1.5 m, $l = 3h$).

Effective Tensile Strength

Effective tensile strength was computed via a uniaxial tensile test. Each sheet was loaded under displacement control using two displacement rates: $v = 0.1$ m/s and 0.01 m/s, as shown in Figure 3. Nodes on the right and on the left boundaries were pulled into the positive and the negative X-directions, respectively, until the sheet was completely fractured. To prevent early fracture near the boundaries, a linearly changing initial velocity field $v_X = 2vX/L$ was applied, as described by Miller et al. (1999). No other constraints were used. Post-fracture contacts were not considered.

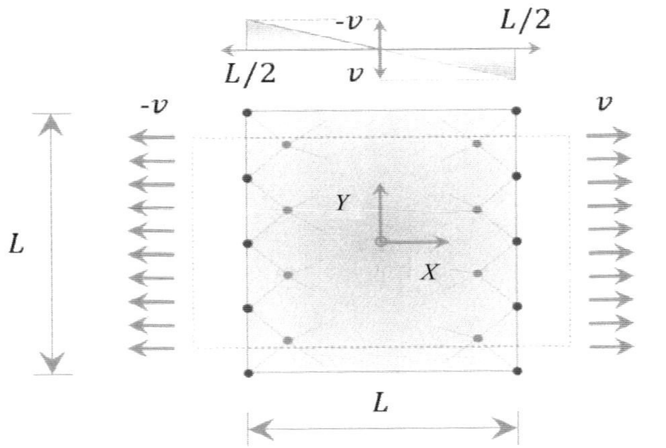

Figure 3. Boundary conditions and the initial velocity field used in the computation of the effective tensile strength.

Effective tensile strength was calculated for each sheet through $\sigma_{TS} = F_X/A_0$, where F_X is the maximum resultant force in the X-direction and $A_0 = Lh$ is the initial cross-sectional area of each sheet. The resultant force F_X was calculated for each sheet by summing up the X-direction components of the internal nodal force vectors of the beams having nodes either on the right or on the left boundary. Table 1 shows the main parameters of the simulations.

Table 1. Main parameters used in the simulations.

Parameter	Symbol	Unit	Value
Young's modulus	E^b	GPa	4
Poisson's coefficient	ν		0.3
Density of ice	ρ_{ice}	kg/m^3	920
Damping constant	c		critical[1]
Time step	Δt	s	$1.0\times10^{-5}...5.0\times10^{-5}$
Critical axial/shear stress	σ_b / τ_b	MPa	0.125
Specific fracture energy	G_f	J/m^2	15

[1] $c = \sqrt{m_{eff}E^b}$, where m_{eff} is the average of the masses of the two discrete elements a beam connects.

Results

Figures 4 and 5 show (with the displacement rates: $v = 0.1$ m/s and $v = 0.01$ m/s, respectively) the computed effective average tensile strengths σ_{TS} and their standard deviations in an order of ascending sheet size L. For each L the results are arranged in an order of ascending sheet thickness h and the discrete element size l (for each h). Results are the averages calculated using ten randomized CVT-meshes, except for the case with L=160 m, h = 0.5 m and $l = 2h$, for which six meshes were used.

With the displacement rate $v = 0.1$ m/s, the effective tensile strength σ_{TS} grows as a function of sheet thickness h and decreases as a function of sheet size L. For the two smallest sheet sizes the differences between the results, whether a discrete element size of either $l = 2h$ or $l = 3h$ is used, are large. Sheets that are smaller, thicker or meshed with a discrete element size of $l = 3h$ have a higher effective tensile strength. Except for the smallest sheet size, the deviations are observed to be small. The ratio $\sigma_{\mathrm{TS}}/\sigma_{\mathrm{b}}$ between the computed effective tensile strength of an ice sheet and the critical axial stress as used with the beams varies from approximately 5.74 for the smallest and the thickest sheets (L = 10 m, h = 1.5 m, $l = 3h$) to an average of 0.97 for the largest sheets.

With the displacement rate $v = 0.01$ m/s, the effective tensile strength σ_{TS} decreases as a function of sheet size L. For the sheets with L = 10 m or L = 20 m, the effective tensile strength grows as a function of sheet thickness h. The sheets with L = 40 m produce (practically) equal results for all of the cases. For the sheets with L = 80 m or L = 160 the tensile strength decreases as a function of sheet thickness h. Except for the smallest sheet size, the deviations are observed to be small. The ratio $\sigma_{\mathrm{TS}}/\sigma_{\mathrm{b}}$ varies from approximately 1.18 for the smallest and the thickest sheets to an average of 0.81 for the largest sheets.

Although a rate-independent cohesive formulation is used and the displacement rates applied are low, a significant rate effect emerges. Ratio between the maximum effective tensile strengths of the smallest and the thickest sheets, for the two displacement rates considered, is approximately 4.9 and reduces to an average value of approximately 1.2 for the largest sheets.

As an example, a fractured ice sheet with L = 160 m, h =0.5 m, $l = 2h$ (29561 discrete elements), and loaded with a displacement rate $v = 0.1$ m/s, is shown in Figure 5. Broad areas become cohesive and are highlighted in purple. Several fully grown cracks appear and are highlighted in gray. Areas remaining elastic are highlighted in blue. Cracks tend to branch, bridge and have a tortuous pattern.

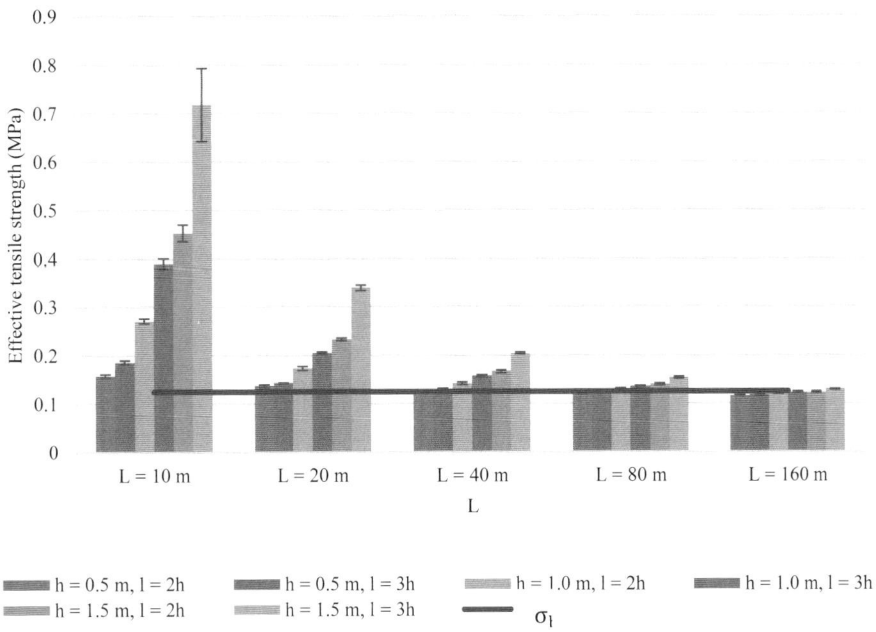

Figure 4. Effective tensile strength σ_{TS} as a function of sheet size L, sheet thickness h and discrete element size l. Displacement rate used: $v = 0.1$ m/s. Critical axial stress σ_{b}, as used with the beams, is highlighted with a red horizontal line.

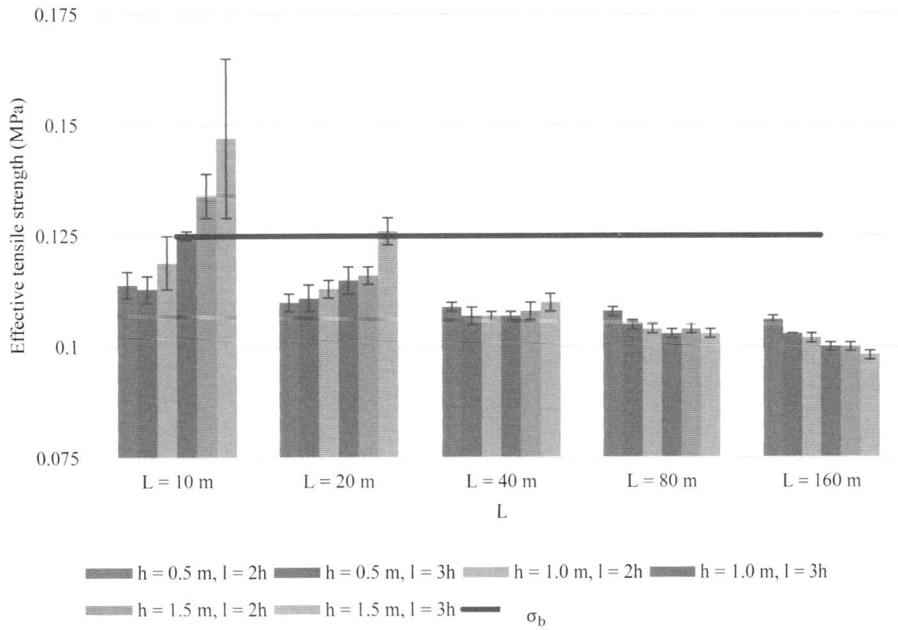

Figure 5. Effective tensile strength σ_{TS} as a function of sheet size L, sheet thickness h and discrete element size l. Displacement rate used: $v = 0.01$ m/s. Critical axial stress σ_{b}, as used with the beams, is highlighted with a red horizontal line.

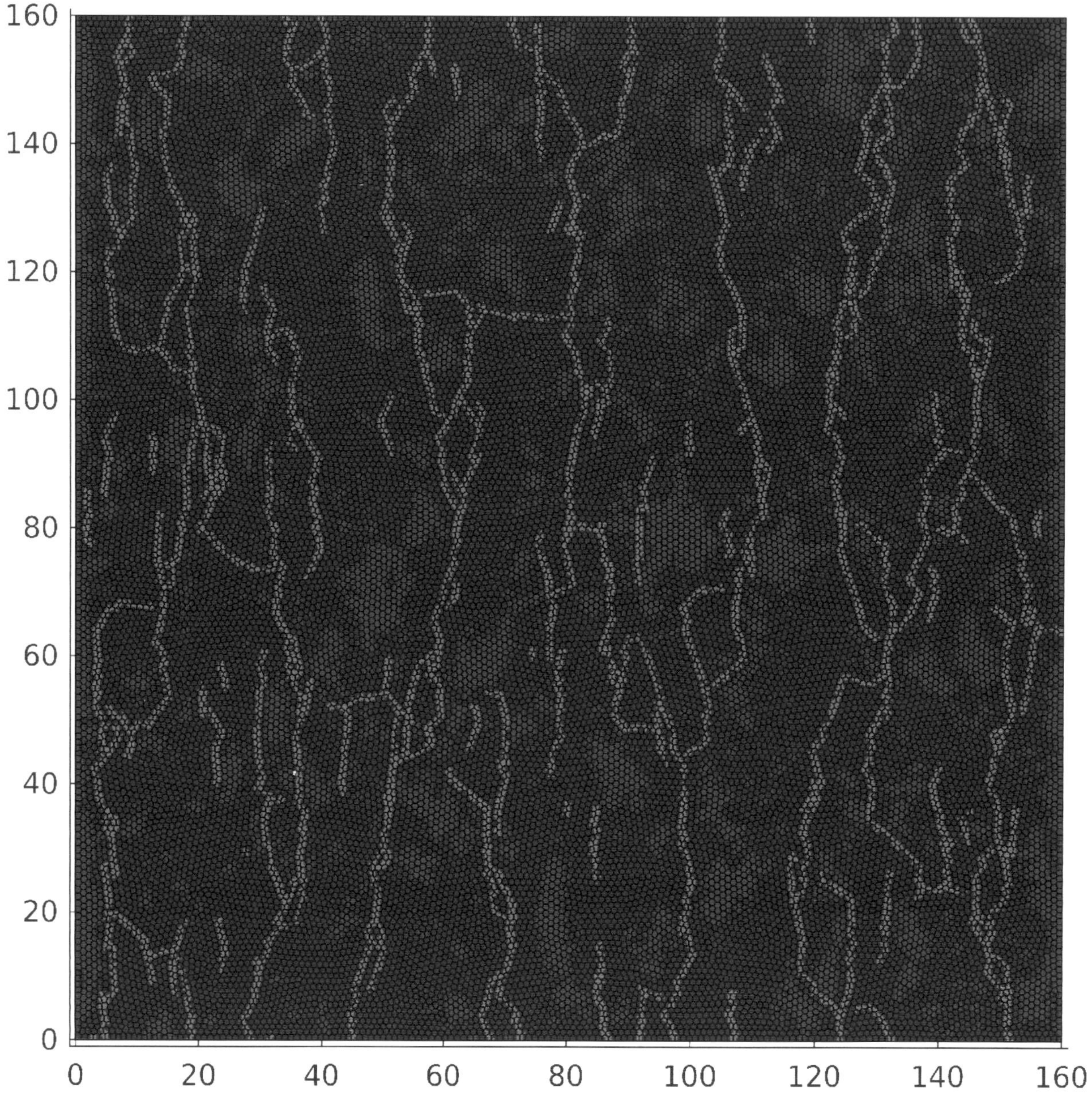

Figure 6. A fractured ice sheet with L = 160 m, h =0.5 m and l = 2h (29561 discrete elements). Completely fractured beams are highlighted in gray, partially fractured (cohesive) beams in purple and elastic beams in blue. Sheet was loaded in tension in the horizontal direction with a displacement rate $v = 0.1$ m/s.

CONCLUSIONS

We computed the effective uniaxial tensile strength of an ice sheet modeled by using a three-dimensional FEM-DEM approach for several different specimen sizes. Model consists of rigid discrete elements combined with co-rotational, viscously damped, cohesive Timoshenko beam elements. Centroidal Voronoi tessellation (CVT) was used to create the meshes. Effect of the loading rate was evaluated by using two displacement rates: $v = 0.1$ m/s and $v = 0.01$ m/s.

With the displacement rate $v = 0.1$ m/s the effective tensile strength σ_{TS} grows as a function of sheet thickness h and decreases as a function of sheet size L. For the smallest sheet size the differences between the results, whether a discrete element size of either $l = 2h$ or $l = 3h$ is used, are large. Sheets that are smaller, thicker or meshed with a discrete element size of $l = 3h$ have a higher effective tensile strength. The ratio σ_{TS}/σ_b between the computed effective tensile strength and the critical axial stress as used with the beams varies from approximately 5.74 for the smallest and the thickest sheets ($L = 10$ m, $h = 1.5$ m, $l = 3h$) to an average of 0.97 for the largest sheets.

With the displacement rate $v = 0.01$ m/s the effective tensile strength σ_{TS} decreases as a function of sheet size L. For the sheets with $L = 10$ m or $L = 20$ m, the effective tensile strength grows as a function of sheet thickness h. The sheets with $L = 40$ m produced (practically) equal results for all of the cases. For the sheets with $L = 80$ m or $L = 160$ the effective tensile strength decreases as a function of the sheet thickness h. The ratio σ_{TS}/σ_b varies from approximately 1.18 for the smallest and the thickest sheets to an average of 0.81 for the largest sheets.

Although a rate-independent cohesive formulation is used and the displacement rates applied are low, a significant rate effect emerges. Ratio between the maximum effective tensile strengths of the smallest and the thickest sheets, with the two displacement rates considered, is approximately 4.9 and reduces to an average value of approximately 1.2 for the largest sheets.

ACKNOWLEDGEMENTS

Authors gratefully acknowledge the financial support by the Academy of Finland through the project: "Discrete Numerical Simulation and Statistical Analysis of the Failure Process of a Non-homogeneous Ice Sheet Against an Offshore Structure."

REFERENCES

Crisfield, M. A., 1990. A consistent co-rotational formulation for non-linear, three-dimensional, beam-elements. *Computer methods in applied mechanics and engineering,* Volume 81, pp. 131-150.

Crisfield, M. A., 1997. chs. 17.1-2. In: *Non-linear Finite Element Analysis of Solids and Structures, volume 2: Advanced Topics.* New York: John Wiley & Sons Ltd., pp. 213-232.

Dorival, O., Metrikine, A. & Simone, A., 2008. *A lattice model to simulate ice-structure interaction.* ASME 2008 27th International Conference on Offshore Mechanics and Arctic Engineering, Estoril, Portugal, pp. 989-996.

Du, Q., Faber, V. & Gunzburger, M., 1999. Centroidal Voronoi Tessellations: Applications and Algorithms. *SIAM Review,* Volume 41, p. 637–676.

Gürtner, A., 2009. *Experimental and Numerical Investigations of Ice-Structure Interaction,* Doctoral thesis, Norwegian University of Science and Technology, NTNU: Trondheim.

Herman, A., 2016. Discrete-Element bonded-particle Sea Ice model DESIgn, version 1.3a – model description and implementation. *Geoscientific Model Development,* Volume 9, p. 1219–1241.

Hocking, G., 1992. The discrete element method for analysis of fragmentation of discontinua. *Engineering Computations,* Volume 9, pp. 145 - 155.

Hopkins, M., 1998. Four stages of pressure ridging. *Journal of geophysical research,* Volume 103, pp. 21883-21891.

Høyland, K. V., 2007. Morphology and small-scale strength of ridges in the North-Western Barents Sea. *Cold Regions Science and Technology,* Volume 48, pp. 169-187.

Jirásek, M. & Bažant, Z. P., 1995. Particle Model for Quasibrittle Fracture and Application to Sea Ice. *Journal of Engineering Mechanics,* Volume 121, pp. 1016-1025.

Kankaanpää, P., 1988. Morphology of sea ice pressure ridges in the Baltic Sea. *Geophysica,* Volume 24, pp. 15-44.

Konuk, I., Gürtner, A. & Yu, S., 2009. *A Cohesive Element Framework for Dynamic Ice-Structure Interaction Problems - Part I: Review and Formulation.* ASME 2009 28th International Conference on Ocean, Offshore and Arctic Engineering, Honolulu, Hawaii, USA, May 31 - June 5, pp. 33-41.

Kulyakhtin, S., 2014. *Distribution of Ice Block Sizes in Sails of Pressure Ice Ridges.* Singapore, August 11 to 15, 22nd IAHR International Symposium on Ice.

Kuutti, J., Kolari, K. & Marjavaara, P., 2013. Simulation of ice crushing experiments with cohesive surface methodology. *Cold Regions Science and Technology,* Volume 92, pp. 17-28.

Leon, S. E., Spring, D. W. & Paulino, G. H., 2014. Reduction in mesh bias for dynamic fracture using adaptive splitting of polygonal finite elements. *International Journal of Numerical Methods in Engineering,* Volume 100, pp. 555-576.

Lilja, V. P., Polojärvi, A., Tuhkuri, J. & Paavilainen, J., 2017. *A three-dimensional FEM-DEM model of an ice sheet.* Busan, Korea, June 11-16, Proceedings of the 24th International Conference on Port and Ocean Engineering under Arctic Conditions.

Lu, W., Lubbad, R. & Løset, S., 2014. Simulating Ice-Sloping Structure Interactions With the Cohesive Element Method. *Journal of Offshore Mechanics and Arctic Engineering,* Volume 136, pp. 1-16.

Miller, O., Freund, L. B. & Needleman, A., 1999. Modeling and simulation of dynamic fragmentation in brittle materials. *International Journal of Fracture,* Volume 96, pp. 101-125.

Paavilainen, J., 2010. Jäälautan murtuminen kartiorakennetta vasten. In: J. Heinonen, ed. *Jatkuvan murtumisprosessin mallinnus jää-rakenne vuorovaikutusprosessissa,* Research - report (in Finnish), project: "STRUTSI". Technical Research Center of Finland, VTT: Espoo, pp. 25-27.

Paavilainen, J., Tuhkuri, J. & Polojärvi, A., 2009. 2D combined finite-discrete element method to model multi-fracture of beam structures. *Engineering computations,* Volume 26, pp. 578-598.

Polojärvi, A., Tuhkuri, J. & Korkalo, O., 2012. Comparison and analysis of experimental and virtual laboratory scale punch through tests. *Cold Regions Science and Technology*, Volume 81, pp. 11-25.

Rimoli, J. J. & Rojas, J. J., 2015. Meshing strategies for the alleviation of mesh-induced effects in cohesive element models. *International Journal of Fracture,* Volume 193, pp. 29-42.

Sand, B., 2008. *Nonlinear finite element simulations of ice forces on offshore structures*, Doctoral thesis, Luleå University of Technology: Luleå.

Sayed, M. & Timco, G., 1998. *A Lattice Model of Ice Failure,* National Research Council of Canada: Ottawa.

Schreyer, H. L. et al., 2006. Elastic-decohesive constitutive model for sea ice. *Journal of geophysical research,* Volume 111, pp. 1-21.

Seagraves, A. N., 2013. *Next generation computational tools for extreme-scale simulation of dynamic fracture and fragmentation in three dimensions*, Ph.D. dissertation, Massachusetts Institute of Technology: Cambridge.

Seagraves, A. & Radovitzky, R., 2010. Advances in Cohesive Zone Modeling of Dynamic Fracture. In: A. Shukla, G. Ravichandran & Y. D. S. Rajapakse, eds. *Dynamic Failure of Materials and Structures.* Springer Science: U.S.A, pp. 349-405.

Selvadurai, A. P. S. & Sepehr, K., 1999. Two-dimensional discrete element simulations of ice-structure interaction. *International Journal of Solids and Structures,* Volume 36, pp. 4919-4940.

Spring, D. W., Leon, S. E. & Paulino, G. H., 2014. Unstructured polygonal meshes with adaptive refinement for the numerical simulation of dynamic cohesive fracture. *International Journal of Fracture,* Volume 189, pp. 33-57.

van den Berg, M., 2016. *A 3-D Random Lattice Model of Sea Ice.* Arctic Technology Conference, St. John's, Newfoundland and Labrador, Canada, 24-26 October.

DEM-FEM Modelling Interaction between Level Ice and Conical Jacket Platform

Shuailin Wang[1], Shunying Ji [1]
[1] State Key Laboratory of Structure Analysis of Industrial Equipment, Dalian University of Technology, Dalian 116023, China

ABSTRACT

In this study, a coupled discrete element method (DEM) and finite element method (FEM) approach is proposed to simulate the interaction between sea ice and conical jacket platform. The DEM with bonding-breaking glue between each element is adopted to describe the breakage of level ice. The FEM consisted of the beam and shell elements can simulate the dynamic response and stress distribution of the jacket platform structure induced by ice load. Additionally, an efficient contact algorithm is developed to get the equivalent nodal force of the element based on the shape function of the triangle element. In this study, the ice-induced vibrations of the platform structure are simulated under different ice conditions. The simulated ice forces of platform structure are compared well with the ISO19906 standard. The numerical results show that the proposed DEM-FEM approach can be aided for the anti-ice design of jacket offshore platforms.

KEY WORDS: sea ice, conical jacket platform, DEM-FEM modelling, structure dynamic response

INTRODUCTION

Jacket platforms have been widely used in ocean oil explorations in the high latitude sea area in China, such as the Bohai Gulf and the northern part of the Yellow sea. The vibrations of these platforms occur during the ice-platform interaction, affecting not only the routine production but also the serviceability and safety of platforms (Liu et al, 2009).

Timco et al. (1992) performed model tests on the four legs jacket platform in the Bohai sea, China, in order to analyze the ice load on the offshore structure (Timco et al. 1992). Huang (2010) found that the ice failures around the cone are related to the ratio of the water line diameter to the ice thickness through series of model tests. When this ratio is larger than 25, the ice failure becomes nonsimultaneous, and the rubble piece size and the ride-up height are found to be proportional to the ratio (Huang 2010). Määttänen et al. (2012) studied the effects of varying ice and structural parameters on the ice-structure interaction, especially on the frequency lock-in vibrations through scale-model tests (Määttänen et al. 2012). Recently, numerical methods have been recognized as playing key roles in the studies of ice-structure interaction (Lu et al. 2014).

Due to the complexity of mechanical properties of sea ice and the interaction between sea ice and platform, numerical simulations of ice-induced vibrations of jacket platforms meet up with great challenges. It is not appropriate to apply traditional continuum mechanics to study the mechanical behaviors of sea ice. Discrete element method(DEM) can effectively simulate the failure process of sea ice transforming from the continuum to bulks (Polojärvi and Tuhkuri 2015), thereby, reasonably reflect the characteristics of the ice load on different types of offshore structures (Polojärvi et al.2011). Sea ice models have been constructed using spheres (Shunying Ji et al. 2015), polyhedra (Lau et al. 2011) and expansion disks (Sun and Shen 2011). The DEM with the bonding-breaking function has also been developed to simulate the breakage characteristics of ice cover and the relative ice load on conical structure.

On the other hand, offshore platforms are continuous structures. Structural analyses under external loads have become quite handy after decades of research and practice. The finite element method (FEM), as a powerful numerical tool, is convenient to analyze the responses of offshore platforms. Therefore, a combined DEM-FEM method is necessary in order to accurately analyze the interaction between discrete ice and continuous platform (Chung and Ooi 2012).

This paper aims to develop a coupled DEM-FEM method to analyze responses of jacket platform under sea ice. A contact algorithm is proposed at the contact surface of ice and platform. The failure mode of sea ice and the platform vibrations are simulated with the proposed DEM-FEM model and compared with field data of the JZ20-2-NW conical platform in the Bohai Gulf, China.

THE DEM MODEL OF SEA ICE

Figure 1 shows the DEM model of sea ice. Ice particles are modeled by spheres, incorporating the bonding-breaking parallel bond between spheres to simulate the breakage of ice cover. The bond strength is the function of the temperature and salinity of sea ice. A force and a moment are developed within the bond induced by the relative motion at the contact. The force and moment can be resolved into normal and shear components F_n and F_s, M_n and M_s respectively. The maximum tensile and shear stresses acting on the bonding region can be calculated based on the beam theory as follows,

$$\sigma_{max} = \frac{F_n}{A} + \frac{M_s R}{I} \qquad \tau_{max} = \frac{F_s}{A} + \frac{M_n R}{J} \tag{1}$$

where the variable A, I and J, denotes the area,radius,polar inertia moment of bonding section,respectively.

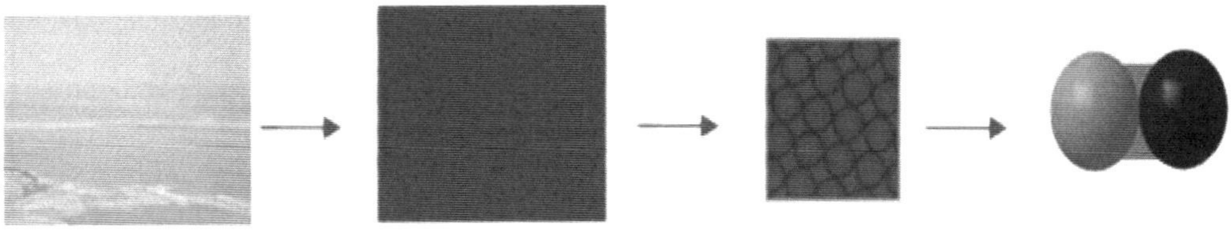

Figure 1. The DEM model of sea ice

THE FEM MODEL OF THE CONICAL JACKET PLATFORM

The JZ20-2 NW conical platform in the Bohai sea, China, as a case study, is used to verify the proposed DEM-FEM method. The platform mainly consists of three parts: the superstructure, the jacket and the pile foundation, as shown in Figure 2. For the platform FEM model, some simplifications are necessarily required for computational efficiency, as shown in Figure 3. In this paper, the ice-anti cone, with some stiffeners, is consist of the flat triangle shell elements and beam elements in the FEM model, which is main part of the platform. The final FEM model of the platform has 1034 elements and 566 nodes.

Figure 2. The JZ20-2 NW conical jacket platform

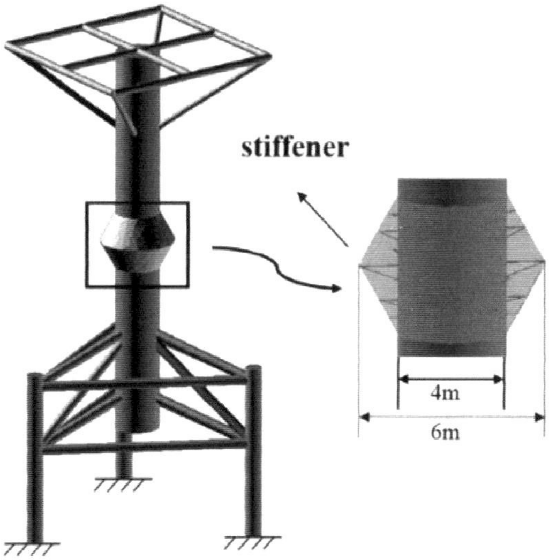

Figure 3. The FEM model of JZ20-2 NW conical jacket platform

THE COMBINED DEM-FEM METHOD

The transmissions of mechanical variables at the interface between DEM and FEM play a criticle role in the coupled DEM-FEM analysis of ice-induced vibrations of jacket platform. Here, the contact forces calculated from the ice DEM are transferred to the structural FEM as external loads, and the deformations of the platform are transferred back to the ice DEM as boundary conditions.

The contact force obtained from the DEM is randomly located on shell elements during the ice-platform interaction process. However, external loads should be appropriately treated so that their action points are located on element nodes in the finite element analyses of the platform. The area coordinate is proposed here to obtain the equivalent nodal ice loads of the FEM in the local coordinate, as shown in Figure 4. The equivalent nodal ice load is expressed as

$$P_{\mathrm{f}}^{\mathrm{e}} = N^{T} f_{\mathrm{dem}} \tag{2}$$

where, $P_{\mathrm{f}}^{\mathrm{e}}$ is the equivalent nodal force; f_{dem} is the force of DEM; N is the shape function expressed by triangle area coordinate:

$$N_{\mathrm{i}} = L_{\mathrm{i}} = A_{\mathrm{i}}/A \tag{3}$$

where A is the triangle element area; A_{i} is sub-triangle area.

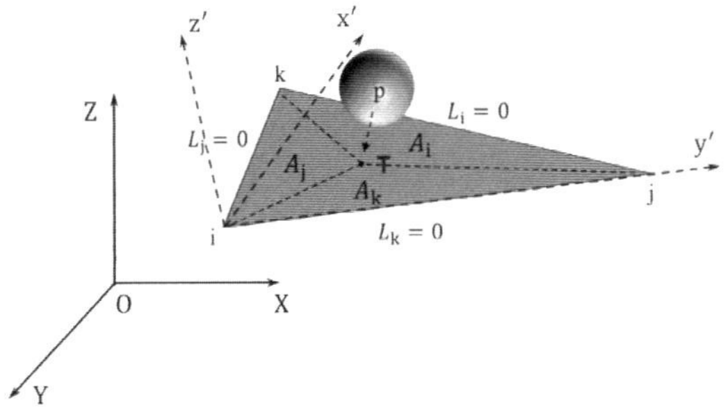

Figure 4.Sketch of contact algorithm between DEM and FEM

DEM-FEM SIMULATION OF INTERACTION BETWEEN SEA ICE AND CONICAL PLATFORM

The parameters of sea ice and platform which are used in the numerical simulation are listed in Table 1. In this paper, four cases with different ice thicknesses are simulated. From Figure 5, it can be found that the phenomenon of bending failure is quite obvious at the area around ice-breaking cones which is also in accordance with field observations (Yue et al.,2007; Tian and Huang, 2013). Figure 6(a) and Figure 6(b) show the close-ups of the sea ice radial failure mode simulated by the DEM-FEM and field monitoring data, which are consistent with each other. The phenomenon of radial failure is quite obvious at the area around ice-breaking cones.

Figure 7 plots the ice loads acting on the ice-breaking cone with $v_{\mathrm{i}} = 0.4\mathrm{m}\cdot\mathrm{s}^{-1}, H_{\mathrm{i}} = 0.2\mathrm{m}$. The force on each pile shows an obvious peak, which is consistent with field monitoring, laboratory model tests and other numerical results (Yue et al., 2007). Meanwhile, the power spectral density (PSD) of the ice load is obtained for analyzing the ice loads in frequency, as shown in Figure 8. It can be seen that the energies of the ice loads are concentrate in a frequency of 0.38Hz.

Table 1. DEM parameters

Parameters	Unit	Value
Sea ice density	kg/m^3	920
Elastic modulus	Gpa	1.0
Ice velocity	m/s	0.4
Ice thickness	m	0.2/0.4/0.6/0.8
Normal stiffness	N/m	8.0e^7
Shear stiffness	N/m	4.0e^7
Normal damping	-	0.36
Particle-particle friction	-	0.1
Particle-wall friction	-	0.3

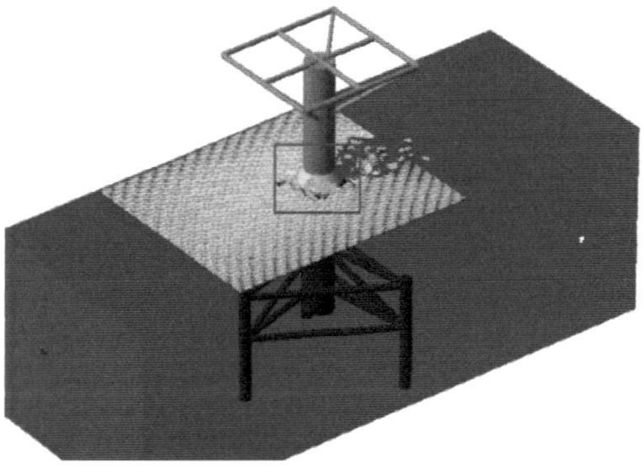

Figure 5. The numerical process of the interaction between sea ice and platform

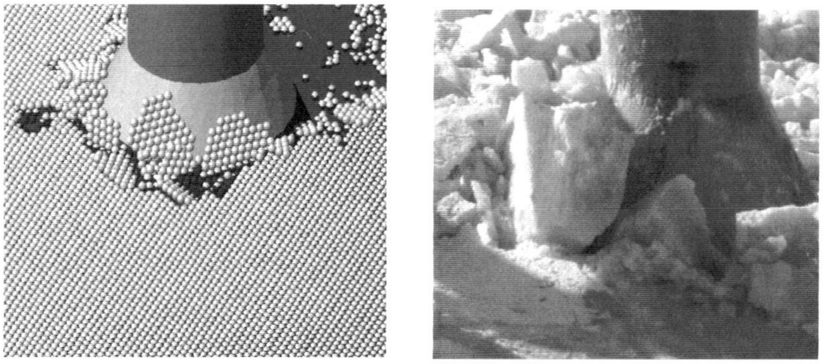

(a)DEM simulation (b)Field monitoring

Figure 6. Close-ups of the sea ice radial failure

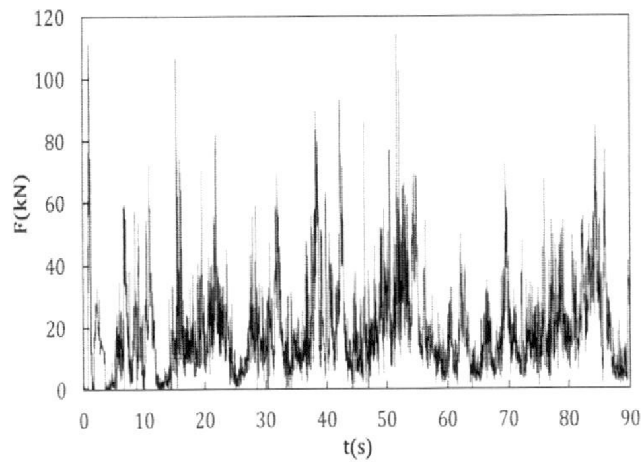

Figure 7. Ice loads obtained by DEM

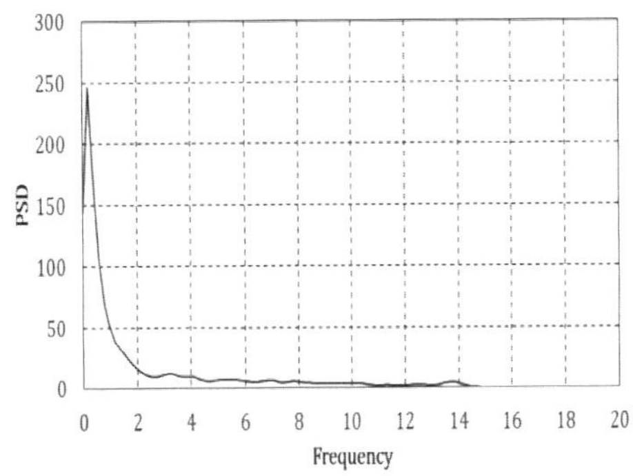

Figure 8. The PSD of the ice force

Figure 9 shows the relationship between the ice thickness and ice force obtained by DEM and ISO19906 standard(ISO19906) with blue triangle and red circle points, respectively. The results indicate the simulated trend is in accordance with the ISO19906. Moreover, the ice force is nonlinearly related to the ice thickness for both the DEM-FEM simulations and ISO19906 standard. The time series of ice-induced vibration acceleration is displayed in Figure 10(a). In Figure 10(b), we plotted the relationship between the ice thickness and the amplitude of vibration acceleration and it indicates that the vibration acceleration of platform is nonlinearly related to the ice thickness.

For analyzing the strength of the conical induced by sea ice, a von Mises stress nephogram of the ice-breaking cone is shown in Figure 11(a) and a time series of the von Mises stress is also plotted in Figure 11(b). As we can see from Figure 11(b), it can be found that the stress curve shows a strong periodicity, which means the fatigue failure of the cone may occur induced by the sea ice.

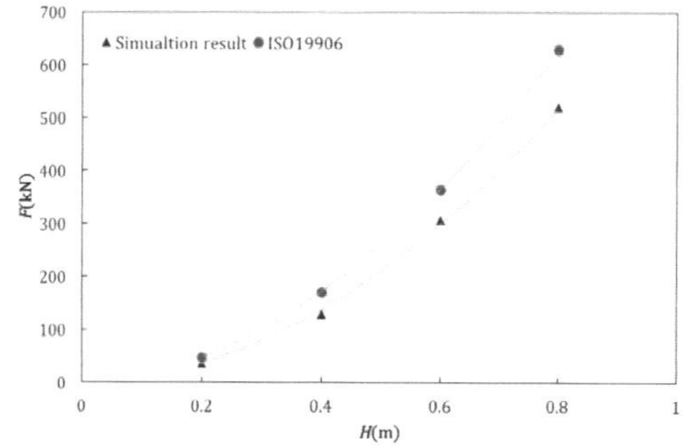

Figure 9. Comparison between numerical results and ISO19906 in ice loads

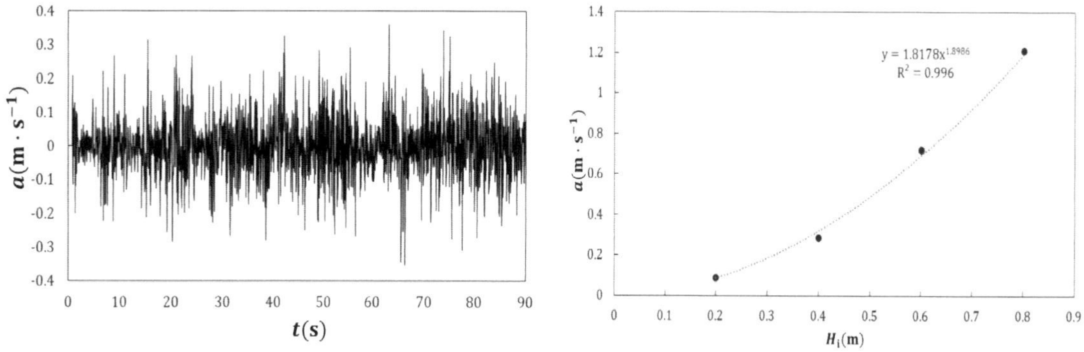

(a) Vibration accelertion under $H_{\mathrm{i}} = 0.2\mathrm{m}$ (b) Response with different ice thickness

Figure 10. The vibration acceleration of the platform

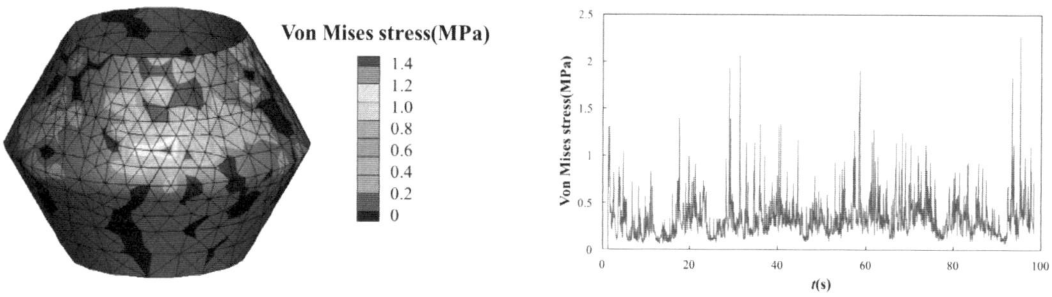

(a) Von Mises stress distribution of the cone (b)Time series curve of the cone stress

Figure 11. Von Mises stress of the cone induced by the sea ice

CONCLUSIONS

A coupled DEM-FEM method is developed to evaluate the dynamic responses and strength of the offshore platform structure. Spherical particles are bonded together to describe the ice cover, and shell-beam elements are used to analyze the dynamic behaviors and stress of the structure. The ice force acting on the platform are calculated by the DEM and transferred to the FEM as equivalent nodal forces, while the responses of the platform are calculated by the FEM and transferred back to the DEM as boundary conditions. Based on the coupled method, the JZ20-2 NW platform is simulated to obtain the ice force, vibration acceleration and stress induced by sea ice. It demonstrates that the approach can provide an effective way to investigate ice-induced vibration and strength of offshore jacket platform structures.

Acknowledgements: The authors gratefully appreciate the financial support provided by the National Natural Science Foundation of China (Grant No. 41576179, 51639004), and the Key Research and Development Program of China (Grant No. 2016YFC1401505).

REFERENCES

A. Polojärvi, J. Tuhkuri, & A. Pustogvar.,2015. DEM simulations of direct shear box experiments of ice rubble: force chains and peak loads. *Cold Regions Science and Technology*, 116, pp, 12-23.

Chung, Y. C., Ooi, Y., 2012. Linking of discrete element modelling with finite element analysis for analysing structures in contact with particulate solid. *Powder Technology*, 217(2), pp,107–120.

Huang, Y., 2010. Model test study of the nonsimultaneous failure of ice before wide conical structures. *Cold Regions Science and Technology*, 63(3), pp, 87-96.

ISO 19906, 2010. Petroleum and natural gas industries - Arctic offshore structures. International Organization for Standardization.

Ji, S., Di, S., & Liu, S., 2015. Analysis of ice load on conical structure with discrete element method. *Engineering Computations*, 32(4), pp, 1121-1134.

Liu, X., Li, G., Yue, Q., & Oberlies, R., 2009. Acceleration-oriented design optimization of ice-resistant jacket platforms in the bohai gulf. *Ocean Engineering*, 36(17-18), pp, 1295-1302.

Lu, W., Løset, S., & Lubbad, R., 2012. Simulating Ice-Sloping Structure Interactions with the Cohesive Element Method. ASME 2012, International Conference on Ocean, *Offshore and Arctic Engineering* ,136, pp,519-528.

Määttänen M., S. Løset S., et al., 2012. Novel ice induced vibration Testing in a Large-scale Facility: Deciphering Ice Induced Vibrations. Part 1. *21st IAHR International Symposium on Ice*, Dalian, China.

Michael Lau, Karl P. Lawrence, & Leo Rothenburg., 2011. Discrete element analysis of ice loads on ships and structures. *Ships and Offshore Structures*, 6(3), pp,211-221.

Paavilainen, J., Tuhkuri, J., & Polojärvi, A.,2011. 2D numerical simulations of ice rubble formation process against an inclined structure. *Cold Regions Science and Technology*, 68(1), pp,20-34.

Sun, S., Shen, H. H., 2012. Simulation of pancake ice load on a circular cylinder in a wave and current field. *Cold Regions Science and Technology*, 78(4), pp,21-4.

Tian Y, Huang Y., 2013.The dynamic ice loads on conical structures. *Ocean Engineering* 59, pp, 37-46.

Timco, G. W., Irani, M. B., et al., 2011. Model tests of dynamic ice loading on the chinese jz-20-2 jacket platform. *Canadian Journal of Civil Engineering*, 19(5), pp, 819-832.

Yue, Q., Qu, Y., Bi, X., & Tuomo, K., 2007. Ice force spectrum on narrow conical structures. *Cold Regions Science and Technology*, 49(2), pp, 161-169.

POAC'17
Busan, KOREA

Proceedings of the 24th International Conference on
Port and Ocean Engineering under Arctic Conditions
June 11-16, 2017, Busan, Korea

ISO 19906 Update - An International Standard for Arctic Offshore Structures

Karen J. Muggeridge[1], Richard F. McKenna[2], Walt Spring[3], Graham A.N. Thomas[4]
[1] K Muggeridge Consulting, Calgary, Alberta, Canada
[2] R. F. McKenna Associates, Wakefield, Quebec, Canada
[3] Bear Ice Technology, Inc., Dallas, Texas, USA
[4] Graham A N Thomas Consulting Limited, Slough, UK

ABSTRACT

In 2014, the International Standards Organisation (ISO) approved the proposal to update ISO 19906, the International Standard for design of Arctic Offshore Structures.

The revision is being developed by an ISO working group (WG8) led by Canada, within ISO Subcommittee TC67/SC7 for offshore structures. WG8 had been established in 2002 with the remit to develop a new standard, now ISO 19906. The first edition (ISO 19906:2010) has been adopted by all "arctic" and many other countries, replacing existing national documents.

Planning for the revision started in 2013. New knowledge gained from front-end engineering design for development projects, joint industry research projects, and activities such as Barents 2020, suggested that ISO 19906 needed to be updated to maintain currency and relevance. The technical work was initiated in 2014, with a review of issues to be addressed and the re-manning of Technical Panels with international experts to address them. A first working draft (WD1) of the revision was completed in October 2016. Publication of the approved second edition is currently scheduled for December 2018.

This paper includes an overview of current status and experiences. It presents the organisation and timeline for the revision work, and a summary of the changes drafted. It also presents some key interfaces, with the others of the ISO 19900 series of International Standards for offshore structures, and with the new ISO Subcommittee TC67/SC8 for arctic operations.

KEY WORDS: Arctic offshore structures; Design; ISO 19906.

INTRODUCTION

Working Group 8 (WG8), within committee ISO/TC67/SC7 Offshore Structures, was responsible for the first edition of International Standard ISO 19906. It was published December 2010, as described in Spring et al, 2011.

After some years of experience in use, and with advances in technical knowledge, the need for an update was recognised. In 2013, a revision to ISO 19906:2010 was proposed and a "new work item proposal" was initiated in February 2014 by WG8, led by Canada.

ISO 19906 SCOPE AND OBJECTIVES

ISO 19906 provides requirements, recommendations, and guidance, for design of offshore structures in the physical environment of arctic and cold regions. It relies on other ISO standards in the 19900 series for common structure types, general requirements, and non-arctic specific requirements. It was written with the assistance of over 150 of the world's leading experts in Arctic and structural design, with particular focus on ice features, ice-structure interaction, ice engineering, ice management, and structural reliability appropriate for arctic and cold regions.

"Arctic and cold regions" is deemed to include both the Arctic and other locations characterized by low ambient temperatures and the presence or possibility of sea ice, icebergs, icing conditions, persistent snow cover, and/or permafrost.

The objective of ISO 19906 is to ensure that complete structures, including substructures, topsides structures, floating production vessel hulls, foundations and mooring systems in arctic and cold regions provide an appropriate level of reliability with respect to personnel safety, environmental protection and asset value. Value includes value to the owner, to the industry and to society in general.

ISO 19906 provides design principles applicable to all phases of the life of the structure, including the successive stages of construction (i.e. fabrication, transportation, and installation), service in-place, and decommissioning.

NATIONAL ADOPTIONS OF ISO 19906:2010

ISO 19906:2010 has been widely adopted, including in all countries with arctic interests, replacing pre-existing national standards and industry technical design documents.

Canada

Canada replaced their offshore structures standard (CSA S471-04) dealing with arctic structures with ISO 19906:2010 and ISO 19900, without deviation. Canada has also adopted the entire ISO 19900 series of standards dealing with offshore structures.

Europe

ISO 19906:2010 was "parallel processed" by the European Committee for Standardization (CEN) under the ISO-CEN Vienna Agreement, resulting in EN ISO 19906:2010 being published at the same time. All CEN Members, the national standards bodies of the European Union and European Economic Area countries, adopted it nationally within the following six months. Significantly for arctic and cold regions, these countries include Denmark and Greenland, Finland, France, Germany, Netherlands, Norway, Sweden, and the United Kingdom.

Russia

A final draft of the Russian version of ISO 19906:2010 was adopted as GOST R ISO 19906 by the responsible national committee. Documentation was sent to the Federal Agency on Technical Regulating and Metrology (Rosstandart) that is responsible for official publishing of Russian standards. The effort to validate the translation and editing continues.

United States

ISO 19906:2010 was adopted in the USA as "ANSI/API RP2N, Third Edition, April 2015 (ISO 19906:2010 modified)". It replaced the previous edition of API RP2N, after an API ballot which resulted in a few minor modifications to ISO 19906:2010. The modifications are being incorporated in the current update work for the second edition of ISO 19906.

Others

ISO 19906:2010 has been adopted or otherwise accepted in many other countries with arctic and cold regions interests, such as China, Japan, Korea, and Singapore.

The European committee responsible for the Structural Eurocodes, the standards for civil and structural engineering practice, has recognized ISO 19906 as providing the definitive provisions for ice actions and arctic offshore structures.

In its design requirements for offshore wind turbines, the International Standard published by the International Electrotechnical Commission (IEC) invokes ISO 19906 for ice actions on turbine supports, for regions such as the Baltic Sea.

EXPERIENCES IN USE

ISO 19906:2010 is used worldwide for arctic design issues such as characterizing ice scenarios and ice actions. Use has ranged from concept development, front-end and detailed engineering design, and completion of projects now in service. ISO 19906 also includes earlier project experiences subsequently incorporated in the content.

Use for Different Regions and Structure Types

Regions in which industry has gained experience for, or has subsequently used, ISO 19906:2010 include the Beaufort Sea, Chukchi Sea, offshore Newfoundland and Labrador, Greenland (East and West offshore), Barents Sea, Pechora Sea, Kara Sea, Ob Bay, offshore Sakhalin Island, and the Caspian Sea.

Some examples of specific use or influence are illustrated in Figures 1 to 5.

Figure 1. Man-made islands: Northstar Island, Alaskan Beaufort Sea. (left, courtesy BP); Kashagan field, Caspian Sea, Kazakhstan (right)

Figure 2. Concrete GBS: Prirazlomnaya, Pechora Sea, Russia (courtesy of GazProm)

Figure 3. Concrete multi-legged GBS: Lunskoya-A, Sakhalin (courtesy of GazProm)

Figure 4. Floating platform: SeaRose FPSO, Grand Banks, Canada (courtesy of Harland and Wolff HI Ltd and Husky Energy)

Figure 5. GBS concepts, Beaufort Sea, Alaska (from Lovecraft and Eicken, 2011)

Hebron, Jeanne d'Arc Basin, Canada

The ISO 19906 standard was used for the specification of iceberg actions on the GBS soon to be installed at the Hebron site in the Jeanne d'Arc Basin 350 km southeast of St. John's, Newfoundland in a water depth of 93 m. The conceptual design is illustrated in Figure 6.

Because iceberg interactions are infrequent at this site, the design of the concrete supporting structure is driven primarily by the abnormal-level iceberg event specified at an annual probability of 10^{-4} (Widianto et al, 2013). Ice pressures were based on a random relationship as presented in ISO 19906:2010, Annex A.8, and capped at a fixed value at large contact areas.

Figure 6. Hebron platform concept, Jeanne d'Arc Basin, Canada (from Widianto et al., 2013)

NEED FOR REVISION AND UPDATE

Issues Deferred from Development of First Edition

During development of the first edition, the focus was on writing provisions that had consensus. If there was more than one approach in common use which could not be reconciled, provisions were written for both approaches with some guidance on applicability.

This gave rise to some unresolved issues when there was insufficient agreement on approaches developed at that time, for example specific provisions for ensuring the reliability of different types of floating structures incorporating operational measures.

And there were a number of unresolved issues related to ice actions as a result of the final review of ISO 19906:2010 which were reported by the relevant Technical Panel.

The final ballot for publication resulted in approximately 120 technical comments which would have required technical revisions to the content. None were considered sufficiently compelling as to require immediate revisions and a second ballot, so these had been archived for the next edition.

Industry Feedback on ISO 19906:2010

After publication, there has been feedback from experts and from industry users, which identified the need to do the following:

- Include other advances in knowledge, from industry activity (see below), and from articles in technical journals and conference proceedings.
- Include further new knowledge on ice actions.
- Clarify some key terms and ensure their consistent use.
- Incorporate issues arising from the API ballot for US adoption of ISO 19906:2010.
- Align with more recent publication of ISO standards in the 19900 series.
- Correct a few editorial errors.

Advances in Knowledge – Barents 2020

In anticipation of imminent developments in the Barents Sea, authorities and industry in Russia and in Norway cooperated on the development of joint offshore standards and guidelines under the Barents 2020 project.

A working group "Design of stationary floating units against ice loads in the Barents Sea" addressed gaps in ISO 19906:2010. In 2012 it published a supplementary guidance document with a number of practical recommendations for updating ISO 19906 related to stationary floating structures and ice management operations. See Moslet et al, 2012.

Other working groups developed technical bases for standards on arctic operations which were later taken forward into ISO and prompted the creation of ISO/TC67/SC8, see below.

Advances in Knowledge – Industry Research Projects

In the 2009-2014 timeframe, several industry research projects contributed advances in knowledge that are being included in the updating of ISO 19906. Some detailed reports are not in the public domain, but key issues can be acknowledged. The projects included:

- IceStruct JIP, a joint industry project managed by DnV, was to produce a document on "Ice Effects on Arctic Offshore Structures", with the intent to supplement ISO 19906 by addressing selected gaps (DnV, 2009). Gaps identified were designer-friendly methods and examples for determining actions and action effects, particularly with respect to floating structures, and to the overall approach to "reliability".
- Ice-Induced Vibrations JIP, a joint industry project was to develop an engineering approach to understanding and assessing ice-induced vibrations for a wide range of structures and ice conditions (Dr. techn. Olav Olsen, 2010). One objective was to provide new content for "dynamic ice actions" in ISO 19906.
- Models for level ice loading on sloping structures and multi-year ridge loading on conical structures were developed as part of a recent research project. These algorithms, presented in Croasdale et al., 2016, supplement those presented in ISO 19906:2010 and

provide for a better characterization of the progressive ice failure process against these types of structures.

ORGANISATION AND RESPONSIBILITIES

ISO/TC67 and ts Arctic Committees

Technical Committee TC67 is responsible for International Standards for oil and gas industry materials and equipment. Its two subcommittees of interest for arctic standards are.

- Subcommittee SC7 "Offshore structures": responsible for the ISO 19900 series for design of offshore structures of all types and in all regions worldwide.
- Subcommittee SC8 "Arctic operations": for ISO Standards for arctic operations.

SC7 Offshore Structures Committee

SC7, under the leadership of the United Kingdom, was created in 1990. The current organisation of WGs is shown in Figure 7. The flags indicate the country of the WG Convenor who provides leadership. SC7/WG8 is responsible for ISO 19906, Arctic offshore structures.

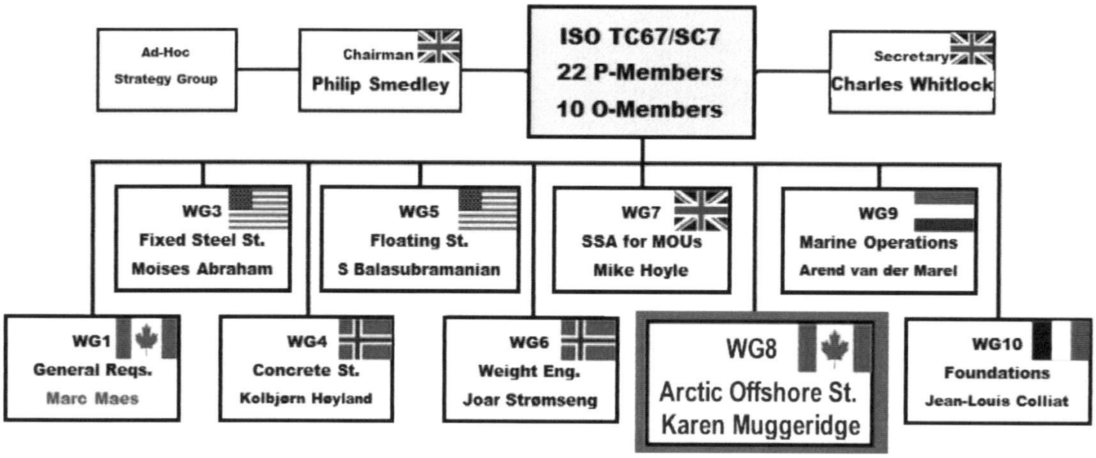

Figure 7. Organisation of ISO/TC67/SC7 Offshore structures

ISO 19900 Series of International Standards

ISO 19906 is one of the ISO 19900 series of International Standards for offshore structures. The full references are provided in ISO 19900 and in ISO 19906. The series is interdependent, and the relationships are illustrated in Figure 8.

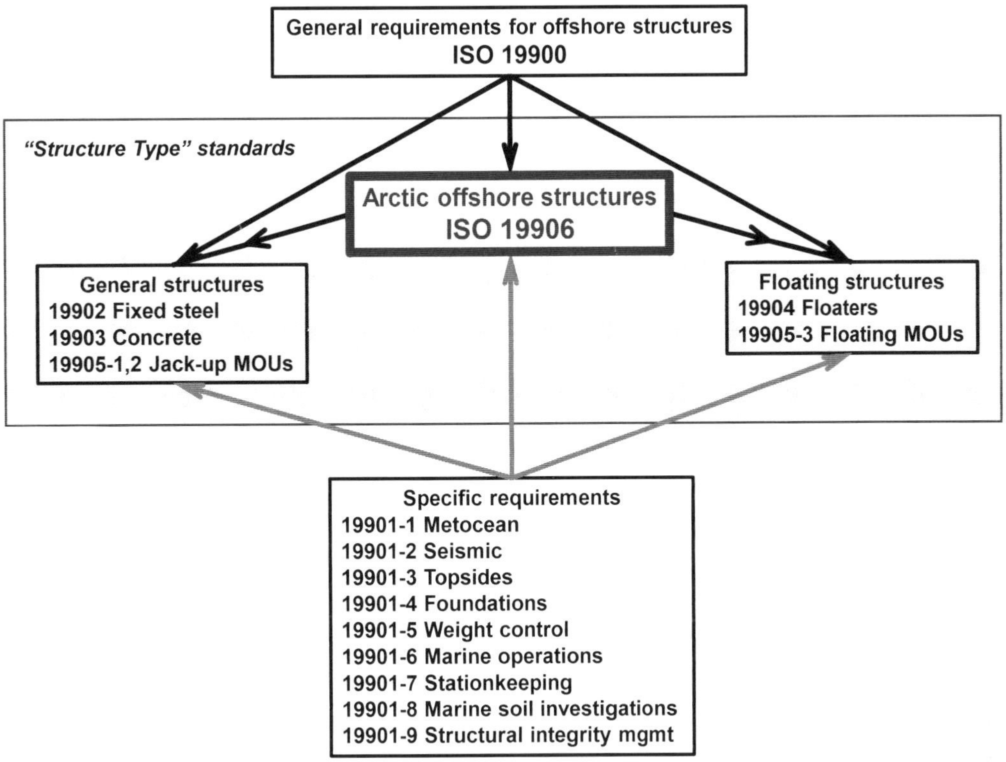

Figure 8. Relationships of ISO 19900 series of Documents for Offshore Structures

General requirements (ISO 19900) and specific requirements (the ISO 19901 series) provide generic input and support to ISO 19906. ISO 19906 adds provisions and guidance necessary for arctic and cold regions. ISO 19906 directs the user to the standard for the relevant "structure type" for generic provisions, without duplicating their content.

WG8 Arctic Offshore Structures Working Group

WG8 is comprised of experts nominated by national bodies as country representatives. The revision of 19906:2010 was initiated by Dan Masterson in 2012. Following Dan's retirement, Karen Muggeridge was selected to lead WG8 and her role as Convenor was approved in June 2014. A meeting of WG8 was held in Hamburg, Germany, in June 2014 to commence the work of revising and updating ISO 19906:2010.

After analysis of the work required, technical panels of recognised experts were organised to focus on the areas requiring significant revisions. TP0 covered minor updates to issues not otherwise covered. Table 1 shows the TPs and their leaders.

Table 1. Organisation of WG8 Technical Panels for revision of ISO 19906

Technical Panel (TP)	**TP Leader**	**Country**
TP0 Coordination/Editing/General	Karen Muggeridge	Canada
TP1 Physical Environment	Anne Barker	Canada
TP2a Reliability and Limit States	Adel Younan	USA

TP2b Ice Actions and Ice Engineering	Robert Frederking	Canada
TP4 Man-made Islands and Foundations	Ken Been	Canada
TP7 Floating Structures and Ice Management	Costas Makrygiannis	Norway

SC8 Arctic Operations Committee

SC8 was created in 2011 under the leadership of the Russian Federation, as recommended by the Barents 2020 project. Its scope includes onshore and offshore operations in arctic and other locations with low ambient temperatures and the presence of ice, snow and/or permafrost.

SC8 is organised with six WGs, each responsible for one new document, see Table 2. Leadership (Convenorship) is indicated by the national flags.

Table 2. ISO/TC67/SC8 Arctic operations: organization and documents

Working group	Title	ISO number	ISO Title: Arctic operations -
SC8/WG1	Working environment	**35101**	Working environment
SC8/WG2	Escape, evacuation and rescue	**35102**	Escape, evacuation and rescue from offshore installations
SC8/WG3	Environmental monitoring	**35103**	Environmental monitoring
SC8/WG4	Ice management	**35104**	Ice management
SC8/WG5	Arctic materials	**35105**	Material requirements for arctic operations
SC8/WG6	Physical environment for arctic operations	**35106**	Metocean, ice and seabed data

There is a close liaison between the WG8 Technical Panels and the SC8 Working Groups. As SC8 developed their documents, relevant content from ISO 19906:2010 was included such that this content could be removed from ISO 19906 and the SC8 document referenced instead.

The work of an original seventh SC8/WG, on man-made islands and land extensions, was transferred to SC7/WG8 and is now incorporated in the revision of ISO 19906, clause 10.

INTERFACES

Interfaces with Other 19900-Series Documents

Interdependencies of the ISO 19900 series were illustrated in Figure 8. Although all these standards have been published, the timelines for revisions are not synchronised due to the different scopes, resources and experts available, and industry needs.

ISO 19900:2013, however, is currently being updated, with publication planned in 2019. A look ahead to some possible changes in ISO 19900 is given in Thomas, 2017. However it is possible that the current update of ISO 19906 will be completed before changes to ISO 19900 are confirmed and can be incorporated.

Interfaces with New Documents from SC8 Arctic Operations

Three of the arctic operations standards have significant interfaces of content with ISO 19906. As of January 2017, they were all at the DIS stage, undergoing ISO member country review.

The new ISO 35102 (Escape, Evacuation, and Rescue) (EER) will incorporate most of the provisions of ISO 19906:2010, clause 18. The revised clause 18 retains only those EER issues that influence the overall design of the offshore structure.

The new ISO 35104 (ice management) will replace some of the previous provisions of ISO 19906:2010, clause 17. The revised clause 17 focuses on the ice management requirements for structural reliability, without any details as to how this is accomplished. ISO 35104 will contain many new provisions on planning and executing ice management for the protection of offshore platforms and for other offshore operations.

The new ISO 35106 (metocean, ice and seabed data requirements) will address collection, analysis, presentation and archival of data for the physical environment for arctic and cold regions operations and design. Its scope includes water depths, winds, waves, currents, air and water temperature, and atmospheric parameters related to visibility. It will also include provisions for data on sea ice, icebergs, and seabed parameters such as ice gouges, strudel scours and permafrost.

KEY AREAS BEING REVISED AND UPDATED

Summary

Significant changes have been made to targeted clauses in this revision to ISO 19906:2010. The changes will be published for public review and comment, as an ISO Draft International Standard (DIS), later in 2017.

The extent of key updates to the clauses is summarized in Table 3.

Table 3. Revisions and updates for ISO 19906 second edition

Subject	Clause	Issues
Terms and definitions	3	Arising from Barents 2020, definitions for "ice event" and "stationary floating structure" have been introduced. Definitions for "broken ice" and "ice management" have been improved.
General	5	Recommendation to consider implementing an ice alert system
Physical environment	6	Thermal ice growth equations have been modified to allow for use in estimating ice thickness.
		Text on temperature specification has been added to distinguish measures used for material and equipment specification, and for ice properties specification.
		Align with and reference ISO 35106.
Regional information	Annex B	Regional data for the physical environment for Western Greenland was combined with that of Baffin Bay. A new section with data for (North) Eastern Greenland was added.
Reliability	7	Alignment with ISO 19900:2013, especially clarifying and referencing descriptions of hazards, exposure levels, and design situations.
		Terminology with respect to ice events and ice actions.

Events and actions	**8**	Terminology with respect to events and actions, ice scenarios, and design situations.
Ice actions	**8**	Global ice pressures, low aspect sea ice interactions, probabilistic analysis.
		Local ice pressures, update with recent work reflecting pressures associated with EL and AL actions.
		More clarity on ice-induced vibrations.
		Revisions to ice properties.
		New algorithm to handle multi-year (MY) ridges.
Foundations	**9**	Alignment with revised foundation and geotechnical standards ISO 19901-4 and ISO 19901-8.
Manmade islands	**10**	New provisions added, including text on "coastal engineering" and "construction", in liaison with ISO/TC67/SC8.
		Experience of islands in the Caspian Sea incorporated.
		Revisions for better alignment with the clauses on foundations (clause 9) and reliability (clause 7).
Floating structures	**13**	Complete revamp of philosophy, emphasizing operational procedures under an ice alert system (activity specific operating guideline).
		Specific requirements have been added to ensure that reliability requirements for abnormal and extreme design situations can be met by means of operational procedures.
		Guidance on the development of ice operating envelopes.
Subsea structures	**14**	Alignment of permafrost and ice gouge content with ISO 35106, as well as transfer of content to that standard.
Other ice engineering topics	**16**	Technical revisions in relation to physical modelling; additional guidance on scaling and limitations of model results.
Ice management	**17**	Significant transfer of content to new ISO 35104.
		Narrowed the focus in 19906 to requirements for ice management systems for structure design aspects.
EER	**18**	Narrowed the focus in 19906 to structure design aspects and moved previous content to new ISO 35102.

PROCESS AND TIMELINE FOR ISO 19906 SECOND EDITION

The Technical Panels were confirmed and started work in 2015, with the target to address all issues assigned and deliver responses and proposed text to the Technical Editor by June 2016. Milestones completed during 2016 and further milestones to achieve the target publication date of December 2018 are detailed in Table 4.

Table 4. ISO 19906 Timeline for second edition

Date	**Milestone**
June 2016	Technical Panels completed technical work.
July – September 2016	Technical Editor merged contributions of TPs to provide first full Working Draft (WD1).
October 2016	TP0 and TP Leaders completed review of WD1.
November 2016	Technical Editor compiled Working Draft 2 (WD2).

December – January 2017	WG8 Country Representatives and all Technical Panels review WD2, and submit comments using the ISO standard form.
February - March 2017	Technical Editor compiles Working Draft 3 (WD3).
Late March, 2017	WD3 and the register of resolved WD2 comments distributed for final review by WG8 and TP0.
Early April, 2017	Make final revisions if any, and prepare Working Draft 4 (WD4) as "ready for DIS".
Mid-April, 2017	Provide WD4 to ISO/TC67/SC7 Secretariat (via IOGP Standard Solution Project Manager and ISO/TC67 Secretariat).
April - May 2017	SC7 Secretariat review. ISO Central Secretariat prepares DIS.
May – July 2017	National standards bodies have two months to translate ISO/DIS 19906 to their national language, if desired.
July – October 2017	Formal twelve-week DIS ballot, supervised by ISO CS. Comments received are sent to the ISO/TC67/SC7 Secretariat.
November 2017 – February 2018	Convenor (with TP0) allocates comments to Technical Panels to resolve. Technical Editor to prepare the full comments register (table of comments and responses), and the "ready for FDIS" version.
March 2018	WG8 review and approval of "ready for FDIS", and comments register with all DIS comments and responses.
April 2018	Convenor sends ISO/FDIS 19906 to ISO/TC 67/SC7 Secretariat, who checks the document and forwards to ISO CS.
May – September 2018	ISO CS perform final editing and formatting, followed by distribution to ISO Members for formal eight-week FDIS vote.
October – November 2018	WG8 final review and approval of editorial corrections. Convenor submits document for publication via ISO/TC67/SC7 Secretariat to ISO CS.
December 2018	**Publication of second edition of ISO 19906.**

CONCLUDING REMARKS

International Standard ISO 19906 specifies requirements and provides recommendations and guidance for offshore structures in arctic and cold regions. The objective is to ensure that these structures provide a level of reliability with respect to personnel safety, environmental protection and asset value, similar to that for offshore platforms in other regions of the world.

ISO 19906:2010 has been widely adopted and recognized as the definitive document for assessment of ice scenarios and ice actions, for design of structures in ice environments and for further developing the science and engineering of ice-structure interaction.

The intent of the revision was to update ISO 19906:2010 with new knowledge accepted by industry and the Arctic engineering community. See Table 5, the work is focused on specific clauses identified from feedback and advances in knowledge, filling gaps in the first edition, and updating the interfaces with other ISO standards.

Table 5. ISO 19906 contents, and clauses being revised or updated

Clause	Clause Title	Clause	Clause Title
1	Scope	**10**	Man-made islands
2	Normative references	**11**	Fixed steel structures
3	Terms and definitions	**12**	Fixed concrete structures
4	Symbols and abbreviated terms	**13**	Floating structures
5	General requirements and	**14**	Subsea production systems
6	Physical environmental	**15**	Topsides
7	Reliability and limit states design	**16**	Other ice engineering topics
8	Events and actions	**17**	Ice management
9	Foundation design	**18**	Escape, evacuation and rescue

Legend	
	White rows: Minor revisions, mostly editorial
	Yellow rows: Some technical revisions and updates
	Orange rows: Significant technical revisions and updates

Details of key technical issues considered during the revision of 19906 and the rationale for the updates will be provided in a suite of technical papers planned to be published in 2018 or 2019, to support the publication of ISO 19906 second edition.

ISO 19906 second edition is expected to be published in December 2018.

ACKNOWLEDGEMENTS

The authors would like to thank the WG8 committee and its leadership, the country members, Technical Panel leaders (past and present), and the many members of the Technical Panels who have worked and continue to work to develop and maintain ISO 19906.

We also thank IOGP (International Association of Oil & Gas Producers) and their members for contributing funding for the technical editor. We also acknowledge the support and efforts of Jonathan Venalainen and the Canadian Standards Association, as Secretariat for WG8.

REFERENCES

Canadian standard CSA S471-04, 2004. *General requirements, design criteria, the environment, and loads,* February 2004. ISBN 1-55397-525-1. Canadian Standards Association.

Croasdale K., Brown T., Fuglem M., Li G., Spring, W., J. Thijssen J., 2016. Improved Equations for the Actions of Thick Level Ice on Sloping Platforms. OTC27385. *Arctic Technology Conference, St John's, NL, Canada.*

DnV, 2009. Project Summary proposal: ICESTRUCT JIP: *Ice effects on Arctic Offshore Structures*, 1-Oct-2009. Det norske Veritas.

Dr. techn. Olav Olsen, 2010. Project Proposal for JIP KO11179: *Ice Induced Vibrations*, Rev.06, 20-Jul-2010. Dr. techn. Olav Olsen.

ISO 19900, 2013, *Petroleum and natural gas industries – General requirements for offshore structures*, International Organization for Standardization.

ISO 19906, 2010, *Petroleum and natural gas industries – Arctic offshore structures*, International Organization for Standardization.

Lovecraft, A.L and H. Eicken, 2011: *North by 2020*. University of Alaska Press, 736 pp.

Moslet, P.O., Eide, L.I., Mansurov, M., Onishchenko, D., Liferov, P., Gürtner, A., Thomas, G.A.N., 2012. Barents 2020 - Working group 02 - Design of floating structures in ice. OTC 23709. *Arctic Technology Conference, Houston, Texas. December 2012.*

Spring, W., McKenna, R.F., Thomas, G.A.N., Blanchet, D., 2011. ISO 19906: An International Standard for Arctic Offshore Structures. *Proceedings of the 21st International Conference on Port and Ocean Engineering under Arctic Conditions (POAC11-K3), Montreal, Canada.*

Thomas, G.A.N., 2017. Design Situations and Limit States in the Design of Arctic Offshore Structures. *Proceedings of the 24th International Conference on Port and Ocean Engineering under Arctic Conditions, Busan, Korea.*

Widianto, Khalifa, J., Younan, A., Karlsson, T., Stuckey, P., Gjorven, A., 2013. Design of Hebron Gravity Based Structure for Iceberg Impact. *Proceedings of the Twenty-third (2013) ISOPE International Ocean and Polar Engineering Conference, Anchorage, Alaska, USA.*

Trends and Developments in Antarctic Icebreaking Ships

Rob Hindley[1], Robert Tustin[2], Chi-Hyun Sung[3]
[1] Lloyd's Register EMEA, Helsinki, Finland
[2] Lloyd's Register EMEA, Southampton, United Kingdom
[3] Lloyd's Register Asia, Busan, South Korea

ABSTRACT

A recent resurgence has been seen in the specification, ordering and building of ice capable ships for Antarctic service. These ships are replacements of existing Antarctic Treaty state ships used to resupply Antarctic research bases. Similar mission demands allow for the fleet of specialist Antarctic icebreaking ships to be studied as a distinct and unique sub-set of the global ice-classed fleet. Trends can be seen between earlier and more recent generations of ships, aligned with changing missions of the ships used in the Antarctic. To draw out these trends a comparison of has been made of ship principal dimensions as well as ice-going performance. A particular emphasis is made on changing mission profiles and an increasing multi-role mission for Antarctic shipping. Trends from procurement and specification practices for recent projects for Antarctic icebreaking ships are described along with insights to inform future procurements.

KEY WORDS: Antarctic; Icebreaker; Specification

INTRODUCTION

Background

The functioning of the Antarctic Treaty System relies to a significant extent on the maintenance of science activities in the Antarctic, including scientific base stations, as well as environmental protection and emergency response capability, all of which require access to (and use of) specialist shipping. The premise of this paper is that because of similar mission demands by Antarctic treaty signatories the fleet of specialist Antarctic icebreaking ships can be categorized, and studied, as a distinct sub-set of the global ice-classed fleet.

Dick & Laframboise (1989) provided an empirical review of the design and performance of icebreakers, which explored characteristics of icebreakers by the use of trend lines, developed from a database of ships existing in 1989. In the first part of this paper we utilize the approaches taken by Dick & Laframboise to look at similar trends for the sub-set of existing Antarctic icebreaking ships and look at the most recent additions to this sub-set fleet in service as well as those currently being designed. For the most recent ships, in addition to characteristics and trends of use to the designer, an overview of trends in procurement processes is offered.

Objective

The objective of this paper is to analyze trends in the design of Antarctic icebreaking ships to understand the implications of evolutions of mission requirements and procurement policies on the design of the latest ships for Antarctic employment.

TREND LINE DEVELOPMENT

Antarctic Ship Data-sets

Data of ships known to be deployed to Antarctic waters for support of Antarctic science has been collated. Ships included have the following characteristics:

- Antarctic research and/or resupply mission
- Ice strengthening for re-supply mission

This screening of ships was undertaken to capture a set of ship data which is consistent, based on an Antarctic mission, and avoids outliers that, in a relatively small data-set, may bias trends unnecessarily. In undertaking this a number of ship types known to have operated in the Antarctic have been excluded, notably: cruise ships operating in the Antarctic – as this ship type is unique in its own right and deserves separate analysis; icebreakers on occasional duty – for example icebreakers such as the *Kapitan Khlebnikov* chartered by tour operators; and finally ice-strengthened general cargo ships used on an ad-hoc basis to supplement resupply activities to Antarctic bases.

Data Sources for the Antarctic Ship Data-set

The majority of data has been collated from two sources:

1. Principal particulars necessary for the trend lines associated with ship proportions have been collated from public domain sources, e.g. Lloyd's Register's Register Book; operator data sheets

2. Data required related to ice performance has been collated primarily from Riska (2013)

Public information sources, such as ship owner webpages have been used to validate these figures. Values associated with bow form (e.g. stem angle) have generally been scaled from drawings in publications and should therefore be considered estimates.

Appendix of Antarctic Ship Data-set

The complete listing of the Antarctic ship data-set analyzed for this paper is included in the appendix. The appendix contains relevant information from which the data used in the graphs have been derived. In order to present the data in a concise manner, derived values are not repeated (e.g. lightweight from deadweight and displacement).

COMMENTARY ON TRENDS AND CHARACTERSTICS

Generations of Antarctic Icebreaking Ship Development

An overview of the data-set allows us to consider three phases of Antarctic icebreaking ship designs, with each phase corresponding to a nominal ship life of 25 to 30 years.

Early Antarctic ships: pre 1970 year of delivery

Access to the Antarctic for science, and strategic purposes, has always required ships. Although early Antarctic ships were often not purpose built – or built for other ice covered environments - the 1950s and 1960s saw the first purpose built ships for Antarctic service to enable access and supply to newly established science research stations. Although history gives useful context we have omitted these early ships from our trend analysis because ship design and ship missions have evolved significantly since these early Antarctic icebreaking ships were introduced.

First Generation: 1970s-2000, peaking in early 1990s

The first generation of Antarctic ships, as we have defined them, began in the 1970s where existing early ships were replaced with purpose built units, aimed at resupplying science bases on the Antarctic content and ensuring countries with claims to the continent were able to fulfill their Antarctic Treaty obligations. The data-set analyzed starts with ships of this first generation (delivered after 1970).

Second Generation: 2000 – present

The second generation is identified as ships delivered after 2000 to the present. Here the line is drawn arbitrarily, but aligns with the replacement of tonnage for first generation ships.

Mission Profiles

Analyzing the Antarctic ship data-set using the generations defined above it is possible to draw conclusions on the changing mission profiles of the ships, and to use this as a way of further analyzing trends.

First generation ships may generally be characterized as *resupply* ships with limited facilities for on-board oceanographic research. Ice going capability is aligned with the need to re-supply Antarctic bases on an itinerary. In particular for ships resupplying more than one station, commencement of resupply is dependent on early-season entry through the Antarctic pack ice.

With the signing of the environment protocol, in the final act of the 11th Antarctic Treaty Special Consultative meeting in 1991, these ships placed more emphasis on missions associated with Protocol (Article 15) provisions for "prompt and effective response to such [environmental] emergencies which might arise in the performance of scientific research programmes, tourism and all other government and non-government activities ...".

The second generation of Antarctic ships may be characterized as multi-role, with a varying emphasis on southern ocean science / research capability in addition to resupply.

To explore this point we have plotted deadweight versus displacement in Figure 1 for the data-set of ships. Figure 1 categorizes ships into *resupply* or *research* primary missions where ships with a deadweight / displacement higher than the trend line slope of 0.38 are those was a *resupply* primary mission, whereas ships with a deadweight / displacement ratio lower than the trend line slope are categorised with a *research* primary mission.

There is a generally consistent ratio for the first generation fleet. This characterizes the cargo carrying nature of these ships. A number of second generation ships are below the trend line, and this indicates a greater emphasis on ocean science compared to cargo carrying. For second generation ships the balance between cargo carrying and research varies with the individual needs of the southern ocean operators with, for example, some national administrations operating two or more vessels, where the role of *resupply* and *research* is distributed between the two ships.

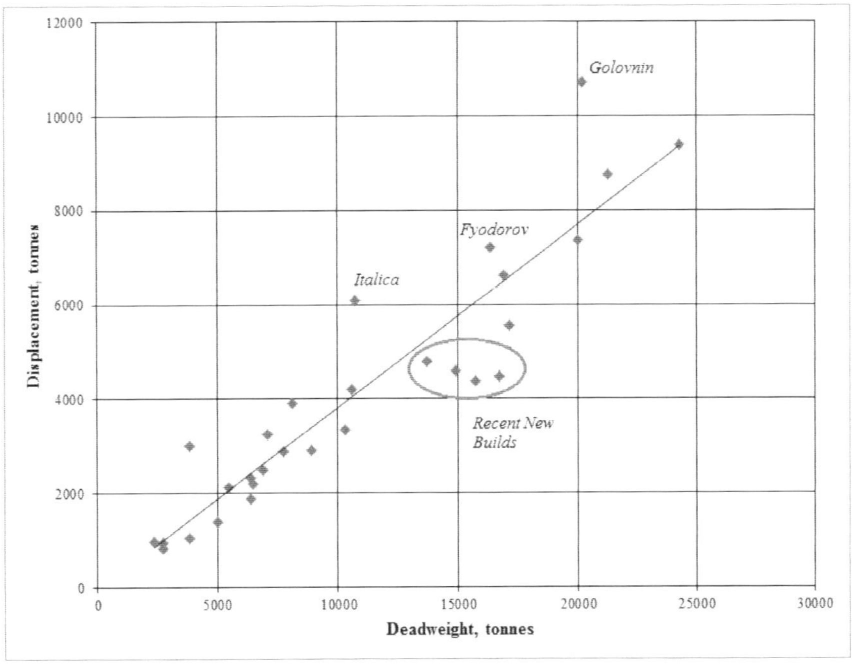

Figure 1. Deadweight vs. Displacement for Antarctic icebreaking fleet

Dimensions

Figure 2 presents a review of ship principal dimensions. Figure 2a shows the entire data-set of Antarctic ships. Linear trends can be seen for ratios of length to breadth, depth and draft albeit with significant data point scatter. Figure 2b presents the same data, but ships are categorised based on their primary mission of *research* or *resupply*. It can be seen that the gradient of the dimension ratios for the *resupply* ships are shallower than those for *research* ships, generally reflecting the increased open water performance characteristics (open water speed) for *research* ships, and the increased cargo carry capacity for *resupply* ships.

Hull Form

Dick & Laframboise discuss hull form, in particular bow form for icebreaking ships in detail. Rather than re-visiting this subject we have focused on the trends between the first and second generation ships and differences in hull form arising from the development of the mission profile for replacement vessels of the second generation associated with an increased emphasis on research missions.

From the data-set it can be seen that stem angle is lower for first generation *resupply* ships when compared with replacement ships of the second generation – which typically have a dual *resupply* and *research* role. A lower stem angle is characteristic of a bow more optimized for ice-performance than open water (*resupply* ships generally have had the need to break ice to carry out their resupply).

Other trends can be derived from two cases where first generation ships have been replaced by second generation ships with increased ice performance, similar stem angles, but with lower deadweight to displacement ratios (*S.A. Agulhas II* replaced *S.A. Agulhas*; *Admiral Tryioshni*kov replaces *Akademik Fyodorov*). This shows the trend that even for second generation ships with a primary *resupply* mission there is more of a consideration, or emphasis, also placed on the *research* role in open water.

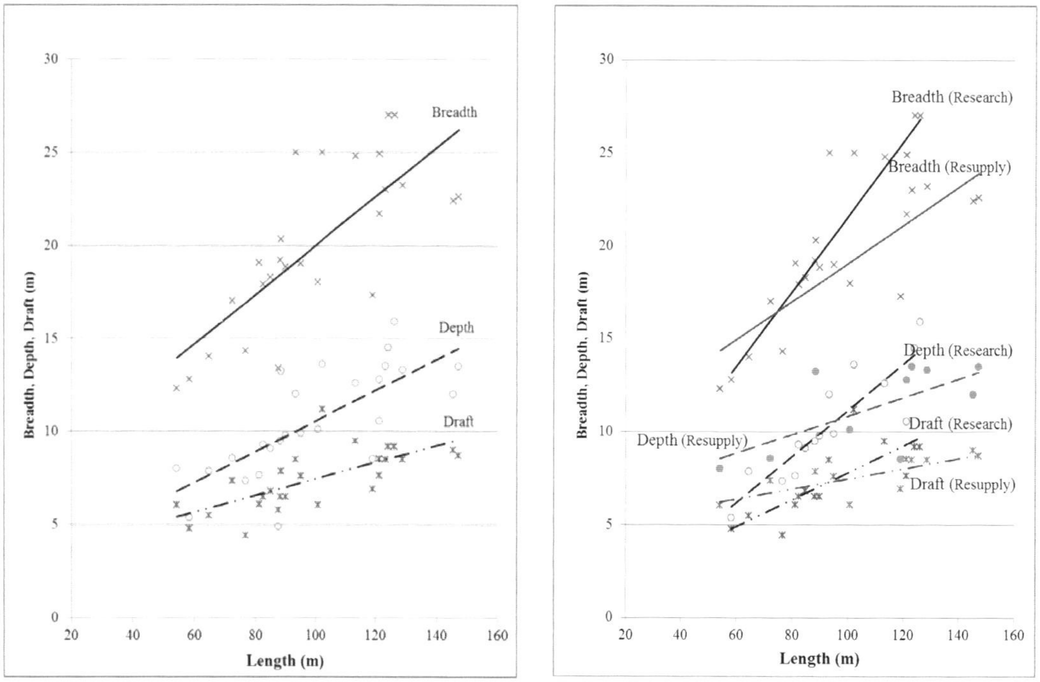

Figure 2. (a) & (b) Antarctic icebreaking ship dimensions

Propulsion

In comparison with the icebreaking fleet analyzed by Dick & Laframboise in 1989 there are no further significant highlights or trends to comment on with respect to propulsion type. We may account for this due to the less extreme mission profile of these Antarctic ships, none of which are dedicated icebreakers.

From the database the distribution between diesel-electric and diesel-mechanical propulsion has changed significantly between the two generations: just 40% of first generation ships are diesel-electric compared to 90% of the later generation ships. This change is probably a result of a number of factors, both from the reduction in cost of electric drives as well as the flexibility such propulsion configurations provide for multi-role ships. There are no significant trends in selection of open and nozzle propulsion units across either generations of ships or primary mission. A small trend in increased use of controllable pitch propellers (CPP) can be seen with the second generation of ships. Flexibility of propulsion that a CPP provides can also be achieved through the combination of fixed pitch propeller and electric propulsion. As such it is suggested that the recent trend to retain CPP even with electric propulsion is related to the requirement to reduce underwater radiated noise for research / science purposes.

Power and Performance

In the data-set principal dimensions and installed power were available for all cases. To determine the bollard pull for all ships the following relationship was used, utilizing installed power and propeller diameter, see Riska (2014):

$$T_B = Ke\,(P_D\,D_P)^{2/3} \qquad [kN] \qquad (1)$$

Where P_D is the installed shaft power and D_P is the propeller diameter. The propulsor efficiency factor, Ke, was determined from Riska (2014), with corrections for propulsion type. These are consistent with the approach taken to the formulations for the powering requirement in the Finnish Swedish ice class rules (2010). Where available the bollard pull values from Riska

(2013), or public domain data sheets, were used for confirmation.

For consistency of presentation we have also normalized the quoted ice performance figures to an ice thickness at a speed of 2 knots. Our approach has been to utilize the ice resistance method presented by Lindquist (1989) to estimate required resistance at the quoted speed and ice thickness and to equate this to the actual thrust available at 2 knots. For the occasional case where ice performance has not been found in the literature, an estimate has been made using the bollard pull and the Lindquist method in combination with an idealized second order polynomial net thrust curve as described in Riska (2014).

While we believe that the values in the data-set for bollard pull and ice-going performance are reliable and are of satisfactory accuracy for determining trends, we would caution against the use of these derived values outside of this context.

Figures 3 and 4 present an evaluation of installed propulsion power and ship performance for the data-set of Antarctic icebreaking ships. The general shape of the trend curve is consistent with that presented by Dick & Laframboise for the wider icebreaking fleet i.e. that propulsion power has a strong dependency on breadth. A small difference is presented between primarily *research* and *resupply* ships and this reflects the trade-off between design points for open water and operation in ice. Figure 4 presents a plot of bollard pull divided by breadth against the quoted ice performance given as a level ice thickness normalized to 2 knot ahead speed. This presentation is consistent with Dick & Laframboise but we have plotted a trend line as opposed to a line of minimum performance. If such a (minimum performance) line would be plotted in this case it would be significantly steeper when compared with the performance line of icebreakers presented by Dick & Laframboise. This steepness indicates that the majority of Antarctic vessels are not "pure" icebreakers and their ice performance is only part of their mission profile (and then to a varying degree in each case).

Capacity and Weight

The trend lines presented in Dick & Laframboise do not, understandably, present a study of weight or capacity: Again being more concerned with ships which primarily, and often solely, perform only icebreaking functions cargo carrying capacity is of little importance. This is not the case for Antarctic ships and consequently we have presented curves of gross tonnage against lightweight for the ships in the data-set. Gross Tonnage (GT) has been used as a value to describe internal volume and has the advantage of being a readily accessible figure for nearly all vessels in the data-set and a straightforward value to develop for a new design.

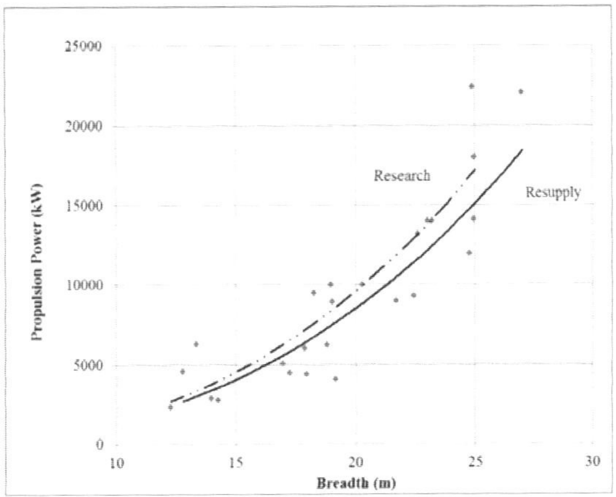

Figure 3. Propulsion power vs. breadth

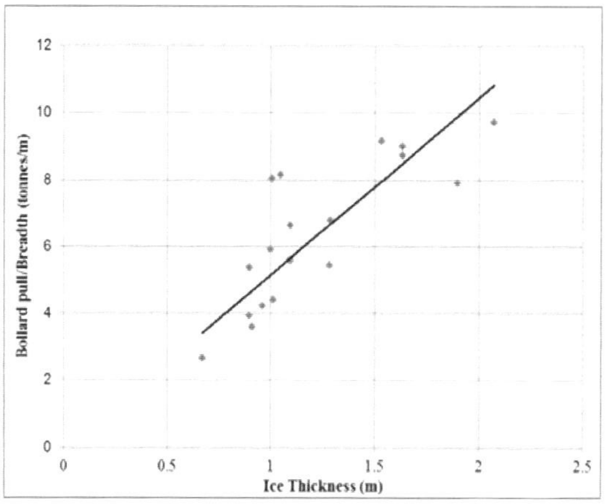

Figure 4. Ice performance

Generally this analysis aligns with Watson's (2002) suggested approaches for estimating steel weight using the old Lloyd's Register Equipment Numeral, which is a function of L, B, T, D and size of superstructure. Lightweight has been plotted, instead of steel weight, due to lack of publically available steel weight data for the majority of ships in the data-set. Figure 5a presents this data. To investigate the effect of ice class on the lightweight, the ships in the database were divided into three strength categories, based on ice class as follows:

- Heavy – ships with an ice class comparable to PC3 and above
- Medium – ships with an ice class comparable to ice class PC5 – PC3
- Light – ships with an ice class below those of "medium"

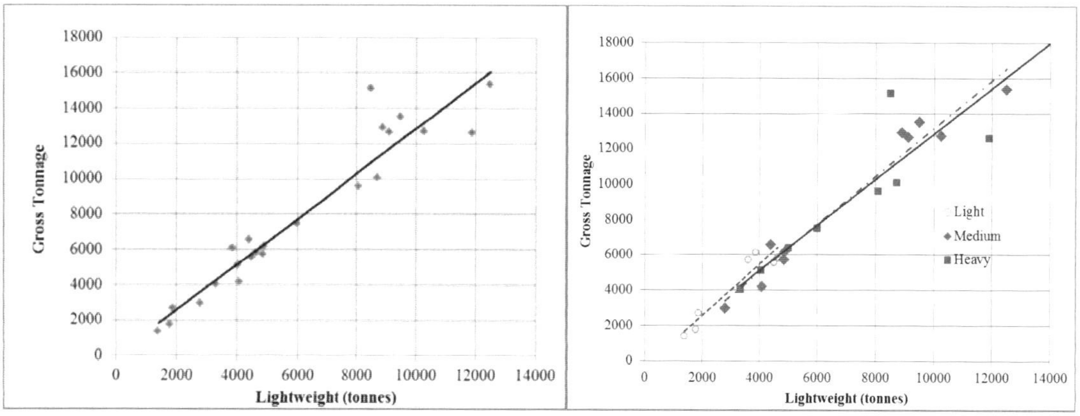

Figure 5. (a) & (b) Antarctic fleet gross tonnage vs. lightweight

The majority of the ships in the data-set were not built to the IACS Polar Class Rules (IACS, 2016). In order to establish a Polar Class for evaluating trends for existing ships built to other ice strengthening standards a comparison of the existing ship bow and midbody structural capacity has been made with that of the capacity of the same structure dimensioned to Polar Class rule requirements. Where Lloyd's Register has had access to the existing ship (or in some cases sister ship) structural plans this has formed the basis of the assessment. Where this has not been available, a standard structural configuration and assigned ice class rule minimum scantlings have been assumed for the comparison.

Trend lines for these ice class (heavy, medium, light) subsets of data are plotted in Figure 5b.

Note that there is not a significant change in slope between the strength categories which in this case is probably due to a reduction in deadweight to accommodate increased steel weight between the categories. Further investigation is warranted here but this relies to access to steel weight data for the ships which is not yet available to be presented in this paper.

PROCUREMENT TRENDS FROM RECENT PROJECTS TO BUILD ANTARCTIC ICEBREAKING SHIPS

Each recent procurement of a second generation Antarctic icebreaking ship has had unique and be-spoke characteristics defined by:

- **Objective for procurement for the new ship:** usually associated with an enhanced mission for supply and research activities to be carried out by a second generation ship, such as, added capability in scientific research aspects and/or enhanced cargo carrying capacity for re-supply missions
- **Procurement processes applied for the new ship:** usually associated with government policies, noting that all projects are for service for an Antarctic Treaty state, including policies for local content as well as requirements for competitive tendering for provision of services by third party providers and suppliers

Three recent examples of procurement objectives for recently contracted Antarctic icebreaking ships are:

Procurement objective – United Kingdom New Polar Research Vessel (NPRV)

- Replacement of two ageing ships with a single, larger, ship with updated, and enhanced, scientific research facilities and sufficient cargo carrying capacity to substitute for two replaced ships

This objective formed part of a science case, NERC (2014), in a public consultation process on the procurement.

Procurement objective – Australian Supply and Research Vessel (ASRV)

- Replacement of a single aged ship with a larger ship with higher cargo carrying capacity and improved operational availability

Procurement objective – Polar Research Institute of China (PRIC) Icebreaking science vessel (IBSV)

- Additional ship with improved ice-going and scientific research capability to support an expected enhanced presence – with additional Antarctic stations and personnel – of the Chinese state in Antarctica

Table 1 presents a summary of characteristics for these second generation ships and compares with existing first generation ships.

PROCUREMENT PROCESSES FOR RECENT ANTARCTIC ICEBREAKING SHIPS

Lloyd's Register, in a consulting capacity, typically focusing on Rules, Regulations and Certification aspects, have been involved in a number of the recent procurement projects for Antarctic icebreaking ships. Each procurement process for these recent projects has been involved three basic phases;

1. **An initial phase involving 3rd parties** – either a concept design and outline specification preparation by third party designers, or alternatively a definition or development of a

statement of requirements by a third party consultant. At this initial phase the third parties involved have been contracted by the owning or chartering Government research organization responsible for the deployment of the ships to the Antarctic.

2. **A validation and tendering phase** – involving science and research body consultations, including, in some cases, public consultations on initial phase work. Following on from this consultation phase a competitive tendering process is launched for bidding parties to be short-listed to:

   a) Design the Antarctic icebreaking ship which typically involves preparation of a basic design package where in-country build is envisaged, as part of government local content policies, with a specialist designer contracted to prepare a design package for bidding in-country shipyards to bid for the build; OR

   b) Design and construct the Antarctic icebreaking ship which may involve specialist designer and shipyard consortia to bid to meet tender qualification requirements; OR

   c) Design, construct, operate and maintain the Antarctic icebreaking ship which typically involves a consortia of a specialist designer, shipyard and owner/operator to bid to meet the tender qualification requirements

3. **A bid evaluation, short-listing and contract award phase** – involving bid evaluations for compliance with tender requirements, including verification of compliance with Government procurement policies. Following from these evaluations there is a short listing of bidders followed by final negotiations and contract award.

Note that the second phase may involve multiple tenders depending upon the scope of work forming part of each tender. Notwithstanding the procurement practices and policies of the countries and organizations involved it should be noted each recent project has adopted the same basic phases.

INSIGHTS TO INFORM PROCUREMENT OF ANTARCTIC ICEBREAKING SHIPS IN FUTURE

Whilst we cannot offer specific examples from individual procurement projects of Antarctic icebreaking ships there are, however, valuable insights that can be generalized from these recent cases that may inform parties involved in future procurements of Antarctic icebreaking ships as follows:

Less detail in requirements definition may be appropriate - in the case of replacements for aged existing ships the requirements definition for a tender document for the second generation replacement ship has often been very much more detailed, extensive and prescribed than the contract technical specification for the first generation ship to be replaced. There may be a case, at least at the initial conceptual and tendering phases, to consider *less* definition for future procurements in order allow bidders to offer design solutions that may possibly represent an improvement to those that might be prescribed in more detailed requirements definitions.

More emphasis on feasibility studies to examine design trade-offs and compromises at very early stages prior to launch of procurement process - in all procurements there appears to have been insufficient emphasis on feasibility studies at very early concept and requirements definition phases and in some instances there was no consideration of funded feasibility studies prior to launching the procurement process.

Where conflicting requirements and standards have been specified without the consideration of any design trade-offs needed this has led, in some cases, to late stage design compromises which possibly could have been avoided with earlier feasibility studies. There may then be a strong justification, prior to launching a tendering process at the very initial conceptual design phase, to fund feasibility studies to consider design trade-offs that may emerge from specification of conflicting requirements and standards.

Pre-procurement funding of conceptual design studies to examine alternatives - Replacement second generation ships have often materialized as a larger sibling of an older ship to be replaced.

	Xue Long	*James Clark Ross*	*Shackleton*	*Aurora Australis*	*NPRV*	*ASRV*	*PRIC*
Antarctic Station (**ice performance case**)	Great Wall, **Zhongshan**	Signey, Rothera, **Halley**	Signey, Rothera, **Halley**	Casey, **Mawson, Davis**	Signey, Rothera, **Halley**	Casey, **Mawson, Davis**	Great Wall, **Zhongshan**
L / B / T (m)	147.2 / 22.6 / 8.7	90 / 18.85 / 6.5	72.4 / 17 / 8.55	88.4 / 20.3 / 7.85	121 / 24 / 7.5	144.2 / 25.6 / 9.2	117 / 22 / 8
Displacement (t)	21 250	7 700	5 455	8 130	16 700 abt	24 300 abt	15 000 abt
Ice performance @2 knots (est.) (m)	1.05m	0.9m	0.93m	1.29m	1.1m	1.79m	1.5m
Non-crew berths	128	31	50	116	60	116	49
Cargo	17 000m3 abt	1 500m3 general cargo 300t cargo AF	3 000m3 general cargo	1 790m3 hold+18 TEU 1000m3 DO	2 200m3 hold 660m3 cargo AF	6 500m3 hold 1 900m3 DO	40 TEU 750t DO
Deadweight (t)	10 225	2 900	1 800	3 893	4 475	ND	ND
Ship's FO	1 800 m3 (abt)	1 200t	1 380 m3	1 200m3	1 800 m3 (abt)	ND	ND
Propulsion Power (kW)	13 200	6 250	5 100	10 000	11 000	26 600	15 000
Propulsion type	Diesel Mech. 1xCPP in Nozzle	Diesel Electric 1XFPP	Diesel Mech. 1xCPP in Nozzle	Diesel Mech. 1xCPP	Diesel Electric 2xCPP	Diesel Mech +PTI 2xCPP in Nozzle	Diesel Electric 2xAzimuth
Station keeping	No	DP1	DP2	DP1	DP2	DP2	DP2
Positioning thrusters	None	2: retractable azimuth	1: retractable azimuth, 4: tunnel	2: retractable azimuths,1: tunnel	4: Gill thrusters	6: tunnel	2: tunnel
Endurance, days Range, nm	ND (No Data) 20,000	57 days 16,500 (abt.)	130 days 40,000	90 days 25,000	60 days 19,000	90 days 16,000	60 days 20,000
Stabilization / ice heeling systems	No / Yes	Intering type tank (passive / active)	Roll reduction tanks (semi passive)	Intering type tank (passive / active)	Intering type tank (passive / active)	Intering type tank (passive / active)	Roll reduction tanks (semi passive)
Science deck area (m^2)	NA	370+150	NA	310	650	ND	500
Science Lab area (m^2)	100 #	330	90	ND	ND	500	580
Labs (+ containers)	3 #	6 (+5)	2	8	7	4 (+24)	6 (+8)
Helideck	Yes + hanger	No	Yes	Yes + hanger	Yes + hanger	Yes	Yes+ hanger
Helideck location	Aft	NA	Aft	Aft	Fwd	Aft	Aft of midship
Tenders / Support craft	NA	NA	NA	10m tender	Cargo tender, workboat	2 x 35t barge, 4 x tender	workboat
Cranes (over 5t)	2x50t cargo cranes (double cranes)	20t / 10t cargo crane 20 t / 30t A frames	1x50t crane 10t telescopic crane	1x25t crane 1x7t crane	1x50t cargo crane 1x30t A frame	2x 55t crane 1x15t crane	1x50t cargo crane A frame & U frame
Winch number	2 #	2	None	6	10 abt	ND	11 abt
Underwater noise	Not Considered	Considered	Not Considered	Not Considered	ICES209 (part)	ICES209 (part)	Considered
Drop keel	No	No	No	No	No	Two	No
Moonpool	No	No	No	No	Yes	Yes	Yes

Table 1. Characteristics of recent Antarctic Icebreaking projects (retrofits marked #)

On the other hand the second generation replacements are incorporating added or enhanced roles as well as new technologies and technological improvements to existing equipment, including, as recent examples, adoption of azimuthing podded propulsions systems and deployment of autonomous vehicles for scientific research tasks.

There may then also be a justification, for adoption of new technologies and improvements to technology that have occurred since the build of a first generation parent, to fund pre-procurement conceptual design studies to identify whether alternative hull forms, layouts and arrangements could offer an improved design and performance in a second generation replacement.

CONCLUSIONS

Because of similar mission demands the fleet of specialist Antarctic icebreaking ships can be categorized, and studied, as a distinct sub-set of the global ice-classed fleet.

In the first part of this paper design trends have been examined. Second generation of Antarctic ships from this analysis may be characterized as multi-role, with more design emphasis on science / research capability in open water and less design emphasis on icebreaking performance for resupply operations.

In the second part of this paper procurement practices for recent projects have been described. Insights to inform future procurements offered include feasibility studies to support study of design trade-offs in multi-role ships, as well as pre-procurement conceptual design studies to address new technologies and alternative design concepts.

REFERENCES

Dick, R.A. & Laframboise, J.E., 1989. An Empirical Review of the Design and Performance of Icebreakers. *Marine Technology*, Vol. 26, No.2 pp.145-149

IACS, 2016 *Unified Requirements for Polar Class*. International Association of Classification Societies

Lindqvist, G., 1989. A straightforward method for calculations of ice resistance of ships. *Proceedings of the 10th International Conference on Port and Ocean Engineering under Arctic Conditions, Lulea, Sweden*, Vol.3 pp.722-736

NERC, 2014. *Science Case for a Replacement Polar Research Ship.* Natural Environment Research Council.

Riska, K., 2011. *The World Icebreaker, Ice Breaking Supply and Research Vessel Fleet.* Baltic Ice Management.

Riska, K., 2014. *Factors Influencing the Power Requirement in the Finnish-Swedish Ice Class Rules.* Winter Navigation Research Board Report No 67, Finnish & Swedish Board of Navigation.

Transport Safety Agency (TRAFI), 2010. *Finnish-Swedish Ice Class Rules 2010.* Finland: TRAFI

Watson, D.G.M., 2002. *Practical Ship Design*, Elsevier.

APPENDIX: DATA-SET OF ANTARCTIC ICEBREAKING SHIPS

Ship	**Admiral Irizar**	**ASRV**	**Aurora Australis**	**Admiral Max'ano**	**Almirante Oscar viel**	**Antarctica I**	**Xue Long**	**PRIC Icebreaker**	**L'Astrolabe**	**L'Astrolabe II**
Country	Argentina	Australia	Australia	Brazil	Chile	Chile	China	China	France	France
Year	1978	In build	1990	1974	1969	In build	1993	In build	1986	2017
Length (m)	113.4	144.2	88.4	87.5	81.15	95	147.2	117	58.32	68
Breadth (m)	24.8	25.6	20.3	13.39	19.05	21	22.6	22	12.8	16
Depth (m)	12.6	13.1	13.25	4.903	7.62	10.6	13.5	11.8	5.36	6.8
Draft, **T** (m)	9.5	9.2	7.85	5.77	6.09	7.2	8.7	8	4.782	5.3
Displacement, **D**, (m)	14900	24300	8130	0	6400	10310	21250	15080	2740	3840
Cb	0.54	0.7	0.56	0	0.66	0.7	0.71	0.72	0.75	0.65
Stem angle, **γ** (deg)	22	16.5	20	ND	53	0	25	20	37	ND
GT (t)	10065	20200	6574	2493	4179	ND	15352	ND	1753	3628
Deadweight, **DW** (t)	4600	9400	3910	1760	2320	3340	8760	ND	950	3000
Endurance (nm)	20160	16000	25000	ND	12000	25900	22000	ND	ND	10080
Compliment	135+45	32+117	24+116	ND	78	64+56	38+128	ND	12+50	60
Shaft **P**ower (kW)	11950	26600	10000	6270	8950	8000	13200	15000	4600	6400
Bollard **Pull** (t)	138	305	138	0	102	101	184	ND	ND	ND
Machinery	DE	DE	DM	DM	DE	DE	DM	DE Th	DM	DM
Propeller **No**	2	2	1	1	2	2	1	2	2	2
Propeller **Dia**meter (m)	4.3	5.65	5.11	ND	3.658	ND	5.7	ND	ND	ND
Propeller **T**ype	FPP	CP	CPP	ND	FPP	FPP	CPP	FPP	CP	CP
Nozzle	Open	Nozzle	Open	ND	Open	Open	Nozzle	Open	Open	Open
OW Speed, **Vow** (kt)	17.2	16	16.8	ND	15	15	17.9	ND	14.2	ND
Ice thick @2kts (m)	1.1	1.79	1.29	ND	0.9	1	1.05	1.5	ND	ND
Ice **Class**	ND	PC3	ASPPR 2/3	Strength'd	Class 1	PC5	ULA	PC3	1A Super	PC5
Polar Class, **PC**	PC3	PC3	PC4	PC7	PC6	PC5	PC5	PC3	PC6	PC5

ND – No Data; DE – Diesel Electric; Th – azimuth thruster; DM – Diesel Mechanical; FPP / CPP – fixed/controllable pitch propeller; In build – preliminary data only

Ship	Polarstern	Maria S Merian	Italica	Shirase I	Shirase II	Lance	Kronprins Haakon	BAP Carrasco	Akademik Fyodorov	Vasiliy Golovnin	Admiral Tryiosh'ov
Co	Germany	Germany	Italy	Japan	Japan	Norway	Norway	Peru	Russia	Russia	Russia
Yr	1982	2005	1981	1982	2009	1978	2017	2017	1987	1988	2012
L	102.2	88.2	119	124	126	54.19	90	84	128.61	145.38	123.24
B	25	19.2	17.3	27	27	12.3	21	18	23.2	22.4	23
D	13.6	9.5	8.54	14.5	15.9	8.01	10.5	9.2	13.31	12	13.5
T	11.2	6.5	6.914	9.2	9.2	6.031	7.8	5.95	8.5	9	8.5
D	15720	6380	10710	17140	20000	2370	10580	5000	16340	20200	16890
Cb	0.53	0.56	0.73	0.54	0.62	0.57	0.7	0.7	0.62	0.67	0.68
Y	25	39	ND	21	19	ND	ND	ND	26	ND	27
GT	12614	5573	5825	ND	ND	1380	10900	5716	12660	13514	12711
DW	4370	1890	6080	5540	7350	980	4200	1400	7200	10700	6630
E	19000	7500	ND	ND	ND	21000	15000	ND	ND	ND	15000
C	44+80	23+23	27	174+60	175+80	13+25	17+35	50+60	90+160	ND	59+80
P	14116	4100	4486	22065	22066	2354	11000	0	14000	9300	14000
BP	197	ND	ND	243	236	28	158	ND	137	ND	152
Mach	DM	DE Th	DM	DE	DE	DM	DE Th	DE Th	DE	DE	DE
PropN	2	2	1	3	2	1	2	2	1	1	2
Dia	4.2	ND	ND	4.9	5.2	2.75	ND	ND	5.1	ND	4.25
PropT	CPP	FPP	FPP	FPP	FPP	CP	FPP	FPP	FPP	FPP	FPP
N	Nozzle	Open	Open	Open	Open	Open	Nozzle	Open	Open	ND	Open
Vow	16	18	14	19.5	19.5	ND	20	ND	16	16	16
Ice	1.8	ND	ND	1.63	1.63	ND	1.29	ND	1	ND	1.1
Class	Arc3	1A	ND	ND	ND	1A	PC3	PC7	ULA	ULA	Arc7
PC	PC2	PC7	ND	ND	ND	PC7	PC3	PC7	PC5	PC5	PC4

Ship	Hesperides	SA Agulhas	SA Agulhas II	Araon	Oden	Ernest Shackleton	James Clark Ross	Endurance	NPRV	Nathaniel B Palmer	Lawrence M Gould
Co	Spain	South Africa	South Africa	South Korea	Sweden	UK	UK	UK	UK	USA	USA
Yr	1991	1977	2012	2009	1989	1995	1991	1990	In build	1992	1997
L	76.75	100.72	121.25	95	93.2	72.4	90	82.5	121	84.94	64.7
B	14.3	18	21.7	19	25	17	18.85	17.9	24	18.288	14.02
D	7.35	10.11	10.55	9.9	12	8.55	9.8	9.3	11	9.1	7.85
T	4.42	6.058	7.65	7.61	8.5	7.359	6.5	6.5	7.5	6.8	5.49
D	2720	7110	13690	8900	12930	5460	7770	6500	16730	6910	3840
Cb	0.54	0.63	0.66	0.63	0.63	0.61	0.68	0.66	0.74	0.63	0.75
Y	ND	35	22	33	21	30	36	30	15	28	35
GT	2682	6112	12897	7507	9605	4028	5732	5129	5000	6174	2966
DW	830	3250	4780	2900	4910	2130	2890	2200	4480	2500	1040
E	12000	15000	15000	20000	ND	40000	16400	ND	19000	15000	12000
C	55+37	44+94	45+100	25+65	22+44	22+50	28+52	35+95	28+62	22+39	16+37
P	2800	4414	9000	10000	18000	5100	6250	6000	11000	9500	2886
BP	ND	48	118	107	243	72	74	79	120	147	50
Mach	DE	DM	DE	DE Th	DM	DM	DE	DE	DE	DM	DM
PropN	1	1	2	2	2	1	1	1	2	2	2
Dia	ND	3.35	4.5	3.5	4.5	3.6	4	3.5	4.5	4	2.65
PropT	FPP	CPP	CPP	FPP	CPP	CPP	FPP	CPP	CP	CPP	CPP
N	Open	Open	Open	Open	Nozzle	Nozzle	Open	Nozzle	Open	Nozzle	Nozzle
Vow	15	14	16	16.65	17	14	15.5	15	14.5	15	12
Ice	ND	0.675	1.285	1.1	2.07	0.963	0.9	1.015	1.1	1.01	0.914
Class	ND	Class 1*	PC5 / Ice10	Polar10	IB Polar 20	Ice05	1AS	Ice05	PC4/PC5	A2	A1
PC	ND	PC6	PC4	PC3	PC2	PC3	PC5	PC3	PC4	PC4	PC5

POAC'17
Busan, KOREA

Proceedings of the 24th International Conference on
Port and Ocean Engineering under Arctic Conditions
June 11-16, 2017, Busan, Korea

Evaluation of the Limit Ice Thickness for the Hull of Various Finnish-Swedish Ice Class Vessels Navigating in the Russian Arctic

Pentti Kujala[1], Mihkel Kõrgesaar[1], Jorma Kämäräinen[2]
[1] Aalto University, Department of Mechanical Engineering, Espoo, Finland
[2] Finnish Transport Safety Agency, Helsinki, Finland

ABSTRACT

Selection of suitable ice class for ships operation is an important but not simple task. The increased exploitation of the Polar waters, both seasonal periods and geographical areas, as well as the introduction of new international design standards such as Polar Code, reduces the relevancy of using existing experience as basis for the selection, and new methods and knowledge have to be developed. This paper will analyse what can be the limiting ice thickness for ships navigating in the Russian Arctic and designed according to the Finnish-Swedish ice class rules. The permanent deformations of ice-strengthened shell structures for various ice classes is determined using MT Uikku as the typical size of a vessel navigating in ice. The ice load in various conditions is determined using the ARCDEV data from the winter 1998 as the basic database. By comparing the measured load in various ice conditions with the serviceability limit state of the structures, the limiting ice thickness for various ice classes is determined. The database for maximum loads includes 3-weeks ice load measurements during April 1998 on the Kara Sea mainly by icebreaker assistance. Gumbel 1 distribution is fitted on the measured 20 min maximum values and the data is divided into various classes using ship speed, ice thickness and ice concentration as the main parameters. Results encouragingly show that present designs are safer than assumed in the Polar Code suggesting that assisted operation in Arctic conditions is feasible in rougher conditions than indicated in the Polar Code.

KEY WORDS: Loads; Serviceability; Limit ice thickness; Polar Code

INTRODUCTION

IMO has adopted the International Code for Ships Operating in Polar Waters (Polar Code) and related amendments to make it mandatory under both the International Convention for the Safety of Life at Sea (SOLAS) and the International Convention for the Prevention of Pollution from Ships (MARPOL). One of the aspects of Polar Code addresses the operational limitations of ships of different categories (A, B and C) according to ice conditions. The approach for evaluating the ice conditions and setting limitations for ships assigned an ice class is called POLARIS – Polar Operational Limit Assessment Risk Indexing System, details of which are given in IMO amendment document (MSC94, 2014). Therein the ice classes are associated with the limiting thickness by combining the experience from three existing approaches used in ice-covered waters: the Canadian Arctic, Baltic (Finnish/Swedish), and Russian Northern Sea Route systems. The assessment given in MSC94 (2014) categorizes ships designed according to Finnish Swedish Ice Class Rules (FSICR) for different operational conditions according to Table 1; assisted operation corresponds to scenario where icebreaker assistance is provided or the ice concentration is less than 100%.

Similarly, the objective of this work is to provide a limiting ice thickness for assisted operation, but through systematic evaluation of structural response to prescribed loads, which are later compared with actual measured loads. This will be achieved using numerical finite element simulations whereby permanent deformations of structures are determined along with the corresponding load level. Analysis are performed for three ice classes according to FSICR notation: IA Super, IA and IB. The permanent deformations comply with DNV serviceability limit state of $s/12$ used by surveyors, Leppik et al. (2010). For the selected structure, long term measured data in different ice thicknesses is available meaning that we can associate the ice thickness with resulting permanent deformation. The latter association lends itself to explicit definition of limiting ice thickness for safe operation.

Table 1. Safe operation of ships in first-year winter ice regime for the Finnish-Swedish ice classes, MSC94 (2014).

	Ice class	WMO description of the ice regime	Thickness of ice floes, h_i
Assisted	IA Super	Medium first-year ice	h_i up to about 100 cm
	IA	Medium first-year ice	h_i up to about 80 cm
	IB	Thin first-year ice	h_i up to about 60 cm
	IC	Thin first-year ice	h_i up to about 40 cm

CASE STUDY

As a case study, we selected M/T Uikku. Ship is classified according to DNV as class +1 A Tanker for Oil corresponding to FSICR as ice class IA Super. Ship was built 1976 in Werft Nobiskrug Gmbh. Ship has a diesel electric propulsion system with four diesel generators. It is important to note that Uikku, in contrast to normal tanker, has a bow shape especially designed for operation in ice. The ship hull and propulsion system was instrumented on 1997 for the EU funded ARCDEV project and the instrumentation was extensive, detail description of the instrumentation can be found in Kotisalo and Kujala (1999).

Description of Full Scale Measurements

The ice load is measured on-board tanker M/T Uikku during one voyage to Ob-estuary in the Russian Arctic waters, from April to May in 1998. The voyage started from the port of Murmansk on 26th of April 1998, where the weather and ice conditions were harsh for the representative area. During the voyage M/T Uikku was always either in convoy or lead by an icebreaker. The route of the convoy is presented in Figure 2 (Kotisalo and Kujala, 1999). 20-minute maximum loads on-board M/T Uikku were measured in the bow, bow-shoulder and stern combined with visual observations. But since visual observations were not conveniently measured with the same 20-minute time interval as the loads, they were later synchronized. As the voyage was quite short only about 500 measured 20-minute loads are available for the analysis. Based on the 500 measurements the time in those ice conditions per year was calculated to be about 7 days.

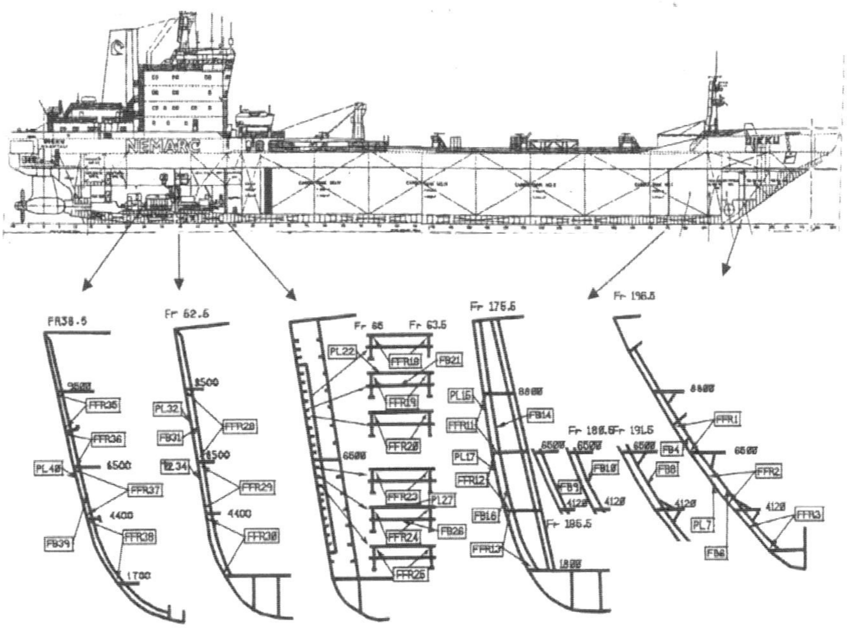

Figure 1. Picture of Uikku showing the instrumented areas (Kotisalo and Kujala, 1999).

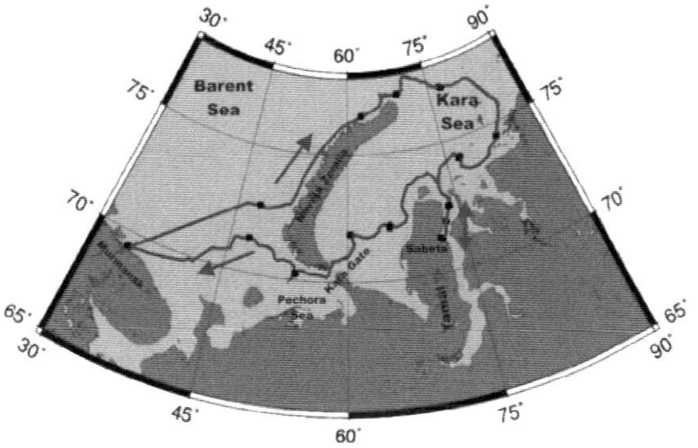

Figure 2. Route of the convoy during ARCDEV voyage.

Observed Ice Conditions

Ice conditions were observed continuously by visual observations on the ship and reported in 20 minute intervals. The ice concentration was observed in tenths during each observation period. The mean ice concentration during each observation period was calculated as a weighted average and the results are presented in Figure 3.

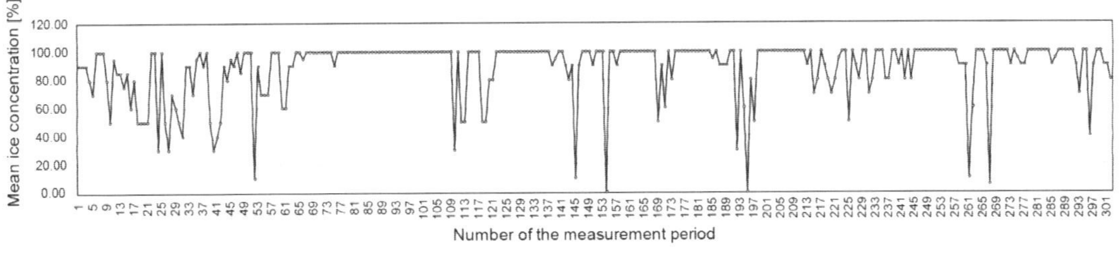

Figure 3. Mean ice concentration in each observation period on-board MT Uikku.

Ice thickness was estimated by observing the thickness of the turning ice floes at the bow of the ship visually. Observations of ice thickness have been reported on five classes: below 10 cm, 10- 30 cm, 30-70 cm, 70-120 cm and above 120 cm. Figure 4 shows the observed mean ice thickness for each period. Figure 5 shows the ship's speed during the voyage – due to the assisted operation the ship could keep a fairly high speed of 10-15 kn most of the time.

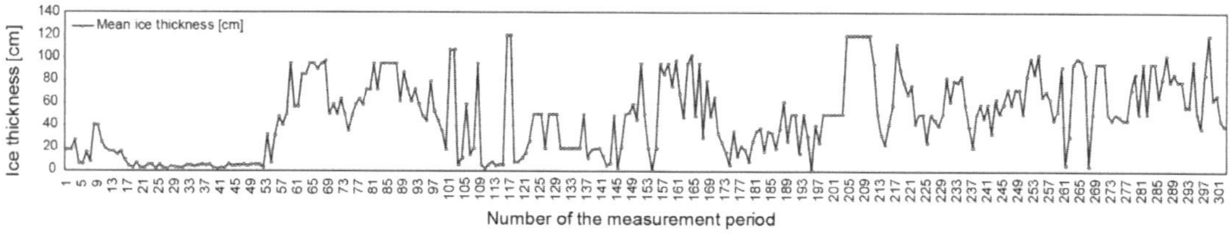

Figure 4. Mean ice thickness in each observation period on-board MT Uikku

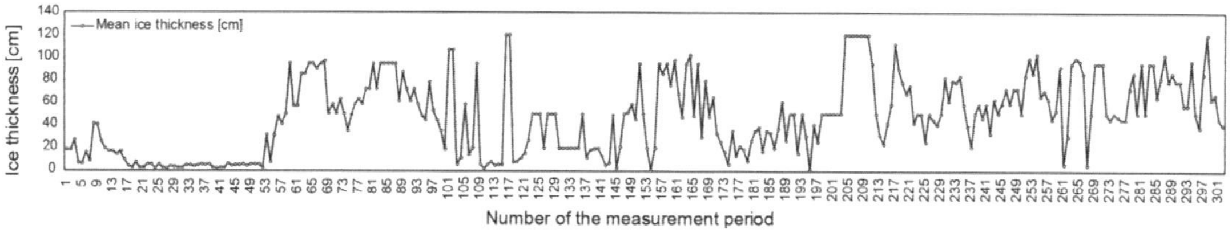

Figure 5. The speed of MT Uikku during each 20-minute period.

Return Period of Loads

The Gumbel I distribution is fitted to the measured ice loads as it has proven to give a good fit to the data, see Kujala (1994), as well as because the initial distributions of the ice loads having an exponentially decaying tail (Gumbel, 1958). The cumulative distribution function of Gumbel I distribution is as follows, Ang and Tang (1984):

$$G(y_n) = \alpha_n e^{-e^{-\alpha_n(y_n-u_n)}} \tag{1}$$

where Gumbel parameter α_n is the inverse measure of dispersion and u_n is the characteristic largest value. These are determined based on the mean μ and standard deviation σ of the measured loads as follows:

$$\alpha_n = \frac{\pi}{\sigma\sqrt{6}} \tag{2}$$

$$u_n = \mu - \frac{\gamma}{\alpha_n} \tag{3}$$

Where $\gamma = 0.577$ is the Euler constant. The measured 20-minute maximum loads at the bow and stern are used in this analysis. The data is dived to three classes using ice thickness as the criteria: smaller than 0.7 m, between 0.7-1.2 m, and between 1.2-2.0 m. Figure 6 summarise the measured load and fitted Gumbel 1 distribution on the data.

Figure 6 shows that the load on the bow shoulder area is almost twice as high as on the bow area. The main reason for this is that the vertical frame angle is much smaller around the frame 176.5 (bow shoulder) than on the frame 196.5 (bow), which will result into higher ice induced loads to break the ice. In FSICR these areas are not separated, therefore in the following analysis, the load measured at the bow shoulder area will be used to evaluate the maximum ice thickness for various ice classes for the bow area. Another reason for the high load at the bow shoulder is that this area encountered more ice hits than the bow itself.

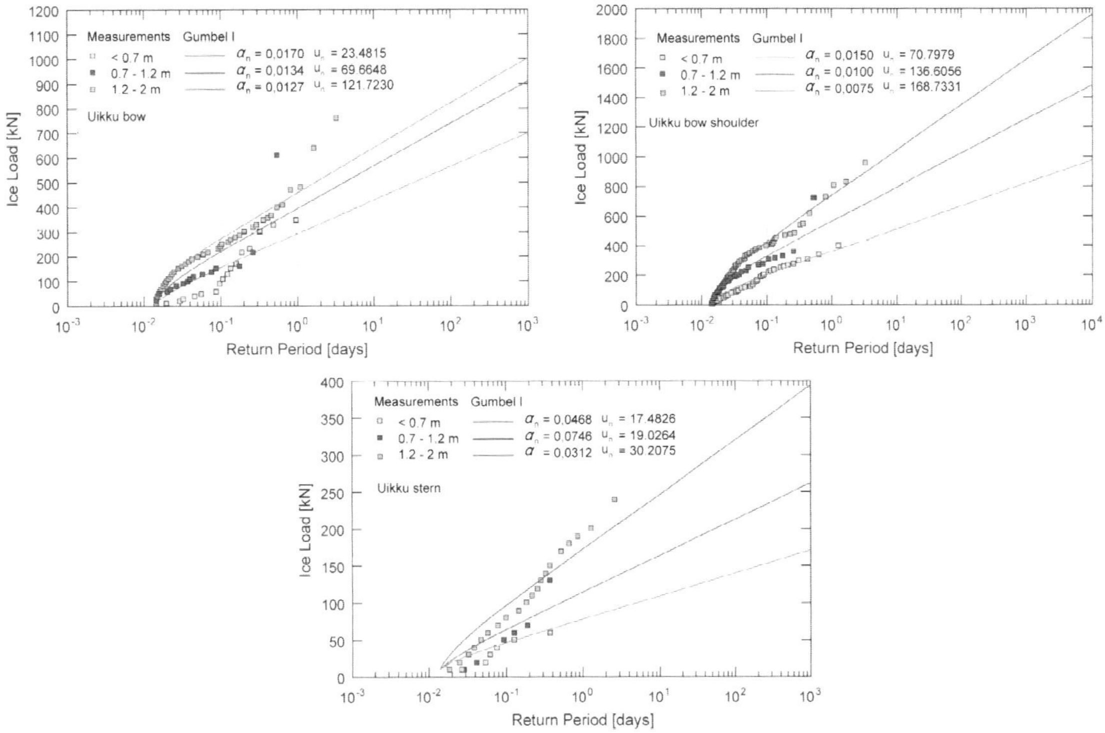

Figure 6. Measure 20-minute maxima with fitted Gumbel 1 distribution for MT Uikku from bow, bow shoulder and stern

Evaluation of Design Loads

The Gumbel fits provide the means to evaluate the extreme values with different probability levels. Therefore, using the Gumbel parameters α_n and u_n the most probable extreme value with the probability of α_s is given:

$$\widehat{y_n} = u - \alpha_n^{-1} \ln\left(-\ln\left(1 - \frac{\alpha_s}{n}\right)\right) \tag{4}$$

Using eq. (4) and the data given in Figure 6, the most probable design load values during one voyage, return period of one day, are given for two safety levels, namely $\alpha_s = 0.1$ and $\alpha_s = 1$. Return period of one day was chosen as a compromise between data availability in lower ice thicknesses, about 3 days, and higher ice thickness, typically less than 1 day. The safety factor of $\alpha_s = 1$ corresponds to one year extreme event which according to FSICR is yield. The factor $\alpha_s = 0.1$ was chosen as it complies with the latest standard for design of Arctic offshore structures (ISO 19906:2010), which defines the three exposure categories and related return period for the probability of occurrence of these:

1. Ultimate limit state (ULS) for all exposure levels (EL's) requires extreme 100-year (10^{-2}) ice event with action factors dependent on the EL. ULS generally correspond to resistance to extreme applied actions.
2. Abnormal/accidental limit state (ALS) for with 1,000-year (10^{-3}) ice events allowing some structural damage, but robustness have to be achieved with no loss of life or harm

to the environment.

3. Serviceability (SLS) ensuring functionality under any 10-year (10^{-1}) ice event. Exceedance of SLS results in the loss of capability of a structure to perform adequately under normal use.

Ensuing probable design load values rising from the analysis of measured data must be associated with the structural response to provide the limiting ice thickness for safe operation. Therefore, structural response is determined in the following Section.

ANALYSIS OF STRUCTURAL RESPONSE

Scantlings

In order to assess the operational restrictions for different ice classes, Uikku's structural scantlings were re-calculated according to FSICR for three ice classes (IA Super, IA, and IB) and two hull regions, bow and stern, see Table 2. In calculations, the original built scantlings for the frame spacing, span and webframe spacing were used. In the bow, frame span is 2 m, 2.92 m in the bow-shoulder and 1.22 m in the stern; the frame spacing is 0.35 m for both regions. Webframe spacing was 2.8 m and 2.1 m, respectively, for bow and stern. In calculations, the minimum required engine power was calculated for IA and IB design according to FSICR requirements, whereas for IA Super, the actual power of 11400 kW was used. All the stiffening members were assumed as T-shaped due to the modelling considerations explained below. The scantling calculations were automatized in the MATLAB routine.

Table 2. Uikku scantlings re-calculated for three ice classes, IA Super, IA and IB.

Bow	Plating	Frames [mm] (spacing=0.35 m)				stringer [mm] (spacing=2 m)				webframe [mm] (spacing=2.8 m)				Power
	t [mm]	h_web	t_web	w_flange	t_flange	h_web	t_web	w_flange	t_flange	h_web	t_web	w_flange	t_flange	P [kW]
IAS	21.5	260	9.5	80	12.5	660	18	140	20	900	15	80	17	11400
IA	20	260	9	70	11.5	580	16	145	21	700	15	60	15	6614
IB	19	240	9	70	11.5	540	17	120	18	650	15	80	15	5300

h_web web height [mm]
t_web web thickness [mm]
w_flange flange width [mm]
t_flange flange thickness [mm]

Stern	Plating	Frames [mm] (spacing=0.35 m)				stringer [mm] (spacing=1.22 m)				webframe [mm] (spacing=2.1 m)				Power
	t [mm]	h_web	t_web	w_flange	t_flange	h_web	t_web	w_flange	t_flange	h_web	t_web	w_flange	t_flange	P [kW]
IAS	14	120	9	40	11.5	450	8	60	14	500	10	80	16	11400
IA	11.5	100	9	50	10	340	8	80	14	380	10	80	16	6614
IB	10	80	9	40	10	290	10	40	10	290	10	40	10	5300

Finite Element Modelling

All FE simulations were performed with FE software Abaqus/Standard 6.13-3. Modelled structures were discretized with reduced integration shell (S4R) elements with 5 integration points through thickness and stiffness based hourglass control. They possess six degrees of freedom on each node. Simulations were set-up using implicit analysis procedure (ABAQUS/Standard) using Static step, number of increments of 500 and minimum time increment of 1e-12; simulation time was 2 seconds. FE model generation based on input dimensions in Table 2 was automated with Abaqus python module.

Material true stress-strain relation is highly non-linear in the plastic region, but can be approximated as linear in the practical ranges of structural deformation of interest here as shown in Kõrgesaar et al. (2017). Therefore, a bi-linear elastic perfectly plastic material was defined with the yield stress of 285 MPa.

Analyses are performed with a large 5-bay grillage structure shown in Figure 7. To simplify the parametric modelling no brackets were modelled and structure was assumed perpendicular, i.e., without curvatures in the side shell. The reasoning behind the large model size was to reduce the effect of boundary conditions on the response. The edges of the model, including the cross-sections of the framing members as shown in Figure 7, are restrained against translations. This is less conservative than fixing all degrees of freedom as stiffness is reduced. Moreover, the uncertainty related with boundary conditions is somewhat relieved by the large chosen model size.

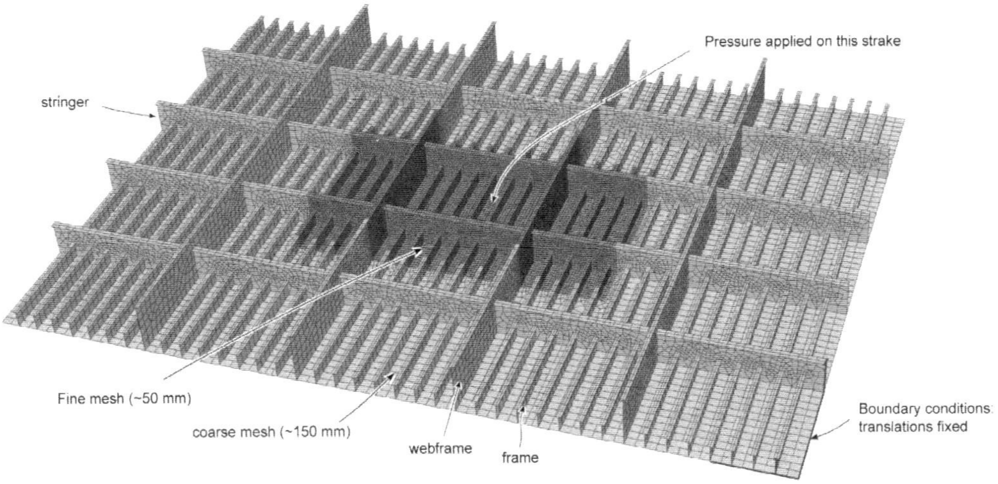

Figure 7. 5-bay grillage model used in the analysis. Scantlings correspond to IA Super bow region.

Load Definition

The definition of the ice load in combination with the corresponding strength criterion is the most important part of the hull rules. The load level and its distribution defines the structural response, which is basis for evaluation of the design point. The ice load in current rules is characterized with uniform pressure distributed over a patch with height depending on the ice class and length depending on the structural component to be dimensioned. In direct analysis using finite element method, the patch length is equal to the webframe spacing. However, a recent study by Kõrgesaar et al. (2017) showed that narrower patch as well as non-uniform distribution of pressure can inflict more damage to the structure. Important perspective considering that recent findings indicate that loads have to be narrow enough (1-4 frame spacings) in order for the maximum load on a frame to occur, Suominen et al. (2017).

Therefore, in this study following non-uniform pressure distribution is used for bow region:

$$p = \begin{cases} 4.35(7.175 - x)^{-0.733} \; if \; x < 6.285 \\ 9.4 \quad if \; 6.825 \leq x \leq 7.175 \\ 4.35(x - 6.825)^{-0.733} \; if \; x > 7.175 \end{cases} \tag{5}$$

where p is pressure in MPa and x is structural coordinate. Pressure on stern model is defined through similar equation:

$$p = \begin{cases} 2.9(5.775 - x)^{-0.733} \; if \; x < 5.075 \\ 6.27 \quad if \; 5.075 \leq x \leq 5.425 \\ 2.9(x - 5.075)^{-0.733} \; if \; x > 5.425 \end{cases} \tag{6}$$

The resulting pressure profiles are shown in Figure 8**오류! 참조 원본을 찾을 수 없습니다.** (a) and they are applied on a large grillage model between two middle webframes and stringers, i.e. on the centre strake as shown in **오류! 참조 원본을 찾을 수 없습니다.** (c). The load is incrementally increased to the maximum value in the loading phase, and then gradually decreased in the unloading phase as shown in Figure 8**오류! 참조 원본을 찾을 수 없습니다.** (b). Similar procedure is advocated by the ABS (2014) guidance notes on ice class. Because of unloading, we could determine the permanent plastic deformation in the structure associated with serviceability limit state. Similar procedure is advocated by the ABS (2014) guidance notes on ice class.

However, to reach the permanent displacement equal to s/12, the maximum pressure had to be chosen large enough so that sufficient permanent deformation would develop, although the load causing s/12 deformation could not be defined *a priori*. Therefore, the unloading portion of the curve is shifted during the post-processing so that the permanent set would be equal to the reference permanent displacement as shown in Figure 9. The underlying assumption is that slope of the unloading portion of the curve remains the same due to the permanent nature of plastic strains, whereas this assumption was validated with simulations with lower peak load in Kõrgesaar et al. (2017) and shown also in Figure 9.

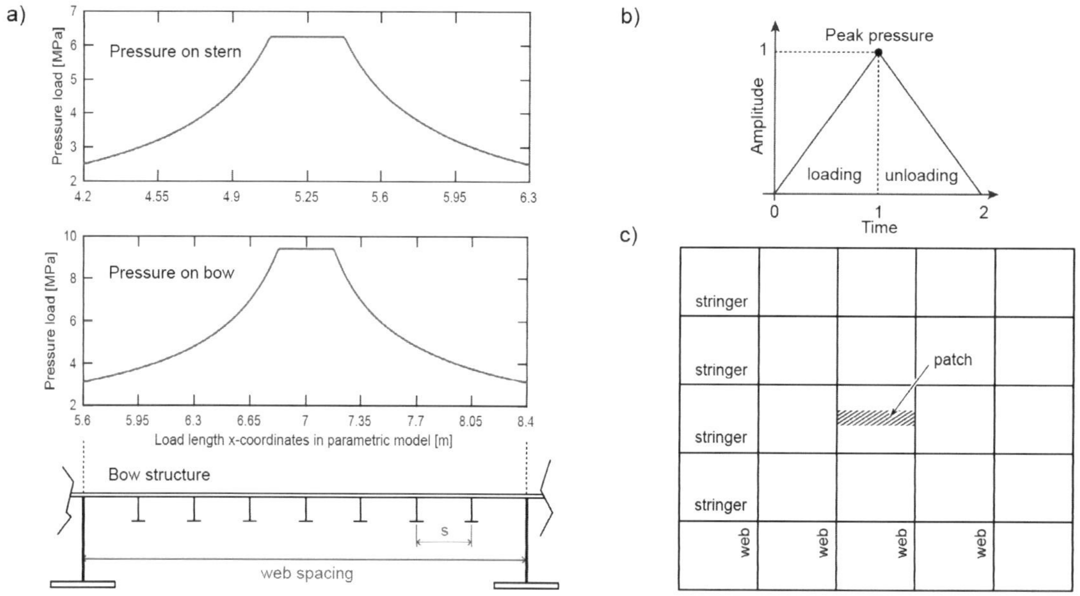

Figure 8. a) Pressure distribution applied on structures showing the peak pressure value. b) The load amplitude with loading and unloading cycle. c) Pressure patch location on large grillage.

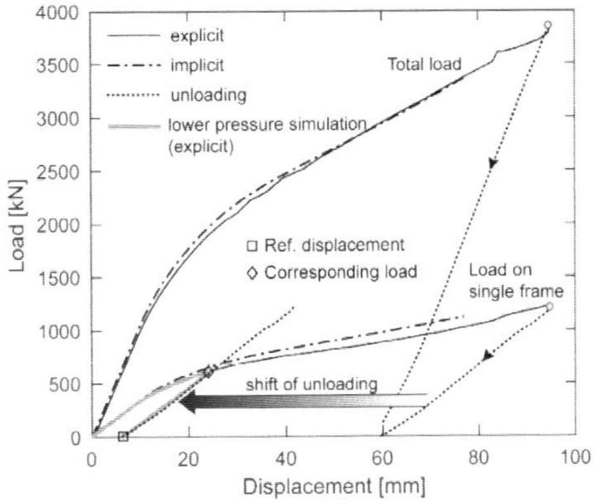

Figure 9. Shifting of the unloading cycle to obtain permanent deformations. Total load corresponds to resultant force in the boundaries. Load on frame is obtained by integrating the pressure over the single frame spacing. Figure from Kõrgesaar et al. (2017).

SIMULATION RESULTS

The analysis results in Figure 10 are presented in the form of load on single frame plotted as a function of displacement. The load on single frame was preferred over a total load as a comparative measure because measurement data also corresponds to single frames facilitating the comparison in the sequel. Load on single frame, that is for the one frame spacing, was found by integrating the applied pressure over the corresponding area. Displacements were measured in the plate field at the frame location.

Stiffness reduction of the frames in Figure 10 correlates with the decrease in ice class as well as hull region – this is consistent with the current design approach. On the curves, three distinctive points are marked: load to permanent deformation *s*/12, design load according to FSICR and three hinge pressure load (Daley, 2002). The obtained results are gathered in Table 3.

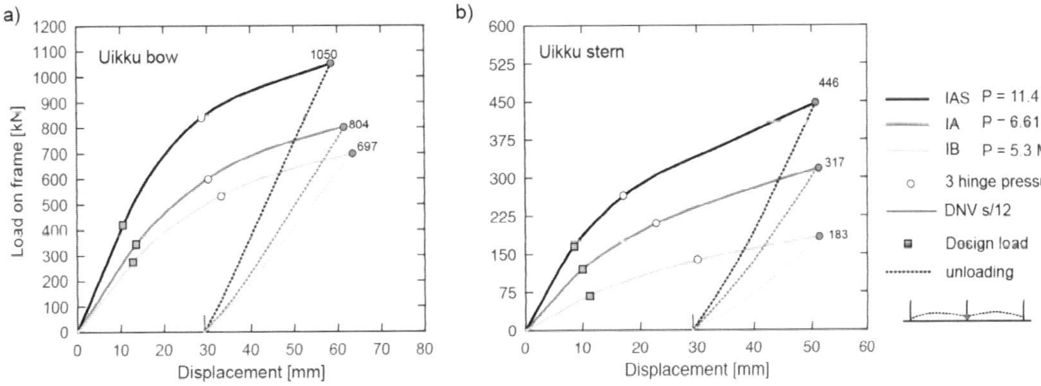

Figure 10. Load-displacement relation for Uikku: a) bow region and b) stern region.

The most captivating results in Table 3 are the safety factors against the yield load and permanent deformation load – for clarity, these are plotted on Figure 11. The closer the safety factor is to one the more successfully the structure follows the design intention of FSICR that is reaching the state of yielding when subjected to design load. Both hull regions follow the design intention surprisingly well, while the stern region of IB class design might be slightly over-dimensioned with safety factor of 1.3 against yield. Similar consistency is observed in safety factors against permanent deformation with the value of ~2.5 for all cases.

Table 3. Detailed analysis results for IA Super, IA and IB for two hull regions.

	BOW			STERN		
	IAS	IA	IB	IAS	IA	IB
Load for permanent deformation [kN]	1050	804	697	446	317	183
Yield limit [kN]	461	316	263	154	105	88
Patch area [m2]	0.123	0.105	0.088	0.123	0.105	0.088
Design pressure [MPa]	3.44	3.31	3.13	1.34	1.13	0.76
Design load [kN]	421	348	274	164	119	67
p3h load [kN]	839	602	532	265	210	137
Safety factor against s/12	2.5	2.3	2.5	2.7	2.7	2.8
Safety factor against yield	1.1	0.9	1.0	0.9	0.9	1.3

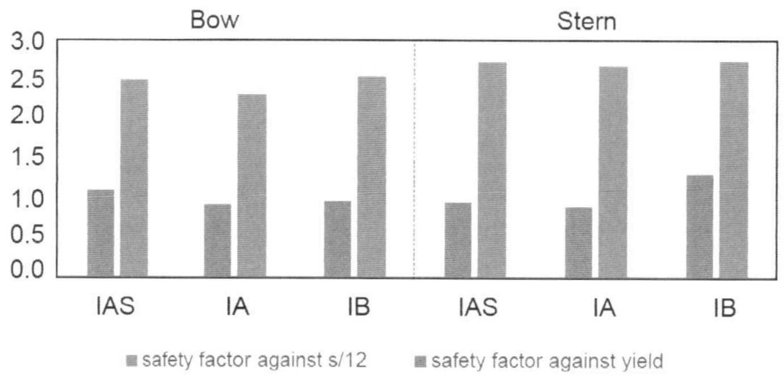

Figure 11. Safety factors against yield and permanent deformation of s/12.

LIMITING ICE THICKNESS

Probable design load values complying with serviceability limit state ($\alpha_S = 0.1$) are determined with eq. (4) and compared with SLS loads from structural analysis (Table 3) in Figure 12. The measured data is plotted with markers at 0.7, 1.2, and 2 m, while dashed lines show the SLS load obtained with FE calculations.

Figure 12 shows that bow shoulder loads are considerably higher than loads on bow. Since FSICR do not differentiate between bow regions, we use bow shoulder loads in comparison with loads obtained from FE analyses.

Therefore, IA Super design can operate in 2 m thick ice safely without sustaining damage equivalent to s/12, i.e. ~30 mm, when icebreaker assistance is available. IA design can operate in 1.2 m and IB in 1 m thick ice.

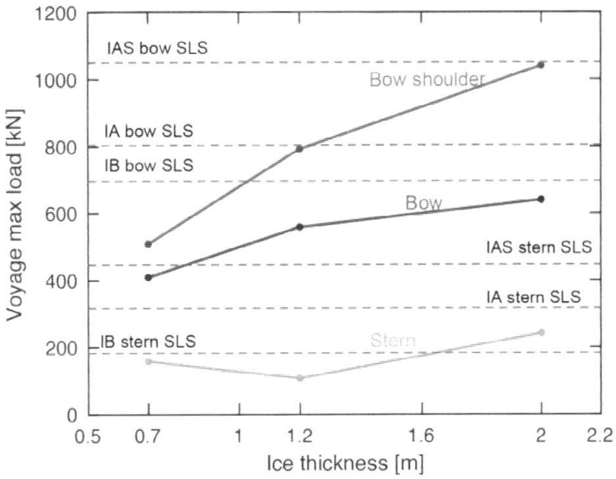

Figure 12. Comparison of measured load with SLS load. Measured data corresponds to safety level of $\alpha_s = 0.1$ (10 years extreme event).

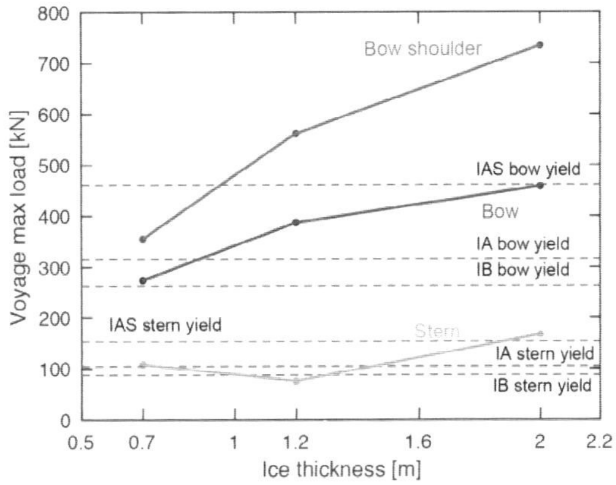

Figure 13. Comparison of measured load with yield load. Measured data corresponds to safety level of 1 (1 years extreme event).

In Figure 13 the curves corresponding to $\alpha_s = 1$ are plotted for reference to show the difference between 10 years and 1 year extreme event. Recall that FSICR design ships to yield once per winter, thus the Figure could be used as a gauge to estimate the compliance between design point and strength according to FSICR. The results are combined into Table 4 where the comparison is also presented with preliminary assessment made in POLARIS (MSC94, 2014).

In POLARIS system, the operational restrictions are based on the definitions given in the Ice Nomenclature of WMO. Comparison reveals that preliminary assessment is more conservative, especially for the highest ice class. Systematic evaluation whereby measured data is compared with structural response analysis shows that IA Super can safely operate with icebreaker assistance or when ice concentration is less than 100% in 2 m ice without sustaining permanent damage of 30 mm – corresponding safe ice thickness according to POLARIS is half of that, 1m. Preliminary assessment for IA and IB is less conservative with current analysis yielding 50% and 67% higher safe ice thickness, respectively, for IA and IB class. Interestingly, the POLARIS assessment is much more consistent with the ice thicknesses corresponding to yield rather than the serviceability criteria.

Table 4. Comparison between preliminary assessment from Polar Code and current results for safe operation of ships in first-year winter ice regime for the Finnish-Swedish ice classes, when icebreaker assistance is provided or the ice concentration is less than 100%.

POLARIS assessment (Polar Code)			Current assessment [cm]	
Ice class	WMO description of the ice regime	Thickness of ice floes, hi	SLS	Yield
IA Super	Medium first-year ice	hi up to about 100 cm	200 cm	100 cm
IA	Medium first-year ice	hi up to about 80 cm	120 cm	70 cm (some yielding)
IB	Thin first-year ice	hi up to about 60 cm	100 cm	50 cm (some yielding)
IC	Thin first-year ice	hi up to about 40 cm	-	-

CONCLUSIONS

Selection of suitable ice class for ships operation is an important, but not simple task. The increased exploitation of the Polar waters, both seasonal periods and geographical areas, as well as the introduction of new international design standards such as Polar Code, reduces the relevancy of using existing experience as basis for the selection, and new methods and knowledge must be developed.

This paper determined the limiting ice thickness for ships designed according to different ice classes of the Finnish-Swedish ice class navigating in the Russian Arctic. MT Uikku was used as the case study as the actual measurement data on-board ship was available. Thereby, the limiting ice thickness was determined by comparison of measured load in various ice conditions with the serviceability limit state of the structures established with FE simulations. Serviceability limit state corresponds to 10-year extreme event that was herein associated with a permanent deformation of $s/12$, i.e., 30 mm.

When serviceability is considered as a limiting condition for safe operation, results encouragingly show that present designs are safer than assumed in the Polar Code when operating under icebreaker assistance or when ice concentration is less than 100%. This extra safety depends on the ice class, with IA Super showing the largest safety margin. When conservatively associating yield with the safe operation, the limiting ice thicknesses provided by POLARIS correlates well with the present findings. Still, it is important to highlight the fact that Uikku has multifunctional icebreaking hull form which possibly decreases the measured loads. Nevertheless, results can be generalized to blunt hull forms as the largest measured loads are measured on the shoulder area, where the frame angle is not substantially different between multifunctional icebreaking hull and blunt hull.

Furthermore, there are two major, but opposing uncertainties in the results. First, pressure patch used in the analysis is a simplification and it excludes the development of line-like contact, which could have deteriorating effect on the capacity as shown in Ehlers et al. (2014), and Kujala and Kõrgesaar (2017). In other words, for the same nominal load, deformations would be higher, meaning that defined deformation limit would be reached earlier. Furthermore, the designed scantlings used in all ice classes were such that stable bending response was obtained during the loading, which implies large overload capacity. The overload capacity and permanent deformation limit could be reached earlier when more slender frames (but with the same section moduli) were adopted because of the consequent shift in deformation mode from bending the buckling (folding or tripping), Daley et al. (2017). Therefore, the effect of frame slenderness and how it affects permanent deformations should be further studied.

Second uncertainty is related to the definition of SLS criteria used in the analysis – permanent set of 30 mm. ISO standard for Arctic offshore structures (ISO 19906:2010) relates exceedance of SLS with the loss of capability of a structure to perform adequately under normal use. Yet, observations of structural response at 30 mm indicate that there is additional reserve capacity before "adequate performance" is lost. Therefore, s/12 of permanent set is believed to be conservative assumption for serviceability. In effect, we consider these two uncertainties to cancel each other rendering the results presented valid for the structures considered.

Conclusively, findings indicate that FSICR designs correlate well with the design point of yield, but large safety margin against permanent deformation limit indicates that further optimization of structures is possible.

ACKNOWLEDGEMENTS

The financial support of the Lloyd's Register Foundation, United Kingdom, is acknowledged with gratitude. The Lloyd's Register Foundation helps to protect life and property by supporting engineering-related education, public engagement and the application of research. The support from Finnish Transport Safety Agency (Trafi) is also acknowledged.

REFERENCES

Ang, A. H. S. & Tang, W. H., 1984. *Probability Concepts in Engineering Planning and Design Volume II*, New York: John Wiley & Sons.

Daley, C.G., 2002. Derivation of plastic framing requirements for polar ships. *Marine Structures*, 15, 543–559.

Daley, C.G., Daley, K.H., Dolny, J., Quinton, B.W.T., 2017. Overload response of flatbar frames to ice loads. Ships and Offshore Structures, 12, S68–S81.

Ehlers, S., Erceg, B., Jordaan, I., & Taylor, R., 2014. Structural analysis under ice loads for ships operating in arctic waters, in: Proceedings of the MARTECH2013.

Gumbel,E., 1958. *Statistics of Extremes*. New York: Columbia University Press.

ISO 19906:2010, Petroleum and natural gas industries - Arctic offshore structures. Section 7: Reliability and limit state design.

Kõrgesaar, M., Kujala, P., Suominen, M., Dastydar, G. S., Kämäräinen, J., 2017. Effect of pressure distribution on the capacity of ship structure frames. Proceedings of the 6th International Conference on Marine Structures, MARSTRUCT 2017.

Kotisalo, K., & Kujala, P. 1999. Ice load measurements onboard MT Uikku during the ARCDEV voyage. In: Proceedings of the 15th International Conference on Port and Ocean Engineering under Arctic Conditions, pp. 974-987., August 23-27, Helsinki, Finland.

Kujala, P. & Kõrgesaar, M., 2017. Application of the Polar Code for the Finnish Swedish Ice Classes, Aalto University publication series SCIENCE + TECHNOLOGY; 02/2017, Aalto University, Finland. *permanent web link will be provided with final submission*

Kujala, P., 1994. *On the Statistics of Ice Loads on Ship Hull in the Baltic.* Doctoral Thesis, Helsinki University of Technology, Espoo.

Lepik, K., Peetsalu, J., Viljakainen, S., & Voog, V., 2010. Allowable plate deflections according to DNV, Internal instructions for surveyors.

MSC 94, 2014. IMO Maritime Safety Committee/INF.13, Technical Backgorund to POLARIS.

Canadian Arctic Shipping Risk Assessment System

Ivana Kubat[1], Lawrence Charlebois[1], Richard Burcher[1], Philippe Lamontagne[1], David Watson[1]
[1] National Research Council of Canada, Ottawa, Canada)

ABSTRACT

The National Research Council of Canada has developed an integrated risk assessment tool called Canadian Arctic Shipping Risk Assessment System (CASRAS). This system has been created in response to the opening of the Arctic and sub-Arctic to industrial exploration and marine operations, which was largely triggered by the decline of sea ice. The increase in northern activities has created the urgent need for rapid access to disparate information from multiple sources, and the ability to plan voyages in compliance with the International Code for Ships Operating in Polar Waters (Polar Code), which entered into force on January 1, 2017. CASRAS brings together the best available environmental and regulatory data, and documents experiences and knowledge of mariners and Northern community members. This information supports decision making by the shipping industry, offshore operators and government departments. Users can easily retrieve data that ensures safe, responsible and economically viable operations.

CASRAS allows for the processing and querying of a wide variety of data types and extensive data sets to obtain required information over any zone and duration through a simple graphical interface. It is a PC based, stand-alone system that allows quick access to essential information in regions where internet access is limited. This paper describes CASRAS and presents examples of information extracted from the system to assess conditions along the route to Milne Inlet from Baffin Bay. Ice conditions, mean daily low and polar surface temperatures, and ice navigability for Type A and PC6 vessels are analyzed. Navigation through shipping safety control zones and sensitive marine areas along the route are also evaluated.

KEY WORDS: Risk Assessment; Shipping; Arctic; Polar Code; Marine Protected Area, Voyage Planning

INTRODUCTION

The Canadian Arctic and Subarctic represent a pristine environment with a sensitive ecosystem and natural beauty, as well as vast natural resources. In the harsh Arctic, the sea is covered by ice most of the year. However, in recent years a significant retreat in sea ice extent has been recorded opening up the region to industrial exploration and marine activities. Tourism, shipping of supplies to Northern communities, and resource development are all increasing. This growth and the associated risks have considerable safety, environmental and economic implications. As a result, ship operators, resource developers, Northern communities and government regulators are increasingly requiring vast amounts of environmental information to ensure safe, responsible and economically viable operations with minimal impact on the life of Northern communities. Rapid access to disparate information from multiple sources is particularly needed. In addition, the mariners and operators have to plan voyages in compliance with the International Code for Ships Operating in Polar Waters (Polar Code), which entered into force on January 1, 2017.

In response to these demands, the National Research Council of Canada (NRC) has developed the Canadian Arctic Shipping Risk Assessment System (CASRAS). This system allows all Arctic stakeholders to compile, assess, and communicate key information relevant to marine safety, transportation efficiency, and regulatory compliance in the Canadian Arctic. Currently, such data are collected and distributed by individual government departments, academia, private consortia, independent consultants, experienced mariners, and Northern residents. The most accessible sources of information are internet-based and facilitate the exploration of data from one or several perspectives (e.g. regulatory, environmental, biological, or climatological). Examples of web-based products include the Canadian Arctic Voyage Planning Guide (Geoportal.gc.ca, 2017), the U.S. Arctic Environmental Response Management Application (ERMA) (Jacobi and al., 2008), the WWF's Mapping the changing Arctic landscape web page (http://wwfarcticmaps.org/), and the View Ice application from Polar View (Polarview.aq, 2017). While comprehensive and user-friendly, such online resources may be of limited value in low bandwidth environments found in many Northern communities and aboard vessels transiting Arctic waters. Other data-driven and innovative projects, such as the Northern Marine Transportation Corridors initiative (Department of Fisheries and Oceans Canada, 2013), and the Passages project (Passages.ie.dal.ca, 2017) also aim to collate and evaluate information that could impact Canadian Arctic shipping. The above generally focus on specific regions, duration, and a specific product.

While the previously mentioned projects provide numerous benefits to Arctic stakeholders, a stand-alone repository of information and a powerful analytical toolset is needed to holistically address shipping safety, efficiency, and effectiveness in Arctic waters. Such a system has to be easy to use and flexible in storing and extracting information relevant to a variety of operators' needs. Therefore, the National Research Council of Canada (NRC) has developed the PC-based CASRAS application as a platform for cataloguing and querying environmental and regulatory Arctic data. CASRAS also documents the experiences and knowledge of professional and local mariners and Northern community members to enable knowledge transfer to new generations and to new operators. The system includes both analytical tools and user-customized reports that facilitate the assessment of shipping risks during voyage planning and operations. CASRAS has been designed to support shipping companies, Northern communities, regulators, government departments, ice navigators and mariners, insurance companies and associations, and academia.

This paper first provides an overview of CASRAS and a discussion of the data and data management philosophy behind the software. Next, a brief analysis of a shipping route is described to illustrate an application of the system. Subsequently, the results are discussed and other questions that can be addressed by CASRAS are identified.

CASRAS APPROACH

Requirements

Collecting all of the information required for assessing risks related to shipping operations in the Canadian Arctic is time consuming. Moreover, successful planning and risk management requires (i) a knowledge of the most reliable data sources available, (ii) an ability to access the data, and (iii) the capacity to manipulate and extract meaning from the data.

Conducting a risk assessment at any stage of an operation (appraisal, planning, or execution) requires assessing Canadian regulations, Polar Code requirements, environmental conditions, impacts on local communities and other considerations. The sources of data and knowledge are diverse. Some sets of information can be accessed from web portals, some must be manually compiled, some require specialized software or programming skills, while other information, like the knowledge compiled from mariners and communities, is not readily available. The absence of a standardized framework to access and analyze data makes it difficult to provide decision-makers with the best information available in order to support a practical and defensible risk assessment methodology. The CASRAS software and approach is designed to address this issue.

Platform Development

CASRAS is built on the application platform EnSim, which has been under active development at the National Research Council since 1997. EnSim is a comprehensive C++ class library that currently supports several applications in the domains of ice dynamics, hydrodynamics, hydrology and wind energy.

CASRAS is an integrated risk assessment system developed for marine operations in the Canadian North. It provides a comprehensive framework for the storage, search and visualization of all key environmental data with specific application to shipping, icebreaking and navigation in Northern waters and Arctic marine corridors. CASRAS is PC-based, easy to learn, easy to use, and allows quick access to essential information at Northern latitudes where internet access is limited. It provides a means for assisting planning, navigation, and knowledge transfer, while improving transparency for the regulatory process.

Over 100 datasets are currently integrated into CASRAS (>300GB). The system contains a range of historical datasets and operational products including: marine weather, marine hydrography, physical oceanography, sea-ice conditions, ice hazards and physical properties, Marine Protected Areas, community information, and mariner knowledge including notes, charts, and digital media. In order to process such diverse information, the platform supports a wide variety of data types including georeferenced static and time-varying rasters, points, polygons, time-series, text, and digital media as part of a centralized workflow.

The granularity of the information is critical for any decision support system, where the user may want to drill down into the raw data to understand the results from a higher-level analysis. CASRAS provides this capability by allowing the user to examine the datasets at multiple levels, depending on the needs of the stakeholder. At the top level, each dataset in the system can be readily queried and visualized, in a similar way to a GIS application. Further to this, data can be extracted along a planned shipping route or a region of interest for a specific time period, providing statistical information and safety compliance reports based on user-defined thresholds. At its deepest level, a multivariate analysis can be carried out to obtain a user-defined risk index or to assess complex risk factors (e.g. survivability time, rescue time).

Applications

The information queried from CASRAS can be used for route planning and optimization, and as a data source for where and when icebreaking is needed. It can be used to evaluate historical information on wind, waves, tides, ice movement, ice deformation, bathymetry and nautical charting, and to provide background information for ice forecasting models and ice movement severity. Information concerning regions of ecological significance and regulated Marine Protected Areas is also included. CASRAS can also be used to identify gaps in important data and to assist in identifying the most beneficial and cost effective methods required to close these gaps.

CASRAS can be used in the offices of shipping companies, government departments, and community administrations that operate in the Arctic. The system produces customizable reports on hazardous conditions along user-defined shipping routes. These reports contain visualizations of historical analyses and identify exceedances of user-defined thresholds for operation. Such information is useful for companies contemplating expanding to new regions, for determining the type of vessel class to be used along a selected route, for complying with the Polar Code requirements, or for determining the historical windows of safe shipping along any route.

CASRAS is also well-suited for use on-board a vessel as it displays information associated with known risks in the operating region. This information includes the level of confidence of the hydrographic information, historical ice conditions including the amount of multi-year ice, and risks identified from local community knowledge and from the experiences of mariners in the region. CASRAS automatically displays this information in a default format or it can be further queried and processed to provide more detailed reports.

Feedback and Continuous Improvement

Throughout the process of the system development, NRC has been closely working with icebreaker Captains, shipping industry contacts, other government departments (Canadian Coast Guard, Canadian Ice Service, Transport Canada, INAC, Government of Nunavut), northern communities and academia to collect relevant data and to ensure that the system is easy to use and aligned with the needs of stakeholders.

EXAMPLE ANALYSIS AND RESULTS

This section presents an example of information extracted from CASRAS to assess conditions along the route to Milne Inlet from Baffin Bay. The route is shown in Figure 1.

To facilitate the voyage planning process, an analysis is undertaken to answer the following set of key questions:

1. What Shipping Safety Control Zones will the vessel be crossing and what will be the permitted window of operations according to the Transport Canada Zone/Date System? Will the vessel encounter any Marine Protected Areas (MPA) along the route?
2. What are the temperature conditions along the route and to what low operating temperature must the vessel and equipment be certified to comply with the Polar Code?
3. What are the general ice conditions along the route for a given time of the year? How variable are conditions from year to year?
4. How much multiyear ice is historically present along the route? How variable is the concentration of multiyear ice?

5. How does ice navigability vary spatially along the route for a Type A vessel? What if a new PC6 vessel is used?
6. How does the ice navigability for a Type A vessel differ between the AIRSS and POLARIS algorithms?

Using CASRAS analyses, these questions are briefly addressed in the following sections.

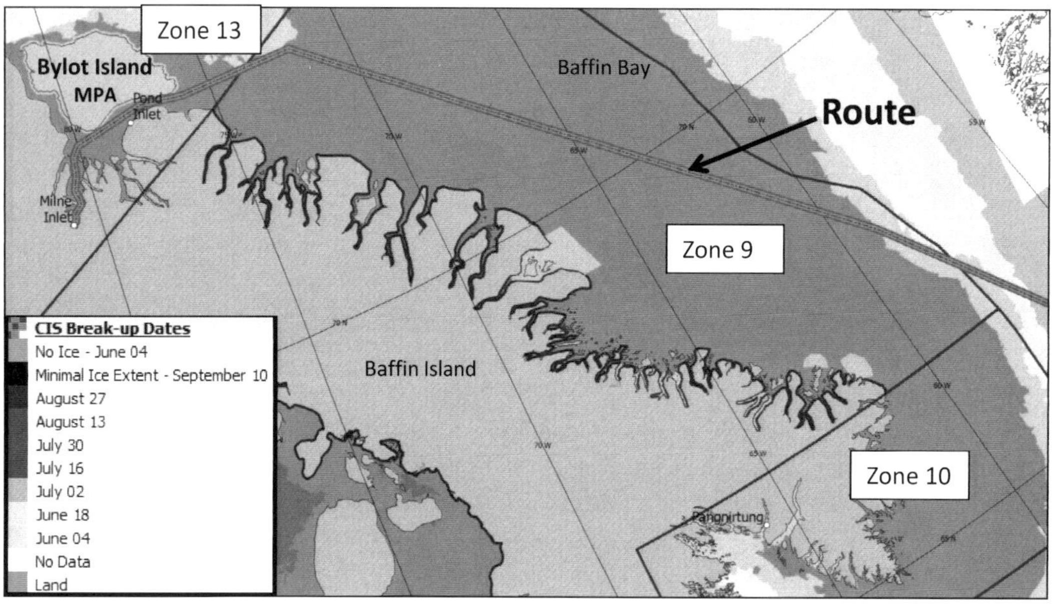

Figure 1. Map showing data exported from CASRAS: route to Milne Inlet, sea ice break-up dates (CIS Ice Atlas 1980-2010), Shipping Safety Control Zones, and the Bylot Island Migratory Bird Sanctuary Marine Protected Area.

Shipping Safety Control Zones and Marine Protected Areas

The Arctic Shipping Pollution Prevention Regulations (ASPPR) regulates navigation in Canadian waters north of 60ºN latitude. These regulations include the Shipping Safety Control Zones and the Date Table made under the Arctic Waters Pollution Prevention Act (ASPPR 1989). Both of these are combined to form the Zone Date System (ZDS) matrix that provides entry and exit dates for various ship types and classes. More information about the Zone Date System including the zones map and dates table can be found on: https://www.tc.gc.ca/eng/marinesafety/tp-tp12819-zone_date-2888.htm

CASRAS can be queried to evaluate what Shipping Safety Control Zones the vessel will cross and what the permitted window of operations will be. As shown in Figure 1, the Milne Inlet voyage passes through Zones 9 and 13. According to Transport Canada regulation, Type A vessels are restricted to operations between August 1 and November 20 and between June 25 and October 22 in Zones 9 and 13, respectively. Figure 1 also illustrates that the route passes by the Bylot Island Migratory Bird Sanctuary and that manoeuvering around ice could bring the vessel into the MPA. This information and any supporting documentation is available for viewing within CASRAS.

Low Temperature Conditions and Vessel Certification

Polar Code

The Polar Code is an international code for ships operating in Polar waters (International Maritime Organization, 2016). It has been developed to supplement existing International Maritime Organization (IMO) instruments in order to protect the two Polar Regions (Arctic and Antarctic) from marine risks and environmental pollution, to increase the safety of ships' operations and to mitigate impacts on the people and environment in these remote, vulnerable and potentially harsh areas of the world. The Polar Code entered into force on January 1, 2017. More information about the Polar Code can be found on the IMO website: http://www.imo.org/en/mediacentre/hottopics/polar/pages/default.aspx

The Polar Code covers the full range of design, construction, equipment, operational, training, search and rescue and environmental protection issues relevant to ships operating in the harsh environment of the Polar Regions. CASRAS is designed to address some components of the code, including operational and voyage planning requirements. In this section the authors present an assessment of a vessel operation in low air temperatures.

CASRAS has been used to compute the minimum Mean Daily Low Temperature (MDLT). MDLT is the mean value of the daily low temperatures for each day of the year over a minimum 10-year period. The computed MDLT results have to comply with the directive of the IMO Polar Code for the assessment of low operating temperature requirements and with the calculation of the required Polar Service Temperature (PST) for any vessel operating in this environment. The required PST can be estimated by subtracting 10°C from the minimum MDLT along the route for the anticipated window of operations. A vessel is compliant if the PST is lower than the lowest calculated MDLT minus 10°C for any portion of the journey.

Figure 2 presents the lowest daily value of the MDLT calculated using the North America Regional Reanalysis and extracted from CASRAS for the Milne Inlet voyage. The analysis of data along the route is based on gridded air temperature data taken from 10, 20, and 30 year historical records for comparison.

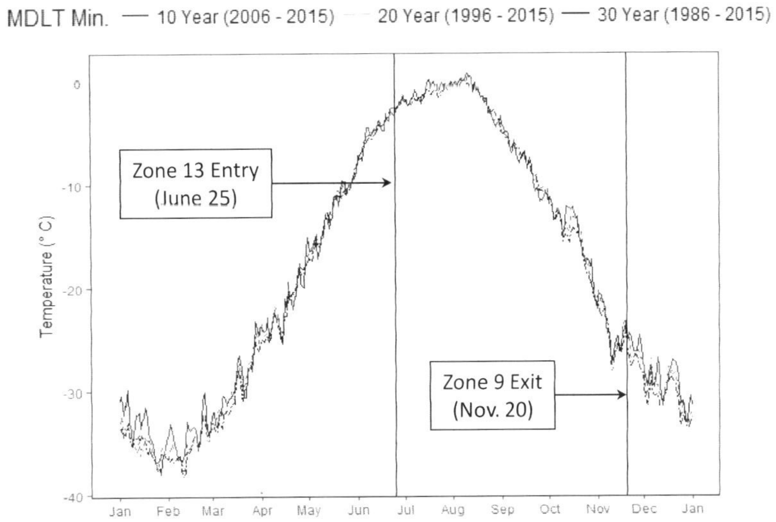

Figure 2. Mean Daily Low Temperature calculated for the Milne Inlet voyage (showing only the lowest MDLT along the route for every day of the year). Vertical solid lines indicate the Zone 9/Zone 13 shipping window.

An indication of Polar Surface Temperature (PST) compliance for a vessel that is certified for -35°C operation (arbitrarily chosen for this analysis) is shown in Figure 3. The analysis was completed using 10 years of climate data (2006 to 2015). The results show that this level of PST certification (-35°C) would be adequate for operations from mid-April through the end of October. The month of November shows frequent variation between the vessel's compliance and non-compliance, therefore additional information was extracted from CASRAS providing the PST calculated for each day of November. These values are presented in Figure 4.

This analysis considered the same date for the entire route, which is a simplification for illustrative purposes. However, the analysis could be modified to account for voyage progression, i.e. changing dates depending on the vessel position along the route.

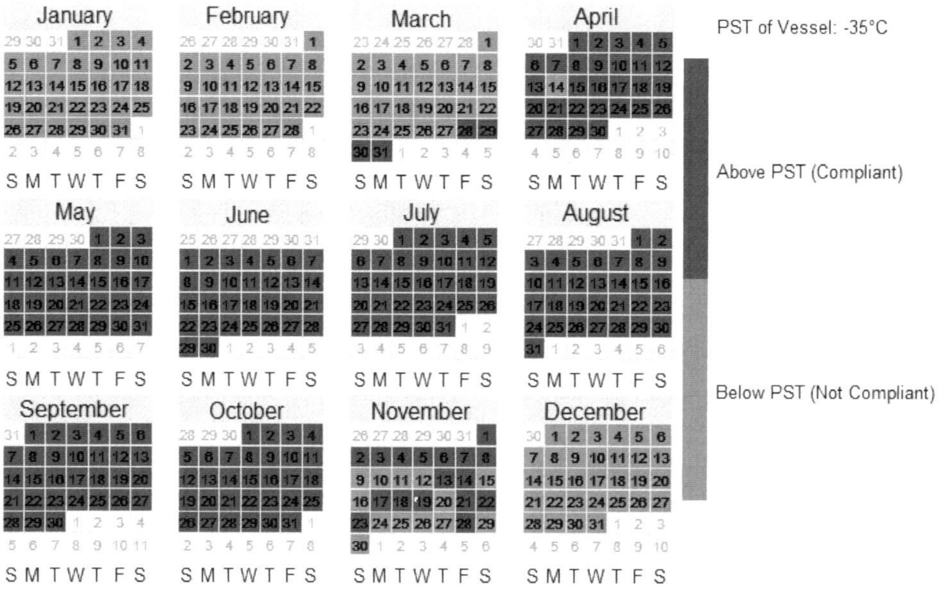

Figure 3. Polar Service Temperature (PST) compliance calculated for a vessel with PST = -35 °C along the Baffin Bay to Milne Inlet route for 10 years (2006-2015).

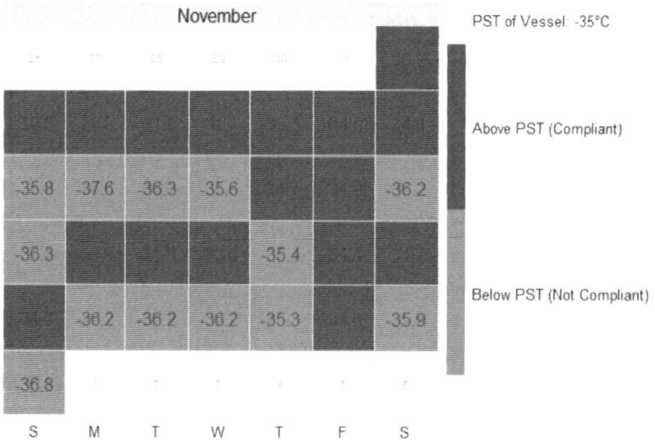

Figure 4. Polar Service Temperature (PST) compliance calculated for a vessel with PST = -35 °C in November along the Baffin Bay to Milne Inlet route; 10 years (2006-2015).

Ice Conditions and Interannual Variability

The presence of open water will positively influence transit times and fuel efficiency. However, it may pose additional risks such as, for example, interaction with icebergs, pieces of floating old ice, rough sea, and marine mammals. Figure 5 depicts a CASRAS analysis showing the presence of open water from 1997 through 2014 for the northern part of the route in question (west of 75°W from the entrance of Eclipse Sound to Milne Inlet).

The analysis shows that some portion of this route was ice-free during the typical shipping season but that the timing is subject to interannual variability.

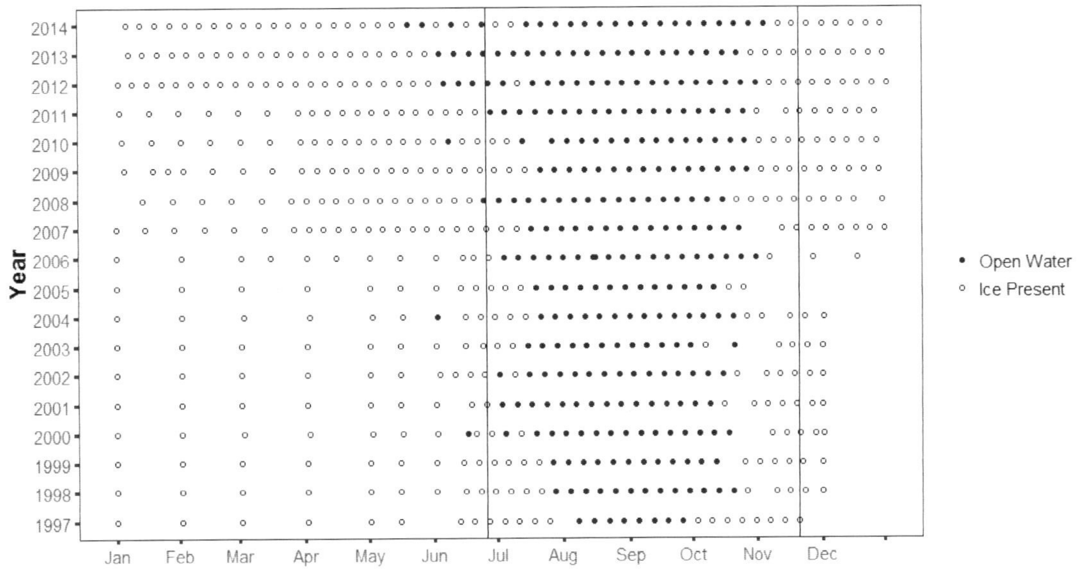

Figure 5. Presence of Open Water west of 75°W (Eclipse Sound approach) based on minimum ice concentration less than 1/10. Vertical solid lines indicate the Zone 13 entry date and Zone 9 exit date shipping.

In the same manner, CASRAS can be used to assess the interannual variability of any ice condition. For example, the database may be queried for the historical presence of user-specified threshold ice concentration and ice type (e.g. 6 tenths of medium first-year ice, 70-120 cm thick)

Old Ice Hazards

The presence and concentration of old ice in shipping lanes needs to be considered for voyage planning. Type A vessels will not be allowed to operate outside the Zone Date System shipping window if the region is covered by high concentrations of old ice. In such cases, the Arctic Ice Regime Shipping System (AIRSS) must be used to determine whether a vessel can proceed. This topic is addressed in depth in Section 5. Figure 6 summarizes the maximum concentrations of old ice encountered along the Baffin Bay to Milne Inlet route in 2006 and 2014. These two years were chosen to compare conditions in a relatively light ice year (2006) and in a relatively heavy ice year (2014).

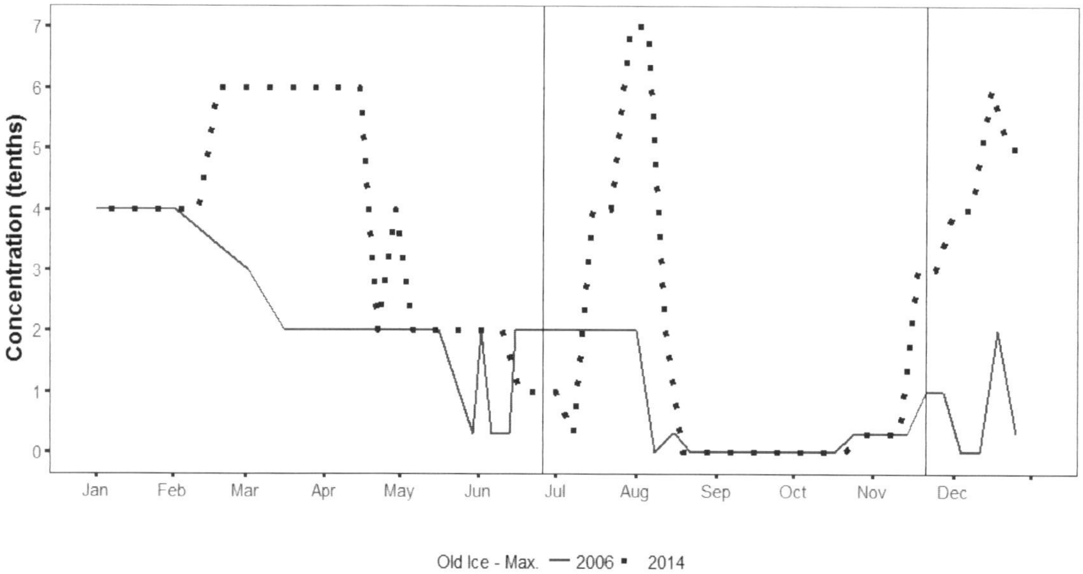

Figure 6. Maximum concentrations of old ice along the Milne Inlet route in years 2006 (relatively light ice year) and 2014 (relatively heavy ice year). Vertical solid lines indicate the Zone 13 entry date and Zone 9 exit date shipping.

The data summarized in Figure 6 suggest that the presence and concentration of old ice can vary considerably depending on annual conditions in the region. Furthermore, as is seen in late July and early August, old ice concentrations can be quite significant (greater than 6 tenths) along the Milne Inlet route during the regulated shipping season (i.e. the window of operation given by the Zone Date System). Note also that the presence of old ice in March and April of 2014 was significantly higher than in 2006. CASRAS has automated the process of manually compiling and interpreting the archived ice charts for this route. In addition to the concentration of old ice, the size of floes, frequency of old ice occurrence, thickness of different types of ice, ice concentrations, and other parameters can be extracted from the system for different legs of the route as well as for different windows of operation.

Ice Navigability

The CASRAS application can be used for evaluating the navigability through ice along the planned route while complying with Canadian regulations.

Arctic Ice Regime Shipping System (AIRSS)

In 1972, Transport Canada, in consultation with stakeholders, made extensive revisions to the Arctic regulations through the introduction of the Ice Regime System (ASPPR 1989; Canadian Gazette 1996; Equivalent Standards 1995; AIRSS 1996). The changes were designed to reduce the risk of structural damage in ships which could lead to the release of pollution into the environment, yet provide the necessary flexibility to ship-owners by making use of actual ice conditions, as seen by the Master, to determine the safest transit.

AIRSS is based on a simple arithmetic calculation, producing an Ice Numeral (IN) that combines the ice regime and the vessel's ability to navigate safely through that ice regime. The "Ice Regime" is a region of generally consistent ice conditions. The Ice Numeral is based on the quantity of hazardous ice with respect to the ASPPR classification of the vessel. Further information about AIRSS and Ice Numeral calculation can be found in Timco and Johnston, 2003.

Polar Operational Limit Assessment Risk Indexing System (POLARIS)

IMO has developed a harmonized methodology for assessing operational limitations in ice called the Polar Operational Limit Assessment Risk Indexing System (POLARIS). This system combines concepts of the AIRSS system, the Russian Ice Certificate, and additional input from coastal administrations that have experience regulating marine traffic in ice conditions. POLARIS assesses ice conditions based on a Risk Index Outcome (RIO). A detailed description of POLARIS and RIO calculation can be found in the IMO Polar Code Advisory, 2016.

Both the AIRSS IN (Ice Numeral) and the POLARIS RIO (Risk Index Outcome) evaluate whether a vessel can safely proceed through the surrounding ice regime. If the values are positive, then a vessel is allowed to proceed; if the values are negative then she cannot. The main difference between the two systems is that AIRSS in the IN calculation considers severity of ice conditions and vessel characteristics, while POLARIS in the RIO calculation accounts for those parameters as well as vessel speed and escort requirements.

In this paper, the authors evaluated the ice navigability of a vessel comparing compliance to both AIRSS and POLARIS. The authors chose to evaluate two approximately equivalent vessel classes, Type A and PC6, that can navigate in the selected region within the June to November shipping season window defined by the Zone Date System. The paper also presents analysis for the entire year.

Ice Navigability for Vessel Type A and PC6

According to Transport Canada (TP 3231 – Ship Safety Bulletin No. 04/2009), an interim equivalent for the Polar Class 6 (PC6) vessel is the Type A vessel. Figure 7 presents the results of the AIRSS calculation of the Ice Numeral for both vessel classes. The month of analysis is March, which is well beyond the typical shipping window, to show the utility of the system in assessing year-round shipping operations. The results are presented as the frequency when navigation is permitted (a "Go" scenario). The calculation was based on Canadian Ice Service (CIS) digital ice charts from 2005 through 2014.

As shown in Figure 7, the navigability through Eclipse Sound (i.e. west of 75°W) would be restricted for a Type A vessel and not recommended for a PC6 vessel. This analysis suggests that the IN calculation for the Type A vessel results in a more conservative evaluation of ice navigability than for the PC6 vessel. This is an important result when considering the performance equivalencies of old (existing) and new (under design) vessels using the AIRSS approach. Closer evaluation of the navigability in the Eclipse Sound for the PC 6 vessel revealed that the frequency of IN > = 0 was in the range between 1 and 3%.

Ice Navigability for Vessel Type A under AIRSS and POLARIS

The CIS Eastern Arctic regional ice chart for March 14, 2016 has been used to calculate the Ice Numerals for the Type A vessel along the route, Figure 8a. The results show that on March 14, 2016 the Type A vessel would not be allowed to proceed through the region because the Ice Numerals values are less than 0.

In Figure 8b, the CASRAS application has been toggled to display the Polar Code equivalent of the Ice Numeral, the POLARIS RIO. Results show that the proposed POLARIS assessment provides more flexibility in operations for March 14, 2016, yet still requires heightened levels of precaution, including speed reductions and escort requirements for the approach to Eclipse Sound.

Other Applications of CASRAS for Voyage Planning

The previous analyses have touched on just a few applications of the environmental data and knowledge that constitutes CASRAS. Additional questions can be addressed using the CASRAS application, for example:

- Which Polar Class vessel is required to navigate year round to approach Milne Inlet?
- What official nautical charts and other bathymetric surveys are available along the route?
- What are the extreme wind conditions and air temperatures in Baffin Bay and Eclipse Sound over the last 30 years?
- Based on the last 30 years, what is the likelihood of having a window of operation of at least 48 hours of ice-free and calm-winds in the first two weeks of May?
- What information is available regarding marine hazards based on community local knowledge or experience from mariners on operating in Eclipse sound

Responding to these questions, however, is beyond the scope of this paper.

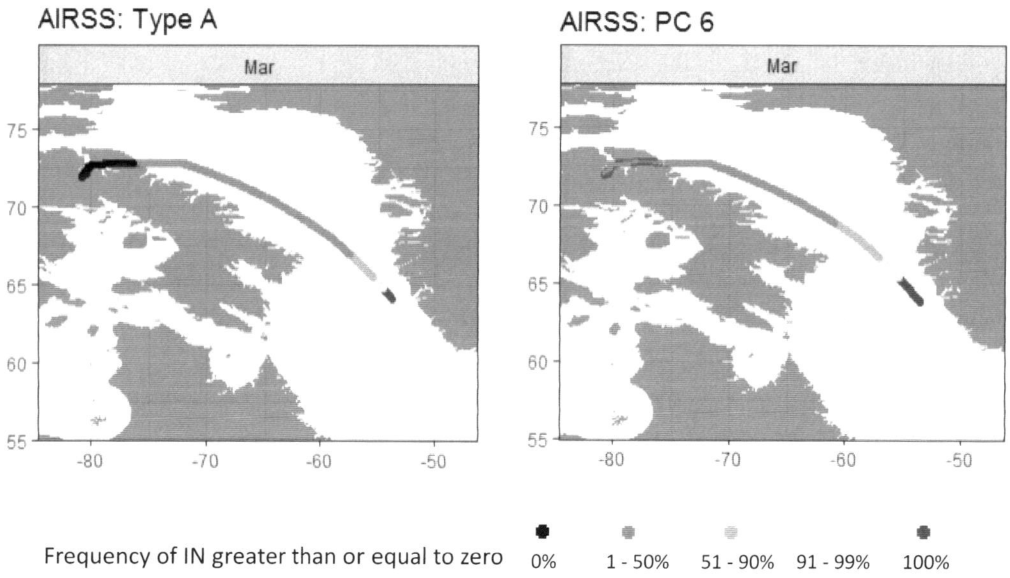

Figure 7. A comparison of the interim 'equivalent' vessel classes Type A and PC6 according to the AIRSS navigability assessment. The frequency of Ice Numeral equal to or greater than zero (i.e. "Go" scenario) is computed for the month of March using data from 2005 through 2014.

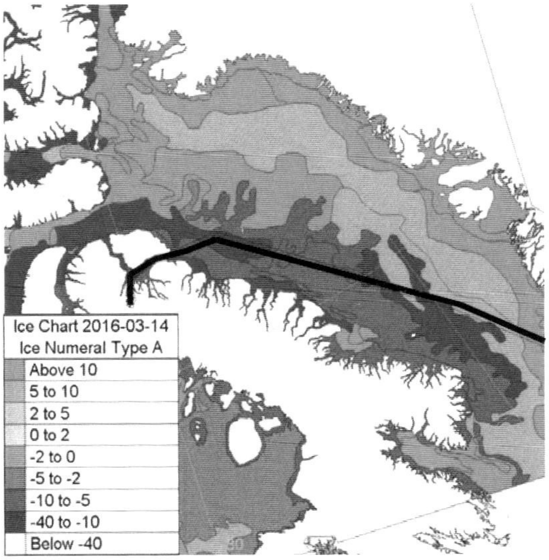

Figure 8. a) Route navigability using the Arctic Ice Regime Shipping System (AIRSS) for a Type A vessel, according to the observed ice conditions on March 14, 2016. Red and orange zones are designated as not navigable by the Type A vessel.

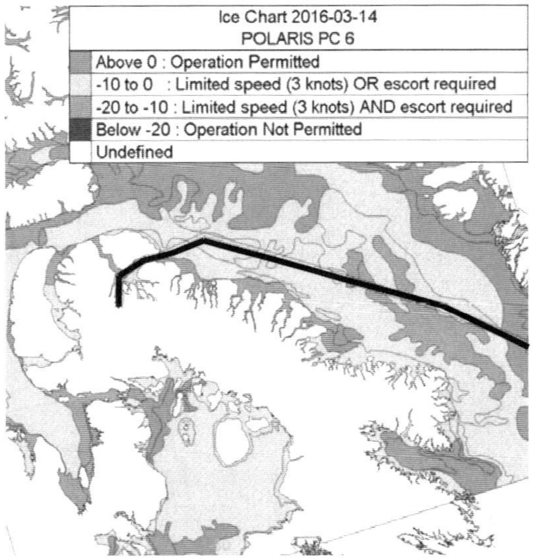

Figure 8. b) Route navigability using the Polar Operational Limit Assessment Risk Indexing System (POLARIS) for a PC6 vessel, according to the observed ice conditions on March 14, 2016. Yellow and orange zone require limiting speed and escorting.

SUMMARY

NRC has developed the Canadian Arctic Shipping Risk Assessment System (CASRAS). It is an integrated risk assessment system for marine operations in the Canadian North that enables sound decision-making, which will lead to safer, more efficient shipping in this harsh environment. It provides a comprehensive framework for the storage, search and visualization of all key environmental data with specific application to shipping, icebreaking, and navigation in Northern waters and Arctic marine corridors. It is PC-based, easy to learn and use, and allows quick access to essential information at Northern latitudes where internet access is limited. It processes large datasets including georeferenced static and time-varying rasters, points, polygons, time-series, text, and digital media as part of a centralized workflow. It is a system in which data can be queried, visualized and extracted, various parameters can be evaluated, and reports can be generated, all for user-defined dates, times and regions of interest, and all on the fly. CASRAS can be used in the office and on-board a vessel. Information extracted from the system is useful for companies contemplating expanding to new regions, for selecting the type of vessel class to be used along the selected route, or for complying with the Polar Code. The system can also be used for determining the historical windows of safe shipping along a route, for evaluating impacts on Northern communities, for the planning of search and rescue operations and for extracting information documented by experienced mariners and Northerners. CASRAS provides a means for assisting in planning, navigation, and knowledge transfer, while improving transparency for the regulatory process.

This paper presented an example of information extracted from CASRAS to assess conditions along the route to Milne Inlet from Baffin Bay. Two vessel classes were considered, Type A and PC6, when evaluating vessel certification based on low air temperature operations, navigation through shipping safety control zones and sensitive marine areas, and ice conditions along the route.

ACKNOWLEDGEMENTS

The authors would like to acknowledge the National Research Council of Canada for funding this project. We thank Captain Mark Taylor for providing waypoints for the route, as well as Captains John Broderick, Terry Camsell, Dave Fowler and Ray Pierce for providing valuable information, comments and feedback throughout the system development.

REFERENCES

AIRSS 1996. *Arctic Ice Regime Shipping System (AIRSS) Standards*, Transport Canada, June 1996, TP 12259E, Ottawa. Ont., Canada.

ASPPR, 1989. Proposals for the Revision of the Arctic Shipping Pollution Prevention Regulations. *Transport Canada Report TP 9981*, Ottawa. Ont., Canada.

Canadian Gazette, 1996. *Regulations Amending the Arctic Shipping Pollution Prevention Regulations. p 1729,* Ottawa, Ont., Canada.

Department of Fisheries and Oceans Canada, (2013). *Report on Plans and Priorities 2013-14.*

Geoportal.gc.ca. (2017). *Arctic Voyage Planning Guide*. [online] Available at: http://geoportal.gc.ca/eng/Gallery/MapProfile/5 [Accessed 14 Feb. 2017].

Equivalent Standards for the Construction of the Arctic Class Ships, 1995. *Transport Canada Report TP-12260*, Ottawa, Ont., Canada.

IMO Polar Code advisory, January 2016, *ABS (American Bureau Shipping*)

International Maritime Organization (2016) International Code for Ships Operating in Polar Waters (Polar Code) MEPC 68/21Add.1 Annex 10, viewed 14 September 2016, http://www.imo.org/en/MediaCentre/HotTopics/polar/Documents/POLAR%20CODE%20TEXT%20AS%20ADOPTED.pdf.

Jacobi, M., Kinner, N.E., Braswell, B.H., Schwehr, K.D., Newman, K.S. and Merten, A.A., 2008, May. Environmental Response Management Application. *International Oil Spill Conference* (Vol. 2008, No. 1, pp. 881-885). American Petroleum Institute.

Passages.ie.dal.ca. (2017). Passages Story Map. [online] Available at: http://passages.ie.dal.ca/Storymap/index.html [Accessed 14 Feb. 2017].

Polarview.aq. (2017). Polar View. [online] Available at: http://www.polarview.aq/ [Accessed 14 Feb. 2017].

Timco, G. and Johnston, M. *Arctic Ice Regime Shipping System: Pictorial Guide*, TP 14044E, January 2003

Transport Canada Ship Safety Bulleting, TP 3231, *Bulletin No.: 04/2009; RDIMS No.: 3908731*, Date: 2009-08-18

POAC'17
Busan, KOREA

Proceedings of the 24th International Conference on
Port and Ocean Engineering under Arctic Conditions
June 11-16, 2017, Busan, Korea

Sustainable Arctic Marine and Coastal Technology

Sveinung Løset[1, 2]

[1]Sustainable Arctic Marine and Coastal Technology (SAMCoT), Centre for Research-based Innovation (CRI), Norwegian University of Science and Technology, Trondheim, Norway

[2] University Centre in Svalbard (UNIS), Norway

ABSTRACT

The Research Council of Norway has established a number of Centres for Research-based Innovation (CRIs). The main objective of these centres is to enhance the capability of the business sector to innovate by focusing on long-term research based on forging close alliances between research-intensive enterprises and prominent research groups. The CRI SAMCoT *(Sustainable Arctic Marine and Coastal Technology*) is hosted by the Norwegian University of Science and Technology (NTNU). SAMCoT started in 2011 and is now entering a phase of harvesting on the research continuously being done. This paper presents some findings of the research at SAMCoT and shows how the clustering of research efforts can lead to innovations in the Arctic offshore industry and a better understanding of the Arctic physical environment.

KEY WORDS: Arctic Marine Technology, Arctic Research

INTRODUCTION

In June 2006, the Executive Board of the Research Council of Norway (RCN) decided to establish Centres for Research-based Innovation (CRIs). The main objective of a CRI is to enhance the capability of the business sector to innovate by focusing on long-term research based on forging close alliances between research-intensive enterprises and prominent research groups. The CRI scheme will

- Encourage enterprises to innovate by placing a stronger emphasis on long-term research and by making it attractive for enterprises that work in the international arena to establish R&D activities in Norway;
- Facilitate active alliances between innovative enterprises and prominent research groups;
- Promote the development of industrially oriented research groups that are on the cutting edge of international research and are part of strong international networks;

- Stimulate researcher training in fields of importance to the business community and encourage the transfer of research-based knowledge and technology.

There have been three waves of CRIs in Norway. Fourteen centres were started in 2007, 7 started in 2011 and 17 started in 2015. The duration of a centre is 8 years assuming that the centre passes a midway evaluation. The funding from RCN is 1.2 million USD per year so long as the industry matches this funding. With the initiation by RCN of waves two and three, the arrangement is considered a great success with high-impact research being conducted.

SAMCoT *(Sustainable Arctic Marine and Coastal Technology*) is one of the centres that started in the 2011 CRI wave. The centre is hosted by the Norwegian University of Science and Technology (NTNU). SAMCoT is now entering a phase of also capitalizing on the on-going research continuously being done at the centre. On average, the total annual turnover of SAMCoT is approximately 4.7 million USD, and the centre has 23 partners. In addition, SAMCoT consistently employs at least 30 PhD students/postdocs. All of them have experience in Arctic surveys in addition to their desk or lab study, providing them with real-world training in handling HSE in the Arctic.

THE STRUCTURE OF SAMCoT

Currently, SAMCoT is divided into five scientific work packages (WPs):

- *WP1 – Data Collection and Process Modelling*: WP1 collects and analyses field data on sea ice, icebergs, currents and coastal permafrost. The primary goal is to provide field data as required by the other WPs.
- *WP3 – Fixed Structures in Ice*: Following the original plans of SAMCoT, WP2 (*Material Modelling*) should merge with WP3. As a result, WP3 comprises today research on constitutive models for ice rubble (ice ridges) and permafrost as well as the development of analytical and numerical models to predict the actions from ice (first- and multi-year, level ice and ridges) on fixed single- and multi-leg structures. Further, the aim is to develop innovative vibration mitigation measures for fixed structures exposed to drifting ice.
- *WP4 – Floating Structures in Ice:* The purpose of WP4 is to develop new knowledge and analytical and numerical models to improve the prediction of loads exerted by first/multi-year ice and ridges as well as icebergs on floating structures. This WP also entails the prediction of the performance of the structures.
- *WP5 – Ice Management and Design Philosophy*: The purpose of WP5 is to establish a philosophy that ensures that the Ultimate Limit State (ULS) and Accidental Limit State (ALS) requirements are fulfilled by Arctic offshore structures without being overly conservative. This philosophy shall consider the use of ice management means such as icebreakers, iceberg towing vessels, structure disconnection and reconnection capabilities.
- *WP6 – Coastal Technology*: This WP develops new knowledge and analytical and numerical models to improve the prediction of coastal erosion in the Arctic. This information is essential for the design of environmentally friendly and sustainable coastal structures and technologies; i.e., the knowledge gained from WP6 will provide bases for design guidelines.

The engineering challenges these WPs focus on are illustrated in Figure 1. The research strategy is very much based on hypothesis and theory development followed by field and laboratory work in combination with numerical modelling (see Figure 2). A high priority is given to collecting and analysing full-scale data when possible.

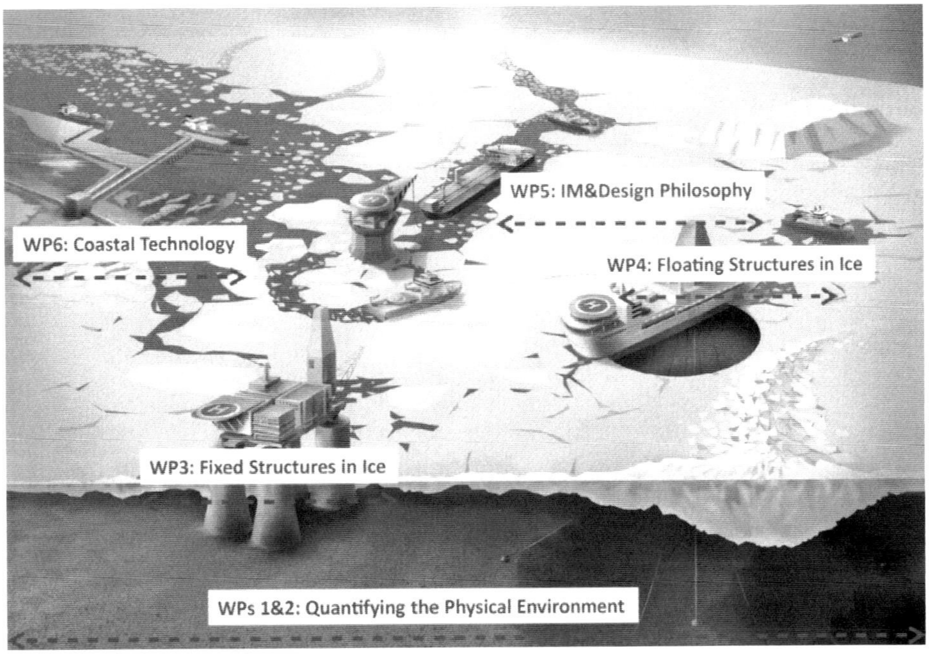

Figure 1. The engineering challenges addressed by SAMCoT.

Figure 2. Research strategy of SAMCoT.

RESEARCH ACTIVITIES

Some highlights from the research being done in the different WPs follow. SAMCoT (2017) is a major source of the text in this Section.

Data Collection and Process Modelling (WP1)

More than 20 years of extensive experimental investigations at Spitsbergen and in the Barents and Kara Seas regions underlie the work done in WP1. Numerous researchers, MSc and PhD students work together with the aim of providing the experimental data required by PhD students over the years. Field studies in the regions of the Russian Arctic are also conducted with the cooperation of Moscow State University (MSU) and Krylov State Research Centre.

The researchers' main goal is to study the following:

- Sea state and ice conditions in the North-West Barents Sea, Barents Sea Opening, Greenland Sea and in the Arctic Ocean to the north of Spitsbergen (CTD); ADCP profiling; and 3D FE modelling of thermodynamic consolidation of ice ridges in the North-West Barents Sea and Barents Sea opening;
- Monitoring of iceberg drift by GPS trackers (Greenland and Barents Sea);
- Investigating and comparing the characteristics of ice drift and surface currents in the North-West Barents Sea and Svalbard region using the data from ice trackers deployed on drift ice and icebergs and the data from floating buoys provided by NOAA (Figure 3);
- Mechanical properties of ice (tensile, compressive and shear strength of floating ice);
- Sea ice actions in the coastal zone (ice actions on shoreline and fixed quays).

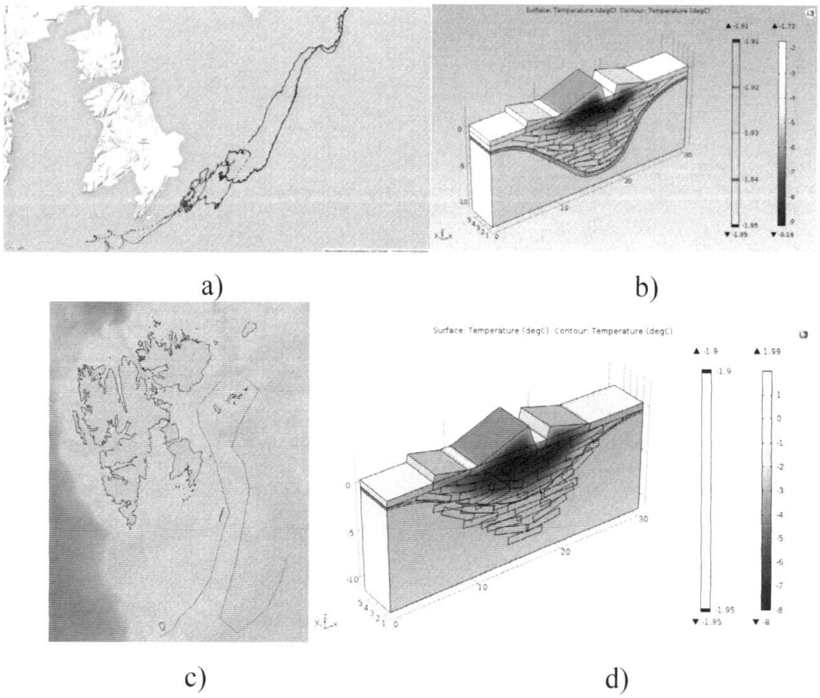

Figure 3. Trajectories of drift ice ridges in the North-West Barents Sea (a) and Barents Sea opening (c); thermodynamic evolution of ice ridges during their 2 month drift along the trajectories (b, d) (SAMCoT, 2016).

The data collected in WP1 are also of high value for assessments of remotely sensed data as provided by several Sea Ice Product Services (e.g., IFREMER, OSI SAF, NSIDC, TOPAZ). The Barents Sea is a very interesting region for SAMCoT's research due to its location and the presence of sea ice that is representative of sea ice trends observed elsewhere. Sea ice is a key factor in marine operations and navigation in the Arctic. An evident decrease of the ice cover both in relation to areal extent and thickness has been scientifically determined. The thinning of the ice cover leads to more broken ice. This broken ice is an important factor to consider in offshore activities in these waters. How this broken ice is affecting these activities is not yet well known. Furthermore, the amount of multi-year ice is reduced significantly, which should be taken into account when planning offshore field developments (SAMCoT, 2017).

The ice conditions in the Barents Sea are representative of the new trends we see in the Arctic ice, including glacial ice (icebergs). The Barents Sea is thus a good study area for SAMCoT as it contains all of these components in addition to being a prosperous area for present and future oil and gas exploitation.

Figure 4 shows a huge rubble field of sea ice accumulated in front of Stonebreen at Edgeøya. The rubble field's extent from the coastline reached 8 km and > 20 km in the N-S direction along the coast (SAMCoT, 2017). The planar, vertical wall was most likely formed by an iceberg drifting by and compressing and shearing the rubble.

Figure 4. An ice rubble field and two icebergs off Stonebreen, Edgeøya (02.05.2016).

Fixed Structures in Ice (WP3)

The research in this work package is twofold:

- Development of constitutive models for ice rubble (ice ridges) and permafrost;
- Studies on vibrations of structures exposed to drifting ice and the development of analytical and numerical models to predict the actions from ice (first- and multi-year, level ice and ridges) on fixed single- and multi-leg structures.

As part of the constitutive modelling of ice rubble, we defined two main directions:

- The application of the ice rubble model developed here is based on breakage mechanics in the ice rubble-structure interaction process;
- The assessment of continuum modelling limitations due to ice rubble heterogeneity.

The use of the developed ice rubble model and its comparison to ice basin experiments showed several interesting results. First, the distribution of breakage in ice rubble did not show a considerable effect on the simulated rubble load (Figures 5 and 6). In contrast, the accumulation of ice rubble appeared to be the governing factor in the load. However, this behaviour may be limited to the case of floating ice rubble and may differ in the case of grounded ice rubble (SAMCoT, 2017).

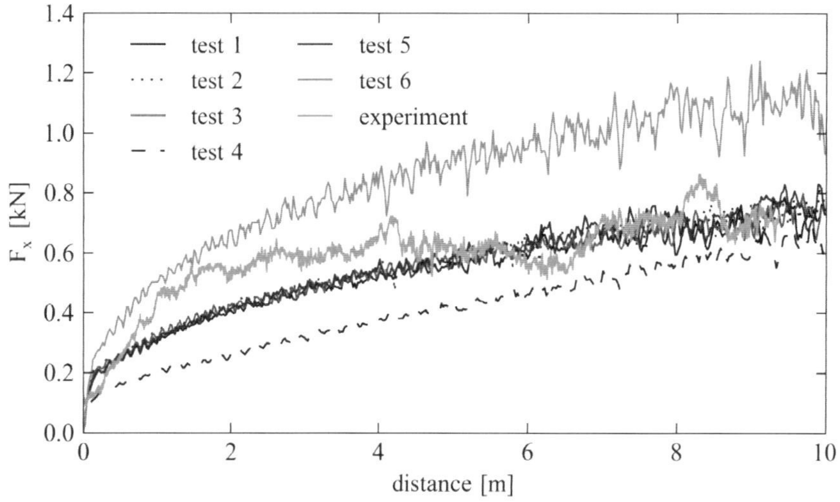

Figure 5. The comparison between simulated and measured ice rubble loads.

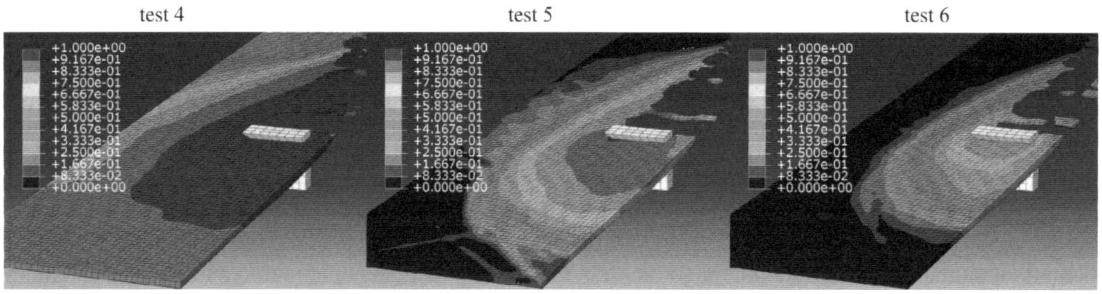

Figure 6. Examples of the distribution of breakage for different model parameters.

A substantial effort is being made to increase the understanding of the causes of ice-induced vibrations exerted by drifting ice on bottom-fixed structures (Nord, 2015; Hendrikse, 2016). One example of these studies is an analysis of the interaction between a laboratory structure and ice in the Hamburg Ship Model Basin, Germany. This analysis was performed by comparing the locally measured pressures at the ice-structure interface to the identified structural responses and forces obtained from a joint input-state estimation algorithm. The framework as it is applied in Nord et al. (2015) consists of the joint input-state estimation algorithm, a model of the structure and a set of response measurements. Displaying the spatially complete structural displacements together with the tactile sensor pressures illustrated how the structural motion influenced the pressure at the ice-structure interface.

This analysis was performed for two regimes of ice-induced vibrations: the intermittent crushing regime and a regime that exhibited the tendencies of frequency lock-in vibrations. An example from intermittent crushing is displayed in Figure 7, in which the actual test structure is displayed on the left, the structural motion is displayed in the middle for three 10 ms time increments, and the corresponding pressure at the ice-structure interface is displayed on the right. An ice failure initiates between frames 7 and 8, which is clearly visible from the pressure changes at the interface. The spring-back deflection at the ice-action point during the intermittent crushing was in the range of 10-15 mm. Both pulverised ice and larger pieces were extruded away from the indenter.

The applied methodology led to the conclusion that the superstructure vibrations influenced the pressure at the ice-structure interface. During intermittent crushing, it was demonstrated that the superstructure vibrations caused cyclic loading and unloading of the ice edge prior to each ice failure.

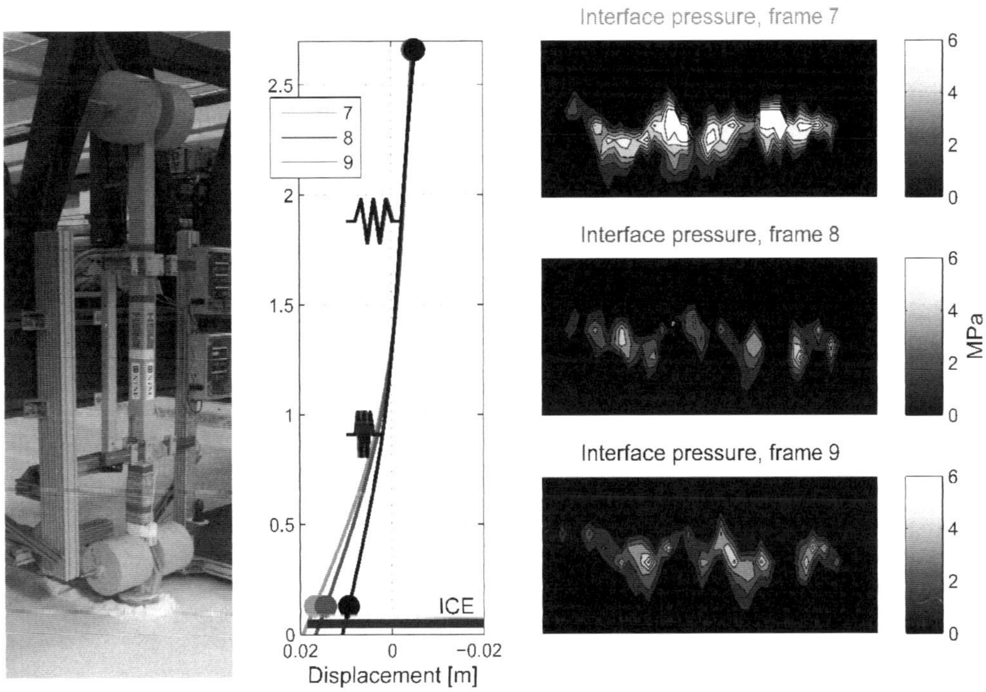

Figure 7. State of the structure and interface pressure during intermittent crushing in Test 4300 at 20 mm/s ice velocity.

Floating Structures in Ice (WP4)

From an engineering point of view, we can split the ice actions on a floating structure into global and local actions. This distinction applies to all floating structures, whether sailing (ship) or station-keeping. Figure 8 illustrates the applicability of such studies.

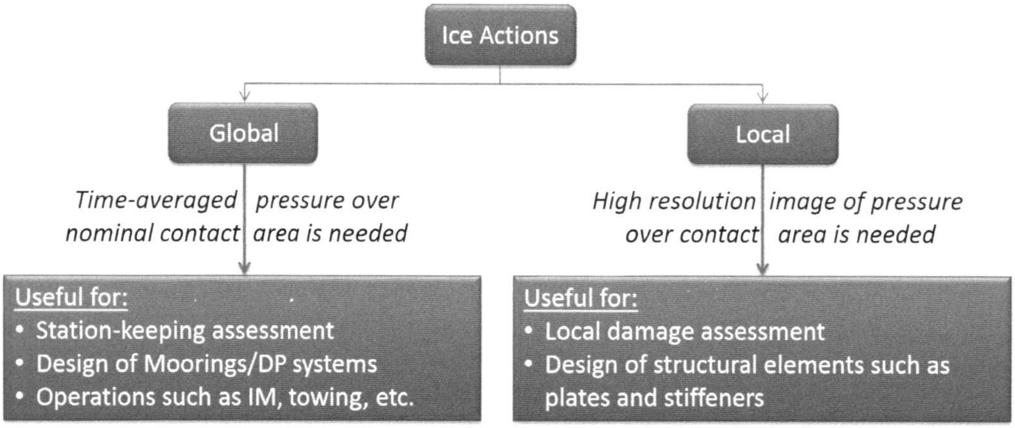

Figure 8. Illustration of needs/relevance for estimation of ice actions.

The success of this work package relies heavily on access to full-scale data in combination with numerical modelling. Figure 9 depicts the participation on some of these research cruises since 2012.

Figure 9. Research cruises in the NE Greenland Sea and the Arctic Ocean where full-scale data were collected.

The importance of these research cruises relates to (see Figure 10) the following:

- Observations of ice features and the interaction processes between structure and ice;
- Studies of ice properties;
- Observations of limiting mechanisms, failure modes and effects of interaction geometry;
- Monitoring of levels of ice actions for the different interaction scenarios;
- Acquisition of full-scale data used in the calibration and validation of numerical models developed in SAMCoT;
- Providing information on scaling from full scale to lab scale.

Sea ice is a key factor affecting marine operations and navigation in the Arctic. An evident decrease of the ice cover with respect to both areal extent and thickness has been scientifically determined. The thinning of the ice cover leads to more broken ice (floe ice). Continuous methods for modelling these conditions are not valid with respect to global ice actions. Thus, SAMCoT has placed a considerable emphasis on developing discrete element methods (DEMs) that enable us to model the interactions between individual ice blocks and the structure of interest. This approach may provide information on relevant processes in ice- structure or ice-ice interactions that cannot be captured by continuous methods. DEMs can broadly be divided into two categories: smooth and non-smooth. While the former is known for its high computational cost, the latter may significantly reduce the calculation time of discrete element simulations by enabling larger time steps (van den Berg et al., 2017).

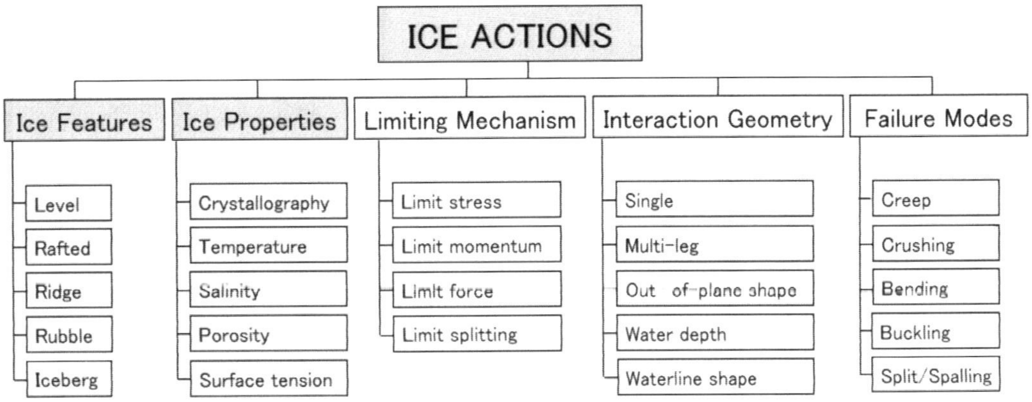

Figure 10. Major parameters influencing the effects of ice actions on structures.

With more broken and dispersed ice present in the Arctic, splitting is a frequently observed limiting mechanism due to the finite size of and reduced confinement in the ice cover (Lu et al., 2015b; Lu et al., 2016a).

During ice management, parallel channels are usually created to facilitate the ice floe size reduction process (Lu et al., 2015c). During such operations, parallel channel fracturing events are continuously observed. Valuable data were obtained during the OATRC'15 expedition in terms of parallel channel fracturing events (Lu et al., 2016b). In 2016, we analysed the test data and programmed a numerical scheme based on XFEM to simulate the crack kinking process to explain the observed fracture events (see Figure 11).

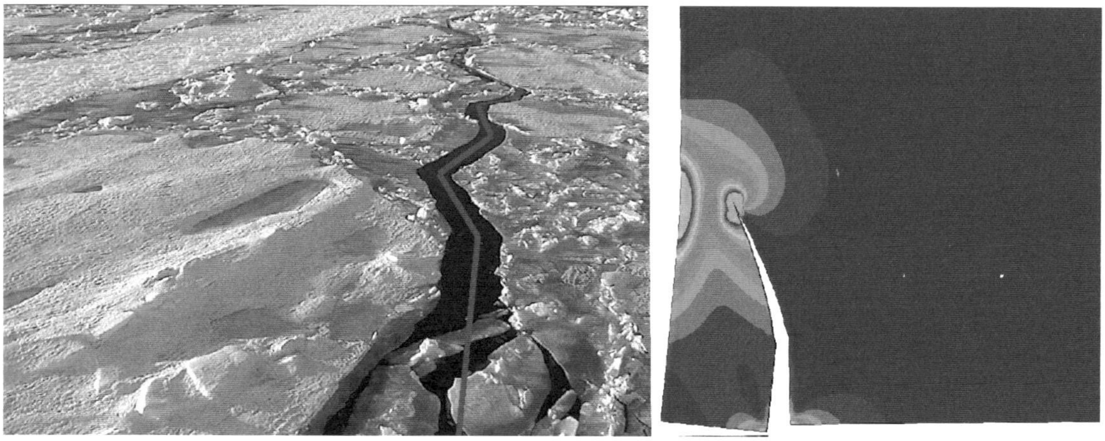

Figure 11. Observed and modelled fracturing of sea ice (Lu et al., 2015c; Lu et al., 2016b).

A question here is the fracture energy associated with the splitting (Lu et al., 2015a). This inquiry led to a large-scale field fracture campaign taking place both in March 2016 and in March 2017 in Svea (Van Mijenfjord, Spitsbergen). After rather extensive preparations, we obtained very valuable data and visually observed for the first time rather stable crack growth within sea ice (see Figure 12).

(a) (b)

Figure 12. The "fracture team" in Svea, Spitsbergen; a) March 2016, b) March 2017.

Ice Management and Design Philosophy (WP5)

The research in WP5 includes a number of different activities, including pursuing the development of next-generation motion control systems for floaters. Such systems will be more proactive in complex environments and be able to "see", predict, and better respond to the surrounding environment. The research necessitates instrumentation of some of the needed surveillance tasks of ice management. Of particular interest is global load monitoring as well as remote ice drift tracking and forecasting with hazard detection. The former enables more reactive motion control of the vessel and improved understanding of the ice conditions. The latter reduces the number of helicopter flights and the disposal of electronic equipment and batteries into the ocean, since physically placing tracking buoys on the ice will not be needed.

Another important activity is the use of inertial measurement units (IMUs) strategically placed on a ship. A time-frequency decomposition method has been developed, and detection algorithms will be used to identify unusual ship-ice interaction events and changes in the ice drift direction.

The activity in this WP is relatively broad, comprising the following research topics:

- A novel design philosophy for Arctic offshore floaters protected by ice management,
- Iceberg drift observations and iceberg towing in broken ice,
- Allison models for harsh environments,
- Spectral analysis of ice-induced acceleration,
- Methods for determining floe size distribution from sea-ice images,
- Dynamic positioning in drifting sea ice,
- Iceberg detection and monitoring using an AUV equipped with multibeam sonar.

As an example of activities in WP5, we here highlight dynamic positioning in drifting sea ice. This is a major issue related to offshore petroleum activities in the Arctic involving station-keeping in ice. In deep water, it can be particularly advantageous for parts of the operations of a vessel to remain in position without being on anchors; this calls for dynamic positioning.

A novel automatic real-time remote sensing algorithm that uses radar images and global positioning satellite system measurements to estimate the ice-drift velocity vector in a region around a free-floating and potentially moving vessel is developed (Figure 13). This algorithm is motivated by the low image frequency of satellite systems together with the impracticality of deploying and retrieving ice trackers (beacons) deployed on the ice. The algorithm developed combines radar image processing with two Kalman filters to produce the estimated local drift vector decoupled from the ship motion. The proposed design is verified using a full-scale data set from an ice management operation north of Svalbard in 2015. The performance of the algorithm is comparable to that of trackers on the ice (Kjerstad et al., 2017).

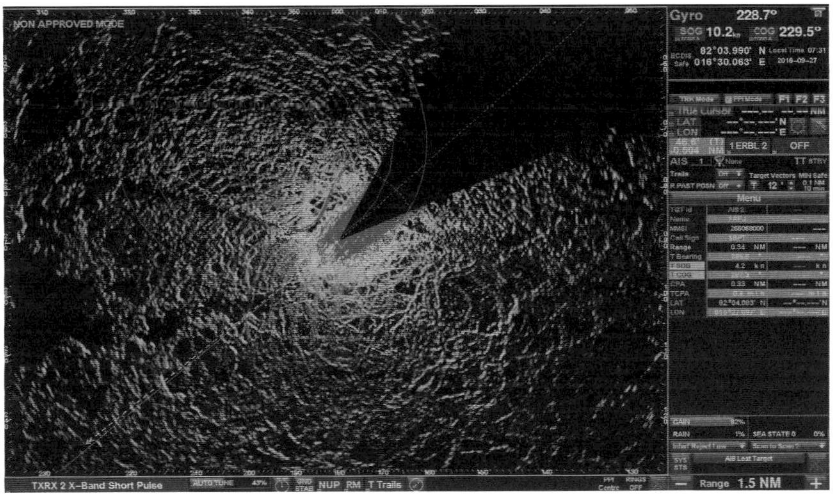

Figure 13. An unprocessed radar image captured from the operator workstation during ice management trials at the OATRC'15 (Kjerstad et al., 2017).

Coastal Technology (WP6)

Coastal zone development in the Arctic is quite demanding. The construction of roads, harbours and other facilities in the Arctic involves several challenges, e.g., exposure to the combined action of waves, currents and ice; high coastal erosion rates; building on permafrost soils; remoteness; and the lack of local materials suitable for construction purposes. Moreover, climate changes may result in a warmer Arctic with less sea-ice cover, leading to higher wave forces on structures, more unstable permafrost soils and increasing rates of coastal erosion during the service lifetime of our structures. Different research projects address these general challenges in response to SAMCoT industry partners' needs for innovation.

The research here comprises the development of/studies on the following:

- Thermo-hydro-mechanical models (THM) constitutive models. An elastic-plastic (and later extended to an elastic-viscoplastic) constitutive model for simulating the mechanical behaviour of saturated frozen soils is developed and implemented in PLAXIS. Efforts are made to extend this work to unsaturated soil, which is quite common in the Arctic.
- Sea ice effects on coastal waves. Coastal erosion in the Arctic is a function of numerous processes, including nearshore hydrodynamics, thermal and mechanical behaviour of frozen and unfrozen soil, atmospheric forcing, and the presence of sea ice. In particular, surface waves play an important role as they force circulation, which can lead to erosion.
- Arctic coastal erosion and protection measures. Here, thermal modelling includes the application of inverse modelling for the interpretation of soil thermal properties from field measurements and the application of passive cooling systems for the preservation of permafrost.
- Effects of climate-warming scenarios on coastal permafrost engineering design. Here, we perform probabilistic analysis utilising the newly developed THM models and taking into account the most recent climate-warming scenarios.

The composition of the Arctic coasts may differ substantially. At Spitsbergen, a majority of the coast contains gravel, as shown in Figure 14a, and the erosion mechanisms are different from those of the more silty Russian Arctic Coast (see Figure 14b).

(a) (b)

Figure 14. Example of coastal erosion: a) Vestpynten site (Spitsbergen) after a severe storm in Nov. 2016; b) low coast of Baydaratskaya Bay with ice-wedge. An MSU student participating in the expedition is sampling the frost grounds for further laboratory research (SAMCoT, 2017).

MSU, a research partner in SAMCoT, is responsible for the test site at Baydaratskaya Bay. The test site is situated on the west coast of the Baydara Gulf of the Kara Sea. Five fieldwork campaigns are being conducted there with the following goals:

- Study the processes and mechanisms behind coastline retreat,
- Investigate the properties of permafrost in the area of interest,
- Examine the interaction between permafrost and existing engineering structures (e.g., cofferdam, pipelines),
- Prepare for in situ validation of some elements of MSU's thermal abrasion modelling.

In previous years, valuable full-scale data were gathered and made available to SAMCoT researchers. These include mapping, measurements of coastline retreat rate and observations of the underlying physical processes, time-lapse photography of slope degradation processes, and continuous temperature measurements of the soil at two typical coastal sites (SAMCoT, 2017).

INNOVATION AT SAMCoT

SAMCoT's goal for innovation is to create opportunities for innovation in any of the following areas:

- Within SAMCoT itself, directly through its own research;
- Within the industry partners, indirectly, through input from SAMCoT research and engagements;
- From collaboration between partners and SAMCoT researchers, leveraging the research for joint industrial partner applications, over and above what was possible through their individual development routes.

An innovation can be a new product, new service, new production process, application or organisation, new methodology or a new tool that is launched in the market or used in production to create economic value.

Innovations can be captured in

- Patents,
- Standards development,
- Spin-off companies,
- Exploitation and innovation templates.

Approximately 30 innovation ideas have been filed. Currently, the most pronounced innovation from SAMCoT is the spin-off company ArcISo AS (Arctic Integrated Solutions) with the software product SAMS (Simulator for Arctic Marine Structures). The vision of this company is to provide pioneering sustainable development of Arctic marine activities. The mission is to secure the integrity of Arctic marine operations. The business idea is to contribute to a safer and more profitable Arctic marine activity through assessments of structure designs and operational procedures and the delivery of decision-support tools using superior sea-ice simulation technology.

CONCLUSIONS

The characteristics of a CRI are 1) high potential for innovation and value creation, 2) active cooperation between innovative companies and prominent research groups, 3) high-quality scientific research, 4) serving as a bridgehead for international cooperation, and 5) recruitment of talented researchers.

As stated in the midway evaluation SAMCoT has met all these milestones as well as bridging with other prominent research centres, in particular AMOS (Autonomous Marine Operations and Systems) at NTNU, Trondheim, and CIRFA (Centre for Remote Sensing and Forecasting for Arctic Operations), UiT, Tromsø, Norway.

ACKNOWLEDGEMENTS

The author would like to acknowledge the support from the SAMCoT CRI through the Research Council of Norway and all of the SAMCoT Partners.

REFERENCES

Kjerstad, Ø. K., Løset, S., Skjetne, R., Skarbø, R. A. (2017): An Ice-Drift Estimation Algorithm Using Radar and Ship Motion Measurements. Submitted to *IEEE Transactions on Geoscience and Remote Sensing.*

Hendrikse, H. (2017): Ice-Induced Vibrations of Vertically Sided Offshore Structures. Doctoral theses at Delft University of Technology, 2017.

Lu, W., Løset, S., Shestov, A. and Lubbad, R. (2015a): Design of a field test for the fracture toughness of sea ice. In *Proc. of the 23rd Int. Conf. on Port and Ocean Eng. under Arctic Conditions (POAC),* Trondheim, Norway.

Lu, W., Lubbad, R., Løset, S. (2015b): In-plane fracture of an ice floe: A theoretical study on the splitting failure mode. *Cold Regions Science and Technology, 110 (2015) 77–101.*

Lu, W., Lubbad, R. and Løset, S. (2015c): Tentative Fracture Mechanisms of the Parallel Channel Effect during Ice Management. In *Proc. of the 23rd Int. Conf. on Port and Ocean Eng. under Arctic Conditions (POAC),* Trondheim, Norway.

Lu, W., Lubbad, R, Løset, S. (2016a): Fracture of an ice floe: Local out-of-plane flexural failures versus global in-plane splitting failure. *Cold Regions Science and Technology, 123 (2016), pp. 1-13.*

Lu, W., Lubbad, R., Løset, S. and Skjetne, R. (2016b): Parallel Channel Tests during Ice Management Operations in the Arctic Ocean. *In Proc. of ATC, Paper OTC-27344.*

Nord, T. S. (2015): Force and response estimation on bottom-founded structures prone to ice-induced vibrations. Doctoral Theses at NTNU, 2015:217.

Nord, T. S, Lourens, E-M., Määtänen, M., Øiseth, O., Høyland, K.V. (2015): Laboratory experiments to study ice-induced vibrations of scaled model structures during their interaction with level ice at different ice velocities. *Cold Regions Science and Technology, 2015;119:1-15.*

SAMCoT (2016). SAMCoT Annual Report 2015, NTNU, Trondheim, *http://www.ntnu.edu/web/samcot/publications*

SAMCoT (2017). SAMCoT Annual Report 2016, NTNU, Trondheim, *http://www.ntnu.edu/web/samcot/publications*

van den Berg, M., Lubbad, R., Løset, S. (2017). Accuracy of a Non-Smooth Time Stepping Scheme with Non-Rigid Contacts for Ice-Structure Interaction. In *Proc. of the 2th Int. Conf. on Port and Ocean Eng. under Arctic Conditions (POAC*), Busan, Korea.

POAC'17
Busan, KOREA

Proceedings of the 24th International Conference on
Port and Ocean Engineering under Arctic Conditions
June 11-16, 2017, Busan, Korea

Confederation Bridge - Lessons and What They Tell Us about the Flexural Failure of Ice

Thomas G Brown[1], Noorma Shrestha[1]
[1] University of Calgary, Calgary, Canada

ABSTRACT

Monitoring systems for the measurement of ice forces and the observation of attendant ice conditions have been in place on several piers of Confederation Bridge, since it opened in 1997. The resulting database of ice interactions is very extensive, amounting to some 20,000 discrete interaction events, many of which have been analyzed in detail. The paper provides a detailed summary of the findings from this monitoring program. These relate to the characteristics of the interaction between ice and the conical piers, the relationship between these characteristics and the resulting ice force, and the relationship between the ice force and the ice physical properties. Because the piers are conical, the ice behavior is flexural. The paper describes the mechanics of the observed ice interactions and how these align and differ from those assumed in accepted ice force algorithms. The paper provides suggestions for improvements to these algorithms, particularly in the treatment of rubble piles and ride-up ice.

KEY WORDS: ice load, flexural failure, ice-conical structure interaction, ice rubble pile.

INTRODUCTION

The monitoring programme at the Confederation Bridge was initiated as a result of uncertainty as to the ice forces on the piers, and the need to understand the behavior of the ice as it interacts with the piers and the forces that result from these interactions. After 18 years of continuous monitoring of ice forces and ice behavior on the two of the main piers of the Confederation Bridge, an extensive ice action database has been created that has provided new insight in flexural failure of ice on conical structures. While the primary emphasis of the monitoring programme has been on the measurement of ice forces, much additional effort has been placed on the measurement of related data and the observation of the behavior of the ice interacting with the piers. These observations includes the behavior of the ice against ice-breaking cones, extracting geometric characteristic of the ice interacting with the piers from ice video, the velocity of the interacting ice, measuring of the pressure and local load distribution resulting from the ice interacting with the pier elements. This paper aims to summarize the findings from the long-term monitoring programme and establish the relationship between ice action and ice physical properties in terms of flexural failure of ice.

ICE LOAD MEASUREMENT

The global ice loads at the Confederation Bridge piers are determined by applying a transfer function from the measured tilts of the pier shaft. The tilts are also corrected for wind effects and baseline shift, detailed in Brown (2007a and 2007b). The tilts at the bottom of the shaft (G3TMY and STMY) are less affected by dynamics due to wind effects than the top tilts and thus are used to determine the global ice load. The tiltmeter data are recorded in two data collection modes, average and trigger. The initial assessment of the ice load is obtained from the average tilt data. In the daily ice load record, ice events, i.e. peak ice load within a predetermined time period that is greater than a preset threshold load (currently set at 0.75 MN), are identified. The time duration is defined by the number of time steps to either side of the peak load and has been set to 10 records of the average file, or 5 records to either side of the peak load, i.e. 170 sec duration for P31 and 150 sec for P32. Trigger data are only collected when pier tilts exceed a threshold and are recorded at intervals of 0.034s for P31 and 0.008s for P32. The number of ice events recorded at the Bridge instrumentation varies every year based on winter condition, availability of both wind and tiltmeter data and reliability of the recorded data. For pier P31 the lowest number of events, 181, was recorded in year 2006 and maximum number of events, 2307, in year 2009. For pier P32, the lowest number of events, 243, was also in year 2006 and maximum number, 1070, was recorded in 1999. For all years from 1998 to 2010, regardless of direction of load, a total of 12,220 ice events at pier P31 and 5,795 ice events at pier P32 had been identified from available average tilt data.

The presence of the Confederation Bridge across the Northumberland Strait has affected the ice conditions in the area. The strong tides, both flood and ebb, and driving forces which move the ice back and forth against the multiple piers, result in either the ice breaking or it lodging against the piers. The ice forces on the piers are a function of a number of different variables (Brown et al., 2010) such as ice growth, ice features, mechanical properties of ice, and ice-structure interaction variables such as rubble height, rubble angle, current velocity, wind speed, tide height etc. The ice concentration upstream (west) of the Bridge is greater than the downstream (east) (Brown et al., 1994), which is also confirmed by the number of events recorded by the instrumentation system at the piers. While not all the variables play a significant role in the final ice load determination, they all have some influence on ice load. In the design process, both the limit stress load (load required to fail the ice feature) and the limit force (driving force) were considered. However, the strong current and prevailing wind regime in the Northumberland Strait ensured that the limit force seldom governed.

The ice load measurement programme on the Confederation Bridge has been in place for the past eighteen years, beginning before the bridge opened in May, 1997. The system was put in place with three objectives: to validate the ice load design criteria, to provide information that could be used in the assessment of the maintenance system on the bridge, with particular emphasis on the ice breaking cones, and to provide a test system for the measurement of ice forces on conical structures.

FLEXURAL FAILURE MODELS

Extensive work to predict ice forces on conical-shaped piers was undertaken in Canada in the mid-1960s for a proposed Confederation Bridge linking Prince Edward Island to the mainland of Canada across the Northumberland Strait (Lavoie, 1966). Some of the earliest flexural failure models for ice-structure interaction include Afanas'ev et al. (1971); Bercha and Danys (1975); Edwards and Croasdale (1976); Pearce and Strickland (1979). In these models, the global ice force is the sum of a breaking term and a ride-up term. The expressions in these models are based on dimensional analysis with coefficients from linear regression or scale experiments. Other early works on flexural failure model include: Ralston (1977) model based on plastic limit analysis of the ice; Croasdale (1980) 2-D model based on the bending of an elastic beam on an elastic foundation. Some of the recent work includes: Määttänen and Hoikkanen's (1990) model which was based on the ice edge crushing effects and ice pile-up on a wedge shaped beam on an elastic foundation; Nevel (1992) model based on the failure of a truncated wedge on an elastic foundation; Croasdale (1994) 3-D model which was specifically modified for the design of the Confederation Bridge to include three-dimensional effects and the presence of a rubble pile; Lau (1999) model for multi-sloped structure with incorporation of rubble pile up loading. The most recent models developed to simulate ice flexural failure on conical structures was completed by Mayne (2007) and Wong (2014). Both Mayne and Wong models have been largely based on full scale ice-structure observations on the Confederation Bridge.

Mayne Model

Mayne developed a flexural failure model based on ice rubble pile behavior observed at the conical piers of the Confederation Bridge. The Mayne model consists of two main components, the breaking load and the load due to the rubble pile. He observed that the profile of the ice rubble pile resting on the cone and ice sheet can be approximated by a bilinear profile. He proposed to use the 2D method of slices to calculate the loads acting on the conical pier and the advancing ice sheet. Mayne focused on the model of the rubble pile formed around the cone. The breaking load of the Mayne model is derived based on the assumption that the ice sheet can be modelled as a semi-infinite beam on an elastic foundation as devised by Hetényi (1946). The model specifically accommodates the difference in deflection of the ice sheet between the apex of the cone and the sides, where no deflection occurs. One consequence of this acknowledgement of the physics of the interaction is that the "circumferential" crack is not circular but elliptic. Observations of this crack, and there are not many due to the presence of the rubble pile, confirm this crack shape. To calculate the breaking load, the rubble pile is treated either as a triangular load or an equivalent point load. A resolution factor was used in converting the 2D horizontal load into a 3D horizontal load based on the assumption that the rubble pile is axisymmetrical around the conical pier.

From the Confederation Bridge Monitoring Program, Mayne observed that due to the complexity of the non-linear profile, the cross-section of a rubble pile at the front end is mostly non-linear, unlike in other flexural failure models where a linear profile is assumed. The dimensions of the bilinear profile (Figure 1) are determined by the following key parameters: upper and lower profile angles (θ_U and θ_L) and horizontal (X_U, X_R) and vertical (Y_R) distances. These key parameters are determined using image analysis software to measure the image pixels and scaling these pixels to the appropriate dimensions. A bilinear rubble pile geometry means two geometric cases are to be considered.

The first geometry case, Case I, is when the bilinear point is above the cone (Figure 1a). The second case is when this point is above the ice sheet (Figure 1b). As the rubble pile profile for a given interaction event, will vary diversely in dimensions and geometry, the calculation of the overall horizontal load for the rubble portion will depend on the geometric case (I or II). The calculation of the total horizontal flexural load due to ice-structure interaction is sum of breaking load and the rubble load.

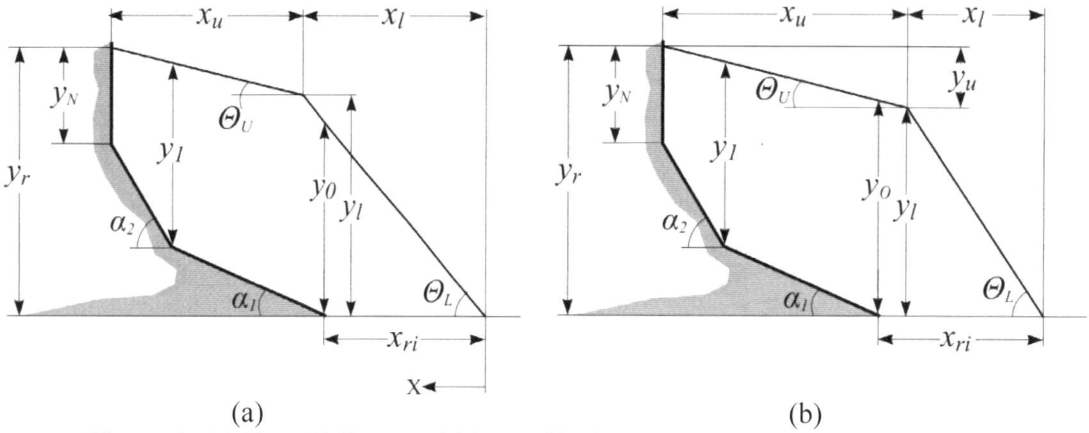

Figure 1. (a) Case I bilinear rubble profile (b) Case II bilinear rubble profile (Mayne, 2007)

Wong Model

One of main drawback of Mayne model was assuming that the rubble pile is axisymmetrical around the conical pier. The rubble pile geometry from video image analysis only provides the estimated dimensions at the middle cross-section of the ice rubble pile where the maximum height of the pile occurs. To overcome this limitation, Wong (2014) proposed a three-dimensional model for ice rubble interaction on conical structure. Wong modeled the geometry of the ice rubble pile around the conical pier through small-scale tests using a cone, and a pile of gravel. The rubble pile around the cone was divided into ten horizontal strips as shown in Figure 2. It was assumed that each section moves at a same velocity and each section acts independent of adjacent sections. The motion of ice rubble in each section aligns with the direction of the advancing ice sheet and is upward which is in contrast with the assumed direction made in most previous flexural failure models (most models assume that all ice rubbles are pushed up converging toward the pier centre). With the measured rubble pile geometry along ten horizontal strip sections, the rubble load around the cone is estimated by up-scaling results from physical model test (Wong, 2014). The ice rubble loads on the conical pier and ice sheet were determined from a generalized Morgenstern–Price method of slices (Fredlund and Krahn, 1976) to simulate the actual ice rubble pile behavior observed in the field. In Wong model, the slices are aligned parallel to the direction of motion, whereas the slices in the Mayne model are aligned perpendicular to the direction of motion. The behavior includes ice rubble formation, ride-up, and motion. A 3D finite element program was used to model the interaction between the ice rubble, ice sheet and conical pier face, and the flexural failure in the ice sheet. The driving forces and the end moments due to the driving force and contact forces are applied to the support contact points between the ice sheet and the conical face. Finally, the total horizontal load is calculated by summing the horizontal loads due to the ice rubble load on the pier and breaking load for flexural failure of the ice sheet.

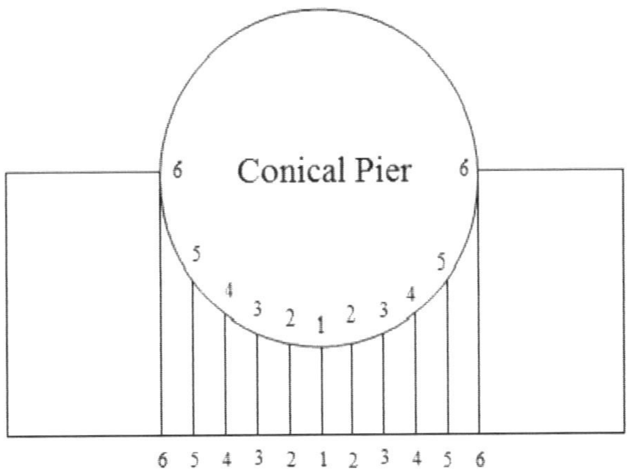

Figure 2. Ice sheet divided into ten horizontal strips with section 1-1 being the middle section (Wong, 2014)

FLEXURAL FAILURE EVENTS

Flexural failure is the predominant ice failure mechanism at the Confederation Bridge. A significant number of limit driving events has also been observed at the Bridge as discussed by Tripathi et al. (2007) and Shrestha et al. (2009). This paper will focus on flexural failure event loads only. The annual maximum flexural loads, based on ice events from 17sec and 15 sec average data, for piers 31 and 32 ranged from 1.29 MN to 6.74 MN and 1.25 MN to 3.6 MN respectively. The annual maximum loads for events failing in flexural are listed in Table 1.

Table 1: Annual maximum flexural loads at either side of both piers, P31 and P32

Year	P31 Load (MN)		P32 Load (MN)	
	NW side	SE side	NW side	SE side
1998	3.8	-3.31	No data	No data
1999	3.5	-3.11	2.77	-1.34
2000	3.45	-1.96	2.33	-1.40
2001	2.26	-1.84	2.12	-1.88
2002	2.42	-2.25	2.26	-1.40
2003	6.74	-2.78	2.27	-2.68
2004	3.09	-1.79	No data	-1.03
2005	3.82	-2.10	3.60	-1.94
2006	1.29	-0.86	1.44	-0.79
2007	3.25	-2.96	2.27	-1.17
2008	6.09	-2.78	No data	No data
2009	3.44	-2.39	2.82	-1.61
2010	1.53	-1.47	1.25	-0.75

Comparison of Measured Load and Model Load

Tibbo (2010) compiled 100 flexural failure events from Confederation Bridge monitoring data and compare the performance of four flexural failure models (Ralston, 1977; Nevel, 1992; Croasdale, 1994; and Mayne, 2007) against full scale trigger load data at pier P31. The predicted loads from the four flexural failure models against measured loads are shown in Figure 3. In her study she concluded that on average all models both underpredicted and overpredicted the ice loads, with Nevel model predicting the highest load. Based on the scatters around 1:1 slope line, more accurate predictions came from the Mayne model (Tibbo, 2010; Wong et al., 2016).

Wong (2014) used a subset of 10 out of 100 flexural failure events used in Tibbo (2010) study to validate his model. Figure 4 shows the predicted load values from Wong model along with loads from Croasdale and Mayne model. For the selected ten events, Wong model gives better load prediction than the Croasdale and Mayne models. Since the Wong rubble pile model considers three dimensional geometry of rubble pile observed at Confederation Bridge, the rubble pile model is more accurate and realistic.

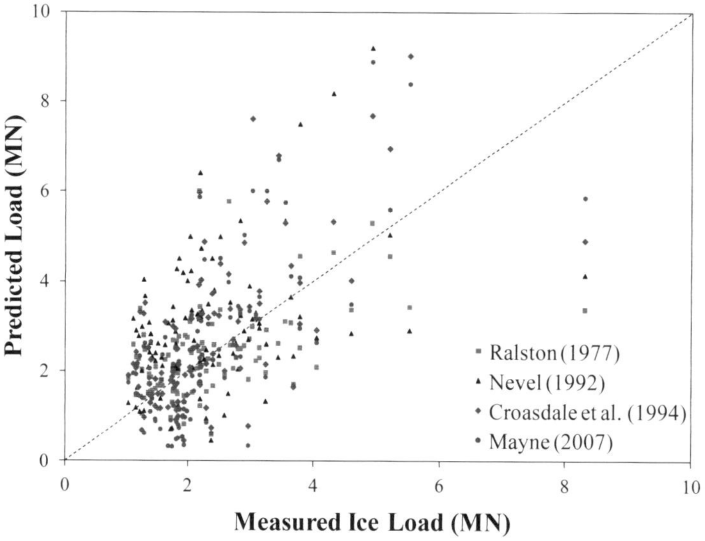

Figure 3. Measured ice load versus predicted ice loads by flexure failure models (modified after Tibbo, 2010)

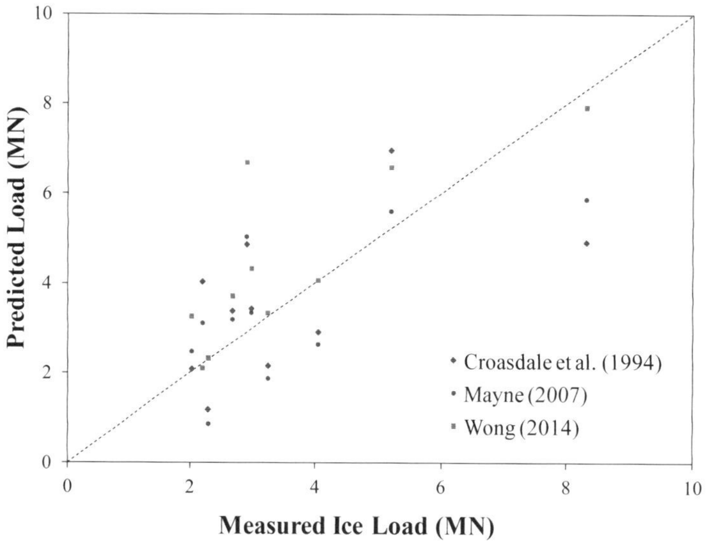

Figure 4. Measured ice load versus predicted ice loads by flexure failure models (modified after Wong, 2014; Wong et al., 2016)

As already discussed, both Mayne model and Wong model were developed based on ice-structure observation at Confederation Bridge and both consider the double conical section and the non-linear rubble pile that is often observed. However, the performance of Wong (2014) model has been validated against only 10 dataset (Wong, 2014; Wong et. al. 2016) and it needs to be expanded for additional full scale data before it could be used for further assessment. Tibbo (2010) flexural failure model assessment suggested Mayne model to be more accurate model for Confederation Bridge data. Hence, for this paper, the model loads for flexural failure events are calculated using Mayne's flexural failure model. The ice interaction data required for the model input is derived from ice video recording. As only daylight data can be extracted from video, the analysis is limited to events occurring during day only. The mechanical parameters such as flexural strength, modulus of elasticity and friction coefficient which are not measurable from video are estimated based on sensitivity analysis (Tibbo, 2010; Shrestha, 2012).

The scatter plot of model load and measured load, both event load and trigger load, are shown in Figure 5 and Figure 6 for piers P31 and P32 respectively. As shown in figures, most of the model loads are within ±1MN of the measured loads. Comparing the model loads with event loads, about 93 % of model loads lies within ±1MN of the measured load. While considering trigger loads, about 71 % of model load for P31 and about 62 % of model load at P32 are within ±1MN of the trigger load. The design load for any structure should never be underpredicted and thus for any model the percentage of underpredicted load are major concern. The model load should be compared with trigger load whenever possible. Considering the trigger loads, about 28% of model loads for P31 and about 38% of model loads for P32 lies outside of lower boundary of 1:1 line and again about 16 % of predicted loads were within ± 10% of trigger loads. In general, when 100% breaking loads is considered, majority of the events loads were overpredicted and the trigger loads were underpredicted. Based on the full scale observation, Mayne incorporated a range of potential failure ice loads: minimum, maximum and average model load. The minimum ice load is sum of the rubble load and 25% of the breaking load and the maximum ice load is sum of the rubble load and 75% of the breaking load. The average load is rubble load plus 50% of the breaking load. Figure 7 and Figure 8 shows the scatter plot of the model load considering 25%, 50% and 75% of breaking load contribution versus measured loads for piers P31 and P32 respectively. The recommended range of model loads (25 % to 75% of breaking load) underpredicted the measured loads especially when comparing it with trigger load, so it is advisable to use 100% of the breaking load for Mayne model.

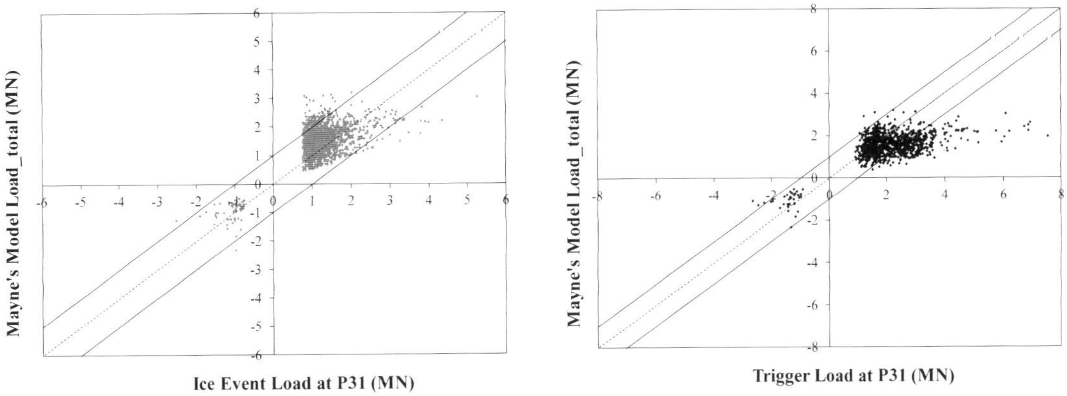

Figure 5. Predicted total load vs. measured ice load: ice events (left) and trigger load (right) at P31 with ±1MN boundary line

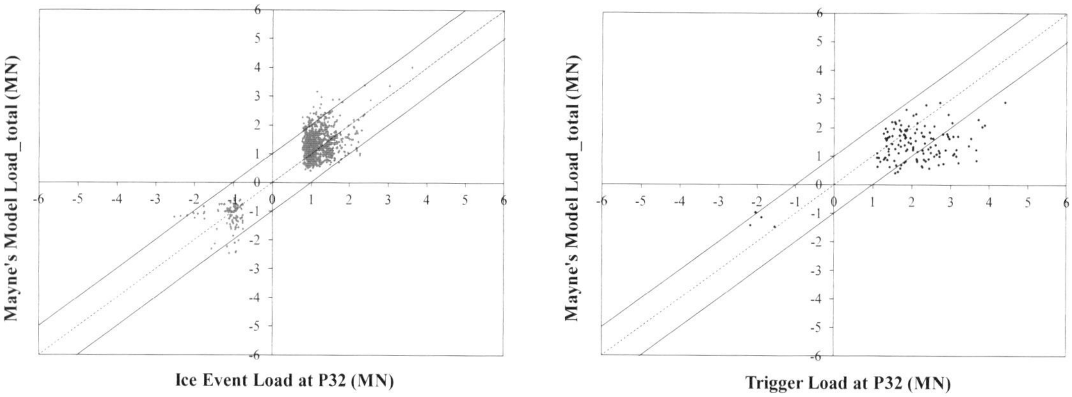

Figure 6. Predicted total load vs. measured ice load: ice events (left) and trigger load (right) at P32 with ±1MN boundary line

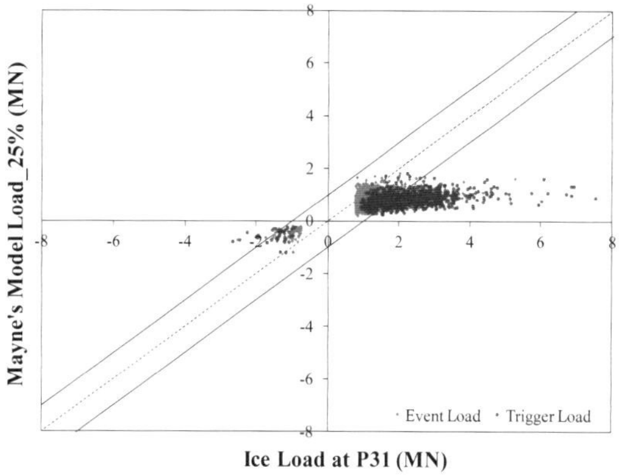

(a)

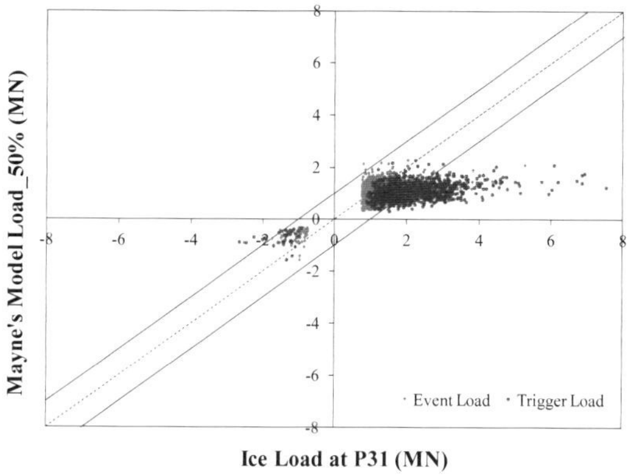

(b)

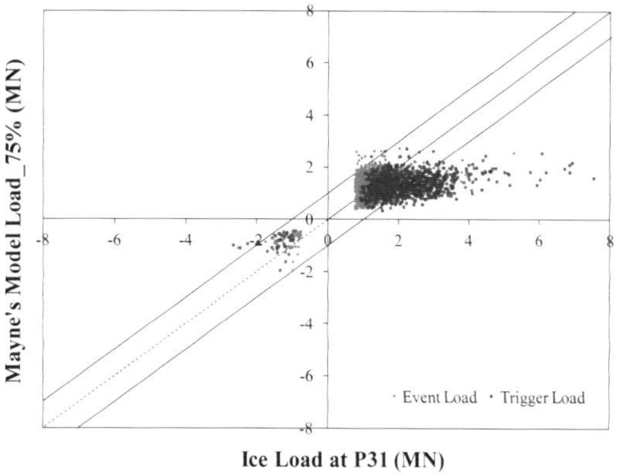

(c)

Figure 7. Predicted load: (a) min (25% of breaking load), (b) mean (50% of breaking load), and (c) max (75% of breaking load) vs. measured load, event and trigger, at P31 with ±1MN boundary line

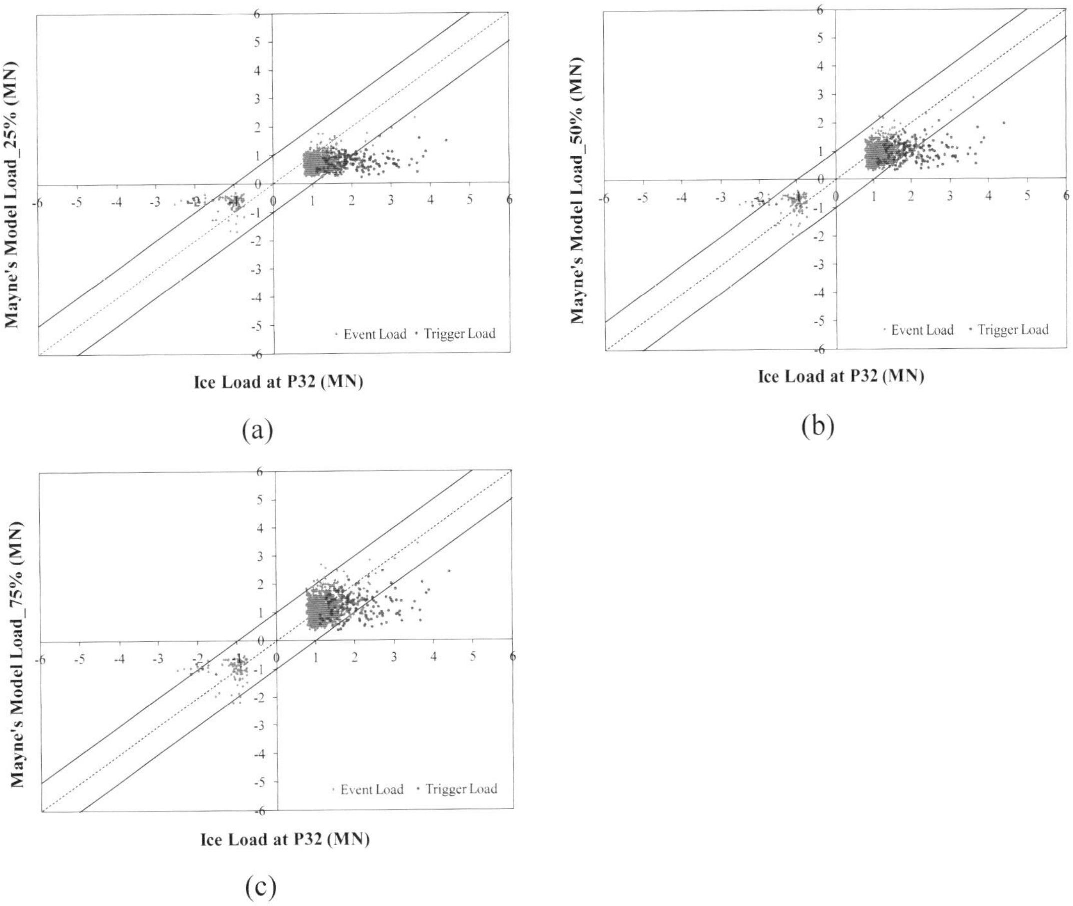

Figure 8. Predicted load: (a) min (25% of breaking load), (b) mean (50% of breaking load), and (c) max (75% of breaking load) vs. measured load, event and trigger, at P32 with ±1MN boundary line

ICE LOAD AND EVENT PARAMETERS

Ice parameters have always been uncertain factors while estimating ice action on the structures. Various studies have been carried out to investigate the relationship between ice geometric parameters and their effects on ice load, individually or combined. However, the conclusions vary between different studies. Conical structures are more effective in reducing the forces associated with interactions with ice features. Still, the failure against the conical structures is a complicated process which involves a number of different parameters affecting ice load on the structure. In this section, three important parameters: ice thickness, ice velocity and rubble pile are analyzed to study their effect on ice load and relation between these parameters.

The event specific parameters for flexural failure events at both piers occurring during daylight hours are extracted from image analysis of the video imagery. These parameters include: ice speed, ice thickness, pileup or ride-up height, rubble angle and upper pile length. The ice thickness is determined from ice pieces within the rubble pile or ride up ice. For each event, multiple ice thickness measurements are taken, if available. To define rubble geometry, pile up height, rubble angle, and upper pile length are measured. The ice sheet velocity is determined from two images taken approximately 10 sec apart and measuring the distance of a particular landmark on the moving ice sheet. Multiple landmarks on the ice sheet surface are identified in each set of pictures and velocity is averaged for a particular event time.

The measured ice thickness for associated events (i.e, loads ≥0.75 MN) varied from 0.17 to 1.6 m. Ice thickness less than 0.17 m was also observed but was not accounted for in the ice events; hence they are not considered in this study. The average ice velocity ranged from 0.012 to 1.64 m/s. The stall events have been excluded in this analysis.

For the ice events, the rubble pile height measured above mean sea level varied from 0.65 m to 6.21 m at P31 and 1.12 m to 6.67 m at P32. The rubble pile-up height data for pier P31 is normally distributed with a mean of 3.4 m and standard deviation of 0.74 m (Figure 9). Similarly, for pier P32, the rubble pile-up height is normally distributed with a mean value of 3.37 m and standard deviation of 0.76 m (Figure 9). For both piers, around 70 % of pile-up height measurements lie within mean ± 1*standard deviation and 95 % of measurements lies within mean ± 2*standard deviation, as shown in Table 1. Combining the information from both piers, the majority of events had rubble height in range of 2.2 to 4.5 m suggesting pile up occurring between cone 1 i.e., the 52° conical section and cone 2, the 78° conical section. The transition from the lower 52° cone to the upper 78° cone occurs at 2.6m above the mean sea level. The observations showed that the vast majority of rubble piles extend above this transition zone.

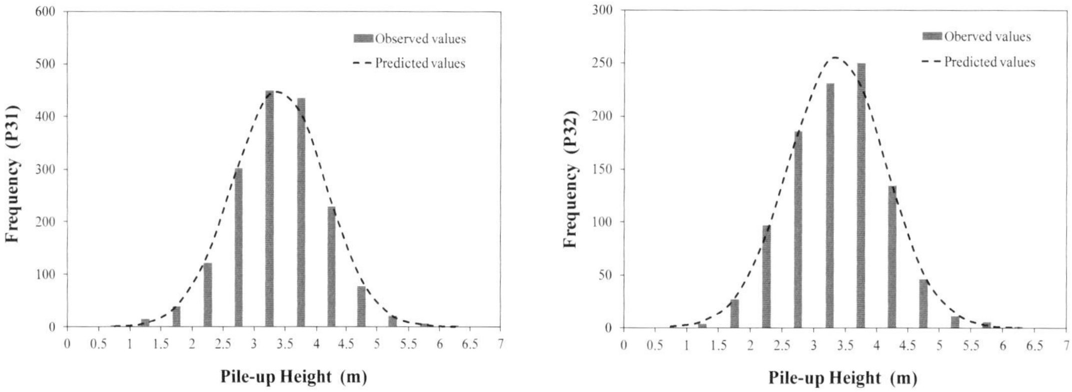

Figure 9. Distribution of pile-up height at P31 (left) and P32 (right)

Table 1. Cumulative distribution of pile-up height measurements

Mean ± X*std dev	P31, Pile-up height		P32, Pile-up height	
	Range	Percentage	Range	Percentage
$\mu \pm 0.5*\sigma$	3.03 – 3.77 m	39.73 %	2.99 – 3.75 m	37.64 %
$\mu \pm 1.0*\sigma$	2.66 – 4.14 m	70.31 %	2.61 – 4.13 m	69.63 %
$\mu \pm 1.5*\sigma$	2.29 – 4.51 m	87.84 %	2.23 – 4.51 m	87.59 %
$\mu \pm 2.0*\sigma$	1.92 – 4.88 m	94.92 %	1.85 – 4.89 m	95.96 %
$\mu \pm 3.0*\sigma$	1.18 – 5.62 m	99.41 %	1.09 – 5.65 m	99.50 %

Effect of Ice Speed on Ice Load

The momentum of the ice is related to the velocity of ice, which affects the load it can impart to a structure during an impact. Brown (2008) study showed a weak relationship between ice loads and ice velocity. He used the Confederation Bridge data from 1998 to 2000 to examine the relationship. This paper incorporates data from 1998 to 2010 to examine any correlation between ice velocity and ice load. Figure 10 illustrates the scatter plot of the measured average ice load and speed of the interacting ice at piers P31 and P32. The plot represents all data for which the speed was measured, and average ice thicknesses ranging from 0.24 m to 0.96 m. Events with zero velocity, representing events from limited driving forces, have been excluded from the analysis. To study the relationship between measured data two approaches are considered, firstly considering all data and secondly taking only upper bound dataset. Considering all data, the relationship between measured load and ice velocity was weak. While using only the upper bound dataset, the envelope curve shows exponential decrease in load with increasing ice velocity (Figure 10). The upper bound relation for measured loads and ice velocity are represented by equation 1 and 2 for two piers. For pier P32, due to lack of data the event load due to limit stress condition is limited to maximum of 3 MN only, thus the envelope curve is much lower than that of pier P31.

$$F_{avg,P31} = 7.85e^{-0.99v} \quad with\ R^2 = 0.87 \tag{1}$$

$$F_{avg,P32} = 4.29e^{-0.69v} \quad with\ R^2 = 0.88 \tag{2}$$

where, F = ice load (MN) and v = ice velocity (m/s)

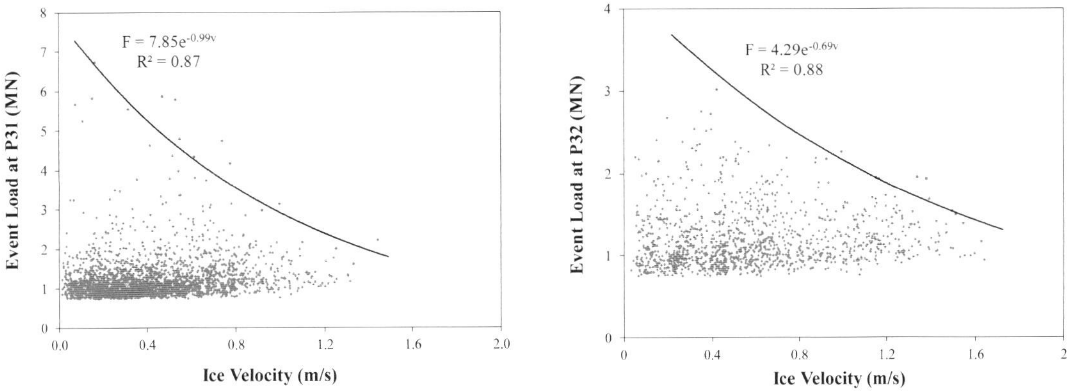

Figure 10. Average ice load against ice velocity at pier P31 (left) and P32 (right)

Effect of Ice Thickness on Ice Load

Ice thickness is one of most important parameters in any flexural failure model. The theoretical flexural failure term depends on the thickness powered by some value (for beam on elastic foundation, the bending term is proportional to the power 1.25); rideup and rubble terms are also related to ice thickness. All flexural failure models (Croasdale, Ralston, Nevel, and Mayne models) reported an increase in load for an increasing ice thickness (Tibbo, 2010). Figure 11 illustrate the plot between the measured average ice load and ice thickness at piers P31 and P32. The stall events have been excluded from the analysis. Taking all data into account, the relationship between measured load and ice thickness is weak. While considering only upper bound dataset, best-fit curve represents a power relationship with increasing ice load for increasing ice thickness (Figure 11). The relationship between measured load and ice thickness is represented by equation 3 and 4 for the two piers. Again due to data limitation the envelope curves for pier P32 is much lower than that of pier P31. As expected the indices in both cases are less than the breaking term index of 1.25, accounting for the effects of the rubble pile.

$$F_{avg,P31} = 6.16\mathrm{h}^{0.95} \qquad with\ R^2 = 0.96 \tag{3}$$

$$F_{avg,P32} = 3.52\mathrm{h}^{0.58} \qquad with\ R^2 = 0.89 \tag{4}$$

where, F = ice load (MN) and h = ice thickness (m)

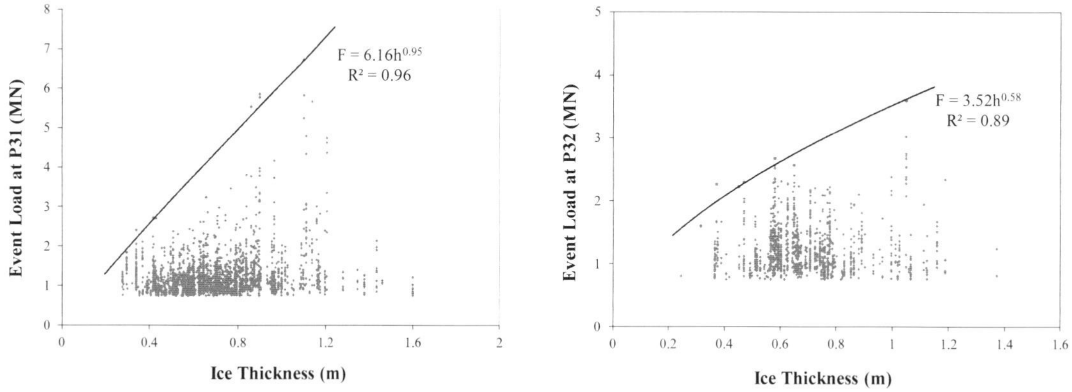

Figure 11. Average ice load against ice thickness at pier P31 (left) and P32 (right)

Effect of Pile-Height on Ice Load

Rubble pile load contribution is the most significant in flexural failure of ice on the conical structures. The shape and size of the rubble pile depend on the surface conditions, ice thickness, ice velocity and piece size (Mayne and Brown, 2000). Figure 12 shows the scatter plots of measured average ice load and rubble pile height at piers P31 and P32 respectively. These plots clearly show that the load increases with increase in rubble pile height. Considering only the maxima, the best-fit trend line suggest power relationship, where equation 5 and 6 represents envelope curve for piers P31 and P32 respectively.

$$F_{avg,P31} = 1.87\mathrm{H}^{0.8} \qquad with\ R^2 = 0.93 \tag{5}$$

$$F_{avg,P32} = 1.15H^{0.65} \qquad with\ R^2 = 0.88 \tag{6}$$

where, F = ice load (MN) and H = rubble pile height (m)

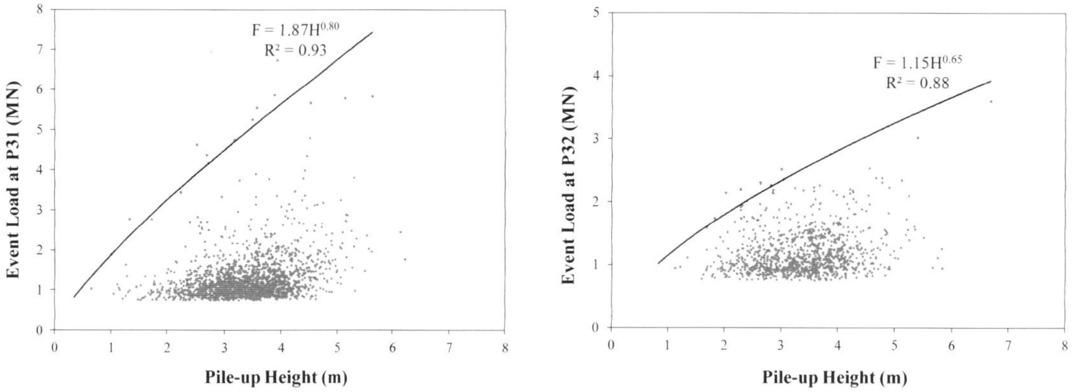

Figure 12. Average ice load against pile-up height at pier P31 (left) and P32 (right)

Relation between Pile-Up Height, Ice Thickness and Ice Velocity

Määttänen and Hoikkanen (1990) and Mayne and Brown (2000) developed relationships to describe the maximum rubble pile height based on ice thickness and ice sheet velocity. The relationships were derived from full scale observations of rubble piles resulting from interaction of level ice with conical structures. Määttänen and Hoikkanen (1990) model is based on field observations made from light piers in the Gulf of Bothnia. The rubble pile is supported by the face of the cone and the advancing ice sheet. Their model considered two cases, (i) when the ice sheet is in contact (supported) with the face of the cone and (ii) when it is not in contact (unsupported) with the face of the cone. The supported or unsupported conditions affect the ability of the ice sheet to support any rubble pile. Using the observations from the Gulf of Bothnia, a relationship is derived between height of rubble pile (H) and ice thickness (h_i) is described by equation 7.

$$H = A.h_i^{0.5} \tag{7}$$

where, the constant (A) is assumed to be 4.5 when rubble pile is unsupported and 6 when the ice sheet is fully supported by the cone.

The Mayne and Brown (2000) envelope curve was based on the full-scale data during period of 1998-2000 at pier P31 from the Confederation Bridge. The relationship between height of rubble pile (H) and ice thickness (h_i) is represented by equation 8. Similarly, the relationship between the height of rubble pile (H) and ice velocity (v) is represented by equation 9.

$$\mathrm{H} = 7.6\mathrm{h}_\mathrm{i}^{0.64} \tag{8}$$

$$\mathrm{H} = -3.12\mathrm{v}^2 + 2.07\mathrm{v} + 7.09 \tag{9}$$

Figure 13 shows the plot of the rubble pile height versus ice thickness for piers P31 and P32 using data from 1998 to 2010. Considering only maxima, the envelope curve for the rubble pile height (H) and ice thickness (h_i) is described using a power function, represented by equation 10 and 11 for piers P31 and P32 respectively. Note, once again, the very similar behavior between the rubble piles at the two piers.

$$H_{P31} = 7.52h_i^{0.51} \qquad with\ R^2 = 0.98 \tag{10}$$

$$H_{P32} = 7.50h_i^{0.58} \qquad with\ R^2 = 0.91 \tag{11}$$

The figures also show the upper envelope curve developed by Mayne and Brown (2000) and Määttänen and Hoikkanen (1990). The envelope curve plotted indicates that Määttänen and Hoikkanen model underestimates the height of rubble pile for the ice regime in the Northumberland Strait. The envelope curve with new datasets agrees well with envelope developed by Mayne and Brown (2000). There is an upper limit to the thickness for which this relationship is no longer valid, otherwise the relationship indicates pile heights continue to grow with thickness, however this maximum ice thickness has not been determined.

Figure 14 shows the relation between rubble height and ice velocity at piers P31 and P32. As speed increases, the maximum rubble pile height decreases as a result of improved clearing of the rubble. Considering only maxima, the relationship between decreasing pile height (H) as velocity (v) increases, fitting second order polynomial trend line is represented by equation 12 and 13 for P31 and P32 respectively. The figure also shows the upper envelope curve developed by Mayne and Brown (2000) indicating similar trends.

$$H = -2.97v^2 + 2.0v + 6.54 \qquad with\, R^2 = 1 \tag{12}$$

$$H = -2.77v^2 + 2.26v + 6.56 \qquad with\, R^2 = 0.93 \tag{13}$$

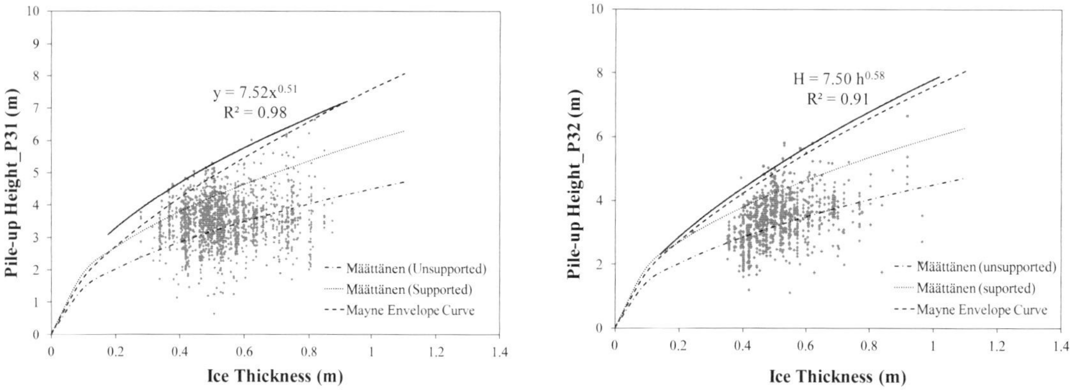

Figure 13. Relationship between rubble pile height vs. ice thickness at pier P31 (left) and P32 (right)

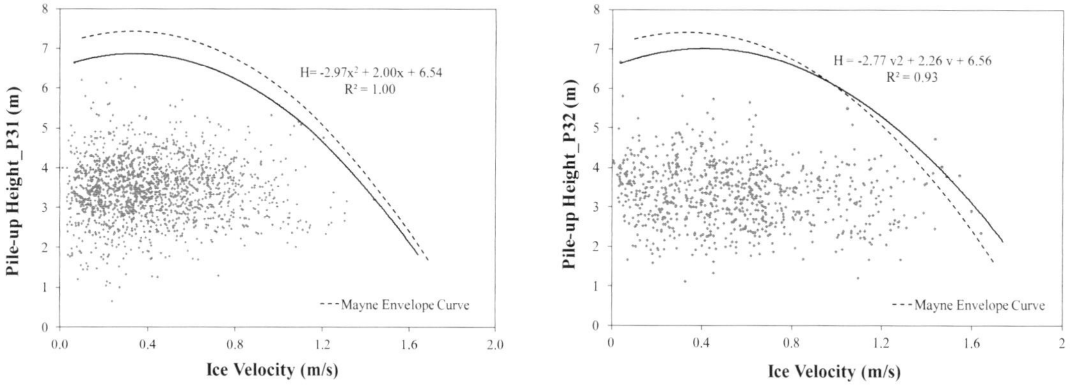

Figure 14. Relationship between rubble pile height vs. ice velocity at pier P31 (left) and P32 (right)

Keel Depth

For the Confederation Bridge, most ridges originate from the Western portion of the Northumberland Strait (Lemeé, 2003). The keel geometric parameters and their effect have been extensively studied for the Confederation Bridge data by Lemée (2003) and Obert (2010). The most recent ridge study for the area have measured maximum keel depths of 7.85 m and 8.49 m for the 2007 and 2008 ice seasons respectively (Obert, 2010). The maximum keel depth recorded between 1996 and 1998 by the Datasonics instrumentation was recorded in 1998 at 15 meters (Brown et. al., 1998). Obert (2010) study has shown that there is no apparent relationship between load and keel depth. Figure 15 shows the plot of load against keel depth and the relationship is very weak. Due to the unique shape of the Bridge pier, no relationship exists between ice load and keel depth as the bottom edge of the ice shield breaks up the keels greater than 4 m, thus reducing the load effect.

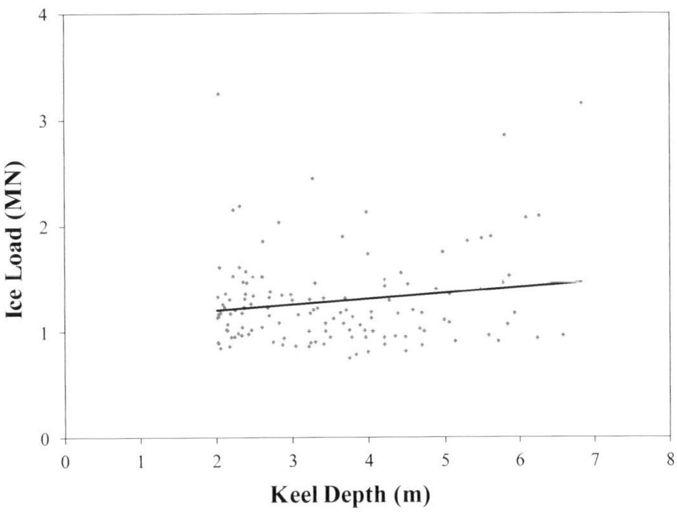

Figure 15. Plot of load against keel depth for P31

PROBABILITY OF EXCEEDANCE

The probability of exceedance plots for the loads for all years from 1998 to 2010 for both piers is shown in Figure 16. The figure indicates that, at the lower levels of probability, negative loads are lower than positive loads for both piers. For pier P32, the curves cross, but this may be an artifact of a limited number of data points. At the 10^{-2} level, the two sets of curves are quite similar with the negative load being about 2.5 MN, and the positive load about 3.0 MN. At higher probabilities of exceedance, the curves from the two piers diverge but, again, this may be a result of few data points. Similarly, Figure 17 shows the probability of exceedance curves for both event and trigger load, irrespective of the loading direction for both piers. Even though the P32 curves are lower than the P31 curves, the probability curves based on event load have a similar pattern but for the trigger load, the curves for the two piers diverge. This is again due to the low number of trigger data at pier P32. For example, at the 10^{-2} level, the event loads are 3.1 MN and 2.8 MN and trigger loads are 4.7 MN and 3.8 MN for piers P31 and P32, respectively.

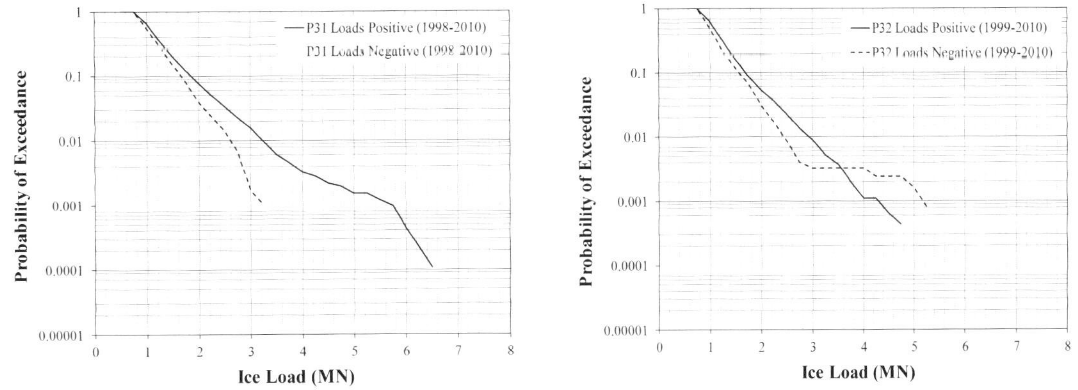

Figure 16. Probability of exceedance curve, P31 (left) and P32 (right)

POAC17-055

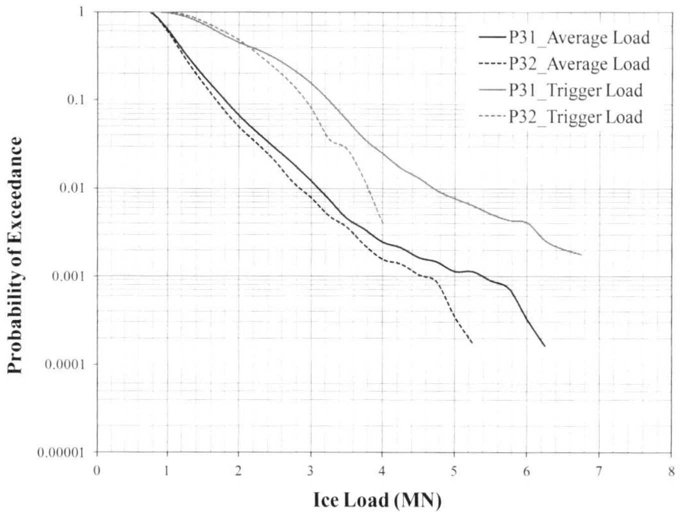

Figure 17. Probability of exceedance curve for average and trigger load for both piers

CONCLUSIONS

This paper presents comprehensive results from the long-term monitoring programme at a Confederation Bridge for both piers P31 and P32. The Mayne's flexural failure model is the most applicable model to assess event based model loads for the Confederation Bridge, as it accounts for both geometries: the double conical section of pier and bilinear rubble profile. The extensive comparison between measured load and model load shows that Mayne's model overpredicted event loads and underpredicted while comparing with trigger loads. About 50% of predicted loads were below 75% of the corresponding trigger load and only about 16 % of loads were within ± 10% of trigger loads. Mayne's model considers 2D rubble pile and use a resolution factor to convert 2D horizontal load into 3D horizontal load. For future study, Wong model should be further investigated to model 3D rubble pile and calculate model loads.

The analysis of the effect of event parameters: ice velocity, ice thickness and rubble height on measured load shows weak correlations between the ice parameters and between load and ice parameters when all events are considered. The weak correlation does not mean they are independent parameters. This suggests that there are many uncertainties involved between ice load and event parameters. A clear relationship exists between event parameters and measured load when only the upper bound data set is considered.

ACKNOWLEDGEMENTS

The authors would like to acknowledge the support of Strait Crossing Bridge Ltd., Public Works and Government Services Canada, the National Science and Engineering Research Council (NSERC) for providing funding. Further, the authors would like to acknowledge the contributions of their colleagues.

REFERENCES

Afanas'ev, D. P., Dolgopolov, Y. V., and Shvaishtein, Z. I., 1971. Ice Pressure on Individual Marine Structures. In Studies in Ice Physics and Ice Engineering, Edited by G.N. Yakovlev, Published by Israel Program for Scientific Translations, Jerusalem, Israel, pp. 50-68.

Bercha, F. G., and Danys, J. V., 1975. Prediction of Ice Forces on Conical Offshore Structures. *Proceeding of 3rd IAHR Ice Symposium*, Hanover, USA, pp. 447-458.

Brown, T. G., Barry, G., Carstens, T., Croasdale, K. R. and Frederking, R., 1994. The Potential Influence of the Prince Edward Island Bridge on the Ice Environment, *IAHR International Symposium on Ice, Trondheim*, Norway, Vol. 2, pp 838-846.

Brown, T., Bruce, J., and Croasdale, K. (1998). *Confederation Bridge Ice Force Monitoring Joint Industry Project Annual Report- 1998*. Calgary: IFN Engineering Ltd.

Brown, T. G., 2007a. Ice Force Monitoring. *Confederation Bridge Engineering Summit*, Charlottetown, PEI: CSCE, pp. 224-254.

Brown, T. G., 2007b. Analysis of Ice Event Loads Derived from Structural Response. *Cold Region Science and Technology*, 47, pp 224-232.

Brown, T., G., 2008. Ice Failure on Conical Structures – Effects of Speed, *Proceeding of the 19th International Symposium on Ice (IAHR)*, 2008, Vancouver, Canada, Vol. II, pp 887-896.

Brown, T. G., Tibbo, S. J., Tripathi, D, Obert, K. and Shrestha, N., 2010. Extreme Ice Load Events on the Confederation Bridge. *Cold Regions Science and Technology*, 60 (2010), pp. 1-14.

Croasdale, K. R., 1980. Ice Forces on Fixed, Rigid Structures. *A State-of-the-Art Report by IAHR Working Group on Ice Forces on Structures*, Edited by T. Carsten, CRREL Special Report 80-26, U.S. Army CRREL, Hanover, N.H., pp. 34-106.

Croasdale, K. R., and Cammaert, A. B., 1994. An Improved Method for the Calculation of Ice Loads on Sloping Structures in First-Year Ice, *Hydrotechnical Construction*, Vol 28, No. 3, pp 174-179.

Edwards, R. Y. and Croasdale, K. R., 1976. Model Experiments to Determine Ice Forces on Conical Structures. *Proceeding of Applied Glaciology Symposium*, Cambridge, U.K., J. of Glaciology, Vol 19, No. 81, pp. 660.

Fredlund, D. G., and Krahn, J., 1976. Comparison of Slope Stability Methods of Analysis. *Canadian Geotech. Journal*, 14(3), pp. 429–439.

Hetényi, M., 1946. Beams on an Elastic Foundation, University of Michigan Studies, *Scientific Series*, Vol. XVI, The University of Michigan Press, Ann Arbor, Michigan, USA.

Lemée, E. M., 2003. *Pressure Ridge Interaction with the Confederation Bridge Piers*. MSc. Thesis, Department of Civil Engineering, University of Calgary, Calgary, Alberta, Canada

Lau, M., 1999. *Ice Forces on A Faceted Cone due to the Passage of a Level Ice Field*. PhD Thesis, Memorial University of Newfoundland, St. John's, NF.

Lavoie, N. Y., 1966. *Ice Effects on Structures in the Northumberland Strait Crossing. In Ice Pressures against Structures*, NRC Technical Memo No. 92 (NRC No. 985), Ottawa.

Määttänen, M., and Hoikkanen, J., 1990. The Effects of Ice Pile Up on the Ice Force of a Conical Structure, *Proceeding 10th International Symposium on Ice*, Vol. II, Espoo, Finland, pp 1010-1021

Mayne, D. C., and Brown, T. G., 2000. Rubble Pile Observations, *Proceeding 10th International Offshore and Polar Engineering Conference*, Seattle, USA, pp 596-599.

Mayne, D. C., 2007. *Level Ice and Rubble Actions on Offshore Conical and Sloping Structures*. PhD. Dissertation, Department of Civil Engineering, University of Calgary,

Canada.

Nevel, D., 1992. Ice Forces on Cone from Floes, *Proceeding IAHR Symposium on Ice 1992*, Vol. 3, Banff Canada, 1391-1404.

Obert, K. M., 2010. *Ridge Keel Geometry and Interaction with the Confederation Bridge*. MSc. Thesis, Department of Civil Engineering, University of Calgary, Calgary, Alberta, Canada.

Pearce, J. C., and Strickland, G. E., 1979. Ice Forces on Conical Structures. *Proceeding of 11th Offshore Technology Conference*, Vol. 4, Paper No. 3635, Houston, USA, pp. 2407-2414.

Ralston, T., 1977. Ice Force Design Considerations for Conical Offshore Structures, *Proceeding of International Conference on Port and Ocean Engineering Under Arctic Conditions (POAC)*, Vol. II, Memorial University of Newfoundland, St. John's Newfoundland, Canada, pp 741-752.

Shrestha, N., Tripathi, D., Mayne, D., and Brown, T. G., 2009. Analysis of Limit Driving Force Events on the Confederation Bridge. *Proceedings of the 20th International Conference on Port and Ocean Engineering under Arctic Conditions (POAC09)*, Luleå, Sweden.

Shrestha, N., 2012. *Ice Action Database for the Confederation Bridge Piers*. PhD. Dissertation, Department of Civil Engineering, University of Calgary, Calgary AB, Canada, 2012.

Tibbo, J.S., 2010. *Flexural Failure of Sea Ice*. MSc. Thesis, Department of Civil Engineering, University of Calgary, Calgary, Alberta, Canada.

Tripathi, D., Shrestha, N., and Brown, T. G., 2008. February 13, 2007 Limit Driving Force Event on the Confederation Bridge. *8th International Conference and Exhibition on Performance of Ships and Structures in Ice (ICETECH),* Banff, AB, pp. 238–242.

Wong, C.K., 2014. *Development of a Three-Dimensional Model for Ice Rubble Interactions on Conical Structures*. M.Sc. Thesis, Department of Civil Engineering, University of Calgary, Calgary, AB, Canada.

Wong, C.K., Brown, T.G., and Robertson, J.S., 2016. Assessment of Flexure Failure Models Using Loads Measured on the Conical Piers of the Confederation Bridge During 1998–2008. *Journal of Offshore Mechanics and Arctic Engineering*, Vol. 139/Issue 1, DOI: 10.1115/1.4034527.

Design Situations and Limit States in the Design of Arctic Offshore Structures

Graham A.N. Thomas[1]
[1] Graham A N Thomas Consulting Limited, Slough, United Kingdom

ABSTRACT

Design of arctic offshore structures, and especially the codification of the design approach as in ISO 19906, the International Standard for the design of arctic offshore structures, requires going beyond the empirical, pragmatic solutions that have been developed for offshore structures in mature, temperate regions. The limit states design philosophy, embodied in ISO 19906, provides an approach for addressing design situations arising from operational, extreme, abnormal, and accidental events so that the desired structural reliability is verified for the relevant limit states. The philosophy also needs to accommodate the effects of operating philosophy and measures (e.g. ice management, disconnection of floating structures).

The Limit States Design procedure accounts for risk and reliability through the use of combinations of actions with specified probabilities, resistance formulations, partial factors, and limit states which arise from each design situation. The limit states for each design situation, such as component failure, local damage, or complete loss of structural integrity, depend on the design situation. Design verification, to confirm the required reliability for each design situation, is key to the design of safe and competitive structures.

This paper discusses issues that are being considered in the current revisions of ISO 19906, and in ISO 19900 which is the overall philosophy and general requirements document.

KEY WORDS: Arctic offshore structures; Design situations; Hazards; ISO 19906; Limit states.

INTRODUCTION

The first edition of International Standard ISO 19906:2010 was published December 2010, as described in Spring et.al, 2011. The second edition of ISO 19900:2013 was published in 2013.

After some years of experience in use, and with advances in technical knowledge, the need for updating both ISO 19906 and ISO 19900 was recognised. There was also a need to improve consistency and interfaces with related standards, including with each other.

Work to revise ISO 19906 started in 2014. The updating is described in Muggeridge et.al., 2017, and is expected to be complete in 2018. The working draft has been in committee as IOGP-ISO/WD 19906, 2016.

Work to revise ISO 19900 started in 2016, and is expected to be complete in 2019. The working draft is in committee as IOGP-ISO/WD 19900, 2017, and is still being revised.

This paper discusses issues and makes proposals which may or may not be incorporated in the final revision work. It focuses on the Ultimate Limit States (ULS) and specifically on the extreme and abnormal design situations, with some discussion of hazards, risk, and reliability for offshore structures. A number of issues related to risk, reliability, limit states, and exposure levels were previously discussed in Thomas, 2015, to which this current paper is a sequel.

BACKGROUND

ISO 19906 provides requirements, recommendations, and guidance, for design of offshore structures in the physical environment of arctic and cold regions. It relies on the other ISO standards in the 19900 series for common structure types, general requirements, and non-arctic specific requirements. Figure 1 shows some typical arctic offshore structures.

Figure 1. Typical Arctic Offshore Structures (left and centre courtesy GazProm)

ISO 19900, first published in 2002, provides general and unifying principles for all types of offshore structures. These principles include the limit states design procedure using the partial factor approach, as well as considerations for hazards, design situations, exposure level, robustness, and limit states.

ISO 19906 is focused on supplementary provisions for offshore structures in arctic and cold regions. ISO 19906 relies on the "structure-type" standards such as ISO 19902 for fixed steel structures for general design requirements for specific types of structure, as explained in Muggeridge et.al., 2017.

The objective of ISO 19906 is to ensure that complete structures, including substructures, topsides structures, floating production vessel hulls, foundations and mooring systems in arctic and cold regions provide an appropriate level of reliability and performance with respect to personnel safety, environmental protection and asset value.

SOME BASIC CONCEPTS

Action and Load

The term "action" is used throughout ISO standards instead of the term "load". ISO 19900:2013 defines action as "external load applied to the structure (direct action) or an imposed deformation or acceleration (indirect action)".

Values of actions termed "representative values" are the actual (unfactored) values, such as actual weights and environmental loads calculated from actual parameters.

Hazard and Hazardous Event

A hazard is a potential source of harm: human injury, damage to the environment, damage to property, or a combination of these. If a hazard can be identified and monitored, such as for an iceberg or a hurricane, planned operational measures such as ice management, or shutdown and evacuation, can be initiated.

A hazardous event occurs when a hazard interacts with a structure, either alone or in combination with other conditions. If the effects are of sufficient magnitude, limit states can be exceeded such that failure occurs.

Ice actions can be plotted against return period (i.e. probability of exceedance) on an action hazard curve, such as shown in Figure 2 which is based on ISO 19900:2013, Figure A.1. Figure 2 shows various types of environmental actions including three ice action curves using data from IOGP, 2010 for 100m wide, vertical-sided GBS structures. The three curves are selected to illustrate that different types of ice can have very different curve shapes and slopes.

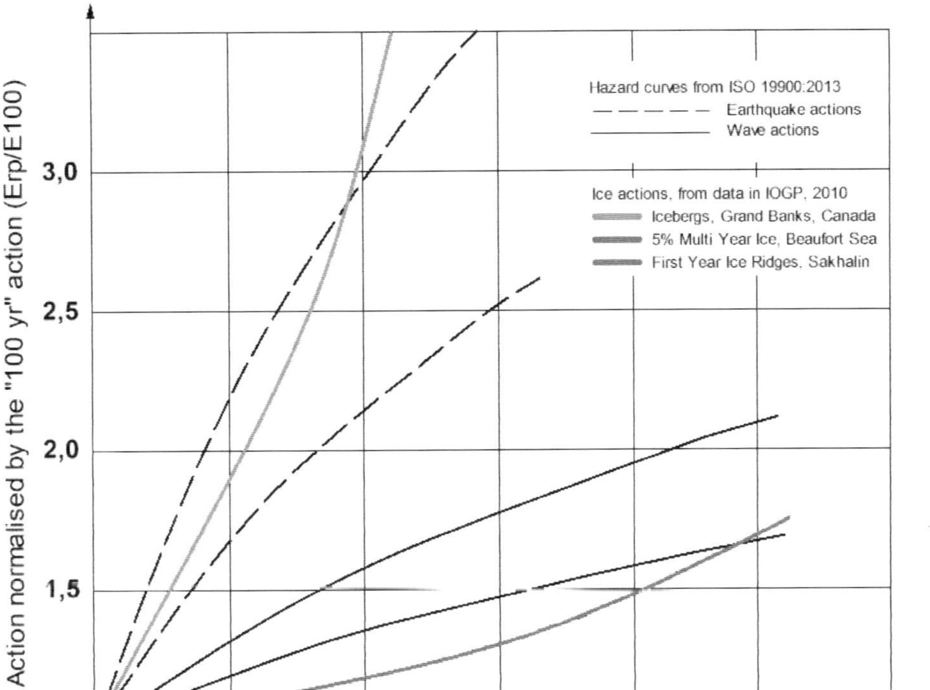

Figure 2. Typical normalised environmental action hazard curves for offshore structures

Extreme and Abnormal

The term "extreme" in ISO 19906 is reserved for events and actions with annual probability (of occurrence or exceedance) of 10^{-2} (100 year return period). This differs from some general use, where, "extreme" can be used as a generic adjective for severe environmental actions beyond an operating level, or as the maximum of a set of values within a defined time period.

Similarly, the term "abnormal" in ISO 19906 is reserved for events and actions used for abnormal design situations. The actual annual probability (of occurrence or exceedance) depends on the exposure level, see below, and is 10^{-4} for exposure level L1.

ISO 19900 explains the relationship between return period and annual probability, in particular that return periods of 100 years and 10,000 years correspond to annual probabilities of 10^{-2} and 10^{-4} respectively. Using "annual probability" terminology helps to avoid fruitless speculation about events 10,000 years in the future.

Limit states

ISO 19900 states that the performance of a structure, in whole or in part, shall be described with reference to a specified set of limit states beyond which the structure no longer satisfies the design requirements and is considered to have failed.

For the revision of ISO 19900, the limit states are divided into the following three categories:

- ultimate limit states (ULS);
- serviceability limit states (SLS);
- fatigue limit states (FLS).

Limit States for ULS design

Limits for ULS design include the following limit states:

a) ULS(a): failure of an individual structural component caused by design action effects exceeding design resistance (in some cases reduced by deterioration), including loss of structural stability (buckling, etc.);

b) ULS(b): loss of static equilibrium of the structure, or of a part of the structure, considered as a rigid body (e.g. overturning, sliding, sinking, or capsizing);

c) ULS(c): complete loss of integrity of the structure or vital parts of the structure when there is no further system ductility or reserve strength, including transformation of the structure into a mechanism (collapse or excessive deformation), loss of stationkeeping (free drifting);

The additional system resistance between failure of the first component with respect to ULS(a), and final collapse with respect to ULS(c), is determined by considerations of ductility, reserve strength, load shedding, energy dissipation, etc., i.e. robustness.

Exposure Level

Exposure level is a categorisation of potential consequences arising from a design situation. The required structural reliability depends on the exposure level.

ISO 19900 describes three exposure levels: L1, L2, and L3. The exposure level to be used in a design situation for design verification is determined by considerations of "life-safety" and of environmental and economic consequences.

L1 applies for a structure which is manned or has significant inventory (of hydrocarbons), and for which evacuation or shutdown is not planned in the event of the design situation under consideration. L2 and L3 can apply if increasing mitigation or elimination of risk to life safety and other consequences is demonstrated. Relevant measures can include hazard detection, monitoring and forecasting, success of ice management, staged shut-in of wells and shut-down of production and export facilities, and evacuation of personnel under prevailing conditions.

DESIGN SITUATIONS

It has been proposed that the concept of design situations be more clearly introduced into the revision of ISO 19906 as part of a more holistic approach to the overall design process, see Thomas, 2015. Now it is expected that the revised ISO 19900 will define different types of design situations, and the characterisation and specification of design situations.

This paper offers the author's view of information and proposals which may or may not be fully realised in the final publication of ISO 19900 third edition, due 2019.

Characterisation of a Design Situation

Design situations can be formulated for both serviceability limit states and ultimate limit states. Each design situation is characterised by relevant parameters and design criteria for which the design is verified, such as:

- Type of design situation;
- Hazardous event or dominating source of action (which provides the principal action);
- Exposure level(s) (includes aspects such as manning and inventory risk);
- Structural configuration (such as under construction, in-place, and damaged);
- Structural system (parts resisting event and actions);
- Operational measures (e.g. ice management, disconnection criteria, evacuation and shutdown criteria);
- Facility operational configuration (e.g. drilling, temporary works, tanker loading);
- Physical conditions and companion actions associated with the principal action;
- Combination of action types, with partial action factors;
- Partial material and/or resistance factors;
- Limit state (specific criteria for serviceability or ultimate limit state).

Choice of structural modelling and method of analysis can also depend on the design situation.

ULS Design Situations

A ULS design situation is specified for each hazardous event and for each dominating source of action. If a hazard curve has been developed, different hazardous events can be selected at different points on the hazard curve.

ULS design situations can be classified as follows:

- Operational
- Extreme
- Abnormal
- Accidental
- Short duration

ULS operational and extreme design situations are intended to ensure that, provided that limit states are not exceeded, there is no after-effect: no consequent reduction in structure reliability, no significant structural damage, no repairs needed for restarting or continuing operation. Limit states ULS(a) and ULS(b) apply. Partial factors, applied to representative values, provide a "safety" margin with respect to the actual limit state design values and structural damage.

ULS abnormal and accidental design situations are intended to ensure that, provided that limit states are not exceeded, there are no life-safety consequences or harm to the environment. There is however the likelihood of structural damage requiring repair or replacement of structure in order to restore adequate in-service reliability.

If structural damage occurs, e.g. after ULS(a) and before ULS(b) or ULS(c), a short duration, after damage design situation can be defined and fitness for continuing service can be assessed.

Extreme Design Situations.

Extreme design situations are defined for each extreme environmental event. The principal action arises from the extreme environmental event. Appropriate partial factors are applied for actions and resistances in order to verify that the design has adequate reliability.

The global behaviour of the structure is essentially elastic, even though local stress concentrations can exceed yield stresses and some nonlinear behaviour (e.g. pile-soil interaction) is expected. Some components may behave inelastically if their capacity to resist action effects is not reduced, e.g. plates in membrane action.

Abnormal Design Situations

Abnormal design situations are defined for each abnormal environmental event. The principal action arises from the abnormal environmental event. Structural reliability is achieved by the specified small probability of exceedance. All partial factors for actions, materials, and resistances are unity.

System and component ductility and reserve capacity may be taken into consideration in determining the ultimate resistance of the structure. Structural analysis may be non-linear and progressed to complete loss of structural integrity or collapse, with design resistance of individual components being exceeded and action effects being redistributed.

For abnormal environmental events when there are life safety implications, the structural integrity at the limit state needs to be assured to enable evacuation following event exposure.

LIMIT STATES DESIGN PROCEDURE

The limit states design procedure verifies that a structure has adequate reliability with respect to defined limit states in relevant design situations. Exceeding the limit state would result in failure. The steps can be outlined as follows:

- Hazardous events and other sources of actions are quantified;
- Design situations are defined for design verification to a limit state;
- Representative values of actions and resistance are derived from basic variables;
- Representative values of action types are derived from combining representative values of principal actions and companion actions;
- Partial factors are quantified, depending on the limit state and design situation;
- Design values for the design situation are calculated by factoring representative values;
- Design actions are applied in combinations of action types dependent on the design situation, and the structure is modelled and analysed to obtain the total design action effect, S_d, for the structure as a whole and for each component;
- Design resistance, R_d, is calculated from formulations using factored representative values of properties of materials and soils and of geometric parameters, or directly by factoring resistance formulations, for the structure as a whole and for each component;

- The design is verified for the relevant limit state by applying Equation (1).

$$S_{\mathrm{d}} \leq R_{\mathrm{d}} \tag{1}$$

The last four bullet points above are illustrated in Figure 3. It also shows the combinations of actions for two design situations. The curves are probability density functions for *S* and *R*.

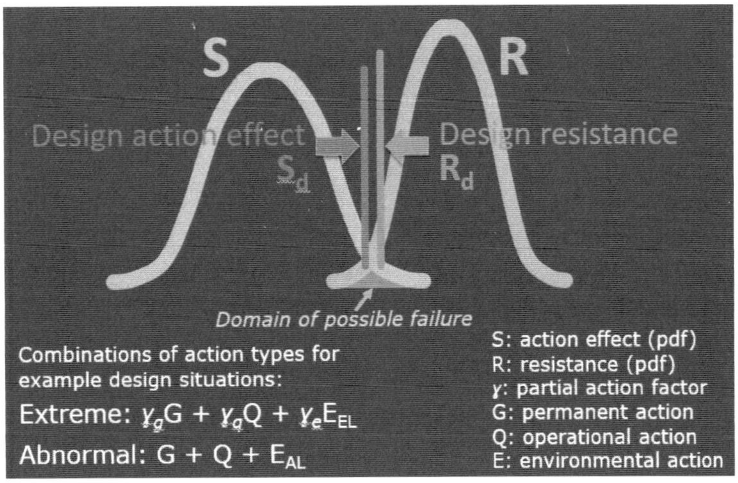

Figure 3. Probabilistic illustration of Equation (1)

RISK AND RELIABILITY

Risk Approach

It is common industry practice, to identify, assess, and manage all risks that can reasonably be foreseen, however unlikely. Risk events are identified, and can be assessed using a risk matrix to plot likelihood and consequence.

In the oil and gas industry, the concept of acceptable level of risk, or tolerable risk, is qualified by the over-riding imperative that no level of risk can be accepted, or tolerated, without a risk management plan. The challenge is what type of risk mitigation or risk management to apply.

The underlying approach of ISO 19906 is that there is a break point at which risks transition from requiring avoidance by design, to being managed by other means such as operational contingency planning. This is illustrated in typical risk matrix form in Figure 4.

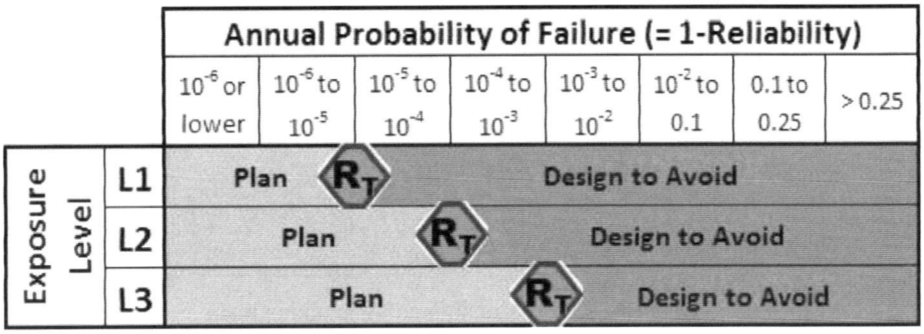

Figure 4. Risk Matrix showing Reliability Targets (R_T) for different Exposure Levels

Reliability Target

The break point between "design to avoid" and other planning, see Figure 4, is expressed as a "reliability target" (R_T) for the structural design verification. R_T is expressed in terms of an annual probability of failure. Failure, in the context of structures, is the risk event associated with exceeding a limit state. Targets can depend upon a variety of factors, see HSE, 2002.

ISO 19906 quantifies ULS reliability targets for each exposure level as $1\text{-}10^{-5}$ for L1, $1\text{-}10^{-4}$ for L2, and $1\text{-}10^{-3}$ for L3, as illustrated in Figure 4. The approach, and the calibration of the ISO 19906 partial action factors to these reliability targets, is explained in Thomas et.al., 2011.

The underlying philosophy of the design approach of both ISO 19900 and ISO 19906 is to ensure that the structure has the appropriate level of structural integrity. This is achieved by demonstrating, by means of structural design and, where applicable, operational measures, that the specified reliability target is satisfied. This is deemed to be achieved by using the limit states design procedure with appropriate partial factors and probabilities.

EXTREME-LEVEL AND ABNORMAL-LEVEL EVENTS AND ACTIONS

Likelihood and Quantification

ISO 19906 requires that representative values for ice actions (and for other environmental actions) are determined both at extreme-level (EL) and at abnormal-level (AL) for all relevant types of event. An ice event is defined as occurrence of ice-structure interaction for which ice actions are calculated. These are hazardous events with specified probabilities which need to be combined with analysis of resistance probabilities, ref. Figure 4, in order to quantify the probability of the risk event (failure). Then it can be plotted on a risk matrix, similar to Figure 4.

EL and AL ice actions represent events in different design situations, and are independently aimed at verifying that, after applying partial action factors and resistance formulations including reserve strength, the structure reliability target is achieved for each design situation.

Figure 5, based on the risk matrix of Figure 4, shows the EL and AL actions (denoted by E and A in the figure) for each exposure level, in a matrix space defined by likelihood of ice event.

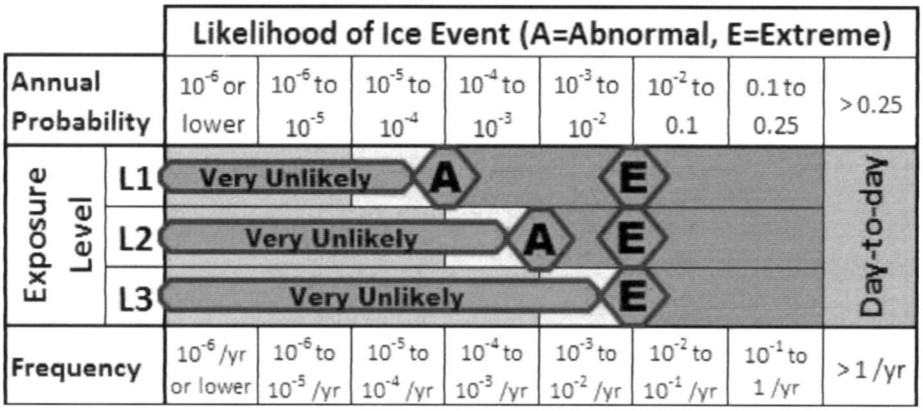

Figure 5. Likelihood Matrix showing EL and AL Ice Events

It is important to not confuse the concept of risk event, which relates to exceeding the relevant limit state in a defined design situation, with the concept of hazardous event, for which actions are calculated for design verification. For the former, various target failure probabilities have been quantified, see HSE, 2002, depending on life safety for the individual, the type of event and, in some approaches, likelihood of other risk events. For the latter, design event occurrence probability is specified, except for earthquakes, as 10^{-2}, 10^{-3}, or 10^{-4}, as indicated in Figure 5.

Action Hazard Curves

The values of actions for the prescribed event probabilities can be read from an action hazard curve for the hazardous event, such as those shown in Figure 2. Figure 6 is an alternative representation of action hazard curves, with axes swapped and not normalised, showing indicative curves for actions for different ice types.

It is also possible for different ice types to be dominant at the different levels. The EL ice action in Figure 6 arises from MY ice ridges, while the critical AL ice action could arise from icebergs.

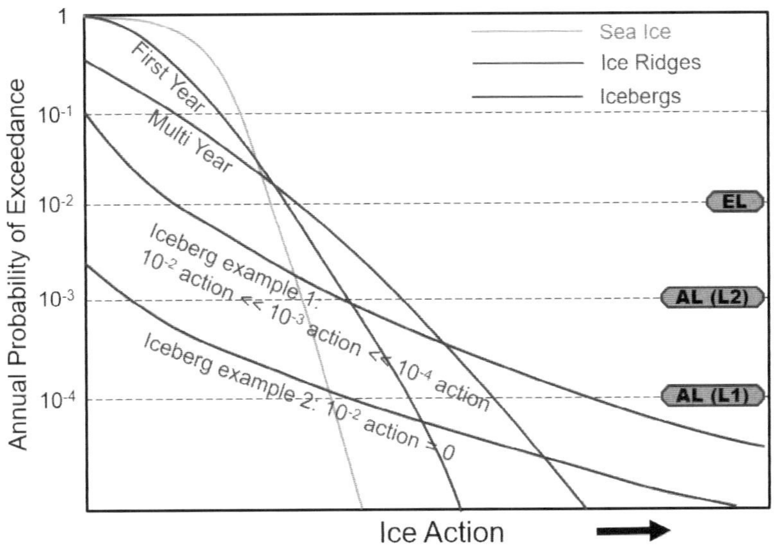

Figure 6. Action-Hazard curves for various types of hazardous ice events

Impact of Non-linear Action Hazard Curves

Few action hazard curves are linear (on a log-probability chart). The "slope" varies with ice type and geographic location. Action hazard curves for icebergs are particularly steep and curved, see Figure 2. Figure 6 illustrates that the ratio of the EL (10^{-2}) value to the AL for L1 (10^{-4}) value of the action hazard arising from a specific hazard is not consistent. The magnitudes of the ice actions also varies with ice type, location, and structural configuration.

Most curves in Figure 6 have both an EL value and an AL (L1) value of characteristic ice action. With reference to IOGP, 2010 typical ratios can be 1.20 for first-year ice ridges and 1.35 for multi-year ice ridges.

For icebergs where the action hazard curve has the form shown for "iceberg example 1", a typical ratio could be 4, as shown for the Labrador Sea. In this example the 10^{-2} p.a. (100-year) value when factored by the partial factor of 1.35 is much less that the value of the 10^{-4} p.a. action. Widianto et al., 2013 showed that the calculated 10^{-4} p.a. iceberg action can be up to 9 times the 10^{-2} p.a. (100 year) iceberg action at the Hebron location, offshore Newfoundland.

It is a myth that the intention of the factored EL ice action is to result in a value of design action with 10^{-4} p.a. probability of exceedance, thus emulating or substituting for the value of the AL ice action. This is one reason for defining separate design situations for extreme and abnormal.

Impact of Action Hazard Curves for Less Likely Events

For icebergs where the action hazard curve has the form shown for "iceberg example 2" in Figure 6, the iceberg encounter probability can be less than 10^{-2} p.a. This could be typical for some areas off the East coast of Canada, off Greenland (Baffin Bay) and for the Barents Sea.

Therefore the 10^{-2} per annum (100-year return period) representative/characteristic ice action value does not exist for "iceberg example 2", it is zero for this design situation, Design verification for this hazard will use only the hazardous event which causes the AL ice action.

There can be significant uncertainties in estimating abnormal ice actions. Discussion of types of uncertainty and approaches for taking this into account can be found in IOGP, 2010.

Operational Measures to Modify Action Hazard Curves

Ice action hazard curves can be modified by operational measures, both for fixed (gravity-based) and for floating structures. If operational procedures such as ice management and/or (for a stationary floating structure) disconnection are a part of the design philosophy to verify structural reliability, all relevant elements of the operational procedures need to be determined, documented, and applied throughout the service life.

For example, the iceberg encounter frequency associated with Grand Banks platforms offshore Newfoundland – Hibernia, Terra Nova, White Rose and soon Hebron – is about 0.02 per annum in the absence of iceberg management and can be about an order of magnitude less with adequate ice management. The corresponding EL ice actions from icebergs are minimal.

If a stationary floating structure is designed to disconnect and move off station, the hazard curve for environmental actions can be limited by the pre-defined conditions and probability of success for disconnection. This can modify or limit the value of the AL ice action and, depending on the criteria for disconnection, the value of the EL ice action. Further discussion can be found in Makrygiannis et al. (2011).

PLANNING FOR VERY UNLIKELY HAZARDOUS EVENTS

General

Very unlikely hazardous events have also been termed "beyond abnormal", see Thomas, 2015, because the upper probability limit of "very unlikely" varies with exposure level, see Figure 5.

The limit states design procedure does not require a design situation to be defined for very unlikely hazardous events. Therefore a very unlikely hazardous event, for which the structure is not designed, could cause collapse. This would be a risk event for which the risk can be assessed and put on a risk matrix, as previously discussed.

Assessing and Planning for Very Unlikely Events.

With reference to Figure 5, the areas in yellow indicate very unlikely hazardous events which occur with an annual probability up to an order of magnitude less than that of the abnormal event. Potentially they can cause a risk event with a probability greater than the reliability target.

The significance with respect to the risk profile of the structure could be an issue that the owner chooses to assess, due to the potential for economic loss and other consequences. The owner in discussion with other stakeholders could develop risk management policy and planning, to determine contingency plans, equipment, processes, and resources to manage the response to the approaching hazard and to mitigate any consequences of a very unlikely hazardous event. However, care should be taken in how reliability targets are interpreted with respect to actual failure probabilities, see HSE. 2002, when assessing the relative risk.

Hazards potentially causing such risk events can include ice islands and large icebergs at some locations. For the Skrugard location, the annual iceberg encounter frequency was assessed as 2.6 x 10^{-5} (Eik et.al., 2013). Therefore, to ISO 19906, the structure design does not need to be verified for resisting icebergs. If an iceberg impact would then cause collapse, the probability of occurrence of the risk event could be similar to that of the hazardous event.

ADVANTAGES OF APPLYING DIFFERENT EXPOSURE LEVELS

A structure, sub-structure, or structural component may have an exposure level that is particular to a specific design situation, if risks and their consequences are managed and mitigated.

A reduction from L1 to L2 for anything larger than the EL event could result in significantly lower AL ice actions. An iceberg action hazard curve presented in Widianto et al., 2013, is annotated in Figure 7 to show that the magnitude of the AL iceberg action at the Hebron location, offshore Newfoundland, would be reduced by 60%.

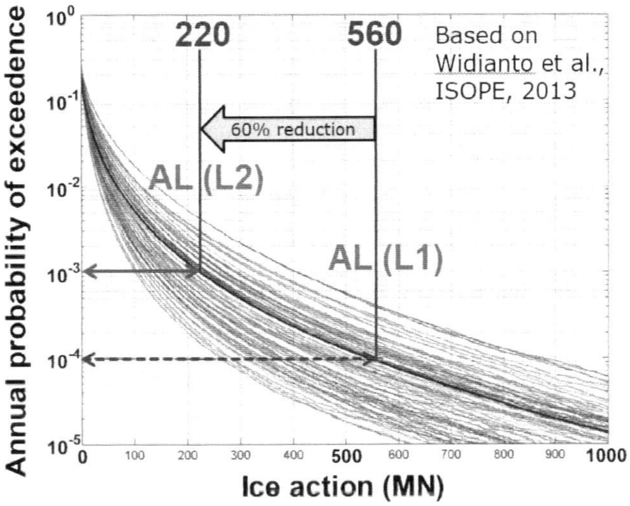

Figure 7. Effect of Exposure Level on AL iceberg action

But there could be significant costs for operational measures and emergency response capability for life safety and environmental protection in accordance with the governing standard (ISO 19900).

Benefits could include cost savings for construction of new structures or enabling an existing structure to continue operating, despite possible deterioration. After establishing the scope and cost of measures and savings, this could essentially be an economic decision by the owner.

CONCLUSIONS

ISO 19906 represents a significant step forward from the traditional approach to the design of offshore structures which has been largely derived from steel jackets and lattice tower structures. Design for all types of structures for arctic and cold regions requires design verification for abnormal design situations, modification of action hazard curves if using operational measures, and consideration of reliability targets,.

Designers need to identify all foreseeable hazards and consider all relevant hazardous events and design situations in the design of arctic offshore structures. The shapes and magnitudes of action hazard curves can vary significantly for different types of ice, for different locations, and for different operational measures.

Defining design situations specifically for extreme and abnormal ice events and actions, with associated ultimate limit states design criteria, will bring more clarity to the design process.

Reliability targets define a break point between risk avoidance by verification of the structural design, and risk management by other means. ISO 19906 encodes the design approach into a limit states design procedure in order to ensure an appropriate level of reliability and acceptable performance in the defined design situations.

ACKNOWLEDGEMENTS

The author thanks many colleagues and contributors in ISO and BSI committees for many instructive conversations, although they would not necessarily agree with all the views expressed here.

REFERENCES

Eik, K.J., Kvingedal, B., and Nygaard, E., 2013, Frequency of sea ice and icebergs at Skrugard - Barents Sea. *Proceedings of the 22nd International Conference on Port and Ocean Engineering under Arctic Conditions (POAC13-219), Espoo, Finland.*

HSE, 2002. OTO 0060/1999. *Target levels for reliability-based assessment of offshore structures during design and operation*. Prepared by Aker Offshore Partner A.S. for the Health & Safety Executive. Available at http://www.hse.gov.uk/research/otohtm/1999/oto99060.htm.

IOGP, 2010, *Calibration of action factors for ISO 19906 Arctic offshore structures*, OGP Report 422. International Association of Oil & Gas Producers. Available at http://www.iogp.org/pubs/422.pdf.

IOGP-ISO/WD 19900, 2017. *Petroleum and natural gas industries – General requirements for offshore structures*, Working Draft. International Association of Oil & Gas Producers.

IOGP-ISO/WD 19906, 2016. *"Petroleum and natural gas industries – Arctic offshore structures*, Working Draft. International Association of Oil & Gas Producers.

ISO 19900, 2013, *Petroleum and natural gas industries – General requirements for offshore structures*. International Organization for Standardization.

ISO 19906, 2010. *Petroleum and natural gas industries – Arctic offshore structures*. International Organization for Standardization.

Makrygiannis, C., McKenna, R.F., Wright, B., Sildnes, T., Jolles, W., Morland. M., Basu, R., Allyn, N.: Ice Management and Operational Strategy for Floaters in Ice. *Proceedings of the 21st International Conference on Port and Ocean Engineering under Arctic Conditions (POAC11-057), Montreal, Canada.*

Muggeridge, K., McKenna, R.F., Spring, W., Thomas, G.A.N., 2017, ISO 19906 Update - An International Standard for Arctic Offshore Structures. *Proceedings of the 24th International Conference on Port and Ocean Engineering under Arctic Conditions, Busan, Korea.*

Spring, W., McKenna, R.F., Thomas, G.A.N., Blanchet, D., 2011, ISO 19906: An International Standard for Arctic Offshore Structures. *Proceedings of the 21st International Conference on Port and Ocean Engineering under Arctic Conditions (POAC11-K3), Montreal, Canada.*

Thomas, G.A.N., 2015, Risk and Reliability in the Design of Arctic Offshore Structures. *Proceedings of the 23rd International Conference on Port and Ocean Engineering under Arctic Conditions (POAC15-241), Trondheim, Norway.*

Thomas, G.A.N., Bercha, F. and Jordaan, I.J., 2011, Reliability, Limit States and Action Factors for ISO 19906. OTC 22072. *Arctic Technology Conference, Houston, Texas.*

Widianto, Khalifa, J., Younan, A., Karlsson, T., Stuckey, P., Gjorven, A. (2013) Design of Hebron Gravity Based Structure for Iceberg Impact. *Proceedings of the Twenty-third (2013) ISOPE International Ocean and Polar Engineering Conference, Anchorage, Alaska, USA.*

Proceedings of the 24th International Conference on
Port and Ocean Engineering under Arctic Conditions
June 11-16, 2017, Busan, Korea

Modelling the Dynamics of Iceberg-Soil Interaction during Seabed Gouging

Jeroen S. Hoving[1], Rutger Marquart[1], Federico Pisanò[2]
[1]Delft University of Technology, Section of Offshore Engineering, Delft, the Netherlands
[2]Delft University of Technology, Section of Geo-Engineering, Delft, the Netherlands

ABSTRACT

Not only in the Arctic, but also well outside the Arctic, icebergs can be a danger to buried offshore pipelines due to seabed gouging. As it is not economically feasible to bury these pipelines at depths where soil deformations are small, these pipelines are buried at sub-gouge depths where the soil experiences significant plastic deformations. Before considering the sub-gouge soil deformations that are vital to determine optimal pipeline burial depths, a better understanding of dynamic iceberg-soil interaction during seabed gouging is required.

In the past, several models have been developed to describe the interaction forces between an iceberg and the soil in front of the iceberg keel based on static soil failure theory or from experimental observations. These models do not fully account for the dynamic behaviour of the iceberg, nor do they consider the corresponding response of the sub-gouge soil. In fact, how the extent of sub-gouge deformations depends on the geotechnical parameters of the seabed is not at all well understood. Therefore, a three-dimensional physics-based model is being developed that describes the transient iceberg-soil interaction (i) based on plasticity theory, (ii) accounting for the dynamic response of the iceberg during gouging and (iii) including the hydrodynamics of the iceberg. This paper presents a two-dimensional version of the iceberg-soil interaction model assuming the iceberg keel to be rigid and plane-strain conditions for the soil. Model predictions are presented and compared for different seabed characteristics.

KEY WORDS: Iceberg dynamics; iceberg-soil interaction; sub-gouge deformation.

INTRODUCTION

Especially in the Arctic and other ice-infested areas, it is attractive to transport hydrocarbons, for example from offshore production locations to shore, or across water bodies between land masses, by means of marine pipelines, thereby largely evading the interaction with sea ice. Nevertheless, icebergs and large ice ridges can be a severe danger to marine pipelines as they ground and gouge the seabed. Generally, marine pipelines in iceberg areas are buried in a trench to protect these pipelines from damage due to iceberg gouging. Figure 1 shows a simplified representation of the interaction between an iceberg keel and the seabed near a pipeline taken from Barrette (2011), where BD is the pipeline burial depth. As Barrette (2011) states, it is not economically feasible to bury pipelines in zone 3, where the soil deformations are negligible and therefore pipelines are buried in zone 2, where the soil experiences significant plastic deformations. Understanding the severity of these sub-gouge deformations is vital for pipeline design and in particular to determine the required depth of pipeline trenches.

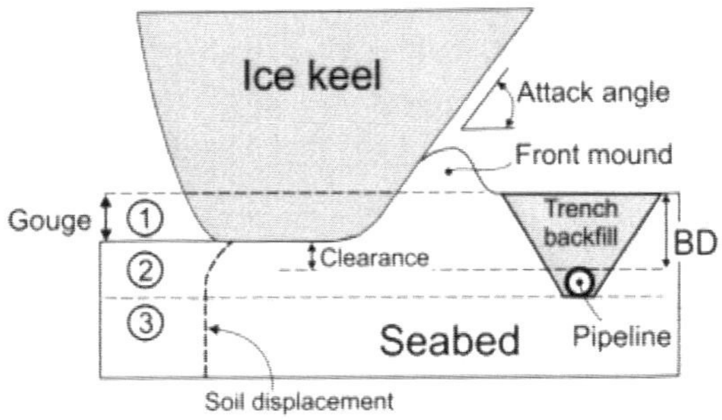

Figure 1. Iceberg keel-seabed interaction near a trenched pipeline (Barrette, 2011).

In the past, several models were developed to describe the interaction between an iceberg keel and the seabed based on static soil failure or from experimental observations. In the two-dimensional gouging models by Kioka and Saeki (1995) and Yoon et al. (1997) for example, assumptions are made that limit the results. The model presented by Kioka and Saeki (1995) assumes that during gouging the iceberg moves with a constant velocity and that the iceberg moves along a so-called pre-described gouging curve, while the gouging model by Yoon et al. (1997) neglects the rotational motion of the iceberg. Clearly, the velocity of an iceberg during gouging is not constant, but decreases due to the soil resistance. In the iceberg-soil interaction model presented in this contribution, the deceleration of the iceberg as it hits and fails the seabed, as well as the rotation of the iceberg, are included to properly account for the dynamic response of the iceberg during seabed gouging. The seabed resistance forces are incorporated into the model quasi-statically, by means of the passive soil-resistance calculated using a slightly adapted version of the theory of Hettiaratchi and Reece (1974).

A PHYSICS-BASED MODEL FOR DYNAMIC ICEBERG-SOIL INTERACTION

Ultimately, it is our aim to develop a three-dimensional physics-based model that describes the transient iceberg-soil interaction that accounts for the plastic deformation of the soil to correctly describe the corresponding soil resistance against iceberg gouging. This paper presents the first stage of this project and regards the two-dimensional version of the iceberg-soil interaction model. The two-dimensional model for dynamic iceberg-soil interaction during seabed gouging is presented here for two separate cases: 1) the free-floating iceberg before grounding, which is aimed at specifically considering the hydrodynamics of the iceberg, and 2) the interaction of the iceberg with the soil as it grounds, specifically focused on the iceberg-soil interaction. In the following, first the general assumptions and modelling constraints, i.e. the assumptions and constraints that apply to both cases, are discussed. Subsequently, the modelling of the two separate cases is discussed and the corresponding model results are presented consecutively. Finally, conclusions regarding the current version of the iceberg gouging model are given and our ideas for future improvements are discussed.

Reference iceberg "Bertha" and the iceberg model

The iceberg-soil interaction model discussed here allows for various iceberg geometries. As a basis for possible iceberg geometries, data is used from the Dynamics of Iceberg Grounding and Scouring (DIGS) field experiments performed in the Eastern Canadian Continental Shelf,

Labrador, in August 1985 (Hodgson et al., 1988). For the purpose of this paper however, from the DIGS experiments, only the iceberg Bertha is considered.

Iceberg Bertha grounded at a water-depth of 107 m on the north-western edge of Makkovik Bank, where water depths range from a minimum of 102 m near the south-eastern end to a maximum of 135 m at the north-western end. At this location, the seabed is relatively smooth and flat with local slopes averaging 1:50. Figure 2a depicts the above- and below-water shape and dimensions of iceberg Bertha that were determined after the iceberg floated free due to a calving event, which additionally caused the iceberg to significantly roll and yaw.

a)

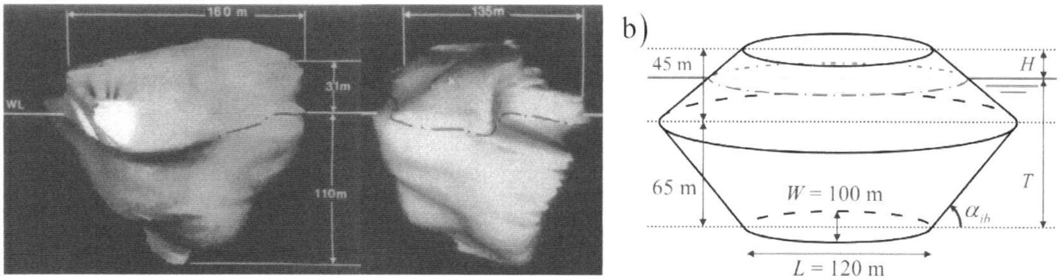

Figure 2. a) Generated perspective of iceberg Bertha (ref Hodgson 1988);
b) Dimensions of iceberg Bertha simplified by two elliptical cone frustums.

The iceberg itself is assumed to be rigid. This means that the iceberg does not deform upon interaction with the water or the soil, nor does it change its shape over time due to thawing or other external influences. To straightforwardly determine the wind and current loads on the iceberg, as well as to calculate the weight and buoyancy of the iceberg, the shape of iceberg Bertha is assumed as depicted in Figure 2b. Maintaining the dimensions of the generated perspective of iceberg Bertha in Figure 2, the ratio between freeboard and keel height, as well as the total volume of the iceberg are too high. The width and length of the elliptical base are therefore respectively chosen as 100 m and 120 m, while the mass and the mass moment of inertia, as well as the freeboard and the keel height of the iceberg model vary depending on the angle α_{ib}. Choosing $\alpha_{ib} = 50°$, the iceberg has a freeboard (H) and a keel height (T) of respectively 18,2 m and 91,8 m, a mass of 2,2 Mt and a mass moment of inertia of 7,0 Gt·m^2.

THE FREE-FLOATING ICEBERG BEFORE GROUNDING

To capture the behaviour of the iceberg during seabed gouging, first it is important to describe the motions of the iceberg before grounding. As long as the iceberg does not interact with the seabed, the iceberg is modelled as a free-floating body, only interacting with the air and the water. This also means that the interaction of the iceberg with other occurrences of ice, such as sea ice or other icebergs, is disregarded.

Figure 3 illustrates the dimensions and the degrees of freedom of the free-floating iceberg, the gravity and buoyancy forces, as well as the aero- and hydrodynamic forces acting on the iceberg. The degrees of freedom of the iceberg, i.e. its surge, heave and pitch, are respectively denoted as x_{ib}, z_{ib} and θ_{ib}. Here, the aero- and hydrodynamic forces applied to the iceberg are all assumed to work in the direction of the x-axis, thus being the direction of forward motion for the iceberg. Additionally, the heave of the iceberg is described in the direction of the z-axis, while θ denotes the pitch of the iceberg.

For large icebergs, the drift of icebergs is governed by ocean currents as the influence of waves on icebergs decreases as the iceberg size increases (Diemand, 2001). Therefore, the contributions of incident, diffracted and radiated waves are assumed negligible, thereby implying that the damping associated with body-induced wave radiation is negligible as well. As a result, only viscous damping is accounted for. Then, as the Coriolis force is a three-dimensional phenomenon and the modelling applied here is two-dimensional, its influence is not included. Additionally, at the scale of an iceberg gouging event, the influence of the Coriolis force is likely to be negligible and iceberg trajectories may be assumed as a straight line.

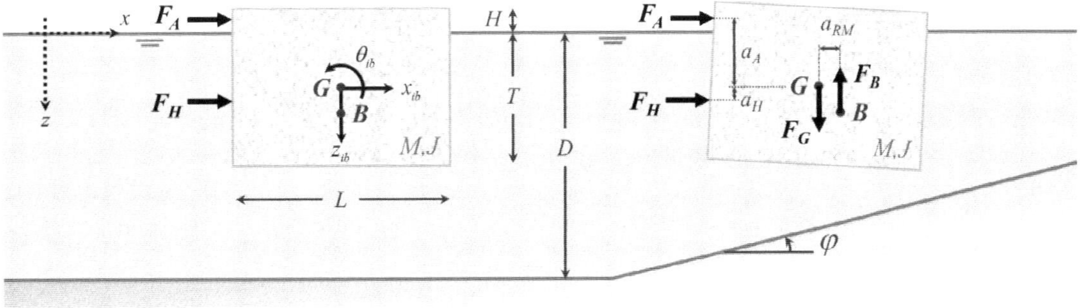

Figure 3. The degrees of freedom of and the forces on a free-floating iceberg.

Thus, the system of equations of motion for the free-floating iceberg reads:

$$\begin{aligned}
&\left(1+A_{xx}\right)Mx_{ib}-F_{A}-F_{H} \quad 0 \quad = \\
&\left(1+A_{zz}\right)Mz_{ib}+B_{zz}z_{ib}-F_{G}+F_{B}=0 \\
&\left(1+A_{\theta\theta}\right)J\theta_{ib}+B_{\theta\theta}\theta_{ib}+F_{A}a_{A}-F_{H}a_{H}-F_{B}a_{RM} \quad \theta
\end{aligned} \tag{1}$$

Here, F_A and F_H are respectively the resultant aero- and hydrodynamic drag forces working on the iceberg, while a_A and a_H are the arms of the corresponding forces with respect to the centre of gravity G of the iceberg. As the iceberg is simplified as an elliptical cylinder and its orientation and dimensions in the two-dimensional plane are assumed to be stable, i.e. it does not roll or yaw, its width W is constant over the height. Consequently, the aero- and hydrodynamic forces working on the iceberg are respectively obtained as:

$$F_{A}=\tfrac{1}{2}\rho_{a}C_{a}W\int_{-H}^{0}\left(u_{a}(z)-x_{ib}\right)\left|u_{a}(z)-x_{ib}\right|dz \tag{2}$$

$$F_{H}=\tfrac{1}{2}\rho_{w}C_{w}W\int_{0}^{T}\left(u_{w}(z)-x_{ib}\right)\left|u_{w}(z)-x_{ib}\right|dz \tag{3}$$

Here, ρ_a, ρ_w, C_a and C_w are respectively the air density, the water density and the air and water drag coefficients. Furthermore, $u_a(z)$ and $u_w(z)$ are respectively the wind (air) and current (water) profiles that are applied over the projected heights H and T of respectively the sail and the keel of the iceberg. The wind and current profiles are described as a function of the z-coordinate as:

$$u_{a}(z)=u_{a}^{10}\left(\frac{-z}{10}\right)^{\frac{1}{7}}; \qquad u_{w}(z)=u_{w}^{\max}\left(\frac{z}{D}\right)^{\frac{1}{7}}. \tag{4}$$

Furthermore in Eq. (1), F_G and F_B respectively denote the weight and the buoyancy of the iceberg, and a_{RM} is the stability lever arm, i.e. the arm between the centre of gravity and the centre of buoyancy due to pitch. The weight of the iceberg is assumed constant throughout and is straightforwardly obtained from the volume of its assumed shape and its density. The submerged volume, and thus the buoyancy of the iceberg, follows from the position of the iceberg in the water and thus depends on the heave and pitch motions of the iceberg. As stated previously, in this model only the viscous damping is accounted for, which in surge direction is already included in the aero- and hydrodynamic drag forces. The hydrodynamic viscous damping coefficients in respectively heave and pitch direction, denoted in Eq. (1) as B_{zz} and $B_{\theta\theta}$, are expressed in terms of the corresponding damping ratios as:

$$B_{zz} = 2\zeta_z\left(1+A_{zz}\right)M\omega_z; \qquad B_{\theta\theta} = 2\zeta_\theta\left(1+A_{\theta\theta}\right)J\omega_\theta. \tag{5}$$

For this model, the heave and pitch damping ratios are respectively chosen as $\zeta_z = 0,04$ and $\zeta_\theta = 0,02$ based on experimental values from free motion decay tests on a cylindrical object in water (Sudhakar and Nallayarasu, 2011). Furthermore, ω_z and ω_θ are the damped natural heave and pitch frequencies that follow from the uncoupled equations of motions for heave and pitch, thus found as:

$$\omega_z = \sqrt{\frac{\rho_w g A_{wl}}{\left(1+A_{zz}\right)M}}\sqrt{1-\zeta_z^2}; \qquad \omega_\theta = \sqrt{\frac{\rho_w g \nabla GM_L}{\left(1+A_{\theta\theta}\right)J}}\sqrt{1-\zeta_\theta^2}. \tag{6}$$

Added Mass of a Free-Floating Iceberg in Variable Water Depth

The hydrodynamic added mass coefficients, respectively denoted in Eq.s (1), (5) and (6) as A_{xx}, A_{zz} and $A_{\theta\theta}$ for respectively surge, heave and pitch, are derived from literature (Bai, 1976, Bass and Sen, 1985). In these contributions, the hydrodynamic coefficients are obtained using the potential flow theory by assuming that the flow is inviscid, irrotational and incompressible. The value of the hydrodynamic added mass coefficients for surge and heave depend on the distance between the bottom of the iceberg and the seabed, or more specifically, on the ratio between the water depth D and the height of the iceberg keel T. Figure 4 depicts the dependency of the added mass coefficients respectively in surge and heave directions as a function of the ratio D/T. The influence of the ratio D/T on the added mass coefficient for pitch however is negligible and therefore assumed constant.

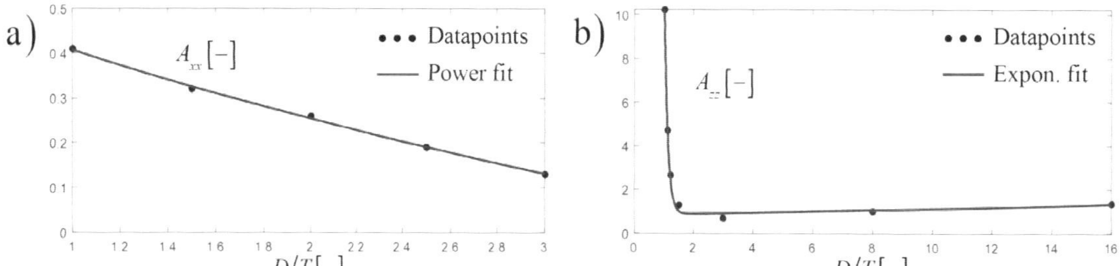

Figure 4. Dependency of the added mass coefficients on the iceberg-seabed distance.

In Figure 4, the added mass coefficients obtained by Bass and Sen (1985) and Bai (1976) for different values of the ratio D/T are given by blue dots, while the continuous red line gives a power fit for surge and an exponential fit for heave. For surge, the influence of the added mass coefficient on the ratio D/T appears to be relatively small. For heave however, the influence of the ratio D/T is significant. In fact, according to Bai (1976), the added mass of the iceberg in heave direction becomes infinitely large as the bottom of the iceberg touches the seabed. In Figure 4b, the data-points have been fitted using a 3rd order polynomial resulting in an added mass coefficient for heave that is 10 times as large as the actual mass of the iceberg. For example, assuming that the free-floating iceberg is in a constant water depth of 150 m, i.e. $D/T = 1,5$, the added mass coefficients for surge, heave and pitch are constant and respectively found as $A_{xx} = 0,32$, $A_{zz} = 0,88$ and $A_{\theta\theta} = 0,04$.

Response of the Free-Floating Iceberg

The response of the free-floating iceberg subjected to winds and currents is obtained in the time domain by solving the system of ordinary differential Eq.s (1) using a Runge-Kutta scheme. To verify the modelling of the free-floating iceberg, the initial conditions for the iceberg are deliberately chosen far from its equilibrium condition. In this case, the initial position of the iceberg is arbitrarily chosen such that its top surface equals the waterline. For the wind and current velocities common average values are chosen as 30 km/h and 1,5 m/s respectively. As depicted by Figure 5 for two different icebergs, upon release from its initial position, the iceberg will ultimately attain a certain steady-state condition that is independent of the starting position, moving at a constant velocity. The iceberg reaches this constant velocity after nearly six hours covering a distance of about 30 km. Additionally, due to its initial vertical position and its own buoyancy, the iceberg oscillates for about 12 minutes before reaching a constant draft. Apparently, the rotation of the iceberg due to the applied constant wind and current is negligible. Note here that during the steady-state orientation, the icebergs restoring moment due to its pitch is equal to the moment caused by winds and currents.

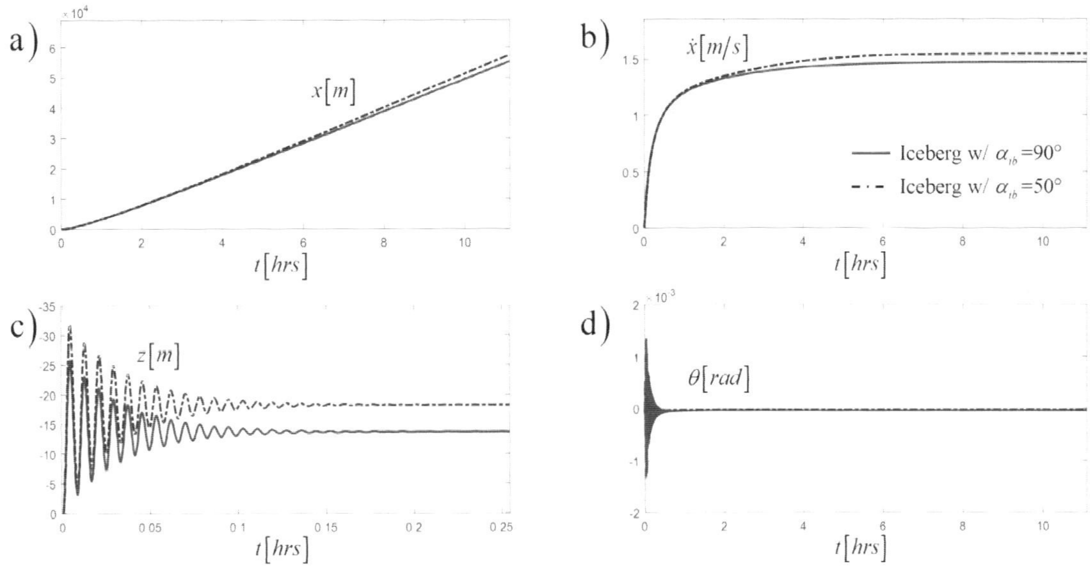

Figure 5. Response of the free-floating iceberg: a) surge; b) surge velocity; c) heave; d) pitch.

THE GROUNDING ICEBERG AND SEABED GOUGING

When an iceberg hits a seabed, the iceberg will penetrate that seabed as depicted in Figure 6b. Because of its inertia, the iceberg continues to move, leaving a seabed gouge in his wake, until the total kinetic energy of the iceberg has been dissipated by the iceberg-seabed interaction. The length and depth of these gouges depend on the mass, the shape and the drift velocity of the iceberg, but also on the bathymetry of the seabed and the mechanical behaviour of the seabed soil (Green et al., 1983). Nevertheless, for a free-floating iceberg to cause seabed gouging, the seabed must at least have a slope somewhere. Although different configurations can be chosen, in this paper, the interaction of an iceberg with a continuous and constant sloping seabed is considered. Before considering the iceberg-seabed interaction in two-dimensional space using plane-strain conditions for the soil, first the interaction of the iceberg with a rigid seabed is addressed as a benchmark for the iceberg-soil interaction problem.

The Interaction of a Grounding Iceberg with a Rigid Seabed

To model the interaction with the rigid seabed, the first contact between iceberg and seabed is assumed to always occur at the corner C of the iceberg as depicted in Figure 6a. The normal and tangential forces that keep the iceberg from penetrating the rigid seabed are denoted as F_N and F_T respectively. Adding the normal and tangential forces to the system of equations of motion (1), previously obtained for the free-floating iceberg, yields the system of equations for the iceberg-rigid-seabed-interaction as:

$$\begin{aligned} &\left(1+A_{xx}\right)Mx_{ib} - F_A - F_H + F_N \sin\varphi + F_T \cos\varphi = 0 \\ &\left(1+A_{zz}\right)Mz_{ib} + B_{zz} z_{ib} - F_G + F_B + F_N \cos\varphi - F_T \sin\varphi = 0 \\ &\left(1+A_{\theta\theta}\right)J\theta_{ib} + B_{\theta\theta}\theta_{ib} + F_A a_A - F_H a_H - F_B a_{RM} - r_N F_N + r_T F_T = 0 \end{aligned} \tag{7}$$

Here, r_N and r_T are the arms of respectively the normal and tangential forces with respect to the centre of gravity of the iceberg. The relation between the normal force F_N and the tangential force F_T is given the Coulomb friction coefficient μ, so that $F_T = \mu F_N$. Consequently, the given system of equations of motion consists of four unknowns, being the degrees of freedom of the iceberg and the normal force F_N and thus, to solve the above system of equations of motion, an additional equation is required. This equation follows from the change of the iceberg momentum due to interaction with the rigid seabed at point C. Accounting for the contributions due to the change both in linear momentum, i.e. the impulse, and in angular momentum, sometimes referred to as the angular impulse, over time, the velocity change of point C in the direction normal to the seabed slope over time is obtained as:

$$\frac{dv_N}{dt} = \frac{dv_N^{lin}}{dt} + \frac{dv_N^{ang}}{dt} = \left(\frac{1}{M} + \frac{r_N^2}{J}\right)F_N - \frac{r_N r_T}{J} F_T \tag{8}$$

Here, v_N denotes the velocity of the iceberg in the direction normal to the seabed slope. Accounting for the friction relation, then yields Eq. (8) as:

$$\left\{\frac{1}{M} + \frac{r_N}{J}\left(r_N - r_T \mu\right)\right\} F_N - \frac{dv_N}{dt} = 0 \tag{9}$$

For the remainder of the iceberg-rigid-seabed-interaction, it is assumed that the corner C remains on the seabed slope. Therefore, after the initial interaction of the iceberg with the rigid seabed, the velocity v_N in the direction normal to the seabed slope, as well as its time derivative, are equal to zero. However, right before the iceberg hits the seabed, the velocity of point C normal to the seabed slope is not equal to zero. In fact, right before the first iceberg-seabed interaction, the normal velocity v_N is known from the system of equations for the free-floating iceberg and the normal force F_N can thus be obtained by solving Eq. (9) numerically.

When the iceberg moves further along the seabed slope and the corner D also hits the seabed, the bottom of the iceberg comes in full contact with the rigid seabed until the iceberg stops moving completely. In this case, the point of application of the normal and tangential forces is shifted to the centre-line of the iceberg. Note here that the above system of equations is only valid as long as the normal force F_N yields soil compression. Furthermore, note that when the iceberg hits the seabed, the iceberg keel is equal to the water depth and the ratio D/T is thus equal to 1. The corresponding added mass coefficients in Eq. (7) for surge, heave and pitch are respectively found as $A_{xx} = 0,41$, $A_{zz} = 10$ and $A_{\theta\theta} = 0,04$.

Iceberg-Soil Interaction for an Iceberg Penetrating the Seabed

When the iceberg interacts with a non-rigid seabed the corner C of the iceberg will penetrate the seabed as depicted in Figure 6b. The soil is assumed to behave as a cohesionless frictional material responding to external loads under fully drained conditions, i.e. with no generation of excess pore pressures (this assumption can be adopted for highly permeable sandy soils). During seabed gouging, the resistance of the soil to the iceberg motion is partly due to the deformation of the soil in front of the iceberg, as well as due to the frictional resistance at the bottom of the iceberg. The soil reaction in front of the iceberg is determined as a quasi-static passive force according to the plasticity approach by Hettiaratchi and Reece (1974). The normal and tangential resistance at the bottom of the iceberg are determined by a traditional bearing capacity equation for shallow foundations under inclined loads (Salgado, 2008).

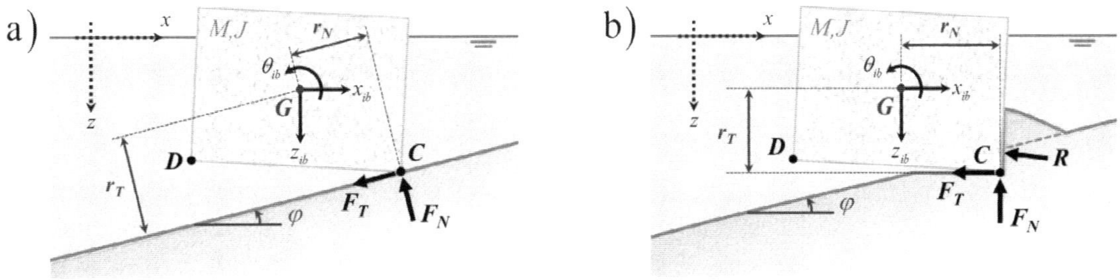

Figure 6. a) The iceberg hitting a rigid seabed; b) The iceberg penetrating the real-life seabed.

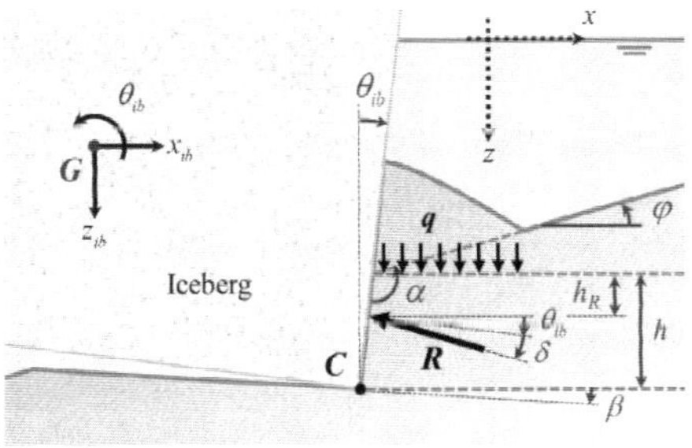

Figure 7. The resultant passive soil resistance force.

The resultant passive soil resistance force R is depicted in Figure 7. According to Hettiaratchi and Reece (1974), the resultant passive soil resistance force for a cohesionless soil reads:

$$R = R_1 + R_2 = qK_q h + \bar{K}_\gamma \gamma' h^2 \qquad (10)$$

Here, q is the surcharge on the seabed surface, i.e. the amount of soil that is removed by the gouging of the iceberg and accumulates at the front and side of the iceberg, thereby yielding additional soil resistance, and γ' is the effective unit weight of the soil, while K_q and K_γ are the corresponding dimensionless soil resistance coefficients taken from the charts by

Hettiaratchi and Reece (1974). Additionally, h is the vertical distance between the corner C of the iceberg and the seabed slope. Following Hettiaratchi and Reece (1975), the angle δ of the resistance force R for cohesionless soils follows from the angle of internal friction of the soil ϕ and the difference between the rake angle α and the angle β that describes the direction of forward motion of the iceberg-seabed interface with respect to the horizontal, so that:

$$\tan\delta = \frac{\sin\phi\cdot\sin\left(270-\phi-2\left(\alpha-\beta\right)\right)}{1+\sin\phi\cdot\cos\left(270-\phi-2\left(\alpha-\beta\right)\right)} \tag{11}$$

The point of application of the force R cannot be determined exactly because the actual distribution of soil stresses along the iceberg edge is unknown. According to Hettiaratchi and Reece (1974), the vertical distance between the seabed surface and the point of application of the force R, denoted as h_R, can be approximated as:

$$h_R = h\left(\frac{3R_1+4R_2}{6R}\right) \tag{12}$$

As the gouging depth of an iceberg is generally smaller than its width, the ultimate bearing capacity of the iceberg during seabed gouging is determined using Terzaghi's bearing capacity equation for shallow foundations in combination with the capacity correction factors by Brinch Hansen (1970). Accordingly, the maximum vertical pressure that the soil is able to provide at the bottom of the iceberg is determined as:

$$q_{ult} = 0,3 s_\gamma d_\gamma i_\gamma b_\gamma g_\gamma \gamma' B N_\gamma \tag{13}$$

Here, a factor 0,3 is used to account for the elliptical shape of the iceberg bottom instead of a factor 0,5 that is used for a strip footing. Furthermore, s_γ and d_γ are respectively shape and depth factors, while i_γ, b_γ and g_γ are load inclination, base inclination and ground inclination factors respectively. The expressions and values for these factors are taken from Salgado (2008). As before, γ' is the effective unit weight of the soil, while N_γ is the corresponding bearing capacity factor and B equals the contact length between the iceberg and the soil during gouging, which varies over time. The normal force F_N is then found at each time step as $F_N = q_{ult} A_c$, where A_c yields the soil-iceberg contact area during seabed gouging. Here, both the normal force F_N and the tangential force F_T are assumed to act in the middle of the contact length. To determine the ratio between the normal and the tangential force, the load inclination factor must be known and vice versa. Since both the ratio and the load inclination factor are unknown prior to each time step, the load inclination factor is found iteratively. Initially, the load inclination factor i_γ is set equal to 1, i.e. a vertical load reaction, or when available, equal to the final load inclination factor from the previous time step. Then, the corresponding motions of the iceberg are determined numerically and the normal and tangential forces are back-calculated from the system of equations of motion. The ratio between the normal and tangential force yields a new load inclination factor as input for the same time step. The iteration is repeated until the load inclination factor i_γ converges to a given threshold. The ratio between F_T and F_N has a lower bound given by the soil-iceberg interface friction angle, relevant to the limit case of horizontal sliding of the bottom iceberg surface.

Accounting for both the resistance of the soil in front and at the bottom of the iceberg, the system of equations of motion for the iceberg follows as:

$$\left(1+A_{xx}\right)Mx_{ib}-F_{A}-F_{H}+F_{T}\cos\beta+F_{N}\sin\beta+R_{x}=0$$
$$\left(1+A_{zz}\right)Mz_{ib}+B_{zz}z_{ib}-F_{G}+F_{B}-F_{T}\sin\beta+F_{N}\cos\beta+R_{z}=0 \qquad (14)$$
$$\left(1+A_{\theta\theta}\right)J\theta_{ib}+B_{\theta\theta}\theta_{ib}+F_{A}a_{A}-F_{H}a_{H}-F_{B}a_{RM}+r_{T}F_{T}-r_{N}F_{N}+r_{Rx}R_{x}-r_{Rz}R_{z} \quad 0$$

Here, R_x and R_z are respectively the horizontal and vertical components of the resistance force R and are thus respectively obtained as $R_x = R\cos\left(\theta_{ib}+\delta_r\right)$ and $R_z = R\sin\left(\theta_{ib}+\delta_r\right)$. It should here be noted that in the model by Hettiaratchi and Reece, the seabed surface is assumed horizontal, while our seabed has a constant positive slope. The slope is accounted for by including the soil volume of the seabed slope in the surcharge. Additionally, Hettiaratchi and Reece assume that the soil-structure interface does not rotate, but translates at a sufficiently slow pace so that inertia loads are negligible. Here, the rotation of the iceberg is incorporated. The influence of the rotation on the passive soil resistance however is neglected by choosing the time step sufficiently small. In addition, this reduces the error by neglecting the influence of the inertia and allows for the quasi-static determination of the soil resistance.

The Iceberg Hitting a Rigid Seabed Versus the Iceberg Penetrating the Seabed

The interaction of an iceberg hitting the rigid seabed, as well as the process of seabed gouging of an iceberg has been simulated solving the corresponding systems of equations in the time domain. The resulting iceberg response is depicted in Figure 8, where the continuous red and dashed blue lines respectively give the interaction with the rigid and the real seabed.

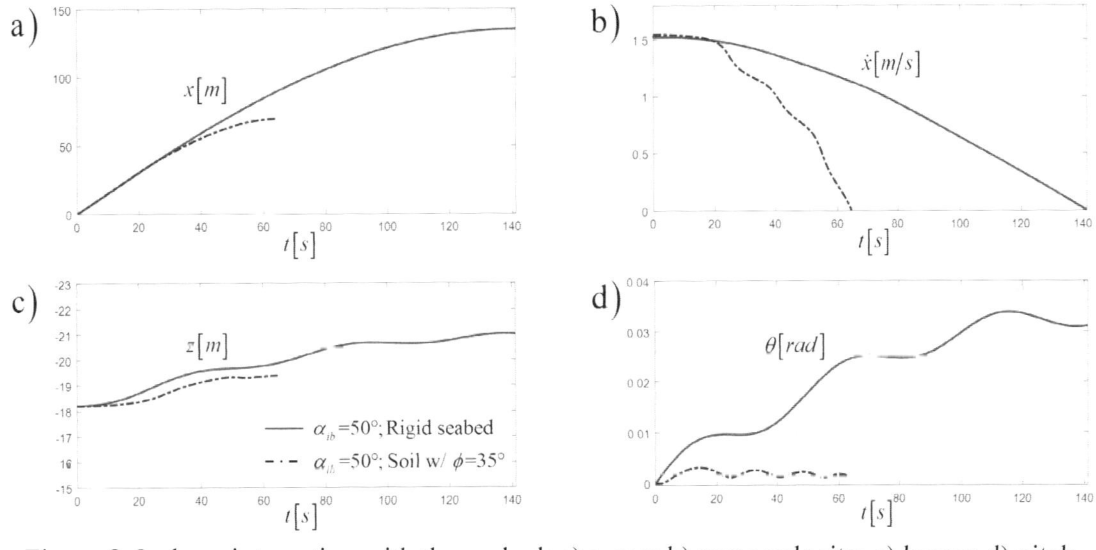

Figure 8. Iceberg interacting with the seabed: a) surge; b) surge velocity; c) heave; d) pitch.

For both considered cases, the initial conditions are assumed equal to the steady-state conditions obtained for the free-floating iceberg response. As the hydrodynamic parameters are chosen equal to those for the free-floating iceberg as well, the iceberg-seabed interaction may be considered to start at the end of the free-floating iceberg response. The iceberg-seabed interaction process consists of two distinct phases, being (i) the phase where the water depth decreases due to the seabed slope until the iceberg touches the seabed, and (ii) the phase of interaction between the iceberg and the seabed until the iceberg comes to a standstill. Evidently, the first phase is identical for the two cases, and the difference in Figure 8 only occurs after the iceberg hits the seabed.

Here, the seabed consists of sand with a density $\rho_s = 2080\,kg/m^3$ and has a constant slope of $\varphi = 2°$. The angle of internal friction of the soil is chosen as $\phi = 35°$ and the cohesion is assumed zero. Furthermore, the iceberg-seabed friction coefficient is chosen as $\mu = 0{,}39$.

Figure 8a and Figure 8b show that the distance travelled by the iceberg interacting with the rigid seabed is significantly larger than that of the iceberg gouging the seabed. As the iceberg cannot penetrate and therefore does not gouge the rigid seabed, the resistance of the rigid seabed against the iceberg motion exclusively consists of the Coulomb friction and thereby highly underestimates the seabed resistance. The distance travelled by the iceberg from the point of first contact with the seabed until it comes to a standstill should therefore be considered as an absolute upper limit for seabed gouging. Additionally, it can be seen from Figure 8c and Figure 8d that the iceberg interacting with the rigid seabed moves up the seabed slope, while it tilts backward until the iceberg rotation coincides with that of the rigid seabed. The iceberg gouging the seabed however, does not follow the sloping seabed at all. Due to the significantly higher seabed resistance forces, the iceberg tilts forward and digs itself into the soil, thereby further increasing the seabed soil resistance causing a fast deceleration of the iceberg.

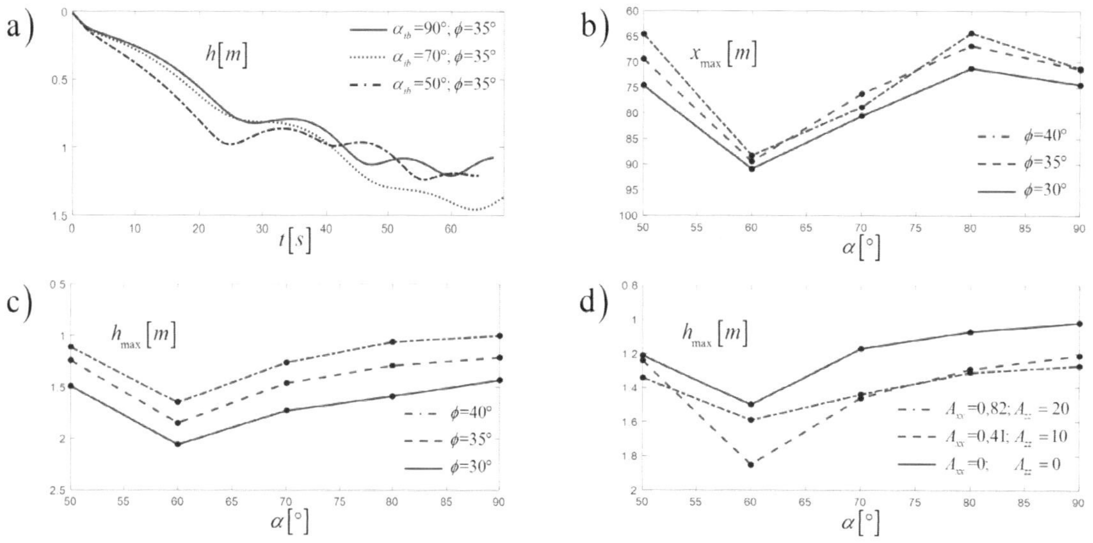

Figure 9. a) gouge depth for several icebergs; b) gouge length versus rake angle for several internal friction angles; c) maximum gouge depth versus rake angle for several internal friction angles; d) maximum gouge depth versus rake angle for varying added mass coefficients.

Figure 9a shows the gouge depth, or seabed penetration, as a function of time for three icebergs with different rake angles. These penetrations show an increasing trend due to the constant seabed slope. It is striking that the iceberg Bertha, with a rake angle $\alpha_{ib} = 50°$ nearly has the smallest maximum penetration and requires the shortest time to come to a standstill, which in fact does not necessarily mean that it has the shortest gouging length. This is verified by Figure 9b that shows the relation between the gouging length and the rake angle for three purely frictional sands with an internal friction angle $\phi = 30°$, $\phi = 35°$ and $\phi = 40°$ respectively. From Figure 9b, it appears that the influence of the rake angle on the gouging length is larger the influence of the soil strength. Figure 9c shows the maximum gouge depth, or maximum seabed penetration, as a function of the iceberg rake angle for the same three sands that have previously been considered in Figure 9b. In this case, the gouge depth is clearly larger for soils with smaller internal friction angle and vice versa. From Figure 9b and

Figure 9c it follows that the shortest gouges are also the deepest gouges and generally occur for icebergs with a rake angle of around 60°. Finally, Figure 9d again shows the maximum gouge as a function of the rake angle, but now for three different sets of added mass coefficients to determine the sensitivity of the iceberg-soil interaction on theses added mass coefficients. As a general case, the added mass coefficients are used that were previously determined for a D/T-ratio equal to 1. In addition, we have here considered added mass coefficients that are twice as large, as well as added mass coefficients that are equal to 0. It should be noted here that the added mass coefficients used in this paper are derived on measurements of the added mass of two-dimensional cylinders heaving in a finite water depth. For frustum-shaped icebergs with small rake angles, the added mass coefficients are therefore slightly overestimated. However, for frustum-shaped icebergs with large rake angles, this difference may be significant and the results may be unreliable.

DISCUSSION, CONCLUSIONS AND FURTHER WORK

This contribution present the results of the first stage of an attempt to dynamically correct model both the free-floating iceberg as well as the dynamic iceberg-soil interaction during seabed gouging. Clearly, as the resulting two-dimensional iceberg model is a work-in-progress, the model is still subject to significant improvements.

First of all, the currently assumed shape of iceberg Bertha is far from realistic. To incorporate the variation of the geometry of the iceberg, and thus to consider more realistic iceberg shapes, the iceberg is considered to be comprised of a set of horizontal layers, where the width, length and side angle may be chosen independently for each layer. This yields a variation of the hydrodynamics over the height of the iceberg, as well as the ability to change the angle of the iceberg-soil interface. Furthermore, to more correctly describe the iceberg-soil interaction during seabed gouging, it is envisaged to introduce a failure criterion that describes the failure of the iceberg keel, thereby allowing the iceberg geometry to change due to its interaction with the soil. Although posing considerable modelling challenges, this will additionally allow us to account for the stability of the iceberg during seabed gouging.

REFERENCES

BAI, K. J. 1976. The added mass of two-dimensional cylinders heaving in water of finite depth. *Journal of Fluid Mechanics,* 81**,** 21.

BARRETTE, P. D. 2011. Offshore pipeline protection against seabed gouging by ice: An overview. *Cold Regions Science and Technology 69***,** 3-20.

BASS, D. W. & SEN, D. 1985. Added mass and damping coefficient for certain 'realistic' iceberg models. *Cold Regions Science and Technology,* 12**,** 13.

BRINCH HANSEN, J. A. 1970. Revised and extended formula for bearing capacity. *Bulletin no. 28.* Copenhagen: Danish Geotechnical Institute Copenhagen.

DIEMAND, D. 2001. Icebergs. *Encyclopedia of Ocean Sciences***,** 1255-1264.

GREEN, H. P., REDDY, A. S. & CHARI, T. R. Iceberg scouring and pipeline burial depth. 7th International Conf. on Port and Ocean Engineering under Arctic Conditions (POAC), 1983 Helsinki, Finland. 280-288.

HETTIARATCHI, D. R. P. & REECE, A. R. 1974. The calculation of passive soil resistance.

Géotechnique, 24, 289-310.
HETTIARATCHI, D. R. P. & REECE, A. R. 1975. Boundary wedges in two-dimensional passive soil failure. *Géotechnique,* 25, 197-219.
HODGSON, G. J., LEVER, J. H., WOODWORTH-LYNAS, C. M. T. & LEWIS, C. F. M. 1988. Dynamics of Iceberg Grounding and Scouring. St. John's, Nfld.
KIOKA, S. & SAEKI, H. Mechanisms of Ice Gouging. 5th International Offshore and Polar Engineering Conf. , June 11-16 1995 The Hague, The Netherlands. 398-402.
SALGADO, R. 2008. *The Engineering of Foundations,* New York, McGraw-Hill.
SUDHAKAR, S. & NALLAYARASU, S. Influence of heave plate on hydrodynamic response of spar. 30th International Conf. on Ocean, Offshore and Arctic Engineering (OMAE), June 19-24 2011 Rotterdam, The Netherlands. 437-447.
YOON, K.-Y., CHOI, K. & PARK, H.-I. 1997. A Numerical Simulation to Determine Ice Scour and Pipeline Burial Depth. *International Offshore and Polar Engineering Conf.* Honolulu, USA.

Comprehensive Monitoring of Ice Gouging Bottom Relief at Key Sites of Oil and Gas Development within the Coastal-Shelf Zone of the Yamal Peninsula, Kara Sea

Stanislav A. Ogorodov[1], Vasiliy V. Arkhipov[1,2], Osip V. Kokin[1,2], Aleksey V. Marchenko[2]
[1] Lomonosov Moscow State University, Moscow, Russia
[2] Zubov State Oceanographic Institute, Moscow, Russia

ABSTRACT

Ice gouging is a dangerous natural process typical for the coastal-shelf zone of the Russian Arctic; because it leads to damaging of the infrastructure it can also be related to the category of catastrophic processes. To lower the risks of occurrence and to prevent emergencies and their consequences, comprehensive monitoring of the dangerous natural processes is necessary. With all lithologic and geomorphologic conditions being equal, the intensity of the ice gouging on the bottom is mostly determined by the changing condition, area and thickness of the ice cover. To assess the real intensity of the ice gouging impact during a given ice season, repeated sounding of the sea bottom topography is necessary; it helps to select the ice gouges which were created in the period between the two consequent observations. At present, the methods and technologies of the monitoring of ice gouging processes are not standardized, and the monitoring, if it is conducted, is often sporadic and lacks systematization. Therefore, the development of a united technology of comprehensive monitoring of ice gouging processes in the coastal and shelf zone is one of the most important tasks of the modern science and practice. Our team was the first one to apply such integrative technology in 2005-2015 in the framework of investigations for the purpose of construction of the underwater gas pipeline at its crossing of the Baydaratskaya Bay, Kara Sea.

KEY WORDS: Sea Ice; Ice Ridges; Seabed; Ice Gouging; Geotechnical Safety; Technology o f Monitoring

INTRODUCTION

Sea ice, as a zonal factor associated with the high latitudes of Arctic seas, plays an important role in the evolution of northern coasts. The ongoing development of oil and gas fields and the construction of relevant engineering facilities in the coastal and shelf areas (including navigation channels, coastal terminals, drilling platforms, submarine pipelines, and artificial islands) require new information on the effects of sea ice on coastal and seabed dynamics. This is one of the most important factors determining the selection of pipeline landfall sites, shore-crossing design, and required depths of burial.

Ice gouging is a destructive mechanical impact of ice on the underlying ground surface (Figure 1). This impact on the shore and the floor of the Arctic seas is driven by the ice cover dynamics and mobility, hummocking (ridging), and formation of grounded hummocks (pressure ridges) controlled by hydrometeorological factors and coastal topography (Ogorodov, 2003). In direct observations from submarines, keels of large hummock formations reaching 50 m depth have been recorded (Lisitzyn, 1994).\

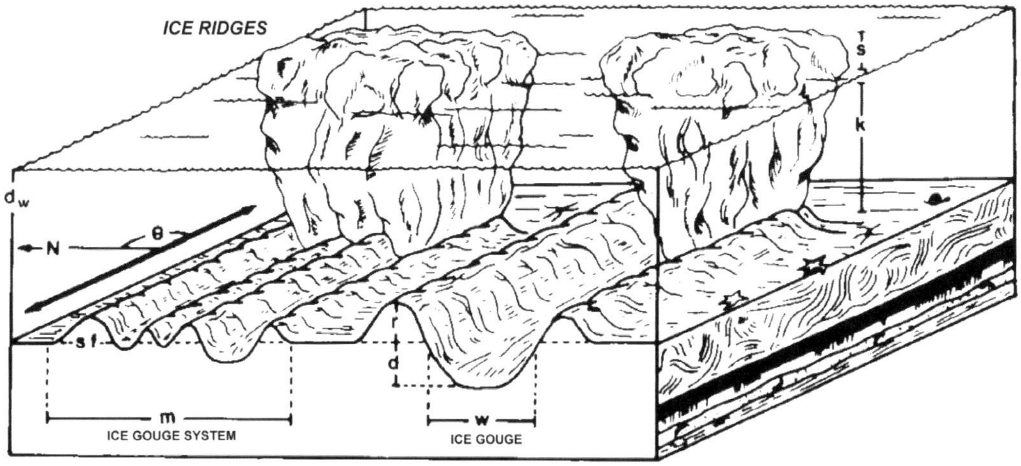

Figure 1. Scheme of formation of systems of parallel ice gouges by hummocky formations with multiple keels, and of separate ice gouges by hummocks with one keel (Barnes, et al., 1984): k – water depth; s – keel height (above sea level); d- depth of the ice gouge; w – width of the ice gouge; r – height of the push ridge; m – width of the system of ice gouges; sf – width of the push ridge; θ – direction of the ice gouge.

Until recently, little work has been done in Russia on the interaction between sea ice and the sea floor in coastal regions, whereas abroad, especially in Canada and the USA, there is a long history of research on this topic. American and Canadian researchers have documented the occurrence and processes of bottom scour and related seabed forms associated with pressure ridges and grounded ridges (stamukhi). Pioneering studies were undertaken on the Alaska shelf (Carsola, 1954; Rex, 1955; Reimnitz & Barnes, 1974; Barnes, 1982; Reimnitz & Kempema, 1984). With the help of echo sounders and diving surveys, later augmented by sidescan sonar, ice gouges were found on the sea floor at depths of 0-65 m (a few deeper), with the highest density in depths of 20-40 m. They were up to 2 km long, incised as deep as 2,5 m into the sea floor, and were up to 67 m wide (Barnes, et al., 1984). In the Canadian Beaufort Sea, seminal work (Shearer & Blasco, 1975; Lewis, 1977; Hnatiuk & Brown, 1977) established that ice scour occurred out to about 55 m depth, was most intense if the depth range of 15-40 m, and that depths of scour penetration were typically <1 m but as deep as 7,6 m.

In Russia, special studies of the sea ice impacts (first of all, the effect of ice gouging) started significantly later and were carried out in the areas of submarine pipeline construction (Dubikov, et al., 1997; Vershinin, et al., 2005; Zubakin, 2006).

In 2011, the so-called "Nord-Stream Gas Pipeline" directly connected Russia and Germany through the seabed of the Baltic Sea. In 2013 this pipeline will reach full capacity. To provide gas for this pipeline, in 2005, the "Yamal-Europe" pipeline project (Dubikov, et al., 1997), the lines of which would cross Baydaratskaya Bay, was revived and by August 2007, the construction of the underwater crossing of the main gas pipeline across the Baydaratskaya

Bay was underway. This was to connect gas fields of the Yamal Peninsula with the pipeline network in the European part of Russia by the shortest route (Figure 2a). The length of the underwater part of the pipeline is approximately 65 km, the maximum bottom depth in the crossing area is 22-23 m. In connection with the renewing of the project, investigations of coastal zone dynamics and sea ice effects continued in 2005 after a 10-year hiatus. In order to assess the impact of the ice on the sea bottom, a wide range of investigations has been carried out, including field studies as well as mathematical modeling. Some of the results and scientific conclusions of the work in which the authors participated directly are presented in this paper.

ICE CONDITIONS

The autumn ice formation in Baydaratskaya Bay usually occurs in the absence of residual ice, i.e. in open water. The climatic mean freeze-up date for the formation of stable ice is about 15-20 October. Year-to-year variations in the hydro-meteorological conditions determine the temporal range of freeze-up. In more seaward parts of the bay remote from the shore, freeze-up occurs 10-15 days later the ice formation close to shore. After the ice cover reaches the grey-white stage, the rate of further growth in ice thickness is directly proportional to the intensity of the underlying surface cooling and poorly correlated to the date of stable ice formation. Ice thickness typically increases every 10 days by 8-10 cm from November to February and then declines to 5-6 cm per 10 days in March, 2-4 cm in April, and 1-2 cm in May. It is important to note that the mean ice thickness along the shallow Yamal coast is generally greater than along the steep Ural coast. By the end of the cold season, the ice thickness along the Yamal coast reaches 140 cm on average, i.e. the ice cover in general fits into the category of thick first-year ice. Across the bay along the Ural coast, the annual maximum ice thickness is typically only 120 cm. The ice cover in that area often does not qualify as thick first-year ice and is categorized as medium first-year ice.

The sea ice of Baydaratskaya Bay consists of drift ice and fast ice (Figure 2b). Close to the open parts of the Yamal and Ural coasts, fast ice forms 20-30 days after the initial freeze-up, with a thickness of approximately 30 cm. This young, relatively thin, fast ice is very unstable in the initial stages and therefore it can be easily broken strong wind shear or sea-level rise. However, with the growth of sea ice thickness (up to 0,5 m) and the formation of ice hummocks and stamukhas, stabilizing ice conditions, the fast ice resistance increases considerably. Stable fast ice forms near the Yamal coast and in the inner part of the bay only. Next to the Ural coast, fast ice is less stable and in the last several years, cases of its break-away have been observed. In the early part of the season, the fast ice spreads seaward quickly, its outer border reaching the 5-7 m isobath in November and the 8-10 m isobath in December. In February-March the seaward border of fast ice corresponds to the isobath of 15 m at the Yamal coast, where its width is 7-9 km, and to the isobaths of 10-12 m off the Ural coast, where its width is approximately 5-7 km.

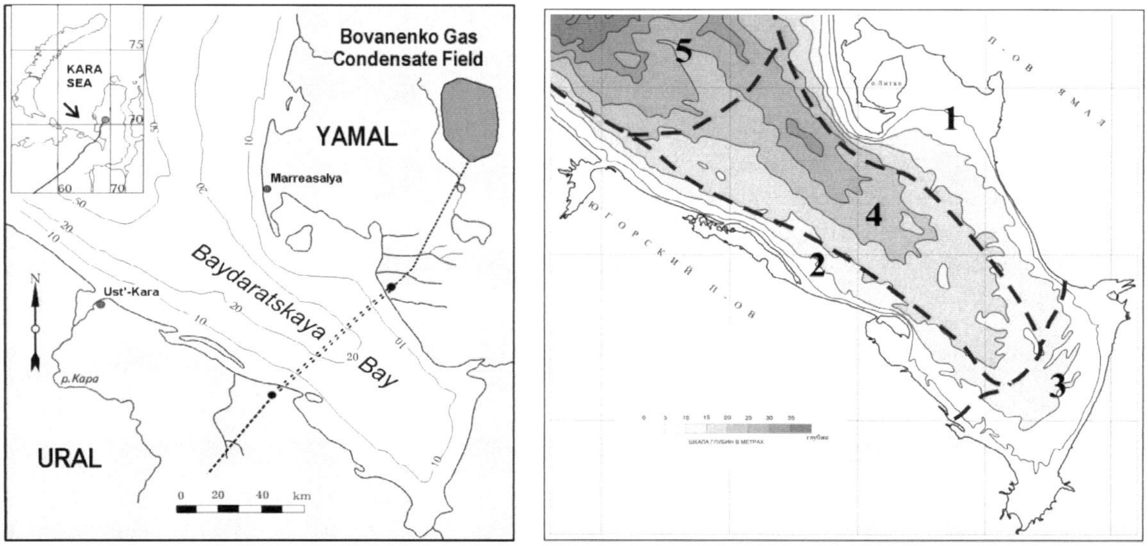

a *b*

Figure 2. (a) - "Yamal-Europe" gas pipelines crossing Baydaratskaya Bay in the Kara Sea; (b) – Ice cover of Baydaratskaya Bay: 1 – fast ice of Yamal Coast, 2 – fast ice of Ural Coast, 3 – fast ice of inner bay, 4 – drift ice of axis part of the bay, 5 – drift ice of open sea

Drift ice consists of ice fields of different sizes. Compared with the immobile fast ice, in the same equal conditions, the thickness of drifting ice is generally less. Along the interface between fast ice and pack ice, as well as on the borders of drifting ice fields, ice hummocks (pressure ridges) form. Ice hummocks freeze into ice fields, building up complex ice formations, which drift together.

Analysis of ice field size for April 2006 based on satellite imagery has shown that most of the ice fields (floes) are less than 2 km in diameter, though several reach sizes of 6-10 km. The maximum vertical dimension of pressure ridges observed in Baydaratskaya Bay, including both the sail and the keel, reaches 30 m, while the length can be up to 300 m. The mass of a 1 km diameter ice field with ice thickness of 1 m is approximately 106 tons. The mass of the largest pressure ridge 30 m × 30 m × 300 m is therefore about 2.7×105 tons, far less than the mass of the 1 km ice field (Marchenko, et al., 2007). Consequently, the ice scour process in Baydarakskaya Bay is dominated by the mass of ice fields with pressure ridges (ice hummocks and their associated keels) frozen in and incorporated within them, rather than by individual ridges.

The driving forces of ice-field drift are the wind and currents and the shear stress they exert on the upper and lower surfaces of the ice field. Currents measured in the ice-free period are practically reversive and aligned long the axis of the bay. Currents are driven by the semidiurnal tide. The maximum speed of the tidal current during the tidal cycle is 0,5, while the measured maximum current speed is 1,0 m per second. The maximum tidal range (spring tides) is 1,1 m, but including storm surges, the water levels have a range of up to 2 m. During the winter season, southwesterly and southerly winds prevail. In the absence of counteracting currents, these winds create conditions for ice drift from the Ural to the Yamal coast of the bay.

METHODS

The most commonly used approach to the assessment of sea-ice impacts on the coast and seabed is statistic analysis of data on the distribution and parameters of the ice scour features and the variability of this parameter over time. Acquiring a wide knowledge of the distribution and penetration depth of ice gouges on an extensive area of seabed is possible only with the use of specialized geophysical equipment – side-scan sonar, multibeam sounding, interferometric sidescan, acoustic sub-bottom profilers and GPS/GLONASS positioning systems. Investigations of the bottom ice gouging are conducted from a specialized research vessel equipped with these devices (Figure 3a). For field surveys of the ice gouging microrelief and determining the morphology and morphometric parameters of ice gouging forms on the sea bottom, we use a complex approach, which allows us to combine the results of several complementary methods. Side-scan sonar and echo-sounding surveys are executed at the same time from the same vessel.

The position of the sounding tracks is determined with high precision using modern GPS and GLONASS receivers, accepting WAAS error corrections. The speed of the vessel during the tracks should not exceed 4 knots. All the information coming from the side-scan sonar, echo sounder, and GPS/GLONASS goes to the computer module and is observed at the monitor in real time. It is registered on the hard drive with the help of the licensed software provided by the sonar equipment suppliers. An echo-sounding profile shows the morphology and morphometry of the forms (depth, width, etc.), as well as whether they are filled with bottom sediments; the side-scan sonar gives the idea of spatial distribution and orientation of the ice gouging forms. Inertial motion units tracking the 3D motion of the survey vessel improve quality by reducing the noise arising from pitch, roll, yaw, or (more commonly) combinations of all three.

As a rule, such investigations are conducted with the aim of assessing ice gouging impacts on the seafloor and oil and gas transportation infrastructure (trenches, pipelines) on the bottom. The spatial resolution of the soundings is determined by proximity to the projected objects and the necessity of repeating previous sounding tracks (Figure 3b). To assess the intensity of the ice scour at the present time, it is necessary to conduct repeated soundings and to distinguish forms which appeared between the two measurements. In the context of climate change, identification of “fresh” gouges is especially important in light of the climate warming of the last few years causing changes in the ice regime and displacing the maximum impact zone towards shallower water.

In this study, the position of the pipeline and the number, depth, width, orientation and bottom sediment fill of the ice scour troughs which have appeared between investigations conducted in different years are documented. For this purpose, the tracks of new surveys should coincide with the previous ones, and the same type of equipment should be used. The field stage is only the first step in the complex geophysical survey processing stream. This work enables statistical processing of morphometric data on the ice gouging forms, distribution, and parameters.

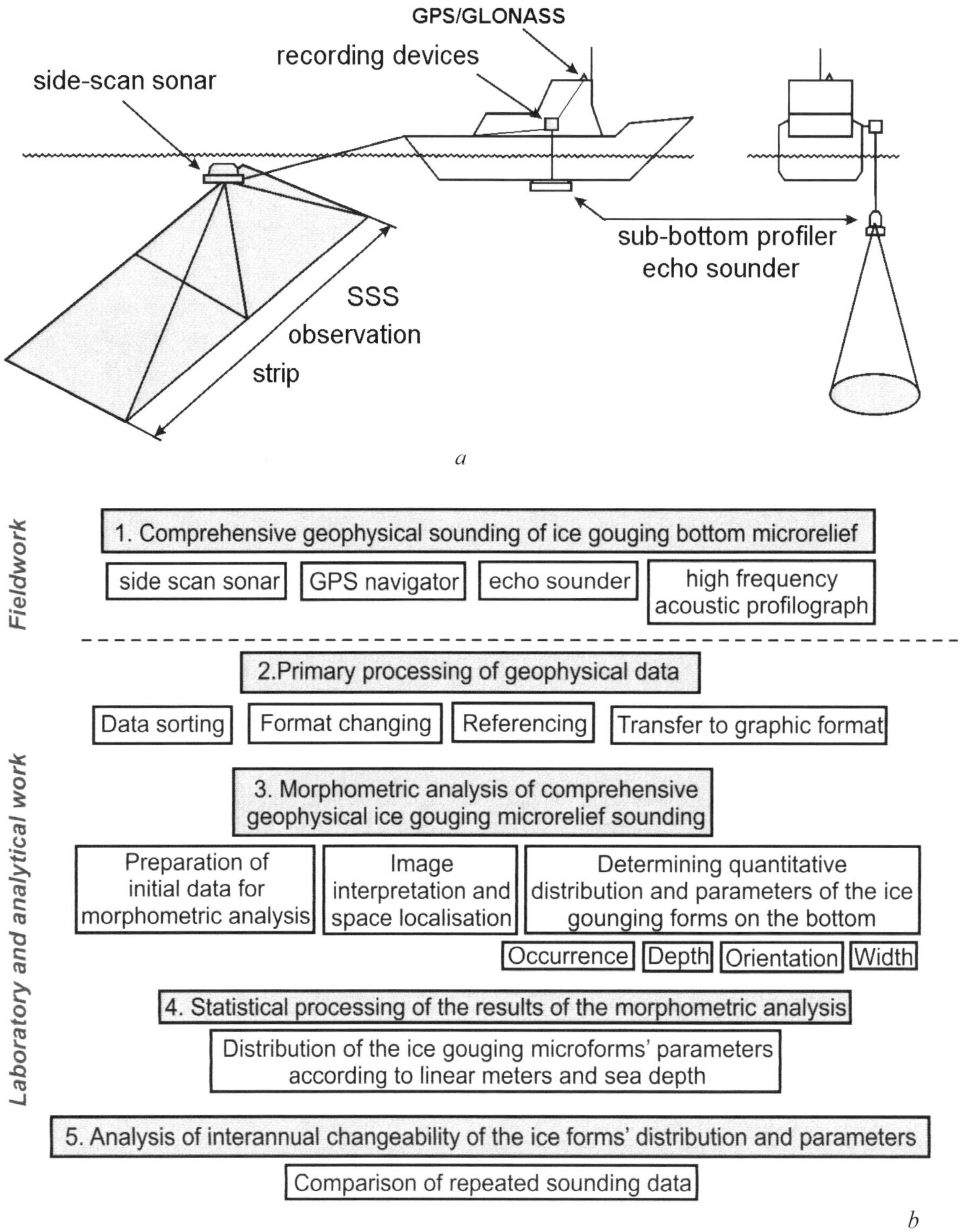

Figure 3. Geophysical survey technology for mapping ice-gouging seabed topography: a – research vessel with geophysical survey devices; b – ice gouging technology intensity assessment.

RESULTS

The first subdivision of the coastal zone of Baydaratskaya Bay by the types of ice formation and their effect on the sea coasts and floor was by V.A. Sovershaev (Dubikov, et al., 1997), who has made a huge contribution to studies of Arctic coastal dynamics. Detailed investigations performed in recent years under the guidance of the lead author, have made it possible to refine Sovershaev's model regarding the characteristics of ice formation as well as understanding of the mechanisms of ice scour (Figure 4).

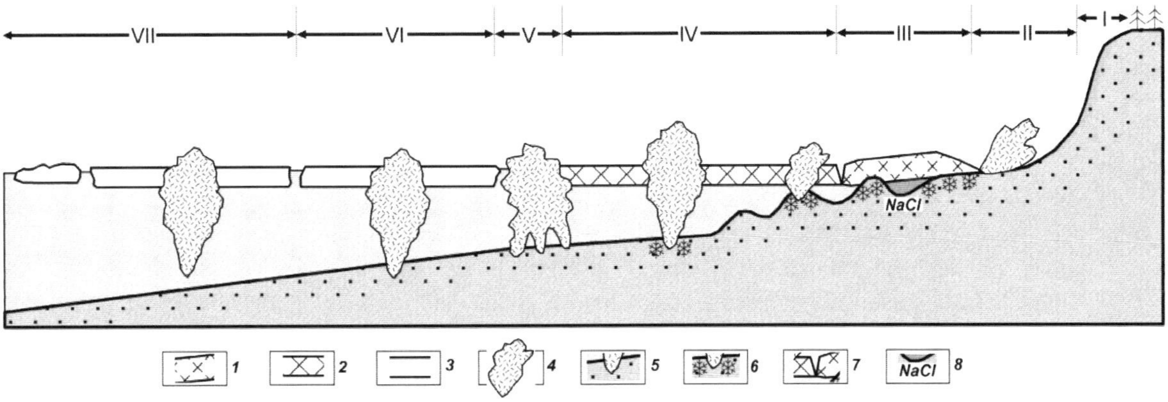

Figure 4. Subdivision of coastal zone by types of ice formations and their effects on coasts and seabed (Ogorodov, S.A., 2011): 1 – fast ice frozen to the bed; 2 – floating fast ice; 3 – drift ice floes; 4 – hummocks ice formations (ice ridges, grounded hummocks and ice dam), ice piles and overthrusts; 5 – hummock keel penetration into the ground; 6 – seasonally frozen ice forming at contact between ice and bed; 7 tidal crack; 8 – high-salinity water in longshore troughs, cryopegs.

Sea coasts are affected by ice during the periods of ice formation in autumn and fast-ice destruction and break-up in spring. On shoreline composed of sand-pebble material, the imprint of ice push is clearly defined. Our surveys reveal the widespread occurrence of ridges formed by ice-push (Figure 4, div. II). In autumn and early winter, young sea ice (20-40 cm thick) can be pushed onshore through wind-driven ice ride-up (unbroken floe) or pile-up (equivalent to pressure-ridge formation). During an ice-push event, the solid ice cover trims off the beach sediments and forms ridges of unsorted material (Barnes, 1982). In spring the ice also produces small-scale forms, such as furrows, striations, wallow depressions and also various ice-pushed ridges, all of which appear on sandy beaches (Forbes & Taylor, 1994). The depth of these forms does not usually exceed 0,5–1,0 m, and the length is typically <50 m. Most ice gouges are oriented transverse to the coastline. These forms, along with most of the ice squeezing forms, are usually preserved until the first considerable storm.

The mechanical action of the ice on the sea bottom lasts from the onset of ice formation until the sea is completely free of ice. After young ice freezes to the seafloor in the nearshore zone, this new strip of ice serves as a protective buffer. The ridges of hummocks closest to the coast develop above submarine bars (Forbes, et al., 2002). Because of a decreased sea depth above these bars, they become the focus of ridging and, thus the number of hummock ridges commonly corresponds to the number of submarine bars. Due to the onshore pushing impact of sea ice, ice gouges in this zone are mostly oriented normal to the coastline. The effect of the coastal hummock ridges and barriers on beaches and in shallow areas (down to a depth of 7-10 m) can be traced only immediately after fast ice is destroyed. The life expectancy of ice-gouged forms developed on sand beaches and shallow areas is very short, until the first summer storm. These forms, the depth of which is mainly <0.5 m, commonly disappear with the first strong waves in summer and autumn.

Further out in the bay, the pattern of hummocky pressure ridges is irregular and controlled by hydrodynamic factors, particularly the location of the fast ice edge during storms. Storm winds can destroy the fast ice edge and form a new ridge of hummocks or single grounded hummocks. The pattern of ice plowing in this case is either chaotic or parallel to the coastline. This is due to the prevailing along-shore drift of hummock formations. After the final onset of the fast ice the stamukhi remain non-mobile, often frozen down to the bottom. After their melting small (up to 1 m deep) holes and shallow, relatively short ice gouges remain. Ice gouges in this area are oriented mostly either chaotically or normally to the coastline, which is enhanced by the pressure of ice from the seaward side. Due to high hydrodynamic activity here, the gouges are quickly smoothed, and the gouge density in shallow areas is lower than in the zone of the fast ice edge.

The fast ice edge (within Baydaratskaya Bay near the 10-15 m isobaths) is the zone where the ice impact on the sea floor is the strongest. Here, pressure ridges reaching the floor ("ice dam") develop throughout the winter (Figure 4, div. V). In this case, ice gouges form a so-called "comb", usually oriented normal to the coastline due to the pressure of ice from the open sea. In 2007, during sonar tracking from the research vessel Ivan Petrov, such a "comb" was observed with dimensions of approximately 70 m wide and 400 m long; it consisted of a system of parallel ice gouges up to 1,5 m deep (Figure 5). Ice gouges are well preserved at this depth. They can be partly smoothed during extreme storms only. At depths of 14–16 m (within Baydaratskaya Bay), the occurrence and density of ice gouges turns out to be lower than at greater depths, though ice gouging is the most intense (most of the mobile systems of hummocks and grounded hummocks are formed here). The above situation is due to more active hydrodynamics at shallow depths, where the wave effect still influences the bottom and the velocities of tidal currents are higher. Due to this, the gouges at depth can exist over several years – first decade years here (as distinct from their short existence in shallow areas), gradually smoothing over and vanishing.

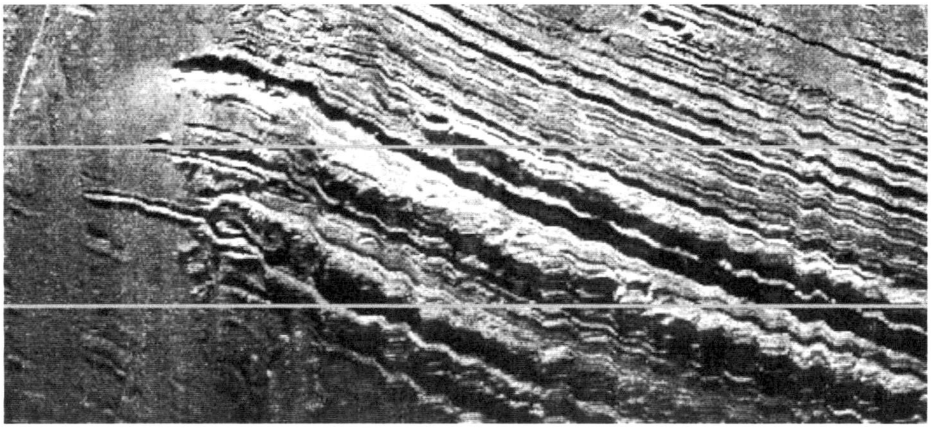

Figure 5. "Comb" ice gouges on Baydaratskaya Bay bottom

The most intensive and deep ice gouges occur in the area of drift ice at the range of depths from 16 to 19 m, next to the Yamal coast fast-ice rim, where ice hummocks is going on during the whole cold season and along which ice fields and hummocks, reaching the bottom frozen into them drift (Figure 6). Given the fact that the mass of the whole ice formation (floe and hummock) participates in the ice gouging, the deepest (up to 2 m), the widest (up to 50 m) and the longest (up to several km) ice gouges have been observed in this area. They are oriented conformably with tidal currents directions – lengthwise the Baydaratskaya Bay. Due to the low hydrodynamic activity at these depths, unaffected by wave action, the rates of sedimentation are very low. As a result, ice gouges are well preserved on the sea bottom and are mostly superimposed. Sometimes they cover up to 100% of the sea bottom (Ogorodov, 2003). Deeper than 19 m, ice gouges occur rather frequently. However, ice formations rarely occur at these depths. This situation is caused by a low hydrodynamic activity and low sedimentation rates. Under such conditions, the gouges, especially large ones, can exist on the bottom surface for decades. Thus, a low intensity of ice gouging is compensated by the long life of gouge forms. This "accumulation" effect gives a false impression of a high intensity of ice gouging here.

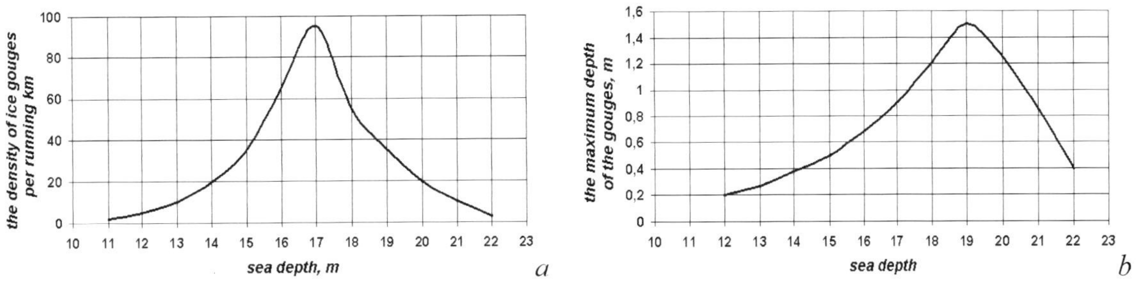

Figure 6. The distribution of the density (a) and depth (b) of the ice gouges with sea depth

At depths exceeding the maximum possible hummock thickness, 23-26 m in this area, bottom ice gouging stops. Ice gouges are rarely observed here. The hypothesis is that ice gouges at these depths are left by icebergs. Icebergs have been documented in Baydaratskaya Bay documented very rarely, including in 1932 and 2007. It is possible that the appearance of icebergs is connected to warm periods and occurs during the time of minimum ice cover.

DISCUSSIONS

With all lithologic and geomorphologic conditions being equal, the intensity of the ice gouging on the bottom is mostly determined by the changing condition, area and thickness of the ice cover. These characteristics can be obtained based on the data of hydrometeorological stations and on satellite observations followed by the analysis of the results of the sounding of ice gouging microrelief for different years.

As a result of the warming, in autumn of 2007 and 2012, a historical minimum of sea ice area since the beginning of satellite observations in 1979 was reached in the Northern Hemisphere. The border of fast ice changes as well. With all other conditions being equal, the greater the sum of temperatures below zero is, the more stable and wide the fast ice can be. So, over the last decade only, the area of fast ice at the moment of its maximal extent decreased by 15% (Frolov, 2015). The ice-free period duration, in its turn, increased. Under the conditions of climate warming, the character and intensity of ice gouging impact of ice formations on the coasts and bottom change considerably. Due to the decrease of the ice period duration and ice thickness, an activation of all the processes connected to sea ice occurs; ice hummocking increases. As a result, the width and stability of fast ice becomes lower; the action of sea ice on the coasts grows.

During the last decades, along with the climate warming in the Arctic basin, not only a general decrease of the ice cover is observed, but also the circulation system of sea water and ice is changing, which leads to a redistribution of areas with thin young and thick perennial pack ice. Among the first-year ice, clear anomalies of the ice cover thickness are observed in the last decade (Frolov, 2015). A decrease of the ice thickness immediately leads to changes in the size of the hummocky ice formations, the deepening of which directly depends on the ice thickness. These two parameters are directly connected. During the last two decades, the duration of the ice period in the Baydaratskaya Bay decreased almost by 1,5 months; the fast ice often breaks and can be carried away during its formation (especially from the side of the Ural coast); the thickness of the ice cover decreased at least by 10-15 sm (Frolov, 2015).

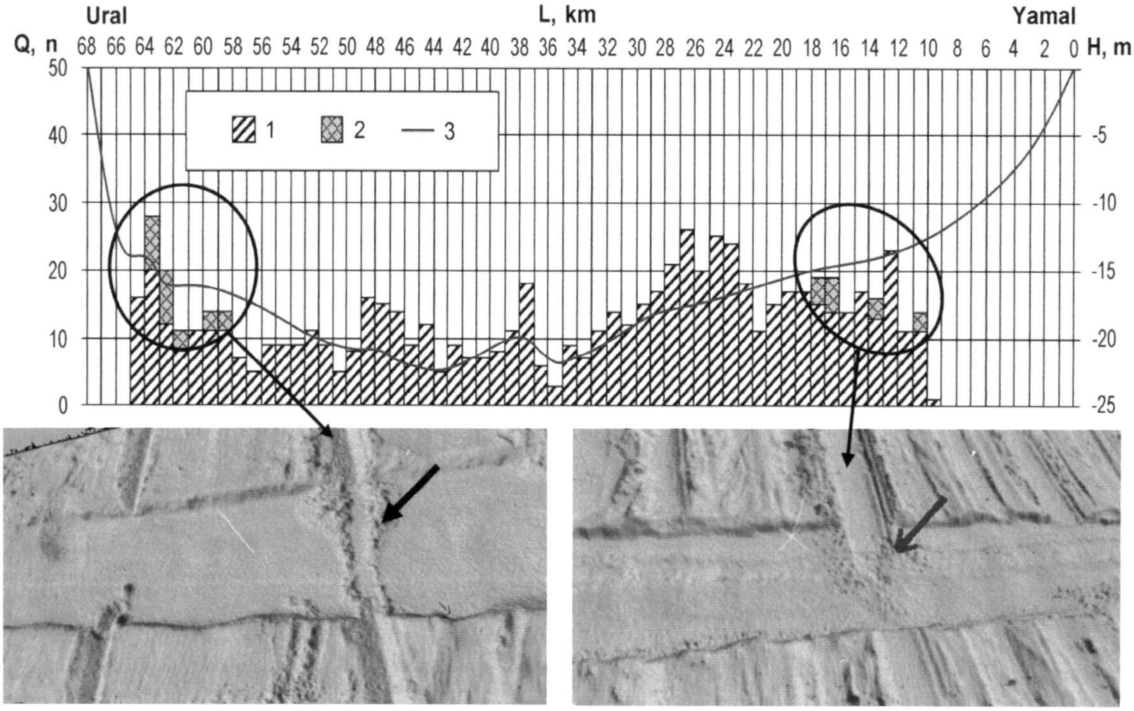

Figure 7. Example of the results of the geophysical sounding data processing for 2011-2012 at the profile of the pipeline crossing "Yamal-Europe" (Ogorodov, et.al., 2015): Q – number of ice gouges ; H – sea depth, m; L – distance from the zero point of the profile, km (Yamal); 1 – number of ice gouges on 1 linear km of the pipeline; 2 – number of fresh gouges formed during the cold season of 2011-2012, 3 – profile of the sea bottom; 4 – fresh ice gouges under filled trench of pipeline underwater crossing

To assess the real intensity of the ice gouging impact during a given ice season, repeated sounding of the sea bottom microrelief is necessary, in order to select the ice gouges which were created in the period between the two consequent observations. Field geophysical observations, including monitoring observations (made for the same profiles repeatedly), were made in the Baydaratskaya Bay of the Kara Sea and within the territory of the gas pipeline underwater crossing in the period from 2005 to 2015, corresponding to a significant climate warming and ice decrease in the Kara Sea.

Young ice gouges, which were formed after the construction of the Baydaratskaya Bay underwater pipeline crossing, are easy to detect above the recently filled trench on the bottom: fresh gouges leave traces on the ground deposited by the constructors, similar to a pencil leaving traces on a clean piece of paper (Figure 7). The executed observations prove that during the last years, a decrease of the ice cover thickness and the ice hummocks' size together with the fast ice width was observed. As a result, the area of the most intensive ice gouging impact moves to smaller water depths. If before, the maximum of the ice gouging impact (by the number and depth of ice gouges) was observed at depths from 17 to 20 m, at present, this zone has moved to 13-17 m (Figure 7). The soil penetration depth of largest ice gouge does not exceed 0,8 meters. Ice gouges have stopped forming in the central, deep part of the bay.

Based on comprehensive monitoring observations, it has been established that in the conditions of climate warming and ice cover decrease, the zone of the most intensive ice impact is moved landwards, to areas of shallower water areas.

ACKNOWLEDGMENT

The present study was prepared with the financial support of the Russian Science Foundation project No.16-17-00034.

REFERENCES

Barnes, P.W., 1982. Marine ice-pushed boulder ridge, Beaufort Sea, Alaska. *Arctic*, 35(2), pp. 312-316.

Barnes, P.W., Rearic, D.M., & Reimnitz, E., 1984. Ice gouging characteristics and processes. In: P.W. Barnes, D.M. Schell, Reimnitz, E., eds., *The Alaskan Beaufort Sea: Ecosystems and Environments*. Orlando: Academic Press, pp. 185-212.

Carsola, A.J., 1954. Extent of glaciation on the continental shelf in the Beaufort Sea. *American Journal of Science*. 252(6), pp. 366-371.

Dubikov, G.I., Sovershaev, V.A. and Tuzhilkin, V.S. (eds.), 1997. *Environmental conditions of the Baydaratskaya Bay. Main results of investigations for construction of submarine passage*. Moscow: GEOS, 432 p. (in Russian).

Forbes D.L. & Taylor, R.B., 1994. Ice in the shore zone and the geomorphology of cold coasts. *Progress in Physical Geography*, 18(1), pp.59-89.

Forbes, D.L., Manson G.K., et al., 2002. Nearshore ice and climate change in the southern Gulf of St. Lawrence. Ice in the Environment: *Proceedings of the 16th International Symposium on Ice. Int. Association of Hydraulic Engineering and Research*, pp.344-351.

Frolov, I.E., 2015. *Review of the hydrometeorologic processes in the Arctic Ocean in 2014*. St. Petersburg: Rotaprint GNC RF AARI, 124 pp. (in Russian).

Hnatiuk J. & Brown, K., 1977. Sea bottom scouring in the Canadian Beaufort Sea. *Proceedings of the 9th Annual Offshore Technology Conference*, v. 3, pp. 519-527.

Lewis, C.F.M., 1977. Estimation of the frequency and magnitude of drift-ice groundings from the ice scour tracks in the Canadian Beaufort Sea. *Proceedings of the 4th International Conference on Port and Ocean Engineering under Arctic Conditions*, v.1, pp. 568-579.

Lisitzyn, A.P., 1994. *Sea-ice sedimentation in the World Ocean.* Moscow: Nauka, 448 p. (in Russian).

Marchenko, A.V., Ogorodov, S.A., Shestov, A.V. & Tsvetsinsky, A.S., 2007. Ice gouging in Baydaratskaya bay of the Kara Sea: field studies and numerical simulations. *Proceeding of the 19th International Conference on Port and Ocean Engineering Under Arctic Conditions, POAC-07*, pp. 747-759.

Ogorodov, S.A., 2003. The Role of Sea Ice in the Coastal Zone Dynamics of the Arctic Seas. *Water Resources*, 30(5), pp.509-518.

Ogorodov, S.A., 2011. *The Role of Sea Ice in Coastal Dynamics*. Moscow: Moscow University Press, 173 p. (in Russian).

Ogorodov, S.A.,. Arkhipov, V.V & Kokin, O.V., 2015. Climate Change Effect on the Intensity of Seabed Gouging by Hummocky Ice Floes, In: Melnikov, V.P. and Drozdov, D.S., eds., Arctic, Subarctic: mosaic, contrast, variability of the Cryosphere, *Proceedings of the International Conference,* pp. 269-271. (in Russian).

Reimnitz, E. & Barnes, P.W., 1974. Sea ice as a geologic agent on the Beaufort Sea shelf of Alaska. In: J.C. Reed and J.E. Sater. *The Coast and Shelf of the Beaufort Sea*. Arlington: Arctic Institute of North America, pp. 301-353.

Reimnitz E. &Kempema, E.W., 1984. Pack ice interaction with Stamukhi Shoal, Beaufort Sea, Alaska. In: P.W. Barnes, D.M. Schell, E. Reimnitz, eds., *The Alaskan Beaufort Sea: Ecosystems and Environments*. Orlando: Academic Press, pp. 159-183.

Rex, R.W., 1955. Microrelief produced by sea ice grounding in the Chukchi Sea near Barrow, Alaska. *Arctic*, 8(3), pp. 177-186.

Shearer, J.M. & Blasco, S.M., 1975. *Further observations of the scouring phenomenon in the Beaufort Sea.* Geological Survey of Canada, 1975, pp. 1-75.

Vershinin, S.A., Truskov, P.A. & Kuzmichev, K.V., 2005. *Ice effect on the constructions at Sakhalin shelf.* Moscow: Giprostroymost Institute, 208 p (in Russian).

Zubakin, G.K., 2006. *Ice formations in the Western Arctic seas.* St. Petersburg: AARI, 272 p. (in Russian).

Salt Domes Influence on Ice Formation Processes in the North Caspian Sea

Piter I. Bukharitsin[1]

[1]Astrakhan State Technical University, Astrakhan, Russia, astrgo@mail.ru

ABSTRACT

During the longstanding ice air-reconnaissance, the observers notice the polynyas in those shallow areas of the North Caspian, where they expect the thick stationary ice (shore-fast ice). Attempts to explain its formation by dynamic factors only (under-ice currents, ice drifting) did not produce any good result. With artificial satellites emerge the data on the Caspian ice increased drastically. Analysis of ice cover state and dynamics data allowed to make a conclusion, that some polynyas form annually in the same locations of the North Caspian Sea. Brusilovsky (1986) supposed, that the polynyas locations may be related to salt domes presence in the water areas of the North Caspian. The Caspian waters dissolve salt domes peaks, leading to local water salinity growth, which in a cold season results in delay in ice forming above the salt domes. Consequently, there form the polynyas. In summer months, local focuses of higher water salinity can hardly be disclosed due to winds, waves and currents. The research should be continued to receive a more detailed hydro-chemical description of the coastal area marine waters in the Caspian Sea (above the salt domes) during a winter season.

KEY WORDS: North Caspian; Ice Cover; Salt Domes; High Salinity; Polynyas.

SATELLITE ICE DATA

Thanks to space technologies development during the 1970-s, radically new and promising methods for hydrological conditions study were developed, including ice processes in seas and oceans. This was facilitated by creation of the satellite information autonomous receivers network. In 1975, such receiver site was created in Astrakhan on the basis of the Astrakhan Zonal Hydro-meteorological Observatory (AZHMO). In winter 1976, for the first time ice cover television pictures were used for precise determination of ice edges and boundaries location acquired from ice air reconnaissance at the Caspian Sea. A simple optical-graphical method was employed for decrypting the satellite television pictures and mapping the ice situation. Informative value and accuracy of these ice maps received was successfully augmented by synchronous shooting. A comparative evaluation between the ice maps accuracy, which were acquired from the Meteor spacecraft, NOAA satellites, etc. and maps of synchronous ice air reconnaissance revealed a satisfactory coincidence of shore-fast ice boundaries, margins of drifting compacted ice, as well as dimensions and location of flaw polynyas, which were averagely 3 miles (5.5 km) across. While the satellite information receiving and processing took a minimum of time (20 to 30 minutes), it sufficiently increased the ice maps operativeness and quality. This offered using the satellite information as a basis for ice reconnaissance pre-flight briefing, coordinating routes for the next air reconnaissance with flight crews, correcting and accurate determining the ice situation in those sea regions, where the air reconnaissance was not carried out due to various reasons, as well, as their fast supplying for scientific purposes and for sea industries of domestic economy [2-5, 9].

REMOTE STUDY OF FLAW POLYNYAS

Satellite data provided an opportunity for monitoring formation, development and extinction process in one of the most important and changeable hydro-meteorological condition elements during winter in the North Caspian - the flaw polynya. The ice air-reconnaissance did not provide such opportunity because of insufficient monitoring frequency and data incompleteness. It was thought, that locations, where polynyas appear as well, as their development processes are determined by wind's speed, direction and duration effect, coupled by under-ice currents' speed and direction. The satellite data helped to reveal that with sustained and offshore winds blowing in the northern Caspian, between the shore-fast ice and the drifting ice, there appear polynyas several hundred metres to 10 miles and more in width, sometimes up to 100 miles or even more (Fig. 1). Formation of these giant polynyas is facilitated by wind-drift currents, which result from wind effect blowing on sea surface free from ice. Ice situation may change rapidly with wind change. Drifting ice will close the existing polynyas, while new ones will form at the windward side [7, 8].

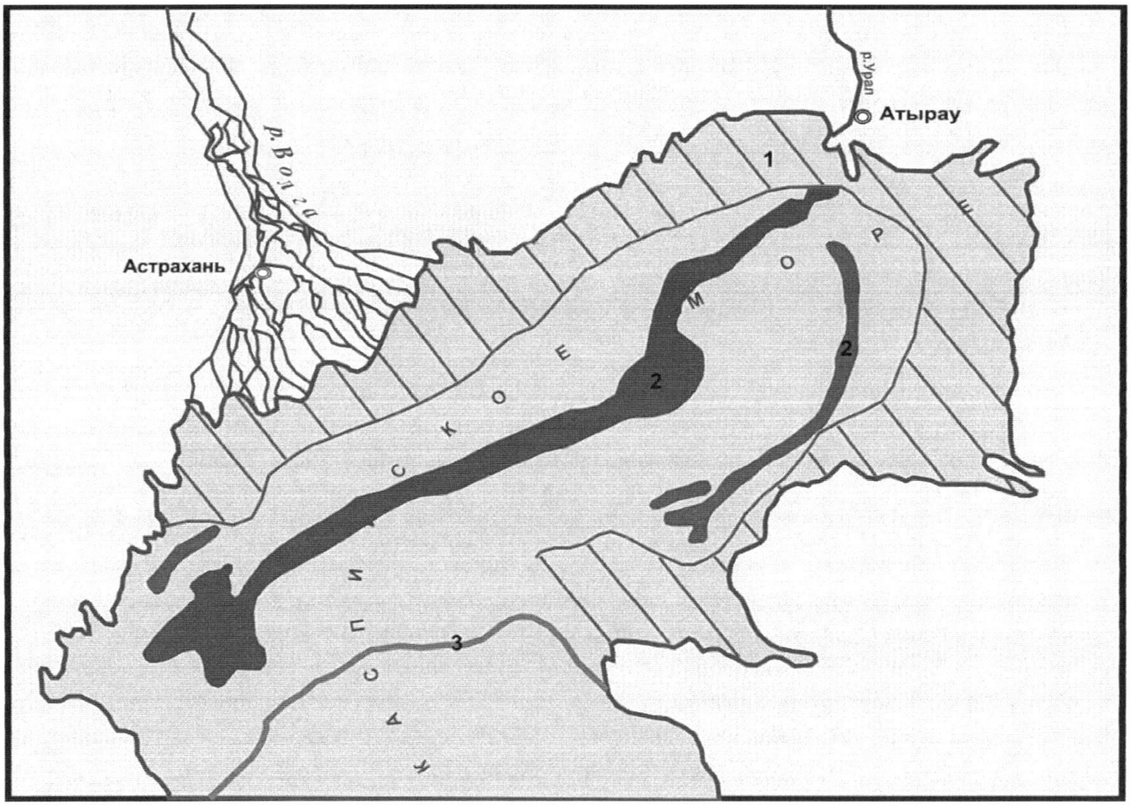

Figure 1. The North Caspian areas, where the polynyas form regularly, are marked in grey-brown

A compiled map shows vast flaw polynyas in shore-fast ice edge, shot by the satellites during 1980-90-s, provided determining the North Caspian areas with a sustained shore-fast ice, i. e. sea areas, where stationary ice remains during the whole ice season from its formation moment to fracturing in spring.

Figure 2. Polynyas in the northern Caspian ice cover
(this photo was found on the Internet df67c592)

It was revealed that during the ice cover formation period, the shore-fast ice movement to the south is not gradual, as it had been thought before, but leaping. Boundaries between older and newer shore-fast ice are often marked by hummock ice, several kilometres in size, which is similar to those forming in drift divides of arctic seas. They may be seen well on satellite pictures and decrypted easily. Thanks to aircraft and satellite data, we managed to trace gradual daily changes in ice distribution on the entire area of the North Caspian, which was not possible with any previously existing traditional methods. As a result, appearance, location and dynamics of ice leads and flaw polynyas in relation to wind direction, speed and duration effect was revealed. A sequence of shore-fast ice fracturing process in spring time was traced as well [6].

Although, for many years during ice air reconnaissance carried out in depth of winter, oceanologists have been noticing vast polynyas in shallow areas of the North Caspian, where, according to their calculations, they could not have been, but there should be a thick stationary ice (shore-fast ice) instead. Attempts to explain its formation by dynamic factors only (under-ice currents, ice drifting) did not produce any good result [1]. There should be some unknown reason, due to which they form (Fig. 2).

NORTHERN CASPIAN SALT DOMES

Brusilovsky (1986) was the first to suppose, that the polynyas locations may be related to salt domes presence in the water areas of the North Caspian (Fig. 3).

In 2002, a Russian book "Regional geology and oil-and-gas-bearing capacity of the Caspian Sea" by Glumov, et al., was published [11]. In this book, the authors draw a conclusion for many years of marine geological and geophysical research conducted by methods of seismic, electrical, magnetic, gravity and geothermal survey, drilling, geochemistry, seismology, etc., which was executed mainly by the VNIIMORGEO (All-Russia Scientific-Research Institute of Marine Geology and Geophysics) in the city of Gelendzhik. This is the former all-Soviet leading scientific institute (renamed many times), which set out hundreds of expeditions across the Caspian Sea, including those under supervision of B.N. Golubov. In their work scientist present a detailed diagram of salt tectonics and adjacent dry land in the North Caspian according to drilling, seismic survey and gravimetric data (Fig. 4). Unfortunately, the authors did not make any mention in their book of the salt domes relation to ice processes in the northern Caspian, nor to the polynyas, in particular.

Marine hydrochemistry results, acquired by Gursky (2003, 2007), are also of an informational value. They are published in his Doctor of Sciences dissertation research, comprising 2 volumes. The diagrams acquired by him resemble much in details salt dome rise contours in the North Caspian.

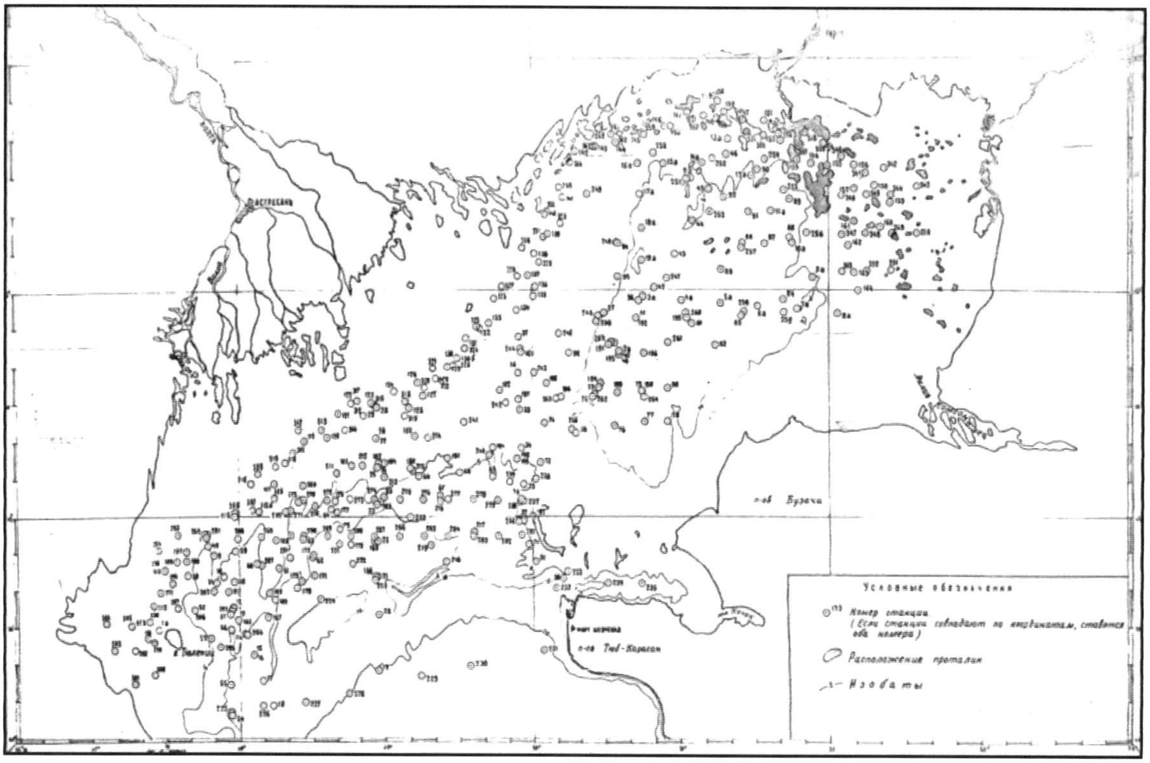

Figure 3. The diagram of salt tectonics and adjacent dry land in the North Caspian (Brusilovsky, 1978)

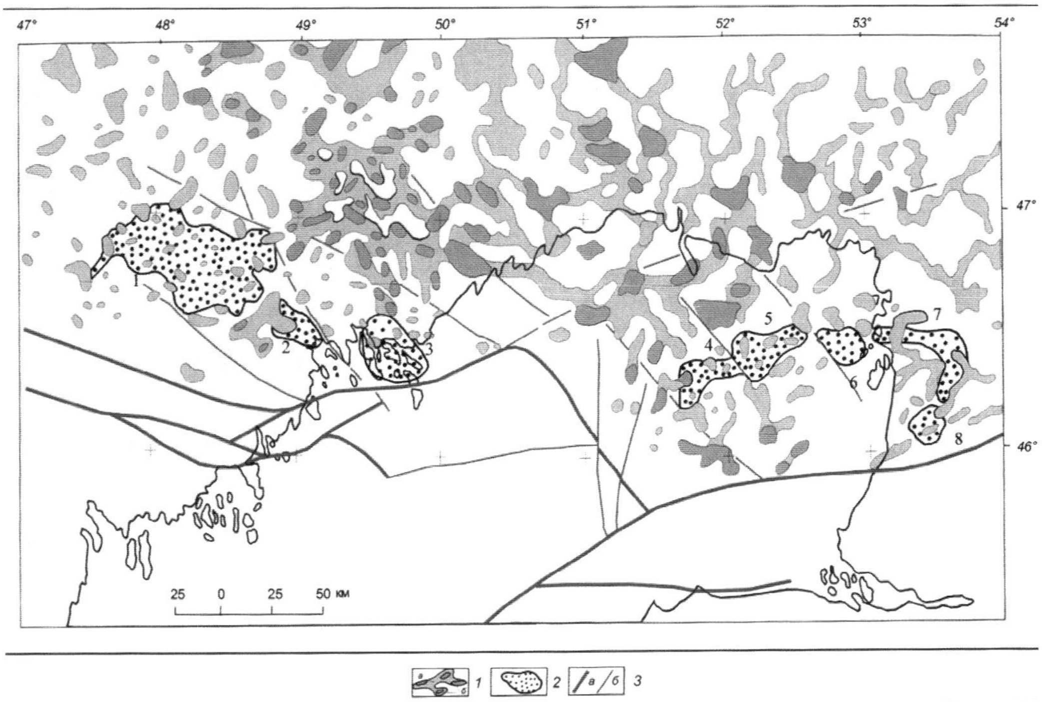

Figure 4. The diagram of salt tectonics and adjacent dry land in the North Caspian (according to drilling, seismic survey and gravimetric data) [11]. Map symbols: 1 - salt beds and masses (a) and local salt cores (b); 2 - subsalt bed protrusions; 3 - subsalt bed fractures: primary (a) and secondary (b). Map figures: subsalt bed protrusions 1 - Astrakhansky, 2 - Imashevsky, 3 - Zhambaisky, 4 - Kerogly-Nubar (western Kashagan), 5 - Kashagan (eastern Kashagan), 6 - Kairan, 7 - Tazhigali-Karaton. 8 - Tengiz.

The author analyses data acquired from many years of observations and proves the supposition made by Brusilovsky: stationary polynyas location practically coincides with salt domes outcropping in the northern Caspian Sea bed (Fig. 5).

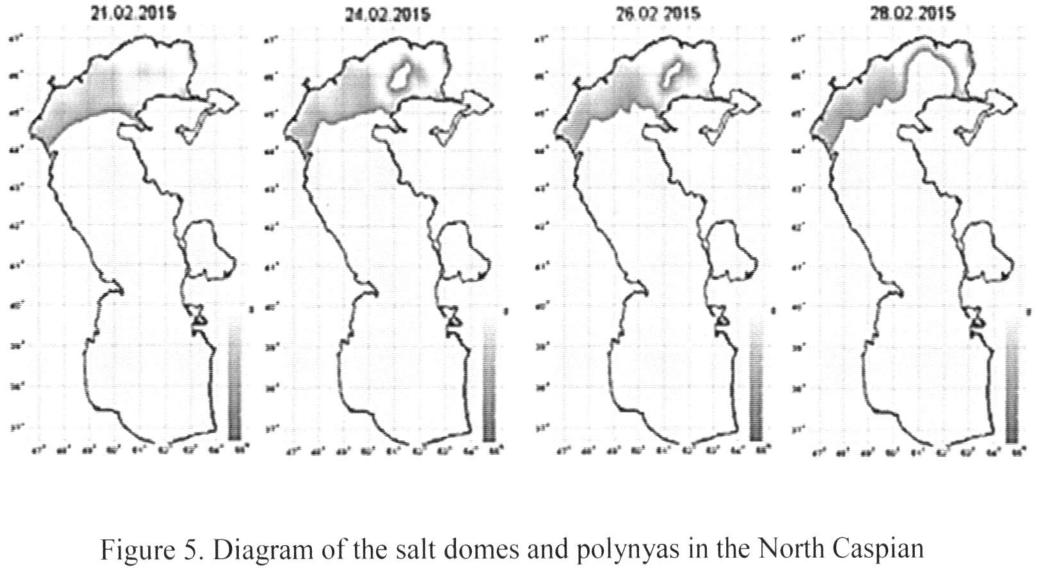

Figure 5. Diagram of the salt domes and polynyas in the North Caspian

The Caspian waters dissolve salt domes peaks, leading to local water salinity growth, which in a cold season results in delay in ice forming above the salt domes. Consequently, there form the polynyas. In summer time, waves and currents hinder findings of local focuses of higher water salinity.

FIELD RESEARCH ON THE CASPIAN ICE STRENGTH PROFILE

Many factors influencing ice strength lead to a large range of strength value, which is d escribed in dedicated literature. Using the same devices and studying methods by the sa me researchers provided grounds to hope for better results accuracy and a much homoge neous data on the strength, which depends only on objective characteristics of the ice co ver. The diagram resulting from many years of study is given in Fig. 6 [10]. Systematic research on the ice strength profile in the eastern part of the North Caspian using a helic opter was conducted in winter season 1970-71. Ice structure and strength study was cond ucted on open cut of Gurievskaya borozdina - Ukatny isle - Karakaisky bank site on 6th and 9th March, 1971. The ice thickness in the cut ranged from 27 to 42 cm. Tests of p lates, obtained from cutting cylinder-shaped ice cores, revealed ice destructive stress valu e at compression related to its salinity (chlorinity). In all plate samples, its average value was 10-16 kg/cm.

On 29 January, 1972 a helicopter-assisted work was carried out to select the ice cores d uring the Caspian ice strength study in: Gogolskaya spit, its continental slope, Balashovs ky, Suendykovsky, Kolkhozny sand islands, totally at 5 locations in the extreme east of t he North Caspian. At the time of the study, the temperature was -19°C, snow cover on i ce 0-5 cm, sea depth in three locations was 60-70 cm, in two locations the ice lay on th e ground 32-38 cm in thickness. The ice was slightly structured, homogeneous across its strata and matted, as a turbid water had been frozen. Lower strata had air bubbles. Snow temperature was -14.5°, under-ice temperature showed -0.4° - (-0.3°). It was found that the temperature in the ice cores increases from surface towards the sea depth. Thus, on t he surface it showed - 13° - (-14°), while in the bottom part only - 5°, with the ice tem perature growing faster from the surface. Difference between first 10 cm of the core was 3°, then 2°, 1.5°and, in the lowest strata 1.0 - 0.5°.

The ice chlorinity (salinity) in the considered samples increased along its depth. At the i ce surface it was 700-900 mg/l and in lower strata - 1300-1500 mg/l. Destructive load f or 100 ice plate samples, as tested by application-interpreted model was: 40-50 kg - for upper strata, 30 kg - for lower strata. The ice strength value varies from 7 to 3-4 kg/cm.

In January, 1973 the aircraft-assisted work for the ice strength study in the eastern part o f the North Caspian was continued. Photography shooting was carried out on January 27 at nearly the same locations, as were in January, 1972. At the time of the study, air tem perature was -7°, snow cover on ice 4–5 cm, the ice temperature -5.5°, upper ice stratu m temperature -4.8°, ice temperature in near-water stratum was -1.5°, water temperature -0.2°. Thus, the temperature distribution in the ice strata measured in January, 1973 diffe red significantly from that of the previous year. Ice temperature increasing gradient did n ot exceed 1.0° (in 1972 it reached 3.0°) across all the ice strata.

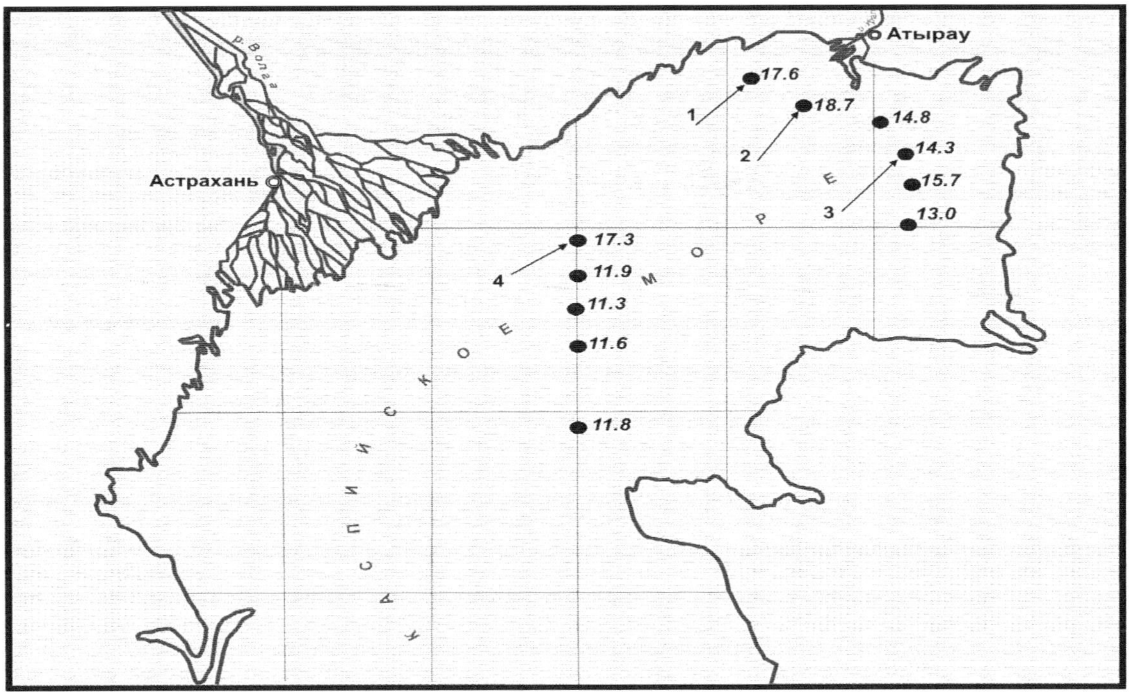

Figure. 6. The ice core sampling points in the east of the North Caspian.
Figures near the points stand for ice average strength in core, kg/cm².
Arrows show the points, data from which was used to draw the ice strength curves.

The chlorinity (salinity) in different ice strata varied from 320 to 1,540 mg/l (Table 1).

Table 1. Comparison of ice chlorinity in eastern areas of the North Caspian in 1972 and 1973, mg/l

Average	Years		Average for 2 years
	1972	1973	
From surface ice samples	826	585	665
From bottom ice samples	1405	890	1062
From all samples	1085	685	835

Such difference in the ice chlorinity (salinity) value for the years given may be explaine d by ice formation peculiarities. So, during the winter seasons 1971-72 ice formation ran quickly. Freezing water captured many salts, which, forming salt brine cells, filled in sp ace between the fresh-water ice crystals, thus increasing the chlorinity (salinity) of the ic e. While in winter 1972-73 decrease in temperature occurred slowly. The ice formation p assed gradually, with the salt brine leaking down into the water and increasing salinity o f the under-ice water, at the same time, the ice chlorinity (salinity) was decreasing.

Like the chlorinity (salinity), average ice strength in 1973 proved to be sufficiently highe r than in 1972 - 14.4 kg/cm. Near the same was an average strength in upper and lower strata.

For the two years (1972-1973), average value of the ice strength in all horizons was 10. 7 kg/cm, and surface stratum average strength was 12.1 kg/cm [10].

The next stage of the ice study was measuring ice maximum strength in different areas of the North Caspian. This work was conducted from December 27, 1973 to February 3, 1974 at 30 equidistant points in the North Caspian water area. Air temperature at the c ore sampling points ranged from -1.5° to -18.1°, ice thickness was 11 to 43 cm, snow c over on the ice was 0 to 4 cm and water temperature under the ice was 0.85° to -0.26°. Water salinity at the surface (under the ice) varied greatly: from almost fresh at estuary seashores of the Volga and the Ural rivers to 20.52 ‰ in coastal area at the extreme eas t of the sea (Fig. 6 and 7) [10].

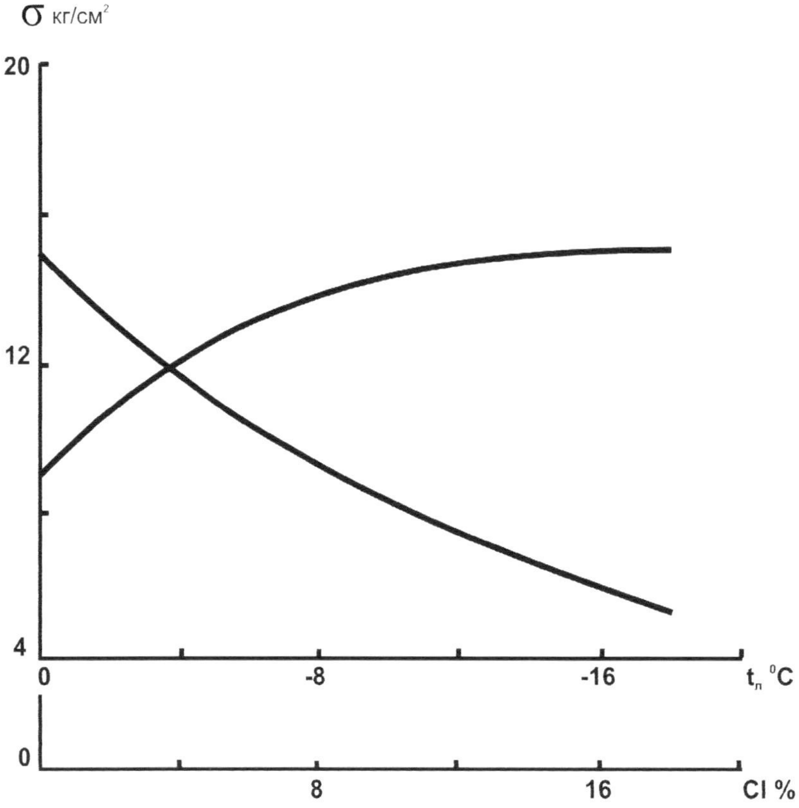

Figure 7. Smoothed ice strength values in the North Caspian related to its temperature a nd chlorinity (salinity).

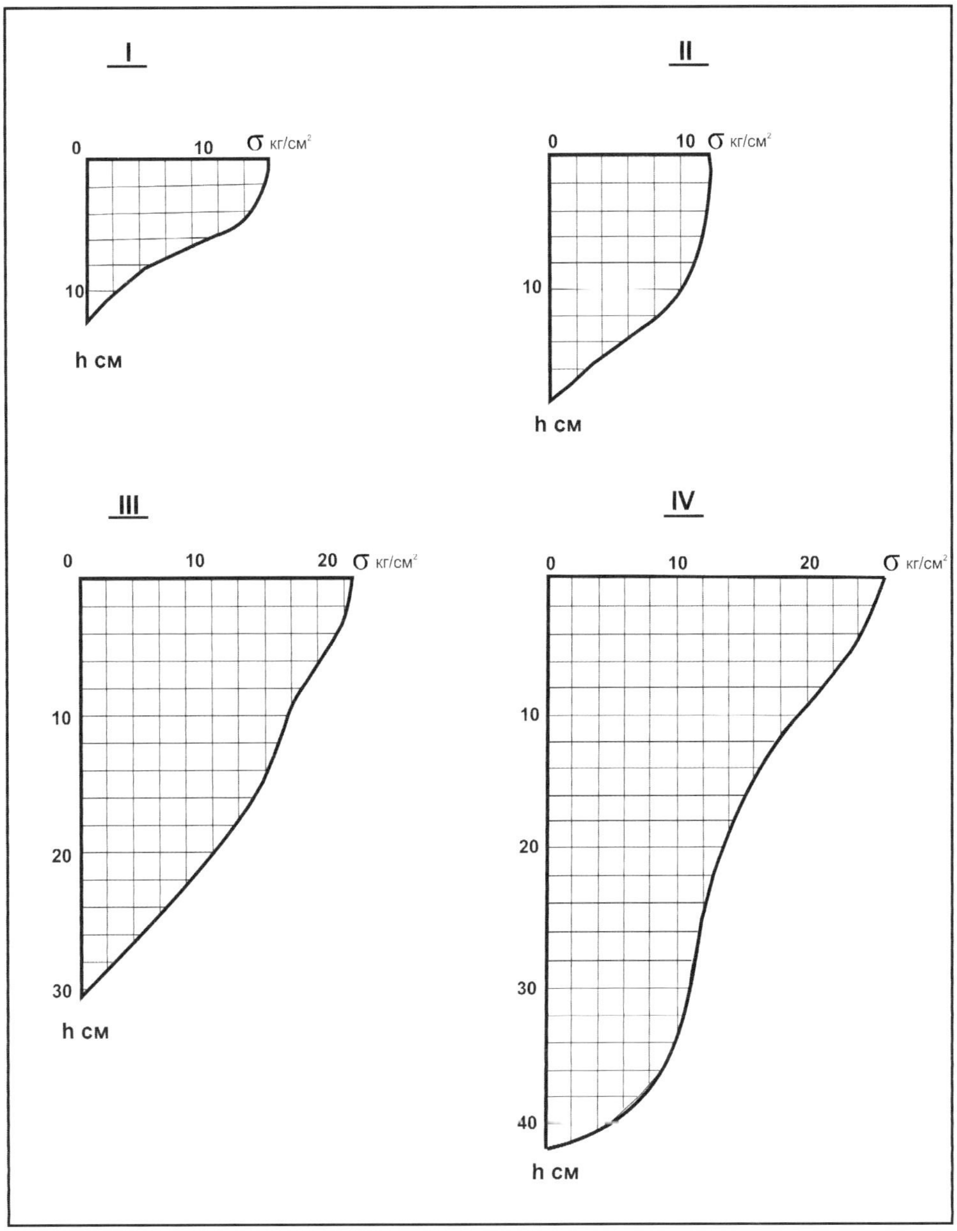

Figure 8. The curves of the ice strength at points located in the eastern part of the North Caspian:

1 - 27. XII.1973; **2** - 05.01.1974; **3** - 10.1.1974; **4** - 20.01.1974.

CONCLUSIONS

Analysis of the gathered aircraft and satellite data proved Brusilovsky's supposition, that some polynyas in freezing shallow areas of the North Caspian Sea form annually in the same locations.

The most probable factor facilitating their local formation is the least strong and thick ice, which forms above the salt domes, and with wind and currents, it is destructed easier than a stronger ice formed from almost fresh water of the Volga and the Ural rivers falling into the northern part of the Caspian. The field research results obtained during wintertime studies 1970-73 prove this supposition completely.

The research should be continued to receive a more detailed hydro-chemical description of the coastal area marine waters in the Caspian Sea (above the salt domes) during a winter season.

REFERENCES

1. Bukharitsin, P.I., 1994. Statsionarnye polyn'i v ledyanom pokrove Severnogo Kaspiya [Immovable polynyas in the ice cover of the North Caspian], *Tezisy dokladov itogovoi nauchnoi konferentsii AGPI im.Kirova*, Astrakhan, 4, p. 73.
2. Bukharitsin, P.I., 1981. Ispol'zovanie dannykh ISZ v izuchenii ledovogo rezhima Severnogo Kaspiya [Employing satellite data for studying ice conditions in the North Caspian]. *Rationalizer's proposition №54 (1439), dated 01.08.1980, adopted for use by North Caucasus Directorate for hydrometeorology and Environment Control (UGKS) on 26.06.1981*, Astrakhan; Bukharitsin, P.I., 1983. Ispol'zovanie televizionnykh snimkov ISZ "Meteor" dlya izucheniya ledovoi obstanovki na Severnom Kaspii [Using television pictures of the Meteor satellite for studying ice conditions in the North Caspian], *Trudy GMTS SSSR, Moscow*, 255, pp. 70-75.
3. Bukharitsin, P.I., 1984. Ispol'zovanie dannykh ISZ pri izuchenii ledovogo rezhima Severnogo Kaspiya [Using satellite data in studying the North Caspian ice regime], *Informatsionnoe pis'mo SK UGKS*, Rostov-on-Don, 1, pp. 48-51.
4. Bukharitsin, P.I., 1987. Opyt ispol'zovaniya dannykh ISZ pri izuchenii ledyanogo pokrova [Experience of using satellite data in studying ice cover in the North Caspian], *Informatsionnoe pis'mo SK UGKS*, Rostov-on-Don, 4(12), pp. 19-23.
5. Bukharitsin, P.I., 1981. Ispol'zovanie dannykh ISZ v izuchenii ledovogo rezhima Severnogo Kaspiya [Employing satellite data for studying ice conditions in the North Caspian]. *Rationalizer's proposition №54 (1439), dated 01.08.1980, adopted for use by North Caucasus Directorate for hydrometeorology and Environment Control (UGKS) on 26.06.1981*, Astrakhan.
6. Bukharitsin, P.I., Vasyanin M.F., Kalinichenko L.A., 1992. Metod kratkosrochnogo prognoza polozheniya kromki splochennykh l'dov na Severnom Kaspii [Short-term forecasting method for packed ice edge location in the North Caspian], *Meteorologiya i gidrologiya*, 4, pp. 74-81.
7. Bukharitsin, P.I., 1996. *Gidrologicheskie protsessy v Severnom Kaspii v zimnii period* [*Hydrological processes during winter season in the North Caspian*], Ph.D. Dissertation in the form of a report. Moscow: Water Problems Institute of RAS (WPI RAS), 61 p.
8. Bukharitsin, P.I., 2005. Mnogoletnyayaizmenchivost' kharakteristik termicheskogo i ledovogo rezhima nizov'ev Volgi i Severnogo Kaspiya [Multi-year changeability of characteristics in temperature and ice regime for the Volga river reach and the North Caspian], *Materialy Mezhdunarodnoi nauchno-prakticheskoi konferentsii «Melioratsiya malykh vodotokov, nerestilishch del'ty r. Volgi i Volgo-Akhtubinskoi poimy»*, Astrakhan, pp. 273-283.
9. Bukharitsin, P.I., 2006. Rabota nad sozdaniem maketa elektronnogo atlasa ledovykh yavlenii i obrazovanii nizov'ev Volgi i Severnogo Kaspiya [Working to create an electronic atlas model for ice events and formations in the Volga river reaches and the North Caspian], *Yuzhno-Rossiiskii vestnik geologii, geografii i global'noi energii*, 1(14), pp. 219-224.

10. Valler, F.I., Prochnost' ledyanogo pokrova Severnogo Kaspiya (po rezul'tatam rabot zimoi 1973/74 g.) [Ice cover strength in the North Caspian (based on 1973-74 yrs. study results)], *Sbornik rabot Astrakhanskoi ZGMO*, 2, pp. 102-108.
11. Glumov, I.F., Malovtskiy, Ya.P., Novikov, A.A., Senin, B.V., 2004. *Regional'naya geologiya i neftegazonosnost' Kaspiiskogo morya* [*Regional geology and oil-and-gas-bearing capacity of the Caspian Sea*], Moscow: Nedra-Businesscentre, 342 p.

POAC'17
Busan, KOREA

Proceedings of the 24th International Conference on
Port and Ocean Engineering under Arctic Conditions
June 11-16, 2017, Busan, Korea

Capability Analysis of Dynamic Positioning for the Arctic FPSO in ICE

Solyoung Han[1], Hyun Joe Kim[1], DongYeon Lee[1], Booki Kim[1]
[1] Central Research Institute (Samsung Heavy Industries Ltd , Korea)

ABSTRACT

Dynamic positioning (DP), which is widely used in a large range of marine operations nowadays, is a computer-controlled system to automatically maintain a vessel's position and heading by using its own thrusters. To evaluate the capability of DP, the standard method proposed by International Marine Contractors Association (IMCA) is widely used at the early design stage. The capability in open water is normally described as capability plots with respect to maximum thrust or maximum environmental condition by considering the given environmental loads. However, in ice-covered water, the ice load is a dominant factor among the other environmental loads. The ice loads are totally stochastic and vary with many parameters such as ice thickness, ice density, ice drift speed, strength of ice. There are some numerical methods to estimate the ice loads have been proposed and developed.

In this study, the global ice loads for a ship-type Arctic FPSO are calculated for various ice parameters with the software GEM (GPU Event Mechanics), developed by Memorial University. Through the DP capability analysis of the Arctic FPSO from the estimated ice loads, the characteristics of DP performance are examined. By this study, the influence of ice parameters are to be identified, which can be used for the design of the capacity and configuration of thrusters and mooring systems .

Keywords: Dynamic positioning (DP), Arctic FPSO, Capability Analysis, Ice load

INTRODUCTION

For offshore operations, it is very important to keep offshore floater at a desired position with adequate heading, nowadays. Operations such as drilling, production, subsea construction, maintenance, pipe-laying, and shuttle offloading are highly dependent on Dynamic Positioning (DP) systems. Environmental forces such as wind, waves and current, tend to move the floater off from the desire position. DP systems play a role to maintain the floater's position and heading by automatically controlling actuators to make balance between thrusts and environmental forces.

At the design stage, it is necessary to identify the DP operational window where the floater can maintain its position and heading because effective operational risk management depends on DP capability. The DP operational window can be found by the static analysis of DP capability. Given environmental conditions, the capability can be examined by quasi-static method to balance the mean environmental loads with the mean thruster forces. The computations and representations of the DP capability have been standardized by the International Marine Contractors Association (IMCA, 2000). The DP capability is generally described as capability plots in polar coordinates with respect to maximum thrust usage or maximum environmental conditions. Currently the capability plots are widely used for floater design in industries due to the advantages of its simplicity.

In ice-covered water, the ice load is a dominant factor among the other environmental loads. Therefore, it is very important to take into account the ice load in analyzing DP performance. The ice loads induced by the interaction between ice and floater are quite different from the other environmental loads in open water. The ice loads are totally stochastic due to the substantial amount of parameters which define ice material such as thickness, elastic modulus, density, flexural strength, etc. Furthermore, it also varies a wide range of ice features, such as level ice, pack ice, ridges, icebergs, etc. Biao et al (2013) carried out a static ice load-based dynamic positioning (DP) capability analysis of the CIVArctic vessel in level ice condition. Kerkeni et al. (2013) reviewed the applicability of the standard method to DP capability analysis in ice-covered water. They suggested the use of dynamic simulation in time-domain to capture stochastic characteristics of the ice field.

This paper presents the DP capability analysis for a ship-type Arctic FPSO in pack ice condition. For the purpose, the ice loads are calculated for various ice parameters with the software GEM (GPU Event Mechanics), developed by Memorial University. In order to capture the stochastic nature of the ice, a series of numerical towing simulation are carried out. Based on the simulated data, the global ice loads are estimated by a numerical analysis of with respect to vessel heading and ice parameters. The characteristics of DP performance are then examined with the estimated global ice loads. Finally, the influence of ice parameters is identified, which can be used for the design of the capacity and configuration of thrusters and mooring systems of the Arctic FPSO.

DP CAPABILITY ANALYSIS

To ensure a standardization in the production of capability plots, IMCA has presented the specification for DP vessels in order to enable a direct comparison of individual vessel's performances and provide an indication of station keeping ability in a common format (IMCA, 2000).

For DP operations of the floaters, the mathematical model at the center of gravity can be described in the body frame as follows:

$$M\dot{v} = F_{coriolis} + F_{wind} + F_{current} + F_{wave} + F_{thrust}. \quad (1)$$

In DP capability analysis, the above model is simplified as quasi-static problem such that the velocity of a floater equal to zero:

$$0 = F_{wind} + F_{current} + F_{wave} + F_{thrust}. \quad (2)$$

The wind load F_{wind} is expressed at the center of gravity of a floater as follows:

$$\begin{aligned} F_{wx} &= \frac{1}{2}\rho_{air} v_w^2 C_{wx}(\alpha_w) A_{wT}, \\ F_{wy} &= \frac{1}{2}\rho_{air} v_w^2 C_{wy}(\alpha_w) A_{wL}, \\ F_{wn} &= \frac{1}{2}\rho_{air} v_w^2 C_{wn}(\alpha_w) A_{wL} \cdot LBP, \end{aligned} \quad (3)$$

where ρ_{air} is density of air, v_w is wind velocity, C_{wx}, C_{wy} and C_{wn} are wind coefficients, A_{wT} and A_{wL} are transverse and longitudinal wind area of floater, LBP is length between perpendiculars. The values for hull wind coefficients are normally determined by wind tunnel test or CFD calculations.

The current load $F_{current}$ is also expressed at the center of gravity of a floater as follows:

$$\begin{aligned} F_{wx} &= \frac{1}{2}\rho_{water} v_c^2 C_{cx}(\alpha_c) A_{cT}, \\ F_{wy} &= \frac{1}{2}\rho_{water} v_c^2 C_{cy}(\alpha_c) A_{cL}, \\ F_{wn} &= \frac{1}{2}\rho_{water} v_c^2 C_{cn}(\alpha_c) A_{cL} \cdot LBP, \end{aligned} \quad (4)$$

where ρ_{water} is density of water, v_c is current velocity, C_{cx}, C_{cy} and C_{cn} are current coefficients, A_{cT} and A_{cL} are transverse and longitudinal wind area of floater. The values for current coefficients are also determined by wind tunnel test or CFD calculations.

The wave load F_{wave} is divided into first and second order term. The first order term is described by vertically oscillating zero mean loads. The second term is described by a low frequency load. In the capability analysis, the first order term normally is not taken into account because the DP control systems filter out the first order motion and only respond to the mean wave drift part among the second order term. The wave load F_{wave} is then expressed as follows:

$$F_{wave} = 2\int_0^\infty S_\zeta(\omega) P(\omega, \omega)\, d\omega, \quad (5)$$

where S_ζ is a wave spectrum such as Pierson-Moskvitz or JONSWAP spectrum, and P is a second order response amplitude operator, which can be computed from the potential-based code such as WAMIT, WADAM, etc.

The thrust load F_{thrust} is a total force delivered from all the actuators of a floater such as tunnel thruster, azimuth thruster or propellers with rudders. The steady-state thrust by an actuator can be expressed as

$$F_{thrust} = k_T \rho n^2 D^4. \tag{6}$$

The power consumption as function of the amount of generated thrust F_{thrust} of an actuator, is typically defined by the non-linear relation:

$$P = (P_{max} - P_{min}) \cdot \left(\frac{F_{thrust}}{F_{thrust}^{max}}\right)^{\eta} + P_{min}, \tag{7}$$

where the parameters P_{max}, P_{min} and η are dependent on actuators, where η is typically the value between 1.3 and 1.7.

If the equation (2) can be solved under inequality constraints of power and limitation of the thrust, the floater is then considered to be capable of maintaining the desired position and heading given the environmental conditions. The DP capability plot can be determined after solving the equation (2) with aforementioned constraints for all headings with a certain heading interval. A sample DP capability plot is shown in the figure 1.

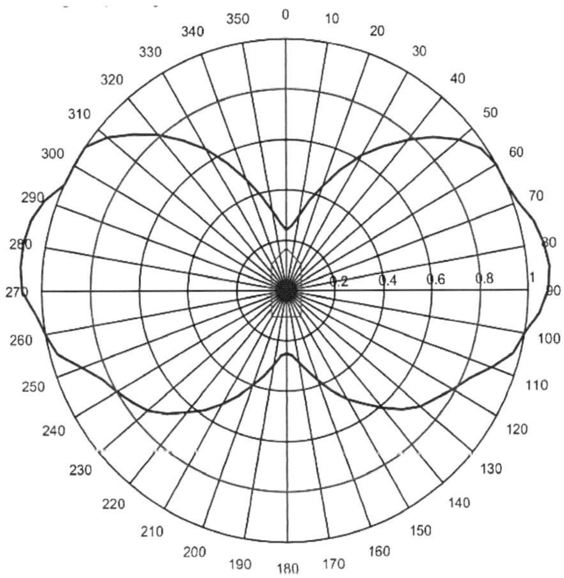

Figure 1. Example of a DP capability plot with respect to maximum thrust usage

ARCTIC FPSO

An Arctic FPSO considered in this study is designed by Samsung Heavy Industries (SHI). This conceptual vessel was designed for a joint collaboration project between KRISO and universities for an initial design of the Arctic FPSO for the government funded project in Korea, started in 2016. Hydrodynamic analysis and a series of experiments will be performed to enhance the way to evaluate the performance of ice capability of an Arctic FPSO .

The Arctic FPSO is 250m long, 50 m breadth and 28 degree as the angle of waterline entrance to have ice capability. The vessel has a turret mooring system, which is integrated into the FPSO's hull, to reduce the ice load on the vessel by ice-vaning to the dominant environment. In the plan view, the vessel's waterline is a polygon as shown in Figure 2. The vessel has a sloped bow section to ensure ice-going capability by reducing the required ice-breaking loads. This is mainly because the bow area experiences the highest loads, while the other of the hull will generally experience the low loads in ice-breaking. The vessel has also an ice skirt to prevent the ice from going under the bottom of the hull. Furthermore, the hull form of the vessel is designed to improve production efficiency in the ship-yard. Figure 3 shows the 3D shape of the vessel.

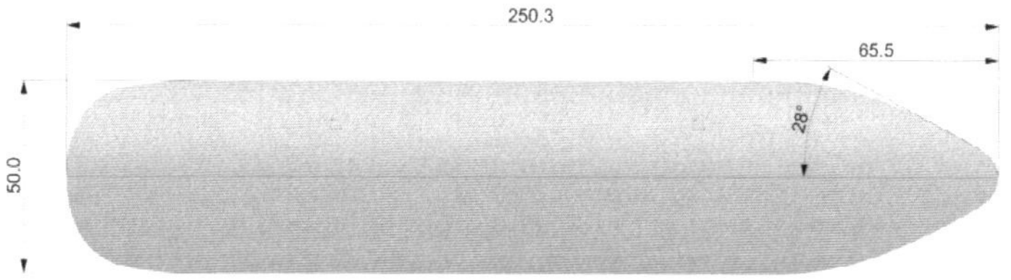

Figure 2. Waterline of the Arctic FPSO

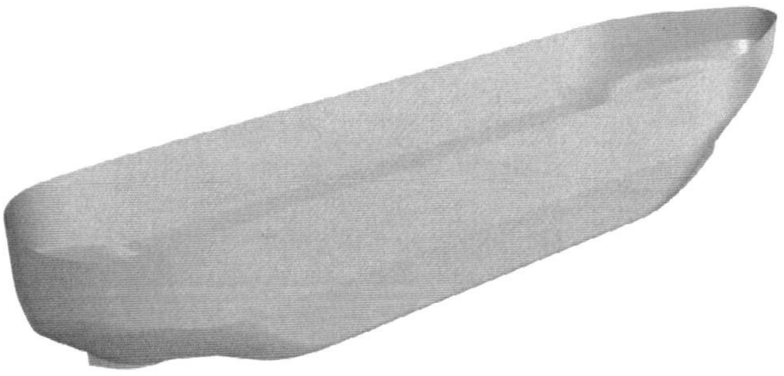

Figure 3. 3D shape of the Arctic FPSO

ICE LOAD ESTIMATION

Ice load estimation on a vessel is very challenging topic. Three different approaches are commonly used for estimating global ice load: i) empirical and statistical models, ii) ice basin tests, and iii) numerical simulations based on physical model. In this paper, the numerical simulator GEM (GPU Event Mechanics) developed by Memorical University (Daley, 2014a and 2014b) is chosen to obtain ice loads in pack ice condition.

The main feature of the GEM is to simulate ice-floater collision and assess global ice loads on the hull operating in pack ice. The GEM can generate ice field that contains hundreds of discrete ice floes. A floater is then modeled to be navigated through the generated ice field. In the simulation, not only ship-ice collision is modeled, but also every ice-ice interaction is modeled. During the simulation, ship-ice collision event is logged and is used later for statistical post-processing. An example of simulation is shown in Figure 4.

For estimating ice loads on the ship-type Arctic FPSO, a series of ice-towing simulations has been carried out to capture stochastic nature of the ice field with some varying parameters of ice thickness, ice density, ice drift speed, crushing strength of ice. The ice represents more than 8/10th ice cover with a constant thickness. During simulation, the Arctic FPSO was towed at varying speeds with fixed heading angles. In the simulations, there were lots of recorded ship-ice collisions, which are used for the basis of the analysis presented. The estimated ice loads are calibrated by multiplying it by a correction factor to reflect the level of ice loads that was measured in model tests with similar size of a vessel.

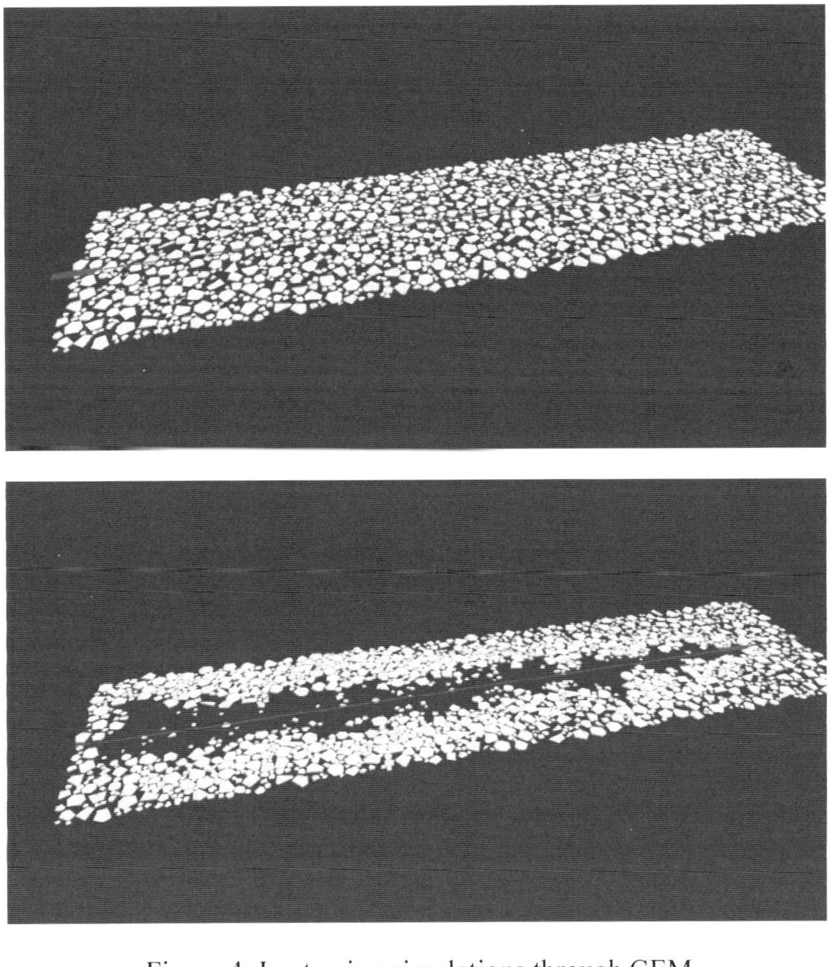

Figure 4. Ice-towing simulations through GEM

Table 1 shows the 60 individual runs that form the data for the capability analysis. A summary of the simulation parameters and results are given. An example of the simulated ice loads is shown in Figure 5. It can be found that the ice loads are gradually increases when the vessel moves through the ice field. This is because the compaction of the ice field leads to gradual force increase. In addition, it can be also found that the signals of the ice loads are significantly oscillated because of the ship-ice interaction, i.e., ice collisions. For DP capability analysis, the data was averaged with a certain period after the data reaching static interval. The list of simulation runs are summarized with summary of the results in Table 1. The results are also presented in Figures 6 - 9.

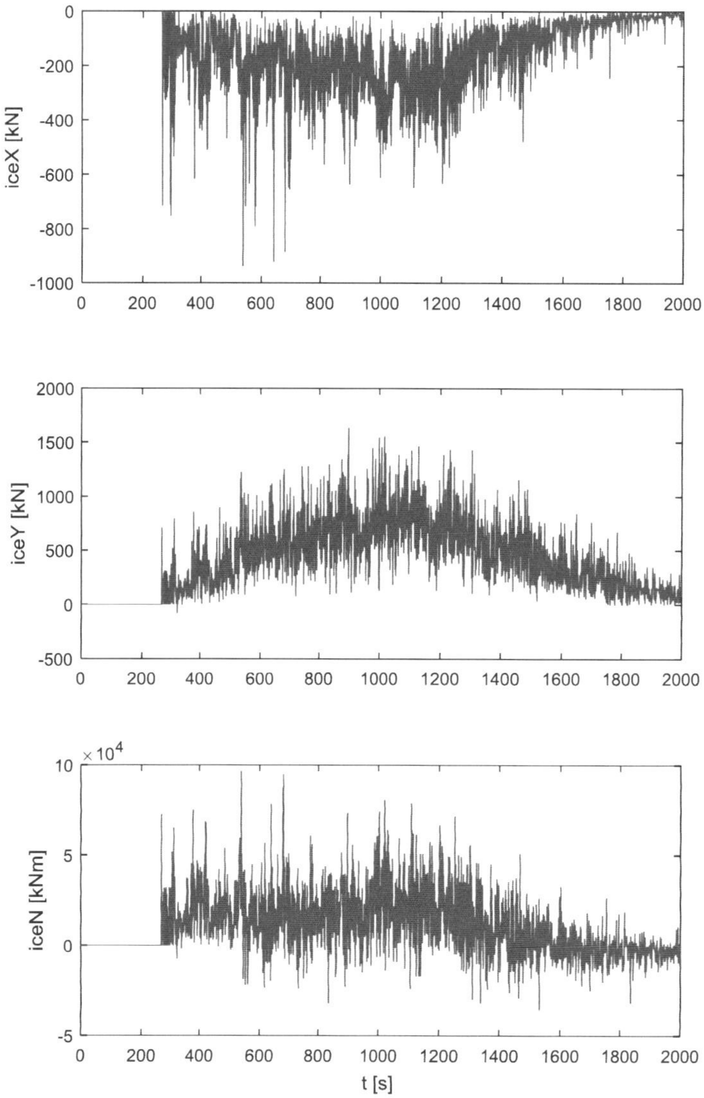

Figure 5. Example of the simulated ice loads

(Run 22: Ice thickness 1 m, Density 850 kg/m^3, Heading 30 deg, Drift speed 0.5 m/s)

Table 1. List of simulation runs with summary of result values

Run No.	Ice thickness	Density	Drift speed	Crushing strength	Heading	Ice Load					
						Xmean	Xmin	Ymean	Ymax	Nmean	Nmax
	[m]	[kg/m^3]	[m/s]	[Mpa]	[deg]	[kN]	[kN]	[kN]	[kN]	[kNm]	[kNm]
1	1	900	1	2	0	-200	-981	-5	432	-346	42,415
2	1	900	1	2	10	-201	-931	132	886	5,093	95,920
3	1	900	1	2	20	-227	-784	456	1,407	17,784	70,885
4	1	900	1	2	30	-263	-939	752	1,854	21,016	75,565
5	1	900	1	2	40	-319	-865	1,200	2,701	26,583	112,510
6	1	900	1	2	50	-354	-874	1,563	3,452	25,292	100,425
7	2	900	1	2	0	-290	-1,830	18	680	1,769	66,045
8	2	900	1	2	10	-314	-1,603	222	1,064	12,523	81,385
9	2	900	1	2	20	-353	-1,702	660	2,040	25,292	156,470
10	2	900	1	2	30	-378	-1,565	1,132	2,745	29,364	133,670
11	2	900	1	2	40	-481	-1,195	1,744	3,478	41,265	169,145
12	2	900	1	2	50	-524	-1,294	2,326	5,149	40,962	197,040
13	2	900	1	2	0	-407	-1,972	-17	1,602	-1,068	180,010
14	2	900	1	2	10	-377	-1,359	335	1,358	16,732	107,625
15	2	900	1	2	20	-467	-2,176	851	2,410	33,955	169,075
16	2	900	1	2	30	-545	-1,456	1,545	3,137	43,683	163,315
17	2	900	1	2	40	-618	-1,753	2,275	5,942	55,900	233,205
18	2	900	1	2	50	-659	-1,864	3,014	6,541	51,835	247,770
19	1	850	1	2	0	-186	-810	-3	353	-216	37,383
20	1	850	1	2	10	-190	-934	128	599	5,942	60,900
21	1	850	1	2	20	-231	-1,255	423	1,061	15,198	73,800
22	1	850	1	2	30	-247	-650	745	1,635	20,468	80,715
23	1	850	1	2	40	-316	-941	1,165	2,730	26,292	103,845
24	1	850	1	2	50	-344	-959	1,500	2,919	27,153	131,450
25	1	950	1	2	0	-196	-883	-9	504	-692	45,344
26	1	950	1	2	10	-229	-1,165	169	742	9,308	72,850
27	1	950	1	2	20	-247	-1,201	486	1,327	17,023	74,845
28	1	950	1	2	30	-279	-777	861	2,032	24,629	97,580
29	1	950	1	2	40	-330	-866	1,285	2,598	26,796	120,065
30	1	950	1	2	50	-364	-998	1,614	3,409	28,285	130,420
31	1	900	1	2	0	-210	-1,212	9	894	1,052	94,510
32	1	900	1	2	10	-245	-2,161	201	1,177	6,048	104,490
33	1	900	1	2	20	-319	-1,487	716	2,142	17,621	126,895
34	1	900	1	2	30	-400	-1,389	1,426	3,499	26,105	167,520
35	1	900	1	2	40	-546	-1,670	2,054	4,663	26,690	201,770
36	1	900	1	2	50	-625	-1,507	3,069	6,143	26,925	185,530
37	1	900	1	2	0	-631	-2,327	13	1,012	1,595	109,270
38	1	900	1	2	10	-690	-2,337	546	1,938	23,604	160,790
39	1	900	1	2	20	-724	-2,270	1,568	3,252	53,250	201,730
40	1	900	1	2	30	-917	-4,471	2,524	8,670	68,985	544,100
41	1	900	1	2	40	-1,112	-4,855	4,191	8,787	76,530	448,630
42	1	900	1	2	50	-1,233	-2,730	5,515	12,583	77,675	512,900
43	1	900	1	2	0	-429	-1,817	19	721	2,346	67,715
44	1	900	1	2	10	-495	-2,389	422	1,690	16,895	165,945
45	1	900	1	2	20	-588	-1,797	1,276	2,543	37,718	164,700
46	1	900	1	2	30	-767	-1,980	2,372	4,309	51,260	275,285
47	1	900	1	2	40	-927	-1,961	3,607	6,981	54,720	212,765
48	1	900	1	2	50	-910	-3,288	4,255	8,730	67,155	356,885
49	1	900	1	2	0	-181	-851	11	539	1,959	48,562
50	1	900	1	2	10	-163	-885	116	724	7,962	68,660
51	1	900	1	2	20	-159	-588	290	932	11,046	64,310
52	1	900	1	2	30	-220	-1,127	602	1,787	18,318	85,940
53	1	900	1	2	40	-264	-823	887	2,253	19,543	105,735
54	1	900	1	2	50	-290	-812	1,269	3,086	20,432	85,565
55	1	900	1	2	0	-218	-1,360	-9	476	-1,126	50,925
56	1	900	1	2	10	-199	-1,151	153	780	6,953	85,165
57	1	900	1	2	20	-240	-1,552	442	1,197	16,814	93,440
58	1	900	1	2	30	-273	-746	792	1,589	23,744	88,380
59	1	900	1	2	40	-328	-1,178	1,154	2,789	25,960	118,650
60	1	900	1	2	50	-348	-843	1,541	3,138	27,882	111,370

It can be found from the results that the global ice loads are quite varying with different ice parameters. Among four parameters, the dominant factors are ice thickness and drift speed. In most cases, the ice loads are increased when the value of ice parameters are increased. Furthermore, the ice loads are increased when the heading of the vessel is increased in the sense of the mean value. However, the above results do not apply in maximum values. The maximum values are highly fluctuated and sometimes have lower values even in larger heading angle with higher ice parameters. This shows the uncertainty of the global ice loads in pack ice condition.

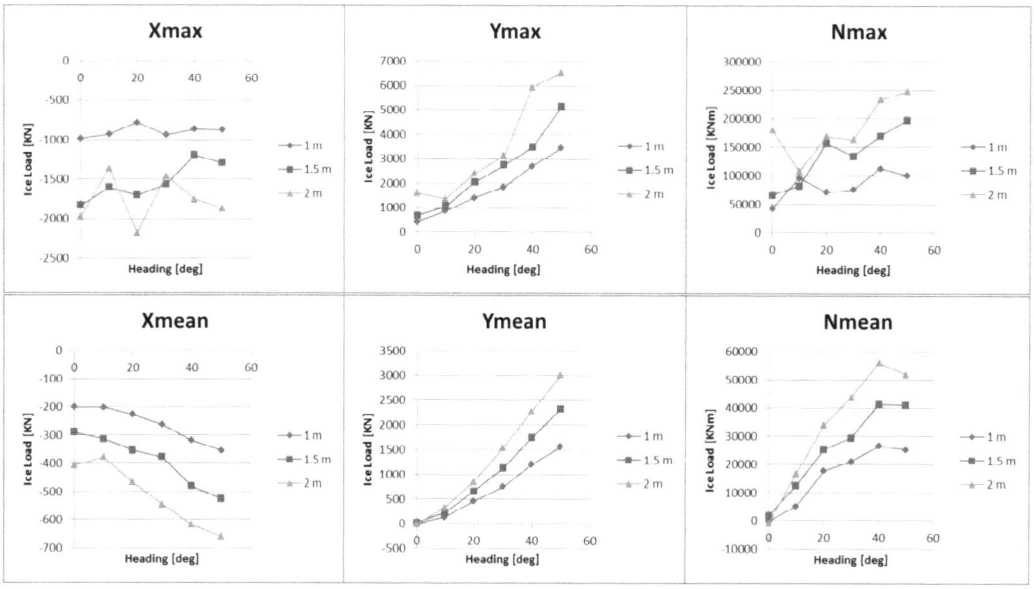

Figure 6. The estimated ice loads according to ice thicknesses

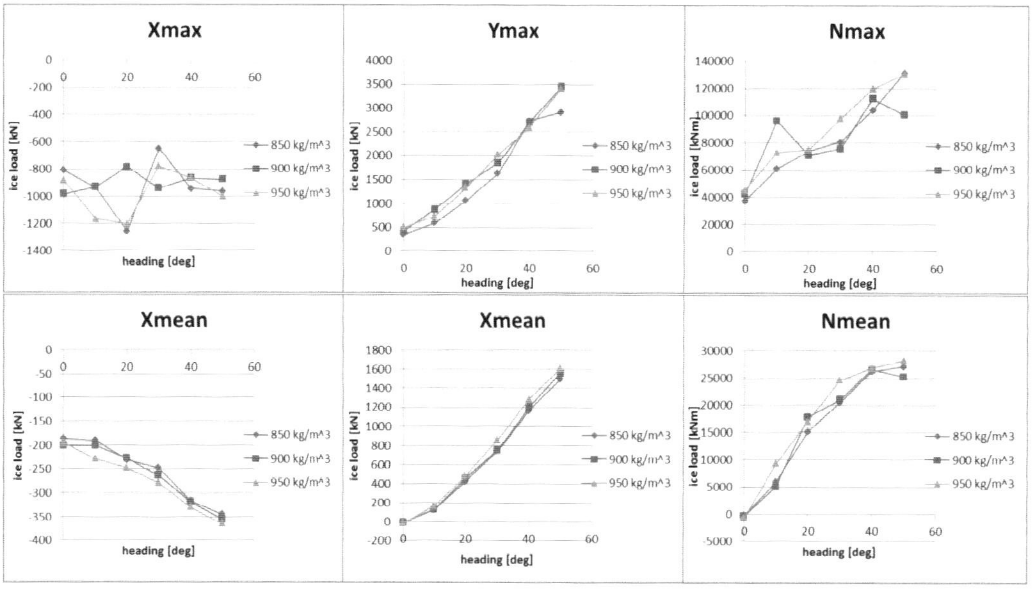

Figure 7. The estimated ice loads according to ice density

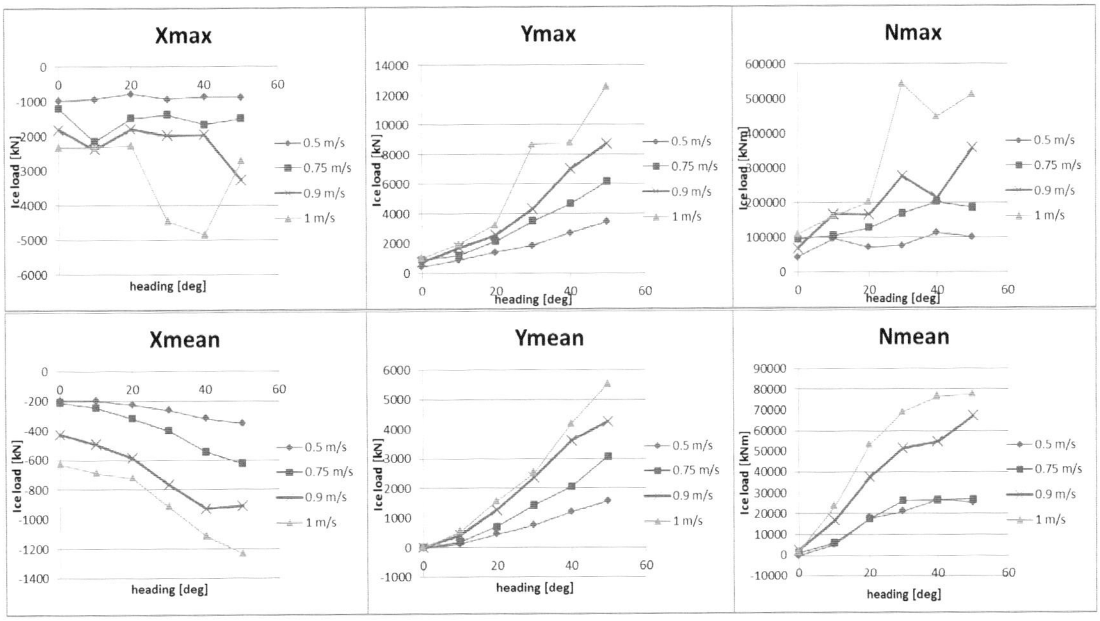

Figure 8. The estimated ice loads according to ice drift speed

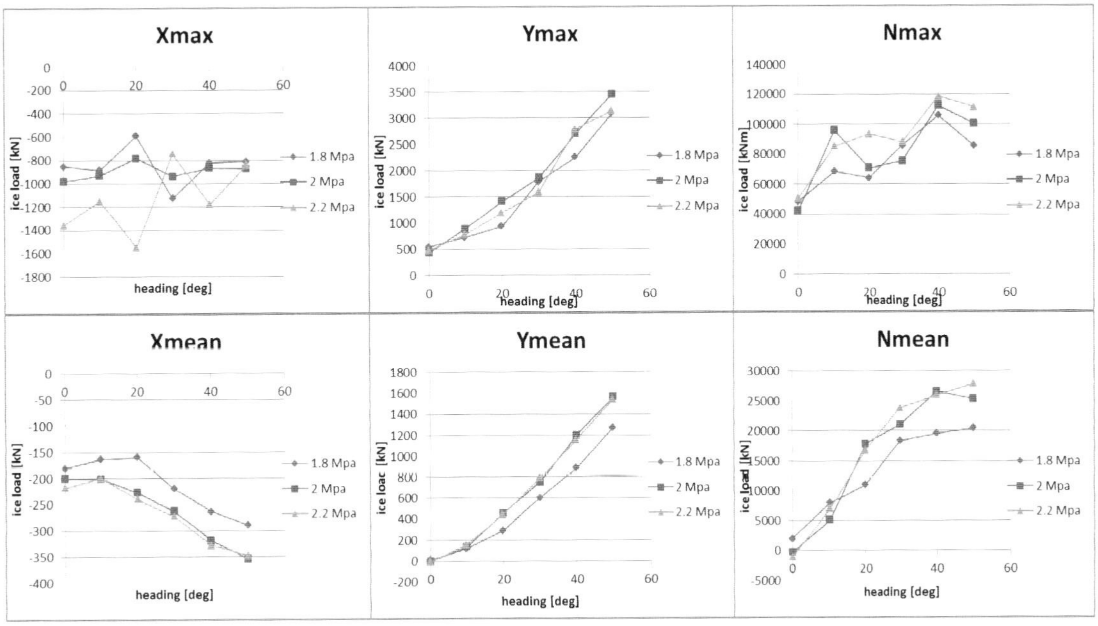

Figure 9. The estimated ice loads according to crushing parameter

DP CAPABILITY OF THE ARCTIC FPSO

The Arctic FPSO adopts a turret mooring system, which provides load-transfer mechanism between mooring system and the vessel. It also provides the mechanism for the ice-vaning capability. Therefore, the vessel can easily maintain the heading into the predominant environment. In this paper, it is assumed that the ice is coming from the same direction which is limited from -50 degree to 50 degree. The current, wind and wave loads are not considered in DP capability analysis.

It is assumed that the DP system of the Arctic FPSO has two bow tunnel thrusters of 3,000 kW capacity and three aft azimuth thrusters of 5,500 kW capacity. The thruster arrangement is summarized in Table 2. In total, the thrusters can give approximately 2.8 MN thrust including 10% of thrust degradation to cope with thruster-hull interaction. In order to investigate the DP capability, we followed the procedures proposed by IMCA with the significant values of the estimated global ice loads.

In the capability analysis, the ice loads are to be counterbalanced by the thrusters for maintaining the desired position. There are three components of ice loads F_x, F_y and M_z. Therefore, the thruster system must be able to provide three independent force components as follows:

$$\begin{aligned} -F_x &= \textstyle\sum_{i=1}^{N} T_i \cos\alpha_i, \\ -F_y &= \textstyle\sum_{i=1}^{N} T_i \sin\alpha_i, \\ -M_z &= \textstyle\sum_{i=1}^{N} (-y_i T_i \cos\alpha_i + x_i T_i \cos\alpha_i), \end{aligned} \tag{9}$$

where T_i is thrust from i th thruster, α_i is the direction of the i th thruster, x_i and y_i are is position from a rotational center.

The equality in Eq. (9) can be achieved by minimizing the sum of relative utilization:

$$J = \textstyle\sum_{i=1}^{N} \left(\frac{1}{W_i} \frac{T_i}{T_{max,i}} \right)^2 \tag{10}$$

where N is the number of thrusters and W_i is a weight.

Table 2. Thruster arrangement

No	Type	x [m] from mid-ship	y [m] from mid-ship	power [kW]	thrust [kN]
1	T/T	91	0	3000	330
2	T/T	86	0	3000	330
3	A/T	-113	-6.5	5500	875
4	A/T	-113	6.5	5500	875
5	A/T	-122	0	5500	875

When the required forces are found, these are allocated to each thruster using the thrust allocation algorithm. If the allocated thrust T_i exceeds its maximum value $T_{max,i}$, the relative utilization becomes greater than 1.0. This means that the DP system is not able to balance the external ice loads and thus DP fails.

In this paper, the DP capability analysis was conducted by using a thrust allocation algorithm that is developed as in-house code in SHI. Four capability plots in Figures 10 - 13 are made based on the DP capability analysis with the varying parameters of the ice thickness, ice density, ice drift speed, crushing strength of ice. It should be noted that regardless of which ice parameter that is varied, the other parameters remain fixed. Figures 10 - 13 show the effect of ice thickness, ice density, ice drift speed, and crushing strength of ice on DP capability, respectively.

It can be found from the results that the DP capability of the Arctic FPSO in pack ice is restricted by a narrow heading range up to 25 degree with current thruster capacity and configuration as in Table 2. The main reason for this is that the hull becomes more exposed to the ice collisions that are severely higher environmental loads when the heading of the vessel becomes large. It can be also found that among four ice parameters that are varied the DP capability is highly dependent on the ice drift speed and ice thickness. Small changes in those two parameters result in large changes in DP performances as in Figures 10 and 12.

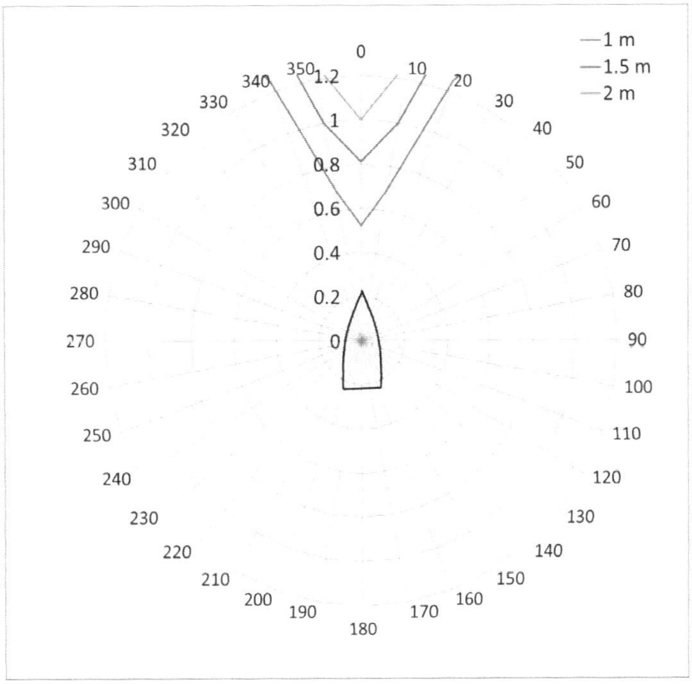

Figure 10. DP capability plot according to varying ice thickness

(Density 900 kg/m^3, Drift speed 0.5 m/s, Crushing strength 2Mpa)

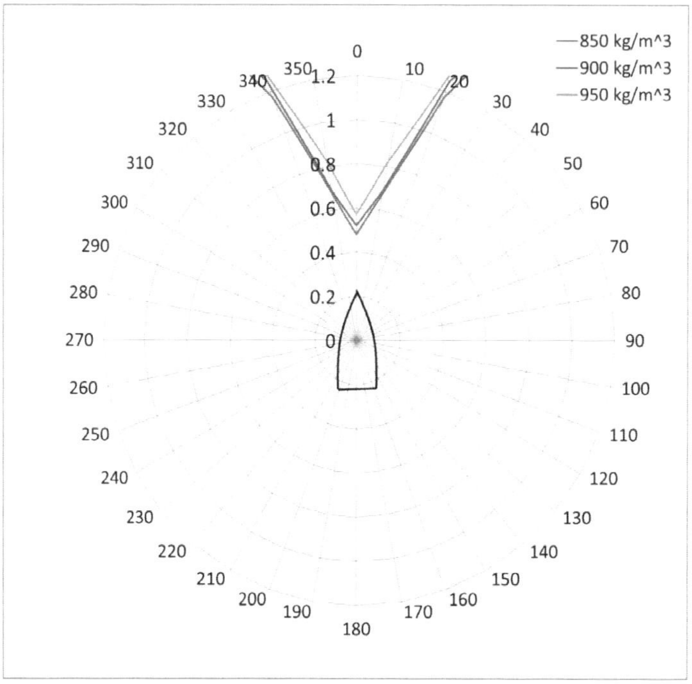

Figure 11. DP capability plot according to varying ice density

(Ice thickness 1 m, Drift speed 0.5 m/s, Crushing strength 2Mpa)

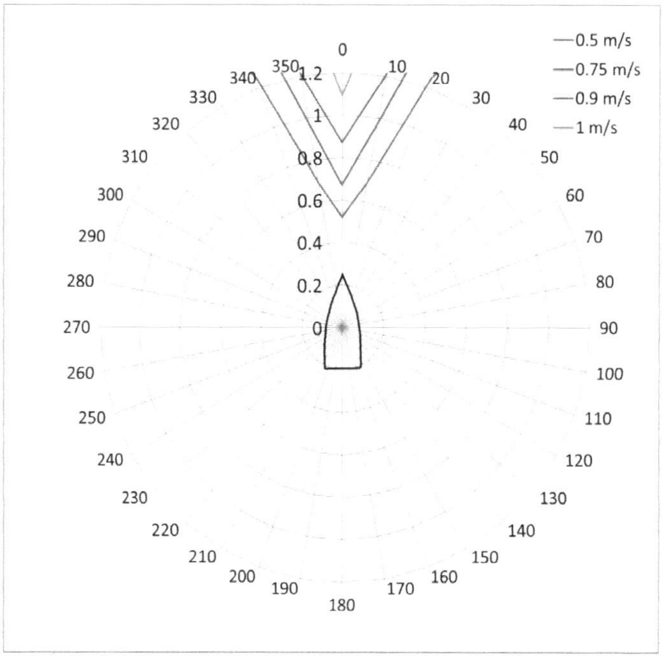

Figure 12. DP capability plot according to varying ice drift speed

(Ice thickness 1 m, Density 900 kg/m^3, Crushing strength 2Mpa)

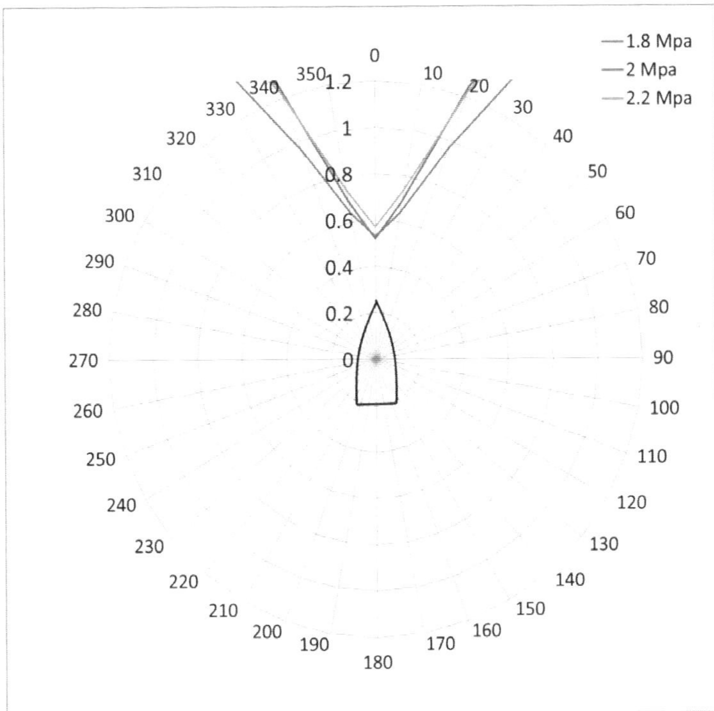

Figure 13. DP capability plot according to crushing strength of ice

(Ice thickness 1 m, Density 900 kg/m^3, Drift speed 0.5 m/s)

CONCLUSIONS

In this paper, the ice load simulator GEM that is developed by Memorial University has been used to investigate the global ice loads on the Arctic FPSO that is conceptually designed by Samsung Heavy Industries. The effects of various ice parameters on the global ice loads have been also examined through a series of ice-towing simulations.

To evaluate the DP performance of the Arctic FPSO in pack ice condition, a static DP-ice capability analysis was conducted by using the estimated ice loads and the in-house thrust allocation algorithm. The results were presented in a DP-ice capability plots according to IMCA that is widely used as a common format. In this paper, four radial parameterizations of the polar plot are presents for ice thickness, density, drift speed, ice strength. Although the results are not validated with model tests or field data, these results can be used qualitatively for the design of the capacity and configuration of thrusters and mooring system.

The precision and validity of the ice load simulator still remain to be solved and verified. It is noted that the results in this paper related to the global ice loads and DP capability should be treated indicatively and illustratively as several numerical challenges. However, it is worthwhile to note that the systematic procedure utilized to create the capability plots, where a numerical simulator is employed to estimate the global ice load, can give insights for analysis of the performance of an Arctic vessel in pack ice conditions.

In addition, uncertainties related with thruster-ice interaction, how to deal with dynamic ice loads shall be investigated in detail with the ice model tests or real-ship measured data with inherent dynamic effects (margin).

ACKNOWLEDGEMENTS

This material is based upon work supported by the Ministry of Trade, Industry & Energy (MOTIE, Korea) under Industrial Technology Innovation Program. No. 10063405, 'Development of hull form of year-round floating –type offshore structure based on the Arctic Ocean in ARC7 condition with dynamic positioning and mooring system'

REFERENCES

Biao Su, Øivind Kåre Kjerstad, Roger Skjetne, Tor Einar Berg, 2013. Ice-Going Capability Assessment and DP-Ice Capability Plot for A Double Acting Intervention Vessel in Level Ice, Proceedings of the 22nd International Conference on Port and Ocean Engineering under Arctic Conditions(POAC`13),

Claude Daley, Shadi Alawneh, Dennis Peters, Bruce Colbourne, 2014a. GPU-Event-Mechanics Evaluation of Ice Impact Load Statistics, Proceedings of Arctic Technology Conference 2014 (ATC2014)

Claude Daley, Shadi Alawneh, Dennis Peters, Bruce Colbourne, 2014b. Simulation of Managed Sea Ice Loads on a Floating Offshore Platform using GPU-Event Mechanics, Proceedings of the 11th International Conference and Exhibition on Ships and Structures in Ice (ICETECH14)

IMCA, 2000. Specification for DP Capability Plots. Technical report, The International Marine Contractors Association (IMCA).

Sofien Kerkeni, Ivan Metrikin & Peter Jochmann, 2013. Capability Plots of Dynamamic Positioning in Ice, Proceedings of the 32nd International Conference on Ocean, Offshore and Arctic Engineering (OMAE2013)

POAC'17
Busan, KOREA

Proceedings of the 24th International Conference on
Port and Ocean Engineering under Arctic Conditions
June 11-16, 2017, Busan, Korea

Environmental Effects on Dynamic Behavior of Moored Turret Ship Based on Numerical Simulations

Evgeny B. Karulin[1], Marina M. Karulina[1]
[1]Krylov State Research Centre, St.-Petersburg, Russia

ABSTRACT

This paper covers further efforts in a series of on-going numerical studies on dynamic behavior of a turret vessel exposed to simultaneous environmental effects like wind, current and broken ice as well loads from the turret mooring lines and own propulsion system. These studies are based on numerical integration of ship's horizontal motion equations. Broken ice impact on the turret vessel is estimated using the two-dimensional discrete element method (DEM), which proved to be an efficient tool in the studies on dynamics of moored structures as seen from earlier publications of the authors. This paper is a further extension of the DEM technique with adaptation of the model to cover new probable scenarios of the turret vessel operation: 1) turning of turret vessel in managed hummock ice; 2) impact of drifting ice on ship under circular tidal current and wind in compressed ice conditions. Broken ice is assumed to be 1 m thick. Typical ice floe sizes vary from 20 m to 40 m, and the initial ice concentration is 72%. The paper also presents the results of studies intended to examine how variations in the roughness of ship hull plating influence the ship motion characteristics, which is also an important operation factor.

KEY WORDS: Turret ship, Managed ice, Numerical simulation, External forces

INTRODUCTION

Moored floaters of various types like conical structures, semi-submersibles and turret vessels are platforms usually operated or planned to be operated for development of deep water oil and/or gas condensate fields, including those located in ice-covered waters. Investigation of dynamic behavior of moored floating structures under various external effects: winds, currents, ice as well as internal effects: own propulsion and mooring system responses is a challenging task. Some specific cases of vessel's behavior in ice are addressed based on analytic models (Onishchenko and Marchenko, 2015), as well as model tests (Bezzubik et al., 2004; Bruun et al., 2009). Numerical simulation of moored platforms in ice is an efficient method for investigations of this kind (Aksnes and Bonnemarie, 2009; Sayed and Barker, 2013; Sayed et al., 2015; Zhou et al., 2012).

In this study the numerical simulation is used as a tool to examine the behavior in ice of one type of the above-mentioned floaters – turret vessel. Distinctive feature of a turret vessel is availability of special structure – anchored turret enabling vessel rotation around a vertical axis. And horizontal ship displacements are dictated by stiffness of turret anchoring system. This structural feature of turret vessel gives it certain advantages over other platforms without turret regarding control of ice, wind and current load levels by rotation along the direction of impact of external force. As a rule, directions of impact of wind, current and ice drift do not coincide. At change of intensity or direction of impact of one of the above mentioned external factors a vessel starts to rotate to position providing it minimum level of total external load. For reducing the level of ice load on hull it is required to make a zone of managed ice around the ship.

The turret vessel under investigation is a FPSO option for the Shtockman field. The behavior of this vessel under some scenarios was examined by the authors earlier using a numerical model (Karulin and Karulina, 2016). The model is based on solution of equations describing vessel motion in horizontal plane under the impact of external forces. For the broken ice loads estimation we take a well-known discrete element model (DEM) approach fundamentals of which are presented in Hopkins (1998). This paper is a further extension of the DEM technique with adaptation of the model to cover new probable scenarios of the turret vessel operation: 1) turning of turret vessel in managed hummock ice; 2) impact of drifting ice on ship under tidal current in compressed ice conditions. Broken ice is assumed to be 1 m thick. Typical ice floe sizes vary from 20 m to 40 m, and the initial ice concentration is 72%.

THE NUMERICAL MODEL DESCRIPTION

Investigation into behaviour of a turret ship under wind, current and ice is based on composition and solution of differential equations for motion of such vessel in the horizontal plane. The problem statement is shown in Figure 1. Here, the following basic assumptions are introduced:

- no impact of roll and pitch on the ship motion in the horizontal plane;
- the ship gravity center is located in the midship frame plane, the vessel is assumed to be symmetrical against the centreline and midship plane;
- motion of the turret ship (horizontal offsets and yawing) is slow and the inertia of mooring system (MS) lines can be disregarded.

Dashed line in Figure 1 shows position of the turret ship at the initial time moment $t = t_0$. We assume that following external forces start acting on the vessel at this moment – current, wind and ice (any combination of them or all together) causing the vessel to move in the horizontal plane. To describe the ship motion we apply two coordinate systems: immobile $\xi O_0 \eta$ and the vessel-fixed XOY. The origin of the vessel-fixed coordinate system lies in point of intersection of center plane and midship section, and axes OX and OY are directed to bow and starboard. At the initial moment of time $t = t_0$ both coordinate systems coincide. Thus, position of the turret ship and its kinematical parameters in an arbitrary moment of time can be found through integration of differential equations of the ship motion (Ship Theory Handbook, 1985) in the horizontal plane. Taking into account symmetry of the turret ship against the midframe plane, we can write the final differential equations in the vessel-fixed coordinate system in the following form:

$$m(1+k_{11})\frac{dV_x}{dt}-m(1+k_{22})V_y\omega=X$$
$$m(1+k_{22})\frac{dV_y}{dt}+m(1+k_{11})V_x\omega=Y \qquad (1)$$
$$I(1+k_{66})\frac{d\omega}{dt}=M$$

where m,L,I are the ship weight, hull length, and inertia against the vertical axis running through the center of gravity (COG), accordingly; k_{11},k_{22},k_{66} are coefficients of the ship added water mass and added moment of inertia; V_x,V_y are projections of linear ship speed to axes of ship-fixed coordinate system; ω is angular velocity of the ship rotation; X,Y,M are projections of total external force and external torque acting on ship hull.

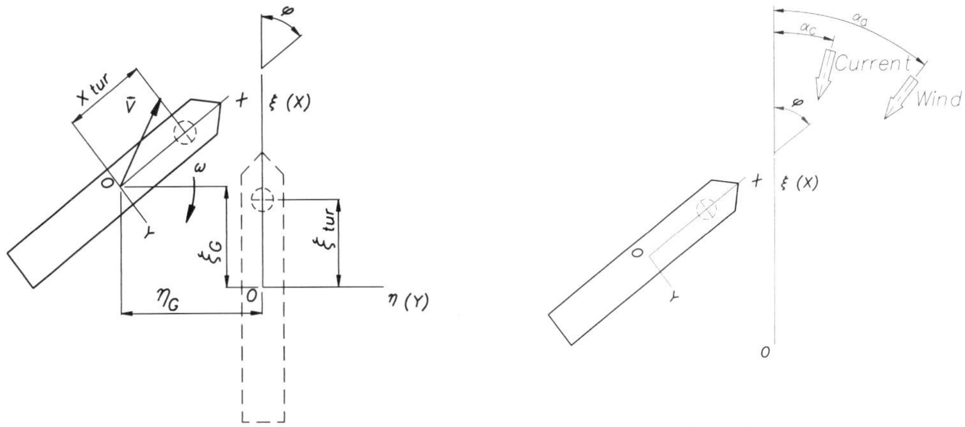

Figure 1. Problem statement. Coordinate systems and directions of external actions

Own ship inertia parameters are known. The total external load contains aero-, hydrodynamic and ice loads, forces and moments produced by the thrusters, and by mooring system. Formulas for determination of aero- and hydrodynamic loads are given in Karulin and Karulina (2011).

Loads caused by the broken ice field acting on floater hull are determined using 2D discrete element method (DEM). According to the method the broken ice is represented by a set of separate (discrete) floating bodies – disks of diameter corresponding to the typical geometrical dimension of separate floes. When the elements are interacting with each other or with the rigid boundaries (ship hull), it is supposed that there are contact forces on the contact zone. The detailed description of the DEM, as well as the contact forces calculation is given in Karulin and Karulina (2011). Additional assumptions have been made:

- disks do not raft – displacement occurs only in horizontal plane, two-dimensional problem is considered;
- ice discs are assumed to be destroyed due crushing when a certain pressure load level is achieved: as a result the ice discs diameters decrease;
- no turning of ice discs: ice blocks move incrementally;
- behaviour of broken ice can be described through the loose medium mechanics and characterized by its inherent features: angle of internal friction φ and cohesion c.

In current calculations, angle of internal friction is equal to 30° and cohesion is 2000 Pa.

These values are based on results of previous studies.

Apart from the described interaction forces each element in horizontal plane is subject to force of hydrodynamic and air resistance, which in general case do not coincide in direction. Force of disk hydrodynamic resistance is calculated under the formula:

$$\vec{F}_{HDdisk} = 0.5c_{wdisk}\rho_w A_{disk}\left|\vec{V}_{disk}\right|\vec{V}_{disk} \quad (2)$$

where $A_{disk}, \vec{V}_{disk}$ are the disk area in plan and velocity in reference to water, accordingly; c_{wdisk} is coefficient of hydrodynamic resistance. Air resistance is determined under similar formula using corresponding values of air density and coefficient of air resistance. Coefficients of hydrodynamic and air resistance of ice floes are assumed c_{wdisk}=6·10-3 and c_{adisk}=3·10-3 according to Timokhov and Kheisin (1987). Added masses of water at disks motion in horizontal plane can be neglected.

After estimation of total force acting on each element, accelerations of each element are found, and differential equation of motion of each element is integrated and velocities and new coordinates of each element are computed. Summation of forces applied from discrete elements to the vessel contour enables obtaining components of ice load in vessel-fixed coordinate system – forces X_{ice}, Y_{ice} and moment M_{ice}

The model was validated by comparing numerical simulations of the Kulluk conical platform with full scale measurements, as well as numerical simulations of moored floater behaviour with ice model tests data (Hansen and Løset, 1999; Karulin and Karulina, 2011).

DESCRIPTION OF THE TURRET VESSEL AND MOORING SYSTEM

Main parameters of a turret vessel assumed to be an object of investigation are given in Table 1. In Figure 2, the vessel waterline with a turret position is shown.

Table 1. Main parameters of a turret vessel.

Parameter	Value
Length at waterline, m	235
Width at waterline, m	38.8
Draft at waterline, m	15.0
Displacement, m3	117000
Inertia, kg·m2	$552.3\cdot10^9$
Turret position fore of midship, m	85.0
Coefficient of water added mass k_{11}	0.0175
Coefficient of water added mass k_{22}	0.360
Coefficient of water added moment of inertia, k_{66}	0.290

In current studies, the turret MS consists of twelve chain lines (R5 114mm) of 2500 m length that are arranged strong radially. Break load of the chain line is 14.5 MN. The fair leads are located symmetrically along a turret circle of 30 m diameter at water depth 300 m. Stiffness of the anchoring system is assumed constant in various directions. Figure 2 illustrates turret MS restoring force depending on the offset.

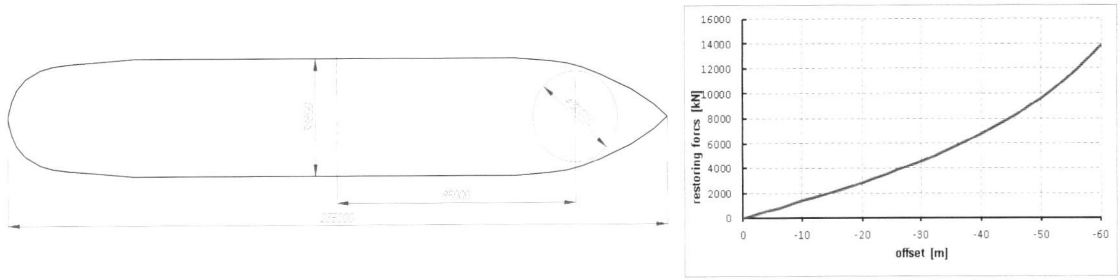

Figure 2. The turret ship waterline and the turret MS restoring force

TURRET SHIP TURNING IN IMMOVABLE MANAGED FIRST-YEAR ICE RIDGE

A zone of managed ice around the vessel is a pre-requisite for her safe operation in ice. If the vessel is operated in immovable ice field and it is predicted that the ice would drift in some direction, the vessel has to turn, under own propulsion or with assistance of ice management vessels, to meet the ice drift with the bow. The earlier studies of the authors for the same moored floater (Karulin and Karulina, 2016) showed that the turning maneuver of the turret vessel under own propulsion depends on ice parameters and managed ice area. At ice thickness 1 m, mean size of ice pieces 20 m and ice concentration 72% the vessel needs a managed ice zone of at least 900 m in diameter for effective turning maneuver. The same values are also assumed in this study, which examines a scenario of the ship turning about the turret using own thrusters in immovable managed first-year ice ridge.

It is assumed that after ice management an ice ridge is broken and some ice blocks from the ridge keel and its consolidated layer have formed a zone (stretch) of broken ice. It is also assumed that the width of this zone is equal to 180 m, while the maximum ice thickness along central axis is 10 m. Ice thickness is linearly reduced to the zone edges reaching the surrounding ice thickness. It is assumed that the propulsion unit generates a thrust of 900 kN applied normally to the ship centreline at the point with the coordinates $x=-102.0$ m, $y=0$ m. The calculation domain is a managed ice area of regular polygon shape inscribed in a circle of 900 m diameter. The calculation domain boundaries are solid walls simulating edges of unbroken ice sheet. In the initial instant the turret center coincides with the circumscribed circle origin. Three initial positions of the ice ridge axis with respect to the ship hull are considered. The results obtained from modeling the ship turn in ice ridge are compared with the ship turn in managed ice of constant thickness. All types of calculation domains at the initial instant and in time after 180° turn are shown in Figure 3. Color gradations reflect different thickness of ice blocks: black color refers to the maximum thickness of 10m.

Some calculation results are shown in Figures 4 and in Table 2. Fluctuations of the restoring force of turret mooring system (Figure 4) are caused by periodical action of ice floes and by compliance of the mooring system itself.

Table 2. Calculation results for the ship turning in managed ice conditions

Ice conditions	Managed ice	Managed ridge 1	Managed ridge 2	Managed ridge 3
Time of turning at 180° [s]	1700	1970	2970	3080
Turret maximal offset [m]	6.9	6.6	5.5	7.3
Maximal restoring force [MN]	1.00	0.95	0.80	1.06

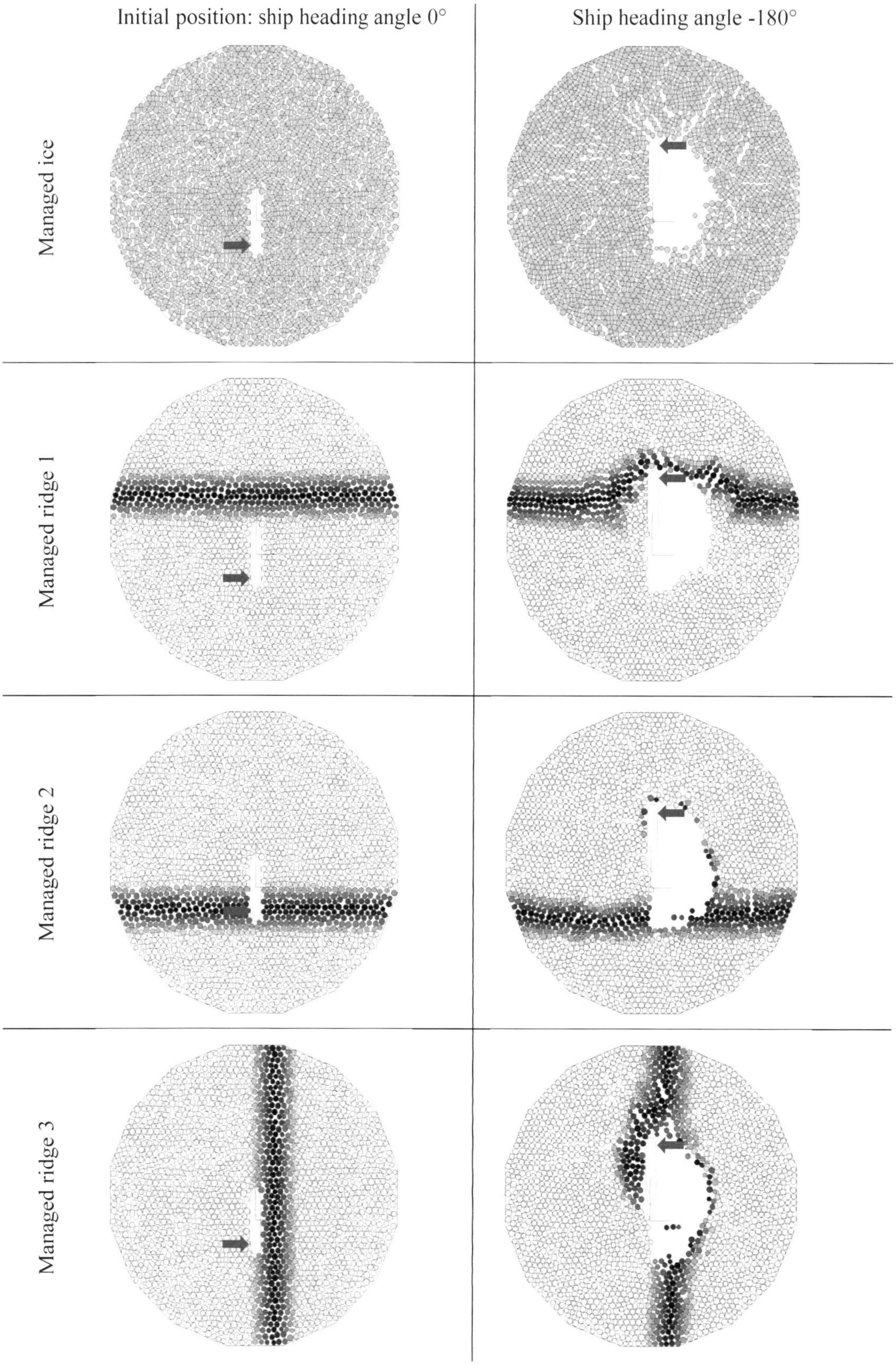

Figure 3. Snapshots of calculation domain (blue arrow shows the thruster action)

POAC17-096

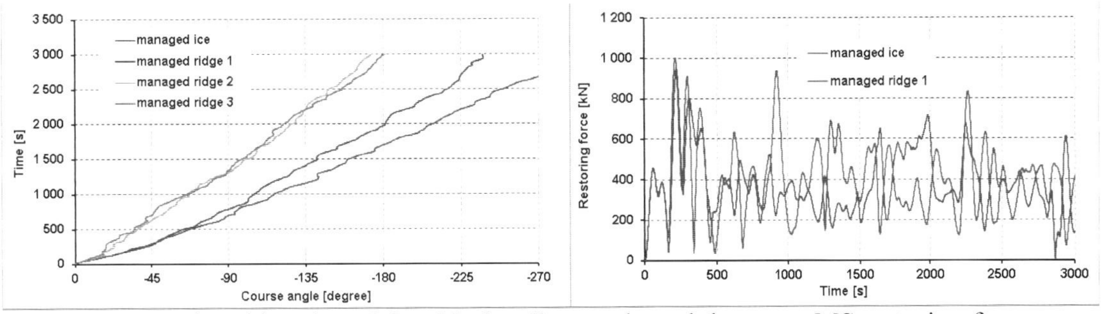

Figure 4. Time histories of the ship heeding angle and the turret MS restoring force

Analysis of computations based on this numerical model indicates the following:

- Turret vessel is able to turn safely under own propulsion in the managed ice ridge of given configuration at different initial orientation of the ridge axis with respect to the hull, while ship's turning speed in this case is lower than that in the managed ice of constant thickness.
- The time required by the ship to change the heading by 180° shows that the turning maneuver in ridge 1 is performed faster than in ridges 2 and 3 with a different initial axis orientation.
- In considered cases of the ship turning the turret excursions and restoring forces are practically the same as in managed ice.

THE TURRET SHIP BEHAVIOUR UNDER ACTION OF SVERDRUP TIDAL WAVE

Earlier the authors studied the turret vessel behavior under Sverdrup tidal wave effects (Karulin and Karulina, 2016). It was shown that the ship slowly turns when there is a zone of managed ice with an average size of ice flows 40 m. In this case the turret excursion and restoring force are small and the ship may safely turn under own propulsion. This result was obtained without ice compression or wind. In this study the Sverdrup wave scenario is consecutively added with two additional external effects: ice compression and wind. In addition, it is examined how the ice/hull friction coefficient influences ice load.

Velocity vector components $U_{S,x}$ and $U_{S,y}$ in the Sverdrup tidal current are as follows:

$$U_{S,x} = \frac{a\omega_S}{k_S H}\cos\theta, \quad U_{S,y} = \frac{af}{k_S H} sin\theta, \qquad \theta = k_S x - \omega_S t, \tag{3}$$

where a– wave amplitude, ω_S and k_S – frequency and wave number of Sverdrup tidal current, respectively, H – water depth and f – Coriolis parameter. In current study, the typical parameters of 74°N were used: a =0.1, k_S = 2.583·10-7, f= 1.398·10-4, ω_S =1.405·10-4, H =3.000·102 (Marchenko, 2015).

The calculation domain of special configuration is generated taking into consideration a circular motion of ice floes drifting due to Sverdrup tidal current. The circle-shaped calculation domain consists of three parts: a central circle and two concentric rings (Figure 5, left). The central circle of 1583 m in diameter is an intact part of the ice field with 1 m thickness. The middle ring, which circles the ship under study, is filled up with ice floes with an average diameter of 40 m and concentration of 72%. The ring outer diameter is 3190 m. For reducing the number of ice floes in calculations the outer ring is represented by a layer of ice floes with an average diameter of 85 m. The calculation domain has free boundary.

Initially the Sverdrup tide wave impact is not accompanied with ice compression and wind effects. The vessel slowly turns under the tide wave action with small spaces of open water being formed around the hull (Figure 5, middle). Approximately 33 minutes after such turning is initiated, the ship is exposed to ice compression. The numerical model simulates ice compressions with additional pressures applied to ice blocks on the periphery of the calculation domain acting towards the center of this domain. The pressure magnitudes depend on the specified compression grade. Calculations are done for ice compression grades 1 and 2 characterized by 10 кPa and 30 kPa, respectively (Timokhov and Kheisin, 1987). Calculations based on the described model indicate that ice compression quickly closes the channel behind the ship: the ice channel was totally closed after about 1.5 min subject to ice compression of grade 1 (Figure 5, right). Exposure to ice compression gives considerable increase in average and amplitude values of ice loads transmitted to the turret mooring system (Table 3 and Figure 6.). Compression level variations within grades 1 and 2 have practically no effect on average ice load values.

It is seen from computations that the ice load significantly depends on hull plating conditions: an increase in ice/hull friction coefficient from 0.09 to 0.36 has increased the average restoring force from 1.7 MN to 2.8 MN under compression of grade 2. (Table 3).

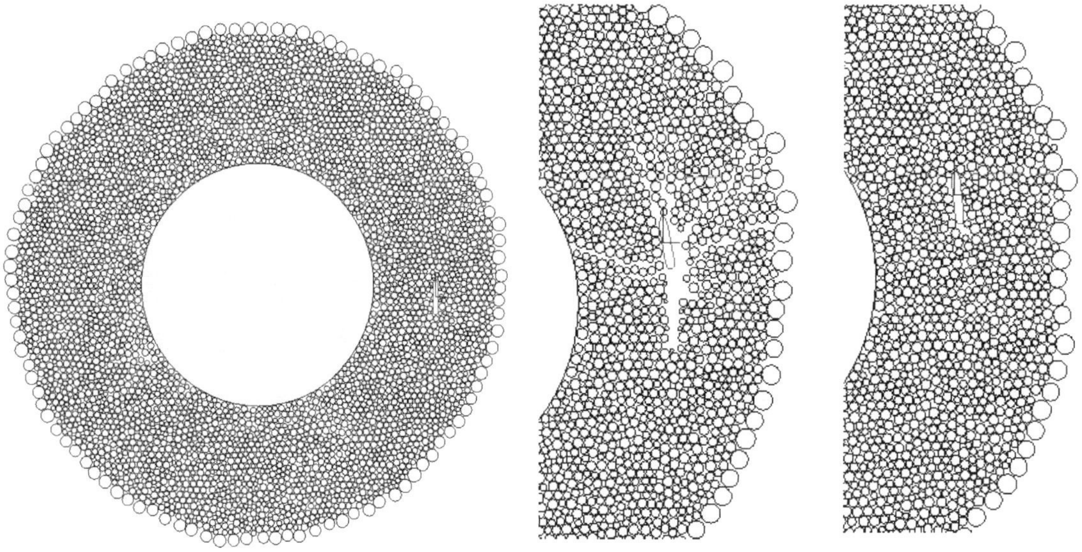

Time step=0 s
initial position

Time step=1980 s
beginning of ice compression

Time step =2070 s
closure of the channel

Figure 5. Snapshots of the calculation domain under ice compression action.

Table 3. Average restoring forces on ship exposed to Sverdrup tidal wave

Friction coefficient ice/hull	Restoring force, MN		
	Without ice compression	Ice compression 1 grade	Ice compression 2 grade
0.09	0.7	1.7	1.7
0.36			2.8

In one of the scenarios, the ship in addition to tide wave and ice compression of grade 1 is subjected to wind of 15 m/s at 135° angle at time instant t= 7020 s (117 min). Trajectories of ice blocks are governed by joint actions of current and wind. Calculations and full scale observations of ice drift in the Pechora Sea suggest that wind speeds higher than 10 m/s govern ice drift directions (Zubakin et all., 2004). Direct wind effects are also observed in view of a large wind exposed area of the vessel under consideration.

Wind onset causes intensive rotation of the vessel, which tends to minimize her windage. Figure 6 shows calculation domains at the wind onset and ship turning instants as well as ship's equilibrium position under joint effects of all factors considered: Sverdrup tide wave, ice compression, wind and managed drifting ice. As it is seen from Figure 7, the equilibrium heading angle is about 196° with small excursions around it. It means that the hull is not aligned exactly into wind due to other environmental factors. Figure 8 shows that the wind onset did not cause any noticeable change in maximal values of the restoring force.

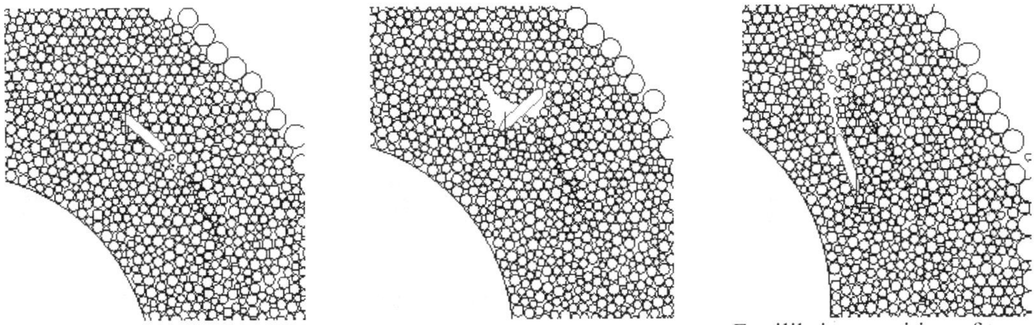

Time step=7020 s beginning of wind action | Ship turning under wind action | Equilibrium position of turret ship

Figure 6. Snapshots of the calculation domain under ice compression and wind action.

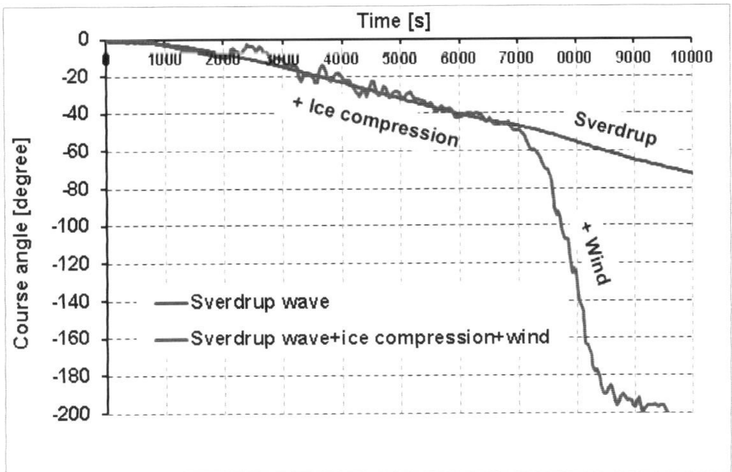

Figure 7. Time histories of the ship heeding angle.

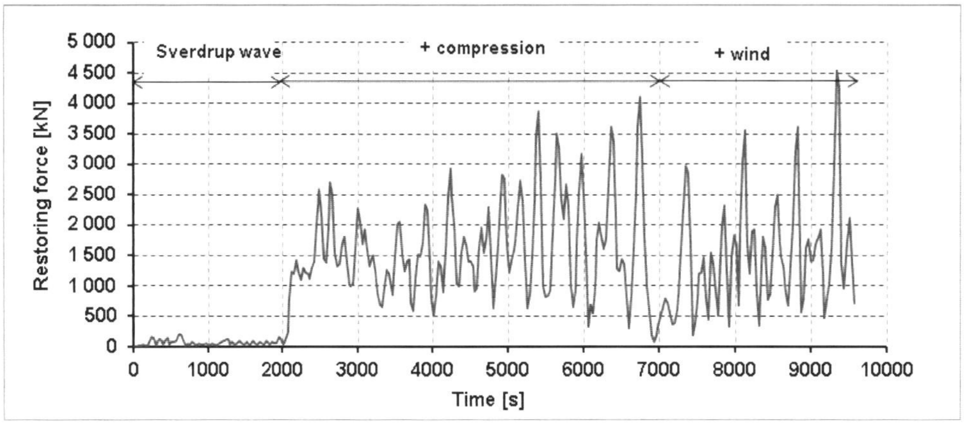

Figure 8. Time histories of the MS restoring force.

CONCLUSIONS

This study examined some scenarios of environmental effects (wind, current and ice) on a turret vessel which may occur during operation of a real vessel in freezing deep-water seas, e.g. at Shtokman field. The numerical model presented in the paper enabled the authors to identify the ship responses under given external effects and determine kinematic parameters of ship's dynamic behavior and forces transmitted to the turret mooring system.

Calculations show that the turret vessel is able to turn safely under own propulsion in the managed first-year ice ridge with a keel up to 10 m. The ship turning speed in the managed ice ridge is slower than in the managed ice of uniform thickness, and it takes more time to perform the turning maneuver. Time estimates are required for operational planning to turn the vessel well in advance into the right position based on weather forecasts to ensure minimum or safe level of environmental loads.

Investigations of ship's dynamics in managed ice under Sverdrup tide wave are complemented with the analysis of ice compression and wind effects. The results of numerical simulations show that ice compressions of grades 1 and 2 have practically no effect on ship dynamics under Sverdrup tide wave effects, but loads on the turret mooring system increase significantly (by an order of magnitude). If an additional wind factor comes into play with a speed higher than 10 m/s, then the circle ice drift trajectory changes, and additional aerodynamic loads are applied to ship hull, which significantly changes the turret ship dynamics during the turning maneuver.

It should be noted that the obtained results are influenced by specific features of the applied numerical model and boundary conditions. The conclusions drawn from these results are valid for the range of external action parameters and specific floater features covered by this study. Therefore, extrapolation of the results and conclusions outside this range should be made with caution and validated by appropriate full-scale or model tests data.

REFERENCES

Aksnes, V., and Bonnemaire, B, 2009. Analysis of the behaviour of a moored ship in variable ice drift. Proc. 20th Int. Conf. POAC, June 9-12, 2009, Lulea, Sweden.

Bezzubik O., Bitsulya A., Karulin E., Karulina M. et al., 2004. Experimental investigation of interaction of moored platforms with drifting ice features. Proceedings of the 17th

International Symposium on Ice IAHR 2004, St. Petersburg, Russia, Volume 1, pp. 76-84.

Bruun P.K., Husvik J., Le-Guennec S., Hellmann J.-H., 2009. Ice Model Test of an Arctic SPAR. Proceedings of the 20th International Conference on Port and Ocean Engineering under Arctic Conditions, POAC 09, June 9-12. Luleå. Sweden, POAC09-136

Hansen, E. and Løset, S. (1999), "Modelling floating offshore units moored in broken ice: comparing simulations with ice tank tests," In Cold Regions Science and Technology, 29, pp. 107-119.

Hopkins, M.A. (1998). "Four stages of pressure ridging," J of Geophysical Research, 103(C10), pp. 21883–21891.

Karulin, E. and Karulina, M. (2011). "Numerical and Physical Simulations of Moored Tanker Behaviour," J Ships and Offshore Structures, 6(3), pp. 179-184.

Karulin, E. and Karulina, M. (2016). " Numerical Simulation of Turret Ship Dynamic Behaviour under External Action", Proceedings of the 23rd International Symposium on Ice IAHR 2016, Ann Arbor, Michigan USA, Paper No. T4869385.

Korotkin, A., 2009. Added masses of ship structures. Springer Science+Business Media BV, 397p.

Marchenko, A., 2015. Influence of added masses on rotation and towing of drifting icebergs. Transactions of KSRC, Issue 86 (370), St.Petersburg, pp.101-116.

Onishchenko, D., Marchenko, A., 2015. Analytical estimation of maneuverability of moored FPU with internal turret in close ice. Proc. of 23rd Int. Conf. POAC, Trondheim, Norway

Sayed, M., Barker, A., 2011. Numerical Simulations of Ice Interaction with a Moored Structure. The Offshore Technology Conference, Houston, Texas, USA, OTC22101.

Sayed, M., Kubat, I., Watson, D., Wright, B., Gash, R. and Millan, J., 2015. Simulations of the StationKeeping of Drillships under Changing Direction of Ice Movement, Arctic Technology Conference, Copenhagen, Denmark, OTC 25565

Ship Theory Handbook (1985), Ed. by Voitkounski, Y., Vol.3, Maneuverability of conventional ships, Leningrad: Sudostroenie.

Timokhov, L. and Kheisin, D (1987), "Dynamics of Sea Ice," Leningrad: Hydrometeoizdat (in Russian).

Zhou, Li, Su, B., Riska, K., Moan, T., 2012. Numerical simulation of moored structure station keeping in level ice. In Cold Regions Science and Technology, 71, pp. 54–66.

Zubakin, G., Buzin, I. and Skutin, A., 2004. Assessment of ice conditions in Shtockman gas-condensate field, Transactions of AARI, Vol. 449. Saint-Petersburg, pp. 124-139.

Development of Platform Motion-Mooring System Coupled Solver for Offshore Plant

Sang Chul Lee[1], Sunho Park[1], Seung Jae Lee[1]
[1] Korea Maritime and Ocean University, Busan, Korea

ABSTRACT

Typically, the commercial softwares for the mooring system analysis are based on the potential flow theory for the analysis of the motion of platform and do not consider the viscous effects around the platform which affect the motion of it. To resolve this problem, in present study, we developed the model for the mooring system analysis with the lumped mass model and coupled it with the open source computational fluid dynamics libraries, OpenFOAM, which can simulate the motion of platform based on the viscous flow theory. Finally, the two-way coupled solver between six-degrees of freedom motion of platform and the tensile loads of the mooring system for offshore plant was developed and validated by comparing the calculation results with the analytic solution and experimental results, and its operability and robustness were confirmed by performing the calculation under various conditions of the platform motion.

KEY WORDS: Open source libraries; 6-DOF platform motion; Mooring system; Lumped mass model; 2-way coupling

INTRODUCTION

Global performance due to station keeping capability is one of the most important factors in the performance of offshore plant, which is determined entirely by the mooring system. Therefore, the mooring system design is essential to ensure the life and performance of the offshore plant. For this purpose, it is necessary to accurately predict the motion of the platform under the given environmental loads and the tensile load of the mooring system accordingly.

However, most commercial codes used for the analysis of mooring systems are based on the potential flow theory which doesn't consider the viscous effects around the platform in analyzing the motion of it, so it is difficult to predict the exact motion of the platform in the actual flow. Additionally, in analyzing the mooring system, a reasonable model is required to accurately predict the tensile load and the overall dynamic characteristics of the mooring line, and the consideration of bidirectional coupled analysis in which the motion of the platform and the tensile load of the mooring system are mutually influenced is also required.

This study was conducted as part of a project to develop a simulation suite for the analysis of polar environmental loads acting on the arctic ships and offshore plants, and the coupled solver between the platform motion and mooring system for offshore plant was developed. In order to analyze the platform motion and mooring system, an open source computational fluid dynamics (CFD) libraries, OpenFOAM, which can simulate the viscous flow analysis and the lumped mass mooring system model which can predict the accurate tensile load and dynamic characteristics of mooring lines were used, respectively.

LUMPED MASS MOORING SYSTEM MODEL

The quasi static model, which has low computation load, easy to use, and easy to apply to open source (Masciola et al., 2013), has been widely used as a model for mooring system analysis. However, since this model does not take into account the hydrodynamic forces and internal forces on the mooring line, there is a difficulty in accurately predicting the platform motion and the tensile load of the mooring system. Therefore, in this study, a lumped mass model which neglects the bending and torsional elasticity of the mooring line was used to reduce the computation load without affecting the calculation results (Hall et al., 2014).

The discretized mooring line is shown in Figure 1. The evenly divided segments have the identical properties, such as length (*l*), diameter (*d*), density (ρ), Young's modulus (*E*), and internal damping coefficient (C_{int}), and the nodes, i.e., lumped mass, have the integer indices then the vector r_i represents the position of node *i*. For the segments which are adjacent to node *i*, the indices *i-1/2* and *i+1/2* were used, respectively. Figure 2 shows internal and external forces acting on the discretized nodes and segments and theses forces are included in the equation of motion at each node (Hall and Goupee, 2015). The details of forces will be described in the following two subsections.

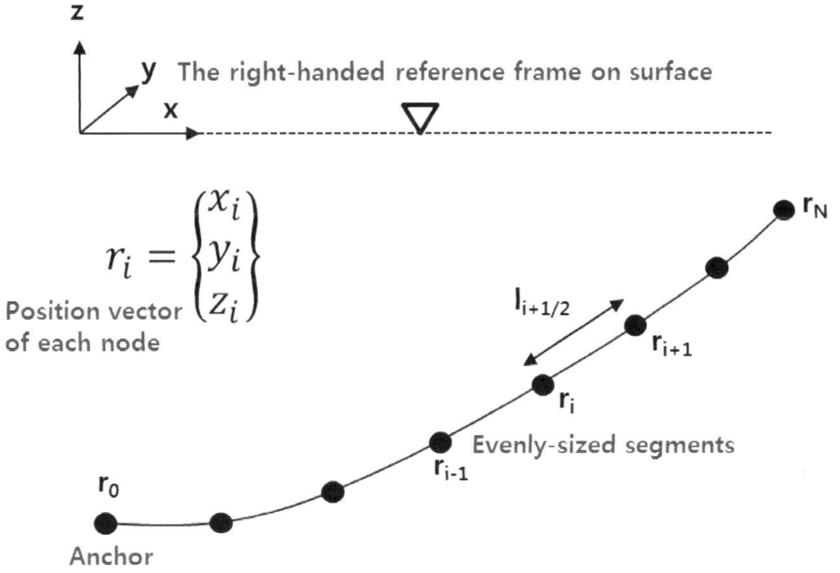

Figure 1. Discretized mooring line and indices

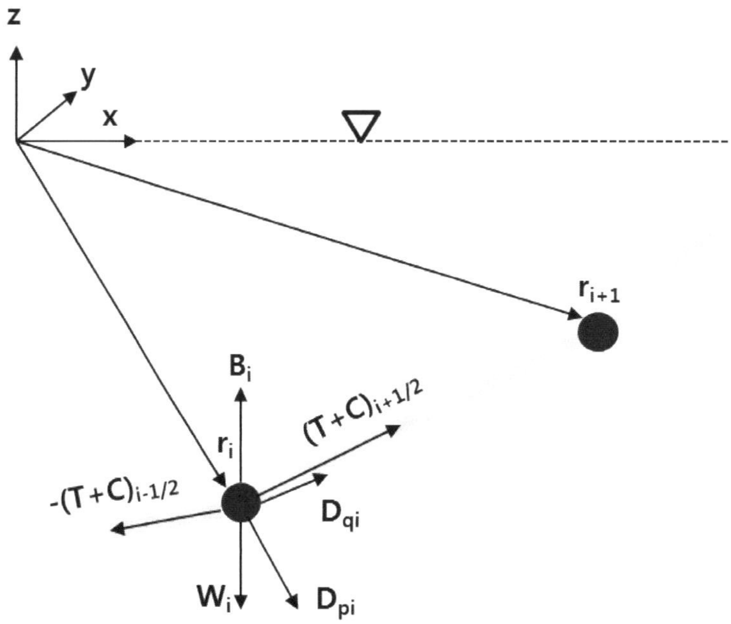

Figure 2. Internal and external forces acting on the nodes and segments

Internal Forces on the Mooring Line

The internal forces consist of axial stiffness (T), axial damping (C), and net buoyancy (W). Axial stiffness and damping are acting on the segments and calculated as

$$T_{i+1/2} = E\frac{\pi}{4}d^2\epsilon_{i+1/2} \tag{1}$$

$$C_{i+1/2} = C_{int}\frac{\pi}{4}d^2\dot{\epsilon}_{i+1/2} \tag{2}$$

where ϵ and $\dot{\epsilon}$ are the strain and strain rate, respectively.

Net buoyancy is acting on the nodes and calculated as

$$W_i = \frac{1}{2}\left[\left\{\frac{\pi}{4}d^2l(\rho_w - \rho)g\right\}_{i+1/2} - \left\{\frac{\pi}{4}d^2l(\rho_w - \rho)g\right\}_{i-1/2}\right]\hat{e}_z \tag{3}$$

where ρ_w is the density of water, g is the gravitational acceleration, and $\hat{e}_z$ is the unit vector in the direction of z-axis.

External Forces on the Mooring Line

The external forces consist of drag forces, added mass forces, and contact force with the seabed, and they are acting on the nodes. The hydrodynamic forces, such as drag forces and added mass forces, are calculated using Morison's equation, and the transverse drag force (D_p) and tangential drag force (D_q) at node i are calculated as

$$D_{p_i} = \frac{1}{2}\rho_w C_{dn} dl \|(\dot{r}_i \cdot \hat{q}_i)\hat{q}_i - \dot{r}_i\| [(\dot{r}_i \cdot \hat{q}_i)\hat{q}_i - \dot{r}_i] \tag{4}$$

$$D_{q_i} = \frac{1}{2}\rho_w C_{dt} \pi dl \|(-\dot{r}_i \cdot \hat{q}_i)\hat{q}_i\| [(-\dot{r}_i \cdot \hat{q}_i)\hat{q}_i] \tag{5}$$

where C_{dn} and C_{dt} are the transverse and tangential drag coefficients, $-\dot{r}_i$ is the relative water velocity due to the motion of nodes, and $\hat{q}_i$ is the tangential unit vector at the nodes.

The transverse added mass force (a_p) and tangential added mass force (a_q) at node i are calculated as

$$a_{p_i}\ddot{r}_i = \rho_w C_{an} \frac{\pi}{4} d^2 l [(\ddot{r}_i \cdot \hat{q}_i)\hat{q}_i - \ddot{r}_i] \tag{6}$$

$$a_{q_i}\ddot{r}_i = \rho_w C_{at} \frac{\pi}{4} d^2 l [(-\ddot{r}_i \cdot \hat{q}_i)\hat{q}_i] \tag{7}$$

where C_{an} and C_{at} are the transverse and tangential added mass coefficients, $-\ddot{r}_i$ is the relative water acceleration due to the motion of nodes.

Finally, the contact force with the seabed is modeled as a linear spring-damper system, and it works only when the node contacts the seabed, i.e., z_i is equal to or less than z_{bot}. The contact force is calculated as

$$B_i = dl[(z_{bot} - z_i)k_b - \dot{z}_i c_b]\hat{e}_z \tag{8}$$

where k_b is the stiffness per unit area of the seabed, c_b is the viscous damping per unit area, z_i and z_{bot} are the z-axis coordinates of the node and the seabed, and $\dot{z}_i$ is the velocity of the node in the direction of z-axis.

MOORING SYSTEM MODEL VALIDATIONS

In order to validate the developed mooring analysis model, two reference cases, i.e., analytical solution and experimental result, were calculated and compared.

Comparison with the Analytical Solution

The first validation case is the analytical solution of line tensions along a single catenary mooring line (Irvine, 1981). The cable properties used are 0.25 m in diameter, 7,000 kg/m^3 in density, 2×10^{11} N/m^2 in Young's modulus, 500 m in mooring line length, 325 m and 350 m in horizontal and vertical distances from anchor to fairlead, 292.8 N/m in net mooring line weight per unit length in the water. The calculated mooring line tensions at nine locations were compared with the analytical solution, and the results were listed in Table 1. The calculation result showed great agreement with the analytical solution within the range of 0.34 %.

Table 1. Comparison of the calculated mooring line tension with the analytical solution

Length from anchor [m]	Analytic solution [N]	Calculated result [N]	Error [%]
55.55	655,966	656,119	0.02
111.10	726,777	726,170	0.08
166.65	822,934	821,526	0.17
222.20	936,663	934,420	0.24
277.75	1,062,335	1,059,250	0.29
333.30	1,196,193	1,192,310	0.32
388.85	1,335,778	1,331,210	0.34
444.40	1,479,469	1,474,390	0.34
500.00	1,626,178	1,620,780	0.33

Comparison with the Experimental Result

The second validation case is the experimental result of node positions along a single catenary mooring line (Nakajima et al., 1982). The cable properties used are 0.599 cm in diameter, 2.15×10^{6} kg/cm^2 in Young's modulus, 9 m in mooring line length, 8.42 m and 3 m in horizontal and vertical distances from anchor to fairlead, 0.1938 kg/m in net mooring line weight per unit length in the water. The calculated mooring line node positions were compared with the experimental result at five points because the reference gives information of only those points. The results were listed in Table 2, and they showed good agreement with the experimental results.

Table 2. Comparison of the calculated mooring line positions with the experimental results

	X-exp [m]	X-cal [m]	X-err [%]	Y-exp [m]	Y-cal [m]	Y-err [%]
Anchor	0.00	0.00	-	0.00	0.00	-
Node 3	3.0021154	2.99761	0.15	0.3605167	0.37579	4.23
Node 6	5.8431810	5.83557	0.13	1.4093504	1.4205	0.79
Node 8	7.6015360	7.58936	0.16	2.4063136	2.423265	0.70
Fairlead	8.42	8.42	-	3.00	3.00	-

COUPLING WITH THE PLATFORM MOTIONS

The developed mooring system analysis model consists of three modules to couple with the platform motion. The first is a module that sets the initial conditions based on the shape of a mooring system and the properties of the mooring line and calculates initial values when the time is zero seconds and is called the initialization module. The second is a module that calculates the tensile load of the mooring system by receiving the information of platform motion such as current time, time step size, position, and velocity as the time progresses and is called the calculation module. The last is a module that frees the variables and memories which were used to calculate the tensile load of the mooring system after the end of calculations and is called the termination module.

After three modules, i.e. functions, that mentioned above are compiled into a single shared library file, the modules can be called as a form of function into the code that generates or analyses the platform motion by the dynamic loading (Beazley et al., 2001), as shown in 3, then the coupling between the mooring system analysis model and the platform motion is completed. Figure 4 shows the calculation procedure. When it starts, the mooring system analysis model gets the initial information of platform motion and calculates the initial tensile load and the mooring system profile until the fairlead tension is converged. Then the platform motion solver goes one time step forward, and the mooring system analysis model repeats receiving the information of platform motion and calculating the tensile load and the mooring system profile at each time step until the platform motion is no more updated.

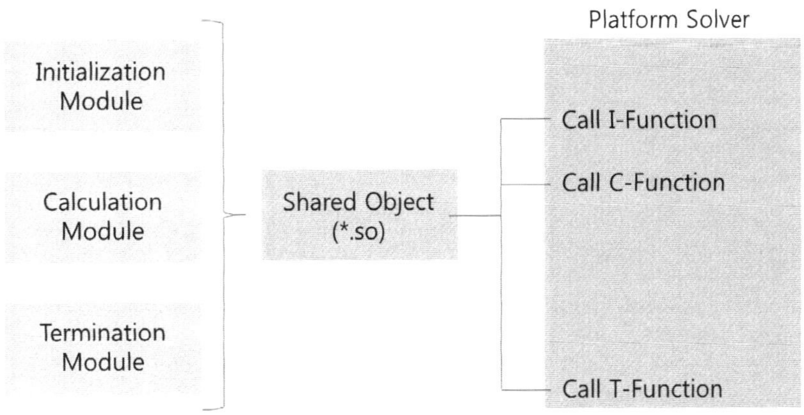

Figure 3. Shared library and dynamic loading of functions

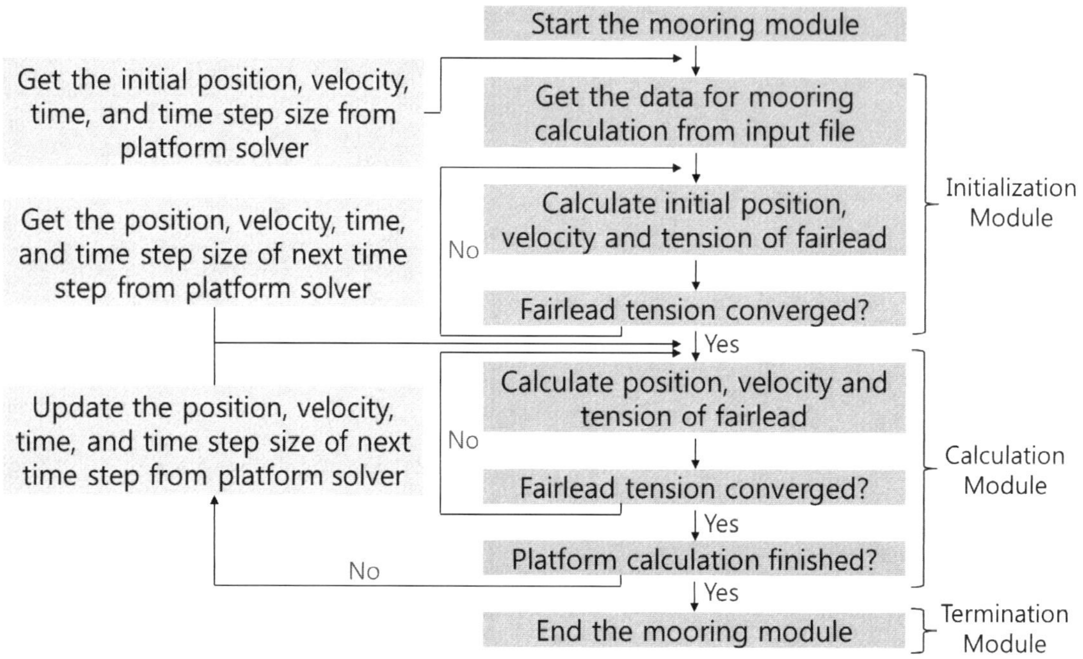

Figure 4. Calculation procedure of the mooring system analysis model

Platform Motion in the Calm Water

Simple calculation was performed to verify that the coupling between the platform behavior and the mooring analysis model through shared library files and dynamic loading works well. In this case, the tensile loads of the mooring system were calculated for a simple platform motion in the calm water. The identical three mooring lines which have arbitrary properties are connected to the SPAR-shaped platform at 120-degree intervals, as shown in Fig. 5, and the platform was set to move at a constant velocity in x-axis direction.

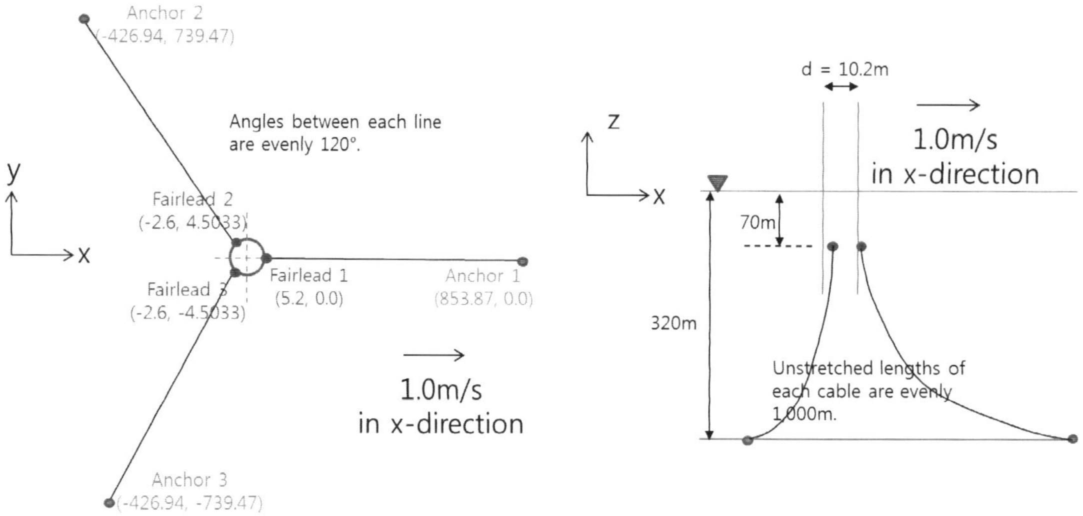

Figure 5. Schematic of the mooring system and the platform motion

Figure 6 shows the calculated tensile loads with the constant velocity in the calm water. The tensile load on the fairlead 1 which was connected to the direction of movement of the platform gradually decreased, while the tensile loads on the fairleads 2 and 3 connected in the opposite direction gradually increased. In addition, since the mooring lines connected to the fairleads 2 and 3 are symmetrical about the x-axis, the tensile loads acting on the fairleads 2 and 3 are identical. These are all physically reasonable results and show that the coupling between the platform motion and the mooring system analysis model works properly.

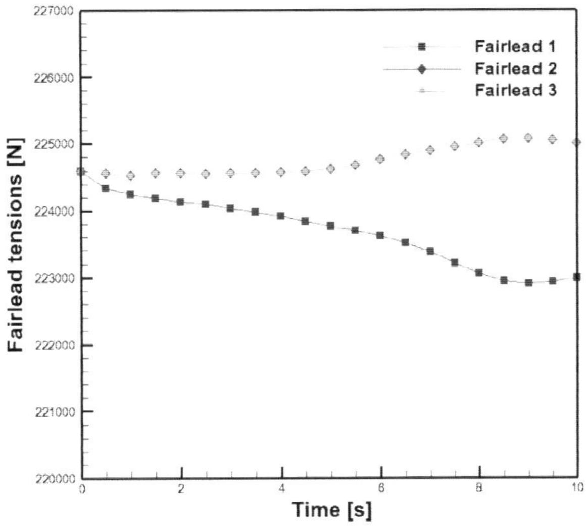

Figure 6. Calculated tensile loads with the constant velocity platform motion in the calm water

Ice Impact Load on the Platform

Another calculation was performed to confirm that the mooring system analysis model works robustly when the impact load, such as a collision with ice or ship, are applied to the platform. In this case, based on the same platform and mooring system as the previous case, the platform was stationary in the calm water at the initial period, then the impact load in the x-axis direction gave a sudden acceleration change to the platform during the short time interval and generated the motion of it, and the platform motion was decaying as time goes by.

Figure 7 shows the calculated tensile loads under the impact load acting on the stationary platform in the calm water. Initially, there is no motion of the platform, so the tensile loads acting on the three fairleads are all the same. When the impact load is applied to the platform after a short time, it can be seen that the tensile load shows the rapid change with a peak and it is stabilized after the external force was disappeared. After that, the tensile loads were increased or decreased with the general tendency by the residual motion. These results show that the developed mooring system analysis model can work robustly under the impact loading conditions, and can capture the abrupt change in the platform motion due to the external force to reflect it in the tensile load calculation.

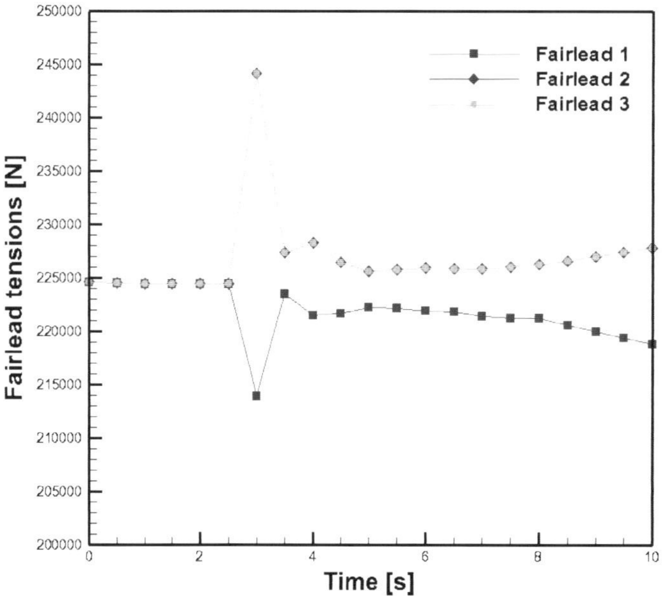

Figure 7. Calculated tensile loads under ice impact load on the stationary platform in the calm water

CONCLUSIONS

The mooring system analysis model based on the lumped mass model was developed to calculate the tensile load of the mooring system as a part of a project to develop a simulation suite for the analysis of polar environmental loads acting on the arctic ships and offshore plants, and the internal forces, such as net buoyancy, internal stiffness, and internal damping, and the external forces, such as drag forces, added mass forces, and contact force with the seabed, were included in it. The validation results of the tensile loads showed very good agreements with the analytical solution and the experimental results.

The developed mooring system analysis model was divided into three modules or functions to couple with the platform motion easily, and was coupled via dynamic loading of shared library. The coupled analyses were performed to confirm its operability and robustness for the cases of platform motion in the calm water, impact load on the platform, and coupling with the open source CFD solver, and it showed good performance and qualitatively reasonable results.

Future studies will be conducted to verify the platform motion and the tensile load of the mooring in the regular and irregular waves with an ice impact load.

ACKNOWLEDGEMENTS

Our research was supported by the National Research Foundation (2015037577) and the IT R&D Program of MOTIE/KEIT (10060329) funded by the Korea government.

REFERENCES

Beazley, D.M., Ward, B.D., & Cooke, I.R., 2001. The Inside Story on Shared Libraries and Dynamic Loading. Computing in Science & Engineering, 3(5), pp.90-97.

Hall, M., & Goupee, A., 2015. Validation of a Lumped-Mass Mooring Line Model with DeepCwind Semisubmersible Model Test Data. Ocean Engineering, 104, pp.590-603.

Hall, M., Buckham, B., & Crawford, C., 2014. Evaluating the Importance of Mooring Line Model Fidelity in Floating Offshore Wind Turbine Simulations. Wind Energy, 17, pp.1835-1853.

Irvine, M., 1981. Cable Structures. Dover Publications.

Masciola, M., Jonkman, J., & Robertson, A., 2013. Implementation of a Multisegmented, Quasi-Static Cable Model, Proceedings of 23rd International Offshore and Polar Engineering Conference, Anchorage, Alaska, USA, 30 June-5 July, 2013.

Nakajima, T., Motora, S., & Fujino, M., 1982. On the Dynamic Analysis of Multi-Component Mooring Lines, Proceedings of 14th Offshore Technology Conference, Houston, Texas, USA, 3-6 May, 1982.

Relation of Maximum Structural Velocity and Ice Drift Speed during Frequency Lock-In

Gesa Ziemer[1], Philipp Hinse[1]
[1] HSVA, Hamburg, Germany

ABSTRACT

Ice model tests with compliant cylindrical structures have been conducted at the Hamburg Ship Model Basin. Ice-induced frequency lock-in vibrations have been observed at different ice drift speeds without variation of other ice parameters. Adapting to different ice drift speeds while vibrating in a frequency close to its natural frequency is possible for the structure mainly due to a variation of the ratio β of maximum structural velocity (V_A) and ice drift speed (U). The paper describes the test setup and main observations from the model test campaign. Boundaries of β are derived based on the model test observations and compared to literature. A formula is presented to estimate the transitional velocities to the possible lock-in range based on β-variation. It is possible to predict the complete range of critical velocities for a given structural configuration based on only one observed lock-in event. A comparison of calculations and measurements shows promising results.

KEY WORDS: Frequency lock-in; Model tests in ice; Ice-induced Vibrations

INTRODUCTION

Ice-induced frequency lock-in vibrations have been subject to extensive research since the late 1960s. Soon after the first observations of lock-in vibrations in full scale, two different mechanisms have been proposed to explain the violent oscillations: Peyton (1968) claims that a characteristic failure length exists, which can correspond to the natural frequency of the structure at certain ice drift speeds and thereby creates a resonant condition. Blenkarn (1970) was the first to propose that lock-in is the result of some source of negative damping. Both concepts have been further developed and most of today's theoretical and numerical models are based on one of them. Despite all research efforts, both explanations are still under debate, and none of the currently existing models is able to predict the occurrence of lock-in with satisfying accuracy (Kärnä et al., 2013). Current research partly concentrates on the role of contact area and high pressure zones, employing rather new measuring techniques that provide better insight into local phenomena (e.g. Ziemer and Deutsch, 2015; O'Rourke et al., 2016; Hendrikse and Metrikine, 2017).

The IVOS project follows this trend and aims at studying the contact area and behavior of high pressure zones during different types of dynamic ice-structure interaction to gain more insight into the physical mechanism leading to lock-in. As a first step, model tests have been performed to create a broad basis for theoretical considerations and model validation. These tests are used in the presented study to formulate a relation of critical ice drift velocities that may lead to lock-in vibrations based on known structural parameters. Note that no scaling was applied, thus all presented results are valid for model scale regime only.

MODEL TESTS

Model tests in ice with compliant cylindrical single-degree of freedom (SDOF) structures have been performed as part of the IVOS project in HSVA's Large Ice Basin in 2015 and 2016. The physical tests were divided into two phases: Phase 1 with a bottom-fixed cylindrical structure, using active loading (structure at rest and moving ice sheet), and Phase 2 with cylindrical and flat structures employing passive loading (structure is pushed through resting ice sheet). Main objectives of the two test series were:

- Development and check of a physical test setup to conduct experiments on ice-induced vibrations with a variety of structures
- Data collection of local ice load measurements to gain insight into dynamic ice-structure interaction and corresponding pressure distribution as well as failure processes

The two test phases are briefly described below.

Phase 1

A compliant basis mounted to the basin's floor was used for Phase 1 tests. The setup bases on previous model test campaigns (see Onken et al. (2013), Ziemer and Evers (2016)) and had been optimized in several steps. The latest version is described by Ziemer and Deutsch (2016) and presented in Figure 1 (left). The compliant basis (yellow) is fixed to a rigid frame (green) by coil springs and linear bearings. A cylindrical model (red) with 830mm diameter was used for the tests. This model was equipped with tactile sensors to monitor local ice loads. Additionally, global loads were recorded by a 6-component load scale connecting the compliant basis and the model, and lasers and accelerometers monitored the ice-induced vibrations of the structure in x- and y-direction (loading direction and perpendicular in-plane motion).

Test conditions are summarized in Table 1. Prior to the tests, the level ice sheet was cut loose from the side walls of the basin. A 10m long stripe of ice was pushed against the structure during each test run by the main carriage. Different failure modes have been observed throughout the test campaign, covering the regimes of intermittent crushing, lock-in and continuous crushing. Due to the relatively high elasticity of the model ice, the ice tended to bend downwards before crushing against the structure. Therefore, the interaction was interrupted frequently by bending failure. This behavior is amplified by the structure's large aspect ratio. At times, release cracks formed that ran from the pushing board of the main carriage towards the structure or the side walls of the basin. Another downside of this setup is the location of the springs at the submerged compliant basis, making all changes of stiffness and natural frequency time consuming. Also, the huge compliant basis built as a steel framework adds considerable hydrodynamic mass and damping.

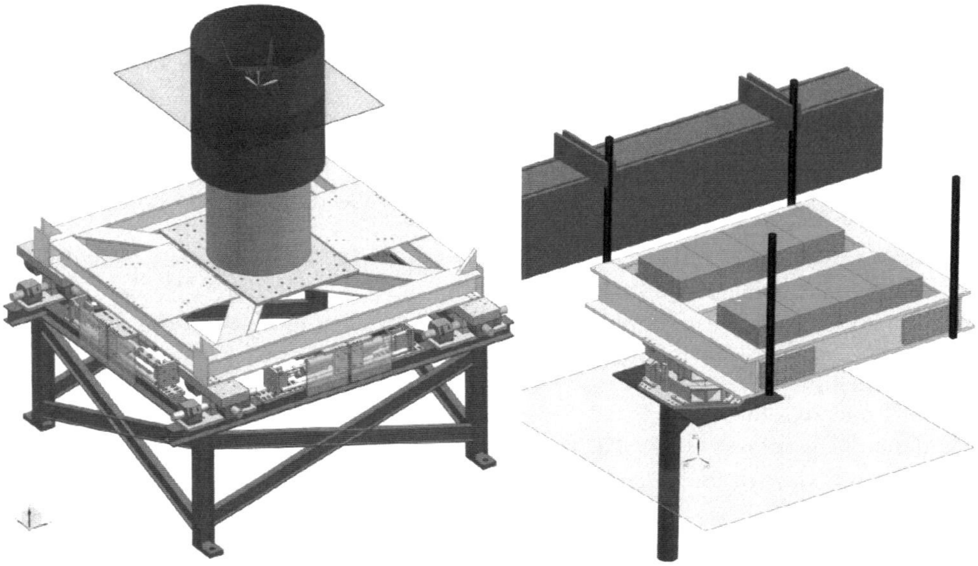

Figure 1. Left: Drawing of the setup used for Phase 1 tests (active loading, mounted to the basin's floor). Right: Drawing of Phase 2 setup (passive loading, mounted to the main carriage)

Table 1. Test conditions Phase 1 (mean values).

Series	Cylinder diameter [mm]	Natural frequency [Hz]	Stiffness [N/mm]	Ice thickness [mm]	Compressive strength [kPa]	Tested ice drift velocity range [m/s]
11000	830	5.5	2230	50	72	0.010 – 0.15
12000	830	5.5	2230	48	95	0.005 – 0.06
13000	830	2.7	490	46	79	0.020 – 0.15
14000	830	2.7	490	46	114	0.010 – 0.044

Phase 2

A new setup was designed for Phase 2 to be more flexible with changes of the model. Passive loading was utilized by attaching a compliant basis to the rigid moonpool of the main carriage, which holds the model and provides the aspired dynamic properties. Aluminum rods with box sections are used as flexible springs. The top mass on the compliant platform can be varied to adjust the natural frequency. The setup is shown in Figure 1 (right).

Passive loading means that the ice sheet can remain frozen to the side walls of the basin. Thus, release cracks are far more unlikely than with passive loading. The acceleration of the main carriage was monitored to check whether control oscillations are transferred to the model. This was not the case; the soft springs are an effective damper for the high frequency oscillations of the main carriage. Shape and diameter of the models were varied: Cylinders were tested with diameters of 120mm, 200mm and 500mm, and a flat indenter with rectangular cross section and width of 200mm was tested additionally in side-first and corner-first configuration. Only results from cylindrical models are considered in this paper. Similar to Phase 1, all models were equipped with tactile sensors and the global loads were measured by a 6-component scale. Lasers and accelerometers monitored the in-plane vibratory motion of the models. The smaller aspect ratios limited the bending failure of the ice sheet at the structure. This occurred only on some of the structures in a certain velocity range. By varying the carriage speed in small steps of 1 mm/s, the transition from intermittent crushing to steady-state response and later to continuous crushing was captured well for all tested structures.

Relevant test conditions studied in Phase 2 are summarized in Table 2.

Table 2. Test conditions Phase 2 (mean values).

Series	Cylinder diameter [mm]	Natural frequency [Hz]	Stiffness [N/mm]	Ice thickness [mm]	Compressive Strength [kPa]	Tested ice drift velocity range [m/s]
21000	500	5.5	2220	31	90	0.005 – 0.10
21000	500	7.6	3020	31	80	0.005 – 0.10
23000	200	5.8	1930	33	82	0.005 – 0.15
23000	200	5.8	1930	42	97	0.005 – 0.15
23000	500	5.5	2030	43	72	0.005 – 0.10
24000	120	5.4	1935	41	136	0.005 – 0.20

RESULTS

The main objective of the tests was to create data sets which cover a large range of ice drift speeds to monitor the transitional velocities limiting the interval where frequency lock-in occurs. A comparison of the conditions leading to lock-in of different structures and in different ice conditions can shed some light onto the still unresolved question of how frequency lock-in develops.

Lock-in has been observed with all tested structural configurations. In most tests, lock-in was monitored along with intermittent and continuous crushing within the same test run, thus without changes of ice parameters except for the simulated ice drift speed. An example of lock-in is given in Figure 2, showing the almost purely sinusoidal response close to the structure's natural frequency, and periodic ice load. Lock-in events had a duration between 7s and 60s.

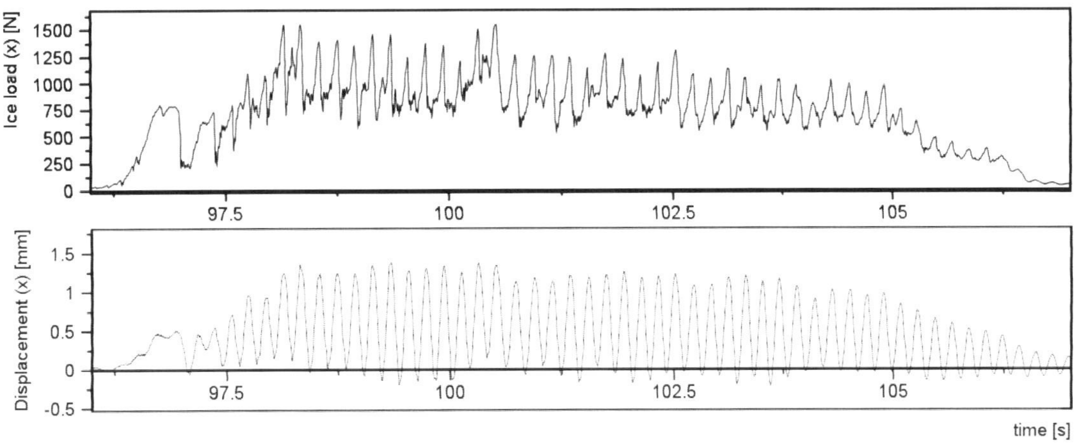

Figure 2. Example of frequency lock-in vibrations observed during Run 12050.

Characterization of a Lock-in Cycle

All time series obtained in the IVOS test phases containing frequency lock-in vibrations show a characteristic time history of load and corresponding response during each loading cycle, which is presented in Figure 3. While a lock-in cycle is often subdivided into a *loading phase* and an *unloading phase* in common literature (compare e.g. ISO 19906:2010), the test results indicate that a further subdivision of the loading phase into a *crushing phase* and a *build-up phase* is reasonable. The unloading phase is referred to as *collapse phase* within this study.

The cycle starts at position ①: After the ice has collapsed in the previous cycle, the deflected structure is released and starts to move back to its equilibrium position. Herein, the equilibrium refers to the mean deflection around which the structure oscillates. When this position is reached, the relative velocity between structure and ice is at its maximum, as structure and ice move in opposite direction. The structure starts to penetrate into the ice and crushes it, which decelerates the structural movement and eventually turns the orientation of movement: When the stored energy is not sufficient to crush the ice any longer, the ice deflects the structure instead. Thus, the relative velocity decreases to almost zero at position ②. This is the case after about one half natural period, and the structure is again close to its equilibrium position at this point in time. This is when the *build-up phase* starts: the ice load increases significantly and deflects the structure while the relative velocity increases as well. When the maximum of deflection is reached (③) the ice fails, and the collapse phase starts. During collapse, the structure is released and returns to its initial position. Then the cycle starts anew.

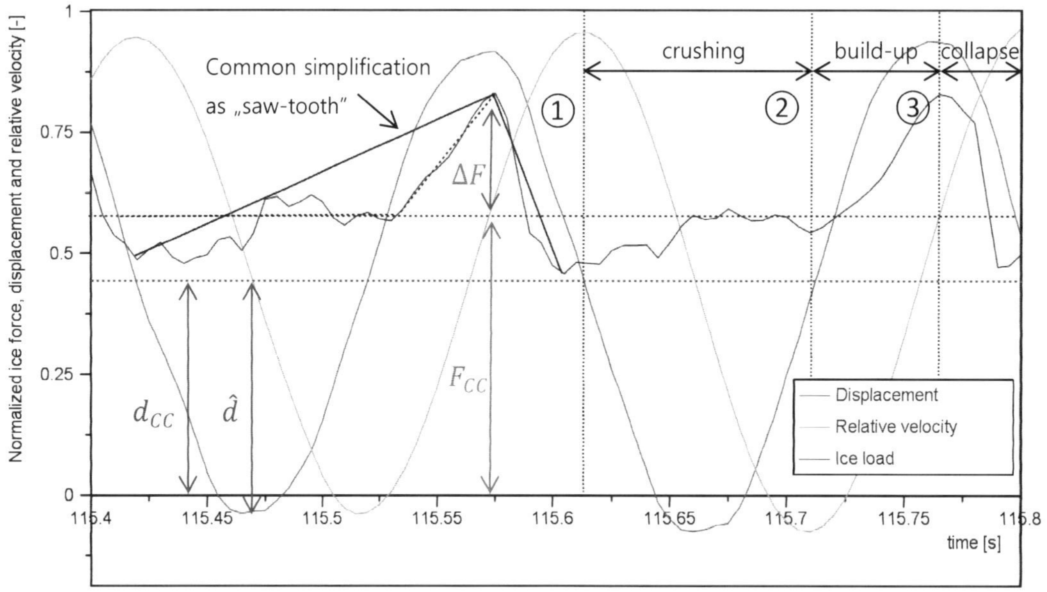

Figure 3. Phases of a lock-in cycle as observed in IVOS tests.

POAC17-071

Calculation of Critical Velocities

It is currently not possible to predict critical velocities for the occurrence of lock-in based on structural and ice parameters with reasonable accuracy, but small calculation efforts. The considerations presented in this section are a step towards such prediction formula.

The previous section has shown that a lock-in cycle has a certain characteristic, mainly prescribed by the initiation and termination of the *build-up phase*. The phase relation of load and response during build-up helps to describe the motion in a simplified way: For further considerations, we simplify the structural response as a purely sinusoidal oscillation which can be described by following function (compare Figure 3):

$$d(t) = d_{CC} + \hat{d}\sin(\omega t) \tag{1}$$

Herein, d_{CC} is the mean deflection at equilibrium position, thus the deflection resulting from the mean crushing force F_{CC} and the structural stiffness k:

$$d(t) = \frac{F_{CC}}{k} + \hat{d}\sin(\omega t) \tag{2}$$

With some simplification, we may define the ice load function as a combination of a time invariant mean crushing force F_{CC} and a time-dependent force $\Delta F(t)$ representing the load increase:

$$F_{ice}(t) = F_{CC} + \Delta F(t) \tag{3}$$

Such description is not correct for collapse and crushing phase, but with the sine function for displacement and a linear load increase during build-up, ice load and structural response are easy to describe for this specific phase of the lock-in cycle.

At position ③, $\Delta F(t)$ becomes ΔF (maximum), and relative velocity between ice and structure becomes almost zero. Hence, we can use the equation of motion to describe the relation of ice load, inertia load and spring force at point ③ which has to be fulfilled to meet a lock-in condition as found in the tests:

$$F_{CC} + \Delta F = -\omega^2 m\hat{d} + k\left(\frac{F_{CC}}{k} + \hat{d}\right) \tag{4}$$

Engelbrektson (1997) was the first to observe that lock-in vibrations occur when the ice drift speed is similar to the maximum velocity of the structure during vibrations, thus when the amplitude of the almost sinusoidal response velocity equals the mean ice drift speed. Subsequently, Kärnä et al. (2007) and others compiled evidence that a certain velocity ratio β between maximum structural velocity V_A and ice drift speed U is required for lock-in:

$$V_A = \beta U = \hat{d}\omega \Rightarrow \hat{d} = \frac{\beta U}{\omega} \tag{5}$$

Kärnä et al. (2007) state that β has to be in a range between 0.9 and 1.4. Inserting eq. (5) in (4) and solving for U yields a formulation for the critical ice drift velocities which allow for lock-in:

$$U = \frac{\omega \Delta F}{\beta(k - \omega^2 m)} \tag{6}$$

It should be noted that this relation of ice drift speed and structural parameters does not imply any physical explanation of lock-in, but rather provides necessary conditions that have to be met in order to maintain a frequency lock-in condition. Although eq. 6 contains several major simplifications, it can be a helpful tool to estimate critical velocities if the unknowns can be predicted in a reasonable range. These predictions are subject to the following sections.

Estimates for Oscillation Frequency ω

During frequency lock-in, the structure oscillates in a frequency close to its natural frequency. IVOS tests indicate a range of $\omega = 0.90 \ldots 0.99\ \omega_n$. This is slightly different from what Kärnä (1994) reported: He found the vibration frequency during lock-in to be 5-15% lower than the natural frequency of the structure ω_n. It should be noted that the vibration frequency can be assumed to be dependent on ice conditions and structural parameters rather than on ice drift speed, as ω is found to be almost constant for each test run (compare example shown in Figure 4).

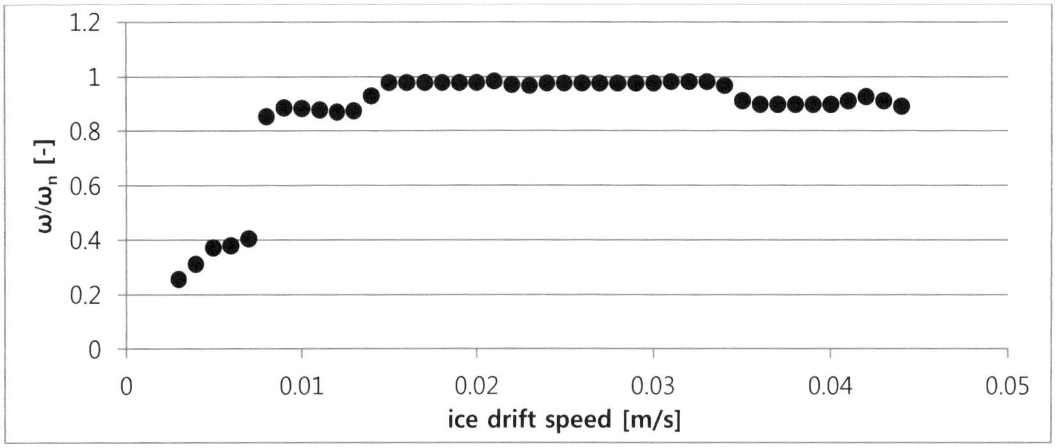

Figure 4. Trend of response frequency over simulated ice drift speed (all values taken from the same test run).

Prediction of ΔF

The increase of global load ΔF during build-up phase is one of the essential parameters to estimate the critical velocities for lock-in. It may depend on ice parameters (i.e. ice thickness, compressive strength) as well as on structural parameters (i.e. diameter of the cylinder, combination of stiffness and natural frequency). The variation of test conditions in the two phases of the IVOS project provides a good basis to study these influences. Such study is still on-going, but first findings are presented below.

Influence of ice parameters

Naturally, the crushing load level depends on the compressive strength of the model ice. Results from Phase 1 tests with invariant structural configurations and ice sheets with similar thickness, but varying strength, confirm a fairly linear increase of crushing load over increasing uniaxial compressive strength (Figure 5). Data indicates that ΔF increases linearly as well. Thus, the ratio of build-up force and mean crushing force remains the same, regardless of ice strength.

The ice thickness variation during the test campaign was too small to allow for significant conclusions about its influence on the build-up force. Detailed analysis of the local loads is required to develop a theory about the influence of the ice thickness on the local pressure distribution which affects ΔF.

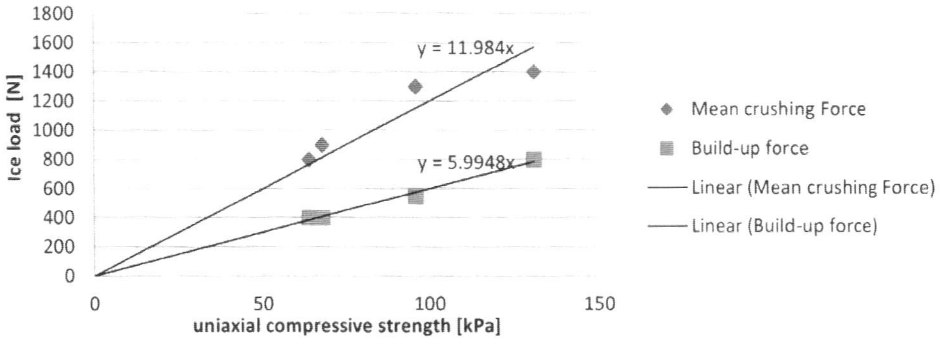

Figure 5. Trend of mean crushing force (F_{CC}) and average build-up force (ΔF) over uniaxial compressive strength of the model ice.

Influence of Structural Parameters

No significant difference between the build-up forces on structures with the same diameter, but varying dynamic properties (mass, stiffness, natural frequency) has been found in the tests. But the mean crushing load depends on the diameter of the tested cylinder, and so does the build-up force. Also, the ratio of build-up force and mean crushing force changes with changing cylinder diameter. Deciphering this relation is subject to current research.

Trend of Maximum Structural Velocity and Ice Drift Speed

Rearranging eq. 6 describes a relation of the build-up force and the inertial and spring forces acting on the structure:

$$\Delta F = -\beta(\omega^2 m + k)U \tag{8}$$

It has been shown above that the build-up force does not change with changing ice drift speed. Also, the term $\omega m + \frac{k}{\omega}$ is not dependent on U as illustrated in Figure 5. Therefore, the expression βU must be nearly constant (Toyama et al., 1983), and the ability of the structure to enter the lock-in condition at different ice drift speeds results from the variability of β. An example of the trend of β over ice drift speed is presented in Figure 6: At low speeds, β is high. As the ice drift speed increases the ratio β becomes smaller and approaches unity. When the ice drift speed increases further, the displacement amplitudes cannot increase accordingly, and β approaches zero.

Kärnä et al. (2007) found β to be in a range between 0.9 and 1.4 during lock-in for a collection of full scale and small scale data. Yap and Palmer (2013) limited this range to 0.9 to 1.1. Thus, the upper bound is somewhat uncertain.

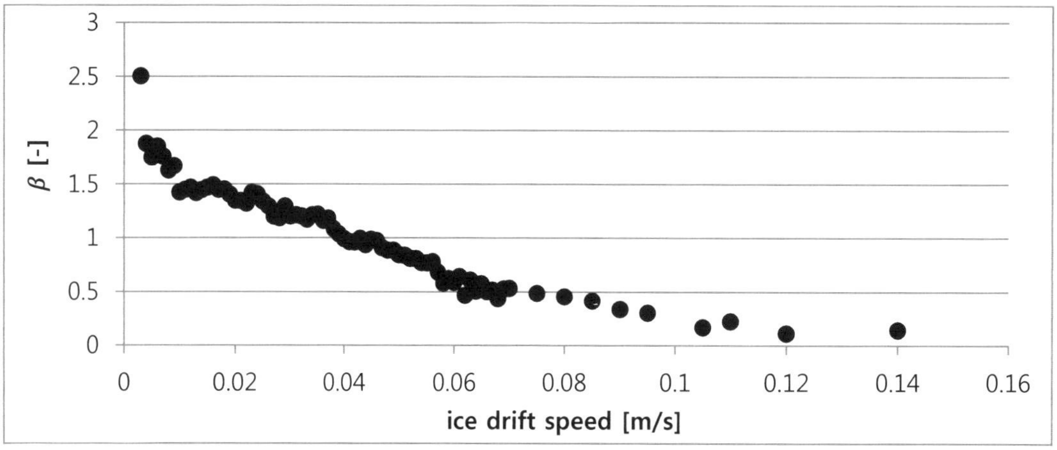

Figure 6. Trend of β over simulated ice drift speed (all values taken from the same test run).

We can find a physical upper limit for β by considering the displacement of the ice sheet during the build-up phase: While the ice edge travels a distance $\Delta d_{ice} = U\Delta t$ within a quarter period ($\Delta t = \frac{\pi}{2\omega}$), the structure deflects by $\hat{d} = \frac{\beta U}{\omega}$ (see eq. 5). We know from the local load measurements that ice and structure never lose contact during build-up phase – otherwise, the load increase would not be possible. Therefore, the displacement amplitude at highest possible β must be smaller than or equal to the distance travelled by the ice edge:

$$\frac{\beta_{max} U}{\omega} \geq U \frac{\pi}{2\omega} \Rightarrow \beta_{max} = \frac{\pi}{2} \tag{7}$$

Eq. 7 corresponds well to the maximum β found in literature. From eqs. 6 and 7 we obtain a formulation for the maximum ice drift speed at which lock-in is theoretically possible:

$$v_{II} = \frac{2\omega\Delta F}{\pi(k - \omega^2 m)}$$

The lower boundary for β depends on the maximum relative velocity at which the build-up phase can be initiated. The value of 0.9 from literature is taken as lower boundary for this study:

$$v_I = \frac{\omega\Delta F}{0.9(k - \omega^2 m)}$$

Comparison of Theoretical and Experimental Results

If the build-up force ΔF is known, critical velocities can be calculated using the derived boundaries for β, known structural parameters, and estimates for ω. At the moment it is not possible to estimate ΔF and ω with sufficient accuracy to use the formula for a prediction of critical conditions for lock-in without experimental results. But if one lock-in event has been monitored on a structure, we can use it to find other conditions which are critical as well. The easiest case is to assess the risk of lock-in based on variation of ice drift speed, but constant ice thickness and strength. This is done for all test runs from the IVOS campaign where lock-in has been observed at different simulated ice drift speeds. For each run, one lock-in event was arbitrarily chosen as a base case. ΔF was calculated as the mean build-up force during the steady-state of the specific event, and ω was found by spectral analysis of its displacement time history. Boundaries for β were set to 0.9 and $\frac{\pi}{2}$. Results are presented in Table 3.

Table 3. Calculated and measured critical velocities for the occurrence of lock-in.

Run	**12050**	**13050**	**21020**	**23010**	**23020**	**24010**
Natural frequency f_n [Hz]	5.45	2.65	7.60	5.80	5.47	5.40
Oscillation frequency during lock-in f [Hz]	5.12	2.52	7.31	5.69	5.42	5.34
f/f_n	0.94	0.95	0.96	0.98	0.99	0.99
Stiffness k [N/m]	2230	490	3020	1930	2030	1935
Ratio of critical damping ξ [%]	1.5	3.0	2.7	2.3	3.1	2.3
Oscillating (modal) mass m [kg]	1712	1591	1192	1308	1547	1513
Average build-up force ΔF based on one arbitrarily chosen event with full ice contact [N]	481	273	264	249	251	131
v_I (calculated) [m/s]	0.021	0.030	0.015	0.022	0.023	0.012
v_I (measured) [m/s]	0.020	0.032	0.020	0.020	0.020	0.009
v_{II} (calculated) [m/s]	0.037	0.053	0.027	0.038	0.040	0.021
v_{II} (measured) [m/s]	0.040	0.050	0.030	0.040	n.a.	0.024

Note that the simulated ice drift speed was incrementally increased using steps of 0.001 m/s (Runs 13050 and 24010), 0.005 m/s (Run 21020) and 0.01 m/s (all other runs). Therefore, the transitional velocities have not been determined with higher accuracy than 0.01 m/s for most runs, and the above fit is satisfactory. In Run 23020, the failure mode changed to buckling instead of continuous crushing; therefore, no experimental upper transitional velocity can be provided.

Although the results show a very good fit, one has to be cautious because of several uncertainties that influence the critical velocities: We assume ice conditions to be homogeneous along the model basin, but in reality there is some variation in thickness and strength which affects the build-up force and the damping during lock-in. Also, the build-up force depends on the contact between structure and ice: After flexural failure, the new ice edge is often asymmetric, and the contact does not fully develop over the structure's circumference. Hence, the build-up force in this case differs from the full contact condition. Furthermore, evaluated events were arbitrarily chosen and only a steady-state part of it was used to determine the mean build-up force. Choosing a different event or a different interval leads to variations of ΔF and thus variation of calculated critical velocities.

SUMMARY AND CONCLUSIONS

Model tests in ice have been performed with a variety of structures as part of the IVOS project. Active and passive loading has been utilized for the test setups. Passive loading seems to be the more reasonable choice for future tests because the setup allowed for more flexibility in changing the models and their structural parameters.

All tested structures were successfully set into lock-in vibrations and the global and local loads as well as the structural response have been measured for different types of ice-structure interaction. The load time histories revealed that the classic definition of a loading and unloading phase, creating a saw-tooth shaped forcing function, may be an oversimplification. This observation was the basis for a simple formula describing the relation of load increase and corresponding structural response during lock-in, mainly influenced by the relation of maximum structural velocity and ice drift speed (β). Although this formula cannot yet be used to predict the occurrence of lock-in for a specific structure without any experiments or full scale observations, it is useful to estimate the range of critical ice conditions based on only one lock-in event. A comparison of experimental and theoretical results shows promising agreement.

The topic of ice-induced lock-in vibrations remains a challenge. The IVOS test results form a valuable basis to shed some more light on unresolved questions. Especially the local load measurements can help to increase the understanding of the phenomenon. More research is needed to decipher the underlying physical mechanisms. Current research concentrates on data analysis with respect to contact area and the behavior of high pressure zones during the different phases of lock-in. This will also help to predict the build-up force theoretically.

ACKNOWLEDGEMENT

The tests have been conducted as part of the SAMCoT-associated project IVOS. Authors are grateful to the Research Council of Norway for the support of the Centre for Research-based Innovation SAMCoT and the support of all SAMCoT partners. In particular, authors wish to thank SAMCoT partners additionally involved in the IVOS project, contributing by valuable feedback and financial support: DNV GL, Engie SA, Kvaerner Concrete Solution AS, Multiconsult AS, Shell Technology Norway AS, Total E&P Norge AS, NTNU and TU Delft.

The authors are also grateful for the financial support of the data analysis by the German Federal Ministry for Economic Affairs and Energy (BMWi) through the project CoPSIS (O3SX408A).

REFERENCES

Blenkarn, K. A., 1970. Measurements and analysis of ice forces on Cook Inlet structures. In Proceedings of the Second Annual Offshore Technology Conference, volume II, pp 365–378, Houston, Texas.

Engelbrektson, A., 1997. A refined ice/structure interaction model based on observations in the Gulf of Bothnia. Proceedings of the Int. Conf. on Port and Ocean Engineering under Arctic Conditions (POAC). Yokohama, Japan, Vol. 4, pp. 373-376.

Hendrikse, H. & Metrikine, A., 2015. Interpretation and prediction of ice induced vibrations based on contact area variation. Int. J. Solids Struct., 75-76:336–348.

Kärnä, T., 1994. Steady-state vibrations of offshore structures. Hydrotechnical Construction, 28(8):446-453.

Kärnä, T., Izumiyama, K., Yue, Q., Qu, Y., Guo, F. & Xu, N., 2007. Ab upper bound model for self-excited vibrations. Proc. of the Int. Conf. on Port and Ocean Engineering under Arctic Conditions (POAC). Dalian, China, Vol. 1, pp. 177-189.

Kärnä, T., Andersen, H., Gürtner, A., Metrikine, A., Sodhi, D., Loo, M., Kuiper, G., Gibson, R., Fenz, D., Wallenburg, C., Wu, J.-F. & Jefferies, M., 2013. Proceedings of the 22nd International Conference on Port and Ocean Engineering under Arctic Conditions (POAC), Espoo, Finland.

Onken, G., Haase, A., Evers, K.-U. & Jochmann, P., 2013. Ice model tests with a cylindrical structure to investigate dynamic ice-structure interaction. Proceedings of the 22nd International Conference on Port and Ocean Engineering under Arctic Conditions (POAC), Espoo, Finland.

O'Rourke, B., Jordaan, I., Taylor, R. & Gürtner, A., 2016. Experimental investigation of oscillation of loads in ice high-pressure zones, part 1: Single indentor system. Cold regions science and technology 124 (2016) pp. 25-39.

Peyton, H. R., 1968. Sea ice forces. In Ice Pressures Against Structures, compiled by L. Gold and G.Williams, NRC Techn.Memo No. 92, Ottawa, Canada.

Toyama, Y., Sensu, T., Minami, M., & Yashima, N., 1983. Model tests on ice-induced self-excited vibration of cylindrical structures. In Proceedings of the Seventh International Conference on Port and Ocean Engineering under Arctic Conditions (POAC), vol 2, pp 834–844, Helsinki, Finland.

Yap, K.T. & Palmer, A., 2013. A model test on ice-induced vibrations: Structure response characteristics and scaling of the lock-in phenomenon. Proceedings of the 22nd International Conference on Port and Ocean Engineering under Arctic Conditions (POAC), Espoo, Finland.

Ziemer, G. & Deutsch, C., 2015. Study of local pressure distribution and synchronization during frequency lock-in. Proceedings of the 23rd International Conference on Port and Ocean Engineering under Arctic Conditions (POAC), Trondheim, Norway.

Ziemer, G. & Evers, K.-U., 2016. Model Tests with a Compliant Cylindrical Structure to Investigate Ice-Induced Vibrations. Journal of Offshore Mechanics and Arctic Engineering 138(4)

The Study on Assessing the Structural Reliability under the Condition of Ice Loads Against a Floating Body of the Ocean Plant Structure in the Polar Regions

Tae Hwan Joung[1], Jong Gil Yum[1], In Sik Nho[2], Young Taek Oh[2], Kuk Jin Kang[1],Su-gil Cho[1]
[1]Korea Research Institute of Ships & Ocean Engineering, Daejeon, Korea
[2]Chungnam National University, Daejeon, Korea

ABSTRACT

The offshore structures should be able to conduct its proper function even while having varied environmental loads on the sea, and thus, when designing the structure, the various environment loads from the ocean should be considered. The environmental loads should be quantitatively measured over a long period of time, in order to be used as load conditions in the structural analysis, but a lot of time and excessive costs are required for the structural analysis. Therefore, it is necessary that the analysis of structural reliability should be conducted by considering the uncertainty of the environmental loads. In addition, even when manufacturing the offshore structure in shipyard, the uncertainties such as tolerance of the structure should be considered, which occurs frequently in the process of producing and manufacturing the structure.

This paper explains an assessment process of structural safety while considering uncertainties under the condition that the floating structure employed in the polar region under the ice loads. The assessment method of the structural safety will be presented by carrying out the structural reliability analysis for the structure that can reasonably assess the safety of a floating structure. Finally, based on the results derived from the analysis of structural reliability, several guidelines in constructing and operating for the structure will be provided to the field.

INTRODUCTION

The offshore plant structures in the polar region operate under different environment loads such as wave and current impact, strong wind, earthquake, extremely low temperature, ice collision etc. Especially, collision with ice influences to the system of the offshore structure, and eventually it can be broken down if the condition are critical. Therefore, it should be taken into consideration in the design stage.

Carney *et al.* (2006) observed strain rates of the ice and made a failure model. Derradji-Aouat*et al.* (2000) also made a mathematical model for the ice with test data under monotonic and cyclic behaviour of fresh water columnar grained. There are, however, a lot of uncertainties in the collision analysis such as dimension of the structure, the speed and the size as well as material properties of ice of the ice bergy bit.

The uncertainties should be systematically considered in the design stage, since the collision analyses contain difficult statistics information. That is, the results of the collision analysis are dependent on the time, and the maximum points are scattered in many places. Accordingly, the results of the collision analysis should be dealt with statistically. The structural reliability analysis can be used for taking uncertainties of design variables into account, which should be dealt with systematically and statically. The safety of structure, therefore, can be assessed by probability of failure as considering the uncertainties. The probability failure of the structure can be computed in the design process.

The structural reliability analysis is required in the offshore structure design for estimating safety under collision with ice, because accurate measurement of the loads in the real sea state is extremely difficult. The uncertainties also exist in the manufacturing process as well. The tolerance or material properties of the structures contain significant uncertainties. As the structural uncertainties impact to safety of the offshore plant, the guideline of the structure in the manufacture process is necessary.

To evaluate structural safety of the offshore plant structure, the structural reliability analysis for computing the probability of failure was employed. In this study, limit state equations were obtained from the response surface method (RSM), and the probability of failure was calculated by the Monte-Carlo Simulation (MCS) method. In order to achieve the target probability of failure, design variables with high sensitivity were chosen, and the statistical properties were adjusted to reach the aimed probability of failure. The obtained statistics characteristics can be used as a guideline when the offshore structures are manufactured.

MECHANICAL CHARACTERISTICS AND MATERIAL PROPERTIES OF THE ICE

The mechanical characteristics of ice were estimated as an elasto-plasticity material for collision of ice and offshore structure. Von-Mises yield condition of the structure was applied for the ice as the commercial software, LS-DYNA contained the material model (Halquist, 2007). Nho *et al.* (2016) used change of hardening curve depends on strain rate speed ($\dot{\varepsilon}_t$) and the same curve was considered in this sturdy. The failure condition of the ice was defined failure stain (ε_f) which is the accumulated equivalent statin.

The material properties of the fresh water ice are shown in the Table 1, which were used for the collision test simulation with LS-DYNA. As shown Table 1, three important random variables (RV[1] ~ RV[3])were considered in the material properties. The 10% of covariance (CoV) value was considered for the random variables, and the statistical properties of the random variables were assumed to have a normal distribution.

Table 1. Mechanical characteristics of fresh water ice

M.P.	Mean value	CoV [%]	Probability distributions	No. of Random Value (RV#)
Elastic modulus (E) [MPa]	7800	10	Normal distribution	RV[1]
Density (ρ) [kg/m^3]	970	10	Normal distribution	RV[2]
Yield strength (σ_y)[MPa]	Figure 1.	-	-	-
Failure strain (ε_f)	7.1	10	Normal distribution	RV[3]

COLLISION ANALYSIS MODEL FOR THE OFFSHORE STRUCTURE AND ICE BERGY BIT

Offshore Structure Model for the Collision Analysis

The structural model for the collision test simulation with ice are shown in Figure 1. The structural model for the simulation was modelled with 1/4 symmetric model as the panel structure are symmetric in x- and y- directions. The thickness of the frame was considered as random variables, and the mean values of the thickness are shown in the Figure 2.
Material properties of the structural model was estimated elastic-perfect plastic material, and they are shown in the Table 2. The 10% of covariance (CoV) value was considered for the random variables(RV[4] ~ RV[6]), and the statistical properties of the random variables were assumed to have a normal distribution.
The thickness of the frame are considered as four random variables (RV[7] ~ RV[10]), and the 5% of covariance (CoV) value was considered for the random variables, and the statistical properties of the random variables were assumed to have a normal distribution.

Table 2. Mechanical characteristics of the target structure

M.P.	Mean value	CoV [%]	Probability distributions	No. of Random Value (RV#)
Elastic modulus (E) [MPa]	210000	10	Normal distribution	RV[4]
Density (ρ) [kg/m^3]	7850	10	Normal distribution	RV[5]
Yield strength (σ_y)[MPa]	285	10	Normal distribution	RV[6]

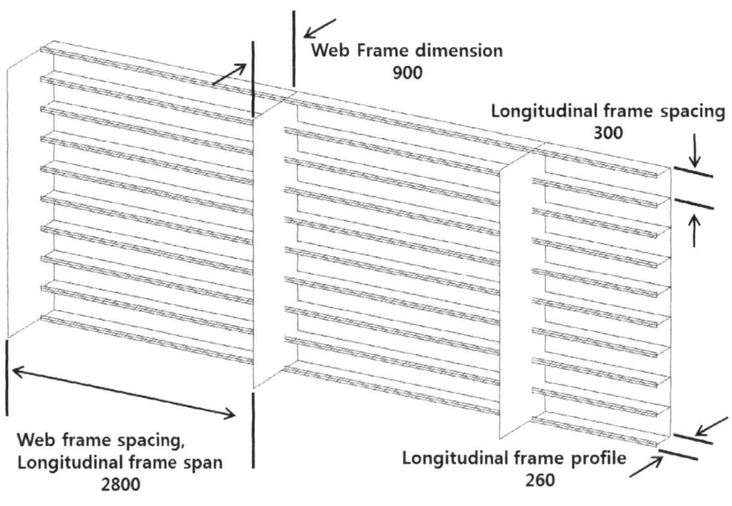

Figure 1. 1/4 symmetric model of offshore panel structure

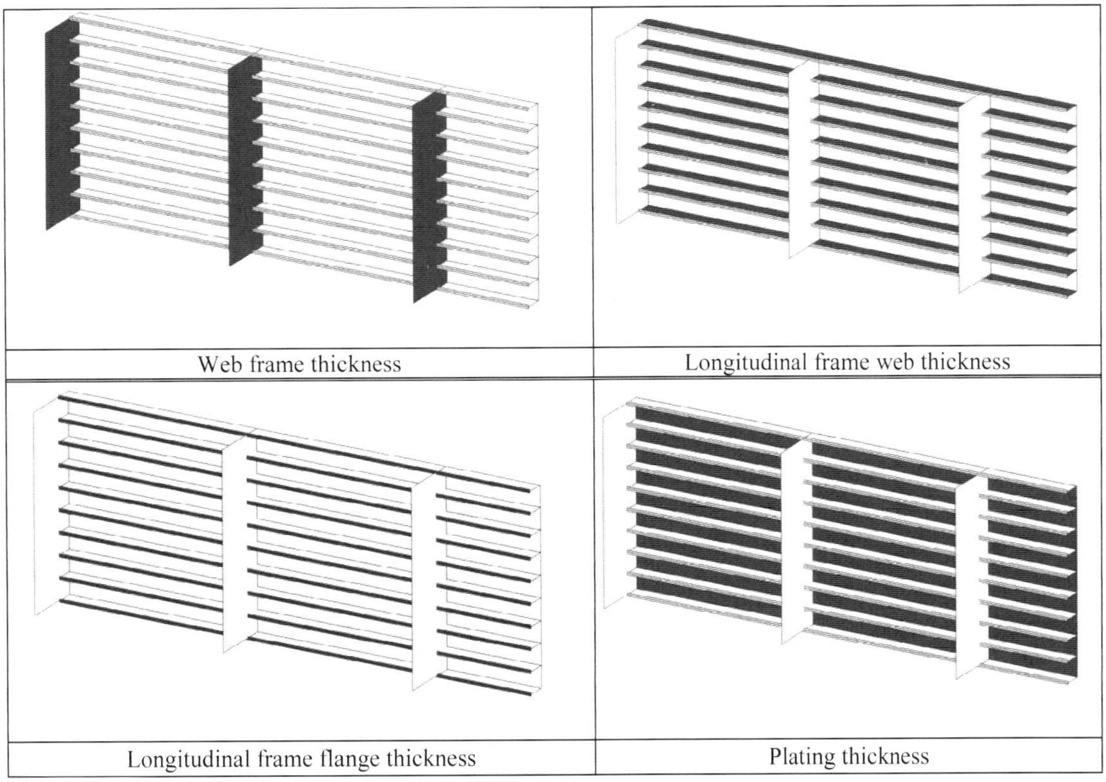

Figure 2. Thickness dimension of the offshore panel structure

Table 3. Dimension of the target structure

Structure member	Dimension (Mean value) [mm]	CoV [%]	Probability distributions	No. of Random Value (RV#)
Web frame thickness	10.0	5	Normal distribution	RV[7]
Longitudinal frame web thickness	10.0	5	Normal distribution	RV[8]
Longitudinal frame flange thickness	27.3	5	Normal distribution	RV[9]
Plate thickness	14.5	5	Normal distribution	RV[10]

Collision Scenario and Modelling for the Ice Bergy Bit

Several cases for the ice collision scenarios were suggested by the class society such as DNV-GL (2006). A typical example case was chosen, in this study, for the collision test simulation between ice bergy bit and structure as shown in Figure 2. There are a lot of things that should be considered for the collision simulation such as collision angle, wave or current speed etc. However, only one case was considered that the sphere ice bergy bit collides to the centre of the structure at 2.0 m/s of speed so that the aims of this study may set up a procedure for the structural reliability analysis. The finite element model of the ice bergy bit is shown in Figure 4, and the size, volume and mass of the ice bergy bit were computed and also shown in Figure 4. Modelling of panel structure and ice bergy bit for the collision test simulation are shown in Figure 5. The 1/4 symmetric model for the panel structure was used and the ice bergy bit collides to the centre of the structure. Simple support condition was applied to the side of the structure, and symmetric condition was applied to the symmetric plane.

The random variables for the structural analysis are shown in Table 4. The diameter and speed of the ice bergy bit is considered as a random variables (RV[11], RV[12]), and the 10% of covariance (CoV) value was considered. The random variables were assumed to have a normal distribution.

In this study, instead of fully coupled hydro-elastic analysis for the collision test, a simple method included added mass of the ice structure for the simulation was considered as Kolari et al. (2009) suggested. That is, added mass of the ice bergy bit was computed (Bishop & Price, 1979), and added on the density of the ice.

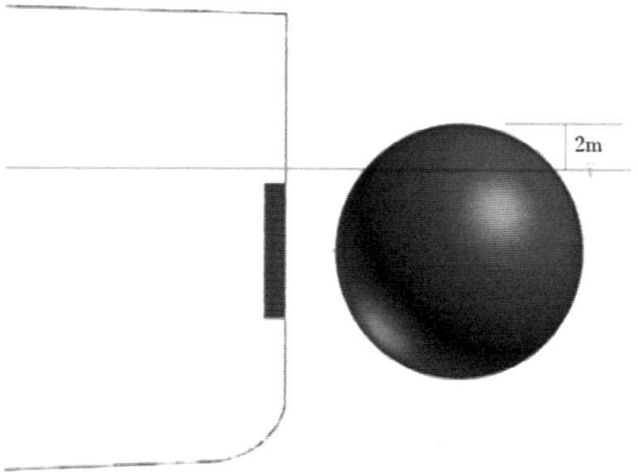

Figure 3. Scenario of ice bergy bit colliding with pannel structure

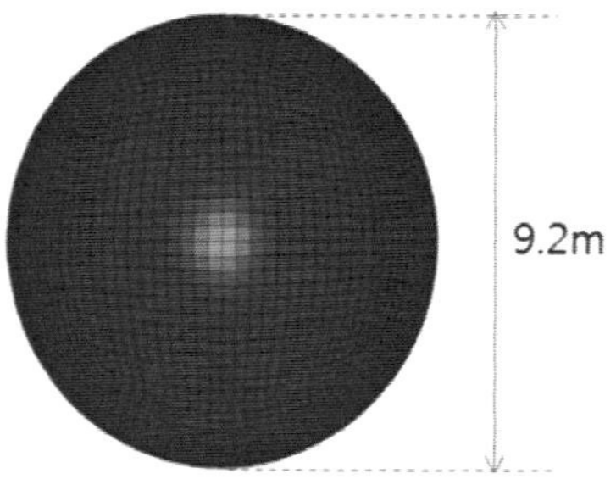

- Radius: 4.6m
- Volume : 584 m^3
- Mass : 526 ton

Figure 4. FE modelling of spherical ice bergy bit

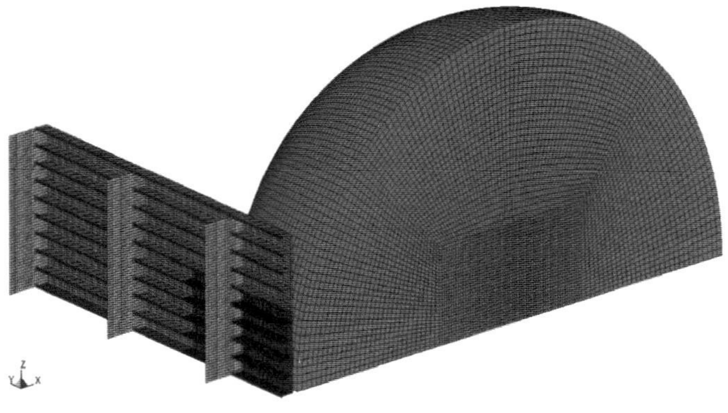

Figure 5. Modelling of panel structure and ice bergy bit

Table 4. Random variables for ice bergy bit

Structure member	Mean value	CoV [%]	Probability distributions	No. of Random Value (RV#)
Diameter [mm]	9200	10	Normal distribution	RV[11]
Velocity [m/s]	2.0	10	Normal distribution	RV[12]

Definition of the Response Value for the RSM Model

The collision results were obtained when all of the random variables are mean values, and ice collision force and effective stress (von-Mises stress) depending on the time were observed as shown in Figure 6 and Figure 7. It is assumed that the failure of the structure occurs when the values of the maximum ice collision force and the maximum effective stress are reaching to the certain figures. Ten of the highest values were selected (round circled values in the Figure 6 and Figure 7), and the averaged values were used for the structural reliability analysis. That is because the values of the maximum ice collision force and the maximum effective stress are changing depends on time. The averaged values were considered as response values for building limit state equation. The limit state equations were obtained by using response surface method (RSM), and probability of failure was computed by using the limit state equations for the two cases, that is, the case for the ice collision force and effective stress (von-Mises stress).

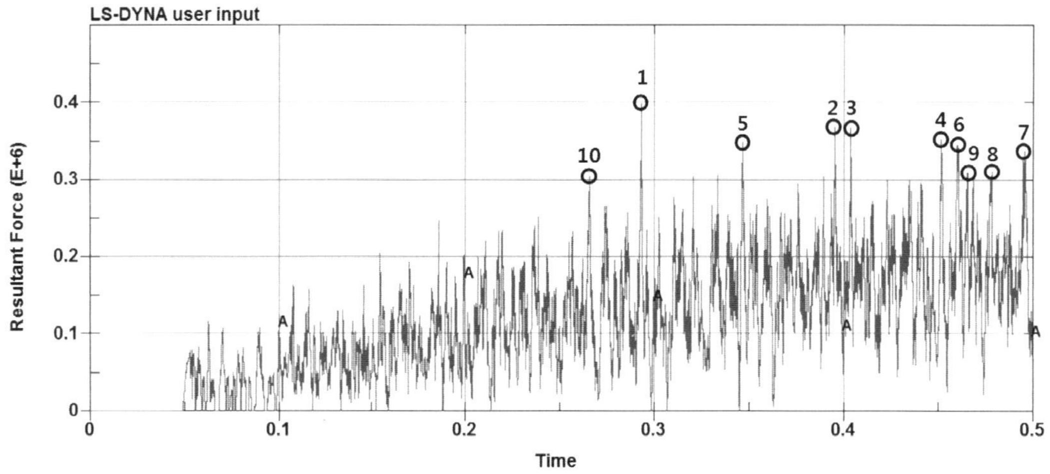

Figure 6. Time history of the collision force

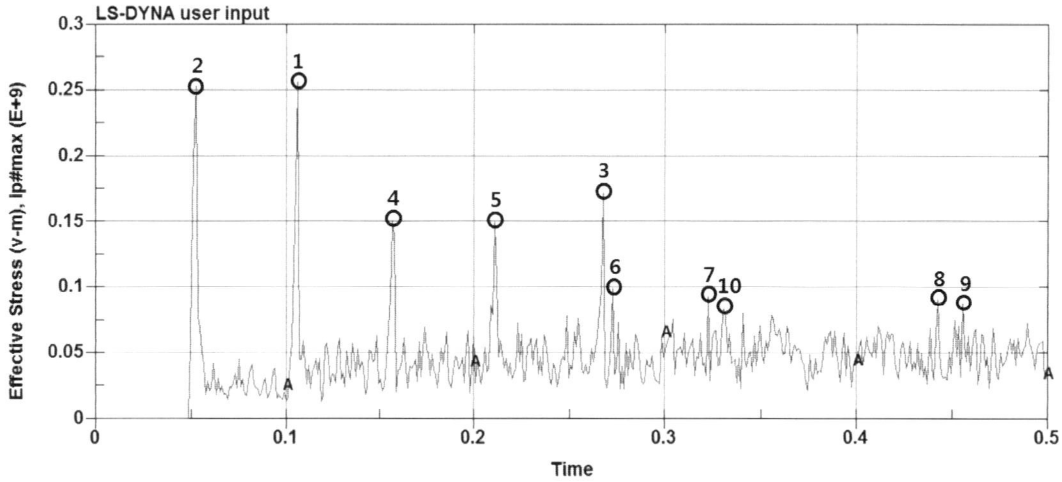

Figure 7. Time history of the effective stress

COMPUTATION FOR THE PROBABILITY FAILURE OF THE STRUCTURE

Limit State Equation and Probability Failure for the Ice Collision Force

Limit state equation was obtained based on the defined 12 random variables in the tables (Table 1 ~4). The response value for the LSE is the averaged values of the ten points among the highest values of the ice collision force. The total number of sampling points for building up LES is 13 and the probability of failure of the structure was computed based on the obtained LSE by using MCS. The sampling points were selected based on orthogonal array method (Lee, 2013). The obtained LSE based on the computed response value at the sampling points is as follows.

$$\begin{aligned} \text{Collision Force}_{ave} &= (-370640) + 15.01 \cdot RV1 + 44.31 \cdot RV2 + 3049225.35 \cdot RV3 \\ &+ (-0.411) \cdot RV4 + (-8.643) \cdot RV5 + 2.278 \cdot RV6 + 2264.95 \cdot RV7 \\ &+ (-10935.05) \cdot RV8 + 1690.46 \cdot RV9 + (-6058.65) \cdot RV10 + 54.91 \\ &\cdot RV11 + 191824.75 \cdot RV12 \end{aligned} \tag{1}$$

where $\text{Collision Force}_{ave}$ represents the averaged values of the ten points among the highest values of the ice collision force, and RV1, .. , RV12 are random variables. The random variables were the same as defined in the tables (Table 1 ~4).

The probability of failure of the structure was computed based on the above LSE by using MCS. The number of sampling points for computing failure probability is 100,000, and the sampling points were generated by using random numbers generator.

It is assumed that the structure is collapsed or failed when the $\text{Collision Force}_{ave}$ reaches to the maximum limit collision force 485kN. The computed probability of failure of the structure is about 26.3% as shown in Table 5.

The optimization process was also carried out by setting the target collision force as the objective function, and the allowable collision force as the constraint function. The optimum values that did not exceed the allowable collision force (of 485kN). Sensitivity values for the analysis were also obtained during the analysis process and the results were shown in Table 5.

Table 5. Optimum points and sensitivity values for the limit collision force and probability failure

RVs	*RV1*	*RV2*	*RV3*	*RV4*	*RV5*	*RV6*	*RV7*	*RV8*	*RV9*	*RV10*	*RV11*	*RV12*
Mean value	7800	1380	7.1E-3	210000	7850	285	10	10	27.3	14.5	9200	2.0
Opt. value	8580	1518	7.8E-3	189286	7065	313.5	11	9	30.03	13.05	10120	2.2
Sens. value	0.03	0.008	0.002	0.018	0.01	0.001	3e-5	0.024	0.006	0.017	0.554	0.314
$\mathbf{P_f}$	0.26308											

As obtained sensitivity values shown in Table 5, the most affected components in terms of sensitivity in the LSE were RV[11], RV[12] and RV[1], but others were insignificant for the response value. Therefore, the top three highest sensitive random variable (RV[11], RV[12], RV[1]) were selected for re-computing the failure probability of the structure for reducing the probability of failure.

The covariance (CoV, or standard deviation) of the selected random variables were modified for reducing the probability of failure less than 10%. As shown in Table 6, the CoV of the selected three random variables needed to be less than 6% to fulfil the requirements of the probability of failure, 10%.

Table 6. Probability failure (P_f) observation, with changes of CoV for the collision force

RVs	Mean value	CoV [%] / Standard Deviation					
		10	9	8	7	6	**5**
RV1	7800	780	702	624	546	468	**390**
RV11	9200	920	828	736	644	552	**460**
RV12	2.0	0.2	0.18	0.16	0.14	0.12	**0.10**
$\mathbf{P_f}$		0.26308	0.22121	0.18427	0.14332	0.10397	**0.06479**

Limit State Equation and Probability Failure for the Effective Stress on the Structure

Limit state equation for the effective stress on the structure was obtained by the same way in the previous section. The response value for the LSE is the averaged values of the ten points of the highest values of the effective stress (von-Mises stress) on the structure.
Total number of the sampling points for building up LES is 13 and the failure probability of the structure was computed based on the obtained LSE by using MCS. The obtained LSE based on the computed response value (for the effective stress on the structure) at the sampling points is as follows.

$$\text{Effective Stress}_{ave} = 147.36 + (-0.01086) \cdot \text{RV1} + 0.02269 \cdot \text{RV2} + 573.24 \cdot \text{RV3} + (-0.00023) \cdot \text{RV4} + 0.0013 \cdot \text{RV5} + (-0.177) \cdot \text{RV6} + (-7.678) \cdot \text{RV7} + (-3.713) \cdot \text{RV8} + 0.412 \cdot \text{RV9} + (-17.43) \cdot \text{RV10} + 0.0466 \cdot \text{RV11} + 39.385 \cdot \text{RV12} \quad (2)$$

where $\text{Effective Stress}_{ave}$ presents the averaged values of the ten points among the highest values of the effective stress (von-Mises stress) on the structure. and RV1, .. , RV12 are random variables. The random variables were defined in the tables (Table 1 ~ 4).

The probability of failure of the structure was computed based on the above LSE by using MCS. The number of sampling points for computing failure probability is 100,000, and it is assumed that the structure is collapsed or failed when the $\text{Effective Stress}_{ave}$ reaches to the maximum limit effective stress185MPa. The computed failure probability of the structure is about 20.26% as shown in Table 6. The optimum values and sensitivity values for the maximum limit effective stress were also obtained in the analysis process, and the results were shown in Table 6.

Table 7. Optimum points and sensitivity values for the limit collision force and probability failure

RVs	*RV1*	*RV2*	*RV3*	*RV4*	*RV5*	*RV6*	*RV7*	*RV8*	*RV9*	*RV10*	*RV11*	*RV12*
Mean value	**7800**	1380	7.1E-3	210000	7850	285	10	10	27.3	**14.5**	**9200**	2.0
Opt. value	**7262**	1312	0.00641	189000	8635	262	9.79	9.00	30.02	**13.94**	**8377**	2.19
Sens. value	**0.027**	0.002	3e-4	0.007	0.001	0.007	0.018	0.006	0.001	**0.225**	**0.68**	0.026
P_f	0.2026											

The most affected components in terms of sensitivity in the LSE of $\text{Effective Stress}_{ave}$ were RV[11], RV[10] and RV[1], but other are trivial. The three highest sensitive random variable

(RV[11], RV[10], RV[1]) were selected for re-computing the failure probability of the structure for reducing the probability of failure.

The covariance (CoV, or standard deviation) of the selected random variables were modified for reducing the probability of failure less than 10%. As shown in Table 8, the CoV of the selected three random variables needed to be less than 7% to fulfil the requirements of the probability of failure, 10%.

Table 8. Probability failure (P_f) observation, with changes of CoV for the collision force

RVs	Mean value	CoV [%] / Standard Deviation					
		10	9	8	**7**	6	5
RV1	7800	780	702	624	**546**	468	390
RV10	9200	920	828	736	**644**	552	460
RV11	2.0	0.2	0.18	0.16	**0.14**	0.12	0.10
P_f		0.2026	0.1910	0.1576	**0.1094**	0.0622	0.280

CONCLUSION

Structural safety of the offshore structure was assessed based on the typical collision scenario with an ice bergy bit. The probability of failure was computed by considering uncertainties of material properties and dimensions of the offshore structure and the ice.

The possible 12 random variables were chosen from the design variables of the offshore structure and ice. The structural reliability analysis, then, was carried out based on the random variables, and the probability of failure was computed.

The highest three sensitive random variables were selected for reducing the probability of failure, and the probability of failure was observed modifying standard deviation (or covariance) of the random variables until the failure probability reaches below 10%.

The limit state equations (LES) for the reliability analysis were obtained with the computed response values which were the averaged value of ten points among the highest values of the ice collision force and effective stress on the structure. The two LSEs were obtained and utilized for computing the probability of failure by using the MCS method.

The obtained statistical properties of the random variables of the structure and ice can be used in the design or manufacturing process as a guideline in near future.

AKNOWLEDGEMENTS

This work was carried out within the scope of the principal R&D program ("Enhancement of the resistance performance and establishment of the station-keeping test assessment method for ships in ice-covered water, "Project No. PES9042), which is supported by Korea Research Institute of Ships & Ocean Engineering (KRISO). All support is gratefully acknowledged.

REFERENCES

Bishop, R.E. and Price, W.G., 1979. Hydroelasticity of ships. Cambridge University Press.

Carney, K.S., Benson, D.J., DuBois, P. and Lee, R., 2006. A phenomenological high strain rate model with failure for ice. International Journal of Solids and Structures, 43(25), pp.7820-7839.

Derradji-Aouat, A., Sinha, N.K. and Evgin, E., 2000. Mathematical modelling of monotonic and cyclic behaviour of fresh water columnar grained S-2 ice. Cold regions science and technology, 31(1), pp.59-81.

DNV, 2006. Iceberg Collision Scenario.Det Norske Veritas Report.

Halquist, J., 2007. LS-DYNA keyword user's manual version 971. Livermore Software Technology Corporation, Livermore, CA.

Kolari, K., Kuutti, J. and Kurkela, J., 2009. Fe-simulation of continuous ice failure based on model update technique. In Proceedings of the International Conference on Port and Ocean Engineering Under Arctic Conditions (No. POAC09-104).

Lee, S.B., 2003. Easy Daguchi Method. Sanjosa, Seoul, Korea

Nho, I.S., Lee, J.M., Oh, Y.T. and Kim, 2016. Analysis Method of Ice Load and Ship Structural Response due to Collision of Ice Bergy Bit and Level Ice. Journal of Ocean Engineering and Technology, 53(2).

Ice Loads Dynamics for Model Scale Cylinders of Various Diameters

P. Zvyagin[1,2,3], A. Dobrodeev[1], K. Sazonov[1]
[1] Krylov State Research Centre, St. Petersburg, RUSSIA
[2] Peter the Great St. Petersburg Polytechnic University, St. Petersburg, RUSSIA

ABSTRACT

The question, if the distribution law of model scale ice load depends on the diameter of a cylindrical indenter and its indentation speed, is still open. In new ice tank facility of Krylov Research Centre (St. Petersburg) a series of experiments with cylindrical indenters of diameters 30, 40 and 96 mm were conducted. In these experiments global ice loads were recorded with high sampling rate. Two ice fields were involved into experiments – with thickness of 55 and 72 mm. The indentation experiments considered in the paper were conducted at speeds of 0.025, 0.1, 0.15, 0.3 m/s.

Tests on distribution law with implementation of D'Agostino's method were performed. Results of statistical processing and ice load process' distribution law evaluation of obtained time series are presented in the paper. The distribution law of global load process appeared to be speed dependent. Probably it also depends on indenter's diameter. Model ice parameters for conducted experiments are provided in the paper. Comparison with results, which were obtained earlier, is performed.

KEY WORDS: Ice; Model; Load; Stochastic; Lognormal.

NOMENCLATURE

In the text all random variables are denoted with bold italic letters, constant values and non-random variables – with regular italic.

$\boldsymbol{X}(t)$ – Stochastic process of ice load.

X – Time series (single record) of process $\boldsymbol{X}(t)$

n – Number of data points in the time series X.

q – Sampling rate of the time series X.

$\boldsymbol{\xi}$ – Statistical population.

μ_i – i-th central moment of the statistical population $\boldsymbol{\xi}$.

γ_1 – Pearson's asymmetry coefficient of the statistical population $\boldsymbol{\xi}$.

$\{\boldsymbol{x}\}$, $\{\boldsymbol{z}\}$ – Random samples taken from statistical populations; random data points from these samples are denoted with $\boldsymbol{x}_i$ and $\boldsymbol{z}_i$. Particular samples with particular data points are denoted with $\{x\}$, $\{z\}$; particular data points are denoted with x_i and z_i.

N, k – Sizes of the samples $\{\boldsymbol{x}\}$ and $\{\boldsymbol{z}\}$ respectively.

l – distance, which separates two almost uncorrelated data points in time series X.

$\overline{\boldsymbol{x}}$ – Sample average. Value of $\overline{\boldsymbol{x}}$ found for the particular sample $\{x\}$ is denoted with $\bar{x}$.

$\boldsymbol{\sigma_z}$ – The point estimator of standard deviation of $\boldsymbol{Z}$. Its particular value is denoted σ_z.

$\boldsymbol{m_i}$ – i-th central moment of sample.

$\boldsymbol{g_1}$ – Pearson's asymmetry coefficient of sample. Value of it found for the particular sample $\{x\}$ is denoted with g_1.

$\boldsymbol{Z}$ – Statistics of D'Agostino's test.

$\boldsymbol{Y, W, \alpha, \delta}$ – Components of expression for $\boldsymbol{Z}$.

H_0 – Hypothesis "Population $\boldsymbol{\xi}$ is normally distributed".

H_0' – Hypothesis "Mean value of $\boldsymbol{Z}$ is 0".

H_0'' – Hypothesis "Standard deviation of $\boldsymbol{Z}$ is 1".

$\chi_p^2(k-1)$ – p-th quantile of χ^2 distribution with $k-1$ dimensions of freedom.

$\boldsymbol{r_{dat}}(\tau)$ – Estimator of the autocorrelation function of the stationary process $X(t)$. Value of it found for the particular sample $\{x\}$ is denoted with $r_{dat}(\tau)$.

τ – Time lag in seconds.

INTRODUCTION

This paper continues the study started earlier by Zvyagin & Sazonov (2015), Zvyagin (2016) and Dobrodeev et al. (2016). Aim of this study is to make reliable stochastic model of ice load dynamics in the case of offshore structures with column supports. For that reason tests with cylindrical indenters of different diameters were performed in the ice tank of Krylov State Research Centre. Global loads caused by the model ice on these cylindrical indenters were recorded and studied.

The stochastic process model was chosen for ice load because of crushing process complexity. Such mathematical model is applied since the paper of Sundararajan and Reddy (1973). According to Bjerkas (2004) stationarity of ice loads is associated with high indentation speed and non-simultaneous brittle ice failure mode. Later Karna et al. (2007) emphasized that the stochastic process approach is applicable to describe loads when sheet of plain ice is failing by continuous crushing, and the structure has vertical walls.

Stationarity and normality or lognormality of ice load stochastic process provides multiple benefits for analysis and simulation. Sodhi (1998) have presented records of loads caused by freshwater ice on vertical T-shaped, 25-mm thick steel plates with 49.5-mm wide contact area at high speed (409.3 mm/s and 401.3 mm/s). These records visually are looking like as time series of stationary processes, though investigation on that was not presented. Example of normal stationary ice load process was presented by Guo (2012). For simulation of a stationary lognormal process the simulation of logarithm of it, which is a normal process, can be implemented. Autocorrelation function of lognormal process and autocorrelation function of its logarithm are related as described by Zvyagin and Sazonov (2014).

Main aims of the previous studies performed by the authors were: verification of ice load process stationarity in wide sense, evaluation of parameters of process and analysis of its autocorrelation function (ACF). Study of loads on indenters of 20 mm diameter at 0.01 m/s indentation speed against 15 and 19 mm thick ice fields have revealed lognormal distribution and features of stationarity (Dobrodeev et. al, 2016). At the same time record of load on 100 mm width cylinder against 30 mm thick ice made with speed of 0.084 m/s has performed normal behavior (Zvyagin and Sazonov, 2015). Appearance of such differences was not explained in previous studies and needs more detailed investigation.

In recent study new records of global ice loads were investigated on their normality/lognormality. Connection of indentation velocity and indenter diameter with load process distribution law is proposed.

The structure of the paper is as following. First Section is introductive. In the second Section parameters of experiments in ice tank are listed. In the third Section details of application of D'Agostino's method to load process normality test are described. In the next three Sections analysis of ice load records measured in experiments in the ice tank is made. After that discussion and conclusions are made.

DESCRIPTION OF EXPERIMENTS IN ICE TANK

Experiments with cylindrical indenter were performed in the new ice tank of Krylov State Research Centre. Detailed description of this facility was earlier made by Timofeev et. al (2015). Three different indenters of 30 mm, 40 mm and 96 mm diameter were used for tests in two ice sheets with thickness 55 mm and 72 mm. Speeds of indentation were 0.025, 0.1, 0.15 and 0.3 m/s.

Tests were performed using FG (Fine Grain) grained model ice. Parameters of ice sheets used in experiments are provided in the Table 1. Ice load on indenter was measured by the dynamometer. Experiments were performed using the principle of inversed motion. The method of indentation experiments is fully described in Dobrodeev et al. (2016). The picture of facility and the example of time history of load measured in test with cylindrical indenter of 30 mm diameter against level ice with thickness of 72 mm are presented in Figure 1.

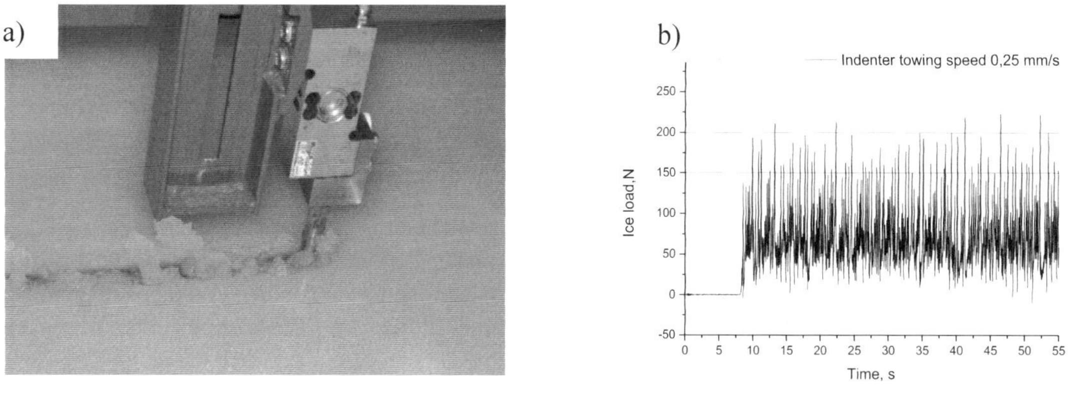

Figure 1. a) Indentation test equipment; b) Time history of ice load measured in indentation test against 72 mm thick ice, diameter of indenter is 30 mm.

Table 1. Parameters of ice fields used in indentation tests (average values)

Number	Thickness, mm	Flexural strength, kPa	Compressive strength, kPa	Elastic modulus, MPa
1	55	22.45	29.1	31.75
2	72	25	47.1	39.25

AN APPLICATION OF D'AGOSTINO METHOD TO THE ICE LOAD RECORD NORMALITY TEST

In this Section we will follow statistical notations accepted in Kendall and Stuart (1977) and D'Agostino and Ralph (1970). As it was said in the Section "Nomenclature", all random variables will be denoted with bold italic letters, and values of those random variables along with constants and non-random variables – with regular italic letters.

Let us consider the model of stochastic process $\boldsymbol{X}(t)$ for the ice load. The particular record of ice load made in the indentation experiment will be denoted with X and considered as time series of the process $\boldsymbol{X}(t)$. To conduct statistical tests we need to get somehow the sample $\{x\}$ of data points from X. Most of statistical tests are designed for samples, elements in which are statistically independent.

To decrease statistical dependency of data in the sample $\{x\}$ we shall take distant points from X for it. The particular method of data points taking will be described later. Then, if probabilistic properties of $\boldsymbol{X}(t)$ are preserved in time, we can consider data points in $\{x\}$ as taken from single statistical population $\boldsymbol{\xi}$.

Let us consider statistical population $\boldsymbol{\xi}$. Let us denote the Pearson's asymmetry coefficient of it with γ_1:

$$\gamma_1 = \mu_3/\mu_2^{3/2}. \qquad (1)$$

Here μ_3 and μ_2 are third and second central moments of population $\boldsymbol{\xi}$. Then corresponding estimator of γ_1, calculated with presence of sample $\{\boldsymbol{x}\}$ of size N taken from that population will be $\boldsymbol{g}_1$:

$$\boldsymbol{g}_1 = \boldsymbol{m}_3/\boldsymbol{m}_2^{3/2} \qquad (2)$$

Value of $\boldsymbol{g}_1$ obtained from the sample $\{x\}$ we shall denote with g_1:

$$g_1 = \frac{\sum_{i=1}^{N}(x_i-\bar{x})^3}{\left(\sum_{i=1}^{N}(x_i-\bar{x})^2\right)^{3/2}}\sqrt{N} \qquad (3)$$

where $\bar{x}$ is the average value of $\{x\}$ and x_i are data points of this sample.

Null hypothesis to test will be the following: $H_0 =$ "Population $\boldsymbol{\xi}$ is normally distributed". To approve or reject H_0 the skewness of sample can be investigated with respect of zero values of μ_3 and γ_1 (1) under the valid hypothesis H_0. In fact, even if H_0 is valid the particular value of random variable (2) calculated with using of particular $\{x\}$ by chance ca n fall rather far from the expected zero value. How large the discrepancy of (3) wi th 0 should be to say, that H_0 is invalid? To answer this question the exact distribution of (2) under the valid hypothesis H_0 should be known.

The problem is that the distribution of $\boldsymbol{g}_1$ with validity of H_0 is depending on the sample size N. With respect of this, D'Agostino (1970) have suggested next transformation of $\boldsymbol{g}_1$, that allows to consider statistics $\boldsymbol{Z}$, which distribution under H_0 is not depending on N:

$$\boldsymbol{Y} = \boldsymbol{g}_1\sqrt{\frac{(N+1)(N+3)}{6(N-2)}}$$

$$W^2 = -1 + \sqrt{2\left(\frac{3(N^2+27N-70)(N+1)(N+3)}{(N-2)(N+5)(N+7)(N+9)} - 1\right)}$$

$$\delta = 1/\sqrt{\ln W} \quad (4)$$

$$\alpha = \sqrt{2/(W^2-1)}$$

$$\boldsymbol{Z} = \delta ln\left[\boldsymbol{Y}/\alpha + \sqrt{(\boldsymbol{Y}/\alpha)^2+1}\right].$$

Under the valid hypothesis H_0 the statistics $\boldsymbol{Z}$ is distributed approximately normally with parameters $(0,1)$ (D'Agostino & Ralph, 1970). Transformation provided above is applicable when $N \geq 8$. Critical area for $\boldsymbol{Z}$, which leads to rejection of H_0, can be found by using of any table of standard normal distribution percentiles.

In our study we shall not rely on single value of $\boldsymbol{Z}$ obtained from single $\{x\}$: simulations have shown that the probability of type II error in the test of H_0 can be rather significant.

Taking in account assumption of $\boldsymbol{X}(t)$ stationarity, we shall take k samples $\{x\}$ and calculate k values of $\boldsymbol{Z}$ according to (4), which will make sample $\{z\}$. For that sample we shall perform Chi-squared goodness-of-fit test on normality and also test two hypotheses: $H_0^{'} =$ "Mean value of $\boldsymbol{Z}$ is 0" and $H_0^{''} =$ "Standard deviation of $\boldsymbol{Z}$ is 1" using corresponding well known confident intervals for normally distributed random variable. If both hypotheses $H_0^{'}$ and $H_0^{''}$ are not rejected, then there are not strong objections against symmetry of distribution of $\boldsymbol{\xi}$. As soon as an alternative to H_0 from our point of view is mostly connected with asymmetric distributions, e.g. lognormal, it means that there are no strong objections against H_0. This way we shall accept hypothesis H_0.

Let us take confidence level 0.95, then significance level will be 0.05. Let the critical area will be two-sided. The critical area for $\bar{\boldsymbol{z}} = \sum_{i=1}^{k} \boldsymbol{z}_i/n$, falling in which leads to rejection of $H_0^{'}$, is the next: $\bar{z} > 1.96/\sqrt{k}$ or $\bar{z} < -1.96/\sqrt{k}$.

Let us denote with $\varepsilon_1 = \chi^2_{0.025}(k-1)$ and $\varepsilon_2 = \chi^2_{0.975}(k-1)$ values of 2.5 and 97.5 percentiles of $\chi^2(k-1)$ distribution. Then the critical area for $\boldsymbol{\sigma}_z = \sqrt{\sum_{i=1}^{k}(\boldsymbol{z}_i - \bar{\boldsymbol{z}})^2/k}$ with significant level 0.05 is $\sigma_z < \sqrt{\varepsilon_1/k}$ or $\sigma_z > \sqrt{\varepsilon_2/k}$. Falling in that critical area leads to rejection of $H_0^{''}$.

In the case of testing of $\boldsymbol{X}(t)$ on lognormality, logarithmic time series should be considered instead of original time series.

LOGNORMAL DISTRIBUTION FOR TESTS AT SPEED 0.025 m/s

At first we shall consider records of global load measured in experiments with cylinders with 30 mm and 40 mm diameter at speed 0.025 m/s against 55 and 72 mm thick model ice. These records were 40 seconds long, which provides $n = 4000$ data points at 100 Hz sampling rate.

Let us study how strong the correlation of neighboring data points in load time series is. For that purpose the autocorrelation values $r_{dat}(\tau)$ were calculated according to the formula:

$$r_{dat}(\tau) = \frac{\sum_{i=1}^{n-q\tau}(x_i-\bar{x})(x_{i+q\tau}-\bar{x})/(n-q\tau)}{\sum_{i=1}^{n}(x_i-\bar{x})^2/n}, \quad (5)$$

here x_i are sequential data points from ice load time series X of size n, $\bar{x} = \sum_{i=1}^{n} x_i/n$ is the average of all data from X, q $[1/s]$ is the sampling rate and τ $[s]$ is the time lag in seconds. This way, $r_{dat}(\tau)$ for every τ is the value which random autocorrelation function estimatior $\boldsymbol{r}_{dat}(\tau)$ took on for particular τ. Plots of these values for considered load time series are presented in Figure 2. From Figure 2 it follows that two data points of X separated by the time lag $\tau \approx 0.1\ s$ might be almost uncorrelated at least for the observed time series.

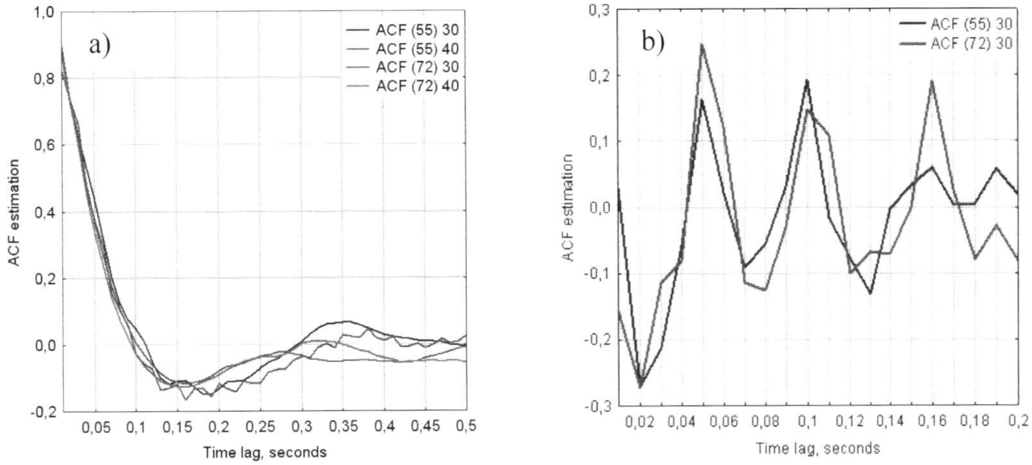

Figure 2. Autocorrelation estimation: a) of global ice loads on 30 and 40 mm wide cylinders at speed 0.025 m/s against 55 mm and 72 mm thick model ice; b) of global ice loads on 30 mm wide cylinder at speed 0.3 m/s against 55 mm and 72 mm thick model ice.

Let us assume that considered ice load records are the time series of lognormally distributed stationary process $\boldsymbol{X}(t)$. The validity of this assumption will be confirmed later. Let us denote the logarithm of time series X with lnX: this way data points of lnX are logarithms of corresponding data points of X.

To study parameters of lnX let us compose a sample $\{x\}$ in the following way: we shall take N data points from lnX separated by $l = 10$ data points from each other. With respect of $q = 100\,Hz$ the distance of 10 data points corresponds 0.1 s, and from Figure 2a it follows that therefore the data in $\{x\}$ might be almost uncorrelated. Taking in account our assumption on lognormal distribution law the selected data points can be assumed as independent.

Having $\{x\}$ let us implement D'Agostino's method described above to calculate one value of $\boldsymbol{Z}$ according to (4). Let us then put this value to the sample $\{z\}$. Let us repeat the procedure of making $\{x\}$ and obtaining values of $\boldsymbol{Z}$ by taking further N data points from lnX separated by l from each other. This way we shall obtain sample $\{z\}$ of size $\left\lfloor\frac{n}{N\cdot l}\right\rfloor$, which is floor of ratio $n/(Nl)$ and equal to 20 in our particular case. This $\{z\}$ is finally should be tested on the centered standard normal distribution law to check validity of our assumption on lognormal distribution of $\boldsymbol{X}(t)$. To perform such a test simple goodness-of-fit Chi squared test was used along with testing hypotheses $H_0^{'}$ and $H_0^{''}$ described in the previous section.

To use D'Agostino's transform the N should be greater than 8. Generally the larger N, the better. But with increasing N and with constant l and constant time series length n, the size of $\{z\}$ can become too small to make confident outcomes. The influence of taken value of N on the properties of sample $\{z\}$ under the valid H_0 should be studied in the future in more details.

For the study described in this section $N = 20$ was taken. Thus with time series of size $n = 4000$ and $l = 10$ size of $\{z\}$ was k =20. For normality goodness-of-fit Chi squared test of $\{z\}$ number of intervals was taken as 4. Here parameters of population are defined as $(0; 1)$, so the number of dimensions of freedom of Chi squared test statistic under the null hypothesis is 3. Resulting parameters of $\{z\}$ obtained for considered ice load time series are presented in Table 2 along with values of Chi squared test statistics $\chi^2_{experim}$.

Table 2. Results of test of global ice load time series on the distribution law.

Ice thickness	55 mm			72 mm				
Trolley speed, m/s	0.025		0.3	0.025		0.1	0.15	0.3
Indenter's diameter	30 mm	40 mm	30 mm	30 mm	40 mm	96 mm	96 mm	30 mm
Type of test	LNormal	LNormal	Normal	LNormal	LNormal	Normal	Normal	Normal
$\bar{z}$	0.18	0.05	-0.01	0.93	0.06	0.17	-0.28	0.22
$\sigma_{\{z\}}$	0.72	1.01	1.01	0.52	1.24	0.99	0.91	0.91
min $\{z\}$	-1.23	-2.15	-1.54	0.16	-2.63	-1.36	-2.39	-0.73
max $\{z\}$	1.61	1.78	1.41	1.86	1.92	2.42	0.82	1.82
$\chi^2_{experim}$	2.1	0.94	0.85	20.39	0.99	0.98	2.57	1.83
N	20	20	10	20	20	10	10	10
k	20	20	7	20	20	13	13	7
Result	Yes	Yes	Yes	No	Yes	Yes	Yes	Yes

From the Table 2 one can see that records of load made for cylinders with diameters of 30 and 40 mm against ice with thickness of 55 mm with indentation speed 0.025 m/s can be considered as of lognormal origin. For the ice thickness of 72 mm, indenter diameter of 30 mm and the same speed, the result of the test on lognormal distribution was negative. For ice thickness of 72 mm, indenter diameter of 40 mm and speed 0.025 m/s the result was positive.

All of 20 samples $\{x\}$ for each of time series X were united to make the sample $\{\{x\}\}$. This way data in $\{\{x\}\}$ can also thought as almost uncorrelated. Histograms of the $\{\{x\}\}$ along with fitted lognormal curves are provided in the Figure 3 a-d. Parameters of process $\boldsymbol{X}(t)$

estimated from samples $\{\{x\}\}$ in those cases when lognormal distribution was confirmed, are provided in corresponding columns of the Table 3.

Table 3. Estimations of ice load process' moments obtained from samples $\{\{x\}\}$.

Ice thickness	55 mm			72 mm			
Trolley speed, m/s	0.025		0.3	0.025	0.1	0.15	0.3
Indenter's diameter	30 mm	40 mm	30 mm	40 mm	96 mm	96 mm	30 mm
Mean	33.21	42.33	58.8	75.7	267	303.4	145.9
Std. deviation (unbiased)	12.04	14.59	24.6	33.1	91.3	113.2	83.2
Number of data points in $\{y\}$	400	400	70	400	130	130	70

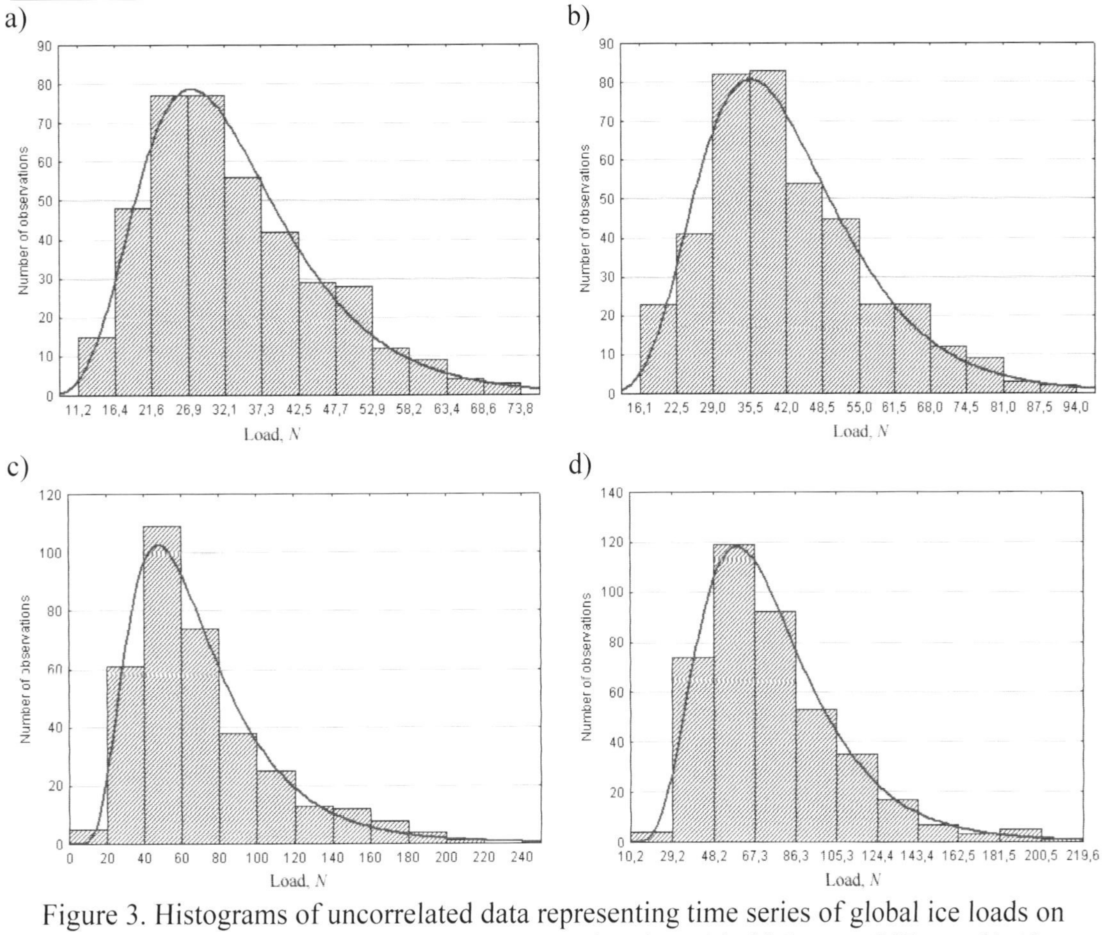

Figure 3. Histograms of uncorrelated data representing time series of global ice loads on cylindrical indenters with diameter: a) 30 mm against ice with thickness of 55 mm; b) 40 mm against ice with thickness of 55 mm; c) 30 mm against ice with thickness of 72 mm; d) 40 mm against ice with thickness of 72 mm. Indentation speed is 0.025 m/s.

Concluding this section we can say, that lognormal distribution of the global load signal has appeared not only in experiments with indenters of 20 mm in diameter, as was found earlier (Dobrodeev et. al., 2016) but also in experiments with indenters of 30 and 40 mm in diameter at moderate speed.

NORMAL DISTRIBUTION OF LOAD AT HIGH INDENTATION SPEED

At high indentation speed the load signal behaves in the other way, than at moderate speed. Case of moderate speed was considered in the previous Section, for 30-40 mm wide indenters lognormal distribution was confirmed.

In opposite, at indentation speed of 0.3 m/s for 30 mm wide indenter normal distribution was discovered for both of ice field thicknesses: 55 and 72 mm. With that speed only 3 seconds of load record were available, thus $n = 300$ data points with $= 100\ 1/s$, but it was quite enough. The plot of autocorrelation function estimation (5) is presented in Figure 2b. From Figure 2b it is clear, that we can take every 4th data point from load time series X to compose the sample $\{x\}$ with low correlation of data points in it. Because of small n the following meaning was taken for N: $N = 10$.

This way with $l = 4$ and $N = 10$, we have for $\{z\}$ the size $k = 7$, which is rather small for conducting Chi squared test. Nevertheless, values of $\chi^2_{experim}$ were found for 3 intervals, and they are provided in the Table 2 along with other parameters for $\{z\}$. Here parameters of population ξ are defined as $(0; 1)$, so the number of dimensions of freedom of Chi squared test statistic under the null hypothesis is 2. From parameters provided in Table 2 it is clear that hypothesis on normal distribution of the process $\boldsymbol{X}(t)$ can be accepted.

All of 7 samples $\{x\}$ for each time series X were united to make the sample $\{\{x\}\}$. The histograms of $\{\{x\}\}$ along with fitted Normal PDF curves are presented in Figure 4a and 4b. Characteristics of $\{\{x\}\}$, which can be considered as estimations of process $\boldsymbol{X}(t)$ parameters are given in the Table 3.

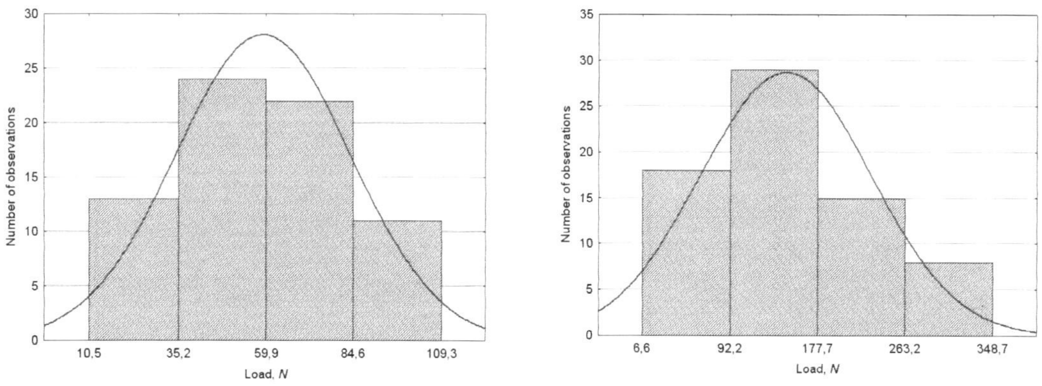

Figure 4. Histograms of uncorrelated data representing time series of global ice loads on cylindrical indenter with diameter of 30 mm at speed 0.3 m/s against ice with thickness a) 55 mm; b) 72 mm.

NORMAL DISTRIBUTION OF LOAD ON INDENTER WITH 96 mm DIAMETER

Another case when normal distribution of the signal was revealed was experiment with relatively wide indenter – cylinder of 96 mm in diameter. Normal distribution appeared in tests with it against 72 mm thick ice at speeds of 0.1 and 0.15 m/s. Records with duration of 16.5 and 9.5 seconds were obtained in these two experiments. Autocorrelation function estimations (4) for these cases are provided in Figure 5. From this Figure it follows that lags l =12 and l =7 respectively seem to be appropriate. For N= 10 this provides $k = 13$ for both of cases.

Parameters of sample $\{z\}$ for both of cases are provided in the Table 2 along with values of $\chi^2_{experim}$ found for 3 intervals. Here the number of dimensions of freedom of Chi squared test statistic under the null hypothesis is 2. From parameters provided in Table 2 it is clear that hypothesis on normal distribution of the process $\boldsymbol{X}(t)$ can be accepted in both of cases.

Histograms of sample $\{\{x\}\}$ are provided in the Figure 6a and 6b. Parameters of $\{\{x\}\}$ are presented in the Table 3.

Results provided in this Section are in agreement with the normal distribution obtained earlier (Zvyagin and Sazonov, 2015) for process of ice load on cylinder with diameter of 100 mm against 30 mm thick ice measured at speed of 0.084 m/s.

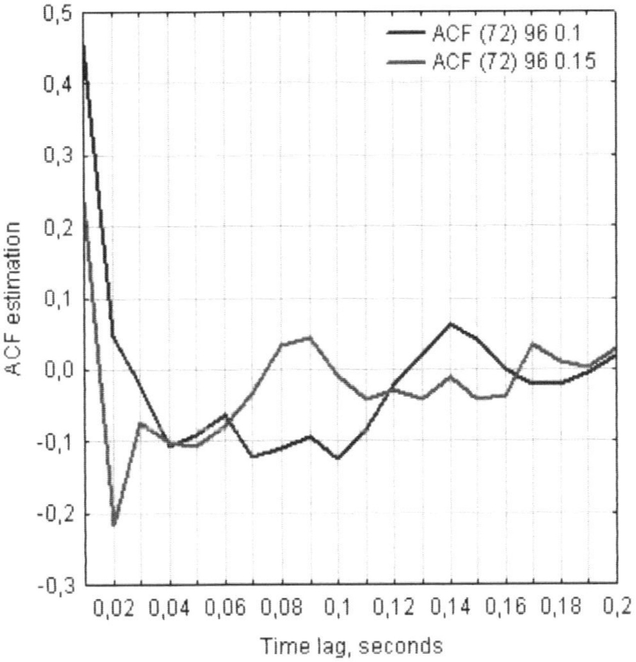

Figure 5. Estimation (5) of autocorrelation of global ice load on cylindrical indenter with diameter of 96 mm against model ice with thickness of 72 mm at speeds 0.1 m/s (blue curve) and 0.15 m/s (red curve).

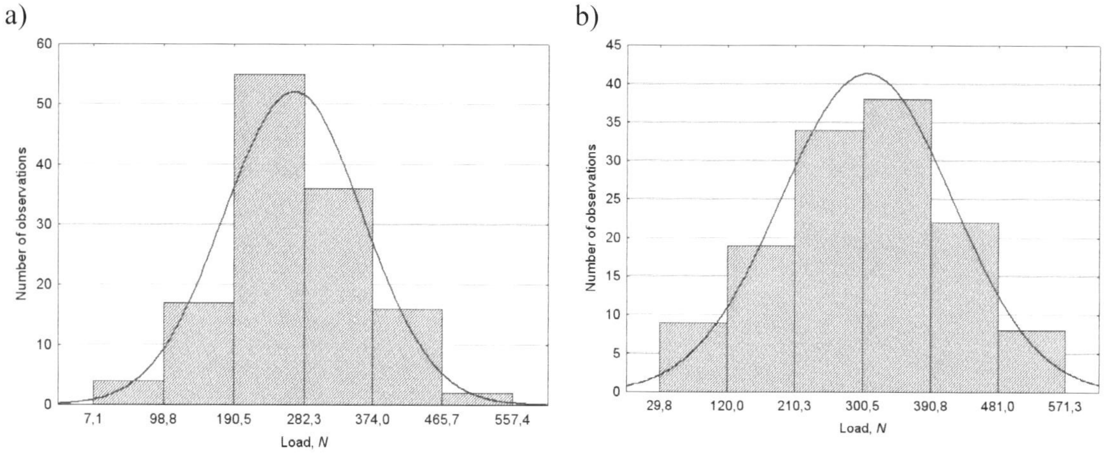

Figure 6. Histograms of uncorrelated data representing time series of ice loads on 96 mm wide indenter against 72 mm model ice at speeds a) 0.1 m/s and b) 0.15 m/s.

DISCUSSION

Dependency of ice load distribution law on indenter's diameter in the case of cylindrical indenters was expected due to previous studies conducted by the authors. But transition of distribution law from lognormal to normal with increasing speed was not revealed before. This phenomenon needs further study. Available length of signals registered at 0.3 m/s was rather short, but due to low autocorrelation of data points in those signals the amount of data was enough to make confident outcomes on the distribution law.

Stationarity, normality (or lognormality) of the distribution law, knowing of autocorrelation function, mean and variance, – these are sufficient conditions for performing successful simulations. As far as authors are aware, there is no software exist by now for simulation of ice loads time series with using of stochastic processes background. For this study authors have used self-written software for data processing and automation of testing procedure. Results in Table 2 were obtained with help of that software.

CONCLUSIONS

Findings of the conducted study are the following:

- The lognormal distribution law was revealed for some of global load signals measured in experiments with cylindrical indenters of 30 and 40 mm diameter at moderate indentation speed of 0.025 m/s and for both studied types of ice thickness – 55 mm and 72 mm. Earlier this distribution law was found for 20 mm wide indenters against ice 15-19 mm at 0.01 m/s.
- Autocorrelation functions of stationary load on cylinders with diameters of 30 and 40 mm at speed 0.025 m/s appeared to be similar for both studied types of ice thickness.
- Distribution law of the load signal is speed dependent. At high speed of 0.3 m/s normal distribution law has appeared.
- Distribution law of the load signal probably also depends on the indenter's diameter. For indenter of 96 mm in diameter the normal distribution law was observed both at indentation speed of 0.1 m/s and of 0.15 m/s. Earlier this distribution law was found for 100 mm wide cylinder against ice 30 mm at indentation speed of 0.084 m/s. However, more detailed investigation is needed on this issue.

The study opens several issues for future investigation. It is important to determine more precisely the indentation speed range and indenter's diameter range at which lognormal or normal distribution is most likely observed. It is important to reveal, how ice thickness influence the distribution law and the stationarity of the load signal. Also further study on the autocorrelation and/or spectral characteristics of the global load signal is desired.

ACKNOWLEDGEMENTS

Authors would like to thank Prof. Oleg Timofeev from St. Petersburg Krylov Research Centre for his support. Authors are grateful to anonymous reviewers of this paper.

REFERENCES

D'Agostino, Ralph B., 1970. Transformation to normality of the null distribution of g1. *Biometrika,* 57 (3), pp.679–681.

Bjerkas, M., 2004. Global Ice Loads' Dependence on Failure Mode. *Int. Journal of Offshore and Polar Engineering*, 14 (3), pp 189-195.

Dobrodeev A, Zvyagin P, Sazonov K., 2016. Study on distribution law and stationarity of global ice loads registered in experiments in ice tank, OMAE2016-54100. In: *Proceedings of 35th OMAE Conference*, Busan, Korea.

Guo, F., 2012, A Spectral Model for Simulating Continuous Crushing Ice Load. In: *Proceedings of 21st IAHR International Symposium on Ice,* Dalian, China, pp.1035-1045.

Karna, T. Qu, Y. Kuhnlein, W. Yue, Q., Bi, X., 2007. A Spectral Model of Ice Forces due to Ice Crushing. *Journal of Offshore Mechanics and Arctic Engineering*, 129, pp.138-145.

Kendall, M.G. and Stuart, A., 1977. *The Advanced Theory of Statistics: Distribution theory.* Vol. 1, Macmillan.

Sodhi, D.S., 1998. Nonsimultaneous Crushing During Edge Indentation of Freshwater Ice Sheets. *Journal Cold Regions Science and Technology*, 27, pp. 179-195.

Sundararajan, C. and Reddy, D.V.,1973. Stochastic analysis of ice-structure interaction. In: *Proceedings of 2nd POAC Conference,* Reykjavik, Iceland, pp.345-353.

Timofeev O.Ya., Sazonov K.E., Dobrodeev A.A., 2015. New ice basin of the Krylov State Research Centre. In: *Proceedings of 23rd POAC Conference,* Trondheim, Norway.

Zvyagin, P., 2016. On autocorrelation of ice loads, measured in ice tank experiments, OMAE2016-54436. In: *Proceedings of 35th OMAE Conference*, Busan, Korea.

Zvyagin, P., 2017. Simulation of normal differentiable stochastic process of ice loads, OMAE2017-62240. In: *Proceedings of 36th OMAE Conference*, Trondheim, Norway.

Zvyagin, P., Sazonov, K., 2014. Analysis and Probabilistic Modeling of the Stationary Ice Loads Stochastic Process with Lognormal Distribution, OMAE2014-24713. In: *Proceedings of 33rd OMAE Conference,* San-Francisco, USA.

Zvyagin, P., Sazonov, K., 2015. Analysis and probabilistic modeling of the unstationary ice loads stochastic process, based on experiments with models of offshore structures, OMAE2015-41619. In: *Proceedings of 34rth OMAE Conference,* St. John's, Canada.

Flexural Strength of Ice Reconstructed from Field Tests with Cantilever Beams and Laboratory Tests with Beams and Disks

Aleksey Marchenko[1], Marina Karulina[2], Evgeny Karulin[2], Peter Chistyakov[3], Alexander Sakharov[3]
[1] The University Centre in Svalbard, Longyearbyen, Norway
[2] Krylov State Research Centre (KSRC), St. Petersburg, Russia
[3] Lomonosov Moscow State University, Moscow, Russia

ABSTRACT

More than 60 in-situ tests have been performed since 2010 to determine flexural strength of sea ice and freshwater ice in North-West Barents Sea, Spitsbergen Fjords, and fresh water lake near Longyearbyen in Spitsbergen. Ice thickness in the tests varied from 20 cm to 80 cm. Dependencies of flexural strength on temperature and salinity, averaged over ice thickness, were found. Results of in-situ, cantilever-beam tests are analyzed and compared with results of laboratory tests of 3-point bending of ice beams and central loading of ice discs.

KEY WORDS: Sea ice, Flexural strength, Cantilever beam, Liquid brine

INTRODUCTION

Flexural strength of ice characterizes capacity of ice to resist pure bending deformations without failure. Flexural strength of materials is measured in standard tests of different types, including most typical 3- and 4-point-bend tests and tests with cantilever beams. In these tests, beam samples have dominant principal stress oriented along the beam axes while absolute values of the other principal stresses are small over the beam volume, excluding the beam edges where the stress concentration occurs. In 3- and 4-point-bend tests, the stress concentration also occurs at the points of loading and beam support. In cantilever-beam tests, the stress concentration occurs near the beam root. Local failure of material in places of stress concentration may not influence global failure of a beam, but its influence on flexural strength and displacements of the beam can be significant. Information about the displacements is used to determine effective elastic characteristics of a beam.

Flexural strength tests with beams are used for concrete (GOST 10180-90), ice (ISO 19906) and other brittle materials. In the case of ceramics, it is shown by comparing tests results with results of finite element modeling that tests with central loading of discs can be used instead of beam tests (Moskvichev et al., 2002). Recommendations in ISO 19906 are “the preferred cantilever beam test, in which the in-situ ice cover is cut along three sides and loaded [at] the free end of the cantilever". Russian documents on Site Investigation of the Continental Shelf for Offshore (2004) accept tests with ice disks for determination of flexural strength of ice. The analytical formula of Timco and Brien (1994) approximating the dependence of flexural strength on the liquid brine content is based on the results of tests with beams. Tests with disks were performed during AARI expeditions in Barents Sea during the time period of 1996-2006 (Krupina and Kubyshkin, 2007).

In the present paper, we describe the results of in-situ tests with ice cantilever beams (FST1) performed since 2009 in Spitsbergen fjords and North-West Barents Sea. These results are compared with the results of laboratory tests performed on 3-point bending of beams (FST2) and tests with central loading of ice discs (FST3).

ORGANIZING OF FIELD AND LABORATORY TESTS

Figure 1 shows sketches of flexural-strength tests to describe the flexural behavior of ice during three types of tests. FST1 is performed on a floating ice in the field and laboratory ice tank. An ice beam is prepared by making three vertical trenches over entire ice thickness (Figure 1a). The trenches should be wide enough to avoid the interaction of the cantilever beam faces with surrounding ice when the beam is loaded by a force F, applied at the free end of the beam. Vertical holes are drilled near the beam roots to avoid stress concentration. Nevertheless, some stress concentrations are created around the holes due to the beam bending, influencing the test results. Length of the beam is approximately 6 times the beam width and ice thickness. Snow is not removed from the surface of the beam to avoid any influence on the beam buoyancy forces.

The following equipment was fabricated and installed for the in-situ tests (Figure 2):

- Steel loading rig transferring force from immovable ice cover to the loaded ice beam;
- Steel loading rig for the mounting of the displacement sensor;
- Steel loading frame transferring force from Single-Acting Cylinder to the ice beam in upward direction;
- Hydrosystem (Single-Acting Cylinder RC-154, Electric Pump ZE-Series);
- Generator (3-phase, 400 V);
- Load cell NTT C8S- 82-1-100kN to measure compressive load;
- PCM 620 - Bi-Directional S-Type Load Cell, 20 kN to measure tensile load;
- Displacement sensor ASM WS 1007287917
- Data Logger Campbell Scientific CR1000.

Records of the load F and the displacement δ of the loaded end of the beam were monitored and recorded during a test at a sampling frequency of 50 Hz. Tests were performed on land-fast ice in Ice Fjord and Van Mijen Fjord in Spitsbergen, on drifting ice in North-West Barents Sea, and on freshwater lake ice near Longyearbyen, Spitsbergen.

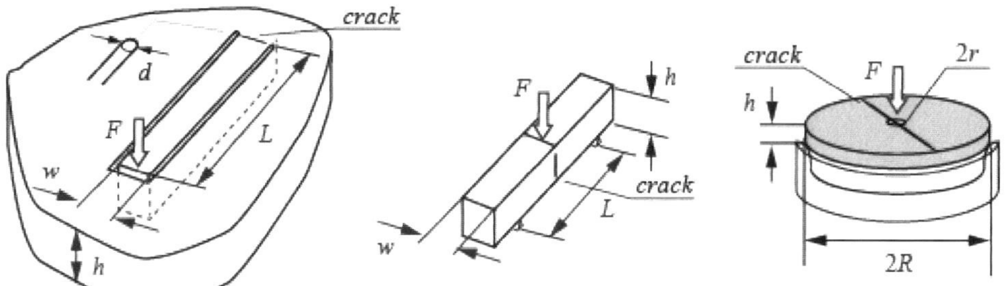

a) b) c)

Figure 1. Schematic sketches of flexural-strength test with a cantilever beam (FST1) (a), 3-point-bend test with a beam (FST2) (b), and test with central loading of a disc (FST3) (c).

The following equipment was used in the laboratory tests FST2 and FST 3 (Figure 3):

- Compression rig Knekkis with maximal load 100 T mounted in the cold laboratory of UNIS;
- Deflectometer Epsilon, Model 3540-004M-ST, travel 4.00 mm;
- HBM C9C Force transducer, 5 kN;
- Bearers for tests FST2 and FST3;

Displacements δ of beams and discs were measured at a sampling frequency of 50 Hz. The load during FST2 was measured at a sampling frequency of 20 Hz. The load in FST3 was measured by HBM Force transducer at a sampling frequency of 50 Hz.

a) b) c)

Figure 2. In-situ flexural strength tests on sea ice with downward (a) and upward (b) loading, Flexural strength test on freshwater lake ice (c).

Several tests FST3 were performed in-situ on drifting ice in North-West Barents Sea (Figure 3c). In this case, square ice plate is placed on a circular support having a diameter of 24 cm, which is hidden below an ice plate in Figure 3c. The load was applied on the ice and measured with the same equipment as in FST1.

STRUCTURE OF ICE SAMPLES IN LABORATORY AND FIELD TESTS

Figure 4 shows the internal structure of ice samples taken from the ice subjected to flexural strength tests in laboratory and in-situ tests. Natural sea ice has columnar structure over almost entire ice thickness. Surface ice layer may have granular structure when it is formed from the mixture of melt snow and seawater. Brine inclusions exist between and inside the grains. Grain boundaries of sea ice are not so sharp as in freshwater lake ice. The grain sizes are similar in the freshwater and seawater ice. Laboratory ice having a salinity less than 3-5 ppt is grown in a 1-m-by-0.5-m ice tank at UNIS from a mixture of seawater and freshwater having a total salinity around 9 ppt. Granular structure of the laboratory ice depends on the distance from the tank walls, because the directions of ice growth are different in different places of the tank.

a) b) c)

Figure 3. Laboratory tests on 3-point-bend test(FST2) of an ice beam (a), central loading of an ice disc (FST3) (b). In-situ test on central loading of ice disc (FST3) (c).

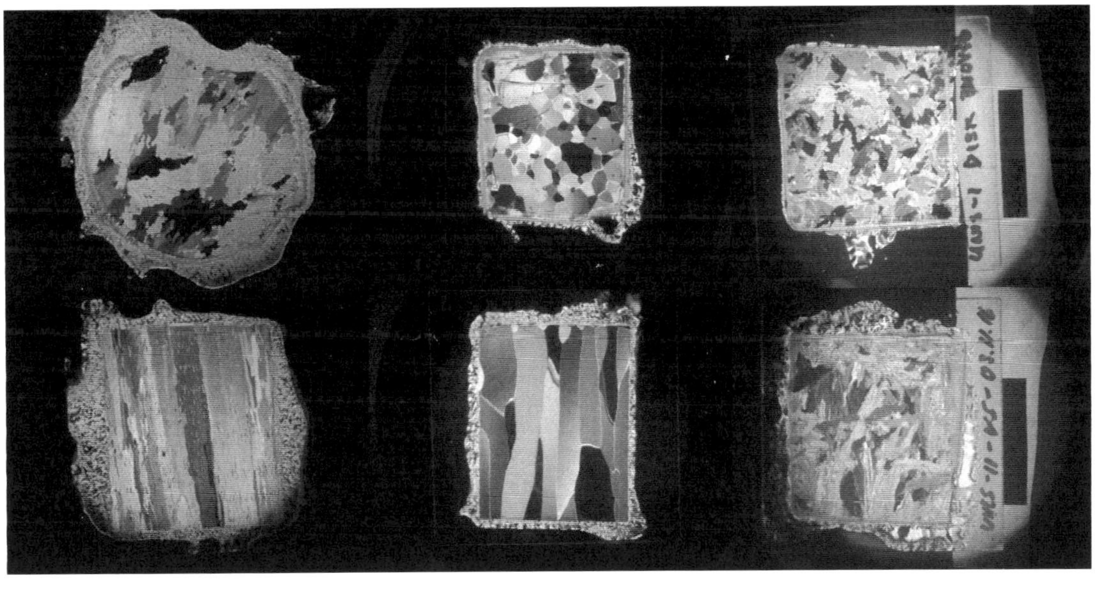

Figure 4. Thin sections of sea ice (left column), lake freshwater ice (middle column) and laboratory ice (right column). Upper and down panels show vertical and horizontal sections respectively.

Sea ice thickness during the FST1 tests was in the range of h=30-75 cm. The thickness of freshwater lake ice during the FST1 tests was in the range of h=30-40 cm. In the present paper, we report the results of 45 FST1 tests performed on sea ice and FST1 14 tests performed on fresh lake ice. Displacement was measured only in 14 FST1 tests performed on sea ice, and in all tests performed on freshwater lake ice. Two FST1 tests were performed in the ice tank on saline ice with cantilever dimensions h=13 cm, w=11 cm and L=55 cm. Geometrical characteristics of the ice beams in the FST2 tests were h=12 cm, w=6.5-12 cm, and L=46-66 cm. In the present paper, we report the results of 6 FST2 tests performed on saline ice beams. The thickness of ice discs in the laboratory FST3 tests was in the range of h=1-1.5 cm, and the diameter of the disks was $2R$=14 cm. The load was applied over the metal disk with diameter $2r$=1 cm. In-situ tests were performed with rectangular plates of sea ice with thickness h=4-5 cm. The load was applied over the plastic disk with diameter $2r\approx4$ cm. In the present paper we report the results of 12 FST3 tests performed on saline ice in the laboratory, and 4 in-situ FST3 tests performed on sea ice.

RESULTS OF TESTS

Figure 5 shows examples of load-displacement curves during tests in the laboratory and the field. All FST1 tests demonstrated brittle behavior of the ice with failure by a vertical crack near the beam root. The linear load-displacement curves demonstrate elastic behavior of ice in these tests (Figure 5a). The flexural strength and effective elastic modulus are calculated using the following formulae

$$\sigma_f = \frac{6F_{max} \cdot L}{wh^2}, \quad E = \frac{F_{max}}{\delta_{max}} \frac{4L^3}{h^3 w}, \tag{1}$$

where F_{max} is the maximal load, and δ_{max} is the displacement of the loading point of beam at the moment of maximal load. The length L is measured between the point of the loading and the crack near the beam root. Representative test duration is t^*=1 s. The strain rate estimated with formula $e=(\delta_{max}/Lt^*)$ varies between $10^{-4}s^{-1}$ and $10^{-3}s^{-1}$.

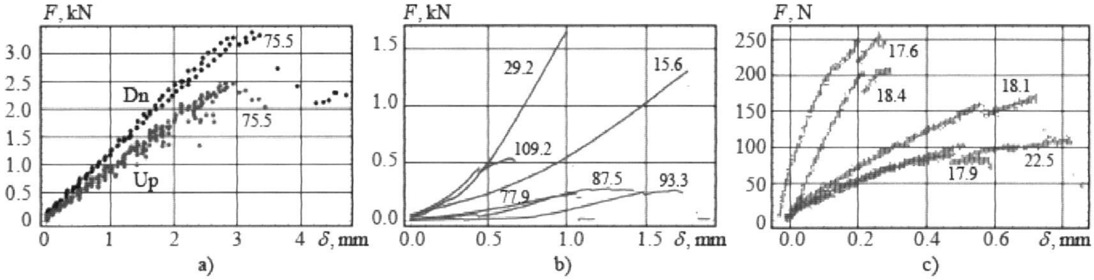

Figure 5. Examples of load-displacement curves during the in-situ tests FST1(a), laboratory tests FST2 (b) and FST3(c). Liquid brine content of ice in ppt is written in the figures.

The dependence of load F with respect to displacement of a beam near the loading point δ is not linear during the tests FST2 (Figure5b). The nonlinear behavior is probably related to local deformations of the ice beam near the loading point and near the contact points of the supports. This introduces error in measurement of the displacement. The flexural strength from a FST2 test is calculated with the following formula:

$$\sigma_f = \frac{3F \cdot L}{2wh^2}, \tag{2}$$

where L is the distance between the supports (Figure 1b). Loading duration is $t^*=15$ s. Maximal measured displacements are about 1 mm. Effective strain rate estimated with formula $e=2\delta_{max}/Lt^*$ is around $2.5\cdot10^{-4}$ s^{-1}.

The dependence of load F with respect to displacement of a disc near the loading point δ is not linear during the tests FST3 (Figure5c). The nonlinear behavior is probably related to creep deformations of the disc due to in-plane compression near the loading point. The function $F(\delta)$ has two local maxima separated by a discontinuity. The first maxima is related to the formation of bending crack which is probably is not going through the disc thickness. The second maxima is probably related to the beam failure due to the compression in surface layer. Similar scenario is observed in the tests with fixed ends beams (Sodhi, 1998; Sakharov et al, 2015; Karulina et al, 2016). Nonlinear dependence $F(\delta)$ demonstrates inelastic behavior of the disc under the load. Therefore the elastic theory can be used for the calculation of the effective elastic modulus of the disc only in the initial period of the loading. The flexural strength is calculated with the formula (Timoshenko and Woinowsky-Krieger, 1959)

$$\sigma_f = \frac{3F}{8\pi h^2}\left[4-(1-\nu)\left(\frac{r}{R}\right)^2 - 4(1+\nu)\ln\left(\frac{r}{R}\right)\right], \quad (3)$$

where ν is the Poisson's ratio. The Poisson's ratio of sea ice depends on the temperature and strain rate and has representative value of 0.33 (Timco and Weeks, 2010). We use the representative value $\nu=0.33$ in formula (3). Loading duration is $t^*=10$ s. Maximal measured displacements are about 0.25 mm. The strain rate estimated with formula $e=\delta_{max}/Rt^*$ is around $3.5\cdot10^{-4}$ s^{-1}.

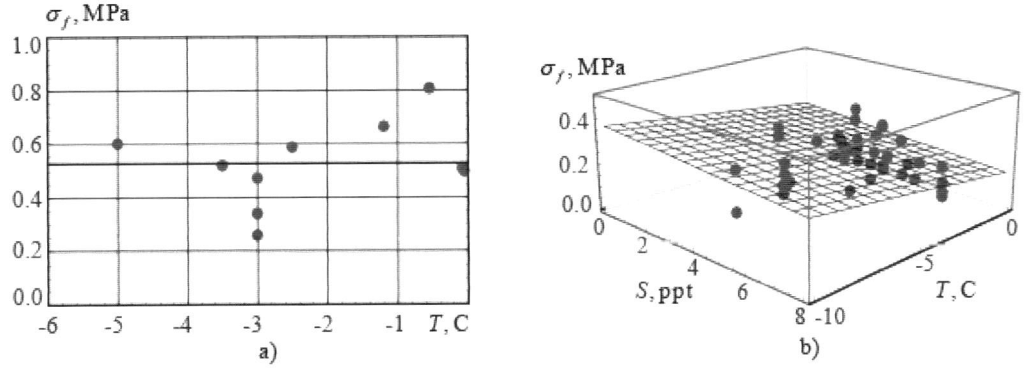

Figure 6. Flexural strength versus temperature of lake freshwater ice (a), and flexural strength versus temperature and salinity of sea ice (b).

Figure 6a shows plots of flexural strength versus temperature of freshwater lake ice as obtained from FST1 test. Mean value of the flexural strength is 0.53 MPa. It is much lower than the values given by Timco and O'Brien (1994) for ice temperatures below -4.5°C. Figure 6b shows plots of flexural strength of sea ice versus the ice temperature and salinity averaged over the ice thickness, as obtained from the FST1 tests. Mean value of the flexural strength is 0.244 MPa. It is very close to the values of the flexural strength of 0.249 MPa and 0.254 MPa obtained from similar in-situ tests performed in the North-East and South-East Barenst Sea (Krupina and Kubyshkin, 2007). Frederking and Hausler (1978) performed 11 in-situ FST1 tests in the Ice Fjord, Spitsbergen. The ice thickness during those tests was around 40 cm. In their tests the ratio L/h varied from 8.6 to 29.4 with mean value 13.64. They found the mean value of the flexural strength of 0.39 MPa. The high value of the flexural strength is attributed to the influence of the buoyancy forces on the deflected shape of long beams. In case when $L/h<6$, the influence of the buoyancy forces on the beam deflected shape is not significant (see, e.g., Karulina et al., 2010).

The plane in Figure 6b is described by the equation

$$\sigma_f=0.236-0.095\cdot S-0.0134\cdot T. \qquad (4)$$

In standard approximation, the flexural strength is expressed as a function of the liquid brine content (Frankenstein and Garner, 1967)

$$\nu_b = S\left(\frac{49.185}{|T|}+0.532\right), \quad -0.5^{\circ}\mathrm{C} \geq T \geq -22.9^{\circ}\mathrm{C}. \qquad (5)$$

Results of in-situ FST1 tests are plotted on a plane ($\sqrt{\nu_b}$,σ_f) in Figure 7a by blue points. Pink and blue lines are described by the equation

$$\sigma_f = A\cdot\exp\left(-B\sqrt{\nu_b}\right). \qquad (6)$$

Timco and O'Brien (1994) proposed to use coefficients A=1.76 and B=5.88 (pink line). The best-fit line for the present data by formula (6) is obtained with coefficients of A=0.45 and B=2.2 (blue line). Standard deviations of the data from the approximations given by formula (4) and formula (6) are of the same order of magnitude.

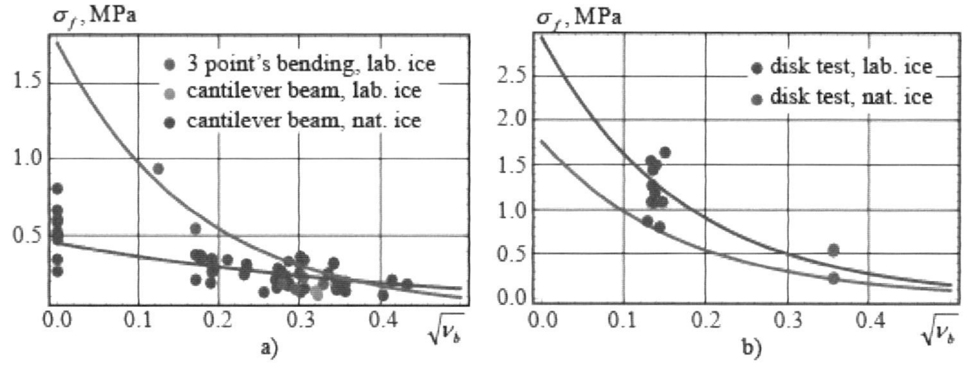

Figure 7. Flexural strength of freshwater and saline ice from in-situ and laboratory tests with beams versus the liquid brine content (a). Flexural strength of saline ice from in-situ and laboratory tests with discs versus the liquid brine content (b).

Pink points in Figure 7a are found from the tests FST2. They fit the approximation given by formula (6) better with the coefficients proposed by Timco and O'brien (1994). Green points in Figure 7a are obtained from the tests FST1 performed in the ice tank.

Blue and pink points in Figure 7b are found, respectively, from the laboratory and in-situ FST3 tests. The pink line in Figure 7b corresponds to the pink line in Figure 7a. The blue line in Figure 7b is described by formula (6) with coefficients A=2.769 and B=5.88. According to this approximation the flexural strength calculated from the tests FST3 is 1.66 times greater than the flexural strength calculated with the formula of Timco and O'Brien (1994). Krupina and Kubyshkin (2007) found that flexural strength obtained from the FST3 tests is greater by more than in 5 times than the flexural strength obtained from the FST1 tests. We think that this difference can be attributed to the difference in temperature: tests with discs were performed at a lower ice temperature than the tests with floating cantilever beams.

Figure 8a shows plots of effective elastic modulus of freshwater ice calculated with the second formula (1) versus ice temperature. There is big spread of the data around the mean value $\langle E_{eff} \rangle \approx 2.94$ GPa. Figure 8b shows plots of effective elastic modulus of sea ice calculated with the second formula (1) versus the liquid brine content. The mean value is equal to $\langle E_{eff} \rangle \approx 1.26$ GPa. Black line shows the approximation of effective elastic modulus as proposed Vaundrey (1977)

$$E_{eff} = 5{,}31 - 0.436\sqrt{\nu_b}\,, \qquad (7)$$

where ν_b is substituted in ppt, and E_{eff} is found in GPa.

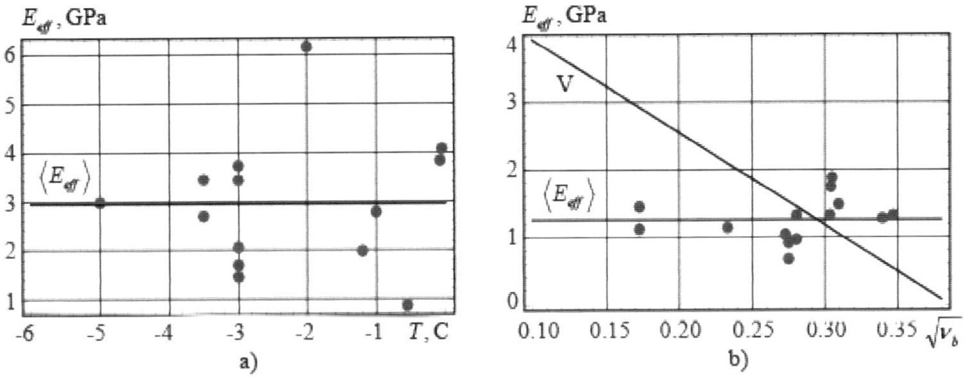

Figure 8. Elastic modulus of lake freshwater ice versus the temperature (a). Elastic modulus of sea ice versus the liquid brine content (b).

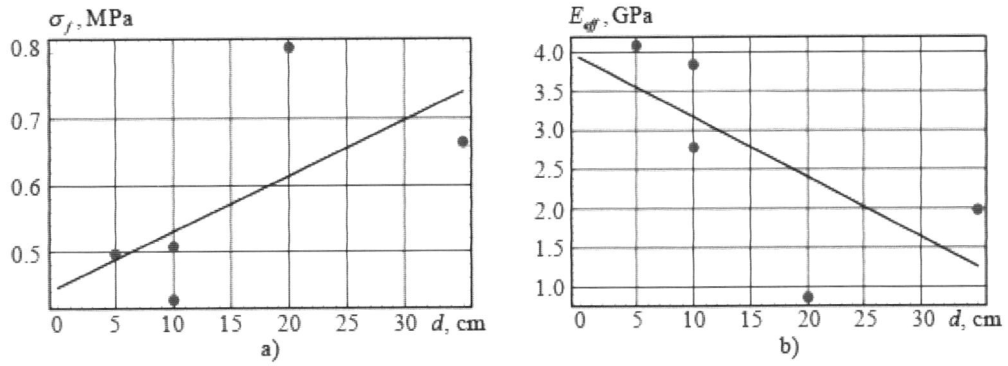

Figure 9. Flexural strength (a) and elastic modulus (b) of lake freshwater ice versus the diameter of the holes near the beam roots.

Figure 9 shows the dependence of flexural strength and effective elastic modulus of fresh ice as obtained from in-situ FST1 tests on diameters d of holes drilled near the beam roots. The tests were performed on the same ice and at the same time. The flexural strength and effective elastic modulus are calculated with formulas (1). We discovered an increase in flexural strength and a decrease of the elastic modulus with an increase in diameter of holes.

$$\sigma_f = 0.447 + 0.835 \cdot d , \ \langle E_{eff} \rangle = 3.927 - 7.615 \cdot d \ , \qquad (8)$$

where d is measured in meters, σ_f is calculated in MPa, and $\langle E_{eff} \rangle$ is calculated in GPa.

CONCLUSIONS

Tests with cantilever FST1 beams provide most useful information about the flexural strength and the effective elastic modulus of floating ice. It is attributed to pure brittle failure of ice cantilevers in the tests and a linear dependence between load and displacement during a test. We discovered that flexural strength of sea ice in North-West Barents Sea and Spitsbergen Fjords is of the same magnitude to those in North-east and South-East regions of the Barents Sea. Flexural strength obtained from in-situ FST1 tests is less than the strength prescribed by the formula of Timco and O'Brien (1994). An effect of ice temperature and salinity on the effective elastic modulus is not found. There is an influence of diameter of holes near the beam roots on the flexural strength and effective elastic modulus as obtained from FST1 tests. An increase in diameter increases the flexural strength and decreases the effective elastic modulus.

ACKNOWLEDGEMENTS

The authors wish to acknowledge the support of the Research Council of Norway through the Center for Sustainable Arctic Marine and Coastal Technology (SAMCoT). Authors thank Dr. Devinder Sodhi for the valuable comments to the paper.

REFERENCES

Frankenstein, G.E., Garner, R., 1967. Equations for determining the brine volume of sea ice from -0.5 to -22.9°C. J. Glaciology, 6(48), 943-944.

Frederking, R.M.W., Hausler, F., 1978. The flexural behavior of ice from in situ cantilever beam tests. Proc. of the IAHR Symposium on Ice Problems, Lulea, Vol. 1, 197-215.

International Standard (ISO 19906), 2010. Petroleum and natural gas industries – Arctic offshore structures. ISO copyright office, Geneva.

GOST 10180-90. Concretes. Methods for strength determination using reference specimens. 1991-01-01. (in Russian)

Karulina, M., Karulin, E., Dmitriev, D.S., Alexeev, Y.N., 2010. Procedural peculiarities of the model ice properties determination. Proc. of the 20the IAHR Symposium on Ice, Lahti, Finland, paper 57.

Karulina, M., Marchenko, A., Sakharov, A., Karulin, E.,Chistyakov, P., 2016. Experimental Studies of Fracture Mechanics for Various Ice Types. Proc. of the 23rd IAHR Symposium on Ice, Ann Arbor, Michigan, paper 4869357.

Krupina, N.A., Kubyshkin, N.V., 2007. Flexural strength of drifting level first-year ice in the Barents Sea. Proceedings of ISOPE, Lisbon, Portugal, 625-632.

Moskvichev, V.V., Makhutov, N.A., Chernyaev, A.P., et al., 2002. Fracture toughness and mechanical properties of construction materials for technical systems. Nauka: Novosibirsk, 334 p.

Sakharov, A., Karulin, E., Marchenko, A., Karulina, M., Sodhi, D., Chistyakov, P., 2015. Failure envelope of the brittle strength of ice in the fixed-ends beam test (two scenarios). POAC15-00230, Trondheim, Norway

Site Investigation on the Continental Shelf for Offshore. СП 11-114-2004.

Sodhi, D.S., 1998. Vertical penetration of floating ice sheets. International Journal of Solids and Structures, 35 (31-32): 4275-4294.

Timco, G.W., O'Brien, S.O., 1994. Flexural strength equation for sea ice. Cold Regions Science and Technology. 22, 285-298.

Timco, G.W., Weeks, W.F., 2010. A review of the engineering properties of sea ice. Cold Regions Science and Technology. 60, 107-129.

Timoshenko, S., Woinowsky-Krieger, S., 1959. Theory of plates and shells. 2nd ed., McGraw-Hill, New York, USA.

Vaudrey, K.D., 1977. Ice engineering – study of related properties of floating sea-ice sheets and summary of elastic and viscoelastic analyses. (Technical Report Naval Construction Battalion Center) Civil Engineering Laboratory, Port Hueneme, CA

Large Scale Simulations of Floe-Ice Fractures and Validation against Full-scale Data

Wenjun Lu[1], Hans-Martin Heyn[1, 2], Raed Lubbad[1], Sveinung Løset[1]

[1]Sustainable Arctic Marine and Coastal Technology (SAMCoT), Centre for Research-based Innovation (CRI) , Norwegian University of Science and Technology, Trondheim, Norway

[2] Centre for Autonomous Marine Operations and Systems (AMOS), Department of Marine Technology, Norwegian University of Science and Technology, Trondheim, Norway

ABSTRACT

While interacting with a sloping structure, an ice floe may fracture in different patterns. For example, it can be local bending failure or global splitting failure depending on the contact properties, geometry and confinement of the ice floe. Modelling these different fracture patterns as a natural outcome of numerical simulations is rather challenging. This is mainly because the effects of crack propagation, crack branching, multi fracturing modes and eventual fragmentation within a solid material are still questions to be answered by the on-going research in the Computational Mechanic community. In order to simulate the fracturing of ice floes with arbitrary geometries and confinement; and also to simulate the fracturing events at such a large scale yet with sufficient efficiency, we propose a semi-analytical/empirical and semi-numerical approach; but with focus on the global splitting failure mode in this paper. The simulation method is validated against data we collected during the Oden Arctic Technology Research Cruise 2015 (OATRC'15). The data include: 1) camera images based on which we specify the exact geometry of ice floes before and after an impact and fracturing event; 2) IMU data based on which the global dynamic force encountered by the icebreaker is extracted for the impact event. It was found that this method presents reasonably accurate results and realistic fracturing patterns upon given ice floes.

KEY WORDS: Icebreaker; Ice; Splitting; XFEM; IMU

INTRODUCTION

Fracturing of sea ice is one of the most visually significant events during ice-structure interactions. However, it is rather challenging to numerically simulate all the fracturing events which spans several scales, from local crushing, bending to the global splitting failure. Instead, we can focus on certain types of fractures and give corresponding appropriate treatment. For example, the local crushing has been treated with a plasticity model taking into account its energy dissipative nature without explicitly modelling the cracks (Liu et al., 2011). Modelling the bending failure is not so straightforward in a three-dimensional space. To the authors' knowledge, there exists not a complete set of numerical scheme to model the initiation and propagation of radial cracks; than the formation of circumferential cracks. These two types of cracks are usually studied separately, e.g., Nevel (1965), Lubbad and Løset (2011) proposed analytical solutions towards the radial crack initiation in a semi-infinite plate on a Winkler type foundation; and Lu et al. (2015b) studied the radial crack propagation within a finite square ice plate before circumferential crack's formation. In comparison, more studies were made on the eventual circumferential crack formation since it governs the final bending failure with a priori assumption of existing radial cracks, e.g., (Nevel, 1958). In terms of the global splitting failure, it is less treated in literatures and engineering practice. However, its importance in nowadays' Arctic marine operation should not be underestimated (Lu et al., 2015a; Lu et al., 2015d). In this paper we adopted a numerical scheme based on which the global splitting failure can be modelled taking into account arbitrary geometries and contact scenarios. The basic numerical method is based on the eXtended Finite Element Method (XFEM), which alleviates us from constant re-meshing procedures to capture the near crack tip field variables (e.g., displacement). This allows us to model a stationary/propagating crack with improved efficiency and accuracy. The numerical scheme is applied upon one interaction case and is compared against field measurement during the Oden Arctic Technology Research Cruise in the Arctic Ocean during September, 2015 (OATRC'15) (Lubbad et al., 2016). The effectiveness of the adopted numerically scheme to simulate the global splitting failure of an ice floe during ice-structure interaction is demonstrated in this paper.

METHODS

In the field expedition of OATRC'15, icebreakers Oden and Frej were extensively instrumented with, e.g., camera systems to monitor the ice environment (ice thickness, ice concentration, and floe sizes), an Electro-Magnetic (EM) inductive device to continuously extract ice thickness, and Inertia Motion Units (IMUs) to get the global impact force during ice-structure interactions. These instrumentations provide necessary inputs for numerical setup and outputs for final comparison. Before we dive into the detailed methodology, it is convenient to first setup the floe-ice structure interaction model.

Interaction Model

For floe ice – structure interactions, the assumed basic interaction model is described in Figure 1. Side views in Figures 1 a), b) and c) described the initial contact between ship hull and the gradually crushing at the ice edge, and eventually the hull-ice contact reached the full ice thickness. During this process, ice fractures at different scales are taking place. Rather small scale fractures are taking place at the crushed zone; relatively larger radial and circumferential cracks are formed due to the dominant vertical force from the sloping structure. Radial cracks are usually formed with relatively small vertical forces and they do not contribute to the final failure of an ice cover (Lu et al., 2015b). As the penetration proceeds to Figure 1f), the ice cover can either fail continuously in local bending failure (i.e., formation of wedges and cusps) or in a global splitting failure (as depicted in the figure). Once the global splitting failure takes place, it is considered as the final limiting mechanism amongst all the fracture events.

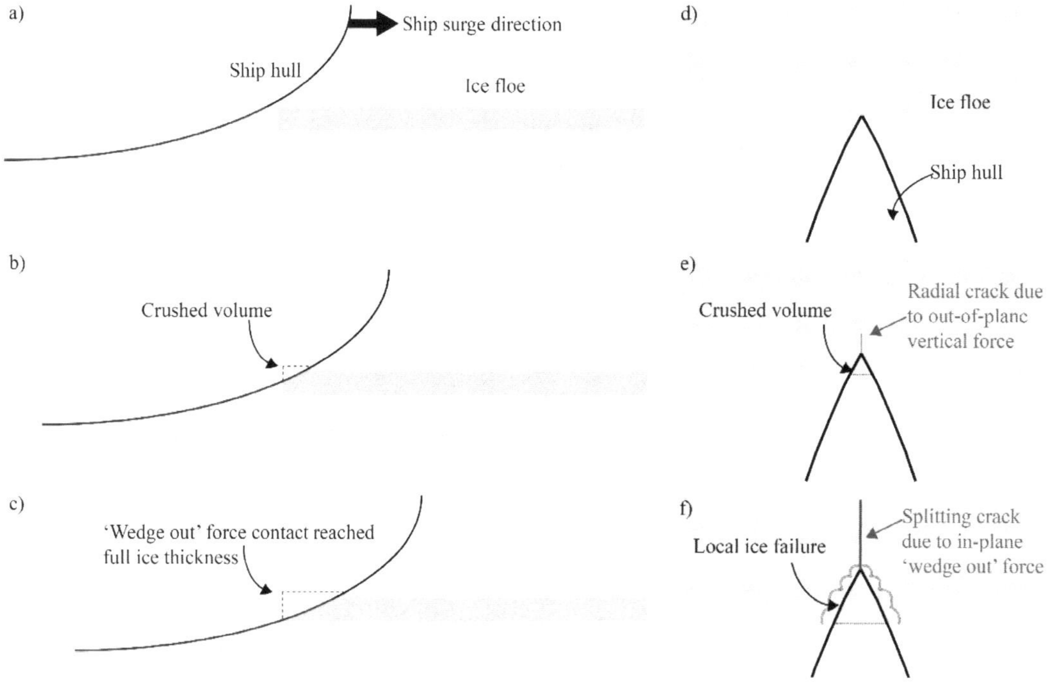

Figure 1. Floe ice – structure interaction model (left: side view; right: plan view).

Instead of simulating all these fracture events with one model from the very beginning, we decide to treat them separately with correspondingly appropriate models. For example, the local crushing has been effectively described by plasticity theories (Liu et al., 2011); by an energy based contact model (Daley, 1999); or the specific energy in a crushed volume approach (Kim and Høyland, 2014; Kinnunen et al., 2016) .The local bending failure can be treated either with theoretical formulas (Nevel, 1958; Nevel, 1972) or empirical formulas (Kerr, 1976) (originally from (Panfilov, 1960)). All these approaches combined together, we are thus analyzing the floe ice – structure interaction in a semi-analytical/empirical and semi-numerical manner. This paper, however, only focuses on the global splitting failure's numerical simulation.

In order to simulate the splitting failure of an ice floe, it is necessary to gain information such as the floe geometry, ice thickness and impact angle as inputs. Afterwards, a numerical scheme based on eXtended Finite Element Method (XFEM) is initiated and output the splitting force history versus splitting crack propagation. In the end, the calculated splitting force is compared with measurements from IMUs. In later sections, the method adopted within each of these steps is described. Particularly, we shall focus on the instrumentations and interaction scenarios on the icebreaker Frej.

Ice Floe Geometry Extraction

The geometry of an ice floe is extracted from images taken by an 180° camera system that has four lens looking in different directions covering the majority of the area on the port and starboard side of Frej. Detailed camera system information, installation location and calibration methods onboard Frej can be found in Lu et al. (2016a; 2016b). In general, chessboards with known geometry were placed within the viewing field of the camera lens to be calibrated (see Figure 2 a and b); afterwards, the method developed by Geetha and Murali (2013) was adopted to rectify 'Figure 2 a and b' into 'c and d', respectively. After perspective rectification, the ice floe's geometry and size information are available.

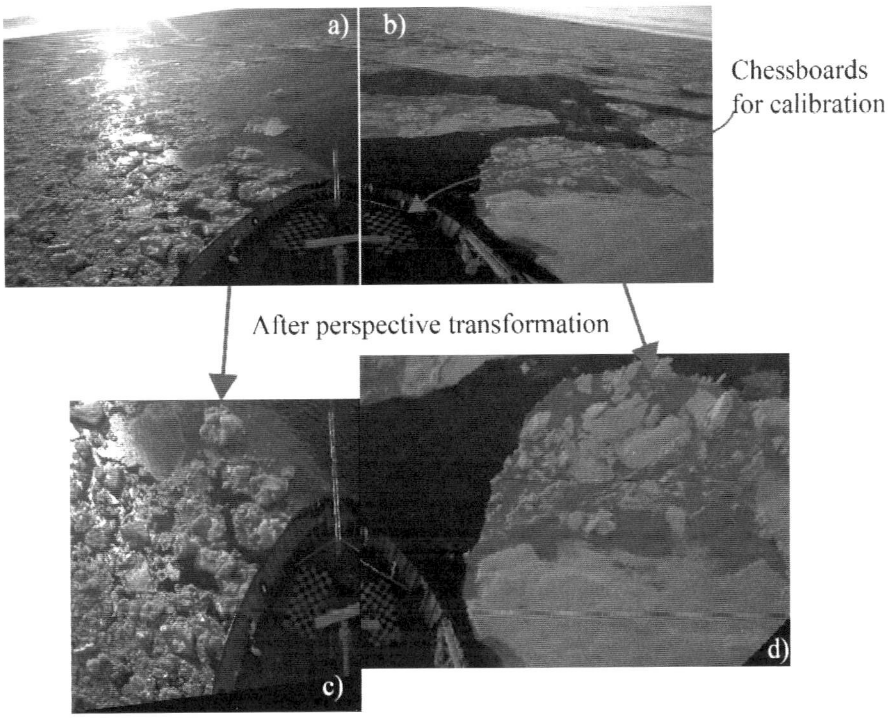

Figure 2. Calibration and perspective rectification of the ice condition images.

In terms of ice thickness, both the EM device and another camera system are available to continuously monitoring the interested ice floe's thickness (Lu et al., 2016b).

Simulating the Splitting Crack with XFEM

Given an ice floe of arbitrary geometry, the splitting crack propagates in a rather nonlinear fashion. In order to capture the crack propagation path correctly, usually a rather refined mesh at the crack tip is required to gain sufficient stress field information and thereby to determine the crack propagation direction (Anderson, 2005). Therefore accordingly, as the crack propagates, continuous mesh refinement at the crack tip would be required. Ingraffea and

Wawrzynek (2004) made extensive reviews on different methods to simulate fractures. Aside from the approach that requires extensive computational power to update the mesh geometry as crack propagates, one can also resolve to use mesh free or meshless methods. The XFEM is one of such methods that minimal mesh refinement is needed to capture arbitrary crack path and crack tip singularity because of the additionally enriched degree of freedoms (Belytschko and Black, 1999). Eq. (1) illustrates the basic formulation of the XFEM, in which, the first term on the Right Hand Side (RHS) represents the conventional FEM with nodal displacement of $\mathbf{u}_i$ multiplying the shape function $N_i(x)$; the second term on the RHS applies to the nodes $\mathbf{a}_i$ of enriched elements that are cut through by the crack and describes the displacement jump over a crack by additionally multiplying the Heaviside function $H(x)$; and the third term applies to the element which encompasses the crack tip describing the near tip displacement by additionally multiplying the function $F_\alpha(r,\theta)$, which is expressed in Eq. (2) with polar coordinate r and θ with origin at the crack tip (see Figure 3). Both $\mathbf{a}_i$ and $\mathbf{b}_i^\alpha$ are nodal enriched degree of freedom vectors.

$$\mathbf{u} = \sum_{i=1}^{n} N_i(x)[\mathbf{u}_i + H(x)\mathbf{a}_i + \sum_{\alpha=1}^{4} F_\alpha(r,\theta)\mathbf{b}_i^\alpha] \tag{1}$$

$$F_\alpha(r,\theta) = [\sqrt{r}\sin(\theta/2) \quad \sqrt{r}\cos(\theta/2) \quad \sqrt{r}\sin(\theta)\sin(\theta/2) \quad \sqrt{r}\sin(\theta)\cos(\theta/2)] \tag{2}$$

This paper employs the XFEM functionality within ABAQUS to evaluate the Stress Intensity Factors (SIFs) so as to determine the crack initiation and propagation. With known SIFs $K_I(a)$ and $K_{II}(a)$ at given crack length a, the crack propagation direction can be identified either based on the maximum energy release rate direction; or the $K_{II}(a+\Delta a)=0$ criteria. For most engineering application purposes, both methods give similar results (Hutchinson and Suo, 1991). This is also confirmed by an earlier study on parallel channel crack propagation in ice (Lu et al., 2015c). In this paper, we adopt the $K_{II}(a+\Delta a)=0$ criteria to identify the propagating direction θ of the crack by solving Eqs. (3) and (4).

$$K_{II}(a+\Delta a) = C_{21}K_I(a) + C_{22}K_{II}(a) \tag{3}$$

$$\begin{aligned} C_{21} &= 1/4[\sin(\theta/2)+\sin(3\theta/2)] \\ C_{22} &= 1/4\cos(\theta/2)+3/4\cos(3\theta/2) \end{aligned} \tag{4}$$

The general flowchart of the calculation is illustrated in Figure 3. The whole procedure in Figure 3 is automated to simulate the crack propagation within an ice floe of arbitrary geometry. For each crack configuration, the same mesh is used so as to minimize computational efforts in remeshing thanks to the elegance of XFEM. The only geometry that needs to be updated during each consecutive simulation is the crack configuration. Specifically, Steps #2 and #3 described in Figure 3 is implemented within Matlab to generate a preprocessing Python script containing updated crack information to feed to ABAQUS. The calculated results are further extracted by another post-processing Python script to feed to Matlab updating the crack geometry for a new loop of simulation. Following the coordinate system defined, the splitting force $F_{splitting}(X=0)$ versus crack extension a can be correlated with the calculated SIFs as in Eq. (5).

$$K_{total}(a)=\sqrt{K_I^2(a)+K_{II}^2(a)}$$
$$F_{splitting}(X=0,a)=\frac{K_{IC}}{K_{total}(a)} \tag{5}$$

In which,

$K_I(a)$ and $K_{II}(a)$	are the Mode I and II SIFs at the crack tip with crack length a under a unit loading at the crack mouth X=0;
$K_{total}(a)$	is the combined SIF effect;
K_{IC}	is the fracture toughness of sea ice;
$F_{splitting}(X=0)$	is the splitting force acting at X=0 pointing towards the positive and negative Y directions.

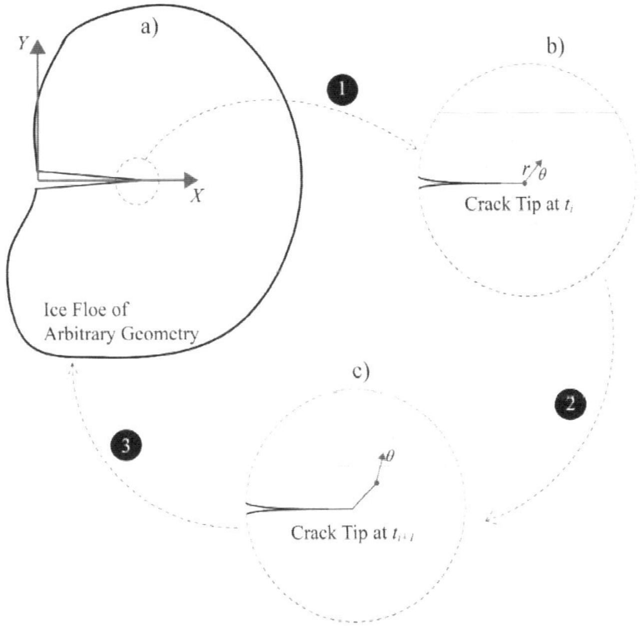

Figure 3. Crack propagation direction and model update.

Back Calculation of Impact Force from IMUs

In order to compare with the calculated results, the IMUs' measurement are utilized. The icebreaker Frej was equipped with four IMUs of type ADIS 16364 and ADIS 16480. IMU 1 was placed directly underneath the bridge level, IMU 2 midship in a compartment in the bow of the vessel; IMU 3 was placed directly on the hull in the bow of the vessel at the port side and IMU 4 on the hull of the vessel on the opposite side (starboard). During the expedition, IMU 2 experienced a malfunction and had to be discarded. IMU 1 was the only ADIS 16480 type sensor and the difference to the other sensor was, that IMU 1 was capable to output the ship's altitude, which is not required for this work. The placement of the sensors is illustrated in Figure 4.

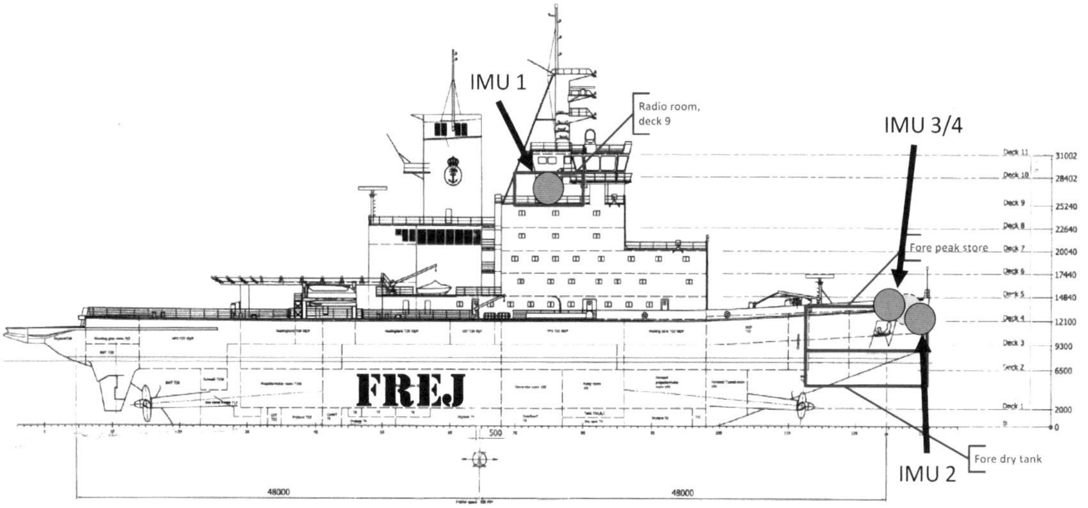

Figure 4. Placement of IMUs on the icebreaker Frej

Besides the altitude from IMU 1, all IMUs measured the acting acceleration on them in all linear directions and the angular rates in all three directions. For this work, we consider only the acceleration output of each sensor, which is given in Eq. (6).

$$\boldsymbol{a}_{m,i}^{s} = \mathbf{R}_{s}^{b}(\psi_i)^{\mathrm{T}}\left(\dot{\mathbf{v}}^{\boldsymbol{b}} - \mathbf{g}^{\boldsymbol{b}}\right) + \mathbf{b}_{i}^{s} + \boldsymbol{\omega}_{i}^{s} \tag{6}$$

$\mathbf{R}_{s}^{b}(\psi_i) \in \mathbb{R}^{3x3}$ is a rotation matrix to correct for the difference between the sensor's inertial frame and the body-fixed reference frame of the vessel, $\dot{\mathbf{v}}^{b} \in \mathbb{R}^3$ is the acceleration acting on the sensor, $\mathbf{g}^{b} = \mathrm{col}(0,0,g)$ contains the earth's gravity, $\boldsymbol{b}_{i}^{s} \in \mathbb{R}^3$ are the accelerometer's biases and $\boldsymbol{\omega}_{i}^{s} \in \mathbb{R}^3$ accounts for measurements noise. For the measurements, we selected days with as little wind as possible to allow for the negligence of wind induced accelerations. Because the ship was travelling in ice, the influence of waves has been neglected as well. The acting acceleration contains under these assumptions

$$\dot{\mathbf{v}}^{b} = \mathbf{a}_{hyd}^{b} + \mathbf{a}_{prop}^{b} + \mathbf{a}_{ice}^{b} \tag{7}$$

with $\mathbf{a}_{hyd}^{b} \in \mathbb{R}^3$ containing the accelerations caused by hydrodynamic and hydrostatic forces, $\mathbf{a}_{prop}^{b} \in \mathbb{R}^3$ the propeller induced accelerations and $\mathbf{a}_{ice}^{b} \in \mathbb{R}^3$ the ice induced accelerations.

Assuming further a constant speed and heading of the vessel before impact with the ice feature and taking only the surge acceleration of the vessel into account, we can assume that the hydrodynamic forces are compensated by the propeller thruster. This means that the measured acceleration upon impact with the ice feature contains the ice induced acceleration as its dominant component

$$\mathrm{a}_{ice,surge}^{b} = \mathbf{B_x}\mathbf{R}_s^b(\psi_i)(\mathbf{a}_{m,i}^s - \mathbf{b}_i^s - \boldsymbol{\omega}_i^s) \tag{8}$$

where $\mathbf{B_x} = (1,0,0)$ is a selection matrix to pick only the surge acceleration. The bias of each sensor is removed by a calibration procedure and the noise of each sensors is filtered out in postprocessing by a Butterworth lowpass filter. Only signal components up to 5 Hz remained in the signal to ensure that only ice induced accelerations are recorded.

The force in surge direction can be evaluated by knowing the total acting mass $m_{11} = m_{ship} + m_{add}$, which contains the ship's displacement mass and the added mass due to the surrounding water masses. Furthermore, an additional damping, with the damping parameter d_{11}, due to the contact of the lower ship hull with water can be assumed. The measured ice induced acceleration then contains

$$\mathrm{a}_{ice,surge}^{b} = m_{11}^{-1} \cdot F_{surge} + d_{11}^{-1} \cdot u_{surge} \tag{9}$$

The velocity in surge direction of the vessel is u_{surge} and the ice induced surge force is F_{surge}. In case the hydrodynamic damping is neglected, the second term of the equation is zero. The force in surge direction is then given as

$$F_{surge} = m_{11} \cdot a_{ice,surge}^{b} \tag{10}$$

The parameters of the icebreaker Frej are given in Table 1.

Table 1. Parameters of the icebreaker Frej

m_{ship} [kg]	780×10^4
m_{add} [% of m_{ship}]	10, 20 and 30
d_{11} [kg·s^{-1}]	2.141×10^4

Relationship between splitting force and impact force

The previous two sections introduced calculation methods of splitting force and impact force separately. However, these two force are not directly comparable. A certain relationship concerning the contact properties has to be constructed to correlate these two terms. This paper assumed a perfectly full contact between the structure and ice floe. During the initial ice-structure interaction before global splitting failure occurs, the stress field within the overlapped area between structure and ice floe is assumed to be averaged upon each hull panel. Based on this assumption, the contact force decomposition between the structure and ice floe reduces to a geometrical problem and the impact force in the structure's surge direction F_{surge} can be correlated to the splitting force $F_{splitting}$ solely based on the structural geometries. Take the icebreaker Frej as an example, the waterline geometry is plotted in Figure 5 together with contact force decomposition in the splitting and surge direction, respectively. The waterline plot is composed of a series of panels with area A^i, angle β_i (see the definition of angle β in Figure 5) for each panel i ; and F_{surge} and $F_{splitting}$ can be correlated as in Eq. (11) if we consider only half part (e.g., $Y \geq 0$ on the port side) of the symmetric ship.

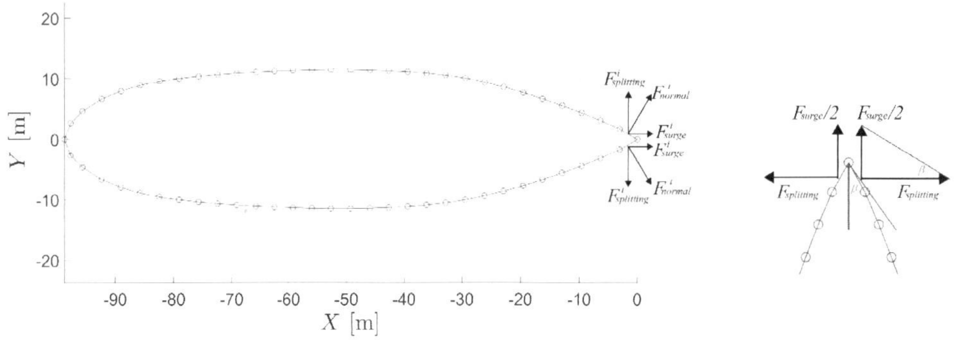

Figure 5. Waterline geometry of Frej and the assumed contact force decomposition.

$$\frac{F_{surge}}{F_{splitting}} = \frac{2\sum_{i=1}^{N} F_{surge}^{i}}{\sum_{i=1}^{N} F_{splitting}^{i}} = \frac{2\sum_{i=1}^{N} A^{i} \sin\beta_i}{\sum_{i=1}^{N} A^{i} \cos\beta_i} \tag{11}$$

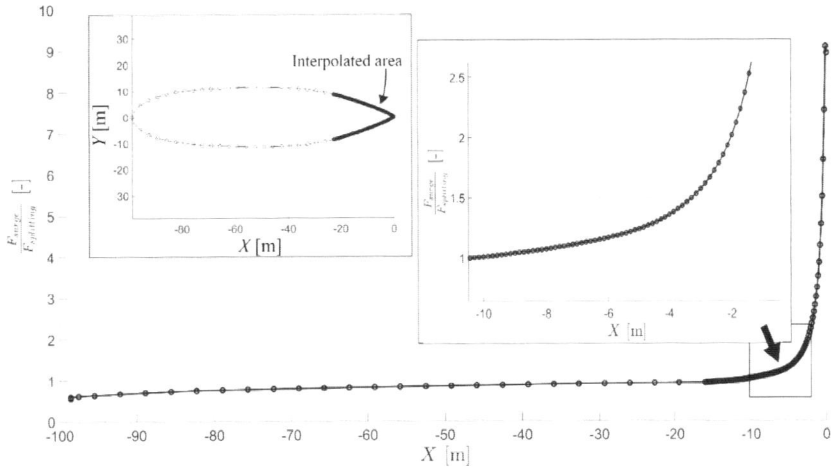

Figure 6. Ratio between the impact force in the surge direction and splitting force based on the waterline geometry of Frej.

Given the geometric properties of Frej, the ratio between surge and splitting forces, i.e. Eq.(11) , is shown in Figure 6 as a function of the penetration. Here we assume that the splitting occurs while the 'wedge out' force reached the full ice thickness as in Figure 1c). One can see from Figure 6 that as the penetration increases, the ratio between the impact force and splitting force decrease gradually as an increasing component in the splitting direction (i.e., Y direction) appears. In the following case study, the ice floe thickness is about 1.0 m based on field measurements. Considering the sloping angle of Frej, the scenario Figure 1c corresponds to a penetration of 2.5 m. A zoomed in sub-figure in Figure 6 illustrates that the ratio $F_{surge} / F_{splitting}$ is around 1.8. This value shall be used for later comparison.

CASE STUDIES

With the above described numerical scheme, a case study is carried out in this paper. The selected case is a splitting scenario of an ice floe while interacting with the icebreaker Frej. It took place from 09:26:20 to 09:26:30 (UTC time) on September 26th, 2015 in the Arctic Ocean. The selected case with the interaction process is shown in Figure 7: after an eccentric impact, a splitting crack eventually took place within the ice floe and kinked sideways.

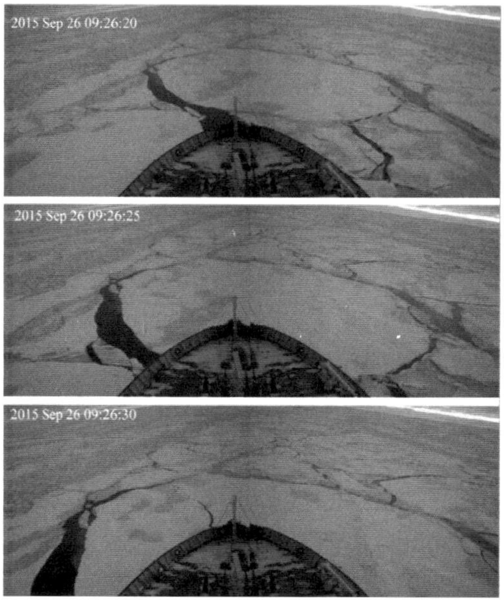

Figure 7. An ice floe's splitting failure scenario chosen for case study.

Based on the perspective rectification process described earlier, the captured images in Figure 7 were utilized to extract the ice floe's pre-failure geometries shown in the 1st row of Figure 8. Then the geometric information together with the loading point (i.e., at $X = 0$ and $Y = 0$) is provided by Matlab and passed over to ABAQUS (via the generated preprocessing Python script) calculating the SIFs at different crack configurations.

Within each of the consecutive simulations, the crack geometry is updated according to the solution from Eqs. (3) and (4). The 2nd row of images in Figure 8 shows the stress field and also the propagating crack. At the crack tip, stress concentration is also shown. The eventual failure of the ice floe is shown in the 3rd row of Figure 8 for both the simulation results and field observations. In addition, the major merit of XFEM, i.e., crack paths' mesh independency, is illustrated in Figure 9. The solution-dependent crack cuts through the pre-defined mesh pattern.

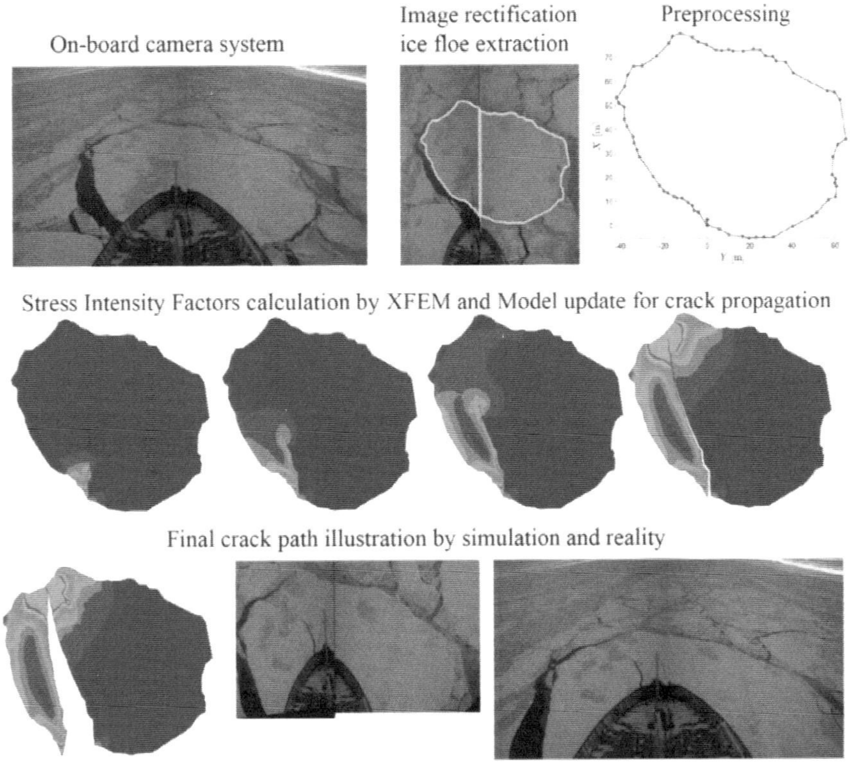

Figure 8. Crack path within the selected ice floe before and after the fracture in both the simulation and reality.

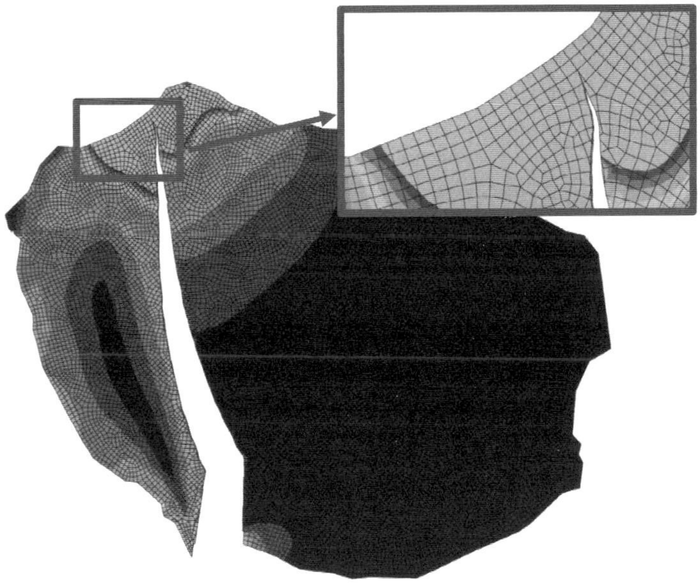

Figure 9. Illustration of crack path's mesh independency.

RESULTS AND DISCUSSIONS

Crack propagation track shown in the 2nd row of Figure 8 visually agrees well with the reality, but not exactly. In both cases, the crack propagates to the upper-left corner of the ice floe.

However, the crack path is not identical. Exact matching the simulated crack path with the field observation is rather challenging for the current case study. During the OATRC 15' expedition, there was one camera installed at the bow area on the starboard side of Frej. The fracture event at the exact time of the chosen case is also captured by the bow camera. Figure 10 qualitatively compares the crack path by intuitively overlaying two images together. Figure 10 shows that the simulation results do not coincide exactly with the actual crack, however, the general crack propagation direction is the same. The discrepancy is considered to be caused by: 1) an 'imperfect' contact exists between the structure and the ice floe; 2) inhomogeneity within the ice floe. In the simulation, a full contact between ice and the structure is assumed. This assumption is different from what Figure 10 shows, in which, local failures create non-smooth ice-structure contact. These additional local fractures lead to an imperfect contact as compared with the numerical simulation input, thereby creating a relatively different boundary condition. One can also see from Figure 10 that the crack kinks into a melt pond which is believed to be a weak zone within the ice floe. However, in the simulation, the ice floe is assumed to be a homogeneous plate.

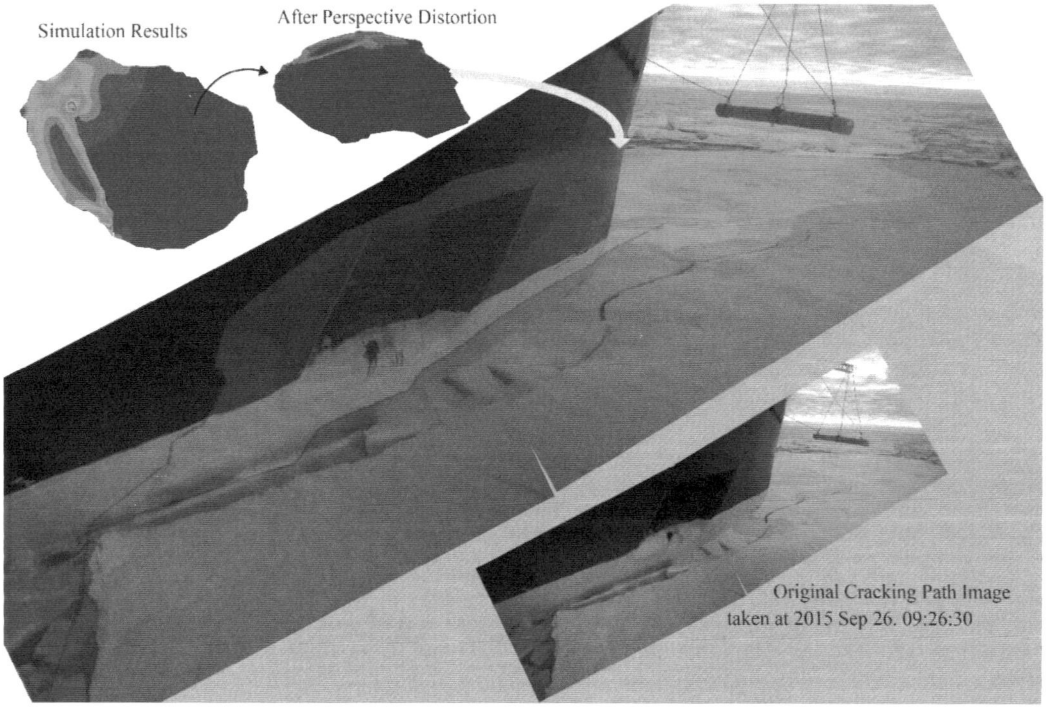

Figure 10. Overlay of the simulation results with actual crack path image.

The splitting force $F_{splitting}(X=0)$ is normalized by dividing $(tK_{IC}\sqrt{L})$, in which, t is the ice thickness and L is the floe size in the surge direction. Figure 11 illustrates the normalized splitting force $F_{splitting}/(tK_{IC}\sqrt{L})$ versus the normalized crack length a (i.e., crack location divided by the total crack length). The results show that the maximum splitting force required to propagate the splitting crack through the entire ice floe occurs at about 22% of the final crack length.

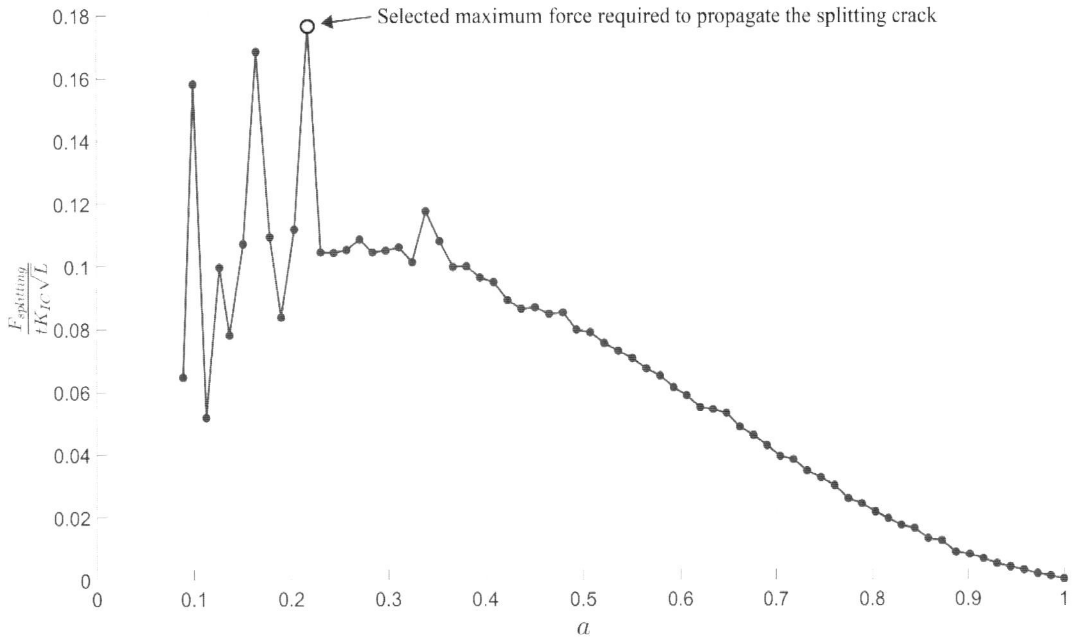

Figure 11. Normalized splitting force versus normalized crack distance by simulation.

Next, the simulated splitting force $F_{splitting}(X=0)$ is converted into the impact force in the surge direction F_{surge} based on hull geometric considerations of Frej (see Figure 5) and further compared with measurement from IMUs. The comparison together with selected material parameters and other coefficients are shown in Tables 2 and 3. Since we did not measure the fracture toughness of sea ice, we choose two existing values in literatures (Dempsey et al., 1999; Schulson and Duval, 2009) representing the lower and upper values, respectively. The final impact force in the surge direction based on ice failure simulation are presented in Table 3 with both the lower and upper values.

Table 2. Inputs for the calculation of splitting force.

$\max(F_{splitting}/(tK_{IC}\sqrt{L}))$	0.1766
t [m]	1
L [m]	75
K_{IC_lower} [$\mathrm{kPa}\sqrt{\mathrm{m}}$]	115
K_{IC_upper} [$\mathrm{kPa}\sqrt{\mathrm{m}}$]	250
$F_{surge}/F_{splitting}$	1.8

Table 3. Impact force in the surge direction calculated based on the ice splitting failure

simulations.

$F_{splitting\_lower}$ [kN]	$F_{splitting\_upper}$ [kN]	F_{surge_lower} [kN]	F_{surge_upper} [kN]
161.81	351.76	291.26	633.17

In the current selected case study, there were three working IMUs. However, a great scatter of impact force were back calculated by the IMUs' measurements. Notably, IMUs 3&4 predicted much larger impact force (around one magnitude larger) than the simulation results in Table 3. On the other hand, back calculated impact force in the surge direction from IMU 1 agrees well with the predicted force in Table 3. This is illustrated in Figure 12. Considering the fact that IMU 1 is installed on the bridge, lots of local vibrations have been damped out such that the global acceleration of the ship during the interaction can be reasonably well extracted. As for IMUs 3 and 4, they were installed near the bow area, which is susceptible to local hull vibrations during the interaction. This may cause too much impact force based on the presented back calculation method. Aside from the IMUs' measurement, strain gauges were also installed at the ship bow area. It would be beneficial to include the strain gauges' measurement for a more complete comparison. On the other hand, correlating F_{surge} and $F_{splitting}$ by Eq. (11) might be oversimplified, especially considering there is a dramatic change on $F_{surge} / F_{splitting}$ at the initial penetration as illustrated by Figure 6. This is speculated to be the reason of the slight overshooting of the IMU 1 based measurement over the simulation based prediction F_{surge_upper} in Figure 12.

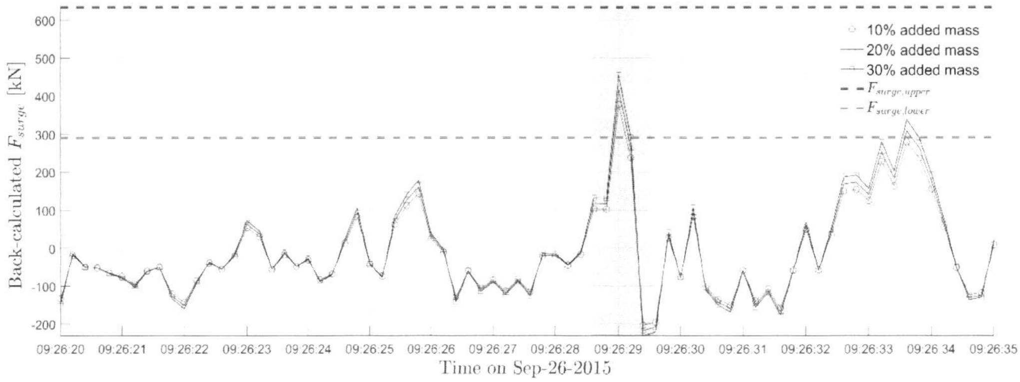

Figure 12. Comparison of F_{surge} obtained by ice floe fracture simulation and IMU 1's back calculation based on $f_s = 5$ Hz

CONCLUSIONS

This paper presents a novel approach to simulate the fracture of an ice floe during ice – structure interactions. The key idea is that different fractures at different scales are treated differently. We focused on the splitting failure mode and proposed a numerical scheme based on the eXtended Finite Element Method to simulate a splitting crack's initiation and propagation. Given the contact properties, and floe geometries, the nonlinear splitting crack propagation path can be numerically predicted together with the required splitting force history. The numerical scheme is applied in one case study with one interaction scenario during the Oden Arctic Technology Research Cruise in September of 2015. During the research cruise, field measurements such as floe geometry and ice thickness were extracted by a camera system and an Electro-Magnetic inductive device, respectively. In addition, the global interaction force can be back calculated by the installed Inertia Motion Units. With the measured floe geometry and ice thickness, the developed numerical scheme predicted the crack path rather effectively. The general trend and propagation direction of the crack in the simulations agree rather well with reality, although an exact path prediction were not achieved, possibly due to inhomogeneity in the ice and imperfect contact between ice and structure as assumed in the numerical scheme. In addition to the favorable visual agreement on crack paths, the simulated impact force based on the floe ice's splitting failure was compared with back calculated results from IMUs. It turned out that not all IMUs give the same impact force, possibly due to local hull vibrations. However, one of the IMU, which was installed at the bridge, measured an impact force; and the value is well within the range of the numerical simulations. The current mutual agreement between simulated results and field measurement is encouraging. However, further assessment on both the numerical scheme and the back calculation method based on IMU measurements are recommended. One good approach is to analyze the strain gauge data installed at the bow area to have a comparison with another set of data from a different source.

ACKNOWLEDGEMENTS

The authors would like to thank the Research Council of Norway through the research center SAMCoT CRI for financial support and all the SAMCoT partners. In addition, Oden Arctic Technology Research Cruise 2015 (OATRC2015) was supported by the ExxonMobil Upstream Research Company and performed by the Norwegian University of Science and Technology (NTNU) in cooperation with the Swedish Polar Research Secretariat (SPRS) and the Swedish Maritime Administration (SMA).

REFERENCES

Anderson, T.L., 2005. Fracture mechanics: fundamentals and applications. CRC press.

Belytschko, T. and Black, T., 1999. Elastic crack growth in finite elements with minimal remeshing. International Journal for Numerical Methods in Engineering, 45(5): 601-620.

Daley, C., 1999. Energy based ice collision forces, Proceedings of the 15th International Conference on Port and Ocean Engineering under Arctic Conditions, Helsinki University of Technology in Espoo, Finland.

Dempsey, J.P., Adamson, R.M. and Mulmule, S.V., 1999. Scale effects on the in-situ tensile strength and fracture of ice. Part II: First-year sea ice at Resolute, NWT. International journal of fracture, 95(1): 347-366.

Geetha, A. and Murali, S., 2013. Automatic rectification of perspective distortion from a single

image using plane homography. IJCSA, 3(5): 47-58.
Hutchinson, J.W. and Suo, Z., 1991. Mixed mode cracking in layered materials. Advances in applied mechanics, 29: 63-191.
Ingraffea,A.R and Wawrzynek, P.A, 2004. Computational fracture mechanics a survey of the field, European Congress on Computational Methods in Applied Science and Engineering.
Kerr, A.D., 1976. The bearing capacity of floating ice plates subjected to static or quasi-static loads. Journal of Glaciology, 17: 229-268.
Kim, E. and Høyland, K., 2014. Experimental Investigations of the Energy Absorption Capacity of Ice During Crushing: Is the Specific Energy Scale Independent?, Proceedings of the 22nd IAHR International Symposium on Ice, Singapore.
Kinnunen, A., Tikanmäki, M. and Heinonen, J., 2016. An energy model for ice crushing in ice-structure impact, 23 r d IAHR International Symposium on Ice, Ann Arbor, Michigan, USA, pp. 1–8.
Liu, Z., Amdahl, J. and Løset, S., 2011. Plasticity based material modelling of ice and its application to ship-iceberg impacts. Cold Regions Science and Technology.
Lu, W., Lubbad, R. and Løset, S., 2015a. In-plane fracture of an ice floe: A theoretical study on the splitting failure mode. Cold Regions Science and Technology, 110(0): 77-101.
Lu, W., Lubbad, R. and Løset, S., 2015b. Out-of-plane failure of an ice floe: Radial-crack-initiation-controlled fracture. Cold Regions Science and Technology, 119: 183-203.
Lu, W., Lubbad, R. and Løset, S., 2015c. Tentative Fracture Mechanisms of the Parallel Channel Effect During Ice Management. In: E. Kim, W. Lu and K. Høyland (Editors), The 23rd International Conference on Port and Ocean Engineering under Arctic Conditions, Trondheim, Norway.
Lu, W., Lubbad, R., Løset, S. and Kashafutdinov, M., 2015d. Fracture of an ice floe: Local out-of-plane flexural failures versus global in-plane splitting failure. Cold Regions Science and Technology.
Lu, W., Lubbad, R., Løset, S. and Skjetne, R., 2016a. Parallel Channel Tests during Ice Management Operations in the Arctic Ocean, Arctic Technology Conference 2016, St. John's, Newfoundland and Labrador.
Lu, W., Zhang, Q., Lubbad, R., Løset, S. and Skjetne, R., 2016b. A Shipborne Measurement System to Acquire Sea Ice Thickness and Concentration at Engineering Scale, Arctic Technology Conference 2016, St. John's, Newfoundland and Labrador.
Lubbad, R. and Løset, S., 2011. A numerical model for real-time simulation of ship-ice interaction. Cold Regions Science and Technology, 65(2): 111-127.
Lubbad, R., Løset, S., Hedman, U., Holub, C. and Matskevitch, D., 2016. Oden Arctic Technology Research Cruise 2015, Proc. of the Arctic Technology Conference (ATC), St. Johns's, Newfoundland and Labardor, Canada.
Nevel, D.E., 1958. The theory of a narrow infinite wedge on an elastic foundation, U. S. Army Snow Ice and Permafrost Research Establishment, Corps of Engineering.
Nevel, D.E., 1965. A semi-infinite plate on an elastic foundation, U. S. Army Snow Ice and Permafrost Research Establishment, Corps of Engineering.
Nevel, D.E., 1972. The ultimate failure of a floating ice sheet, International Association for Hydraulic Research, Ice Symposium, pp. 17-22.
Panfilov, D., 1960. Experimental investigation of the carrying capacity of a floating ice plate. Izvestia Vsesojuznogo Nauchno-Issledovatelsskogo Instituta Gidratekhniki, 64: 80-8.
Schulson, E.M. and Duval, P., 2009. Creep and fracture of ice, 1. Cambridge University Press, Cambridge, 190-211 pp.

A Study of Ice Oblique Drift Action onto the Structure Partially Sheltered by another Object

Marina M. Karulina[1,4], Alexander N. Sakharov[3,4], Aleksey V. Marchenko[2,4], Evgeny B. Karulin[1,4], Peter V. Chistyakov[3,4]
[1]Krylov State Research Centre, St.-Petersburg, Russia
[2]The University Centre in Svalbard, Longyearbyen, Norway
[3]Moscow State University, Moscow, Russia
[4]Sustainable Arctic Marine and Coastal Technology (SAMCoT), Centre for Research-based Innovations (CRI), Norwegian University of Science and Technology, Trondheim, Norway

ABSTRACT

Ice action on a structure may follow different scenarios primarily depending on ice environment (conditions) and type (shape) of the structure itself. The study examines a case of oblique drift ice acting on a structure partially sheltered by another object (offshore structure) or Ice Protection Structure (IPS). This scenario was investigated in a model experiment whose purpose was to study ice effect on support beams (foundations) placed near each other. The supports had extended side walls and rectangular plan form. The paper contains main results of the model experiment showing that the rear support partly sheltered by the forward support was subject to impulse ice loads. The magnitude of ice load impulse is determined by the splitting force acting on the ice sheet section enclosed between the two units, and such magnitudes can be quite significant. In March 2016 two mesoscale tests were conducted in sea ice of about 0.6 m thickness in a fjord of Svalbard archipelago to study this process. The tests consisted in loading of short ice cantilevers by horizontal force distributed over one of the specimen side. The cantilevers had triangular form in plan. The test procedure and load measurements are described. The ice sheet failure analysis is performed using both a simple engineering approach and an analytical model.

KEY WORDS: Ice loads, Ice failure, Splitting, Tensile stress, Compressive stress

INTRODUCTION

For specification of design ice loads on engineering structures it is required to investigate and analyze all potential ice exposure scenarios and identify the worst cases of ice impact levels. As a rule, the global ice load on a structure is governed by a so-called "limit stress scenario" when ice cover is broken in interaction with the structure. It is assumed that external driving forces acting on the ice cover are sufficient to break it, while the failure modes can be of various types. In accordance with ISO 19906 (2011) the "limit stress scenario" includes ice floe splitting cases when ice may interact with part rather than total width of the structure. Some specific cases of ice cover splitting are addressed in Bhat (1988), Sodhi and Chin (1995) where formulas for estimation of splitting forces are given for the cases under consideration. Ice floe splitting (when part of the ice cover is broken off) can be expected to occur at an "oblique" ice drift at a glance ice impact on a structure which is sheltered by another object. This kind of situations may happen, e.g., in case of ice protection barriers or system of offshore field facilities (Figure 1).

An interesting case of the scenario under consideration is ice interaction with the substructure blocks installed on the Philanovsky field in the northern part of the Caspian Sea. The substructure consists of two extended parallel structures (blocks) of pentagonal planform (Figure 1, right). At some directions of ice drift the rear block is partly sheltered by the front one, and, as shown by earlier published model ice test data (Karulina et al., 2012), the rear support will periodically break off quite large ice floes. In these scenarios the maximum global force on this block will be governed by ice splitting forces.

Ice protection barriers near the drilling structure (Schiff & Hafen, 2008) | Platforms on Philanovsky field, Caspian Sea (www.lukoil.ru)

Figure 1. Examples of the offshore structures arrangement

It the purpose of this study to investigate the ice cover failure mechanism and to estimate ice loads on the structure sheltered by another object. The study is based on some results of model tests in ice basin (Karulina et al., 2012), as well as on the data of special-purpose meso scale tests conducted in sea ice with subsequent analysis of these tests. The tests were carried out under the international project SAMCoT, WP1, and partly reproduced the ice/structure interaction scenario under study here.

INVESTIGATION OF OBLIQUE ICE DRIFT IN ICE BASIN

Ice interactions with the platforms designed for the Philanovsky field were investigated on models at different ice drift directions at the ice basin of Krylov State Research Centre (KSRC), St.Petersburg. As it is seen in Figure 1, right, the substructure consists of two parallel extended blocks. The internal and end side surfaces of these structures are made as vertical walls, while the external sides have inclined flat surfaces in way of interaction with ice. This study examines a scenario when the rear structure is partly sheltered by the front structure and analyzes global ice loads on the rear structure. This model test case is illustrated in Figure 2.

a) Ice crushing on internal and end side of rear block
b) Formation of macrocrack

Figure 2. Ice model tests of platform foundation consisting of two blocks

Figure 3 shows step-wise failure of a level ice field drifting towards the substructure at an angle β:

- Initially the ice edge comes into contact and interacts with the front structure, then the ice breaks against the vertical end surface and inclined external side surface of the structure (Figure 3a)
- When the ice edge comes into contact with the rear block it crumbles against its vertical internal and end surfaces (Figure 3b). This process can also be observed from a photo taken during the model tests (Figure 2a).
- The crushed ice area at the internal surface is gradually increasing, which cause the global ice load to grow. Once a certain load is achieved to initiate a macro crack in ice between two blocks, an ice floe is splitted (Figure 3c and 2b). The crack inception occurs at the stress raiser – edge where the end side and internal surface of the block intersect.
- Drifting ice is pushing the split ice floe between the blocks into a channel behind the facility.

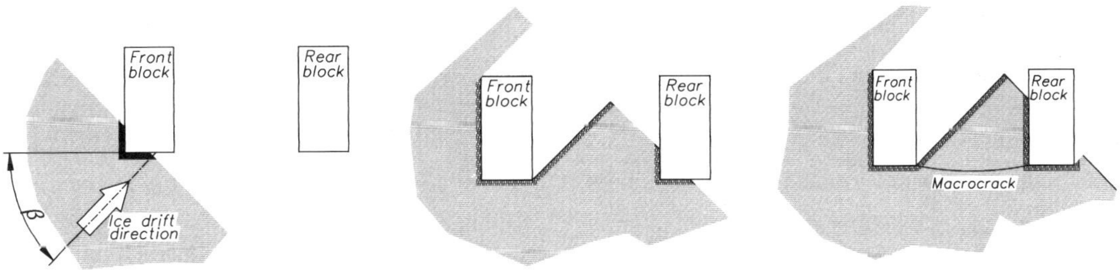

a) Front block interaction with ice
b) Ice cruching against the rear block
c) Formation of macrocrack

Figure 3. The ice action stages

Then the above-described processes are repeated. A cyclic pattern of interaction between the rear block and ice governs the global ice load behaviour. Figure 4 shows variation of the horizontal ice force component acting normally to the internal side surface of the rear structure. The force was recorded in the ice model tests (Karulina et al., 2012) and extrapolated to full-scale condition. The ice drift angle was 45°, ice thickness 1.2 m, drift speed 0.5 m/s. The peak load values at the time of splitting (macro crack inception) are in the range of 10.2 – 12.8 MN (average peak value is 11.4 MN), then the load falls to practically zero values. The average peak load period is 70 s.

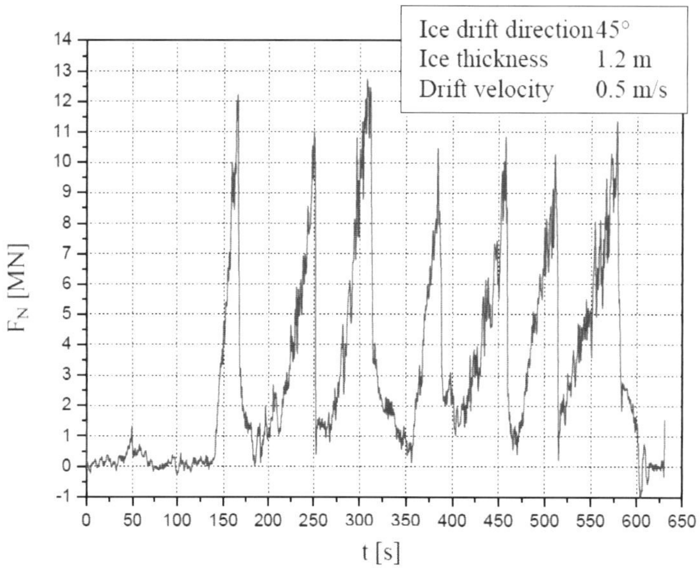

Figure 4. Normal ice force acting on rear block.

Apart from the cyclic pattern of ice action on the rear block, it can be expected that the rear block will be subject to quite a large turning moment about the vertical axis due to asymmetrical application of ice force (with only one peripheral side of the structure interacting with ice). This factor affects the ability of the structure to resist shearing force and should be taken into consideration for design, however in this study it is left out of consideration.

One of the causes for macro cracking and ice floe splitting can be a bending moments generated by the forces F_N and F_T about the point B – crack tip on the front block edge (Figure 5). F_N and F_T are normal and tangential forces acting on the exposed part of ice field from rear block, accordingly. Tensile stresses at the point A – crack base can be approximately estimated based on the theory of cantilever beams using a simple formula of ultimate bending strength $\sigma_f = \frac{6M}{hW^2}$, where M is the total moment of forces F_N and F_T about the point B, h is ice thickness, W is crack length, which is determined by the distance between blocks in this case. If thus obtained value $\sigma_f = 0.412$ MPa exceeds the ice bending strength in horizontal plane, then bending moments acting on the exposed part of ice field may cause cracking.

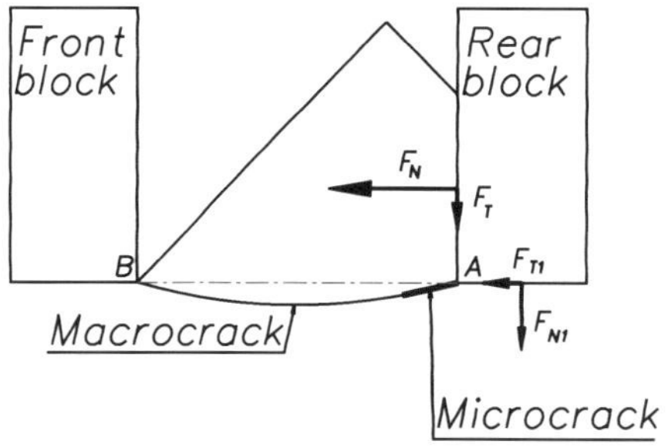

Figure 5. Forces acting from rear block on ice cover.

Another approach to analyzing the causes of ice floe splitting is based on representation of the rear block as a rectangular indenter penetrating ice with its corner edge. The process of radial

POAC17-158

cracks initiation and propagation at penetration of a structure into ice that cause ice floe splitting is described by Bhat (1988). In the case under consideration the forces acting on ice from the internal (F_N, F_T) and end (F_{N1}, F_{T1}) faces of the rear block contribute to transformation of a micro crack at the corner (point A) into the splitting macro crack (Figure 5).

For further investigation of macro cracking mechanisms mesoscale in-situ tests were performed in full scale ice to reproduce a similar scenario of ice cover loading and breaking. The tests are described in more detail below.

IN-SITU TESTS WITH HORIZONTAL LOADING OF SHORT ICE CONSOLES

Test Set-up and Equipment

In March 2016 two in-situ tests were performed in a fjord of Svalbard archipelago using floating ice specimens of right-angle triangle and rectangular trapezium planforms.

Figures 6 and 7 show the test set-up and dimensions of test specimens. The specimens were cut from ice so that one of their edges remained attached to the ice field. For the purposes of these experiments the length of uncut edge was important rather than the shape of specimens. It was 0.7 m for the triangular specimen and 1.2 m for the trapezium. In the following these are referred to as the "narrow" and the "wide" specimen, respectively.

A horizontal force was applied to one of the free side surfaces using a 600×800 mm vertical plate driven by a rig of two hydraulic cylinders capable to exert a total horizontal force of about 600 kN (Figure 6). The rig is equipped with displacement and pressure gages whose indications were recorded by computer.

Loading of triangle ("narrow") specimen

Loading of trapezium ("wide") specimen

Figure 6. Test set-up.

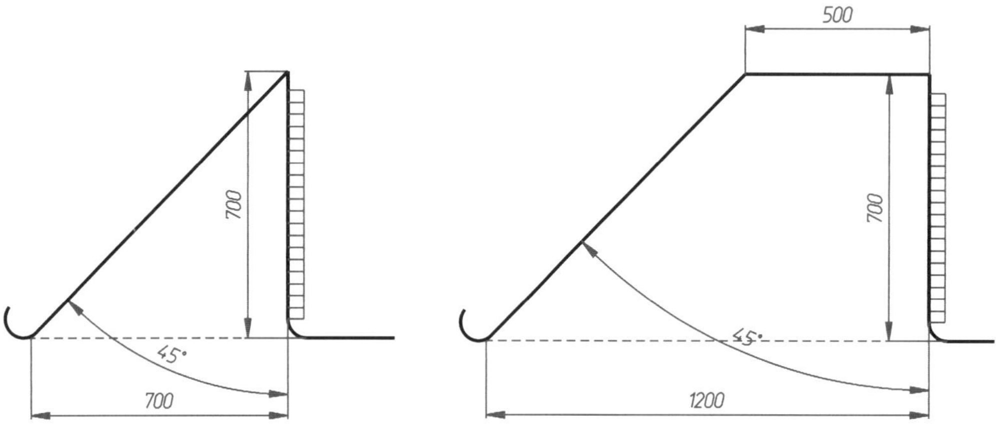

Figure 7. Dimensions of ice specimens

The tests were conducted at ice thickness of 0.6 m, ice temperature –3.5°C and salinity 4.5 ppt averaged over thickness. The horizontal plate speed is 1.5 mm/s. The tests were performed in sea ice and have some characteristic differences from the model tests in ice basin with simulation of ice loads on structures:

- The size and thickness of loaded ice section have the same order of magnitude, while in model experiments the linear dimensions of split ice section were by an order of magnitude greater than its thickness;
- The in-situ test setup prevented normal displacement of the loading plate with respect to the loading force on block, which excluded any friction force shown in Figure 5.

Description of Ice Failure Processes

Failure modes of ice cantilevers were analyzed based on synchronized records of forces (Figure 8) and test videos. Figures 9 and 10 show crack patterns obtained during testing of specimens.

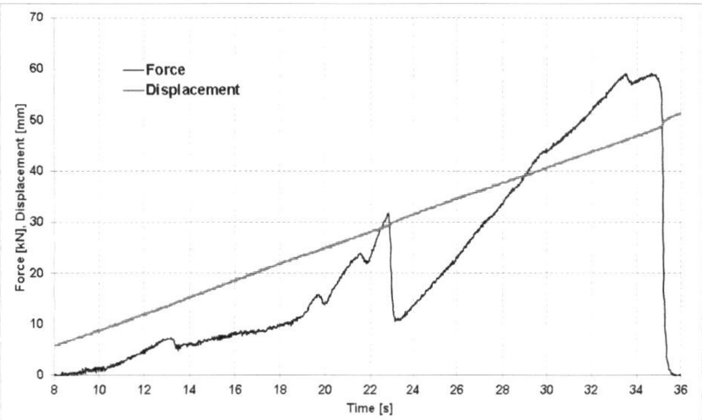

a) Loading of triangle ("narrow") specimen

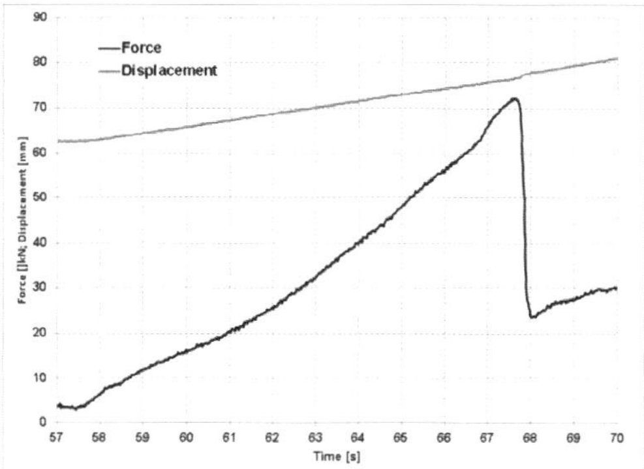

b) Loading of trapezium ("wide") specimen

Figure 8. Time records of force and plate displacement

a) formation of first crack due to bending
b) opening of the first crack
c) specimen is fully separated from ambient ice due to shear

Figure 9. Testing of triangle ("narrow") specimen

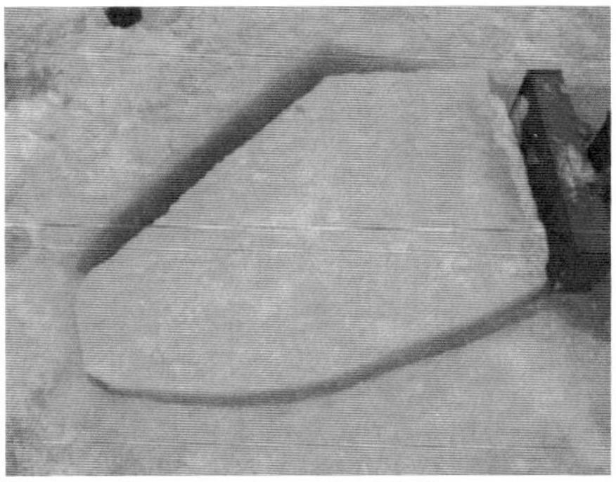

Figure 10. View of trapezium ("wide") specimen after testing.

Two phases of macro crack development were observed in the loading tests of the narrow (triangular) specimen:

– Indentation of vertical plate into ice causing ice crushing and load growth; initiation of a macro crack in the root section of the specimen caused by bending moment (peak load $F_1 = 32$ kN at instant $t_1 = 22.8$ s Figure 9a). A simple formula from the beam bending theory gives the ice bending strength in horizontal plane of $\sigma_f = 0.229$ MPa. This value practically coincides with the value $\sigma_f = 0.231$ MPa obtained by the authors from bending tests of cantilever ice beams in horizontal plane in the same ice. From Figures 9a and 9b it is seen that a bending crack does not propagate across the entire root section: its length is about 0.56 m, while the root section length is 0.7 m.

– Under further loading of the specimen the force is growing monotonously for almost 11 seconds (till $t_2 = 33.6$ s). At $F_2 = 60$ kN, corresponding to the second peak value, the specimen is fully splitted. Limit stresses at failure in this case can be estimated from: $\sigma_{lim} = \frac{F_2}{h \cdot a}$, where $h = 0.6$ m – ice thickness, $a = 0.14$ m – length of root section subject to failure. The obtained limit stress $\sigma_{lim} = 0.714$ MPa is in good agreement with other tests performed by the authors in the same ice using a similar loading setup with simulation of shearing failure. The average stress value in these tests was $\sigma_{lim} = 0.750$ MPa.

In loading test of the wide (trapezium) specimen the load monotonously increased as the plate penetrated the ice specimen. When the applied horizontal load reached 72 kN, a macro crack was initiated, which broke the specimen away from the ice field. In comparison with the failure (split) of the narrow specimen it should be note that in this case:

– The specimen was split in one step, i.e. the load gradually increased up to the limit value and then dropped practically at once after the macro crack was made.

–Unlike the triangular specimen, when the crack practically ran across the root section, the wide specimen had an L-shaped crack. It is an evidence of a complex stress state of the split ice section and a simplified analysis of the crack initiation and propagations will not work in this case.

Analytical stress/strain studies for ice cantilever specimens of triangular shape (wedges) under a horizontal force are described below.

LIMIT STRESS ANALYSIS OF ICE WEDGES UNDER HORIZONTAL LOADING

Let us consider a two-dimensional problem of an ice cantilever wedge-shaped specimen being loaded with a horizontal force uniformly distributed over one edge of the specimen, assuming that ice is a homogeneous material. A narrow wedge AOB (Figure 11a) will fail under the load when the stress in extreme tensioned fibers reaches the tensile ice strength limit. In this case ice stressed state is close to tension-compression and the bending theory provides a rather good approximation of the limit load p causing failure by propagation of a splitting crack AB. For larger angles $2\alpha > 45°$ (in case of a wide wedge base) a 2D stressed state can be analyzed using the limit curve shown in Figure 11b (Schulson and Duval, 2009).

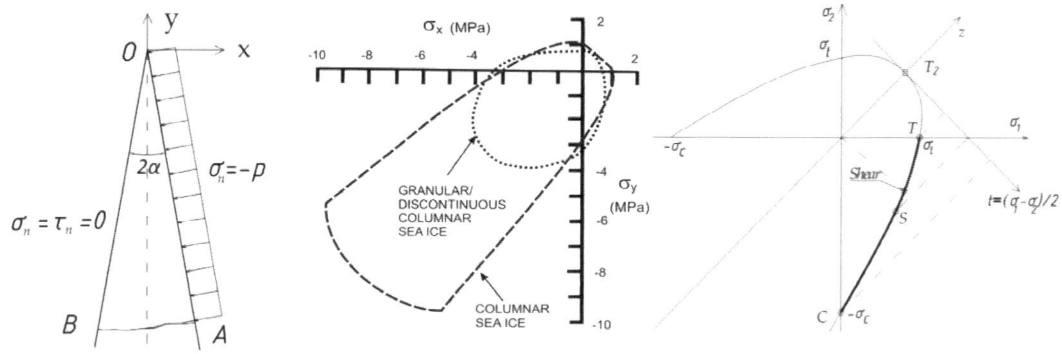

a) Narrow wedge loading with uniformly distributed over one edge pressure

b) Failure envelope of sea ice

c) Representation of limit curve segment as a parabola

Figure 11. Scheme of narrow wedge loading and ice strength diagrams

The stressed state of wedge-shaped ice specimen is located in the fourth quadrant of the principle stress plane. Let us represent the limit curve as a parabola (Figure 11c) passing through two points: C $(0; -\sigma_c)$ – ice compression strength and T $(\sigma_t; 0)$ – ice tensile strength. The parabola axis lies on the bisectrix of the first quadrant: $\sigma_1 = \sigma_2$. The parabolic equation in terms of new variables $t = (\sigma_1 - \sigma_2)/2$ and $z = (\sigma_1 + \sigma_2)/2 - \sigma_0$ is written as

$$f(\sigma_1, \sigma_2) = z + At^2 = 0 \tag{1}$$

where $A = 2/(\sigma_c - \sigma_t)$. Other characteristic points of the parabola are as follows:

$T_2 = (\sigma_0, \sigma_0)$ – parabola apex corresponding to the strength under bi-axial tensioning $\sigma_0 = \dfrac{\sigma_c \sigma_t}{2(\sigma_c - \sigma_t)}$, and point $(\tau_s; -\tau_s)$, where $2\tau_s = \sqrt{\sigma_c \sigma_t}$ – uniaxial shear strength.

The statement of problem on limiting wedge equilibrium is reduced to equilibrium equations in the polar coordinate system

$$\begin{cases} \sigma_r - \sigma_\theta + \dfrac{d\tau_{r\theta}}{d\theta} = 0 \\ \dfrac{d\sigma_\theta}{d\theta} + 2\tau_{r\theta} = 0 \end{cases} \tag{2}$$

where stress components are dependent only on the angle Θ and related by the principal stresses

$$\begin{cases} \sigma_r = \dfrac{\sigma_1 + \sigma_2}{2} + \dfrac{\sigma_1 - \sigma_2}{2}\cos 2\psi = z + \sigma_0 + t\cos 2\psi \\ \sigma_\theta = \dfrac{\sigma_1 + \sigma_2}{2} - \dfrac{\sigma_1 - \sigma_2}{2}\cos 2\psi = z + \sigma_0 - t\cos 2\psi \\ \tau_{r\theta} = \dfrac{\sigma_1 - \sigma_2}{2}\sin 2\psi = t\sin 2\psi \end{cases} \tag{3}$$

Equation (2) at limit conditions (1) may have two solutions – discontinuous stress field and continuous one (Figure 12).

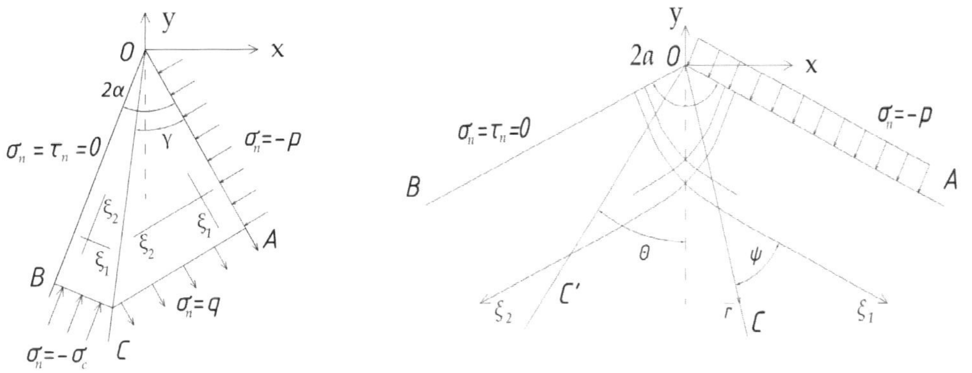

a) Discontinuous stress field b) Continuous stress field

Figure 12. Two solutions for a wedge loading

Let us find the solution based on the discontinuous stress field for a wedge with angle 2α (Figure 12a). Considering that stresses on the boundary OB are equal to zero, the stressed state within the triangle OBC is uniform $(0; -\sigma_c)$. At the section OA stresses are equal to $\sigma_n = -p$, $\tau_n = 0$, the stressed state within the triangle OAC is also uniform $(q; -p)$, $q \le \sigma_t$, and corresponds to the parabola arc CT (Figure 11c). After reducing the stresses into non-dimensional form, dividing the same by σ_c and introducing $\mu = \sigma_t / \sigma_c < 1$, $s_1 = q/\sigma_c$; $s_2 = -p/\sigma_c$, we can re-write the parabolic equation as follows

$$(s_1 - s_2)^2 + (1-\mu)(s_1 + s_2) - \mu = 0 \tag{4}$$

For matching the stress fields on the section OC let us write the equilibrium equation $[\sigma_n] = [\tau_n] = 0$. The equilibrium conditions and limit curve equation (4) are equivalent to the transcendental equation

$$\frac{\sin^2 2(2\alpha - \gamma)}{\sin^2 2\gamma} + 2\cdot(1-\mu)\frac{\sin(2\alpha - \gamma)\cos(2\alpha + \gamma)}{\sin 2\gamma} - \mu = 0 \tag{5}$$

where γ – angle between the sections OA and OC. Eq. (5) can be solved either numerically or graphically.

If the ratio between tensile-to-compression ice strength is assumed $\mu = \sigma_t / \sigma_c = 0.25$, then from (5) for a wedge with angle $2\alpha = 45°$ we obtain $\gamma = 30.3°$. The limit load on OA in this case is $p = 0.84\sigma_t$, and tension on AC is $q = 1.04\sigma_t$, i.e. the stressed state in the triangle OAC is very close to the uniaxial shear, while the limit load on the wedge is just a little bit higher than the ice tensile strength.

The performed theoretical analysis examines the stressed state of ice wedge prior to initiation of the first crack. The theory is not extended to stage after first crack formation in the case when the ice wedge is broken in two (or more) steps as during the tests on the "narrow" specimen in sea ice (Figures 8a and 9).

CONCLUSIONS

Ice field splitting is quite a probable scenario of ice interaction with an engineering structure partly sheltered by another structure. The global ice load on the structure in this case is governed by the force required to break off a large ice floe being a cantilever beam of wedge or truncated wedge shape. The results obtained in model tests show that the maximum ice loads correspond to the values calculated from the simple beam bending theory. Two meso scale tests in sea ice with horizontal loading of short cantilever ice specimens indicate that the ice failure patterns depend on the specimen geometry. Based on the theoretical analysis of 2D stressed state of wedge-shaped ice specimens possible mechanisms of ice failure are identified: for relatively narrow specimens the breaking force may be estimated using the theory of cantilever beam bending, while the wide specimens may be subject to shear failure.

ACKNOWLEDGEMENTS

The authors wish to acknowledge the support of the Research Council of Norway through the Center for Sustainable Arctic Marine and Coastal Technology (SAMCoT).

REFERENCES

Berger, J., 2008. Ice protection structures. Journal Schiff&Hafen, No.6, pp.16-18.

Bhat, S.U., 1988. Analysis for splitting of ice floes during summer impact. In Cold Regions Science and Technology, 15, pp. 53-63.

International Standard ISO 19906, 2011. Petroleum and natural gas industries – Arctic offshore structures.

Karulina M., Karulin E., Blagovidova I., Bardin I., 2012. Experience of enhancing ice resistance of platforms for exploring North Caspian oil and gas fields. Transactions of KSRC, Issue 70 (354), St.Petersburg, pp.83-98.

Schulson, E.M., Duval, P., 2009. Creep and Fracture of Ice. Cambridge University Press.

Prager W. and Hodge Ph.G., 1951. Theory of perfectly plastic solids. N.Y.

Sodhi D.S. and Chin S.N., 1995. Indentation and splitting of freshwater ice floes. Journ. of Offshore Mechanics and Arctic Engineering. Vol. 117, pp. 63-69

POAC'17
Busan, KOREA

Proceedings of the 24th International Conference on
Port and Ocean Engineering under Arctic Conditions
June 11-16, 2017, Busan, Korea

Variation of Stress in Virtual Biaxial Compression Test of Ice Rubble

Sergey Kulyakhtin1 and Arttu Polojärvi[1,2]
[1] Sustainable Arctic Marine and Coastal Technology (SAMCoT), Centre for Research-based Innovation (CRI), Norwegian University of Science and Technology, Trondheim, Norway
[2] Aalto University, School of Engineering, Department of Mechanical Engineering
P.O. Box 14300, FI-00076 Aalto, Finland, e-mail: sergey.kulyakhtin@ntnu.no

ABSTRACT

Ice rubble constitutes ice ridges and rubble fields, which can exert high loads on offshore structures. To characterize the ice rubble, its average properties are often desired. The properties, such as, the critical state friction angle, rely on the continuum notion of stress. This study uses discrete element model for the ice rubble. We use an average stress tensor from the contact forces between the ice blocks and compare it to the macroscopic stress tensor yielded by the boundary contact forces. These stress definitions are used to evaluate the uncertainty in defining the stress tensor for the ice rubble, and the related uncertainty in defining the critical state friction angle for the rubble.

INTRODUCTION

To describe ice rubble properties in both natural and laboratory conditions, bulk parameters are often used. These parameters, such as the internal friction angle or cohesion (ISO, 2010), rely on the continuum notion of stress and displacement fields, or in other words, continuum approach. The ice rubble, however, is a granular material (consists of individual ice blocks), and it remains unclear, when the continuum approach can be applied for it. Here we will use the results from a 2D discrete element method (DEM) simulations of bi-axial compression tests on ice rubble, and study the error in the assumed stress state of a an ice rubble sample of given size.

The average stress tensor is a stress measure that can be applied for granular materials having a large number of individual particles. This is not necessarily the case for ice rubble, as an ice rubble sample of interest may only consist of tens, or hundreds, of ice blocks. Here we use a virtual bi-axial compression test on ice rubble to study the effect of averaging-volume size on the averaged stress tensor. First, we will describe the stress measures and the simulation set-up. Second, we will investigate how accurately the averaged stress tensor inside the sample represents the stress tensor derived from the boundary forces. We will also discuss the uncertainties in defining the critical state friction angle from the test results.

STRESS MEASURES

Several definitions for the averaged stress tensor for granular media exist. Here we use a definition that uses a weighting function. This allows us to vary the size of averaging volume independently of the configuration of the particles in the sample, which further allows us to study the applicability of the stress tensor itself on describing the stress state of rubble samples of various sizes. Babic (1997) defines such an average stress tensor $\bar{\sigma}$ at point x in the quasi-static case a

$$\bar{\sigma}_{ij} = \frac{1}{2}\sum_{\alpha}\sum_{\beta>\alpha} f_i^{\alpha\beta} x_j^{\alpha\beta} \int_0^1 w(\mathbf{x} - \mathbf{x}^{\alpha} + s\mathbf{x}^{\alpha\beta}, a)ds, \tag{1}$$

Where $f_i^{\alpha\beta}$ is the ith component of the contact force between blocks α and β, $x_j^{\alpha\beta}$ is the jth component of the vector $x^{\alpha\beta}$ connecting their centroids, $\mathrm{w(x, a)}$ is a weighting function, x^{α} is the position vector of the centroid of block α, and a is the side length of a square shaped AV. The above equation is not defined for the particles contacting the sample boundary, because it uses the distance between the centroids of blocks in contact (boundaries do not have one). For w(x, a), we use the Heaviside function that is normalized over a square domain

$$w(\mathbf{x}, a) = \left(\frac{1}{a^2}\right) H(a + x_i^0 - x_i)H(a + x_j^0 - x_j). \tag{2}$$

For equilibrium under uniform stress state, the average stress inside the sample can be always computed from the forces acting on the boundaries of the sample. These forces can be used to define a boundary stress tensor σ^b as (Drescher and de Josselin de Jong, 1972)

$$\sigma_{ij}^b = \frac{1}{V}\sum_{\alpha \in boundary} f_i^{\alpha} x_j^{\alpha}, \tag{3}$$

where V is sample volume and summation is over the sample boundary. In continuum sense, under uniform boundary conditions, the stress defined by Eq. 3 should be the same everywhere in the sample. Therefore, in our bi-axial compression tests, $\bar{\sigma}$ should be equal to σ^b, if the side length a of the averaging volume is large enough.

SIMULATION SET-UP

Our simulation tool has been validated against model- and full-scale experiments on ice-structure interaction in Paavilainen et al. (2009) and Paavilainen et al. (2011), respectively. In addition it was used to model laboratory scale shear box experiments on ice rubble in Polojärvi et al. (2015). In the model, the ice blocks are rigid and the contact forces are calculated using an elastic-viscous-plastic contact force model together with an incremental coulomb friction model (see Hopkins (1992) and Paavilainen et al. (2009) for details). Further, there is no cohesion between the blocks. The block fracture is not modelled. Table 1 gives the main parameters of the simulations.

Figures 1a and b illustrate the simulation set-up for our virtual biaxial compression experiment. All simulated samples consisted of 5000 uniformly sized blocks (block size was 0.3 x 0.9 m). We generated the initial configuration the figures illustrate by starting with a very loose and random blocks configuration, and by compressing it to an initial confinement $\sigma_{xx}^b = \sigma_{yy}^b =$ 5kPa. The boundaries were rigid and frictionless.

During the actual virtual compression experiments, the cover of the box moved downwards with a strain rate of $\dot{\epsilon}_{yy}^{b} = 5 \cdot 10^{-4} s^{-1}$ (we defined vertical component of the strain as $\epsilon_{yy}^{b} = \Delta h / h_0$ where Δh is the change in the specimen height and h0 is the initial height of the sample). Chosen strain rate was low enough to ensure a quasistatic loading as was indicated by the total load on the top and bottom of the box remaining virtually equal throughout the simulations. The location of the vertical boundaries was iterated on every simulation time step so that σ_{xx}^{b} remained constant. We used the confining stress
$\sigma_{xx}^{b} = 5$kPa as it is an approximate average confinement on the top of a ten meters deep ridge keel (if continuous rubble mass is assumed). We ran simulations with five different ice block configurations to account for the potential effect of the initial ice rubble configuration.

RESULTS AND ANALYSIS

Boundary Stress

Figures 2a and b show the diagonal and off-diagonal components of the boundary stress σ^{b} (Eq. 3) as functions of strain ϵ_{yy}^{b}. The data in each figure are from five simulations with different initial configurations. Additionally, the figures show the mean of five records. As the figures show, σ_{xx}^{b} records are virtually equal to all configurations, but σ_{yy}^{b} records show some scatter. All configurations show similar behaviour: σ_{yy}^{b} first increases with high rate up to $\epsilon_{yy}^{b} \approx 0.01$, then continues to increase with slower rate, and finally reaches a plateau after $\epsilon_{yy}^{b} \approx 0.05$.
The off-diagonal stress components show random oscillations, while the mean value for different configurations is approximately zero (Figure 2b). The nonzero off-diagonal components are due to the external moments, which arise from the asymmetric boundary force distributions. To describe these stresses, the couple stress tensor, such as described in (Malvern, 1969), would be needed. This is however outside of the scope of this paper.

Table 1. Main simulation Parameters. Sample width and height depended on the initial configuration and confinement. Parameters were chosen as in Paavilainen et al. (2011)

Normal contact stiffness	4 GPa	
Tangential contact stiffness	1.5 GPa	
Normal damping factor	0.5	
Tangential damping factor	0.25	
Poisson coefficient	0.3	
Bulk material density	900 kgm^{-3}	
Plastic limit in contact (σ_p)	2 MPa	
Ice-ice friction coefficient	0.3	
Ice-boundary friction coefficient	0.0	
Block length (l)	0.9 m	
Block thickness	0.3 m	
Initial sample width	≈40 m	
Initial sample height (h_0)	≈40 m	
Strain rate		$5 \cdot 10^{-4}$
Confining stress	5 kPPa	

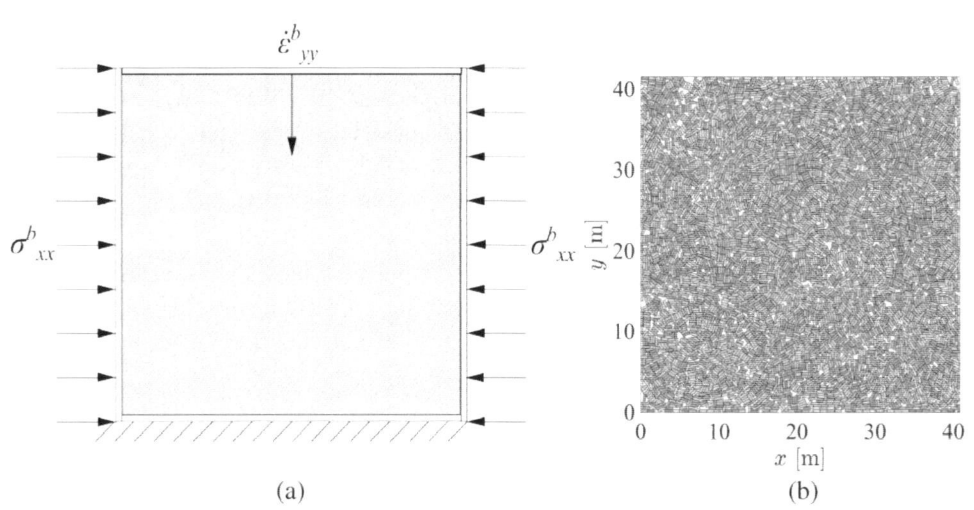

Figure 1. Virtual compression experiments: (a) simulation set-up and (b) a 5000 block sample in its initial configuration. The confining stress σ^b_{xx} was kept equal to 5 kPa and the cover moved down at the constant strain rate of $5\times10^{-4}s^{-1}$.

due to the external moments, which arise from the asymmetric boundary force distributions. To describe these stresses, the couple stress tensor, such as described in (Malvern, 1969), would be needed. This is however outside of the scope of this paper.

Convergence of averaged stress to boundary stress

In a continuum approximation, a uniform boundary stress imposes a uniform stress throughout the sample. This means that the average stress given by Eq. 1 should be equal to the boundary stress given by Eq. 3 at all material points of the sample. This will not hold if the averaging volume is small and only few blocks are used to compute $\bar{\sigma}$. However, $\bar{\sigma}$ should approach σ^b as the averaging volume size increases.

To study the convergence of $\bar{\sigma}$ we first chose to study the stress state of the sample at strain $\epsilon^b_{yy} = 0.15$ where the component σ^b_{yy} is approaching an approximately constant value. (We note that the stress $\bar{\sigma}$ can naturally be computed for any given strain.) Figures 3a and b show $\bar{\sigma}_{xx}$ and $\bar{\sigma}_{yy}$ values, which are here normalized by the corresponding boundary stress components, as a function of the averaging volume side length a divided by by the block length l. The figures clearly show that the diagonal components $\bar{\sigma}_{ii}$ of the average stress approach the corresponding components of σ^b_{ii}.

However, in some cases, the average stress for even the largest averaging volume differs by $>10\%$. This can be explained by non-uniform contact force distribution and the applicability of Eq. 1 and is described by Figure 4. The figure illustrates the contact force distributions for two samples at the initial state of the test $\epsilon_{yy} = 0$ (only initial confinement of 5 kPa) and at strain $\epsilon_{yy} = 0.15$ on its left and right columns, respectively. For clarity, only the forces that have a value of at least 20% of the maximum contact force are shown in each figure.

In both cases, the contact force distribution is seen to be initially uniform (left column of the figure). For the configuration in Figure 4a, the distribution also remains somewhat uniform up to $\epsilon = 0.15$, as the top right figure shows. The configuration in Figure 4b, however, has the majority of the load carried by the blocks close to the left boundary at $\epsilon = 0.15$. The average stress $\bar{\sigma}$ (see Eq. 1) does not apply for these blocks, since they are in contact with the boundaries. This leads to σ not converging to σ^b. We studied the convergence of $\bar{\sigma}_{ij}$ for all time instances of our simulations, and found that about 7% of the cases behaved like this. As this behavior depends on the boundaries, these cases are excluded from the following analysis.

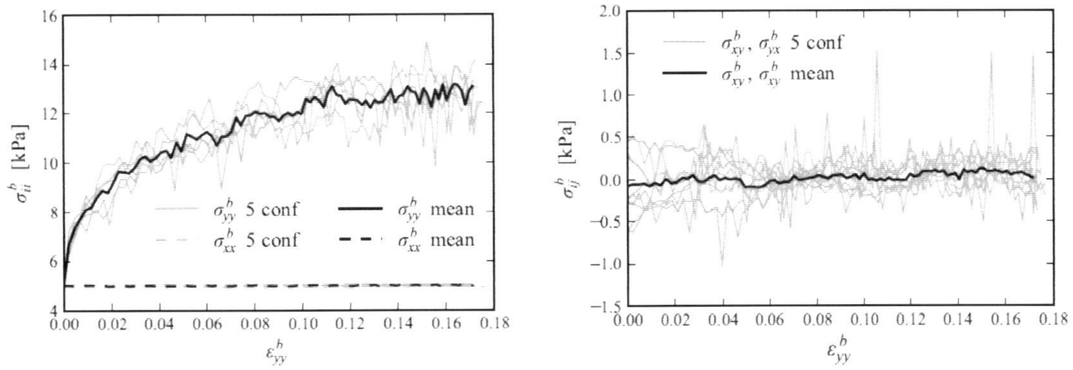

(a) (b)

Figure 2. The diagonal (a) and off-diagonal (b) components of boundary stress tensor σ_{ij}^b as a function of strain ϵ_{yy}^b for five configurations and means of these configurations

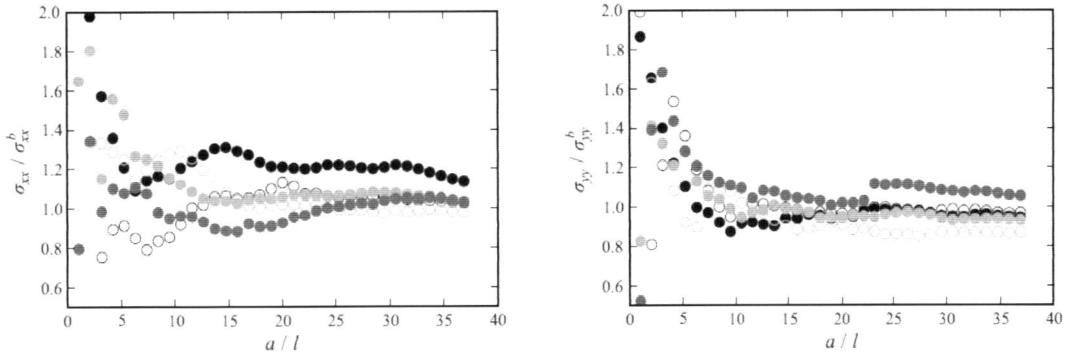

(a) (b)

Figure 3. The diagonal stress components nomalised by the corresponding boundary stress components ($\bar{\sigma}_{ii}/\sigma_{ii}^b$) as a function of AV size normalised by block length (a/l) for five different configurations

We also calculated the $\bar{\sigma}_{ij}$ values for random points within our sample using a number of AV side lengths α. The aim in this study was two-fold: (1) to study the probability and the magnitude of error, if the sample size is chosen after some given value, and (2) to provide a proof-of-concept for the applicability of our approach on approximating the macroscopic stress state of the sample. In addition, the study indicates whether our analysis is sensitive to the choice of the stress averaging point. (Above, the $\bar{\sigma}_{ij}$ was always calculated at the center point of the sample.) We calculated $\bar{\sigma}_{ij}$ for 100 random points for all of our samples at every 0.01 strain increments.

Figures 5a and b show histograms that respectively illustrate the distribution of the normalised stresses $\bar{\sigma}_{xx}/\sigma_{xx}^b$ and $\bar{\sigma}_{yy}/\sigma_{yy}^b$ for random points of the sample. The histograms illustrate that in all of the cases the normalized stress values are distributed around 1.0, and that their variation gets smaller as averaging volume size increases. The relative standard deviations for $\alpha = 5l$ are 28 % and 27 % for σ_{xx} and σ_{yy}, respectively. For averaging volumes having $\alpha = 20\,l$, the relative standard deviations are only 7 % and 9 % for σ_{xx} and σ_{yy}, respectively.

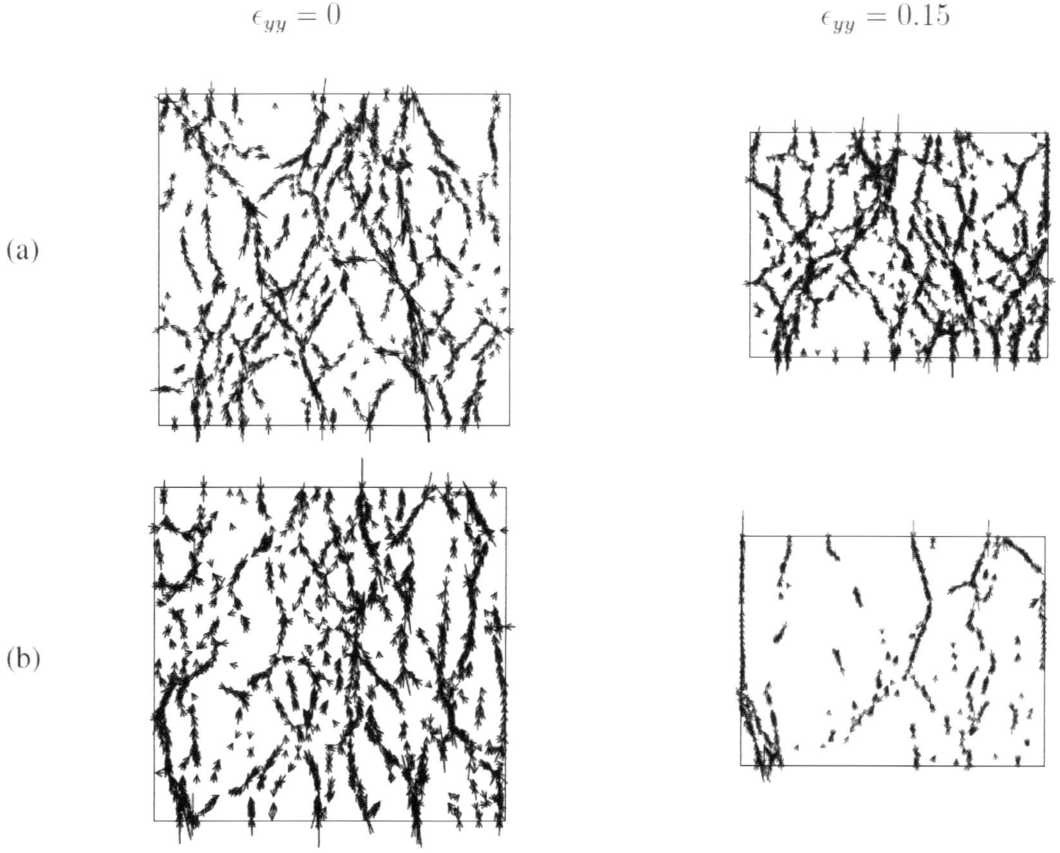

Figure 4. Two different initial configurations (a) and (b); and two instances of biaxial compression test $\varepsilon = 0$ and $\epsilon = 0.15$; the contact force vectors are shown that are ≥20% of the maximum contact force in the given instance

Variation in critical state friction angle

There is also uncertainty in the material parameters that are measured from the bi-axial test on insufficiently large sample. We demonstrate this by using the critical state friction angle ϕ_{cs} In the bi-axial compression test, ϕ_{cs} is calculated using the ratio of the major (here σ_{yy}) and minor (σ_{xx}) stress components as the ratio reaches a constant value (here this occurred at about $\epsilon = 0.15$). The critical state friction angle is given by

$$\phi_{cs} = \arcsin\left(\frac{\sigma_{yy}/\sigma_{xx} - 1}{\sigma_{yy}/\sigma_{xx} + 1}\right). \tag{4}$$

Since there is a functional dependence of ϕ_{cs} on σ_{yy}/σ_{xx}, the relative standard deviation of the latter, std[σ_{yy}/σ_{xx}]/mean[σ_{yy}/σ_{xx}], can be used to calculate the relative standard deviation of ϕ_{cs}. This is given by

$$\frac{\text{std}[\phi_{cs}(\sigma_{yy}/\sigma_{xx})]}{\phi_{cs}(\text{mean}[\sigma_{yy}/\sigma_{xx}])} = \sqrt{\left(\frac{\partial\phi_{cs}}{\partial\left(\sigma_{yy}/\sigma_{xx}\right)}\bigg|_{\frac{\sigma_{yy}}{\sigma_{xx}}=\text{mean}\left[\frac{\sigma_{yy}}{\sigma_{xx}}\right]}\right)^2} \frac{\text{std}[\sigma_{yy}/\sigma_{xx}]}{\phi_{cs}(\text{mean}[\sigma_{yy}/\sigma_{xx}])}$$
$$= \frac{1}{\sqrt{\text{mean}[\sigma_{yy}/\sigma_{xx}]\left(\text{mean}[\sigma_{yy}/\sigma_{xx}]+1\right)^2}} \frac{\text{std}[\sigma_{yy}/\sigma_{xx}]}{\phi_{cs}(\text{mean}[\sigma_{yy}/\sigma_{xx}])} \quad (5)$$

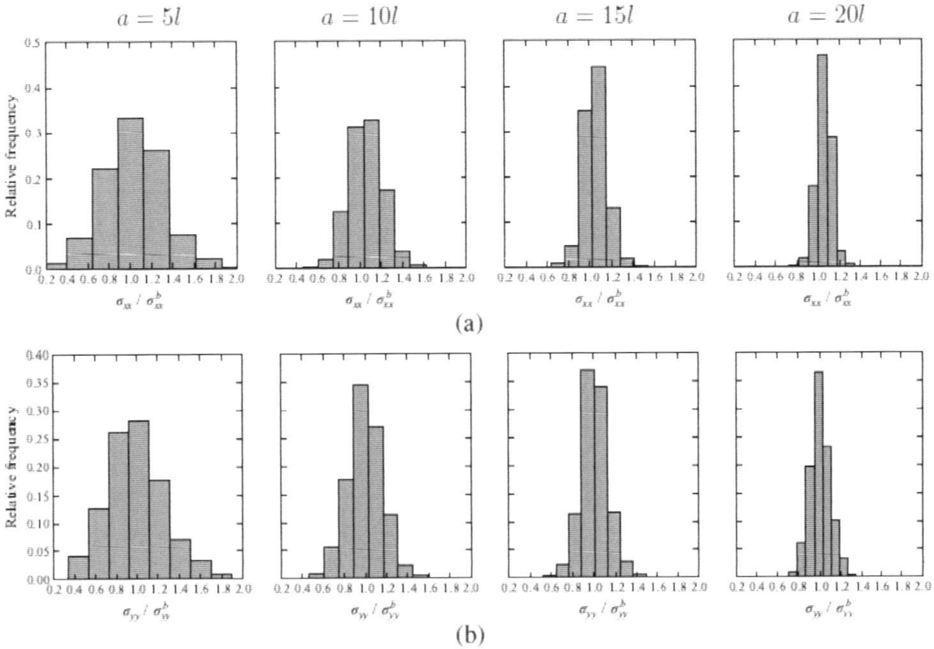

Figure 5. Histograms describing the distributions of the ratios of averaged stress to boundary stress ($\bar{\sigma}_{ij}/\sigma_{ij}^b$) calculated about 100 random points for all samples at intervals of 0.01 strain increment: graphs of (a) show $\bar{\sigma}_{xx}/\sigma_{xx}^b$ and (b) $\bar{\sigma}_{yy}/\sigma_{yy}^b$. The calculation was done using four different AV side lengths α (here a multiple of block length l) as indicated above each column.

We again computed the stress components in 100 random points to study the variation of σ_{yy}/σ_{xx} for averaging volumes of different sizes with the results shown in Figure 6. Here we only use data points with $\epsilon_{yy}^b \geq 0.15$, since only they should be used to compute ϕ_{cs} (approximately constant stress ratio was reached with $\epsilon_{yy}^b \geq 0.15$). The figure shows that, similarly to normalized stress, the variation in σ_{yy}/σ_{xx} gets continuously smaller as the averaging volume size increases.

Table 2. The relative standard deviations of the critical state friction angle (std[$\phi_{cs}(\sigma_{yy}/\sigma_{xx})$/ ϕ_{cs}(mean[σ_{yy}/σ_{xx}])]

	AV size			
	5 l	10 l	15 l	20 l
std[$\phi_{cs}(\sigma_{yy}/\sigma_{xx})$/ ϕ_{cs}(mean[σ_{yy}/σ_{xx}])	30%	18%	13%	10%

Table 2 summarises the relative standard deviations of ϕ_{cs} calculated by Eq. 5. These are 30% and 10% for the smallest (α/l=5) and largest (α/l= 20) averaging volumes, respectively, which is similar to the relative standard deviations of σ_{xx} and σ_{yy}. Table 2 also indicates that if a sample of only 10 block lengths is used to measure the critical state friction angle, even for the ice blocks with identical shape/size and properties, the variation in measured ϕ_{cs} of 18% should be expected.

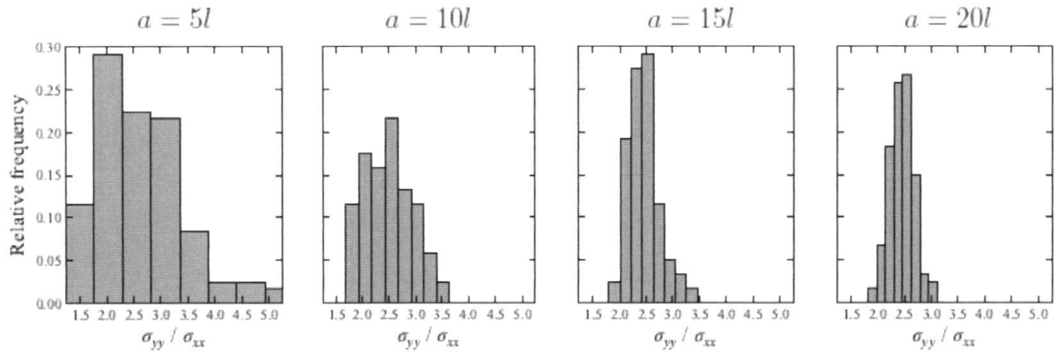

Figure 6. Histograms of σ_{xx}/σ_{yy} ratio computed in 100 random points for all samples at intervals of 0.01 strain increment starting from ϵ_{yy} = 0.15. The calculation was done using four different AV side lengths α (here a multiple of block length l) as indicated above each column.

CONCLUSIONS

We studied the applicability of the continuum approach in measuring the stress state of an ice rubble sample in bi-axial numerical experiment. Study was based on 2D DEM. The averaged stresses usually converged towards the boundary stress of the sample. In about 7% of cases this did not occur due to boundary conditions. For the rubble samples of size 5 block lengths, the ratio of averaged to boundary stress had a relative standard deviation of approximately 30%. This fell to about 10% with samples sizes of 20 block lengths. We also studied the effect of stress variation on the critical state friction angle. Its relative standard deviation was similar to those of the ratio of average to boundary stress.

ACKNOWLEGMENTS

The authors wish to acknowledge the support from the Research Council of Norway through the Centre for Research-based Innovation SAMCoT and the support from all SAMCoT partners.

REFERENCES

Babic, M. (1997). Average balance equations for granular materials. International journal of engineering science, 35:523–548.

Drescher, A. and de Josselin de Jong, G. (1972). Photoelastic verification of a mechanical model for the flow of a granular material. Journal of the Mechanics and Physics of Solids, 20(5):337–340.

Hopkins, M. (1992). Numerical simulation of systems of multitudinous polygonal blocks.

Technical
Report 92-22, Cold Regions Research and Engineering Laboratory, CRREL. 69 p.
ISO (2010). Petroleum and natural gas industries - Arctic offshore structures. Standard, International
Organization for Standardization, ISO 19906, Geneva, Switzerland.
Malvern, L. (1969). Introduction to the Mechanics of a Continuous Medium. Prentice-Hall, Inc.
Paavilainen, J., Tuhkuri, J., and Polojärvi, A. (2009). 2D combined finite–discrete element method to model multi-fracture of beam structures. Engineering Computations, 26(6):578–598.
Paavilainen, J., Tuhkuri, J., and Polojärvi, A. (2011). 2D numerical simulations of ice rubble formation process against an inclined structure. Cold Regions Science and Technology, 68(1-2):20–34.
Polojärvi, A., Pustogvar, A., and Tuhkuri, J. (2015). DEM simulations of direct shear box experiments
of ice rubble: Force chains and peak loads. Cold Regions Science and Technology, (116):12–23.

POAC'17
Busan, KOREA

Proceedings of the 24th International Conference on
Port and Ocean Engineering under Arctic Conditions
June 11-16, 2017, Busan, Korea

Experimental and Numerical Investigation of Ship–Ice Interactions in Pack Ice

Hyoil Kim[1], Namkyun Im[2], Junji Sawamura[1]
[1] Department of Naval Architecture and Ocean Engineering, Osaka University, JAPAN
[2] Division of Marine Transportation System, Mokpo Maritime University, Jeonnam, KOREA

ABSTRACT

Ship–ice interaction in pack ice is important issue for safe ship operation in ice-covered water. This paper presents an investigation of ship–ice interaction when an ice-going ship turns and moves straight in pack ice. Free-running model tests were conducted to elucidate ship–ice interaction in pack ice. A 1/100 scale ship model and disk-shaped synthetic ice floes were used in a free-running model test. The model ship, propelled with a constant engine RPM at full loading condition, had a GPS and a motion sensor installed onboard to measure the ship's trajectory and motion. Ice removal was measured using a vision inspection technique. In free-running tests, the model ship passed through pack ice of three ice concentrations (10%, 20%, and 30%) with no rudder action. In turning circle tests, the model ship passed through pack ice with 7.2% ice concentration. The authors proposed a numerical model to calculate ice loads and ship motions in pack ice based on Discrete Element Method. To validate the numerical model, ship motions and ice removal in the model tests were compared with the simulations. Numerical results showed good agreement with experimental test results. The effects of ice concentration on ship–ice interactions were investigated experimentally and numerically.

KEY WORDS: Ship–ice interaction; Pack ice; Discrete element method; Free-running model test; Turning circle test

INTRODUCTION

Sea ice in the Arctic region has been decreasing rapidly because of global warming, which has led to opening of Arctic routes in summer. Cho et al. (2013) anticipated that operations in level ice will decrease, but the operation in brash ice, pack ice, and re-frozen ice will increase dramatically. They suggested more effective model test methods for pack ice conditions in a square-type ice tank to reduce the test time and cost. Woolgar and Colbourne (2010) conducted towing tests in an ice basin using an offshore model ship to investigate factors affecting pack ice loading. Results show that the hull–ice friction coefficient is a main factor influencing the pack ice load. In addition, the relations between the ice load and various factors such as the ship speed, ice concentration, and ice floe size were studied. Huang et al. (2016) also conducted towing tests in the ice tank to investigate the ice failure processes and the resistance of a large transport vessel navigating in pack ice. Kim et al. (2013) measured the resistances of an icebreaking cargo model ship in a towing tank using wax-type synthetic ice floes. Their results show that the towing tank test using synthetic ice can provide a means of comparison with results obtained from ice basin tests and simulations. Sawamura et al. (2016) conducted the model test with synthetic ice in a towing tank, and demonstrated that the ice resistance of the model test with synthetic ice shows good agreement with the ice resistance formula. A model test with synthetic ice in a towing tank can be alternated with model ice tests conducted in an ice tank.

Simulation studies using Discrete Element Modeling (DEM) have been used to calculate the ice loads acting on ice-going ships in pack ice (Hopkins, M. A., and Tuhkuri, J., 1999; Ji et al., 2013). The maneuverability of an ice-going ship in pack ice has been studied numerically by carrying out turning circle and zigzag tests. Full-scale ice trials and simulations done by Zhou and Peng (2013, 2016) revealed that the turning diameter in level ice is approximately three times greater than that in open water. Oppositely, Zhan et al. (2010) found from simulations that the turning diameter in pack ice was less than that in open water. Especially, the turning diameter in a larger ice concentration (90%) was smaller than that in the lower ice concentration (70%) (Zhan and Molyneux, 2012). Kim and Sawamura (2016) also conducted turning circle tests in six ice concentrations to elucidate the relation between ice concentration and turning ability numerically. They concluded that turning capabilities such as Advance, Transfer, and Tactical diameter became smaller as the ice concentration increased. As described above, ship–ice interactions in pack ice have not been understood yet.

For this study, free-running model tests using a KVLCC1 model ship and synthetic ice floes were conducted to elucidate ship–ice interaction and turning capabilities in pack ice. A numerical model of an ice-going ship in pack ice was developed by coupling the MMG model for KVLCC1 with Discrete Element Modeling (DEM) for ice loads acting on the vessel. Free-running tests were conducted with three ice concentrations (10%, 20%, and 30%). Based on the test results, contact stiffness and hull-ice friction coefficient affecting ship–ice contact forces in DEM were ascertained through comparison of ship speed reduction rates in model test and simulation. In addition, ice floe trajectories in free-running tests were analyzed using a vision inspection to clarify and quantify ship–ice interactions in pack ice as well as simulations. Results related to ice floe trajectories provide a better understanding of physical phenomena related to ship–ice collisions depending on the ice concentration and hull contact position. Turning circle tests were performed in open water and pack ice conditions. Turning ability (Advance and Transfer) was calculated using GPS and compared. In addition, turning circle tests were simulated in various ice concentrations to investigate the relation between ice concentration and turning capabilities.

FREE-RUNNING MODEL TEST USING SYNTHETIC ICE

On October 19–20, 2016, free-running model tests for an ice-going ship navigating in pack ice conditions were conducted at the outdoor pool at Mokpo Maritime University, Korea. According to the Korea Meteorological Agency, maximum and average wind speeds were reported as 4.0 m/s and 2.38 m/s during 09:00–18:00. The pool dimensions are 50 m long, 20 m wide and 2.0 m deep. The pool was filled to 1.2 m depth with seawater, which is sufficient to prevent the squat effect. A 1/100 scale KVLCC1 model ship was used for the free-running model test, which consists of various systems such as device control systems, sensor and data acquisition systems, and a wireless communication system for remote control and real-time monitoring (Im, 2010). In the sensor system, a GPS with a 10 Hz sampling rate and an IMU sensor with a 50 Hz sampling rate were installed on board to measure the ship trajectory and motion. Disk-shaped synthetic polypropylene (955 kg/m^3 density) ice floes of 40 cm diameter and 2 cm thickness were used for tests. A video camera was installed vertically over the water surface to observe the ice floe motions. Commercial software (NI Vision Builder AI 2015) was used to calculate the ice floe motion from camera images. The model ship was propelled with a constant engine revolutions per minute at a full loading condition (20.8 cm draft), which let her speed reach the target speed (approximately 0.5 m/s) in open water. Free-running and turning circle tests were conducted. In free-running tests, the pack ice area was 12 m long and 3 m wide. It was comparted by a floating fence and course ropes as depicted in Figure 1(a). Cho et al. (2013) and Kim et al. (2013) reported that the ice channel width should be at least six times greater than the ship's breadth. For this study, the pack ice channel width was approximately 5.2 times the ship breadth. Using no rudder action, the model ship passed through pack ice of three given ice concentrations. In the turning circle test, the pack ice area was 12 m long and 12 m wide. When the model ship enters the pack ice area, a steering order (hard port) was commanded. The Advance and Transfer were measured and compared with those achieved in open water. All test conditions are presented in Table 1.

NUMERICAL MODELING OF SHIP–ICE INTERACTION

Motion Equations in Ice-Covered Water

The vessel and ice floes are assumed as rigid bodies. Neglecting wave loads and vertical motions, the 3DOF surge–sway–yaw motion equations described in the ship-fixed coordinate system are expressed in vectorial form (Fossen, 2011).

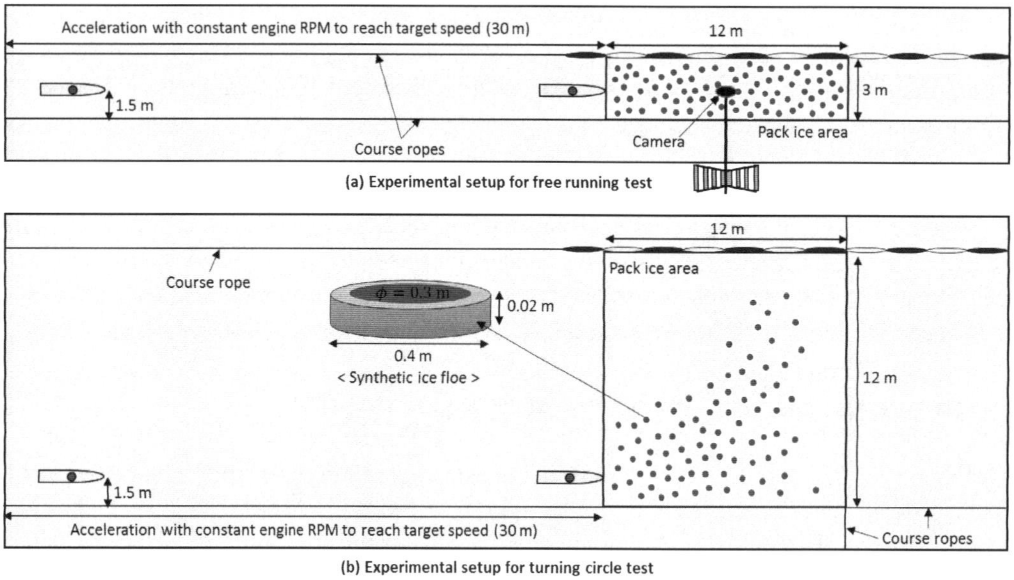

Figure 1. Experimental setup.

Table 1. Summary of test conditions

Test type	Steering order (°)	Test No.	Ice concentration (%)	No. of ice floes
Free-running	0	Case 01, 02, 03	Open water	
		Case 04, 05, 06	10.1	29
		Case 07, 08, 09	20.2	58
		Case 10, 11, 12	29.0	83
Turning circle	-35	Case 13, 14, 15	Open water	
		Case 16, 17, 18, 19	7.2	83

$$\mathbf{M}_{\mathrm{RB}}\dot{\mathbf{v}} + \mathbf{C}_{\mathrm{RB}}(\mathbf{v})\mathbf{v} = \begin{bmatrix} X_H \\ Y_H \\ N_H \end{bmatrix} + \begin{bmatrix} X_P \\ 0 \\ 0 \end{bmatrix} + \begin{bmatrix} X_R \\ Y_R \\ N_R \end{bmatrix} + \begin{bmatrix} X_I \\ Y_I \\ N_I \end{bmatrix} \tag{1}$$

Therein, $\mathbf{M}_{\mathrm{RB}}$ is the inertia matrix of the ship; $\mathbf{C}_{\mathrm{RB}}(\mathbf{v})$ is the Coriolis and Centripctal matrix.

$$\mathbf{M}_{\mathrm{RB}} = \begin{bmatrix} m+m_x & 0 & 0 \\ 0 & m+m_y & mx_G \\ 0 & mx_G & I_{ZZ} \end{bmatrix}, \quad \mathbf{C}_{\mathrm{RB}} = \begin{bmatrix} 0 & -(m+m_y)r & -mx_G r \\ (m+m_x)r & 0 & 0 \\ mx_G r & 0 & 0 \end{bmatrix} \tag{2}$$

Therein, X_H, Y_H, and N_H are hydrodynamic forces and moment acting on the hull; X_P, X_R, Y_R and N_R are the hydrodynamic forces and moment generated by the propeller and rudder; X_I, Y_I, and N_I respectively denote the ice-induced forces and the moment, which are obtained from the sum of all ship–ice contact. The hydrodynamic forces and moments in Eq. (1) attributable to the hull, propeller, and rudder are expressed by the MMG model (Yasukawa et al., 2015) as follows.

$$X_H = (1/2)\rho L_{pp} dU^2(-R'_0 + X'_{vv}v'^2_m + X'_{vr}v'_m r' + X'_{rr}r'^2 + X'_{vvvv}v'^4_m) \tag{3a}$$

$$Y_H = (1/2)\rho L_{pp} dU^2(Y'_v v'_m + Y'_R r' + Y'_{vvv}v'^3_m + Y'_{vvr}v'^2_m r' + Y'_{vrr}v'_m r'^2 + Y'_{rrr}r'^3) \tag{3b}$$

$$N_H = (1/2)\rho L^2_{pp} dU^2(N'_v v'_m + N'_R r' + N'_{vvv}v'^3_m + N'_{vvr}v'^2_m r' + N'_{vrr}v'_m r'^2 + N'_{rrr}r'^3) \tag{3c}$$

$$X_P = (1-t_P)\rho n_P^2 D_P^2 K_T(J_P) \tag{3d}$$

$$X_R = -(1-t_R)F_N \sin\delta \tag{3e}$$

$$Y_R = -(1+a_H)F_N \cos\delta \tag{3f}$$

$$N_R = -(x_R + a_H x_H)F_N \cos\delta \tag{3g}$$

Therein, prime ′ attached to the symbol signifies a non-dimensionalized value. t_P is the thrust deduction factor. n_P and D_P respectively denote the revolutions and propeller diameter. K_T is the thrust coefficient. t_R is the steering resistance deduction factor. a_H and x_H are the coefficients of lateral force because of rudder deflection. Also, x_R represents the longitudinal position of the rudder. The rudder normal force F_N is given as presented below.

Table 2. Characteristics and coefficients of KVLCC1 model ship

Model ship		Propeller / Rudder	
Scale ratio	1/100	Thrust deduction factor, t_P	0.22
L_{pp} (m)	3.20	Engine, n_P (RPM)	64.955
Breadth (m)	0.58	Propeller diameter, D_P (m)	0.0986
Draft (m)	0.208	Rudder area, A_R (m^2)	0.01125
Ship mass, m (kg)	320.415	Rudder aspect ratio, Λ	1.827
Added mass, m_x (kg)	24.015	Steering resistance deduction factor, t_R	0.387
Added mass, m_y (kg)	243.423	Rudder force increase factor, a_H	0.312
Moment of inertia, I_{zz} (kgm^2)	36.918	Longitudinal position of acting point, x_H (m)	-1.485
Longitudinal position of COG, x_G (m)	0.3584	Longitudinal position, x_R (m)	-1.60

$$F_N = \frac{1}{2}\rho \frac{6.13\Lambda}{\Lambda + 2.25} A_R U_R^2 \sin \alpha_R \tag{4}$$

Therein, Λ and A_R respectively denote the aspect ratio of the rudder and profile area of the movable part of the rudder; U_R and α_R stand for the effective rudder inflow velocity and angle. Table 2 presents the main particulars of the model ship and coefficients for the MMG model.

Ice Loads Acting on a Ship in Pack Ice

Ice loads acting on the vessel in pack ice were calculated using 2D Discrete Element Method (Hopkins, 1999; Ji et al., 2013). Contact forces induced by collisions between two DEM elements are based on linear force-displacement law. The contact force has normal and tangential components at the contact point as below.

$$\begin{aligned} F_n^n &= k_n \delta_{is} - \eta \mathbf{V}_{is} \cdot \mathbf{n} \\ F_t^n &= \min\left[F_t^{n-1} - k_t \Delta t (\mathbf{V}_{is} \cdot \mathbf{t}),\ \mu F_n^n\right] \end{aligned} \tag{5}$$

where, F_n^n signifies the normal contact force, superscript n denotes the current time step, k_n denotes the normal contact stiffness, η represents the normal contact viscosity, and $\mathbf{V}_{is}$ and $\mathbf{n}$ respectively denote the relative velocity of floe i with respect to the ship and unit normal vector at the contact point. F_t^n and F_t^{n-1} represent the tangential forces at time step n and $n-1$, k_t stands for the tangential contact stiffness, and μ is the hull–ice friction coefficient. The tangential force F_t is scaled as $|F_t| = \mu F_n$ when it exceeds its limit. The elements for the ice floe and ship hull are modeled, respectively, as 2D circle and 2D polygonal hull form as shown in Figure 2. The overlap distance between two elements (ship–ice and ice–ice) is determined as below.

$$\begin{aligned} \delta_{is} &= |\Delta_{is}| - R_i, && \text{(Ship – ice)} \\ \delta_{ij} &= |\Delta_{ij}| - R_i - R_j, && \text{(Ice – ice)} \end{aligned} \tag{6}$$

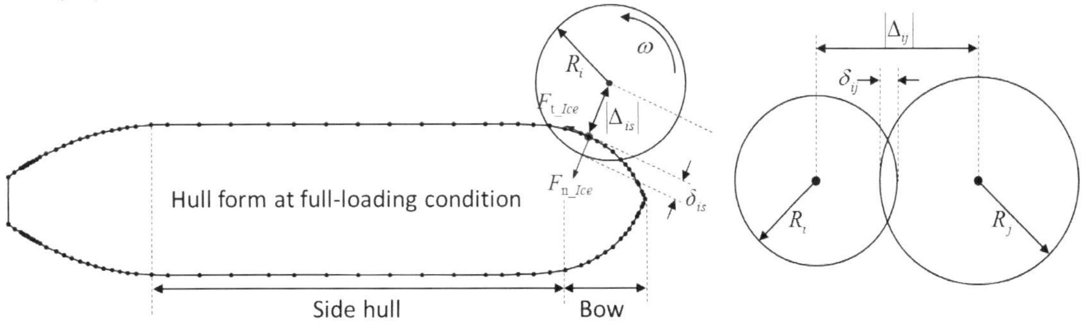

Figure 2. Scenarios of ship–ice and ice–ice contacts.

Therein, δ_{is} stands for the overlap distance between floe i and the ship. In those equations, $|\Delta_{is}|$ signifies the distance between the ice floe's center and hull, δ_{ij} denotes the overlap distance between floe i and floe j, and $|\Delta_{ij}|$ stands for the distance between two ice floes' centers. R_i and R_j respectively represent the radii of floe i and floe j. When values δ_{is} and δ_{ij} are negative, the two elements are mutually contacting.

RESULTS AND DISCUSSION

Comparison of Numerical Simulation and Model Test

A numerical model of an ice-going ship in pack ice was developed by coupling the MMG model for KVLCC1 with Discrete Element Modeling (DEM) for ice loads acting on the vessel. The contact stiffness and hull–ice friction coefficient affecting ship–ice contact forces in DEM were ascertained through comparison of ship speed reductions in model tests and simulations. Figures 3(a) and 3(b) present time histories while passing through the pack ice region. Figure 3(c) presents a comparison of speed reductions in the model test and simulation. In this numerical model, the normal contact stiffness k_n and hull–ice friction coefficient μ were chosen as 5.87 kN/m and 0.2, respectively. As shown in Figures 3(a), 3(b), and 3(c), the numerical results show good agreement with those of the model test. The ship speed decreased linearly while passing through the pack ice. The ship speed gradient became steeper depending on the ice concentration.

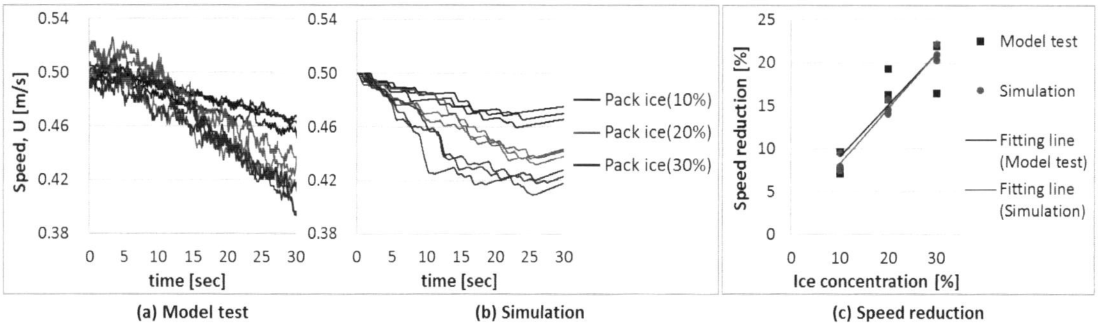

Figure 3. Time histories and speed reductions of ship speed in free-running test.

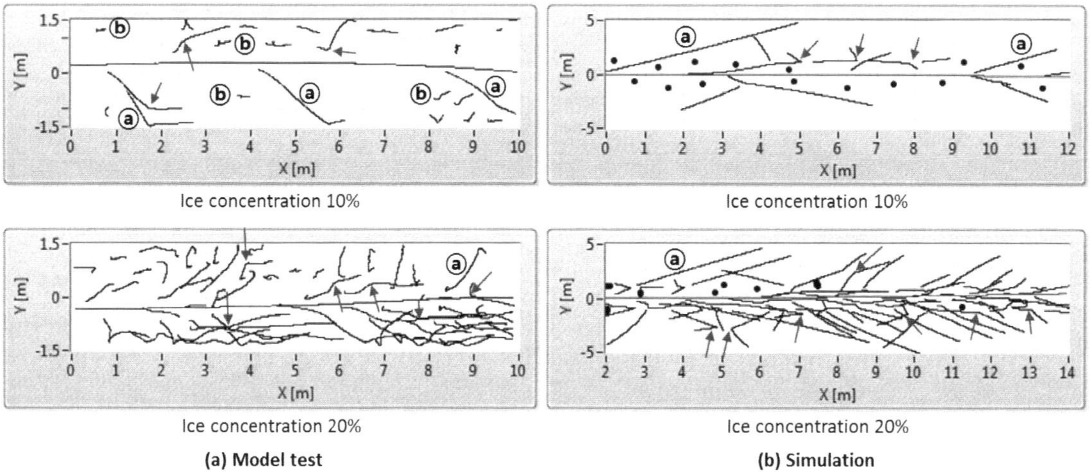

Figure 4. Comparison of ice floes' trajectories in experiment and simulation.

Ship–ice Interactions in Pack Ice

(1) Ice floe trajectories

Figure 4(a) shows ice floe trajectories from the free-running tests, as measured using a vision inspection. A long centerline represents the ship trajectory. Figure 4(b) shows the simulated results for the same ice conditions with the model test. The red solid line shows the ship trajectory. The black lines are ice floe trajectories. The black dots represent the ice floes that have never contacted with the ship. A few straight lines pointed as ⓐ represent that the ice floes were pushed aside and that they have not interacted with other ice floes after ship–ice collisions at the bow. Some short lines pointed as ⓑ represent that ice floes stationed in their own positions or drifted by the wake flow generated by ship motion in spite of the lack of ship–ice contact.

It was observed from the model test and simulation that the number of ship–ice collisions increased with the ice concentration, which is one of the largest contributions of the increment of ice resistance and the decrease of ship speed in pack ice. At low ice concentrations, ship–ice interactions mostly occurred at the bow; they seldom occurred at the side hull. In addition, ice–ice collisions occurred only slightly at low ice concentrations. Once ice in low ice concentration collided with the bow, it only slightly interacted with other ice and the bow again. However, as the ice concentration increased, ice–ice collisions were increased along with ship–ice interactions. These ice–ice interactions are apparent in the curved lines pointed as red arrows in Figure 4. Moving directions of ice floes were changed suddenly by the ice–ice interaction with neighboring ice floes. Ice that collided with the bow at high ice concentrations more frequently interacted with other ice and the bow after the first ship–ice collision at the bow. Moreover, some ice interacted with the side hull. This increase of ice–ice collisions induced ship–ice contacts at the side hull, which engenders increased friction force. The friction force causes the increase of ice resistance and the decrease of ship speed in pack ice.

(2) Ship–ice interactions

To investigate details regarding ship–ice interactions at the bow and side hull, the model ship was divided into a bow and side hull, as depicted in Figure 2(a). Free-running tests (constant engine RPM and no rudder action) were simulated with six ice concentrations: 10%, 20%, 30%, 40%, 60%, and 80%. Three random pack ices were made in the respective ice concentration. Consequently, 18 simulations were conducted. Figure 5 presents one result obtained for ice concentration of 80%. At the beginning of simulation for 3–4 s, the ice resistance caused by ship–ice interactions at the bow mainly appeared. However, the ice resistance attributable to ship–ice interaction at the side hull gradually increased and then fluctuated depending on the ship–ice interaction at the side hull.

Figure 6 portrays the relation between ice concentration and ice resistance in pack ice. The ice resistance at the bow increased with the ice concentration, which means that the number of ship–ice collisions at the bow also increased with the ice concentration. The ice resistance generated by ship–ice interactions at the side hull was increased with the ice concentration, which means that the number of ship–ice collisions at the side hull also increases with the ice concentration. The ice resistance at the side hull was extremely low (negligible) at low concentrations. However, the portion of ice resistance at the side hull with respect to the total ice resistance grew with increased ice concentration. Consequently, in high ice concentrations, the ice resistance (ice friction) at the side hull is also an important component to evaluate the ice resistance in pack ice.

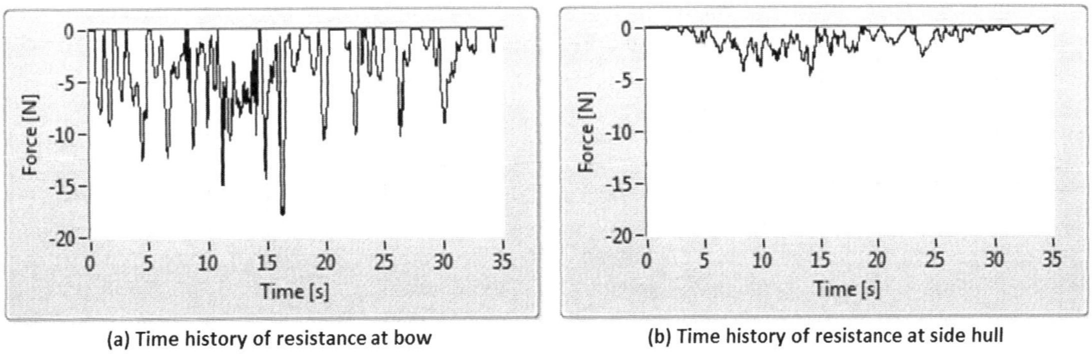

Figure 5. Time histories of ice resistances at bow and side hull.

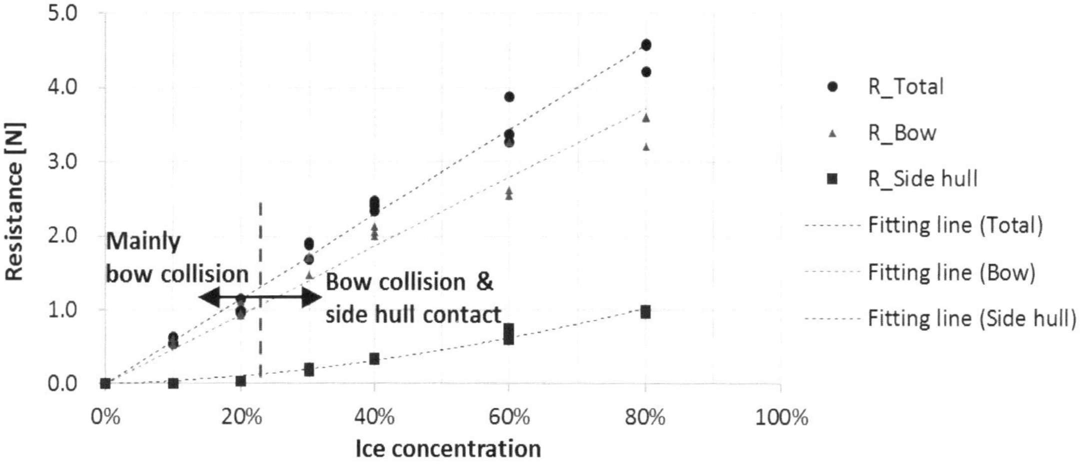

Figure 6. Relation between ice concentration and ice resistance in pack ice.

Turning Ability in Pack Ice

In this study, turning circle tests using a free-running model ship were conducted to investigate the turning ability (Advance and Transfer). The pack ice area was 12 m × 12 m. In all, 83 synthetic ice floes were used for the test and were arranged intentionally on the expected ship trajectory to increase ship–ice interactions, as shown in Figure 1(b). Consequently, even though the ice concentration of the entire pack ice region was 7.2% (= Sum of the ice floes' area / Pack ice area 12 m × 12 m), the turning circle test was locally conducted at higher ice concentrations. The ice concentrations in the model test were approximately 20%–30%. Figure 7 presents ship trajectories acquired from the model test (a) and simulation (b). In Figure 7(a), red lines show the trajectories in open water; black lines show those in pack ice. The average turning radius in pack ice was smaller than that in open water. Figure 7(b) portrays turning circle tests simulated in three ice concentrations as well as in open water. The turning radius became smaller as the ice concentration increased. Table 3 presents turning abilities (Advance and Transfer) obtained from the model tests and simulations. The turning abilities become smaller as the ice concentration increases. The reduction of turning ability in pack ice might be related to speed reduction induced by the increase of the ship–ice contact. However, further investigations must be conducted to elucidate the relation between the ice concentration and turning ability.

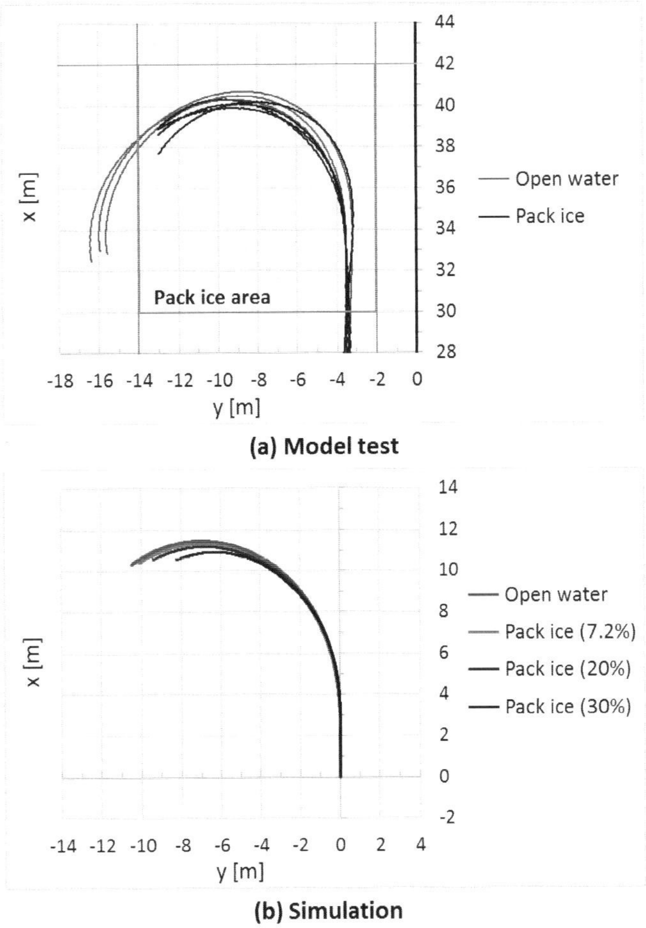

Figure 7. Turning circle plots with data from free-running model tests and simulation.

Table 3. Comparison of turning abilities

Ice condition	Model test		Ice condition	Simulation	
	Advance (m/L_{pp}) (Reduction ratio)	Transfer (m/L_{pp}) (Reduction ratio)		Advance (m/L_{pp}) (Reduction ratio)	Transfer (m/L_{pp}) (Reduction ratio)
Open water	3.746	2.048	Open water	3.506	1.650
Pack ice	3.645 (2.71%)	1.946 (4.96%)	7.2%	3.463 (1.24%)	1.653 (-0.16%)
			20%	3.419 (2.49%)	1.592 (3.56%)
			30%	3.342 (4.69%)	1.498 (9.23%)

CONCLUSIONS

For this study, free-running model tests using a free-running model ship (1/100 scale KVLCC 1) and synthetic ice floes were conducted at an outdoor pool to investigate the ship–ice interaction of an ice-going ship in pack ice. In addition, a numerical model was developed by coupling the MMG model of KVLCC1 with DEM method. Main results from the free-running tests and simulations include the following:

(1) In our developed numerical model, contact stiffness and the hull–ice friction coefficient affecting ship–ice contact forces in DEM were ascertained from comparisons of ship speed reduction rates in free-running tests. The time history of ship speed in the numerical simulations shows good agreement with those of model tests.

(2) At low ice concentrations, ship–ice interactions mostly occurred at the bow and seldom at the side hull. As the ice concentration increased, ice resistance attributable to ship–ice interaction at the bow increased. Ice resistance attributable to ship–ice interaction at the side hull also increased with the ice concentration. Ice resistance at the side hull was much lower than the ice resistance at the bow. However, the portion of ice resistance at the side hull with respect to the total ice resistance became larger with the ice concentration.

(3) The turning circles by numerical simulation show good agreement with those of model tests. Both model tests and numerical simulation showed that the turning radius in pack ice is smaller than that in open water. In the simulations, the turning ability becomes smaller concomitantly with the ice concentration.

ACKNOWLEDGEMENTS

This research was supported by JSPS KAKENHI Grant Number 26630455 and the Shipbuilders' Association of Japan (SAJ). The support of both organizations is greatly acknowledged. This research was also supported by "Development of Ship-handling and Passenger Evacuation Support System" funded by the Ministry of Oceans and Fisheries (MOF, South Korea).

REFERENCES

Cho, S.R., Jeong, S.Y., & Lee, S., 2013. Development of effective model test in pack ice conditions of square-type ice model basin. *Ocean Engineering*, 67, pp.35-44.

Fossen, T.I., 2011. *Handbook of marine craft hydrodynamics and motion control*. John Wiley & Sons.

Hopkins, M.A. & Tuhkuri, J., 1999. Compression of floating ice fields. *Journal of Geophysical Research: Oceans*, 104(C7), pp.15815-15825.

Huang, Y., Li, W., Wang, Y., & Wu, B., 2016. Experiments on the resistance of a large transport vessel navigating in the Arctic region in pack ice conditions. *Journal of Marine Science and Application*, 15(3), pp.269-274.

Im, N., & Seo, J.H., 2010. Ship maneuvering performance experiments using a free-running model ship. TransNav-International Journal on Marine Navigation and Safety of Sea Transportation, 4(1), pp.29-33.

Ji, S., Li, Z., Li, C., & Shang, J., 2013. Discrete element modeling of ice loads on ship hulls in broken ice fields. *Acta Oceanologica Sinica*, 32(11), pp.50-58.

Kim, M.C., Lee, S.K., Lee, W.J., & Wang, J.Y., 2013. Numerical and experimental investigation of the resistance performance of an icebreaking cargo vessel in pack ice conditions. *International Journal of Naval Architecture and Ocean Engineering*, 5(1), pp.116-131.

Kim, H.I., & Sawamura, J., 2016. A simulation study of the turning ability of ice-going ship navigating in pack ice. *23rd IAHR International Symposium on Ice.*

Sawamura, J., Senga, H., Imaki, K., Suga, K., & Kim, H.I., 2016. Ice resistance test using synthetic ice for a ship advancing in ice-covered water. *23rd IAHR International Symposium on Ice.*

Woolgar, R.C. & Colbourne, D.B., 2010. Effects of hull-ice friction coefficient on predictions of pack ice forces for moored offshore vessels. *Ocean Engineering*, 37(2), pp.296-303.

Yasukawa, H. & Yoshimura, Y., 2015. Introduction of MMG standard method for ship maneuvering predictions. *Journal of Marine Science and Technology*, 20(1), pp.37-52.

Zhan, D., Agar, D., He, M., Spencer, D., & Molyneux, D., 2010. Numerical simulation of ship maneuvering in pack ice. In *ASME 2010 29th International Conference on Ocean, Offshore and Arctic Engineering*. pp.855-862.

Zhan, D. & Molyneux, D., 2012. Three-dimensional Numerical Simulation of Ship Motion in Pack Ice. In *ASME 2012 31st International Conference on Ocean, Offshore and Arctic Engineering*. pp.407-414.

Zhou, Q. & Peng, H., 2013. A numerical simulation for operating a dynamic positioned vessel in level ice. *Proceedings of the Twenty-third International Offshore and Polar Engineering*, pp.1300-1307.

Zhou, Q., Peng, H., & Qiu, W., 2016. Numerical investigations of ship–ice interaction and maneuvering performance in level ice. *Cold Regions Science and Technology*, 122, pp.36-49.

POAC'17
Busan, KOREA

Proceedings of the 24th International Conference on
Port and Ocean Engineering under Arctic Conditions
June 11-16, 2017, Busan, Korea

New Ice Resistance Evaluation Technique Considering Hull form and Icebreaking Pattern for Arctic Vessels

Kyung-Duk Park[1] and Donghyeong Ko[1]
[1] Hyundai Heavy Industries Co., Ltd.

ABSTRACT

This study deals with a new method of predicting the ice resistance for arctic vessels considering their hull forms and icebreaking patterns. The previous version of ice resistance calculation program (Park et al., 2015) combined three resistance components such as breaking, clearing and buoyancy resistances from the empirical formulas considering the hull form geometry only. However, the breaking empirical formula based on Shimanskii's theory (1938) does not allow for ship speed dependency relevant to a change on the breaking patterns due to variation of ship speed. To remedy this problem in this study, the breaking resistance is estimated by time domain numerical simulation which enables considering the ship speed dependency on icebreaking patterns, but the clearing and buoyancy resistances are estimated by the same empirical formulas as before. For verification, this scheme is applied to both bow-first and stern-first icebreaking models and the calculated results are compared with the model test results.

KEY WORDS: Ice resistance, Icebreaking pattern, Empirical formulas, Stern-first icebreaking vessels

INTRODUCTION

Starting from the hull form development for a 190,000 dwt class arctic ore carrier (Park et al., 2011), HHI (Hyundai Heavy Industries) has involved in many projects relating to natural resources transportation by icebreaking vessels from or through the arctic region. When designing hull forms for icebreaking vessels, ship designers need to estimate ice resistance induced by the contact between hull and ice because it's a starting point in calculating the required engine power. For this reason, HHI developed an ice resistance prediction program called HIceRes taking hull form geometry into consideration (Park et al., 2015). It is a hybrid method combined by three kinds of empirical formulas, in which total ice resistance decomposes into three components: breaking, clearing and buoyancy. In detail, the Shimanskii formula for breaking resistance (1938), the Enkvist formula for buoyancy resistance (1972) and the newly modified formula for clearing resistance from Ionov's one (Poznyak and Ionov, 1981) were combined in HIceRes.

After using the hybrid method in many projects, a critical weakness was found in that the breaking empirical formula based on the Shimanskii's theory does not allow for ship speed dependency relevant to a change on the breaking pattern induced by variation of ship speed.

Furthermore, it can't reliably reflect local change in icebreaking pattern induced by the local modification of hull form. For example, local change on the shoulder part of icebreaking ship can give a different breaking resistance due to changes on breaking pattern although the other parts are kept in her shape. Generally as the shoulder shape is well designed to break ice easily, the ice channel produced by icebreaking vessel will become wide and thus the resulting ice resistance goes low. Moreover, an increase of icebreaking ability around the shoulder will give rise to better turning ability in level ice.

On the other hand, a time domain numerical simulation technique was proposed by Biao Su in his doctoral thesis (2011). In his method, icebreaking resistance is estimated by averaging the time signals induced by ice-hull contacts and its interactions, but clearing and buoyancy resistances are calculated by Lindqvist's empirical formula (1989). To estimate the icebreaking patterns and the turning ability of icebreaking ships during icebreaking procedure, he suggested the time domain numerical simulation with constant crushing stress and his results show a good agreement with full scale measurements for small-sized icebreakers by Riska et al. (2001).

Inspired by Biao Su's theory, HHI developed the similar simulation tool to consider ship speed dependent icebreaking pattern and to estimate the turning ability of icebreaking ships, and its availability was verified by applying to a few bow-first icebreaking models only (Ko et al., 2016). By comparing the simulation results with model test results for large-sized commercial icebreaking ships, it was checked whether it is suitable for the estimation of icebreaking and turning capability or not. However, it was found that the estimated contact forces between hull and ice dramatically increased around long parallel middle body due to the assumption of constant crushing stress. To avoid the problem, a modified method with varying crushing stress was proposed by assuming that the crushing stress varies with not only the contact area between hull and ice (Ashby et al., 1986) but also the velocity vector of ship.

In this study, a new scheme for ice resistance prediction considering hull form and icebreaking pattern for arctic vessels is introduced. To remedy the principal defect of Shimanskii's empirical formula where ship speed dependent icebreaking patterns are ignored, it was modified by combining breaking resistance from the time domain numerical simulation and clearing and buoyancy resistances from the hybrid empirical formulas. For verification, this scheme is applied to both bow-first and stern-first icebreaking models and the calculated results are compared with the model test results.

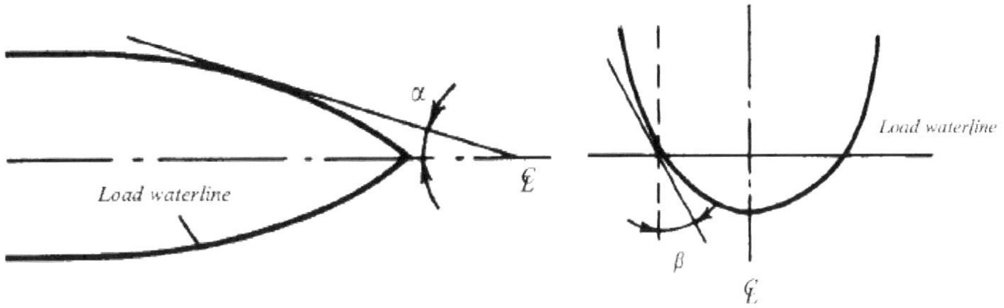

Figure 1. Definition of hull angles

CONVENTIONAL HICERES WITH HYBRID EMPIRICAL FORMULAS

Definition of Hull Shape and its Angles

The direction of x axis is followed by length of ship from After Perpendicular (AP) to Fore Perpendicular (FP), the direction of y axis is heading from center line to port side and the direction of z axis is from bottom of ship to deck. Figure 1 shows definition of angles. Angle α is defined as the angle between the x axis and the tangent of waterline at an arbitrary point in the x-y plane, angle β is the angle between the z axis and the tangent of section line in the y-z plane.

Definition of Stern-First Icebreaking Vessels

After new stern-icebreaking concept (Juurmaa et al., 1995) was introduced, the concept of Double Acting Ship (DAS) has widely come into general use for especially Russian arctic. Unlike conventional bow-first icebreaking ships, the stern shape of stern-first icebreaking ships normally comprises of two or three separate icebreaking parts according to the sectional shape as shown in Figure 2. In this paper, the icebreaking shape with one icebreaking part is named 'single V shape' and the stern-first icebreaking shapes are classified by 'double V shape' or 'triple V shape' according to the number of icebreaking parts for DAS.

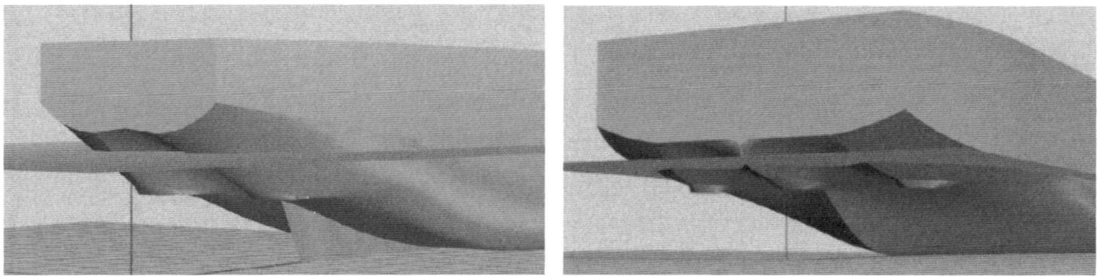

Figure 2.Stern views of double V shape (left) and triple V shape (right)

Definition of Ship Resistance in Ice

According to the model test procedure for icebreaking vessels, ship resistance in ice (ITTC, 2005) give the equation for the total resistance in ice R_{total} as the sum of 4 individual components:

$$R_{total} = R_{br} + R_{cl} + R_{bu} + R_{ow} \tag{1}$$

where, R_{br} is the resistance component due to breaking the ice, R_{cl} is the component due to clearing the ice, R_{bu} is the component due to buoyancy of ice, and R_{ow} is the resistance component in open water. In this study, R_{ow} was estimated by extrapolating the model test data in open water.

Estimation of Breaking Resistance

Because icebreaking mainly occurs on the plane composed with the DLWL (Design Load Water Line) of hull form geometry and thus breaking resistance is closely related to the shape of DLWL, this study decided to use Shimanskii empirical formula (1938) to estimate breaking resistance.

Assuming that the unit external force along the DLWL of icebreaking vessel is applied on the hull surface with unit beam, he derived the resulting forces acting on hull surface (F_x, F_y, F_z) as follows:

$$F_x = \int_0^{L_E} \frac{tan^2 \alpha \sqrt{1+tan^2 \alpha}}{1+tan^2 \alpha+tan^2 \beta} dx \quad (2)$$

$$F_y = \int_0^{L_E} \frac{tan \alpha \sqrt{1+tan^2 \alpha}}{1+tan^2 \alpha+tan^2 \beta} dx \quad (3)$$

$$F_z = \int_0^{L_E} \frac{tan \alpha \, tan \beta \sqrt{1+tan^2 \alpha}}{1+tan^2 \alpha+tan^2 \beta} dx \quad (4)$$

where, L_E is the entrance length between the fore end and the starting point of the parallel part of the icebreaking bow with single V shape, as shown in Figure 3.

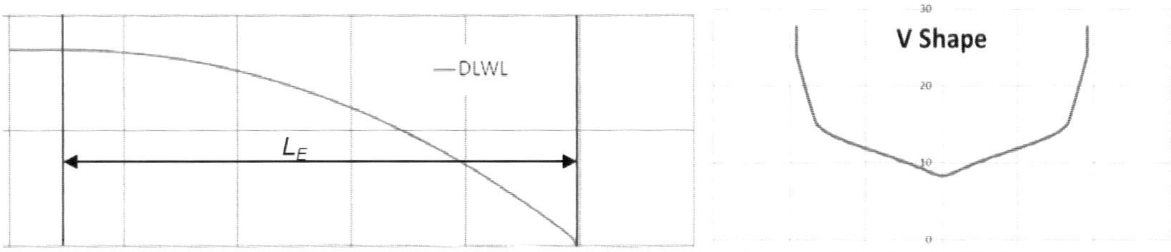

Figure 3. Integral part of traditional icebreaking bow with single V shape

When a ship with unit beam breaks the ice, he also defined icebreaking parameter as F_Z/F_X (ratio of the force induced by ice in *z* direction to that in *x* direction). Assuming that the resulting breaking resistance is proportional to this ratio and her beam, he derived:

$$R_{br} = \frac{\lambda \sigma_f h^2 B}{1.93} \frac{F_X}{F_Z} \quad (5)$$

where, $\lambda = \sqrt[4]{\frac{3\rho_w g}{E h^3}}$ (6)

In Eqs. (5) and (6), *g* is acceleration due to gravity, *h* is the thickness of the ice, *E* is the Young's modulus of the sea-ice and ρ_w is the density of water.

For application to stern-first icebreaking vessels with 'double or triple V shape', the calculations are carried out separately after dividing the DLWL into two or three parts based on the tip point with an angle α of ±90°, as shown in Figure 4. Then, the resulting breaking resistance is defined as the sum of individual forces.

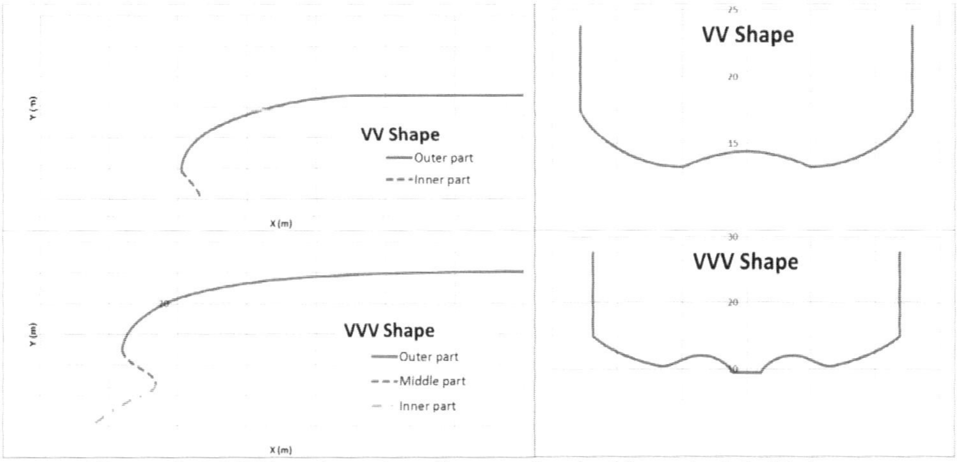

Figure 4. DLWL and section diagram for double and triple V shapes

Estimation of Clearing Rresistance

To estimate the clearing resistance, the Ionov method (Poznyak and Ionov, 1981) was used initially. He assumed that ice clearing happens only on the *x-y* plane along the DLWL as shown in Figure 5. Then clearing speed in *y* direction is derived:

$$V_y = \frac{dy}{dt} = V_S \cdot y'(x) \tag{7}$$

where, V_S is ship speed in ice and $y'(x)$ is the slope of the tangential line along the DLWL. Because clearing resistance is a function of a ship's speed, Ionov described the following formula:

$$R_{cl} = 2\rho_i hB \cdot Fr_B \cdot \left[k_3' \int_0^{L/2} \frac{\{y'(x)\}^2}{\cos\alpha} dx + k_3'' \cdot f_g \int_0^{L/2} \frac{y'(x)}{\cos\alpha} dx\right] \tag{8}$$

where, L is ship length, k_3' and k_3'' are empirical constants, f_g is the friction coefficient, and ρ_i is the density of ice. In Eq. (8), Froude number Fr_B is a dimensionless number defined as the speed-beam ratio of a ship as follows:

$$\mathrm{Fr_B} = \frac{\mathrm{V_S}}{\sqrt{\mathrm{gB}}} \tag{9}$$

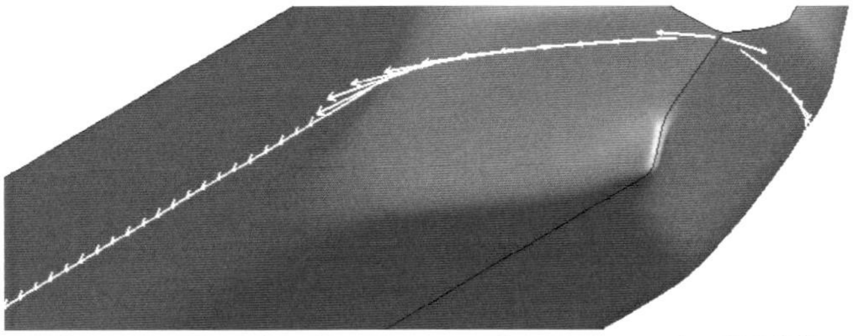

Figure 5. Ice clearing on x-y plane along the DLWL

However, the authors found out that Eq. (8) has certain irrational issues as follows:

- The integrand in Eq. (8) can be divergent when $\cos\alpha$ goes to zero, especially at the end point of the DLWL of icebreaking bow. For DAS with double or triple V shape, the value of the integrand cannot also be determined when $y'(x)$ and $\cos\alpha$ go to zero simultaneously, at the inflection point of the DLWL as shown in Figure 4.
- It is unrealistic to assume that the actual ice clearing happens only according to the clearing angle α and on the *x-y* plane along the DLWL. Ice model tests generally show that ice clearing happens over the overall underwater hull surface covered ice not only along the DLWL.

To solve these problems in this study, a new formula was proposed by modifying the Ionov formula. For defining a new realistic ice clearing direction over the underwater hull surface, a new vector is introduced as shown in Figure 6. A normal vector $\vec{n}$ at an arbitrary point on the hull surface is defined:

$$\vec{n} = n_x\vec{i} + n_y\vec{j} + n_z\vec{k} \tag{10}$$

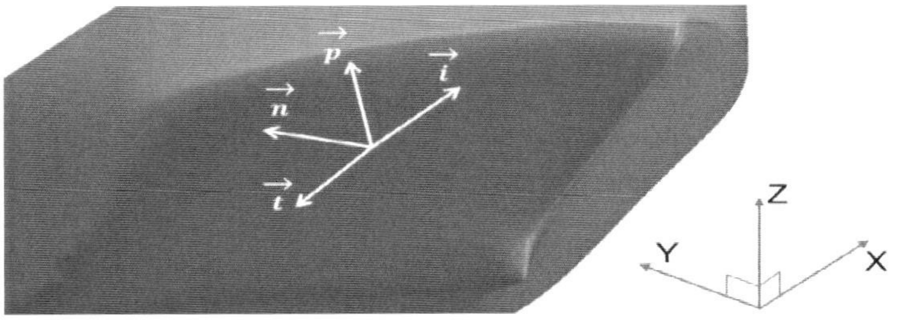

Figure 6. Diagram of clearing direction

The vector $\vec{\imath}$ indicates the direction of advancing ship. If we define the vector $\vec{p}$ as $\vec{\imath}\times\vec{n}$, the vector $\vec{p}$ is normal to the vector $\vec{n}$ and $\vec{p}$ is derived:

$$\vec{p} = \vec{\imath}\times\vec{n} = -n_z\vec{\jmath} + n_y\vec{k} \tag{11}$$

When a curl calculation is done between $\vec{p}$ and $\vec{n}$, a new vector $\vec{t}$ is defined as an ice clearing direction vector at an arbitrary point on the underwater hull surface surrounded by ice as follows:

$$\vec{t} = \frac{\vec{p}\times\vec{n}}{|\vec{p}\times\vec{n}|} = t_x\vec{\imath} + t_y\vec{\jmath} + t_z\vec{k} \tag{12}$$

In this study, all broken ice pieces surrounding hull surface are assumed to flow down along the ice clearing direction vector defined by Eq. (12). For reference, Figure 7 shows the comparison of ice flow over the underwater hull surface of an icebreaking vessel between simulated by ice clearing direction vectors and observed from ice model test. The simulated result matches well with the observed one.

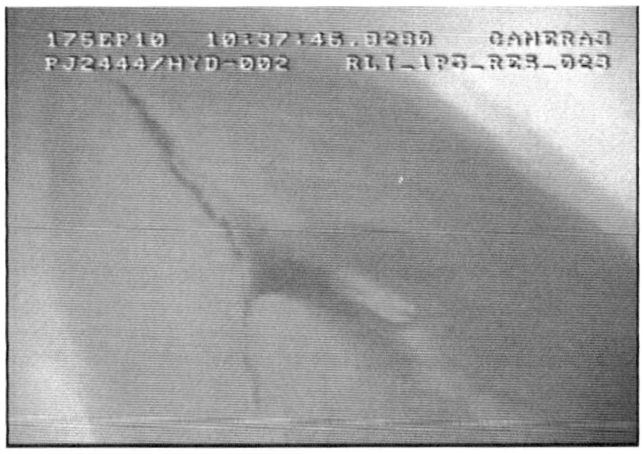

↑ Simulated by ice clearing direction vector ↑ Observed from ice model test

Figure 7. Comparison of underwater ice flow pattern

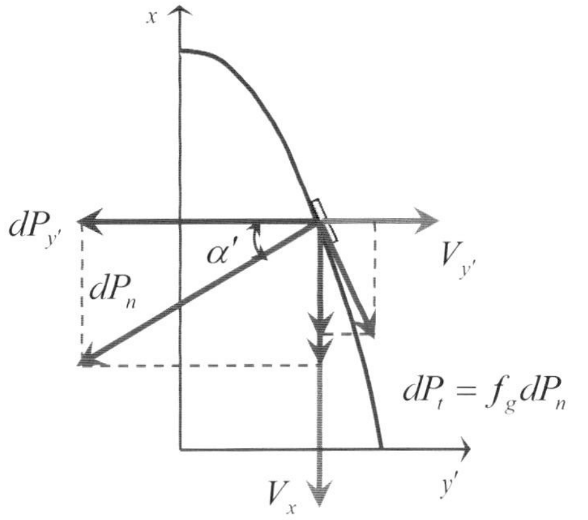

Figure 8. Force components induced by ice clearing

If we consider that the force components induced by ice clearing along the arbitrary line on x-y' plane is shown in Figure 8, the ice clearing speed and force toward y' direction can be assumed:

$$V_{y'} = \frac{dy'}{dt} = V_S\sqrt{\{y'(x)\}^2 + \{z'(x)\}^2} \tag{13}$$

where, $y'(x)$ is the tangential slope in y direction and $z'(x)$ is in z direction over hull surface. Then the ice clearing force acting on the infinitesimal length ds and the resulting normal and tangential forces can be defined as Eqs. (14) to (16).

$$dP_{y'} = c \cdot V_{y'}ds \tag{14}$$

$$dP_n = dP_{y'}/\cos\alpha' \tag{15}$$

$$dP_t = f_g \cdot dP_n \tag{16}$$

where c in Eq. (14) is proportionality constant and α' in Eq. (15) is the angle between $\overrightarrow{dP_{y'}}$ and $\overrightarrow{dP_n}$ on x-y' plane. Resultantly, the clearing force in x direction can be derived using the following relation:

$$dP_x = dP_n \cdot \sin\alpha' + dP_t \cdot \cos\alpha' \tag{17}$$

By integrating Eq. (17) along the underwater ice flow line covered with ice, the clearing resistance is newly derived as follows:

$$R_{cl} = 2cV_S \cdot \left[\int_0^{L/2} \sqrt{\{y'(x)\}^2 + \{z'(x)\}^2}\left(k_3' \cdot \sin\alpha' + k_3'' f_g \cdot \cos\alpha'\right) dx\right] \tag{18}$$

Considering that the clearing resistance of icebreaking vessels is normally proportional to $B^{1.5}$ and h, the proportional constant c is defined as:

$$c = k_3 \frac{\rho_i g h B}{\sqrt{gB}} \tag{19}$$

In addition, if we assume that $y'(x) = t_y/t_x$ and $z'(x) = t_z/t_x$, then the new formula of the clearing resistance can be obtained as follows:

$$R_{cl} = 2\rho_i hgB \cdot Fr_B \cdot \left[\int_0^{L/2} \sqrt{\left(\frac{t_y}{t_x}\right)^2 + \left(\frac{t_z}{t_x}\right)^2}\left(k_3' \cdot \sin\alpha' + k_3'' f_g \cdot \cos\alpha'\right) dx\right] \tag{20}$$

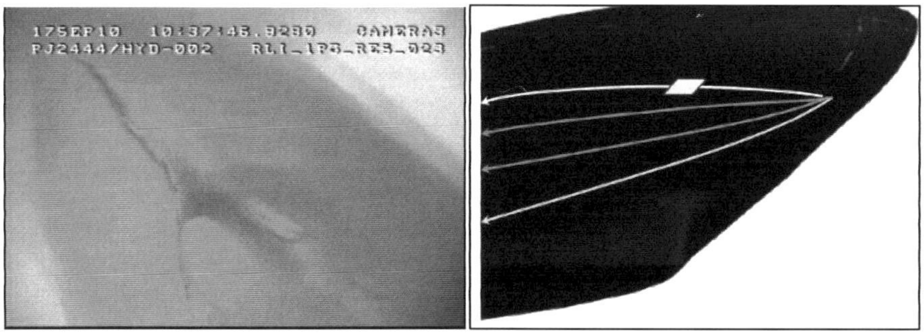

Figure 9. Definition of ice covered area and ice flow lines

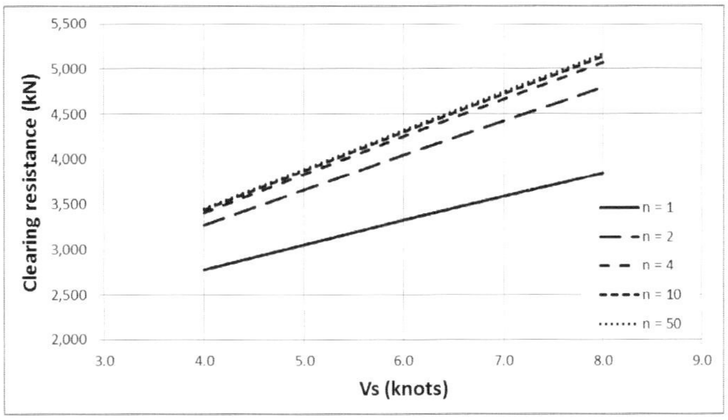

Figure 10. Convergence test for number of ice flow lines

For practical application of the new clearing resistance formula in this study, ice covered area and ice flow lines are defined as shown in Figure 7. Assuming that the hull surface is evenly covered with the broken ice pieces and then the arc length of an arbitrary sectional curve is identical with the half width of ship, a group consisting of 'n' ice flow lines can be defined by cutting the arc length of the sectional curve with same interval and connecting their nodes as shown in Figure 9. Then the line integral of Eq. (20) is applied along each of ice flow lines and their average values gives resultant clearing resistance. For reference, Figure 10 shows the convergence of resulting clearing resistance according to the number of ice flow lines 'n' for a bow-first icebreaking model. It means the result with one integral equation along the DLWL for $n = 1$ and two integral equations along the two border lines of underwater ice covered area for $n = 2$. For $n > 3$, the result comprises of multiple integral equations and its average values. From the convergence test, the resulting clearing resistance was approximately converged for $n = 4$ and its resulting value was about 25% bigger than one for $n = 1$.

Estimation of Buoyancy Resistance

According to Enkvist (1938), buoyancy resistance can be divided into the pressure component induced by submerged ice in water, and the friction component caused by broken ice pieces. In the calculation of the buoyancy force, the geometry of the hull under the DLWL serves as important information. Enkvist assumed that the underwater hull surface covered with ice can be defined with the arc length of sectional curves equal to half width of the ship, as shown in Figure 11. Then he derived the buoyancy resistance as follows:

$$R_{bu} = \frac{\sum_0^{L/2} L_i \sum_0^B \rho_\Delta g h \bar{s} b_i}{L/2} + \sum_0^{L/2} L_i \sum_0^B f_g \rho_\Delta g h s b_i \sin\beta \qquad (21)$$

where b_i is a beam of ice piece in any section, L_i is the length between neighboring sections, ρ_Δ is the difference between the density of water and that of ice, and $\bar{s}$ is averaged depth of submergence.

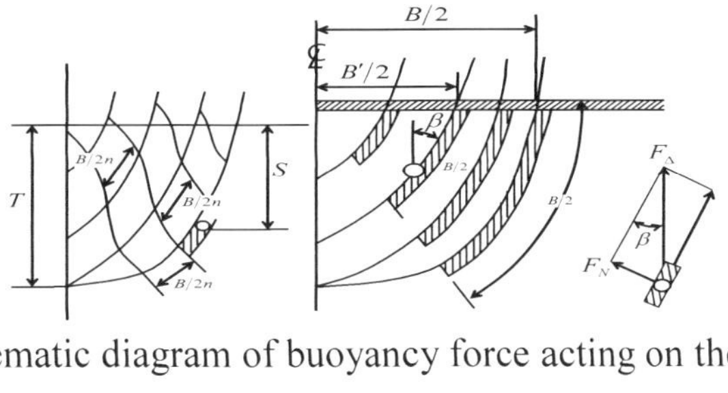

Figure 11. Schematic diagram of buoyancy force acting on the hull surface

NUMERICAL CALCULATION FOR BREAKING RESISTANCE

Ice-Hull Interaction

Instead of Shimanskii method for breaking resistance, the time-domain numerical simulation proposed by Biao Su (2011) and its availability on the application to icebreaking pattern analysis was dealt with in this study. As a ship moves forward in level ice with a constant speed as shown in Figure 12, the hull contacts the ice sheet. Then the crushing force will happen along the DLWL and keep growing with the increase of contact areas between hull and ice until its vertical load exceeds the bending failure of ice.

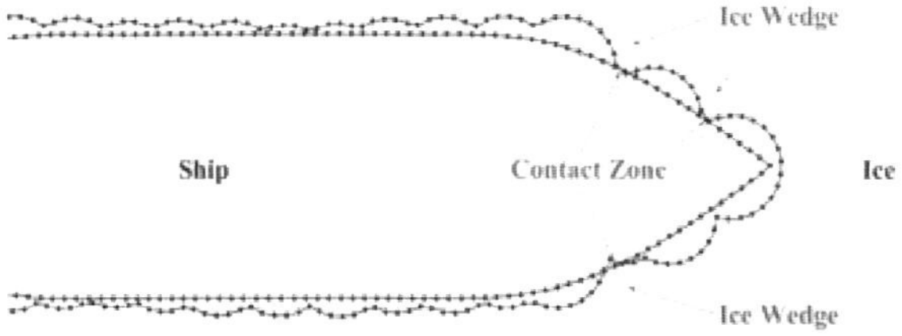

Figure 12. Description of node generation (Biao Su, 2011)

From the viewpoint of numerical calculation, the contact area between hull and ice is calculated at each time step with the initial inputs of hull and ice information (waterline shape of hull and ice properties) and then it is checked whether its vertical load exceeds the bending failure load of idealized ice wedges or not. For each time step, the ice crushing force $\overrightarrow{F_n}$ is normal to an arbitrary contact area A_c and it is defined as:

$$\overrightarrow{F_n} = (\sigma_c \cdot A_c)\vec{n} \tag{22}$$

where, σ_c is ice crushing strength from the measured ice crushing pressure on ship hull. For the contact area, as shown in Figure 13, two cases can be considered and they are simply determined by the contact length L_h and the indentation depth L_d as follows:

$$A_c = \frac{1}{2} L_h \frac{L_d}{\cos\varphi} \qquad Case\ 1:\ L_d \cdot \tan\varphi \le h_i \tag{23}$$

$$A_c = \frac{1}{2}\left(L_h + L_h \frac{L_d - h_i/\tan\varphi}{L_d}\right)\frac{L_d}{\sin\varphi} \qquad Case\ 2:\ L_d \cdot \tan\varphi > h_i \tag{24}$$

where, h_i is the ice thickness and φ is a contact angle between hull and ice.

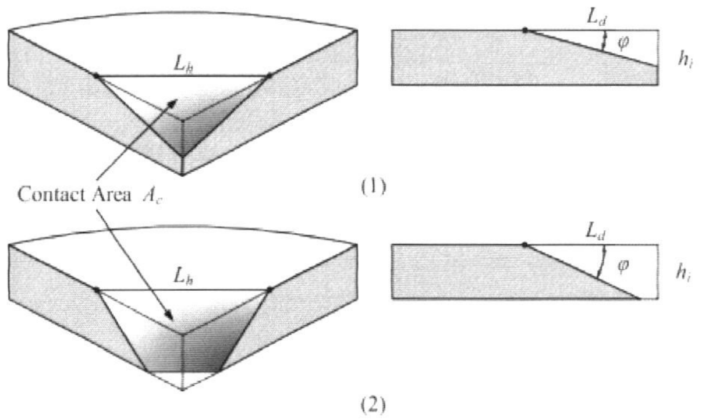

Figure 13. Two cases for the calculation of contact area (Biao Su, 2011)

On the other hand, all crushing forces induced by hull-ice contacts are not necessarily proportional to their contact areas. For example, if the hull surface drifts apart from the ice sheet then there can be no crushing force applied to the ship despite the hull-ice contact as shown in Figure 14. Furthermore, the crushing force can be increased or decreased due to the mutual movement between hull and ice. Thus, once the contact areas are determined, the resulting crushing stress is in this study assumed to be divided into two components: planar component σ_u in the ice sheet and vertical component σ_v to the ice sheet as shown in Figures 14 and 15. Especially it is noted that $\overrightarrow{\sigma_u}$ always appears to be normal to the hull surface from the top view. Then, the resulting crushing stresses are decomposed as follows:

$$\sigma_u = \sigma_c \cdot c_1 \cdot sin\,\varphi \quad (25)$$

$$\sigma_v = \sigma_c \cdot c_2 \cdot cos\,\varphi \quad (26)$$

In Eqs. (25) and (26), c_1 and c_2 are computed by the inner product between the following two vectors.

$$c_1 = \begin{cases} \frac{\vec{U}\cdot\overrightarrow{\sigma_u}}{|\vec{U}|\cdot|\overrightarrow{\sigma_u}|} & if\ \frac{\vec{U}\cdot\overrightarrow{\sigma_u}}{|\vec{U}|\cdot|\overrightarrow{\sigma_u}|} < 0 \\ 0 & if\ \frac{\vec{U}\cdot\overrightarrow{\sigma_u}}{|\vec{U}|\cdot|\overrightarrow{\sigma_u}|} \geq 0 \end{cases}$$
$$c_2 = \begin{cases} 1 & if\ \frac{\vec{U}\cdot\overrightarrow{\sigma_u}}{|\vec{U}|\cdot|\overrightarrow{\sigma_u}|} < 0 \\ 0 & if\ \frac{\vec{U}\cdot\overrightarrow{\sigma_u}}{|\vec{U}|\cdot|\overrightarrow{\sigma_u}|} \geq 0 \end{cases} \quad (27)$$

Meanwhile, the frictional force tangential to the crushing force is induced by the relative motion between hull and ice. In this study, the frictional force $\overrightarrow{F_t}$ is assumed to follow the ice clearing direction vector $\vec{t}$ in Eq. (12) and then it can be defined as:

$$\overrightarrow{F_t} = f_g \cdot |\overrightarrow{F_n}|\vec{t} \quad (25)$$

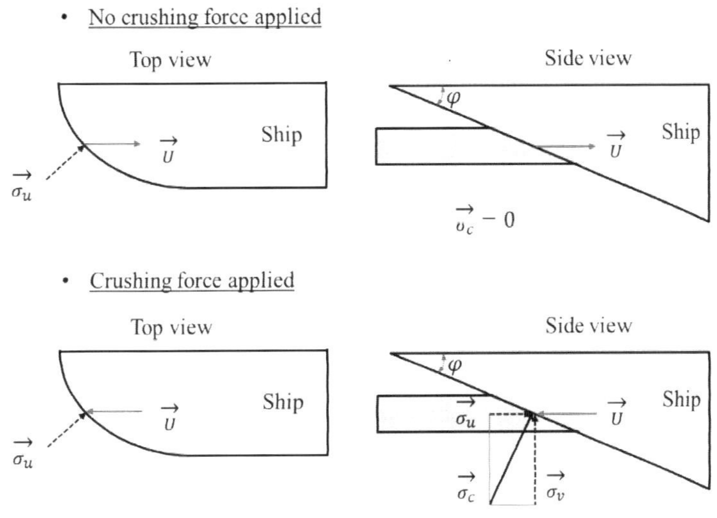

Figure 14. Two cases for crushing force and no crushing force applied to the ship

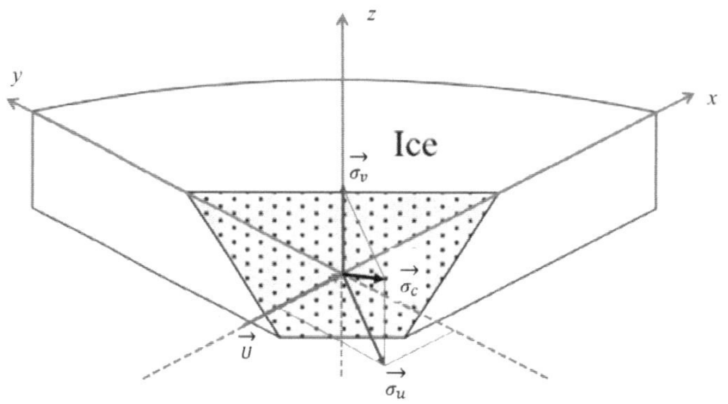

Figure 15. Description of contact area and crushing stress

Varying Crushing Stress

Biao Su (2011) proposed using the constant crushing stress to calculate ice load in level ice, but it was found that this constant value makes the ice load increase sharply around the long parallel middle body of ship. To solve this problem, the concept of varying crushing stress defined by the contact area between hull and ice is newly introduced in this study. According to Ashby et al. (1986), the crushing stress σ_c can be defined as a function of contact area as follows:

$$\sigma_c = \frac{P}{A_c} = \frac{P_L \Delta_L}{L_i^3}\left[1 + 3\left(\frac{L_i^3}{A_c \Delta_L}\right)^{\frac{1}{2}}\right] \tag{26}$$

where, L_i is an idealized size of cubical independent cells, Δ_L is a moved distance before the cell fails and P_L is the average force during the period of contact. The relation between crushing stress and contact areas is given as a theoretical curve in Figure 16, where $L_i = 1$, $P_L = 15\ MN$ and $\Delta_L = 0.02\ m$ are used. In addition, the peak stresses measured in the field are assumed to be approximately 1 to 2 MPa for contact areas of $A_c \approx 100\ m^2$. By slightly moving the theoretical curve to the left using the assumption that the peak stress is approximately 1 MPa for contact areas of $A_c = 100\ m^2$, the resulting theoretical curve is defined:

$$\sigma_c = 0.33\left(1 + 20.3/\sqrt{A}\right) \tag{27}$$

which is valid for $A_c \gg L_i^2$ and higher than the minimum level of 0.33 MPa for very large contact areas. On the other hand, for very small contact areas, the varying crushing stress curve is constrained to be bounded by the observed data of the maximum stress, 10.5 MPa to avoid the unrealistic situation in which the stress soars infinitely as contact areas decreases to zero in this study. (Sanderson, 2014)

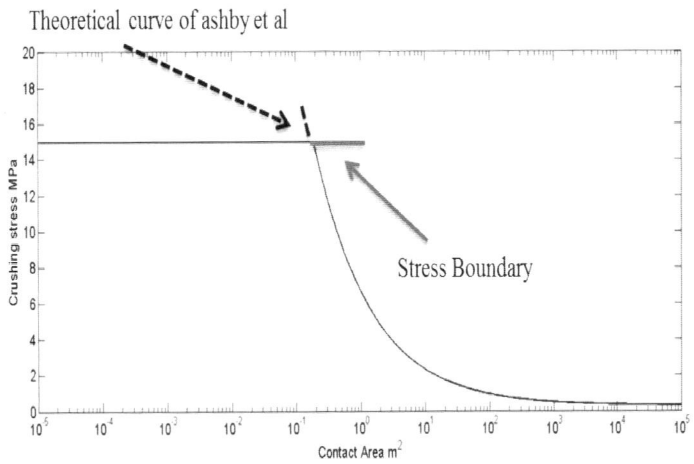

Figure 16. Theoretical curve of Ashby et al. (1986) bounded by the collected data

Ice bending failure

Ice bending failure arises from the vertical component of the contact force between ice and hull. The horizontal and vertical components (F_u and F_v) of the total contact force are then decomposed as:

$$F_u = |\vec{F_n}| \sin\varphi + |\vec{F_t}| \cos\varphi \tag{28}$$

$$F_v = |\vec{F_n}| \cos\varphi + |\vec{F_t}| \sin\varphi \tag{29}$$

In point of numerical calculation, the idealized ice wedge with the icebreaking radius (R) can be defined as shown in Figure 17 and it will be broken from the edge of the ice sheet if the vertical force in Eq. (29) exceeds the bending failure load of ice sheet (P_f). The bending failure load and the icebreaking radius are respectively calculated as:

$$P_f = C_f \left(\frac{\theta}{\pi}\right)^2 \sigma_f h_i^2 \tag{30}$$

$$R = C_l \cdot l \cdot \left(1 + C_v \cdot |\vec{U}| \cdot \cos\alpha \cdot \sin\varphi\right) \tag{31}$$

where, θ in Eq. (30) is the opening angle of the idealized ice wedge shown in Figure 17 and l in Eq. (31) is the characteristic length of the ice defined as:

$$l - \left(\frac{Eh^3}{12(1-\nu^2)\rho_w g}\right)^{\frac{1}{4}} \tag{32}$$

where, E is the elastic modulus of ice, ν is the Poisson's ratio, ρ_w is the density of water and g is the acceleration of gravity.

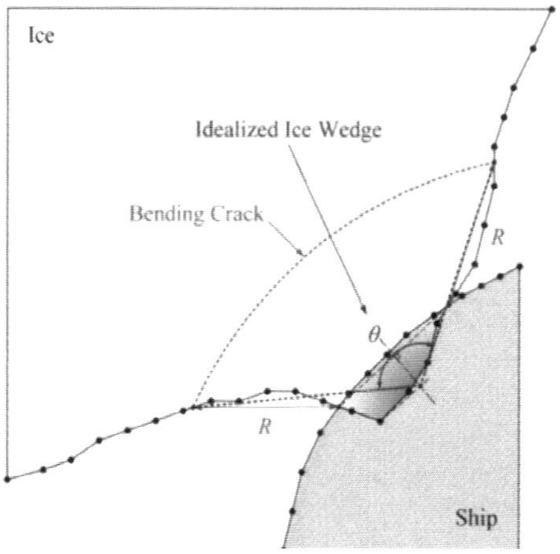

Figure 17. Geometrical idealization of the ice wedge (Biao Su, 2011)

If the bending failure occurs, the geometry of ice sheet is regenerated and then the forces and moment working on the ship is calculated. During the process, the nodes not contacted with hull are excluded from the calculation to enhance the computational efficiency and the number of ice nodes in the far field from hull is drastically reduced to avoid a waste of memory as shown in Figure 18.

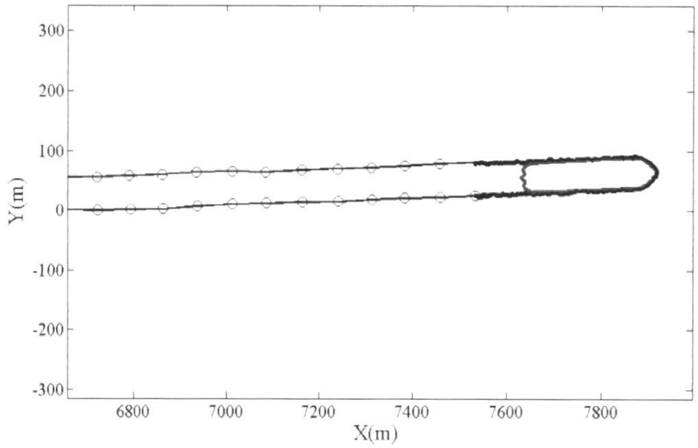

Figure 18. Ice node generation highlighted by blue circles

Numerical Breaking Resistance

Based on the logical strategy introduced in Sections 3.1 to 3.3, a numerical code has been developed to simulate ice-hull interactions along the waterline shape of an icebreaking vessel with a constant speed in this study. As the simulation goes, the normal and tangential forces induced by ice-hull interactions are recorded in succession. Then, breaking resistance is numerically defined by the sum of x components of both forces as in the following equation:

$$R_{br} = \frac{1}{t_1 - t_0} \sum_{t_0}^{t_1} \left[\left(\overrightarrow{F_n} + \overrightarrow{F_t} \right)_x \right] \tag{32}$$

NUMERICAL RESULTS

Numerical Ship Models

To verify the ice resistance estimation scheme suggested in this study, ice resistance estimations including numerical simulations are carried out for two different ships (Arctic

LNG Carrier and Arctic Bulk Carrier) and their results are compared with model test results. Arctic LNG Carrier is designed with the concept of double acting ship whose bow shape is optimized for light ice condition and whose stern shape is for heavy ice condition respectively as shown in Figure 19, and Arctic Bulk Carrier has traditional icebreaking bow with an inclination of 20 degree to ice sheet as shown in Figure 20. For reference, ice model tests for Arctic LNG carrier are conducted at AARC ice model basin in Finland, but those for Arctic Bulk Carrier at IOT ice model basin in Canada. Table 1 shows the characteristics of the ships and model testing programs.

Figure 19. Arctic LNG Carrier model tested in AARC ice model basin

Figure 20. Arctic Bulk Carrier model tested in IOT ice model basin

Table 1. Numerical ship models

Items	Arctic LNG Carrier		Arctic Bulk Carrier
Ship characteristics	$L/B = 5.5$, $B/T = 4.3$ Double acting ship with 3 POD propellers		$L/B = 6.1$, $B/T = 2.6$ Ducted propellers
Model test facility	AARC in Finland		IOT in Canada
Test methods	Free running		Towed resistance
Icebreaking function	Bow-first (light ice condition)	Stern-first (heavy ice condition)	Bow-first only
Tested ice thickness	*1.5 m*	*1.5 m, 2.1 m*	*1.5 m, 2.4 m*
Tested flexural strength	*500 kPa*	*500 kPa*	*600 kPa*

Verification for Bow-First Icebreaking Models

To begin with, numerical simulations for bow-first icebreaking models are verified comparing the numerical results with test ones. The numerical simulation runs 1,000 second computation times with time step of 0.01 second assuming that the ship advances with a constant speed. Figure 21 shows the simulated breaking resistance of Arctic LNG Carrier with 5 *knots* of advancing speed.

For the verification for Arctic LNG Carrier, the numerical results of three typical ship speeds (1, 3 and 5 *knots*) were compared with test results to check ship speed dependency from quantitative and qualitative perspective. In Figure 22, HIceRes_0 indicates the initial version of ice resistance prediction, in which total resistance is composed of the Shimanskii's solution, Eq. (5) plus the Ionov's one, Eq. (5) plus Enkvist's one, Eq. (21) as listed in Table 2.

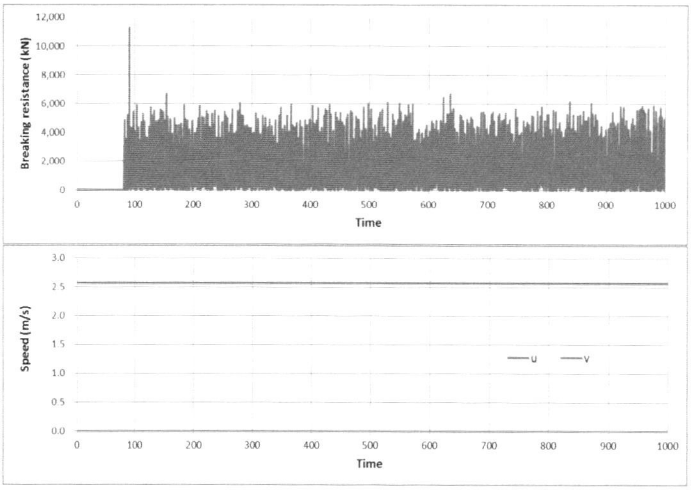

Figure 21. Simulated results for Arctic LNG Carrier model (Ahead)

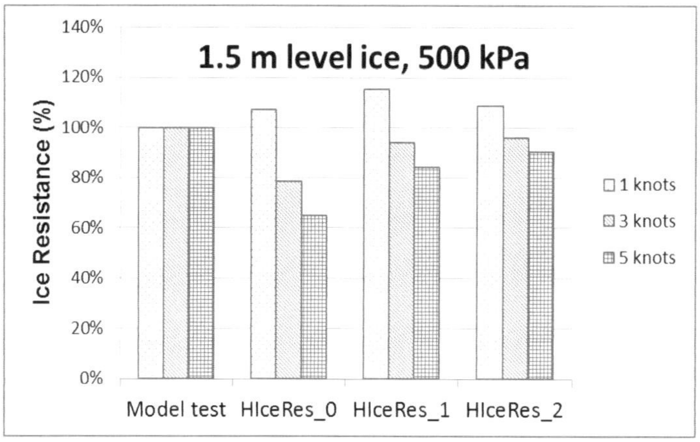

Figure 22. Ship speed dependency of estimated ice resistances for Arctic LNG Carrier (Ahead)

Table 2. Ice resistance estimation schemes

Estimation methods	Resistance components in ice		
	Breaking	**Clearing**	**Buoyancy**
HIceRes_0	Eq. (5)	Eq. (8)	Eq. (21)
HIceRes_1	Eq. (5)	Eq. (20)	Eq. (21)
HIceRes_2	Eq. (32)	Eq. (20)	Eq. (21)

HIceRes_1 means that total resistance is composed of Shimanskii's and Enkvist's solutions but the clearing component is only replaced into the modified equation, Eq. (20) instead of Eq. (8). Lastly, HIceRes_2 is the most updated solution in estimating ice resistance, where the breaking pattern of icebreaking vessels is taken into consideration through the numerical simulation and thus total ice resistance is composed of the following solutions: Eq. (32) plus Eq. (20) plus Eq. (21). Comparing with test results (assumed to be 100% for each ship speeds), the accuracy and consistency of HIceRes_1 is better with approaching to the test results than HIceRes_0. This indicates that the underwater ice clearing effect proposed in this study was an effective method in estimating ice resistance for arctic vessels. In the same way, the accuracy and consistency of HIceRes_2 is better than HIceRes_1. For estimating breaking resistance, Shimanskii's formula was used in HIceRes_1 while the numerical simulation in HIceRes_2. This is implying that Shimanskii's breaking resistance is independent of ship speed variation as shown in Eq. (5), while real icebreaking phenomena is closely affected by ship speed.

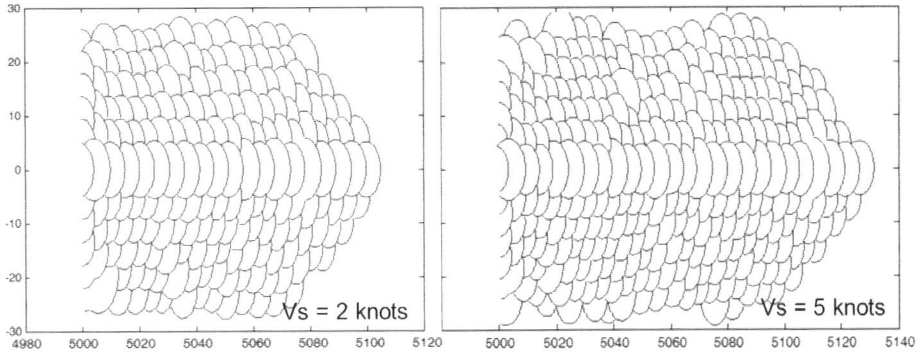

Figure 23. Simulated icebreaking patterns for different ship speeds

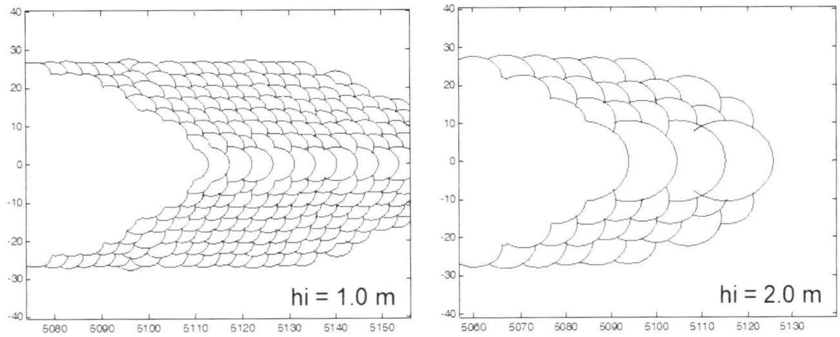

Figure 24. Simulated icebreaking patterns for different ice thickness

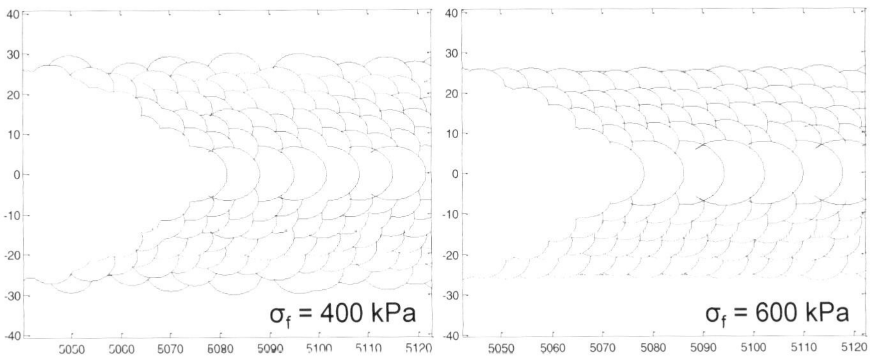

Figure 25. Simulated icebreaking patterns for different flexural strength of ice

For the same hull form, some simulated icebreaking patterns are compared in Figures 23 to 25 to check on dependencies of ship speed, ice thickness and flexural strength of ice. In case of ship speed variation (Figure 23), overall icebreaking patterns are quite similar between two different speeds. The size of ice wedges becomes slightly smaller but the channel width becomes bigger as the ship goes faster. In case of ice thickness variation (Figure 24), there is a big difference in ice wedge sizes between two different ice thicknesses. As in Eq. (32), the wedge size becomes bigger as the ice thickness get thicker. In case of flexural strength of ice (Figure 25), the bending failure load increases as in Eq. (30) and then the wedge size and angle becomes slightly bigger but the channel width becomes narrower as the strength get stronger.

Lastly, numerical simulations for Arctic Bulk Carrier model are carried out for verification of accuracy and consistency. The comparison between the numerical and test results is shown in Figure 26. The ice model tests for Arctic Bulk Carrier were carried out for two different ice thicknesses at the ice model basin of IOT in Canada. As in Arctic LNG Carrier, HIceRes_1 is more consistent with test results than HIceRes_0 for Arctic Bulk Carrier. HIceRes_2 is also the best for all the ship speeds from the accuracy and consistency point of view even though the ice thickness changes drastically.

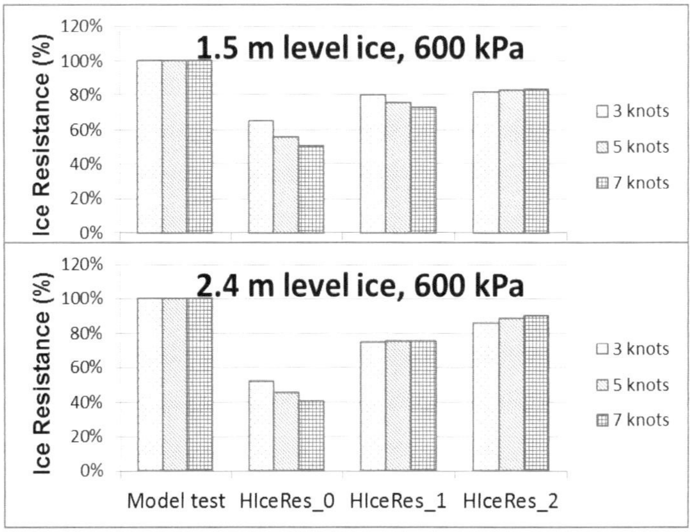

Figure 26. Ship speed dependency of estimated ice resistances for Arctic Bulk Carrier (Ahead)

Verification for Stern-First Icebreaking Models

In this study, the same method used to simulate icebreaking patterns of bow-first icebreaking vessels was newly applied to a stern-first icebreaking model. The numerical simulations for Arctic LNG carrier were carried out for two different ice thicknesses for verification of accuracy and consistency. For reference, Arctic LNG carrier was tested at the ice model basin of AARC in Finland.

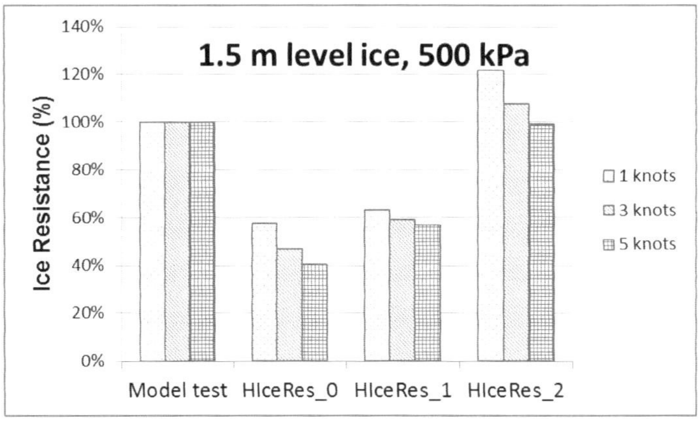

Figure 27. Estimated ice resistances in 1.5 *m* level ice for Arctic LNG Carrier (Astern)

The estimated values of ice resistance in 1.5 *m* level ice are compared with test results in Figure 27 and those in 2.1 *m* level ice are compared in Figure 28. Both figures show that the results estimated by HIceRes_2 are well consistent with test results. Unlike bow-first icebreaking models, the results estimated by HIceRes_1 are drastically smaller than test results in stern-first icebreaking mode. This means that the breaking component comprises a large portion of total ice resistance in case of stern-first icebreaking model. That is, Shimanskii's formula gives quite a small value so that it results in a large amount of error, while the numerical simulation gives rise to a reasonable value for stern-first icebreaking models.

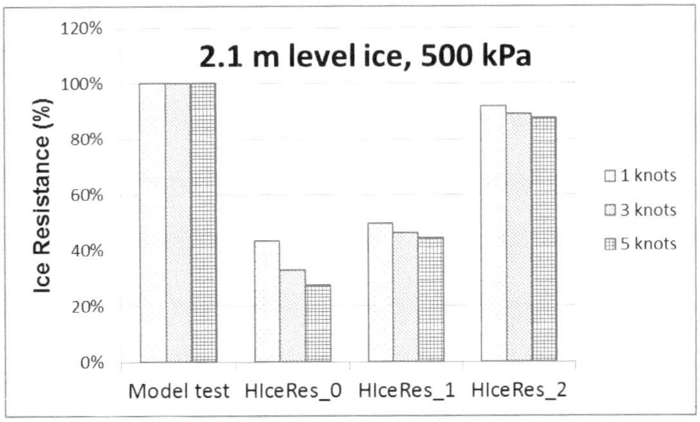

Figure 28. Estimated ice resistances in 2.1 *m* level ice for Arctic LNG Carrier (Astern)

Icebreaking patterns are compared between two different ice thicknesses and ship speeds as shown in Figure 29. Bow-first icebreaking vessels normally have single V shape in their cross-sectional views, while this stern-first icebreaking model has triple V shape as shown in Figure 2. For this reason, the overall breaking patterns are characterized by three ice wedges parallelized in a row. On the other hand, the ice wedge sizes of 2.1 *m* level ice are bigger than those of 1.5 *m* level ice as was the case in bow-first icebreaking mode.

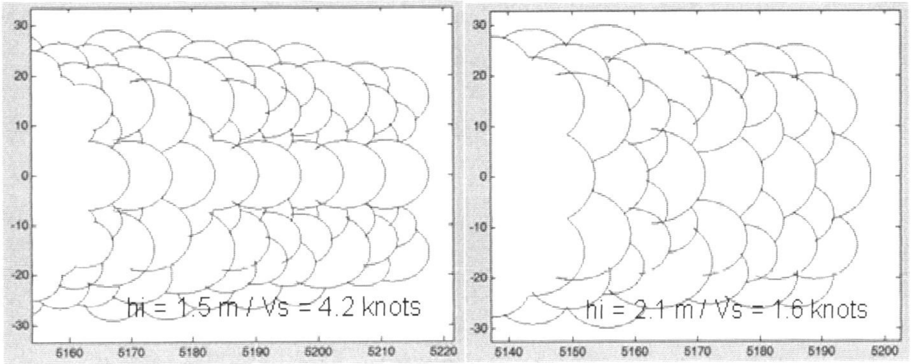

Figure 29. Icebreaking patterns in astern mode for different ice thicknesses and ship speeds

CONCLUSIONS

In this study, new ice resistance evaluation technique considering hull form and icebreaking pattern for arctic vessels was proposed and it was verified through the application to bow-first and stern-first icebreaking models. To improve the accuracy and consistency in estimating ice resistance, a new combination of empirical formulas and numerical simulation was introduced as follows:

(1) Enkvist's empirical formula for buoyancy component of ice resistance
(2) Modified empirical formula from Ionov's theory for underwater ice clearing effects
(3) Numerical simulation for icebreaking pattern analysis for breaking component

With comparing with model test results, it was found that the estimated results by the new method give rise to better ship speed dependency than those of the previous version considering hull form geometry only.

REFERENCES

Park, K.D., Chung, Y.K., Jang, Y.S., Kim, H.S. and Molyneux, D., 2011. Development of hull forms for a 190,000 dwt icebreaking ore carrier. Preceedings of the ASME 2011 30th International Conferences on Ocean, Offshore and Arctic Engineering (OMAE2011), Rotterdam, The Netherlands.

Park, K.D., Kim, M.C. and Kim, H.S., 2015. Calculation of ice clearing resistance using normal vector of hull form and direct calculation of buoyancy force under the hull. International Journal of Naval Architecture and Ocean Engineering (IJNAOE) Vol. 7, pp. 699-707

Shimanskii, Y.A., 1938. Conditional standards of ice qualities of a ship. Trans. Arctic Research Institute, Northern Sea Route Administration Publishing House, Vol. 130, Leningrad. Translation T-381-01 by Engineering Consulting and Translation Center (ECTC), P.O. Box 1377, Jackson Heights, New York

Enkvist, E., 1972. On the Ice Resistance Encountered by Ships Operating in the Continuous Mode of Icebreaking. Report No.24, The Swedish Academy of Engineering Science in Finlande, Helsinki, Finland.

Poznyak, I.I. and Ionov, B.P., 1981. The division of icebreaking resistance into components. Preceedings 6th STAR Symposium. SNAME, New York, pp. 249-252

Su, B, 2011. Numerical Predictions of Global and Local Ice Loads on Ships. PH.D. Thesis. Department of Marine Technology, Norwegian University of Science and Technology, Norway.

Lindqvist, G., 1989. A straightforward method for calclation of ice resistance of ships. Proceedings of 10th International Conference on Port and Ocean Engineering under Arctic Conditions (POAC), Lulea, Sweden.

Riska, K.,Leiviska, T., Nyman, T., Fransson, L.,Lehtonen, J., Eronen, H. and Backman, A.,2001. Ice Performance of the Swedish Multi-Purpose Icebreaker Tor Viking II. Proceedings of PAC 2001, pp.849-865.

Ko, D.H., Park, K.D. and Ahn, K.S., 2016. Time domain simulation for icebreaking and turning capability of bow-first icebreaking models in level ice. International Journal of Naval Architecture and Ocean Engineering (IJNAOE) Vol. 8, pp. 228-234.

Ashby, M.F., Palmer,A.C., Thouless, M., Goodman, D.J., Howard, M.W., Hallam, S.D., Murrell, S.A.F., Jones, N., Sanderson, T.J.O. and Ponter, A.R.S, 1986. Nonsimultaneous failure and ice loads on Arctic structures. Offshore Technology Conference 1986 NO. OTC 5127, pp. 399-404, Houston, USA.

Sanderson T.J.O, 2014, Ice Mechanics. Springer, Lexington, USA.

Study on the Procedure to Obtain Attainable Speed in Pack Ice

Hyun Soo Kim[1], Seong-Yeob Jeong[2], Sunhong Woo[3], Donghwa Han[3]
[1] Department of Naval Architecture and Ocean Engineering, Inha Technical College, Incheon, Korea(Institution, City, Country)
[2] Korea Research Institute of Ships & Ocean Engineering (KRISO), Deajeon, Korea
[3] Department of Naval Architecture and Ocean Engineering, Inha University, Incheon, Korea

ABSTRACT

Cost evaluation for voyage route planning in ice covered sea is one of the major topics among ship owners. Information of ice properties, such as ice type, concentration of ice, ice thickness, flexural strength of ice and speed-power relation in given ice properties, are very important to decide optimum route in cost perspective.

To find procedure of attainable speed evaluation at any designated pack ice condition model test results of resistance, self-propulsion and overload test in ice and ice free water were used. Also calculation of net thrust and estimation of ice resistance by empirical formula were performed. All model tests were carried out in KRISO ice tank and towing tank.

Combination of model test results and careful observation of underwater videos shows physical phenomena of underwater situation and suitable buoyancy and clearing resistance of empirical formula. Detailed description of determination of attainable speed in certain power and pack ice condition will be presented in this paper.

KEY WORDS: Pack Ice; Pack ice resistance; Model test; Attainable speed;

INTRODUCTION

In available literature there are not so many published articles regarding safe speed in an operation prospective. John Dolny et al. (2013) presented good research about safe speed of structure prospective and Hyun Soo Kim et al. (2016) presented new approach for evaluation of attainable speed in pack ice. They separated clearing and buoyancy components in pack ice resistance and made a regression formula for the Araon vessel based on the model test results. Another approaching technique to know performance in pack ice is studied and will be described in this paper.

Attainable speed is one of speed features that are used for operation in ice covered sea. From an operational point of view several definitions of speed should be compared. First one is attainable speed in ice is defined as the speed which is possible to achieve in certain ice type and properties. To know attainable speed, relationships between thruster and resistance in ice and ice free water are compared carefully in propulsion performance prospective. Second speed definition is safe speed. Hull is not safe if vessel is passing through in heavy ice condition with critical speed. Hence, to know safe speed of structure hull stress and collision effect of ice should be evaluated. Third one is limitation speed in ice passport. The authorities permit limitation of operating speed in ice passport based on vessel's speed-power performance in ice, operating method, routes, season and ice thickness. Three different speeds effect voyage route system and the lowest one will be operating speed by authority.

PROCEDURE TO GET ATTAINABLE SPEED

Previous study, which was presented by first author in IAHR 2016, is based on precise observation of underwater video and finding coverage of vessel's underwater surface area. But separation of clearing and buoyancy resistance in pack ice resistance is not easy because relationship among ice concentration, velocity of vessel and pack ice resistance is complicated. Total pack ice resistance was used instead of separation of clearing and buoyancy resistance. The flow chart of attainable speed evaluation used in this study is shown in Figure 1.

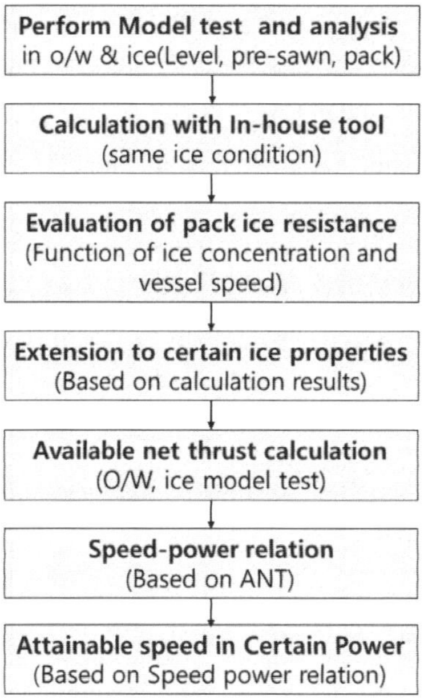

Figure 1. Flow chart to find attainable speed

First Korean ice breaking Research vessel Araon is used for the model test in this paper. Table 1 shows short information of Araon. Hyun Soo Kim (2011) presented more detailed information about Araon's specification in his paper.

Table 1. Principle dimension of Araon

Length, B. P. (m).	Length, B. P. (m)	Design Draft (m)	Propulsion Motors (kW)
95.0	19.0	6.8	2 set of 5,000

All model tests in ice and ice free water are carried out in KRISO's (Korea Research Institute of Ships and Ocean Engineering) ice tank and towing tank. Analysis software, which is explained in Cheol ho Ryu (2013) and Kyung Duk Park (2014)'s articles, is used for calculation of ice resistance.

MODEL TEST IN KRISO ICE TANK AND TOWING TANK

Model tests were carried out with three ice thicknesses which is categorized case 1 (41cm), case 2 (56cm) and case 3 (90cm). Figure 2 and 3 are shown the photos of model at level ice and pre-sawn ice (underwater view) respectively.

Figure 2. Photo of level ice test

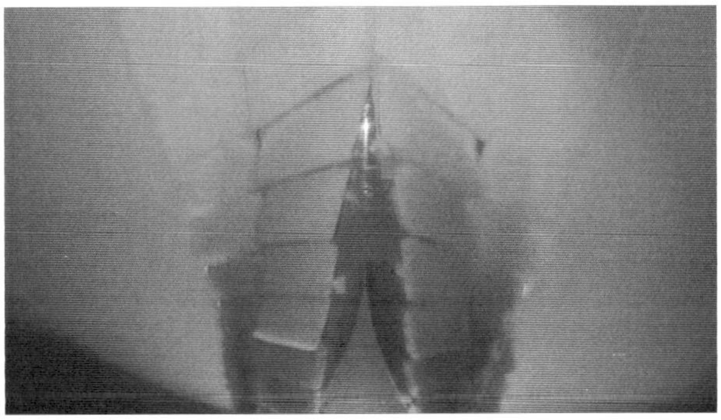

Figure 3. Photo of Pre-sawn ice (underwater view)

Model test for pack ice condition were performed in three different ice thickness (0.5m, 1.0m, 1.5m) and three different concentration of pack ice (60%, 80%, 90%) per each ice thickness.

Figure 4. Photo of pack ice at 6/10 concentration

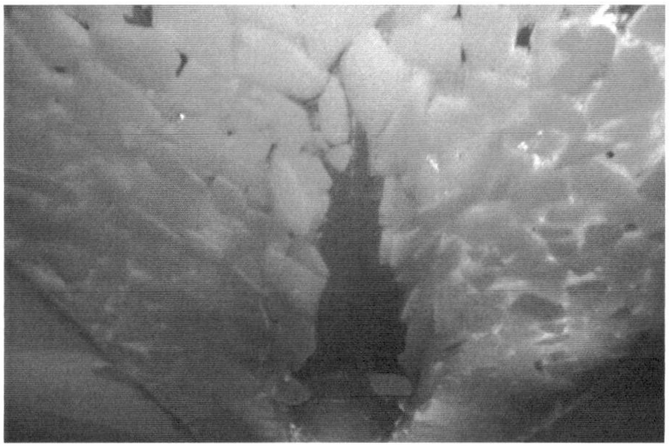

Figure 5. Photo of pack ice at 8/10 concentration (under water view)

COMPARISON BETWEEN MODEL TEST RESULTS AND CALCULATION RESULTS

Accuracy of In-house code for ice resistance calculation was verified by comparison of code results with model test ones and published in several papers, for example Woo jin Kim(2015), Cheol ho Ryu (2013) and Kyung Duk Park (2014). This comparison showed that difference of ice resistance evaluated by code and measured by experiment is less than 10%. The accuracy of In-house code is checked one more time in this study and Figure 6 shows the comparison of in-house code results with model test results. It's less than ±8% but case 2 has 13% difference when speed is 5 knots

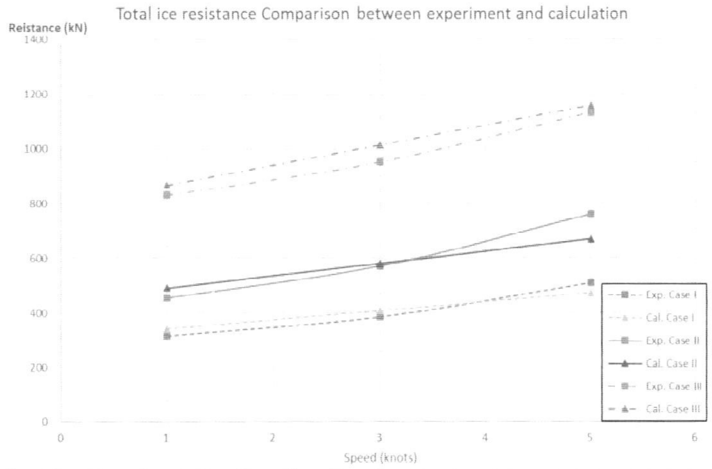

Figure 6. Total ice resistance comparison between experiment and calculation

In In-house code pre-sawn ice resistance is calculated by following formula.

$$R_{pre-sawn} = R_{Buoyancy} + R_{Clearing} + R_{open\ water} \tag{1}$$

$$R_{pre-sawn} = R_{pack\ ice\ at\ 100\%\ concentration} \tag{2}$$

Where R means Resistance and all subscripts are the meanings of resistance

Open water resistance is come from open water model test results.

Figure 7 shows pre-sawn ice resistance comparison between experiment and calculation. The difference is less than ±8% but in case 1 difference reaches 9.7% when speed is 5 knots. Figure 7 shows that in Case 2 coincidence of calculated result and experimental one is the best among three cases.

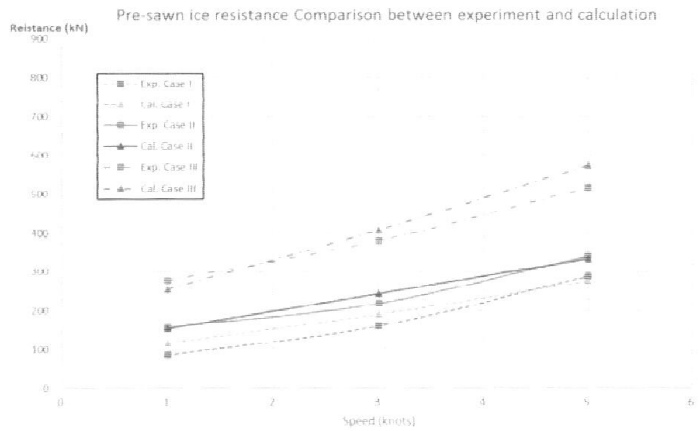

Figure 7. Pre-sawn ice resistance comparison between experiment and calculation

Figure 8 shows model test results of pack ice (three kinds of ice concentration) and pre-sawn (100% concentration) results when ice thickness is 1.06m. The relation according to variation of speed, ice concentration and pack ice resistance show reasonable tendency.

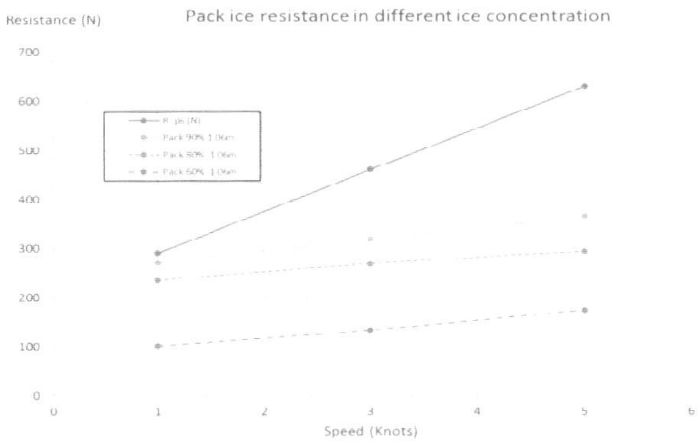

Figure 8. Pack ice resistance in different ice concentration

Based on above results authors tried to figure out relationship among ice concentration, ice resistance and ship speed. MATLAB was used to find plane of those relationship. MATLAB result is depicted in Figure 9. The merit of MATLAB is ability to generate relation formula and find RMS (Root Mean Square) value among the variables. RMS was 0.9359 and formula is shown in equation (3).

$$f(x, y) = a + bx + cy + dx^2 + exy \tag{3}$$

Where, a, b, c, d, e are constant value and x, y are ice concentration and velocity respectively.

The ranges of speed in formula (3) are applied between zero and maximum speed of the vessel from the open water results. This formula can calculate ice resistance in given ice properties using correlation factor.

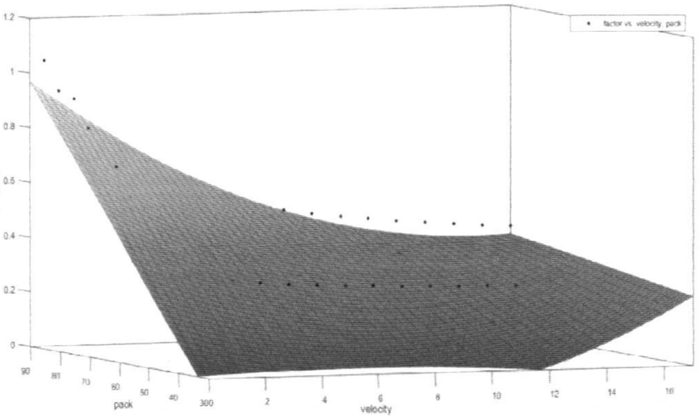

Figure 9. 3D plane of velocity, ice concentration and ice resistance

ANALYSIS OF ATTAINABLE SPEED

Bollard pull test results and available net thrust are shown in Figure 10 & 11. Detailed procedure for available net thrust is explained in Finnish Maritime Administration (FMA) and Hyun soo Kim (2015).

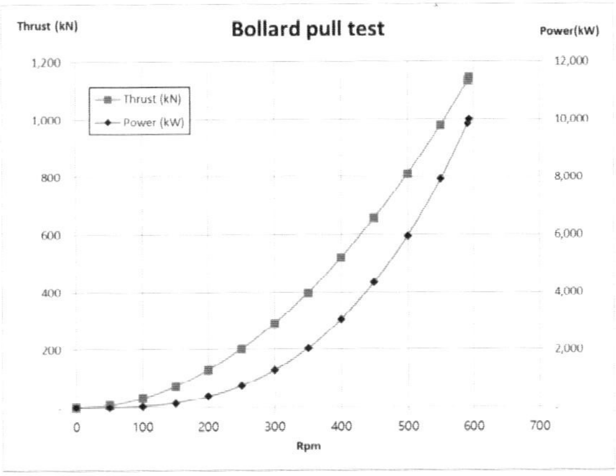

Figure 10. Bollard Pull test results

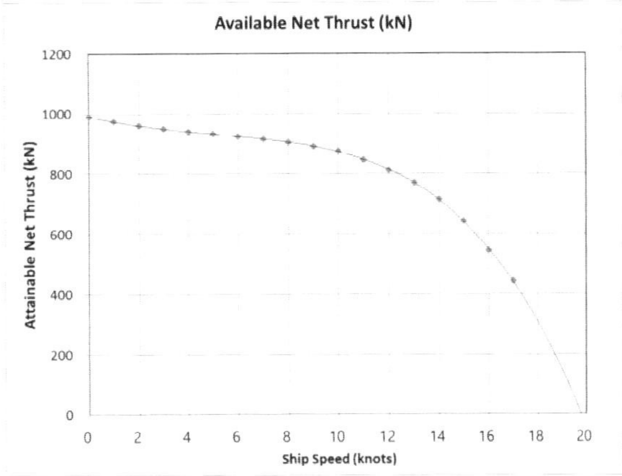

Figure 11. Available net thrust for two thrust

Based on pack ice model test results which are performed with different ice properties (concentration, thickness and flexural strength), analysis results of available net thrust and bollard pull test results, so called over-thrust, can be calculated. Over-thrust means it still have available thrust after overcome in certain ice resistance at maximum engine power of 10MW. It is possible to calculate pack ice resistance in variable pack ice properties (concentration: 6 to 9, thickness, strength and velocity) using In-house code. Ice thickness is changed from 0.5 to 1.4m, flexural strength is varied from 540kPa to 800kPa and velocity of vessel is varied from 1 knot to 16 knots. Figure 12 shows ice thickness and over-thrust relation when speed is at 5 knots pack ice and flexural ice strength of 540kPa. It's plotted with different pack ice concentration. The results show that over-thrust at 1.4m pack ice is about 350kN at 90% concentration. It means that 10MW powered vessel can operate of 5 knots at 1.4m, 90% pack ice and have a margin of 350kN thrust and 4.5MW power.

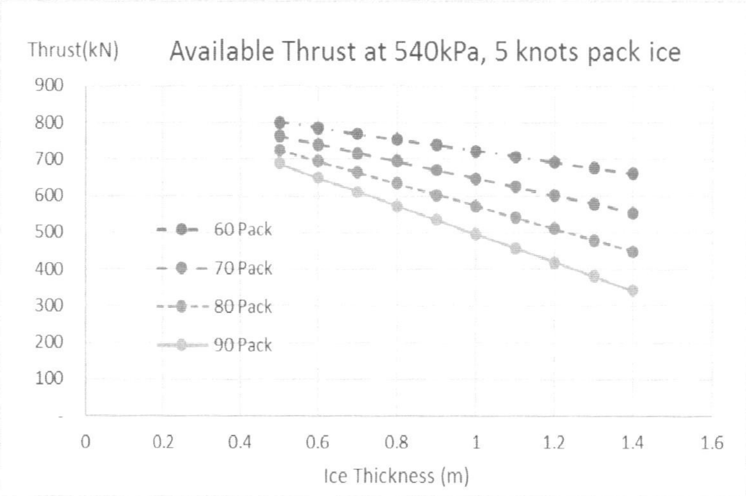

Figure 12. Ice thickness vs thrust relation

Attainable speed can obtain from Figure 13 & 14. This figures show residual thrust after overcome of pack ice resistance including open water resistance. Different ice concentrations are considered in calculation and it is found out the attainable speed in Figure 13 and 14 at 0.5m and 1.0m ice thickness respectively.

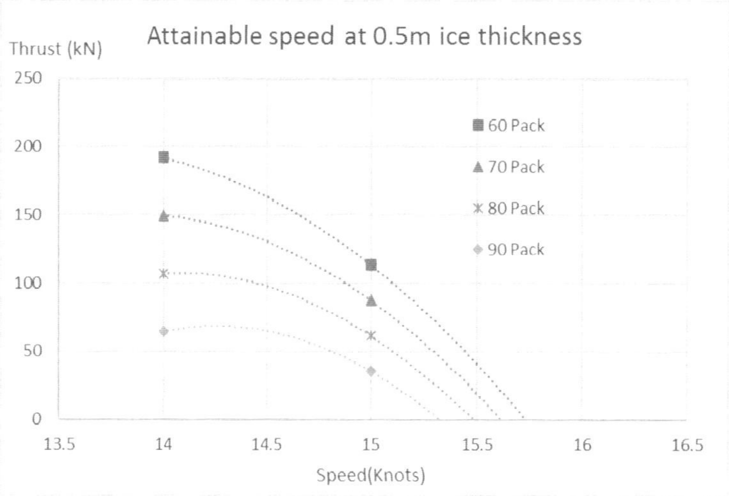

Figure 13 Attainable speed calculation at 0.5m ice thickness

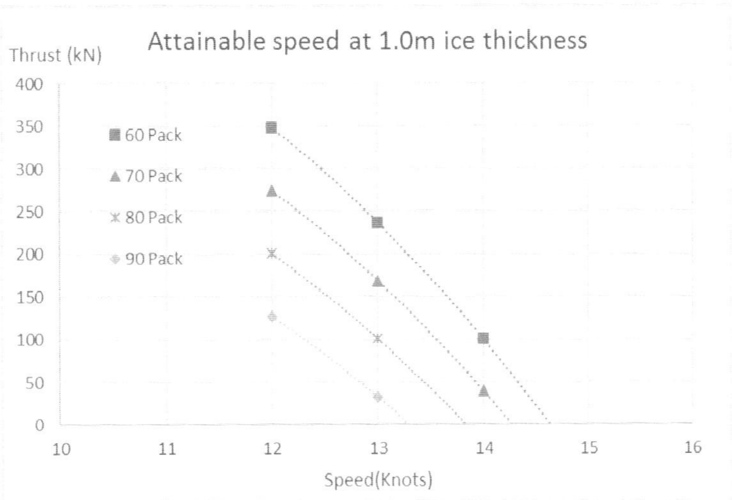

Figure 14. Attainable speed calculation at 1.0m ice thickness

Final attainable speed at 0.5m and 1.0m ice thickness, 540kPa ice strength and 10MW power is same as Table 2. It was calculated along different ice concentration.

Table 2. Attainable speed at 0.5 & 1.0m depend on ice concentration

Pack ice concentration (%)	60	70	80	90
Attainable speed (knots) at 0.5m	15.73	15.61	15.49	15.32
Attainable speed (knots) at 1.0m	14.63	14.27	13.84	13.30

Attainable speed in any pack ice condition can be obtained using interpolation method if ice properties (flexural strength, thickness, concentration) are given but real ice field is not categorized by ice properties (concentration, strength, strength). In natural ice field it's mixed with different ice properties (type, thickness, strength, concentration).

Anyway, this procedure is very useful to know speed and estimated arrival time based on pack ice condition and these results are also helpful to make voyage route planning. The merit of this procedure is possibility to calculate attainable speed if operator decides certain power level.

CONCLUSIONS

This paper is focused on attainable speed calculation. For this open water resistance and self-propulsion performance, characteristic of propulsion system, performance in different pack ice concentration and calculation of In-house code were used. The procedure is developed and will be useful in voyage route planning system.

Araon's performance in pack ice is evaluated in this paper. Based on the model test and analysis results it can be concluded that Araon is able to operate at 13.3 knots in pack ice with full power in next conditions: ice thickness is 1.0m, ice strength is 540kPa and ice concentration is 90%. This attainable speed should be not final operating speed. Operator must check safe speed in structure prospective and limitation of ice passport carefully. Lowest speed among these three kinds of speed will be operating speed.

For the future work, brash ice performance will be checked again and attainable speed in brash ice will be studied using similar approach that was presented in this research.

ACKNOWLEDGEMENTS

The research is part of the project titled "Development of Safe Voyage Planning System for Vessels Operating in Northern Sea Route (PJT200642)" which is supported by the Ministry of Oceans and Fisheries, Korea and also part of the project titled "Development of hullform of year-round floating-type offshore structure based on the Arctic ocean in ARC7 condition with dynamic positioning and mooring system (Project number 10063405)" which is supported by the Ministry of Trade, Industry & Energy(MOTIE, Korea).

REFERENCES

Cheol ho Ryu, 2013, Development of ice resistance estimation system using empirical formula, *Proceedings of the POAC 2013,* Espoo, Finland

Enkvist, E.,1972, *On the Ice Resistance Encountered by Ships Operating in the Continuous Mode of Ice breaking*, Report No.24. The Swedish Academy of Engineering Sciences in Finland, Helsinki.

Finnish Maritime Administration(FMA), 2002, *Finnish-Swedish ice class rules*

Hyun Soo Kim H.S., 2011, Study on icebreaking performance of the KOREA Icebreaker ARAON in the ARCTIC sea, *International Journal Naval Architecture and Ocean Engineering*, 3, pp.208-215.

Hyun Soo Kim, Cheol Ho Ryu, Kyung Duk Park, June Lee, 2014, Development of estimation system of ice resistance with surface information of hull form, *Ocean Engineering* 92, pp. 12-19

Hyun Soo Kim, Cheol-Ho Ryu, Dong-Hyun Ahn, SuHeon Ju, Kuk-jin Kang, Jong Gye Shin, 2015, Development of software which is calculating ice resistance using 2-dimensional Hull form design program(HCAD), *Proceedings of Society of Naval Architects of Korea*, Je-Ju, Korea

Hyun Soo Kim, 2015, Study on the equivalent model test and analysis method of brash ice at KRISO ice tank, Trondheim, Norway

Hyun Soo Kim, 2016, Investigation of the Extension of an Ice-Resistance System for a Various Pack Ice Condition, *Proceedings of the 23nd IAHR International Symposium on Ice Conference*, Ann Arbor, Michigan USA.

Hyun Soo Kim, Dong-Hyun Ahn, SuHeon Ju, Kuk-Jin Kang, Seong-Yeob Jeong, 2016, An Investigation of the Extension of an Ice-Resistance System for a Various Pack Ice Condition, *The 31st International Symposium On Okhotsk Sea & Sea Ice*, Monbetsu, Japan

John Dolny, 2013, Developing a technical methodology for the evaluation of safe operating speeds in various ice conditions, *Proceedings of the POAC 2013,* Espoo, Finland

Kyung Duk Park, 2014, Study on the ship ice resistance estimation using empirical formulas, *Proceedings of the OMAE Conference*, 2014-23971, San Francisco, California USA

Naude, P. & Holland, C., 1998. Marketing in the information domain. In: Halinen-Kaila, A. and Nummela, N. eds., Interaction, Relationships and Networks: Visions for the Future, *Proceedings of the 14th Annual IMP Conference*, pp.245-262.

Poznyak,I.I., Ionov,B.P., 1981, The division of ice breaking resistance into components. *Proceedings of the Sixth STAR Symposium*. SNAME, NewYork, pp. 249-252.

Shimanskii, Yu.A, 1938, *Conditional Standards of Ice Qualities of a Ship, Northern Sea Route Administration Publishing House Leningrad*, vol.130. Translation T-381-01 by Engineering Consulting and Translation Center (ECTC), New-York.

Spencer, D., 1992, A Standard Method for the Conduct and Analysis of Ice Resistance Model Tests, *Proceedings of the 23rd ATTC Symposium*, pp.301-307.

Woo Jin Kim, 2015, Ice resistance estimation and comparison the results with model test, *Proceedings of the ISOPE Conference*, Hawaii, USA

POAC'17
Busan, KOREA

Proceedings of the 24th International Conference on
Port and Ocean Engineering under Arctic Conditions
June 11-16, 2017, Busan, Korea

2D Numerical Modeling of Icebreaker Advancing in Ice-covered Water

Junji Sawamura[1]
[1] Department of Naval Architecture and Ocean Engineering Osaka University, Osaka, Japan

ABSTRACT

This paper presents a 2D numerical modeling to calculate removal and breaking of ice floes when an icebreaker advances into ice-covered water. The numerical simulation calculates repeatable ice-breaking and ice-removal in pack ice. The ice-breaking is calculated by a ship-ice contact detection and fluid-structural interaction of bending behavior of a plate ice. The ice-removal are calculated by a physically based modeling using 3DOF rigid body equations. A ship advances with a constant speed or constant thrust. A plate ice is broken by bending and splitting mode. The ice foes drift with constant force in which ice floes are assumed to be drifting by wind and current. An open channel caused by the ice-breaking and ice-removal by icebreaker are numerically obtained. A simulated ice channel after icebreaker's advancing into ice-covered water depends on the floe size as well as the ship's rigid motion and ice drifting force. The numerical results show the proposed numerical model can be useful in order to identify an efficient way of ship handling in ice-covered water.

KEY WORDS: Ice management; Pack ice; Icebreaking; Numerical simulation

INTROCUDTION

Ice management using an icebreaker is vital for safe and efficient operations for vessels and offshore structures in ice-covered waters. During an ice management operations, an icebreaker opens a cannel behind her by removing and breaking of ice floes. An escorted vessel can reduces the ice load in the managed ice channels. For successful ice management operations, the estimation of the managed ice channels opened by an icebreaker, and planning of the management vessels and fleet deployment configuration are needed.

Hamilton et al. (2011) developed a numerical simulator and quantified the ice management performance using real ice conditions data. They numerically study the effects of the ice management strategies such as the number of icebreaker and icebreaker speed performance on the ice floe size in the managed ice channel. Farid et al. (2014) investigated the sea ice breaking pattern of several short-term ice management activities during the research cruise (OATRC, 2013), and proposed a preliminary analysis. The maximum floe size resulting from the numerical simulations was fairly equivalent to that of real ship trial. Lu et al. (2015, 2016) investigated fracturing phenomena of ice floe during icebreaker's ice management. They proposed the theoretical models of in-plane and out-plan ice failure for implementing into the numerical simulator of ship-ice interaction (e.g., Lubbad and Løset, 2011). The numerical simulation, as shown above, has the potential to evaluate of ice management strategy efficiently.

This paper develops a 2D numerical modeling to calculate removal and breaking of ice floes when an icebreaker advances into ice-covered water. The numerical model can predict repetitive ice floe removal and plate ice breaking by icebreaker. The motion of ice floes and icebreaker are described by 3DOF rigid body equations. The ice floes is drifting with constant drifting force which assumed to be ice floe's drifting by wind or current. Example results show the distributions of the ice channels and time history of ice force in different ice conditions. The effect of ice drifting and ice floe size on the ice channel made by an icebreaker are investigated.

NUMERICAL MODELING

Interaction between Ship and Plate Ice

The numerical simulation of ship advancing into level ice has been developed by Sawamura et al. (2009). In the relative large ice floe, a ship repetitively contacts with the ice edge and breaks the plate ice in various ice failure mode such as bending, crushing and splitting. The repetitive ship–ice contacts are represented using circle contact detection (e.g. Dimgliana et al., 2000). Figure 1 shows a schematic whereby circle contact detection is applied to ship–ice contacts. A plate ice and ship waterline are represented by contact circles. When the ship waterline contacts with the ice edge, the distance of both contact circles between the ship and ice becomes zero. When the plate ice is broken, contact circles within the ice cusp are removed from the plate ice. Contact detection and removal of contact circles within the broken cusp are progressing as the ship advances. The circle contact detection allows us to detect the contact points between the two complicated objects in less computational time. The accuracy of ship–ice contact and icebreaking depends on the contact circle radius.

The ice edge at the ship–ice contact is crushed as a ship advancing distance. The ice force increases proportionally with increased crushing area of the ice edge. When an ice pressure on the crushing area is assumed to be a constant, the ice force induced by a ship–ice contact is given as:

$$F_{\mathrm{n}} = A_{\mathrm{c}}\sigma_{\mathrm{c}} \tag{1}$$

where A_c represents the crushing area induced by ship–ice contact. σ_c is the compressive strength of ice. F_n acts in the direction normal to the ice crushing surface. Crushing area A_c increases with the icebreaker's advancing distance into ice edge. As the deflection of the plate ice is assume to be neglected and the geometrical relation between ship and ice are considered (Figure 2), the contact area A_c is given as:

$$A_{\mathrm{c}} = \frac{1}{2\sin\theta_{\mathrm{s}}}\left(\tan\theta_{\mathrm{wf}} + \tan\theta_{\mathrm{wb}}\right)\cdot(vt\cos\theta_{\mathrm{c}})^2, \qquad 0 < t < \frac{h_{\mathrm{i}}\tan\theta_{\mathrm{s}}}{v\cos\theta_{\mathrm{c}}}$$

$$= \frac{1}{2\cos\theta_{\mathrm{s}}}\left(\tan\theta_{\mathrm{wf}} + \tan\theta_{\mathrm{wb}}\right)\cdot(2vth_i\cos\theta_{\mathrm{c}} - h_i^2\tan\theta_{\mathrm{s}}), \qquad t > \frac{h_{\mathrm{i}}\tan\theta_{\mathrm{s}}}{v\cos\theta_{\mathrm{c}}} \tag{2}$$

where v denotes the ship velocity, t is the ship advancing time into the ice edge, h_i is the ice thickness, θ_c and θ_s are, respectively, the ship's entrance angle and the ship hull's inclined angle. θ_{wf} and θ_{wb} are the forward and backward edge angles at the contact point.

Kinematic friction takes into account in direction horizontal to the surface of ice crushing. Coulomb type friction is represented as:

$$F_f = \mu F_n \tag{3}$$

where μ denotes the friction coefficient between the ship and ice. A direction of frictional force is opposite to the relative velocity between ship and ice. The total ice force attributable to ship–ice interaction is related to ice breaking of a plate ice that is calculated by the sum of collision force F_n and friction force F_f.

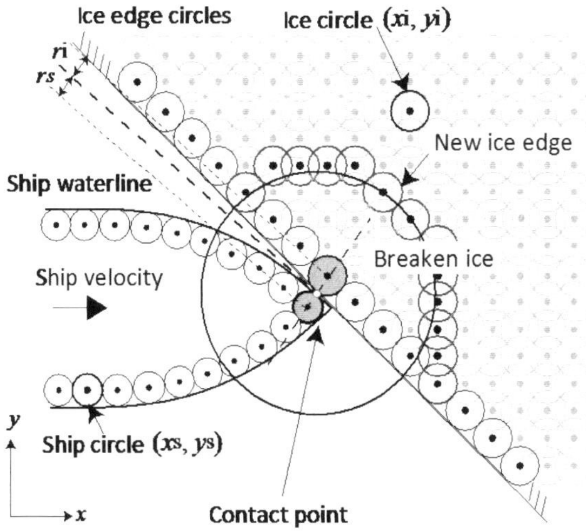

Figure 1. Circle contact detection applied to ship–ice interactions.

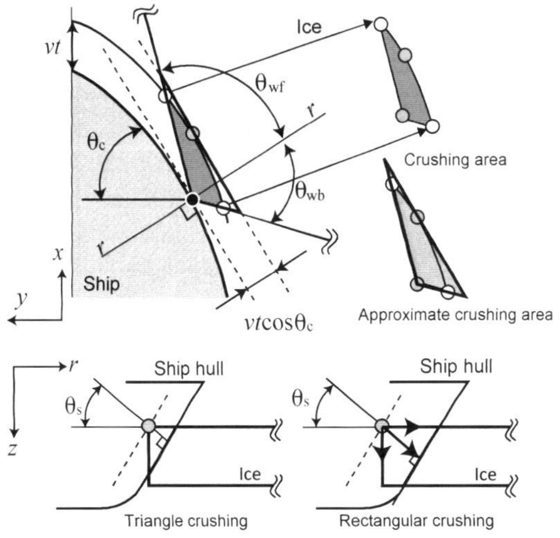

Figure 2. Ship-ice relation of ship-ice contact, ice crushing and ice contact force.

Plate Ice Breaking

A bending failure of plate ice occurs when the bending stress in the plate ice σ_b increases as the ship advances and exceeds the flexural strength of the ice, σ_f. The bending stress in the plate ice σ_b is calculated using a fluid–structure interaction (FSI) in which the dynamic effect of fluid underneath the plate ice is calculated as well as a plate ice bending (Sawamura et al., 2008). After bending failure, a circler ice cusp is broken off from the plate ice. The broken cusp radius is calculated by the distance from the contact point and maximum stress point when a plate ice is broken. The icebreaking time is calculated by a time from the ship–ice contact to the ice bending failure. The bending stress in the plate ice σ_b in various ship–ice contact conditions are calculated by using FSI. A database of an icebreaking force, area and time was made. In numerical simulations for different ship–ice contact conditions, the icebreaking force, area, and time at each contact point can be derived from this database.

An alternative mode of ice failure is splitting. A splitting failure occurs because of normal tensile stresses. The splitting force is obtainable as (Michel, 1978 and Cammaert, A.B. et al., 1988)

$$F_s = nLh\sigma_t \tag{4}$$

where σ_t stands for the tensile strength of the ice and n signifies a shape factor for the structure which varies as the wedge angle. L and h respectively denote the length and thickness of ice floe. As described in this paper, the shape factor n is 0.25 (Cammaert, A.B. et al., 1988). The crack of the splitting is assumed to propagate normal to the ship waterline.

Motions Equations and Contact Force

Physically based modeling (Baraff, D., 1997) is applied to calculate the ice floe and ship motions. Ship and ice floe are assumed to be a rigid body. A circle contact detection is applied for ship–ice and ice–ice contacts as described above section. The motions of ship and ice floe are described by 3DOF rigid body equations. The position and orientation of ice floe are solved by Newton's second law. A collision response between two objects (ship–ice and ice–ice) is calculated using an instantaneous impulse. The impulse vector **J** in a normal direction at the contact surface can be written as;

$$\mathbf{J} = j \cdot \mathbf{n}_{ik} \text{ where, } j = \frac{-(1+\varepsilon)v_{ref}}{\dfrac{1}{m_i} + \dfrac{1}{m_k} + \mathbf{n}_{ik}\left\{\left(\mathbf{I}_i^{-1}(\mathbf{r}_i \times \mathbf{n}_{ik}) \times \mathbf{r}_k\right) + \left(\mathbf{I}_k^{-1}(\mathbf{r}_k \times \mathbf{n}_{ik}) \times \mathbf{r}_i\right)\right\}}. \tag{5}$$

Therein, j signifies the magnitude of an impulse, $\mathbf{n}_{ik}$ denotes the direction of impulse force of object i received from the object k, ε is a coefficient of restitution, v_{ref} stands for the relative velocity between two objects, m_i represents mass of an object i, and $\mathbf{I}_i$ is the inertia of an object i. Also, r_i is the displacement vector representing the displacement from a center of mass of object i to the center of the contact circle. The mechanical friction force of the Coulomb model described by Eq. (3) is presented in the direction of horizontal to a surface of contact. In the numerical simulation, the time interval of an impulsive response is assumed to be 0.5 s, which is used to transfer from impulsive force [N·s] to peak force [N].A fluid force of surrounding water acting on an ice floe is represented as:

$$\mathbf{F}_w = -C_D A \frac{1}{2} \rho |\mathbf{v}| \mathbf{v} \tag{6}$$

where C_D signifies the drag coefficient, A denotes a projected area, ρ represents water density, and **v** stands for the ice floe velocity. The fluid force is opposite to the ice floe velocity. In the numerical simulation, the drag coefficient C_D is 0.5. A buoyancy force at the gravity center of the ice floe is considered. For simplicity, other fluid forces related to the ship motion are neglected.

Ice Removal and Ice Breaking

During ship–ice interaction, the ice floe size is major variable for icebreaking scenario. When a ship collide with ice floe, a small ice floe is simply removed. Splitting and bending failure occur in a large-sized ice floe. As described in this paper, it is simply determined that the direct rotation of a small ice floe is occurred when an ice floe size L is smaller than a characteristic length of ice l (Lu et al., 2016). A characteristic length of ice is represented as:

$$l = \sqrt{Eh_i^3 / 12(1-\nu^2)/(\rho_w g)} \tag{7}$$

where, E signifies Young's modulus of an ice, h_i denotes the plate ice thickness, ν represent Poisson's ratio ρ_w, stands water density, g is gravitational acceleration. In the larger ice floes than the characteristic length of ice, splitting occurs when an impulse between the ship and ice floe calculated using Eq. (5) is larger than a splitting force obtained by Eq. (4). Bending failure occurs in all remaining ship–ice contacts when a collision force is less than splitting force. An ice floe alternates with ice breaking (bending or splitting) and ice removal (motion of ridged body) within the time interval Δt. During the ice removal contact, ice crushing occurs when the impulse calculated using Eq. (5) is larger than ice-ice contact force calculated using Eq. (1).

RESULTS AND DISCUSSIONS

Numerical Model

Numerical simulations that calculate repetitive ice breaking and ice removal of pack ice have been conducted. The icebreaker waterline and the pack ice model for simulations are shown in Figure 3. A virtual icebreaker (Length × Breadth = 120×24 m) is modeled. The water entry angle and bow angle of the model ship are, respectively, 27° and 30°. The displacement and moment of inertia of the icebreaker are, respectively, 10000 t and 5×10^9 kgm^2. Three different sizes of squared ice floes are arranged within 550×300 m area of the pack ice model. In Figure 3, the sizes of ice floes are, respectively, 15×15 m (110 floes), 40×40 m (21 floes), and 110×200 m (2 floes). Free boundary conditions are given in the numerical domain. The ice floes are 0.5 m thick. Young's modulus of the plate ice, E, is 5.4 GPa. The flexural strength of the ice, σ_f, is 1.0 MPa. The compressive strength of the ice, σ_c, is 2.8 MPa. The tensile strength of the ice, σ_t, is 0.8 MPa. The friction coefficients μ between the ship–ice and ice–ice are 0.2. The coefficient of a restitution ε between the ship–ice and ice–ice are 0.1. The density of ice and water are, respectively, 900 kg/m^3 and 1000 kg/m^3. The ship and ice floe motions are described by 3DOF rigid body equations. The radius contact circle that is used to detect the collision points between the ship and ice is 1.0 m. The simulation time step Δt is 0.005 s.

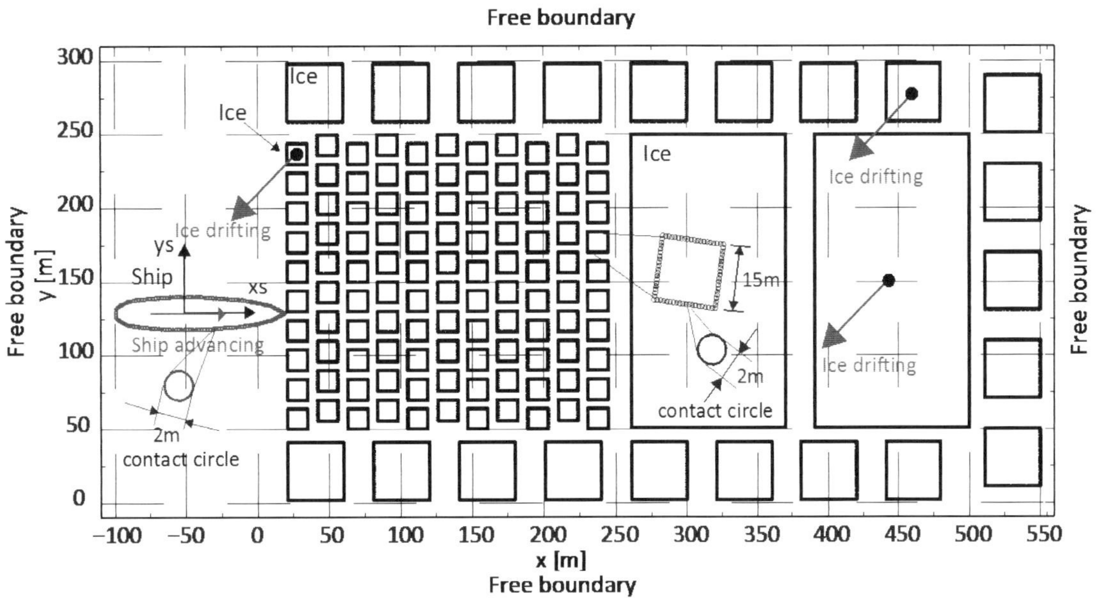

Figure 3. Numerical model when icebreaker advances into ice–covered water with pack ice (The sizes of smallest ice floes = 15×15 m, the number of smallest ice floe = 110 floes).

Effect of Ship Motion and Ice Floe Drifting

Figure 4 presents distributions of ice floes (ice channels) when an icebreaker advances in pack ice. In Figure 4(a), the icebreaker advances with constant ship speed: v_s = 1.0 m/s. In Figure 4(b), the icebreaker advances with constant thrust: $F_t = 0.2\times10^5$ N. In Figure 4(c), the icebreaker advances with constant thrust, and the pack ice floes move with constant drift speed: v_i = 0.071 m/s (v_{ix} = -0.05 m/s and v_{iy} = -0.05 m/s), in which all of the ice floes are drifting in the directions from the portside of icebreaker. The icebreaker removes small ice floes, and makes an ice channel behind the ship. The large ice floe (plate ice) is broken by the bending failures and drifted by the repetitive ship–ice contacts. The ice splitting takes place when the ship–ice collision force acquires the ice splitting force described by Eq. 4. The plate ices divided into 2 or 3 pieces are drifted and removed by ship-ice collision force and icebreaker's advancing. In the comparison with Figure 4 (a), (b) and (c), the ice removals when the icebreaker advances into the small ice floes with different ship–ice conditions show the similar ice channel. On the other hand, the icebreaking by the icebreaker advancing into the large ice (plate ice) show different ice channels. In Figure 4(c), the ship course is changed by the effect of the ice floes drifting from the portside of the icebreaker.

Figure 5 shows the time histories of the peak ice force acting on the icebreaker. The x-force represents surge force in ship coordinate system. The y-force represents sway force. The ice collision force when the icebreaker collides with small ice floes is smaller than the ice breaking force when the icebreaker collides with a large ice and breaks the plate ice by bending failure. The icebreaking force when an icebreaker advances with constant ship thrust (Figure 5(b) and 5(c)) becomes larger than those with the constant ship speed (Figure 5(a)). Two splitting failure are observed in all of ship–ice interactions in Figures 5(a), (b) and (c). After the second ice splitting, the time history of ice force become more complicated. When an icebreaker advances with the constant ship thrust and ice drifting speed (Figure 5(c)), the ice force induced by both ice breaking and ice contact are observed. These results show that ship motion and ice drifting have large influence on the icebreaking of a large ice floe (plate ice), but small influence on the ice removal of small ice floes.

Effect of Ice Floes Size

Figure 6 shows the ice floes distribution in different ice floe size. The size of smallest ice floes (15×15 m, 110 floes) in Figure 3 are changed. In Figure 6(a), the size of ice floes are 10×10 m and the number of ice floes are 225 ices. In Figure 6(b), the size and the number of ice floes are, respectively, 20×20 m and 64 ices. In Figure 3, the ice concentration in which smallest ice floes are distributed is 53%, and in both of Figure 6(a) and (b) are 55 %. In Figure 6, the icebreaker advances with constant thrust: $F_t = 0.2\times10^5$ N. The pack ice drift with constant speed: v_i = 0.071 m/s. The ice channel width becomes wider depending on the ice floe size. The ship route of icebreaker removes small ice floes is changed by the effect of the ice floe size.

Figure 7 shows time histories of the peak ice force in different floe size. The peak values of the ice collision force when an icebreaker removes the small ice floes depend on the ice floe size. The peak value of the ice removal force colliding with the small ice floes (Figure 7(a)) is smaller than those with large ice floes (Figure 7(b)). The peak value of the ice breaking forces when an icebreaker breaks the plate ice are almost same in Figure 7(a) and (b). In Figure 7(a), the ice breaking starts at 231 s, and ice splitting occurs at 231 s and 366 s. In Figure 7 (b), the ice breaking starts at 293 s, and the ice splitting occurs at 367 s and 454 s. The delay of the

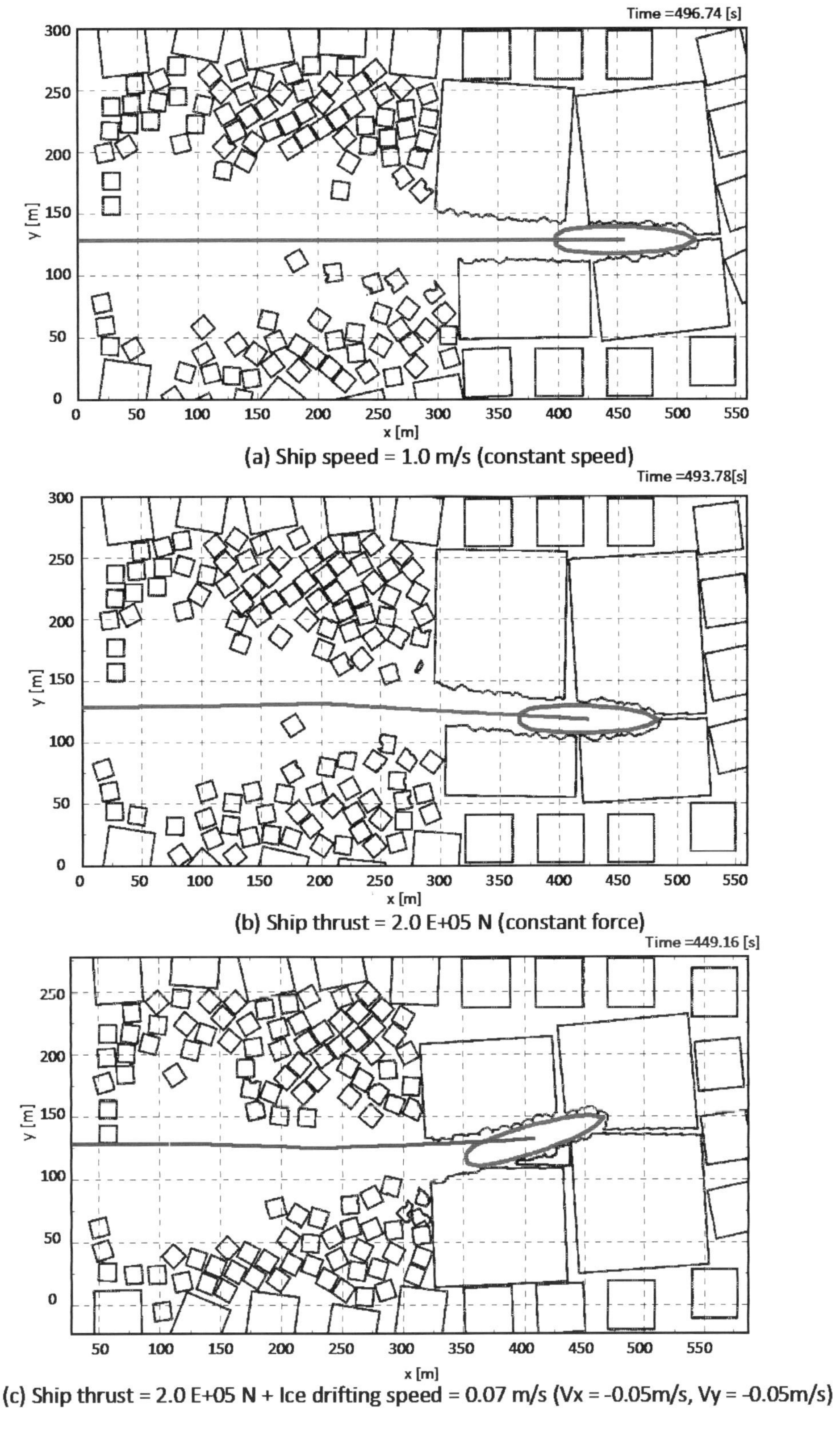

Figure 4. Distributions of ice floes (ice channels) (The sizes of smallest ice floes = 15×15 m, the number of smallest ice floe = 110 floes).

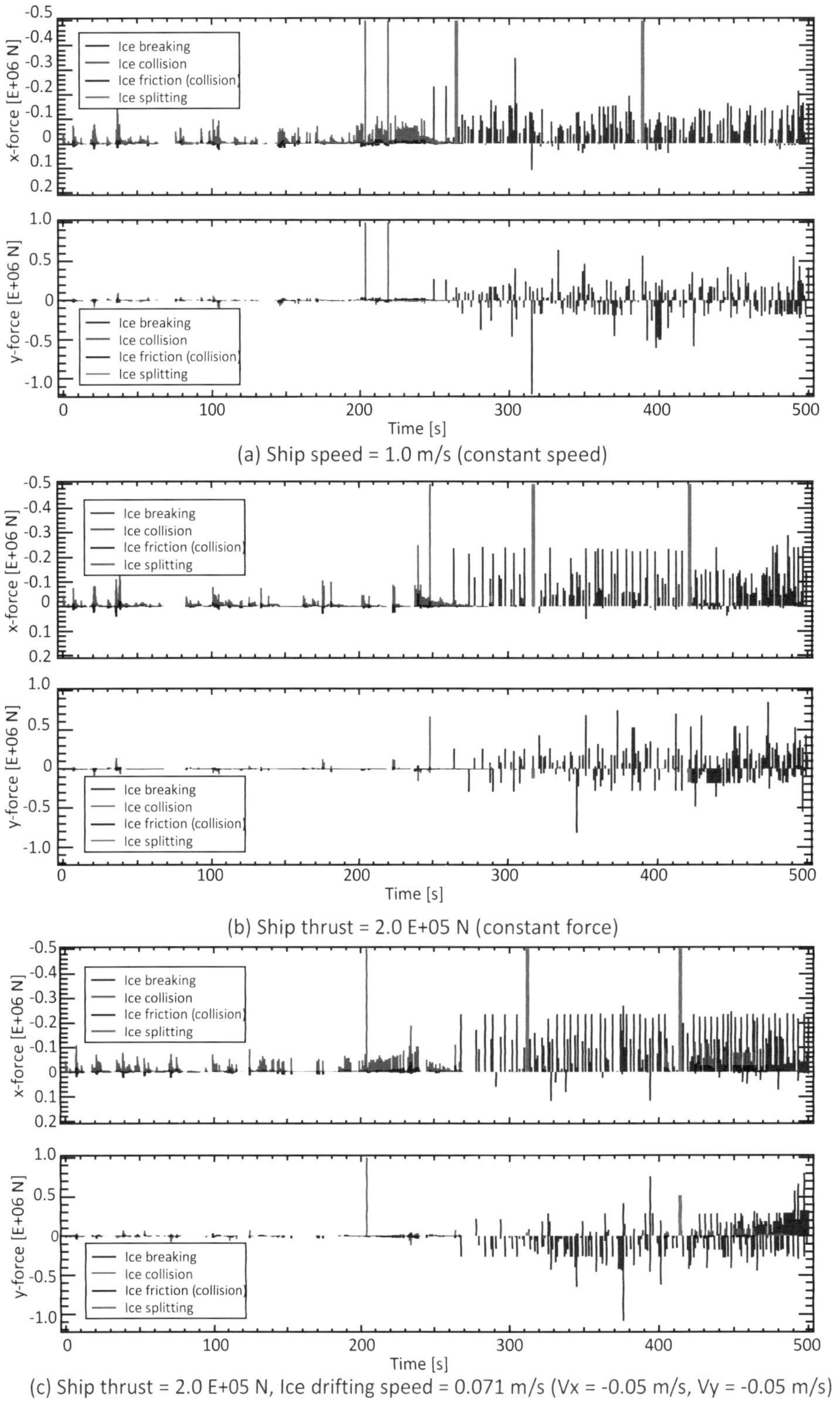

Figure 5. Time history of peak ice force (The sizes of smallest ice floes = 15×15 m).

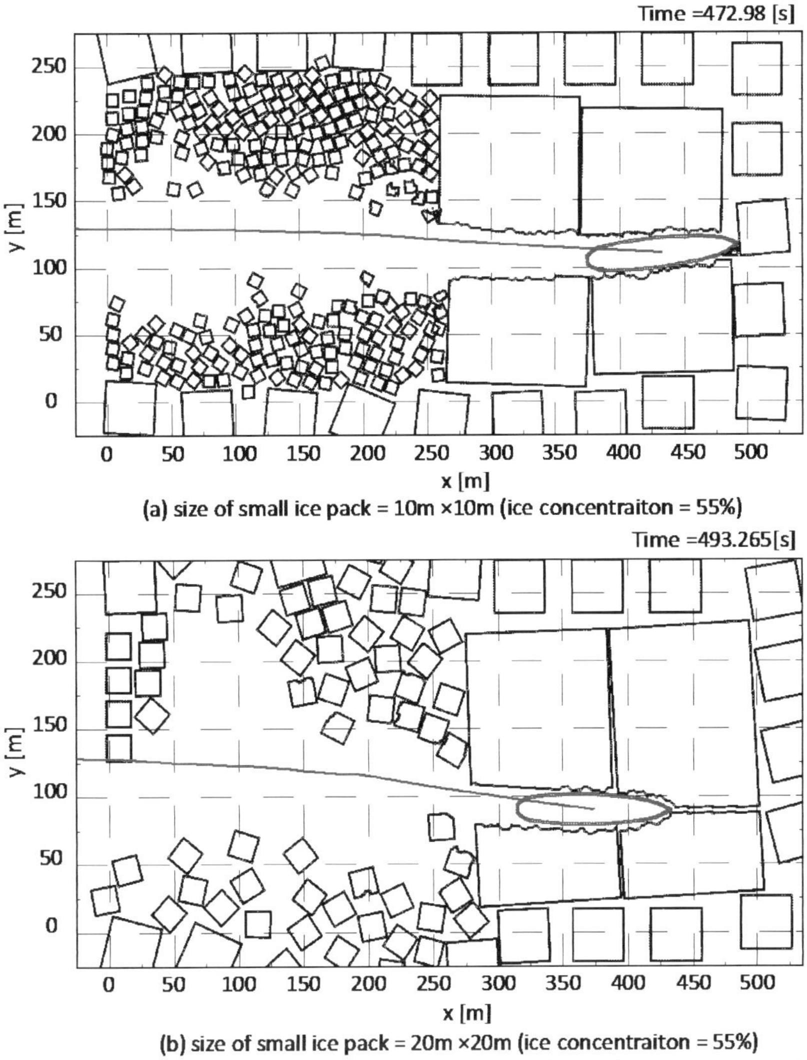

Figure 6. Comparison of ice channel in different ice floes size (The sizes of smallest ice floes = (a) 10×10 m and (b) 20×20 m).

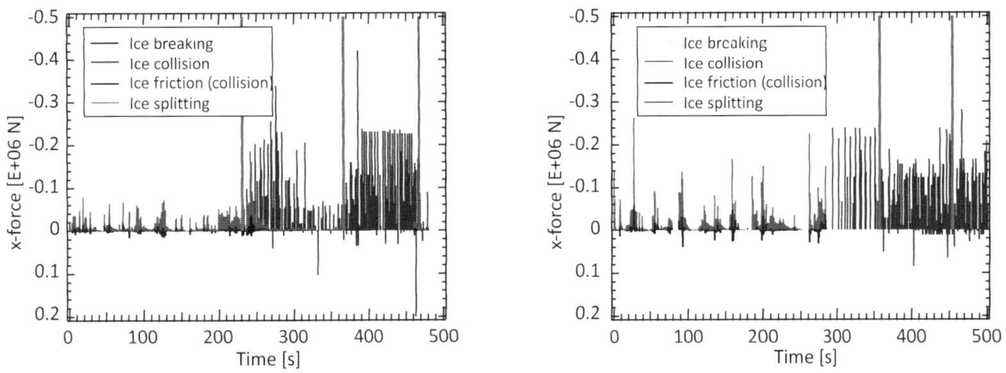

(a) size of small ice pack = 10m ×10m (ice concentraiton = 55%)

(b) size of small ice pack = 20m ×20m (ice concentraiton = 55%)

Figure 7. Comparison of peak ice force in different ice floes size (The sizes of smallest ice floes = (a) 10×10 m and (b) 20×20 m).

icebreaking and splitting can be explained by the speed reduction due to the large contact force when an icebreaker removes the large ice floes. These results show that the ice floe size removed by icebreaker has large influence on both ice breaking of a plate ice and the ship–ice contacts of small ice floe.

CONCLUSIONS

This paper presents a numerical model to calculate ice breaking and ice removal when an icebreaker advances into ice-covered water with pack ice. The ice removal and ice crushing are considered when a small ice collides with a ship or ice floe. The bending failure and splitting are introduced as an ice failure mode of the plate ice. The effect of the ship motion, ice floe drifting and ice floe seize on the ship–ice interaction in pack ice are investigated. The ship motion and ice drifting have large influence on the icebreaking of a plate ice, but small influence on the ice removal of small ice floes. The ice floe size removed by icebreaker has large influence on both ice breaking of a plate ice and the ship–ice contacts of small ice floe. The numerical results show that more realistic numerical modeling including a ship performance and ice conditions is needed for the estimation of managed ice channels. A numerical simulation, however, has the potential to evaluate of ice management strategy. Verification of the numerical model is necessary to identify an efficient mode of shiphandling in actual ice sea.

ACKNOWLEDGEMENTS

This research was supported by JSPS KAKENHI Grant Number 26630455 and 15kk0206. The support from JSPS is greatly appreciated.

REFERENCES

Baraff, D., 1997. Rigid Body Simulation I and II. Siggraph 97, Course notes.

Cammaert, A.B. and Muggeridge, D.B., 1988. Ice Interaction with Offshore Structures, Chapter 7, Static Global Ice Forces on Vertical Structures, Section 7.3, Splitting Mode, pp 232-233.

Dimgliana, J. and O'Sullivan, C., 2000. Graceful degradation of collision handling in physically based animation, *Computer Graphics Forum*, Vol. 19, Issue 3, 239-248.

Farid, F., Scibilia, F., Lubbad, R. and Løset, S., 2014. Sea Ice Management Trials during Oden Arctic Technology Research Cruise 2013 Offshore North East Greenland. *Proceedings of the 22nd IAHR International Symposium on ICE*, pp.518-525.

Hamilton, J., Holub, C.J. and Blunt, J., 2011. Simulation of Ice Management Fleet Operations using Two Decades of Beaufort Sea Ice Drift and Thickness Time Histories. *Proceedings of International Society of Offshore and Polar Engineers*, pp 1100-1107.

Lu, W., Lubbad, R. and Løset, S., 2015a. Tentative Fracture Mechanics of the Parallel Channel Effect during Ice Management. *Proceeding of 23th International Conference on Port and Ocean Engineering under Arctic Conditions*, Paper No.253.

Lu, W., Lubbad, R., Løset, S. and Kashafutdinov, M., 2016. Fracture of an ice floe: Local out-of-plane flexural failures versus global in-plane splitting failure. *Cold Regions Science and Technology*. 123. pp 1-13.

Lubbad, R. and Løset, S., 2011. A numerical model for real-time simulation of ship-ice interaction. Cold Regions Science and Technology. 65(2). pp 111-127.

Michel, B., 1978. Ice Mechanics. Les Presses de l'universite Laval, Quebec.

Sawamura J., Riska K., Moan T., 2008. Finite Element Analysis of Fluid-Ice Interaction during Ice Bending. *19th IAHR International Symposium on Ice,* pp.191-132.

Sawamura J., Riska K., Moan T., 2009. Numerical Simulation of Breaking Pattern in Level Ice at Ship's Bow. *The proceeding of 19th International Offshore and Polar Engineering Conference*, pp.600-607

Proceedings of the 24th International Conference on
Port and Ocean Engineering under Arctic Conditions
June 11-16, 2017, Busan, Korea

Experimental Studies of Ice Loads on Pod Propulsors of Ice-Going Support Ships

Alexey Dobrodeev[1], Kirill Sazonov[1],
Alexander Andryushin[2], Sergey Fedoseev[2], Sergey Gavrilov[3]
[1] Krylov State Research Centre, St. Petersburg, Russian Federation
[2] Central Marine Research and Design Institute, St. Petersburg, Russian Federation
[3] Sapphire Center, Russian Federation

ABSTRACT

Intense developments of offshore oil & gas fields in the Russian Arctic call for a more capable fleet of icebreakers. Today, ice-resistant oil rigs are already operating on the Russian shelf in harsh ice conditions, regularly offloading their production to shuttle tankers. One of the main challenges in operation, for example, of Prirazlomnaya platform is a nearby buildup of drifting ice pieces noted in Zaykin (2016). To ensure the safety of oil offload to tankers, special ice-breaking ships are being designed and constructed. These ships are intended to ensure safe mooring & loading operations at sea in ice conditions.

Direction and speed of ice drift at the locations of offshore production facilities might vary significantly, so state-of-the-art ice-breaking ships are fitted with 360° pod propulsion units (PPUs) that significantly improve their maneuvrability (Sazonov, 2006). An important issue is investigation of ice loads on PPUs to ensure their safe operation.

This paper provides the test results obtained on the dummies of PPUs in Ice Basin of KSRC. The model tests were performed under the scope of design activities regarding the rudder & propulsion system of an Icebreaker 7 ship. The results of these tests allow determination of ice loads on PPUs, as well as analyse their distribution for the ship moving in different ice conditions and with different turning angles of PPUs.

KEY WORDS: Ice loads; POD; Icebreaker supply vessel; Level ice; Ice management.

Wide-scale Arctic developments call for a more capable marine logistical system in the Arctic, ice-breaking supply ships being one of its core elements. Ice-breaking ships are mainly intended to pilot cargo carriers, to perform emergency & rescue operations in ice conditions, to support drilling and production oil rigs, to perform ice management operations in order to safeguard the tanker while she is being loaded with oil, as well as to make sure that the ice near the tanker will be brash ice in small pieces; and also, in certain cases, to reduce the global ice load on the production rig (Kostylev, 2016) by means of constant maneuvers serving to break drifting ice formations, including ridges. Accordingly, an ice-breaking vessel must have a good maneuvrability and be able to gyrate in ice both ahead and astern.

For this reason, one of the most important aspects of an ice-going ship design is the development of a propulsion system intended for operation in ice conditions. Currently, pod propulsion units (PPUs), being efficient propulsion and maneuvrability tools of the ships operating in ice conditions, are seriously investigated. The point is that operation in ice conditions involves significant ice loads on both hull of an ice going ship (Sazonov, 2010) and her PPUs. The hull has flat areas at the PPU location, and the frames in the transition area between the bottom and the side are rounded. Due to this hull shape, significant amounts of ice come to the propulsors (see Figure 1. below). Russian Maritime Register of Shipping states certain requirements to the loads induced by propeller interaction with ice. However, these requirements do not cover the ice loads induced by the ice hits against the PPU body, so investigation of ice loads on PPUs of ice-going ships is an important issue.

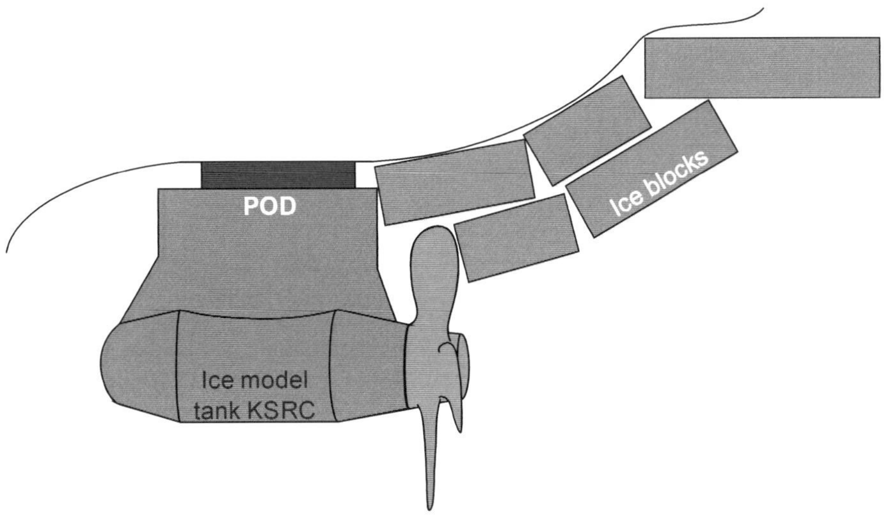

Figure 1. – Ice interaction with the PPU of an ice-going ship moving astern

Currently, electric PPUs (Azipod) are the most common. They operate in pulling mode, have a good flow around their strut, and their torpedo-shaped body accommodates the drive and the propulsion shaft. Examples of ships with Azipod propulsion are Vasily Dinkov-class Arctic tankers, Norilsky Nikel-class container ships, LK-25 icebreakers (Project 22600), etc. Another type of pod propulsors installed on modern ice-breaking vessels are Arctic PPUs with mechanical gearing. PPUs of this type are applied on Projects 21900 and 21900M icebreakers, ice-breaking support ships Varandey, Toboy, etc. It shall be noted that the PPUs mentioned above successfully operate on the ships with high ice class.

Thus, a growing fleet of icebreakers and ice-going carriers calls for a design of a Russian 360° electric PPU capable of operating in ice conditions. Such PPU must meet very tough reliability requirements, so its development is impossible without profound studies. This paper discusses the studies of ice loads on the ice-class PPU being designed by Russian engineers.

TEST RIG FOR STUDIES

The study of ice loads on the PPU under design was performed in Ice Basin of KSRC. The ice sheet generated in this test tank is 80 m long and 10 m wide (Timofeev, 2015). Such dimensions allow maneuvrability studies of icebreakers, e.g. exit of the ship from her own channel, determination of gyration radius, “star” turning maneuver, and also the studies of icebreaker behavior during ice management operations.

During the studies of ice loads on PPU, an experimental rig was developed, consisting of a self-propelled icebreaker model with two PPUs. One of its PPUs is installed with the instrumentation system specially developed for these tests.

The instrumentation system includes two 3-component dynamometers installed between the foundation rigidly fixed in the aft of the icebreaker model, and the flange of the PPU (see Figure 2.). In this arrangement, the PPU would only have two points of support, so to make it stable without prejudice to its ability to move due to ice load, two flexible elastic lines were installed additionally.

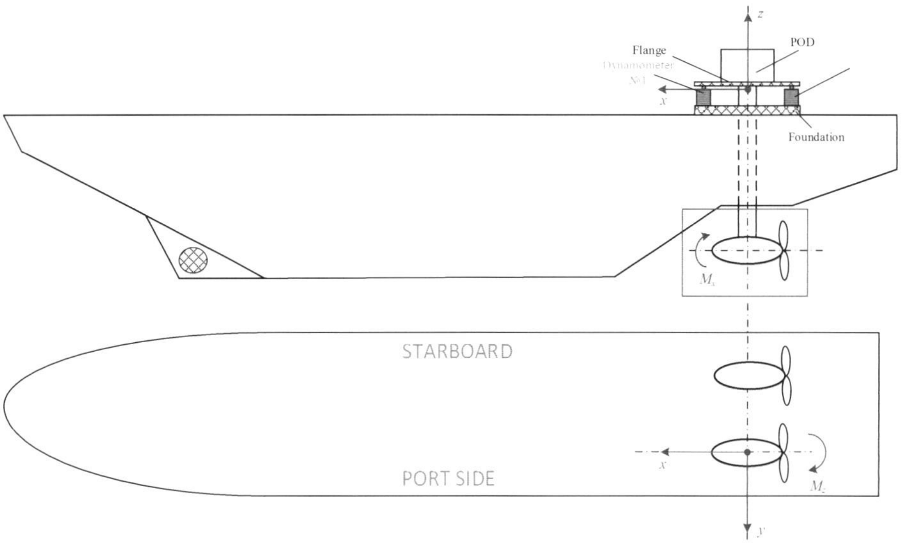

Figure 2. – Arrangement of PPU with instrumentation on the model of the ship

The developed instrumentation system is intended to experimentally determine longitudinal Px and transverse Py component of ice-induced force, bending moment Mx and twisting moment Mz. Positive directions of ice load along coordinate axes Ox, Oy, Oz are shown in red. The elastic membrane of vertical force gauge Pz installed in the dynamometer used in the tests lies within 0XY plane.

To measure the ice loads on PPUS by means of Gauges 1 and 2 shown in Figure 1. above, the following calculation formulas for forces and moments were suggested:

$$P_x = \frac{(U_{xz} - U_0)}{K_{x1}} + \frac{(U_{x2} - U_0)}{K_{x2}} \text{ (N)} \tag{1}$$

$$P_y = \frac{(U_{yz} - U_0)}{K_{y1}} + \frac{(U_{y2} - U_0)}{K_{y2}} \text{ (N)} \tag{2}$$

$$M_z = \left[\frac{(U_{xz} - U_0)}{K_{x1}} - \frac{(U_{x2} - U_0)}{K_{x2}}\right] \cdot \frac{L}{2} \text{ (Nm)} \tag{3}$$

$$M_x = \left[\frac{(U_{zz} - U_0)}{K_{z1}} - \frac{(U_{z2} - U_0)}{K_{z2}} \right] \cdot \frac{L}{2} \text{ (Nm)} \tag{4}$$

where subscripts 1 and 2 correspond to the number of the gauge used; F_x is the longitudinal component of force, N; F_y – transverse component of force, N; M_x – bending moment, N·m; M_z – twisting moment, N·m; U_0 – initial readings of corresponding measurement channel for Gauge 1, mV; $Ux(y,z)_1$ – current readings of corresponding measurement channel for Gauge 1, mV; $Ux(y,z)_2$ – current readings of corresponding measurement channel for Gauge 2, mV; $Kx(y,z)1$ – conversion coefficient of corresponding measurement channel for Gauge 1, mV/N; $Kx(y,z)2$ – conversion coefficient of corresponding measurement channel for Gauge 2, mV/N; L – distance between vertical axes of Gauges 1 and 2.

The calculations always considered the polarity sign of the output voltage for corresponding measurement channel, $Ux(y,z)1$ and $Ux(y,z)2$. Prior to tests, calibration loadings of dynamometers were performed, where the maximum measurement error for each component did not exceed 3%.

The experimental studies were performed in the model level ice sheet, thickness 1.3-1.8 m full-scale, as well as in broken ice and in ridges. The strength of level ice also varied within the range of 500-800 kPa full-scale. The experiments was performed with different PPU turning angles, propeller rotation rate being adjusted so as to ensure the thrust corresponding to 100% propulsion power of the ship for which these PPUs are being designed. The PPUs were operating in the pulling mode, and the tests were performed with the ship running astern (Figure 3.).

Figure 3. – Fragment of the tests in Ice Basin

To ensure the simulation of the specified speed in ice, the experiments were performed with the model fastened to the planar motion mechanism of the towing carriage. The planar motion mechanism moves the model across the towing tank, perpendicularly to the direction of towing carriage movement.

RESULTS OF EXPERIMENTAL STUDIES

Currently, there exist various approaches to determination of ice loads on PPUs, however, the variety of ice classes and ship types requires comparison of calculation results versus model and ful-scale test data. This paper is an attempt to compare the calculated loads obtained by the method described in DNV Rules (Classification Notes No. 51.1, January 2011) versus the results of the model tests performed in KSRC Ice Basin.

The method described in DNV Rules is a simplified one. Still, it allows sufficiently correct estimate of ice loads on PPUs for the majority of existing ice classes of ships. This procedure discusses ice classes PC1-PC7 as defined in the IACS document issued in 2006 (ice classes Arc4 – Icebreaker8 as per the Rules of Russian Maritime Register of Shipping).

The ice load on the specified area of the PPU can be determined as follows:

$$F = p_0^{\;0.8} \cdot (S \cdot C^{0.3})^t K_1 K_2 K_3 K_4 \quad [\text{MN}] \tag{5}$$

where p_0 is ice pressure, MPa, determined as per the table of correspondence between ice thicknesses and ice pressures provided in DNV Rules; S – area of ice impact, m^2; $C = S/2h_{ice}^2$, where h_{ice} is ice thickness; K_1 – location and propeller type factor; K_2 factor reduces POD ice loads by 1/3 for under bottom mounted units; K_3 – ship type factor; K_4 – statistical factor for expected maximum load.

Figure 4. compares calculation results obtained as per the procedure described above versus the model test data. The comparison regards the total ice load on the PPU turned in advance by the angle required. Turning angle of 0 degrees corresponds to the position of the PPU working in the pulling mode as the ship moves astern. It shall be noted that the test data regarding the ice loads on the PPUs have a good correlation with the calculation results obtained as per DNV procedure, and the analysis of the plot has detected some peculiarities:

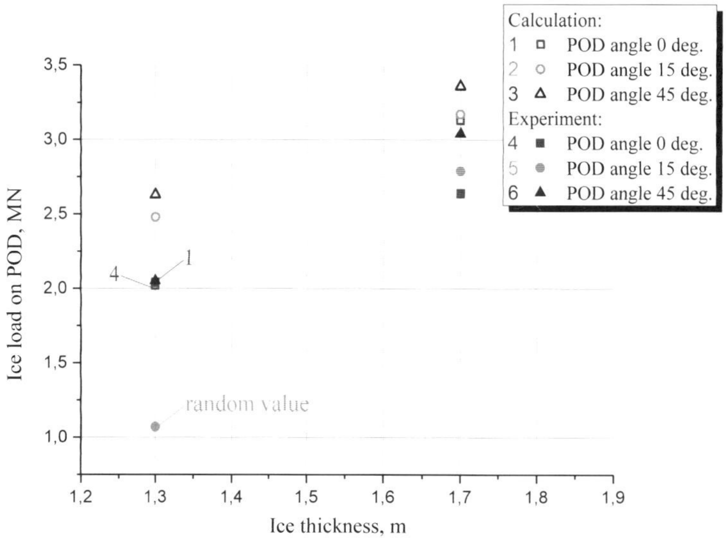

Figure 4. – Ice load: calculation results vs test data

The plot shows that calculation procedure yields the results slightly higher than the test data.

POAC17-031

However it shall be noted that the procedure cannot consider the speed of the ship in the calculation of loads on strut and pod, whereas the test data show that this relationship exists. Figure 5. shows the change in relative ice load $R_{V_N} - R_{V_0} / R_{V_0}$, where R_{V_N} is the ice load at specified ship speed V_N, R_{V_0} – ice load at initial ship speed V_0 as the speed increases by a certain amount, ΔV, when the ship moves in different conditions: movement in continuous level ice and exit of the channel in the ice of different thickness. According to these data, the growth of ship speed by 2 knots might result in ~30% higher load on the PPUs.

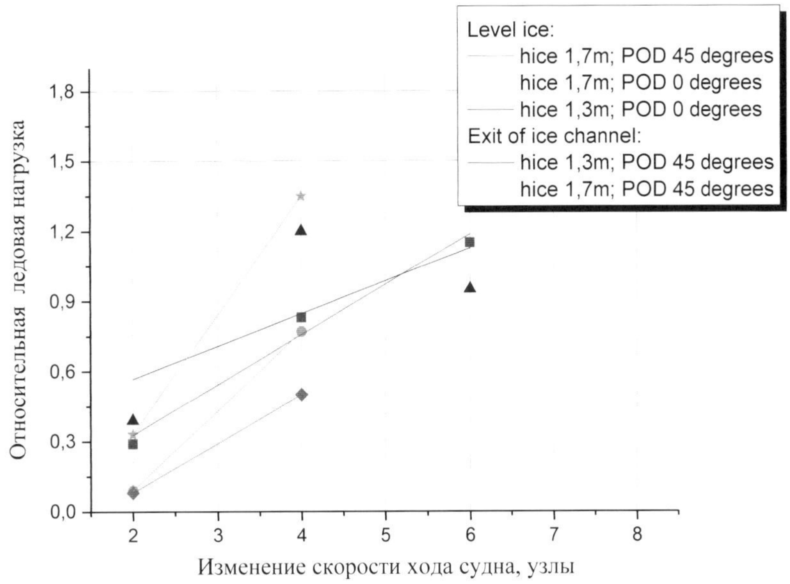

Figure 5. – Load on PPUs versus ship speed

Calculating the ice load on the PPU turned at the specified angle, it is hard to determine the area of ice impact. If the contact area is less than $2h^2_{ice}$, the actual contact area has to be determined to perform the calculation. However the procedure does not give any reference to how the actual area of contact between ice and PPU parts can be determined. Thus, this factor could also contribute to the increase of the total load on the PPU obtained as per the calculation procedure.

Andryushin (2009) extensively discusses analytical and experimental methods for prediction of ice loads on the elements of ship propulsion system. An important aspect of these studies is not only the development of analytical methods for determination of loads on PPU parts, but also the development of the methods to extrapolate model test data to the full scale. For fine-grain (FG) model ice, it is necessary to consider the relationship for the simulated crushing strength and the full-scale value of this parameter. To this effect, the present work included determination of the ice load coefficients required to extrapolate the ice-induced forces and moments acting on the PPUs. These coefficients were derived from the indentor test data obtained in the Ice Basin that allowed determination of ice crushing force for the horizontal loading, as well as calculation of ice strength under vertical uniaxial compression. The main approaches to calculation of ice load on PPUs are discussed in *Andryushin* (2013).

The tests performed in KSRC Ice Basin allowed the assessment of the ice load on the PPU arising when the podded propulsor is hit by an ice piece of specified size. The ratio between the maximum ice load and the average one gives an idea about the level of the extreme load applied to the PPU as it is hit by ice formations (see Figure 6.). These shock loads generate significant bending stresses in the PPU, thus deteriorating its operability.

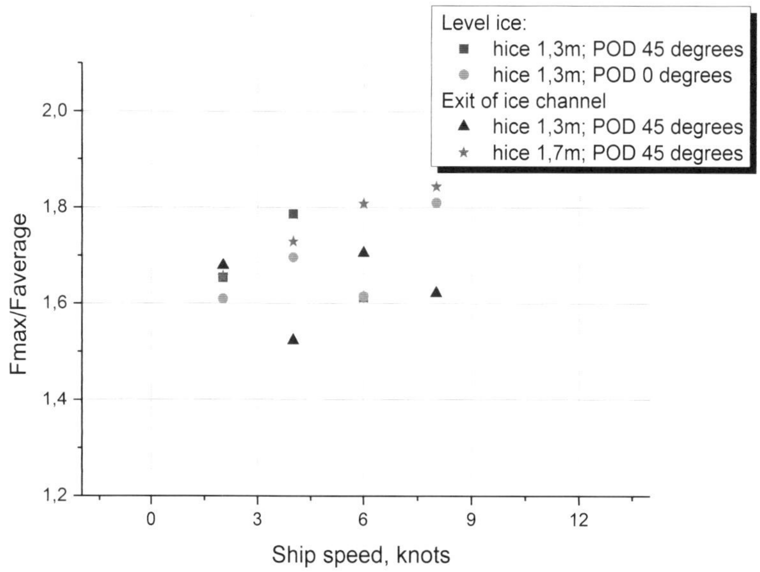

Figure 6. – Growth of maximum ice load (F_{max} / $F_{average}$) versus ship speed

Model tests in Ice Basin yield a large amount of the statistical data on the ice loads on PPUs, and these data can be used to update the existing calculation procedures for ice loads, as well as to develop new ones.

CONCLUSIONS

The results of this work show that the instrumentation system developed to experimentally study the ice loads on the PPUs under design is truly efficient. The test data show a good correlation with the analytical loads obtained by means of DNV procedures and show that to update the existing ice load calculation procedures, it is necessary to introduce the corrections taking into account the speed of the ship fitted with the PPU at question, as well as to develop the recommendations on how to determine the area of contact between an ice piece and the surface of a PPU element at different angles of the PPU turning. The test data show that the growth of ship speed by 2 knots results in ~30% increase of the ice load, and at the moment of contact between an ice piece with the PPU, the ice load jumps 1.6-1.8 times above its average level.

Thus, optimization of calculation procedures for the ice loads on PPUs is an important task that can be solved by obtaining a large amount of statistical data based on the results of ice model tests.

REFERENCES

Sazonov K.E., 2006. Ship maneuverability in ice. Krylov Shipbuilding Research Institute, St. Petersburg, 251 pp.

Sazonov K.E., 2010. Theoretical principles of ship navigation in ice. Krylov Shipbuilding Research Institute, Saint Petersburg, 278 pp.

Zaykin D., Karulin E., Pronyashkin A., 2016. Method of destruction o hummocked ice applied in practice near ice-resistant marine platforms in «Prirazlomnoe» oilfield. *Proceedings of the third International Scientific Conference «Polar Mechanics»*, Vladivostok, Russia, 16 pp.

Timofeev O.Ya., Sazonov K.E., Dobrodeev A.A., 2015. New ice basin of the Krylov State Research Centre. *Proceedings of the International Conference on Port and Ocean Engineering under Arctic Conditions*, POAC 23. Cep. "POAC 2015 - Proceedings of the 23rd International Conference on Port and Ocean Engineering under Arctic Conditions", 8 pp.

Kostylev A.I., Sazonov K.E., 2016. Study of world experience of ice conditions control. *Arctic: ecology and economy*, № 3 (23), pp. 86-97.

Ice strengthening of propulsion machinery, 2010. Classification notes, Det Norske Veritas, Norway.

Andryushin A.V., Evenko V.I., Sergeev A.V. et al., 2009.Современные методы отработки пропульсивных комплексов судов ледового плаванию ледовые нагрузки. Научно-технический сборник РМРС, Issue №32, pp. 182-204.

Andryushin A.V., Hänninen S., Heideman T., 2013. «Azipod» Azimuth Thruster for large capacity arctic transport ship with high ice category Arc7. Ensuring of operability and operating strength under severe ice conditions. P*roceedings of the 22nd International Conference on Port and Ocean Engineerin under Arctic Conditions, Espoo, Finland, 10 pp.*

Full-scale Ice Impact to an Azimuthing Thruster in Laboratory Conditions

Ikka Perälä[1], Aki Kinnunen[2], Pekka Koskinen[1], Jaakko Heinonen[1]
[1] VTT Technical Research Center of Finland, Espoo, Finland
[2] Vibrol Oy, Tuusula, Finland

ABSTRACT

Ships navigating in winter conditions frequently come in contact with ice blocks floating on the waterways. The ice blocks can collide with the propulsion machinery of the ship. Especially the interaction between ice and ship azimuthing propulsion call for greater understanding.

Ice blocks can collide with the azimuthing propulsion units in many different ways, one of them being the propulsion unit hitting the ice block as the ship advances. This is an impact type loading case and the dynamics of the thruster and ship hull affect to the load response level obtained in this case. In a dynamic system, the damping is greatly affecting to the response level.

To study the dynamics and damping in an ice impact event, a full-scale test opportunity was found in an azimuthing thruster test bench. An impact test with a full-scale thruster and ice blocks was carried out in the laboratory with velocities corresponding to realistic ship operating speeds. In addition, numerical analyses of the ice-structure interaction were carried out and structural response was compared with the impact test measurements. Comparison between the simulations and the tests gave valuable information about the thruster dynamics during the ice impact.

KEY WORDS: Ice; Ice impact; Azimuthing propulsion; Simulation; Full-Scale measurement;

THE THRUSTER TEST STAND STRUCTURE

The Ice impact tests were conducted in a test facility where a full scale thruster can be run. The thruster is attached to a test stand structure where it is mounted like in ship bottom. Thruster is driven by an electric motor and the propeller shaft is linked to a generator. Thus, there is no real propeller rotating. The purpose of the thruster test stand is in general research of thrusters.

For the ice impact tests, a structure with rails, a wagon, weight, cable and pulleys was constructed. The ice impact with the thruster was carried out by placing the ice blocks into the wagon and then accelerating the wagon with a weight to a desired impact speed. By varying the masses of the ice and wagon and the weight, the impact speed could be set as desired. The cable pulling the wagon was attached to a hook at the bottom of the wagon and released just before impact. This way, only the mass of the wagon and ice block took part in the impact. A drawing showing the thruster and the ice impact structure can be seen in Figure 1.

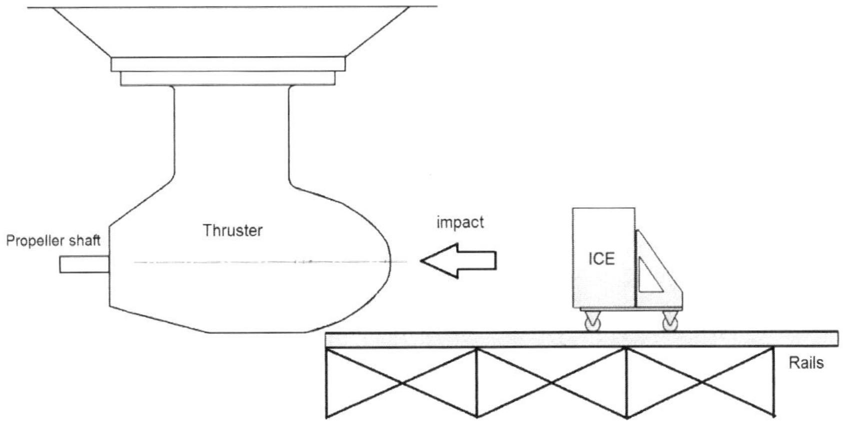

Figure 1. Impact test structure

ICE IMPACTS TO THRUSTER - LABORATORY TESTS

All together 12 impact tests were carried out. The number of tests was limited by the amount of available ice blocks and time. The most interesting test results with respect to ice impact are discussed here.

Ice blocks were frozen in a freezing container in insulated boxes. One of the ice blocks had temperature sensors to monitor the freezing process. First, the boxes were filled with tap water and the container temperature was se to -30 degrees Celsius. After about 10 days when it was monitored that the ice blocks have completely frozen, the container temperature was gradually brought to -2 degrees. During the freezing, all of the ice blocks developed some cracks, however these were not located directly at the contact area. Before the impact tests, the ice blocks were cut to correct size with a saw. See Figure 2 for ice block just before test.

Figure 2. Ice block before test. The impact area is in the middle of the block. One larger crack is visible at the top of the block

Measurement setup in the thruster included the following sensors: Strain gauges in half bridge configuration to measure the bending force in the direction of impact on the thruster strut. Strain gauges in half bridge configuration to measure the shear force on the thruster strut in the direction of impact force. Three triaxial acceleration sensors on the thruster body. One triaxial acceleration sensor on the wagon to measure the accelerations, and thus the contact force, of the impacting ice. Two pulse sensors on the rails to measure the speed of the wagon just before impact. And finally, a high-speed camera to record the impact with speed of 2000 FPS. In addition, a laser distance sensor was used for final tests to see the amount of movement on the bottom of the thruster.

To calculate the response force of the thruster, a calibration was carried out by pulling the thruster from the impact area with a force sensor equipped winch. This way the measured strain gauge data from the strut could be transformed to response force.

Various thruster and test stand specific impact tests were done, but only the validation tests which could be compared to simulated results are discussed here. Table 1 lists these tests. The first test was a low speed test, which is interesting because in this test the ice block developed only minor cracks. In the second test the wagon had much higher speed and the ice block developed major cracks. The last test saw the use of a sharper hubcap, which purpose was to allow a larger indentation and thus lower the contact and the response forces in the impact.

Table 1. Test parameters

Test number	Impact velocity [m/s]	Impact mass [kg]	Notes
1	2	139	Low speed test
2	5,5	136	High speed test
3	5,1	137	Test with sharper hubcap

In Figure 3 The post-impact state of ice blocks can be seen.

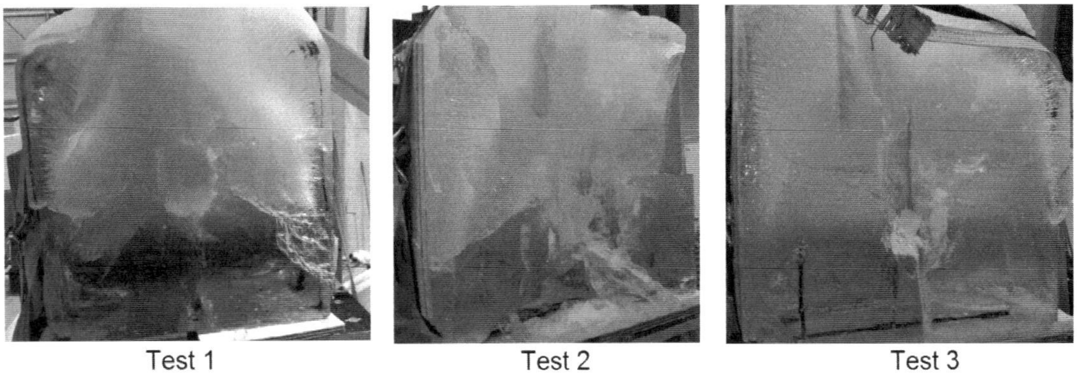

Figure 3. Ice blocks after impact

With the high-speed camera, the impact duration and ice crushing and cracking was possible to see. The duration of events in the video could be determined with an accuracy of 0,005 seconds. Below in figure 4, video screen captures from the test 3 are shown. During the impact in can be seen that the transparency of the ice block changes, which indicates cracks opening in the ice.

Figure 4. Screen captures of the high-speed camera from test 3.

An example of measured load during test 1 is presented in Figure 5. The curve on the left shows the overall timescale, that is five seconds after impact, and it is clear that the dynamic structure did not vibrate in this case after impact. On the right, a detailed view of the impact response shows how the load is varying during the impact event. A cyclic crushing process is assumed to take place and the structure local dynamics is contributing to this result.

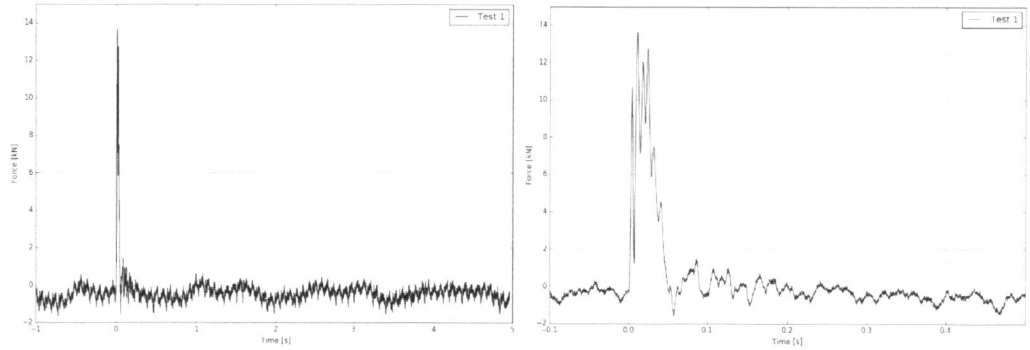

Figure 5. Measured response force time history for test 1. On the left, timescale of 5 seconds, showing impact and no vibratory behavior. On the right, detail of the impact event, showing that there is variation of response load during the impact event.

The thruster indentation into ice is determined for test 1 based on the size of the impact mark on the ice. Impact mark can be seen in Figure 6. The diameter of the marking was on average 110 mm, and the radius of the impacting steel part was 270 mm. When the maximum projection contact area with indentor radius r and indentation depth x is

$$A_{max} = \pi(r_i^2 - (r_i - x_i)^2) \tag{1}$$

It can be seen that with the maximum indentation projection area defined with the radius r_m

$$\pi r_m^2 = \pi(r_i^2 - (r_i - x_i)^2) \tag{2}$$

$$x_i = r_i - \sqrt{r_i^2 - r_m^2} \tag{3}$$

This results in 5,66 mm depth of indentation when the mark has 110 mm diameter and indentor has a radius of 270 mm.

For tests 2 and 3 the indentation could be measured directly because of much larger indentation. The indentation for test 2 was 85 mm and for test 3 175 mm.

Figure 6. Indentation mark on the ice after impact test no 1. Diameter on average is 110 mm.

Response force and contact force measurement results for tests 1-3 can be seen in Figure 7. It is clear that the test 2 has the highest contact and response forces, and that the thruster shows vibratory behavior after the impact. In comparison, in test 3, with even higher initial kinetic energy, the maximum response force is only 1/3 of the maximum response force of test 2 and the thruster does not oscillate notably after the impact. This difference is due to the sharper hubcap, which penetrates the ice block, and thus the dissipation of impact energy is divided to longer period. It should be noted here that for test 1 the thruster was equipped with force cylinders which essentially oppose the impact force. Thus, these cylinders affect the response results. For tests 2 and 3 these cylinders were removed.

The contact forces shown in the Figure 7 are calculated from the ice block acceleration measurements. For force levels this is not very accurate method, but gives some indication of force levels involved. Duration of the contact could be calculated accurately and was found out to be well in line with measurements from the high-speed camera footage.

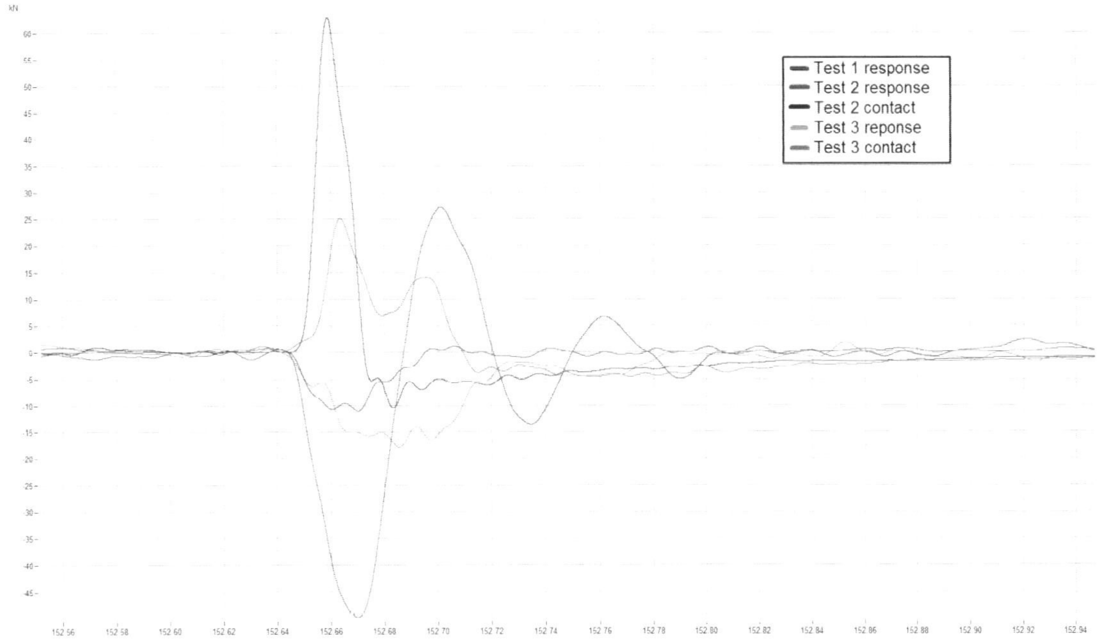

Figure 7. Time histories of all the tests. All signals are filtered with low pass filter. For test 1 contact measurement data was not available. Contact and response force have opposite directions in the graph.

For test 2, a comparison test with identical input parameters was carried out. This test showed very similar response to test 2. Figure 8 Shows results of these tests.

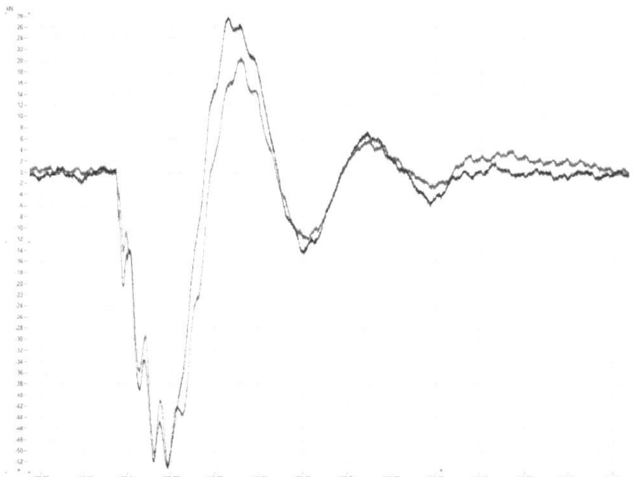

Figure 8. Time history of response force signals for test 2 and comparison test with nearly identical mass and impact speed

ICE IMPACTS TO THRUSTER - NUMERICAL SIMULATIONS

One of the main objectives of the ice impact tests was to compare the measurements to numerical simulation results. Simulation code for ice block to thruster impacts was developed at VTT and described in detail in references (Kinnunen et al 2016, Kinnunen et al 2014, Kinnunen et al 2013, Kinnunen et al 2012, Tikanmäki et al. 2010.)

The code models the ice impact with an average contact pressure over the contact area, combined with basic dynamics of the thruster structure. The contact area is based on the geometry of the structure. A simple spherical geometry can be used in normal ice block to thruster impact, but also other geometries are possible. The code also predicts the dynamic behavior of the thruster when thruster parameters are given. The system is solved with difference method in the time domain.

The main input parameters for the simulation are:

- Impact speed
- Impacting ice mass
- Mass of the thruster
- Thruster natural frequency
- Ice compressive strength

Impact speed and impacting ice mass were measured for the impact tests. The mass of the thruster means the effective mass of the first mode. Initially the mass was not known, and thus it was estimated, based on a modal test and confirmed by the impact test acceleration measurements. For ice compressive strength a value of 3 MPa was used.

Before ice impact tests, the first bending mode of the thruster was determined with experimental modal analysis. The modal analysis was done with impact excitation. The analysis gives the natural frequency, corresponding mode shape and damping of the mode. However, during the actual impact tests it was observed that the ice impacts caused vibrations, which were occurring at much lower frequency than the measured thruster natural frequency.

This implies that the dynamic behavior of the thruster stand structure is more complex. A likely reason for the change in the natural frequency is the splined coupling joining the propeller shaft and the loading generator. The splined coupling axial load carrying capacity depends on the torque of the shaft. Due to complexity of the structure, the exact behavior of the splined coupling and the influence on the dynamics of the thruster could not be investigated thoroughly.

With the measured input parameters, the code calculates time histories for the contact force at the impact and the subsequent response of the thruster structure. One such result can be seen in Figure 9.

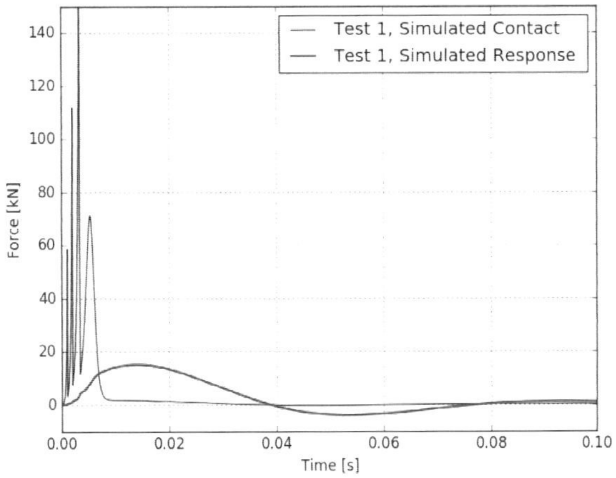

Figure 9. Simulated time history of ice block impact, test 1

The energy consumed in the ice indentation process is estimated by the integral

$$W = \int_0^{u(t)} F du \tag{4}$$

where F is the contact force and u is the indentation to ice. Figure 10 shows the energy balance and indentation into ice for test 1 simulation. The E_{kin} limit is the initial kinetic energy. The gap between the E_{crush} and E_{kin} goes mostly to damping of the structure.

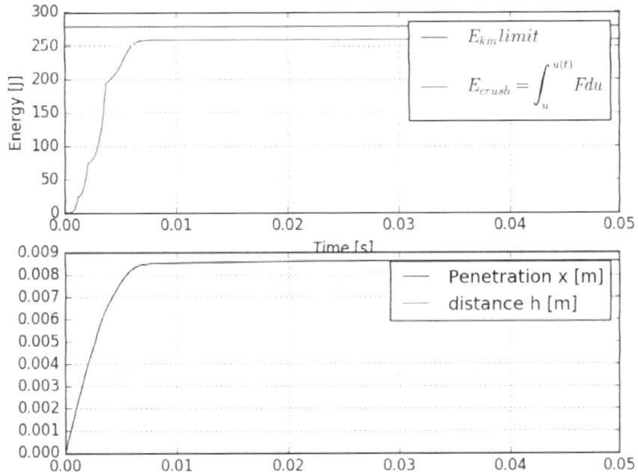

Figure 10. Energy balance and indentation to ice. Test 1 simulation

COMPARISON OF NUMERICAL SIMULATIONS TO MEASUREMENTS

Next, the comparison of simulations and measurements is discussed.

In Figure 11 a comparison is shown for test 1. Here it is notable that the simulated response force initially follows the measurement quite well, but predicts higher force. The simulated maximum response force is slightly over 15 kN and the measured maximum response force was around 10 kN. The vibratory behavior predicted by the code is not present. The smaller force and lack of vibratory behavior is most likely due to the attached force cylinders discussed earlier. It is likely that the cylinders initially have some room to move until they start to oppose the impact and thus cut the response force. 14 Hz natural frequency was used in the simulation, as this was measured in the modal analysis with force cylinders attached. Simulated indentation to ice was 8 mm, a 40% more than measured indentation.

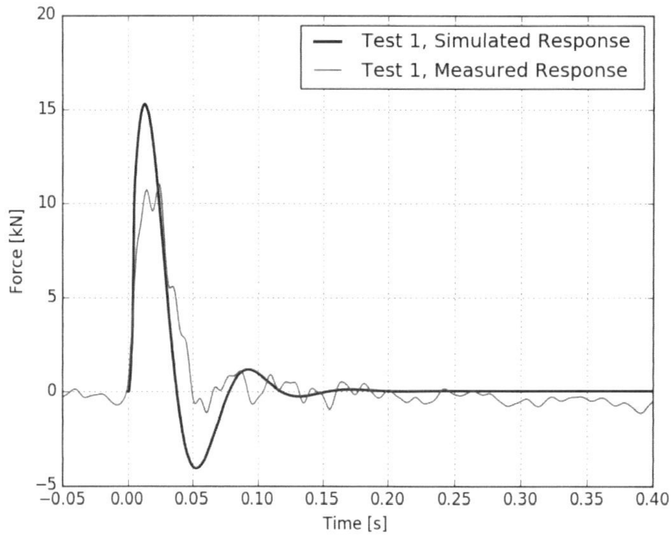

Figure 11. Test 1, comparison of simulated response and measurement results. Parameters: ice mass 139 kg, impact velocity 2,0 m/s and 14 Hz natural frequency of thruster.

In Figure 12 the comparison of simulation and measurements for test 2 is shown. It is noticeable that here the simulation and measurements give similar response force. Measurements and simulation show clear vibratory behavior after the impact. Indentation to ice in the simulation was 23 mm, 73% less than measured.

In Figure 13 The comparison of simulation and measurements for test 3 is shown. There is quite a large difference in the simulated and measured response. Simulated response force is now more than twice as high as the measured response. In addition, the response duration is longer. Simulation shows oscillatory movement of the thruster which is missing from the real test. It would appear that the longer contact, seen in Figure 7, does not excite the thruster natural frequency. Indentation to ice in the simulation was 110 mm, 37% less than measured in the tests.

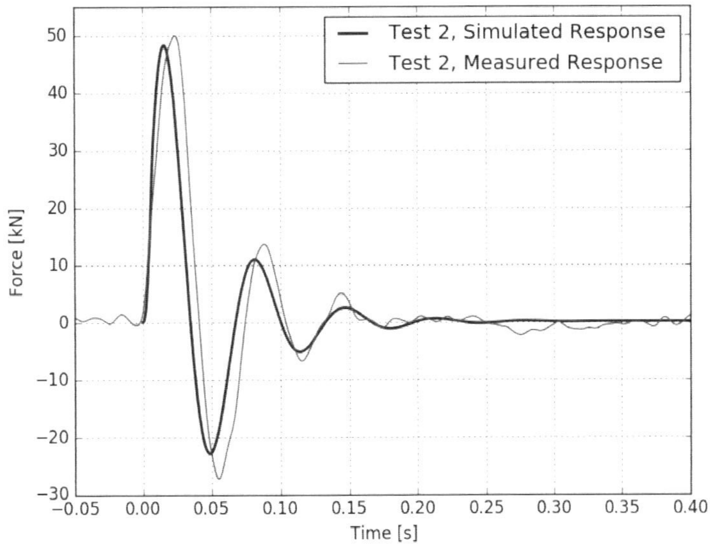

Figure 12. Test 2, comparison of simulated response and measurement results. Parameters: ice mass 136 kg, impact velocity 5,1 m/s and 16 Hz natural frequency of thruster.

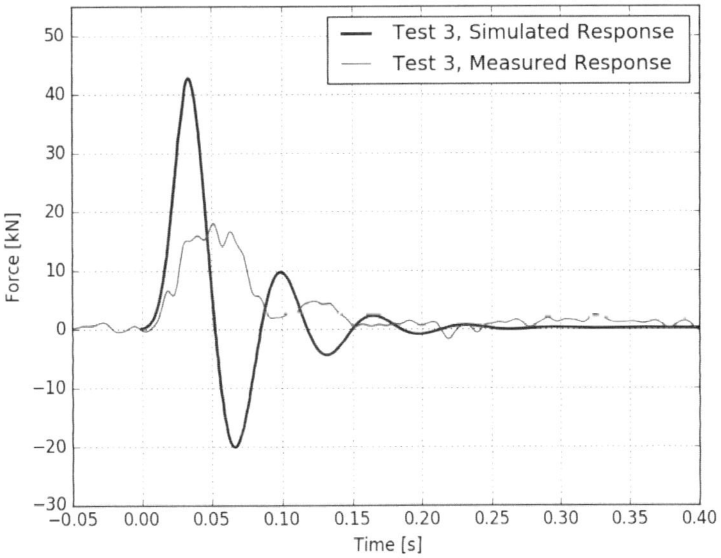

Figure 13. Test 3, comparison of simulated response and measurement results. Parameters: ice mass 137 kg, impact velocity 5,5 m/s and 16 Hz natural frequency of thruster, sharper hubcap.

CONCLUSIONS

The results from the ice impact tests reveal some interesting points. It was observed that the thruster test stand used in these tests showed complex dynamical behavior. Due to this behavior some test results are hard to interpret and require more investigation. For example, it would appear that, in addition to thruster natural frequency, the natural frequency of the whole test stand also plays part in the dynamics of the system.

Some clear results were gained, such as the use of a sharper thruster hubcap which resulted in 70% lower response force levels when compared to standard hubcap. It was also observed that with nearly the same initial impact energy in tests 2 and 3, the thruster showed very different dynamic behavior because of different geometry of the indentor. In test 2 with standard hubcap the impact was short and force leves high, whereas in test 3 with the sharper hubcap the impact duration was longer and the response force levels lower. In test 2 the thruster showed clear vibratory behavior whereas in test 3 there was hardly any oscillation after the impact. This shows that even with same initial impact energy the response of the thruster can vary depending on the contact geometry, thruster natural frequency and other factors contributing to impact duration.

The comparison of test measurements to simulated results proved in some parts challenging. Test 1 results were influenced by the presence of force cylinders, which were believed to have affected the measured response levels. There was no major cracking or splitting of ice block in test 1. On the other hand, test 2 showed very good resemblance between simulation and measured results. Due to high impact forces, there was noticeable cracking in the ice blocks in tests 2 and 3. Ice blocks were partly or completely splitted in the tests. For test 3 it would seem that the sharper hubcap penetrated the ice block and splitted it completely which greatly affected the measured response force. The simulation does not take into account the cracking of ice blocks. It was observed by Kim et al (2012) that unconfined ice blocks developed large cracks and failed by splitting during indentation experiments. After failure the ice was not able to absorb more energy.

ACKNOWLEDGEMENTS

"This work was carried out under the Maritime Technologies II Research Programme in the framework of the ERA-NET MARTEC II project CA 266111 ArTEco 'Arctic Thruster Ecosystem'"

REFERENCES

Kinnunen, A., Tikanmäki, M., Heinonen, J., 2016. An energy model for ice crushing in ice-structure impact. Proceedings of the 23nd IAHR International Symposium on Ice, Ann Arbor, May 31 to June 3, 2016.

Kinnunen, A., Tikanmäki, M., Koskinen, P., 2014. Ice-structure impact contact load test setup and impact contact load calculation. Proceedings of the 22nd IAHR International Symposium on Ice, Singapore, August 11 to 15, 2014.

Kinnunen, A., Lämsä, V., Jussila, M., Koskinen, P. 2013. Determining of ice loads for azimuthing propulsion units. Proceedings of the Dresdner Maschinenelemente Kolloquium – DMK 2013, 3. and 4. December 2013, Dresden, Germany

Kinnunen, A., Tikanmäki, M., Heinonen, J., Kurkela, J., Koskinen, P., Jussila, M., 2012. Azimuthing thruster ice load calculation. VTT Research Report VTT-R-08842-12.

Tikanmäki, M., Heinonen, J., Kinnunen, A., 2010, Comparison of ice load models for azimuthing thruster ice load calculation, VTT Research report VTT-R-10310-10, 25pp.

Kim, E., Golding, N., Schulson, E.M., Loset, S., Renshaw, C.E., 2012, Mechanisms governing failure of ice beneath a spherically-shaped indenter, Cold Regions Science and Technology , vol 78, p 46-63., http://dx.doi.org/10.1016/j.coldregions.2012.01.011

POAC'17
Busan, KOREA

Proceedings of the 24th International Conference on
Port and Ocean Engineering under Arctic Conditions
June 11-16, 2017, Busan, Korea

Ice Load Characteristics on a Model Ship Hull Installed with Tactile Sensor Panels

Seong-Yeob Jeong [1], Jeong-Gil Yum[1] ,Eun-Jee Cheon [2], Kyungsik Choi [2],
[1] Korea Research Institute of Ship and Ocean Engineering, Daejeon, Korea
[2] Korea Maritime and Ocean University, Busan, Korea

ABSTRACT

The ice load characteristics are described through the analysis of measured level ice load distribution on a model ship hull at the KRISO ice model basin. A 1:18.667 scale model of the icebreaking research vessel ARAON was used for model test. To measure the spatial distribution of ice pressures on model ship hull, six tactile sensor panels were installed along the waterline sections of the model ship. Three different ship speeds were selected for the model test simulating ice-ship interaction process under level ice conditions. By analyzing model test data, the correlation between ice resistance force in horizontal direction and the ice load acting on ship hull in normal direction is discussed in this paper. The results show that the resistance force calculated from the measured ice load component is in good agreement with the ice resistance force directly measured from model test.

KEY WORDS: Ice Load-Resistance Correlation; Model Ship Test; Level Ice; Tactile Sensor.

INTRODUCTION

Ice load plays most dominant role for ice-going ships and it is important to estimate actual ice load on ships in ice. Field measurement is often carried out to obtain full-scale ship-ice interaction data because it gives a basis for the structural design of ice-going vessel. However the field tests require extensive equipment and techniques to obtain correct ice loads and a model-scale test in an ice tank could be an alternative for obtaining valuable information such as the ice load distribution over the ship hull under various ice conditions. Contrary to ice resistance model tests, experiences of model test to measure the ice loads are very rare to find mainly because of the complicated nature of ice load distribution along the surface of the ship hull. Izumiyama et al.(2001) conducted a series of resistance tests in ice, in which tactile sensors were employed to measure local ice loads. They compared the tactile sensor data to ship resistance in ice. Frederking(2004) reported the use of tactile sensors to measure ice pressure distribution around a fixed offshore structure.

Present study aims to analyze ice load characteristics during a model ship-ice interaction process and to figure out the correlation between ice resistance force in horizontal direction and the ice load acting on ship hull in normal direction. It is described through the analysis of measured level ice load distribution on a model ship hull tested at the KRISO ice model basin.

Figure 1. Ice load model test in the KRISO ice model basin

ICE MODEL TEST WITH TACTILE SENSOR PANELS

Model tests were conducted in an ice tank at the Korea Research Institute of Ships and Ocean engineering (KRISO). The model was manufactured with a scale of $\lambda = 18.667$. The specific dimensions of the model ship and the test ice conditions are summarized in Tables 1 and 2, respectively. Three different ship speeds, 0.119, 0.357, 0.595m/s were selected for the model test simulating ice-ship interaction process under two different level ice conditions. The thickness of model ice was selected as 25mm and 50mm which correspond to 47cm and 94cm in full-scale respectively. The model ship speed were recorded when the model ship achieved a steady speed in ice. The model test was performed along a straight line and maintained the design draft of model ship.

Table 1. Principal particulars of the IBRV ARAON

Scale (λ = 18.667)	Full-scale	Model
Length between perpendiculars (m)	93.5	5.01
Maximum beam (m)	19.0	1.02
Design waterline (m)	6.8	0.36
Stem angle (°)	35.0	35.0
Waterline entrance angle (°)	34.0	34.0

Table 2. Ice model test conditions

	Scale (λ = 18.667)	Ice thickness (mm)	Flexural strength (kPa)	Speed (m/s)
Test No.1	Model	25	30	0.119, 0.357. 0.595 (0.514, 1.54, 2.57 m/s in full scale)
	Full-scale	470	560	
Test No.2	Model	50	30	
	Full-scale	940	560	

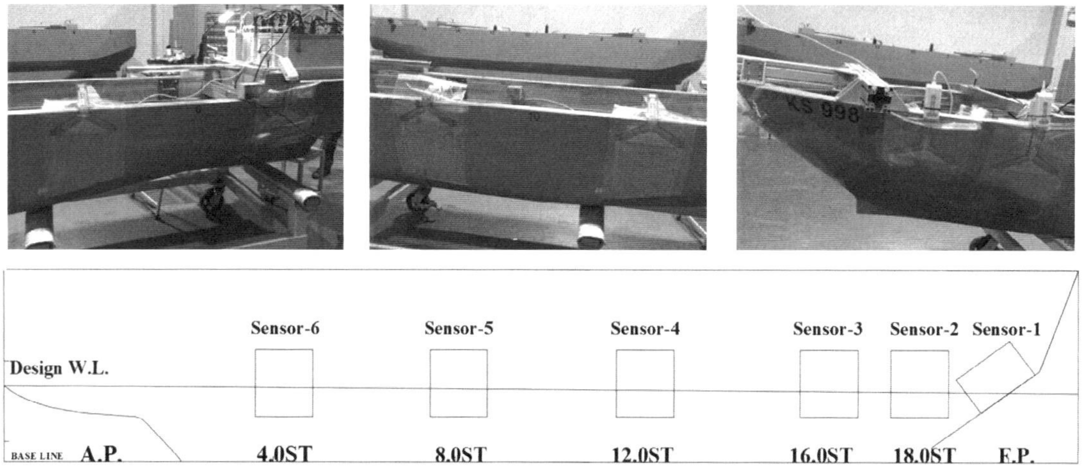

Figure 2. Locations of installed tactile sensor panels

To measure the spatial distribution of ice pressures on model ship hull, six tactile sensor panels were installed along the waterline sections of the model ship. Tactile sensor panels called as I-SCAN system were designed to add up whole ice pressures acting on each panel. One tactile sensor panel has a total of 1,936 sensing spots arranged in 44x44 grids. The inter-row and -column spacing is 0.54 cm and one sensing spot area is about $0.29cm^2$. Ice load acts on model ship hull close to the waterline and is dominant in the forward region during the straight-going test. Therefore, six tactile sensor panels were installed along the waterline sections as shown Figure 2.

In previous studies, Jeong et al.(2015) performed model tests with the same model ship. Using three tactile sensor panels in the bow, shoulder and midship areas (Station No.20, No. 17 and No.11), the ice pressure distribution around the ship hull were measured. Later in this study, an improved procedure was adopted with six tactile sensors, installed in the bow area (FP, Sensor-1), shoulder areas (Station No.18, Sensor-2; Station No.16, Sensor-3), midship area (Station No.12, Sensor-4) and stern areas (Station No.8, Sensor-5; Station No.4, Sensor-6).

ANALYSIS OF MEASURED ICE LOAD DATA

Figure 3 shows a typical load time history recorded during icebreaking processes in model test. As seen in the figures, so many saw-tooth type peaks were recorded. They represent the typical ship-ice interaction process. As the model ship penetrates into the ice sheet, vertical downward force acting on the ice sheet becomes larger. When this force becomes greater than the flexural strength of ice sheet, ice failure occurs and then next interaction process repeats. By adopting a concept of interaction 'event', a section marked '(a)' between 20sec and 30sec, in Figure 3 is enlarged for determining peak ice load during each ship-ice interaction event. In model scale, the icebreaking cycle occurs approximately during 0.1~0.4s and this is equivalent to 0.4~1.7s in full scale.

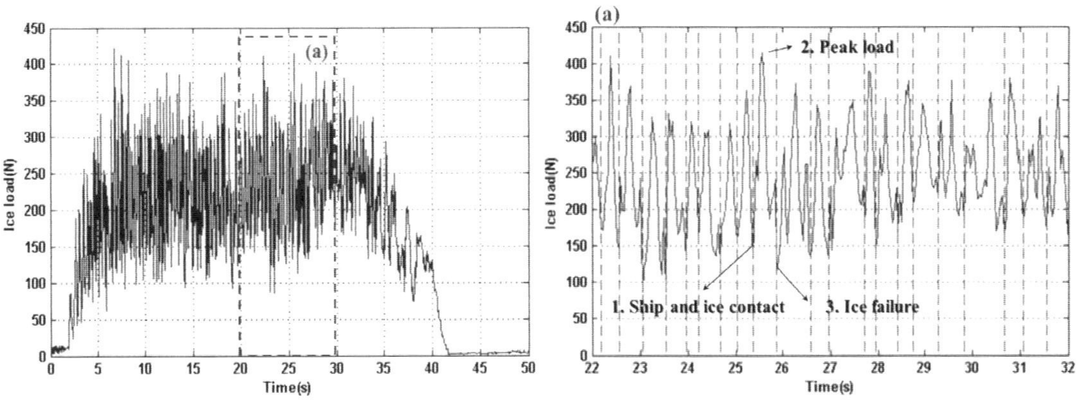

Figure 3. Typical time history of ice load measured from the model test. The section marked '(a)' between 20sec and 30sec in left figure is enlarged for selecting the peak ice loads.

For three different speeds and two different ice thicknesses in the model tests, peak ice loads were collected and the average values of peak ice loads were calculated for each tactile sensor panels. In Figure 4 and 5, the average values of peak ice loads were denoted with bar charts and it is clear to see the ice load distribution around the ship hull. Because of straight-going test, ice load is dominant in the stem and shoulder areas where three sensor panels (FP, St.18 and St.16) were located. It is unexpected that the mid shoulder area (St.16) recorded somewhat large peak ice load comparable to those on bow area. This can be explained by that broken ice pieces clogged, instead of passing by underneath the ship hull, in the vicinity of mid shoulder area especially in narrow channels. Clogging phenomena seemed to be clear at lower speed than at higher speed. In the stern areas (St.8 and St.4) the peak ice loads were less significant and it did not seem to show any correlation between the model ship speed and peak ice loads.

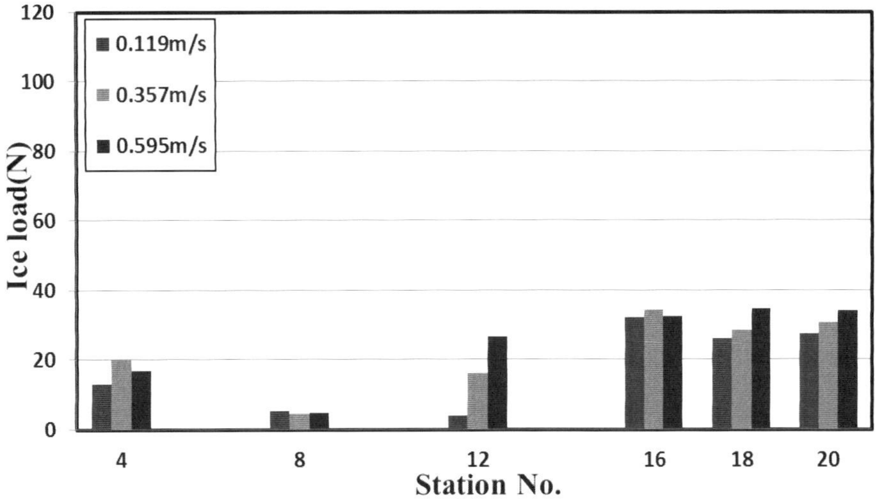

Figure 4. Average value of peak ice loads measured from tactile sensor panels

(ice thickness 25mm)

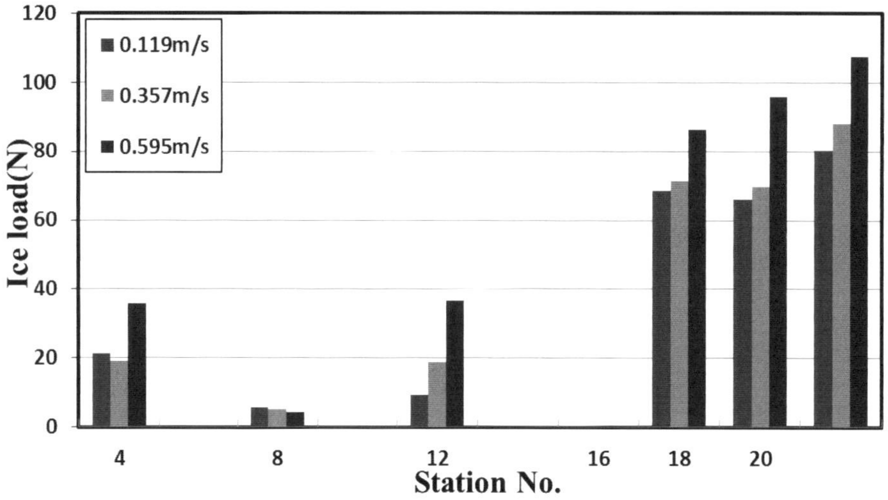

Figure 5. Average value of peak ice loads measured from tactile sensor panels

(ice thickness 50mm)

CORRELATION BETWEEN ICE LOADS AND ICE RESISTANCE

In the previous research by Izumiyama et al.(2001), they used tactile sensors covering most of the hull area forward and they calculated the ice loads and ice resistance of model ship by summing the force components measured from tactile sensors. Most of front hull area was covered by tactile sensor panels in their model test. The contact force on the hull is divided into components in horizontal and vertical directions as shown in Figure 6 and they derived a correlation factor C_R which is simply defined to relate ice loads and ice resistance forces as in Eq.(1).

$$C_R = F_X / F_N = \sin\alpha \cos\gamma + \mu \cos\alpha \quad (1)$$

where α and γ are the waterline and normal frame angles shown in Figure 6 and μ is the friction coefficient, hence the correlation factor C_R depends on hull shapes of the ship. The correlation factor C_R can be used to estimate ice loads from measured ice resistance force or can be used to estimate ice resistance force from measured ice loads data in reverse way.

In the present study, correlation factor C_R for the model ship of the ARAON was derived as in Figure 7. The ice load components in forward direction were calculated by summing up average peak ice loads measured from all tactile sensor panels. Then equivalent ice resistance force can be estimated multiplying by the correlation factor.

As shown in Figure 8 and Figure 9, the equivalent ice resistance calculated from peak ice loads components in forward horizontal direction was compared to actual ice resistance force measured from model tests. The difference was within about 10% for both of two different ice thicknesses. In this model test, the correlation between ice resistance force in horizontal direction and the ice loads acting on ship hull was in good agreement.

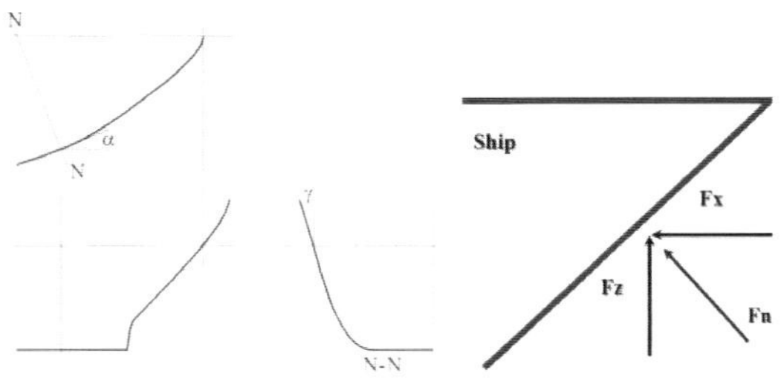

Figure 6. Definition of angles of the ship hull form

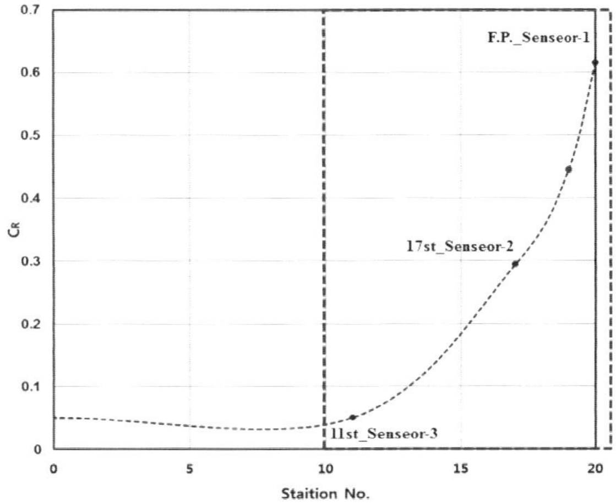

Figure 7. Correlation factor C_R derived for the model ship

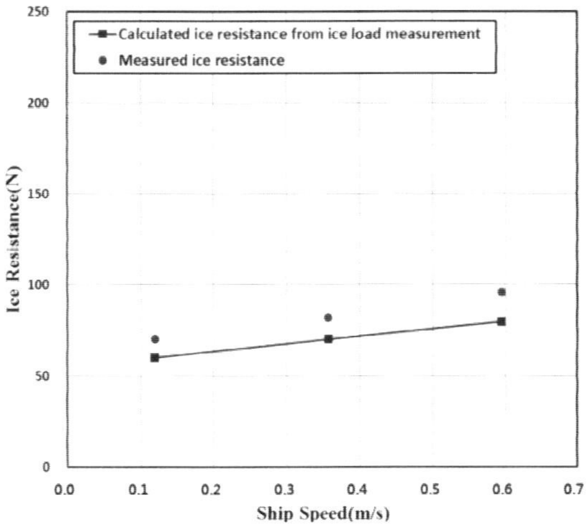

Figure 8. Ice resistance(square) calculated from the measured ice loads and ice resistance (circle) directly measured from model test (ice thickness 25mm)

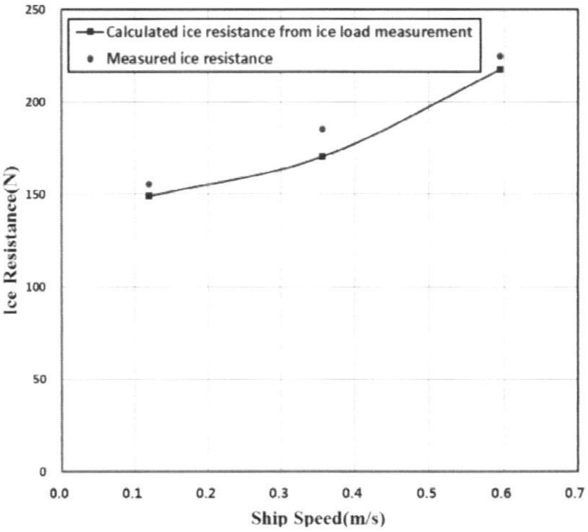

Figure 9. Ice resistance(square) calculated from the measured ice loads and ice resistance (circle) directly measured from model test (ice thickness 50mm)

CONCLUSIONS

Present study aims to analyze ice load characteristics on a model ship hull and to find out the correlation between ice resistance force in horizontal direction and the ice loads acting on ship hull in normal direction. The model test was performed in KRISO ice model basin for the scale model of the Korean IBRV ARAON with six tactile sensor panels. Following conclusions can be drawn:

1) In general, ice load is dominant in the stem and shoulder areas. Large peak ice load at mid shoulder area may be explained by that broken ice pieces clogged in the vicinity of mid shoulder area, instead of passing by underneath the ship hull. In the stern areas, the peak ice loads were less significant and it did not seem to show any correlation between the model ship speed and peak ice loads.
2) The resistance force calculated from the measured ice load component is in good agreement with the ice resistance force directly measured from model test. The difference is within 10~20% and the use of correlation factor C_R is recommended.

ACKNOWLEDGEMENTS

Research funds supported by the Korea Research Institute of Ships & Ocean Engineering (Project No.PES3550) and also supported by Ministry of Trade, Industry and Energy of Korea (Project No.10063417) are greatly acknowledged.

REFERENCES

Cheon, E.J., Choi, K. and Jeong, S.Y., 2016. An analysis of ice load characteristics on a model ship hull. *Proceedings of the 31st International Symposium on Okhotsk Sea & Sea Ice.* pp.88-91, Mombetsu, Japan.

Frederking, R., 2004. Ice pressure variations during indentation. *Proceedings of the 17th International Association of Hydraulic Engineering and Research (IAHR)*, Vol.2, pp.307-314, Saint Petersburg, Russia.

Izumiyama, K., Wako, D. and Uto, S., 2001. Ice pressure acting over a model ship hull. *Proceedings of the 16th International Conference on Port and Ocean Engineering under Arctic Conditions (POAC)*, Vol.2, pp.793-802, Ottawa, Canada.

Jeong, S.Y., Choi, K., Ha, J.S., Kang, K.J. and Cheon, E.J., 2015. Distribution of ice load acting on model hull due to ship–ice interaction. *Proceedings of the 23th International Conference on Port and Ocean Engineering under Arctic Conditions (POAC)*, POAC15-093, Trondheim, Norway.

Lu, W., Serre, N., Høyland, K. and Evers, K., 2013. Rubble ice transport on Arctic offshore structure (RITAS), Part IV: Tactile sensor measurement of the level ice load on inclined plate. *Proceedings of the 22nd International Conference on Port and Ocean Engineering under Arctic Conditions (POAC)*, POAC13-087, Espoo, Finland.

Sodhi, D., Takeuchi, T., Kawamura, M., Nakazawa, N. and Akagawa, S., 2001. Measurement of ice forces and interfacial pressure during medium-scale indentation tests. *Proceedings of the16th International Conference on Port and Ocean Engineering under Arctic Conditions (POAC)*, Vol. 2, pp.617–626, Ottawa, Canada.

Proceedings of the 24th International Conference on
Port and Ocean Engineering under Arctic Conditions
June 11-16, 2017, Busan, Korea

Friction Correction for Model Ship Resistance and Propulsion Tests in Ice

Michael Lau[1]
[1] National Research Council, St. John's, Canada

ABSTRACT

This paper documents the result of a preliminary analysis on the influence of hull-ice friction coefficient on model resistance and power predictions and their correlation to full-scale measurements. The study is based on previous model-scale/full-scale correlations performed on the National Research Council - Ocean, Coastal, and River Engineering's (NRC-OCRE) model test data. There are two objectives for the current study: (1) to validate NRC-OCRE's modeling standards in regarding to its practice of specifying a *CFC* (Correlation Friction Coefficient) of 0.05 for all its ship models; and (2) to develop a correction methodology for its resistance and propulsion predictions when the model is prepared with an ice friction coefficient slightly deviated from the *CFC* of 0.05. The mean *CFC* of 0.057 and 0.050 for perfect correlation as computed from the resistance and power analysis, respectively, have justified NRC-OCRE's selection of 0.05 for the *CFC* of all its models. Furthermore, a procedure for minor friction corrections is developed.

KEY WORDS: Model Test; Ice Resistance; Power; Friction Correction; Correlation Friction Coefficient

INTRODUCTION

Measurements of ice-hull friction coefficient at full-scale (Mäkinen et al., 1994; Schwarz, 1986) result in a very wide range of values from 0.01 to 0.6; therefore, at early days most ice tanks tested a model at two or more hull-ice friction coefficients, and the friction coefficient that gives the best correlation with full-scale data was found. This is called the "Correlation Friction Coefficient" (*CFC*). This standard friction is then used for other models. This concept is analogous to the correlation allowance used in clear-water testing. If the *CFC* falls in a narrow band of values for all ships of similar hull surface condition, it is assumed that full-scale performance for another similar hull can be predicted from model tests based on the same *CFC*.

The validity of *CFC* concept has been studied using good-quality full-scale information

(Edwards et al., 1979; Williams et al., 1991 & 1992; Noble and Comfort, 1979; Cowper, 1991; Sodhi et al., 2001) and model test data for five icebreaker models in the NRC-OCRE's model test database (Newbury and Williams, 1986; Spencer, 1987; Spencer et al., 1988; NORDCO Ltd, 1989; Spencer et al., 1992; Jones and Moores, 2002; Jones and Lau, 2006). The *CFC* obtained from each of these correlation studies (Spencer, 1992; Spencer and Jones, 2002; Jones and Lau, 2006; Wang and Jones, 2008) suggest a good correlation between model test data and full-scale measurements by painting the models with a hull-ice friction coefficient of 0.05-0.07. Lau (2015) performed additional assessment by correcting model ice resistance to a hull ice friction to 0.05 prior to comparison with full-scale measurements. For details of the NRC-OCRE's *CFC* methodology, the reader is referred to Spencer and Jones (2002).

NRC has adopted a *CFC* of 0.05 as the standard frictional preparation for their ship model surfaces. Sometimes, a poor control of the frictional preparation may result in a friction coefficient deviating from the target, and hence, affecting the test result. A reliable correction methodology is needed for correcting resistance and propulsion predictions when the model is prepared with an ice friction coefficient slightly deviated from the expected friction target.

This paper documents the result of a preliminary analysis on the influence of model hull-ice friction coefficient on model resistance and power predictions and their correlation to full-scale measurements with the objective of developing such correction methodology. The study is based on 5 sets of full scale trials and their respective model tests conducted at the National Research Council. The full-scale trial and model test data are summarized, and the analysis procedure presented. The mean *CFC* of 0.057 and 0.050 for correlation as computed from the resistance and power analysis, respectively, are found to fit the dataset, and a correction procedure is established for minor friction correction.

CURRENT FRICTION CORRECTION PROCEDURE

NRC-OCRE has been painting its model ships with its old paint formula to 0.05 *CFC* with a close tolerance and there was no need for friction correction for its full-scale prediction. When the vender replaced the old paint with a new formula, the repeatability and control of the frictional preparation were lost, resulting in a friction coefficient deviating from the target as much as 0.02. These required a friction correction procedure to be developed at NRC-OCRE, while it explores and develops a new surface treatment technique. In this study, we adopted ITTC friction correction method as given in the ITTC recommended procedures and guideline for the testing and extrapolation methods for resistance test in level ice 7.5-02-04-02.1 (ITTC, 2002). The relevant paragraphs are excerpted as follow for quick reference:

> "If the surface treatment of the model does not result in a desired friction between the model and the ice, some further corrections are needed. The best correction is to repaint the model to the desired surface roughness. However, if the difference between the measured friction coefficient and the target value is small, instead of a new surface treatment, a friction correction coefficient may be applied. The friction coefficient may be taken into account by the following equation:
>
> $$R_{i,corr} = C_{\mu} R_{i,p} \quad (1)$$
>
> The friction correction coefficient can be based on the coefficient of the measured model ice friction and a comparison between full-scale tests results

and model tests with different model/ice-friction coefficients. Actual test results (Figure 1) can be used to derive the friction correction coefficient. If the correction is assumed to be linearly proportional to the coefficient of friction, then from Figure 1 the following formula is obtained:

$$\frac{\lambda^3 R_{i.m}}{R_{i,p}} = \frac{1}{C_\mu} = a + f_{ID} b \qquad (2)$$

where f_{ID} is the dynamic model ice friction coefficient and a and b are empirical coefficients (From Figure 1, a = 0.8 and b = 5.8). This formula is valid for a new ship with a hull surface in good condition.

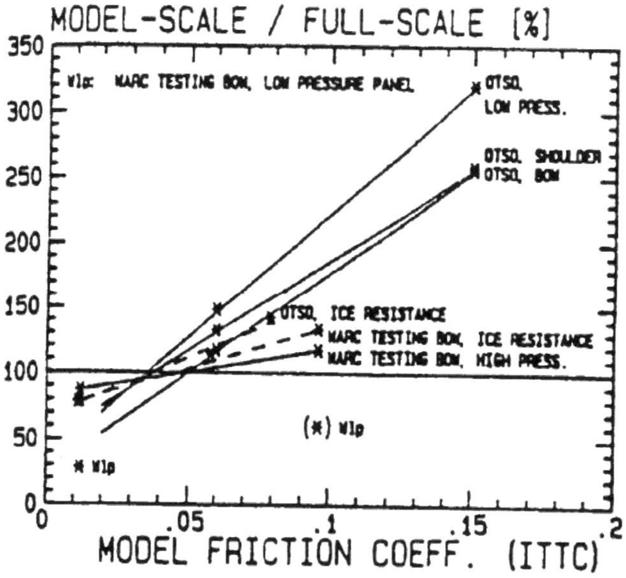

Figure 1. The ratio between extrapolated model scale ice resistance and measured full scale ice resistance plotted versus the coefficient of friction in model scale to justify correction of minor friction derivation (taken from ITTC, 2002)

NRC-OCRE testing standard require a nominal value of 0.05 for the *CFC*. A friction correction equation is used to correct the ship resistance to that of 0.05.

FULL-SCALE TRIAL AND MODEL TEST DATA

The model tests and associated sea trials used in this study are summarized in Table 1 and they are briefly described as follows. For details, please refer to the cited publications.

CCGS R-Class Icebreakers

Several full-scale speed-power icebreaking trials have been conducted with the R-class icebreakers CCGS Pierre Radisson and CCGS Sir John Franklin.

The CCGS icebreaker Pierre Radisson was tested soon after delivery in 1978 during transit in the Central Arctic with ice thickness in the range of 0.8 to 1.3 m and ice flexural strength estimated at 200 kPa. In February 1979, further tests were conducted in thinner but stronger ice in the St. Lawrence River where ice thickness ranged from 0.2 to 0.7 m and ice flexural strength varied from 400 to 500 kPa (Edwards et al., 1979).

Tests were conducted on the CCGS icebreaker Sir John Franklin in Notre Dame Bay, Newfoundland in February 1991, where ice thickness was between 0.5 and 0.6 m, and flexural strength ranged between 200 and 300 kPa (Williams et al., 1991 & 1992).

Model level ice resistance tests were conducted at NRC-OCRE on a 1:20 scale model covering a wide range of speed, ice thickness and strength with three hull-ice friction coefficients of 0.03 & 0.09 (Spencer et al., 1992) and 0.071 (Newbury, 1988). Newbury and Williams (1986) summarized R-Class icebreaker model test results.

CCGS Icebreaker Terry-Fox

The full-scale trial for the CCGS icebreaker Terry-Fox used in this study was conducted in 1990 by Fleet Technology Limited (Cowper, 1991). The ice thickness was 1.55 m and the flexural strength was 150 kPa. In this trial, the Terry-Fox was towed by the MV Ikaluk to measure resistance directly. Self-propulsion tests were also conducted to measure power and thrust.

A 1:21.8 scale model has been tested with three different hull-ice friction coefficients (0.11, 0.045 and 0.005) and several ice conditions over a period of more than 20 years. The datasets used in this review were from Spencer et al. (1988) with a hull-ice friction of 0.11, NORDCO Ltd. (1989) with a hull-ice friction coefficient of 0.045 and Spencer (1990) with a hull-ice friction coefficient of 0.005.

CCGS Icebreaker Louis S. St. Laurent

Full-scale data was obtained from ARCTEC Canada (Noble and Comfort, 1979) in its July 1979 Strathcona Sound trials. Immediately prior to these trials, a new INERTA 160 coating was applied over the ship's bow and waterline regions; thus the hull condition during these trials may be similar to a new ship on its maiden voyage. Ice thickness varied from 0.76 to 1.36 m, and mean flexural strength was only 150 kPa averaged over five cantilever beam tests. Snow cover was very light and was not a factor in these trials. Shipboard measurements were shaft speed (*RPM*) and ship speed. Ship resistance was estimated from *RPM* only in model overload tests.

Resistance of the 1:15 scale model of Louis S. St. Laurent was extensively measured at NRC-OCRE (Spencer, 1987) with a hull-ice friction coefficient of 0.13 and ice thicknesses of 47, 67 and 100 mm corresponding to 0.7, 1.0 and 1.5 m full-scale for the ship. The target ice strength was 30 kPa corresponding to 450 kPa full-scale. It was subsequently refinished and tested with a 0.075 hull-ice friction coefficient (Fleet Technology Ltd., 1992) and the same target ice thickness and ice strength.

Table 1. Summary of model-scale/full-scale studies.

	Sea Trial					Model Test	
Ship	**Date**	**Location**	**Thickness (m)**	**Flexural Strength (kPa)**	**Reference**	**Friction Coefficient**	**Reference**
CCGS Pierre Radisson (R-Class)	July & August 1978	Central Arctic	0.8 - 1.3	200	Edwards et al., 1979	0.03, 0.071 & 0.09	Spencer and Jones, 2002
CCGS Pierre Radisson (R-Class)	Feb. 1979	St. Lawrence River	0.2 - 0.7	400 – 500	Edwards et al., 1979	0.03, 0.071 & 0.09	Spencer and Jones, 2002
CCGS Sir John Franklin (R-Class)	Feb. 1991	Notre Dame Bay, NL	0.5 - 0.6	200 – 300	Williams et al., 1991 & 1992	0.03, 0.071 & 0.09	Spencer and Jones, 2002
CCGS Louis S. St. Laurent	1979	Strathcona Sound	0.8 - 1.4	150	Noble and Comfort, 1979	0.07 & 0.13	Spencer, 1992
CCGS Terry-Fox	1991	Central Arctic	1.55	150	Cowper, 1991	0.005, 0.045 & 0.11	Wang and Jones, 2008

RESULTS

Preliminary analysis on 5 sets of full scale trials and their respective model tests is given in this paper. More detailed analysis on the influence of hull ice friction on ice resistance and power as a function of relevant parameters, e.g., ice thickness and ice strength, will be explored in next phase of the study. The confidence level of the correction procedure as proposed in this report will also be assessed and further datasets may be added to the analysis.

Since the model test set-up uses a towed resistance and propulsion arrangement, and full-scale trial was conducted in free propulsion, a brief procedure to compute resistance from full-scale measurement, and the power estimated from the model towed propulsion tests were described before presenting the comparison.

Resistance

IOT standard method for analyzing resistance tests in ice (Jones and Moores, 2002) was followed to compute model resistance in ice and then it was scaled to full-scale according to Froude's law prior to comparison.

The resistance correlation was performed for 5 sets of full-scale data, including the trials for the CCGS icebreaker Louis S. St. Laurent, CCGS icebreaker Terry-Fox, CCGS icebreaker Pierre Radisson and CCGS icebreaker Sir John Franklin (See Table 1).

The correlation between full-scale and model scale tests in ice is shown in the following manner. First, full-scale open water bollard and speed-power trials are compared with corresponding model tests to confirm that the model tests accurately predict the ship performance in open water. In general, a ship moving through an ice sheet is in an overload condition. Thus, a thrust deduction factor is determined from open water and overload model tests. While it would be preferable to measure the thrust deduction factor at full-scale, these tests are expensive. An alternative is to use thrust deduction factors derived from model overload tests. The full-scale resistance in ice for the ship is then determined from the measured trials data and the thrust deduction factor mentioned above. Detailed procedure is given in Spencer and Jones (2002).

The result of the ship resistance correlation is given in Figures 2 to 4. In these figures, the Ratio of Model to Full Scale Resistances (*RMFSR*) was plotted against the model's friction coefficient. In Figure 2 data from individual dataset were fitted with a linear relationship to examine the trend exhibited from individual dataset. In Figures 3 and 4, all data were fitted with a linear relationship and a second degree polynomial, respectively, to give overall trends. The level line at *RMFSR* equal to1 indicates perfect correlation.

From Figure 2, assuming a linear relationship between the *RMFSR* ratio and ice friction, the following formula is obtained:

$$\frac{R_{i.mp}}{R_{i,f}} = b + f_{ID}m \quad (3)$$

Where $R_{i.mp}$ and $R_{i,f}$ are the model prediction and full-scale measurement for ice resistance, f_{ID} is the ice-hull friction of the model, m is the slope and b is the y-intercept for the linear fit. The equation for the individual dataset is summarized in Table 2.

The targeted ice friction, i.e., *CFC* for perfect correlation, for the individual dataset can be computed as follow:

$$CFC = \frac{1-b}{m} \tag{4}$$

If the *CFC* is known, the model prediction at *CFC*, can be corrected from model prediction with the same model of any arbitrary friction preparation, i.e.,

$$R_{i,f} = \frac{R_{i,mp}}{(f_{ID}-CFC)m+1} \tag{5}$$

Figure 2 shows a good correlation between the full-scale prediction and the sea trial measurement for a *CFC* within a narrow range of 0.048 to 0.063. On average, the *CFC* is at 0.056 with a standard derivation of 0.0064.

If we assume the *CFC* equals to 0.05, the corresponding $R_{i.mp}/R_{i,f}$ ranges from 0.86 to 1.03 for individual dataset with an average of 0.951. Hence, there were a maximum of 3% under-prediction and 14 % over-prediction among the datasets at the adopted *CFC* of 0.05, which is acceptable within modeling uncertainty. The result is summarized in Table 2.

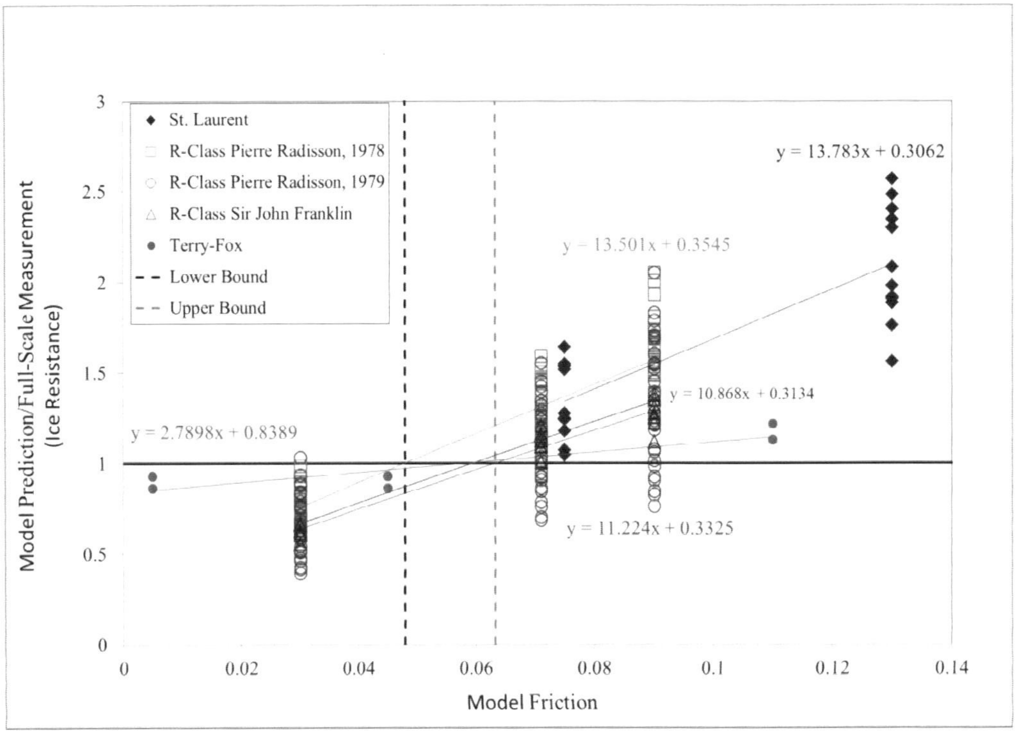

Figure 2. The ratio between full-scale predicted and measured ice resistance versus the model's coefficient of friction: Linear fit to individual dataset

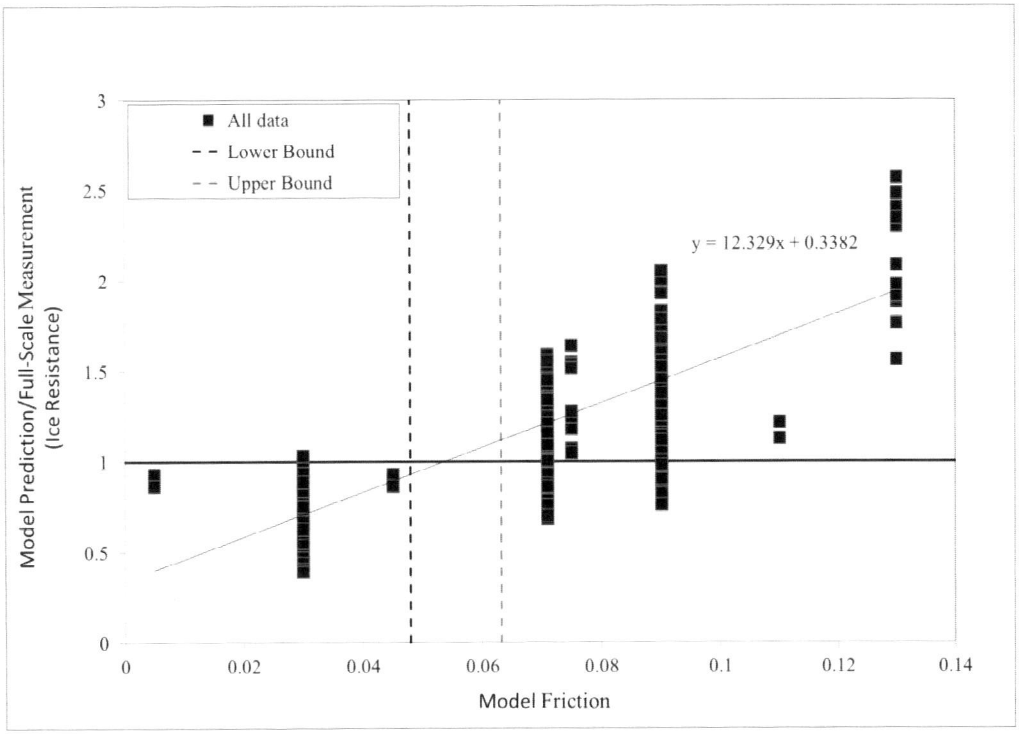

Figure 3. The ratio between full-scale predicted and measured ice resistance versus the model's coefficient of friction: Linear fit to all datasets

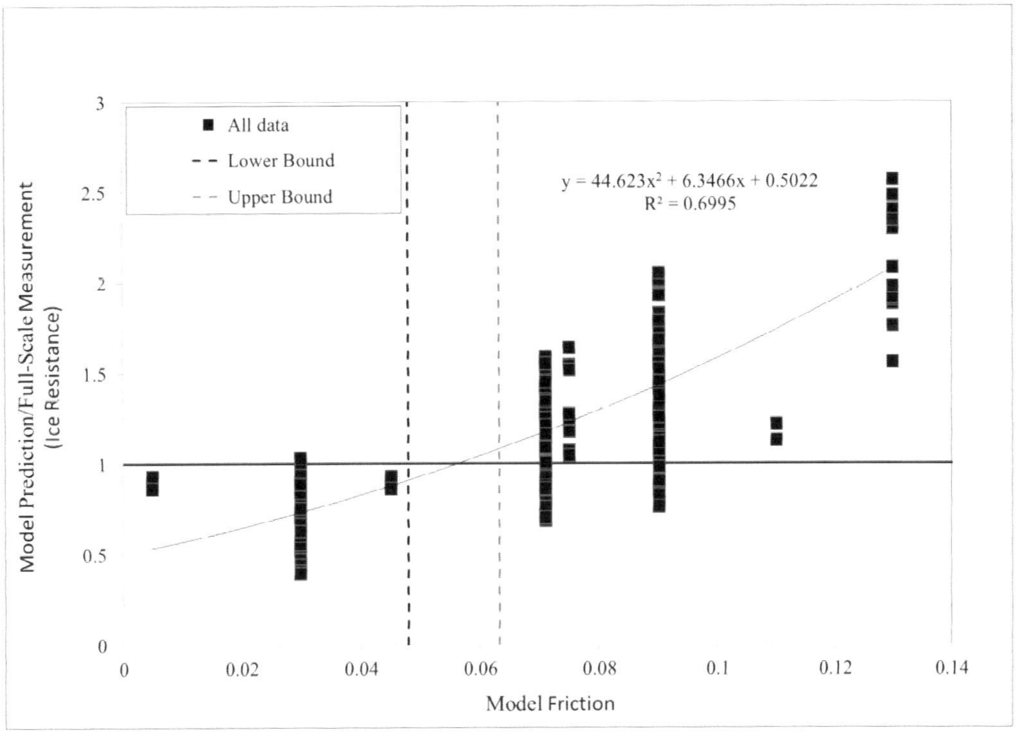

Figure 4. The ratio between full-scale predicted and measured ice resistance versus the model's coefficient of friction: Second order polynomial fit to all datasets

Table 2. Result of resistance analysis: Summary of the equation for the individual dataset

Dataset	m	b	***CFC***	$R_{i.mp}/R_{i,f}$@f_{ID}=**0.05**
CCGS Louis S. St. Laurent	13.78	0.3062	0.050	1.00
CCGS Pierre Radisson, 1978	13.50	0.3545	0.046	1.03
CCGS Pierre Radisson, 1979	11.22	0.3325	0.058	0.89
CCGS Sir John Franklin	10.87	0.3134	0.063	0.86
CCGS Terry-Fox	2.790	0.8389	0.057	0.98

As shown in Figures 3 and 4, linear and second order polynomial fits give the following equations, respectively:

$$\frac{R_{i.mp}}{R_{i,f}} = 0.3382 + 12.329 f_{ID} \quad (6)$$

$$\frac{R_{i.mp}}{R_{i,f}} = 0.5022 + 6.3466 f_{ID} + 44.623 {f_{ID}}^2 \quad (7)$$

The *CFC*s at perfect correlation are 0.054 and 0.056 for the linear and second order polynomial fits, respectively. Since the second order polynomial fits better the data, we recommend adopting Equation 7 for the friction correction to the *CFC* of 0.056. Assuming all ships following the same trend, the ratio $R_{i.mp}/R_{i,f}$ corresponding to a particular model friction can be obtained directly from the equation, with which the model prediction $R_{i.mp}$ obtained from a particular model friction finish f_{ID} can be corrected to the expected full-scale measurement $R_{i,f}$. For convenience, the ratio $R_{i.mp}/R_{i,f}$ for a range of f_{ID} is tabulated in Table 3.

Table 3. The ratio $R_{i,mp}/R_{i,f}$ for a particular model's friction coefficient

f_{ID}	$R_{i.mp}/R_{i,f}$ @f_{ID}	f_{ID}	$R_{i.mp}/R_{i,f}$ @f_{ID}	f_{ID}	$R_{i.mp}/R_{i,f}$ @f_{ID}
0.000	0.502	0.045	0.882	0.090	1.443
0.005	0.535	0.050	0.936	0.095	1.516
0.010	0.571	0.055	0.991	0.100	1.592
0.015	0.609	0.060	1.049	0.105	1.670
0.020	0.649	0.065	1.109	0.110	1.750
0.025	0.691	0.070	1.171	0.115	1.833
0.030	0.735	0.075	1.236	0.120	1.917
0.035	0.782	0.080	1.303	0.125	2.004
0.040	0.831	0.085	1.372	0.130	2.093

The mean of 0.054 or 0.056 for perfect correlation has justified NRC-OCRE's current practice in regarding to its adoption of a value of 0.05 for its *CFC*. It should be noted that all datasets used in this preliminary analysis are from mid- to large-size icebreakers, and in the absence of full-scale trial data, vessels from other classes and sizes are assumed to follow similar trend. Possible inclusion of vessels from other classes would further refine the proposed correction procedure, including the selection of the *CFC* and the associated correction method.

In a previous analysis of the same datasets, Lau (2015) pointed out that the correlation was better with thicker ice, i.e., the range of interest, which corresponds to higher resistance. There was substantial uncertainty on ice thickness and ice strength within the full scale data sets that contributes to data scattering as shown in Figures 2-4. Lau suggested a conservative estimate can be obtained to address this uncertainty reasonably by increasing the model prediction by 15%.

Ship Power

Similar analysis was performed with the power data. At the present time, the power correlation was performed on 3 sets of full-scale measurements including two trials for the CCGS icebreaker Pierre Radisson and one for the CCGS icebreaker Sir John Franklin (see Table 1).

IOT standard method for analyzing propulsion tests in ice includes propulsion overload tests in open water combined with limited ice tests. The principle of the method is that overload experiments in open water are used to predict the hydrodynamic torque required to develop a thrust sufficient to move the hull against a force equal to the hull resistance in ice. Because such open water tests cannot take account of any ice-propeller interaction, it is necessary to conduct a corresponding experiment in ice to determine the increase in torque due to propeller-ice interaction. It is assumed that propeller-ice interaction has a negligible effect on the thrust developed by the propulsion system. This has been shown to be true for small values of h/D where h is the ice thickness and D is the diameter of the propeller (Molyneux, 1989). The background and theory of the method is given in Jones and Lau (2006).

The result of the ship power correlation is given in Figures 5 and 6. In these figures, the ratio of Model to Full Scale Power (*RMFSP*), $R_{i.mp}/R_{i,f}$, was plotted against the model's friction coefficient. Again, data from individual dataset were fitted with a linear relationship in Figure 5 to examine the trend exhibited from individual dataset. In Figure 6, all data were fitted with a linear relationship to give overall trends.

Figure 5 shows a good correlation between the full scale model prediction and the sea trial measurement for a *CFC* within a narrow range of 0.045 to 0.062. On average, the *CFC* is at 0.056 with a standard derivation (*STDEV*) of 0.0098. The result is summarized in Table 4.

As shown in Figures 6, linear fit for all data give the following equation with the *CFC* equal to 0.0505:

$$\frac{P_{i.mp}}{P_{i,f}} = 0.1186 + 17.455 f_{ID} \tag{8}$$

Again, the mean of 0.0505 for perfect correlation has justified NRC-OCRE's practice in regarding to specifying a standard value of 0.05 for the *CFC*.

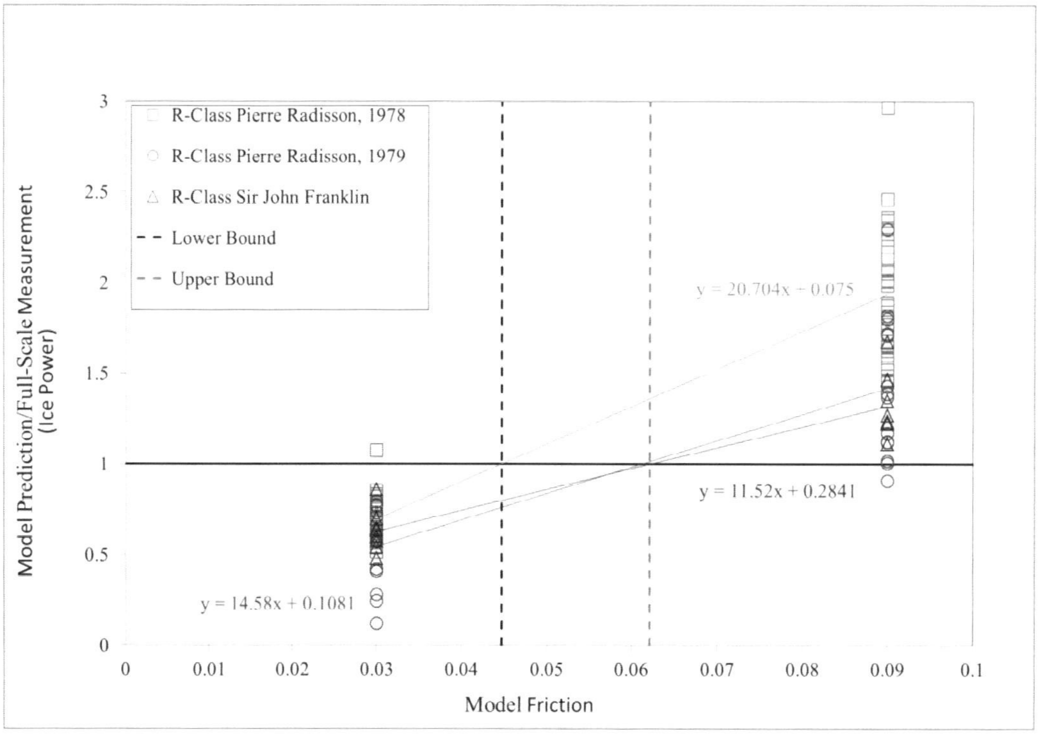

Figure 5. The ratio between model prediction and measured full-scale ice power $P_{i.mp}/P_{i,f}$ versus the model's coefficient of friction: linear fit to individual dataset

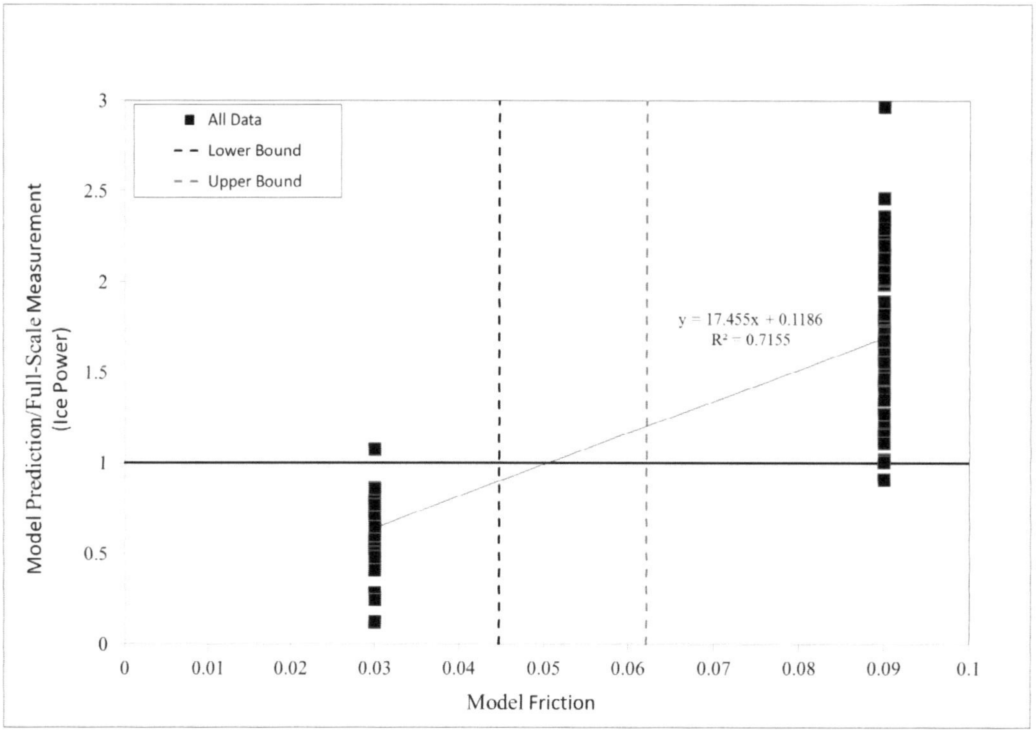

Figure 6. The ratio between model prediction and measured full-scale ice power $P_{i.mp}/P_{i,f}$ versus the model's coefficient of friction: linear fit to all datasets

Table 4. Result of power analysis: Summary of the equation for the individual dataset

Dataset	m	b	***CFC***	$P_{i.mp}/P_{i,f}$@f_{ID}**=0.05**
CCGS Pierre Radisson, 1978	20.70	0.0750	0.045	1.1102
CCGS Pierre Radisson, 1979	14.58	0.1081	0.061	0.8371
CCGS Sir John Franklin	11.52	0.2841	0.062	0.8601

Assuming all ships following the same trend, the power ratio $P_{i.mp}/P_{i,f}$ corresponding to a particular model friction can be obtained directly from Equation 8, with which the model prediction $P_{i.mp}$ obtained from a particular model friction finish f_{ID} can be corrected to the expected full-scale measurement $P_{i,f}$. For convenience, the ratio $P_{i.mp}/P_{i,f}$ is tabulated in Table 5.

Again, in the previous analysis by Lau (2015) of the same datasets, the data suggested a conservative estimate can be obtained to address test uncertainty reasonably by increasing the model prediction by 15%.

Table 5. The ratio $P_{i.mp}/P_{i,f}$ for a particular model's friction coefficient

f_{ID}	$P_{i.mp}/P_{i,f}$ @f_{ID}	f_{ID}	$P_{i.mp}/P_{i,f}$ @f_{ID}	f_{ID}	$P_{i.mp}/P_{i,f}$ @f_{ID}
0.000	0.119	0.045	0.817	0.090	1.602
0.005	0.206	0.050	0.904	0.095	1.690
0.010	0.293	0.055	0.991	0.100	1.777
0.015	0.380	0.060	1.079	0.105	1.864
0.020	0.468	0.065	1.166	0.110	1.951
0.025	0.555	0.070	1.253	0.115	2.039
0.030	0.642	0.075	1.340	0.120	2.126
0.035	0.730	0.080	1.428	0.125	2.213
0.040	0.817	0.085	1.515	0.130	2.300

Summary of Friction Correction Procedure

Friction correction to full-scale ice resistance prediction at *CFC*, i.e., 0.057, from any arbitrary model friction f_{ID}

The full-scale ice resistance predicted from a model test measurement $R_{i.mp}$ with a particular model friction f_{ID} can be corrected to that corresponding to the *CFC*, i.e., 0.057, by dividing the full-scale ice resistance prediction with the ratio $R_{i.mp}/R_{i,f}$ @ f_{ID} as given in Table 3. For example, the ice resistance at *CFC* is 1/0.831 or 20.3% higher that obtained from a model with a model ice friction of 0.04.

Friction correction to full-scale ice resistance prediction for ships with any friction coefficient from *CFC*, i.e., 0.057

There are limited data in full-scale to extract the dependency of ice resistance on the hull-ice friction. Until further full-scale data are available with ships tested in difference friction, we assume the trends as observed in model tests, i.e., the regression equation given in Figure 4, is representative of that at full-scale; hence, the full-scale ice resistance corrected from *CFC* can be corrected to ice resistance corresponding to any arbitrary hull-ice friction by multiplying the full-scale ice resistance prediction @ *CFC* with the ratio $R_{i.mp}/R_{i,f}$ @ f_{ID} as given in Table 3. For example, the ice resistance for an ice hull friction of 0.1 is 59.2% higher than the hull with the *CFC*.

Friction correction of towed propulsion data

Based on the limited data used, the ice power correction is of lesser confidence although the data show similar trend observed for the resistance data. Assuming all ships following the same trend, the ratio $P_{i.mp}/P_{i,f}$ corresponding to a particular model friction, i.e., *CFC* or an arbitrary ship hull friction can be obtained directly from equation 8. Please note the *CFC* from the power correlation was 0.05. For convenience, the ratio $R_{i.mp}/R_{i,f}$ is tabulated in Table 5. For example, the delivered power measured at *CFC* is expected to be 1/0.817 or 22% higher than that from a model with an ice friction of 0.04, and on the other hand the delivered power is 77.8% higher for a hull furnished with a friction of 0.1 rather than the *CFC*.

CONCLUSIONS

A re-analysis of NRC-OCRE's model-scale/full-scale correlation studies on the CCGS icebreaker Terry-Fox, CCGS R-Class icebreakers Pierre Radisson and Sir John Franklin and CCGS icebreaker Louis S. St. Laurent has confirmed a good agreement between NRC-OCRE model test predictions and full-scale measurements in resistance and power using its standard *CFC* friction coating of 0.05.

Equations 7 and 8 for resistance and power adjustment, respectively, are recommended for minor friction corrections, i.e., f_{ID} within the range of 0.04 to 0.06. In a previous analysis of the similar datasets, Lau (2015) suggested a conservative estimate can be obtained to address data scattering by increasing the model prediction by 15%.

It should be noted that all datasets used in this preliminary analysis are from mid- to large size icebreakers. In the absence of full-scale trial data, vessels from other classes and sizes are assumed to follow similar trend. Possible inclusion of vessel in next phase of analysis from other classes would further refine the proposed correction procedure, including the selection of the *CFC* and the associated correction method.

REFERENCES

Cowper, B., 1991. *Resistance and propulsive performance trials of the MV Terry-Fox and MV Ikaluk in level ice*. Transport Canada, Report Number TP10845E.

Edwards, R.Y. Jr., et al., 1979. *Result of trials in ice and open water of CCGS Pierre Radisson*, Report 245C-8, 2 vols, ARCTEC Canada Ltd.

Fleet Technology Ltd., 1992. *Low friction coefficient CCG Louis St. Laurent – Test report*. NRC-OCRE report CR 1992-01, Ocean, Coastal and River Engineering, National Research Council Canada, St. John's, Newfoundland and Labrador.

ITTC, 2002. ITTC – *Recommended procedures and guidelines 7.5-02-04-02.1, testing and extrapolation methods - ice testing - resistance test in level ice*. ITTC, 2002, Rev 01.

Jones, S.J. & Lau, M., 2006. Propulsion and maneuvering model tests of the USCGC Healy in ice and correlation with full-scale. *Proceedings of the 7th International Conference and Exhibition on Performance of Ships and Structures in Ice*, 16-19 July 2006, Banff, Alberta. Paper 104.

Jones, S.J. & Moores, C., 2002. Resistance tests in ice with the USCGC Healy. *Proceedings of the IAHR 16th International Symposium on Ice*, Dunedin, New Zealand, Vol. 1, p. 410-415.

Lau, M., 2015. Model-scale/full-scale correlation of NRC-OCRE's model resistance, propulsion and maneuvering test results. *Proceedings of the 34th International Conference on Ocean, Offshore and Arctic Engineering*, Volume 8: Ian Jordaan Honoring Symposium on Ice Engineering, St. John's, Newfoundland, Canada.

Mäkinen, E., Liukkonen, S., Nortala-Hoikkanen, A. & Harjula, A., 1994. Friction and hull coatings in ice operations. *Proceedings of the IceTech '94*, SNAME E1–E22.

Molyneux, W.D., 1989. Self-propulsion experiments for icebreakers in ice and open water. *Proceedings of the 22nd American Towing Tank Conference, ATTC, National Research Council, Ottawa*, p.326-331.

Newbury, S., 1988. *Icebreaking experiments with 1:20 and 1:40 scale models of the CCG R-class vessels*. NRC-OCRE report TR AVR-10, Ocean, Coastal and River Engineering, National Research Council Canada, St. John's, Newfoundland and Labrador.

Newbury, G.S. & Williams, F.M., 1986. "R-class" icebreaker model experiment results. *Ice Technology*, Computational Mechanics, Springer-Verlag, Berlin, p. 333-348.

Noble, P. & Comfort, G., 1979. *Ice trials to evaluate low friction coating on C.C.G.S. Louis S. St. Laurent*. ARCTEC Canada report FR357C-4.

NORDCO Ltd, 1989. *Report on resistance tests of Model 417 (Terry-Fox/Kalvik) in the ice tank*. NRC-OCRE report CR-1989-25, Ocean, Coastal and River Engineering, National Research Council Canada, St. John's, Newfoundland and Labrador.

Schwarz, J., 1986. Advances in icebreaker technology in West Germany. *Proceedings of the International Polar Transportation Conference*, Vancouver, D.F. Dickins Assoc., 1, 201–219.

Sodhi, D.S., Griggs, D.B. & Tucker, W.B., 2001. Healy ice trials: performance tests in ice. *Proceedings of the 16th International Conference on Port and Ocean Engineering under Arctic Conditions*, Vol. 2, pp.893-907.

Spencer, D., 1992. *Model-scale/Full-scale ice resistance correlation for the CCGS Louis St. Laurent - Part II*. NRC-OCRE report LM-1992-05, Ocean, Coastal and River Engineering,

National Research Council Canada, St. John's, Newfoundland and Labrador.

Spencer, D., 1990. *Ice resistance tests of Model 417 M.V.'s Terry Fox/Kalvik hull-ice friction = 0.005*. Report Number TR-1990-09, Ocean, Coastal and River Engineering, National Research Council Canada, St. John's, Newfoundland and Labrador.

Spencer, D.S., 1987. *Ice model tests on CCG Louis St. Laurent – Phase I: Models M395A, M395B, and M395C*, NRC-OCRE report TR AVR-01, Ocean, Coastal and River Engineering, National Research Council Canada, St. John's, Newfoundland and Labrador.

Spencer, D. & Jones, S.J., 2002. *Model-scale/Full-scale correlation in open water and ice for Canadian Coast Guard 'R-class' icebreakers*. NRC-OCRE report IR 2002 05, Ocean, Coastal and River Engineering, National Research Council Canada, St. John's, Newfoundland and Labrador.

Spencer, D., Molynuex, D. & Keinonen, A., 1988. *Model tests of M.V.'s "Terry-Fox"/"Kalvik" and their correlation to full scale*. NRC-OCRE report TR AVR-12, Ocean, Coastal and River Engineering, National Research Council Canada, St. John's, Newfoundland and Labrador.

Spencer, D., Williams, F.M. & Newbury, S., 1992. *Model-scale/full-scale ice resistance correlation for the CCG R-class icebreakers*. NRC-OCRE report LM-1992-20, Ocean, Coastal and River Engineering, National Research Council Canada, St. John's, Newfoundland and Labrador.

Wang, J. & Jones, S.J., 2008. Resistance and propulsion of CCGS Terry Fox in ice from model tests to full scale correlation. *Proceedings of the 8th International Conference and Exhibition on Performance of Ships and Structures in Ice*, 20-23, Banff, AB.

Williams, F.M. et al., 1992. *Full-scale trials in level ice with Canadian R-class icebreaker*. Trans. SNAME, 100, 293.

Williams, F.M. et al., 1991. *Full-scale icebreaker trials CCGS Sir John Franklin Indian Arm/Little Burnt Bay 1991*. NRC-OCRE report TR-1991-03, Ocean, Coastal and River Engineering, National Research Council Canada, St. John's, Newfoundland and Labrador.

Proceedings of the 24th International Conference on
Port and Ocean Engineering under Arctic Conditions
June 11-16, 2017, Busan, Korea

Load-Varying Methods for Ship Power Estimation in Brash Ice Channel by Ice Tank Model Test

Takatoshi Matsuzawa[1], Haruhito Shimoda[1], Daisuke Wako[1], Shotaro Uto[1], Qing He[2], Shinpei Watanabe[3]

[1] National Maritime Research Institute, Tokyo, Japan
[2] Oshima Shipbuilding Co., Ltd., Nagasaki, Japan
[3] Nippon Kaiji Kyokai (Class NK), Tokyo, Japan

ABSTRACT

A series of ice tank tests of a bulk carrier was conducted at National Maritime Research Institute. According to the Finnish-Swedish Ice Class Rules, 1m thick brash ice channel was assumed as actual condition for ice class IA vessel and simulated in the ice tank. For the propulsion tests, two kinds of load-varying methods i.e. in brash ice and in open water, were employed. The difference in thrust at self propulsion point between these two methods was 8.1% at most, while RMSE of measurement in brash ice was 15.1%. The power required for 5knots in brash ice were estimated and compared with the rule-based calculation.

KEY WORDS: Ice tank test; Brash ice channel; Load-varying method

NOMENCLATURE

R	Resistance
V	Ship speed
F	Towing force
T	Thrust
Q	Torque
N	Propeller revolution
ρ	Density
g	Gravity acceleration
h_B	Mean thickness of brash ice

Subscriptions I and W represent ice and water, respectively.

INTRODUCTION

New buildings of ice-strengthen vessel seem to be increasing as a consequence of recent positive use of the Arctic Ocean. As for large vessels such as tankers and bulkers, the main engine power is facing a paradox as needs to reduce power for energy-saving and at the same time needs to fulfill the power required for navigability in ice. The fair estimation of propulsion power of ice-strengthen vessels is thus important to meet all requirements by such as ice class rules and the authority's regulations. Ice tank model test is a major tool for power estimation of ice-strengthen vessels. The Finnish-Swedish Ice Class Rules (FSICR) allows to prove sufficiency of main engine power by ice tank tests instead of provided formulae calculation (Trafi, 2010).

ITTC Recommended Procedures gives standards of general propulsion tests in ice (ITTC, 2011) which is similar way to that of in open water. However, obtaining propulsion factors and self propulsion point in the ice tank test needs to correct many parameters due to non-uniform nature of ice condition. Load-varying method is a convenient way to determine required thrust regardless of detail propulsion factors.

A series of resistance tests and towing propulsion tests in ice tank was conducted to estimate power in ice. The ice condition was assumed as brash ice channel of 1m thick, which corresponds to the thickness for ice class IA of FSICR. As applying load-varying method, two procedures for testing and analysis to obtain required thrust were considered such as by towed propulsion tests in ice; by a set of resistance tests in ice and towed propulsion tests in open water. The thrusts and the power derived from these two methods were compared to each other in order to evaluate the differences.

METHODS

Load-Varying Method

The procedures of load-varying method for open water test is briefly introduced in ITTC Recommended Procedures. When towed propulsion test is performed, the parameters such as F, T, Q and N are measured. These parameters can be obtained normally in brash ice directly (Method-1). As a replacement, another method that does not require towed propulsion test is proposed (Method-2). The overall flow of test and analysis is shown in Figure 1.

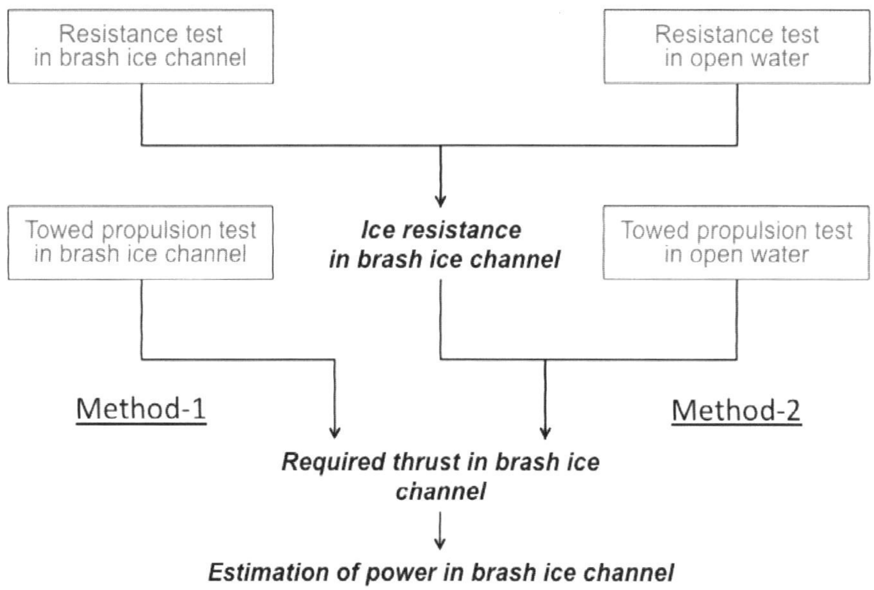

Figure 1. Flow chart of the test program and analysis for load-varying methods.

Method-1 is a direct method to obtain propulsion parameters in ice. Measured T and F are plotted in *F-T* diagram like Figure 2. Their regression line should cross with the *F* axis at all resistance in ice (R_I+R_W), and should cross with the *T* axis at the thrust required in ice, respectively.

Method-2 requires towed propulsion test only in open water. This can reduce both of the number of ice sheet to be used and the parameters to be measured in ice. The resistance of ice is analytically added after the measurement so that the *T* at self propulsion point balances in ice. Resistance tests in both of ice and open water must separately be performed to determine R_I to calculate required *T*.

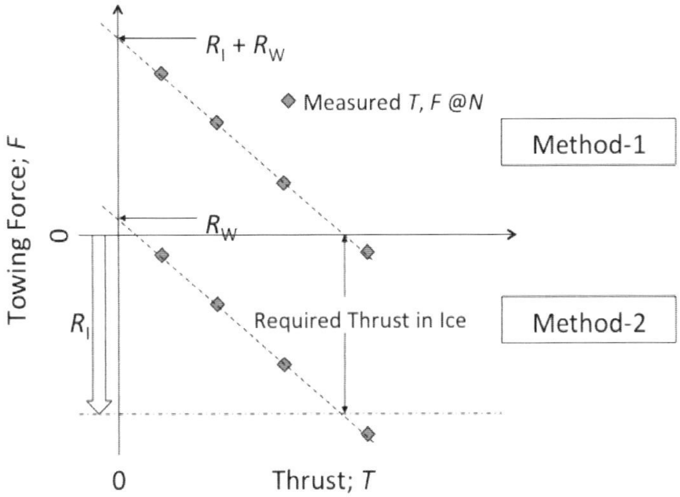

Figure 2. Supposed *F-T* diagram of towed propulsion tests.

Model Ship

The tested ship is a Panamax bulk carrier of 75,000DWT with 1 propeller designed by Oshima Shipbuilding Co., Ltd. (OSY) as shown in Figure 3. It is considered as one of the typical ships in the NSR having ice-strengthen hull. Ice breaking ability is not important for those ships; therefore they have rather normal shape to keep open water performance and cargo capacity.

The model length *L* was 5.7m and the width *B* was 0.84m. The main body of the model was made of wood and coated by FRP. The surface was painted by a fine-grained epoxy paint to control hull-ice friction coefficient to 0.1 which was confirmed by the measurement using model ice and painted test specimen, according to ITTC Recommended Procedures.

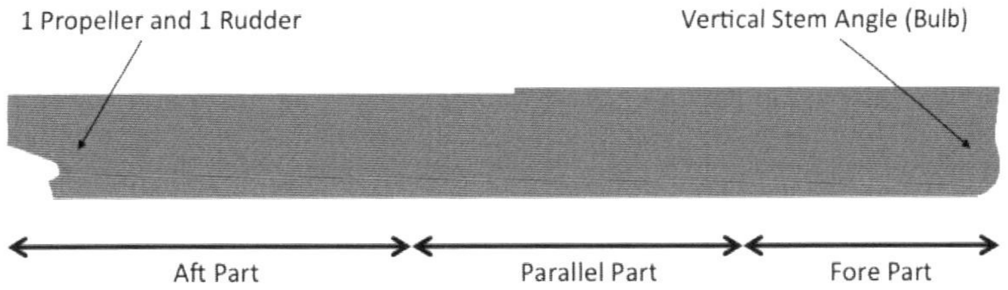

Figure 3. Side view of the model ship.

EXPERIMENTS

Ice Tank and Model Ice

The ice tank of National Maritime Research Institute (NMRI) was used in this study (Figure 4 left). The length of the tank is 35m and the width is 6m. The capacity of the towing carriage is 50kN and the speed can be varied in the range of 5 to 1500mm/s. The water contains Propylene Glycol as a dopant. Model ice with columnar crystal structure, shown in Figure 4 right, is produced for normal model test. Wet seeding is performed at the first stage of each model ice production. Density of the ice can be adjusted to a value required for the test by controlling air bubble concentration in the ice.

Figure 4. NMRI ice tank and model ice.

Left: ice chamber and model ice. Right: typical crystal structure of the model ice.

Preparation of Brash Ice Channel

According to FSICR Guidelines (Trafi, 2010), brash ice channel for ice tank test should be filled by broken ice pieces. In each brash ice test, mother ice sheet was crashed by ice chisels as shown in Figure 5 left, and compressed to control ice concentration and to obtain desired thickness.

Figure 5. Making brash ice channel.

Ice Tank Test

The model ship was connected to the carriage via a resistance dynamometer to measure towing force F. During towed propulsion test, it equipped with a model propeller so that thrust T, torque Q and propeller revolution N were measured. The system of measurement was shown in Figure 6.

The propeller rotates at a rate of N while the carriage tows the model at a constant speed in each run. The test was repeated varying N to obtain several set of F, T, Q and N. The speed was set to model scaled value equivalent to 5 knots for ship.

For Method-2, resistance tests were also conducted in both conditions of ice and open water respectively to distinguish between the channel ice resistance R_I and water resistance R_W. The system of measurement was almost the same as towed propulsion test except for dismounting the model propeller.

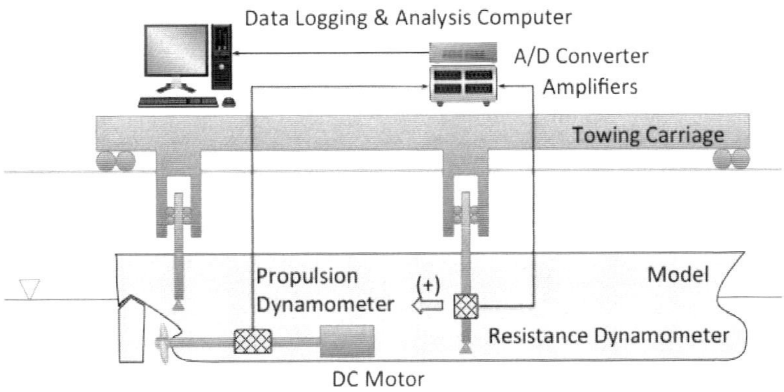

Figure 6. Measurement system for towed propulsion test.

RESULTS AND DISCUSSION

Results of Towed Propulsion Tests

Towed propulsion tests were carried out for both of full and ballast conditions. Derived from measured towing force F and thrust T through all test runs, F-T diagrams were made as shown in Figure 7. The values were extrapolated to full scale, taking into account skin friction correction. Brash ice thickness and density of ice were also corrected using regression formula as following:

$$\frac{R_I}{\rho_I g B_I h_B^2} = a_0 + a_1 \frac{V}{\sqrt{g h_B}} \tag{1}$$

where a_0 and a_1 are regression coefficients obtained from measurement data.

Towing force at no thrust means resistance, thus the intercept of each regression line was set to $R_I + R_W$ which were obtained from resistance tests. Note that measured thrust by Method-2 was raised by R_I because it was obtained in open water. In the figure, thrust is normalized by T_{SP} which means thrust at ship self propulsion point in brash ice. Towing force is also normalized by R_I.

Plots of thrust and torque on the basis of propeller revolution are shown in Figure 8.

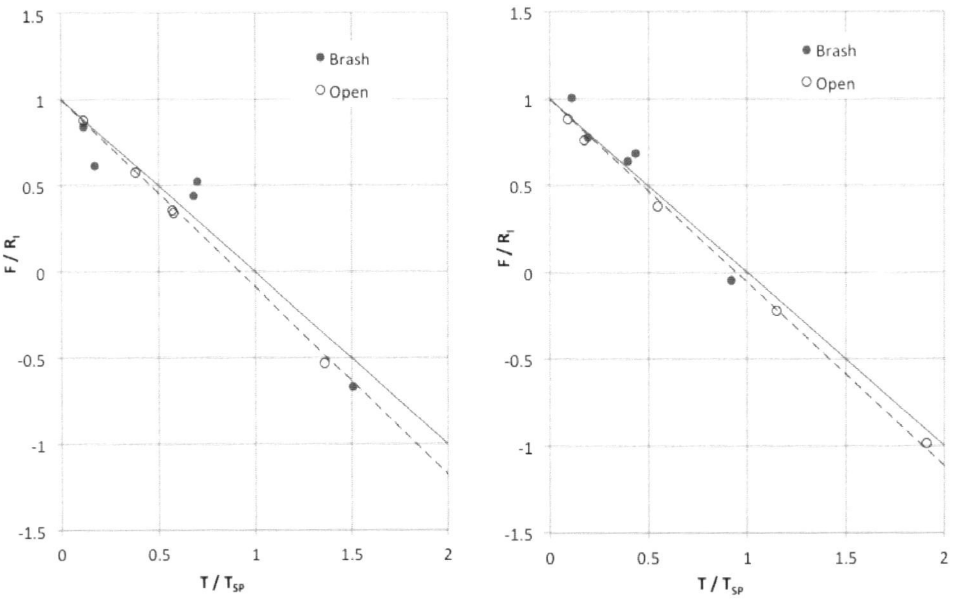

Figure 7. *F-T* diagrams in full scale derived from towed propulsion tests.
Left: Full. Right: Ballast.

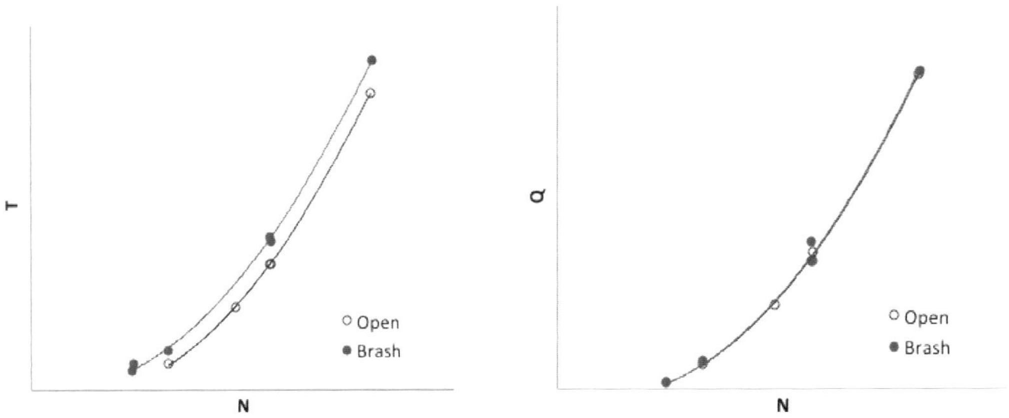

(a) Full

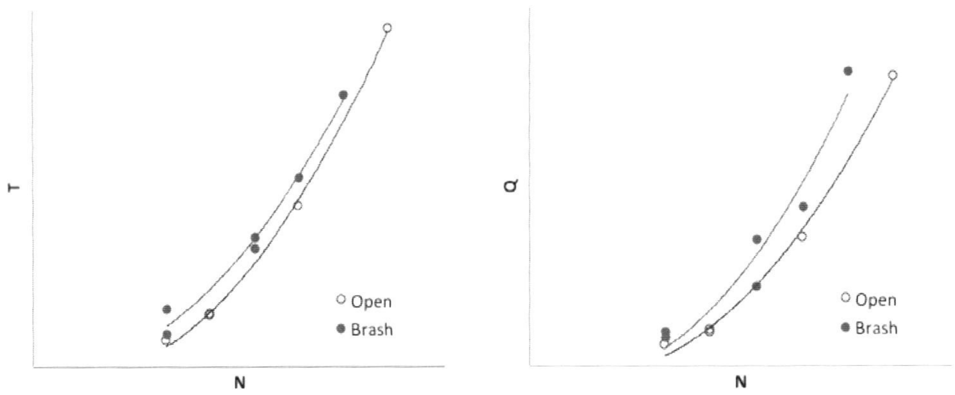

(b) Ballast

Figure 8. Thrust and torque measured in towed propulsion tests.

Discussion

The RMSEs of measured data against the regression lines from the *F-T* diagrams are shown in Table 1. Although the deviation of towing force (or thrust) of Method-1 are significant, those of Method-2 agree well with the lines. Differences between thrusts at T_{SP} of each method were 8.1% for full and 5.5% for ballast which are less than RMSEs of Method-1. Therefore, in this case, Method-2 was reliable enough as a replacement of Method-1.

Table 1. RMSEs of thrust against regression lines.

Condition	Method-1	Method-2
Full (UIWL)	0.151	0.033
Ballast (LIWL)	0.085	0.030

In order to estimate power from towed propulsion tests, T_{SP} should be obtained from *F-T* diagram at first. Then *N* must be determined by *T-N* curve, and *Q* at *N* is derived. As shown in Figure 8, some differences of these parameters clearly exist between two methods and they should be accumulated in calculation of power. Estimated powers normalized by the value calculated by FSICR formulae are shown in Table 2. The powers of full-load condition agreed with each other well, however in ballast condition power of Method-1 are greater than that of Method-2. Torque in ice possibly increases due to propeller-ice interaction and it happens with more frequency when the draft is small. Therefore Method-2 should be used with care for ballast condition.

Table 2. Estimated power of each method against required value by FSICR formulae.

Condition	FSICR	Method-1	Method-2
Full (UIWL)	1	0.46	0.48
Ballast (LIWL)	1	0.37	0.28

CONCLUSION

Towed propulsion tests in ice tank were carried out to estimate power in brash ice channel. The tests in brash ice could obtain thrust at self propulsion point; T_{SP} directly, however the data showed large deviation. As for the tests in open water, in spite of separately requires resistance test in brash ice, the data was robust and reasonable. The difference of T_{SP} between two methods were 8.1% for full-load condition and 5.5% in ballast condition. Both were within RMSE of measurement in brash ice channel.

Thrust, torque and propeller revolution were measured in the tests and used for estimation of required power. The powers of full-load condition agreed with each other well. However, in ballast condition, power in brash ice are greater than that derived from open water tests. Higher torque in brash ice tests due to propeller-ice interaction possibly resulted in greater power, thus care should be paid for applying open water load-varying method to small draft condition.

ACKNOWLEDGEMENT

This research was carried out jointly by NMRI, OSY, Kogakuin University and ClassNK as part of the ClassNK Joint R&D for Industry Program.

REFERENCES

ITTC; International Towing Tank Committee, 2011. Recommended Procedures and Guidelines 7.5-02-03-01, Testing and Extrapolation Methods, Propulsion, Performance.

ITTC; International Towing Tank Committee, 2011. Recommended Procedures and Guidelines 7.5-02-04-02, Testing and Extrapolation Methods, Ice Testing.

TraFi; Finnish Transport Safety Agency, 2010. Finnish -Swedish Ice Class Rules 2010. TRAFI/31298/03.04.01.00/2010.

TraFi; Finnish Transport Safety Agency, 2011. Guidelines for the application of the Finnish-Swedish Ice Class Rules. TRAFI/21816/03.04.01.01/2011.

POAC'17
Busan, KOREA

Proceedings of the 24th International Conference on
Port and Ocean Engineering under Arctic Conditions
June 11-16, 2017, Busan, Korea

Field Observations and Preliminary Investigations of a Wave Event in Solid Drift Ice in the Barents Sea

Aleksey Marchenko[1], Jean Rabault[2], Graigory Sutherland[2], Clarence O. Collins III[3], Peter Wadhams[4], Mikhail Chumakov[5]
[1] The University Centre in Svalbard, Longyearbyen, Norway
[2] University of Oslo, Oslo, Norway
[3] Naval Research Laboratory, Stennis Space Center, Hancock County, Mississippi, USA
[4] Univetrsity of Cambridge, Cambridge, UK
[5] Gazprom VNIIGAZ LLC, Moscow, Russia

ABSTRACT

Waves propagating below drift ice were observed in the North-West Barents Sea. In-situ measurements of ice and water characteristics were performed with the equipment deployed on the ice and in the water below the ice during 16 hours from 16:00, May 1, to 08:00, May 2, 2016. Peak frequencies of the observed waves were around 0.35 rad/s, 0.5 rad/s, 0.6 rad/s and 0.8 rad/s. It is found that the eddy viscosity in the boundary layer below the ice has a mean value of 140 cm^2/s with significant variations correlated with the floe acceleration. Spectral properties of the observed waves are compared with the results of the WaveWatch III model. The origin of the observed waves is discussed in the paper.

KEY WORDS: waves, ice, turbulence, drift

INTRODUCTION

Propagation of wind waves and swell in ice covered regions can influence the break up of drift ice into small floes in a span of a few hours over relatively large areas. At the same time, drift ice influences damping of surface waves and prevents their propagation over long distances in ice covered regions. In regions of offshore activities, broken ice and icebergs accelerated by waves create risks to human activities. Investigation of wave interaction with ice is of interest for the development of human activity in the Barents Sea. There are regions where the water is consistently covered by solid drift ice in the winter time and other regions which are exposed to stronger wave action with rare occurrences of drift ice. Collins et al (2015) described details of ice break up in the Barents Sea due to the penetration of storm waves from open waters.

An analysis of some previous in-situ measurements of wave characteristics in ice covered regions of the Barents Sea is presented by Marchenko et al, 2015. Measurements were performed with an Acoustic Doppler Velocimeter (ADV, SonTek 5MHz Ocean Probe) deployed on the drift ice in a downward looking position and a Directional Waverider buoy deployed on top of the ice. Typical periods of waves observed below the drift ice were 10-12 s. The results of ADV measurements gave the possibility to calculate the eddy viscosity in the under-ice boundary layer. According to Liu and Mollo-Christensen (1988) the eddy viscosity influences wave damping. Field measurements have shown great variability of the eddy viscosity from very small values of about 0.002 cm^2/s to 100 cm^2/s (Marchenko et al, 2015). The high values of the eddy viscosity were found from the analysis of field data collected near Edgeøya. The increase of the eddy viscosity is explained by the influence of non-stationary ice drift on the surface water layer. Representative damping distance for a swell with period of ten seconds is 10-20 km when the eddy viscosity is about 100 cm^2/s. Wind waves with periods of 5-6 s dampen over shorter distance and, therefore, were never observed in solid ice.

Previously, Marchenko et al (2013) introduced a method for wave measurement from the ice with two pressure recorders SBE-39 mounted on the same wire at different depths. The method was used to register waves propagating from the front of Tuna glacier due to calving events. Rabault et al (2016) constructed instruments based on Inertial Motion Units (IMUs) and used them for the registration of surface waves penetrating below the ice from the open water in Templefjorden in Spitsbergen. The IMUs were configured to output acceleration and angular rate at 10Hz. In May 2016 all sensors previously cited, including the ADV, SBE-39 and 10 IMUs were deployed on the drift ice near Edgeøya and registered wave characteristics synchronously during 16 hours. In the present paper we described and analyze the results of those field measurements, and compare the output from the different processing methods.

LOCATION AND ORGANIZATION OF THE FIELDWORK

Fieldwork was performed on the drift ice near Edgeøya in the North-West Barents Sea (Figure 1a). The geographical coordinates are 77.76°N, 25.5°E. Sea depth in the region was measured around 160 m. RV Lance was moored to a floe of thickness around 30 cm and diameter around 2 km (Figure 1b). The floe mass is estimated between $7{\cdot}10^5$ tons and $8{\cdot}10^5$ tons. The mass of the fully loaded Lance is M_L=2370 tons, and her length and breadth are around L_L=60m and w_L=12 m. Therefore the influence of the Lance on the floe dynamics is small. The frequency of natural oscillations of the Lance is estimated from the formula $\omega_L^2=\rho_w g S_L/M_L$, where ρ_w is the water density, g is the gravity acceleration, and $S_L= w_L L_L$. We find a natural frequency of $\omega_L\approx 1.7$ rad/s, i.e. a period $T_L\approx 3.5$ sec. This is outside of the range of wave frequencies, and therefore cannot be mistaken for incoming water waves.

The equipment was deployed on the floe in the afternoon of May 01, 2016, and wave measurements started from 16:00, May 01 (here and further UTC is used), and extended until 09:00, May 02. The equipment consists of the pressure and temperature recorders SBE 39, Acoustic Doppler Velocimeter SonTek Ocean Probe 5 MHz (ADV), Ice tracker Oceanetic Measurement (model 703 equipped with anemometer) and Inertial Motion Units (IMUs) from Vectornav (VN100). SBE 39 were fixed on a steel wire on the depths 3.6 m and 11.4 m, and the wire was mounted on the ice (Figure 2a). The ADV probe is equipped with tiltmeter and compass. Magnetic inclination is about 15° in the region and is taken into account during data processing. The ADV was mounted on a vertical wooden pole fixed on the ice with a tripod (Figure 2a). Velocity measurements were performed at a depth of 80 cm below the ice. The depth is measured by the ADV pressure sensor. SBEs were deployed around 16:00, May 01, and recovered around 09:00, May 02. The ADV sensor was deployed on 19:00, May 01, and around 04:30, May 02, it was disconnected from the electric power source on the Lance board because of the floe motion relatively the ship. The ice tracker was deployed on the floe around 11:00 and sent data via Iridium during one month. The positions of the 10 IMUs deployed are presented in Figure 2b. One IMU was deployed alone, while all other IMUs were grouped into arrays of three sensors.

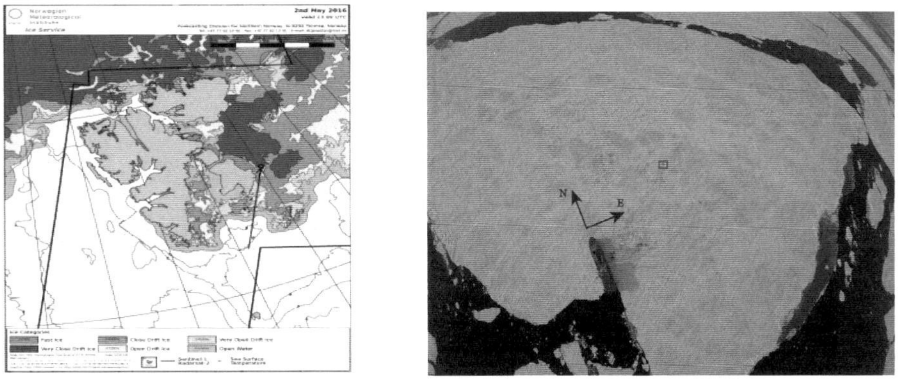

a) b)

Figure 1. (a) Location of the field works near Edgeøya on May 1-2, 2016, is shown on the ice map by black circle and arrow. (b) View of RV Lance from unmanned helicopter. Red square shows location of ADV and SBE

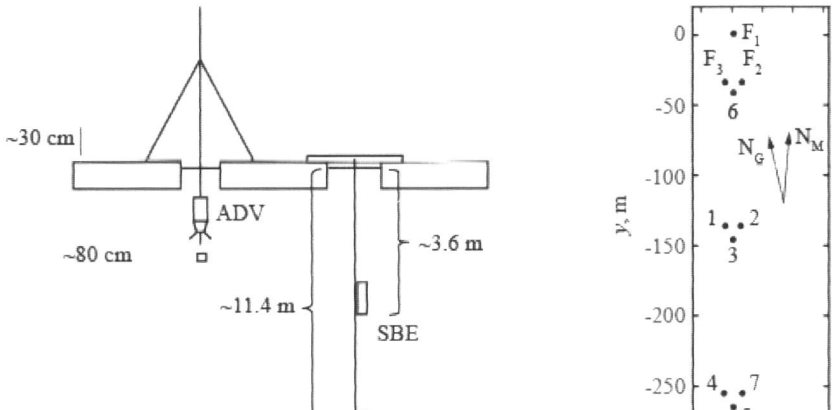

Figure 2. (a) Schematic of the deployment of ADV and SBE sensors. (b) Positions of the IMUs on the ice floe. N_M and N_G indicate the position of the Magnetic and Geographic North, respectively.

User setup for the ADV measurements is given in Table 1. SBE sensors provided continue record with sampling frequency 2 Hz. Ice tracker provided data on GPS location of the tracker, its yaw angle and local wind velocity with sampling period of 10 min. The trajectory and the drift velocity of the ice tracker are shown in Figure. Wind velocity (Figure 3a) didn't change significantly 16:00 UTC, May 01, to 04:00 UTC, May 02. The shape of the floe trajectory (Figure 3a) and the evolution with time of the flow velocity are explained by the influence of sea current modulated by semidiurnal tide. Evolution of the floe acceleration and yaw angle shown in Figure 4 are explained by the floe interaction with surrounding ice.

Table 1. User setups of ADV.

Sensor	Sampling frequency, Hz	Burst Interval, s	Samples per burst	Number of bursts	Depth, cm
ADV	10	360	2400	94	80

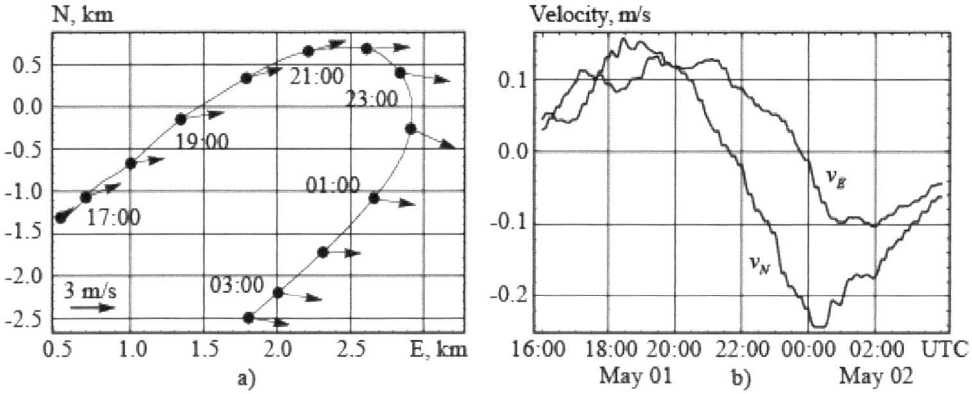

Figure 3. Trajectory (a) and drift velocity components versus the time (b) of the ice tracker deployed on the ice floe. Vectors of the wind velocity measured along the drift trajectory are shown by arrows (a).

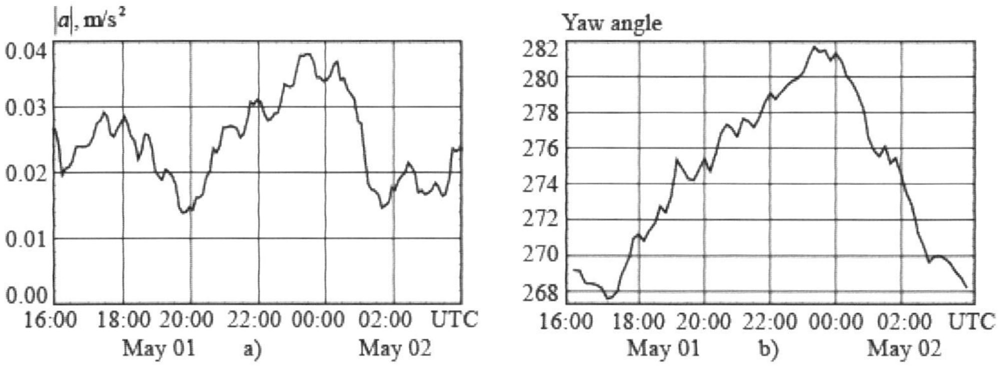

Figure 4. Absolute value of the acceleration (a) and yaw angle (b) of ice tracker deployed on the drift floe versus time.

ANALYSIS OF SBE DATA

Records of the water pressure measured by both SBE39s are shown in Figure 5. The quality of the data recorded at 3.6 m depth before 21:00 is not sufficient to distinguish wave induced oscillations of the water pressure, but after 21:00 the oscillations are well visible (Figure 5a). The oscillations are well visible in the pressure data recorded at 11.4 m depth from 16:00. Spectral analysis was performed with pressure fluctuations δp calculated as a difference between the pressure averaged over each consequent 5 min interval and actual pressure measured in this interval. Spectrograms (Figure 6a) and spectrums (Figure 7) constructed in Mathematica software show that wave spectrum has four local maxima at frequencies around 0.35 rad/s, 0.5 rad/s, 0.6 rad/s and 0.8 rad/s before 23:00, May 01, and there are only two local maxima at frequencies around 0.35 rad/s and 0.6 rad/s after 00:00, May 02. The frequency of waves with maximal energy was 0.8 rad/s before 23:00, May 01, and 0.6 rad/s after 00:00, May 02. Spectrums of the pressure fluctuations recorded at 3.6 m depth are very similar to shown in Figure 7.

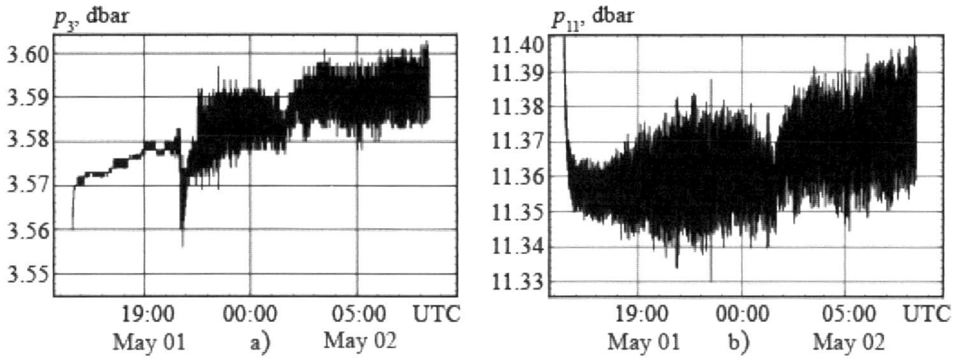

Figure 5. Records of the water pressure measured by recorders SBE39 deployed at 3.6 m (a) and 11.4 m (b) depths.

It was shown (Marchenko et al, 2013) that the ratio of wave induced pressure amplitudes measured at depths $z=z_1$ and $z=z_2$ is equal to:

$$\gamma \equiv \frac{p_{1,dyn}}{p_{2,dyn}} = \frac{\omega^2 \cosh[k(z_1+H)] - gk\sinh[kH]}{\omega^2 \cosh[k(z_2+H)] - gk\sinh[kH]}. \qquad (1)$$

Eqn. (1) is used for the calculation of the wave number k when the wave frequency is known. We consider the Fourier transforms $\delta p_{3,f}(\omega)$ and $\delta p_{11,f}(\omega)$ of the pressure fluctuations $\delta p_3(t)$ and $\delta p_{11}(t)$ recorded at depths 3.6 m and 11.4 m. It is assumed that the ratio γ is equal to the ratio $|\delta p_{3,f}|/|\delta p_{11,f}|$, where the values of $|\delta p_{3,f}|$ and $|\delta p_{11,f}|$ are taken in the points of their local maxima shown in Figure 6 by grey and black circles. The values of $|\delta p_{3,f}|$ and $|\delta p_{11,f}|$ in their local maxima are given in Table 2. Black dots in Figure 8a have coordinates (k, ω), where the values of ω are taken from Table 2, and the values of k are calculated from Equn. (1) with the values of ω and γ calculated using Table 2. Solid line in Figure 8a shows the dispersion relation of gravity waves propagating in the water with free surface and described by the formula $\omega^2=gk\cdot \tanh(kH)$ with H=160 m. One can see that experimental points sit on the dispersion curve. It means that the influence of ice on dispersion properties of observed waves is very small.

The last conclusion follows from the dispersion equation of flexural-gravity waves $\omega^2=gk\cdot \tanh(kH)(1+Dk^4)$, when $Dk^4<<1$. Here $D\approx Eh^3/(12\rho_w g)$, where E is the effective elastic modulus of the ice and h is the ice thickness. In-situ tests of flexural strength of floating cantilever beams performed on the drift ice on May 01 shown that E=1.2-1.9 GPa. Assuming h=0.3 m we find that $Dk^4<0.01$ when k=0.07, i.e. the influence of ice elasticity on the waves with registered frequencies is expected to be very small, in good agreement with observations.

Wave amplitude is calculated using water pressure records at depth z below the ice and the formulae:

$$\delta p_{z,f}(\omega)=\rho_w g\eta_f(\omega)P(k,z),\quad P(k,z)=\frac{\cosh[k(z+H)]}{\cosh[kH]}-1, \tag{2}$$

where $\delta p_{z,f}$ and η_f are the Fourier images of the water pressure fluctuations $\delta p_z(t)$ at the depth z and water surface elevation $\eta(t)$ caused by waves. We used records of the water pressure at 11 m depth. Values of $\delta p_{z,f}$ were calculated using the discrete Fourier transform in Mathematica. Then the inverse Fourier transform was used to calculate water surface elevation as a function of the time separately. The procedure was realized independently for each hour of the record. Significant wave height (SWH) is calculated using the formula:

$$SWH=4\cdot \text{StandardDeviation}[\eta]. \tag{3}$$

Mean value of SWH was found around 10 cm.

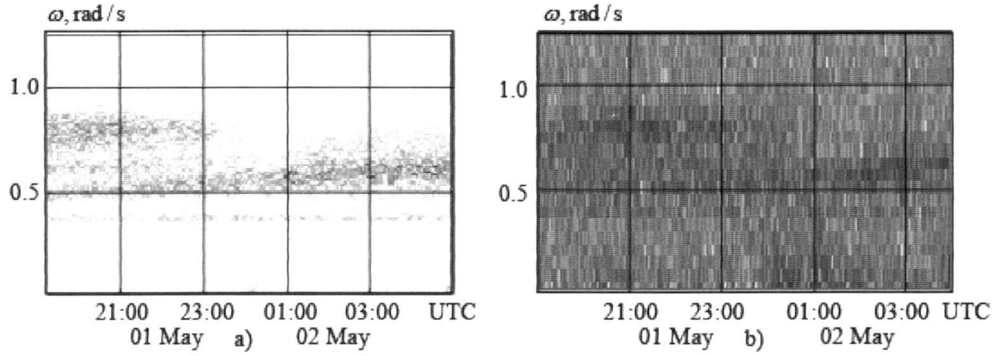

Figure 6. Spectrogram of pressure fluctuations reconstructed with SBE data recorded at 11.4 m depth (a). Spectrogram of the fluctuations of the North component of water velocity recorded by ADV (b).

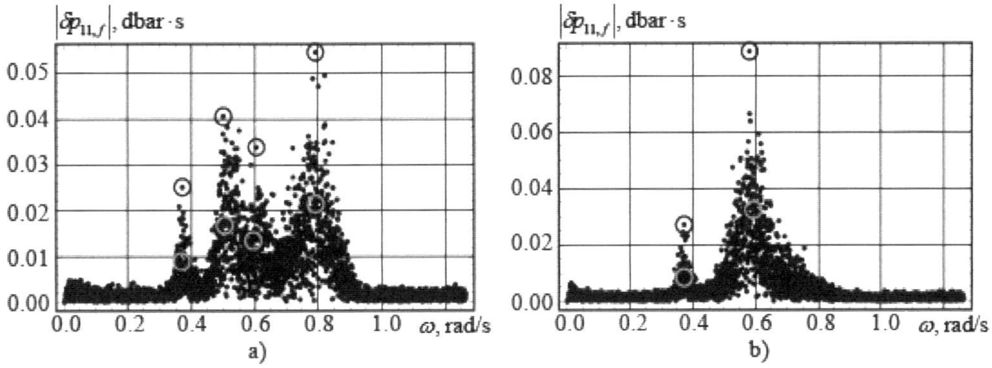

Figure 7. Spectra of water pressure fluctuations recorded at 11.4 m depth from 19:00 to 23:00, May 01, (a) and from 23:00, May 01, to 03:00, May 02, (b). Black and gray circles show spectral maxima of pressure fluctuations records at 11.4 m and 3.6 m depths.

Table 2. Local maxima of spectrums $|\delta p_{3,f}|$ and $|\delta p_{11,f}|$, dbar·s.

	19:00 to 23:00, May 01				23:00, May 01, to 03:00, May 02	
ω, rad/s	0.35	0.5	0.6	0.8	0.35	0.6
3.6 m	0.009	0.017	0.013	0.021	0.009	0.032
11.4 m	0.026	0.04	0.034	0.054	0.025	0.09

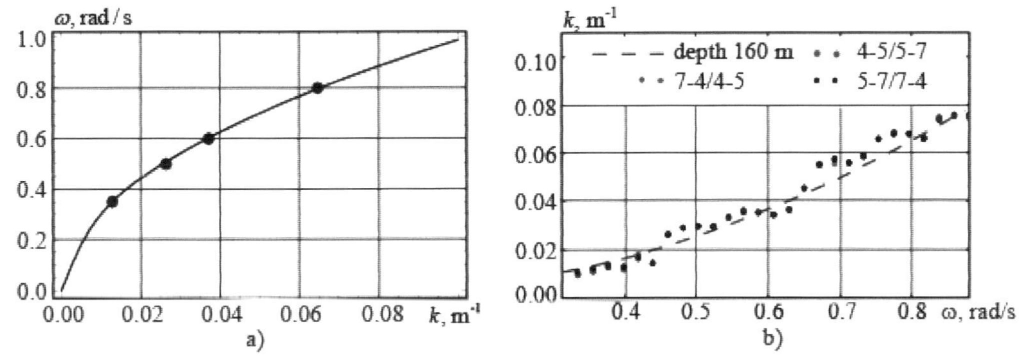

Figure 8. Dispersion relation of surface gravity waves in the water of 160 m depth. Black dots are constructed using SBE records at 3.6 m and 11.4 m depths (a). Dispersion relation obtained from the analysis of the cross correlation between IMU signals 4, 5 and 7 using Eqn. (11), at 20:00 UTC on 01/05/2016 (b).

ANALYSIS OF ADV DATA

Records of the East (v_E) and North (v_N) velocity components at 80 cm depth below the ice were used for the calculation of the eddy viscosity and the analysis of spectrums and directions of wave propagation. The velocity fluctuations in the East (δv_E) and North (δv_N)

directions were calculated for each burst of the ADV record as follows:

$$\delta v_E = v_E - \langle v_E \rangle, \quad \delta v_N = v_N - \langle v_N \rangle, \tag{4}$$

where $\langle v_E \rangle$ and $\langle v_N \rangle$ are the mean values of the East and North velocities averaged over the burst. All fluctuations were grouped in one file and used for the construction of spectrogram. The spectrogram of the fluctuations of the North velocity component is shown in Figure 6b. It looks similar to the spectrogram constructed with the SBE data. Spectrums of δv_N are shown in Figure 9. Spectral maxima are well visible in the spectrums of δv_N and are absent in the spectrum of δv_E from 19:00 to 23:00, May 01. The δv_N spectrum looks very similar to the spectrum shown in Figure 7a. Figure 9b shows spectral maxima at the frequency of 0.6 rad/s in the spectrum δv_N. There is spectral maxima in the spectrum of δv_E at the same frequency. Spectral maxima at the frequency of 0.35 rad/s appears only in the spectrum of δv_N. The velocity directions should be corrected by clockwise shift on 15° because of the magnetic inclination.

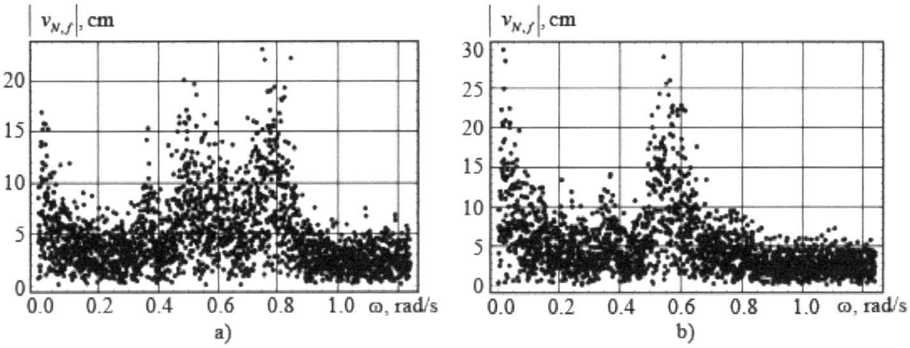

Figure 9. Spectrums of fluctuations of the North velocity component recorded from 19:00 to 23:00, May 01, (a) and from 23:00, May 01, to 03:00, May 02, (c).

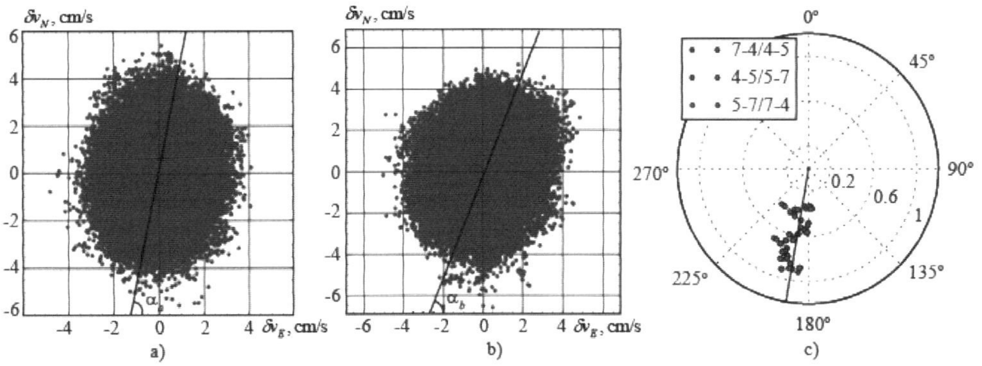

Figure 10. Hodograph of the velocity fluctuations recorded from 19:00 to 23:00, May 01, (a) and from 23:00, May 01, to 03:00, May 02, (b). (c): direction of propagation of the waves obtained from the analysis of the cross correlation between IMU signals 4, 5 and 7 using Eqn. (11), at 20:00 UTC on 01/05/2016, concentric circles indicate wave frequency rad/s). The black line indicates the direction of azimuth 190 degrees.

Each point in Figure 10a,b corresponds to the vector of the velocity fluctuations with corrected direction. The sampling frequency is 10 Hz. From Figure 10a it follows that dominant direction of wave propagation from 19:00 to 23:00, May 01, was from the South-West with azimuth angle $\alpha_a \approx 189.6^o$. Later from 23:00, May 01, to 03:00, May 02, the azimuth angle of wave propagation increased to $\alpha_b \approx 199.8^o$ (Figure 10b).

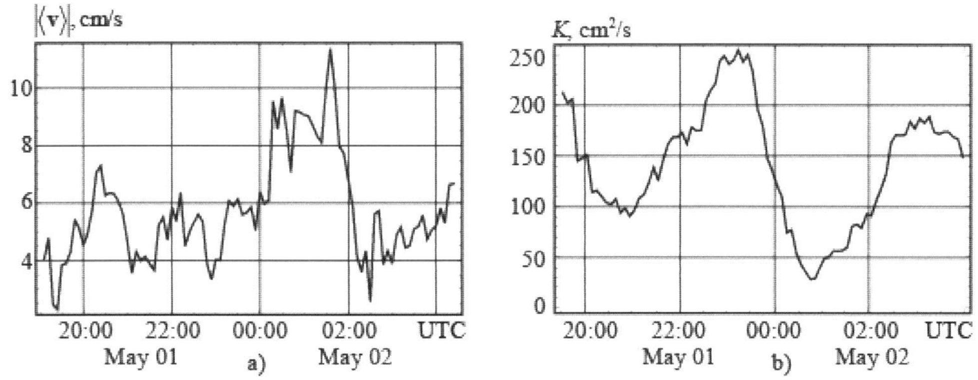

Figure 11. The mean horizontal water velocity relatively the ice (a) and the eddy viscosity (b) versus the time.

Figure 11a shows the mean horizontal velocity of the water below the ice measured with the ADV versus the time. Figure 11b shows the eddy viscosity versus the time. The eddy viscosity is calculated with the formula:

$$\tau = -\rho_w K \frac{\partial \langle v_h \rangle}{\partial z}, \tag{5}$$

where τ is the turbulent shear stress applied to the ice at the ice-water interface, ρ_w is the water density, $\langle v_h \rangle$ is the mean horizontal velocity of the water relative to the ice, z is the vertical coordinate, and K is the eddy viscosity.

The turbulent shear stresses applied to the ice bottom in the East and North directions are calculated with the formula of Reynolds stresses:

$$\tau_E = \rho_w \langle \delta v_E \delta w \rangle, \quad \tau_N = \rho_w \langle \delta v_N \delta w \rangle, \tag{6}$$

where δw is the fluctuation of the vertical velocity, and symbol $\langle .. \rangle$ means the averaging over the burst. The turbulent shear stress τ in Eqn. (4) is calculated from the formula:

$$\tau = \sqrt{\tau_E^2 + \tau_N^2} . \tag{7}$$

The vertical gradient of the mean horizontal velocity of the water relatively the ice is approximated as $\partial \langle v_h \rangle / \partial z \approx -\langle v_h \rangle / h_{ADV}$, where h_{ADV} =80 cm. The mean horizontal velocity is calculated from the formula:

$$\langle v_h \rangle = \sqrt{\langle v_E \rangle^2 + \langle v_N \rangle^2} \,. \tag{8}$$

From Figure 10 it follows that K reaches a local maximum in time when $\langle v_h \rangle$ is minimal, and when the floe acceleration shown in Figure 3a. The mean value of the eddy viscosity over is found to be 140cm^2/s.

ANALYSIS OF IMU DATA

The signals recorded by the IMUs are of good quality, as confirmed by the example of raw data shown in Figure 12. The PSD of the wave elevation is computed from the PSD of the wave vertical acceleration following the formula: $PSD[\eta] = \omega^{-4} PSD[\eta_{tt}]$, where η is the wave elevation. Excellent agreement is found between the shape of the PSD reported in Figure 7a and the one reported in Figure 12 (right), which validates both methods. Significant wave height can be obtained from the zeroth order moment of the PSD, and the results are very similar to what is obtained from Eqn. (3), with a maximum value of around 12 cm.

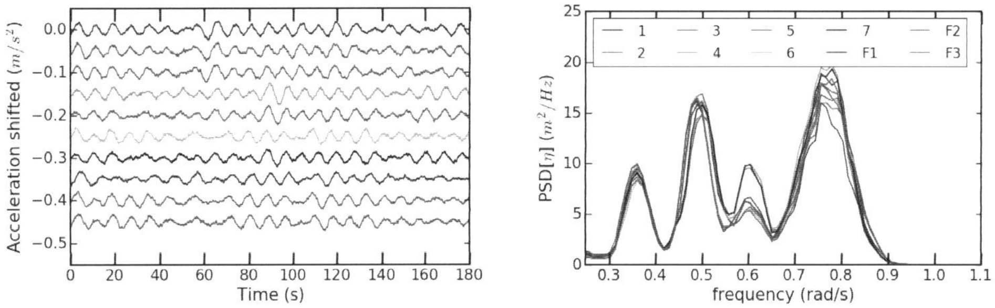

Figure 12. Left: sample of the raw signal for vertical acceleration recorded by all 10 IMUs starting at UTC time 20:00:00.000 on 01/05/2016 (vertical offsets are used for ease of visualization). Right: PSD of wave elevation computed for the 10 IMUs at the same time as presented in Figure 7 (a), using Welch method on 30 minutes intervals with 75% overlap.

The signals obtained by each set of 3 sensors grouped together can be used to perform a 2D cross-correlation analysis, extending what was presented by Sutherland and Rabault (2016). The normalized cross-spectral density is computed between adjacent sensors as:

$$C_{xy} = \frac{|S_{xy}|^2}{S_{xx} S_{yy}}, \text{ with } S_{xy}(f) = \frac{1}{2\pi} \sum_{n=-N/2}^{n=N/2} R_{yx}(n) e^{-j2\pi f n}, \tag{9}$$

where $R_{xy}(n)$ is the cross-correlation between the discrete signals $x(n)$ and $y(n)$. The phase shift of the waves between two adjacent sensors is then computed at all frequencies for which the coherence is significant with a probability higher or equal to 95%, following the formula:

$$\phi_{xy}(f) = \arctan 2\left(\frac{\mathrm{Im}[S_{yx}(f)]}{\mathrm{Re}[S_{yx}(f)]}\right) \quad (10)$$

Typical outputs from this method are presented in Figure 13.

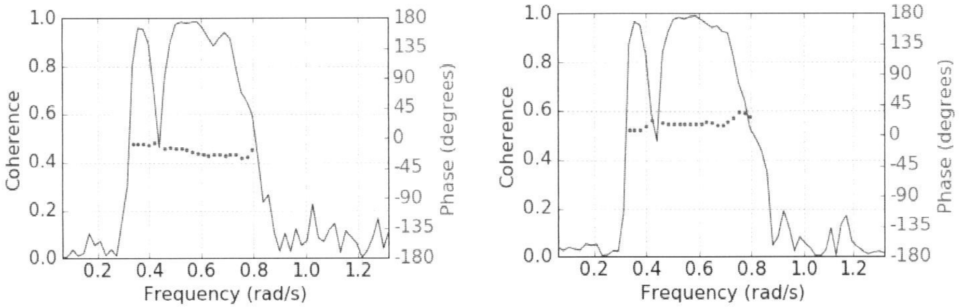

Figure 13: coherence and phase shift of the wave acceleration recorded between sensors F2 and F3 (left) and F3 and 6 (right). Coherence is computed based on a 45 minute interval, with 75 percent overlap between segments.

The phase shift between adjacent sensors is related to the incoming waves through the relation: $\phi_{ij}(f) = \vec{k}(f)\cdot\vec{x}_{if}$(Eqn. 11), where $\vec{k}(f)$ is the incoming wave vector at frequency f and $\vec{x}_{ij}$ is the vector from IMU i to IMU j. Therefore, the two dimensional wave vector can be obtained at each frequency using the phase shift from two sides of a triangle of sensors and elementary trigonometry. This gives access to both the direction of the incoming waves and the dispersion relation. An example of the results of such analysis is presented in Figure 8b and Figure 10c. Similar to what is presented in Figure 8a,b, the waves are found to follow the deep water dispersion relation and the effect of the ice is negligible. The direction of propagation of the waves is also in very good agreement with the results presented in Figure 10a.

COMPARISON WITH NUMERICAL SIMULATIONS

Wave frequency and wave direction of propagation were obtained from the web (https://earth.nullschool.net), where the results of modeling with WaveWatch III are provided. Figure 14 shows that there are two wave systems coming to Spitsbergen from South-West and from South-East. The white line in Figure 14 marks the boundary between them. Waves coming from the South-West have periods around 10 sec, and waves coming from the South-East have periods around 6.3 sec near Spitsbergen. According to WaveWatch III, the dominant wave period and direction of propagation at the point (76.93 N, 21.60 E) is shown in Figure 14 by a white circle were 8.4 s and 210° on 22:00, May 01, and 10.1 s and 205° on 02:00, May 2. Frequencies of waves in the region extended to the East from the marked location are around 5-6 sec. Spectra in Figure 7 show the dominant wave frequency to be 0.8 rad/s (wave period 7.8 sec) before 23:00, May 01, and 0.6 rad/s (wave period 10.4 s) after 23:00, May 01. Thus the waves observed propagate from the South-West along the shore of

Edgeøya. The ice map in Figure 1a shows green and yellow regions of very open drift ice which extended along Edgeøya. The waves arrive at the observation site following this route.

CONCLUSIONS

In-situ records of water and ice characteristics are performed on the drift floe in the North-West Barents Sea during wave propagation event of 16 hours duration. The ice floe of 30 cm thickness was cyclically bended by waves with an amplitude of several centimeters and was not broken. Spectral analysis has shown the existence of local spectral maxima at the frequencies 0.35 rad/s, 0.5 rad/s, 0.6 rad/s and 0.8 rad/s. Dominant waves with frequencies 0.6 rad/s and 0.8 rad/s propagated from South-West along the Edgeøya shoreline. Several measurement methods are presented and validated against each other, including direct measurement of the dispersion relation and of the direction of propagation of waves.

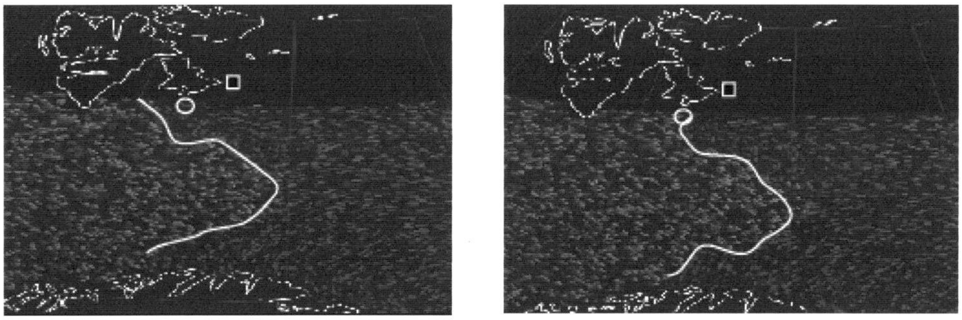

a) b)

Figure 14. Results from simulations performed by WaveWatch III. Direction of wave propagation are indicated at 22:00, May 01, (a), and 02:00, May 02 (b).

ACKNOWLEDGEMENTS

The authors wish to acknowledge the support of the Research Council of Norway through the Petromaks2 project Experiments on waves in oil and ice, and NORRUS project Field studies and modelling of sea state, drift ice, ice actions and methods of icebergs management on the Arctic shelf. The help of Masters Student Benjamin Hergoualch in deploying the IMUs is gratefully acknowledged.

REFERENCES

1. Collins, C,O., Rogers, W.A., Marchenko, A., Babanin, A.V., 2015. In situ measurements of an energetic wave event in the Arctic marginal ice zone. Geoph. Res. Letters, 42, 6, 1863-1870.
2. Liu, A. K. and E. Mollo-Christensen, 1988. Wave propagation in a solid ice pack, J. Phys. Oceanogr., 18(11), 1702-1712.

3. Marchenko, A.V., Morozov, E.G., Muzylev, S.V., 2013. Measurements of sea-ice flexural stiffness by pressure characteristics of flexural-gravity waves. Annals of Glaciology, 54(64), 51-60.
4. Marchenko, A.V., Gorbatsky, V.V., Turnbull, I.D., 2015. Characteristics of under-ice ocean currents measured during wave propagation events in the Barents Sea. POAC15-00171, Trondheim, Norway.
5. Rabault, J., Sutherland, G., Ward, B., Christensen, K.H., Halsne, T., Jensen, A., 2016. Measurements of Waves in Landfast Ice using Inertial Motion Units. IEEE Transactions on Geoscience and Remote Sensing, 54(11), 6399-6408.
6. Sutherland, G., Rabault, J., 2016. Observation of wave dispersion and attenuation in land fast ice. Journal of Geophysical Research, Oceans, 121, 3, 1984-1997.

Experimental Researching of the Specific Energy Mechanical Fracture of Ice by Method of Uniaxial Compression of Samples

Vladimir G. Tsuprik[1], Aleksandr T. Bekker[1], Egor E. Pomnikov[1], Evgeny S. Ivolgin[1]
[1] Far Eastern Federal University, Vladivostok, Russia

ABSTRACT

Long known for the method of determining the specific energy breaking the ice using falling hard body on the surface of the ice (DBT method). Earlier studies have shown that in such experiments the ice is destroyed by layers, and a specific energy mechanical fracture of ice is fairly stable. It have the expressed commitment of the temperature dependence of ice and not dependent on mass and velocity. But the fracture process in the ice array is different from the development such fracture processes in the edge of ice field limited thickness. There are also difficulties in integrating energy spent on fracturing only the amount of ice that is counted as a result of the experience. It is known that the ice sheet dissected by vertical radial cracks and horizontal cracks her cleavage on the trapeze-shaped and wedge-shaped blocks. But such blocks not separated from ice array in far from contact area, as large branches from a tree trunk. This process is fundamentally different from the testing process of samples on uniaxial compression, as well as from the testing interaction process of rigid sphere in half-space of ice. The objective of the laboratory's experiments is to improve the method of determining the specific energy of mechanical destruction of ice. In this study for testing not adopted prismatic shape of ice samples, but research trapeze-shaped forms of beams like at the edge of the ice field. In the lab are simulated the deformation and destruction of edges sea ice blocks-samples that have such special forms, like the destruction of the trapeze-shaped and wedge-shaped ice blocks on contact with the surface of the interacted structure leg into edge ice cover in nature.

KEY WORDS: Ice; Fracture; Compression test; Specific energy fracture.

INTRODUCTION

Many studies are known, where basic goal of contemporary authors in the field of research the mechanics of interaction of drifting ice formations with feet of drilling platforms is to create a mathematical model of the phenomenon that would adequately describe the process of the formation of cyclic ice load on the structure. Not a stationary process of breaking the ice on contact drifting ice field and structure can lead to dangerous vibrations and potentially dangerous dynamic loads on offshore structures. Extreme resonant oscillations of the platforms fundaments of can cause not only violations of the regular functioning of the object, but also significantly reduce the reliability of the facility and its durability, causing fatigue fracture in Foundation design elements of the platform or its equipment.

The solution of these tasks has a close link with the mechanism of breaking the ice, as a way to transfer the kinetic energy of the ice field into energy of elastic displacement of structures and energy spent on ice fracture. Therefore, on the way to the development of mathematical descriptions one or other the fracture mechanism for ice, there was a need to examine patterns of the laws of inception and development of all physical phenomena, which together constitute the process of breaking the ice. In this connection, on first plan standing the tasks of research problems of physics and mechanics of breaking the ice, as the materially medium, which creating a pressure on structure. Such a model has not yet been established, it would require to take into account the phenomenological features of destruction of sea ice is as a mechanism for converting the kinetic energy of the ice field into energy elastic deviations structures and energy spent on fracture of ice.

Sea ice is a very complex material that consists of separate various shape and size crystals, inclusive of brine areas, air impurities and mechanical fractions. The mechanical properties of ice depend upon its composition, physical state and temperature. The ice fracture is a complex, integrated process, and describe any changes occurring destruction from reaching limits stresses and deformations stretching, longitudinal and transverse shear for a compressed volume portion of the ice using power strength criterion is impossible. But you can take into account the energy costs for the development of all fracture processes specific volume of ice, using the energetical concept of fracture.

It was assumed that the critical state of stress volume of ice is a condition of the limit equilibrium the total elastic surface energy of cracking in this volume and input external energy. In the Griffith's concept the development the fracture crack for one-dimensional problem as the failure criteria is adopted threshold values the specific surface energy G_c, which "propagating a crack", the corresponding critical value velocity of release of elastic energy stored in the body. A similar approach was developed by Kheisin et al. (1966), Khrapaty & Tsuprik (1976a). They offered to take as the fracture criterion in a tensile single amount of ice in this case the criterion of ε_{cr} – as a moment of the beginning increase and merge of all types of cracks formed in this volume from of the many defects in the crystal lattice, long cells with brine, inclusions and defects on the borders of the crystals. Therefore during the investigation of sea ice strength, the integral strength properties shall be mentioned inclusive of structural strength (links between its components) and fresh ice crystal lattice strength.

Khrapaty & Tsuprik (1976a) proposed ε_{cr} as a criterion for calculation of ice loads on structures. In this time a mathematical model that describes the implementing of this process was developed by Tsuprik (1978, 2012). Determination of ε_{cr} as the fracture criteria was considered as a necessary and sufficient condition for beginning the process of fracture in a unit volume of ice. The execution of this requirement follows from the conservation equations and boundary condition (in relation to the equations of the theory of elasticity) when deciding on the ultimate energy balance of the array with the micro-cracks and defects. With this in mind, and based on the hypothesis above, the decision of model regarding this criterion, written in such form

$$\Delta U/\delta W \geq \varepsilon_{cr}, \quad (1)$$

where ΔU is changing the energy of the body due to elastic volume change on δW.

The verification of the ice fracture model with using the value of ε_{cr} by use comparisons of calculated and DBT spheres with masses from 70 to 300 kg was carried out and gave a good convergence (Khrapaty & Tsuprik, 1976b).

In the work of Afanasyev et al. (1978) in materials international symposium IAHR was published the first dependence of values ε_{cr} versus temperature of the ice received according to results of a large number of DBT experiments on the sea ice of the Amur Bay, Sea of Japan. The authors proposed to consider this value as the strength characteristic of ice for calculation of ice load. Ayer (1978) noted that ice is a very brittle material with low destruction energy and to calculate the sharp changes of load through use in calculations of the force criterion is very difficult. Therefore, Ayer (1978) concluded that the unit volume of ice in contact zone is correlated with the energy of its destruction. Sodhi & Morris (1984) held a large number of laboratory experiments on the interaction of ice with models structure in ice pool where also determine the value ε_{cr} and they also noted that "*the concept of specific energy may be useful in estimating the ice forces during impact of ice floes on structures...* ". Vershinin et al. (2006) offered his version of the calculation method of ice load from impacting the ice at the structures using "specific crushing energy of ice" based on the energy balance equation. The consumption of kinetic energy of ice floe at ice-structure interaction can be expressed as the sum of energy consumption for ice destruction in a volume of formed area (Bekker et al., 2016). In this case the energy for ice volume destruction, i.e. the specific energy of destruction ε_{cr}, shall be considered as a calculation unit of sea ice strength. Therefore, as we believe, this criterion must be defined in the integrated fracture process of ice in the array of ice field or by test of the large specimens, if use in the test such specimens the condition that identical with condition fracture of the edge facets of ice field.

The purpose of this investigation is a complex task for verification of hypothesis that the volume of elastic energy being released due to the brittle failure of ice volume ε_{cr}, is a constant value for this material. The verification of this hypothesis provides the application of energy criterion for identification of ice load calculation values, i.e. an actual ice pressure acting upon the structure. The achievement of this complex purpose is possible subject to the development of method for experimental identification of the specific index of energy consumption of ice floe against the mechanical destruction of ice due to the interaction with structure support, i.e. the specific energy of mechanical ice fracture ε_{cr}.

DEVELOPMENT OF ICE FRACTURE PROCESSES ON CONTACT WITH ICE COVER

The change of global load per time acting upon the structure is of an accidental character. Furthermore, according to the observations by different scientists, variation of contact force values depends on the following parameters of ice field penetration by the structural support: velocity of ice filed and its thickness; physical parameters of ice; dimensional parameters of support, etc. The combination of these factors has defined ice destruction process: spalling of free surface of cold ice subject to high velocity of loading due to significant variation of contact forces values; crushing of "warm" ice subject to relatively low velocity of loading due to smooth changing of load. Accounting for experimental research of ice penetration of many authors (Peyton, 1966; Afanasyev et al., 1974; Hirayama et al., 1975; Croasdale, et al., 1975, 1977; Schwarz, 1975; Saeki et al., 1976; Maattanen,1978; Khrapaty & Tsuprik, 1979; Frederking et al., 1999) we concluded that the contact force have high changeability due to the differences of mechanical ice destruction.

The study of ice destruction mechanisms in large-scale tests with natural ice in the Amur Bay (Khrapaty & Tsuprik, 1979) have shown that the edge of ice field under local loading was destructed by development of the vertical and horizontal cracks at the contact area (Fig. 1). Such picture of the edge destruction were studied by many authors in that years, e.g. Schwarz (1970,1975), Croasdale (1970, 1975), Tryde P. (1974), Hyrayama et al. (1975), Kivisild H.R. and Iyer S.H. (1976), Saeki et al. (1976, 1977), Nevel et al., (1977), Michel and Toussaint (1977), Kry (1980), Taylor (1981), Ojima et al. (1985), Yamashita M. et al. (1985) and others.

All researchers have noted the same picture of the destruction, depending on the thickness of the ice, temperature and velocity of the load. Researchers noted that in addition to the development of the initial radial cracks and full crushing ice throughout its height, in front of the structural support is formed by a limited volume of the heavily compressed ice in the form of a prism, inside which the ice crushed. In addition, in front of the surface of structural support occurs: the crushing and crumbling of ice; the crumbly products of ice extrude the outside; spalling of wedge-shaped wreckages; shifting all products of ice fracture of towards the upper and lower surfaces of the ice cover; the periodic repeat all these processes.

Later studies of Palmer et al. (1983), Sodhi & Morris (1984), Sanderson (1988), Joensuu et al. (1988), Daley (1991), Muhonen (1992), Kujala (1994), Gagnon (1994), Kärnä et al. (1999) and others show that the primary role in breaking down the edges of the ice cover in front of the structural support play a crack of different nature. Further investigations by different scientists were dedicated to the mechanics of destruction of the central block by the structural support (Fig. 1), after the initial fracture of upper and lower surface of the field limited by two radial vertical cracks in front of the support with central angle . Several scientists have focused their research projects on theoretical and experimental investigation of block destruction mechanisms, inclusive of Gagnon et al. (2006), Kärnä (2006), Jordaan and Frederking (2006), Vershinin & Truskov (2010), Taylor & Jordaan (2011), Karulin et al. (2014) and other researchers.

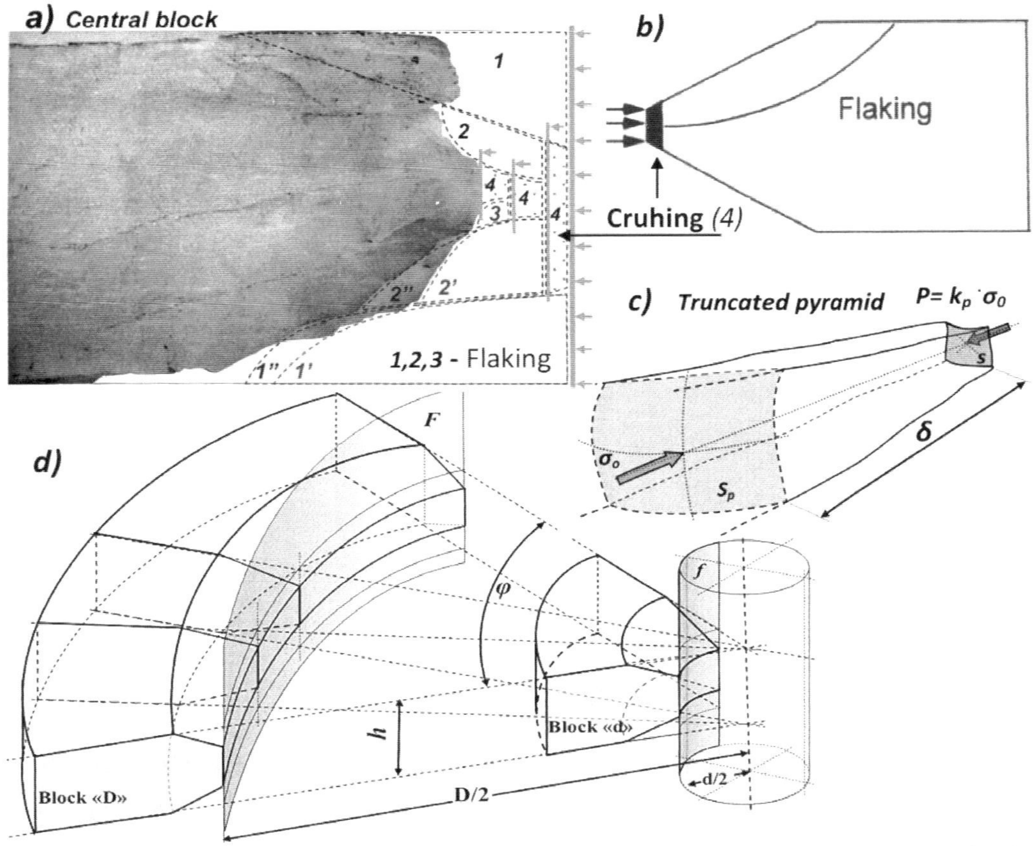

Figure 1. The theoretical and experimental investigation of ice field edge fracture mechanisms: a) - the picture of natural ice edge fracture as per Khrapaty & Tsuprik, (1979), b) - the failure process central V-area of field as per Kujala (1994); c) - scheme of destructed element within Tsuprik, (2012); d) - the hypothesis of field destruction by cracks in front of large (***F***) and small (***f***) supports and the form of central ice block

In compliance with Fig. 3, the central block of edge area of ice field is divided by the system of radial and horizontal cracks into separate trapezoidal and V-shaped elements due to the interaction with hard surface of the support. It can be supposed that the global ice load upon the structural elements consists of local loads providing the destruction of separate elements of the central block. This destruction process was studied in laboratory conditions by Joensuu & Riska (1988), Gagnon (1994) (Fig. 2) and results are close to ones of Khrapaty & Tsuprik (1979) (Fig. 1).

Large-scale and significant investigations of ice block destruction processes featured with V-shaped central block of the ice field were performed by Habib et al., 2014). Fig. 3, a, b show the results of series of experiments of specific energy consumptions. But the authors of this work had the other goal in it study. Fig. 3, c, d show the results of Storheim et al. (2015) of impact ice-ship and ice-structure. Unfortunately, there is no measurement data on ice destruction volume W_{cr} due to the contact (Fig. 3, a, c). Therefore, it is not possible to correlate the energy consumption on to destructed of ice volume U_{cr} (Fig. 3, b, d) for calculation the specific energy mechanical fracture of ice εcr. Besides, the external holder provides the high level of compression that increases the ice floe resistance level against cracks formation. It is obvious that most energy is used for compression of ice subject to closed volume. In reality the volume of destruction can be much higher.

As a result of investigations specified herein, it can be concluded that there is a large theoretical experience and high number of different experiments. Subject to the complex investigation, it provides the basis for distinct and clear sequence of ice brittle destruction process development due to the contact with hard support. So, it is important to describe the typical development of ice destruction per time.

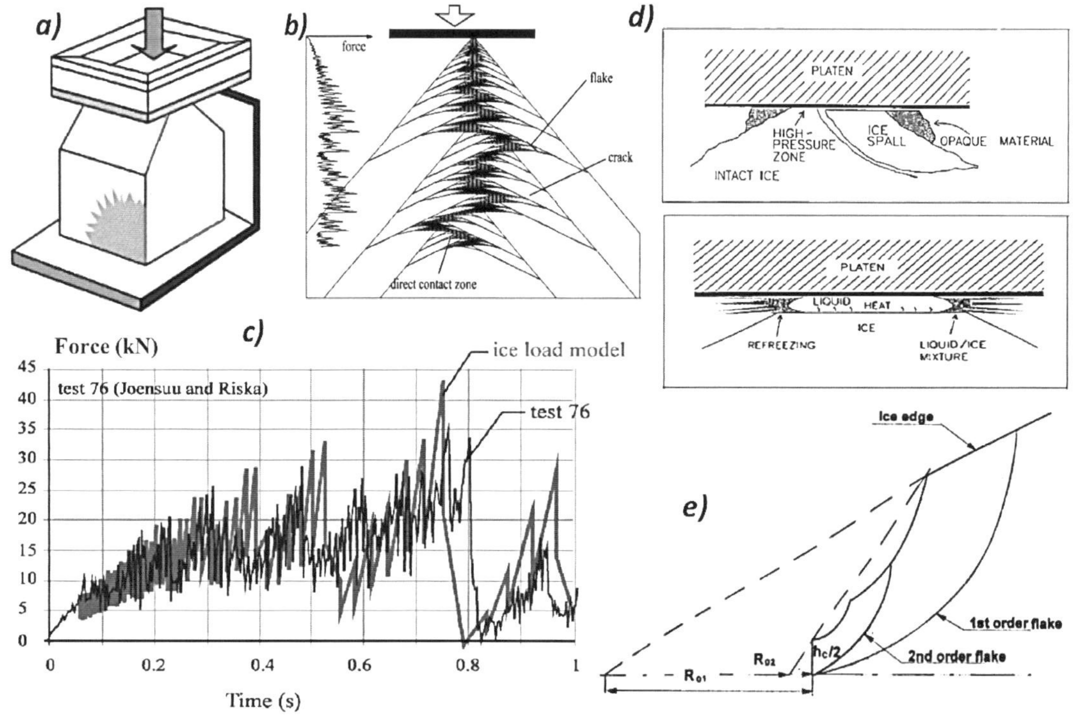

Figure 2. Failure of central block of ice sheet: a) load records (Joensuu et al., 1988); b) model of unsymmetrical fracture (Daley, 1991); c) comparison of contact force of model Daley versus "test 76" of Joensuu et al., 1988); d) destruction of cone (Gagnon, 1994); e) calculation scheme of ice volumes (Kujala, 1994)

ENERGY TRANSFER MECHANISM FROM ICE FIELD TO STRUCTURE

This section of article provides the integrated description of brittle destruction of ice within the central V- shape area of ice floe. Several details herein are especially emphasized as this description is intended for verification of experimental identification method selection related to the numerical parameters of specific energy of ice destruction, i.e. these details are of a high importance. It is wildly known that the process of ice destruction due to the contact of ice and support surface has cycled character. Each cycle of load variation acting upon the structure consists of repeated stages with complex process. Let's consider one cycle of ice destruction process consisting of three basic stages and different independent continuous processes.

The first stage of typical cycle is featured with accumulation of elastic strain within the ice field. The exceeding of definite tension limit at the contact area provides the initial fracture of the upper and lower surface of field resulting in vertical radial cracks. At this an active zone of compaction can be defined within the ice massive (central V- shaped area). As a result of several contact processes the kinetic energy of ice field is transferred into the elastic energy of ice massive. The size of active area depends as follows: support diameter; thickness of ice field and its velocity; ice temperature and its structure.

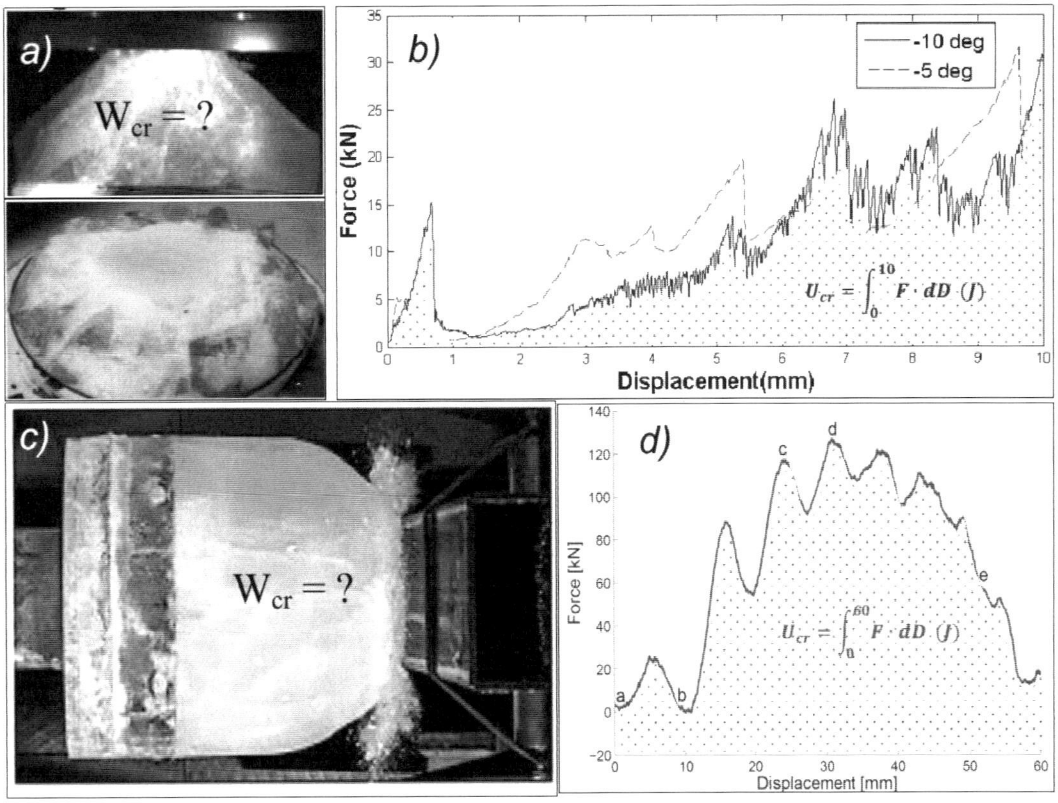

Figure 3. The cone ice sample destruction and the result of destruction:
a, b) pictures and force-deformation curve as per Habib et al. (2014) for low velocity (0.1 mm/s);
c, d) pictures and force-deformation curve (Storheim et al., 2015) for impact (8.0 m/s)

Then the horizontal crack appears at the centre of strained volume of ice field. This crack is developed in both sides. As a result, two plates act with the same scheme of tension accumulation and get fractured into smaller pieces, i.e. field is fractured into secondary pieces. Some previously formed cracks continue developing into ice massive due to energy transfer to its upper sections; the other cracks get stopped due to gaps or other cracks; several new cracks are formed at the points of maximum tension concentration within the crystals border or foreign inclusions. Considering the volume feature of this event, it shall be noted that cracks are developed within all radial areas. The volume of non-destructed ice is located between cracks. These ice blocks are featured with extended shape due to compacted volume of ice destruction products subject to the previous cycle of destruction. Therefore, V- shaped and trapezoidal ice blocks are formed within the local contact zone of ice field. These blocks are extended into ice massive by the ends located against the contact surface (Fig. 1).

The second stage of destruction is featured with strained volumes of contact zone of large ice blocks, appeared from ice massive. The complex process is developed with these blocks: a part of ice kinetic energy is accumulated in the form of elastic energy providing the significant change due to small side and internal fractures inclusive of multiple radial cracks within active areas of single large ice blocks. The "layers of initial fracture" will be accidentally destructed along different blocks forming the edge of ice field, crashing the most rigid areas of ice (high pressure zones – HPZ, Fig. 2, d). At the same time, ice destruction products (small blocks, crystals and ice fractures) are pressed out of the large blocks contact areas. At this, accidental new rigid areas (HPZ) appear. The contact force is higher in comparison with the other sections of contact area. These sections are destructed and replaced with new sections that are subject to further destruction.

Intermediate processes of layer-by-layer destruction are featured with accumulation of elastic energy by the trapezoidal blocks due to increase of their contact areas. The diagram of contact pressure is equalized. The area of block compaction appears subject to standard cracking that subdivides the block into two pieces. Certain energy is released via enlarged master cracks forming continuously extended large blocks.

The third stage is an avalanched type development of large side fractures and significant decrease of blocks cross section at the contact area. Increase of contact pressure provides development of layer-by-layer destruction of edge contact areas. The process of field edge destruction will continue till the moment when the velocity of energy transfer to ice massive via destruction surface is less than the total velocity of energy dissipation within the ice field due to formation of multiple cracks, crashing of blocks, etc. The process of cycle destruction of ice and load transfer upon structure will continue until the kinetic energy is less than energy for destruction of ice. This logical description of sea ice destruction subject to penetration with solid body is based on the results of theoretical and experimental studies, considered in this article (Fig. 1-3). This conceptual description of ice destruction mechanism was considered as a basis for development of method for identification of specific energy of fractured ice due to dynamic loading of **large samples** using test equipment with constant velocity of loading plate Displacement.

SPECIFIC ENERGY MECHANICAL FRACTURE OF ICE AND COMPRESSION OF LARGE SAMPLES OF NON-PRISMATIC SHAPE

In this work, the new non-tested method of ε_{cr} value investigation was applied. The method of trapezoidal or V- shape ice sample loading is featured with dynamic compression by testing machine plate that moves with actual velocity of the ice field. This method of physical modelling considers both physical and phenomenological particulars of ice - inclusive of all process parameters due to destruction and solid ice surface impact. The prototype of this method can be considered a similar method that was applied in Sodhi & Morris (1984) to identify specific energy fracture of the modeled ice in the ice basin.

The purpose of this suggested method is its possible application for identification of specific energy of ice destruction ε_{cr} under the laboratory conditions similar to ice field central section destruction conditions. This purpose can be achieved by compliance of dimensional and kinematic similarity of ice destruction process under natural and laboratory conditions. It can be obtained via establishment of the dynamic multi-cycled loading of trapezoidal shape ice block sample formed under the natural conditions between vertical and horizontal cracks at the middle of ice field (see Fig. 1). Another requirement was the approximation of sample size to ice block size under field conditions.

In order to obtain the kinematic conformity of testing conditions against the real ice destruction conditions, velocity of press plate shall be equal to ice field Displacement velocity under the natural conditions related to investigation of ice strength parameter. Compliance of the above-mentioned experimental conditions shall provide high accuracy of obtained data, necessary for structures design under freezing seas shelf conditions.

So, the suggested method provide the definition of specific energy of mechanical ice destruction ε_{cr} as per the velocity of its loading corresponding to ice field drift velocity and related to realization of multi-cycle destruction of ice sample section, as the sum of specific integral consumption of energy for shear and compressed deformation used for displacement of ice destruction products out of the contact area and displacement of fractures

$$\boldsymbol{\varepsilon^{o}_{cr}} = \boldsymbol{U_{cr}/W_{cr} \cdot \rho} \tag{2}$$

where $\boldsymbol{\varepsilon^{o}_{cr}}$ is specific energy of mechanical destruction of ice, defined as per the dynamic destruction of samples; $\boldsymbol{\rho}$ – ice density.

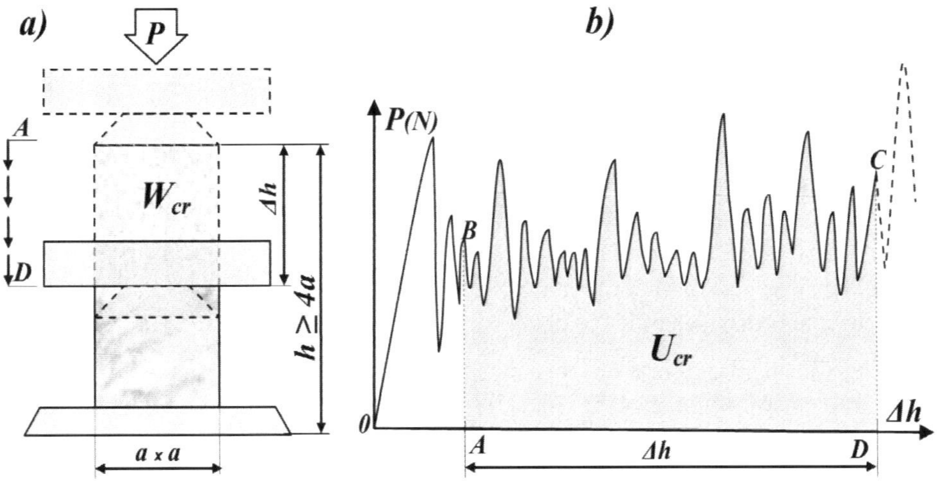

Figure 4. Calculation scheme of specific destruction energy and force-deformation diagram:
a) scheme of "high" ice sample; b) scheme for calculation specific energy required to fracture ice

Thus, the main difference of the suggested method from the standard method is the establishment of conditions for the beginning of continuous destruction of ice sample upper area. Normally it is achieved by application of samples with shape truncated pyramid or cone shape (Fig. 5). It is the first and basic condition for multi-cycle ice destruction. The area of upper basis of sample located on movable press support shall be less than area of bottom foundation area.

The second condition is preparation of relatively high sample that is essential to establish conditions for long-term process of ice destruction. There are two basic reasons for ice destruction: firstly, the most tensile condition inside sample occurs at its upper section, as its upper surface area is larger than its lower surface area. Secondly, due to significant height of the sample against its foundation, high tensile area of sample surface in contact with upper and lower surface shall provide no areas of high tensile within the central area of sample.

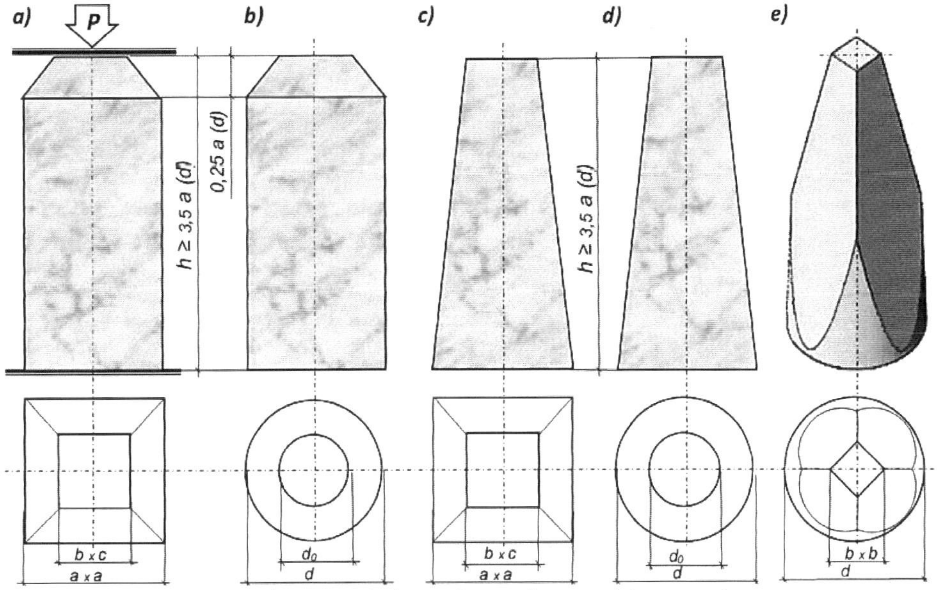

Figure 5. Shape and size of samples used in experiments

Considering these conditions subject to selection of sample size and shape, inclusive of press plate Displacement velocity shall provide the achievement of specified purpose, i.e. the dynamic multi-cycle process of ice sample loading. In this case, the specific energy of destruction becomes the *integral feature* inclusive of crack formation of ice massive, ice fracture energy at the surface of ice field and the compression energy of ice central block, inclusive of energy for displacement of ice fractures within the contact area of ice and structure.

EXPERIMENTAL RESULTS

The suggested method has been realized in compliance with the procedure of maximum ice strength identification as per the standard samples investigation method for uniaxial compression. Some samples of artificial fresh ice and nature sea ice were tested for a method adjustment. The significant preliminary results were obtained presented here, but studding all events of fracture ice process and values of parameter $\boldsymbol{\varepsilon}^{o}_{cr}$ dependences from different factors, are continuing.

1. Force-Displacement diagram of ice sample testing by new method (Fig. 6) is similar to presented earlier (Peyton, 1966; Afanasyev et al., 1971; Croasdale, 1975; Hyrayama et al., 1975; Schwarz, 1975; Maattanen, 1978; Yue et al., 1998; Frederking et al., 1999; Kärnä, et al., 2003, 2006; Jordaan et al., 2006; Karulin et al., 2014).

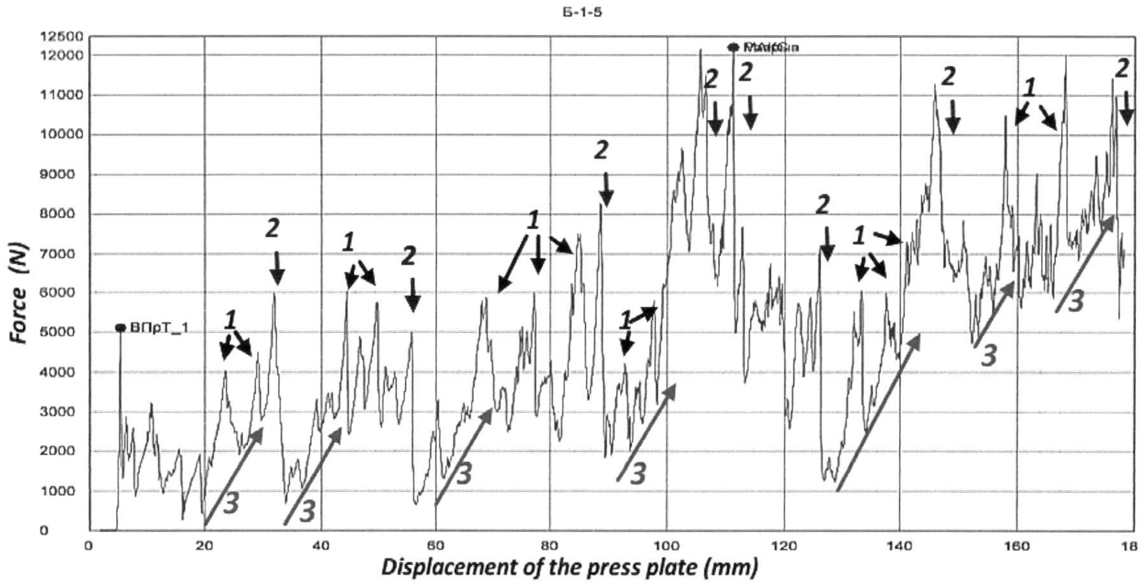

Figure 6. The Force-Displacement curve destruction of sample showed cyclical process where occurring: 1 - the fracture ice in alternating of high pressure zones (HPZ); 2 - spalls of side edges and 3 – ejection of the crushed ice layer and the shards of ice out

2. The process of continuous destruction of sample upper surface under the compression show accidentally repeated deformations of side sections at the contact area of sample and surface of moving plate, i.e. the middle area of its working contact is compressed and ice destruction products are moved out of the contact area (Fig. 7). The process of destruction is accompanied with expansion of side surface and occurrence of pulsating sizzle sound. The wreckages from spalls of ice have curve surface (Fig. 7).

3. As the moving press plate approaches to the lower immovable plate to the distance equal to vertical size of cubic sample, the remained section of sample is crushed with one of the standard types of cubic ice sample destruction: longitudinal vertical cracks 1-2 or displacement of side cones.

According to testing results of some samples with different temperature (no less than -5^{0}C however) subject to change of loading velocity, the authors have concluded that the experimental purpose was achieved: the observed destruction of sample is similar to real parameters of ice field and structural support interaction due to destruction type and parameters of this event. The obtained experimental curves of Force-Deformation (Fig. 6) was processed in order to obtain the total value work of elastic deformations of ice that spent on destruction of ice on fixed sample height as per selected method.

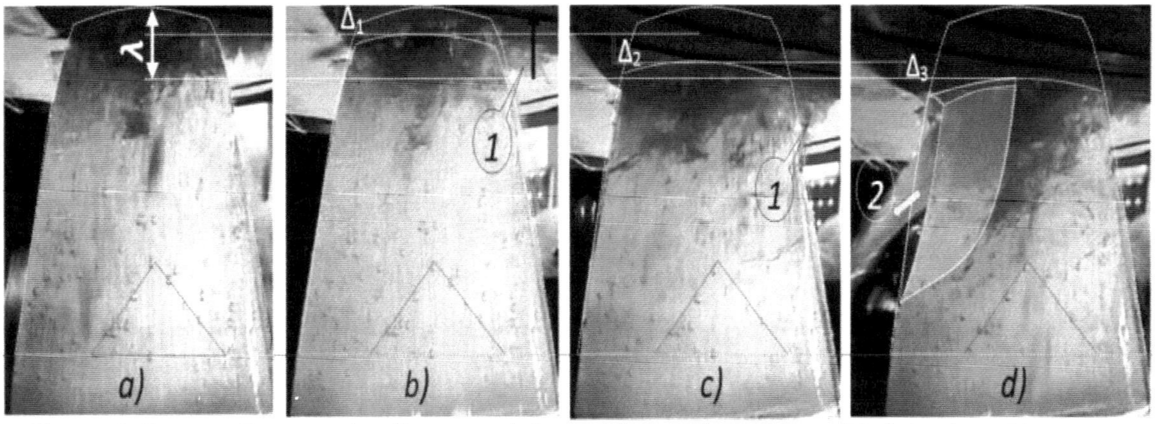

Figure 7. Stages of process ice fracture: a) formed layer of crushing with λ depth; a, b) process of compression of crushed ice and displacement products of ice fracture (1) out of the contact area; c, d) rigid contact and spalling (2) of side edges of ice block

This work is equal to released energy of elastic deformation of ice, U_{cr} per the fractured volume W_{cr}. It was calculated via the height of destructed area Δh of sample using trapezium method. Dependency of the specific energy of mechanical fracture of ice has been obtained. This dependency has been obtained during ice samples testing under the dynamic multi-cycle load from temperature (Fig. 8), but not dependence from velocity of press plate and from time testing of specimen (Fig. 9, a). When increasing the velocity of interaction Indenter with ice, as is well known, the contact force decreases. This phenomenon can be seen and for crushing process of samples in the graph in Fig. 9, b.

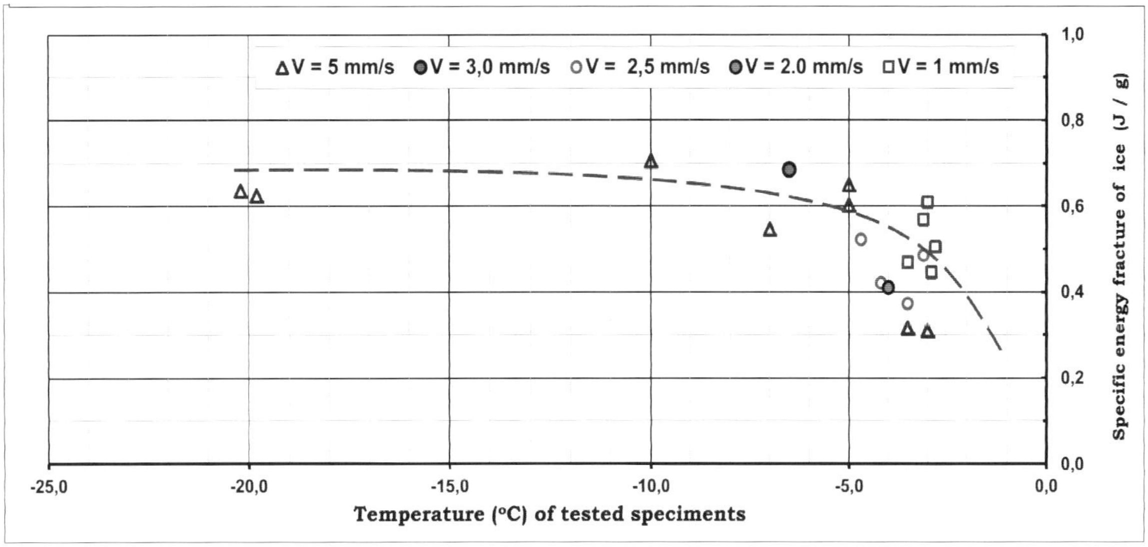

Figure 8. Dependency of specific energy of mechanical fracture of ice ε^{o}_{cr} from temperature

The obtained dependency ε^{o}_{cr} from temperature is identical by the nature of the curve with such dependency for this magnitude obtained by DBT. However its numerical values are less than obtained data related to drop ball test more than 10 times (Tsuprik, 2012, 2013). It is said that in the DBT method virtually impossible to properly account for all energy loss, which is not spent on local failure of ice. Given the importance of this physical magnitude as the energetical of ice strength criterion this study gives justification for spending in further very careful review these two methods.

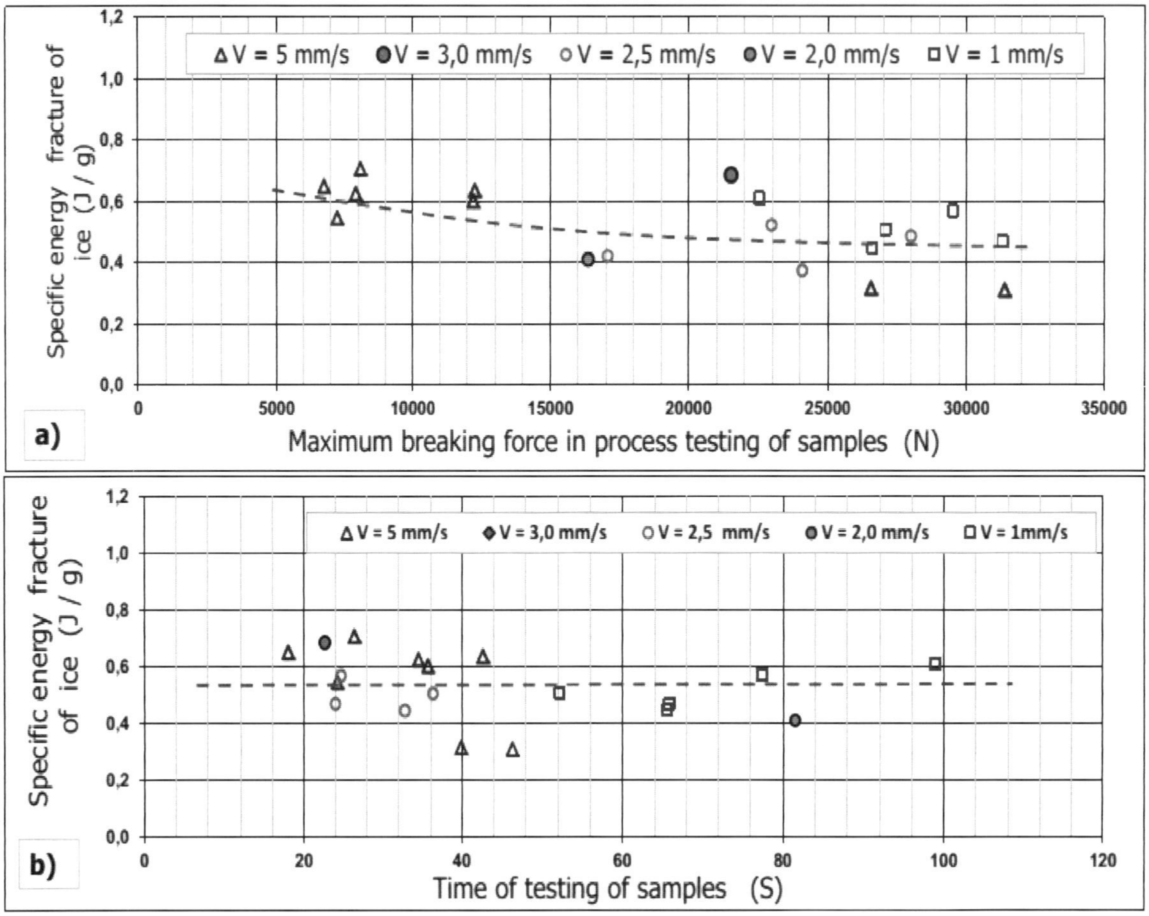

Figure 9. Dependency of specific energy of mechanical fracture of ice $\boldsymbol{\varepsilon}_{cr}^{o}$:
a) - from contact force; b) - from speed of press plate and from testing time of specimens

CONCLUSIONS

The objective of this laboratory experiments is the approbation of the testing method uniaxial compression of high ice samples to determine the specific energy of mechanical fracture of ice. This new method of determining the specific energy of mechanical destruction of ice considered as an alternative method of Drop Test Boll (DBT). In this study adopted a samples form of trapezoidal and wedge-shaped beams such like as the fracture edges of blocks in contact zone with the surface of penetration into the ice cover of the leg of structure as in nature.

During testing of samples by of dynamic loading conditions were obtained different mechanisms breaking the ice depending on its temperature and speed the movement of plates press. The increase of the speed of interaction plate with ice leads to decreases of contact force, this phenomenon corresponds the processes that occur naturally in interoperating ice with real constructions. This results are fully supported by the results of studies by other authors.

The performed investigation of energy criterion of ice destruction provides the simplicity of its experimental identification, independence from experiment conditions, stability of values and flexibility. The specific energy of mechanical fracture of ice $\boldsymbol{\varepsilon}_{cr}^{o}$ does not depend from speed of moving press plates and from testing time of sample.

These all facts provide the possibility of method development for the calculation of ice load acting upon the structure on the basis of application of the specific energy of mechanical fracture of ice $\boldsymbol{\varepsilon}_{cr}^{o}$ as an alternative to the existing method.

REFERENCES

Afanasyev, V., Dolgopolov, Yu.V., Shmeleva, L.A., Khrapaty, N.G., Tsuprik, V.G., 1978. Effect of ice impact forces on the support of structures in estuaries and shelf zones, *Proceedings IAHR Symposium on Ice Problems*, pp. 453-464.

Afanasyev, V.P., Dolgopolov, Yu.V., Shvainshtein, Z.I., 1971. *Pressure ice on marine separately-standing legs*. Gidrometeoizdat: Leningrad.

Ayer, S.H., 1978. Existing ice codes and suggested criteria, *Proceedings IAHR Symposium on Ice Problems,* pp. 363-375.

Bekker, A.T. & Tsuprik, V.G., 2016. Energy concept for determining ice strength in calculation of Ice load on vertical offshore structures, *Proceedings International Ocean and Polar Engineering conference,* pp. 1282-1289.

Croasdale, K.E., 1970. The Nutcracker ice strength tester and its operation in the Beaufort sea, *Proceedings IAHR Symposium on Ice Problems,* pp. 4-6.

Croasdale, K.R., Morgenstern, N.R., Nuttal, J.B., 1977. Medium-scale indentation tests on sea ice at various speeds. *J Glaciology,* 19(81), pp. 309-12.

Daley, C.G., 1991. *Ice Edge Contact - A Brittle Failure Process Model*. Ph.D. Helsinki: Finnish Academy of Technology.

Frederking, R., Timco, G.W., Wright, B., 1999. Ice Pressure Distributions from First-Year Sea Ice Features Interacting with the Molikpaq in the Beaufort Sea, *Proceedings International Ocean and Polar Engineering conference*, pp. 541-548.

Gagnon, R.E. & Daley, C., 2006. Apparatus design evolution and supporting experiments for a novel technique to study ice crushing, pp. 289-298.

Gagnon, R.E., 1994. Melt-layer thickness measurements during crushing experiments on fresh-water ice. *J Glaciology*, 40(134), pp. 119-124.

Habib, K.B., Taylor, R.S., Jordaan, I., Bruneau, S., 2014. Experimental Investigation Of Compressive Failure Of Truncated Conical Ice Specimens, *Proceedings Offshore Mechanics and Arctic Engineering conference*, pp. 1-10.

Hyrayama, K., Schwarz, J., Wu, H.C., 1975. Ice forces of vertical indentation and penetration. *Proceedings IAHR Symposium on Ice Problems*, pp. 442-445.

Joensuu, A., Riska, K., 1988. *Contact Between Ice and Structure.* Rept M-88, Otaniemi: Helsinki University of Technology.

Jordaan, I.J., Frederking, R., Li, C., 2006. Mechanics of ice compressive failure, probabilistic averaging and design load, *Proceedings IAHR Symposium on Ice Problems*, pp. 223-230.

Kärnä, T. & Järvinen, E., 1999. Symmetric and asymmetric flaking presses, *Proceedings International conference on port and ocean engineering under Arctic conditions,* pp. 988-1000.

Kärnä, T., Kolari, K., Jochmann, P. et al., 2003. Tests On Dynamic Ice-Structure Interaction., *Proceedings Offshore Mechanics and Arctic Engineering conference*, pp. 233-239.

Kärnä, T., Qu, Y., Yue, Q.J., 2006. An extreme value analysis of local pressures, *Proceedings International conference exhibition on performance of ships and structures in ice*, pp. 230-236.

Karulin, E., Marchenko, A., Karulina, M. et al., 2014. Field Indentation Tests of Vertical Semi-Cylinder on First-Year Ice, *Proceedings IAHR Symposium on Ice Problems*, pp. 179-186.

Kheisin, D.,E., 1966. About determine of contact pressure active in the strike zone of the vessel on the ice. *Problem of Arctic Region and Antarctic*, 22, pp. 99-102.

Khrapaty, N.G. & Tsuprik, V.G., 1976a. Experimental study of impact the solid on to ice, *Proceedings Hydroengineering Meetings,* Leningrad, pp. 166-169.

Khrapaty, N.G. & Tsuprik, V.G., 1976b. To the issue of calculating infinite plates on elastic Foundation with consideration of contact phenomena. *Proceedings of Far Eastern State University: Hydrotechnic Engineering Structures"*, Vladivostok, pp. 41-54.

Khrapaty, N.G., Tsuprik, V.G., 1979. The Semi-Natural Experiments to Study the Fracture of Ice Floe Edges in the Area of Contact with the Cylindrical Leg of Facilities. *Proceedings Hydroengineering Meetings,* Leningrad, pp. 101-104.

Kivisild, H.R. & Iyer S.H., 1976. In city test for ice strength measurements. *Ocean Engineering*, 3, 5, pp. 329-342.

Kry, P.R., 1980. Ice forces on wide structures. *Canadian Geotechnical J*, 17, pp. 97-13.

Kujala, P., 1994. Modelling of the ice-edge failure process with curved failure surfaces. *Annals Glaciology,* 19, pp. 158-164.

Määttänen, M., 1978. Ice-force measurements at the Gulf of Bothniaby the Instrumented Keti lighthouse, *Proceedings International conference port and ocean engineering in Arctic conditions,* pp. 730-740.

Michel, B., Toussaint, N., 1977. Mechanics and theory of indentation of ice plates. *J Glaciology*, 19(8), pp. 285–300.

Muhonen A., Kärnä T., Eranti E. et al., 1992. *Laboratory indentation tests with thick freshwater ice.* Espoo: VTT Research Notes, 1, 1370.

Nevel, D.E., Perham, R.E., Hogue, G.B., 1977. *Ice forces on vertical piles.* Hanover: USA CRREL Rept.

Ojima, T., Matsushima, Y., Yamashita, S., 1985. Some considerations on the designing of Arctic structures, *Proceedings Offshore Mechanics and Arctic Engineering conference*, pp. 128-134.

Palmer, A. C., Goodman, D. J., Ashby, M.F. et al., 1983. Fracture and its role in determining ice forces on offshore structures. *Annals of Glaciology,* 4, pp. 216-221.

Peyton, H.R., 1966. *Sea ice strength.* Rept NNR 307-247, Alaska: Geophys Inst. Univ.

Saeki, H., Hamanaka, K., Ozaki, A., 1977. Experimental on ice forces on pile, *Proceedings International Ocean and Polar Engineering conference,* 2, pp. 195-706.

Saeki, H., Saito, S., Hamanaka, K., Ozaki, A., 1976. Experimental study on the compressive strength of sea ice and the ice forces on a circular pile. *Coastal Engineering, Tokyo*, 19, pp. 73-87.

Sanderson, T.J.O., 1988. *Ice Mechanics*. London: Graham and Trotman.

Sodhi, D.S. and Morris, C.E., 1984. *Ice Forces on Rigid, Vertical, Cylindrical Structure*. Hanover: USA CRREL Rept. 84-33. p. 47.

Schwarz J., 1970. The pressure of floating ice fields on piles, *Proceedings IAHR Symposium on Ice Problems*, pp. 12-23.

Schwarz, J., 1975. Action of floating ice fields on piles, *Proceedings IAHR Symposium on Ice Problems*, pp. 373-386.

Storheim, M., Nord, T., Kim, E. et al., 2015. Pilot study of ice-structure interaction in a pendulum accelerator, *International conference on port and ocean engineering under Arctic conditions,* pp. 1-10.

Taylor, R.S., & Jordaan, I.J., 2012. Probabilistic fracture mechanics applied to compressive ice failure, *Proceedings International Offshore and Polar Engineering Conference,* pp. 233-239.

Taylor, T.P., 1981. An experimental investigation of the crushing strength of ice, *International conference on port and ocean engineering under Arctic conditions,* pp. 332-345.

Tryde P., 1974. Ice research in Denmark, *Symposium Krig Isfragor,* Stockholm, pp. 233-239.

Tsuprik, V.G., 1978. The model of dynamic ice sheet interaction with a single pier, *Proceedings Far Eastern State University: Hydro Engineering Structures,* Vladivostok, pp. 82-89.

Tsuprik, V.G., 1980. On the influence of the diameter of the structure and the thickness of the ice for determination of ice load on the constructions of hydraulic structures for shelf. *Proceedings Far Eastern Polytechnic Institute,* pp. 84-93.

Tsuprik, V.G., 2012. Theoretical and Experimental Studies of Specific Energy of Mechanical Failure of Sea Ice, *Proceedings International Ocean and Polar Engineering conf.*, pp. 1242-1246.

Tsuprik, V.G., 2013. The method of experimental study of the energy criterion of the sea ice failure requirements to the method and to the criterion, *Proceedings International Ocean and Polar Engineering conference*, pp. 1098-1103.

Vershinin, S.A., & Truskov, P.A., 2010. *Friction and abrasion effects of ice on the Offshore Platform Structures*. Moscow: Athlete Press.

Vershinin S.A., Truskov P.A., Kuzmichev K.V. 2006. *Sakhalin Island - Offshore Platforms Structures – The Impact and Influence of Ice.* Moscow: Institute Giprostroymost, 208 p.

Yamashita, M., et al., 1985. Model test and analytical simulation on fracture mechanism of ice, *Proceedings International Ocean and Polar Engineering conference*, pp. 195-204.

Yue, Q.J. and Bi, X.J., 1998. Full-scale tests and analysis of dynamic interaction between ice sheet and conical structures, *Proceedings IAHR Symposium on Ice Problems,* pp. 233-239.

Proceedings of the 24th International Conference on
Port and Ocean Engineering under Arctic Conditions
June 11-16, 2017, Busan, Korea

Arctic LNG Carrier Structural Risk Analysis for Iceberg Collisions

R. Gagnon[1], J. Wang[1], D. Seo[1], H. Ki[2], J. Choi[2] and S. Park[2]
[1] Ocean, Coastal and River Engineering, National Research Council Canada, St. John's, NL, Canada
[2] Daewoo Shipbuilding & Marine Engineering (DSME), Seoul, Korea

ABSTRACT

Structural risk analysis for a new 172,600 m^3 Arctic LNG Carrier was carried out. Finite element software, LS-DYNA, was used for the analysis. Three different iceberg masses were used: 3,320 ton, 6,640 ton and 10,000 ton. Six impact simulations were conducted for the condition where no water was present where the vessel forward speed was ~19.5 kt and the iceberg speed was 5 kt perpendicular to the tracking line of the ship. Impacts were targeted on specific areas of the vessel's bow section. Bell-shaped icebergs, specified by DSME, and more realistic vase-shaped icebergs, developed by NRC, were used for the simulations. The maximum contact force that was measured was in the 80 - 90 MN range for the 10,000 ton iceberg for either shape. The maximum deflections of the outer and inner hulls for these cases were -263.9 mm and -29.5 mm respectively. Two simulations using the 10,000 ton NRC vase-shaped iceberg and DSME bell-shaped iceberg were conducted where water, and associated hydrodynamics, was included. For these wet-case simulations the vessel speed was ~19.5 kt and the maximum impact force was in the same approximate range as the dry-case simulations. The outer and inner hull deflections for the wet-case simulations were significantly higher than those for the dry case because the deformable hull section was less constrained and consequently more flexible than the actual case corresponding to the dry-case simulations. Ice contact areas and average pressures were determined for seven cases. All of the simulations generated sliding-load impacts. No rupturing/tearing of the outer hull was observed for any case.

KEY WORDS: Ship-iceberg collisions; Numerical simulations of ice impacts; Ice impact damage to ships; Ice impact loads and pressures.

INTRODUCTION

The objective of this study was to perform numerical simulations to evaluate the structural integrity of the new 172,600 m^3 Arctic LNG Carrier for various iceberg collision scenarios. Most of the simulations did not include water (dry case) because the results were of value to the client for comparative reasons since similar simulations had been conducted before. Two more realistic simulations were conducted where water was included (wet case). The validated ice model used for all the simulations in this study was developed by NRC. The simulations presented here involving the highly detailed ship mesh, provided by DSME, for the cases where water and hydrodynamics are included represent the most advanced simulations ever conducted of a ship sustaining damage from a collision with an iceberg. The methods used in these simulations can be applied to a wide range of scenarios.

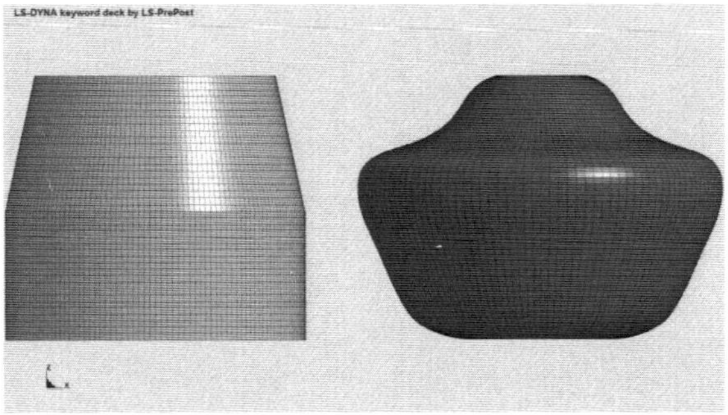

Figure 1. Iceberg shapes: (left) DSME bell-type; (right) NRC vase-type.

SIMULATION SCENARIOS

Two different-shaped icebergs were used: one had a bell shape (specified by DSME) and the other had a vase shape (generated by NRC). Figure 1. shows images of the two iceberg shapes. Three different iceberg masses (3,320 ton, 6,640 ton and 10,000 ton) were used for each shape for the dry-case simulations. Only the 10,000 ton icebergs were used for the two wet-case simulations.

Simulation scenarios are shown in Table 1. There are two contact locations of interest on the vessel, i.e. at FR. 131 and FR. 125 + 1400 mm as shown in Figure 2., where the intention is that the maximum load for any particular impact simulation occurs at either of these locations. It is important to note that the scenario Cases 1- 6 are not realistic since these scenarios could not occur under real circumstances where water is present. These simulations were conducted at the request of the client so that comparisons could be made with earlier simulation results that they had acquired.

HARDWARE, SOFTWARE AND ICE MODEL

The simulations were run on a HP Z820 Workstation that has two Intel Xeon Processors E5-2660 v2, where each has 10 cores running at 2.20 GHz (up to 3 GHz). The system has 28 GB of DDR3 RAM. 8 CPU's with SMP (single memory processing) were used for each simulation. The software used was LS-Dyna™ version Ls971d Dev, Revision 101132, double precision. It uses LS-Dyna's ALE formulation to handle fluid hydrodynamics and it has a number of contact algorithms and a large suite of material types that can be chosen for the interacting structures. ANSYS Workbench V12.1 with ANSYS DesignModeler™ was used for the modeling and generation of meshes for the study.

A crushable foam model (Gagnon and Derradji, 2006; Gagnon and Wang, 2012) was used for the ice that made contact with the vessel during the simulations. The isotropic foam model crushes one-dimensionally with a Poisson's ratio that is essentially zero, as described by Hallquist (1998). Unloading is elastic to the tension cutoff stress. Subsequent reloading follows the unloading curve. Application of a pressure, prescribed by the user, on a block or layer of such foam that is rigidly supported from below causes it to irreversibly compress in the direction of the applied load.

Table 1. Simulation Scenarios.

Case Number	Contact Location	Iceberg Mass (metric ton)	Iceberg Speed (kt)	Iceberg Shape	Ship Speed (kt)
Case 1	FR. 131	10,000	5	Bell type	19.44
Case 2	FR. 131	10,000	5	Vase type	19.44
Case 3	FR. 125 + 1400 mm	6,640	5	Bell type	19.44
Case 4	FR. 125 + 1400 mm	6,640	5	Vase type	19.44
Case 5	FR. 131	3,320	5	Bell type	19.44
Case 6	FR. 131	3,320	5	Vase type	19.44
Case 7-ALE*	FR. 131	10,000	0	Bell type	19.44
Case 8-ALE*	FR. 131	10,000	0	Vase type	19.44

* Note that Cases 7-8 use ALE (Arbitrary Lagrangian Eulerian) methods to simulate hydrodynamic effects with the presence of water.

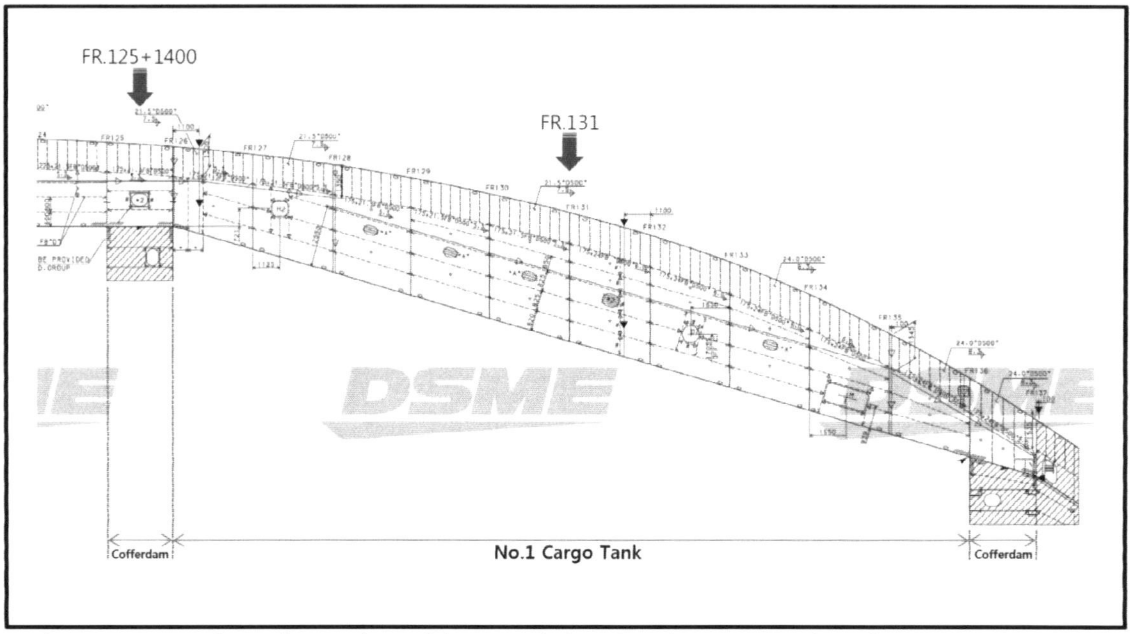

Figure 2. Top view of a portion of the vessel showing the cargo tank region. Two contact locations on the vessel hull are indicated by red arrows where, during any particular simulation, the ice impact load will reach its approximate maximum value at one or the other location.

The ice model has been validated using data from growler impact tests conducted in NRC's Ice Tank facility (Gagnon, 2004) and from full-scale measurements (Gagnon et al., 2008). Properties of the icebergs used in the simulations are given below in Table 2. As mentioned above, for the present study iceberg shape was investigated by using two shapes. The bell-shaped iceberg model was supplied by DSME and the vase-shaped iceberg model, deemed to be more realistic than the bell-shaped one, was generated by NRC. The vase shape is very roughly based on observed shapes of icebergs where underwater protrusions are common and the protrusions are generally rounded and smooth due to melting.

SHIP MODEL

The ship model used for the simulations was supplied by DSME. The full meshed model is shown in Figure 3. The portion of the cargo section of the vessel where the simulated ice impacts occurred is shown in Figure 4. The double-hull design structure of the vessel is evident.

ICEBERG SHAPE EFFECT ON LOCAL PRESSURE, GLOBAL LOAD AND DAMAGE

Table 2. Iceberg Properties.

Density	870 kg/m^3
Young's modulus	9.0 GPa (for deformable portion)
Poisson's ratio	0.003 (for deformable portion)
Element type	*SECTION_SOLID (brick)
Typical element dimension in contact region	0.48 m – 0.50 m (DSME Iceberg) 0.50 m – 0.78 m (NRC Iceberg)
Material properties	*MAT_CRUSHABLE_FOAM (for the deformable portion)
Number of elements for Cases 1-8	DSME 10,000 ton: 88,640 DSME 6,640 ton: 66,700 DSME 3,320 ton: 32,880 NRC 10,000 ton: 186,714 NRC 6,640 ton: 132,120 NRC 3,320 ton: 70,128

Generally speaking, in the results below, the more rounded and protrusive aspect of the vase-type iceberg leads to more concentrated forces in the contact region that result in higher pressures and ultimately more local damage (plastic strain) than in the case of the bell-type iceberg. A related consequence of this is that the peak loads will be somewhat less for the vase-type iceberg impact because more energy is absorbed by the vessel as plastic deformation, as opposed to elastic deformation, than the case of the bell-type iceberg. One can appreciate this issue more clearly by imagining the extremely unlikely scenario of an iceberg shape virtually matching the shape of an extensive area of the hull of a vessel during a collision and see that the load would be distributed over a very large area leading to low local pressures and very little or no plastic damage on the one hand and a high resultant force on the other hand because the vessel response would be fully elastic. Similarly one would expect lower pressures and less damage to the ice itself when the iceberg shape conforms to the shape of the vessel in the contact region.

RESULTANT IMPACT FORCES AND HULL DEFLECTIONS (DRY-CASE)

In order to run a simulation where the maximum load would occur at the specified frame, that is, either frame 131 or frame 125 +1400 mm, it was necessary to run a few trial simulations that involved adjusting/offsetting the initial collision contact point by a suitable amount. Figure 5. shows the resultant forces for the simulations where the offset adjustments were made. One can see

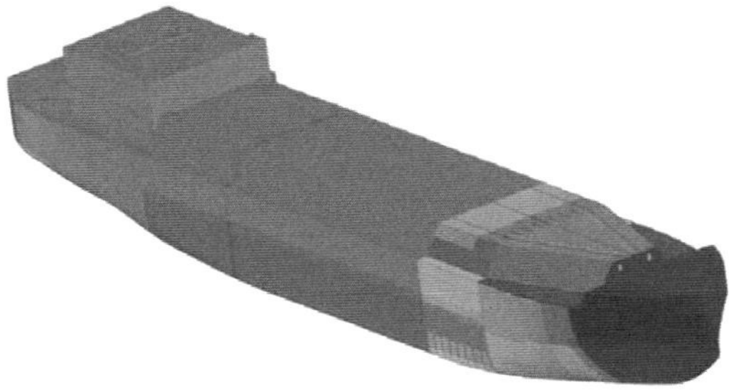

Figure 3. Full FE model of the ship. The multicoloured length segment between the red forward-bow section and the large blue main-body section had deformable properties during the dry-case simulations.

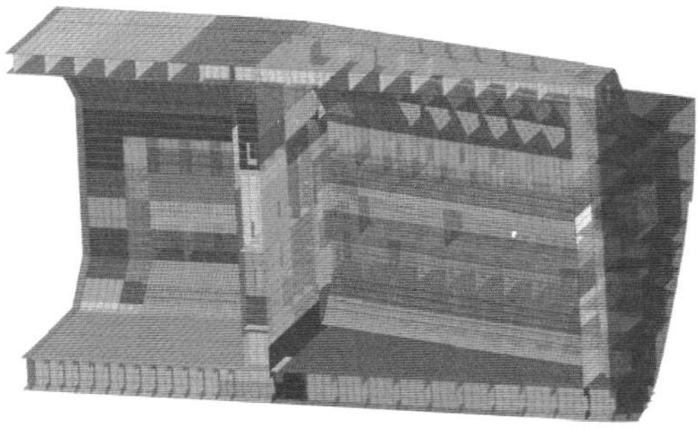

Figure 4. Sectional view of the FE ship model showing inner portions of cargo holds 1 and 2 where details of the mesh structure of the outer and inner hulls are evident. The simulated ice impacts occurred on the outer hull of this portion of the vessel.

that the peak loads for the DSME and NRC icebergs were fairly close to one another for the more massive ones, that is, the differences were about 10% for the 10,000 ton iceberg, 15% for the 6,640 ton iceberg. Since the bergs exhibited similar degrees of rotation during the impacts we might attribute the higher load exerted by the bell-type iceberg to the fact that it had less curvature (i.e. less protrusiveness) than the NRC iceberg and therefore presented a larger contact area, thereby resulting in greater loads due to the wider load distribution and less plastic damage, as previously discussed.

For the case of the 3,320 ton icebergs the DSME iceberg was subject to considerable rotation

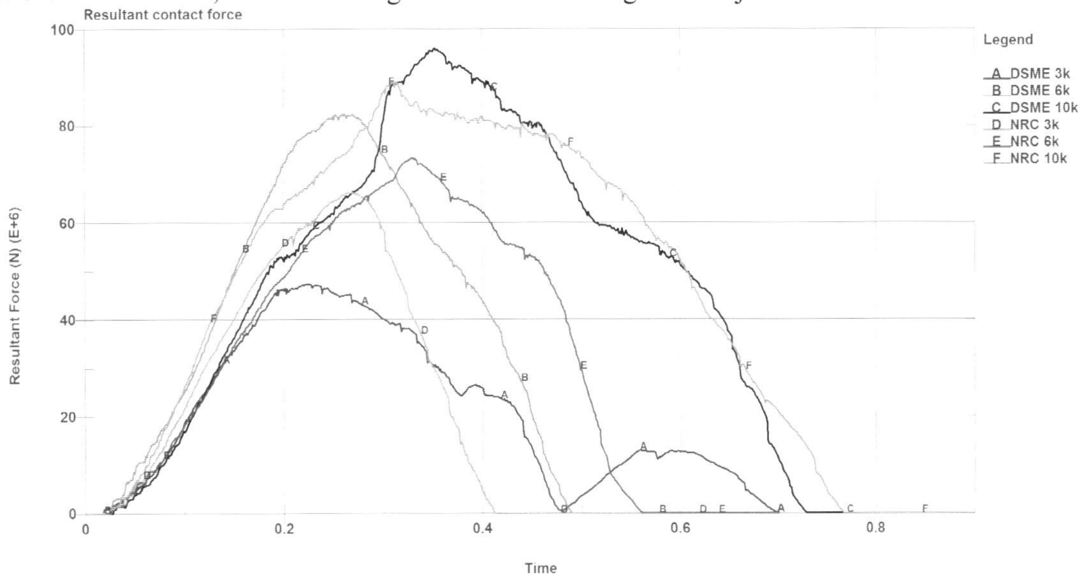

Figure 5. Resultant force time series for the impacts where load reaches its approximate maximum value at FR.131 for the 3,320 ton iceberg cases (traces A and D) and the 10,000 ton cases (traces C and F), and at FR. 125+1400 for the 6,640 ton case (traces B and E). The unit on the X axis is 'second'.

when the impact occurred (i.e. the force vector passed far from the iceberg's center of mass) because of where the vessel contacted it. The NRC iceberg experienced less rotation during the impact and consequently exerted more force on the vessel. We also note that the DSME iceberg experienced a secondary hit.

Figure 6. shows the maximum hull deflections for all the dry cases (Cases 1- 6). The higher outer hull deflections exhibited by all three NRC icebergs support the earlier conjecture that the NRC iceberg's protrusiveness leads to greater local damage than the DSME icebergs. The inner hull deflections are similarly greater for the NRC icebergs for the two lower-mass cases, however the pattern switches for the largest iceberg mass. The implication is that much more crumpling of the webbing occurs in the NRC 10,000 ton iceberg case than for the DSME 10,000 ton iceberg case.

NRC 10,000 TON ICEBERG WET-CASE (ALE) SIMULATIONS

Some of the simulations where water was included were for the NRC 10,000 iceberg. The methods used to set up and run the simulations are similar to those described by Gagnon and Wang (2012). These simulations took considerably longer to run than the dry cases (Cases 1-6, Table 1) because of the inclusion of the large number of water elements in the mesh. The full-run simulation using the NRC iceberg took 167 hr. The dry-case simulations, on the other hand, typically took approximately 23 hr to run. Furthermore, a number of preliminary simulations had to be run in order to locate the correct initial position of the iceberg, relative to the tracking line of the vessel as it moved forward, to facilitate an impact occurring near frame 131.

To continue, a series of short and simple simulations were run where the vessel was given a certain speed (19.44 kt) to determine the required initial position of the ice mass in the water in order to get a 'hit' at the desired location on the bow of the vessel. For these simulations all solid components (the vessel and the ice mass) had rigid-body designations and the contact definitions for both objects applied only to the water so that the simulations could run quickly. That is, during the relatively long real-time portion of the simulation where the ship is transiting over a distance of ~130 m, in order to generate a realistic bow wave, the simulation runs with a reasonably long time step. This procedure took about 4 simulations to perform, where each took 8 hr to run, corresponding to 13 s of simulated time.

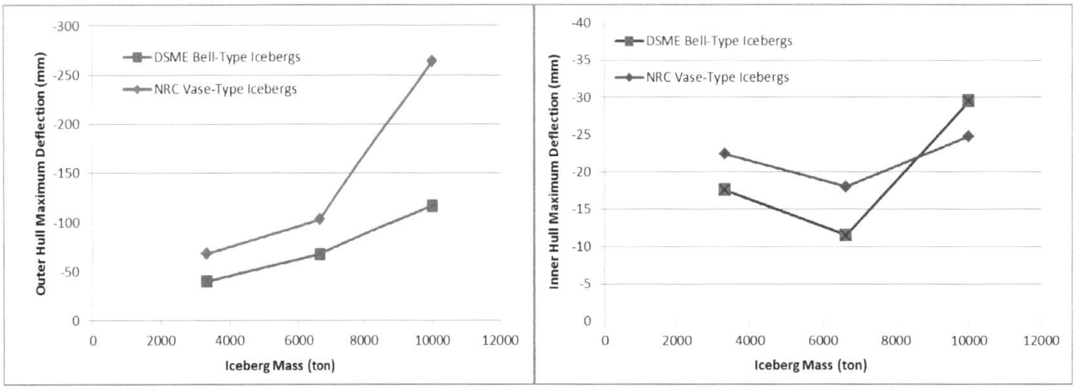

Figure 6. Maximum outer hull (left) and inner hull (right) deflections.

So far what we have described will get a wet-case simulation to the point where the vessel is at a close approach to the ice mass and the mass has begun to respond significantly to the hydrodynamics associated with the ship's bow wave. This takes about 8 hr of run time. To this point, as previously mentioned, the ice mass and vessel have been treated as rigid bodies interacting only with the water. But just before the actual collision occurs it is necessary for these components to be activated, that is, their properties need to be changed to realistic ones rather than simple rigid body ones. In LS-Dyna this can be done 'on the fly' during a simulation. This is facilitated by using the LS-Dyna command '*RIGID_DEFORMABLE_R2D' to activate the ice mass portion (i.e. the half of the mass facing the vessel) and vessel portion (the large section of the hull encompassing the impacted region) as deformable parts. This command enables one to specify the activation time, i.e. the time just before contact occurs (at time = 13 s), prior to running the simulation. Once activated, the formerly rigid components now have realistic deformable properties that they maintain throughout the rest of the simulation. Now, however, the computations are much more intensive due to the large numbers of deformable elements of portions of the ship and ice mass, where the portions interact with the water and with each other, so the simulation runs much more slowly, i.e. for the NRC iceberg case it takes ~160 hr for the remaining simulation time segment from 13 s to 14.35 s. Note that the simulations ran on only 8 CPU's. Workstations are presently available with several times this number of CPU's, and Beowulf computer clusters have even more. Consequently simulations like the present ones can be run in much shorter time periods.

Figure 7. is an image from the simulation showing the full view of the ship/ice collision scenario that includes water waves associated with the ship and iceberg motions. The time series load record for the simulated collision is shown in Figure 8. The time series deflections of the inner and outer hull are shown in Figure 9.

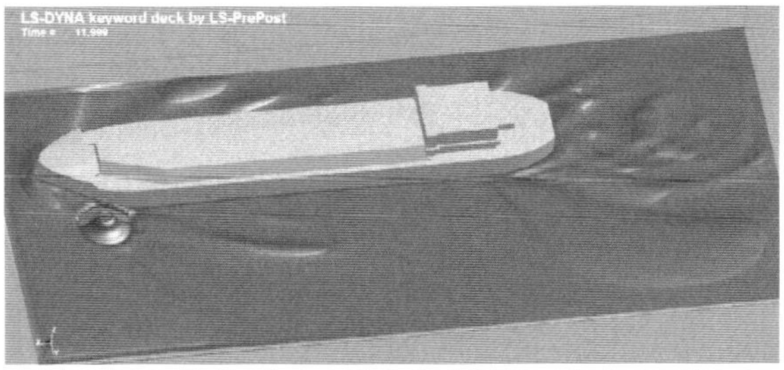

Figure 7. An image showing the full view of the ship/ice impending collision (~ 1.3 s before impact) that includes water waves associated with the ship and iceberg motions for the case of the NRC 10,000 ton iceberg. The deformable portion of the iceberg is the half (light blue in color) that faces the side of the ship. The ship speed was 19.44 kt. (Case 8, Table 1)

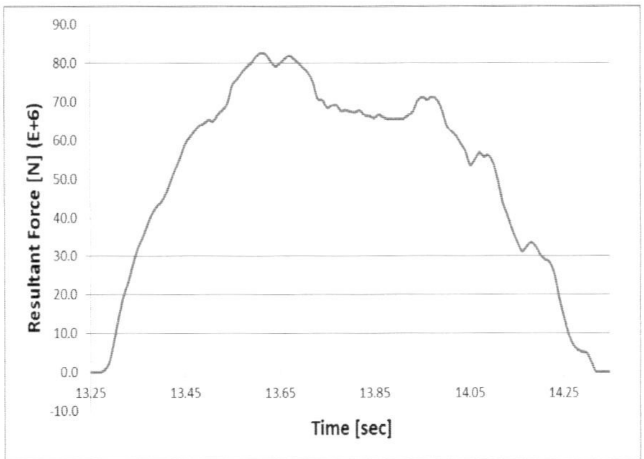

Figure 8. Resultant force time series for the NRC 10,000 ton iceberg (wet-case). (Case 8, Table 1.)

Now it is important to draw the reader's attention to certain aspects of the ALE wet simulations that signal caution in terms of reliability of the results. During the dry-case simulations (Cases 1-6) the deformable segment of the vessel, as shown in Figures 3. and 4. above, was large and included all the strengthening components in that length segment. This was somewhat intensive computationally but was handled within reasonable amounts of time for the simulations to run. At that earlier stage of the work it was understood that wet-case simulations would take a relatively long time to run if that whole section of the vessel was given deformable properties. So it was decided to use a smaller deformable section than what was used for the dry-case simulations, as shown in Figure 10. This was attached to the rest of the vessel at the fore and aft ends of the segment. That implied that the structures at the top and bottom of the segment, such as the deck, did not impart their strength to it. Consequently the segment was more flexible elastically than would normally have been the case.

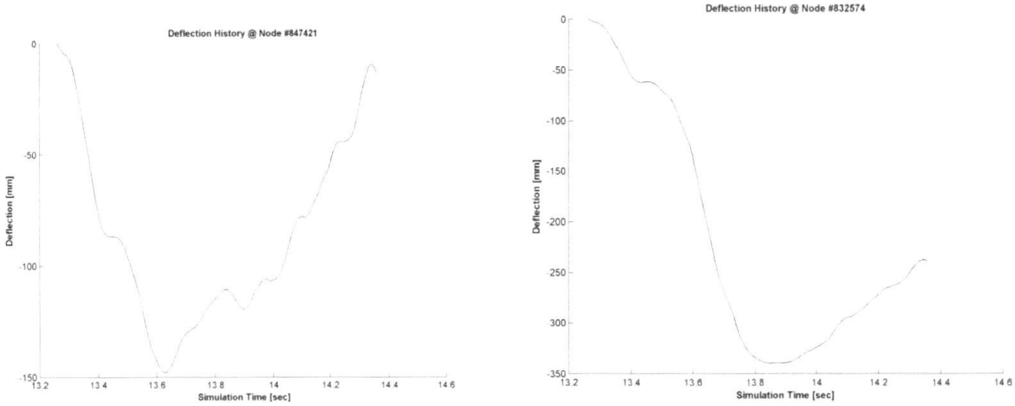

Figure 9. Deflection history for the node showing the maximum deflection of the inner hull (left) and outer hull (right) where the NRC vase-shaped 10,000 ton iceberg (wet-case) interacted with the ship during a sliding-type impact. The small bumps on the plots, most prominent at the right of the left plot, likely correspond to resonant oscillations (~ 9 Hz) of the hull. (Case 8, Table 1.)

Hence, it is our opinion that the wave patterns and iceberg motions during the wet-case simulations were not affected much by this issue. Similarly the impact load data should still be reasonably accurate, indeed they seem to be in the same range as the dry-case values. The deflection data for both the outer and inner hulls would be inaccurate, that is, the deflections would be greater than what they should be due to the extra flexibility of the affected deformable hull section, i.e. for comparison the dry-case and wet-case maximum deflections of the outer hull are -263.9 mm and -339.4 mm respectively for the NRC 10,000 ton iceberg. The maximum deflection values of the inner hull suffer a much greater relative discrepancy since those deflections are normally considerably smaller than those of the outer hull, i.e. for comparison the dry-case and wet-case maximum deflections of the inner hull are -24.8 mm and -148.1 mm respectively. The issues discussed above can fairly easily be remedied by either using the full deformable segment of the hull that was used in the dry-case simulations or by judiciously attaching the top and bottom edges of the smaller deformable section to the rest of the rigid vessel. Time ran out under the present contract to perform such alterations.

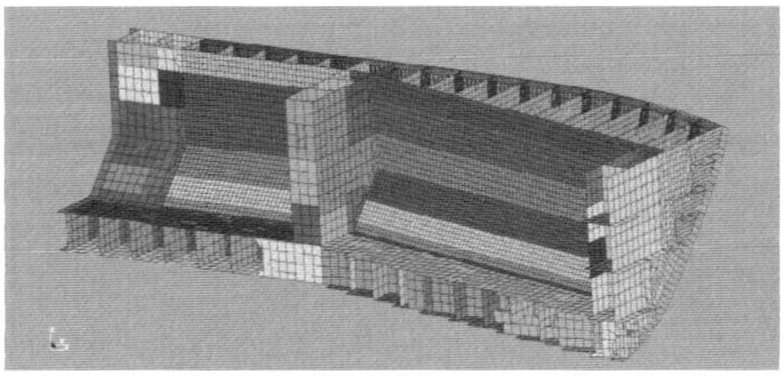

Figure 10. Hull portion that was given deformable properties for the NRC 10,000 ton iceberg (wet-case) simulation. (Case 8, Table 1.)

INTERFACE CONTACT PRESSURE AND AREA

The following discussion primarily regards the simulation results from the wet-case vase-type 10,000 ton iceberg. Figure 7. shows the whole vessel and iceberg to give the reader some perspective of the scale of the ice/ship interaction for the NRC 10,000 ton case. Figure 11. is an expanded view of the inside of the hull and shows detail of a typical interface pressure pattern during the impact. The 'qualitative' aspect of the pressure patterns is discussed below.

One thing that is noteworthy is that the pressure is highest on the plating that is backed by the frames. This characteristic is intuitively sensible and is similar to the results of an earlier study of a tanker collision with a bergy bit (Gagnon and Wang, 2012). Another aspect is that the pressure distribution in the ice contact zones is generally characterized by high pressures in the central contact regions and considerably lower pressures surrounding the high-pressure zones. Both these characteristics are visible in the expanded view shown in Figure 11. These characteristics arise because the NRC ice model was designed to reflect the pressure distributions observed when real ice is crushed at strain rates in the brittle regime. During crushing the high-pressure zones in real ice correspond to relatively intact ice whereas the low-pressure zones correspond to softer crushed ice. Details of the physics involved in ice crushing have been discussed extensively before (Gagnon, 1999).

Figure 11. Image showing the iceberg-ship 'qualitative' interface contact pressure where the NRC vase-shaped 10,000 ton iceberg interacted with the ship during a sliding-type impact in water. Similar patterns occurred for dry-case simulations. The view is from the inside of the outer hull so that structural members are visible. The scale for the colored contours of pressure is at the right of the image. Pressure is elevated in contact regions where the ice contacts hull plating that is supported by structural members on the inside of the hull. (Case 8, Table 1.)

We may also determine the contact area and average pressure. This was done by first counting the elements that experience contact at the particular instant in time during the interaction in order to get the contact area. An element was considered to be experiencing ice contact if it registered at least 0.1 MPa for the interface pressure. The average contact area was determined by dividing the resultant force by the measured contact area. The time series contact area and average contact pressure are shown in Figure 12. The contact areas and average pressures determined in this way are considered to be reasonably accurate. The same method was used by Gagnon and Wang (2012).

Here we note that while the correct patterns of pressure are qualitatively presented in images such as Figure 11, the magnitudes of the actual interface pressures are higher than what is reflected in the scale at the right because not all of the resultant force that generates the pressure patterns is accounted for in this LS-Dyna rendering. This occurred because the shell-plating elements of the outer hull share some nodes with the structural members behind the plating. One can roughly estimate to what extent the pressure scale should be upwardly adjusted by comparing the approximate perceived average pressure from any image with the actual average contact pressure in the plot below in Figure 12.

It is noteworthy that for the majority of the impact duration the average pressure tends to level off at around 9 MPa. This relative lack of dependence on contact area has been observed before in other cases of ice/structure interaction. Gagnon (2014) reported a fairly constant average pressure for a wide range of contact area for collisions of the CCGS Louis St. Laurent with sea ice. Frederking and Sudom (2008) and Gagnon (2014) reported similar results for indentation experiments at Hobson's Choice Ice Island. Furthermore, numerical simulations by Gagnon and Wang (2012) and Gagnon (2007) showed qualitatively similar trends.

In all the simulations of this study a specific strain criteria for failure had been assigned to the deformable ship elements, depending on the type of metal. For the larger icebergs, in particular, there were generally a few elements that reached the plastic strain criteria (e.g. effective plastic strain = 0.1) during a simulation so that those elements eroded.

CONCLUSIONS

Eight scenarios of iceberg collisions with the DSME 172,600 m^3 Arctic LNG Carrier have been conducted using LS-Dyna software that involved two iceberg shapes and three sizes for each shape. Six of the simulations did not include water and associated hydrodynamics. For the largest ice mass a full-run simulation was conducted for the NRC 10,000 ton iceberg where water was included using LS-Dyna's ALE formulation. For the wet-case simulations motions of the icebergs prior to contact with the vessel (data not shown here), due to hydrodynamic effects, appeared to be realistic. Ice/ship contact area and average contact pressure were determined for seven impact simulations. Average pressure appeared to be independent of contact area for significant portions of the time series records and was in the approximate range of 6-11 MPa.

Impact loads for the largest icebergs were in the ~ 80-90 MN range. Deflections of the outer and inner hulls in the impacted areas of the vessel's bow were in the ranges -116.6 mm to -263.9 mm and -24.8 mm to -29.5 mm respectively for the largest ice masses for the dry-case simulations. No ruptures of the outer hull (side shell) were observed.

The wet-case simulations are the most sophisticated ship/iceberg impact simulations ever conducted. The techniques and methodology utilized have been shown to be robust, suggesting the usefulness of such simulations for vessel design and for testing/improving of codes in general.

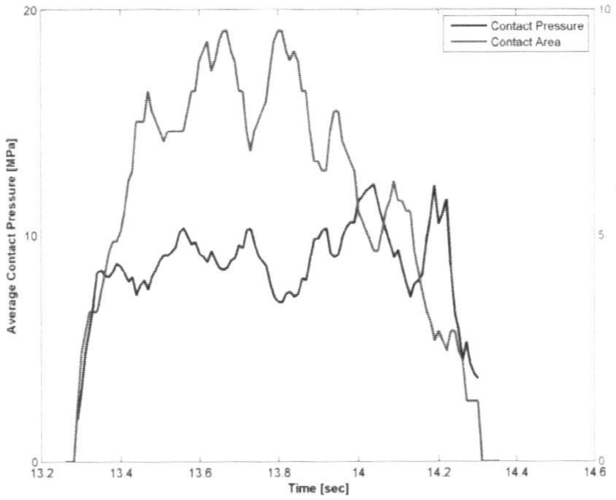

Figure 12. Average contact pressure and contact area for the case of the NRC vase-shaped 10,000 ton iceberg interacting with the ship during a sliding-type impact (Wet-Case). The average contact pressure was calculated by dividing the resultant force by the measured contact area (determined by counting contacted elements) assuming that the surface is flat, whereas the surface is usually deforming, i.e. becoming concave due to damage to the hull. Hence the actual average contact pressure is somewhat higher depending on the degree of concavity of the surface. However, the discrepancy amounts to < 5% even in the most extreme cases. (Case 8, Table 1.)

REFERENCES

Frederking, R. and Sudom, D., 2008. Local Ice Pressure Distributions during 1990 Hobson's Choice Ice Island Multi-Year Ice Indentation Tests. Proc. 19th IAHR International Symposium on Ice, Vol. 2, 815-827.

Gagnon, R.E., 1999. Consistent Observations of Ice Crushing in Laboratory Tests and Field Experiments Covering Three Orders of Magnitude in Scale. Proceedings of the 15th International Conference on Port and Ocean Engineering under Arctic Conditions, 1999, POAC-99, Helsinki, Finland, Vol. 2, 858-869.

Gagnon, R., 2004. Analysis of laboratory growler impact tests. Cold Regions Science and Technology 39, 1–17.

Gagnon, R., 2007. Results of Numerical Simulations of Growler Impact Tests. Cold Regions Science and Technology 49 (2007) 206–214.

Gagnon, R., 2014. Re-analysis of Load and Pressure Data Acquired from Ice Impacts During the CCGS Louis St-Laurent 1994 Arctic Voyage. Proceedings of the 22nd International Symposium on Ice, IAHR 2014, Singapore.

Gagnon, R.E. and Derradji-Aouat, A., 2006. First Results of Numerical Simulations of Bergy Bit Collisions with the CCGS Terry Fox Icebreaker. Proceedings of IAHR 2006, Sapporo, Japan.

Gagnon, R., Cumming, D., Ritch, A., Browne, R., Johnston, M., Frederking, R., McKenna, R., Ralph, F., 2008. Overview accompaniment for papers on bergy bit impact trials. Cold Regions Science and technology 52, 1–6.

Gagnon, R.E. and Wang, J., 2012. Numerical simulations of a tanker collision with a bergy bit incorporating hydrodynamics, a validated ice model and damage to the vessel. Cold Regions Science and Technology 81 (2012) 26–35.

Hallquist, J.O., 1998. LS-Dyna Theoretical Manual. Livermore Software Technology Corporation. May 1998.

POAC'17
Busan, KOREA

Proceedings of the 24th International Conference on
Port and Ocean Engineering under Arctic Conditions
June 11-16, 2017, Busan, Korea

Accuracy of a Non-Smooth Time Stepping Scheme with Non-Rigid Contacts for Ice-Structure Interaction

Marnix van den Berg[1], Raed Lubbad[1], Sveinung Løset[1]
[1] Sustainable Arctic Marine and Coastal Technology (SAMCoT), Centre for Research-based Innovation (CRI), Norwegian University of Science and Technology, Trondheim, Norway

ABSTRACT

Discrete element methods enable us to model the interactions between individual ice blocks and the structure of interest. This may give information on relevant processes in ice- structure or ice-ice interaction that cannot be captured by continuous methods. Discrete element methods can broadly be divided into two categories; smooth and non-smooth. While the former is known for high computational cost, the latter may significantly reduce the calculation time of discrete element simulations by enabling larger time steps. This paper investigates the applicability bounds of a novel non-smooth discrete element time stepping scheme that includes contact non-rigidity. Contact non-rigidity allows us to compute ice-structure or ice-ice contact forces based on crushing assumptions. Additionally, it enables the modelling of deformable ice features and failure. A major advantage of the applied method is that it maintains stability for a much wider range of time step sizes. This enables discrete element simulations with a much reduced calculation time compared to smooth methods. The accuracy of the applied method is investigated for two case studies. Results show that accurate results can be generated with significantly larger time steps than used in other methods.

KEY WORDS: Lattice model; Non-Smooth DEM; Ice-Structure interaction

INTRODUCTION

Full-scale ice-structure interaction is of interest in the design of structures and ships, and in the planning of ice management operations. Several numerical approaches are possible to model these interactions. A distinction can be made between continuum methods and discrete methods. The latter models the individual ice blocks as discrete bodies, while continuum methods take the influence of broken ice into account by approximating the bulk material properties of ice and ice rubble. Some phenomena can only be studied using discrete methods, such as the formation of force chains within broken ice. A downside of discrete methods is the computational demand.

Discrete element methods can broadly be divided into two categories; smooth and non-smooth. The difference between smooth and non-smooth discrete element methods can be understood as the difference between explicit and implicit time integration (Servin et al. 2014). Smooth- or explicit methods require only simple evaluations per time step, but need very small time steps to maintain stability. Non-smooth- or implicit methods, on the other hand, require the solution of a system of equations for each time step, but they maintain accuracy for larger time step sizes. The use of non-smooth discrete element methods may significantly reduce the calculation time of discrete ice-structure interaction simulations. A downside of many existing non-smooth discrete element methods is that contacts are assumed infinitely rigid. This makes the contact forces predicted by such methods unphysical for most ice structure interaction scenarios.

This paper presents a novel method that extends the 'classical' non-smooth formulation to account for contact non-rigidity. This results in a time stepping scheme that still has all advantages of the non-smooth methods, namely 1) the stability is independent of time step size, 2) it can handle rigid (infinitely stiff) contacts, and 3) it gives accurate results for larger time steps. In addition, the new method is capable of accounting for contact non-rigidity, and it predicts physically correct contact forces. With the proper choice of model parameters, the current scheme can be as accurate as smooth schemes. The time stepping scheme used in this paper differs from schemes published by others (Lacoursiere 2007; Servin et al. 2014; Krabbenhoft et al. 2012) in that it is derived based on the second-order accurate Newmark-Beta method. In contrast to other schemes, it does not introduce (artificial) numerical damping.

Although the proposed method maintains stability for a wide range of time step sizes, the accuracy of the results may be compromised before instability occurs. This paper investigates the influence of time step size, as well as mesh size on the accuracy of the results for two different interaction scenarios; the interaction between a free floating ship and a single square ice floe, and the interaction between a fixed, vertical walled cylindrical structure and a confined floe field.

The paper starts with a short description of the used model. Two case studies are presented. Thereafter, the results are described and discussed. Lastly, the results are summarized and conclusions are drawn.

MODEL DESCRIPTION

The proposed model consists of four parts:

- A *material model*, used to account for the influence of ice floe deformation on the results.
- A *contact model*, used to determine the contact parameters for in ice-ice and ice-structure interactions.
- A *time stepping scheme* used for stable and accurate time integration.
- A *hydrodynamic model*, which accounts for hydrostatic forces, form drag and skin friction.

Ice failure (except local crushing) is not taken into account in the model and results presented in this paper. Therefore, the results should be seen as describing trends in the contact behavior, and should not be perceived as loads that would occur in actual full-scale interaction.

Material Model

A random lattice model based on Voronoi Tesselation is used as a material model for the ice. This model is presented in (van den Berg 2016). The model parameters are determined as described in (Yip et al. 2005) and are adapted to ice plates as described in (van den Berg 2016). Lattice models based on Voronoi tessellation give accurate results in fracture modelling of brittle, inhomogeneous, and/or polycrystalline materials such as concrete and rock.

Contact Model

The contact model is used to derive contact parameters in ice-ice and ice-structure contacts. The model is based on the assumption that local crushing will occur at ice-ice or ice-structure contacts when a maximum crushing pressure is exceeded. Full-scale observations show that local crushing does indeed occur in most ice-ice and ice-structure interactions. An example of local crushing is shown in Figure 1.

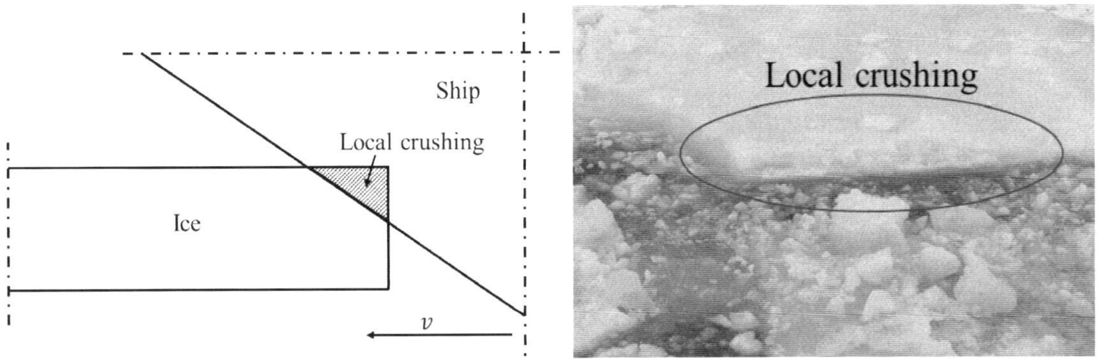

Figure 1. Local crushing at ship-ice contact. Sketch of interaction scenario (left) and full-scale observation (right, Oden 2016).

We find that it is essential to take this local crushing into account for an accurate contact force determination. The contact parameters are determined assuming constant energy dissipation per crushed volume. This is similar to (Kinnunen et al. 2016) and is supported by experiments described in (Kim & Høyland 2014). We note that the assumption of a constant energy absorption during crushing is a simplification of the actual phenomena that occur during crushing. Such a simplification is needed because of the large difference in scale between the phenomena that occur at the ice-structure contact and the global ice sheet behavior. To our knowledge, a similar simplification is made in any global ice structure interaction model, the only difference being the assumptions made in deriving the force-penetration curve. Models that attempt to physically model the crushing process are currently limited to small scales.

We derive an expression for the contact force as a function of penetration in the form:

$$F(\delta) = p_{cr1}\delta^2 + p_{cr2}\delta + p_{cr3} \quad (1)$$

In which the constants p_{cr1}, p_{cr2} and p_{cr3} are determined based on the contact geometry and the specific energy absorption during crushing. F is the contact force in the normal direction of the contact and δ is the crushing penetration in the normal direction. Crushing contacts are fully dissipative; the energy 'consumed' in the crushing process cannot be recovered.

Time Stepping Scheme

The time stepping scheme utilizes the contact parameters derived from the lattice model and the contact model to propagate the simulation in an implicit manner. The time stepping scheme used for the simulations in this paper is a relatively new way of modelling, and shows similarities with the methods described in (Lacoursiere 2007; Servin et al. 2014; Krabbenhoft et al. 2012). However, the derivation of the current method is based on a different implicit integration method, resulting in different integration factors and model properties. It is derived based on the second-order accurate Newmark-Beta method. In contrast to other schemes, it does not introduce (artificial) numerical damping.

Each time step, the velocity change of the bodies $\boldsymbol{\Delta v}$ and the contact force integral at each contact $\boldsymbol{\lambda}$ is calculated by solving the system of equations shown in formula 2:

$$\begin{bmatrix} \mathbf{M} & -\mathbf{J}^T \\ -\boldsymbol{J} & \boldsymbol{\Sigma} \end{bmatrix} \begin{bmatrix} \boldsymbol{\Delta v} \\ \boldsymbol{\lambda} \end{bmatrix} = \begin{bmatrix} \mathbf{0} \\ \mathbf{G} \end{bmatrix} \tag{2}$$

Here, $\mathbf{M}$ is the mass matrix, $\boldsymbol{J}$ is a matrix containing the constraint Jacobians, $\boldsymbol{\Delta v}$ contains the velocity change of each body and $\boldsymbol{\lambda}$ are the Lagrange multipliers at each contact, which have the dimension of impulse in this formulation. The diagonal matrices $\boldsymbol{\Sigma}$ and $\mathbf{G}$ describe the contact non-rigidity. The terms in $\boldsymbol{\Sigma}$ and $\mathbf{G}$ are derived by ensuring energy balance at each contact and utilizing an implicit integration scheme similar to the Newmark-Beta method. The terms in $\boldsymbol{\Sigma}$ and $\mathbf{G}$ can be used to represent elastic contacts, like the lattice elements representing the ice deformation, or they can represent dissipative contacts for contact between separate ice floes or between ice and structure, in which local crushing occurs.

At ice-ice or ice-structure contacts, there cannot be any tension and the frictional force is limited based on the normal force. Coulomb friction is assumed, resulting in the following constraints for loose contacts;

$$\lambda_n \geq 0 \text{ (no tension)} \tag{3}$$

$$\lambda_{t,i} \leq \mu \lambda_{n,i} \tag{4}$$

In which μ is the friction coefficient. λ_n is the integral of normal force over the time step and λ_t is the integral of frictional forces over a time step;

$$\lambda_n = \int_{t_n}^{t_{n+1}} F_n(t)dt \tag{5}$$

$$\lambda_t = \int_{t_n}^{t_{n+1}} F_t(t)dt \tag{6}$$

Hydrodynamic Model

Hydrodynamic drag and hydrostatic forces are included in the model. The drag is calculated based on a quadratic drag equation and skin friction. More details can be found in (Tsarau 2015). The effect of added mass is not taken into account. This may influence the results at higher interaction velocities.

DESCRIPTION OF CASE STUDIES

We investigate the influence of time step, mesh size and interaction velocity on the contact force, contact area and sheet deformation in the time domain. The influence of these parameters is investigated for two cases:

- Case #1: Interaction between a ship-shaped structure (sloping at the waterline) and a single 100x100 m ice floe.
- Case #2: Interaction between a bottom-fixed circular-shaped structure (vertical at the waterline) and a confined floe field of 100 x 100 m. The floes are randomly generated according to the method described in (Yulmetov et al. 2014) and have areas between 100 and 400 m^2.

Figure 2 and Figure 3 show the two case studies. In Case #1, a ship interacts with a free-floating ice floe. In the initial condition, the ice has no velocity and the ship propagates in the direction indicated in the figure. The velocity of the ship is not constant, as it interacts with the ice, the ship slows down and the ice floe accelerates.

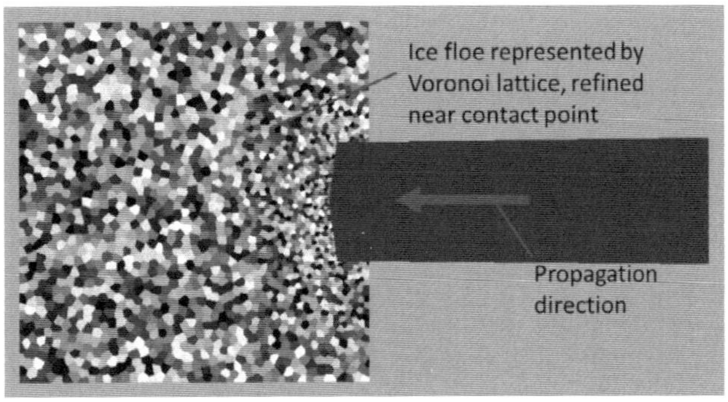

Figure 2. Ship-ice interaction scenario, top view.

In Case #2, the ice floes are confined by the domain boundaries, as they would be in an ice tank. In full-scale, the confinement could come from larger ice floes or ice pressure situations. The cylindrical structure is propagated with a constant velocity of 2.0 m/s in the direction indicated in the figure.

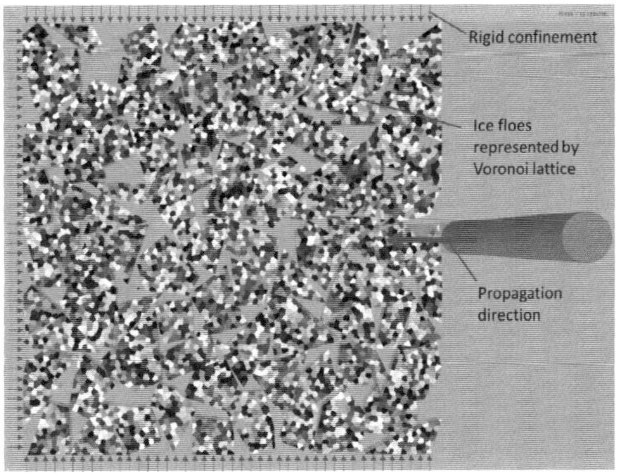

Figure 3. Fixed structure interaction scenario, top view.

Other simulation parameters are shown in Table 1.

Table 1. Simulation parameters

Parameter	Value
Ice thickness	1.0 m
Time step	variable, see Table 2
Ice-structure friction coefficient	0.3
Water density	1025 kg/m^3
Ice density	910 kg/m^3
Young's modulus of ice	$5.0 \cdot 10^9$ N/m^2
Poisson's ratio*	0
Specific energy absorption during crushing**	$2.0 \cdot 10^6$ J/m^3
Mesh size	variable, see Table 2
Interaction velocity	variable, see Table 2

* The current method of lattice properties determination does not allow for accurate modelling of Poisson effects. Therefore, a value of 0 is used.
** This corresponds to a constant crushing pressure of 2.0 MPa.

The ship dimensions are roughly similar to that of existing icebreakers, however it does not match any existing icebreaker in particular. The ship has a stem angle of 33.7°. The middle of the ship bow protrudes compared to the sides of the bow. This ensures initial point contact between the ship hull and the ice, which makes the analysis of contact model behavior and sheet deformation more convenient. The diameter of the fixed cylindrical structure matches the diameter of the Nordstrømsgrund lighthouse. The geometry of the ship and the fixed structure are shown in Figure 4.

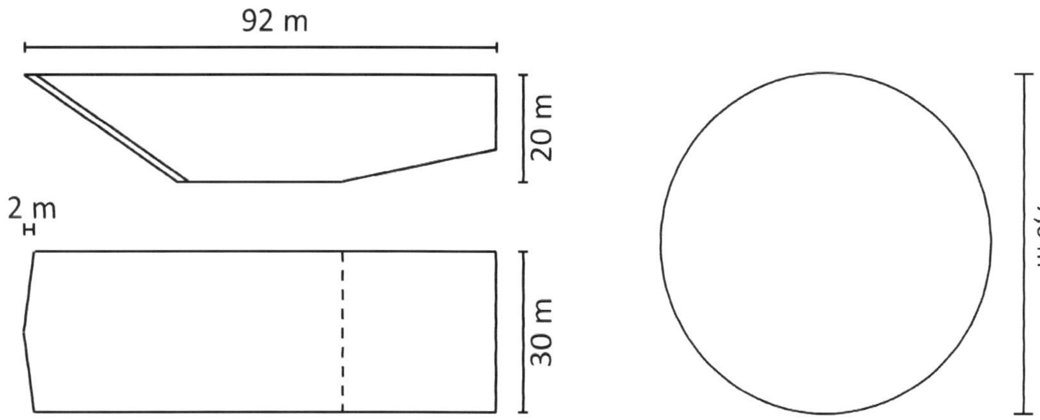

Figure 4. Geometry of ship and fixed cylindrical structure.

RESULTS

In all ice-structure interaction simulations presented below, the following phenomena occur:

- Ice and structure come in contact. Because of roughness of the ice surface area and/or the contact geometry, the contacting area at initial contact between the ice and the structure is close to zero.
- Local crushing occurs at the ice-structure contact, which results in a higher contact area. This higher contact area is needed to transfer a load from the structure to the ice.
- As the contact area and the total contact force between ice and structure increases, the ice sheet starts to deform. This deformation influences the temporal evolution of the contact force.

We consider the above described phenomena a reasonably good approximation of the phenomena that occur during actual ice- structure interaction. During the simulations, the stress distributions in the ice were similar to those displayed in Figure 5. Bending waves were observed in Case #1 (the ship interaction scenario) and force chains formed in Case #2 (the fixed structure scenario).

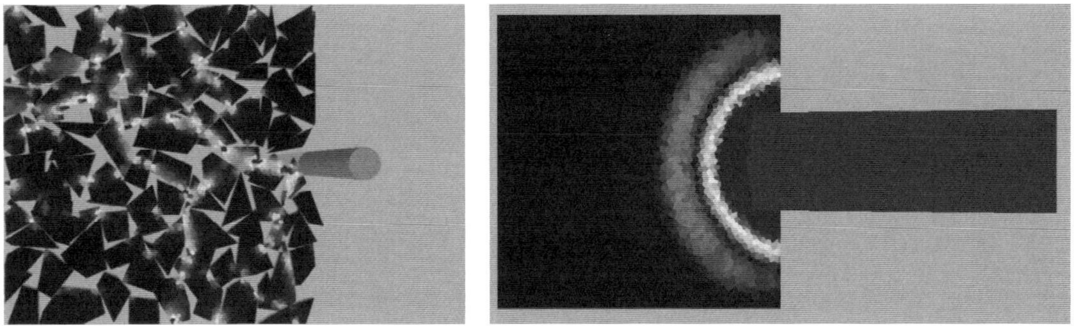

Figure 5. Stress in ice sheets during interaction with fixed structure and ship

We investigate the effect of mesh size (d_s), time step (dt), and interaction velocity (v) on the results of Case #1. For Case #2, we study the effect of mesh size and time step on the results. The combinations of the tested parameters are shown in Table 2.

Table 2. Numerical test matrix

Case number #	Time step (dt) [s]	Mesh size (d_s) [m]	Interaction velocity (v) [m/s]	Structure type
1.1	$1.0 \cdot 10^{-4}$, $5.0 \cdot 10^{-4}$, $1.0 \cdot 10^{-3}$, $5.0 \cdot 10^{-3}$, $1.0 \cdot 10^{-2}$, $5.0 \cdot 10^{-2}$, $1.0 \cdot 10^{-1}$	[0.5 / 2.0]*	5.0	Ship-shaped floater
1.2	$1.0 \cdot 10^{-4}$	[0.5 / 2.0], [1.0 / 4.0], [1.5 / 6.0], [2.0 / 8.0], [2.5 / 10.0], [3.0 / 12.0], [3.5 / 14.0]*	5.0	Ship-shaped floater
1.3	$1.0 \cdot 10^{-2}$	[0.5 / 2.0]*	0.05, 0.1, 0.5, 1.0, 5.0	Ship-shaped floater
2.1	$1.0 \cdot 10^{-3}$, $5.0 \cdot 10^{-3}$, $1.0 \cdot 10^{-2}$, $5.0 \cdot 10^{-2}$, $1.0 \cdot 10^{-1}$	1.0	2.0	Fixed cylindrical
2.2	$1 \cdot 10^{-2}$ s	1.0, 2.0, 5.0, 10.0, ∞	2.0	Fixed cylindrical

*[d_{smin} / d_{smax}] The mesh changes from fine near the contact to a coarser mesh further away.

In the following sections, selected results are presented. First, we present the results for ship-ice interaction. Then we present the results for the fixed structure scenario.

Case #1.1; Influence of Rime Step in Ship-Ice Interaction

In the ship simulation case, the influence of time step size on the results is minimal for $dt \leq 0.01$ s, and is significant for the presented higher step sizes. The results are shown in Figure 6. An interesting observation in the below figures is that while the contact force is lower at $t = 0.1$ s for the large time step cases, the vertical sheet deformation at the contact point is higher at this time. This indicates that the sheet dynamics are no longer accurate at these time steps.

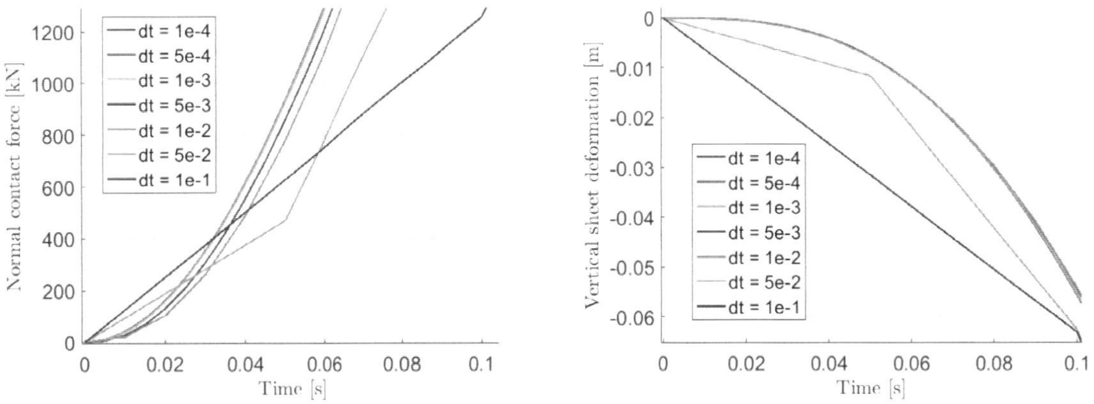

Figure 6. Contact force as a function of time (left) and vertical sheet deformation as a function of time (right), for a range of time step sizes *dt*.

Case #1.2; Influence of Mesh Size on Ship-Ice Interaction

The influence of mesh size on the contact force and the sheet deformation is shown in Figure 7. Note that the contact force is hardly effected by the mesh size in this simulation case. This may be explained by the result presented in Figure 9. At an interaction velocity $v = 5.0$ m/s, bending deformation is much smaller than the local crushing deformation. This reduces the effect of sheet deformation on contact force. Therefore, the inaccuracy in sheet deformation behavior that occurs for coarser meshes has little effect on the contact force build up. The right plot of Figure 7 shows that the local mesh deformation stiffness is increased as the mesh size increases. This is in line with the results presented in (ref ATC 2016) and can be attributed to the inability of coarse meshes to capture the local sheet deformation near the contact point.

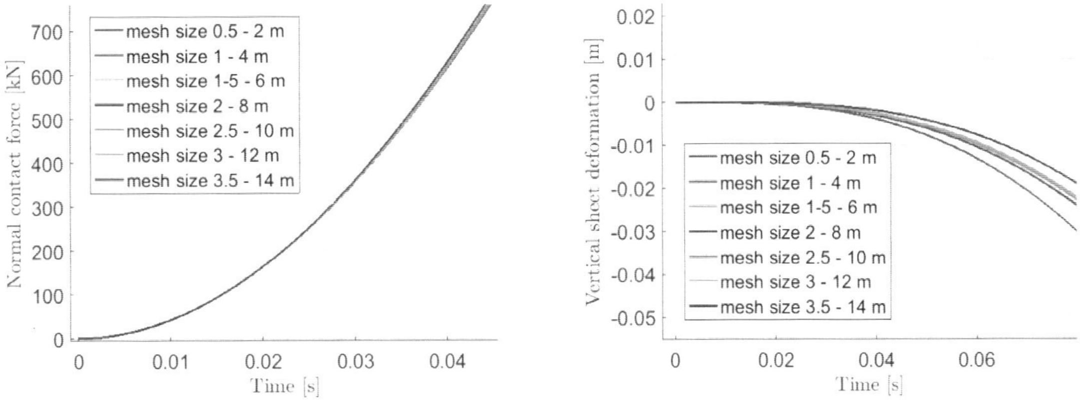

Figure 7. Contact force as a function of time (left) and vertical sheet deformation as a function of time (right), for a range of time step sizes.

Case #1.3; Influence of Interaction Velocity in Ship-Ice Interaction

In contrast to mesh size and time step, which are purely numerical features, the interaction velocity is a parameter of the actual interaction. We choose to investigate and demonstrate the influence of interaction velocity on the model results because it demonstrates the importance of sheet dynamics on the results. The influence of interaction velocity is apparent in Figure 8. Here, the vertical sheet deformation at the loading point is plotted against the contact normal force. At higher interaction velocities, the sheet deformation increases much slower with an increasing normal force. This can be attributed to ice sheet inertia and hydrodynamic damping. Since strain is often used as a failure criterion, different deformation behavior will also result in a different failure pattern. In case of bending failure, this will most likely mean a shorter failure length.

Note that the contact normal force is limited for the lowest interaction velocities, 0.05 m/s and 0.1 m/s. At these velocities, the sheet reaches the same velocity as the ship before the cut-off normal force is reached.

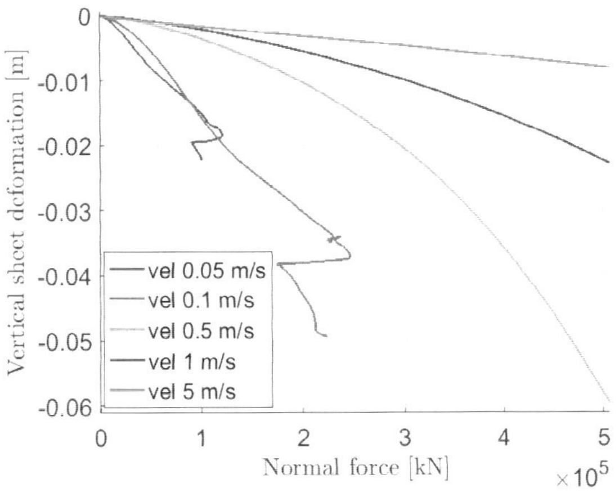

Figure 8. Contact force against sheet deformation for different interaction velocities

Another interesting observation can be made when looking at the difference between local crushing deformation and sheet bending deformation. This is shown in Figure 9. At lower interaction velocities, there is both bending deformation in the ice sheet and local crushing deformation. At high interaction velocities, the sheet deformation is much smaller compared to the crushing deformation, which may result in a steeper increase of the contact force.

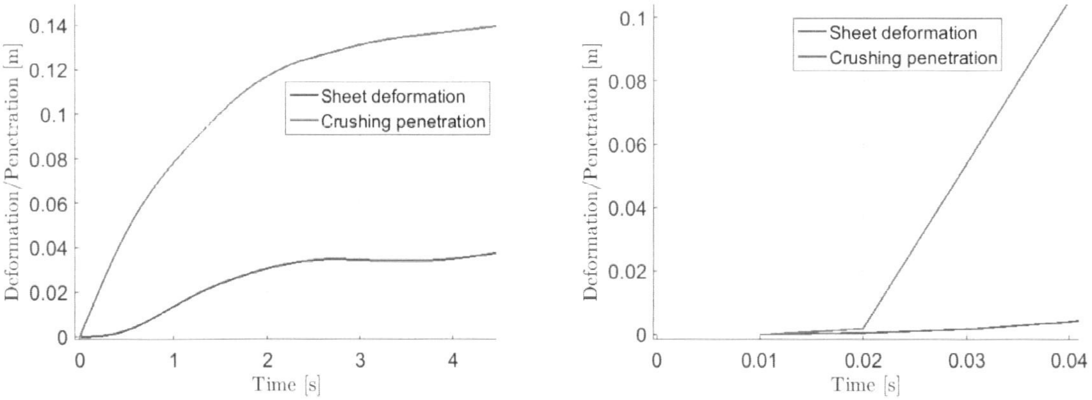

Figure 9. Sheet deformation and crushing penetration at an interaction velocity of 0.1 m/s (left) and 5.0 m/s (right).

Case #2.1; Influence of Time Step on Fixed Structure – Ice Interaction

Figure 10 and Figure 11 shows the load from the ice on the fixed structure in the direction opposite to the propagation direction. Any change in time step causes a clear difference in the load signal. We may wonder if a complete correlation between load-time signals should be expected. Due to the nature of the modelled case, the load at any moment in time depends on the ice floe displacements in the previous simulation steps. Because of this, minimal (numerical) differences in the beginning of the simulation may lead to a completely different load signal later in the simulation. Therefore, the properties of the signals over the duration of the simulation should be compared, rather than comparing the exact load values at any moment in time. The current simulations are considered too short to say anything statistically significant about the load-time signals, but from visual comparison we conclude that the signals for time steps $5.0 \cdot 10^{-2}\ s$ to $1.0 \cdot 10^{-3}\ s$ show similar behavior. Even the load-time signal for $1.0 \cdot 10^{-1}$ s has peak loads and an average load value of comparable size.

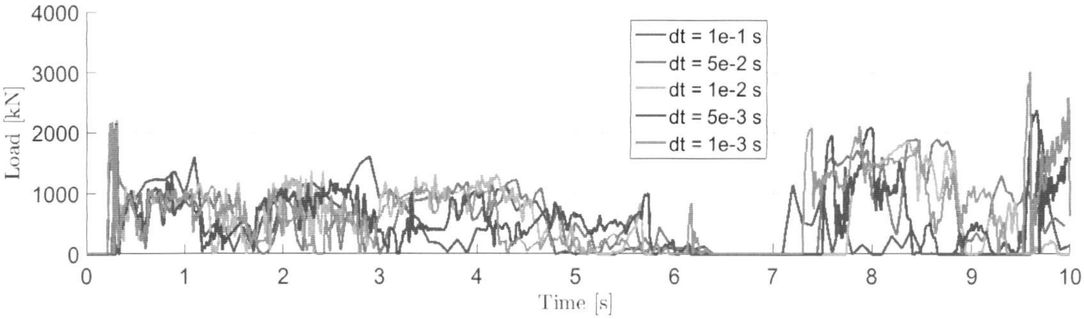

Figure 10. Unfiltered load-time signals, time steps $1 \cdot 10^{-3}\ s$ to $1 \cdot 10^{-1}$ s.

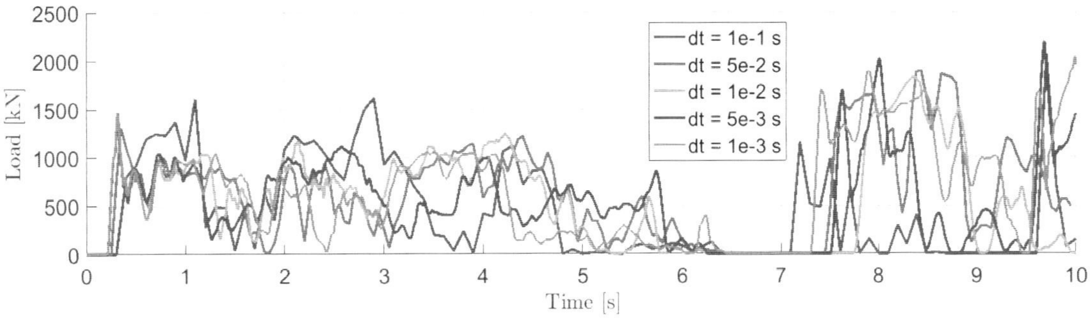

Figure 11. Filtered load-time signals, time steps $1 \cdot 10^{-3}\ s$ to $1 \cdot 10^{-1}\ s$.

Case #2.2; Influence of Mesh Size on Fixed Structure – Ice Interaction

Figure 12 and Figure 13 shows the load from the ice on the fixed structure in propagation direction, in a comparison for different mesh sizes. As discussed in the previous section, we should not expect a correlation over the whole time signal because of the nonlinear nature of the simulation. The results are similar for the first second of the simulation, after which the results start to differ. Comparing the load-time signals over the complete duration, we see that the signals for mesh sizes 5.0 – 1.0 m have broadly similar characteristics. The load-time signals for no mesh and for a mesh size of 10.0 m are similar to each other, but are different from the signals obtained with finer mesh sizes.

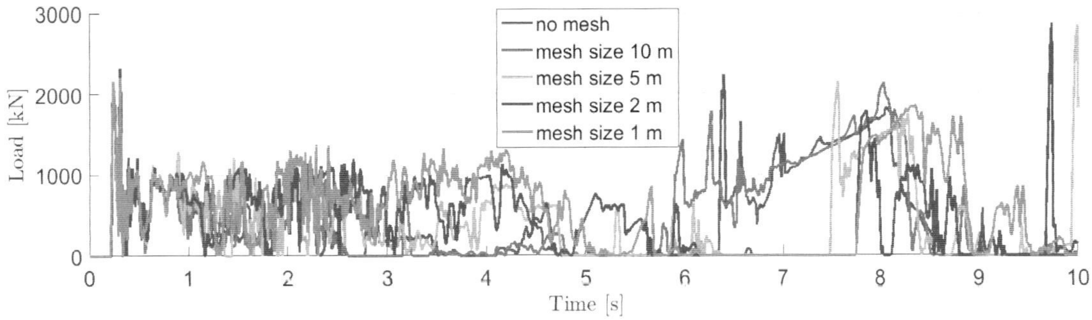

Figure 12. Unfiltered load-time signals, no mesh to mesh size 1 m.

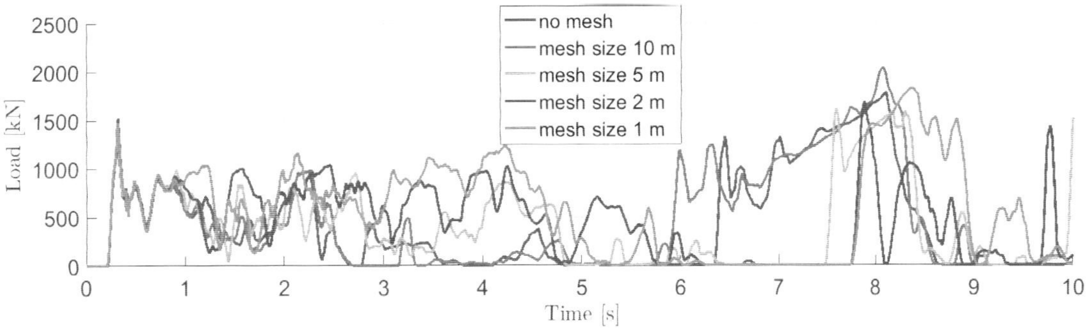

Figure 13. Filtered load-time signals, no mesh to mesh size 1 m.

CONCLUSIONS

The accuracy limits of a novel non-smooth discrete element time stepping scheme that includes contact non-rigidity were investigated. The sensitivity of model results to mesh density, time step and interaction velocity were compared. Based on the results of this analysis, we draw the following conclusions:

Case #1: Ship – Single Floe Interaction

- The ship-ice interaction results can be considered accurate for time steps ≤ 0.01 s. For larger time steps, the model results start to differ significantly.
- The mesh size has little influence on the results in the performed simulations. This is due to the high interaction velocity, which reduces the effect of ice floe deformation on the results.
- At low velocity interaction, the ice floe deformation was significant compared to the local crushing deformation. For high interaction velocities, (> 1 m/s) the sheet deformation is much smaller than the local crushing deformation. Hydrodynamic damping and inertia of the ice floe cause the difference in deformation behavior. This will most likely result in a different failure pattern for high velocity interactions.

Case #2: Fixed Structure Floe Field Interaction

- Due to the nonlinear nature of the simulation, we should not expect a correlation between results over the full time signal. A minimal (numerical) difference in the beginning of the simulation may lead to completely different results later on.
- Comparing the load-time signals for different time step sizes, we find broadly similar results for time steps $5.0 \cdot 10^{-2}\ s$ to $1.0 \cdot 10^{-3}\ s$.
- Comparing the load-time signals for different mesh sizes, the results for mesh sizes 5.0 -1.0 m are broadly similar.

In the fixed structure simulation case #2, the local crushing at ice-ice and ice structure contacts is important for a correct contact force prediction, but the elastic ice floe deformation (taken into account by the lattice model) does not have much influence on the load signal. In the ship simulation case #1, the bending deformation of the ice floe influences the contact force at low interaction velocities. At higher interaction velocities, local crushing is much more important. Since the sheet deformation as a function of contact force is different at high ice velocities, we expect that the failure pattern will also change as a function of interaction velocities. These results were obtained while neglecting the added mass of the water. Including added mass will most likely exacerbate the velocity effects.

ACKNOWLEDGEMENTS

The authors would like to acknowledge the support from the SAMCoT CRI through the Research Council of Norway and all of the SAMCoT Partners.

REFERENCES

van den Berg, M., 2016. A 3-D Random Lattice Model of Sea Ice. In *Arctic Technology Conference. St. Johns, Newfoundland and Labrador.*

Kim, E. & Høyland, K. V., 2014. Experimental Investigations of the Energy Absorption Capacity of Ice During Crushing: Is the Specific Energy Scale Independent? *Proceedings of the 22nd IAHR International Symposium on Ice*, pp.163–170.

Kinnunen, A., Tikanmäki, M. & Heinonen, J., 2016. An energy model for ice crushing in ice-structure impact. In *23 r d IAHR International Symposium on Ice*. Ann Arbor, Michigan, USA, pp. 1–8.

Krabbenhoft, K. et al., 2012. Granular contact dynamics with particle elasticity. *Granular Matter*, 14(5), pp.607–619.

Lacoursiere, C., 2007. *Ghosts and machines: regularized variational methods for interactive simulations of multibodies with dry frictional contacts*. Umeå University.

Servin, M. et al., 2014. Examining the smooth and nonsmooth discrete element approaches to granular matter. *International Journal for Numerical Methods in Engineering*, (March), pp.1885–1891.

Tsarau, A., 2015. *Numerical Modelling of the Hydrodynamic Effects of Marine Operations in Broken Ice*. Norwegian University of Science and Technology.

Yip, M., Mohle, J. & Bolander, J.E., 2005. Automated Modeling of Three-Dimensional Structural Components Using Irregular Lattices. *Computer-Aided Civil and Infrastructure Engineering*, 20, pp.393–407.

Yulmetov, R., Løset, S. & Lubbad, R., 2014. An Effective Numerical Method for Generation of Broken Ice Fields, Consisting of a Large Number of Polygon-Shaped Distinct Floes. In *Proceedings of the 22nd IAHR International Symposium on Ice*.

POAC'17
Busan, KOREA

Proceedings of the 24th International Conference on
Port and Ocean Engineering under Arctic Conditions
June 11-16, 2017, Busan, Korea

Sources of Stochasticity in Ice-Structure Interaction Process

Janne Ranta[1,2], Arttu Polojärvi[1,2] and Jukka Tuhkuri[1,2]

[1]Aalto University, School of Engineering, Department of Mechanical Engineering P.O. Box 14300, FI-00076 Aalto, Finland
[2]Sustainable Arctic Marine and Coastal Technology (SAMCoT), Centre for Research-based Innovation (CRI), Norwegian University of Science and Technology, Trondheim, Norway

ABSTRACT

We simulated the ice-structure interaction process using 2D FEM-DEM. The simulations are deterministic, but sensitive to initial conditions. Ice loads statistics from three sets of 50 simulations were considered. The sets differed by ice crushing strength. In two set the ice sheet was homogeneous and in one set the crushing strength of ice varied. In simulations with homogeneous ice sheets the initial condition varied causing different load records. In the case of the non-homogenous ice sheets, variation in load records came directly through the non-homogeneity of the sheet. Our core result is that the stochastic nature of ice loads could be due to the non-homogeneity of sea ice, but virtually as much variation in the data was observed with homogenous ice sheets. This suggests that the variability in the ice loads was mainly caused by the ice-structure interaction process instead of the non-homogeneity of ice.

INTRODUCTION

Arctic marine operations increase continuously. This includes developments in Northern sea transportation, offshore drilling operations, and offshore wind energy. One of the key factors in developing safe Arctic operations is a reliable prediction of sea ice loads, which are induced by ice-structure interaction process. Here we focus on on the statistics of the ice loads induced by an ice-inclined structure interaction processes by studying these processes using 2D combined finite-discrete element method simulations (2D FEM-DEM). We investigate where the well- known stochastic nature of ice loads (Daley et al., 1998; Jordaan, 2001) arises from. Often this stochasticity is assumed to be due to the inhomogeneity of ice.

Our 2D FEM-DEM simulations of the ice-structure interaction process are deterministic, but we know that the process is very sensitive to its initial conditions (Ranta et al., 2017a). Here we used this sensitivity to create sets of simulations using the same simulation parameters, but different initial conditions. For a set of simulations, we conducted 50 simulations with constant parameters, but varying vertical initial velocity v_0 at the free end of the ice sheet (see Figure 1). This v_0 was given a random value of the order of 10^{-9} mms^{-1}. This was enough to cause different ice failure processes. After this we ran one more set of 50 simulations where the ice sheet was non-homogenous. This gave us data on the processes in the case of non-homogenous ice (vertical initial velocity v_0 was zero in these simulations).

The core finding here is that, the stochastic nature of ice loads could be due to non-homogeneity of sea ice, but it can as well simply arise from the ice-structure interaction process itself. The

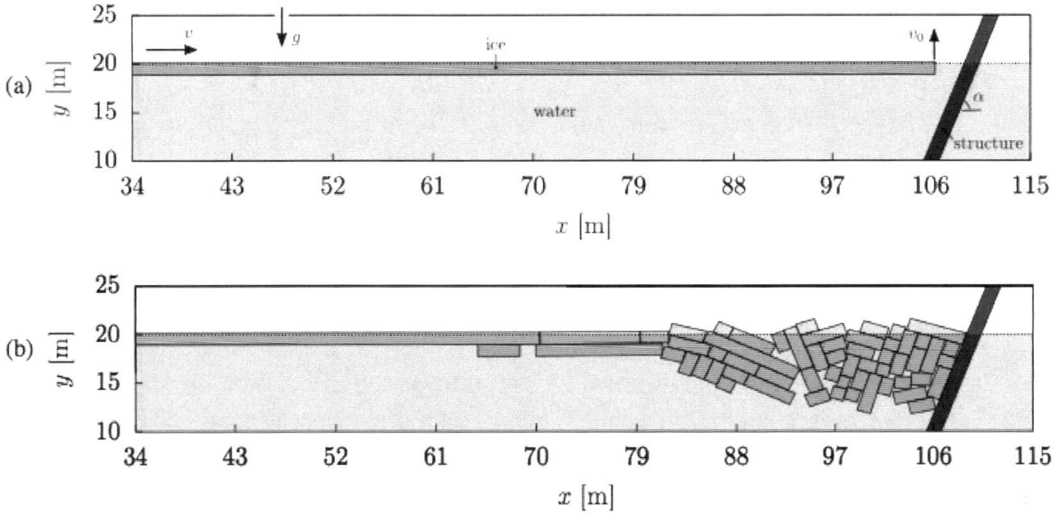

Figure 1: (a) Simulation set-up: A floating ice sheet (dark grey) is pushed with a velocity of v = 50 mms^{-1} towards an inclined ($\alpha = 70^{\circ}$) structure (black). The water is colored light gray. A perturbation for the initial conditions was created by giving a low vertical velocity v_0 for the free end of the ice sheet. v_0 was of the order of 10^{-9} mms^{-1} and varied randomly between all simulations. (b) A snapshot from a simulation at a stage where the length of the pushed ice L = 125 m. Ice thickness h = 1.25 m and plastic limit σ_p = 1 MPa.

ice-structure interaction process is itself a stochastic process. We will first describe the simulation tools. After this we present our results, and analyze and discuss them. We look into the peak load statistics and compare the processes that have homogenous ice sheets to those which have a non-homogenous sheets. We conclude the paper with some remarks on our work.

SIMULATIONS

The 2D FEM-DEM simulation tools we used are the same as used in Paavilainen et al. (2009, 2011), Paavilainen and Tuhkuri (2012, 2013), and Ranta et al. (2017a). We use the 2D FEM-DEM code of Aalto University Ice Mechanics Group. The DEM part of the code follows the model by Hopkins (1992) while Paavilainen et al. (2009) developed the FEM-DEM Timoshenko beam model for the ice sheet. The model was validated in Paavilainen et al. (2009, 2011) against laboratory and full-scale measurements on ice loads. Polojärvi et al. (2015) successfully used the DEM part for modeling direct shear box tests on ice rubble.

Figure 1 illustrates the simulations. We have a floating, homogenous and continuous ice sheet pushed against an inclined rigid structure with constant velocity v. The initially intact ice sheet breaks into ice blocks, which then interact with each other and the structure. Modeling of the ice-structure interaction process requires a numerical model, where the formation of the rubble pile and its interaction with both, the advancing ice sheet and the structure, are accounted for. Table 1 lists the simulation parameters, which we mostly chose after Timco and Weeks (2010).

We ran three sets S1, S2 and S3 of 50 simulations summarized in Table 2. Within sets S1 and S2, all parameters, except the initial velocity perturbation v_0, were constant. These simulation sets had a different value for the ice plastic limit σ_p in contact. Plastic limit σ_p describes the local ice crushing in contacts, but the model is not able to describe the continuous crushing of intact ice nor the extrusion of crushed ice. The simulations of set S3 were ran with a non-homogeneous ice sheet. In these simulations σ_p randomly varied between values 1 and 2 MPa within the sheet. All of the ice sheets of the simulations in S3 were unique, with σ_p varying after uniform distribution throughout the sheet. This was done by giving each discrete element of the ice sheet a random σ_p value. Simulations of S3 did not have an initial velocity perturbation; The potential stochasticity in the results of simulations in S3 thus arises from the non-homogeneity of the modeled ice.

Figures 2a and b illustrate how the initial velocity perturbation affects the load records from the ice failure process in simulations of set S1. Here F is plotted as a function of L, the length of the ice sheet pushed against the structure. The load records show similar features, but they diverge after the first load peaks at around L = 10 . . . 15 m. Figure 2b shows the maximum ice loads F^p event for both of the simulations in Figure 2b. The F records between the simulated processes differ, which causes the values of F^p and L at their occurrence to differ. Simulations of sets S2 and S3 yielded similar load records, and the discussion of this paragraph applies for them as well.

Table 1: Summary of the main simulation parameters used in the 300 simulations of this paper. Ice thickness h and plastic limit σ_p were varied. Other parameters had the values given by the table in all simulations. The parameter values were mostly chosen after Timco and Weeks (2010).

	Description and symbol		Unit	Value or Range
General	Gravitational acceleration	g	m/s^2	9.81
	Ice Sheet velocity	v	m/s	0.05
	Drag coefficient	c_d		2.0
Ice	Thickness	h	m	1.25
	Effective modulus	E	GPa	4
	Poisson's ratio	ν		0.3
	Density	ρ_i	kg/m^3	900
	Tensile strength	σ_f	MPa	0.6
	Shear strength	τ_f	MPa	0.6
Contact	Plastic limit	σ_P	MPa	1.0-2.0
	Ice-ice friction coefficient	μ_{ii}		0.1
	Ice-structure friction coefficient	μ_{iw}		0.1
Water	Density	ρ_w	kg/m^3	1010
Structure	Slope angle	α	deg	70

Table 2: Summary of simulations sets S1-S3. Sets differed of simulations was 150. by the values of σ_p. Total number

Set	IDs	n	h	σ_P
			[m]	[MPa]
S1	1-50	50	1.25	1
S2	51-100	50	1.25	2
S3	101-150	50	1.25	$U(1,2)$

The data analyzed here consisted of above-described three sets of 50 simulations. If data from the 50 simulations were combined into a single graph, we would get a similar graph as in Figure 2a for 2 simulations, but such a graph would not be informative. Since we have 50 load values for each time step from the 50 simulations in each set, we could calculate the mean ice load, standard deviation and maximum load for each time step. Figure 3 illustrates how these statistical measures for a given time step were obtained from the load records. In this way we define the mean load, standard deviation of the load, and the maximum load for each time step.

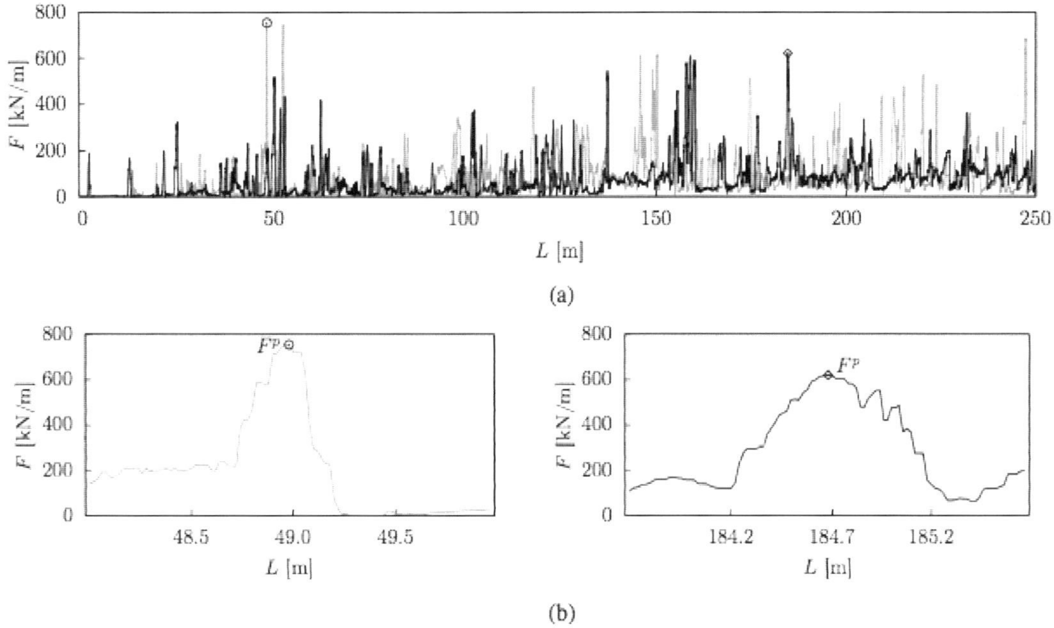

Figure 2: Two ice load F records from simulations of set S1: (a) F plotted against length L of pushed ice and (b) two close-ups of the peak ice load F^p events from the same two simulations. The F records diverged after the L = 10 . . . 15 meters of pushed ice.

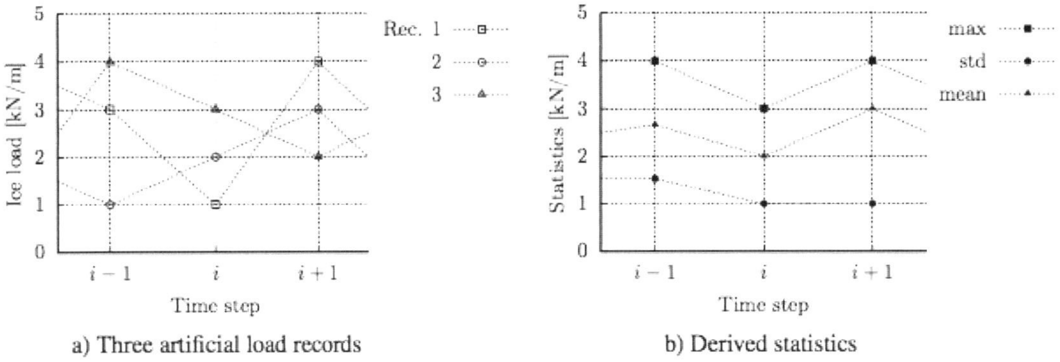

Figure 3: Illustration on how the mean (mean), standard deviation (std) and maximum (max) ice load records were obtained: (a) Data from three time steps $i-1$, i and $i+1$ from three artificial load records. (b) The derived statistics for these time steps.

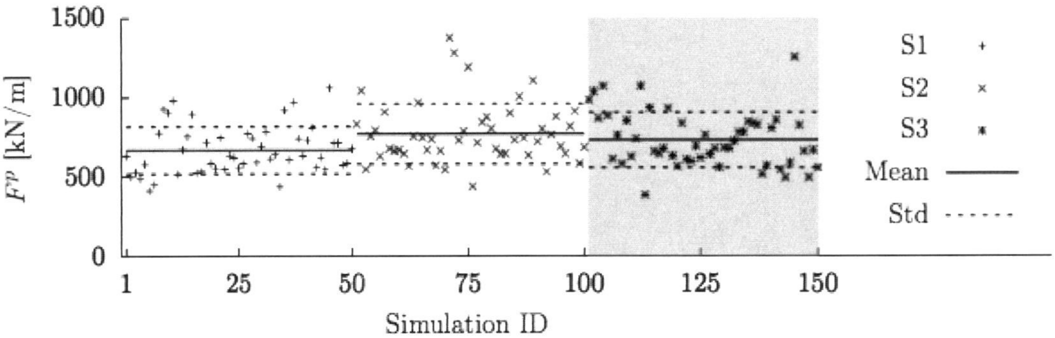

Figure 4: Peak ice load observations from simulation sets 1–3. Solid horizontal lines indicate set mean values and dashed lines are ± one standard deviation away from the mean. Results from set 3 with non-homogenous ice sheets are highlighted by gray background color.

RESULTS AND ANALYSIS

Figure 4 shows global peak ice load F_P values from all simulations used in the analysis. In the figure, the horizontal axis indicates the simulation ID and the vertical axis shows F_P value for each observation. In addition, the figure shows the mean values and standard deviations for the data points belonging to each set S1-S3. These mean values and the standard deviations are also summarized in Table 3, which additionally gives the coefficient of variation c_v (the standard deviation of a set divided by the corresponding mean value) values for all sets. Value of c_v describes the variability in the underlying data set.

As seen from Figure 4, the data points from the different sets fall onto same regime. The mean F_P value for S2 (for which σ_p = 2 MPa) is slightly higher than for S1 (σ_p = 1 MPa), and the mean value for S3 falls between the means of sets S1 and S2. The figure together with the standard deviations of Table 3 indicate, the data set pair S1 and S3 are very similar. Same applies to pair S2 and S3. The similarity in the data sets is also indicated by the values of the c_v, also given in the table. They all are about equal. These statistical figures do not allow us to distinguish the data set yielded by the simulations with non-homogenous ice sheet from those yielded by homogenous sheet.

In addition to these features, also the distributions of the peak load data were very similar as shown by Figures 5a and b. The figures, respectively, show the relative frequencies of the peak load F_P observations, sorted here into 10 equally sized bins, and the empirical cumulative distribution functions for F_P values. The figures show that the distributions for sets S1-S3 clearly follow very similar trends, and in fact, the histograms of Figure 5a are even difficult to tell apart. We did not attempt to look for the best fitting distribution for the data sets in the histogram of Figures 5a, but the shape of the histogram indicates them to be non-normal (Ranta et al., 2017b).

Table 3: The mean of global peak loads F_P values, their standard deviations (std), and the coefficients of variation (c_v) for the simulations of all sets S1-S3.

Set	Mean(F^P)[kN/m]	std(F^P)[kN/m]	c_v[-]
S1	670	150	0.22
S2	770	190	0.25
S3	730	170	0.23

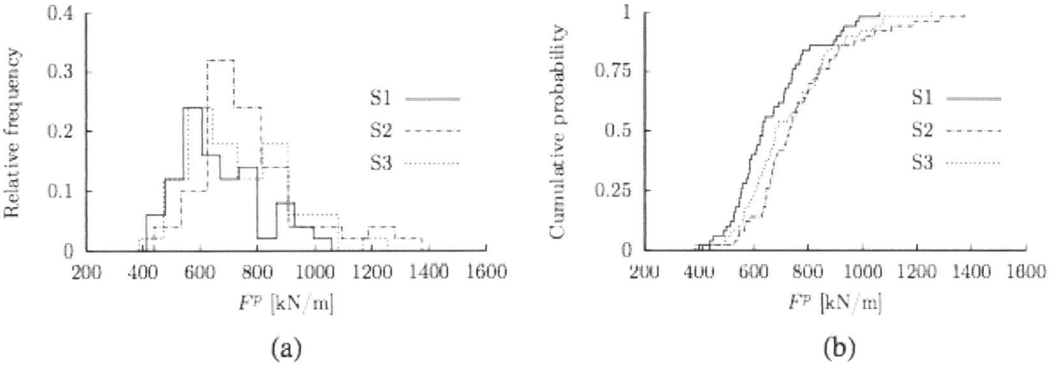

Figure 5: Relative frequencies (a) of peak loads from simulation data sets 1–3 and empirical cumulative distribution functions (b) of the same data sets.

We can also compare the so-called concurrent statistical values from the different processes, that is, the statistics of load F values from the same time step of the different simulations. The derivation of these time-step-wise statistical figures was described above (see Figure 3). Figures 6a-c show comparisons for the mean load, for the standard deviation, and for the maximum load, respectively. Left column of the figure depicts comparisons of loads between simulation S1 and S3 and the right column between simulation sets S2 and S3. The x- and y-coordinates of each data point are, respectively, defined by the value of a given statistical figure from the sets S1-S3 given as the label of x- and y-axis. If the data sets matched one-to-one, all of the data points would fall on the diagonal also shown in the plots.

Figures 6a-c illustrate that the concurrent load values from the different sets of 50 simulations do not show any dependency of the set: all of the point clouds are virtually symmetrical about the diagonal. The same is shown by the coefficient of determination R^2 values, which the graphs of the figures also show. While R^2 has a high value in the case of the mean load (the points lye close to the diagonal) the more spread out data in rest of the plots do not indicate any set-depended correlation in the data R^2 (very low R^2).

Different gray-scale levels in graphs of Figures 6a-c indicate the process stage from where the data point is from. Lightest color stands for the early process stage (small L) and the black color denotes the final stage (L = 250 m). Gray-scale levels in the figure demonstrate that mean, standard deviation and maximum (of concurrent load values) tend to be increase throughout the simulations. Plots in Figure 6c also demonstrate the earlier finding (Kujala et al., 2009; Suominen and Kujala, 2014; Ranta et al., 2017a) on the variation in the data increasing with the increasing load values. This is indicated by the wider spread in the cloud of data points for maximum load with the high than with the low maximum load values.

CONCLUSIONS

We performed 2D combined finite-discrete element method (FEM-DEM) simulations on ice-structure interaction process for studying the reasons behind the stochasticity of ice loads. The study based on three sets of 50 simulations, two sets with homogenous ice and one with nonhomogeneous ice. No clear set-dependency was found. In all of the cases load values were stochastic throughout the process. The stochastic nature of ice loads can be thus partly due to non-homogeneity of sea ice, but it can as well arise from the ice-structure interaction process itself; The ice-structure interaction process is itself a stochastic process. Understanding the sources for the stochasticity can be used to increase the accuracy of ice load estimates.

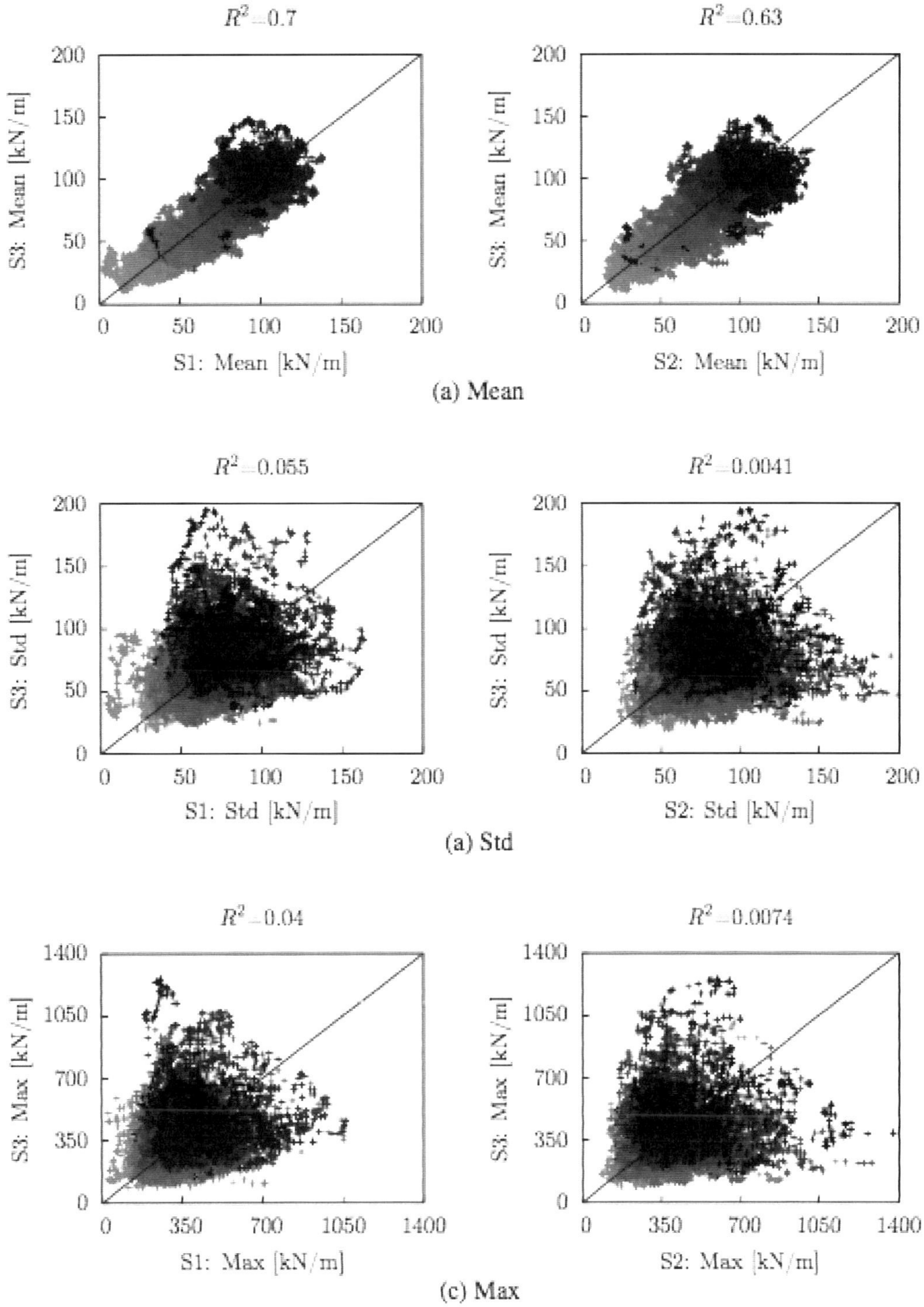

Figure 6: Comparison of concurrent mean (a), standard deviation (b) and maximum (c) load observations between simulations sets 1 and 2 (left column) and between simulation sets 2 and 3 (right column). Coefficients of determination R^2 are shown with 2 significant digits. Lightest colors stand for the early process stage (small L) and the black color the final stage (L = 250 m).

ACKNOWLEDGEMENTS

The authors wish to acknowledge the support from the Research Council of Norway through the Centre for Research-based Innovation SAMCoT and the support from all SAMCoT partners. Also the financial support from the Finnish Academy through research project Discrete Numerical Simulation and Statistical Analysis of the Failure Process of a Non-Homogenous Ice Sheet Against an Offshore Structure (DICE) is acknowledged.

REFERENCES

Daley, C., Tuhkuri, J., and Riska, K. (1998). The role of discrete failures in local ice loads. *Cold Regions Science and Technology*, 27(3):197 – 211.

Hopkins, M. (1992). Numerical simulation of systems of multitudinous polygonal blocks. Tech- nical Report 92-22, Cold Regions Research and Engineering Laboratory, CRREL. 69 p.

Jordaan, I. (2001). Mechanics of ice-structure interaction. *Engineering Fracture Mechanics*, 68:1923–1960.

Kujala, P., Suominen, M., and Riska, K. (2009). Statistics of ice loads measured on MT Uikku in the Baltic. In *Proceedings of the 20th International Conference on Port and Ocean Engineering under Arctic Conditions, POAC'09 (electronic publication)*.

Paavilainen, J. and Tuhkuri, J. (2012). Parameter effects on simulated ice rubbling forces on a wide sloping structure. *Cold Regions Science and Technology*, 81:1 – 10.

Paavilainen, J. and Tuhkuri, J. (2013). Pressure distributions and force chains during simulated ice rubbling against sloped structures. *Cold Regions Science and Technology*, 85:157 – 174.

Paavilainen, J., Tuhkuri, J., and Polojärvi, A. (2009). 2D combined finite–discrete element method to model multi-fracture of beam structures. *Engineering Computations*, 26(6):578–598.

Paavilainen, J., Tuhkuri, J., and Polojärvi, A. (2011). 2D numerical simulations of ice rubble formation process against an inclined structure. *Cold Regions Science and Technology*, 68(1-2):20–34.

Polojärvi, A., Pustogvar, A., and Tuhkuri, J. (2015). DEM simulations of direct shear box experiments of ice rubble: Force chains and peak loads. *Cold Regions Science and Technology*, 116:12–23.

Ranta, J., Polojärvi, A., and Tuhkuri, J. (2017a). The statistical analysis of peak ice loads in a simulated ice-structure interaction process. *Cold Regions Science and Technology*, 133:46– 55.

Ranta, J., Polojärvi, A., and Tuhkuri, J. (2017b). Statistics of peak loads in ice-structure interaction process: A study based on virtual experiments. *Submitted to Cold Regions Science and Technology*.

Suominen, M. and Kujala, P. (2014). Variation in short-term ice-induced load amplitudes on a ship's hull and related probability distributions. *Cold Regions Science and Technology*, 106-107:131–140.

Timco, G. and Weeks, W. (2010). A review of the engineering properties of sea ice. *Cold Regions Science and Technology*, 60(2):107–129.

Model Tests with Ships and Offshore Structures in HSVA's Ice Tanks

Karl-Ulrich Evers[1]

[1] Arctic Technology Department, Hamburgische Schiffbau-Versuchsanstalt GmbH, Hamburg Ship Model Basin (HSVA), Bramfelder Straße 164, D-22305 Hamburg, Germany

ABSTRACT

There is a long history of conducting experimental tests of the interactions between water or ice and structures. For a century, the private and independent Hamburg Ship Model Basin (HSVA) founded in 1913 has been at the forefront of hydrodynamic research. HSVA operates various test facilities such as: Large Towing Tank, Small Towing Tank, Hydrodynamic Cavitation Tunnel (HYKAT), Arctic Environmental Test Basin (AETB) and Large Ice Model Basin (LIMB). This paper describes the historical development of ice technology in the Hamburg Ship Model Basin (HSVA) with its test facilities Large Ice Model Basin (LIMB) and Arctic Environmental Test Basin (AETB). The procedure of model ice production is shown in detail and the methods for determining the mechanical model properties (flexural strength, crushing strength, elastic modulus, etc.) are described. The simulation of various ice conditions (level ice, brash ice, pancakes, ridges, rubble fields, ice floes, etc.) in which model tests with ice breaking ships and offshore structures are carried out are explained. A selection of different projects is presented to demonstrate the versatility and the complex requirements for the execution of model tests.

KEY WORDS: Ice testing facility; ice model tests; model ice properties; offshore structures; ice breaking ship models.

INTRODUCTION

There is a long history of conducting experimental tests of the interactions between water or ice and structures, from Archimedes, through Leonardo da Vinci to the great 19th century pioneers such as Froude and Reynolds. The 20th century saw the development of many of the great hydrodynamics laboratories that remain in use to this day.
For a century, the private and independent Hamburg Ship Model Basin (HSVA) founded in 1913 has been at the forefront of hydrodynamic research. HSVA has influenced and led developments of testing technology, methods, standardisation and numerical procedures to solve complex problems.
HSVA operates various test facilities such as: Large Towing Tank, Small Towing Tank, Hydrodynamic Cavitation Tunnel (HYKAT), Arctic Environmental Test Basin (AETB), Large Ice Model Basin (LIMB) and workshops (Figure 1).
HSVA offers the full range of hydrodynamic services for ships and marine structures. From initial design to detailed investigations of propulsion improvement devices, from concept studies to full scale CFD predictions, all can be done in-house or in close collaboration with

our partners. Model testing with scaled models are still irreplaceable to demonstrate the behavior and the performance of ships and offshore structures, both in open water and in ice covered waters.

This paper gives an overview of the ice testing facilities at HSVA and the production of the model ice, the determination of its mechanical properties are discussed. Selected projects with ships and offshore structures in ice are described.

Figure 1. Oblique aerial view of HSVA's facilities

In 1972 an ice tank was built at HSVA, which with a length of 30 m, a width of 6 m and a depth of 1.2 was at that time one of the most modern in the world. First, sodium chloride and later urea was used as a dopant. These sodium chloride and urea solutions, together with scaled down ice crystals, made it possible to scale the properties of the model ice (e.g. bending strength, compressive strength, and elasticity) corresponding to Froude and Cauchy (Valanto and Schwarz, 1999).

Advanced model testing techniques (e.g. self-propelled ice breaking ships, fixed and floating offshore structures, etc.) required a larger ice tank. The Large Ice Model Basin (LIMB) was put ceremoniously into operation in August 1984 during the 7th IAHR Ice Symposium in Hamburg, Germany.

In the new ice tank "LIMB" with a length of 78 m, 10 m width and 2.5 m / 5 m depth (Figure 2 and Figure3) the HSVA model ice has been further developed.

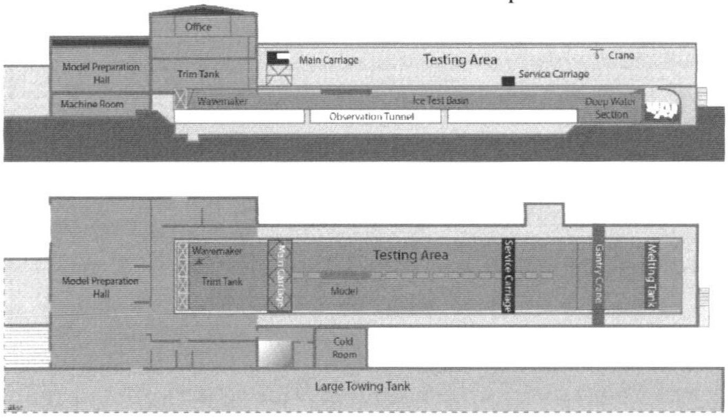

Figure 2. Elevation view (top) and plan view (bottom) of the LIMB.

Figure 3. Large Ice Model Basin (LIMB)

Until late 1980's HSVA has produced a refrigerated model ice grown from an aqueous solution containing urea as a chemical dopant by application of a seeding technique. This type of model ice is strictly columnar structured but has a thin upper granular layer. The brine is entrapped along the boundaries of columnar ice crystals.

Investigations with respect to the improvement of ice properties were made by using various dopants like EG/AD/S and glycerol instead of urea. However, on one hand these chemical dopants reduce the translucence in the water and glycerol is rather costly. The aspect of excellent water translucence is important, because the observation of ice failure processes, propeller-ice interaction and ice clearing by means of underwater video recording. On the other hand the content of sugar in the EG/AD/S dopant may cause bacterial pollution and the chemical stability of the dopant with sugar additive cannot be easily maintained (Timco, 1986; Evers, 2016; Kim and Choi, 2011; Lee, et al. 2010).

Thus, today an aqueous solution containing sodium chloride (~6.8 ppt concentration) is used as dopant at HSVA in combination with air bubble entrapment, in order to get a stable chemical condition in the tank water and to improve the mechanical properties in particular ice brittleness and cracking replication. An air bubble system allows the production of various ice densities. To control ice density and ice porosity the pressure necessary to obtain air saturated water and flow volume is varied (Hellmann, et al.,1992; Evers and Jochmann, 1993).

Transverse carriage

In 2003 the test capabilities in the LIMB were extended. A transverse carriage is permanently installed as a sub-carriage at the rear side of the main carriage. Both carriages together make it possible to run offshore structures or floating vessels in a combined and computer-controlled x-y-motion (planar motion) through the ice sheet. The device gives us the opportunity to simulate, for instance, ice drift scenarios with slow or rapid ice drift direction changes or manoeuvring tests, whereby the model is guided by main carriage and transverse carriage simultaneously while the ice sheet is kept stationary.

The transverse carriage has a maximum static load capacity of 5 kN in any horizontal direction and a load capacity of nearly 10 kN in vertical direction. The horizontal load can be applied on a vertical lever of up to 1.2 m length. A maximum driving force of about 3 kN is applied to the transverse carriage at speeds of up to 0.5 m/s by a geared electric motor (Figure 4).

Figure 4. Transverse carriage installed as a sub-carriage on rear side of towing carriage.

For free running tests model and carriage are connected only by the cables, which are needed to transfer electrical power into and the measuring signals out of the model. Cables are kept above the model during manoeuvring operations in order to avoid any force application. The model itself is self-propelled while the rpm of each propeller may be remote controlled by throttles from the carriage. For manoeuvring tests, the rudder or azimuth angle can also be controlled by the operator.
A Qualisys™ motion capture system is used to detect the rigid body motions in all six degrees of freedom (6-DOF). The system uses four cameras installed at the main carriage to detect markers which are located at different positions on the model. Thereby the relative motion between carriage and model are recorded during the free running tests (Figure 5).

Figure 5. Manoeuvering test of a free running model, model motions in 6-DOF are detected by motion capture system

Removable wave generators

The experiences and results obtained in various projects in the past with "waves in ice" experiments carried out in the Arctic Environmental Test Basin (AETB) have revealed that a larger basin than the AETB is required to perform complex investigations on the behavior of waves in the ice. Therefore HSVA has decided in 2014 to enhance its testing equipment portfolio in the Large Ice Model Basin (LIMB) facility with a wave generator. The system was successfully installed in December 2014. The generator consists of four flap type wave making modules and covers the entire 10 m width of the basin. The flaps of 1.3 m height can be attached to a basement structure such that the hinge position is set 1.2 m above bottom. All four modules can be installed and removed from the basin within two hours while the basin is filled up with water (Figure 6).

The system is able to produce regular and irregular waves while the maximum wave height is limited to 0.25 m and the maximum deep water wave period is 1.8 seconds with respect to the basin depth of 2.5 m. The system is equipped with an active absorption to reduce reflections. The actual wave signal input is done via user friendly software *AwaSys™* which transmits the signal to the control software. Figure 7 shows for example the ice floe distribution along the entire basin after wave propagate through an intact ice sheet.

Figure 6. Removable flap type wave generator in Large Ice Model Basin.

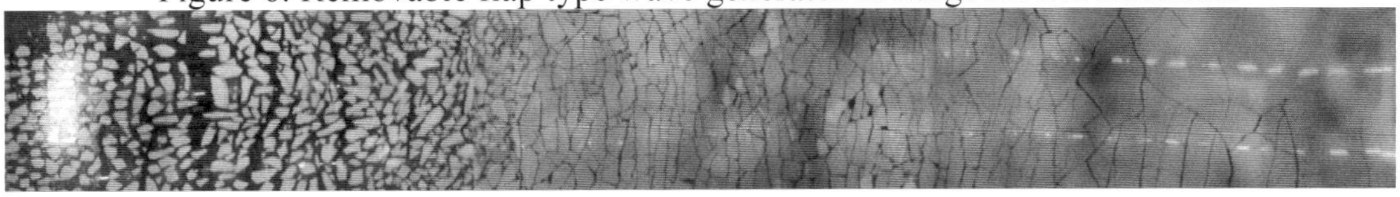

Figure 7. Aerial view of ice floe size distribution along the entire ice tank, the waves propagated from left hand side into undisturbed level ice sheet

Arctic Environmental Test Basin (AETB)

In 1989 after the EXXON VALDEZ oil spill in Prince William Sound, Alaska occurred, the previous "small ice tank" has been renovated and converted to the Arctic Environmental Test Basin (AETB). After the renovation, numerous experiments and investigations were carried out with regard to "oil in ice", testing of mechanical oil recovery systems, geophysical experiments like frazil and pancake formation and research projects in the area of artic marine biology and chemistry.

In this context it is mentioned that since 1996 possibility is given for international research groups to make use of HSVA's ice tanks (LIMB and AETB) for research projects financially supported by the European Union. Young engineers and researchers have carried out numerous projects in these facilities which were financially supported by the European Commission (www.hydralab.eu).

In 2016 the oil-resistant coating of the basin was completely renewed and a movable crane platform was designed and built (Figure 8).

Figure 8. Arctic Environmental Test Basin (AETB) with movable working platforms

MODEL ICE AT HSVA

General

By using a single dopant system the grown model ice sheet consists of a three-layer system:

- thin hard upper granular layer
- transition layer with higher dopant concentration
- weak columnar crystal layer

Figure 9 shows a thin section of first-year sea ice sample taken in the Arctic. The structure of this ice type is columnar with a thin hard granular layer on top. Since the model ice is denser than in full scale , the criteria for buoyancy similitude may not be met. Thus methods were developed at NRC Canada for EG/AD/S CD-ice (Spencer and Timco, 1990) and HSVA, Germany, for sodium chloride doped model ice to incorporate air bubbles (Figure 10) into a growing ice sheet in order to control its overall density (Evers and Jochmann, 1993).

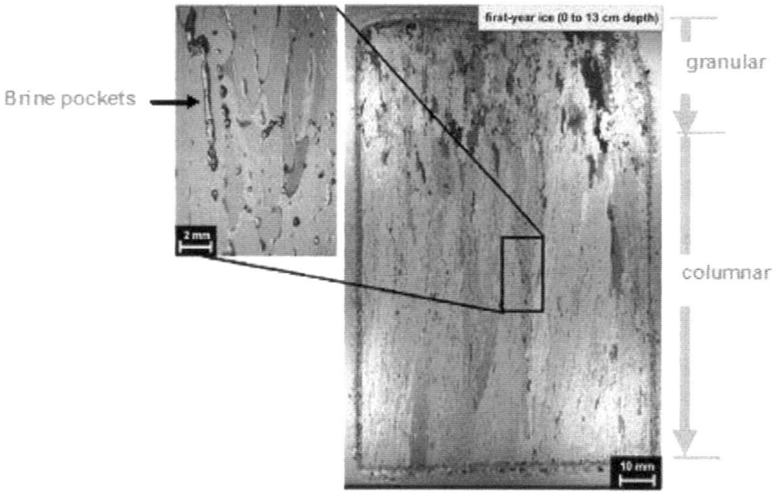

Figure 9. Thin section of first-year ice showing the ice platelets and the brine pockets long the grain boundaries (photo courtesy of M. Johnston; Timco and Weeks, 2010).

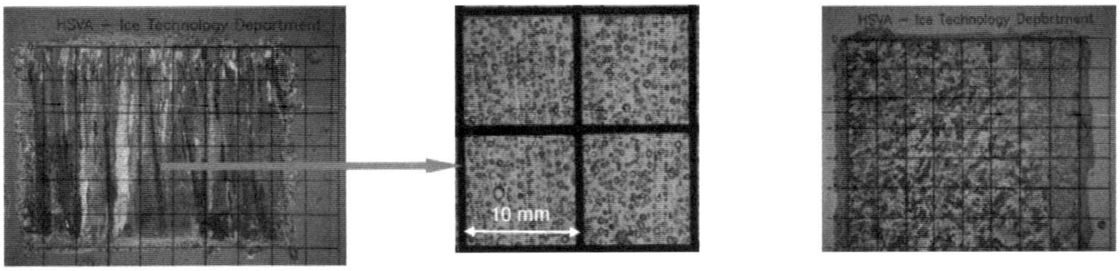

Vertical thin section
(grid 10x10 mm)

Air bubbles
200 – 500 μm

Horizontal thin section
(horizon 13 mm from top)

Figure 10. Vertical and horizontal thin sections with micro bubble inclusions in HSVA's model ice.

For this purpose a patented technique (Hellmann, et al.,1992) was developed at HSVA and is applied by pressing air saturated water under high pressure through perforated tubes fixed at the bottom of the ice tank during ice growth process at about – 20°C air temperature. When the water leaves the tubes the pressure drops. Tiny air bubbles of 200-500 μm diameter rise upwards and are embedded in the growing ice sheet. The content of micro bubbles affects also the elasticity/ strength ratio E/σ_f which increases significantly and the ice behaves more brittle when it fails. A secondary effect of air inclusions is the opaque colour of the ice giving a better contrast between model and ice in images taken from underwater (Evers and Jochmann, 1993).

Model ice production technique

The preparation of the ice sheet is started by a seeding procedure. For this purpose fresh water is sprayed under pressure into the cold air of the ice tank (Figure 11). The droplets freeze in the air forming small ice crystals which settle on the cold water surface. By this method the growth of a fine-grained ice of primarily columnar crystal structure is initiated. The water in the tank is fresh water doped with 6.8 ppt salt. After a thin closed icy skin has formed on the surface of the water the ice tank is cooled to -20°C and the ice sheet continues to grow.

During the ice growth continuously air-supersaturated water is discharged in the ice tank through the perforated pipe system fixed at the bottom of the ice tank. As a result of the pressure rundown the air dissolved in water is released and air bubbles rise upward to the top where they are embedded between the growing ice crystals as shown in Figure 10.

This impure ice sheet is inherently softer than pure-water ice but may be harder than the scaled strength desired. Once a desired thickness is achieved, the air temperature is raised to a tempering temperature, in order to weaken the ice according target strength.

As the temperature of the ice rises the dopants come out of frozen solution and form liquid brine pockets. These brine pockets slowly drain out of the ice sheet thus weakening it. The strength of the ice continues to decrease approaching an asymptotic value. This softening is often referred to as "warming-up".

Figure 11. Water is sprayed by means of nozzles under high pressure into the air of the pre-cooled ice tank (left photo) where crystal nuclei are formed which settle down on the water surface (right photo).

Due to the embedded air the model ice fails brittle, i.e. the E/σ_f-ratio increases and the size of broken ice pieces becomes more realistic compared to model ice without air bubble treatment. Another advantage of the air content in the ice is the possibility to adjust the ice density so that the density difference between ice and water is within the natural range (Evers and Jochmann, 1993).

Before reaching the target ice thickness the cooling system is switched off and the air temperature is raised slightly above the freezing point. This temperature is kept until the properly scaled strength is reached. Then, the heat transfer into the tank is reduced to a minimum in order to keep the strength nearly constant for the duration of the test. A typical "freezing-warming cycle" is shown as an example in Figure 12.

In a first step the ice tank is precooled [1]. During the seeding process the cooling process is interrupted and the air temperature rises [2]. When the seeding procedure is finished the cooling process starts again and the air temperature in the ice tank decreases to about –20°C. The duration of cooling depends on the required target ice thickness and has to be finished in due time taking a certain additional ice growth into account [3]. The heating process is initiated by heated air blown into the tank instead of cold air. The tempering of the ice causes a decrease of ice strength. During the tempering ice properties (e.g. ice strength, ice thickness, etc.) at different locations in the ice tank are measured in order to determine the point of time when the model test run can start [4]. During the test runs the air temperature is kept constant slightly above 0°C [5] until the last test run is executed and the pre-cooling procedure for production of the next ice sheet starts [6].

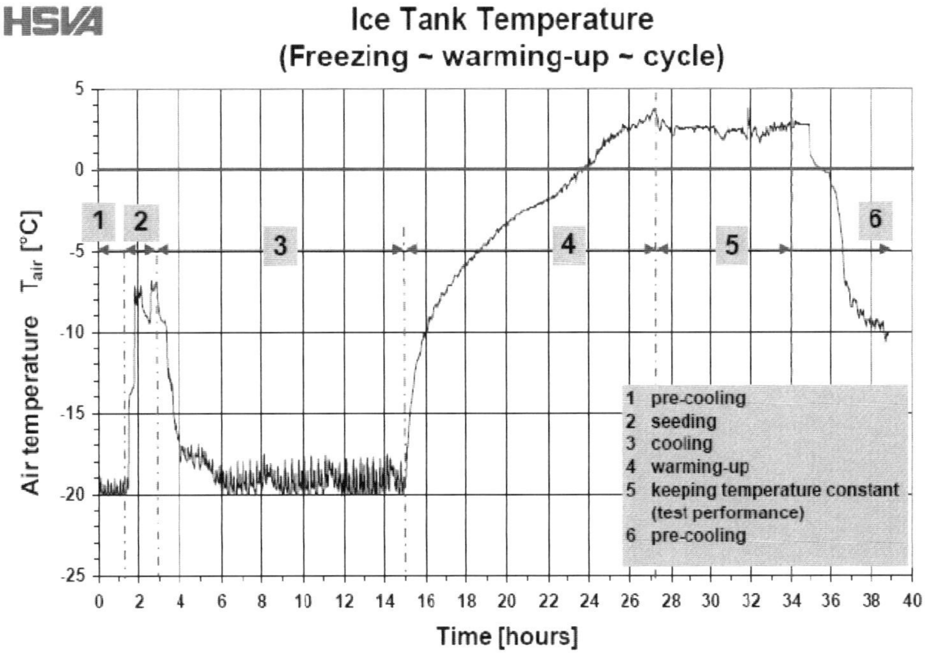

Figure 12. Example of the temporal temperature profile during a freeze-thaw-cycle.

ICE PROPERTIES

Since sea ice, in general, is an inhomogeneous, anisotropic, and nonlinear viscoelastic material, the mechanical properties of sea ice can be characterized as:

1) short-term, or generally brittle behavior, or
2) long-term, or generally ductile behavior.

The short-term properties usually considered are compressive, tensile, flexural, shear and adhesive strengths together with elastic modulus, Poisson's ratio and friction coefficient. These properties exist to a certain extent as a function of grain size, and crystallographic orientation, porosity and brine content, temperature, strain or stress rate, confinement and scale. The long-term properties are also a function of the same parameters but depend on time and the stress or strain rate that is usually held constant during testing (Sand, 2008).

Ice thickness

The thickness of sea ice is one of the most important ice parameters, because the way that ice fails is a function of the ice thickness. The ship performance of icebreaking vessels and ice forces on offshore structures exposed to ice depend on the ice thickness. Also the bearing capacity of ice sheets depends mainly on ice thickness, which has to be considered when trucks are moving or heavy loads are stored on the ice surface. Ice loads on offshore structures increase significantly with increasing ice thickness and the velocity at which vessels move through the ice is directly related to ice thickness (Timco and Weeks, 2010).

As a standard procedure in ice model testing at HSVA the ice thickness of a level ice sheet is measured at equal intervals along the center line of the basin prior to the test runs. These data are used to draw an ice thickness profile of the undisturbed ice sheet. After the test run with a ship model or offshore structure model, ice thickness measurements are carried out at equal distances on both sides along the channel broken by the model.

Flexural ice strength

Although the flexural strength of ice is not a basic material property it is an important parameter which is used in ice engineering for design of icebreaking ships and offshore structures. Many real sea ice failures occur in flexure, e.g. pressure ice ridge formation, bending ice failure by icebreaking vessels and bending of ice on sloped and conical shaped offshore structures. Therefore the performance of flexural ice strength tests is justified.

The flexural ice strength is determined by creating a bending failure in the ice and measuring the failure load and the geometry of the specimen. It is most common to do this in-situ by cutting a cantilever beam and calculating the strength by means of the formulae derived from elastic beam theory $\sigma_f = 6\,PL/bh^2$, where σ_f is the flexural strength, P is the failure load, L, b and h are the length, breadth and thickness of the cantilever beam.

When using this equation it is assumed that the ice is isotropic and homogeneous, that the root of the beam is rigidly fixed and that there is no buoyant support of the beam by the water. However none of these assumptions is completely valid. Consequently the flexural ice strength is considered as a strength index.

Since the beam geometry can significantly influence the apparent strength (Timco, 1981; Tatinclaux and Hirayama, 1982) it is important to specify the beam dimensions. The relative size and geometry of the beam is also important since it is known for both sea ice (Frederking and Häusler, 1978) and freshwater ice (Svec and Frederking, 1981) that the apparent flexural strength depends upon the length to thickness ratio of the beam; the same is true for model ice.

It is important that a standard size and geometry is adopted in order to have a meaningful comparison of the flexural strength from one ice tank to another. If the beam is short in relation to its thickness, the failure mode involves shearing as well as bending. If the beam is very long in relation to its thickness, the buoyancy of the beam affects the apparent strength (Sutherland and Evers, 2013).

The IAHR Working Group on Standardizing Testing Methods in Ice recommends a beam length to thickness ratio of 7 to10 (Schwarz et al., 1981), however tests carried out by Timco indicate that, for model ice, the length to thickness ratio should be less than this in order that the strength is not a function of this ratio. Preliminary tests on the influence of the width to thickness ratio on the apparent flexural strength of the ice indicate that strength is not a strong function of this ratio (Timco, 1981). It seems that for tests on model ice, the length to thickness ratio of the cantilever beams should be of the order of 5 to7, and the width to thickness ratio should be 1 to 2.

To load the beams a screw-driven load cell can be used to determine the breaking load of the beam. The load cell gives information on the loading rate and time. A loading time of 1 to 2 seconds to failure is considered to be realistic to represent the time in which a model icebreaker would ride up and break the model ice. For most tests, the cantilever beams are pushed downwards. In the cases where ice fails against an inclined structure the strength should be determined by pull-up tests (Timco, 1981).

There are different approaches to determine the flexural ice strength as cantilever beam tests and simple beam tests. When doing in situ cantilever beam tests, the floating ice sheet is cut to form three sides of a beam. The fourth side is uncut and connected to the floating ice sheet (root of the cantilever beam). At the free end of the beam an increasing load is applied until the beam breaks. The test procedure is easy to conduct and the results are analyzed in terms of simple elastic beam theory (Timco and Weeks, 2010). Figure 13 shows the set-up of cantilever beam test.

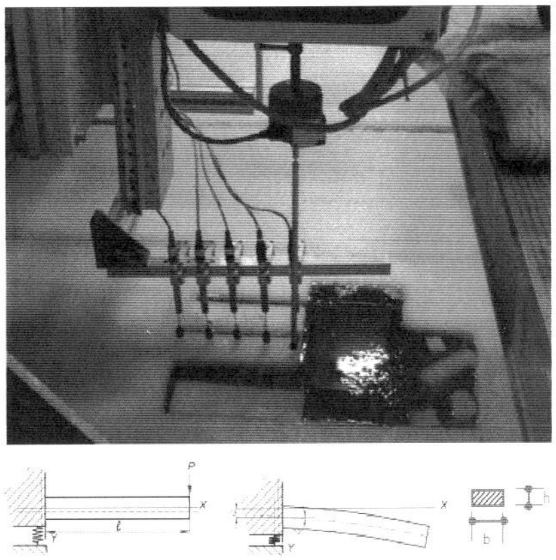

Figure 13. In situ cantilever beam test set-up with Linear Variable Differential Transformers (LVDT).

Effective modulus (elastic strain modulus)

The determination of the elastic strain modulus is based on the measurement of load and deflection of either a beam or a plate. Taking the geometry involved the strain modulus can be calculated by means of equations which are derived from theory. It is important to specify the technique used for the measurements, because various measurement techniques result in different E values in the same ice. Currently the "plate technique" as proposed by Sodhi et al. (1982) and the beam methods as proposed by Tatinclaux and Hirayama (1982) are used in ice tanks. The latter one may lead to a better method for comparing the model and full scale modulus, because more information and comparisons have been made both in laboratories and field tests.
Similar to the ice strength, the elastic strain modulus E is scaled by the geometric scaling factor of the tests. Since the proper scaling of this parameter in model tests is important, an acceptable test must be defined to measure this property. Basically, it can be measured using either static or dynamic techniques (Timco, 1981).

In the static method, the strain modulus can be determined using either a beam or plate approach. In the former, the modulus is determined by measuring both the load and deflection characteristics of a cantilever beam during loading to failure, whereas in the latter the modulus is determined by monitoring the deflection of the ice sheet as it is loaded with a known increasing load. In ice tanks which measure the strain modulus, the latter technique is usually preferred since it integrates over a large area of the ice sheet and is both simpler and quicker to perform.

In the flexural strength test, the ice can be loaded with either with a screw-driven apparatus in series with a load cell (Figure 14a) or dead weights in discrete increments (Figure 14b), or. During loading, the deflection can be measured either adjacent to or slightly away from the load (within one characteristic length) by either dial indicator gauge, LVDT or LASER-sensor which gives a continuous recording. On the other hand, the dead weights and dial indicator are much simpler and faster to use and may serve as a method of quickly determining the modulus prior to a model test.

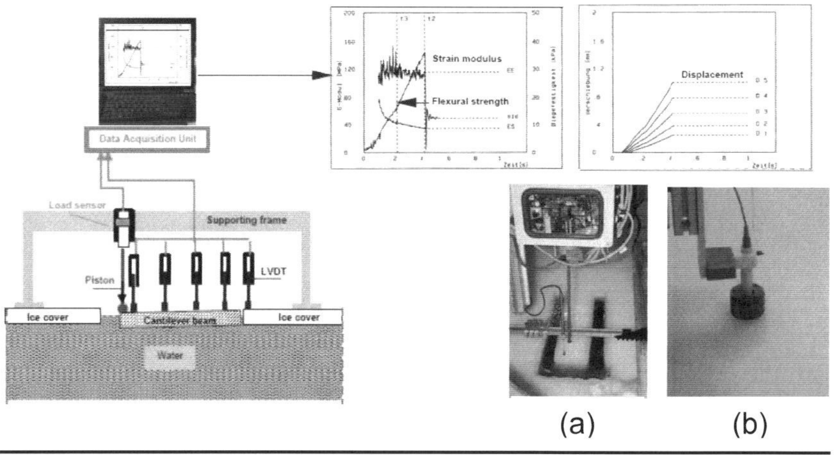

Figure 14. Determination of elastic strain modulus: cantilever beam with LVDTs (a) and plate method (b).

Uniaxial compressive strength

Measurement of the compressive strength of model ice is important, especially for the horizontal loading direction, when ice impacts rigid vertical structures it often fails in a crushing mode.

After the ice sample has been collected from the field or grown under laboratory conditions, rough cutting of prismatic ice blocks or cylindrical cores is advisable in order to obtain manageable sample size. Cubes have also been used conventionally, but the results are unreliable hence not recommended for uni-axial compression strength tests.

In the conventional uni-axial compression test, axial force is applied to the ends of a prismatic sample or circular cylinder through compliant platen that make direct contact with the test specimen. Friction between platen and specimen produces radial restraint, so that there is a triaxial state of stress near the end planes; the triaxial field is significant over an axial distance from the end planes of about one specimen radius. Placing a highly compliant sheet (elastic or plastic) between platen and specimen often changes the sign of radial end forces, but does not eliminate the triaxial stress state. The usual procedure for testing ice in uni-axial compression is to accept positive frictional restraint at the specimen end planes and to use a specimen that is long enough to provide a mid-section that is reasonably free from end-effect stress perturbations (length/diameter ratio of 2 to 3) (Schwarz et al. 1981).

At HSVA uni-axial compression tests are carried out on a rigid closed loop controlled testing machine by a hydraulic piston for load application. The specimen for compression tests should be sampled from the same location in the ice tank, where in situ flexural strength tests are being carried out at the same time and the ratio between flexural strength and compressive strength of the model ice can be determined.

Prismatic ice samples having a size of $1h \times 2h \times 4h$, with h= ice thickness, are cut and placed in between the two loading plates of the test frame. Compliant platens or thin sheet of compressible material is used in order to avoid sliding of the specimen and to apply uniform axial load to the specimen. The applied load and crosshead movement are recorded and the results can be presented as load deflection curves. Figure 15 shows the testing machine and the hydraulic aggregate in the refrigerated laboratory.

Figure 15. Uni-axial compressive strength test: (a) test set-up (b) prismatic ice sample placed between load platens (c) failure of ice sample during compressive failure.

Crushing strength (indenter tests)

One practical method is the indenter test. The indentation test is done in a place where the indenter can be put into the ice sheet through an existing hole (holes for beam or density tests). The device is placed vertically as shown in Figure 16. The tests are typically performed at several speeds, such as 30, 10, 3, and 1 mm/s.

The compressive strength is calculated using equation (1):

$$\sigma_c = \frac{F}{ci * mkbh} \qquad (1)$$

where

F = force (measured)
m = shape factor (round structure 0.9)
k = contact factor (0.4 - 0.7)
h = ice thickness
b = diameter of indenter
ci = constant, function of b/h

Figure 16. Indenter test with large indenter (diameter 250 mm) outfitted with tactile sensors to measure local ice pressure distribution along the perimeter (a) and small indenter (diameter 60 mm (right).

Ice density

The determination of the density of model ice using the standard technique - cutting a sample block, measuring its dimensions (i.e. volume) and weighing it - has several inaccuracies, the primary one being that a large number of the liquid impurity pockets in the ice drain out when it is removed from model ice sheet. Nevertheless, the technique can be used to obtain semi-quantitative density values, although measurements of the freeboard of model ice may prove to be a better method of ice density determination (Timco, 1981).

A more advanced method is the density measurement by the principle of displacement. Ice samples of 120 mm times 120 mm size are cut from the existing ice sheet at locations where cantilever beam tests are carried out and stored in a plastic bowl filled with water from the ice tank in order to avoid any drainage.

The specimen is submerged and the displaced water volume is collected and weighed. The buoyancy force is measured by means of a load sensor. The test set-up for the density measurement is illustrated in Figure 17.

Figure 17. Set-up for determination of ice density

The ice density is calculated by equation (2):

$$\rho_i = \rho_w - [\, F_b \,/\, (V_d * g)\,] \qquad (2)$$

where

ρ_i = ice density [kgm^{-3}]
ρ_w = water density [kgm^{-3}]]
F_b = buoyancy force [N]
V_d = displaced water volume [m^3]
g = acceleration due to gravity is 9.814 [ms^{-2}]

Friction coefficient

A testing apparatus developed and designed at HSVA is used to determine the dynamic ice friction coefficient. During the varnishing process of the model a plate of same material as the model is painted at the same time to ensure same surface characteristics. The plate is fixed on the carriage of the friction apparatus (Figure 18a) which can be moved back and forth. The ice specimen is adopted in a frame (Figure 18b), where two slightly pre-stressed load cells are mounted between this frame and the support and remains stationary while the carriage moves back and forth. A schematic diagram of the friction device used at HSVA is illustrated in Figure 18.

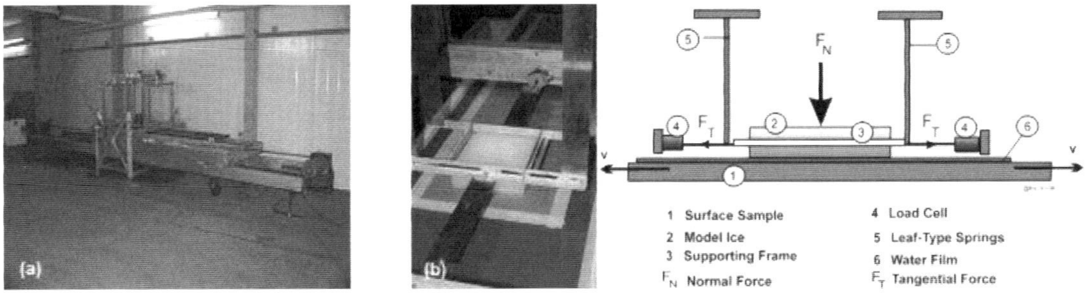

Figure 18. Ice friction testing device (a) and ice specimen adopted in a frame (b) and schematic diagram (right).

The tangential force is measured by two load cells. Two different normal loads, e.g. 50 and 100 N, are applied by adding ballast weights to the weight of the ice specimen. For each normal load 10 test runs are carried out. An average value is calculated giving the dynamic friction coefficient between the ice and the tested (hull) surface of the model.
The dynamic friction coefficient is calculated according to Coulomb's friction law:

$$C_{if} = F_t / F_n \tag{3}$$

where

C_{if} is the dynamic friction coefficient
F_t is the mean value of measured tangential force in [N]
and F_n is the normal load in [N]

The ice friction coefficient is inversely proportional to the normal load applied to a specimen, i.e. the higher the normal load, the less the ice friction coefficient C_{if}. Typical values are $0.04 < C_{if} < 0.14$.

It is recommended to carry out friction tests at least once during a testing campaign. The friction coefficient should be determined for the case that the top surface of ice specimen is in contact with the plate with the same surface property of the model. The tests should be conducted under dry and wet conditions.

MODELLING OF VARIOUS ICE CONDITIONS

Level ice

Level ice is a homogeneous ice sheet unaffected by any deformation with consistent ice thickness. The production is described exemplarily for HSVA's current model ice. It is frozen from a 0.7% sodium chloride solution in the natural way, i.e., the ice surface is exposed to cooled air. The preparation of the ice sheet is started by a seeding procedure.

Main parameters to be measured in level ice sheets are:

- flexural strength (when bending failure occurs predominantly)
- uniaxial unconfined compressive strength (when crushing failure occurs predominantly)
- ice thickness along the broken channel
- width of broken channel
- ice density
- modulus of elasticity (optional)

Figure 19 shows as an example the ice-vaning scenario of a moored Floating Production Unit model (FPU) in level ice.

Figure 19. Ice-vaning test with a moored FPU in level ice.

Broken ice floes

For ship navigation in broken ice (ice floes) conditions it is essential to consider the floe size and the ice coverage or ice concentration. Usually the two terms "ice coverage" and "ice concentration" are used to describe the ratio of ice floes and the area of open water referenced to a unit area (e.g. 1 km^2), thus both terms have the same denotation. The unit of ice coverage or ice concentration is given as tenth or in percentage. As an example six-tenths or 60% coverage means that 60% of a unit area (100%) is covered with ice.

Also for ice management around fixed or floating structures broken ice conditions have to be taken into account. Thus the performance of tests in broken ice is in many cases part of the model test program.

When modeling broken ice conditions a predefined parental level ice sheet has to be prepared first and then cut in pieces corresponding to the required size and shape, which is illustrated in Figure 20.

Figure 20. Ice floes in full scale (left) and ice floes in model scale produced in the ice tank (right).

The preparation of ice floes in the ice tank is done manually by cutting the parental level ice sheet in pieces of required size and shape according to model scale factor and distributing them to achieve the required different ice concentration (e.g. 60% to 90%). Figure 21 shows the preparation of ice floes with different sizes and the ice floe arrangement at the end of the ice tank (about 90% ice concentration).

The shape of ice floes could be quadratic-, rectangular- or rhombus shaped. Circular shape floes seldom occur except when pancake ice is taken into account. It should be noted that the most realistic movement of ice floes in the tank can be achieved by rhombus shaped floes, while quadratic shaped floes mostly do not rotate when the vessel encounters an ice floe field.

When the model encounters quadratic shaped ice floes they are pushed together and usually do not rotate, when the ice concentration is high. At high ice concentrations, e.g. c = 90% and more, the sidewalls of the basin may affect the flow of ice floes (ice clearing around the model) due to the confinement.

Figure 21. Production of ice floes in the ice tank (left photo) and 90% ice concentration (right photo).

Parameters to be measured:

- ice thickness
- ice coverage respectively ice concentration (before test run starts)
- ice floe size

Method to determine parameters:

- calliper gauge (ice thickness)
- documentation by photography (ice concentration, ice floe size and shape)

The ice concentration can be determined by means of photogrammetry techniques and image processing software. The processing of images is done with the main focus on the static ice concentration in advance of the model tests over the total ice tank length.

Before each test in broken ice, several pictures are taken by a digital still camera from a certain level above water surface along the ice tank. If the camera lens does not capture the entire tank width, two pictures or more, depending on the width of the ice tank, have to be taken along the width.

By image processing all of those pictures are stitched together using the overlap that is among them. The resulting images are evaluated to obtain the ice concentration. The results of ice concentration are based on a technique using a pixel analysis. To evaluate the image pixel by pixel it has first to be converted to grey scale. Then each pixel has a value between 0 (pure black) and 255 (pure white). A set threshold value divides the pixels into two groups; grey scale values below the threshold become black and those above become white. In this way a black and white (or binary) image is created. An example is given in Figure 22. With the ratio between the black and white pixels the coverage of ice on the water surface can be determined. The results of the pixel analysis from the overall pictures from top of the ice tank are given as global ice concentrations. A more detailed description of the method is given by van der Werff (2012), van der Werff et al. (2012), Haase et al. (2012), Zhang et al. (2012) and Sutherland and Evers (2013). Image editing software ImageJ ™ is used to analyse the images and gives information about the amount of ice floes, average ice floe size [L^2] and ice concentration [%].

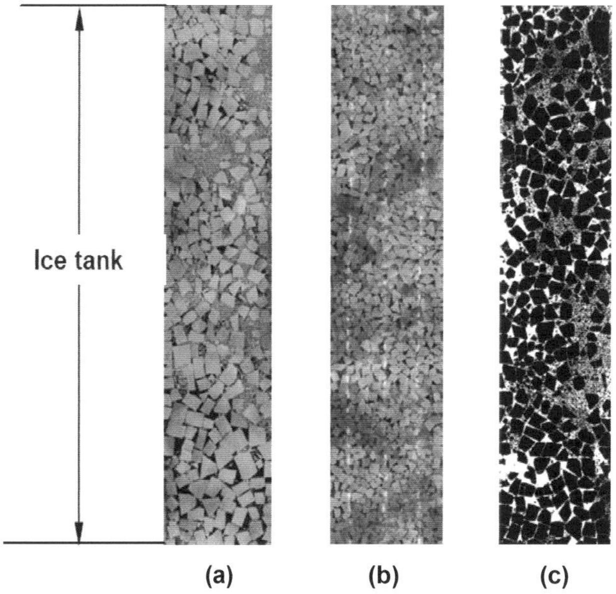

Figure 22. Examples of stitched pictures (a-b), (c) example of binary image. The pictures show the ice coverage of the entire ice tank area.

Figure 23 shows examples of a moored conical drilling unit and a ship model during dynamic positioning in broken ice conditions the ice tank.

Figure 23. Model of a moored conical drilling platform (left) and ship model (right) in broken ice conditions.

Brash ice

According to the *World Meteorological Organisation* (WMO) the term "brash ice" is defined as accumulations of floating ice made up of fragments not more than two meters across. Brash ice forms quickly during frequent passages in an ice channel filling the channel with ice fragments behind the vessel. As the broken ice consolidates between each passage, new brash ice is successively produced resulting in increased brash ice thickness which leads to higher vessel resistance.

Model tests in brash ice are required to demonstrate that a ship has sufficient propulsion power to navigate safely along a route in the ice. This is necessary for issuance of the appropriate ice class of the ship. Figure 24 presents a typical brash ice channel from the northern Baltic.

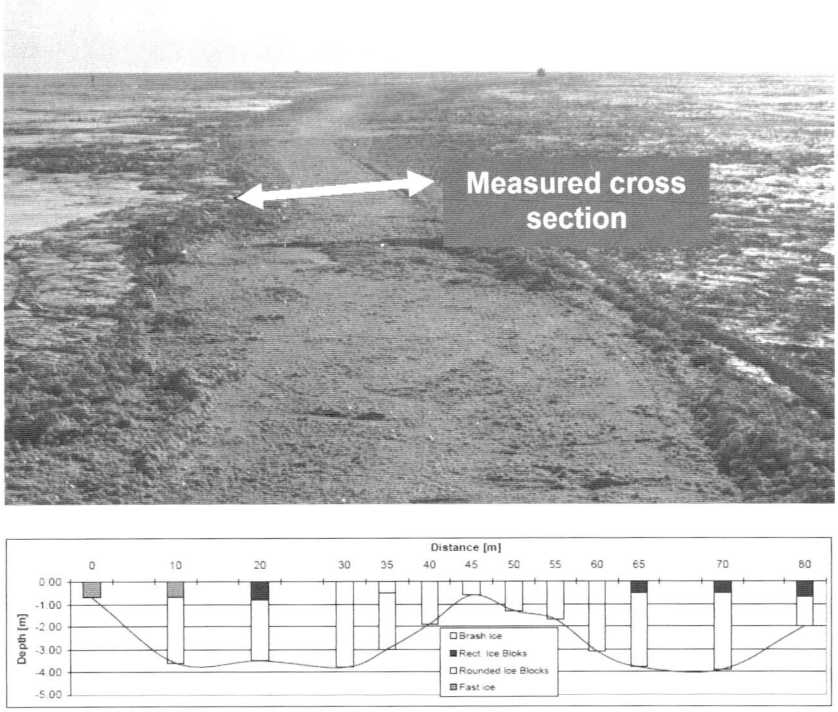

Figure 24. A photograph of a brash ice channel leading to the Finnish port Kemi and the measured cross section of the channel (after Riska, 2014).

Preparation of brash ice channels in model ice sheets

At HSVA a parental level ice sheet of adequate thickness is produced according to the standard preparation procedure. After a predefined level ice thickness has been achieved, the air temperature in the ice tank is raised. When the air temperature is close to -3°C, an ice channel with straight edges is cut into the ice sheet. This is done by means of two ice knifes being pushed by an auxiliary carriage. Thereafter, the ice stripe between the two cuts is manually broken up into relatively small ice fragments using special ice chisels (Figure 25).

In order to obtain a brash ice channel with desired brash ice thickness according to the Ice Class guidelines, the ice fragments in the channel have to be compacted. The compacting process is done in sections by means of a channel-wide grid, starting from the rear end of the channel.

For a second test run with larger brash ice thickness the ice fragments will be rearranged in the channel and furthermore compacted. Since the channel filled with brash ice becomes shorter due to the compacting procedure, some of the level ice from beside the forward part of the channel is additionally broken up and refilled into the remaining part ice-free section of the channel.

Figure 25. Preparation of brash ice fragments from level ice by means of special ice chisels (a) and channel filled with brash ice fragments (b).

Typically, the brash ice thickness is measured before the test run starts at 7 locations across and in 1m distances over the whole length of the brash ice channel. A special device is used for the measurements. This device consists of a slide gauge with a perforated plate mounted under a right angle at the lower end of the gauge. With the slide gauge the thickness of the brash ice is measured whereby the ice is floating in the water. This is done by slightly pulling the lower plate up against the bottom of the brash ice fragments. The top plate is lowered in such way that it just get in contact with the top of the brash ice. The porosity of the brash ice is calculated based on the total area and the thickness of the level ice being used to generate the brash ice. Figure 26 shows an example of a model in a brash ice channel.

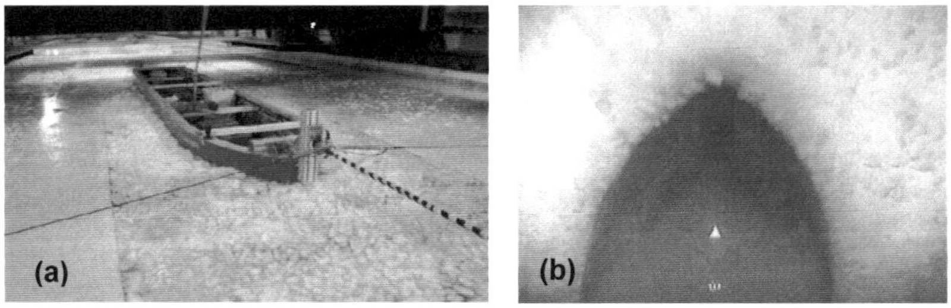

Figure 26. Model test in a brash ice channel (a) and view from underwater towards the bow of the vessel in brash ice channel (b).

Pressure ice ridges

The formation of pressure ice ridges in an ice cover is a result of changing stress regime within the plane of the ice sheet. The ridges originate from the interaction between ice floes as they collide with each other cause by wind and/or current driving forces. Pressure ridges consist of angular ice blocks of different sizes which pile up on the floes (sail) or submerge below the floes (keel). Pressure ice ridges which are grounded as a result from interaction between fast ice and drifting pack ice are called Stamukhi or grounded hummocks, according to *WMO Sea Ice Nomenclature*.

The main geometrical characteristics of pressure ice ridges are keel depth, sail height and thickness of consolidated layer as illustrated in the schematic diagram (Figure 27). In addition the parameters ice block size, porosity and bulk density characterize the ice ridge.

During the summer the ridge can undergo a substantial amount of weathering. During the weathering process the salt drains out of the ice blocks (brine drainage). When a ridge undergoes freezing a consolidated layer is formed, i.e. at cold air temperatures the water between the individual ice blocks is frozen. As a result the porosity is reduced and the bulk density increases. This has an impact on the mechanical properties and is associated with an increase in ice strength. The keel depth of a ridge is much larger than its sail height and can reach up to more than 40 m. Typically the the keel depth is about 4 - 5 times the sail height and the keel width is also 2 - 3 times wider than the sail.

Figure 28 shows an area of first year ridges in the Kara Sea investigated during an Arctic expedition in 1998. From offshore engineering and ice navigation perspective, pressure ice ridges are of particular interest and subject of investigation.

Firstly, drifting ice with embedded ridges, in particular when they are consolidated, causes high loads on offshore structures.

Secondly, when pressure ice ridges drift into shallow water areas, the ridge keel may contact the seabed and represents a risk of scouring for pipelines installed on the sea bottom.

Thirdly, when taking into account that ridged ice makes up to 40% of the overall mass of artic sea ice, ice ridges have a significant impact on ship navigation in cold regions, because these ice features may impede the navigation in ice.

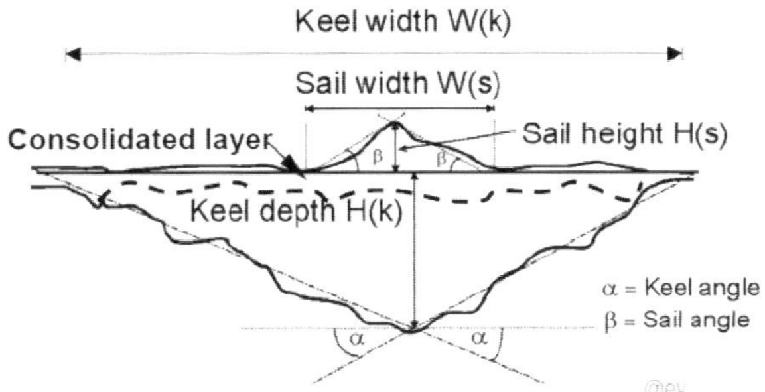

Figure 27. Main characteristic ridge dimensions.

Figure 28. First year pressure ice ridges in the Kara Sea, photo was taken during the ARCDEV expedition 1998.

Ridge production techniques

There are various methods for the production of model pressure ice ridges, which are used depending on the available test devices in the various ice tanks. The method described below is used at HSVA.

Prior to the ridge building process ice and air temperature as well as ice thickness in level ice are measured along the mid-line of the tank. When a specific ice temperature, which might be different in the various ice tanks, is achieved the ridge building process starts. The thickness and strength of the consolidated layer in the ridge strongly depends on the temperature and strength of the initial ice sheet used to build the ridge. The ice temperature is measured in the level ice sheet 2 cm below the ice surface, while the air-temperature is measured 2 cm and 1 m above the ice surface.

Depending on the required ridge geometry parameters (e.g. length, width, sail height and keel depth) the required volume of ridge fragments taking into account the ridge porosity is calculated. If the volume and porosity is known the length for cutting the ice sheet into stripes is calculated.

The width of the stripes is adjustable depending on the required size of ridge fragments. Thereafter the ice is cut along the sidewalls, and a cut transverse to the longitudinal axis of the tank is made at the spot where the ridge is to be built. At this spot a box girder with inclined face is put across the tank supporting the ridging process (Figure 29).

The free floating ice sheet with stripes is pushed against the ridge bar by means of the towing carriage's pushing board. During the ridge formation the ridge beam is successively stepwise moved forward (step 1, 2,…4 in Figure 30). The length and thickness of the floating ice sheet, and the bulk density determine the volume of an ice ridge. By this way a ridge of defined ice mass can be generated. The size of the ridge is controlled by the number of dislocations of the ridge beam. Smaller dislocations lead to a narrow ridge with larger keel depth. Wider ridges with smaller keel depth can be produced when the ridge beam is dislocated several times.

The ridge has a typical natural underwater profile (triangular or trapezoidal) and is embedded in level ice. If desired the upper layer of the ridge can be consolidated. Figure 30 shows the schematic principle of ridge preparation procedure.

Figure 29. Principal ridge preparation procedure.

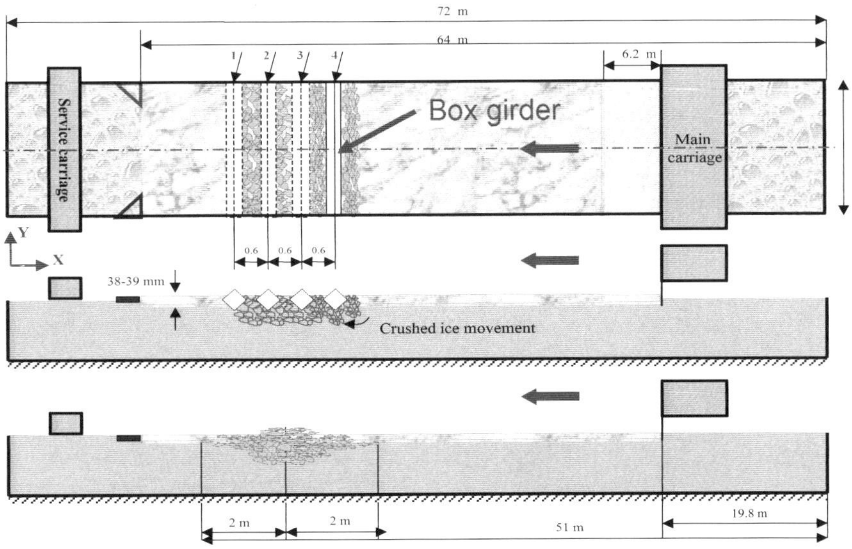

Figure 30. Schematic diagram of model ridge preparation procedure at HSVA (example).

The number of ridges to be built in one ice sheet depends on the dimension of ridges and the length of the ice tank. It is recommended that the distance between two ridges should be larger than 2.5 times the ship length to ensure that the vessel can pass the ridge completely before penetrating the second one. Figure 31 shows a model ice ridge embedded in a level ice sheet after it was built (a) and ad-frozen ridge fragments from the consolidated part of the ridge (b-c).

Figure 31. Ice ridge embedded in level ice (a) and ad-frozen ridge fragments from the consolidated layer (b-c).

After a ridge has been built the keel depth and sail height are determined by profiling. In general three profiles are taken preferably in the area of the model trace (portside – center –

starboard). This is done by pressing a stick in equidistant intervals through the ridge. At the lower end of the stick a cross-bar (Figure 32) is activated and the stick can be lifted upwards until a certain resistance indicates the bottom of the ridge. The keel depth is indicated at the reading line (yellow circle) by a marker in the upper part of the stick (Figure 32). From the measured data a profile of the ridge sail and keel can be plotted (Figure 33).

Alternatively the underwater contours of the ridge can be profiled with acoustic echo sounder and the sail topography above water by laser level (Sutherland and Evers, 2013).

Ice rubble fields are produced according to the same procedure as ice ridges are built as described above. An example of an unconsolidated rubble ice field in the ice tank is shown in Figure 34.

Figure 32. Ridge profiling device to determine keel depth and sail height.

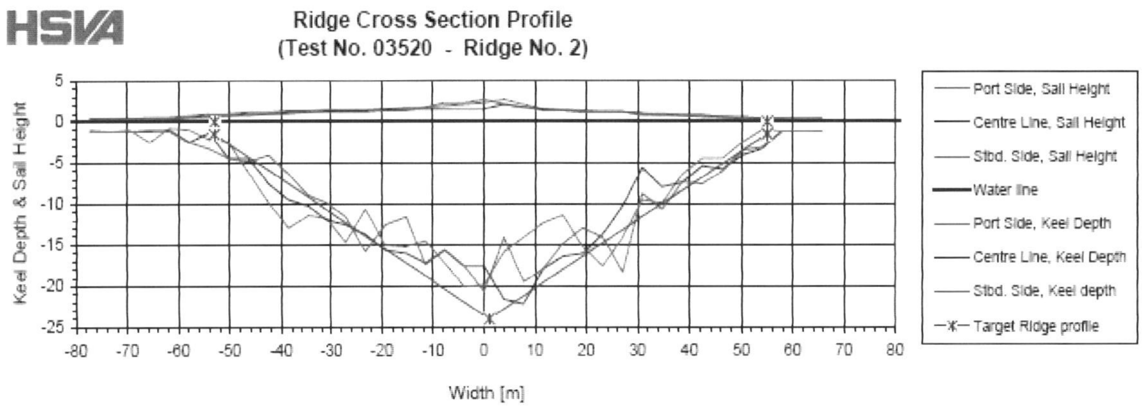

Figure 33. Model ridge cross-section (data are given as full scale values in meters).

Figure 34. Ice rubble field in the Kara Sea (left) and ruble field modeled in the ice tank.

SELECTED PROJECTS

Project: Model testing of moored structures in ice

Numerous ice model test projects with fixed and moored floating structures (e.g. (i.e. FPU, FPSO, shallow draft buoy and Spar-platform, etc.) to support the design with respect to the operability in various ice conditions (i.e. level ice, pressure ice ridges, ice floes, ice rubble fields and managed ice) have been carried out at HSVA.

The choice of a proper scale plays an important role, because the environmental conditions (ice , water depth, current, etc.), dimensions of the structure/vessel, the dimensions of the ice tank, and the configuration of the mooring system are to be considered. For the model test campaign, in general, the full scale situation has to be modelled as accurate as possible, Deviations of modelled test set-up from full scale set-up cannot always eliminated, because some features have to be simplified and need adaption to the available infrastructure.

The dynamic behaviour of the moored vessel and its interaction with the mooring system is of great importance and has to be simulated as accurate as possible in the model tests. For the conceptual design of the vessel and the mooring system the ice forces, restoring forces, mooring line forces, vessel displacement, vessel motions (heave, pitch, roll and yaw) have to be determined. Also the interaction between submerged broken ice floes and the turret respectively mooring lines is of interest (Evers and Jochmann, 2011).

The first moored floating structure in ice was tested at HSVA about 35 years ago. HSVA was contracted by Gulf Canada Ltd. to develop the design of the hull shape of the well-known conical drilling unit "Kulluk" in order to deflect submerged broken ice. Subsequently comprehensive ice model tests were conducted in level ice and ridges. Within the following years other concepts (Figure 35) like Arctic shuttle tanker, STL, SPM, offloading buoys, shallow draft buoys, moored FPU and FPSO were subject of ice model tests at HSVA (Evers and Jochmann, 2011).

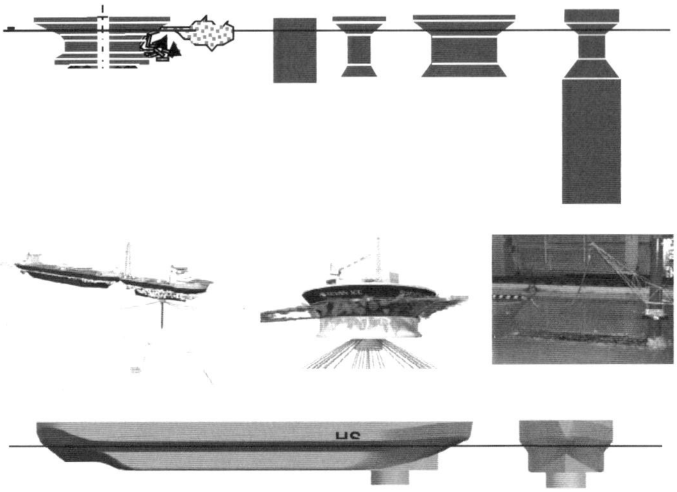

Figure 35. Different concepts of moored floating structures and vessels tested in ice at HSVA

In general the moored floating structure should be capable to break level ice sheets associated with ridges and ice rubble fields and should also be able to disconnect from the mooring system in the case of iceberg collision risk. A reconnection at a later date should be possible.

Ice management by means of purpose built ice breaking supply and standby vessels may be necessary to ensure maximum operability in case of extreme ice conditions.

The hull at the waterline should be designed to allow the level ice and the consolidated layer of the ridge to fail by bending. Even more important is the hull design with respect to ice clearing, i.e. the transportation of broken ice floes and ridge fragments around the floating structures respectively vessels. In particular for buoys the hull diameter below the waterline should be reduced as much as possible to avoid serious accumulation of broken ice pieces around the structure, which consequently increases the effective diameter and generates higher ice loads.

Set-up of a moored Arctic Tandem Offloading Terminal (ATOT)

The Arctic Tandem Offloading Terminal (ATOT) is a concept for offshore offloading in ice-covered waters . The concept exists of two units: a turret moored Offloading Icebreaker (OIB) and a shuttle tanker moored in tandem at the OIB aft as shown in Figure 36a-b (Bonnemaire, et al., 2008).

The OIB design basis is that it should be able to withstand the interaction from severe ice conditions, such as unmanaged deep ice ridges in straight drift. The operational strategy is that the tanker may disconnect and reconnect to the OIB if required, however the OIB should be able to remain moored at the location as long as possible. The OIB prototype is thus moored on mooring lines fixed on a sub-surface loading buoy connected inside a turret. The OIB is optimised for icebreaking with an icebreaking spoon bow and reamers on the side to increase the manoeuvrability in ice and break a wider channel to reduce the ice forces on the tanker. (Bonnemaire, et al., 2010)

(a) (b)

Figure 36. The OIB with 4 pods penetrating a ridge, with the tanker moored in close tow in the background (model scale 1:24) (a) and artist impression of OIB (b).

For the tests a "dry" mooring system as shown in Figure 38 and Figure 39 was designed and connected to the transverse carriage at the rear of main towing carriage facilitating the execution of manoeuvring or dynamic positioning tests in the ice tank with free sailing icebreaking vessels in level ice, ridges and ice rubble fields.

This device is very suitable for the execution of tests regarding slow or rapid ice drift change scenarios. In this case the turning radius of the vessel and the ice drift speed can be varied, when the model is guided through the ice by the towing carriages and the ice sheet is kept stationary (Figure 37).

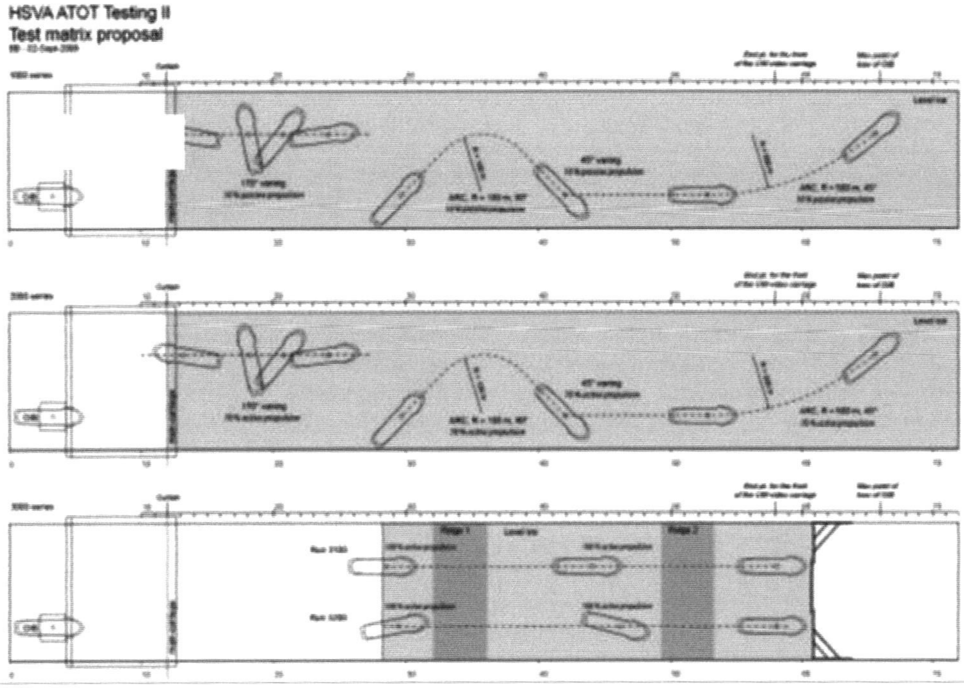

Figure 37. Schematic diagram of slow or rapid ice drift change scenarios, the turning radius of the vessel and the ice drift speed is varied.

For practical reasons, a "dry" mooring system mounted "upside down" above the model is more convenient to handle, i.e. easy access to the mooring system without discharge of tank water (Figure 38). In the test campaign the model was tested in different ice conditions (e.g. level ice, ridges, ice rubble and managed ice).

Figure 38. OIB model with "dry" mooring system (a); FPU model in fixed mode connected to the transverse carriage (b).

Figure 39 presents a schematic diagram of the dry mooring device. A vertical column is fixed to the bottom of the vessel by means of a universal joint and a triaxial load cell. The column is thus fixed in surge, sway and heave to the vessel. However the column is free to heave in the mooring frame. The column is mounted on the transverse carriage which can move simultaneously with the main carriage in x and y-directions. On each beam a coil spring - mooring line and load cell system is mounted, that controls the force in surge and sway between the vessel and the towing carriage.

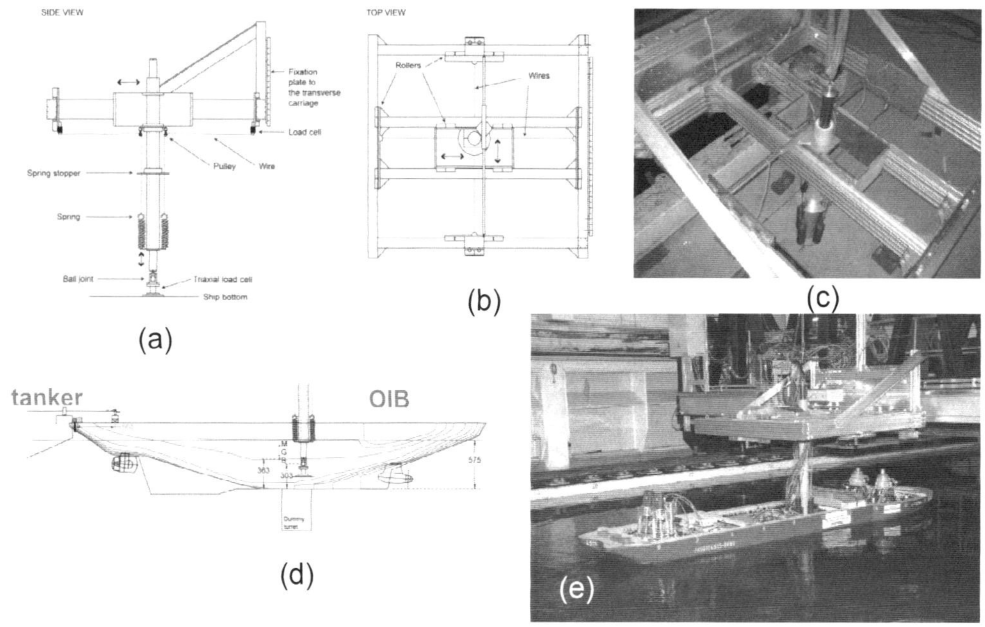

Figure 39. Schematic diagram of the dry mooring device (Evers and Jochmann, 2011).

Project: Investigation of influence of structural compliance and slope angle on ice loads and dynamic response of cylindrical and conical structures

Slender cylindrical structures are often found in offshore applications such as wind turbines, legs of GBS etc. Their simple, cost-efficient shape provides reasonable open water properties. However, ice loads on vertical structures are generally high, and flexible cylindrical structures may be set into lock-in vibrations due to structural feedback to the ice load. Such vibrations should be avoided to minimize the risk of severe structural damage (Ziemer, et al., 2015)

Model tests in ice have been conducted with cylindrical and conical, compliant structures exposed to drifting level ice to investigate the influence of slope and compliance on the ice load and its breaking frequency. Main goal of the test campaign was to study the importance of structural feedback during ice-cone interaction. Four shapes were tested: 50°, 60°, 80° and 90° slope angle. The cylinder was tested in order to define the worst case scenario regarding magnitude of ice load and severity of ice-induced vibrations. Stiffness and natural frequency of the structure were chosen similar to typical values for offshore wind turbine support structures. All shapes were tested both in a compliant and fixed configuration.

The breaking frequency was found to be more pronounced for the lower slope angles where the ice failed in flexural failure only, while a transition to crushing failure as observed on a cylindrical structure takes place at 80° cone angle already. This results in significantly higher ice loads on the 80° cone than on those with lower angles, but a reduced risk of severe ice-structure interaction due to the unsteady nature of the mixed mode breaking process. Although the breaking frequency is rather constant e.g. during ice impact on the 60° cone, it was not possible during the model tests to match the ice drift speed and the dynamics of the structure in a way that causes resonance. However, model test results prove that there is a risk of conical structures with low natural frequencies and low stiffness in ice plane being excited by periodic ice failure in their natural frequency, thus response amplification may take place and pose a risk to the structural integrity of conical offshore structures exposed to sea ice (Ziemer, et al., 2015).

Figure 40 shows a schematic diagram of the test set-up and upward bending ice failure on a conical structure.

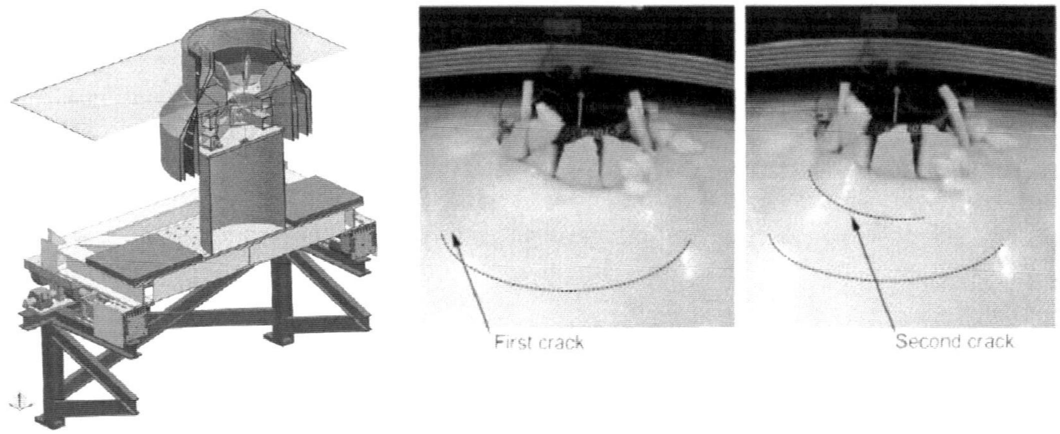

Figure 40. Schematic diagram of the model set-up (left); photographs showing the two-time cracking during upward bending ice failure on the cone (right).

Load and response on the cylindrical reference structure in case of lock-in

Only short events of lock-in have been observed, all interrupted by bending failure, eliminating the chance of persistent lock-in vibrations. However, several short events can be used as reference case to define the worst case scenario. All events show similar high forces and response amplitudes, i.e. 3500 N to 4000 N (model scale) maximum ice load and 4 mm periodic amplitude of oscillation. Figure 41 shows FFT of load and response of a lock-in like event. Next to a frequency close to the natural frequency (3.15 Hz, i.e. 10% lower than natural frequency), the first harmonic is very pronounced in the frequency spectrum of the ice load (Ziemer, et al., 2015).

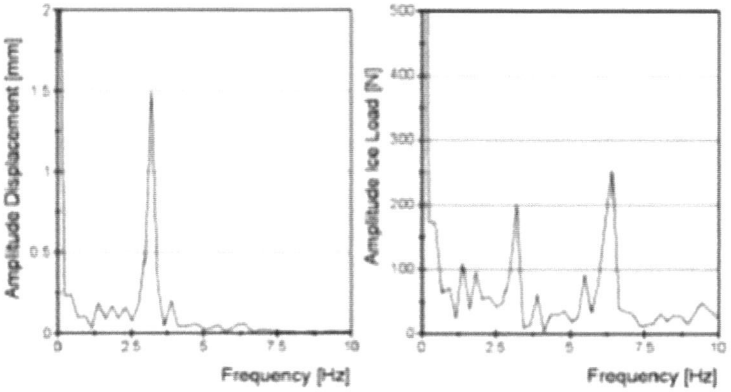

Figure 41. FFT of response (left) and load (right) during lock-in like event of 5 seconds duration (Ziemer, et al.,2015)

EU-Hydralab-Plus project: Loads on Structure and Waves in Ice (LS-WICE)

A multi-group research project was conducted recently under the EU-Hydralab+ Transnational Access project: Loads on Structure and Waves in Ice (LS-WICE). There are three parts to this investigation: (a) ice fracture under wave actions, (b) wave attenuation/dispersion in broken ice covers, and (c) ice-structure interaction under wave conditions. This paper focuses on the wave attenuation/dispersion part of the investigation. A level ice sheet was produced first. The physical and mechanical properties were measured in the Large Ice Model Basin (LIMB) before passing through a range of waves to obtain the attenuation/dispersion relation. Wave measurements were monitored with pressure transducers and ultrasound sensors. The test sequence began with a continuous ice sheet followed by broken ice sheet. The broken ice sheet was produced by cutting the continuous ice sheet into uniform size floes. A range of floe sizes and wave periods were tested (Cheng, et al., 2017; Herman, et al., 2017; Tsarau, et al., 2017). The test program was executed in open water and in 4 ice sheets. A cylindrical structure was stationary anchored in ice sheet no. 4 in order to measure the ice and wave forces acting on the cylinder as shown in Figure 44 (Tsarau, et al., 2017). Each of these tests had different purposes. The open water case was performed to characterize the wave tank behavior and to calibrate all sensors. Ice sheet no. 1 was produced to perform the break-up test. Ice sheets no. 2 and no. 3 were used to test wave attenuation and dispersion under different monochromatic waves. Ice sheet no. 4 was produced to test ice-structure interaction in the presence of a wave field.

Ice fracture under wave actions

The effect of ice floe size relative to wavelength on wave propagation was investigated. Parameters to be determined from these two inputs include the wavelength and the attenuation rate.

A level ice sheet was formed by seeding method first to create a granular layer. Sustained cooling thickened this top layer by columnar growth until the whole sheet reaches about 0.035 m thickness. The Young's modulus, flexural strength, ice thickness, density and salinity were then measured.

A sequence of wave tests with different wave periods was first applied to the intact level ice cover. The intact level ice was then cut longitudinally into six 1.5 m wide strips parallel to the wave tank wall. Transverse cuts were applied to create floes of uniform sizes. Prior to the tests the ice sheet was cut into 6 m long and 1.5 m wide ice floes. After applying a set of wave tests with different periods, these floes were further cut into 3 m, then 1.5 m, and finally 0.5 m floes. Each time a pattern of uniform size floes was produced. Figure 42 shows the floe field for the 6 m case and the 0.5 m case (Cheng et al., 2017).

Figure 42. The regular width of ice floes tested were 1.5 m wide. The length of the floes started with 6 m, as shown in the left side photo and ended with 0.5 m, as shown in the right side photo, both taken just after the cutting prior to the wave tests. The whole ice cover was refrained from spreading by the boom at the leading edge. The first row of the 6 m long floes was broken by waves as evident from the right side photo (Cheng, et al., 2017).

The wave tests included wave periods between 0.9 to 2 s, starting from long wave to minimize the risk of breaking floes. However, the 6 m floes were proven to be too long to keep their integrity. The first row of these floes did break, as can be seen from Figure 42 (right side photo).

A stop-go wave test was performed. Each test run lasted between 80 to 90 s, sufficient for each of the wave tested to travel the entire wave tank and return to ice edge. However, because no beach could perfectly absorb wave energy, to avoid effects from reflection, only a portion of the wave record before any wave reflection will be used for further analysis (Cheng et al., 2017).

Wave attenuation/dispersion in broken ice covers

The test sequence began with a continuous ice sheet followed by broken ice sheet. A few markers were placed on the ice sheet tracking the 3D motion with infrared camera system (Qualisys™ Motion Capture System). Prior to the tests the physical and mechanical properties were measured before passing through a range of waves to obtain the attenuation/dispersion relation. Pressure transducers and ultrasound sensors located at different positions along the ice tank were used to monitor the wave propagation through the ice. The progress of breaking was recorded with cameras placed above as well as sideways from the tank. The measurements consisted of two groups of tests, with a constant wave period in each group. The wave amplitude was increased stepwise, starting from a value too low to break the ice, until a major fragmentation of ice occurred. After each group of tests, the locations of cracks were determined and the floe-size distribution was estimated (Herman et al., 2017).

Figure 43. Snapshots of the ice sheet from the GoPro Silver camera after test 1440 (a), 1450 (b) and 1510 (c), after removal of fish-eye effects. The Qualisys markers are highlighted with yellow circles for better visibility. In (b) and (c), the location of the initial crack is marked with a dashed blue line. Additionally in (b), the time sequence of the crack formation in the initial phases of test 1450 is shown with lines of different colors (according to the color code in the upper left corner of the image); (Herman, et al., 2017).

Ice-structure interaction under wave conditions

Another experiment in this project focused on studying combined wave and ice actions on a fixed structure represented by a cylinder with a diameter of 0.69 m. The structure was located approximately in the middle of the ice tank at 43.7m from the wave maker as shown in Figure 44. The ice concentration near the structure was approximately 100%. Apart from the structure and the ice field, Figure 45 shows a number of sensors and video cameras that were used in the tests: 12 pressure sensors and 2 ultrasound sensors to measure wave height; Qualisys cameras and markers to monitor ice deflection; an Inertial Measurement Unit (IMU) to record the accelerations of an ice floe; video cameras on the ceiling, side walls, above the structure and under the water surface to monitor the ice cover and the ice-structure interaction area. A set of load cells that were installed inside the structure and fixed to a rigid carriage was used to measure loads on the structure. As shown in Figure 44 the IMU was installed at the center of an ice floe adjacent to the structure to obtain the response in waves and impact accelerations (Tsarau, et al., 2017).

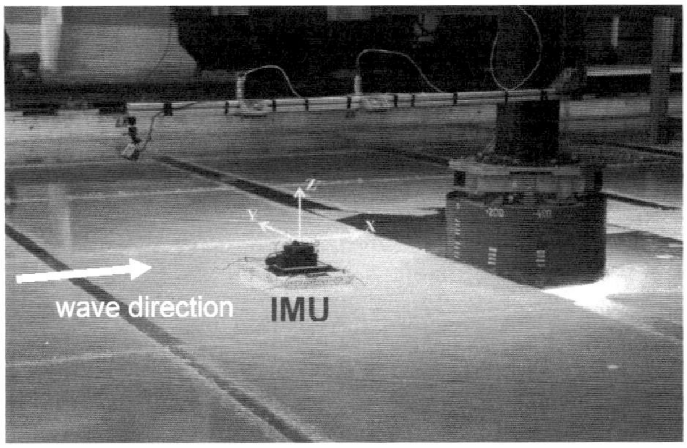

Figure 44. Structure and an adjacent ice floe equipped with an IMU.

The layout of the locations of sensors used in the ice-structure interaction tests is shown in Figure 45.

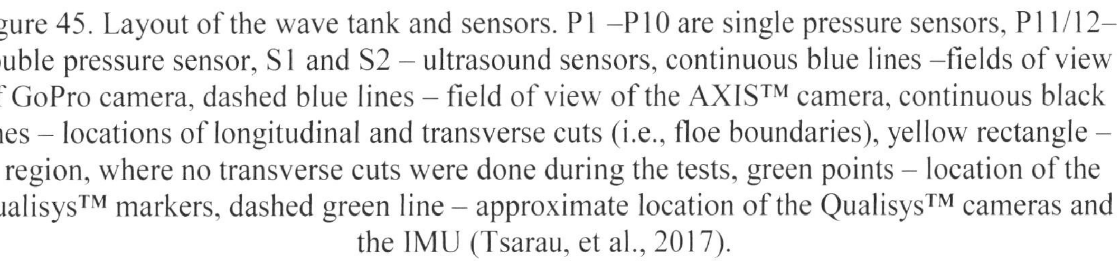

Figure 45. Layout of the wave tank and sensors. P1 –P10 are single pressure sensors, P11/12– double pressure sensor, S1 and S2 – ultrasound sensors, continuous blue lines –fields of view of GoPro camera, dashed blue lines – field of view of the AXIS™ camera, continuous black lines – locations of longitudinal and transverse cuts (i.e., floe boundaries), yellow rectangle – a region, where no transverse cuts were done during the tests, green points – location of the Qualisys™ markers, dashed green line – approximate location of the Qualisys™ cameras and the IMU (Tsarau, et al., 2017).

The global loads from both only waves and combined ice/wave action were measured. Waves with periods T = 1.5 s – 2.0 s and heights 25 mm – 75 mm were used in the tests. Two additional test runs were performed with wave heights H = 100 mm and 200 mm and wave period T =1.6 s; major fractures of the ice floes were observed in these test runs. (Tsarau, et al., 2017).

Among the parameters influencing the impact force, following was identified:

- wave height and period (also wave length for shallow water);
- ice-floe kinematics in waves, including momentum exchange due to floe-floe interaction;
- Interaction area between the structure and the ice floe.

A major difference of this experiment compared to previous experiments on impacts of a wave-driven ice mass on a structure is the utilization of a wave tank fully covered with ice floes instead of considering only one isolated ice mass and a structure. Full ice coverage ensures a better representation of MIZ conditions and allows taking into account floe-floe interactions and the wave dispersion effects in ice-covered water.

CONCLUSIONS

The increase in offshore activities in Arctic regions as well as the further development of icebreaking ships with regard to performance in ice, strength and safety over the past 50 years has stimulated to further develop hydrodynamic test facilities in order to meet the extended requirements. Due to the diversity of the projects, especially in the offshore sector and development of the Northern Sea Route, HSVA responded with the expansion of its facilities. A new type of model has been developed and introduced. Test procedures, measurement technology and data analysis were standardized and extended to simulate complex test scenarios. Increasingly, physical experiments are being performed to validate numerical simulations. Currently, a trend is observed that due to climate change the ice tanks are not only used by commercial clients from shipbuilding and oil industry but more often used for investigations by researchers from universities, because it is possible to model different scenarios under controlled environmental conditions, which is a cost-effective alternative compared to expensive field tests in the Arctic.
Nowadays numerical models are becoming increasingly important in the field of ice engineering and Arctic technology. However numerical simulations always have to be verified. Consequently the execution of physical ice model tests is furthermore essential and the ice testing facilities have to adapt to the future challenges.

ACKNOWLEDGEMENTS

The work described in this publication was partly financially supported by the European Community's Horizon2020 Research and Innovation Programme through the grant to HYDRALAB-PLUS, Contract no. 654110.

REFERENCES

Bonnemaire, B., Lundamo, T., Evers, K.-U., Løset, S., and Jensen, A.,2008. Model Testing of the Arctic Tandem Offloading Terminal Mooring Ice Ridge Loads. *19th IAHR International Symposium on Ice "Using New Technology to Understand Water-Ice Interaction"*, Vancouver, British Columbia, Canada, July 6 to 11, 2008.

Bonnemaire, B., Aksnes, V., Lundamo, T., Evers, K.-U., Sveinung Løset, S., and Ravndal, O., 2010. Ice Basin Testing of a Moored Offloading Icebreaker in Variable Ice Drift: Innovations and New Findings. *Proceedings of the HYDRALAB III Joint User Meeting*, Hannover, February 2010.

Cheng, S., Tsarau, A., Li, H., Herman, A., Evers, K.-U., and Shen, H. (2017). Loads on Structure and Waves in Ice (LS-WICE) project, Part 1: Wave attenuation and dispersion in broken ice fields, *Proceedings of the 24th International Conference on Port and Ocean Engineering under Arctic Conditions*, June 11-16, 2017, Busan, Korea

Evers, K.-U. (2016), Modeling ice processes in laboratories and determination of model ice properties, in *Cold Regions Science and Marine Technology* , [Eds. Pat Langhorne, Hayley Shen, Matti Leppäranta, Knut Høyland], in *Encyclopedia of Life Support Systems (EOLSS)*, Developed under the Auspices of the UNESCO, Eolss Publishers, Oxford, UK, [http://www.eolss.net]

Evers, K.-U. and Jochmann, P. (2011). Experiences at HSVA with model testing of moored structures in ice-covered waters, *Proceedings of the 21st International Conference on Port and Ocean Engineering under Arctic Conditions (POAC)*, July 10-14, 2011 Montréal, Canada

Evers, K.-U. and Jochmann, P., (1993) .An Advanced Technique to Improve the Mechanical Properties of Model Ice Developed at the HSVA Ice Tank, *Proceedings of the 12th International Conference on Port and Ocean Engineering under Arctic Conditions (POAC)*, 17- 20 Aug. 1993, Hamburg, Vol. 2, pp. 877-888

Frederking, R. and Häusler, F.U., (1978). The flexural behaviour of ice from *in situ* cantilever beam tests, *Proc. IAHR Symp. on Ice Problems*, Part I, Lulea, Sweden, pp. 197-215.

Haase, A., van der Werff, S. and Jochmann, P., (2012). DYPIC - Dynamic Positioning in Ice: First Phase of Model Testing, *Proceedings of the 31st International Conference on Ocean, Offshore and Arctic Engineering OMAE 2012*, July 1-6, 2012, Rio de Janeiro, Brazil, Paper No. OMAE2012-83455.

Hellmann, J.-H., Rupp, K.-H., Evers, K.-U. and Häusler, F.U., (1992). *Method and devices for forming a sheet of ice for, in particular, tests with ship models or models of marine structures*, Patent No. PCT/EP1992/000305; DE 4106930 A1.

Herman, A., Tsarau, A., Evers, K.-U., Li, H. and Shen, H. (2017). Loads on Structure and Waves in Ice (LS-WICE) project, Part 2: Sea ice breaking by waves, *Proceedings of the 24th International Conference on Port and Ocean Engineering under Arctic Conditions (POAC'17)*, June 11-16, 2017, Busan, Korea

Kim, J.-H. and Choi, K., (2011). Comparison of EG/AD/S and EG/AD model ice properties, *International Journal of Ocean System Engineering* ,1(1) (2011) 32-36 DOI 10.5574/IJOSE.2011.1.1.32

Lee, C.-J., Cho, S.-R., Jeong, S.-Y., Chun, E.-J., (2010). MOERI's Ice Model Test for Korean Icebreaking Research Vessel, *Proc. Intern. Conference and Exhebition on Performance of Ships and Structures in Ice, ICETECH 2010* , September 20-23, 2010, Anchorage, Alaska, USA

Riska, K.(2014). *Factors influencing the power requirement in the Finnish-Swedish Ice Class Rules*, Research Report Finnish Transport Safety Agency - Swedish Maritime Administration, Research Report No. 67, 2014

Sand, B., (2008). *Nonlinear finite element simulations of ice forces on offshore structures*, Doctoral Thesis, Luleå University of Technology, Department of Civil, Mining and Environmental Engineering Division of Structural Engineering

Schwarz, J. *et al*. (1981). Standardized testing methods for measuring mechanical properties of ice, *Cold Regions Science and Technology*, 4 (1981) 245-253 245, Elsevier, Amsterdam - Printed in The Netherlands

Sodhi, D.S., Kato, K., Haynes, F.D. and Hirayama, K. (1982). Determining the Characteristic Length of Model Ice Sheets, *Cold Regions Science and Technology*, 6, 1982, pp.99-104.

Spencer, D.S. and Timco, G.W., (1990). CD Model ice: A process to produce correct density (CD) model ice. *In Proceedings of the IAHR 10th International Symposium on Ice*, August, Espoo, Finland. International Association for Hydraulic Research, vol. 2, p. 745 755.

Sutherland, J. and Evers, K.-U. (editors), (2013). *Foresight study on laboratory modelling of wave and ice loads on coastal and marine structures*, Deliverable D2.3 , EC contract no. 261520, HYDRALAB-IV.

Svec, O.J. and Frederking, R.M.W., (1981). Cantilever beam tests in an ice cover: Influence of plate effects at the root, *Cold Regions Science and Technology* 4(2): 93-101

Tatinclaux, J.C. and Hirayama, K. (1982). Determination of the Flexural Strength and Elastic Modulus of Ice from *In Situ* Cantilever Beam tests, *Cold Regions Science and Technology,* 6, 1982, pp.37-47.

Timco, G.W., (1981). Invited commentary: on the test methods for model ice. *Cold Regions Science and Technology*., 4: 269-274.

Timco, G.W., (1986). EG/AD/S: A New Type of Model Ice for Refrigerated Towing Tanks, *Cold Regions Science and Technology*, 12 (1986) 175-195, Elsevier Science Publishers B.V.,Amsterdam - Printed in The Netherlands

Timco, G.W. and Weeks, W.F. (2010). A review of the engineering properties of sea ice, *Cold Regions Science and Technology* 60 (2010) 107 129

Tsarau, A., Sukhorukov, S., Herman, A., Evers, K.-U., Li, H. and Løset, S. (2017). Loads on Structure and Waves in Ice (LS-WICE) project, Part 3: Ice-structure interaction under wave conditions, *Proceedings of the 24th International Conference on Port and Ocean Engineering under Arctic Conditions (POAC'17)*, June 11-16, 2017, Busan, Korea

Valanto, P. and Schwarz, J., (1999). Eisbrechtechnik in Deutschland, *100 Jahre Schiffbautechnische Gesellschaft*, Berlin-Heidelberg- New York, Springer, 2001, pp 148-156, DOI: 10.1007/978-3-642-93390-5

van der Werff, S. (2012). *Dynamic positioning in Ice – Assessment of hull forces induced by managed ice fields*, TU Delft, Master thesis, June 2012.

van der Werff, S. and Haase, A., (2012). Influence of the Ice Concentration on the Ice Loads on the Hull of a Ship in a Managed Ice Field, *Proceedings of 31st International Conference on Ocean, Offshore and Arctic Engineering OMAE 2012*, July 1-6, 2012, Rio de Janeiro, Brazil, Paper No. OMAE2012-83927.

Ziemer, G., Evers, K.-U. und Voosen, C., 2015. Influence of structural compliance and slope angle on ice loads and dynamic response of conical structures, *Proceedings of the ASME 2015 34th International Conference on Ocean, Offshore and Arctic Engineering OMAE2015*, May 31-June 5, 2015, St. John's, Newfoundland, Canada, [OMAE2015-41769].

Zhang, Q. and van der Werff, S. (2012). Image Processing for the Analysis of an Evolving Broken-Ice Field in Model Testing, *Proceedings of the 31st International Conference on Ocean, Offshore and Arctic Engineering OMAE 2012*, July 1-6, 2012, Rio de Janeiro, Brazil, [OMAE2012-84117].

Proceedings of the 24th International Conference on
Port and Ocean Engineering under Arctic Conditions
June 11-16, 2017, Busan, Korea

Analysis of Ship Structural Response due to Ice Collision

In Sik Nho[1] , Seung Jae Lim[1], Young-Taek Oh[1], Jaewon Ryu[1] and Sung-Chan Kim[2]
[1] Chungnam National University, Daejeon, Korea
[2] Inha Technical College, Incheon, Korea

ABSTRACT

The ice impact load is considered as the most significant factor in the structural design of ships and offshore structures operating in arctic sea. It results from the mutual interactions during the ice-structure collision process and which is closely related to mechanical characteristics of ice. The strength and failure properties of ice, however, show very complicated aspect varying with temperature, volume fraction of brine, grain size, strain rate and so on. So it is nearly impossible to establish a perfect unified material model of ice satisfying all the mechanical characteristics completely. Therefore, in general, ice collision analysis was carried out by relatively simple material models considering only specific aspects of mechanical characteristics of ice and it would be the major cause of inevitable errors in the analysis. Especially, it is well-known that the most distinctive mechanical property of ice is high dependency on strain rate. Ice shows brittle attribute in higher strain rate while it becomes relatively ductile in lower strain rate range. In this study, the simulation method of ice collision to ship hull using the nonlinear dynamic FE analysis was dealt with. To consider the strain rate effects of ice during ice-structural interaction, strain rate dependent constitutive model in which yield stress and hardening behaviors vary with strain rate was adopted. To reduce the huge amount of computing time, the modeling range of ice and ship structure were restricted to the confined region of interest. Under the various scenario of ice-ship hull collision, the structural behavior of hull panels and failure modes of ice were examined by nonlinear FE analysis technique.

KEY WORDS : Ice collision simulation; Fresh water ice; Sea ice; Material property; Strain rate dependency; Nonlinear FE analysis; LS-DYNA

INTRODUCTION

Along with the development of oil and natural gas resources in the arctic region, recently many shipping companies and shipbuilders have interest in the construction of various types of ships with ice grades as an attempt to reduce the transportation cost through exploiting the arctic sea route. However, since the arctic sea is always covered with thick ice except during the short summer season, there is a growing need for arctic engineering to overcome the cold ice sea in order for ships to operate in the area.

Ice loads, which are considered to be the most important in the structural design of vessels and offshore plants operating in the arctic sea, take place as interactions due to collision of ice and structures. However, the collision phenomenon between ice and structures can be varied very diversely, and the analysis method of ice load can be changed according to occasions. In addition, since magnitude and duration time of the ice loads show various aspects depending on the strength of the structure and the material characteristics of the ice during the collision, there are still many difficulties in accurate estimation of the ice collision loads. In order to secure the safety of ships operating in the arctic region, therefore, it is essential to establish reasonable estimation methods of ice load, and apply them to the structural analysis and design process.

As mentioned above, ice loads due to ice-structure interaction are very sensitive to the collision conditions and material properties of ice. The mechanical property of ice is, however, so diverse and complicated depending on the various environmental conditions different from common structural materials. Therefore, it is very difficult to propose an unified material model for ice strength that can be applied to the all cases of ice-structure interaction. Moreover, since the ice-hull collision condition is also very diverse, the design of arctic structures should be carried out considering the rational ice collision conditions. The ice colliding scenarios proposed by DNV(2006) and ABS(2009) are typical examples of them.

Recently, studies for the ice collision problem by numerical approach have been attempted (Kujala, et al, 2007). Han et al(2008) and Lee et al(2009) carried out the ice collision analysis using nonlinear finite element analysis codes in order to evaluate the structural safety of LNG carriers operating in the arctic region.

In the numerical analysis, however, the obtained ice collision loads are also very irregular due to the diversity of material characteristics of ice and the collision conditions and they are the main source of inevitable uncertainty in ice load estimation. Therefore, it would be necessary to set up a rational procedure for the numerical simulation to evaluate ice collision loads.

This study dealt with the simulation method of ice-hull collision using the nonlinear dynamic FE analysis. In order to show the standard structural safety evaluation procedure for the ice-hull collision in the arctic region, 2 simple ice collision scenarios for the level ice and ice bergy bits colliding a ship's hull were simulated and results were considered. The structural modeling range were restricted to minimum level to reduce the simulation time within reliable accuracy for evaluating the hull structural response. In addition, the constitutive model considering strain rate effect, which is the most typical characteristic of ice was adopted.

MATERIAL PROPERTIES OF ICE

The mechanical properties of ice are complicatedly varying with many factors such as temperature, brine volume, size of crystalline, strain rate and so on. In particular, as shown in Figure 1, the deformation behavior of ice tends to be brittle at high strain rate while it becomes relatively ductile at low strain rate. And the overall shape of the stress-strain curves varies sensitively according to the strain rates.

Such complexity of ice deformation behavior is caused by microstructural changes in the crystalline structure of ice including micro crack growth and propagation. For the reasonable estimation of the ice-structure interaction and ice load due to ice collision by applying the theoretical and numerical approach, it is necessary to set up the rational constitutive equation, which can represent the actual stress-strain relationship of ice deformation. Although it is not easy to model the complex strength characteristics of ice closely, the basic mathematical form of the constitutive equation is assumed considering the physical behavior of the ice in various mechanical aspects related to the purpose of analysis. And the related coefficients of constitutive equation can be determined by experiments or field measurement.

Ice generally shows typical brittle material characteristics, it is much stronger for compressive stress than tensile stress with high strain rate dependency. In this study, ice was basically regarded as an elasto-plastic material. Considering the limitations of the material library included in the nonlinear structural analysis code (LS-DYNA), an piecewise linear plasticity material model with the von-Mises yield condition was adopted, which can define arbitrary stress versus strain curves according to the strain rates considering the strain rate dependency of the hardening curves. Also, the failure condition of ice was defined as failure strain expressed by cumulative equivalent plastic strain.

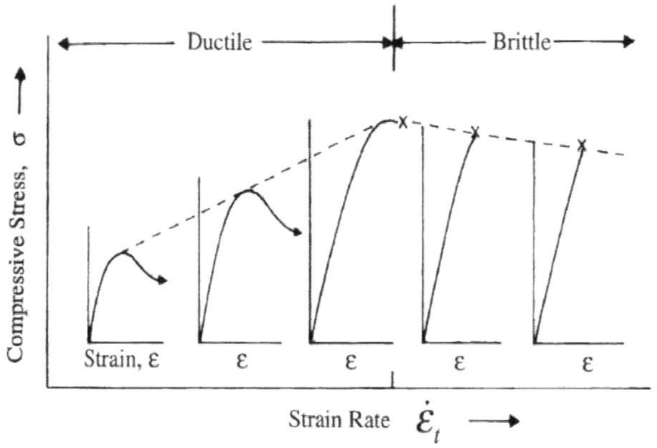

Figure 1. Compressive failure modes in ice as a function of strain rate (Carney, 2006)

Characteristics of Sea Ice

A lot of research has been carried out about the properties of sea ice in various aspects and viewpoints. Wang(1982) showed the different shapes of stress-strain curves of sea ice depending on the strain rates as shown in Figure 2(a). The present study also used this experimental results for the level ice collision problem.

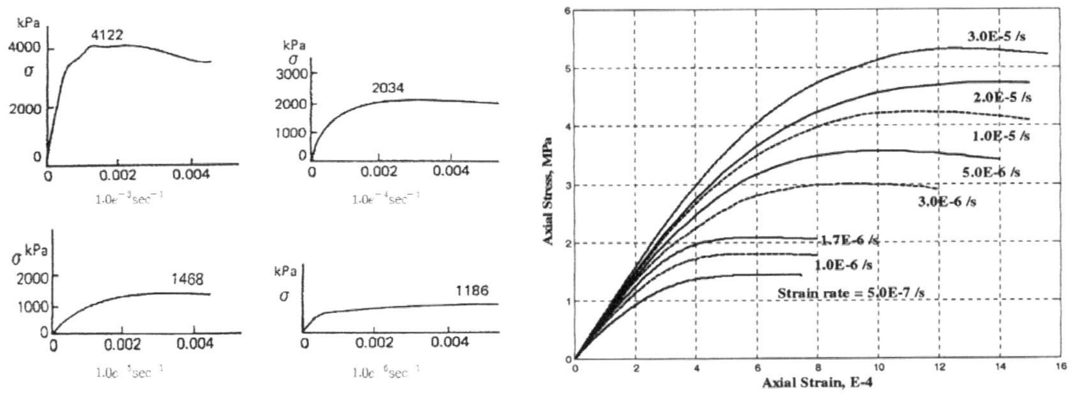

a) sea ice(Wang,1982) b) fresh water ice(Derradji-Aouat, 2000)

Figure 2. Stress-strain curves of ice depending on strain rates obtained by experiments

Characteristics of Iceberg

Iceberg generally falls out of land glaciers, so it is basically similar to fresh water ice. Johns(2006) showed that the strength of iceberg is slightly weaker than freshwater ice in high strain rate range, but the difference is not significant.

Therefore, in the present study, the stress-strain curves for fresh water ice, as shown in Figure 2(b) (Derradji-Aouat, 2000) was adopted for the collision analysis of iceberg. The shapes of the hardening curve vary considerably depending on the strain rates.

ICE SHIP COLLISION SCENARIO

There can be so many different situations of ice collision to ship's hull. In this study, the collisions of level ice sheet and iceberg was dealt with.

Considering the hydrodynamic and buoyancy forces of a mass of ice drifting in water, it could be a more accurate way to include the surrounding fluid region in the analysis modeling range for the collision simulation.

Even present day, however, it is not easy to provide a quantitatively reliable results for the FSI problems based on the CFD technique in spite of a great computation time. Therefore, a simplified practical method proposed by Kolari et al (2009) was applied, in which the added mass of drifting ice in water was calculated roughly (Bishop & Price, 1979) and modified the overall density of ice considering it.

Collision of Level Ice

The collisions of level ice sheet should be considered when ships are directly breaking the ice covered sea or pass through the ice channels provided by ice breakers. When passing through the ice channels, the ice sheets or the ice floes may collide the bow or shoulder of the hull. The collision speed can be determined by considering the velocity and turning speed of ship.

Level ice is generally composed of 1-year-old sea ice. The speed of the ship in this situation is usually about 3 ~ 5 knots, and thickness of the level ice was assumed to be about 0.8m ~ 1.0m.

Collision of Iceberg

Han et al(2008) mentioned that icebergs of 2m high above water line could be easily detected by deployed ice monitoring systems of ships and avoided easily by change of ship speed and course. Hence, the icebergs with sail heights less than 2m can be bigger threat to the safety of ships and should be fully investigated with possibility of collision to ships voyaging in their full speed. So it would be better to call them ice bergy bits from now on.

Therefore, it is rational to consider situations in which ice bergy bits may collide with bow, shoulder, or side of ships. DNV (2006) proposed some typical ice collision scenarios and shapes of ice bergy bits. For the iceberg collision analysis, not only the relative speed of the ship and ice but also the collision angle should be considered to simulate the actual phenomena accurately, However, this study aimed to establish a standard interpretation procedure, it was assumed that the spherical ice bergy bit collides to the side structure of a membrane type LNG carrier in vertical direction with a speed of 2 m/s as a typical example.

ICE COLLISION ANALYSIS AND CONSIDERATIONS

As shown in Figure 3 and 6, the colliding process of iceberg and level ice to ship's hull were simulated precisely using a commercial nonlinear structural dynamic FEA code LS-DYNA.

Material Properties

Even though nonlinear FE structure analysis codes, including LS-DYNA, provide hundreds of material models, it is not easy to find an adequate material model to implement the complicated mechanical properties of ice as described above. In this study, MAT_ PLASTIC_KINEMATIC model of LS-DYNA which can consider the strain rate dependency of ice was adopted. In this material model, elastic modulus is thought to be constant during deformation and yield stress and hardening curve can be defined differently according to the strain rate by piecewise linear curves. The isotropic strain hardening model was used neglecting the Bauschinger effect. The mild steel used for hull structure was regarded as an elastic-perfect plastic material for simplicity.

Table 1 summarizes the mechanical properties of freshwater ice, sea ice, and mild steel for ship structure. The failure condition of ice was defined by failure strain given by a representative constant value regardless of the strain rate because of the limitation of material model.

Table 1. Mechanical characteristics of materials

	Mild steel	Sea ice	Fresh water ice
Elastic modulus, GPa	210	7.3	7.8
Density, kg/m^3	7850	920	920
Poisson's ratio	0.3	0.34	0.34
Yield strength, MPa	285	Figure 2(a)	Figure 2(b)
Hardening curves	Elastic- perfect plastic material		

Collision Analysis of Level Ice

As an example of level ice collision analysis, a side structure of arctic cargo ship with a longitudinally framed system was selected. The target ship structure was assumed to collide to an 1 m thick ice sheet located in an ice channel with relative speed of 5 kts(2.572 m/sec) vertically.

Considering the geometrical symmetry of the target structure, the only 1/4 part was modeled and analyzed. In order to reproduce a realistic situation, the colliding ice was modeled as an arc shaped sheet for gradual contact and failure. The FE modeling of collision scenario and boundary condition was described in Figure 3. As shown in the figure, the mesh of ice contact part of the panel was divided more precisely to enhance the analysis accuracy. The material properties of level ice are shown in Figure 2(a).

In Figure 4, the equivalent stress distribution and time histories are shown at the 3 positions of panel around the collision point including the midpoints of long and short side(position A, B) and the corner position C of the plating, in which the stress levels are expected to be high.

The stress level of point A was relatively higher than other positions, however, the possibility of plastic deformation on the shell panel seems to be low since the average stress level was not so high and the duration times of peak value are very short.

The time series of the stresses are very randomly distributed because the failure of ice sheet takes place very irregularly during collision. In general, the actual ice load data obtained from onboard measurement shows typical regularly periodic triangular shaped response during ice collision process and random high frequency behaviors are overlapped on it. The high frequency behavior may take place due to the impact load generated during the partial corruption and removal process of contacting ice.

The periodic triangular shape response is caused by the bending failure at a distance from the loading point of ice, untill now, however, the phenomenon is not easy to implement by FE simulation technique and it is thought to be a challenging field of the future.

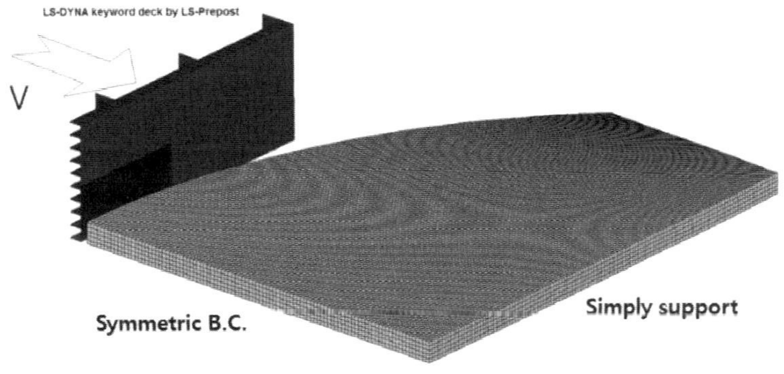

Figure 3. Modeling of ship structure, level ice and boundary conditions

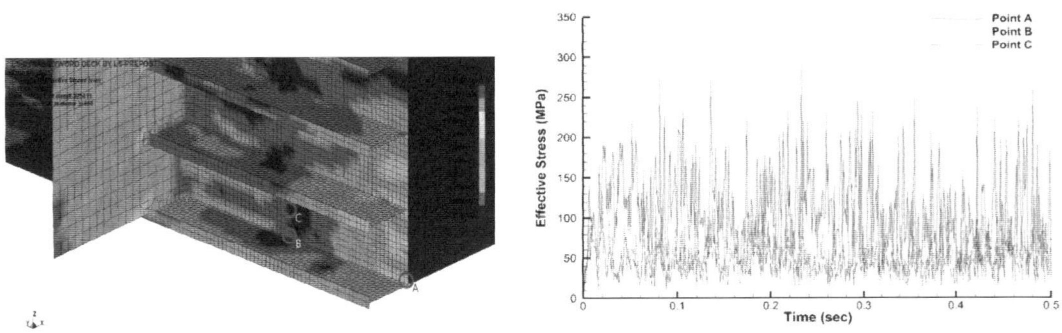

Figure 4. Distribution and time histories of effective stresses in ship structure

Collision Analysis of Ice Bergy bit

The side structure of the Mark III membrane type LNG carrier was analyzes under the situation that an ice bergy bit is colliding at the speed of 1, 2 and 3 m/sec to it as a scenario shown in Figure 5. 4 different ice bergy bits with sphere, bell or cubic types represented in Figure 6 were considered as recommended by DNV(2006).

The ice bergy bits were considered to collide at the center of the side structure and an insulation panel located at the collision position was included in the structural analysis model (Nho et al, 2014). As shown in Figure 7, the R-PUF, mastic and plywood in the insulation panel were modeled precisely to investigate the structural integrity of CCS. The side hull of LNG carrier was modeled by 2-D shell elements and the ice bergy bits and insulation panel were modeled by 3-D solid elements. For the material properties of the ice bergy bits, the characteristics of fresh water ice by Derradji-Aouat(2000) shown in Figure 2(b) was applied and ABS(2006) data were used for those of insulation panel.

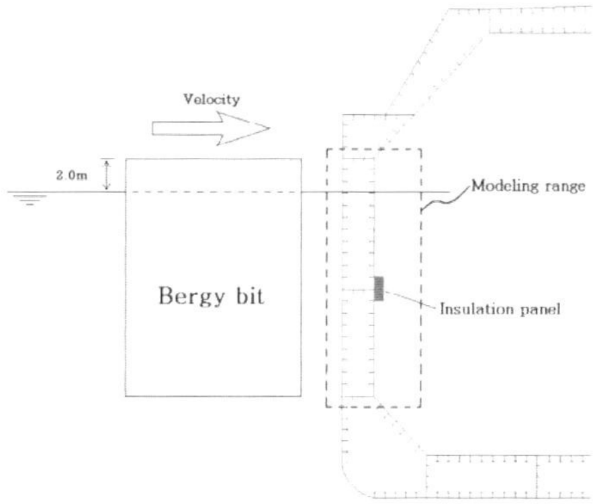

Figure 5. Ice bergy bit colliding scenario and modeling range of side structure

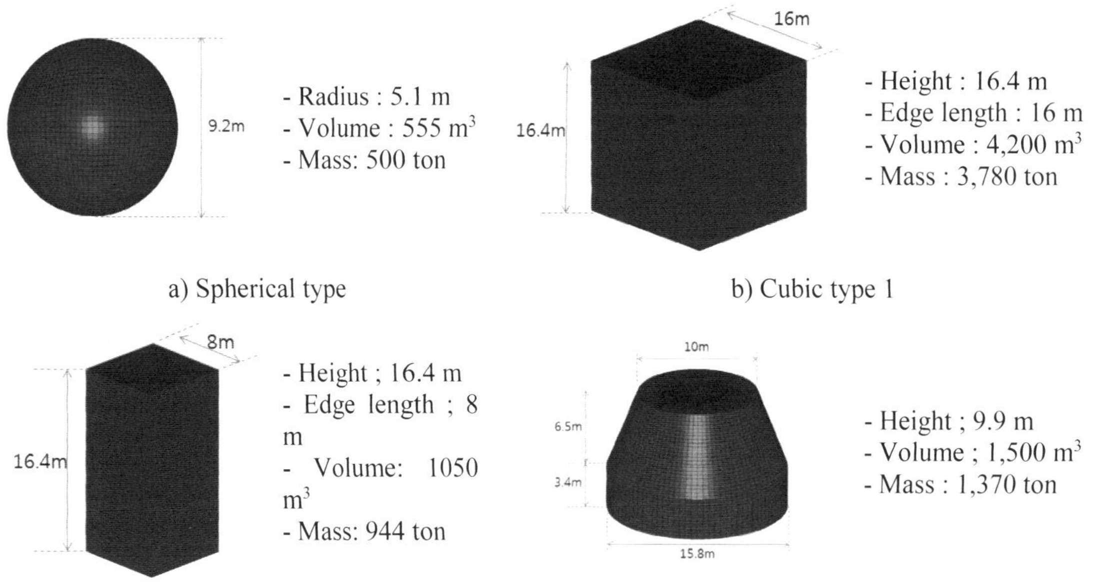

a) Spherical type

b) Cubic type 1

c) Cubic type 2

d) Bell type

Figure 6. Modeling details of ice bergy bits

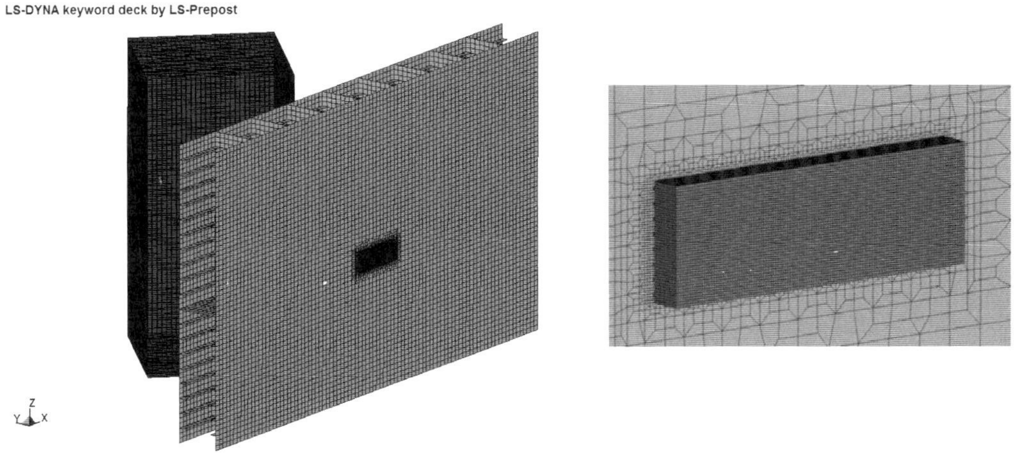

Figure 7. Structural modeling of side hull structure and insulation panel

The maximum collision loads according to the collision velocities and shapes of ice bergy bits were compared in Figure 8 respectively. It was obvious that the iceberg collision load was depends on the weight of iceberg. So, the analysis results for the cuvic 1 which is the heaviest model are dealt with from here.

The obtained equivalent stresses time histories of the side hull and insulation panel according to the variation of collision velocity were shown in Figure 9. And maximum von Mises equivalent stress distributions in hull structure and R-PUB were presented in Figure 10.

In the side hull structure of LNG carrier, the collision position suffered relatively high stress level and slight plastic deformations were taken place, however, most of the rest parts including the insulation panel showed low stress level within elastic limit. Therefore, it was concluded that the structural integrity of the CCS of this membrane type LNG ship against iceberg collision can be validated.

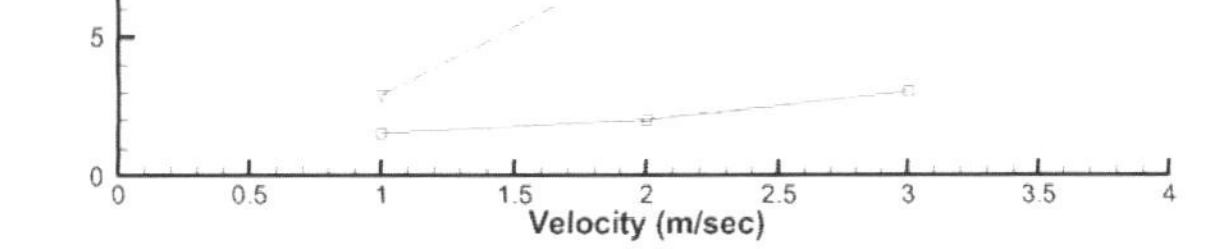

Figure 8. Maximum collision forces according to collision velocity

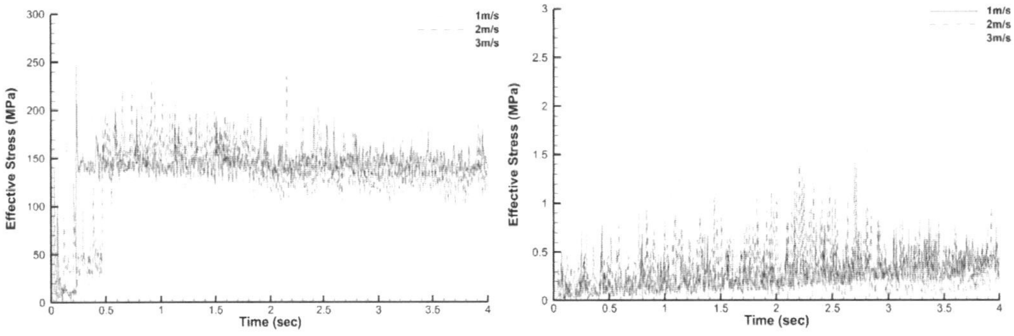

a) in ship's side structure b) in R-PUF

Figure 9. Time history of effective stress

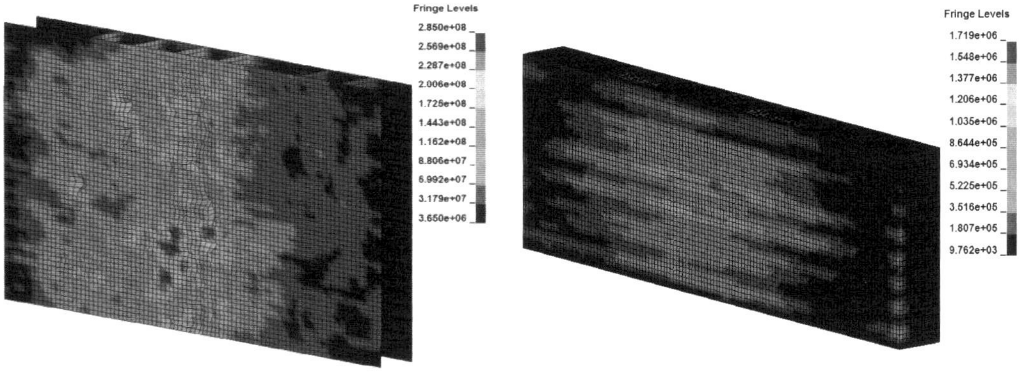

a) in ship's side structure b) in R-PUF

Figure 10. Distribution of effective stress

CONCLUSIONS

In this study, the simulation method of ice-hull collision process using the nonlinear finite element method (LS-DYNA) was dealt with to evaluate the structural safety of the hull. The estimation of material properties of ice is also a challenging field to be studied in the future, however, the strain rate dependency of material properties of ice was considered. And material test data for sea ice(Wang, 1982) and fresh water ice(Derradji-Aouat, 2000) were adopted.

Though the conditions of ice-hull collision are very diverse, in this study, a few relatively simple scenarios were assumed for the example analysis to verify the suggested simulation methods. Based on the consideration for the example analyses, the suggested simulation method of ice collision was considered to be a selectable means to evaluate the ice load and structural safety of ship's hull.

In future, it would be necessary to develop more accurate ice collision simulation technology for the rational structural design of arctic ships and more precise interpretation would be possible if the detailed basic researches on material characteristics and failure mechanism of ice are supported.

ACKNOWLEDGEMENTS

This work was supported by the National Research Program ("Development of Safe Voyage Planning System for Vessels Operating in Northern Sea Route," PMS3360) funded by the Ministry of Oceans and Fisheries, Korea and the Industrial Convergence Strategic technology development program ("Development of basic design technology for ARC7 class Arctic offshore structures," 10063417) funded by the Ministry of Trade, Industry & Energy, Korea.

REFERENCES

ABS, 2006. *Guidance note on Strength Assesment of Membrane-Type LNG Containment Systems under Sloshing Loads*.

Bishop, R., and Price, W., 1979. *Hydroelasticity of ships*, Cambridge University Press.

Carney, Kelly S., Benson, David J., DuBois, Paul and Lee, Ryan, 2006. "A phenomenological high strain rate model with failure for ice," *International Journal of Solids and Structures*, Vol. 43, pp.7820-7839.

Derradji-Aouat, A., Sinha, N.K. and Evgin, E., 2000. "Mathematical Modelling of Monotonic and Behaviour of Fresh Water Columnar Grained S-2 ice", *Cold Regions Science and Technology*, 31, pp.59-81.

DNV, 2006. "Iceberg Collision Scenario," *DNV Report No. 2006-0672*, pp.8-14.

Han, S.K., Park, Y.I., and Che, J.S., 2008. "Structure Risk Analysis of a GTT NO96 Membrane Type LNG Carrier in Baltic Ice Operations", *Proceedings of IMechE 2008*, Vol. 222, Part M, pp.179.

Jones, S., 2006. "Comparison of the Strength of Iceberg and Other Freshwater Ice and the Effect of Temperature," *PERD/CHC Report 20-83 TR-2006-07*, NRC Canada.

Kolari, K., Kuutti, J. and Kurkela, J., 2009. "FE-simulation of Continuous Ice Failure based on Model Update Technique," *Proceedings of the 20th International Conference on Port and Ocean Engineering under Arctic Conditions*, June 9-12, 2009, Luleå, Sweden.

Kujala, P., Suominen, M. and Jalonen, R., 2007a. "Increasing The Safety of Icebound Shipping," *Final Scientific Report: Volume 1&2*, Helsinki University of Technology, Ship Laboratory.

Lee, S.G., Lee, J.S., Baek, Y.H., Paik, J.K., 2009. "Structural Safety Assessment in Cargo Containment System in Membrane Type LNGC under Iceberg Collisions," *Proceeding of the Annual Autumn Meeting, the Society of Naval Architecture of Korea*, Mungueong, pp.29-30, October.

LSTC, 2007. *LS-DYNA keyword user's manual, Volume 1*, Version 971.

Nho, I.S., Yun, Y.-M., Park, M.-J., Oh, Y.-T. and Kim, S.-C., 2014. "Structural Safety Assessment of Mark III Membrane type LNG CCS under Ice Collision," *Journal of Ocean Engineering and Technology*, 28(2), pp.126-132.

Wang, Y.S., 1982. "A Rate-Dependent Stress-Strain Relationship for Sea Ice," *Proceedings of The 1st International Offshore Mechanics And Arctic Engineering Symposium*, Vol. Ⅱ.

Propulsion Shaft Line Ice-Induced Dynamic Torque Response Calculation and Comparison to Full Scale

Aki Kinnunen[1], Tuomas Turunen[2], Pekka Koskinen[1], Jaakko Heinonen[1]
[1] VTT Technical Research Centre of Finland, Finland
[2] Tevo Ltd, Finlad

ABSTRACT

Winter and arctic conditions set specific requirements for ship design in the form of ice-covered water. One major aspect is propeller and propulsion design for ice conditions. The propeller-ice interaction has been widely studied over the years and there are current regulations among classification societies to ensure safety of ice-going vessel propulsion design. The design process involves propeller dimensioning for ice loads as well as rest of the propulsion line. The propeller-ice interaction causes torque loading to the propulsion system. This is dynamic in nature and is somehow to be dealt with in design phase. In this paper, the principle of dynamic propulsion shaft line model is presented, the model is validated with ship design torsional vibration analysis calculations and the time domain response calculation is performed with ice class requirements, and these results are compared with full scale measurements.
In full scale tests, the propulsion shaft line input torque from propulsion motor was measured simultaneously with rotating speed. By comparing the measured and simulated dynamic response, it is possible to backward determine the magnitude of ice torque excitation.
In conclusion, the propulsion shaft line dynamic ice-induced torque response in the measurements was significant, even so that the ice class required excitation level for simulation was underestimating the torque response level encountered in measurements.

KEY WORDS: Propeller ice torque; response simulation; full-scale measurements

DYNAMIC TORQUE MODEL FOR PROPULSION SHAFT LINE

Model Principle

The model is built on simplification of the ship propulsion shaft system. The shafting is simplified to lumped masses, stiffnesses and dampings. The approach is well used in the shipbuilding industry for the torsional vibration analysis of the propulsion systems. In general it is used for estimating the torsional natural frequencies and also for placing a correct flexible coupling in the system.
The simplification is presented in schematic way in Figure *1*, where the J denotes rotational inertia, k refers to torsional stiffness, c means torsional damping and T is referring to external torque to the system, like motor torque or propeller resistance.

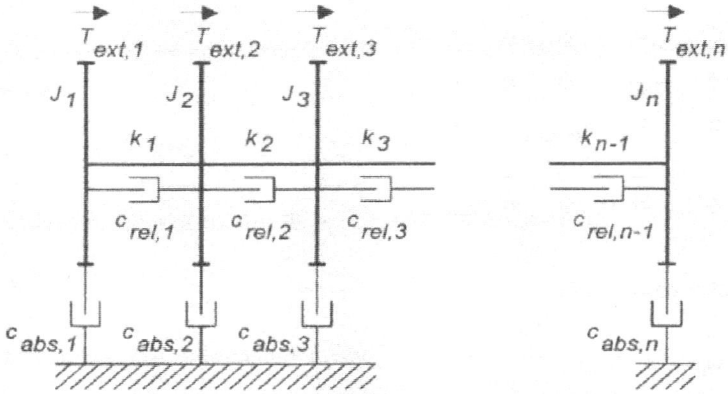

Figure 1. The system simplification approach.

The equation of motion for this type of system is simply written in matrix form

$$[J]\{\ddot{\varphi}\} + [C]\{\dot{\varphi}\} + [K]\{\varphi\} = \{T_{ext}\} \tag{1}$$

The angular displacement is described with the variable f. When this type of system includes gears, they must be taken into account for deriving the system matrixes.
The propeller hydrodynamic model is well-known torque-rotation speed dependency from propeller theory (Carlton, 2012):

$$Q = KQ\rho n^2 D^5 \tag{2}$$

The propeller torque *Q* is described with torque coefficient *KQ*, water density r, propeller rotation speed *n* and propeller diameter *D*. The torque coefficient is generally function of the propeller advance coefficient *J*, but in this simulation case the *KQ* is assumed constant, because ship speed is assumed to remain constant throughout the short-duration propeller-ice contact event.
Engines or electric drives used in ship propulsion systems are modelled as torque sources, with PID controller or similar means to provide functionality corresponding to the actual motor response.
The practical implementation was done with MATLAB® Simscape, and the appearance of model can be observed in Figure *2*.

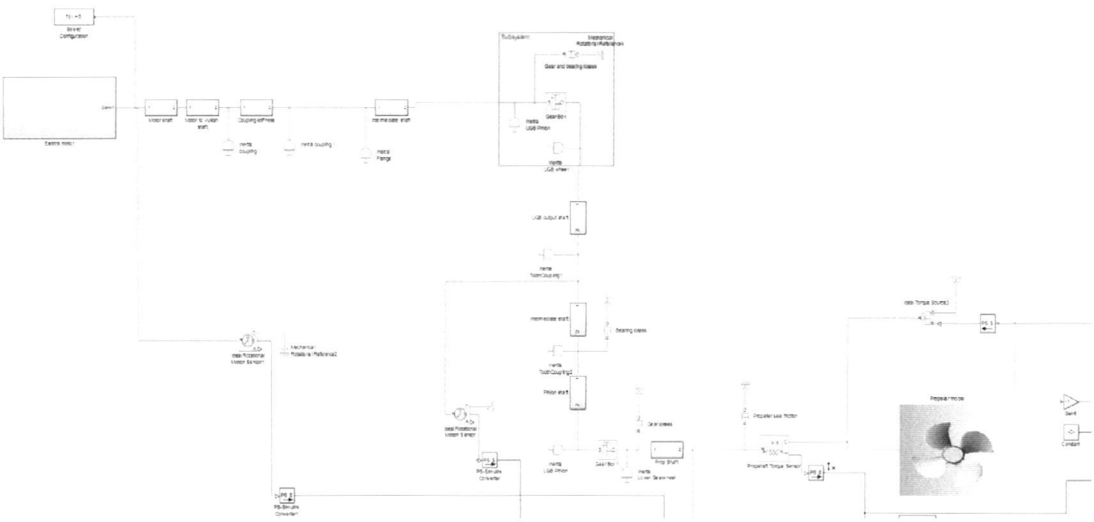

Figure 2. The modelling view: Electric motor (on top left) connected to Z drive propulsion unit model. Shafts are represented with stiffness parameters. Coupligns, gears and propellers include inertias as well.

Dynamic Shaftline Model Validation

The model dynamic properties are usually checked to be the same as are used in the ship design torsional vibrational analysis (TVA). The model set-up for TVA is used to calculate the first few natural frequencies and the MATLAB model for response calculation is checked to reproduce the same first natural frequency. This is sufficient because the propulsion shaft line dynamics related to propeller blade frequency excitation is usually dominated by the first torsional mode.

For the modelling method validation as such, a comparison of model results with three full scale cases was made and good agreement was found. (Kinnunen, 2017)

Modelling Ice Excitation: Ice Class Rule Requirements

The propeller-ice interaction resulting into propeller torque is addressed in IACS and Finnish-Swedish ice class rules in the form of additional torque on top of the hydrodynamic torque. The representation of the torque is based on full scale experience and propeller-ice interaction simulation. The propeller-ice interaction simulation (e.g. Veitch 1995, Veitch 1997, Soininen 1998, Kinnunen, 2015) is laborious task and requires rather extensive set of input data.
The ice torque calculation, as presented in the ice class rules, relies on determining the maximum ice torque for the propeller in question and using that together with a predetermined torque load pattern to the propeller. The load patterns used in this study are from Finnish-Swedish ice class rules 2010 version.
The definition of ice torque is expressed as a function of propeller shaft rotational degree of freedom in following way:

$$Q(\varphi) = C_q Q_{max} \sin\left(\varphi\left(\frac{180}{\alpha_i}\right)\right), \varphi \in [0..\alpha_i] \quad (3)$$

$$Q(\varphi) = 0, \varphi \in (\alpha_i \ldots 360] \quad (4)$$

Here the parameter Q_{max} is the maximum ice torque in ice contact, C_q is amplitude scaling factor relating to ice torque case, α_i is the ice torque impulse length in propeller shaft rotation angle. Angular variables in these formulas are in degrees. The torque load cases in the rules are schematically presented in Figure *3*.

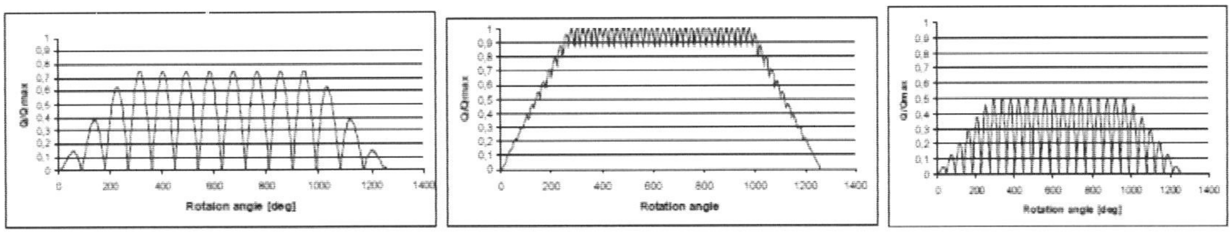

Figure 3. Ice torque load sequences according to Finish-Swedish ice class rules. On the left; excitation case 1, in middle excitation case 2, on the right excitation case 3.

Simulation Result Example

The simulation results provide the time-domain estimates for the propulsion system response to ice excitation. The propeller shaft response to Finnish-Swedish ice class rule ice torque *load excitation case 1*, together with the ice excitation is presented in Figure *5*.

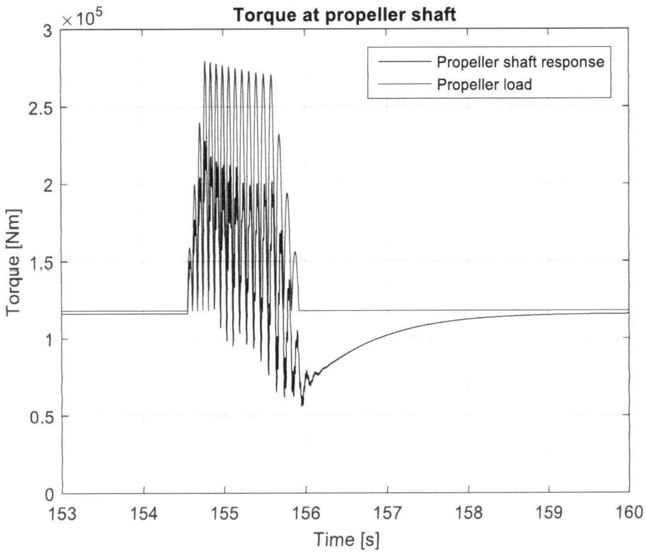

Figure 4. Propeller shaft response to ice torque excitation.

THE FULL SCALE MEASUREMENTS

The full scale measurements were done during winter 2016, in Bothnian Bay, the northernmost region of Baltic Sea. The main purpose of the ice trials was to verify the performance of the propeller in ice conditions. The ship was equipped with two Z-drive type azimuthing propulsion units, that is, the propeller is driven with motor inboard the vessel via two bewel gears The machinery arrangement was diesel-electric, i.e. main engines drive generators, and propulsion units are driven by electric motors. Propellers are fixed pitch propeller. The measurement setup consisted of measuring input shaft torque and RPM with the steering angle of both propulsion units simultaneously. Ship speed was recorded with GPS. In Figure *5* is the propulsion system principle and measurement location.

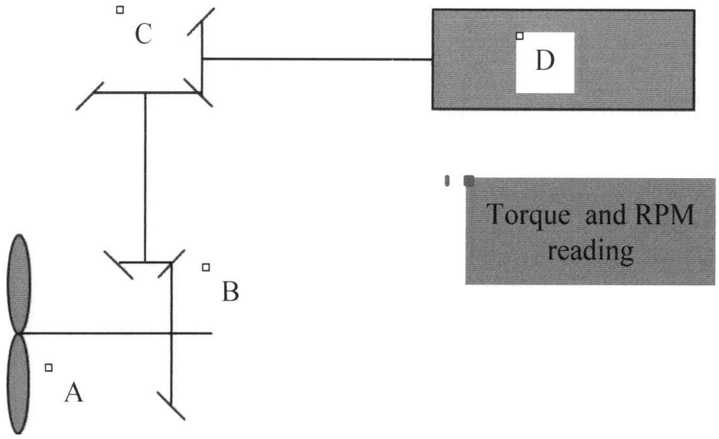

Figure 5. Z-drive type azimuthing propulsion unit principle. A: propeller, B: lower gearbox, C: upper gearbox, D: electric drive. Measurement of torque and RPM on the input shaft of propulsion unit.

The ice conditions were good for ice trials, at least 0.8 m solid ice with lots of ice ridges. The conditions are at least comparable to the IA Super ice class in Finnish-Swedish context. The ice channels on the area were in normal condition. The ice properties were studied with compression tests in-situ and the ridges were observed by drilling. The compressive strength samples were taken from solid ice, values are presented in Table *1*. The two checked ridges were 4,25m and 2,55m total height. Temperature on the area was between -2 .. 0°C.

Table 1. Compressive strength of ice observed on site.

Sample id	Uniaxial compression strength [MPa]
4	3,61
6	3,61
7	2,00
8	2,19
9	3,38
10	0,81
11	3,63
12	1,68
13	2,94
14	2,67

15	1,88
16	1,30
17	1,57
18	1,63
19	2,70
21	2,02
Mean value	**2,35**
St dev	0,87

COMPARING MEASUREMENT AND SIMULATION

The comparison of measurement and simulation is done for torque and RPM on the input shaft of the propulsion unit. In Figure *6* is compared the maximum measurement result and the simulation result for Finnish-Swedish ice class IA Super excitation case 1.

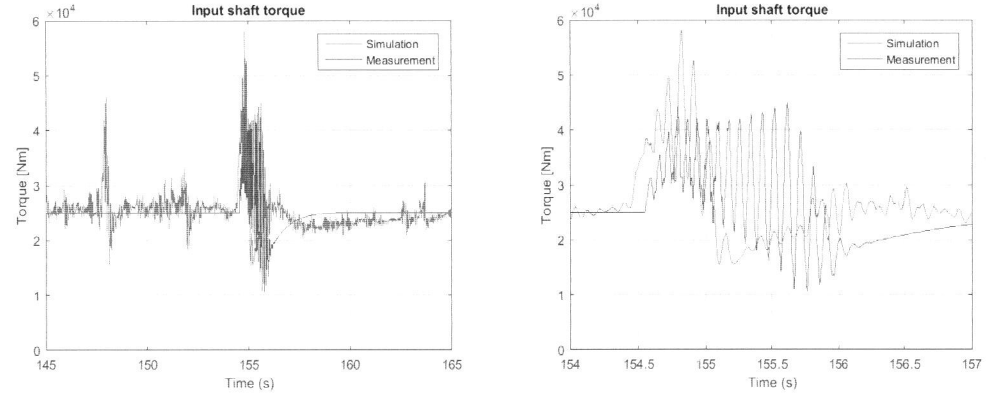

Figure 6. Comparing measured and calculated response from input shaft. Overview on left and detail on the right. Calculated torque clearly lower and longer ice-contact than measured.

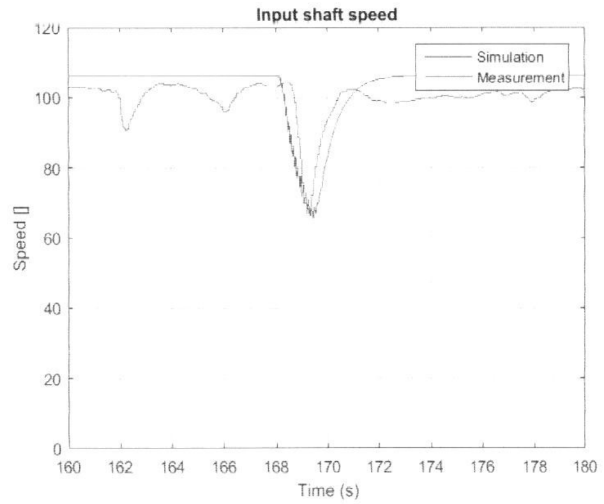

Figure 7. Input shaft speed comparison. Simulation with FSCIR ice class IA Super excitation.

The measured response is clearly higher in amplitude and shorter in duration than what the IA Super ice class excitation causes in simulation. This leads to question: how much ice torque excitation the simulation needs to match simulated response to the measurement? The simulation ice load sequence duration is scaled to 50% of the original and amplitude increased by 75%, then the simulation result is closer to the measured, see Figure *8*.

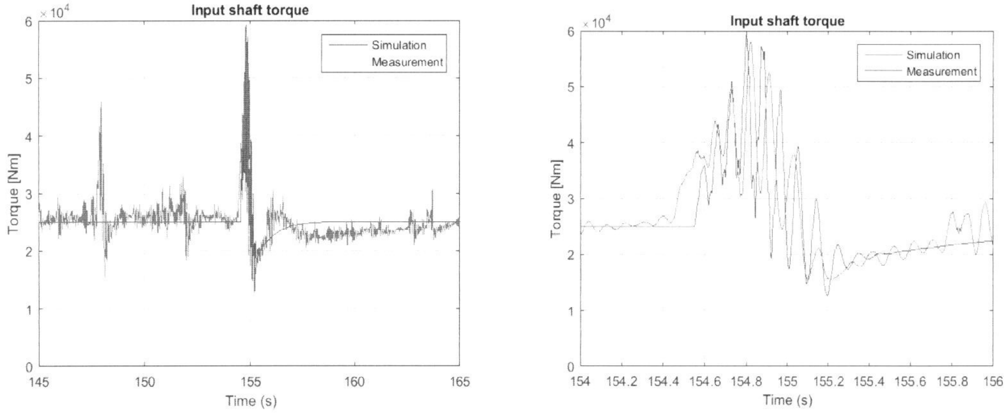

Figure 8. Comparing measured and calculated response (overview on the left, detail on the right) from input shaft. Excitation duration shortened by 50% and amplitude increased by 75% compared to the Finnish-Swedish ice class rule IA Super requirement.

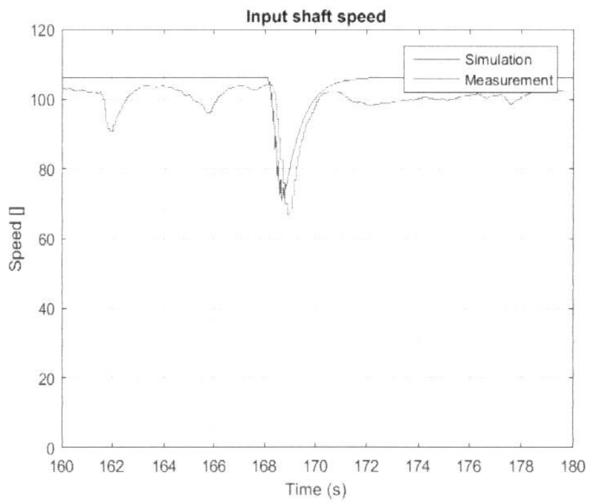

Figure 9. Input shaft speed comparison. Excitation duration shortened by 50% and amplitude increased by 75% compared to the Finnish-Swedish ice class rule IA Super requirement.

CONCLUSIONS

The dynamic torsional model of the propulsion shaft line was constructed with a lumped mass method. The electric drive in the model was a controlled torque source and the ice torque excitation was taken from ice class rules (FSICR IA Super). The dynamic model properties were checked to match the ship design time torsional vibration analysis results. The time domain response calculation was done with MATLAB® Simscape environment.
The full scale measurements provided data for propulsion shaft line input speed and torque. The measured torque is considered as a system response to the ice torque load at the propeller. The measurements were compared to the ice class required simulations and it was noted that the measurement gave higher response than the simulation suggested. The simulation

response was scaled by scaling the ice torque load at the propeller, and good match of simulated response was found to the measured response by this method.
In conclusion, it is reasonable to say that the dynamics in the shaft line system is in significant role in determining the actual torque level in operation, and that the ice class-required dynamic ice torque load for the propeller is not conservative in this case.
It is also worth to remember that the full scale test was short, two ice days, and such a good ice load was encountered during that time. The simulation based on the ice class excitation is trying to represent the maximum loading over service life of the vessel was indicating lower values.

ACKNOWLEDGEMENTS

The full-scale measurements were funded and arranged by Tevo Ltd. The expert support from VTT colleagues is highly appreciated.

REFERENCES

Belyashov, V.A., and Spakov, V.S., On Mechanics of Ice Crushing by Propeller Blades', Ice Mechanics and physics, Nauka, Academy of Sciences of the USSR, Moscow.

Carlton, J.S., Marine Propellers and Propulsion, Elsevier, 2012. ISBN: 978-0-08-097123-0

Finnish-Swedish ice class rules, 2010, Trafi, Finland.

IACS UR I3, unified requirements, 2010.

Kinnunen, A., Lämsä, V., Koskinen, P., Jussila, M., Turunen, T., MARINE PROPELLER-ICE INTERACTION SIMULATION AND BLADE FLEXIBILITY EFFECT ON CONTACT LOAD. POAC 2015.

Kinnunen, A., Dynamic response of propulsion shaft line systems to propeller ice torque excitation, VTT Reserch Report, 2017

Koskinen, P., Jussila, M. and Soininen, H. 1996. Propeller ice load models.

Espoo: Technical Research Centre of Finland, VTT Tiedotteita - Meddelanden - Research Notes 1739. 82 p. + app. 1 p

Soininen, H., 1998, A Propeller-Ice contact model, VTT Publications 343, Technical Research Centre of Finland, Espoo, 119 p.

Veitch, B., 1995. Predictions of ice Contact Forces on a Marine Screw Propeller During the Propeller-Ice Cutting Process, Acta Polytechnica Scandinavia, Mechanical Engineering Series No. 118, Helsinki 1995, 140p.

Veitch, B., 1997. Predictions of propeller loads due to ice contact. Int. Shipbuild. Progr., 44, no 439 (1997) pp. 221 – 239

POAC'17
Busan, KOREA

Proceedings of the 24th International Conference on
Port and Ocean Engineering under Arctic Conditions
June 11-16, 2017, Busan, Korea

A Parametric Study on Turning Ability of Icebreaking Ships in Level Ice

Donghyeong Ko[1], Kyung-Duk Park[1], Youngjae Sung[1]
Hyundai Heavy Industries Co., Ltd

ABSTRACT

A parametric study is carried out to determine major design parameters in turning ability of ice breaking vessels in level ice. Turning ability in ice is becoming an important part of ship design along with the interest in ice resistance. Thus, HHI developed the time-domain simulation to analyze turning ability of the icebreaking ships. This simulation considers icebreaking process and pattern induced by the ice-hull interactions. The usefulness of the scheme was validated by the application of a few bow-first icebreaking models (Ko et al. 2016). In this study, the application of the simulation is extended to various icebreaking models including conventional twin-screw ship and pod-driven ships. To understand which design parameter of icebreaking ships is dominant on turning ability, the effects of length-breadth ratio, rudder area ratio, stem angle and shoulder angle on ship design parameters were studied through many numerical simulations.

KEY WORDS: Turning ability, Icebreaking ships, Pod-driven ships, Twin-screw ships

INTRODUCTION

HHI has designed several icebreaking ships during the arctic projects to meet the growing demand for the natural resource transportation through the northern sea route. In the early stage of icebreaking ship design, the ship performance, such as speed and manoeuvrability in level ice is mostly considered because these items are directly related to owner's requirement.

The ship performance of icebreaking ships is influenced by the interaction between ship and ice. To understand the ship-ice interaction, the mechanism for icebreaking in level ice divides into three parts: breaking, buoyancy, clearing by many researchers (Shimanskii, 1938; Enkvist, 1972; Inonov, 1981; Lindqvist, 1989). The researchers also suggested a couple of empirical formulas based on principal hull dimension, such as length, breadth, depth, stem angle around waterline, etc. The formulas give a certain level of accuracy, but the formulas cannot reasonably reflect the detailed design variation of candidate hull forms which have different design characteristics despite same principal dimension. For this reason, HHI developed the new ice resistance prediction program which is based on modified empirical formulas to take hull form variation into account (Park et al., 2015). The method showed some improvement in estimation of ice resistance for icebreaking ships. However, because the method cannot handle yaw moment and side force on the hull, it has a limit to estimating turning ability in level ice.

Biao Su (2011) has introduced semi-empirical time domain simulation to understand the icebreaking performance of small sized ice breakers in level ice. In the simulation, breaking process is numerically considered by adopting crushing-bending failure while buoyancy and clearing parts are computed by the empirical formula of Lindqvist (1989). This method well represents the icebreaking patterns and the turning ability, and it shows good agreement with full scale measurements for an icebreaker by Riska et al. (2001).

HHI has suggested modified semi-empirical time domain simulation for icebreaking commercial vessels which has long parallel middle body. (Ko et al., 2016) To release excessive ice loads on the long parallel middle body, they proposed varying crushing stress which is varied according to the velocity vector and theoretical curve of Ashby et al. (1986). For the simulation of ship motions, the modular type mathematical model with multiple pod propulsors (Kim et al., 2006) was used. The simulation results are compared with the model tests.

In this study, the application of HHI's modified semi-empirical time domain simulation is extended to various icebreaking models including conventional twin-screw ship and pod-driven ship. To understand which design parameter of icebreaking ships is dominant on turning ability in level ice, the effects of length-breadth ratio, rudder area ratio, stem angle at bow and flare angle at shoulder on ship design parameters were studied through many numerical simulations.

NUMERICAL SIMULATION

Procedure

Basic procedure of the modified semi-empirical time domain simulation is derived from the Biao Su's thesis (2011). Among the ice induced force by ship-ice interaction, breaking component is numerically simulated and the other ones (buoyancy and clearing components) are simply computed from Lindqvist's empirical formula (1989).

The flow chart of the numerical simulation is briefly summarized in figure 1. The detailed procedures were also described in the paper of Ko et al. (2016). At first, it is assumed that a ship moves ahead in the level ice with given initial conditions, such as dimension of the ship, hydrodynamic coefficient in open water, propeller open water characteristic and ice data. As the ship goes on, it detects contact areas around the hull at each time step. If the contact area was found, the code checks crushing-bending failure and calculates the force and yaw moment acting on the ship. After the bending failure occurs, then ice nodes are regenerated.

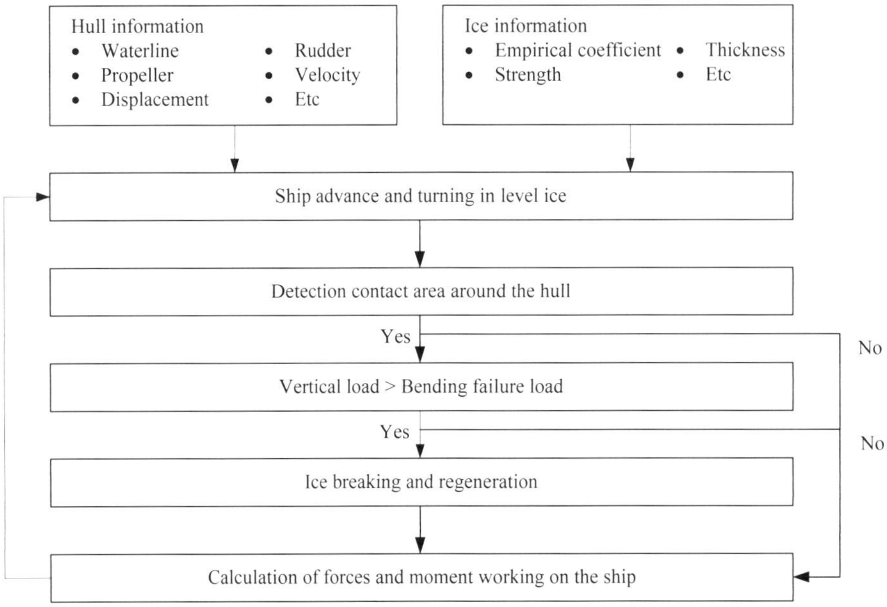

Figure 1. Flow chart of numerical simulation

Varying crushing stress

This procedure adopts a new concept of varying crushing stress. The new concept considers the influence of the contact area and the direction of impact due to a ship-ice interaction. If the ship and ice are fully crushed, crushing stress normal to the contact area is generated. According to Ashby et al. (1986), the crushing stress σ_{area} can be defined as a function of contact area A as follows.

$$\sigma_{\text{area}} = \frac{P}{A} = \frac{P_L \Delta_L}{L_i^3}\left[1 + 3\left(\frac{L_i^3}{A\Delta_L}\right)^{\frac{1}{2}}\right] \tag{1}$$

where, L_i is an idealized size of cubical independent cells, Δ_L is a moved distance before the cell fails and P_L is the average force during the period of contact.

On the other hand, the Eq. (1) shows that the stress extremely soars as contact area decreases. Thus, for small contact area, the varying crushing stress curve is bounded by the experimental results of the maximum stress, 10.5 MPa to avoid the unrealistic situation. (Sanderson, 2014) The overall crushing stress curve is described in figure 2.

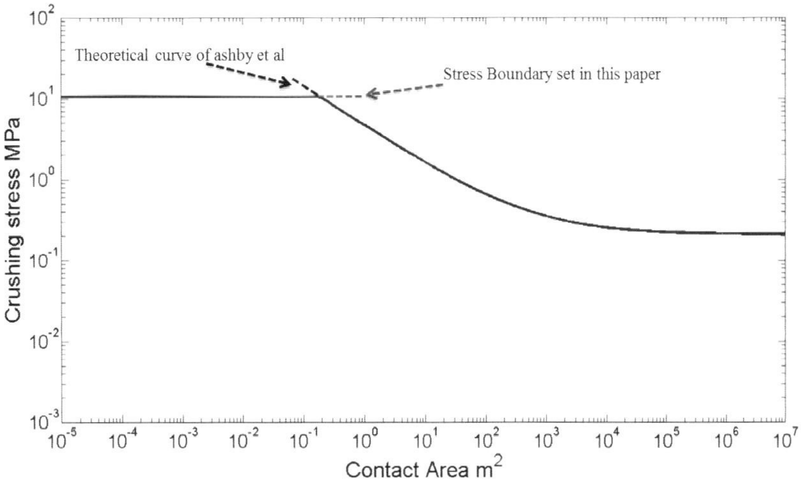

Figure 2. Theoretical curve of Ashby et al. (1986) bounded by the collected data

The relation between the direction of advancing ship and the direction of the crushing force defined by the contact area is also considered as well as the varying crushing stress. If the ship crushes the ice diagonally, due to the direction of ship advance, the ice is not fully crushed. To implement this effect, the velocity vector of the ship is newly considered. Velocity vector of ship ($\vec{U}$) is assumed to comprise of two components, linear velocity ($\overrightarrow{U_l}$) and angular velocity ($\overrightarrow{U_r}$) and defined as

$$\vec{U} = \overrightarrow{U_l} + \overrightarrow{U_r} \tag{2}$$

Once the contact areas are determined, the resulting crushing stress is calculated according to the curve in figure 2 and divided into two components: normal and vertical to the ice sheet, considering the slope angles between hull and ice such as α and β in figure 3. Then, vertical crushing stress σ_v and normal crushing stress σ_n are generated as the ship contacts ice. However, they are zero if the ship moves far from the ice although such a contact between hull and ice happens. Therefore, the resulting crushing stresses are given as in Eq.s (3) and (4), where the normal stress is the cosine α component and the vertical stress is the cosine β component of σ_{area}.

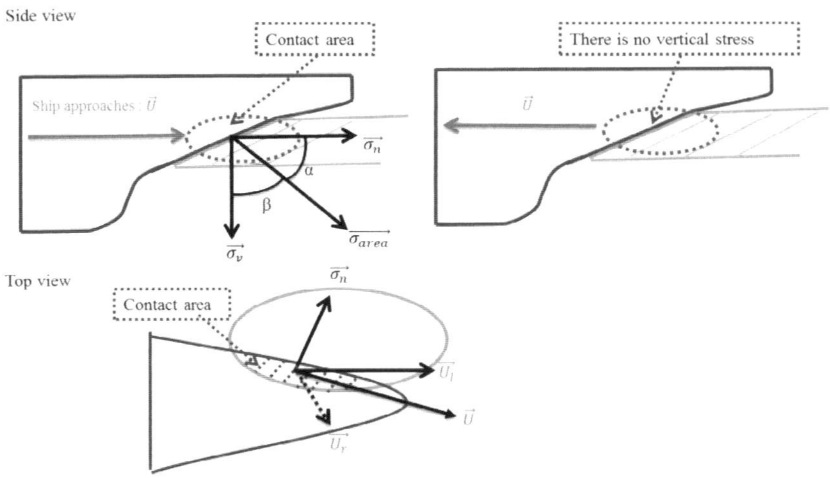

Figure 3. Description of contact area and crushing stress

$$\begin{aligned} \sigma_n &= \sigma_{area} \cdot t_1 \cdot \cos\alpha \\ \sigma_v &= \sigma_{area} \cdot t_2 \cdot \cos\beta \end{aligned} \tag{3}$$

$$\begin{aligned} t_1 &= \begin{cases} \frac{\vec{U}\cdot\overrightarrow{\sigma_n}}{|\vec{U}|\cdot|\overrightarrow{\sigma_n}|} & if \ \frac{\vec{U}\cdot\overrightarrow{\sigma_n}}{|\vec{U}|\cdot|\overrightarrow{\sigma_n}|} > 0 \\ 0 & if \ \frac{\vec{U}\cdot\overrightarrow{\sigma_n}}{|\vec{U}|\cdot|\overrightarrow{\sigma_n}|} \le 0 \end{cases} \\ t_2 &= \begin{cases} 1 & if \ \frac{\vec{U}\cdot\overrightarrow{\sigma_n}}{|\vec{U}|\cdot|\overrightarrow{\sigma_n}|} > 0 \\ 0 & if \ \frac{\vec{U}\cdot\overrightarrow{\sigma_n}}{|\vec{U}|\cdot|\overrightarrow{\sigma_n}|} \le 0 \end{cases} \end{aligned} \tag{4}$$

In Eq. (4), t_1 and t_2 represent the effect of the ship advance and their values are determined according to the inner product between velocity vector ($\vec{U}$) and inward normal vector of ice sheet ($\overrightarrow{\sigma_n}$).

Simulation of ship performance based on modular model

In the study, the modular type mathematical model is and two different types of ships are considered for the simulation. One is a pod-driven ship which has three pod propulsors and the other is a conventional twin-screw ship. This mathematical model is based on twin pod propulsor model (Kim et al., 2006), and then slightly modified for the pod-driven ship and twin-screw ship. Right-handed coordinate system with Z vertically downward is used and its origin is located at the center of gravity. Prior to the simulation, by the Newton's second law, the ship's equations of motion (u, v, r) can be described in Eq. (5).

$$\begin{aligned} M \cdot (\dot{u} - vr) &= X_H + X_P + X_R + X_{ice} \\ M \cdot (\dot{v} + ur) &= Y_H + X_P + Y_R + Y_{ice} \\ I_{zz} \cdot \dot{r} &= N_H + N_P + N_R + N_{ice} \end{aligned} \tag{5}$$

The terms (X, Y and N) with subscript (H, P, R, ice) represent the forces and yaw moment acting on hull induced by hydrodynamic effect, propeller (or pod propeller), rudder (or pod struts) and ice respectively. Prior to the simulation, the test results from open water propeller test and PMM (Planar Motion Mechanism) tests were conducted for icebreakers to obtain relative coefficient induced by propeller hydrodynamic effect.

Derivatives of hydrodynamic effect are introduced below.

$$\begin{aligned} X_H &= -M_x\dot{u} + \left(M_y + X_{vr}\right)\cdot vr + X_{vv}\cdot v^2 + X_{rr}\cdot r^2 + X(u) \\ Y_H &= -M_y\dot{v} + Y_v\cdot v + Y_{vvv}\cdot v^3 + Y_r\cdot r + Y_{rrr}\cdot r^3 + Y_{vrr}\cdot vr^2 + Y_{rvv}\cdot rv^2 \\ N_H &= -J_{zz}\cdot\dot{r} + N_v\cdot v + N_{vvv}\cdot v^3 + N_r\cdot r + N_{rrr}\cdot r^3 + N_{vrr}\cdot vr^2 + N_{rvv}\cdot rv^2 \end{aligned} \tag{6}$$

Forces and moment induced by propeller (pod propellers) and rudders (or pod struts) are formulated in Eq. (7) and (8) respectively.

$$\begin{aligned} X_P &= (1-t)(T^P + T^C + T^S)\cos\delta_p \\ Y_P &= (1-t)(T^P + T^C + T^S)\sin\delta_p \\ N_P &= x_p\cdot(1-t)(T^P + T^C + T^S)\sin\delta_p - y_p\cdot(1-t)(T^P - T^S)\cos\delta_p \end{aligned} \tag{7}$$

$$\begin{aligned} X_R &= -(1-t_R)(F_N^P + F_N^C + F_N^S)\sin\delta_r \\ Y_R &= (1+a_H)(F_N^P + F_N^C + F_N^S)\cos\delta_r \\ N_R &= (x_R + a_H x_H)(F_N^P + F_N^C + F_N^S)\cos\delta_r - y_R\cdot(1-t_R)(F_N^P - F_N^S)\sin\delta_r \end{aligned} \tag{8}$$

where, *P, C, S* (port, center, starboard) represent the location of each device.

In the equation, δ_p and δ_r mean turning angle of propeller and rudder respectively. Since the twin-screw ship has two stationary propellers (δ_p=0), rudder is assumed to be a single device for the ship manoeuvering.

SIMULATION RESULTS

Comparison with Model Tests

In the design phase, various ship designs are compared for the decision-making. There are a few ways to determine optimum design of a ship, such as experience-based design, numerical simulation and model test. For the non-ice class ships, experience-based design is useful approach in the early phase. Accumulated data from a number of model tests and sea trial tests help to evaluate maneuverability of the ship in a relatively short period of time. For icebreaking ships, however, experience-based design is inadequate due to the lack of data. Although the model test is the most accurate method, a lot of time and cost are needed to refreeze the ice repeatedly.

On the other hand, there is no restriction in cost and time required for numerical simulation. Even if the uncertainty of numerical simulation is much more than model test, its cost and time are cheaper and shorter. Thus, if we have a numerical simulation which corresponds with the model test result it will be a good alternative for the small design change.

In this study, the model tests for two different types of mother ships are compared with their numerical simulations before a parametric study. The principal dimensions of the ships and ice model test conditions are summarized in table 1. The model ships are tested in the same ice condition but they have slightly different slenderness ratio and bow angle. The ships tested in the ice model basin are presented in figure 4.

Table 1. Principal dimension of ships and ice model test conditions

	Pod-driven ship	Twin screw ship
L/B	5.5	5.6
Bow angle (degree)	20	25
Ice thickness (m)	1.7	1.7
Flexural strength (kPa)	500	500
Propulsion unit	Azipod	Propeller
Number of propulsion unit	3	2

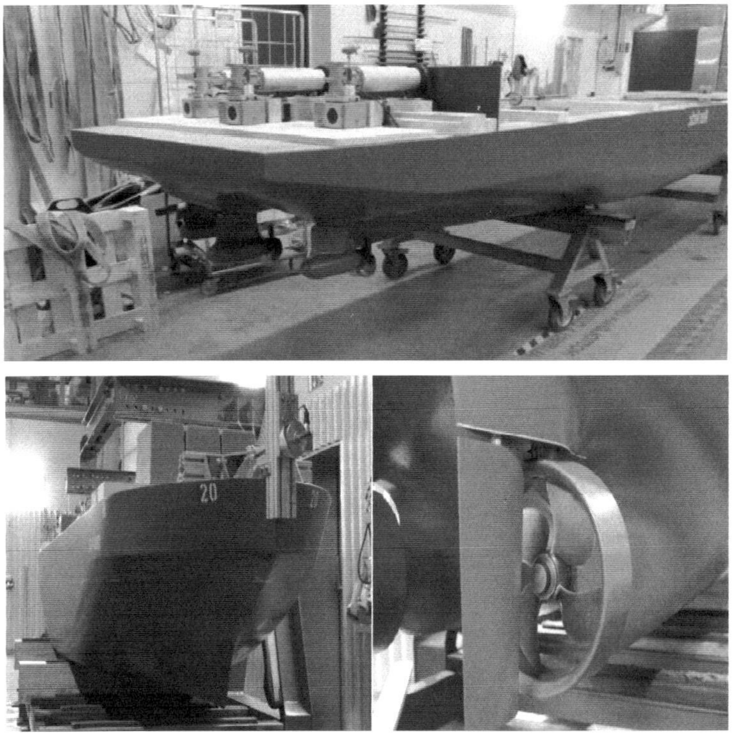

Figure 4. Stern of the model ships (Pod-driven ship: Top, Twin-screw ship: Bottom)

Comparison results between the simulation and model tests are shown in table 2. The full turning diameter of model tests is extrapolated from initial turning track because ice model basin has long and relatively narrow channel in general. The resulting mean velocity and turning diameter from the numerical simulations well agree with the model tests for both types of ships. From the turning ability point of view, the most distinctive characteristic of the pod-driven ship is that pod body is able to rotate for turning and this makes the thrust force generate additional yaw moment directly. For this reason, pod-driven ship has a great advantage in its turning ability. This tendency is shown in both model tests and numerical simulations. Dimensionless turning diameter (Turning diameter / Length) of the both ships are significantly different. In the model tests, dimensionless turning diameter of the Twin screw ship marked 112 but the pod-driven ship 34. For reference, turning tracks of both simulations and model tests are shown in figure 5 and figure 6 respectively.

Table 2. Turning data from model test and numerical simulation

	Mean velocity [m /s]	Turning diameter [Turing diameter / L]
Pod-driven ship [model test]	0.4	33.6
Pod driven ship [simulation]	0.5	33.3
Twin-screw ship [model test]	1.2	112.1
Twin-screw ship [simulation]	1.3	103.4

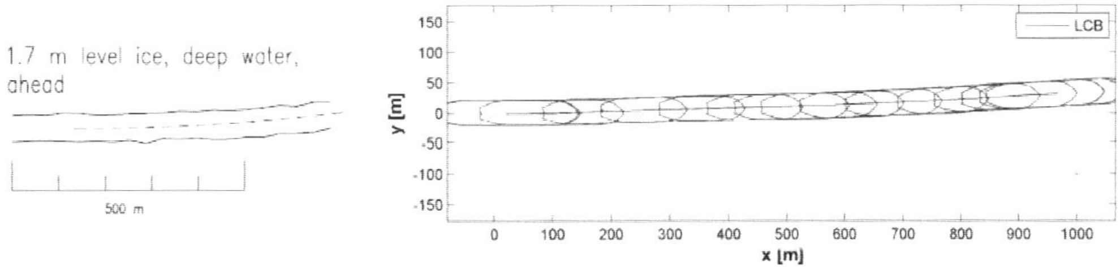

Figure 5. Turning tracks of the model tests (Pod-driven ship: Left, Twin-screw ship: Right)

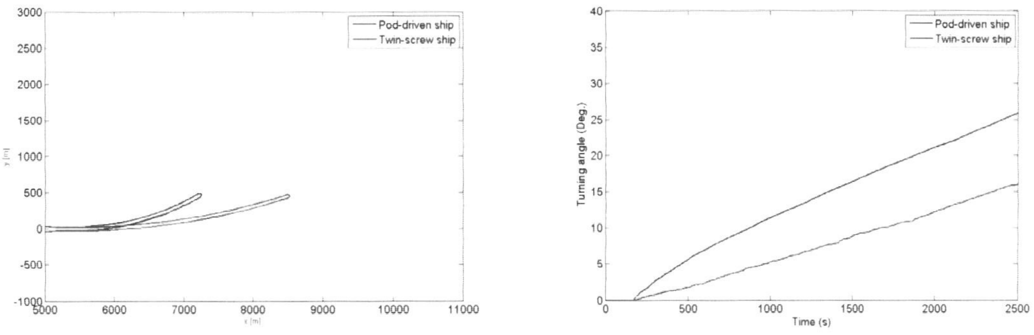

Figure 6. Turning track and angle of the numerical simulations

Parametric Study

For many decades, many different kinds of ships has been tested to verify their turning ability in open water. Through the tests, it was found that some significant dimensionless coefficients related to the turning ability were C_bB/L (C_b: block coefficient, B: Breadth, L: Length), A/LT (A: rudder area, L: length, T: depth), stem angle at bow region and flare angle at shoulder region. Higher C_bB/L normally has an influence on the smaller turning diameter. Also, higher A/LT (rudder area ratio) has same characteristic as C_bB/L.

For dimensionless coefficients, numerical simulations are conducted to monitor what the turning ability in level ice is like. In the study, some design parameters such as slenderness, block coefficient and rudder area are changed to check which parameter of a ship is dominant on the turning ability. Figure 7 shows the tendency of the dimensionless turning diameter with respect to slenderness ratio (B/L). Turning diameter decreases as the ratio increases regardless of types of ships and this tendency is similar to open water test result. As a result, slender ship is unfavorable to turning ability in ice as it is in open water. In addition, the impact of block coefficient (C_b) is studied in figure 8. In general, C_b is directly related to inertial mass as a factor of course stability in open water turning ability. However, parallel shifts are observed with the change of C_b in the simulation in level ice. This indicates C_b has no effect on the turning ability in level ice. In other words, the effect of ice induced force is much larger than that of the inertial mass induced by C_b variation for turning ability.

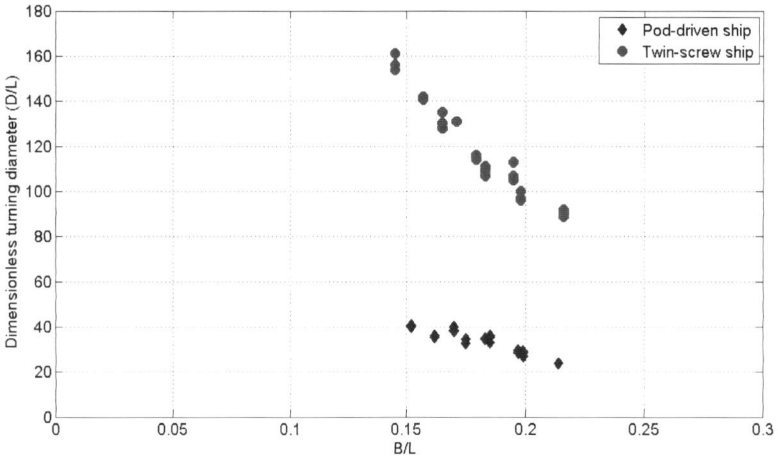

Figure 7. Dimensionless turning diameter with respect to slenderness ratio (B/L)

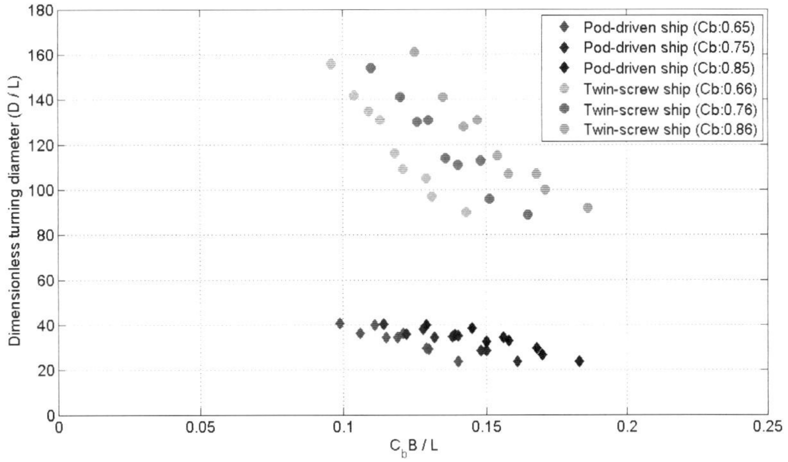

Figure 8. Dimensionless turning diameter with respect to C_bB/L

The rudder area ratio effect in turning ability was studied. Length and rudder area of the ships altered to see the aspect of the relationship between the rudder area ratio and turning diameter. Figure 9 indicates that rudder area ratio effectively impacts on the turning diameter for both ships but the gradient is rather different. Since pod propeller is turning along with the pod strut, pod propeller activates as an auxiliary device for ship manoeuvering. On the other hand, twin-screw ship has conventional propeller-rudder system. In the system, rudder only acts as a manoeuvering device. For that reason, variation of the rudder area ratio directly affects the gradient of the turning diameter on the twin-screw ship. The gradient of the twin-screw ship is approximately 4 times larger than the pod-driven ship.

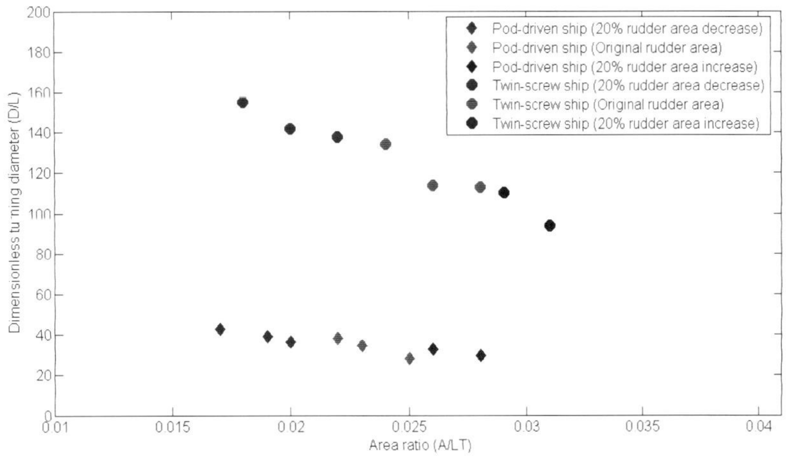

Figure 9. Dimensionless turning diameter with respect to rudder area ratio (A/LT)

To see the impact of hull form variation on the turning ability, stem angles at bow region and flare angles at shoulder region are changed in the simulation. The meaning of bow region and shoulder region illustrates in figure 10. Bow region is defined from the position of forward perpendicular to a width of 0.1 B. In addition, shoulder region is within the region not exceeding 0.1 B from the start position of parallel middle body.

Figure 10. Definition of bow and flare region

For both ships, stem angle variation at bow region was studied. The relation between turning diameter and stem angle is represented in figure 11. With reduced stem angles, turning ability in level ice is not noticeably improved. Reduced stem angle helps to relieve the stress level at bow region and it makes the average velocity of the ships increase as shown in table 3. Thus, it is thought that the increased ship velocity leads to slightly bad turning ability in level ice. On the other hand, ~~However,~~ the direction of crushing force at bow region is almost aligned with the x-axis and thus the crushing stress variation with nearly zero moment arm do not much contributed to the turning ability in level ice.

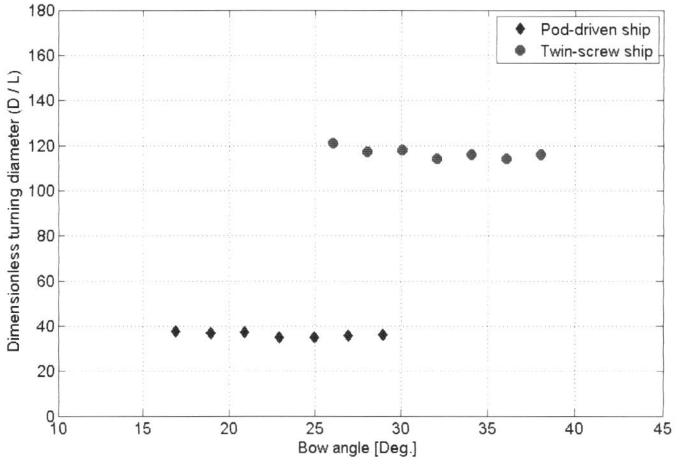

Figure 11. Dimensionless turning diameter with respect to bow angle

Table 3. Average velocity and yaw rate of both ships *(In this table, velocity ratio and yaw rate ratio are represented by the ratio of the corresponding model to reference model.)*

Stem angle variation	Velocity ratio (Pod-driven)	Velocity ratio (Twin-screw)	Yaw rate ratio (Pod-driven)	Yaw rate ratio (Twin-screw)
Model 1(-6 degree)	111.4 %	103.5 %	104.9%	103.6%
Model 2(-4 degree)	107.5 %	101.5 %	108.4%	98.9%
Model 3(-2 degree)	103.7 %	101.1 %	99.1%	101.3%
Original ship (reference point)	100.0 %	100.0 %	100.0%	100.0%
Model 4(+2 degree)	98.1 %	97.6 %	101.6%	98.7%
Model 5(+4 degree)	94.1 %	98.5 %	99.4%	106.7%
Model 6(+6 degree)	93.6 %	97.9 %	93.8%	100.4%

Flare angle variation at the shoulder part of a ship is also studied. Figure 12 shows the effect of flare angle variation about turning diameter in level ice. With reduced flare angles, the turning ability is conspicuously improved in level ice. By contrast with the stem angle variation, the variation of crushing occurred at the shoulder area affect the yaw moment of a ship because of relatively long moment arm.

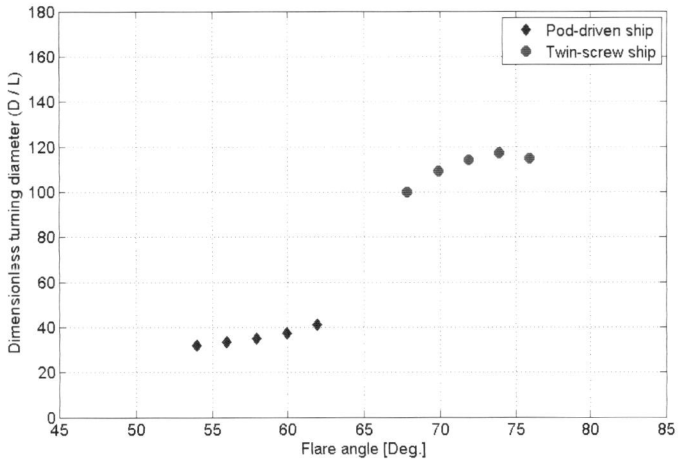

Figure 12. Dimensionless turning diameter with respect to flare angle

CONCLUSIONS

Using the modified semi-empirical time domain simulation, a parametric study is carried out to find out what design parameters of ice breaking vessels are dominant on turning ability in level ice. Prior to a parametric study, comparison study between numerical simulation and model test for pod-driven and twin-screw models is conducted to see the usefulness for the simulation technique suggested in this study. Tendency of dimensionless turning diameter is monitored with variation of ship design parameters, such as slenderness ratio, block coefficient, rudder area ratio, stem angle at bow and flare angle at shoulder. The results of the parametric study are summarized as follows.

- Slenderness ratio is quite related to the turning ability in level ice as with in open water. Higher slenderness is a bad influence on the turning ability in ice.
- While block coefficient normally affects the turning ability in open water, it doesn't in level ice because the effect of ice induced force is much larger than the inertial mass.
- Higher rudder area ratio is helpful for small radius of turning trajectory. This effect emerged as a key parameter in twin-screw ship. However, it is less remarkable for pod-driven ship because pod strut is not a main device for turning.
- Flare angle variation at shoulder region is much more related to the turning ability compared with stem angle at bow for both ships. In bow region, crushing stress generally acts on the ship in the same direction of x-axis. Thus, crushing stress around the bow region do not much contributes to the yaw moment in the simulation. On the other hand, acting point of crushing force around the shoulder region has relatively long moment arm and it is an effective uptake to increase yaw moment.

Even though this study enables us to analyze some features of turning ability in level ice, there are two typical shortcomings in the simulation. First, there is no guarantee that the breaking patterns estimated by this simulation correspond well with that of model tests. Second, both buoyancy and clearing components are estimated by using Lindqvist's empirical formula suggested to calculate ice resistance in x direction only and thus this formula does not reflect an additional influence on yaw moment in practice. It means that turning operation in level ice leads to the unbalanced flow patterns of the fractured ices on the sides of a ship after icebreaking process, and thus it needs to be considered to include secondary effect due to the unbalanced buoyancy and clearing forces. These problems will be dealt with in future works.

REFERENCES

Ashby, M.F., Palmer,A.C., Thouless, M., Goodman, D.J., Howard, M.W., Hallam, S.D., Murrell, S.A.F., Jones, N., Sanderson, T.J.O. and Ponter, A.R.S, 1986. Nonsimultaneous failure and ice loads on Arctic structures. Offshore Technology Conference 1986 NO. OTC 5127, pp. 399-404, Houston, USA.

Enkvist, E., 1972. On the Ice Resistance Encountered by Ships Operating in the Continuous Mode of Icebreaking. Report No.24, The Swedish Academy of Engineering Science in Finlande, Helsinki, Finland.

Izumiyama, K., Wako, D., Shimoda, H. and Uto, S., 2005. Ice load measurement on a model ship hull. Proceedings of 18TH International Conference on Port and Ocean Engineering under Arctic Conditions (POAC), New York, USA.

Ko, D.H, Park, K.D., Ahn, K.S., 2016. Time domain simulation for icebreaking and turning capability of bow-first icebreaking models in level ice. International Journal of Naval Architecture and Ocean Engineering. Vol. 8, Issue 3, pp 228-234

Park, K.D. and Kim, H.S., 2014. Study on the Ship Ice Resistance Estimation Using Empirical Formulas. Proceedings of the ASME 2014 33rd International Conference on Ocean, Offshore and Arctic Engineering (OMAE), San Francisco, USA.

Lindqvist, G., 1989. A straightforward method for calclation of ice resistance of ships. Proceedings of 10th International Conference on Port and Ocean Engineering under Arctic Conditions (POAC), Lulea, Sweden.

Riska, K.,Leiviska, T., Nyman, T., Fransson, L.,Lehtonen, J., Eronen, H. and Backman, A.,2001. Ice Performance of the Swedish Multi-Purpose Icebreaker Tor Viking Ⅱ. Proceedings of PAC 2001, pp.849-865.

Sanderson T.J.O, 2014, Ice Mechanics. Springer, Lexington, USA.

Schneider, P.J. and Eberly, D.H., 2002. Geometric Tools for Computer Graphics. Morgan Kaufmann Publishers, San Francisco, USA.

Shimanskii, Y.A., 1938. Conditional standards of ice qualities of a ship. Trans. Arctic Research Institute, Northern Sea Route Administration Publishing House, Vol. 130, Leningrad. Translation T-381-01 by Engineering Consulting and Translation Center (ECTC), P.O. Box 1377, Jackson Heights, New York, NY 11372.

Su, B, 2011. Numerical Predictions of Global and Local Ice Loads on Ships. PH.D. Thesis. Department of Marine Technology, Norwegian University of Science and Technology, Norway.

Yeon-gyu K., Sunyoung K., Youngha P., Byeongseok Y. and Sukwon L., Prediction of Maneuverability of a Ship with POD Propulsion System. Journal of the Society of Naval Architects of Korea Vol. 43, No. 2, pp. 164-170

Visual Antarctic Sea Ice Condition Observations during Austral Summers 2012-2016

Mikko Suominen[1], Anriëtte Bekker[2], Pentti Kujala[1], Keith Soal[2], Mikko Lensu[3]
[1] Aalto University, Dep. of Mechanical Engineering, Marine Technology, Espoo, Finland
[2] Stellenbosch University, Sound and Vibration Research Group, Stellenbosch, South Africa
[3] Finnish Meteorological Institute, Helsinki, Finland

ABSTRACT

Due to Antarctica's remoteness, sea ice condition observations and measurements on board ships are rare. However, ship based annual expeditions are conducted from Cape Town to Antarctica within the South African National Antarctic Program. The S.A. Agulhas II replaced the S.A. Agulhas in 2012 and departed Cape Town on December 2012 for her maiden voyage to Antarctica. Intensive sea ice observations have been performed since by researchers during annual relief voyages from December to February.

In this paper, we present the visual observations on board the S.A. Agulhas II from the annual expeditions of four austral summers 2012 to 2016. The sea ice condition observations include ice thickness, concentration, and floe size. The changes in these parameters are studied between different years and during the voyages. The observations show that the thickness, concentration and floe size increase when moving from north towards south.

KEY WORDS: Ice conditions; Ice thickness; Ice concentration; Floe Size; Antarctica.

INTRODUCTION

Several research stations have been established in Antarctica. As most of the stations are located relatively close to the ice shelf edge the main material logistics are handled through sea transportation. Thus, several ships annually visit Antarctica. Despite the retreat of the Arctic sea ice, Antarctic sea ice extent has not shown indications of shrinking by the onset of 2016. As such, ships necessarily have to navigate through sea ice in order to resupply research stations. Although several ships sail to the Antarctic in the austral summer, there is a lack of sea ice observations.

Ice thickness and its measurement is a central problem in sea ice geophysics. Although progress in satellite altimetry is promising, there is still no means to remotely sense Antarctic ice thickness with sufficient accuracy. Both the basic understanding and the development of altimetry require thorough data sets. Current data sets are lacking both in the spatial and temporal domains. Current studies include temporal data series from bottom mounted sonars (Worby et al., 2001) and fast ice thickness records (Murphy et al., 1995). Electromagnetic (EM) campaigns have provided regional data (Haas, 1998, Reid et al., 2003). Drilling and observations during various campaigns (e.g. Worby et al., 1996) have resulted in a rough understanding of typical thicknesses in frequently visited Antarctic seas. A large number of observations from ships have been analysed by Worby et al. (2008) for circumpolar coverage. However, winter season thickness data is almost nonexistent, except for a few dedicated campaigns (Wadhams et al., 1987) and ship observations.

In order to enlarge the data pool, this paper presents the visual sea ice observations on board the S.A. Agulhas II (SAA II) from the annual South African National Antarctic Expeditions during austral summers 2012 to 2016. Earlier, a part of the visual observations and stereo camera measurements have been reported by Suominen et al. (2016) and EM measurements by Lensu et al. (2015) and Suominen et al. (2016). The observed sea ice parameters include ice thickness, concentration, and floe size. The changes in these parameters are studied during the summer seasons and between the years.

DESCRIPTION OF VOYAGES

Voyage during Austral Summer 2012 – 2013

The ship departed Cape Town on December 6, 2012 and headed to the zero Meridian, which she followed to Antarctica. She encountered ice for the first time on December 13. The Antarctic ice shelf was reached on December 15, which she followed towards Akta Bukta close to Neumayer III (the German Antarctic research station). The ship operated next to the ice shelf between Penguin Bukta, the closest point to the SANAE IV (the South African Antarctic research station), and Neumayer III between December 15, 2012 and January 10, 2013. The sea ice in the area mainly comprised ice floes, with diameters which did not exceed the length of the ship, with brash ice between. In some locations the thickness of the ice floes was in excess of two meters. The ice concentration varied from open water to 100 % ice concentration. The bay ice next to Neumayer III consisted of the broken ice shelf and thick ice floes that had frozen together.

On January 13, the ship initiated her voyage towards South Georgia and the South Sandwich Islands along the zero Meridian exiting the ice conditions on January 14. The vessel returned from the voyage to the ice infested waters adjacent to Antarctica on January 28. Between January 28 and February 10, the ship operated in the same area as before. Generally, the ice conditions in the area were the same as before. An iceberg, which had stuck close to Neumayer III, had formed a level ice sheet in the bay next to Neumayer III. The cargo loading of the vessel was completed on February 10, and the ship headed towards Cape Town. The last ice was encountered on the same day.

Voyage during Austral Summer 2013 – 2014

The SAA II departed from Cape Town on November 28, 2013, and headed to Antarctica following the zero Meridian. The first ice was encountered at a latitude of 62.2° on December 7. The ice concentration was typically less than 90 % until December 10. More difficult ice conditions were encountered around 68° latitude, when the concentration increased to 100 % and a maximum ice thickness 1.6 m. From 69° latitude (December 9, 20:30) to the Akta Bukta, latitude of 70.3°, it took 12 days and 19 hours. In open water the voyage would have taken approximately 6 hours. During this period, the maximum observed ice thickness was 3 m and heavy ridges were encountered. The ship arrived to the Akta Bukta on December 23. On December 24, the ship navigated to the Penguin Bukta in fairly easy ice conditions with a lot of open water. On December 30, the SAA II headed towards South Georgia and the South Sandwich Islands. The ice conditions were more challenging until December 31. Thereafter the ice conditions eased with intermittent navigation through open water. The vessel arrived at the South Sandwich Islands on January 4 and exited the ice field at this point.

Whale observations were performed after reaching South Georgia, during which the ship was navigating at the ice edge between 68° and 65° latitude and 0° and 20° longitude without ice passage. The ship was back in ice on January 23 and broke through the pack ice to Penguin Bukta which she reached on January 25. The ice conditions were fairly similar as before. The final phase close to Penguin Bukta proved challenging due to a particularly resistant ridge on the afternoon of 23 December. Several hours were spent to navigate through the ridge. On January 26, the ship started the voyage to Akta Bukta arriving there January 28. From Penguin Bukta to Akta Bukta the ice conditions encountered were undemanding (January 26 to 28). The thickness of the bay ice was fairly constant at an estimated 1.8 m. The return voyage to Cape Town started on January 31 and the last ice was observed February 1. The final phase of leaving Akta Bukta to Cape Town was again fairly easy with typical ice concentrations below 90 %.

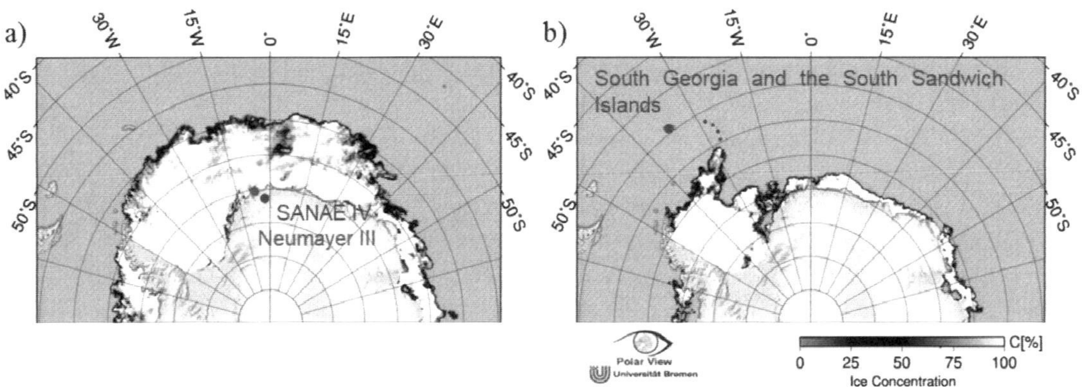

Figure 1. The sea-ice extent in Antarctica on (a) December 6, 2013, and (b) February 2, 2014. Pictures reproduced from Suominen et al. (2015) and the concentration maps produced by Spreen et al. (2008).

Voyage during Austral Summer 2014 – 2015

The ship headed from Cape Town to the zero Meridian on December 5, which she followed to Antarctica visiting Bouvet Island on the way. The first ice berg was observed on December 13. The first ice was encountered in the morning of December 14 and the ice shelf was reached on December 14. After reaching the shelf, the ship headed towards Akta Bukta arriving there on December 18. She was beset a few times on the way, but was able to circumnavigate the most challenging ice conditions. She started carving bay ice, but the carving was ended after a while as the thickness of the bay ice in Akta Bukta was too great to penetrate. Thus, the cargo off- and on-loading was performed on bay ice between December 18 and 19. Subsequently the ship headed towards Penguin Bukta. The ship was able to navigate in a lead close to the shelf and Penguin Bukta was reached in the morning of December 20.

The SAA II left for South Georgia at the daytime on December 29 after which the number of ice floes and the ice concentration increased. The SAA II attempted to avoid challenging ice navigation and exited icy waters on December 30, 2014. She arrived to Thule Island on January 4, where some ice was encountered, but not for long because of the wind situation. The ship returned to ice conditions on January 19, 2015. The next day the encountered sea ice was increasingly compact and thicker. On January 20 the ship had to ram for one day before reaching open water on January 21. She approached the ice shelf and passenger transportation and cargo handling activities ensued. On January 29, the SAA II approached Neumayer III to reach her destination on January 30 where cargo was collected. The ship left the ice shelf and headed back to Cape Town on February 6. On February 7, the ship entered open water.

Voyage during Austral Summer 2015 – 2016

The SAA II departed Cape Town harbour on December 5 first navigating south west and then due south on the Greenwich Meridian. Ice was first encountered on December 11 and continued until arriving at the Antarctic ice shelf at Penguin Bukta on December 16. On December 19 the vessel navigated in a lead with some light ice conditions to Akta Bukta. After cargo operations the ship left the shelf for South Thule and soon encountered thick pack ice, requiring heavy ramming maneuvers. The vessel became heavily beset and after repeated ramming remained stationary from December 22 to December 23. When the wind conditions changed, the pack ice started opening allowing further ice navigation. The vessel reached South Thule on December 24 and left the pack ice on December 28 on route to South Georgia, arriving on December 30.

The vessel re-entered ice on January 12 on route to Penguin Bukta and conducted seal tagging in light to medium floe ice conditions until January 24. The ship performed cargo operations at Penguin Bukta with intermittent short runs into the pack ice for further seal tagging. The vessel reached Akta Bukta on January 26 and began carving the fast ice at the shelf, however unfavorable wind conditions ended this navigation after only a few hours. The vessel continued short runs into the ice for seal tagging, while the Polarstern finished the carving once the wind conditions became favorable once more. After cargo back loading the SAA II started her return voyage on Februray 1 and left the sea ice on February 2.

VISUAL OBSERVATIONS

The visual observations were conducted from the bridge of the ship when the ship was operating in ice. As the sun does not set during the austral summer in Antarctic waters, enough light was available also during the nights for observations. The observations were executed in 2 to 3 hour shifts with 5 to 7 observers in the group to avoid fatigue strain on observers. Observations were reported for 15 minute periods in the earlier voyages of 2012-2013 and 2013-2014, and for 10 minute intervals during further voyages. The observations included ice thickness, ice concentration, snow thickness, brash ice amount, floe size as well as general comments. The thickness, concentration, and floe size were estimated as occurrence percentages (in tenths) for given categories during the measurement period.

The ice thickness was estimated by comparing the thickness of the cross-section of the upturning pieces with a scaled yard stick which was suspended overboard from the main deck. The stick was marked with 10 cm wide black and white painted stripes to calibrate the estimations of observers, see Figure 2. Observers estimated the ice thickness with the stick as it was seen from the bridge and the real ice thickness was obtained by scaling the observations with a factor of 1.5. This was done in order to correct the parallax error. The factor was determined based on the distances to the sea surface and measurement stick from the bridge. The thickness classes during the 2012-2013 voyage included the classification of ice in categories with 20 cm increments between 0 m to 2 m with a final category for ice thickness in excess of 2 m (see Figure 3). Additional thickness classes 2.0-2.5 m and 2.5-

3.0 m were added to observation classes for the following voyages. The voyages 2013-2014 and 2015-2016 also included a class of >3.0 m. However, as the measurement stick is only 1.5 m long the uncertainty increases significantly for the thickness classes exceeding 2.0 m. The average ice thickness for a 10 minute period was determined by calculating a weighted average from the thickness observation periods.

The concentration of the ice field was estimated from inboard observations. Thus, the concentration was estimated from conditions experienced in the close vicinity of the ship. As the crew preferably navigated in open water instead of ice, the ship followed open water conditions when possible. In this case, the concentration was marked as zero, although floes of ice could be seen. A range from 0 to 100 % was divided into bins with a size of 10 % each, i.e. 0-10 %, 10-20 %, etc., which formed the observation classes. Here, zero denotes open water and one full ice cover. Figure 3 presents an example from a part of the visual observation sheet.

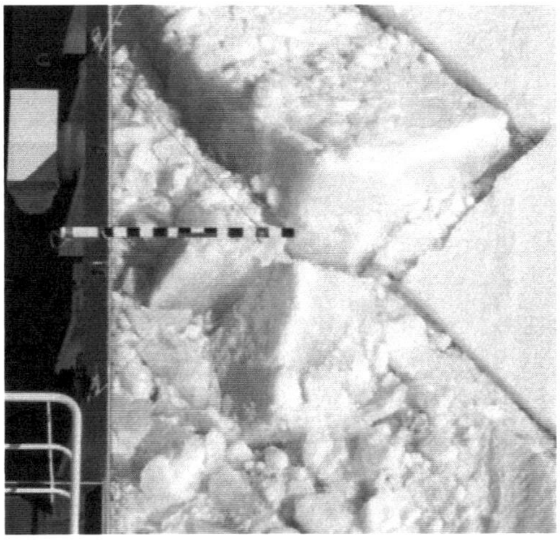

Figure 2. The measurement stick for ice thickness estimation. Each stripe is 10 cm wide.

Time UTC+0									snow [cm]		Ice concentration in tenths									
			start		end				min	max	0-	10-	20-	30-	40-	50-	60-	70-	80-	90-
year	mm	dd	hh	mm	hh	mm	Lat	Lon			10	20	30	40	50	60	70	80	90	100
2013	12	22	8	0	8	10	-70.46	-8.426		50	6			3	1					
2013	12	22	8	10	8	20	-70.45	-8.379		60	0				1	4		3	2	
2013	12	22	8	20	8	30	-70.45	-8.377			4							2	1	3
2013	12	22	8	30	8	40	-70.45	-8.376		60	0						3	4		3

Figure 3. An example from a part of the visual observation sheet.

The classes for observed ice floe diameter were <20 m, 20-100 m, 100-500 m, 500-2000m, 2-5 km, >5 km. The classes were selected based on the egg code (WMO, 2014) used, for example, in the Baltic Sea. The floe diameters were estimated with the help of the main dimensions of the ship. If the floes were smaller than the breadth of the ship (21.7 m), those belonged to the first class. If the floes were larger than the breadth, but smaller than the overall length of the ship (134 m), those belonged to the class 20-100 m. If the floes were larger than ship length, the diameter was estimated as multiples of the ship length.

ICE CONDITIONS

Figure 4 presents the 10 minutes mean ice thickness for observation periods during the four annual voyages in 2012-2016 and Figure 5 shows the mean ice thickness in latitudes for

December, January, and February. The onset of ice navigation is typically around 60 degrees latitude when the ship started her southerly voyage to Antarctica. Generally, the thickness increased towards the south and the most challenging conditions were encountered close the ice shelf. However, thick ice floes have also been encountered in lower latitudes. Three meter thick ice has been observed in 60 degree latitude in the voyage 2015-2016. However, the concentration of the ice field was generally lower in 60 degree latitude and increases southwards, see Figure 6 and Figure 7. Thus, the ship was commonly able to navigate relatively easily through the ice by navigating between the floes or pushing floes aside at the onset of the voyage. Difficulties were encountered closer to the ice shelf when the concentration and thickness increased. Furthermore, floe size increased southwards, which hinders maritime operations, see Figure 8 and Figure 9. Ships are able to push smaller floes aside and navigate between the floes, However, in a case of large floes, the ship is forced to break through the ice, if the floe cannot be circumnavigated.

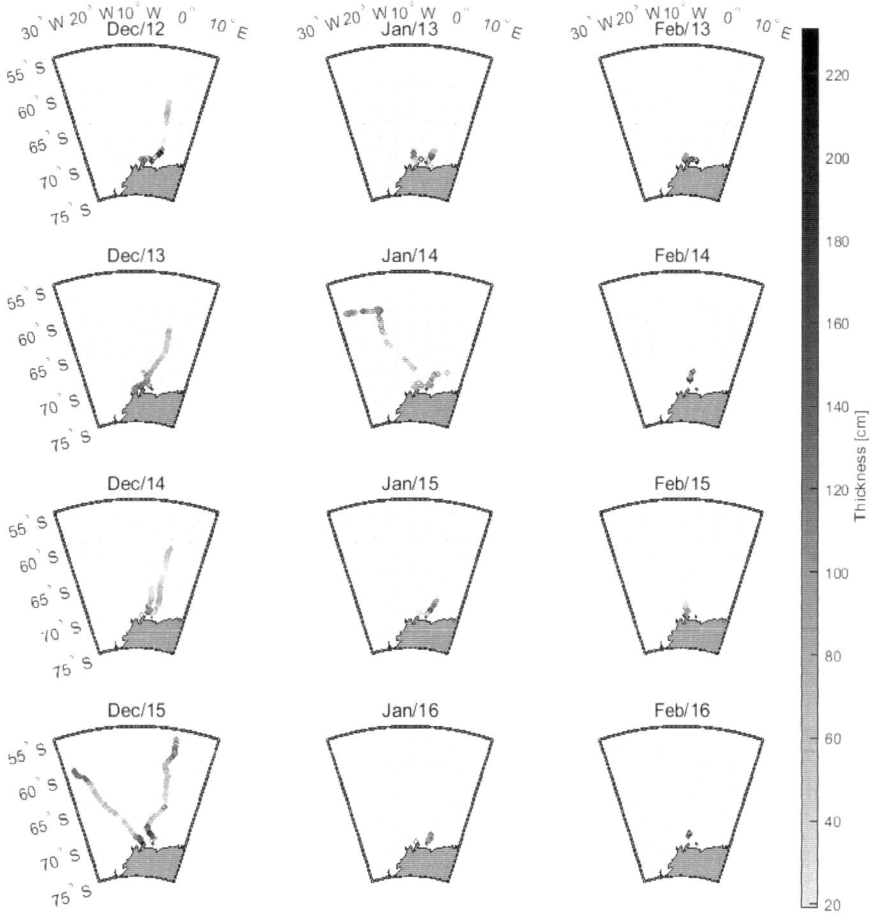

Figure 4. 10 minutes mean ice thicknesses during the four voyages.

As can be noted from Figure 4 and Figure 6, in the voyages 2012-2013 and 2014-2015 the ship was basically able to avoid the ice conditions when she headed for South Georgia. However, in the voyages 2013-2014 and 2015-2016, the ice conditions lasted up to the South Sandwich islands at 60 degree latitude and two meter thick floes were encountered in these latitudes. During the summer season, the sea ice extension around Antarctica shrinks significantly and the ice is encountered only near the continent, see Figure 1. The shrinking can also be observed from Figure 4 to Figure 9 by comparing the December and February in different years. The only ice in January and February has been encountered in the vicinity of the ice shelf. However, as the surviving ice is generally multiyear ice, the thickness can be significant despite the late season, see Figure 4 and Figure 5.

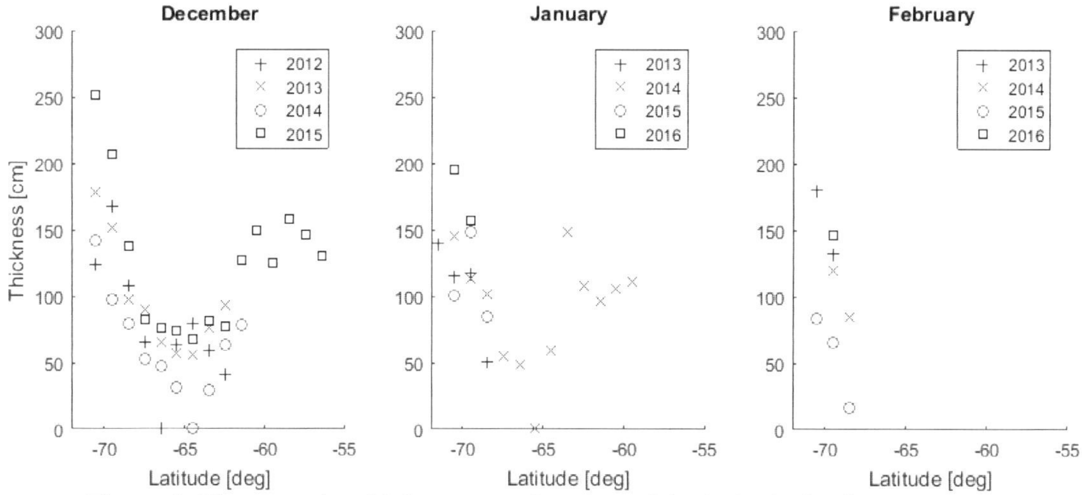

Figure 5. The mean ice thickness as a function of the latitude for four voyages.

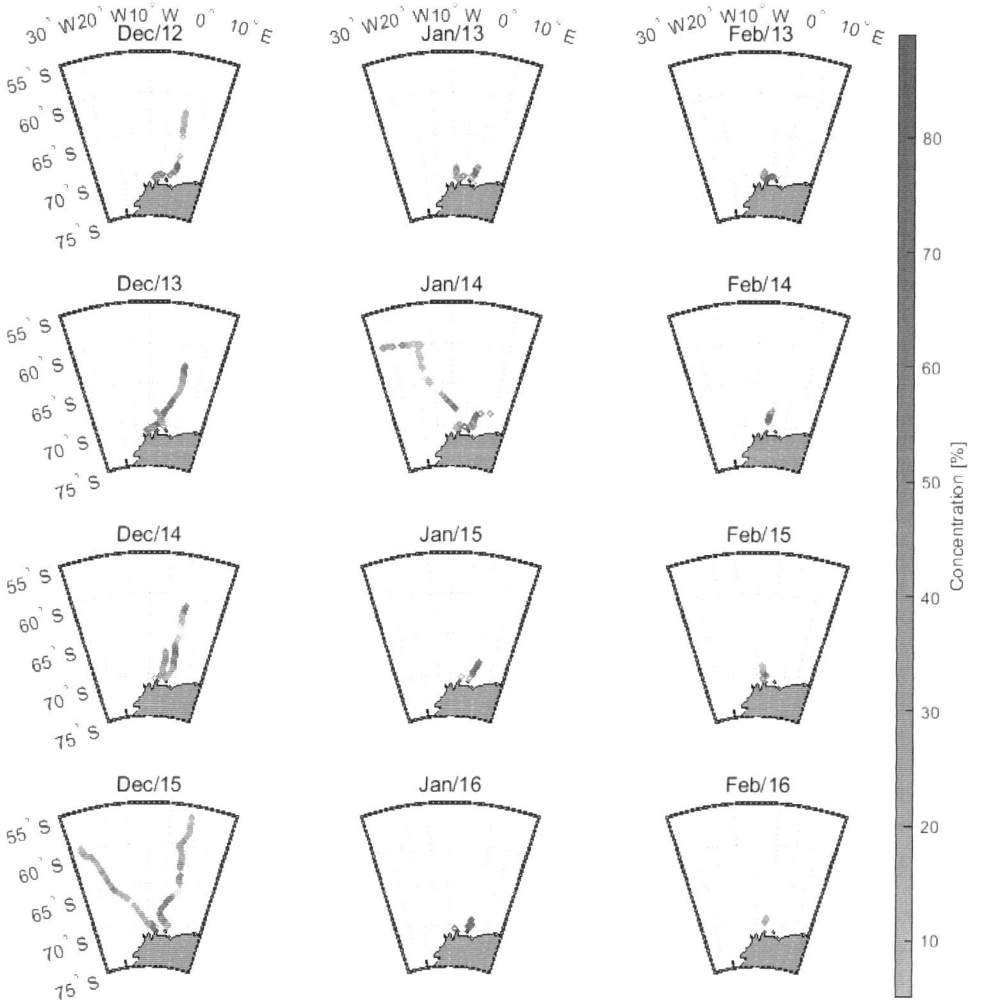

Figure 6. 10 minutes mean ice concentration during the four voyages.

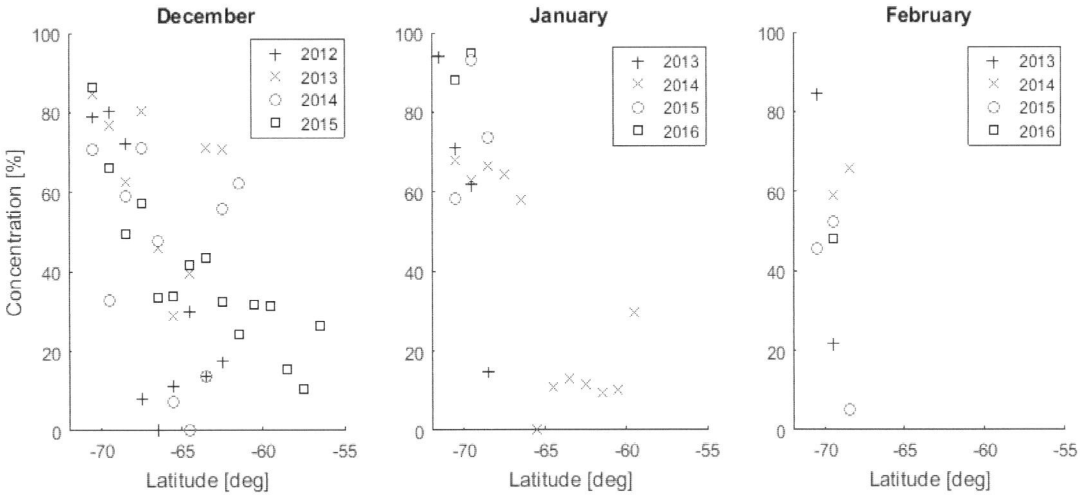

Figure 7. The mean ice thickness as a function of the latitude for the four voyages.

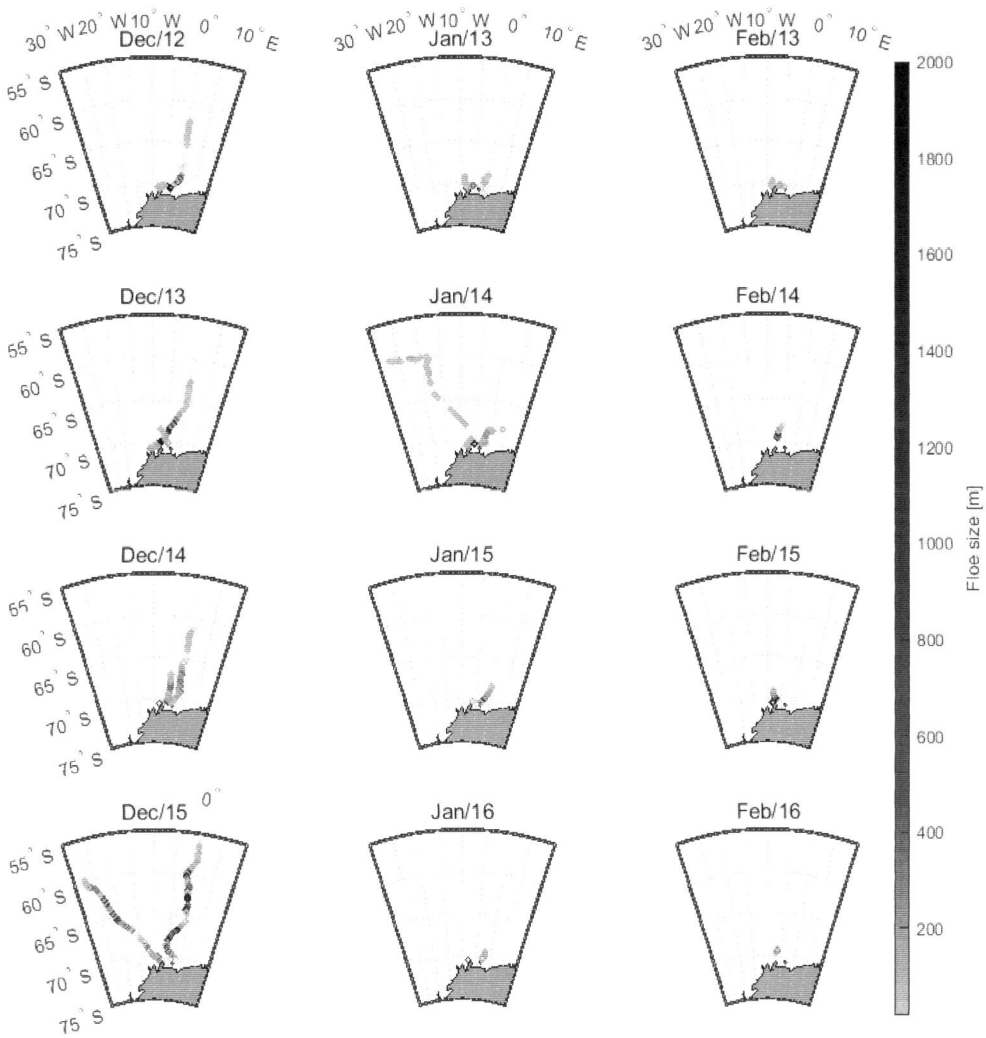

Figure 8. 10 minutes maximum floe size during the four voyages.

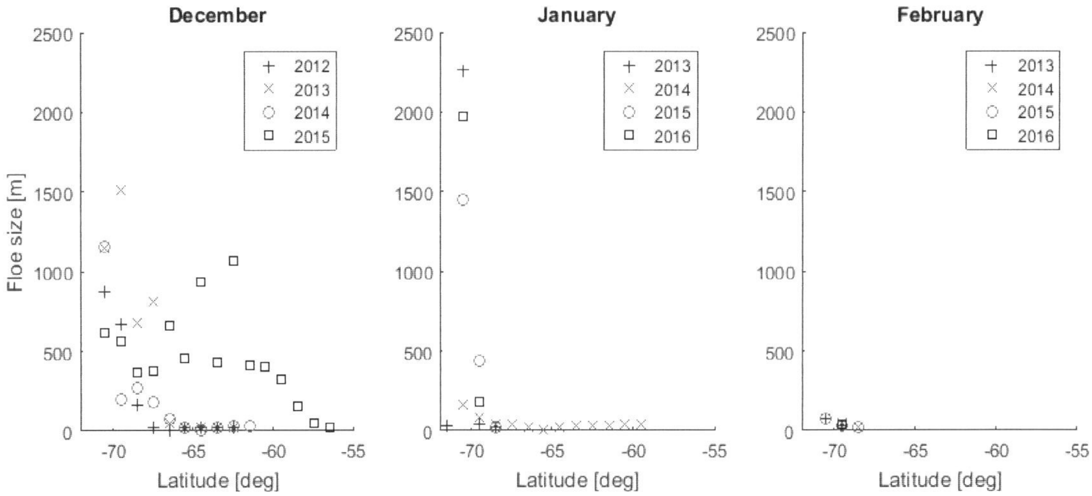

Figure 9. The mean floe size as a function of the latitude for four voyages.

CONCLUSIONS

Visual observations on ice thickness, ice concentration and floe size are presented from four annual Antarctic voyages on board the SAA II in 2012-2016. The observations show that the ice conditions commonly extend to approximately 60°S latitude in austral summers. Generally, the thickness, concentration and floe size increases towards the ice shelf. Furthermore, the observations show that the sea ice extent significantly shrinks during a year and multiyear ice only exists close to the ice shelf in early February. Similar observations have been reported in earlier studies (see e.g. Worby et al., 2008).

ACKNOWLEDGEMENTS

The work of Aalto University was funded by Tekes (Finnish Funding Agency for Technology and Innovaitons), Academy of Finland, and Lloyd's Register Foundation through the projects NB1369 Full-scale ice trial, ANTLOAD, and CEARCTIC. Their financial support is gratefully acknowledged. The observations would not have been possible without the access to S.A. Agulhas II and help of the crew of the vessel. The Department of Environmental Affairs (South Africa) and South African National Antarctic Program (SANAP) are acknowledged for enabling the research and crew of the vessel for enabling the observations during the voyage. Authors would also like to express their gratitude for the researchers participating to the observations Håkan Enlund, Jani Hoikkala, Jakke Kulovesi, Brendan Boulle, Jonni Lehtiranta, Kim McMahon, Fourie Gildenhuys, Rosca De Waal, Hamza Omer, Jukka Tuhkuri, Katriina Veijola, Clinton Saunders, Ryan Cloete.

CONCLUSIONS

Haas, C., 1998. Evaluation of ship-based electromagnetic-inductive thickness measurements of summer sea-ice in the Bellingshausen and Amundsen Seas, Antarctica. *Cold Regions Science and Technology*, 27 (1), 1-16.

Lensu, M., Kujala, P., Kulovesi, J., Lehtiranta, J., Suominen, M., 2015. Measurements of Antarctic sea ice thickness during the ice transit of S.A. Agulhas II. *Proceedings of the 23rd International Conference on Port and Ocean Engineering under Arctic conditions*.

Murphy, E., Clarke, A., Symon, C. and Priddle, J., 1995. Temporal variation in Antarctic sea-ice: analysis of a long term fast-ice record from the South Orkney Islands. *Deep Sea Research Part I*: 42(7), pp. 1045-1062.

Reid, J.E., Pfaffling, A., Worby, A.P. and Bishop, J.R, 2003. Shipborne electromagnetic measurements of Antarctic sea-ice thickness. *Geophysics*, 68 (5).

Spreen, G., Kaleschke, L., Heygster, G., 2008. Sea ice remote sensing using AMSR-E 89 GHz channels. *Journal of Geophysical Research*, 113, C02S03, doi:10.1029/2005JC003384.

Suominen, M., Aalto, H., Kulovesi, J., Lensu, M., Kujala, P., Soal, K., Lehtiranta, J., 2016. Ship-based visual Antarctic sea ice thickness measurements. *Proceedings of the 23rd IAHR International Symposium on Ice.*

Wadhams, P., Lange, M. A, Ackley S. F., 1987. The Ice Thickness Distribution Across the Atlantic Sector of the Antarctic Ocean in Midwinter. *Journal of Geophysical Research*, 92(C13), pp.14535–14552.

WMO, 2014. WMO Sea Ice Nomenclature, WMO No. 259, volume I-III.

Worby, A.P., Bush, G.M, and Allison, I., 2001. Antarctic sea ice thickness distribution as determined from moored Upward Looking Sonar. *Annals of Glaciology*, 33 (2001), pp. 177–180.

Worby, A. P., Geiger, C. A., Paget, M. J., Van Woert, M. L, Ackley S. F., DeLiberty T. L., 2008. Thickness distribution of Antarctic sea ice. *Journal of Geophysical Research*, 113, C05S92.

Worby, A. P., Jeffries, M. O., Weeks, W. F., Morris, K., Jaña R., 1996. The thickness distribution of sea ice and snow cover during late winter in the Bellingshausen and Amundsen Seas, Antarctica. *Journal of Geophysical Research*, 101(C12), pp. 28441–28455.

Investigation of Large Ice Rubble Field in the Barents Sea

N.A.Marchenko[1], A.V.Marchenko[1]
[1] The University Centre in Svalbard, Longyearbyen, Norway

ABSTRACT

Characteristics of ice ridges and rubble fields in the Barents Sea are of great interest for the formulation of design criteria for offshore structures. But there are quite a few field observations and measurements. On 2 of May 2016 UNIS researchers and students, due to unusual little ice coverage, reached on RV Lance Northern-East end of the Edge Island (East of Spitsbergen) and observed large ice rubble field with frozen in icebergs stretching on several km along the coast. Laser scanning from the Lance board was performed to estimate spatial variations of vertical elevation of above water part of the rubble. 2 CTD profiles performed near the rubble identified that entire water column of 40 m depth was around the freezing point with couple of fresher water intrusions on different depths.

Further investigation of high-resolution satellite images since 2009 gave information about structure, size, formation and evolution of the rubble every year and revealing possible cause as grounded on the shallow icebergs. Analysis of ice drift data from IFREMER and own ice trackers allowed to describe possible origins of the ice rubble and its future development.

KEY WORDS: Ice; Rubble; Barents Sea

INTRODUCTION

A characteristic of ice-covered waters is the presence of ice ridges and rubbles. Pressure ridges are composed of ice fragments that are piled up along a line, with the steep-sloped ridge rising up as much as 5 to 10 feet or more above the adjacent stretches of level ice. Rubble ice, by comparison, is a jumble of ice fragments or small pieces of ice (such as pancake ice) that covers a larger expanse of area without any particular order to it. The height of surface features in rubble ice is often lower than in ridges. Rubble ice can cover large expanses, such as in Antarctica where the ocean swells breaks up newly forming ice and herds it together into rubble ice field (NOAA, 2017).

The investigations devoted to ice ridges (Timco and Burden, 1997) and ice rubble (Barker and Timco, 2017) describes their shapes and sizes. Studies of ice rubble was performed by laboratorial experiments (Liferov and Bonnemaire, 2005), field observation and modeling (Marchenko et al., 2016). Observations in the Caspian and Beaufort Seas show maximum rubble heights of about 15m above the ice line can be achieved, even with relatively thin ice

(McKenna et al., 2011)(Croasdale, 2012). Mathematical models of ice ridge formation were formulated by R.R.Parmeter and M.D.Coon (Parmeter and Coon, 1972) and A.V. Marchenko (Marchenko and Makshtas, 2005). Discrete elements model of ice ridge build was elaborated by M.Hopkins (Hopkins, 1998) Formulas for the estimates of upper bounds of vertical sizes of ice ridges and rubble sails and keels were considered in (Hopkins, 1998) and (Marchenko, 2016a).

T.Vinje and Å.Kvambekk describing drifting ice in the Barents Sea wrote: "Ridging events may take place with onshore winds. Rubble fields, possibly due to this effect, are often observed in the NW part of the Barents Sea and ice piles 10-15m high have been observed on the shore of the islands. The height of the ridging is also illustrated by the fact that fairly large accumulations of ice blocks may be observed on top of 7-15 m high grounded icebergs (Løset et al., 1988, Vinje et al., 1989). Ice ridging also occurs in convergences caused by wind, currents and land constraints. The most frequent ridge height observed from the meteorological station on Hopen is 2 m. The most frequent ridge density in the ice fields of the Barents Sea as observed from a long term project using aircraft is about 10 km-l. The frequency distribution is fairly broad, with a maximum of 26 km-' (Vinje, 1985)" (Vinje and Kvambekk, 1991). But these characteristics are related to cold 1980-s. Multiple Norwegian and Russian expeditions performed since 2003 reported about maximal draft of ice ridge registered by drilling studies about 15 m (Marchenko et al., 2009, Hoyland, 2007, Zubakin, 2006). Origins of big ice ridges are usually related to ice drift from Arctic Ocean into the Barents Sea.

On the background of "global warming" and Arctic Ice shrinking in last years, it was very interesting to discover the huge ice rubble in the western part of the Barents Sea and make investigation, using modern equipment. Knowledge about ice rubble is very important for many engineering problems, and relevant for future hydrocarbon exploration in the Barents Sea. That's why our research of the rubble can be interesting both from a purely scientific and practical point of view.

DISCOVERING AND LOCATION

The rubble field had been discovered during cruise on RV Lance in the frame of UNIS course AT-211 and research work. The weather conditions this spring was unusually warm and Barest Sea was almost ice free, with ice not going south longer than 76°N. On the 2 of May 2016 UNIS researchers and students finished the investigation on the first station south in Ryke Yse Island and moved in the NW direction looking for large enough and stable ice floe. Ice in the area was presented by drifting ice floes average size 300 m with ice thickness 40-50 cm, smaller floes and ice free water. Near the north-east end of Edge Island, the large ice rubble field with frozen in icebergs stretching on several km along the coast was observed. (Figure 1,2) The edge of rubble was high (3-3,5 m) surprisingly vertical and sharp/even, as it was cut by huge knife. Offshore of rubble field was ice free water (Figure 1).

Figure 1. Ice rubble field discovering. Photo: N.Marchenko

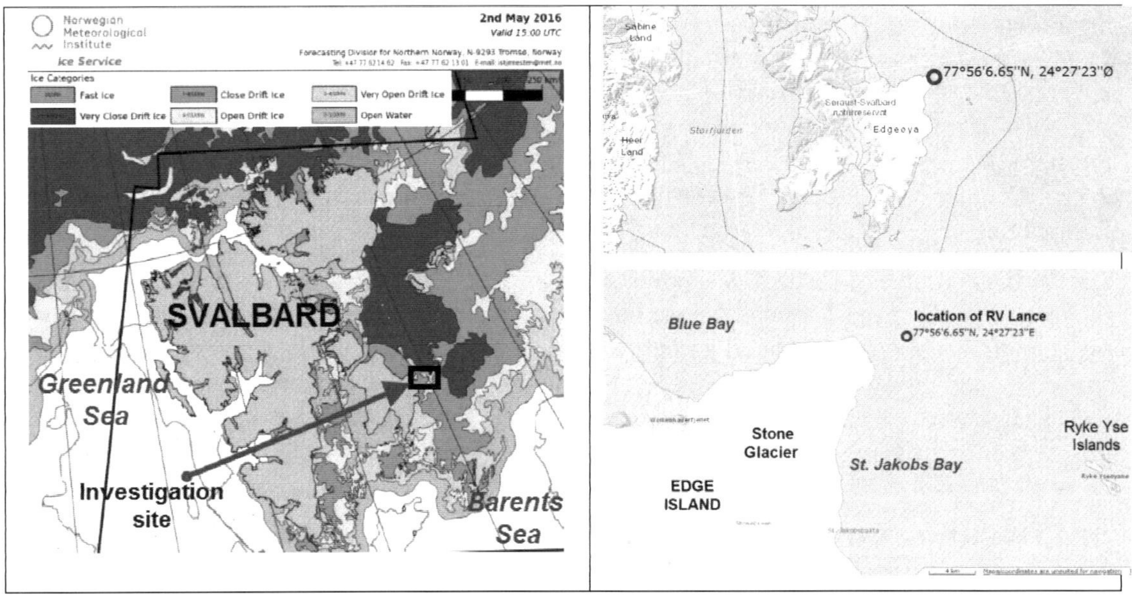

Figure 2. Geographical location of the discovered ice rubble on the base map (Norwegian Polar Institute, 2017) and sea ice map for the day of observation (Norwegian Meteorological Institute, 2017)

LASER SCANNING AND RUBBLE SPATIAL CHARACTIRISTICS

Laser scanning from the Lance board was performed to estimate spatial variations of vertical elevation of above water part of the rubble with scanner REIGL VZ-1000 from the board of Lance. Scanning from several positions (program: 950m (150kHz) and resolution 0,03 deg) had been done, but due to short distance between them it was not possible to create full 3D image. So there is a shadow from high ice objects on scan. Foggy weather created significant noise on the point cloud around the scanner and the quality of the picture decreases considerably farther than the distance 300-400 m. Nevertheless the obtained point cloud clearly shows the structure of the rubble and dimension. Scans were processed in RiScan program: The high of ice wall near the ship was more than 3 m, high of nearest iceberg – more than 9,5 m.

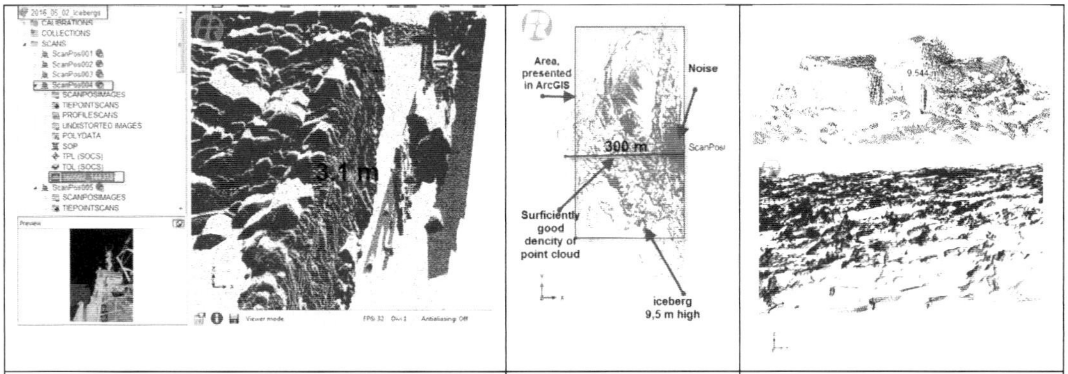

Rubble scan in RiScan program	Scan orientation	Iceberg (above) and rubble (below)

Figure 3. Point cloud images of ice rubble in RiScan software.

Point cloud was exported to ArcGIS software, where elevation distribution was mapped and estimated (Figure 3-4). Among 160 thousand measured points min is 0 (water level, near the RV Lance), max – 11,96 m (iceberg). Mean value is 1,61m. Taking into consideration that considerable part of points was in the shadow of large ice pieces, mean vertical size of the rubble is estimated as above two m. Histogram in Figure 4a shows smooth distribution of measured values of the rubble high in the points of laser scanning. Pink histogram in Figure 4b is constructed by the multiplication of the rubble sail high on 4.4 according to the sail-keel proportion discovered by Timco and Burden (1997). Blue histogram in Figure 4b is constructed by the multiplication of the rubble sail high on 9 according to the sail-keel proportion of tabular icebergs. Gray area in Figure 4a is bounded by ice thicknesses 5.5 m and 15 m calculated with the formulas specifying the upper and lower boundaries of the rubble draft (Marchenko, 2016a)

$$h_k = 4.6 + 2.88 \cdot h(\mathrm{m}),\ h = 0.3\,\mathrm{m};\ h_k \approx 15\sqrt{h}(\mathrm{m}),\ h = 1\,\mathrm{m}.$$

These dependencies were formulated with the assumption that level ice is broken by bending when it forms the floating rubble. The observed rubble is not completely floating, and can be partially grounded. It may explain high estimated values of the draft up to 30 m (Figure 4b).

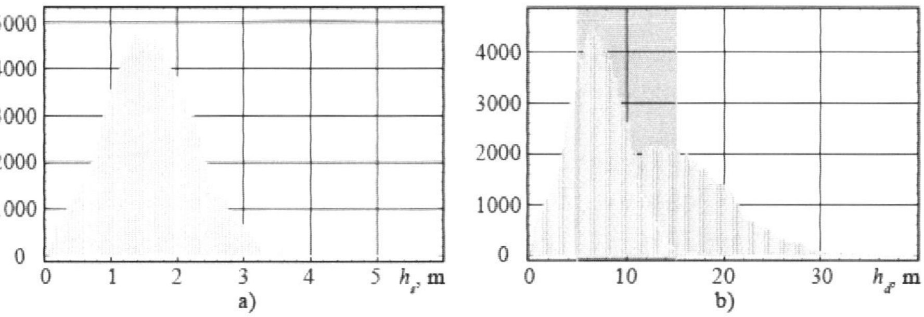

Figure 4. Frequency Distribution of point elevation above sea level (a) and below sea level calculated with

On Figure 5 the space distribution of point elevation in ice rubble and profiles across rubble are presented on the map in ArcGIS software.

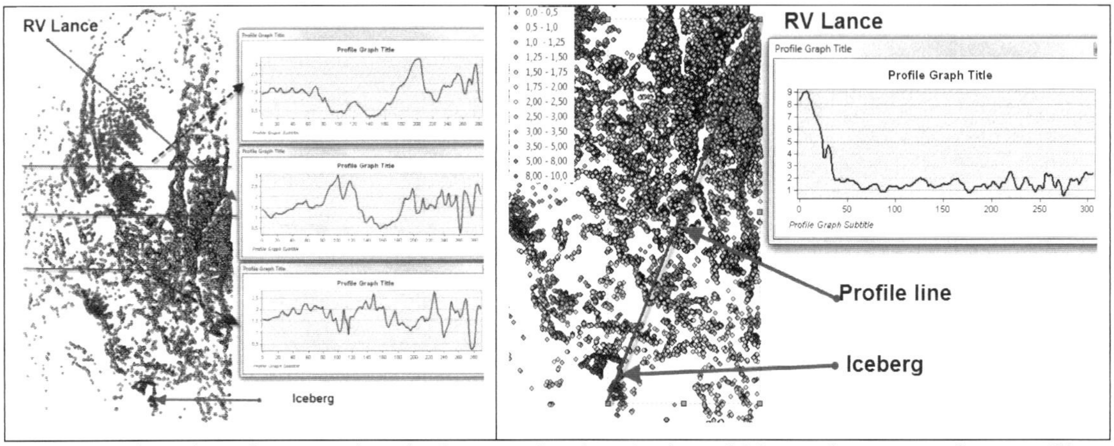

Figure 5. Elevation above sea level of scanned points in the rubble on the map and on profiles in ArcGIS software

HIGH RESOLUTION IMAGES. RUBBLE STRACTURE AND ORIGIN

Fortunately, our location was captured by satellite high resolution (50 cm) image in suitable time and two images from 25 April (Sensor WV3, Off-nadir 25 °, sun elevation 10 , sun azimuth – 297,4) was bought from Apollo mapping service (Figure 6).

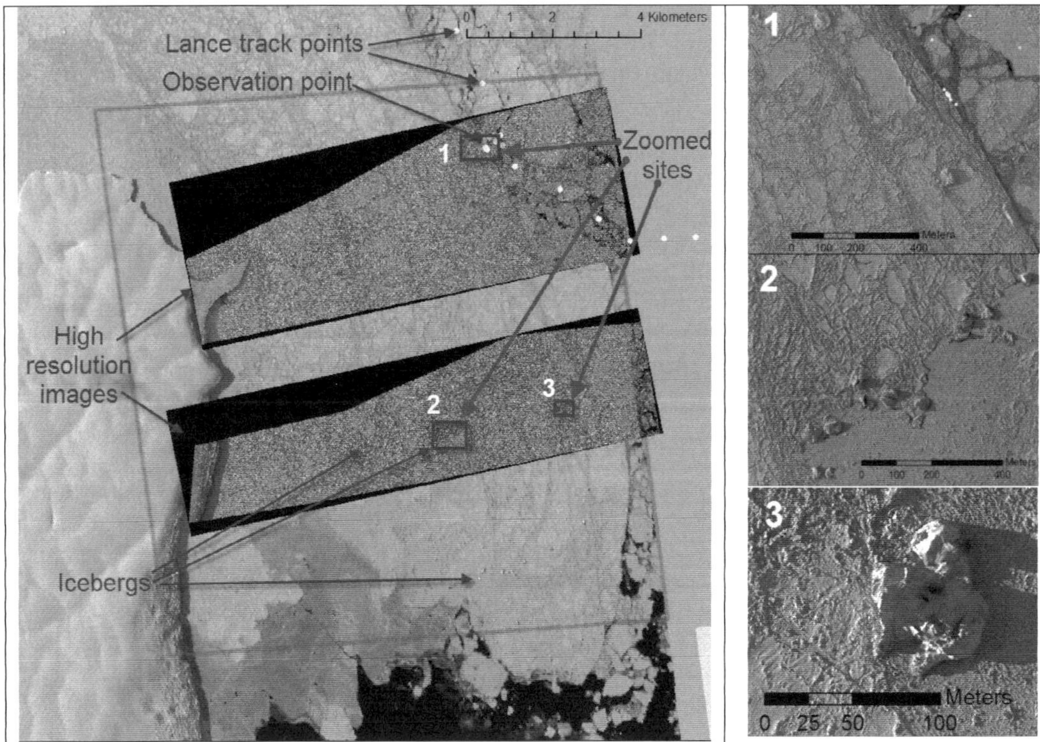

Location of high resolution images on the back ground of low resolution images	Zoom to 1)the RV Lance place, 2)icebergs, 3)one of the biggest iceberd.

Figure 6. High resolution images (location and zoom to particular featres)

On the satellite images the width of rubble field can be estimated as 9 km. It consists of different size flat ice floe (small 50-150 m and large – 400-500 m) connected by compressed ridges, stretching along the floes. Some of ridges reach up to 1 km length.

We can assume also the origin and forming process of the rubble field as stopping of drifting from the North ice by aground icebergs and further accumulation and compression of ice on the shallow. Such icebergs are visible on the south image (number 2, 3 on Figure 6). Icebergs gather in the groups of 2-5 ones along the line and hold the ice flow (Figure 6. 2). The size of small icebergs is 10-50 m, the large icebergs reach up to 50 m (Figure 6.3). Over time, the ice fills the entire space to the bottom.

LOW RESOLUTION SATELITE IMAGES. YEARLY AND INTERYEAR DYNAMICS AND LARGE SPATIAL CHARACTIRISTICS

The set of satellite images was taken from Apollo Mapping web-site (Apollo Mapping). They present on site as preview of high resolution images and give nearly 50 m resolution). 28 images such captured from 2009 in light time show that the rubble field starts to form in this place every year at the end of March, reaches max at the end of April and exists in some years till the end of June (Figure 7). In 2015 rubble was still on the place on the 2 of July (Figure 8). The remnants of the destroying rubble drift to the south, following the general scheme of drift

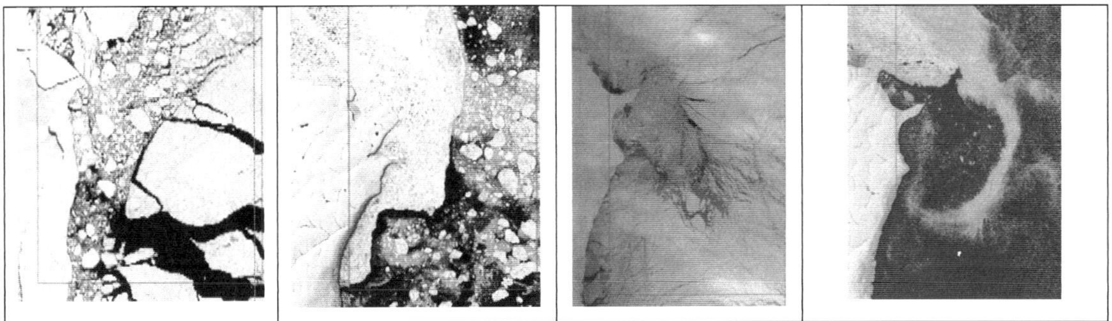

1 April 2013	28 June 2009	1 March 2016	21 June 2016

Figure 7. Satellite images of rubble for different years and dates.

Modis image of 2 July 2015 shows huge rubble (80 km long) fasted to the Edge Island - The width changes from 10 to 20 km. The rubble is separated from the main body of floating ice massive with width 100-120 km by clear polynya.

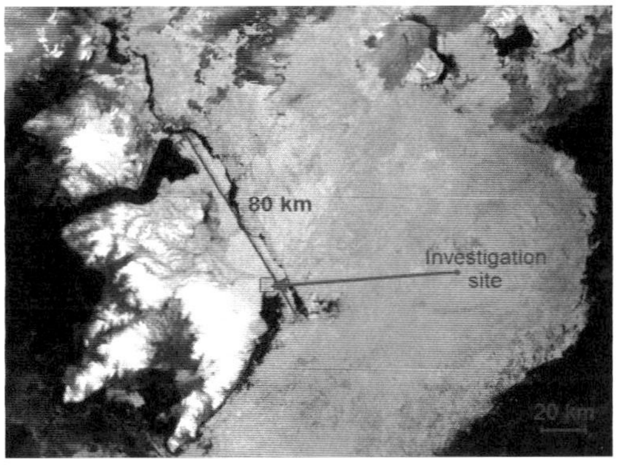

Figure 8.MODIS image 2 July 2015

(NASA, 2017)

SEA ICE PRODUCTS AND DRIFTING BUOYS. ICE MOVEMENT TO AND FROM RUBBLE PLACE

Information about actual ice drift can be obtained using drifting buoys and by interpretation consequent satellite images. Several institutions and agencies provide so called "Sea ice products", showing mean ice motion vector, derived from passive microwave sensors, visible and infrared sensors, and other sources. Among them as in was found in (Marchenko, 2016c) , IFREMER (Ifremer, 2017) is the most relevant for our region and purpose. We took IFREMER data for 2003 and 2010 with 6 days interval and constructed drifting lines, which show south-west direction of drift (Figure 9). The distance of 250 km was passed in 55 days, so near Bear Island (the new hydrocarbon exploration area), locating in 400 km on South-West, such ice creature would be in two and half month in case of average drift. But in case of polar cyclones and strong wind gust, the drift can be much faster. As it was shown in (Marchenko et al., 2016) ice fields in our region can drift with speed 1,7 m/s and reach Bear Island in a month. The significant consolidation of ice rubble occurs due to the melting from below if the rubble drifts from the region of Kong Karls Land to the Bear Island. On the satellite images Terra the ice tongues reaching Bear Island was visible both in March 2015 and March 2016 (Marchenko et al., 2016).

Our ice trackers, installed on ice floe show more detail picture of the drift, as they transmit coordinated each 10 minutes. Tidal loops and short time changing of the direction are visible on the plot (Figure 9). In case of strong wind with combination with current, which goes along the south east coasts of Spitsbergen and Edge Island, ice floe can make several tenths km per day.

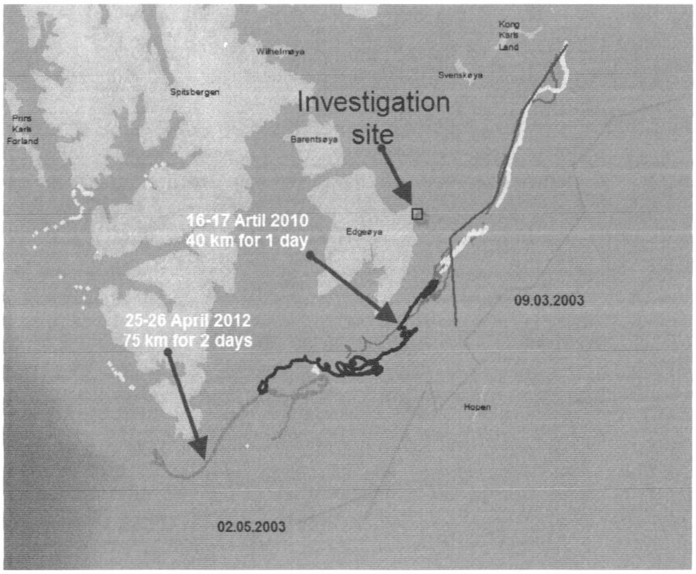

Figure 9. Drifting line of ice trackers (installed in 2010 – green, and 2012 – violet) and constructed from (Ifremer, 2017) data (dark blue – 2010, light blue – 2003)

CTD PROFILING

CTD profiling was performed with SBE 19 from the Lance board near the rubble wall in two locations. The horizontal distance between the locations was 20-30 m. Sea depth about 40 m was registered in both places. The ship hull may influence the profiling within the ship draft of 5.5 m. Dependencies of the temperature, salinity and density of the water from the depth are shown in Figure 10 and Figure 11. In both of the locations the entire water column is above the freezing point. The difference between the actual temperature and the freezing point increases with the approaching to the sea surface. Temperature and salinity profiles are not monotonic and include 4 local maxima in the location 1 and 2 local maxima in the location 2 at depth below 10 m. Density profiles shown in Figure 11b are very stable, but also include similar local maxima. The origin of the observed vertical water structure can relate to the convection and double diffusion of salt water caused by the rubble melting. Similar structures were observed in the experiment (Josberger and Martin, 1981) and obtained in numerical simulations (Gayen et al, 2015).

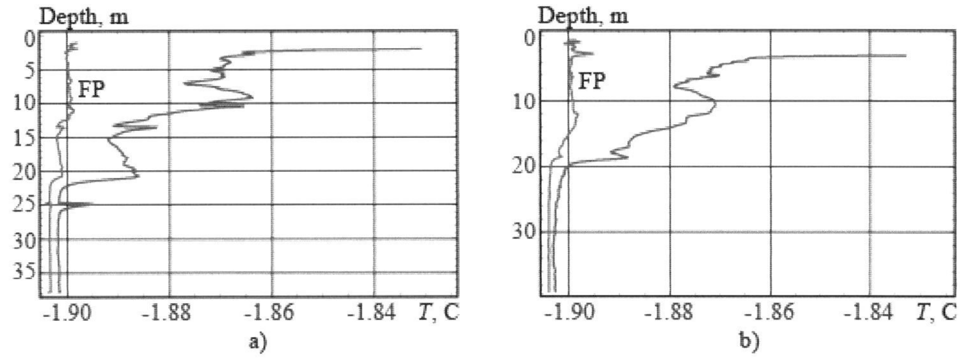

Figure 10. Vertical profiles of the temperature and freezing point measured near the rubble in two locations 1 (a) and 2 (b).

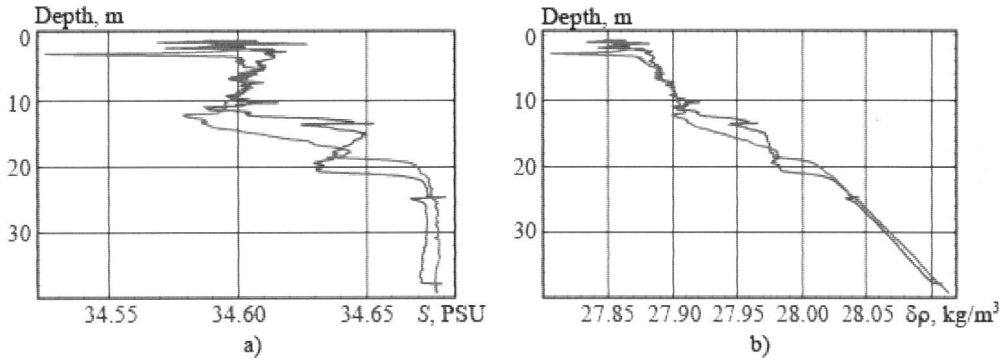

Figure 11. Vertical profiles of the salinity (a) and density fluctuations (b) measured near the rubble in two locations 1 (blue lines) and 2 (pink lines).

CONCLUSIONS

Large ice rubble, unusual for such warm period and location was discovered on 2 of May near North East point of Edge Island (NW Barents Sea). Investigations using laser scanning, and high resolution satellite images shown that it is a part of the rubble forming regularly in spring by stopping of ice drifted from the North by grounded icebergs. The average sizes of icebergs in 2016 were 20x40 m and concentration 20 icebergs per 1 km^2. In May 2016 the rubble had vertical size of edge wall – 3 m, average elevation above water level 2 m (that gives us draft 18 m), width from the coastal line 8 km and stretching in N-S direction along s coast was more than 20 km. Rubble formed at the beginning of April and existed till the beginning of July.

In July 2015 the rubble of 80 km long and 10-20 km width was fasted to the NE of Edge Island. Estimation on the base of IFREMER and ice trackers data show that such ice formation can potentially reach Bear Island region (the new hydrocarbon exploration site) in two and half month in case of average drift and in a month in case of polar cyclones and strong wind gust.

ACKNOWLEDGEMENTS

The authors would like to thank the FIMA and SAMCoT project (funded by the Research Council of Norway), the SITRA project (funded by the Norwegian Centre for International Cooperation in Education) and the colleagues for supporting this study.

REFERENCES

APOLLO MAPPING. *Image hunters* [Online]. Boulder, Colorado, USA. Available: https://apollomapping.com/, https://imagehunter.apollomapping.com/.

BARKER, A. & TIMCO, G. W. 2017. Maximum pile-up heights for grounded ice rubble. *Cold Regions Science and Technology,* 135, 62-75.

CROASDALE, K. R. 2012. Ice rubbling and ice interaction with offshore facilities. *Cold Regions Science and Technology,* 76-77, 37-43.

HOPKINS, M. A. 1998. Four stages of pressure ridging. *Journal of Geophysical Research-Oceans,* 103, 21883-21891.

HOYLAND, K. V. 2007. Morphology and small-scale strength of ridges in the North-western Barents Sea. *Cold Regions Science and Technology,* 48, 169-187.

IFREMER 2017. psi-drift/quicklooks/arctic/. *In:* IFREMER (ed.) *Index of /ifremer/cersat/products/gridded.*

LIFEROV, P. & BONNEMAIRE, B. 2005. Ice rubble behaviour and strength: Part I. Review of testing and interpretation of results. *Cold Regions Science and Technology,* 41, 135-151.

LØSET, S., VINJE, T., LØVÅS, S. M., JOHNSEN, A. S., JENSEN, H., ERLINGSSON, B. & VEFSOMO, S. 1988. IDAP 88 Vessel deployment. Volume 2. Field observations and analysis. Trondheim: Norwegian Hydrotechnical Laboratory.

MARCHENKO, A. 2016a. Physical mechanisms limiting sizes of pressure ice ridges and model of ice rubble formation. *23nd IAHR International Symposium on Ice.* Ann Arbor, Michigan USA,.

MARCHENKO, A., DIANSKY, N., FOMIN, V., MARCHENKO, N. & KSENOFONTOVA, D. 2016. Consolidation of Drifting Ice Rubble in the North-West Barents Sea. *23nd IAHR International Symposium on Ice.* Ann Arbor, Michigan USA,.

MARCHENKO, A. V., DIANSKY, N. A., ONISHCHENKO, D. A., CHUMAKOV, M. M., NIKITIN, M. A., FOMIN, V.V.. & MARCHENKO, N. A. 2016. Studies of ice drift and ridge consolidated layer evolution in the North-West Barents Sea. *Proceedings of Hydrometcentre of Russia*. - 2016. - Vol. 361. - P. 231–260.

MARCHENKO, A. B. & MAKSHTAS, A. 2005. A dynamic model of ice ridge buildup. *Cold Regions Science and Technology,* 41, 175-188.

MARCHENKO, N. A. 2016c. Comparison of Sea Ice Products and data of drifting buyos. *23rd IARH International Symposium on Ice (IAHR-ICE 2016).* AnnArbor, Michigan-US.

MARCHENKO, A.V., OGORODOV, S.A., TSVETCINSKY, S.A., SHESTOV,A.S.,2009. D angerous ice phenomena on the shelf of the Barents and Kara Seas. *Proc. RAO/CIS Offshore 2009,* St.-Petersburg, HIMIZDAT, V.1, pp. 91-96.

NASA. 2017. *MODIS. Moderate resolution imaging spectroradiometer* [Online]. Available: https://modis.gsfc.nasa.gov/.

NOAA. 2017. *What is the difference between rubble ice and pressure ridges?* [Online]. Available: http://www.pmel.noaa.gov/arctic-zone/ice-rubble-pressure.html.

NORWEGIAN METEOROLOGICAL INSTITUTE. 2017. *Icechart and Ice information* [Online]. Available: http://polarview.met.no/index_HI.html.

NORWEGIAN POLAR INSTITUTE. 2017. *TopoSvalbard* [Online]. Available: http://toposvalbard.npolar.no/ [Accessed 2017.

PARMETER, R. & COON, M. 1972. Model of pressure ridge formation in sea ice. *J. Geophys. Res.,* 77, 6565– 6575.

TIMCO, G. W. & BURDEN, R. P. 1997. An analysis of the shapes of sea ice ridges. *Cold Regions Science and Technology,* 25, 65-77.

VINJE, T. 1985. Drift, composition, morphology and distribution of the sea ice fields in the Barents Sea. *Norsk Polarinstitrutt Skrifrer.* Oslo.

VINJE, T., JENSEN, H., JOHNSEN, A. S., LØSET, S., HAMRAN, S., LØVÅS, S. M. & ERLINGSSON, B. 1989. IDAP-89 R/V LANCE deployment. Volume 2: Field observations and analysis. Oslo: Norsk Polarinstitutt.

VINJE, T. & KVAMBEKK, A. S. 1991. Barents Sea Drift Ice Characteristics. *Polar Research,* 10, 59-68.

ZUBAKIN, G. K. 2006. Ice formation of the seas of Western Arctic. Saint Petersburg, AARI.

Thermodynamic Consolidation of Ice Rubble in the Small Scale in-situ Experiment. Van Mijen Fiord, Spitsbergen, March 2016

Aleksey Shestov[1], Evgenii Salganik[2]
[1] The University Centre in Svalbard, Longyearbyen, Norway
[2] Sustainable Arctic Marine and Coastal Technology (SAMCoT), Centre for Research-based Innovations (CRI), Norwegian University of Science and Technology, Trondheim, Norway

ABSTRACT

In March 2016 ice rubble thermodynamic consolidation experiment was performed in saline Lake Vallunden connected through a narrow straight to Van Mijen Fiord, Spitsbergen. Ice blocks of 23 x 23 x 5 cm were produced from the level ice section 1 x 1 x 0.6 m and placed back into the appeared basin. Ice rubble field was equipped with thermistor string probe through it and three one-point measurement conductivity and temperature sensors. Two of them were placed inside cavities between ice blocks in the upper and lower parts of the rubble, and the third one was located in sea water below the rubble. Part of the main phase of ice rubble thermodynamic consolidation due to atmospheric cooling was observed during 14 days. The developing of the consolidated layer and salt balance processes inside rubble cavities were observed and measured. Water inside cavities was found to be supercooled up to 0.3 °C. The final thickness of the consolidated layer, as defined by isoline corresponding to sea water freezing temperature and zero temperature gradient, by mechanical drilling and by Stefan's equation, was found to be 23, 30 and 29 cm correspondingly.

KEY WORDS: Ice Rubble, Thermodynamics, Consolidation, In-situ Experiment

INTRODUCTION

Arctic ice cover is rapidly changing, and it is most obviously seen in the summer ice extent. Ice is losing in thickness too. For example, Renner et al. (2014) report thinning by over 50% during 2003-2012. Such changes affect dynamic and thermodynamic models of ice cover – ocean interaction. Ice cover becomes more fragile and exposed to air and ocean interactions. The rate of deformation and speed of drift are increasing (Spreen et al., 2011). Ice ridges constitute a significant part of the ice volume (see e.g. Hansen et al. (2013)). As more of the Arctic sea ice becomes first-year, a larger fraction of the total volume will be first-year ridges. An essential difference between first-year and old ridges is that the former melt more easily, so the Arctic ice cover as a whole becomes more sensitive to global warming. Ice ridges are key feature both in geophysical context and in the engineering context. They often give the quasi-static design forces on marine structures when icebergs are not a threat (CAN/CSA-S471-92, R2001; ISO/FDIS/19906, 2010; SNiP2.01.07-85, 1987).

The role of ice ridges in the global heat and mass balance has not been studied extensively, and in particularly their role in a thinning Arctic sea ice cover needs to be better quantified. Strub-Klein and Sudom (2012) and Timco and Burden (1997) give comprehensive literature reviews of ridge studies from 1976 and onwards in the engineering context. Very few of the ridge studies investigated the seasonal development of ridges, exceptions are, to the authors' knowledge, Leppäranta et al. (1995), Blanchet (1998), Høyland (2002) and Strub-Klein and Høyland (2011). Thermodynamic processes in ice ridges define their morphology structure and, as a result, mechanical properties too. These properties are essential for estimating ridge loads. The three last papers mostly deal with the consolidated layer, whereas the campaign of Leppäranta et al. (1995) measured the season development of the consolidated layer, temperature profiles, the keel depth and volume and the ridge and rubble macro-porosities and extended into the decay phase. Shestov and Ervik (2016) report on ice ridges studies in the Arctic Ocean during at the end of the main phase and the beginning of the decay phase.

Usually, the life cycle of first-year ice ridges from their formation to break up is divided into three phases: initial, main and decay. In case first-year ice ridge survives the summer and its decay phase, it becomes a second-year ice ridge. The initial phase is relatively short and characterized by formation and thermodynamic processes mostly related to extraction of reserved cold from submerged ice blocks, though oceanic heat flux and atmospheric cooling are present at the same time. The main phase is related to atmospheric cooling, and it is the phase when most of the consolidated layer growth occurs. The decay phase starts when ridge begin to be heated both from the ocean and the atmosphere. Summer transformation processes of ice ridges during the decay phase are not well understood yet but include both mechanical erosion and thermal effects. Shestov and Marchenko (2016b) and Shestov and Marchenko (2016a) considered thermodynamics of ice rubble in the water with the varying freezing point, due to freshening during the decay phase and showed that it could cause further consolidation (reduction of macro-porosity) to the keel. The presence of the brine pockets in the sea ice and the conditions of thermodynamic equilibrium explain a strongly non-linear specific heat capacity of sea ice. In the conditions of water freshening, this results in transferring energy from brine pockets to rubble cavities and reduction in macro-porosity of rubble due to an increase of its micro-porosity.

In both aspects, geophysics and engineering, ice ridges are the ice features of the interest for studying. Authors are involved in laboratory size, small field size and full-scale size studies and consider also performing basin and mesoscale size studies. This paper is dedicated to the small-scale field studies performed during March 2016 in Van Mijen Fjord, Spitsbergen. Due to seasonal ice conditions, the location of the experiment was chosen at the Lake Vallunden, connected with the fiord through a strait. Thus, the lake is filled with salt water and covered by saline ice. Ice rubble field was produced in a basin of 1 by 1 m and temperature and conductivity measurements in the rubble were organized. The aim was put on tracking initial phase and development of the consolidated layer both through the change in temperature pattern and salinity in rubble cavities.

EXPERIMENTAL SETUP AND METHODS

The setup and preparation stages are illustrated in Figure 1. Chainsaw operation was involved in opening up the basin and producing ice blocks. First, four large ice cubes were lifted up and left in the air overnight to accumulate more reserved cold. Next day 188 ice blocks of average dimensions 23 cm x 23 cm x 5 cm were prepared from large ice cubes. To avoid contacts between ice blocks and basin walls rope net was set inside the basin. This simplified lifting of the rubbles at the end of the experiment. Ice blocks were one by one placed into the basin and sensors were deployed in parallel. Thermistor string (TSP) probe was deployed through the rubble field, and three conductivity and temperature sensors (CTS) were deployed: two inside the rubble (CTS1 and CTS2) and one under rubble (CTS3). Development of the rubble and its consolidated layer was monitored in such setup during two weeks from the beginning of March. At the end of the experiment newly frozen ice was cut through along the walls of the basin and consolidated ice rubble was lifted up in the net using engine crane with the help of a snowmobile. Once it was lifted, it was observed, pictured, sensors were recovered and samples were cored through the consolidated layer.

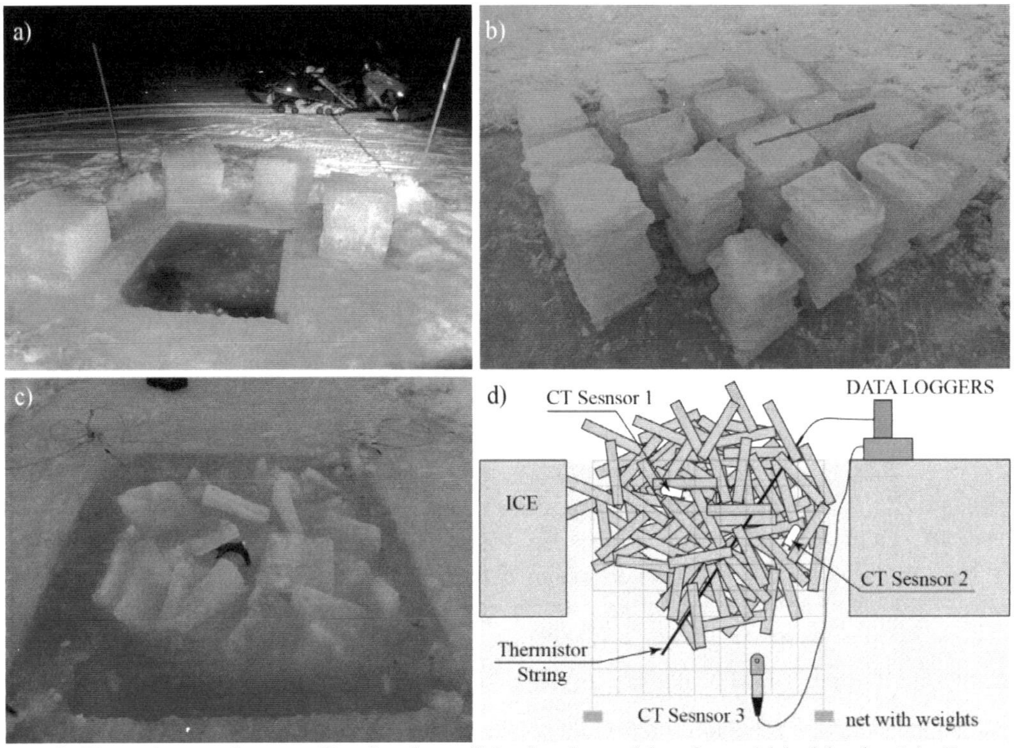

Figure 1. Experimental setup. Production of the basin and ice for rubble blocks (a). Prepared ice rubble blocks (b). The process of submerging ice rubble blocks into the basin and deployed CTS2 sensor (c). Schematic view of the whole experimental setup.

CTS1 and CTS2 sensors are autonomous sensors DST CT Type I (3-37 mS/cm) conductivity and temperature loggers (Star-Oddi, 2017). CTS3 is the Aanderaa conductivity/temperature sensor 4120 with analog output and external data logger. CTS1 and CTS2 provide conductivity measurements transferred in practical salinity units PSU. CTS3 provide conductivity measurements in mS/cm. Using a corresponding sequence of function from the TEOS-10, all salinities further are reported on the TEOS-10 absolute scale, S_a, (IOC et al., 2010) with salinity anomaly taken from version 3.0 of the McDougall et al. (2012) database. Conservative temperature freezing point is found using the corresponding function of TEOS-10 (IOC et al., 2010). TSP is the thermistor string from GeoPrecision GmbH company (GeoPrecision, 2017) with a sequence of measuring points spaced 4 cm apart from each other and external data logger. Readings from TSP were used to calculate vertical conductive heat flux according to Fourier's law, where temperature gradient was calculated between each two sequent TSP measuring points and the thermal conductivity was chosen to be 2.1 $Wm^{-1}K^{-1}$. Based on physical properties of sea ice measured in situ and using the model of thermal conductivity proposed by Schwerdtfeger (1963), Shestov et al. (2017) found the coefficient of thermal conductivity to be varying through the depth of ridge keel from 2.0 to 2.2 $Wm^{-1}K^{-1}$. Thus average value was chosen to be used in current estimates of conductive heat flux through the rubble.

RESULTS

During the experiment, there were two major atmospheric cooling events (Figure 2a). Though at the very beginning of the experiment there were mild temperatures and even above zero for a short period. Between first and second cooling events there was a period when the air temperature was above zero more than one day. Temperature profile through the rubble responded to the cooling events accordingly with almost no delay and caused propagation of cold front down inside the rubble (Figure 2b). In the beginning there was no snow insulation and temperature of the top surface of the rubble keel followed air temperature, while after March 7th snow came and total effective atmospheric cooling at the keel top surface during 14 days was considerably lower: FDD_{ice} = 42 °C days against FDD_{air} = 104 °C days in the air (Figure 2a). Vertical heat flux due to conduction in the rubble varied from 20 to 80 Wm^{-2} during first two cooling events and exceeded 120 Wm^{-2} during the third one (Figure 2c).

Both conductivity sensors placed inside the rubble showed lower salinity, in the beginning, compare to the salinity of surrounding waters (Figure 3a). Readings at CTS1 and CTS3 began at the levels of about 26 and 29 gkg^{-1} respectively and gradually increased. In the case of CTS1, it reached about 31 gkg^{-1} and rapidly dropped to 0, where stayed until the end of the experiment. This moment was interpreted as a freezing event – CTS1 event (Figure 3a). In the case of CTS2 absolute salinity in the cavity was growing until it established finally at the level of 33.4 gkg^{-1}, where it stayed until the end of the experiment. The rapid increase of the absolute salinity, prior it established at the final level, was interpreted as penetration of the surrounding water inside the cavity – CTS2 event. The absolute salinity of seawater measured below the rubble and level sea ice varied at the level 33.7 gkg^{-1} through the experiment. Based on registered absolute salinity freezing point temperature of water at different points (CTS1, CTS2, CTS3) was calculated and compared with registered temperatures in those points respectively (Figure 3b, c, d). Inside the rubble (CTS1 and CTS2) the registered temperature of the water was below the freezing point temperature, while below the rubble (CTS3) registered temperature of water was below its freezing point temperature. In the upper part of the rubble (CTS1) (Figure 3a) difference between freezing temperature and registered temperature was about 0.3 °C in the beginning and then freezing temperature was decreasing, and the difference became 0.15 °C at the moment of the CTS1 freezing event. In the bottom part of the rubble difference between freezing temperature and the registered temperature began with up to 0.5 °C and reduced to less than 0.1 °C due to a decrease of freezing temperature. Sea water temperature below the rubble (CTS3) was above freezing point temperature all time during the experiment (Figure 3d). The difference between registered temperature and freezing temperature was growing during the experiment from the initial value of 0.15 °C to final 0.2 °C. Freezing point temperature was varied with salinity fluctuations in the range between -1.85 to -1.84 °C with a noticeable increase towards the end of the observation period.

Comparison of temperature and salinity profiles in level ice before and in rubble field after the experiment is shown in Figure 4. Ice cores were taken respectively in the level ice and rubble field. Before coring samples, ice rubble was lifted up, and the consolidated layer of up to 30 cm thickness was observed (Figure 5). For the level ice temperature profile was measured in situ, while for the rubble field, readings from thermistor string probe are shown inside the keel part. Salinity profiles were constructed by melting respective sections of the ice cores in both cases.

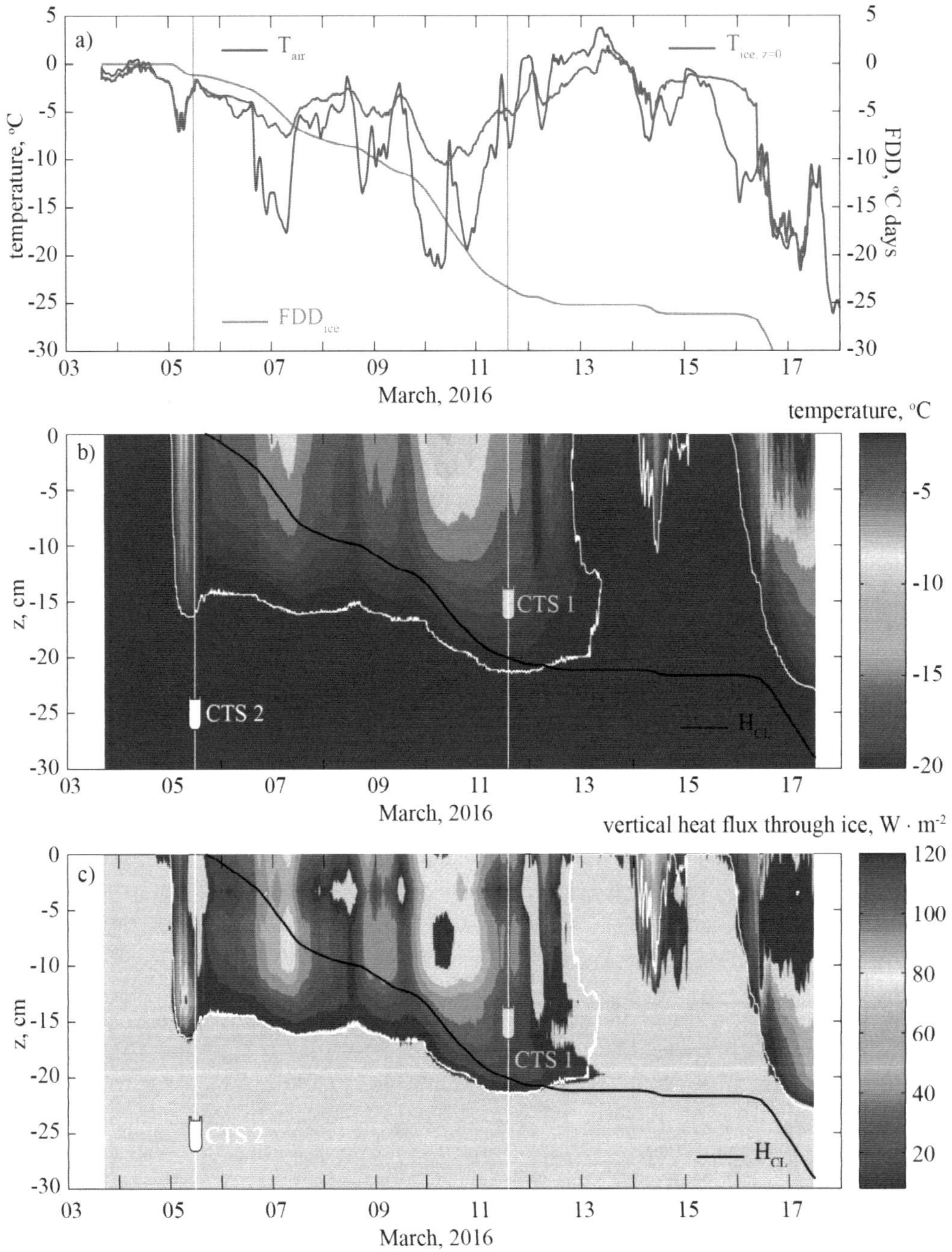

Figure 2. Air temperature close to the ice surface, ice temperature and freezing degree days at the top surface of the keel (a). The temperature profile of ice rubble measured by thermistor string and plotted in time-height space (white line shows isotherm T=-1.8 °C) (b). Vertical heat flux through the rubble, calculated using thermistor string data according to Fourier's law (c) (thermal conductivity 2.1 $Wm^{-1}K^{-1}$, temperature gradient is calculated correspondingly between each two following temperature measuring points). Black curves show consolidated layer thickness of ice rubble calculated using Stefan's equation. Vertical lines on all plots and white signs of CTS sensors are related to the time moments of two respective events happened at two different CTS sensors. Note that positions of CTS sensors are rather qualitative while time moments logged in data series.

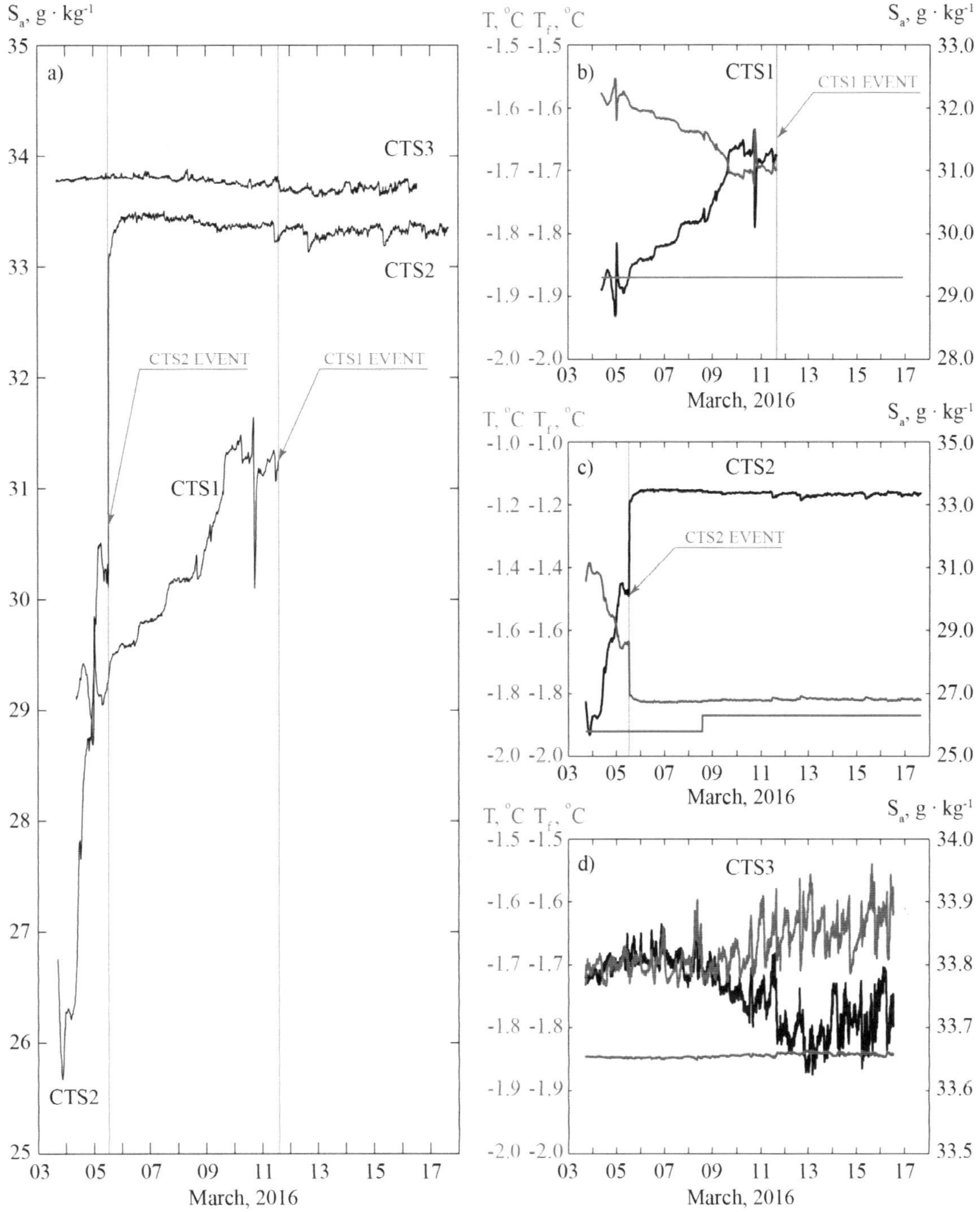

Figure 3. The absolute salinity of water measured in all three points: CTS1, CTS2, and CTS3 respectively (a). Actual temperature, absolute salinity and calculated freezing point temperature for CTS1, CTS2 and CTS3 in (b), (c) and (d).

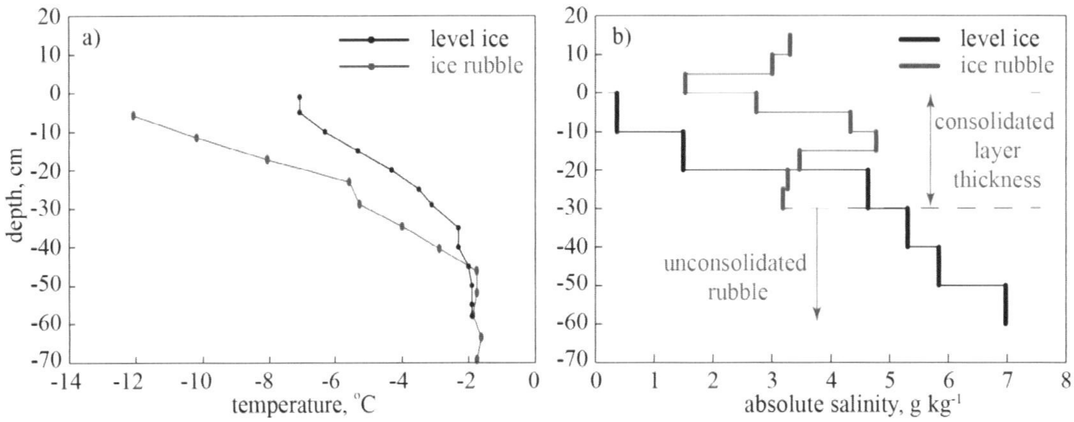

Figure 4. Temperature (a) and salinity (b) profiles in the level ice at the beginning of the experiment (black curves) and in the rubble field (red lines) at the end of the experiment.

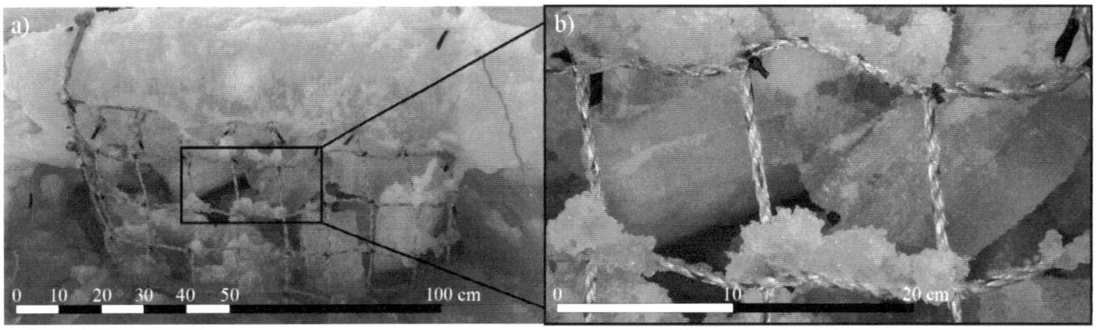

Figure 5. Consolidated layer of rubble and unconsolidated rubble below it at the end of experiment (a); zoomed view of the bottom part of the consolidated layer, where single blocks are frozen, while cavities between them are not (b).

DISCUSSIONS

Estimated ice rubble consolidated layer thickness calculated from Stefan's equation (Stefan, 1891) using temperature at the top surface of the keel predicts the final thickness of consolidated layer defined by the thermal isoline or zero temperature gradient line (Figure 2b, c). Note that increase in temperature of ice rubble during June 13th – 16th (Figure 2b) does not cause its melting. Sea ice can be heated above freezing point temperature of saline water from which it was formed; any change in temperature of saline ice causes the adjustment of brine volume fraction on the micro-porosity scale. During mentioned period of no atmospheric cooling, the thickness of consolidated layer has to be considered kept at the same level until the new event of atmospheric cooling will occur. The difference (Figure 2b, c) in consolidated layer thickness between one defined by the isothermal line (and zero temperature gradient line) and one calculated using Stefan's equation can be caused by reserved cold and process during the initial phase of consolidation; and also by affected temperature measurements. The used thermistor string probe consists of sensors placed into the metal tube and filled with some thermal isolation material between sensors. Metal coater has a higher thermal conductivity compare to ice, what cause cold to propagate locally down into rubble along the probe (see the beginning of the cooling events on 5th, 14th and 16th of June, Figure 2b). One can see that cold temperature front first rapidly penetrates down and then in several hours it lifts back up again. In particular, such strikes are the consequence of the warm temperatures in general and probe not being frozen properly (not in full contact) into ice. This is short time phenomena, but on the long run, it is not excluded that conduction along the probe affects results noticeably. That has to be studied.

As shown in results (Figure 3a), CTS1 and CTS2 inside the rubble experienced lower salinity, in the beginning, compare to a salinity of surrounding water measured by CTS3. This must be due to the melting of ice blocks to provide the source of the fresher water. However, what was the exact reason of melting is remaining an open question. Surrounding water was above the freezing point temperature (Figure 3d). Thus we can expect considerable oceanic heat flux towards the rubble from below. The fact that salinity in the cavity of CTS2 (deeper in rubble) lower than in cavity with CTS1 (higher in rubble), is for the hypothesis that melting is caused by the high oceanic turbulent heat flux. Since the lower part of the rubble exposed more to the oceanic flux, there is more melting occurs and lower salinity drops. Further, in the experiment, both sensors CTS1 and CTS2 showed an increase of water salinity in cavities where sensors were measuring. There were two events noticed related to salinity change: CTS1 event and CTS2 event. First, in time, CTS2 event occurred when salinity in cavity rapidly jumped to the value close to surrounding water salinity (CTS3) and further stayed at that level (Figure 3a, c). In total salinity has grown from 25.7 to 33.5 g kg^{-1} during two days. As we think, while melting of the rubble from below was developing, at the same time it was increasing permeability of the upper rubble to the sea water. At the moment when cavity with sensor became accessible to sea water, salinity in the cavity instantly jumped (CTS2 event). Second, in time, the CTS1 event occurred, when, as we believe, conductivity measuring cell got frozen and as a result salinity readings instantly dropped to zero. Whether only measurement cell or the entire cavity got frozen is an open question. In the case of CTS1, the rate of salinity increase was lower, during six days salinity has grown from 29 to 31 g kg^{-1} (Figure 3a, b). Temperature readings of CTS1 and CTS2 show that both cavities were supercooled (Figure 3b, c) with temperatures close to the freezing point temperature of sea water (Figure 3b, c, d). In the case of CTS1 water in the cavity was supercooled up to 0.3 °C in the beginning, while conductivity measurement cell got frozen when water was supercooled by 0.2 °C. Since we do not observe temperature decrease in CTS1 data (Figure 3b) after conductivity measurement cell got frozen, the conclusion pending to be that only measuring cell got frozen, while the cavity remained unfrozen. Otherwise, we would observe temperature drop, due to cold temperature front propagation down once ice cavity is frozen. The facts that we cannot see a drop in the temperature at CTS1 and that its value is higher compared to the thermistor probe temperature value at the same level, are consistent with the suggestion made above that metal coating of the thermistor probe can affect the measurements and bring to the observed situation. One should understand that larger spacing between sensors in thermistor string probe makes results very discrete and hard to observe thermal processes with high spatial resolution, while CTS1 and CTS2 measure directly inside the cavities between ice blocks.

Due to a different sequence of submerging blocks and their orientation in the rubble field, we cannot expect similarity in salinity profiles between level ice and ice rubble. At the end of the experiment salinity profile in the rubble varies between 2 and 5 g kg-1 compare to 0 and 7 g kg-1 for level ice respectively (Figure 4b). Since the upper bound of the range is becoming less, there are no new salt sources in the ice rubble, i.e. cavities between ice blocks have possibilities to expel salts out below it while freezing. This probably can explain a maximum of salinity in ice rubble at the level of 10-20 cm depth.

At the end of the experiment keel of the rubble field consisted of three parts: a) fully consolidated layer with frozen cavities; b) consolidated layer part where ice blocks frozen together while cavities were not frozen and c) unconsolidated loose ice blocks (Figure 5).

CONCLUSIONS

The small-scale field experiment on ice rubble thermodynamic consolidation was conducted in Van Mijen Fiord, Spitsbergen in March 2016. One of the most important aspects of studies was put on the processes related to salt balance inside the cavities between ice blocks. In conditions of mild atmospheric cooling (FDD_{air} = 104 °C days, which with snow and sail insulation resulted in FDD_{ice} = 42 °C days) and sea water being above freezing temperature by up to 0.2 °C, at the end of the experiment (after 14 days) we observed a consolidated layer up to 30 cm thick.

Water inside both cavities between ice blocks at two different levels was all time supercooled by up to 0.3 °C. It was indicated that oceanic heat flux continuously melts bottom part of the unconsolidated part of the keel; what results in two processes: it provides fresher water and at the same time makes the bottom part of the keel more permeable to the seawater of high salinity. At the same time ice cavities in the upper layers of the rubble, which are subjected to freezing, have possibilities to expel salt further below them as they freeze.

Temperature data interpretation using thermal isoline (or zero temperature gradient isoline corresponding to the sea water freezing temperature gives a reasonable estimate of the final thickness of consolidated layer compared to the thickness calculated according to Stefan's equation using FDD_{ice} = 42 °C days (Figure 2). At the same time, it was seen that metal coating of the thermistor string probe should preferably be avoided to reduce the thermal conductivity of the probe itself. Moreover, at the scale experiment was performed, higher spatial resolution of thermistor string would be beneficial to catch more local processes in the cavities.

At the end of the experiment all three expected parts of rubble keel were observed: a) fully consolidated layer with frozen cavities; b) consolidated layer part where ice blocks are frozen together while cavities were not frozen and c) unconsolidated loose ice blocks (Figure 5). Consolidated layer thickness as defined by isoline corresponding to sea water freezing temperature and zero temperature gradient, by mechanical drilling and by Stefan's equation is 23, 30 and 29 cm correspondingly.

ACKNOWLEDGEMENTS

The author wishes to acknowledge the support of the Research Council of Norway through the Centre for Research-based Innovation SAMCoT (project number 203471) and the support of all SAMCoT partners.

REFERENCES

Blanchet, D., 1998. Ice loads from first-year ice ridges and rubble fields. Canadian Journal of Civil Engineering, 25(2): 206-219.

CAN/CSA-S471-92, R2001. General Requirements, Design Criteria, the Environment, and Loads. Standards Council of Canada.

GeoPrecision, 2017. Environmental Technology for Industry, Science and Research (http://www.geoprecision.com/en/).

Hansen, E., Gerland, S., Granskog, M.A., Pavlova, O., Renner, A.H.H., Haapala, J., Løyning, T.B. and Tschudi, M., 2013. Thinning of Arctic sea ice observed in Fram Strait: 1990–2011. Journal of Geophysical Research: Oceans, 118(10): 5202-5221.

Høyland, K.V., 2002. Consolidation of first-year sea ice ridges. Journal of Geophysical Research: Oceans, 107(C6): 15-1-15-16.

IOC, SCOR and IAPSO, 2010. The international thermodynamic equation of seawater – 2010: Calculation and use of thermodynamic properties. Intergovernmental Oceanographic Commission, Manuals and Guides, UNESCO (English), 56(56): 196.

ISO/FDIS/19906, 2010. Petroleum and natural gas industries - Arctic offshore structures. International Standarization orginaization, Geneva, Switzerland.

Leppäranta, M., Lensu, M., Kosloff, P. and Veitch, B., 1995. The life story of a first-year sea ice ridge. Cold Regions Science and Technology, 23(3): 279-290.

McDougall, T.J., Jackett, D.R., Millero, F.J., Pawlowicz, R. and Barker, P.M., 2012. A global algorithm for estimating Absolute Salinity. Ocean Sci., 8(6): 1123-1134.

Renner, A.H.H., Gerland, S., Haas, C., Spreen, G., Beckers, J.F., Hansen, E., Nicolaus, M. and Goodwin, H., 2014. Evidence of Arctic sea ice thinning from direct observations. Geophysical Research Letters, 41(14): 5029-5036.

Schwerdtfeger, P., 1963. The thermal properties of sea ice. Journal of Glaciology, 4: 789-807.

Shestov, A. and Ervik, Å., 2016. Studies of Drifting Ice Ridges in the Arctic Ocean during May-June 2015. Part II. Thermodynamic properties and melting rate, Proceedings of 23rd IAHR International Symposium on ICE 2016, Ann Arbor, Michigan, USA.

Shestov, A., Hoyland, K.V. and Ervik, Å., 2017. Thermodynamics of Ice Ridges during Spring Season in the Arctic Ocean. Journal of Geophysical Research: Oceans (submitted).

Shestov, A.S. and Marchenko, A.V., 2016a. The consolidation of saline ice blocks in water of varying freezing points: Laboratory experiments and computer simulations. Cold Regions Science and Technology, 122: 71-79.

Shestov, A.S. and Marchenko, A.V., 2016b. Thermodynamic consolidation of ice ridge keels in water at varying freezing points. Cold Regions Science and Technology, 121: 1-10.

SNiP2.01.07-85, 1987. Construction Codes and Regulations. Loads and Forces. Techincal Standards Used in Russia and Kazakhstan, pp. 34.

Spreen, G., Kwok, R. and Menemenlis, D., 2011. Trends in Arctic sea ice drift and role of wind forcing: 1992–2009. Geophysical Research Letters, 38(19): n/a-n/a.

Star-Oddi, 2017. Fish tagging & data logging with data storage tags (http://www.star-oddi.com/).

Stefan, J., 1891. Theory of ice formation, especially in the Arctic Ocean (German). Ann. Physik, N.f: 269-286.

Strub-Klein, L. and Høyland, K.V., 2011. One Season of a 1st Year Sea Ice Ridge Investigation - Winter 2009, Proceedings of 21th International Conference on Port and Ocean under Arctic Conditions (POAC) Montreal, Canada.

Strub-Klein, L. and Sudom, D., 2012. A comprehensive analysis of the morphology of first-year sea ice ridges. Cold Regions Science and Technology, 82(0): 94-109.

Timco, G.W. and Burden, R.P., 1997. An analysis of the shapes of sea ice ridges. Cold Regions Science and Technology, 25(1): 65-77.

POAC'17
Busan, KOREA

Proceedings of the 24th International Conference on
Port and Ocean Engineering under Arctic Conditions
June 11-16, 2017, Busan, Korea

Calculation of Ice Abrasion Induced by Unbreakable Ice Floes using Discrete Element Method

Jeong-Hwan Kim[1], Yooil Kim[2]

[1] Ph.D student (Department of Naval Architecture and Ocean Engineering, INHA University, Incheon, South Korea)

[2] Associate professor (Department of Naval Architecture and Ocean Engineering, INHA University, Incheon, South Korea)

ABSTRACT

In this study, a calculation of ice abrasion induced by unbreakable ice floe using discrete element method (DEM) has been carried out. Under the assumption that unbreakable floes behave as rigid body, DEM was applied to simulate the interaction between a fixed structure and ice floes. In DEM simulation, individual ice floe was treated as single rigid element which interacts with each other following the given interaction rules. Interactions between the ice floes and structure were defined by soft contact and viscous Coulomb friction laws. To derive the details of the interactions in terms of interaction parameters, the finite element method (FEM) was employed. An abrasion process between a structure and an ice floe was simulated by FEM, and the parameters in DEM such as contact stiffness, contact damping coefficient, etc. were calibrated based on the FEM result. Resultantly, contact pressure and contact path length, which are the most important factors in ice abrasion prediction, were calculated from both DEM and FEM and compared with each other. The results showed good correspondence between the two results, providing superior numerical efficiency of DEM.

KEY WORDS : Ice abrasion; Unbreakable ice floe; Discrete element method; Finite element method; Contact pressure; Contact path length.

INTRODUCTION

Main factors in the calculation of ice abrasion are contact pressure, length of abrasion path and resistance to ice abrasion of particular material (Saeki et al., 1987). While evaluation of material resistance to ice abrasion requires experimental studies, an appropriate mathematical tool is needed for contact pressure and abrasion path length during service life. The joint use of both numerical model for ice abrasion load and empirical model for material resistance enables to calculate ice abrasion depth (Bekker et al., 2011).

The research on material resistance is relatively large, but the study to obtain ice abrasion load is insufficient. Since there is no standardized calculation method with regard to ice abrasion yet, it is still difficult to find an appropriate model for the calculation.

Bekker et al. (2012) developed a probabilistic macro-scopic approach on the ice-concrete abrasion process. In this model, contact force is defined by the equations predetermined by experiments and field study, and contact path length is calculated by simple equilibrium equations. Although it skips some physical representation of the ice abrasion processes and relies on many assumptions, it is evaluated as the most complete model of its kind(Jacobsen et al., 2014). Dorival et al. (2008) developed a lattice model for ice crushing against a rigid structure. The lattice model can describe ice behavior in high detail so it is expected to be suitable to implement abrasion(Tijsen, 2015). However, it is still under development for abrasion case and has a drawback that it is demanding heavy computational intensity. Consequently, it is required to develop a new calculation model reflecting the strengths of the both models.

On the other hand, the first thing to do before developing a numerical model for interaction between a structure and an ice floe is to define if the floe is treated as breakable or unbreakable. The criterion is of importance since the unbreakable ice floe can be regarded as a rigid body. In other words, it is not necessary to spend heavy computational expense on calculating internal stress of the floe different from the breakable ice floe whose failure mode such as local bending and global splitting fracture should be considered critically. In this study, a simple criterion for determination of unbreakable ice floe developed by Lindseth (2013) was employed.

This paper focuses on the numerical simulation of ice abrasion induced by unbreakable ice floe. Under the assumption that unbreakable floes behave as rigid body, it is possible to apply the discrete element method (DEM) to simulate the interaction between a structure and ice floes. In DEM simulation, individual ice floe was treated as single rigid element which interacts with each other following the given interaction rules. Interactions between the ice floes and structure were defined by soft contact and viscous Coulomb friction laws. However, when each floe is regarded as one particle in DEM there is a concern for loss of accuracy since it simplifies some physical mechanisms such as local failure mode and contact behavior. In this study, to overcome this drawback the finite element method (FEM) was employed. An abrasion process between a structure and an ice floe was simulated by FEM and the parameters in DEM such as contact stiffness, contact damping coefficient, etc. were calibrated based on the FEM result.

In this paper, efforts have been made on the numerical simulation of ice-induced abrasion using the DEM. In order to set up the interaction rule between ice and structure, the nonlinear finite element analysis was carried out in Eulerian domain. The failure of ice colliding with the fixed structure was realized using Crushable foam material model, whose detail parameters were tuned using experimental results of ice cone crushing test. Also, an example of abrasion calculation using DEM was presented based on the actual environmental condition of Gulf of Bothnia. The simulation was performed for 60 seconds, and the resulting contact pressure, path length and abrasion depth were calculated.

DISCRIPTION OF TEST CONDITION

Test conditions used in this chapter are summarized in Table 1. A target structure is Raahe lighthouse in the Gulf of Bothnia as motivated by the research of Bekker et al. (2011). Ice floes having various thickness and diameter were tested to verify the simulation results. Based on the test conditions, numerical simulations using FEM and DEM were carried out, and the results were compared with each other.

Table 1. Test conditions

Diameter of structure	7.4m
Thickness of ice floe	0.5m / 1.0m
Diameter of ice floe	2.5m / 5.0m / 10m
Ice-concrete friction coefficient	0.1
Current velocity	2.0m/sec.
Test case 1	Diameter: 2.5m / Thickness:0.5m
Test case 2	Diameter: 5.0m / Thickness:0.5m
Test case 3	Diameter: 5.0m / Thickness:1.0m
Test case 4	Diameter: 10.m / Thickness:1.0m

NUMERICAL SIMULATION OF ICE ABRASION USING FEM

FE analyses were performed to calibrate the major coefficients such as contact normal stiffness, contact damping coefficient and drag coefficient used in the DEM calculation. To investigate the difference according to the thickness and diameter of the ice floe, four different models were used considering the diameter and thickness presented in Table 1. Figure 1 shows the FE model for ice-structure collision using Coupled Eulerian-Lagrangian method. The structure was simulated using rigid elements in Lagrangian domain, and the ice floe and sea water were modeled in Eulerian domain. The failure of ice colliding with the structure was realized using Crushable foam material model, whose detail parameters were tuned using experimental results of ice cone crushing test performed by Kim et al (2015). The results of the FEM analysis are shown in Figures 2, 3 and 4. The details of the FEM analysis results are explained along with the DEM analysis results in the next chapter.

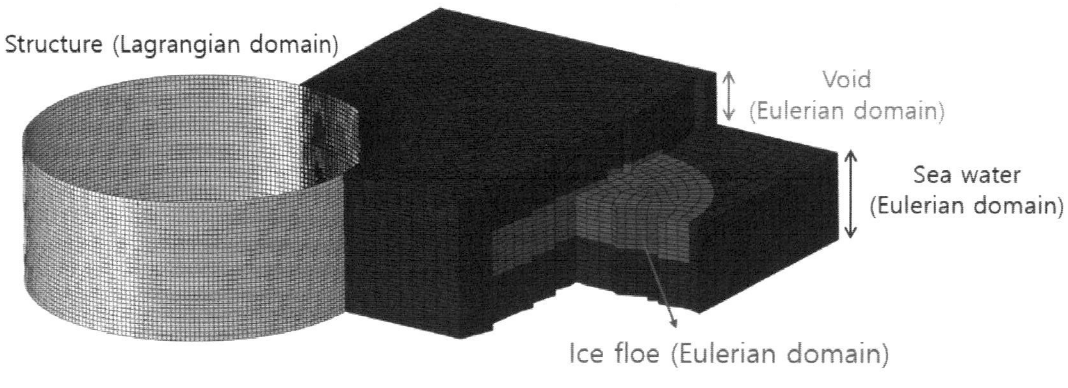

Figure 1. Ice-structure collision simulation using Coupled Eulerian-Lagrangian method

NUMERICAL SIMULATION OF ICE ABRASION USING DEM

Numerical simulation using DEM was carried out on the same conditions as FEM analysis. First, after determining the contact stiffness which has the greatest effect on the contact force, damping coefficient and drag force coefficient were adjusted to match the behavior of the ice floe obtained in the FEM as shown in Figure 2, 3 and 4. Contact stiffness was determined based on the relationship between penetration and contact force obtained from the cone shape test results by Kim et al (2015). However, since the specimen used in the cone shape test is different in shape from the ice floe and it diameter is only 10 cm, instead of applying the values obtained in the experiment directly to DEM calculation as the contact stiffness, it was determined as the value that gives the most similar result to the FEM result by changing little by little based on the experimental value. In other words, the experimental results show the contact stiffness distribution between $1.0 * 10^6$ and $4.0 * 10^6$ depending on the penetration stage, so that an optimized value was found based on that range assuming the contact stiffness of the unbreakable ice floe have a range similar to that. The contact damping coefficient is the same as the critical damping coefficient considering the plasticity of the ice. In the case of the drag force coefficient, it was adjusted so that the behaviors of particles (the change in the x and y coordinates or the duration of the collision) in DEM and FEM results match well each other. The determined major coefficients are presented in Table 2. Considering that the range of ice floe thickness distribution found in Okhosk is between 0.3m and 1.3m (Bekker, 2011), applying the coefficients obtained by case studies using the thickness 0.5m and 1.0m in this study to the whole of unbreakable ice seems to be reasonable.

Table 2. Parameters used in DEM calculation

Contact normal stiffness (kN/m)	2000
Contact damping coefficient (kNs/m)	Critical damping coefficient
Drag force coefficient	10

Figure 2, 3 and 4 show the comparisons of the numerical simulation results using FEM and DEM for a measuring point on the surface of a structure. In Figure 2, the changes in contact pressure over time for each case were compared. As a result, each line shows a very similar shape in both results although there is a slight difference in maximum value. In case of DEM model, since the contact area calculated at the moment of first contact is very small, the corresponding pressure shows a pointed shape at the moment. However, this does not seem to affect the overall result because it is the pressure that occurs for very short contact lengths in a very short time. In Figure 3 and 4, to compare the movement of the ice floe the changes of the coordinate with time for each case were presented. It is best to directly compare the contact lengths, but since the contact length is difficult to know in case of the FEM model, the coordinate changes of the origin of the ice floe were used for comparison. It is also reasonable because the change of the particle coordinate after contact determines the contact length. As a result, it can be said that there is small difference in Figure 3 (b) but both results showed good agreement overall.

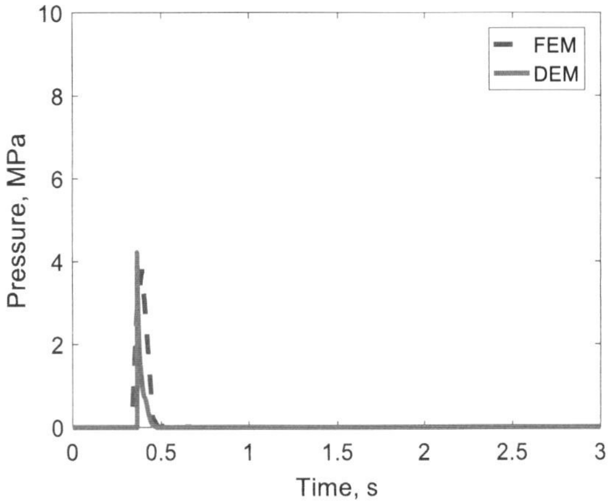

(a) Diameter: 2.5m / Thickness:0.5m

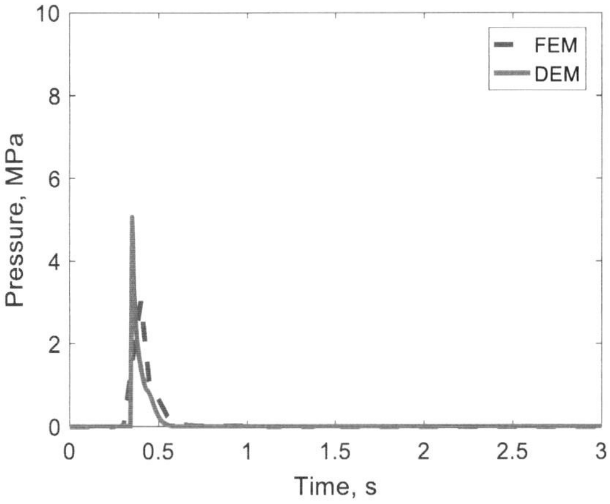

(b) Diameter: 5m / Thickness: 0.5m

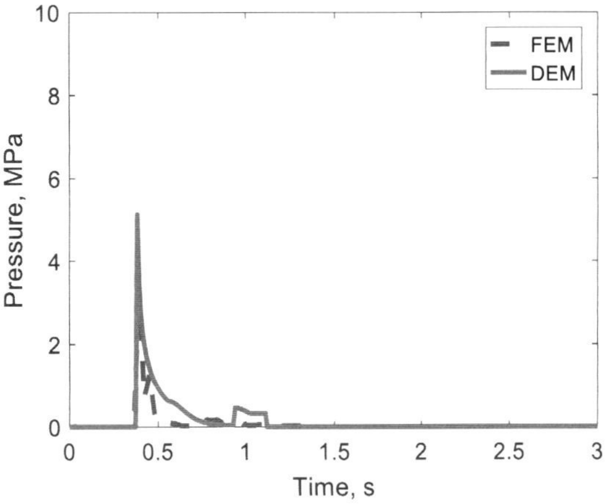

(c) Diameter: 5m / Thickness: 1m

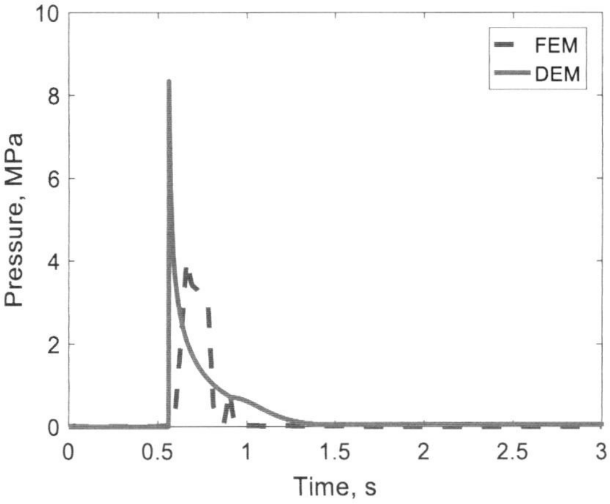

(d) Diameter: 10m / Thickness: 1m

Figure 2. Comparison of contact pressures between FEM and DEM models

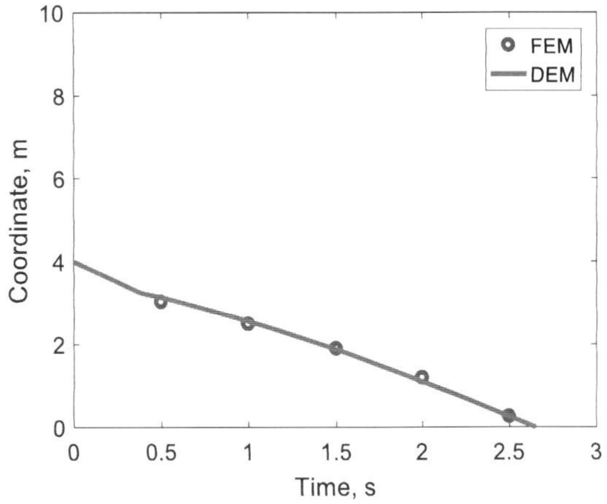

(a) Diameter: 2.5m / Thickness:0.5m

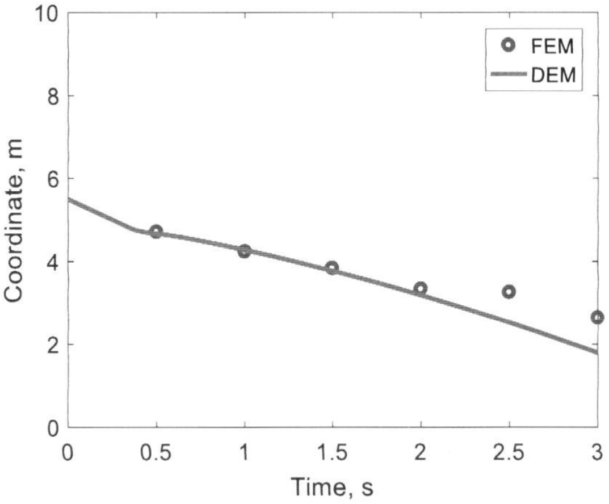

(b) Diameter: 5m / Thickness: 0.5m

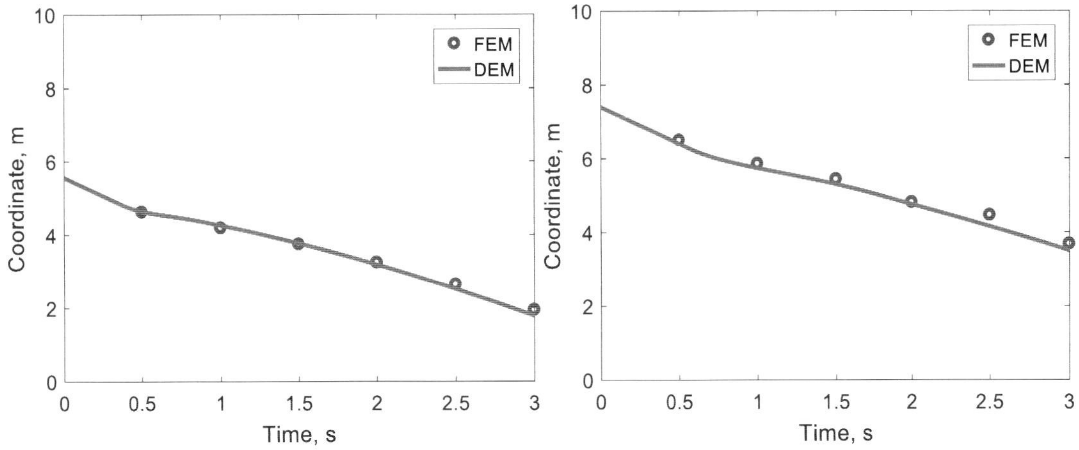

(c) Diameter: 5m / Thickness: 1m (d) Diameter: 10m / Thickness: 1m

Figure 3. Comparison of ice floe movements in x-direction between FEM and DEM models

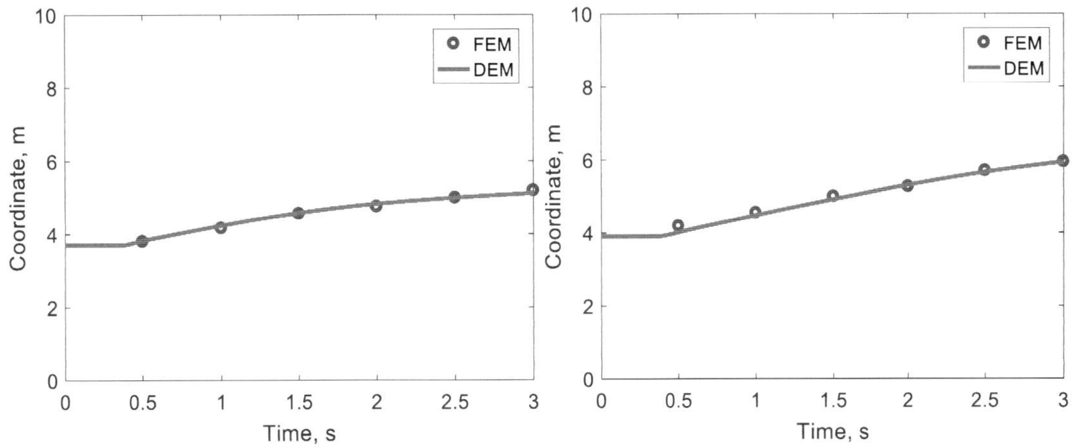

(a) Diameter: 2.5m / Thickness:0.5m (b) Diameter: 5m / Thickness: 0.5m

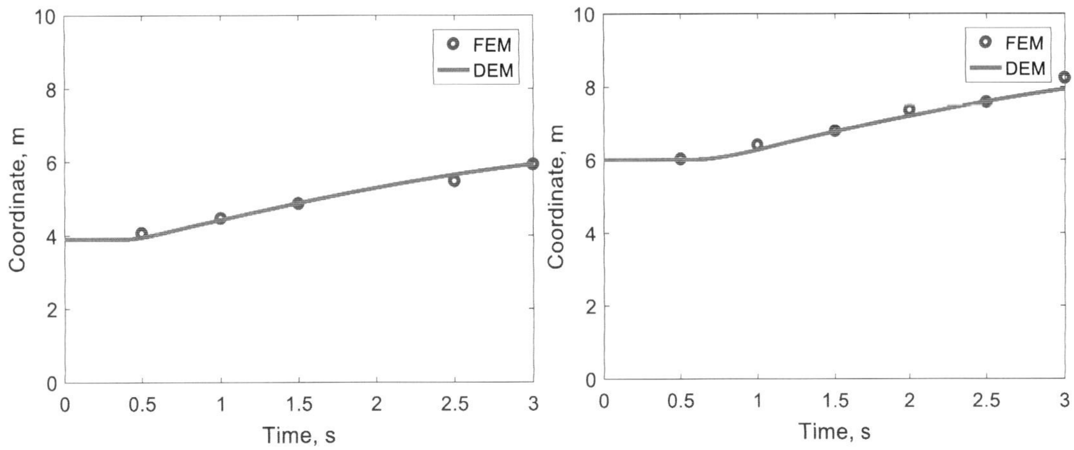

(c) Diameter: 5m / Thickness: 1m (d) Diameter: 10m / Thickness: 1m

Figure 4. Comparison of ice floe movements in y-direction between FEM and DEM models

EXAMPLE OF ICE ABRASION CALCULATION

In this chapter, an example of ice abrasion depth calculation using DEM is presented. Raahe lighthouse in Gulf of Bothnia was targeted, and a numerical simulation for 60 seconds was carried out based on the actual environmental conditions (Bekker, 2011) The analysis conditions are shown in Table 2. Figure 5 shows the measuring points on the surface of structure, and Figure 6 shows the initial setting of particle through random distribution. After setting measuring points on the surface of structure at 15 degree intervals, the contact pressures and the corresponding contact length at those points were calculated. Unbreakable ice floes were randomly distributed with ice concentration of 40%, which was selected for the example because it is relatively common concentration in Sakhalin offshore area. (Bekker, 2004) Considering the simulation for 60 seconds, the area was set at 120m × 100m, and no boundary was set. Figure 7 shown the relation between contact pressure and contact length measured at point 1 and point 6, where minimum and maximum abrasion depth are expected respectively. It can be observed that the middle part of the structure has a larger contact pressure and length than the edge. Abrasion rate can be obtained by substituting the contact pressures and contact lengths into Eq. (1), which was derived by the material resistance test (Bekker, 2011). Finally, the total abrasion depth is obtained by summing the product of the abrasion rate and contact length calculated at each time step. Figure 8 shows the final abrasion depth distribution for each measuring point. In actual calculations, it should include all possible concentrations, directions and current velocities. Also, the final abrasion depth should be calculated by simulation for 3 hours or more and include an appropriate statistical processing additionally.

$$\delta = 0.116\times\left(\frac{T}{\sigma}\right)^{-1.121} \tag{1}$$

$$D = \sum_{i}^{n} \delta_i \cdot L_i$$

where δ is the abrasion rate (mm/km). T is the temperature (°C), and σ is the contact pressure (MPa). D is the total abrasion depth (mm), and δ_i and L_i are the abrasion rate and the contact length calculated at a specific time.

Table 2. Environmental condition for DEM simulation

Diameter of structure	7.4m
Ice concentration	40%
Thickness distribution of ice floe	0.5m / 0.7m / 0.9m / 1.1m / 1.3m (Evenly distributed by one-fifth)
Diameter of ice floe	Randomly distributed between max. and min. diameter • Min. diameter: 4 × h • Max. diameter: $\sqrt{130h - 11}$ where h is the thickness of ice floe
Ice-concrete friction coefficient	0.1
Current velocity	2.0m/sec.

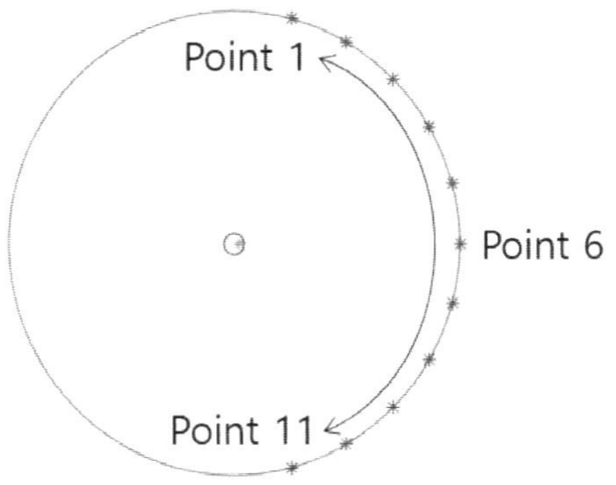

Figure 5. Measuring points

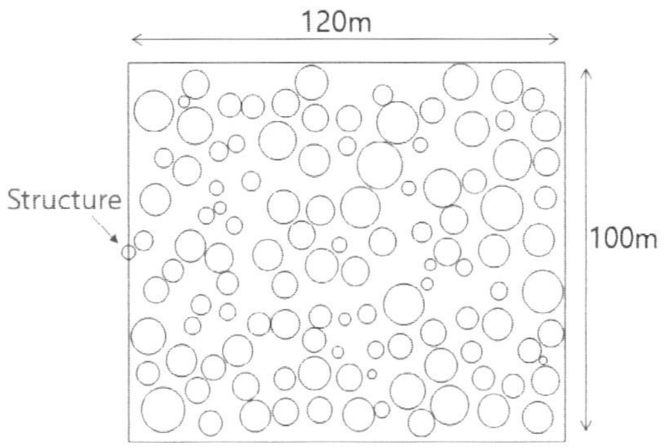

Figure 6. Initial setting of particle through random distribution (Concentration 40%)

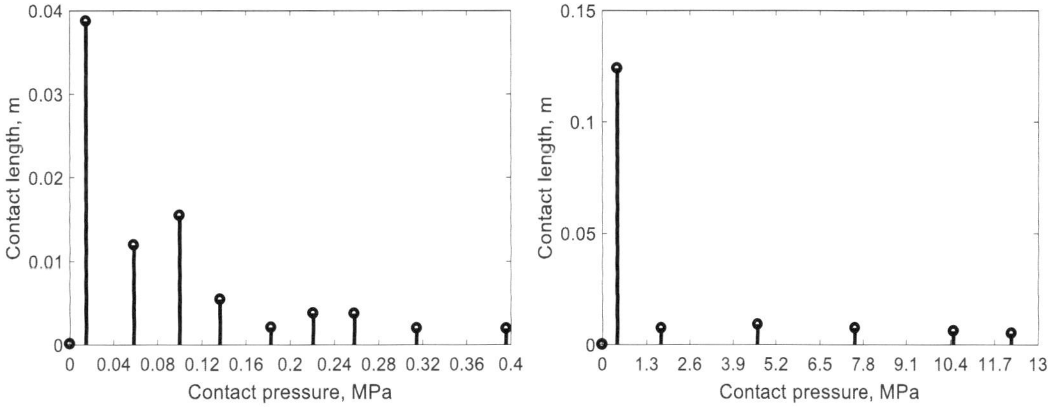

(a) Point 1 (b) Point 6

Figure 7. Contact pressure vs. contact length

POAC17-130

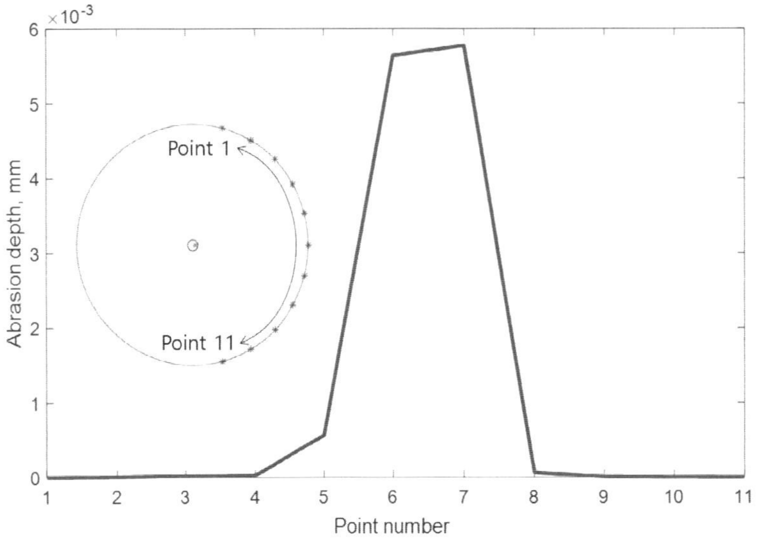

Figure 8. Ice abrasion depth calculated using DEM for 60 seconds

CONCLUSIONS

In this study, DEM was used as a numerical simulation method to calculate ice abrasion induced by unbreakable ice floes. Through the application of DEM, a rapid numerical analysis is expected, and the accuracy problem was solved by calibration with FEM results. Based on the study results described so far, the following conclusions are drawn.

- A coupled Eulerian-Lagrangian method of FEM was employed to grasp the ice abrasion processes that occur during the collision between the structure and the ice floe. The material properties of the ice were determined based on the results of recent material ice tests, and the Crushable foam model was used as the ice material model. Through this analysis, the characteristics of the contact stiffness between the structure and ice floe were found, and the result has become an important basis for the determination of various coefficient used in DEM model.
- Collision of a structure and an ice floe was simulated using DEM. The major coefficients in DEM such as contact stiffness, contact damping coefficient, and drag force coefficient were adjusted based on the pressure distribution through FEM.
- The resultant contact pressure and movement of ice floe calculated using the two methods were compared at the same conditions and locations. As a result, there is a slight difference from the maximum, but it can be said that the overall results are very similar. Also in terms of movement, although there is a little difference in one case, most of the results show good correspondence.

An example of ice abrasion depth calculation using DEM is presented in the final chapter. A numerical simulation for 60 seconds was carried out based on the actual environmental conditions, and contact pressures and the corresponding contact length at measuring points were calculated. Finally, the abrasion depth was obtained by combining contact pressures, contact lengths and material resistance test results.

ACKNOWLEDGEMENTS

This work was supported by the Industrial Convergence Strategic technology development program (10063417, Development of basic design technology for ARC7 class Arctic offshore structures) funded by the Ministry of Trade, industry & Energy (MI, Korea).

REFERENCES

Bekker, A., Uvarova, T., Pomnikov, E., 2011. Calculation of ice abrasion for the lighthouses installed in the Gulf of Bothnia, Proc. Of Port and Ocean Eng. under Arctic Conditions (POAC) Conference, Canada, pp.155-160

Bekker, A., Uvarova, T., Pomnikov, E., Shamsutdinova, G., 2012. Numerical simulation of ice abrasion on offshore structures, Proc. Of 21st IAHR International Symposium on Ice, China, pp.897-906

Bekker, A., Uvarova, B., Kim, S., 2004. Numerical simulation of the process of interaction between drifting ice fields and structure support, Proc. of The Sixth ISOPE Pacific/Asia Offshore Mechanics Symposium, 123-128

Dorival, A., Metrikine, A., Simone, A., 2008. A lattice model to simulate ice-structure interaction, International conference on offshore mechanics and arctic engineering, Estoril, pp.989-996

Jacobsen, S., Scherer, G., Schulson, E., 2015. Concrete-ice abrasion mechanics, Cement and Concrete Research 73, pp.79-95

Kim, H., Daley, C., Colbourne, B., 2015. A numerical model for ice crushing on concave surfaces, Ocean Engineering 106, 289-297

Lindseth, S., 2013. Splitting as a load releasing mechanism for a floater in ice, Norwegian University of Science and Technology, Master thesis

Saeki, H., Takeuchi, T., Yoshida, A., Asai, Y., Suenaga, E., 1987. Abrasion test for concrete due to sea ice, Proc. Of Port and Ocean Eng. under Arctic Conditions (POAC) Conference, Alaska

Tijsen, J., 2015. Experimental study on the development of abrasion at offshore concrete structures in ice conditions, Delft University of Technology, Master thesis

POAC'17
Busan, Korea

Proceedings of the 25th International Conference on
Port and Ocean Engineering under Arctic Conditions
June 11-16, 2017
Busan, Korea

Systems Performance Measurement in the Arctic Using Internet of Things (IoT)

Zaki, R.[1], Ayele, Y. Z. [1], Barabadi, A. [1]

[1] Department of Engineering and Safety, UiT The Arctic University of Norway, NO-9037 Tromsø, Norway

ABSTRACT

The Arctic has unique and challenging operational conditions with strict regulations and requirement for safety and environment. A study of performance terminologies and standards revels that the performance elements measured such as reliability, maintainability, quality, though important, are not sufficient to describe the performance of a component for the Arctic. Some important elements like those that financial performance and sustainability are absent which, we mean are essential to reflect the performance of a system. This paper attempted to identify the shortcomings in the integrated performance frameworks for the Arctic. Hence, it begins with a review of the literature and evolution of performance measurement systems. Thereafter, the paper will illustrate how the concept of financial performance and sustainability should be integrated in performance measurement and suggest calculating process.

Moreover, due to the importance of availability and accuracy of data in calculation of the performance measurement, automated data-collection methods, quality of information being collected and processed are vital. Without available of accurate data, performance measuring can easily lead to a lack of credibility. Various indicators developed by processing raw data and other related information characterize such information. However, considering the high complexity of equipment and dynamicity of working environment collecting and analyzing of such data is a challenging task. It requires a platform to the generation of enormous amounts of data, which have to be stored, processed and presented in an efficient form. This paper will discuss the application of the concept of Internet of thing (IOT) to facilities such process.

KEY WORDS: Arctic, performance measurement, economic performance, IOT

POAC17-121

INTRODUCTION

Performance measurement (PM) of the system, in general, is an essential tool that enables companies to achieve and control their desired objectives(Simons, 2013). The field of PM has been evolved over a long period of time; and, has been defined from different perspectives by different researchers. For instance,(Neely et al., 1995) defined, PM as the set of metrics used to quantify both the efficiency and effectiveness of actions. Effectiveness refers to the extent to which customer requirements are met, while efficiency is a measure of how economically the firm's resources are utilized when providing a given level of customer satisfaction. Since 1880s different explanations and perspective used within the performance measuring concept including internal and external structure, financial, customer demand, effectiveness, efficiency, learning perspective, growth, stakeholder satisfaction, stakeholder contribution, productivity, Capacity, etc.

A study of the performance terminologies and standards reveals that the performance elements measured by them, though important, but not enough and still should be improve due to the unique and challenging Arctic operational conditions with strictly regulations and requirement for safety and environment. The operating environment of the Arctic, such as low temperatures, sea and atmospheric icing, polar low pressures, poor visibility and seasonal darkness, etc. has significant effects on the performance of system in various ways, including increasing failure rate and repair time; consequently, leads to production losses (Barabadi, 2014). Moreover, Arctic operational condition can increase the power losses, life cycle costs, and safety hazards. Furthermore, less developed infrastructure in the Arctic create several challenges such as limitations to the logistics of supplies, material and personnel required for the operation and maintenance activities (Barabadi et al., 2013). Hence, the Arctic region due to its peculiar operational conditions coupled with stringent environment regulation and safety requirement, requires new criteria for judging performance. However, most of the available literatures and methods lacks the ability to capture the effect of the dynamic nature of the Arctic on the performance of the system. Some of the measures has been used already to measuring performance in the Arctic but this paper will bring some new and important measures, such as sustainability and financial performance. Moreover, show how to calculate the suggest performance measure system.

The main purpose of this paper is, thus to modify the performance measurement system for Arctic conditions. Further, the concept of Internet of Things (IOT), for efficient ways of storing and processing data has been integrated in the modified PM concept. The rest of the paper is organized as follows: after a litreture review in section 2, the performability measurement concept for Arctic environment are discussed in Section 3. Section 4 presents formulations for quantifying the PM. The required data and information for PM estimation and the application of Internet of things (IOT) for this aim is described in Section 5. Finally, the concluding remarks is presented in Section 6.

LITRETURE REVIEW

Over the years, several methods, frameworks, and models have been developed, in order to quantify performance measurement (See e.g. (Nakajima, 1988), (Hayes and Abernathy, 2007) , (Kaplan, 1984), (Bourne et al., 2003), (Ghalayini and Noble, 1996), and (Khan and Shah, 2011)). Ghalayini and Noble (1996) categorized the performance measurement literatures into two phases: i) traditional phase, which is regarded as the first phase, was cost accounting orientation, and was started in the late 1880s; ii) the second phase, which is started after 1980, and attempted to present a balanced and integrated view of PM.

In general, the traditional phase literatures – cost accounting orientation approaches – have tried to quantify the system performance and, other improvement efforts in financial terms, by quantifying the return on investment (ROI), return on sales (ROS), etc. (Bourne et al., 2003), (Khan and Shah, 2011). There are several drawbacks of the traditional phase approaches. Most of the approaches, for instance, ignore clients and their needs (Ghalayini and Noble, 1996), are internal rather than externally focused (Hayes and Abernathy, 2007), (Kaplan, 1984), are backward looking and is historically focused therefore is not predictive measures and also criticized for not providing adequate information for a productivity measurement. To overcome these challenges, (Nakajima, 1988) introduced overall equipment effectiveness (OEE), by categorizing the major losses or reasons for poor performance, which is based on its availability, performance, and quality rate of the output. However, OEE and its measures, availability, performance speed, and quality rate only reflect the internal effectiveness of a system and financial performance and external effectiveness, which is characterized by customer satisfaction and measures that have long-term effect on company's profitability, was missing.

The end-1980 was a turning point in the performance measurement literatures, as it marked the beginning of the second phase (Hayes and Abernathy, 2007), As markets became competitive and customers became more demanding, due to globalization, companies attempts to find more balanced, multi-criteria, and integrated PM frameworks, by considering financial, non-financial, internal, and external performance perspectives. Consequently, several frameworks has been modified, suggested, and developed, see e.g. (Neely et al., 1995), (Keegan et al., 1989), (Kaplan and Norton, 2005), (Neely et al., 2002). Table 1 summarized the relative measures and/or criteria of various frameworks.

The applicability and suitability of second phase approaches vary from field to field, business-to-business, and even by time and location. Furthermore, these approaches are not well suited for operation and facilities with new and unique structures, functions, and business positions, such as the Arctic region.

Table 1. The relative measures and/or criteria of various frameworks

Framework	**Proposed in:**	**Measure/Criteria**	**Reference**
Performance measurement matrix	1989	Cost, non-cost, external, and internal factors	(Keegan et al., 1989)
SMART (performance) pyramid	1991	Quality, delivery, process time, cost, customer satisfaction, flexibility, productivity, marketing, and financial measures	(Lynch and Cross, 1995)
Results and determinants matrix	1991	Financial performance, competitiveness, quality, flexibility, resource utilization, and innovation	(Brignall et al., 1991)

Balanced Scorecard (BSC)	1992	Financial, customer, internal process, learning and growth	(Kaplan and Norton, 2005)
Integrated dynamic PM system	1997	Timeliness, finance, customer satisfaction, human factors, quality, and flexibility	(Ghalayini and Noble, 1996)
NORSOK Z-016	1998	Reliability, maintainability, and supportability	(Standard, 1998)
Performance prism	2001	Stakeholder satisfaction, strategies, processes, capabilities, stakeholders contribution	(Neely et al., 2002)
BSC of advanced information. Services Inc (AISBSC)	2003	Financial perspective, customer perspective, process, people, infrastructure and innovation	(Abran and Buglione, 2003)
System perfomability	2008	Survivability, dependability, and sustainability	(Misra, 2008a)
ISO 20815	2008	Item availability, production availability, and deliverability	(Standardization, 2008)
Production assurance performance	2010	Capacity, dependability, customer demand	(Barabady et al., 2010)
Production performance	2010	Economical, functional, and HSE	(Markeset, 2010)

PERFORMANCE MEASURMENT FOR THE ARCTIC

During the critical review of the available literatures, it can be concluded that for a comprehensive PM the following measurement should be considered:

- Financial performance
- HSE performance
- Overall Equipment Effectiveness,
- Sustainability performance

Figure 1 illustrates the concept of the performance measuring system and its related concepts. For instance, the Overall Equipment Effectiveness include functional and quality performance. The functional performance of a system basically, is expressed as a function of the availability and capacity of the system. Moreover, availability is expressed as a function of the system reliability, maintainability and supportability.

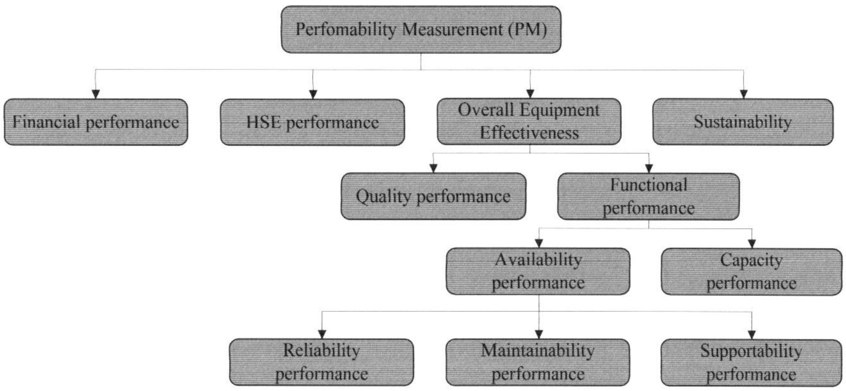

Figure 1. The concept of the performance measuring system

Sustainability Performance

Products and systems implied, especially in the Arctic should be on compliance with the principles of sustainable to increase the energy and material efficiencies, preserve ecosystem integrity, and promote human health, which in turn result in minimum life-cycle costs (Hallstedt et al., 2010). Without sustainable analysis, which include economic, environment and social aspects (Figure 2) an overall performance evaluation cannot be comprehensive particularly in the Arctic with strictly regulations and requirement for safety and environment.

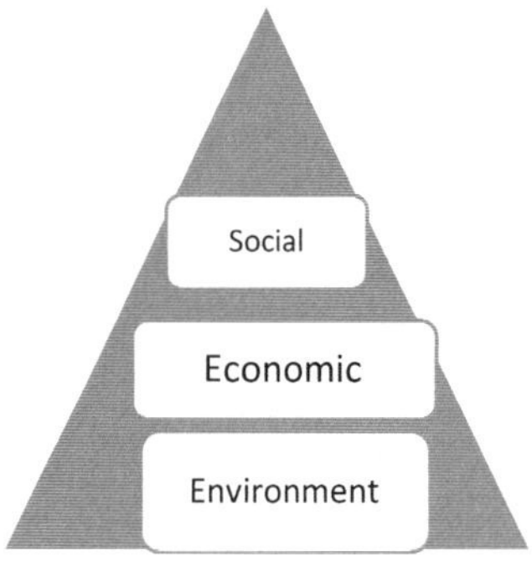

Figure 2. Sustainability aspects

For environment impact assessment, indicators such as air pollution, generated waste, waste recycling rates, use of non-hazardous materials, etc. should be included. In response, a vast range of methods, tools and concepts have been developed, each focusing on certain part of environment aspects of sustainability concept. These include Environmental Management Systems (EMSs), cleaner production and eco-design and various kinds of ecological indicators, such as ecological foot printing and Factor (Markeset and Kumar, 2000).

Social impact assessment should be included how company contributes to better health, education and safety of employees and the impact the company have on the local community. Economic impact assessment also need to be evaluated, economic activity can reduce the environmental and social capital. A company need environmental and social capital - alongside economic capital to create value in the future (Jan and Petra, 2016). A sustainable company can improve company reputation, brand value and can increase shareholder value or cost savings due to minimize use of material and energy. Moreover, sales may increase or strengthen of customer loyalty due to there is growing number of people who prioritize environmentally friendly product and services (Jan and Petra, 2016).

Figge and Hahn (2008) have developed a value-oriented methodology to calculate the sustainability performance of companies, called Sustainable Value. Sustainable Value allows assessing sustainable performance in a value-orientated way. Sustainable Value can have calculated by:

$$SV = \frac{\sum_1^n vc_n}{n} \quad (1)$$

VC = Return value – OC (2)

where:

- VC is Value Contribution for all environmental, economic and social resources for the company.
- OC is Opportunity Cost; is how much return would be created, if the resources in the company were used by the benchmark, and
- Return value is how much return the company creates with its resources.
- Finally, n is number of indicators for economic, environment and social aspects.

Financial Performance

In general, LCC analysis combined with a LCA (Life Cycle Assessment) analysis can be used for financial performance (Misra, 2008b). Several LCC tools have been developed, for cost benefit evaluation of a system, which describes the costs of system from the early planning stages to the end of use and, gives the decision makers information to find the correct balance or best solution considering cost and benefit. For instance, Norwegian Standard NS3454 provides a standardized methodology for carrying out the LCC. Design for the harsh climate condition, increases the LCC of the systems, and consequently, increases business risk. This is due to the lack of infrastructure in the Arctic, limitations to the logistics of supplies, material, and personnel required for the operation and maintenance activities, etc.

In general, the financial performance of a system FP_S, is a measure of efficiency of an investment and, can be used to express the measure of relative effect the operating environment of the Arctic has compared to that of an 'ideal' or designed operating environment. It is the ratio of the rate of investment (ROI) of a system under the Arctic condition and the ideal' operating environment. This ratio shows how the FP_S will be changed (increased or decreased) based on the effect of the operating environment.

Suppose our system is installed or is planned to be install in the Arctic region, and we wish to estimate the financial performance of the system under Arctic and 'ideal'-operating conditions. For the two operating environments with ROIs, ROI_{AR} and ROI_{ID}, the financial performance of a given system FP_S, can be expressed as follows:

$$FP_S = \frac{ROI_{AR}}{ROI_{ID}} = \frac{\frac{IG_{AR}-CI_{AR}}{CI_{AR}}}{\frac{IG_{ID}-CI_{ID}}{CI_{ID}}} \qquad (3)$$

where:

- IG_{AR} and IG_{ID} are the gain from the system investment under Arctic and 'ideal' operating environment, respectively.
- CI_{AR} and CI_{ID} are the cost of the system investment under Arctic and 'ideal' operating environment, respectively.

The FP_S are useful to describe the multiplication of the cost that occurs due to the peculiar Arctic operating environment. For instance, an FP_S of 1.5 for a particular system implies that the operating environment of the Arctic increases the cost of investment by $100 \times (FP_S - 1)\% = 50\%$. In similar manner, an FP_S of 0.5 is interpreted as the financial performance of the system installed in the Arctic region being more than half of that in the 'ideal' operating environment. The numerical value of the estimated FP_S must always be between zero and $\frac{1}{ROI_{ID}}$.

Health, Safety, and Environment (HSE) Performance

Engineering products and systems can cause hazards during operation or maintenance. Hence, performable system must seek to minimize the possibility of hazards (Misra, 2008b). The Arctic has great resources of different fish species, planktonic organisms, and bird habitats, which makes the area vulnerable and in the public perception Arctic is an icon for undisturbed nature, and the ''last'' wilderness (Paulsen et al., 2005). Hence, hazard reduction measures should be implemented to improve the HSE performance of the system. In addition, at very low temperature, electrical insulation can be cracked and, eventually exposed the conductors to the environment, and this can lead to a serious hazard for personnel. Further, salt water ice on antennas can bridge the insulators, then causing arcing and loss of communication. Moreover, low temperature generates static electricity, which destroys computer and can cause loss of data. Icing on stairs, deck and other surfaces can further cause slippery hazards that can result in accidents. Further, icing can damage structures and equipment, can hinder access to equipment and prevents work (Ryerson, 2011). There are several tools for consequence evaluation as well as quantifying probabilities of failure, such as failure mode and event analysis (FMECA), event tree analysis (ETA), cause consequence analysis (CCA) and fault tree analysis (FTA).

Risk matrix can be used to estimate the HSE performance. After calculating the associate risk for Health, Safety, and Environment the highest associate risk with these elements can be considered as the HSE performance of the production facilities. For example, if the risk of specific activity is high regarding the Health, medium regarding the safety and low with respect to the Environment the HSE performance will be High in this case.

Capacity Performance and Availability Performance

Functional performance of a system can be expressed as a function of availability and capacity. Capacity performance can be defined as an item's ability to deliver according to design capacity and/or current demands (requirements/needs). Capacity can calculate as:

$$\text{Capacity} = (\text{Ideal Cycle Time} \times \text{Total Count}) / \text{Run Time} \quad (4)$$

where, Ideal cycle time is minimum time to produce one part and run time is calculated by subtracting down time from production time and includes time when the process could be experiencing small stope, reduced speed and making reject part. The availability performance is also expressed, as dependability, which is a function of reliability, maintainability, and maintenance supportability, for details see (Barabady et al., 2010).

Quality is another important term to measure the performance of a system. The quality of an equipment is a measure of its degree of conformance to applicable design specifications and workmanship standards. It can realize that the degree of perfection is inversely proportional to the variability present in the equipment. Production facilities are made up of complex subsystems and components and they employ materials and machines. Hence, variation of parameters and dimensions of components leads to weakening, component mismatch, incorrect fits, vibration, etc. These issues can increase the failure (Misra, 2008b). Quality can calculate by dividing produced part that meet the quality standard by total of produced part including defect parts.

Quality = Acceptable count / Total count (5)

ESTIMATION OF PERFORMABILITY MEASURING INDEX

To quantify the PM, firstly the concept of performance measuring index (PMI) needs to be developed. In general, quantifying the PMI is a bottom–to–top process, which starts by quantifying the Availability Performance (AP). This constitutes an analysis of the reliability, maintainability and supportability performance of the system. Different form of the availability have been developed for example in the form of the Steady State Availability when the reliability, maintainability ad supportability is model by the Exponential Distributions it can be express as:

$$AP = \frac{MTTF}{MTTF+MTTR+MTTS} \tag{6}$$

where, MTTF represent the men time to failure, MTTR is the mean time to repair and MTTS is the men time which need to bring the spare parts and logistic support to the failure location.

Having the AP, the quality performance (QP) and the capacity performance (CP) of the production facility the OEE can be calculated by:

$$OEE = AP \times QP \times CP \tag{7}$$

Finally, by considering the value of the OEE, the Performability Measuring Index (PMI) can expressed as:

$$PMI = Fi^{\alpha_{Fi}} \times HSE^{\alpha_{HSE}} \times OEE^{\alpha_{OEEI}} \times Sus^{\alpha_{Sus}} \tag{8}$$

where,

- Fi, HSE and Sus are financial, HSE, and sustainability performance measures of the equipment, respectively
- α_{Fi} , α_{HSE} , α_{OEEI}, and α_{Sus} are weight vectors for financial, HSE, OEE and sustainability performance measures, respectively
- $\alpha_{Fi} \geq 0, \alpha_{HSE} \geq 0, \alpha_{OEEI} \geq 0$ and $\alpha_{Sus} \geq 0$ and $\alpha_{Fi} + \alpha_{HSE} + \alpha_{OEEI} + \alpha_{Sus} = 1$.

The basic assumption embedded in the estimation of AMI is that the performability measures, i.e. the financial, HSE, and sustainability performance have a different importance or weight on the overall performability of the system. Hence, when estimating the AMI, the weight of each of the performability measures needs to be quantified. Typically, the criticality of the selcted will decide the weight of each of the performability measures. For instance, for a critical component, which needs to be repaired as soon as it fails, maintainability is an important factor that has a significant impact on its availability; and, consequently on its overall performability. Hence, its maintainability has a high weight compared to a system, which has several redundancies, and it can be repaired later.

PM DATA COLELCTION AND SHARING UISNG IOT

Performance measurement is an information system and a reporting process through which the employees are given feedback on the outcome of their actions (Bititci et al., 1997). To obtain such feedback, availability, accuracy and quality of relevant data is a key factor. To have an effective PM analysis both currently reported data as well as historical data available in corporate databases should be used. Different type of the data should be collect and used in PM analysis including: data from manufacturing, experts, local information about climate, life cycle data of equipment , event and accident data, reliability data, financial data and products life cycle costs etc. Moreover, information about the behavior equipment under a given operational setting, their functional characteristics, and the technical faults and failures need to be collected precisely.

However, considering the high complexity of equipment and dynamicity of working environment in the Arctic, generation of enormous amounts of data, which have to be stored, processed and presented in an efficient form to calculate PM is a challenging task. Moreover, considering this fact that there is no sufficient data, information regarding the industrial activity in this area and conventional O&M practices may not be suffocation, thus some new technical solution should be developed. More specifically the proposed practice should enable differ department and even different industries in the region to achieve the timely critical PM data. The growth of implementation of advanced information and communication technology ICT solutions have generated new opportunities to data management of the complex, high-risk assets and capital-intensive industrial plants and facilities such as oil and gas industry in the Arctic.

Internet of Things (IOT) is a novel paradigm, which makes a data exchange and communication platforms for Business-to-Business (B2B) communication and technical solutions where data acquisition, processing and interpretation, and decision support components are integrated (Stojkoska and Trivodaliev, 2017), (Sundmaeker et al., 2010), (Gubbi et al., 2013). The connection of physical things to the Internet makes it possible to access remote sensor data and to control the physical world from a distance. A thing can be any real/physical object like simple autonomous sensor nodes, actuators, a virtual/ digital entity, or machines (Mihailovic, 2017).

IOT can brought unique capabilities to share information, knowledge, and experience to optimize decisions and actions. In the industry, integrated e-operation and e-maintenance are some example of IOT. In Norway from 2004 the term Integrated e-Operation is used in offshore industry. The Norwegian Oil and Energy Department support the Oljeindustriens Landsforening initiative and OLF is now coordinator for Norwegian efforts to improve and implement e-Operation at the NCS. There are examples of implementation of IO in relation to the O&G activity on the NCS. An example of integrated e-Operation framework modeling is the 'Secure Oil Information Link' (so-called SOIL) network available in the North Sea for oil & gas exploration and production industry, operated and administrated by the network service provider Oil Camp. SOIL was introduced to the Norwegian E&P industry in 1998, which consists of a number of application services actively connecting almost all the business sectors of the Norwegian O&G industry. This network, through the use of fiber-optic cables and wireless communications helps establishing the connectivity and interactivity between different parties, for instance, offshore O&M teams, operator's onshore O&M support groups, third- party CBM experts, logistic contractors, etc. Real-time equipment data can be acquired, jointly analysed and results can be exchanged online between these parties, enhancing the ability for shared interpretation and decision-making (Liyanage and Langeland, 2009). Another example is the onshore support center OSC in SKF-Norway, a CBM expert center that has remote diagnostic and prognostic capabilities and serves various operators in the Norwegian and Danish O&G sectors. Over the past, few years provide expert assistance, in logistics, and it has carried out online remote vibration monitoring of critical machinery of offshore production platforms in its OSC (Liyanage, 2008). Moreover, ConocoPhillips also as the operator of the Ekofisk asset has two such onshore centers OSC onshore support center. One of them is called onshore operational center (OOC) and has built-in integrated solutions for O&M planning, logistics, and other production and operation related activities.

Industry activities in the Arctic is not limited to oil and gas exploration. Mining operations, hydropower development, power lines, windmill parks, military activities, shipping activities and tourist traffic, particularly around Svalbard and Greenland have also been arisen across the past decades. These industries have varies experience and data about operation and maintenance in the Arctic. Connectivity and interactivity enhancing decisions and work processes regardless of the geographical location. This helps industries for manage the challenges due to less developed infrastructure in the Arctic. Limitations to the logistics of supplies, material and personnel required for the operation and maintenance activities in the Arctic create several challenges. Thus, IOT can be contributing to implement new business solutions between these industries by sharing the expertise between different industries in the Arctic, particularly between shipping and O&G industry. The connectivity and the interactivity between offshore and onshore, as well as between different onshore-based competence groups and other sectors of the industry (e.g. engineering contractors, equipment suppliers, technical expert centers, spare-part vendors, logistics, etc.), allows more effective PM analysis and consequently more effective decision loops as well as more coordinated planning and execution of O&M.

CONCLUSION

This work has introduced a modified performance measurement (PM) analysis of a given system, by considering the impact of the arduous Arctic operating environment. The proposed PM formulations comprises the estimation of the financial performance (to measure of efficiency of an investment), HSE performance (to measure the risk profile of the system), Overall Equipment Effectiveness (to measures the availability and quality rate of the system), and sustainability performance (to measure the material and energies used to build, run and dispose a specific system).

The proposed concept of performance measuring index (PMI) is particularly important in the Arctic operating environment since it expresses the measure of relative effect the operating environment of the Arctic has compared to that of an 'ideal' or designed operating environment. Moreover, the paper introduced the concept of measuring the sustainability and financial performance and, integrate it with the overall PM estimation process. Further, the paper illustrates the application of the concept of Internet of Thing (IOT) for measuring the performance of a given system. Incorporating IOT with PM analysis will be especially vital in the Arctic, due to the dynamicity of working environment as well as high complexity of equipment in the region. Our conclusion is that the use of a modified PM estimation technique has a vital role for ensuring that a system has high-level performance index, when operating in cold Arctic environment.

REFERENCES

ABRAN, A. & BUGLIONE, L. 2003. A multidimensional performance model for consolidating balanced scorecards. *Advances in Engineering Software,* 34, 339-349.

BARABADI, A. 2014. Reliability analysis of offshore production facilities under arctic conditions using reliability data from other areas. *Journal of Offshore Mechanics and Arctic Engineering,* 136, 021601.

BARABADI, A., NASERI, M. & RATNAYAKE, R. C. Design for Arctic Conditions: Safety and Performance Issues. ASME 2013 32nd International Conference on Ocean, Offshore and Arctic Engineering, 2013. American Society of Mechanical Engineers, V02AT02A024-V02AT02A024.

BARABADY, J., MARKESET, T. & KUMAR, U. 2010. Review and discussion of production assurance program. *International Journal of Quality & Reliability Management,* 27, 702-720.

BITITCI, U. S., CARRIE, A. S. & MCDEVITT, L. 1997. Integrated performance measurement systems: a development guide. *International journal of operations & production management,* 17, 522-534.

BOURNE, M., NEELY, A., MILLS, J. & PLATTS, K. 2003. Implementing performance measurement systems: a literature review. *International Journal of Business Performance Management,* 5, 1-24.

BRIGNALL, T., FITZGERALD, L., JOHNSTON, R. & SILVESTRO, R. 1991. Performance measurement in service businesses. *Management Accounting,* 69, 34.

FIGGE, F. & HAHN, T. 2008. Sustainable investment analysis with the sustainable value approach? a plea and a methodology to overcome the instrumental bias in socially responsible investment research. *Progress in Industrial Ecology, An International Journal,* 5, 255-272.

GHALAYINI, A. M. & NOBLE, J. S. 1996. The changing basis of performance measurement. *International Journal of Operations & Production Management,* 16, 63-80.

GUBBI, J., BUYYA, R., MARUSIC, S. & PALANISWAMI, M. 2013. Internet of Things (IoT): A vision, architectural elements, and future directions. *Future generation computer systems,* 29, 1645-1660.

HALLSTEDT, S., NY, H., ROBÈRT, K.-H. & BROMAN, G. 2010. An approach to assessing sustainability integration in strategic decision systems for product development. *Journal of Cleaner Production,* 18, 703-712.

HAYES, R. H. & ABERNATHY, W. J. 2007. Managing our way to economic decline. *Harvard business review,* 85.

JAN, F. & PETRA, L. 2016. Sustainable Value as a Tool for Corporate Performance Management within New Public Management Framework. *WSEAS Transactions on Environment and Development, volume 12, issue: 12, 2016.*

KAPLAN, R. S. 1984. Yesterdays accounting undermines production. *Harvard business review,* 62, 95-101.

KAPLAN, R. S. & NORTON, D. P. 2005. The balanced scorecard: measures that drive performance. *Harvard business review,* 83, 172.

KEEGAN, D. P., EILER, R. G. & JONES, C. R. 1989. Are your performance measures obsolete? *Strategic Finance,* 70, 45.

KHAN, K. & SHAH, A. 2011. Understanding performance measurement through the literature.

LIYANAGE, J. P. 2008. Integrated e-operations-e-maintenance: Applications in north sea offshore assets. *Complex system maintenance handbook.* Springer.

LIYANAGE, J. P. & LANGELAND, T. 2009. Smart assets through digital capabilities. *Encyclopedia of Information Science and Technology, Second Edition.* IGI Global.

LYNCH, R. L. & CROSS, K. F. 1995. Measure up!: Yardsticks for continuous improvement.

MARKESET, T. 2010. Design for performance: review of current research in Norway. *The Proceedings of Condition Monitoring and Diagnostic Engineering Management (COMADEM2010). June.*

MARKESET, T. & KUMAR, U. Application of LCC techniques in selection of mining equipment and technology. Mine planning and equipment selection 2000: Proceedings of the Ninth International Symposium on Mine Planning and Equipment Selection/Athens/Greece/6-9 November, 2000. 635-640.

MIHAILOVIC, A. 2017. Liberalising Deployment of Internet of Things Devices and Services in Large Scale Environments. *Wireless Personal Communications,* 92, 33-49.

MISRA, K. B. 2008a. Maintenance engineering and maintainability: An introduction. *Handbook of performability engineering.* Springer.

MISRA, K. B. 2008b. Performability Engineering: An Essential Concept in the 21st Century. *Handbook of Performability Engineering.* Springer.

NAKAJIMA, S. 1988. Introduction to TPM: Total Productive Maintenance.(Translation). *Productivity Press, Inc., 1988,* 129.

NEELY, A., GREGORY, M. & PLATTS, K. 1995. Performance measurement system design: a literature review and research agenda. *International journal of operations & production management,* 15, 80-116.

NEELY, A. D., ADAMS, C. & KENNERLEY, M. 2002. *The performance prism: The scorecard for measuring and managing business success*, Prentice Hall Financial Times London.

PAULSEN, J. E., HOSET, H., RØRHUUS, T., LARSEN, V., ALM, D., BIRKELAND, O. & MARKER, R. Exploration Drilling In The Barents Sea; Prevailing Zero Discharge Regime, Challenges And Learning From Two Recent Exploration Wells. SPE Asia Pacific Health, Safety and Environment Conference and Exhibition, 2005. Society of Petroleum Engineers.

RYERSON, C. C. 2011. Ice protection of offshore platforms. *Cold Regions Science and Technology,* 65, 97-110.

SIMONS, R. 2013. *Performance Measurement and Control Systems for Implementing Strategy Text and Cases: Pearson New International Edition*, Pearson Higher Ed.

STANDARD, N. 1998. Regularity management & reliability technology. *Norwegian Technology Standards Institution, Oslo, Norway*.

STANDARDIZATION, I. O. F. 2008. Petroleum, Petrochemical and Natural Gas Industries– Production Assurance and Reliability Management. International Organization for Standardization Geneva.

STOJKOSKA, B. L. R. & TRIVODALIEV, K. V. 2017. A review of Internet of Things for smart home: Challenges and solutions. *Journal of Cleaner Production,* 140, 1454-1464.

SUNDMAEKER, H., GUILLEMIN, P., FRIESS, P. & WOELFFLÉ, S. 2010. Vision and challenges for realising the Internet of Things. *Cluster of European Research Projects on the Internet of Things, European Commision*.

Cold Stress: Safety Issues when Working in Cold Environments

Yonas Zewdu Ayele[1], Abbas Barabadi[1]
[1] Department of Engineering and Safety, UiT The Arctic University of Norway, Tromsø, Norway

ABSTRACT

Exposure to cold environments can be a significant challenge for a person's comfort, for performing a given task in a safe and efficient manner and in the most extreme cases, also for survival. Understanding of the impact of cold exposure to human body is important to make the necessary preventive measures. Further, investigating the negative adverse impact help us to prepare personal protective clothing and equipment (PPC/PPE) and to layout the procedure how to use the PPC/PPE in optimal manner. The purpose of this paper is thus to propose a simplified cold stress analysis methodology, by investigating the risk factors that contribute to cold stress. The central thrust of the paper is to understand the work-related hazards and risks in cold operating environments; and, propose control measures to eliminate or reduce the identified risks.

KEY WORDS: Cold exposure; Cold stress; Risk management; Personal protective equipment

INTRODUCTION

Humans are "designed" to operate in very narrow temperature range. Low temperatures, wind, snow, and darkness reduces cognitive and reasoning abilities and, cognitive errors are more likely to occur. Further, cold environment significantly reduced the effectiveness of the workers, and possibility of mistakes or being inaccurate increases. For instance, since 1994, human error causes 9 out of 10 fatal accidents on the Norwegian continental shelf (NCS) and the main reason is the lack of concentration, which is caused by cold working environment (Aalipour et al., 2016). Moreover, the direct economic costs of work-related (occupational) injuries and illnesses, due to climatic and environmental conditions, on the NCS is estimated between NOK 710 and NOK 1,415 million per year (approximately EUR 91.5 to 172.5 million) (ECON, 2005).

In general, humans who are exposed to extreme cold, or work in cold environments maybe at risk of "Cold Stress". Cold stress is the response of the body to cold temperatures resulting from heat loss and the opposite of heat stress and, kills 20 times as many people as hot weather. (Risikko et al., 2003). Various factors can affect the body's ability to tolerate or maintain internal core body temperatures, which is the operating temperature of human beings. Factors include, the heat generated within the body from the work activity, how we lose heat to the environment (conduction, convection, radiation, evaporative heat loss), environmental conditions (e.g. air temperature, humidity, air movement), and the clothing worn (Ayele et al., 2015, Risikko et al., 2003). There are two main types of cold stress: whole-body cooling, and local cooling, including extremity cooling, convective skin cooling (wind chill), conductive skin cooling (contact cooling), and cooling of respiratory tract.

Understanding the impact of cold exposure to human body is important to make the necessary preventive measures. In general, decision makers must be able to recognize, how cold weather affects human performance and, consequently adjusts management and operational tools and approaches. The evaluations of these factors can help to have an effective working environment with high level of safety in the harsh, remote and cold environment. Further, investigating the negative adverse impact help us to prepare personal protective clothing /equipment (PPC/PPE) and to layout the procedure how to use the PPC/PPE in optimal manner.

Over the years, a number of work-related (occupational) health and safety hazards assessment models and approaches have been developed, to measure the relative effect of the cold operating environment on the human performance and, evaluate the risk of human error; see e.g. Aalipour et al. (2016), Ayele et al. (2015), Risikko et al. (2003), Jussila et al. (2010), and Mearns et al. (2003). For instance, Ayele et al. (2015) proposed a risk-based approach to eliminate and manage occupational hazards when working in the cold and harsh operational condition and, offered a step-by-step risk reduction measures. Risikko et al. (2003) developed a model for managing cold-related health and safety risks at workplaces. Further, there are several regulations and standards, which define the methods and procedures for evaluating and managing work-related health and performance risks in cold workplaces. One of such standards, for instance, is ISO 15743:2008, Ergonomics of the thermal environment – Cold workplaces – Risk assessment and management.

However, there is a lack of implementation of simplified as well as systematic cold stress analysing methodology. Furthermore, the traditional available tools or methodologies lacks an explicit causal model linking observed behaviour to personal and situational factors, and most of the methodologies are too cumbersome, time-consuming and generalised. Hence, developing a simplified methodology that supports and, facilitates the decision-making process, by measuring the relative effect of the cold operating environment, can offer the solution to filling the gaps that exist in the present cold stress assessment methodologies. The purpose of this paper is thus to propose a simplified cold stress analysis methodology, by analyzing the unique characteristics of cold operating environment and, their effect on human performance. Further, the paper comprise the assessment of potential job hazards, risks of work-related (occupational) injury, and health risks stemming from exposure to harsh environmental factors, such as cold, icing of work surfaces, ice falling from heights, darkness, etc.

The rest of the paper is organized as follows: The simplified cold stress analysis methodology is presented in Section 2. Thereafter, in Section 3, the direct and indirect cold related hazards are summarized. Afterwards, the control measures that can be employed for reducing cold exposure, are depicted in Section 4. Finally, the concluding remarks is presented in Section 5.

PROPOSED COLD STRESS ANALYSIS METHODOLOGY

Cooling influences muscle function greatly, resulting in: decreased muscle strength, slower reaction, poorer endurance, poorer muscle coordination, reduced manual skill (dexterity), and lower rate of muscle contraction (Risikko et al., 2003). Moreover, heat loss and cold experience affects the mental function in a human and maybe a significant distraction factor. Since, humans have a narrow temperature range in which it works optimally and, at higher and lower temperatures, we act poor. This leads among other things to the increasing number of errors. Figure 1 illustrates the proposed cold stress analysis methodology for investigating work-related hazards in the cold operating environment. The proposed methodology consists of four main steps: *i*) understanding body heat balance concept, *ii*) evaluating direct and indirect cold related hazards, *iii*) analyzing control measures, which can be employed to reduce the negative impact of cold exposure, and *iv*) establishing cold exposure monitoring programme.

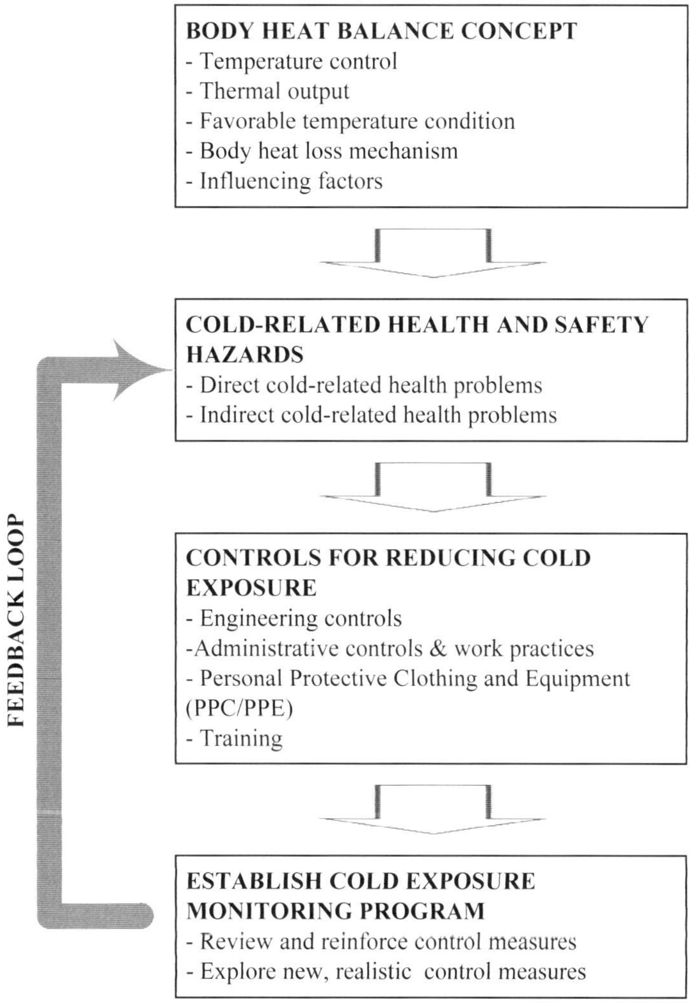

Figure 1. A simplified cold stress analysis methodology

The key purpose of the first stage is to understand the body heat balance concept, which is depicted in Figure 2. Understanding the effects of cold stress on the body, and how the body responds, can help to make life-saving decisions. The body, in general, attempts to maintain a constant core temperature (homeostasis) through a balance of heat loss and heat gain. For instance, body heat is normally gained by the application of external heat sources such as PPC/PPEs and, in normal situations through activities such as exercise and shivering. In general, body temperature is held nearly constant at 37°C at the heat production and heat dissipation are balanced accurately, under changing conditions, such as during work and rest as well as in cold and hot weather.

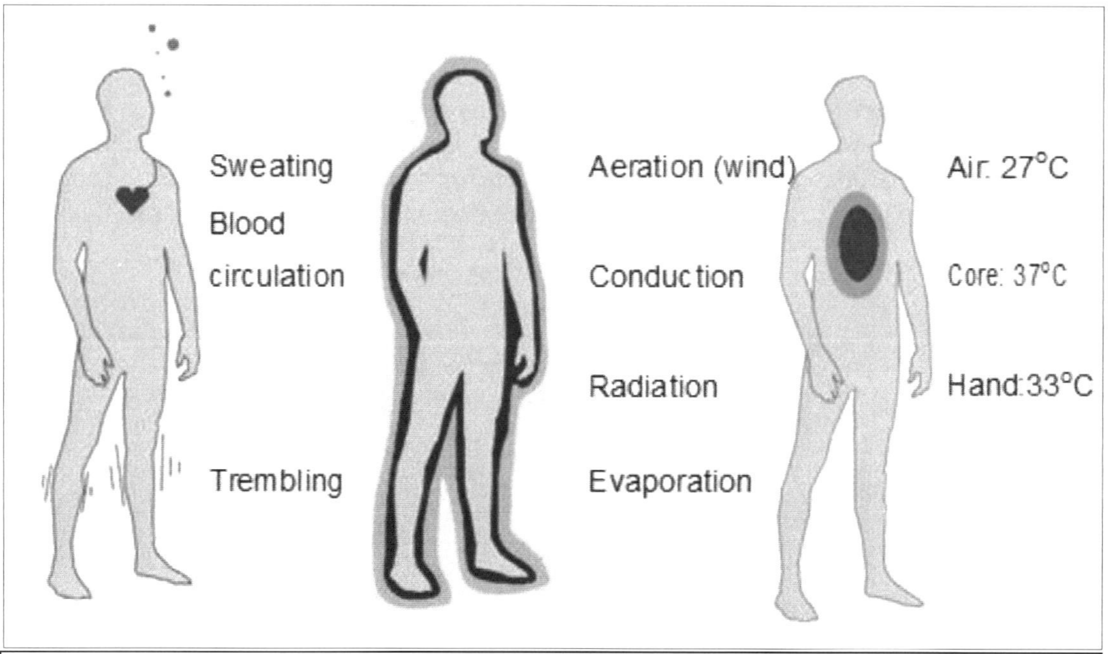

A) Temperature control B) Thermal output C) Favorable temperature

Figure 2. Body heat balance concept, modified from Norskoljeoggass (2012)

Body heat is lost in several ways, and there are four major mechanisms (Canadian Safe Boating Council, 2011):

- *Radiation*, which is a mechanism where the heat generated from within the body, is given-off to the surrounding atmosphere.
- *Evaporation,* is the second heat loss mechanism, and it occurs when you sweat or when your skin or clothing gets wet, the evaporation of that liquid (i.e., the change from liquid to vapor form) promotes heat loss, and the natural result is a cooling effect.
- *Convection:* is the third heat loss mechanism, where air or water flowing by the skin and, carrying away body heat. For instance, in case of cold-water contact, a heat loss via convection is the heat loss, which needs to be prevented by staying as still as possible in the water and, consequently the heated boundary layer remains undisturbed (OSHA, 2016). The reason behind staying still is that, if you move around in the water, you disrupt that boundary layer of warmer water, and that increases heat loss.
- *Conduction:* is the other heat loss mechanism, and it occurs due to temperature difference between the body and the operating environment. Normally, the body conducts heat to whatever the skin is in direct contact with. Conductive heat loss, in general, occurs when the skin is subjected to either cold air or water; however, it is especially critical in water, as body loses heat about 25 times faster in water than in air of the same temperature.

DIRECT & INDIRECT COLD-RELATED HEALTH AND SAFETY PROBLEMS

Several factors influence the cold injuries, such as negative air and sea temperature, icing, and their negative synergy effect on the human performance. The focal point of the second stage in the proposed methodology is to investigate the direct and indirect cold-related hazards, which are mainly due to predominant factors. Table 1 depicts the main factors that increase danger from the cold. In order to work safely, challenges have to be counterbalanced by proper insulation (e.g. layered protective clothing), buddy system, physical activity, diet and nutrition, acclimatization, and controlled exposure to cold (work/rest schedule) (Jussila et al., 2010). For instance, to protect heat loss in cold water, dry and insulated clothing between the human body and the cold water is critical. This is very important, both to avoid the dangerous immediate cold shock, maintain muscle function, and maintain the body's core temperature. The inside lining is the most important part of the insulation of the rescue equipment. Moreover, it is vital to avoid insulated clothing from getting wet inside. If a liter of water gets inside of the insulated cloth, for instance, the insulation will be reduced by about 40% (Killian, 2012).

Table 1. Factors influence the cold injuries, modified from OSHA (2016)

Factor	Description
Air movement	High wind exposure, strong ventilation, moving through air - as in an open vehicle.
Humidity (wetness)	Water conducts heat away from the body 25x faster than dry air. Wet skin, clothing or shoes from being wet with water, gasoline, alcohol, solvent, or other liquid that evaporates.
Physical Activity	The production of body heat by physical activity (metabolic rate) is difficult to measure. In most case, Tables are used to show metabolic rates for a variety of activities. Metabolic heat production is in general measured in kilocalories (kcal) per hour. For instance, one kilocalorie is the amount of heat needed to raise the temperature of one kilogram of water by 1°C.
Work/rest schedule	For safe working environment in cold climate, The American Conference of Governmental Industrial Hygienists (ACGIH) has adopted work schedule as Threshold Limit Values (TLVs) for cold stress.
Protective clothing	For work at or below 4°C, adequate protective clothing is vital. Clothing should be selected to suit the temperature, weather conditions (e.g., wind speed, rain), the level and duration of activity, and job design. - *Under-dressing:* Exposed fingers, cheeks, nose, ears; uncovered head. - *Over-dressing:* Too-thick clothing rather than layers; tight-fitting belt, clothing or shoes that restrict circulation; waterproof clothing that restricts evaporation.
Medical conditions	Especially those affecting circulation, such as diabetes, an underactive thyroid, heart disease, history of frostbite.
Medications	Some medications such as tranquillizers and beta-blockers cause drowsiness or decrease vasoconstriction.
Alcohol, caffeine, nicotine	Alcohol impairs judgement and reduces shivering. Caffeine increases urine production and blood circulation; both lead to a loss of body heat. Nicotine decreases blood flow to the extremities and raises the risk of cold injury.

Direct cold-related health problems

Effective cold stress analysis requires managers to draw together the technical expertise to evaluate the cold-related hazards and, the risks these hazards pose to the workers. Further, effective safety decision-making involves selecting appropriate protective clothing and equipment for workers. As part of controlling cold risk efforts, it is essential to put in place a coordinated, multiagency plan for monitoring hazards, which arises due to cold exposure.

Hypothermia

Hypothermia is one of the most common direct cold-related health problems and, it is in general a drop in the body's internal temperature below 35°C and, it is a threat in frigid weather; it can also strike at moderate temperatures (Ward et al., 2012). In most of the cases, hypothermia is considered as a medical emergency with a high fatality rate and, Table 2 summarized the main conditions of hypothermia. In general, hypothermia can be generally categorized into five subcategories (Ward et al., 2012, Brown et al., 2012):

- *Acute immersion hypothermia:* this type of hypothermia in general occurs when a person falls into cold water.
- *Sub-acute exhaustion hypothermia:* can happen to any worker in a cold environment as well as to skiers, climbers and walkers in the mountains.
- *Accidental hypothermia:* this type of hypothermia is associated with significant morbidity and mortality and, it is due to a drop in core body temperature to <35°C (Brown et al., 2012).
- *Hypothermia in trauma:* a type of hypothermia, which disrupt virtually all major body systems, and is a risk for all trauma patients (Ruffolo, 2002).
- *Sub-clinical chronic hypothermia:* this type of hypothermia poses a serious challenge because the condition may negatively impact the plan of care for very old and frail patients (Campbell and Travis, 1997).

Table 2. Conditions of hypothermia, modified from Group (2013).

Mild Hypothermia	Moderate Hypothermia	Severe Hypothermia
- Core temperature 95°F (35°C) – 93.2°F (34°C) - Uncontrolled, intense shivering - Unable to solve problems - Mild depression, confusion - Loss of fine motor skills - Cool, pale skin - Coldness is creating some pain and discomfort - The 'umbles' – stumble, mumble, fumble	- Core temperature 91.4°F (33°C) – 87.8°F (31°C) - Uncontrolled shivering - Muscles begin to stiffen - Mental confusion and apathy sets in - Speech becomes slow, vague and slurred - Breathing becomes slower and shallow - Drowsiness and strange behaviour may occur - Shivering stops. Lethargic.	- Core temperature below 87.8°F (31°C) - Skin is cold, may be bluish-gray in colour - Eyes may be dilated - Muscle rigidity - Gradual loss of consciousness - No apparent breathing - Unconscious - May appear dead.

Frostbite

A frostbite is a condition caused by extreme cold exposure, in which your extremities such as fingers, toes, etc. freeze, so that they swell, become darker, and sometimes drop off (Killian, 2012). In most of cases, the feet, hands, ears, nose, cheeks, and penis are the most frequently injured sites (Jussila et al., 2010). In general, in case of frostbite condition, skin, muscle, blood vessels, and nerves freeze and form ice crystals and, blood vessels become blocked with tissue debris that causes more damage (OSHA, 2016). Frostbite is often irreversible and amputation is sometimes required. On the other hand, in the case, the injured site heals; the victim may suffer chronic pain or numbness, excessive sweating, abnormal skin color, and joint pain. There are some common sign and symptoms related to frostbite, such as: sensations of coldness, tingling, stinging, or aching sensation, skin feels numb and cold to the touch, skin is waxy and appears yellowish/white, and blistering, turning red, then black (Group, 2013).

Immersion foot/ Trench foot

Immersion or trench foot is the case when the body is covered with water or wet mud, which is just above freezing, the area may become chronically swollen, weak, and sensitive to the cold (Group, 2013). In general, trench foot is mainly due to prolonged exposure (12 hours or longer) of the feet to wet, cold conditions. It is very serious non-freezing cold injury and, occurs when cold exposure softens skin, causing tissue loss and, often, infection. If untreated, in most cases, can eventually require amputation (OSHA, 2016). Early signs for immersion foot case includes, itching, numbness, or tingling pain and, eventually swelling of feet. Trench foot is characterized by: yellowish, smelly feet, possible numb, and sloughing of skin tissue/itching (Group, 2013).

Chilblain and snow blindness

Chilblain, also known as pernio, are the painful inflammation of small blood vessels in your skin, which occurs in response to prolonged exposure to cold but not freezing air (Group, 2013). Redness, swollen skin, usually on hands and feet, that feels hot, tender to touch, and itchy after cold exposure are the most common symptoms and signs of chilblain (Ward et al., 2012). In general, the snow blindness symptoms occur within 8 to 12 hours of cold exposure; at first eyes feel dry and irritated, thereafter, it feels as if they are full of sand, moving or blinking becomes extremely painful, exposure to light hurts the eyes, eyelids may swell, eye redness, and excessive tearing (Group, 2013).

Indirect cold-related health problems

Cold exposure can also cause various types of indirect cold related hazards, in addition to the direct cold-related hazard and safety problems, these problems include (Ayele et al., 2015, OSHA, 2016, Killian, 2012, Mearns et al., 2003):

- *Disease flare-ups*: Conditions like arthritis and asthma can be made worse in cold environments.
- *Increase in injuries*: Cold weather can decrease dexterity, mental skills, coordination, and cause a general decline in performance that affects safety.
- *Strains and sprains*: Working in cold weather can increase the risk of injuries to muscles and tendons, for instance, back strain.
- *Other health effects*: Sinus irritation, viral infections, chronic lung disease, arthritis.
- *Increased ear and eye infections*, which in turn can impair hearing and vision, which may increase the risk of accidents.
- Worsening of existing rheumatic conditions, and
- *Occurrence of Raynaud's Syndrome*, a vascular disease in which constricts blood vessels in the extremities such as the hands and feet, causing the blue coloring of the affected areas accompanied by on-going pain and discomfort and cold stress.

CONTROL MEASURES TO PREVENT COLD-RELATED INJURIES

The emphasis of the third stage in the proposed methodology is to evaluate control measures that can be employed to reduce the negative impacts of cold exposure. In addition, effective cold stress analysis involves understanding the available options or control measures and, how to choose among them. Analyzing control measure can help to determine what additional safety resources maybe needed when operating in cold environment. Control measures for reducing cold exposure, in general, fall into four categories: engineering controls, administrative controls and work practices, personal protective clothing and equipment (PPC/PPE), and training.

Engineering controls

Engineering controls are one of the vital controls, which can be employed to reduce the impact of the cold-related injuries (Ayele et al., 2015). The first step in implementing engineering controls is reviewing equipment design. For instance, metal handles and bars should be covered by thermal insulating material and, unprotected metal chair seats should not be used. In addition, machines and tools should be designed so that they can be operated without having to remove mittens or gloves (Group, 2013). In addition, reducing the cold weather exposure of the work task through mechanical assistance (such as snow-blowers, etc.), is the other engineering control measure (Risikko et al., 2003). Moreover, providing warm areas for break, such as portable heating units, heated shelters, access to warm buildings, etc., and protecting the worker from drafts to the greatest extent where possible (e.g. shielding) are the other control measures (OSHA, 2016, Group, 2013).

Administrative controls & work practices

In most cases, implementing the administrative controls and work practices, will help to assess the demands of all jobs, have monitoring, and control strategies in place for cold days and cold workplaces (Killian, 2012). For instance, for outdoor workplaces, triggers for applying control strategies could include, wind chill (equivalent chill temperature), "Work Warm-Up Schedule", and environment weather reports (OSHA, 2016). In addition, it is also helpful to increase the frequency and length of rest breaks and, allowing a period of adjustment to the cold before assigning a full work schedule. Moreover, allowing workers to set their own pace, and take extra work breaks when needed and, educating new or newly transferred workers on the hazards of working in a cold environment is vital (Group, 2013). Further, carrying out as many tasks as possible indoors and reduce timeframe people must work outdoors and, working outside during the warmer hours of the day (mid-day/early afternoon) will reduced the negative impact of cold exposure (Group, 2013).

The other vital administrative measure, is training workers to recognize the signs and symptoms of cold stress and start a ''buddy system'' since people are not likely to notice their own symptoms (Jussila et al., 2010). This system will help take appropriate preventive steps, if symptoms arise in a co-worker. This is especially essential, for older workers, or those with certain medical problems, which need careful attention in regards to cold stress (Group, 2013). Where appropriate, developing an emergency response plan should be in place in the event of a cold related illness. Furthermore, keeping track of the temperature and air movement and, linking it to an action program is the other key administrative measure. For instance, for indoors activities, readings of temperature and air movement should be taken in all cold work areas at the start, middle, and end of each shift, at least every four hours (Group, 2013, Risikko et al., 2003). On the other hand, for outdoors activities, the weather report can be used. For instance, where there is air movement from wind, ventilation or travel in an open vehicle like a forklift, the wind chill index should be used to evaluate the hazard, instead of the air temperature (Jussila et al., 2010). The reason is that wind chill index takes into account the wind blowing the heat away from the body. In general, wind-chill effect is the perceived decrease in air temperature felt by the body on exposed skin due to the flow of cold air. The air temperature combines with wind speed to describe the wind chill factor, which is the effect of these two parameters on exposed skin (Ayele et al., 2015).

Personal Protective Clothing and Equipment (PPC/PPE)

When working at or below 4°C personal protective clothing are essential, and Table 3 elucidates typical personal protective clothing. In general, selection of cold weather protection should be done in compliance with international and national standards, such as OSHA (US Occupational Safety & Health Administration) and, PEOSH (Public Employees Occupational Safety and Health) standards on personal protective equipment can be useful for selecting appropriate PPC/PPE. These are PEOSH 1910.132 for general requirements, PEOSH 1910.136 for foot protection and PEOSH 1910.138 for hand protection. Moreover, for the indoor environment based on air temperature and relative humidity in air, ISO 7720 and ASHRAE 55-2010 standards can be used. Furthermore, PPC/PPE must be properly selected to suit the temperature, weather conditions, duration of activity and job design:

- Multiple layers of clothing, inner layer to 'wick' moisture away from body to keep it dry (OSHA, 2016). Wear, in general, at least three layers: outer layer, which is windproof and that allows some ventilation; middle layer of wool, quilted fibers, or synthetic fleece to create an insulating layer; and, inner layer of synthetic fabric or wool to provide ventilation and allow moisture to escape (Group, 2013).
- Appropriate wool knit cap or a liner under a hard hat and appropriate gloves or mittens. In general, use gloves for work below 4°C for light work and below -7°C for moderate work; and, mittens for work below -17°C (Group, 2013). It is recommended that to wear mittens instead of un-insulated gloves, since mittens are warmer because your fingers stay together or use thermal insulated gloves and, nylon overmitts can be easily put on and taken off if gloved hands are necessary (OSHA, 2016, Group, 2013).
- Appropriate boots, such as felt-lined, rubber-bottomed, leather-topped boots with removable felt insoles for heavy work are necessary to reduce the trench foot hazard (OSHA, 2016). Further, when the work involves standing on the water, then waterproof boots are required. Moreover, boots with linings are preferred as they can be taken apart to dry more easily, or new linings can be used (Group, 2013).
- Eye and face protection, such as glasses/goggles, scarfs, to reduce the hazard from wind, blowing snow, glare, UV, etc. are vital(OSHA, 2016). In extremely cold conditions, eye protection must be separated from the nose and mouth to prevent exhaled moisture from fogging and frosting eye shields or glasses (Jussila et al., 2010, Group, 2013).

Training

Workers who are exposed to cold as part of their job and, their supervisors should take the health, safety, and environment (HSE) training course regarding working in cold environments. In general, the training should discuss various controlling measure for reducing cold exposure, such as engineering controls, work practices, and protective equipment to reduce the risk of cold stress. Moreover, supervisors and workers should be trained to watch for signs of cold stress and allow workers to interrupt work when uncomfortable as well as managing work schedules to allow rest periods (Group, 2013). Furthermore, supervisors and managers should also take the course on Job Safety Analysis (JSA). JSA generally, assists supervisors and managers in planning a job safely from beginning to end. In addition to formal training, supervisors can also take the opportunity to review: department procedures or work-specific procedures for cold stress (or other health and safety requirements) in other forums such as toolbox talks, operations meetings, etc. (OSHA, 2016).

CONCLUSION

This work introduced a simplified methodology for cold stress analysis, by considering the effect of the extreme cold operational conditions. The proposed methodology consists of four main stages: *i*) understanding body heat balance concept (to recognize the temperature control process, body heat loss mechanisms, factors that increase danger from the cold), *ii*) evaluating cold-related hazards (to investigate the direct and indirect hazards and risks due to cold exposure), *iii*) analyzing control measures (to evaluate measures that can be employed to reduce the negative impact of cold exposure), and *iv*) establishing cold exposure monitoring programme (to review, explore, and reinforce control measures).

The proposed methodology is particularly important in the cold operating environment since it is easy to implement and, enforce a course of the proposed actions during a cold-related incident, which can prove to be highly useful. Further, enforcing the simplified steps can protect the health and safety of the workforce by managing the cold stress related issues. Our conclusion is that the use of a simplified cold stress analysis methodology has a vital role for ensuring that an organization do not accepts unnecessary risks, makes risk decisions in a way that guarantees safe operation during cold exposure, and manages risk by planning and, establishing cold exposure monitoring program.

REFERENCES

Aalipour, M., Ayele, Y. Z. & Barabadi, A. 2016. Human reliability assessment (HRA) in maintenance of production process: a case study. *International Journal of System Assurance Engineering and Management,* 7, 229-238.

Ayele, Y., Barabadi, A. & Barabady, J. A risk-based approach to manage the Occupational Hazards in the Arctic drilling waste handling practices. Safety and Reliability: Methodology and Applications, European Safety and Reliability Conference, ESREL, 2015. 1329-1334.

Brown, D. J., Brugger, H., Boyd, J. & Paal, P. 2012. Accidental hypothermia. *New England Journal of Medicine,* 367, 1930-1938.

Campbell, D. & Travis, S. S. 1997. Chronic subclinical hypothermia: home care alert. *Home Healthcare Now,* 15, 727-732.

Canadian Safe Boating Council. 2011. *Mechanism of heat loss* [Online]. Available: http://www.beyondcoldwaterbootcamp.com/mechanisms-of-heat-loss.

ECON 2005. Cost due to work-related injuries. ECON- Report No.:2005-052. Project No.: 44670.

Group, T. W. C. 2013. *Working in cold environment - How to keep workers safe and productive when the mercury plunges* [Online]. Available: http://www.slideshare.net/windsgroup/asse-presentation-on-cold-stress-28296102.

Jussila, K., Valkama, A., Remes, J., Anttonen, H. & Peitso, A. 2010. The effect of cold protective clothing on comfort and perception of performance. *International Journal of Occupational Safety and Ergonomics,* 16, 185-197.

Killian, H. 2012. *Cold and Frost Injuries—Rewarming Damages Biological, Angiological, and Clinical Aspects: Biological, Angiological, and Clinical Aspects*, Springer Science & Business Media.

Mearns, K., Whitaker, S. M. & Flin, R. 2003. Safety climate, safety management practice and safety performance in offshore environments. *Safety science,* 41, 641-680.
Norskoljeoggass. 2012. *Work in cold climate: Personal protection against heat loss in cold weather* [Online]. Available:https://www.norskoljeoggass.no/Global/HMS-utfordringer%20i%20nordomr%C3%A5dene/Underlagsmateriale/Helse%20og%20arbeidsmiljø%C3%B8/Norsk%20Olje%20og%20gass_Arbeid%20i%20kaldt%20klima%202012.pdf.
OSHA. 2016. *New Jersey Public Employee Occupational Safety and Health (PEOSH) Program's Cold Weather Safety Initiative* [Online]. Available: https://www.osha.gov/dcsp/osp/njpeoshcoldsafety.pdf.
Risikko, T., Makinen, T. M., Påsche, A., Toivonen, L. & Hassi, J. 2003. A model for managing cold-related health and safety risks at workplaces. *International journal of circumpolar health,* 62, 204-215.
Ruffolo, D. C. 2002. Hypothermia in trauma: The cold, hard facts.
Ward, D., Foley, T. & Prabhakar, G. 2012. *Temperature, Humidity, Winds, and Human Comfort* [Online]. Available: http://www.atmo.arizona.edu/students/courselinks/fall12/atmo336/lectures/sec1/comfort.html.

Proceedings of the 24th International Conference on
Port and Ocean Engineering under Arctic Conditions
June 11-16, 2017, Busan, Korea

Winterization Design based on Weather Occurrence Probability for Arctic Ship and Offshore Plant

Daehwan Ahn[1], Yooseok Song[1], Youngdae Ko[2]
[1] Corporate Research Center, Hyundai Heavy Industry (Ulsan, South Korea)
[2] Shipbuilding Division, Hyundai Heavy Industry (Ulsan, South Korea)

ABSTRACT

The ship or offshore plant operating in the Arctic region needs the proper winterization to secure the safe working environment and normal operation of the topside facilities in harsh weather condition. In this paper, winterization design methodology for a drilling platform is established based on the weather occurrence probability in the operation site eventually inducing the local flow and temperature around the topside equipment. Each heat loss from the topside equipment is calculated by this probability basis synchronizing the weather database with the local flow database corresponding to various wind speeds and directions. Heat loss from the pipelines, deck plates and other equipment according to weather conditions is estimated and then the total required heat and electric load can be added up to the considered facility availability, which is applicable to the design of the electric power system. This methodology can be utilized for the energy saving design of the Arctic ship and offshore plant.

KEY WORDS : Winterization; Weather database; CFD; Heat loss; Heat tracing.

INTRODUCTION

The ship or offshore plant operating in the cold environment needs proper winterization provisions to secure the safe working environment and normal operation of the topside facilities. Electrical heat tracing (EHT) as a heat source has been widely used to prevent freezing or icing of the topside facilities. Heating capacity of the EHT has to be designed under the considerations of wind chill effects and its heat loss for each component in the exposed space. DNV-GL (2016) shows the winterization class notation prescribing how much air temperature has to be reflected to the heat loss calculation and it is rated by the ranges of air temperature. However, the heat loss is not calculated by the air temperature only. An additional engineering procedure is therefore necessary to get an exact value of the heat loss by combining air temperature with the effects of external wind chill around the heated parts. Overestimation of the heat loss causes overheating of the parts and the energy loss. Recently, a ship builder has been requested to design heating systems with efficient and economic perspectives. Risk-based winterization (Yang, et al., 2013) has been proposed for determination of winterization levels or requirements on a case-by-case basis. In here, estimation of probability of the failure was calculated by using the limit state function based on the temperature. Risk assessments of the designed winterization systems were performed. However, fixed wind speed condition without the consideration of flow distribution on the topside was used and it was limited to an evaluation method in order to check whether heating systems meet the need and what condition they may fail.

In this paper, winterization design methodology has established based on the weather occurrence probability analyzed from the past data over 20 years in the operation site eventually inducing the air flow and temperature around the topside equipment. The occurrence probability of the heat loss from the topside equipment is calculated by synchronizing weather occurrence database with local flow database corresponding to various wind speeds and directions. Maximum heat loss for the each topside equipment can be defined by just selecting the satisfaction criterion in the respect of the safety and availability. The proper heating capacity of the EHT is automatically applied to the design procedure and total electrical loads for the winterization are also provided. For a demonstration of the methodology, a case study applying to a semi-submersible drilling platform.is described.

WINTERIZATION METHODOLOGY

The heat loss of topside equipment due to wind chill effects, which are governed by weather conditions such as air flow and temperature, is the key value to design winterization system and its heating capacity. Weather characteristics at the operation site or local position along the ship route can be quantified with analysis of past weather data. Local flow features around the topside equipment induced by external wind are varied with the position of target equipment and external wind conditions like wind speed and wind direction. Flow analysis based on CFD techniques can be useful to define local flow characteristics with consideration of complex topside structure. Each quantitative heat loss by wind chill effects is calculated by using empirical equations expressed with Nusselts number for the forced convection heat transfer. Consequently, occurrence rate of the certain heat loss is defined by probability calculation synchronized weather database with local flow database. Details are as bellows.

Weather Analysis

Weather data like air temperature, wind speed and wind direction for the operation site have been openly supplied on the website of NOAA (National Oceanic and Atmosphere Administration). Past weather data collected 4 times a day for 20 years are used. Within each 6 hours, mean values for wind speed and wind direction and minimum value for air temperature are taken for this work. For example, occurrence distribution diagrams for wind speed and air temperature and wind roses are shown in Figure 1. Data of air temperature and wind speed are divided with intervals of 1°C and 1 m/s, respectively. Wind directions are separated with the angle of 45° and thus 8 intervals in the wind rose are applied.

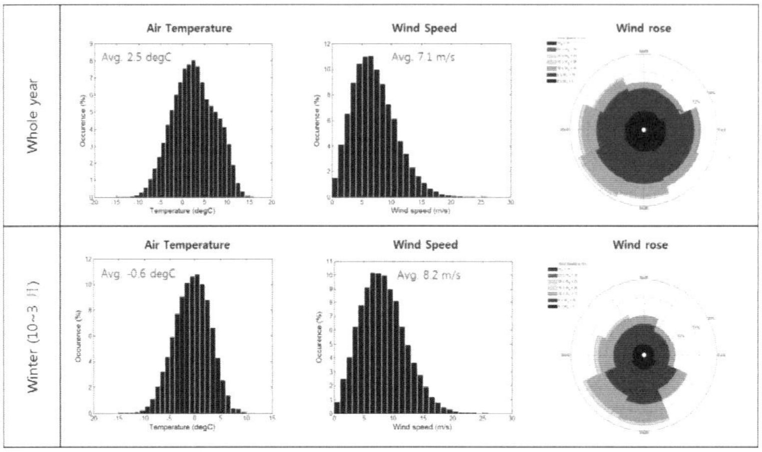

Figure 1. Example of diagnosis of weather probability

Flow Analysis

Local air flows in the exposed space of the ship or offshore plant are analyzed with a conventional CFD package. External wind conditions varied with several combinations of wind speed and direction, 5 external wind speeds (2, 5, 10, 20, 30 m/s) and 8 wind directions (N ~ NW), are used for input boundary conditions of the CFD simulation. 40 cases are calculated, and from the results local flow speed under the certain external speed is defined with spline interpolation method. Topside structure of the ship or the offshore plant disturbs external wind flow and induced local flow speed around the topside facilities can be down to about 1/10 of the external wind speed as the flow contours are shown in Figure 2(a). Moreover, Figure 2(b) shows that local flow speed is not linearly proportional to the external wind speed due to the turbulence flow phenomenon.

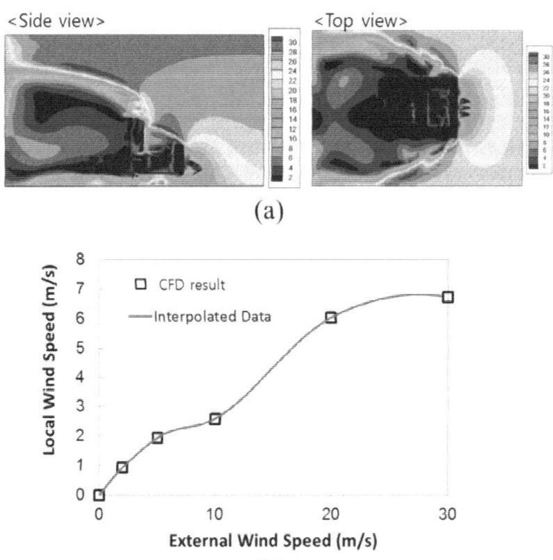

Figure 2. (a) Wind speed contour for external wind of 20 m/s and (b) local flow speed distribution under various external wind speeds

Occurrence Probability of Heat Loss from Topside Equipment

Empirical equations (Bergman, et al., 2011) for the forced convection heat transfer are used to get how much heat loss occurs on the plate or cylinder, representing the deck surfaces, passage ways and pipes on the topside. The equations and heat flux are as follows,

$$\text{Plate: } \overline{Nu_L} = \left(0.037Re_L^{4/5} - A\right)Pr^{1/3} \tag{1}$$

$$\text{Cylinder: } \overline{Nu_D} = 0.3 + \frac{0.62Re_D^{1/2}Pr^{1/3}}{\left[1+(0.4/Pr)^{2/3}\right]^{1/4}}\left[1+\left(\frac{Re_D}{28200}\right)^{5/8}\right]^{4/5} \tag{2}$$

$$h = \overline{Nu_L} \times \frac{k}{L} \ \ or \ \ \overline{Nu_D} \times \frac{k}{D} \tag{3}$$

$$q'' = h(T_s - T_\infty) \tag{4}$$

where q'' is the heat flux of wind chill, T_s is surface temperature, T_∞ is air temperature, L is the length of plate, D is the diameter of pipe, Re_L and Re_D are Reynolds number, Pr is Prandtl number and k is the thermal conductivity.

By using interpolated CFD results, an external weather combination, i.e. wind speed (V) and wind direction (θ), specifies the local flow speed (V_{local}) at the point of a target topside facility and thus the probability of the local flow speed (P_{local}) can be obtained by using a summation of the probabilities corresponding to the weather combination like Eq.(5~8). Finally, the local flow speed and air temperature can be translated to the specific heat loss and its probability [Eq.(9)].

$$P_{ws}(V),\ P_{wd}(\theta),\ P_{temp}(T) \leftarrow Weather\ DB \tag{5}$$

$$V_{local}(V,\theta) \leftarrow Local\ Flow\ DB \tag{6}$$

$$q''(V_{local},T) \leftarrow \text{Eq. (1~4)} \tag{7}$$

$$P_{local}(V_{local}) = \sum P_{ws} \times P_{wd} \tag{8}$$

$$P_{heat\ loss}(q'') = \sum P_{local} \times P_{temp} \tag{9}$$

Winterization of a Drilling Platform

For an example of winterization methodology, a case study is prepared by applying to a semi-submersible drilling platform operated at the Barents Sea. From the weather analysis of the operation site and local flow database, probability distribution of the local flow speed (P_{local}) at the certain location of the topside is calculated as illustrated in Figure 3(a). Probability distribution of the heat loss from the 100A pipe with thermal insulation is shown in Figure 3(b). Below 12.4 m/s of local flow speed satisfies 90% of a year. For satisfying 99.999% of a year, 30.4 m/s of local flow speed should be considered to calculate the maximum heat loss. 16.3 W/m^2 of heat loss for the 90% basis and 35.9 W/m^2 for 99.999% are provided at the final. 0.001% of the time is about 5 minutes a year and it could be negligible compared to icing and freezing phenomena. If the satisfaction criteria are defined for each of the topside facilities, the proper heating capacity of EHT are automatically selected with consideration of their heat loss, and moreover total electric loads for the winterization can be summarized. It is helpful to design the electric power system. The winterization system can be optimized by utilizing an adjustable satisfaction criterion for each of topside facilities reflecting its safety aspects and operation availability.

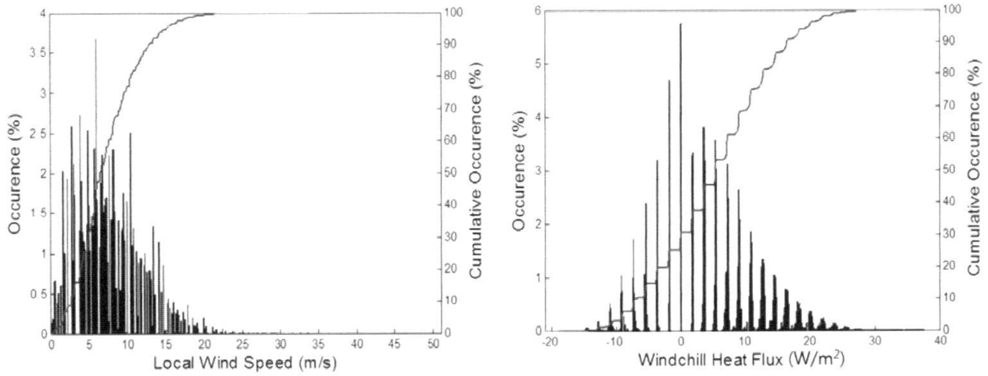

Figure 3. Occurrence distributions of (a) local flow speed and (b) heat flux due to wind chill effects

CONCLUSIONS

Winterization design methodology has been established based on the weather occurrence probability in the operation site eventually inducing the external air flow and temperature around the topside equipment. For the winterization system, the proper heating capacity is designed on the basis of occurrence probability of the heat loss. It is applicable for optimized heating system and saving energy of ship or offshore plant operating in the Arctic region.

REFERENCES

Bergman, T. L., Lavine, A. S., Incropera, F. P., & Dewitt, D. P., 2011, *Introduction to Heat Transfer*, John Wiley & Sons:Hoboken.

DNV-GL, 2016, *Ships for Navigation in Ice*, DNV-GL:Norway.

Yang, M., Khan, F, Lye, L., Sulistiyono, H., Dolny, J., & Oldford, D., 2013, Risk-based Winterization for Vessels Operations in Arctic Environments, Journal of Ship Production and Design, 29(4), pp.199-210

Risk-based Offshore Facility Design (RB-OFD) under Arctic Operational Conditions

Yonas Zewdu Ayele[1], Javad Barabady[1]
[1] Department of Engineering and Safety, UiT The Arctic University of Norway, Tromsø, Norway

ABSTRACT

Offshore facility design is a complex activity involving technical challenges from a wide range of sources, a considerable degree of risks, and substantial financial resources. Design strategies, especially in the Arctic offshore, must take account of the risk of structural failure due to ice impact and ice loading, in addition to the 'conventional' structural failure; that is, both the probability of failure and its consequences have to be considered. Further, offshore structures in Arctic have the added complication of being placed in harsh environment where winterization, ice loading, icing, hydrodynamic interaction effects, and dynamic response become major considerations in their design. Hence, when designing offshore facilities for Arctic conditions, it is imperative to identify and quantify all risks. The purpose of this paper is to suggest a simplified risk–based offshore facility design (RB–OFD) methodology, by considering the complex and fast-changing nature of the Arctic. The main goal is to highlight the fact that risk-based approach is particularly well suited to offshore project designs in the Arctic since there is less experience and data in the region. Further, a design strategy based on risk avoids the inadequacies of the traditional approach, maximizes the value of each offshore project, and increases the effectiveness of risk mitigation strategies.

KEY WORDS: Arctic; FPSO; Risk-based; Offshore facility design; Risk mitigation

INTRODUCTION

The offshore industries are drawing upon the lead set by other industries (nuclear, aircraft, etc.), in the application of risk–based approaches for design as well as in-service inspection (Serratella et al., 2007). Risk–based methodologies for design and inspection plan optimization originated in the nuclear industry in the 1970's and, over the years have migrated into other industries such as the downstream petrochemical industry in the 1980's and 1990's; these approaches are now moving into the upstream sector of the oil and gas industry. The risk-based methods include aspects of the conditions–based methods using trending techniques to estimate likelihood, but also factors in an estimation of the consequence of the structure's degradation and potential failure, enabling the program resources to be optimized and focused towards inspecting those items which have a greater overall risk weight (Serratella et al., 2007). Once those items are identified, optimum methods of design are the selected (Serratella et al., 2007).

Offshore projects, especially in the Arctic oil and gas industry, involve great technical challenges, considerable risks and massive financial resources (Ayele et al., 2016c). Moreover, as the offshore industry expands into the Arctic: remote locations, harsher environments, and deeper water, the hazards associated with these projects is expected to increase significantly compared to the well-established practices of exploration and production in other region, such as North Sea (Sember, 2004, Ayele et al., 2016a). Further, the potential for economic loss, environmental degradation, equipment damage, and work-force injuries is expected to be magnified (Ayele et al., 2013). In addition, oil and gas production project in the Arctic region could face unforeseen uncertainties and challenges. These challenges are mainly due to the new and advanced technological innovations related to the production platforms, stringent requirements to robustness against cold, little expertise on the use of international and national regulations and standards for Arctic offshore operation.

Given the inherent obstacles required to make any project successful, this industry has to serve as a proving ground for cutting edge project valuation methodologies and tools in the region (Hartke, 2012, Ayele et al., 2015). Further, as oil and gas companies operating in the Arctic offshore attempt to maximize the value of each project and optimize their portfolio of investment opportunities, it is imperative that all risks are properly identified and quantified, in order to maximize the value and increase the effectiveness of mitigation strategies (Hartke, 2012, Ayele et al., 2016c). Hence, to smooth the progress of the expansion of the offshore industry further north and, to implement new design aspects in the Arctic offshore, risk–based approach (RBA) have a key role to play. RBA in general allows for a more technical approach to the consideration of innovative designs, an appropriate methodology for assessing alternative designs to a high level of safety, and a better understanding of the hazards, mitigation measures, and risks inherent in the proposed designs (Sember, 2004). Moreover, RBA ensures the safety standards and regulation associated with design of offshore facilities by taking into consideration the peculiar Arctic challenges. In particular, minimizing the overall risk profile of the complex offshore projects under the Arctic operating environment is the main objective of risk-based approach.

In general, when offshore facility design are based on risk assessment, several factors have to be taken into consideration, such as choice of material, equipment, support strategies, physical environment, human factors, HSE (health, safety, and environment) aspects. Further, for the Arctic offshore operation, beyond the traditional risk factors, additional factors have to be taken into consideration. These peculiar risk factors are due to the stringent HSE requirements in terms of emissions to the atmosphere and discharges to the sea, and stringent design procedures with respect to risk reduction measures – prevention of risk incident, control of risk incident, and mitigation measures. To provide a basis for comparing alternative ways of achieving a certain benefit, aid safety design, and offer a fair basis for evaluating alternative design practices, a number of quantitative risk assessment models have been developed; see e.g. (Abimbola et al., 2014, Guo et al., 2005, Ayele et al., 2015, Khakzad et al., 2013, Abimbola et al., 2015, Øien, 2013). For instance, Khakzad et al. (2013) demonstrated the application of bow-tie and Bayesian network methods in conducting quantitative risk analyses during offshore facility design process. Abimbola et al. (2014) has proposed a dynamic safety assessment approach based on bow-tie analysis and a real-time barriers' failure probability assessment for offshore facility design.

However, in most of the available risk analysis literatures, risk factor is the only factor considered; and there is a lack of consideration of the impact of the operating environment on the risk profile. This is considered as a big shortcoming, particularly in the arduous Arctic operating environment (Barabadi et al., 2015). This is especially vital in the Arctic offshore operation because of its slow, non-linear, and potentially irreversible ecological and physical process (Ayele et al., 2016b). Moreover, and most of the methodologies are too cumbersome, time-consuming and generalised. Further, the unique environmental and climatic factors of the Arctic, such as large variations in temperature within a short period of time, sudden wind increase and large changes in wind direction, snow, and inadequate weather forecasting, increases the risk of system failure. The purpose of this paper is thus to proposed a simplified risk–based offshore facility design methodology, by considering the complex and fast-changing nature of the Arctic. This is to have better understanding of the hazards, mitigation measures, and risks inherent in the proposed designs.

The rest of the paper is organized as follows: In Section 2 a bird's eye view of offshore facility design concepts, procedures, standards, etc. are presented. Thereafter, the proposed risk–based offshore facility design (RB-OFD) methodology is described in Section 3. Finally, the concluding remarks is presented in Section 4. For the purpose of this paper, 'the Arctic' is taken simply to mean the Norwegian Arctic and the starting point of our discussion is the Barents Sea. Moreover, in this paper, risk is taken to mean the probability times the consequence of an adverse or hazardous event.

OFFSHORE FACILITY DESIGN: A BIRD'S EYE VIEW

Offshore facilities in general, are floating, production and mooring systems, such as FPSO (Floating Production, Storage and Offloading), FSU (floating and storage unit), FSO (Floating, Storage & Offloading), etc. Figure 1 illustrates the typical offshore productions systems and FPSO layout. The design and construction of an FPSO, for instance is divided into two parts: one for the hull system and the other for the topside system. The topside, like chemical plants, produces and off-loads crude oil and gas, and the hull, like a big tank, stores the produced oil (Hwang et al., 2009).

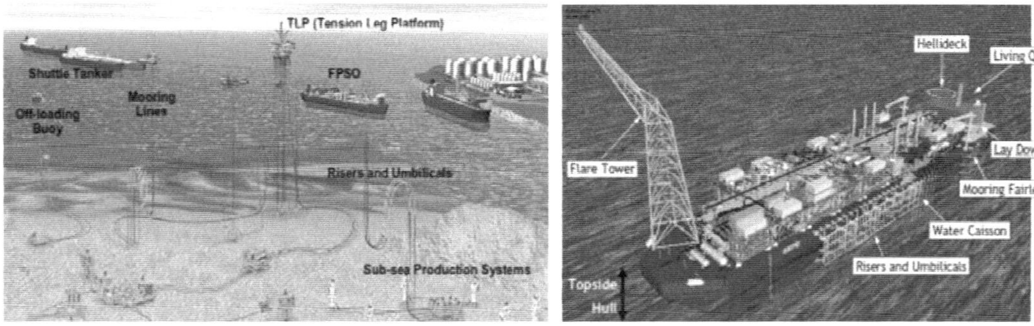

Figure 1. A) Offshore productions systems. B) A typical FPSO layout, adapted from Hwang et al. (2009).

A typical FPSO produces and processes crude oil and gas on the topside, and stores the stabilized oil in cargo tanks of the hull. The FPSO also off-loads the stabilized oil and gas to a shuttle tanker through the oil export/metering pump (Mather, 2000). However, the FPSO alone cannot produce oil and gas in the oil field. It requires many offshore production systems such as the mooring lines, riser structure, accommodation, process (oil, gas, and water) facilities, gas and water injection facilities, flaring facilities, oil storage facilities, export facilities (pumping and metering), gas export facilities, and control systems (Hwang et al., 2009). Thus, the design and construction of an FPSO includes many other offshore production systems (Shimamura, 2002).

Characteristically, when developing an oil and/or gas field, an exploration and feasibility study must first be performed. Thereafter, based on the result of the study, FEED (Front End Engineering Design) of the offshore facilities is commenced if the economic value of the field development is positive, taking the circumstances of offshore market into account(Hwang et al., 2009). Afterwards, based on the initial data of FEED, the design of the hull should be performed (Shimamura, 2002). At this time, IFR (Issued For Review) P&ID (Pipe & Instrumentation Diagram) should be prepared for submission to the client. Main activities are, specification incorporation, FEED clarification, and incorporation of lessons learned from previous project. In going from IFR P&ID to IFA (Issued For Approval) P&ID, the most important activity is the design of various systems in the hull (Hwang et al., 2009). The detailed design procedure of the offshore facilities, by taking an FPSO as an example, is depicted in Figure 2. In general, the interface data between the topside and the hull of FPSO must first be incorporated, and the utility balance, which was fixed at FEED stage, must be verified at the detailed design stage. After that, the design of the systems, including hydraulic calculations, should be performed based on the updated utility balance (Hwang et al., 2009). During this procedure, preparing equipment and instrument datasheets and overall line size calculation are very important activities in achieving a balance of all systems of the hull (Mather, 2000).

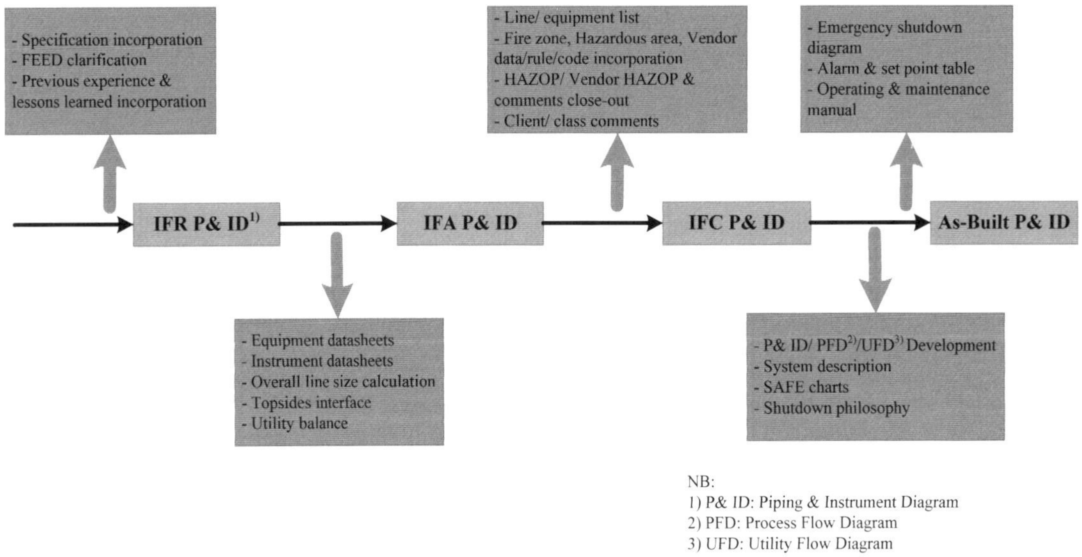

Figure 2. Detailed design procedure for P&ID of the hull of FPSO, adapted from Hwang et al. (2009)

Furthermore, in the process of constructing IFC (Issued For Construction) P&ID from IFA P&ID, the P&ID should include actual construction information. In this stage, the most important activity is the incorporation of safety factors and client/class comments. Line and equipment lists, which are very important items at the construction stage should be updated based on IFC P&ID. The design should be performed considering safety factors, such as the fire zone and hazardous area. The Hazard and Operability Analysis (HAZOP) evaluation is vital for safety in offshore design, and needs to be executed to remove potential risks during the site operation of FPSO (Hwang et al., 2009). Moreover, stakeholder's comments on consideration items of the safety, operation, and maintenance of each system should be incorporated in the IFC P&ID through technical review (Hwang et al., 2009). The interface data of each vendor and related rules/codes needs to be incorporated in IFC P&ID. In the final stage, in preparing the As-Built P&ID, the principal action is the integration of changes in the modelling or construction designs (Shimamura, 2002). Moreover, safety work and items on operation and maintenance are also important. In detail, preparing system descriptions, safety charts, shutdown philosophy, an emergency shutdown diagram, alarm, a set point table, and the operating and maintenance manual should be included at this stage (Mather, 2000).

A SIMPLIFIED RISK-BASED OFFSHORE FACILITY DESIGN (RB-OFD) METHODOLOGY

A risk-based design approach is an alternative to the prescriptive or performance-based design approaches. Under risk-based design approach, the potential hazards and consequences of undesired events related with an offshore facility design are firstly identified and, presented in a structured format. The expected frequency of each event is then determined based on historical data and expert judgment. For offshore facility design in Arctic offshore, for instance the historical data from NCS (Norwegian Continental Shelf) can be extrapolated in order to consider the ''additional'' hazards due to the Arctic operational conditions. With this information, the data can then be analysed, to identify those risks, which need to be mitigated and, to select the most appropriate risk-reduction approach. The steps involved in the risk–based offshore facility design (RB–OFD) for Arctic offshore facilities are outlined in Figure 3.

The initial step in the simplified RB-OFD methodology is to select components for further analysis. By analysing case-specific design features/components and, identifying and quantifying the unique risks involved, the industry can take appropriate measures to mitigate those risks (Sember, 2004). For instance, the range of possible design solutions pose their own peculiar demands in terms of failure modes, response to ice loading, application of winterization measures, hydrodynamic loading effects, foundation support conditions, and character of the dynamic response of not only the structure itself but also of the riser systems for oil extraction adopted by them.

Thereafter, performing evaluation of the peculiar Arctic risk influencing factors (RIFs) is the next stage. RIFs are factors that potentially affect the barriers and barrier performance (Vinnem, 2007). In the Arctic offshore, the predominant RIFs are the climatic and environmental conditions, such as snowstorms, atmospheric and sea spray icing conditions. These factors depend on various variables and they interact with each other. Their interaction is very complex and has a cumulative negative synergy effect on the offshore facilities. In combination, these factors will determine the performance of the offshore facilities or the suitability of new floating, production and mooring technologies in the region. In this stage, a considerable background knowledge and expertise is required from those involved in the engineering design and approval process, in order capture the impact of the Arctic operating environment and, to meet the defined companies and governing bodies requirements (Gareth and Luiz, 2008).

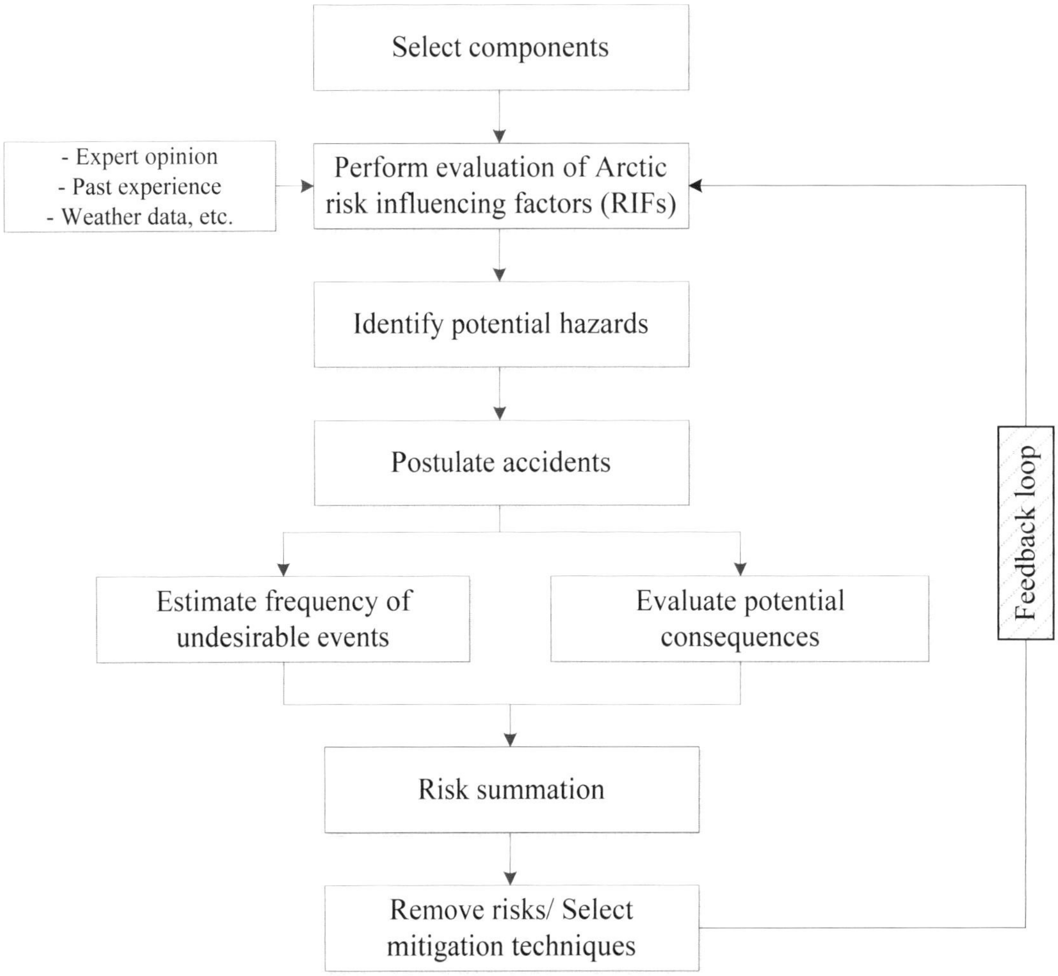

Figure 3. A simplified risk–based offshore facility design (RB–OFD) methodology for Arctic operating conditions

The next stage, requiring the identification of potential hazards of selected components and postulating accidents, while seemingly simple requires considerable professional judgment. All hazards need to be recognized during this step. This consists of those which are known, but their effects are not totally understood and those which have not yet been identified (Gareth and Luiz, 2008). In particular, for Arctic offshore facility design, risk assessment must take account of the risk of structural failure due to ice impact and ice loading, in addition to the ''conventional'' structural failure; that is, both the probability of failure and its consequences have to be considered. Further, a probabilistic consideration of 'unforeseen' challenges, due to the Arctic operating condition, as additional risk, should be carried out to determine the most probable levels of damage, and to check the adequacy of the design loads and resistance values.

Moreover, for Arctic offshore operations, since complete information is not available for frequency of events and potential consequences, appropriate tolerances need to be considered (Gareth and Luiz, 2008). In general, definition of scenario is a very important part of the consequence analysis, and it is related to the frequency estimates. The elements in consequence evaluation are illustrated in Figure 4. Overall, risk-based design provide a proactive systematic approach to assist the identification and categorization of the potential risks. Sound management and decision-making permits this information to be used as a tool in the design and regulatory compliance process (Gareth and Luiz, 2008).

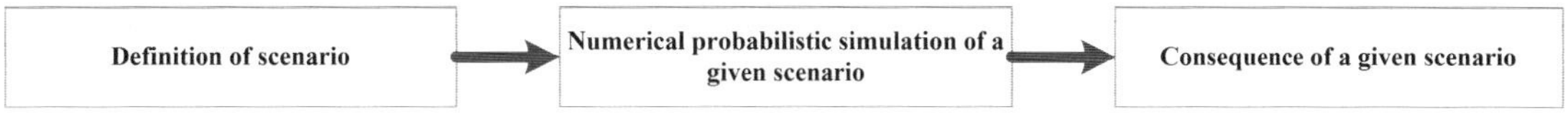

Figure 4. Consequence Evaluation, based on Bjerketvedt et al. (1997)

After evaluating the frequency of the undesirable events as well as the consequence of a given scenario, then, the summation of the risk need to be performed. In the case of the summed risk are above the acceptance criteria, the next step will be removing the risk or putting risk mitigation measures. Acceptance criteria are criteria used as a basis for decisions about acceptable risk – a risk that is accepted in a given context based on the current standards, regulations and values of the company (Rausand, 2013). The mitigation measures will include reinforcing the already identified barriers, exploring and/or evaluating new (realistic) barriers and evaluating the hazards/ risk reduction measures. Furthermore, there should be a feedback loop where the recommendations should help to review the risk assessment and, the choice of the alternative risk reduction measures. This loop helps to detect any changes in the operating environment and their effect on the risk profile.

An illustrative example: Application of the RB–OFD methodology

The main steps of the simplified RB–OFD methodology are briefly discussed, by using FPSO as an example.

Subdivision

The entire hull of the FPSO can be split into three main zones: submerged, above water and internal. The main advantages of breakdown the hull structure in such way are (Biasotto and Rouhan, 2004):

- To logically structure the risk assessment, the consequences, barriers, inspections and repair actions are specified for each zone, and
- To adapt the sub-groups of components in accordance with the asset definition.

Risk screening

The main objective of this step is to identify those components to be included in the detailed risk analysis and, determine the suitable methodology for risk assessment for each selected component: quantitative or qualitative (Biasotto and Rouhan, 2004). In this stage, components are ranked according to their criticality; and thus the number of components selected for the quantitative analysis is reduced to the potentially most critical ones. The remaining components are analysed separately, either by qualitative analysis, semi-quantitative or using class requirements for example (Biasotto and Rouhan, 2004). Therefore, the first screening can be conducted through HAZID (hazard identification) sessions, performed by a group of different experts including the consultant and the operator (Hwang et al., 2009). In general, the work group includes risk experts, naval architects and structural engineers. The HAZID is a brainstorming exercise where each hazard or threat is identified in terms of possible causes for its occurrence, its possible effects and mitigation actions to prevent the hazard and its effects and consequences (Rausand, 2013). Each scenario is also assessed regarding the likelihood or frequency of its occurrence regarding the severity of possible effects and consequences (Biasotto and Rouhan, 2004).

Investigating the peculiar Arctic risk influencing factors

Another important challenge in the RB–OFD process is the need to employ risk management practices during the operational phase of an offshore facility. In particular, the complexities of risk management in the Arctic offshore arise from the peculiar Arctic RIFs – low sea and air temperature, sea spray and atmospheric icing, strong wind. Further, the following challenges also play an important role:

- Stringent design procedure, with respect to risk reduction measures,
- Stringent HSE requirements,
- ''Untested'' knowledge in Arctic offshore, especially Norwegian Arctic, regarding to risk management of major hazards, such as Fires, Explosions and Blow-outs,
- Winterization measures,
- Sensitive environment to disruption, on one hand, but harsh and unforgiving on the other, and
- Environmental impacts take longer to heal and cost more to remediate.

Quantitative risk assessment

Structural risks may be assessed using various models, such as quantitative degradation models and so on. This leads to compute explicitly the annual probability of failure versus time, given a quantitative probabilistic degradation model and, a set of limit state functions (Biasotto and Rouhan, 2004). For instance, for crack propagation, a fracture mechanics model can be used. Then, the consequences of events can be used to compute the total expected cost of failure. One advantage of this approach is the analysis of a high number of components with a limited amount of computational time. The main steps in quantitative risk assessment of FPSO, basically are: risk acceptance criteria, damage computations (fatigue, corrosion, etc.), and detailed risk analysis (Biasotto and Rouhan, 2004).

Qualitative risk assessment

The purpose of the qualitative assessment is to establish plans for those components where degradation models are not available or for those elements disregarded in the quantitative assessment (Biasotto and Rouhan, 2004). This is attained through the risk rating obtained after assessment of the likelihood and consequences of failure for each component. Each hazard identified is analysed in terms of its functional failure, failure mode, consequences (including the possible different scenarios), existing barriers, control methods and repair strategies (Biasotto and Rouhan, 2004). Once the possible hazards are analysed, the Risk Ranking Matrix is used to evaluate the risk of each hazard and ranks its risk level based on the Risk Acceptance Criteria for the project (Rausand, 2013). In general, the hazards identified are qualitatively classified on the basis of the likelihood and the related consequences regarding risks to personnel, to environment and to asset and production (Biasotto and Rouhan, 2004).

CONCLUSION

Offshore structures in Arctic have the added complication of being placed in harsh environment where winterization, ice loading, icing, hydrodynamic interaction effects, and dynamic response become major considerations in their design. To have a better perceptive of the hazards, mitigation measures, and risks inbuilt in planned designs of offshore facilities, this paper proposed a simplified risk-based approach to offshore facility design (RB-OFD) methodology for Arctic offshore operation. The proposed RB-OFD methodology is particularly well suited to offshore facilities in the Arctic offshore where oil and gas industry faces with unpredictable and tough challenges.

Our conclusion is that the use of risk-based approach as base of design tool of offshore facilities, by considering the peculiar Arctic challenges, is vital, for the expansion of the petroleum industry in the region. Employing the risk-based approach helps the industry to look for other design/modification alternatives with better risk reduction measures, by considering unforeseen uncertainties and challenges. However, a shortage of valid system performance data can be a challenge associated with a risk-based regulatory regime. Since such data is critical to the decision making process, the absence of valid data can be a significant restriction to this approach. Hence, the proposed approach should be updated as new data/evidence becomes available, preferably in the form of field (hard) data reflecting the actual operational experience in this Arctic region.

REFERENCES

Abimbola, M., Khan, F. & Khakzad, N. 2014. Dynamic safety risk analysis of offshore drilling. *Journal of Loss Prevention in the process industries,* 30, 74-85.

bimbola, M., Khan, F., Khakzad, N. & Butt, S. 2015. Safety and risk analysis of managed pressure drilling operation using Bayesian network. *Safety Science,* 76, 133-144.

Ayele, Y. Z., Barabadi, A. & Barabady, J. 2016a. Dynamic spare parts transportation model for Arctic production facility. *International Journal of System Assurance Engineering and Management,* 7, 84-98.

Ayele, Y. Z., Barabadi, A. & Barabady, J. 2016b. A methodology for identification of a suitable drilling waste handling system in the Arctic region. *International Journal of Environment and Waste Management,* 17, 227-250.

Ayele, Y. Z., Barabadi, A. & Droguett, E. L. 2016c. Risk-Based Cost-Effectiveness Analysis of Waste Handling Practices in the Arctic Drilling Operation. *Journal of Offshore Mechanics and Arctic Engineering,* 138, 031301.

Ayele, Y. Z., Barabadi, A. & Markeset, T. 2013. Spare Part Transportation Management in High North.

Ayele, Y. Z., Barabady, J. & Droguett, E. L. 2015. Risk assessment of Arctic drilling waste management operations based on Bayesian Networks. Safety and Reliability of Complex Engineered Systems: ESREL 2015, Zurich, Switzerland. CRC Press 1907-1915.

Barabadi, A., Gudmestad, O. T. & Barabady, J. 2015. RAMS data collection under Arctic conditions. *Reliability Engineering & System Safety,* 135, 92-99.

Biasotto, P. & Rouhan, A. Survey and Inspection management for FPSOs. Proc., 23rd International Conference on Offshore Mechanics and Arctic Engineering, 2004.

Gareth, C. B. & Luiz, P. F. 2008. Practical Role of Classification Societies in Verifying Safe Design on Offshore Oil and Gas Facilities. *In:* SHIPPING, A. B. O. (ed.). Houston, Texas, USA.
Guo, Q., Geehan, T. & Pincock, M. 2005. Managing Risks and Uncertainties in Drill Cuttings Re-Injection in Challenging Environments–Field Experience from Sakhalin Island. SPE 93781, paper presented at the SPE/EPA/DOE E&P Environmental Conference, Galveston, March 2005.
Hartke, R. 2012. *The Application of Quantitative Risk Analysis Techniques in the Oil and Gas Industry* [Online]. Oil and Gas Monitor. Available: http://www.oilgasmonitor.com/application-quantitative-risk-analysis-techniques-oil-gas-industry/1966.
Hwang, J. K., Bang, G. J., Roh, M. I. & Lee, K. Y. 2009. Detailed Design and Construction of the Hull of an FPSO (Floating, Production, Storage, and Off-loading Unit).
Khakzad, N., Khan, F. & Amyotte, P. 2013. Quantitative risk analysis of offshore drilling operations: A Bayesian approach. *Safety Science,* 57, 108-117.
Mather, A. 2000. Offshore Engineering – An Introduction. *Witherby.*
Rausand, M. 2013. *Risk assessment: theory, methods, and applications*, John Wiley & Sons.
Sember, W. J. 2004. A Risk-Based Approach to Offshore Classification.
Serratella, C. M., Wang, G. & Conachey, R. 2007. Risk-based strategies for the next generation of maintenance and inspection programs. 20-21.
Sshimamura, Y., 2002. FPSO/FSO: State of the Art, *Journal of Marine Science and Technology,* VOL. 7, NO. 2, PP. 59-70.
Vinnem, J.-E. 2007. *Offshore risk assessment: principles, modelling and applications of QRA studies*, Springer Science & Business Media.
Øien, K. 2013. Remote operation in environmentally sensitive areas: development of early warning indicators. *Journal of Risk Research,* 16, 323-336.

AUTHOR INDEX

AUTHOR INDEX

AUTHOR INDEX